ENCYCLOPEDIA OF

ENVIRONMENTAL POLLUTION AND CLEANUP

VOLUME 1

ENCYCLOPEDIA OF

ENVIRONMENTAL POLLUTION AND CLEANUP

VOLUME 1

Robert A. Meyers

Editor in Chief

Diane Kender Dittrick

Editor

A Wiley-Interscience Publication

John Wiley & Sons, Inc.

New York / Chichester / Weinheim / Brisbane / Singapore / Toronto

This book is printed on acid-free paper.

Library of Congress Cataloging-in-Publication Data:

The Wiley encyclopedia of environmental pollution and cleanup / Robert A. Meyers.
p. cm.
Abridged ed. of: Encyclopedia of environmental analysis and remediation. c1998.
"A Wiley-Interscience publication."
Includes bibliographical references and index.
ISBN 0-471-31612-1 (cloth : alk. paper)
1. Pollution—Encyclopedias. 2. Environmental protection—Encyclopedias.
TD173.M48 1999
628.5 21—dc20 99-17884

Printed in the United States of America

10 9 8 7 6 5 4 3 2 1

EDITORIAL STAFF

CONTRIBUTORS

Isam Mohammed Abdel-Magid, *Sudan University for Science & Technology,* Wastewater Reclamation and Reuse

Fred Adams, *University of Antwerp, Wilrijk, Belgium,* Mass Spectrometry, Inorganic

Marc Adams, *University of Waterloo, Faculty of Science, Waterloo, Ontario, Canada,* Solid-Phase Microextraction

John Addison, *John Addison Consultancy, Edinburgh, Scotland,* Asbestos, Analysis of

Diana S. Aga, *Swiss Federal Institute for Environmental Science and Technology, Dubendorf, Switzerland,* Immunoassay, Environmental

Ann-Christine Albertsson, *Royal Institute of Technology, Stockholm, Sweden,* Biodegradable Polymers

D. G. Allen, *University of Toronto, Toronto, Ontario, Canada,* Pulp Mill Effluent Remediation

J. Angerer, *University of Erlangen-Nuremberg, Erlangen, Germany,* Biological Materials, Analysis of Hazardous Substances

J. M. Azcue, *National Water Research Institute, Burlington, Ontario, Canada,* Aquatic Sediment Sampling

Sergey Babichenko, *Institute of Ecology, Tallinn, Estonia,* Fluorescent Signatures in Environmental Analysis

Amalendu Bagchi, *Consultant, Madison, Wisconsin,* Landfills

Eddie G. Baker, *Battelle, Pacific Northwest Division, Richland, Washington,* Petroleum Sludge Treatment

Ralph S. Baker, *ENSR, Acton, Massachusetts,* Soil Vapor Extraction

Brandon R. Ball, *Parametrix, Inc., Sumner, Washington,* Ozonation, Wastewater Treatment

Shankha K. Banerji, *University of Missouri, Columbia, Missouri,* Carbonaceous BOD and Suspended Solids, Removal from Wastewater; Nitrogen Removal from Municipal Wastewater

Ian Barnes, *Bergische Universitat, Wuppertal, Germany,* Hydrocarbons in the Atmosphere

Rod Barratt, *The Open University, Milton Keynes, United Kingdom,* Dust, Measurement of Trace Elements in

Rajat S. Basu, *AlliedSignal, Inc., Buffalo, New York,* Chlorofluorocarbon Replacements, Applications of

Charles E. Baukal, *Air Products and Chemicals, Inc., Allentown, Pennsylvania,* Oxygen-Enhanced Waste Incineration

Richard L. Bechtold, *Consultant, Columbia, Maryland,* Ethanol Motor Fuel; Methanol Vehicles

David J. Becker, *U.S. Army Corps of Engineers, Omaha, Nebraska,* Soil Vapor Extraction

Roger Beckie, *University of British Columbia, Vancouver, British Columbia, Canada,* Hydrogeology

David J. Belson, *Consultant, Nottingham, England,* Earth Surface Monitoring from Space

Alain Berthod, *Université Claude Bernard, Villeurbanne, France,* Countercurrent Chromatography

Don Betowski, *U.S. Environmental Protection Agency, Las Vegas, Nevada,* Dyes and Pigments, Environmental Chemistry

Leon D. Betowski, *Lockheed Martin Environmental Services, Las Vegas, Nevada,* Polycyclic Aromatic Hydrocarbons, Determination

Vladislav Bevc, *Consultant, Danville, California,* Electric Cars, Environmental Effects

A. Bevolo, *Iowa State University, Ames, Iowa,* Site Characterization, Expedited

Dibakar Bhattacharyya, *University of Kentucky, Lexington, Kentucky,* Membranes for Hazardous Wastewater Treatment; Reverse Osmosis

Michael Biddle, *MBA Polymers and American Plastics Council, Washington, D.C.,* Automotive Plastics, Environmental Compatibility and Recycling

J. R. Bidwell, *Virginia Polytechnic Institute and State University, Blacksburg, Virginia,* Ecological Toxicity Testing

Richard J. Bigda, *The Technotreat Corporation, Tulsa, Arizona,* Fenton's Chemistry in Wastewater Treatment

Gary A. Bishop, *University of Denver, Denver, Colorado,* Automobile Emissions Control; Vehicle Emission, On-Road

Gabriel Bitton, *University of Florida, Gainesville, Florida,* Wastewater Microbiology

Dianna Blair, *Sandia National Laboratories, Albuquerque, New Mexico,* Volatile Organic Compounds, On-Line Monitoring Techniques

Peter Bode, *Interfaculty Reactor Institute, Delft University of Technology, Delft, the Netherlands,* Activation Analysis

Gordon B. Bonan, *National Center for Atmospheric Research, Boulder, Colorado,* Climate System Models, Land–Atmosphere Interactions

Bruce Boomer, *Consultant, Prarie Village, Kansas,* Air Permitting

Abdolhamid Borazjani, *Mississippi State University, Mississippi State, Mississippi,* Pentachlorophenol-Contaminated Soils, Bioremediation

John K. Borchardt, *Shell Chemical Company, Houston, Texas,* Paper Recycling; Petroleum, Enhanced Oil Recovery

Darrell J. Bosch, *Virginia Polytechnic Institute and State University, Blacksburg, Virginia,* Soil and Water Quality in Agriculture, Assessing and Improving

Andrew Bouie, *BDM International, Gaithersburg, Maryland,* Buried Radioactive Waste

Jeffrey H. Boy, *U.S. Army Corps of Engineers, Champaign, Illinois,* Lead-Based Paint Hazard Mitigation

Jacob I. Bregman, *Bregman Company Inc., Bethesda, Maryland,* Environmental Impact Statements

Udo A. Th. Brinkman, *Free University, Amsterdam, the Netherlands,* Two-Dimensional Gas Chromatography

N.F. Brockmeier, *Argonne National Laboratory, Argonne, Illinois,* Carbon Dioxide Recovery from Fossil-Fueled Power Plants

Carl E. Brown, *Environment Canada, Ontario, Ottawa, Canada,* Oil Spill Remote Sensors

Clifford J. Bruell, *University of Massachusetts, Lowell, Massachusetts,* Soil Remediation by Air Sparging

Douglas Bryant, *McMaster University, Hamilton, Ontario, Canada,* Tires, Open Burning

Jens Bublitz, *Universität Kiel, Kiel, Germany,* Laser-Induced Fluorescence Spectroscopy

Verónica Bucalá, *Planta Piloto de Ingenieria Química, 12 de Octubre 1842, Bahía Blanca, Argentina,* Soil Thermal Decontamination: Fundamentals

Daniel B. Bullen, *Iowa State University, Ames, Iowa,* Nuclear Fuel Reserves and Production

Amar C. Bumb, *Fluor Daniel GTI, Inc., Greenville, South Carolina,* Pump-and-Treat Technology, Aquifer Remediation

Nigel J. Bunce, *University of Guelph, Guelph, Ontario, Canada,* Dioxin-Like Compounds, Screening Assays

Lufgarde Buydens, *Optical Measurement Systems, Wageningen, the Netherlands,* Plastics Among Nonplastics, Identification in Mixed Waste

John Cairns, Jr., *Virginia Polytechnic Institute and State University, Blacksburg, Virginia,* Ecological Toxicity Testing

Anne Camper, *Montana State University, Bozeman, Montana,* Microbial Quality of Drinking Water

Kirk J. Cantrell, *Pacific Northwest National Laboratory, Richland, Washington,* Sorptive Barriers for Groundwater Remediation

Andrew M. Carleton, *The Pennsylvania State University, University Park, Pennsylvania,* Ocean–Atmosphere Interactions

Martin R. Carter, *Agriculture & Agri-Food Canada Research Centre, Charlottetown, Prince Edward Island, Canada,* Soil Quality Analytical Methods

Joseph A. Caruso, *University of Cincinnati, Cincinnati, Ohio,* Chromtography/Plasma Mass Spectrometry

Rick Cee, *U. S. Department of Labor, Salt Lake City, Utah,* OSHA Standards, Sampling, and Measurements

Ernesto R. Cespedes, *U.S. Army Corps of Engineers, Vicksburg, Mississippi,* Cone Penetrometer-Deployed Chemical Sensors

A. J. Cessna, *National Hydrology Research Institute, Saskatoon, Saskatchewan, Canada,* Herbicide Residues in Biota, Determination

Wing-Hong Chan, *Hong Kong Baptist University, Kowloon, Hong Kong,* Formaldehyde in the Environment

Ronald Chaney, *University of Connecticut, Storrs, Connecticut,* Sediments, Management and Remediation of Contamination

Louis W. Chang, *University of Arkansas for Medical Sciences, Little Rock, Arkansas,* Metals, Toxicology

Prosenjit D. Chaudhuri, *The BOC Group Technical Center, Murray Hill, New Jersey,* Pollution Prevention Design

Baoshu Chen, *TDA Research, Inc., Wheat Ridge, Colorado,* Chlorinated Hydrocarbons, Destruction of

Cary Chiou, *U.S. Department of the Interior, Denver, Colorado,* Soil Sorption of Organic Pollutants and Pesticides

Chen-Hwa Chiu, *Bechtel Corporation, Houston, Texas,* Chlorofluorocarbon Replacements, Applications of

Ming Fat Choi, *Baptist University of Hong Kong, Kowloon, Hong Kong,* Oxygen and Carbon Dioxide Detection

Chien Chung, *National Tsing Hua University, Hsinchu, Taiwan Republic of China,* Gamma-Ray Activation Analysis

Roy Cohen, *RJC Consultants, Fairfield, Ohio,* Validation of Environmental Data

Hilary Coller, *Massachusetts Institute of Technology, Cambridge, Massachusetts,* Mutational Spectra Technology

K. L. Compton, *Clemson University, Anderson, South Carolina,* Risk Assessment

Ronald L. Cook, *TDA Research, Inc., Wheat Ridge, Colorado,* Chlorinated Hydrocarbons, Destruction of

Anthony Cooper, *APV Anhydro, Copenhagen, Denmark,* Volatile Organic Compounds in Wastewater, Removal and Recovery

Robert M. Counce, *University of Tennessee, Knoxville, Tennessee,* Gas Cleanup, Absorption

David A. Cremers, *Los Alamos National Laboratory, Los Alamos, New Mexico,* Laser-Induced Breakdown Spectroscopy

Burton B. Crocker, *St. Louis, Missouri,* Air Pollution Control for Stationary Sources

M. John Cullinane, *U.S. Army, Research Triangle Park, North Carolina,* Military Toxic Materials, Destruction of

E.J. Daniels, *Argonne National Laboratory, Argonne, Illinois,* Carbon Dioxide Recovery from Fossil-Fueled Power Plants

W. Lee Daniels, *Virginia Polytechnic Institute and State University, Blacksburg, Virginia,* Reclamation of Disturbed Lands

B. E. Davies, *Clemson University, Clemson, South Carolina,* Lead Levels in Soil

J. Eric D. Davies, *Lancaster University, Lancaster, England,* Raman Spectroscopy in Water Analysis

S. Michael Davis, *U.S. Environmental Protection Agency, Research Triangle Park, North Carolina,* Oxygenated Fuel Additives, Health Effects of

Jacob de Boer, *DLO-Netherlands Institute for Fisheries Research, IJmuiden, the Netherlands,* Two-Dimensional Gas Chromatography

David DeMarini, *U.S. Environmental Protection Agency, Research Triangle Park, North Carolina,* Tires, Open Burning

Scott F. DeMuth, *Oak Ridge National Laboratory, Oak Ridge, Tennessee,* Nuclear Materials Processing, Waste Reduction

Thomas N. Debo, *Georgia Institute of Technology, Atlanta, Georgia,* Stormwater Management, Municipal

Jiří Dědina, *Academy of Sciences of the Czech Republic, Prague, Czech Republic,* Atomic Absorption Spectrometry, Hydride Generation

Henk-Jan de Geus, *DLO-Netherlands Institute for Fisheries Research, IJmuiden, the Netherlands,* Two-Dimensional Gas Chromatography

Jeroen J.M. de Goeij, *Interfaculty Reactor Institute, Delft University of Technology, Delft, the Netherlands,* Activation Analysis

Kenneth Demars, *University of Connecticut, Storrs, Connecticut,* Sediments, Management and Remediation of Contamination

Lloyd E. Deuel, Jr., *Soil Analytics Services, Inc., College Station, Texas,* Salt Waste in Oil and Gas Production, Remediation

Joseph S. Devinny, *University of Southern California, Los Angeles, California,* Biofiltration

Antonio DiCorcia, *Universita La Sapienza, Rome, Italy,* Pesticides, Trace Analysis

L. C. Dickson, *National Hydrology Research Institute, Saskatoon, Saskatchewan, Canada,* Herbicide Residues in Biota, Determination

Susan V. Diehl, *Mississippi State University, Mississippi State, Mississippi,* Pentachlorophenol-Contaminated Soils, Bioremediation

R. Bertrum Diemer, Jr., *E. I. du Pont de Nemours & Co., Inc., Wilmington, Delaware,* Hazardous Waste Incineration

Urmila M. Diwekar, *Carnegie Mellon University, Pittsburgh, Pennsylvania,* Pollution Prevention Design

Yassine Djebbar, *University of Ottawa, Ottawa, Ontario, Canada,* Volatile Organic Compounds, Air Stripping from Contaminated Groundwater

Joseph R. Donnelly, *Lockheed Martin Environmental Services, Las Vegas, Nevada,* Chemical Warfare Agents; Polycyclic Aromatic Hydrocarbons, Determination

Paul V. Doskey, *Argonne National Laboratory, Argonne, Illinois,* Cone Penetrometer-Deployed Chemical Sensors

Lisa G. Dowden, *Spiegel & McDiarmid, Washington, D.C.,* Municipal Environmental Compliance

Donald E. Drinkwater, *Texaco, Inc., Beacon, New York,* Petroleum Hydrocarbons, Environmental Analysis of

James A. Dyer, *Consultant, Newark, Delaware,* Volatile Organic Compounds, Control at Industrial Plants

Timothy M. Dyess, *U.S. Environmental Protection Agency, Research Triangle Park, North Carolina,* Radon In Buildings

W. Wesley Eckenfelder, *Eckenfelder, Inc., Nashville, Tennessee,* Refinery Wastewater, Toxicity Reduction; Wastes, Industrial

Robert P. Eganhouse, *U.S. Geological Survey, Reston, Virginia,* Molecular Markers of Anthropogenic Waste

B. Eklund, *Radian International LLC, Austin, Texas,* Modeling Atmospheric Dispersion

Douglas C. Elliott, *Battelle, Pacific Northwest Division, Richland, Washington,* Petroleum Sludge Treatment

Thomas D. Ellis, *E. I. du Pont de Nemours & Co., Inc., Wilmington, Delaware,* Hazardous Waste Incineration

M. Eltgroth, *Radian International LLC, Austin, Texas,* Modeling Atmospheric Dispersion

Evan J. Englund, *U.S. Environmental Protection Agency, Las Vegas, Nevada,* Soil Sampling at Hazardous Waste Sites

V. P. Evangelou, *University of Kentucky, Lexington, Kentucky,* Acid Mine Drainage

Anna M. Fan, *California Environmental Protection Agency,* Metals, Toxicology

Robert J. Farrauto, *Engelhard Corporation Research Center, Iselin, New Jersey,* Air Pollution Control, Catalytic

Antonio Febo, *CNR Istituto Inquinomento Atmosferico, Roma, Italy,* Nitrous Acid Generation, Fate and Detection

Yaping Feng, *The Open University, Milton Keynes, United Kingdom,* Dust, Measurement of Trace Elements in

Mervin F. Fingas, *Environment Canada Emergencies Science Division, Ottawa, Ontario, Canada,* BTEX Quantitation in Oils by GC/MS; Oil Spill Remote Sensors

Robert A. Fjeld, *Clemson University, Anderson, South Carolina,* Risk Assessment

Jerry Fosnaugh, *The Dow Chemical Company, Midland, Michigan,* Automotive Plastics, Environmental Compatibility and Recycling

Harold S. Freeman, *U.S. Environmental Protection Agency, Las Vegas, Nevada,* Dyes and Pigments, Environmental Chemistry

Gregory C. Frye, *Sandia National Laboratories, Albuquerque, New Mexico,* Volatile Organic Compounds, On-Line Monitoring Techniques

Arthur Furst, *University of San Francisco, San Francisco, California,* Metals, Toxicology

Yuri V. Gankin, *Tufts University, Medford, Massachusetts,* Polychlorinated Biphenyls Congeners Identification

Chris G. Ganotis, *Wheelabrator Environmental Systems. Inc., Hampton, New Hampshire,* Municipal Waste-to-Energy Technology

R. Gijbels, *University of Antwerp, Wilrijk, Belgium,* Mass Spectrometry, Inorganic

Todd D. Giorgio, *Vanderbilt University, Nashville, Tennessee,* Supercritical Fluid Extraction for Soil Remediation

Earnest F. Gloyna, *University of Texas at Austin, Austin, Texas,* Supercritical Water Oxidation for Wastewater and Sludge Remediation

Michel G. Goedert, *Hewlett-Packard Laboratories, Palo Alto, California,* Field-Portable Instrumentation

Norbert Gonzalez-Flesca, *Parc Technologique ALATA, Verneuil en halatte, France,* Polluted Atmosphere Generation, Validation of

Tadeusz Gorecki, *University of Waterloo, Waterloo, Ontario, Canada,* Solid-Phase Microextraction

Judith A. Graham, *U.S. Environmental Protection Agency, Research Triangle Park, North Carolina,* Oxygenated Fuel Additives, Health Effects of

Michal Green, *Israel Institute of Technology, Haifa, Israel,* Groundwater Treatment, Nitrate Contamination

Gary Grieco, *Stone & Webster, New York, New York,* Air Toxic Control Strategies for Utilities

Zoe A. Grosser, *The Perkin-Elmer Corporation, Norwalk, Connecticut,* Inductively Coupled Plasma/Atomic Emission Spectrometry; Inductively Coupled Plasma Mass Spectrometry

Robert A. Haas, *California Environmental Protection Agency, Berkeley, California,* Polychlorinated Biphenyls, Detection by Immunoassay

Richard A. Haimann, *CH2MHILL, Los Angeles, California,* Fungal Technologies for Hazardous Waste Treatment

Martin A. Hamilton, *Montana State University, Bozeman, Montana,* Microbial Quality of Drinking Water

D. Alan Hansen, *Electric Power Research Institute, Palo Alto, California,* Air Quality Modeling

Thomas M. Harris, *University of Tulsa, Tulsa, Oklahoma,* Brine-Impacted Soils, Remediation

Ronald G. Harvey, *University of Chicago, Chicago, Illinois,* Polycyclic Aromatic Hydrocarbons in the Environment

John M. Haschke, *Los Alamos National Laboratory, Los Alamos, New Mexico,* Plutonium Storage

Peter Hawkins, *University of the West of England, Bristol, England,* Oxygen and Carbon Dioxide Detection

J. V. Headley, *National Hydrology Research Institute, Saskatoon, Saskatchewan, Canada,* Herbicide Residues in Biota, Determination

Ronald M. Heck, *Engelhard Corporation Research Center, Iselin, New Jersey,* Air Pollution Control, Catalytic

John H. Heiser, *Brookhaven National Laboratory, Upton, New York,* Subsurface Containment Barrier Integrity, Verification of

Douglas T. Heitkemper, *U.S. Food & Drug Administration, Cincinnati, Ohio,* Chromatography/Plasma Mass Spectrometry

M.-C. Hennion, *Ecole Supérieure de Physique et de Chimie Industrielles, Paris, France,* Bioanalytical Methods

J. A. Hestekin, *University of Kentucky, Lexington, Kentucky,* Membranes for Hazardous Wastewater Treatment

Douglas Hileman, *Pilko & Associates, Inc., Woodland Hills, California,* ISO

Byron Y. Hill, *Union Carbide Corporation, South Charleston, West Virginia,* Gas Cleanup, Absorption

Jack B. Howard, *Massachusetts Institute of Technology, Cambridge, Massachusetts,* Soil Thermal Decontamination: Fundamentals

Hann S. Huang, *Argonne National Laboratory, Argonne, Illinois,* Hazardous Waste, Solidification and Stabilization

Charles K. Huston, *Varian Chromatography Systems, Walnut Creek, California,* Mass Spectrometry, Quadrupole Ion Trap

Marvin H. Hyman, *Bechtel, Richland, Washington,* Groundwater and Soil Remediation

Yuri Orlov Igo, *Moscow State University, Moscow, Russia,* Fluorescent Signatures in Environmental Analysis

K. I. Irgolic, *Karl-Franzens Universitat, Österreich, Austria,* Aresenic and Arsenic Compounds, Electrochemical Determination

Trevor Jackson, *SAIC, Idaho Falls, Idaho,* Vitrification, Plasma-Arc

William A. Jacoby, *National Renewable Energy Laboratory, Golden, Colorado,* Trichloroethylene or Perchloroethylene Contamination, Remediation of

W. Jambers, *University of Antwerp, Wilrijk, Belgium,* Mass Spectrometry, Inorganic

Bruce R. James, *University of Maryland, College Park, Maryland,* Chromium(VI) in Soils

R.K.M. Jayanty, *Research Triangle Institute, Research Triangle Park, North Carolina,* Carcinogens, Monitoring of Indoor Air

B.J. Jody, *Argonne National Laboratory, Argonne, Illinois,* Carbon Dioxide Recovery from Fossil-Fueled Power Plants

Mitchell E. Johnson, *Duquesne University, Pittsburgh, Pennsylvania,* Fluorescence Spectroscopy

Alan S. Kao, *ENVIRON, Arlington, Virginia,* Air Sampling Instruments; Atmospheric Aerosols, Size Distributions

Daniel I. Kaplan, *Pacific Northwest National Laboratory, Richland, Washington,* Sorptive Barriers for Groundwater Remediation

S. Karlsson, *Royal Institute of Technology, Stockholm, Sweden,* Biodegradable Polymers

S. V. Karmarkar, *Zellweger Analytics, Inc., Milwaukee, Wisconsin,* Ion Chromatography

Alexander E. Karu, *University of California, Berkeley, California,* Polychlorinated Biphenyls, Detection by Immunoassay

Sidney A. Katz, *Rutgers University, Camden, New Jersey,* Chromium, Environmental Analytical Chemistry

Michael J. Kennish, *Rutgers University, New Brunswick, New Jersey,* Marine Pollution

B. M. Kim, *GE Corporate Research and Development,* Membranes for Hazardous Wastewater Treatment

David E. Kimbrough, *Castaic Lake Water Agency, Santa Clarita, California,* Elemental Pollution in Air

B. H. Kjartanson, *Iowa State University, Ames, Iowa,* Site Characterization, Expedited

Mark S. Klima, *The Pennsylvania State University, University Park, Pennsylvania,* Coal Preparation

Brian S. Kocher, *Dover Chemical Corporation, Dover, Ohio,* Soil Decontamination by Supercritical Water Oxidation; Supercritical Water Oxidation

Joseph Kovach, *Cornell University, Ithaca, New York,* Pesticides, Ranking by Environmental Impact

Konrad B. Krauskopf, *Stanford University, Stanford, California,* Geochemistry, Environmental

Carolien Kroeze, *Wageningen Institute for Environment and Climate Research, the Netherlands,* Nitrous Oxide Emissions and Control

Paul H. Krumrine, *Waste Policy Institute, Gaithersburg, Maryland,* Minimum Additive Waste Stabilization

James C. Ku, *U.S. Department of Labor, Salt Lake City, Utah,* OSHA Standards, Sampling, and Measurements

William L. Kuhn, *Pacific Northwest National Laboratory, Richland, Washington,* Radioactive Wastes, Separation Technologies

Ashok Kumar, *U.S. Army Corps of Engineers, Champaign, Illinois,* Lead-Based Paint Hazard Mitigation

Herbert Lacayo, *University of Maryland, Baltimore, Maryland,* Sample Size for Site Investigation

Wendy S. Lader, *Spiegel & McDiarmid, Washington, D.C.,* Municipal Environmental Compliance

Kate Lajtha, *Oregon State University, Corvallis, Oregon,* Forest Ecosystems in a Changing Global Context

Janusz S. Laskowski, *University of British Columbia, Vancouver, British Columbia, Canada,* Coal Preparation

Barry K. Lavine, *Clarkson University, Potsdam, New York,* Underground Fuel Spills, Source Identification

Edward Laws, *University of Hawaii at Manoa, Honolulu, Hawaii,* Water Pollution Toxicology

Gary Leatherman, *SAIC, Idaho Falls, Idaho,* Vitrification, Plasma-Arc

Frank S.C. Lee, *Hong Kong Baptist University, Kowloon, Hong Kong,* Formaldehyde in the Environment

Tim LeGore, *University of Maryland, Baltimore, Maryland,* Sample Size for Site Investigation

Paul Lemieux, *U.S. Environmental Protection Agency, Research Triangle Park, North Carolina,* Tires, Open Burning

Barry Lesnik, *U.S. Environmental Protection Agency, Washington, D.C.,* Polychlorinated Biphenyls, Detection by Immunoassay

Lois Levitan, *Cornell University, Ithaca, New York, Integrated Pest Management Program, Geneva, New York,* Pesticides, Ranking by Environmental Impact

Lixiong Li, *University of Texas at Austin, Austin, Texas,* Supercritical Water Oxidation for Wastewater and Sludge Remediation

Steven Licht, *APV Anhydro, Copenhagen, Denmark,* Volatile Organic Compounds in Wastewater, Removal and Recovery

JoAnn S. Lighty, *University of Utah, Salt Lake City, Utah,* Hazardous Waste Incineration

H. W. Liu, *University of Toronto, Toronto, Ontario, Canada,* Pulp Mill Effluent Remediation

Garrick E. Louis, *University of Virginia, Charlottesville, Virginia,* Recycling Municipal Solid Waste

George Lunn, *Consultant, Baltimore, Maryland,* Laboratory Hazardous Waste, Destruction of

Milton Manes, *Kent State University, Kent, Ohio,* Activated Carbon Adsorption Fundamentals

W. C. Mangum, *University of Kentucky, Lexington, Kentucky,* Reverse Osmosis

Bruce A. Manning, *U.S. Department of Agriculture, Riverside, California,* Selenium Immobilization in Wastewater Remediation

Bernd Markert, *Zittau, Germany,* Industrial Waste Dumps, Sampling and Analysis

Joseph C. Martz, *Los Alamos National Laboratory, Los Alamos, New Mexico,* Plutonium Storage

Paul M. Mather, *University of Nottingham, Nottingham, United Kingdom,* Remote Sensing Data Systems and Networks

Tadashi Matsunaga, *Tokyo University of Agriculture & Technology, Tokyo, Japan,* Bacterial Pathogens in Drinking Water

William H. McBride, *UCLA Medical Center, Los Angeles, California,* Radioprotective Agents

Brian McCarry, *McMaster University, Hamilton, Ontario, Canada,* Tires, Open Burning

Louis M. McDonald, Jr., *University of Kentucky, Lexington, Kentucky,* Acid Mine Drainage

Lisa McKenzie, *University of California, Irvine, California,* Smoldering Biomass Fuels, Measuring and Modeling

Francis C. McMichael, *University of Virginia, Charlottesville, Virginia,* Recycling Municipal Solid Waste

Walt W. McNab, Jr., *Lawrence Livermore National Laboratory, Livermore, California,* Organic Contaminants in Fresh Water, Distribution and Fate

John McNurney, *R.W. Beck, Denver, Colorado,* Municipal Environmental Compliance

James W. Mercer, *HSI GeoTrans, Inc., Sterling, Virginia,* Groundwater Transport and Remediation of Chemicals in

Markku Mesilaakso, *University of Helsinki, Helsinki, Finland,* Chemical Weapons Convention, NMR Analysis

Eline P. Meulenberg, *ELTI Support, Nijmegen, the Netherlands ,* Molecular Biology, Environmental Applications; Pesticides, Immunoassay

Carol J. Miller, *Wayne State University, Detroit, Michigan,* Groundwater Monitoring

Fred P. Miller, *The Ohio State University, Columbus, Ohio,* Soil as an Ecosystem

Jaffer Mohiuddin, *U.S. Department of Energy, Germantown, Maryland,* Buried Radioactive Waste

Gerardo A. Montero, *North Carolina State University, Raleigh, North Carolina,* Supercritical Fluid Extraction for Soil Remediation

Kenneth Mopper, *Washington State University, Pullman, Washington,* Organic Carbon, Analysis Methods for Aqueous Samples

Lidia Morawska, *Queensland University of Technology, Brisbane, Australia,* Indoor Air Risk Assessment and Management

Edward C. Morettl, *Baker Environmental, Inc., Coraopolis, Pennsylvania,* Clean Air Act, Regulations and Implementation

Alena Mudroch, *National Water Research Institute, Burlington, Ontario, Canada,* Aquatic Sediment Sampling

Kenneth L. Mulholland, *Consultant, Newark, Delaware,* Volatile Organic Compounds, Control at Industrial Plants

David Murray, *University of Alberta, Edmonton, Alberta, Canada,* Radioprotective Agents

Raymond L. Murray, *Consultant, Raleigh, North Carolina,* Nuclear Power; Nuclear Reactor Types; Nuclear Waste Management

George R. Mussoline, *Environmental Standards, Inc., Valley Forge, Pennsylvania,* Chromium(VI) in Soils

T. N. Narasimhan, *Lawrence Livermore National Laboratory, Livermore, California,* Organic Contaminants in Fresh Water, Distribution and Fate

Roberto M. Narbaitz, *University of Ottawa, Ottawa, Ontario, Canada,* Volatile Organic Compounds, Air Stripping from Contaminated Groundwater

Nagaraj K. Neerchal, *University of Maryland, Baltimore, Maryland,* Sample Size for Site Investigation

Vincent P. Nero, *Texaco, Inc., Beacon, New York,* Petroleum Hydrocarbons, Environmental Analysis of

Peter A. Nick, *Consultant, Yorba Linda, California ,* Petroleum Fuels Analysis for Meeting Regulatory Limits

B. R. Niederlehner, *Virginia Polytechnic Institute and State University, Blacksburg, Virginia,* Ecological Toxicity Testing

David E. B. Nightingale, *EMCON, Inc., Olympia, Washington,* Household Hazardous Material Management

Scott Noesen, *The Dow Chemical Company, Midland, Michigan,* Automotive Plastics, Environmental Compatibility and Recycling

Mina Okochi, *Tokyo University of Agriculture & Technology, Tokyo, Japan,* Bacterial Pathogens in Drinking Water

John E. Oliver, *Indiana State University, Terre Haute, Indiana,* Climate–Biosphere Interactions

Hans Oonk, *TNO Institute of Environmental Sciences, Apeldoorn, the Netherlands,* Nitrous Oxide Emissions and Control

Chi-Shiu Ordaz, *U.S. Department of Energy, Germantown, Maryland,* Minimum Additive Waste Stabilization

Yuri Orlov, *Moscow State University, Moscow, Russia,* Fluorescent Signatures in Environmental Analysis

Janusz Pawliszyn, *University of Waterloo Faculty of Science, Waterloo, Ontario, Canada,* Solid-Phase Microextraction

David Peakall, *King's College, London, England,* Biomarkers

William E. Pepelko, *U.S. Environmental Protection Agency, Washington, D.C.,* Carcinogens, Identification and Risk Assessment

Joseph J. Perona, *University of Tennessee, Knoxville, Tennessee,* Gas Cleanup, Absorption

Cinzia Perrino, *CNR Istituto Inquinomento Atmosferico, Roma, Italy,* Nitrous Acid Generation, Fate and Detection

David W. Pershing, *University of Utah, Salt Lake City, Utah,* Hazardous Waste Incineration

Igor Persiantsev, *Moscow State University, Moscow, Russia,* Fluorescent Signatures in Environmental Analysis

William Peters, *Massachusetts Institute of Technology, Cambridge, Massachusetts,* Soil Thermal Decontamination: Fundamentals

Max R. Peterson, *Research Triangle Institute, Research Triangle Park, North Carolina,* Carcinogens, Monitoring of Indoor Air

David M. Petrovski, *U.S. Environmental Protection Agency, Chicago, Illinois,* Sediments, Management and Remediation of Contamination

John R. Petrulis, *University of Guelph, Guelph, Ontario, Canada,* Dioxin-Like Compounds, Screening Assays

John C. Petura, *Applied Environmental Management, Inc., Malvern, Pennsylvania,* Chromium(VI) in Soils

Russell W. Phifer, *Environmental Assets, Inc., West Chester, Pennsylvania,* Laboratory Hazardous Waste, Management

Alice E. Pichahchy, *Los Alamos National Laboratory, Los Alamos, New Mexico,* Laser-Induced Breakdown Spectroscopy

Valerie Pichon, *Ecole Supérieure de Physique et de Chimie Industrielles, Paris, France,* Bioanalytical Methods

David K. Pijawka, *Arizona State University, Tempe, Arizona,* Hazardous Waste Planning

William A. Piper, *Amoco Corporation, Houston, Texas,* Drilling Waste Management

Larisa Poryvkina, *Institute of Ecology, Tallinn, Estonia,* Fluorescent Signatures in Environmental Analysis

Patricia D. Powell, *Powell & Associates Science Services, Las Vegas, Nevada,* Subsurface Remediation, Iron Metal for

Robert M. Powell, *Powell & Associates Science Services, Las Vegas, Nevada,* Subsurface Remediation, Iron Metal for

Roger C. Prince, *Exxon Research and Engineering Company, Annandale, New Jersey,* Crude Oil Biodegradation

Jian-quo Qian, *Washington State University, Pullman, Washington,* Organic Carbon, Analysis Methods for Aqueous Samples

Jerzy M. Rajtar, *New Mexico Institute of Mining and Technology, Socorro, New Mexico,* Radioactive Materials, Naturally Occurring

John D. Randolph, *Oak Ridge National Laboratory, Oak Ridge, Tennessee,* Nuclear Materials Processing, Waste Reduction

W. Rasemann, *Academy of Mining & Technology, Freiberg, Germany,* Industrial Waste Dumps, Sampling and Analysis

Sergei Rebrik, *University of San Francisco, San Francisco, California,* Fluorescent Signatures in Environmental Analysis

Katta J. Reddy, *University of Wyoming, Laramie, Wyoming,* Selenium Speciation in Natural Waters

Andrew J. Reese, *Georgia Institute of Technology, Atlanta, Georgia,* Stormwater Management, Municipal

Abraham Reife, *U.S. Environmental Protection Agency, Las Vegas, Nevada,* Dyes and Pigments, Environmental Chemistry

Richard Reiss, *OGDEN, Nashville, Tennessee,* Ozone Chemistry in Buildings

Gregory Richardson, *University of Connecticut, Storrs, Connecticut,* Sediments, Management and Remediation of Contamination

Susan D. Richardson, *U.S. Environmental Protection Agency, Athens, Georgia,* Drinking Water Disinfection By-Products

Otto Richter, *Institut fur Geographie und Geookologie, Braunschweig, Germany,* Pesticides, Environmental Fate

Albert Robbat, Jr., *Tufts University, Medford, Massachusetts,* Field Analytics, Dynamic Work Plans; Polychlorinated Biphenyls Congeners Identification

Kim R. Rogers, *U.S. Environmental Protection Agency, Las Vegas, Nevada,* Biosensor Technology for Environmental Measurement

Brian E. Rood, *Mercer University, Macon, Georgia,* Mercury, Determination in the Environment

Gregory L. Rorrer, *Oregon State University, Corvallis, Oregon,* Heavy-Metal Ions, Removal from Wastewater

Mike C. Rose, *U.S. Department of Labor, Salt Lake City, Utah,* OSHA Standards, Sampling, and Measurements

Rolf J. Rosenberg, *VIT Chemical Technology, Espoo, Finland,* Radionuclides, Methods of Analysis

N. Phillip Ross, *University of Maryland, Baltimore, Maryland,* Sample Size for Site Investigation

Donald R. Rowe, *D. R. Rowe Engineering Services, Inc., Bowling Green, Kentucky,* Wastewater Reclamation and Reuse

James F. Rusling, *University of Connecticut, Storrs, Connecticut,* Electrochemical Remediation of Soils

Hugh H. Russell, *CHR_2 Environmental Services, Oilton, Oklahoma,* Chlorinated Ethenes, Anaerobic Bioremediation

Jeffrey Ryan, *U.S. Environmental Protection Agency, Research Triangle Park, North Carolina,* Tires, Open Burning

Hiroshi H. Saito, *Massachusetts Institute of Technology, Cambridge, Massachusetts,* Soil Thermal Decontamination: Fundamentals

Caryn L. Sanford, *Mercer University, Macon, Georgia,* Mercury, Determination in the Environment

Wolfgang Schade, *Universität Kiel, Kiel, Germany,* Laser-Induced Fluorescence Spectroscopy

K. H. Schaller, *University of Erlangen-Nuremberg Institute, Erlangen, Germany,* Biological Materials, Analysis of Hazardous Substances

George Schneider, *U.S. Department of Energy, Idaho Falls, Idaho,* Buried Radioactive Waste

Karl B. Schnelle, Jr., *Vanderbilt University, Nashville, Tennessee,* Air Pollution Control for Stationary Sources; Supercritical Fluid Extraction for Soil Remediation

Jüergen Schultz, *Geographisches Institut, Aachen, Germany,* Ecozones, Global

Brian Schumacher, *U.S. Environmental Protection Agency, Las Vegas, Nevada,* Soil Sampling at Hazardous Waste Sites

John Schwemberger, *Consultant, Rockville, Maryland,* Lead Paint Testing

Glenn T. Seaborg, *Ernest Orlando Lawrence Berkely National Laboratory, Berkeley, California,* Plutonium Management and Disposition

L. John Sealock, *Battelle, Pacific Northwest Division, Richland, Washington,* Petroleum Sludge Treatment

Richard A. Selg, *British Nuclear Fuels Limited—Savannah River Corporation, Aiken, South Carolina,* Hazardous Waste Cost Control

Ananto K. Seta, *Universitas Bengkulu, Bengkulu, Indonesia,* Acid Mine Drainage

Guy W. Sewell, *U.S. Environmental Protection Agency Risk Management Research Laboratory, Tulsa, Oklahoma,* Chlorinated Ethenes, Anaerobic Bioremediation

Robert W. Shaw, *U.S. Army, Research Triangle Park, North Carolina,* Military Toxic Materials, Destruction of

Joseph Sherma, *Lafayette College, Easton, Pennsylvania,* Thin-Layer Chromatography

Kusiel S. Shifrin, *Oregon State University, Corvallis, Oregon,* Atmospheric Aerosols, Methods of Analysis

Lee Shugart, *L. R. Shugart & Associates, Inc., Oak Ridge, Tennessee,* Biomarkers

Hiba Shukairy, *U.S. Environmental Protection Agency, Cincinnati, Ohio,* Ozonation in Drinking Water Treatment

Markus W. Sigrist, *Institute of Quantum Electronics, ETH Zurich, Zurich, Switzerland,* Air Monitoring, Optical Spectroscopic Methods

S. K. Sikdar, *U.S. Environmental Protection Agency, Cincinnatti, Ohio,* Membranes for Hazardous Wastewater Treatment

Geoff Silcox, *University of Utah, Salt Lake City, Utah,* Hazardous Waste Incineration

Laura J. Silva, *Battelle, Pacific Northwest Division, Richland, Washington,* Petroleum Sludge Treatment

Surinder P. Singh, *Oak Ridge National Laboratory, Oak Ridge, Tennessee,* Nuclear Materials Processing, Waste Reduction

Walter Slavin, *Bonaire Technologies, Ridgefield, Connecticut,* Inductively Coupled Plasma/Atomic Emission Spectrometry; Inductively Coupled Plasma Mass Spectrometry

Leslie Smith, *University of British Columbia, Vancouver, British Columbia, Canada,* Hydrogeology

William Smith, *Yale University, New Haven, Connecticut,* Acid Rain

J. Andy Soesilo, *Western International University, Phoenix, Arizona,* Hazardous Waste Planning

O. David Sparkman, *O. David Sparkman & Associates, Antioch, California,* Mass Spectrometry

Thomas Sprio, *Princeton University, Princeton, New Jersey,* Dioxin Emissions Inventory

Donald H. Stedman, *University of Denver, Denver, Colorado,* Automobile Emissions Control; Vehicle Emission, On-Road

G. A. Stenback, *Iowa State University, Ames, Iowa,* Site Characterization, Expedited

Ronald G. Stillman, *Roy F. Weston, Inc., West Chester, Pennsylvania,* Hazardous Waste Cost Control

Barbara H. Stuart, *University of Technology, Sydney, Australia,* Infrared Spectroscopy

G. Subramanian, *Consultant, Kent, England,* Quality Assurance in Environmental Monitoring

Joan D. Sulzberg, *Science Applications International Corporation, McLean, Virginia,* Household Hazardous Material Management

Rao Y. Surampalli, *U.S. Environmental Protection Agency, Kansas City, Kansas,* Carbonaceous BOD and Suspended Solids, Removal from Municipal Wastewater; Nitrogen Removal from Municipal Wastewater

Robert W. Sylvester, *Consultant, Newark, Delaware,* Volatile Organic Compounds, Control at Industrial Plants

Hideshige Takada, *Tokyo University of Agriculture and Technology, Tokyo, Japan,* Molecular Markers of Anthropogenic Waste

John J. Taylor, *Electric Power Research Institute, Palo Alto, California,* Nuclear Power Facilities Safety

Larry Taylor, *Nova Tech, Westtown, Pennsylvania,* Electroanalysis in Environmental Analysis

Malcolm Taylor, *Mintek, Randburg, South Africa,* Vanadium Speciation Analytical Methods

S. Templeman, *Radian International LLC, Austin, Texas,* Modeling Atmospheric Dispersion

William Thilly, *Massachusetts Institute of Technology, Cambridge, Massachusetts,* Mutational Spectra Technology

V. M. Thomas, *Princeton University, Princeton, New Jersey,* Dioxin Emissions Inventory

David A. Tillman, *Foster Wheeler Environmental Corporation, Sacramento, California,* Trace Metal Emissions from Coal Utilization

Terrence Toy, *University of Denver, Denver, Colorado,* Reclamation of Disturbed Lands

Dimiter L. Tsalev, *University of Sofia, Sofia, Bulgaria,* Electrothermal Atomic Absorption Spectrometry

A. Tsechanski, *Ben-Gurion University of the Negev, Beer Sheva, Israel,* Nuclear Detectors and Methods for Environmental Analysis

Wayne L. Turnberg, *Washington Department of Health and University of Washington, Seattle, Washington,* Biohazardous Waste

Lawrence Valcovic, *U.S. Environmental Protection Agency, Washington, D.C.,* Carcinogens, Identification and Risk Assessment

Willie van den Broek, *Optical Measurement Systems, Wageningen, the Netherlands,* Plastics Among Nonplastics, Identification in Mixed Waste

J. Jeffrey van Ee, *U.S. Environmental Protection Agency, Las Vegas, Nevada,* Soil Sampling at Hazardous Waste Sites

R. Van Grieken, *University of Antwerp, Wilrijk, Belgium,* Mass Spectrometry, Inorganic

Philip Vella, *Carus Chemical Company, LaSalle, Illinois,* Permanganate Wastewater Treatment

Rock J. Vitale, *Environmental Standards, Inc., Valley Forge, Pennsylvania,* Chromium(VI) in Soils

Karl S. Vorres, *Argonne National Laboratory, Argonne, Illinois,* Coal

Arun S. Wagh, *Argonne National Laboratory, Argonne, Illinois,* Hazardous Waste, Solidification and Stabilization; Waste Generation

Mohan K. Wali, *School of Natural Resources, The Ohio State University, Columbus, Ohio,* Soil as an Ecosystem

John M. Walker, *U.S. Environmental Protection Agency, Washington, D.C.,* Biosolids Management, Use and Disposal

Zhendi Wang, *Environment Canada, Ottawa, Ontario, Canada,* BTEX Quantitation in Oils by GC/MS

Carl J. Watras, *Wisconsin Department of Natural Resources, Boulder Junction, Wisconsin,* Mercury Deposition, Atmospheric Monitoring

James C. Weaver, *Massachusetts Institute of Technology, Cambridge, Massachusetts,* Electric and Magnetic Fields

Todd S. Webster, *Envirogen, Inc., Lawrenceville, New Jersey,* Biofiltration

Chris Wedig, *Stone & Webster, New York, New York,* Air Toxic Control Strategies for Utilities

Gregory J. Wells, *Varian Chromatography Systems, Walnut Creek, California,* Mass Spectrometry, Quadrupole Ion Trap

Dietrich Wienke, *Optical Measurement Systems, Wageningen, the Netherlands,* Plastics Among Nonplastics, Identification in Mixed Waste

Ebtisam S. Wilkins, *University of New Mexico, Albuquerque, New Mexico,* Oil Spill Containment, Terrestrial

M. E. Williams, *EET Corporation, Knoxville, Tennessee,* Reverse Osmosis

William R. Williams, *Praxair, Inc., Tarrytown, New York,* Chlorofluorocarbons, Destruction of

Neal E. Wilson, *Minnesota Pollution Control Agency, St. Paul, Minnesota,* Soil Water and Groundwater Sampling

Richard Wilson, *Harvard University, Cambridge, Massachusetts,* Plutonium Burning for Weapons Disposal

Ryan D. Wilson, *University of Waterloo, Waterloo, Ontario, Canada,* Partitioning Tracers for Soil Characterization

Mary Leigh Wolfe, *Virginia Polytechnic Institute and State University, Blacksburg, Virginia,* Soil and Water Quality in Agriculture, Assessing and Improving

George Wolff, *GM Public Policy Center, Detroit, Michigan,* Air Pollution

Karen A. Wolnick, *U.S. Food & Drug Administration, Cincinnati, Ohio,* Chromatography/Plasma Mass Spectrometry

Alan M. Wolsky, *Argonne National Laboratory, Argonne, Illinois,* Carbon Dioxide Recovery from Fossil-Fueled Power Plants

J. D. Wonder, *Iowa State University, Ames, Iowa,* Site Characterization, Expedited

John D. Wright, *TDA Research, Inc., Wheat Ridge, Colorado,* Chlorinated Hydrocarbons, Destruction of

Craig Wyman, *Liskow & Lewis, New Orleans, Louisianna,* Hazardous Materials Transportation

Yuk Yee Yan, *Hong Kong Baptist University, Hong Kong,* Climate–Biosphere Interactions

Min Yang, *University of Waterloo Faculty of Science, Waterloo, Ontario, Canada,* Solid-Phase Microextraction

Jehuda Yinon, *Weizmann Institute of Science, Rehovot, Israel,* Explosives Contamination, Determination in Water and Soil

Markus Zahn, *Massachusetts Institute of Technology, Cambridge, Massachusetts,* Electric and Magnetic Fields

Mark E. Zappi, *Mississippi State University, Mississippi State, Mississippi,* Groundwater Treatment, Explosives Contamination

John P. Zelina, *University of Connecticut, Storrs, Connecticut,* Electrochemical Remediation of Soils

Brenda S. Zimmer, *U.S. Food & Drug Administration, Cincinnati, Ohio,* Chromatography/Plasma Mass Spectrometry

Shmuel Zitrin, *Israel Police Headquarters, Jerusalem, Israel,* Explosives Contamination, Determination in Water and Soil

FOREWORD

John Wiley and Sons has recently provided the environmental community with an eight-volume encyclopedia dealing with environmental analysis and remediation at the level of the working professional. This work—*The Encyclopedia of Environmental Analysis and Remediation*—contains 5400 pages and 280 articles by 500 authors. The Encyclopedia fills a significant gap in the literature; its articles summarize in detail, and with professional clarity, procedures associated with the treatment of numerous practical environmental problems. Each article is written by one or more experts in the given area. Working professionals will find this large, unabridged encyclopedia appropriate for their needs. Many individuals who do not specialize in environmental research, but have an appropriate background in science or engineering, will also find the large encyclopedia informative and useful for them.

However, members of the public who have an interest in environmental issues, including many high-school students, would like to have material more nearly at their educational level. The present, two-volume, condensed, and carefully edited version of the eight-volume set is intended to fill this need. This two-volume set is called *The Wiley Encyclopedia of Pollution and Cleanup*. It should earn a place on many bookshelves, particularly those in public libraries and in high schools, as a useful and handy reference on the environment.

One of the remarkable features of the last half of the twentieth century, the period since the end of World War II, has been the gradual but steadfast rise of general interest in environmental science, technology, and practice. The driving force, which has been strongest in the more developed countries but is evident all over the world, stems from an interest in improving health, aesthetics, and, in many cases, the economics of individual operations. In fact, in many countries, environmental concerns rank at the same level as those related to political and economic stability. *The Wiley Encyclopedia of Pollution and Cleanup*, and the more extensive eight-volume encyclopedia from which it is derived, should provide general information and practical guidance at the working level for all who wish to have it.

Professor Frederick Seitz
President Emeritus, Rockefeller University
Past President, National Academy of Sciences

PREFACE

The Wiley Encyclopedia of Environmental Pollution and Cleanup is a two-volume, 1800-page set that provides an introduction to environmental science, environmental technology, and environmental issues in a finely crafted collection of some 230 articles.

This work is based on the highly regarded eight-volume, 5400-page *Encyclopedia of Environmental Analysis and Remediation*, edited by Dr. Robert A. Meyers and published by John Wiley in 1998. The condensed set aims to maintain the high quality of articles on environmental topics and retain the information of greatest interest, while bringing this information to the widest possible audience, including readers having limited scientific training. In his role as its Editor, Dr. Meyers served as the guiding force for the "Unabridged Encyclopedia." He continued to lend his expertise to the series by planning, selecting, and shaping material for this condensed work.

The critical operation of the condensation of the material and the fine-tuning of the selection process were admirably carried out by the Editor of this work, Diane Kender Dittrick, Laboratory Director, Department of Environmental Science, Barnard College. Her extensive experience includes university teaching, science writing, and various positions in the reference-book publishing industry.

This Concise encyclopedia focuses more on describing pollution and cleanup and less on specialized specific analytical techniques. By doing so, the work—even though it is condensed, compact, and economic—can emphasize areas of general interest such as pollution in the biosphere, air pollution and its control, the cleanup of hazardous waste, water reclamation, and environmental regulations. This condensed work contains lengthy survey articles on many important topics: coal; ecozones; environmental geochemistry; field-portable instrumentation; groundwater and soil remediation; groundwater transport; hydrocarbons in the atmosphere; nuclear reactors; ocean–atmosphere interactions; petroleum fuels analysis; risk assessment; sediments, management and remediation of contamination; soil and water quality in agriculture; soil as an ecosystem; and wastewater reclamation and reuse.

Features of this Encyclopedia include a large number of illustrations and tables, sizeable bibliographies for each article, with these bibliographic entries keyed to the text; extensive cross-referencing, both within the article text and as separate cross-reference entries to direct the reader to information on the indicated subject; and an extensive subject index to allow the reader to find appropriate information no matter what initial term is sought. A total of 14 color photographs have been added to the Encyclopedia to explain, emphasize, and illustrate key concepts more fully.

As Professor Frederick Seitz indicates in the Foreword, "One of the remarkable features of the last half of the twentieth century has been the gradual but steadfast rise of general interest in environmental science, technology, and practice.

The Wiley Encyclopedia of Environmental Pollution and Cleanup should earn a place on many bookshelves, particularly those in public, high school, and college libraries, where it can serve as a useful and handy reference on the environment."

Arthur Biderman
Executive Editor

ABBREVIATIONS AND ACRONYMS

AA	Atomic absorption
AAS	Atomic absorption spectrometry
AE	Atomic emission
AES	Atomic emission spectrometry
BDAT	Best Demonstrated Available Technology
BMP	Best Management Practices
CAAA	Clean Air Act Amendments
CEQ	Council on Environmental Quality
CERCLA	Comprehensive Environmental Response, Compensation and Liability Act of 1980 (as amended)
CWA	Clean Water Act
DNAPL	Dense nonaqueous-phase liquids (soil)
DOD	U.S. Dept. of Defense
DOE	U.S. Dept. of Energy
DOI	U.S. Dept. of the Interior
EA	Environmental assessment
EIS	Environmental impact statement
EPA	U.S. Environmental Protection Agency
EXAFS	Extended X-ray absorption fine structure spectroscopy
FFCA	Federal Facilities Compliance Act
GC	Gas chromatography/gas chromatograph
GC-MS	Gas chromatograph-mass spectrometer
HAPs	Hazardous air pollutants
HPLC	High performance liquid chromatography
ICP	Inductively coupled plasma
ICP/AES	Inductively coupled plasma/atomic emission spectrometry
ICP-AES	Inductively coupled plasma-atomic emission spectrometer
ICP-MS	Inductively coupled plasma mass spectrometer
ISO	International Organization for Standardization
IUPAC	International Union of Pure and Applied Chemistry
LC-MS	Liquid chromatograph-mass spectrometer
MACT	Maximum achieveable control technology
MS	Mass spectrometry/mass spectrometer
NAA	Neutron activation analysis
NAAQS	National Air Quality Standards (a section of CAAA)
NASA	National Aeronautics and Space Administration
NEPA	National Environmental Policy Act of 1969 (as amended)
NESHAP	National Emission Standards for Hazardous Air Pollutants
NIST	National Institute of Standards and Technology
NMR	Nuclear magnetic resonance
NOAA	National Oceanic and Atmospheric Administration
NORM	Naturally occurring radioactive materials
NPDES	National Pollution Discharge Elimination System
NPL	National Priority List
PAHs	Polycyclic aromatic hydrocarbons
PCBs	Polychlorinated biphenyls
PCDD	Polychlorinated dibenzo-*p*-dioxins
PCDD/F	Mixed PCDD/PCDF
PCDF	Polychlorinated dibenzofurane
PICs	Products of incomplete combustion
PIXE	Particle induced X-ray emission
PM	Particulate matter
POTWs	Publicly Owned Treatment Works
RCRA	Resource Conservation and Recovery Act of 1976 (as amended)

SARA	Superfund Amendments and Reauthorization Act of 1986
SDWA	Safe Drinking Water Act
SFC	Supercritical fluid chromatography
Superfund	An extension of RCRA enacted in 1988 (formal name is CERCLA)
SVE	Soil vapor extraction
TRI	Toxics release inventory
TSCA	Toxic Substances Control Act
TSD	Treatment, storage, and disposal of military munitions
USDA	U.S. Dept. of Agriculture
UV	Ultraviolet
VOCs	Volatile organic compounds
XRF	X-ray fluorescence

CONVERSION FACTORS

The following tables are provided:

1. Common Energy Conversion Factors
2. Aggregate Energy Equivalents
3. Weights of Typical Petroleum Products
4. Equivalence of Mass and Energy
5. Energy Factors
6. Power Factors
7. Fluid Flow Factors
8. Useful Quantities for Global Climate Change

Table 1. Common Energy Conversion Factors

	Joule	Quadrillion BTU	Kilogram Calorie	Metric Ton of Coal Equivalent
1 Joule	1	947.9×10^{-21}	239×10^{-6}	34.1×10^{-12}
1 Quadrillion Btu	1055×10^{15}	1	252×10^{12}	36.0×10^{6}
1 Kilogram Calorie	4184	3966×10^{-18}	1	142.9×10^{-9}
1 Metric Ton of Coal Equivalent	29.3×10^{9}	27.8×10^{-9}	7×10^{6}	1
1 Barrel of Oil Equivalent	6119×10^{6}	5.8×10^{-9}	1462×10^{3}	0.21
1 Metric Ton of Oil Equivalent	44.8×10^{9}	42.4×10^{-9}	10.7×10^{6}	1.53
1 Cubic Meter of Natural Gas	37.3×10^{6}	35.3×10^{-12}	8905	1272×10^{-6}
1 Terawatt Year	31.5×10^{18}	29.9	7537×10^{12}	1076×10^{6}

Table 2. Aggregate Energy Equivalents

1 MBDOE = 1 million barrels per day of oil equivalent
= 50 million tons of oil equivalent per year
= 76 million metric tons of oil equivalent per year
= 57 billion cubic meters of natural gas per year
= 2.2 10^{18} joules per year
= 530 × 10^{12} kilocalories per year
= 2.1 × 10^{15} Btus per year = 2.1 quads
= 620 10^{9} kwh per year

1 QUAD = 1 quadrillion Btus = 10^{15} Btus
= 500,000 petroleum barrels a day
182,500,000 barrels per year
= 40,000,000 short tons of bituminous coal = 36,363,636 metric tons
= 1 trillion (10^{12}) cubic feet of natural gas
= 100 billion (10^{11}) kwh (based on 10,000 Btu/kwh heat rate)

1 kilowatt-hour of hydropower
= 10 × 10^{3} Btus
= 0.88 lb coal
= 0.076 gallon crude oil
= 10.4 cubic feet of natural gas

1 MTCE = one million short tons of coal equivalent
= 4.48 × 10^{6} barrels of crude oil
= 67 tons of crude oil
= 25.19 × 10^{12} cubic feet of natural gas

Table 3. Weights of Typical Petroleum Products[a]

	Pounds per U.S. Gallon	Pounds per 55-gal Drum	Kilograms per Cubic Meter	Barrels (42-gal) per Short Ton	Barrels (42-gal) per Metric Ton
LP-Gas	4.52	248	541.6	10.5	11.6
Aviation gasoline	5.90	325	707.0	8.2	8.9
Motor gasoline	6.17	339	739.3	7.7	8.5
Kerosene	6.76	372	810.0	7.0	7.8
Distillate fuel oils	7.05	388	845.8	6.8	7.5
Lubricating oils	7.50	413	898.7	6.3	7.0
Residual fuel oils	7.88	434	944.2	6.0	6.7
Paraffin wax		367	800.1	7.1	7.9
Grease		458	998.8	5.7	6.3
Asphalt		477	1039.2	5.5	6.1

[a] Source: U.S. Energy Information Administration, *Monthly Energy Review*.

Table 4. Equivalence of Mass and Energy

	1 Electron Mass	1 Atomic Mass Unit	1 Gram
Million electron volts	0.511	931.5	5.61×10^{26}
Joules	8.19×10^{-14}	1.49×10^{-10}	8.99×10^{13}
Btu	7.76×10^{-17}	1.42×10^{-13}	8.52×10^{10}
Kilowatt hours	2.27×10^{-20}	4.15×10^{-17}	2.50×10^{7}
Quads	7.76×10^{-35}	1.42×10^{-31}	8.52×10^{-8}

Table 5. Energy Factors

	1 Electron-Volt	1 Joule	1 British Thermal Unit	1 Kilocalorie	1 Kilowatt-Hour
Electron volt	1	6.24×10^{18}	6.58×10^{21}	2.61×10^{22}	2.25×10^{25}
Joule	1.60×10^{-19}	1	1054	4184	3.6×10^{6}
Calorie	3.83×10^{-20}	0.24	252	1000	860.4×10^{5}
Btu	1.52×10^{-22}	9.48×10^{-4}	1	3.97	3413
Kilocalorie	3.83×10^{-23}	2.39×10^{-4}	0.252	1	860.4
Kilowatt-hour	4.45×10^{-26}	2.78×10^{-7}	2.93×10^{-4}	1.16×10^{-3}	1
Megawatt-day	1.85×10^{-30}	1.16×10^{-11}	1.22×10^{-8}	4.84×10^{-8}	4.17×10^{-5}
Quad	1.52×10^{-40}	9.48×10^{-22}	10^{-18}	3.97×10^{-18}	3.41×10^{-15}

Table 6. Power Factors

	1 Btu per Day	1 Kilowatt-Hour per Year	1 Watt (W)	1 Kilowatt	1 Megawatt	1 Gigawatt	1 Quad per Year
Btu/day	1	9.35	81.95	8.2×10^{4}	8.2×10^{7}	8.2×10^{10}	2.74×10^{15}
Kilowatt/year	0.11	1	8.77	8766	8.8×10^{6}	8.77×10^{9}	2.93×10^{14}
Watts	0.012	0.114	1	1000	10^{6}	10^{9}	3.34×10^{13}
Kilowatts	1.22×10^{-5}	1.14×10^{-4}	0.001	1	1000	10^{6}	3.34×10^{10}
Megawatts	1.22×10^{-8}	1.14×10^{-7}	10^{-6}	0.001	1	1000	3.34×10^{7}
Gigawatts	1.22×10^{-11}	1.14×10^{-10}	10^{-9}	10^{-6}	0.001	1	3.34×10^{4}
Quad/yr	3.65×10^{-16}	3.41×10^{-15}	2.99×10^{-14}	2.99×10^{-11}	2.99×10^{-8}	2.99×10^{-5}	1

Table 7. Fluid Flow Factors

	1 Gallon per Day	1 Acre-Foot per Year	1 Cubic Foot per Minute	1 Cubic Meter per Second	1 Billion Gallons per Day
Gallon/day	1	892	1.077×10^{4}	2.28×10^{7}	10^{9}
Acre-cubic feet/yr	1.12×10^{-3}	1	12.07	2.56×10^{4}	1.12×10^{6}
Cubic feet/min	9.28×10^{-5}	8.28×10^{-2}	1	2119	9.28×10^{4}
Cubic feet/s	1.55×10^{-6}	1.38×10^{-3}	1.67×10^{-2}	35.31	1547
Cubic meters/s	4.38×10^{-8}	3.91×10^{-5}	4.72×10^{-4}	1	43.8
Billion gallons/day	10^{-9}	8.92×10^{-7}	1.08×10^{-5}	2.28×10^{-2}	1

Table 8. Useful Quantities for Global Climate Change

Quantity	Value	Ref.[a]
Solar constant	1.375 kilowatts square meters	1
Earth mass	5.976×10^{24} kilogram	2
Equatorial radius	6.378×10^{6} meters	2
Polar radius	6.357×10^{6} meters	2
Mean radius	6.371×10^{6} meters	
Surface area	5.101×10^{14} square meters	
Land area	1.481×10^{14} square meters	3
Ocean area	3.620×10^{14} square meters	4
Ice sheets and glaciers area	0.14×10^{14} square meters	5
Mean land elevation	840 meters	4
Mean ocean depth	3730 meters	
Mean ocean volume	1.350×10^{18} cubic meters	4
Ocean mass	1.384×10^{21} kilograms	
Mass of atmosphere	5.137×10^{18} kilograms	6
Equatorial surface gravity	9.780 meters/second	2

[a] Sources

1. D. V. Hoyt, *Rev. Geophys. Space Physics* **17,** 427–458 (1979).
2. F. Press and R. Siever, *Earth,* W. H. Freeman and Co., San Francisco, Calif., 1974.
3. B. K. Ridley, *The Physical Environment*, Ellis Horwood, Ltd., West Sussex, U.K., 1979.
4. H. W. Menard and S. M. Smith, *J. Geophys. Res.* **71,** 4305–4325 (1966).
5. M. F. Meier, ed., *Glaciers, Ice Sheets, and Sea Level: Effect of a CO_2^- Induced Climatic Change*. DOE/ER-60235-1, U.S. Dept. of Energy, Carbon Dioxide Research Division, Office of Basic Energy Sciences, Washington, D.C., 1985.
6. K. E. Trenberth, *J. Geophys. Res.* **86**, 5238–5246 (1981).

ENCYCLOPEDIA OF

ENVIRONMENTAL POLLUTION AND CLEANUP

VOLUME 1

ACID MINE DRAINAGE

V.P. (BILL) EVANGELOU
LOUIS M. MCDONALD, JR.
University of Kentucky
Lexington, Kentucky

ANANTO K. SETA
Universitas Bengkulu
Bengkulu, Indonesia

As a consequence of mining, a variety of sulfide minerals, primarily pyrite, are brought to the earth's surface. When this pyrite oxidizes after exposure to air and water, acidity and high concentrations of iron and sulfate are released into the environment. Depending on what other minerals are present, the acidity may react with these minerals to release manganese, zinc, copper, lead, mercury, aluminum, and cadmium. These low pH, metal contaminated waters, called acid mine drainage (AMD), degrade surface and groundwater quality and upset the ecosystems of streams, lakes, and estuaries.

The discharge of AMD into streams, rivers, and lakes has affected several thousand kilometers of natural fresh water systems in the United States and Canada. The U.S. Bureau of mines estimates that abandoned coal and metal mines and the associated piles of mine wastes alone adversely affect over 19,000 kilometers of rivers and streams and over 73,000 hectares of lakes and reservoirs in the United States. Mines in Canada, excluding gold, uranium and coal mines, and abandoned mines under the control of the Canadian Government, have generated at least 12,000 hectares of tailings and 320 Gkg of waste mine rock (1). Ontario alone has over 2,000 abandoned mines, many of which can produce AMD (1). Large sums of money are spent yearly to prevent and treat AMD. West Virginia alone, one of the largest coal-producing states in the country, spends roughly $350 million per year to treat AMD. Aside from the direct costs, there are additional costs due to diminishing land and water quality.

Handling the environmental aspects of a coal mining operation is just as costly as the actual coal extraction, and cleaning up a contaminated site is far more difficult than planning to avoid the problem (2). Technologies exist to prevent and control AMD. However, to implement an effective AMD control strategy, one must predict AMD production and, to predict AMD production, one must understand the oxidation chemistry of pyrite. What follows is an overview of the current understanding of pyrite oxidation chemistry and a review of AMD predictive methods and control strategies (see COAL).

ACID MINE DRAINAGE CHEMISTRY

Pyritic Acid Mine Drainage

Pyrite (FeS_2) is the most widespread metal sulfide mineral in coal fields and ore deposits including zinc, copper, uranium, gold, and silver. Pyrite occurs as euhedral, octahedral, hexahedral, and pyritohedral framboids and polyframboids, acicular conglomerates, coarse-grained masses, and cubes, depending on the environment of formation (Fig. 1). Because of its abundance in the environment, pyrite is recognized as the major source of AMD (3,4).

Pyrite oxidation takes place when the mineral is exposed to air and water. The process involves chemical, biological, and electrochemical reactions and varies with environmental conditions. Therefore, there is no single rate law available to describe the overall kinetics of pyrite oxidation for all cases. The effects of oxygen (O_2), dissolved ferric iron (Fe^{3+}), pH, microorganisms, carbon dioxide (CO_2), galvanic cell, temperature, and pyrite specific surface morphology are discussed below. Details of pyrite oxidation chemistry are given by in several studies (3, and references therein; and 4).

O_2 and Fe^{3+} Effects. Oxidation of pyrite mediated by O_2 and Fe^{3+} is the major cause for production of AMD. The reaction is surface-controlled and involves an electron transfer, via a S_2^{2-} bridge at the mineral's surface, to an electron acceptor, either O_2 or Fe^{3+} (5,6). Initially, a surface disulfide is oxidized by O_2 and water producing dissolved ferrous iron (Fe^{2+}) and acidity (H^+) according to the following reaction (7):

$$FeS_2 + \frac{7}{2}O_2 + H_2O \rightarrow Fe^{2+} + 2\ SO_4^{2-} + 2\ H^+ \tag{1}$$

Several studies (8,9) report a linear relationship between pyrite oxidation rate and O_2 concentration (Fig. 2). The rate law proposed herein is

$$\text{Rate} = 10^{-6.77}[O_2]^{0.5} \tag{2}$$

where $[O_2]$ is the molar concentration of dissolved oxygen. It was found that the rate of pyrite oxidation is nonlinear with respect to oxygen concentration and the following rate law based on oxygen adsorption was proposed (11):

$$\text{Rate} = wSKA\exp\left[\frac{-E_a}{RT(1 + KC)}\right] \tag{3}$$

where w represents mole ratio of SO_4 to O_2, S is the surface area, C is the gas phase O_2 concentration, K is the oxygen adsorption constant (1.36 $m^3\ mol^{-1}$), A is the Arrehenius factor and E_a is the activation energy.

The Fe^{2+} produced in reaction 1 is further oxidized to Fe^{3+} which then is hydrolyzed to iron hydroxide [$Fe(OH)_3$] releasing additional acidity into the environment (7), as follows:

$$Fe^{2+} + \frac{1}{4}O_2 + H^+ \rightarrow Fe^{3+} + \frac{1}{2}H_2O \tag{4}$$

and

$$Fe^{3+} + 3\ H_2O \leftrightarrow Fe(OH)_3(s) + 3\ H^+ \tag{5}$$

The nature of the iron hydroxide [$Fe(OH)_3$] precipitate formed depends on solution pH. At low pH, schwertmannite [$Fe(OH)_3$] forms (12,13), whereas, at medium and high pH, lepidocrosite (g-FeOOH) and goethite (a-FeOOH) form.

The Fe^{3+} produced by reaction 4 is reduced at the pyrite surface, releasing acidity according to reaction 6:

$$FeS_2 + 14\ Fe^{3+} + 8\ H_2O \rightarrow 15\ Fe^{2+} + 2\ SO_4^{2-} + 16\ H^+ \tag{6}$$

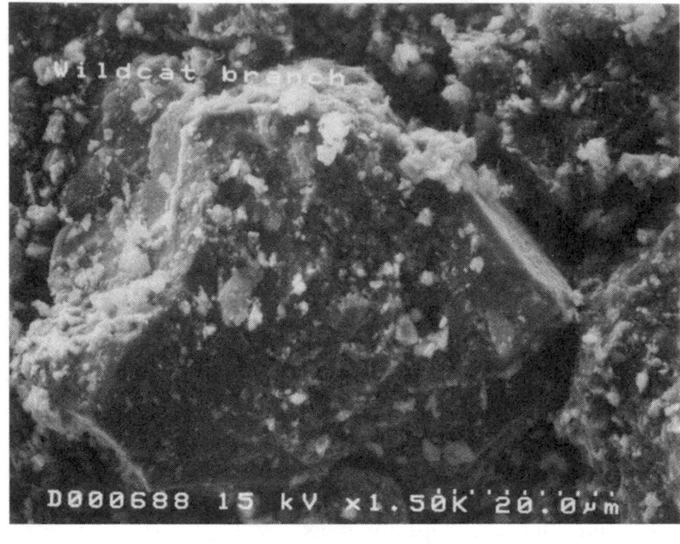

(a)

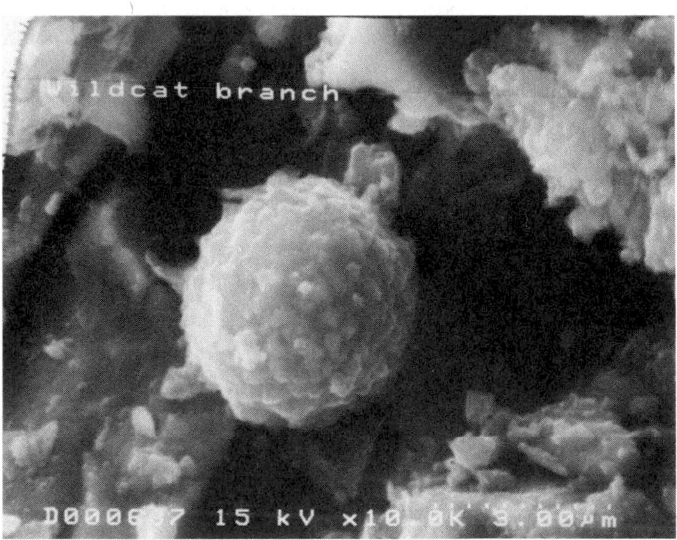

(b)

Figure 1. Examples of pyrite morphology: (**a**) massive pyrite (magnification is 1500×); (**b**) framboidal pyrite (magnification is 10,000×).

In abiotic low pH conditions, reaction 4 is very slow whereas reaction 6 is very fast. For example, at pH 3, the half-life for reaction 4 is on the order of 1000 days; the half-life for reaction 6 is on the order of 20–1000 min (11). Therefore, reaction 4 is the rate-limiting step in pyrite oxidation. The rate of pyrite oxidation rate in the absence of O_2 is described by

$$\text{Rate} = 10^{-9.74}\frac{[Fe^{3+}]^{0.5}}{[H^+]^{0.5}} \quad (7)$$

where $[Fe^{3+}]$ and $[H^+]$ represent molar concentrations (9). The oxidation rate is largely independent of Fe^{2+} concentration (9). Fe^{3+} oxidizes pyrite faster than O_2 because Fe^{3+} binds chemically to the pyrite surface whereas O_2 does not. The mechanism of this reaction is described in detail based on molecular orbital theory (5).

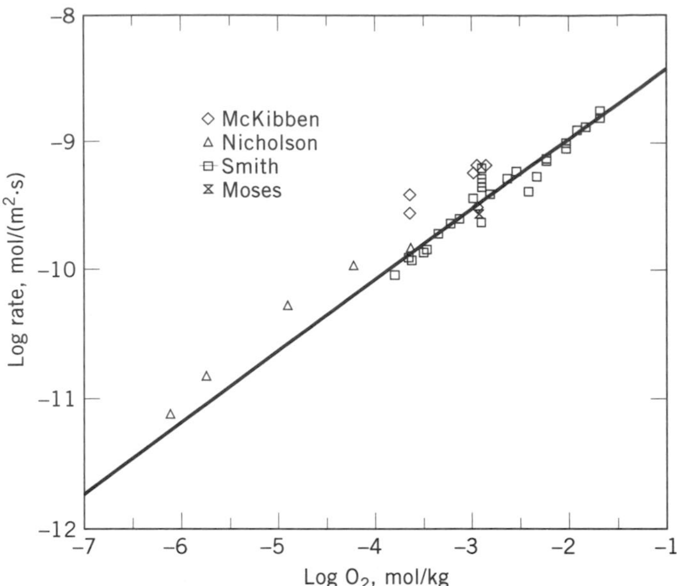

Figure 2. Rates of pyrite oxidation as a function of oxygen concentration (Ref. 10; with permission).

Microbial Effects. Abiotic oxidation of Fe^{2+} (reaction 3) at relatively low pH (pH <5) is much slower than reduction of Fe^{3+} by pyrite (reaction 6). However, Fe^{2+} oxidizing bacteria, especially *Thiobacillus ferrooxidans,* accelerate the rate of Fe^{2+} oxidation by a factor of 10^6 (14,15). *Thiobacillus ferrooxidans* is an acidophilic, chemolithotrophic organism that is ubiquitous in geologic environments containing pyrite and oxidizes Fe^{2+}, S^0, metal sulfides, and other reduced inorganic sulfur compounds (15). In the presence of *T. ferrooxidans* and at low pH, pyrite oxidation is described by the following (3):

$$FeS_2 + 14Fe^{3+} + 8H_2O \longrightarrow 15Fe^{2+} + 2SO_4^{2-} + 16H^+$$

(Regeneration loop: O_2, Iron oxidizing bacteria) (8)

Pyrite oxidation kinetics, then, are strongly linked to Fe^{2+} oxidation by bacteria. Therefore, the environmental conditions that affect *T. ferrooxidans* activity affect the pyrite oxidation rate. The optimum temperature and pH for *T. ferrooxidans* activity is around 30°C and pH 3.2 (16,17). *T. ferrooxidans* (obligate aerobes) activity ceases when O_2 is depleted. However, bacterial activity resumes even with extremely small increases in O_2 content and reaches a maximum at approximately 1% O_2.

CO_2 Effects. Laboratory data show that nonmicrobial pyrite oxidation increases with increasing CO_3^{2-} concentration (18). The mechanism of this phenomenon and direct molecular evidence are reported by using Fourier transform infrared (FTIR) spectroscopy (19). It was proposed that formation of a ferrous-carbonate complex on the surface of pyrite is responsible for the increase in the pyrite oxidation rate (19). The proposed reactions are as follows:

$$Fe - S_2 + Fe^{2+} \leftrightarrow Fe - S - S : Fe^{2+} \quad (9)$$

$$Fe - S - S : Fe^{2+} \overset{H_2O}{\leftrightarrow} Fe - S - S : Fe(OH)_2 \quad (10)$$

$$\mathrm{Fe-S-S:Fe(OH)_2 + CO_2 \leftrightarrow Fe-S-S:FeCO_3} \tag{11}$$

or

$$\mathrm{Fe-S-S:Fe(OH)_2 + CO_2 \leftrightarrow Fe-S-S:FeHCO_3^+ + OH^-} \tag{12}$$

Formation of the ferrous-bicarbonate complex on the surface of pyrite increases the basicity of Fe^{2+} leading to its rapid oxidation (20):

$$\mathrm{Fe-S-S:FeHCO_3^+ + \frac{1}{4}O_2 + H^+ \rightarrow Fe-S-S:Fe(OH)^{2+} + CO_2 + \frac{1}{2}H_2O} \tag{13}$$

Reaction 13 spontaneously forms a persulfido bridge by decarboxylation and pyrite oxidation initiates as shown below:

$$\mathrm{Fe-S-S:Fe^{3+} \rightarrow Fe-S-\underbrace{S^{\cdot}:Fe^{2+}}_{\textit{Free radical}}} \tag{14}$$

Formation of a pyrite free radical (reaction 14) leads to a continuous loss of electrons until thiosulfate is produced:

$$\underbrace{\mathrm{Fe-S-S^{\cdot}:(Fe^{3+})_5}}_{\textit{Persulfido bridge}} \mathrm{+ 3\,H_2O \rightarrow Fe-\underbrace{S-S-O_3}_{\textit{Thiosulfate}} + 6\,H^+ + 5\,Fe^{2+}} \tag{15}$$

In reaction 15, the bracketed surface undergoes five continuous redox reactions to produce thiosulfate. Thiosulfate is rapidly oxidized by Fe^{3+} to SO_4^{2-}, yielding eight more H^+ (3).

Galvanic Cell Effects. A galvanic cell results from physical contact between two different metal disulfide minerals in an acid/ferric sulfate solution. Such a system is caused by physical mixing of strata at the Earth's surface. If two metal disulfide minerals with different electrical rest potentials are in contact, the mineral with the lower electrical rest potential, acting as anode, is dissolved, whereas the mineral with higher rest potential, acting as cathode, is galvanically protected (21,22). For example, in the case of the chalcopyrite/pyrite galvanic cell ($CuFeS_2$/FeS_2), chalcopyrite, with a lower rest potential, dissolves rapidly whereas pyrite remains practically intact. In the galvanic cell reaction, bacteria play a role by continuously oxidizing the elemental sulfur (produced from the anodic oxidation reaction) to sulfate. Thus, formation of a sulfur barrier on the surface of chalcopyrite is eliminated. Details of these reactions are available (3 and 4).

Little is known about the contribution of galvanic cells to pyrite oxidation in mining or other disturbed geologic environments. However, one may assume that galvanic cells are responsible for pyrite oxidation because pyrite is usually associated with other sulfide minerals, such as those of lead, copper, nickel, and zinc. It is most likely that in such systems various mechanisms of oxidative dissolution operate simultaneously. These various mechanisms include galvanic dissolution, electrochemical oxidation, and oxidation by Fe^{3+}, O_2, and bacteria.

Temperature Effects. The effect of temperature on the rate k of pyrite oxidation is explained by the Arrhenius equation (12):

$$k = A\exp\frac{-Ea}{RT} \tag{16}$$

where E_a is the activation energy (J/mol), A is the preexponential factor called the Arrhenius factor (units of k), R is the universal gas constant (J/(mol · K)), and T is the temperature in K. An Arrhenius plot of log k versus $1/T$ yields a line with slope of $-E_a/2.303R$ from which E_a is derived (12). Several studies (9,12,23,24) suggest that the activation energy of pyrite oxidation by O_2 is about 60–80 kJ/mol. It is suggested that E_a is a means of predicting temperature effects on the rate constant and also indicates the controlling factors in kinetic reactions. For example, the high E_a values reported for pyrite oxidation suggest a surface chemical control. An E_a value of about 20 kJ/mol indicates a diffusion-controlled reaction rate.

Surface Area Effects. Because pyrite oxidation is a surface-mediated reaction, the oxidation rate is proportional to the available surface area. This finding is important in predicting pyrite oxidation rate under laboratory or field conditions. For example, improper washing pulverized pyrite results in a bimodal grain-size distribution which produces inaccurate rate constants. In addition, when significant pyrite mass is lost during the oxidation process the surface area is changed, resulting in "shrinking particles" and, consequently, increasing oxidation rates (25). However, oxidation products accumulate on the surfaces. Therefore, the magnitude of pyrite surface area is also related to the accumulation of a product layer on the surface (26). Figure 3 shows that the diameter of pyrite particles decreases with time and the thickness of the product layer (ie, ferric hydroxide) increases. Consequently, pyrite oxidation rates are limited. This mechanism provides a significant control on long-term rates of pyrite oxidation in near-neutral pH.

Nonpyritic Acid Mine Drainage

Three mechanisms are responsible for acid mine drainage produced by nonpyritic geologic material. The first mechanism

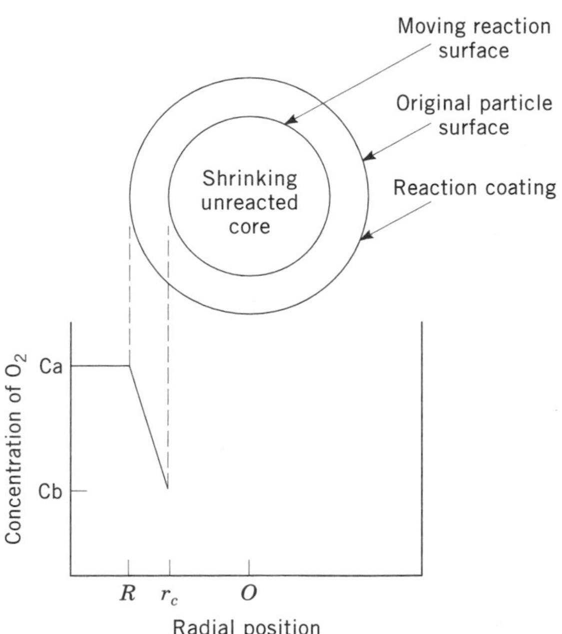

Figure 3. Schematic diagram of reacting particle with increasing product layer (Ref. 27; with permission).

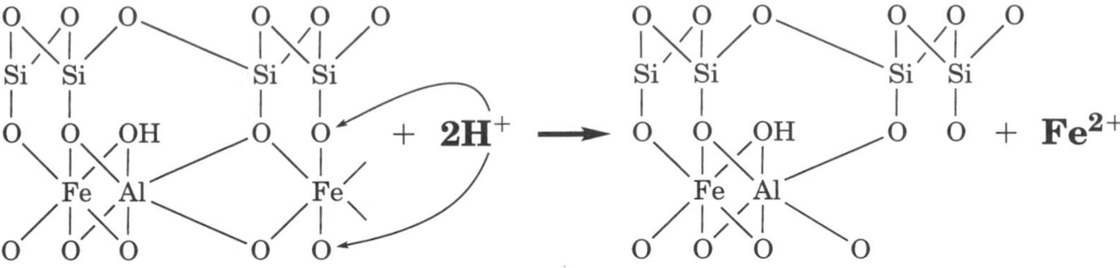

Figure 4. Schematic of acid attack of octahedral mineral structures and release of Fe^{2+}.

is the hydrolysis of exchangeable acid metals, for example, exchange of Al and Mn for Ca and Mg (3). A second mechanism involves a change in the oxidation state of the coordinating cations (eg, Fe^{2+} to Fe^{3+} or Mn^{2+} to Mn^{3+} or Mn^{4+}) which destabilizes the mineral structure. An index of such behavior is a change in color of geologic material upon exposure to the atmosphere. Ferrolysis is the surface-induced oxidation and hydrolysis of exchangeable iron. Ferrolysis produces leachates with pH as low as 3 and is a well-known process in iron-rich soils with high rainfall and alternating reducing/oxidizing conditions. A third mechanism responsible for producing non-pyritic AMD is release of structural Fe^{2+}. It is hypothesized that two steps are involved in this process. One step is acidification of the mineral edge surfaces, illustrated in Figure 4. It shows protonation of two oxygen atoms coordinated by Fe^{2+} in an octahedral mode (octahedral is defined as an eight-sided structure of six oxygens coordinated by Fe^{2+}). Protonation of the coordinated oxygens destabilizes the mineral structure and Fe^{2+} is released. The remaining silica structure possesses a negative charge (7) and retains exchangeable Fe^{2+}. The exchangeable Fe^{2+} is displaced by another cation, e.g., Ca^{2+}, or undergoes ferrolysis to release additional acid. These processes are often apparent in weathering sandstone.

An example of structural iron release from sandstone minerals is described below. The data in Table 1 show total iron concentration in the drain of a sandstone-dominated mine site that did not contain pyrite. The drainage becomes acidic when the ferrous iron oxidizes. Samples of unweathered, partially weathered, and deeply weathered sandstone, and a subsoil from the area were pulverized and then extracted with either distilled water, 0.1 *M* NaCl, 0.1 *M* HCl, or 6 *M* HCl (Table 2). The differences in iron content among the samples are probably caused by sample heterogeneity. The important point is that the sandstone samples at all stages of weathering contained significant levels of soluble iron (as shown by the results of extraction with 0.1 *M* NaCl and 0.1 *M* HCl) and that

Table 1. Quality of Water Emanating from a Nonpyritic, Sandstone-Dominated, Mine Drainage Site

Component	Drainage
pH	6.20
SO_4^{2-}, mg/L	78.4
Total Fe, mg/L	18.0
Fe^{2+}, mg/L	15.3
Mn, mg/L	4.65
Ca, mg/L	19.64
Mg, mg/L	11.95

most of the iron extracted with 0.1 *M* NaCl and 0.1 *M* HCl was ferrous iron. The iron content extracted with distilled water is much lower than that extracted with 0.1 *M* NaCl, indicating that some of the soluble Fe^{2+} in the sandstone was potentially exchangeable.

The data in Table 2 also show that the soil sample contains very low levels of Fe^{2+}, although it does contain a high level of total iron. Most iron in soils and deeply weathered sandstone is expected to be some form of ferric oxide which does not contribute to iron release in underground waters. Thus, it seems from the results (shown in Table 2) that release of Fe^{2+} from the weathering sandstone is the main source of the ferrous iron in the drainage. Principal minerals in the samples are chlorite, goethite, illite, kaolinite, and quartz.

ACID MINE DRAINAGE PREDICTION

Acid–Base Accounting

Currently, the most acceptable method for predicting AMD is acid-base accounting (ABA). In acid-base accounting, all potential alkalinity (neutralization potential = NP) and all potential acidity (maximum potential acidity = MPA) are determined in a given mine spoil. The value of NP and MPA are subtracted to obtain a net neutralization potential (NNP):

$$NNP = NP - MPA \tag{17}$$

Generally, if acidity is greater than alkalinity, AMD is produced. On the other hand, if alkalinity is greater than acidity, AMD is not produced. These conclusions are drawn due to (27):

$$FeS_2 + 2\,CaCO_3 + 3.75\,O_2 + 1.5\,H_2O \Leftrightarrow 2\,SO_4^{2-} + Fe(OH)_3 + 2\,Ca^{2+} + 2\,CO_2 \tag{18}$$

Based on reaction 18, two moles of calcite are required to neutralize all of the acidity produced by oxidizing one mole of pyrite. In other words, for each 10 kg of sulfur present in geologic material (1% pyritic sulfur), 31.25 kg of calcite are needed to neutralize all potential acidity.

In the pH range of 6 to approximately 8.4, a solution system in equilibrium with $CaCO_3$ obeys the following relationship (7):

$$Ca^{2+} \cong \frac{HCO^{3-}}{2} \tag{19}$$

assuming that no other Ca salt is present. Note that HCO_3^- is the main species determining alkalinity in the pH range of 4.2 to 8.4. Maximum alkalinity levels produced in mining waste material by calcite or dolomite depend partially on equilibrium considerations and partially on kinetic

Table 2. Iron Content in Unweathered, Partially and Deeply Weathered Sandstone and Soil Derived from the Sandstone[a]

Component	Unweathered Sandstone, ppm	Partially Weathered Sandstone, ppm	Deeply Weathered Sandstone, ppm	Soil, ppm
Distilled Water				
Ferrous iron	3	4	3	2
0.1 *M* NaCl				
Total iron	74	43	53	1
Ferrous iron	74	41	53	1
0.1 *M* HCl				
Total iron	282	128	183	48
Ferrous iron	240	93	84	10
6 *M* HCl				
Total iron	705	920	9,800	36,400

[a] Measured by extracting with distilled water, 0.1 *M* NaCl, 0.1 *M* HCl, and 6 *M* HCl. Extraction was conducted by shaking 10 g ground sandstone in 50 mL of solution for 16 h. Total iron was determined by atomic absorption spectroscopy, ferrous iron with 0.5% dipyridyl.

considerations. The following equation describes $CaCO_3$ solubility at equilibrium (3):

$$\log Ca^{2+} \cong 9.76 - pH - \log P_{CO_2} \tag{20}$$

where P_{CO2} denotes the partial pressure of CO_2. By substituting equation 19 in equation 20, alkalinity is expressed as a function of P_{CO2} and pH (note however, that P_{CO2} also controls pH):

$$\log\left(\frac{HCO_3^-}{2}\right) \cong 9.76 - pH - \log P_{CO_2} \tag{21}$$

Equation 21 reveals that alkalinity released by $CaCO_3$ depends on P_{CO2}, which also controls pH. In general, as P_{CO2} increases, pH decreases and alkalinity or bicarbonate increases. At equilibrium, for any given P_{CO2}, only a given concentration of HCO_3^- is present. On the other hand, pyrite oxidation is an electrochemical process, and as long as the redox potential is high enough (3), pyrite continues to oxidize, and oxidation products accumulate.

Various statistical correlations were carried out on ABA data obtained from the Pennsylvania Department of Environmental Resources (28). The data were segregated into "with thresholds" (strata with greater than 0.5% sulfur, or NP greater than 30 t 1000 t^{-1} with carbonates) and "without thresholds" (all reported values for NP and % sulfur for all strata). Data from these correlations are shown in Figures 10 to 14. There is a linear correlation between net alkalinity (NA) in the water and NP or NNP. However, a "gray zone" appears at NNP or NP values between approximately −10 and 10 t 1000 t^{-1} where net alkalinity is positive or negative. In this "gray zone", samples predicted to form AMD may or may not. It was concluded that for samples in the gray zone, additional information is needed, such as "results of previous and adjacent mining, paleoenvironmental interpretations, baseline hydrologic data, proposed mining and operational plans, and other analytical procedures (28)."

Proposed Empirical Models

There is no doubt that predicting AMD, especially from ABA data, is very difficult. Pyrite oxidation and/or neutralization are surface-controlled reactions, and unless parameters, such as total surface area of pyrite and carbonate and rates of pyrite oxidation and alkalinity release are known, AMD prediction remains elusive. Below, an empirical approach is proposed that can be tested in leaching columns and then extended to the field. This AMD predictive approach is based on total surface area of pyrite and carbonates at a NNP ≥ 0.

The proposed AMD predictive model is based on the hypothesis that the ratio of total surface area of acid-producing to acid-neutralizing material determines the acidity of mine drainage. A relationship between specific surface (cm^2/g) to particle diameter (cm) can be established for calcite and pyrite. This relationship for spherical particles is given in Figures 5 and 6. The relationship is curvilinear, and the maximum change in specific surface is attained for this choice of units as the size of the particle becomes less than 1 cm. Small decreases in particle size below 1 cm exhibit large increases in specific surface (cm^2/g). Taking logarithms on both axes, a linear relationship between specific surface and particle size is apparent (Fig. 6). Based on Figure 6, a relationship between total surface area (m^2) of the material (pyrite, calcite or dolomite) in a mine site (t 1000 t^{-1}) is plotted against the mass (pyrite, calcite or dolomite) assuming various particle sizes (Fig. 7). Based on Figure 7, the total surface area of the minerals (pyrites and carbonates) present in the mine spoil can be predicted. Finally, an empirical relationship can be established between net acidity or alkalinity in the leachate (mg/L) and the ratio of the total pyrite surface area relative to the carbonate surface area. An ideal form of this empirical relationship is shown in Figure 8.

The relationship in Figure 8 reveals three important points. First, there should be a linear relationship between acidity/alkalinity released in the drainage and the ratio of total surfaces. Second, for a particular mine waste, there should be a different relationship between calcite and dolomite, because the kinetics of dissolution of the two minerals differ (38,39). Third, for a particular mine spoil system, there should be a single unique crossover point on the x-axis at $y = 0$. This point reflects the smallest amount of limestone (calcite or dolomite) needed to produce drainage neither alkaline or acidic.

The relationships in Figure 8 are unique to a particular system. The relationship, however, is independent of total quantities of carbonates and pyrite. It is assumed, though, that to produce alkaline drainage there must be at least enough carbonate to equal potential acidity. The relationship in Figure 8 cannot universally apply because the slope of the line is system-dependent. In other words, it depends on the type of carbonate, temperature, reaction rates, hydraulic conductivity, etc. The relationship in Figure 8 can be established for a

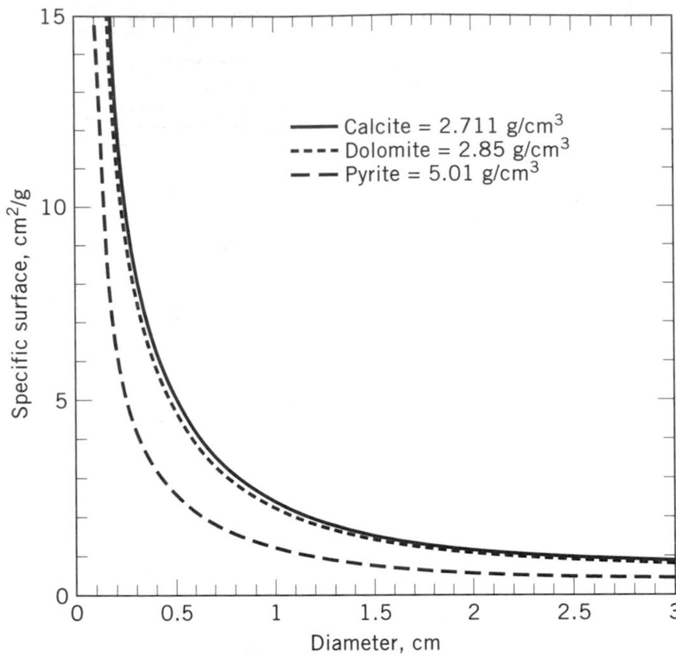

Figure 5. Relationship between specific surface and particle diameter.

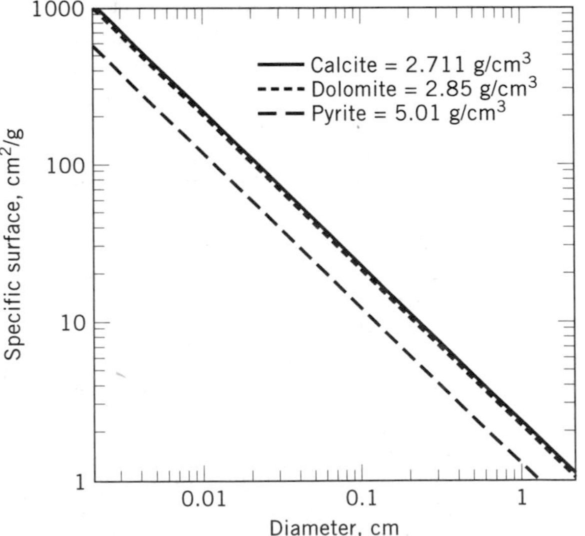

Figure 6. Linearized relationship between specific surface and particle diameter.

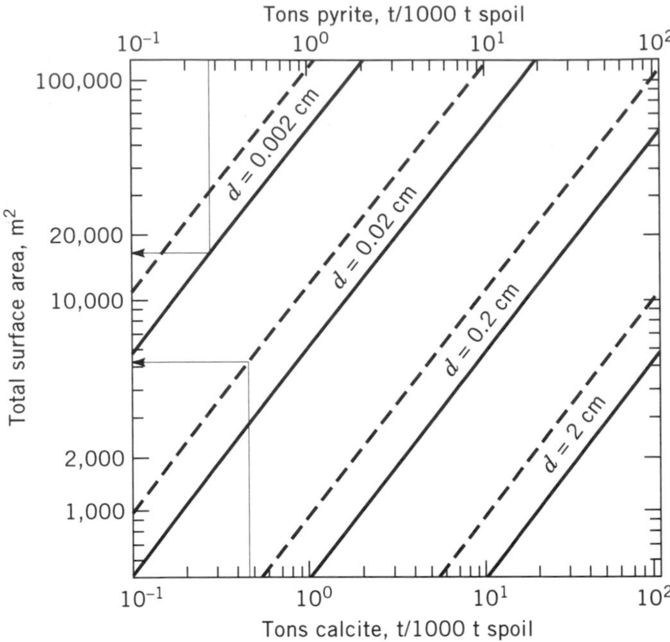

Figure 7. Mass–surface area relationships for spherical calcite and pyrite particles.

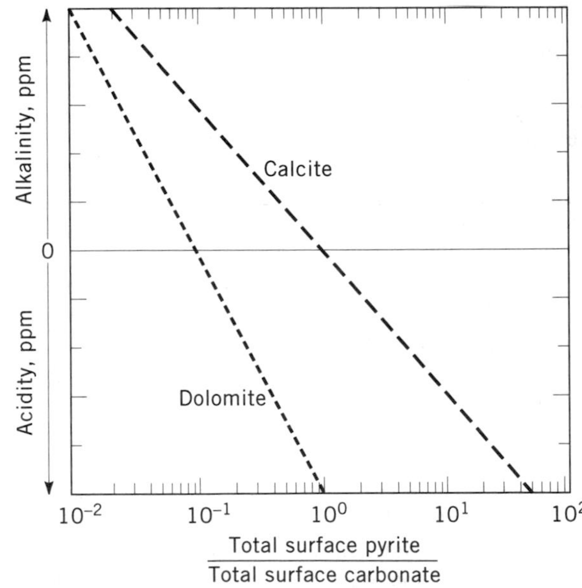

Figure 8. Relationship between acidity or alkalinity and ratio of total surface of the two reactants (carbonate and pyrite).

particular mine site by column leaching studies. In contrast, a correlation between acidity and mass (27), instead of surfaces, always exhibits a "gray zone" because the reactions of carbonates and pyrite(s) are surface-dependent. In the case of carbonates, this is clearly illustrated in Figure 9. Note that the acid dissolution rate constant k of calcite is much greater than the rate constant k of dolomite. Additionally, for particle sizes less than 0.125 mm, the specific surface and k values increase dramatically. It is important to note that the point at which there is an abrupt change in the slope of the line, relating specific surface and particle size, closely represents the particle size considered ideal for agricultural limestone.

In the proposed model it was assumed that pyrite and limestone particles are blended and therefore in close proximity to each other in the mine spoil. Thus, alkalinity or acidity release is strictly surface-controlled. However, where pyrite and carbonates represent two different geologic strata, alkalinity release is described by chemical equilibria whereas acidity release (pyrite oxidation) is described by kinetics. The idea here is that released alkalinity intercepts released acidity and neutralization takes effect assuming that alkalinity is equal to or greater than acidity. A simple but eloquent study demonstrated that, in the

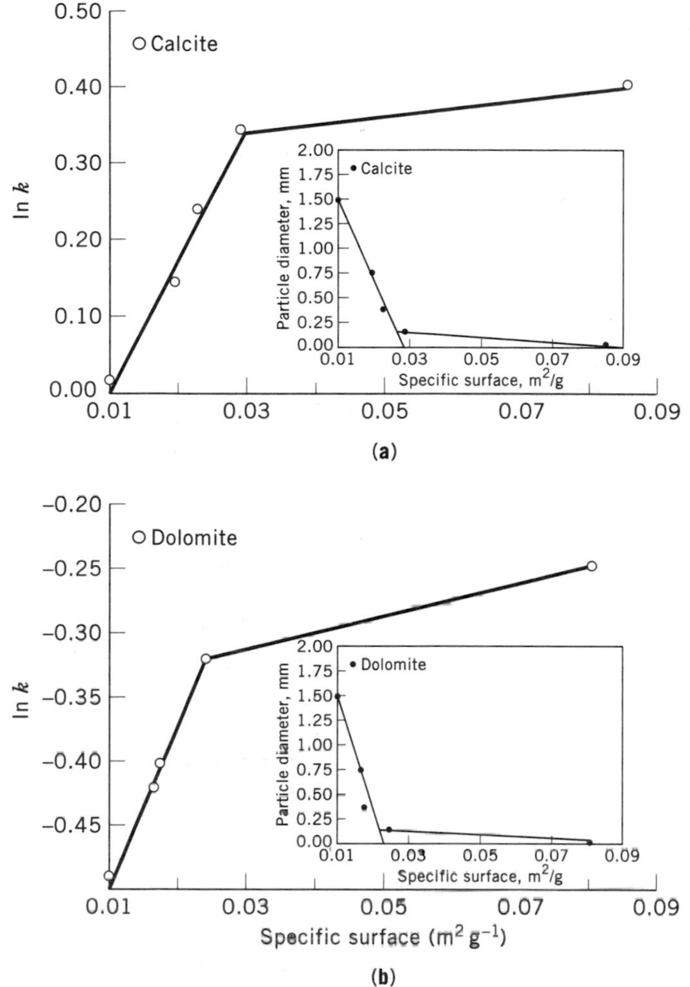

Figure 9. Relationships of the natural logarithm of the first-order rate constant (lnk) of acid dissolution of (**a**) calcite specific surface of calcite and mean particle diameter vs specific surface of calcite, and (**b**) dolomite vs specific surface of dolomite and mean particle diameter vs specific surface of dolomite (Ref. 29; with permission).

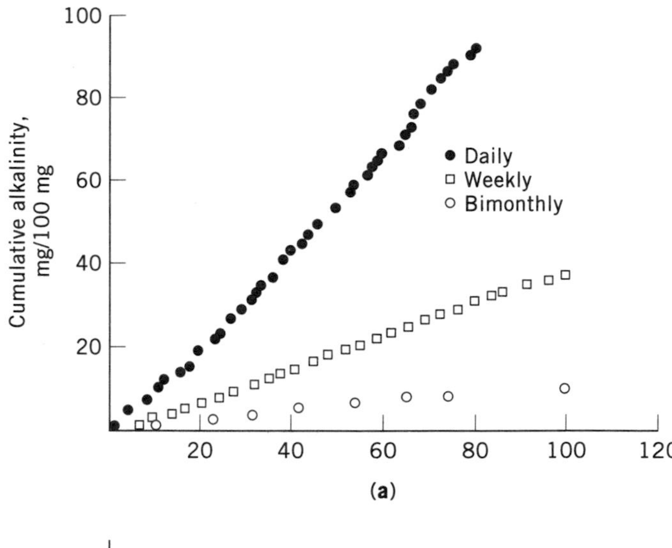

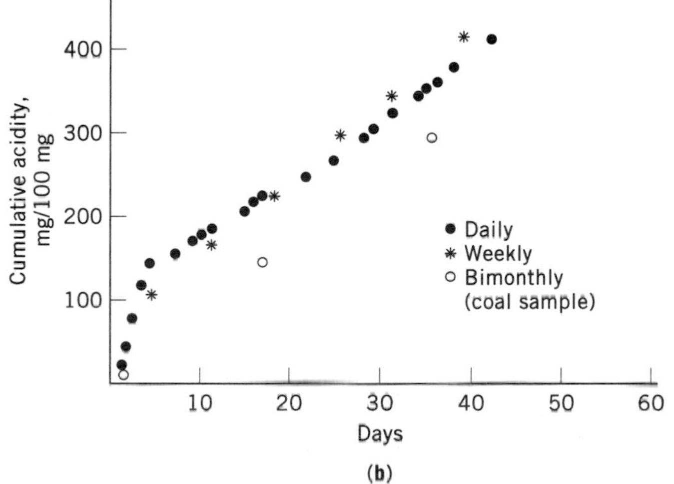

Figure 10. Cumulative (**a**) alkalinity for a limestone sample for various leaching intervals and (**b**) acidity for a pyritic coal sample for various leaching intervals (Ref. 30; with permission).

latter case, frequency of rainfall controls net acidity or net alkalinity (30). The data in Figure 10 show that alkalinity release from limestone under chemical equilibrium considerations depends on leaching intervals. On the other hand, acidity released by pyrite oxidation is virtually independent of leaching intervals. A good example supporting the case where AMD prevention depends on equilibrium and kinetic control is the old groundwater–new groundwater concept.

In the old groundwater–new groundwater concept, AMD accumulates in groundpools. As rainwater infiltrates (new groundwater), the old groundwater (AMD) is displaced in a piston mode, and AMD is discharged. The concept of old groundwater–new groundwater has been observed for pesticide runoff, heavy metal runoff, etc. A schematic representation of the concept is shown in Figure 11.

Deciding which model is most appropriate for predicting and controlling AMD depends on knowing the mechanisms of AMD production. For example, in the case where old-groundwater–new groundwater is the controlling mechanism of AMD production, a large quantity of alkalinity is required to control AMD. Limestone may not necessarily provide this much soluble alkalinity, therefore more soluble bases are needed, eg, $KHCO_3$.

Conventional AMD Prevention Technologies

Inundation. Underwater disposal of pyritic materials has been used by metal mining industries (31). This technology is based on the fact that the diffusion coefficient of O_2 through the covering water table is only 10^{-4} of that through the atmosphere (32) and, therefore, it reduces the rate of pyrite oxidation. In addition, in this reduced environment, the activity of *T. ferrooxidans* is almost nil (33).

The effectiveness of inundation, however, as a practical technology is questionable. For example, field and laboratory research (34,35) show incomplete reduction of acidity in flooded pyritic waste because of the availability of Fe^{3+} as an alternate oxidant (reaction 5). In addition, water table

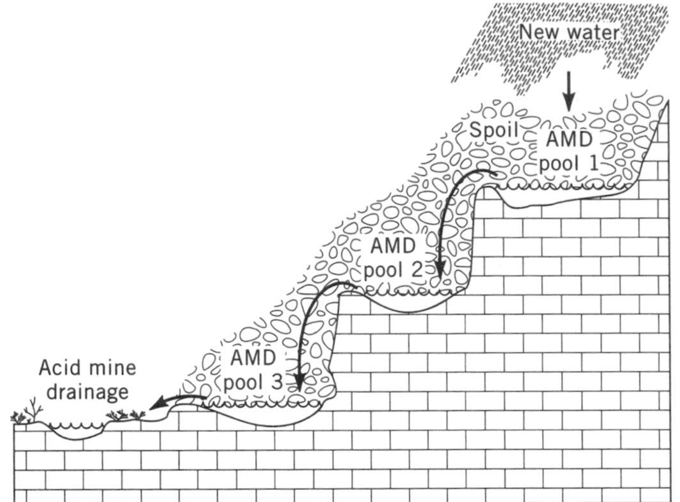

Figure 11. Schematic of old groundwater–new groundwater (infiltrating rainwater) concept explaining AMD production. The schematic shows that as "new water" reaches the AMD pools, a piston-like action pushes AMD out the drain and into the stream. This explains the occurrence of AMD during heavy rain.

fluctuations cause incomplete inundation. A detailed discussion of inundation technology has been presented (3,4).

Alkaline Amendments. The effects of alkaline amendments in controlling AMD in both active and abandoned mines have been discussed in detail (36). Almost all alkaline amendment technologies rely on ABA. There are several materials used as alkaline amendments, such as fluidized-bed combustion (FBC) ash, kiln dust, steelmaking slags, phosphate rock, and limestone.

Limestone is the least expensive and most readily available source of alkalinity. It has a neutralization potential between 75 and 100% and is safe and easy to handle. Alkalinity derived from limestone acts as a pH buffer and AMD neutralizer; it also hydrolyzes most heavy metals, precipitating them as metal hydroxides.

An effective way of utilizing limestone to control AMD is through anoxic drains (37–39). When AMD contacts limestone in an anoxic environment, limestone dissolution and subsequent production of alkalinity proceed readily. Construction details of an anoxic limestone drain are given (3,4).

Another approach to controlling AMD production is alkaline recharge trenches (40,41). This technology does not require physical mixing of the strata with the ameliorator. Effectiveness, however, lasts only as long as neutralizers are sufficient.

In abandoned mines, alkaline injection can be an effective method in preventing AMD. This is relatively new method and requires long-term observations before a conclusion about effectiveness is reached. Preliminary results indicate small improvements in water quality after injecting coal combustion residues into buried pods of pyritic materials in abandoned mines (42,43).

Constructed Wetlands. Constructed wetlands are man-made ecosystems specifically designed for wastewater treatment and located at sites where natural wetlands did not exist, at least at the time of construction. Constructed wetlands are used successfully to remove metals from AMD (44). Mechanisms of metal removal within wetlands include (*1*) precipitation of metals as oxides and oxyhydroxides or sulfides; (*2*) complexation of metals by organic materials; (*3*) direct uptake by living organisms and plants; (*4*) adsorption of metals by ferric oxyhydroxides; and (*5*) physical filtration of suspended metal colloids onto algal mats. Other beneficial reactions in wetlands include generation of alkalinity by microbial mineralization of organic matter, microbial dissimilatory reduction of iron oxyhydroxides and/or sulfate, and chemical dissolution of carbonates.

There are two types of constructed wetlands: (*1*) *surface flow* or *aerobic* wetlands, and (*2*) *subsurface flow* or *anaerobic* wetlands. In surface flow wetlands, AMD treatment results primarily from metal oxidation which causes metal precipitation and physical retention. In subsurface flow wetlands, AMD treatment results from metal oxidation in the shallow aerobic surface layer and from reduction reactions in the anaerobic subsurface which induces metal removal as metal sulfide precipitates. In addition, the capacity for alkalinity generation is greater in subsurface flow wetlands. For these reasons, subsurface flow wetlands present a much greater metal retentive capacity.

Bactericides. Anionic surfactants, such as sodium dodecylbenzene sulfonate and sodium lauryl sulfate, are used to control bacteria that catalyze the conversion of Fe^{2+} to Fe^{3+}, thereby controlling pyrite oxidation. An application rate of 225 kg/ha on a refuse site reduced the acidity and iron by about 79 and 82%, respectively (36). Bactericides have also been successful in reducing acidity in silver mining (45).

Although bactericides are relatively easy to handle and apply, wide use of such substances is limited for two specific reasons. First, anionic surfactants are very soluble and move with water, thus repeated treatments are required to prevent bacterial repopulation. Second, anionic surfactants are adsorbed on the surfaces of minerals and may not reach the pyrite-bacteria interface (46,47).

ACKNOWLEDGMENTS

The authors wish to thank E. Portig and M. Vandiviere for their help in producing this manuscript. Some work in this manuscript was funded by the U.S. Dept. of Energy (Grant Number DE-FG22-95PC95226) and the Kentucky Agricultural Experimental Station and published with the approval of the Director. The views and conclusions contained in this document do not necessarily represent the official policies or recommendations of the U.S. Dept. of Energy or of the U.S. Government.

BIBLIOGRAPHY

1. *CIM Bull.* **86,** 29 (1993).
2. T.R. Wildeman, "Drainage from Coal Mines" in *Geology in Coal Resource Utilization,* Techbooks, Fairfax, Va., 1991, p. 499.
3. V.P. Evangelou, *Pyrite Oxidation and Its Control,* CRC/Lewis Press, Boca Raton, Fla., 1995, and references therein.
4. V.P. Evangelou and Y.L. Zhang, *Crit. Rev. Environ. Sci. Technol.* **25,** 141–199 (1995), and references therein.
5. G.W. Luther III, *Geochim. Cosmochim. Acta* **51,** 3193–3199 (1987).
6. C.O. Moses, D.K. Nordstrom, J.S. Herman, and A.L. Mills, *Geochim, Cosmochim. Acta* **51,** 1561–1571 (1987).

7. W. Stumm and J.J. Morgan, *Aquatic Chemistry,* 2nd ed., John Wiley & Sons, Inc., New York, 1981, pp. 469–471.
8. E.E. Smith, K. Svanks, and K. Shumate, "Sulfide to Sulfate Reaction Studies" in *Proceedings Second Symposium on Coal Mine Drainage Research,* Bituminous Coal Research Institute, Monroeville, Pa., 1968, pp. 1–11.
9. M.A. McKibben and H.L. Barnes, *Geochim. Cosmochim. Acta* **50,** 1509–1520 (1986).
10. R.V. Nicholson, "Iron-Sulfide Oxidation Mechanisms: Laboratory Studies" in J.L. Jambor and D.W. Blowes, eds., *Short Course Handbook on Environmental Geochemistry of Sulfide Mine Wastes,* Mining Association Canada, Waterloo, Ontario, Canada, 1994, Vol. 22, pp. 163–183.
11. R.V. Nicholson, R.W. Gillham, and E.J. Reardon, *Geochim. Cosmochim. Acta* **52,** 1077–1085 (1988).
12. J.M. Bigham, U. Schwertmann, L. Carlson, and E. Murad, *Geochim. Cosmochim. Acta* **54,** 2743–2758 (1990).
13. J.M. Bigham, U Schwertmann, and M. Wolf, *Geochim. Cosmochim. Acta* **60,** 2111–2121.
14. P.C. Singer and W. Stumm, *Science* **167,** 1121–1123 (1970).
15. P.R. Dugan, *Ohio J. Sci.* **75,** 266–279 (1975).
16. D.K. Nordstrom, *Geochim. Cosmochim. Acta* **46,** 681–692 (1982).
17. D.B. Jaynes, A.S. Rogowski, and H.B. Pionke, *Water Resour. Res.* **20,** 233–242 (1984).
18. T.A. Hood, *The Kinetics of Pyrite Oxidation in Marine Systems,* Ph. D. Thesis, University of Miami, Coral Gables, Fla., 1991.
19. V.P. Evangelou, "A Silica/Fly Ash-Based Technology for Controlling Pyrite Oxidation", *Semi-Annual Technical Progress Report* submitted to the U.S. Dept. of Energy, Pittsburgh, Pa., 1996.
20. G.W. Luther III, J.E. Kostka, T.M. Church, B. Sulzberger, and W. Stumm, *Marine Chem.* **40,** 81–103 (1992).
21. A.E. Torma, "Leaching of Metals" in H.J. Rehm and G. Reed, eds., *Biotechnology,* VCH Verlagsgesellschaft, Weinheim, Germany, 1988, Vol. 6B, pp. 367–399.
22. A.P. Mehta and L.E. Murr, *Hydrometallurgy* **9,** 235–256 (1983).
23. R.T. Lowson, *Chem. Rev.* **82,** 461–497 (1982).
24. C.L. Wiersma and J.D. Rimstidt, *Geochim. Cosmochim. Acta* **48,** 85–92 (1984).
25. B. Elberling, R.V. Nicholson, and D. David, *Nordic Hydrol.* **24,** 323–338 (1983).
26. R.V. Nicholson, R.W. Gillham, and E.J. Reardon, *Geochim. Cosmochim. Acta* **54,** 395–402 (1990).
27. A.A. Sobek, W.A. Schuller, J.R. Freeman, and R.M. Smith, *Field and Laboratory Methods Applicable to Overburdens and Mine Soils,* U.S. Environmental Protection Agency, Cincinnati, Ohio, 1978.
28. K.B. Brady, E.F. Perry, R.L. Beam, D.C. Bisko, M.D. Gardner, and J.M. Tarantino, "Evaluation of Acid-Base Accounting to Predict the Quality of Drainage at Surface Coal Mines in Pennsylvania" in *Proceedings of the International Land Reclamation and Mine Drainage Conference on the Abatement of Acidic Drainage,* Pittsburgh, Pa., 1994, pp. 138–147.
29. K.S. Sajwan, V.P. Evangelou, and J. Lumbanraja, *Soil Sci.* **151,** 444–451 (1991).
30. G. Geidel, *Reclam. Rev.* **3,** 101–107 (1980).
31. G.M. Ritcey, "Deep Water Disposal of Pyritic Tailings" in *Proceedings of the 2nd International Conference on the Abatement of Acidic Drainage,* Montreal, Quebec, 1991, Tome 1 pp. 421–442.
32. Ohio State University Research Foundation, "Acid Mine Drainage Formation and Abatement" in *Water Pollution Control Research,* Series DAST-42-14210 FPR-04/71. USEPA, Washington, D.C., 1971.
33. R.L.P. Kleinmann and D.A. Crerar, *Geomicrobiol. J.* **1,** 373–388 (1979).
34. J.W. Foreman, "Evaluation of Mine Sealing in Butler County, Pennsylvania" in *4th Symposium Coal Mine Drainage Research,* Louisville, Ky., 1972, pp. 83–95.
35. G.R. Watzlaf, "Pyrite Oxidation in Saturated and Unsaturated Coal Waste" in *Proceedings of 9th National Meeting of the American Society for Surface Mining and Reclamation,* Duluth, Minn., 1992, pp. 191–205.
36. P. Ziemkiewicz and J. Skousen, *Green Lands,* **26,** 45–52 (1996).
37. R.W. Nairn, R.S. Hedin, and G.R. Watzlaf, "Generation of Alkalinity in an Anoxic Limestone Drain," in *Proceedings of the 9th Annual Meeting of the American Society for Surface Mining and Reclamation,* Duluth, Minn., 1992.
38. G.R. Watzlaf and R.S. Hedin, "A Method for Predicting Alkalinity Generated by Anoxic Limestone Drains," in *Proceedings of the 1993 West Virginia Surface Mine Drainage Task Force Symposium,* Morgantown, W. Va., 1993.
39. G.A. Brodie, C.R. Britt, T.M. Tomaszewski, and H.N. Taylor, "Use of the Passive Anoxic Limestone Drains to Enhance Performance of Acid Drainage Treatment Wetlands," in W. Oaks and J. Bowden, eds., *Proceedings Reclamation 2000: Technologies for Success,* Durango, Colo., 1991, pp. 211–222.
40. K.J. Ladwig, P.M. Erickson, and R.L.P. Kleinmann, "Alkaline Injection: An Overview of Recent Work," in *Control of Acid Mine Drainage,* Bur. of Mines IC 9027. USDA, Bureau of Mines, Pittsburgh, Pa., 1985, p. 35–40.
41. F.T. Caruccio and G. Geidel, "The Prediction of Acid Mine Drainage from Coal Strip Mines" in Proceeding on the Reclamation of Abandonment Acid Spoils, Missouri Department of Natural Resources and Land Reclamation Commission, 1985.
42. A.G. Kim and T.E. Ackman, "Disposing of Coal Combustion Residues in Inactive Surface Mines: Effects on Water Quality," in International Land Reclamation and Mine Drainage Conference, USDI, Bureau of Mines SP 06B-94, Pittsburgh, Pa., 1994, pp. 228–236.
43. J. Schueck, T.E. Ackman, and B. Scheetz, "Acid Mine Drainage Abatement Using Fluidized Bed Combustion Ash Grout After Geophysical Site Characterization," in International Land Reclamation and Mine Drainage Conference, USDI, Bureau of Mines SP 06B-94, Pittsburgh, Pa., 1994, pp. 218–227.
44. D.A. Hammer, ed., *Constructed Wetlands for Wastewater Treatment: Municipal, Industrial and Agricultural,* Lewis Publishers, Chelsea, Mich., 1989.
45. D. Parisi, J. Horneman, and V. Rastogi, Use of Bactericides to Control Acid Mine Drainage from Surface Operations," in International Land Reclamation and Mine Drainage Conference, USDI, Bureau of Mines SP 06B-94, Pittsburgh, Pa., 1994, pp. 319–325.
46. P.M. Erickson and K.J. Ladwig, "Control of Acid Formation by Inhibition of bacteria and by Coating Pyritic Surfaces", Final Report to the West Virginia Department of Energy, Division of Reclamation, Charleston, W. Va., 1985.
47. M. Shellhorn and V. Rastogi, "Practical Control of Acid Mine Drainage Using Bactericides," in *Proceedings of 6th West Virginia Surface Mine Drainage Task Force Symposium,* Morgantown, W. Va., 1985.

ACID RAIN

WILLIAM SMITH
Yale University
New Haven, Connecticut

Acid rain is appropriately described as an old environmental problem with a new image. Acid rain, more than any other

Plate 1. Effect of acid rain on forests.

environmental contaminant, has focused societal concern on ecosystem toxicology. Robert Angus Smith (1) is credited with first using the term acid rain in 1872. During the first half of the twentieth century numerous European investigators added insight to this emerging environmental challenge. Oden, a Swedish scientist, is appropriately credited with detailing the regional scale nature (rain-out hundreds of kilometers downwind from the source) of the acid rain problem through a series of publications and lectures between 1967 and 1977. North American interest in acid rain was significantly advanced by the studies of the Sudbury area smelters during the 1960s (2,3). The longest continuous record of rain chemistry in the United States has been maintained at the Hubbard Brook Experimental Forest (U.S. Department of Agriculture, Forest Service) in central New Hampshire since 1963 (4). The federal governments of both Canada and the United States initiated systematic rain chemistry monitoring in 1976 (Canadian Network for Sampling Precipitation) and 1978 (National Atmospheric Deposition Program), respectively. Results of the latter program reveal that acidity of precipitation in the eastern portion of the United States is approximately 10 times more acid than the western portion (Fig. 1).

Over the 150-year history of acid rain research, concerns have been raised in regard to the ability of acid rain to erode and corrode buildings, sculpture, monuments, and other structures; to reduce visibility; to impair human health; and to impact adversely agricultural as well as stream, lake, and forest ecosystems. Acid rain can result in significant long-term deterioration of limestone, marble, bronze, unpainted galvanized steel, and numerous paints. In regard to visibility, it has been concluded that more than 50% of the light extinction in rural areas of the eastern United States, on an annual average basis, is due to sulfate particles in the atmosphere. A significant portion of these sulfate particles are the result of acid precursors in the atmosphere. Evidence generated from epidemiological, human clinical, and animal toxicological studies suggest that a high enough exposure to acidic aerosols may produce a number of direct and indirect adverse health effects in sensitive human populations. High risk populations include the elderly and the very young and anyone with chronic lung dysfunction such as chronic bronchitis, emphysema, or asthmatic disease. Agricultural ecosystems are thought to be generally resistant to acid rain damage due to the high buffering capacity typical of agricultural soils. Known effects on less well-buffered natural ecosystems such as streams, lakes, and forests, however, may range from beneficial (fertilization from nitrogen, phosphorus, calcium, potassium, and sulfur) to adverse (toxicity from hydrogen, aluminum, nitrate, sulfate, or heavy metals) (5,6).

In recognition of the expansive size of freshwater and forest ecosystems, the multiple values of these systems to people, and the potential for adverse interaction of regional-scale air pollutants on ecosystem health, the current understanding of the relationship between forests, lakes, streams, and acid rain are described herein.

CHARACTERISTICS OF ACID RAIN

At room temperature, pure water dissociates to produce equal amounts (0.0000001 g/L) of hydrogen (H^+) and hydroxyl (OH^-) ions. When these ionic concentrations are equal, the solution is said to be neutral. The concentration of hydrogen ions in a solution is represented by the pH scale. The pH number reflects the negative logarithm of the hydrogen ion concentration. At neutrality, when hydrogen and hydroxyl concentrations are

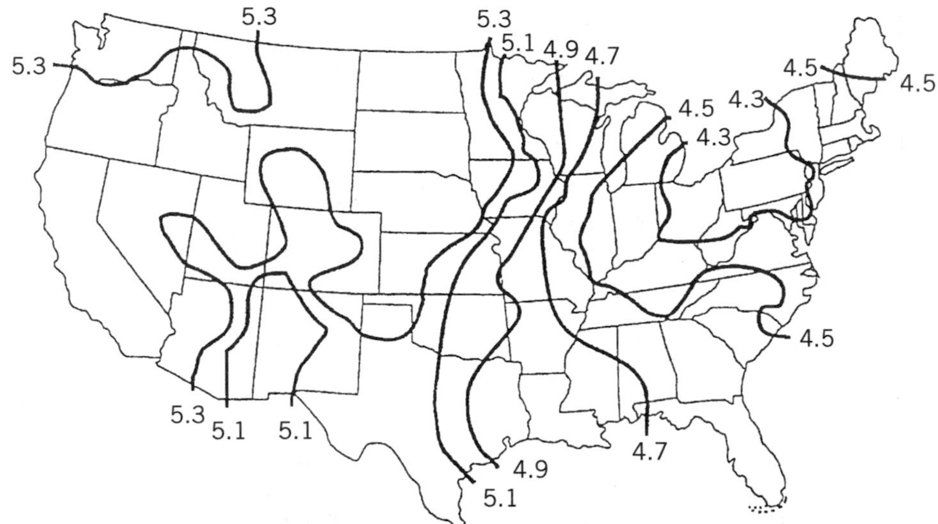

Figure 1. Average 1990 annual hydrogen ion concentrations in precipitation, expressed as pH and adjusted for amount of precipitation (5).

equal, pH is 7.0. Acid solutions in water are defined as those solutions that have hydrogen ion concentrations greater than the hydroxyl concentration (pH < 7). Keep in mind that the pH scale is logarithmic and that a change from pH 7.0 to pH 6.0 represents a 10-fold increase in hydrogen ion concentration and, therefore, a 10-fold increase in acidity.

Natural rain, including precipitation in relatively clean or unpolluted regions, is naturally acidic, with a pH in the range of 5.0 to 6.0. This natural acidity results from the oxidation of carbon oxides and the subsequent formation of carbonic acid. Formic and acetic acids, originating primarily from natural sources, may also contribute minor amounts of acidity to precipitation.

In regions downwind from electric-generating power stations employing fossil fuels, industrial regions, or major urban centers, precipitation can be acidified below pH 5.0. Precipitation with a pH less than 5.0 is designated acid rain. This human-caused acidification of precipitation results primarily from the release of sulfur dioxide (SO_2) and nitrogen oxides (NO and NO_2) from smokestacks and tailpipes. The sulfur and nitrogen oxides are subsequently oxidized to sulfate (SO_4^{2-}) and nitrate (NO_3^-), hydrolyzed, and returned to earth as sulfuric (H_2SO_4) and nitric (HNO_3) acids. This additional acidification of precipitation by human activities can readily reduce the pH of precipitation in downwind regions to between 4.0 and 5.0 on an annual average basis. Individual precipitation events that have a pH in the 3.0 to 4.0 range are not uncommon.

The atmosphere deposits acidity onto the landscape both during and in between precipitation events. In the latter case, termed dry deposition in contrast to wet deposition, the acids are delivered in the gas phase or in association with fine particles (aerosols). Acid deposition is a term that includes acid delivery in the form of precipitation (rain, snow, fog, and cloud moisture) plus dry deposition. In view of the importance of both wet and dry deposition in acid transfer from the atmosphere to the biosphere, acid deposition is a much more appropriate descriptor than acid rain.

The Hubbard Brook Experimental Forest, which is part of the White Mountain National Forest in central New Hampshire (45°56′ N, 71°45′ W), has been a focal point for North American study of acid deposition. Despite the fact that Hubbard Brook is more than 100 km from any large urban-industrial area, the average annual pH since 1963 has generally fallen in the 4.0–4.4 range. The lowest annual average pH recorded was 4.03 and occurred in 1971. Individual storm pH values have ranged from 3.0 to 5.95.

Precipitation solutions are electrically neutral, requiring the sum of cations (eg, H^+) to be balanced by the sum of anions (eg, NO_3^-). Because of the initiation of precipitation chemistry measurements at Hubbard Brook in 1964, precipitation cations have been dominated by hydrogen (H^+) and ammonium (NH_4^+), and anions dominated by sulfate (SO_4^{2-}) and nitrate (NO_3^-). Decreases in the concentration of sulfate in the precipitation at Hubbard Brook since 1964 have been attributed to decreases in sulfur dioxide emissions in the upwind areas of the U.S. Northeast and Midwest.

It is important to recognize that both the positive and negative effects of acid deposition on ecosystems involve the consideration of hydrogen ions, sulfate ions, nitrate ions, and heavy-metal ions in acid deposition. Selected forests are judged to be at a special risk to the adverse effects of acid deposition. These forests receive especially high exposure to hydrogen, sulfate, or nitrate ions due to their proximity to primary sources of sulfuric and nitric acids or because of their elevation. High elevation, or montane, forest ecosystems, eg, the high elevation coniferous forests along the Appalachian Mountain chain in the eastern United States, are exposed to high levels of acid deposition because of frequent occurrence of fog immersion and cloud-water deposition. Concentrations of ionic species in fog may be 10 to 15 times that in low elevation rain. The duration of exposure to fog or cloud water events may be longer than rain events and thus result in greater deposition of ionic species to high elevation forests. Repeated alteration of high elevation precipitation events with periods of evaporation, facilitated by high elevation winds, may further act to concentrate ionic species on the surfaces of forests or forest soils. In August 1984, a wide-area cloud-fog event was simultaneously recorded in several northern U.S. forest regions. This event recorded an extremely low pH (2.80–3.09) and concentrations of sulfate and nitrate 7 to 43 times greater than previously recorded in cloud or fog water in the eastern United States (7).

ADVERSE IMPACT OF ACID DEPOSITION ON FOREST SYSTEMS

Assessment of air pollution impact on forest systems is extremely challenging for a variety of reasons; three of the most important are forest system variability, deficiency of understanding of ecosystem- and landscape-scale phenomena, and large variation in system exposure to acid deposition. In general, forest disturbance from air pollutants is exposure related, and dose-response thresholds for a specific pollutant are different among the various organisms of the ecosystem. Ecosystem response is, therefore, a complex process. In response to low exposure to air pollution, the vegetation and soils of an ecosystem function as a sink or receptor. When exposed to intermediate loads, individual plant species or individual members of a given species may be subtly and harmfully affected by nutrient stress, impaired metabolism, predisposition to entomological or pathological stress, or direct induction of disease. Exposure to high deposition may induce acute morbidity or mortality of specific plants. At the ecosystem level, the impact of these various interactions would be highly variable. In the first situation, the pollutant would be transferred from the atmosphere to the various elements of the biota and to the soil. With minimal physiological effect, the impact of this transfer on the ecosystem could be undetectable (innocuous effect) or stimulatory (fertilizing effect). If the effect of the pollutant dose on some component of the biota is harmful, then a subtle adverse response may occur. The ecosystem impact in this case could include reduced productivity or biomass, alterations in species composition or community structure, or increased morbidity. Under conditions of high dose, ecosystem impacts may include gross simplification, impaired energy flow and biogeochemical cycling, changes in hydrology and erosion, climatic alteration, and major impacts on associated ecosystems.

For North American forests, it is generally concluded that acid deposition influences on forest systems are neutral, ie, no adverse effects can be discriminated from natural forest dynamics. Actual effects may be slightly stimulatory (eg, nitrogen fertilization via nitrate input) or contributory to multiple-factor forest stress (eg, in high elevation, high risk (montane) forest ecosystems). In the latter case, acid deposition is presumed to be highly interactive with other stresses, subtle in manifestation, and long-term (several decades) in development. The primary hypotheses for these subtle effects are listed below. Not all of these hypotheses are supported by equal scientific evidence. The first six hypotheses are best understood and supported by the greatest evidence; they are summarized in the following sections.

Tree Population Interaction	Forest Ecosystem Perturbation
Increased rate of soil acidification causes altered nutrient availability and root disease.	Population dynamics, tree competition, and species composition.
Cation nutrients are leached from foliage to throughfall and stem flow.	Biogeochemical cycle rates.
Cation nutrients are leached below soil horizons of active root uptake.	Biogeochemical cycle rates.
Increased available soil aluminum results in fine-root morbidity.	Population dynamics, tree competition, and species composition.
Increased available heavy-metal and hydrogen ion concentrations in soil result in enhanced root uptake or impact on soil microbiota.	Decomposer impact, biogeochemical cycle rates, and species composition.
Increased nitrogen (sulfur) deposition alters nitrogen (sulfur) cycle dynamics.	Biogeochemical cycle rates.
Deposition causes alteration of carbon allocation to maintenance respiration or repair or to aboveground instead of belowground tissues.	Productivity and energy storage.
Deposition increases or decreases phytophagous arthropod activity.	Consumer impact and insect population dynamics.
Deposition increases or decreases microbial pathogen activity.	Consumer impact and pathogen population dynamics.
Deposition increases or decreases abiotic stress influence (temperature, moisture, wind, and nutrient stresses).	Population dynamics, tree competition, and species composition.
Increased soil weathering alters cation availability.	Biogeochemical cycle rates.
Deposition increases or decreases microbial symbioses.	Productivity and energy storage.
Deposition impacts one or more processes of reproductive or seedling metabolism.	Population dynamics, tree competition, and species composition.
Deposition impacts a critical metabolic process, eg, photosynthesis, respiration, water uptake, translocation, or evapotranspiration.	Population dynamics, tree competition, and species composition.

Table 1. Soil Acidity Designations by pH

Descriptor	pH
Extremely acid[a]	<4.5
Very strongly acid[a]	4.5–5.0
Strongly acid[a]	5.1–5.5
Medium acid[a]	5.6–6.0
Slightly acid[a]	6.1–6.5
Neutral	6.6–7.3
Mildly alkaline	7.4–7.8
Moderately alkaline	7.9–8.4
Strongly alkaline	8.5–9.0
Very strongly alkaline	>9.1

[a] Forest soils.

SOIL ACIDIFICATION

Natural processes make forest soils acid (Table 1). These processes act by controlling the chemistry of the soil cation-exchange complex. This complex consists of negative charges located on clay minerals in the soil or on soil organic matter. On clay minerals, negative charges generally result from substitution, within the mineral lattice, of a cation of lower positive charge for one of higher charge. On organic matter, negative charges result from the ionization of hydrogen ions from carboxyl, phenol, and phenol groups. In acid mineral soils the cation-exchange complex is typically dominated by aluminum species, eg, $Al^{3+}Al(OH)^{2+}$, formed by the dissolution of soil minerals. In acid organic soils, the hydrogen ion may dominate the cation-exchange complex.

Processes that acidify forest soils include those that increase the number of negative charges, such as organic matter accumulation or clay formation, and those that remove basic cations, such as leaching of bases in association with an acid anion. Weathering by carbonic acid, organic acids (podzolization), humification, and cation uptake by roots all increase the negative charge of forest soils.

It has been noted that soil acidification is a complex of processes that cannot be quantitatively described by any single index (8). There has been an emphasis on the utility of using capacity and intensity factors. Capacity refers to the storage of hydrogen ions, aluminum ions, or base cations on the soil-exchange complex or in weatherable minerals. Intensity refers to the soil solution concentration at any point in time; in the case of the hydrogen ion, it refers to the soil solution pH. In forest regions receiving acid deposition from the atmosphere, the most likely effect on the capacity will be an increase in the exchange acidity and a reduction of exchangeable bases. The former is increased directly by hydrogen ion input or, more likely, by increasing exchangeable aluminum through the reaction of hydrogen ions with soil minerals. The latter (reduction of exchangeable bases) results via replacement of base cations on the exchange complex by aluminum species. The base cations are subsequently leached from upper soil horizons in association with strong acid anions.

The evidence for soil acidification (ie, lowering of pH) of forest soils at the present rates of acid deposition in North America is not great. Forest soils at greatest risk to pH reduction from acid deposition are restricted to those limited soil types characterized by no renewal by fresh soil deposits, low cation-exchange capacity, low clay and organic matter content, low sulfate absorption capacity, high input of acidic deposition without significant base cation deposition, high present pH (5.5–6.5), and deficiency of easily weatherable materials to a 1-m depth. In addition, these high risk forest systems would need to be exposed to significant levels of acid deposition for decades. In Sweden, where some forests have been exposed to acid deposition for over six decades, acidification of forest soils has been reported.

CATION DEPLETION FROM FOLIAGE

Vegetative leaching refers to the removal of substances from plants by the action of aqueous solutions such as rain, dew, mist, and fog. Precipitation washout of chemicals from trees has been appreciated for some time. Inorganic chemicals leached from plants include all the essential macroelements. Potassium, calcium, magnesium, and manganese are typically leached in the greatest quantities. A variety of organic compounds, including sugars, amino acids, organic acids, hormones, vitamins, and pectic and phenolic substances, are also leached from vegetation. As the maturity of leaves increases, susceptibility to nutrient loss via leaching also increases and peaks at senescence. Leaves from healthy plants are more resistant to leaching than leaves that are injured, infected with microbes, infested with insects, or otherwise under stress.

Deciduous trees lose more nutrients from foliage than do coniferous species during the growing season. Conifers, however, continue to lose nutrients throughout the dormant season. The stems and branches of all woody plants lose nutrients during both the growing and dormant seasons.

The mechanism of leaching is presumed to be primarily a passive process. Cations are lost from free-space areas within the plant. Under uncontaminated, natural environmental conditions, little if any cations are thought to be lost from within cells or cell walls. It has been demonstrated that leaching of cations on the leaf surface involves exchange reactions in which cations on exchange sites of the cuticle are exchanged by hydrogen from leaching solutions. Cations may move directly from the translocation stream within the leaf into the leaching solution by diffusion and mass flow through areas devoid of cuticle.

Pollution exposure may predispose foliage to leaching loss by cuticular erosion, membrane dysfunction, or metabolic abnormality (Fig. 2). Epicuticular wax, the outermost layer of plant leaves, consists of a complex and variable mixture of long-chain alkanes, alkenes, aromatic hydrocarbons, fatty acids, ketones, aldehydes, alcohols, and esters. The composition and integrity of this layer is strongly controlled by climate, foliar age, and air contaminants. Because acid precipitation may increase hydrogen ion activity by one or two orders of magnitude (pH 5.6–3.6), due to increasing concentrations of sulfuric and nitric acids in precipitation and because damage to cuticles and epidermal cells may also result from exposure to acid precipitation, the potential for accelerated leaching under this stress may be important.

The movement of nutrient cations via leaching in forest soil profiles is an extremely important component of forest nutrient cycling. The evidence that has been provided by numerous experiments subjecting soil lysimeters to natural or artificially acidified precipitation indicates a potential for a meaningful acid precipitation

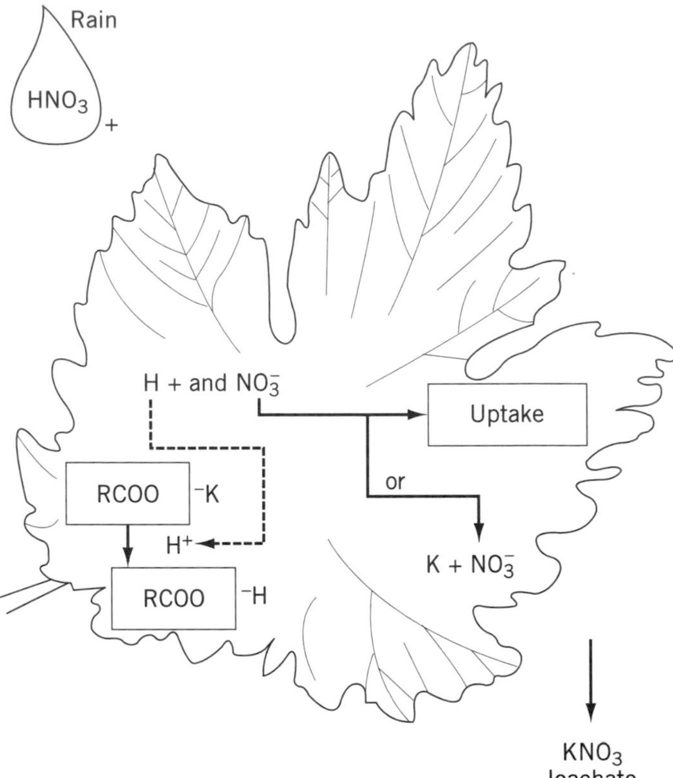

Figure 2. Foliar leaching can result from the deposition of nitric acid. Hydrogen ions from nitric acid may displace nutrient cations held on leaf wall exchange sites. Nitrate ions from the acid, if not taken up by the leaf microorganisms, may be available to combine with potassium and to remove this nutrient from the leaf. If potassium ions are not resupplied by the roots, forest trees may suffer from nutrient stress (9).

influence on the soil leaching process. The threshold for significant increases in the rate of movement of calcium, potassium, and magnesium appears to require precipitation in the pH range of 3–4 for most systems examined. For certain forests, eg, subalpine balsam fir in New England, the threshold of increased leaching may be higher and in the range of pH 4.0–4.5.

Despite the efficacy of SO_4^{2-} leaching in selected soils, it must be realized that forest soils vary greatly in their susceptibility to it (depending on chemistry, depth, and parent and bedrock material). Mineral weathering and deep rooting constantly replenish base cations for tree growth, and even for sensitive soils, leaching losses affect cation reserves only on an extended (multiple decade) time scale.

Forest soils have been judged more vulnerable to leaching influence by acid precipitation than are agricultural soils. Forest soils supporting early regeneration following harvest or severe natural disturbance events may be especially vulnerable to an adverse impact on nutrient cycling by acid rain, because the system controls on nutrient conservation are weakest at such times. The bulk of evidence, however, remains consistent with the conclusion that unless the acidity of precipitation increases substantially or the buffering capacity of forest soil declines significantly, acid rain influence will not quickly or dramatically alter the productivity of most temperate forest soils via cation depletion. Important declines in the total calcium (Ca^{2+}) and magnesium (Mg^{2+}) concentrations or surface organic soil at the Hubbard Brook Experimental Forest, however, have been measured since 1970. In addition, annual forest biomass has declined since 1987. It is possible that the loss of these important forest tree nutrients is causing, or contributing to, the reduced growth rate of this northern hardwood forest.

FRESHWATER SYSTEMS

The pH range for undisturbed lakes and streams with full compliments of native biota, especially selected finfish species important to humans, typically is within the range of 6–8. Lakes and streams with a pH <6 do occur naturally. Land use practices, for example, conversion from agriculture to forest in the watershed of a lake can, over time, also cause lake acidity to increase. Aquatic resources acidified from the input of acid rain are also well documented, particularly in the northern United States, eastern Canada, and Scandinavia.

Fortunately, most surface waters in the United States have sufficient buffering (acid-neutralizing) capacity to be resistant to acidification via atmospheric deposition. In the instance of high acid deposition inputs and poorly buffered aquatic systems, however, declines in water column pH can be caused by acid rain. In the mid-1980s a National Surface Water Survey was conducted in areas of the United States known to contain lakes and streams with little capacity to neutralize acids. This survey identified the Adirondacks and mid-Atlantic Highlands region as having both sensitive aquatic resources and high levels of acid deposition (5).

The adverse response of aquatic biota to acidification is very well documented (10,11). Sensitive species may be stressed or lost at small increases in acidity. Acid-sensitive mayfly and stonefly species can be impacted at pH levels near 6.0 and sensitive finfish species, for example fathead minnows, can be lost at pH 5.6–5.9. Acid-sensitive phytoplankton, zooplankton, and benthic macroinvertibrate species may also be lost. Acid-tolerant species may increase significantly with acidification. Survival of fish in acidic waters is primarily related to pH level, inorganic chemistry of aluminum, and the concentration of calcium. Other relevant considerations include food web effects, spatial and temporal variation in exposure to acidity and aluminum at various life stages, and behavioral avoidance.

PROGNOSIS

Across a landscape as large as the United States, some forests undoubtedly benefit from fertilization by nutrients contained in acid deposition. Most forests are not detectably influenced positively or negatively by acid deposition due to low input or inability to separate acid rain responses from the background noise of forest processes, and some high risk forests, eg, those receiving cloud moisture acidity along the Appalachian Mountain chain, may be subjected to chronic and interactive stress associated with nutrient depletion and aluminum toxicity. Lakes and streams in areas where geologic substrates and associated soils do not provide acid-neutralizing materials can be made more acidic by acid rain. Fish losses attributable to acidification have been documented for some acidic surface waters in the Adirondacks, New England, and the mid-Atlantic Highlands.

As the twenty-first century nears, it is increasingly clear that humans have the ability to influence the global environment, especially the global atmosphere. Emphasis on the atmosphere is not surprising when it is recognized that this medium integrates human activities quite independently of political boundaries. Increasing evidence emphasizes the ability of human beings to influence specific global-scale and regional-scale atmospheric processes. At the global scale, some of the most important include trace gas loading of the troposphere with the radiatively active species carbon dioxide, methane, nitrous oxide, and halocarbons, resulting in the hypothesis of global warming; chlorine loading of the stratosphere via chlorofluorocarbons, resulting in the hypothesis of stratospheric ozone depletion and potential for increased UV-B radiation at the surface of the earth; and global circulation and ultimate deposition of toxic pollutants, including chlorinated pesticides (DDT), polychlorinated biphenyls (PCBs), polynuclear aromatic hydrocarbons (PAHs), and heavy metals.

At the regional scale, the two most important air pollutants influencing terrestrial ecosystems are acid rain and oxidant (ozone) pollutants (9). Research indicates that regional air pollution is one of the significant contemporary stresses imposed on some temperate forest ecosystems. Gradual and subtle change in forest metabolism and composition over high risk areas of the temperate zone over extended time, rather than dramatic destruction of forests in the immediate vicinity of point sources over short periods, must be recognized as the primary consequence of regional air pollution stress. Global air pollution, with its associated capability to cause rapid climate change, has the potential to alter dramatically forest ecosystems in the next century. The integrity, productivity, and value of forest and lake systems is intimately linked to air quality. Failure to give careful consideration to forest and aquatic resources in societal considerations of energy technologies and in management and regulation of air resources is unthinkable.

BIBLIOGRAPHY

1. E. B. Cowling, *Environ. Sci. Technol.* **16,** 110A–124A (1982).
2. E. Gorham and A. G. Gordon, *Can. J. Botany* **38,** 307–312 (1960).
3. E. Gorham and A. G. Gordon, *Can. J. Botany* **41,** 371–378 (1963).
4. G. E. Likens, F. H. Bormann, R. S. Pierce, J. S. Eaton, and N. M. Johnson, *Biogeochemistry of a Forested Ecosystem,* Springer-Verlag, New York, 1977.
5. National Acid Precipitation Assessment Program, *1990 Integrated Assessment Report,* Washington, D.C., 1991.
6. W. H. Smith, in D. C. Adriano and A. H. Johnson, eds., *Acidic Precipitation,* Vol. II, *Biological and Ecological Effects,* Springer-Verlag, New York, 1989, pp. 165–188.
7. K. C. Weathers, G. E. Likens, F. H. Bormann, J. S. Eaton, W. B. Bowden, J. L. Andersen, D. A. Cass, J. N. Galloway, W. C. Keene, K. D. Kimball, P. Huth, and D. Smiley, *Nature,* **319,** 657–658 (1986).
8. J. O. Reuss and D. W. Johnson, *Acid Deposition and the Acidification of the Soils and Waters, Ecological Studies,* Vol. 59, Springer-Verlag, New York, 1986.
9. W. H. Smith, *Air Pollution and Forests,* Springer-Verlag, New York, 1990.
10. J. P. Baker and S. W. Christensen, in D. F. Charles and S. Christie, eds., *Acidic Deposition and Aquatic Ecosystems,* Springer-Verlag, New York, 1991, pp. 83–106.
11. D. W. Schindler, *Oikos* **57,** 25–41 (1990).

ACTIVATED CARBON ADSORPTION FUNDAMENTALS

MILTON MANES
Kent State University
Kent, Ohio

Adsorption on activated carbon is of environmental significance because in many cases it is the method of choice for the removal of contaminants from air and water. Adsorption is a surface process that is exemplified by the concentration of gases, vapors, or solutes from solution at or near the surfaces of solids, or within surface cracks, pores, or crevices of nanometer dimensions. Activated carbons have extremely high surface areas and porosity; the high surface areas magnify the extent of adsorption, and the porosity increases the (exothermic) energy of adsorption. The amount of adsorption of a given single component on a given activated carbon at equilibrium may vary from near zero to about 50% by weight; it depends on the nature of the substance adsorbed (adsorbate), the partial pressure (or concentration), the temperature, and the presence of any other possibly competing substances. For a given single component on a given carbon, a plot of the amount adsorbed against the partial pressure or concentration at constant temperature is known as an adsorption isotherm. Knowledge of the isotherms for any given system enables the calculation of the amount of carbon that is needed for any specified purification task. In the absence of direct knowledge, it is important to be able to make reasonable estimates from other available data.

ACTIVATED CARBONS

Description

Activated carbon is a generic term for a variety of products consisting principally of elemental carbon that have been treated (activated) to enhance their surface areas, ie, areas that are external to carbon but internal to individual particles. A variety of activated carbons are available as articles of commerce; they are produced from a variety of carbonaceous raw materials, eg, bituminous coal, lignite, wood, anthracite, petroleum coke, coconut shells, waste matter from paper manufacture, pitch, and textile fibers. They differ from one another in a number of properties, such as shape, hardness, ash content, density, particle size, and surface area. Their distinguishing characteristic is an extremely high surface area; for example, a typical commercially available activated carbon has a surface area of 900–1000 m^2 g. At this area over half of the carbon atoms are exposed at a surface, and a layer of nitrogen molecules one molecular layer thick has a mass of about one-fourth the mass of the carbon. Surface phenomena that are difficult to detect on bulk crystals of graphite become greatly enhanced on activated carbons, and it is this enhancement that accounts for the utility of activated carbons in a wide variety of industrial and, in particular, of environmental applications. In such applications a wide variety of contaminants are removed from gases and liquids, particularly from air and water. Typical applications include removal of odorous components from circulating air, such as in homes, submarines, and automobiles; retention of gasoline vapors from automobile motors and from gasoline pumps; removal of hydrogen sulfide from vented sewer air; removal of trace taste and odor components and other contaminants from drinking water and gross contaminants from waste waters; dechlorination of drinking water; and acting as a broad-spectrum antidote for a wide variety of ingested poisons. Some contaminants in air that are not effectively removed by unmodified carbons are removed by activated carbons that have been impregnated with appropriate chemical reagents. For example, mercury vapors can be removed from air and other gases by activated carbons containing either elemental sulfur or silver, and cyanogen can be removed from air by carbons containing copper chromate.

Activated carbons are commonly available as granules, pellets, powders, and most recently as fibers. Carbons made from bituminous coal, char, lignite, wood, or coconut shells are usually sold as irregular granules, although in special cases they may be sold in pulverized form. Carbons from petroleum coke are usually available as cylindrical pellets; carbons made as by-products of paper manufacture are available only in pulverized form. All pristine activated carbons, ie, those that are not spent in use, may be readily distinguished from other forms of carbon by a sharp rise in temperature (of the order of 50°C) when most organic liquids are poured into a vessel filled with the activated carbon. In fact, the extent of the temperature rise can be used as an index of the activity of the carbon, as measured by its ability to remove contaminants from a particular stream of liquid or gas.

Production Methods

The activation process is the development of surface area (concomitant with fine porosity) in an appropriate carbonaceous precursor, and reactivation is the term for the regeneration of surface area in a spent carbon, ie, one that has been loaded with impurities in use. The most common activation process is slow oxidation with a mixture of carbon dioxide and water vapor at a dull red heat in a kiln or a multiple hearth furnace. The (endothermic) chemical reactions are

$$CO_2 + C = 2\,CO$$

and

$$H_2O + C = CO + H_2$$

The process usually takes several hours. The oxidation process must be relatively slow in order to oxidize the interior of the particles without excessive burnout of the exterior. In all likelihood, surface area is developed by oxidative attack at the ends of graphite-like planes, producing nonuniform submicroscopic slits in which, for example, vapors can condense.

For some raw materials a preliminary treatment is necessary before exposure to the high temperature of the activation furnace. For example, bituminous coal, which is the raw material for most of the commercial activated carbon, cannot be directly activated because it forms inactive coke on direct heating. Pretreatment of the bituminous coal consists of grinding the coal to a fine powder; compacting the powder with a binder to form pellets, which are then broken to granules; and oxidizing and devolatilizing the granules in air at about 300°C. This pretreatment destroys the coking properties of the coal, and the pretreated product can then go to the activating furnace. The interparticle space within the granules provides microscopic channels which enhance the availability of the interior of the granules to an external liquid or gas stream.

For any given feed to the activation process, different grades of carbon are produced by variation of the severity of activation, in particular by variation of the temperature and contact time of activation. More severe activation results in the formation of larger pores in the adsorption region or surface, which pores may be necessary for the removal of impurities with relatively bulky molecular structures, such as the color bodies in colored sucrose solutions. These larger pores may be of little or no consequence, however, in the removal of taste and odor compounds from potable waters or for the removal of odorous or toxic components from air. Increased severity of activation results in lower carbon yields, higher ash contents, and softer products. The choice of activation conditions is therefore a compromise between different demands on the product. The choice between different grades of carbon depends on the kind of use to which it is put; some require a particular grade of activated carbon, whereas others are well served by a wide variety of carbons.

Environmental Applications

Activated carbons are widely used in the removal of undesired impurities from air and from water. Applications for air purification include removal and recovery of solvent vapors in dry-cleaning plants, ductless fume hoods in kitchens, odor-removing filters in homes and workplaces, carbon-filled canisters in automobiles to prevent the escape of gasoline vapors, military and civilian gas masks which frequently use impregnated carbons, and protective clothing in military applications, in which activated carbon is incorporated into textiles. Other applications include the removal of hydrogen sulfide from vented sewer air by manhole covers containing activated carbon.

A major application of activated carbon-to-water purification is in filter beds in water treatment plants, where it serves as a final polishing agent. Another major application is in the treatment of plant effluents before they are discharged to sewers or to bodies of water. In these applications the spent carbons are usually regenerated by reactivation. Activated carbons also act to dechlorinate waters; potable waters that are treated in carbon filter beds may be subsequently rechlorinated before they are dispensed to the water supplies. Water slurries of activated carbon that are used as broad-spectrum antidotes in poison ingestions, particularly by children, are very effective for such poisons as aspirin and barbiturates, but are not effective for either ethanol or methanol. When the fundamentals of adsorption are considered, this will be clarified.

Activated carbon is used in gold-mining operations, in which gold is recovered from cyanide solutions. Although this is not, strictly speaking, an environmental application, the overall process poses some environmental problems.

Properties and Specifications

The three properties of activated carbons that are critical for most applications are high surface area and nanometer porosity, strong attraction of carbon for organic substrates, and weak attraction of carbon for water.

Consider surface area. The attractive forces between the carbon and the substances it attracts do not extend significantly beyond three molecular diameters from a carbon surface. If significant amounts are to be attracted it is necessary that a significant fraction of the carbon atoms be at the surface. In typical commercial activated carbons about half of the carbon atoms are surface carbons. At one (average) monolayer of coverage of nitrogen, for example, the nitrogen capacity of a typical commercial carbon is about 25 wt %. Although there are better ways of characterizing an activated carbon than by its surface area, this is a commonly used specification for activated carbons.

The most widely recognized method for surface area determination is the method of Brunauer, Emmett, and Teller (1), generally known as the BET method. The experimental determination consists in measuring the uptake of an appropriate gas or vapor at constant temperature, most commonly nitrogen at liquid nitrogen temperatures, as a function of its partial pressure. The range of relative pressure, x, ie, the ratio of the partial to saturation pressures, usually extends from about 0.05 to near saturation. The resulting data (adsorption isotherm) results in a curve when plotted on a linear scale as uptake (in mass or standard volume units) against partial or relative pressure. According to the model on which the method is based, the plot becomes linear when the function $x/V(1 - x)$ is plotted against the relative pressure and the model leads to the following equation:

$$x/V(1 - x) = 1/CV_m + (C - 1)x/CV_m \qquad (1)$$

where V_m is the volume of gas corresponding to monolayer coverage. The constant C, which is usually not computed, is approximately equal in the model, to $\exp((Q_1 - Q_V)/RT)$, where Q_1 is the heat of condensation of the first monolayer and Q_V the heat of condensation of the vapor to bulk liquid. According to Equation 1, $1/V_m$ is equal to the sum of the slope and the ordinate intercept of the linear plot. Given the number of molecules in a monolayer, the surface area per molecule is the only determination needed to calculate the (assumed smooth) surface area. The molecular surface area is estimated from the liquid density of the adsorbate, eg, 16.2×10^{-20} m^2 for N_2.

Although the BET method appears to be the best available, it nevertheless has a number of limitations. Perhaps the most serious is that the model is not well suited for a surface with a high degree of energetic heterogeneity (the reasons for which will become apparent later). Although the BET plot for activated carbons is sufficiently linear at relatively high capacities (or relative pressures from about 0.1 to near saturation), it deviates widely from linear at much lower relative pressures, where the capacities are quite significant. For example, at liquid nitrogen temperatures the nitrogen capacity of a typical activated carbon may exceed 5 wt % at a relative pressure of the order of 10^{-8}. Moreover, some specialized carbons of unusually high activities have reported surface areas of the order of 3000 m^2/g, which is higher than the surface area would be if all the carbon atoms were at a surface. Finally, there is the conceptual difficulty of deciding whether the walls of a slit one or two molecular layers in width should be considered to be independent surfaces. The BET surface area is useful in the sense that a carbon with an unusually low surface area is likely to be a poor carbon, at least for most purposes. However, the surface area is a better index of the activity of the carbon for high capacity applications, eg, recovery of dry-cleaning solvent vapors, than for low-capacity operations, eg, final polishing of drinking waters. Finally, the carbon in a group of samples that has the highest surface area is not necessarily the best carbon for a specific purpose.

Pore-Size Distributions. Activated carbons have been traditionally and frequently characterized by their pore-size distributions, although it should be stated at the outset that these characterizations are of limited utility. In granular carbons one recognizes two distinct kinds of porosity: macropores, which result from the interparticle spaces between the particles that were originally agglomerated to form briquettes and then granules; and micropores, which are of the order of a nanometer in width and which account for practically all of the adsorptive activity of a carbon. Typically a plot of the pore-size distribution of a carbon has as its ordinate the cumulative volume of all pores having a diameter or width that is equal to or less than the abscissa value.

Macropores provide diffusion channels for the transport of vapors or solutes to the micropores in the interior of the carbon granules. Macropore distributions are commonly determined by mercury intrusion. Because mercury neither spreads nor wicks into porous carbon (these terms will be defined more precisely in a later section), it does not enter a carbon pore unless forced to do so by hydrostatic pressure, and the pressure required to force mercury into an assumed cylindrical pore is inversely proportional to the pore diameter. The determination consists essentially of submerging a weighed sample of carbon granules under mercury in a vessel that can withstand high pressures, and measuring the volume of mercury forced into the carbon as a function of the imposed hydrostatic pressure. The pressures are then related to pore diameters, and the result is a pore-size distribution. The relation between pressure and pore size is at best approximate because the pore geometry is unknown. The method is not suited for measuring micropore distributions because the required pressures would be inconveniently high.

Micropore size distributions are determined from adsorption measurements of gases and vapors, such as nitrogen, argon, and carbon dioxide at or near their boiling points. The measurements are quite similar to the measurements in the BET determination of surface areas, but over a wider range of relative pressures. The translation of the adsorption data into a pore-size distribution requires a model. Several such models have been proposed (2) and all appear to give significantly different results. Although differences in distributions between different carbons have some significance, the absolute values of pore sizes should not be taken too literally, if only because the shape of the pores remains both complex and undetermined.

The Polanyi model gives a distribution of adsorption energies which does not depend on any assumption of pore size or shape, and from which one may derive adsorption isotherms of practical value. Although considerable effort and numerous publications have been devoted to micropore distributions (2) the practical results have been quite meager. In fact the author has spent nearly 25 years on the theory and practice of adsorption on activated carbon without finding them of significant use. They are noted herein so that readers will neither be surprised on hearing about them nor overly impressed with presented data on them.

Additional Adsorptive Specification Numbers. Although there are much better ways to characterize the adsorptive activity of an activated carbon, several figures of merit, in addition to the surface area, have become well entrenched. The most popular of these are the iodine number and carbon tetrachloride activity (3).

The iodine number is the number of milligrams of elemental iodine adsorbed per gram of dry carbon when it is in equilibrium with a 0.0100 M solution of I_2 in a 0.15 M water solution of KI. In general the iodine numbers are similar in magnitude to the BET surface areas and are a measure of the carbon activity in the high capacity range. The iodine number is a rough measure of the quality of the carbon and is useful in determining the extent to which a spent carbon has been reactivated. However, it is not a good index of the activity of a carbon in low-capacity applications such as polishing drinking water. Small differences in iodine number between competing samples should not be a basis for choice between them.

The carbon tetrachloride activity (or number) is the carbon tetrachloride capacity, expressed as percent of the mass of dry carbon when the carbon is in equilibrium with carbon tetrachloride vapor at 25°C and a relative pressure of 0.275. Like the iodine number, it is a measure of the adsorptive activity at relatively high capacities. It is likely to be superseded by some similar method that does not use the toxic carbon tetrachloride.

Modes of Action of Activated Carbon

Most applications of activated carbon may be accounted for by what is commonly termed physical adsorption, in which the attractive force of the carbon for the substances that stick to it are chiefly London forces (4). Adsorption by London forces is amenable to extensive theoretical treatment and leads to results of considerable practical value. However, there are a few applications of activated carbon that are not primarily the result of London forces. It is important to recognize them as being outside of the scope of the theory (Polanyi adsorption potential theory) that will later be presented. It is also important to note that a number of the common adsorption specifications, eg, surface area and iodine number, have little relevance to the performance of an activated carbon in these applications.

Consider the dechlorination of water by activated carbon, in which the carbon appears to act as a reagent that reduces the chlorine and is consumed in the process. Unlike graphite, activated carbon can act as a reducing agent that is not quite as powerful (at room temperatures) as metallic copper. For example, activated carbon precipitates silver crystals from silver nitrate and silver–ammonia solutions, but not from silver thiosulfate solutions, whereas metallic copper plates silver out of silver thiosulfate. A slurry of pulverized activated carbon yields a brown solution with no visible particles when treated with a solution of bleaching powder. Because of the low concentrations of chlorine in drinking water, it is difficult to detect any carbon loss from reaction with chlorine, but it seems reasonable to expect that it is consumed by the chlorine even at low concentrations.

Another mode of action of activated carbon is as a catalyst. This is exemplified by the removal of hydrogen sulfide from air, in which elemental sulfur remains on the carbon. The presumed reaction is as follows:

$$2\ H_2S + O_2 \rightarrow 2\ H_2O + 2\ S$$

More concentrated mixtures of H_2S in air, eg, a stoichiometric mixture, ignites on passage over activated carbon. Moreover, H_2S is relatively weakly adsorbed, either by itself or in an inert gas such as nitrogen. Activated carbon also catalyzes the oxidation of sulfur dioxide to sulfates in the presence of air and moisture. More recently, it has been found that some phenols in water solution are oxidatively polymerized on activated carbon in the presence of oxygen (5).

A further mode of action is by chemisorption, ie, by bonding with specific organic groups on the carbon surface. Activated carbons may be weakly acidic or basic to strong acids and bases (6). The adsorption of acetic acid on activated carbon appears to be partially by London forces and partly by chemisorptive forces.

Finally, activated carbon can serve as a relatively inert support for a variety of reagents and catalytic agents with which it is impregnated. Impregnation on activated carbon is an effective means of increasing the surface area, and therefore the chemical reactivity, of the impregnants. For example, carbons for military gas masks have been impregnated with copper chromate to oxidize cyanogen, and with a trace of silver to catalyze the decomposition of arsine, AsH_3, to its elements. Traces of mercury vapor in air and other gases may be removed by passage through activated carbon that has been impregnated with either elemental sulfur or metallic silver. The sulfur reacts to form mercury sulfide; the silver either dissolves or amalgamates the mercury.

Having recognized what may be considered the secondary modes of action of activated carbon, the most important mode, ie, London force (or physical) adsorption will be considered in some detail after reviewing some background information.

ADSORPTION ON SOLIDS: GENERAL

Adsorption as a Surface Phenomenon

Adsorption onto solids from the gas or vapor phase is exemplified by the ability of solids of high surface area to condense on their surfaces considerable amounts of gases or vapors. Adsorption onto solids from liquids is exemplified by the removal from solution of considerable amounts of impurities from water. The prefix ad- is used to indicate that this is a surface process, ie, that the adsorbed substances (adsorbates or adsorptives) are exterior to the bulk of the solid. The attractive forces of solids for adsorbates fall off rapidly with increasing distance from the surface, with the result that the adsorbed material is not likely to be more than several molecular layers thick on the solid. The mass ratio of adsorbate to solid becomes of significant magnitude when a significant fraction of the atoms of the solid is at a surface.

Surface chemists distinguish between chemisorption, in which the attractive forces come from specific chemical bonding, and physical or London-force adsorption, in which the attractive forces are essentially London forces. In this adsorption the attractive forces per molecule increase with molecular size and decreasing volatility and are not strongly dependent on chemical specificity.

Examples of chemisorption are the adsorption of hydrogen on tungsten and of carbon dioxide on potassium oxide in synthetic ammonia catalysts. In these examples the high volatility of the adsorbates does not hinder adsorption because it takes place by specific chemical bonding. By contrast, inert substances boiling much higher than hydrogen would not be significantly adsorbed. An example of London-force adsorption is the adsorption on activated carbon of hydrogen, nitrogen, and butane. Hydrogen is strongly adsorbed in the region of its boiling point (21 K), but quite weakly at the boiling point of nitrogen (78 K). Nitrogen is strongly adsorbed near its boiling point, but hardly at all at ambient temperatures, eg, 300 K. Butane is quite strongly adsorbed at ambient temperatures. These examples illustrate that physical adsorption on activated carbons tends to be relatively strong at temperatures around normal boiling points, and tends to be stronger for larger molecules.

Adsorption Isotherms

For any given gas or vapor adsorbing on any given sample of activated carbon, the amount adsorbed on the sample at equilibrium, typically expressed as grams or standard cm^3 per gram of carbon, increases with increasing pressure and decreases with increasing temperature, the latter because adsorption is exothermic. In adsorption from solution the amount adsorbed at equilibrium increases with increasing concentration of adsorbate solute and usually decreases with increasing temperature, although adsorption from water solution is usually less temperature-sensitive than vapor-phase adsorption.

Adsorption data are conveniently plotted as some measure of capacity or loading against pressure or concentration at constant temperature. In many cases it is convenient to plot relative concentrations or pressure, ie, the ratios of the pressures or concentrations to the saturation values. These plots, or their mathematical equivalents, are called adsorption isotherms; they may differ very widely for different adsorbates. For data extending to very low pressures or concentrations, it is frequently more convenient to use logarithmic units for both capacities and concentrations. The resulting plots are referred to as Freundlich plots or isotherms, after the name of their attributed originator.

Because the adsorption isotherm for a given system gives the capacity as a function of the partial pressure or concentration at a given temperature, knowledge of the isotherm for a given system is critical for calculating the amount of an activated carbon that would be required, for example, to remove a given amount of contaminant from a liquid or gas stream at a given temperature, or, alternatively, to purify a given amount of a gas or liquid stream from a given impurity, given its partial pressure or concentration. The determination of such isotherms, or their estimation when they cannot be conveniently determined, is critical for feasibility studies and for plant design, and will be the major preoccupation of this article.

Characteristic shapes for a wide variety of adsorption isotherms for a wide variety of both adsorbates and solids have been classified (7,8) into five different categories. Only the types that are found on activated carbons, ie, Types IV and V, are considered here.

Type IV. This isotherm (on a linear scale) is concave downward at low concentrations and goes to a finite upper limit at saturation pressures (Fig. 1). On a linear scale it may appear to coincide with the ordinate scale for a significant length, so that this part of the isotherm would be better plotted on a logarithmic scale. This type of isotherm, although it may vary widely in scale, is typical for the adsorption of a wide variety ofcontaminants from air or water solution onto activated carbon. Alternative representation of this type of isotherm will be dealt with later.

Type V. This isotherm is concave upward at low pressures and goes to a finite maximum at saturation. It is exemplified by the adsorption of water vapor on an ash-free carbon. As in the preceding example, the finite limit at saturation is attributed to the adsorption taking place by condensation in pores, which adsorption comes to a limit when the pores are filled. The upward concavity is typical for the adsorption of vapors of liquids that do not spread on smooth graphite.

Linear Isotherms. Linear isotherms are found for adsorbates that are very weakly adsorbed, such as nitrogen at room tem-

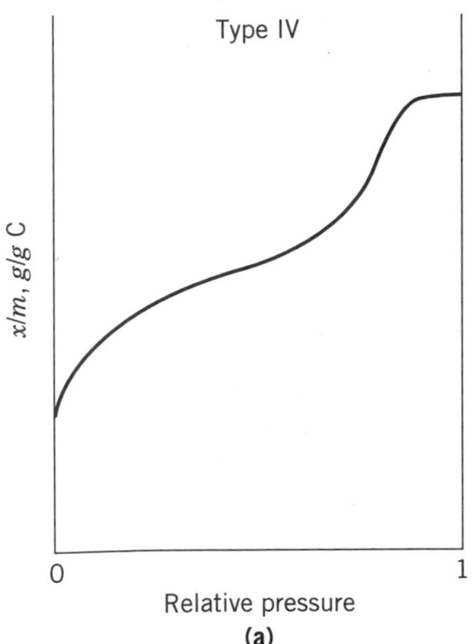

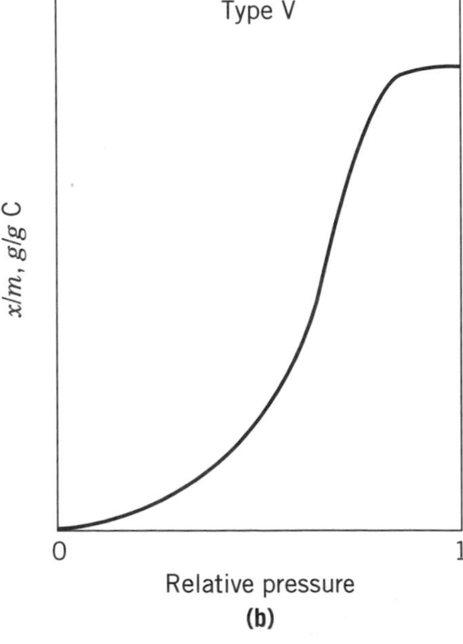

Figure 1. Two types of isotherms (on linear scales) that occur on activated carbons: (**a**) is characterized by appreciable adsorption at relative pressures that are too low to be represented accurately on a linear scale, and by a limit on the amount of adsorption at saturation, which is characteristic for the filling of pores; (**b**) is characteristic for water adsorption on an ash-free carbon, and is not followed by most organic vapors.

peratures, relatively volatile adsorbate vapors at high temperatures, very dilute miscible solutes from water, and, as will be seen, for trace adsorbates in the presence of a dominant component in multicomponent adsorption. A special case for the adsorption of trace gases is the adsorption of radon at extreme dilutions (9). With rare exceptions, such as the adsorption of radon for air testing, linear isotherms are of much more interest to the regeneration of activated carbons, where linear isotherms are desirable, than for the adsorption of trace contaminants, where highly concave (downward) isotherms are desirable.

Log–Log Plots. Plots of adsorption isotherms, like the Type IV isotherms, on a linear scale cannot be used to depict adsorption in the trace concentration region, eg, parts per million, which is frequently of most interest. For such cases it is more convenient to plot the data as the (common) logarithm of capacity against the logarithm of partial pressure (or relative pressure) or concentration (or relative concentration). Such plots have the advantage that they can depict adsorption over orders of magnitude of concentration, for example, to parts per billion and lower. Moreover, on most activated carbons these plots tend to be approximately (logarithmically) linear over a wide range of partial pressures or concentrations, although in general not quite up to saturation conditions. Unit slope on a log–log (Freundlich) plot (8) corresponds to a linear isotherm on a linear plot, whereas isotherms that are concave downward on a linear plot have a flatter slope. In general, for reasons that will be seen, isotherms of larger molecules tend to exhibit flatter slopes.

One form of the mathematical representation of a log–log plot over the linear range is as follows:

$$\log x/m = \text{constant} + (1/n)\log(p/p_s) \quad (2)$$

$$x/m = A(p/p_s)^{1/n} \quad (3)$$

where x/m is the adsorbate mass per unit mass of carbon, p/p_s is the relative partial pressure, and n and A are constants, with n usually greater than 1.

For illustrative calculations with adsorption isotherms, consider, for example, two equations that represent approximately the respective adsorption of n-butane and n-hexane at 300 K on a typical activated carbon:

$$x/m\ (\text{butane}) = 0.47p^{0.31} \quad (4)$$

$$x/m\ (\text{hexane}) = 0.85p^{0.23} \quad (5)$$

where the partial (absolute) pressures, p, are in atmospheres (using absolute rather than relative pressures). The equations may be presumed to be reasonably accurate except in the limit of saturation pressures (2.3 atm for butane, 0.237 atm for hexane), where the calculated values of approximately 0.6 g/g carbon are ~1.6 times the actual values.

The equations have been cast in a form that permits ready estimation of activated carbon capacities for removing butane and hexane (separately) from atmospheric air at 300 K and at, for example, 1 and 1000 ppm in partial pressure units. The respective partial pressures are 10^{-6} and 10^{-3} atm. Substitution into the above equations gives the following capacities in g/g carbon at equilibrium at the given partial pressures:

	1000 ppm	1 ppm
x/m (butane)	0.055	0.0065
x/m (hexane)	0.17	0.035

The capacity for hexane at both partial pressures is higher than for butane. In addition, the reduction of the partial pressure by three orders of magnitude reduces the capacity for hexane only fivefold, whereas the butane capacity is reduced by a factor of about 8.5, which is still a low dependence on partial pressure, as can also be seen from the magnitude or the exponents in the equations.

The practical significance of these results can be illustrated by calculating the volume of contaminated air that can be cleaned per gram of carbon, assuming that all of the impurity goes onto the carbon, which can be approximately attained in carbon columns. To accomplish this, concentrations in ppm must be converted to g/L. Assuming ideal gas behavior, the respective densities, d, of butane and hexane vapors per atm of partial pressure are as follow:

$$d_{\text{butane}} = (58\ \text{g/mol})/22.4\ \text{L/mol} \times 273/300 = 2.36\ \text{g/L}$$

$$d_{\text{hexane}} = (86/22.4) \times (273/300) = 3.49\ \text{g/L}$$

The volume (liters) cleaned per gram of carbon is simply the capacity divided by the gas density at the given partial pressure. For 1 ppm butane the calculation is as follows:

$$\text{liters} = (0.0065\ \text{g/g C})/(2.36 \times 10^{-6}\ \text{g/L}) = 2750\ \text{L}$$

The results of all of the calculations are as follows (volume in L/g carbon):

	1000 ppm	1 ppm
butane	23	2750
hexane	49	10000

This calculation illustrates how the power of adsorption as a purification process increases with decreasing concentration of the adsorbable contaminant when the slope of the log–log plot ia relatively flat, ie, when on a linear scale the adsorption isotherm is highly concave.

This can be expressed more generally using concentrations as the variable, because the log–log plots apply to adsorption of solutes as well as to vapors. If the Freundlich equation is written as follows,

$$x/m = KC^{1/n} \quad (6)$$

and the volume of solution containing the mass x is written as V, then equation 6 becomes

$$Vc/m = KC^{1/n} \quad (7)$$

$$V/m = KC^{(1/n-1)} \quad (8)$$

These equations show how the volume containing any given adsorbate varies to an inverse power of the concentration (or partial pressure) for any value of n greater than 1.

Returning to the equations for the adsorption of butane and of hexane, the amount adsorbed can be written in terms of the relative pressure p/p_s, where p_s is the saturation pressure, in which case the equations become:

Butane $$x/m = 0.61(p/p_s)^{0.31} \quad (9)$$

Hexane $$x/m = 0.61(p/p_s)^{0.23} \quad (10)$$

When adsorption is considered in further detail, it will be found that the exponents tend to be inversely proportional to the (liquid) molar volumes for many vapors and that the multiplicative constants do not vary by much when pressures are expressed as relative pressures, although they are not as constant as in the present example. Some of these generalizations will appear to become obvious when the Polanyi model for adsorption on activated carbons is considered.

Adsorption Energy (Thermicity) and Temperature Dependence

Adsorption from the vapor phase is generally exothermic, in keeping with the reduction of entropy on condensation; capacities at any given partial pressure therefore decrease with increasing temperature. Adsorption of solutes from water usually decrease also with increasing temperature, although usually less sharply than in vapor-phase adsorption. There are exceptions, however. For example, if a low-melting solute in water can liquefy as the temperature rises, the capacity may go through a stepwise increase at the temperature of the phase change, although it will continue to decrease as the temperature is raised further. This stepwise increase, which has been erroneously attributed to a reversal of the thermicity of adsorption (10), has been better accounted for by increased efficiency of packing of liquids in the carbon pores when the solid adsorbate melts (11). The Polanyi model provides methods for making quantitative estimates of the temperature dependence of adsorption, as will be shown later.

In considering the enthalpy of adsorption, or the energy of adsorption, which is of the same magnitude, it is first necessary to distinguish between the differential and the integral heats of adsorption. The differential heat is the heat liberated per unit mole or mass by each small increment of adsorption, whereas the integral heat is the total heat liberated when a given amount is adsorbed. The integral heat is therefore the limit of the sum of the differential heats, starting from zero adsorption to the total amount adsorbed, as the increments become infinitesimally small. In general the (exothermic) differential heats of adsorption of most organic liquids on a typical activated carbon decrease from a maximum of about 100 cal/cc (liquid) at the lowest capacities, to about 50 cal/cc at a capacity of 0.1 cc (liquid)/g carbon, and to zero at the limiting capacity, which is of the order of 0.5 cc/g carbon. For the adsorption of vapors, the exothermic heat of condensation must be added, which for typical organic vapors is of the order of 50 cal/cc. Therefore, as a rule of thumb, the differential excess adsorption energy of a typical organic vapor or the differential heat of adsorption of a typical organic liquid on typical activated carbons ranges from about twice the energy of vaporization to zero. The distribution of the differential adsorption energies to the adsorbate volume is related to the different individual adsorption isotherms on a given carbon, and is a characteristic of the individual carbon sample. Later it will be shown how this distribution may be derived from adsorption isotherms, how it may be used to predict other isotherms, and how much more useful it is than the conventional pore size distributions.

The maximum excess enthalpy of adsorption of a liquid on an activated carbon is its heat of immersion. When activated carbon is immersed in water the temperature change is not immediately obvious. However, when activated carbon is immersed in an organic liquid, in amounts to cover the carbon, there is an immediate temperature rise, of the order of 50°C. This temperature rise can in fact be used as an index of the activity of a carbon.

The thermicity of adsorption is important in the static determination of vapor-phase adsorption isotherms, where the mass increase of a carbon is determined at constant temperature and at the desired range of partial pressures. In such experiments it is necessary to wait for dissipation of the heat of adsorption before taking measurements. The heat of adsorption can have some dramatic effects; for example, when glycerine is passed into a column of dry activated carbon, the interior may get hot enough to produce charring. In this case the heat of adsorption is concentrated in a relatively small space by the preheating that takes place upstream of the hot zone. The charring occurs because the glycerine does not boil before charring. In other systems, the thermicity is of no consequence, as in the removal of trace impurities from air or water, where the heat generated per unit of mass flow is too small to be readily detected.

Mathematical Representation of Adsorption Isotherms

In addition to the Freundlich equation, a wide variety of mathematical expressions have been used to represent adsorption isotherms, of which some of the more prominent have been given by Langmuir Halsey-Hill, Frenkel, Dubinin, and others. They are extensively reported in the literature (8). They are not discussed here in further detail because they are of minimum utility as compared to the Polanyi model.

Competitive Adsorption

Consider the removal by activated carbon of trace color bodies from concentrated solutions of sucrose. Because water and sucrose both adsorb on the carbon, and because both are much more concentrated than the color bodies, this is clearly a case of competitive adsorption, in which traces of one component can adsorb in the presence of high concentrations of others. If the adsorption is taking place in a column, most of the carbon is saturated with sucrose before the color bodies reach it, in which case the color bodies must displace sucrose from the carbon. Similar cases arise in the adsorption from water of a series of trace impurities. Later, the issues of why one adsorbate can displace another and, specifically, the criteria for a displacement series will be discussed. First, consider the following review of some of the macroscopic behaviors of surfaces.

Surface Energy (Surface Tension), Capillarity, and Adhesion

A later discussion of adsorption in pores will be related to some macroscopic surface phenomena. First, consider the Gibbs free energy (or work) that is required to increase the surface of a (macroscopic) liquid. This free energy is best expressed in units of energy per unit of surface, eg, joules/m^2 or ergs/cm^2. It is more commonly expressed in units of force per unit length, eg, newtons/m or dynes/cm and referred to as surface tension. Although the alternative methods of expression are equivalent and give identical magnitudes, it is preferable for the purposes herein to think in terms of surface energies rather than tensions.

When comparing the molecules at a liquid surface with similar ones in the interior, an obvious distinction between them is that the interior molecules have neighboring molecules all around them, whereas surface molecules have lost a significant fraction (approximately less than half) of their neighbors. Because the loss of all neighbors would amount to vaporization, surface formation may be viewed as partial vaporization. This admittedly simplistic point of view accounts qualitatively for the observation that liquids with abnormally high heats of vaporization per unit volume (or cohesive energy density) also have anomalously high surface energies; it also gives a better intuitive feel for the energetics of surface formation than the concept of surface tension. More sophisticated models that give surface energy estimates of the correct approximate magnitudes are available (12).

Capillarity, spreading, and wicking are considered herein in accordance with the Harkins approach (13). The cohesive work of a liquid or solid is defined as the work required to separate a column of 1 cm^2 cross-sectional area so that 2 cm^2 of new surface is thereby created. This work is twice the surface energy per cm^2 because it creates two surfaces of 1 cm^2 each. Consider a column of 1 cm^2 cross section that consists of a column of liquid and one of solid that are joined together, the joint also having the area of 1 cm^2. The work required to separate the column at the joint is defined as the adhesive work, W_{ad}. The same definition applies to the junction of two solids. If the separation of the column is carried out in practice as, for example, with some solid and an adhesive, the joint surface breaks cleanly (adhesively) at the junction if the adhesive work is less than the cohesive work of either of the adhering components, usually the adhesive. By contrast, if the adhesive work exceeds the cohesive work, then the joint breaks cohesively, ie, within the component with the lesser cohesive work.

Consider now the behavior of a liquid drop on the horizontal, plane, and smooth surface of a solid. First consider the case where the adhesive work, W_{ad}, exceeds the cohesive work, W_c, of the liquid. The driving force for the spreading of a film is W_{ad} per unit of interfacial area. The restraining force is equal to twice the surface energy of the liquid because the spreading of a film of macroscopic thickness (the duplex film of Harkins) requires the formation both of the contact surface and the upper free surface of the film. The condition for the spreading of the liquid film is therefore $W_{ad} > 2\gamma$ or $W_{ad} > W_c$. For most adsorbates (water and mercury being the most frequent exceptions), the fact that their liquid phases spread on graphite can be related to their adsorptive behavior.

Consider now a liquid drop on a horizontal, plane, and smooth solid surface, that does not spread, but comes to equilibrium at a contact angle θ, where the contact angle is zero for a spreading liquid, positive and less than 90° for a liquid that is close to spreading, and greater than 90° for a liquid with a reentering contact angle, such as mercury on glass (Fig. 2). Consider now the energy balance at the line of contact between solid and liquid. For each increment of surface contact area, dA, the driving force is $W_a dA$. The corresponding increase in liquid surface is also dA at the contact surface, but is $dA\cos\theta$ at the upper surface. The energy balance for equilibrium is therefore

$$(W_{ad} - \gamma - \gamma\cos\theta)da = 0 \qquad (11)$$

whence

$$\cos\theta = 2(W_{ad}/W_c) - 1 \qquad (12)$$

If the adhesive and cohesive works are equal, then $\cos\theta = 1$ and the contact angle is zero, which is the spreading condition. If $W_c > W_{ad} > \gamma$, then $\cos\theta$ will be positive and less than 1, so that the contact angle will be less than 90°. At $W_{ad} =$

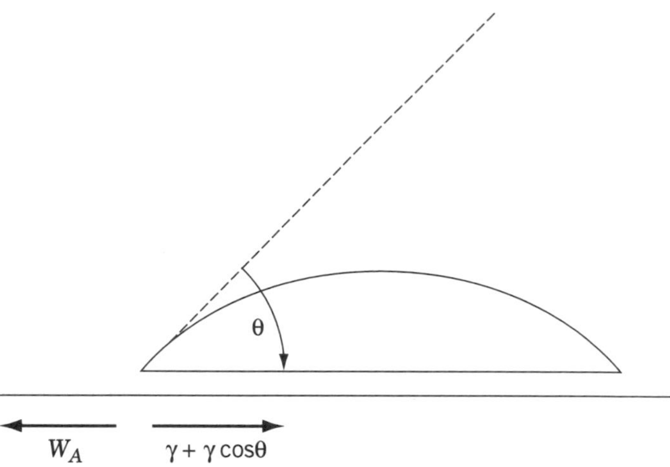

Figure 2. Diagram showing the balance of forces at the edge of a sessile drop of liquid on the horizontal, plane, and smooth surface of a solid (14). Reprinted courtesy of Federation of Societies for Coatings Technology (Blue Bell, Pa.).

$W_c/2$ ($= \gamma$), $\cos\theta$ becomes zero and $\theta = 90°$; for $W_{ad} < \gamma$, the contact angle will exceed 90°.

An alternative method for expressing the contact angle is in terms of the surface tension of the solid and the interfacial tension of the solid–liquid interface. This derivation is commonly found in treatises on surface chemistry (8) and the resulting equation is known as the Young equation. However, neither the surface tension of the solid nor the interfacial tension of the solid–liquid interface are amenable to direct measurement. The Harkins approach is more useful for the present purposes.

Before considering the adsorption of vapors and of liquids in micropores, it is of interest to consider the behavior of liquids in macroscopic pores, if only because some adsorption has been attributed to capillary condensation. In considering the behavior of liquids in capillaries or pores of fixed geometry, it is customary to regard the forces exerted by a liquid meniscus as a function of its curvature. The much more general approach of Schwartz is adopted herein (15) for dealing with the behavior of a liquid in a porous solid of as yet undefined geometry. For a liquid in contact with a column of porous solid, the driving force for increasing the contact surface is $W_{ad}dA$ and the restraining force is simply γdA, where dA is the differential increment in the contact area. It is important to emphasize the distinction between capillary rise and the spreading of a surface drop, namely that in capillary rise there is no concomitant increase in the area of a free surface. The criterion for the horizontal travel of liquid in a porous solid or in a capillary is therefore $W_{ad} \geq \gamma$ or $W_{ad} \geq W_c/2$, and the energetic driving force per unit of contact area is

$$W_{ad} - W_c/2 = \gamma\cos\theta \qquad (13)$$

Note that the $\cos\theta$ refers to the contact angle of a drop of liquid on a plane horizontal surface of the smooth solid, and not to any contact angles in the porous solid or capillary, which are usually much more difficult to measure.

The criterion for what may be called wicking, as exemplified by horizontal travel of liquid in a porous solid or by capillary rise, is less severe than for spreading of a film on a smooth surface, again because there is no energy cost for generating a free surface. The term wetting, which has been applied to both phenomena, is therefore less precise than spreading or wicking. Water, for example, does not spread on graphite, but wicks into carbon pores or crevices. This fact will turn out to relate to the adsorption of water on activated carbon and the adsorption of solutes from water solution.

A relation (Kelvin equation) can now be derived between the size of a well-defined pore or crevice and the effect on the vapor pressure of a liquid in it. Consider a capillary or crevice or other pore that is closed at one end and partially filled with a liquid. The driving force for increasing (or decreasing) the contact surface is $\gamma\cos\theta$, where θ refers to the contact angle on a smooth surface. If θ is less than 90°, the driving force for extension of the capillary will produce a negative hydrostatic pressure on the liquid, ie, there will be a pull on the liquid. If $\theta > 90°$, the result will be compression of the liquid. These hydrostatic pressures can affect the vapor pressure of the liquid by well-known thermdynamic relationships. In the general case the hydrostatic pressure can be calculated from the surface–volume relationships even in irregular pores (15). Herein the discussion is limited to the special cases of cylindrical pores and of slits of known width and infinite lengths. For the general case for a virtual expansion of length dl, the hydrostatic pressure P is given by the energy balance:

$$-PdV/dx = \gamma \cos\theta dA/dx \qquad (14)$$

$$-P = \gamma \cos\theta dA/dV \qquad (15)$$

For a cylindrical pore,

$$dA/dV = 2\pi r/\pi r^2 = 2/r \qquad (16)$$

where r is the pore radius. For a slit of distance d and length l, neglecting the ends,

$$dA/dV = 2l/dl = 2/d \qquad (17)$$

Therefore for a cylindrical pore,

$$-P = 2\gamma \cos\theta/r \qquad (18)$$

and for a slit,

$$-P = 2\gamma \cos\theta/d \qquad (19)$$

The dependence of vapor pressure on hydrostatic pressure is given by equation 20.

$$V_m dP = \mathrm{RT} d \ln p \qquad (20)$$

At zero hydrostatic pressure the partial pressure of the vapor in equilibrium with liquid is p_s, the saturation pressure. Integrating from $P = 0$ to P (at assumed constant V_m for an assumed incompressible liquid) gives equation 21.

$$\mathrm{RT} \ln p/p_s = V_m P \qquad (21)$$

whence for cylindrical pores,

$$\mathrm{RT} \ln p/p_s = -2\gamma V_m \cos\theta/r \qquad (22)$$

and for plane parallel slits,

$$\mathrm{RT} \ln p/p_s = -2\gamma V_m \cos\theta/d \qquad (23)$$

These are the corresponding Kelvin equations.

Note that the derivation based on energetics does not make any assumptions about the curvature of the meniscus. It makes much more sense to reason that the shape of the meniscus is the result rather than the cause of the adhesive driving forces.

Whereas the Kelvin equation has been derived for macroscopic pores or capillaries, it cannot be expected to be accurate where the pore dimensions are of molecular size. One reason is that the surface free energy cannot be expected to remain constant down to molecular dimensions. The second problem is that the maximum adhesive driving force in the derivation of the Kelvin equation is one-half of the cohesive work of the liquid, regardless of the magnitude of the adhesive work. This assumption, which is both reasonable and realistic for macroscopic capillaries, turns out to be grossly inaccurate for adsorption on activated carbon, particularly in the low-capacity region.

The problem may be illustrated for the adsorption of hexane on a typical activated carbon at 20°C (293 K), where the ratio p/p_s can easily get below 10^{-8} at low capacities. Given γ for hexane as 18.4 ergs/cm^2 and V_m as 131 cm^3/mol, and taking R as 8.3×10^7 ergs/deg mol, and assuming that liquid hexane is between parallel slits, the slit width calculates to 1 A°. This is clearly an impossibly small number because the separation of the planes in solid graphite is over 3 A°.

The problem of extrapolating macroscopically derived relationships to molecular dimensions is again illustrated by the derivation of the well-known equation for the rise of liquids in capillaries with solid walls. For a capillary of cylindrical cross section the energy loss per unit advance, dh, of the meniscus height is $(W_{ad} - \gamma)2\pi r dh$, and the potential energy gain of the liquid of density ρ is $\rho g h \pi r^2 dh$. At equilibrium the two become equal, whence

$$\rho g h \pi r^2 = (W_{ad} - \gamma)2\pi r \qquad (24)$$

$$h = 2(W_{ad} - \gamma)/\rho g r \qquad (25)$$

When considering the magnitude of $W_{ad} - \gamma$, first consider the special case where the liquid forms a stable contact angle on the solid, in which case, as has been seen, $W_{ad} - \gamma = \gamma\cos\theta$. Substitution of this value into equation 25 gives the well-known equation for capillary rise:

$$h = 2\gamma \cos\theta/\rho g r \qquad (26)$$

Consider now the case of a strongly spreading liquid, where there is no stable contact angle and where $(W_{ad} - \gamma) > \gamma$. In this case a capillary rise may be found that is higher than that calculated from equation 26 with $\cos\theta = 1$. As a matter of experience, a capillary rise any higher is never found. The apparent reason for this (16) is that no matter how great the adhesive work, the joint between liquid and capillary wall cannot sustain the corresponding stress without breaking cohesively. In effect, if $W_{ad} > W_c$, there is a multilayer of liquid (or condensed vapor) on the capillary wall, so that in effect the liquid rises in contact with its own film rather than with the wall. If $W_{ad} < W_c$, the liquid in the capillary rises in contact with the wall. If the capillary is of molecular dimensions, there is no room for a multilayer on the sides, and the model for deriving the capillary rise equation does not apply for strongly spreading liquids, nor, as has been seen, does the assumption of a size-independent surface free energy.

For a physical understanding of physical (London-force) adsorption it is useful to consider the conditions under which a liquid-like multilayer of vapor would form on the smooth surface of a bulk solid. If the solid may be thought of as a solid and nonvolatile replica of the liquid phase of the adsorbing vapor (as polyethylene might be an approximate replica for a paraffin hydrocarbon), then the attractive force of the solid for vapor molecules would be about the same as for the liquid phase of the vapor. Under these circumstances multilayer formation could be expected at the saturation pressure of the vapor, or at a relative pressure of 1. Similarly, for a solid for which $W_{ad} > W_c$ for the liquid phase, multilayer formation would be expected below saturation pressures. For $W_{ad} < W_c$ no multilayer formation would be expected on the solid even at saturation pressures. This leads to the expectation that if a liquid forms a stable contact angle on a smooth solid, the contact angle would not be sensitive to the presence of the vapor phase. For adsorption on activated carbon it is worth remembering that most organic liquids spread on graphite and that water does not. However, water can wick into carbon crevices or pores. More specificity will be provided by the Polanyi model (17), which in effect incorporates most of the foregoing ideas, but which avoids any assumptions on the size or shapes of presumed pores.

LONDON-FORCE (PHYSICAL) ADSORPTION ON ACTIVATED CARBON: THE POLANYI MODEL

The adsorption surfaces (or pore volumes) of activated carbons are energetically heterogeneous because of the existence of multiple crevices, cracks, or pores of differing shapes and sizes, which affect the extent of approach to carbon, and therefore the energy of adsorption. For example, a molecule in a deep pore or crevice of molecular dimensions has more closely neighboring carbon atoms than a molecule on a block of carbon. The Polanyi model has long (but not universally) been recognized as the most powerful available model for dealing with adsorption on energetically heterogeneous surfaces (7). It relates a wide variety of both vapor-phase and liquid-phase isotherms to each other, and, in particular, it relates vapor-phase to liquid-phase adsorption. In addition, it can be readily adapted to estimate adsorption in multicomponent vapor-phase adsorption with no limit on the number of components, to multicomponent adsorption of organic liquids from water, and, with more limited success to the adsorption of solids from water and from organic solvents. The extended model also accounts for competitive adsorption, takes into account nonuniformity of a multicomponent adsorbate in pores, and provides criteria both for the recognition of chemisorption in cases where it occurs and for the recognition of steric factors in the adsorption of molecules that cannot be made to fit a planar surface. It may be stated that no other known model comes close to handling such a variety of systems, and that those interested in adsorption on activated carbon would be well advised to become familiar with it, both for an intuitive physical understanding and for practical applications.

Concentration of Molecules in a Force Field

Consider a gas, for the moment assumed ideal, in some kind of attractive force field. For example, the gas could be in a gravitational or centrifugal force field or, alternatively, it could be attracted to a nearby solid by London forces. It turns out that resulting modifications to the otherwise uniform concentration distribution are all governed by the same equations.

In general, a system tends to minimize its energy and to maximize its entropy. In the case of an attractive force field the two tendencies are in opposite directions. Energy is minimized by concentration of the molecules in the region of lowest potential energy, but the resulting concentration reduces the entropy. The balance between these opposing tendencies is given by thermodynamics as follows:

$$-dE = VdP \qquad (27)$$

where $-dE$ is the differential loss of potential energy per mole, for example, as a function of distance or height in gravitation, V is the molar volume, and dP is the differential increase in pressure. For ideal gases equation 27 becomes

$$-dE = \mathrm{RT}d \ln P \qquad (28)$$

In a gravitational field, $dE = Mgdh$, where g is the acceleration of gravity, M is the molecular weight, and h is the height. Substitution into equation 28 and integration gives the familiar Boltzmann equation for atmospheric concentration due to gravity:

$$-Mg(h_2 - h_1) = \mathrm{RT}\ln(p_2/p_1) \qquad (29)$$

For gas-phase adsorption the Polanyi model postulates a loss in potential energy resulting from the attraction of the solid, and this loss of potential energy (taken as positive) is called ϵ, the adsorption potential. The model postulates that the difference between the ambient pressure (far from solid) of a gas or vapor and local pressure in the vicinity of the solid is given by the Boltzmann equation, in the following general form:

$$\epsilon = \int VdP \qquad (30)$$

which is valid for nonideal gases and vapors, such as gases at high pressures and supercritical temperatures. When the ideal gas assumption applies, equation 30 becomes equation 31:

$$\epsilon_{\mathrm{local}} = \mathrm{RT}\ln(p_{\mathrm{local}}/p_{\mathrm{ambient}}) \qquad (31)$$

This equation is one of a number of analogies that can be drawn between adsorptive and gravitational energies, and which facilitate a physical understanding of London-force adsorption.

Magnitudes of Adsorption Energies

The (earlier considered) adsorption energies per unit of liquid volume will be found to be useful in considering the competition between different adsorbates in multicomponent adsorption. However, the adsorption energy per mole is the important variable for determining adsorption of single components from the vapor phases or from solution. The reasons for the variation of adsorption energy with loading are now considered.

Consider the adsorption energy or potential of a single molecule between two carbon planes that touch the molecule from both sides. Here the potential energy loss is expected to be about twice what it would be for the same molecule on a single planar surface. The *deep well,* or maximum adsorption energy, would be about two and one-half times the potential for the closest approach to a plane. This provides some idea of the range of adsorption energies to be expected.

It may be anticipated that in the Polanyi model the physical adsorption of gases and vapors takes place primarily by the concentration of the vapors to their saturation pressures, at which point they condense to liquid-like adsorbates. Incidentally, in adsorption from solution the solutes are concentrated to saturation with similar results. It is useful to estimate the extent of concentration that can be achieved by the available adsorption energies. From the Boltzmann equation, and taking common logarithms, equation 32 is derived:

$$\log_{10} p_2/p_1 = \epsilon/2.3RT \qquad (32)$$

At 300 K, for example, taking R as 1.985 cal/deg mol, equation 32 becomes equation 33

$$\log p_2/p_1 = \epsilon(\text{cal/mol})/1370 = \epsilon(\text{J/mol})/5733 \qquad (33)$$

For n-hexane, for example, ϵ is about equal to its heat of vaporization of 7200 cal/mol or 30,000 J/mol at a capacity of about 5% by weight. The adsorption energy can therefore concentrate the hexane vapor by about five decades. At very low capacities the adsorption potential is about doubled, in which case about 10 decades of concentration can be achieved. These magnitudes are smaller for smaller molecules and larger for larger molecules. It is this variation of molar adsorption energies with molecular size and with capacity, together with the wide ranges of volatility (and for solutes, of solubility), that is responsible for the wide range of effectiveness of adsorption on activated carbon for different systems.

BIBLIOGRAPHY

1. S. Brunauer, P.H. Emmett, and E. Teller, *J. Am. Chem. Soc.* **60,** 309 (1938).
2. S.J. Gregg and K.S. Sing, *Adsorption, Surface Area, and Porosity,* Academic Press, Inc., New York (1967).
3. *1994 Annual Book of ASTM Standards,* Section 115, Vol. 15.01, American Society for Testing Materials, Philadelphia, Pa., 1994.
4. F. London, *Z. Phys. Chem.* **B11,** 222 (1930).
5. R. Vidic, M. Suidan, and R. Brenner, *Environ Sci. Technol.* **27,** 2079 (1993).
6. H.P. Boehm, *Advan. Catal.* **16,** 79 (1966).
7. S. Brunauer, *The Adsorption of Gases and Vapors,* Vol. I, Princeton University Press, Princeton, N.J., 1945.
8. A.W. Adamson, *Physical Chemistry of Surfaces,* John Wiley and Sons, Inc., New York, 1976.
9. B. Cohen, *Rad. Protec. Man.* **5,** 47 (1989).
10. J.H. Mattson, H.B. Mark, Jr., M.D. Malbin, W.J. Weber, Jr., and J. C. Crittenden, *J. Colloid Interface Sci.* **31,** 116 (1969).
11. C.T. Chiou, *J. Phys. Chem.* **78,** 622 (1974).
12. R. Defay, I. Prigogine, A. Bellemans, and D.H. Everett, *Surface Tension and Adsorption,* John Wiley and Sons, Inc., New York, 1966.
13. W.D. Harkins, *The Physical Chemistry of Surface Films,* Reinhold Publishing Co., New York, 1952.
14. M. Manes and C.J. Zahradnik, *J. Paint Tech.* **47,** 44 (1975).
15. A.M. Schwartz, *Ind. Eng. Chem.* **61,** 10 (1969).
16. M. Manes and C.J. Zahradnik, *J. Paint Technol.* **47,** 43 (1975).
17. M. Polanyi, *Verh. deut. physik. Ges.* **16,** 1012 (1914); **18,** 55 (1916).

ACTIVATION ANALYSIS

PETER BODE
JEROEN J. M. DE GOEIJ
Interfaculty Reactor Institute, Delft University of Technology
Delft, the Netherlands

Activation analysis is a method for qualitative and quantitative element determination. The method is based on conversion of stable nuclei of atoms into radioactive ones and subsequent measurement of characteristic nuclear radiation emitted by the radioactive nuclei. The conversion into radioactive nuclei is performed by bombardment with neutrons, energetic charged particles, or high energy photons.

NUCLEAR STRUCTURE, RADIOACTIVITY, NUCLEAR REACTIONS

Nuclear Structure

In a simple representation an atom is built up from a nucleus and electrons, organized into electron shells. The nucleus contains protons and neutrons. In a neutral atom, the number of protons equals the number of electrons, and is denoted with Z, the atomic number. The number of electrons determine the chemical characteristics of the atom; therefore, Z is related to these characteristics. The total number of protons and neutrons is called the mass number, indicated by A. Atoms with the same number of protons but with different numbers of neutrons are called isotopes. Though their mass numbers are different, their chemical behaviour is identical. Some elements have only one stable isotope, eg, fluorine with 9 protons and 10 neutrons in the nucleus. Most of the elements have two or more stable isotopes (Fig. 1). For instance, chlorine has two stable isotopes, each of them with 17 protons, but one with 18 neutrons and one with 20 neutrons. The mass number is used to distinguish different isotopes of one element. To this end, the mass number is written as a superscript to the left of the chemical symbol of the element, eg, ^{35}Cl and ^{37}Cl. The fraction in which a particular stable isotope occurs in the mixture of isotopes of an element is called isotopic abundance. Not all combinations of protons and neutrons are stable: some are unstable in that they lose their excess energy by radioactive decay. An unstable isotope is called a radioactive isotope or radioisotope.

Radioactivity

It has been described that a combination of protons and neutrons in an atomic nucleus may have an excess of energy, which makes the nucleus unstable. In such a case it releases its energy by radioactive decay under emission of nuclear radiation. This is also denoted as nuclear disintegration; the unstable nucleus is called radionuclide. Four types of nuclear radiation can be distinguished:

1. α-Radiation: emission of a helium nucleus, containing 2 protons and 2 neutrons; the new nucleus gets a mass number 4 units lower.
2. β^--Radiation: emission of (negative) electrons. In β^--decay effectively a neutron transforms into a proton in the atomic nucleus. The mass number remains the same, but the atomic charge changes, and thus an isotope of a different element results from the decay.
3. β^+-Radiation: positive electrons or positrons. In β^+ decay, effectively a proton transforms into a neutron. As an alternative to β^+-decay, the nucleus may capture an electron from one of the electrons shells. (electron capture (EC)). The vacancy in the shell is filled via electron transitions from outermost shells and in this

Protons	114	115	116	117	118	119	120	121		
79	Au 193	Au 194	Au 195	Au 196	Au 197 7.8 s γ279 | 100 σ 98.8	Au 198 2.3 d Iγ215 | 269 d β^- 1.0 γ412	Au 199 3.14 d β^- 0.3 γ158	Au 200 18.7 h β^- 0.6 γ498 | 2.69 d β^- 2.3 γ368	Au 201 26.4 m	Au 202
78	Pt 192 0.79 σ 2.2 + 14	Pt 193 4.33 d Iγ136 e^- | 2.5 h β^- 1.0 γ99	Pt 194 32.9 σ 0.09 + 1.2	Pt 195 4.02 d Iγ99 | 33.8 σ 27	Pt 196 25.3 σ 0.05 + 0.7	Pt 197 94.4 m Iγ346 e^- | 18.3 h β^- 0.6 γ77;191	Pt 198 7.2 σ 0.027 + 3.7	Pt 199 13.6 s Iγ392 e^- | 30.8 m β^- 1.7 γ543	Pt 200 12.5 h β^- 0.6; 0.7 γ76; 137	Pt 201 2.5 m
77	Ir 191 4.94 s Iγ129 e^- | 37.3 σ 300 + 624	Ir 192 241 yr Iγ161 | 1.4 m γ317 e^- β^- | 74 d β^- 0.7 γ317	Ir 193 10.6 d Iγ80 e^- | 62.7 σ?+110	Ir 194 171 d β^- γ484 | 19.1 h β^- 2.2 γ328	Ir 195 3.8 h β^- 0.4 γ320 | 2.5 h β^- 1.0 γ99	Ir 196 1.4 h β^- 1.2 γ394 | 52 s β^- 3.2 γ356	Ir 197 8.9 m β^- 2.0 | 52 s β^- γ470; 431	Ir 198 8 s β^- γ407; 507		

Neutons

Figure 1. Part of the chart of radionuclides showing stable and radioactive isotopes of the elements: ■ = stable isotopes; ■ = radioisotopes; □ = metastable radioisotopes.

process characteristic X-radiation is emitted. In both processes, the mass number remains the same, but the atomic charge changes, and thus an isotope of a different element results from the decay.

4. γ-Radiation: emission of high energetic electromagnetic radiation. In this case the nucleus is at a so-called excited level and decays to a less excited level or even direct to the ground level, without changes in either the number of protons and neutrons, and thus in the atomic number. The latter process is also denoted as isomeric transition. Excited levels of an atomic nucleus often result from preceding α, β^-, β^+, and/or EC decay, and incidentally these excited levels may result directly upon activation. Often, the decay via excited levels takes place by subsequent emission of two or more γ's. This is called a γ-cascade, since often the excited level is populated and depopulated in a very short time (eg, in μs). The related radiations are in coincidence. In a few cases, the excited level has a measurable lifetime from seconds (s) to hours (h) or even longer. Such a level is considered a special radionuclide, and denoted with the symbol m (metastable), eg, ^{116m}In.

In radioactive decay generally more than one type of radiation is emitted, such as β^-- and γ-radiation. The process of radioactive decay can be derived from a decay scheme (Fig. 2). Not every transition in γ-decay results in γ-radiation. Some γ-rays are converted, ie, instead of the emission of γ-rays, electrons are thrown out of the shells. The fraction of the number of disintegrations that results in γ-radiation is called the γ-yield. This is also given in the decay scheme.

Gamma-radiation is the radiation of choice in activation analysis since it is monoenergetic and in most cases characteristic for the emitting nucleus. The other advantage of γ-radiation is that it has a high penetrating power, so that it is hardly absorbed in the radioactive material itself.

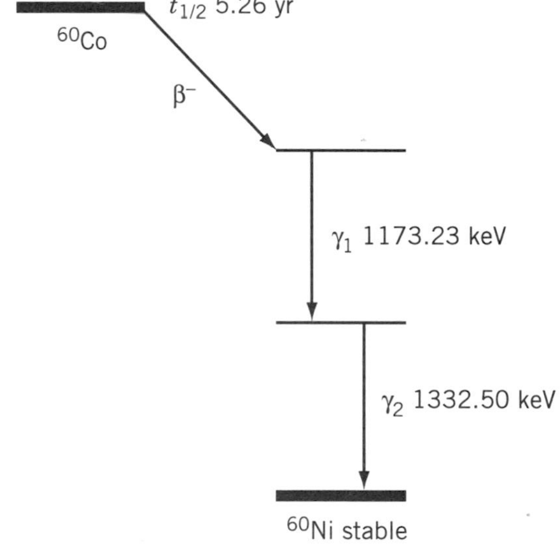

Figure 2. Simplified decay scheme of the radionuclide ^{60}Co.

Radioactive decay is a stochastic process, so it cannot be predicted exactly when a radioactive nucleus will decay. However, which fraction of the original number of radioactive nuclei remains after a given time is measurable. The decay rate is proportional to the number of radioactive nuclei:

$$-\frac{dN}{dt} = \lambda N \tag{1}$$

where $-dN/dt$ = the decrease of the number of radioactive nuclei per unit of time, often denoted with the term activity or disintegration rate (s^{-1}), N = the number of radioactive nuclei, and λ = the decay constant(s^{-1}).

The disintegration rate is denoted with Becquerel (Bq), which is the equivalent of 1 disintegration per second.

Instead of the decay constant, the half-life of a radionuclide is the preferred term in characterizing radionuclides. The half-life ($t_{1/2}$) is the time in which the number N_0 of radioactive nuclei has been reduced by a factor of 2:

$$N_{t=t_{1/2}} = 0.5\ N_0 = N_0 \exp(-\lambda t) \quad (2)$$

from which it follows that $\lambda = \ln 2/t_{1/2}$. The half-life is characteristic for a radionuclide and may range from a fraction of a second to millions of years.

Nuclear Reactions

Activation analysis requires the conversion of stable nuclei into radioactive nuclei or radionuclides. This conversion is attained via a nuclear reaction, resulting from irradiation or bombardment of the target nuclei with neutrons, energetic charged particles (eg, protons), or high energy photons. The nuclear reactions also often lead to the production of additional charged particles, neutrons, and/or photons. Examples of nuclear reactions are shown in equations 3 and 4:

$$^{18}\mathrm{O}(Z = 8) + {}^{1}\mathrm{H}(Z = 1) \rightarrow {}^{18}F(Z = 9) + {}^{1}\mathrm{neutron}(Z = 0) \quad (3)$$

and

$$^{23}\mathrm{Na}(Z = 11) + {}^{1}\mathrm{neutron}(Z = 0) \rightarrow {}^{24}\mathrm{Na}(Z = 11) + {}^{0}\mathrm{photon}(Z = 0) \quad (4)$$

These equations show that the total number of protons and neutrons does not change during the reaction.

The usual notation of nuclear reactions, as an example for these two reactions, is as follows:

$$^{18}\mathrm{O}(p, n)^{18}\mathrm{F} \qquad \mathrm{Na}(n, \gamma)^{24}\mathrm{Na}$$

or, in general,

$$A'(x, y)B$$

where A' is the isotope of element A; x, bombarding particle or photon; y, particle and/or photon emitted upon reaction (also denoted as prompt radiation); and B, produced nuclide (usually radioactive). For x and y the following notations are used: γ = gamma, p = proton, α = alpha particle, n' = neutron, and f = fission which implies that the neutron captured splits the nucleus into two or more fragments.

The probability that a nuclear reaction between a bombarding particle or photon with an atomic nucleus will take place is expressed as cross section, σ, a fictitious surface area of the atomic nucleus. The value of the cross section depends on the atomic nucleus, the reaction under consideration, the type of bombarding particle, and its energy. For many nuclear reactions, except for the (n, γ) reactions, a lower threshold energy exists below which no reaction takes place. The cross section is often expressed in the unit barn, b, corresponding to 10^{-28} m^2.

ACTIVATION ANALYSIS: PRINCIPLES

When a nuclear reaction results at a radioactive nucleus, the process is denoted as activation; the bombardment is irradiation. In the reaction $A'(x,y)B$, the produced radionuclide B decays to a stable atomic nucleus C under emission of characteristic radiation, as described in previously. Often, characteristic γ-radiation is emitted. By determining the energy of the γ-radiation and using the decay schemes, the radionuclide B may be identified. Eventually, this identification may be even more selective by examining the half-life of radionuclide B. When B has been identified, the nature of the target isotope A' can be derived from knowledge of the nuclear reaction that has taken place. It all forms the basis for qualitative activation analysis. Quantitative activation analysis is based on measurement of the intensity of the radiation, ie, measurement of the activity of radionuclide B. This radioactivity is proportional to the number of target nuclei A' of element A in the irradiated sample. Thus, in activation analysis the following typical steps can be distinguished: activation, decay, measurement, and interpretation.

Activation analysis is not a method that can simply be purchased as "push-button" equipment as in atomic absorption spectroscopy, X-ray fluorescence spectrometry, or inductively coupled plasma spectrometry, although the necessary equipment and software for activation analysis are largely commercially available.

ACTIVATION

Sources for Activation

The particles or photons required for activation can be obtained by various means. Energetic charged particles, such as protons, deuterons, or α-particles are generated with a cyclotron. The disadvantage of charged particle activation is the limited penetration depth (only several hundreds μm), thus it is actually a surface analysis technique. Moreover, high energy dissipation upon bombardment places demands on radiation and temperature stability of the samples and requires cooling during irradiation. High energetic photons with energies of several tens of MeV can be produced as Bremsstrahlung with an electron accelerator. High energetic photons have a good penetration power, thus enabling bulk analysis. However, problems may also arise in respect to heat generation. Furthermore, the cross sections for the related nuclear reactions are rather small. Charged particle activation analysis and photon activation analysis have only advantages compared to neutron activation analysis especially with low Z, for particular elements such as H, C, N, and O.

Fast neutrons with energies of several MeV can be produced with a neutron generator or in an isotopic neutron source. In a neutron generator a nuclear reaction between deuterium nuclei, accelerated up to about 250 keV, and tritium nuclei is evoked, in which neutrons with an energy of about 14 MeV are produced. Typically the yields are between $10^{10}-10^{12}$ neutrons s^{-1}. In isotopic neutron sources a nuclear reaction takes place between an α-particle and a low Z-element (eg, Be) and neutrons with energies of several MeV are produced. The α-particle is provided by radioactive decay of ^{226}Ra or ^{238}Pu. Alternatively, the radionuclide ^{252}Cf decays by spontaneous fission under emission of neutrons of several MeV. The neutron yields of isotopic neutron sources vary typically between 10^4 and 10^9 neutrons s^{-1}, depending on the amount of, eg, ^{252}Cf.

It is not so much the neutron yield as the neutron fluence rate (or neutron flux) that is an important parameter for activation analysis. The neutron fluence rate is the number of neutrons per second per square meter. With neutron generators and isotopic neutron sources the ratio between neutron yield and neutron fluence rate depends strongly on type and construction. The high energy of the neutrons implies that mainly activation reactions of the (n,p), (n,α), and $(n,2n)$ type will take place. Fast neutrons have a high penetration depth but in samples with high concentrations of light elements, in particular hydrogen, moderation takes place. The energy of the neutron reduces, which may result in a final energy lower than the threshold energies for the reactions.

The nuclear research reactor is the most common source of neutrons for activation analysis, and typical neutron fluence rates range from 10^{16} to 10^{18} $m^{-2}s^{-1}$. The majority of the neutrons in a nuclear reactor is in thermal equilibrium with its environment, and therefore denoted as thermal neutrons. With thermal neutrons usually reactions of the (n,γ) type result. The gamma radiation emitted during the reaction is called a prompt gamma. Many nuclei have a high cross section for (n,γ) reactions, whereas others nuclei have a low cross section, eg, the low Z-elements H, C, N, O, and Si. In many matrices these are the major elements; because of their low cross sections neutrons are hardly absorbed in many materials. Since the thermal neutrons are already in thermal equilibrium, no heat is generated upon irradiation. A few elements, such as Li, B, and Cd, have such high cross sections that if they are present in large quantities their capture of neutrons may result in a local neutron attenuation or neutron self-shielding.

The advantage of isotopic neutron sources and neutron generators is that they are relatively inexpensive, easily transported, and generators can be switched off when not in use. Their main application lies with in-field activation (eg, borehole logging) or industrial analysis of bulk samples. In most cases only the major elements can be determined.

Neutron activation analysis (NAA) (1–3) has some attractive aspects. First, the cross sections are generally high and high neutron fluence rates are available, so that good detection limits may be achieved for a sizeable number of elements. Second, neutrons penetrate well in materials and relatively homogenous neutron fluence rates may be obtained, so that bulk analyses may be performed. Third, the heat production in the sample is small. Finally, irradiation may be performed as a parasitic activity in a nuclear research reactor and ample irradiation facilities are available or accessible. Activation analysis with thermal reactor neutrons has become the most commonly applied method of activation analysis.

INSTRUMENTAL NEUTRON ACTIVATION ANALYSIS

The term nondestructive or instrumental neutron activation analysis (INAA) is used when the sample is not chemically processed after irradiation, and only measure a mixture of radionuclides may be expected and, consequently, also a range of gamma-ray peaks in the spectrum. Simultaneous determination (in one measurement) of many gamma-ray peaks of various radionuclides make INAA a multielement technique. It emphasizes the importance of detectors with high energy resolution properties. But even with the state-of-the-art semiconductor detectors overlap of peaks may occur. However, the resulting doublets or multiplets can often be resolved mathematically without too many difficulties.

The Compton continuum in the gamma-ray spectrum reduces the signal-to-noise ratio of the peaks. As a consequence, small photopeaks may be difficult to detect. This problem can be partly encompassed by recognizing that the various radionuclides decay with different half-lives. A long irradiation time, a long decay time, and a long counting time are advantageous for the measurement of the radionuclide with the longest half-life, without many interferences from the other radionuclide produced. Similarly, short irradiation, short decay time, and short counting time are advantageous for the measurement of the short half-life radionuclide. If the interfering Compton continuum, therefore, is due to radionuclides with a shorter half-life than the radionuclide of which the gamma-ray peak is located at this continuum, a longer waiting time may improve the detectability of this peak. When the situation is reversed, however, the irradiation time and decay time have to be shortened to attain a better detectability.

In many cases these principles are combined with the phenomenon that a number of elements have more than one activation product, often with different decay times. For instance, the element zinc emits gamma-ray radionuclides ^{71}Zn ($t_{1/2}$ 2.4 min, $E_\gamma = 511.7$ keV), ^{69m}Zn ($t_{1/2}$ 14.1 h, $E_\gamma = 438.6$ keV), and ^{65}Zn ($t_{1/2}$ 243 d, $E_\gamma = 1115.5$ keV) after neutron activation. When, for instance, the interfering Compton continuum in a spectrum is primarily caused by ^{24}Na ($t_{1/2} = 14.9$ h, $E_\gamma = 1368.4$ keV), instead of determining zinc via measurement of ^{69m}Zn, eg, five days after irradiation, a better detection limit may be attained via measurement of ^{65}Zn after a decay time of eg, three weeks, after which the ^{24}Na activity has completely decayed. The shape of the gamma-ray spectrum thus changes continuously as a consequence of the differences in half-lives of the radionuclides produced.

In INAA samples are irradiated and/or counted more than once, all during different times. Sample sizes are usually kept limited to a maximum of approximately 0.1–1 g, so the neutron attenuation and gamma-ray attenuation effects can be neglected. Since the contribution by the blank (if any) can be well controlled and kept small and generally negligible, INAA is also a suitable method for handling very minute quantities of material, like air particulate matter or collected cosmic dust.

A typical analysis protocol may include a first-irradiation during 30 s, a 300 s decay, and a 600 s counting. This is followed by a second irradiation during 4 h, and measurements at 5 d and 20 d after irradiation during 1 h and 4 h, respectively. The results of the three measurements are combined to the final analysis report which may contain concentrations and detection limits on 50–60 elements. Sometimes use can be made of radionuclides with very short half-lives, eg, ^{77m}Se ($t_{1/2} = 17.5$ s) for the determination of selenium in biological material. The entire analysis can be completed in only 1–2 minutes. For a full multielement analysis, however, or for determination of certain elements, the total (passage) time may be on the order of 3–4 weeks due to the long decay times necessary to attain the required detection limit. The choice of the analysis protocol depends on the element(s) to be determined and required detection limit, and on the expectation and experience with interferences, typical for the type of material analyzed (eg, biological or geological).

Many materials of organic–biological and geological origin consist primarily of the elements H, C, N, O, Al, Si, and/or Ca. Upon irradiation with reactor neutrons no substantial or only short-lived radioactivity is produced. As a result, the matrix elements are apparently absent and the sample becomes transparent for radioactivity of the trace elements.

In principle, materials can be activated in any physical state, ie, solid, liquid, or gaseous. There is no fundamental necessity to convert solid material into a solution prior to activation. During activation the material is heated by the absorption of gamma-rays present in the reactor because of the fission of uranium and to a lesser extent to processes of activation of the various reactor materials. Radiolysis is another effect caused by the gamma-rays. In aqueous samples H_2, O_2, and H_2O are formed. The gases cause pressure buildup in the sample's container. Because of the explosion, hazard samples are preferentially dried prior to activation. Radiolysis may also lead to decomposition of proteins into gaseous compounds.

Because of the limited number of operations, INAA can relatively easily be automated and computerized. The latter may also include bookkeeping, control of spectrometers, and sample changers and quality control.

Advantages and Disadvantages of INAA

The most attractive analytical characteristics of INAA are included here.

1. The method is nondestructive. There is no need to convert the sample into a solution prior to analysis. This is advantageous for analysis of, eg, sediments, rocks, catalysts, plastics, etc. The nondestructive nature of the technique, however, is also important when precious or unique materials must be analyzed, such as archaeological and art objects, cosmic dust, or material from forensic research (no loss of evidence).
2. There are various adjustable experimental parameters that can be exploited so that the best detection limit for the desired element is achieved. The type, energy, and fluence rate of the neutrons may be varied which allows for selective activation. The irradiation time can be varied resulting in enhanced activation of elements with short half-life radionuclides. The differences in half-life between the various radionuclides is a tool to discriminate between them. Spectral interferences can be overcome by using multiple gamma-ray lines and multiple radionuclides to identify the elements.
3. The method provides simultaneous multielement data. Analysis reports may contain concentrations and detection limits on 50–65 elements.
4. For many elements detection limits can be obtained at the $\mu g kg^{-1}$ to $mg kg^{-1}$ level.
5. NAA is a method based on processes that take place in the nuclei of the elements. The chemical form of the elements does not influence the final result.
6. The physical nature of the technique, by which each individual step can be described in a parametric form, leads to a basis for accurate results independently of the concentration over a large dynamic range. In addition, the method has self-validating character: concentrations should agree, within limits of uncertainty, when different peaks are considered of one or more radionuclides produced from a given element.
7. Standardization in INAA is done "once and for all". There is no need to determine new standardization parameters for every new type of material to be analyzed.
8. The absence of any sample pretreatment steps (with the possible risk of introducing contaminants or losing elements) makes the technique suitable for measurement of low concentrations in, eg, ultrapure materials, and for handling very small quantities of materials such as cosmic dust and air particulate matter.
9. Of the major components present in many samples, the elements C, H, N, and O are almost insensitive to neutron activation, whereas from many other major elements (eg, Si or Al) only short half-life radionuclides are formed. As such, they are virtually absent, making the sample by the long-lived radionuclides transparent for its trace elements.

The following disadvantages can be noted, when compared to other analytical techniques:

1. Several elements cannot be determined, at least not at a low level. For instance, activation of Pb results only at milligram quantities to measurable activities; for applications related to environmental research such a detection limit is inadequate.
2. INAA is not suitable for a sensitive analysis of water samples without extensive pretreatment.
3. The method, like most other methods of elemental analysis, only yields information on total element concentrations, not specified to their chemical form and/or physical state.
4. For elements involving radionuclides with long half-lives, the turn-around time of an analysis may be on the order of 2–4 weeks.
5. INAA is not available as push-button equipment with a complete software package to be operated on-the-spot at any random analytical laboratory.
6. The INAA laboratory has to meet the legal requirements for radiological safety. This may imply high investment costs when a laboratory has to be adapted. A radiological health officer may be required and provisions exist for storage and disposal of radioactive waste. Employees must be trained in radiological safety and practical aspects of handling radioactivity.
7. The merge of a physical method and chemical analysis requires simultaneous expertise from both disciplines for an effective operation. Not all NAA laboratories have such an interdisciplinary composition.
8. The requirement of the availability of or access to a nuclear research reactor.

Applications

Neutron activation analysis (instrumental and radiochemical) has been applied in a vast range of applied sciences. Examples of applications of conventional, reactor based INAA are included here (4):

Archaeological: ceramics, pottery, glass, coins

Biological: aquatic and marine biota, medicinal plants, snake venom

Environmental: atmospheric samples, aerosols, volcanic gases, sediments, plants, wastes, industrial raw products, lichens, mosses, tree leaves, water

Forensic: hair, skin tissue, finger/toenails, bomb debris, paint, glass fragments, drugs, bullet lead

Geological: cosmic spheres, fossil fuels, ores, separated minerals, sand, sediments

Human: bone, brain, breast tissue, blood, hair, kidney, liver, teeth, finger/toenails, seman, urine

Materials: alloys, ceramics, refractory materials, plastics, quartz, semiconductors, superconductors, ultrapure materials

Nutritional: composite diets, cigarettes, corn, cereals

Pharmaceutical: drugs, medicine

Neutron activation analysis is not always selected as a method of elemental analysis due to the developments of other techniques, particularly ICP MS, which are not only commercially available as standalone apparatus, but which also provide data at a much shorter turnaround time. It is generally accepted that the various methods of elemental analysis each have their own specific niches although they may be usable for almost any analytical quest. For NAA these areas deal with (*1*) samples that are difficult to dissolve completely; (*2*) samples in which the trace element concentrations are so low that there is a risk of substantial contamination during sample destruction, dissolution, or other processing; (*3*) analyses in which a high degree of accuracy has to be obtained, and (*4*) very small and very large analytical portions. Many of the examples given above are closely related to these areas.

BIBLIOGRAPHY

1. D. de Soete, R. Gijbels, and J. J. Hoste, *Neutron Activation Analysis,* John Wiley & Sons, Inc., New York, 1972.
2. S. J. Parry, *Activation Spectrometry in Chemical Analysis,* John Wiley & Sons, Inc., New York, 1991.
3. W. D. Ehmann and D. Vance, *Radiochemistry and Nuclear Methods of Analysis,* John Wiley & Sons, Inc., New York, 1991.
4. N. M. Trahey, ed., *NIST Standard Reference Materials Catalog 1995–1996,* NIST Special Publication 260, NIST, Gaithersburg, Md., p. 84.

ADSORPTION ON SOILS. See SOIL SORPTION OF ORGANIC POLLUTANTS AND PESTICIDES.

AIR MONITORING, OPTICAL SPECTROSCOPIC METHODS

MARKUS W. SIGRIST
Institute of Quantum Electronics, ETH Zurich
Zurich, Switzerland

This article discusses various optical spectroscopic methods employing ultraviolet, visible, and infrared wavelengths and their applications to detecting atmospheric trace gases in different environments. Although lasers are important in diverse areas including environmental sensing, nonlaser-based spectroscopic systems also have great potential and the most promising schemes presently are treated here. Because *the* optical scheme that fulfils all of the requirements of an air monitoring system does not exist, this chapter tries to provide an overview of presently used spectroscopic methods in air monitoring and of their potentials and limitations.

The first technique to be discussed is differential optical absorption spectroscopy (DOAS). The Lidar scheme yielding range-resolved information on air pollutants is treated next, followed by conventional infrared spectroscopy. The tunable diode laser absorption spectroscopy (TDLAS) is discussed in the next section. Finally, the features of photoacoustic spectroscopy (PAS) are outlined in the last section.

The enormous variety and concentration ranges of atmospheric trace gases and the complex chemical interactions among them set a challenging task to detecting them. Ideally, a detection technique should fulfill the following requirements:

1. High sensitivity to enable detecting very low concentrations.
2. High selectivity to differentiate among different species in a multicomponent mixture; this permits quantitative analysis or at least selective monitoring of specific compounds in mixtures.
3. Feasibility of detecting numerous compounds with a single instrument (multicomponent capability).
4. Large dynamic range to monitor low and high concentrations with the same instrument.
5. Good temporal resolution to enable on-line monitoring.
6. Good portability for *in situ* measurements or even capability for remote sensing.
7. Easy sampling procedure eventually with pretreatment of the sample or contactless "open-air" measurement.

Depending on the situation, the importance of these requirements varies. For example, it may suffice to know the total hydrocarbon content of a sample, eg, of car exhausts, and selectivity would thus be a minor concern. On the other hand, selectivity is crucial for understanding and modelling photochemical smog formation. In some cases it is desirable to have local data (point monitoring), whereas in other cases the integrated pollution situation over a certain area in the sense of pollution load is of primary interest. Finally, sometimes range-resolved information, eg, vertical profiles of the pollutant distribution, is desirable.

During the past few decades various different air monitoring schemes have been developed and extensively discussed in the literature. Continuous impetus to novel developments has arisen from progress in modern physical and chemical research, especially in surface physics and optics and in electronics and computerization.

SPECTROSCOPIC SCHEMES

This article focuses on modern optical techniques applied to air monitoring. Although known for quite some time, optical methods have gained importance because they offer some unique features that render such schemes very versatile for trace gas monitoring. The progress is discussed in some selected reviews (1–7). Furthermore, still novel developments are expected from such techniques combined with present and future laser technologies. Today, there is a great variety of optical schemes. Figure 1 separates them into those with conventional light sources, such as Xe lamps, and laser-based devices. The latter

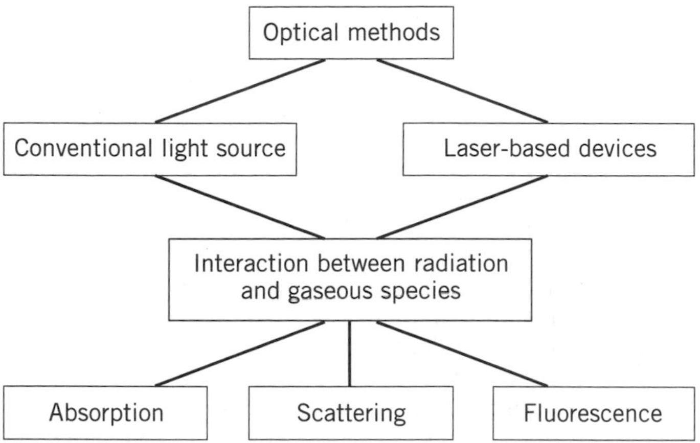

Figure 1. Schematic diagram of sources and interaction processes used in air monitoring by optical techniques.

essentially rely on lasers tunable in appropriate wavelength ranges. The advantage of lasers is the extremely high spectral brightness not achievable with conventional sources combined with filters or gratings. This characteristic of lasers results in high spectral resolution which is crucial for detection selectivity. On the other hand, conventional light sources often offer a larger tuning range to compensate for the lack of spectral brightness which is also important for the desired multicomponent capability. Furthermore, broadly tunable lasers are still often complex and expensive devices which hinder their wide distribution. However, this situation may change in the future with the development of compact diode-laser-based systems.

As Figure 1 implies, the basis of all optical schemes is an interaction process between the radiation and the air mass being monitored. It is thus convenient to differentiate between the essential processes involved, namely absorption, scattering, and fluorescence. Hence, these methods rely on the spectroscopic properties of the gases under study. However, an important category of optical detectors called fiber optic sensors (8) usually uses other interaction processes. In the case where the fiber only acts as a lightguide to a sensor head spectroscopic properties are still used for detection. But often other sensing principles are applied, such as the probing of the evanescent field in specifically sensitized films on the fiber surface or the direct interaction of the evanescent field with the investigated substance. Hence, the transfer of the chemical interaction into a measurable signal is the key. Apart from their important role as sensors in liquids, eg, for detecting oil traces in water and pH, p_{CO2} and p_{O2} values in blood, their use in ambient air monitoring is still limited. So far fiber optic sensors are installed in mines and gas wells for detecting methane. They are sometimes also used for ambient air sensing, mainly for CO and CO_2. However, these sensors are specifically designed for these specific gases and hence lack the multicomponent capability inherent in spectroscopic devices. Therefore they are not discussed in this article.

The different types of scattering processes employed in true spectroscopic schemes are illustrated in Figure 2. The solid lines indicate real atomic or molecular energy levels. Initial and final levels are denoted by i and f, respectively, whereas ν_O and ν_S represent the frequency of the incident and scattered radiation, respectively. Figure 2**a** refers to elastic scattering (Rayleigh and Mie), where no frequency shift occurs ($\nu_S = \nu_O$). Elastic scattering is important for atmospheric studies on aerosols, clouds, and particulates, eg, by light detection and ranging (lidar) techniques (see the later section), but not for trace gas detection because the scattering process per se is not molecule-specific. The situation is, however, different

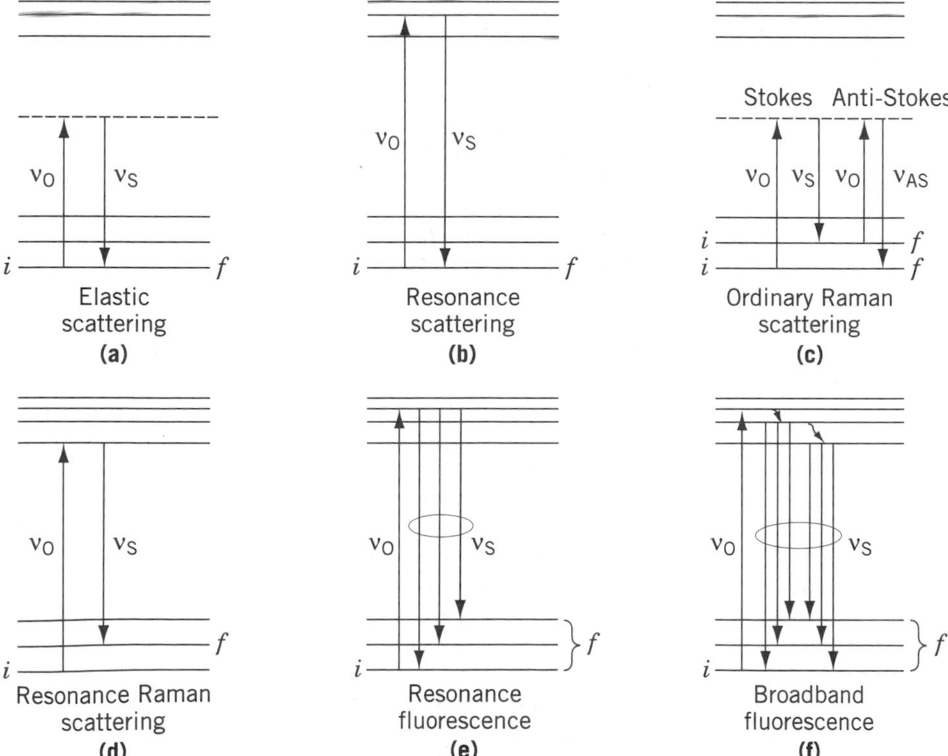

Figure 2. (**a–f**) Schematic atomic or molecular energy level diagrams with real (—) and virtual (- - -) energy levels. Various interaction processes associated with scattering and fluorescence are shown.

for all the other scattering schemes which involve specific molecular or atomic levels and hence offer the potential for species-selective detection and for quantitative analyses of multicomponent gas mixtures. Except for the ordinary (nonresonant) Raman scattering (Fig. **2c**), all schemes involve transitions between specific energy levels and thus require tunable radiation sources. The resonant scattering (Fig. **2b**) is applied successfully for observations in the mesosphere (9). The intriguing feature of Raman scattering (Figs. **2c** and **2d**) is the fact that with a laser, emission at only one wavelength (untuned for nonresonant and tuned for resonant) is necessary to detect numerous gases because the selectivity is given by the frequency shifts $\pm\nu_r$ with $\nu_r = \nu_O - \nu_S$ for the Stokes and $\nu_r = \nu_{AS} - \nu_O$ for the anti-Stokes process. The shifts ν_r correspond to vibrational-rotational frequencies characteristic of the molecules under study. The main disadvantages are the low scattering cross sections that are typically 10^3 times lower for the nonresonant case than Rayleigh scattering cross sections and possible interferences with fluorescence. This restricts Raman scattering methods to recording rather high molecular concentrations and/or short ranges. Furthermore, because the cross section varies with $\nu_O{}^4$, as in Rayleigh scattering, studies must be performed at short wavelengths (UV to visible). The application of resonant Raman scattering yields an increase in the cross section of 3–6 orders of magnitude compared to the ordinary Raman process but requires radiation tunable close to a proper molecular or atomic resonance. Raman scattering is used in short-range lidar systems as discussed later on.

The remaining scattering processes relate to resonance fluorescence (Fig. **2e**) and broadband fluorescence (Fig. **2f**). This phenomenon involves spontaneous emission of a photon following excitation into an excited state by absorption of the incident radiation at ν_O. The excited level decays by reemitting photons via transitions to lower levels, resulting in discrete emission peaks (Fig. **2e**), or by a combination of broad fluorescence and radiationless transitions caused by collisions with other atoms or molecules (Fig. **2f**). The fluorescence schemes require a tunable excitation source but yield specific information about the atomic and molecular species responsible for the fluorescence. The cross sections for fluorescence emission are small and comparable to those for Raman scattering (2). Furthermore, fluorescence is quenched in the atmosphere by collisions with air molecules. The elegant technique of laser-induced fluorescence thus plays a minor role in monitoring pollution in the troposphere. However, sodium atoms in the mesosphere at an altitude of 90 km are monitored by resonant scattering (also referred to as resonance fluorescence) as mentioned above. Other examples include hydrospheric pollution monitoring with airborne sensors or combustion diagnostics.

The most important tool for the spectroscopically detecting gaseous air pollutants is based on measuring the absorption of radiation by trace species. This approach inherently permits identifying and determining the concentrations of different species because the absorption features and strengths are molecule-specific. The spectra occur over a broad portion of the electromagnetic spectrum, from the microwave region, where rotational transitions are responsible for absorption, to the vacuum ultraviolet (VUV) where outer shell electronic transitions cause the characteristic absorption (2). However, the main wavelength ranges for chemical analysis and spectroscopic detection of gases are the mid- or fundamental infrared between, say, 2.5 and 25 μm and the visible-to-ultraviolet (700–250 nm). The middle part of Figure 3 shows the main absorption ranges of some selected molecules of environmental concern. In the mid-IR range, the so-called "fingerprint" region, absorption is caused by fundamental (as well as some overtone) and combination vibration-rotational bands, whereas electronic transitions with vibrational-rotational structure contribute to absorption in the visible-to-UV region. Molecular absorption cross sections in the mid-infrared at atmospheric pressure, typically on the order of 10^{-17} to 10^{-19} cm^2 are thus 6 to 8 orders of magnituted higher than Raman cross sections. This results in a considerably increased detection sensitivity which further enhances the attractiveness of absorption schemes for trace gas monitoring, particularly in the infrared range.

Absorption spectroscopy of *trace* gases involves measuring small absorption coefficients resulting from low concentrations. This often requires long path lengths, anyhow, for remote detection in the free atmosphere and restricts the applicable wavelength ranges to atmospheric windows. The main windows occur for wavelengths shorter than 2.5 μm, from 3 to 5 μm, and 8 to 14 μm. Atmospheric H_2O vapor and CO_2 drastically reduce transmission beyond these spectral windows. The bottom part of Figure 3 illustrates atmospheric transmission for a horizontal path of 500 m at an altitude of 500 m for standard conditions (total pressure 950 mbar, temperature 20°C, relative humidity 50%) based on calculation by the LOWTRAN computer code. Long-path measurements and Lidar schemes are obviously restricted to atmospheric windows whereas short-path measurements are also performed outside those regions. The quantitative identification of air constituents is limited by the spectral resolution and by the minimum detectable absorption coefficient α_{min}, ie, by the sensitivity of the particular technique. A high spectral resolution is of primary concern for the proper identification of species, ie, for detection selectivity. On the other hand, the sensitivity determines the minimum concentrations that are measured reliably. Both aspects, sensitivity and selectivity, relate to the radiation source and detector, particularly to the tunability and spectral brightness of the source and to the spectral resolution and responsivity of the detector.

On the one hand, incoherent radiation sources, such as incandescent and arc lamps, are successfully used for air monitoring pollution. On the other hand, the development of tunable lasers has impelled the application of optical methods in air pollution control thanks to their high spectral brightness and low divergence. Various laser types are of interest including solid state, diode, dye, and gas lasers or laser systems combined with nonlinear optical processes such as optical parametric oscillators (OPO) and difference frequency generation (DFG). The tuning ranges of some of these systems are depicted at the top of Figure 3. Different measurement schemes are applied to measure the small absorptions encountered in trace detection. Most are based on the well-known Beer–Lambert absorption law relating the transmitted radiation power $P(\lambda)$ at wavelength λ to the incident power $P_0(\lambda)$:

$$P(\lambda) = P_0(\lambda)\exp[-\alpha(\lambda)\cdot L] \quad (1)$$

where $\alpha(\lambda)$ indicates the common absorption coefficient (in cm^{-1}) at wavelength λ and L is the absorption path length. The absorption coefficient α is related to the molecular

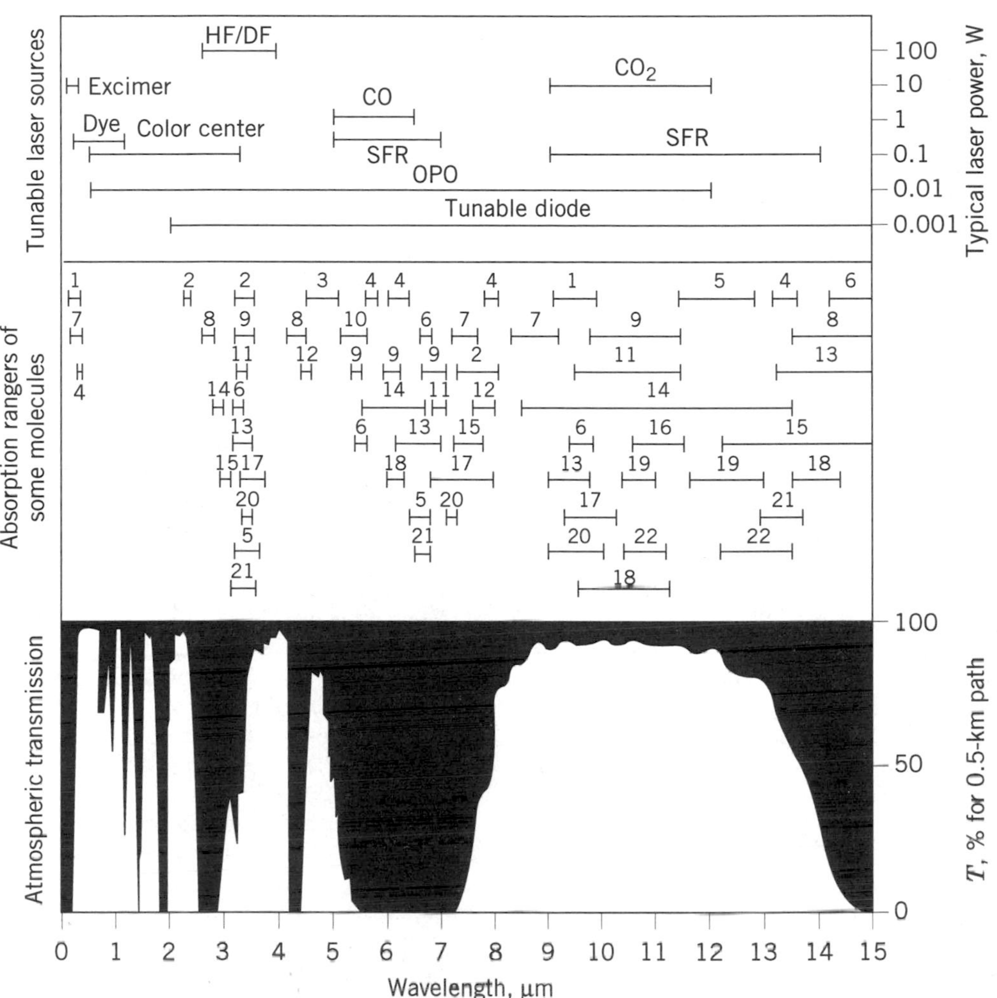

Figure 3. Emission ranges of available laser sources (SFR: spin flip Raman, OPO: optical parametric oscillator) including typical powers, absorption regions of various molecules of environmental concern (see list below), and transmission of the atmosphere for wavelengths between 0.25 and 15 μm.1, O_3; 2, CH_4; 3, CO; 4, NO_2; 5, C_2H_6; 6, C_6H_6; 7, SO_2; 8, CO_2; 9, C_3H_6; 10, NO; 11, C_2H_4; 12, N_2O; 13, C_7H_8; 14, NH_3; 15, C_2H_2; 16, C_4H_6; 17, CH_3OH; 18, C_2H_3Cl; 19, C_2HCl_3; 20, C_2H_5OH; 21, C_3H_8; and 22, C_2Cl_4.

absorption cross section $\sigma(\lambda)$ in cm^2 by

$$\alpha(\lambda) = N\sigma(\lambda), \tag{2}$$

where N indicates the density of the absorbing molecules in molecules per cm^3. Straightforward differential absorption measurements are limited to absorbances αL of $\approx 10^{-4}$ to 10^{-5}. Based on the Beer-Lambert absorption law (eq. 1) and a minimum detectable relative power loss $\Delta P_{\min}/P_0 \approx \alpha L$, a minimum detectable absorption coefficient $\alpha_{\min}$, the minimum detectable number of molecules in the optical beam $n_{\min}$, the minimum detectable density $N_{\min}$, and a minimum relative concentration $c_{\min}$ of molecular impurities in a mixture is evaluated. The corresponding expressions and numbers for a representative example are listed in Table 1. Here, A denotes the cross section of the optical beam and N_{tot} is the total density of molecules in the mixture. The numerical example is calculated for a typical molecular absorption cross section $\sigma = 10^{-18}$ cm^2, for $A = 1$ cm^2, an optical pathlength $L = 1$ m and for $N_{tot} = 2.5 \times 10^{19}$ cm^{-3}. This concentration corresponds to the total number of molecules present in 1 cm^3 of air at a total pressure of 1 bar and a temperature of 25°C. The numbers in Table 1 imply that the achievable detection limits permit monitoring ppb concentrations under the assumptions made for deriving these numbers. This concerns particularly the challenging limit of $\Delta P_{\min}/P_0 = 10^{-5}$ for measuring absorption. Furthermore, the data are derived for interference-free conditions. To some extent long path lengths L are employed to achieve sufficient sensitivity. This is accomplished by choosing an appropriate path through the (transparent) atmosphere with a separate sender and receiver yielding path-integrated data or by using a special cell design, such as a White cell combined with a spectrometer. The former approach is used in DOAS systems (see later section) whereas the latter is encountered, eg, in diode laser spectrometers (see that section).

Table 1. Main Parameters Used to Characterize Absorption Detection Limits

Parameter	Expression	Example[a]
Sensitivity limit in absorption	$\Delta P_{min}/P_0$	10^{-5}
Minimum detectable absorption coefficient	$\alpha_{min} = \frac{\Delta P_{min}}{P_0}\frac{1}{L}$	10^{-7} cm^{-1}
Minimum detectable number of molecules	$n_{min} = \frac{\Delta P_{min}}{P_0}\frac{A}{\sigma}$	10^{13}
Minimum detectable density of molecules	$N_{min} = \frac{\Delta P_{min}}{P_0}\frac{1}{\sigma L}$	10^{11} cm^{-3}
Minimum relative concentration of impurity (analyte) in mixture	$c_{min} = \frac{\Delta P_{min}}{P_0}\frac{1}{\sigma L N_{tot}}$	4 ppb

[a] Example calculated for $\Delta P_{min}/P_0 = 10^{-5}$, $L = 100$ cm, $A = 1$ cm^2, $\sigma = 10^{-18}$ cm^2 and $N_{tot} = 2.5 \times 10^{19}$ cm^{-3}.

Most commercial spectrometers operate with conventional broadband sources in conjunction with monochromators or interferometers to establish the desired spectral resolution. The latter device is known as a Fourier transform (FT) spectrometer. Compared with dispersive spectrometers, FT systems possess higher optical efficiency and greater efficiency in observation time because all spectral elements are simultaneously observed and measured. This unique feature and the available broad infrared range renders FTIR spectrometers very attractive for chemical analysis of multicomponent mixtures (10,11). However, the sensitivity is not always sufficient, eg, for measuring trace gases in ambient air.

The transmitted intensity in *differential* absorption measurements is recorded as the radiation wavelength is tuned across an absorption line of the species of interest or as it is switched from a wavelength with peak absorption ("on" position) to a nearby wavelength with negligible absorption ("off" position).

Commercial instruments based on differential optical absorption spectroscopy (DOAS) are available today. They consist of a broadband radiation source such as a Xe lamp, and a receiver combined with a high-resolution spectrometer located some 100 m or km apart. Two adjacent wavelengths are used to detect a specific compound. Hitherto, these devices mainly operated in the UV and visible range and were thus somewhat limited in the number of measurable species. However, they are successfully applied to monitoring various gases, including highly reactive species such as the important OH radicals with detection limits below the ppb range.

Long-path absorption spectroscopy with *lasers* has profited from recent developments of tunable IR lasers, particularly lead salt diode lasers (6,12). This technique of tunable diode laser absorption spectroscopy (TDLAS) offers very high spectral resolution, sensitivity, and time response. TDLAS is performed both with open atmospheric paths and with confined air samples in long-path cells. The tuning range of a single diode laser is limited, and the technique is thus not as universal as FTIR spectroscopy because generally a different diode must be used for each compound being measured. Nevertheless, the method is applied to *in situ* measurements of a great variety of atmospheric trace gases and detection limits in the ppt range are achieved.

The sensitivity of long-path absorption measurements is further improved by several orders of magnitude if the straightforward differential measurement is replaced by a first- or second-order *derivative* absorption measurement (2). In this scheme the wavelength of the incident radiation, generally of a tunable laser, is periodically modulated at a frequency ω_m across a small fraction of the line width of the molecular transition being measured. The absorbed and the transmitted power thus contain a time-varying component at both ω_m and $2\omega_m$. Hence, the problem arising from amplitude fluctuations, which limit the sensitivity of the straightforward technique of differential absorption, is avoided.

An interesting scheme that uses both scattering and absorption is differential absorption Lidar called DIAL. Its operation principle and versatile application for air monitoring are presented below. The double-ended version uses a topographic target or a retroreflector and thus yields an average concentration of specific substances along the laser beam path between the source and the reflecting target. The single-ended version relies on effective backscattering in the atmosphere and is thus most useful in the UV and visible spectral range where Rayleigh and Mie scattering are strong. The unique feature is the range-resolved monitoring, ie, the remote three-dimensional profiling of specific atmospheric trace species.

Finally, another method for *in situ* absorption measurements that differs in various aspects from the previously discussed schemes is *Photoacoustic* Spectroscopy (PAS). It is primarily a calorimetric technique whereby the absorbed energy is determined directly not by measuring transmitted or backscattered radiation. Instead of radiation detectors, acoustic sensors are usually employed which measure the pressure modulation in the sample caused by the absorption of modulated radiation. Combined with a tunable laser as a radiation source, this scheme measures absorbances $\alpha L \approx 10^{-9}$ or even lower. Thus, small absorption, ie, small gas concentrations, are measured even at short path lengths. Therefore, radiation wavelengths beyond atmospheric windows, eg, in regions of strong water vapor absorption, can be employed for measurements. A further aspect is the wide dynamic range of PAS which spans at least five orders of magnitude and is thus of special interest for air monitoring because of the large gamut of pollutant concentrations. Photoacoustics is a rather universal technique that is successfully applied to many other fields (13).

DIFFERENTIAL OPTICAL ABSORPTION SPECTROSCOPY (DOAS)

Principle

As the name implies, DOAS systems rely on absorption measurements, often performed along long, open paths between a light source and a receiver in the atmosphere. The attenuation of a light beam by absorption by atmospheric constituents is in principle described by equation 1. For practical measurements in the open atmosphere, however, additional causes of light extinction must be taken into account. These include Rayleigh and Mie scattering. These scattering processes are treated like absorption processes because light scattered out of the beam does not normally reach the detector. The Rayleigh scattering process describes scattering by air molecules.

Experimental Arrangement

A DOAS system involves a light source, an open path in the atmosphere, and a receiver consisting of a spectrometer and a radiation detector. A typical setup is shown in Figure 4. The light source should emit radiation which varies only slowly with wavelength, ideally "white" light. The key parameters are the spectral brightness (in watts per unit area) and the wavelength interval. Thermal light sources, such as incandescent lamps or arc lamps, are often employed. In this case, the emission spectrum is similar to that of a black body radiator and is essentially given by Planck's law with a temperature of the radiating area of approximately 3000 K for incandescent lamps and of 6,000–10,000 K for a xenon arc lamp. In the latter case, a smooth spectrum in the range of 200 nm to beyond 400 nm is emitted, superimposed by a group of emission lines between 400 and 450 nm. Lasers yield a much higher spectral brightness and a low beam divergence but their emission bandwidths or tuning range usually makes it difficult to observe molecular vibrational bands. Furthermore, lasers are more complex devices than lamps. Measurements are also performed with natural light sources outside the atmosphere, ie, with sun-, moon-, or even starlight. The largest problem with these light sources relates to their structured spectra that contain many Fraunhofer lines.

The common design of a DOAS system uses an open path through the atmosphere between a searchlight-type light source and the receiver. Appropriate optical elements are thus introduced to allow a path from a few hundred meters to one of possibly several km. The light source is thus designed to emit a beam of light with minimum divergence, and the light is collected by a telescope at the other end of the path. The instrument thus averages trace gas concentrations over this path. A variation of this design uses a reflector in the field to return the light from the source located next to the receiver, thus doubling the light path. Alternatively, the path is folded in a (closed) multiple reflection cell, such as a White cell (15), that contains the air under study. In this case measurements are also performed at reduced gas pressures which can be advantageous for improving the detection selectivity in multicomponent mixtures. Finally, range-resolved Lidar techniques (see section below) are also employed.

After being collected the received light is dispersed in a spectrometer (also shown in Fig. 4) to measure the spectra $I(\lambda)$ and $I_0(\lambda)$. Different detector designs are used depending on the wavelength and sensitivity requirements. They range from nondispersive semiconductor or photomultiplier detectors to spectrograph-photodiode array combinations. The most commonly are Czerny–Turner, flat field, and Fourier transform spectrometers. Although the popular Czerny–Turner spectrometers require at least two additional reflecting surfaces (collimating and camera mirror) besides the grating, modern holographic grating technology allows the design of flat-field spectrometers consisting only of an entrance slit, grating, and exit slit and detector, or detector array. The present lower efficiency of the holographic grating compared with a ruled grating is approximately compensated for by the smaller number of optical surfaces. Finally, a Fourier transform spectrometer has recently been combined with DOAS in the UV-visible. The most important issue in the design of the spectrometer concerns the stray light. Stray light is composed of light scattered from optical elements, reflection of unused diffraction orders off the spectrometer walls, reflection of unused portions of the spectrum from walls near the focal plane, and reflections from the detector surface. A typical stray light level for a Czerny–Turner spectrometer for a very narrow spectral width (single line definition) is 10^{-5}. In DOAS applications, however, with comparatively low spectral resolution of 0.3 nm, stray light levels I_{SL}/I are closer to 10^{-2} than to 10^{-5}. An additional source of stray light is sunlight

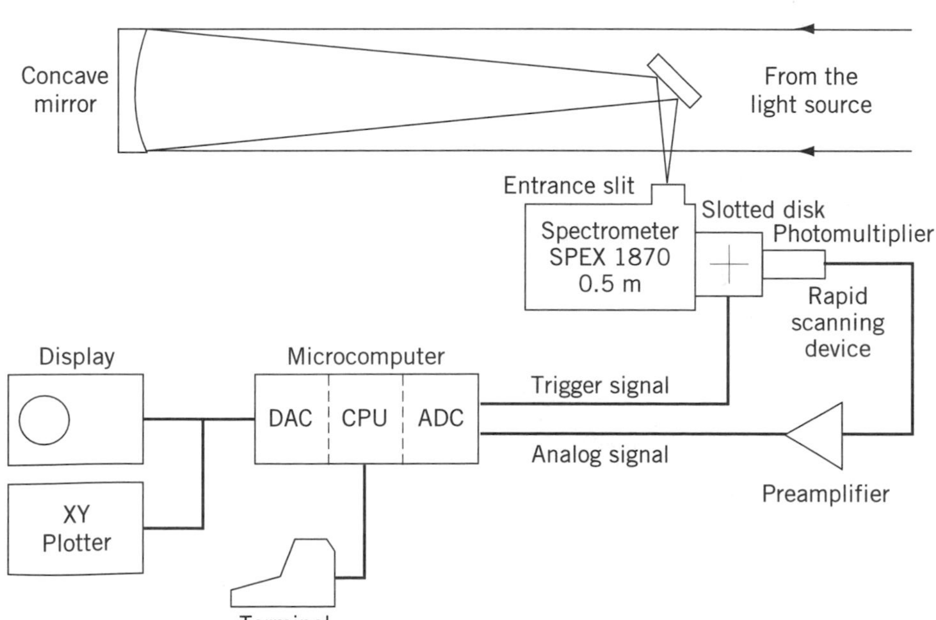

Figure 4. Typical setup of a low-resolution differential optical absorption spectrometer (DOAS); the searchlight-type light source is not shown. ADC, analog-to-digital converter; DAC, digital-to-analog converter; CPU, central processing unit (14).

scattered within the instrument volume which also reaches the spectrometer. However, because of the very narrow field of view of the instrument, levels of externally scattered sunlight are extremely low, except when very hazy conditions prevail.

Figure 4 shows that the spectrum projected onto the focal plane of the spectrometer is recorded by a mechanical scanning device combined with a photomultiplier tube or by a photodiode array. Finally the recorded spectrum is digitized and transferred to a computer which analyzes the spectral features of trace gas absorptions along the light path and calculates the corresponding concentrations. A more advanced design uses a coaxial merge of a transmitting and a receiving telescope in conjunction with an array of corner-cube retroreflectors. This approach has the advantage of cancelling the effect of atmospheric turbulence on the light beam because these retroreflectors return the incident light exactly (although with some lateral offset) in the direction of incidence.

Finally, DOAS systems are commercially available (16) and are routinely applied to air monitoring at many locations.

Examples of DOAS Studies

DOAS measurements of atmospheric trace species range from air pollution control to detecting very rare species. The measurements with open paths render the technique particularly attractive for monitoring unstable species which would be difficult in the presence of inner surfaces of tubes and cells. The first reliable studies on nitrous acid HONO and on NO_3 and OH radicals were made by DOAS. With respect to air pollution, DOAS systems have been applied to monitoring nitric oxides (NO and NO_2), SO_2, NH_3, Hg, H_2O, CO_2, HCl, HF, O_3, and VOCs, such as benzene, toluene, formaldehyde, etc. Applications include air measurements in urban areas including street level air quality studies, background measurements in rural air, and monitoring along the surrounding fence of industrial areas. In general, detection limits at the few $\mu g/m^3$ range, corresponding to ppb concentrations, are achieved with typical path lengths of 500 m.

Conclusions

Long-path DOAS is a very useful technique for monitoring atmospheric trace gases and radicals. The main strengths of DOAS are the absence of any sampling and hence of any wall losses (which permits measurements of radicals), good selectivity, and the potential for real-time measurements. Furthermore, as with any spectroscopic technique, field calibration is obsolete because the absorptive cross sections of the molecules of interest are measurable independently in the laboratory. However, many species of interest in atmospheric chemistry and pollution control are not detectable by DOAS because of lack of absorption. Furthermore, it should be pointed out that the technique is capable of multicomponent monitoring but, because optimum paths and path lengths for different species differ, simultaneous multicomponent monitoring may require installing several paths and receivers. Setting up instrumentation at two sites separated by several kilometers but in sight of each other is sometimes also cumbersome. In the future, measurements will probably be extended to shorter wavelengths below 300 nm. Finally, folded path arrangements and diode array detectors will most likely gain in importance.

LIGHT DETECTION AND RANGING (LIDAR)

Principle

Whereas DOAS is based on transmission measurements to sense the average absorption over an atmospheric path, the Lidar scheme provides active remote sensing in backscattering. Lidar is the acronym for *l*ight *d*etection *a*nd *r*anging. It uses a pulsed-laser transmitter, and light, back-scattered from molecules and particles, is collected by an optical telescope and detected as a function of time by an optical receiver in a radar-like mode. Then the backscattered light signal yields information on the presence, distance, and concentration of atmospheric scatterers and absorbers. Hence, the technique is applied to study aerosol distributions, clouds, and also atmospheric trace gases. If the laser wavelength is varied from an absorption line of a pollutant gas to an adjacent wavelength, the detected changes in backscattered light intensity are used to evaluate range-resolved concentration profiles. This is the differential absorption version of the lidar scheme known as DIAL. The general schematic of a lidar system is depicted in Figure 5.

Different types of lidar schemes exist which are called topographic target lidar, Mie scattering lidar, fluorescence lidar, Raman scattering lidar and differential absorption lidar (DIAL).

Experimental Arrangements

The main features of a good DIAL system suitable for detecting low gas concentrations even at large distances are the available laser intensity, telescope area, and detector efficiency. A schematic coaxial setup is shown in Figure 6. The laser beam is expanded and directed into the free atmosphere via a telescope of typically 30 to 60 cm in diameter. The backscattered signal is collected and fed to a detector, eg, a photomultiplier tube, via a narrowband filter to suppress the daylight. The pulsed lasers for DIAL have to offer tunability, high pulse energy, and sufficiently short pulse length. In view of field measurements they must be sufficiently rugged and practical

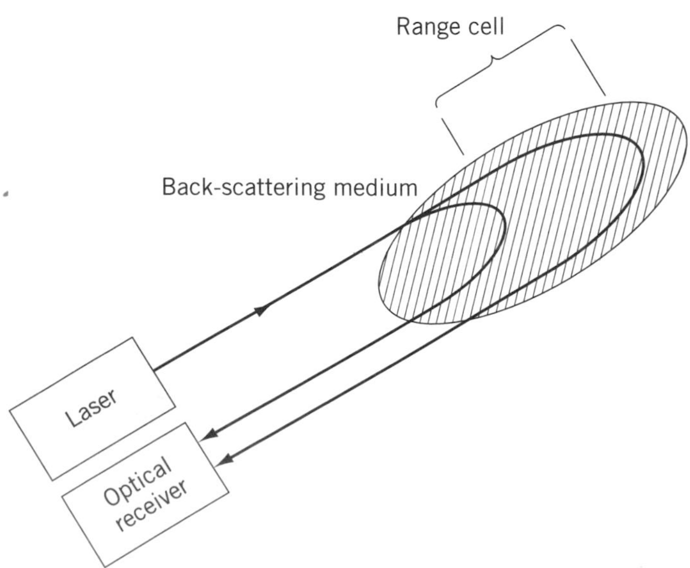

Figure 5. Principle of lidar scheme.

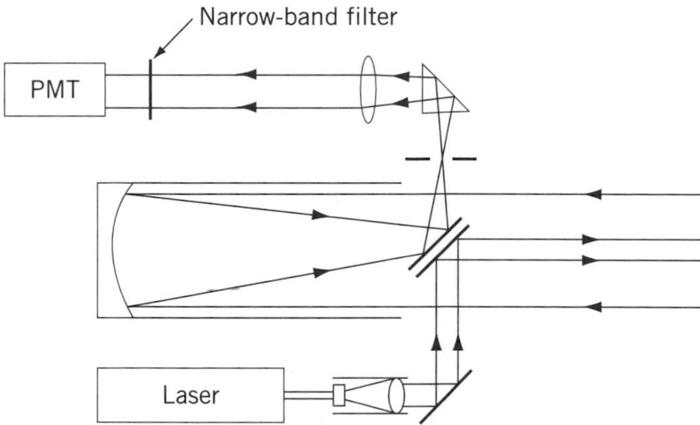

Figure 6. Schematic coaxial Lidar setup (PMT = photomultiplier tube).

to operate. Lasers of choice are Nd:YAG, excimer, dye, titanium sapphire, CO_2, and diode lasers. The fixed-frequency Nd:YAG and excimer lasers are normally used as pump lasers for the widely tunable dye and Ti:sapphire lasers. Flashlamp-pumped Q-switched Nd:YAG lasers with pulse repetition rates of 10 to 30 Hz easily achieve pulse energies above 1 Joule and pulse durations of a few ns at a wavelength of 1064 nm. Efficient frequency-doubling, -tripling and -quadrupling with nonlinear optical crystals to 532 nm, 355 nm and 266 nm, respectively, is readily obtained. In the future, very compact, reliable high-energy, diode-pumped Nd:YAG lasers will play an important role in DIAL systems. *Excimer* lasers emit in the UV and typically yield pulse energies of hundreds of mJ, pulse durations in the 10 ns range, and high repetition rates up to several hundred Hz. The most common types are the KrF laser emitting at 249 nm used for tropospheric ozone Lidar systems. The second wavelength required for DIAL measurements is often obtained by stimulated Raman scattering in high-pressure deuterium or hydrogen, On the other hand, the less absorbed longer wavelengths of 308 nm of XeCl or of 351 nm of XeF are often employed for stratospheric ozone monitoring. *Dye* lasers have long been the standard source in DIAL systems. These are pumped by the harmonics of a Nd:YAG laser or by a XeCl excimer laser. Instead of the normal (slow) grating tuning of the dye laser wavelength, a dual-wavelength option with alternate λ_{on} and λ_{off} operation is frequently employed in these lasers. By using dye amplifier cells, pulse energies up to 200 mJ are achieved. A promising source is the Ti:sapphire laser which offers tunability between 670 nm and about 1 μm. It is pumped by flashlamps or (with higher efficiency) by a frequency-doubled Nd:YAG laser. Output pulse energies for flashlamp-pumped systems reach more than 1 J at either μs or at 20–50 ns (Q-switch) pulse durations. The appropriate wavelengths for NO_2 detection are achieved by frequency doubling, for SO_2, O_3, and Hg by frequency tripling, and for NO by frequency quadrupling. The CO_2 laser is tunable in the 9 to 11 μm range. However, the tunability is not continuous but limited to some 80 vibrational–rotational transitions. One has thus to rely on accidental coincidences of pollutant molecular absorption lines and laser transitions. Nevertheless, many pollutants are accessible because of the richness of molecular absorption lines in this IR region. However, DIAL applications are somewhat limited by the strongly decreased backscattering efficiency at long wavelengths. In some cases, *copper vapor* lasers at 511 and 578 nm, *gold vapor* lasers at 628 nm, and *lead vapor* lasers at 723 nm are employed. These lasers yield average powers in the watt range with high pulse-repetition rates of 25 kHz. Finally, *diode* lasers are very compact, highly efficient tunable solid state lasers. They are used for solid-state laser pumping in certain cloud height meters and range finders, for seeding pulsed IR lasers, and as local oscillators in heterodyne detection systems. In more complex DIAL systems, several laser sources are combined to form a multifrequency system.

DIAL systems are arranged as fixed, mobile, air or even spaceborne devices, with different degrees of versatility and complexity. For atmospheric trace gas monitoring, mobile systems on the ground (or eventually on a ship) are most suitable. Over the years, various systems have been constructed and also commercialized. Mobile systems are installed in a truck and often equipped with a vertical telescope arrangement with a dome on the roof carrying a mirror which is rotated around two axes under computer control. This permits two-dimensional scans of the spatial distribution of pollutants with a full 360° horizontal scanning capability and eye-safe near field operation with the laser beam leaving the system high above the ground. Airborne systems are very useful for providing data over wide areas. Such systems are normally looking downward or upward in a fixed position. Because of the swift movement, special requirements for a high repetition rate and minimum temporal spacing between the on- and off-wavelength must be considered.

Conclusions

Apart from aerosol, wind, and cloud measurements, the lidar scheme offers the unique possibility of three-dimensional mapping of various atmospheric trace gases. Among the different versions of lidar schemes, the differential absorption lidar called DIAL is best suited for pollution monitoring and also for measuring total fluxes from industrial and urban areas. Most DIAL studies are performed in the UV and visible spectral ranges where sufficient backscattering and atmospheric transmission exist. Detection ranges in the km range and trace gas concentrations in the ppb range are often accessible. To achieve the desired sensitivity, path-integrated measurements, rather than range-resolved studies, must be performed. Most DIAL studies so far have concentrated on trace species such as NO, NO_2, SO_2, and O_3. Fewer investigations have been performed on the hydrocarbons benzene, toluene and ethylene and on HCl. Furthermore, atomic mercury, an important natural and man-made pollutant is addressed in some studies. In this case an extraordinary detection limit in the ppt range is achieved because of the huge absorption cross section. Despite the lower back-scattering in the IR wavelength range, the potential of IR DIAL systems is promising because of the characteristic absorption features of pollutants. However, the applications are still limited because there are only few tunable, high power lasers available and the detection techniques are not as well developed as in the UV and visible. In general, the accuracy and detection limits in DIAL schemes could be further improved by measuring the lidar signals not only at two wavelengths as is usually done today but at several wavelengths of an absorption profile. The future applicability of such a technique is coupled to the advances made in laser and detector technology. Already, increased compactness possible with new technology allows mobile DIAL systems in small vehicles and also enables the construction of realistic DIAL airborne systems. The ultimate application of the DIAL technique is global monitoring of meteorological parameters and monitoring atmospheric pollution from space. Several projects on space Lidar systems are presently being explored.

INFRARED SPECTROSCOPY

Principle

The spectral region between 3700 and 500 cm^{-1} (2.7–20 μm) includes almost all the important absorptive bands of gaseous molecules. Hence, this fundamental or often called middle IR or fingerprint region, is very rich in molecular absorption features and therefore is the favorable region for spectroscopic measurements. At higher frequencies there are only overtones with correspondingly small absorption coefficients whereas at lower frequencies there are strong rotational lines but because of extremely strong absorption by water vapor they are quite inaccessible. The absorption pattern depends on the physical properties of the molecule, such as the number and type of atoms, the bond angles, and bond strengths. Each spectrum differs from all the others and is considered the molecular "signature" or "fingerprint". Diamotic molecules, such NO, CO or HCl have a single major band that is an array of individual lines, each with a width of about 0.2 cm^{-1}. Linear polyatomic molecules, such as CO_2, N_2O and C_2H_2, also show arrays of individual lines. Nonlinear polyatomic molecules, such as O_2, SO_2, and NH_3 have many absorption bands with widths varying from 0.2 cm^{-1} to many cm^{-1}. For larger polyatomic molecules at atmospheric pressure there are so many lines overlapping each other that the spectral features are broad and smooth, except for occasional "spikes". Figure 7 shows the principal absorption bands of five molecules (from bottom to top): nitric oxide (heteronuclear diatomic molecule), acetylene (four-atom linear molecule), formaldehyde (four-atom nonlinear molecule), benzene (12-atom symmetrical molecule) and methyl ethyl ketone (13-atom nonsymmetrical molecule). These spectra illustrate the different shapes and center frequencies of these bands depending on the molecular size and geometry (17).

The measurement principle in infrared spectroscopy is based on the Beer-Lambert absorption law (see eq. 1) and therefore relies on the accurate determination of the absorbed radiation power along a given path length. In addition, the spectral line shapes depending on the special situation must also be considered (17,18). In both laboratory and atmospheric studies, observed line shapes are rarely caused solely by natural, Doppler or collisional broadening but are often mixed in shape. Temperature and pressure regions of interest often cover a transition from one line shape to another, eg, from the Doppler regime with a Gaussian line shape where the line width is constant and the peak absorption varies with pressure, to the collision-broadened Lorentzian regime, where the peak absorption remains constant but the line width increases proportional to pressure. It should be mentioned that the Beer-Lambert absorption law strictly holds only for isolated molecules, interacting with light. This requirement is practically best met with gases or vapors at low partial

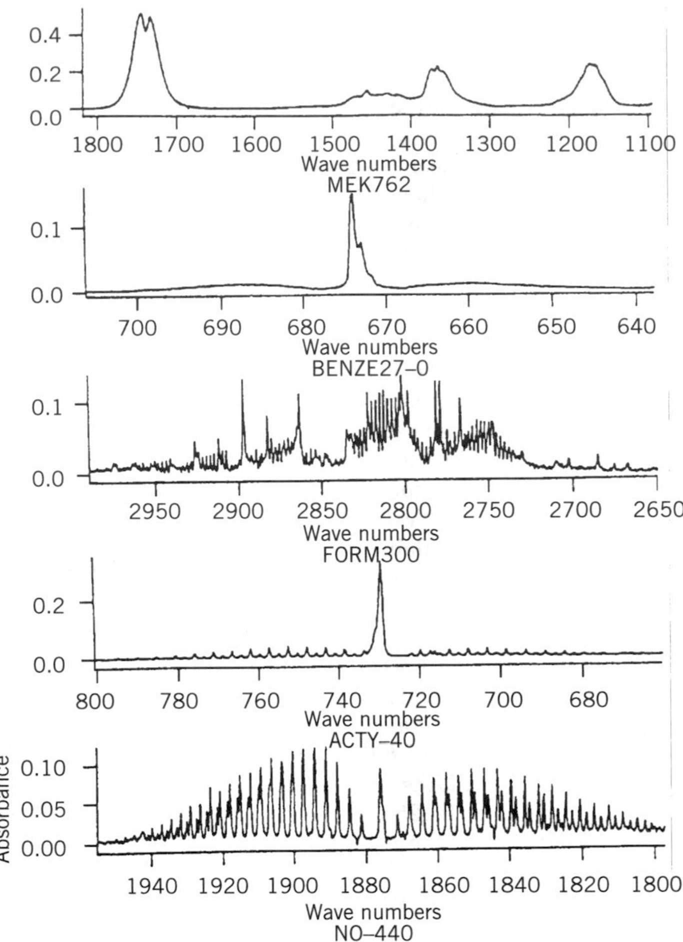

Figure 7. Molecular signatures. From bottom to top: nitric oxide, a heteronuclear diatomic molecule; acetylene, a four-atom linear molecule; formaldehyde, a four-atom nonlinear molecule; benzene, a 12-atom symmetrical molecule; and methyl ethyl ketone, a 13-atom nonsymmetrical molecule (17).

pressures diluted in a nonabsorbing (inert) buffer gas, eg, ppm concentrations of substances in air, as usual in air pollution monitoring. Deviations from ideal behaviour arise from optical interferences, nonlinearities at high absorbance values, or differences in total pressure or temperature between the sample and possible reference spectra taken for the analysis, rather than from intermolecular interactions in the case of liquid-phase spectra.

The detection limit is determined by the path length and the signal-to-noise ratio in the spectrum. An important aspect involves spectral resolution. The fine structure in the spectra of gaseous molecules is of great value in identification, in discrimination against interferences, and in quantitative measurements. High enough resolution should be used to take advantage of this fine structure. The widths of individual absorption lines at atmospheric pressure and room temperature are about 0.2 cm^{-1} (FWHM, full width at half maximum). Therefore, to see all spectral details requires a resolving power on the order of 0.1 cm^{-1}. Such a resolution is easily obtained with a laser source. Infrared laser-spectroscopic applications to trace gas detection are discussed separately in the following sections. However, despite the great success of laser spectroscopy in this field, considerable effort has also been put into conventional IR sources owing to the still limited availability and/or complexity of widely tunable IR lasers. In this section we restrict ourselves to the discussion of IR spectrometers employing incoherent light sources. Infrared spectrometers have been widely used for the last 40 years and many improvements have been introduced so that it is now feasible to measure gases over the range of partial pressures from above 1 bar to less than 10^{-9} bar.

Experimental Arrangements

A Nernst glower is the best available broadband source for IR absorption spectroscopy. Nernst glowers operate at higher temperatures than globars or coilforms, and therefore they emit more energy in the infrared. In addition to bringing a maximum number of photons to the detector while avoiding its saturation, the noise level is lowered by choosing a low-noise detector. The best available detectors today are the photoconductors that operate at liquid nitrogen temperature, such as mercury–cadmium–telluride (MCT, HgCdTe). These detectors usually produce a noise level in the spectrum about 100 times lower than the noise level produced by detectors that operate at room temperature.

For maximum spectral detail and maximum line heights, the spectrometer should fully resolve the spectral lines. Considering the line widths at a working pressure of 1 bar, a resolving power of 0.125 cm^{-1} is appropriate. For most instruments, however, the noise level increases at a rate faster than the gain in line height when the resolution is increased. A value of 0.5 cm^{-1}, which is easily achieved with a grating spectrometer, represents a good choice.

According to the Beer-Lambert absorption law in equation 1, the detection sensitivity is improved by lengthening the optical path L. However, maximum path lengths should be kept at 100 m or less because of the loss of photons reaching the detector and absorption by water vapor and CO_2. The lengthening of the optical path is achieved by simply increasing the distance between the light source and the detector or by employing a multiple-pass cell. In the latter case, reflecting mirrors are used at either end of a cell of fixed length, and the light beam is reflected back and forth between the mirrors many times, giving an effective absorbing path length for the pollutant which is some multiple of the cell length. Different cell designs including the so called White cell (15) have been introduced over the years. The basic three-mirror system of the White cell displayed in Figure 8 is still widely used. The radiation from the source is focused in the entrance aperture. From there the diverging beam passes to the first objective mirror on the left which focuses an image on the field mirror located on the right. From there the beam is returned to the second objective mirror on the left which directs the beam either out of the cell after four passes or back to the field mirror for additional multiple passing. The images line up on the field mirror in two rows, as shown in Figure 9 an example of 28 passes. Modifications of this cell type allow 100 or more passes through the cell resulting in path lengths up to kilometers.

A major improvement over conventional IR dispersive spectrometers with gratings was achieved with the development of *F*ourier *t*ransform *IR* spectrometry, called FTIR, in the early 1970s (10,11,19). This technique offers a number of significant advantages at once including sensitivity, speed, and improved data processing. The basis of a FTIR spectrometer, shown in Figure 10, is a Michelson interferometer consisting of a fixed mirror, a moving mirror, and a beam splitter. The light beam (incident from the left) strikes the beam splitter at point O. After the beam splitter the partial beams are reflected off the fixed mirror A and the moving mirror B and after recombination are directed out to the detector. For a monochromatic beam, a regular interference pattern is obtained after recombination. With a broadband IR source such as a Nernst glower containing a larger number of wavelengths, a different interferogram is obtained with a maximum signal for OA = OB. This interferogram is transferred into an infrared spectrum by applying a Fourier transform. If a sample that selectively absorbs certain wavelengths is present in the path before the beam reaches the detector, the interferogram changes in a characteristic way and thereby yields information on the concentrations of pollutants present.

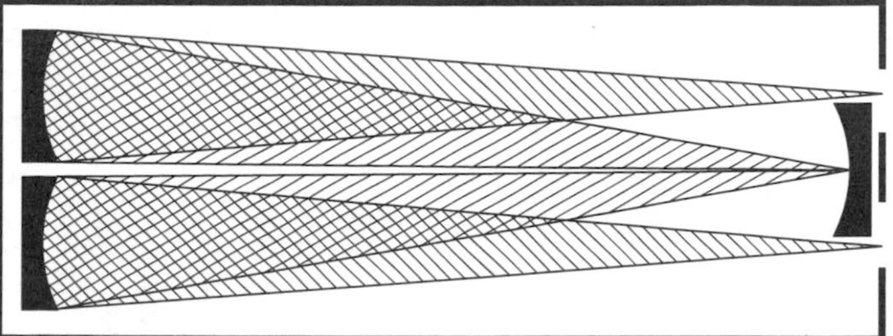

Figure 8. White cell with basic set of four passes (17).

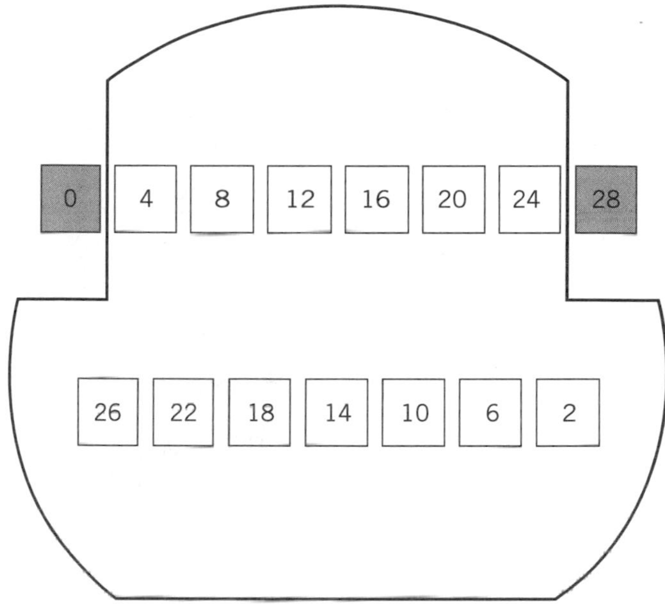

Figure 9. Placement of images on the field mirror of 28-pass White cell (17).

Various IR and FTIR spectrometers and multipass cells are commercially available today which have increased the variety of applications of IR absorption spectroscopy to atmospheric studies.

Example of Atmospheric Studies

The potential of IR spectroscopy is illustrated by a representative example involving the analysis of ambient air in Los Angeles in 1980. Figure 11 shows the portion of an

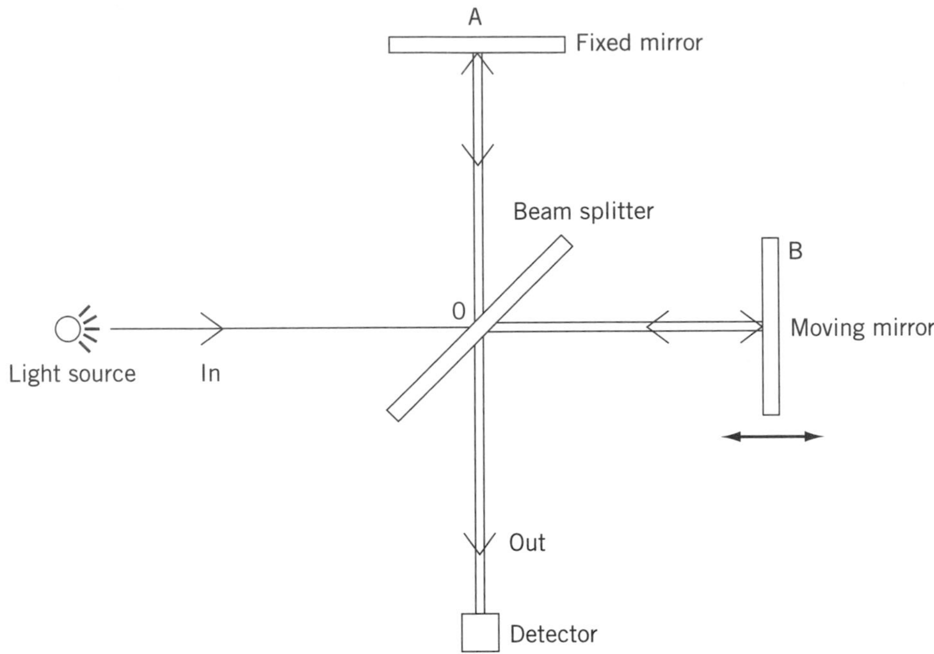

Figure 10. Schematic diagram of Michelson interferometer as often used in commercial FTIR spectrometers.

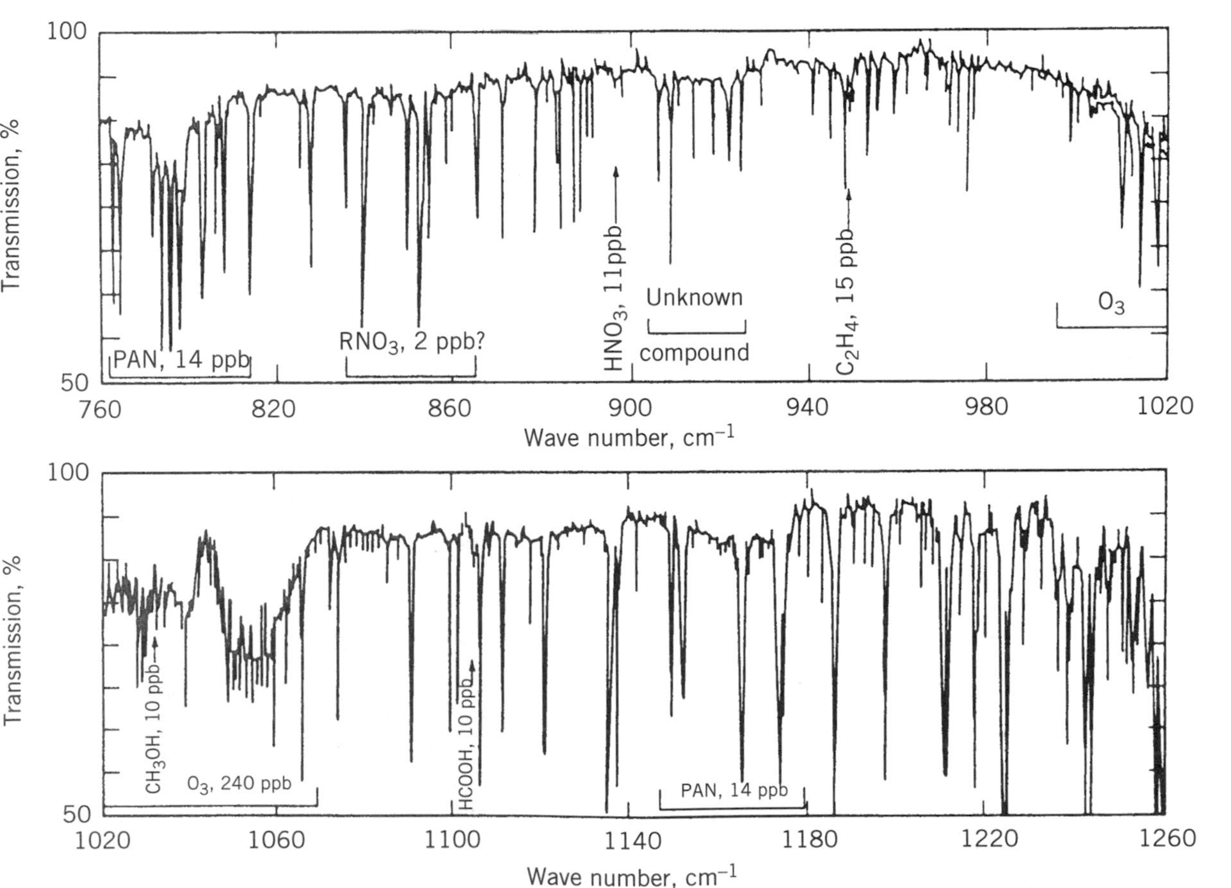

Figure 11. Ambient air FTIR spectrum from 780 to 1260 cm^{-1} in Los Angeles on June 26, 1980 at 2:45 p.m. at a path length of 1.26 km and a spectral resolution of 0.25 cm^{-1}. The identified compounds and the derived concentrations are indicated (20).

infrared spectrum between 760 and 1260 cm^{-1} with a resolution of 0.25 cm^{-1} (20). The spectrum was recorded with an FTIR spectrometer with a pathlength of 1.26 km. The figure does not represent the direct spectrum recorded but a spectrum ratioed to that of reference air made up of 1 atm of N_2 with CO_2, CH_4, and N_2O equivalent to the amounts in clean

Table 2. Molecules Detected and Measured in Air by Infrared Absorption[a]

Molecule	Location of Bands or Lines, cm^{-1}	Approximate Atmospheric Mixing Ratio[b]
H_2O	Everywhere in spectrum	10^{-6} in S, 10^{-2} in *T*
CO_2	2380, 670	3×10^{-4}
CH_4	3020, 1305	1.5×10^{-6}
N_2O	2220, 1280	3×10^{-7}
CO	2146	2×10^{-6}
O_3	1050	10^{-6} in S, 10^{-7} in lower T
NO	1900	10^{-6} in S, 10^{-10} in clean T, 10^{-7} in urban smog
NO_2	2920, 1620	10^{-6} in S, 10^{-9} in clean T, 10^{-7} in urban air
N_2O_5	1240	10^{-9} in S(?)
HNO_2	791	10^{-9} in urban smog
HNO_3	879, 967	10^{-9} in clean T, 5×10^{-9} in S, 10^{-8} in urban areas
NH_3	932, 967	$10^{-8}-10^{-9}$
$CINO_3$	780, 809, 1293	10^{-9}
HCl	2924, 2926	10^{-9} in S, not seen in T
HF	4029, 4174	10^{-10} in S, 10^{-8} near fertilizer plant
HCOOH	1105	10^{-8} in urban smog, 10^{-9} in upper T
H_2CO	2780, 2870	10^{-8} in urban smog, 10^{-10} in upper atmosphere
HCN	3270, 3290	10^{-10}
Chlorofluorocarbons (CFCs)	Various	10^{-9}
Nonmethane hydrocarbons	2850–3000	10^{-7} in smog, 10^{-9} in S

[a] Ref. 17.
[b] S = stratosphere; T = troposphere.

air and H_2O equal to approximately 30% relative humidity. The peaks due from peroxyacetyl nitrate (PAN), HNO_3, C_2H_4, O_3, CH_3OH, HCOOH, and alkyl nitrates (RNO_3) are indicated in the spectrum. The remaining peaks are primarily from atmospheric H_2O and CO_2 which were not completely cancelled when the ratio was obtained. Clearly a wide variety of organic and inorganic pollutants are monitored with detection sensitivities down to the ppb level. The problems and uncertainties in determining quantitative data are obviously related to the occurrence of absorption lines, to the limited spectral resolution, to the long path length necessary (which may cause problems with extinction from scattering and absorptions of major air compounds), and to the "clean air" reference spectrum used for ratioing.

Table 2 summarizes the molecules monitored *in situ* in ambient air up to 1984 by IR absorption spectroscopy. The locations of the absorption bands and lines employed are given as well as approximate mixing ratios (ratio of the number of molecules of the trace gas to the number of molecules of all gases) encountered in the troposphere and stratosphere. Many additional molecules, mainly industrial pollutants, have been identified by IR spectroscopy after being captured and taken into the laboratory.

Conclusions

IR spectroscopy is a successful and versatile tool for trace gas monitoring in ambient air and in emission studies. The development of FTIR spectroscopy, with its significant advantages over dispersive systems, has provided substantial impetus for the increased application of IR absorption spectrometry to air monitoring. Today, a wide variety of compounds are detected with sufficient sensitivity for many ambient air studies depending on the absorptivity, path length used, the presence of interfering gases, and the spectral resolution chosen. This versatility is one of the significant advantages of FTIR compared with UV spectroscopy. The major disadvantage of FTIR is the sensitivity limit for some important species, including H_2O_2, whose concentrations are lower than the current FTIR detection limit. A further noteworthy application of IR spectroscopy gaining in importance is the area coverage and perimeter protection of plants. In this case, various open IR measurement paths are arranged so that any gaseous emissions above a certain concentration that leave or enter the area trigger an alarm. A wavelength modulation scheme monitors a gas in the presence of interfering gases with sufficient sensitivity over distances of some 100 m. Typical sensitivities between 100 ppm · m and 10′000 ppm · m, depending on the gas, are achieved. Such applications are of interest for flammability sensing at chemical and petroleum plants, for fire and combustion sensing (CO emissions) in buildings, tunnels, etc, for fugitive emissions of flammable hydrocarbons at landfill sites and storage areas, etc, and for vehicle emissions on roads or airfields.

TUNABLE DIODE LASER ABSORPTION SPECTROSCOPY

The fundamental or middle infrared between 2 and 15 μm is a rich spectral region for atmospheric measurements where almost all species of interest have strong fundamental absorptions but the major constituents, nitrogen and oxygen, do not. To avoid any absorption interferences between species and to provide unequivocal identification of constituents of polluted air samples or even of clean samples where high sensitivities are required, a high spectral resolution is mandatory and facilitates the spectral analysis substantially. Spectrometers with broadband light sources like FTIR are generally incapable of such high resolution. Hence, rather than using a continuous wavelength light source and scanning the entire IR spectrum, laser sources are employed with very narrow line widths which are tunable over a limited wavelength range. In this respect tunable diode lasers have undergone substantial development and today tunable diode laser absorption spectroscopy (TDLAS) represents a second established technique based on infrared absorption. The major theoretical advantage of TDLAS over FTIR spectroscopy is increased sensitivity. The major disadvantage, on the other hand, is that scanning the entire IR spectrum quickly is not possible. Thus TDLAS is more appropriate for following specific pollutants known to be present in an air sample than searching for previously unidentified species.

Diode Lasers

The diode lasers used in TDLAS are not the common diode lasers operating in the 800 nm or 1300–1500 nm range because only high vibrational overtone bands fall into their emission range. Instead the tunable diode lasers for TDLAS are made of lead salts (PbS, PbSe, PbTe) and their mixed alloys with themselves, with tin salts (eg, SnS or SnSe), and other materials (eg, Cd, Eu, or Yb). These semiconductors form *p-n* junctions with small energy bandgaps that emit in the 3–30 μm spectral region. The exact wavelength range depends on the composition of the alloy. These lasers therefore are manufactured to operate over the chosen wavelength by selecting the alloy materials and the composition. The bandgap energy defines the peak wavelength of the gain envelope within which laser emission occurs. Unfortunately, these diode lasers must be operated at cryogenic temperatures to permit continuous wave (CW) emission. The homojunction types are generally operated at temperatures below 50 K. The output power is typically 10 to 500 μW. The line width is extremely narrow and, even taking into account the influence of temperature, diode current instabilities, mechanical vibrations, and some optical effects, line widths below 10^{-4} cm^{-1} are achieved. Because Doppler half widths of absorbing molecules are on the order of 10^{-3} cm^{-1}, tunable diode laser systems obtain absorption spectra limited by Doppler or pressure broadening, not by the radiation source itself. The most useful feature of these diode lasers, namely the wavelength tunability, is most easily accomplished by changing the temperature of the diode. This permits coarse wavelength variations over a range of about 200 cm^{-1}. Fine tuning is done by varying the current through the diode. However, due to mode hopping while tuning, the tuning curve is not smooth but consists of a series of continuous portions 1–2 cm^{-1} wide, separated by gaps where no emission occurs. Fortunately, this is not a serious limitation for atmospheric measurements where coincidence of the laser output is required with only a single, well-chosen, rotational line of a specific compound. Furthermore, mode hopping is circumvented with new diode laser designs, such as distributed feedback (DFB), cleaved-coupled cavity or external cavity configurations. Finally, it should be mentioned that new fabrication techniques, such as molecular beam epitaxy (MBE) and new diode types, particularly quantum-well lasers, have steadily improved the performance of tunable diode lasers with respect to higher efficiency, lower threshold currents, higher output power and operating temperatures, and dominant single-mode operation.

Experimental Arrangement

A TDLAS system applied to atmospheric measurements requires a number of essential components (21). A schematic of a simple laboratory arrangement is displayed in Figure 12.

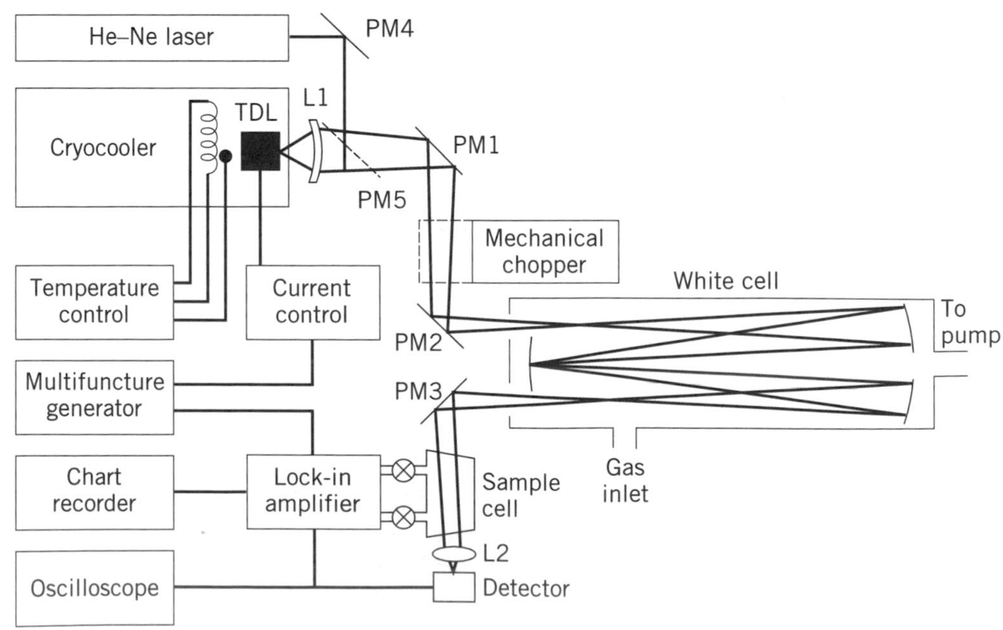

Figure 12. Schematic of a laboratory tunable diode laser system: TDL is the tunable diode laser, L1 and L2 are the lenses used to focus the beam onto the White cell entrance and the detector, respectively, and PM1–PM5 are gold-coated mirrors used for beam alignment (21).

The key part is of course a properly chosen tunable diode laser (TDL) that emits radiation in the desired wavelength region. For multicomponent monitoring, several diode lasers must be implemented. The TDL must be operated at temperatures below 120 K and therefore is mounted on a cold-finger in a cryocooler which holds up to four diodes. Both closed-cycle refrigerators and Dewar systems containing liquid cryogens (helium or nitrogen) are used. Recently, a miniature integral Stirling cooler has become available which reduces the system complexity. The increased availability of TDLs operating near room temperature should eventually even allow the use of thermoelectric coolers. As indicated in Figure 12, the temperature control and laser tuning is provided by a heating element. The modulation is achieved by a mechanical chopper (AM) or by current control in combination with a multifunction generator (WM). An optical system consisting of several lenses and mirrors is necessary to collect the multilobe pattern of the TDL radiation, to direct the radiation through a multipass gas cell (with sufficient path length to attain the desired sensitivity), and finally to focus the exiting beam onto the IR detector. This detector is usually a nitrogen-cooled, mercury cadmium telluride (HgCdTe, MCT) detector whose output is fed into an oscilloscope for visual representation and then to a lock-in amplifier for processing. Data acquisition and handling and spectrometer control are done by computer. The system calibration is often performed by measuring the signal when the beam is passed through a short cell of known length (indicated as sample cell in Fig. 12) containing a known concentration of the gas to be measured. In recent years, complete TDLAS atmospheric measurement systems have been developed (6). A commercial system has been introduced that is designed to operate from a mobile ground-based laboratory, on a meteorological tower, or even from an aircraft (22).

Examples of Atmospheric Studies

An example of a ground-based study on the ambient air is the *in situ* measurement of hydrogen peroxide (H_2O_2) by TDLAS. Hydrogen peroxide is an important atmospheric species, a so-called reservoir gas. Its detection represents a challenging task because of the small ppb to subppb concentrations in tropospheric air.

A second example involves measurements of formaldehyde (HCHO). Earlier studies indicate that there are two distinct types of temporal concentration profiles for HCHO. In the absence of local pollution sources, the HCHO diurnal variation is weak and the HCHO concentration drops to zero at night. At sites downwind of pollution sources, the HCHO concentration maximizes near solar maximum and declines in the afternoon as a result of photochemical formation from primary air pollutants. Minimum concentrations at night are in the low ppb range. An extensive evaluation of techniques for measuring ambient HCHO was carried out in the Los Angeles area in 1986. The HCHO concentration profiles recorded with these spectroscopic schemes during an 8-day measurement period are compared in Figure 13. Agreement is within 15 % of the mean of the three measurements, with the exception of the DOAS data at night when the concentration is at or below the detection limit of that system. Two maximum concentrations are usually observed each day. The early morning peak is attributed to traffic emission because it correlates in time with maxima in particle carbon. On the other hand, the second peak in the early afternoon originates from photochemical HCHO formation. The measurements demonstrate that the contributions to air pollution from primary and secondary HCHO production are comparable at this location. Because the spectroscopic data agree within 25% with

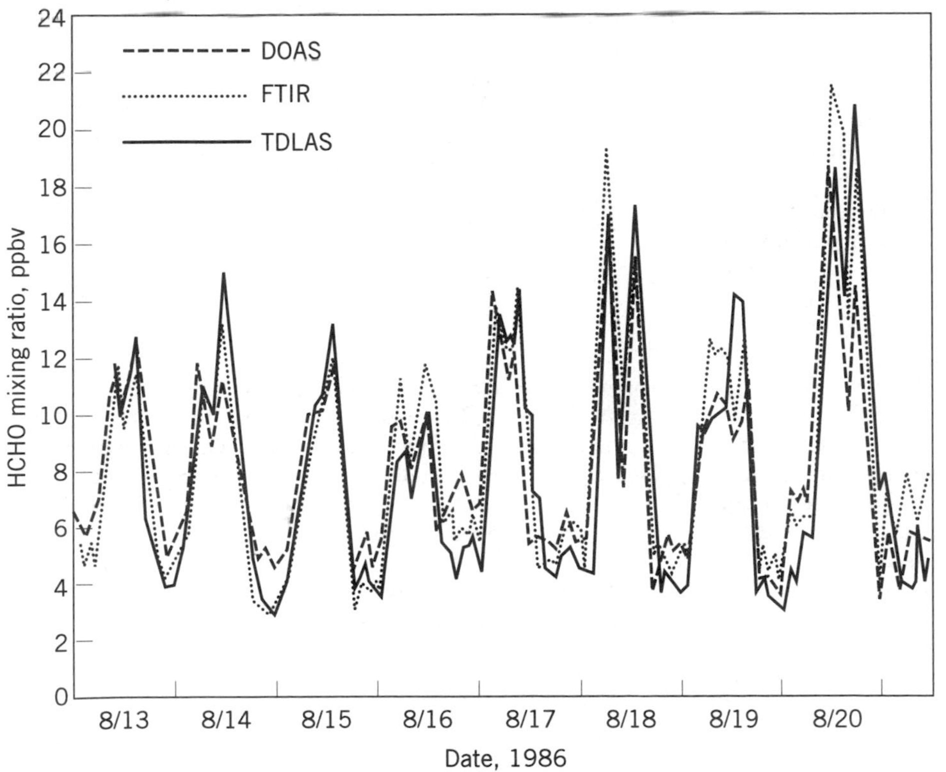

Figure 13. Time series plot for formaldehyde concentrations by differential optical absorption spectroscopy (DOAS), Fourier transform infrared spectroscopy (FTIR), and tunable diode laser absorption spectroscopy (TDLAS) at Glendora, California, during Aug. 1986 (23).

Table 3. Detection Limit of TDLAS for Some Atmospheric Trace Gases[a]

Species	Wave Number, cm^{-1}	Typical Atmospheric Concentration, ppb	Detection Limit Measured, ppb	Path Length, m
CH_4	1301.4	1700	0.25	12
C_2H_2	743.3	<30		
C_2H_4	950.0	~10		
C_2H_6	822.3	<50	10	12
C_6H_6	686.4	1–10		
H_2CO	2781	0.5–75	0.05	150
CH_3OH	1030	~1	1.0	150
CH_3OOH			2.5	
HCOOH	1107	<1	1.0	150
COS	2050.9	<0.1	0.005	120
CO	2111.5	50–1000	0.1	40
CO_2	2339.4	3×10^5		
N_2O	1250.6	310	0.05	150
NO	1890.9	0.01–100	0.04	150
NO_2	1600	0.02–1000	0.025	150
HNO_3	1720	0.1–50	0.1	150
H_2O_2	1285.7	<1	0.1	150
SO_2	1360.7	1–100	0.5	150
O_3	1050	10–500		
NH_3	1065	~1	0.025	150
HCl	2900		0.2	17

[a] Ref. 21.

independent measurements performed with (DNPH)-impregnated cartridges, an enzymatic technique and a diffusion scrubber, this study demonstrates the excellent performance of spectroscopic methods for monitoring pollution at low ppb concentrations in ambient air.

Table 3 lists some of the gases that are monitored with TDLAS in ambient air together with the wavelength used for the measurement, the typical atmospheric concentration, and the minimum detectable concentrations achieved. The table reflects the excellent sensitivity that is realized with TDLAS.

Conclusions

Lead salt diode lasers emit in the 3 to 30 μm region where most gases of atmospheric interest have their strong fundamental rotational–vibrational absorptions. Such lasers thus provide ideal sources for IR spectrometers. The high monochromaticity, ie, extremely narrow bandwidth, of these diodes makes TDLAS the most specific method available for unequivocal identification of trace atmospheric species. This property is particularly useful for detecting specific compounds in complex multicomponent gas mixtures such as polluted air or automobile exhausts. The combination of modulation (AM, WM, or FM) and long path lengths enables detection limits in the subppb or even ppt range. The rapid wavelength tuning of these diodes permits real-time measurements in the ms to second range. This is exploited for measurements from fast-flying aircraft, for eddy correlation flux measurements, and for study of transients in automobile exhaust.

The main disadvantage of TDLAS for atmospheric studies is the rather limited tuning range within a single mode for a single diode. This limits to 1 to 3 the number of species that are detected with a single diode. Therefore TDLAS is mostly applied to monitor a specific compound in a multicomponent mixture and not to the analysis of mixtures. Multiplexing techniques are of course employed that permit the simultaneous operation of a number of diodes. The high detection selectivity of TDLAS depends on the resolved spectral structure of the molecule. This essentially limits the method to molecules with less than 10 atoms or to symmetric molecules. Best results are obtained with measurements performed at subatmospheric pressures. However, this requires the use of sampling cells that introduce adverse effects like adhesion, surface reactions, etc. Finally, the necessity to use cryogenic temperatures and the rather limited available laser power have increased the interest in using diode lasers made of III–V compounds for gas studies. Despite the 100 to 1000 times smaller absorption, the higher quality and power of these lasers and very high frequency modulation techniques are used to regain some of the lost detection sensitivity. However, the number of detectable molecules in this spectral range is considerably smaller than in the middle infrared region. Therefore, it is most likely that TDLAS systems with both III–V and Pb-salt diode lasers will be applied in the future to trace gas studies.

PHOTOACOUSTIC SPECTROSCOPY (PAS)

Principle

In conventional absorption spectroscopy, the measurement of absorption is transferred to a measurement of the radiation power transmitted through the sample. On the contrary, in photoacoustic spectroscopy, the absorbed power is determined directly via its heat and hence the sound produced in the sample. Photoacoustics, also known as optoacoustics, was pioneered more than a century ago (24). The photoacoustic (PA) effect involves transforming modulated radiation energy, represented by photons, into sound.

Experimental Arrangements

The typical arrangement for PA air monitoring studies is schematically shown in Figure 14. The beam of a tunable laser

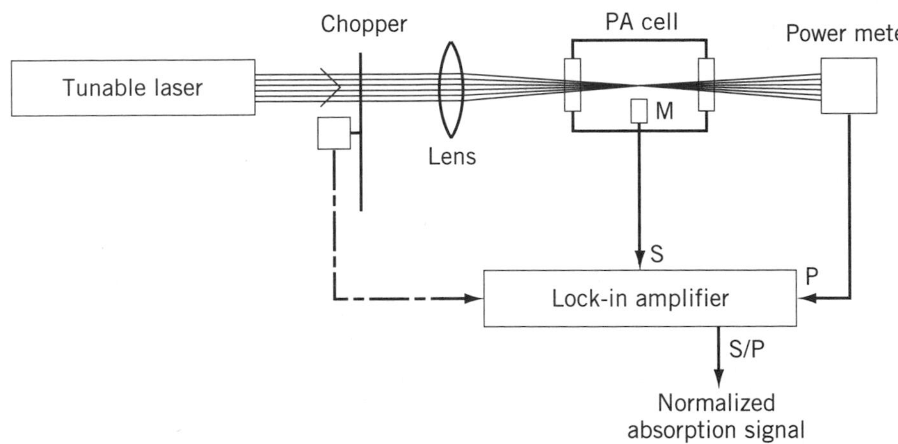

Figure 14. Schematic setup for gas-phase photoacoustic spectroscopy with tunable laser and microphone detection; M represents the microphone yielding signal S at laser power P (25).

used as radiation source is modulated by some means, eg, mechanically by a chopper or directly pulsed. The radiation is directed through the PA cell that either contains the gas under study or through which the air is pumped continuously. Then the generated acoustic waves are detected by an appropriate microphone or some other device whose signal is further processed. In the example shown in Figure 14, the microphone signal is fed to a lock-in amplifier locked to the modulation frequency. The average laser power P is recorded simultaneously by a power meter to normalize the microphone signal S according to equation 19. Numerous versions and modifications of this general scheme have been presented in the literature.

Conclusions

The feasibility and versatility of laser-photoacoustic spectroscopy (PAS) for air monitoring is demonstrated by numerous examples. Minimum detectable concentrations of gaseous pollutants are often in the ppb or subppb range depending on the molecular absorption cross section and on possible absorption interferences. In some cases, excellent selectivity is achieved enabling, eg, differentiation between the isomers of dichlorobenzene. Another important parameter is the temporal resolution to be achieved. Almost simultaneous and continuous monitoring of several components of interest is obtained by sequential measurements at appropriate laser wavelengths. The latter are selected to yield maximum sensitivity at minimum interference for selectively detecting the individual components. The excellent time resolution is important for surveillance of toxic gases, process control in chemical industry, etc. A further aspect concerns the large dynamic range that permits monitoring trace gas concentrations exceeding a range of five orders of magnitude. Thus low concentrations of pollutants in rural air and high concentrations at or near emission sources are detected with the same instrument.

It should be emphasized that PA monitoring is not restricted to laboratory measurements but is performed *in situ* with mobile systems as demonstrated by various examples involving stack emission, ambient air and even stratospheric air. However, unlike long-path or lidar measurements performed in the atmosphere, PA studies are restricted to point monitoring, and remote sensing is excluded. On the other hand, PAS investigations are carried out in the entire IR wavelength range, notably also outside atmospheric windows, because high sensitivity is already achieved with short absorption path lengths. This represents advantages with respect to both accessible wavelength ranges and required sample volumes.

The great advantage that a large variety of organic and inorganic species are detected also reflects the main problem of present PA systems which is related to absorption interferences in multicomponent mixtures. Detection thresholds for a specific substance in a given mixture may therefore be considerably higher than for interference-free conditions. This phenomenon often aggravates a quantitative analysis of unknown gas mixtures with numerous species of similar absorptivities. Hence, PA studies so far have concentrated on the monitoring of specific species in gas or air samples with at least roughly known composition, and PAS was used to quantify them. However, as examples have demonstrated, real analyses of multicomponent samples and the identification of unknown species is feasible on the basis of spectral fits. In this respect, the potential of a PA system strongly depends on the availability of tunable IR sources. Because the PA signal is proportional to radiation power, high detection sensitivity is achieved only with lasers with CW power in the mW to W range.

Like other spectroscopic schemes laser-PAS will also benefit from novel laser development with respect to wide continuous tunability. For practical purposes, the compactness of the laser source and of the entire system, easy operation, and automation and portability of the apparatus are important. One can imagine the development of a small laser PA instrument with restricted applications, eg, with lower sensitivity for emission control, etc.

Based on the great potential of laser - PAS, an increasing number of successful applications of *in situ* trace gas monitoring for air pollution surveillance and many other areas of current and future interest are expected.

BIBLIOGRAPHY

1. P.L. Hanst, "Spectroscopic Methods for Air Pollution Measurement" in J. N. Pitts, Jr. and R.L. Metcalf, eds., *Advances in Environmental Science and Technology,* John Wiley & Sons, Inc., New York, 1971, Vol. 2, pp. 91–213.
2. E.D. Hinkley, ed., *Laser Monitoring of the Atmosphere,* Topics in Applied Physics, Springer-Verlag, Berlin, 1976, Vol. 14.

3. D.A. Killinger and A. Mooradian, eds., *Optical and Laser Remote Sensing,* Springer Series in Optical Sciences, Springer-Verlag, Berlin, 1983, Vol. 39.
4. G. Christian and J.B. Callis, ed., *Trace Analysis: Spectroscopic Methods for Molecules,* Chemical Analysis, John Wiley & Sons, Inc., New York, 1986, Vol. 84.
5. R.M. Measures, ed., *Laser Remote Chemical Analysis,* Chemical Analysis, John Wiley & Sons, Inc., New York, 1988, Vol. 94.
6. R. Grisar, H. Böttner, M. Tacke, and G. Restelli, eds., *Monitoring of Gaseous Pollutants by Tunable Diode Lasers,* Proc. Int. Symposium, Freiburg/Germany, Oct. 17–18, 1991, Kluwer Academic, Dordrecht, the Netherlands, 1992.
7. M.W. Sigrist, ed., *Air Monitoring by Spectroscopic Techniques,* Chemical Analysis, John Wiley & Sons, Inc., New York, 1994, Vol. 127.
8. D.A. Krohn, *Fiber Optic Sensors: Fundamentals and Applications,* 2nd ed., Instrument Society of America, Research Triangle Park, N.C., 1992.
9. R.D. Hake, Jr., D.E. Arnold, D.W. Jackson, W.E. Evans, B.P. Ficklin, and R.A. Long, *J. Geophys. Res.* **77,** 6839–6848 (1972).
10. P.R. Griffiths, and J. A. de Haseth, *Fourier Transform Infrared Spectrometry,* Chemical Analysis, John Wiley & Sons, Inc., New York, 1986, Vol. 83.
11. T. Theophanides, ed., *Fourier Transform Infrared Spectroscopy,* Reidel, Dordrecht, the Netherlands, 1984.
12. Yu.A. Kuritsyn, "Infrared Absorption Spectroscopy with Tunable Diode Lasers" in V.S. Letokhov, ed., *Laser Analytical Spectrochemistry,* Adam Hilger, Bristol, U.K., 1985, Chap. 4.
13. A. Mandelis, ed., *Principles and Perspectives of Photothermal and Photoacoustic Phenomena,* Progress in Photothermal and Photoacoustic Science and Technology, Elsevier, New York, 192, Vol 1.
14. U. Platt, "Differential Optical Absorption Spectroscopy (DOAS)" in M.W. Sigrist, ed., *Air Monitoring by Spectroscopic Techniques,* Chem. Analysis, John Wiley & Sons, Inc., New York, 1994, Vol. 127, Chap. 2.
15. J.U. White, *J. Opt. Soc. Am.* **32,** 285–288 (1942).
16. Firm OPSIS AB, Furulund, Sweden.
17. P.L. Hanst and S.T. Hanst, "Gas Measurement in the Fundamental Infrared Region" in M.W. Sigrist, ed., *Air Monitoring by Spectroscopic Techniques,* Chem. Analysis, John Wiley, & Sons, Inc., New York, 1994, Vol. 127, Chap. 6.
18. C.R. Webster, R.T. Menzies, and E.D. Hinkley, "Infrared Laser Absorption: Theory and Applications" in R. Measures, ed., *Laser Remote Chemical Analysis,* Chem. Analysis, John Wiley & Sons, Inc., New York, 1988, Vol. 94, Chap. 3.
19. S. F. Johnston, *Fourier Transform Infrared: A Constantly Evolving Technology,* Ellis Horwood Ltd., Chichester, U.K., 1991.
20. P.L. Hanst, N.W. Wong, and J. Bragin, *Atmos. Environ.* **16,** 969–981 (1982).
21. H.I. Schiff, G.I. Mackay, and J. Bechara, "The Use of Tunable Diode Laser Absorption Spectroscopy for Atmospheric Measurements" in M.W. Sigrist, ed., *Air Monitoring by Spectroscopic Techniques,* Chem. Analysis, John Wiley & Sons, Inc., New York, 1994, Vol. 127, Chapt. 5.
22. Technical data, Unisearch Associates, Inc., Concord, Ontario, Canada.
23. D.R. Lawson, H.W. Biermann, E.C. Tuazon, A.M. Winer, G.I. Mackay, H.I. Schiff, G.L. Kok, P.K. Dasgupta, and K. Fung, *Aerosol Sci. Technol.* **12,** 64–76 (1990).
24. A.G. Bell, *Am. J. Sci.* **20,** 305–324 (1880).
25. M.W. Sigrist, "Air Monitoring by Laser Photoacoustic Spectroscopy" in M.W. Sigrist, ed., *Air Monitoring by Spectroscopic Techniques,* Chem. Analysis, John Wiley & Sons, Inc., New York, 1994, Vol. 127, Chap. 4.

AIR PERMITTING

BRUCE BOOMER
Prairie Village, Kansas

The term air permitting essentially refers to an agreement between the owner/operator of a source of air pollution, ie, a factory or other industrial facility, and the governmental air pollution control agency with jurisdiction. Newer air permits tend to be very specific in listing allowable industrial operations and the extent of allowable air emissions. The approach for obtaining a Title V operating permit (also called a Part 70 permit) (1) under the Clean Air Act Amendments of 1990 (CAAA) in the mid-1990s is routine. Please note the following limitations:

1. Each state and many regional or urban agencies have their own very specific air permit programs and regulations. It is advisable to contact the local agency to obtain specific regulations and instructions.
2. The information in this article is intentionally general. Each agency has its own requirements. Also, the Title V permitting program is currently undergoing significant scrutiny and suggested change.

EVOLUTION OF AIR PERMITTING PROGRAMS

A growing public interest in cleaner air and protection of health resulted in the evolution of air quality regulations and air quality control agencies at the federal, state, and local levels to enforce those regulations. State and local agencies needed to catalog the air pollutant emissions under their jurisdiction and place some type of limitations on allowable rates of pollution. Agencies started issuing permits to catalog activities and define accountabilities. Many of the early permits were vague, eg, "This refinery is allowed to operate", with perhaps some limitations placed on extent of operation or air quality issues. Some permit programs became more specific, requiring limits on the type of operations allowed, the timing of operations, total emissions allowed (in tons per year), and perhaps imposed an emission standard or limit in terms of opacity limits or contaminant concentration limits in stack gas (see AIR QUALITY MODELING).

The following types of programs also evolved:

1. The *prevention of significant deterioration (PSD)* permitting program is a federal and state program that protects local air quality. It requires close scrutiny of new industrial activities that would cause deterioration of air quality even if local air quality is very good.
2. The *new source review (NSR)* program evaluates newly proposed polluting activities to regulate compliance with all pertinent regulations.
3. Various *toxic air pollutant* programs in selected localities cataloged or regulated to some extent a variety of chemical emissions that were not regulated otherwise (the federally regulated pollutants had been limited to particulate matter, carbon monoxide, sulfur oxides, ni-

trogen oxides, volatile organic compounds as a group, and lead compounds.)

Facilities were also subjected to a variety of national, state, and local programs that set emission standards for specific types of industrial facilities or set specific air quality standards based on local or regional air quality.The CAAA included a new air permit program intended to bring together all existing air quality and air pollution regulatory programs into one facility operating permit. This Title V operating permit would clearly define all obligations under existing laws and regulations and take the confusion out of air quality compliance. The regulations defining the permit program were *introduced* by U.S. Environmental Protection Agency regulation (CFR 1995, Part 70) so the program is also known as the Part 70 permit program. In 1995, the state of Missouri noted "Operating permits represent a "contract" between citizens and industry on what represents compliance on air matters" (2). The Title V permit program itself did not create new air quality standards or emission limits, but incorporated any new regulations such as the broader visions included throughout the CAAA and the Act's subsequent regulations. Some of these broader issues include the consideration of 189 hazardous air pollutants (HAPs), looking at perhaps a broader range of air pollution (qv) sources (including smaller facilities), and greater public involvement in the permitting process. The CAAA also set the stage for more intensive enforcement of requirements and greater accountability for each permittee. A permit fee program was outlined for states and local agencies to adapt to help pay for the expanded programs.

The Title V program started with significant national and local controversy. Many of the points of controversy and suggested changes have not currently been resolved and several years of finetuning are expected for the future. Accordingly, this article is general in its approach. The most recent information should be obtained from local/state permitting agencies before embarking on a major permit development or modification activity.

OVERVIEW OF MAJOR STEPS

The air quality agency needs a number of specific details about a facility and how it pollutes (on an actual or potential basis) to issue a permit. The facility owner/operator must provide this information, but also perform strategic planning in proposing limits that are at least tolerable to the facility over the typical five-year life of the permit. This includes the following steps:

- Characterize the facility and its emissions in detail in an emissions inventory document.
- Collect supplementary emissions data as appropriate to improve the accuracy/relevance of the emissions inventory.
- Evaluate compliance with all applicable requirements.
- Submit the permit application/obtain the permit.
- Live with the permit and all of its requirements.

EVALUATING COMPLIANCE

One intent of the Title V air permitting program was to fold together all of a facility's existing air quality and air pollution control requirements into a single, specific document. The diversity of these requirements had been growing since the 1970s with the development of industry-based emission standards, general emission standards, local air quality requirements/goals, and other local, state, and federal requirements. Many facility owners/operators were unclear concerning how all requirements fit together and how the facility could demonstrate compliance. Additional health/risk concerns could be an issue also as many agencies began to evaluate the emission of toxic materials under other new programs. Inconsistencies in regulatory interpretations over the years have added to the confusion.

Unfortunately, all of these requirements must be sorted out and clarified to develop a facility's first Title V permit. For existing facilities, this can involve an extensive search through old permits, reports, and file communications combined with a review of all current regulations and air program requirements. Definitions and requirements of the earlier air programs tended to be more lenient and general. Many early Title V permit applicants discovered that their facility had been operating out of compliance due to changes in rules and agency interpretations over the years. A typical example includes failure to modify an old permit when equipment had been replaced, modified, or added to the facility (eg, replacing a boiler or changing its standby fuel). Each permit applicant must be in compliance with all applicable requirements prior to obtaining a Title V permit. Although each agency has its own policies on coming back into compliance, some states have required facilities to apply for construction permits for previously nonpermitted long-time existing operations prior to receiving a new operating permit. Other states tend to ignore past issues and focus only on processing the new Title V permit based on complete and accurate information.

DEVELOPING COMPLIANCE STRATEGIES

The information presented herein has primarily addressed the collection of information. This section suggests how to use that information. The results of the emission inventory process are added together to determine the facility's total annual emissions. The emission values may be too high, bringing up one of the following concerns:

1. If emissions exceed the values stated earlier that define a major source (eg, 10 t/yr of a single HAP) the facility will be required to meet all of the requirements associated with a major source.
2. Total emissions or emission rates may be high enough to trigger an emission limit or control requirement contained in an applicable regulation or program (such as limits in SIPs as mentioned earlier).

SUBMITTING THE PERMIT APPLICATION AND OBTAINING THE PERMIT

Each permitting agency will provide the permit applicant a standard set of instructions and application forms to complete for obtaining the air permit. Typical information requested will include:

- The information comprising the emission inventory (including general facility information)

- Summary information of total emissions (probably actual emissions and potential emissions)
- Statements concerning compliance with all regulations or plans to achieve compliance
- Optional information to propose operational flexibility within the permit, limit emissions, etc

The permitting agency typically requests very detailed information to allow the agency staff to understand the extent of operations at the facilities, the type and quantity of emissions, and the specific needs/strategies of the permit applicant in obtaining a permit. The applicant should try to lead the agency through the permitting process to obtain a permit that is most favorable to the facility. This involves satisfying all needs of the agency for specific information with an appropriate level of detail. Applicants should try to anticipate all of the types of questions the agency reviewer may have. Some applicants prepare a draft permit as part of the permit application to make their requested terms as clear as possible.

The agency will likely review the application and request additional information. A typical problem includes an incomplete or unclear application for certain specific points. After resolving these issues, the agency will draft a permit. Following any negotiations and any public review, the agency will issue the permit for generally a five-year term. The entire permitting process may require several months from first application to final issuance, depending on the flow of work at the agency, the quality of the information, the size and complexity of the facility, the degree of controversy associated with the facility and the requested permit, as well as a host of possible contributing issues. Some agencies may offer special expedited permit issuance under certain circumstances. Delays in permitting can be quite costly, for example, by delaying the start of production at a new facility. Maintaining close communications with the agency concerning issues and expectations is highly beneficial from the first inquiry through the issuance of the final permit.

LIVING WITH THE PERMIT

Once the permit has been obtained, the facility owner/operator must be prepared to demonstrate continuing compliance with *all* terms of the permit. This involves some combination of operational documentation and emissions verification (perhaps simply through keeping the emission inventory information up to date), documenting any required monitoring, and meeting any administrative requirements in the permit. Examples of administrative requirements include:

- Providing notice to the control agency concerning facility modifications/new equipment, etc
- Providing notice of emergency releases, malfunctioning control equipment, etc (although the permit may specify that such activities are a permit violation)
- Maintaining maintenance records for selected monitoring equipment and control devices

Agencies may add standard provisions to each permit that specify a number of specific accountabilities.

Modifications to an existing permit may require significant effort for the applicant. Likewise, *renewals* of permits may create a major effort if the regulatory programs continue to change significantly through the coming years. Many of the goals of the CAAA will require future implementation, depending on the pace of regulatory development and funding. As a minimum, the permit renewal process will require an updating of facility information and regulatory requirements. Improved guidance, such as expanded emission factor information or test databases, could simplify the renewal process.

BIBLIOGRAPHY

1. *Code of Federal Regulations,* Vol. 40 (1995), Parts 60, 70, and 71, U.S. EPA Office of Air Quality Planning and Standards, Washington, D.C., 1995.
2. State of Missouri, *Operating Permit Application and Forms,* Missouri Dept. of Natural Resources, Air Pollution Control Program, Jefferson City, Mo., 1995.

AIR POLLUTION

GEORGE WOLFF
GM Public Policy Center
Detroit, Michigan

The concept of air pollution has changed significantly during the past several decades. Many years ago, air pollution was only associated with smoke, soot, and odors. Textbooks published since the 1970s define air pollution as any atmospheric condition in which substances are present in concentrations high enough above their normal ambient levels to produce a measurable effect on humans, animals, vegetation, or materials. Even this definition, however, is deficient today because it would not include "greenhouse" or ozone-depleting gases since the effects of these gases on humans, animals, vegetation, or materials have not been, and may never be, observed. Still, there is a need to include them because of their potential to alter the global climate and, hence, the global ecosystem. Therefore, in an attempt to be more comprehensive and timely, the following definition is offered: air pollution is the presence of any substance in the atmosphere at a concentration high enough to produce an objectionable effect on humans, animals, vegetation, or materials, or to alter the natural balance of any ecosystem significantly. These substances can be solids, liquids, or gases, and can be produced by anthropogenic activities or natural sources. In this article, however, only nonbiological material will be considered. Airborne pathogens and pollens, molds, and spores will not be discussed. Airborne radioactive contaminants will not be discussed either, except for radon, which will be discussed in the context of an indoor air pollutant.

As the definition of air pollution changed, the breadth of air pollution also expanded. Initial perceptions of objectionable effects of air pollutants were limited to those easily detected: odors, soiling of surfaces, and smoke-belching stacks. Three rare meteorological events, however, made it clear that air pollutants can be hazardous to human health, and can even cause death at high enough concentrations. The first event was in the Meuse Valley in Belgium in 1930. Meteorological conditions produced a week-long air stagnation which caused pollutants to accumulate to presumably extremely high concentrations. As a result, 60 people died and a large number of people experienced respiratory problems. In 1948, similar meteorological conditions in Donora, Pennsylvania, resulted in

(a)

(b)

Plate 1. Top view shows how fumes would pour from the basic oxygen furnace stack at Bethlehem Steel Corporation's Bethlehem, Pa., plant if an electrostatic precipitator control system had not been installed. Bottom view indicates successful control with the precipitators at work. Courtesy of Bethlehem Steel Corp.

nearly 7000 illnesses and 20 deaths. In 1952, 4000 deaths were attributed to a four-day "killer fog" in London. While these episodes dramatized the acute health effects of air pollution at high concentrations, it was the concern over longer term, chronic effects that led to the initiation of National Ambient Air Quality Standards (NAAQS) for six Criteria Pollutants (so named because EPA is required to summarize published information on each pollutant—these summaries are called Criteria Documents) in the United States in the early 1970s. The six criteria pollutants were: sulfur dioxide (SO_2), carbon monoxide (CO), nitrogen dioxide (NO_2), ozone (O_3), suspended particulates, and nonmethane hydrocarbons (NMHC, now referred to as volatile organic compounds or VOC). These Criteria Pollutants captured the attention of regulators because they were ubiquitous, there was substantial evidence linking them to health effects at high concentrations, three of them (O_3, SO_2, and NO_2) were also known phytotoxicants (toxic to vegetation), and they were fairly easy to measure. The NMHC were listed as Criteria Pollutants because NMHC are precursors to O_3. However, shortly after NMHC were so designated, it became obvious that the simple empirical relationship that the U.S. Environmental Protection Agency (EPA) used to relate NMHC concentrations to O_3 concentrations was not valid; consequently, NMHC were dropped from the criteria list. In the late 1970s, EPA added lead (Pb) to the list. Particulate matter with an aerodynamic diameter of less than or equal to 10 μm, PM_{10}, was added to the list in 1987.

Several developments since the establishment of the Criteria Pollutants greatly expanded the geographic scale of concern. Until the 1970s, one air pollution axiom was "dilution is the solution to pollution," and the result was that tall smoke stacks were built. In the mid-1970s, however, it was shown that high concentrations of O_3 and sulfate haze could be transported hundreds of miles across state and international borders. The acid deposition studies of the 1980s clearly illustrated the international and global aspects of the issue and, finally, when stratospheric O_3 depletion and global warming became issues, air pollution became viewed in a global context. At the same time that the geographic scale of the issue was expanding, the number of pollutants of concern also increased. In the 1970s, it was realized that hundreds of potentially toxic chemicals were being released to the atmosphere. As detection capabilities improved, these chemicals were, indeed, measured in the air. This led to the establishment of a "hazardous air pollutant" category which included any potentially toxic substance in the air that was not a criteria pollutant.

AIR POLLUTION COMPONENTS

Air pollution can be considered to have three components: sources, transport and transformations in the atmosphere, and receptors. The sources are any process, device, or activity that emits airborne substances. When the substances are released, they are transported through the atmosphere. Some of the substances interact with sunlight, or other substances in the atmosphere, and are transformed into different substances. Air pollutants that are emitted directly to the atmosphere are called primary pollutants. Pollutants that are formed in the atmosphere as a result of transformations are called secondary pollutants. The reactants that undergo the transformation are referred to as precursors. An example of a secondary pollutant is O_3, and its precursors are nitrogen oxides (NO_x = nitric oxide[NO] + NO_2) and NMHC. The receptor is the person, animal, plant, material, or ecosystem affected by the emissions.

Sources

There are three types of air pollution sources: point, area, and line sources. A point source is a single facility that has one or more emissions points. An area source is a collection of smaller sources within a particular geographic area. For example, the emissions from residential heating would be treated collectively as an area source. A line source is a one-dimensional, horizontal configuration. Roadways are an example. Most, but not all, emissions emanate from a specific stack or vent. Emissions emanating from sources other than stacks, such as storage piles or unpaved lots, are classified as fugitive emissions.

Table 1. Nationwide Air Pollutant Emissions Estimates for the United States in 1991[a]

	Pollutant Emissions, 10^6 t/yr					
Source Category	PM_{10}	SO_x	NO_x	VOC	CO	Pb[b]
Transportation	1.5	1.0	7.3	5.1	43.5	1.6
Stationary fuel combustion	1.1	16.6	10.6	0.7	4.7	0.5
Industrial processes	1.8	3.2	0.6	7.9	4.7	2.2
Solid waste	0.3	0.02	0.1	0.7	2.1	0.7
Miscellaneous	0.7	0.01	0.2	2.6	7.1	0.0
Totals[c]	*5.4*	*20.7*	*18.8*	*16.9*	*62.1*	*5.0*

[a] Ref. 4.
[b] 10^3 t/yr.
[c] The sums of the subcategories may not equal totals due to rounding.

EPA requires that individual states develop emissions inventories for all primary pollutants and precursors to secondary pollutants that are classified as criteria or hazardous air pollutants. The degree of geographical resolution and detail required in the inventory depends on the severity of the local air pollution problem. In clean rural areas, countrywide emissions totals for individual pollutant species may be all that is required. Emissions from large point sources are inventoried separately, however. For urban areas with severe air pollution problems, gridded emissions inventories are required. In a gridded inventory, the area is divided into squares which are typically 5 to 10 kilometers on a side; area- and line-source emissions are calculated for each grid square. Small point-sources are included in area sources, but the large point-sources are listed individually. Such detailed inventories are required as inputs to sophisticated air-quality models, which are used to develop air pollution control strategies. Details on the construction of emissions inventories have been presented by EPA (1).

Emissions rates for a specific source can be measured directly by inserting sampling probes into the stack or vent. While this has been done for most large point sources, it would be an impossible task to do for every individual source included in an area inventory. Instead, emission factors (based on measurements from similar sources or engineering mass-balance calculations) are applied to the multitude of sources. An emission factor is a statistical average or quantitative estimate of the amount of a pollutant emitted from a specific source-type as a function of the amount of raw material processed, the amount of product produced, or the amount of fuel consumed. Emission factors for most sources have been compiled (2).

To obtain emission factors for motor vehicles, the vehicles are operated using various driving patterns on a chassis dynamometer. Emissions are determined as a function of vehicle model year, speed, temperature, etc. Dynamometer-based emissions data are used in EPA MOBILE4 model (3) to calculate total fleet emissions for a given roadway system.

Each year, EPA publishes a summary of air pollution emissions and air quality trends for the Criteria Pollutants (4). Table 1 contains the summary for 1991. United States emissions estimates for these pollutants are available back to 1940 (5).

Transport and Transformation

Once emitted into the atmosphere, air pollutants are transported and may be transformed. The fate of a particular pollutant depends on the stability of the atmosphere and the stability of the pollutant in the atmosphere. The former determines the concentration of the species, initially in the atmosphere, while the latter determines the persistence of the substance in the atmosphere. The stability of the atmosphere depends on the ventilation. The stability of a pollutant depends on the presence or absence of clouds or fog, the presence or absence of precipitation, the pollutant's solubility in water, the pollutant's reactivity with other atmospheric constituents (which may be a function of temperature), the concentrations of other atmospheric constituents, the pollutant's stability in the presence of sunlight, and the deposition velocity of the pollutant.

As the ventilation increases the concentration of the pollutants decrease for a given emission rate. The atmospheric stability is determined by comparing the actual lapse rate to the dry adiabatic lapse rate. An air parcel warmer than the surrounding air will rise and cool at the dry adiabatic lapse rate of 9.8°C/1000 m, the atmosphere is unstable, and the concentrations of pollutants are lower. As the lapse rate becomes smaller, the dispersive capacity of the atmosphere declines and reaches a minimum when the lapse rate becomes positive, at which point a temperature inversion exists. Temperature inversions form every evening in most places as the heat from the earth's surface is radiated upward, and the air in the lower layers of the atmosphere is cooled. However, these inversions are usually destroyed the next morning as the sun heats the earth's surface, which in turn heats the air adjacent to the surface, and convective activity (mixing) is initiated. Most episodes of high pollution concentrations are associated with multiday inversions.

The stability or persistence of a pollutant in the atmosphere depends on the pollutants atmospheric residence time, which depends on the pollutant's reactivity with other atmospheric constituents, surfaces, or water. Mean atmospheric residence times and principal atmospheric sinks for a variety of species are given in Table 2. Species like SO_2, NO_x (NO and NO_2), and coarse particles have lifetimes less than a day; thus important environmental impacts from these pollutants are usually within close proximity to the emissions sources. Secondary reaction products may have a larger zone of influence, however, depending on the residence times of the products. The principal sink for the SO_2 and NO_2 is the reaction with the hydroxyl radical (OH). Sources of OH and the nature of the secondary reaction products are discussed in subsequent sections. The residence time of O_3 varies considerably, depending on the presence of other constituents. In the presence of high concentrations of NO, the lifetime is on the order of hours to seconds. In the relatively nonpolluted environment of the free troposphere (from approximately 1500 m to the top of the troposphere (~12 km)), the maximum lifetime applies.

Table 2. Mean Residence Time (τ) of Species in the Atmosphere

Species	CAS Registry Number	τ	Dominant Sink[a]	Location[b]	Refs.
SO_2	*[7446-09-5]*	0.5 days	OH	T	6
NO_x	*[10102-44-0]* *[10102-43-9]*	0.5 days	OH	T	6
Coarse particles (>2.5 μm dia)		<1 day	S, P	T	7
O_3 (tropospheric)	*[10028-15-6]*	90 days[c]	NO, UV, Sr, O	T	6
Fine particles (<2.5 μm dia)		5 days[d]	P	T	7
CO	*[630-08-0]*	100 days	OH	T	6
CO_2	*[124-38-9]*	120 yr[e]	O	T	8
CH_4	*[74-82-8]*	7–10 yr	OH	T	6
CFC-11	*[75-69-4]*	65–75 yr	UV	St	6,8
CFC-12	*[75-71-8]*	110–130 yr	UV	St	6,8
N_2O	*[10024-97-2]*	120–150 yr	UV	St	6,8
CFC-113	*[76-13-1]*	90 yr	UV	St	6

[a] Sinks: (OH), reaction with OH; S, sedimentation; P, precipitation scavenging; NO, reaction with NO; UV, photolysis by ultraviolet radiation; Sr, destruction at surfaces; O, adsorption or destruction at oceanic surface.
[b] Location of sink: T, troposphere; St, stratosphere.
[c] Tropospheric residence time only; shorter lifetime applies to urban areas where NO quickly destroys O_3; upper limit applies to the remote troposphere.
[d] Applies to particles released in the lower troposphere only; the most important sink is scavenging by precipitation, so in the absence of precipitation, these particles remain suspended longer.
[e] Combined lifetime for atmosphere, biosphere, and upper ocean.

Consequently, under the right conditions, O_3 could have important environmental impacts far downwind from the O_3 precursor sources. In fact, concentrations of O_3 near the NAAQS have been transported from the Gulf Coast of the United States to the Northeast over a several-day period (9).

Particles with diameters less than 2.5 μm have negligible settling velocities and, therefore, have residence times that are considerably longer than those of larger particles. As a result, observations have shown multiday transport of haze produced by fine particles over distances of more than a thousand kilometers (10). The longer life-times of the greenhouse gases, those listed below CO in Table 2, result in the accumulation and relatively even distribution of these gases around the globe. Chlorofluorocarbons (CFCs) and nitrous oxide (N_2O) are essentially inert in the troposphere and are only destroyed in the stratosphere by ultraviolet (UV) solar radiation. Unfortunately, the photolysis products of CFCs are the reactants which are responsible for stratospheric O_3 depletion.

To determine the fate of a pollutant after it is released, two approaches, monitoring and modeling, are available. Monitoring of the criteria pollutants is done routinely by state and local air pollution agencies in most large urban areas and in some other areas as well. The recommended techniques for measuring the criteria and many other pollutants are found in Reference 11. Monitoring is expensive and time consuming, however, and even the most extensive urban networks are insufficient to assess the geographic distribution of pollutants accurately.

Receptors

The receptor is the person, animal, plant, material, or ecosystem affected by the pollutant emission. The criteria pollutants and the hazardous air pollutants were so designated because, at sufficient concentrations, they can cause adverse health effects to human receptors. Also, some of the criteria pollutants cause damage to plant receptors. For each criteria pollutant, an "Air Quality Criteria Document" exists and is updated periodically (see *General References*). These documents summarize the most current published literature concerning the effects of these pollutants on human health, animals, vegetation, and materials.

With respect to acid deposition, the receptors that have generated the most concern are certain aquatic ecosystems, although there is also some concern that acid deposition adversely affects some materials. For visibility-reducing air pollutants, CFCs, and greenhouse gases, the receptor is the atmosphere. Visibility-reducing species alter the optical properties of the atmosphere, and CFCs alter the natural chemical composition of the atmosphere in such a way that it becomes more transparent to potentially harmful UV solar radiation. The greenhouse gases alter the radiative properties of the atmosphere and, consequently, have the potential to alter the global heat budget.

AIR QUALITY MANAGEMENT

In the United States, the framework for air quality management was established by the 1965 Clean Air Act (CAA) and subsequent amendments in 1970 and 1977, and the comprehensive amendments of 1990. The CCA defines two categories of pollutants: criteria pollutants and hazardous air pollutants. For the criteria pollutants, the CAA required that EPA establish NAAQS and emissions standards for some large new sources and for motor vehicles, and gave the primary responsibility for designing and implementing air quality improvement programs to the individual states. For the hazardous air pollutants, only emissions standards for some sources are required. The NAAQS apply uniformly across the United States whereas emissions standards for criteria pollutants can vary somewhat, depending on the severity of the local air pollution problem and whether an affected source already exists or is proposed as a new source. In addition, individual states have the right to set their own ambient air quality and emissions standards (which must be at least as stringent as federal standards) for all pollutants and all sources except motor vehicles. With respect to motor vehicles, the CAA allows the states to choose between two sets of emissions standards: federal standards or the more stringent California standards.

Table 3. National Ambient Air Quality Standards

Pollutant	Primary[a] $\mu g/m^3$	Primary[a] ppm	Secondary[a] $\mu g/m^3$	Secondary[a] ppm	Averaging Time
PM_{10}	50		50		Annual arithmetic mean
	150		150		24-h[b]
SO_2	80	(0.03)			Annual arithmetic mean
	365	(0.14)			24-h[b]
			1300	(0.50)	3-h[b]
CO	(10)	9			8-h[b]
	(40)	35			1-h[b]
NO_2	(100)	0.053	(100)	0.053	Annual arithmetic mean
Pb	1.5		1.5		Maximum quarterly average
O_3	(235)	0.12	(235)	0.12	Maximum daily[c] 1-h average

[a] Parenthetical value is an approximately equivalent concentration.
[b] Not to be exceeded more than once per year.
[c] Not to be exceeded on more than three days in three years.

The two levels of NAAQS, primary and secondary, are listed in Table 3. Primary pollutant standards were set to protect public health with an adequate margin of safety. Secondary standards, where applicable, were chosen to protect public welfare, including vegetation. The pollutant PM_{10} refers to particulate matter with an aerodynamic diameter less than or equal to 10 μm. Originally, there was a nonsize-selective NAAQS for total suspended particulates (TSP), but it was revised to the PM_{10} standard in 1987. According to the CAA, the scientific bases for the NAAQS are to be reviewed every five years so NAAQS levels reflect current knowledge. In practice, however, the review cycle takes considerably longer. In order to analyze trends in Criteria Pollutants nationwide, the EPA has established three types of monitoring systems. The first is a network of 98 National Air Monitoring sites (NAMS), located in areas with high pollutant concentrations and high population exposures. The system was established by regulations promulgated in May 1979 to provide accurate and timely data on the national air quality. In addition, EPA also regularly evaluates data from the State and Local Monitoring system (SLAMS) and from Specific Purpose Monitors (SPM). These three types of stations comprise the 274-site national monitoring system, which is required to meet rigid quality assurance criteria. To determine if an area meets the NAAQS, the states are required to monitor the concentrations of the Criteria Pollutants in areas that are likely to be near or exceed the NAAQS. If an area exceeds a NAAQS for a given pollutant, it is designated as a nonattainment area for that pollutant, and the state is required to develop and implement a State Implementation Plan (SIP). The SIP is a strategy designed to achieve emissions reductions sufficient to meet the NAAQS within a specific deadline. The deadline is determined by the severity of the local pollution problem. Areas that receive long deadlines (six years or more) must show continuous progress by reducing emissions by a specified percentage each year. For SO_2 and NO_2, the initial SIPs were very successful in achieving the NAAQS in most areas. However, for other criteria pollutants, particularly O_3 and to a lesser extent CO, many areas are starting a third round of SIP preparations with little hope of meeting the NAAQS in the near future. If a state misses an attainment deadline, fails to revise an inadequate SIP, or fails to implement SIP requirements, EPA has the authority to enforce sanctions such as banning construction of new stationary sources and the withholding of federal grants for highways.

In nonattainment areas, the degree of control on small sources is left to the discretion of the state, and it is largely determined by the degree of required emissions reductions. Large existing sources must be retrofitted with reasonable available control technology (RACT) to minimize emissions. All large new sources and existing sources that undergo major modifications must meet EPA's new source performance standards at a minimum; and in nonattainment areas they must be designed with lowest achievable emission rate (LAER) technology, and emissions offsets must be obtained. Offsets require that emissions from existing sources within the area must be reduced below legally allowable levels so that the amount of the reduction is greater than or equal to the emissions expected from the new source. RACT usually is less stringent than LAER because it may not be feasible to retrofit certain sources with the LAER technology.

In attainment areas, new large facilities must be designed to incorporate the best available control technology (BACT). Generally, BACT is more stringent than RACT, and equal to or less stringent than LAER. In addition, there are also rules that specify how much deterioration in baseline air quality a new facility can cause in an attainment area, and in no situation can the facility cause a new violation in the NAAQS.

Large sources of SO_2 and NO_x may also require additional emission reductions because of the 1990 Clean Air Act Amendments. To reduce acid deposition, the amendments require that nationwide emissions of SO_2 and NO_x be reduced on an annual basis by 10 million and 2 million tons, respectively, by the year 2000.

Once a substance is designated by EPA as a Hazardous Air Pollutant (HAP), EPA has to promulgate a National Emission Standard for Hazardous Air Pollutants (NESHAP), which is designed to protect public health with an ample margin of safety. The 1990 Clean Air Act Amendments identify 189 HAPs. These are further discussed in the Air Toxics section.

AIR POLLUTION ISSUES

Photochemical Smog

Photochemical smog is a complex mixture of constituents formed when VOCs and NO_x are irradiated by sunlight. From an effects perspective, O_3 is the primary concern and it is the most abundant species formed in photochemical smog.

Extensive studies have shown that O_3 is a lung irritant and a phytotoxicant. It is responsible for crop damage and it is suspected of being a contributor to forest decline in Europe and in parts of the United States. There are, however, a multitude of other photochemical smog species that have significant environmental consequences. The most important of these additional pollutant species–particles are hydrogen peroxide (H_2O_2), peroxyacetyl nitrate (PAN), aldehydes, and nitric acid.

Photochemical smog is a summertime phenomenon for most parts of the United States because temperatures are too low and sunlight is insufficient during the other seasons. In the warmer parts of the country, especially in southern California, the smog season begins earlier and lasts into the fall. Despite almost two decades of reducing VOC emissions from stationary and mobile sources and NO_x emissions from mobile sources, progress had been slow in reducing the number of areas in the United States designated as nonattainment for O_3. For the 1985–1987 period, there were 64 areas, mostly urban, that experience O_3 concentrations above the NAAQS. This number increased to 101 areas after the anomalously hot summer of 1988, but then gradually decreased to 56 by 1991. The recent decrease is due primarily to decreased VOC emissions resulting from reducing the Reid vapor pressure of gasoline and the replacement of older vehicles with new, cleaner ones. An area is classified nonattainment if the O_3 design value (the design value is equal to the fourth highest maximum daily 1-h O_3 concentration within a 3-yr period) exceeds 0.12 ppm. For most of the nonattainment areas, the design value falls between 0.12 to 0.15 ppm. The three areas with the highest design values for the 1989–1991 period were the Los Angeles (0.31 ppm), Houston (0.22), and New York (0.17) metropolitan areas.

There is a significant clean air background O_3 concentration that varies with season and latitude. The clean air background is defined as the concentrations measured at pristine areas of the globe. It consists of natural sources of O_3, but it undoubtedly contains some anthropogenic contribution because it may have increased since the last century (12). In the summertime in the United States, the average background is about 0.04 ppm (13). This background O_3 has four sources: intrusions of O_3-rich stratospheric air, *in situ* O_3 production from methane (CH_4) oxidation, the photooxidation of naturally emitted VOCs from vegetation, and the long-range transport of O_3 formed from the photooxidation of anthropogenic VOCs and NO_x emissions. Although there are several mechanisms which will transport O_3-rich air from the stratosphere into the lower troposphere, the most important appears to be associated with large-scale eddy transport that occurs in the vicinity of upper air troughs of low pressure associated with the jet stream (14). This is an intermittent mechanism, so the contribution of stratospheric O_3 to surface O_3 will have a considerable temporal variation. On very rare occasions, this mechanism has produced brief ground level concentrations exceeding 0.12 ppm (15). The other three mechanisms will be described below.

In the presence of sunlight ($h\nu$), NO_2 photolyzes and produces O_3:

$$NO_2 + h\nu \rightarrow NO + O \quad (1)$$

$$O + O_2 + M \rightarrow O_3 + M \quad (2)$$

$$NO + O_3 \rightarrow NO_2 + O_2 \quad (3)$$

where M is any third-body molecule (most likely N_2 or O_2 in the atmosphere) that remains unchanged in the reaction. This process produces some steady-state concentration of O_3 that is a function of the initial concentrations of NO and NO_2, the solar intensity, and the temperature. Although these reactions are extremely important in the atmosphere, the steady-state O_3 produced is much lower than the concentrations usually observed, even in clean air. In order for O_3 to accumulate, there must be a mechanism that converts NO to NO_2 without consuming a molecule of O_3, as does reaction 3. Reactions among hydroxyl radicals (OH) and hydrocarbons or VOC constitute such a mechanism. In clean air, OH may be generated by:

$$O_3 + h\nu \rightarrow O_2 + O(1\,D) \quad (4)$$

$$O(1\,D) + H_2O \rightarrow 2\,OH \quad (5)$$

where $O(^1D)$ is an excited form of an O atom that is produced from a photon at a wavelength between 280 and 310 nm. This produces a "seed" OH which can produce the following chain reactions:

$$OH + CH_4 \rightarrow H_2O + CH_3 \quad (6)$$

$$CH_3 + O_2 + M \rightarrow CH_3O_2 + M \quad (7)$$

$$CH_3O_2 + NO \rightarrow CH_3O + NO_2 \quad (8)$$

The NO_2 then photolyzes and produces O_3 (eqs. 1 and 2). The CH_3O radical continues to react:

$$CH_3O + O_2 \rightarrow HCHO + HO_2 \quad (9)$$

and the HO_2 radical also forms more NO_2:

$$HO_2 + NO \rightarrow NO_2 + OH \quad (10)$$

which result in more O_3. In addition, OH is regenerated, and it can begin the cycle again by reacting with another CH_4 molecule. Further, the formaldehyde photodissociates

$$HCHO + h\nu \xrightarrow{a} H_2 + CO \quad (11)$$

$$\xrightarrow{b} HCO + H \quad (12)$$

$$HCO + O_2 \rightarrow HO_2 + CO \quad (13)$$

$$H + O_2 \rightarrow HO_2 \quad (14)$$

and the HO_2 from both equations 13 and 14 will form additional NO_2. Furthermore, the CO is oxidized:

$$CO + OH \rightarrow CO_2 + H \quad (15)$$

and the H radical can form another NO_2 (eqs. 14 and 9). Thus, the oxidation of one CH_4 molecule is capable of producing three O_3 molecules and two OH radicals by the reaction sequence. However, this chain reaction is less than 100% efficient because there are many competing chain-terminating reactions. Two examples are equations 16 and 17.

$$HO_2 + HO_2 \rightarrow H_2O_2 + O_2 \quad (16)$$

$$OH + NO_2 \xrightarrow{M} HNO_3 \quad (17)$$

Both of these reactions terminate the chain by scavenging a free radical. On the average, however, the chain results in a net production of O_3 and OH.

Table 4. Median Concentration of the 10 Most Abundant Ambient Air Hydrocarbons in 39 U.S. Cities and their Relative Reactivity with OH

Compound	CAS Registry Number	Median Concentration[a] ppbC	Relative Reactivity with OH[b,c]
Isopentane	[463-82-1]	45.3	494
n-Butane	[106-97-8]	40.3	351
Toluene	[108-88-3]	33.8	831
Propane	[74-98-6]	23.5	143
Ethane	[74-84-0]	23.3	36
n-Pentane	[109-66-0]	22.0	480
Ethylene	[74-85-1]	21.4	1013
m-Xylene	[108-38-3]	18.1	3117
p-Xylene	[106-42-3]	18.1	1818
2-Methylpentane	[107-83-5]	14.9	
Isobutane	[75-28-5]	14.8	325
		Biogenic species	
α-Pinene	[80-56-8]		7792
Isoprene	[78-79-5]		12078

[a] Ref. 16.
[b] Ref. 17.
[c] Relative to CH_4 + OH reaction at 298°C.

In a polluted or urban atmosphere, O_3 formation by the CH_4 oxidation mechanism is overshadowed by the oxidation of other VOCs. The "seed" OH can be produced from equations 5 and 6, but the photodisassociation of carbonyls and nitrous acid (HNO_2, formed from the reaction of OH + NO and other reactions) are also important sources of OH in polluted environments. An imperfect, but nevertheless useful, measure of the rate of O_3 formation by VOC oxidation is the rate of the initial OH–VOC reaction. The rate of the OH-CH_4 reaction is much slower than any other OH–VOC reaction. Table 4 contains the reaction rates with OH relative to the OH–CH_4 rate for some commonly occurring VOCs and their median concentrations from 39 cities. Also shown for comparison are the relative reaction rates between OH and two VOC species emitted by vegetation: isoprene and α-pinene. It is obvious from the data in Table 4 that there is a wide range of reactivities. In general, internally bonded olefins are the most reactive, followed in decreasing order by terminally bonded olefins, multialkyl aromatics, monoalkyl aromatics, C5 and greater paraffins, C2–C4 paraffins, C2–C4 paraffins, benzene, acetylene, and ethane.

Once the sun sets, O_3 formation ceases and, in an urban area, it is rapidly scavenged by freshly emitted NO by equation 3. On a typical summer night, however, a nocturnal inversion begins to form around sunset, usually below a few hundred meters. Consequently, the surface-based NO emissions are trapped below the top of the inversion. Above the inversion to the top of the mixed layer (usually about 1500 m), O_3 is depleted at a much slower rate. The next morning, the inversion dissipates and the O_3-rich air aloft is mixed down into the O_3-depleted air near the surface. This process, in combination with the onset of photochemistry as the sun rises, produces the sharp increase in surface O_3, shown in Figure 1. Notice in Figure 1 that the overnight O_3 depletion is less in the more rural areas than in a large urban area (ie, New York City). This is due to lower overnight levels of NO in rural areas. Even in the absence of NO or other O_3 scavengers (olefins, for example), O_3 will still decrease at night near the ground faster than aloft due to the destruction of O_3 at any surface (ie, the ground, buildings, trees, etc). At the remote mountaintop sites, Whiteface and Utsayantha, there is no overnight decrease in O_3 concentrations.

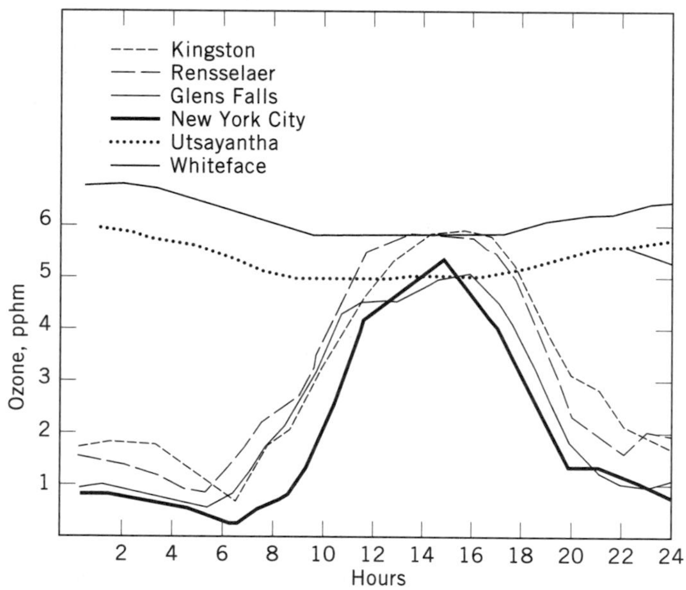

Figure 1. Hourly average ozone concentrations during August 1–17, 1973 for selected sites in New York State (18). Courtesy of Air and Waste Management Association.

Although photochemical smog is a complex mixture of many primary and secondary pollutants and involves a myriad of atmospheric reactions, there are characteristic pollutant concentration versus time profiles that are generally observed within and downwind of an urban area during a photochemical smog episode. In particular, the highest O_3 concentrations are generally found 10–100 km downwind of the urban emissions areas, unless the air is completely stagnant. This fact, in conjunction with the long lifetime of O_3 in the absence of high concentrations of NO, means that O_3 is a regional problem. In the

Los Angelos basin, high concentrations of O_3 are transported throughout the basin, and multiday episodes are exacerbated by the accumulation of O_3 aloft, which is mixed to the surface daily. On the East Coast, a typical O_3 episode is associated with a high pressure system anchored offshore, producing a southwesterly flow across the region. As a result, emissions from Washington, D.C., travel and mix with emissions from Baltimore and over a period of a few days will continue traveling northeastward through Philadelphia, New York City, and Boston. Under these conditions, the highest O_3 concentrations typically occur in central Connecticut (19).

It is obvious from the above discussion that in order to reduce O_3 in a polluted atmosphere, reductions in the precursors (VOC and NO_x) are required. However, the choice of whether to control VOC or NO_x or both as the optimum control strategy depends on the local VOC/NO_x ratio. At low VOC/NO_x ratios for example, O_3 formation is suppressed principally by equations 3 and 17. Consequently, reducing NO_x emissions (which are emitted mainly as NO) in this case will reduce the amount of O_3 scavenged by equation 3 and the amount of OH scavenged by equation 17. The consequence of both will be an increase in O_3 concentrations. The 1990 Clean Air Act Amendments require that O_3 nonattainment areas reduce both VOC and NO_x from big stationary sources unless the air quality benefits are greater in the absence of NO_x reductions. Large cities in the Northeast tend to have ratios significantly less than 10:1, while cities in the South (Texas and east) tend to have ratios greater than 10:1. Determining a workable control strategy is further complicated by the transport issue. For example, on high O_3 days in the Northeast, the upwind air entering Philadelphia and New York City frequently contains O_3 already near or over the NAAQS as a result of emissions from areas to the west and south (20). Consequently, control strategies must be developed on a coordinated, multistate, regional basis.

Because of the complex mixture of VOCs in the atmosphere, the composition of the reaction products and intermediate species is even more complex. Some of the more important species produced in the smog process include hydrogen peroxide (H_2O_2), peroxyacetyl nitrate (PAN), aldehydes (particularly formaldehyde, HCHO [*50-00-0*]), nitric acid (HNO_3), and particles. The H_2O_2 is formed and dissolved in cloud droplets; equation 17 is an important oxidant, responsible for oxidizing SO_2 to H_2SO_4, the primary cause of acid precipitation. The oxidation of many VOCs produce acetyl radicals, CH_3CO, which can react with O_2 to produce peroxyacetyl radicals, $CH_3(CO)O_2$, which in turn can react:

$$CH_3(CO)O_2 + NO_2 \rightleftharpoons CH_3C(O)O_2NO_2(PAN) \qquad (18)$$

At high enough concentrations, PAN is a potent eye irritant and a phytotoxicant. On a smoggy day in the Los Angeles area, PAN concentrations are typically 5–10 ppb. PAN concentrations in the rest of the United States are generally a fraction of a ppb. An important formation route for HCHO is equation 9. However, ozonolysis of olefinic compounds and some other reactions of VOCs can also produce HCHO and other aldehydes. Aldehydes are important because they are temporary reservoirs of free radicals (see eqs. 11 and 12), and HCHO is a known carcinogen. Nitric acid is formed by OH attack on NO_2 and by a dark-phase series of reactions initiated by $O_3 + NO_2$. Nitric acid is important because, next to H_2SO_4, it is the second most abundant acid in precipitation. In addition, in southern California it is the main cause of acid fog.

Particles are the principal cause of the haze and the associated brown color that is often apparent with smog. The three most important types of particles produced in smog are composed of organics, sulfates, and nitrates. Organic particles are formed when large VOC molecules, especially aromatics and cyclic alkenes, react with each other and form condensable products. Sulfate particles (formed initially as H_2SO_4) are formed by a series of reactions initiated by the attack of OH on SO_2 in the gas phase or by liquid-phase reactions. Nitrate particles are formed by

$$HNO_3\ (g) + NH_3(g) \rightleftharpoons NH_4NO_3(s) \qquad (19)$$

or by the reactions of HNO_3 with salt (NaCl) or alkaline soil dust.

Volatile Organic Compounds (VOC). VOCs include any organic carbon compound that exists in the gaseous state in the ambient air. In some of the older air pollution literature the term VOCs was used interchangeably with the term nonmethane hydrocarbons (NMHC). Strictly speaking, this is incorrect organic chemistry nomenclature because hydrocarbons include only those compounds that contain carbon and hydrogen exclusively; whereas organic carbon includes any compound that contains organic carbon (ie, HCHO). Sources of VOCs include any process or activity that uses organic solvents, coatings, or fuel. Emissions of VOCs are important because some are toxic by themselves and most are precursors of O_3 and other species associated with photochemical smog. As a result of control measures designed to reduce O_3, VOC emissions are declining in the United States. Between 1979–1988, nationwide VOC emissions declined 17% (4). Trends in ambient VOC concentrations cannot be determined, however, because of the lack of measurements.

Nitrogen Oxides (NO_x). In air pollution terminology, nitrogen oxides include the gases NO and NO_2. Most of the NO_x is emitted as NO, which is oxidized to NO_2 in the atmosphere (see eqs. 3 and 8). All combustion processes are sources of NO_x. At the high temperatures generated in the combustion process, some N_2 is converted to NO in the presence of O_2. In general, the higher the combustion temperature, the more NO_x produced. Since NO_2 is one of the original Criteria Pollutants and it is a precursor to O_3, it has been the target of successful emissions reduction strategies for two decades in the United States. As a result, in 1987, all areas of the United States, with the exception of the Los Angeles/Long Beach area, were in compliance with the NAAQS for NO_2. From 1979 to 1988, nationwide NO_x emissions declined 8%, while ambient concentrations declined 7% (4).

Throughout the United States, however, NO_x remains an important issue because it is an essential ingredient of photochemical smog and some of the NO_x is oxidized to HNO_3, an essential ingredient of acid precipitation and fog. In addition, NO_2 is the only important gaseous species in the atmosphere that absorbs visible light, and in high enough concentrations can contribute to a brownish discoloration of the atmosphere. Additional information on NO_x appears in the sections on Acid Deposition, Particulate Matter, and Visibility.

Sulfur Oxides (SO_x). The combustion of sulfur-containing fossil fuels, especially coal, is the primary source of SO_x. Between 97 and 99% of the SO_x emitted from combustion sources is in the form of SO_2, which is a Criteria Pollutant. The remainder is mostly SO_3, which in the presence of atmospheric water vapor is immediately transformed into H_2SO_4 [*7664-93-9*], a liquid particulate. Both SO_2 and H_2SO_4 at sufficient concentrations produce deleterious effects on the respiratory system. In addition, SO_2 is a phytotoxicant. As with NO_2, control strategies designed to reduce the ambient levels of SO_2 have been highly successful. In the 1960s, most industrialized urban areas in the eastern United States had an SO_2 air quality problem. By 1991, only Steubenville, Ohio, exceeded the annual NAAQS. Over the past 10 years, nationwide emissions declined 2%, and ambient concentrations decreased about 20% (4). However, the 1990 Clean Air Act Amendments require additional SO_2 reductions because of the role that SO_2 plays in acid deposition. Further discussion is found in the Acid Deposition, PM_{10}, and Visibility sections. In addition, there is some concern over the health effects of H_2SO_4 particles, which are not only emitted directly from some sources, but are also formed in the atmosphere from oxidation of SO_2 (21).

Carbon Monoxide (CO). Carbon monoxide is a colorless, odorless gas emitted during the incomplete combustion of fuels. CO is emitted during any combustion process, and transportation sources account for about two-thirds of the CO emissions nationally. However, in certain areas, woodburning fireplaces and stoves contribute most of the observed CO. CO is absorbed through the lungs into the blood stream and reacts with hemoglobin to form carboxyhemoglobin, which reduces the oxygen carrying capacity of the blood.

Emissions of CO in the United States peaked in the late 1960s, but have decreased consistently since that time as transportation sector emissions significantly decreased. Between 1968 and 1983, CO emissions from new passenger cars were reduced by 96%. This has been partially offset by an increase in the number of vehicle miles traveled annually. Even so, there has been a steady decline in the CO concentrations across the United States, and the decline is expected to continue until the late 1990s without the implementation of any additional emissions–reduction measures. In 1991, there were still 41 urban areas in the United States that were classified as nonattainment for CO, but the number of exceedances declined by about 90% from 1982 to 1991. Over the same time period, nationwide CO emissions decreased 31%, and ambient concentrations declined by 30% (4).

Particulate Matter

In the air pollution field, the terms particulate matter, particulates, particles, and aerosols are used interchangeably and all refer to finely divided solids and liquids dispersed in the air. The original EPA primary standards for total suspended particulates (TSP) were 75 $\mu g/m^3$ as an annual geometric mean, and 260 $\mu g/m^3$ as 24-h mean. TSP is the weight of any particulate matter collected on the filter of a high volume air sampler. On the average, the high volume sampler collects particles that are less than about 30–40 μm in diameter. However, because of the unsymmetrical design of the sampler's inlet, the collection efficiencies for particles varies with both wind direction and speed. In 1987, the EPA promulgated new standards for ambient particulate matter using a new indicator, PM_{10}, rather than TSP. PM_{10} is particulate matter with an aerodynamic diameter of 10 μm or less. The 10-μm diameter was chosen because 50% of the 10 μm particles deposit in the respiratory tract below the larynx during oral breathing. The fraction deposited decreases with increasing particle diameter above 10 μm. The primary and secondary PM_{10} NAAQS are given in Table 3. Because the standard was only enacted in 1987, currently available PM_{10} data are insufficient to determine trends. However, in 1991, EPA designated 70 areas in the United States as nonattainment for PM_{10}% (4) (see ATMOSPHERIC AEROSOLS, METHODS OF ANALYSIS; ATMOSPHERIC AEROSOLS, SIZE DISTRIBUTIONS).

Lead

Lead [*7439-92-1*] (Pb) is of concern because of its tendency to be retained and accumulated by living organisms. When excessive amounts accumulate in the human body, lead can inhibit the formation of hemoglobin and produce life-threatening lead poisoning. In smaller doses, lead is also suspected of causing learning disabilities in children. From 1982 to 1991, nationwide lead emissions decreased 90%, with the primary source, transportation, showing a 97% reduction (4). The most important reason for this dramatic reduction was the removal of lead compounds from fuels, primarily gasoline. Tetraethyllead [*78-00-2*] was added to gasoline to improve the octane rating to prevent engine knock. With the introduction of the catalytic converter, the lead compound had to be eliminated because it would poison the catalyst. Trends of lead in the ambient air have responded to the emissions reductions. In 1991, only a few isolated monitoring sites that were dominated by industrial sources of lead experienced violations of the NAAQS (4).

Air Toxics

There are thousands of commercial chemicals used in the United States. Hundreds of these substances are emitted into the atmosphere and have some potential to affect human health adversely at certain concentrations. Some are known or suspected carcinogens. Identifying all of these substances and promulgating emissions standards is beyond the present capabilities of existing air quality management programs. Consequently, toxic air pollutants (TAPs) need to be prioritized based on risk analysis, so that those posing the greatest threats to health can be regulated. Although the Criteria Pollutants were so designated because they can have significant public health impacts, the Criteria Pollutants are not considered TAPs because the Criteria Pollutants are regulated elsewhere in the CAA. A distinguishing feature between TAPs and Criteria Pollutants is that Criteria Pollutants are considered national issues whereas TAPs are most often isolated issues, localized near the source of the TAP emissions. For example, O_3 is likely to be an issue in all large U.S. metropolitan areas, whereas TAPs usually are of concern only in areas with certain types of sources.

There are three types of TAP emissions: continuous, intermittent, and accidental. Intermittent sources can be routine emissions associated with a batch process or a continuous process operated only occasionally. An accidental release is an inadvertent emission. A dramatic example of this type was the release of methyl isocyanate in Bhopal, which was responsible for over 2000 deaths. As a result of this accident, the U.S. Congress created Title III, a free-standing statute included in the Superfund Amendments and Reauthorization Act (SARA) of 1986. Title III provides a mechanism by which the public can be informed of the existence, quantities, and releases of toxic substances, and requires the states to develop plans to respond to accidental releases of these substances. Further, it requires anyone releasing specific toxic chemicals above a certain threshold amount to annually submit a toxic chemical release form to EPA. At present, there are 308 specific chemicals subject to Title III regulation (22).

A valuable resource that contains a listing of many potential TAPs is the American Conference of Governmental Industrial Hygienists' *Threshold Limit Values* (TLV) (23). This booklet lists the workplace air standards for over 700 substances, many of which would be considered a TAP if present in sufficient quantity in the ambient air. The exceptions would be those listed as simple asphyxiants, or classified as Criteria Pollutants. Toxicological data for these substances are available (24).

The 1970 Clean Air Act required that EPA provide an ample margin of safety to protect against Hazardous Air Pollutants (HAPs) by establishing national emissions standards (NESHAPs) for certain sources. From 1970 to 1990, over 50 chemicals were considered for designation as HAPs, but EPA's review process was completed for only 28 chemicals. Of the 28, NESHAPs were promulgated for only eight substances: beryllium [*7440-41-7]*, mercury [*7436-97-6*], vinyl chloride [*75-01-4*], asbestos [*1332-21-4*], benzene [*71-43-2*], radionuclides, inorganic arsenic [*7440-38-2*], and coke-oven emissions. EPA decided not to list 10 of the substances and intended to list the other 10 substances as HAPs (25). However, in the 1990 Clean Air Act Amendments, 189 substances are listed (Table 5) that EPA must regulate by enforcing maximum achievable control technology (MACT). The Amendments mandate that EPA issue MACT standards for all sources of the 189 substances by the year 2000. In addition, EPA must determine the risk remaining after MACT is in place and develop health-based standards that would limit the cancer risk to one case in one million exposures. EPA may add or delete substances from the list.

Because EPA was so slow in promulgating standards for HAPs prior to the 1990 Amendments, most states developed and implemented their own TAP control programs. Such programs, as well as the pollutants they regulate, differ widely from state to state. Some states have emissions and/or ambient standards. The ambient standards for a given substance are usually selected to be some small fraction of the TLV for that substance.

Odors

The 1977 Clean Air Act Amendments directed EPA to study the effects, sources, and control feasibility of odors. Although no federal legislation has been established to regulate odors, individual states have responded to odor complaints by enforcing common nuisance laws. Because about 50% of all citizen air pollution complaints concern odors, it is clear that a disagreeable odor is perceived as an indication of air pollution. However, many substances can be detected by the human olfactory system at concentrations well below those considered harmful. For example, hydrogen sulfide can be detected by most people at 0.0047 ppm, whereas the occupation health 8-h TLV is 10 ppm. Although exposures to such odors in low concentrations may not in itself cause apparent physical harm, the exposure can lead to nausea, loss of appetite, etc. On the other hand, the absence of an odor is no indication of healthy air. For example, CO is an odorless gas.

Odors are characterized by quality and intensity. Descriptive qualities such as sour, sweet, pungent, fishy, and spicy, to name a few, are commonly used. The strength intensity is determined by how much the concentration of the odoriferous substance exceeds its detection threshold (the concentration at which most people can detect an odor). Odor intensity is approximately proportional to the logarithm of the concentration. However, several factors affect the ability of an individual to detect an odor: the sensitivity of a subject's olfactory system, the presence of other masking odors, and olfactory fatigue (ie, reduced olfactory sensitivity during continued exposure to the odorous substance). In addition, the average person's sensitivity to odor decreases with age.

Visibility

Although there is no NAAQS designed to protect visual air quality, the 1977 Clean Air Act Amendments set as a national goal "the remedying of existing and prevention of future impairment of visibility in mandatory Class I Federal areas which impairment results from man-made pollution." Class I areas are certain national parks and wildernesses that were in existence in 1977. The 1977 Amendments also directed EPA to promulgate appropriate regulations to protect against visibility impairment in these areas. In 1981, EPA directed 36 states to amend their State Implementation Plans to develop control programs for visual impairment that could be traced to particular sources. This type of impairment is called plume bright, and it was the initial focus of EPA's effort because it involved easily identifiable sources. The 1990 Clean Air Act Amendments direct EPA to promulgate appropriate regulations to address regional haze in affected Class I areas. EPA has not dealt with a third type of visibility impairment, urban-scale haze, because the source-receptor relationships are extremely complex (26).

Visibility or visual range is the maximum distance at which a black object can be distinguished from the horizon. In other words, the ability to distinguish an object from the background horizon depends on the contrast between the target and the background. In general, light scattering is dominated by particles, primarily fine particles. The most efficient light-scattering particles are those that are the same size as the wavelengths of visible light (0.4–0.7 μm). A peak in the mass distribution occurs in the size range comparable to the wavelength range of visible light. The particles in this size range, therefore, almost always dominate b_{sp}. Exceptions to this occur during fog, precipitation, and dust storms. On a per mass basis, the most efficient light-scattering fine particles are hygroscopic particles, such as sulfate, nitrate, and ammonium particles, which sorb significant amounts of water at moderate to high relative humidities. As the particles sorb water, they become more efficient light scatterers. Light absorption by particles in the atmosphere is almost exclusively due to elemental carbon which also scatters light. The only common light-absorbing gaseous pollutant is NO_2, which usually accounts for a few percent or less of the total extinction.

Air pollutants can also cause discolorations of the atmosphere. The most common are brownish discolorations (eg, the "brown Los Angeles haze" and the "brown clouds" observed

Table 5. Substances Listed as Hazardous Air Pollutants in the 1990 Clean Air Act Amendments

Substance	CAS Registry Number	Substance	CAS Registry Number	Substance	CAS Registry Number
Acetaldehyde	*[75-07-0]*	Dimethyl carbamoyl chloride	*[79-44-7]*	Parathion	*[56-38-2]*
Acetamide	*[60-35-5]*	Dimethyl formamide	*[68-12-2]*	Pentachloronitrobenzene	*[82-68-8]*
Acetonitrile	*[75-05-8]*	1,1-Dimethyl hydrazine	*[54-14-7]*	Pentachlorophenol	*[87-86-5]*
Acetophenone	*[98-86-2]*	Dimethyl phthalate	*[131-11-3]*	Phenol	*[108-95-2]*
2-Acetylaminofluorene	*[53-96-3]*	Dimethyl sulfate	*[77-78-1]*	*p*-Phenylenediamine	*[106-50-3]*
Acrolein	*[107-02-8]*	4,6-Dinitro-*o*-cresol, and salts	*[534-52-1]*	Phosgene	*[75-44-5]*
Acrylamide	*[79-06-1]*	2,4-Dinitrophenol	*[51-28-5]*	Phosphine	*[7803-51-2]*
Acrylic acid	*[79-10-7]*	2,4-Dinitrotoluene	*[121-14-2]*	Phosphorus	*[7723-14-0]*
Acrylonitrile	*[107-13-1]*	1,4-Dioxane	*[123-91-1]*	Phthalic anhydride	*[85-44-9]*
Allyl chloride	*[107-05-1]*	1,2-Diphenylhydrazine	*[122-66-7]*	Polychlorinated biphenyls	*[1336-36-3]*
4-Aminobiphenyl	*[92-67-1]*	Epichlorohydrin	*[106-89-8]*	1,3-Propane sultone	*[1120-74-4]*
Aniline	*[62-53-3]*	1,2-Epoxybutane	*[106-88-7]*	β-Propiolactone	*[57-57-8]*
o-Anisidine	*[90-04-0]*	Ethyl acrylate	*[140-88-5]*	Propionaldehyde	*[123-38-6]*
Asbestos	*[1332-21-4]*	Ethyl benzene	*[100-41-4]*	Propoxur (Baygon)	*[114-26-1]*
Benzene	*[71-43-2]*	Ethyl carbamate	*[51-79-6]*	Propylene dichloride	*[78-87-5]*
Benzidine	*[92-87-5]*	Ethyl chloride	*[75-00-3]*	Propylene oxide	*[75-56-9]*
Benzotrichloride	*[98-07-7]*	Ethylene dibromide	*[106-93-4]*	1,2-Propylenimine	*[75-55-8]*
Benzyl chloride	*[100-44-7]*	Ethylene dichloride	*[107-06-2]*	Quinoline	*[91-22-5]*
Biphenyl	*[92-52-4]*	Ethylene glycol	*[107-21-1]*	Quinone	*[106-51-4]*
Bis(2-ethylhexyl)phthalate	*[117-81-7]*	Ethylene imine	*[151-56-4]*	Styrene	*[100-42-5]*
Bis(chloromethyl)ether	*[542-88-1]*	Ethylene oxide	*[75-21-8]*	Styrene oxide	*[96-09-3]*
Bromoform	*[75-25-2]*	Ethylene thiourea	*[96-45-7]*	2,3,7,8-Tetrachlorodibenzo-*p*-dioxin	*[1746-01-6]*
1,3-Butadiene	*[106-99-0]*	Ethylidene dichloride	*[75-34-3]*	1,1,2,2-Tetrachloroethane	*[79-34-5]*
Calcium cyanamide	*[156-62-7]*	Formaldehyde	*[50-00-0]*	Tetrachloroethylene	*[127-18-4]*
Caprolactam	*[105-60-2]*	Heptachlor	*[76-44-8]*	Titanium tetrachloride	*[7550-45-0]*
Captan	*[133-06-2]*	Hexachlorobenzene	*[118-74-1]*	Toluene	*[108-88-3]*
Carbaryl	*[63-25-2]*	Hexachlorobutadiene	*[87-68-3]*	2,4-Toluene diamine	*[95-80-7]*
Carbon disulfide	*[75-15-0]*	Hexachlorocyclopentadiene	*[77-47-4]*	2,4-Toluene diisocyanate	*[584-84-9]*
Carbon tetrachloride	*[56-23-5]*	Hexachloroethane	*[67-72-1]*	*o*-Toluidine	*[95-53-4]*
Carbonyl sulfide	*[463-58-1]*	Hexamethyl-1,6-diisocyanate	*[822-06-0]*	Toxaphene	*[8001-35-2]*
Catechol	*[120-80-9]*	Hexamethylphosphoroamide	*[680-31-9]*	1,2,4-trichlorobenzene	*[120-82-1]*
Chloramben	*[133-90-4]*	Hexane	*[110-54-3]*	1,1,2-Trichloroethane	*[79-00-5]*
Chlordane	*[57-74-9]*	Hydrazine	*[302-01-2]*	Trichloroethylene	*[79-01-6]*
Chlorine	*[7782-50-5]*	Hydrochloric acid	*[7647-01-0]*	2,4,5-Trichlorophenol	*[95-95-4]*
Chloroacetic acid	*[79-11-8]*	Hydrogen fluoride	*[7664-39-3]*	2,4,6-Trichlorophenol	*[88-06-2]*
2-Chloroacetophenone	*[532-27-4]*	Hydroquinone	*[123-31-9]*	Trimethylamine	*[121-44-8]*
Chlorobenzene	*[108-90-7]*	Isophorone	*[78-59-1]*	Trifluralin	*[1582-09-8]*
Chlorobenzilate	*[510-15-6]*	Lindane (all isomers)	*[58-89-9]*	2,2,4-Trimethylpentane	*[540-84-1]*
Chloroform	*[67-66-3]*	Maleic anhydride	*[108-31-6]*	Vinyl acetate	*[108-05-4]*
Chloromethyl methyl ether	*[107-30-2]*	Methanol	*[67-56-1]*	Vinyl bromide	*[593-60-2]*
Chloroprene	*[126-99-8]*	Methoxychlor	*[72-43-5]*	Vinyl chloride	*[75-01-4]*
Cresols/cresylic acid	*[1319-77-3]*	Methyl bromide	*[74-83-9]*	Vinylidene chloride	*[75-35-4]*
o-Cresol	*[95-48-7]*	Methyl chloride	*[74-87-3]*	Xylenes (isomers and mixture)	*[1330-20-7]*
m-Cresol	*[108-39-4]*	Methyl chloroform	*[71-55-6]*	*o*-Xylenes	*[95-47-6]*
p-Cresol	*[106-44-5]*	Methyl ethyl ketone	*[78-93-3]*	*m*-Xylenes	*[108-38-3]*
Cumene	*[98-82-8]*	Methyl hydrazine	*[60-34-4]*	*p*-Xylenes	*[106-42-3]*
2,4-D, salts and esters	*[94-75-7]*	Methyl iodide	*[74-88-4]*	Antimony compounds	
DDE	*[3547-04-4]*	Methyl isobutyl ketone	*[108-10-1]*	Arsenic compounds	
Diazomethane	*[334-88-3]*	Methyl isocyanate	*[624-83-9]*	Beryllium compounds	
Dibenzofurans	*[132-64-9]*	Methyl methacrylate	*[80-62-6]*	Cadmium compounds	
1,2-Dibromo-3-chloropropane	*[96-12-8]*	Methyl-*tert*-butyl ether	*[1634-04-4]*	Chromium compounds	
Dibutylphthalate	*[84-74-2]*	4,4-Methylene bis(2-chloroaniline)	*[101-14-4]*	Cobalt compounds	
1,4-Dichlorobenzene(p)	*[106-46-7]*	Methylene chloride	*[75-09-2]*	Coke oven emissions	
3,3-Dichlorobenzidene	*[91-94-1]*	Methylene diphenyl diisocyanate	*[101-68-8]*	Cyanide compounds	
Dichloroethyl ether	*[111-44-4]*	4,4′-Methylenedianiline	*[101-77-9]*	Glycol ethers	
1,3-Dichloropropene	*[542-75-6]*	Naphthalene	*[91-20-3]*	Lead compounds	

Table 5. *(continued)*

Substance	CAS Registry Number	Substance	CAS Registry Number	Substance	CAS Registry Number
Dichlorvos	*[62-73-7]*	Nitrobenzene	*[98-95-3]*	Manganese compounds	
Diethanolamine	*[111-42-2]*	4-Nitrobiphenyl	*[92-93-3]*	Mercury compounds	
N,N-Diethyl aniline	*[121-69-7]*	4-Nitrophenol	*[100-02-7]*	Fine mineral fibers	
Diethyl sulfate	*[64-67-5]*	2-Nitropropane	*[79-46-9]*	Nickel compounds	
3,3-Dimethoxybenzidine	*[119-90-4]*	*N*-Nitroso-*N*-methylurea	*[684-93-5]*	Polycyclic organic matter	
Dimethyl aminoazobenzene	*[60-11-7]*	*N*-Nitrosodimethylamine	*[62-75-9]*	Radionuclides (including radon)	
3,3′-Dimethyl benzidine	*[119-93-7]*	*N*-Nitrosomorpholine	*[59-89-2]*	Selenium compounds	

in Denver and elsewhere). Three factors can contribute to the brown tint. The first is nitrogen dioxide which is a brownish gas. This is most commonly viewed in a plume of NO_2. In the urban hazes, the effect of NO_2 is usually overwhelmed by the effects caused by particles. Because fine particles preferentially scatter blue light in the forward direction, the light viewed through an optically thin cloud with the sun behind the observer is deficient in the blue wavelengths and appears brown. In dense haze clouds, the preferential scattering is masked by multiple scattering effects and the haze is seen as white. However, a dense haze cloud can appear brown along its edges where it is optically thin. If the cloud is between the observer and the sun, it will appear as white. Also, a dense cloud in the distance against a bright blue sky background can appear brown through a process called chromatic adaptation. With chromatic adaptation, the blue receptors in the human eye are desensitized by the bright blue background; as a result, the white light from the haze appears to be brown.

Acid Deposition

Acid deposition is the deposition of acids from the atmosphere to the surface of the earth. The deposition can be dry or wet. Dry deposition refers to the process whereby acid gases or their precursors or acid particles come in contact with the earth's surface and are retained by the surface. The principal species associated with dry acid-deposition are SO_2 (g), acid sulfate particles (H_2SO_4 and NH_4HSO_4), and HNO_3 (g). Measurements of dry deposition are quite sparse, however, and usually only speciated as total SO_4^{2-} and total NO_3^-. In general, the dry acid deposition is estimated to be a small fraction of total acid deposition because most of the dry deposited SO_4^{2-} and NO_3^- has been neutralized by basic gases and particle in the atmosphere. However, the sulfate and nitrate deposited from dry deposition is estimated to be a significant fraction of the total sulfate and nitrate deposition. More specific estimates are not possible because current spatial and temporal dry deposition data are insufficient. On the other hand, there are abundant data on wet-acid deposition. Wet-acid deposition or acid precipitation is the process by which acids are deposited by the rain or snow. The principal dissolved acids are H_2SO_4 and HNO_3. Other acids, such as HCl and organic acids, usually account for only a minor part of the acidity, although organic acids can be significant contributors in remote areas.

Both acid particles and gases can be incorporated into cloud droplets. Particles are incorporated into droplets by nucleation, Brownian diffusion, impaction, diffusiophoresis (transport into the droplet induced by the flux of water vapor to the same surface), thermophoresis (thermally induced transport to a cooler surface), and electrostatic transport. Advective and diffusive attachment dominate all other mechanisms for pollutant gas uptake by cloud droplets. Modeling and experimental evidence suggest that most of the H_2SO_4 is formed in cloud water droplets. SO_2 diffuses into the droplet and is oxidized to H_2SO_4 by one of several mechanisms. At pHs greater than about 5.5, oxidation of SO_2 by dissolved O_3 is the dominant reaction. At lower pHs, SO_2 oxidation is dominated by the reaction with H_2O_2. Under some conditions, oxidation by O_2, catalyzed by metals or soot, may contribute to the formation of H_2SO_4. Most of the HNO_3 in precipitation is due to the diffusion of HNO_3 into the droplet. However, there is observational evidence that some HNO_3 is formed in the droplets, but the mechanism has not been identified.

The pH of rainwater in equilibrium with atmospheric CO_2 is 5.6, a value that was frequently cited as the natural background pH of rainwater. However, in the presence of other naturally occurring species such as SO_2, SO_4^{2-}, NH_3, organic acids, sea salt, and alkaline crustal dust, the "natural" values of unpolluted rainwater will vary between 4.9 and 6.5 depending on time and location. Across the United States, the mean annual average pH varies from 4.2 in western Pennsylvania to 5.7 in the West (Fig. 2). Precipitation pH is generally lowest in the eastern United States within and downwind of the largest SO_2 and NO_x emissions areas. In general, the lowest pH precipitation occurs in the summer. In the East, SO_4^{2-} concentrations in precipitation are 1.5 to 2.5 times higher during the summer than winter, but the NO_3^- values are about the same year round. Consequently, the lower pHs in the summer are mostly due to the higher SO_4^{2-} concentrations. The equivalent ratio of sulfate to nitrate in precipitation is an often used (but inexact) measure of the relative contribution of these two species to the acidity. An equivalent ratio of 1.0 would mean equal contributions by both. In the East during the winter, this ratio ranges from 1.0 to 2.5, while during the summer it ranges from 2.0 to 3.0. On the average in the eastern United States, about 60% of the wet-deposited acidity can be attributed to SO_4^{2-} and 40% to NO_3^- (27).

Since SO_2 and NO_2 are Criteria Pollutants, their emissions are regulated. In addition, for the purposes of abating acid deposition in the United States, the 1990 Clean Air Act Amendments require that nationwide SO_2 and NO_x emissions be reduced by 10 million t/yr and 2 million t/yr, respectively, by the year 2000. The reasons for these reductions are based on concerns which include acidification of lakes and streams,

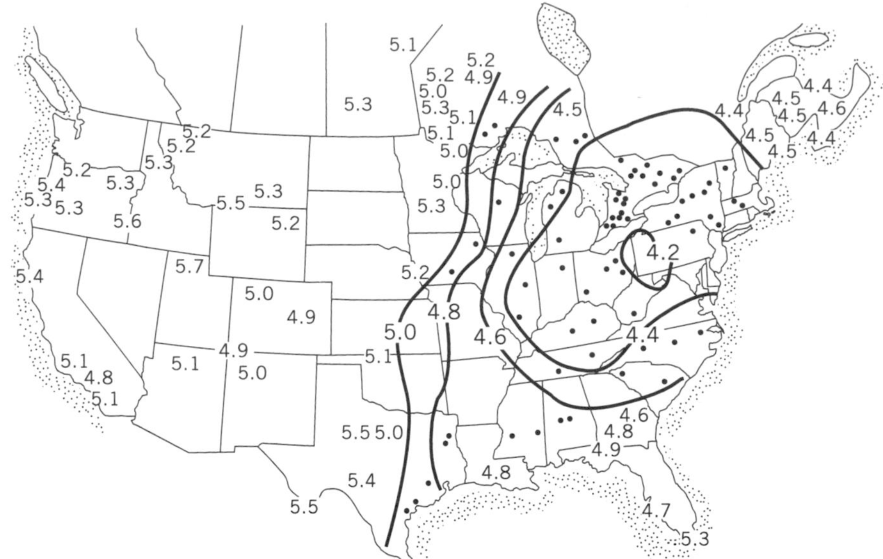

Figure 2. 1985 annual precipitation-weighted pH (27).

acidification of poorly buffered soils, and acid damage to materials. An additional concern was that acid deposition was causing the die-back of forests at high elevations in the eastern United States and Europe. Although a contributing role of acid deposition cannot be dismissed, the primary pollutant suspected in the forest decline issue is now O_3.

Global Warming (The Greenhouse Effect)

Solar energy, mostly in the form of visible light, is absorbed by the earth's surface and reemitted as long-wavelength infrared (IR) radiation. Certain gases in the atmosphere, primarily water vapor and to a lesser degree CO_2, have the ability to absorb the outgoing ir radiation which is translated to heat. The result is a higher atmospheric equilibrium temperature than would occur in the absence of water vapor and CO_2. This temperature enhancement is called the greenhouse effect, and gases that have the ability to absorb ir and produce this effect are called greenhouse gases. Without the naturally occurring concentrations of water vapor and CO_2, the earth's mean surface temperature would be $-18°C$ instead of the present $17°C$. There is concern that increasing concentrations of CO_2 and other trace greenhouse gases due to human activities will enhance the greenhouse effect and cause global warming. Speculated scenarios based on global warming include the following: an alteration in existing precipitation patterns, an increase in the severity of storms, the dislocation of suitable land for agriculture, the dislocation and possible extinction of certain biological species and ecosystems, and the flooding of many coastal areas due to rising sea levels resulting from the thermal expansion of the oceans, the melting of glaciers, and, probably less so, from the melting of polar ice caps.

Measurements of CO_2 since 1958 clearly show that CO_2 concentrations in the atmosphere are increasing at the rate of about 0.3% per year. The present concentration of ~350 ppm compares to the preindustrial revolution (1800) value of 285 ppm (estimated from ice cores). Projections based on the current rate of the CO_2 increase and future energy uses show that the CO_2 concentration will approach 600 ppm some time in the middle of the 21st century. Within the past two decades, it was discovered that the concentrations of other trace greenhouse gases are also increasing. These gases include methane (CH_4), nitrous oxide (N_2O), O_3 (tropospheric), and a variety of chlorofluorocarbons (CFCs). A list of the greenhouse gases (not including water), present concentrations, current rates of increase in the atmosphere, and estimates of relative greenhouse efficiencies and atmospheric residence times are presented in Table 6. Ozone is not included in this table because there is insufficient information available for quantifying the global radiative influence of O_3 (29). CFCs are included in the table, but their warming potentials should be used with caution because there presently is a controversy over whether the warming due to the presence of CFCs is partially or completely offset by the cooling resulting from O_3 losses in the lower stratosphere (29). Three important features are evident from Table 5. First, today CO_2 is by far the most abundant anthropogenic greenhouse gas in the atmosphere. Second, all of the other trace greenhouse gases are much more efficient absorbers of ir radiation than CO_2. Third, most of the gases have very long atmospheric residence times so that even if emissions were to cease, the gases would remain in the atmosphere for decades (some for centuries). The principal sources of greenhouse gases, as they are understood today, are summarized in Table 7.

From the analyses of air trapped in Antarctic and Greenland ice, the concentrations of greenhouse gases (except for O_3) in the preindustrial atmosphere (averaged over ~1000 years) can be estimated quite accurately. The enhancement of the greenhouse effect due to current concentrations of greenhouse gases relative to preindustrial concentrations is called the enhanced greenhouse effect or radiative forcing. Using a radiative convective model, the contributions from the various greenhouse gases to the radiative forcing in the 1980s can be estimated (6). Such estimates are shown in Figure 3. At present, CO_2 accounts for about half of the radiative forcing. The relative contribution from CO_2 has been shrinking and will continue to do so because other species that are much more efficient ir absorbers are increasing in concentration at a faster relative rate than CO_2.

Table 6. Summary of Important Greenhouse Gases

Gas	CAS Registry Number	Present Concentrations[a]	Concentration Increase,[a] %/yr	Warming Potential [b,c]	Atmospheric Residence Times,[b,d] yr
CO_2		350 ppm	0.3	1	120[e]
CH_4		1.68 ppm	0.8–1	11	10.5
N_2O		340 ppb	0.2	270	132
$CFCl_3$ (CFC-11)	[*75-69-4*]	226 ppt	4	3400	55
CF_2Cl_2 (CFC-12)	[*75-71-8*]	392 ppt	4	7100	116
$CHClF_2$ (HCFC-22)	[*75-46-6*]	100 ppt	7	1600	16
$C_2H_2F_4$ (HFC-134a)	[*811-97-2*]			1200	16
$C_2Cl_3F_3$ (CFC-113)	[*76-14-2*]	30–70 ppt	11	4500	110
CH_3CCl_3 (Methychloroform)	[*71-55-6*]	125 ppt	7	100	6
CCl_4 (Carbon tetrachloride)	[*56-23-5*]	75–100 ppt	1	1300	47

[a] Ref. 28.
[b] Ref. 29.
[c] Relative to CO_2, based on a 100-yr time horizon (29).
[d] Residence times may be slightly different from those reported in Table 2 because the primary source is different.
[e] Ref. 8.

Table 7. Principal Sources of Greenhouse Gases[a]

Gases	Principal Sources
CO_2	Fossil fuel combustion, deforestation oceans, respiration
CH_4	Wetlands, rice paddies, enteric fermentation (animals), biomass burning, termites
N_2O	Natural soils, cultivated and fertilized soils, oceans, fossil fuel combustion
O_3	Photochemical reactions in the troposphere, transport from stratosphere
CFC-11	Manufacturing of foam, aerosol propellant
CFC-12	Refrigerant, aerosol propellant, manufacturing of foams
HCFC-22	Refrigerant, production of fluoropolymers
CFC-113	Electronics solvent
CH_3CCl_3	Industrial degreasing solvent
CCl_4	Intermediate in production of CFC-11, -12, solvent

[a] Ref. 28.

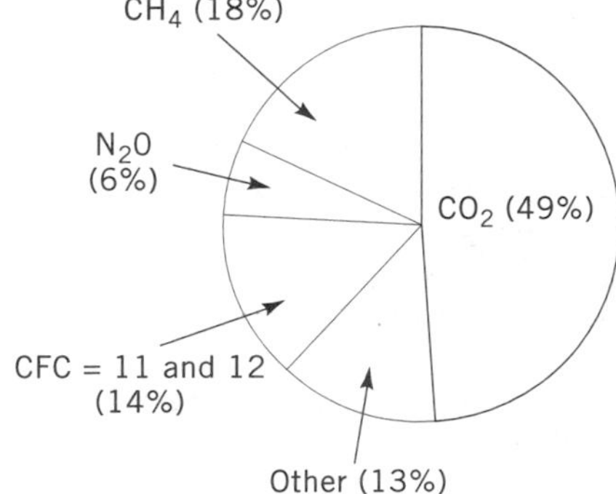

Figure 3. Estimates of greenhouse gas contributions to global warming in the 1980s.

Although there is no doubt that greenhouse gas concentrations and the radiative forcing are increasing, there is no unequivocal evidence to suggest that the forcing is actually causing a net warming of the earth. Analyses of global temperature trends since the 1860s show that the global temperature has increased about 0.5–0.7°C (30,31), but this number decreases to ~0.5°C when corrections for heat-island effects are considered (32). This is in reasonable agreement with modeling results which predict a temperature increase of ~1°C (33). However, it has been shown that most of the temperature increase occurred prior to the big increase in CO_2 (34). A detailed analysis of global temperature and CO_2 concentration time-series over the period of atmospheric measurements of CO_2 (1958–1988) shows an excellent positive correlation between the two variables, but the CO_2 changes lag temperature change by an average of five months (35). Thus, although there is strong evidence linking temperature and CO_2 changes, the cause and effect not only has not been demonstrated, but it is not clear which is the cause and which is the effect. The lack of a definitive relationship may also be obscured by changes in other factors that affect the earth's heat budget, such as increased atmospheric aerosols or cloud cover as well as natural climatic cycles.

Predictions of future temperature changes due to increased greenhouse gas forcing are made using global circulation models (GCMs). The GCMs are sophisticated but incomplete models that incorporate expressions for the basic physical processes that govern the dynamics of the atmosphere and allow for some atmosphere–ocean interactions. Depending on which model is used and the various assumptions incorporated into the model, the models predict that the average global temperature will increase from 2.0 to 5.5°C with a doubling of CO_2 concentrations from preindustrialized levels (36,37). These temperature increases will not be uniformly distributed. Although temperature change predictions from the models are in fairly good agreement, the uncertainties are considered to be large because many of the important feedback processes involving the oceans and clouds are not adequately incorporated into the models because many of them are poorly understood. Consequently, the models are an indication that global warming could occur, but the results are not definitive (see OCEAN–ATMOSPHERE INTERACTIONS).

Stratospheric O_3 Depletion. In the stratosphere, O_3 is formed naturally when O_2 is dissociated by ultraviolet (UV) solar radiation in the wavelength (λ) region 180–240 nm:

$$O_2 + UV \rightarrow O + O \quad (20)$$

$$O + O_2 + M \rightarrow O_3 + M \quad (21)$$

where M is any third-body molecule (most likely N_2 or O_2 in the atmosphere) that remains unchanged in the reaction. UV radiation in the 200–300 nm λ region can also dissociate the O_3:

$$O_3 + UV \rightarrow O_2 + O \quad (22)$$

In this equation 22, O_3 is responsible for the removal of UV-B radiation ($\lambda = 280-330$ nm) that would otherwise reach the earth's surface. The concern is that any process that depletes stratospheric ozone increases the UV-B (in the 293–320 nm region) reaching the surface. Increased UV-B will lead to increased incidence of skin cancer and could have deleterious effects on certain ecosystems. The first concern over O_3 depletion was from NO_x emissions from a fleet of supersonic transport aircraft that would fly through the stratosphere and cause (38):

$$NO + O_3 \rightarrow NO_2 + O_2 \quad (23)$$

$$NO_2 + O \rightarrow NO + O_2 \quad (24)$$

The net effect of this sequence is the destruction of 2 molecules of O_3 since the O would have combined with O_2 to form O_3. In addition, the NO acts as a catalyst because it is not consumed, and therefore can participate in the reaction sequence many times.

In the mid-1970s, it was realized that the chlorofluorocarbons (CFCs) in widespread use because of their chemical inertness, would diffuse unaltered through the troposphere into the mid-stratosphere where they would be photolyzed by UV ($\lambda < 240$ nm) radiation. For example, CFC-12 would photolyze forming Cl and ClO radicals:

$$CF_2Cl_2 + UV \rightarrow CF_2Cl + Cl \quad (25)$$

$$CF_2Cl + O_2 \rightarrow CF_2O + ClO \quad (26)$$

The following reactions would then occur:

$$Cl + O_3 \rightarrow ClO + O_2 \quad (27)$$

$$ClO + O \rightarrow Cl + O_2 \quad (28)$$

In this sequence the Cl also acts as a catalyst, and two O_3 molecules are destroyed. Before the Cl is finally removed from the atmosphere in 1–2 years by precipitation, each Cl atom will have destroyed approximately 100,000 O_3 molecules (39). The O_3-depletion potential is defined as the ratio of the emission rate of a compound required to produce a steady-state O_3 depletion of 1% to the amount of CFC-11 required to produce a 1% depletion. Another class of compounds, halons, are also ozone-depleting compounds. The halons are bromochlorofluorocarbons or bromofluorocarbons that are widely used in fire extinguishers. Although their emissions and thus their atmospheric concentrations are much lower than the most common CFCs, they are of concern because they are 3 to 10 times more destructive to O_3 than the CFCs.

Evidence that stratospheric O_3 depletion is occurring comes from the discovery of the Antarctic O_3 hole. In recent years during the spring, O_3 depletions of 60% integrated over all altitudes and 95% in some layers have been observed over Antarctica. During winter in the southern hemisphere, a polar vortex develops which prevents the air from outside of the vortex from mixing with air inside the vortex. The depletion begins in August, as the approaching spring sun penetrates into the polar atmosphere, and extends into October. When the hole was first observed, existing chemical models could not account for the rapid O_3 loss. Attention was soon focused on stable reservoir species for chlorine, namely HCl and $ClNO_3$. These species are formed in competing reactions involving Cl and ClO that temporarily or permanently remove Cl and ClO from participating in the O_3 destruction reactions. Two important reactions are the following:

$$Cl + CH_4 \rightarrow HCl + CH_3 \quad (29)$$

$$ClO + NO_2 + M \rightarrow ClNO_3 + M \quad (30)$$

Within the polar vortex, temperatures as low as −90°C allow the formation of polar stratospheric ice clouds. On the surfaces of the ice particles that compose these clouds, heterogeneous reactions occur which break down the reservoir species HCl and $ClNO_3$. Two important reactions are the following:

$$ClNO_3 + HCl\,(s) \rightarrow Cl_2 + HNO_3\,(s) \quad (31)$$

$$H_2O\,(s) + ClNO_3 \rightarrow HOCl + HNO_3\,(s) \quad (32)$$

During the polar winter night, Cl_2, HOCl and $HNO_3(s)$ accumulate. When sunlight returns to the polar regions, the chlorine compounds are photolyzed, producing Cl and ClO. Nitrogen oxides remain sequestered, and without NO_2 to deplete ClO, massive O_3 destruction occurs until the polar vortex dissipates later in the spring and the mixing of air from lower latitudes occurs.

Other data show that globally, stratospheric O_3 concentrations have declined during the winter, spring, and summer in both the northern and southern hemispheres at middle and high latitudes (29). Declines were most evident during winter months (40,41).

In 1976, the United States banned the use of CFCs as aerosol propellants. No further steps were taken until 1987 when the United States and some 50 other countries adopted the Montreal Protocol which specifies a 50% reduction of fully halogenated CFCs by 1999. Since then, however, because of the Antarctic ozone hole and the observed global decreases in stratospheric ozone, there has been increased support for a faster phaseout. In 1990, an agreement was reached among 93 nations to accelerate the phaseout and completely eliminate the production of CFCs by the year 2000. The 1990 Clean Air Act Amendments contain a phaseout schedule for CFCs, halons, carbon tetrachloride, and methyl chloroform. Such steps will stop the increase of CFCs in the atmosphere but, because of their long lifetimes, they will remain in the atmosphere for centuries.

Indoor Air Pollution

Indoor air pollution is simply the presence of air pollutants in indoor air. The focus of this section is on air in residential buildings as opposed to the industrial environment which would be covered under industrial hygiene. The concentrations of indoor pollutants depend on the strength of the indoor sources of the pollutants as well as the ventilation rate of the building and the outdoor concentrations of the pollutants. In response to the energy crisis in the early 1970s, new buildings were constructed to be more airtight. Unfortunately, airtight structures created a setting conducive to the accumulation of indoor air pollutants. Numerous sources and types of pollutants found indoor can be classified into seven categories: tobacco smoke, radon, emissions from building materials, combustion products from inside the building, pollutants which infiltrate from outside the building, emissions from products used within the home, and biological pollutants (see CARCINOGENS, MONITORING OF INDOOR AIR).

Tobacco smoke contains a variety of air pollutants. In a survey of 80 homes in an area where the outdoor TSP varied between 10–30 $\mu g/m^3$, the indoor TSP was the same or less in homes with no smokers. In homes with one smoker, the TSP levels were between 30–60 $\mu g/m^3$, in homes with two or more smokers, the levels were between 60–120 $\mu g/m^3$ (42). In other studies, indoor TSP levels exceeding 1000 $\mu g/m^3$ have been found in homes with numerous smokers. In addition to TSP, burning tobacco emits CO, NO^x, formaldehyde, benzopyrene, nicotine, phenols, and some metals such as cadmium and arsenic (43).

Radon-222 (Rn) is a naturally occurring, inert, radioactive gas formed from the decay of radium-226 (Ra). Because Ra is a ubiquitous, water-soluble component of the earth's crust, its daughter product, Rn, is found everywhere. Although Rn receives all of the notoriety, the principal health concern is not with Rn itself, but with its alpha (α) particle-emitting daughters (radioactive decay products). Because Rn is an inert gas, inhaled Rn is not retained in the lungs. With a half-life of four days, Rn decays to polonium-218 (Po-218) with the emission of an α-particle. It is Po-218, an α-emitter with a half-life of three minutes, and Po-214, also an α-emitter with a half-life of 1.6×10^{-4} seconds, that are of most concern. Po-218 decays to lead-214 (a β-emitter with a $t_{1/2} = 27$ minutes), which decays to bismuth-214 (a β-emitter with a $t_{1/2} = 20$ minutes), which decays to Po-214. When inhaled, the Rn daughters, either by themselves or attached to an airborne particle, are retained in the lung and the subsequent α-emission irradiate the surrounding lung tissue. Rn can enter buildings through emissions from soil, water, or construction materials. By far, the soil route is the most common, and construction material the least common source of Rn contamination (there have been isolated incidents where construction materials contained high levels of Ra). The emission rate of Rn depends on the concentration of Ra in the soil, the porosity of the soil, and the permeability of the building's foundation. For example, Rn transports faster through cracks and sumps in the basement floor than through concrete. In the ambient air, Rn concentrations are typically 0.25–1.0 picoCuries per liter (pC/L), whereas the mean concentration in U.S. residences is about 1.2 pC/L (44). However, it is estimated that there are 1 million residences that have concentrations exceeding 8 pC/L, which is the action level for remedial action recommended by the National Council on Radiation Protection and Measurements (44). The highest values ever measured in U.S. homes exceeded 1000 pC/L (45). Remedial action consists of (*1*) reducing the transport of Rn into the building by sealing cracks with impervious fillers and installing plastic or other barriers that have been proven effective; (*2*) removing the daughters from the air by filtration; and (*3*) increasing the infiltration of outside air with an air-exchanger system.

Of the pollutants emitted from construction materials within the home, asbestos [*1332-21-4*] has received the most attention. Asbestos is a generic term for a number of naturally occurring fibrous, hydrated silicates. By EPA's definition, a fiber is a particle that possess a 3:1 or greater aspect ratio (length:diameter). The family of asbestos minerals is divided into two types: serpentine and amphibole. One type of serpentine, chrysotile [$Mg_6Si_4O_{10}(OH)_8$], accounts for 90% of the world's asbestos production. The balance of the world's production is accounted for by two of the amphiboles: amosite [$Fe_5Mg_2(Si_8O_{22})(OH)_2$] and crocidolite [$Na_2(Fe^{+3})_2(Fe^{+2})_2Si_8O_{22}(OH)_2$]. Three other amphiboles, anthophyllite [$(Mg,Fe)_7Si_8O_{22}(OH)_2$], tremolite [$Ca_2Mg_5Si_8O_{22}(OH)_2$], and actinolite [$Ca_2(Mg,Fe)_5Si_8O_{22}(OH)_2$], have been only rarely mined. The asbestos minerals differ in morphology, durability, range of fiber diameters, surface properties, and other attributes that determine uses and biological effects. Known by ancients as the magic mineral because of its ability to be woven into cloth, its physical strength, and its resistance to fire, enormous heat, and chemical attack, asbestos was incorporated into many common products including roofing materials, wallboard, insulation, spray-on fireproofing and insulating material, floor tiles, pipes, filters, draperies, pot holders, brake linings, etc (46). In the 1940s and 1950s, however, evidence accumulated linking exposure to the airborne fibers with asbestosis (pulmonary interstitial fibrosis), lung cancer, and mesothelioma (a rare form of cancer of the lung or abdomen). Although all forms of asbestos were implicated in the early studies, studies indicate that most of the asbestos-related diseases are due to exposure to airborne amphiboles rather than the most common type, chrysotile, and to fibers greater than or equal to 5 μm in length (47). In the 1970s, the spray-on application of asbestos was banned and substitutes were found for many products. Nevertheless, asbestos was used liberally in buildings for several decades and many of them are still standing. Asbestos in building materials does not spontaneously shed fibers, but when the materials become damaged by normal decay, renovation, or demolition, the fibers can become airborne. When such situations arise, specific procedures should be followed to contain and remove the damaged materials.

Another important pollutant emanating from building material is formaldehyde (HCHO). Formaldehyde is important because of its irritant effects and its suspected carcinogenicity. Although traces of formaldehyde can be found in the air in virtually every modern home, mobile homes and homes insulated with urea–formaldehyde foam have the highest concentrations. Higher emissions can occur in mobile homes using particle boards which are held together with a urea–formaldehyde resin. This can also be a problem in a conventional home, but it is exacerbated in a mobile home because of the usually low rate of air exchange in a mobile home. Plywood is also a source of formaldehyde as the layers of wood are held together in a similar urea–formaldehyde resin adhesive. In general, however, particle board contains more adhesive per unit mass, so the emissions are greater. Urea foam is an efficient insulation material that can be injected into the sidewalls of conventional homes. Production of this foam peaked in 1977, when about 170,000 homes were insulated. When improperly formulated or installed, this foam can emit significant amounts of formaldehyde. In 1982, the use of the foam was banned in the United States. Other sources of formaldehyde indoors are paper products, carpet backing, and some fabrics.

Whenever unvented combustion occurs indoors or when venting systems attached to combustion units malfunction, a variety of combustion products are released to the indoor environment. Indoor combustion units include stoves and ovens (except electric), furnaces, hot-water heaters, space heaters, and wood-burning fireplaces or stoves. Products of combustion include CO, NO, NO_2, fine particles, aldehydes, polynuclear aromatics, and other organic compounds. Especially dangerous sources are unvented gas and kerosene space heaters which discharge pollutants directly into the living space. The best way to prevent the accumulation of combustion products indoors is to make sure all units are properly vented and properly maintained.

Pollutants from outdoors can also be drawn inside under certain circumstances. Incorrectly locating an air intake vent downwind of a combustion exhaust stack can cause this condition. High outdoor pollutant concentrations can infiltrate buildings. Unreactive pollutants such as CO diffuse through any openings in the building and pass unaltered through any air-intake system. Given sufficient time, the indoor/outdoor ratio for CO will approach 1.0 if outside air is the only source of CO. For reactive species such as ozone, which is destroyed on contact with most surfaces, the indoor/outdoor ratio is usually around 0.5, but this ratio varies considerably depending on the ventilation rate and the internal surface area within the building (48).

Air contaminants are emitted to the indoor air from a wide variety of activities and consumer products. It is obvious from this list that most indoor activities produce some types of pollutants. When working with these products or engaging in these activities, care should be exercised to minimize exposures by proper use of the products and by providing adequate ventilation.

Biological air pollutants found indoors include airborne bacteria, viruses, fungi, spores, molds, algae, actinomycetes, and insect and plant parts. Many of the microorganisms multiply in the presence of high humidity. The microorganisms can produce infections, disease, or allergic reactions, whereas the nonviable biological pollutants can produce allergic reactions. The most notable episode was the 1976 outbreak of *Legionella* (Legionnaires') disease in Philadelphia, where American Legion convention attendees inhaled *Legionella* virus from a contaminated central air-conditioning system. A similar incident in an industrial environment occurred in 1981 when more than 300 workers came down with "Pontiac fever" as a result of inhalation exposure to a similar aerosolized virus from contaminated machining fluids (49). Better preventative maintenance of air management systems and increased ventilation rates reduce the concentrations of all species, and this should reduce the incidence of adverse affects.

BIBLIOGRAPHY

1. *Procedures for Emission Inventory Preparation,* Volumes I–IV, Publication No. EPA 450/4-81-026A-E, U.S. Environmental Protection Agency, Research Triangle Park, N.C., 1981.
2. *Compilation of Air Pollution Emission Factors,* Publication No. AP-42, 5th ed., U.S. Environmental Protection Agency, Research Triangle Park, N.C., 1989.
3. *Compilation of Air Pollution Emission Factors,* Vol. II, Mobile Sources, Publication No. AP-42, 5th ed., U.S. Environmental Protection Agency, Research Triangle Park, N.C., 1989.
4. *National Air Quality & Emissions Trends Report, 1991,* Publication No. EPA-450-R-92-001, U.S. Environmental Protection Agency, Research Triangle Park, N.C., 1992.
5. *National Air Pollutant Emissions Estimates 1940–1986,* Publication No. EPA-450/4-87-024, U.S. Environmental Protection Agency, Research Triangle Park, N.C., 1988.
6. V. Ramanathan and co-workers, *Review Geophys.* **25,** 1441 (1987).
7. P. Warneck, *Chemistry of Natural Atmospheres,* Academic Press, New York, 1988, p. 367.
8. Intergovernmental Panel on Climate Change, *Scientific Assessment of Climate Change, Section 2, Radiative Forcing of Climate,* United Nations, New York, 1990, p. 14.
9. G. T. Wolff and P. J. Lioy, *Environ. Sci. Technol.* **14,** 1257 (1980).
10. G. T. Wolff, N. A. Kelly, and M. A. Ferman, *Science* **311,** 703 (1981).
11. J. P. Lodge, ed., *Methods of Air Sampling and Analysis,* Lewis Publishers, Chelsea, Mich., 1989, 763 pp.
12. A. M. Hough and R. G. Derwent, *Nature* **344,** 645 (1990).
13. N. A. Kelly, G. T. Wolff, M. A. Ferman, *Atmos. Environ.* **16,** 1077 (1978).
14. W. Johnson and W. Viezee, *Atmos. Environ.* **15,** 1309 (1981).
15. W. Attmannspacher and R. Hartmannsgruber, *Pure Appl. Geophysics.* **106–108,** 1091 (1973).
16. R. L. Seila, W. A. Lonneman, and S. A. Meeks, *Determination of C_2 to C_{12} Ambient Air Hydrocarbons in 39 U.S. Cities from 1984 through 1986,* Publication No. EPA/600/3-89/058, U.S. Environmental Protection Agency, Research Triangle Park, N.C., (1989).
17. P. Warneck, *Chemistry of Natural Atmospheres,* Academic Press, New York, 1988, pp. 721–729.
18. W. N. Stasiuk, Jr. and P. E. Coffey, *J. Air Pollut. Control Assoc.* **24,** 564 (1974).
19. G. T. Wolff and co-workers, *Environ. Sci. Technol.* **11,** 506 (1977).
20. G. T. Wolff and co-workers, *J. Air Pollut. Control Assoc.* **27,** 460 (1977).
21. *An Acid Aerosols Issue Paper,* Publication No. EPA/600/8-88/005F, U.S. Environmental Protection Agency, Washington, D.C., 1989.
22. P. W. Fisher, R. M. Currie, and R. J. Churchill, *J. Air Pollut. Control Assoc.* **38,** 1376 (1988).
23. *Threshold Limit Values and Biological Exposure Indices,* American Conference of Governmental and Industrial Hygienists, Cincinnati, Ohio, 1989, p. 124.
24. N. I. Sax, *Dangerous Properties of Industrial Materials,* Van Nostrand Reinhold, New York, 1979, p. 1108.
25. J. A. Cannon, *J. Air Pollut. Control Assoc.* **36,** 562 (1986).
26. J. C. Mesta, in P. S. Bhardwaja, ed., *Visibility Protection Research and Policy Aspects,* Air and Waste Management Association, Pittsburgh, Pa. 1987, pp. 1–8.
27. D. Albritton and co-workers, *NAPAP Interim Assessment: Atmosphere Process and Deposition,* Vol. 2, National Acid Precipitation Assessment Program, Washington, D.C., 1987.
28. *Policy Options for Stabilizing Global Climate,* U.S. Environmental Protection Agency, Washington, D.C., 1990.
29. *Scientific Assessment of Ozone Depletion: 1991,* World Meteorological Organization/United Nations Environment Programme, New York, 1991.
30. P. D. Jones, T. M. L. Wigley, and P. B. Wright, *Nature* **322,** 430 (1986).
31. J. Hansen and S. Lebedeff, *J. Geophys. Res.* **D11,** 13, 345 (1987).
32. T. R. Karl and P. D. Jones, *Bull. Amer. Meteor Soc.* **70,** 265 (1989).
33. V. Ramanathan, *Science* **240,** 293 (1988).
34. R. S. Lindzen, *Bull. Amer. Meteorol Soc.* **71,** 288 (1990).
35. C. Kuo, C. Lindberg and D. J. Thomson, *Nature* **343,** 709 (1990).
36. S. H. Schneider, *Sci. Amer.* **261,** 70 (1989).
37. J. F. B. Mitchell, C. A. Senior, and W. J. Ingram, *Nature* **341,** 132 (1989).
38. P. J. Crutzen, *Quart J. Royal Meteorol Soc.* **96,** 320 (1970).
39. M. J. Molina and F. S. Rowland, *Nature* **249,** 810 (1974).
40. M. B. McElroy and R. J. Salawitch, *Science* **243,** 763 (1989).

41. F. S. Rowland, *Amer. Sci.* **77,** 36 (1989).
42. J. D. Spengler and co-workers, *Atmos. Environ.* **15,** 23 (1981).
43. California Department of Consumer Affairs, *Clean Your Room, Compendium on Indoor Air Pollution,* Sacramento, Calif., 1982, pp. III.Ei–III.E.II.
44. Mueller Associates, Inc., Syscon Corp. and Brookhaven National Laboratory, *Handbook of Radon in Buildings,* Hemisphere Publishing Corp., New York, 1988, p. 95.
45. H. W. Alter and R. A. Oswald, *J. Air Pollut. Control Assoc.* **37,** 227 (1987).
46. P. Brodeur, *New Yorker* **44,** 117 (1968).
47. B. T. Mossman and co-workers, *Science* **247,** 294 (1990).
48. C. J. Weschler, H. C. Shields, and D. V. Naik, *J. Air Waste Manage. Assoc.* **39,** 1562 (1989).
49. L. A. Herwaldt and co-workers, *Ann. Intern. Med.* **100,** 333 (1984).

General References

References 4, 5, 6, 7, 8, 11, 28, 29, 36, 40, and 43, and the following books and reports constitute an excellent list for additional study. Reference 7 is an especially useful resource for global atmospheric chemistry.

J. H. Seinfeld, *Atmospheric Chemistry and Physics of Air Pollution,* John Wiley & Sons, Inc., New York, 1986.

B. J. Finlayson-Pitts and J. N. Pitts, Jr., *Atmospheric Chemistry Fundamentals and Experimental Techniques,* John Wiley & Sons, Inc., New York, 1986.

T. E. Graedel, D. T. Hawkins, and L. D. Claxton, *Atmospheric Chemical Compounds Sources, Occurrence and Bioassay,* Academic Press, New York, 1986.

Air-Quality Criteria for Ozone and Other Photochemical Oxidants, Publication No. EPA/600/8-84-020F (5 Vols.), U.S. Environmental Protection Agency, Research Triangle Park, N.C., 1986.

EPA publishes separate Criteria Documents for all the Criteria Pollutants, which are updated about every five years.

Atmospheric Ozone 1985, World Meteorological Organization, Geneva, Switzerland (3 Vols.), an excellent compendium on tropospheric and stratospheric processes.

G. T. Wolff, J. L. Hanisch, and K. Schere, eds., *The Scientific and Technical Issues Facing Post-1987 Ozone Control Strategies.* Air and Waste Management Association, Pittsburgh, Pa., 1988.

J. H. Seinfeld, "Urban Air Pollution: State of the Sciences," *Science* **243,** 745 (1989).

S. H. Schneider, "The Greenhouse Effect: Science and Policy." *Science,* **243,** 771 (1989).

Indoor Air Pollutants, National Academy Press, Washington, D.C., 1981, p. 101.

AIR POLLUTION CONTROL FOR STATIONARY SOURCES

BURTON B. CROCKER
St Louis, Missouri

KARL B. SCHNELLE, Jr.
Vanderbilt University
Nashville, Tennessee

Air pollution has been defined as the presence in ambient air of one or more contaminants of such quantity and time duration as to be injurious to human, plant, or animal life, to property, to the conduct of business (1,2), or to alter significantly the natural balance of an ecosystem (3). Because of more recent concerns about the effect of the release of greenhouse gases and of stratospheric ozone-depleting gases to the atmosphere, the last phrase has been added to the definition above (3). The effect of a time–dosage relationship has been used by the U.S. Environmental Protection Agency (EPA) in the establishment of the National Ambient Air Quality Standards (NAAQS) for Criteria Pollutants (Table 1).

Increasing concerns about the health of sensitive groups that react to ozone and fine particulate matter have caused the EPA to propose new and somewhat controversial standards for these two pollutants. People with asthma, emphysema, children, and elderly persons are adversely affected by high ozone levels. Thus, the new ozone standard is designed to protect these people. This standard has been set to be 0.09 ppm to be measured by averaging hourly ozone readings over 8 hr. The standard would be compared to the average of the third highest 8-hr average per year for three years. Background for this standard is provided elsewhere (4).

Concern about exposure to fine particles led to the establishment of the particulate (PM_{10}) standard in 1987. Operationally this standard is defined in terms of the sampler collection efficiency as the 10-μm cut point that represents the size of the particle collected with 50% efficiency. However, it was also recognized that particle sizes of 1.0 μm in diameter are more easily breathed into the deepest recess of the lung. Moreover, it has also been established that the sources of particles greater than 2.5 μm in diameter are formed from mechanical action, and particles less than 2.5 μm in diameter come from combustion or secondary chemical reaction in the atmosphere. Thus, different methods of control are required for both sizes, and it seems logical to establish a standard that would recognize the differences. Therefore, the EPA set a 24-h PM_{10} standard of 150 $\mu g/m^3$ along with a $PM_{2.5}$ standard of 50 $\mu g/m^3$ and an annual PM_{10} standard of 50 $\mu g/m^3$ with a $PM_{2.5}$ standard of 20 $\mu g/m^3$. Background information for this standard has been published (5).

Within a local area, the definition suggests that the harmfulness of air pollutants can be reduced by reducing the downwind concentration of pollutants or the exposure time to them or both. However, atmospheric dispersion studies have indicated that high concentrations of sulfur dioxide, sulfite-sulfate complex, and of NO_x can be transported hundreds of kilometers downwind from tall stacks and deposited as acid rain, killing fish in lakes and affecting the health and growth of forests greatly removed from the source of the emission. Acid rain caused by SO_x and NO_x emissions from large-scale release of fossil-fuel combustion flue gases has been a matter of concern to the electric power industry. Large-scale, tall-stack releases of SO_x from smelting of sulfide-containing nonferrous ores can also contribute to the problem. The combination of such releases has resulted in extensive ecological damage in eastern Canada. Releases from fossil fuel combustion in western and northern Europe have been channeled by wind currents to impact the Black Forest of Germany with extensive damage to fir trees. Likewise, in the U.S. Clean Air Reauthorization Act of 1990, Title IV covers acid deposition control, or acid rain. Its purpose is to cut SO_x and NO_x emissions by 90% from 111 fossil-fuel power plants that emit major quantities of these pollutants in the central and midwestern portions of the United States. Beginning in 1995, an annual reduction of

Table 1. National Ambient Air Quality Standards

Pollutant	Maximum Permissible Concentration[a]			Measurement Method
	Averaging Time	Primary Standard	Secondary Standard	
Sulfur oxides	Annual arithmetic mean	80 $\mu g/m^3$ (0.3 ppmv)		West-Gaeke Pararosaniline
	24 h max	365 $\mu g/m^3$ (0.14 ppmv)		
	3 h max		1300 $\mu g/m^3$ (0.5 ppmv)	
Particulates (PM_{10})	Annual arithmetic mean	50 $\mu g/m^3$	Same	Gravimetric 24-h high volume sample with PM_{10} classifying head
	24 h max	150 $\mu g/m^3$	Same	
Carbon monoxide	8 h max	10 mg/m^3 (9 ppmv)	Same	Nondispersive infrared analyzer
	1 h max	40 mg/m^3 (35 ppmv)	Same	
Ozone	1 h max	235 $\mu g/m^3$	Same	Gas-phase chemiluminescence analyzer
Hydrocarbons	3 h max (6–9 a.m.)	160 $\mu g/m^3$ (0.24 ppmv as CH_4)	Same	Flame ionization detector
Nitrogen oxides	Annual arithmetic mean	100 $\mu g/m^3$ (0.05 ppmv as NO_2)	Same	Chemiluminescence analyzer
Lead	Calendar quarter arithmetic mean	1.5 $\mu g/m^3$	Same	Lead analysis by atomic absorption spectrometry on extract from high volume sample catch

[a] Standards for periods shorter than annual average may be exceeded once per year.

9 million tons of SO_2 and 1.8 million tons of NO_x per year is mandated as phase 1 in 21 states, extending from Minnesota and Wisconsin to New York, New Jersey, and Pennsylvania, and from the Great Lakes to the Gulf of Mexico to include Tennessee, Mississippi, Alabama, Georgia, and Florida. Beginning in year 2000, the acid rain regulations will impact an additional 200 fossil-fueled power plants. The impetus for these regulations is concern that large-scale SO_2 and NO_x emissions from tall stacks of power plants in the Midwest travel northeastward for long distances and damage forests and lakes in the United States and Canada.

In 1995 there was a total of 445 boilers in the program, including the 111 major fossil-fueled power plants noted previously. In 1996 these boilers emitted only 5.3 million tons of SO_2 compared to the 1980 levels of 10.9 million tons (6). Thus, there is an indication that the program is working.

To control ozone depletion of the stratosphere, the 1990 Clean Air Reauthorization Act contains a phase-out schedule for chlorofluorocarbons (CFCs), halons, CCl_4, and methylchloroform. It includes new requirements for collecting and recovering refrigerant gases during air-conditioner maintenance in fixed installations as well as in automobiles.

Selection of pollution-control method is generally based on the need to control ambient air quality to achieve compliance with NAAQS standards for criteria pollutants or, in the case of nonregulated contaminants, to protect human health and welfare. There are three elements to a pollution problem: a source of pollutants, a receptor affected by the pollutants, and the transport of the pollutants from source to receptor. Modification or elimination of any one of these elements can change the nature of the pollution problem. For instance, tall stacks that disperse the effluent that modifies the transport of pollutants can reduce nearby pollution levels. If done from numerous sources, better dispersal aloft can solve a local problem, however, it can cause a regional problem such as the acid rain problems mentioned above. Atmospheric dilution as a control measure has been discussed (7–19). Today's regulations do not permit tall stacks to be used as a control measure. Instead emissions must be controlled at the source. It should be noted that limitations have been set to the stack height that can be used in atmospheric modeling, but not on the height of stack that can be built (20).

There are three main classes of pollutants: gases, particulates, and odors. Particulates may be either liquid or solid or a combination. Odors may originate as gases or particulates. Although odors are controlled as are other pollutants, they often require different methods for their sensing and measurement. Many effluents contain several contaminants; some present as gases and the others often exist as liquid or solid particulates of various sizes. The possibility that effluent pollutants may be present in more than one physical state must be taken into account in sampling, analysis, and control. To achieve air pollution control, reliable measurements are needed to quantify both the pollutant concentration and the contribution of individual sources. These data are necessary for designing control equipment, monitoring emissions, and maintaining acceptable ambient air quality.

PRINCIPAL ENERGY SOURCES OF AIR POLLUTANTS

Principal energy sources producing and emitting air pollutants to the atmosphere are fossil-fuel combustion for electric power generation, steam production, and space heating. Another large source is fuel combustion for transportation. On a mass-emission basis, the EPA estimates show the following percentage of total U.S. emissions by source: transportation 43%; stationary fuel combustion 29%; industrial processes, 16%; solid waste disposal (incineration) 4%; and miscellaneous 8%. Of the transportation emissions, by far the largest component is carbon monoxide (78%). The second highest component is NO_x (10%), followed by hydrocarbons (9%), particulates (2%), and SO_x (1%). This article discusses air pollution control techniques for stationary sources only. Stationary consumption of fossil fuels takes place in boilers (approximately 33% of the total U.S. fossil-fuel consumption), residential use, and direct and indirect heating of processes such as steel production and rolling, nonferrous metallurgical processes, and many types of chemical and petroleum process industry operations.

Boilers can be classified industrially as utility boilers, industrial boilers, and commercial or institutional boilers. Utility boilers are generally large (averaging approximately 10.5×10^{12} J/h of heat release), and used for producing steam for electric-power generation. On a combustion heat release capacity basis, the total U.S. utility boiler capacity is approximately 80% of the combined U.S. capacity for all other classes of boilers. More than 50% of the U.S. utility boiler capacity burns coal or residual oil. These two fuels present the most significant air pollution problems, because both contain ash and often sulfur. In the past, these boilers have also been designed for high flame temperatures, which maximizes the fixation of NO_x by reaction between N_2 and O_2 in combustion air. It is estimated that there are 4000 to 5000 utility boilers in the United States, 700,000 to 800,000 industrial boilers, and more than 1.5 million commercial or institutional boilers.

Industrial boilers average approximately 6.3×10^9 J/h heat release. Commercial boilers are even smaller. Approximately 33% of these boilers are natural gas or distillate oil fired. The commercial and institutional boilers present complicated pollution-control problems because of their large number, their small size, and their proximity to population centers. In addition, many of these boilers have fluctuating load swings and unsteady operating conditions, which influence the rate and type of emissions released.

For boilers burning coal, the largest specific release, uncontrolled, is typically SO_2, (in tons released per ton of fuel fired). This can be appreciably reduced by burning low sulfur coal. NO_x is the second most-released-pollutant, with particulate emissions close behind. Hydrocarbon emissions are generally quite low, but they are important in the overall atmospheric pollution situation. Although even lower, there are small quantities of aldehydes released that are important as precursers of photochemical irritants. Emissions and their control from fossil-fuel combustion has been discussed (20), and EPA estimates of emissions from most pollution sources have been given (21). Emission release and control from sources burning unconventional fuels (22) and from municipal solid waste incinerator sources (including incineration of industrial waste) have been discussed (23). Emissions and control in petroleum refining have been reviewed (24).

MEASUREMENT OF AIR POLLUTION

Measurement techniques are divided into two categories: ambient and source measurement. Ambient air samples often require detection and measurement in the parts per billion of volume to parts per million of volume range, whereas source concentrations can range from a few hundred parts per million of volume to tenths of a volume percent. Federal regulations (25) require periodic ambient air monitoring at strategic locations in a designated air quality control region. The number of required locations and complexity of monitoring increases with region population and with the normal concentration level of pollutants. Continuous monitoring is preferable, but for particulates the usual sampling interval is one 24-h sample every 6 days. In some extensive metropolitan sampling networks, averaged results from continuous monitors are telemetered to a single data-processing center. Special problems have been investigated using portable, vehicle-carried, or airborne, ambient-sampling equipment. The use of remote-guided miniature aircraft has been reported as a practical, cost-effective ambient-sampling method (26). Ambient sampling may fulfill one or more of the following objectives: (*1*) establishing and operating a pollution alert network, (*2*) monitoring the effect of an emission source, (*3*) predicting the effect of a proposed installation (compliance with prevention of significant deterioration (PSD) regulations requires 1 year of background ambient air monitoring at the proposed installation site before filing an application for a construction permit for a new installation), (*4*) establishing seasonal or yearly trends, (*5*) locating the source of an undesirable pollutant, (*6*) obtaining permanent sampling records for legal action or for modifying regulations, and (*7*) correlating pollutant dispersion with meteorological, climatological, or topographic data and with changes in societal activities.

The problems of source sampling are distinct from those of ambient sampling. The source gas may have high temperature or contain high concentrations of water vapor or entrained mist, dust, or other interfering substances, so that particulates or gases may be deposited on or absorbed into the grain structure of the gas-extractive sampling probes. Depending on the objective or regulations, source sampling may be infrequent, occasional, intermittent, or continuous. Typical objectives are (*1*) demonstrating compliance with regulations; (*2*) obtaining emission data; (*3*) measuring product loss or optimizing process operating variables; (*4*) obtaining data for engineering design, such as for control equipment; (*5*) determining collector efficiency or acceptance testing purchased equipment; and (*6*) determining need for maintenance of process or control equipment.

Sampling of Gaseous Pollutants

Gaseous pollutant detection depends on the chemistry of the material involved. Reference methods for criteria (25) and hazardous (25) pollutants established by the EPA include sulfur dioxide by the West-Gaeke method, carbon monoxide by nondispersive infrared analysis, ozone and nitrogen dioxide by chemiluminescence, and hydrocarbons by gas chromatography coupled with flame-ionization detection. Gas chromatography coupled with a suitable detector can also be used to measure ambient concentrations of vinyl chloride monomer, halogenated hydrocarbons and aromatics, and polyacrylonitrile (27,28). Methods of sampling and analysis for ambient gases have been reviewed (29).

Automated analyzers may be used for continuous monitoring of ambient pollutants. Therefore, the EPA has developed continuous procedures (30) as alternatives to the referenced methods. For source sampling, the EPA has specified extractive sampling trains and analytical methods for pollutants such as SO_2 and SO_3, sulfuric acid mists, NO_x mercury, beryllium, vinyl chloride, and volatile organic compounds (VOCs). Some EPA new-source performance standards require continuous monitors on specified sources.

Sampling of Particulates

Ambient air suspended particulate concentration was traditionally measured gravimetrically over a 24-h period with a *hi-vol* (high volume) sampler. However, in 1987 the EPA changed ambient particulate control to the PM_{10} reference method (31). In the PM_{10} method, a particle size classification head is attached to a Hi-Vol sampler so that particulates finer than an aerodynamic 10 μm are collected on the filter. Although tape samplers, used for more frequent determination of suspended particulates, have been tied into the EPA Alert Warning System, it is not yet apparent how they will be correlated with PM_{10} monitoring. In the tape method, particulate quantity is measured automatically by light transmittance or β-ray attenuation and converted to an electronic signal for transmission and data processing.

The EPA presented a draft method for the measurement of PM_{25} (32). This method calls for daily sampling of ambient mass concentration of fine particulate matter having an aerodynamic diameter less than or equal to a nominal 2.5 μm in air over a 24-h period. This method and a device for carrying out the measurement have been discussed further (33).

Source sampling of particulates requires isokinetic removal of a composite sample from the stack or vent effluent to determine representative emission rates. Table 2 lists 50 specific source sampling procedures developed and promulgated by the EPA as specific regulatory compliance methods in specific situations (34,35). Samples are collected either extractively or using an in-stack filter. From Table 2, EPA method 5 is representative of extractive sampling and EPA method 17 of in-stack filtration. Other means of source sampling have been used, but they have largely been supplanted by EPA methods. Continuous in-stack monitors of opacity use attenuation of radiation across the effluent. These opacity measurements are affected by the particle size, shape, size distribution, refractive index, and the wavelength of the radiation (36). Detailed procedures for these methods have been given (37).

Particle size measurements for particulates extracted by filtration, electrostatic or thermal precipitation, or impaction may be performed using microscopy, sieve analysis, gas or liquid sedimentation, centrifugal classification, or electrical or optical counters. For aerosol particulate size determination, however, questions arise such as whether the collected particles agglomerate after capture or whether they are redispersed to the same degree in the measuring media as they were originally. These problems can be avoided mainly by performing particle size measurements on the original aerosol by using devices such as cascade impactors (38), virtual impactors (39), and diffusion batteries and mobility analyzers.

AIR POLLUTION AND CONTROL REGULATIONS

There has been considerable improvement, especially in industrial areas, in U.S. air quality since the adoption of the Clean Air Act of 1972. Appreciable reductions in particulate emissions and in SO_2 levels are especially evident. In 1990, however, almost every metropolitan area was in nonattainment on ozone air-quality standards; 50 metropolitan areas exceeded the carbon monoxide standard, and between 50 and 100 exceeded the PM_{10} standard for particulate level (40,41).

The U.S. Congress adopted a new clean air act in 1990 that has three areas of emphasis: acid rain reduction in the northeastern United States; severe limitation on atmospheric emissions of 189 chemicals on the EPA hazardous or toxic substance list; and ozone and smog reduction by tightened regulations on vehicular exhaust, reformulated vehicular fuels, and vehicles capable of using alternative fuels. Regulations associated with acid rain prevention emphasize reductions in sulfur oxide and nitrogen oxide emissions from combustion processes,

Table 2. EPA Reference Methods for Source Sampling

Method 1—Sample and velocity traverses for stationary sources
Method 2—Determination of stack gas velocity and volumetric flow rate (Type S pitot tube)
Method 2A—Direct measurement of gas volume through pipes and small ducts
Method 2B—Determination of exhaust gas volume flow rate from gasoline vapor incinerators
Method 3—Gas analysis for carbon dioxide, oxygen, excess air, and dry molecular weight
Method 3A—Determination of oxygen and carbon dioxide concentrations in emissions from stationary sources (instrumental analyzer procedure)
Method 4—Determination of moisture content in stack gases
Method 5—Determination of particulate emissions from stationary sources
Method 5A—Determination of particulate emissions from the asphalt processing and asphalt roofing industry
Method 5B—Determination of nonsulfuric acid particulate matter from stationary sources
Method 5D—Determination of particulate matter emissions from positive pressure fabric filters
Method 5E—Determination of particulate emissions from the wool fiberglass insulation manufacturing industry
Method 5F—Determination of nonsulfate particulate matter from stationary sources
Method 6—Determination of sulfur dioxide emissions from stationary sources
Method 6A—Determination of sulfur dioxide, moisture, and carbon dioxide emissions from fossil fuel combustion sources
Method 6B—Determination of sulfur dioxide and carbon dioxide daily average emissions from fossil fuel combustion sources
Method 6C—Determination of sulfur dioxide emissions from stationary sources (instrumental analyzer procedure)
Method 7—Determination of nitrogen oxide emissions from stationary sources
Method 7A—Determination of nitrogen oxide emissions from stationary sources
Method 7B—Determination of nitrogen oxide emissions from stationary sources (ultraviolet spectrophotometry)
Method 7C—Determination of nitrogen oxide emissions from stationary sources
Method 7D—Determination of nitrogen oxide emissions from stationary sources
Method 7E—Determination of nitrogen oxides emissions from stationary sources (instrumental analyzer procedure)
Method 8—Determination of sulfuric acid mist and sulfur dioxide emissions from stationary sources
Method 9—Visual determination of the opacity of emissions from stationary sources
Method 10—Determination of carbon monoxide emissions from stationary sources
Method 10A—Determination of carbon monoxide emissions in certifying continuous emission monitoring systems at petroleum refineries
Method 11—Determination of hydrogen sulfide content of fuel gas streams in petroleum refineries
Method 12—Determination of inorganic lead emissions from stationary sources
Method 13A—Determination of total fluoride emissions from stationary sources—SPADNS zirconium lake method
Method 13B—Determination of total fluoride emissions from stationary sources—specific ion electrode method
Method 14—Determination of fluoride emissions from potroom roof monitors of primary aluminum plants
Method 15—Determination of hydrogen sulfide, carbonyl sulfide, and carbon disulfide emissions from stationary sources

Table 2. *(continued)*

Method 15A—Determination of total reduced sulfur emissions from sulfur recovery plants in petroleum refineries
Method 16—Semicontinuous determination of sulfur emissions from stationary sources
Method 16A—Determination of total reduced sulfur emissions from stationary sources (impinger technique)
Method 16B—Determination of total reduced sulfur emissions from stationary sources
Method 17—Determination of particulate emissions from stationary sources (instack filtration method)
Method 18—Measurement of gaseous organic compound emissions by gas chromatography
Method 19—Determination of sulfur dioxide removal efficiency and particulate matter, sulfur dioxide and nitrogen oxides emission rates
Method 20—Determination of nitrogen oxides, sulfur dioxide, and oxygen emissions from stationary gas turbines
Method 21—Determination of volatile organic compounds leaks
Method 22—Visual determination of fugitive emissions from material sources and smoke emissions from flares
Method 24—Determination of volatile matter content, water content, density, volume solids, and weight solids of surface coating
Method 24A—Determination of volatile matter content and density of printing inks and related coatings
Method 25—Determination of total gaseous nonmethane organic emissions as carbon
Method 25A—Determination of total gaseous organic concentration using a flame ionization analyzer
Method 25B—Determination of total gaseous organic concentration using a nondispersive infrared analyzer
Method 27—Determination of vapor tightness of gasoline delivery tank using pressure-vacuum test
Appendix B—Performance Specifications
Performance Specification 1—Performance specifications and specification test procedures for transmissometer systems for continuous measurement of the opacity of stack emissions
Performance Specification 2—Specifications and test procedures for SO_2 and NO_2 continuous emission monitoring systems in stationary sources
Performance Specification 3—Specifications and test procedures for O_2 and CO_2 continuous emission monitoring systems in stationary sources [a]

[a] Ref. 36. Courtesy of McGraw-Hill.

especially coal-fired power boilers in the Midwest. The chemical process industry and its customers are increasingly under pressure to eliminate atmospheric releases of VOCs and carcinogenic-suspect compounds.

Minimizing Pollution Control Cost

Although the first impulse for emission reduction is often to add a control device, this may not be the environmentally best or least costly approach. Process examination may reveal changes or alternatives that can eliminate or reduce pollutants, decrease the gas quantity to be treated, or render pollutants more amenable to collection. The following principles should be considered for controlling pollutants without the addition of specific treatment devices (42).

1. Eliminate the source of the pollutant.
   - Seal the system to prevent interchanges between the system and the atmosphere.
   - Use vacuum vessels.
   - Interconnect vents on receiving and discharging containers.
   - Provide seals on rotating shafts and other necessary openings.
   - Change raw materials, fuels, and so on to eliminate the pollutant from the process.
   - Change the manner of process operation to prevent or reduce formation or air entraninment of a pollutant.
   - Change the type of process step to eliminate the pollutant.
   - Use a recycled gas or recycle the pollutants rather than using fresh air or venting.
2. Reduce the quantity of pollutant released or the quantity of carrier gas to be treated.
   - Minimize entrainment of pollutants into a gas stream.
   - Reduce the number of points in a system in which materials can become airborne. Recycle a portion of process gas.
   - Design hoods to exhaust the minimum quantity air necessary to ensure pollutant capture.
3. Use equipment for dual purposes, such as a fuel combustion furnace to serve as a pollutant incinerator.

Steps such as the substitution of low sulfur fuels or nonvolatile solvents, change of raw materials, lowering of operation temperatures to reduce NO_x formation or volatilization of process material, and installation of well-designed hoods (43–50) at emission points to reduce effectively the air quantity needed for pollutant capture are illustrations of the above principles.

Selection of Control Equipment

Engineering information (51) needed for the design and selection of pollution-control equipment include the following properties of the pollutants: chemical species, physical state, particle size, concentration, and quantity of conveying gas, and effects of pollutant on the surrounding environment. The design must consider future collection requirements and alternative collection techniques. The following factors should be taken into account:

1. Collection efficiency.
2. Ease of reuse or disposal of recovered material.
3. Ability of the collector to handle variations in gas flow and loads at required collection efficiencies.
4. Equipment reliability and freedom from operational and maintenance attention.
5. Initial investment and operating cost.
6. Possibility of recovery or conversion of contaminant into a saleable product.

The physical state of a pollutant is obviously important; for example, a device designed to collect particulate matter only cannot remove vapor. Pollutant concentration and carrier gas quantity are necessary to estimate collector size and required efficiency, and knowledge of a pollutant's chemistry may suggest alternative approaches to treatment. Emission standards

Table 3. Checklist of Applicable Devices for Control of Pollutants

	Pollutant Classification			
			Particulate	
Equipment Type	Gas	Odor	Liquid	Solid
Absorption	•	•		
Aqueous solution				
Nonaqueous				
Adsorption	•	•		
Throw-away canisters				
Regenerable stationary beds				
Regenerable traveling beds				
Chromatographic adsorption				
Air dispersion (stacks)	•	•	•	•
Condensation	•	•		
Centrifugal separation (dry)			•	•
Chemical reaction	•	•		
Coagulation and particle growth			•	•
Filtration				
Fabric and felt bags				•
Granular beds			•	•
Fine fibers			•	•
Gravitational settling				•
Impingement (dry)				•
Incineration	•	•	•	•
Precipitation, electrical				
Dry			•	•
Wet	•	•	•	•
Precipitation, thermal			•	•
Wet collection[a]	•	•	•	•

[a] Includes cyclonic, dynamic, filtration, inertial impaction (wetted targets, packed towers, turbulent targets), spray chambers, and venturi.

may set collection efficiency, but specific regulations do not exist for many trace emissions. In such cases, emission targets must be set by dose-exposure time relationships obtained from effects on vegetation, animals, and humans. With this information, a list of possible treatment methods can be made as seen in Table 3.

Control devices that are too inefficient for a particular pollutant or too expensive can then be stricken from the list. For example, condensation may require costly refrigeration to give adequate collection. Although both absorption and adsorption devices for contaminant gases can be designed for almost any efficiency, cost generally dictates the choice between the two. Grade-efficiency curves should be consulted in evaluating particulate collection devices, and the desirability of dry or wet particulate collection should be considered, especially with respect to material recycle or disposal.

Other factors to be evaluated are capital investment and operating cost, material reuse or alternative disposal economics, relative ruggedness and reliability of alternative control devices, and the ability to retain desired efficiency under all probable operating conditions. Control equipment needs to be both rugged and reliable, in part to avoid shuting a process down for control equipment repair. Efficiency of control devices varies with processing conditions, flow rate, temperature, emission concentration, and particle size. Control devices should be designed to handle these variations. Combinations of gaseous and particulate pollutants can be especially troublesome as gaseous removal devices are often unsuitable for heavy loadings of insoluble solids. Concentrations of soluble particulates up to 11 g/m^3 have been handled in gas absorption equipment with some success. However, to ensure rapid particle solution, special consideration must be given to wet–dry interfaces and adequate liquid quantities.

CONTROL OF GASEOUS EMISSIONS

Five methods are available for controlling gaseous emissions: absorption, adsorption, condensation, chemical reaction, and incineration. Atmospheric dispersion from a tall stack, considered as an alternative in the past, is now not a permitted method of control. Absorption is particularly attractive for pollutants in appreciable concentration; it is also applicable to dilute concentrations of gases having high solvent solubility. Adsorption is desirable for contaminant removal down to extremely low levels (less than 1 ppmv) and for handling large gas volumes that have quite dilute contaminant levels. Condensation is best for substances having rather high vapor pressures. Where refrigeration is needed for the final step, elimination of noncondensible diluents is beneficial. Incineration, suitable only for combustibles, is used to remove organic pollutants and small quantities of H_2S, CO, and NH_3. Specific problem gases such as sulfur and nitrogen oxides require combinations of methods and are discussed separately.

Applicability of the control techniques depends more on the individual emission stream under consideration than on the particular source category. Thus, selection of applicable control devices for point sources is made on the basis of stream-specific characteristics and desired control efficiency. The EPA Handbook *Control Technologies for Hazardous Air Pollutants* (52) provides guidance in selection of devices to control hazardous air pollutants (HAP) in Table 4 and an estimation of efficiency that might be expected in Figure 1. The course of action for VOC control in the future is discussed elsewhere (53).

Absorption

Absorption is a diffusional mass-transfer operation by which a soluble (or semisoluble) gaseous component can be removed from a gas stream by causing the absorbable component to dissolve in a solvent liquid through gas–liquid contact. The driving force for absorption is the difference between the partial pressure of the soluble gas in the gas mixture and the vapor pressure of the solute gas in the liquid film in contact with the gas. If the driving force is not positive, no absorption will occur. If it is negative, desorption or stripping will occur and pollution of the gas being treated will actually be enhanced. Absorption systems can be divided into those that use water as the primary absorbing liquid and those that use a low volatility organic liquid. The system can be a simple absorption in which the liquid (usually water) is used in a single pass and then disposed of, while still containing the absorbed pollutant. Alternatively, the pollutant can be separated from the absorbing liquid and recovered in a pure, concentrated form by stripping or desorption. The absorbing liquid is then used in a closed circuit and is continuously regenerated and recycled. Regeneration alternatives to stripping are pollutant removal through precipitation and settling; chemical destruction through neutralization, oxidation, or reduction; hydrolysis; solvent extraction; pollutant liquid adsorption; and so on. Absorption is one of

Table 4. Key Emission Stream and HAP Characteristics for Selecting Control Technologies for Organic Vapors from Point Sources

	Emission Stream Characteristics					HAP Characteristics[a]			
Control Device	HAP/Organics Contents[b], ppmv	Heat Content, BTU/scf	Moisture Content, %	Flow Rate scfm	Temp, °C	Molecular Weight, lb/lb · mole^{-1}	Solubility	Vapor Pressure, mm Hg	Adsorptive Properties
Thermal incinerator	>20; (<25% of LEL[c])			<50,000[d]					
Catalytic incinerator	50–10,000; (<25% of LEL[c])			<50,000					
Flare		>300[e]		<2,000,000[f]					
Boiler/process heater[g]		>150[h]		Steady					
Carbon adsorber	700–10,000 (<25% of LEL[c])		<50%[i]	300–200,000	≤54	45–130			Must be able to absorb on/desorb from available adsorbents
Absorber	250–10,000			1,000–100,000			Must be readily soluble in water or other solvents		
Condenser	>5,000–10,000			<2,000				>10 (at room temp)	

[a] Refers to the characteristics of the individual HAP if a single HAP is present and to that of the HAP mixture if a mixture of HAPs is present.
[b] Determined from HAP/hydrocarbon content.
[c] For emission streams that are mixtures of air and VOC; in some cases, the LEL can be increased to 40 to 50% with proper monitoring and control.
[d] For packaged units; multiple-package or custom-made units can handle larger flows.
[e] Based on the EPA's guidelines for 98% destruction efficiency.
[f] Units: lb/h. Source: Ref. 12.
[g] Applicable if such a unit is already available on-site.
[h] Total heat content.
[i] Relative humidity. Applicable for HAP concentration less than about 1000 ppmv.

the most frequently used methods for removal of water-soluble gases. Acidic gases such as HCI, HF, and SiF_{m4} can be absorbed in water efficiently and readily, especially if the last contact is made with water that has an alkaline pH. Less soluble acidic gases such as SO_2, $C1_2$, and H_2S can be absorbed more readily in a dilute caustic solution. The scrubbing liquid may be made alkaline with dissolved soda ash or sodium bicarbonate or with NaOH (usually no higher a concentration in the scrubbing liquid than 5–10%). Lime is a cheaper and more plentiful alkali, but its use directly in the absorber may lead to plugging or coating problems if the calcium salts produced have only limited solubility. A technique often used, such as in the two-step flue gas desulfurization process, is to have a NaOH solution inside the absorption tower, and then to lime the tower effluent externally, precipitating the absorbed component as a slightly soluble calcium salt. The precipitate may be removed by thickening, and the regenerated sodium alkali solution is recycled to the absorber. Scrubbing with an ammonium salt solution is also employed. In such cases, the gas is often first contacted with the more alkaline solution and then with the neutral or slightly acid contact to prevent stripping losses of NH_3 to the atmosphere. When flue gases containing CO_2 are being scrubbed with an alkaline solution to remove other acidic components, the caustic consumption can be inordinately high if CO_2 is absorbed. However, if the pH of the scrubbing liquid entering the absorber is kept below 9.0, the amount of CO_2 absorbed can be kept low. Conversely, alkaline gases, such as NH_3, can be removed from the main gas stream with acidic water solutions such as dilute H_2SO_4, H_3PO_4, or HNO_3. Single-pass scrubbing solutions so used can often be disposed of as fertilizer ingredients. Alternatives are to remove the absorbed component by concentration and crystallization. The absorbing gas must have adequate solubility in the scrubbing liquid at the resulting temperature of the gas–liquid system.

For pollutant gases with limited water solubility, such as SO_2 or benzene vapors, the large quantities of water that would be required are generally impractical on a single-pass basis, but may be used in unusual circumstances.

Although water is the most common liquid used for absorbing acidic gases, amines (monoethanol-, diethanol-, and triethanolamine; methyldiethanolamine; and dimethylanaline) have been used for absorbing SO_2 and H_2S from hydrocarbon gas streams. Such absorbents are generally limited to solid particulate-free systems, because solids can produce difficult-to-handle sludges as well as use up valuable

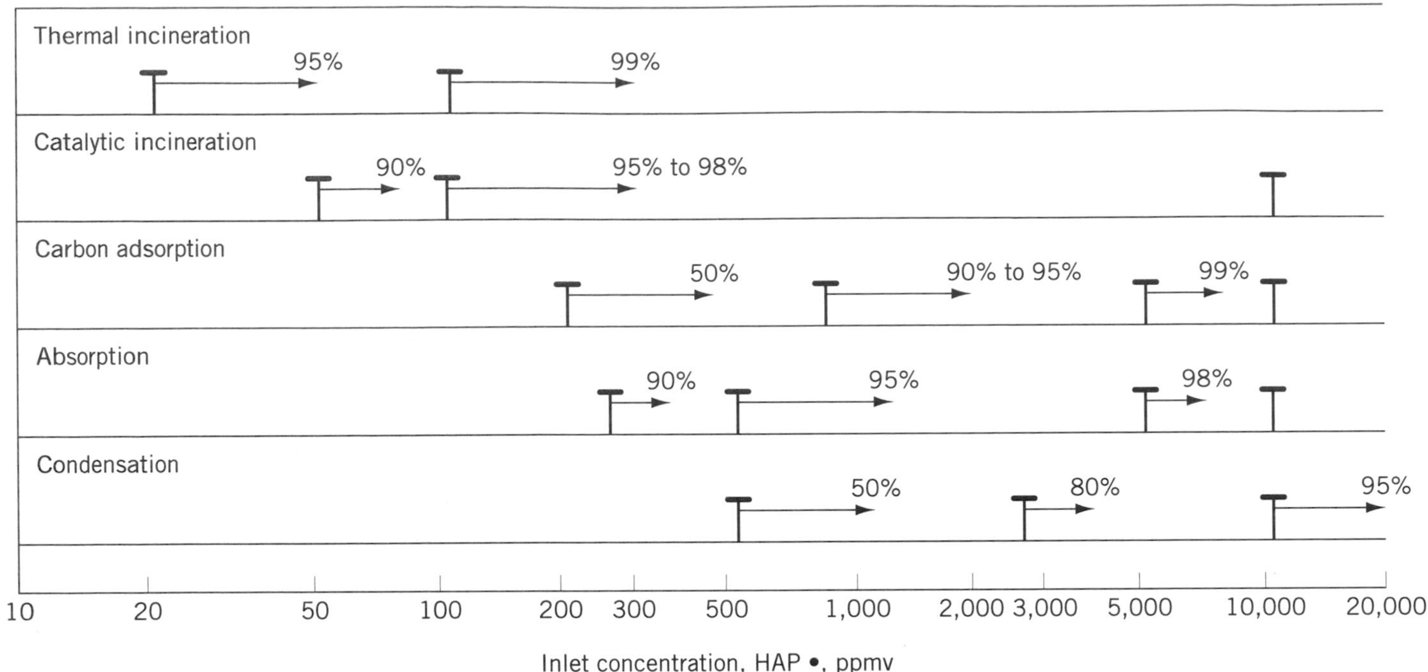

Figure 1. Approximate percent reduction ranges for add-on equipment.

organic absorbents. Furthermore, because of absorbent cost, absorbent regeneration must be practiced in almost all cases.

At first glance, an organic liquid appears to be the preferred solvent for absorbing hydrocarbon and organic vapors from a gas stream because of improved solubility and miscibility. The lower heat of vaporization of organic liquids is an energy conservation plus when solvent regeneration must occur by stripping. Many heavy oils, hexacane (No. 2 fuel oil) or heavier, and other solvents with low vapor pressure can do extremely well in reducing organic vapor concentrations to low levels. Care must be exercised in picking a solvent that will have sufficiently low vapor pressure, so that the solvent itself will not become a source of VOC pollution. Obviously, the treated gas will be saturated with the absorbing solvent. An absorber-stripper system for recovery of benzene vapors has been described (54). Other aspects of organic solvent absorbtion requiring consideration are stability of the solvent in the gas-solvent system (such as its resistance to oxidation) and possible fire and explosion hazards.

Adsorption

The attractive forces in a solid that exist between atoms, molecules, and ions, holding the solid together, are unsatisfied at the surface and thus are available for holding other materials such as liquids and gases. This phenomenon is known as adsorption. If the solid is produced in a highly porous form with extensive pores and microstructure, its adsorptive capacity can be greatly enhanced. Adorption can be used as a pollution-control measure to remove pollutant gases from an otherwise harmless gas when released to the atmosphere and is quite adaptable for removing such contaminants, especially VOCs, down to extremely low levels (less than 1 ppmv). Adsorption's best applications are (*1*) handling large volumes with dilute pollution levels and (*2*) removing the contamination level, regardless of gas quantity, down to only trace pollutant levels. Removal of solvent losses from large quantities of ventilation air is an example of the former, and recovery of toxic and hazardous vapors to extremely low concentrations exemplifies the latter. In any case, absorption may be used alone for the entire control requirement or in combination with other removal methods. In the latter case, adsorption typically becomes the final cleanup step because of its capability of achieving low emission concentrations. Recently a review of adsorption applied to the recovery of VOCs has been published (55).

In being adsorbed, a gas molecule travels to an adsorption site on the surface of the solid, where it is held by attractive forces and loses much of its molecular motion. This loss of kinetic energy is released as heat. The heat of adsorption is often close in magnitude to the heat of condensation for the species being adsorbed. Thus adsorption is always exothermic. Desorption is a reversal of the adsorption process, and heat must be supplied to cause desorption to take place. Hence temperature rise tends to reverse the process or cause a loss in capacity of the sorbent. Cooling of the sorption bed (which is often difficult because of poor heat transfer within the bed) or precooling of the gas stream to be treated is desirable to provide a sink for the heat of adsorption being released. Some sorption processes can occur so strongly that they are irreversible. That is, the adsorbed material can only be desorbed by removal of some of the solid substrate. Such a process is referred to as chemisorption. For example, under certain circumstances oxygen can be adsorbed so strongly on activated carbon that it can be removed from the solid only in the form of CO or CO_2. The Reinluft process (56), developed as a means for removing SO_2 from flue gas, adsorbed a portion of the SO_2 on the carbon as sulfuric acid. Desorption was achieved by heating the bed to 370°C, causing the acid and carbon to react, producing CO_2 and SO_2. Similarly, in the U.S. Bureau of Mines alkalized alumina process (57) the SO_2 became oxidized to sulfates, which could be removed only by reacting the adsorbent with hydrogen or reformed natural gas at 650°C.

The adsorbing solid is called the adsorbent or sorbent; the adsorbed material, the adsorbate or sorbate. A thorough discussion of adsorption processes for air pollution and design equations have been given (58), and other more general references are also available (59–69).

Types of Adsorbents. Commercially important adsorbents are activated carbon, other simple or complex metallic oxides, and impregnated sorbents. Activated carbon is a general adsorbent. It is composed primarily of a single species of neutral atoms with no electrical gradients between molecules. Thus there are no significant potential gradients to attract or orient polar molecules in preference to nonpolar molecules. For this reason, carbon has less selectivity than other sorbents and is one of the few that will work in absorbing organics from a humid gas stream. Because the polar water molecules attract each other as strongly as the neutral carbon, the latter tends to be slightly selective for organic molecules. However, some water is adsorbed, especially if its partial pressure is greater than that of the organic molecules. The water being adsorbed must be taken into consideration in selecting the sorptive capacity to be provided in the design. Typical sources of activated carbon are coconut and other nut shells, fruit pits, bituminous coal, hardwoods, and petroleum coke and residues.

Simple and complex metal oxides are polar and have a greater degree of selectivity than carbon and a great preference for polar molecules. They can be useful for removal of a particular species from the gas stream but are ineffective when moisture is present, because most of these adsorbents are excellent desiccants. Siliceous adsorbents include materials such as silica gel, Fuller's and diatomaceous earth, synthetic zeolites, and molecular sieves. They are available in a wide range of capacities, with the best equaling the capacity of the best activated carbons. Synthetic zeolites can be prepared with specific and uniform pore sizes to give sorptive specificity based on the shape and size of sorbate molecules. Even this property tailored to specific organic molecules will not overcome their chemical preference for polar molecules such as water vapor. Activated alumina and other metallic oxides are even more polar than the silicas and are seldom used directly for pollution control adsorption.

Impregnated sorbents fall into three general classes (*1*) those impregnated with a chemical reactant or reagent, (*2*) those in which the impregnant acts as a continuous catalyst for pollutant oxidation or decomposition, and (*3*) those in which the impregnant acts only intermitantly as a catalyst. Reagent impregnants chemically convert the pollutant into a harmless or adsorbable pollutant. Carbon may be impregnated with 10–20% of its mass with bromine, which is used to react with olefins. Thus, ethylene, which is poorly adsorbed from an air stream because of its low molecular weight, is converted at the brominated surface to 1,2-dibromoethane, which is readily adsorbed. Other impregnant reagents are iodine for collecting mercury vapor, lead acetate for collecting H_2S, and sodium silicate for collecting HF. Other applications of impregnated sorbents and continuous and intermittent catalytic sorbents have been discussed elsewhere (67,68).

Condensation

Condensation has been discussed extensively as a pollutant control method (70–72). It is most applicable for vapors fairly close to their dew points. As such, it is suitable for hydrocarbons and organic compounds that have reasonably higher boiling points than ambient conditions and are present in the effluent gas in appreciable concentrations. Pollutants having reasonably low vapor pressures at ambient temperatures may be controlled satisfactorily in water-cooled or even air-cooled condensers. For more volatile solvents, two-stage condensation may be required, using cooling water in the first stage, and refrigeration in the second. Refrigeration to extremely low temperature levels is seldom attractive for pollution control alone, unless dictated by process considerations, because alternative control methods will generally be more attractive. Minimizing the presence of inert diluent gases in the condenser will reduce the need to cool to very low dewpoints and enhance the economic attractiveness of condensation as a control method. Condensation is not a practical method of total control for reasonably volatile toxic or hazardous organics in appreciable concentrations in streams of noncondensables if the effluent concentration must be reduced to a few parts per million of volume. Condensation may still be useful as a preliminary treatment method to recover valuable solvents or to reduce the capacity required for the final treatment method. Partial condensation can be useful when the stripped gas can be recycled to the process (rather than vented) or when it can be used as primary combustion air. (The remaining pollutants are incinerated.) Condensation can be an attractive pretreatment method when it can serve to cool the gas before final control by adsorption.

Chemical Reaction

Removal of an objectional or hazardous gaseous pollutant by chemical reaction has interesting possibilities. It is difficult to generalize about such means, because they are so specific to the chemistry of the species of concern. In addition, the process suitability also can vary with the pollutant concentration in the gas stream as well as the temperature and composition of the carrier gas. The unit operations of absorption and wet scrubbing provide opportunities to carry on chemical reactions by adding a chemical reactant to the absorbing or scrubbing liquid, such as an alkali to enhance the absorption of an acidic gas (discussed above.) If reactive in nature, the chemical nature of the contacting liquid could change or destroy the pollutant vapor by the presence of an oxidizing agent, such as potassium permanganate, hydrogen peroxide, ozone, strong and hot nitric acid, or hypochlorites and hypochlorates. In addition, agents can be added to remove the absorbed vapors from the liquid by precipitation, forming insoluble compounds with the gaseous pollutant. Likewise, with adsorption, adsorbents can be impregnated with bromine, iodine, lead acetate, and sodium silicate. Another category is that of catalytic adsorbents, which are oxcar catalysts resulting in the oxidation of organic pollutants. More process development research has been carried out on chemical methods for removal of SO_x and NO_x, probably because of their widespread occurrence in flue gases. These specific chemical solutions are discussed below. Gas–solid reactions are feasible for removal of specific gaseous pollutants by injection of dry solids into the conveying gas steam. Hydrated lime injection to remove SO_2 is a prime

Table 5. Gases Classified According to Degradability

Rapidly Degradable VOCs	Rapidly Reactive VOCs	Slowly Degradable VOCs	Very Slowly Degradable VOCs
Alcohols	H_2S	Hydrocarbons	Halogenated hydrocarbons
Aldehydes	NO_x (not N_2O)	Phenols	Polyaromatic hydrocarbons
Ketones		Methylene	CS_2
Ethers	SO_2	Chloride	
Esters	HCl		
Organic acids	NH_3		
Amines	PH_3		
Thiols	SiH_4		
Other molecules with O, N, or S functional groups	HF		[a]

[a] Reprinted from Ref. 75.

example of such applications. Dry injection of sodium bicarbonate to remove both SO_2 and NO_x from flue gas is another. Generally, dry solid injection falls considerably short of 100% total removal, but recent research with gas humidification and the addition of activity enhancers to the injected solids have been shown to help. As would be expected, increasing solid surface area through fine particle grinding and increasing particle surface porosity are beneficial.

Some fundamental objectives to consider for chemical reactions are (*1*) convert the pollutant into a different material with a lower vapor pressure (ideally into a liquid or solid particle) or into another chemical species that is more easily collectible, (*2*) convert the pollutant molecule into a harmless or at least less harmful molecule, (*3*) destroy the pollutant. Simple examples of conversion would be to increase the molecular weight of the pollutant through gas-phase reaction, causing the particle to condense so that it can be captured in a particulate collector. Another example is modification of an organic molecule by substitution, rendering it highly water soluble for absorption. A simple example of changing a gaseous pollution problem to a particulate one is the addition of ammonia to a gas stream containing HCl, to produce NH_4Cl smoke. An example of destruction is oxidation of a pollutant to simple harmless species such as CO_2 and water.

Control of Volatile Organic Compounds and Hazardous Air Pollutants by Biofiltration

A recent treatment process for VOCs and HAPs that has come into use is biofiltration (73,74). A biofilter consists of a bed of soil or compost beneath which is a network of perforated pipe. Contaminated air flows through the pipe and out the many holes in the sides of the pipe thereby being distributed throughout the bed. A biofilter works by providing an environment in which microorganisms thrive. The organic substrate provides the salts and trace elements for the bacteria, and the VOC provides the food source. This action is an adaptation of biodegradation in which the air cleanses itself naturally. The microorganisms are the same that degrade organic wastes in nature and in wastewater treatment plants. These microorganisms in a moist environment oxidize organic compounds to CO_2 and water. The soil or compost beds provide a network of fine pores with large surface areas. In soils the pores are smaller and less permeable. Therefore, soil requires larger areas for biofiltration. Table 5 lists compounds in order of their degradability.

Incineration

Incineration of gaseous contaminants is also known as thermal oxidation and fume incineration. It is primarily applicable to gaseous impurities that can be oxidized or burned to decompose the original molecules to simpler nonhazardous compounds such as CO_2 and water. This process is simply controlled combustion. As such, it may be used to destroy airborne or air-mixed hydrocarbon gases, other organic vapors, and similar blowdown gases; mercaptans; and undesirable inorganic gases such as H_2S, HCN, CO, H_2, and NH_3. Volatile organic compound emissions are frequently destroyed by incineration (76–79) as are gases evolving from landfills.

Substituted organic vapors can also be decomposed by incineration, but it is important to consider the complications imposed when passing the substituted radical through the combustion process. For instance, halogen substituted organics will result in the release of the corresponding acidic hydrogen halide gas. Sulfur-containing groups will generally be oxidized to SO_2. These acidic gases generally must be scrubbed or otherwise removed from the incinerator effluent before the combustion products are released to the atmosphere. Cooling of the effluent gas in the scrubber may require adding reheat for good atmospheric dispersion. Saturating the effluent with water may also produce condensing plumes in cold weather. Recovery of the acidic products, often in dilute aqueous form, may constitute a disposal problem.

Incineration also may be used to destroy odors in those cases in which the odor substance can be decomposed by combustion. With proper design, aerosol incineration could destroy combustible airborne liquid or solid particles using a burner much like the one used to burn pulverized coal or activated sludge from sewage treatment. However, literature references for aerosol incineration are essentially nonexistent. In such an incinerator, rapid ignition of particles by quickly heating them above their kindling temperature is required as well as providing adequate residence time and flame space for complete combustion. Particulates have a considerably slower burning rate than combustible gases. Means to reduce the particulate size to subsieve size would greatly enhance the combustion speed. The possible presence of noncombustible residue that can produce a solid or molten ash and its subsequent handling must also be considered.

Thermal oxidizers can be designed to yield from 95% to more than 99% destruction of all combustible compounds. They can be designed with a capacity to handle from 0.5 to

236 m^3/s and for inlet combustible pollutant concentrations of 100 to 2000 ppmv.

CONTROL OF PARTICULATE EMISSIONS

The removal of particles (liquids, solids, or mixtures) from a gas stream requires deposition and attachment to a surface. The surface may be continuous, such as the wall and cone of a cyclone or the collecting plates of an electrostatic precipitator, or it may be discontinuous, such as spray droplets in a scrubbing tower. Once deposited on a surface, the collected particles must be removed at intervals, without appreciable reentrainment in the gas stream. One or more of seven physical principles (gravity settling, centrifugal deposition, flowline interception, inertial impaction, diffusional deposition, electrostatic precipitation, thermal precipitation, or thermophoresis) are frequently employed to move particles from the bulk gas stream to the collecting surface. In some instances, a few other principles such as diffusiophoresis and methods of particle growth and agglomeration have also been used. The magnitude of the force developed to move a particle toward a collecting surface is influenced markedly by the size and shape of the particle.

DEVELOPING PARTICULATE CONTROL TECHNOLOGY

Present control methods for particulates are least efficient in the size range from 0.2 to 2.5 μm; this range is the most costly to collect and the most energy intensive. Health studies indicate that particles in this size range are also those that penetrate most deeply into the human respiratory system and often become deposited there. This is the principal reason for the EPA change in the ambient air quality standard from total suspended particulate (TSP) concentration to the $PM_{2.5}$ standard (ambient air particles equal to or smaller than 2.5 μm aerodynamic diameter). The new standard will undoubtedly place even more emphasis on the need to collect particulates in this difficult-to-control fine-particle range, and therefore, collection of this size range is most in need of improvements in technology. Improved collection requires the use of a separating force that is independent of gas velocity or of the growth of particles that can be more readily collected. Particle growth can be accomplished through coagulation (agglomeration), chemical reaction, condensation, and electrostatic attraction. Promising separation forces include, the *flux forces* involving diffusiophoresis, thermophoresis, electrophoresis, and Stefan flow. Although particle growth techniques and flux-force collection theoretically can be considered independently, both phenomena are applied in many practical devices.

Thermophoresis may be considered a special form of thermal precipitation. If a hot submicrometer particle is close to a large cold particle or droplet of such relative size that it resembles a wall to the small particle, the kinetic motion of the hot gas molecules opposite the cold particle will bombard and propel the submicrometer particle toward the cold droplet. If cold droplets are introduced into a warmer saturated gas stream, water vapor will condense on the cold droplets reducing the water vapor pressure near its surface and produce a water vapor pressure gradient. The hydrodynamic flow of vapor toward the condensing surface is known as Stefan flow. If the molecular mass of the diffusing vapor is different from the molecular mass of the carrier gas, the motion of small particles is further affected by a density differential. Both of these forces influence the movement of submicrometer particles, and their algebraic sum is known as the force of diffusiophoresis. In condensation, diffusiophoresis tends to move submicrometer particles toward cold droplets; in droplet evaporation, the action is reversed, moving the particles away from the droplet. Therefore, if the bulk gas stream is water saturated before its entry to the scrubber and the scrubbing liquid is cooled below the gas dewpoint, wet scrubbing collection is enhanced. Electrophoresis is the movement of a small particle toward a charged particle. Mathematical description of these forces has been presented (80). Diffusiophoresis has also been discussed (81).

In time, submicrometer particles will coagulate into chains or agglomerates thorough Brownian motion. Increasing turbulence during coagulation will increase the frequency of collisions and coagulation rate. The addition of fans to stir the gas, or gas flow motion through tortuous passages such as those of a packed bed, will aid coagulation. Sonic energy is also known as an aid to the coagulation process. Production of standing waves in the confines of long narrow tubes can bring about concentration and coagulation of aerosols in band zones in the tube. The addition of water and oil mists to the treated aerosol can improve the effectiveness of sonic agglomeration by improving the tendency of the colliding particles to stick together. Sulfuric acid mist (82) and carbon black (83) have been sufficiently agglomerated so that the coagulated product can be collected in a cyclone. Sonic agglomeration has been tested in the past for many metallurgical fumes. However, it has generally been found too power intensive for practical consideration. One problem is the low energy efficiency of transforming other energy sources to sonic energy in available sonic generators. The development of more energy-efficient sonic generators, coupled with improved knowledge of the phenomenon of sonic coagulation, might justify further investigation. The combination of sonic agglomeration and electrostatic precipitation could result in considerable reduction of precipitator size and perhaps capital cost (84).

Particle growth may be brought about (or the charge on particles modified) by the introduction of a gas to react with the particles. Another procedure used is chemical modification of the particle to render it hygroscopic. This can be particularly beneficial for particle growth if the particle continues to absorb moisture to form hydrates. It may also be possible to control the size of aerosols initially formed by chemical means. In studies of chemical reactions producing aerosols (85), it was found that reactions having a large chemical driving force and hence releasing large quantities of energy tended to produce very fine particles (high surface energy); conversely, reactions that occur without release of large amounts of energy tend to produce larger particles (2–6 μm). Therefore, limitation of the chemical driving force may often be beneficial to prevent formation of fine aerosols. If steam or other condensable vapors can be condensed while cooling an aerosol, particle growth will occur through vapor condensation on the existing aerosol nuclei. A given mass of smaller particles will present more surface area for condensation than the same mass of larger particles. Thus the smaller particles selectively grow faster in size than the larger particles. Also, the addition to the system of particles opposite in charge results in mutual attraction and aids in particle growth.

Of these various possibilities, those involving electrostatic forces and condensation have received the most interest. Flux-force condensation scrubbing may be desirable for hot gases needing treatment when there is no attractive alternative available to recover the energy. Sizable amounts of low pressure waste steam are also useful. Reports have appeared on condensation scrubbing in multiple sieve-plate towers (85,86), aspirative condensation (87), and some of the parameters that affect particle growth and collection in a conventional orifice scrubber with and without condensation (88). Fine-particle collection is appreciably improved by scrubbing a hot saturated gas stream with cold water rather than using recirculated hot water. Little collection improvement resulted in cooling below a 50°C dewpoint, but much better collection was achieved when the hot gas was introduced already saturated close to its initial dewpoint (rather than admitting it unsaturated with substantial superheat). A decrease in scrubbing efficiency of an evaporative scrubber was also found. Addition of an adiabatic presaturator for hot gases ahead of a scrubber should be quite beneficial.

ODOR CONTROL

Odor is a subjective preception of the sense of smell. Its study is still in a developmental stage: information including a patent index has been compiled (89), 124 rules of odor preferences have been listed (90), detection and recognition threshold values have been published (91), and odor technology as of 1975 has been assessed (92). Odor control involves any process that gives a more acceptable perception of smell. Whether as a result of dilution, removal of the offending substance, or counteraction (masking).

Odor Measurement

Both static and dynamic measurement techniques exist for odor. The objective is to measure odor intensity by determining the dilution necessary so that the odor is imperceptible or doubtful to a human test panel, that is, to reach the detection threshold, the lowest concentration at which an odor stimulus may be detected. The recognition threshold is a higher value at which the chemical entity is recognized. An odor unit (ou) has been widely defined in terms of 0.0283 m^3 of air at the odor detection threshold. It is a dimensionless unit representing the quantity of odor that when dispersed in 28.3% of odor-free air produces a positive response by 50% of panel members. Odor concentration is the number of cubic meters that one cubic meter of odorous gas will occupy when diluted to the odor threshold. Selection of people to participate in an odor panel should reflect the type of information or measurement required, for example, for evaluation of an alleged neighborhood odor nuisance, the test subjects should be representative of the entire neighborhood. However, threshold determinations may be done with a carefully screened panel of two or three people (93). A general population test panel of 35 people has been described (94). Odor measurement in general has been surveyed (95).

Odor Control Methods

Absorption, adsorption, and incineration are all typical control methods for gaseous odors; odorous particulates are controlled by the usual particulate control methods. However, carrier gas, odorized by particulates, may require gaseous odor-control treatment even after the particulates have been removed. For oxidizable odors, treatment with oxidants such as hydrogen peroxide, ozone, and $KMnO_4$, may sometimes be practical. Catalytic oxidation has also been employed. Odor control as used in rendering plants (96), spent grain dryers (97), pharmaceutical plants (98–99), and cellulose pulping (100) has been reviewed (101–103); some reviews are presented in two symposium volumes from the Air and Waste Management Association (AWMA) (formerly the Air Pollution Control Association, APCA) specialty conferences (104,105). The odor-control performances of activated carbon and permanganate-alumina for reducing odor level of air streams containing olefins, esters, aldehydes, ketones, amines, sulfide. mercaptan, vapor from decomposed crustacean shells, and stale tobacco smoke have been compared (106). Activated carbon produced faster deodorization in all cases. Activated carbon adsorbers have been used to concentrate odors and organic compounds from emission streams, producing fuels suitable for incineration (107). Both air-pollution control and energy recovery were accomplished. A summary of odor-control technology is available (108).

ACKNOWLEDGMENT

Much of this article has been taken from J. Kroschwitz and M. Howe-Grant, eds., *Kirk-Othmer Encyclopedia of Chemical Technology,* 4th ed. (*ECT*4), Vol. 1, John Wiley & Sons, Inc., New York, 1991, and M. Grayson, ed., *Kirk-Othmer Encyclopedia of Chemical Technology,* 3rd ed. (*ECT*3), John Wiley & Sons, Inc., New York, 1978.

BIBLIOGRAPHY

1. *Guiding Principles of State Air Pollution Legislation,* U.S. Department of Health, Education, and Welfare, Washington, D.C., 1965.
2. Sect. 1420, Chapt. 111, *General Laws,* Chapt. 836, *Acts of 1969,* the Commonwealth of Massachusetts, Dept. of Public Health, Division of Environment, Health, Bureau of Air Use Management.
3. G. T. Wolff, in "Air Pollution," *The Encyclopedia of Chemical Technology,* Vol. 1, John Wiley & Sons, Inc., New York, (1978).
4. G. T. Wolff, *EM,* 27–32 (Sept. 1996).
5. G. T. Wolff, *EM,* 26–31 (Oct. 1996).
6. News Focus, *EM,* 35–36 (Oct. 1996).
7. D. H. Slade, ed., *Meteorology and Atomic Energy 1968,* U.S. Atomic Energy Commission, July 1968; available as *TID-24190,* Clearinghouse for Federal Scientific and Technical Information National Bureau of Standards, U.S. Dept. of Commerce, Springfield, Va.
8. D. B. Turner, *Workbook of Atmospheric Dispersion Estimates,* U.S. EPA, OAP, Pub. AP26, Research Triangle Park, N.C., revised 1970, U.S. Dept. Printing Office Stock No. 5503-0015, Washington, D.C.
9. A. D. Busse and J. R. Zimmerman, *User's Guide for the Climatological Dispersion Model,* U.S. EPA Pub. No. EPA-R4-73-024, Research Triangle Park, N.C., Dec. 1973.
10. M. Smith, ed., *Recommended Guide for the Prediction of the Dispersion of Airborne Effluents,* American Society of Mechanical Engineers, New York, 1968.
11. G. A. Briggs, *USAEC Critical Review Series* TID-25075, NTIS, Springfield, Va., 1969.

12. G. A. Briggs and D. B. Turner in *Effective Stack Height: Plume Rise,* US EPA Air Pollution Training Institute Pub. SI:406, 1974.
13. J. E. Carson and H. Moses, *J. APCA* **19,** 862 (1969).
14. H. Moses and M. R. Kraimer, *J. APCA* **22,** 621 (1972).
15. G. A. Briggs, *Plume Rise Predictions, Lectures on Air Pollution and Environmental Impact Analyses,* American Meteorological Society, Boston, Mass., 1975.
16. *Guideline on Air Quality Models,* EPA Research Triangle Park, N.C., 1980.
17. G. A. Schmel, *Atmos. Environ.* **14,** 983–1011 (1980).
18. N. E. Bowne, R. J. Londergan, R. J. Minott, D. R. Murray. *Preliminary Results from the EPRI Plume Model Validation Project—Plains Site,* Report EPRI EA-1788, Electric Power Research Institute, Palo Alto, Calif., 1981.
19. N. E. Bowne, in S. Calvert and H. M. Englund, eds., *Handbook of Air Pollution and Technology,* John Wiley & Sons, New York, 1984, pp. 859–891.
20. T. W. Devitt in S. Calvert and H. M. Englunds, eds., *Handbook of Air Pollution and Technology,* John Wiley & Sons, Inc., New York, 1984, pp. 375–417.
21. *Compilation of Air Pollutant Emission Factors.*Publication No. AP-42, 3rd ed., EPA, Research Triangle Park, N.C., Apr. 1980 (or latest).
22. N. J. Weinstein, in S. Calvert and H. M. Englunds, eds., *Handbook of Air Pollution and Technology,* John Wiley & Sons, Inc., New York, 1984, pp. 419–434.
23. A. J. Klee, in S. Calvert and H. M. Englunds, eds., *Handbook of Air Pollution and Technology,* John Wiley & Sons, Inc., New York, 1984, pp. 513–550.
24. L. Short, in Ref. 19, pp. 673–695.
25. *Code of Federal Regulations* 40 (CFR 40). *Fed. Reg.,* C-50–99.
26. W. W. Lund and R. Starkey, *J. Air Waste Manage Assn.* **40**(6), 896–897 (1990).
27. T. A. *Gosink. Environ. Sci. Technol.* **9,** 630–634 (1975).
28. S. R. Heller, J. M. McGuire, and W. L. Budde, *Environ. Sci. Technol.* **9,** 210–213 (1975).
29. H. H. Westberg, in S. Calvert and H. M. Englunds, eds., *Handbook of Air Pollution and Technology,* John Wiley & Sons, Inc., New York, 1984 771–783.
30. *Fed. Reg.* **40,** 46250 (1975).
31. *Fed. Reg.* **52,** 24634–24750 (1987); *Fed. Reg.* **54,** 4121S–41232 (1989).
32. 40 CFR, Part 50, Appendix L, "Reference Method for the Determination of Fine Particle Matter as $PM_{2.5}$ in the Atmosphere" (Draft).
33. T. M. Merrifield, *EM,* 35–36, (1997).
34. D. S. Ensor and M. J. Pilot, *J. APCA* **21,** 496–501 (1971).
35. B. B. Crocker, *Chem. Eng. Prog.* **71,** 83–89 (1975).
36. Ref. 21, part 60 as compiled in R. A. Corbett, *Standard Handbook of Environmental Engineering,* McGraw-Hill, New York, 1989.
37. Appendix to *Code of Federal Regulations,* CFR 40, Part 60, U.S. Government Printing Office, Washington, D.C.
38. L. E. Sparks, in S. Calvert and H. M. Englunds, eds., *Handbook of Air Pollution and Technology,* John Wiley & Sons, Inc., New York, 1984 pp. 800–818.
39. B. W. Loo, J. M. Jaklevic, and F. S. Goulding, in B. Y. H. Liu, ed., *Fine Particles—Aerosol Generation, Measurement, Sampling and Analysis,* Academic Press, New York, 1976, pp. 311–350.
40. Report of 18th Government Affairs Seminar, *J. Air and Waste Management Assn.* **40**(6), 855–860 (1990).
41. *Proceedings of the 18th Air and Waste Management Assn. Govt. Affairs Seminar,* Air and Waste Management Assn., Pittsburgh, Pa., 1990.
42. R. H. Perry, ed., *Engineering Manual,* 3rd ed., McGraw-Hill, New York, 1976.
43. American Conference of Governmental Industrial Hygienists, Committee on Industrial Ventilation, *Industrial Ventilation,* 15th ed., Sect. 4, Lansing, Mich., 1978.
44. R. Jorgensen, *Fan Engineering,* 7th ed., Buffalo Forge Co., Buffalo, N.Y., 1970, pp. 471–480.
45. W. E. L. Hemeon, *Plant and Process Ventilation,* Industrial Press, New York, 1954.
46. J. M. Dalla Valle, *Exhaust Hoods,* Industrial Press, New York, 1952.
47. J. L. Alden, *Design of Industrial Exhaust Systems for Dust and Fume Removal,* 3rd ed., Industrial Press, New York, 1959.
48. J. A. Danielson, ed., *Air Pollution Engineering Manual.* Pub. No. 999-AP-40, U.S. Dept. of Health, Education, and Welfare, Cincinnati, Ohio, 1973, Chapt. 3.
49. H. D. Goodfellow, in J. A. Buonicore and T. Davis, eds., *Air Pollution Engineering Manual,* Von Nostrand Reinhold, New York, 1992, pp. 155–206.
50. B. B. Crocker, in *Proceedings, Control of Specific (Toxic) Pollutants* Air Pollution Control Assn., Pittsburgh, Pa, 1979, pp. 415–433.
51. B. B. Crocker. *Chem. Eng. Prog.* **64,** 79 (1968).
52. EPA*Handbook, Control Technologies for Hazardous Air Pollutants,* EPA/625/6-91/014, Cincinnati, Ohio, 1991.
53. E. C. Moretti and N. Mukhopadhyay, *Chem. Eng. Prog.,* **89**(7), 20–26, (1993).
54. B. B. Crocker, in *Proceedings: Control of Specific Toxic Pollutants,* Air Pollution Control Assn., Pittsburgh, Pa., 1979, pp. 360–376.
55. M. J. Ruhl, *Chem. Eng. Prog.,* **89**(7), 37–41 (1993).
56. D. Bienstock, J. H. Field, S. Katell, and K. D. Plants, *J. APCA* **15,** 459 (1965).
57. D. Bienstock, J. H. Field, and J. G. Myers, *J. Eng. Power* **86**(3), 353 (1964).
58. B. B. Crocker and K. B. Schnelle, Jr., in S. Calvert and H. M. Englunds, eds., *Handbook of Air Pollution and Technology,* John Wiley & Sons, Inc., New York, 1984, pp. 135–192.
59. D. M. Ruthven, in Ref. 3, pp. 493–528.
60. J. D. Sherman and C. M. Yon, in Ref. 3, pp. 529–573.
61. M. Suzuki, *Adsorption Engineering,* Kodansba-Elsevier, Tokyo, 1990.
62. A. E. Rodrigues, M. D. LeVan, and D. Tondeur, *Adsorption, Science and Technology,* NATO ASI E158, Kluwer, Amsterdam, 1989.
63. R. T. Yang, *Gas Separation by Adsorption Processes,* Butterworths, Stoneham, Mass., 1987.
64. P. Wankat, *Large Scale Adsorption and Chromatography,* CRC Press, Boca Raton, Fla., 1986.
65. T. Vermeulen, M. D. LeVan, N. K. Hiester, and G. Klein, in R. H. Perry and D. Green, eds., *Perry's Chemical Engineers Handbook,* 6th ed., McGraw-Hill, New York, 1984, Sec. 16.
66. D. M. Ruthven, *Principles of Adsorption and Adsorption Processes,* Wiley-Interscience, New York, 1984.
67. A. Turk, in A. C. Stern, ed., *Air Pollution,* 3rd ed., Vol. 5, Academic Press, New York, 1977, pp. 329–363.
68. *Ibid.,* pp. 337–339.
69. R. J. Buonicore, in A. J. Buonicore and T. Davis, eds., *Air Pollution Engineering Manual,* Van Nostrand Reinhold, New York, 1992, pp. 31–52.

70. J. J. Spivey, *Environ. Progress* 7(1), 31–40 (1988).
71. S. M. Hall, *J. Air Waste Manage Assn.* **40**(3), 404–407 (1990).
72. A. J. Buonicore, A. J. Buonicore and T. Davis, eds., *Air Pollution Engineering Manual,* Van Nostrand Reinhold, New York, 1992, pp. 52–58.
73. H. Bohn, *Chem. Eng. Prog.,* **88**(4), 34–40 (1992).
74. A. Rozich, *Env. Eng. World,* **1**(4), 32–34 (1995).
75. J. L. Humphrey, *Chem. Eng. Prog.,* **91**(10), 38 (1996).
76. V. S. Katari, W. M. Vatavuk, and A. H. Wehe, Part I, *J. APCA* **37**(1), 91 (1987); Part II, *J. APCA* **37**(2), 19S–201 (1987).
77. M. Kosusko and C. M. Nunez, *J. Air Waste Manage Assn.* **40**(2), 254–255 (1990).
78. M. A. Palazzolo and B. A. Tichenor, *Environ. Progress* **6,** 172–176 (1987).
79. E. N. Ruddy and L. A. Carroll, *Chem. Eng. Progress* **89**(7), 28–35 (1993).
80. R. V. Kleinschmidt and A. W. Anthony in L. C. McCabe, ed., *U.S. Technical Conference on Air Pollution,* McGraw-Hill, Inc., New York, 1952, p. 310.
81. L. E. Sparks and M. J. Pilat, *Atmos. Environ.* **4,** 651 (1970).
82. H. W. Danser, Jr., *Chem. Eng.* **57,** 158 (1950).
83. C. A. Stokes, *Chem. Eng. Prog.* **46,** 423 (1950).
84. E. P. Mednikov, *Acoustic Coagulation and Precipitation of Aerosols,* USSR Academy of Science, Moscow, 1963, translated by C. V. Larrick, Consultants Bureau, 1965.
85. G. R. Gillespie and H. F. Johnstone, *Chem. Eng. Prog.* **51,** 74F (1955).
86. S. Calvert and N. C. Jhaveri, *J. APCA* **24,** 946 (1974); S. Calvert, J. Goldshmid, D. Leith, and N. Jhaveri, *Feasibility of Flux Force/Condensation Scrubbing for Fine Particle Collection,* NTIS Pub. PB-227307, Oct. 1973.
87. S. R. Rich and T. G. Pantazelos, *J. APCA* **24,** 952 (1974).
88. K. T. Semrau and C. L. Witham, Preprint 75-30.1, *68th APCA Annual Meeting,* Boston, Mass., June 15–20, 1975.
89. J. P. Cox, *Odor Control and Olfaction,* Pollution Sciences, Lynden, Wash., 1975.
90. R. W. Moncrieff, *Odour Preferences,* John Wiley & Sons, Inc., New York, 1966.
91. W. H. Stahl, ed., *Compilation of Odor and Taste Threshold Values Data,* American Society for Testing and Materials Data Series 48, Philadelphia, Pa., 1973.
92. P. N. Cheremisinoff and R. A. Young, eds., *Industrial Odor Technology Assessment,* Ann Arbor Science, Ann Arbor, Mich., 1975.
93. J. Wittes and A. Turk, *Correlation of Subjective-Objective Methods in the Study of Odors and Taste,* American Society for Testing and Materials Special Technical Publication 440, Philadelphia, Pa., 1968, pp. 49–70.
94. F. V. Wilby, *J. APCA* **19,** 96 (1969).
95. H. Van Langenhove and N. Schamp, in P. N. Cheremisinoff, ed., *Encyclopedia of Environmental Control Technology,* Vol. 2, Gulf Publishing, Houston, Tex., 1989, pp. 935–965.
96. R. M. Bethea and co-workers, *Environ. Sci. Technol.* **7,** 504 (1973).
97. M. W. First and co-workers, *J. APCA* **24,** 653 (1974).
98. D. E. Quane, *Chem. Eng. Prog.* **70,** 51 (1974).
99. D. J. Eisenfelder and J. W. Dolen, *Chem. Eng. Prog.* **70,** 48 (1974).
100. J. E. Paul, *J. APCA* **25,** 158 (1975).
101. J. E. Yocom and R. A. Duffee, *Chem. Eng.* **77**(13), 160 (1970).
102. M. Beltran, *Chem. Eng. Prog.* **70,** 57 (1974).
103. R. M. Bethea, *Engineering Analysis and Odor Control.* P. N. Cheremisinoff and R. A. Young, eds., *Industrial Odor Technology Assessment,* Ann Arbor Science, Ann Arbor, Mich., 1975, pp. 203–214.
104. *Proceedings, State of the Art of Odor Control Technology Specialty Conference,* Air Pollution Control Association, Pittsburgh, Pa., 1974.
105. *Proceedings, State of the Art of Odor Control Technology Specialty Conference,* Air Pollution Control Association, Pittsburgh, Pa., 1977.
106. A. Turk, S. Mehlman, and E. Levine, *Atmos. Environ.* **7,** 1139 (1973).
107. W. D. Lovett and F. T. Cunniff, *Chem. Eng. Prog.* **70,** 43 (1974).
108. W. H. Prokop, in A. J. Buonicore and T. Davis, eds. *Air Pollution Engineering Manual,* Van Nostrand Reinhold, New York, 1992, pp. 147–154.

General References

R. G. Bond and C. P. Straub, eds., *Handbook of Environmental Control,* Vol. 1, CRC Press, Cleveland, Ohio, 1972.

A. J. Buonicore and W. T. Davis, eds., *Air Pollution Engineering Manual,* Van Nostrand Reinhold, New York, 1992.

S. Calvert and H. M. Englund, eds., *Handbook of Air Pollution Technology,* John Wiley & Sons, Inc., New York, 1984.

P. N. Cheremisinoff, ed., *Encyclopedia of Environmental Control Technology,* Vol. 2, Gulf Publishing, Houston, Tex., 1989.

P. N. Cheremisinoff and R. A. Young, *Air Pollution Control and Design Handbook,* Parts 1 and 2, Marcel Dekker, New York, 1977.

R. A. Corbett, *Standard Handbook of Environmental Engineering,* McGraw-Hill, New York, 1989.

C. N. Davies, ed., *Aerosol Science,* Academic Press, New York, 1966.

H. E. Hesketh, *Air Pollution Control,* Ann Arbor Science, Ann Arbor, Mich., 1979.

L. H. Keith, *Compilation of EPA's Sampling and Analysis Methods,* Lewis Publishers, CRC Press, Boca Raton, Fla., 1991.

W. Licht, *Air Pollution Control Engineering: Basic Calculations for Particulate Collection,* Marcel Dekker, New York, 1980.

B. Y. H. Liu, ed., *Fine Particles—Aerosol Generation, Measurement, Sampling and Analysis,* Academic Press, New York, 1976.

J. J. McKetta, *Unit Operations Handbook,* Vols. 1 and 2, Marcel Dekker, New York, 1992.

H. C. Perkins, *Air Pollution,* McGraw-Hill, New York, 1974.

R. H. Perry and D. Green, eds., *Perry's Chemical Engineers' Handbook,* 6th ed., McGraw-Hill, New York, 1984.

W. Ruch, ed., *Chemical Detection of Gaseous Pollutants,* Ann Arbor Science, Ann Arbor, Mich., 1966.

J. H. Seinfeld. *Air Pollution: Physical and Chemical Fundamentals,* McGraw-Hill, New York, 1975.

A. C. Stern, ed., *Air Pollution,* 3rd ed., Vols. 1–5, Academic Press, Inc., New York, Vols. 1–3, 1976; Vols. 4–5, 1977.

W. Strauss, *Industrial Gas Cleaning,* 2nd ed., Pergamon Press, Oxford, U.K., 1975.

L. Theodore and A. J. Buonicore, *Air Pollution Control Equipment—Selection. Design, Operation and Maintenance,* Prentice-Hall, Englewood Cliffs, N.J., 1982.

L. K. Wang and N. C. Pereira, eds., *Handbook of Environmental Engineering,* Vol. 1—*Air and Noise Pollution Control,* The Humana Press, Clifton, N.J., 1979.

K. Wark and C. F. Warner. *Air Pollution: Its Origin and Control,* 2nd ed., Harper & Row, New York, 1981.

P. O. Warner, *Analysis of Air Pollutants,* John Wiley & Sons, Inc., New York, 1976.

AIR POLLUTION CONTROL, CATALYTIC

RONALD M. HECK
ROBERT J. FARRAUTO
Engelhard Corporation
Iselin, New Jersey

GENERAL CATALYSIS

Concerns for the preservation of the environment can now be heard all over the world, independent of political philosophies. Pollution has no boundaries and thus must be treated at its point of origin. To meet these challenges countries are establishing and enforcing regulations to minimize the emission of specific air pollutants such as carbon monoxide, hydrocarbons, ozone, nitric oxides, and gas-borne particulates. Gasoline- and diesel-fueled vehicles are a major source of these toxins; however, stationary power plants, chemical and petroleum industries, food processes, etc are contributors as well. Small engines, such as lawn mowers, snow blowers, weed whackers, and chain saws, are also coming under careful scrutiny as sources of pollution. The primary solution for controlling gas-borne emissions is the catalytic converter, which provides in many cases an inexpensive, durable, and somewhat passive solution to converting pollutants to harmless products.

Catalysis and Catalysts

Basic Concepts. A catalyst accelerates the rate (molecules converted per unit time) of a chemical reaction, but does not undergo any permanent change. Reactants (or just one reactant) adsorb on to active catalytic sites and are converted to products at a lower temperature and more quickly than those would convert in the absence of a catalyst. An example of the oxidation of propane using a Pd catalyst follows:

$$C_3H_8 + 5O_2 \xrightarrow{\text{Catalyst(Pd)}} 3\ CO_2 + 4\ H_2O \quad (1)$$

The catalytic reaction occurs at <200°C compared to 500°C in the absence of the catalyst. Furthermore, the mass throughput is so much greater for the catalytic reaction that smaller reactors can be used. These properties make catalysts ideal for environmental applications.

A catalyst lowers the energy barrier or activation energy in the transformation of reactants to products, which increases the net rate of reaction. The difference between the energy states of the reactants and products is the exothermic heat of reaction.

It is of great importance to recognize that both the net enthalpy and net free energy are unaffected by the presence of the catalyst. A catalyst enhances the kinetics of reaching equilibrium. Thus the reaction enthalpy ΔH and reaction free energy ΔG (and, consequently, the equilibrium constant K) are not altered by the presence of a catalyst. The rate of reaction is inversely proportional to the activation energy. The activation energy is a measure of the slowest step for the entire process of converting reactants to products. This is called the rate-limiting step. The overall rate can never be greater than the slowest step (1).

The catalyst directs the reactants to desirable products, ie, selectivity, by preferentially lowering the activation energy for a particular step in the reaction sequence, and increases the rate at which this step proceeds. For catalytic air pollution control reactions, it is obvious that carbon-containing reactants must be oxidized completely to carbon dioxide instead of to partially oxidized products such as aldehydes. The nature of the catalyst and reaction conditions affect the product distribution. In air pollution control, all catalytic processes use solid heterogeneous catalysts through which gaseous reactants pass.

Dispersed Catalyst Model. The number of reactant molecules converted to products in a given time is directly related to the number of catalytic sites available to the reactants. It is, therefore, essential to maximize the number of active sites by dispersing the catalytic components onto the surface of a substrate or carrier. Maximizing the surface area of the catalytic components, such as Pt, Rh, Pd, and V_2O_5, increases the number of sites on which chemisorption and catalytic reaction can occur.

The active catalytic components are dispersed on high surface area porous oxide particles (carriers) such as Al_2O_3, SiO_2, TiO_2, and $SiO_2-Al_2O_3$ combinations (zeolites). The carriers are themselves not usually catalytically active but do play a major role in maintaining the overall stability and durability of the finished catalyst.

High surface area gamma alumina (γ-Al_2O_3) is the most commonly used carrier in environmental applications, and it will be used as a model to describe a catalytic metal (Pd) dispersed heterogeneous catalyst.

Figure 1 shows a few select pores of a high surface area Al_2O_3, which has pores 30, 60, and 120 Å into which Pd has been deposited by solution impregnation. The Pd particles, or crystallites, are represented as dots. The internal surface of the Al_2O_3 is rich in surface OH groups (not shown), depending on the type of Al_2O_3 and its thermal history. These OH species, which cover the entire surface and are part of the walls of each pore, represent sites on which one can chemically or physically bond a catalytic substance. The physical surface area of the Al_2O_3 is the sum of all internal areas of the oxide from all the walls of each and every pore. It is on these internal walls and at the OH sites that the catalytic components are bound. The catalytic surface area is the sum of all the areas of the active catalytic components, in this example Pd. The smaller the individual size of the crystallites of the Pd, the higher the catalytic surface area. As a rough approximation, one assumes the higher the catalytic surface area, the higher the rate of reaction for a process controlled by kinetics and not mass transfer.

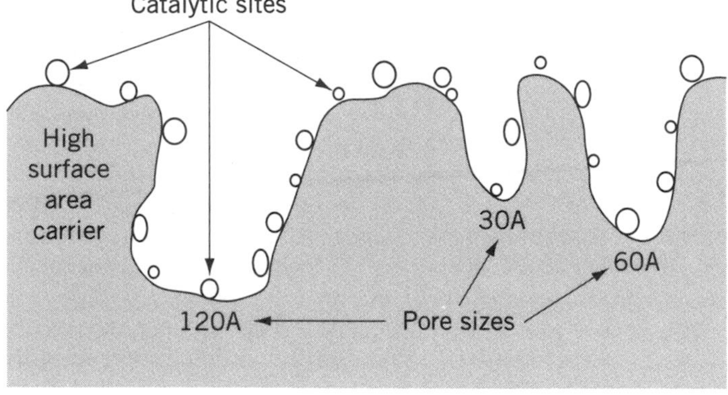

Figure 1. Model of a supported catalyst dispersed on a high surface area carrier.

The tiny Pd-containing particles are dispersed throughout the porous Al_2O_3 carrier network and generate a high Pd surface area. This procedure maximizes the catalytic area but does introduce other resistances for the reactants and products. They must now diffuse through the porous network to arrive at the Pd sites. The rate of pore diffusion is influenced by the pore size of the carrier and the molecular dimensions of the diffusing molecule (1).

The Steps in Heterogeneous Catalysis. To maximize reaction rates it is essential to ensure accessibility of all reactants to the active catalytic sites dispersed within the internal pore network of the carrier. Consider the oxidation of propane, with the reactants flowing through a bed of a heterogeneous catalyst. As a simple model, it can be assumed that the following physical and chemical steps must occur for the propane to convert catalytically to products.

1. The propane and oxygen must diffuse through a stagnant thin layer of gas in close contact with the outer surface of the Pd/Al_2O_3. Bulk molecular diffusion rates have little sensitivities to temperature and typically have activation energies, $E \sim 2$ kcal/mol (8.3 kJ/mol).
2. The Pd is dispersed primarily within the Al_2O_3 pore structure so the majority of the reactant molecules must diffuse through the porous network to the Pd. This pore diffusion process has some sensitivity to temperature. Its activation energy is, approximately half that of a chemical reaction, or 6–10 kcal/mol (25–42 kJ/mol).
3. The reactant (or reactants) must chemisorb onto the Pd. The kinetics generally follow exponential dependence on temperature ie, $\exp(-E/RT)$, where E is the activation energy, which for chemisorption is typically $>$ 10 kcal/mol (42 kJ/mol).
4. Catalytically controlled surface reactions involve chemisorbed reactants that convert to adsorbed products. Kinetics also follow exponential dependence on temperature. Activation energies are typically $>$10 kcal/mol (42 kJ/mol).
5. Products must desorb from the Pd site also following exponential kinetics. Activation energies are typically $>$10 kcal/mol (42 kJ/mol).
6. Products must then diffuse through the porous network with activation energies and kinetics similar to step 2.
7. The products must diffuse through the stagnant layer and finally into the bulk gas. Activation energies are also similar to step 1, or $<$2 kcal/mol (8.3 kJ/mol).

Steps 1 and 7 represent bulk mass transfer, which is a function of the specific molecules, the dynamics of the flow conditions, and the geometric surface area (external) of the catalyst. Pore diffusion (steps 2 and 6) depends primarily on the size and shape of both the pore and the diffusing reactant and product. Steps 3, 4, and 5 are related to the chemical interactions at the catalytic site between reactants and products.

Any of the steps listed above can be rate limiting and control the overall rate of reaction. Chemical steps, being exponentially dependent on temperature and having higher activation energies, are generally not difficult to determine when they are rate limiting (1). Take, for example, the conversion of gaseous reactant propane in equation 1 to its completely combusted products, CO_2 and H_2O, as the temperature is increased (Fig. 2).

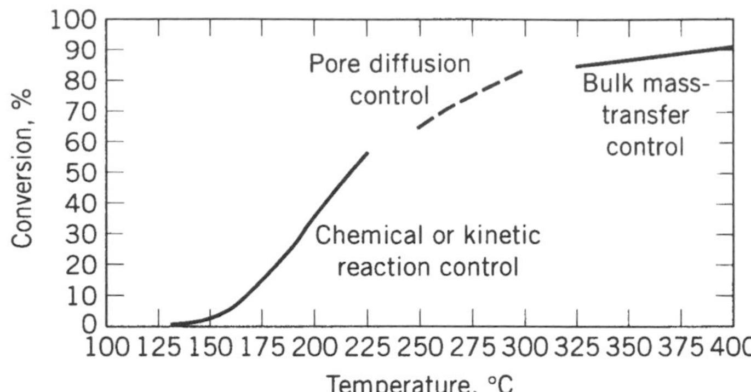

Figure 2. Model of a typical conversion–temperature profile.

At low temperatures the chemical reaction rate steps are slow relative to diffusion. As the temperature is increased steps with higher activation energies and exponential dependence increase the fastest, and control of the overall rate will shift from chemical to pore diffusion. Finally, at higher temperatures both the chemical and pore diffusion rates become sufficiently fast that bulk mass transfer, having a small relative temperature dependence, becomes rate limiting.

The slope of the conversion–temperature curve gives a qualitative picture of the rate-controlling steps (1). The steeply rising lower part of the curve is indicative of chemical or kinetic control (either Step 3, 4, or 5). The relatively flat temperature-insensitive portion reflects bulk mass transfer control (either step 1 or 7), while the intermediate portion is characteristic of pore diffusion control (either step 2 or 7).

For a simple first-order reaction, the rate is proportional to the concentration of the reactant C and its rate constant k:

$$dC/dt = -kC \qquad (2)$$

The rate constant can be described by the Arrhenius equation, which relates k to the activation energy E:

$$k = k_o \exp(-E/RT) \qquad (3)$$

The preexponential factor k_o is related to a number of mechanistic terms but is also proportional to the number of active sites on the catalyst; T and R are the absolute temperature and universal gas constant, respectively. The natural log of equation 3 generates a convenient equation that allows the determination of a series of straight lines whose slopes are directly related to the activation energy of the rate-controlling step (Fig. 3):

$$\ln k = \ln k_o - E/R(1/T) \qquad (4)$$

Figure 3 shows three distinctly different slopes, each related to a possible rate-limiting step. The largest slope is for chemical, intermediate for pore diffusion, and the smallest for bulk mass transfer (1).

When the reaction is controlled by one of the chemical steps, the diffusion of reactants to the active sites is fast. The concentration of reactants within the catalyzed carrier is essentially uniform. With pore diffusion control the concentration of reactants decreases from the outer periphery of the catalytic surface to the center. Finally, with mass transfer control, the concentration of reactants is zero immediately at the outside surface of the catalyst.

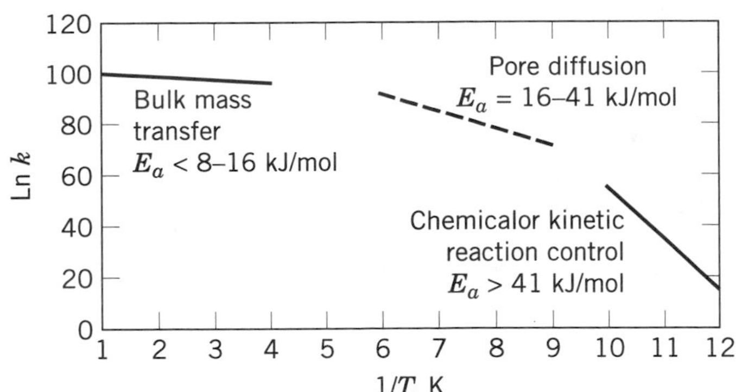

Figure 3. Arrhenius–activation energy plot. To convert kJ/mol to kcal/mol, divide by 4.184.

Catalyst Characterization

The characterization of a heterogeneous catalyst is the quantitative measure of its physical and chemical properties assumed to be responsible for its performance in a given reaction. These measurements have value in the preparation and optimization of a catalyst and even more important in elucidating mechanisms of deactivation and subsequent catalyst design to minimize such deactivation. Physical properties such as pore size, surface area, and morphology of the carrier as well as the geometry and strength of the monolithic catalyst support must be well defined for the given end use application. Similarly, the composition, structure, and nature of the carrier and the active catalytic components and their changes during the catalysis process are critical goals in characterization.

During the conversion process, a number of physical and chemical properties must be well defined to produce an optimized catalytic system. The most important for environmental monolithic catalysts are as follows (1):

- Nitrogen Brunauer-Emmett-Teller (BET) surface area for the internal surface area of carriers.
- Mercury penetration for pore size distribution of carriers.
- Laser methods for particle size distribution of carriers.
- Classical and instrumental analysis for chemical composition.
- X-ray diffraction (XRD) for structural and crystallite size analysis of carriers, catalytic metals (oxides), and substrate supports.
- Selective chemisorption for metal dispersion.
- Scanning electron microscopy for morphology examination of carriers, catalytic components, and substrates.
- Electron microprobe for catalytic metal and contaminant profiles within the washcoat.
- X-ray photoelectron (XPS) spectroscopy for surface analysis of a catalyst.
- Thermal analysis for measuring small weight changes as a function of gaseous environment and temperature.

Catalyst Deactivation

General Aspects. One of the major sources of catalyst deactivation occurs due to high temperature exposure. This is especially true in the automobile catalytic converter where temperatures can reach 1000°C. Other sources of deactivation such as poisoning or masking (fouling) can occur due to exhaust or process contaminants adsorbing onto or blocking active catalytic sites. It is essential to understand the modes of poisoning before developing resistant materials and methods of regeneration when possible. Since environmental catalytic applications use washcoated monoliths, deactivation by attrition and erosion of the washcoat from the support must also be considered (1).

A convenient laboratory tool for studying deactivation is the model reaction. Here one takes a representative model compound and conditions under which the catalyst is expected to function and develops a conversion–temperature curve, similar to Figure 2. Shifts in the profile give great insight into the mechanism of deactivation.

Poisoning. A common cause of catalyst deactivation results from contaminants present in the feedstock or in the process equipment depositing onto the catalyst surface. There are two basic mechanisms by which poisoning occurs: *selective poisoning,* in which a chemical directly reacts with the active site or the carrier, rendering it less or completely inactive; and *nonselective poisoning,* such as deposition of fouling agents onto or into the catalyst carrier, masking sites and pores resulting in a loss in performance due to a decrease in accessibility of reactants to active sites.

Attrition. Loss of the catalytic washcoat due to attrition or erosion is a serious source of irreversible deactivation. Washcoat loss from a monolithic environmental catalyst could likely be a problem, since the gases, often containing particulates, are flowing at high linear velocities. Furthermore, due to expansion differences between the washcoat and the monolith, process and temperature changes can lead to attrition. This has been a problem for washcoated metal substrates. Pretreatment techniques have been developed by which the surface of the metal monolith is modified to produce an adherent washcoat. Occasionally, binders such as SiO_2 can be added to the washcoat formulation to improve the chemical bond between the washcoat and the SiO_2-containing substrate to ensure an adherent washcoat.

Washcoat loss is observed by preparing a cross-section of the honeycomb catalyst and scanning the wall of the channel with either an optical or a scanning electron microscope (1).

HYDROCARBON AND CARBON MONOXIDE

Commercial manufacturing and service industries are the major sources of emissions of hydrocarbons (HC) and carbon monoxide (CO). In these industries either in the raw materials, the intermediates, or the finished product, organic materials are present as chemicals, solvents, release agents, coatings, decomposition products, pigments, etc that end up in stack gases. The CO is usually a by-product, through either partial oxidation or combustion. When the gaseous effluent contains low concentrations of HC and CO (ie, <5000 vppm), catalytic abatement is usually the preferred method. Examples of commercial process having HC and CO emissions are (1):

- Chemical plants
- Petroleum refineries
- Pharmaceutical plants

- Automobile manufacturers
- Airplane manufacturers
- Food processors
- Fiber manufacturers
- Textiles manufacturers
- Printing plants
- Can coating plants
- Wire enameling plants
- Electronic component plants
- Painting facilities
- Wood stoves
- Power generation
- Restaurants

Because the manufacturing processes vary for all these stationary sources of pollution, catalysts must be developed to react to a broad range of hydrocarbons, which can be present either as vapors or as aerosols.

The basic exothermic catalytic oxidation reaction of HC and CO is shown below:

$$HC + air(O_2) \xrightarrow{catalyst} CO_2 + H_2O \quad (5)$$

$$CO + air(O_2) \xrightarrow{catalyst} CO_2 + H_2O \quad (6)$$

Actual operating conditions are variable and depend on the organic molecule, contaminants, water vapor, etc. The majority of the catalytic abatement applications can be classified as hydrocarbons, carbon monoxide, halogenated hydrocarbon, and hydrocarbon aerosols. Hydrocarbon abatement is the classic application of catalysts for abatement of volatile organic compounds (VOCs) and organic chemicals.

NITROGEN OXIDES

Nitrogen oxides—NO, NO_2 (or NO_x)—are formed in the high temperature combustion of fuels in utility boilers, industrial boilers, gas turbines, and stationary engines and in the chemical industry from production of nitric acid. N_2O is found in the exhaust emissions in chemical plant operations associated synthetic nylon fiber production. NO and NO_2 participate in the formation of ozone (smog) through photochemical reaction with hydrocarbons and are components of acid rain along with SO_x, since when mixed with water vapor in the clouds they form nitric and sulfuric acid. Nitrous oxides (N_2O), although much smaller in total yearly mass emissions, are stable and contribute to global warming. Advances in combustion technology have substantially lowered the NO and NO_2 emissions; however, catalytic after-treatment is required to meet federal regulations (2).

AUTOMOTIVE CATALYST

Emissions and Regulations

It is estimated that in 1995 there were 195 million light-duty vehicles on the road in the United States (>400 million worldwide). The majority of these vehicles (automobiles and small trucks) use a spark-ignited gasoline engine to provide power and have become the most common form of transportation. Gasoline is a mixture of paraffins and aromatic hydrocarbons that combust in air at a high efficiency. The simplified reaction is

$$\text{Gasoline} + O_2(\text{in air}) \rightarrow CO_2 + H_2O + \text{heat} \quad (7)$$

Since the efficiency of the combustion is <100%, there are incomplete combustion products. A typical exhaust gas composition from a automobile consists of:

- Carbon monoxide 1–2 vol %
- Unburned hydrocarbons 500–1000 vppm
- Nitrogen oxides 100–3000 vppm
- Hydrogen 0.3–0.6 vol %
- Water 10–15 vol %
- Carbon dioxide 10–15 vol %

HC, CO, and NO_x are the major exhaust pollutants. HC and CO occur because the combustion efficiency is <100%; while NO_x is formed during the high temperatures (>1500°C) of the combustion process, resulting in thermal fixation of the nitrogen in the air, which forms NO_x (3). Typical distribution of total HC out of the engine are as follows (4).

Hydrocarbon Group	Mass, %
Aromatics	46.2
Olefins	27.5
Paraffins (C_5^+)	11.5
Methane	7.0
Paraffins (C_2–C_4)	4.5
Other	3.3

The predominant influence on the quantity of pollutants is the air:fuel ratio in the combustion cylinder. Figure 4 shows the engine emissions as a function of the air:fuel ratio (5).

Since large amounts of air are processed in this enormous vehicle population, a significant amount of HC, CO, and NO_x

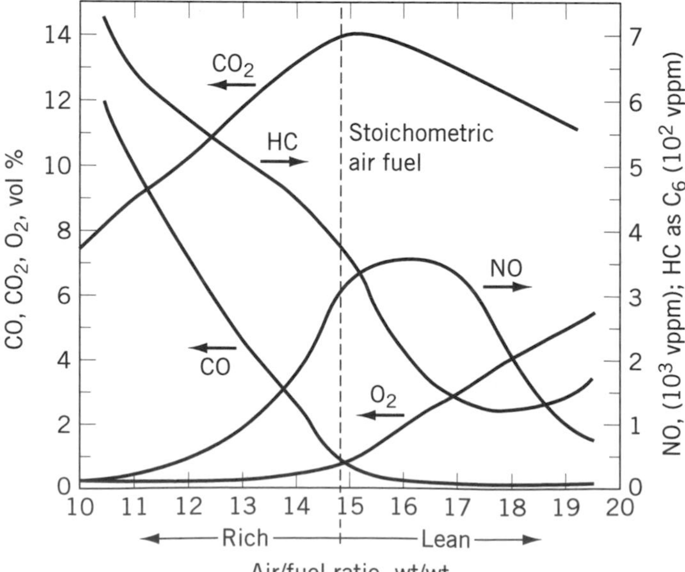

Figure 4. Engine air/fuel ratio determines vehicle emissions (51).

is emitted to the atmosphere. The consequences of these emissions has been well documented (6); but briefly, CO is a direct poison to humans, while HC and NO_x undergo photochemical reactions in sunlight that lead to the generation of smog and ozone. Through 1994 statistics accumulated by a major catalyst supplier showed that for the 150 million catalysts it sold an estimated 14 million t of HC, 180 million t of CO, and 17 million t of NO_x have been abated and prevented from entering the atmosphere (7).

Automobile emission controls began in the United States in 1970 with the passage of the Clean Air Act, which required a 90% reduction (based on raw emissions from a 1970 vehicle) in CO and HC emissions by 1976, and further reductions in NO_x were to follow at a later date. The Clean Air Act also required the 90% reductions be maintained for at least 80,465 km (50,000 mi).

To implement measurement of vehicle tailpipe emissions, the Environmental Protection Agency (EPA) established the Federal Test Procedure (FTP), which simulated on-the-road average driving conditions. Overall, the total FTP cycle generated 21–26 km (13–16 g/mi) of HC, 83–90 g/mi of CO, and 3.5–7.0 g/mi of NO_x on uncontrolled vehicles (8). Reductions of 90% for CO, nonmethane hydrocarbons (NMHC) and NO_x were required by 1976 for all gasoline-powered cars sold in the United States.

Recent amendments to the Clean Air Act have set up more stringent requirements for automotive emissions (9). The catalysts will be required to last 160,930 km (100,000 mi) for new automobiles after 1996. These amendments reduce tailpipe emissions as follows:

	Emissions, g/mi		
Year	HC	CO	NO_x
1991	0.41	3.4	1.0
2004	0.125	1.7	0.2

California has set more stringent regulations and NMHC emissions; must be reduced to 0.075 g/mi by 2000 for 96% of all passenger cars. By 2003, 10% of these must have emissions ≤0.04 g/mi, and 10% must emit no NMHCs at all.

A summary of the California emission standards for passenger cars, including low emission vehicles (LEVs), transitional LEVs (TLEVs) ultra-LEVs (ULEVs), and zero-emission vehicles (ZEVs) is as follows:

	Emissions, g/mi			
Category	Nonmethane Organics	CO	NO_x	Aldehydes
TLEV	0.125	3.4	0.4	0.015
LEV	0.075	3.4	0.2	0.015
ULEV	0.040	1.7	0.2	0.008
ZEV	0	0	0	0

Presently, 49 states are considering adopting the LEV regulations. The ZEV vehicles will probably be battery-operated electric vehicles. At the time of this writing, a proposed amendment is being considered to the California regulations, which involves inclusion of an equivalent zero emissions vehicle (EZEV), which would have the following emissions requirements: nonmethane organics, 0.04 g/mi; CO, 0.17 g/mi; and NO_x, 0.02 g/mi.

Automobile manufacturers have explored a wide variety of engine modifications and technologies to meet the Clean Air Act requirements. Catalysis has proven to be the most effective passive system. As the emission requirements became more stringent, the engine-control devices and combustion modifications have proven to be synergistic with catalyst technology; and today the emission control unit and operating strategy for each vehicle incorporates catalyst in the engineering design. The major worldwide suppliers of automotive catalysts are Engelhard Corporation (Iselin, New Jersey), Johnson Matthey (London), Degussa (Hanau, Germany), Nippon Shokubai (Osaka, Japan), and Allied Signal (Tulsa, Oklahoma).

The Catalytic Reactions for Pollution Abatement

The catalyst promotes the following desired reactions for HC, CO, and NO_x in the exhaust of the automobile. For oxidation:

$$C_yH_n + (1 + n/4)O_2 \rightarrow yCO_2 + n/2\ H_2O \tag{8}$$

$$CO + 1/2\ O_2 \rightarrow CO_2 \tag{9}$$

$$CO + H_2O \rightarrow CO_2 + H_2 \tag{10}$$

For reduction:

$$NO(\text{or } NO_2) + CO \rightarrow 1/2\ N_2 + CO_2 \tag{11}$$

$$NO(\text{or } NO_2) + H_2 \rightarrow 1/2\ N_2 + H_2O \tag{12}$$

$$(2 + n/2)NO(\text{or } NO_2) + C_yH_n \rightarrow (1 + n/4)N_2 + yCO_2 + n/2\ H_2O \tag{13}$$

When a driver first starts the automobile, both the engine and catalyst are cold. As the engine and exhaust piping heats up, a temperature is reached within the catalyst that initiates the catalytic reactions. This light off temperature and the concurrent reaction rate is kinetically controlled, ie, depends on the chemistry of the catalyst, since the transport reactions are fast. Typically, the CO reaction begins first followed by the HC and NO_x reaction. Upon further heating, the chemical reaction rates become fast, and the overall conversions are controlled by pore diffusion and/or bulk mass transfer.

The Monolithic Catalyst

Both beaded (or particulate) and monolithic or honeycomb structures (Fig. 5), have been used as a catalyst support for control of automotive emissions for passenger vehicles. Because of the advances in monolith technology, catalyst mounting methods, flexibility in reactor design, low pressure drop, high heat, and mass transfer rates, the monolithic support dominates the entire market as the preferred catalyst support (1).

Ceramic materials were chosen for the initial monolith. The ceramic monolithic catalyst had to have sufficient physical and chemical properties to withstand the environment encountered in the engine exhaust and provide sufficient strength and resistance to cracking under thermal shock conditions experienced during rapid accelerations and decelerations. Thermal shock was eventually satisfied by mechanical design coupled with the use of a low thermal expansion ceramic material called cordierite (synthetic cordierite has a composition

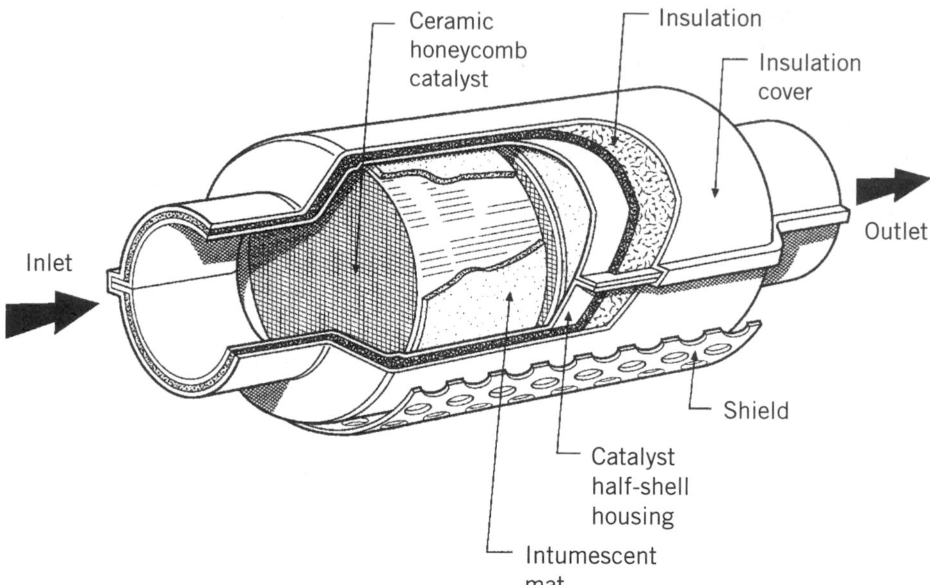

Figure 5. Automotive three-way catalyst structural design, including honeycomb support and mounting can.

approximating 2MgO, $5SiO_2$, and $2Al_2O_3$ and a softening point >1300°C). In preparing the catalyst, this desirable property must be matched by the thermal expansion properties of the catalyst carrier or washcoat to prevent a mismatch in thermal properties.

Monolithic structures were ultimately produced by a novel extrusion technique that allowed mass production to be cost effective. The first honeycombs used in the auto exhaust had 300 cells per square inch with wall thicknesses of about 0.030 cm (0.012 in.) and open frontal area of about 63%. These dimensions were finalized based on mechanical specifications and activity performance requirements to ensure a high degree of contact between the reactants and the catalyst washcoat (high mass transfer) and the lowest possible light off temperature. Later developments in extrusion technology resulted in a 400 cpsi honeycomb with a wall thickness of 0.006 inch and open frontal area of 71%. This increased the geometric surface area for the mass transfer controlled reactions. Today, cordierite monoliths of 400 cpsi and 0.010 cm (0.004 in.) wall thickness and 600 cpsi and 0.010 cm wall thickness are available (1).

Catalyst companies began to explore these new structures as catalyst supports. They developed slurries of the catalytic coating that could be deposited onto the walls of the honeycomb, producing adherent washcoats. The washcoat thickness—about 20 and 60 μm on the walls and corners (fillets), respectively—could be kept at a minimum to decrease pore diffusional effects while allowing sufficient thickness for anticipated aging due to deposition of contaminants. One method of preparing a washcoated honeycomb is to submerge it in a slightly acidified slurry (slip) containing the γ-Al_2O_3 already impregnated with stabilizers (10) and precious metals. The washcoat chemically and physically bonds to the honeycomb surface, where some of the washcoat fills the large pores of the ceramic. The slurry must have the proper particle size distribution to be compatible with the pores of the ceramic wall. Another method involves first washcoating the honeycomb with the alumina slurry, drying and calcining, and then dipping it into the impregnating solutions. The coated honeycomb is air dried, and calcined to 450–500°C, which ensures good adhesion. Typically, the catalyst contains 0.1–0.15% precious metals. For the oxidation catalysts of the first generation, the weight ratio of Pt:Pd was 2.5:1, while the second generation contained a weight ratio of Pt:Rh (1). The ceramic honeycomb catalyst is mounted in a steel container with a resilient matting material wrapped around it to ensure vibration resistance and retention (11). Usually, the catalyst is prepared at the catalyst manufacturer ("coater") and shipped to a metal fabricating company to be canned ("canner"). A typical ceramic honeycomb supported catalyst is shown in Figure 5.

Although the early honeycombs were ceramic, recently metallic monolith structures are being used in certain niche markets for exhaust control, because they can be made with thinner walls and have open frontal areas close to 90%, allowing lower pressure drops. Cell densities >400 cpsi can be used, which permits smaller catalyst volumes and smaller converters. The base material of construction is an ferritic stainless steel alloy with iron, chrome, aluminum, and rare earths. Typical monoliths have 400–600 cpsi with a 0.5 m (0.002-in.) wall thickness. With some metal substrate suppliers, the catalyst is first coated onto the sheet metal and then fabricated into the honeycomb structure (12). This has the advantage of producing uniform coating thicknesses, thus eliminating the fillets. Other catalysts are prepared similarly to the ceramic counterparts (13).

Catalyst Technology Development

Over the past 25 years, the technology development of the automotive exhaust catalyst has been dictated by the quest for lower emissions and the full integration of the catalyst into the emission control system. During the early phase of development, the catalyst was thought of as an add-on device to the tailpipe to reduce emissions. However, more recently, the catalyst has become integral to the design strategy for vehicle operation. The automobile catalyst has progressed through the following development phases: oxidation catalyst (bead and monolith support, HC and CO emissions) to three-way catalyst (HC, CO and NO_x emissions) to a high temperature three-way catalyst (approaching 950°C) to all-palladium, three-way catalyst to meet LEVs (high temperature, close-coupled catalyst, approaching 1050°C). These approaches for automobile emission control will meet the majority of emission regulations through the year 2000.

Emerging Technology for Low Emission Vehicles

The requirements for cleaner air will continue to place great demands on the catalyst. With the present generation TWCs, the majority of hydrocarbon emissions (60–80% of the total emitted) are produced in the cold start portion of the FTP, in the first 2 min of operation. Typical composition of the hydrocarbons during cold start are as follows (14):

Hydrocarbon Type	Approximate Hydrocarbon Composition, % (after cold start)	
	3 s after Cold Start	30 s after Cold Start
Paraffins	20	35
Olefins	45	20
Aromatics		
C_6, C_7	20	20
$\geq C_8$	15	25

As LEV regulations become effective, strategies to further reduce these hydrocarbons had to be developed. Hill 1 of the cold start portion of the FTP includes transients of temperature, exhaust flow rate, emissions (type and content), perturbation, and air:fuel ratio. The vehicle control strategy is feed forward, since the oxygen sensor cannot respond to these transients at the low temperatures during this cold start portion. A number of approaches are being considered to meet the requirements of 0.075 g/mi (LEV) and 0.040 g/mi (ULEV) HC emission standards, which require increased destruction rates during the first 2 min of the cycle.

Noncatalytic Approaches

Three system design approaches have been proposed and studied in combination with catalysts to reduce the cold start emissions. The preheat burner uses the gasoline fuel in a small burner placed in front of the catalyst. The burner is turned on during cold start and the heat generated warms up the catalyst so the catalyst is hot when the cold exhaust from the manifold reaches the catalyst (15).

The exhaust gas igniter involves placing an ignition source (eg, glow plug) in between two catalysts. During cold start, the engine is run rich and a small amount of air is injected to make the mixture flammable. This is then ignited and heats the catalyst.

The chemically heated catalyst uses a highly reactive species, usually H_2, which is generated in a device onboard the vehicle. Since this reacts at room temperature over the catalyst, the heat of reaction warms up the catalyst to react during cold start (16).

These system approaches rapidly heat the catalyst during cold start, resulting in low emission operation; however, little is known of the system durability and they are complex and expensive (17).

Next Generation Technologies

Operating the spark-ignited engine lean of stoichiometric conditions results in a fuel economy benefit. Unfortunately, the TWC catalyst has poor NO_x reduction activity. Reduction of NO_x in lean environments is under investigation. A new catalyst must be developed to promote the following reaction:

$$HC + NO_x + O_2 \rightarrow N_2 + CO_2 + H_2O \qquad (14)$$

A lean NO_x reduction system must be integrated with the engine so the exhaust stream will have the type and amount of hydrocarbons needed to reduce these oxides at the optimum temperature for the particular hydrocarbon (18). This area will undoubtedly continue to receive a great deal of attention and is covered more thoroughly below.

Another approach is to operate the vehicle in a partial lean burn operation by which the engine operates near stoichiometric conditions and lean under cruise conditions. In these studies, the NO_x is trapped during the lean mode of the engine operation and released during the rich mode of the engine operation to then react over a standard TWC located downstream. Adsorber or trap components being considered are based on alkaline earths, rare earth metal oxides, and carbonates (19–21).

Alternative fuels are another area of active study (22). Fuels such as compressed natural gas (23,24) liquid petroleum gas and alcohols (25) are attractive alternatives to gasoline because they are potentially less polluting, although they still produce gaseous emissions (see ALCOHOL FUELS). Natural gas–fueled vehicles are discussed below. Engines that burn them could be run lean with an oxidation catalyst for emissions (26). Emission systems have been developed for methanol fuel vehicles (27,28).

CATALYST FOR THE REDUCTION OF DIESEL EMISSIONS

The popularity of diesel engines is derived primarily from their fuel efficiency relative to the gasoline engine, because they operate in lean stoichiometric air:fuel ratios. Another advantage is their longevity, since it is not uncommon for a diesel engine to have a life of 1 million miles or about 10 times that of the gasoline counterpart. Furthermore, in Europe diesel fuel is less heavily taxed than gasoline and thus is less expensive. It is for these reason that commercial transporters as well as many passenger cars, especially in Europe, prefer this type of engine. The sale of diesel vehicles is expected to increase dramatically into the next century due to economy of operation and decrease in emission of greenhouse gases, most notably CO_2.

Diesel fuel is injected into a highly compressed charge of air where the temperature is sufficiently high that combustion occurs. Thus unlike the spark-ignited engine the diesel is a compression–ignited process. Its lean nature results in a cooler combustion with less NO_x formation. The liberal design of the combustion process does, however, result in larger quantities of unburned fuel, lube emissions, and dry soot. Some characteristics of diesel fuel are shown in Table 1. A large cetane number is indicative of a high combustibility of the fuel under compression.

Diesel emissions represent all three phases of matter: (*1*) solids, dry carbon or soot; (*2*) liquids, unburned fuel, lubricat-

Table 1. Chemistry of Typical Diesel Fuel

Characteristic	Measure
Sulfur, ppm	500
Cetane number	50
Total aromatics, vol %	35
Mono-aromatics	30
Diaromatics	5
Paraffins, vol %	65
Boiling point, °C	
Initial distillation, °C	185
10, vol %	225
95	345
Final	360

ing oil, and liquid sulfates (primarily sulfuric acid, H_2O associated with the SO_3 is included as a component of the liquid particulates); (*3*) gases, carbon monoxide, hydrocarbons (derived primarily from partially burned fuel), and oxides of nitrogen and sulfur. The percent composition of the particulate depends on the mode of engine operation and the sulfur content of the fuel (29). For example, for a pre-1994 diesel fuel containing 0.3–1% sulfur at high temperatures (high engine load), the sulfuric acid portion of the particulates represents almost 50%; dry carbon, 40%; and the soluble organic fraction (SOF), only about 10%. In contrast, at lower temperatures (low load operation) the SOF can be as high as 65% and dry carbon and sulfuric acid, both <20%.

Since 1994, in the United States and Europe, the sulfur content has been mandated to <0.05% (500 ppm), dramatically decreasing the engine out sulfuric acid emissions. The role of the catalyst was to reduce the SOF and gaseous CO and HC emissions but specifically to minimize the oxidation of SO_2 to SO_3 (H_2SO_4).

The incorporation of a diesel oxidation catalyst into the exhaust of heavy-duty trucks required sophisticated combustion engineering and catalysis (30). Engine manufacturers have made great progress in redesigning the combustion process to minimize dry carbon emissions. Through the use of sophisticated flow models (31), the use of a flow through monolith is possible. The catalytic reactions are as follows:

$$SOF + O_2 \rightarrow CO_2 + H_2O \tag{15}$$

$$HC + O_2 \rightarrow CO_2 + H_2O \tag{16}$$

$$CO + O_2 \rightarrow CO_2 \tag{17}$$

Undesirable:

$$SO_2 + O_2 \rightarrow SO_3 \tag{18}$$

Reaction 15 is the oxidation of the soluble organic fraction (SOF) (unburned fuel and lubricating oil) portion of the particulates. This coupled with reductions in dry carbon or soot result in compliance with particulate emission standards.

Reactions 16 and 17 are typical of the gasoline vehicle in that the HC and CO are oxidized. The reduction of the gas phase HC positively impacts smog-generating photochemical reactions and also decreases odor-bearing compound concentrations.

Reaction 18 gives rise to an increase in liquid particulates in the form of sulfuric acid and counters the benefits obtained from the reduction in SOF. This becomes a critical problem in catalyst selectivity: the catalyst must perform only the desirable reactions; ie, reactions 15–17.

The units of measurement for particulates are grams per brake-horsepower generated in an hour (g/bhp · h) in the United States for heavy-duty truck transient test cycles; grams per kilometer (g/km) is used for passenger cars. In Europe, Japan, and Korea the standards are expressed in grams per kilowatt hour (g/kWh) for heavy-duty trucks. Countries have developed standardized transient test cycles that reflect the driving habits and terrain of each respective country. Analysis details are available (32).

Table 2 lists the respective standards for the United States and Europe. What should be especially noted is the dramatic decreases required for NO_x and HC plus NO_x proposed for the year 2003.

Alternate Fuels

There is a growing use of lean-burn natural gas–fueled vehicles currently for service operations such as delivery trucks and buses. Natural gas is used as a replacement for diesel fuel to more easily meet emission standards (33), especially for urban bus applications. These vehicles still need catalysts to control particulate emissions, primarily derived from lubricating oil, and specific gaseous hydrocarbon emissions, called reactive organic gases (ROGs), which participate with NO_x and sunlight in the generation of smog. Catalysts for lean-burn engines are typically Pd and/or Pt supported on CeO_2 plus γ-Al_2O_3 deposited on a ceramic monolith. The diameter can be as large as 25 cm with a length of 18 cm for an engine with a 10-L displacement.

Methane is the major hydrocarbon in natural gas. It is unreactive and does not participate in photochemical smog reactions; thus in the United States it is unregulated. If it were to be required (CH_4 is a greenhouse gas), a major breakthrough in catalyst technology would be needed (26,34) to pass the U.S. test, due to inhibition by sulfur oxides. For the European bus market, a catalyst is successful for abating methane emissions because the 13-mode test favors high temperature modes where the deactivation by fuel and oil sulfur have a minimum effect on the performance of the Pd catalyst.

Diesel Particulate Traps and Fuel Additives

One of the earliest attempts to treat diesel particulate emissions was the ceramic wall flow (cordierite) filter with each channel alternately plugged at each end to force flow through the highly porous wall that filters the particulate matter. Although this approach is still considered a possibility (29), with and without a catalyst (35,36), problems with reliable regeneration at modest exhaust temperatures and/or runaway temperatures during regeneration leading to substrate melting have plagued this technology.

The addition of metalloorganic additives to the diesel fuel uses the same wall flow filter but provides more reliable regeneration of the particulate matter accumulated. Copper-, iron-, or cerium-containing fuel additives can catalytically reduce the temperature for particulate burnoff to considerably lower temperatures than the 600°C needed for thermal regeneration. For example, copper can initiate oxidation of the particulates at 250°C, allowing for more frequent regeneration with less chance of particulate buildup. The cerium-organometallic fuel-bound additive (29,37) coupled with exhaust gas recirculation (EGR) claims to simultaneously decrease particulates and

Table 2. Some U.S. and European Diesel Standards

Vehicle and Year	Particulate	Carbon Monoxide	Hydrocarbons	Nitric Oxides	Hydrocarbons plus Nitric Oxide
United States, g/bhp · h					
Truck, 1991	0.25	15.5	1.3	5	—
Truck, 1994	0.1	15.5	1.3	5	—
Truck, 1998	0.1	15.5	1.3	4	—
Truck, 2003 (proposed)	0.1	—	—	2	—
Europe, g/km					
Car, 1996	0.1	1	—	—	0.9
Car, 1999	0.04	0.5	—	—	0.5
Car, 2003 (proposed)	0.025	—	—	—	0.19

NO_x. Such systems are in an advanced state of testing but are still not commercially available.

Reducing NO_x in a Diesel Environment Using Onboard Derived Hydrocarbons

Future diesel emission standards will require significant reductions in NO_x (NO and NO_2) (see Table 2) for which an ideal solution is a catalyst capable of decomposing it according to the following reaction:

$$NO \rightarrow N_2 + O_2 \tag{19}$$

No such catalyst exists today, so an alternative approach is to catalytically reduce NO_x selectivity using hydrocarbons derived from onboard diesel fuel, according to reaction 20 and Figure 6.

$$HC + NO_x + O_2 \rightarrow CO_2 + H_2O + N_2 \tag{20}$$

$$HC + O_2 \rightarrow CO_2 + H_2O \tag{21}$$

The undesirable reaction 21 (Fig. 6) consumes the hydrocarbon by reaction with the O_2 present in the exhaust. This reaction becomes dominate as the temperature is increased, resulting in the maximum conversion shown in Figure 6.

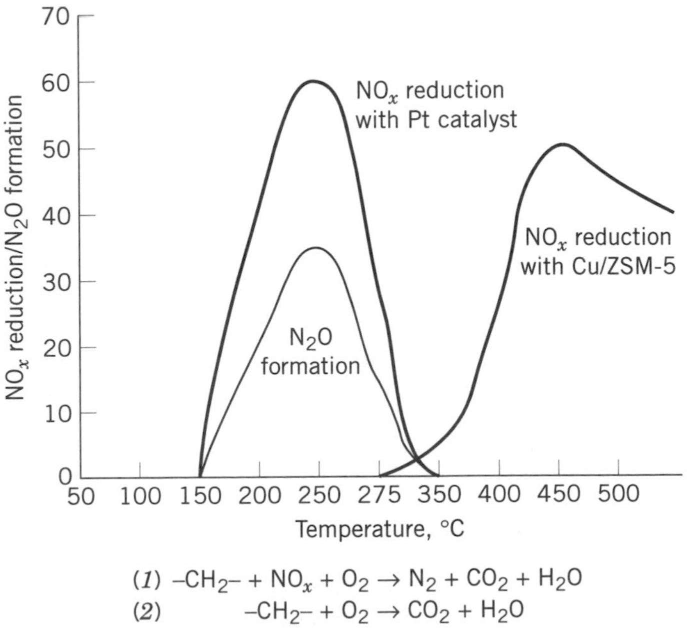

Figure 6. Catalysts for NO_x reduction in a lean environment using hydrocarbons.

Selective catalytic reduction in the presence of large excesses of O_2 is commonly practiced in power plants and other stationary engines, using NH_3 as the reductant (1). When hydrocarbons such as diesel fuel are used in place of NH_3, the activity and selectivity decrease significantly. To date, the only two candidate materials are Pt supported on an Al_2O_3 or a zeolite and Cu/ZSM-5, where the ZSM-5 is a pentasil zeolite with a Si:Al ratio of about 20 and pore size openings of about 5.5 Å.

Literally thousands of materials have been investigated, but a better one has not yet been found (38,39). Pt is active in a narrow temperature window of 180–275°C (40) while the Cu catalyst is active >350°C, as shown in Figure 6 (41,42). Furthermore, the Pt reduces most of the NO_x to N_2O, which is a powerful greenhouse gas. It does have the advantage of being generally insensitive to the presence of SO_2, and it possesses good thermal stability. Cu/ZSM-5 catalyzes the NO_x reduction with certain hydrocarbons to N_2 but is poisoned by SO_2 (43) and lacks thermal stability above about 600°C.

Therefore, neither will be sufficient to meet emission standards for 2003. Some benefit for NO_x reduction is being realized by the activity of Pt present in diesel passenger car oxidation catalysts, because it does convert 10–15% NO_x without added hydrocarbon. This technology is called four-way catalysis (47). The engine is tuned to permit sufficient slippage of HC exiting the combustion chamber to bring about the reduction. Maximum possible conversions for a Pt-based catalyst with diesel fuel injection in the exhaust are placed at 25% in the U.S. FTP test (44).

OZONE ABATEMENT—AIRCRAFT APPLICATIONS

Modern-day aircraft commonly fly above 40,000 ft, where the ambient ozone (O_3) concentration becomes significant. This ozone-laden air is brought into the passenger cabin and is used to pressurize the cabin atmosphere. FAA regulations now require that the airplane cabin ozone concentration (on a time-weighted average) cannot exceed 0.1 vppm (sea level equivalent) (45).

Ozone can be removed through a number of processes, such as thermal degradation and carbon adsorption, but the most economical method has proven to be catalytic abatement (1).

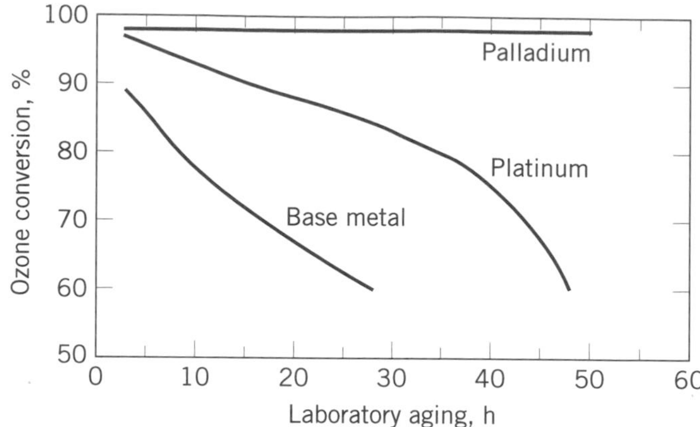

Figure 7. Palladium is the best catalyst for ozone removal, even though many materials have good initial activity (46).

Figure 8. Installation of ozone catalytic abater in underbody of jet aircraft.

Table 3. Design Conditions for Ozone Catalysts for Wide-Body Aircraft

Condition	Value
Air flow conditions, kg/s	0.68–1.4
Temperature, °C	120–200
Pressure, atm	1.6–4.0
Allowable pressure drop, atm	0.034–0.102
Vessel Proof Pressure, atm	5–30
Required conversion, %	83–93
Housing	
diameter, cm	20–28
Length, cm	40–60
Maximum weight, kg	4.5–16
Shock and vibration	Individual aircraft manufacturer's specification

The catalytic reaction proceeds as follows:

$$2\ O_3 \rightarrow 3\ O_2 \quad (22)$$

In the actual application, the reaction requires some preheat of the ambient air, which is accomplished through the isoentropic compression by removing the air off the appropriate compression stage of the jet engine. Using a catalyst permits the decomposition reaction to occur at much lower temperatures than does thermal decomposition; thus the air can be removed at a lower stage of air compression. The higher the stage of compression for using the preheated air, the higher the parasitic power loss and the less efficient the jet engine.

Ozone is a reactive chemical species; and many catalytic materials, both precious metals and base metals, have been explored for this application. Figure 7 shows some candidate catalysts after short-term testing in the laboratory. Initial activity is apparent for many different catalytic materials, and the issue in developing a suitable catalyst was that ozone itself is a powerful oxidant and will deactivate many catalytic materials after a short time. This caused considerable problems in early testing, since many materials looked good on bench tests but failed in the aircraft after a relatively short time (1).

Some typical ranges of operating conditions for wide-body jet aircraft is shown in Table 3. The most effective catalytic material is about 1% Pd on γ-Al_2O_3 supported on a high cell density (300–400 cpsi) ceramic or metallic monolith; the catalyst operates at 200,000–500,000 h^{-1} space velocity. Figure 8 shows an ozone abater being installed in the underbody of a commercial jet aircraft.

Since the environment that the ozone catalysts sees depends on the type of aircraft, airline maintenance procedure, the flight routes, and time spent awaiting takeoff, the long-term performance of the ozone abatement catalyst will vary.

BIBLIOGRAPHY

1. R. M. Heck and R. J. Farrauto, *Catalytic Air Pollution Control: Commercial Technology,* Van Nostrand Reinhold, New York, 1995.
2. *U.S. Code of Federal Regulations,* Title 40, Washington, D.C., U.S. Government Printing Office.
3. J. Zeldoovich, *Acta Phsiochim. USSR* **21,** 577 (1946).
4. R. Impens in A Crucq and A. Frennet, eds., *Catalysis and Automobile Pollution Control,* Elsevier, Amsterdam, the Netherlands, 1987.
5. J. Kummer, *Prog. Energy Combust. Sci.* **6,** 177–199 (1980).
6. A. Viala, *Petrol. Technol.* **351,** 25–27 (1993).
7. R. Heck and R. Farrauto, *Auto. Eng.,* **75,** 93–96 (Feb. 1996).
8. *Hydrocarbon Proc.,* **50,** 85–88 (May 1971).
9. J. Calvert, J. Heywood, R. Sawyer, and J. Seinfeld, *Science,* **261,** 37–45 (1993).
10. B. Beguin, E. Garbowski, and M. Primet *J. Catalysis* **127,** 595–604 (1991).
11. U.S. Pat. 3,441,381 (Apr. 29, 1969), C. Keith, T. Schreuders, and C. Cunningham (to Engelhard).
12. J. Adomaitis, M. Galligan, J. Kubsh, and W. Whittenberger, *SAE 962080,* SAE International, Warrendale, Pa., Oct. 1996.
13. H. Harkonen and co-workers, *SAE 960560,* SAE International, Warrendale, Pa., Feb. 1996.
14. K. Kumitake, T. Wanatanbe, K. Uchida, T. Yaegashi, and H. Ito, *SAE 961954,* SAE International, Warrendale, Pa., Oct. 1996.

15. D. Eade, R. Hurley, B. Rulter, G. Inman, and R. Bakshi, *Auto. Eng.,* 70–73 (Apr. 1996).
16. Y. Kamada, M. Hayashi, M. Akaki, S. Tsuchikawa, and A. Isomura, *SAE 960346,* SAE International, Warrendale, Pa., Feb. 1996.
17. P. Langen, M. Theissen, J. Malloy, and R. Zielinski, *Auto. Eng.,* 31–35 (Sept. 1994).
18. H. Iwamoto and H. Hamada, *Catal. Today,* **10,** 57–71 (1991).
19. M. Brgan, R. Brisley, J. Moore, and A. Clark, *SAE 962045,* SAE International, Warrendale, Pa., Oct. 1996.
20. W. Strehlau and co-workers, *SAE 962047,* SAE International, Warrendale, Pa., Oct. 1996.
21. G. Lutkemeyer and co-workers, *SAE 962046,* SAE International, Warrendale, Pa., Oct. 1996.
22. R. Farrauto, R. Heck, and B. Speronello, *Chem. Eng. News,* **70**(36), 34–44 (1992).
23. J. White, J. Carroll, M. Brady, W. Burkmyre, W. Liss, and M. Church, *SAE 930222,* SAE International, Warrendale, Pa., Mar. 1993.
24. S. Subramanian, R. Kudia, and M. Chattha, *SAE 930223,* SAE International, Warrendale, Pa., Mar. 1993.
25. R. McCabe and P. Mitchell, *Appl. Catal.* **44,** 73–93 (1988).
26. J. Lambert, M. Kazi, and R. Farrauto, *SAE 961971,* SAE International, Warrendale, Pa., Oct. 1996.
27. R. McCabe, P. Mitchell, F. Liapri, W. Scruggs, and R. Warburton, *SAE 872138,* SAE International, Warrendale, Pa., Nov. 1987.
28. J. Hochmuth and J. Mooney, *SAE 930219,* SAE International, Warrendale, Pa., Mar. 1993.
29. P. Zelenka, W. Cartellieri, and P. Herog, *Appl. Catal. B,* **10**(1–3), 3 (1996).
30. J. Clerk, *Appl. Cataly. B,* **10**(1–3), 99 (1996).
31. J. Johnson and D. Kittelson, *Appl. Cataly. B,* **10**(1–3), 117 (1996).
32. R. D. Cuthbertson and P. R. Shore, *J. Chromatogr. Sci.* **26,** 106–112 (1988).
33. M. Samsa, *GRI Report-91/0183,* GRI, Chicago, Ill., June 1991.
34. R. McCormick, A. Newlin, D. Mowery, and M. Grabowski, *SAE 961967,* SAE International, Warrendale, Pa., 1996.
35. K. Voss and G. Rice, *SAE 950156,* SAE International, Warrendale, Pa., 1995.
36. V. Rao, H. Cikanek, and R. Horrocks, *SAE 940458,* SAE International, Warrendale, Pa., 1994.
37. J. Summers, S. Van Houtte, and D. Psaras, *Appl. Catal. B,* **10**(1–3), 139 (1996).
38. M. Amiridis, T. Zhang, and R. J. Farrauto, *Appl. Catal. B,* **10**(1–3), 203 (1996).
39. M. Iwamoto, *Cataly. Today,* **28,** 29 (1996).
40. G. Ansell and co-workers, *Appl. Catal. B,* **10**(1–3), 183 (1996).
41. K. Adams, J. Cavataio, and R. Hammerle, *Appl. Catal. B,* **10**(1–3), 157 (1996).
42. A. Konig and co-workers, *SAE 963041,* SAE International, Warrendale, Pa., 1996.
43. J. Feeley, M. Deeba, and R. J. Farruto, *SAE 950747,* SAE International, Warrendale, Pa., 1995.
44. J. Hochheim and co-workers, *SAE 962042,* SAE International, Warrendale, Pa., 1996.
45. *Federal Register,* 14 CFR, Parts 25 and 121 (1980).
46. R. Heck, R. Farrauto, and H. Lee, *Catal. Today,* **13,** 43–85 (1992).
47. Y. Lui and co-workers, *SAE 962048,* SAE International, Warrendale, Pa., 1996.

AIR QUALITY MODELING

D. ALAN HANSEN
Electric Power Research Institute
Palo Alto, California

IMPETUS FOR MODELING

A host of natural processes put into the air materials called pollutants. Since human beings discovered fire and started working the soil, they have added to the burden. Even before the Industrial Revolution and the coming of Europeans, Los Angeles, California, had air pollution. In 1542, during a voyage of discovery, Juan Rodriguez Cabrillo noted in his diary that what is now the Los Angeles Basin was covered with a smoky haze, leading him to name the area *Bahia de los Fumos,* or Bay of Smokes (1,2). He speculated that the cause was brush fires started by the local Native Americans to drive small game toward their hunters.

In view of the then prevailing societal values and the relatively primitive state of medical knowledge, such emissions were likely considered a natural consequence of day-to-day living and at most a nuisance rather than a threat to human health. But with the growth of cities and the reliance on coal for heating, and then with the Industrial Revolution, combustion and smelting emissions led to such high ambient pollutant concentrations in some locations that people clearly suffered ill affects, sometimes fatally (3). In the 1950s a new type of air pollution, photochemical smog, was identified. It was traced to the production of oxidants from the sunlight-induced reactions of nitrogen oxides (mainly from combustion) and gaseous organic compounds (mainly from hydrocarbon fuels and their combustion products) (4). Ill effects, not only to people, but also to plants, were also associated with exposure to photochemical smog (5).

Industrial societies suffer from yet other manifestations of air pollution than photochemical smog, including acid deposition, respirable aerosols, so-called air toxics, and visibility degradation. Reactions of gaseous organic, sulfur, and nitrogen compounds in the atmosphere are major players in the production of these phenomena. The aerosols and gases involved can have harmful effects on plants, aquatic environments, soils, and building materials. They also reduce visibility.

In addition to airborne materials resulting from anthropogenic combustion and industrial process emissions, gases and volatile liquids can be released accidentally, intentionally, or incidentally in war, for example, and naturally from biological and geophysical sources such as vegetative and soil emissions, volcanoes, and lightning.

The preferred approach to mitigating air pollution effects is to reduce emissions that contribute to them. In situations where systematic and rational strategies are necessary for responding to or mitigating the release of these airborne substance at the least cost to society, the following must be known:

- The rates and locations of emissions to establish a baseline from which changes are intended
- The chemical and physical reactions transforming emissions to gaseous, aerosol, and dissolved products, both in the air and in clouds and precipitation

- Locations where emitted and transformed materials will be dispersed by atmospheric motions (winds)
- Rates of removal by deposition
- Downwind exposures
- Health, environmental, and economic consequences of exposures
- Available technologies for reducing emissions and their costs
- The process by which exposures will respond to emissions changes

Also necessary are the identification of reasonable alternatives strategies for mitigating impacts and ways to distinguish those that are most likely from those less likely to meet the objectives, be they health- and welfare-related or economic.

Because of the complexity of the chemical and physical processes involved, only numerical models can simulate all the atmospheric phenomena of interest with any skill. Further, only models that capture the essence of all the important chemical and physical tropospheric dynamics, including their interactions and feedbacks, allow multiple issues to be assessed simultaneously. Models are the only means of exploring how pollutant concentrations will change if emissions are changed into the atmosphere. The desired model might be characterized as a high-fidelity tropospheric simulator. Simpler approaches, such as assuming proportional relations between emissions and exposures, have little likelihood of finding realistic, least-cost answers because the relations between sources and receptor exposures to atmospheric pollutants are often nonlinear.

Such models do not now exist and represent a departure from the past practice of using different types of models to assess different air pollution issues, eg, one type for photochemical oxidants and another for acidic deposition. This was a consequence of a relatively primitive state both of the understanding of the relevant atmospheric processes and of computational science and technology. However, these constraints are diminishing rapidly and serious efforts are now underway to develop modeling systems that not only simulate all important tropospheric phenomena but also link the simulated results with decision analysis systems that take into account uncertainty and economic factors. These developments are a reflection of the growing appreciation that air quality issues related to common precursor emissions cannot be treated in isolation and that it is in our best interest to examine issues holistically, within a scientifically sound framework.

The type of highly sophisticated tropospheric simulator and linked decision analysis tool currently under development is referred to herein as a comprehensive modeling system (CMS) (6). However, the focus here is primarily on the atmospheric modeling aspects and much less so on those related to decision analysis.

NEED FOR CMS

A CMS for air quality research and management would fulfill a societal need not satisfiable by any other mechanism: a systematic integrator of knowledge about the response of the atmosphere to natural and artificial perturbations, such as weather, changes in land use, and emissions, and the social and economic ramification of those perturbations.

To fully appreciate the need for a CMS, rather than continuing the traditional piecemeal approach to modeling, it helps to review some of the existing impediments to the facile use of models in air quality management, to the realistic simulation of the atmosphere, and to the distribution and use of air quality models. To set the stage, we can make a summary statement that few would challenge: urban- and regional-scale air quality models are hard to use.

Considerable time, effort, and engineering knowledge are required to assemble modeling emissions inventories for the more complex air quality models. Not only is information on emissions factors and activity levels required, as for regulatory inventories, but also on temporal and spatial variations in biogenic as well as anthropogenic emissions on scales comparable to those in the models and in a format that can be assimilated by them. The number of species whose emissions must be estimated can be huge, especially if the air quality model to be used follows the chemical evolution of organic gases and aerosols. Developers of biogenic emissions inventories must have information on vegetation types and cover, allocated spatially and temporally. To explore cost-effective air quality management strategies also requires the capability to project future emissions, to change them according to options explored, to estimate the costs associated with any changes, and to estimate the uncertainty in all derived data.

Converting the emissions factors and activity data to emissions strengths, spatially and temporally allocating the emissions strengths to the appropriate grid cells, changing the strengths and locations of the emissions according to hypothetical air quality management strategies, estimating costs and uncertainties, and using the accumulated information to support decision analysis all require mathematical models. In general, many of these activities are handled by separate models requiring many levels of data manipulation by separate teams of specialists. In principle, most of these activities could be automated and conducted through a single user interface, thus streamlining the process enormously.

Meteorological model output is needed by both advanced emissions models that use temperature, wind speed, and insolation to estimate some biogenic and volatile organic compound emissions, and air quality models as input to the chemistry, transport, diffusion, and deposition algorithms. Meteorological models themselves need observational data as input either to be interpolated or in nudging algorithms to control error growth. Observational data can fall in the categories of meteorological, land use, soil moisture, and albedo. Accessing and formatting both input data, either historical or, in some cases, real-time, used by the model itself and output data for input to emissions and air quality models consume more than an order of magnitude more time and effort than the actual model execution. Many of these activities, as well as interfacing the meteorological model with the emissions and air quality models and visualizing computed meteorological fields, could be automated, substantially easing the data processing burden.

To represent realistically the forcing of wind flows and pollutant transport by nonuniform surface elevation, terrain data must be input to both meteorological and air quality models. The grid structure will determine how well terrain feature are resolved, which can influence the accuracy of simulated transport, among other things.

Regional- and urban-scale air quality models, in addition to meteorological and emissions input, require air quality data for initialization and for boundary conditions. They also need land-use information for input to their dry deposition algorithms. As with meteorological models, processing the data for initialization and execution and then postprocessing and displaying outputs in interpretable forms take much more time and effort, and therefore money, than the actual model execution. The lion's share of this effort could be streamlined and made less costly.

Uncertainty of all input and derived data must be estimated, if model users are to be able to associate levels of confidence with the model outputs. Uncertainty estimation techniques for geophysical modeling are currently the subject of intense research. Some of them involve tortuous propagation of error analyses. But methods for evaluating model performance through sensitivity analyses and diagnostic tests and then expressing the results in a meaningful form to users of model outputs are still relatively primitive in the air quality modeling community.

The complexity of modern society is reflected in the number of issues or applications addressable with air quality models. One of the most fundamental applications of models is in their role as integrators of our knowledge of atmospheric processes. In this role they allow investigation on a detailed level how the processes interact to produce what is physically and chemically observed in the atmosphere. This could be termed the scientific application. Other applications address the general issue of remediation. These have been laid out as follows (7):

- Permitting of new sources or modifications of existing sources, including (in the United States) new source review (NSR), prevention of significant deterioration (PSD) determination, and Clean Air Act Title V permits.
- Ozone control: estimation of the effects of alternative volatile organic compound (VOC)/NO_x emission control strategies on ozone concentrations and ozone attainment demonstration modeling.
- Acid deposition issues: estimation of the sources of acid deposition and identification of potential remedial control strategies.
- Particulate matter: air quality, deposition, and health effect impacts; regional-scale and local particulate matter (PM) concentration impacts at coarser (PM_{10}) and finer ($PM_{2.5}$) size distributions; visibility impairment.
- Toxics: impacts of toxic compounds, eg, formaldehyde, benzene, 1,3-butadiene, polycyclic organic materials (POMs), dioxin, mercury, and cadmium, on health, both through direct exposure to ambient air and through indirect sources, eg, bioaccumulation of toxics in fish, and on ecosystems, eg, deposition of toxics into the Great Lakes.
- Source Apportionment: the assessment of the fractional contribution of a source (or a group of sources) to the air pollution concentrations measured in a certain region.
- Emergency planning and response through estimation of the range and level of impact of accidental releases (real or hypothetical) of toxic compounds.
- Exposure of humans to harmful air pollutants through inhalation, eg, ozone, PM_{10}, and toxics, and other pathways.
- Risk assessment of different mitigation measures, including cost-benefits analysis of alternative emission control strategies.

Currently, each of these issues is treated with different sets of modeling tools with varying levels of scientific detail and complexity.

Historically, each of the models needed to cover this suite of air quality assessments has been developed by groups working independently, using a variety of programming languages, for execution often on specific and different types of computers, adhering to varying levels of coding efficiency, modularity, and quality control. Different coordinate systems are employed according to individual preferences. This has resulted in an unhealthy inconsistency among models, especially in view of the great pains that must be taken to interface models when the results from one must feed another. There is a clear need for modernizing and standardizing procedures, facilitating data access, manipulation, and visualization, and for a means to estimate modeling uncertainty and express it in a syntax that decision makers can use.

Although not every user must address all issues and applications, accessibility of all software for doing so through a single user interface would accommodate the broadest range of user needs and encourage multiple-issue assessments. For example, if a user wants to assess multiple issues associated with a particular pollutant, eg, ecosystem acidification, acid aerosols, ozone, and visibility impacts of NO_x (NO plus NO_2), interpretation of results would be vastly simplified and their credibility enhanced, if consistent chemistry and physics were applied uniformly throughout. Further, the time to gain familiarity and ultimate mastery of more than one part of the modeling system would be reduced by ensuring consistency in the look and feel across components and applications.

Need for Coding Standards

The air quality community has typically not applied the most up-to-date software engineering principles in its models. To enhance efficiency, speed, flexibility, and the overall quality of model codes, the community must modernize its approach to software development. Part of this modernization would be the establishment of coding standards. Such standards would maximize compatibility of software developed by different groups and facilitate its execution on a broad variety of computers. Their implementation would encourage the exchange of modeling technology among researchers for testing, evaluation, and incorporation into their own modeling systems. Progress in air quality modeling would accelerate with easier code-debugging, more efficient execution, and reduction of redundant developmental efforts.

Need for Convenient Code Modification

As new and better process representations and numerical solution techniques evolve, the need arises to incorporate them in models. Traditionally, model improvement cycles have been on the order of years. This was a consequence not of the pace of new knowledge but of the difficulty in modifying specific blocks of the code and the resistance by model custodians to adopting new code that had not been adequately evaluated. The models were written as continuous, interdependent sequences of operations, so that if a change was made in one place, it often required changes in other places. Often the linkages were so obscure, and tracing them so difficult, that some parts of the code requiring change eluded detection. This could result in seemingly inexplicable model misbehavior. An insidious aspect of such a situation is where resulting errors do not lead to obvious misbehavior, but cause the model to depart subtly from the intended and expected accuracy of simulation. These hidden errors can yield a model that provides false guidance.

The problem of convenient and controlled code modification is now being successfully addressed by modularizing models and writing code in easily interpretable formats. A particular process representation or numerical solution algorithm is coded as a single module. The model then calls these modules on demand. Computational efficiency is maintained by compilers that create continuous strings of machine-executable code, apportioned to the engaged computers according to a computational optimization scheme. An excellent example of how these principles have been put into practice is the Advanced Regional Prediction System (8), which facilitates model upgrades in two ways: by isolating the code to be modified in manageable packages and by making the code structure intuitive.

Need for Reducing Turnaround Time

Existing urban- and regional-scale emissions as well as meteorological and air quality models are very computationally and labor-intensive. As noted above, they require hours to weeks to compile and format data for their input and for evaluating them. They also require supercomputers or high-end workstations to execute them. If run on supercomputers, they often need to be submitted in batch mode and a day or more is necessary before the output is available for examination. The cost of CPU time on supercomputers is prohibitively expensive for many model users. On workstations, present meteorological or air quality models generally run in near real-time, ie, each hour of simulated time takes about an hour of central processing unit (CPU) time. Relatively long turnaround time between job initiation and completion (logistical delay), cost, and relative slowness of execution (*1*) make it impractical to simulate more than a few days at a time; (*2*) discourage sensitivity analyses involving multiple runs where parameters are varied from run to run; (*3*) constrain the number of grid cells in the model domain, thus limiting either the size of the domain or the spatial and temporal resolution of the model; (*4*) may encourage oversimplification of atmospheric process representations or numerical solvers at the expense of accuracy; (*5*) discourage exploration for optimal outcomes through the simulation of a large number of candidate situations; and (*6*) lead to frustration in those cases where inadvertent errors have been made and the modelers must pause in their work before the erroneous output is available for examination and error detection.

Computing Platforms. To minimize logistical delays, access to supercomputers should be more direct so that runs can be made interactively. This would allow early error detection and changes to be made in the model setup without having to complete the entire simulation. Generalized approaches to coding should be developed that exploit whatever computing environment is available, should it include the vector and parallel architectures of the most advanced supercomputers or clusters of workstations. Cost can be minimized by including clusters of top-end microcomputers for example, PCs, which are rapidly blurring the distinction between PCs and workstations. Distributing the computing problem, ie, dividing it up among a variety of networked PCs, workstations, and supercomputers, in a way that optimally matches the code structure and timing with the computer architectures has the greatest potential for balancing cost, speed, and efficiency with off-the-shelf computers. Even greater gains might be realized with the development of processors customized to solve specific parts of the computing problem, as is done in arcade video games, computer-assisted design modules, and some military applications. The penalty paid with customized processors, however, may be the loss of flexibility and ease of upgrading as new science and computational technology emerges.

Process Algorithms and Numerical Techniques. Although not inactive, research on numerical algorithms for simulating atmospheric processes should be stimulated to address more effectively the problem of slow throughput. Examples of processes that consume inordinate amounts of CPU time are advection, cloud microphysics, aerosol dynamics, and aqueous- and gas-phase chemistry. Algorithms should be more computationally efficient while maintaining accuracy. Some improvement could be achieved by tailoring the solution technique closer to the particular capabilities of the computers being used. Other improvements would rely, however, on the development of totally new numerical techniques for the solution of differential equations.

Need for Conducting Sensitivity Tests

Complex geophysical modeling systems simulate many processes that may interact nonlinearly and lead to unexpected outcomes. Many of the processes are necessarily represented by numerical approximations because they are not amenable to analytical solution. Others are highly parameterized (simplified) either to make them less computationally intensive or because their details are not understood. The principal tool for gaining diagnostic insights into model performance is sensitivity analysis. It has been applied to studying process interactions (9), identifying which parts of the models have the strongest influence on outcomes (10,11), studying error propagation (12), and testing for simulation reasonableness (8). Reflecting the importance of each of these capabilities to definitive modeling, a CMS attribute must be provision for real-time (automatic differentiation) as well as retrospective sensitivity analyses.

Need for Established Procedures for Evaluating Many Types of Models

The enthusiasm among model developers for incorporating new science and technology into their models has not been matched in their efforts to characterize their models' ability to simulate physical and chemical reality. More often than not, modelers make relatively superficial comparisons of model output with observational data and find "reasonable agreement" between the two. It is remarkable how large a range of disagreement is considered reasonable by many in the geophysical modeling community; it easily spans an order of magnitude.

The large number of processes simulated, the complexity of their independent and interactive dynamical representations, and the absence of observational data sufficient for illuminating diagnostic studies have typically discouraged in-depth performance evaluations of modern urban- and regional-scale air quality models. Definitive evaluations require large, multivariate three-dimensional data sets and cleverly designed diagnostic evaluation procedures. The difficulty of the problem, however, does not excuse inappropriate use of the models, such as basing policy decisions on models that have only been superficially evaluated. Using them in this way may lead to misguided policies that are ineffective, overly costly, or both. It is now well-established (13) that these models can ostensibly simulate similar ground-level ozone concentrations with different combinations of input parameter choices covering a wide range of values. Example parameters are boundary conditions, NO_x and VOC emissions, depth of the planetary boundary layer, diffusivity, cloud cover, and wind speeds. By judiciously selecting parameter values within their uncertainty ranges the models can serve as multidimensional curve fitters, where the data to be fitted are often sparse surface observations. The problem is exacerbated by the often spatial and temporal nonrepresentativeness of the observational data. Such a fitting exercise does not ensure that the model is dynamically correct and that it will accurately simulate an inherently nonlinear system for different input conditions, such as those with changed emissions.

The U.S. EPA guidelines for regulatory photochemical modeling (14) illustrate how endemic the problem is. Although they specify up to ±30% target ranges for statistical measures of agreement between models and observations, they do so only for ozone, not for ozone precursors or any other indicator of the fidelity of the numerical simulation to the actual physical and chemical dynamics of this very nonlinear system. Further, they do not specify the extent or quality of the observational data that should be used for the comparisons. The unfortunate consequence is that models are being used for regulatory purposes without any tests to ensure that compensating errors are not responsible for ostensible agreement within the specified tolerances.

What is needed is a general recognition that complex geophysical models are not inherently good simulators of natural phenomena. In fact, students of these models have posited that beyond a certain level modeling uncertainty grows with model complexity, although we would like to believe that the more explicitly the atmospheric processes are represented, the more realistic the simulation. Following recognition of these intrinsic model limitations must be the formulation and application of rigorous diagnostic evaluation procedures that demonstrate how well models perform and why they perform as they do. This exercise will rely on developing an understanding of the error structure, eg, probability distribution functions, for both the input variability and uncertainty and the model formulation, ie, grid structure and numerical algorithms, uncertainty. The objective should be to establish confidence bounds on simulated outcomes. Once confidence bounds are available, then a basis can be established for (*1*) better defining the precise role that models should play in air quality management, (*2*) identifying what modeling components contribute most to modeling uncertainty, and (*3*) deciding how much effort should be expended on improving model reliability.

Need for Accessing, Storing, and Manipulating Large Data Sets

The data sets input to and generated by air quality modeling systems are by any measure huge. They include literally dozens of types of data. Taken together, the gridded, typically hourly averaged emissions, meteorological, and air quality input data number in the hundreds of megabytes, while the model outputs for simulation of regional-scale domains, at relatively fine resolution and for time periods of a week or more, are in the gigabyte range. Only recently have database management systems (DBMSs) been developed which have the potential to make the problem of storing, accessing, and manipulating such large and diverse data sets tractable. The problem is complicated by the fact that the individual types of data required are often only subsets of even larger data sets maintained by a variety of organizations at widely separated locations. The DBMSs used at these locations may differ.

A CMS must be able to access remotely these data on demand and reformat them according to its needs. It must be able to manipulate and display them in visually interpretable forms. Thus, the problem extends the need beyond that just for a sophisticated and versatile DBMS to that for similarly capable geographical information and visualization systems. Because the problem has yet to be adequately resolved, research continues on adapting and enhancing these systems for CMS applications.

Need to Facilitate Assembly of Input Data Sets

Accessing, formatting, and compiling input data sets is perhaps the most time-consuming aspect of air quality modeling. The needs here are for (*1*) creating data archives in locations and formats that are easily accessible with adequate documentation to allow the user to ascertain the appropriateness of the data for the application, (*2*) routines for screening data quality, and (*3*) software for creating gridded data fields at the spatial and temporal resolution required by the models. The whole process should be made faster, more convenient, and accurate.

Need for Assessing All Impacts of Airborne Materials

A common use for air quality modeling systems is to assess the air quality impacts of changes in emissions. Because of the way regulations are written, regulators have historically assessed single-issue impacts such as ozone or acid deposition. Recognizing that emissions changes usually influence more than one issue (because issues are largely subjective constructs whereas emissions alter the chemical and physical state of atmospheric constituents that do not respect issue boundaries), model users are beginning to be more interested in looking at all the atmospheric manifestations of some emitted materials. An often-cited example of an emitted substance that is associated with many issues is NO_x. It catalyzes formation of tropospheric ozone, plays a role in determining the oxidizing power of the atmosphere, is capable of acidifying precipitation, is a precursor of suspended particles, and can influence visibility. If studied from the perspective of a single issue, its mitigation may exacerbate another, as when NO_x reductions actually increase the rate of production of ozone. Another example is provided by implementation of the provisions of Title IV of the 1990 Clean Air Act Amendments in the United States regarding reduction of SO_2 emissions from large, elevated point sources. The resulting lower ambient particulate sulfate has resulted in a larger absolute and proportional concentration of nitrate particles. This is a consequence of the fact that, in the competition for available ammonia to form ammonium salts, ammonium sulfate is thermodynamically favored over ammonium nitrate.

A strong motivation for this trend toward a broader perspective is the growing appreciation that effective air quality managers should seek to optimize mitigation across all relevant impacts. Of course, there will probably be a continuing demand for single-issue assessments until regulations are written from a broader perspective; nevertheless, a CMS must provide a capability for studying both the troposphere as a whole and more limited sets of its chemical and physical responses to emissions changes.

Need to Make Models Accessible and Intelligible

Only highly trained specialists are able to set up and run today's regional- and urban-scale emissions, meteorological, and air quality models. This severely constrains the number of people who can do the modeling. Sometimes these specialists are the clients themselves, ie, the ones who intend to use the modeling results. More often, however, the clients and modelers are distinct. This arrangement discourages experimentation and fosters ignorance on the part of the clients of models' limitations and strengths, because at best they only see the models at arms' length. It also prevents the beneficial cross-fertilization that accrues when many users provide feedback on the strengths and weaknesses of the models. One consequence of the current situation that is good for some and bad for others is that a few entrepreneurial specialists have been rewarded with a virtual monopoly on the exercise of some regulatory models, causing the clients to pay more than they would have to in an environment more conducive to hands-on model use.

Efforts are now underway to alleviate the situation by automating the model setup and application process. A description of one example of a capable early prototype, the Environmental Decision Support System (EDSS), developed by the North Carolina Supercomputing Center, can be found on the World Wide Web at [http://www.mcnc.org]. An example of an alternative approach is the first prototype CMS from the Consortium for Advanced Modeling of Regional Air Quality (CAMRAQ) that can be accessed at [http://camraq.owt.com]. Both of these systems are described below. The idea of letting the computer do most of the work in an intuitive but rigorous manner, so that credible and affordable modeling is available to a wider community of users, is one of the principal objectives of CMS developers.

Need for Working Collaboratively

Only recently have the communications and computing abilities of networked computers been exploited to allow multiple users to work interactively on air quality model development, application, and evaluation. Such collaborations can lead to large productivity increases and could be fostered by making the process easy, flexible, convenient, and broadly functional. Much work is needed to incorporate these attributes into the modeling environment of a broader community of users.

Need for Analyzing Policy Options

There has been an unfortunate tendency in the policy community over the last decade or so to ignore model uncertainty and to treat model outcomes as deterministic, reliable predictions of how the real world would respond to changes in initial conditions such as emissions and temperature. Avoiding the subject of modeling uncertainty has probably contributed to misguided air quality management policy and public misconception of the probable efficacy of that policy. Two factors that have contributed to this tendency are the absence of rigorous methods for estimating uncertainty in modeling outcomes and the lack of formal procedures for explicitly taking into account uncertainty in analyzing air quality management options. Although both research areas, ie, modeling uncertainty estimation and decision analysis in the face of uncertainty, are enjoying increased attention in the air quality management arena (15), their essential nature needs to be more widely recognized. Such recognition dictates that a CMS must include a capability for uncertainty estimation and decision analysis.

Need for Effectively Communicating Results

In most instances the primary reason for applying air quality models is to explore the outcome of some change that has been made to the model's input or formulation. The outcome in its raw form will be voluminous. It is the result of complex computations with uncertainty that may or may not be known or easily quantifiable. The volume of output must be organized and reduced to a more manageable and interpretable form. If the results are to be presented to nonmodelers unfamiliar with model limitations, some statement of confidence in the results should accompany them to ensure they are used in the proper context. Confidence levels must also accompany outputs from air quality modeling systems that are to be input to decision support or risk assessment systems. Modeling system capabilities that are required to accomplish these tasks include not only those referred to above for data manipulation, sensitivity analysis, and model evaluation, ie, a DBMS, a geographical information system (GIS), and a visualization system, but also an uncertainty estimation system that synthesizes the sensitivity and model evaluation results into estimated confidence levels in the modeled outcomes and a report generation system that organizes and displays the results in appropriate media such as maps, charts, animations, and summary tables.

MODEL TYPES

A distinction should be made here between models and modeling systems. A model is something that simulates one of the many aspects of the air quality assessment problem. These modeled aspects are emissions, meteorology, air quality, and decision analysis or risk management. Collectively, together with the hardware and ancillary software that facilitates their use and the data required to apply and evaluate them, they comprise a modeling system. Less complex modeling systems treat fewer aspects or execute them in less detail. A schematic representation of the relationships among the major components (aspects) of an air quality modeling system is given in Figure 1. Comprehensive modeling systems are able to treat all aspects at whatever level of detail that can be formulated and afforded. One rendition of the major components of comprehensive modeling system is shown in Figure 2.

WHAT MUST A CMS PROVIDE?

Although we have described why we need a CMS, what some of its parts should be, and existing systems and concepts from which those parts can evolve, this section attempts to answer in greater detail the question, what must a CMS provide? Simply stated, a CMS must provide an infrastructure that helps

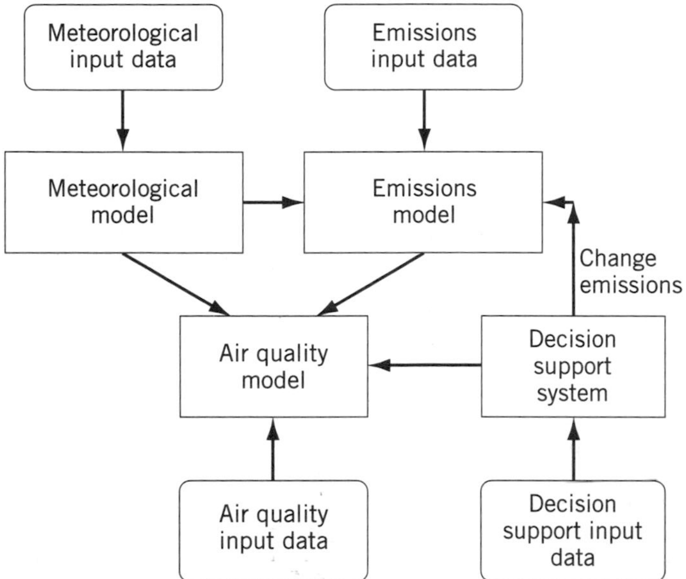

Figure 1. Relations among the major components of a modeling system for air quality management.

its users to do their jobs better, faster, and more economically, whether those jobs be regulatory and policy analysis, source impact assessment, understanding atmospheric chemistry and physics, or performing atmospheric research studies. As such, a CMS should provide the following: (*1*) a platform for modeling pollutant emissions, atmospheric physics and chemistry, and the impact of pollution in as scientifically sound a fashion as is desired or possible; (*2*) a readily accessible and user-friendly interface, so that its use is a benefit, not a distraction; (*3*) a powerful set of decision support and analysis tools, be they graphical, visual, economic, or scientific, including report preparation; and (*4*) a method to make optimal use of the available computational resources, including central processing unit (CPU) power, disk storage, and communication systems.

Many options should be available in a CMS for modeling air quality, ranging from inert dispersion models through reduced-form Eulerian models to fully comprehensive 3-D representations of tropospheric chemistry and physics, with optional process algorithms available for selection. A CMS should be designed in a way that facilitates its continuous evolution with that of science, computer capabilities, and user needs.

PRECURSORS OF A CMS

Recently, a number of groups have undertaken the task of building or designing systems of models that incorporate a growing number of the features desired in a CMS. The following are descriptions of a few of these, starting with a modeling system that combines meteorological and global dispersion modeling in one package, and proceeding to modeling systems that facilitate the exploration of alternative emissions control strategies by making model setup and input changes more convenient than with traditional models. This latter category is known as decision support systems.

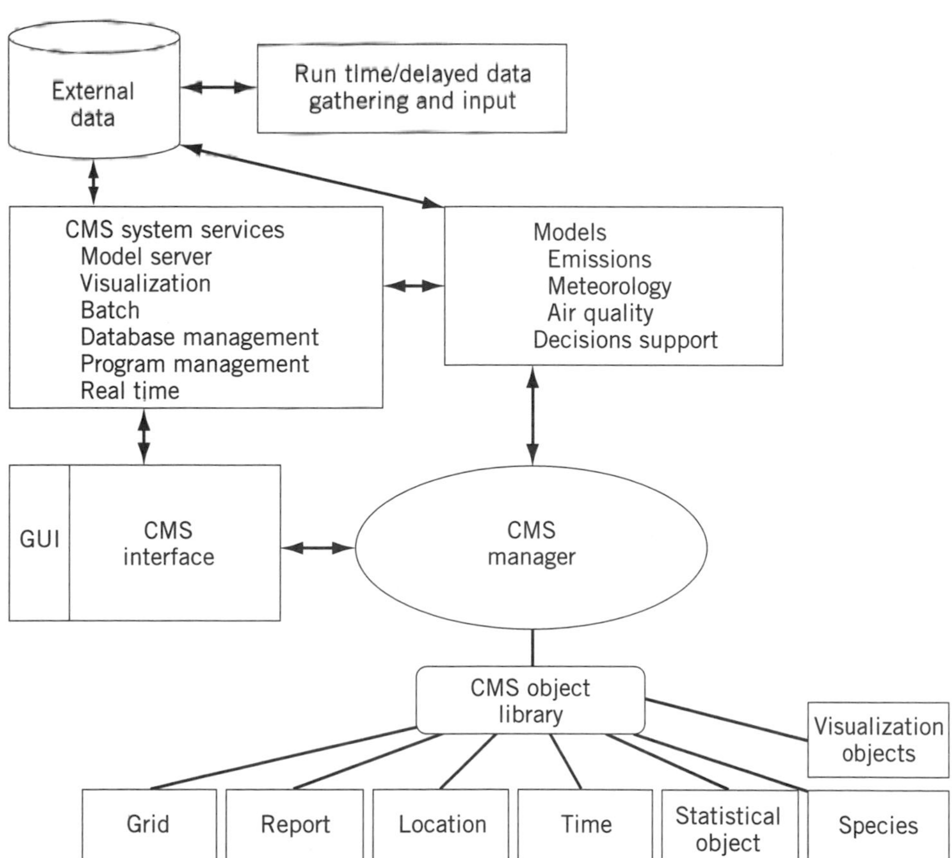

Figure 2. One rendition of a CMS (7). The external data required to support the broad range of CMS applications come from distributed archives and real-time sources. The graphic user interface (GUI) is the medium for communications between the user and the CMS. The CMS manager directs (as objects) all of the programs, modules, subprograms, and libraries that appear in the system.

ACTUAL CMS DEVELOPMENT

Two efforts of significantly different scale have been mounted in the United States explicitly to develop comprehensive modeling systems, initially with the focus on air quality applications, with all the features and attributes described above. One is the U.S. EPA's program under the auspices of the Federal High Performance Computing and Communications (HPCC) initiative, Models-3, which is funded annually at a level of millions of dollars. The other is CAMRAQ's program, funded at an annual level of a few hundred thousand dollars. Reflecting the disparate funding levels, each program is following a different developmental philosophy. EPA is investigating design and then structural components in great depth and has a strong emphasis on developing technology for many components inhouse. CAMRAQ, or Consortium for Advanced Modeling of Regional Air Quality, has opted to benefit from EPA's in-depth analyses through coordination with Models-3 developers in the selection of CMS elements and to rely more heavily on adaptation of legacy air quality models and quick-prototyping to test out concepts and gain user feedback. CAMRAQ, being an international consortium of public and private organizations, views itself as having potentially to serve a wider user community than does EPA.

FUTURE EXPECTATIONS

Those with a stake or interest in comprehensive modeling hope that the introduction of early prototypes will stimulate sufficient enthusiasm in the air quality modeling and management community that backing and progress will accelerate, or at least continue at present levels, on all the fronts required to realize the dream of a fully functional system with all the attributes and advantages discussed above. Once the tangible benefits of a CMS can be demonstrated, support for its further development should grow. But what can be expected beyond what has been described so far? To conclude, let us consider four possibilities.

Multimedia Assessments

Many of the same system functions that are developed to support air quality research and assessment could be easily adapted to other media such as ground and surface water, soils, and ecosystems. Because the air is coupled to surface water and terrestrial systems, multimedia modeling is a natural extension of the emissions and deposition components of an air quality CMS to exposure and effects models.

Virtual Reality

Zannetti's vision of the future is a CMS in which the user is a participant and even the graphic user interface (GUI) is transparent to the user (16). Virtual reality could speed up problem diagnosis by allowing the expert participant to compare details of the simulated experience with knowledge of those expected in the actual experience. In principle, much more information could be simultaneously processed and its internal consistency assessed in this mode than through a conventional graphic user interface. Although the full benefits of virtual reality applied to environmental research and assessment are not abundantly apparent, an open mind should be kept in case opportunities for exploitation should present themselves in the future.

Desktop Systems

Desktop systems already play a prominent role in the CAMRAQ/CMS/GUI. If the power of these systems continues to increase at anything approaching its present rate, more and more CMS functions that were formerly handled in the distributed computing environment can and will be shifted to the desktop.

Megacomputing

Existing air quality modeling systems already run on supercomputers, although there is a distinct trend toward porting them to high-level work stations as these increase in power. This trend and that toward more widespread use of PCs as GUIs will undoubtedly continue. However, as modelers and managers demand more explicit process representations across more media (perhaps in a virtual reality setting), with more advanced decision support capabilities, only the most powerful computing environments are expected to be up to this grand challenge. These cutting edge computing environments will probably include a variety of massively parallel and vector computers, workstations, and PCs, with architectures particularly suited for subsets of the overall computational task, networked by very high bandwidth ($\gg$1 Gigabit/s) links, and managed by an operating system optimized for the type of application. This would be the world of megacomputing, which would allow applications for a CMS which can only be speculated about today.

BIBLIOGRAPHY

1. J. R. Cabrillo, *Diary,* 1542.
2. R. J. Pierson, personal communication, 1995.
3. J. Evelyn, *Fumifugium: or, The Inconvenience of the Aer and Smoak of London Dissipated, Together with Some Remedies Humbly Proposed,* Bedel and Collins, London, 1661.
4. A. J. Haagen-Smit, C. E. Bradley, and M. M. Fox, *Indust. Eng. Chem.* **45,** 2086 (1953).
5. A. J. Haagen-Smit, *Indust. Eng. Chem.* **44,** 1342 (1952).
6. D. A. Hansen and co-workers, *Environ. Sci. Technol.* **28,** 71A–77A (1994).
7. P. Zannetti and co-workers, *Framework Design: Design and Development of a Comprehensive Modeling System for Air Pollution,* Consortium for Advanced Modeling of Regional Air Quality, Palo Alto, Calif., 1996.
8. K. K. Droegemeier and co-workers, in A. Tentner, ed., *High Performance Computing,* Addison-Wesley Publishing Co., Inc., Rending, Mass., 1995.
9. H. E. Jeffries and S. Tonnesen, *Atmos. Environ.* **28,** 2991–3003 (1994).
10. J. B. Milford and co-workers, *Environ. Sci. Technol.* **26,** 1179–1189 (1992).
11. D. Hwang and D. W. Byun, in Ref. 8, pp. 52–59.
12. M. T. Odman and co-workers, *An Investigation of Error Propagation in the California Air Resources Board Air Quality Model,* California Air Resources Board, Sacramento, Calif., 1992.
13. S. Reynolds and co-workers, *The Importance of Using Alternative Base Cases in Photochemical Modeling,* API Publication Number 4616, American Petroleum Institute, Washington, D.C., 1994.

14. U.S. EPA, *Guidelines for Regulatory Applications of the Urban Airshed Model,* EPA-450/4-91-013, U.S. Environmental Protection Agency, Research Triangle Park, N.C., 1991.
15. S. J. Roselle, *Atmos. Environ.* **28,** 1757–1772 (1994).
16. P. Zannetti, in H. Power, N. Moussiopoulos, and C. A. Brebbia, eds. *Air Pollution Theory and Simulation,* Computational Mechanics Publications, Southampton, U.K., 1995, pp. 3–10.

AIR SAMPLING INSTRUMENTS

ALAN S. KAO
ENVIRON
Arlington, Virginia

Considerable advances have been made in the development of instruments for gas- and particulate-phase air contaminants over the past several decades. Whereas early efforts in the early to mid-1900s to measure air pollutants focused on optical observations of black smoke from industrial stacks or "British Smoke Shade" measurements, air sampling instruments currently are capable of collecting data on a wide range of physical and chemical properties of air constituents in both the gas and particulate phases. Many of the different types of instruments available for sampling both gas- and particulate-phase air contaminants, including methods that have been approved by the United States Environmental Protection Agency (U.S. EPA), are discussed herein. The scope is primarily limited to methods that capture samples of air pollutants on a collection substrate for laboratory analysis; continuous monitoring techniques that do not involve sample collection are not discussed in depth.

SAMPLING CONSIDERATIONS

When sampling gaseous species, the principal considerations include collecting representative samples, ie, identifying appropriate sampling locations, and selecting the proper analytical technique. The property of concern usually is the concentration of an individual chemical compound, eg, ozone, or the total amount of a class of chemical compounds, eg, total nonmethane organic compounds (NMOC).

Atmospheric particulate matter is especially complex, consisting of solid or liquid particles ranging in size from a few nanometers (nm) to greater than fifty micrometers (μm). Each particle may contain a broad mixture of chemical species, including organic compounds, trace metals, and water-soluble compounds, eg, hydrogen (H^+), sulfate (SO_4^{2-}), nitrate (NO_3^-), and ammonium (NH_4^+) ions. Airborne particles are of interest for several reasons. Small particles can be inhaled deeply into the respiratory system, potentially resulting in adverse health effects, eg, pulmonary inflammation or mechanical changes in lung function. These health effects may be associated with chemical components of the particles, with reactive species, eg, free radicals, that may be associated with fine particles, or with the physical properties of the particles themselves (1). Particles are also responsible for visibility problems and haze in national parks and for the soiling of statues and buildings, which results from dry deposition.

As a result, airborne particles can be characterized in several different manners, including size, chemical composition, physical properties, and formation mechanism or origin. Examples of different chemical and physical properties of interest for different applications include the following:

- Quantification of the exposure of a population to inhalable particulate matter (IPM) involves knowledge of the total mass of particles small enough to be inhaled deeply into the respiratory system; parameters of interest in this case are total mass and particle size.
- Epidemiologists seeking to link a chemical constituent of IPM with observed health effects are interested in characterizing the chemical composition as well as particle size.
- Physical properties such as light scattering potential is of interest in the investigation of regional haze or visibility impairment problems in a national park or urban airshed. Other physical properties such as deposition velocity can be used to assess potential soiling or erosion of a building or statue.
- A regulator or risk manager may classify the aerosol in terms of its sources in order to develop strategies and policies to reduce ambient concentrations. The aerosol particles may also be characterized as primary (directly emitted or generated) or secondary (formed by gas–particle conversion).

The type of instrument, or combination of instruments, selected for air sampling depends on the properties of interest and several additional considerations.

Compliance Monitoring

Pursuant to the Clean Air Act, EPA has established and periodically updates National Ambient Air Quality Standards (NAAQS) for six criteria pollutants: ozone (O_3), nitrogen dioxide (NO_2), sulfur dioxide (SO_2), carbon monoxide (CO), particulate matter, and lead (Pb). States and local air quality management districts are required to meet these standards. The current NAAQS for the criteria pollutants are summarized in Table 1. Depending on the chronic or acute nature of the health effects for each pollutant, various averaging times (ranging from one hour to one year) have been specified as the basis for the NAAQS. Areas that do not meet the NAAQS for any of the criteria pollutants are designated as nonattainment areas.

Among the six criteria pollutants only particulate matter is not chemical-specific. The first standard for particulate matter, promulgated in 1971, applied to total suspended particulate (TSP) mass. The measurement of TSP was based on a specific sampling method, the high-volume (hivol) sampler, which has been found to collect particles with aerodynamic diameters $<$25–50 μm, depending on wind speed and direction. The current particulate standard is based on total mass of particulate matter with aerodynamic diameters $<$10 micrometer (PM_{10}). The 10-μm cutpoint was selected in recognition of the fact that smaller particles are more likely to cause respiratory problems. Most particles larger than 10 μm are removed in the mouth or nose prior to entering the respiratory tract; as much as 60% of particles with aerodynamic diameters $<$10 μm pass the trachea and reach the lung and alveolar sacs, where they might cause harm. As of this writing (1996), the NAAQS for particulate matter was in the process of being reviewed, and expected to be revised in 1997. The review of the particulate

Table 1. National Ambient Air Quality Standards[a]

Criteria Pollutant	Primary Standard	Secondary Standard
Sulfur oxides[b]	80 μg/m^3 (0.03 ppm), annual arithmetic mean; and 365 μg/m^3 (0.14 ppm), 24-h (max) concentration not >1/yr	1300 μg/m^3 (0.5 ppm), 3-h (max) concentration not >1/yr
Particulate matter[c,d]	150 μg/m^3, 24-h (max) PM_{10} concentration not >1/yr and 50 μg/m^3, annual arithmetic mean (PM_{10})	same as primary standard
Carbon monoxide[e]	10 μg/m^3 (9 ppm), 8-h (max) average and 40 μg/m^3 (35 ppm), 1-h (max) average concentration not >1/yr	none
Ozone[f,g]	235 μg/m^3 (0.12 ppm), hourly average (max) concentration not >1/yr (averaged over 3 consecutive years)	same as primary standard
Nitrogen dioxide[h]	100μg/m^3 (0.053 ppm), annual arithmetic mean	same as primary standard
Lead[i]	1.5 μg/m^3, arithmetic mean (max) averaged over calendar quarter	same as primary standard

[a] Ref. 2.
[b] Ref. 3.
[c] Ref. 4.
[d] On July 18, 1997, the NAAQS for PM was revised, and is scheduled to be implemented after 2005. The revised standard includes an annual average of 15 μg/m^3 and a 24-h maximum of 65 μg/m^3 for $PM_{2.5}$ (2,4,5).
[e] Ref. 6.
[f] Ref. 7.
[g] On July 18, 1997, the NAAQS for ozone was revised, and is scheduled to be implemented after 2004. The revised standard replaces the one-hour standard with an eight-hour standard of 0.08 ppm for the fourth highest daily maximum (2,7,8).
[h] Ref. 9.
[i] Ref. 10.

standard focuses on finer particles with aerodynamic diameters $\leq$2.5 μm, ie, $PM_{2.5}$ (2,4,5).

According to 1995 data, the following number of regions in the United States did not meet the NAAQS (11): ozone, 68; nitrogen dioxide, 1; sulfur oxides, 43; carbon monoxide, 31; particulate matter (PM_{10}), 81; and lead, 10.

Measurements for determining an area's attainment status are collected from a national monitoring network comprised of both national air monitoring stations(NAMS) and state and local air monitoring stations (SLAMS). NAMS were established to ensure a long-term national network for urban air-oriented ambient monitoring, and to provide a systematic, consistent database for air quality comparisons and trends analysis. These monitoring stations are supplemented with SLAMS, which allow state and local governments to develop networks tailored for their immediate monitoring needs. All data collected from the NAMS and SLAMS conform to uniform criteria for monitor siting, instrumentation, and quality assurance, and are available from USEPA's Aerometric Information Retrieval System (AIRS) on-line database. Data are available from over 1200 NAMS and SLAMS locations.

Measurement methodologies used to collect data for compliance monitoring purposes must meet certain performance standards (2). Performance standards for particulate matter, for example, specify the particle properties that are to be determined and the permissible tolerances for accuracy and precision in quantifying those properties. Any PM_{10} sampler that meets these criteria can be applied to determine compliance with the NAAQS. The performance standards for PM_{10} samplers include (12) the following.

Sampling effectiveness: for liquid particles, the expected mass concentration must be within $\pm$10% of that predicted for an "ideal" sampler (12). For solid particles, the sampling effectiveness obtained using 25-μm solid particles at two wind speeds (8 and 24 km/h) must be no more than 5% above that obtained for liquid particles of the same size.

Cutpoint: the 50% cutpoint must fall within 10 $\pm$ 0.5-μm aerodynamic diameter, as determined in a wind tunnel using 10 liquid particle sizes ($3 \pm 0.5, 5 \pm 0.5, 7 \pm 0.5, 9 \pm 0.5, 10 \pm 0.5, 11 \pm 0.5, 13 \pm 1, 15 \pm 1, 20 \pm 1$, and 25 ± 1 μm) and three wind speeds (2, 8, and 24 km/h).

Precision: measurement precision must be within 5 μg/m^3 (7%) for three collocated samplers for 10 periods of 24 h.

Flow rate stability: average flow rate over each of ten 24-h sampling periods must be within $\pm$5% of initial flow rate; all measured flow rates over each 24-h period must be within $\pm$10% of initial flow rate.

Sampling methods that meet these criteria and are based on the principle of inertial separation of the PM_{10} size fraction followed by filtration and gravimetric mass determination can be designated as reference methods. Sampling methods based on other principles that achieve equivalent performance when collocated with reference samplers under certain test conditions can be designated as equivalent methods. Either reference or equivalent sampling methods can be used for demonstrating compliance with the NAAQS.

Currently, seven reference and three equivalent methods have been approved by USEPA (13) for sampling PM_{10}. These methods, described in the most recent PM_{10} criteria document (14), are summarized in Table 2. Similarly, numerous reference and equivalent methods have been approved by USEPA for sampling gas-phase criteria pollutants. These methods are summarized in Tables 3 through 7.

Source Apportionment

States and local air quality management districts with nonattainment areas are required to develop or revise state implementation plans (SIPs) that demonstrate how emissions that contribute to ambient levels of criteria pollutants will be

Table 2. U.S. EPA-Approved Reference and Equivalent Methods for PM_{10}[a]

Method Number	Identification	CFR Reference
	Manual Reference Methods	
RFPS-1087-062	Wedding & Associates PM_{10} Critical Flow High-Volume Sampler	52 FR 37366 (10/6/87)
RFPS-1287-063	Sierra-Anderson or General Metal Works (GMW) Model 1200 PM_{10} High-Volume Air Sampler	52 FR 45684 (12/1/87); 53 FR 1062 (1/15/88)
RFPS-1287-064	Sierra-Anderson or GMW Model 321-B PM_{10} High-Volume Air Sampler	52 FR 45684 (12/1/87); 53 FR 1062 (1/15/88)
RFPS-1287-065	Sierra-Anderson or GMW Model 321-C PM_{10} High-Volume Air Sampler	52 FR 45684 (12/1/87); 53 FR 1062 (1/15/88)
RFPS-0389-071	Sierra-Anderson Model SA-254 Medium-Volume PM_{10} Sampler (Oregon Dept. of Environmental Quality)	54 FR 12273 (3/24/89)
RFPS-0789-073	Sierra-Anderson Models SA241 and SA241M or GMW Models G241 and G241M PM_{10} Dichotomous Samplers (virtual impactor)	54 FR 31247 (7/27/89)
	Automated Reference Methods	
RFPS-0694-098	Rupprecht & Patashnick Partisol Model 2000 Low-Volume Air Sampler	59 FR 35338 (7/11/94)
	Automated Equivalent Methods	
EQPM-0990-076	Anderson Instruments Model FH621-N PM_{10} Beta Attenuation Monitor	55 FR 38387 (9/18/90)
EQPM-1090-079	Rupprecht & Patashnick TEOM Series 1400/1400a PM_{10} Monitors	55 FR 43406 (10/29/90)

[a] Refs. 13 and 14.

Table 3. USEPA-Approved Reference and Equivalent Methods for Nitrogen Dioxide[a]

Method Number	Identification	Reference
	Automated Reference Methods	
RFNA-0691-082	Advanced Pollution Instrumentation Model 200 Nitrogen Oxides Analyzer (NOA)	56 FR 27014 (6/12/91)
RFNA-1194-099	Advanced Pollution Instrumentation Model 200A NOA	59 FR 601892 (12/2/94)
RFNA-0179-034	Beckman Model 952-A $NO/NO_2/NO_x$ Analyzer	44 FR 7806 (2/7/79)
RFNA-0470-038	Bendix Model 8101-B Oxides of Nitrogen Analyzer	44 FR 26792 (5/7/79)
RFNA-0777-022	Bendix or Combustion Engineering Model 8101-C Oxides of Nitrogen Analyzer	42 FR 37435 (7/21/77)
RFNA-0977-025	Columbia Scientific Industries (CSI) Model 1600 and 5600 Oxides of Nitrogen Analyzers	42 FR 46574 (9/16/77)
RFNA-1192-089	Dasibi Model 2108 Oxides of Nitrogen Analyzer	57 FR 55530 (11/25/92)
RFNA-0795-104	Environment SA Model AC31M Chemiluminescent NOA	60 FR 38326 (7/26/95)
RFNA-1292-090	Lear Siegler Measurement Controls Corporation or Monitor Labs Model ML9841 or ML9841A NOA	57 FR 60198 (12/18/92)
RFNA-1078-031	Meloy Model NA530R NOA	43 FR 50733 (10/31/78); 44 FR 8327 (2/9/79)
RFNA-0677-021	Monitor Labs Model 8440E NOA	42 FR 37434 (7/21/77); 42 FR 46575 (9/16/77); 46 FR 29986 (6/4/81)
RFNA-0280-042	Monitor Labs or Lear Siegler Model 8840 NOA	45 FR 9100 (2/11/80); 46 FR 29986 (6/4/81)
RFNA-0991-083	Monitor Labs or Lear Siegler 9941 NOA	45 FR 47473 (9/19/91)
RFNA-0879-040	Philips Model PW9762/02 $NO/NO_2/NO_x$ Analyzer	44 FR 51683 (9/4/79)
RFNA-0179-035	Thermo Electron or Thermo Environmental Instruments Model 14 B/E Chemiluminescent $NO/NO_2/NO_x$ Analyzer	44 FR 7805 (2/7/79); 44 FR 54545 (9/20/79)
RFNA-0279-037	Thermo Electron or Thermo Environmental Instruments Model 14 D/E Chemiluminescent $NO/NO_2/NO_x$ Analyzer	44 FR 10429 (2/20/79)
RFNA-1289-074	Thermo Electron or Thermo Environmental Instruments Model 42 or Model 42C $NO/NO_2/NO_x$ Analyzer	54 FR 50820 (12/11/89)
	Manual Equivalent Methods	
EQN-1277-026	Sodium Arsenite Method for NO_2	42 FR 62971 (12/14/77)
EQN-1277-027	Sodium Arsenite Method for NO_2, Technicon II	42 FR 62971 (12/14/77)
EQN-1277-028	TGS–ANSA Method for NO_2	42 FR 62971 (12/14/77)
	Automated Equivalent Methods	
EQNA-0495-102	Opsis Model AR 500 System Open (Long) Path Ambient Air Monitoring System for NO_2	60 FR 21518 (5/2/95)

[a] Ref. 13.

reduced. The levels of pollutants measured at ambient monitoring stations generally originate from numerous sources. When developing strategies for reducing ambient levels of air pollutants, it is important to quantify the contributions from the sources, thereby allowing the greatest resources to be allocated to addressing the largest contributors.

Receptor models are data analysis techniques that use ambient measurements of chemical species to apportion the

Table 4. U.S. EPA-Approved Reference and Equivalent Methods for Sulfur Dioxide[a]

Method Number	Identification	Reference
	Manual Reference Method	
40 CFR Part 50, Appendix A	Reference Method for SO_2 (Pararosaniline method)	47 FR 54899 (12/6/82); 48 FR 17355 (4/22/83)
	Manual Equivalent Methods	
EQS-0775-001	Pararosaniline Method for SO_2, Technicon I	40 FR 34024 (8/13/75)
EQS-0775-002	Pararosaniline Method for SO_2, Technicon II	40 FR 34024 (8/13/75)
	Automated Equivalent Methods	
EQSA-0990-077	Advanced Pollution Instrumentation Model 100 Fluorescent SO_2 Analyzer	55 FR 38149 (9/17/90)
EQSA-0495-100	Advanced Pollution Instrumentation Model 100A SO_2 Analyzer	60 FR 17061 (4/4/95)
EQSA-0877-024	ASARCO Model 500 SO_2 Monitor[b]	42 FR 44264 (9/2/77); 44 FR 67522 (11/26/79)
EQSA-0678-029	Beckman Model 953 Fluorescent Ambient SO_2 Analyzer	43 FR 35995 (8/14/78)
EQSA-1078-030	Bendix Model 8303 Sulfur Analyzer	43 FR 50733 (10/31/78)
EQSA-0494-095	Columbia Scientific Industries Model 5700 SO_2 Analyzer	59 FR 18818 (4/20/94)
EQSA-1086-061	Dasibi Model 4108 UV Fluorescence SO_2 Analyzer	51 FR 32244 (9/10/86)
EQSA-0292-084	Environnement SA Model AF21M SO_2 Analyzer	57 FR 5444 (2/14/92)
EQSA-0486-049	Lear Siegler Model AM2020 SO_2 Monitor	45 FR 79574 (12/1/80); 46 FR 9997 (1/30/81)
EQSA-1275-005	Lear Siegler Model SM1000 Ambient SO_2 Monitor	41 FR 3893 (1/27/76); 41 FR 32946 (8/6/76); 42 FR 13044 (3/8/77); 45 FR 1147 (1/4/80)
EQSA-0193-092	Lear Siegler Measurement Controls Corporation or Monitor Labs Model ML9850 SO_2 Analyzer	58 FR 6964 (2/3/93)
EQSA-1275-006	Meloy Model SA185-2A SO_2 Analyzer	41 FR 3893 (1/27/76); 4 FR 38088 (8/25/78)
EQSA-1078-032	Meloy Model SA285E SO_2 Analyzer	43 FR 50733 (10/31/78)
EQSA-0580-046	Meloy Model SA 700 Fluorescence SO_2 Analyzer	45 FR 31488 (5/13/80)
EQSA-0876-013	Monitor Labs Model 8450 Sulfur Monitor	41 FR 36245 (8/27/76); 44 FR 33476 (6/11/79)
EQSA-0779-039	Monitor Labs or Lear Siegler Model 8850 Fluorescent SO_2 Analyzer	44 FR 44616 (7/30/79)
EQSA-0390-075	Monitor Labs or Lear Siegler Model 8850S SO_2 Analyzer	55 FR 5264 (2/14/90)
EQSA-0495-101	Opsis Model AR 500 Open (Long Path) Ambient Air Monitoring System for SO_2	60 FR 21518 (5/2/95)
EQSA-0676-010	Philips PW9755 SO_2 Analyzer	41 FR 26252 (6/25/76); 41 FR 46019 (10/19/76); 42 FR 28571 (6/3/77)
EQSA-0876-011	Philips PW9700 SO_2 Analyzer	41 FR 34105 (8/12/76)
EQSA-0276-009	Thermo Electron Model 43 Pulsed Fluorescent SO_2 Analyzer	41 FR 8531 (2/27/76); 41 FR 15363 (4/12/76); 42 FR 20490 (4/20/77); 44 FR 21861 (4/12/79); 45 FR 2700 (1/14/80); 45 FR 32419 (5/16/80)U
EQSA-0486-060	Thermo Electron or Thermo Environmental Instruments Model 43A or 43B Pulsed Fluorescence SO_2 Analyzers	51 FR 12390 (4/10/86)

[a] Ref. 13.
[b] No longer commercially available.

contributions of various sources. These models assume that emissions from different sources have characteristic chemical profiles. The differences in chemical composition among source emissions allow the contributions from each source type to be inferred. The most widely used receptor model is the chemical mass balance (CMB) (15). The primary applications of the CMB have been for particulate matter and volatile organic compounds (VOCs). In addition to the CMB, a number of multivariate techniques have been developed that evaluate similarities in temporal variations among chemical components that occur with emissions from individual sources (14).

In order to obtain data suitable for use with the CMB or other receptor models, several types of chemical analysis are usually used. For example, typical source profiles for particulate matter include measurements of trace elements, water-soluble ions, and organic and elemental carbon. The chemical compositions measured at a particular sampling location can vary significantly depending on the size fraction collected. These differences in elemental, organic, and ionic composition among sources in different size fractions allow the contributions from each source to be inferred. Therefore, the samples collected must be amenable to analysis by several different analytical techniques, such as x-ray fluorescence analysis,

Table 5. U.S. EPA-Approved Reference and Equivalent Methods for Carbon Monoxide[a]

Method Number	Identification	CFR Reference
	Automated Reference Methods	
RFCA-1093-093	Advanced Pollution Instrumentation Model 300 Gas Filter Correlation CO Analyzer	58 FR 58166 (10/29/93)
RFCA-0876-012	Beckman Model 866 Ambient CO Monitoring System	41 FR 36245 (8/27/76)
RFCA-0276-008	Bendix or Combustion Engineering Model 8501-5CA Infrared CO Analyzer	41 FR 7450 (2/18/76)
RFCA-0381-051	Dasibi Model 3003 Gas Filter Correlation Environmental CO Analyzer	46 FR 20773 (4/7/81)
RFCA-0488-067	Dasibi Model 3008 Gas Filter Correlation CO Analyzer	53 FR 12073 (4/12/88)
RFCA-0995-108	Environment SA Model CO11M Ambient Carbon Monoxide Analyzer	60 FR 54684 (10/25/95)
RFCA-1278-033	Horiba Models AQM-10, AQM-11, and AQMR Ambient CO Monitoring Systems	43 FR 58429 (12/14/78)
RFCA-1180-048	Horiba Model APMA-300E Ambient CO Monitoring System	45 FR 72774 (11/3/80)
RFCA-0895-106	Horiba Model APMA-360 Ambient CO Monitor	60 FR 39382 (8/2/95)
RFCA-0992-088	Lear Siegler Measurement Controls Corp. or Monitor Labs Model ML9830 CO Analyzer	57 FR 44565 (9/28/92)
RFCA-1280-050	MASS-CO Model 1 CO Analyzer	45 FR 81650 (12/11/80)
RFCA-0979-041	Monitor Labs Model 8310 CO Analyzer	44 FR 54545 (9/20/79); 45 FR 2700 (1/14/80)
RFCA-0388-066	Monitor Labs or Lear Siegler Model 8830 CO Analyzer	53 FR 7233 (3/7/88)
RFCA-0177-018	MSA/LIRA Model 202S Air Quality CO Analyzer System	42 FR 5748 (1/31/77)
RFCA-0981-054	Thermo Electron or Thermo Environmental Instruments Model 48 Gas Filter Correlation Ambient CO Analyzer	46 FR 47002 (9/23/81)

[a] Ref. 13.

Table 6. U.S. EPA-Approved Reference and Equivalent Methods for Ozone[a]

Method Number	Identification	Reference
	Automated Reference Methods	
RFOA-0577-020	Beckman Model 950A Ozone Analyzer	42 FR 28571 (6/3/77)
RFOA-0176-007	Bendix or Combustion Engineering Model 8002 Ozone Analyzer	41 FR 5145 (2/4/76); 45 FR 18474 (3/21/80)
RFOA-0279-036	Columbia Scientific Industries Model 2000 Ozone Meter	44 FR 10429 (2/20/79)
RFOA-1076-014 RFOA-1076-015 RFOA-1076-016	McMillan (MEC) Models 1100-1, 1100-2, and 1100-3 Ozone Meters	41 FR 46647 (10/22/76); 42 FR 30235 (6/13/77)
RFOA-1075-003	Meloy Model OA325-2R Ozone Analyzer	40 FR 54856 (11/26/75)
RFOA-1075-004	Meloy Model OA350-2R Ozone Analyzer	40 FR 54856 (11/26/75)
RFOA-1176-017	Monitor Labs Model 8410E Ozone Analyzer	41 FR 53684 (12/8/76)
	Automated Equivalent Methods	
EQOA-0992-087	Advanced Pollution Instrumentation Model 400 Ozone Analyzer	57 FR 44565 (9/28/92)
EQOA-0577-019	Dasibi Models 1003-AH, 1003-PC, or 1003-RS Ozone Analyzers	42 FR 28571 (6/3/77)
EQOA-0383-356	Dasibi Models 1008-AH, 1008-PC, or 1008-RS Ozone Analyzers	48 FR 10126 (3/10/83)
EQOA-0990-078	Environics Series 300 Computerized Ozone Analyzer	55 FR 38386 (9/18/90)
EQOA-0895-105	Environnement SA Model $O_3$41M UV Photometric Ozone Analyzer	60 FR 39382 (8/2/95)
EQOA-0881-053	Monitor Labs or Lear Siegler Model 8810 Photometric Ozone Analyzer	46 FR 52224 (10/26/81)
EQOA-0193-091	Lear Siegler Measurement Controls Corp. Model ML9810 or Monitor Labs Models ML9810, ML9811, and ML9812 Ozone Analyzers	58 FR 6964 (2/3/93)
EQSA-0495-103	Opsis Model AR 500 Open (Long) Path Ambient Air Monitoring System	60 FR 21518 (5/2/95)
EQOA-0382-055	PCI Ozone Corp. Model LC-12 Ozone Analyzer	47 FR 13572 (3/31/82)
EQOA-0777-23	Philips PW9771 O_3 Analyzer	42 FR 38931 (8/1/77); 42 FR 57156 (11/1/77)
EQOA-0880-047	Thermo Electron or Thermo Environmental Instruments Models 49 UV Photometric Ambient O_3 Analyzer	45 FR 57168 (8/27/80)

[a] Ref. 13.

thermal/optical reflectance, ion chromatography, and automated colorimetry. This need for multiple chemical analyses often results in combinations of air sampling instruments being operated in parallel or in series.

Exposure Assessment

A key component of risk assessment is quantification of the levels of substances of potential concern to which receptors are exposed. For determining exposure, samples must be collected that are representative of the exposed population. Depending on the purpose of the exposure assessment, different methods for collecting samples are utilized. For estimating personal exposure, a portable sampler is typically used. This type of sampler provides an estimate of human exposure to particles as individuals conduct their daily activities, moving from location to location, eg, at home, at work, or during recreational activities. Because of the wide variability among the activities of diverse individuals within a population, large amounts of personal sampling data are required to evaluate personal

Table 7. U.S. EPA-Approved Reference and Equivalent Methods for Lead[a]

Method Number	Identification	CFR Reference
	Manual Reference Method	
40 CFR Part 50, Appendix G	Reference Method for the Determination of Lead in Suspended Particulate Matter Collected from Ambient Air	43 FR 46258 (10/5/78)
	Manual Equivalent Methods	
EQL-0783-058	Energy-Dispersive X-Ray Fluorescence Spectometry	48 FR 29742 (6/28/83);
EQL-0589-072		54 FR 20193 (5/10/89)
EQL-0380-043	Flame Atomic Absorption Spectrometry	45 FR 14648 (3/6/80)
EQL-0380-044	Flameless Atomic Absorption Spectrometry	45 FR 14648 (3/6/80);
EQL-0895-107		60 FR 39383 (8/2/95);
EQL-0785-059		50 FR 37909 (9/18/85)
EQL-0380-045	Inductively Coupled Argon Plasma-Optical Emission Spectrometry	45 FR 14648 (3/6/80);
EQL-1193-094		58 FR 61902 (11/23/93);
EQL-0592-085		57 FR 20823 (5/15/92);
EQL-0483-057		48 FR 14748 (4/5/83);
EQL-1188-069		53 FR 44947 (11/7/88);
EQL-1290-080		55 FR 49119 (11/26/90);
EQL-0592-086		57 FR 20823 (5/15/92);
EQL-0995-109		60 FR 54684 (10/25/95);
EQL-0995-110		53 FR 30866 (8/16/88);
EQL-0888-068		53 FR 48974 (12/5/88);
EQL-1288-070		59 FR 29429 (6/7/94)
EQL-0694-096		
EQL-0581-052	Wavelength Dispersive X-Ray Fluorescence Spectrometry	46 FR 29986 (6/4/81)

[a] Ref. 13.

exposures for the population. If, however, the purpose of the exposure assessment is to estimate potential exposures at a particular indoor or outdoor location, a stationary air sampler is more appropriate.

Analytical Methods

The type of air sampling instrument or sample collection substrate used is dependent on the type of analytical method to be used with the collected material. Care must be taken to eliminate any sampling artifacts specific to the chemical components of interest.

BIBLIOGRAPHY

1. A. S. Kao and S. K. Friedlander, *Inhal. Toxicol.* **7,** 149–156 (1995).
2. United States Environmental Protection Agency (U.S. EPA), National Primary and Secondary Ambient Air Quality Standards, *Code of Federal Regulations,* 40 CFR Part 50 and Appendices, U.S. Government Printing Office, Washington, D.C., 1996.
3. Ref. 2, § 50.4 and 50.5.
4. Ref. 2, § 50.6.
5. United States Environmental Protection Agency (U.S. EPA), *Review of the National Ambient Air Quality Standards for Particulate Matter: Policy Assessment of Scientific and Technical Information*, OAQPS Staff Paper, EPA-452/R-96-013, Office of Air Quality Planning and Standards, Research Triangle Park, N.C., July 1996.
6. Ref. 2, § 50.8.
7. Ref. 2, § 50.9.
8. United States Environmental Protection Agency (U.S. EPA), *Review of the National Ambient Air Quality Standards for Ozone: Assessment of Scientific and Technical Information, OAQPS Staff Paper,* EPA-452/R-96-007, Office of Air Quality Planning and Standards, Research Triangle Park, N.C., June 1996.
9. Ref. 2, § 50.11.
10. Ref. 2, § 50.12.
11. United States Environmental Protection Agency (U.S. EPA), National Air Quality and Emissions Trends Report, 1995, EPA 454/R-96-095, Office of Air Quality Planning and Standards, Research Triangle Park, N.C., Oct. 1996.
12. United States Environmental Protection Agency (U.S. EPA) Ambient Air Monitoring Reference and Equivalent Methods: Procedures for Testing Performance Characteristics of Methods for PM_{10}, *Code of Federal Regulations,* 40 CFR Part 53, Subpart D, U.S. Government Printing Office, Washington, D.C., 1996.
13. United States Environmental Protection Agency (U.S. EPA), *List of Designated Reference and Equivalent Methods*, Office of Research and Development, Research Triangle Park, N.C., Mar. 10, 1996.
14. United States Environmental Protection Agency (U.S. EPA), *Air Quality Criteria for Particulate Matter*, EPA/600/P-95/001aF-cF, Office of Research and Development, Washington, D.C., Apr. 1996.
15. S. K. Friedlander, *Environ. Sci. Technol.* **7,** 235–240 (1973).

AIR TOXIC CONTROL STRATEGIES FOR UTILITIES

GARY J. GRIECO
CHRIS WEDIG
Stone & Webster
New York, New York

Title III of the Clean Air Act Amendments (CAAA) of 1990 identifies 189 chemicals subject to control as hazardous air pollutants (HAPs). Many HAP-emitting sources will be required to implement Maximum Achievable Control Technology

(MACT). The CAAA established a schedule for EPA to create source categories and applicable emissions standards for Maximum Achievable Control Technology. The CAAA also mandated studies to regulate HAPs for electric utilities. These studies are currently in progress.

Studies have already been performed on the potential site-specific and global impacts of anticipated HAP legislation on the electric generating industry. Typical HAP emission studies being performed include (1)

- measurement of air HAPs emissions from utility generating units.
- evaluation of the HAP emission differential from an individual generating unit attributable to fuel-switching scenarios.
- placing an emission bubble over an existing generating station and evaluating HAP emission differentials due to one or more of the following: fuel switching; repowering one or more units; or retiring one or more units in conjunction with either repowering a unit or the construction of a new generating unit.
- industrywide impacts of various anticipated HAP legislation scenarios on emissions levels and control costs.

With the exception of certain advanced mercury control techniques, the technologies needed to reduce HAPs presently exist and are generally the same equipment currently used to reduce particulate, acid gases, and organic emissions.

BACKGROUND

The CAAA of 1990 were signed into law by President Bush on November 15, 1990. These amendments have generally been described as the most complex and comprehensive piece of environmental legislation ever enacted. Under Title III hazardous air pollutants (HAPs) are defined as pollutants that cause serious, irreversible or acute health effects in humans. These health effects are in the form of cancer, development effects, reproductive dysfunctions, neurological disorders, inheritable genetic mutations, or chronic toxicity, and are caused by ambient concentrations, bioaccumulation, or deposition. Only those pollutants not regulated by National Ambient Air Quality Standards (NAAQS) are considered HAPs under Title III. Therefore, ozone, carbon monoxide, particulates that are less than 10 microns in mean aerodynamic diameter (PM10), sulfur dioxide, and nitrogen dioxide are considered HAPs. In addition, pollutants regulated under Title VI of the CAAA (Stratospheric Ozone Protection) are not subject to regulation under Title III (2).

Under Title III, EPA published an initial list of 174 source categories required to implement controls for 189 HAPs. Rulemaking for all categories will take over 10 years. Existing sources affected by the rules will then have approximately three years to comply with standards once promulgated.

Other potential sources of HAPs not currently required to be regulated but must be studied under the CAAA include electric utility emissions, hydrogen sulfide emissions from oil and gas wells, mercury emissions, and accidental releases of hydrogen fluoride from chemical industries.

Title III of the CAAA virtually replaces an existing program for the control of HAPs known as NESHAP. Title III shifts the focus from risk-based control to a technology-based program of source category control. There is still a risk management component to the title which requires the EPA to conduct a study of the effectiveness of the technology-based program in reducing cancer risks, but this will result in further control only if the risks are found to be unacceptable.

As previously stated, EPA has published an initial list of 174 source categories required to implement MACT controls for 189 HAPs. MACT standards will apply to sources that emit more than 10 tons per year of an individual HAP or more than 25 tons per year of all HAPs combined. For new sources, MACT standards will be based on the control technology used by the best controlled source in the same source category. Existing source MACT will be based on the best performing 12 percent of existing sources in the same source category. The EPA is to take into account the cost, nonair quality environmental impact, and energy usage of control options in developing MACT standards. These standards will include control technologies and also consider process changes, enclosures, controls, and work practices that help prevent emission.

The CAAA are complex and contain many provisions with potential to impact the future business practices of the utility industry (3). The uncertainty associated with HAP regulations complicates long-term environmental compliance planning. As electric utilities consider expansion of existing facilities and/or construction of new facilities, the cost associated with environmental compliance remains uncertain as the HAP rule-making process develops. As an example, electric utilities are confronted with conducting generation planning that must consider CAAA requirements imposed by Title IV (Acid Deposition), Title I (Nonattainment) and Title III (Hazardous Air Pollutants).

A decision on whether to list utility generating units as a MACT-regulated source category is not planned until specific studies, identified in the CAAA for this source category, are completed. The five studies listed below are mandated by the CAAA and potentially affect electric utilities (4):

1. A three-year study of the hazards to public health reasonably anticipated to result from emissions from utility steam generating units after the imposition of the CAAA requirements.
2. A four-year study of mercury emissions from electric utility steam generating units, municipal waste combustion units and other sources, including area sources.
3. A three-year study by the National Institute of Environmental Health Sciences to determine the threshold level of mercury exposure below which adverse human health effects are not expected.
4. A three-year study by the National Academy of Sciences to review the risk assessment methodology used by EPA.
5. A three-year study by EPA in cooperation with the Under Secretary of Commerce for Oceans and Atmosphere concerning deposition of HAPs in the Great Lakes, the Chesapeake Bay, Lake Champlain, and coastal waters.

The EPA is directed by the CAAA to regulate utility generating units under the MACT program if the EPA finds such regulation appropriate and necessary considering the results of the first study listed above. The utility HAP study and the study of mercury emissions have been delayed until later in this decade. EPA's delay in completing the utility air toxic studies will allow the agency more time for data gathering and analysis, such as ongoing field tests being conducted by the Electric Power Research Institute (EPRI) and the DOE (5,6).

Many electric utilities have completed their plans for complying with the requirements of Phase I of Title IV of the CAAA. The Phase I sulfur dioxide (SO_2) reduction requirements were to be implemented by 1995. During the Phase I compliance planning process, these utilities were confronted with the possibility of future HAP regulations. Certain utilities have indicated that HAPs affected their Phase I compliance planning (7). For example, the selection of fabric filter collectors for a dry scrubbing application and the selection of wet flue gas desulfurization (wet FGD) were influenced by the ability of these technologies to reduce HAP emissions. Other utilities have indicated that they did not change their acid rain strategies because of pending legislation (8). Another utility planner identified the uncertainty associated with receiving credit for mercury reductions from Phase I wet FGD installations if enacted as a compliance planning consideration prior to HAP regulations (9).

Potential HAP regulations also complicate the long-range Phase II compliance planning process (10). Until the CAAA-mandated studies on utility HAPs are completed, compliance planners must consider the "robustness" of selected plans to possible environmental regulatory scenarios such as HAPs regulations. Additional environmental requirements, such as the possibility of postcombustion nitrogen oxides (NO_x) controls in certain ozone nonattainment areas or the possibility of a carbon tax, provide added incentives to consider repowering options. In some instances the collective capital investments and incremental operating costs required to comply with Titles I, III and IV of the CAAA may justify early retirement or repowering of certain generating units.

In some instances, recent coal-fired power plant projects have had to address HAP emissions to comply with State HAP programs. Also, other source categories such as waste incineration have had to address HAP emissions as part of their new source permitting process. Experience in these areas can provide useful insights into possible air toxic control strategies for electric utilities.

Table 1 presents a list of the 189 HAPs identified in the CAAA. Table 2 presents the definition of release category, and also the division of major groupings of HAPs among the three release categories. Table 3 presents suggested sampling methods for flue gas stream HAP field testing. Table 4 shows some of the possible laboratory analytical methods for analyzing HAP samples (12). Reference (14) provides a detailed description of the sampling methods, analytical procedures, and properties of the EPA's list of 189 HAPs. It has been found that some of the sampling and analytical procedures for HAPs need continued development and validation efforts to improve accuracy and precision. For additional information concerning the toxicological and safety issues of HAP species see Reference (15).

The EPA, DOE, and EPRI have instituted programs to collect data concerning HAPs emissions from coal-fired power plants. These programs will provide data to better understand HAP removal efficiencies of air pollution control devices installed on coal-fired power plants. For example, in 1989 EPRI introduced the Power Plant Integrated Systems Chemical Emissions Studies (PISCES) project. The project involves three components: one, a data base of existing literature; two, a computerized systems model; and three, a field monitoring program to supplement the database, as necessary. In collaboration with EPRI and the Utility Air Regulatory Group (UARG), the Pittsburgh Energy Technology Center (PETC) has initiated field studies under its Flue Gas Cleanup Program to supplement the PISCES database and to provide input to the EPA studies previously identified. Various companies are conducting studies to determine how effectively different pollution control subsystems remove trace elements under varying operating conditions and to determine the concentration of pollutants in flue gas streams as a function of particle size. The DOE has included an initiative in their Clean Coal Technology (CCT) Program to address HAP monitoring in the individual CCT projects managed by PETC and the Morgantown Energy Technology Center (METC). The CCT programs include the two low-NO_x combustion system projects conducted by Southern Company Services Inc. (SCS), the Pure Air Advanced Flue Gas Desulfurization project, and the Chiyoda CT-121 project. PETC is also sponsoring a study which compares HAP emissions from two full-scale coal-fired boilers with those from a laboratory combustor. Other programs are being conducted to collect data concerning HAP emissions from coal-fired power plants (12,16).

Information concerning trace elements within a coal-fired power plant is available for use. For example, material balances on a pulverized coal-fired power plant were performed in a little known thesis given in 1974 for nineteen trace elements (17). In this study, atomic absorption, X-ray fluorescence, and wet chemistry were used to determine concentrations of these elements. Closures of less than ten percent were obtained on nine of these elements. Material balances were determined for potassium, calcium, manganese, iron, copper, zinc, gallium, arsenic, selenium, mercury, rubidium, strontium, yttrium, zirconium, niobium, molybdenum, tin, antimony, and lead. Samples for the material balances were obtained from the raw coal, bottom ash, mechanical collector, electrostatic precipitator, wet FGD blowdown, and chimney inlet. Results of the material balances offer clues about the nature and distribution of the elements in the various effluents. Aluminum, iron, rubidium, strontium, yttrium, and niobium were at constant concentrations in all of the effluents. Copper, zinc, arsenic, molybdenum, tin, antimony, and lead had progressively higher concentrations in the successive downstream ashes. Zirconium had the highest concentration in the bottom ash. Mercury and selenium had progressively higher concentrations in the successive downstream effluents; however, mercury and selenium also had the poorest closures. These results indicate that particulate removal devices capable of collecting very fine ashes successfully capture most trace elements, whereas highly volatile elements, such as mercury and selenium, are not well captured.

Consol Inc. has prepared a summary of the available information concerning the trace element contents of U.S. coals (18). Consol performed a study of commercial coals routinely used by electric utilities and found the following (see COAL):

- Trace element contents vary by an order of magnitude. No coals are predictably lower in trace elements.
- Individual trace element contents are weakly to moderately correlated to ash content, but no correlation was noted for sulfur content.

Table 1. Clean Air Act Title III List of Hazardous Air Pollutants (HAPs)[a]

CAS Number	Chemical Number	CAS Number	Chemical Number
75070	Acetaldehyde	133062	Captan
60355	Acetamide	63252	Carbaryl
75058	Acetonitrile	75150	Carbon disulfide
98862	Acetophenone	56235	Carbon tetrachloride
53963	2-Acetylaminofluorene	463581	Carbonyl sulfide
107028	Acrolein	120809	Catechol
79061	Acrylamide	133904	Chloramben
79107	Acrylic acid	57749	Chlordane
107131	Acrylonitrile	7782505	Chlorine
107051	Allyl chloride	79118	Chloroacetic acid
92671	4-Aminobiphenyl	532274	2-Chloroacetophenone
62533	Aniline	108907	Chlorobenzene
90040	*o*-Anisidine	510156	Chlorobenzilate
1332214	Asbestos	67663	Chloroform
71432	Benzene (including benzene from gasoline)	107302	Chloromethyl methyl ether
92875	Benzidine	126998	Chloroprene
98077	Benzotrichloride	1319773	Cresols/Cresylic acid (isomers and mixture)
100447	Benzyl chloride	95487	*o*-Cresol
92524	Biphenyl	108394	*m*-Cresol
117817	Bis (2-ethylhexyl)phthalate (DEHP)	106445	*p*-Cresol
542881	Bis(chloromethyl)ether	98828	Cumene
75252	Bromoform	94757	2,4-D, salts and esters
106990	1,3-Butadiene	3547044	DDE
156627	Calcium cyanamide	334883	Diazomethane
105602	Caprolactam	132649	Dibenzofurans
96128	1,2-Dibromo 3-chloropropane	104885	Ethyl acrylate
84742	Dibutylphthalate	100414	Ethyl benzene
106467	1,4-Dichlorobenzene(*p*)	51796	Ethyl carbamate (Urethane)
91941	3,3-Dichlorobenzidene	75003	Ethyl chloride (Chloroethane)
111444	Dichloroethyl ether (bis(2-chloroethyl)ether)	106934	Ethylene dibromide (Dibromoethane)
542756	1,3-Dichloropropene	107062	Ethylene dichloride (1,2-Dichloroethane)
62737	Dichlorvos	107211	Ethylene glycol
111422	Diethanolamine	151564	Ethylene imine (Aziridine)
121697	*N,N*-Diethyl aniline (*N,N*-Dimethylaniline)	75218	Ethylene oxide
64675	Diethyl sulfate	96457	Ethylene thiourea
119904	3,3-Dimethoxybenzidine	75343	Ethylidene dichloride (1,1-Dichloroethane)
60117	Dimethyl aminoazobenzene	50000	Formaldehyde
119937	3,3′-Dimethyl benzidine	76448	Heptachlor
79447	Dimethyl carbamoyl chloride	118741	Hexachlorobenzene
68122	Dimethyl formamide	87683	Hexachlorobutadiene
57147	1,1-Dimethyl hydrazine	77474	Hexachlorocyclopentadiene
131113	Dimethyl phthalate	67721	Hexachloroethane
77781	Dimethyl sulfate	822060	Hexamethylene-1,6-diisocyanate
534521	4,6-Dinitro-*o*-cresol, and salts	680319	Hexamethylphosphoramide
51285	2,4-Dinitrophenol	110543	Hexane
121142	2,4-Dinitrotoluene	302012	Hydrazine
123911	1,4-Dioxane (1,4-Diethyleneoxide)	7647010	Hydrochloric acid
122667	1,2-Diphenylhydrazine	7664393	Hydrogen fluoride (Hydrofluoric acid)
106898	Epichlorohydrin (1-Chloro-2,3-epoxypropane)	123319	Hydroquinone
106887	1,2-Epoxybutane	78591	Isophorone
108316	Maleic anhydride	58899	Lindane (all isomers)
67561	Methanol	82688	Penthacloronitrobenzene (Quintobenzene)
72435	Methoxychlor	87865	Pentachlorophenol
74839	Methyl bromide (Bromomethane)	108952	Phenol
74873	Methyl chloride (Chloromethane)	106503	*p*-Phenylenediamine
71556	Methyl chloroform (1,1,1-Trichloroethane)	75445	Phosgene
78933	Methyl ethyl ketone (2-Butanone)	7803512	Phosphine
60344	Methyl hydrazine	7723140	Phosphorus
74884	Methyl iodide (Iodomethane)	85449	Phthalic anhydride
108101	Methyl isobutyl ketone (Hexone)	1336363	Polychlorinated byphenyls (Arochlors)
624839	Methyl isocyanate	1120714	1,3-Propane sultone
80626	Methyl methacrylate	57578	β-Propiolactone
1634044	Methyl *tert*-butyl ether	123386	Propionaldehyde

Table 1. Clean Air Act Title III List of Hazardous Air Pollutants (HAPs)[a] *(continued)*

CAS Number	Chemical Number	CAS Number	Chemical Number
101144	4,4-Methylene bis(2-chloroaniline)	114261	Propoxur (Baygon)
75092	Methylene chloride (Dichloromethane)	78875	Propylene dichloride (1,2-Dichloropropane)
101688	Methylene diphenyl diisocyanate (MDI)	5569	Propylene oxide
101779	4,4′-Methylenedianiline	15558	1,2-Propylenimine (2-Methyl aziridine)
91203	Naphthalene	91225	Quinoline
98953	Nitrobenzene	106514	Quinone
92933	4-Nitrobiphenyl	106423	*p*-Xylene
100027	4-Nitrophenol	0	Antimony compounds
79469	2-Nitropropane	0	Arsenic compounds (inorganic including arsine)
684935	*N*-Nitroso-*N*-methylurea	0	Beryllium compounds
62759	*N*-Nitrosodimethylamine	0	Cadmium compounds
59892	*N*-Nitrosomorpholine	0	Chromium compounds
56382	Parathion	0	Cobalt compounds
100425	Styrene	0	Coke oven emissions
96093	Styrene oxide	0	Cyanide compounds[b]
1746016	2,3,7,8-Tetrachlorodibenzo-*p*-dioxin	0	Glycol ethers[c]
79345	1,1,2,2-Tetrachloroethane	0	Lead compounds
127184	Tatrachloroethylene (Perchloroethylene)	0	Manganese compounds
7550450	Titanium tetrachloride	0	Mercury compounds
108883	Toluene	0	Fine mineral fibers[d]
95807	2,4-Toluene diamine	0	Nickel compounds
584849	2,4-Toluene diisocyanate	0	Polycyclic organic matter[e]
95534	*o*-Toluidine	0	Radionuclides (including radon)[f]
8001352	Toxaphene (chlorinated camphene)	0	Selenium compounds
120821	1,2,4-Trichlorbenzene		
79005	1,1,2-Trichloroethane		
79016	Trichloroethylene		
95954	2,4,5-Trichlophenol		
88062	2,4,6-Trichlorophenol		
121448	Triethylamine		
1582098	Trifluralin		
540841	2,2,4-Trimethylpentane		
108054	Vinyl acetate		
593602	Vinyl bromide		
75014	Vinyl chloride		
75354	Vinylidene chloride (1,1-Dichloroethylene)		
1330207	Xylenes (isomers and mixture)		
95476	*o*-Xylene		
108383	*m*-Xylene		

[a] For all listings above which contain the word compounds and for glycol ethers, the following applies: unless otherwise specified, these listings are defined to include any unique chemical substance that contains the named chemical (ie, antimony, arsenic, etc) as part of that chemical's infrastructure.
[b] X′CN where $X = H'$ or any other group where a formal dissociation may occur, for example KCN or $Ca(CN)_2$.
[c] Includes mono- and diethers of ethylene glycol, diethylene glycol, and triethylene glycol $R(OCH_2CH_2)_n{-}OR'$ where $n = 1$, 2, or 3; R = alkyl or aryl groups, and R′ = R, H, or groups which, when removed, yield glycol ethers with the structure: $R(OCH_2CH)_n{-}OH$. Polymers are excluded from the glycol category.
[d] Includes mineral fiber emissions from facilities manufacturing or processing glass, rock, or slag fibers (or other mineral derived fibers) of average diameter one micrometer or less.
[e] Includes organic compounds with more than one benzene ring and which have a boiling point greater than or equal to 100°C.
[f] A type of atom which spontaneously undergoes radioactive decay.

- Trace element variabilities ranged from 20 to 40 percent (percent relative to standard deviation) for coals from a single mine, from coals within the same seam, and even from geographically proximate coals.
- Coal cleaning to remove ash is effective in removing a portion of many of the trace elements. The degree of trace element removal is often similar to the overall ash reduction.
- Estimates of trace element inputs and emissions based on most published data (usually compilation of core or channel sample analyses or hypothetical washabilities) overestimate actual values.

Of the 189 substances now designated as HAPs, EPRI recognizes that approximately 37 are emitted from utility generating units and are of concern to the industry. Included are compounds containing antimony, arsenic, beryllium, cadmium, chromium, cobalt, lead, manganese, mercury, nickel, and selenium.

The flue gas cleaning technologies to abate HAP emissions from utility fossil-fuel-fired generating units currently exist and are the same types of devices presently utilized to control particulates, acid gases, and organics. Site-specific economics dictate which technology or combinations of technologies is needed. The risk of noncompliance is usually lower when utilizing proven technologies. Control technologies

Table 2. HAP Release Categories[a]

Element	Abbreviation	Element	Abbreviation
Release Category 1: Equally Distributed in Bottom Ash and Fly Ash			
Barium	Ba	Rubidium	Rb
Cerium	Ce	Scandium	Sc
Cobalt	Co	Samarium	Sm
Europium	Eu	Strontium	Sr
Hafnium	Hf	Tantalum	Ta
Lanthanum	La	Thorium	Th
Manganese	Mn		
Release Category 2: Preferentially Concentrated in Fly Ash			
Antimony	Sb	Lead	Pd
Arsenic	As	Molybdenum	Mo
Beryllium	Be	Nickel	Ni
Cadmium	Cd	Thallium	Tl
Copper	Cu	Uranium	U
Chromium	Cr	Vanadium	V
Gallium	Ga	Zinc	Zn
Release Category 3: Emitted as Vapors			
Chlorine	Cl	Mercury	Hg
Fluorine	F	Selenium	Se

[a] Ref. 11.

Table 3. Summary of Flue Gas Stream Sampling Methods for HAPs[a]

Gas Stream Component Analyzed	Suggested Sampling Method
Volatile organics	Volatile Organics Sampling Train (VOST) (SW-846 Method 0010 or 0030)
Semivolatile organics	Modified Method 5 Sampling Train (MM5) (SE-846 Method 0010) (XAD)
Acid gases and aldehydes	Impinger Sampling, Method 5
Vapor-phase elements	Method 5 Metals Train
Entrained particulate	
Particle loading	EPA Method 5 or 17
Bulk	Filtration (in-stack filter or thimble)
Size fractionated	Impactor sampling (PSD)

[a] Ref. 12.

Table 4. Summary of Laboratory Analytical Determinations for HAPs[a]

Analyte	Suggested Method[b]
Trace elements	
Antimony	ICAPES
Arsenic	HGAAS
Barium	ICAPES
Beryllium	ICAPES
Boron	ICAPES
Cadmium	GFAAS
Chromium[c]	ICAPES
Cobalt	ICAPES
Copper	ICAPES
Lead	GFAAS
Manganese	ICAPES
Mercury	CVAAS
Molybdenum	ICAPES
Nickel	ICAPES
Selenium	HGAAS
Vanadium	ICAPES
Radionuclides[d]	Gamma ray counting
Ions	
Phosphates (PO_4^{-3})	Spectrophotometric
Sulfates (SO_4^{-2})	Ion chromatography
Inorganics	
Ammonia	Ion-selective electrode
Cyanide	Ion-selective electrode
Hydrogen chloride	Ion chromatography
Hydrogen fluoride	Specific ion electrode
Organic	
Benzene[e]	GC/MS (Method 8240)
Toluene[e]	GC/MS (Method 8240)
Formaldehyde	HPLC
Polycyclic organic matter[f]	GC/MS (Method 8270)
Dioxins[g]	Method 23 coupled with GC/HRMS
Furans[g]	Method 23 coupled with GC/HRMS

[a] Ref. 13.
[b] Index of abbreviations: HGAAS = hydride generation atomic absorption spectrophotometry, GFAAS = graphite furnace atomic absorption spectrophotometry, CVAAS = cold vapor atomic absorption spectrophotometry, ICAPES = inductively coupled argon plasma emission spectroscopy, GC/MS = gas chromatography/mass spectrometry, GC/HRMS = gas chromatography/high resolution mass spectrometry, and HPLC = high performance liquid chromatography.
[c] Reported as total chromium.
[d] Atoms that undergo spontaneous radioactive decay.
[e] Other volatiles associated with GC/MS (Method 8240).
[f] All organic compounds with more than one aromatic ring that are associated with GC/MS (Method 8270).
[g] All polychorinated dibenzo-*p*-dioxins (PCDDs) and polychlorinated dibenzofurans (PCDFs) associated with GC/HRMS.

which control HAPs emissions in an efficient and cost-effective manner are

- conventional electrostatic precipitator;
- gas-conditioned electrostatic precipitator;
- pulse-jet fabric filter collector;
- reverse air fabric filter collector;
- wet electrostatic precipitator;
- wet flue gas desulfurization system (FGD); and
- dry scrubbing system with activated carbon injection for mercury removal.

Electrostatic precipitators (ESPs), fabric filter collectors, and gas-conditioned ESPs remove fine particles of heavy metals. A conventional wet FGD system scrubs out substantial amounts of residual fines and condensable organics and also achieves partial removal of ionic mercury. Additives, such as activated carbon and alkali chlorides, enable mercury removal at higher efficiency levels in dry scrubbing systems. Wet ESPs remove submicron solid particulates, acid mists, and condensable organic materials at very high efficiencies. NO_x removal technologies, such as Selective Catalytic Reduction (SCR), Selective Noncatalytic Reduction (SNCR), and combustion modifications can also be integrated into the overall air quality control system.

New electric generating units have several viable solutions for removing HAPs. A fabric filter collector or a dry ESP followed by a wet FGD system, with the possibility for future retrofitting an advanced type of sorbent injection system for additional mercury removal, is a promising course of action.

Table 5. Typical Fuel Ultimate Analyses for No. 6 Fuel Oil, Orimulsion®, and Bituminous Coal

Parameter	No. 6 Fuel Oil	Orimulsion®	Bituminous Coal
Carbon, %	85.0	59.0	68.5
Hydrogen, %	9.9	7.4	4.0
Nitrogen, %	0.4	0.5	1.5
Sulfur, %	1.8	2.7	0.6, 0.75, 2.0
Oxygen, %	2.2	0.2	6.0
Ash, %	0.35	0.25	9.0
Moisture, %	0.4	30.0	9.0
HHV, Btu/lb	18,420	12,680	12,070

Table 6. Assumed (90th Percentile) HAP Emission Factors

	Uncontrolled Emission, lb/10^{12} Btu		
Parameter	No. 6 Fuel Oil	Orimulsion®	Bituminous Coal
Antimony (Sb)	46.0	2.6	87.0
Arsenic (As)	114.0	90.0	911.4
Beryllium (Be)	4.2	11.0	251.0
Cadmium (Cd)	211.0	20.0	44.4
Cobalt (Co)	121.0	35.0	1,209.6
Chromium (Cr)	128.0	44.0	1,910.0
Lead (Pb)	194.0	59.0	744.0
Mercury (Hg)	1.4	1.6	11.6
Manganese (Mn)	74.0	36.0	1,963.5
Nickel (Ni)	2,330.0	7,300.0	1,590.7
Selenium (Se)	38.0	12.60	435.8
Total heavy metals (without Se, As, Hg)	3,108.2	7,507.6	7,800.2

Depending on site conditions, coal characteristics and applicable HAP emissions standards, a dry scrubbing system with activated carbon injection upstream of a fabric filter is more economical. As an alternative to a pulverized coal-fired boiler, a circulating fluidized-bed boiler can be utilized. Then the emissions control system consist of a mechanical cyclone followed by a dry ESP or a fabric filter.

Existing electric generating units are constrained by equipment retrofit issues, primarily, the type and condition of the existing emissions control equipment, space constraints, constructibility, and unit downtime penalty. As such, less straightforward combinations of technologies may prove to be the most economical, with a high degree of variation from one site to the next.

Typical fuel properties of two major utility fossil fuels, coal and fuel oil, and a new fuel named Orimulsion® which is expected to be important in the near future, are listed in Table 5, and uncontrolled HAP emission factors are compiled in Table 6. The emission factors shown on this table are indicative of high end (90th percentile) release rates—hence it should be expected that in most instances a site-specific study would yield lower uncontrolled HAPs average emission rates.

BIBLIOGRAPHY

1. U.S. Environmental Protection Agency, *Initial List of Categories of Sources Under Section 122(C)(1) of the Clean Air Act Amendments of 1990, Fed. Reg.* **57,** 137, July 16, 1992.
2. A.C.D. Leslie, *Compliance Strategies Rev.* (Feb. 14, 1994).
3. S.A. Vigeant, "Is Your Plant Ready to Comply with the Clean Air Act?," presented at the *21st New England Plant Engineers Conference and Show,* Newport, R.I., May 6–8, 1992.
4. W. Chow, L. Levin, and M.J. Miller, "Air Toxics and the 1990 Clean Air Act: Managing Trace Element Emissions," presented at the *1991 SO_2 Control Symposium,* Washington, D.C., Dec. 3–6, 1991.
5. Electric Power Research Institute, "New Focus on Air Toxics," *EPRI J.* (Mar. 1991).
6. T.D. Brown, C.E. Schmidt, and A.D. Radziwon, "Comprehensive Assessment of Toxic Emissions from Coal-Fired Power Plants," presented at *Managing Hazardous Air Pollutants: State of the Art,* Nov. 4–6, 1991.
7. *PRECIP Newsletter,* McIlvaine Co., Northbrook, Ill., Oct. 1991.
8. M. Leone, "Air Toxics Studies Key to Emissions Rules," *Electrical World* (Feb. 1992).
9. "Compliance Planning, What's Next?," *Electric Perspec.* (Jan–Feb. 1992).
10. F.E. Depenbrock and W.F. Frazier, "Compliance Planning for Title IV, Phase II of the Clean Air Act Amendments of 1990," presented at the *AIChE 1992 Summer National Meeting,* Minneapolis, Minn., Aug. 12, 1992.
11. K. Felsvang, R. Gleiser, G. Juip, and K. Nielsen, "Control of Air Toxics by Dry FGD Systems," *Power-Gen '92,* Orlando, Fla., Nov. 17–19, 1992.
12. DOE, *Comprehensive Assessment of Toxic Emissions from Coal-Fired Power Plants,* PETC Request for Proposal Number DE-RP22-92PC91349, Feb. 10, 1992.
13. J.J. Cramer, and R.L. Gordon, "The Role of the New Clean Air Act in the Control of Accidental Releases of HAPs," presented at the *1990 Annual Meeting of the American Institute of Chemical Engineers,* Chicago, Ill., Nov. 12, 1990.
14. L.H. Keith, and M.M. Walker, *Handbook of Air Toxics, Sampling, Analysis, and Properties,* 1995.
15. R.J. Lewis, *Hazardous Chemicals Desk Reference,* 1991.
16. DOE, Gathering Data on Air Toxics Emissions, published in *PETC Review,* **6,** 18–24 (Summer 1992).
17. J.W. Kaakinen, Ph.D. Thesis, University of Colorado, 1974.
18. P.R. Tumati, and M.S. DeVito, "Trace Element Emmissions from Coal Combustion - A Comparison of Baghouse and ESP Collection Efficiency," *EPRI Conference on the Effects of Coal Quality on Power Plants,* San Diego, Calif., Aug. 25–27, 1992.

ALCOHOL FUELS

Conventional fuels such as gasoline and diesel fuel are derived essentially from crude oil processed in refineries. Concerns about the extent and reliability of petroleum sources and about air quality have led to interest in alternative fuels, such as various alcohols. Ethanol and methanol are two alcohols that are leading candidates for alternative fuels. See the following two articles: ALCOHOL FUELS: ETHANOL; ALCOHOL FUELS: METHANOL. Also see AUTOMOTIVE ENGINES AND VEHICLES.

ALCOHOL FUELS: ETHANOL

RICHARD L. BECHTOLD
Columbia, Maryland

Ethanol has been known to humankind for thousands of years. It is believed that ethanol was discovered when fruits or grains were allowed to overripen in the presence of water under favorably warm conditions. Ethanol is the alcohol in alcoholic beverages such as beer and wine, and most ethanol today is made from the fermentation of agricultural crops.

The attraction of using ethanol as a fuel is multifaceted. Farmers like it because it creates greater demand for their crops. Environmentalists like it because it is less reactive in the atmosphere than many other fuels, and under the right conditions, it can reduce greenhouse gas emissions in comparison to those produced using petroleum fuels. Those concerned about energy security like it because it is a domestically produced fuel. The primary disadvantage of using ethanol as a fuel is that it is currently about three times as expensive as gasoline on an energy basis.

Ethanol, C_2H_5OH, is the second lightest alcohol, methanol being the lightest. There are three primary ways in which ethanol can be used as a transportation fuel: (*1*) as a blend with gasoline, typically 10 vol %, commonly known as Gasohol; (*2*) as a component of reformulated gasoline, ie, gasoline refined to specifications set by the U.S. Environmental Protection Agency (EPA), properties of which result in fewer emissions of nonmethane organic gases and carbon monoxide (EPA requires that reformulated gasoline be used in urban areas around the United States that are in nonattainment for ozone), with the ethanol both directly used as a component or transformed into a compound such as ethyl tertiary butyl ether (ETBE); or (*3*) used directly as a fuel, with 15% or more of gasoline known as "E85." Ethanol can also be used directly in diesel engines specially configured for alcohol fuels, such as the Detroit Diesel Corporation 6V-92TA model (no longer in production). Using ethanol to make Gasohol or in reformulated gasoline, or transformed into ETBE for use in reformulated gasoline, does not require specially configured vehicles; almost all current vehicles will tolerate these fuels without problems and likely with advantageous emissions benefits. For example, ethanol and ETBE blends in gasoline are approved for use by the EPA for mandated wintertime oxygenated fuel programs whose objective is to lower vehicle carbon monoxide emissions.

ETHANOL PRODUCTION

Brazil is the largest producer of ethanol in the world, with production levels nearly three times as high as those in the United States, which is the second largest producing country. Europe produces about half the ethanol production of the United States, with other countries producing far less (1).

Production Processes and Resources

Ethanol is produced through two processes: (*1*) the fermentation of sugars extracted from crops such as corn and sugarcane (although many others are possible), referred to as fermentation ethanol; or (*2*) it is made from the catalytic hydration of ethylene, ethanol made this way being referred to as synthetic ethanol. All of the ethanol used as fuel in vehicles in the United States is produced via fermentation, because a $0.54 per gallon Federal tax credit, with some states giving additional tax credits, makes it less expensive than synthetic ethanol and competitive with petroleum fuels. Synthetic ethanol is used to produce other chemicals.

Fermentation Production Process. Figure 1 illustrates the basic components of a fermentation ethanol plant. The first step is to prepare the feedstock. Cleaning is usually required, followed by cutting or grinding (starch crops) or pressing or steaming (sugar crops). For starch crops, the preparation step exposes the starch and provides an opportunity to extract valuable components such as germ, oil, meal, and vitamins. For sugar crops, the preparation step is used to extract the sugar juice from the crop. Following preparation, a "mash" is created by adding water to the ground starch or the sugar juice. The mash is then cooked, with starch mash having to go through a few additional steps to convert it to fermentable sugars. Cooking the starch mash first creates a dispersion of starch in the mash and sanitizes the mash. Then, an enzyme called fungal amylase is added, which converts the starch to sugars. This process of converting starch to fermentable sugars is called saccharification. In comparison, sugar crops need only be cooked to sanitize them (2).

The cooked mash is cooled to about 27°C (80°F) before the beginning of the fermentation process. Yeast is added to the mash which feeds on the sugars, converting them to ethanol, carbon dioxide, and small quantities of liquid co-products, including higher alcohols, aldehydes, etc. The process is completed when all the sugars are converted by the yeast. The resulting liquid typically contains between 7 and 10% ethanol and is called dilute beer.

In the alcohol recovery step, the dilute beer is distilled into an alcohol-rich product containing up to 96% ethanol, 4% water, and a co-product stream known as whole stillage. The ethanol product is then further distilled using a process called azeotropic distillation to remove most of the remaining water.

Ethanol intended for uses other than consumption in alcoholic beverages must be denatured (made unfit for human consumption). For ethanol intended for use as a motor fuel, 5 vol % unleaded gasoline is used as the denaturant.

Feedstocks. There are three types of feedstocks used to produce fermentation ethanol: sugar crops, starch crops, and cellulose. The sugar crops for producing ethanol that are typically considered include sugar cane, sugar beets, sweet sorghum, fodder beets, sugar beets, Jerusalem artichokes, and by-products of fruit processing such as citrus molasses. Of these, sugar cane and sugar beets are the two that are most

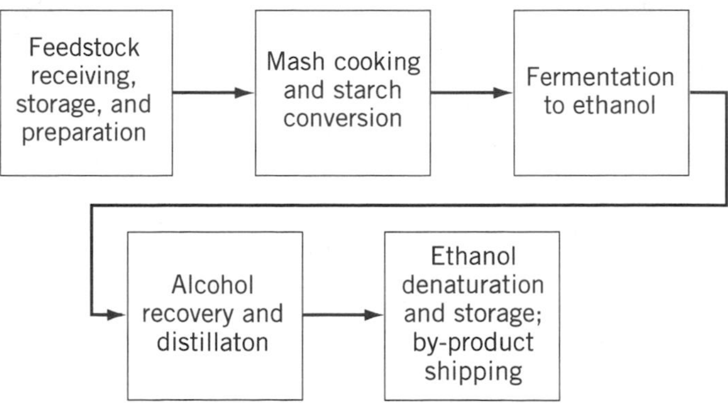

Figure 1. Simplified ethanol production process flow chart (2).

widely available in the United States. About 57 L (15 gal) of ethanol can be produced from one ton (0.9 Mg) of sugar cane, while 83 L (22 gal) can be produced from one ton (0.9 Mg) of sugar beets (2). The primary advantage of sugar crops is that less processing is required compared to starch and cellulose feedstocks. The primary disadvantage is that sugar crops are difficult to store for long periods of time.

The starch crops for producing ethanol that are typically considered include grains (corn, wheat, barley, oats, rice, grain sorghum, and others) and tubers (potatoes and sweet potatoes). Of these, corn is by far the one used in largest quantity to produce ethanol. Up to 90 gallons of ethanol can be produced from one ton (0.9 Mg) of corn, which is the most of any crop. Most of the other grains will yield nearly as much ethanol per ton (2). The primary advantage to using starch crops to produce ethanol is that they are easy to store for long periods of time without spoilage. The primary disadvantage is that they need more processing than sugar crops.

Making ethanol from cellulose is not commercially viable at present, but cellulose feedstocks are in plentiful supply. Examples of cellulose feedstocks include corn stover, straw, sugar cane bagasse, forest residues, and hybrid strains of fast-growing trees. Cellulose feedstocks are inexpensive but the processing required to release the fermentable sugars is much more extensive than that required for starch crops.

Ethanol production plants, regardless of the feedstock used, are all similar once fermentable sugars are obtained. The differences in the production plants have to do primarily with the front-end processing steps to extract fermentable sugars from the feedstocks.

ENVIRONMENTAL REASONS FOR USING ETHANOL

Emissions

Ethanol by itself has a very low vapor pressure, but if it is blended in small amounts with gasoline as in making Gasohol, it causes the resulting blend to have a disproportionate increase in vapor pressure. (Fuels with high vapor pressure tend to have high refueling, evaporative, and running losses.) For this reason, there is considerable interest in using fuels such as ETBE, which is made using ethanol, as reformulated gasoline components, because ETBE has a very small or even negative effect on blend vapor pressure. The primary emission advantage of using ethanol blends is that CO emissions are reduced through the "blend-leaning" effect that is caused by the oxygen content of ethanol. The oxygen in the fuel contributes to combustion much the same as adding additional air. Because this additional oxygen is being added through the fuel, the engine fuel and emission systems are fooled into operating leaner than designed, with the result being lower CO emissions and typically slightly higher NO_x emissions. The blend-leaning effect is most pronounced in older vehicles that do not have feedback control systems; however, even the newest technology vehicles typically show some reduction in CO emissions.

The vehicle technology to use E85 is virtually the same as that to use M85, which has contributed to the rapid development of E85 vehicles. The emission characteristics of E85 vehicles have not been extensively documented, but it is expected that they will be comparable to the latest vehicles using reformulated gasoline, with the exception that E85 produces more acetaldehyde, as compared to M85, which produces more formaldehyde. An advantage of acetaldehyde over formaldehyde is that it is less reactive in the atmosphere, which offers less of a potential contribution to ground-level ozone formation.

The Ford Motor Company has developed an E85 version of their flexible/fuel Taurus passenger car and has presented results of their emissions testing (3). It was found that engine-out emissions using E85 were slightly lower than when using gasoline, but tailpipe emissions (downstream of the exhaust catalyst) were insignificantly different. The postulated reason for these findings is that ethanol creates exhaust gases that are lower in temperature because it burns with a lower flame temperature than gasoline. The lower temperature exhaust gases take a longer time to heat up the catalyst to operating temperature and in the meantime, more emissions are released. After the catalyst is warmed up, Ford found that the lower engine-out emissions were also reflected in lower tailpipe emissions. Ford demonstrated with their M85 flexible/fuel vehicle that the increase in emissions owing to cold-start (using M85 in this case) can be eliminated by adding small light-off catalysts just downstream of the engine before the main catalyst. It is likely that similar results are possible using E85.

The National Renewable Energy Laboratory (NREL), as of 1997, is conducting a project to determine whether E85 vehicles can be made to meet the California Ultra Low Emissions Vehicle regulations (the most stringent in the United States except for Zero Emissions Vehicles). Significant progress has been made toward achieving this goal, and it is believed to be achievable (4).

Ozone Production

As stated, flexible fuel vehicles using E85 have mass emissions similar to those using gasoline. Unburned hydrocarbons from using E85 and evaporative, running loss, and refueling emissions will have a significant portion represented by unburned ethanol, and likely will have more acetaldehyde in exhaust gases than when using gasoline. Ford has estimated that ethanol has about 63% of the reactivity of a composite of gasoline hydrocarbons (3). Overall, it is expected that using E85 would significantly lower the reactivity of hydrocarbon emissions in the air from vehicles, and thus would have a positive impact on the amount of ozone formed.

Advanced technology vehicles using neat ethanol as a fuel (E100) would be expected to have lower evaporative, running loss, and refueling emissions than gasoline, and because of the lower mass of emissions, ozone production might be lowered. While the technology to build such advanced technology vehicles has been demonstrated, as of ca 1997 plans for production are not yet on the horizon.

Greenhouse Gases

The largest single contribution to greenhouse gases (GHGs) from vehicles is carbon dioxide, CO_2, from combustion of the fuel. Where fermentation ethanol is used as a fuel, it can be assumed that all the CO_2 produced from combustion in the vehicle is recycled by the plants grown from which the ethanol was produced. However, combustion of ethanol also produces GHGs such as methane, CO, and NO_x that are not recycled. In addition, the production of crops from which ethanol is produced and the fermentation process itself use significant amounts of energy, which results in production of GHGs. The type of energy used in the production of ethanol crops and for the fermentation process has a large influence on the GHGs produced. For example, the amount of fertilizer used is very important and the type of energy used to supply power and steam for fermentation is also very important. Use of coal to supply fermentation power and steam produces high GHG emissions, whereas lesser amounts are produced if natural gas is used, and much lower amounts if renewable sources such as wood or agricultural wastes are used, because the CO_2 produced is recycled.

In contrast, whereas all the CO_2 produced from combustion of gasoline or diesel fuel increases GHGs, the GHGs produced from refining of crude oil are much less per equivalent gallon than from production of fermentation ethanol. Overall, production and use of fermentation ethanol as a transportation fuel is projected to cause fewer GHG emissions than using conventional petroleum fuels (5). Gasoline produces by far the largest amount of GHG emissions from combustion in the vehicle, but producing the fuel creates about the same amount of GHGs as E85 made using fermentation ethanol, and less than E85 made using cellulosic ethanol. The GHG emissions from vehicles using E100 are assumed to be zero because of increased engine efficiency, which offsets the small net increase in GHGs from ethanol combustion. The GHG emissions from feedstock preparation are largest for fermentation ethanol, followed by cellulosic ethanol, and smallest for crude oil used to make gasoline. Based on these data, it appears that using ethanol as a transportation fuel could reduce GHG emissions in the near term by about 25% and in the future, when E100 vehicles are developed, by about 40%. It should be noted, however, that significant variations in these estimates can occur, depending on the amount of fertilizer used for producing the ethanol crops and the energy sources used to produce fermentation and cellulosic ethanol.

HEALTH AND SAFETY

Ingestion

Ingestion of ethanol, eg, from alcoholic beverages, causes progressively more severe physiological changes as the amount of ethanol ingested increases. At low levels of ingestion (50–150 mg/100 mL of blood), loss of coordination, slow reaction time, and blurred vision are typical. At higher levels of ingestion (150–300 mg/100 mL of blood), visual impairment, staggering, and slurred speech occur. In the next level of ingestion (300–500 mg/100 mL of blood), marked incoordination, stupor, and convulsions occur. At levels of ingestion beyond 500 mg/100 mL of blood, coma and death are likely. Absorption of ethanol from the gastrointestinal tract is rapid, with peak blood levels achieved within 30–60 min following ingestion. The metabolism of the typical adult will eliminate from 7–11 g of ethanol/h. The physiological changes described previously are dependent on the experience of the individual with ingestion. Chronic alcoholics have been known to ingest over 500 mg/100 mL of blood without injury (6).

The greatest danger from ethanol ingestion occurs to the fetuses of pregnant women. Abnormalities that can occur to the fetus from exposure to ethanol include interference with neuronal migration, facial features that are not typical, and mental retardation. Fetal alcohol syndrome has been demonstrated in several animal species and is believed caused by acetaldehyde, which is produced as ethanol is metabolized by the body (6).

A very real personnel safety hazard of fuel ethanol is ingestion. The 5-vol % gasoline added to fuel ethanol as a denaturant gives it an unpalatable odor but does not completely eliminate the odor of ethanol. Recognizing the syphoning hazard presented by alcohol alternative fuels, the National Highway Traffic Safety Administration (NHTSA) has promulgated regulations requiring the incorporation of anti-syphoning devices on all new vehicles using fuels containing 20 volume percent or more alcohol (7). The larger amount of gasoline contained in E85 increases the odor of gasoline and makes it less likely that anyone would attempt knowingly or willingly to ingest it. As with other fuels, syphoning for fuel transfer purposes outside of the vehicle should never be initiated orally.

Whereas gasoline is not particularly toxic, ingestion will cause irritation of mucous membranes, vomiting, and central nervous system depression. Hydrocarbon pneumonia often develops within 24 h of ingestion, and may take several weeks to be resolved completely. Doses as low as 10 mL have been reported to be potentially fatal, though cases of survival from ingestion of as much as 60 mL have been reported (6). Ingestion of ethanol containing gasoline is doubly hazardous for this reason.

Spills

Whereas ethanol is biodegradable, the presence of other components such as gasoline results in the classification of spills of fuel ethanol as hazardous wastes. Fuel ethanol that is spilled and enters the water table represents a hazard because of its water solubility, which facilitates rapid transport in solution. Because of this solubility, spilled fuel ethanol can be dispersed to a much wider area than spilled gasoline, which has only very limited solubility in water.

ETHANOL VEHICLE TECHNOLOGY

Dedicated Vehicles

As of ca 1997, there are no vehicles being produced to use ethanol as the only fuel. If there were, these vehicles could make use of ethanol's high octane relative to gasoline to improve the energy efficiency of these vehicles. Depending on the engine and emission standards, when using ethanol, the engine compression ratio can be increased by 1–2 numbers, with an estimated corresponding thermal efficiency increase of about 6% compared to using gasoline (4). Some heavy duty diesel engines have been built to use ethanol as the only fuel.

Flexible Fuel Vehicles

As of ca 1997, there is only one vehicle model available that can use E85 as a fuel—the Ford Taurus Flexible Fuel Vehicle (FFV). Ford has steadily improved its FFV technology over time, and offers a similar model that uses M85 as fuel (3). Ford has been selling Ethanol FFV Tauruses in the United States since the 1993 model year. In 1995, Ford sold about 1,500 Ethanol FFV Tauruses and in 1996, it sold about 3,600 (8). Ford has offered to sell the first 12,000 of its 1997 Ethanol FFV Tauruses for $345 less than the comparable gasoline Taurus (9). (The retail option price of the Ford Taurus Ethanol FFV is $1,165.)

Chevrolet developed an ethanol FFV model of their Lumina sedan and sold 320 of them in 1993 (10). Chevrolet is working on an ethanol FFV version of their S-10 pickup truck, to be introduced in 1998 or 1999. Chrysler has announced that they will be building up to 180,000 minivans starting in the 1998 model year, capable of running on E85. Ford has announced that they will build up to 250,000 E85 Ranger pickup trucks and Taurus passenger cars in the 1999 model year.

In the past there have been some abortive attempts to import ethanol vehicles from Brazil. Although Brazil does have a large number of ethanol vehicles, there are several reasons why they cannot and should not be used in the United States: they do not meet U.S. safety regulations; they do not meet U.S. emissions regulations; and they will not be supported by U.S. manufacturers for warranty coverage or parts availability.

E85 has 71% of the heating value of an equal volume of straight gasoline, assuming the ethanol is denatured using 5-vol % gasoline. As a consequence, flexible fuel vehicles will have 71% of the driving range using E85 compared to gasoline. Conversely, if a vehicle uses only E85, the fuel tank capacity must be increased by about 40% for the vehicle to have the same range as when using gasoline. Advanced engines developed to take advantage of fuel ethanol's properties should be more efficient than gasoline engines and should narrow the driving range difference.

The only significant differences drivers should notice about flexible fuel E85 vehicles is reduced range and the need for a specially formulated engine oil. No other differences should be apparent. Diesel vehicles with fuel ethanol diesel engines should have the same characteristics as their diesel counterparts, and the same driving range if larger fuel tanks (they would need to be nearly 50% larger) are specified.

The power of E85 engines in flexible fuel vehicles should be the same or slightly higher when operating on E85 vs gasoline. If engines are optimized for E85 operation, significant increases in power and efficiency are possible.

Maintenance and Durability

E85 vehicles should present maintenance requirements similar to those of their conventional petroleum fuel counterparts. Up until mid-1997, Ford recommended that their E85 flexible fuel vehicles should use a special synthetic oil that cost about $3.00 per quart ($2.83 per liter). Based on the field experience Ford collected, they decided that regular engine oil is acceptable for use in their E85 flexible fuel vehicles.

There is no reason to believe that E85 vehicles should not last as long as gasoline vehicles. Long-term tests of M85 vehicles have shown similar engine wear to gasoline vehicles. Tests of engine wear indicate that ethanol causes lower wear rates than methanol, and similar wear rates to gasoline if there is no water in the ethanol (11). If significant amounts of water are present in the ethanol, engine wear rates are increased compared to using gasoline.

Ethanol Diesel Engines

Ethanol's low cetane number (<20) and lower energy density (ethanol contains only about 60% of the energy of an equal volume of diesel fuel) does not allow it to be used directly in diesel engines as a replacement for diesel fuel. Several methods have been employed to use ethanol in diesel engines, including emulsions, fumigation, dual injection, spark ignition, ignition improvers, and hot-surface ignition (12). Of these methods, the first three only displace some of the diesel fuel used by the diesel engine, while the latter three displace all the diesel fuel used by a diesel engine.

ETHANOL STORAGE, DISPENSION, AND FIRE SAFETY

Recommended Practices

Fuel ethanol must be denatured before it may be transported for use as a fuel, with 5-vol % gasoline being the preferred denaturant. General Motors (GM) pioneered a fuel ethanol specification for E85 which is 85 vol % fuel ethanol and 15 vol % gasoline, which results in an overall blend of 79 vol % ethanol and 21 % gasoline, taking into account the gasoline used as a denaturant. During those times of the year when typically the temperature is below freezing, GM recommends that the gasoline portion of the blend be increased to 25 vol % (13). Ethanol has very low volatility, and the gasoline is needed to provide sufficient fuel vapor to achieve reliable cold-starts for spark-ignited light duty engines. The American Society for Testing and Materials (ASTM) developed specifications for fuel ethanol based largely on the work performed by GM and others (14). The specification defines three classes of fuel that vary in gasoline content and vapor pressure. The classes are designed to assure that ethanol fuels have sufficient vapor pressure to achieve acceptable cold-starts in all areas of the country during all months of the year. The specifications recommend which fuel classes should be used throughout the country around the year. While gasoline varies in vapor pressure throughout the year, ethanol fuel would vary in both gasoline content and vapor pressure.

The technology for storing and dispensing fuel ethanol is similar to that for gasoline, except for changes in materials to make them compatible with ethanol. Since fuel ethanol is hygroscopic, ie, it readily absorbs water, storage tanks should include conservation (pressure/vacuum) vents to minimize breathing and water ingress. The National Fire Protection Association (NFPA) does not specifically address fuel ethanol storage and dispensing for vehicle use. Fire suppression systems should use dry chemicals or foams specifically formulated for ethanol.

The distribution system for fuel ethanol consists of transport from the production facility by rail or tank truck to existing petroleum fuel terminals, where it is stored in dedicated storage tanks (15). Pipeline transport of ethanol does not occur yet (ca 1997) because of water content and fungibility issues. (Fungibility of fuel means that one producer's fuel can be exchanged for another at the pipeline and distributor level, since they are both made to the same specifications. Fuel additives are added before the gasoline is delivered, making it different from other brands of gasoline. As of ca 1997, no system to assure fungibility of ethanol for fuel purposes has been developed.) Current petroleum pipelines (ca 1997) contain water from condensation and leakage into storage facilities, which is readily absorbed by fuel ethanol. Thus, if fuel ethanol is put through a pipeline with water in it, it will absorb all the water, which will contaminate the fuel. In addition, the solvency properties of fuel ethanol will remove hydrocarbon residues and pipeline scale, which will further degrade the fuel ethanol quality. If fuel ethanol were transported frequently, it would keep the pipeline clean and no significant degradation of fuel quality would occur. Pipeline operators will be reluctant to transport fuel ethanol until demand is sufficient to justify committing storage volume and if acceptable specifications for pipeline transport can be established.

From fuel terminals, fuel ethanol is delivered via tank truck to user storage tanks. Fuel ethanol deliveries can and should make use of Stage I vapor recovery, which is the practice for gasoline. The same practices for transferring petroleum products apply to fuel ethanol.

Fire Safety

Fuel ethanol poses hazards similar to those of gasoline in terms of fire safety. The low volatility of ethanol results in flammable mixtures present in the ullage space of vehicle fuel tanks, ie, the space not occupied by liquid fuel, under some conditions of cold weather. (The same thing occurs with gasoline but only at very low temperatures.) Diesel vehicles with fuel ethanol engines should have fuel systems with vapor control and flame arresters because fuel ethanol blends with <15 vol % gasoline will be prone to formation of flammable mixtures in the ullage space of the fuel tanks.

E85 Refueling Facilities

As of July 1997, there are 67 refueling stations in the United States offering E85 for sale, of which 38 are public stations (16). These stations are located primarily in the Midwest states where ethanol is predominately produced and where most of the E85 flexible fuel vehicles produced to date are located. An additional 112 stations are planned to be established by the year 2000 (16).

E85 can be stored in most existing steel underground storage tanks (fiber glass tanks may not be compatible). Gasoline dispensers can be used if they are suitably modified. Most modifications involve replacing incompatible elastomers. Modification of existing gasoline storage tanks and dispensers to store and dispense E85 can be completed for a cost of about $2000 (17).

Fire Characteristics

Ethanol is classified as a flammable liquid the same as gasoline, and the same regulations apply concerning transport, storage, and use with respect to risks for fire. Ethanol burns with a visible flame but it is not as prone to soot production as gasoline or diesel fuel when they burn.

The flash point of ethanol is +13°C (55°F), ie, much higher than that of gasoline (−43°C; − 45°F). Ethanol's lower flammability limit (LFL) in air is 4.3 vol % compared to gasoline at 1.4 vol % in air. The vapor pressure of ethanol is 15.9 kPa (2.3 psi) compared with gasoline, which has a range of 48–103 kPa (7–15 psi). These characteristics indicate that ethanol is less prone to produce vapors, and where it does, more of these vapors at a higher temperature are needed to create a flammable mixture in air. However, where gasoline is added to ethanol, eg, to make E85, these characteristics become more like those of gasoline than those of ethanol.

Once ethanol does ignite, its lower heat of combustion and higher heat of vaporization cause it to burn more slowly and less violently than gasoline. Under equivalent conditions, one would expect ethanol to be less likely than gasoline to ignite in open-air situations such as a fuel spill. This is not necessarily the case for E85, however, because of the addition of gasoline.

Explosion Potential

Ethanol's combustion properties are such that a flammable mixture in air above the liquid is likely inside fuel storage tanks under normal ambient temperatures (7–42°C; 45–108°F), whereas gasoline is virtually always too rich to ignite. Thus a partially full tank of ethanol presents a hazard. However, the electrical conductivity of ethanol is higher than for gasoline, so there is less danger of spark ignition by static discharge. Also, precautions can be taken to prevent either flammable vapor/air mixtures from forming in storage tanks, through use of eg, nitrogen blanketing, bladder tanks, or floating roof tanks, or to prevent ignition sources from entering the tanks, eg, through use of flame arresters, or by removing or modifying in-tank electrical devices, thereby mitigating any additional risk.

Firefighting Techniques

Water is a poor extinguisher of ethanol fires. Dry chemicals, CO_2, or alcohol-resistant foam concentrates are the desirable extinguishers (18). Alcohol-resistant foams are needed for pure ethanol fires and function equally well on gasoline fires. First aid extinguishers with CO_2 or powders seem to function equally well on methanol and gasoline fires (19). Service stations typically use dry chemical extinguishers for fire suppression.

BIBLIOGRAPHY

1. *Assessment of Costs and Benefits of Flexible and Alternative Fuel Use in the U.S. Transportation Sector Technical Progress Report One: Context and Analytical Framework,* DOE/PE-0080, United States Department of Energy, Washington, D.C., Jan. 1988.
2. Solar Energy Research Institute, *A Guide to Commercial-Scale Ethanol Production and Financing,* Report No. SERI/SP-751-877, National Technical Information Service, Springfield, Va., Mar. 1981.
3. J. S. Cowart and co-workers, *Powertrain Development of the 1996 Ford Flexible Fuel Taurus,* SAE Paper No. 952751, SAE International, Warrendale, Pa., 1996.
4. G. Bourn and co-workers, *Development of a Dedicated Ethanol Ultra-Low Emission Vehicle (ULEV) System Design,* National Renewable Energy Laboratory Report No. NREL/TP-425-6722, Alexandria, Va., Feb. 1995.
5. M. Q. Wang, *Development and Use of the GREET Model to Estimate Fuel-Cycle Energy Use and Emissions of Various Transportation Technologies and Fuels,* Argonne National Laboratory Report No. ANL/ESD-31, Argonne National Laboratory, Argonne, Ill., Mar. 1996.
6. M. O. Amdur, J. D. Doull, and C. D. Klaassen, *Toxicology—The Basic Science of Poisons,* 4th ed., Pergamon Press, Elmsford, N.Y., 1991.
7. "Federal Motor Vehicle Safety Standards; Fuel System Integrity; Alcohol Fuels," *Fed. Reg.* **57** (10) (Jan. 15, 1992).
8. *Alternatives to Traditional Transportation Fuels 1995,* U.S. Department of Energy, Energy Information Administration, internet site (http//www.doe.eia.gov), Dec. 2, 1996.
9. *Natl. Ethanol Vehicle Coalition Newsltr,* **I** (2), (Fall 1996).
10. *Altern. Fuel Insider,* **1** (9) (July 30, 1993).
11. D. W. Naegeli and E. C. Owens, *Final Report on Investigation of Alcohol Combustion Associated Wear in Spark Ignition Engines: Mechanisms and Lubricant Effects,* Interim Report AFLRL No. 176, U.S. Army Fuels and Lubricants Research Laboratory, Southwest Research Institute, San Antonio, Tex., Dec. 1984.
12. R. L. Bechtold, *Status of Alcohol Fuels Utilization Technology for Highway Transportation: A 1986 Perspective, Volume II—Compression Ignition Engines,* Report No. ORNL/Sub/85-22007/

3, National Technical Information Service, Springfield, Va., Jan. 1988.

13. GM Specification for Fuel Ethanol-E85 (GM4729M), General Motors Co., Aug. 10, 1992.
14. ASTM D5798—*Specification for Fuel Ethanol Ed70–Ed85 for Automotive Spark-Ignition Engines,* American Society for Testing and Materials, Philadelphia, Pa., 1996.
15. *Assessment of Ethanol Infrastructure for Transportation Use,* EA Engineering, Science and Technology, Inc., Silver Spring, Md., Apr. 1991.
16. S. Hentges, personal communication, National Ethanol Vehicle Coalition, Jefferson City, Mo., Aug. 4, 1997.
17. P. Lambert, personal communication, National Ethanol Vehicle Coalition, Oct. 10, 1996.
18. *Assessment of Costs and Benefits of Flexible and Alternative Fuel Use in the U.S. Transportation Sector, Technical Report Seven: Environmental, Health, and Safety Concerns,* DOE/PE-0100P, U.S. Department of Energy, Washington, D.C., 1991, p. 13.
19. *Alcohols as Alternative Fuels for Road Vehicles,* Report EUCO-Cost 304/I/85, VII/40/85 Brussels, EN, European Cooperation in the Field of Scientific and Technical Research, Belgium, 1984, pp. 19–21.

ALCOHOL FUELS: METHANOL

RICHARD BECHTOLD
Columbia, Maryland

Methanol has many characteristics that make it suitable as a fuel for vehicles: it has a high octane value, it burns without soot, and it is less reactive in the atmosphere than typical hydrocarbons from gasoline. Methanol can be used neat (ie, undiluted or pure with no additives, or 100%) or near-neat in specially configured vehicles or blended with gasoline in low concentrations. Consideration of methanol as motor fuel did not emerge until it became a common industrial chemical. It was used as an automotive fuel during the 1930s to replace or supplement gasoline supplies, in high performance engines in Grand Prix racing vehicles, and since the mid-1960s as the only fuel allowed in competing vehicles at the Indianapolis 500 (1,2). For general transportation vehicle use, serious research attention started in the late 1960s based on emissions advantages and was greatly expanded when energy security problems developed in the 1970s. Air quality has become the near-term catalyst for alternative fuel vehicle expansion, but in the long-term the United States and the rest of the world will likely have to rely on alternative energy as petroleum reserves diminish, become more expensive, or societal pressures demand use of fuels with lower environmental impact (3).

With the government mandated phase-out of lead as a gasoline octane additive, the high octane blending value of methanol made it a prime nonmetallic candidate for use in low concentrations. Methanol requires incorporation of higher order (C_3–C_8) alcohols to obviate phase separation deficiencies when blended with gasoline in small percentages, and a 50/50 mixture with tertiary butyl alcohol (TBA) was initially found to be more advantageous than other organics (4). Soon thereafter, methyl tertiary butyl ether (MTBE; made by reacting methanol with isobutylene) was found to have more favorable characteristics than methanol as an octane blending agent for use in existing gasoline vehicle models (5). Today, MTBE use in gasoline represents approximately 2.5% by volume of the total gasoline supply, and methanol is not used directly as a gasoline blending component (6).

Researchers in industrialized and developing countries throughout the world have been seriously interested in alcohol fuels since the 1960s, and Canada has an active commercially oriented program paralleling that in the United States (7). The cost of methanol produced from chemical processes has long been lower than that for both chemical and biological production of ethanol, and this shows little promise of changing in the near term.

Methanol has several major advantages in vehicular use: it is a convenient, familiar liquid fuel that can be produced readily using well-proven technology. It is a fuel for which vehicle manufacturers can, with relative ease, design a vehicle that will outperform an equivalent gasoline vehicle while having the same or lower emissions and efficiency (8). Next to hydrogen, methanol is the preferred fuel for fuel cells, which convert fuel into electricity with efficiency higher than internal combustion engines with the only emissions being water vapor and carbon dioxide. Fuel cell vehicles are currently in the research prototype stage, with only limited examples being tested. Fuel cell vehicles would be more practical than all-electric vehicles (no limitation on range) and have many of their benefits.

Major disadvantages of methanol are its higher cost than gasoline, the impact of reduced energy density on driving range or the need for a larger fuel tank; and the need for user and handler education on toxicity and safety (8).

Provisions to operate vehicles on either petroleum or an alternative fuel (bifuel) provide the means for a smooth transition from conventional petroleum to alternative fuels as the fuel supply infrastructure keeps pace. This offers a solution to "the chicken or egg" problem of industry not having the incentive to make either fuel or vehicles until the other is readily available (2). Early bifuel experience with carburated engines and gaseous fuels resulted in compromises in performance and emissions from both fuels. With the advent of fuel injection and electronic engine control, a new technique was developed to identify the fuel composition of mixed liquids and instantaneously adjust parameters for optimal operation (9). This concept, called the flexible fuel vehicle (FFV), also called the variable fuel vehicle (VFV), can provide operation on any combination of methanol or gasoline. (FFVs have also been developed to operate on ethanol and gasoline. The fuel sensor responds to both methanol and ethanol, and with the proper engine software development, an FFV that can operate on methanol, ethanol, or gasoline is possible.) When operating on methanol, FFVs typically exhibit slightly faster acceleration and increased thermal efficiency, without any degradation of emissions (2). An engine optimized for methanol alone would demonstrate even larger advantages, but the FFV is limited in that it will always represent a compromise between what is best for gasoline and what is best for methanol (10).

Methanol has two characteristics that make its use difficult in spark ignition engines: its low vapor pressure does not allow it to support cold starts below approximately 7°C (45°F); and it burns without a luminous flame, which makes methanol flames virtually impossible to see in bright sunlight. Adding 15% gasoline to methanol adequately solves both these problems, although it reduces the exhaust emissions benefits of straight methanol somewhat, and causes methanol fuel to have similar evaporative emissions as gasoline. Such blends of gasoline and methanol are generically known as M85.

All U.S. and most foreign automobile manufacturers have FFV designs. The U.S. manufacturers have produced several thousand FFVs per year since 1991. There were 18,319 such vehicles on the road at year-end 1995, primarily in federal government and the State of California fleets (11).

Although methanol has a very low cetane rating, which makes it difficult to autoignite in a compression ignition (CI; diesel) engine, methods have been developed to apply it in this use. The primary advantages are reduced oxides of nitrogen and particulate emissions while generally retaining other operational features. One of the several techniques for using methanol in place of diesel fuel has seen limited production in hundreds of urban buses and a number of specialty trucks.

Rules on emissions standards and test procedures for methanol-fueled vehicles including light-duty methanol cars, light-duty methanol trucks, heavy-duty methanol vehicles and engines, and methanol motorcycles have been established by the Environmental Protection Agency (EPA), and minimum range requirements for FFVs have been set by the National Highway Traffic Safety Administration (NHTSA) (12).

METHANOL PRODUCTION

Methanol is produced by the reaction over a catalyst of synthesis gas (ie, hydrogen and carbon oxides) generated from natural gas, liquefied petroleum gas (LPG), naphtha, vacuum residue, wood or other lignocellulose material, peat, or coal. Large-scale synthesis of methanol was first carried out in 1924, and the technologies for converting natural gas and coal into methanol have improved since then (14). Today, most methanol is made from natural gas, of which world resources are plentiful. Although methanol for motor fuel use would likely be imported at the outset, it could play a positive U.S. energy security role because of the nature of the suppliers or differences between the oil and methanol markets (8).

In 1995, methanol production in the United States totaled 6.5 billion liters (1.7 billion gal) which made it the 21st ranked chemical in terms of use (15). One of the largest single uses of methanol is to make MTBE, the preferred oxygenate for addition to gasoline for winter-time oxygenate programs (3) and to make reformulated gasoline. (The U.S. Environmental Protection Agency requires that MTBE or other oxygenates be added to gasoline in certain areas of the United States during the winter to reduce emissions of carbon monoxide.) In 1995, 10.0 billion liters (2.64 billion gallons) of MTBE was produced in the United States, making it the 12th largest used chemical (15).

Methanol was once referred to as wood alcohol, because it originally was made from the destructive distillation of wood. The technology for large-scale production of methanol was developed by Badische Anilin und Soda Fabrik (BASF) in Germany in 1924 (14). The currently preferred process for producing methanol is steam reforming natural gas. In this process, any sulfur present in natural gas is first removed. Next, the natural gas is reacted with steam in the presence of a catalyst under high heat and pressure to form carbon monoxide and hydrogen.

These elements are then put through the methanol production catalyst to make methanol. There are many variations of the basic steam reforming process, all aimed at increasing the overall thermal efficiency. Figure 1 shows a typical methanol plant in use today.

Figure 1. Methanol production plant, Trinidad, West Indies. Courtesy of the Caribbean Methanol Co.

Larger methanol production plants are more efficient than smaller ones. The size of a large (called world-scale) methanol plant is in the range of 2000 to 2500 metric tons/day. If methanol were to become a widely used alternative fuel, many more methanol production plants would be required. Plants as large as 10,000 metric tons/day have been postulated to serve the demand created by transportation vehicles.

In the United States, the primary methanol production location is in the Gulf coast area. Methanol is also produced in Canada, South America, Europe, and the Middle East. Methanol production and price is not controlled by any single country or consortium of countries. (However, most of the merchant methanol is currently produced by one company, Methanex.) Any country with remote natural gas reserves is a candidate for methanol production since production of methanol can be a cost-effective means of developing those reserves.

ENVIRONMENTAL REASONS FOR USING METHANOL

Emissions

Methanol has superior emissions properties to petroleum fuels. Because it is less photochemically reactive than gasoline, its evaporative emissions contribute less to smog and ozone-forming chemicals; also because it contains oxygen and has good lean combustion capability, it can reduce formation of carbon monoxide in exhaust emissions (16). The higher latent heat of vaporization results in lower combustion temperatures, reducing generation of NO_x (11). Unburned hydrocarbons (HCs) from vehicles using M85 have been measured to have less than half of the ozone-forming potential as the same mass of HCs from conventional or reformulated gasoline (17). (Reformulated gasoline is gasoline that is specially refined to reduce reactive and toxic hydrocarbons, and is required in areas of the country that are in nonattainment of air quality standards.)

Evaporative emissions of methanol during transport, storage, dispensing and use fall about midway between gasoline and diesel fuel, but increase with use of gasoline/methanol blends. Even though nearly twice as much methanol by volume is required to achieve the same operating range as gasoline, evaporative losses from M100 distribution could be about two-thirds those of gasoline (13). Methanol's low reactivity means that unburned methanol and evaporated methanol emissions have less smog- and ozone-forming potential than an equal weight of organic emissions from gasoline-fueled vehicles and infrastructure (8).

The combustion of methanol creates a higher percentage of aldehydes (almost totally formaldehyde) of total HCs as compared to gasoline. This was first identified in 1968 by Steadman and given great attention since then (18). Fortunately, formaldehyde is reduced appreciably along with other HCs by catalytic converters which are used on all new vehicles in the United States. Also, the HC emissions from methanol will result in decreased levels of secondary formaldehyde, which is formed in the ambient air from photochemical oxidation of HCs (18). This is the primary reason why combustion of methanol results in lower ozone-forming potential than gasoline (19).

Methanol contains no sulfur, so it does not contribute to atmospheric sulfur dioxide (SO_2). Since SO_x and NO_x emissions lead to acidic deposition, use of methanol would make a minor contribution to reducing acid rain.

Methanol-fueled vehicles will emit lower carbon dioxide emissions[5] (approximately 94%) than gasoline-fueled vehicles having the same fuel efficiency measured in energy per mile (20). Carbon dioxide emissions are attributed to be the primary cause of global warming. In addition, several studies have shown that FFVs using M85 and gasoline have from 4% to 7% higher energy efficiency using M85 than gasoline, which further reduces carbon dioxide emissions (21). However, when methanol production is taken into account, the net production of carbon dioxide emissions is the same or slightly higher for M85 compared to gasoline.

The need to achieve clean air has resulted in California requirements for progressive reduction in vehicle emissions. Methanol-fueled vehicles have the potential to achieve all but the zero-emission-vehicle (ZEV) level.

Ozone Production

Ozone (O_3) is formed in the lower atmosphere by photochemical reactions between volatile organic compounds (VOC), nitrogen oxides (NO_x), and to a smaller extent, carbon monoxide (CO). The ozone-forming potential of automotive emissions depends on the amount of VOC, NO_x, and CO emitted, on the composition of the VOC emissions, and on atmospheric conditions including VOC/NO_x ratios and the concentrations of the reactive species present from other sources (22).

The effectiveness of methanol fuels as an ozone control measure will vary considerably from area to area, with some areas benefiting significantly and some not benefiting at all. In particular, methanol's effectiveness will tend to be high in areas that characteristically have low ratios of reactive organic gas (ROG) levels to NO_2 levels, and will tend to be low in areas with high ratios. Other area variables affecting methanol effectiveness include average temperatures and mixing heights of the atmosphere. Low mixing heights (low dilution) are most characteristic of ozone episodes in California cities; high mixing heights (high dilution) are characteristic of summertime conditions in the Eastern United States (23). Industry and government analysts are not in agreement regarding the benefits of methanol emissions on ozone.

Greenhouse Gases

Greenhouse gases in the atmosphere have an important influence on the earth's climate. These gases intercept infrared radiation from the earth (turning it into heat), and as a consequence the temperature of the atmosphere is increased. Growth in the concentration of greenhouse gases is believed to create a net increase in the earth's temperature; this effect is called *global warming*.

Although some amount of greenhouse gases (natural carbon cycle) are necessary to maintain the earth's temperature, human activity has contributed to increased atmospheric concentration of greenhouse gases. The burning of fossil fuels for electricity generation (ie, coal, natural gas, and petroleum) and in transportation vehicles (petroleum) is a major contributor to greenhouse gases.

Transportation plays a major role in the emission of greenhouse gases. The transportation sector accounts for 31% of the fossil fuel and 69% of the petroleum consumed in the United States (24). The greenhouse gases produced by burning transportation fuels include carbon dioxide, methane, nitrous oxide, and ozone (created in sunlight-driven reactions involving NO_x and VOCs). When compared on an equivalent energy basis, methanol produces only 94% of the carbon dioxide of gasoline and no methane. However, when the production processes for making methanol from natural gas and gasoline from crude oil are included, making methanol from natural gas creates sufficient additional carbon dioxide to overwhelm this benefit (25). If methanol is used in dedicated engines with higher efficiency than similar gasoline engines, or in fuel cells, the increase in efficiency would proportionately decrease the amount of greenhouse gases produced per mile of vehicle travel. However, even if methanol facilitates significant increases in fuel economy, the net overall effect on greenhouse gases is likely to be minimal relative to using conventional petroleum fuels.

Impact of Spills

It would be expected that there would be a larger number of methanol spills because of the larger quantities of fuel that would have to be transported (26). However, the road slip/skid hazard from spilled methanol fuel is expected to be as low for M85 as for gasoline (27). Also, aquatic and terrestrial spills have minimal and brief consequences (28). Spills are of special concern because of methanol's solubility in water.

Once in groundwater, methanol may spread without detection unless it has been tagged with an odorant or denaturant (28). However, as a result of methanol's inherent properties of water solubility, biodegradability, and relative ease of complete evaporation, it could quickly dilute to nontoxic concentrations, disperse downstream, and decompose if spilled into large bodies of water, and evaporate or decompose if spilled on land areas (28).

Cleanup. In general, cleanup of methanol spills requires less extensive efforts and costs than cleanups associated with spills of water-insoluble petroleum fuels. Small methanol spills usually do not require any cleanup efforts, because of the effectiveness of natural biodegradation; whereas large methanol spills may require aeration of the water (to supply depleted oxygen to marine life and speed biodegradation) and/or use of methanol-destroying bacteria. Cleanup of a methanol fuel spill into a moving body of water would be handled similarly to that of a spill into the ocean (29).

Reversible Effects. Chronic leaks or major spills from methanol fuel systems are not expected to have major irreversible effects. Although the effect of methanol–gasoline blends may be somewhat stronger, no long-term or permanent damage is expected (29). The EPA has concluded that, like all fuels, methanol has certain characteristics that justify protective regulatory safeguards, but that overall the use of methanol would be less hazardous than the use of current petroleum fuels (30).

HUMAN HEALTH EFFECTS

Ingestion

Methanol is toxic to humans. Generally, it is not better or worse than gasoline or diesel fuel, just different (29). Methanol is odorless, colorless, and tasteless, but will be used with additives that will give warning and/or otherwise discourage use (29). From an overall safety and human health perspective, methanol represents some new dangers, but the EPA has concluded that its use would be less hazardous than the use of current petroleum fuels (26,31).

Methanol, like all combustible fuels such as gasoline, poses a potential human safety risk. Because of the differences in the physical and chemical properties of methanol and gasoline, the human safety risks of neat methanol are dramatically different than those of gasoline. Based on what is currently known, methanol would appear to offer safety benefits compared to gasoline (31). Experience with gasoline-fueled vehicles has not shown any significant risk from inhaling fuel vapors during driving, and the same is probable with methanol. During refueling, maintenance, and repair, the risk is also considered to be about the same as for gasoline vehicles (27).

Besides oral ingestion, methanol is readily absorbed by the lungs when present in the air (60%–85% of the methanol will be absorbed), and liquid methanol will be absorbed through contact with the skin. Once in the body, methanol is distributed to body tissues in proportion to their water content. More than 90% of the methanol in the body is metabolized in the liver and exhaled as carbon dioxide with the remainder removed through the lungs and the urine (30).

The minimum lethal dose of methanol (in the absence of medical assistance) varies between 0.3 and 1.0 g/kg of body weight (32). Following ingestion of a lethal dose, a transient, mild depression occurs. A latent period then follows that may last several hours to two days or more. Methanol in the body is first metabolized in the liver to formaldehyde, which is then quickly metabolized to formate. The formate is converted to carbon dioxide and water. The entire process of metabolism is as short as a few hours for low doses and a day or more for high doses. It is the accumulation of formate in the body that causes the toxic effects, and not that of methanol alone. High levels of formate in the body will cause damage to the visual system and can result in blindness, and in more severe cases, coma and then death. Low levels of methanol are metabolized harmlessly. In intermediate cases where some damage to the visual system has occurred, the effects sometimes reversed over time. The susceptibility to methanol depends on coexposure to ethanol and the level of folate in the liver, which impacts the ability to oxidize the formate that is present in the body (32). It is important that medical attention be given immediately to anyone who has ingested significant amounts of methanol.

In a clinical study, volunteers where exposed to a concentration of 192 ppm of methanol in the air for 75 min. (This concentration is the most that individuals might experience from normal use of methanol vehicles.) The volunteers were tested using 20 commonly used tests of sensory, behavioral, and reasoning performance, before, during, and after exposure. The presence of methanol had no detectable effect on the subjects' performance for most tests. Performance was impaired in one test measuring memory and concentration, but the effects were minor and within the range of the responses of volunteers exposed only to air (33).

Antisiphoning Rules. Industry specialists feel that a single device installed in the filler tank can be designed having both antisiphoning and antispitback functions in addition to being a flame arrester. The NHTSA estimated that 90% of the potential ingestion injuries and fatalities would be prevented by antisiphoning protection and promulgated regulations requiring the incorporation of antisiphoning devices on all new vehicles using fuels containing 20 vol% or more alcohol in 1992 (34).

METHANOL VEHICLE TECHNOLOGY

The Otto cycle engine, generally known as the spark ignition (SI) engine, operates on gasoline and offers the greatest opportunity for use of methanol. From the mid-1970s to the mid-1980s, laboratory research and development activities sponsored by the federal government built a technology base on SI engine use of neat methanol and ethanol and low (<30 vol%) concentration blends of these in gasoline, including laboratory, road, and fleet blend tests (35). Foreign and U.S. vehicle original equipment manufacturers (OEMs) added to the know-how, and information was freely exchanged in domestic and international forums (7,36,37). Small quantities of modified production vehicles kept pace with technology development, and helped identify problems and data gaps that were promptly filled. This work on fuel alcohol was almost totally devoted to neat methanol (M100), with occasional action on ethanol (neat E100 or hydrated E96). Following Brazil's initial production of E96-fueled automobiles, the California Energy Commission (CEC) launched fleet tests in 1981 on approximately 100 near-neat methanol and ethanol vehicles primarily supplied by Ford and Volkswagen. These vehicles incorporated then current technology, adding performance and emission provisions to characteristics otherwise similar to the Brazilian designs. The features of the two alcohols are similar, with those of methanol deviating the most from those of gasoline, whether positive or negative. Thus methanol posed the worst-case conditions for engine/vehicle designers. The CEC fleet program continued as technology advanced, successively incorporating pentane and gasoline additives up to 15 vol%; early flexible fuel vehicles that ran on gasoline, M85, E85, or any combination thereof; and preproduction FFVs with performance and emissions comparable to or better than gasoline counterparts even in cold weather. The primary features of the SI engine are its use of a homogeneous mixture of air and fuel typically in stoichiometric proportions (ie, a volume of air just sufficient to burn all of the fuel).

Materials Changes

Methanol attacks many metals, notably zinc, lead, aluminum, and magnesium. Exposure to methanol results in swelling, shrinking, hardening, and softening of elastomers, plastics, cork (gaskets), leather, polyurethane, and Viton in different ways than when applied to gasoline. Elastomers are particularly affected by blends of methanol and gasoline. Thus, many components in a methanol engine must be replaced with stainless steel, high-fluorine-content elastomers, or other methanol- and gasoline-compatible materials. These components include all those in the fuel delivery system (ie, the tank, fuel pump, fuel lines, and fuel injection rail in the engine), the injectors and pressure regulator in the fuel metering system, and piston rings (and possibly valve guide materials) in the engine (38). Electroless nickel plating has been extensively used on such items as carburetor parts and fuel dispensing nozzles. Galvanic corrosion also occurs in electric fuel pumps, fuel injectors, and fuel-level sending units. Anode dissolution is caused by leakage current flow due to the high conductivity of methanol (29).

Fuel Contamination

Fuel contamination must be avoided. Undesirable materials include chlorides, sodium formate, organic peroxides, phosphorus, sulfur, water, and reactive metals. Trouble has been encountered in vehicle use where the fuel was delivered in drums or tank trailers cleaned out by a chloride-containing solvent prior to shipment and where methanol fuel was stored in a tank where it came in contact with a calcium chloride desiccant.

METHANOL DIESEL ENGINES

The use of methanol in heavy-duty diesels is a more complex technical problem than using it in light-duty SI engines, because methanol's characteristics are poorly suited to diesel engine's CI requirements (39). In CI engines, diesel fuel ignites spontaneously (autoignites) when it comes into contact with air at high pressure and high temperature in the combustion chamber. A fuel's autoignition capability is measured by cetane number or ignition delay. If the ignition is delayed too much (low cetane number), excess fuel will be injected before ignition. Engine pressure will then increase rapidly, producing diesel knock (40).

Diesel fuels usually have cetane ratings between 40 and 50. Unfortunately, alcohols have poor autoignition characteristics and a much lower cetane number than diesel fuel. Methanol may have a rating as low as zero or as high as ten and is unsuitable in pure form (M100) in unmodified CI engines (41).

A variety of methods to use methanol in CI engines was investigated. Some require minimal changes to diesel engines, whereas others require significant modifications to fuel, combustion, and associated system parameters (42). These are alcohols mixed with diesel fuel (solutions; emulsions), diesel fuel with ignition additive (cetane improver), bifuels (diesel fuel ignition plus methanol fumigation; dual fuel injection), spark assist, and hot surface (active, eg, glow plug; passive) (43,44).

METHANOL STORAGE, DISPENSION, AND FIRE SAFETY

Methanol Infrastructure

A methanol infrastructure will not look or operate differently than the present gasoline-based system. The technology and equipment to adapt the existing gasoline system to methanol is readily available. Materials changes and some capacity addition will be required at fuel terminals, in transport equipment, and at retail outlets (45). If initial bulk distribution were to be conducted solely by water to existing petroleum fuel terminals and then by tank truck to retail outlets within a 100 mile radius, 75% of the present gasoline market would be served. Now, there are approximately 187,000 retail fuel outlets with 676,000 underground tanks in the United States. Approximately 50,000 new tanks are being installed annually, of which 30,000 to 35,000 are replacements. If this gasoline infrastructure were to be upgraded for methanol to replace 1 million gal/day of gasoline, the cost would be about $4.8 billion. This is one-half that of a comparable natural gas system and one-quarter that for electric vehicles (46).

There are currently 80 methanol refueling facilities in the United States (as of October 1996; not all commercial). These stations were established to serve the methanol FFVs that have been purchased by the state and Federal governments. Figure 2 illustrates the location of these stations. Most are in California because California has been proactive in promoting use of methanol as a measure to alleviate their air-quality problems. California has established a methanol fuel reserve that assures methanol availability without the price swings typical of the spot market for methanol.

Approximately 10 petroleum fuel marketers have participated in establishing the existing methanol service stations, most of which were majors corporations. Many equipment manufacturers also participated. Most required changes have been identified and shaken out. A typical methanol-dispensing system serving urban bus fleets includes a compatible submersible pump, steel double-walled underground tank, 40-gal/min island dispenser, dry-brake nozzle, appropriate piping and hose, filters, and vacuum monitoring with alarm system (47–49).

Service station installations are similar, with choice of dispenser/pump brand, and steel or fiberglass tank. Nozzles are nickel-plated aluminum, and hoses are cross-linked polyethylene with nickel-plated swivels (47–49). Dispensers resemble conventional fuel pumps, but typically operate in conjunction with a computerized cardlock system to prevent use by

Figure 2. Methanol fuel stations in the United States in 1996. Courtesy of the Alternative Fuels Data Center.

motorists that have conventional vehicles (48). A blending pump and electronic blending dispenser are also available to provide M85 from separate neat fuel storage tanks (47). Private vehicle refueling takes slightly longer than with gasoline, as reflected by the larger volume of methanol required. The Canadian Oxygenated Fuels Association has published a *Guide for Underground Storage and Dispensing Systems* that includes installation procedure and a list of equipment manufacturers. Participants in the CEC have similar information (48). Complete portable refueling systems are also available (47).

Fires

Methanol is classified as a flammable liquid with risk equal to gasoline, and the same regulations apply concerning transport, storage, and use with respect to risks for fire, explosions, and damage to the environment.

The methanol flash point of +11°C (52°F) is much higher than that of gasoline (−40°C; −40°F). The lower flammability limit (LFL) of methanol in air is 10°C (50°F). This determines the minimum concentration of fuel vapor in air that is required for ignition. The higher the LFL, the less likely it is that ignition will occur (50). The flammability temperature range is between 10°C (50°F) and 40°C (104°F) for methanol, and below −20°C(−4°F) for volatile gasoline (51). Addition of 5 to 10% gasoline will lower the flashpoint to less than −30°C(−22°F) and move the upper flammability limit to 0°C (32°F) or below (52).

Open Air Fires. In ventilated areas, the ignitability of neat methanol is between that of gasoline and diesel fuel. Methanol's low volatility, relatively high LFL, and low vapor density cause it to be much less likely to ignite in an open area resulting from a spill of fuel or release of vapor. However, spontaneous ignition can occur if methanol vapor comes in contact with a hot surface. In addition, once it does ignite, methanol's low heat of combustion and high heat of vaporization cause it to burn much slower and less violently, releasing heat at roughly one-fifth the rate of gasoline. When gasoline is added to methanol, the high volatility results in behavior more like gasoline (50). The volatility and LFL can be combined and expressed as the flammability index. The flammability index at common ambient temperatures for M100 is roughly 10% of that for gasoline. The flammability index for M85 is roughly 60% of that for gasoline (50).

Under equivalent conditions, one would expect methanol (M100) to be less likely than gasoline to ignite in open-air situations such as a fuel spill. This is not necessarily the case for M85, however, because the gasoline component would be volatile (52).

Explosion Potential. Methanol's combustion properties are such that a flammable mixture in the air above the liquid is likely inside fuel storage tanks under normal ambient temperatures (7°C to 42°C or 45°F to 108°F), whereas gasoline is virtually always too rich to ignite. Thus, a partially full tank of methanol presents a hazard. However, the electrical conductivity of methanol is higher than for gasoline, so there is less danger of spark ignition by static discharge. Also, precautions can be taken to prevent either flammable vapor–air mixtures from forming in storage tanks (eg, nitrogen blanketing, bladder tanks, floating roof tanks) or to prevent ignition sources from entering the tanks (eg, flame arresters, removing or modifying in-tank electrical devices), thereby mitigating any additional risk (50). The high volatility of blends make them less of a fire risk in enclosed spaces (53). Fire in an enclosed area is likely to start in open air and migrate to the enclosure.

Methanol Fire Characteristics. Because of the lower combustion heats and higher vaporization heats, M100 and M85 fires tend to burn less intensely and with a cooler flame than gasoline or diesel fuel fires. More heat is absorbed by the fuel as it burns, releasing less to the environment (52,53). Due to the lack of any large carbonaceous particles in its products of combustion, pure methanol burns with a light blue flame that is essentially invisible to the human eye in bright daylight. The only means of detecting the burning methanol in such situations is by feeling the heat generated or seeing the heat waves (26).

Gasoline addition (also useful for cold start) can provide the necessary flame luminosity. Addition of 15% gasoline having approximately 40% aromatics (predominately toluene) will provide luminosity until all the liquid is burned. Addition of 10% gives visibility until 50% is burned, and 5% toluene or 5% to 10% pentane works until 50% to 70% is consumed (29).

Fire Fighting. Water is a poor extinguisher of methanol fires. Dry chemicals, CO_2, or alcohol-resistant foam concentrates are the best extinguishers (54). Alcohol-resistant foams are needed for pure methanol fires and function equally well on gasoline fires. First aid extinguishers with CO_2 or powders seem to function equally well on methanol and gasoline fires (51).

Spill Fire Hazards. Comparing methanol (M100) with diesel fuel, fuel system leaks are equally likely, risk of fire is slightly higher, and damages will be lower (31). Comparing M85 with gasoline, fuel system leaks are equally likely, ignition of fuel leaks is somewhat less likely, and fires that do occur will tend to be less damaging (31).

The risk of a fire involving spilled M100 during repair and maintenance is greater than that for diesel fuel, although less than the currently acceptable risk level for gasoline (31). Spill fires involving M85 are likely to be less frequent and less damaging than gasoline fuel fires, whereas the overall risk for fires resulting from the spread of nonfuel-related vehicle fires is less for M85 than for gasoline (31).

Because of the lack of ignition sources, the risk of spilled M85 fuel fires during refueling is low and comparable to the risk of fires associated with gasoline fueling. In addition, the risk of a fire involving spilled M85 during maintenance and repair is the same or less than that for gasoline, as is the risk of storing M85 on site (31).

Field Experience. The large-scale industrial handling and transportation of methanol has a very good record (51). In addition, the NHTSA has established a standard for methanol vehicle storage tanks (54).

BIBLIOGRAPHY

1. A.W. Nash and D.A. Hawes, *Principles of Motor Fuel Preparation and Application,* John Wiley & Sons, Inc., New York, 1938.
2. U.S. Dept. of Energy, "Assessment of Costs and Benefits of Flexible and Alternative Fuel Use in the U.S. Transportation Sector Technical Progress Report One: Context and Analytical Framework," DOE/PE-0080, Washington, D.C., 1988.
3. R. Nichols, *Oxy-Fuel News,* 3–4 (1990).
4. W.H. Douthit, *Effects of Oxygenates and Fuel Volatility on Vehicle Emissions at Seasonal Temperatures,* SAE Paper No. 902180, SAE International, Warrendale, Pa., 1980.
5. E.E. Ecklund, *Options for and Recent Trends in Use of Alternative Transportation Fuels,* U.N. Center for Human Settlements Ad Hoc Expert-Group Meeting in Human Settlements, New York, 1986.
6. U.S. Dept. of Energy, *Alternatives to Traditional Transportation Fuels—1994,* Vol. 1, Report No. DOE/EIA-0585(94)/1, Washington, D.C., 1996.
7. *Proceedings of International Symposium on Alcohol Fuels* I, Stockholm (1976); II, Wolfsburg (1977); III, Asilomar (1979); IV, Guaraja Brasil (1980); V, Auckland (1982); VI, Ottawa (1984); VII, Paris (1986); VIII, Tokyo (1988); IX, Florence (1991); X, Colorado Springs (1993).
8. *Replacing Gasoline Alternative Fuels for Light-Duty Vehicles,* Congress of the United States Office of Technology Assessment, Washington, D.C., 1990, p. 13.
9. J.V.D. Wiede and R.J. Wineland, "Vehicle Operation with Variable Methanol/Gasoline Mixtures," *Proceedings of the VI International Symposium on Alcohol Fuels,* Ottawa, 1984.
10. U.S. Dept. of *Energy, Final Report of the Interagency Commission on Alternative Motor Fuels,* DOE/EP-0002, Washington, D.C., 1992.
11. U.S. Dept. of Energy, *Alternatives to Traditional Transportation Fuels—1995,* Vol 1, Report No. DOE/EIA-0585(94)/1, Washington, D.C., 1997.
12. *Driving Ranges for Dual Energy and Natural Gas Dual Energy Passenger Automobiles,* 55 FR 17611, Apr. 26, 1990.
13. U.S. Dept. of Energy, *Assessment of Costs and Benefits of Flexible and Alternative Fuel Use in the U.S. Transportation Sector, Technical Report Seven: Environmental, Health, and Safety Concerns,* DOE/PE-0100P, Washington, D.C., 1991, p. 1.
14. U.S. Dept. of Energy, *Assessment of Costs and Benefits of Flexible and Alternative Fuel Use in the U.S. Transportation Sector Technical Report Three: Methanol Production and Transportation Costs,* DOE/PE-0093, Washington, D.C., 1989, p. vii.
15. E. M. Kirschner, *Chem Eng News,* **74**(15), 16–20 1996.
16. U.S. EPA, *Clean Fuels Rept.,* 11–14, (1993).
17. V.R. Burns, and co-workers, *Emissions with Reformulated Gasoline and Methanol Blends in 1992 and 1993 Model Year Vehicles,* SAE Paper No. 941969, SAE International, Warrendale, Pa., 1994.
18. R. Pefley, *Alcohols as Extenders and Supplanters of Petroleum in the Transportation Sector,* SAE Paper No. 801378, SAE International, Warrendale, Pa., 1980.
19. B. M. Bertilsson, in *Proceedings of the III International Symposium on Alcohol Fuels,* Asilomar, Calif., 1979.
20. Ref. 13, p. ix.
21. R.L. Bechtold, *Alternative Fuels for Vehicles Fleet Demonstration Program - Final Report,* New York State Energy Research and Development Authority, Report 97-4, Albany, N.Y., Mar. 1997.
22. Mobil Research and Development Corp., *Clean Fuels Rept.,* **3**(2), 83 (1991).
23. *Replacing Gasoline Alternative Fuels for Light-Duty Vehicles,* Congress of the United States Office of Technology Assessment, Washington, D.C., 1990, pp. 67–69.
24. *Annual Energy Outlook 1996,* U.S. Dept. of Energy, 1996.
25. U.S. Dept. of Energy, *Alternatives to Traditional Transportation Fuels—1994,* Vol. 2, *Greenhouse Gas Emissions,* DOE/EIA-0585(94)/2, 1996.
26. *Office of Mobile Sources Analysis of the Economic and Environmental Effects of Methanol as an Automotive Fuel,* (Draft), U.S. Environmental Protection Agency, Ann Arbor, Mich., 1989.
27. "Battelle 1990 Report to the Trucking Research Institute" *Clean Fuels Rept.,* **3**(1), (1991).
28. Ref. 13, pp. 10–11.

29. *Status of Alcohol Fuels Utilization Technology for Highway Transportation: A 1986 Perspective,* Vol. 1, *Spark Ignition Engines,* ORNL/Sub/85-22007/4, Oak Ridge, Tenn., 1986.

30. M.G. Costantini, *Environ. Health Persp. Suppl.* **101** (Suppl. 6), 151–160 (1993).

31. *Replacing Gasoline Alternative Fuels for Light-Duty Vehicles,* Congress of the United States Office of Technology Assessment, Washington, D.C., 1990, pp. 72–73.

32. Health Research Committee, *Automotive Methanol Vapors and Human Health: An Evaluation of Existing Scientific Information And Issues for Future Research,* Health Effects Institute, Cambridge, Mass., 1987.

33. M.R. Cook, F.J. Bergman, H.D. Cohen, M.M. Gerkovich, C. Graham, R.K. Harris, and L.G. Siemann, *Effects of Methanol Vapor on Human Neurobehavioral Measures,* Research Report No. 42, Health Effects Institute, Cambridge, Mass., 1991.

34. "Federal Motor Vehicle Safety Standards; Fuel System Integrity; Alcohol Fuels," *Fed. Reg.* **57** (10), 1992.

35. U.S. Dept. of Energy, *Program Planning Document - Highway Vehicle Alternative Fuels Utilization Program (AFUP),* DOE/CS-0029, 1978.

36. *Proceedings of Advanced Power Systems Contractor's Coordination Meeting, 1–27 spanning EPA, ERDA and DOE,* National Technical Information Services, Springfield, Va., 1987–1993.

37. *ERDA / DOE / EMR Roundtables on Alcohol Fuels Research,* U.S. Contractor's Coordination Meetings and Windsor Workshops, 1975–1989.

38. U.S. Dept. of Energy, *Assessment of Costs and Benefits of Flexible and Alternative Fuel Use in the U.S. Transportation Sector, Technical Report Four: Vehicle and Fuel Distribution Requirements,* DOE/PE-0096P, 1990.

39. U.S. Dept. of Energy, *Assessment of Costs and Benefits of Flexible and Alternative Fuel Use in the U.S. Transportation Sector, Technical Progress Report One: Study Objectives and Methodologies,* DOE/PE-0086, Washington, D.C., 1988, p. 16.

40. *Substitute Fuels for Road Transport,* International Energy Agency, Paris, France, 1990, pp. 51–59.

41. R. L. Bechtold, *Status of Alcohol Fuels Utilization Technology for Highway Transportation: A 1986 Perspective,* Vol 2, *Compression Ignition Engines,* ORNL/Sub/85-22007/3, Oak Ridge, Tenn.

42. Ref. 13, p. 5.

43. New York City Emissions Report, *Oxy-Fuel News,* **2**(25), 7, 12 (1990).

44. Ref. 13, p. 6.

45. B. McNutt, J. Dowd, and J. Holmes, *The Cost of Making Methanol Available to a National Market,* SAE Paper No. 872063, SAE International, Warrendale, Pa., 1987.

46. U.S. Dept. of Energy, *Assessment of Costs and Benefits of Flexible and Alternative Fuel Use in the U.S. Transportation Sector, Technical Report Four: Vehicle and Fuel Distribution Requirements,* DOE/PE-0095P, Washington, D.C., 1990, p. vii.

47. J. Spacek, "Infrastructure Issues from the Retailer's Perspective–Panel Discussion," *1993 Windsor Workshop,* NTIS, Springfield, Va., 1993, pp. 511–529.

48. A. Argentine, "Maintaining Fuel Quality: California's Methanol Experience," *1993 Windsor Workshop,* NTIS, Springfield, Va., 1993, pp. 465–488.

49. J. Wiens, "Methanol Fuel Formulation Issues," *X International Symposium on Alcohol Fuels,* U.S. Dept. of Energy Report, DOE/CH10093-245, Colorado Springs, Colo., pp. 159–170.

50. P. A. Machiele, *Flammability and Toxicity Tradeoffs with Methanol Fuels,* SAE Paper No. 872064, SAE International, Warrendale, Pa., 1987.

51. *Alcohols as Alternative Fuels for Road Vehicles,* Report EUCO-Cost 304/I/85, VII/40/85-EN, European Cooperation in the Field of Scientific and Technical Research, Brussels, 1984, pp. 19–21.

52. U.S. Dept. of Energy, *First Interim Report of the Interagency Commission on Alternative Motor Fuels,* Washington, D.C., 1990, pp. 6-33–6-34.

53. Ref. 13, p. 12.

54. Ref. 13, p. 13.

AQUATIC SEDIMENT SAMPLING

ALENA MUDROCH
J. M. AZCUE
National Water Research Institute
Burlington, Ontario, Canada

Materials on the lake, stream, and ocean floor, known as bottom sediments, have been studied for many years to characterize their nature and properties. The characterization of bottom sediments is an important part of geological and geotechnical surveys of stream, lake, or ocean floors before construction or oil and mineral exploration. It has been proven that bottom sediments are a sink for as well as a source of contaminants and that sediment-associated contaminants can be transported by resuspension of sediment particles, may accumulate in the food chain, or may affect the health of biota and water quality in aquatic ecosystems. Consequently, the knowledge of the physicochemical properties of bottom sediments is important to the studies of the fate and pathways of contaminants in aquatic ecosystems. Furthermore, the quality of the bottom sediments needs to be assessed in the evaluation of the health of aquatic ecosystems.

The knowledge of the character of the sediments and the biogeochemical processes at the sediment–water interface as well as the identification of quality and quantity of different contaminants in the sediments are necessary in the development of techniques for remediation of aquatic ecosystems. Sediment quality affects the selection of equipment and the methods for the disposal of the sediments in the dredging of harbors and waterways.

Since the mid-1980s, sediment toxicity has been intensively studied. Many methods were developed and tested for the determination of the effects of contaminated sediments on aquatic biota. Laboratory bioassays were designed to test whole sediments and sediment pore water and elutriates under different conditions. In addition, research studies indicated that the toxicity of some trace elements in sediments strongly depends on their chemical forms under specific environmental conditions. Therefore, different methods were developed to determine bioavailable forms of the trace elements in sediments. Extensive information on assessing toxicity of freshwater sediments is available (1).

Adequate definition of the physical, chemical, and biological character of sediments and all analyses and tests involved in the evaluation of the sediment quality require collection of sediment samples. The samples must be collected, stored, and handled by proper techniques, which ensure the preservation of sediment integrity requested in different analysis and tests.

To process nonrepresentative or incorrectly collected or stored sediment samples may lead to a waste of money and human effort and to erroneous conclusions.

PLANNING OF SEDIMENT SAMPLING PROGRAM

Study Objectives

A clear definition of the problem and a statement of the objective(s) or the reason for the characterization of sediments are necessary before planning an adequate design of the sediment sampling program. The objectives for the characterization of sediments can, be eg, the scientific interest in the nature of stream, lake, or sea floor in an area for which no information exists, for the assessment of geotechnical properties of sediments in gas or mining exploration, or for laying underwater cables. The objectives for sediment characterization in environmental studies are, in most cases, the determination of the extent to which the sediments are either a source or a sink of contaminants and the evaluation of the effects of sediment-associated contaminants on the aquatic ecosystem of interest. Sediment characterization studies may have regulatory implications (such as dredging and disposal of the dredged material or remediation of the contaminated area) or be carried out to assess risk to human and environmental health through research of different sediment–water interaction processes.

Regardless of the type of study, the objectives should clearly outline the physical extent of the study area. For example, the monitoring of sediment quality in a lake may be carried out with the objective of evaluating the changes of inputs of different contaminants into the lake over the past 50 years. In this case the study area is the lake. On the other hand, an investigation of environmental effects of industrial wastes disposed of near a river may involve a study of a large terrestrial area, river, lake, groundwater, etc. The character and quantity of data that need to be collected to draw conclusions should be accurately identified in the study plans. In studies dealing with the assessment of the effects of contaminated sediments, different actions, which must be applied as a result of study conclusions, should be outlined together with the action level (eg, criteria or guidelines for disposal of dredged sediments).

Collection and Review of Available Data

The next step following the clear definition of the problem and statement of objective(s) should be the collection of historical information and data relevant to the study. The collection of historical data with a comprehensive review of literature, reports, and all available previously published data generated by surveys and studies (including the characterization of the sediments) should be completed before planning the sediment sampling method. Even old or incomplete data are useful, eg, they can provide a first estimate of the concentrations of a parameter of interest in the sediments or information on the deposition of the sediments. Sometimes old data provide sufficient information to warrant additional sampling at the study area. Data from regional reconnaissance surveys could provide a wide range of information, such as concentrations of different metals and trace elements in stream and lake systems predicted from the known geology and mineralogy of soils and bedrock in the drainage basin; the geochemistry of the bottom sediments; and general background concentrations of different chemicals in soils that, through weathering or erosion within the drainage basin, contribute material to the stream and lake sediments. Material could enter from the drainage basin in either dissolved form or associated with eroded soil materials and could include pesticides or fertilizers from agricultural practices, metals and metalloids for mining wastes or excavated materials, or different by-products from industrial and mining processing effluents.

Data that are relevant to the planning of sediment sampling include general information on the drainage basin of the studied water body, such as land use and type of industries; and the quality and quantity of runoffs and effluents. Information on the water body to be studied includes distribution, thickness, and types of sediments, particularly fine-grained sediments; horizontal and vertical profiles of physical and chemical characteristics of the sediments; structure of the biological community; composition and diversity of the species; bioaccumulation of contaminants; and results of different bioassays.

Collected old data and information must be carefully reviewed, particularly with respect to the overall objectives of the project for which they were collected. The completeness and quality of the data should be assessed, as indicated by quality assurance and quality control in the data collection. Furthermore, it is important to visit the study site to evaluate the completeness of the collected information and to identify any significant changes that have occurred since the old data were collected.

Planning Sediment Sampling

Sediment sampling is the operation of removing a portion of sediments distributed over the study area in such a way that the samples are representative of the physical, chemical, and biological properties of the study area as a whole. From a statistical point of view, sampling is expected to provide analytical data from which the sediment character may be determined with known and controlled errors and at the lowest cost. In most cases, the cost of the physical, chemical, and biological analyses of collected sediment samples exceeds that of the collection of the samples. The funds spent on the analyses are wasted if the samples are collected at inappropriate locations or if they do not represent the study area.

The proper selection and use of sediment sampling equipment and the sample handling, storage, and transport are all equally important to the selection of sampling locations. Considering the above, it is desirable to spend more than a half of the time allocated to sediment sampling on detailed planning of where and how to collect the samples. The plan should also include all logistics associated with the travel of personnel involved in the sampling; shipping the equipment to the sampling location; and handling, preservation, storage, and transport of collected samples.

The sediment sampling plan needs to be designed to achieve the stated objectives for the characterization of the sediments. The sampling plan should describe in detail which objectives of the study will be selected and how these objectives will be achieved within a given time frame and budget for the study. Furthermore, the sediment sampling plan should contain the following: a detailed work plan, manual and standard operating procedures, and the rationale for the proposed techniques that will be used to achieve the objectives. The plan should include an identification of individuals responsible for each operation, the equipment, and special requirements to perform the work. The plan may be outlined as a flow chart; and in cases in which the study must be carried out in a given time frame, the timetable schedule should also be outlined. An adequate quality assurance (QA) program for sediment sampling and sediment analyses should be included in the sediment sampling plan. For example, under the established U.S. Environment Protection Agency (EPA) mandatory quality assurance program (2), data quality acceptance criteria and QA project plans are prepared for all data collection projects. These project plans clearly describe which operations will be performed at each stage of data collection (eg, sediment sampling site selection; techniques for sediment collection, sample handling, and analyses; and data handling and analyses) and include instructions of standard operating procedures for each field and laboratory activity. Properly described and standardized methods for sediment sampling, analyses, and data processing allow for the comparison of results from different studies and support the confidence in study conclusions and, in projects dealing with pollution, in selection of proper remedial actions.

Sediment sampling must be planned in such a way that its execution will meet the stated objectives within the given cost constrains. A detailed work plan needs to be prepared that will outline the individual steps of the sampling. The sediment sampling budget should closely follow the sediment sampling plan by estimating the cost of each step of the work plan. The estimated cost of individual steps should be obtained from the personnel responsible for a particular task. The responsibility of the study manager is to ensure that sediment sampling is accurate and cost-effective. The cost of sediment sampling and processing of the samples should include all expenditures associated with the collection of the sediments into proper containers, measurements and sample handling in the field, transportation and storage of the samples, sample preparation, sediment tests and analyses, quality control procedures, data processing, and preparation of reports. The most important factors affecting the cost of sediment sampling include the number and location of sampling stations; selected, and in many cases expensive, sampling procedures; the number of samples to be collected at each station and/or the amount of sediment material required for analyses and different experiments; and sample handling and preservation procedures. The location and size of the study area, the number of sampling stations, and the selected sampling procedures greatly affect the cost of the sediment collection. For example, there is a difference between the cost for sediment sampling in a small, shallow lake using a small vessel operated by two or three workers and the cost for sediment sampling in a large, deep lake or ocean, which requires a large vessel with an experienced crew. In cases for which the estimated cost for sediment sampling exceeds available funds, alternatives must be considered. Extensive discussions of the costs associated with sediment sampling have been published (3,4).

SELECTION OF SAMPLING STATIONS AND PREPARATION OF SAMPLING PROTOCOL

The sediment sampling plan should start with the selection of the sampling stations and the preparation of a sampling protocol that contains all the details of the sampling within the study area and, if desirable, in a selected control site (the control site is considered a water body within the general vicinity of the study area, relatively free of contaminants and with sediments of similar geochemical character to that of the sediments in the study area). In selection of the number and location of sediment sampling stations, the following factors should be considered: objective(s) of the study, purpose of characterization of the sediments, historical data and other available information, bottom dynamics and bathymetry of the sampling area, size of the sampling area, and available funds for the sediment sampling program.

The study objectives will further define the type of samples to be collected, such as whole sediments and sediment pore water, and designate individual samples for different analysis and tests, such as determination of physicochemical character of sediments, concentrations of bioavailable contaminants in the sediments, extraction of pore water, and elutriation for geochemical analysis and bioassays. Methods for collecting and handling individual samples with respect to their designation need to be outlined in detail in the sampling protocol. Collection, storage, and handling of samples for toxicity using different bioassays need special attention to preserve the integrity of sediments. Only a few standard test methods are available for the evaluation of sediment toxicity. However, several published standard guidelines contain descriptions of the procedures for collecting and handling sediment samples for testing sediment toxicity (5–10). Recently, guidelines for the collection and preparation of sediments for physicochemical characterization and biological testing have been prepared (11). Some guidance documents attempt to standardize sediment collection and handling methods by presenting limited options with their associated strengths and weaknesses. However, it is apparent that the field of sediment assessment is in its infancy, and the guidance information is of a provisional nature (1).

Sediment Heterogeneity

An ideal study area would have a uniform distribution of geochemically homogeneous bottom sediments. However, such places rarely occur. It is recognized that sediment sampling operates in an environment of uncertainty. In most cases, the knowledge of spatial heterogeneity of sediment chemistry within the study area does not exist before the sediment sampling. In studies of the distribution of contaminants, sediment samples featuring the most suitable grain size for different physicochemical and biological analyses are preferred. It is well known that fine-grained sediments contain the greatest concentrations of contaminants (12,13). Consequently, in environmental studies, samples of fine-grained sediments are usually selected to determine the extent and effects of contamination, unless there is an indication that certain contaminants are associated with coarse-grained sediments at the study area. The general size classes of sediment particles are gravel, sand, silt, and clay. Different nomenclatures of particle size were proposed and have been used in the characterization of sediments (14,15). A review of the historical development of the particle size classification has been published (16). In most studies of aquatic sediments, the particles are classified by their sizes into the following groups: <4 μm clay, 4–63 μm silt, 63 μm–2.00 mm sand, and 2.0–64 mm gravel. Fine-grained sediments are usually referred to as silt- and clay-size particles, ie, <63 μm.

Considering the association of contaminants with the silt- and clay-size particles, sediment samples should be collected from sites with deposition of fine-grained sediments in the study area. However, fine-grained sediments are not uniformly distributed in lakes, oceans, and particularly, rivers. The distribution of the sediments is affected by energy-controlled processes. The sorting of sediments by texture and particle size is mainly affected by current and wave actions. Fine-grained sediments usually accumulate in areas of low energy zones, such as bays or the outer side of the main channel of a meandering river. Coarse particles, such as sand and gravel, and hard glaciolacustrine clay are found on the bottom of fast-flowing rivers. In large lakes, the deposition of fine-grained sediments is affected mostly by current circulation and wave action, and fine-grained sediments are deposited in deep, quiet waters in the offshore areas. In small lakes, fine-grained sediments usually accumulate in the greatest water depth.

Personnel involved in the selection of the location of sediment sampling stations should have a basic knowledge of the distribution of fine-grained sediments at the study area. If such information is not available, a preliminary study should be carried out to define the distribution of the fine-grained sediments in the study area. In addition to the size of the sediment particles, it is important to know the origin of the fine-grained sediments. Glaciolacustrine fine-grained sediments deposited in lakes and rivers of glacial origin are typically clay size, firm, sticky materials containing approximately 25% water in contrast to fine-grained, soft, recent sediments containing 70–95% water. Glaciolacustrine sediments do not contain anthropogenic contaminants, and their consistency prevents the penetration of most frequently used sediment samplers.

Physicochemical heterogeneity of fine-grained sediments is another important factor to be considered in the selection of the location of sediment sampling stations. Great variations in the chemistry of the fine-grained sediments, even within a small study area, are usually the result of different distances from pollution sources, bathymetry of the water body, and sediment resuspension and redeposition, etc. The above examples indicate that the selection of sediment sampling stations is a complex task and requires local and detailed information on the characteristics of the water body from which the sediments will be sampled. More details on the bottom dynamics and energy-controlled processes that affect the distribution of sediments on lake and river bottoms are available (17–19). Furthermore, examples of sediment heterogeneity in large rivers and lakes have been discussed (20); and examples of heterogeneity of different physical, chemical, and biological characteristics of sediments have been published (19).

Preliminary Sampling

In studies of sediment contamination, it is necessary to obtain information on the distribution of the sediments in the study area before the selection of the location of sampling stations. It is particularly important to locate the deposition of fine-grained sediments. Generally, there are two methods that can be used to obtain the information. The first is an acoustic survey of the bottom of the water body to be sampled. The second is a preliminary sediment sampling at a few selected locations in the study area. The information obtained by one or both of these methods will provide guidance in the selection of the locations of sediment sampling stations in the study area.

Echo sounding, seismic reflections, and refraction are acoustic survey techniques suitable for detecting the type of surficial sediment layers (such as sand; gravel; or soft, silty clay) and the subsurface sediment layers. Acoustic penetration is most effective in unconsolidated, soft, silty clays with high water content. Firm, compacted sediments, such as sand and glaciolacustrine clays, display minimal acoustic penetration. More information on acoustic techniques is available (21,22). However, acoustic survey techniques require special equipment and a good positioning capability. Furthermore, the interpretation of records obtained by these techniques often requires the skills of a geophysicist. In addition to the acoustic survey, the information on the physical characteristics of sediments may be obtained by the fall cone technique (23) carried out at the preliminary sampling stations. A sediment penetrometer, for *in situ* measurement of sediment physical characteristics, can rather simply and rapidly determine sediment types and prevailing bottom dynamics. The penetrometer has a few different cones. The depth of the penetration of these cones reveals the type of sediments. A calibration table is supplied with commercially sold sediment penetrometers to convert the penetration depth into several different sediment types ranging from very soft to hard sediments.

In the second method, the design of the preliminary sediment sampling carried out to obtain information on sediment distribution in the study area generally depends on the size and shape of the water body to be sampled. Sampling along transects with minimum sampling stations along each transect may give sufficient information on the distribution of the sediments in lakes and rivers (24). Sediment samples obtained at each of the preliminary sampling stations should be visually inspected, and the type of sediments and water depth recorded. The results of an acoustic survey, description of sediment texture, and a bathymetric chart should be sufficient for prediction of the distribution of sediments in the study area. The information is further used in the selection of the location and number of sediment sampling stations in the preparation of the final sampling plan. The preliminary information on the distribution of the sediments is also necessary for the preparation of a complete list of sampling equipment, logistics of the field work, estimation of the quantities of sediments to be sampled, etc., which all need to be included in the sampling plan.

Location and Number of Sampling Stations

Information on statistical methods of sampling and application of the methods of sampling in the environmental studies are useful in the selection of locations and number of sediment sampling stations in freshwater and marine systems (25–30). At a study area where some information exists on the chemical composition of the sediments, the estimation of variability and mean concentrations of chemicals of interest should be the first step in the selection of the locations and number of sampling stations. The estimation should be followed by a statistical method of calculating the number of stations needed to achieve the acceptable confidence levels and to meet the objectives of the determination of the character of sediments in the study area. At study areas with no information on sediment chemistry, the number of required sampling stations can usually be estimated only after statistical analyses of different parameters have been determined in sediments collected in a preliminary sampling. The preliminary sampling should determine the heterogeneity of sediment chemistry and concentration ranges of the chemicals of interest in the study area. Additional statistical analyses following these determinations can help select the number of sampling stations for the final sampling plan. However, only a few examples are available for dealing with statistical methods for small data sets, which can be obtained usually by costly and time-consuming preliminary sampling.

In addition to the application of the statistical approach to the sediment sampling design, different conditions in the study area (described below) need to be considered. The physical size of the study area must be considered in addition to the distribution of the sediment in the selection of the locations and number of sampling stations. In a study area with a relatively homogeneous sediment character, such as the particle size, sampling stations can be widely spaced. On the other hand, a denser sampling grid will be required in areas with relatively heterogeneous sediment character. Examples of spacing sampling stations in studies of marine sediments and sediments in large lakes can be found in different reports (31–34). In studies of environmental pollution of small areas, sediment sampling stations usually need to be located fairly close, in particular in areas with many contaminant sources, such as effluents from different industries and municipal water treatment plants. Common sense, good judgment, knowledge of the historical background of the sampling area, and experience in sediment sampling are the best help in the selection of the locations and number of sampling stations in the study area.

In most cases, rivers and creeks have more complex bottom dynamics than lakes. Therefore, it is important to consider the following conditions in the selection of the sampling stations in a river or creek. Only a few areas of fine-grained bottom sediments occur at a high-speed water flow. In addition, some of these areas are transitional only, with fine-grained particles settling on the bottom becoming resuspended and transported downstream. Natural fine-grained sediment traps, such as the inside of channel bends, isolated pockets along the margins of the channel, and areas sheltered from the main flow, are usually good locations for sediment sampling in a river or creek. In meandering rivers and creeks with a moderate or slow flow, a preliminary sediment sampling along a few transects across the river or creek will reveal the depositional areas of fine-grained sediments. Seasonal changes in flow and discharge can affect the deposition of the sediments, such as more intensive erosion and transport of fine-grained bottom sediments during the spring or fall runoff.

The conditions described above indicate that the selection of sediment sampling stations in rivers and creeks can be more difficult than that in lakes where depositional areas of the fine-grained sediments can usually be easily defined and remain unchanged. A physical inspection of the area to be sampled should be the first step in the selection of the sampling stations in a river or creek. A preliminary sediment sampling along one or a few transects across the river or creek should follow the physical inspection to help determine the depositional areas of fine-grained sediments. The number of sampling points along the transect(s) should be based on the flow speed, with more sampling points being located in areas with low flow speed, water depth, and water channel width. An example of sediment sampling in rivers and creeks in a reconnaissance survey carried out on a grid with a density of one to several samples per square kilometer is available (35).

The number of sampling stations depends on the objectives of the study and heterogeneity of the parameters of interest in the sediments in the study area. Clearly stated objectives and a knowledge of sediment distribution over the area to be sampled help determine the number of sampling stations. Sediment sampling in studies of the distribution of sediment types and geochemistry in freshwater and marine environments is usually carried out on a regular grid system. Examples of using a regular grid have been given (36,37). One of the most common objectives in determing the number of sediment sampling stations in large lakes and oceans is to cover the largest possible area in the least amount of time and, consequently, at minimum cost.

Description of the physicochemical character of sediments has often been based on samples collected in a haphazard pattern. Using this pattern, a disproportionate number of samples is collected from areas easiest to access or easiest to sample. The implications of such sampling have been discussed (38). Examples of the number of sampling stations necessary for producing maps of distribution of different contaminants in lake sediments have been shown (19). A pilot sample formula to provide an even cover of sampling stations over an entire lake has been proposed (39). This formula is based on two morphometric standard parameters: the lake area (eg, more samples should be collected in large lakes than in small lakes) and the development of the shoreline, which is used as an indirect measure of the bottom roughness. Statistical concepts used in soil sampling can be applied to determine the number of stations in bottom sediment sampling. These concepts have been described (40). Further examples of the selection of locations and number of sediment sampling stations in different studies of sediments, including dredging projects, have been discussed (19,20,38).

Positioning of Sampling Stations

The positions of selected sampling stations need to be marked on good-quality navigation charts of the study area. Sediment samples must be collected as close by as possible to these marked positions. In studies in which the sampling is to be repeated, accurate positioning of the stations is extremely important. The accuracy and precision of the position of each sampling station depend on the nature of the study. For example, a 5–10% error in the distance between sampling stations can be acceptable in a baseline survey carried out in a large study area (eg, >10 km^2). However, in studies where the sediment sampling must be subsequently repeated, accuracy of positioning within a few meters may be necessary.

Many different positioning methods are presently available. Some of them will require topographic maps of the shore adjacent to the water body to be sampled. The selection of a suitable positioning method depends on the character of the study area, the distance between the sampling stations, and the shore of the sampled water body. Furthermore, a proper positioning method needs to be selected and identified in the work plan before beginning the sediment sampling. There are vessels that have a good positioning system and personnel trained with its operation. When the owner or operator of a sampling vessel does not have the necessary expertise and/or positioning equipment, a competent surveyor, a survey company, and/or government agencies should be contacted for advice to assist in positioning. Regardless of the type of system selected, the positioning equipment must be properly set up and calibrated, and standard operating procedures must be followed by trained personnel to achieve maximum accuracy of the positioning system.

Simple techniques for the positioning of the sediment sampling stations (such as sextant observations and measurements of distances by line or taut wire) can be used in small study areas. Positioning of sampling stations in large study areas, such as large harbors, lakes, and oceans, with a large number of sampling stations requires an electronic positioning system or radio navigation and satellite navigation systems. Underwater positioning methods use different types of underwater acoustic beacons at known positions on the lake or sea floor and sensors, such as echo sounders, sonars, and TV cameras. The characteristics, advantages, and disadvantages of the different positioning methods have been described (21,22,41). Government agencies may provide information on a large-scale positioning system available within the country. Information on small-scale positioning systems may be obtained from different private companies. Many low-frequency, long-range navigation systems are permanently established and operated by national and international agencies.

Sediment Sampling Protocol

The sediment sampling protocol should be prepared before selecting the sediment sampling equipment and the determining the quantity of sediment samples to be collected. The sampling protocol should contain the following information:

- Character of the samples to be collected to meet the study objectives as well as the objectives of the sampling program, including information on the number and volume of the surface sediment samples and the number and length of the sediment cores.
- Compilation of all analyses and tests (eg, physical and chemical analyses and bioassays) that will be carried out on each of the samples; list of all tests and observations that will be carried out in the field during the sediment sampling program.
- Estimation of the quantity of wet and dry sediment required for each of the above analyses and tests, including the quantity of sediment necessary to meet the quality assurance and quality control program. Some studies require collection of an additional volume of sediment for banking for future analysis; this should be included in the estimation of the required quantity of sediments.
- Morphometry, water depth, hydrological conditions, distribution of sediments on the bottom (eg, depositional areas of fine-grained sediments), climatic conditions, and other features of the study area that are relevant to the sediment sampling program.
- List of areas that may be considered as a control site.
- Available funds and number and availability of trained and nontrained personnel needed to carry out the sampling program.

In addition, the following information should be added to the sampling protocol:

- Chart of the sampling area with plotted positions of sediment sampling stations (after the number and location of the stations are selected); the sampling stations should be numbered in the most logical sequence relevant to the sediment sampling program and objectives.
- Outline of the time frame of the sediment sampling program with consideration of the optimal use of the time spent on the sediment sampling in the study area.
- Information on the weather conditions expected during the sampling period and severity of contamination of sediments and water to which the sampling personnel will be exposed, to ensure the safety of the personnel carrying out the sampling program.
- Selection and listing of sampling equipment and other materials necessary to support the sampling program, including tools and spare parts for emergency repairs and maintenance of the sampling equipment in the field; maps and charts relevant to the study area also considering safe access to the area; notebooks with sediment logging sheets; equipment for field measurements, homogenization, and subsampling of collected sediments; equipment for extruding sediment cores; sample containers and material for labeling the containers; storage boxes; and other equipment relevant to the sampling program.
- The last date the sediment sampling equipment was tested together with any problems encountered (and repaired) during the testing, including a list of all necessary spare parts and tools that *must* accompany the sediment sampling equipment for emergency repairs in the field.
- Typed report of the complete sediment sampling and subsampling procedures, sample handling and preservation methods, field storage procedure, method of transport of samples from the sampling site to the laboratory, and method of storage after the samples are delivered to the laboratory (eg, required storage temperature, freezing, and light conditions).

A copy of the sediment sampling protocol should be distributed to the personnel participating in the sediment sampling program. A meeting among the personnel to discuss the program before leaving for the field is usually beneficial.

DESCRIPTION OF METHODS AND EQUIPMENT FOR BOTTOM SEDIMENT SAMPLING

The objective of sampling is to collect a representative, undisturbed sample of the sediment to be investigated. There are many factors that need to be considered in the selection of suitable equipment for bottom sediment sampling. These factors include the objective of the sediment characterization, type of available sampling platform (such as vessel or ice surface), location and access to the sampling site, physical character of the sediments, number of sites to be sampled, weather, number and experience of personnel who will carry out the sampling, and budget. The number of factors involved make standardization of sampling techniques difficult. Generally, the selected sampling equipment should recover an undisturbed sediment sample.

Several excellent comprehensive reviews are available on bottom sediment sampling devices. These reviews generally describe different equipment for particular purposes, eg, grab samplers and corers suitable for sampling benthic organisms and a wide variety of bottom sediment samplers designed for biological and geological work mainly in the marine environment, which may also be used in pollution studies in the marine or freshwater environments and sediment sampling techniques in studies of sedimentary structure (42,43). There are many reviews that discuss the limitations of different sediment sampling equipment, eg, theoretical and practical aspects and advantages and disadvantages of various types of sediment samplers (44,45). Other reviews consider sediment sampling for different purposes in freshwater and marine environments (46–49). A summary of many reports on sediment sampling has been given (50).

Many samplers described in the literature are only variations of a few early models modified to overcome observed deficiencies or to be used for specific objectives and for different operating conditions. Many different names used for sediment samplers can be confusing to those who need to choose one suitable for a specific project. In addition to the above reviews, commercially available or custom-made sediment samplers have been described in studies involving bottom sediments in marine and freshwater systems, eg, the investigation of benthic fauna, the collection of samples at sites with different water depths and sediment textures for geological interpretation, or the comparison of the performance of newly designed samplers to fulfill a specific task.

QUALITY CONTROL IN SEDIMENT SAMPLING

Information on the concentrations of contaminants in sediments is often the basis for important environmental decisions. Due to the improvements in analytical techniques, the present inaccuracies in the assessment of sediment quality are usually the result of an improper sediment sampling design and sampling procedures instead of errors in analysis of collected sediments. The degree of uncertainty in the final result of a sampling program depends on the performance of each sediment sampling step, such as sample design, sample collection, sample preservation, transport, and storage. Consequently, the data obtained in all the intermediate steps of the sampling program must be reliable. Errors can occur in all steps, and the objective of quality control (QC) is to identify and quantify them. The goal is to correct and minimize individual errors and their cumulative effects. Quality control procedures reduce and maintain random and systematic errors within tolerable limits, while quality assurance (QA) is the management system that ensures an effective QC system is in place and working as intended (75). In the laboratory, QC consists of the technical, day-to-day activities, such as the use of reference materials, spikes, and blanks, to control and assess the quality of the measurements. It should be noted that QC in the laboratory applies only to possible errors introduced after the collection of sediment samples and their transport to the laboratory. Therefore, sediment sampling procedures are required to have their own QC.

Much attention has been focused on QA and QC in chemical analyses of sediment samples. However, it is more complex to measure the accuracy of sampling the sediments, which are, in most cases, heterogeneous. The following two techniques can be used for QC in sediment sampling. One technique consists of the collection of more than one sediment sample at selected sampling sites with identical sampling equipment, such as multicorers, as well as using identical procedures for field subsampling, methods for handling and storage of the samples, and methods for sediment analyses. The results of the analyses will show variations that are due to the sampling and subsampling techniques, but the heterogeneity of the sediment at the sampling site will still affect the tests. The sediment sampler must be selected to suit the sediment texture at the sampling site.

In another QC technique, the collected sample is divided into a few subsamples and each subsample is treated as an individual sample. The results of chemical analyses of all subsamples indicate the variability due to the sampling and analytical techniques and sediment heterogeneity within a single collected sample.

A complete sequence of sediment sampling, such as sample collection, preservation, storage, and transport, must be evaluated to measure and minimize the sources of systematic and accidental errors. Typically, 10% of the samples should be collected in duplicate as a component of the QA–QC plan. Basic requirements to ensure the quality of sediment sampling are detail in planning, consistent use of qualified personnel, and adequate and well-maintained equipment. A few control sites should be included in a sampling program for investigation of sediment contamination. The risk of contamination is much greater when sampling sediment pore waters, due to considerably lower concentrations of different parameters in the pore water than in the sediments. There is no universal procedure for cleaning sampling containers; a specific method can be effective for one element but not sufficient for another. There are standard cleaning procedures for containers that hold sediment destined for organic and inorganic constituents analyses (76). Handling sediments in the field (for some physical and chemical measurements) and preservation of subsamples should be kept to a minimum. There are potential sources of cross-contamination and alteration of the nature of sediment samples, including deterioration and chemical changes of different compounds and elements. Due to the complex matrix of sediment samples, the determination of representative field blanks is much more complex than in water sampling.

HANDLING, STORAGE, AND PRESERVATION OF SEDIMENT SAMPLES

Sediment samples can be collected for the determination of inorganic or organic substances, particle size distribution, biological tests, etc. The containers, the handling of the samples, and the preservation of the samples depend on the objective of the sample collection.

Preservation of Sediment and Pore Water Samples

Clearly, it is impossible to prescribe absolute rules for preventing all possible changes during the storage of sediment and pore water samples. Preservation techniques are usually intended to retard microbial degradation, oxidation, and/or loss of volatile components. Methods are limited to pH control, poisoning, drying, refrigeration, freezing, and isolation from the atmosphere. No single preservation method is applicable to all constituents. Therefore, it is often necessary to take replicate samples, or subsamples, and preserve them by different methods when the determination of a variety of parameters is

Table 1. Sample Containers and Preservation for Different Parameters Measured in Sediments[a]

Parameter	Container[b]	Preservation	Maximum Storage	Comments
Particle size	P, G, or M	Wet, 4°C, tightly sealed	14 d	Drying, freezing, and thawing cause aggregation of particles
Stratigraphy	Core	Wet, 4°C	Several months	Preserve original consistency
Metal-contaminated sediments for toxicity testing	P or G	4 ± 2°C, dark	<6 wks, preferably <2 wks	Mixing and sieving recommended before testing sediment toxicity
Organic compound–contaminated sediments for toxicity testing	P, T, or M	4 ± 2°C, dark	<6 wks preferably <2 wks	
Bacteriological testing	Sterile G	Wet, 4°C	Processed within 6 h	Clumping/adsorption of bacteria to particles makes accurate enumeration difficult
pH, Eh, CEC	Bucket or core	Wet, undisturbed, and untreated	Determined in field	Difficult and problematic temperature corrections
Major elements (total Ca, Mg, Si, Na, K, Al, Fe, P, S, Ti, Mn)	T or P	4 ± 2°Ca	<2 wks	Dried samples stored at room temperature for 1 yr
Trace elements (total Ag, Cd, Co, Cr, Cu, Mo, Pb, V, Zn, etc.)	P or T	4 ± 2°C; −20°C	<2 wks; <6 months	Dried samples stored at room temperature for 1 yr
Oil and grease	M or G	Wet, 4 ± 2°C	24 h	Wet sample stored for up to 1 month at -10°C with 1–2 mL concentrated H_2SO_4 80 g
Nutrients (NH_4–N, NO_2–N, NO_3–N, TKN, TC, TOC)	P, T, or G	4 ± 2°C	<48 h	If possible analyze within 24 h
Mercury	G or T	Freeze, −20°C	1 month	Mercury analysis is performed with wet samples
Volatile organics	G vials with Teflon septums	Freeze, −20°C	1 month	No preservatives added; possible loss of compounds
Cyanides	P	Freeze, −20°C	Up to 1 month	Sulfides interfere with colorimetry
Pesticides and PCB	M or G covered with aluminum foil	Freeze, −20°C, dark	7 d until extraction	If samples are not analyzed within 48 h, freeze-dried at −20°C up to 6 months

[a] Data from Refs. 3, 11, and 51.
[b] P, polyethylene or polypropylene; G, glass; T, Teflon; M, metal.

required. In general, sample containers should be tightly sealed and head space should be minimized as soon as the samples are taken. If acid digestion or chemical sequential extractions are required, these procedures should be carried out as soon as possible. Table 1 outlines the preservation requirements for the analyses of different parameters in sediment samples.

The most common changes in sediments and pore water samples are oxidation, reduction, volatile losses, and biodegradation of different elements and compounds. Two parameters are important for preserving the integrity of samples: temperature and storage time. Temperature can affect the collected sediment samples in a variety of ways from the time of recovery through handling and processing to the final analyses. Sediment samples intended for analyses or experiments after air or oven drying can be stored in containers such as cans and plastic bags at ambient or room temperature. However, sediments collected for the determination of organic contaminants and mercury should be stored frozen (approximately −20°C). The higher the temperature, the higher the risk of losses or changes of volatile compounds. Preservation at low temperatures reduces biodegradation and sometimes loss of volatile elements and compounds. However, freezing water-containing samples can cause degassing, can fracture the sample, or can cause a slightly immiscible phase to separate. Considerable changes in sediment toxicity were found after storage time between <7 days and 12 months (51–55). Therefore, it was suggested that the storage of sediments collected for toxicity tests should be minimized (51,54). Sediments may be stored for longer periods, such as up to 6 weeks, in monitoring and assessment studies, if it can be demonstrated that significant changes in toxicity or chemistry will not occur over the storage period (54).

Table 1 outlines the preservation requirements for the analysis of different elements in sediment pore water samples. Significant changes were observed within 24 h in samples stored at 4°C. Immediately after retrieval, pore water samples should be pretreated with the appropriate type and quantity of preservative based on the analysis to be performed. It is recommended to use preservatives only when they are shown not to interfere with the analysis being made. When used, the preservative should be added to the sample container initially so that all sample portions are preserved as soon as collected. It has been shown that temperature and pressure exert a strong influence on the composition of pore water. Temperature changes quickly; pH may change significantly in a matter of minutes; dissolved gases (oxygen, carbon dioxide) may be lost. Therefore, these parameters should be determined in the field immediately after the collection of the samples. During changes in the pH–carbon dioxide equilibrium, calcium carbonate may precipitate and cause a decrease in the concentrations of dissolved calcium and total hardness. To avoid loss of volatile species, collected samples should completely fill, even overflow, the storage container before capping or sealing. For nonvolatile species, pore water samples should be collected in preacidified vials to a final pH of 2.5–3.5, usually ultrapure HNO_3, about 10 μL/5 mL. The samples should be stored at 4°C, and analyzed within 2 weeks. Pore water samples collected for the determination of dissolved organic carbon and methane should be preserved with $HgCl_2$ and analyzed within 3 days. Freezing of pore water did not affect toxicity in some studies; however, studies with waste water showed that freezing and thawing can cause changes in toxicity (56,57). Therefore, toxicity tests should be carried out immediately after pore water collection.

Biological changes taking place in a sediment sample may change the oxidation state of some constituents. For example, iron and manganese are readily soluble in their lower oxidation states but relatively insoluble in their higher oxidation states. Therefore, they may dissolve from the sediment or precipitate, depending on the redox potential of the sample. Microbial activity may be responsible for changes in the nitrate-nitrite-ammonia content, for decreases in phenol concentration and in biochemical oxygen demand (BOD), or for reducing sulfate to sulfide (58). To minimize the potential for volatilization or biodegradation, samples should be kept as cool as possible without freezing until analysis. Preferably pack samples in crushed ice before shipment; however, the use of dry ice should be avoided because it will freeze the samples. If immediate analysis is not possible, storage at 4°C is generally recommended for most samples (Table 1). As a general rule, the shorter the time between collection of a sample and its analysis, the more reliable will be the analytical results. Determination of certain physicochemical parameters, such as pH, Eh, temperature, and cation exchange capacity, must be made in the field immediately after sample collection.

Storage of Sediment Samples

Careful planning and selection of containers and utensils involved in sediment handling before analysis are extremely important. Containers and other equipment used in handling sediment samples after retrieval can be a significant source of contamination. Containers should neither contaminate the sample nor promote loss of parameters of interest through adsorption. Synthetic polymeric plastics have found widespread application in trace metal work. Among these materials, Teflon, polypropylene, and high pressure polyethylene are the most desirable. Borosilicate glass has been found useful for the storage of samples for analysis of organic compounds, mercury, and the volatile organometallic forms of several elements, such as lead, bismuth, and selenium. Already cleaned containers for water and sediment samples have recently become commercially available. Table 1 summarizes the recommended sample containers for collecting sediments before the determination of different parameters.

Generally, containers used for drying should be made of material resistant to corrosion and not subject to change in weight or disintegration on repeated heating and cooling. Crucibles, dishes, and trays made of aluminum, nickel, glass, and porcelain are recommended for drying sediments in an oven. The selection of the material depends on the subsequent analyses of sediment samples. Therefore, considerations in the selection of containers also apply to the choice of containers for drying. Shipment of collected samples must be planned, specified, and defined before any sediment sampling. It should be general procedure to ship the containers with the samples in coolers filled with ice cubes, as soon as possible for processing, preservation, and analyses.

Because of operational considerations, it may not be possible to store samples in the field in the same way that they will be ultimately stored. A temporary storage method may have to be adopted. One should strive to compromise the samples as little as possible and implement the best storage conditions as soon as possible. Samples collected for bioassays should be refrigerated and transported or stored in coolers filled with ice. Although freezing is considered appropriate for sediment samples collected for chemical analyses, it is not recommended for toxicity tests (59) because it can affect the toxin levels. After transporting the samples from the field, storage in refrigerators (usually at 4°C) and/or freezers (usually at −20°C) is essential to preserve the integrity of the collected sediments.

PREPARATION OF SAMPLES BEFORE ANALYSES

Drying of Sediment Samples

Three types of drying are commonly used to prepare solid samples before analysis: air, oven, and freeze-drying. Sediments, particularly the topmost 10 cm, typically contain up to 95% water. To permit the comparison of obtained data, sediments are dried and analyses are carried out on dry material or a subsample is taken for drying to determine the water content while the analyses are performed on the wet sediment. In either case, results of analyses are usually presented on a "dry weight" basis.

Air drying is commonly used in sedimentology (60) but it is only rarely used for the preparation of sediments for pollution studies. Air drying may generate undesirable changes in sediment properties and may increase the risks of contamination. For example, changes in metal availability and complexation for samples that were air dried have been demonstrated. On the other hand, air drying has been used in some studies to avoid the loss of components, such as mercury, which are volatile at temperatures greater than 50–60°C (61).

Oven drying of sediments at approximately 100°C is usually carried out on samples collected for the determination of inorganic components, such as major and trace elements. Lower heating temperatures (<60°C) are essential when preparing sediment to determine the levels volatile trace elements, such as mercury, or for determination of grain size.

In the freeze-drying process water in the frozen or solid state is sublimated and is removed from the material as a vapor. Freeze-drying (also called lyophilization) can be used for drying sediments collected for to determine the presence of most organic pollutants as well as for analyses of inorganic components. However, it should not be used for particle size analyses and toxicity tests.

Particle Size Distribution

Particle size distribution analyses should be carried out on wet sediments. Samples for the analyses should be stored at 4°C and never frozen. Tightly sealed plastic bags, glass jars, or other containers can be used to store samples before particle size analyses (Table 1). Sediments with a high iron content should be stored in airtight containers to avoid precipitation of iron oxides on particle surfaces and should be analyzed as soon as possible after collection. Drying, freezing, and thawing of the sediments can cause nonreversible aggregation of particles and should be avoided. Size analyses of fine-grained sediments should be carried out on only completely dispersed samples that have been treated for the removal of organic matter and carbonate and iron coatings. Sediment samples collected for stratigraphic or geotechnical studies can be stored at 4°C in a humidity-controlled room, without any large changes in sediment properties for several months (Table 1). Long cores, such as those collected by piston coring, can be cut into lengths suitable for storage, sectioned longitudinally, described, labeled, wrapped to preserve their original consistency, and stored in a refrigerated room.

Separation of Sediment into Different Granular Fractions

Sediments are often used to monitor metal pollution in natural waters. One of several methods to investigate the significance of the granular composition in the concentration of metals or other contaminants in sediments is based on the isolation of individual granular fractions and determination of metal concentrations in each fraction.

Wet sieving has been used to separate different particle size fractions in sediments. However, wet sieving involves resuspension of sediments in water and may change their original size distribution. Furthermore, because the water used for wet sieving usually does not have the *in situ* ionic composition, resuspension may even break particles that were originally agglomerated. This effect may be reduced by using water collected at the sediment sampling site. Another system for separating particles into size ranges is based on the principle of elutriation. A wet sediment sample is separated into specific size fractions by a process that depends on the forces present in a moving fluid. The Cyclosizer is a commercially available instrument (62) that separates particles according to their relative size and density, in a series of hydraulic cyclones where the centrifugal force produces the elutriating action. This unit is capable of separating silt-sized particles into six standard fractions: >44; <33–44; <23–33; <15–23; <11–15 and <11 μm.

Sediment Sieving

Sieving is an effective and economical process for dividing sediment samples into different fractions containing particles of more or less the same size. Wet sieving is particularly useful for processing fine-grained sediments. Wet sieving of a small quantity of sediment is generally carried out manually and requires one or more sieves and two types of containers, such as buckets, dishes, beakers, bowls, and jars. In one method, all material in the container is transferred to the sieve and washed with the smallest volume of running water (63). To facilitate passing the material through the sieve, a nylon brush can be used. Particles retained in the sieve (the coarse fraction) are examined and, if they are of interest to the study, retained and redried. The sediment washed through the sieve (the fines) is redried and weighed. After use, all sieves should be thoroughly cleaned under running water, with special care taken to remove the material caught by the screen. Sediments are usually classified simply as gravel (>2 mm), sand (<2 mm–63 μm), silt (<63–2 μm), and clay (<2 μm). These definitions are based on arbitrary divisions between the median diameters of the soils; there are several different classification systems (60).

For sediment toxicity tests, it may be desirable to remove indigenous organisms that interfere directly (eg, predators) or indirectly (eg, competitors) with the test organisms (65,66). Pressure or wet sieving were used in testing sediment toxicity. Pressure sieving involves mechanically pressing the sediments through a sieve of a specific mesh size (66,67). However, pressure sieving of sediments containing debris, vegetation, or high clay content through a single mesh size of 1 mm is difficult, and a series of sieves might be required. Wet sieving used in sediment toxicity tests has also been described (58,68). Generally, sediments have been sieved through stainless-steel; brass; or plastic woven polymer sieves, such as polyethylene, polypropylene, nylon, and Teflon, with mesh sizes from 0.24 to 2.0 mm (53,66–74). The most frequently used mesh size is 1.0 mm. However, it must be recognized that sediment sieving can change the sediment geochemistry and has the potential to ameliorate its toxicity.

Sediment Grinding

Sediment samples are commonly reduced to powder before analysis. The required final size is usually between 149 and 44 μm (100 and 325 mesh). The choice of equipment generally depends on the quantity of sediment to be ground, the hardness of the particular mineral particles, and contamination considerations. An efficient method for obtaining particle uniformity is alternate grinding and sieving. Mortars and pestles of suitable size are commonly used manually or operated mechanically to grind small samples to the desired particle size. Manual grinding is time-consuming and, consequently, relatively costly. There are several commercially available grinders able to reduce sediment samples to 100 μm (150 mesh) and smaller grains. Ball and disc mills are effective for disintegration, but there is less control over the final particle size. Hammer-type mills are designed for capacities from 2 kg/h to several thousand kilograms per hour. Small hammer-type mills are suitable for reducing small quantities of fragments that are <6 mm in size to various degrees of fineness.

BIBLIOGRAPHY

1. G. A. Burton Jr., *Environ. Toxicol. Chem.,* **10**(12), 1585–1627 (1991).
2. "Sediment Sampling Quality Assurance User's Guide," *EPA/600/S4-85/048,* Environmental Protection Agency, Center for Environmental Research Information, Cincinnati, Ohio, 1985.
3. A. Mudroch and S. D. MacKnight in A. Mudroch and S. D. MacKnight, eds., *Handbook of Techniques for Aquatic Sediments Sampling,* 2nd ed., Lewis Publishers, Boca Raton, Fla., 1994, pp. 5–16.

4. J. M. Azcue, J. R. D. Guimaraes, A. Mudroch, P. Mudroch and O. Malm in Ref. 3, pp. 203–228.
5. "Standard Practice for Conducting Acute Toxicity Tests with Fishes, Macroinvertebrates, and Amphibians," in ASTM, *Annual Book of ASTM Standards,* Philadelphia, Pa., 1988, pp. 1–20.
6. "Standard Test Methods for Measuring the Toxicity of Sediment-Associated Contaminants with Freshwater Invertebrates," in ASTM, Annual Book of Standards, Vol. 11.05, E 1706-95a, Philadelphia, Pa., 1995, pp. 1204–1285.
7. U.S. Environmental Protection Agency, "Assessment and Remediation of Contaminated Sediments (ARCS) Program, Biological and Chemical Assessment of Contaminated Great Lakes Sediments," *905-R93-006,* Chicago, Ill., 1993.
8. U.S. Environmental Protection Agency, "Methods for Measuring the Toxicity and Bioaccumulation of Sediment-associated Contaminants with Freshwater Invertebrates," *600-R24-024,* Duluth, Minn., 1994.
9. U.S. Environmental Protection Agency, "Assessment Guidance Document, Assessment and Remediation of Contaminated Sediments (ARCS) Program," *905-B94-002,* Great Lakes National Program Office, Chicago, Ill., 1994.
10. U.S. Environmental Protection Agency and Army Corps of Engineers, "Evaluation of Dredged Material Proposed for Discharge in Inland and Near Coastal Waters (Draft), *823-B94-002,* Washington, D.C., 1994.
11. Environment Canada, "Guidance Document on Collection and Preparation of Sediments for Physico-Chemical Characterization and Biological Testing, Environmental Protection Series," *Report EPS 1/RM/29,* Ottawa, Ont., Dec. 1994, p. 166.
12. A. J. De Groot, W. Salomons, and E. Allersma in J. D. Burton and P. S. Liss, eds., *Estuarine Chemistry*, Academic Press, Inc., New York, 1976, pp. 131–157.
13. F. Ackermann, H. Bergmann, and U. Schleichert, *Environ. Technol. Lett.* 4(7), 317–328 (July 1983).
14. F. P. Sheppard, *J. Sed. Petrol.* **24**(3), 151–158 (Sept. 1954).
15. R. L. Folk, *Sedimentology*, **6**(2), 73–93 (Mar. 1966).
16. W. F. G. Tanner in R. W. Fairbridge and J. Bourgeois, eds., *The Encyclopedia of Sedimentology*, Dowde, Hutchinson & Ross, Inc., Stroudsburg, U.K., 1978, pp. 376–382.
17. P. G. Sly and R. L. Thomas, *J. Fish Res. Board Can.* **31**(5), 795–825 (May 1974).
18. L. Håkanson, *Can. J. Earth Sci.* **14**(3), 397–412 (Mar. 1977).
19. R. Baudo in R. Baudo, J. Giesy, and H. Muntau, eds., *Sediments: Chemistry and Toxicity of In-Place Pollutants*, Lewis Publishers, Ann Arbor, Mich., 1990, pp. 15–60.
20. A. Mudroch and J. M. Azcue, *Manual of Aquatic Sediment Sampling*, Lewis Publishers, Boca Raton, Fla., 1995, pp. 5–43.
21. Ref. 20, pp. 22–25.
22. U.S. Environmental Protection Agency, *Guidance for Sampling and Analyzing for Organic Contaminants in Sediments*, Office of Water, Regulations and Standards, Criteria and Standards Division, Washington, D.C., 1987, p. 49.
23. L. Håkanson and M. Jansson, *Principles of Lake Sedimentology*, Springer-Verlag, Berlin, Germany, 1983, p. 32.
24. Ref. 20, pp. 16–18.
25. W. G. Cochran, *Sampling Techniques*, 2nd ed., Wiley Interscience, New York, 1963, p. 330.
26. R. H. Green, *Sampling Design and Statistical Methods for Environmental Biologists*, Wiley-Interscience, New York, 1979, p. 257.
27. R. J. Howarth and I. Thornton in I. Thornton, ed., *Applied Environmental Geochemistry*, Academic Press, Inc., New York, 1983, p. 41.
28. L. P. Provost in E. G. Schweitzer and J. A. Santolucito, eds., *Environmental Sampling for Hazardous Wastes*, American Chemical Society, Washington, D.C., 1984, p. 79.
29. D. F. Parkhurst, *Environ. Sci. Technol.* **18**(7), 521–523 (July 1984).
30. R. O. Gilbert, *Statistical Methods for Environmental Pollution Monitoring*, Van Nostrand Reinhold, New York, 1987, p. 320.
31. T. C. Moore Jr. and G. R. Heath in J. P. Riley and R. Chester, eds., *Chemical Oceanography*, Vol. 7, Academic Press, Ltd., London, 1978, pp. 75–126.
32. R. L. Thomas, A. L. W. Kemp, and C. F. M. Lewis, *J. Sed. Petrol.* **42**(1), 66–84 (Mar. 1972).
33. R. L. Thomas, A. L. W. Kemp, and C. F. M. Lewis, *Can. J. Earth Sci.* **10**(2), 226–271 (Feb. 1973).
34. R. L. Thomas and A. Mudroch, Small Craft Harbours—Sediment Survey, Lakes Ontario, Erie and St. Clair 1978: Dredging Summary and Protocol, *Report to Small Craft Harbours,* Ontario Region, Great Lakes Biolimnology Laboratory, Dept. of Fisheries and Oceans, Burlington, Ontario, Canada, 1979, p. 149.
35. J.-B. Chaussier, *Mineral Prospecting Manual*, North Oxford Academic Publishers, London, 1987.
36. R. A. Cahill, "Geochemistry of Recent Lake Michigan Sediments," *Circular 517,* Illinois State Geological Survey, Champaign, 1981, p. 94.
37. R. L. Thomas, J.-M. Jaquet, A. L. W. Kemp, and C. F. M. Lewis, *J. Fish Res. Board Can.* **33**(3), 385–403 (Mar. 1976).
38. Ref. 3, pp. 17–28.
39. Ref. 23, p. 39.
40. J. Crépin and R. L. Johnson in M. R. Carter, ed., *Soil Sampling and Methods of Analysis*, Lewis Publishers, Boca Raton, Fla., pp. 5–18.
41. Radio Technical Commission for Maritime Services, *RTCM Recommended Standards for Differential NAVSTAR GPS Services*, Washington, D.C., 1990, pp. 3–8.
42. N. A. Holme in F. S. Russel, ed., *Advances in Marine Biology*, Vol. 2, Academic Press, Inc., New York, 1964, p. 171.
43. T. L. Hopkins in M. Sears, ed., *Progress in Oceanography*, Vol. 2, Pergamon-Macmillan, New York, 1964, p. 213.
44. L. Håkanson and M. Jansson in Ref. 23, p. 316.
45. S. Blomqvist, *Hydrobiologia,* **206**(3), 245–254 (Oct. 1990).
46. Z. Kajak in W. T. Edmonson, ed., *A Manual on Methods for the Assessment of Secondary Productivity in Fresh Water,* IBP Handbook No. 17, Blackwell Scientific, Oxford, U.K., 1969, p. 25.
47. P. G. Sly in *Proceedings of the 12th Conference of the International Association of Great Lakes Research,* International Association of Great Lakes Research, Ann Arbor, Mich., 1969, pp. 883–898.
48. J. M. Elliot and P. A. Tullet, *Occ. Publ. Freshwater Biol. Assoc.,* (4), 61 (1978).
49. J. M. Elliot and P. A. Tullet, *Occ. Publ. Freshwater Biol. Assoc.,*(20), 27 (1983).
50. A. Mudroch and S. D. MacKnight in Ref. 3, pp. 29–95.
51. "Standard Guide for Collection, Storage, Characterization, and Manipulation of Sediments for Toxicological Testing" in ASTM, *Annual Book of ASTM Standards,* Vol. 11.04, E 1391-90, Philadelphia, Pa., 1990, pp. 1105–1119.
52. T. H. DeWitt, R. C. Swartz, and J. O. Lamberson, *Environ. Toxicol. Chem.* **8**(7), 1035–1048 (1989).
53. B. L. Stemmer, G. A. Burton Jr., and G. Sasson-Brickson, *Environ. Toxicol. Chem.* **9**(8), 1035–1044 (1990).
54. R. A. Othoudt and co-workers, *Chemosphere,* **22**(9–10), 801–807 (1991).

55. P. F. Landrum, B. J. Eadie, and W. R. Faust, *Environ. Toxicol. Chem.* **11**(8), 1197–1208 (1992).
56. J. T. Hardy, S. Kiesser, L. Antrim, A. Stubin, R. Kocan, and J. Strand, *Mar. Environ. Res.* **23,** 227–249 (1987).
57. R. S. Carr, J. W. Williams, and C. T. B. Fragata, *Environ. Toxicol. Chem.* **8**(6), 533–543 (1989).
58. American Public Health Association, American Water Works Association, and Water Environment Federation, *Standard Methods for the Examination of Water and Wastewater,* 18th ed., APHA, AWWA, WEF, Washington, D.C., 1992.
59. K. W. Malueg, G. S. Schuytema and D. F. Krawczyk, *Environ. Toxicol. Chem.* **5** (3), 248–253 (1986).
60. R. L. Folk, *Petrology of Sedimentary Rocks,* Hemphill, Austin, Tex., 1974, pp. 159.
61. U. Förstner and W. Salomons, *Environ. Technol. Lett.* **1**(11), 494–505 (1980).
62. "Particle Size Analysis in the Subsieve Range," *Bulletin WCS/2,* Warman International Ltd., Sydney, Australia, 1981.
63. *General Methods and Instrumentation,* American Society for Testing and Materials, Philadelphia, Pa., 1986, p. 5.
64. M. S. Redmont and K. J. Scott, *Estuaries,* **12**(3), 205–207 (1989).
65. C. G. Ingersol and M. K. Nelson in W. G. Landis and W. H. van der Schalie, eds., *Aquatic Toxicology and Risk Assessment,* Vol. 13, ASTM STP 1096, ASTM, Philadelphia, Pa., 1990, pp. 93–109.
66. J. P. Giesy, C. J. Rosiu, R. L. Graney, and M. G. Henry, *Environ. Toxicol. Chem.* **9**(2), 233–248 (1990).
67. D. M. Johns, R. A. Pastorak and T. C. Ginn in M. A. Mayers and M. G. Barron, eds., *Aquatic Toxicology and Risk Assessment,* Vol. 14, ASTM STP 1124, ASTM, Philadelphia, Pa., 1991, pp. 280–293.
68. R. A. Pastorok and D. S. Becker in Ref. 136, pp. 123–139.
69. B. L. Stemmer, G. A. Burton Jr., and S. Leibfritz-Frederick, *Environ. Toxicol. Chem.* **9**(3), 381–389 (1990).
70. T. J. Keilty, D. S. White, and P. F. Landrum, *Aquat. Toxicol.* **13,** 251–270 (1988).
71. T. J. Keilty, D. S. White, and P. F. Landrum, *Aquat. Toxicol.* **13,** 227–250 (1988).
72. T. J. Keilty, D. S. White, and P. F. Landrum, *Arch. Environ. Contam. Toxicol.* **17**(1), 95–101 (1988).
73. M. J. Lydy, K. A. Bruner, K. A. Fry, and S. W. Fisher in Ref. 136, pp. 140–164.
74. P. F. Landrum and W. R. Faust in Ref. 138, pp. 263–279.
75. L. H. Keith, *Environmental Samples and Analyses: A Practical Guide,* Lewis Publishers, Chelsea, U.K., 1991, p. 143.
76. A. Mudroch and J. M. Azcue in Ref. 3, pp. 186–188.

AQUIFER REMEDIATION. See PUMP-AND-TREAT TECHNOLOGY, AQUIFER REMEDIATION.

ASBESTOS, ANALYSIS OF

JOHN ADDISON
John Addison Consultancy
Edinburgh, Scotland

The asbestos minerals consist of a group of naturally occurring fibrous silicate minerals that have been exploited industrially for their useful properties such as thermal insulation, chemical and thermal stability, and high tensile strength. They include chrysotile, a member of the serpentine mineral group, and the amphibole minerals amosite, crocidolite, tremolite asbestos, actinolite asbestos, and anthophyllite asbestos (1–3). Industrial use of the minerals over the twentieth century, in addition to natural erosion of rocks containing asbestos, has led to a widespread dispersion of asbestos fibers throughout the atmosphere, hydrosphere, and soils of many areas (4). Concerns about the health risks associated with exposure to asbestos minerals have encouraged the development of increasingly sensitive techniques for the detection, identification, and quantification of asbestos.

ASBESTOS MINERALS

Definition of Asbestos

The simple definition used in the introduction for asbestos is adequate, but only for the most practical purposes, and finding a better, more comprehensive and exclusive definition has been controversial and difficult (5). There are a variety of reasons for this. The fact that the term asbestos includes members of two fundamentally different mineral families, the serpentines and the amphiboles, makes a simple mineralogical definition immediately impossible. In addition, there are distinctions between a normal amphibole mineral and an asbestos variety of the same amphibole that are technically complex but already implicit in the definition in the use of the term fibrous (6). For example, tremolite is a common mineral in the amphibole family that sometimes occurs in a fibrous habit and when it does it is called tremolite asbestos; when the amphibole riebeckite occurs in its fibrous form it is called crocidolite or riebeckite asbestos. A distinction must also be made between a fibrous particle and an asbestos fiber: a particle matching the dimensions in the various asbestos regulations (usually length >5 μm, diameter <3 μm, aspect ratio $>3:1$) is defined as a fiber and counted as asbestos for the purposes of those regulations. However, the shape of any particle does not identify it as asbestos in any way other than for compliance with the regulations and in the course of a specific test unless that test involves an additional identification process.

There is a further distinction to be made between a fibrous mineral, meaning one that has crystallized in the fibrous mineral habit and an asbestos mineral. Possession of a fibrous or asbestiform habit does not identify a mineral as asbestos. Therefore, tremolite that has formed in the asbestos habit is asbestos; gypsum that has formed in the asbestos habit is not asbestos; fibrous erionite is not asbestos even though it may have formed in the asbestos habit (7), and in those circumstances may be more hazardous than asbestos. Ultimately there is no real alternative to a complicated mineralogical definition despite the desire of some legislators and biological scientists for a simple hazard-related definition.

U.K. regulations (2) refer to the fibrous forms of a number of naturally occurring silicate minerals which have been exploited commercially for their useful properties of flexibility, high tensile strength, incombustibility, low thermal conductivity, and resistance to chemical attack. The six commonly accepted asbestos minerals are listed in Table 1. In the United States, OSHA relies on a similar commercial and

Table 1. Chemical Formulas of Main Asbestos Varieties

Chrysotile	$Mg_3Si_2O_5(OH)_4$
Amosite	$(FeMg)_7Si_8O_{22}(OH)_2$, Fe : Mg 5 : 2
Crocidolite	$Na_2(MgFe)_3(Fe^{3+})_2Si_8O_{22}(OH)_2$, Fe : Mg 4 : 1
Anthophyllite	$(MgFe)_7Si_8O_{22}(OH)_2$, Fe : Mg 1 : 6
Tremolite	$Ca_2Mg_5Si_8O_{22}(OH)_2$
Actinolite	$Ca_2(MgFe)_5Si_8O_{22}(OH)_2$, Fe : Mg < 4 : 1
Ferro-actinolite	$Ca_2Fe_5Si_8O_{22}(OH)_2$

morphological definition (1). The physical properties of asbestos minerals that distinguish them from other minerals are described by the Committee on Nonoccupational Health Risks of Asbestiform Fibers as fiber-like morphology and dimensions, enhanced strength and flexibility, diameter dependent strength, increased physical strength and chemical durability, and improved surface structure (8,9). Many of these properties appear to be related to the absence of transverse crystal defects or cracks across the fiber crystals (9).

Further properties considered to be important in the definition of asbestos are high aspect ratio (>20 : 1 to ≥1000 : 1); very thin fibers (mostly <1 μm); curved, or matted masses of fibers; and splayed ends to fibers and split fibers.

Environmental Contamination by Asbestos

World asbestos production was probably at its peak throughout the 1970s and 1980s when on average about 5 million metric tons were produced every year (10). A conservative estimate of the total amount of asbestos used since the industry started is ~200 million t. More than 95% of the asbestos used throughout the world was chrysotile. Amosite and crocidolite asbestos usage constituted most of the rest, whereas the use of tremolite asbestos, actinolite asbestos, and anthophyllite asbestos was comparatively rare.

Uses of Asbestos. There has been an enormous diversity of products using asbestos (11) although increasing restrictions are now confining its use in certain countries to asbestos cement and friction products. Existing buildings and other environments may contain materials that can be considered in the following broad categories.

Low density materials (<500 kg/m^3), such as sprayed asbestos coatings, lagging, quilts, blankets, ropes, yarns, cloths, and lightweight, preformed, sectional insulation made from calcium silicates or magnesia, all contain loosely bonded asbestos that are easily damaged and readily release fibers. Often the asbestos contents are very high and disturbance of such materials can lead to high airborne fiber concentrations. The use of many of these materials is now prohibited but they may still be found in older buildings or in contaminated industrial sites.

Medium density insulating boards in the 500–770 kg/m^3 were very common in construction industry use in the 1950s through the mid-1970s. They contained mostly amosite, and sometimes chrysotile, in a matrix of Portland cement or hydrated lime and silica. They are found in all types of industrial, commercial, public, and private buildings as wall panels, ceiling tiles, and acoustic panels. Even careful handling of these products can produce high airborne concentrations, and machine working with power tools is likely to produce very high concentrations.

Asbestos cement (AC) products are high density materials in the range of 1200–1500 kg/m^3. All three main asbestos types have been mixed with Portland cement, usually at about 10–15% of the mixture, although AC sheet manufactured today is exclusively made with chrysotile. AC is found as flat, pressed, and semipressed sheets, corrugated sheets, tiles and slates, flues, pipes, drains, water tanks, cores of steel sheets, etc. The asbestos fibers are usually firmly bonded into the cement and the materials are unlikely to lead to high airborne emissions unless they are cut or sanded with power tools or become weathered.

Friction products constituted an almost universal use of chrysotile for much of the twentieth century. Brake shoes, brake pads, and clutch plates were ubiquitous throughout the automotive industry and are still used in many countries. Their use is probably the main reason for the higher ambient atmospheric concentrations of chrysotile in urban environments although most of the fibers released are extremely fine and usually short.

Other assorted materials have also used major amounts of asbestos. Bitumen, plastics, mastics, sealants, paints, and putties have all used asbestos, mainly chrysotile, as fillers. As with the AC the rate at which asbestos may be released to the environment is very low unless the materials are corroded or are worked mechanically.

Asbestos in the Industrial Environment. Almost any derelict industrial site is likely to contain asbestos contamination from chrysotile asbestos cement material, and depending on the nature of the industry, the amphibole asbestos varieties might also be present. Derelict industrial sites such as ship-building, repair, and ship breaking therefore often have a higher risk of being of contaminated by crocidolite, as do railway rolling-stock factories and repair works. Chemical industries, power stations, gas-works, and coke-works also are more likely to have used crocidolite asbestos for insulation purposes. Many large public and commercial buildings erected in the 1960s and 1970s had amosite or crocidolite asbestos sprayed onto the surfaces of structural steel or concrete girders in order to provide improved and extended structural integrity during a major fire. Although many or all of these have been subject to asbestos abatement programs in the United States and U.K. there are many countries that still have extensive amounts of sprayed asbestos cement in place.

General industrial buildings were frequently constructed using asbestos cement sheeting for roofs, wall cladding, and panels. While the advice for removal of such installations would normally have been decontamination before dismantling rather than immediate demolition, the latter option was frequently taken out of ignorance or on the basis of lower costs so that many derelict industrial sites may now contain asbestos cement debris as widespread contamination. Second-hand problems have also arisen where "clean" demolition wastes used as landfill have been found later to contain asbestos materials because insufficient checks were made on the quality of the waste.

The asbestos industries themselves constitute special cases of high levels of asbestos contamination. During their working lives these industries obviously produced major emissions of asbestos to the atmosphere and created major waste disposal problems in the form of spoil heaps at mines, highly contaminated crushing and milling plants, and cement waste scrap at asbestos cement factories. Mining regions have significant problems if required to remediate the mine sites, but the remediation problems of the hundreds of derelict asbestos cement factories throughout the industrial world may be even greater as the industry continues to decline in the face of public pressure. The question that might be asked is if this amount of concern is justified, especially in the case of chrysotile and chrysotile products.

Ambient Atmospheric Pollution. The continued exposure and weathering of chrysotile cement sheets of roofs and walls of industrial agricultural and domestic buildings, as well as the continuing abrasion of chrysotile-containing friction products (still used in many countries) are contributing to the levels of fibers in the ambient atmosphere. However, these might be considered to be relatively small when compared to the natural sources of chrysotile in areas such as southern California (4). Chrysotile fibers have been present in the ambient atmosphere since before the time of any commercial exploitation, as borne out by the presence of chrysotile in old Antarctic snow and ice (12), but it is not yet clear how much the ambient concentrations might have increased since that exploitation began. On the other hand, airborne asbestos fiber concentrations in mining regions such as Quebec have been recognized as being much higher than normal background for many years. Similarly, urban atmospheres have higher airborne chrysotile concentration as a result of abrasion from friction products.

Degradation of AC pipes used in public water supplies has led to increased asbestos concentration of potable water, as has the run-off from the surface exposure of mine tailings and AC roofing and sheeting (13). The full scale of these additions to natural waterborne asbestos in the environment is not yet established, and given the amount of natural exposure and weathering that must be taking place it is by no means certain that the anthropogenic contribution to potable waters other than from the supply pipes would be significant outside mining areas. It does seem fairly obvious that natural erosion of geological outcrops of asbestos has made a much greater contribution to the oceanic content of asbestos than would any human use of the materials, including the possible disposal at deep ocean sites of all of the asbestos that was ever used. For example, it is estimated that 6×10^{12} tons of crocidolite has been eroded from the Hammersley deposits of Australia over the last 150 million years (4).

The amphibole asbestos minerals may pose a more serious problem than those posed by chrysolite since the amphibole minerals are considerably more durable than chrysotile and persist longer throughout the cycles of erosion and weathering. They are the among the few minerals to be associated with endemic disease (mesothelioma in Greece and Turkey, etc), even though these occurrences are more properly be ascribed to para-occupational exposure rather than ambient environmental exposure.

DETERMINATION OF ASBESTOS IN THE ENVIRONMENT

Sampling Methods

Airborne Dust Sampling. Airborne dust samples are collected by drawing measured volumes of air through fine, pore-size filters using current methods (14–16). The pore size and filter type used depends on the analytical technique used for evaluation. Tortuous-pore cellulose nitrate, acetate, or mixed ester membrane filters with pore sizes of 0.8 or 1.2 μm are usually recommended for optical microscopy (see Plate 1). Capillary-pore polycarbonate filters of 0.4 or 0.8 mm pore size are usually used for direct electron microscope analysis.

The volume of air that may be collected is often limited by the presence of background dust in the air, but sample volumes of 2 m^3 are useful, where possible, to achieve appropriate low detection limits for ambient air.

Water Sampling. Water samples are normally collected in clean, sterile, pretreated plastic bottles and transported quickly to the analytical laboratory for filtration according to ISO methods (17–21). Great care must be taken to ensure that the containers used are themselves free from asbestos contamination. Filtration of the water is carried out using standard laboratory glassware, usually through 25-mm diameter polycarbonate or cellulose ester filters. Usually 0.1 or 0.2 μm pore-size filters are required since water samples are conventionally analyzed by transmission electron microscopy and the concentrations of the finest chrysotile fibrils may be of interest. Careful cleaning of any glassware used is very important, as is the preparation of suitable blank control filters, since there may be serious social or economic consequences from the finding of a few fine fibrils. Pretreatment of the water samples using ultraviolet light and an ozone bubbler or using sodium hypochlorite may be necessary if there are excessive amounts of suspended organic matter in the water (22).

Sampling Bulk Materials. Sampling of bulk materials or of contaminated land should normally be undertaken using procedures described in BSI and ISO draft methods (23,24). Visual inspection, carried carefully and systematically, with collection of small samples of suspect materials, is usually the first step in the assessment of a contaminated land site (after any desktop studies). On the basis of this a sampling strategy can be developed. This normally involves a series of sampling points that may be anything from surface scrapings using garden tools, hand augurs using simple percussion methods, or test coring using small drilling rigs or the digging of trial pits to some meters depth using mechanical diggers. In the case of the latter it is important to follow the safety guidelines for entry into excavated pits. The sample size is effectively determined by the particle size of the material (BS 1377, 1990) although there may be other constraints that prevent the ideal amounts from being sampled. Sieving or size separation on-site during sampling may also allow smaller, more easily handled samples to be collected. Simple or weighted grid sampling patterns should be used, taking samples of a few kilograms as a minimum in the case of contaminated land. There may be many different strategies to the sampling program for a contaminated land site, but the simplest is probably a regular (square) sampling grid or a stratified systematic unaligned grid. Both will provide a 95% confidence probability of detecting a "hot spot" concentration high that constitutes 5% of the total area of a site of any given area with a total sample number of around 30 (25). Other statistical models may be necessary if different contamination targets or strategies are required. Sample splitting should be carried out carefully with due precautions to minimize any losses of asbestos to the atmosphere. Simple coning and quartering techniques or riffle boxes are suitable for the purpose.

Detection and analysis of trace amounts ($<0.01\%$ m/m) of asbestos in soils, aggregate bulk materials, or in soluble substances can be carried out by preparation of a suspension or a solution of the sample. This is then filtered through a fine-pore filter using the same processes as for the simple water samples and the filters are prepared and evaluated using the techniques described below. Alternatively, airborne dusts may be generated from the bulk materials and airborne fiber concentrations derived from them, related to the dust concentrations and bulk proportions (26).

Survey and Sampling of Buildings. Before any asbestos abatement work or any demolition is started it is essential that any structure that potentially contains asbestos materials is carefully and systematically surveyed for those materials and that suspect materials are sampled for analysis in the laboratory. An alternative procedure is to presume that suspect materials actually contain asbestos and to proceed accordingly with all necessary precautions.

Any sampling should only be carried out by suitably qualified persons, and strict attention must be paid to the relevant asbestos regulations to ensure that no hazards are caused by the survey and sampling process. Visual walk-around inspection by an experienced, qualified person is usually the first step in a survey, and a sampling program is established on the basis of these results. The sampling scheme should allow the detection of all possible installations within the building, including not only the obvious laggings on boilers and pipes but also areas between walls, above suspended ceilings, and below floors. It may not be acceptable to owners or occupants to sample all of the possible situations; it is therefore important to record any areas that could not be accessed at the time. Simple tools such as pliers, hand drills, and saws may be used with care and cleaned carefully between uses. Samples of a minimum size of around 5 cm square are suitable for most homogeneous building materials containing asbestos, but larger samples may be necessary for ad hoc materials such as old laggings and insulations. All samples must be sealed individually in containers and fully labeled and documented at the time of sampling. All sampling sites must be recorded and may be marked for future reference. All damage must be repaired and any exposed surfaces should be resealed. Care must be taken to ensure that no cross-contamination occurs between samples.

There is extensive guidance available from government departments on the use of asbestos in buildings and a range of protocols for building surveys have been developed. The use of an appropriate method is advisable. Various point scoring systems have been devised to decide the eventual advice as a result of an asbestos survey, but they must be used with care as they may not be the best predictor of asbestos health risks in a building. These surveys are usually based on factors such as types of asbestos, nature and amounts of asbestos-containing materials, their state of repair or degradation, accessibility and potential for damage, and the potential health risks to building occupants (27). Airborne asbestos concentrations in buildings are probably the best guides to the advisability of remedial action (28).

ANALYTICAL METHODS

Optical Microscopy

Polarized Light Microscopy. The simplest and most widely used methods of asbestos identification are those using optical microscopy. Examples of published methods are available, and these are broadly similar in approach (29–31). A high quality polarized light microscope is recommended for the tests and usually such a microscope has strain-free optics to provide the best light interference properties for the observed minerals. However, modern phase contrast microscopes have excellent optic systems for fiber counting at 400–600× magnification and the addition of simple polarizing and analyzing lenses with suitable accessory plate slots provides an adequate system for asbestos identification.

The whole sample from whatever source is usually examined visually to assess its homogeneity. Small subsamples of two to three grams are selected from the whole and these may be prepared in a variety of ways before they are then searched thoroughly using a low power stereo binocular microscope. A variety of preparation steps may be necessary before examination of the sample; acid treatment may be used to remove carbonates from cement matrices, or organic solvents may be necessary to remove plastic binder from materials such as floor tiles. The main purpose of any procedure is to present clean examples of the fiber types present for analysis, although if successful, they will also improve detection limits.

Suspected asbestos types are characterized and selected for further examination on the basis of their fibrous nature, color, lustre, elasticity, flexibility, and other handling properties. The selected fibers are mounted on glass microscope slides in refractive index liquids chosen to match the refractive index (RI) of the suspected identity of the fiber, eg, 1.55 for chrysotile, 1.67 for amosite, 1.7 for crocidolite, 1.605 for tremolite and anthophyllite, and 1.64 for actinolite. Polarized light microscope methods are then used to assess optical properties of each selected fiber, including birefringence, pleochroism, and extinction orientation; dispersion-stain microscopy or phase-contrast microscopy are used for the assessment of refractive index of the selected fibers in comparison with the refractive index of the liquids in which the fibers are mounted. Fibers mounted in matching RI liquids display characteristic optical color stains when appropriate systems of lenses and lens stops are used. The optical properties of the asbestos minerals are well documented and fully tabulated in the published methods. Differentiation between the asbestos and more normal prismatic forms of the amphibole asbestos minerals is usually possible by observation of extinction angles and coarser morphology, although properties such as refractive indices may be exactly the same. Individual fibers of anthophyllite may still be indeterminate in type since both asbestos and nonasbestos forms have straight extinction.

The detection and identification of asbestos in suitable bulk materials down to levels about 10 mg/kg has been demonstrated (29). The detection limits of the method are determined by the amount of effort put into the low power microscope search and also by the nature of the matrix of the test material. Great care is needed in searching bulk samples of materials such as contaminated soils to maintain low levels of detection, and with such a sensitive method care must also be taken to avoid errors from cross-contamination of samples.

Standard reference asbestos minerals are used for comparison with unknown samples and with the refractive index liquids for quality assurance purposes. Suitable materials are obtainable from NIST in the United States and from the Institute of Occupational Medicine in the U.K. (32).

Phase Contrast Optical Microscopy

The procedures for PCOM fiber counting have been fully documented (14,15).

Scanning Electron Microscopy

SEM methods are fully documented (33,34) and are the subject of further work for the International Standardization Organization.

Correct analysis of samples by SEM requires a well-adjusted, good quality instrument in addition to considerable operator skills and experience. It is important to ensure that all the operating parameters are regularly checked and provide the appropriate conditions for fiber visibility and identification using energy dispersive X-ray spectrometry (EDXS). SEM methods have the advantage over TEM methods in having more direct preparation with the least amount of sample manipulation before analysis. On the other hand, the degree of confidence with which fibers can be identified is limited by the lack of a fully quantitative chemical analysis, and by the absence of crystal structure data. Also there is a tendency for fibers with diameters less than about 0.1 μm to be missed during the search.

Transmission Electron Microscopy

Analysis of environmental samples for asbestos by TEM requires the highest levels of training and skill. The costs, in terms of both equipment and time, are by far the greatest of any of the analytical techniques applied to asbestos. Even so the value of the analysis with its high degree of certainty and sensitivity is often sufficient to justify the expense, particularly with the present climate of public opinion about the health risks of asbestos. Fully detailed methods for transmission electron microscopy are available (35,36).

LEGISLATION

The asbestos minerals are classified as carcinogens by the International Agency for Research on Cancer in Lyons (IARC) (37) and as a result almost all national and international agencies have established bodies of legislation that control their use, transport, disposal, packaging and labeling, and handling. Many countries have introduced total prohibitions on the use of all asbestos types; others have prohibitions of the import or use of amphibole asbestos types and a ban on new uses of chrysotile while permitting the continued use of chrysotile in products such as asbestos cement and friction products. Almost all countries have legislation in place controlling the ways in which asbestos may be handled in the workplace, the ways that the workplace environment is monitored, and the protection and health surveillance of the workers, imposing duties of care upon both the management and workforces of the asbestos industries. Many countries also have environmental protection legislation controlling the extent of emissions permitted from workplaces using asbestos and regulations concerning notification of asbestos in waste materials to authorities for landfill disposal. Many countries classify asbestos as a special waste requiring the most careful control on disposal.

ASBESTOS REMEDIATION

Asbestos Abatement in Buildings

There are a number of options open to the proprietors of buildings that contain known installations of asbestos containing materials:

- leaving the material in place but instituting a management system to ensure its continuing safety
- leaving it in place but sealing or enclosing it and introducing a management system
- removing and disposing of the asbestos-containing material

The first two might involve minor repairs that should only be undertaken within the relevant regulations, involve recording of all asbestos sites, labeling of materials, and notification to all interested parties and provision for regular reinspection. Any asbestos removal should only be carried out by appropriately trained operatives and according to all regulations. In the U.K. the Control of Asbestos at Work Act (38) and the Asbestos (Licensing) Regulations (39) apply to any removal work. In the United States the OSHA Final Rule on Occupational Exposure to Asbestos applies (1). Asbestos abatement processes usually require complete containment of the work-site in a negative pressure enclosure with suitable air-locks for access and decontamination units for the workers. Wet-stripping methods for sprayed asbestos or asbestos insulation materials are normally recommended in the U.K. and Europe because they have been shown to reduce workers exposure to airborne asbestos fibers. This may involve pressurized injection of wetting agents into the sealed layers of insulation. All workers have to be fully trained in the processes of asbestos removal, educated in the risks of asbestos exposure, and trained in the safe use of personal protective equipment. As asbestos workers they will also have regular health surveillance. Any airborne or aqueous emissions from the site during work have to be carefully controlled, and any material removed has to be taken from the site in a sealed container to be disposed of as a special waste at an appropriately licensed site.

Demolition of entire buildings is likely to create large amounts of airborne dust so it is important that all asbestos-containing materials are carefully removed in a controlled way before demolition to minimize risks to the health of the public and the work-force. Asbestos cement components should be carefully dismantled by cutting bolts, and lowered to the ground rather than simply knocking them down.

Asbestos in Fires

Asbestos cement sheeting materials explode violently in very severe fires and debris is often dispersed over wide areas. The heating process may destroy some of the asbestos, but the AC will be made more friable and likely to release fibers.

All of the asbestos minerals are susceptible to thermal degradation to varying degrees (40). Chrysotile starts to dehydrate and turn brown at comparatively low temperatures (200–400°C). It is almost entirely dehydrated by about 400°C, and by about 650°C the chrysotile structure is destroyed and the mineral is converted to a mixture of forsterite (olivine) and silica. At higher temperatures the mineral enstatite may be formed. The amphibole asbestos minerals are more thermally resistant but also start to turn brown and dehydrate at relatively low temperatures (250°C). Crocidolite is stable up to temperatures around 930–970°C, above which the iron-rich pyroxene acmite forms with cristobalite and spinel. Amosite is slightly less stable and decomposes into pyroxene, cristobalite, and spinel at around 800°C.

All of the main commercial asbestos varieties show signs of thermal alteration in normal use and are frequently turned brown by oxidation and dehydration. However, they may still maintain their high tensile strength fiber form and some parts of the thicker fiber bundles often retain the necessary optical properties to allow unambiguous identification by optical microscopy.

Contaminated Land Remediation

Decontamination of large contaminated land sites is covered by all of the national asbestos regulations and rules. Decisions about remediation options depend on factors such as the nature and amounts of asbestos present, asbestos concentrations present, the lateral extent and depth of the contamination, the local availability of disposal sites, and not least, the proposed final use of the site. Large deposits are obviously more difficult to deal with than more limited quantities. Some of the options available for decontamination are as described below.

Removal of the contamination and disposal in some suitable site is the simplest option. Successive layers can be scraped up using mechanical diggers and deposited into earth-moving trucks for dumping elsewhere. If diggers work backward they can leave behind a cleared area free of asbestos or at least with asbestos below an appropriate action level. Care must be taken to ensure that the contamination is not spread beyond the site by careless handling, and trucks leaving the site must themselves be decontaminated before they move off the site.

Smaller amounts of asbestos contamination may be rendered safe by vitrification of the material. Purpose-built furnaces are capable of heating the waste to temperatures high enough to melt the asbestos minerals in mixtures with glass cullet (scrap glass) and to fuse the materials. The dense glassy solids formed on cooling contain no fibers and can then be disposed of without concern for the asbestos content. This process may be more expensive than simple disposal and the rates at which material can be processed is limited (about 50 t/d), but it may be appropriate in some situations.

Similar treatment using hydrofluoric acid or other acids in combination can also destroy the asbestos, leaving a nontoxic residue that can be disposed of safely. Alternatively it may be possible to use cement to stabilize asbestos soil mixtures thus minimizing the release of fibers and allowing safer disposal.

Simple coverage of the asbestos-contaminated land by a two-meter thick layer of hard core and clean soil, possibly on top of geotextile sheeting, may allow the development of recreational grassland. Well-established vegetation can in such circumstances provide adequate protection. Such treatment requires that full records of the contamination are maintained in perpetuity, and that no other development is allowed in the future without remediation.

Retention of asbestos beneath permanent hard cover in the form of buildings, roads, and paved parking areas may be a long-term method of dealing with contaminated land that allows it to be used for beneficial purposes. Industrial or commercial buildings rather than residential developments may be considered, but whatever the purpose it is essential that the construction workers are properly protected by observance of all asbestos occupational legislation, and that site clearance is conducted in a way that minimizes release of asbestos to the atmosphere. As before, it is essential that full records are maintained and that no further redevelopment is carried out without due consideration for the asbestos.

BIBLIOGRAPHY

1. Occupational Safety and Health Administration, *Occupational Exposure to Asbestos; Final Rule,* Department of Labor 29 CFR Parts 1910, *Fed. Reg.* (Aug. 10, 1994).
2. *The Control of Asbestos at Work Regulations,* Statutory Instruments No. 2115, Her Majesty's Stationery Office, London, 1987.
3. M. Ross, R. A. Kuntz, and R. A. Clifton, "A Definition for Asbestos," in B. Levadie, ed., *Definitions for Asbestos and Other Health-Related Silicates,* ASTM STP834, American Society for Testing and Materials, Philadelphia, Pa., 1984, pp. 139–147.
4. C. Klein, "Rocks, Minerals, and a Dusty World," in G. D. Guthrie, Jr., and B. T. Mossman, eds., *Health Effects of Mineral Dusts,* Mineralogical Society of America, Washington D.C. *Rev. Mineral.* **28,** 7–59 (1993).
5. Occupational Safety and Health Administration, *Occupational Exposure to Asbestos, Tremolite, Anthophyllite and Actinolite; Final Rule,* Department of Labor 29 CFR Parts 1910 and 1926, *Fed. Reg.* (June 8, 1992).
6. D. R. Veblen and A. G. Wylie, "Mineralogy of Amphiboles and 1:1 Layer Silicates," in Ref. 4; *Rev. Mineral.* **28,** 61–137 (1993).
7. A. J. Gude III and R. A. Sheppard, *Clay and Clay Miner.* **29,** 5, 378–384 (1981).
8. *The Control of Asbestos in the Air Regulations,* Statutory Instruments No. 556, Her Majesty's Stationery Office, London, 1990.
9. Committee on Nonoccupational Health Risks of Asbestiform Fibers, *Asbestiform Fibers—Nonoccupational Health Risks,* National Academy Press, Washington D.C., 1984.
10. G. Clarke, *Indus.* Mineral. (Mar. 1982).
11. *Asbestos Materials in Buildings,* 3rd ed., Dept. of the Environment, London, Her Majesty's Stationery Office, 1990.
12. N. Kohyama, "Airborne Asbestos Levels in Non-Occupational Environments in Japan," in J. Bignon, J. Peto, and R. Saracci, eds., *Non-Occupational Exposure to Mineral Fibres,* IARC Scientific Publications No. 90, International Agency for Research on Cancer, Lyon, 1989.
13. H. M. Cunningham and R. D. Pontefract, *Nature (London)* **232,** 322 (July 30, 1971).
14. Health and Safety Executive, *Asbestos Fibres in Air—Light Microscope Methods for Use with the Control of Asbestos at Work Regulations, Methods for the Determination of Hazardous Substances,* MDHS 39/4, Health and Safety Executive Books, London, 1994 (rev).
15. National Institute for Occupational Safety and Health, *NIOSH Manual of Analytical Methods,* 3d ed., DHSS/NIOSH Pub. No. 84-100, Method No. 7400, Government Printing Office, Washington D.C., 1984.
16. International Organisation for Standardisation, *Air Quality—Determination of the Number Concentration of Airborne Inorganic Fibres by Phase Contrast Optical Microscopy—Membrane Filter Method,* ISO 8672, Geneva, Switzerland, 1993.
17. "Guidance on the Design of Sampling Programmes," *Water Quality, Sampling,* Part 1, ISO 5667-1, 1980.
18. "Guidance on Sampling Techniques," *Water Quality, Sampling,* Part 2, ISO 5667-2, 1982.

19. "Guidance on the Preservation and Handling of Samples," *Water Quality, Sampling,* Part 3, ISO 5667-3, 1985.
20. "Guidance on Sampling from Lakes, Natural and Man-Made," *Water Quality, Sampling,* Part 4, ISO 5667-4, 1987.
21. "Guidance on Sampling from Rivers and Streams," *Water Quality, Sampling,* Part 6, ISO 5667-6, 1990.
22. E. J. Chatfield, "Measurement of Asbestos Fibers in the Workplace and in the General Environment," in R. L. Ledoux, ed., *Mineralogical Techniques of Asbestos Determination,* Short Course Handbook, Vol. 4, Mineralogical Society of Canada, Toronto, 1979.
23. *Code of Practice for the Identification of Potentially Contaminated Land and its Investigation,* Draft for Development, DD 175, British Standards Institution, 1988.
24. International Organisation for Standardisation, *Soil Quality, Sampling,* Part 2, Committee Draft ISO/CD 10381-2, 1992.
25. P. Haggett, A. D. Cliff, and A. Frey, *Locational Methods,* Edward Arnold, London, 1977.
26. D. W. Berman and A. Kolk, *Superfund Method for the Determination of Releasible Asbestos in Soils and Bulk Materials,* EPA 68-W9-0059, U.S. EPA, San Francisco, Calif., July 1995.
27. *Asbestos-Containing Materials in Schools: Final Rule and Notice,* 40 CFR Part 763, U.S. EPA, Washington, D.C., Oct. 1987.
28. R. Wilson and co-workers, *Reg. Tox. Pharm.* **20** (1994).
29. *Interim Method for the Determination of Asbestos in Bulk Insulation Samples,* 600/M4-82-020, U.S. EPA, Washington, D.C., 1982.
30. Health and Safety Executive, *Asbestos in Bulk Materials: Sampling and Identification by Polarised Light Microscopy (PLM)—Methods for the Determination of Hazardous Substances,* MDHS 77, Health and Safety Executive Books, London, 1994.
31. American Society for Testing Materials, "Proposed Test Method for Asbestos—Containing Materials by Polarized Light Microscopy," *Annual Book of ASTM Standards,* D22 Proposal, 236, ASTM, Philadelphia, Pa., 1993.
32. B. E. Tylee, L. S. T. Davies, and J. Addison, *Ann. Occup Hyg.* **40,** 711–714 (1996).
33. Asbestos International Association, Health and Safety Publication, *Recommended Technical Method No. 2 (RTM2); Method for the determination of Airborne Asbestos Fibres and Other Inorganic Fibres by Scanning Electron Microscopy,* Asbestos International Association, London, 1984.
34. Verein Deutscher Ingenieure, *Measurement of Inorganic Fibrous Particulates in Outside Air—Scanning Electron Microscopy Procedure,* VDI 3492, Leaflet 1, Germany Apr. 1989.
35. International Organisation for Standardisation, *Ambient Air: Determination of Asbestos Fibres—Direct-Transfer Transmission Electron Microscopy Method,* ISO 10312, 1995.
36. International Organisation for Standardisation, *Ambient Air: Determination of Asbestos Fibres—Indirect-Transfer Transmission Electron Microscopy Procedure,* ISO/CD 13794, 1993.
37. International Agency for Research on Cancer (IARC), *IARC Monographs on the evaluation of the Caecinogenic Risks of Chemicals to Humans,* Vol. 42, *Silica and Some Silicates,* World Health Organization International Agency for Research on Cancer, Lyon, France, 1987.
38. *The Control of Asbestos at Work (Amendment) Regulations,* SI No. 3068, Her Majesty's Stationery Office, London, 1992.
39. *The Asbestos (Licensing) Regulations,* Her Majesty's Stationery Office, London, 1983.
40. J. B. Griffiths, "The Identification of Partially Oxidised Asbestos," *Fourth International Conference on Asbestos,* Vol. 1, Torino, 1980, pp. 29–248.

ATMOSPHERIC AEROSOLS: METHODS OF ANALYSIS

KUSIEL S. SHIFRIN
Oregon State University
Corvallis, Oregon

This article is devoted to methods of analyzing atmospheric aerosols. The lively interest in studies of atmospheric aerosols has quickened in the past years, resulting in the development of many new methods and devices. The intent is not to describe all of them within the framework of this relatively short article. The main concern is to provide an account of existing methods.

It is impossible to give a complete list of subjects treated in this article, much less possible to exhaust the range of subjects which belong to this theme and could be included. The condensed form presents what is considered most important, with the aim of striking a sensible compromise among the interests of different readers.

LIST OF SYMBOLS

The principal symbols used in this article are listed in the Nomenclature, except general physical symbols, like ω, λ, t, k, ..; the latter have their usual meanings. Symbols used in only one place are not listed.

Units of measurements are given in the MKS system. To find factors for converting MKS units into units used in aerosol science, consult the Tables 1 and 2. Table 1 is for units that differ from MKS only by a decimal factor. For instance, one can find there that the MKS density unit 1 kg m^{-3} is related to 1 g/m^3 by a factor of 10^{-3}. Table 2 gives conversion factors for off-system units commonly used in American literature. For example, the same density unit 1 kg m^{-3} is equal to 1.940.10^{-3} slug foot^{-3}. When calculating Table 2, 2.999×10^{10} cm s^{-1} was taken as the speed of light in a vacuum, and 9.806 m/s^2 for gravitational acceleration.

Sometimes the same symbol designates quite different quantities. If the meaning of a symbol is unclear from the context, consult the Nomenclature.

PRELIMINARY NOTES

Aerosols are small solid or liquid particles suspended in air, for example, drops of water, dissolved acids, salts and their bases,

Table 1. Recommended Unit Prefixes

Multiples and Submultiples	Prefix	Symbols
10^{12}	tera	T
10^{9}	giga	G
10^{6}	mega	M
10^{3}	kilo	K
10^{2}	hecto	h
10	deka	da
10^{-1}	deci	d
10^{-2}	centi	c
10^{-3}	milli	m
10^{-6}	micro	μ
10^{-9}	nano	n
10^{-12}	pico	p

Table 2. Selected Conversion Factors

Quantity	MKS Units	Conversion Factors to Other Units			
Length	Meter	10^6 μm	10^{10} Å	39.37 in.	3.281 ft
Time	Seconds	$1.667 \cdot 10^{-2}$ min	$2.778 \cdot 10^{-4}$ h	$1.157\ 10^{-5}$ day	$3.169 \cdot 10^{-8}$ yr
Plane angle	1 degree	$2.778 \cdot 10^{-3}$ rw	$1.745 \cdot 10^{-2}$ radian	3600″	60′
Solid angle	Steradian	$(4\pi)^{-1}$ sphere	$7.955 \cdot 10^{-2}$ sphere		
Mass	Kg		2.205 lb	$6.852.10^{-2}$ slug	35.27 oz
Density	Kg m^{-3}		$1.9070 \cdot 10^{-3}$ slug ft^{-3}	$3.613.10^{-5}$ lb $in.^{-3}$	
Speed	Meter s^{-1}	3.6 km hr^{-1}	3.281 ft s^{-1}	2.237 mi h^{-1}	1.944 knot
Force	Newton	10^5 dyn	0.2278 lb	7.233 pdl	102.0 gf
Pressure	Newton m^{-2}	$9.869 \cdot 10^{-6}$ atm	10 dyn sm^{-2}	$7.501.10^{-4}$ cm · Hg	5.353 in of water
Energy, heat	Joule	10^7 erg	0.2389 cal	1 watt s	0.102 m kgf
Power	Watt	0.2389 cal · s^{-1}	10^{-3} kw	0.7376 ft lb s^{-1}	$1.341 \cdot 10^{-3}$ hp
Energy flow	Watt m^{-2}	joule $(s \cdot m^2)^{-1}$	$1.433 \cdot 10^{-3}$ cal $cm^{-2}min^{-1}$	langley	

ice crystals, dust particles (of mineral and organic origin), and pollen and spores. The structure of aerosol particles is very diverse. Side by side with dense homogeneous particles, there often are complex inhomogeneous ones, sometimes chains, sometimes particles of peculiar configuration. Particle dimensions are very small, so they practically do not precipitate and travel along with the air element in which they are suspended.

The smallest aerosol particles are referred to as clusters. They make up formations of several molecules with dimensions of about $3-5 \times 10^{-4}$ μm. The cluster is a temporary combination and exists a very short time. The largest particles belong to coarse-dispersive aerosols. The limit is 50–100 μm. Large particles (radii > 20 μm), similarly to small particles, disappear rapidly due to precipitation, and are observable only near their sources. Generally, boundaries between small aerosol particles and molecules and between large particles and macroscopic pieces of a substance are quite indistinct. For example, some authors assign the term clusters to molecular formations. It was supposed earlier that particles with macroscopic properties ought to be defined as aerosols. However, it transpired that properties of even large aerosols differ from those of macroscopic particles, and this definition lost its meaning.

Atmospheric aerosols is the general name for dispersive systems suspended in the air. Different fields of science also use specific names, such as cloud, dust, fog (or mist), fume, smog, smoke, and haze. These names are connected with the origin of aerosols and do not have an exact definition. Their meaning is approximately as follows (1):

Cloud: aerosol clouds in which the mass of contained particles is >1% of the mass of contained gases.

Dust: solid particles formed by crushing or other mechanical breakage of a parent material. These particles generally have irregular shapes and are >0.5 μm.

Fog or mist: liquid-particle aerosol. These can be formed by condensation of supersaturated vapors or by physical shearing of liquids, such as in nebulization, spraying, or bubbling.

Fume: particles are usually the result of condensed vapor with subsequent agglomeration. Solid fume particles typically consist of complex chains of submicrometer-sized particles (usually <0.05 μm) of similar dimension. Fumes often result from combustion and other high temperature processes.

Smog: aaerosol consisting of solid and liquid particles, created, at least in part, by the action of sunlight on vapors. The term smog is a combination of the words smoke and fog and refers to the entire range of such pollutants, including the gaseous constituents.

Smoke: a solid or liquid aerosol resulting from incomplete combustion or condensation of supersaturated vapor. Most smoke particles are submicrometer in size.

Haze: consists of liquid and solid particles of a size of ~0.1 μm appearing in the process of the condensation of the water vapor in the atmosphere. The term is common in meteorology in the theory of oblique visibility.

Herein the general term *atmospheric aerosols* will be used. The science of aerosols, being at the crossroads of a number of scientific disciplines, sometimes uses several terms for the same quantity in addition to some special terms (1):

Primary particle: a particle introduced into the air in solid or liquid form. A primary particle is often contrasted with a secondary particle.

Secondary particle: usually, a particle formed in the air by gas-to-particle conversion. Secondary particle is sometimes used to describe agglomerated or redispersed particles.

Aggregate: a heterogeneous particle in which the various components are not easily broken up. The term *heterogeneous* indicates that the individual components may differ from each other in size, shape, and chemical composition.

Agglomerate: a group of particles held together by van der Waals forces or surface tension.

Flocculate: a group of particles very loosely held together, usually by electrostatic forces. Flocculates can easily be broken up by shear forces within the air.

The curve of density of particle size distribution equals aerosol particle spectrum which equals aerosol microstructure.

A number of classifications of the atmospheric aerosols have been offered:

By origin: natural and artificial aerosols

By mechanism of origin: sea water bubbles (Woodcock-Blanchard's mechanism); separation of soil particles (Bagnold's mechanism); condensation of water vapor

(Aitken's mechanism); gas-to-particle conversion; biogenic–biological processes

By particle size: fine-dispersive and coarse-dispersive aerosols

By level at which aerosols are situated: boundary layer aerosols (0–2 km); aerosols of the rest of the troposphere (2–12 km); stratospheric aerosols (12–35 km); ionospheric aerosols (>35 km)

Indoor and outdoor aerosols

By radioactivity, and chemical hazard, etc

The scientific history of aerosol studies falls into two periods: (*1*) the initial stage and (*2*) the contemporary stage. During the initial stage, investigations were of relatively modest character. They dealt mostly with the condensation of water vapor in the atmosphere and with problems of air purification from industrial stack pollutants, which became a great concern of industrially developed countries by the mid-1930s. England found itself in a particularly difficult position, being snowed under by approximately 2.5 million metric tons of soot yearly.

The contemporary period started in the 1940s. It was initiated by the development of efficient smoke screens for masking military operations, followed by other scientific and technical problems that appeared. The solutions to these problems rested on the lack of knowledge of aerosols. Some of these problems follow: (*1*) the increase of environmental pollution resulted in the origination of ecology, for which aerosols are among its main interests; (*2*) the appearance of a number of industries where a material goes through the aerosol stage, such as, powder metallurgy, the production of some ceramic materials, pigment production, some food industries, etc; and (*3*) the appearance of some technologies with especially strict requirements for air purity on industrial premises, such as, the production of electronic integrated circuits, of medicines, and of particularly pure materials.

In the process of solving each problem, a number of investigative methods for aerosols were created, and a great body of new information was obtained. Of all these results, only those concerned with the methods of analyzing atmospheric aerosols will be considered. The methods employed in ecological studies will be discussed in greater detail.

It is common to divide aerosol characteristics into two groups: (*1*) physicochemical characteristics (chemical composition, particle shape, etc) and (*2*) dispersive characteristics by which are meant the moments and particle size distributions. The best-accepted distribution curves and their features are presented. The descriptive methods for finding an analytical formula and determining the parameters involved from empirical data are described.

The so-called standard methods for analyzing aerosols are also considered. These methods include direct air sampling, particle measurement, and counting of particles in air samples. Different sampling methods are described, such as precipitation, centrifuging, filtration, and others, and the principal sources of errors inherent in these methods are discussed. Samples obtained by these methods are studied under the microscope, ultramicroscope, or electron microscope. Problems associated with these observations are noted.

Counters of individual particles are described, (such as the Owl, Coulter counter, and others). In all of these counters, particles pass a counting device one after another. The particle size is determined from the magnitude of the response of a measured characteristic. It can be, for example, the variation of the conductivity in the counting chamber when a particle enters it, or the light radiance scattered at some angle, etc. The errors in each method are also noted. These methods comprise the so-called direct methods. They supply basic information on aerosols now available to us.

The essence of the inverse problem (indirect methods) consists in different ways of retrieving the particle size distribution of atmospheric aerosols from their physical characteristics. The limited inverse problem, the complete problem (analytical methods), and the complete problem (numerical methods) are considerd.

Classifying investigative methods as direct and indirect is a matter of convention. For example, the study of particles under the microscope, when the light scattered by a particle is observed, falls in the category of direct method, whereas, strictly speaking, it is indirect. However, certain common practices have been established.

Indirect methods are currently being developed vigorously. Most often these methods are based on the inversion of light-scattering characteristics, though other physical characteristics of aerosol particle systems can also be inverted. In contrast to the direct methods, indirect methods are presently in the development stage of principles and laboratory tests. The principal problem associated with indirect methods lies in the fact that the inversion of a measured optical characteristic into a sought for physical one is unstable. Often small measurement or calculation errors result in large oscillations of the solution. Nonetheless, it is clear that indirect methods are quite promising.

PHYSICOCHEMICAL CHARACTERISTICS OF AEROSOLS

Elementary and Chemical Composition: The State of Particulate Material

The physicochemical characteristics of aerosols are (*1*) the elementary and chemical composition (state of the material constituting a particle); (*2*) the particle shape, and (*3*) the particle optical properties.

The chemical and elementary composition of aerosol particles, the state of their substance, and their optical properties can differ significantly from the corresponding properties of the original substance. This applies, for example, to the elementary and chemical composition of marine aerosol particles and sea water generating these aerosols. The state of the same substance may also differ noticeably for the aerosol and macroscopic conditions. For example, small water fog drops remain liquid even at temperatures well below the freezing point (down to $-12°C$), whereas a large volume of water solidifies at this temperature in a moment. The density of aerosol particles can also differ from that of a lump of the same substance. In particular, such is, the case with smoke particles whose density is usually many times less than the density of their substance.

The Shape of Aerosol Particles

Several sets of characteristics are used to describe the particle shape (5). The basic tool for examining particles is microscopy, either optical or electronic. It reveals that solid particles making up most dusts and fumes are usually very irregular in shape.

Liquid particles present in mists and sprays tend to be spherical, although large ones might be more or less irregularly shaped. Examination of a particle under the microscope reveals only a two-dimensional outline randomly oriented in the plane of observation.

The observed irregular shapes tend to fall into one of three general classes:

Isometric (granular, modular) in which all three overall dimensions of the particle are roughly of the same magnitude

Flat (flaky) in which the lengths in two dimensions are much greater than in the third, eg, platelets, scales, leaves, etc

Needle-like (acicular, fibrous) in which there is a much greater length in one dimension than in the other two, for instance, as in prisms or fibers.

When describing irregularly shaped particles, various equivalent sizes are introduced. The size is then determined based on some sort of equivalent behavior, usually equivalent to the behavior of a spherical particle under similar conditions (1,5). For example, the rate of settling in air by gravity may be measured for some irregular particles. Then a size number may be assigned to each particle based on calculating the diameter of a spherical particle of the same density which would settle at the same speed as that measured. The diameter so calculated may be referred to as the effective size of the particle.

With these concepts in mind, some definitions of particle size in common use may be stated:

Projected area diameter: diameter of a circle having the same area as the image of the particle projected parallel to the plane of the microscope view.

Equivalent volume diameter: diameter of a sphere having a volume equal to that of the particle.

Sedimentation diameter (free-falling diameter): diameter of a sphere of equal density, having the same settling velocity as the particle in a specified fluid.

Aerodynamic diameter: diameter of a sphere of unit density (1 g/cm^3) having the same settling velocity in air as the particle.

Drag diameter: diameter of a sphere having the same resistance to motion (drag force) as the particle in a fluid of the same viscosity and at the same velocity.

Other diameters are defined with reference to the particle image under the microscope as indicated in Figure 1. All of these possible diameters depend on the application for which the particle behavior is being considered. In most aerosol work, it is customary to assume, for simplicity, that particles are spheres. For different types of particle collectors, however, different diameters may be most appropriate.

Another method of describing a real irregular particle is in terms of the size of some simple geometrical shape which behaves equivalently. The simple geometrical shapes employed are sphere diameter (diameter = d), cube (side = d), cylin-

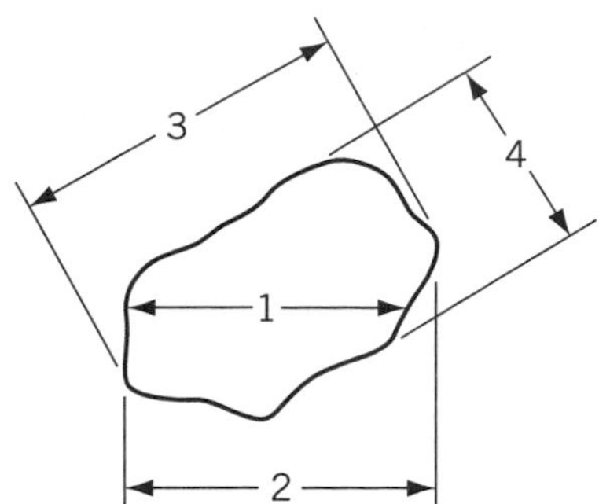

Figure 1. Other diameters: 1, Martin's diameter, divides projected area into equal parts along a line parallel to the base of field of view; 2, Feret's diameter, must be the longest dimension along a line parallel to the base of field of view; 3, longest diameter, is without regard to direction; and 4, shortest diameter, without regard to direction (5).

der (length = l, diameter = d), and ellipsoid of revolution (polar radius = p, equatorial radius = r).

$$\beta = \frac{p}{r} \begin{cases} \text{prolate} & \beta > 1 \\ \text{oblate} & \beta < 1 \\ \text{sphere} & \beta = 1 \end{cases}$$

An isometric particle is approximated by a sphere or a cube, a disk-like particle by a cylinder with a very small l/d or an oblate ellipsoid with $\beta \rightarrow 0$, and a needle-like particle by a cylinder with a large l/d or prolate ellipsoid with $\beta \rightarrow \infty$. The system of ellipsoids is used to cover many cases.

The particle silhouette shape as seen in two dimensions is represented by the method of Fourier transforms. This consists of finding the particle outline and its center of gravity, then measuring the distance from the centroid to the circumference at equal angular intervals of rotation around the centroid. Then the relationship between the radial distance to the periphery and the angle of rotation is expressed in terms of Fourier coefficients.

Optical Properties

The optical properties of a substance are described by three pairs of characteristics: n and k, or n and κ, or ε' and ε'', where n is the refractive index, κ is the absorption index, k is the dimensionless electrodynamic absorption coefficient, and ε' and ε'' are the real and imaginary parts of the dielectric constant ε. These three parts are related in the following way:

$$\kappa = \frac{k}{n}$$
$$\varepsilon' = n^2 - k^2$$
$$\varepsilon'' = 2\,nk \tag{1}$$

Usually n and k are used as basic characteristics, and the remaining ones are derived from them. Such a choice is explained by the fact that, when describing electromagnetic wave propagation, the equations involving n and k are the simplest. In this case, Maxwell's equations for an infinite homogeneous medium have the solution of the following form:

$$\exp\left(-\frac{2\pi}{\lambda}kz\right)\exp i\left(\omega t - \frac{2\pi}{\lambda}nz\right) \tag{2}$$

where λ is the wavelength in vacuum. This describes an attenuating plane wave propagating along the z-axis. Its length $\lambda' = \lambda/n$, and its amplitude is lowered by the factor $\exp(-2\pi/\lambda \cdot kz)$ when the wave travels the length z.

From equation 2 it is seen that n and k play different roles. The refractive index n determines the phase advance, and the electrodynamic absorption coefficient k determines the reduction in wave amplitude. One should distinguish characteristics relating to the field from those dealing with the radiance.

Light radiance is equal to the square of the field module. For the wave in equation 2 it has the following form, where $a = 4\pi k/\lambda$:

$$\exp(-az) \tag{3}$$

In courses of optics, the quantity a is called the absorption coefficient. It has the dimensionality L^{-1}. This is the standard Bouguer absorption characteristic. The quantities n and k depend on the frequency.

Values of $n(\omega)$ and $k(\omega)$ are commonly obtained from experiments with macroscopic quantities of a substance. The experiments usually involve determining k directly from the attenuation of light after passing through a layer of z thickness and simultaneous determination of n from the reflectance R. The value of n is determined at normal reflection by equation 4:

$$n = \frac{1+R}{1-R} + \sqrt{\left(\frac{1+R}{1-R}\right)^2 - 1 - k^2} \tag{4}$$

Reference books (6) give values of n and k for various substances. Similar data for atmospheric aerosols are presented (7), where it is assumed that 12 different types of atmospheric aerosols (urban aerosols, rural aerosols, marine aerosols, and others) consist of 22 original components. Values of n and k for these components are given (7) for different λ in the range 0.3–40 μm, and for different humidities in the interval 0–99%. However, the quantities $n(\omega)$ and $k(\omega)$ are not independent but are related to each other by integral relationships. These relationships, found in 1927, are called the dispersion relationships, and make it possible to find $n(\omega_0)$ at some frequency ω_0 from values of $k(\omega)$ given over the entire axis ω, from 0 to ∞. And vice versa. The dispersion relationship is used in optics for practically determining $n(\omega)$ and $k(\omega)$ for various substances. The functions $n(\omega)$ and $k(\omega)$ over the entire range of frequencies from zero to infinity must be known. The values of the optical constants $n(\omega)$ and $k(\omega)$ presented in Tables 1 and 2 refer to macroscopic quantities of a substance. Their application to small particles is erroneous in some cases. For example, the conductivity of metal σ is proportional to the mean free path l of electrons in it. In a large fragment, the mean free path is affected by thermal wave scattering. In a particle whose linear dimensions a are considerably less than the quantity l for a large fragment, the mean free path is determined by the particle dimensions. In this case, conductivity is proportional to the particle dimensions $\sigma \sim a$. When the particle dimensions increase, the linear relationship between σ and a is disturbed, and σ changes into a constant value ie, the conductivity of a massive fragment (8).

The current theory of metals assumes that l is 100 times greater than the lattice constant. Thus, for example, l is of the order of 500 Å for silver at room temperatures. Consequently, this value indicates the limit of the particle dimensions for which we may employ the macroscopic value of σ in the given case.

The general theory for n and k, with consideration for particle sizes, has been treated (9). It follows that the size effect can be neglected with assurance for particles larger than 0.1 μm, and in many cases, even for particles larger than 0.01 μm. As the particle size diminishes, the size effect, naturally, increases (9).

DISPERSIVE AEROSOL CHARACTERISTICS

The most important feature of dispersive systems is their huge surface. The surface S of the aerosol mass M is the larger, the smaller the particle sizes. For a monodispersive system of spherical particles of the radius a,

$$S = 3M/\delta a \tag{5}$$

where δ is the density of a particle substance. Aerosol optical properties, their absorption capacity, and much more are determined by their surface. This is why determining the particle size a (or, in the case of a polydispersive system, of a mean particle size a_1) is an important problem of aerosol studies. Of course, the aerosol effect depends on the particle size and also on the particle concentration N/cm^3. Usually these two quantities are not enough for an accurate estimate of the aerosol effect. It is known, for example, that different parts of the human and animal respiratory system retain various sized particles differently. Most large particles ($a > 1$ μm) are stopped in the nasopharyngeal area, whereas smaller particles ($0.1 \le a \le 1$ μm) are entrapped deeper in the trachea and bronchi. This means that to estimate the danger of atmospheric aerosol pollution, the aerosol particle size distribution must be known. Determining methods for aerosol particle size distribution $n(a)$ (cm^{-4}) is the essence of our work. Assume that the measurements are performed by some means. The measurement results are usually presented as a table for the number of particles ΔN whose radii fall within the interval $a + \Delta a$. The choice of the intervals Δa depends on consideration and the measurement accuracy. Often, this table is represented as a step-shaped curve called the distribution histogram. The table and histogram are inconvenient to use and scientists usually try to replace a table or histogram by an analytical formula. The choice of the formula is suggested by the histogram shape.

STANDARD METHODS

Sampling

Inertial and Gravitational Precipitation. Standard methods of sampling are most common in everyday aerosol measurements. The measurement process starts with sampling of aerosol particles out of the air. In some cases, sampling proceeds simultaneously with filtration.

The principle of inertial precipitation is very simple. A backing in a form of strip or cylinder, sometimes covered with a specific oil (to increase adhesiveness), is placed into the air flow. The gas flows around the backing while aerosol particles inertially continue their forward motion and hit the backing. The inertia of particles depends on their mass. Very small particles flow around the backing with the air. Large ones strike the backing. For the quantitative description of this phenomenon, a collision efficiency cross section of particles with the backing is used, expressed by the following formula (eq. 6):

$$E = \frac{y_c^2}{(R + r)^2} \tag{6}$$

The meanings of y_c, R, and r are clear from Figure 2. If $\pi(R + r)^2$ is designated as s, and t is the exposure time, then the numerical N or mass q aerosol particle concentration in the undisturbed flow is described by the following formulas (eq. 7):

$$N = \frac{N_d}{\pi y_c^2 tV} = \frac{N_d}{EstV} \ (\text{cm}^{-3}) \tag{7}$$

$$q = \frac{q_d}{\pi y_c^2 tV} = \frac{q_d}{EstV} \ (g \cdot \text{cm}^{-3})$$

where N_d is the number of particles deposited on the backing, q_d is the mass of deposited particles, V is the modulus of the flow velocity vector, and E is the collision efficiency cross section.

Regrettably, this simple procedure has a number of drawbacks: (*1*) the collision efficiency cross section E depends on the ratio of the particle radius r to the backing radius R. For small values of the parameter $\eta = r/R, E = 0$. E increases with η (10,11). Precipitation depends on the regime of the flow around the backing. It is possible to calculate E for two regimes, the potential flow (E_p) and viscous flow (E_v). For each regime, E depends on the dimensionless parameters η and κ. The parameter η determines the geometry of the flow, and the parameter κ describes the inertia of particles (12). If the particle inertia is low ($\kappa \rightarrow 0$), the particle is entrained by a flow, and $E = 0$. If the inertia is high, the particle hits the backing. The table of values of E_p and E_v is presented (12). With potential flow, for very small particles ($\eta = 0$) at $\kappa < 0.0833$, $E_p = 0$. E_p increases with κ(for $\eta = 0$) and becomes equal to 1 at $\kappa = \infty$. For the viscous regime, $E_v = 0$ at $\kappa < 2$. This is natural, because the more viscous the flow, the stronger is the bond between the flow and the particle. In reality, the flow differs greatly from theoretical models. This is why theoretical values of E calculated for a specific model may be considered only a qualitative estimate of the entrainment effect. Accurate values of E for a specific device can be obtained from a model experiment with a set of monodispersive aërosols. (*2*) As particles pass through the collector, some particles deposit at the entrance and on inside walls. (*3*) In the process of transport, particles coagulate. (*4*) In the process of transport, evaporation, and condensation take place at the particle surface. (*5*) A certain number of particles already deposited are entrapped and carried out with the flow. (*6*) When hitting the backing, some large particles get broken.

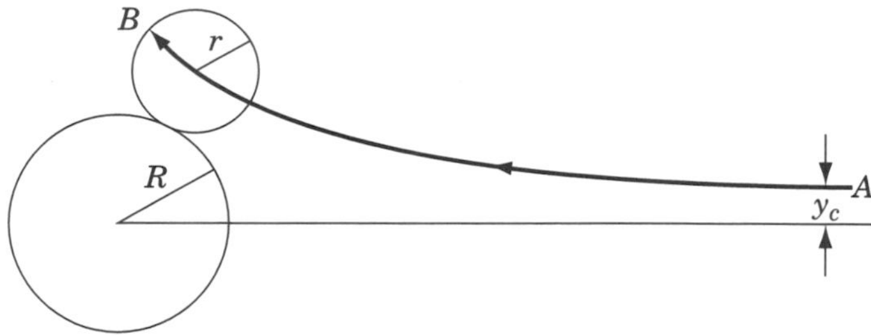

Figure 2. Diagrammatic representation of the trajectory of a small drop relative to a larger one (12).

As a result, particle size distributions observed on a backing in the undisturbed atmosphere vary significantly. Because it is impossible to account for all these factors, it is preferable to calibrate collectors with a set of homogeneous aerosols. The similarity of flow, evaporation, and other conditions must be ensured between calibration and measurement.

When the concentration of measured aerosols is high, it is sometimes necessary to resort to pure air dilution, to attain a minimum overlap of particles in the process of transit or ensuing particle counting under the microscope. Dilution is also advantageous because particle coagulation slackens when particle concentration is low. If a sample has to be kept, it must remain at the same temperature at which it was collected.

One should also take into account the inhomogeneity of air when taking external samples. In particular, this concerns rooms, shafts, or pipes where high temperature and humidity gradients occur. The inhomogeneity is especially noticeable when the aerosol concentration is low, for example, on particularly clean premises.

The method of inertial precipitation is used in many devices. Diagrams of the four most common types of classifiers are shown in Figure 3 (13). These are body impactor, conventional impactor, virtual impactor, and cyclone. The cascade im-

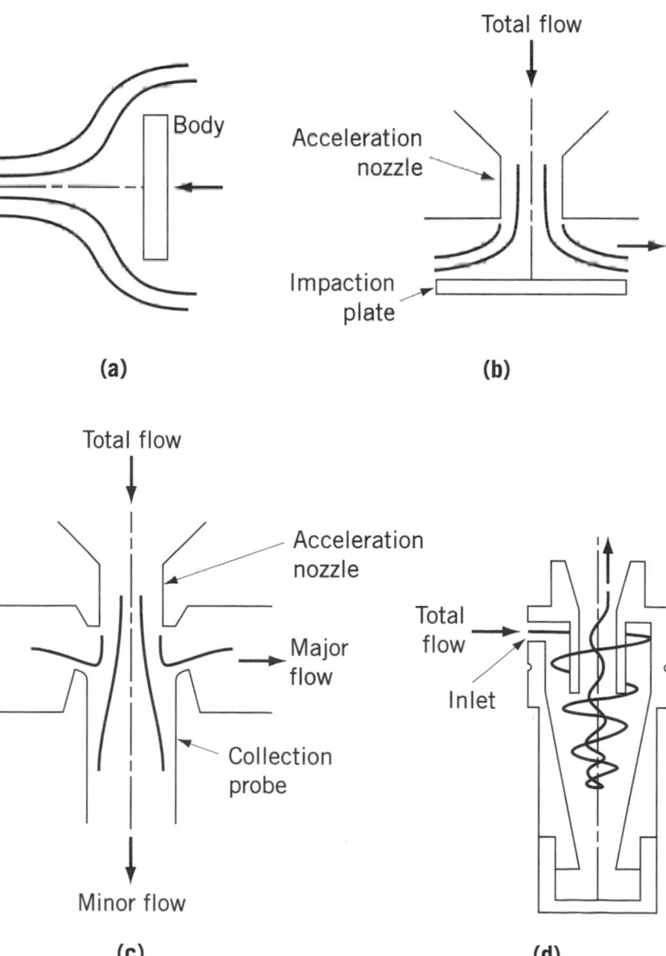

Figure 3. Four types of inertial classifiers (**a**) body impactor, (**b**) conventional impactor, (**c**) vitual impactor, and (**d**) cyclone (32).

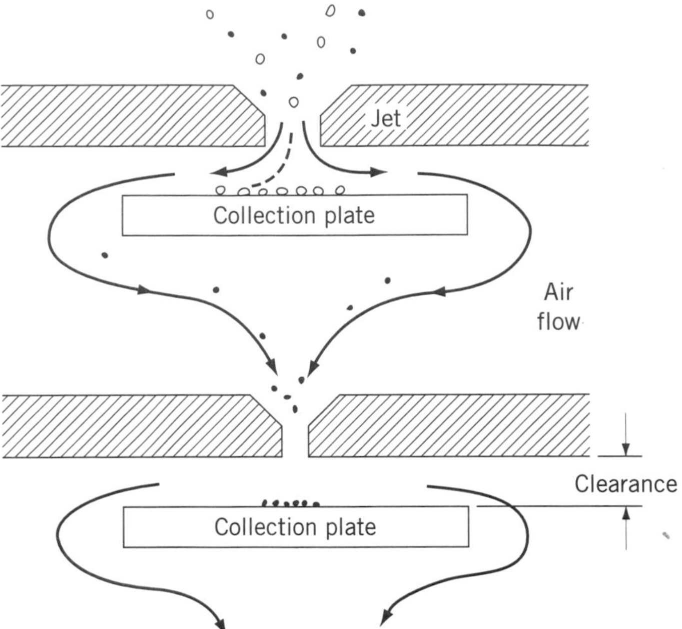

Figure 4. Schematic diagram showing two stages of a cascade impactor. The last stage is often followed by a filter (14).

pactor should also be mentioned. It is most commonly used for classifying aerosol particles by size for subsequent chemical analysis. The device consists of a series of stages, each composed of an orifice through which the aerosol flows normal to a collecting surface (Fig. 4) (14). The orifice may be rectangular or circular. The air flows over the collecting surface and on to the next stage, whereas particles too large to follow the air motion deposit on the surface. The basic mechanism of collection is inertial impaction. The orifice diameter (or width) is largest at the first stage, and the gas velocity is lowest. The largest suspended particles deposit at the first stage; smaller ones pass on to succeeding stages of smaller orifice diameter and progressively increasing efficiency of removal. The drawbacks of the method listed above are inherent, to a variable degree, in all devices based on the inertial precipitation. More details on the construction and operational features of specific devices are available (13).

Centrifugation. The precipitation method described previously is suitable only for particles whose diameter d is larger than 2 μm. The sedimentation rate of smaller particles is very low, and it takes too much time to observe the process. The lower limit of particle diameter can be decreased if centrifugal instead of gravitational acceleration is employed. The acceleration in contemporary ultracentrifuges is nearly one million times g. In common ultracentrifuges the acceleration is, of course, not so high. Those can be used for particles of $d \geq 0.05$ μm. The principle of centrifugation is simple. A particle moves curvilinearly in a device. The centrifugal forces affect the particle. The acceleration generated by these forces is directed along the radius outward from the center of the trajectory curvature. Particles deposit on a backing placed on the inner surface of the device. The momentary centrifugal acceleration is equal to $\omega^2 r$, where ω is the momentary angular velocity, and r is the distance from the center of curvature. Two principles are used to create curvilinear acceleration: (*1*) particles are positioned at the inner rapidly rotating part of the centrifuge and (*2*) gas flows together with particles along a spiral path. The pattern of motion resembles that in atmospheric cyclones which is why such devices are called cyclones. They are much simpler than common centrifuges, because there are no fast rotating parts. A device of this type, shown in Figure 5 (15), consists of a chamber in which the gas stream is given a swirling motion. The resulting centrifugal acceleration moves the airborne particles to the wall.

There are several types of cyclone collectors. Each has a tangential inlet near the top of a vertical cylindrical chamber. The gas outlet is a cylinder mounted concentrically at the top of the chamber and extends about one-third of the way down. The particles approaching the wall of the chamber are let into a cone-shaped extension of the chamber, which in turn leads the collected particles down to an oval discharge port. The entering aerosols spiral downward below the gas outlet and, near the bottom of the cone, reverse direction to form an inner vortex which is generally smaller in diameter than the gas outlet. A layer of dust-rich gas spirals down to the discharge port.

Because of their simplicity, ease of operation, and high size resolution, cyclones of various forms are used for sampling aerosol particles, but they are more useful for industrial applications. They are nearly useless for collecting submicrometer-diameter particles but have the great advantage over static sedimentation chambers of being able to sample a large amount of air in a short time. The use of conventional ultracentrifuges introduces a number of complications, and some quite elaborate techniques have been developed to overcome them. In particular, because particles are not moving in a vacuum but in the air, their speed is governed by the so-called fall size which depends on the air resistance; that is, not only on the particle size but on its shape.

Filtration. Filtration ensures sampling particles out of air and also selecting them in size. This is why filtration is the most commonly used method for aerosol analysis. The original literature and reviews on filtration are quite extensive

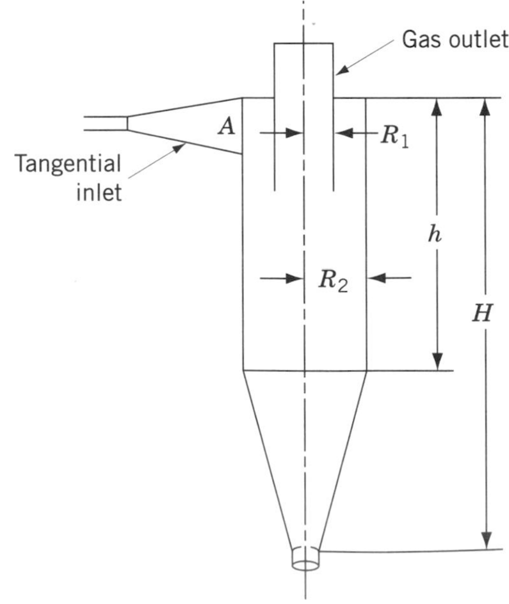

Figure 5. Schematic diagram of common cyclone collector (15).

(14). Fundamental problems connected with filtration, the disadvantages of the filtration method, and the methodology for evaluating its validity are discussed herein.

Sampling by filtration starts with the suction of aerosol particles by a sampling probe. The process depends on the pressure differential, on the shape of entrance, on the type of aerodynamic process, on electrostatic phenomena at the entrance and exit, etc. The efficiency of the process is estimated by aspiration efficiency η_a. Next, particles travel through the canal of the filtering device and arrive at the filter frame. The efficiency of this stage of the process is estimated by the transport efficiency η_t. The full inlet efficiency is $\eta_f = \eta_a\eta_t$. After this, particles pass through the filter itself. At this point, a number of factors affect the working quality of the filter: (*1*) colmatage, or clogging of the filter; (*2*) chemical and moisture interaction of the filter walls with the particles, in particular, with hygroscopic ones; (*3*) deformation of "soft" biological particles in the process of suction; (*4*) deformation of particles in the process of stripping for microscopic analysis; (*5*) ingress of particles through the filter (to avoid this, several filters are usually installed); and (*6*) penetration of particles bypassing the filter because of improper contact of the filter frame with holders.

Different types of filters may be used. (*1*) Fibrous filters are composed of separate fibers. Their thickness varies from about 1 μm to several hundredths of a micrometer. Sometimes a small amount (~10%) of binder is added to the filter medium. Filters are manufactured from cellulose, glass, quartz, and various plastic materials. Cellulose (paper) filters, (eg, Whatman) are the most common. (*2*) Porous-membrane filters are manufactured from Teflon and synthesized metals. Usually they consist of a number of layers. These filters are finer than fibrous ones, and the filtration through them proceeds much more slowly. (*3*) Nucleipore filters are manufactured by the neutron bombardment of a polycarbonate membrane. Neutrons make pores in the membrane of exactly the same size. The number of pores depends on the exposure time. The pores are continuous canals piercing the membrane. (*4*) Granular-bed filters: in the process of filtration, aerosols pass through a bed containing granules that stop aerosol particles. They are removed from the filter by a special extraction procedure. The main advantage of these filters is that, given the proper choice of bed material, both particle and gaseous pollutants can be collected.

This list is, of course, incomplete. The industry produces many filters differing in membrane pore size, intensity of pore clogging, etc. Exhaustive information is presented in booklets of manufacturers, such as Poretics (Livermore, California) a leading American firm in filter manufacturing (16).

When choosing a filter for the aerosol analysis, it is advisable to try using coarse filters. The use of fine filters takes much more time and requires more complex instruments for analysis. Sampling methods discussed above are unreliable because they distort the particle size distribution of the suspension. The fine fraction of the distribution is the most deformed.

Thermal Precipitation. In this method, thermophoretic forces are employed for sampling. The phenomenon of thermophoresis was discovered by J. Tyndall in 1870 who noticed that if aerosols are placed in a vessel where a thermal gradient is created (for example, with a heated wire) and the aerosols then are illuminated, a dark area is observable around the wire. The nature of the phenomenon is quite clear. The gas pressure on the particle is stronger on the hot side than on the cold side. As a consequence, the hot gas pushes the particles out of the heated area to the cold area, which depletes the particles in the hot area around the wire, hence darkening occurs. The method is easily realized instrumentally. In the simplest case, this can be just a heated wire between two cold plates. Cover glasses are placed on the plates opposite the wire. They collect the particles. The schematic diagram of such a device is presented in Figure 6 (13).

The phenomenon depends on the ratio between the particle diameter D and the mean free path length λ of gas molecules called the Knudsen number $Kn = \lambda/D$. If Kn is very small, the gas behaves as a continuous medium, and the phenomenon can be estimated according to hydrodynamic laws. If the Knudsen number is high, the phenomenon develops as in ultrararefied gases. For air under the standard conditions, $\lambda = 0.2\mu$m. This means that the complete theory of thermal precipitation of aerosol particles should cover the complete interval of values of the Knudsen number.

The thermophoric force is obviously proportional to the thermal gradient affecting the aerosol particle. Because the temperature differential is usually preset, the area through which aerosols pass should be as narrow as possible. This creates the problem of aerosol transit through a narrow vessel. It should be also remembered that the temperature differential cannot be too high, because this results in fast transformation of the aerosols in vessels. The heat conduction in the aerosol substance is much higher than in the gas in which the particles are suspended. It follows that the thermal gradient immediately near the particle, is significantly lower than in the vessel as a whole. Accordingly, the thermophoric force will be very weak.

A considerably thorough theory treatment of different cases for various size particles, different thermal gradients in a vessel, and different heat conductivities of aerosol substances, is available (15).

Electrostatic Precipitation. In electrostatic precipitators, particles are first allowed to pass through a corona dis-

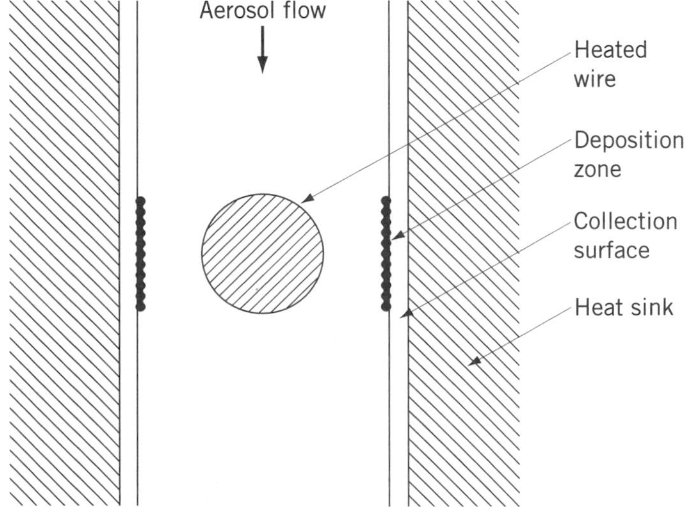

Figure 6. Cross-sectional view of a heated wire and plate thermal precipitator (13).

charge where they acquire a charge (8,17,18). After this, they go through a capacitor where they precipitate on plates of opposite charge. A cylindrical capacitor is often used as a precipitator. A wire in the center of the capacitor serves as a cathode. The outer enclosure, which, conventionally is, an aluminum cylinder concentric around the cathode, works as an anode. Sometimes just a metal plate is used as an anode. The device frequently consists of two parts. Of course, one can change the charges and make the central wire an anode and the external part (cylinder or plate) a cathode.

Aerosol particles are charged in the corona discharge by two mechanisms: (*1*) collisional charging, and (*2*) diffusive charging. When collisional charging takes place, an ion collides with a particle, couples with it, and imparts its charge to the particle. In diffusive charging, the particle acquires the charge as a result of the ion's Brownian motion to the particle and subsequent precipitation on it. The first mechanism is particularly efficient for large particles of radii $r > 0.25\ \mu$m; the second one fits small particles of radii $r < 0.1\ \mu$m. For intermediate-sized particles, both mechanisms work. Under the collisional charging, the process stops after the particle acquires some maximum charge $q_{\max}$. The value of $q_{\max}$ is determined by the following formula (eq. 8):

$$q_{\max} = \left(1 + \frac{\varepsilon - 1}{\varepsilon - 2}\right)E_0 r^2 \tag{8}$$

where E_0 is the external field surrounding the particle, r is the particle radius, and ε is the dielectric constant of the particle substance.

When diffusive charging occurs, the charge q of the particle depends on the time t. It is described by the following formula (eq. 9):

$$q = \frac{rkT}{e^2}\log\left(1 + \frac{\pi rcN_0e^2t}{kT}\right) \tag{9}$$

where T is the absolute temperature, c is the root-mean-square velocity of ions, e is the charge of an electron, and N_0 is the gas ion concentration.

Numerous comparisons have been made of theoretical predictions with experimental results. Unfortunately, the results of different authors are contradictory. After the particles are charged, they are collected by passing the aerosol through a continuation of the corona field in single-stage precipitators or by allowing the aerosol to pass through the field produced by the nondischarging plates of a second stage.

According to the Stokes equation, the velocity v of the particle moving to the collecting surface in single-state collectors is given by equation 10:

$$v = \frac{qE_0}{6\pi\eta r} \tag{10}$$

where η is the gas viscosity coefficient. These equations are based on the assumption that air is a continuous medium and that there is no influence of a turbulent wake, in other words, that Stokes' law would apply. If the particles are small relative to the mean free path, the Cunningham correction must be applied; the equations do not yield useful results for Reynolds numbers much greater than unity. Also, of course, these equations apply only when aerosol flow through the collection region is laminar.

Much of the theory of electrostatic precipitators has been developed for application to large-scale precipitators designed to collect industrial effluents. Airflow through such precipitators is usually turbulent. For such conditions, the following equation is suggested (18):

$$E = 100\left(1 - \exp\frac{-Av}{V}\right) \tag{11}$$

where E is the efficiency in percent, A is the total electrode collecting surface in square meters, and V is the volumetric flow rate in cubic meters per second. Equation 11 has been applied to various particle size distributions, such as normal and lognormal distributions (18).

The process of electrostatic precipitation usually has two stages, one for charging and the other for collection. The charging section has the common arrangement of an axial wire in a tube. However, the corona current is very high, permitting a high sampling rate. The collector zone reduces the interelectrode spacing by splitting the flow stream into two annuli, one inside and one outside the sampling tube. This arrangement reduces the interelectrode spacing without reducing the free cross-sectional area.

The principles described above are realized in many devices. One of the oldest and best known electrostatic precipitator for aerosol sampling is manufactured by the Mine Safety Appliance Co. (MSA). The sampling rate of 85 l/min is kept constant by a voltage stabilizer in the power supply, which operates on power between 105 and 120 volts, 60-cycle ac. The precipitator is the wire and cylinder variety, and the air is drawn through by a fan.

The particle size distribution obtained by electrostatic precipitators was compared with distributions obtained for the same aerosols with thermal precipitators. The results for aerosols of different natures were also compared. The coincidence of results from both methods was quite satisfactory. Part of the oxygen in electrostatic generators turns into ozone through the corona discharge. This is extremely undesirable because ozone is chemically active and affects aerosol particles, especially biological substances (15).

Wet Scrubbing. Wet scrubbing is probably the oldest and most commonly used method of aerosol sampling. More specialists are preoccupied with estimating its efficiency than all other methods combined. In those methods, the purpose of particle sampling is further analysis. The goal of wet scrubbing is to purify air by removing pollutant particles. This is one of the most important ecological problems, although this method is rarely employed for aerosol studies because it is difficult to extract aerosol characteristics from the deposits obtained. In industrial practice, particles collected from spent gases are usually recycled in an industrial process. Wet scrubbing devices are also used for removing harmful gaseous admixtures. With an appropriate choice of liquid, settling drops absorb the gases and eliminate them from the air. In addition, specially moistened surfaces are sometimes installed to collect the purified gas.

The idea of wet scrubbing is simple. A current of large liquid drops is run through aerosols. The drops collect aerosols on their surface and carry them to the collector. This purification is similar to the process of precipitation in the atmosphere. The mechanism of capturing aerosol particles by a separate drop is the same as the mechanism of inertial precipitation discussed above. However, in wet scrubbing, not one but a multitude of randomly arranged drops precipitates. A picture of the aerodynamic motion of a group of drops differs significantly from the simple superposition of pictures of individual falling drops. The hydrodynamic interaction of falling drops depends on their spacing and sizes. Simultaneously falling neighboring drops give additional speed to an individual drop, because the neighbors open the way for it. The dynamics of the collective fall of an ensemble of drops is a subject of interest in colloidal and aerosol physics.

The efficiency η, introduced for estimating the performance of this or other devices, is expressed by the following formula:

$$\eta = \frac{M_0 - M}{M_0} \tag{12}$$

where M_0 is the dust content at the scrubber entry, and M is the dust content at its exit. It is usually interesting to evaluate the integral efficiency η and also efficiencies of individual particle-capturing fractions $\eta(a)$. The concentration, density, and the dust particle size distribution at the scrubber entry are denoted as N_0, σ_0, and $f_0(a)$, respectively, and the same quantities at the exit as N_1, σ_1, and $f_1(a)$. Obviously,

$$\eta(a) = \frac{f_0(a) - \xi f_1(a)}{f_0(a)} \qquad \xi = \frac{N_1\sigma_1}{N_0\sigma_0} \tag{13}$$

It is clear from equation 13 that $\xi(a)$ (and η) are expressed in terms of the particle size distribution function.

Because the curves $f_0(a)$ and $f_1(a)$ are usually unknown, the dust is often assumed to be homogeneous when estimating η. The error of this assumption depends on the distribution width $\Delta\varepsilon$. For the Gaussian distribution (size distributions of large particles are usually close to Gaussian), $\Delta\varepsilon = 2.355\sigma/a$, where σ and a are the distribution parameters (see the section above on the distribution function). The monodispersive approximation means that $a_3 = a_1^3$. However, it is easy to show that $a_3 = a^3 + 3a\sigma^2$. For presenting particle size distributions in industrial smoke pollutants, $a \simeq 15$ μm (19). Thus, the mass of suspended particles is underestimated by the coefficient 4 when using this approximation.

Of different devices using the wet scrubbing method, the most used is the Venturi scrubber. It consists of a Venturi tube which has a smooth contraction at the inlet, a cylindrical neck, and a smoothly extending outlet (diffuser); a device for atomizing a liquid, either a hydrodynamic (atomizer) or an aerodynamic (central nozzle); and a device for removing drops, with dust particles deposited on them (drip pen). The inertial capture of dust particles by liquid drops takes place in the Venture tube. The drops also coagulate into larger compounds which makes it easier to separate them in the drip pen. The Venturi scrubber is often used as a primary stage for gas-cleaning systems and for removing large particles. In this way, gases are prepared for electrostatic filtration where finer cleaning takes place (5,20).

Problems of Particle Observation and Counting

Optical Microscope. The final stage of standard methods is the observation and counting of particles, for which microscopes, ultramicroscopes, and electron microscopes are used. The specific method applied depends on the size and nature of the particles, on what kind of data we wish to obtain, and on the material of the backing. For example, a microscope is used for particles whose size exceeds its resolution; an ultramicroscope is appropriate only for counting the particles, and so on. Particles collected on fibrous filters are usually difficult to remove. In this case, one measures only the increase of the weight of a filter and the chemical composition of the matter deposited on the filter (to do this, the particles settled on the filter are usually dissolved in a liquid; next, the liquid is evaporated, and the precipitate is analyzed).

Consider the methods for observing and counting particles after they collect on a backing or filter. The schematic diagram of an optical microscope is shown in Figure 7. The major parts of the microscope are the light source, the objective, and the eyepiece. Light from a source, either reflected from an object or transmitted through it, is focused by the objective. The objective creates the image of the object. The eyepiece enlarges this image and transmits it to the eye or to photographic film. The image of the object is usually positioned behind the plane on which the object is situated, as shown in Figure 7. The following features of a microscope are usually significant: (*1*) magnification (the ratio of the image to object, dimensions); (*2*) the depth of field (distance between the plane of the object and the plane on which the object is in focus); (*3*) the numerical aperture N describing the microscope's maximum capacity for collecting light from the object; (*4*) resolution indicating the minimum size of the object discernible under the microscope. The linear resolution of a microscope is described by the Rayleigh–Abbe formula:

$$h = \frac{0.6\lambda_0}{N} \tag{14}$$

where λ_0 is the wavelength of light in a vacuum and N is the numerical aperture. To increase the resolution, immersion systems are often used. In these, the object is placed in an immersion oil applied to a cover glass. The oil is usually

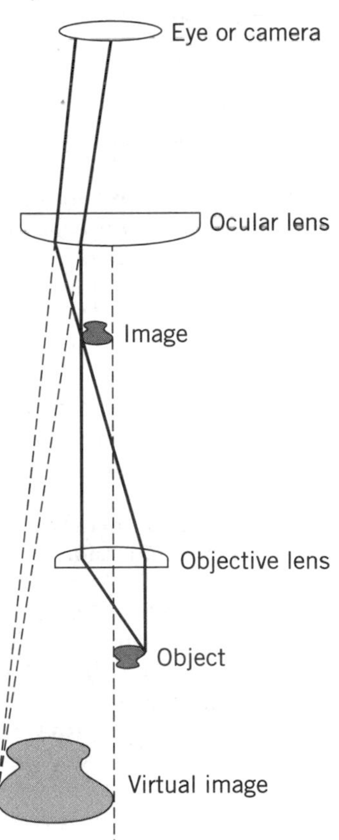

Figure 7. Schematic diagram of a light microscope. Light is transmitted through the sample and focused by an objective lens. The intermediate image is then enlarged and transmitted to the eye or the detector (21).

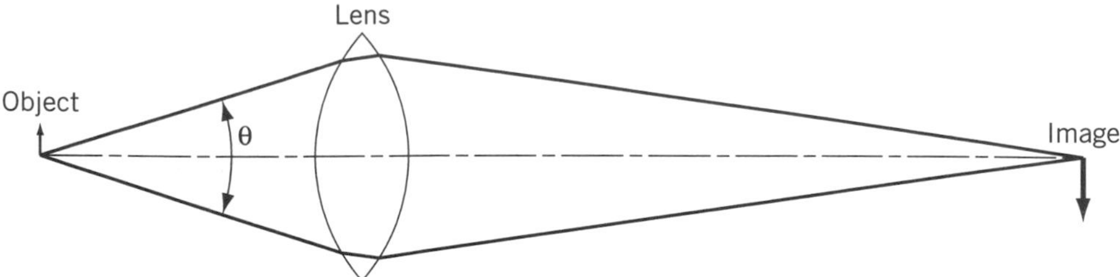

Figure 8. Light from a single point on the object is intercepted over a θ-angle (aperture) by the lens (14).

transparent and has a high refraction index n. The numerical aperture of the immersion system is $N = nN_0$, where N_0 is the numerical aperture of the dry object. N_0 is equal to the half-angle formed by rays coming from the extreme points of an object (Fig. 8). As seen from the formula, to increase the resolution, one has to choose small λ_0 and possibly great ϑ and n. At present, microscopes are manufactured with $\vartheta = 144°$, and hence, $N_0 = 0.95$. Immersion oils are found with $n = 1.5623$ which gives $N = 1.45$. For $\lambda_0 = 0.550$ μm, the minimum particle radius discernible under the dry microscope is 0.35 μm; under the immersion microscope, it is 0.25 μm. By comparison, the resolution h of the naked eye is $\simeq$200 μm, of a magnifying glass $\sim$100–25 μm, and of an average standard microscope $\simeq$10–1 μm. Present-day microscopy is an essential part of technical optics in which many important improvements are developed. To improve resolution (to decrease h), ultraviolet illumination is coming into use. In this case, the image is received on a photographic plate or on a luminescent or television screen. These systems are called ultraviolet microscopes. They use quartz lenses which are transparent down to $\lambda_0 = 0.275$ μm. Unfortunately, it is impossible to work with shorter wavelengths, because air is highly absorbing. When observing transparent objects, the so-called phase contrast is used to achieve greater contrast. This instrument features coherent illumination, which makes it possible to observe even a small difference in phase resulting from light passing through a transparent object. The same way of increasing contrast is used in interference microscopes. The principle of these microscopes is the same as for interferometers. The light beam from a source is split in two. The object under study is placed in one of the beams, and its interference is observed under the second beam. In this way, the optical thickness of the object is determined. Enhanced contrast is also achieved with the help of polarizing microscope, oblique illumination of the field of vision, tinting the backing or the object, and other methods.

Microscope observations and particle counting are tedious processes, especially when a multitude of small particles are analyzed. However, this is inevitable, particularly so, in cases of inertial or thermal precipitation, when particles have been already collected on a slide. A variety of automatic scanning attachments are employed to make the task easier.

Except for the investigation of atmospheric aerosols, all these devices are only a slight improvement, because they do not solve the major problem of visual microscopy, namely, its limited resolution. The observer just does not see a great number of small particles whose size is beyond the resolution limit. This usually leads to underestimating the numerical concentration. Coalescence of particles and shading of small particles by large ones also adds to underestimating the concentration.

Ultramicroscope. The ultramicroscopefor registering very small particles was proposed in 1903 (23). The idea is based on counting light flashes. This makes it possible to observe particles whose sizes are much smaller than the resolution of a conventional microscope. The instrument has to be constructed so that particles cross the light beam individually. Flashes are observed against a dark background. The number N_c of flashes is counted from a given weight of the mass Δm. The quantity $\Delta m/N_c$ gives the mean mass of an aerosol particle. If the particle substance density σ is known, one also obtains the mean particle volume. The ultramicroscope was first used for studying particles collected in a glass vessel or on a backing. However, the same method was applied to particles suspended in air (22). The axle shaft of the microscope was positioned horizontally, and aerosols flowed freely through. Flashes were counted when particles crossed the light beam. The advantage of this method was that particles were not removed from their natural medium. However, the method was not widely accepted because it took a great deal of time and involved many inevitable inconveniences and measurement errors (including systematic errors).

Most of these inconveniences and sources of error disappear if, instead of the number of particles in a stationary volume of the disperse system, the number of particles N_t is counted, traversing a field of view on passage through a definite volume V. The particle concentration will then be given by the following:

$$N = \frac{N_t}{V}$$

Such a flow ultramicroscope was developed in the 1960s (23). In this device, the aerosol flow moved along the optical axis, which allowed an increased counting rate approximately by a factor of hundred. The increase of the number of observed flashes made it possible to diminish essentially the fluctuation counting error.

The other important accomplishment is as follows (23). In the classic version of the Zsigmondy's microscope, the scattered light was observed at an angle of 90°. The instrument constructed employed small scattering angles improving the resolution significantly, because the scattering function of small particles is noticeably elongated. As a result, the lower limit of particle sizes was $a = 0.05$ μm in (23).

Electron Microscope. The electron microscope is an important step in the development of microscopy. It is known that an electron traversing a potential difference V (in volts) is similar to an electromagnetic wave of a wavelength of $\lambda = \frac{12.3}{\sqrt{V}}$ (Ångströms). Electron microscopes use the potential difference V of an order of 50 kV. This means that electron beams created under these conditions are similar to electromagnetic

Table 3. Properties of Objectives

Numerical Aperture of Objectives	Theoretical Limit of Resolution, μm	Depth of Field, μm
0.10	2.8	28
0.15	1.8	12
0.25	1.1	4.2
0.50	0.6	1.0
0.85	0.3	0.2

waves of wavelength $\lambda = 0.055$ Å. This is 20 times less than the atom, and 10^3 times less than the visible wavelength. Because electron beams are strongly absorbed in a substance, a vacuum is created in electron microscopes. Electrostatic and magnetic fields serve as lenses. The functions of these lenses are the same as those in objectives and oculars of conventional microscopes. The same problems of spherical and chromatic aberration arise which are inherent in the optics of glass and microscopes. The electron beam is generated by an electron gun. It consists of a tungsten filament (cathode), Wehnelt's cylinder, and anode. Electron lenses are created by nonuniform electric and magnetic fields, which are symmetric about the axle of the microscope. The transfer of an electron beam through the magnetic lens resembles the transfer of a photon through the glass lens, but the trajectory shapes are different. The photon moves along a straight line; the electron follows a spiral path. When observed under the electron microscope, the particles are placed on a thin membrane, commonly of a metal or silicone.

***In Situ* Photomicrography.** The microscopic examination of particles usually requires a preliminary collection step. However, photomicrographs of particles can be taken while particles are airborne (1,8).

The requirements for the direct photomicrography of airborne particles are similar to conventional photomicrography. However, two characteristics of airborne particles add complications: (*1*) the particles are in motion; (*2*) the number of particles in focus at any given time is generally very small when a microscope or similar objective is used. There may be only a small chance that an aerosol particle will be in focus at any given time. The limitations with regard to the volume in focus are given in Table 3.

If a film size of 5×7 in.^{2} and a magnification of 250× the numerical aperture is assumed, the volume in focus for numerical apertures of 0.85 will be 0.0001 mm^3. If the aerosol number concentration is 10^4 particles cm^{-3}, there will be only one chance in 1000 that a particle is in focus at any given moment.

Two general methods have been used to stop the particles photographically. If the aerosol is moving rapidly past the photographic equipment, the motion of the image across the photographic film can be slowed or stopped with a rotating mirror or prism, as in the Michelson–More experiment. If the particle motion is more or less random, very short exposure times with high intensity flashes are required, especially at high magnifications.

When photographing airborne particles, usually, high-intensity, very short flashes (~1 μs) are common practice. An appropriate optical system obtains a number of images of individual particles on each negative.

Suspensions of transparent droplets produce somewhat misleading images. When such a droplet is illuminated from the side, two small sections of the droplet, at opposite ends of the diameter parallel to the illuminating beam, were very much brighter than the rest. Thus the image of each droplet consisted essentially of two small dots, and the faint image of the rest of the droplet was easily overlooked.

Photomicrographic equipment for water droplets and ice crystals was developed (24). A modification for mounting on a sailplane produces images of droplets and particles larger than about 4 μm diameter. The airfoil housing was mounted above the glider.

Holography. Holography is closely related to *in situ* photomicrography (25). One of the first practical applications of holography was a device to investigate rapidly moving airborne particles in the ranging from 3–3000 μm in diameter. It was designed to provide images of all the particles in a single volume simultaneously and with equal sharpness. Thus it was especially useful for recording the images of snowflakes, raindrops, and cloud droplets. A modulated ruby laser of 10-mW power illuminated the particles in a volume as large as 5000 cm^3 for 20 μs. The hologram recorded the three-dimensional distribution of particles that later could be viewed with a continuous-wave laser.

Equipment for pulse holography for measuring aerosol particles emerging from a nozzle has been described (26) and shown schematically in Figures 9 and 10. The laser beam pulse generated by a ruby laser with $\lambda = 0.6943$ Å is divided into an object beam and reference beam. The former is directed through a window into a chamber containing the aerosol and out of a window on the opposite side. The light scattered by the droplets in a forward direction produces an interference pattern with the reference beam which is recorded on a photographic plate to form the hologram. The image reconstructed in three-dimensional space is magnified with a microscope and

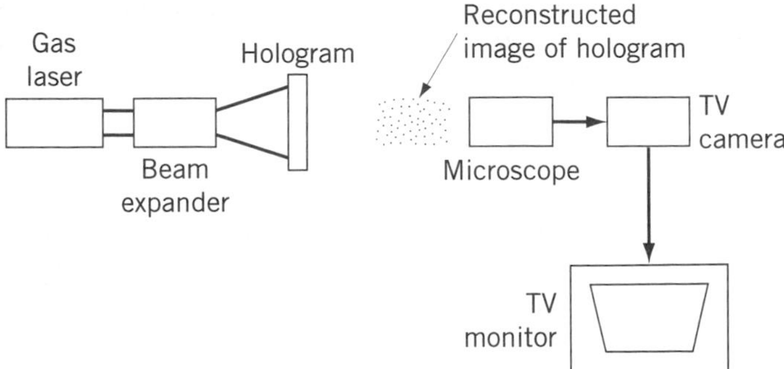

Figure 9. Equipment for preparing holograms of aerosol droplets emerging from a nozzle (15).

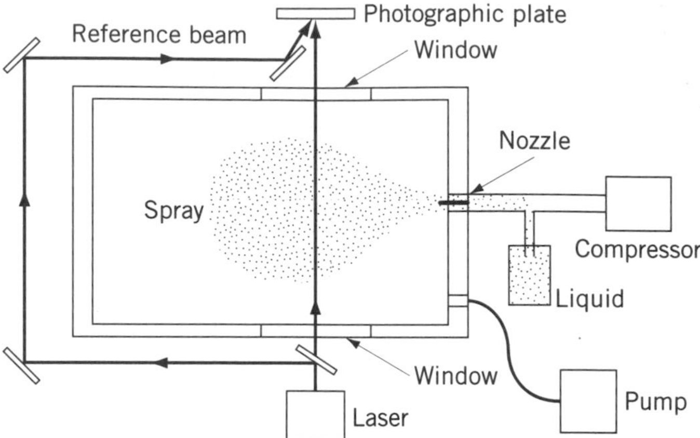

Figure 10. Equipment for reconstructing a hologram of airborne droplets and studying them with a microscope and TV monitor.

displayed by closed circuit television. By moving the hologram, the entire volume sampled is studied.

Pulse holography has some essential advantages over stationary holography: (*1*) to freeze the image of a moving particle, during its exposure, its travel must be no longer than some fraction ε of its radius. Suppose that the particle velocity is governed by its thermal equilibrium with surrounding gas. The mean-square-root of the velocity $\sqrt{v^2}$ is equal in this case to $\sqrt{3kT/m}$. The path traversed by the particle in a time Δt $\Delta l = \Delta t\sqrt{3kT/m}$. Assuming that $\Delta l = \varepsilon a$, Δt is related to the particle radius a by the expression

$$a^5 = (\Delta t)^2 \frac{9kT}{4\pi\sigma\varepsilon^2} \tag{15}$$

After specifying $\varepsilon = 0.1$, $\sigma = 1\ g/\text{cm}^{-3}$, and $T = 300$ K, equation 16 will be found:

$$\Delta t^2 = 3.32 \times 10^7 a^5 s^2 \tag{16}$$

Calculated Δt values from equation 16 for different aerosol particles suspended in air are presented in Table 4.

As seen from Table 4, it is most difficult to obtain a distinct image of small particles. Ultrashort picosecond pulses must be used for small particles. It is also important to observe a particle illuminated from the side against a dark background. When meeting these conditions, the particle speed can be two orders higher than when illuminating it along the receiver axle (26). When illuminating along the axle, the maximum permissible speed was 3 m/s. This meant that the limiting resolution in radius was 0.21 μm, whereas it was 0.05 μm by side illumination. The second advantage of pulse holography is that it can be used even with outside illumination.

Table 4. Δt Values for Aerosols in Air

a, μm	$10^6\ \Delta t \times s$
0.05	4.2×10^{-6}
0.1	2.4×10^{-5}
0.2	1.4×10^{-4}
1.0	7.6×10^{-2}
5.0	7.2×10^{-1}
10	2.4

Additional Remarks

Some simple methods which do not fall into the format of the chapter are presented in this review.

Measurements of Aerosol Concentration Using Webs. The simplest way of measuring aerosol particles in the most difficult size interval from 0.1 to 0.5 μm radius is with spider webs. The method was first proposed in 1940s (27). The measurements under electron microscope show that the radius of spider fibers is ~0.05 μm, and thus the capturing efficiency E will be zero only for very small particles of a radius $r << 0.05\ \mu$m (see the section above on inertial precipitation). This method was previously used quite actively, but now it is called on rarely, because it is not very accurate. Small particles seized by the spider fibers are not seen in conventional microscopes. Nonetheless, in its time (1952–1955), spider webs were used by Junge when he discovered his famous law for the particle size distribution of natural aerosols described by the formula $f \sim a^{-4}$.

Nets are commonly used to measure the aerosol concentration in a marine atmosphere. A tightly stretched, finely-structured net is set up at the windward side of a boat. Knowing the net area, boat velocity, exposure time, and gain of the net's weight from captured aerosols, it is easy to determine the mass concentration of atmospheric aerosols. Of course, in doing so, the efficiency E is assumed to be equal to 1.

Ringenlmann charts are used in the United States for estimating the density c(mg m^{-3}) of air pollutants. A black net looks like a grey one against a white background. A gradual change of the black-to-white ratio gives a continuous gray chart usually calibrated in five gray gradations. The grayness number and the corresponding density of a smoke cloud are determined by comparing its colour with the Ringenlmann chart. Translucent slides viewed against the sky are also used for the same purpose. In this case, one can use optically neutral filters. The measurements are performed even at night with a night viewing device. In its current modification, the device consists of a tube and a sensitive photometer which increases the accuracy significantly, because it excludes outside light.

Determination of Elementary Composition of Aerosol Particles. A variety of microanalytical techniques (such as wet chemical methods, atomic absorption spectroscopy, neutron activation analysis, and X-ray fluorescence), are used for chemical analysis of atmospheric particle matter collected on filters or on impactor slides (14). In all cases, it is necessary to sample for a more or less long period of time and then wait for the results of the chemical analysis. Two methods for specific application to aerosols will be considered, both based on the analysis of X-rays emitted by excited particle matter. The first method is based on observations by X-ray fluorescence.

The X-ray fluorescence detector consists of a monochromatic X-ray source, sample holder, and semiconductor detector (28). A sample of particulate matter collected on a filter is excited by the X-ray source. The excited sample produces an X-ray spectrum characteristic of the atomic species present in the sample.

Enough material can be conveniently collected on a filter in a period of one or two hours for analysis of many of the species in polluted urban air. It is difficult to detect elements lighter than potassium because of their low fluorescence yields and the strong absorption of fluorescent X-rays by other matter including the particles.

The second method is referred to as electron microprobes. It unites the features of electron microscopy and X-ray spectrochemical analysis. When a particle is irradiated by a hard electron beam, X-rays characteristic of the sample are generated. By measuring the wavelength and intensity of the emitted X-rays, the concentrations of various elemental species can be determined.

A schematic diagram of the electron microprobe is shown in Figure 11. The particles are mounted on a metal substrate, evaporated carbon film, or electron microscope grid. They must be sufficiently far apart to avoid interference, but the concentration must be high enough to permit easy identification in the scanning process. As an example of the scale of operation of such systems, an electron beam with a diameter of 1 μm, when passed through an acceleration field of 15 kV, will penetrate a volume of about 3 μm^3 in the sample. This means that the limit of resolution for this instrument is only a little more than 1 μm.

Elemental composition is determined by X-ray spectrometry. The size of the smallest particle that can be analyzed is presently limited by spectrometric techniques. A computer connected to the system is programmed to calculate the elemental concentrations. In this way, data are obtained in a few minutes on the composition of each area on which the beam is focused.

COUNTERS OF INDIVIDUAL PARTICLES

The standard methods described in the previous section have a number of drawbacks in the sampling methods and analysis of samples obtained. Consequently, the development of stable methods free of the shortcomings of standard methods started in the middle of this century. These devices count individual

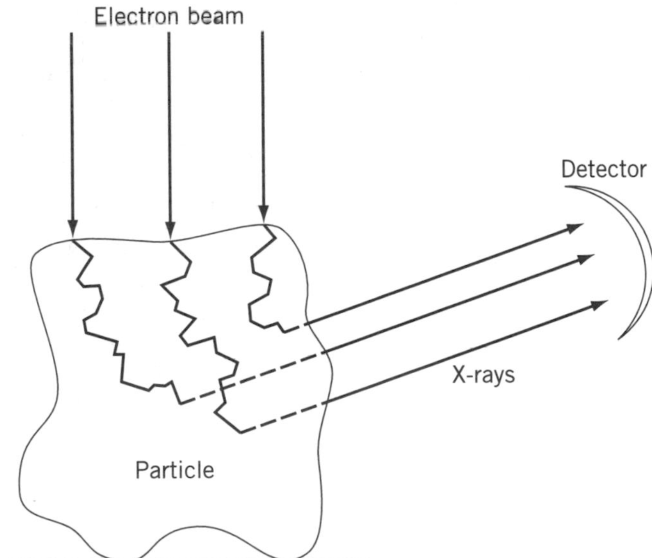

Figure 11. Principle of the electron microprobe. Schematic diagram of the interaction of an electron beam with a deposited particle. Irregular lines represent the path of an electron before it produces ionization and the generation of x-rays characteristic of the emitting element. Straight lines represent the paths of x-rays headed in the direction of the detector. The irregular shape of the particle considerably complicates the interpretation of results (14).

particles, which pass singly through a sizing facility. The responses of this facility depend on the particle size. The count is performed for responses related to a specific size interval. Next, the particle size distribution is easily determined. Unfortunately, all counters of individual particles, to a certain degree, have common shortcomings. The most important of them are the following: (*1*) for nonspherical particles or for particles of a complex shape, the response of the system depends on the particle size, how the particle is situated in the flow, and also on the particle structure; (*2*) the response intensity is ambiguously related to the particle size. So, for example, when the light intensity $I(\vartheta)$ scattered at the angle ϑ is measured, it is easy to confuse particles of different sizes, because $I(\vartheta)$ is related to the particle size in a nonunique way; and (*3*) When two small particles pass through the counter simultaneously, they can mistakenly be taken for a single large particle.

Some designs for counters of individual particles are examined here.

INDIRECT METHODS: RESTRICTED PROBLEM

General Notes

The direct methods described above are commonly used in observing atmospheric aerosols. Most of them are also employed in hydrosol studies. Unfortunately, the drawbacks and limitations inherent in these methods impair their efficiencies and sometimes make them unfit for use. This is why indirect methods are under development.

Among various indirect methods, optical methods are the most widely used. They are based on the inversion of optical into microphysical characteristics, such as the concentration and sizes of particles and their shape.

In the general case, the method for setting up a problem is as follows. Assume that a parallel light beam illuminates an elementary volume of a medium and that interference and multiple scattering are absent from the scattered light. In this case, the measured optical characteristic $S(x)$ is the sum of the same characteristics for all observed particles. Let us temporarily denote the measured characteristic as $S(x)$. This can be the volume extinction coefficient $\sigma(\lambda)$ or the intensity of the scattered light, etc. Let the minimal particle size in the system be a_1 and the maximum one be a_2; the number of particles of sizes from a to $a + da$ in a unit volume is $Nf(a)da$, where N is the total number of particles in the unit volume:

$$S(x) = N \int_{a_1}^{a_2} s(x, a) f(a) da \tag{17}$$

$S(x)$ is usually measured within some interval (x_1, x_2). The problem of inversion theory is to indicate a calculation method for an unknown function $f(a)$, limiting values of a_1 and a_2 from the measured $S(x)$, and known optical characteristics $s(x, a)$ for individual particles. It is also necessary to determine the minimal interval (x_1, x_2), where $S(x)$ has to be measured, and the number and position of measured points within this interval. In short, one should determine the minimum of information on $S(x)$ needed for retrieving $f(a)$ with desirable accuracy. The solution is based on the minimum information on $S(x)$ needed for retrieving $f(a)$ with desirable accuracy. The solution to the problem stated so completely presents considerable difficulties. The fact is that the inversion of the integral equation 17 is an incorrect problem. This means that the calculation procedure for such problems is very sensitive to small errors in the initial data or calculations. A small, usually uncontrollable variation in the initial data or calculation error leads to a significant change in the answer, and a rigorous solution to the problem might lead to physically absurd results. In other words, incorrect problems are practically ambiguous for the reason that optical data are satisfied by a great number of solutions. One ought to impose some additional conditions to obtain a unique solution. These conditions must narrow the set of possible solutions. For this, the sought for function $f(a)$ is usually assumed to possess some other features in addition to those satisfying equation 17. For example, $f(a)$ can be supposed to be smooth, etc. Entering this information into that contained in equation 17 is called regularization of the problem. The choice of the regularization method is difficult because additional information must not contradict the basic data. To circumvent this difficulty, the so-called restricted problems received wide acceptance in aerosol optics. There are two kinds of such problems: (*1*) determining specific characteristics of the system (for example, the total particle concentration N or the moments of particle size distribution); and (*2*) determining the parameters within some specified distribution family. The determination is performed by comparing measured characteristics with calculated ones. One should differ such restricted problems from complete problems, where no conditions are imposed on N and $f(a)$; they are determined entirely from optical experimental data.

Concluding Remarks

Basic concepts used in aerosol physics were defined. Methods for studying the structure of aerosol systems are classified as direct and indirect. The direct methods are those dealing with aerosol samples collected from air with subsequent examination of individual particles under a microscope (or a similar device). Indirect methods deal with the light scattered by particles. The particle concentration and their properties are estimated from light-scattering characteristics. Of indirect methods, those associated with the solution of the complete problem are particularly promising. These methods can be used in diluted systems, where there is no need to account for multiple scattering and interference; such conditions are typical of atmospheric and oceanic suspensions. It was also assumed that particles are spherical. In reality, this assumption is not critical, because one retrieves a distribution of equivalent spherical particles; we only recall that the meaning of equivalent radius varies with the different light-scattering experiments. In the analytical methods, the inversion is very similar to the Fourier inversion problem; such inversions on a limited range are connected with the rise of false oscillations in the result. We described some methods for quelling them.

The three indirect analytical methods have different fields of applicability. The total scattering function method (TSFM) has to do with small particles close to Rayleigh particles (the range of applicability of the Rayleigh–Debye approximation). The spectral transmittance method (STM) handles somewhat larger particles (the range for which the van de Hulst formula is suitable), and the small-angle method (SAM) is suited for large particles, the range of applicability of the Kirchhoff formula, possibly with some correction. Because the intervals of applicability are partially overlapping, it is useful to set up combined experimental plans that use more than one method, so that information can be obtained concerning different parts of the particle size distribution. In practice, techniques and methods are often used together.

For each of the inversion methods treated above, the same general considerations are applicable. It is necessary to have a sufficient amount of experimental information and to know the minimum which provides the desired result with the required accuracy. Often we encounter situations wherein optical information is clearly insufficient; in this case it is necessary to add extra information, as, for instance, in the fitting method. It is necessary to account for these criteria to perform the optical experiment in the best way.

NOMENCLATURE

Symbol	Measurement Units	Definition
a	m	Radius of spherical particle
a	m^{-1}	Optical absorption coefficient
a	m	Parameter of normal distribution
a_m	m	Mode of distribution
a_1	m	Mean particle radius, first initial moment
a_2	m^2	Second initial moment
$a_{min} = a_1$	m	Minimal particle radius, parameter of power distribution
$a_{max} = a_2$	m	Maximum particle radius, parameter of power distribution
$a_{2,3}$	m	Sauter's (mean cubic-square) of particle radius
$a_{n,m}^{n-m}$	m^{n-m}	$(n-m)$th particle radius
a_2	m^2	Variance or dispersion of distribution
a_{eff}	m	Effective particle radius
B		Turbidity coefficient
c	m^{-1}	Root-mean-square velocity
c	kg/m^3	Mass concentration of the suspension
C_ν	$m^{\nu-1}$	Parameter of power distribution
E		Collision efficiency cross section
f	m^{-1}	Density of particle size distribution
F		Integral distribution function
g,g^*	m^{-1}	Spectral transmittance of dispersive system
i_1, i_2		Mie sums
I	watt $m^{-2}sr^{-1}$	Light radiance
I_0	watt m^{-2}	Light irradiance
k		Electromagnetic absorption coefficient
n		Refractive index
N	m^{-3}	Numerical particle concentration

Symbol	Measurement Units	Definition
N	m^{-3}	Numerical particle concentration
Q		Attenuation efficiency factor
r		Dimensionless particle radius
R		Ratio a_{max}/a_{min} of power distribution
S	m^{-1} or $watt(m^2 sr)^{-1}$	Measured optical characteristic
α		Diffraction (Mie) parameter
α		Ångstrom parameter
β	m^{-1}	Parameter of gamma and generalized gamma distributions
$B(z,w)$		Beta function of z and w
γ		Parameter of generalized gamma distribution
$\Gamma(x)$		Gamma function of x
Δa	m	Width of distribution
$\Delta \varepsilon$		Relative width of distribution
δ	kg/m^3	Density of particle substance
δ		Phase shift of light ray when it passes particle
ε, ε', ε''		Dielectric constant (real and imaginary parts)
η	$kg \cdot m^{-1}s^{-1}$	Gas viscosity coefficient
θ	deg, radian	Scattering angle
x		Absorption index
μ		Parameter of gamma distribution
ν		Parameter of power distribution
σ	kg/m^3	Density of particle substance
σ	m	Parameter of normal and lognormal distributions
σ	m^{-1}	Volume extinction coefficient
τ		Optical thickness
$\Phi_0(z)$		Laplace function of z
ψ	deg, radian	Angle, scattering angle

ACKNOWLEDGMENTS

This work was supported by the Office of Naval Research (ONR) under Grant No. 9410064, and is dedicated to the memory of my wife Lily.

BIBLIOGRAPHY

1. K. Willene and P. A. Baron, (eds). *Aerosol measurement. Principles, Techniques and Application.* Van Nostrand, New York, 1993.
2. M. Kerker, *The Scattering of Light and Other Electromagnetic Radiation.* Academic, New York, 1969.
3. K. S. Shifrin, *Physical Optics of Ocean Water.* American Institute of Physics, New York, 1988; *Introduction to Ocean Optics.* Gidrometeoizdat, Leningrad, 1983.
4. K. S. Shifrin and G. Tonna, Inverse problems related to light scattering in the atmosphere and ocean. In Vol. 34 of *Advances in Geophysics* Academic Press, New York, 1993, pp. 175–252.
5. W. Light, *Air Pollution Control Engineering,* 2nd ed. Marcel Dekk, New York and Basel, 1988.
6. *Landolt-Börnstein Zalenwerte und Functione aus Physik, Chemie, Astronomie, Geophysik und Technik,* Secte Aflage, Bond 1 and 2, Teil 8, Optische Constants, Springer-Verlag, Berlin, 1962.
7. G. d'Almeida, P. Koepke, and E. Shettle, *Atmospheric Aerosols.* Deepak, Hampton, Va., 1991.
8. K. S. Shifrin, *Scattering of Light in a Turbid Medium.* Gostechteoretizdat, Moscow-Leningrad, 1951; NASA Technical Translation, TT F-477, Washington D.C., 1968.
9. C. Bohren and D. Huffman, *Absorption and Scattering of Light by Small Particles.* John Wiley & Sons, Inc., New York, 1983.
10. I. Langmuir. *J. Meteorol.* **5,** 175, 286, 318, 328, 571 (1948).
11. A. Fonde and H. Herne, *The Aerodynamic Capture of Particles by Spheres,* National Coal Board Mineral Research Establishment Rep. No 2068, 572–3, 1957.
12. B. Mason, *The Physics of Clouds,* 2nd ed. Clarendon Press, Oxford U.K., 1971.
13. V. Marple, K. Rubow, and B. Olson, in Ref. 1, Chapt. 11.
14. S. Friedlander, *Smoke, Dust, and Haze.* John Wiley & Sons, Inc., New York, 1977.
15. R. D. Cadle, *The Measurements of Airborne Particles.* John Wiley & Sons, Inc., New York, 1975.
16. *Microfiltration and Laboratory Products Catalog,* Osmonies, Livermore, Calif., 1996.
17. H. White, *Industrial Electrostatic Precipitation,* Wesley Publ Co., Palo Alto, Calif., 1963.
18. H. Rose and A. Wood, *An Introduction to Electrostatic Precipitation in Theoric and Practice,* Constable, London, 1956.
19. L. Bayvel and A. Lagunov, *Measurement and Verification of the Dispersion of Particles by the Method of Light Scattering at Small Angles,* Energiya, Moscow, 1977.
20. S. Calvert, J. Goldschmid, D. Leith, and D. Mente, *Scrubber Handbook,* NTIS, PB 213–016, 1972.
21. R. Ditchburn, *Light,* 3rd ed. Academic, London, New York, San Francisco, 1976.
22. R. Whytlaw–Gray and K. Patterson, *Smoke.* Edward Arnold & Co., London, 1932.
23. B. Derjagin and G. Vlasenko. *J. Colloid Sci.* **17,** 605–627 (1962).
24. T. Cannon. *J. Appl. Meteorol.* **9,** 104 (1970).
25. G. Stroke. *An Introduction to Coherent Optics and Holography,* 2nd ed. Academic, New York, 1969.
26. W. Koechner. *Ind. Res.,* 44–48, (Apr. 1973).
27. H. Dessens. *Q. J. R. Meteorol. Soc.* **75,** 23 (1949).
28. F. Goulding and J. Jaklevich, *X-ray Fluorescent Spectrometer for Airborne Particulate Monitoring,* Office of Research and Monitoring Report EPA, R-2-73-182, NTIS PB-225038, 1973.

ATMOSPHERIC AEROSOLS: SIZE DISTRIBUTIONS

ALAN S. KAO
ENVIRON
Arlington, Virginia

Atmospheric aerosols consist of solid or liquid particles ranging in size from a few nanometers (nm) to greater than fifty microns (μm). Each particle contains a broad mixture of chemical species, including organic compounds, trace metals, and water-soluble compounds (eg, hydrogen [H^+], sulfate

[SO_4^{2-}], nitrate [NO_3^-], and ammonium [NH_4^+] ions). Exposure to airborne particles is associated with adverse health effects (eg, pulmonary inflammation and mechanical changes in lung function), and the United States Environmental Protection Agency (U.S. EPA) has developed National Ambient Air Quality Standards (NAAQS) for particulate matter to protect human health and welfare. Particles also have a strong effect on atmospheric visibility and haze in national parks and the soiling of statues and buildings that results from dry deposition. All of these properties are functions of the size distribution of atmospheric aerosols.

PHYSICAL AND CHEMICAL PROPERTIES

Particle Size

Significant portions of atmospheric aerosols are generated by human activities, which contribute to local, regional, and global air pollution. Coal combustion, diesel exhaust, and resuspended soil dust are among the numerous sources of direct emission of particles. However, much of the aerosols results from secondary sources, ie, the conversion of gaseous precursors (eg, sulfur dioxide, nitrogen oxides, and organic vapors) to particles.

Because its influence on the physical behavior of suspended particles is large, particle size is one of the most important properties of aerosols. Atmospheric particles are often irregularly shaped, and their sizes generally are characterized by an "equivalent" diameter, which relates a specific measurable physical property of an irregularly shaped particle to that of a spherical particle. The two most commonly used equivalent diameters are the *aerodynamic diameter* and the *Stokes diameter.* These two parameters describe a particle in terms of a sphere with the same physical behavior:

- Aerodynamic diameter (d_{aero}) is defined as the diameter of a spherical particle of unit density ($\rho_0 = 1\ \text{g cm}^{-3}$) with the same settling velocity as the particle.
- Stokes' diameter (d_{Stk}) is defined as the diameter of a spherical particle with the same aerodynamic resistance as the particle. For a smooth, spherically shaped particle, the Stokes' diameter equals the physical particle diameter.

The aerodynamic diameter and the Stokes diameter are related as follows:

$$d_{aero} = d_{Stk}\sqrt{\rho\frac{C(d_{Stk})}{C(d_{aero})}} \qquad (1)$$

where ρ is the particle density (g/cm^3) and $C\ (d_p)$ is the Cunningham slip correction factor for the particle diameter d_p (μm). The Cunningham slip factor is an empirical correction parameter that accounts for the reduction in the drag force on particles small enough to "slip" by gas molecules, thereby decreasing the number of molecular collisions near the particle surface. This correction factor is important for particles smaller than 1 μm in diameter and is determined by the following:

$$C(d_p) = 1 + \frac{\lambda}{d_p}\left[2.514 + 0.800\exp\left(-0.55\frac{d_p}{\lambda}\right)\right] \qquad (2)$$

where λ is the mean free path of the air (μm). At normal temperature (20°C) and pressure (1 atm), the mean free path of air λ is 0.0665 μm. Because $C\ (d_{aero})$ is a function of the aerodynamic particle diameter, equation 1 must be solved iteratively for d_{aero}. However, for particles with diameters greater than approximately 5 μm, $C = 1$ and equation 1 reduces to the following:

$$d_{aero} = \sqrt{\rho}\, d_{Stk} \qquad d_p \gg \lambda \qquad (3)$$

The simplified equation 3 is often used for particles as small as 0.5 μm (1).

Particles with the same physical size and shape, but different densities, have the same Stokes diameter, but different aerodynamic diameters. For particles greater than about 0.5 μm, the aerodynamic diameter is generally the quantity of interest, because it is the parameter that is important to particle deposition in the respiratory tract and collection in air sampling instruments. U.S. EPA has established NAAQs for particulate matter with aerodynamic diameters less than 10 μm (PM_{10}) and 2.5 μm ($PM_{2.5}$).

Modes in the Size Distribution

The mass distribution of atmospheric aerosols is generally represented by a trimodal size distribution (2,3). The three modes and typical median diameters for each mode are as follows:

- *Nuclei mode* (0.018 μm): Particles in this mode (diameters below about 0.1 μm) are formed by condensation of atmospheric gases into primary particles, which then coagulate into aggregates. The nuclei mode is observed as a separate mode only in clean or remote areas or near sources of new particle formation by nucleation. Particles grow out of this range primarily by coagulation into larger aggregates and are removed by interception because of their rapid Brownian motion. This mode does not contribute appreciably to the mass of a typical urban aerosol.
- *Accumulation mode* (0.21 μm): Particles in this mode are formed by gas-to-particle conversion through chemical reactions and condensation. Particles from the nuclei mode move into the accumulation mode by coagulating with accumulation mode particles. Once these particles reach 0.1 μm in diameter, removal by diffusion is negligible, and the particles grow slowly by coagulation until they exceed 2.0 μm, where sedimentation and impaction are significant. These particles at the upper end of the accumulation mode size range deposit to the ground before growing much larger.
- *Coarse mode* (4.9 μm): This mode consists primarily of particles generated by mechanical processes and contains windblown dust, sea salt spray, and plant material. Coagulation models show that there is little mass exchange between the accumulation and coarse modes, resulting in stable and sharply separated modes.

These modes are described by lognormal distributions, as shown in Figure 1. Each mode is described by the following probability distribution:

$$\frac{dc}{d(\ln d_p)} = \frac{1}{\sqrt{2\pi}\ln\sigma_g}\exp\left[-\frac{(\ln d_p - \ln GM)^2}{2(\ln\sigma_g)^2}\right] \qquad (4)$$

where c is the total particle mass concentration of one of the modes, GM is the mass median diameter, and σ_g is the geometric standard deviation (GSD) of the size distribution.

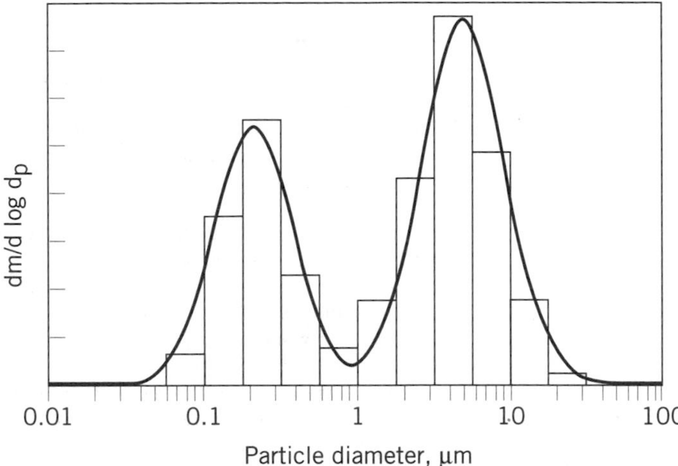

Figure 1. Typical histogram and fitted bimodal size distribution of atmospheric aerosols. Fitted fine particle distribution has a geometric mean (GM) of 0.21 μm and a geometric standard deviation (GSD) of 1.8; coarse particle distribution has GM = 4.9 μm and GSD = 1.9.

Assuming that the mass of the nuclei mode is negligible, ambient aerosols typically are modeled by a bimodal lognormal distribution function as follows:

$$\frac{dc}{d\ln d_p} = \frac{\alpha_F}{\sqrt{2\pi}\ln\sigma_{gF}}\exp\left[-\frac{(\ln d_p - \ln GM_F)^2}{2\ln^2\sigma_{gF}}\right] + \frac{1-\alpha_F}{\sqrt{2\pi}\ln\sigma_{gC}}\exp\left[-\frac{(\ln d_p - \ln GM_C)^2}{2\ln^2\sigma_{gC}}\right] \quad (5)$$

where GM_F and GM_C are the mass median diameters of the fine mode (ie, accumulation mode) and coarse mode particles, respectively; σ_{gF} and σ_{gC} are the GSDs of the fine and coarse particle modes, respectively; and α_F is the fraction of the total mass in the fine particle mode. GSDs for fine and coarse particle size distributions typically range from 1.5 to 1.9 (4). Fine particles represent approximately 40 to 80% of total PM_{10} mass (5,6) and PM_{10} represents approximately 45 to 64% of total suspended particulate (TSP) mass (7). Thus, values of α_F range from 0.18 to 0.51.

Fine and coarse mode particles differ from each other in many respects, including formation mechanism, composition, solubility, sources, and atmospheric lifetimes. Some of these differences are summarized in Table 1 and are discussed in greater detail later in this article.

PARTICLE FORMATION AND GROWTH

Although some combustion sources emit particles in the fine particle size range, the fine particle fraction is dominated by particles formed by gas-to-particle reactions. As fine particles are transported through the atmosphere, mass transfer to or from the particles continuously occurs. Thus, the size distribution and chemical composition of the particles are constantly changing with time. These changes occur because of a number of concurrent chemical and physical processes, including

- nucleation
- condensation
- evaporation
- chemical reaction
- coagulation

Once these particles reach 0.1 μm in diameter, removal by diffusion is negligible, and the particles grow slowly by coagulation until they exceed 2.0 μm, where sedimentation and impaction are significant. These particles at the upper end of the accumulation mode size range deposit to the ground before growing into the coarse particle size range, and there is little mass exchange between the fine and coarse modes, resulting in stable and sharply separated modes. In this section, the physical aspects of these gas-to-particle conversion mechanisms and their effects on the particle size distribution are discussed. For detailed discussions of particle formation and growth mechanisms, see (1,8,9).

Two important properties for understanding gas–particle interactions are partial pressure and equilibrium vapor pressure. The *partial pressure p* of a vapor is the pressure the vapor would exert if it were the only component present. The *equilibrium vapor pressure* p_s, (or saturation pressure) of a liquid, is the minimum partial pressure of a liquid's vapor required at the gas–liquid interface to maintain equilibrium. The equilibrium vapor pressure (in units of mm Hg) is often estimated by Antoine's equation:

$$\log p_s = A - \frac{B}{T - C} \quad (6)$$

where T is the temperature (K) and A, B, and C are chemical-specific constants. For water, $A = 8.11$, $B = 1750$, and $C = 38$ for the temperature range 0 to 333 K.

The *saturation ratio* S_R is defined as the ratio of the partial pressure of a species to its equilibrium vapor pressure above a flat liquid surface;

$$S_R = \frac{p}{p_s} \quad (7)$$

When the saturation ratio is unity, the rate of evaporation is equal to the rate of condensation, and the mixture is described as saturated. Values of S_R greater or less than unity correspond to supersaturated and unsaturated mixtures, respectively. For either condensation or nucleation, the species vapor pressure must exceed its equilibrium vapor pressure (ie, supersaturated mixture).

The equilibrium vapor pressure is defined above as the partial pressure required for mass equilibrium for a flat liquid surface. However, liquid aerosol particles have a curved surface. The curvature of the particle surface modifies the attractive forces between surface molecules so that, the smaller the droplet, the easier it is for molecules to leave the surface. Therefore, a greater partial pressure is required to maintain mass equilibrium at a given temperature for a curved surface than for a flat liquid surface, a phenomenon known as the Kelvin effect. The Kelvin effect is significant only for particle sizes smaller than 0.1 μm. The saturation ratio required for mass equilibrium for a droplet of diameter d_p is given by the Kelvin equation:

$$S_{RK} = \frac{p_d}{p_s} = \exp\left[\frac{4\gamma M}{\rho_p RT d_p}\right] \quad (8)$$

where S_{RK} is the Kelvin equation saturation ratio, p_d is the partial pressure of vapor at the droplet surface, γ is the surface tension of the liquid, M is the molecular weight, ρ_p is the density of the liquid particle, R is the gas constant, and T is the

Table 1. Comparison of Ambient Fine and Coarse Mode Particles[a]

Property	Fine Mode	Coarse Mode
Formed from	Gases	Large solids/droplets
Formation mechanism	Chemical reaction; nucleation; condensation; coagulation; evaporation of fog and cloud droplets in which gases have dissolved and reacted.	Mechanical disruption (eg, crushing, grinding, abrasion of surfaces); evaporation of sprays; suspension of dusts.
Composition	Sulfate (SO_4^{2-}); nitrate (NO_3^-); ammonium (NH_4^+); hydrogen ion (H^+); elemental carbon; organic compounds (eg, PAHs); metals (eg, Pb, Cd, V, Ni, Cu, Zn, Mn, Fe); particle-bound water.	Resuspended dusts (eg, soil dust, street dust); coal and oil fly ash; metal oxides of crustal elements (Si, Al, Ti, Fe); $CaCO_3$, NaCl, sea salt; pollen, mold spores; plant/animal fragments; tire wear debris.
Solubility	Largely soluble, hygroscopic, and deliquescent.	Largely insoluble and nonhygroscopic.
Sources	Combustion of coal, oil, gasoline, diesel, wood; atmospheric transformation products of NO_x, SO_2, and organic compounds including biogenic species (eg, terpenes); high-temperature processes, smelters, steel mills, etc	Resuspension of industrial dust and soil tracked onto roads; suspension from disturbed soil (eg, farming, mining, unpaved roads); biological sources; construction and demolition; coal and oil combustion; ocean spray.
Atmospheric lifetimes	Days to weeks	Minutes to hours
Travel distance	100s to 1000s of kilometers	<1 to 10s of kilometers

[a] Ref. 1.

temperature. Note that the increase in the partial pressure of vapor required for equilibrium increases with decreasing particle size. Evaporation occurs if S_R is less than S_{RK}, even if the saturation ratio is greater than unity. Similarly, if the saturation ratio is greater than that required by the Kelvin equation, condensation and growth occur.

Nucleation

Nucleation is the initial formation of new particles from a vapor. This initial droplet formation process occurs in the absence (homogeneous) or presence (heterogeneous) of condensation nuclei. Homogeneous gas-phase chemical reactions form a condensable species, such as the oxidation of sulfur dioxide to form sulfuric acid. Condensable species nucleate to form a new particles (nucleation) or condense onto the surface of existing particles (condensation).

Homogeneous nucleation, or self-nucleation, requires large saturation ratios, usually in the range of 2–10, which normally occur in only special laboratory or chemical process situations. The more common new particle formation mechanism is heterogeneous nucleation, or nucleated condensation. Heterogeneous nucleation, which relies on the existence of submicron condensation nuclei that serve as sites for the addition of vapor molecules, is the primary mechanism for cloud formation in the atmosphere. New particle formation via nucleation usually is not important except near sources of condensable species (eg, in traffic areas and in cleaner, remote regions).

Condensation and Evaporation

Condensation is the process of net particle growth in which the rate of arrival of vapor molecules at a particle's surface exceeds the rate of molecules leaving the surface (evaporation). Growth by condensation occurs when a droplet of pure liquid is in a supersaturated environment that exceeds the requirement given by the Kelvin equation. The rate of growth depends on the saturation ratio and the particle size. Although nucleation reactions lead to increases in particle number and particle mass, condensation reactions result in an increase of aerosol mass and surface area, but do not affect the total number of particles.

In condensational growth processes, particle growth rate is limited by the rate of transport of condensable species to the particle surface. In the free molecular regime, where particles are much smaller than the mean free path of air (λ), transport occurs by single molecular bombardment of the surface, and the volume (or mass) of these particles grows in proportion to the surface area. Therefore, particle density is independent of particle size and is governed by the following growth rate:

$$\frac{d(d_p)}{dt} = \frac{2(p - p_d)}{\rho_p\sqrt{2\pi RT/M}} \qquad d_p < \lambda \tag{9}$$

where p is the partial pressure of vapor near the droplet (calculated from the product of S_R and the vapor pressure p_s), p_d is the partial pressure of vapor at the droplet surface (as given by the Kelvin equation), M is the molecular weight, ρ_p is the density of the liquid particle (g/cm^3), R is the gas constant, and T is the temperature in K.

In the continuum regime, where particles are larger than the mean free path, transport is governed by the rate of molecular diffusion. In this regime, the loss of diffusing species at the surface of the particle causes a gradient in the concentration of the diffusing species near the surface of the particle, so that the volume of the particle grows in proportion to particle diameter rather than surface area. Thus, the diameter is inversely proportional to particle diameter, and the growth rate is governed by the Maxwell equation:

$$\frac{d(d_p)}{dt} = \frac{4DM(p - p_s)}{\rho_p d_p RT} \qquad d_p > \lambda \tag{10}$$

where D is the diffusion coefficient of the vapor molecules, as given by the Stokes–Einstein relationship:

$$D = \frac{kTC}{3\pi\mu d_p} \tag{11}$$

where k Boltzmann's constant (1.38×10^{-16} dyne-cm/K), C is the Cunningham slip correction factor, and μ is the gas viscosity (g/cm/s).

In the transition regime, where the particle size is on the order of the mean free path, the growth law is expressed as

$$\frac{d(d_p)}{dt} \propto \frac{1}{d_p}\left(\frac{1.51 + 2.08Kn}{1.51 + Kn(1.08 + 6.04\beta) + 8.32\beta Kn^2}\right) \quad d_p \sim \lambda \qquad (12)$$

where $Kn = 2\lambda/d_p$ is the Knudsen number and β is proportionality constant relating the mean free path to the diffusion coefficient for the condensing species (10).

The process of *evaporation* of a pure liquid droplet is similar to the process of condensational growth, except that it proceeds in the opposite direction. Evaporation occurs when the partial pressure of vapor is less than the saturated vapor pressure (ie, $p < p_0$), and is predicted by equation 10 with proper corrections. Evaporation is discussed in greater detail in a review (8).

Coagulation

Thermal *coagulation* is an aerosol growth process in which aerosol particles collide with each other as a result of Brownian motion, adhering to one another to produce larger particles. The net result is an increase in particle size and a decrease in number. In the absence of any loss or removal mechanisms, there is no change in mass concentration as resulting from coagulation.

The net effect of coagulation over some period of time is generally the quantity of interest, rather than the rate of coagulation. The change in particle size from coagulation over a period of time t is given by the following:

$$d_p(t) = d_{p0}(1 + N_0Kt)^{1/3} \qquad (13)$$

where d_{p0} and N_0 are the initial count median diameter and number concentration, respectively, and K is the coagulation coefficient (cm^3/s). For large particles with negligible slip correction, K is independent of particle size. However, as particle size decreases, the value of K increases because of the effect of d_p on C. For a monodisperse aerosol of particle size larger than the mean free path, the coagulation coefficient is expressed as

$$K = 4\pi d_p D = \frac{4kTC}{3\mu} \qquad d_p > \lambda \qquad (14)$$

where D is the particle diffusion coefficient, μ is the gas viscosity (g/cm/s), k is Boltzmann's constant (1.38×10^{-16} dyne · cm/K), and T is the temperature. For a polydisperse aerosol with a lognormal size distribution, a count mean diameter of CMD and a geometric standard deviation of σ_g, the average coagulation coefficient is given by:

$$K = \frac{2kT}{3\mu}\left[1 + e^{\ln^2\sigma_g} + \left(\frac{2.49\lambda}{CMD}\right)\left(e^{0.5\ln^2\sigma_g} + e^{2.5\ln^2\sigma_g}\right)\right] \qquad (15)$$

The average coagulation coefficient calculated by equation 15 assumes that there is only a modest change in *CMD* over the period of time t. For larger changes in particle size, equation 13 is used in a series of steps, each time with a constant but different value of K.

Chemical Relations

Another important class of gas-to-particle conversion mechanisms is heterogeneous *chemical reactions,* which involve both gas-phase and particle-phase constituents. Chemical reactions occur via surface reactions or droplet- or aqueous-phase reactions. An example of surface reactions is the uptake of nitric acid on the surface of calcium carbonate particles to produce calcium nitrate. A common aqueous-phase chemical reaction is the dissolution of sulfur dioxide into a hygroscopic particle or fog or cloud droplet, followed by oxidation of the dissolved sulfur dioxide to sulfate. For heterogeneous chemical reactions, the rate limiting step for growth is not the transport of the reacting species to the particle, but the rate of reaction on or within the particle. For reactions at the surface of the particle, the rate of growth is controlled by the particle surface area; for droplet-phase reactions, the rate of growth depends on the volume of the particle.

When material is added to the particle phase by condensation or by particle-phase chemical reactions, particles of different sizes grow at different rates, depending on the mechanism involved. Condensational growth has a different effect on the size distribution of aerosols than that of heterogeneous conversion through chemical reactions within a droplet. The relative rates at which the sizes of particles change depend on whether the rate-limiting step in the growth process is transport to the particle, chemical reactions at the surface of the particle, or chemical reactions within the particle. These are called transport-limited, surface-reaction rate-limited, or volume-reaction rate-limited, respectively. The different physical mechanisms give rise to a different form of the growth law for the particle.

DEPENDENCE OF AEROSOL PROPERTIES ON PARTICLE SIZE

Chemical Composition and Sources

Atmospheric aerosols are normally composed of mixtures of chemical compounds, which are distributed with respect to particle size depending on their sources and past history. As a result, different substances accumulate in different parts of the particle size distribution. Fine mode particles are composed primarily of varying proportions of several major components: sulfates (SO_4^{2-}), nitrates (NO_3^-), ammonium ions (NH_4^+), carbonaceous compounds (both elemental and organic carbon compounds), metal compounds, and water. The chemical composition of the fine particle fraction largely reflects its formation from gas-to-particle conversion involving organic and inorganic gases (eg, sulfur dioxide and nitrogen oxides). Coarse mode particles consist mainly of crustal constituents, such as Si, Al, Fe, and K. Biological material, such as bacteria, pollen, and spores are also found in the coarse mode. As a result of the fundamentally different chemical compositions and sources of fine and coarse fraction particles, the chemical composition of the sum of these two fractions (ie, PM_{10}) is more heterogeneous than either mode alone.

In most locations, a variety of diverse activities contribute significantly to particulate matter levels, including fuel combustion (from vehicles, power generation, and industrial facilities), residential fireplaces, agricultural burning, and atmospheric formation from gaseous precursors (largely produced from fuel combustion). Other sources include construction and demolition activities, traffic or agricultural operations, windblown dust, and resuspended road dust. The amounts of these components vary from place to place and over

Table 2. Chemical Composition and Sources of Fine and Coarse Mode Particles[a]

Chemical Component	Primary Sources		Secondary Sources	
	Natural	Anthropogenic	Natural	Anthropogenic
		Fine Mode Particles		
SO_4^{2-}	Sea spray	Fossil fuel combustion	Oxidation of reduced sulfur gases emitted by the oceans and wetlands and SO_2 and H_2S emitted by volcanism and forest fires	Oxidation of SO_2 emitted from fossil fuel combustion
NO_3^-		Motor vehicle exhaust	Oxidation of NO_x produced by soils, forest fires, and lighting	Oxidation of NO_x emitted from fossil fuel combustion and in motor vehicle exhaust
NH_4^+		Motor vehicle exhaust	Emissions of NH_3 from wild animals; undisturbed soil	Emission of NH_3 from animal husbandry, sewage, fertilized land
Crustal elements	Erosion; soil reentrainment	Fugitive dust; paved and unpaved roads; agriculture and forestry		
Organic carbon (OC)	Wild fires	Open burning; wood burning; cooking; motor vehicle exhaust; tire wear	Oxidation of hydrocarbons emitted by vegetation (terpenes, waxes); wild fires	Oxidation of hydrocarbons emitted by motor vehicles, open burning, wood burning
Elemental carbon (EC)	Wild fires	Motor vehicle exhaust; wood burning; cooking		
Metals	Volcanic activity	Fossil fuel combustion; smelting; brake wear		
Bioaerosols	Viruses; bacteria	–		
		Coarse Mode Particles		
Crustal elements	Erosion; soil reentrainment	Fugitive dust; paved and unpaved road dust; agriculture and forestry		
Metals	Erosion; soil reentrainment; organic debris	–		
Miscellaneous ions	Sea spray	Road salting		
Organic carbon		Tire and asphalt wear		
Organic debris	Plant and insect fragments	–		
Bioaerosols	Pollen; fungal spores; bacterial agglomerates			

[a] Ref. 1.

time. The chemical composition and sources of fine and coarse particles are summarized in Table 2.

In general, fine and coarse particles exhibit different degrees of solubility and acidity. With the exception of carbon and some organic compounds, fine particle mass is largely soluble in water and hygroscopic (ie, fine particles readily take up and retain water). The fine particle mode also contains the acidic fraction. By contrast, coarse particles are mostly insoluble, nonhygroscopic, and generally basic.

Removal Mechanisms

Fine accumulation mode particles typically have longer atmospheric lifetimes (ie, days to weeks) than coarse particles (minutes to hours). As discussed previously, secondary fine particles are formed by gas-to-particle conversion. Such transformation processes take place locally during atmospheric stagnation or during transport over long distances. Once formed, the low, dry, deposition velocities of fine particles contribute to their atmospheric persistence.

On the other hand, larger particles generally deposit more rapidly than small particles. Larger particles ($d_p > 10$ μm) rapidly fall out of the air by gravitational settling, and have atmospheric lifetimes of only minutes to hours. Such large particles are not readily transported across urban or broader areas because they are generally too large to follow air streams and they are easily removed by impaction on surfaces. The atmospheric behavior of smaller coarse mode particles is intermediate between that of the larger particles and smaller fine particles. Thus, coarse particles have lifetimes on the order of days and travel distances of up to 100 km or more.

Nevertheless, because fine particles remain suspended for longer times (typically on the order of days to weeks as opposed to days for coarse fraction particles) and travel much farther (ie, hundreds to thousands of kilometers) than coarse fraction particles (ie, tens to hundreds of kilometers), all else being equal, fine particles are theoretically likely to be more uniformly dispersed across urban and regional scales than coarse fraction particles. In contrast, coarse particles are be less evenly dispersed spatially around urban areas and exhibit more localized elevated concentrations near sources.

METHODS FOR MEASURING SIZE DISTRIBUTIONS

In general, it is difficult to obtain particle size information on the entire five-decade size range of 0.001 to 100 μm with a single instrument. For example, inertial measurement techniques (eg, cascade impactors) are inefficient for particle sizes below about 0.5 μm under normal conditions. Most aerosol sizing instruments effectively measure over a size range no larger than one and a half orders of magnitude. Thus, different instrumental techniques are used to measure different ranges of particle sizes. Combinations of sizing instruments are used to develop a more complete aerosol size distribution. These instruments are discussed in detail in other reviews (11,12). These instruments either use inertial measurement techniques or make direct readings. Inertial-measurement instruments include the cascade impactor, low-pressure impactor, and microorifice uniform deposit impactor (MOUDI). Direct-reading instruments include the optical particle counter, particle relaxation size analyzer, electrical aerosol analyzer, and diffusion battery. These instruments are discussed in detail in other reviews (11,12).

BIBLIOGRAPHY

1. United States Environmental Protection Agency, Air Quality Criteria for Particulate Matter, EPA/600/P-95/001aF-cF, Office of Research and Development, Washington, D.C., April 1996.
2. K. Willeke, and K. T. Whitby, "Atmospheric Aerosols: Size Distribution Interpretation," *JAPCA* **25**, 529–534 (1975).
3. K. T. Whitby "The Physical Characteristics of Sulfur Aerosols," *Atmos. Environ.* **12**, 135–159 (1978).
4. W. E. Wilson, L. L. Spiller, T. G. Ellestad, P. J. Lamothe, T. G. Dzubay, R. K. Stevens, E. S. Macias, R. A. Fletcher, J. D. Husar, R. B. Husar, K. T. Whitby, D. B. Kittelson, and B. K. Cantrell, "General Motors Sulfate Dispersion Experiment: Summary of EPA Measurements," *JAPCA* **27**:46–51 (1979).
5. R. A. Eldred, T. A. Cahill, and R. G. Flocchini, "Composition of $PM_{2.5}$ and PM_{10} Aerosols in the IMPROVE Network," *J. Air Waste Manage. Assoc.* **47**, 194–203 (1997).
6. J. C. Chow, J. G. Watson, Z. Lu, D. H. Lowenthal, C. A. Frazier, P. A. Solomon, R. H. Thuillier, and K. Magliano, "Descriptive Analysis of $PM_{2.5}$ and PM_{10} at Regionally Representative Locations During SJVAQS/AUSPEX," *Atmos. Environ.*, **30**(12):2079–2112 (1996).
7. United States Environmental Protection Agency (USEPA), "Revisions to the National Ambient Air Quality Standards for Particulate Matter," *Fed. Regist.* **52**(126), 24634–24663 (1987).
8. W. C. Hinds, "Physical and Chemical Changes in the Particulate Phase," in K. Willeke and P. A. Baron, eds., *Aerosol Measurement: Principles, Techniques, and Applications,* Van Nostrand Reinhold, New York, 1993, pp. 41–53.
9. S. K. Friedlander, *Smoke, Dust and Haze,* John Wiley & Sons, New York, 1977.
10. P. H. McMurry, and J. C. Wilson, "Growth Laws for the Formation of Secondary Ambient Aerosols: Implications for Chemical Conversion Mechanisms," *Atmos. Environ.* **16**(1), 121–134 (1982).
11. B. S. Cohen and S. V. Hering, eds., *Air Sampling Instruments,* 8th ed., American Conference of Governmental Industrial Hygienists, Inc., Cincinnati, Ohio, 1995.
12. K. Willeke and P. A. Baron, eds., *Aerosol Measurement: Principles, Techniques, and Applications,* Van Nostrand Reinhold, New York, 1993.

ATMOSPHERIC DISPERSION MODELING. See MODELING ATMOSPHERIC DISPERSION.

ATOMIC ABSORPTION SPECTROMETRY, HYDRIDE GENERATION

JIŘÍ DĚDINA
Academy of Sciences of the Czech Republic
Prague, Czech Republic

Hydride generation (HG) has been employed for over a 100 years for determining arsenic by the Marsh reaction or the Gutzeit test. The introduction of atomic absorption spectrometry (AAS) in laboratories throughout the world was incentive for a massive application of HG to AAS and also to other methods of analytical atomic spectrometry. Beginning around 1970, HG was used to overcome problems associated with flame AAS determinations of arsenic and selenium. In the following years, the unquestionable advantages of the method led to its application to virtually all elements forming volatile hydrides, namely arsenic, antimony, bismuth, germanium, lead, selenium, tellurium, and tin. Hydride generation of indium and thallium was even described. Most hydride-forming elements are of significant environmental importance. Due to instrumental and methodological progress in the last decade, this technique is widely applied to analyses of diverse environmental samples. The popularity of HG arises for several reasons. One is its relative simplicity and the low cost of the apparatus. However, the main reason lies with the principle of the method. It involves analyte preconcentration and separation from the sample matrix. This results in a superior sensitivity and a striking suppression of interference during atomization.

SCOPE

In principle, a sample can be introduced to an atomizer/detector in a condensed or a gaseous phase. Techniques for introducing samples in the gaseous phase are usually based on *chemical vapor generation,* a selective conversion of the analyte from the liquid sample to the gaseous phase via an appropriate chemical reaction resulting in a volatile compound of the analyte. Covalent hydrides are the most popular volatile compounds generated; this article is devoted exclusively to generation of hydrides. The process of determining volatile hydride-forming elements by any atomic spectrometric method involves three independent steps: (*1*) sample preparation, (*2*) hydride generation (performed in generators), and (*3*) atomization/detection. An exhaustive and critical treatment of the theory, instrumentation, methodology, and analytical applications of HG AAS (ie, all aspects of this article) are exhaustively covered until 1992–1993 in a recent monograph (1). Additional recent reviews cover applications of HG to AAS or to other atomic spectroscopy methods (2–5).

Besides hydrides, other useful *volatile compounds* can be tentatively classified into classes: chelates, alkyls, carbonyls, oxides, and halides. The subject has previously been reviewed (1,6). Among the other volatile compounds, alkyl generation has recently grown in significance, because it is a useful approach for determination and speciation mainly of tin, lead, mercury, cadmium, and selenium in environmental samples (1,6–11). A specific kind of chemical vapor generation is the *cold vapor technique,* which is a generation of monoatomic vapors of mercury (6,12–14) and cadmium (15–17).

In addition to chemical vapor generation, gaseous phase sampling occurs when atomic spectroscopy methods are employed as detectors for speciation of volatile metal compounds in various matrices. These applications are not covered here (for more information see Refs. 18–23). Also electrothermal vaporization (5,24,25), which can be regarded as a kind of gaseous-phase sampling, is not discussed here.

BACKGROUND

The principle of AAS and a basic classification and assessment of individual AAS techniques is treated elsewhere (see ELECTROTHERMAL ATOMIC ABSORPTION SPECTROMETRY).

One of the most important analytical characteristics is sensitivity. The term sensitivity invariably means the slope of the calibration graph. The dependence of peak height absorbance (A_p) or peak area (A_{int}) (integrated absorbance) either on analyte mass or on analyte concentration in the sample can be employed to construct the calibration graph. The performance of the whole analytical process is usually expressed as sensitivity defined by relating peak height absorbance to analyte concentration in the sample (A_p/c_0). The dimension is absorbance/(mass × volume^{-1}). The alternative is to define sensitivity as the ratio of either the peak height absorbance or integrated absorbance to the analyte mass in the sample (A_p/c_0V_s or A_{int}/c_0V_s). Corresponding dimensions are: absorbance/mass or (absorbance × time)/mass. Sensitivity in AAS is conveniently expressed as characteristic concentration or characteristic mass. For example, instead of sensitivity (A_p/c_0) the characteristic concentration (C_0), defined as analyte concentration corresponding to a peak height absorbance of 0.0044, can be used:

$$\frac{A_p}{c_0} = \frac{0.0044}{C_0}$$

Characteristic mass for peak height absorbance (m_p) and for integrated absorbance (m_0), respectively, are defined analogously: as analyte mass corresponding to peak height absorbance 0.0044 and to integrated absorbance 0.0044 s. In contrast to sensitivity, limit of detection is influenced by contaminations and/or by performance (noise) of the detector.

HYDRIDE GENERATION

Hydride generation consists of *hydride release* from the mineralized sample solution, that is, conversion of an analyte in sample to hydride and its transfer to the gaseous phase, and *transport of the released hydride* by a flow of the purge gas to an atomizer. The hydride release must be preceded by a proper *sample preparation.* See Table 1 for the list of analytically important hydride-forming elements, optimum conditions for their conversion to corresponding binary hydrides, and basic physical and spectroscopic constants of the hydrides and elements, respectively.

Sample Preparation

The purpose of sample preparation is to convert a sample into a solution without analyte losses and contamination. The biological or organic matrix should be completely decomposed. In reality, sample preparation is a crucial step, because it is prone to major volatilization losses, incomplete decomposition of the biological matrix (increased risk of matrix interferences), significant reagent blanks, and incomplete conversion of the analyte to the optimum valency (compound interferences) and so on. With the exception of speciation studies (see below), the analyte should finish up in the valency optimum for HG (see Table 1). In the case of antimony, arsenic, selenium, and tellurium, which are often present in the higher valency, a prereduction is required since the analytes in higher valency either are not converted to hydrides at all (selenium, tellurium) or the conversion may be less efficient (antimony, arsenic). Numerous procedures for sample preparation, often specific for a given matrix/analyte combination, and for prereduction can be found in the literature. There is a large variety of described procedures for prereduction, although the results of existing reports are not always consistent. In essence, the procedures for prereduction of As(V) and Sb(V) are rather similar. A widely used prereductant is potassium iodide. Also the prereduction procedures for Se(VI) and Te(VI) are mutually similar but different from those used for As(V) and Sb(V). It should be noted that selenium and tellurium species are reduced to their elemental forms (and therefore lost for HG) by most common prereductants for As(V) and Sb(V). A typical procedure for Se(VI) or Te(VI) consists of heating sample solutions with HCl (1 + 1) for 25–30 min on a boiling water bath at 95–100°C.

Hydride Release

Several *wet chemistry* reactions have been used to convert analyte present in the solution to hydride. Presently, reduction of analyte by tetrahydroborate in acidic (most often HCl) medium is employed almost universally for HG:

$$BH_4^- + 3\,H_2O + H^+ \rightarrow H_3BO_3 + 8\,H$$

$$A^{m+} + (m + n)H \rightarrow AH_n + mH^+$$

where m and n, respectively, are valence of the analyte A in the sample solution and in the hydride. The tetrahydroborate decomposition is complete within a few microseconds in an acidic medium. The resulting nascent hydrogen reduces analyte to hydride according to the above reaction. Sodium or (more rarely) potassium salt of tetrahydroborate is employed. Initial applications used pellets, but currently a water solution stabilized by potassium or sodium hydroxide is the most popular and most convenient agent. Tetrahydroborate concentrations vary by several orders of magnitude depending on the analyte element and on the type of hydride generator used. It should be underlined that even alkaline tetrahydroborate solutions slowly decompose (and hydrogen is formed); the decomposition is probably surface catalyzed and is slower in

Table 1. Volatile Hydride Forming Elements

Analyte	Optimum Valency	Optimum HCl Concentration	Hydride Formula	Hydride Name	Hydride Melting Point, °C	Hydride Boiling Point, °C	Hydride Solubility in Water, μg/mL	Useful Wavelengths, nm
Antimony	Sb(III)	pH9–9M	SbH_3	Stibine	−88	−18	4100	217.6, 231.1
Arsenic	As(III)	pH6–9M	AsH_3	Arsine	−117	−62	696	193.7, 197.2
Bismuth	Bi(III)	pH6–9M	BiH_3	Bismuthine	−67	17	-	222.825, 223.061, 306.8
Germanium	Ge(IV)	pH1–pH7	GeH_4	Germane	−165	−88	Insoluble	265.118[a], 265.158[a]
Lead	Pb(II)	pH1–pH2[b]	PbH_4	Plumbane	−135	−13	-	217.0, 283.3
Selenium	Se(IV)	0.2M–9M	SeH_2	Selenium hydride	−66	−41	38,000	196.0[a], 204.0[a], 206.3[a]
Tellurium	Te(IV)	0.2M–7M	TeH_2	Tellurium hydride	−51	−4	Very soluble	214.3[a], 225.9[a]
Tin	Sn(II), Sn(IV)	0.1M–1M	SnH_4	Stannane	−146	−52	-	224.6[a], 286.3[a], 235.5[a]

[a] These lines may have a relative population of their lower energy states markedly lower than unity at temperatures below 2200°C (see Ref. 1 for more complete spectral information).
[b] Reaction modifiers required (32).

more concentrated hydroxide solutions. Filtration, or preferably, membrane ultrafiltration is therefore advisable. A total of 0.05–2% m/V of sodium hydroxide is typically employed for stabilization. Large excesses of hydroxide could be detrimental because of the higher blanks entailed and because more acidic samples are thus required. Although reductant solutions are prepared fresh daily in many laboratories, it is not generally necessary. The stability is generally better in filtered and more alkaline solutions. The lifetime of the reductant could be extended considerably when stored under refrigeration or even frozen.

Technically, the reducing reaction is performed almost exclusively by mixing an excess of acidic sample volume with tetrahydroborate so that the reacted out mixture is still acidic. As discussed below, various chemical forms of individual analyte elements can be converted to hydrides. Table 1 presents the most convenient chemical forms of individual analyte elements to be converted to binary hydrides and a rough estimate of optimum HCl concentration for hydride release. The change of the reaction mixture acidity during the reaction should be taken into account. However, optimum conditions for hydride release involving the effect of various acids as reaction media, their concentration, sometimes presence of reaction controllers (such as peroxodisulfate for lead) or buffers for tin or germanium, depend, beside analyte identity and chemical form, also on the method of HG employed.

There is always an excess, usually of many orders of magnitude, of the reducing agent over the analyte under typical conditions of HG. Resulting hydrogen obviously drives the hydride from the reaction mixture to the gaseous phase, because the concentration of hydride in the reaction mixture is typically much lower than the hydride solubility (see Table 1). This is the additional function of the reducing agent. In some designs of hydride generators (see below) the purge gas is mixed with or bubbled through the reaction mixture. This supports the release of reduced hydride from the solution and can partially reduce the demand of the reducing agent. Under the optimum conditions and in the absence of a matrix, the hydride release efficiency approaches unity.

Every hydride generator can be characterized by two parameters: (*1*) a hydride supply rate that is the number of analyte atoms delivered to the atomizer in the form of hydride per unit time and (*2*) a total gas flow rate.

Hydride Transport

The hydride released from sample solution is transported by a flow of the purge gas to an atomizer. Purge gas may also assist in hydride release by stripping the hydride out of the sample solution. A large number of gasses satisfies the essential requirement on the purge gas: that it must be inert towards the hydride. The purge gas is chosen with respect to optimum economy and atomizer performance. Only in the case of cryogenic trapping it is not convenient to employ a gas having the boiling point above the trapping bath temperature. The most popular purge gas is argon; nitrogen is utilized often, too. Helium and hydrogen have been used as well. Principally, hydrogen released during HG can be used as the purge gas as well. Lower dilution of analyte thus contributes to enhanced sensitivity.

Losses of hydride on glass or plastic surfaces during its transport can be serious. Losses or delay of transported hydrides are probably due to their decomposition or sorption. Water traces, which are typically present in hydride generators, can also enhance the losses. For example, solubility of selenium hydride in solutions at a pH of 2–3 is very high due to its acid dissociation. These effects can reduce hydride transport efficiency (i.e., the observed sensitivity) but mainly deteriorate reproducibility of measurements. An efficient way to prevent losses of hydrides on glass surfaces is their silanization to block the surface-active sites.

In conclusion, hydride transport losses can be made negligible in an optimized experimental setup. The optimization includes the use of a reasonably high gas flow rate, minimization of surfaces coming into contact with transported hydride, and, if necessary, silanization of glass surfaces.

The tubing serving for the transport of hydrides to the atomizer should be as short as possible to avoid the transport losses. The interaction of hydride with the tubing surface is obviously reduced in narrower tubings; however, tubing that is too narrow increases the risk of overpressure problems. A certain amount of the reaction mixture is sprayed to the gaseous phase and transported together with generated hydride. Condensation of the reagent mist and/or water vapor in the transfer line should be avoided because of the above-mentioned losses or delay of hydrides. The spray can also negatively influence performance of hydride atomizers, namely quartz-tube atomizers. Various desiccants have thus been employed, such as sulfuric acid, calcium chloride, silica gel, calcium sulfate (Drierite), magnesium perchlorate, or a glass or quartz tube cooled to temperatures below zero down to −80°C. Welz and Schubert-Jakobs (26) successfully eliminated the spray carry over by installing a liquid trap on the gas outlet from the hydride generator. The trap was made of GoreTex membrane tubing fitted to the outlet, closed at the other end, and surrounded by a silicone tubing. Liquid particles were retained, and the gas phase could penetrate through the wall of the tubing quantitatively so that no loss of sensitivity was observed (26). A different possibility is to dry the gases by a hygroscopic-membrane dryer tube. This device might be a very useful component of hydride generators; however, it has to be extensively tested whether there are not hydride losses in the membrane. Another means to avoid carryover of liquid drops to the transfer line is to employ membrane gas–liquid separators.

Methods of Hydride Generation

There are two basic modes of HG (Fig. 1). In the direct-transfer mode, hydride released from a sample solution is directly transported to an atomizer. Three direct-transfer methods are currently employed: continuous flow (CF), flow injection (FI) and batch. In the collection mode, the hydride is trapped in a collection device until the evolution is completed and then is transported to an atomizer all at once. Collection methods were employed much more frequently in early years of the application of HG when relatively slow approaches to hydride release were employed. Obviously, it was advantageous to collect the released hydride and then to sweep it into the atomizer in the shortest possible time. It should be underlined that, when using the present BH_4^{-1}/acid system as the reducing agent, the need to collect released hydride is much less pressing, so that the simpler direct-transfer methods are more often utilized now. A collection generator consists of a reaction vessel, which is most often actually a batch generator (but a CF generator can be employed as well), and of a collection device. After hydride evolution from the reaction vessel has been finished, the collected hydride is either released from the collection device and transported to the atomizer on line or it can be stored and atomized off line. The collection may be done either by bubbling the hydride released from the reaction mixture into an absorbing solution or the hydride can be collected in a closed vessel under pressure or in a cold trap. Methods employing *in situ* trapping of hydride in a graphite furnace are not classified in our description among collection methods, since the hydride leaves the generator uncollected. The *in situ* trapping is therefore considered as a first stage of the atomization and is treated among atomization methods.

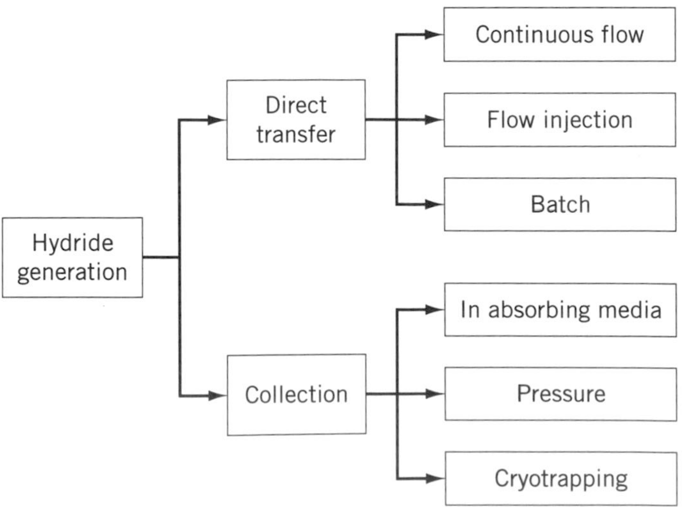

Figure 1. Methods of hydride generation. Reproduced with permission (1). Copyright John Wiley & Sons, Ltd.

HYDRIDE ATOMIZATION

In a hydride atomizer, a distribution of free analyte atoms is formed that is reflected in the absorbance value. It is important to note that the chemical composition of the system in current hydride atomizers is usually far from thermodynamical equilibrium: Free analyte atoms are thermodynamically forbidden at conditions prevailing in most hydride atomizers. For example, there are no selenium free atoms under equilibrium conditions at temperatures below 1300°C, and a temperature above 2200°C is required to expect all selenium to be in the free atom state. Arsenic and antimony behave similarly to selenium. Bismuth, lead, tellurium, and tin are present as free atoms, under equilibrium conditions, at temperatures above 900°C, 1100°C, 1300°C, and 1900°C, respectively. Consequently, the hydride atomization is typically not a thermal dissociation (for a detailed discussion, see Ref. 1). The following atomizers are presently in use for AAS: (*1*) inert gas-hydrogen diffusion flames, (*2*) graphite furnaces, and (*3*) quartz-tube atomizers.

INTERFERENCES

Two classes of interferences are observed in AAS: spectral and nonspectral. Spectral interferences are due to radiation absorbed by other species than free atoms of the analyte. Nonspectral interferences are due to the influence of sample constituents on the analyte signal.

Because of the separation of the analyte from the matrix, spectral interferences in HG AAS are much less serious than for liquid sampling AAS (see ELECTROTHERMAL ATOMIC ABSORPTION SPECTROMETRY). Line interferences do not occur at all; nonselective background absorption can be observed in the presence of an excess of other hydride forming elements in the sample matrix. Changes in gas flow composition or rate from hydride generators of the batch or collection type may also alter transparency (ie, background absorption of diffusion flames). Additionally, oxygen produces background absorption at low wavelengths. This has been shown to be from the structured absorbance by molecular oxygen in this region. The observed absorbance increases with temperature. Generally, background absorption takes place in flame atomizers and in graphite furnaces rather than in externally heated quartz tubes. Spectral interferences are additive in nature, that is, the relative influence of the interferent on the signal depends on analyte concentration. The method of standard additions obviously cannot help.

Nonspectral interferences may principally occur either in the liquid phase during hydride formation and its transfer from solution (liquid phase interferences) or they can affect the analyte in the gaseous phase (gaseous phase interferences). Liquid-phase interferences are changes in the rate of hydride release from the liquid phase (kinetics interferences) and/or by a decreased efficiency of hydride release caused by interfering species present in sample solutions (efficiency interferences). Gaseous phase interferences are obviously caused by volatile species, most often by hydrides but also by other compounds, or by liquid spray, produced in hydride generator. These interferences can take place on the surface or in the dead volume of generator, the connective tubing and/or the atomizer.

With the exception of compound interferences, species responsible for nonspectral interferences are usually present in great excess to the analyte. Then the extent of the interference should not depend on the analyte concentration but only on the interferent concentration. A consequence of the independence of interference magnitude on analyte concentration is that the method of standard additions may be, in principle, employed to alleviate the interferences; however, it is usable only when the interference does not reduce the observed signal too much. The other generally applicable method is to dilute the sample. However, sensitivity is reduced accordingly when employing the direct transfer HG methods. This approach is obviously possible only for samples with higher analyte content or when employing collection methods of HG or *in situ* trapping in graphite furnaces. The extent of feasible dilution is thus limited mainly by the analyte blank in reagents.

Naturally, the risk of an interference is enhanced if the efficiency of the process where the interference takes place is low. For example, a relatively slight change in sample matrix composition is probable to change the efficiency of hydride release (ie, to cause the interference) when the generation conditions are poorly optimized so that the hydride release is less than 100% even in standard solutions. Therefore, optimum experimental conditions should be observed.

SPECIATION

Element speciation analysis of environmental samples consists of identifying and quantifying the various forms of an inorganic analyte. The most sensitive speciation techniques utilize separation of the various analyte forms and their subsequent detection by an atomic spectrometry method. In principle, there are three approaches to using HG for speciation analysis; all of them are characterized briefly in subsequent sections. Detailed information on applications of HG for speciation analysis (for all atomic spectrometry detectors) can be found (1,2,5,10,27–31).

APPLICATIONS

The popularity of HG can be illustrated by the many application articles on individual hydride-forming elements that appeared in the literature prior to 1996 (over 400 on antimony almost 1000 on arsenic; around 250 on bismuth; over 100 on germanium; around 150 on lead; over 700 on selenium; around 150 on tellurium; over 300 on tin). The share of AAS in these papers for individual elements is approximately two thirds. The rest of the papers are devoted to other atomic spectroscopy methods—such as emission spectrometry employing plasma excitation (most often inductively coupled plasma, direct current plasma and microwave induced plasma), atomic fluorescence spectrometry employing either plasmas or flame for atomization, and inductively coupled plasma mass spectrometry.

Environmental applications of HG for AAS and also for other methods of analytical atomic spectrometry published until 1992/1993 have been exhaustively discussed (1). Since then, new articles on environmental applications of HG appear roughly at a rate of 100 per year. Some of these are covered in review articles devoted to determination of antimony (29), arsenic (27), lead (32) and selenium (28) and to various aspects of HG (2–5,30,31,33–37), (see ELECTROTHERMAL ATOMIC ABSORPTION SPECTROMETRY).

GENERAL ASSESSMENT OF HYDRIDE GENERATION AND ATOMIZATION METHODS

Accuracy and sensitivity are the main criteria to assess hydride generators and atomizers. There are also other criteria such as difficulty, speed of analysis, and cost of the instrumentation. Although the optimum choice for a given laboratory is always a compromise, it is useful to rank HG and atomization methods according to the individual criteria.

Accuracy

The accuracy of an analytical process must be controlled by employing standard analytical procedures (independent parallel determinations, standard reference materials, etc). However, the individual HG and atomization methods offer various resistance to interferences that can critically influence the accuracy. Obviously, the freedom to control matrix interferences and atomization interferences, respectively, is given by the choice of the method of HG and hydride atomization. As discussed above, the extent of atomization interferences is much lower in graphite furnaces or in diffusion flames than in quartz-tube atomizers. Regarding HG methods, direct introduction as well as collection generation modes are performed either in a batch or in a flow (CF or FI) arrangement. The batch arrangement is generally more susceptible to matrix interferences in the liquid phase than both flow arrangements, which offer generally the same freedom to control the matrix interferences in the liquid phase. Another disadvantage of batch generation performed in the direct introduction mode is that the kinetics interferences influence the peak height. Collection methods of HG and *in situ* trapping allow suppression of the matrix interferences by the sample dilution. The extent of feasible dilution is limited mainly by the analyte blank in the reagents.

Sensitivity

Sensitivity of the whole analytical process is usually expressed as peak-height absorbance related to analyte concentration in the sample. The peak-height absorbance depends on atomizer parameters, that is, on its length, L, temperature, T_a, and on the integral atomization efficiency, B_a, and on generator performance, that is, on the peak concentration of hydride in the gas delivered to the atomizer, C_{max}:

$$A_{peak} \propto \frac{LB_aC_{max}}{T_a} \qquad (1)$$

This formula applies for atomization in quartz-tube atomizers and for on-line atomization in graphite furnaces. Thus it can be used as a general guide to optimize the experimental arrangement for reaching maximum sensitivity when employing the above-mentioned atomizers.

The optimum choice of hydride generator and atomizer for reaching maximum sensitivity depends also on the sample volume at disposal. Two extreme cases can be considered: microanalysis when only a very small sample volume is available, and trace analysis characterized by sufficient volume sample with a very low analyte concentration. The sensitivity in the case of trace analysis is typically expressed as (peak-height) absorbance related to analyte concentration in the sample, whereas in microanalysis the analyte mass is the relevant criterion.

Microanalysis. In the case of microanalysis, quartz-tube atomizers should be recommended for the atomization step since they are more sensitive than flames or graphite furnaces employed in the on-line atomization mode (*in situ* collection does not offer any advantage for microanalysis). Hydride generation modes to be considered are batch and FI, because collection methods and the CF generation do not make sense for small sample volumes. According to the treatment above, fully optimized FI generation provides inferior sensitivity compared to fully optimized batch generation. The sensitivity reachable by FI is that reachable by CF mode reduced by the actual dispersion, that is, its sensitivity could be made equal to that of the batch mode only if sufficiently high sample volumes were injected. However, high sample volumes are not typical for microanalysis. This suggests batch generation as the method of choice. It should be underlined that this is valid only for generators optimized to provide maximum hydride concentration in the gas leaving the generator (1). The batch generators that are used typically are not optimized in this respect for small sample volumes so that a typical FI generator will perform better for microanalysis than a standard batch generator. To take full advantage of the batch generation, the generator design should be optimized for small sample volumes: the generator and connecting tubing should be miniaturized and tetrahydroborate mass required to release hydride from the sample should be added all at once. Anyway, HG, even optimized, cannot principally offer substantially better sensitivity for microanalysis than the classic electrothermal AAS employing the liquid sampling.

Trace Analysis. The method of choice for trace analysis is *in situ* trapping in a graphite furnace or cryogenic trapping followed by atomization in a quartz-tube atomizer. These methods make possible an efficient collection of analyte from large sample volumes, as discussed in the corresponding sections above. When employing other approaches to HG and/or atomization, equation 1 can be used to assess their potential to reach maximum sensitivity. For the atomization step, a maximum value of $(LB_a)/T_a$, and consequently the maximum sensitivity, is provided by quartz-tube atomizers. For the hydride generation step, pressure collection, batch modes, and CF modes, when optimized, reach the same C_{max} and consequently the same sensitivity. The FI mode can reach the same sensitivity if the sample injection volume is large enough to reach the steady-state hydride concentration in the gas delivered to the atomizer (dispersion coefficient equals unity). This is valid for generators optimized to provide maximum hydride concentration in the gas leaving the generator (1). The CF generators that are typically used are not optimized in this respect so that a typical batch generator will provide better sensitivity for trace analysis than a standard CF generator.

Limits of Detection

As shown above, the observed sensitivity is critically influenced by the method and by actual experimental parameters of hydride generation and atomization. Taking into account the whole analytical process, the sample preparation step is critical as well, because of the analyte dilution, the extent of which is given by a matrix and by the mineralization procedure adopted. However, the relevant parameter for analysis of real samples is not the sensitivity but the limit of detection. Besides sensitivity, it is usually controlled by contaminations arising in the sample preparation step but also in the process of HG. This is obviously the reason for wide divergences of limits of detection reported in literature—the values are often within two to three orders of magnitude even for the same sample, analyte, method of HG, and detector. This is illustrated in Table 2 on the example of *in situ* trapping in graphite furnaces for AAS. The table also shows trends in the limits of detection of various hydride-forming elements for the most frequently used HG methods (CF and FI) combined with the most frequently used detection approaches—that is, atomization in quartz-tube atomizers for AAS, inductively coupled plasma atomic emission spectrometry, and inductively coupled plasma mass spectrometry.

Other Criteria

Degree of Difficulty. Laboriousness is significantly influenced by the choice of HG mode. The methods of choice are CF and FI modes, because they can be easily automatized. Automation of *in situ* trapping in a graphite furnace is also possible. The batch mode and collection methods of HG require, in principle, cumbersome manual operation.

Speed of Analysis. With the exception of inherently slow *in situ* trapping in a graphite furnace, the speed of analysis is controlled by the HG mode employed. The FI mode clearly allows highest sample throughput—approximately 180 samples per hour which is six times more than for the batch mode (38). The CF mode is slower because of the time required to achieve the steady-state concentration of hydride in the gas leaving the generator. Collection methods are the most time consuming.

Cost. The cost of the instrumentation is influenced by the choice of both the generator and the atomizer. The advantage of using the batch mode is that it is done with very simple equipment. Because they usually employ the FI valve, FI generators

Table 2. Limits of Detection, pg/mL, for Flow Modes of HG

Analyte	*In situ* Trapping[a]	QT[b] AAS[c]	ICP/AES[c,d]	ICP/MS[c,e]
Antimony	2–1000	80	100	0.3
Arsenic	3–2000	11	32	1.5
Bismuth	0.1–500	20	200	0.7
Germanium	3–18	500,000	150, 63[f]	12[g]
Lead	440–4000, 250[h]	100	100	10
Selenium	1–120	20	55	1.4
Tellurium	40–80	100	1000	1
Tin	2–10	30	30	210

[a] Range of values reported in the literature (3); also batch methods of HG taken into account.
[b] Quartz-tube atomizers
[c] The lowest values reported in the literature (1).
[d] Inductively coupled plasma atomic emission spectrometry (qv).
[e] Inductively coupled plasma mass spectrometry (qv).
[f] See Ref. 37.
[g] See Ref. 39; the value of 0.01 pg/mL reported for cryogenic trapping (40).
[h] See Ref. 41.

are more complicated than CF ones. The cost of hydride generators naturally depends on the level of sophistication. Quartz-tube atomizers can be made relatively cheaply compared to flame atomizers or even graphite furnaces. However, the burners for diffusion flames and graphite furnaces often belong to standard equipment of atomic absorption spectrometers so that they do not require additional expenses.

Conclusions

For HG, the CF and FI modes are very convenient methods. They are simple to operate, easy to automate, accurate, and reasonably sensitive. The only disadvantage of the FI mode compared to CF is its lower sensitivity and slightly more complicated apparatus. However, it can be converted to the CF mode simply by increasing the sample coil volume, and it has lower sample consumption and higher sample throughput. The only significant advantage of the batch mode is the simplest equipment. The pressure collection mode is inferior, because it is complicated and does not provide better sensitivity than batch or CF modes. Cryogenic trapping provides very high sensitivity; however, it is time consuming and much labor is involved. *In situ* trapping in graphite furnaces could substitute for cryogenic trapping, because it offers at least the same sensitivity and it is simpler and can be automated. Its only disadvantage is that, in contrast to cryogenic trapping, it does not allow speciation of individual volatile analyte species.

Currently the most often employed hydride atomizers are quartz-tube atomizers. They provide higher sensitivity than flames or on-line atomization in graphite furnaces. Their inherent disadvantage is the adverse influence of their inner surface on atomizer performance and relatively pronounced atomization interferences. Diffusion flame atomizers in the current design provide much lower sensitivity; however, they are much more resistant against atomization interferences. The weak point of on-line atomization in graphite furnaces is the interface between the hydride generator and the atomizer. On the other hand, sensitivity is not much lower than in quartz-tube atomizers and atomization interferences are strongly reduced. *In situ* trapping in graphite furnaces is a very promising approach that offers high sensitivity and resistance to atomization interferences.

BIBLIOGRAPHY

1. J. Dědina and D. L. Tsalev, *Hydride Generation Atomic Absorption Spectrometry,* John Wiley & Sons, Ltd., Chichester, U.K., 1995.
2. A. G. Howard, *J. Anal. Atom. Spectrom.,* **12,** 267–272 (1997).
3. H. Matusiewicz and R. E. Sturgeon, *Spectrochim. Acta B,* **51,** 377–397 (1996).
4. T. Nakahara, *Advanc. Atom. Spectrom.,* **2,** 139–178 (1995).
5. L. K. Olson, N. P. Vela, and J. A. Caruso, *Spectrochim. Acta B,* **50,** 355–368 (1995).
6. W. C. Story and J. A. Caruso, in Z. B. Alfassi, C. M. Wai, eds., *Preconcentration Techniques for Trace Elements,* CRC Press, Boca Raton, Fla., 1992, pp. 333–362.
7. S. Rapsomanikis, *Analyst,* **119,** 1429–1439 (1994).
8. W. M. R. Dirkx, R. Lobinski and F. C. Adams, *Anal. Chim. Acta,* **286,** 309–318 (1994).
9. R. Lobinski, W. M. R. Dirkx, J. Szpunar-Lobinska, and F. C. Adams, *Anal. Chim. Acta,* **286,** 381–390 (1994).
10. R. M. Olivas, O. F. X. Donard, C. Camara, and P. Quevauviller, *Anal. Chim. Acta,* **286,** 357–370 (1994).
11. P. B. Stockwell and W. T. Corns, *Analyst,* **119,** 1641–1645 (1994).
12. D. L. Tsalev, *Atomic Absorption Spectrometry in Occupational and Environmental Health Practice, Vol. II: Determination of Individual Elements,* CRC Press, Boca Raton, Fla., 1984.
13. D. L. Tsalev and Z. K. Zaprianov, *Atomic Absorption Spectrometry in Occupational and Environmental Health Practice, Vol. I: Analytical Aspects and Health Significance,* CRC Press, Boca Raton, Fla., 1983.
14. D. L. Tsalev, *Atomic Absorption Spectrometry in Occupational and Environmental Health Practice, Vol. III: Progress in Analytical Methodology,* CRC Press, Boca Raton, Fla., 1995.
15. P. Bermejo-Barrera, J. Moreda-Pineiro, A. Moreda-Pineiro, and A. Bermejo-Barrera, *J. Anal. Atom. Spectrom.,* **11,** 1081–1086 (1996).
16. X. W. Guo and X. M. Guo, *J. Anal. Atom. Spectrom.,* **10,** 987–991 (1995).
17. A. Sanz-Medel, M. C. Temprano, N. B. Garcia, and M. R. F. de la Campa, *Anal. Chem.,* **67,** 2216–2223 (1995).
18. R. M. Harrison, S. Rapsomanikis, eds., *Environmental Analysis Using Chromatography Interfaced with Atomic Spectroscopy,* Ellis Horwood, Chichester, U.K., 1989.

19. L. Ebdon, S. J. Hill, and R. W. Ward, *Analyst,* **111,** 1113–1138 (1986).
20. F. J. Fernandez, *At. Abs. Newslett.,* **16,** 33–35 (1977).
21. F. J. Fernandez, *Chromatogr. Newslett.,* **5,** 17–21 (1977).
22. P. C. Uden, in J. Sneddon, ed., *Sample Introduction in Atomic Spectrometry,* Elsevier, Amsterdam, 1990, pp. 195–252.
23. P. C. Uden, in R. M. Harrison and S. Rapsomanikis, eds., *Environmental Analysis Using Chromatography Interfaced with Atomic Spectroscopy,* Ellis Horwood, Chichester, U.K., 1989, pp. 96–126.
24. K. C. Ng and J. A. Caruso, in J. Sneddon, ed., *Sample Introduction in Atomic Spectrometry,* Elsevier, Amsterdam, 1990, pp. 165–193.
25. J. M. Carey and J. A. Caruso, *Crit. Rev. Anal. Chem.,* **23,** 397–439 (1992).
26. B. Welz and M. Schubert-Jacobs, *Atom. Spectrosc.,* **12,** 91–104 (1991).
27. B. Amran, F. Lagarde, M. J. F. Leroy, A. Lamotte, C. Demesmay, M. Olle, M. Albert, G. Rauret, and J. F. Lopezsanchez, in P. Quevauviller, E. A. Maier, and B. Griepnik, eds., *Quality Assurance for Environmental Analysis,* Elsevier, Amsterdam, 1995, pp. 285–304.
28. C. Camara, M. G. Cobo, M. A. Palacios, R. Munoz, and O. F. X. Donard, in P. Quevauviller, E. A. Maier, and B. Griepnik, eds., *Quality Assurance for Environmental Analysis,* Elsevier, Amsterdam, 1995, pp. 235–262.
29. M. B. Delacalleguntinas, Y. Madrid, and C. Camara, in P. Quevauviller, E. A. Maier, B. Griepnik, eds., *Quality Assurance for Environmental Analysis,* Elsevier, Amsterdam, 1995, pp. 263–283.
30. R. Ritsema, F. M. Martin, and P. Quevauviller, in P. Quevauviller, E. A. Maier, and B. Griepnik, eds., *Quality Assurance for Environmental Analysis,* Elsevier, Amsterdam, 1995, pp. 489–503.
31. K. Pyrzynska, *Chem. Analyticzna,* **40,** 677–686 (1995).
32. Y. Madrid and C. Camara, *Analyst,* **119,** 1647–1658 (1994).
33. M. R. F. de la Campa, E. S. Garcia, M. C. Temprano, and B. A. Fernandez, *Spectrochim. Acta B,* **50,** 377–391 (1995).
34. Q. Deren, *Trends Anal. Chem.,* **14,** 76–82 (1995).
35. J. D. Hwang and W. J. Wang, *Appl. Spectrosc. Rev.,* **30,** 231–350 (1995).
36. B. L. Huang, *Can. J. Appl. Spectrosc.,* 39, 117–122 (1994).
37. T. Nakahara and T. Wasa, *Microchem. J.,* 49, 202–212 (1994).
38. T. Inui, S. Terada, H. Tamura, and N. Ichinose, *Fresenius' J. Anal. Chem.,* **318,** 502–504 (1984).
39. L. S. Zhang and S. M. Combs, *J. Anal. Atom. Spectrom.,* **11,** 1043–1048 (1996).
40. K. Jin, Y. Shibata, and M. Morita, *Anal. Chem.,* **63,** 986–989 (1991).
41. H. O. Haug, *Spectrochim. Acta B,* **51,** 1425–1433 (1996).

AUTOMOBILE EMISSIONS, ON-ROAD

GARY A. BISHOP
DONALD H. STEDMAN
University of Denver
Denver, Colorado

Picture for a moment any major industrialized city in the world on a hot summer day. Perhaps Los Angeles, Mexico City, or London come to mind. Few residents can ignore the ever-present signs of humankind that linger in the air, many of which are directly attributable to the internal combustion engine and the automobile. Carbon monoxide (CO), hydrocarbons (HC), nitrogen oxides (NO_x), fine particles, and lead are some of the compounds emitted by automobiles that foul the air humans breathe. See Plate 2.

In the United States, air pollution control measures to mitigate mobile source emissions in nonattainment areas (as defined by the Clean Air Act) include inspection and maintenance (I/M) programs, oxygenated fuel mandates, and transportation control measures. Nonetheless, many areas remain in nonattainment past the 1987 deadline for compliance with federal standards, and some are projected to remain in nonattainment for many more years, despite the measures currently undertaken. A major emphasis in controlling mobile source emissions has centered around periodic testing for identifying vehicles with excessive emissions.

Current federal and state governments employ a variety of testing methods and protocols for certifying new and in-use vehicle emissions. These range from the extremely comprehensive new car certification test (the Federal Test Procedure), to which all newly manufactured vehicles sold in the United States must comply, to the more common one- or two-speed idle measurements performed in many state I/M programs (1–4). The former requires at least 12 h to complete and costs in excess of $700. The later requires less than 15 min and costs around $10. Both tests are designed as surrogates for evaluating the emissions of vehicles under actual operating conditions. The limitations of each type of testing have been hotly debated. A review of the Federal Motor Vehicle Control Program (5) outlined the following criteria for an ideal test: it should evaluate the vehicle under real-life conditions; be reproducible, accurate, quick, and inexpensive; and measure all pollutants of concern.

In 1987, with support from the Colorado Office of Energy Conservation, the University of Denver developed an infrared (IR) remote monitoring system for automobile carbon monoxide (CO) exhaust emissions (6). Significant fuel economy improvements result if rich-burning (high CO and HC emissions) or misfiring (high HC emissions) vehicles are tuned to a more stoichiometric and more efficient air:fuel (A:F) ratio. Therefore, the University of Denver CO/HC remote sensor is named Fuel Efficiency Automobile Test (FEAT). The basic instrument measures the carbon monoxide:carbon dioxide ratio ($CO:CO_2$) and the hydrocarbon:carbon dioxide ratio ($HC:CO_2$) in the exhaust of any vehicle passing through an IR light beam that is transmitted across a single lane of roadway. Figure 1 shows a schematic diagram of the instrument (7).

THEORY OF OPERATION

The FEAT instrument was designed to emulate the results that would be obtained using a conventional nondispersive infrared (NDIR) exhaust gas analyzer; thus FEAT is also based on NDIR principles. An IR source sends a horizontal beam of radiation across a single traffic lane, approximately 25.4 cm (10 in.) above the road surface. This beam is directed into the detector on the opposite side of the road and is divided between four individual detectors: CO, CO_2, HC, and reference. An optical filter that transmits infrared light of a wavelength known to be uniquely absorbed by the molecule of interest is placed in front of each detector, determining its specificity. Reduction in the signal caused by absorption of light by the molecules of interest reduces the voltage output. One way of conceptualizing the instrument is to imagine a typical garage-type NDIR instrument in which the separation of the IR source

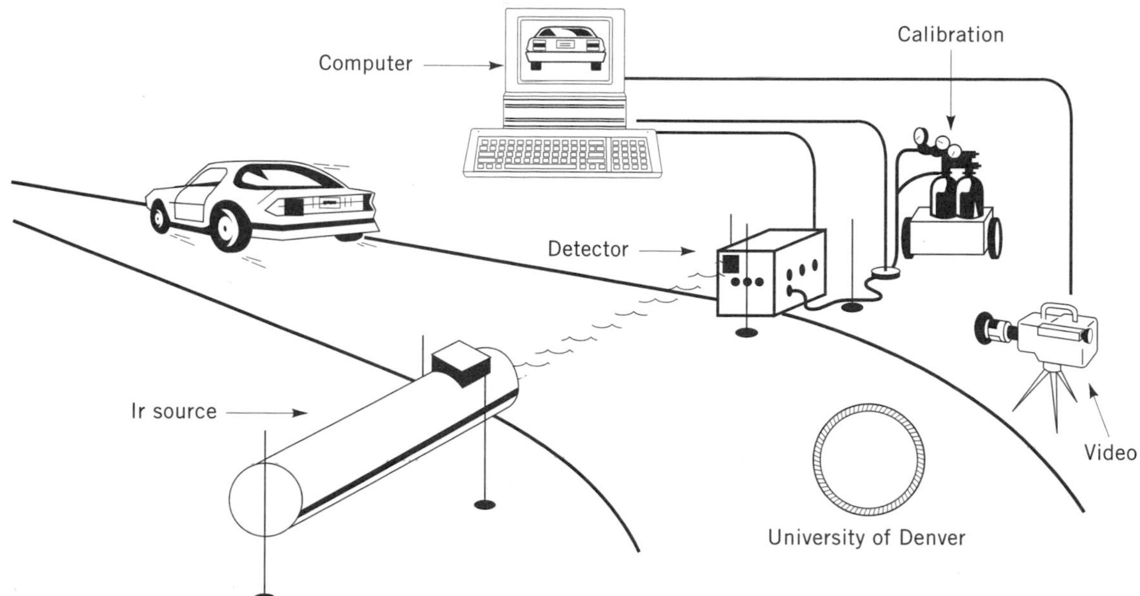

Figure 1. A schematic diagram of the University of Denver on-road emissions monitor, capable of monitoring emissions at vehicle speeds between 2.5 and 150 mph in under 1 s per vehicle.

and detector is increased from 8 cm to 6–12 m (20–40 ft). Instead of pumping exhaust gas through a flow cell, a car now drives between the source and the detector.

Because the effective plume path length and amount of plume seen depend on turbulence and wind, FEAT can directly measure only ratios of $CO:CO_2$ and $HC:CO_2$. These ratios are constant for a given exhaust plume and by themselves are useful parameters to describe the combustion system. However, with a fundamental knowledge of combustion chemistry, it is possible to determine many other parameters of the vehicle's operating characteristics from these ratios, including the instantaneous air:fuel ratio, grams of CO or HC emitted per liter of gasoline (g CO/L or g HC/L) burned, and the percent of CO or HC in the exhaust gas. Most vehicles show ratios of zero, since they emit little to no CO and HC. To produce a $CO:CO_2$ ratio greater than zero, the engine must have a fuel-rich air:fuel ratio and the emission control system, if present, must not be fully operational. A high $HC:CO_2$ ratio can be associated with either a fuel-rich or a fuel-lean air:fuel ratio coupled with a missing or malfunctioning emission control system. A lean air:fuel ratio, while impairing driveability, does not produce CO in the engine. If the air:fuel ratio is lean enough to induce misfire, then a large amount of unburned fuel is present in the exhaust manifold. If the catalyst is absent or nonfunctional, then high HC will be observed in the exhaust without the presence of high CO. To the extent that the exhaust system of the misfiring vehicle contains some residual catalytic activity, the HC may be partially or totally converted to a $CO:CO_2$ mixture.

REAL-WORLD VEHICLE EMISSION CHARACTERISTICS

Through the use of FEAT, it is now possible to collect data quickly and easily on a large number of vehicles. This enables the study of important questions concerning the automobile's contribution to urban air quality and what mitigation actions might be taken.

Not all cars have equal emissions. Data from the measurements show that a small fraction of the passing vehicles are responsible for half or more of the emissions in any given area. In Denver, half the emissions come from only 7% of the vehicles. In Kathmandu, half the emissions come from 25% of the vehicles. Figure 2 shows this for data collected at a single site in the Los Angeles area. At this location, half of the CO is emitted by 7% of the vehicles and half of the HC is emitted by 11%. The few vehicles emitting half of the CO and HC are referred to as "gross polluters." For automobile emissions, the adage that the Tail Wags the Dog holds true.

The overall characteristics of these fleets are similar regardless of age, location, or the presence of I/M programs and can be mathematically described by a γ-distribution (16). Most vehicles show mean emissions of 1% CO and 0.1% HC (as propane) or less in the exhaust. The newer the fleet, the more skewed the emissions. This is because more of the vehicles have near zero emissions, and thus a smaller number of gross polluters dominate the total emissions. Table 1 compares the results for many of the locations that have been sampled (8–15).

The good news is that for the U.S. fleet, 50% of the vehicles produce only 4% of the CO emissions and 16% of HC using current gasoline formulas as fuel. This shows that alternative and reformulated fuels are not likely to solve the problem, which apparently arises due to a lack of maintenance.

Not all gross polluters are old vehicles (only about 25% of pre-1975 vehicles in the United States). In fact, the majority of even precatalyst vehicles are relatively low emitting. There is a strong correlation between fleet age and fleet emissions, as shown in Figure 3**a**. However, this correlation has less to do with emissions-control technology than it does with vehicle maintenance. Any well-maintained vehicle regardless of age

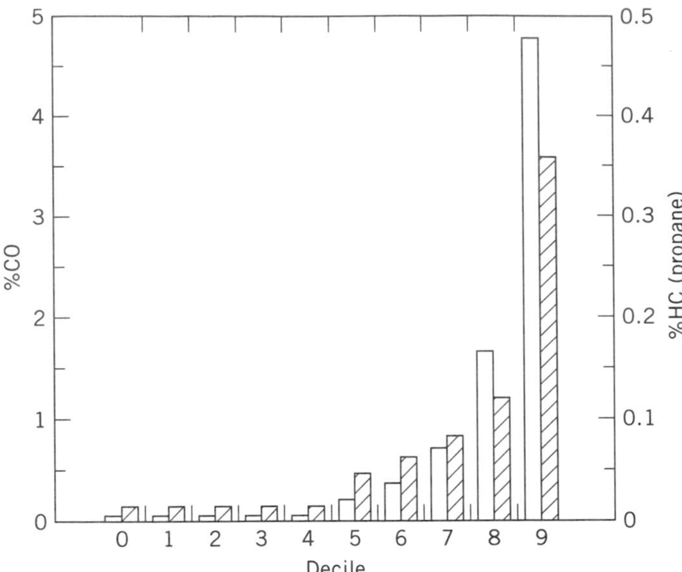

Figure 2. FEAT data collected in 1991 in El Monte, California (Rosemead Blvd. site), showing that automobile emissions are not normally distributed. The fleet is rank-ordered and divided into deciles; the average percent of CO and HC for each decile is plotted (▨ represents CO; □, HC). The first five deciles are displayed as an average of all five.

can be relatively low emitting as seen by the mean emissions for all model years in the lowest emitting quintile. It can also be seen that the most rapid deterioration in emissions occurs during the first 11 years of ownership. It is interesting to note that the vehicles with the most rapidly deteriorating emissions are all newer technology vehicles with computer-controlled fuel-delivery systems and three-way catalysts (installed in the United States after 1982). All of the studies point to the fact that these vehicles have negligible emissions when first purchased, thus emphasizing the need for proper maintenance.

If the problem then is not the old clunkers, as is usually assumed, then which vehicles are responsible? Figure 3**c,** which is the product of the average emissions shown in Figure 3**a** and the fleet distribution data shown in Figure 3**b,** represents the overall contribution for each model year. What is obvious is that although older vehicles (pre-1980) have higher average emissions it is the newer and more prevalent (post-1980) vehicles that actually contribute the most to the total. This has important implications for regulatory agencies that design programs around age or emission rates only.

For HC, the fleet emissions tend to be less skewed than for CO, with a larger percentage of vehicles being responsible for the majority of the fleet emissions. Only four cities (Bangkok, Hong Kong, Kathmandu, and Taipei) have half of the HC emissions produced by >15% of the fleet. Most of the same conclusions that are drawn regarding CO emissions and fleet characteristics hold true for HC emissions, because HC emissions increase as engine combustion gets richer and produces more CO. Bangkok and Kathmandu stand out for HC emissions, in large part because of the high percentage of two-cycle engines (necessarily high HC emitters by design), many of which also appear to be poorly tuned and maintained.

CONCLUSION

On-road remote sensing can now be used to analyze the emissions of a large fleet of vehicles in a cost-effective manner without inconveniencing the driving public. Statistical analyses of the data can be used to plan and evaluate emission-control programs. The identification of gross polluters can be used

Table 1. Worldwide Data Summary

Location	Year	Number of Measurements	Percent CO_2 Mean	Percent CO_2 Median	Percent HC Mean	Percent HC Median
Denver, CO	1991–1992	35,945	0.74	0.11	0.057	0.033
	1993	58,894	0.58	0.13	0.022	0.013
Chicago, IL	1990	13,640	1.10	0.37	0.139	0.087
	1992	8,733	1.04	0.25	0.088	0.064
California	1991	91,679	0.82	0.14	0.076	0.042
Provo, UT	1991–1992	12,066	1.17	0.45	0.220	0.127
El Paso, TX	1993	15,986	1.22	0.37	0.073	0.044
Juarez, Mexico	1993	7,640	2.96	2.18	0.170	0.091
Mexico City, Mexico	1991	31,838	4.30	3.81	0.214	0.113
Göteborg, Sweden	1991	10,285	0.71	0.14	0.058	0.046
Denmark	1992	9,038	1.71	0.67	0.177	0.058
Melbourne, Australia	1992	15,908	1.42	0.57	0.107	0.058
Thessaloniki, Greece	1992	10,536	1.40	0.55	0.155	0.082
London, UK	1992	11,666	0.96	0.17	0.136	0.071
Leicester, UK	1992	4,992	2.32	1.61	0.212	0.131
Edinburgh, UK	1992	4,524	1.48	0.69	0.129	0.084
Bangkok, Thailand	1993	5,260	3.04	2.54	0.948	0.567
Hong Kong	1993	5,891	0.96	0.18	0.054	0.037
Kathmandu, Nepal	1993	11,227	3.85	3.69	0.757	0.363
Seoul, Korea	1993	3,104	0.82	0.26	0.044	0.019
Taipei, Taiwan	1993	12,062	1.49	0.88	0.062	0.050

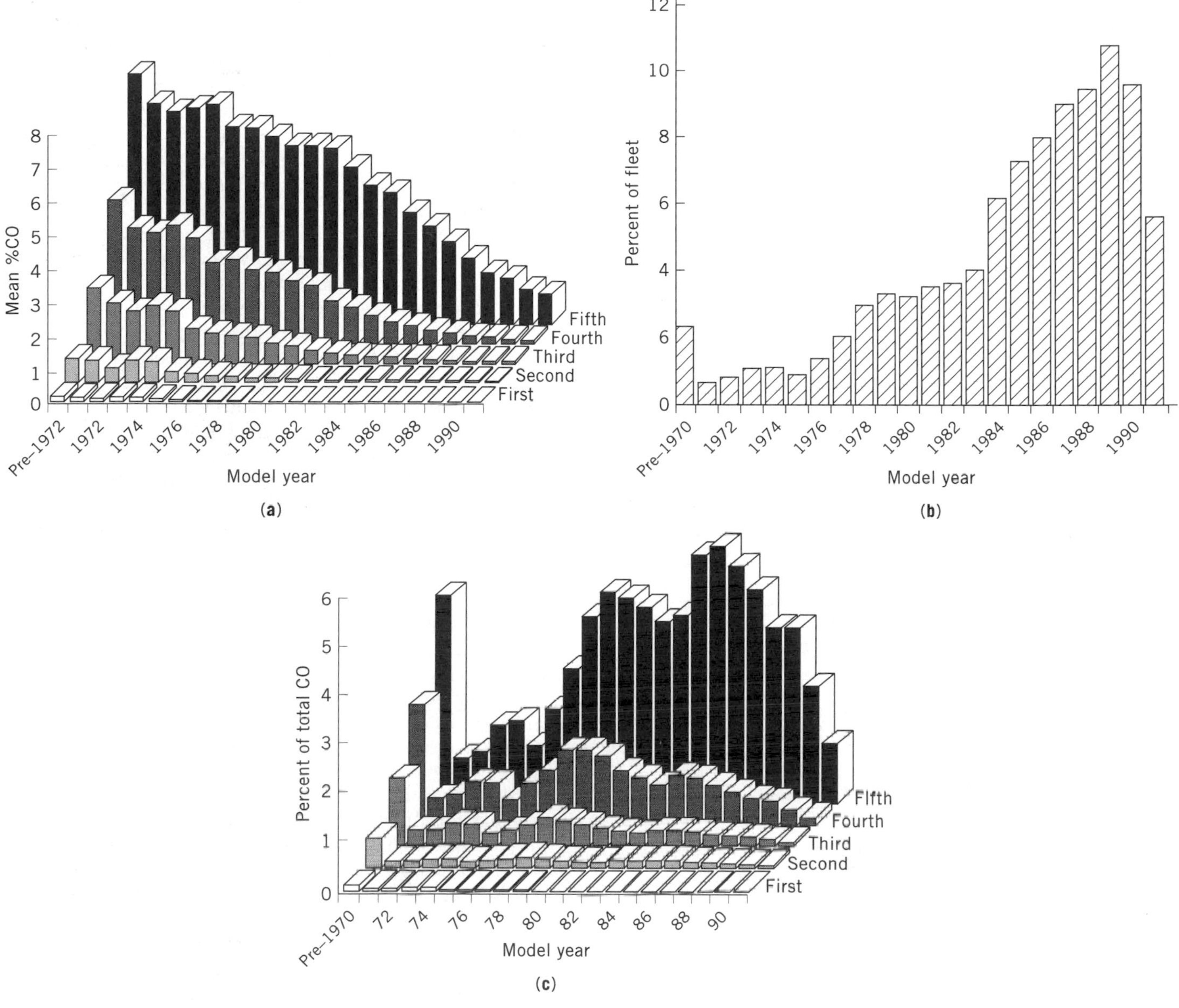

Figure 3. (**a**) The 1991 California data for CO presented as emission factors by model year for each of the five quintiles of data; (**b**) distribution of vehicles by model year for data in Figure 3**a;** (**c**) product of Figures 3**a** and 3**b,** showing average contributions of CO emissions by model year. The majority of emissions are contributed by vehicles in the fifth quintile that are post-1980 models.

as a component of a program designed to ensure that those vehicles receive effective repair (an inspection and maintenance program). Excessive on-road readings have been used to pull vehicles over to the side of the road to check for tampering with the emissions-control system; a high incidence of such behavior has been found.

Remote sensing can be periodically applied at the same site under the same conditions to determine how well efforts to reduce emissions are working. Since the correlation between low emissions and proper maintenance is high, most people operate vehicles that do not contribute significantly to pollution. These people could be rewarded for their socially responsible behavior based on low on-road emissions readings.

On-road remote sensing of motor vehicle emissions is a new tool that has progressed from a university prototype to a commercially available system. Air pollution officials who have used this system have suggested new ways in which the tool could be applied to solving mobile source emissions problems.

BIBLIOGRAPHY

1. *Fed. Reg.* **21**(60), part II (1966).
2. *Fed. Reg.* **33**(108), part II (1968).
3. *Fed. Reg.* **35**(214), part II (1970).

4. *Fed. Reg.* **36**(128), part II (1971).
5. M. P. Walsh, *Critical Analysis of the Federal Motor Vehicle Control Program,* NESCAUM, Albany, N.Y., 1988.
6. G. A. Bishop, J. R. Starkey, A. Ihlenfeldt, W. J. Williams, and D. H. Stedman, *Anal. Chem.,* **61,** 671A–676A (1989).
7. U.S. Pat. 5,210,702 (May 11, 1993), G. Bishop and D. H. Stedman (to Colorado Seminary).
8. *Performance Audit of Colorado's Oxygenated Fuels Program,* Final Report to the Colorado State Auditor Legislative Services, PRC Environmental Management, Inc., Denver, Colo., Dec. 1992.
9. G. A. Bishop and D. H. Stedman, *Environ. Sci. Technol.* **24,** 843–847 (1990).
10. D. H. Stedman, G. A. Bishop, J. E. Peterson, P. L. Guenther, I. F. McVey, and S. P. Beaton, "On-Road Carbon Monoxide and Hydrocarbon Remote Sensing in the Chicago Area," *ILENR/RE-AQ-91/14,* Illinois Department of Energy and Natural Resources, Springfield, Ill., 1991.
11. D. H. Stedman, G. Bishop, J. E. Peterson, and P. L. Guenther, "On-Road CO Remote Sensing in the Los Angeles Basin," *Contract No. A932-189,* California Air Resources Board, Sacramento, Calif., 1991.
12. D. H. Stedman and co-workers, "On-Road Remote Sensing of CO and HC Emissions in California," *Contract No. A032-093,* California Air Resources Board, Sacramento, Calif., 1994.
13. J. E. Peterson, D. H. Stedman, and G. A. Bishop, *Currents,* Ontario Air Waste Manage. Assoc., 1991.
14. Å. Sjödin, *Rena och Smutsiga Bilar: En Pilotstudie av Avgasutsläpp Från Svenska Fordon i Verklig Trafik,* Swedish Environmental Research Institute, Göteborg, Sweden, 1991.
15. S. P. Beaton, G. A. Bishop, and D. H. Stedman, *J. Air Waste Manage. Assoc.* **42,** 1424–1429 (1992).
16. Y. Zhang, G. A. Bishop, and D. H. Stedman, *Environ. Sci. Technol.* **28,** 1370–1374 (1994).

AUTOMOTIVE ENGINES AND VEHICLES

JEFFREY W. HODGSON
DAVID K. IRICK
University of Tennessee
Knoxville, Tennessee

LORI M. FUSSELL
Central Wyoming College
Jackson, Wyoming

QUALITATIVE ROLE OF MOBILE SOURCES ON AIR QUALITY

Emissions are extremely complex and consist of hundreds of compounds in the form of gases, aerosols, and particles. Additionally, many of these emissions undergo transformations in the atmosphere forming secondary pollutants such as ozone, acid aerosols, and carcinogenic hydrocarbons. Regulated emissions from automobiles include carbon monoxide (CO), volatile organic compounds (VOC), oxides of nitrogen (NO_x), and particulates. In addition to the emissions that leave the engine tailpipe, fuel vapors can be released into the atmosphere. Additionally, recent attention has been focused on carcinogen and carbon dioxide (CO_2) emissions from automobiles. These pollutants, their effects, and the relative contribution which motor vehicles make to the national inventory are discussed in the following sections.

Exhaust Emissions

Emissions from the tailpipes of motor vehicles represent the products of combustion created as fuel and air burn in the engine. These products include the following.

Carbon Monoxide. Carbon monoxide (CO) is a colorless and odorless gas which is the result of incomplete combustion of hydrocarbon fuels. It is considered dangerous because it binds to the hemoglobin in blood (forming carboxyhemoglobin) and renders the hemoglobin incapable of transporting oxygen. Health effects range from neural–behavioral effects (at 2–3% carboxyhemoglobin) to headaches and fatigue (at 10% carboxyhemoglobin) to respiratory failure and death. Reduced blood oxygen levels from CO exposure are particularly dangerous to the elderly, people with cardiovascular disease or other circulation problems, people with anemia, fetuses, young infants, and pregnant women. Carbon monoxide most strongly affects populations exposed to exhaust emissions that are concentrated in limited areas, ie, parking garages, tunnels, and toll booths, but also may reach significant levels in metropolitan areas (1). The major sources of carbon monoxide in the United States are shown in Figure 1. Approximately 69% of the CO emitted in the United States originates from the transportation sector, primarily motor vehicles (2).

Volatile Organic Compounds. Volatile organic compounds (VOC) are the product of incomplete combustion or the loss of fuel vapors into the atmosphere. They are defined by the U.S. Environmental Protection Agency (EPA) as organic compounds which, once released, can remain in the atmosphere long enough to participate in photochemical reactions. Typically, they have a vapor pressure greater than 13 Pa at 20°C and 101.3 kPa (1 atm). Some VOCs are classified as air toxics. The primary risk associated with VOCs (except for acute exposure to specific compounds) is that they react with oxides of nitrogen in the presence of sunlight to form photochemical oxidants, such as ozone, peroxyacetyl nitrate, and peroxybenzoyl nitrate (1). These oxidants irritate mucus membranes, inflame the eyes and upper respiratory tract, cause chest constriction, and at high concentrations cause severe coughing and the inability to concentrate. The major sources of volatile organic compounds in the United States are shown in Figure 1.

Oxides of Nitrogen. There are seven different oxides of nitrogen. Of these, nitric oxide (NO) and nitrogen dioxide (NO_2) are the major air pollutants. Oxides of nitrogen (NO_x) can be produced when a mixture containing oxygen and nitrogen (such as air) is heated to a high temperature (greater than 1100°C). If the cooling process is slow, the reaction is reversible. However, most internal combustion engines cool the combustion products rapidly, and the oxides of nitrogen do not have a chance to decompose. Although acute exposure to NO is nontoxic (even though it may fix itself to hemoglobin much like CO), NO emissions are of concern because NO is an NO_2 precursor and acute exposure to NO_2 may irritate eyes, and damage the lungs, liver, and heart. Chronic lung damage can occur with long-term exposure at low concentrations. The major effects of NO_x, however, are indirect because, as mentioned above, NO_x reacts with VOCs to form photochemical oxidants. Additionally, NO_2 can react with moisture present in the atmosphere to form nitric acid, which causes corrosion of metal surfaces, turns lakes acidic, damages forests and crops, alters

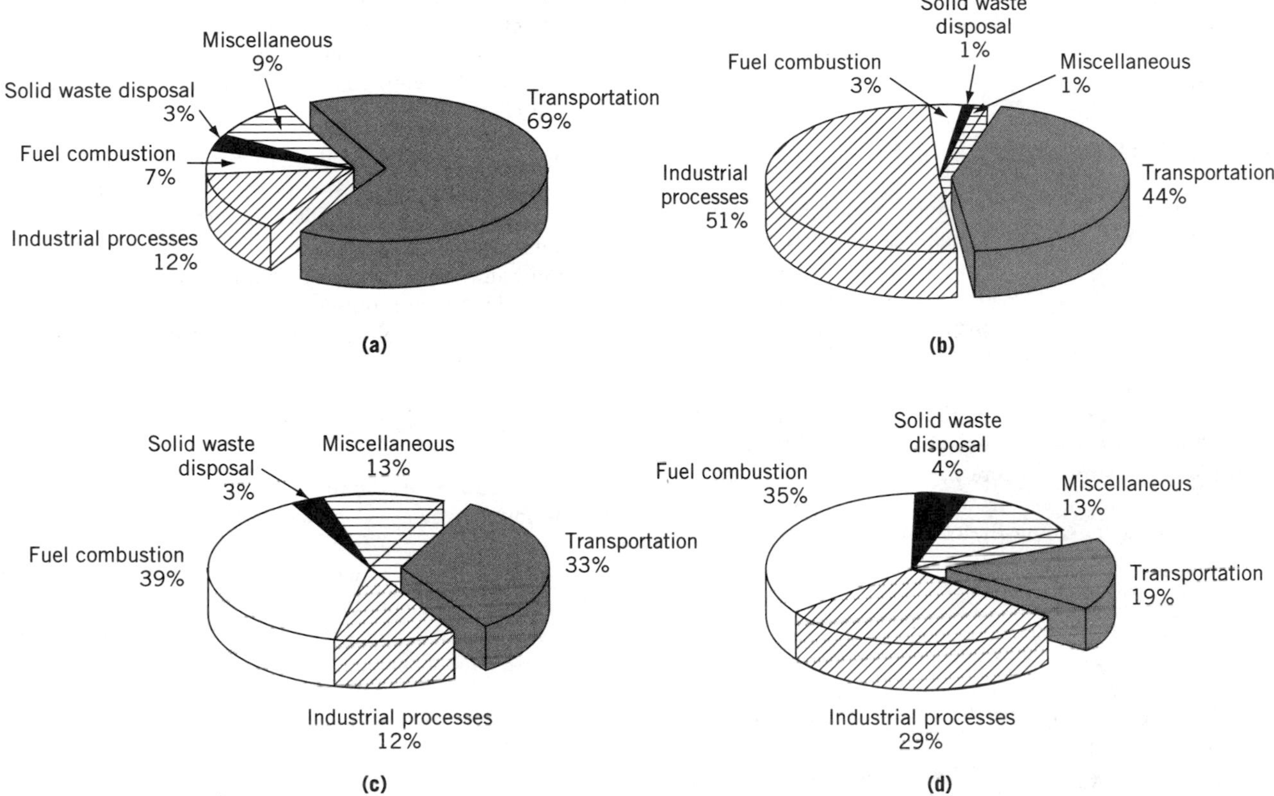

Figure 1. Major sources of pollutants in the United States where (**a**) represents carbon monoxide, (**b**) oxides of nitrogen, (**c**) volatile organic compounds, and (**d**) total suspended particulates.

soil fertility, and destroys sandstone and marble structures. (see ACID RAIN). The major sources of oxides of nitrogen in the United States are illustrated in Figure 1.

Particulates. Particulates are dispersed airborne solid and liquid particles that are larger than single molecules and range in size from 0.0002 to 500 μm. Larger particles are readily removed from the air by inertial processes (including gravity) whereas smaller particles are much more persistent. The smaller the particle, the longer the lifetime suspension of the particle. One of the main health threats posed by particulates results from their deposition in the lungs. Particles in the size range of 0.01 to 2.5 μm are particularly effective in overcoming the body's natural protection against the inhalation of foreign matter. The major sources of particulates in the United States are illustrated in Figure 1.

Carcinogens. Carcinogenicity is an extremely complex form of slow poisoning. Cancer experts generally cannot determine with high confidence the specific cause of a cancer in an individual. The best estimates of the causes of cancer were published in 1981 (3), where it was estimated that 2% of cancer deaths in the United States were attributable to environmental pollution. (Diet was the number one cause of cancer deaths at 35%, with tobacco a close second at 30%).

Evaporative Emissions

Emissions from motor vehicles include those that result when fuel evaporates into the atmosphere. Evaporation of fuel occurs when the fuel is introduced into the vehicle (refueling losses), when the fuel tank is heated and cooled due to variations in ambient temperature (diurnal losses), when fuel in the tank is heated as the vehicle is operated (running losses), and when the vehicle is parked after being operated (hot soak losses). Since each of these is associated with the volatility of the fuel, they are of concern for vehicles fueled with liquids such as gasoline.

Refueling Losses. As the vehicle uses fuel from its fuel tank, the space above the liquid contains fuel vapors. The mass of the fuel vapor in this space depends on the volume of this space, the temperature, and the composition of the fuel being used. When the vehicle is refueled, the incoming liquid displaces the fuel vapors in the tank and, if provisions are not made to capture these vapors, they are pushed out into the atmosphere.

Running Losses. As the vehicle is operated the fuel usually heats up. In some cases this is due to heat transfer to the fuel tank from the atmosphere, the road, or from hot surfaces (exhaust system) of the vehicle itself. The situation is exacerbated by fuel system designs that use the fuel tank as a thermal sink to keep the fuel in the rest of the fuel system from getting too hot. This is accomplished by allowing fuel to circulate continuously through the fuel system components before it is returned to the fuel tank. If the fuel temperature increases, some of the liquid fuel vaporizes and if the fuel tank is vented directly to the atmosphere, some of these vapors are pushed into the surrounding air. The solution for this problem has been to seal the fuel tank (except during refueling) from the atmosphere and allow venting (or breathing) of the tank to occur through a canister filled with activated carbon where the fuel vapors are adsorbed on the charcoal.

CONTRIBUTIONS TO CLIMATE CHANGE

There is significant interest in how humankind may be contributing to global climate change (global warming) by releasing gases that absorb infrared radiation in the atmosphere (4,5). The primary gas of interest from motor vehicles fueled with hydrocarbon fuels is carbon dioxide. Ways to reduce the emissions of carbon dioxide from motor vehicles are to reduce fuel consumption of vehicles either by reducing vehicle use, by improving fuel economy of vehicles that are used, or by using fuels that produce less carbon dioxide (6). For equivalent vehicles, the superior fuel efficiency of diesel engines results in a lower global warming effect from the diesel-powered vehicle when compared to its gasoline-fueled counterpart (7).

EMISSION MEASUREMENTS

It is necessary to measure exhaust emissions in order to meet legislative requirements and to gain insight into engine performance. The most common methods of exhaust gas analysis are the nondispersive infrared (NDIR) analyzer, the flame ionization detector (FID), the photochemiluminescent analyzer, the Fourier transform infrared (FTIR) analyzer, and various smoke meters.

The Federal Test Procedure

Presently, automobile emissions in the United States are measured in accordance with the Federal Test Procedure (FTP). For light-duty vehicles (LDVs), this procedure involves the measurement of exhaust emissions while the vehicle is driven according a prescribed driving schedule on a chassis dynamometer. Evaporative emissions are measured using the Sealed Housing for Evaporative Determination (SHED) method. The procedure used for exhaust emission testing involves the Urban Dynamometer Driving Cycle (UDDC). The UDDC was originally derived from traffic studies in Los Angeles and simulates an urban drive, approximately seven and a half miles (12.06 km) long. The cycle has frequent stops, an average speed of 19.7 mph (31.7 km/h), a maximum speed of 56.7 mph (91.2 km/h), and the vehicle engine idles for approximately 18% of the cycle.

The test is organized into four phases: the cold transient phase (phase 1), the cold stabilized phase (phase 2), the hot transient phase (phase 3), and the hot stabilized phase. The cold transient phase is comprised of the first 505 seconds of the test. During this phase, exhaust emissions are typically at their worst because of cold-start enrichment strategies and catalysts which have not lit off. The rest of the seven-mile drive is termed the cold stabilized phase. After the cold stabilized phase is completed, the vehicle is turned off and soaked for 10 minutes. The cold transient and cold stabilized phases represent an urban trip starting with a "cold" engine. The vehicle is then restarted and the first 505 seconds of the UDDC are repeated (hot transient phase). The hot stabilized phase is the remainder of the drive, but the results are assumed to be the same as in the cold stabilized phase. The hot transient and the hot stabilized phases represent an urban trip starting with a "hot" engine. The emissions corresponding to the first three phases are measured separately and the overall mass emissions of the vehicle are calculated using the following weighting factors.

Chassis Dynamometer. Vehicle emissions are tested while driving on a chassis dynamometer. This allows ambient conditions to be closely controlled and exhaust emissions to be sampled without disturbing the engine. Essentially, the chassis dynamometer consists of rollers upon which the vehicle is driven. The dynamometer absorbs the mechanical energy transmitted by the vehicle's tires and simulates the inertia of and road-load power required by the vehicle.

ENGINE FUNDAMENTALS

In order to gain a better understanding of the emissions from combustion engines, it is useful to examine what is going on inside the engine. Because the combustion process is so complex, it is often worthwhile to model the engine using whatever tools are available. Various excellent introductory reference books dealing with internal combustion engines are available (8–11).

Engine Types

Internal combustion engines typically compress fuel and air to create a high pressure mixture that is burned to create a high temperature, high pressure gas (products of combustion). This is subsequently expanded to deliver work; ie, the products of combustion constitute the working fluid. Essentially all current internal combustion engines used in transportation use what is termed a slider-crank mechanism in which a piston moves with a reciprocating motion along the axis of a cylinder. The only other type of engine currently (ca 1997) in production for motor vehicles is the rotary (Wankel) engine. Reciprocating internal combustion engines can be further classified as either four-stroke cycle ("four-stroke") engines or two-stroke cycle ("two-stroke") engines.

Four-Stroke Cycle Engines. In the four-stroke cycle engine, the piston traverses the cylinder four times (thus, four strokes) for each complete cycle that the engine executes. This requires two revolutions of the crankshaft. The sequence of events is shown in Figure 2 and consists of the following five steps.

1. *Intake stroke:* the piston travels from "top dead" center (TDC) to "bottom dead" center (BDC). Air (and in the case of the conventional spark ignition engine, fuel) is drawn into the engine past the intake valve(s).
2. *Compression stroke:* the fresh charge is compressed as the intake valve closes and the piston moves toward TDC.
3. *Ignition:* initiated in the case of a spark ignition engine by the occurrence of a spark between the electrodes of a spark plug protruding into the combustion chamber. In the case of a compression ignition engine the combustion process is initiated by injecting fuel into the hot compressed air.

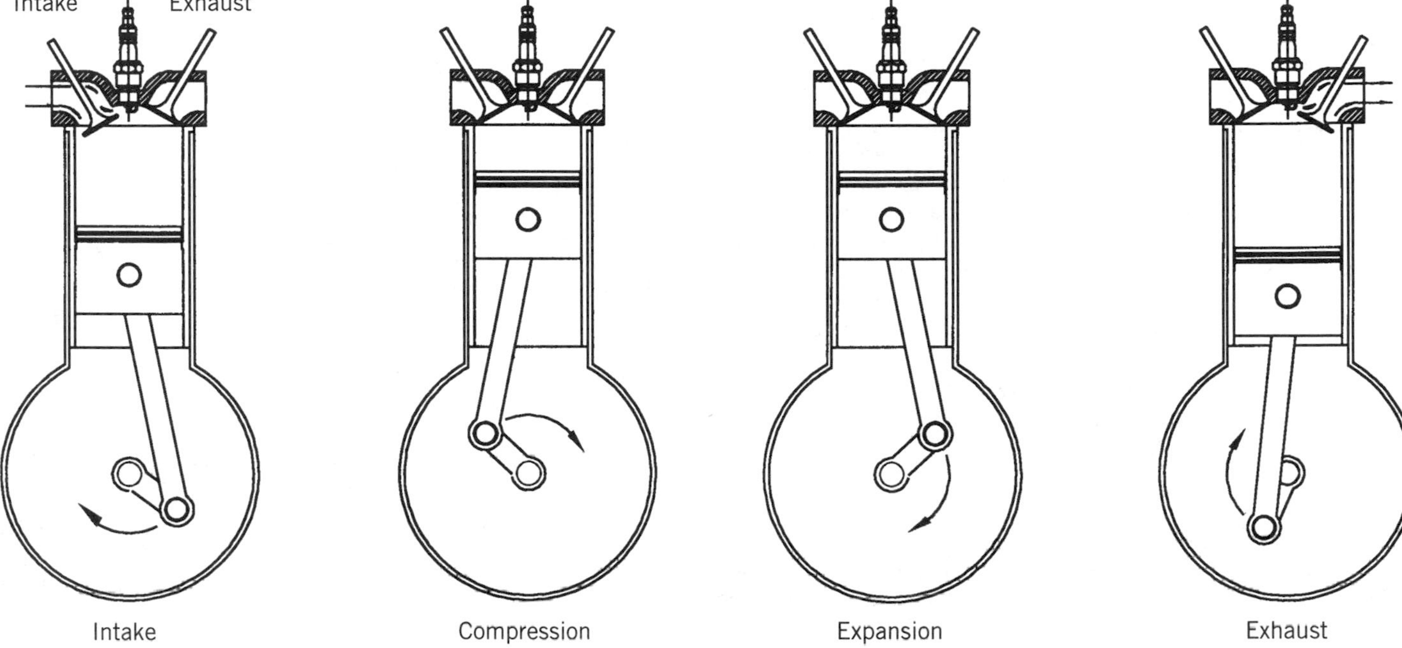

Figure 2. Four-stroke cycle engine schematic.

4. *Expansion stroke:* during this stroke the combustion products expand as the piston moves toward BDC and work is done by gases on the piston. Near BDC the exhaust valve opens and the pressure in the cylinder drops as the gases rush out of the cylinder.
5. *Exhaust stroke:* during this stroke the piston moves toward TDC and as it does it pushes gases out of the cylinder past the open exhaust valve. As the piston approaches TDC the intake valve starts to open and the exhaust valve starts to close. The cycle repeats itself.

Two-Stroke Cycle Engines. The two-stroke cycle engine has a compression stroke and an expansion stroke, but the intake and exhaust processes are combined into events that occur as the piston approaches BDC on the expansion stroke. As the piston approaches BDC it uncovers exhaust ports that allow the combustion products to flow out of the cylinder. As the piston moves further toward BDC it uncovers intake ports. The intake charge is pressurized either by crankcase compression or by an externally driven compressor so that the intake charge helps push out the combustion products that are in the cylinder.

ENGINE OPERATING VARIABLES

There are a number of important variables in engine operation. They include fuel–air stoichiometry; carburetion, including carburetors and fuel injectors; spark timing; and valve actuation.

ENGINE FUELS

Spark Ignition Engines

In the conventional homogeneous charge spark ignition engine the fuel is introduced into the air outside the cylinder with the expectation that the fuel and air will form a homogeneous mixture through which the flame will propagate in a predictable manner. In order for this to occur the fuel must have certain characteristics.

Volatility. In order for the engine to start under cold conditions, a combustible mixture must be present around the spark plug and throughout the cylinder when the engine is being cranked over and after the engine starts. For gaseous fuels (natural gas, dimethyl ether, liquefied petroleum gas) this occurs naturally, but in the case of fuels that are liquid under normal temperatures and pressures some of the fuel must evaporate to form the vapor. Rather than have all the fuel capable of vaporizing, gasolines are blended to create a mixture of hydrocarbons, some of which vaporize readily under cold temperatures.

Octane Rating. In homogeneous charge engines the unburned fuel is subjected to increasing pressures and temperatures during the compression stroke and as the flame travels across the combustion chamber. Because successful operation of the engine depends on having the combustion process occur in a predictable manner, it is essential that the fuel have the ability to exist at these temperatures and pressures without autoigniting until the flame reaches it.

The autoignition process can be simplified by considering that in order for a fuel–air mixture to autoignite, the mixture must be held at a temperature higher than its self-ignition temperature for a period of time longer than its ignition delay period. The concept of the self-ignition temperature is fairly self-explanatory; in order for a gasoline–air mixture to autoignite its temperature must be raised above some threshold value.

One measure of a fuel's ability to resist autoignition is determined by running standard tests on the fuel in a standard engine. The results are expressed in terms of the octane rating of the fuel. This indicates that under the test conditions encountered, the fuel exhibits the same ability to resist autoignition as a mixture of the primary reference fuels isooctane and normal heptane. Isooctane is arbitrarily assigned an octane number of 100 and normal heptane is arbitrarily assigned an octane number of 0. Thus, a fuel having an octane rating of 90 would have the "same" ability to resist autoignition as a mixture of 90% isooctane and 10% normal heptane. Since there are several different tests that can be run, it is necessary that the test methodology also be specified when reporting the octane rating. The two most common test methods used for gasoline are the research method and the motor method, as specified by the American Society for Testing Materials (ASTM). Typical gasolines have a higher octane rating by the research method than by the motor method. Octane ratings posted on gasoline dispenser pumps in the United States are the average of the two results.

Compression Ignition Engines

In the case of compression ignition engines the fuel is introduced into the hot air toward the end of the compression stroke. The entire combustion process is based on the presumption that the fuel will autoignite as it is injected into the air. If the fuel does not promptly autoignite, it may accumulate in the air until it autoignites simultaneously. If enough fuel simultaneously autoignites, knock can result. The tests for compression ignition engine fuels are different from those of spark ignition engine fuels and the comparable quantity reported is termed the cetane rating of the fuel.

EMISSIONS CONTROLS

Carbon Monoxide

The emissions of carbon monoxide from an engine result primarily from fuel-rich combustion. In the case of the homogeneous charge engine, rich combustion can result from the overall fuel–air ratio being rich (greater than stoichiometric), from poor fuel or air distribution among the cylinders of a multicylinder engine, or from poor fuel distribution within a given cylinder. Carbon monoxide also appears in the exhaust of engines because it is formed due to the dissociation of carbon dioxide at the very high temperatures following combustion.

The primary control strategy for reducing engine-out CO is to avoid rich combustion. Aftertreatment of the exhaust consists of using either an oxidizing catalyst (with air injection sometimes incorporated) or a three-way catalyst.

Volatile Organic Compounds

Vehicles emit volatile organic compounds (VOC) through evaporation of fuel and in the exhaust gases. The VOC exhaust emissions can be thought of as representing unburned fuel, but the speciation of typical exhaust gas VOC shows that many different organic compounds are present as the fuel (and lubricating oil) are exposed to the combustion process. There are several commonly accepted explanations for VOC emissions (12).

Control of VOC emissions from the vehicle involves placing limits on the volatility of the fuel and installation of an effective control system for evaporative emissions, taking steps to reduce the engine-out VOC emissions and treating the exhaust catalytically.

Oxides of Nitrogen

At the high temperatures created during the combustion process, nitrogen present can react to form various oxides of nitrogen (NO_x). If the expansion process could take place slowly so that chemical equilibrium could be maintained (shifting equilibrium) the oxides of nitrogen would disappear and be negligible at temperatures associated with the exhaust gas. However, the expansion process occurs so rapidly that the mixture composition, although not frozen at levels found near TDC, are somewhere between the values at the start of the expansion process and those at the end of this process. This is similar to the situation that occurs with carbon monoxide.

There are several steps that can be taken to reduce NO_x emissions. Control of the stoichiometry is important. NO_x emissions are low under fuel-rich combustion due to the relatively low combustion temperature and the relatively low abundance of oxygen. As the air–fuel ratio increases, the NO_x increases as the combustion temperature and availability of oxygen increase. At higher air–fuel ratios the combustion temperature starts to decrease, the lower temperature prevails over the greater abundancy of oxygen, and the NO_x emissions decrease.

The Three-Way Catalyst

As pointed out previously, the catalytic treatment of CO, VOC, and NO_x emissions is one strategy that results in low exhaust emissions. The catalyst material is typically incorporated on the surface of a substrate material which is often a monolith of either ceramic or metallic material (13). The actual catalyst materials used include platinum, palladium, rhodium, and other materials used to increase the effectiveness of the catalyst.

Stoichiometry. Figure 3 shows how catalyst efficiency (percentage of incoming specie eliminated by the catalyst) varies with the air–fuel ratio supplied to the engine. For fuel-rich combustion the catalyst is ineffective for CO and VOC conversion, and for fuel-lean combustion the catalyst is ineffective for NO_x conversion. The catalyst must also be hot in order to be effective.

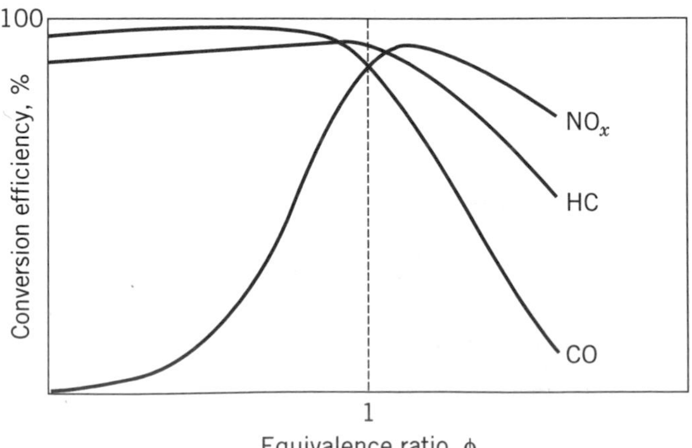

Figure 3. Catalyst efficiency.

Other Controls. These include exhaust gas oxygen feedback, electronic engine management (26), on-board diagnostics (28), and fuel metering strategies.

Vehicle Technologies

Traditional vehicles will most likely be supplemented by other vehicles, some of which are already in use.

Electric Vehicles. Currently electric vehicles and small trucks are available from major U.S. and Japanese manufacturers and a great deal of work is going on to develop more effective batteries. Although advanced lead–acid batteries are being used in the base General Motors EV-1 electric car, nickel metal hydride batteries are also available, and the Honda electric car uses nickel metal hydride batteries. It appears that the lithium ion battery will be a strong contender for batteries of the future (14).

Hybrid Electric Vehicles. Another technology is the hybrid electric vehicle which uses two energy sources: electricity and some other fuel (15). The ability to use conventional fuel with its high energy density means that the hybrid vehicle can have a significantly higher range than its pure electric counterparts (see ELECTRIC CARS, ENVIRONMENTAL EFFECTS).

Fuel Cell Hybrids. Another type of hybrid is the fuel cell hybrid in which a fuel cell (which generates electricity directly from a fuel) is used to keep on-board batteries charged so that they can provide electricity to the vehicle's electric motor(s). The fuel cell can be thought of as electrolysis in reverse in that it combines hydrogen and oxygen to produce water and electricity. The source of the hydrogen can be either from hydrogen stored on board or by an on-board reformer that generates hydrogen from other fuels.

ALTERNATIVE FUELS

The characteristics of the emissions from internal combustion engines are inexorably tied to the characteristics of the fuel that is burned in the engine (16). Conventional gasolines and diesel fuels are engineered fuels derived essentially from crude oil processed in refineries. As such they are not pure substances, but are mixtures of hydrocarbons with carefully tailored properties to meet the requirements of engine manufacturers and governmental regulations. The fuels are also treated with additives to provide various attributes that may not be present in the base fuel. The oil industry has responded to concerns about mobile sources contributions to air quality and gasolines sold today are different from their predecessors (17). Concerns about the reliability of petroleum sources and air quality have led to a significant interest in alternative fuels.

Reformulated Gasoline

The oil industry has made changes to gasoline to reduce its impact on the environment. Specifically, tetraethyllead has been eliminated from gasoline sold in the United States. The term *reformulated gasoline* has been introduced to designate gasoline that has the following attributes: reduced vapor pressure to reduce evaporative emissions, lower benzene content to reduce the benzene emissions, and normally some sort of oxygenate, typically methyl tertiary butyl–ether (MTBE). The MTBE also serves to raise the octane rating of the fuel. Reformulated gasolines are being mandated in various parts of the United States.

Reformulated Diesel Fuel

Although conventional diesel fuel is a product of refineries, concern over air pollution has led to the introduction of diesel fuel that produces reduced levels of particulates, primarily through reduced sulfur content of the fuel.

Biodiesel Fuel

The use of vegetable oils as replacements for petroleum-based diesel fuel has been the subject of many investigations. The fuel has been shown to give reduced particulate emissions, but slightly higher NO_x emissions than conventional diesel fuel (18). The disadvantages of this fuel include higher cost and a greater tendency to form wax at low temperatures.

Natural Gas

Natural gas is a mixture of various gases, but in most instances the main constituent is methane. It occurs naturally and is extracted directly from the earth or as a coproduct during the production of crude oil. In some cases where there is no economical means of transporting it, it is flared (burned) at the sources. It is also transported in pipelines, liquefied and transported in cryogenic vessels, and can also be converted into methanol or ammonia. Both light- and heavy-duty engines can use natural gas as fuel (19,20).

In light-duty engines, the natural gas is either metered into the intake air using a gas carburetor or a special fuel injection system designed to meter gas rather than liquid.

Natural gas is most commonly stored on-board the vehicle as a highly compressed gas (pressures of 25 MPa are common). In order to accommodate this pressure, special tanks are required.

Liquefied Petroleum Gas

Liquefied petroleum gas (LPG) is a mixtures of hydrocarbons, but the primary component is propane. Often a significant amount of butane is also present. LPG is a by-product of natural gas production and petroleum refinery operations. Because the fuel is primarily saturated paraffins, the exhaust emissions are relatively nonreactive as far as ozone is concerned. LPG must be pressurized at room temperature to keep it liquefied, but the pressures are not as high as those under which natural gas is typically stored. It is the most widely used alternative fuel in the United States and in many countries of the world it is commonly used as a transportation fuel. LPG has a high octane rating and offers the engine designer an opportunity to use higher compression ratios than those considered to be prudent for gasoline.

Methanol

Methanol (sometimes referred to as wood alcohol) is currently produced from natural gas. It is a liquid at room temperature and pressure and has a very high octane rating. Its energy content is about one-half that of gasoline per unit volume and the stoichiometric air–fuel ratio is significantly lower (6 vs 15). Because of this difference in the stoichiometry, one cannot simply pour methanol into a vehicle that is designed to use gasoline because the fuel metering system will not meter enough methanol into the air. Methanol is also corrosive to some materials typically used in gasoline-fueled vehicles, so vehicles that use methanol as a fuel must be engineered to use this fuel (see ALCOHOL FUELS: METHANOL).

Ethanol

Ethanol is sometimes referred to as grain alcohol and has properties similar to methanol. It is currently made by fermenting various feedstocks, most commonly corn. Like methanol it is a liquid at room temperature and pressure and has a lower energy content (heating value) than gasoline and diesel fuel (approximately two-thirds that of gasoline on a volumetric basis). It also has a high octane rating (see ALCOHOL FUELS: ETHANOL).

Hydrogen

Hydrogen is considered by many to be the ultimate transportation fuel. The hydrogen energy cycle has been discussed in the literature and engines have been designed and operated successfully using hydrogen as the fuel. As discussed earlier, hydrogen is the fuel on which fuel cells operate and this adds to the fuel's versatility. The biggest problem with hydrogen is that of on-board storage. The density of hydrogen is low, even when compressed or liquefied. Concerns about the safety of hydrogen (sometimes referred to as the Hindenburg syndrome) also have to be addressed.

BIBLIOGRAPHY

1. R. McGill, A. Compere, B. Griffith, J. Hodgson, K. Lenox, and C. Pilo, *Comparison of Relative Environmental Impacts of Alternative and Conventional Motor Fuels,* International Energy Agency, 1995.
2. A. Watson, R. Bates, D. Kennedy, *Air Pollution, the Automobile, and Public Health,* National Academy Press, Washington, D.C., 1988, p. 20.
3. R. Doll and R. Peto, *J. Nat. Cancer Inst.* (1981).
4. M. Benarde, *Global Warming...Global Warning,* John Wiley & Sons, Inc., New York, 1992.
5. F. Krause, W. Back, and L. Kooney, *Energy Policy in the Greenhouse,* John Wiley & Sons, Inc., New York, 1992.
6. C. Amann, *The Passenger Car and the Greenhouse Effect,* SAE Paper No. 902099, Society of Automotive Engineers, Warrendale, Pa., 1990.
7. M. DeLuchi, *Emissions of Greenhouse Gases from the Use of Transportation Fuels and Electricity,* Technical Memo No. ANL/ESD/TM-22, Argonne National Laboratory, Argonne, Ill., 1991.
8. J. Heywood, *Internal Combustion Engine Fundamentals,* McGraw-Hill Book Co., Inc., New York, 1988.
9. R. Stone, *Introduction to Internal Combustion Engines,* 2nd ed., Society of Automotive Engineers, Warrendale, Pa., 1992.
10. C. Ferguson, *Internal Combustion Engines, Applied Thermosciences,* John Wiley & Sons, Inc., New York, 1986.
11. E. Obert, *Internal Combustion Engines and Air Pollution,* Harper and Row, New York, 1973.
12. W. Cheng, D. Hamrin, J. Heywood, S. Hochgreb, K. Min, and M. Norris, *An Overview of Hydrocarbon Emissions Mechanisms in Spark-Ignition Engines,* SAE Paper No. 932708, Society of Automotive Engineers, Warrendale, Pa., 1993.
13. E. Becker, R. Watson, and K. Zygourakis, *Advanced Substrate Design for Low Emission Vehicles,* SAE Paper No. 961899, Society of Automotive Engineers, Warrendale, Pa., 1989.
14. M. Origuchi, T. Miyamoto, and H. Horie, *Development of a Lithium-ion Battery System for EVs,* SAE Paper No. 970238, Society of Automotive Engineers, Warrendale, Pa., 1997.
15. M. Cuddy and K. Wipke, *Analysis of the Fuel Economy Benefit of Drivetrain Hybridization,* SAE Paper No. 970289, Society of Automotive Engineers, Warrendale, Pa., 1997.
16. P. Gabele, *Ozone Precursor Emissions from Alternatively Fueled Vehicles,* SAE Paper No. 941905, Society of Automotive Engineers, Warrendale, Pa., 1994.
17. L. Gibbs, *How Gasoline has Changed II—The Impact of Air Pollution Regulations,* SAE Paper No. 961950, Society of Automotive Engineers, Warrendale, Pa., 1996.
18. C. Choi, G. Bower, and R. Reitz, *Effects of Biodiesel Blended Fuels and Multiple Injections on D. I. Diesel Engines,* SAE Paper No. 970218, Society of Automotive Engineers, Warrendale, Pa., 1997.
19. *Methane as a Motor Fuel,* The International Energy Agency, 1992.
20. C. Weaver, *Natural Gas Vehicles—A Review of the State of the Art,* SAE Paper No. 89133, Society of Automotive Engineers, Warrendale, Pa., 1989.

AUTOMOTIVE PLASTICS, ENVIRONMENTAL COMPATIBILITY AND RECYCLING

JERRY FOSNAUGH
SCOTT NOESEN
The Dow Chemical Company
Midland, Michigan

MICHAEL BIDDLE
MBA Polymers and American Plastics Council
Washington, D.C.

The use of plastics in vehicles has increased over the last 20 years to about 8 wt % of the vehicle, an average of 120 kg (265 lbs) per vehicle. The reason for this increase is due to the benefits that plastics provide the automotive companies and their customers in appearance, affordability, durability, safety, and fuel economy. The environment also benefits. Because less material (mass) is required for the average part on the vehicle, fuel economy is improved, resulting in lower emissions over the life cycle of the vehicle. Today's automotive scrapping process is designed mainly to recycle only the metals in the vehicle, and very little of the 8% plastics is recycled. However, progress in automotive plastics recycling technology is significantly advancing.

PLASTICS IN THE AUTOMOBILE

Although synthetic plastics are considered to be new materials, they date back to the 1800s. In 1869, John Hyatt invented celluloid, the first synthetic plastic material used commercially, in the search for an ivory replacement for billiard balls. In 1872, Leo Baekeland in the United States discovered Bakelite, a reactive plastic of phenolic formaldehyde resin that could be molded into parts (1). However, it was not until the 1920s and 1930s that many more familiar plastics were discovered:

Year	Plastic Resin
1929	Cellulose acetate
1934	Epoxies
1935	Nylon-6
1937	Polystyrene
1937	Polyurethanes
1938	Nylon-6,6
1940s	Polyethylene and silicones

After World War II plastics manufacturing and use became popular, and the automobile industry quickly found advantages with this new material. During the 1970s, when the world experienced the oil crisis, and an increase in fuel economy for vehicles became paramount, automotive engineers turned to plastics to lighten the vehicle and improve the aerodynamics of the design. Since then, the technical advancement of plastic resins has focused on improving their performance in vehicles, and plastics use has increased from an average of 32 kg (70 lbs) per vehicle in 1970 to 120 kg (265 lbs) per vehicle in 1995. Figure 1 lists the average weight of plastics in U.S. vehicles for 1990–1995 (2,3).

In North America, over 95% of all end-of-life vehicles are economically scrapped. Approximately 75% of each vehicle is recycled for a profit, and the remainder goes to the landfill, making the automobile one of the most recycled products today. This scrapping process is possible because of efficient auto recyclers (dismantlers) and scrap processors (shredders) (Fig. 2). Automobile recyclers have perfected the process of removing valuable metal, glass, and plastic parts from the scrapped vehicles to sell for reuse to repair shops. They also remove parts that harm the environment later in the scrapping process, such as batteries, catalytic converters, gas tanks, fluids, tires, etc. The auto recyclers sell the hulk to shredders, who process it through a large hammermill that reduces the car to small fist-size chunks, followed by separation of the metals for recycling. This automotive scrapping process is discussed in more detail later.

The increased use of plastics in vehicles has created concern in the vehicle scrapping industry. The concern involves the speculation that increased amounts of plastics in vehicles and

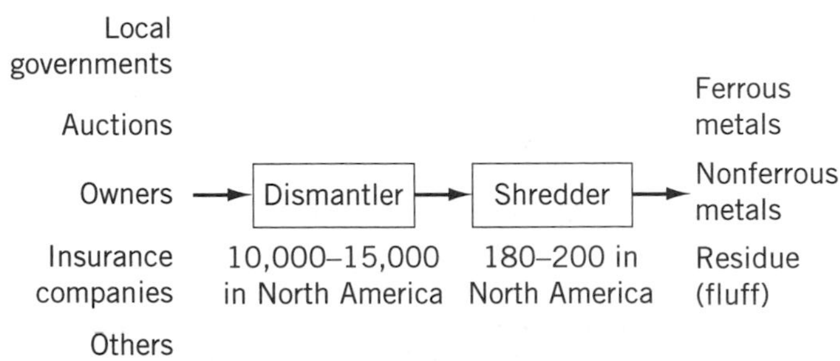

Figure 2. Vehicle scrapping infrastructure in North America. Total process is profitable and economically self-sustaining.

the corresponding decrease in metals content, will break down currently self-sustaining and profitable vehicle salvaging. If true, this could create a dilemma for the automotive original equipment manufacturers (OEMs) as they turn to plastics to improve fuel efficiency, cost, safety, and manufacturability of the vehicle.

However, plastics recycling and recovery technology is evolving rapidly, and as more plastics in vehicles are recovered, other benefits may also be realized by the OEMs, who predict that a more recoverable and recyclable vehicle will, in the future, be differentiated in the showroom; by the communities, which prefer to conserve resources and minimize the need for more landfills; and, by the repair and recycling companies, which are developing processes to profitably recover and recycle plastics.

The terms plastics and polymers are used to describe many different materials. Certainly plastics are polymers, but not all polymers are plastics, as generally described in the automotive industry. The terms plastics and plastic resins in the automobile are used only for those applications that include films, such as vinyl skins on instrument panel topper pads; foams, in seat cushioning; and plastic parts, such as center consoles in the interior or bumper fascia. The applications for polymers not considered plastics are textiles, such as floor liners or cloth on seats, and rubber parts (in hoses and tires).

Defining a plastic resin application is difficult. For instance, some floors on vehicles have a sandwich construction

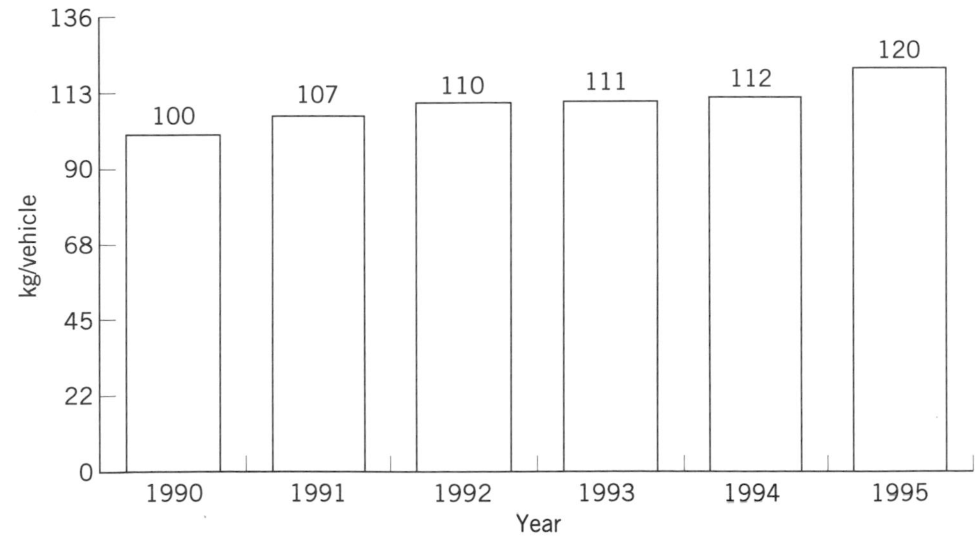

Figure 1. Average weight of plastic resin in automobiles and light trucks for 1990–1994 and 1995 calculated (3).

with polyurethane foam between textiles. Also, some elastomeric materials used for hoses and belts are either rubber or plastic resin.

Plastics Applications In The Automobile

The following is a list of common applications for plastics in vehicles (3).

Cowl vent grille	Energy absorber	Spare tire cover
Doors	Bumper trim	Steering wheel
Lift gate/Hatch	Intake manifold	Interior trim
Fender (front)	Fuel tank	Trunk liner
Fender (rear)	Heater/AC housing	Radiator support
Front end panel	Center console	Spring
Hood	Cowl vent leaf catch	Deck lid reflector
Quarter panel	Fan shroud	Molding side trim
Rear deck	Fender liner	Radiator grille
Roof	Gas tank shield	Seat cushion (front)
Spoiler	Headliner	Seat cushion (rear)
Bumper air dam	Instrument panel	Upholstery trim
Bumper beam	Package shelf	
Fascia	Seat load floor	

In this article, the term vehicles and the term automobiles are used interchangeably and refer to automobiles and light trucks sold in the United States. Light trucks include sport utility vehicles (SUVs) and vans. Approximately 13,293,000 automobiles and light trucks were produced in the United States in 1994. Table 1 identifies the larger plastic parts found on some vehicles, and the typical weight of those parts as reported in Market Search's *Automotive Plastics Report '95*. The *Automotive Plastics Report*, published annually by Market Search, Inc. (Toledo, Ohio), is the source of information that the automotive and plastics industries rely on for key facts related to plastics use in automobiles and small trucks (3).

In each of the applications listed above, a plastics material is used because it is more cost effective, provides the required functionality, and is processed better than competing materials. Many of the applications, such as seat cushioning and interior trim have been made of plastics for many years. Others, such as body panels, bumper beams, gas tanks, and air intake manifolds, have only recently switched to plastics from more traditional materials.

Table 1. Plastics Applications[a]

Application	Kilograms	Pounds
Lift gate/hatch	13.6–18.2	30–40
Quarter panel	13.6–18.2	30–40
Hood	7.7–15.9	17–35
Roof	6.4–18.2	14–40
Doors	11.3–13.2	25–29
Fender (rear)	8.2–15.9	18–35
Instrument panel	4.5–5.4	10–12
Rear deck	7.7–9.5	17–21
Fuel tank	7.3–8.2	16–18
Front end panel	3.2–6.8	7–15
Bumper beam	4.5–6.4	10–14
Seat cushion (rear)	5.9–6.4	13–14
Seat cushion (front)	5.4–5.9	12–13
Bumper fascia	3.2–6.8	7–15
Seat load floor	4.5–12.3	10–27

[a] Ref. 3.

Plastic Resin Types

Table 2 lists the most commonly used plastic resins in the automobile and light trucks (3). In each case, a specific resin is chosen because of a unique benefit it provided at an optimal cost. Approximately 20 base resins are used to formulate the plastic resins required to meet all of the needs in the vehicle. For instance, polycarbonate (PC) and acrylonitrile–butadiene–styrene (ABS) are each used to make plastic components in the vehicle, and they are also blended together, or alloyed, to form a new plastic resin (PC/ABS) which has desirable properties not provided by the individual resins. Polyvinyl chloride (PVC or vinyl) is used by itself and it is alloyed with ABS to modify its properties for applications, such as the skin on the instrument panel.

Many of these resins and blends of resins are also filled with finely pulverized glass, wood, or talc to meet the dimensional stability required for specific applications. In other cases glass mat or glass strands are required for increased strength. Additives are also used in some cases to make the plastic resin more resistant to sunlight, more impact resistant, more flexible, or to provide better aging performance. To create color, the parts are painted or the plastic resin is manufactured with a color pigment in the resin prior to molding added as a concentrate or during the molding process. The plastic part is also covered with a vinyl skin to create the desired color and texture. Generally, plastic resins are divided into two categories, thermosets and thermoplastics. The thermosets are unique because they are formed by a reaction that sets the polymer. Examples of thermoset resins are polyurethane foam and glass-reinforced thermosetting polyester resin (sheet molding compound SMC). Thermoplastic resins can be reformed by

Table 2. Typical Plastic Resins in Automobile, 1995[a,b]

Resin	Kg (lb)
ABS	7.0 (15.5)
Acetal	0.9 (2.0)
Acrylic	1.6 (3.6)
Butyral	0.5 (1.1)
PC/ABS	1.0 (2.2)
PC/PBT	1.4 (3.2)
PPO/nylon	1.0 (2.2)
PPO/styrene	2.8 (6.2)
Ionomer	0.7 (1.5)
Nylon	7.3 (16.0)
Phenolic	1.7 (3.8)
Polycarbonate	3.2 (7.0)
Polyester thermoplastic	3.0 (6.6)
Polyester thermoset	6.2 (13.6)
Polyethylene	7.2 (15.9)
Polypropylene	17.7 (39.1)
PP/EPDM	6.0 (13.2)
Polyurea	0.7 (1.6)
PVC	9.8 (21.5)
SMA	2.0 (4.5)
Polyurethane (foam and RIM)	18.2 (40.0)
Vinyl ester	0.5 (1.0)

[a] Ref. 3.
[b] Values are estimated.

melting and cooling to form a new shape. Examples of thermoplastic resins used in the auto industry are ABS, nylon, polycarbonate, and polypropylene. Both thermosets and thermoplastic resins can be recycled.

Features, Benefits, and Environmental Attributes

Plastic resins increase the safety of the occupants in the vehicle. Examples are protective films in the windshield; the airbag; the plastic knee bolster; the pad on the instrument panel; and the energy management foam behind the bumper. Specially designed plastic parts are also utilized to meet new head and side impact requirements.

Comfort and aesthetics are improved when plastic resins are used to cushion the occupants in seating and to reduce noise, vibration, and harshness (NVH), inside and outside the vehicle.

Plastic resins create design flexibility that contributes to the affordability of the vehicle, primarily because they are molded into "net shapes", or three dimensional parts, that give parts high strength using a minimal amount of material. Also, the moldability of plastic resins allows making parts with unique designs that can be manufactured quickly, inexpensively, and better meet requirements to minimize the coefficient of drag. These unique designs allow the designer to reduce the total number of parts in an assembly, saving labor costs. Finally, capital costs to manufacture plastic parts are relatively low compared to more traditional materials.

The durability of plastics increases the environmental benefits and the affordability of the vehicle. Most plastic parts are designed to last the lifetime of the vehicle, thereby reducing the material and energy demands of the vehicle. For instance, the polyurethane foam used for seat cushioning generally performs well for the lifetime of the vehicle, despite constantly applied compression to which the cushioning is exposed for many years. Plastic parts also withstand significant impact with no damage. Finally, technology is developing to repair plastic parts that have been damaged, decreasing demand on landfill and reducing material usage. The American Plastics Council, an industry association of plastic resin manufacturers, reports that an average of 30% of all plastic parts removed during the dismantling or repair of a damaged vehicle are reused, as is, or are repaired and reused in a similar application (5). Member companies of the American Plastics Council include the following:

Amoco Corporation	Huntsman Chemical Corporation
ARCO Chemical Company	ICI Americas Inc.
Ashland Chemical Company	Lyondell Petrochemical Company
BASF Corporation	Mobil Chemical Company
Bayer Corporation	Monsanto Company
The BF Goodrich Company	Montell North America, Inc.
Chevron Chemical Company	NOVA Corporation
The Dow Chemical Company	Occidental Chemical Corporation
DuPont	Quantum Chemical Company
Eastman Chemical Company	Rohm and Haas Company
Exxon Chemical Company	Shell Chemical Company
FINA, Inc.	Solvay America, Inc.
GE Plastics	Union Carbide Corporation
Hoeschst Celanese Corporation	

Plastics use in automobile manufacturing provides significant environmental benefits, including the following.

Feature	Benefit
Light weight	Reduced fuel consumption
	Reduced emissions
	Reduced material usage
	Longer tire wear
	Smaller engine requirements
	Reduced suspension requirements
	Reduced fuel consumption for parts delivery
Durability	Longer, more reliable vehicle life
	Better economy to the owner
	Reduced landfill demand
	Reduced material usage
Manufacturability	"Net shape" engineering minimizes material usage
	Efficient part manufacturing
	Consolidation of parts, conserves resources

The relatively *low specific gravity* of plastics reduces the weight of the vehicle, which allows the OEM to meet federal Corporate Average Fuel Economy (CAFE) requirements without reducing the vehicle size. It has been estimated that constructing vehicle fenders with plastics versus conventional materials permits a saving of 7.5×10^6L (2×10^6 gal) of gasoline for each model year's production (4).

Life cycle analysis, as described later, defines the energy requirements and emissions from each part of the life cycle of the vehicle: the manufacturing phase, the use phase, and the scrapping phase. Not surprisingly, the use phase of the vehicle's life cycle has the most potential to impact the environment. As will be discussed, reducing the weight of the vehicle affect the environment over the life cycle of the vehicle more than modifications in manufacturing processes or increasing the recyclability of the scrapped vehicle.

RECYCLING AND RESOURCE CONSERVATION

The recycling of automotive plastic parts is not new. For several years battery cases, made with polypropylene, have been recycled. The Battery Council estimates that at least 98% of all scrapped batteries were recycled in 1994 (6). Assuming that 70% of all batteries are made with polypropylene resin and there is an average of two pounds of plastic per battery, over 45,400 metric tons (100 million lbs) of polypropylene (PP) are recycled each year from automobiles and trucks. An efficient and sustainable collection system has developed to recover the lead in the battery and the PP resin. Applications for recycled PP resin include new battery cases. Thus a closed-loop recycling program is accomplished. More recently, post-consumer plastic resin bumper fascias are being recycled from scrapped cars, especially those made with thermoplastic olefin (TPO) resin and polycarbonate—polybutylene terephthalate (PC/PBT) resin. These resins are recycled into similar or new applications in the vehicle. The process scrap from bumper fascia made with reaction injection molded (RIM) polyurethane resin is also recycled back into the original application using a relatively new "third stream" technology (7). A more thorough account of current recycling programs is listed later on.

In addition to recycling of plastics from scrapped vehicles, there is a long history of resale and reuse of plastic parts which are dismantled and re-sold as used parts by automotive recyclers. The general practice is to remove only those parts with a high probability of selling in the shortest period of time. The demand for such parts is predicted by experience or by an "on-line" computer system shared by many auto recyclers (8). It is estimated that as much as 45,400 metric tons/yr (100MM lbs/yr) of plastics are recovered in this manner (9).

Benefits of Recycling

There are several benefits resulting from increased recycling of plastics from scrapped vehicles. The benefit most referred to may be less significant than originally estimated. Landfill avoidance is actually quite small. Currently, it is estimated that 1 million tons/year of plastics (hard plastics, foams and films) from scrapped vehicles are landfilled. Although this is a large amount, it represents only about 0.5% of the 207×10^6 t/yr of municipal waste that is landfilled each year (10). Therefore, the automotive industry must recognize the relative importance of benefiting the environment by increased recycling of the automobile versus other environmental benefits, such as increased fuel economy and lower emissions.

A second benefit resulting from increased automotive plastic parts recycling is lower cost disposal and improved economics for the auto shredder. It is unclear now how significant this is. A study at the Massachusetts Institute of Technology (MIT) analyzed the impact on automotive shredder operation economics, if plastics content significantly increased or if the cost of landfilling the Automotive Shredded Residue (ASR) significantly increased (11). As can be seen from Figure 3 increasing tipping fees from \$40 to \$60/t is equivalent to scrap steel prices fluctuating from about \$55 to \$65/t. It is not uncommon for scrap steel prices to fluctuate by as much as \$50/t over a few years. However, it is unlikely that a \$20/t increase in tipping fees will occur soon. Therefore, it seems unlikely that the impact of tipping fees will seriously threaten the viability of efficient auto shredders.

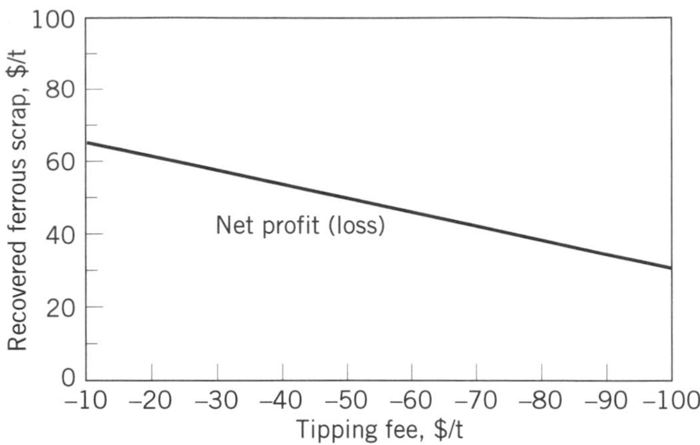

Figure 3. Auto shredder's net profit (loss) equivalency per ton of recovered ferrous scrap with changes in landfill tipping fees for the automotive shredder residue (11).

The MIT study also indicates that a doubling of the amount of plastics in the vehicle from the current 113–225 kg/vehicle (250–500 lbs/vehicle), will have less impact on shredder economics than a \$10/t decrease in scrap steel prices (Fig. 4). Again, the fluctuation in scrap steel prices is significantly more likely to occur than a doubling of the amount of plastics in the average vehicle (11), inferring that an increase in plastics content will not significantly impact the viability of the efficient shredder operator.

The ASR without plastics may be less environmentally friendly than the current mixture in ASR which contains approximately 25–35% plastics. The National Research Council of Canada has demonstrated that automotive shredder residue actually provides a benefit to the landfill by locking-up some problem materials, such as lead and cadmium. Additional studies have implied a correlation between the plastics-rich, large particle size fraction of ASR and improved performance in reducing the leachability of ASR that had been spiked with various materials, such as lead, zinc, and cadmium. While this research is ongoing, and inconclusive at this time, plastics in landfill may be a necessary component that keeps ASR safe to landfill in cost-effective municipal landfills (12).

In addition, using ASR in the landfill makes excellent "day cover", ie the cover spread over that day's addition of garbage to the landfill to prevent paper from blowing into the neighbors' properties and to limit a bird and rodent problem. Current soil or dirt used for day cover costs an average of \$2.50–\$3.25/t and is considerably higher if the landfill operator needs to truck the soil from other locations to meet the quality required for day cover. The soil used for day cover also occupies 20–25% of the volume in the average landfill. Therefore, it is beneficial to carefully managed landfills to use ASR as a day cover and reduce the volume of soil required each day. Considering that new landfills built in full regulatory compliance cost as high

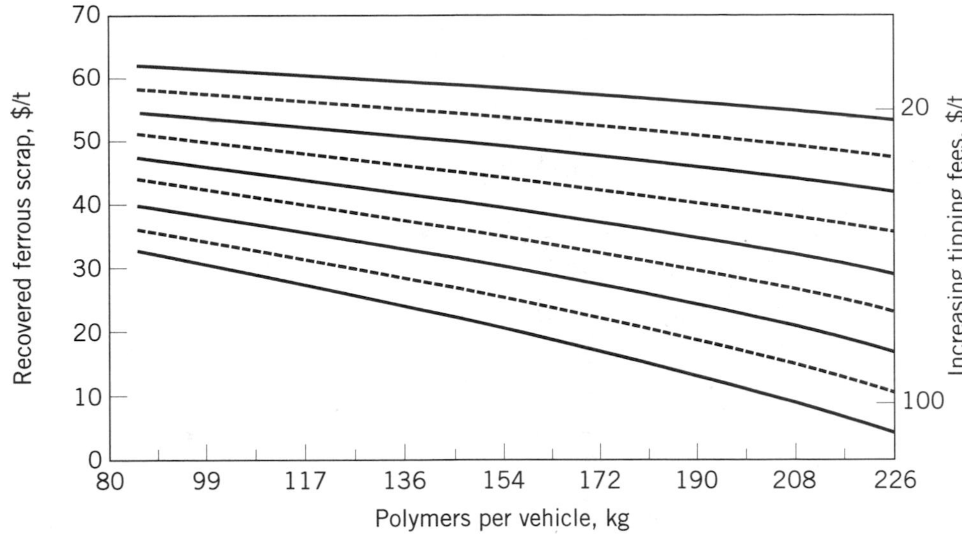

Figure 4. Auto shredder's net profit (loss) equivalency per ton of recovered scrap with increasing tipping fees and changes in amount of plastics per vehicle. Ferrous scrap (—) price = $100/t; nonferrous scrap (- - -) price = $900/t (11).

as $400,000, or $40/t of capacity, according to the Solid Waste Association of North America, it is important to evaluate the advantages of using ASR as a landfill day cover (13).

Another benefit of improved recyclability of automotive plastic components is that the public prefers increased recycling, which may improve the marketability of a more environmentally friendly vehicle. Several OEMs have advertised the recyclability and recycled content of their vehicles, but it is inconclusive if this has created showroom appeal for today's buyers. Some feel that future generations will value the recyclability of the vehicle more than the current generation of buyers, because younger generations are exposed to recycling at an earlier age (14).

It is also assumed that increased automotive recyclability will avert future recycling legislation. European debate about the scrapping of end-of-life vehicles has continued since 1992. European Union legislation which places the responsibility for the proper disposition of the vehicle on the shoulders of the vehicle manufacturer, is predicted by the end of 1998, and has already passed in a few countries. Recycling targets vary by country, and the final outcome of regulation is still unknown.

Finally, a long-term benefit of increased end-of-life vehicle recycling is a lower cost raw material source for plastic resins. Today, some plastics are economically recycled from the electronics industry, such as computer housings and telephones. Some plastics such as milk jugs, soda bottles, and large water bottles, are also recycled from the consumer packaging industry. In these specific cases, the recycled materials are less expensive than the virgin plastic resins. As the auto scrapping infrastructure and recycling technology is tailored to automotive plastics, the automotive industry is optimistic that lower priced recycled plastic resins will become available.

Recycle Rate vs Recycled Content

Recycling occurs only when scrapped material is reprocessed into a new material and sold for an application. Therefore, it is impossible to discuss recycling rates (the percentage of the plastics in the vehicle that are recycled), without also talking about "recycled content" (the percentage of a new part that is made from post- or preconsumer recycled resin). Of course, recycled content resin can originate from nonautomotive sources, such as scrapped telephones or soda bottles. Likewise, recycled battery cases or polyurethane foam seats from automobiles can be recycled into consumer products or carpet underlayment. Nevertheless, the two concepts go hand-in-hand and initiatives in both recycle rate and recycled content are needed for a successful recycle program.

The Federal Trade Commission (FTC) has published guidelines for the terms "recyclable" and "recycled content" for content calculations, titled *Guides for the Use of Environmental Marketing Claims* (15). It states that a recycled content claim can be made only when certain requirements are met, including the diversion from the waste stream of pre- and postconsumer scrap that otherwise would have gone into the waste stream. Also, the content of preconsumer and postconsumer recycled resin must be clearly identified, if not 100% postconsumer scrap. The American Automotive Manufacturers Association (AAMA) has also published an AAMA Recycling Definitions document on the same subject (16). The definitions for recycle content and recyclability are described in more detail later in the article.

The Ford Motor Company issued a goal of using 25% recycled content in its plastic parts in all newly designed vehicles after 1997, thus creating a market for recycled material (17). Several OEMs have established recyclability guidelines for their engineers solely to modify designs which improve the economics of recycling the vehicle once it is scrapped. The details of these guidelines will be discussed later.

VEHICLE SCRAPPING INFRASTRUCTURE

Because of the development of the auto shredding and electric furnace technologies in the 1960s, end-of-life vehicles are economically recycled and the vehicle recycling infrastructure is self-sustaining and profitable without government mandates. Even in the late 1960s and early 1970s, as much as 20% of end-of-life vehicles were abandoned, causing an unsightly problem and a place for rodents and insects to breed. However, by 1994, 95.2% of all end-of-life vehicles were recovered for recycling. The auto shredder economically reduces the vehicle to fist size pieces and separates the steel from other materials (Fig. 5). The electric furnace requires relatively low capital and

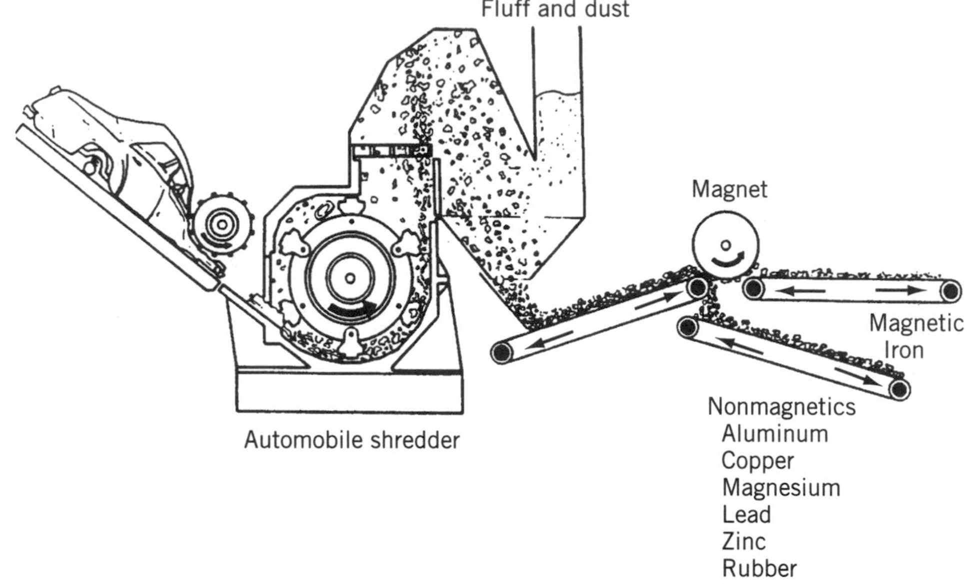

Figure 5. Simplified illustration of a typical auto shredder hammermill and ASR separation process. Alternative processes separate the fluff and dust by blasting air across the conveyor belt (18).

accommodates higher levels of scrapped steel in the feed than possible with basic oxygen furnaces, thus using steel scrap from end-of-life vehicles as feed, despite some level of contamination (18).

ASR from a typical shredder contains approximately one-quarter to one-third plastics by weight (19). Although exact figures are not available, in 1996 about 77 kg (170 lbs) of plastic parts, foam, and skins were landfilled from each scrapped vehicle, originally manufactured about 10 years earlier.

INDUSTRY INITIATIVES

Recycling technology is being developed by several segments in the industry. In addition, the OEMs are developing design guidelines to ultimately reduce the cost of removing plastic parts and separating the plastics from other materials at the end of the vehicle's life. The resin manufacturers are developing grades of plastic resins that contain post-consumer recycled material. Plastic resin recyclers are developing processes that can cost effectively separate plastics from contaminants, such as paint, incompatible resins, and metal parts.

American Plastics Council

Significant progress toward plastics recycling is being made by industry groups focused on determining the cost and value of recovering plastics from scrapped vehicles. The American Plastics Council (APC) represents 27 major plastics manufacturers. Since the early 1990s it has focused on developing and better understanding the viability of various recovery technologies for automotive plastics.

The work of the APC can be divided into four phases: Phase I, information gathering; Phase II, technology assessment; Phase III, technology optimization; and Phase IV, technology demonstration (20). Following Phase IV, a pro forma business plan can be developed by plastic recyclers. This information also allows the industry to properly evaluate which resource recovery option (recycling, energy recovery, feedstock recovery, ASR re-use, etc) is best suited for various automotive plastic parts. In addition, this information is being used by the OEMs to develop design guidelines. Finally, this information will assist with life cycle analyses of the automobile, to ensure that overall environmental impact, energy efficiency, and functional performance are properly balanced.

The APC has designed its projects around the technologies required to recover the plastic components. Table 3 lists the projects according to the vehicle scrapping infrastructure. The results of these studies will be described.

Polyurethane Recycle and Recovery Council

Another industry group focused on plastics recovery is the Polyurethane Recycle and Recovery Council (PURRC), a group organized by the Polyurethanes Division of the Society of the Plastics Industry, Inc., shown below (members as of Mar. 1996).

Air Products and Chemicals, Inc.	The Dow Chemical Company
AlliedSignal, Inc.	Eastman Chemical Company
ARCO Chemical Company	ICI Polyurethanes
BASF Corporation	Olin Corporation
	Bayer Corporation

Its mission is to develop and demonstrate polyurethane recycling and recovery from all markets and applications, including automotive seating and instrument panel topper pads. Figure 6 (21) lists the most popular technologies for recycling and recovering polyurethane-based components, such as auto seat cushioning and RIM fascia. Polyurethane foam seat cushioning may be one of the first large-scale plastic products to be recycled from scrapped vehicles. Relatively large quantities can be recovered from the average automobile (about 10 kgs/car (22 lbs/car)), it is easy to identify, it is used

Table 3. APC Progress in Four-Phase Approach

Progress	Phase I, Generating Information	Phase II, Evaluate Technology	Phase III, Develop Technology	Phase IV, Demonstrate Technology
Repair and reuse	+			+
Advanced dismantling	+	+	+	
Identification and sorting	+	+	+	+
Paint and coatings removal	+	+	+	
Size reduction	+	+	+	+
Separation	+	+		+
Advanced separation	+	+	+	
ASR feedstock recovery	+	+		
ASR energy recovery	+	+		1995–1996
ASR application development	+	+	+	

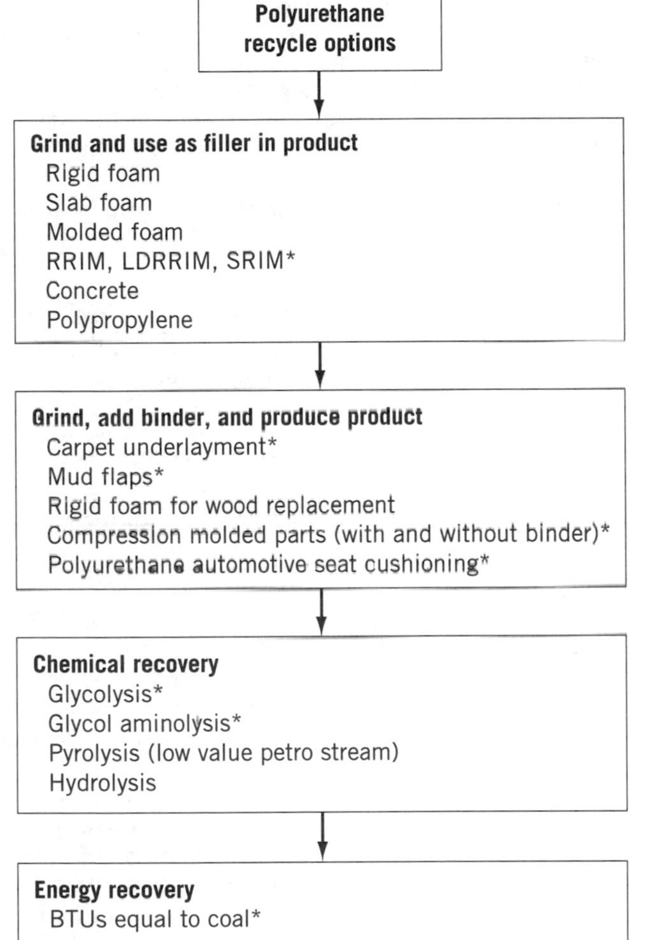

Figure 6. Processes that can be used to recycle and recover polyurethane products. Processes with an * have been commercialized (21).

almost universally in every car, the recycling process is relatively inexpensive (separate from cloth and grind into chunks (crumbs) about 1.3 cm (1/2 in.) thick), and it can be recycled into a relatively high valued application (rebond for carpet underlay) (22).

SMC Automotive Alliance

The SMC Automotive Alliance is composed of raw material suppliers and molders of sheet molding compound products that are sold to the automotive industry. An example of an SMC product is glass filled polyester thermosetting resins used in the body panels of the Chevrolet Corvette. The recycling committee of the SMC AA has helped implement a program whereby SMC process scrap, including the recovered glass fibers, is recycled into new automotive and nonautomotive parts (23–25).

The Vehicle Recycling Partnership

The Vehicle Recycling Partnership (VRP) was organized by Chrysler Corporation, General Motors Corporation, and the Ford Motor Company in North America. Together they have sponsored the Vehicle Recycling Development Center (VRDC) in Highland Park, Michigan, to understand and advance the knowledge of design for dismantling and recycling by full-scale automotive dismantling experience. The VRDC also is a meeting place for various recycling committees organized by the VRP and attended by personnel from plastics resin manufacturers, auto recyclers (dismantlers), plastics recyclers, automotive shredder operators, and automotive component manufacturers. The committees are focused on interior trim, bumper systems, seats, instrument panels, and sound management products made from recycled seats (26).

DESIGN CONSIDERATIONS

In the mid-1980s the Institute of Scrap Recyclers (ISRI) introduced a program trademarked Design For Recycling, whose goal is to encourage preproduction planning for safe and efficient recycling (27). This concept represents the most important contribution that can be made to advance the recyclability of the vehicle. However, the compromises required to advance recycling design may be in conflict with other environmental and economical considerations. Indeed, as illustrated in the section on Life Cycle, whether the end-of-life vehicle is recycled or landfilled is of little consequence to the environment when compared to the efficiency of the vehicle during use. On the other hand, improving the recyclability of the plastics has other benefits, as presented earlier, as long as the recycling can be done sustainably. For the majority of the plastics to become economically recycled, automotive design, the auto scrapping infrastructure, and recycling technology need to advance in a consistent and synergistic manner. For instance, the designs to optimize recycling by manual separation differ from the designs to optimize recycling by density-based mechanical separation. Likewise, designs to optimize recycling where the plastics recyclate is to be used by the sponsoring party (resin compatibility) differ from the designs preferred if the plastics recyclate is to be sold on the open market (resin separability).

Designing to optimize manual recycling requires common screws or fasteners, a minimum number of parts with different resins, large parts, well-marked parts, or a rapid identification instrument. On the other hand, designing to optimize density-based mechanical separation is not concerned about common fasteners, although magnetic materials are preferred if metal fasteners are required. Adhesives need to be compatible. All the different resins in a component need to have distinct densities, so that pure resins can be generated. Paints should be compatible with the resin or avoided. Designing to optimize recycling, where the recycling sponsor is going to reuse the plastic recyclate, can incorporate multiple resins as long as the resins are compatible in the correct proportions to generate the physical properties the sponsor requires in its application. The value of the recyclate is equal to the cost to the sponsor of the virgin resin normally used in that application. On the other hand, if not used by the sponsor of the recycling program, mixtures of resins need to be sold commercially. Historically, a mixture of resins does not sell for a value reflective of its properties because of the high degree of requalification, testing, and processing changes required. Therefore, to optimize for recycling where the recyclate is to be sold commercially, it is best if the resins in the part be easily separated, mechanically or by density.

Recycling Guidelines

Each OEM has developed a list of proper design criteria. The Vehicle Recycling Partnership is focusing on proper design to aid recycling by taking vehicles apart at its Vehicle Recycling Development Center. The American Plastics Council has conducted studies to better understand the impact of design on the dismantling of plastic parts. Finally, several plastic manufacturers and Tier 1 integrators have promoted designs that aid in separability, compatibility, or parts consolidation. Recycling guidelines generally include the components shown below (28).

Materials selection: *Where possible, use recyclable materials and materials having recycled content.* Select materials for which recycling technologies are currently practiced. *Minimize substances of concern.* Control or eliminate the use of substances that are potentially hazardous in the manufacture and recycling of vehicles.

Design: *Integrate parts.* Reduce the number of materials and parts used within an assembly. *Facilitate disassembly.* Select fastener systems that facilitate disassembly by various methods, including destruction, after the vehicle reaches the end of its useful life. *Reduce fasteners.* Reduce the number and types of fasteners used. Select fasteners that do not require separation for recycling. *Mark plastic parts.* Use the SAE/ISO recommended practices for marking parts to facilitate recycling and repair.

The emphasis on recyclability has generated the concept that the ideal component system is made of only one resin type. Examples are the "all polypropylene", "all styrenics", "all polyurethane", and "all PET" component system, which may be the seating system, instrument panel, door panel, headliner, or front/rear end bumper systems. The commercial value and sustainability of recycling these integrated systems is still unknown, because, to create the desired properties in the component, the resins are filled with different materials and reacted into different forms that may not produce an acceptable resin when recycled. For instance, the all polypropylene (PP) instrument panel contains PP parts with several different filler levels and also PP foam for cushioning and PP skin on the surface of the topper pad. The value of these mixed PP types, all recycled together, may not be as great as the individual resins and insufficient to cover the cost of recycling the instrument panel. Likewise, it is unknown if the design and material selection required to build the "all compatible system" is justified, if cost or performance of even one part in the component is compromised. It is generally agreed that the best solution is a reasonable emphasis on fewer resin types and parts consolidation, which aid recycling and also add efficiency to the manufacturability of the component (26,29).

In addition to establishing design guidelines, OEMs are developing criteria to measure the recyclability of the complete vehicle. For example, General Motors currently counts the materials listed below as recyclable. Other plastic materials will be added to the list as the scrapping infrastructure for plastics recycling is developed (30).

Materials:	All metals
Components:	Catalytic converters Batteries Polypropylene battery cases
Fluids:	Engine coolant Engine oil CFC

Eco-Efficiency Criteria

The broader perspective for automotive design is to consider the eco-efficiency of the vehicle and its individual components. The term eco-efficiency is defined by The Eco-Efficiency Workshop (organized by the Business Council for Sustainable Development under the United Nations Environmental Program and The Commission of the European Communities), as being "reached by the delivery of competitively priced goods and services that satisfy human needs and bring quality of life, while progressively reducing ecological impacts and resource intensity throughout the life cycle, to a level at least in line with the earth's estimated carrying capacity" (31). Six criteria are evaluated to create an eco-efficiency quotient that predicts the vehicle's compatibility with the environment, as shown below (32). A qualitative or quantitative scoring system is developed to determine the quotient for comparison with multiple design options.

1. Minimum material consumption during *total manufacturing,* including oil, gas, ores, landscape, trees, virgin and recycled materials, etc.
   - Basic materials manufacturing
   - Component manufacturing
   - Automotive assembly
2. Minimum air and water *emissions* during life cycle.

Regulated materials emissions
Health and safety of workers and occupants
Paperwork and permitting

3. Minimum *fuel consumption* during use.
Light weighting
Aerodynamic designs

4. Maximum *long-term performance* of components.
Durability
Reusability
Repairability

5. Maximum *recycling.*
Recycle content
Recyclability of process and end-of-life scrap

VISION FOR THE FUTURE: "SUSTAINABLE VALUE"

The final analysis highlights the need for sustainable value. The concept of sustainable value emphasizes the desirability and the need for future products and services offered by our society to contribute to the long term sustainability of the economic well being of industry and its employees and also to the long-term sustainability of our environment and its resources (33).

A vehicle that uses minimum fuel and creates minimum emissions is the preferred vehicle of the future as our society becomes more conscious of limited resources and ability to manage emissions, a society that prefers a high mileage vehicle, because it saves money on fuel and also because products are preferred that contribute to sustainability of resources. Recycling will be judged successful only if, by recycling, the recycler is profitable and fewer total resources are consumed. Resource management selection, whether it be landfilling, material recycling, parts reuse, feedstock recovery or energy recovery, will be based on the total long-term contribution to conserving resources.

Plastic resins provide many of the necessary requirements for sustainable value of the vehicle. They are light weight and can be molded efficiently into high strength shapes. They can be used effectively in the design of an aerodynamic vehicle. They can be manufactured and used safely. They require less than 5% of the oil our society consumes for energy creation and can be safely converted back to energy at the end of their lives. Finally, because of their durability, they last the lifetime of the vehicle, resist breakage, can be reused, and, in a growing number of cases, recycled back into other products. For these reasons, plastic resins will continue to be one of the materials of choice in our more environmentally conscious society.

The Challenge

Recycling is certainly not new to plastics, as molders and processors have recycled and reused these valuable raw materials in-house since the beginning of their commercial use. This internally generated and reused material is commonly called regrind and represents a very controlled and well understood source of material.

Recycling post-consumer plastics packaging presents a greater challenge than internally recycled regrind because packaging is mixed with other types of plastics, contains all sorts of foreign material (food or other product residues, paper labels, other plastics, dirt, etc) and has experienced some degradation during its life cycle. Nevertheless, plastics packaging recylers have commercialized the post-consumer recycling of some plastic bottles and film by overcoming many technical challenges.

Most plastic parts coming from durable goods streams contain unique challenges not met by the recycling technologies developed for packaging materials. The principal practice today for recovering highly commingled scrap is hand separation, which is frequently not done because it is cheaper to buy new plastic than to recover the scrap. In addition, conventional automated plastics cleaning and sorting processes (eg, as used for bottle recycling) cannot handle these streams because of technical challenges, such as

- The plastics used in the durable goods are more specialized than those used in packaging. Whereas the majority of plastic packaging can be categorized in approximately five commodity types of plastic resin, more than fifty plastic resin *grades* might be required to comprise a similar fraction of the durables market. For example, whereas polyethylene terephthalate (PET) plastic used to make soda bottles may also be appropriate for water bottles, acrylonitrile–butadiene–styrene (ABS) plastic used to make a computer housing is very different from the type used in a refrigerator door, which differs again from that used in an automobile. This broad variety of materials increases the separation challenge.
- Different plastics are frequently attached to one another in durable goods.
- In addition to different plastic types, many parts contain a wide variety of *reinforcements, fillers,* and *pigments,* which change their performance and physical properties, such as density.
- The use of *structural foam* results in moldings with material density that varies even within the same part.
- Durable plastic parts often contain high levels of different types of metals, including wiring, brackets, structural pieces, and molded-in screw inserts, which represent a size reduction and melt processing hazard.
- Durable plastic parts often contain other attached materials, such as rubber, other elastomers, glass, foam, fabrics, and/or plastic films made of different materials.
- Paint and metallic *coatings* on some parts make identification, sorting, and the melt reprocessing much more difficult. Some coatings on the engineering thermoplastics used for durable products cause a decrease in properties if recycled with the plastic.
- Larger and more variable *thickness* wall sections increase the challenges associated with size reduction and particle size and shape control, which is extremely important to effective subsequent material separation.
- *Composites* and *thermosets,* which represent additional separation and reprocessing challenges, are used to a much larger extent in durable goods.
- Collection costs for plastics in many durable applications are higher than those for packaging, where curbside collection programs already exist.

Although these challenges are significant, some of the plastics used in durable products are more highly engineered and more valuable per pound than packaging plastics. By definition, plastics used in durable goods are required to perform for long periods of time and at much higher levels than plastics used in packaging. Therefore, some durable goods streams offer greater potential for recovery in an economically sustainable manner.

As with other types of materials, such as metals and glasses, different types of post-use plastics must be separated from foreign materials and from one another to achieve good performance and acceptable market values. Most plastics are not highly compatible with one another, and although some commingled applications have been demonstrated, particularly with compatibilization technology, they typically are of much lower value than virgin plastic.

In summary, the primary reasons for segregating plastics by type are

- Most foreign material contaminants decrease the performance of the host material, and most plastics are not uniformly compatible.
- Properties are consistent and understood. Even if compatible blends are produced, it would be difficult to ensure consistent composition of a blend made from a mixed recycle stream.
- The maximum formulation/upgrading flexibility is with pure materials.

Because there are many different types of plastics used in durable products and the prelabeling of plastics to assist with identification will not impact most end-of-life durables streams for many years (reflecting the long life cycle of many of these products), some method to sort plastics by type is required. One approach is to depend on an intrinsic material property, such as density, to sort during the recycling operation. This has limitations with highly mixed streams of plastics which frequently have density ranges that overlap one another. Another complementary approach is first to identify the plastic and then sort.

SUMMARY

The technologies of mechanically recycling plastic components from end-of-life vehicles are still evolving. Until these technologies and the associated infrastructure are optimized, it is not possible to know if mechanical recycling is the preferred resource management process. Other recycling and recovery methods, such as feedstock recovery (thermal or chemical regeneration of the resin building blocks), energy recovery, long-term applications, such as addition to high-performance concrete and landfill day-cover, and repair and reuse of parts are all acceptable and are also being optimized.

As shown by life cycle considerations, cost effectively manufacturing lightweight and aerodynamically designed vehicles with plastic resins improves the overall environmental performance of the vehicle, regardless of the preferred end-of-life treatment options. In addition, plastics provide economical solutions for improved safety, comfort, and aesthetics.

The next 100 years will place greater demands on the environmental and safe performance of automobiles, including efficient and environmentally sound manufacturing, use, and end-of-life treatment. The technologies developed today to advance the environmental performance of plastics are key to the role of plastics in the vehicle of the future.

BIBLIOGRAPHY

1. W. F. Gum, Jr., W. Riese, and H. Ulrich, *Reaction Polymers,* Hanser Publishers, Munich, Vienna, New York, Barcelona, 1992, pp. 1–4.
2. D. Winter, *WARD'S Auto World,* Sept. 1993, p. 34.
3. *Automotive Plastics Report '95*, Market Search, Inc., Toledo, Ohio, 1995.
4. *Green Wheels,* American Plastics Council, Washington D.C., 1993, p. 3.
5. "Repair and Reuse of Automotive Plastic Parts," *Automotive Report Series,* American Plastics Council, Washington, D.C., Oct. 1994.
6. *BCI Recycling Rate Study,* Battery Council International, Chicago, Il., 1994.
7. R. E. Morgan, L. Nemedy, S. G. Yester, D. Peterson, and B. Armstrong, *Automotive Eng.,* 25–28 (Aug. 1995).
8. *The Hollander,* Minneapolis, Minn., 1992, pp. 3–4.
9. "Disposal Practices for Post-Use Automotive Plastics," *Automotive Report Series,* American Plastics Council, Washington, D.C., Oct. 1994.
10. *Characterization Of Municipal Solid Waste In The United States: 1994 Update,* EPA530-S-94-042, Environmental Protection Agency, Washington, D.C., Nov. 1994.
11. F. R. Field III, A. C. Chen, and J. P. Clark, "Automotive Recycling: Environmental Policy In A Technologically Complex Market," *World Economic Forum-Automobile Industry Summit,* Massachusetts Institute of Technology, Cambridge, Mass., 1993.
12. M. Day and F. T. Awadalla, *Auto Shredder Residue: A Fixation Medium For Lead,* Special Report No. EC-1270-93S, National Research Council Canada, Ottowa, Ontario, Jan. 1993.
13. T. Arrandale, *Governing,* 70 (Aug. 1995).
14. D. P. Levin, "Imperatives of Recycling are Gaining on Detroit," *The New York Times,* Sept. 6, 1992.
15. *Fed. Reg.* **57**,(157), pp. 36366–36367 (Aug. 13, 1992).
16. *AAMA Recycling Definitions,* American Automobile Manufacturers Association, Detroit, Mich. Dec. 22, 1993.
17. "Ford Initiative Has Placed New Urgency on Automotive Plastics Recycling," *Reinforced Plastics Newsletter,* **XVIII**(10), 1–2 (Mar. 7, 1994).
18. *The Automobile Cycle: An Environmental and Resource Reclamation Problem,* U.S Environmental Protection Agency, Washington, D.C., SW-80ts. 1, 1972; *The Impacts of Material Substitution on the Recyclability of Automobiles,* American Society of Mechanical Engineers, New York, 1994, p. 50.
19. *Inside Automotives,* 42–45 (Jan. 1994).
20. J. L. Fosnaugh, *Int. Autom. Tech. Update,* 41–44 (1994–1995).
21. R. J. Cain, *Body Eng. J.,* 58–64 (Fall 1995).
22. J. Fosnaugh, *Autom. Trans. Interiors,* 76 (May 1994).
23. C. L. Smith and F. J. Good, *Plastics Eng.,* 27 (Jan. 1996).
24. E. Millerschin and D. Graham, *Body Eng. J.,* 31 (Fall 1993).
25. W. R. Diem, *Automotive News,* 24B (June 12, 1995).
26. B. Miller, *Plast. World,* 32–37 (Oct. 1993).
27. *It's Time To Design For Recycling,* Institute of Scrap Recycling Industries, Inc., Washington, D.C., 1991.
28. *Backgrounder,* American Automobile Manufacturers Association, Washington, D.C., Dec. 1993, p. 3.

29. *Fortune,* 104 (Feb. 6, 1995).
30. W. K. Lange, *Method To Calculate Percent Recyclable,* Technical Paper, SAE International Congress and Exposition, Feb. 27–Mar. 2, 1995.
31. N. Robins and C. Fussler, "Getting Eco-Efficient—Competitive Advantages Through Eco-Efficiency," *Report of the Business Council for Sustainable Development,* UNEP, Antwerp Eco-Efficient Workshop, Nov. 1993, p. 9.
32. J. Fosnaugh, technical data, The Dow Chemical Company, Midland, Mich., 1995.
33. J. H. Benson and J. L. Fosnaugh, "A Business View of the Recycling of Plastics from Durables Goods," *The Society of Automotive Engineers International Congress and Exposition,* SAE, Warrendale, Pa., Feb. 1996, p. 169.

B

BENEFICIATION OF COAL. See COAL PREPARATION.

BIOANALYTICAL METHODS

VALÉRIE PICHON
M.-C. HENNION
Ecole Supérieure de Physique et de Chimie Industrielles (ESPCI)
Paris, France

Today there is an increasing effort to monitor the environment for compounds that may pose a risk to human and ecosystem health. The monitoring also provides a better knowledge of fate and transport of pollutants in the environment. Thus analytical methods that can be applied to different types of matrices (waters, soils, sediments, food) are needed. Classical analytical methods are time-consuming, expensive, and require specialized instrumentation such as liquid or gas chromatography and associated pretreatment methods. The emergence of biochemical methods as an alternative to these conventional techniques have proven to be sensitive, reliable, cost-effective, and rapid. They also allow a high sample throughput analysis. Biochemical methods are based on the use of biological substances such as enzymes, antibodies, DNA, and microorganisms. These biological entities should respond selectively to the presence of one or several analytes. However the reaction between the target compounds and the biological element cannot be directly measured. A special device is used to measure the signal; this device can be an enzymatic reaction or a fluorescent or radiolabeled compound. It can also be an electrochemical, optical, or acoustic transducer as for biosensors. These biological entities can then be used in a classical analytical technique, as in the case of the immunosample pretreatment.

The most important biological techniques are the immunochemical methods, which are based on the use of antibody, or a mixture of antibodies—the analytical reagent—synthesized against an antigen—the target analyte. Antibodies are the key components of all immunochemical methods. They are characterized by specific recognition sites in their structure, which enables a highly specific interaction with the antigen. The high affinity toward the antigen and the specificity of the antigen–antibody interaction allow a sensitive and selective determination of the pollutants in various matrices. Immunochemical methods in environmental analysis include mainly immunoassays and the use of antibodies for sample preparation, for detection in flow injection analysis, and for immunosensors.

Antibodies have been used for a quantitative immunological assay of human insulin in body fluid (1). Since that time, the main applications were in the clinical field for the detection of hormones, drugs, and viruses. Although in 1980, this technique was used in immunochemistry for environmental analysis (2), few immunoassays (IAs) were commercially available for pesticides in Europe in the early 1990s. Recently the use of IAs for environmental analysis has dramatically increased, and many examples have demonstrated the reliability of this technique and its advantages over conventional analytical methods.

Sample preparation based on immunochemical methods is also an increasing area of research. This technique is used to simplify the sample pretreatment required before the analytical separation step to obtain more reliable results.

IMMUNOASSAYS

Immunoassay techniques were first developed in the clinical field to detect and quantify organic compounds at low concentration, down to the fentogram level (10^{-15} g). The generation of antibodies for small molecules, such as pesticides and many organic contaminants, is difficult because small molecules are unable to produce an immune response. This is certainly the major reason for the later development of IAs for environmental analysis and for their clinical application. Moreover, the recent progress in the strategies for the rational design of haptens for the synthesis of antibodies to small molecules partly explains the strong increase in recent interest (3–15) (see IMMUNOASSAY, ENVIRONMENTAL).

Antigen–Antibody Interaction

The antibodies (ABS) used in IAs belong to the immunoglobulin gamma (IgG) fraction and are produced by mammalian lymphocyte B cells as part of the immune system response to foreign substances. An IgG molecule consists of two identical heavy chains of 50,000 Da linked by disulfide bonds and two light chains (25,000 Da) (Fig. 1). One heavy chain is connected to a light one by a disulfide bond. Both heavy and light chains are divided onto constant and variable domains based on their aminoacid sequence variability. In the variable region, there is a considerable variability of the amino acid arrangement from one Ab to another one that confers the specificity of the

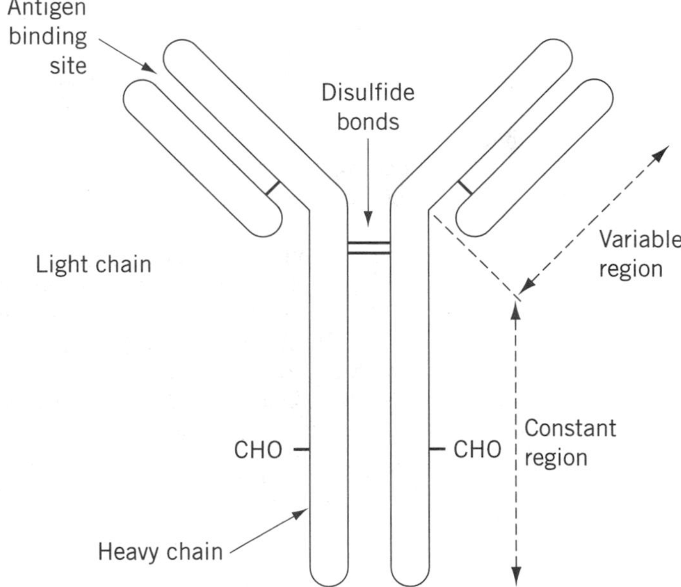

Figure 1. Structure of an IgG antibody molecule.

recognition site for a given Ab. The antigen–antibody (Ag–Ab) interaction involves only noncovalent bonds such as Van der Waals interactions, hydrogen bonds, electrostatic bonds, and hydrophobic bonds. Since these interactions occur over short distances, a close steric fit combine with an accurate match of oppositely charged ions will promote binding. All these stringent binding requirements make the recognition of Ags by Abs very selective (3).

The antigen–antibody interaction obeys to the law of mass action:

$$\mathrm{Ag} + \mathrm{Ab} \leftrightarrows \mathrm{Ag{-}Ab}$$

$$K = \frac{[\mathrm{Ag{-}Ab}]}{[\mathrm{Ag}][\mathrm{Ab}]}$$

where K is the affinity of the Ab to the Ag. K values are often in the range of 10^9-10^{12} L · mol^{-1}. These high values explain that the complex Ag–Ab can be formed with low concentration of antigens or antibodies.

Antibody Production

Technologies for developing specific antibodies is an active area, and some methods are quite sophisticated. To date, commercial Abs have used mainly polyclonal or monoclonal antibodies. In both cases, the different steps required to produce Abs are similar up to the immunization. Low molecular mass compounds, such as most of the pollutants in the environment, are unable to evoke an immune response. Production of antibodies toward a small molecule (hapten) involves the attachment of this molecule (or an analogue of it) to a larger molecule (a carrier protein) to form an immunogenic molecule that can be used for the immunization.

Choice of the Hapten. The design of hapten is the most crucial step in the development of an immunochemical technique for small molecules. Many reviews have shown that an appropriate hapten design determines the features of the resulting antibodies (16–20). The farther the conjugation site is from the target part of the molecule, the more specific the antibodies are. Thus the choice of conjugation site determines if the Ab will recognize a compound or a class of related compounds. The optimum hapten for a selected target analyte must be a near-perfect mimic of that molecule, in structure, geometry, electronic density, H bonding capabilities, and hydrophobic properties (21). However, characteristic portions of the molecule may sometimes be sufficient to generate valuable antibodies. Often, the target compound contains a functional group that can be covalently coupled to the carrier molecule (NH_2, COOH, OH, SH, HCO). If not, an active group must be introduced in the hapten molecule without masking its specific parts used as recognition sites (22). Once the characteristic site has been chosen, the handle should be attached as far as possible from this site, because its exposure to the immune system will be then maximized. Moreover, the strategy will be different if a single compound within a class is targeted than if the whole class is. For the class-selective assay, the handle will be best located at a position that leaves the common sites exposed to the immune system. For small molecules, the use of a spacer arm in the linker is required to favor recognition by the immune system. The length and the chemical structure of the linker should be selected to maximize the exposure of the target molecules, and the spacer itself should not be selectively recognized. A spacer of three to six atoms has been shown to be optimal.

Synthesis of the Immunogen. The selection of the conjugation method depends mainly of the functional group of the hapten. The stability of the hapten during this synthesis is important and can require some functional protection before protein conjugation. The carrier protein must be from a different species than is the host animal to obtain a strong immune response. The carrier proteins most frequently used in IAs are bovine serum albumin (BSA), β-lactoglobuline (β-LG), keyhole limpet hemocyne (KLH), conalbumine (CONA), ovalbumine (OVA), and tyroglobuline (TG). It is worthwhile to verify the coupling reaction and to characterize the immunogen structure before immunization, ie, to determine the epitopic density corresponding to the number of haptens linked to the protein. A good immune response seems to be obtain with 10–30 haptenic groups per 100 kDa carrier protein (23). The measure of this epitopic density depends on the carrier molecule but is mainly based on spectrophotometric measurements. Nevertheless, labeled haptens used in association with spectrophotometric and electrophoretic methods or mass spectrometry with different interfaces have also been employed.

Polyclonal Antibodies. Polyclonal antibodies (PABs) are obtained by immunizing the host animal with the immunogen. In the immune system, a single B lymphocyte produces a single type of antibody. After the injection of the immunogen B lymphocytes produce several different Abs that are specific to different part of the immunogenic molecule. Then the obtained serum contains a mixture of antibodies with various affinities called polyclonal Abs.

They are often produced in rabbits but much larger amounts of pAbs can be obtained using sheep, goats, or cows. After each boost (every four or five weeks and in two or three rabbits to minimize the animal variability) the Ab titer is measured to check the progress of the immunization. After three to six months, the Ab titer no longer increases. Then the collected serum must be purified to obtain the IgG fraction.

If the polyclonal antibodies are proven to be effective, they are used in many commercial kits. They have some limitations because polyclonal antiserum can vary from one animal to another and the supply of Abs ends when the animal dies. A constant supply of Abs can be provided by the hybridoma technique generating homogeneous antibodies.

Monoclonal Antibodies. The technique of monoclonal Abs first developed in 1975 was applied to antipesticide Abs in 1985. Because B lymphocytes cannot grow in culture, only a finite production of polyclonal Abs can be expected. But this can be overcome by the use of this hybridoma technique based on the fusion of spleen lymphocytes from immunized mice that are isolated and fused with myeloma cells (6). The resultant hybridomas are purified, cloned, and screened for the ability to produce high affinity Abs of the desired specificity. This screening is the critical step of the production of monoclonal antibodies (mAbs). Suitable clones are propagated on a larger scale in culture flasks or in larger bioreactors. This can also be made *in vivo* via ascetic fluids in mice but with low yields. The advantage of this technology is an unlimited production of mAbs with constant affinities. However, in contrast to pAbs production, this technique is difficult, laborious, time-consuming, and expensive. Nevertheless, mAbs are now used in commercial immunoassays for their long-term supply. Because they have constant characteristics, the performances of the kits are expected to be reproducible.

Other techniques for cloning and engineering antibodies are evolving rapidly (24,25). For example, DNA technology is an active area, even though the development of recombinant antibodies is expensive and time-consuming.

Description of Immunoassays

As the antigen–antibody reaction cannot be directly measured, immunoassays are based on the use of labels to detect the immunological reaction. This label can be a radioisotope in radioimmunoassay (RIA), fluorescent (FIA), or chemiluminescent species. However, enzyme labels such as horseradish peroxidase or alkaline phosphatase have gained in popularity in the last few years (8,12,14). Among the enzyme immunoassay (EIA), those based on heterogeneous conditions, with either Abs or Ags immobilized on a solid phase, are the most commonly employed. IAs can be based on a competitive or on a noncompetitive system. Competitive IAs involve the measurement of unoccupied sites when using limiting Abs concentration. Figure 2 describes IAs with either Abs or Ags immobilized. In the direct enzyme-linked immunosorbent assay (ELISA), the sample and labeled analytes compete for Ab binding sites. After a separation method, which allows the removal of the free analytes and free labeled analytes, the amount of bound labeled analytes is measured by the signal tracer. In the indirect ELISA, Ags are immobilized on the solid support, and immobilized analytes compete with sample analytes for the nonlabeled antibodies. The amount of Abs linked to the immobilized Ags is determined by the addition of labeled antibodies specific to the nonlabeled Abs. In the noncompetitive ELISAs, unlabeled Abs are immobilized while an excess of labeled Abs is used to determine the amount of analyte recognized by the immobilized Abs. This configuration can be performed only when the analyte possesses at least two binding sites and, therefore, is often not appropriate for small molecules.

ELISA Formats

Two ELISA formats are commercially available; they differ only in the solid phase used for the immobilization. One common format consists of antibody-coated tubes or wells. In another format, Abs are covalently bound to magnetic particles as solid supports. In the tube or well format, the sample (100–200 μL) and a known amount of labeled antigen are added and compete for the limited number of binding sites. In the particle-based format, the sample (250 μL) and known amounts of enzyme conjugate and particles are added in a tube. After an incubation step, lasting 10–60 min depending on the tests, nonbonded reagents are removed by a washing step. According to the law mass action, the more analyte there is in the sample, the more enzyme conjugate it will remove from the binding sites. The analyte concentration in

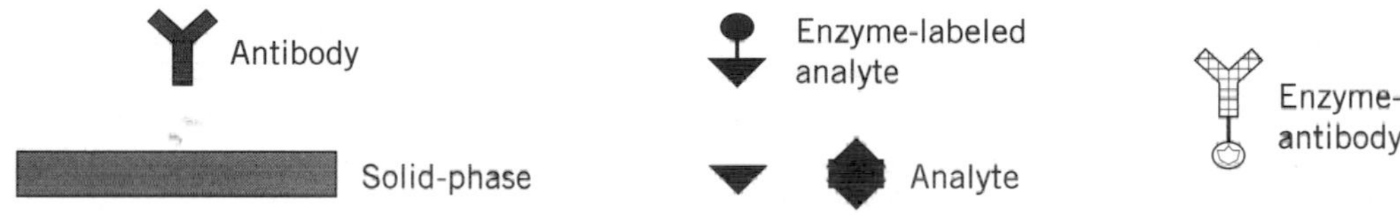

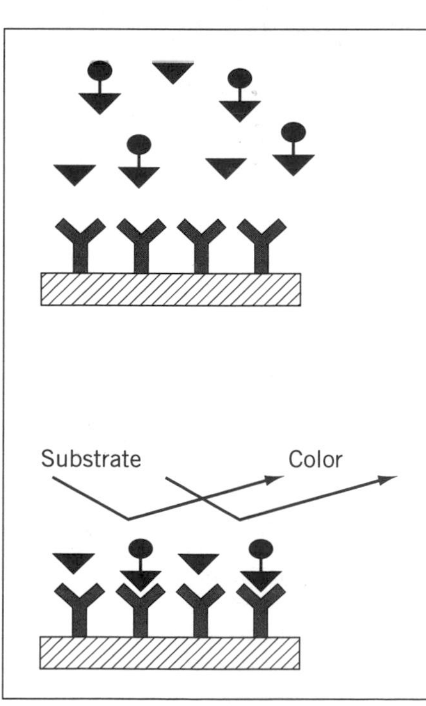

Direct competitive ELISA

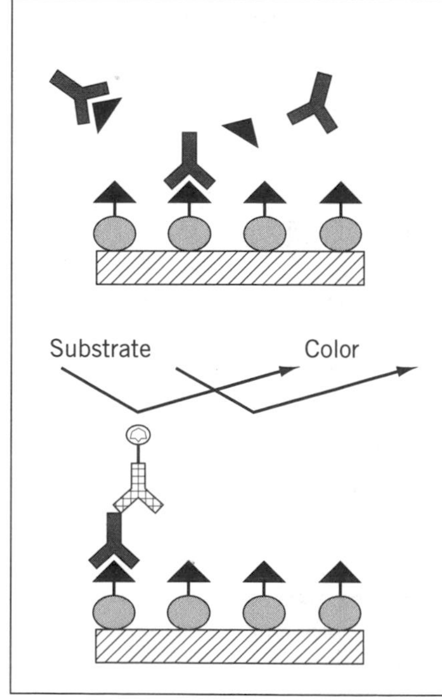

Indirect competitive ELISA

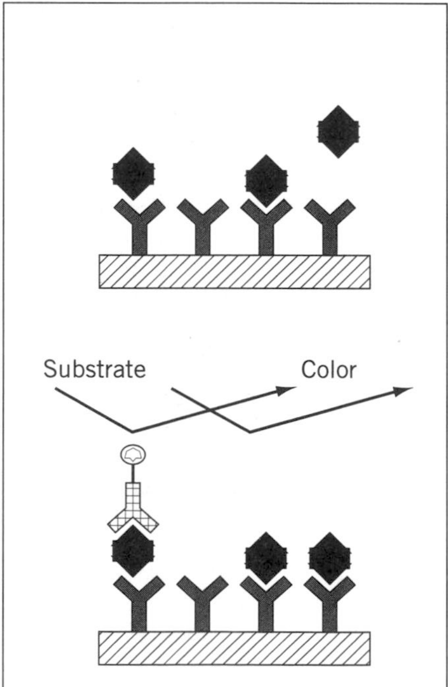

Sandwich ELISA

Figure 2. Competitive and noncompetitive immunoassays.

the sample is determined by measuring the amount of bonded enzyme conjugate. This is measured by adding a substrate and a chromogen: the enzyme reacts with the substrate, which in turn causes the chromogen to produce a colored compound. The color intensity is, therefore, inversely proportional to the amount of analytes present in the samples.

Competitive Immunoassays

Immunoassays, whatever their format, contain a target analyte (or antigen), a specific antibody, and a hapten conjugated to a protein–enzyme that is an analog of the analyte.

Use of ELISAs: Advantages and Drawbacks

Immunoassay acceptance depends on the demonstration of quality and validity compared to more traditional methods. It has been noted that both the strength and the weaknesses of immunoassays must be considered to effectively and efficiently use the techniques.

Appropriate use of immunoassays is first screening and, when possible, quantification in environmental fate and residue studies, because there are many samples to analyze. Conventional methods cannot compete with speed and cost of IAs. Immunoassays can be also selected because analyses can be performed in the field. Immunoassays are appropriate when they provide a sensitivity not attainable by other methods and/or when traditional methods are too cumbersome at the required limit of detection. Their use is inappropriate when there is a small number of samples, when the matrix effects are too important, and when the purpose of the study is to identify transformation products. They cannot be used for quantitative purposes when cross-reactants are present in the samples.

The frequency for confirming positive and negative samples using other analytical methods depends mainly on the type of study and on the intended use of the data. For enforcement applications, such as water monitoring, ideally, all positive samples should be confirmed. However, when most levels are low, the expense of residue verification is not justified and analyses for confirmation should be performed to levels of regulatory significance. The regulatory cut-off level should be determined for each compounds in accordance with applicable laws. To ensure that false-negative results do not occur, a fraction of the negative samples should be re-analyzed by another method.

Other Immunoassay Formats

There have been several attempts to simplify ELISA formats, such as reducing the number of washing steps, avoiding the separation between the free and bound analyte, and accelerating the equilibrium step. Field formats based on the use of strips have been also described. Another attempt was to automate IAs on site with flow-injection techniques.

Polarization Fluoroimmunoassays. Polarization fluoroimmunoassay (PFIA) is a homogeneous immunochemical method that does not require washing or separation of the free and bound analyte. It is a competition method based on detection of the difference of fluorescence polarization between a small fluorescent-labeled antigen and its immunocomplex with specific antibody. It depends on the difference in the signal given by a relatively small fluorescing-labeled hapten in the free form compared to the much higher polarization values when it has been bound to its specific antibody. The polarization or fluorescence is determined by exciting the mixture of antibody, sample, and tracer with vertically polarized light and measuring the intensity of both the vertically and horizontally polarized components of the emitted fluorescence. PFIAs are used in clinical chemistry because of their simplicity, precision, and possible automation.

Flow-Injection Immunoassays. On-site automated flow-injection monitors are a low cost option for obtaining continuous and quantitative data of dissolved aquatic chemical parameters (26). Much research is being devoted toward the automation of immunoassays through the use of continuous-flow systems (27).

In the simplest homogeneous flow-injection immunoassays (FIIAs), no separation of bound and free species is required. The sample is injected in the flow stream containing reagents. Usually, a labeled antibody present in the flow stream produces a signal that changes when the antibody–antigen complex is formed, due to a quenching of a fluorescent label or inhibition of an enzyme label by the binding reaction.

Dipstick Immunoassays. Immunoassays in the dipstick format have been studied (28,29) and will certainly become commercially available for environmental analysis of pesticides in the near future. The main advantage is that they are easy to use in the field: simply dip the strips in water, with no sample preparation. The dipstick techniques involve, first, competitive immunoreaction and, second, color development.

Liposome-Amplified Immunoanalysis. The advantages of liposome-encapsulated markers as signal enhancers of the competitive binding reaction of small molecule immunoassays instead of enzymatically produced color have been demonstrated (30,31). Liposomes provide instantaneous rather than time-dependent enhancement and offer considerable potential for both automated and field assays. Single-use field immunomigration strips for determining alachlor and PCBs have been performed (32–34).

Liposomes are bilayered vesicles that are formed spontaneously when lipids are dispersed in water. During this formation they encapsulate a portion of the aqueous solution in which they are dispersed; and if this solution contains a marker molecule such a dye, it will be present in the aqueous core of the liposome. If the analyte of interest is conjugated to a lipid, this can be also incorporated into the liposome surface. In a competitive liposome immunoassay, the tagged liposomes and the sample containing the analyte are passed over a solid surface onto which the antibodies have been immobilized. Competition occurs between the free analyte molecules and the analyte molecules conjugated to the liposomes. The number of liposomes that bind the antibodies is inversely proportional to the amount of free analyte in the sample. Unbound liposomes move out from the antibody region and can be measured by an appropriate downstream detector. Alternatively, the bound liposomes can be measured *in situ* or a detergent can be added in a flowing stream to release the marker, which is then measured downstream.

Conclusion

Immunoassay techniques are simple, powerful, and inexpensive screening methods that have enormous potential, including the generation of quantitative data. They are gaining acceptance with and the confidence of analytical chemists. They are competing successfully with traditional analytical methods because they are now evaluated by the same criteria according to a well-defined quality assurance plan. But users must know and recognize the methods' limitations. In particular, they must be aware of the data interpretation and know that, depending on the selected application, some immunoassays can be quantitative whereas some other cannot. Given the widespread adoption and the ease with which these assays are performed, many people, manufacturers, academic and industrial researchers, and end users should be concerned with correct data generation and interpretation. Development of guidelines will promote consistent validation and data reporting, and application will certainly aid in a wider acceptance of this technology.

IMMUNOCHEMICAL SAMPLE PREPARATION METHODS

The increasing development of antibodies tailored for immunoassays has induced their use in other fields, especially in the sample preparation area. Generally, analytical methods used for the trace-level determination of pollutants in environmental matrices (surface and groundwaters, soils, sediments) are time-consuming and often not reliable because the sample pretreatment procedures require many steps and are nonselective. Many other compounds are co-extracted, which render difficult the analysis of the target analytes. This problem can be overcome by using immunochemical sample preparation methods based on the use of the antigen–antibody reaction for the selective extraction of an analyte or a group of structurally related analytes from complex environmental matrices. Antibodies are covalently bonded onto an appropriate sorbent to form a so-called immunosorbent (IS), to be packed into a solid-phase extraction cartridge or precolumn. In contrast to IAs, ISs provide individual quantification of each analyte, because, once the compounds have been extracted onto the IS, they are desorbed, separated, and analyzed.

Immunochemistry has been used for a long time for analysis and/or sample pretreatment in medical and biological fields, but its introduction in the environmental field is relatively recent, likely because of the difficulty of making antibodies for small molecules. In the medical field, ISs have been often developed for the analysis of a single analyte and its metabolites that are structurally related to the parent molecule.

The first commercial ISs used in environmental analysis were introduced for the cleanup of samples before determining the aflatoxins (35,36). Other ISs have been described in the literature for the analysis of single pesticides such as carbendazim, chlortoluron, atrazine, and terbuthylazine (37–40). Because of the unavoidable cross-reactivity of antibodies against atrazine, ISs were also developed for trapping atrazine and its major degradation products (41). Some studies have taken advantage of the cross-reactivities of antibodies to develop ISs that were able to selectively extract a whole group of structurally related compounds, such as the triazine and phenylurea pesticides (42–47). Class-selective trapping is now recognized as the key feature of ISs to be tailored for environmental analysis of pesticides. There is also interest in developing IS for single analytes, which are particularly difficult to analyze at trace levels because of the lack of available extraction methods to extract them from water.

Synthesis of Immunosorbents

The overall process for the preparation of an immunosorbent first involves the production of antibodies. All the different steps of this production were described above. After the antibody purification, the immunosorbents are obtained by binding antibodies onto a solid support. Immobilization of antibodies can be done by different means, such as physical adsorption, affinity adsorption, and covalent binding. Physical adsorption, which has been widely used for coating of microtiter plates in immunoassays, is based on hydrophobic interactions and thus cannot be used when the desorption of the trapped compounds requires an organic modifier. The affinity adsorption, using protein A or G immobilized on a stationary phase allows a good orientation of the antibodies, but this process suffers from the risk of lingand leakage and steric problems due to the size of the immobilized protein. Covalent binding appears to be the best process of immobilization to obtain a good stability between the sorbent and the antibodies.

The selected sorbent should have large pores because antibodies are large molecules, be hydrophilic to avoid any nonspecific interactions, and be pressure resistant to be used in on-line techniques. The third requirement is not obligatory for off-line disposable cartridges, and agarose is often used for antibody immobilization (48). Only silica-based sorbents meet the three features, and to reduce the preparation steps, it is faster to use commercial silicas that have already been modified by appropriate functional groups. Early ISs were made from silica containing active aldehyde groups and had a pore size of 300 nm (42). But results have shown a rather low capacity. A silica-based sorbent with a higher specific area, but lower pore size (50 nm), is now selected, although there is a compromise between large pore sizes and high specific surface areas.

Many have attempted to use hydrophilic polymers because they allow an oriented bonding of the antibodies and they have higher capacities. The bonding procedure using silica-based sorbents gives rise to antibodies grafted onto the surface in a random way, so some antibodies are not well oriented toward the analytes. However, hydrophilic polymers all contain π-bonds, which necessarily give raise to nonspecific hydrophobic interactions and dramatically decrease the selectivity of the IS, which is their primary interest in the sample preparation methods.

PERSPECTIVES IN BIOANALYTICAL TECHNIQUES

The area of environmental chemical analysis and monitoring is expanding worldwide. Field analytical methods have been shown to reduce the time and the cost of environmental monitoring. In this regard, immunochemical methods will certainly be pursued as simple, cheap, and efficient ways to monitor pesticides (and other pollutants) in the environment. Much effort has been made in the last decade to develop biosensors in light of the need for fast, sensitive, cost-effective, continuous, and *in situ* monitoring methods. A biosensor is defined as a miniaturized device that integrates an immobilized biological-sensing element attached to an appropriate transducer (14, 49–51). The sensing element can be enzymes, antibodies, DNA, or microorganisms. Advances in biochemistry, molecular biology, and immunochemistry have expanded the range of biological recognition elements, whereas development in fiber optics and microelectronics have expanded the capabilities of signal transducers. Although they are still in the development stage, biosensors show many advantages for the monitoring of micropollutants in the environment. They have been developed for the detection of compounds within different classes such as phenols, PAHs, pesticides, and VOCs; and their potential has been demonstrated in a variety of field applications. These techniques still must be improved, but recent advances in the biotechnology and in the development of transducers should allow to this technique to be rapidly competitive with widely accepted methods. Biosensors are an active area of research and are beginning to move from the laboratory, concept stage to field testing and commercialization. And, as evidenced by the number of reported biosensors for pesticides, it appears probable that biosensors will be among the future tools for monitoring purposes.

Immunoassays are now at an advanced stage. There is an increasing availability of immunoassays now developed for monitoring many pesticides. However, the situation is different in North America (where immunoassays are now accepted as a valuable analytical method) than it is in Europe (where they have still to be accepted and validated). The dipstick format will certainly be the field format of the near future.

The first IA developments dealt with single pesticides, but there is clearly now a trend for developing class-specific immunoassays, especially for the monitoring of herbicides in water because several active ingredients are applied in the same areas and detected together. For the same reason, the simultaneous detection of a plurality of analytes by immunochemical techniques would meet many of the requirements of pesticide pollution monitoring. In this direction, it has been reported that recent advances in nonenvironment target application areas indicate that immunochemically based, simultaneous, multianalysis capabilities are possible, with emphasis on simplicity, avoidance of sample treatment, speed, sensitivity, automation, and low cost (52).

Class-specific immunosorbents are expected to be developed and become available in the near future for immunochemical sample preparation, as a consequence of the wider acceptance of immunoassays by analytical chemists. Commercialization of these immunosorbents will require the availability of reproducible antibodies, which can be better guaranteed by monoclonal antibodies. Our group has compared the class-specific properties of ISs using monoclonal antibodies against isoproturon, atrazine, and simazine. First results indicate that higher capacities are obtained because of the higher concentration of serum containing the monoclonal antibodies; but similar recoveries have been measured for the two groups of pesticides using similar conditions, including the bonding procedure of the antibodies onto silica. This is a promising result for development of class-specific IS. Other classes, such as phenoxyacetic acids and carbamates, are under development.

Moreover, for all the ISs obtained by a covalent immobilization of antibodies on silica, the elution conditions are the same and consist in a mixture of water and organic solvent. Then, by mixing several ISs, it is possible to carry out in one run a selective multiresidue analysis.

Finally, considering all these techniques, future advances in the development of immunochemical methods and biosensors will require scientists of different discipline to join their research efforts. Progresses in genetic engineering techniques, newly discovered microorganisms such as hyperthermophilic bacteria, and new synthetic biomaterials may also bring new developments in this area.

BIBLIOGRAPHY

1. R. S. Yalow and S. A. Berson, *Nature,* **184,** 1648 (1959).
2. B. D. Hammock and R. O. Mumma in J. Harvey Jr. and G. Zewig, eds, *Recent Advances in Pesticide Analytical Methodology;* ACS Publications, Washington, D.C., 1980, p. 321.
3. J. P. Sherry, *Crit. Rev. Anal. Chem.,* **23,** 217–300 (1992).
4. M. Schwalbe-Fehl, *Intern. J. Environ. Anal. Chem.,* **26,** 295–304 (1986).
5. J. M. Van Emmon and V. Lopez-Avila, *Anal. Chem.,* **64,** 79A–88A (1992).
6. B. Hock, T. Giersh, and K. Kramer, *Analysis,* **20**(8), M29–M33 (1992).
7. B. S. Ferguson, D. E. Kelsey, T. S. Fan, and R. J. Bushway, *Sci. Total Environ.,* **132,** 415–428 (1993).
8. R. Niessner, *Anal. Method Instrum.,* **1,** 134–144 (1993).
9. D. Knopp, *Anal. Chim. Acta,* **311,** 383–392 (1995).
10. K. Rubach, *Chem. Plant Protection,* **11,** 185–217 (1995).
11. J. A. van Emmon and C. L. Gerlach, *Environ. Sci. Technol.,* **29**(3), 312A–317A (1995).
12. E. P. Meulenberg, W. H. Mulder, and P. G. Stoks, *Environ. Sci. Technol.* **29**(3), 553–561 (1995).
13. A. D. Lucas, S. G. Gee, and B. D. Hammock, *J. AOAC Intern.,* **78**(3), 585–591 (1995).
14. M. P. Marco, S. G. Gee, and B. D. Hammock, *Trends Anal. Chem.,* **14**(7), 341–350 (1995).
15. M. P. Marco, S. G. Gee, and B. D. Hammock, *Trends Anal. Chem.,* **14**(7), 415–425 (1995).
16. J. M. Schlaeppi, W. Föry, and K. Ramsteiner, *J. Agric. Food Chem.,* **37,** 1532–1538 (1989).
17. M. H. Goodrow, R. O. Harrison, and B. D. Hammock, *J. Agric. Food Chem.,* **38,** 990–996 (1990).
18. R. O. Harrison, M. H. Goodrow, and B. D. Hammock, *J. Agric. Food Chem.,* **39,** 122–128 (1991).
19. M. T. Muldoon, R. N. Huang, C. J. Hapemen, G. F. Freis, M. C. Ma, and J. O. Nelson *J. Agric. Food Chem.,* **42,** 747–755 (1994).
20. S. A. Eremin, Z. V. Samsonova, and A. M. Egorov, *Russian Chem. Rev.,* **63**(7), 611–622 (1994).
21. M. H. Goodrow, J. R. Sandborn, D. W. Stoutamire, S. J. Gee, and B. D. Hammock in J. O. Nelson, A. E. Karu, and R. B. Wong, eds., "Immunoanalysis of Agrochemicals, Emerging Technologies," *ACS Symposium Series, Vol. 586,* American Chemical Society, Washington, D.C., 1995, pp. 119–139.
22. F. Jung, A. Szekacs, Q. Li, and B. D. Hammock, *J. Agric. Food Chem.,* **39**(1), 129–136 (1991).
23. M. Worterg, K. Camman, K. Strupat, and F. Hillenkamp, *Fresenius J. Anal. Chem.,* **348,** 240–245 (1994).
24. H. A. Lee, G. Wyatt, S. D. Garrett, M. C. Yanguela, and M. R. A. Morgan in Ref. 21, pp. 22–30.
25. S. B. Kreissig, V. K. Ward, B. D. Hammock, and P. V. Choudary in Ref. 21, pp. 31–49.
26. K. N. Andrew, N. J. Blundell, D. Price, and P. J. Worsfold, *Anal. Chem.,* **66,** 916A–922A (1994).
27. J. Emneus and G. Marko-Varga, *J. Chromatogr. A,* **703,** 191–243 (1995).
28. C. Wittman, S. Loeffler, A. Zell, and R. D. Schmid, paper presented at the *211th ACS National Meeting,* New Orleans, Mar. 24–28, 1996.
29. C. Wittman, *Int. J. Environ. Anal. Chem.,* **65,** 113–126 (1996).

30. R. A. Durst, L. Locasio-Brown, and A. Plant in R. D. Schmid, ed., "Flow Injection Analysis Based on Enzymes or Antibodies," *GBF Monograph Series, Vol. 14,* VCH Publishers, Weiheim, 1990, pp. 181–190.
31. W. T. Yap, L. Locasio-Brown, A. L. Plant, S. J. Choquette, V. Horvath, and R. A. Durst, *Anal. Chem.,* **63,** 2007–2011 (1991).
32. A. Siebert, S. G. Reeves, M. A. Roberts, and R. A. Durst, *Anal. Chim. Acta,* **311,** 309–318 (1995).
33. S. G. Reeves, S. A. Siebert, M. A. Roberts, and R. A. Durst, *Trends Anal. Chem.,* **14,** 351–355 (1995).
34. S. G. Reeves, S. A. Siebert and R. A. Durst in Ref. 21, pp. 210–222.
35. J. F. Lawrence and P. M. Scott in D. Barcelo, ed., *Environmental Analysis: Techniques, Applications and Quality Assurance,* Elsevier, Amsterdam, 1993, pp. 273–309.
36. M. W. Trucksess and D. W. Koeltzow in Ref. 21, pp. 326–334.
37. A. Marx, T. Giersch, and B. Hock, *Anal. Lett.,* **28,** 267 (1995).
38. G. S. Rule, A. V. Mordehal, and J. Henion, *Anal Chem.,* **66,** 230–235 (1994).
39. D. H. Thomas, M. Beck-Westermeyer, and D. S. Hage, *Anal. Chem.,* **66,** 3823–3829 (1994).
40. S. J. Shahtaheri, M. F. Katmeh, P. Kwasowski, and D. Stevenson, *J. Chromatogr. A,* **697,** 131–136 (1995).
41. D. S. Hage, J. G. Rollag, and D. H. Thomas, *paper presented at the 211th ACS National Meeting,* New Orleans, Mar. 24–28, 1996.
42. V. Pichon, L. Chen, R. Daniel, A. Martel, F. Le Goffic, J. Abian, and D. Barcelo, *Anal. Chem.,* **67,** 2451–2460 (1995).
43. V. Pichon, L. Chen, and M.-C. Hennion, *Anal. Chim. Acta,* **311,** 429–436 (1995).
44. V. Pichon, L. Chen, N. Durand, F. le Goffic, and M.-C. Hennion, *J. Chromatogr. A,* **725,** 107–119 (1996).
45. J. F. Lawrence, C. Menard, M.-C. Hennion, V. Pichon, F. Le Goffic, and N. Durand, *J. Chromatogr. A,* **732,** 277–281 (1996).
46. J. F. Lawrence, C. Menard, M.-C. Hennion, V. Pichon, F. Le Goffic, and N. Durand, *J. Chromatogr. A,* **752,** 147–154 (1996).
47. V. Pichon, H. Rogniaux, N. Fischer-Durand, S. Ben Rejeb, F. Le Goffic, and M.-C. Hennion, *Chromatographia,* **45,** 289–295 (1997).
48. A. Farjam, *The Use of Immobilized Antibodies for the Selective On-Line Sample Pre-Treatment in Liquid- and Gas Chromatography,* Ph.D dissertation, Free University of Amsterdam, Amsterdam, the Netherlands, 1991.
49. M.-P. Marco and D. Barcelo, *Maes. Sci. Technol.,* **7,** 1–16 (1996).
50. K. R. Rogers and L. R. Williams, *Trends Anal. Chem.,* **14,** 289–294 (1995).
51. K. R. Rogers and C. Gerlach, *Environ. Sci. Technol.,* **30,** 468A–491A (1996).
52. A. Brecht and R. Abuknesha, *Trends Anal. Chem.,* **14,** 361–371 (1995).

BIODEGRADABLE POLYMERS

A.-C. ALBERTSSON
S. KARLSSON
Royal Institute of Technology (KTH)
Stockholm, Sweden

Biodegradable polymers cover a vast area of high-molecular-weight compounds. Usually, it is valuable to distinguish between biodegradable polymers of natural and of synthetic origin. Natural biodegradable polymers are a result of a synthesis developed during millions of years of evolution, leading to tailor-made materials for different applications in nature. These biopolymers include proteins, polysaccarides, nucleic acids, or lipids that show completely different characteristics depending on the situation in which they are used. Synthetic polymers, on the other hand, are the result of a mere century of research and development. Synthetic polymers susceptible to biodegradation can be of different types, for example, polymers containing hydrolyzable backbone polyesters. Recent research activity on biodegradable synthetic polymers has often focused on the simulation of different biopolymers or polymers with different backbones, for example, polyanhydrides, polycarbonates, polylactones, etc. Other concepts in the search for new biodegradable materials include the use of microorganisms that can produce polymers, for example, poly(β-hydroxybutyrate) (PHB) and copolymers of PHB. Difficult synthesis and purification steps leading to low yields are avoided by using microorganisms. In packaging applications, a biodegradable additive is often included as a way to promote environmental degradation, for example, starch in polythylene (PE).

BIODEGRADATION

The term *biodegradation* has no clear definition. Several different definitions are being discussed and the question of a general definition is still under debate. The term biodegradation has been used to encompass events taking place in both the natural environment and the living human body, although environmental degradation by microorganisms like bacteria and fungi is probably not comparable with that taking place in mammals. In the field of biomaterials (eg, sutures, bone, reconstruction, and drug delivery), the definition may simply be hydrolysis. Environmentally speaking, however, biodegradation may be fragmentation (eg, fungi penetrating the materials) or loss of mechanical properties of degradation through the action of living organisms (ie, endo- and/or exo-enzymes degrade the materials).

The accessibility of a polymer to degradative attack by living organisms is dependent on not its origin, but its molecular composition and architecture. Complex macromolecules such as lignin and asphalt show great inertness despite being biopolymers. On the other hand, synthetic polymers with intermittent ester linkages (eg, polyesters and polyurethanes) are readily accessible to the biodegradative action of esterases despite their usual enzymatic specificity. In addition, solitary examples of an extremely uniform chain, characteristic of a synthetic polymer molecule (for instance, polyethylene with 100–1000 or more carbon atoms), certainly occur sporadically in nature as an artefact, intermixed throughout with countless types of other aliphatic and aromatic macromolecules of higher petroleum derivatives.

Fungi and bacteria may also use plasticizers and fillers as a source of nutrient and by this way accelerate the aging of plastics. The growth of a pure culture of a single, defining microbial strain on a specified accessory substance in a plastic material is unlikely to to induce the ribosomal production of an entirely different enzyme in the same strain, directed toward the main molecular species in this particular plastic product. Such a mechanism is not in agreement with existing genes to produce a number of molecules of a corresponding enzyme.

It is even more futile to hope that the induction of biosynthesis of yet unknown enzymes might eventually occur in the case of polyethylenes. There is instead a synergism between biodegradation and environmental degradation.

BIODEGRADABLE POLYMERS OF NATURAL ORIGIN

Biopolymers form the basis of intelligence on earth. Besides being sources of nutrients, many natural polymers have been used, for example, in the fabrication of clothes for thousand of years.

The polypeptides (polyamides of α-amino acids) make up fibers and occur in proteins extracted from nonfibrous natural products: Important examples are wool and silk. Several of these natural, biodegradable materials have now been replaced by synthetic ones, although a material such as silk remains a product with very competitive characteristics. Marine cuticle collagen has been investigated for use in adhesives, especially bioadhesives in aqueous environments (1). Bioadhesives have been studied in order to prepare synthetic proteins of the same type (2). Several research activities have attempted to imitate natural proteins and obtain materials with novel properties (3). Caseins are biodegradable proteins whose main application is in the food industry, but they are also used as glues in plywood, for example. An undesirable side effect of the biodegradability of caseins was observed when caseins were used in self-leveling concretes. The biodegradation of the materials caused malodorous low-molecular-weight compounds and this was one phenomenon observed in connection with the "sick building syndrome" (4–9).

Polysaccharides are a class of widely used compounds, for example, dextrans, cellulose, or chitins. Starch, another important polysaccharide, is used not only in the food industry but also in the manufacture of paper, adhesives, etc. Recently, starch, of which amylose and amylopecetin are the basic components, has become important as a cheap biodegradable additive to inert synthetic polymers. Since starch is an energy-storing biopolymer in nature, it is not surprising that including pure starch in, for example, LDPE reduces the mechanical properties in the material. The most important properties of polysaccharides are their ability to alter the flow characteristics of fluids and their ability to act as absorbants, gel formers, ion-exchange agents, etc. In addition, all these polysaccharides individually perform biological functions.

Polysaccharides have been synthesized via the ring-opening polymerization of anhydro sugar derivatives, which has been given materials with novel properties (10). Polysaccharides are generally degraded by various enzymes and microorganisms and are thus truly biodegradable materials. They are also used in encapsulation, gelling or thickening and suspending agents. Xanthan is the one major bacterial polysaccharide commercially available (11). The bacterial polysaccharide gellan, a gelling agent, is obtained by the aerobic fermentation of *Pseudomonas elodea* (12). Several other types of bacteria have been investigated for the production of gelation polymers (13). *Lactid acid bacteria* and *Acetobacter* are examples of bacteria capable of synthesizing polymers with gelling characteristics.

Polymer chemistry had its beginning with the characterization of cellulose. Cellulose was isolated for the first time some 150 years ago. It differs in some respects from other polysaccharides produced by plants, the molecular chain being very long and consisting of only one repeating unit (the cellobiose). Naturally, it occurs in a crystalline state. From the cell walls, cellulose is isolated in microfibrils by chemical extraction. Nonglucose sugars range from 15–50% in the cell wall (14). When cellulose is treated with sulfuric acid and titrated back to pH 3, the microfibrils fall into short rods, still giving the X-ray diagram of cellulose (15). The biodegradation of cellulose is complicated, partly because cellulose exists together with lignin in wood cell walls. White rot fungi (attacking lignin preferentially) and brown rot fungi (attacking cellulose) are well-known microorganisms responsible for the deterioration of wood.

The degradation rate depends on both structure and chemical composition. The biodegradation of modified cellulose depends on both the type of substituents and the degree of substitution. Cellulose triacetate is resistant to microorganisms, whereas cellulose with a low degree of acetylation readily degrades. One study (16) reports on different synthetic routes to well-defined, tailored cellulose/starch synthetic graft copolymers. These graft copolymers function as compatibilizing/interfacial agents for allowing cellulose and starch with synthetic polymers to yield a biodegradable/biocompatible material (16–17).

Combining cellulose with chitosan results in various kinds of strong, gas-barrier, and water-resistant composite films (18). Films, nonwoven textiles, and various extruded products yielding biodegradable materials are planned (19). Chitin is a macromolecule formed in the shell of crabs, lobsters, shrimps, and insects, among others. The chemical treatment of chitin produces chitosan. The materials are biocompatible and exhibit antimicrobial activity as well as the ability to absorb heavy metal ions. Because of their water-retaining and moisturizing ability, they also find application in the cosmetic industry. With chitin and chitosan as carriers, a water-soluble prodrug has been synthesized (20).

Modified chitosan has been prepared with various chemical and biological properties (21). *N*-carboxymethylchitosan and *N*-carboxybutylchitosan have been prepared for use in cosmetics and wound treatment (22).

Chitin derivatives may also be used as drug carriers, and a report on the use of chitin in absorbable sutures shows that chitin has the lowest elongation among suture materials consisting of chitin, PGA (poly(glycolic) acid), plain catgut, and chromic catgut (23–24). The tissue reaction of chitin is similar to that of PGA.

Alginate, the structural biopolymer of brown kelp, is used because of its ability to thicken aqueous solutions and form gels under mild conditions (25). Pullulan, a bacterial polysaccharide produced by *Aeurobasidium pullulans,* is a water soluble polymer with a viscosity comparable to that of arabic gum (the repeating unit is maltobiose where glucose is the monomer). It has application in materials such as molding resins and packaging films, since it forms excellent films.

Microbial polyesters have been the subject of considerable research over many years. Different bacteria produce polyesters as an energy and carbon reserve just like plants produce starch. Poly(β-hydroxybutyrate) (PHB) was first isolated in 1925 by Lemoigne in France. PHB belongs to the family of poly(hydroxyalkanoates) (PHA). PHB is produced by several bacteria and fungi during starvation, that is, nutrient limitation, lack of essential nutrients such as nitrogen, oxygen, phosphorus, etc. PHB is an interesting material due to its biodegradability that, together with its biocompatibility, makes it useful in applications such as surgical pins, sutures, wound dressings, bone replacement, stimulation of bone growth, and healing by piezoelectric properties (26–27). PHB also exhibits such interesting properties that it is used in a packaging in spite of its relatively high cost. Parallel to starch, PHB also is an energy reserve for cells and not a fiber that has any influence on the mechanical properties obtained.

Considerable interest arose recently when Holmes (28) developed a large-scale, controlled fermentation process for the production of copolymers of PHB. Feeding bacteria with a variety of carbon sources led to the production of different copolymers and a material was obtained with better mechanical properties than natural PHB (29–38).

The biodegradation of PHB and its copolymers has been studied using environments such as soil, activated sludge, and sea water (33). Films (0.07 mm thick) of the homopolymer PHB, a copolymer of 91% 3HB and 95 4HB and a copolymer of 50% 3HB and 50% 3HV were subject to biodegration in soil. The fastest biodegration rate was obtained for P(3HB-*co*-9% 4HB). In activated sludge, P(3HB-*co*-9% 4HB) was completely decomposed after 2 weeks (33).

The natural polyesters are also hydrolyzed in water at a very slow rate. *In vivo,* this is the main mechanism, involving chain scission of the polymer. The hydrolytic degradation of hydroxybutyrate–hydroxyvalerate copolymers in vitro begins with a surface modification, accompanied by water diffusion into the matrix (39). A progressive increase in porosity facilitates the diffusion by removal of degradation products (39). The hydrolytic degradation of microbial polyesters occurs by homogeneous erosion over two stages: random hydrolytic chain scission of the ester group, leading to a decrease in molecular weight, is followed by a second step (Mn 13,000) in which weight loss occurs (40). The microbial polyesters are degradable in soil, sludge, and seawater. Bacterial polyesters have also been blended with PE and PS. The goal was to expand their physical properties while retaining biodegradability (41–42).

BIODEGRADABLE SYNTHETIC POLYMERS

Biodegradable synthetic materials are useful for temporary medical devices where the original function is supposed to be recovered with time, for example, soft-tissue injuries, bone fractures, or perturbed biological processes. Typical products are surgical sutures, bone bins and plates, staples, surgical adhesion, drug-delivery systems, and artificial plasma (43–45). The degradation products should be nontoxic, biocompatible, and preferably water-soluble.

The first of the biodegradable synthetic polymers that became commercially available was poly(glycolic acid) (PGA) chosen on the basis of the results of screening of potential materials subjected to degradation *in vivo* and in physiological solution over a period of 90 days (46–47).

The glycolide has also been copolymerized with lactide, with the result that a stiff polymer, poly(glycolic-*co*-lactic acid) (PGA/LA:3) was obtained (48). Sutures have been produced from a 90:10 glycolide:lactide monomer composition (49) and the suture retains its tensile strength somewhat longer than the poly(glycolic acid) homopolymer (50) and has a lower melting point. This line of application also includes artificial tendons made of a carbon fiber composite with poly(L-lactic acid) as the matrix (51–53) and bone plates of poly(L-lactic acid) (54–56). Stereoblock copolymers of L- and D-lactides are synthesized by living ring-opening polymerization in order to prepare biodegradable materials (57).

The degradation of poly(glycolic acid) is also used in the medical field to obtain slow-release systems for pharmaceutical drugs. By synthesizing block copolymers, increased elasticity is obtained, and this is amplified by aliphatic block copolymers with which poly(glycolic acid) and poly(ethylene glycol) have been synthesized (58).

Polylactide is perhaps the most frequently used polyester in biomedical applications due to its many favorable characteristics, for example, high strength, biocompatibility, and biodegradability. Several commercial products made of homo- and copolymers of polylactide are available. There has been much interest in the polymerization and degradation behavior of homo- and copolymers of lactide (59–61). The degradation rate of poly(agr;-hydroxyacid)s is dependent on chemical structure, molecular weight and molecular weight distribution, presence of residual compounds (solvent, monomers, oligomers, initiators), morphology, ionic strength, and pH. Polylactide with a low degree of purity initially degraded very rapidly, while a polymer with a high degree of purity degradation went through a lag phase before rapid degradation occurred (62).

Polylactide and polyglycolide hydrolyze in the presence of moisture, whereas bacterially produced poly(hydrooxyalkanoates) are more resistant to moisture, but are rapidly hydrolyzed by enzymes (63).

Polylactide meets many requirements as a packaging thermoplastic and may be used as a commodity resin for general packaging applications. Its physical properties and melt processing are similar to those of conventional packaging resins. When plasticized by its own monomer LA, polylactide becomes increasingly flexible so that products can be prepared that mimic PVC, LDPE, LLDPE, PP, and PS (64). The degradation rate increases with the increasing amount of plasticizer.

Degradable block copolyesters of the aromatic polyester poly(ethylene terephthalate) with poly(oxyalkenes), poly(ethylene glycol) (65–66) or poly(tetramethylene glycol) (67) are used in biodegradable materials in surgical applications. The use of poly(tetramethylene glycol) instead of poly(ethylene glycol) increases elasticity and hydrophobicity, which in turn leads to slower hydrolytic degradation.

Aliphatic homopolyesters, for example, poly(tetramethylene adipate) (PTMA), and block copolymers, such as poly(ethylene succinate)-*b*-poly(ethylene glycol) (PES/PEG) and poly(ethylene succinate)-*b*-poly(tetramethylene glycol) (PES/PTMG; 4), were synthesized and the subsequent degradation studied in a pseudoextracellular fluid (PSE) buffered at pH 7.3 and maintained at 37°C (68). The materials obtained exhibited thermoplastic elastomer behavior, the degradation rate depending on the polyether's composition.

The polymerization of 1,5-dioxepan-2-one (DXO) yielded a poly(ether ester) with amorphous properties, implying its usefulness as an amorphous block in copolymers possessing elastic properties (69). Recently, degradable cross-linked polymers were prepared based on DXO and a cross-linker of bis-ϵ-caprolactone (70). The cross-linked polymers exhibit a degradation time of about 1 year (71).

Polycaprolactones (PCL) aliphatic polyethers without branches, have been thoroughly studied, especially in biodegradation tests. They are generally prepared by the ring-opening polymerization of ϵ-caprolactones. Researchers have discussed the hydrolysis of PCL and biodegradation of PCL by fungus, and shown that PCL can be degraded enzymatically (72–74). The fungus assimilates various polyesters, but in general the assimilation of aliphatic polyesters is better the greater the number of carbon atoms between the ester bonds.

Another approach has been made using enzymes capable of degrading PCL and blends of PCL and polymers such as polythylene (PE), nylon-6, polystyrene (PS), and poly(β-hydroxybutyrate) (PHB) (75). PCL has also been blended with low density polyethylene (LDPE) and the biodegradability of this blend was monitored (76) showing that the degradability could be controlled by combining the melt viscosity of PCL with that of LDPE. Blends of poly(ϵ-caprolactones) (PCL) and poly(L-lactic acid) (PLA) with poly(glycolic acid-*co*-L-lactic acid) yield a material with better permeability and form stability, but at the same time a higher degradation rate, suitable for drug delivery systems (77).

The biodegradability of polyurethanes was shown to be dependent on whether or not the prepolymer is a polyester or polyether (78). The polyether-polyurethanes are resistant to biodegradation, whereas the polyesterurethanes are readily attacked. The influence of diisocyanates is also important to the biodegradability of polyurethanes, with the linear diisocyanates being less resistant than cyclic diisocyanates. One study (79) has monitored biodegradability as a function of structure for polyesters and shown that linear structure has higher biodegradability than the corresponding branched structure. Another study also monitored the biodegradability of polymers with different structures (80). Important factors determining biodegradability are the presence of hydrolyzable and/or oxidazable linkages. The presence of hydrophobic and hydrophilic segments in synthetic polymers improves degradation (80).

Poly(α-amino acid)s are often susceptible to hydrolysis and the amino acids and their copolymers are potentially useful as biodegradable biomaterials. Copolymers of L-aspartic acid and L-glutamic acid have been prepared and their degradation in vitro by papain in pseudoextracellular fluids followed (81). They degrade by random chain scission and the nature of the side chains is important to the rate of degradation. The synthetic poly(α-amino acid)s resembling native proteins can be degraded by proteolytic enzymes and the tissue reaction of these materials is similar to that associated with native proteins.

Polyamides-nylons are generally reported to be quite resistant to microorganisms (82). It has been proposed, however, that nylon-6,6 can be modified by *N*-acetyl substitution to yield materials that are biodegradable (83). Copolymers of ethylene containing 10% ester units have been shown to be highly biodegradable, whereas copolymers containing only 2% are only slowly biodegradable (84). Researchers have reported on biodegradable polymers of poly(amide enamine) (85). These materials are similar to proteins in that they produce both acidic and basic products during degradation. Poly(amide enamine)s are promising materials for controlled-release applications.

Polyanhydrides are a group of polymers with two sites in the repeating unit susceptible to hydrolysis. These are interesting materials due to their good biocompatibility (86). The bioerodible polymers, especially polyanhydrides, are useful materials for drug delivery. Polyanhydrides are useful in biomedical applications due to their fiber-forming properties. The type of degradation products from the degradation of poly(trimethylene carbonate) has been reported (87). In this case, the main product is trimethylene carbonate, whereas other workers report several different degradation products besides the six-membered molecule (the monomer) (88,89). Poly(trimethylene carbonate) was synthesized using cationic and anionic initiators, yielding high-molecular-weight compounds with rubbery character at room temperature (90). Since the material obtained is an aliphatic polycarbonate, it could be useful as a biodegradable polymer for medical applications.

Poly(vinyl alcohol) (PVA) is a water-soluble polymer with applications as a thickening agent for emulsions and suspensions and as packaging film where water solubility is desired. By chemical treatment, the final form of the polymer can be insoluble. Polyethylene glycol (PEG) and PVA are readily degraded by microorganisms, and it has been shown that isolated cultures cannot grow on liquid cultures containing PVA as the sole carbon source, but mixed cultures (*Pseudomonas* sp *Alcaliqenes* sp, and *Pseudomonas putida*) can grow (91).

Polycarboxylates containing biodegradable blocks have been prepared where the blocks consist of polysaccharides and PVA and the block polymers obtained have been shown to exhibit improved biodegradability (92–93). Water-soluble poly(sodium carboxylate) has been subjected to biodegradation tests in activated sludge, The polymer containing more than 60% glycopyranose groups shows enhanced biodegradability (94).

A review of water-soluble polymers (95) contains information on the biodegradability of PVA, poly(carboxylic acid)s, poly-(hydroxycarboxylic acid) and poly(acrylic acid), among others. PVA is the only carbon–carbon backbone polymer that is biodegradable. For water-soluble polymers, branching on the polymer backbone is detrimental to biodegradation, whereas heteroatomic backbone polymers (C—O—C) have enhanced biodegradability (95).

BIODEGRADATION IN NATURE

The degradation of polyethylene in different environments, such as landfills, composts, and seawater has been well-studied. It is not realistic to expect plastics to degrade in landfills, where available sunlight, oxygen, water, and microbial activity are very limited. Even food has been found to be essentially undegraded after many years in landfill. This is also positive as no leaching of the degradation products into water sources and little methane formation will occur. Composting is a natural way of disposing of biodegradable solid waste. It offers also more realistic possibilities for degradation; microorganisms, moisture, and oxygen are usually present and temperatures as high as 70°C may be obtained.

The degradation of polyethylene during composting has been studied by several authors. Of the starch content, 6% was lost when polyethylene films containing 0–20% starch were composted for 60 days (96). Polyolefin/starch blends without prooxidants did not show any oxidation of the polyethylene matrix during composting (97). In addition, not more than 30% of the starch was removed. Starch-based materials (52 and 67%) demonstrated rapid and almost complete starch removal during soil burial for 40 days, but at 29% starch content, only the starch from the surface of the films was removed during

8 months (98). At the same time, no oxidation of the polyethylene matrix was observed. Already in 1978, IR analysis gave evidence for PE matrix oxidation in simulated composts (99). Chiellini et al. uncovered, also, evidence of polyethylene matrix oxidation and the formation of ester bonds during exposure to prolonged composting (96). Recently, soil burial for 32 years yielded signs of degradation in pure polyethylene films (100).

The initiation of thermal oxidation of polyethylene films by transition metal catalysts during composting proceeds slowly compared to oxidation at the same temperature in an oven (60–70°C). Possible reasons are the low oxygen concentration in the wet compost and the leaching out or deactivation of the catalysts (101). The starch-filled polyethylene bags exposed on the surface of the compost broke down into small pieces, while the buried bags remained intact after 49 days of exposure (102–103).

Several authors have compared the natural and artificial weathering of polyethylene (104–107). Starch-polyethylene films deteriorated rapidly in the strawline of a marsh, whereas less degradation was observed in the films placed in compost, landfill, soil, and seawater (108). Polyethylene films on the surface of water showed a more rapid loss of properties than those samples partially or completely submerged (109). This can be attributed to decreased light intensity and the lack of heat buildup. The degradation of LDPE placed in water was slower than when the films were kept dry. In polyethylene containing from 0–30% starch and metal catalysts, very little degradation, as measured by molecular weight and tensile strength, was observed during 12 weeks in seawater (110). Without metal catalysts, no changes were observed. Of the starch content, 8–15% was lost during 13 months in a marine environment (108).

DEFINITION OF BIODEGRADABLE POLYMERS

Various definition of biodegradation have been given and discussed. Biodegradation may be defined as a degradation caused by the action of microorganisms and/or enzymes (111). International standards generally state that biodegradation should yield metabolic end-products and a plastic material is called biodegradable if all components undergo total biodegradation (112). Many scientists also declare that degradation should reduce molecular weight, giving metabolic end-products in a very short period of time (eg, weeks). Generally, standard biodegradation tests are specified to 28 days and during that time 60% degradation shall have been achieved as measured by, for example, BOD or CO_2 evolution (113). The time for total biodegradation and the generation of natural metabolities are not as important as the role small molecules play in nature. The type of (bio)degradation products and the possibility of their taking part in the cycles of nature are the crucial point. This interaction of biodegradable polymers with the environment is ultimately the key question in understanding biodegradable polymers.

BIBLIOGRAPHY

1. T. Takimoto and H. Yamamoto, *Polym. Prepr. Jpn.* **39,** E1087 (1990), English translation, Tokyo.
2. H. Yamamoto, A. Nishida, T. Takimoto, K. Ikeda, and S. Yamaguchi, *Polym. Prepr. Jpn.* **39,** E1088 (1990), English translation.
3. J. H. Richards, in M.J. Epton, ed., *Macramolecular Preprints,* 2nd Euro- American Conference on Functional Polymers and Biopolymers, Oxford University Press, 1989, p. 1.
4. S. Karlsson, E. Banhidi. Z. G. Banhidi, and A.-C. Albertsson, in B. Berglund, et al., eds., *Procedings of 3rd International Conference Indoor Air,* Vol. 3, Swedish Council for Building Research, Stockholm, 1984, p. 247.
5. S. Karlsson, Z. G. Banhidi, and A.-C. Albertsson, *Appl. Microbiol. Biotechnol.* **28,** 305 (1988).
6. S. Karlsson, Z. G. Banhidi, and A.-C. Albertsson, *J. Chromatogr.* **442,** 267 (1988).
7. *Proceedings of the 33rd IUPAC, International Symposium on Macromolecules.*
8. S. Karlsson, Z. G. Banhidi, and A.-C. Albertsson, *Mater. Constr. (Paris)* **22,** 163 (1989).
9. S. Karlsson and A.-C. Albertsson, *Mater. Constr. (Paris)* **23,** 352 (1990).
10. K. Kobayashi, *Proc. ACS. Div. Polym. Mater. Sci. Eng.* **62,** 477 (1990).
11. G. T. Colegrove, *Ind. Eng. Chem. Prod. Res. Dev.* **22,** 456 (1983).
12. K. S. Kang, G. T. Veeder, P. J. Mirrasoul, T. Kanecko, and W. Cottrell, *Appl. Environ. Microbiol.* **43,** 1086 (1982).
13. V. J. Morris, *Proc. ACS Div. Polym. Mater. Sci. Eng.* **62,** 462 (1990).
14. J. Cronhaw, A. Myers, and R. D. Preston, *Biochim. Biophys. Acta* **27,** 89 (1958).
15. B. G. Rånby and G. Ribi, *Experientia* **6,** 2 (1950).
16. R. Narayan, *Appl. Biochem. Biotechnol.* **17,** 7 (1988).
17. C. J. Biermann, J. B. Chug, and R. Narayan, *Macromolecules* **20,** 954 (1987).
18. J. Hosokawa and M. Nishiyama, in Y. Kishimoko, ed., *Proceedings of International Symposium on Biodegradable Polymers,* Biodegradable Plastic Society, Tokyo, Japan, 1990, p. 144.
19. J. Hosokawa, M. Nishiyama, K. Yoshihara, and T. Kubo, *Ind. Eng. Chem. Res.* **29,** 800 (1990).
20. K. Inosaka, Y. Ohya, and T. Ouchi, *Polym. Prepr. Jpn.* **39,** E975 (1990), English translation.
21. R. A. A. Muzzarelli, *Chitin in Nature and Technology,* Plenum Press, New York, 1986.
22. R. A. A. Muzzarelli, *Polym. Prepr., ACS Div. Polym. Chem.* **31,** 626 (1990).
23. S. Tokura, Y. Miura, Y. Uraki, K. Watanabe, I. Saiki, and I. Azuma, *Polym. Prepr., ACS, Div. Polym. Chem.* **31,** 627 (1990).
24. M. Tashibana, A. Yaita, H. Tamiura, K. Fukasawa, N. Nagasue, and T. Nakanura, *Jpn. J. Surg.* **18,** 533 (1988).
25. P. A. Sandford, *Polym. Prepr., ACS Div. Polym. Chem.* **31,** 628 (1990).
26. R. Lenz, R. A. Gross, H. Brandl, and R. C. Fuller, *Chin. J. Polym. Sci.* **7,** 289 (1989).
27. P. J. Hocking and R. H. Marchessault, *Biopolyesters,* 1989.
28. P. A. Holmes. L. F. Wright, and S. H. Collins (ICI), European Patent 0052 459, 1982; European Patent 0069 497, 1983.
29. H. Preusting, A. Nijenhuis, and B. Witholt, *Macromolecules* **23,** 4220 (1990).
30. S. Bloembergen, D. A. Holden, T. L. Bluhm, G. K. Hamer, and R. H. Marchessault, *Macromolecules* **22,** 1656 (1989).
31. S. Bloembergen, D. A. Holden, T. L. Bluhm, G. K. Hamer, and R. H. Marchessault, *Macromolecules* **22,** 1663 (1989).
32. Y. Doi and C. Abe, *Macromolecules* **23,** 3705 (1990).
33. M. Kunioka, Y. Kawaguchi, and Y. Doi, *Appl. Microbiol. Biotechnol.* **30,** 569 (1989).
34. M. Kunioka, A. Tamaki, and Y. Doi, *Macromolecules* **22,** 694 (1989).

35. M. Kunioka, Y. Nakamura, and Y. Doi, *Polym. Commun.* **29,** 174 (1988).
36. Y. Doi, A. Tamaki, M. Kunioka, and K. Soga, *J. Chem. Soc., Chem. Commun.* **5,** 1635 (1987).
37. Y. Doi, M. Kunioka, Y. Nakamura, and K. Soga, *Macromolecules* **20,** 1988 (1987).
38. Y. Doi, M. Kunioka, Y. Nakamura, and K. Soga, *Macromolecules* **19,** 2860 (1986).
39. S. J. Holland, M. Yasin, and B. J. Tighe, *Biomaterials* **11,** 206 (1990).
40. Y. Doi, Y. Kanesawa, Y. Kawaguchi, and M. Kunioka. *Macromol. Chem., Rapid Commun.* **10,** 227 (1989).
41. S. N. Bhalakia, T. Patel, R. A. Gross, and S. P. McCarthy, *Polym. Prepr. ACS Div. Polym. Chem.* **31,** 441 (1990).
42. P. B. Dave, N. J. Ashar, R. A. Gross, and S. P. McCarthy, *Polym. Prepr. ACS Div. Polym. Chem.* **31,** 442 (1990).
43. W. Leenslag, A. J. Pennings, R. R. M. Bos, F. R. Rozema, and G. Boering, *Biomaterials* **8,** 70 (1987).
44. S. Vainionpää, P. Rokkanen, and P. Törmälä, *Prog. Polym. Sci.* **14,** 679 (1989).
45. M. Vert, *Makromol. Chem.* **166167,** 155 (1989).
46. U.S. Pat. 3297 033, (1967), E. E. Schmitt and R. A. Polistina.
47. E. J. Frazza and E. E. Schmitt, *Biomed. Mater. Symp.* **I,** 43 (1971).
48. U.S. Pat. 3 636 956, 1972, A. K. Schneider.
49. Can. Pat. 950 308, 1974, D. Wasserman and A. Levi.
50. R. L. Kronenthal, in R. L. Kronenthal, Z. Oser, and E. Martin, eds., *Polymers in Medicine and Surgery,* Plenum Press, New York, 1975, p. 10.
51. J. Aragona, J. R. Parsons, H. Alexander, and A. B. Weiss, *Clin. Orthop. Relat. Res.* **160,** 268 (1981).
52. H. Alexander, A. B. Weiss, and J. R. Parsons, *Actuel. Probl. Chir. Orthop.* **26,** 78 (1983).
53. J. R. Parsons, H. Alexander, and A. B. Weiss, *Surg. Sci. Ser.* **2,** 417 (1994).
54. M. Vert, F. Chabot, J. Leray, and P. Christel, *Macromol. Chem. Suppl.* **5,** 30 (1981).
55. P. Christel, F. Chabol, J. Leray, C. Morin, and M. Vert, *Adv. Biomater.* **3,** 271 (1982).
56. F. Chabot, M. Vert, S. Chapelle, and P. Granger, *Polymer* **24,** 53 (1983).
57. N. Yui, P. J. Dijkstra, and J. Feigen, *Makromol. Chem.* **191,** 481 (1990).
58. D. J. Casey and M. S. Robys, European Patent 108933, 1984.
59. D. K. Gilding and A. M. Reed, *Polymer* **20,** 1459 (1979).
60. C. C. Chu, *J. Biomed. Mater. Res.* **16,** 117 (1982).
61. D. W. Grijpma, A. J. Nijenhuis, and A. J. Pennings, *Polymer* **31,** 2201 (1990).
62. X. Zhang, U. P. Wyss, D. Pichora, and M. F. A. Goosen, *Bioact. Compat. Polym.* **9,** 80 (1994).
63. Y. Kanesawa, N. Tanahashi, Y. Doi, and T. Saito, *Polym. Degr. Stab.* **45,** 179 (1994).
64. R. G. Sinclair, *J. Macromol. Sci., Pure Appl. Chem.* **A33,** 585 (1996).
65. D. K. Gilding and A. M. Reed, *Polymer* **20,** 1454 (1979).
66. A. M. Reed and D. K. Gilding, *Polymer* **22,** 499 (1981).
67. V. V. Shevchenko, A. S. Chegolya, and E. M. Aisenshtern, *Vysokomol. Soedin.* **A27,** 2333 (1985). *Chem. Abstr.* **104,** 149, 525 (1985).
68. A.-C. Albertsson and O. Ljungquist, *Acta Polym.* **39,** 95 (1988).
69. T. Mathisen, K. Masus, and A.-C. Albertsson, *Macromolecules* **22,** 3842 (1989).
70. R. Palmgren, S. Karlsson, and A.-C. Albertsson, *J. Polym. Sci., Polym. Chem.,* **35,** 1635 (1997).
71. R. Palmgren, S. Karlsson, and A.-C. Albertsson, (in press) (1997).
72. Y. Tokiwa and T. Suzuki, *Nature* **270,** 76 (1977).
73. Y. Tokiwa, T. Ando, and T. Suzuki, *J. Ferment. Technol.* **54,** 603 (1976).
74. Y. Tokiwa, T. Suzuki, and T. Ando, *J. Appl. Polym. Sci.* **24,** 1701 (1979).
75. Y. Tokiwa, in *33rd IUPAC, International Symposium on Macromolecules,* Book of Abstracts, Montreal, Canada, 1990, p. 2.6.6.
76. Y. Tokiwa, T. Ando, K. Tatkeda, A. Iwamoto, and M. Koyama, in Y. Kishimoto, ed., *International Symposium on Biodegradable Polymers,* Biodegradable Plastic Society, Tokyo, 1990, p. 69.
77. Y. Cha and C. G. Pitt, *Biomaterials,* Vol. 11, Guildford, England, 1990, p. 108.
78. R. T. Carby and A. M. Kaplan, *Appl. Microbiol.* **16,** 900 (1968).
79. J. E. Potts, R. A. Clendinning, and W. B. Ackart, *Polym. Prepr., ACS Div. Polym. Chem.* **13,** 629 (1972).
80. S. J. Huang, *Polym. Prepr. ACS Div. Polym. Chem.* **31,** 54 (1990).
81. T. Hayashi and M. Iwatsuki, *Biopolymers* **29,** 549 (1990).
82. J. E. Potts, in M. Grayson, ed., *Kirk-Othmer Encyclopedia of Chemical Technology,* Wiley-Interscience, New York, 1984, suppl. vol., p. 626.
83. D. M. Ennis, A. Kramer, C. W. Jamesson, P. H. Mazzocchi, and W. J. Bailey, *Appl. Environ. Microbiol.* **35,** 51 (1978).
84. W. J. Bailey and B. Gapaud, *ACS Symp. Ser.* **280,** 423 (1985).
85. S. J. Huang and O. Kitchen, *Polym. Prepr., ACS Div. Polym. Chem.* **31,** 207 (1990).
86. K. W. Leong, B. C. Brott, and R. Langer, *J. Biomed. Mater. Res.* **19,** 941 (1985).
87. I. C. McNeill and A. Rincon, *Polym. Degrad. Stab.* **24,** 59 (1989).
88. I. C. McNeill and H. A. Leiper, *Polym. Degrad. Stab.* **11,** 267 (1985).
89. I. C. McNeill and H. A. Leiper, *Polym. Degrad. Stab.* **12,** 373 (1995).
90. A.-C. Albertsson and M. Sjöling, *J. Macromol. Sci. Chem.* **A29,** 43 (1992).
91. M. Shimao and N. Kato, in Y. Kishimoto, ed., *Proceedings of International Symposium on Biodegradable Polymers,* Biodegrable Plastic Society, Tokyo, Japan, 1990, p. 80.
92. S. Matsumura, M. Nishioka, H. Shigeno, and S. Yoshikawa, in *Proceedings of International Symposium on Biodegradable Polymers,* Biodegradable Plastic Society, Tokyo, 1990, p. 75.
93. S. Matsumura, S. Maeda, and Y. Yoshikawa, *Macromol. Chem.* **191,** 1269 (1990).
94. S. Matsumura, M. Nishioka, and S. Yoshikawa, *Polym. Prepr. Jpn.* **39,** E969 (1990), English translation.
95. G. Swift. *Proc. ACS, Div. Polym. Mater. Sci. Eng.* **63,** 848 (1990).
96. E. Chiellini, F. Cioni, R. Solaro, G. Vallini, A. Corti, and A. Pera, *J. Environ. Polym. Degrad.* **1,** 167 (1993).
97. D. F. Gilmore, S. Antoun, R. W. Lenz, S. Goodwin, R. Austin, and R. C. Fuller, *J. Ind. Microbiol.* **10,** 199 (1992).
98. S. M. Goheen and R. P. Wool, *J. Appl. Polym. Sci.* **42,** 2691 (1991).
99. A.-C. Albertsson, *J. Appl. Polym. Sci.* **22,** 3419 (1978).
100. Y. Otake, T. Kobayashi, H. Asabe, N. Murakami, and K. Ono, *J. Appl. Polym. Sci.* **56,** 1789 (1995).
101. M. Weiland and C. David, *Polym. Degrad. Stab.* **45,** 371 (1994).
102. H. B. Greizerstein, J. A. Syracuse, and P. I. Kostyniak, *Polym. Degrad. Stab.* **39,** 251 (1993).

103. K. E. Johnson, A. L. Pometto III, and Z. L. Nikolov, *Appl. Environ. Microbiol.* **59,** 1155 (1993).
104. C. David, M. Trojan, A. Daro, and W. Demarteau, *Polym. Degrad. Stab.* **37,** 233 (1992).
105. A. L. Andrady, J. E. Pegram, and Y. Song, *J. Environ. Polym. Degrad.* **1,** 117 (1993).
106. A. Tidjani and R. Arnaud, *Polym. Degrad. Stab.* **39,** 285 (1993).
107. S. H. Hamid and M. B. Amin, *J. Appl. Polym. Sci.* **55,** 1385 (1995).
108. V. T. Breslin and B. Li, *J. Appl. Polym. Sci.* **48,** 2063 (1993).
109. K. K. Leonas and R. W. Gorden, *J. Environ. Polym. Degrad.* **1,** 45 (1993).
110. K. E. Gonsalves, S. H. Patel, and X. Chen, *J. Appl. Polym. Sci.* **43,** 405 (1991).
111. A.-C. Albertsson and S. Karlsson, in S. A. Barenberg et al., eds., *Degradable Materials,* CRC Press, Boca Raton, Fla., 1990, p. 263.
112. DIN Standard Test, FIVK 103.2.3 (BRG/BPG 194/05).
113. R. J. Muller, J. Augusta, and M. Dantke, *Material und Organismen* **27,** 179 (1992).

BIODEGRADATION OF CRUDE OIL. See CRUDE OIL BIODEGRADATION.

BIOFILTRATION

TODD S. WEBSTER
Envirogen, Inc.
Lawrenceville, New Jersey

JOSEPH S. DEVINNY
University of Southern California
Los Angeles, California

Since the 1960s, air pollution regulations have required the commercial and industrial sectors to control their emissions through the use of traditional technologies such as carbon adsorption, incineration, or scrubbing. As the regulations become ever more stringent and encompassing, many companies have started to look past traditional technologies to alternative cost-effective control systems. A system that has attracted significant interest is biofiltration. In biofiltration, a humid, contaminated air stream is passed through a porous filter bed supporting pollutant-degrading microorganisms. Like wastewater biological treatment processes, biofiltration relies on microbial catabolic reactions for the degradation of specific waste compounds. Under optimal conditions, the pollutants can be converted completely to harmless products and biomass. Biofilters have found most of their success in treatment of low concentration odors and volatile organic compounds (VOCs), where traditional control technologies are less cost-effective. They are now widely used for odor control at wastewater treatment facilities and industrial installations. Air from food-processing plants, such as slaughterhouses and cocoa-roasting ovens, is commonly treated by biofilters. As research advances the technology, biofilters are being applied to difficult contaminants, such as ketones and aldehydes, and to air toxics, such as the gasoline vapors from soil-venting projects. A detailed treatment of the field is available in the literature (1).

TRADITIONAL CONTROL TECHNOLOGIES

There are two distinct means of air pollution emissions control. Primary control reduces emissions at their source through raw product substitution, reduction, or recycling methods. However, these changes may reduce the quality of the product or increase its cost. Secondary control involves treating the waste gas after it has been produced. Various technologies exist for this, but the choice is often dictated by economic constraints. Such constraints reflect the nature of the compound being treated, the concentration, the flow rate, and the mode of emission of the gaseous waste stream. Combinations of the techniques are often required to meet regulatory standards. The most common traditional control technologies are adsorption, incineration, and absorption.

Adsorption

Adsorption is the transfer of a contaminated gas to the surface of a solid. Adsorption occurs on a fixed or fluidized bed of porous material, such as activated carbon or zeolite, and is highly efficient for the treatment of low concentration contaminant vapors. Once the activated bed has reached adsorptive capacity, the material can be regenerated and the pollutant recovered by steam desorption. Disposal or incineration of the spent adsorbant, however, is often more economical when mixed contaminant waste streams have been treated. A carbon adsorption system has a high capital cost because of the cost of the carbon, but is appropriate for lower organic loadings and when solvents are recoverable. However, this technology transfers the waste from the air to a solid phase that still requires further treatment. Regeneration requires specially trained personnel and additional equipment, incurring additional costs.

Incineration

Thermal and catalytic incineration are widely used and effective treatments for concentrated waste gases. Thermal incineration involves the combustion of pollutants at temperatures from 700 to 1400°C. Catalytic incineration allows process temperatures to be reduced to between 200 and 700°C. Incineration is the most widely used secondary technique, because it can easily handle multicomponent pollutant streams at large flow rates. However, because of high energy costs, incinerators are best applied to effluents with higher concentrations of organics. Incinerators require large fuel inputs when the waste gas concentration is low and provides little heat of its own. Incomplete combustion by incineration may produce harmful byproducts, like dioxins or NO_x.

Absorption

Absorption involves the interaction of a pollutant, often an odorous compound, with a scrubbing solution. The gas enters a large contactor, such as a packed or bubble column, where the gaseous pollutants are transferred to a liquid phase. Success is dictated by the affinity of the pollutant for the liquid phase. Water is the most frequently used scrubbing solution, with the pH adjusted to increase the solubility of acidic or basic gases. Once pollutant transfer has occurred, treatment of the liquid phase is necessary. This may be achieved by desorbing the pollutant at high temperatures, followed by distillation or burning. This technology requires no media disposal, but incurs high chemical operating costs.

DESCRIPTION OF BIOLOGICAL REACTORS

Gas-phase biological reactors use microbial metabolic reactions to treat contaminated airstreams. They are economical for large quantities of air with low concentrations of contaminant vapors. As these vapors pass through a biological reactor, the contaminants are sorbed to the aqueous phase, where microbial attack occurs. Through oxidative and occasionally reductive reactions, the contaminants are degraded to water vapor, carbon dioxide, mineral salts, and organic biomass. Both organic and inorganic contaminants can be used as an energy source for maintenance and growth by the microbial population. The contaminants of interest must be nontoxic to the microorganisms at the treated concentrations.

The most successful removal occurs with low molecular weight, highly soluble substances. Examples of such compounds include odorous materials (sulfides, mercaptans, etc), alcohols, aldehydes, ketones, and some simple aromatics. Other compounds that show moderate to slow degradation include phenols, chlorinated hydrocarbons, polyaromatic hydrocarbons, and highly halogenated hydrocarbons. Certain anthropogenic compounds may not biodegrade at all. Compounds that are biologically treatable come from a wide array of sources (Table 1).

The three types of biological reactors are biofilters, biotrickling filters, and bioscrubbers. The basic contaminant removal mechanisms are similar for all reactor types. However, the microbes may be suspended or immobilized, and the liquid may be mobile or stationary.

Biofilters

In a biofilter, the microorganisms grow in a biofilm on the surface of the medium or in the water layer surrounding the medium particles (Fig. 1). The filter-bed medium (or packing) consists of organic materials (compost, peat, etc) or structured inorganic materials that ensure large surface areas and porosity. As the air passes through the media, the contaminants in the air phase are transferred into the biofilm and onto the filter medium, where they are degraded.

Biofilter reactors may be open- or closed-bed systems, and include six major components: air distribution system, humidification system, water irrigation system, reactor vessel, filter-bed medium, and computer-control system (Fig. 2).

Table 1. Possible Industrial Sources for Biological Air Pollution Control

Adhesive production	Landfill gas extraction
Animal husbandry	Petrochemical manufacturing
Chemical manufacturing	Petroleum industry
Chemical storage	Printing industry
Coatings industry	Pulp and paper
Composting	Rendering
Food processing	Sewage treatment
Furniture manufacturing	Site remediation
Investment foundries	Wood products production
Iron foundries	

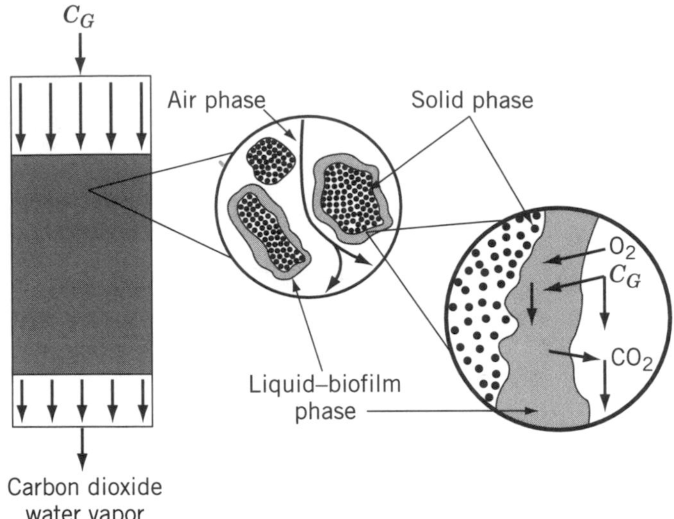

Figure 1. Internal mechanisms of a biofilter. Contaminated air (C_G) passes through the filter-bed medium (compost, peat, soil, etc) with oxygen and sorbs into a microbial biofilm–liquid phase attached to the filter medium. Microbes degrade the contaminant into carbon dioxide and water.

The overall effectiveness of a biofilter is largely governed by the properties and characteristics of the support medium, including porosity, tendency for compaction, water-retention capabilities, and the ability to host microbial populations. Critical biofilter performance parameters include the medium moisture content, medium pH, microbial inoculum, and nutrient content.

Biofilters have low operating costs and produce little secondary pollution. However, they may be inappropriate when high concentrations of organics or poorly degradable compounds are present in the waste gases. In general, biofilter technology is most cost-effective for flows of 1,000–50,000 m^3/h and pollutant concentrations up to 1 g/m^3 (3,4).

Biotrickling Filters and Bioscrubbers

Biological trickling filters and bioscrubbers are biological reactor units in which the contaminants are absorbed by a flowing liquid phase and biodegraded by either suspended or immobilized microorganisms. The liquid is collected at the bottom of the reactor and recirculated. Biotrickling filters operate with either countercurrent or cocurrent movement of the air and water phases, depending on the specific operation. As the water is recirculated over the packing material, nutrients, acids, or bases may be added by the operator to control environmental conditions for optimal pollutant removal. Biotrickling filters are governed by many of the same phenomena as biofilters. Most important, the system must host a thriving microbiological population while avoiding clogging.

In a bioscrubber, absorption occurs in a packed or unpacked washing tower or a bubble column. Degradation of the contaminants is performed by a suspended consortium of microbes in a separate vessel. The system is aerated to ensure maximum microbial degradation.

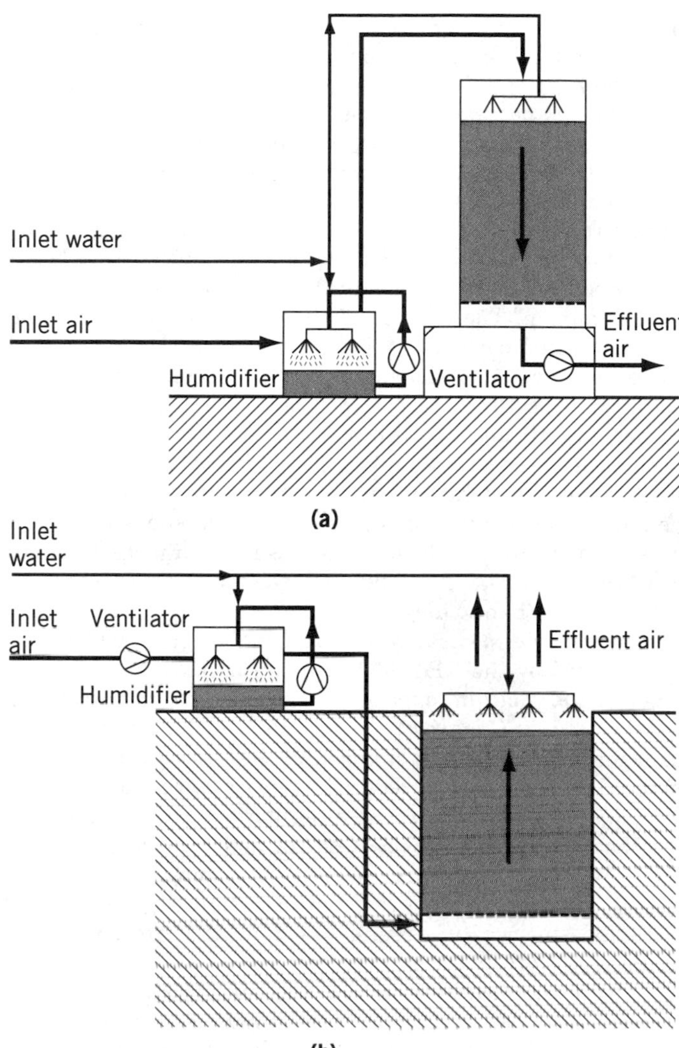

Figure 2. Schematic diagram of (**a**) a closed biofilter and (**b**) an open-air biofilter (2).

The flowing water benefits both biotrickling filters and bioscrubbers by providing a continuous supply of nutrients, removing toxic degradation by-products, suspending biomass for continual reseeding of the system, and aiding in the diffusion of hydrophilic pollutants into the biofilm. Both systems can operate on waste streams similar to those treated by biofilters but may be better for more soluble compounds. However, complex computer monitoring and control are required, raising costs.

HISTORY OF BIOFILTERS

With the increasingly stringent air pollution control regulations developed since the 1960s, biofiltration technology has slowly become an accepted alternative to traditional air pollution control measures. The idea of using microorganisms to treat organic waste streams is not new. Since the late 1890s, microbial reactions have been used in engineered systems for wastewater and solid waste treatment. In the last 40 years, microbial remediation techniques have been used to treat waste gases. Some of the earliest biofilter systems, built in the mid-1950s, were open, excavated pits filled with soil (5). These pits had a simple air-distribution system of perforated pipes beneath the soil, and odorous air was cleaned as it passed upward through the filter bed. However, problems with homogeneous air distribution and clogging of filter material occurred.

In the 1970s, as industry and its pollutant releases grew, so did the need for biofilters capable of handling larger loads of odors and volatile organic compounds. Primarily in Germany and the Netherlands, more advanced systems were developed. These biofilters were open to the atmosphere and had better air-distribution systems. Structural support media (bark, wood chips, polystyrene balls, etc) were added to reduce bed compaction and to evenly distribute the air.

During the 1980s and 1990s, biofiltration progressed rapidly in Europe and slowly in North America. Computer-operated, enclosed systems have been designed to treat odors, volatile organic compounds, and mixtures (6). Experiments with inorganic filter bed media have been performed (7,8). Granular activated carbon, polystyrene coated with powdered activated carbon, and ceramics have been used. Mixed with organic materials, they improve porosity while increasing the life expectancy of the bed. The development of advanced design techniques and a mathematical model (9) began the transformation of biofiltration research from a black-box approach to a science-based effort.

As biofilter technology has been demonstrated to be a cost-effective, reliable means of controlling biodegradable waste gases and has been accepted by regulators, new biofilter development and research companies have emerged. Research is now being directed at understanding pollution biodegradation pathways, mixed pollutant treatment, transient behavior, nutrient limitation, degradation inhibitors, suppression of excessive growth, and process modeling.

DESIGN OF BIOFILTERS

Biofilter design has become more sophisticated since the 1960s. Initial designs were developed by trial and error, but biofilter design has progressed dramatically. However, because waste gases differ in contaminant type, concentration, flow rate, temperature, and humidity, the ability to design an effective biofilter still involves fundamental science coupled with practical experience from past designs. With this dual approach, cost-effective systems can be designed without the mistakes that have plagued past efforts. A detailed explanation of biofilter design is available (1).

CONCLUSIONS

Increased production by industry will lead to enactment of further waste gas control regulations. The severity of the standards will increase as more harmful effects of air pollution are discovered. Such stringent conditions are creating the need for better, cost-effective emission control technologies for low concentration waste streams. Traditional technologies such as incineration and absorption are effective at higher loading rates, but they are expensive for dilute wastes. Biofiltration is an alternative to these traditional technologies. Since the 1970s, the biofiltration field has been slowly growing and becoming more prominent in the waste gas control market. Reasons for its rise in prominence include its simplicity, effectiveness on dilute waste streams, and low costs. The technology has advanced from treating odors with simple, open soil beds to treating volatile organic compounds in enclosed, multibed systems. Advancements in the technology have come from better control of operational conditions and a better understanding of the removal mechanisms of pollutants. Through the next century, further research will be implemented toward transforming the field from an empirical approach to one deeply rooted in scientific principles.

BIBLIOGRAPHY

1. J. S. Devinny, M. Deshusses, and T. S. Webster, *Biofiltration for Air Pollution Control,* CRC Lewis, New York, in press.
2. M. A. Deshusses, *Biodegradation of Mixtures of Ketone Vapours in Biofilters for the Treatment of Waste Air,* Ph.D. dissertation, Swiss Federal Institute of Technology, Zurich, Switzerland, 1994.
3. A. J. Dragt in A. J. Dragt and J. Van Ham, eds., *Biotechniques for Air Pollution Abatement and Odour Control Policies,* Elsevier Science Publishers, Amsterdam, 1992, pp. 3–9.
4. H. J. G. Kok in Ref. 3, pp. 77–82.
5. U.S. Pat. 2,793,096 (Oct. 5, 1953) R. D. Pomeroy (to R. D. Pomeroy).
6. T. S. Webster, J. S. Devinny, E. M. Torres, and S. S. Basrai, *Environ. Prog.,* **15**(3), 141–147 (1996).
7. V. F. Medina, T. Webster, M. Ramaratnam, and J. S. Devinny, *J. Environ. Sci. Health,* **A30**(2), 407–422 (1995).
8. J. R. Graham, *Proceedings of the 1996 Conference on Biofiltration (an Air Pollution Control Technology),* University of Southern California and The Reynolds Group, Tustin, Calif., 1996, pp. 85–93.
9. S. P. P. Ottengraf in H. J. Rehm and G. Reed, eds., *Biotechnology,* 8th ed., VCH Verlagsgesellschaft, Weinheim, Germany, 1986, pp. 426–452.

BIOHAZARDOUS WASTE

WAYNE L. TURNBERG
Washington Department of Health and University of Washington
Seattle, Washington

On November 1, 1988, the Medical Waste Tracking Act (MWTA) was signed into law (1). This action, in part, required the Administrator of the U.S. Environmental Protection Agency (EPA) to publish an interim final rule for a two-year demonstration medical waste management and tracking program. The MWTA was enacted in an attempt to end the washups of medical waste along beaches of the Atlantic Coast and Great Lakes which had occurred during the summers of 1987 and 1988. The Act established a cradle-to-grave medical waste tracking program which was ultimately implemented in four states and one territory of the United States. Of note, the beach washups from which this issue arose consisted primarily of garbage and other debris. Only a very small part of the waste consisted of syringes, medical vials, or other wastes of medical origin. Nevertheless, the perception of medical waste littering the beaches raised public fears about the spread of the human immunodeficiency virus (HIV) through contact with such waste. The presence of garbage on the beaches and the fear that much of that waste was medical in origin resulted in beach closures and a fearful public often unwilling to use the beaches when open. As a result, public confidence was shaken and local shoreline economies suffered.

In the heat of public concern, the United States Congress enacted quick passage of House Bill 3515, the Medical Waste Tracking Act of 1988 (MWTA), codified at 42 U.S.C. 6992 and amending the Resource Conservation and Recovery Act of 1976 by adding a new Subtitle J. In its enactment, Congress decried the appearance of such waste on the beaches to be "repugnant, intolerable and unacceptable," which was being attributed by Congress to medical mismanagement and "midnight dumping." The sponsor of the bill, Representative Tom Luken of Ohio, declared: "Medical waste pollution is lethal and grotesque. It is sickening and frightening the public, fouling our coastlines, and crippling local tourist economies. We have a critically ill medical waste disposal system—and the hypodermic needles and vials of infected blood washing ashore from Massachusetts to Florida and New York to Ohio are the leading symptoms of this illness." Of note, subsequent investigation found the beach washups to be largely attributed to a poorly functioning solid and wastewater management system rather than to illegal dumping.

Although the medical waste tracking program which was implemented by the EPA had only limited participation, and although the program expired in June of 1991 without being reauthorized by Congress, the course of biohazardous waste management was forever changed in the United States. Prior to the MWTA of 1988, federal agencies with authority to promulgate biohazardous waste regulations such as the EPA and the Centers for Disease Control and Prevention (CDC), often directed their efforts toward developing guidelines or "best management practices," rather than regulations for managing this waste stream, based on the viewpoint that regulations were not warranted or necessary. However, the public's growing fear of the AIDS epidemic in conjunction with the washups of medical waste on the nation's beaches led to a national shift in biohazardous waste policy in response to perceived risks associated with this waste stream.

Biohazardous Waste in the Environment

The EPA Standards for the Tracking and Management of Medical Waste, published at 40 *CFR* Part 259, was promulgated on March 24, 1989 as an interim final rule (1). The preamble to the EPA rule quoted from the findings of a study conducted by the New York Department of Environmental Conservation on beach washups in 1988 along the Atlantic Ocean, which concluded:

> *With few exceptions, floatable debris cannot be traced to any specific source. Most of these wastes are likely to come from: (1) the improper transport and handling of solid waste destined for disposal at the Fresh Kills landfill, (2) inadequate handling procedures, supervision, and maintenance at the marine transfer stations, (3) combined sewer overflows, (4) raw sewage discharges caused by occasional breakdowns at one or more of New York City's sewage treatment plants, and (5) storm water outlets. Other activities that were judged less likely to contribute included: (1) litter deposited by beach users, (2) recreational boating, and (3) commercial shipping. Other findings included: (1) floatables that become stranded on sandbars are sometimes refloated by the tides and washed ashore, (2) weather conditions contributed to the volume and persistence of washups; and (3) illegal disposal appears to account for some of the wastes (ie, the blood vials).*

Table 1. Estimated U.S. Medical Facilities and Annual Waste Generated[a]

Generator Category	Number of Facilities	Annual Infectious Waste Generated, t	Annual Total Waste Generated, t
Hospitals	7,000	360,000	2,400,000
Laboratories			
Medical	4,900	17,600	117,500
Research	2,300	8,300	55,500
Total	*7,200*	*25,900*	*173,000*
Clinics (Outpatient)	41,300	26,300	175,000
Physicians' offices	180,000	35,200	235,000
Dentists' offices	98,000	8,700	58,000
Veterinarians	38,000	4,600	31,000
Long-term care facilities			
Nursing homes	18,800	29,700	198,000
Residential care	23,900	1,400	9,000
Total	*42,700*	*31,100*	*207,000*
Free-standing blood banks	900	4,900	33,000
Funeral homes	21,000	900	6,000
Health units in industry	22,1700	1,400	9,000
Fire and rescue	7,200	1,600	11,000
Corrections	4,300	3,300	22,000
Police	13,100	>100	>1,000
Total	*682,400*	*504,000*	*3,361,000*

[a] Courtesy of the U.S. Environmental Protection Agency, EPA-453/R-94-042a.

Problems associated with the presence of syringes on the beaches of the Great Lakes states, particularly Lake Erie, have been generally attributed to improper disposal by intravenous drug users or individuals with diabetes, rather than illegal dumping by medical facilities (2).

Before the interim final rules were implemented, the EPA identified its skepticism about the success the program might have on medical waste beach washups. Nevertheless the EPA noted that the program would result in the proper management of medical waste, and would also allow for the collection of information that could be used by Congress when considering additional legislative decisions. See Plate 3.

Sources and Volumes of Biohazardous Waste

Biohazardous waste is most simply defined as waste which is capable of producing an infectious disease. It represents that subset of the overall medical waste stream with infectious characteristics and is often referred to both in regulations and in the scientific literature by a number of different terms, such as infectious waste, red bag waste, biomedical waste, and regulated medical waste. Depending on how it is defined, it may include such wastes as (*1*) cultures and stocks of infectious agents and associated biologicals; (*2*) human blood and blood products; (*3*) pathological wastes; (*4*) contaminated sharps; (*5*) contaminated animal carcasses, body parts, and bedding; (*6*) isolation waste; (*7*) wastes from surgery and autopsy; (*8*) contaminated laboratory wastes; (*9*) dialysis unit wastes; or (*10*) contaminated equipment.

Because there is no nationally mandated definition for biohazardous waste, significant definitional changes in the waste stream routinely occur as one crosses borders between state and local jurisdictions. However termed or defined, proper and safe management of biohazardous waste must be an integral part of any healthcare strategy to protect the safety and health of healthcare providers and support staff, patients and their families, waste industry workers, and the general public.

Biohazardous waste is generated primarily in the course of healthcare or research by both medical institutions and home healthcare activities and to a lesser extent by illegal drug users. The primary medical institutions generating biohazardous waste in the United States include hospitals, laboratories, physicians, dentists, veterinarians, long-term healthcare facilities, clinics, blood establishments, and funeral homes. The estimated number of generators in each of these categories and volumes of waste generated are presented in Table 1.

EVALUATING HUMAN HEALTH RISK

Biohazardous Waste and Infectious Disease Transmission

During the past few years, a significant effort has been directed toward proper and safe management of biohazardous waste streams by government agencies and the regulated industry, but without a clear understanding of the infectious disease transmission risks associated with improper disposal practices.

To date, infection in the community resulting from exposure to biohazardous waste has not been identified. Furthermore, only two incidents of infectious disease have been reported in the scientific literature associated with exposure to biohazardous waste within the medical environment: one involving possible infection to a hospital employee from a chute-hydropulping waste system in a hospital (3); the other involving a hospital housekeeper who developed staphylococcal bacteraemia and endocarditis following injury by a waste hypodermic needle (4).

Elements of Infection. The term *infection* is defined to mean the entry and development (of many parasites) or multiplication of an infectious agent in the body of persons or animals (5). Infection of a host may occur without recognizable clinical signs or symptoms (inapparent infection), or may lead to a clinically manifest disease (infectious disease).

Infection requires a specific series of events to occur. There must be (*1*) a reservoir of an infecting organism (human, animal, or inanimate); (*2*) a susceptible host; and (*3*) a mode of transmission to the host (contact, vehicle, airborne, and vectorborne). There must also be (*4*) an infective dose of the pathogen to cause infection in the susceptible host, and (*5*) a portal of entry for the pathogen to enter the host (eg, inhalation, ingestion, breaks in the skin and mucous membranes, or direct bloodstream inoculation by sharp objects such as hypodermic needles).

Estimating Infectious Disease Transmission—ATSDR Report. A comprehensive examination of the role that biohazardous waste may play in human infection and disease was published by the Agency for Toxic Substances and Disease Registry (ATSDR) in November 1990 (6). The study was mandated under Section 11009 of the Medical Waste Tracking Act of 1988 which required the Administrator of ATSDR to prepare a report on the health effects of medical waste that included:

> *(1) a description of the potential for infection or injury from the segregation, handling, storage, treatment, or disposal of medical wastes; (2) an estimate of the number of people injured or infected annually by sharps, and the nature and seriousness of those injuries or infections; (3) an estimate of the number of people infected annually by other means related to waste segregation, handling, storage, treatment, or disposal, and the nature and seriousness of those infections; and (4) for diseases possibly spread by medical waste, including Acquired Immune Deficiency Syndrome and hepatitis B, an estimate of what percentage of the total number of cases nationally may be traceable to medical wastes.*

Populations under consideration in the study for estimating medical waste injuries, infection, and disease included healthcare providers (those involved with direct patient care such as nurses or laboratory technicians, and nonpatient care workers such as janitors and housekeeping staff); waste handling workers (eg, trash collectors and handlers, landfill operators, and incinerator operators); and the general public. The strategy for estimating injury, infection, and disease related to medical waste exposure involved estimating the probabilities of occurrence of each step in the chain of events necessary for injury, infection, and disease to occur. These events require that an injury occur (eg, physical injury or mucous membrane contamination allowing entry of infectious agents), that the mechanism of infectious disease be completed, and that the infection result in a state of disease. Formulas used for determining estimates for the number of medical waste related injuries, infections, or disease are presented as follows:

$$\text{Estimated number of medical waste-related injuries} = (\text{number of persons in each occupational subgroup}) \times (\text{medical waste-related injury rate})$$

$$\text{Estimated number of medical waste-related infections} = (\text{number of medical waste injuries}) \times (\text{prevalence of contaminated material}) \times (\text{sero-conversion rate})$$

$$\text{Estimated number of medical waste-related diseases} = (\text{number of medical waste-related infections}) \times (\text{clinical disease rate among sero-converted individuals})$$

Estimates of the number of persons suffering injury and infection related to contact with medical waste had not been made prior to the ATSDR effort. Injury data estimates were based on information derived from the medical literature or special surveys. When information was not available, estimates were made from similar situations. Because of the lack of information and methods to collect or calculate data, certain biases and inaccuracies were recognized. Data collected from multiple sources may not reflect identical degrees of involvement with medical waste. In addition, individual studies always contain biases, which will most likely remain when the studies are combined. Therefore, ATSDR chose to base estimates on extremes, such as worst-case situations that reflect the highest estimates in the available information and best-case situations that reflect the lowest estimates in the available information.

Origins, types, and estimated quantities of medical waste, as defined by the MWTA, were identified in the report based on estimates derived from various sources. The sources of waste were identified to be (*1*) hospitals; (*2*) physician's offices; (*3*) dentist's offices; (*4*) biomedical research facilities; (*5*) clinical laboratories; (*6*) manufacturing facilities; (*7*) veterinary offices and clinics; (*8*) funeral homes; (*9*) in-home medical care; (*10*) other healthcare and residential care facilities; (*11*) illicit intravenous drug use; (*12*) other sources (eg, cruise ships and naval vessels).

Nonsharps Medical Waste. ATSDR addressed questions posed by Congress in the MWTA by categorizing medical waste as being either "nonsharp" (representing all medical waste categories except Category 4 waste) or "sharp" (representing only Category 4 waste) in nature. When conducting its analysis, ATSDR distinguished between those injuries and infections related to medical waste from those occurring in the course of patient care that are unrelated to medical waste. Additionally, injuries and infections related to medical waste before it is discarded are identified in the report as not related to medical waste.

ATSDR described conditions necessary for infection to occur and noted that nonsharp medical waste generally does not create a portal of entry upon contact. Therefore, for infection to occur, a portal of entry would have to already exist at the time of contact for transmission to occur. Mechanisms for entry into the body by infectious agents associated with nonsharp waste include transmission via contact with breaks in the skin (eg, cuts and abrasions) or contact with mucous membranes (eg, contact with eyes or inner lining of the nose). This could occur for instance if infectious agents are suspended in blood.

In its review, ATSDR noted that several cases suggested contact with blood, other than via percutaneous routes (eg, needlestick or other injury producing entry) may produce HIV infection during patient care. However, on examination of eight longitudinal studies that evaluated healthcare workers who had been exposed to blood or blood containing body fluids through skin or mucous membrane contact from an HIV positive patient, no HIV infections were observed. This is based on evaluations of 538 healthcare workers that had experienced up to 921 mucous membrane contacts with HIV-positive blood or blood contaminated body fluids.

Based on this information, ATSDR estimated the probability for transmission following mucous membrane exposure (eg, splash to the eyes or other mucous membrane) to blood or blood contaminated body fluids from an HIV positive patient of <0.1%, and the upper bound of the 95% confidence interval for the rate of transmission is approximately 0.33%. Based on literature reports, ATSDR noted that the probability of transmission following a percutaneous contact (eg, needlestick or other skin breaking injury) with HIV contaminated blood is approximately 0.36%, and that it seems reasonable that the probability following a mucous membrane contact would be less. The report also noted that there were no HIV or hepatitis B viral (HBV) infections reported in the scientific literature associated with medical waste and that no injury data for medical waste management had ever been reported in the scientific literature, except as related to sharps waste. The report concluded:

> *Based on the principles of infectious disease transmission, the potential for infection as the result of contact with non-sharp medical waste is likely to be significantly less than that for injuries related to medical waste sharps. The primary reason for the lower potential is, in contrast to sharps, a portal of entry must exist prior to contact with non-sharp medical waste.*

Sharps Medical Waste. In contrast to nonsharps medical waste, sharps (ie, Category 7 wastes) have the ability to create a portal of entry through the skin. As required by the MWTA, ATSDR estimated the number of medical waste injuries that occur for various occupational subgroups based on available data sources. Limiting factors included (*1*) the general unavailability of overall injury data in specific occupations at the national level, (*2*) the use of workers' compensation claims in selected states to calculate overall injury rates, and (*3*) the fact that needlestick injuries generally are not reported.

Transmissions of infectious agents have been reported in the literature by contaminated sharps, although almost all transmissions occurred in the course of patient care or laboratory procedures, and were not associated with medical waste. ATSDR identified only one medical waste related occurrence which involved a hospital housekeeper reported earlier in this article (4). Other than this one case, no other information was identified in the scientific literature that involved transmission from exposure to a contaminated waste needle.

Calculating Theoretical Transmission Rates. To calculate a theoretical infectious disease transmission rate, ATSDR estimated the probability that any medical waste sharp would be contaminated with an infective dose (eg, sufficient number of infectious agents to cause infection in a susceptible host), that infection will occur following injury, and that the infection will result in disease.

The formula used by ATSDR to estimate the number of medical waste-related infections is:

(Number of medical waste injuries)

× (prevalence of contaminated material)

× (sero-conversion rate)

The formula to estimate the number of medical waste-related diseases is:

(Number of medical waste-related infections)

× (clinical disease rate among sero-converted individuals)

In its study, ATSDR developed infection and clinical disease estimates for HIV and HBV only. Information needed to calculate estimated transmission rates for other infectious diseases is not available.

Estimating Injury Numbers. To determine estimates of the number of medical waste injuries, ATSDR relied on information that reported injuries resulting from improperly disposed medical sharps in the waste stream for several occupational groups. Four sources of information were used: (*1*) a medical waste-related injury survey conducted by 17 health departments for the ATSDR report; (*2*) data provided by the solid waste industry; (*3*) data provided by the U.S. Department of Defense; and (*4*) the scientific literature. Recognizing the limitations of these data, and adjusting data as necessary, the injury rates were used by ATSDR to calculate the number of medical waste injuries for each occupational category.

Implications of Infectious Agents in Solid Waste

Microbial agents infectious to human are present in general (7,8). Pathogenic microorganisms routinely enter the waste stream via common contaminated materials such as human excreta from disposable diapers, animal excrement, blood, exudates or secretions from dressings, bandages, sanitary napkins, facial tissue, condoms, bandages, home used syringes, or other inanimate objects that have come into contact with human body substances. Any waste source, whether of medical or residential origin, may contain pathogenic microorganisms capable of causing infection in a susceptible host if all elements required for infection to occur are met.

Infectious Disease and the Waste Industry. As a result of the AIDS epidemic, medical waste disposal has emerged as a concern for waste industry workers, although such disease transmission has never been epidemiologically demonstrated. Few studies have been conducted to examine waste industry worker exposure to potentially infectious agents in the waste stream. Researchers examined 97 waste workers employed by the City of Cincinnati during January and February of 1968 for skin disorders (9). Forty-one cases of bacterial, viral, or fungal dermatitis were observed in this group, but all were classified as nonoccupational in origin. The authors reported that no systemic infectious diseases had been diagnosed in the Cincinnati Division of Waste Collection at the time of their study and that only one claim had been filed for occupational skin disease that was later judged to be nonoccupational in origin.

New York City Sanitation Department health records of waste workers employed between 1968 and 1969 have been examined (10). Needlestick injuries were reported due to the presence of uncontained needles in waste collected from hospitals, doctor's and dentist's offices, and discarded needles from drug addicts. All workers reporting needlestick injuries were given gamma globulin prophylaxis and no cases of hepatitis were reported.

A subsequent 1987 publication of occupational hazards from New York City sanitation workers examined death records between 1975 and 1984 for those employed as solid waste collectors as of January 1973 (11). Of those 10,565 individuals, 511 died during that period. The author did not report any deaths or illness due to infectious disease.

A 1979 report examined the incidence of viral infection among 43 waste collection workers (12). Sera antibody levels for 18 viruses were examined from blood samples collected during the spring and fall. The authors found no evidence for an increased occupational risk to bloodborne viral infections.

In a 1990 report, 940 waste industry workers were surveyed to evaluate occupational exposure to potentially infectious materials in the municipal waste stream (13). Responses were received from 438 (47%) of the 940 workers surveyed. Waste worker employee job safety training rates were ascertained as well as occurrences of occupationally incurred cuts and scratches. The prevalence of exposure to blood contaminated waste and injury from hypodermic needles were also estimated. Sixty-nine percent of respondents reported having received job safety training but only 26% were trained specifically to deal with safety hazards associated with medical waste. Seventy-four percent of respondents reported having received cuts and scratches on-the-job and 32% of respondents reported direct contact with waste blood on their clothing or shoes. Thirteen percent of respondents reported blood exposure on their skin and 5% reported blood exposure on their face or eyes. Occupational needlestick injuries were reported by 21% of respondents overall, with 10% of 240 responding waste collectors reporting having sustained a needlestick injury in the year preceding the survey. Needlesticks were reported from both residential and commercial waste collectors as well as by landfill/transfer station operators. Although injuries were reported, none was linked to infectious disease transmission.

Despite any clear evidence implicating waste industry employment with a higher incidence of infectious disease, advisories have been developed to safeguard against such potential occurrences. For example, in 1982, the National Institute for Occupational Safety and Health (NIOSH) reported on the potential for infectious disease transmission to residential waste collectors from direct contact with the waste stream (14). The report stressed personal hygiene as a health matter of considerable importance to this group and recommended that employees bathe daily and wash before eating during the day and before going home. The report recommended that proper techniques for cleaning and covering wounds be addressed by employers and that clean gloves and coveralls be provided daily by employers. In a 1987 publication, the Health Division of the Oregon Department of Human Resources further recommended that safety protective equipment be employed by waste workers to add further protection against potential infectious disease transmission (15). Protective measures include use of safety glasses, hard hats with chin straps, coveralls, waterproof gauntlet gloves, boots with sufficient thickness and strength to protect the wearer from injury from sharp objects, and NIOSH approved dust masks when working indoors (eg, at a transfer station) or whenever necessary. The report also urged waste worker employees to report all injuries and illness to the person responsible for employee health.

Surface and Groundwater Contamination. A modest level of research has been conducted to determine the ability of microorganisms to migrate from waste disposed in landfills to surface or groundwater. The health significance of pathogenic microorganisms in landfills has been described as related to (*1*) the concentration and nature of the pathogen, (*2*) the pathogen's ability to survive and retain its infectious properties in the landfill environment, and (*3*) the pathogen's ability to migrate through the landfill into the surrounding environment and be a potential human hazard (16).

Although studies examining direct measurement of human pathogens in landfill environments are limited, several studies have been undertaken that have quantified indicator organisms of fecal pollution (total coliform, fecal coliform, and fecal streptococcus) from leachate derived from field or simulated sources (17). The substantial presence of fecal waste in landfills from human and animal sources has been suggested based upon indicator bacterial recoveries (18,19).

Enteric pathogenic bacteria have been identified from leachate sources, but not enumerated (17). Enteric virus particles, including polio virus, have also been isolated from landfill leachates, although one study that examined leachate samples collected from 22 landfill sites isolated enteric virus particles (polio types 1 and 3) from only one site which was described by the author as having deficient sanitary landfill practices (20).

Several studies have been published in the literature that examine the survival of microorganisms in landfills and landfill leachates. Significant decreases in microbial populations have been observed both in the field and during laboratory lysimeter testing. For example, a significant decrease of bacteria and an absence of virus were observed from leachate derived from a laboratory lysimeter simulating a landfill environment. These results indicated that the harsh conditions of landfill leachate have an inactivating effect on certain microorganisms (16). However, the inactivating constituents could not be correlated. In a field examination of leachates from 22 landfills, viruses could be isolated from only one site which was described as poorly operated (20). It was suggested that leachates from properly run sanitary landfills pose little threat to the public with regard to infection by enteric virus.

Although research indicates that the chemical and physical characteristics of the landfill environment produces an inactivating effect on viruses and bacteria, it is still possible that microbial pathogens may survive. However, movement of microorganisms through soils of the landfill would be dependent on many factors, including soil texture/composition, soil moisture, salt concentrations, pH, climate (rainfall and temperature), nutrient availability, and antagonisms (21). Absorption of viruses onto fill material is likely and may also explain the low recovery of viruses from landfill leachate studies.

Landfilling of solid waste remains the most widely used and accepted method for waste disposal. In 1991, the EPA promulgated minimum landfill standards that must be adopted and implemented by each state (22). These requirements specify how landfills are located and operated. The standards address how municipal solid waste landfills are sited, constructed, operated, closed, and cared for during a 30-year post-closure period to ensure that the environment is protected. The standards were largely written to protect ground and surface water from leachate pollutants. In doing so, EPA established requirements for landfill liners, leachate collection systems, compaction and covering of waste, and environmental monitoring. The operation of sanitary landfills has become far more sophisticated in terms of protecting the health of the public and environment when compared to the open dump sites of only 20 years ago. In conclusion, the risk associated with ground and surface water contamination by infectious microorganisms in a properly operated landfill appears to be low.

This premise is supported in the literature although further examination of this issue may be warranted (see LANDFILLS).

Infectious Aerosols. Exposure to aerosols containing pathogenic microorganisms can result in human infection, particularly in laboratory settings (23). The ability of an aerosol to cause infection by inhalation is related to the susceptibility of the host and the infective dose and virulence of the pathogen.

The role of infectious aerosols originating from the solid waste stream in terms of human morbidity is less understood and supporting studies are limited. Two studies have been identified that focus on the subject. Researchers conducted an examination of aerosols associated with waste collection and their effects on the workers (24). It was concluded that reports of the start and continuation of chronic bronchitis could be associated with exposure to airborne infectious agents.

Airborne levels of bacteria and viruses were examined at a refuse processing plant, a municipal incinerator, a waste transfer station, a landfill, and a wastewater treatment plant (25). No viruses were isolated during the study, which may actually reflect more on the limitations of the assay recovery system than on the absence of airborne virus particles. The highest levels of airborne bacteria colonies were observed at the refuse derived fuel facility, both within the facility and at the property line. The health significance of any of the observed levels could not be determined. The study also involved a comprehensive literature review to determine existing information regarding bacteria and virus emissions from waste handling facilities (25). The authors identified studies observing airborne bacterial colonies ranging from 200/cubic meter in a laboratory, up to 700,000/cubic meter in a sewage treatment plant and between 2,000 and 4,000/cubic meter in offices, factories, and streets. Again, the health significance for any of these levels could not be judged, based on the existing literature.

Implications of Infectious Agents in Wastewater

Attention has been focused on the practice of disposing of bulk blood into the sanitary sewage system and the potential for this practice to spread bloodborne pathogens to plumbers, sewage treatment plant operators, or others who may come into contact with such liquid waste.

BIOHAZARDOUS WASTE POLICY IN THE UNITED STATES

A principal consideration in managing biohazardous waste is understanding the applicable federal, state, and local regulations, guidance policies, and industry standards that influence or directly affect how this waste stream is managed. To date there is no national regulation that comprehensively addresses how biohazardous waste is to be managed from cradle to grave, nor does it appear that such a mandate is forthcoming from the U.S. Congress. Nevertheless, at least five federal agencies, the (OSHA) (bloodborne pathogen standards), the Public Health Service (PHS) (transportation standards), the Department of Transportation (DOT) (transportation standards), the U.S. Postal Service (USPS) (mail system standards), and the EPA (proposed medical waste incineration standards), have developed or are in the process of developing requirements that address biohazardous waste that are specific to their respective jurisdictional responsibilities. Other federal agencies such as the EPA, the National Institutes of Health (NIH), and the CDC have also developed guidance policies specific to biohazardous waste. Other guidance comes in the form of industry standards, such as those developed by the Joint Commission on Accreditation of Healthcare Organizations (JCAHO) and the American Society for Testing and Materials (ASTM). In addition, the EPA's interim medical waste tracking program, which expired in 1991, continues to influence biohazardous waste management practices. Each of these policies is summarized.

In addition to the federal activities, most states and many local governments have also developed biohazardous waste regulations during the past few years. However, without a federal mandate or clearly understood regulatory need, state and local regulations are typically inconsistent and widely variable in approach.

MANAGING BIOHAZARDOUS WASTE IN A MEDICAL FACILITY

Source to Endpoint Management

This section addresses the management of biohazardous waste when generated by medical facilities such as hospitals, laboratories, physicians offices, dental clinics, veterinary clinics, long-term healthcare facilities, clinics, blood establishments, and funeral homes. Although biohazardous waste types and volumes vary significantly between the various potential generators, common basic principles of management can be applied. The management approach presented is provided primarily as an overview of commonly observed practices which can be incorporated into a facility's biohazardous waste management program. Nevertheless, the reader should be aware that waste management programs are first of all based on the federal, state, and local requirements established by government agencies with which the biohazardous waste generator must comply.

As a first step, the biohazardous waste management system developed by each generating facility should address those individuals in the facility responsible for developing the program and ensuring that it is followed. Those responsible may consist of a waste management team comprised of key department heads (eg, directors of safety, environmental services, engineering, and infection control) for larger generators such as hospitals or of a single individual for small generators such as physician's offices. Whatever approach is taken, those given this responsibility to develop and implement the program must also have the full backing by the facility management to carry out their duties.

Biohazardous waste management requires an understanding of all aspects of the waste stream within a given medical facility from the point of initial generation through final treatment and disposal. Proper management would take into account mechanisms to protect the health of the healthcare provider, housekeeping and custodial staff within a facility, waste collectors/workers, and the public once the waste has left a facility for final disposal.

The program should address protocols to meet all applicable federal, state, and local management requirements. For a program to be effective, protocols for each element must be recorded in a written biohazardous waste management plan which is maintained at a location where it can be easily accessed by facility users and government or accreditation agencies should it become necessary. Protocols should be established to meet all applicable federal, state, and local requirements and accreditation agency standards if applicable. When developing protocols, the health and safety of facility staff, visitors, waste industry workers, and the public must also be taken into account.

Biohazardous Waste Management Plan

Central to a facility's biohazardous waste management program is the biohazardous waste management plan. The plan should document information describing all aspects of the waste management program from the point of waste generation to final disposal. Written plans should be prepared by both larger generators such as hospitals as well as smaller generators such as dental practices. The exercise of writing a plan, large or small, focuses attention on how waste is managed by a facility, and documents step-by-step how waste management is to be carried out. A management plan also provides a written record of a facility's disposal policies.

Once written, the plan should be periodically reviewed and updated as necessary, and should be made readily available to all employees conducting waste management activities. The plan should be written in a format that is easily understood and should be maintained as a working document by the facility, and should be available for inspection by government regulators and the public if ever requested. The plan should include: (*1*) compliance with all applicable regulations; (*2*) responsibilities of all involved staff (eg, infection control members, environmental control and housekeeping personnel, department, and individual responsibilities); (*3*) definition of all biohazardous wastes handled within a facility; (*4*) procedures for biohazardous waste management, including biohazardous waste: (*a*) definition; (*b*) segregation; (*c*) containment; (*d*) labeling; (*e*) storage; (*f*) treatment; (*g*) transport (both in-house and off-site); (*h*) disposal; (*i*) monitoring and recordkeeping; (*j*) contingency planning; and (*k*) training for all involved staff (professional, medical, environmental services, housekeeping, etc). Each of these elements is addressed.

Definition. This element requires knowledge of the specific definition for the wastes within a facility to be categorized as *biohazardous waste.* Because there is no national standard defining what wastes compose the biohazardous waste stream, each facility's waste management team must develop its in-house definition based on definitions established in all applicable federal, state, and local biohazardous waste-related regulations, and based on the need to protect health and safety. Regulatory definitional inconsistencies are common, and the waste management team will be expected to deal with these. Inconsistencies can often be resolved through discussions (which should always be documented in writing) with the regulatory agencies, although a facility may ultimately have to establish an expanded in-house definition that includes all of the inconsistencies established by the various governmental bodies to insure compliance with each.

Identification. Once the biohazardous waste stream has been defined, a waste stream audit should be conducted. Before a management strategy can be developed, the biohazardous waste team should identify the location of waste generated and the types and amounts (volume or weight) of waste generated. Lists identifying types and amounts (either in volume or weight depending on how disposal costs are based) of wastes generated should be developed for each location where biohazardous waste is generated. By documenting location of waste generation, types, and amounts generated, informed management decisions can be made.

Segregation, Containment, and Labeling. In general, biohazardous waste should be segregated from the general waste stream at the point of origin by the generator of the waste into clearly marked containers that take into consideration the waste type (eg, liquid wastes, nonsharp/nonliquid wastes, and sharp wastes). To contain costs, facility staff should be trained to only segregate that which has been specifically defined as biohazardous waste by the facility. Casual disposal of nonbiohazardous waste materials can dramatically increase a facility's waste disposal costs.

Nonsharp Biohazardous Wastes. Solid/semisolid wastes should be segregated into disposable leak-proof containers or plastic bags that meet specific performance standards. A suggested performance standard would be the *Standard Methods of Test for Impact Resistance of Polyethylene Film by the Free Falling Dart Method,* Standard of the *ASTM*'s Designation D1709-67, Method B, which has been used by the State of California biohazardous waste rule. The bags should also be constructed in such a manner that precludes ripping, tearing, or bursting under normal use. These bags should be tagged or effectively marked by the generator as containing biohazardous waste. The bags should be secured to prevent expulsion of contents during handling, storage, or transport. Liquid biohazardous wastes should be prevented from entering these bags.

Sharps Waste. Sharps waste such as hypodermic needles should be contained in rigid, leak-proof, puncture-resistant, break-resistant containers that can be tightly lidded during storage, handling, or transport. The most suitable container material would be plastic.

Liquid Biohazardous Wastes. These should be segregated into leak-proof containers that are capable of transporting the waste without spillage.

When storing, handling, or transporting wastes that have been packaged as described, additional protection should be provided to the original containers (plastic bags, sharps containers, and liquid containers) by placing them into other durable containers such as disposable or reusable pails, cartons, boxes, drums, or portable bins.

If containers are to be reused for biohazardous waste storage, handling, or transport, they should be thoroughly washed and decontaminated by an approved method each time they are emptied unless the surfaces of the containers had been protected by disposable liners, plastic bags, or other means. The process of cleaning should include agitation (scrubbing) to remove any visible solid residue, followed by disinfection. Disinfection could be accomplished using chemical disinfectants. A minimum standard should require that the disinfectant be used in accordance with manufacturer's recommendations for tuberculocidal and viricidal (Polio type 1 or 2, SA Rotovirus) killing capacities or by other comparable process. These containers should not be used for any other purpose, unless they have been properly disinfected and have had biohazardous waste symbols and labels removed.

Specific segregation, containment, and labelling requirements are specified in rules published by OSHA for occupational waste handling, by PHS when wastes are transported over the highways, and by the USPS when wastes are mailed through the USPS mail system. Regulations have also been published by the DOT. Additionally, state and local rules must be consulted to insure compliance with all requirements.

Storage. Facilities that store biohazardous waste should have a specific storage area for that purpose. The storage area should be inaccessible to unauthorized entry. The area should offer protection from animals, the elements (eg, rain, and wind) and should not provide a breeding place or a food source for insects or rodents. Storage time and temperature should be considered because of putrefaction of the waste with time. Microorganisms will grow and decompose the waste in storage, creating the unpleasant odors associated with putrefaction or rotting garbage.

Treatment. Biohazardous waste treatment options are described later in this article. In general, treatment options are available either on- or off-site, depending on the facility's needs and budget. Options include incineration, steam sterilization, and various alternative treatment technologies including microwave irradiation, electrothermal deactivation, chemical treatment, or ionizing radiation.

Transportation. Transport of biohazardous waste must be considered as the waste moves through the facility to storage areas, and if the waste is to be transported off-site to a treatment/disposal facility. Carts used to transport waste within a facility from the point of generation to the storage site should only be used for that purpose, and not for other purposes (eg, food carts or miscellaneous equipment transfers). Carts should be routinely cleaned and disinfected. Off-site transportation must meet requirements established by the PHS, the DOT, and all state and local transportation requirements relating to the inter- and intrastate transport of biohazardous waste. In general, the waste destined for off-site transport should be transported only in leak-proof and fully enclosed containers or vehicle compartments. Biohazardous waste should not be transported in the same vehicle with other waste or medical specimens unless separately contained. Biohazardous waste spills should be promptly decontaminated. The waste should only be transported to a treatment facility that meets all local, state, and federal environmental regulations.

Contingency Planning. Generators of biohazardous waste must be prepared for unexpected events occurring from unexpected situations such as equipment failure, spills, exposures, or any other event that would interrupt the day-to-day management of the waste stream. Unexpected events must be anticipated, and procedures developed to address them should it ever become necessary. Procedures to follow should be documented and maintained in the facility's biohazardous waste management plan, and periodically updated as necessary.

Biohazardous Waste Minimization

Minimizing wastes generated by a medical facility that will require special handling, treatment, and disposal as biohazardous waste is an important cost containment step. For example, biohazardous waste haulage is at least five times more expensive than general waste haulage (26). This discussion is limited to minimizing only the biohazardous waste fraction of all waste generated by a medical facility. Techniques for minimizing biohazardous waste include proper segregation, waste reduction, product reuse, and recycling.

Segregation. The primary mechanism for reducing the volume of biohazardous waste generated by a medical facility involves proper waste segregation. For example, in an effort to reduce the volume of regulated medical waste (biohazardous waste) in its facility after the implementation of EPA's Medical Waste Tracking Program, the Beth Israel Medical Center located on Manhattan's lower East Side implemented a program "regulation interpretation, education, training, placement of appropriate waste receptacles, and monitoring" that resulted in a significant decrease in the volumes of biohazardous waste generated by the facility (26). In 1990 the medical center generated 1,040,558 kg (2,294,000 lbs) of regulated medical waste, with an annual cost of $858,550 for disposal. By 1992, this volume had been reduced to 545,045 kg (1,201,600 lbs) with a total disposal cost of $311,087 as a result of implementing its biohazardous waste reduction program.

A similar program of "aggressive waste segregation and behavior modification" was conducted by the Mount Sinai Medical Center in New York City in response to skyrocketing biohazardous waste disposal costs resulting from increased regulatory policies and the Medical Waste Tracking Act of 1988 (27). During this period the facility observed a red bag waste reduction of from 5,443 kg/d (12,000 lbs/d), which occurred immediately after implementation of the Medical Waste Tracking Act rules, to 1,814 kg/d (4,000 lbs/d) one year later. This translated to a cost reduction of from $110,000 per month to $40,000 per month.

The percentage of a medical facility's waste that is designated as biohazardous is largely determined by how the waste stream is defined by prevailing regulations. Because of the high costs associated with biohazardous waste handling, treatment, and disposal, it is important to insure that staff involved with segregating biohazardous waste from general waste have been adequately educated and trained to make this determination. Far to often, waste audits in hospitals have revealed general wastes such as soda cans or office wastes being disposed into biohazardous waste red bags. Such practices result in significant but completely unnecessary waste disposal cost increases to the facility.

Techniques to improve segregation efficiency begin with the waste audit. Audits may be conducted by observing staff waste segregation practices, and by examining contents of biohazardous waste red bags and containers. One purpose of the audit is to identify the consistency by staff of segregating only wastes that meet the biohazardous waste definition, and excluding all other nonbiohazardous waste types. Another purpose is to identify the locations of red bags that may be used for general waste disposal.

The audit should include a survey of biohazardous waste disposal container locations. Clearly marked biohazardous waste disposal containers should be conveniently located within all areas in which biohazardous waste is generated. However, experience has found that improperly located red bags will result in their improper and costly use for general waste disposal (26,27). Red bags in patient rooms should be either placed in locations to minimize use by visitors for general waste disposal or removed from patient rooms altogether. A clear bag (general) waste container should be located next to all red bag containers for the biohazardous waste segregation program to be successful. In addition, each facility should insure that enough clear bags have been ordered, and that they are conveniently stocked in all supply closets so that red bags will not be substituted by housekeeping staff for general waste collection.

Each facility should develop a plan for biohazardous waste segregation and insure that all staff involved with waste segregation have received adequate training and education and are committed to following the plan. The Mount Sinai Medical Center (New York) experience noted the single most important factor in achieving success in reducing the volume of red bag waste is cooperation by the nursing department. For this reason, the bulk of early educational efforts focused on nursing.

Reduction. Once proper segregation techniques have been implemented, the biohazardous waste stream can be further reduced by using less of a product that will require special handling as biohazardous waste when disposed. In a 1991 report by the Minnesota Hospital Association entitled "The MHA Recycling and Conservation Guide," one hospital estimated up to 10 tons of plastic per year would be removed from the biohazardous waste stream by requiring that biohazardous waste be only single bagged instead of double or triple bagged as had been the previous policy (28). Another hospital identified in the report estimated a plastic disposal cost savings of $7740 through an 86-ton reduction in plastic waste by converting to lighter, stronger plastic bags for containing infectious wastes. That hospital also projected a plastic purchase cost savings of $90,000 a year by changing plastic bags. As stated by a spokesperson for the hospital: "It is extremely important that a hospital make clear to the vendor its specifications for durable, lightweight bags, and make sure the bags provide protection. We had our vendor perform the ASTM D1709 Dart test on the bags to test durability, and we weighed the cases they were shipped to us in to make sure they met our specifications." The Beth Israel Medical Center switched from disposable to reusable sharps containers throughout its facility by contracting with a company offering a full-service sharps collection system. Since the switch, that medical center reports that the incidence of over-filled containers has dropped significantly, needle sticks associated with disposal are rare, and the facility has eliminated the need to incinerate 2700 disposable containers each month at an off-site location.

Product Reuse. Another biohazardous waste reduction technique involves replacing disposable with reusable materials when feasible. During the past two decades, disposable products have come into wide use by medical institutions because of their infection control benefits. However, in many cases the benefits of reusable over disposable products are being reexamined in part because of mounting costs of waste disposal. For example, the Methodist Hospital in St. Louis Park, Minnesota, switched to a reusable, cost-effective isolation gown that meets OSHA requirements to reduce the volume of biohazardous waste requiring incineration by the facility (28).

Decisions regarding the return to reusable medical items should be made on a case-by-case basis, addressing potential effects on staff safety and patient care. In its 1990 Report to Congress, the Office of Technology Assessment states (29), "The use (and reuse) of disposables can be considered on an item-by-item basis, in light of how they will be used, including consideration of infection risks and other factors associated with those risks."

Recycling. Recycling involves the diversion of a used product to a future use rather than to disposal. Options for recycling untreated biohazardous waste are extremely limited due to the infectious nature and inherent hazards associated with this waste stream. However, options for recycling biohazardous waste that has been treated and rendered noninfectious are available, although from a technical standpoint, treated biohazardous waste would no longer be considered part of the biohazardous waste stream. For example, plastics or metals that have been rendered noninfectious may be removed for future processing. Plastics recycling is being commercially conducted by Stericycle, Inc. (Deerfield, Illinois), a company that offers a regional biohazardous waste treatment service based on an electrothermal deactivation treatment process. As part of the process, treated plastics are diverted for recycling into medical sharps containers and other products by the Sage Company. Other plastics are used for their high Btu value as refuse-derived fuel, which is packaged into bales and burned as fuel in cement kilns.

MANAGING NONMEDICAL FACILITY BIOHAZARDOUS WASTE

Home Healthcare Management Practices

Managing biohazardous waste generated in the course of home healthcare or self-care activities brings additional challenges to regulatory agencies, home healthcare providers, and healthcare educators. Although the home care biohazardous waste stream shares many similarities to its institutional counterpart, it is handled outside of the institutional context often by untrained, unskilled individuals or family members whose primary interest is on the administration of healthcare and not on the disposal of waste. Home care agencies are often unclear about their role in waste management and how they should approach waste management decisions.

Home healthcare has been expanding rapidly during the past decade as a means of reducing healthcare costs in response to changes in Medicare reimbursement (30). Home healthcare is broadly defined as consisting of medical, nursing, pharmaceutical, and other care provided to patients in their place of residence (31). The burgeoning home healthcare industry is evidenced by the increasing number of home healthcare agencies offering client services. Often unaccounted for is the patient care provided by family and friends, who also play a significant role in the wastes that are placed into the trash.

Biohazardous waste generated in the course of home healthcare activities bears similarities to institutionally generated biohazardous waste. Such procedures are not unlike those conducted in the institutional setting and generate similar biohazardous waste.

Developing a Waste-Management Strategy

Biohazardous waste management in the home healthcare setting presents challenges because such activities are less standardized and controlled than in their institutional counterparts. Nevertheless, poor waste management practices increase the risk of injury or infection to others both in and outside the home (eg, waste industry workers). It is the responsibility of the home healthcare agency to develop proper waste management systems and insure that those responsible for following them, including family and friend care givers, have been adequately educated.

When developing systems for managing home care biohazardous waste programs, home care agencies should (*1*) identify all applicable regulations, accreditation standards and guidelines, and the agencies and contacts responsible for their enforcement or implementation; (*2*) based on regulations, accreditation standards, and guidelines, develop protocols for biohazardous waste management that addresses as applicable: (*a*) identification; (*b*) segregation; (*c*) containerization; (*d*) labelling (when necessary); (*e*) storage; (*f*) transportation (only as necessary); (*g*) disposal; (*h*) education; and (*i*) recordkeeping; (*3*) develop and conduct an educational program for both the skilled agency home care providers and the family and friends as day-to-day care providers; and (*4*) implement the program through agency staff and the team of family and friend home care givers.

Syringe Disposal Outside of the Healthcare Facility

Used syringes, when improperly disposed by home users (eg, diabetes, allergy care, and intravenous drug users), pose a physical injury risk and a potential infection risk to waste industry workers and possibly to the public. Improperly disposed syringes from sources outside of the formal medical setting have been reported in the scientific literature, which indicates that proper syringe disposal techniques and options are often either unknown or misunderstood (32,33).

Options for safely disposing both used syringes have been explored by various government and academic groups in the United States. For example, the Washington State Department of Ecology conducted an education program targeting injectable insulin using populations which identified seven disposal options: Option 1, needle clipping with a hand-held needle clipper manufactured by the Becton-Dickinson Corporation followed by disposal of the needleless syringe in the trash; Option 2, syringe containment by placement into a common 2-l PETE (polyethylene terephthalate plastic resin) soda bottle affixed with a bright orange warning label prior to disposal in the trash; Option 3, return to a pharmacy participating in a syringe return program; Option 4, return to physician willing to accept used syringes for disposal; Option 5, disposal at a municipal disposal site equipped to accept used syringes; Option 6, collection by a participating garbage hauler equipped to collect and dispose of used syringes; and Option 7, return of used syringe to medical waste disposal companies through the USPS via syringe mailback programs. Each option shared advantages and disadvantages (34). A concern for Option 1 (needle clipper) was that a 1/4-in. portion of the needle remained on the syringe following clipping which presented an injury risk. Option 2 became a concern for the plastics recycling industry due to problems with syringe containing plastic soda bottles entering recycling facilities, putting workers at risk of needlestick injuries and possibly contaminating the recycled plastic. Options 3–6 were observed to have a very low rate of participation by pharmacies, physicians, municipal waste collection facilities, and garbage haulers. The seventh option, mailback programs, was found to be too costly for most home syringe users.

In the late 1980s, a community-wide needle disposal program was implemented in Florida by the De Soto County Public Health Unit and the Desoto Memorial Hospital in an effort to stem the presence of stray syringes that were appearing in public locations with greater frequency (35). A needlestick injury sustained by a child picking up litter near a public school prompted the response. The program involved locating a secure, highly visible, and easily accessible red needle collection container within all city and county fire departments and police/sheriff stations, for needle disposal by the public. The program also was designed to provide needle storage containers to the public for home use. Old mailboxes mounted to the ground were used for the collection containers. The program has been successful, with new pick-up sites each year, and is recognized as a practical, inexpensive solution.

BIOHAZARDOUS WASTE TREATMENT

This discussion describes technologies for treating biohazardous waste prior to disposal. Such technologies include incineration, steam sterilization, and alternative treatment technologies. Biohazardous waste is treated prior to disposal primarily to reduce or eliminate the presence of pathogenic microorganisms, thereby reducing or eliminating risk of infection to humans. Technologies for processing biohazardous waste may also include mechanical grinding or shredding to render the waste unrecognizable or to facilitate the treatment process, or compaction to reduce final waste volume.

ALTERNATIVE BIOHAZARDOUS WASTE TREATMENT TECHNOLOGIES

During the past few years, many treatment technology alternatives to incineration and steam sterilization have become available for treating biohazardous waste prior to disposal. Such systems are becoming increasingly relied upon as public and regulatory pressures continue to move the medical industry away from its historical reliance on incineration as the preferred method for treating biohazardous waste. Yet challenges remain for those responsible for determining the ability of these emerging technologies to effectively treat biohazardous waste in a manner that is not detrimental to human health or the environment.

State and Territorial Association on Alternate Treatment Technologies. In April of 1994, a coalition of US state and territorial government medical waste specialists under the name of the State and Territorial Association on Alternate Treatment Technologies (STAATT) published a report entitled *Technical Assistance Manual: State Regulatory Oversight of Medical Waste Treatment Technologies.* The purpose of this document was to help standardize the process used by state agencies when reviewing alternative medical waste treatment or destruction technologies within the scope of the regulations that such agencies are bound by. A secondary purpose was to standardize an approach to these systems to assist manufacturers who are developing new technologies which ultimately require governmental approval.

Copies of the STAATT report may be obtained by contacting the Rhode Island Department of Environmental Management, 9 Hayes Street, Providence, Rhode Island 02908-5003,

Tel: (401)277-2771/Fax: (401) 277-6802. Key points in the report are summarized subsequently.

The STAATT report recognized emerging medical waste treatment technologies as following into one of four treatment modes for inactivating microorganisms: thermal (wet and dry heat, microwaving, infrared, laser, plasma pyrolysis); chemical (chlorine, chlorine derivatives, ozone, enzymes); irradiation (UV, cobalt 60); and other treatment mechanisms designed for specific medical waste categories generated in small volumes (thermal/electrical). For some technologies, a combination of modes may apply (eg, chemical/thermal, or chemical/irradiation). In addition to these treatment modes, some technologies rely on a post-treatment process such as grinding, shredding, or compaction to destroy the waste, render it unrecognizable, or otherwise facilitate the treatment method.

Absent a uniform approach from federal or state government, the STAATT guidance document provided an approach to evaluate new treatment technologies. In that approach the report recommended a level of microbial inactivation that all medical waste treatment process should be capable of achieving, which is defined as:

> *Inactivation of vegetative bacteria, fungi, lipophilic/hydrophilic viruses, parasites, and mycobacteria at a 6 Log_{10} reduction or greater; and inactivation of Bacillus stearothermophilus spores or B. subtilis spores at a 4 Log_{10} reduction or greater.*

Criteria were established for defining surrogate pathogen selection for vegetative microorganisms used during the efficacy testing of new technologies. Such surrogates should not affect healthy individuals, be easily obtainable, be an ATCC registered strain, as available, be easily cultured and maintained, and meet quality control requirements.

Mechanisms to quantify levels of microbial inactivation are based on the formula:

$$\log_{10}\text{kill} = \log_{10}(\text{cfu/g introduced}) - \log_{10}(\text{cfu/g recovered})$$

where $\log_{10}$ kill is equivalent to the term $\log_{10}$ reduction, introduced is the number of viable test microorganisms introduced into the treatment unit, recovered is the number of viable test microorganisms recovered after treatment; and cfu/g are colony forming units per gram of waste solids. Procedures for conducting efficacy testing of new equipment are described in the report. In addition, the STAATT report presents a suggested checklist for information that should accompany an application for treatment technology approval by a state agency.

Alternative Treatment Technologies. The following alternative treatment technologies are summarized: (*1*) chemical disinfection; (*2*) microwave technology; (*3*) dry heat technology; (*4*) electrothermal deactivation; (*5*) plasma arc technology; (*6*) electron beam sterilization; and (*7*) cobalt-60 irradiation. The reader should be aware that this cross section of treatment technologies does not represent all technologies available today. The reader should also be aware that product examples are primarily based on information supplied by the manufacturers which was not necessarily substantiated by independent third-party review.

Disclaimer

The views presented in this article represent those of the author and do not necessarily reflect the views or the policies of the Washington Department of Health or the University of Washington, Seattle, Washington.

BIBLIOGRAPHY

1. U.S. Environmental Protection Agency, 40 *CFR* Parts 22 and 259, *Fed. Reg.* **54**(56), 12326–12395 (Mar. 24, 1989).
2. A. Shockley, Ohio Environmental Protection Agency, Columbus, OH, personal communication, Sept. 1996.
3. H. G. Grieble, T. J. Bird, H. M. Nidea, and C. A. Miller, *J. Infect. Dis.,* **130**(6), 602–607 (Dec. 1974).
4. J. T. Jacobson, J. P. Burke, and M. T. Conti, *Inf. Control,* **4**(2), 100–102 (1983).
5. A. S. Benenson, ed., *Control of Communicable Diseases Manual,* 16th ed., American Public Health Association, 1995.
6. U.S. Dept. of Health and Human Services, Agency for Toxic Substances and Disease Registry, Public Health Service, *The Public Health Implications of Medical Waste: A Report to Congress,* Atlanta, Ga., 1990.
7. M. Peterson, *The Occurrence and Survival of Viruses in Municipal Solid Waste,* Ph.D. Thesis, University of Michigan, Ann Arbor, Mich., 1972.
8. J. Donnelly, P. Scarpino, and D. Brunner, "Recovery of Fecal Indicator and Pathogenic Microbes from Landfill Leachate," in: *Land Disposal: Municipal Solid Waste,* EPA-600/9-81-002a, U.S. Environmental Protection Agency, Cincinnati, Ohio, 1981, pp. 37–54.
9. G. A. Gellin and Z. R. Mitchell, *Arch. Environ. Health,* **20,** 510–515 (1970).
10. J. A. Cimino, *Am. J. Public Health* **65**(1), 38–46 (Jan. 1975).
11. J. A. Cimino, and R. Mamtani, *J. Environ. Health* **50**(1), 8–12 (Jul/Aug. 1987).
12. C. Clark, G. Van Meer, A. Bjornsen, C. Linneman, G. Schiff, and P. Gartside, *Incidence of Viral Infections Among Waste Collection Workers,* University of Cincinnati Medical Center, Institute of Environmental Health Kettering Laboratory, Cincinnati, Ohio, Jan. 19, 1979.
13. W. L. Turnberg and F. Frost, *Am. J. Public Health* **80**(10), 1262–1264 (1990).
14. National Institute for Occupational Safety and Health, *Residential Waste Collection: Hazard Recognition and Prevention,* U.S. Department of Health and Human Services, Centers for Disease Control, Atlanta, Ga., Mar. 1982.
15. Oregon Health Division, *Recommendations for Preventing Disease Transmission while Handling Solid Waste in Oregon,* Oregon Dept. of Human Resources, Sept. 1988.
16. R. Englebrecht, M. Weber, P. Amirhor, D. Foster, and D. La Rossa, in *Virus Survival in Water and Waste Water Systems,* J. F. Malina and B. P. Sagick, eds., Water Resources Symposium No. 7, Center for Research in Water Resources, University of Texas, Austin, Tex., 1974, pp. 201–217.
17. F. Donnelley and P. Scarpino, *Isolation, Characterization and Identification of Microorganisms from Laboratory and Full-Scale Landfills,* EPA-600/2-84-119, PB84-212 737, Environmental Protection Agency, Washington, D.C., July 1984.
18. R. Cooper, S. Klein, C. Leong, J. Potter, and C. Golueke, SERL Report No. 74-3, Sanitary Engineering Research Laboratory, College of Engineering and School of Public Health, University of California, Berkeley, Calif., 1974, p. 93.
19. R. Cooper, J. Potter, and C. Leong, in J. F. Malina and B. P. Sagick, eds., *Virus Survival in Water and Waste Water Systems,* Water Resources Symposium No. 7, Center for Research in Water Resources, University of Texas, Austin, Tex., 1974, pp. 218–232.
20. M. Sobsey, *Am. J. Public Health* **68**(9), 858–863 (1978).

21. S. Ware, *A Survey of Pathogen Survival During Municipal Solid Waste and Manure Treatment Processes,* EPA-600/8-80-034, U.S. Environmental Protection Agency, Cincinnati, Ohio, Aug. 1980.
22. U.S. Environmental Protection Agency, 40 *CFR* Part 258, "Municipal Solid Waste Landfill Standards," *Fed. Reg.* (Oct. 9, 1991).
23. R. M. Pike, *Ann. Rev. Microbiol.* **33,** 41–66 (1979).
24. G. Ducel, J. Pitteloud, M. Rufener-Press, M. Bly, and P. Rey, *Sozial und Praventivmedizin* **21,** 136–138 (1976).
25. D. Fiscus, P. Gorman, M. Schrag, and L. Shannon, *Assessment of Bacteria and Virus Emissions at a Refuse Derived Fuel Plant and Other Waste Handling Facilities,* EPA-600/2-78-152, Environmental Protection Agency, Washington, D.C., Aug. 1978.
26. J. Brown, *Reg. Anal.* **1**(10), 1–8 (July 1993).
27. M. J. Connelly, *Med. Waste Anal.* **2**(10), 1–7 (July 1994).
28. J. Paprock, *The MHA Recycling & Conservation Guide,* Minnesota Hospital Association, Minneapolis, Mn., fourth printing (1991).
29. U.S. Congress, Office of Technology Assessment, *Finding the Rx for Managing Medical Wastes,* OTA-O-459, U.S. Government Printing Office, Washington, D.C., Sept. 1990.
30. A. N. Lorenzen, and D. J. Itkin, *Am. J. Infect. Control* **20**(6), 326–329 (Dec. 1992).
31. P. N. Catania and M. M. Rosner, eds., *Home Health Care Practice,* 2nd ed., Health Markets Research, Palo Alto, Calif., 1994.
32. D. W. Satterfield, J. Kling, and D. L. Gallina, *Diabetes* **39** (Suppl. 1), 51A (1990).
33. A. Swislocki and co-workers, *Diabetes Care* **14**, 930–932 (Oct. 1991).
34. W. L. Turnberg and L. D. Lowen, *Diabetes Educat.* **20**(6), 489–492 (Nov./Dec. 1994).
35. D. W. Toews, *Am. J. Public Health* **85**(10), 1447–1448 (Oct. 1995).

BIOLOGICAL MATERIALS: ANALYSES OF HAZARDOUS SUBSTANCES

J. ANGERER
K.H. SCHALLER
University of Erlangen-Nuremberg
Erlangen, Germany

BIOLOGICAL MONITORING OF EXPOSURE TO CHEMICAL AGENTS

Definitions

In 1980, the Commission of the European Community (CEC), the U.S. National Institute for Occupational Safety and Health (NIOSH) and the Occupational Safety and Health Administration (OSHA) organized a seminar on Assessment of Toxic Agents at the Workplace. The participants agreed upon the following definition for biological monitoring (1): "Biological monitoring is the measurement and assessment of workplace agents or their metabolites either in tissues, secreta, excreta, expired air or any combination of these to evaluate exposure and health risk compared to an appropriate reference".

By this definition biological monitoring is restricted to measuring internal dosage (2). In other countries, especially in Europe the concept of biological monitoring is extended to include the detection of early reversible nonadverse effects (biological effect monitoring) (eg, aminolevulinic acid in urine or zinc protoporphyrine in blood to assess exposure to lead). An adverse effect (eg, increased microproteinuria and cadmium exposure) indicates that exposure is or has been excessive, and, therefore, such a measurement is more logically included in a program for early detection of health impairment than in a biological monitoring program for evaluating exposure (2–4). Biological monitoring aims at preventing chemically induced health effects by providing a basis for assessing the uptake of chemicals in the workplace and the related risk. Therefore biological monitoring data are fundamentally important within preventive medical examinations.

Therefore, the following definition of biological monitoring is usually used in Germany, also providing a good reflection of the view of other countries in the European Union.

> Biological monitoring is the directed systematic continuous or repetitive health-related activity for collection of biological samples for the measurement and assessment of hazardous chemical compounds, their metabolites or their specific non-adverse biochemical effect parameters. The objective is to evaluate the exposure and health risk of exposed persons by comparing the obtained data with appropriate reference values (slightly modified from (5)).

That means:

1. Biological monitoring in the field of occupational and environmental health is a medical approach to preventing diseases by hazardous chemical substances. It is an integral part of medical health surveillance of persons exposed to chemical agents.
2. To appreciate the individual health risk, data from biological monitoring should be evaluated along with other clinical data of the exposed person.
3. Data from biological monitoring are governed by the rules of medical secrecy because they concern individuals.

Advantages of Biological Monitoring

Biological monitoring has a long history. Elevated urinary lead concentrations were reported in the 1920s, and the first evaluation of lead in blood and urine in diagnostics and occupational health care was published more than 60 years ago (6).

Biological monitoring has expanded rapidly in the last decades, from measuring lead, mercury, cadmium, phenol, and hippuric acid to measuring more than 100 chemicals or groups of related chemicals (2). Biological monitoring has also been included in national guidelines and legislation (6).

Biological monitoring of exposure offers several advantages over environmental monitoring (7).

1. Many industrial chemicals enter the body by routes other than inhalation. Absorption through the skin or the gastrointestinal tract is an important way of intake. Personal hygiene varies from one person to another. The lack of care in personal hygiene leads to significant ingestion of the substance. If dermal exposure is important, then airborne monitoring is an inadequate measure of total exposure.
2. Environmental monitoring does not adequately assess exposure. This may be the case when respirators or other personal protection is in use. Absorption is uncertain (particle size, solubility), individual variability

is extreme (respiratory volume, and work practices) or exposure fluctuates rapidly over time. A biological parameter takes all of these factors into consideration.

3. Cumulative toxicants are present, such as metals bound to tissues (eg, lead, cadmium, mercury) or organic chemicals that are highly fat soluble and poorly metabolized (eg, polychlorinated biphenyls (PCB), hexachlorocyclohexane (HCH); dioxins). The common characteristic of these toxicants relevant for biological monitoring is a long half-life of elimination.
4. Biological monitoring also reflects nonoccupational background exposure (leisure activity, residency, dietary habits, smoking, etc), which is also be expressed at the biological level. The organism integrates the total external (environmental and occupational) exposure into one internal load.
5. In some cases the measurement of biological effect parameters allows the assessment of total exposure, such as the use of acetylcholinesterase monitoring for exposure to many organophosphate pesticides, met-hemoglobine for exposure to multiple inducers, and DNA adducts for exposure to multiple carcinogens.

Requirements for Biological Monitoring

The following conditions must be present to consider biological monitoring (7):

The determinant must be present in media (namely blood, urine, exhaled air) and must be suitable for sampling.

The proposed monitoring must be acceptable to the population being monitored.

The strategy of sample collection must produce representative samples.

A method of analysis must be available, that is practical and yields reliable results over the existing range of concentration.

The analytical determination must be carried out under a quality control scheme.

The results must be interpretable by comparing them with reference limits.

Biological Media. The majority of the available biological tests rely on analyzing blood, urine, or exhaled air. Considerations for using these different sample media have been reviewed in detail (8–10).

Selection of sample media depends on several factors, such as the kinetics, the convenience of sample collection, and the possibility of sample contamination or loss of analytes.

Blood constitutes the main vehicle for transporting and distributing chemicals in the body. Therefore, most systemically active substances or their metabolites are found in blood. Blood is used for measuring most inorganic chemicals and organic substances which are poorly biotransformed and have a sufficient half-life. Moreover determining an unchanged substance in blood has greater specificity than determining its metabolites in urine. Blood is also useful for measuring substances that bind to macromolecules, for example, to hemoglobine, DNA, or albumin (11). Blood levels usually reflect a shorter period of exposure for a particular substance than urine, eg, cadmium in urine reflects, primarily, body burden for exposure over years, whereas blood cadmium reflects, primarily, recent exposure. Blood is less affected than urine by variation in composition.

On the other hand, blood sampling is invasive. Proper sample storage is critical especially for analyzing volatile solvents. For substances with extremely fast elimination rates, true of many organic solvents, the rapid fluctuation of blood concentrations during and immediately after exposure make the interpretation of results difficult.

Urine is easy to collect, the procedure is noninvasive, and large volumes can be collected. Spot urine specimens are subject to a large degree of variability in urine concentration because of variability in urine output. Longer collections, particularly 24-hour sampling, improves interpretability but is impractical for most persons and leads to contamination. The concentration of analytes in spot samples may be adjusted for creatinine concentration, specific gravity or osmolality to correct for urine output, but with some chemicals this procedure produces more variability than it corrects for (7,9).

In the case of exposure to substances with short biological half-times or with fluctuating airborne concentrations, the level of a metabolite in urine, collected at the end of exposure, usually indicates the average exposure better than the concentration of the substance itself in blood samples. The danger of contamination of urine during sampling is high compared to blood sampling. Care must be taken to avoid contamination from clothes and from container walls. Renal alterations of glomerular filtrations, tubular secretion, and tubular reabsorption greatly influence elimination of the analyte.

Alveolar air analysis has not gained any importance in environmental health. The low concentrations of the analyte in exhaled air and the need for specialized sampling equipment are the main limitations of breath analysis. Moreover the concentration of the analyte in exhaled air fluctuates very rapidly with the changes in exposure intensity. The time of sampling is very critical. Furthermore there is a risk of external contamination during collection.

Media other than urine, blood and exhaled air have not become routine for monitoring any substance. Hair and nails are not appropriate for most substances because several factors, such as contamination, lack of standardization of sampling, analytical methods and quality control, influences of washing procedures and so on, which make the results uninterpretable under routine conditions (12).

Analysis of milk or adipose tissue has sometimes been performed to assess the body burden of lipophilic compounds (such as organochlorine pesticides) or to assess the risk of transferring toxic substances to the newborn (11).

Techniques are being developed for measuring *in vivo,* the concentration of lead in bone and cadmium in liver and kidney. At present these techniques are used only for research and are not generally applied to routine monitoring (11).

Indicators of Exposure and Effect. The biological tests currently used for the biological monitoring of exposure to chemicals can be classified into 3 categories (13):

Determination of the chemical or its metabolites in biological media

Quantification of nonadverse biological effects related to the internal dose

Direct measurement of the amount of active chemical interacting with the target molecules

The great majority of the biological tests available for monitoring exposure belong to the first category. Most of the tests in the second category are nonspecific. Unlike the preceding exposure test, those belonging to the third category directly estimate the amount of chemical interacting with the size of action. The determination of protein adducts, hemoglobin adducts, or DNA adducts assesses the health risk more accurately than other monitoring procedures. This is relevant especially for mutagens and carcinogens. Much research is still needed before this test can be introduced in routine biological monitoring of exposed persons (11).

The parameters used for estimating internal exposure and health risks should sensitively and specifically indicate the exposure to a given substance. Furthermore the parameter should reflect the toxic effect as close as possible. The benzene content in blood, for example, is the most sensitive and yet specific parameter for benzene exposure down to the concentration range caused by environmental exposure. Using dynamic headspace analysis, it is possible to distinguish nonsmokers, passive smokers, and active smokers from one another using benzene blood levels as a criterion. The unchanged benzene molecule, however, is relatively removed from the effect. In the case of benzene leukaemia, the determination of *t,t*-muconic acid, *S*-phenylmercapturic acid, or adducts of benzene to proteins or DNA are more closely related to the biological response.

Today the possibilities for biological monitoring can be roughly categorized according to specificity and sensitivity. Specificity must be subdivided into specificity for the substance and specificity for its effect. Determining the unchanged substance in body fluids is, of course, highly specific for exposure to a given substance, but in many cases unspecific for the effect. Metabolites in body fluids take a medium position more or less specific for substance and effect. The sensitivity of the detection of chemicals and their metabolites in biological media is, however, unsurpassed by all other possibilities of biological monitoring.

Biological effect parameters, such as δ-aminolaevulinic acid, cholinesterase, zinc protoporphyrin, and biochemical effect parameters, such as adducts to proteins or to DNA, are almost ideal for biological monitoring. These parameters are specific for both the substance and the effects. However, from a diagnostic viewpoint, they do not indicate internal exposure as sensitively as the determining concentrations of chemicals in blood or urine.

Cytogenetic parameters, such as sister chromatid exchanges, micronuclei, and chromosomal aberrations, are highly specific for the effect but are not specific with regard to the exposure to a given chemical. The same is true for tests on mutagenicity in urine which, moreover, are rather insensitive. When cytogenetic parameters, mutagenicity tests, or certain metabolites are used for biological monitoring or health surveillance, the unchanged substance should be determined in parallel. Only in this way is the observed effect traced back to a particular exposure: a particular substance is responsible for adverse health effects only if at least one of the parameters used for biological monitoring is a specific indicator for that substance (4).

Table 1 is an overview of the substances which are routinely determined in biological materials. This list contains practically all metals from aluminium to zinc, which are of occupational and environmental interest. These substances are determined in both blood and urine. As already mentioned, urine offers the advantage that lower detection limits are obtained. Among the organic substances, solvents and their metabolites play a major role in occupational and environmental medicine. Organochlorine compounds are important from an environmental point of view and in specific work places, such as those where decontamination of polluted soil and combustion processes for incinerating municipal waste take place.

Table 2 presents the possibilities of biological monitoring of an exposure to pesticides. The pesticide and the corresponding parameter of biological monitoring and the biological material are listed. Table 2 gives the current scientific knowledge for monitoring pesticide exposure, preferably assessed by biological monitoring.

Analytical Methods. Specific and sensitive analytical methods with defined and tested reliability criteria are indispensable for carrying out the analyses necessary for biological monitoring. They must be reproducible, lead to reliable and comparable results, and also be practicable, so they can be used in routine investigations. For the terminology of the analytical reliability criteria, refer to the International Organization for Standardization (ISO, 1990 and 1992) and to the International Union of Pure and Applied Chemistry (IUPAC) and the Association of Official Analytical Chemists (AOAC) (14). The terms used in this contribution are defined in a later section. When planning the use of analytical methods, one must consider its practicability, ie, the speed of the method, the equipment, the technical skill required, the precautions and procedures required for safety, the workload, specimen handling, the cost, and the space needed (11).

The complex matrix of biological media and the extremely low concentrations of the parameters usually demand complicated instrumental analysis. The state of the art is represented by the determination of polychlorinated dibenzodioxins as examples of organic substances and platinum as an example of inorganic substances. These chemicals can be determined in body fluids at concentrations as low as 1 pg/L (15) and 1 ng/L (16), respectively. Mass spectrometric analysis, combined with gas chromatography, liquid chromatography, and ICP will gain further use in biological monitoring. There also indications that immunologic methods will become routine in biological monitoring (17).

Therefore, biological monitoring can be carried out only in specialized laboratories which are responsible for the analytical reliability of the results.

The working group "Analyses of Hazardous Substances in Biological Materials" of the Commission for the Investigation of Health Hazards of Chemical Compounds in the Work Area of the Deutsche Forschungsgemeinschaft (DFG) has published a collection of tested methods and analytical procedures for toxicological analysis of human biological materials.

Table 3 shows the analytical procedures contained in this collection of methods, listed according to analytical technique and class of substance. About 120 methods are available for determining about 180 parameters (18,19).

HAZARDOUS SUBSTANCES ANALYSES IN ENVIRONMENTAL MEDICINE

Preanalytical Phase (Sample Collection, Transport, and Storage)

In common with other medical disciplines, the object of any examination in environmental medicine is to obtain findings

Table 1. Overview of Substances Determined Routinely in Biological Materials

Substance	Whole Blood	Serum/Plasma	Urine
	Metals and Relevant Intermediary Metabolites		
Aluminum		•	•
Antimony			•
Arsenic			•
δ-Aminolevulinic acid			•
Barium			•
Beryllium			•
Cadmium	•		•
Calcium		•	
Chromium	•	•	•
Cobalt	•		•
Copper		•	•
Iron		•	•
Lead	•		•
Manganese	•	•	•
Mercury	•		•
Nickel	•		•
Platinum		•	•
Potassium		•	
Selenium		•	•
Silver		•	•
Sodium		•	
Tellurium			•
Thallium			•
Vanadium			•
Zinc			•
	Organic Solvents and Their Metabolites		
Acetone	•	•	•
Aliphatic hydrocarbons-screening	•		
Formic acid			•
Aromatic hydrocarbons-screening	•		
Benzene-Environment	•		
Benzene-Workplace	•		
Butanol	•		
Butanone (Methylethylketone)			•
Butoxyacetic acid			•
Butyl acetate (as Butanol)	•		
Butoxypropanol	•		•
Chloroform	•		•
Chlorophenols			•
Cyclohexane	•		
Cyclohexanone	•		
1,1- and 1,2-Dichloroethane	•		
1,1- and 1,2-Dichloroethene	•		
Dichloromethane	•		
1-Ethoxy-2-propanol	•		•
Ethoxyacetic acid			•
Ethylbenzene	•		
Frigenes	•		
Halogen hydrocarbons-screening	•		
n-Heptane	•		
Heptanone	•		
n-Hexane	•		
Hexanedione			•
Hexanol	•		
Hexanone	•		
Hippuric acid			•
Isopropanol (as Acetone)			•
o-Cresol			•
Mandelic acid			•
Methanol			•
1-Methoxypropanol-2	•		•
2-Methoxypropionic acid			•
Methoxyacetic acid			•

Table 1. Overview of Substances Determined Routinely in Biological Materials *(continued)*

Substance	Whole Blood	Serum/Plasma	Urine
Methyl acetate (as methanol)			•
Methylhippuric acid			•
Methylisobutylketone	•		•
Methylpentane	•		
t,t-Muconic acid			•
Octane	•		
Phenol			•
Phenylglyoxylic acid			•
Propanol	•		•
Propylbenzene	•		
S-Phenylmercapturic acid			•
Carbon disulfide (as TTCA)			•
Styrene	•		
Tetrachloromethane	•		
Tetrachloromethane (as Per)	•		
Tetrahydrofurane			•
Toluene	•		
Toluric acids			•
(Methylhippuric acids)			
1,1,1-Trichloroethane	•		
Trichloroethene	•		
Trichloroacetic acid			•
Trichloroethanol	•		
Xylenes	•		
Others			
Acrylamide		•	•
CO-Hemoglobin	•		
Fluoride		•	•
Hydroxybenzo(*a*)pyrene			•
(PAH-metabolites)			•
Hydroxyphenanthrenes			
(PAH-metabolites)			
1-Hydroxypyrene			•
(PAH-metabolite)			
Hydroxyethylvaline	•		
(Ethylene oxide-metabolite)			
Met-Hemoglobin	•		
N-Methylformamide			•
(Dimethylformamide)			
Thiodiglycolic acid			•
2-Thio-thiazolidine-4-carboxylic acid			•
Aromatic Amines, Nitro Compounds and Their Metabolites			
Aminodinitrotoluenes (DNT)			•
Aminonitrotoluenes (TNT)			•
Aromatic amines-screening			•
Aniline			•
Chloraniline			•
Diaminodiphenylmethane			•
Dichlorobenzidine			•
Dimethylbenzidine			•
4,4′-Methylene-bis-(2-chloroaniline)			•
Nitroaromatic compounds			•
o-Nitrophenol			•
p-Nitrophenol			•
o-Toluidine			•
Toluylendiamine			•

related to the individual patient. The analytical finding is the result of two stages. The preanalytical phase and the analytical determination. Both stages of analysis and thus the results are associated with specific errors.

In all steps of the preanalytical phase from the time of sampling the material until removal of a sample from the matrix, disturbing influences and interference factors can falsify the results and make interpretation more difficult or even

Table 2. Possibilities of Biological Monitoring to Assess Exposure to Pesticides

Pesticides	Analytical Parameter	Matrix
	Pyrethroids	
Cyfluthrine	Cl_2CA, 4-Fluoro-3-phenoxic benzoic acid (F-PBA)	Urine
β-Cyhalothrine	3-Phenoxybenzoic acid	Urine
λCypermethrine	Cl_2CA, 3-Phenoxybenzoic acid (PBA)	Urine
αDeltamethrine	Br_2CA, 3-Phenoxybenzoic acid (PBA)	Urine
Fenpropathrine	3-Phenoxybenzoic acid (PBA)	Urine
Fenvalerate	3-Phenoxybenzoic acid (PBA)	Urine
Permethrine	Cl_2CA, 3-Phenoxybenzoic acid (PBA)	Urine
	Chlorophenoxycarboxylic Acids and Derivatives	
Bifenox	2,4-Dichlorophenol	Urine
2,4-Dichlorophenoxyacetic acid (2,4-D)	2,4-Dichlorophenol	Urine
Dichloroprop	Dichloroprop	Urine
MCPA	MCPA	Urine
Mecoprop	Mecoprop	Urine
	Organophosphates	
Dichlofenthion	2,4-Dichlorophenol	Urine
Fenchlorphos	2,4,5-Trichlorophenol	Urine
Parathion	*p*-Nitrophenol	Urine
Parathion-methylene	*p*-Nitrophenol	Urine
Prothiofos	2,4-Dichlorophenol	Urine
Trichloronat	2,4,5-Trichlorophenol	Urine
	Organochloro Compounds	
Hexachlorobenzene (HCB)	Hexachlorobenzene (HCB)	Serum
Lindane (HCH)	Lindane/hexachlorocyclohexane (HCH)	Serum
Pentachlorophenol (PCP)	Pentachlorophenol (PCP)	Serum
Polychlorinated biphenyls (PCB)	Polychlorinated biphenyls	Serum
Carbamates		
Propoxur	2-Isopropoxyphenol	Urine
	Other Pesticides and Metabolites	
Acetylcholinesterase	Acetylcholinesterase	Erythrocytes
Chloroaromatic compounds	Chlorophenols	Urine
DDT	*p,p′*-DDE	Urine
Parathion	*p*-Nitrophenol	Urine

Table 3. Overview of Parameters and the Corresponding Analytical Techniques

	Inorganic Substances	Organic Ccompounds	Effect Parameter
Analytes	Metals: Al, As, Ba, Be, Cd, Cr, Co, Cu, Hg, Mn, Ni, Pb, Pt, Sb, Se, Sr, Ti, V, Zn Anions: F^-, Br^-, CN^-	Adducts (eg, hydroxyethylvaline, solvents, organochlorine compounds, aromatic amines, nitro compounds, phenols, PAH-metabolites, pesticides (pyrethroids, carbamates, phenoxy carboxylic acids, organophosphates)	Acetylcholinesterase, ALA-U, ALA-D, cholinesterase, CO-Hb, coproporphyrins, Met-hemoglobin, β2-microglobuline albumine, RBP
Analytical techniques	F-AAS, GF-AAS, hydride AAS, cold vapor-AAS, ICP-OES, ICP-MS, inverse voltammetry, ion-selective electrode	GC-FID, GC-ECD, GC-TID, GC-MS, head space technique, HPLC, fluorimetry	Photometry, IR spectroscopy, GC, immunologic methods

impossible. The main factors that influence the results *in vivo* are the time of sampling and the great variations in the composition of diuresis. The most important interference factors *in vitro* are

- exogenous contamination of the material (at the site of sampling);
- contamination of the material from the sampling utensils or the vessels used;
- evaporation of volatile components from the material;
- absorption of the components to be analyzed through the walls of the vessels used; and
- changes in the samples during storage and transport (eg, coagulation of blood samples, sedimentation of urinary components, chemical changes in the parameters).

Therefore the preanalytical phase must be coordinated with the laboratory before beginning an investigation. Taking advantage of the services generally offered by the analytical laboratories and ordering from them sampling utensils, transport vessels, and information on sampling is recommended. This is particularly important if specially cleaned vessels or special containers are necessary. It is useful for the laboratory to set up a consulting service to provide information about possible sources of errors.

In the case of biological monitoring, analytical determination is almost exclusively trace analysis and ultratrace analysis. Therefore preparation of the samples and the choice of sampling utensils and vessels must be carried out with possible exogenous contamination in mind. All materials which come into contact with the body fluids must meet high standards of cleanliness. To prevent carryover, all sampling utensils and vessels should be discarded after use. Contamination of these disposable articles must be prevented during production, transport, and storage. Random checks should be carried out to ensure that these vessels are free of contamination. For special purposes these vessels should be washed out before use with 1 *M* nitric acid followed by distilled water.

Analyte losses occur as a result of evaporation or absorption through the walls of the vessels and in sediments. Therefore sealable head space vials should be used for highly volatile substances. Absorption losses, in particular of metals, are counteracted by acidifying urine samples (1 mL glacial acetic acid/ 100 mL urine).

Description of Analytical Methods

The reliability of analytical results is an essential requirement of biological monitoring in occupational and environmental medicine. Only results from carefully elaborated and validated methods can bear comparison among themselves, with different laboratories, and can be used to check adherence to the threshold limit values of biological materials. The analytical methods used to analyze biological samples must be sensitive enough to detect determinant concentrations well below levels of concern, must be specific to the degree that most of the results truly reflect exposure or effect of the substance being monitored, and must be practical. Some national organizations, such as the DFG and NIOSH describe analytical methods controlled for reliability. These methods are published in a form useful as a guide to the analytical technician. These so-called standard operating procedures (SOP) are provided by a working group Analyses in Biological Materials of the Commission for the Investigation of Health Hazards of Chemical Compounds in the Work Area of the DFG.

In selecting suitable analytical methods, the working group is guided mainly by the relevant scientific literature and the expertise of the members. If appropriate analytical methods are not available they are worked out within the working group. The chairman designates an author, who assumes the task of developing and formulating a method. The proposed method is examined experimentally by at least one other member of the group who then submits a written report of the results of the examination. As a matter of principle, the examination must encompass all phases of the proposed analytical procedures. Only methods, for which criteria of analytical reliability are explicitly assigned, are released for publication. The values for the detection limit, sensitivity, accuracy, and specificity must fulfil the requirements of statistical quality control and the specific standards set by occupational and environmental health. This procedure guarantees that only reliable methods are published, which are reproducible within the framework of the given reliability criteria in different laboratories and can also be monitored over the course of time.

Methods released by the working group are published in German by the DFG as a loose-leaf collection entitled "Analysen in biologischem Material." The German edition has been appended 12 times, and about 120 methods for measuring about 180 parameters have been published (19).

In response to the worldwide demand for chemical methods for biological monitoring, the DFG has decided to make the contributions available to an international audience by bringing out an English edition entitled "Analyses of Hazardous Substances in Biological Materials." So far five volumes of this work with 62 selected methods for determining 34 inorganic and 30 organic substances or groups of substances and 7 parameters of effect have been published (18).

Quality Assessment

Quality assessment refers to the quality of the analytical results. It has two components: internal quality control, a set of procedures used by the laboratory staff to continuously assess results as they are produced and to decide whether they are reliable enough to be released, and external quality assessment, a system of objectively checking the laboratory performance by an external agency (19,20). The precision and accuracy of the values determined in the environmental-medical-toxicological laboratory must be continually checked by adequate internal and external quality control.

Internal Quality Control. Internal laboratory quality control is systematic monitoring of precision under repeated conditions to determine random errors and the accuracy of quantitative laboratory investigations. It includes a check of precision for every series of analyses and accuracy after every fourth series. In practice it is carried out with a control sample system. The control materials used should be as similar as possible to the investigation material and commercially available. These are standard reference materials with a certified analyte content and control materials with a constant analyte content for "daily use" during precision and accuracy checks.

Standard reference materials are samples whose quantitative composition of certain components is determined by various methods and by qualified laboratories. A certificate is provided which states the concentration of these components. Reference materials with certified values are offered, for example, by the Community Bureau of Reference (CBR) of the Commission of the European Union, by the American National Institute of Standards and Technology (NIST), and by the International Atomic Energy Agency (IAEA). There are only a few materials with parameters relevant to occupational medicine and environmental medicine in the programs of these institutions. In addition reference materials are very expensive because of the complicated certification procedure. Therefore they are not used for routine quality control but during validation of an analytical procedure.

Usually, therefore, commercially available control samples are used for routine internal quality control.

For toxicological parameters not contained in these commercially available control samples, control materials are used for internal quality control that have been produced by the laboratory itself. Animal or human blood and human serum are spiked with known concentrations of the analytes. These samples are in liquid form and aliquots are stored deep-frozen until analysis (21).

The results of daily internal quality control are registered on control cards and compared with the standard deviation for the method. The standard deviation of the method is determined in a so-called preperiod where control samples are analyzed on 20 different working days. A result is correct if it does not deviate from the actual value by more than three times this relative standard deviation and if there are no trends towards higher or lower values over a longer period.

The possibilities and results of internal quality control for medical-toxicological analyses in biological materials are described elsewhere (22).

External Quality Control. External quality control based on intercomparison programs objectively monitors the accuracy of quantitative laboratory investigations under comparable conditions. The aim of this is to guarantee the comparability of results from laboratory to laboratory and with the valid tolerable threshold limit values (eg, Biological Tolerance Value for Occupational Exposures/Exposure Equivalents for Carcinogenic Substances (BAT/EKA) values, biological exposure indices (BEIs), reference values, human biomonitoring (HBM) values). Intercomparison programs for occupational–medical–toxicological investigations are organized by the following institutions:

Deutsche Gesellschaft für Arbeitsmedizin und Umweltmedizin eV; round robins organized by Institut und Poliklinik für Arbeits-, Sozial- und Umweltmedizin der Universität Erlangen-Nürnberg, Erlangen, Germany

Institute of Occupational Health, Helsinki, Finland

Danish National Institute of Occupational Health, Arbejolsunljøiustitüttet (AMJ), Copenhagen, Denmark

Centre de Toxicologie du Québec, Toxicology Laboratory Québec, Canada

Deutsche Gesellschaft für Klinische Chemie eV, Bonn, Germany

Instand eV Ringversuche, Düsseldorf, Germany

Except for the intercomparison programs organized by the German Society for Occupational and Environmental Medicine (Deutsche Gesellschaft für Arbeitsmedizin und Umweltmedizin), usually only metals or a limited number of parameters are taken into consideration. The external quality control program of the German Society for Occupational and Environmental Medicine has been organized since 1982. In the meantime the nineteenth intercomparison program has been carried out by this society. These intercomparison programs are now highly accepted nationally and internationally. About 150 laboratories from home and abroad take part in these intercomparison programs. The high degree of acceptance is understandable considering that a range of 96 parameters for 47 substances is offered, the most extensive quality control program in toxicological analysis of biological materials currently offered worldwide.

Evaluation of the results of analysis of the laboratories which participated in the intercomparison program is carried out using reference values and tolerance ranges. These values are determined by reference laboratories. At present 19 national and international laboratories, selected according to certain criteria, participate in evaluating these reference values. Participants in the intercomparison program have fulfilled the analytical requirements when both results for one parameter are within the tolerance range, ie, in the ±3 SD range determined by the reference laboratories. The participants receive a certificate for this parameter which is valid for one year. There are numerous descriptions of the spectrum, organization, and results of these intercomparison programs (22,23).

Evaluation of the intercomparison programs has shown that the percentage of correct analyses for the laboratories taking part is between 50 and 60%. Improvement of this unsatisfactory result must be achieved by critical review of the analytical methods currently used, consistent use of internal quality control, and improvement in the performance of these laboratories.

Although quality control is primarily the responsibility of toxicological laboratories, the physician contacting the laboratory should find out about the quality of the laboratory. The laboratory receives a certificate for successful participation in the intercomparison program of the German Society for Occupational and Environmental Medicine. Physicians carrying out occupational and environmental medical health checks should use only laboratories which carry out their toxicological analyses under quality control because of liability.

BIBLIOGRAPHY

1. A. Berlin, R. Yodaiken, and B. Henman, eds., *Assessment of Toxic Agents at the Workplace. Roles of Ambient and Biological Monitoring,* Proceedings of NIOSH-OSHA-CEC-Seminar, Martinus Nijhoff Publishers, Boston, Mass., 1984.
2. R. Lauwerys and P. Hoet, eds., *Industrial Chemical Exposure: Guidelines for Baological Monitoring,* Lewis Pub., Boca Raton, Fla., 1993.
3. G. Lehnert and K.H. Schaller, *Israel J. Med. Sci.* **31,** 549–557 (1995).
4. J. Angerer and J. Gündel, *Ann. Jst. Super. Sanita* **32,** 199–206 (1996).
5. R. Zielhuis, "Recent and Potential Advances Applicable to the Protection of Worker's Health-Biological Monitoring II," in A. Berlin, R. Yodaiken, and B. Henman, eds., *Assessment of Toxic Agents at the Workplace. Roles of Ambient and Biological Monitoring,* Martinus Nijhoff Publishers, Boston, Mass., 1984.
6. A. Aitio, *Scand. J. Work Environ. Health* **20,** 46–58 (1994).
7. J. Rosenberg and D. Rempel, *State of the Art Rev.,* **5,** 491–498 (1990).
8. L. Lowry, J. Rosenberg, and V. Fiserova–Bergerova, *Appl. Ind. Hyg.* **4,** F11–13 (1989).
9. J. Rosenberg, V. Fiserova–Bergerova, and L.K. Lowry, *Appl. Ind. Hyg.* **4,** F16–20 (1989).
10. V. Fiserova–Bergerova, L.K. Lowry, and J. Rosenberg, *Appl. Ind. Hyg.* **4,** F10–13 (1989).
11. P. Hoet, *Biological Monitoring of Chemical Exposure in the Workplace,* Vol. 1, Contribution to the International Programme on Chemical Safety (IPCS) World Health Organization, Geneva, 1996, pp. 1–19.
12. K.H. Schaller, W. Arnold, K. Bencze, and P. Schramel, "Die Bedeutung der Haaranalyse in der arbeitsmedizinischen Toxikologie zur Beurteilung exogen aufgenommener Gefahrstoffe" *Arbeitsmed. Sozialmed. Präventivmed.* **26,** 225–227 (1991).
13. R. Lauwerys and A. Bernard, *Ann. Am. Conf. Ind. Hyg.* **12,** 327–329 (1985).

14. M. Thompson and R. Wood, *Pure Appl. Chem.,* **65,** 2123–1244 (1993).
15. P. Schrey, J. Wittsiepe, U. Ewers, M. Exner, and F. Selenka, *Bundesgesundheitsblatt* **36,** 455–463 (1993).
16. J. Messerschmidt, F. Alt, G. Tölg, J. Angerer, und K.H. Schaller, *Fresenius J. Anal. Chem.* **343,** 391–394 (1992).
17. D. Knopp and S. Glass, *Int. Arch. Occup. Environ. Health* **63,** 329–333 (1991).
18. J. Angerer, and K.H. Schaller, eds., *Analyses of Hazardous Substances in Biological Materials,* VCH Verlagsgesellschaft, Weinheim, Germany, Vol. 1, 1985, Vol. 2, 1988, Vol. 3, 1991, Vol. 4 1994, Vol. 5 1997.
19. World Health Organization, *External Quality Assessment of Health Laboratories,* Euro reports and studies 36, World Health Organization, Regional Office Europe, Copenhagen, 1981.
20. A. Aitio, *Biological Monitoring of Chemical Exposure in the Workplace,* Vol. 1, Contribution to the International Programme on Chemical Safety (IPCS) World Health Organization, Geneva, 1996, pp. 20–51.
21. J. Angerer and K.H. Schaller, *Arbeitsmed. Sozialmed. Präventivmed.* **11,** 311–12 (1976); *Arbeitsmed. Sozialmed. Präventivmed.* **12,** 33–35 (1977).
22. K.H. Schaller, J. Angerer, and G. Lehnert, *Int. Arch. Occup. Environ. Health* **62,** 537–542 (1991).
23. J. Angerer and co-workers, *Arbeitsmed. Sozialmed. Umweltmed.* **30,** 191–203 (1995).

BIOMARKERS

DAVID B. PEAKALL
King's College
London, England

LEE R. SHUGART
L.R. Shugart and Associates, Inc.
Oak Ridge, Tennessee

Ecotoxicology is a relatively new scientific discipline. Only within the past 10–15 years has it been regarded as a true science. It is a branch of toxicology that studies the subtle toxic effects that pollution exert on living organisms (eg, fish, birds, and plants) by investigating the fate and effects of chemicals and natural substances in ecosystems. Ecotoxicology is sometimes used synonymously with environmental toxicology, however, the latter also encompasses the effects of environmental pollution on humans.

The term ecotoxicology merges the fields of ecology and toxicology. A major difficulty with the merging of these two fields is that each focuses on a different organizational level. Toxicology is concerned with adverse effects of chemicals on living organisms, whereas ecology is focussed at the population, community or even ecosystem level. In their book *Ecotoxicology in Theory and Practice* (1) Forbes and Forbes put forward the following definition: "The field of study which integrates the ecological and toxicological effects of chemical pollutants on populations, communities and ecosystems with the fate (transport, transformation and breakdown) of such pollutants in the environment."

Here we are concerned with the use of ecotoxicology tests in analyzing environmental problems and in assisting with remedial actions. Human toxicology is based on the extrapolation of data from many species to one species, humans. Ecotoxicology, on the other hand, has to extrapolated data from a few species to many and from laboratory or limited field tests to effects on natural communities and ecosystems.

OFFICIAL ECOTOXICOLOGY TESTS

European Economic Community Tests

The testing of industrial chemicals within the EEC is based on a tiered approach. These levels are as follows: level 0 for all new industrial chemicals, level 1 when production exceeds 100 t/yr and level 2 when production exceeds 1000 t/yr. The ecotoxicological data requirements for these various levels are summarized in Table 1.

U.S. Environmental Protection Agency Tests

For most toxic substances generated by industrial, agricultural, and domestic activities, the consequences to human health and to the ecological integrity of the environment is not known *a priori.* Ecosystems can adapt through a variety of physical, chemical, and biological mechanisms on exposure to these toxic substances. However, when the assimilating capacity of the receiving ecosystem is surpassed for these toxic substance, the development, growth, reproduction, behavior, and ultimately the survival of the organisms present will be affected. Toxicity testing and laboratory-derived data can be useful for the identification and control of toxic substances entering the environment; they can be used to assess the potential for damage. The emphasis on pollution control activities, such as toxicity testing, have been driven primarily by various federal regulations for the protection of human and environmental health. The U.S. EPA was formed in December of 1970 with the responsibility, under one administrator, for setting standards and a compliance timetable for air and water qualities improvement. The regulatory authority exercised by the EPA comes mainly from three acts: (*1*) Federal Insecticide, Fungicide, and Rodenticide Act (FIFRA) of 1947 and reauthorized in 1972; (*2*) Toxic Substance Control Act (TSCA) of 1976; and (*3*) Clean Water Act (CWA) of 1984.

Table 1. Ecotoxicological data Required by Directive 79/831/EEC

Level	Description
0	Physicochemical properties (vapor pressure, water solubility, octanol-water partition coefficient)
	Abiotic and biotic degradation (minimum BOD and BOD/COD ratio)
	Acute toxicity for *Daphnia* and fish
1	Algal growth inhibition test
	Prolonged toxicity study of *Daphnia* (21 days) and fish (14 days)
	Test on a higher plant and earthworm
	Test for bioaccumulation
	Prolonged biodegradation study
2	Additional tests for accumulation, degradation, and mobility
	Prolonged toxicity study with fish including reproduction
	Avian toxicity study
	Other tests may be required based on likely movement within the environment

FIFRA requires that any pesticide distributed or sold in the United States be registered with the U.S. EPA. TSCA gives the U.S. government broad authority to control the production, distribution and use of all potentially hazardous industrial chemicals. The CWA established a national policy for the development of water-quality-based permit limitations for toxic pollutants in effluent discharges. Under these acts, the U.S. EPA must regulate chemicals so as to protect the environment and yet reap the benefits that derive from their use. An integrated approach using chemical-specific analysis and acute and chronic aquatic toxicity tests has been employed by the U.S. EPA to assess the degree of risk to the environment and to enforce these regulations.

The ultimate recipients of most toxic substances are the natural waters of the environment. Therefore, many of the toxicity tests are orientated toward the aquatic environment. As stated, government authorization for the manufacture of new chemicals is regulated under TSCA, which requires that a premanufacturing notification (PMN) be submitted for the U.S. EPA review. For chemicals expected to be used, transported or otherwise released into the aquatic environment, the PMN should include information on their potential toxicity to aquatic organisms. For biocides such as insecticides, rodenticides, fungicides and herbicides, the U.S. EPA under FIFRA requires that specific aquatic toxicity tests be conducted to evaluate the potential hazard to nontarget aquatic species and that resulting data be submitted in support of registration permits for sale and use of these chemicals. Aquatic toxicity test data are also used to evaluate the potential hazards and risks resulting from the discharge of municipal and industrial wastes (effluents) into natural waters. The U.S. EPA, under the National Pollutant Discharge Elimination System (NPDES) permit program, as authorized by the CWA, regulates industrial discharges. It should be noted that for pharmaceuticals, the Food and Drug Administration also requires aquatic toxicity test data for some new drug applications.

In the aquatic environment, toxicity test data have a variety of applications (2), including:

- Corporate industrial decisions on product development, manufacture, and commercialization
- Registration of products to satisfy regulatory requirements
- Permitting for the discharge of municipal and industrial wastes
- Environmental hazard-risk assessments
- Prosecution and defense of chemical-related activities in environmental litigation
- Derivation of numerical water and sediment quality criteria for protection of aquatic organisms

In addition to the governmental regulatory requirements listed previously, aquatic toxicity tests with indigenous species and site-specific conditions can be used to evaluate the potential environmental hazards and risks of an unanticipated, accidental release of a chemical into the environment. This evaluation includes the potential human health hazard posed by contamination of commercially important species with a chemical. Also, with the trend in the past toward use of litigation to resolve concerns about environmental pollutants, toxicity tests have been used to evaluate the effects of long-term chemical releases. Toxicity tests can be used to confirm or refute a relationship between observed effects and the concentration of chemicals shown to exist or to have existed in the environment.

The creation of the U.S. EPA and the enactment of FIFRA, TSCA and CWA in the United States have been a powerful stimulus to the development of standardized toxicity testing. Adverse effects on ecosystems may result from toxicant exposure that directly causes death of organisms present or produces sublethal effects, ie, may result in an organism's inability to grow an reproduce normally. Human health may be affected either through direct exposure to the toxicants or through uptake of contaminated fish, game, etc. To control and identify toxic inputs into the environment, data from toxicity tests that causes detrimental effects on species under control experimental conditions in the laboratory are extrapolated to determine appropriate safe conditions for the environment. Thus the evaluation of the biological effects of chemicals has been based on the results of single-species laboratory toxicity tests. Standardization of these laboratory test procedures has provided, in most instances, a means for assessing the potential adverse effects of chemicals (2).

The U.S. EPA alone has been mandated under TSCA to assess the hazard potential of about 70,000 chemicals already in production in the United States with about 1,000 new chemicals being added to the inventory each year. Even though the state of the art of toxicological testing has advanced to meet the regulatory demands, a large number of chemicals in use to date have not been tested. Furthermore, to cover this gap in knowledge, the unknown toxicity of these many chemicals are often approximately predicted by extrapolation from measured toxicity of a few tested chemicals. In addition, the prediction of the unknown toxicity toward many species in the environment is often made on the basis of measured toxicity in a few laboratory species. These deficiencies in toxicity testing procedures highlight the urgent need for development of new and more appropriate tests. Detailed description of the many biological test using numerous and well-studied laboratory species, on various environmental samples (air, fresh and salt water, sediment, etc) as well as commercially produced chemicals and the theoretical basis for the use of these tests is accessible in the scientific literature.

Comparison of EEC and U.S. EPA Tests

Some international (ie, Europe, Canada, and South America) organizations are following the lead of the United States by increasing their focus on controlling toxicants in the environment. The OECD was formed to ensure better health conditions as new products reach the international marketplace and the current testing system in the EEC came out of expert committees appointed by the ODEC in the late 1970s and early 1980s. However the OECD recommendations did not become binding on the member states, which is virtually the entire western industrialized world. Thus, the opportunity for a unified approach to testing, which would have accounted for some 90% of chemical production, was lost. In view of the ever increasing international aspects of the trade in chemicals, this lack of uniformity in available safety data is a serious loss. If such a system had been put in place 15 years ago, it is likely that it would have been adopted by the rapidly developing countries of the Pacific Rim.

Reducing Animal Testing

When the current testing system was devised in the late 1970s, virtually no concern was expressed for the numbers of animals used in testing. Although this is not the place to debate the ethics of the issue of animal rights, there is no doubt that it has become a political force throughout the western world. Although testing procedures have not been altered yet, there is considerable pressure to do so in the near future. Here, the scientific consequences of such changes are examined.

The two main avenues available, short of banning all testing, are to reduce the number of animals used in a specific test and to use alternative tests. The growing public demand for greater safety regarding the effects of chemicals on human health and the environment coupled at the same time with a demand for fewer or even no animal experiments has produced a paradox.

One of the basic tests is the LD_{50} (dose that produces death in 50% of the population being tested). Here pilot experiments can be used to decrease the numbers of animals. A method based on small numbers of animals has been validated (3). This method is based on using three rats or mice of each sex at each dosage level. The initial dose used was 2000 mg/kg. If no animals died, the tests were suspended; if deaths occurred, then toxicity tests at 200 and 20 mg/kg were carried out. These workers examined 10 chemicals using this method and compared their results to literature values of LD_{50}s. Reasonable agreement was found. For dichlorvos, for example a value of 20–200 mg/kg was obtained using this method, whereas literature values ranged from 133–275 mg/kg. In view of the uncertainty in extrapolating from rodents to man, a range value could be considered acceptable. For example, the LD_{50} for 2,3,7,8-tetrachlorodioxin ranges from 0.6 mg/kg for the guinea pig to 115 mg/kg for the rabbit, with the rat given an intermediate value of 20 mg/kg.

From the point of view of wildlife toxicology, range values are acceptable in view of interspecies variation. Assuming that the LD_{50} is available for the target species, what is the value of this information?

Even though LD_{50}s are not the be all and end all that their high profile in regulations suggest, they are often a useful starting point. A comparison of the LD_{50} value with environmental levels gives some indication of the degree of hazard to which a species is exposed. Many experimental studies are based on LD_{50}s. Sublethal effects are often studied at various fractions of the LD_{50}. Other approaches that use fewer animals, or at least more acceptable animals, move toward the use of lower organisms and *in vitro* testings.

The LD_{50} tests on *Daphnia* and a fish (the guppy, *Poecilia reticulata*) have been compared (4) and a good correlation between the two tests was demonstrated. This suggests that preliminary tests could be undertaken on *Daphnia* and additional testing on fish carried out only under specific circumstances: high toxicity of substances being tested or high likelihood of substance getting into the aquatic food chain. Another possibility is the Microtox test which uses phosphorescent bacteria as the test organism. This test has been used satisfactorily to assess the toxicity of industrial effluent (5). Again, as in the *Daphnia* test, good correlations have been found with other tests, including tests with mammals. Although it is unlikely that a regulatory authority would be prepared to rely solely on tests on lower organisms such tests are likely to increase in importance in the near future.

For ecotoxicology studies outside the regulatory framework, there is increasing interest in nondestructive methods and *in vitro* testing. In addition to being nondestructive, sampling of blood has the advantage that serial studies can be made on the same individual. This can decrease the amount of variability of the data by allowing the use of an individual as its own control in experimental work. It also makes it possible to follow the time course of exposure. This can be important, particularly in cases where the time of onset of an effect is variable.

A great deal of progress has been made over the last two decades in the technique of *in vitro* experiments, a term meaning literally "in glass". Early *in vitro* experiments on the inhibition of enzymes, in which a variety of compounds were tested, are very limited. Frequently, there was a lack of relationship with that observed *in vivo*. The more sophisticated approach of using *in vitro* experiments based on tissue cultures has much more promise. Experiments can run the entire gamut from those on intact animals to those using established cell lines. Whether a particular feature is an advantage or a disadvantage depends on the question being asked. The brief discussion of this major field here will focus on the comparison between experiments using primary cell cultures and those on intact animals.

An important difference between *in vivo* and *in vitro* experiments is that the latter are designed to reduce the number of variables by eliminating the influence of biochemical regulation, such as those imposed by the endocrine and nervous systems. This enables individual components of the system to be examined separately and in detail, but ultimately they require integration. For example, the metabolism of a xenobiotic by the P450 system can be examined in one series of experiments, but another series of experiments is needed to examine for the effect of altered steroid levels caused by P450 induction. Well-integrated studies encompassing all of the *in vitro* approaches will be required if we are ever going to understand fully the biological mechanisms associated with chemical toxicity.

Another important feature of tissue culture is the fact that large numbers of experiments that can be run. For example, much of the tedious work on the effect of specific polychlorinated biphenyl (PCB) congeners on induction of porphyria was carried out in tissue culture experiments because the number of animals that would have been required for intact animal studies would have been prohibitive. An additional factor is the cost of the congeners because the amount of chemical needed per experiment is much less than that required for testing with whole animals. Finally, it will be necessary to take these types of laboratory studies into intact animals and into the field situation. For both practical and moral reasons, it is important that the two main streams of toxicology, *in vivo* and *in vitro* studies, do not remain two solitudes.

ADDITIONAL ECOTOXICOLOGICAL TESTS

The tests considered so far are those devised for regulatory purposes. These tests are only a small number of those that have been devised for ecotoxicology studies. There is no possibility of completeness, but a representative sample of tests is given.

Tests at the Organism Level

The Biomarker Approach. The use of biological responses (biomarkers) in organisms present in a polluted environment is an approach that has caught the attention of the ecotoxicologist. One of the primary reasons to conduct tests on organisms in situ, is to provide a more realistic exposure regime. Chemicals and physical agents are known to elicit measurable and characteristic biological responses in exposed organisms. Such evidence can provide a link between exposure and effect. Recent books and symposia on the subject of biomarkers are available (6–9).

There have been several attempts to define the term biomarker, but the definition of biomarker used here is "a biological response to a chemical or chemicals that gives a measure of exposure and sometimes also of toxic effect in an organism." It is proposed that biological responses measured above the organism level be considered as "ecological indicators." Even though ecological indicators are obviously important, they are much less well developed than biomarkers and are not yet to the stage where they can be used in formal risk assessment. Therefore, they will not be considered in this article.

There have been a number of attempts to subdivide biomarkers. Four classes of biomarkers have been proposed (10): exposure biomarkers, effect biomarkers, exposure/effect biomarkers, and latent effect biomarkers. A new journal, *Biomarkers,* makes the following division: biomarkers of exposure, biomarkers of effect, and biomarkers of susceptibility. The latter category is considered to include genetic factors and changes in biological receptors which alter the susceptibility of an organism to exposure to a chemical substance. Although such biomarkers are of considerable interest, they have not yet reached the point where they can be considered at a policy level.

However, these subdivisions seem artificial. All biomarkers are, by definition, biomarkers of exposure. Whether they are also biomarkers of effect depends on the state of knowledge. For example, the induction of mixed function oxidases (MFOs) is well-known as a biomarker of exposure to a variety of chemicals. There are many other examples that show chemicals in the environment cause induction of various biological processes, although the consequences of the induction is unknown. In a few cases, such as for the fish-eating birds in the North American Great Lakes, enough data are available to relate this induction, expressed as dioxin equivalents, to embryonic death. This concept is considered in more detail in a later section.

Specific and Nonspecific Biomarkers. The specificity of biomarkers to chemicals varies greatly. The current status of measuring biomarkers, given in order of decreasing specificity, is summarized in Table 2. At one end of the spectrum the enzyme amino levulinic acid dehydratase (ALAD) is highly specific, being inhibited only by lead. At the other end, mixed function oxidase enzymes and stress proteins are induced by many different classes of chemicals. Both specific and nonspecific biomarkers have their place in environmental assessment. A nonspecific biomarker can tell one that a pollutant is present in a meaningful concentration but does not tell one which chemical is present. Based on this information a more detailed chemical investigation can be justified. In contrast, a specific biomarker tells one which chemical is present, but gives no information on the presence of other chemicals.

Concept of Meaningful Exposure. Biomarkers have the advantage over chemical analysis in that they can demonstrate whether or not an organism is meaningfully exposed. For some classes of persistent organic chemicals, especially the organochlorines (OCs), detection limits are now down to parts per trillion. Thus, in almost all samples these human-made chemicals can be detected, but the physiological significance is rarely known.

With biomarkers it is possible to determine if the physiology of the organism is significantly different from normal. If it is, then the organism can be considered to be meaningfully exposed; equally important, if the physiology is not significantly different, then the organism can be considered not to be meaningfully exposed even though the chemicals can be detected. The ability to determine whether or not an organism is meaningfully exposed is important in making the decision whether regulatory action should be taken and also in making the decision whether or not remedial action has been successful. The following criteria need to be meet before the concept of meaningful exposure can be used:

1. Data must be available on what is normality for each biomarker. Because of the diversity of species involved, this is a good deal more complex than in the case where biochemical levels are used in the diagnosis of human health. Obviously it is impossible to have data on all species; the question of sentinel species will be discussed in more detail in a later section. Although there is a great deal of data available, there is a need for a centralized database to collect, verify, and validate this baseline information.
2. To adequately assess the impact of the major classes of chemicals of concern, biomarkers are needed that indicate the status of the important functions of the organism. Even though this point has not yet been reached, the rate of progress toward the goal is encouraging.

It is important to note that there is a need to relate biomarker responses to changes in Darwinian fitness parameters in individuals so that population and community level effects can be predicted (9,11). The biochemical and physiological consequences of toxicant exposure of most concern to ecotoxicologists are those which might affect reproductive health. There is considerable evidence that some toxicants may potentiate any number of biological mechanisms and eventually manifested this toxicity at the population level (12). Included are such occurrences as mutational events in embryonic tissue, impaired growth, change in genetic diversity, etc.

Criteria for Evaluating Biomarkers. A list of criteria for evaluating biomarkers, in addition to those on specificity (7), is considered below:

1. *Biological specificity:* It is important to know which classes of organisms the biomarker may be used on.

Table 2. Some Biomarkers and Their Status (In Order of Decreasing Specificity)[a]

Biomarker	Measure of Toxic Effect	Pollutant	Status
Inhibition of ALAD	Yes	Lead	Sufficiently reliable to replace chemical analysis; can be related to mortality
Induction of metallothionein	No	Cadmium	No advantage over chemical analysis; not related to mechanism of action
Eggshell thinning	Yes	DDT, DDE Dicofol	Wide variation in sensitivity Related to reproductive success
Anticoagulant; clotting proteins	Yes	Rodenticides	Has been related to mortality, risk assessed from blood protein levels
Porphyrin profiles	No	Several OCs	Levels of porphyrins found in environmental samples are well below those causing adverse effects
Depression of plasma, retinol, and thyroxine	Yes	3,4,3′,4′TCB	Dermal and epithelia lesions; binding to specific protein has been shown
Inhibition of AChE	Yes	Ops and carbamates	More reliable than chemical analysis Directly related to mortality
Induction of vitellogenin	No	Alkylphenols	Mechanism related to environmental estrogens
Induction of MFOs	Yes	OCs, PAHs	Analysis of TCDD-EQ has been related to reproductive success Induction of P450s related to specific chemicals
DNA and hemoglobin adducts	No	PAHs	Good monitor of exposure, especially for PAHs
DNA integrity	Yes	Metals, OCs, PAHs, and genotoxicants	DNA damage is serious indication of harm; relationship to effects tenuous
Other serum enzymes	No	Metals, OCs, OPs	A considerable number of enzymes are altered by pollutants, but relationship to effects are not clear
Immune response	Yes	Metals, OCs, PAHs	Proper functioning of immune system is critical to health, but system has considerable reserve
Stress proteins	No	Metals and OCs	Difficult to separate effects of chemicals from other stressors

[a] AChE: acetylcholinesterase; ALAD: aminolevulinic acid dehydratase; DDE: 2,2-bis(*p*-phenyl)-1,1-dichloroethylene; DDT: 1,1,1-trichloro-2,2-bis (chlorophenyl) ethane; Dicofol: 2,2,2-trichloro-1,1-di-(4-chlorophenyl) ethanol; MFOs: mixed function oxidases; OCs: organochlorines; PS: organophosphates; PAHs: polycyclic aromatic hydrocarbons; 3,4,3′,4′TCB: 3,4,3′,4′-tetrachlorobiphenyl; TCDD-EQ: dioxin equivalents.

The inhibition of the enzyme acetyl choline esterase (AChE) can be applied throughout the animal kingdom, whereas the induction of vitellogenin is confined to those vertebrates that lay eggs.

2. *Clarity of interpretation:* How clearcut is the endpoint as an indicator of exposure to anthropogenic stress? Can the endpoints be clearly distinguished from natural stresses? It is valuable to know the mechanism of response to the chemical in assessing this point.
3. *Time of responses:* The response temporal expression of different biomarkers can vary widely from nearly instantaneous to years. Depending on the type of study, slow or rapid manifestation maybe desirable.
4. *Permanence of response:* Similarly, it is important to know how long the response lasts. If it is transient, it may readily be missed. The inhibition of AChE, especially in blood, is a transient response and thus it is necessary to know when the exposure occurred to assess the importance of the degree of inhibition. In contrast, the inhibition of ALAD is only slowly reversed.
5. *Reliability:* This can be considered under two headings: (*1*) environmental influences that modulate the organism's response to a chemical, and (*2*) inherent variation in the biological response to a given exposure. It is important to know the extent of all variation in order to have a reliable biomarker.
6. *Methodological considerations:* Important considerations here are precision (analytical reproducibility of the method) and cost and ease of the analysis. Although many reliable assays have been developed there is a need for standardization, along the lines used in analytical chemistry, so results from different laboratories are comparable.
7. *Relative sensitivity:* It is important that the biomarker be sensitive when compared to other endpoints, such as

mortality or reproductive impairment, and it is important to know the relative sensitivity of this comparison.

8. *Validation in the field:* For a biomarker to be useful in environmental assessment, it must be validated in the field. Organisms in the field are subjected to a wide range of variables that are usually accounted for or controlled in laboratory experimentation.
9. *Linkage to higher level effects:* A biomarker is more useful if there is clear linkage to effects at higher levels of organization. Studies on invertebrates have been particularly fruitful as population changes occur more rapidly than in higher species.

Two specific tests, DNA structural damage and the induction of mixed function oxidases, are considered in more detail to illustrate the strengths and weaknesses of the biomarker approach.

Genotoxic Tests. One of the more serious environmental concerns about pollution is the potential for exposure to substances that are genotoxic. This problem arises because some of these pollutants are carcinogens and mutagens with the capacity to affect both the structural integrity of DNA and the fidelity of its biological expression (13).

Genetic toxicology is an area of science in which the interaction of DNA-damaging agents with the cell's genetic material is studied in relation to subsequent effects on the health of the organism. Structural changes to the integrity of DNA caused by DNA-damaging agents are useful endpoints for assessing exposure to hazardous environmental pollutants on human health (14,15) and biota (16,17). The organism functions as an integrator of exposure, accounting for abiotic and physiological factors that modulate the dose of toxicant taken up. The resulting magnitude of the change in DNA structure provides an estimate of the severity of exposure, hopefully in time to take preventive or remedial measures.

In most instances, studies have been concerned with documenting exposure of environmental species to genotoxic agents via the detection of DNA structural damage. DNA can be analyzed for specific modifications such as chemical adducts (covalent attachment of a specific chemical to DNA) and photoproducts (dimerization of bases due to ultraviolet light) or generalized structural damage (ie, DNA strand breakage) that is induced from exposure to any of a number of genotoxicants. Three specific examples are considered: DNA adducts in Belgula whales, DNA strands breaks in turtles and fish, and UVB-induced photoproducts in the DNA of plants. Each example contains a brief description about the environmental issue/concern being addressed, the approach used (ie, species sampled and methodology employed to detect DNA structural damage), and results obtained.

Tests at the Population Level

Ecotoxicology is more concerned with populations than with the effects on individuals. Although population changes can be determined by a wide variety of techniques it is difficult to assign a specific chemical cause to these changes.

Tests at the Community Level

Effects on the structure of communities have been studied largely with invertebrates following the use of pesticides or in heavily contaminated aquatic systems. Invertebrates account for 95% of all species and are thus important components of all ecosystems. Populations of invertebrates are often large and thus studies can be made without serious impact on population dynamics. The short life cycle of many invertebrates means that multigeneration studies can be carried in a short period of time. The extrapolation to populations and community structure has been considered (18). The most successful studies relating effects to the individual to population and community changes has been studies on resistance to pesticides. Here it has been possible to relate enzyme changes to marked changes at the population and community structure level.

Tests at the Ecosystem Level

The ultimate concern of ecologists is the integrity of the structure and function of ecosystems. Thus ecotoxicologists are concerned with the possibility that pollutants can affect ecosystems. The problem is that while the importance of the effects increases as we move up the organizational levels so does the time of response and the difficulty of linking specific effects with specific chemicals. Chemical effects on ecosystems can be difficult to distinguish from natural fluctuations. Nevertheless, despite these difficulties, there have been a number of possible ecosystem effects that have been investigated. These include the effects of acid rain on aquatic and terrestrial ecosystems, the effects of chloroflurocarbons on the ozone layer and global warming (see ACID RAIN).

FUTURE DIRECTION OF ECOTOXICOLOGICAL TESTING

The direction in ecotoxicological testing is to move from merely protecting against toxic damage at a local level to ensuring robust ecosystem health. Although few ecotoxicologists would dispute this objective, especially those strongly wedded to the eco- side of the question, the concern raised by those more firmly wedded to the toxicology side is: How this can be achieved?

The difficulties with the standard toxicological approach of assessing exposure and assessing environmental toxicity are well known. The problem of prediction of environmental exposure (PEC) of living organisms moving within the environment, eating a variety of foods and, thus, exposed to a wide variety of contaminant levels, is clear enough. It was difficult when we were dealing with highly persistent compounds such as the organochlorines which are now banned in most countries. The difficulties with less persistent materials which are used persistently is now even more challenging. Ecotoxicology testing on a small number of species to assess environmental toxicity and to determine a predicted no-effect concentration (PNEC) also has severe limitations.

One of the most exciting areas of molecular biology that is beginning to become important in toxicology is the field of receptors. Much detailed work on the pharmacodynamics of the interaction of environmentally important compounds with receptors and characterization of receptors is needed. It can be expected that major advances will be made in the next few years.

Numerous books are available that address not only the growing interest in ecotoxicology as a scientific endeavor, but also the many new and unique scientific approaches that are accessible to ecotoxicologists to help with their studies.

BIBLIOGRAPHY

1. V. E. Forbes and T. L. Forbes, *Ecotoxicology in Theory and Practice,* Chapman and Hall, London, 1994.
2. U.S. NRC, *Toxicity Testing,* National Academy Press, Washington, D.C., 1984.
3. S. Yamanaka, S. M. Hashimoto, M. Tobe, K. Kobayashi, J. Sekizawa, and M. Nishimura, *Arch. Toxicol.* **64,** 262–268 (1990).
4. C. H. Walker, P. W. Grieg-Smith, N. O. Crossland, and R. Brown, in M. Balls, J. Bridges, and J. Souphee, eds., *Animals as Alternatives in Toxicology,* MacMillian, New York, 1991, pp. 223–241.
5. M. Richardson, *Ecotoxicology Monitoring,* VCH, New York, 1993.
6. J. F. McCarthy and L. R. Shugart, *Biomarkers of Environmental Contamination,* Lewis Publishers, Boca Raton, Fla., 1990.
7. R. J. Huggett, R. A. Kimerle, P. M. Mehrle, Jr. and H. L. Bergman, *Biomarkers. Biochemical, Physiological, and Histological Markers of Anthropogenic Stress,* Lewis Publishers, Boca Raton, Fla., 1992.
8. D. B. Peakall, *Animal Biomarkers as Pollution Indicators,* Chapman and Hall, London, 1992.
9. D. B. Peakall and L. R. Shugart, *Research and Application in the Assessment of Environmental Health,* Springer-Verlag, Berlin, 1993.
10. M. C. Fossi and C. Leonzio, *Nondestructive Biomarkers in Vertebrates,* Lewis Publishers, Boca Raton, Fla., 1994.
11. M. H. Depledge, in Ref. 13, pp. 271–295.
12. S. Anderson, W. Sadinski, L. R. Shugart, P. Brussard, M. Depledge, T. Ford, J. Hose, J. Stegeman, W. Suk, I. Wirgin and G. Wogan, *Environ. Health Perspec.* **102,** 3–8 (1994).
13. G. N. Wogan and N. J. Gorelick, *Environ. Health Perspec.* **62,** 5–18 (1985).
14. H. W. Kohn, *Ann. NY Acad. Sci.* **407,** 106–118 (1983).
15. Committee on Biological Markers of the National Research Council, *Environ. Health Perspec.* **74,** 3–9 (1987).
16. L. R. Shugart, in J. F. McCarthy and L. R. Shugart, eds., *Biological Markers of Environmental Contaminants,* Lewis Publishers, Boca Raton, Fla., 1990, pp. 205–216.
17. L. R. Shugart, J. F. McCarthy, and R. S. Halbrook, *Risk Anal.* **12,** 353–360 (1992).
18. L. Lagadic, T. Caquet, and F. Ramade, *Ecotoxicol.* **3,** 193–208 (1994).

BIOREMEDIATION, INDUSTRIAL. See BIOFILTRATION.

BIOSENSOR TECHNOLOGY FOR ENVIRONMENTAL MEASUREMENT

KIM R. ROGERS
U.S. Environmental Protection Agency
Las Vegas, Nevada

Biosensors are analytical devices that show potential for development in the clinical, diagnostic, food analysis, process control, and environmental areas. Examples of commercially available biosensors include enzyme electrode biosensors for detection of glucose in blood and single analytes in food, as well as biological oxygen demand (BOD) in waste water. Although biosensors show a significant potential for commercial applications, relatively few of these devices have reached that stage of development. There are a number of reasons for the slow movement of reported biosensor configurations into commercial markets. The primary reasons are that these devices have not met the performance requirements or that they are not economically competitive with other available methods for potential applications. This is particularly evident in the environmental area, where relatively few of the numerous biosensor configurations that have been reported are commercially available (1).

INTRODUCTION

Biosensors historically have been viewed as a class of chemical sensors that use biological recognition rather than chemical reactions to measure small organic compounds such as glucose. The use of biological recognition coupled to a signal transducer (eg, electrochemical, optical, thermal, or acoustic) has required a certain degree of multidisciplinary approaches. Most biosensor research, however, has focused on innovative biochemistry or signal transducer technology (ie, physics and engineering) rather than the overall analytical problem. This is particularly true for areas such as environmental monitoring. These versatile analytical devices have been used to measure a wide variety of environmentally related chemical and biological parameters including organics, heavy metals, proteins, bacteria, biological oxygen demand, toxicity, and DNA hybridization. Nevertheless, biosensor technology developed in the environmental area seems to represent potential solutions looking for specific analytical problems.

Considering the need for field analytical methods in the environmental area, the wide variety of biosensors reported for potential environmental applications provides an interesting opportunity. More specifically, in response to the increasing time and expense associated with environmental monitoring, there exists a critical and growing need for cost-effective field methods. Driven by this need, a variety of field screening methods have been reported and are currently under investigation (eg, miniaturized laboratory techniques, field test kits, and chemical sensors). Biosensors, as a result of their unique characteristics, may complement these field analytical technologies. In order for biosensors to be competitive, however, these characteristics must be exploited to fill specific technology gaps. For example, biosensors show a great deal of analytical diversity (eg, potential range of analytes) and format diversity (eg, potential for flow-through and *in situ* operation). Nevertheless, in the area of environmental monitoring, the development of commercial biosensors faces a number of significant obstacles. This may be due, in part, to the enormous scope of the environmental monitoring task. As a consequence of such issues as compound diversity, matrix complexity, variety of data quality requirements, and diversity of potential monitoring applications, the development and commercialization of biosensors for the environmental area will be extremely challenging. This article will outline some of the most commonly reported biosensors for potential environmental applications as well as discuss some of the relevant environmental issues and how they are related to biosensor development.

PRINCIPLES OF OPERATION

The definition of a biosensor is by no means static and continues to change with the introduction of new methods and technologies. Because it is relevant to the application of these devices to environmental monitoring, a brief discussion concerning several definition-related concepts will be presented. The concept of the *sensor* as being an analytical device which selectively and reversibly responds to the concentration of a chemical species has been generally accepted in the literature (2) and has been adopted under the International Union of Pure and Applied Chemistry (IUPAC) convention. Nevertheless, many chemical sensors and biosensors respond selectively and quantitatively but not reversibly. The concept of a probe has been suggested for this type of biosensor. For the purpose of the following discussion, however, both sensors and probes will be considered to be biosensors, which will be defined as analytical devices composed of a biological recognition element (eg, enzyme, receptor, DNA, antibody, or microbe) in intimate contact with a signal transducer (eg, electrochemical, optical, thermal, or acoustic), which together relate the concentration or chemical property of an analyte to a measurable electronic signal (Fig. 1).

The recent introduction of a variety of applied methods and technologies such as flow injection analysis and fluorescent techniques as well as the use of genetically engineered microorganisms (GEMs) has further blurred the classical concept of a biosensor as an enzyme electrode. Although many of the biosensor-type techniques depend on biomolecular recognition, for many of these methods the target analyte is interrogated by a sensor that is not necessarily in intimate contact with the indicator biochemistry. These techniques are referred to here as bioanalytical methods. Because many of the biochemical indicators involved in these methods have been (or potentially could be) monitored using an integrated signal transducer, and because this differentiation has not been routinely or uniformly made in the literature, many of these bioanalytical methods as related to biosensors are discussed also.

Biosensors are classified or grouped in several different ways. When considered as a class of chemical sensors, biosensors are typically grouped by their physical transducers, as is evident in comprehensive chemical sensor reviews (4) and physical science-oriented reviews (5,6). In contrast, when viewed as a separate field of study, biosensors have been grouped according to the operational mechanism of their biological recognition element (ie, biocatalytic, bioaffinity, or microbial), as in the format adopted by the World Congress on Biosensors. Yet another view of biosensor classification has been suggested with respect to biosensor characteristics and the targeted market (7). This classification places biosensors in two groups: those which are portable and those which serve as detectors for laboratory instruments. For this practical approach, the portable biosensors are characterized by a simple assay format, low capital investment, and a disposable test format. The laboratory-based biosensors are suggested to require more complex assay formats, higher capital investments, and reusable assay formats. Regardless of how they are grouped, however, biosensors show a wide diversity in the combinations of biological recognition elements, signal transducers, and operating formats that have been reported.

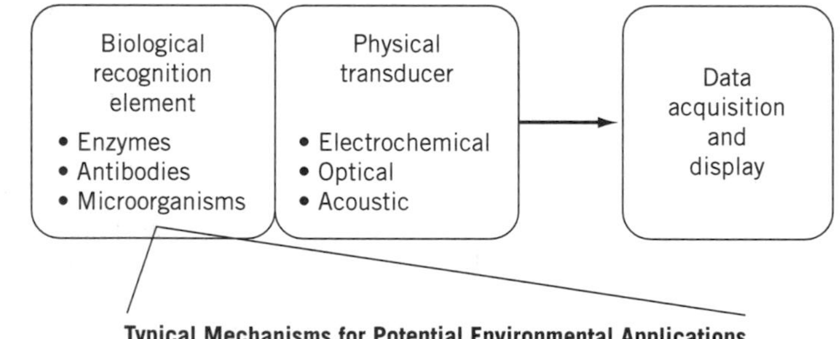

Typical Mechanisms for Potential Environmental Applications

Enzymes	Catalytic transformation of the pollutant (eg, analyte, phenolics; enzyme, tyrosinase)
	Specific inhibition of enzyme activity by the pollutant (eg, analyte, organophosphates; enzyme, acetylcholinesterase)
	Modification of enzyme activity by a pollutant that acts as a modulator or cofactor (eg, analyte, Mn(II); enzyme, horseradish peroxidase)
Antibodies	Compound or class specific affinity toward the pollutant (eg, analyte, pesticides or benzo(*a*)pyrene ; antibody, monoclonal or polyclonal)
Microorganisms	General inhibition of cellular respiration by pollutant (eg, analyte, respiratory toxicants; test organism, *P. phosphoreum*; response, luminescence)
	Promotor recognition by specific pollutant followed by gene expression, enzyme synthesis, and catalytic activity (eg, analyte, Hg(II); expressed enzyme, luciferase; response, luminescence)

Figure 1. Biosensor components and typical mechanisms for environmental applications. Reprinted with permission (3).

In addition to differences in the terminology associated with biosensors themselves, approaches to issues such as sensitivity, detection limits, and assay interferences show considerable diversity in the biosensors literature. Consequently, several terms will be defined with respect to environmental biosensor methods. For the purpose of this article, the lowest analyte concentration that can be measured will be referred to as the *detection limit* and the smallest change in concentration that can be measured will be referred to as *sensitivity*. In addition, the concentration range over which the biosensor can reliably and unambiguously measure the analyte will be referred to as the *dynamic operating range*.

Molecular Recognition

A considerable number of biosensor formats have been reported using a variety of biological recognition elements. Nevertheless, general mechanisms for biological recognition typically fall into three basic mechanistic classes. These classes include:

1. the measurement of products, inhibitors, or modulators in a biocatalytic process (eg, using enzymes);
2. the measurement of a stoichiometric binding event (eg, using antibodies, DNA, or receptors) either directly or through catalytic amplification; and
3. the use of isolated tissues, organs, or microorganisms that can respond to an initial recognition event followed by a cascade of other events resulting in a measurable response (eg, luminescence of a microorganism in response to a specific analyte).

Biocatalytic. Enzymes have certain advantages and limitations for use as biological recognition elements. Historically, these proteins were incorporated into the first biosensors, most likely due to their ability to catalytically convert specific substrates to products and thus amplify the sensor response (8). The range of analytes that can be detected by enzyme-based biosensors is limited to substrates, inhibitors, or modulators of the enzyme. This would seem to limit the use of enzymes as recognition elements for biosensors in potential environmental applications. Nevertheless, the toxic properties of many environmental pollutants primarily arise from their effect on enzyme-catalyzed reactions in the affected organisms. In addition, because many organisms have adapted to toxins by developing systems for the detoxification of a variety of environmental contaminants, enzymes responsible for this decontamination may be used as sensing elements for biosensors that target these compounds. Other advantages, particularly for potential environmental applications, are the reversible nature of enzymes and the extensive research reported on immobilization and stabilization of these proteins.

Detection principles for enzyme-based biosensors rely on three basic mechanisms; those for which the analyte of interest is the enzyme substrate (Fig. 2**a**), those for which the analyte is an enzyme inhibitor (Fig. 2**b**), and those for which the analyte is an enzyme, modulator, or cofactor (Fig. 2**c**).

Bioaffinity. In contrast to biocatalytic-type biosensors, bioaffinity-based biosensors rely on a stoichiometric binding event between the recognition element and the analyte. These biosensors rely on the use of antibodies, receptors, or DNA as bioaffinity elements (Fig. 3). Although the stoichiometric nature (ie, one receptor, with a specific number of binding sites) of bioaffinity-based biosensors limits the amount of analyte that can be detected to a finite number (based on the number of receptors present on the sensor surface), signal transducers are usually sensitive enough to measure a relative percentage of sites occupied by the analyte.

Due primarily to their versatility, high binding affinities, and commercial availability, antibodies are the most widely reported affinity-type biological recognition element. This is true for both general applications and potential environmental applications. These proteins are generated in response to an animal having been immunized with an antigen (immunogen). Because small molecular weight compounds such as pesticides are not themselves immunogenic, they are typically bonded to larger proteins, which are then used as immunogens. Antibodies derived from serum of an immunized animal (polyclonal) form a consortium of proteins that recognize various regions of the immunogen. By contrast, monoclonal antibodies are derived from an immunized animal's spleen cells, which are fused with myeloma cells resulting in hybridoma cells that secrete a specific antibody that recognizes a single portion of the

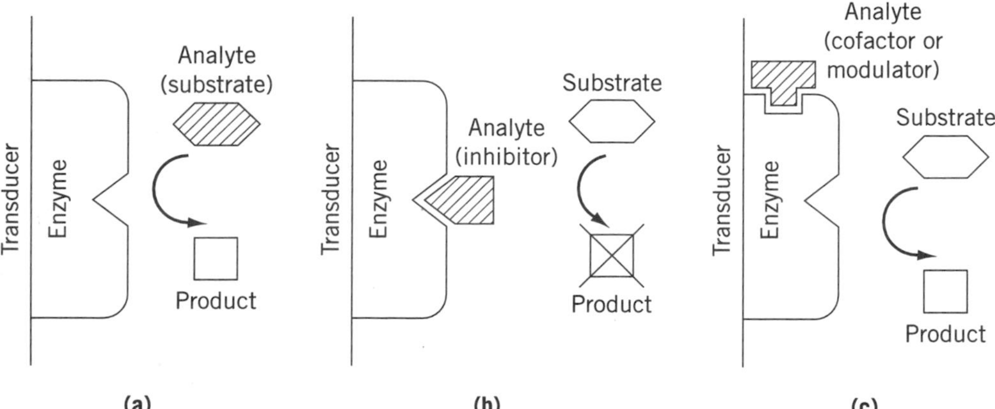

Figure 2. Principles of biosensor operation. Biocatalytic (enzyme) mechanisms in which the analyte of interest is a (**a**) substrate, (**b**) competitive inhibitor, or (**c**) cofactor of the enzyme used as the biological recognition element.

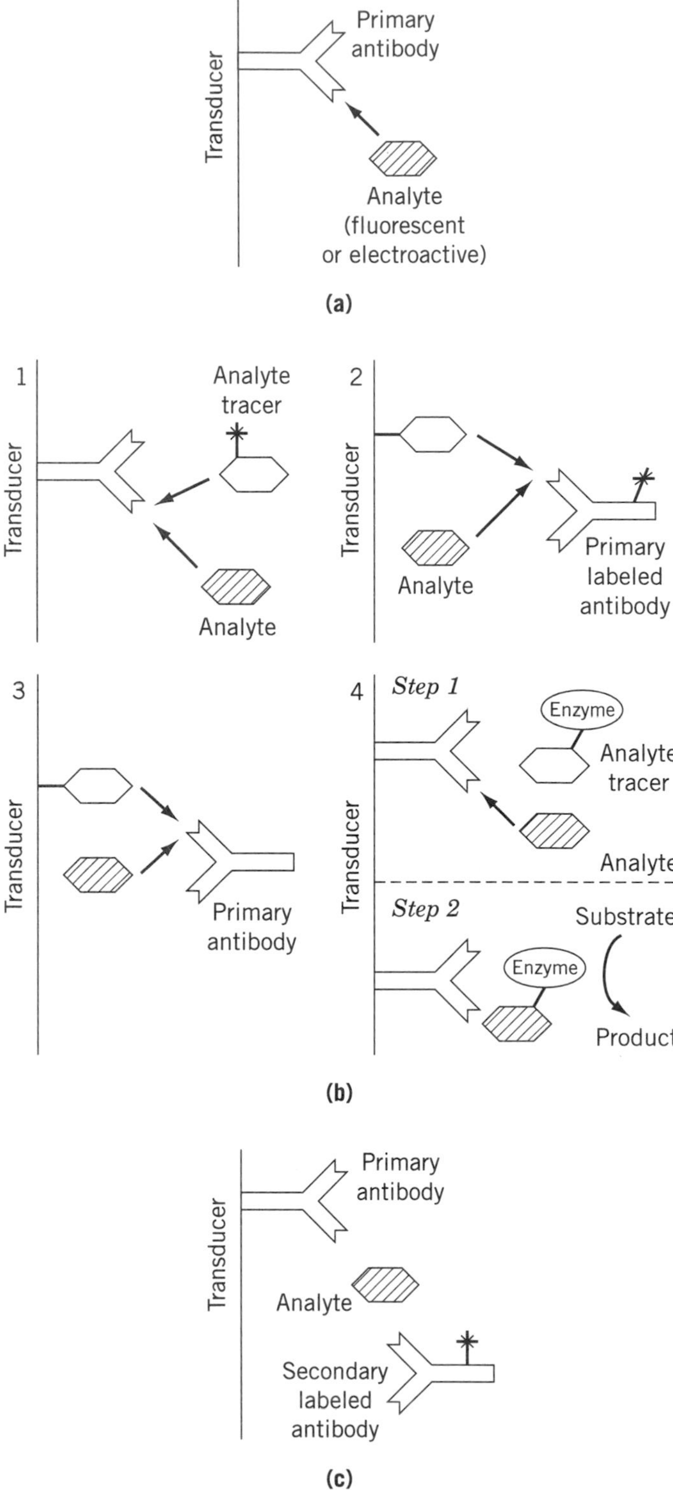

Figure 3. Principles of biosensor operation. Bioaffinity (antibody, receptor, and enzyme) mechanisms. Schematic representation of (**a**) direct nonlabel format, (**b**, 1) competitive format in which the analyte is labeled, (**b**, 2) competitive format where the primary antibody is labeled, (**b**, 3) competitive format for which the antigen is immobilized and the nonlabeled primary antibody binding is detected, (**b**, 4) competitive format where the label on the analyte is an enzyme, and (**c**) sandwich format requiring the use of two antibodies directed toward different portions of the antigen.

immunogen. Numerous cell lines are then screened for an antibody clone with the desired affinity characteristics. Antibodies show certain ranges of specificity in their ability to recognize specific or structurally related groups of compounds. In the case of monoclonal antibodies, individual clones show differing degrees of cross-reactivity and may be selected for applications that require the detection of a single compound or a group of structurally related compounds.

Although there are a wide variety of reported immunoassay formats, they can be grouped into several main categories (9). These immunoassay formats include radioimmunoassay (RIA), fluorescence immunoassay (FIA), and enzyme-linked immunosorbant assay (ELISA). These formats are typically competitive in nature and can be described by the following equation:

$$Ab + Ag + Ag^* \rightarrow AbAg + AbAg^*$$

where Ab is the antibody, Ag is the antigen (analyte of interest), and Ag^* is the analyte tracer. Various assays under this paradigm differ from each other, primarily on the basis of the tag used to label the analyte. For example, in RIA, the antigen is labeled with ^3H, ^{125}I, or ^{14}C; for FIA, the antigen is labeled with a fluorescent dye; and for ELISA, the antigen is labeled with an enzyme that amplifies the assay response for a given amount of analyte tracer bound to the antibody. This last format requires that the free tracer be separated from that which is bound to the antibody (ie, heterogenous assay format). In the case of antibody-based biosensors, the selectivity of the system for structurally related compounds is usually reported in terms of percent cross-reactivity, or EC_{50}. As is typically the case for competitive immunoassay formats, the EC_{50} is the concentration of analyte that causes a 50% reduction in the measured response. Other interferences for these assays may involve pH, ions, heavy metals, and organic solvents.

In many respects, antibody-based biosensor formats are similar to the previously mentioned immunoassay methods. The most frequently reported assay formats used for bioaffinity-based sensors include direct binding, competition, and sandwich assay formats. In the case of the direct binding format, a unique optical or electrochemical property of the analyte of interest is observed as the target compound binds to the recognition site of the bioaffinity element and accumulates at the sensor surface (see Fig. 3**a**). An example of this format is the fiber optic biosensor for the environmental pollutant benzo(*a*)pyrene. This compound is one of the few environmental contaminants that can be directly measured by an inherent optical property (ie, fluorescence). In this particular biosensor, the selectivity is provided by the antibody, the sensitivity of the assay is influenced by both the affinity of the antibody and the fluorescence properties of the analyte, and the application versatility is provided by a portable fiber optic fluorimeter.

Most analytes of environmental interest do not possess optical or electrochemical properties that can be uniquely distinguished by the sensor. Consequently, competition or sandwich assay formats are often used in conjunction with optical or electrochemical tracers. Because of the availability of antibodies for low molecular weight environmental contaminants such as pesticides, antibody-based biosensors using competitive assay formats have been frequently reported. There are several types of formats typically used in biosensor competition assays. In one format, the antibody is immobilized on the sensor surface, and the analyte and the labeled analyte-probe compete for available binding sites (see Fig. 3**b**, 1). In another, the antigen is immobilized on the sensor and the free analyte (in solution) competes with the analyte immobilized on the sensor surface for binding sites on the antibody (see Fig. 3**b**, 2). In both cases, the sensor signal is inversely proportional to the analyte concentration.

There are also several signal transducers that do not require the use of optical or electrochemical tracers. These methods, however, typically require the use of an immobilized antigen format (see Fig. 3**b**, 3) so that binding of the high molecular weight antibody can be observed.

There have also been a number of reported biosensor-type assays that require multi-step enzyme tracer formats (see Fig. 3**b**, 4). In these formats, the analyte and the enzyme-labeled analyte-tracer compete for a limited number of binding sites on the immobilized antibodies. After separation of the unbound antibody-tracer, enzyme substrate is added and the product is optically or electrochemically detected. Although a significant amplification can be realized by the use of assay formats in which enzymes are used as the analyte of antibody tracers, the primary disadvantages include the requirement for removing unbound tracer, the addition of substrate, and the incubation time required for enzyme-catalyzed product formation.

Sandwich assay formats are typically used for measuring large molecular weight molecules (ie, >10, 000 Da). For this assay format, the analyte must have two separate recognition sites. The analyte is then "sandwiched" between an antibody that is immobilized to the sensor surface and a second tracer antibody (labeled with an enzyme or fluorescent tracer) directed toward a second site on the analyte of interest (see Fig. 3**c**). Most environmental chemical pollutants are too small to have multiple bioaffinity recognition sites; however, this format has been used to detect microorganisms such as pathogens found in pre-processed drinking water or bacteria used for bioremediation.

Given the wide range of bioaffinity-type biological recognition elements and the versatility of how these may be configured into biosensors, it is important to address several issues relevant to applying these devices to the range of analytes, matrices, and potential uses required for environmental monitoring. These issues include reversible versus irreversible formats, direct binding versus competitive formats, homogeneous versus heterogeneous assay formats, potential continuous and *in situ* assay formats, and potential dynamic operating range. For example, reversible or regenerable bioaffinity-based formats (eg, immunosensors) would be of considerable value for continuous assay formats such as flow injection analysis, as detectors for chromatographic systems, or as continuous sentinels. Another issue involves homogeneous versus heterogeneous assay formats. Homogeneous formats are inherently simpler than heterogeneous but are not of much advantage for use in flow systems where the separation of bound and free analyte tracer can be easily accomplished.

For remote and *in situ* capabilities with respect to potential environmental applications, there are two strategies. One strategy is to place the biosensor in the sample and bring the signal back to the monitoring instrument. The other is to bring the sample to a biosensor located in or near the instrument. Although the strategy involving the remotely placed sensor would be better for some potential monitoring applications, it would require reversible, homogeneous, and reagentless operation.

Because of the wide range of pollutant concentrations encountered in the environment, it is, in some cases, advantageous for a field analytical method to show a wide dynamic operating range. This is not always possible, however, because biological recognition elements impose certain constraints in terms of optimal operating conditions. Several approaches in biosensor development have been reported to address the issue of dynamic operating range for potential environmental applications. For example, in enzyme-or microbial-based biosensors, access of the substrate to the biological recognition element can be controlled through the use of permeability-selective membranes or coatings. Additionally, this approach can be used to exclude co-contaminating interferences. Another approach that has been used to control assay characteristics in bioaffinity (competition) formats involves the use of rational synthetic design of the analyte tracer. One of the advantages of biosensors for potential environmental applications is the ability to select, modify, and design the molecular recognition elements used in these devices.

Although the most common affinity-based biosensor and biosensor-related formats have been outlined here, it should be noted that numerous formats that are potentially adaptable to biosensors (10) (eg, liposome-based assays or micro-particle-based assays) are beyond the scope of this chapter and are not discussed.

Microbial-Based Biosensors. The use of microorganisms to monitor biological oxygen demand, toxicity, and bioavailability of environmental pollutants as well as to detect specific compounds has been extensively investigated (11). Compound-specific or general toxicant-induced responses have been measured using a variety of methods including changes in doubling times, general respiratory indicators (eg, pH changes or oxygen consumption), and bioluminescence. Furthermore, development of these microorganism-based toxicity and genotoxicity assays has resulted in several commercially available products. Several microbial-based biosensors have also been reported. The most mature microbial biosensors are based on the measurement of respiration. Microbial respiration may be detected electrochemically from the consumption of oxygen, using a Clark electrode, or from the extrusion of protons using a pH electrode (Fig. 4**a**).

Because of improvements in assay speed, simplicity, and format versatility over existing microbial assays, the use of microorganisms as biological recognition elements in biosensors is being actively investigated by a number of research groups. For example, BOD represents the total amount of biodegradable organic material measured in terms of oxygen consumption by microorganisms and represents an important indicator of organic materials in waste water. Several microbial-based biosensors are commercially available and have been used to measure BOD in support of a Japanese industrial regulation for secondary effluent from waste water treatment plants. BOD-5, the classical standard method, measures the total oxygen consumed by an inoculation of native organisms after 5 days (12). Because the BOD biosensor, which

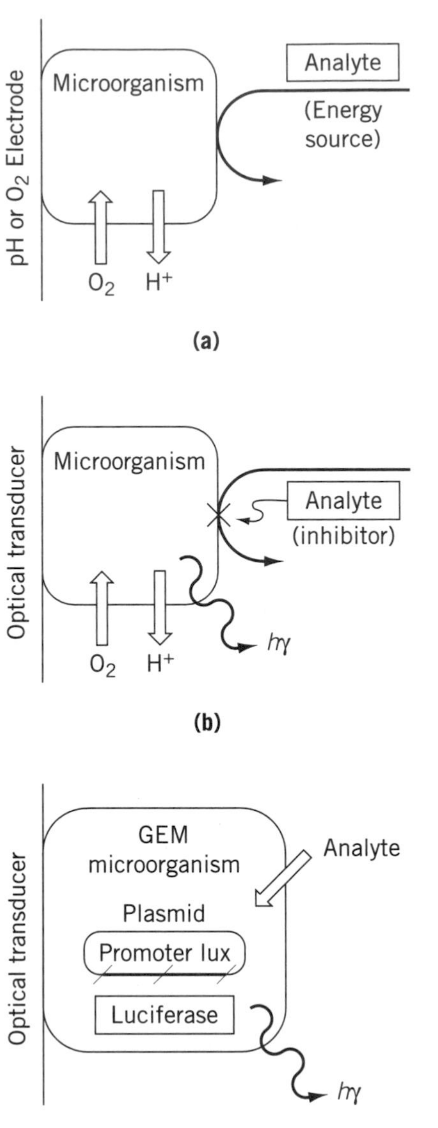

Figure 4. Principles of biosensor operation. Microbial (bacteria) mechanisms. Schematic representation of (**a**) detection of analytes that serve as energy sources using cellular respiration, (**b**) detection of compounds that inhibit cellular respiration using bioluminescence as an indicator, and (**c**) detection of specific compounds that interact with gene expression (resulting in changes in bioluminescence) in GEMs.

uses yeasts immobilized to an oxygen electrode, can measure BOD in several minutes rather than 5 days, it can be used in a feedback monitoring system to improve the process control of waste water treatment.

Both naturally occurring microorganisms and GEMs have been used in microorganism-based biosensors. For biosensors that use these organisms, respiration may be measured optically or electrochemically. The analyte of interest may be an energy source or a respiratory inhibitor. For example, the respiration of luminescent bacteria such as *Photobacterium phosphoreum* can be optically monitored (13). Because bioluminescence involves electron transfer processes, which are an integral part of the organism's metabolism, compounds that adversely affect this respiration can be measured as a decrease in luminescence (Fig. 4**b**). These biosensor formats have been used to measure respiratory substrates such as ammonia and benzene, as well as respiratory inhibitors such as chromium(VI).

Biosensors using GEMs that recognize specific environmental pollutants or groups of pollutants, such as heavy metals found in fly ash, have also been reported (14). The microorganisms used in these biosensors are typically produced by constructing a plasmid in which genes that code for luciferase or β-galactosidase are placed under the control of a promoter that recognizes the analyte of interest (Fig. 4**c**). For some of these GEM-based biosensors, the microorganisms protect themselves by metabolizing the analytes of interest, such as in the case of benzene, or by chemical reduction as in the case of mercuric ion (15). In other cases, these organisms initiate transport systems to remove the toxic substance or chelate it into a biologically inactive form. Because the organism's biological recognition system is linked to the reporter system, the presence of analyte results in the synthesis of the inducible enzymes, which then catalyze reactions resulting in detectable products. A number of these systems have been engineered to detect environmental pollutants such as benzene, Hb, Ni, Co, Sb, As, Cr, Cu, Pb, Ni, Zn, PCBs, and naphthalene.

Signal Transduction

Signal transducers that have been reported for biosensors include electrochemical, thermal, optical, optical-electronic, and acoustic. For environmental applications, most reports involve electrochemical, optical, optical-electronic, or acoustic devices. Theoretically, enzymes, receptors, antibodies, DNA, and microorganisms may be interfaced to any of these classes of transducers, provided an appropriate reaction product or tracer can be devised and measured. Successful biosensors for environmental applications, however, will require simple, rugged, sensitive, and inexpensive configurations which lend themselves easily to miniaturization and mass production.

ENVIRONMENTAL MONITORING

The ability to define and manage risk to humans and ecosystems resulting from chemical contamination of the environment depends in large part on our ability to monitor the air, water, and terrestrial zones for compounds of concern. In many cases, however, the ability to measure these environmental pollutants is limited by the time and expense required by currently employed analytical methods. Consequently, the development of rapid, cost-effective, and field portable methods may play a key role in the improvement of the risk assessment process. For field analytical methods (such as biosensors) to have a significant impact on this monitoring task, however, they must address the following challenges associated with environmental applications:

- pollutant diversity
- matrix diversity
- variety of data quality requirements
- environmental monitoring applications

Pollutant Diversity

One of the obstacles in the development of biosensors for environmental applications is the large number of pollutants that must be monitored and the broad range of chemical classes from which they arise. For example, over 700 chemical species have been identified at hazardous waste sites, and unidentified compounds may number in the thousands. Numerous agricultural and industrial compounds are regulated under waste disposal and treatment regulations as well as under the Toxics Release Inventory (TRI). Because these chemical species pose differing risks to human health and ecosystems, the Agency for Toxic Substances and Disease Registry (ATSDR) has ranked the most common priority hazardous substances (16). This ranking is based on the frequency of occurrence at sites present on the National Priorities List, available toxicity data, and the potential for human exposure. The ATSDR has further grouped these compounds by chemical class with respect to overall public concern (Fig. 5). Biosensors have been reported that detect compounds in several classes of concern including phenols/phenoxy compounds, polyaromatic compounds, halogenated pesticides, volatile organic compounds, and inorganic substances.

Another major challenge in the design, development, and commercialization of field analytical methods such as biosensors with respect to compound diversity involves the development and production of product lines that are adaptable to multi-analyte environmental monitoring scenarios. Several approaches have been used to increase the versatility and adaptability of biosensors. For example, in an assay system that shows potential for biosensor development, the use of a suite of enzymes (each of which responds differently to specific compounds) has been used to identify up to five of the most commonly encountered environmental pollutants (17). In addition, the range of target pollutants that can be analyzed by enzyme-array type biosensors may be expanded through neural network methods of data processing. Another approach, which has been investigated to increase the versatility of biosensors with respect to environmental pollutants, is the use of a base sensor platform with interchangeable immunoassay cartridges. This would allow the rapid and cost-effective development of new immunosensor-based assays for compounds of environmental concern.

Environmental Matrix Diversity

The range of contaminated matrices also contributes to the overall environmental monitoring challenge faced by field analytical methods. Typical examples of environmental matrices include air, drinking water, ground water, soil (of many types and characteristics), and sludge. The matrix problem for environmental monitoring is made more difficult by the complex nature, the heterogeneous distributions, the wide concentration range of target pollutants, and the presence of co-contaminants within these matrices. For example, soil samples often contain humic substances and nitrates, while Superfund sites are often contaminated with mixtures of heavy metals, chlorinated hydrocarbons, and reactive organics. Such mixtures may pose a particular challenge to biosensors if elements of their biological recognition systems (eg, enzymes, antibodies, or microorganisms) are inactivated by these co-contaminants.

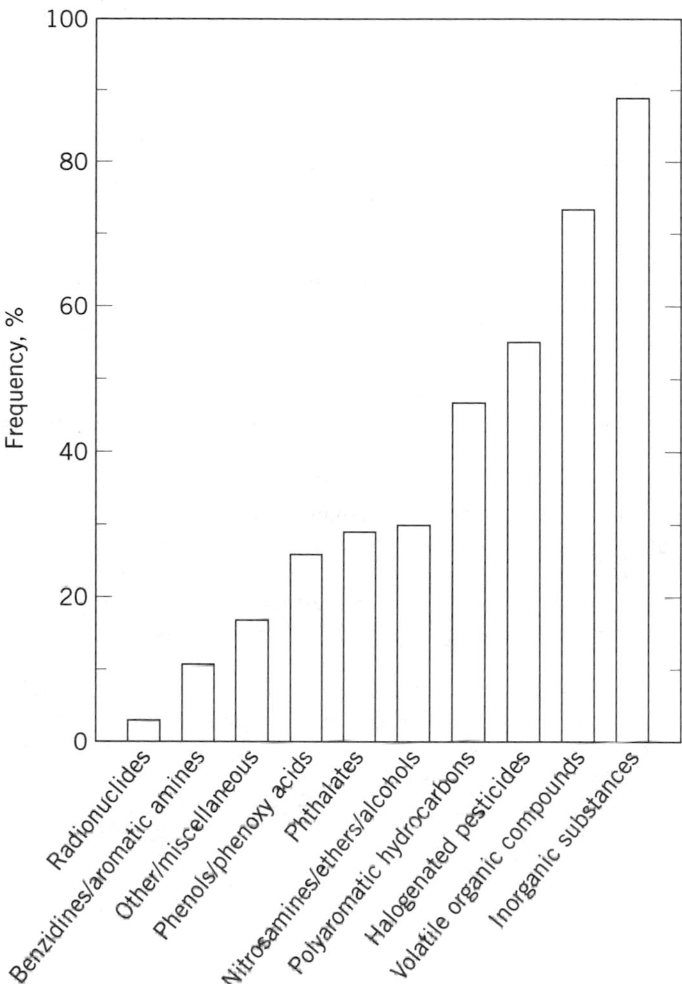

Figure 5. Contaminants of concern most often identified in public health assessments and advisories (3).

Biological recognition elements classically have been isolated and assayed in aqueous media of controlled pH and ionic strength. It has been shown in a number of cases, however, that enzymes and antibodies can operate in a variety of nonaqueous media. This is particularly relevant to the detection of environmental pollutants in the variety of matrices in which they are located. One interesting approach to biosensor development recently was reported for measuring formaldehyde vapor (18). This enzyme electrode employs formaldehyde dehydrogenase which catalyzes the oxidation of formaldehyde. Gas exchange is facilitated and enzyme electrode desiccation prevented by using a series of gas- and liquid-permeable membranes.

A major challenge in the development of biosensors for environmental applications involves interfacing these devices with relevant matrices (eg, measuring pollutants in matrices that range from drinking water to sewage sludge). For laboratory-based analytical methods, the problem of diverse environmental matrices is typically addressed by using a variety of complex and time-consuming extraction techniques. For a number of field analytical methods, substantial progress has been made to simplify these extraction methods (eg, the use of single solvent short-duration agitation with metal or glass beads followed by a filtration step). In addition, supercritical fluid extraction (SFE) methods pioneered for environmental applications in the laboratory also show the potential for miniaturization for field use. These methods essentially rely on the use of CO_2 maintained at temperatures and pressures that confine it to the super-critical state and allow it to extract both nonpolar and polar environmental contaminants from matrices such as soils. Because the solvent (ie, CO_2) is easily removed by bringing the sample back to standard temperature and pressure,this method shows a number of advantages over organic solvent extraction methods.

Although these extraction methods are relatively fast and efficient, they still require multi-step procedures using, in many cases, organic solvents and multiple dilutions prior to the assay. The requirement for extraction procedures limits many of the potential advantages (eg, single-step flow injection or continuous formats) that might be realized using biosensors. Application of technological advances in the clinical and diagnostic areas may, however, help to solve some of these problems inherent in environmental applications. For example, microdialysis techniques, which have been used to extract μL quantities of subcutaneous fluids for glucose analysis, could conceivably be used to extract liquids from saturated soils. Although many biosensors show the ability to operate in environmental matrices such as drinking water or groundwater, significant challenges remain to apply these methods to more difficult matrices commonly encountered in environmental settings (eg, surface water, sludge, and saturated soils).

Data Quality Requirements

Another challenge facing environmental regulators, contractors, and methods developers is compliance with data quality requirements specified by environmental regulations. The data quality objectives for sampling and analysis should theoretically be derived from the environmental management decisions to be made on the basis of the expected data. In practice, however, the data quality requirements, for many projects, are prescribed under environmental regulations or dictated by the performance characteristics of currently used laboratory-based methods. One of the strategies in adopting field analytical methods for environmental monitoring is their use for field screening followed by laboratory conformation for positive samples. A number of field analytical methods (particularly bioanalytical methods) have been configured to yield a yes-or-no response to contamination at a specific threshold or action concentration. As a result, these assays have been widely used in screening applications. Nevertheless, cost-effective field analytical methods, such as biosensors that are configured to report analyte concentrations over a dynamic range, may also be of considerable value even though the error rates may be higher than for standard laboratory methods. In modeling studies of cost-effective sampling design where specific variables (sample size, sampling patterns, and analysis error) were examined for their effect on cost and uncertainty, the size of the sample set was more significant than either sampling pattern or sample error. These modeling studies suggested that, for the same overall cost, the use of less expensive methods to generate a greater number of analyses (even with an increased error per sample), should yield a more accurate overall site characterization (19).

Environmental Monitoring Applications Diversity

Field Screening and Monitoring. Environmental monitoring covers a diverse group of potential applications. These applications range from the classical regulation-driven characterization of environmental matrices for listed compounds through emerging areas such as biological screening for endocrine disrupter function. With such a broad array of potential applications, biosensor technology, which has been developed primarily for the clinical and diagnostic areas, might be expected to find a variety of niches in the environmental area. Nevertheless, despite advancements in field analytical methods in general and in biosensors specifically, regulators, contractors, and developers are struggling to define their acceptable uses in environmental monitoring.

There are several issues that must be considered when deciding whether or not to use a field analytical method for an environmental monitoring project. Field analytical methods that rely on bioanalytical mechanisms (such as immunoassay kits) typically have been configured to give yes-or-no answers at a threshold contaminant level for use as screening tests. These tests should accurately differentiate samples with concentrations of a specific analyte or defined analyte class above or below a given level of concern, should have a known and stable sensitivity, and should be relatively free from (or well characterized with respect to) positive and negative interferences. Analytical tasks for which *screening* technologies are best suited include the spatial characterization of specific environmental contaminants (eg, identification of hot spots in contaminated soils or delineation of contaminant plumes). Although positive samples (ie, those containing the analyte of interest at a concentration above the threshold value) are routinely confirmed by laboratory analysis, considerable analytical cost savings can be realized because only a small portion of the negative samples are re-run in the laboratory.

The use of screening methods in conjunction with laboratory confirmation appears to be compatible with the analytical requirements of current environmental regulations. As increasingly versatile analytical techniques continue to advance, however, additional options for environmental monitoring are expected to be demonstrated. As a consequence of this increasing body of information, it is also anticipated that regulations will be adapted to reflect these new analytical options.

One of the format capabilities for which certain biosensors may prove useful is the field monitoring assay. These monitoring assays (as compared to field screening assays which are typically designed for operator-assisted discrete sampling) are expected to be capable of continuous and *in situ* operation under defined conditions for a given period of time. Tasks for which field *monitoring* technologies might best be suited include the temporal characterization of specific environmental contaminants that may change on a short time scale (eg, min) or *in situ* monitoring where sample removal is inconvenient, difficult, or dangerous. Specific examples might include *in situ* monitoring of a process control stream to determine the efficiency of hazardous waste site remediation; monitoring agricultural runoff during peak application periods; or continuous monitoring of wells to determine whether concentrations of an analyte of interest (during remediation procedures or after site closure) are in compliance. Although the continuous monitoring format does not, at present, appear to be as compatible with environmental monitoring regulations as screening formats, a number of chemical sensors are routinely used for continuous monitoring of water quality parameters such as temperature, pH, clarity, and salinity.

Another potential dimension of environmental monitoring for which biosensors may provide solutions is in the area of monitoring biological parameters. These might include measurements of toxicity, BOD, bioavailability, and the presence of either harmful or useful microorganisms. Although methods for measuring these parameters have been established, biosensor methods in these areas have been shown to have some advantages in terms of time, expense, and simplicity over currently used methods. Among environmental monitoring applications, the determination of biological parameters may be most significant for preprocessed and finished product in waste water treatment processes.

ENVIRONMENTAL APPLICATIONS

Recently reported biosensors for potential environmental applications measure a fairly broad spectrum of environmental pollutants including pesticides, organic compounds, metals, and biological parameters (Table 1). Although these prototype biosensors have been demonstrated primarily using laboratory standards in buffer solutions, a number have been tested using matrices such as waste water, surface water, and mixed organics. In addition, several of these devices are undergoing field trials in environmental settings. Many of the compounds targeted by these biosensors reflect environmental pollutants of national concern. For example, a wide range of organophosphate and carbamate insecticides have been measured using cholinesterase-based biosensors. For herbicide detection, antibody-base biosensors that measure triazines, imadazolinones, and 2,4-D have been reported. Biosensors based on inhibition of Photo System II have been used to measure a wide array of herbicides, and biosensors using GEMs have been used to measure pesticides such as meturon and propanil.

A variety of organic compounds shown to contaminate Superfund sites throughout the United States can be detected by a number of reported biosensors. These include enzyme-based biosensors for detection of phenolics, organophosphates, and cyanide; antibody-based biosensors for PCBs, potent carcinogens such as benzo(a)pyrene, and explosives such as TNT and RDX; and microbial biosensors for toxicants such as benzene and ammonia.

GEMs have been incorporated into biosensors that are specific for particular heavy metals such as mercury and copper, or that respond nonspecifically to metals such as lead, cadmium, chromium, and manganese, as well as to other pollutants such as benzalkonium ions, laurel sulfate, and 3-chlorobenzene.

In addition to specific compounds, biosensors have been reported to measure a variety of biological parameters, including BOD, biomarkers of human exposure, potential carcinogens, bioremediation efficiency, and bacterial identification or

Table 1. Biosensors for Environmental Applications

General Compound Class	Biochemical Mechanism
Pesticides	
Insecticides	Antibody
Herbicides	GEM Photosystem II (inhibition)
Organic compounds	
Benzene	GEM
Benzo(a)pyrene	Antibody
Ammonia	GEM, enzyme (substrate)
Formaldehyde	Enzyme (substrate)
Trinitrotoluene (TNT) or Cyclonite (RDX)	Antibody
Polychlorinated Biphenyls (PCBs)	Antibody
Alcohols	Enzyme (substrate)
Organonitriles	Enzyme (substrate)
Phenolics	Enzyme (substrate)
Cyanide	Enzyme (inhibitor)
Metals	
Zn	Enzyme (cofactor)
Hg, Cu	GEM
Biological parameters	
Toxicity[a]	GEM, indigenous microorganisms
Biological oxygen	GEM, indigenous microorganisms
Demand (BOD)	Enzyme (substrate)
Phosphates	Enzyme (substrate)
Potential carcinogens	DNA (intercalation)
Bioremediation efficiency	GEM
Bacterial identification or enumeration	Antibody, DNA (sequence)

[a] Compounds that have been measured using these devices include benzene, mercury, nickel, cobalt, chromium, PCBs, naphthalene, and ammonia.

enumeration. The BOD biosensors have been developed and tested primarily in Japan and Europe. The short response time (ie, 15 min compared to 5 d for classical determinations), make these devices particularly useful in process control applications for waste water treatment in which rapid analyses are required.

Biomarkers of human exposure is another area currently gaining attention (see BIOMARKERS). Fiber optic antibody-based biosensor technology recently was developed for detection of DNA adducts such as benzo(a)pyrene tetrol. This compound is a potent carcinogen and has been shown to be a frequent co-contaminant with other polycyclic aromatic hydrocarbons (PAHs). Biosensor methods also have been developed to measure potential carcinogens through their ability to intercalate into DNA. This DNA-based fiber optic biosensor detects a wide range of potentially carcinogenic polyaromatic hydrocarbons.

Although bacteria are not typically considered to be environmental pollutants, there are a number of circumstances where field monitoring for these microorganisms is important and biosensor methods may prove to be a cost-effective alternative to classical methods. For example, when bioremediation is used for decontamination of toxic compounds in the environment, it is crucial (especially in the case of GEMs) that the fate and transport processes of these organisms is well defined. Biosensors demonstrated to detect organisms of clinical interest potentially could be developed for environmental applications.

FUTURE OPPORTUNITIES

Biosensors can be thought of as essentially biological or biochemical assays that have been interfaced to signal transducers to yield fast, simple, and cost-effective analytical methods. Because biosensor characteristics such as sensitivity and selectivity are, to a great extent, determined by the biological recognition element, many of the major advances in biosensor technology are expected to occur in this area. The most active of the current biosensor research areas are the development of new biological recognition elements (eg, the application of techniques developed through molecular biology and genetic engineering, use of enzyme arrays, development of new antibodies, and construction of new GEMs); environmental interfaces (eg, the use of membranes and coatings that maintain biological activity, remove interferants, and buffer analyte concentrations); and signal transducers (eg, development of smaller and more easily manufactured devices). One area that shows considerable promise for biosensor development is in molecular biology. Due to their versatility, sensitivity, afd speed, biosensors may provide the ability to quickly measure sequence-specific DNA hybridization (20), oxidation-reduction properties, and the binding and intercalation of genotoxicants.

Although the development of biosensors for potential environmental applications appears to be an active and growing area of research, commercial development and widespread use of biosensors for environmental applications will require progress in areas such as:

- improved performance characteristics (eg, sensitivity, dynamic range, reproducibility, and stability) under field conditions
- versatile environmental interfaces (eg, multimatrix)
- simple sensor design and operating formats
- field demonstrations
- regulatory acceptance

Developing biosensors for environmental applications is not a trivial task; however, there appears to be sufficient evidence that biosensors can be configured to be selective, sensitive, and inexpensive to manufacture. It is likely that with appropriate development, testing, and commercialization, biosensors can have a significant impact on reducing the cost and increasing the efficiency of specific environmental monitoring applications.

ACKNOWLEDGMENT

The U.S. Environmental Protection Agency (EPA), through its office of research and development, funded the work involved in preparing this article. It has been subject to the EPA's peer and administrative review and has been approved for publication. Mention of trade names or commercial products does not constitute endorsement or recommendation by the EPA for use. This document contains copyrighted material (3) in Figures 1 and 5, reprinted with kind permission of Elsevier Science-NL, S. Burgerhartataat 25, 1055 KV Amsterdam, the Netherlands.

BIBLIOGRAPHY

1. K. R. Rogers and C. L. Gerlach, *Environ. Sci. Technol.* **30,** 486A–491A (1996).
2. P. M. Kramer, *J. AOAC, Int.* **79,** 1245–1254 (1996).
3. K. R. Rogers and L. R. Williams, *Trends in Analytical Chemistry,* **14,** 289–294 (1995).
4. J. Janata and co-workers, *Anal. Chem.* **66,** 207R–228R (1994).
5. A. Sharma and K. R. Rogers, *J. Meas. Sci. Technol.* 1994 Book Series **5,** 461–472 (1994).
6. A. Sharma and K. R. Rogers, *Food Technol. Biotechnol.* **34,** 113–123 (1996).
7. R. F. Taylor, in R. F. Taylor and J. S. Schultz, eds., *Handbook of Chemical and Biological Sensors,* Institute of Physics Publishing, Bristol, 1996, pp. 553–577.
8. G. Bardeletti and co-workers, in L. J. Blum, and P. R. Coulet, eds., *Biosensor Principles and Applications,* Marcel Dekker, Inc., New York, 1991, pp. 7–46.
9. P. Tijssen, *Practice and Theory of Enzyme Immunoassays,* Elsevier, New York, 1987.
10. S. G. Reeves and co-workers, *Trends Anal. Chem.* **14,** 351–355 (1995).
11. I. Karube and co-workers, *Trends Anal. Chem.* **14,** 295–299 (1995).
12. H. Tanaka and co-workers, *Wat. Sci. Tech.* **30,** 215–227 (1994).
13. S. Lee and co-workers, *Biosens. Bioelectron.* **7,** 273–277 (1992).
14. P. Corbisier and co-workers, *Environ. Toxicol. Water Qual.* **11,** 171–177 (1996).
15. L. Tescione and G. Belfort, *Biotechnol. Bioeng.* **42,** 945–952 (1993).
16. Agency for Toxic Substances and Disease Registry, *FY 1993 Agency Profile and Annual Report,* U.S. Department of Health and Human Services, Atlanta, Ga., 1993.
17. D. C. Cowell and co-workers, *Biosens. Bioelectron.* **10,** 509–516 (1995).
18. M. Hammerle and co-workers, *Biosens. Bioelectron.* **11,** 239–246 (1996).
19. E. J. Englund and co-workers, *Mathemetical Geology* **22,** 417–455 (1992).
20. J. Wang and co-workers, *Anal. Chem.* **68,** 2629–2634 (1996).

BIOSOLIDS MANAGEMENT: USE AND DISPOSAL

JOHN M. WALKER
U.S. Environmental Protection Agency
Washington, D.C.

Human, animal, and plant life are invariably accompanied by wastes (see WASTE GENERATION). These wastes are part of the life cycle and are recycled back to the earth. New animal and plant life result from the nutrients contained in these wastes. This recycling process, if done naturally and wisely, supports life on earth (Fig. 1). The challenge is to recycle naturally occurring waste materials so that life is sustainable.

GENERATION

During the Stone Age, life was hard but brutally simple. From a human perspective, virtually all time was spent gathering

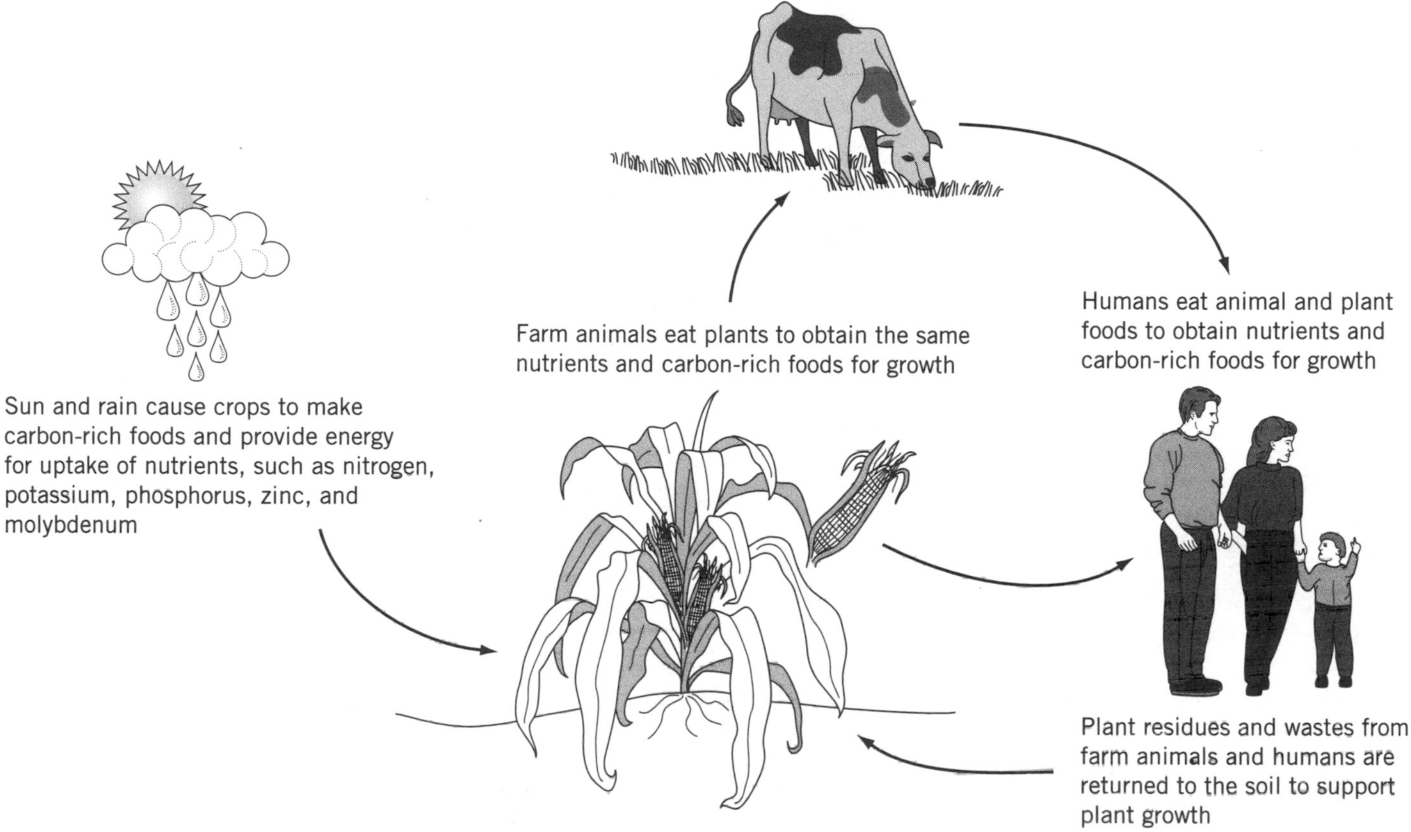

Figure 1. Natural cycling of nutrients (1).

food and water and trying to avoid becoming the food of a larger predator. A place to live away from the elements was often in hollowed out spaces in the ground or rock. At some point in the evolutionary process humans learned that naturally accumulating human wastes enriched the soil to produce additional food. Manually collected liquid and solid human excreta, often collected at night, have been used for centuries even in some places today, where it is called "night soil."

As populations became more concentrated, so did the potential for disease, odor, and ground and surface water pollution from improperly collected and treated human waste. The role of human waste in disease and other environmental problems became more apparent. Simultaneously, human tolerance of malodorous systems for managing wastes lessened. With the advent of indoor plumbing and water closets and underground sewers for collecting and discharging sewage, chamber pots and outhouses were replaced. During the first half of the twentieth century, the waters from sewers carrying these human wastes were often discharged directly into water bodies without treatment. However, this practice was damaging to the quality of the receiving waters.

Humans began to separate solid materials from the sewage, first by gravity sedimentation. The solids separated out by gravity are called primary wastewater solids. More complex commercial and industrial wastes were also discharged into the sanitary sewers. Primary treatment was not enough to clean up the wastewater, and the effluents discharged into water bodies after primary treatment were still damaging to streams and other water bodies. Additional or secondary treatment of the primary wastewater effluent came into practice. An early form of secondary treatment was the use of trickling filters. A trickling filter consists of a bed of rocks onto which effluent water from primary treatment of wastewater is sprayed. The trickling filter provides a surface on which aerobic microorganisms grow and further decompose the solids remaining in the wastewater. Microbial growth converts entrained wastewater solids into microbial biomass instead of fecal material and pathogenic organisms. The microbial biomass also transforms the previously entrained fecal solids into a form more readily separated from the wastewater. The microbial biomass is also very competitive with and less likely to support the growth of fecal coliform and disease causing organisms, such as the ubiquitous *Salmonella* bacteria that survive and multiply outside of the human body. The resultant solids are called secondary wastewater solids. A more advanced form of secondary treatment is known as activated sludge treatment in which large quantities of air are forced through the wastewater effluent after primary solids are removed. The air supports a highly active mass of aerobic microorganisms that transform the entrained wastewater solids into microbial residues called biomass.

In some countries, more advanced treatment of the wastewater is being achieved. The effluent from primary and secondary wastewater treatment is further subjected to chemical and biological treatment, called tertiary treatment. The additional treatment removes more pollutants and nutrients, so that, when the effluent is discharged into streams, the receiving waters can absorb the remaining entrained pollutants

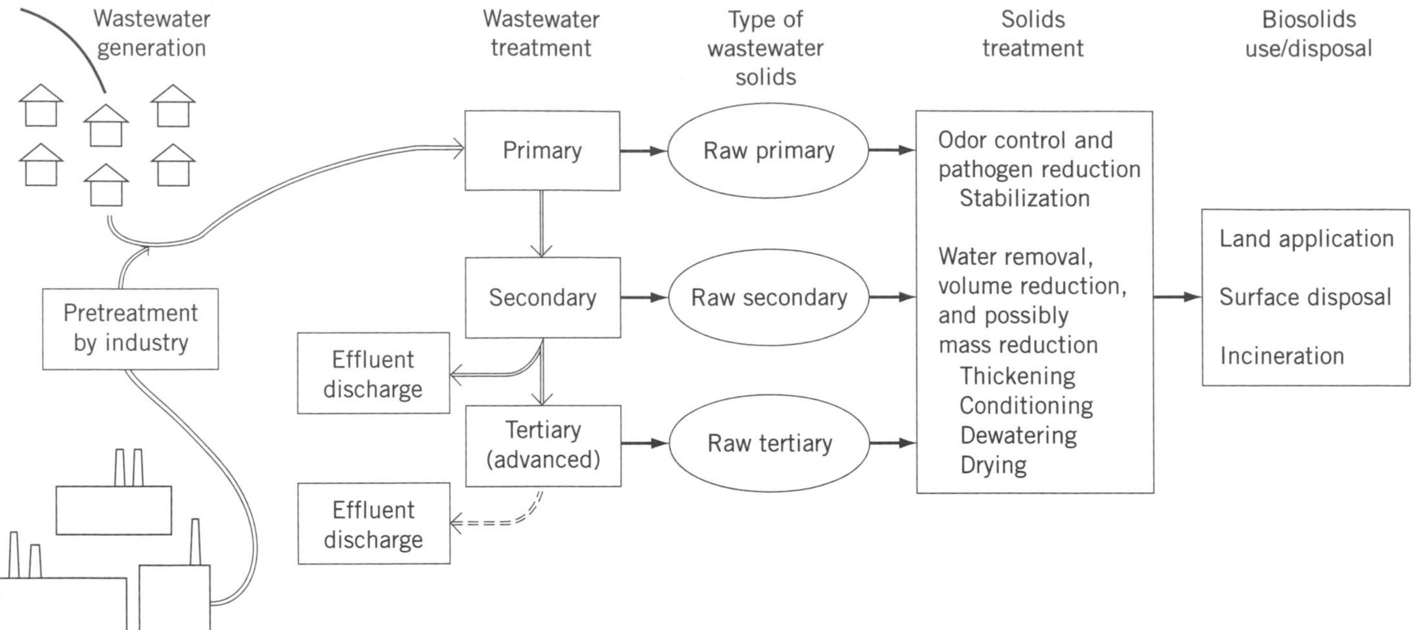

Figure 2. Generation, treatment, and disposal of municipal wastewater solids, where (⇒) represents wastewater and (→), biosolids (3).

and nutrients. Solids separated from this process are called tertiary wastewater solids and are combined with the primary and secondary wastewater solids for treatment and use or disposal. Figure 2 is a schematic diagram of a centralized wastewater treatment system.

In the years prior to the discharge of large amounts of industrial waste into sewage collection and treatment systems, domestic wastewater and its by-products were land-applied as advocated by Sir Edwin Chadwick in the mid-1800s with the slogan "rain to the river and sewage to the soil." In some cases by-products were processed to improve hygienic safety and sensory acceptability and to reduce the volume and weight. The application to land was acceptable raising little public concern. Disease was generally not a problem because of biosolids treatment prior to the land application or because the pathogens died in the soil after land application prior to consumption of the crops grown in those soils (4).

The environmental awakening of the 1960s and early 1970s generated new issues over the management of wastes. The images created by toxic industrial wastes buried in places such as Love Canal (New York) and the horror stories associated with those situations were transferred to all wastes including the solid residuals from wastewater treatment which in the past were called sewage sludges. The term sludge became associated with toxic industrial wastes and oily residues like oil sludge drained from an automobile crankcase. At the same time, as a result of industries discharging many toxic chemicals into the sewer systems, concern over properly managed municipal wastewater solids also grew. Testing of some of these wastewater solids showed high levels of metals. People were warned not to use any wastewater solids. What had been successful for many years was now suddenly a health concern for the use of all solids from wastewater treatment (4).

Agencies like the U.S. Department of Agriculture, Land Grant universities, and the U.S. Environmental Protection Agency launched extensive studies to learn more about the proper treatment and safe use and disposal of solids from wastewater treatment. Research concerning the use of wastewater and wastewater solids on land increased dramatically in the 1970s and 1980s. The need to reduce the level of certain pollutants, especially heavy metals, in these wastewater solids, became increasingly apparent from this research. Thus a program of industrial pretreatment began to limit the release of pollutants into wastewaters prior to their discharge into sanitary sewers. Over a period of about 20 years, this program of industrial pretreatment, along with source control, resulted in drastically reduced levels of metals and other pollutants going into treatment plants and hence into wastewater solids.

TREATMENT OF SEPARATED WASTEWATER RESIDUALS INTO BIOSOLIDS

Biosolids are produced by municipal wastewater treatment and, hence, contain organic and inorganic solids and soluble substances entrained or dissolved in the incoming wastewater or added or generated during wastewater treatment (3,6). Several processes used to treat biosolids before utilization or disposal include thickening, stabilization, conditioning, and dewatering. The treatment process greatly affects how the product is used.

CONSTITUENTS OF BIOSOLIDS

The characteristics and volumes of biosolids depend on the initial wastewater composition and the subsequent wastewater treatment processes. Pollutant levels in biosolids from a given wastewater treatment works vary less over time than levels in the incoming raw wastewater because of the considerable mixing and relatively long retention time prior to ultimate use or disposal.

Organic Content

The organic content (3) in biosolids is expressed as the percent of total solids that are volatile. The organic or volatile solids content as used in this context is defined as the naturally occurring carbonaceous materials of which plants and animals are comprised and not synthetically produced compounds that may also be potentially toxic. Some unstabilized biosolids contain as much as 75–80% volatile solids on a dry weight basis but more often the volatile solids content is not more than about 50% after biological treatment. Organic content is important in determining the amount of auxiliary fuel needed for incineration, the potential for odor problems during storage and land application, value as a soil conditioner, and the potential for methane gas generation during anaerobic digestion.

Pathogens

A significant proportion of bacteria, viruses, protozoa, and the eggs of parasitic worms in wastewater concentrate in biosolids during treatment of the wastewater. A small percentage of these organisms are pathogenic to humans (cause human disease) (3,5,7). Pathogens of concern are shown in Table 1. Pathogen abundance is difficult to measure directly, but pathogen reduction is estimated from the reduced concentration of organisms, such as total fecal coliform or *Salmonella* sp. Treatment to reduce pathogens to a safe level is based on knowledge of exposure routes and of environmental and other treatment mechanisms which restrict exposure and/or cause their reduction.

Macroinorganic Nutrients

Biosolids contain macronutrients (nutrients required in relatively large quantities), such as nitrogen, phosphorus, potassium, sulfur, calcium, and magnesium (1,3). Biosolids are applied at a rate that supplies the agronomic need for the nutrients nitrogen or phosphorus. Other sources of fertilizer are required to supply enough of the crop's needs for potassium.

Microinorganic Nutrients Including Metals

Biosolids contain varying amounts of heavy metals, such as cadmium, copper, lead, nickel, mercury, and zinc (1,2,5,7–12). Some of these elements are essential for plant, animal, and human nutrition and are referred to as micronutrients, that is, they are required for plant and animal life in relative small quantities (1). Ranges and median concentrations of trace metals in digested biosolids are shown in Table 2. Some of these metals are added to commercial fertilizers to provide a more complete nutrient source. If biosolids are utilized that contain these metals in sufficient but not excessive quantity, plants and animals benefit, and these elements then do not have to be added commercially. Further discussions about properties, usefulness, and need for control are discussed later.

Toxic Organic Compounds

Biosolids contain synthetic organic chemicals from industrial wastes, household chemicals, and pesticides (10,11,12–15). Most biosolids contain low levels of these substances and do not pose a human or environmental health problem. For these reasons, no toxic organic compounds are currently regulated by the EPA's Part 503 biosolids rule. Further discussions about reasons for this approach to toxic organic compounds appear later.

Chemical Form

A nutrient may be present in biosolids in one of several chemical forms (1,3,12,15). For example, nitrogen occurs as organic nitrogen and/or as inorganic ammonium and/or nitrate, and phosphorus occurs as inorganic phosphate or as organically bound phosphate. Both the form and the concentration of nitrogen and phosphorus, and the many other nutrients and pollutants in biosolids affect the fertilizer value or potential to cause harm if a pollutant is present in too great a concentration, is too highly mobile, and/or is too readily bioavailable.

Consider the impact of chemical form on the bioavailability of lead as an example of the potential for risk to human health. Lead contained in high quality crystal drinking glasses (>50% lead) poses no real risk to children who drink from them because the lead in such crystal is tightly bound and not bioavailable. Similarly, lead in biosolids, that does not exceed levels permitted by the federal Part 503 biosolids rule, poses virtually no risk if children ingest the biosolids themselves or even the soils to which the biosolids had been applied because the lead in biosolids is tightly bound in a stable nonbioavailable form by oxides of iron, aluminum and manganese and this inorganic matrix persists whether or not the biosolids are mixed into soils and whether or not the organic part of the biosolids degrades with time after being land applied. On the other hand, lead is not tightly bound in soils from homes painted with lead-containing paint or with residues from past automotive exhausts. Lead in soils from these two sources is much more bioavailable and poses a risk to health especially when ingested by children.

Biosolids Use and Disposal

The characteristics of biosolids greatly affect their suitability for various use and disposal (16).

CONTROLLING BIOSOLIDS USE AND DISPOSAL WITH APPLICABLE FEDERAL REGULATORY REQUIREMENTS

Need for Control

Appropriate control is needed for safe agricultural use of all fertilizers and soil conditioners, whether in the form of biosolids, other organic amendments, or chemical fertilizers, to ensure that the proper amounts of essential elements are provided. Controls are also needed with all fertilizers and soil conditioners to avoid contaminating groundwater with excess leachable nitrogen. Controls are needed with biosolids and animal wastes, because, depending on the level of treatment, disease-causing organisms (pathogens) may be present and vectors, such as flies and rodents, are attracted that transmit disease. These controls come from many sources. Some control comes from following state fertilizer recommendations and sound agricultural practices. Additional control is obtained by wastewater treatment required to reduce pathogens to levels that are not harmful. Pretreatment by industry, mandated by law, is another primary control that prevents excessive

Table 1. Principal Pathogens in Municipal Wastewater and Biosolids[a]

Organism	Disease/Symptoms
Bacteria	
Salmonella sp	Salmonellosis (food poisoning), typhoid fever
Shigella sp	Bacillary dysentery
Yersinia sp	Acute gastroenteritis (including diarrhea, abdominal pain)
Vibrio cholerae	Cholera
Campylobacter jejuni	Gastroenteritis
Escherichia coli (pathogenic strains)	Gastroenteritis
Enteric Viruses	
Hepatitis A virus	Infectious hepatitis
Norwalk and Norwalk-like viruses	Epidemic gastroenteritis with severe diarrhea
Rotaviruses	Acute gastroenteritis with severe diarrhea
Enteroviruses	
Polioviruses	Poliomyelitis
Coxsackie viruses	Meningitis, pneumonia, hepatitis, fever, cold-like symptoms, etc
Echoviruses	Meningitis, paralysis, encephalitis, fever, cold-like symptoms, diarrhea, etc
Reovirus	Respiratory infections, gastroenteritis
Astroviruses	Epidemic gastroenteritis
Caliciviruses	Epidemic gastroenteritis
Protozoa	
Cryptosporidium	Gastroenteritis
Entamoeba histolytica	Acute enteritis
Giardia lamblia	Giardiasis (including diarrhea, abdominal cramps, weight loss)
Balantidium coli	Diarrhea and dysentery
Toxoplasma gondii	Toxoplasmosis
Helminth Worms	
Ascaris lumbricoides	Digestive and nutritional disturbances, abdominal pain, vomiting, restlessness
Ascaris suum	Coughing, chest pain, fever
Trichuris trichiura	Abdominal pain, diarrhea, anemia, weight loss
Toxocara canis	Fever, abdominal discomfort, muscle aches, neurological symptoms
Taenia saginata	Nervousness, insomnia, anorexia, abdominal pain, digestive disturbances
Taenia solium	Nervousness, insomnia, anorexia, abdominal pain, digestive disturbances
Necator americanus	Hookworm disease
Hymenolepis nana	Taeniasis

[a] Ref. 7.

levels of unwanted pollutants in wastewater and resultant biosolids. And finally, compliance with the new federal and existing state regulations requires the carefully implementing management practices and biosolids application rates based on crop needs.

The EPA Part 503 Biosolids Rule

The EPA developed a comprehensive, risk-based rule to protect public health and the environment from reasonably anticipated adverse effects of pollutants present in biosolids used or disposed of (5). The regulation (40 CFR Part 503), known as the Part 503 rule, was published in the *Federal Register* on February 19, 1993. Much of the rule was based on the results of risk assessments scientifically conducted to identify what, if any, risks were associated with the use or disposal of biosolids. Those parts of the rule not based on risk assessment were based on performance or technology standards or on management, monitoring, and record keeping practices which protect human health and the environment.

The EPA's scientific risk assessment methods and policy decisions were used to establish pollutant limits for biosolids that

Table 2. Ranges and Median Concentrations of Trace Elements in Dry Digested Biosolids[a]

	Reported Range, mg/kg		
Element	Minimum	Maximum	Median, mg/kg
As	1.1	230	10
Cd	1	3,410	10
Co	11.3	2,490	30
Cu	84	17,000	800
Cr	10	99,000	500
F	80	33,500	260
Fe	1000	154,000	17,000
Hg	0.6	56	6
Mn	32	9,870	260
Mo	0.1	214	4
Ni	2	5,300	80
Pb	13	26,000	500
Sn	2.6	329	14
Se	1.7	17.2	5
Zn	101	49,000	1,700

[a] Ref. 9.

protect *highly exposed individuals.* The EPA also relied on the best professional judgment based on research and operational data to determine appropriate site restrictions (eg, requiring waiting periods before harvesting crops grown on soils where biosolids have been applied) and other requirements to ensure the safe use or disposal of biosolids. The result was the Part 503 rule, which imposes general requirements, pollutant limits, management practices, operational standards (such as technology-based requirements for pathogen reduction and vector control), and frequency of monitoring, record keeping, and reporting requirements. These elements of the Part 503 rule are presented graphically in Figure 3.

PUBLIC CONCERN

Concerns about the processing, use, and disposal of waste-derived materials like biosolids are entirely natural and legitimate. Gathering information to address these concerns may lead to improved designs and operations for facilities or may provide sound evidence that the current designs and operations are safe. When concern is expressed by an individual or group, logic and reasoning to address the concern may not make much difference (17).

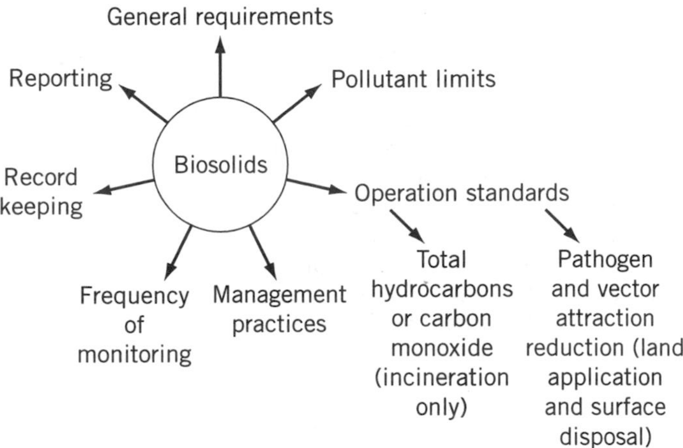

Figure 3. Elements of the Part 503 Rule (2).

It is difficult to decide when a concern has been legitimately addressed. For example, not everyone agrees that the specific management practices used to manage biosolids are sound and based on good scientific information and practical experience. It is also common for some uncertainty to remain even when factual scientific information has been obtained. Especially confusing to the public is the doubt raised by a small part of the scientific community. Such doubt may be based on a desire for more and better data and sometimes on personal opinions not based on careful analysis of benefits, risks and comparative costs of alternate approaches. What level of uncertainty should be tolerated? This level varies for different individuals. Such factors challenge those who responsibly manage biosolids. A middle ground is needed because these waste by-products are a natural consequence of life itself and must be managed prudently. This section recognizes public concerns, explores their importance, and discusses their resolution.

What Standard of Protectiveness Is Acceptable?

Should the standard of protectiveness from waste-borne pollutants be based on scientific risk assessment or a no-change policy? Consider the case when biosolids are added to the soil as fertilizers or soil conditioners. A risk-based approach permits an increase in pollutants from biosolids provided a scientifically determined acceptable level for pollutants is not exceeded. On the other hand, a no-change or minimal change policy limits the increase of each pollutant from biosolids to some fraction of its background concentration in soil. Some European countries, provinces in Canada, and scientists in the United States have adopted or urge, as a matter of policy, a protectiveness standard that permits minimal to zero change in soil metals from waste addition to soils. A lengthy discussion of arguments about the adoption of risk- and policy-based approaches for setting pollutant limits concludes with recommendations for more research (18). The requirements in the Part 503 rule and its predecessors in EPA have not been not overly stringent. Moreover, the Part 503 rule includes research and risk-based incentives and management alternatives that stimulate the production and use of better quality biosolids. The risk-based alternative in the Part 503 rule that defines quality biosolids used with minimal controls has resulted in continual improvement in the quality of biosolids in the United States.

A minimal change policy-based approach is thought to be more protective of public health and the environment than an acceptable change approach. However, difficulties in meeting policy-based limits cause shifts to disposal alternatives which pose even greater risks. Therefore, regulatory requirements for biosolids use and disposal that are more stringent than they need to be, just to be safe, cost more and therefore deprive society of funding for higher priority public health and environmental needs. The ultimate decision of what level of risk and uncertainty to accept is most certainly a policy decision, and of course more and better information is desirable to assist in that decision process. The EPA's biosolids rule (40 CFR Part 503) established scientific risk-based standards that permit an acceptable change in soil metal concentrations (2). The EPA Part 503 rule took nearly eight years to produce; was very carefully thought out; had relevant data, models, and algorithms carefully selected; and underwent repeated review by groups with divergent points of view. The final Part 503 rule even underwent careful review by the National Research council (NRC) of the National Academy of Sciences several years after promulgation of the regulation (19). The NRC endorsed the risk-based approach used in its development.

Studies at the University of Texas in Austin endorse standards based on acceptable environmental change. The view that nature is not fragile and that a standard of zero risk followed to its ultimate end means very limited production of wealth has been expressed. The wealth, which comes from producing quality foods and goods, enables research to be conducted and environmental controls that enhance the environment to be instituted (20).

How to Deal with Concerns about Possibilities of Harm

Many concerns are expressed about the use and disposal of biosolids. Odor is perhaps the single most important factor triggering concern. Once odor has triggered concern, many other fears are expressed. Given the odor trigger and the public knowledge that biosolids are a combined human, municipal, and industrial waste, is it no surprise that people expect that the soil, groundwater, and surface water will be polluted and diseases spread. Other nuisance issues, such as increased traffic, congestion, dust and noise heighten environmental and public health fears and also cause people to fear losing their quality of life and the value of their property. Further exacerbating concerns are the ever changing pro and con views each of us keeps hearing from the media about the risks from alar, drinking coffee, eating margarine vs butter, taking aspirin, drinking alcoholic beverages in moderation, undergoing hormone replacement therapy, using saccharin, and so on.

Key ingredients for addressing public concerns are proactiveness, honesty, fairness, adherence to facts, and a framework by which the concerns can be anticipated and addressed. This framework should involve dedicated professionals at the local level who have responsible for wastewater and biosolids management supported by people from across the county. These support people should have widely varied backgrounds and experience and should work together and encourage each other. It is essential that public concerns be anticipated before they become an overwhelming issue. The framework should anticipate segments of the public to be dealt with at any one time and should continually and credibly provide information that addresses new and existing concerns. The information provided should fairly balance the pros and cons of the concerns and the benefits and risks of the alternative used to manage the biosolids. A communication plan is needed for the following seven public audiences (21) who are concerned about biosolids:

academics/agricultural scientists
water quality professionals
public health officials
agricultural groups/farming representatives
environmentalists
regulatory officials
media

The communication plan should contain vehicles for obtaining the best scientific knowledge available to address concerns about biosolids management and for disseminating information knowledgeable and trained spokespersons credible to the concerned segments of the public. The spokespersons should be continually provided with valid technical and scientifically based information. In turn, they can explain it as accurately and understandably as possible, giving good perspectives on what is known and not known. The spokespersons should be backed up by scientists and other experts who can explain the procedures being used, the science and policy-based decisions that have been made, and any uncertainties that still exist. Because we are all human, we all have our own agendas and preferences, those of us in the biosolids field should be ready to acknowledge our agendas and preferences and express our position as clearly as possible. We should be able to explain various distinctions among biosolids rules and management practices in terms of science, policy, and cost. We should also recognize that science and logic may not satisfactorily address some concerns. In these instances, previously established alliances with members of the seven listed public audiences can be very helpful. More details about important aspects of the ingredients of a communication plan and alliances for gaining public acceptance are described elsewhere (21,22).

When some determined individuals or groups oppose biosolids processing, use, or disposal nearby, a never-ending list of new issues and concerns is raised even as previous concerns are carefully addressed. The EPA has sponsored an effort to develop fact sheets that carefully explain the facts about such incidents and concerns.

Some individuals and groups have criticized the EPA for collecting and disseminating information to encourage the beneficial use of biosolids. It is good public policy to safely recycle by-products like biosolids, and to do so also follows the mandate of Congress. Section 405 (g) of the Clean Water Act of 1986 as amended authorizes the EPA's Administrator "to conduct or initiate scientific studies, demonstration projects and public information and educational projects which are designed to promote the safe and beneficial management and use of sewage sludge (biosolids) for such purposes as aiding the restoration of abandoned mine sites, conditioning soil for parks and recreation areas, agricultural and horticultural uses and other beneficial purposes. The Administrator may make grants" (to a wide variety of private and public agencies and organizations) and "is authorized to collect and disseminate information pertaining to the safe and beneficial use of biosolids."

QUESTIONS AND ANSWERS ON PART 503 RISK ASSESSMENTS

The following questions are about the Part 503 rule. The questions and answers should enhance the reader's understanding of the risk assessment process. They help define terms and further explain the reason for assumptions used and policy decisions made. Finally, the questions indicate a number of issues that arose during the development of the rule, and the answers indicate how they were addressed. A number of the following issues are also discussed previously and in greater detail in (2).

Risk Assessment

What do the Part 503 risk assessments accomplish? They assess the potential for risk to humans, other animals, and plants from pollutants in biosolids. The risk assessments evaluated exposure via 14 pathways for land application, two for surface disposal, and one for incineration.

Risk Level of 1 in 10^{-4} or 1 in 10^{-6}

What does a risk level of 1 in 10^{-4} *mean?* For carcinogenic compounds (compounds that induce or cause cancer), a 1 in 10^{-4} risk level means there is 1 chance in 10,000 of the highly exposed individual getting cancer.

Does this 1 in 10^{-4} *risk level mean that, as a result of the Part 503 biosolids rule, 2,500 of the 2.5 million persons living in the United States (1 person for each 10,000) could get cancer from exposure to that pollutant in biosolids?* The risk of getting cancer is related only to the population exposed to that risk. In the United States, the number of persons highly exposed to risks from biosolids is actually very small. For example, if 10,000 individuals were in the highly exposed population, then there might be one case of cancer arising in the United States from exposure to that pollutant in biosolids. However, if the population of highly exposed individuals was 10, then there might be 0.001 cases of cancer arising in the United States from that pollutant.

Were the limits for metals in the Part 503 rule established based on a 1 in 10^{-4} *risk?* The Part 503 metals are noncarcinogens (they do not cause or induce cancer) in the exposure pathways evaluated.

If metals were not regulated on a 1 in 10^{-4} *risk basis, then on what basis?* The pollutant limits for each of the Part 503 metals in biosolids are based on threshold limits, such as risk reference doses (RfDs) which represent the daily intake of a particular noncancer causing substance not expected to cause adverse effects; the RfD is a conservative determination of the upper level of the acceptable intake. The RfD (or other threshold limit) was then combined with pollutant intake information (eg, the amount of a particular food consumed) to derive a pollutant limit. Each pollutant limit is set to protect a highly exposed individual (plant or animal) from the reasonably anticipated adverse effects of a pollutant in biosolids.

Understanding now that the limits for metal pollutants in biosolids used or disposed were not based on a 1 in 10^{-4} *risk in the Part 503 rule, were any pollutant limits established on the basis of a 1 in* 10^{-4} *risk?* Yes, in that pollutant limits based on 1 in 10^{-4} cancer risk level were determined for potentially toxic organic pollutants that occur in biosolids; and no, because the pollutant limits determined in this way were not included in the final rule as discussed below:

- Land Application: Thirteen pollutant limits were determined for toxic organic pollutants using the 1 in 10^{-4} approach, but they were not included in the final Part 503 rule. It was decided to drop these organic pollutants from the final Part 503 rule because (*1*) the pollutant has been banned, restricted for use, or is no longer manufactured in the United States; (*2*) the pollutant is not in biosolids at significantly high frequencies of detection, based on data gathered from the National Sewage Sludge Survey (NSSS); or (*3*) the limit for the pollutant identified in the biosolids risk assessments is not expected to be exceeded in biosolids that are used or disposed of, based on data from the NSSS (25).
- Surface Disposal: Pollutant limits also were determined for toxic organic pollutants in surface-disposed biosolids based on a 1 in 10^{-4} cancer risk. None of the organics were retained in the final Part 503 rule and three inorganics were deleted from the regulation because each of these organic and inorganic pollutants met one of the three criteria described in the previous paragraph.
- Incineration: Pollutant limits were also determined for toxic organic pollutants associated with incinerated biosolids for which q_1^*s (cancer potency values) exist, based on a 1 in 10^{-4} cancer risk. Because of the limitations of the risk assessment process in reflecting all of the individual toxic organic pollutants emitted from biosolids incinerators, the EPA's Science Advisory Board recommended using an operational standard rather than pollutant limits. The recommended operational standard involves monitoring the emission of total hydrocarbons from incinerators to ensure that the levels from stacks not exceed 100 ppm. This standard is believed to be protect of public health for the spectrum of toxic organic pollutants emitted from biosolids incinerators.

Why was a risk limit of 1 in 10^{-4} *chosen as a basis for the pollutant limit for carcinogens instead of a 1 in* 10^{-6} *risk (1 in 1 million chance of getting cancer)?* The less restrictive 1 in 10^{-4} risk limit was a policy decision. The aggregate (overall) risk from biosolids use or disposal in the United States is especially low (ie, from a fraction to several persons at risk out of the total U.S. population). Because the risk is especially low, the less restrictive risk limit still provides adequate protection.

If a risk limit of 1 in 10^{-4} *is sufficient, then why not apply a more protective risk limit just to be safe? After all, a 1 in* 10^{-6} *risk limit is only 100 times more restrictive than the 1 in* 10^{-4} *risk limit?* In addition to the fact that cancer risk from the use of biosolids is very low, a 1 in 10^{-6} cancer risk target level was not chosen to be more protective because

Use of more conservative levels in risk assessment calculations has sometimes led to predictions that the levels of certain substances in the environment are more hazardous than relevant research indicates. A good discussion of the way risk assessment methodology erroneously predicts that the levels of certain substances in the environment are too high is available (29).

Although not used as a determining factor during the development of the Part 503 rule, a more stringent risk level would require thousands of facilities to achieve the stricter limit of 1 in 10^{-6} for a given substance rather than 1 in 10^{-4}, even though the limit is only one hundred times more stringent. It is difficult to justify such an expense for little or no actual difference in risk to the highly exposed organism.

Selection of the Part 503 Pollutant Limits

How were the pollutant limits chosen for the Part 503 rule? For all pollutants evaluated, first the highly exposed individual was identified for each of the applicable pathways of exposure. For example, for land application practices, a different highly exposed individual was identified for each of the 14 different applicable exposure pathways that were applicable. The risk assessment limit for each pollutant was selected from the pathway with the highest exposure and lowest permitted dose. For example, the pollutant limit for copper was set at 1500 kg copper/hectare of land based on the Pathway

8 pollutant limit being the most stringent (lowest) 1500 kg copper/ha land; the pollutant limits were higher for all other copper exposure pathways. For land application, pollutant concentration limits and annual pollutant loading rate limits were derived from cumulative pollutant loading limit values and incorporated into the rule.

Were all pathways evaluated for each pollutant? Risk assessments were not conducted for all pathways for each pollutant. A risk assessment is made up of several components, including hazard identification and exposure assessment. Where the exposure assessment indicated that exposure to the pollutant was not significant via a certain pathway or where the EPA lacked data, that pathway was not evaluated for a particular pollutant.

Most Exposed Individual vs Highly Exposed Individual

Was the final Part 503 rule designed to protect the most exposed individual (MEI) or the highly exposed individual (HEI)? The MEI is a hypothetical (imaginary) individual that experts do not believe could exist. Protecting an individual who does not even exist was believed unrealistic. The Agency's risk assessment policy states that the individual who should be protected is an HEI. In contrast to the MEI, the HEI does exist, although in small numbers.

The MEI was used as the target organism to be protected in the proposed Part 503 rule which was developed with very conservative assumptions and overly stringent models. As an example, one of the MEIs in the proposed Part 503 rule for land application was the hypothetical home gardener:

- Who raised and consumed essentially all of his or her own food for 70 years in a home garden amended with biosolids.
- Whose biosolids-amended garden soil contained the maximum cumulative permitted application of each of the evaluated pollutants for that 70-yr period.
- Whose food harvested from the garden had the highest plant uptake rate for the 70-yr period for each of the pollutants, as calculated using data from pot/salt studies.
- Who for 70 years consumed foods grown in that garden, with the gardener always at the age, sex, and physiological state for maximum absorption and/or ingestion (eg, simultaneously male and female, pregnant, an infant, and a teen-age male).

In contrast, the use of an HEI combines high end and midrange assumptions in models and algorithms (descriptive mathematical equations). The HEI attempts to represent a real individual. This is indicated by the data, models, and assumptions used for protecting the highly exposed home gardener during the revised risk assessment and development of the final Part 503 rule. In this risk assessment,

- The home gardener HEI raised and consumed up to 59% of his or her own food (depending on the food group) for a 70-yr period in a biosolids-amended garden
- The biosolids-amended garden soil contained the maximum cumulative permitted application of each of the evaluated pollutants for the 70-yr period
- The food harvested from the garden had plant uptake slopes for biosolids pollutants determined by the geometric mean of relevant data from field studies with both acid and neutral biosolids-amended soils
- The food consumption was apportioned among several different age periods during the 70-yr life of the HEI gardener.

Soil pH

Why wasn't soil pH management included as a biosolids land application requirement in the Part 503 rule, especially because it was a requirement in the former Part 257 rule? The Part 503 rule was designed to be self-implementing and to cover all practices involving the use of biosolids. Hence, the plant uptake data for establishing the regulatory limits for land application pathways in the Part 503 rule included data from acidic, neutral, and alkaline soils (pH below 6.0 to greater than 7.0).

Some sensitive plant species exhibit phytotoxicity when grown in soils amended with biosolids containing high concentrations of zinc, nickel, or copper at low soil pH and near the cumulative pollutant loading rates. At the recommendation of experts who assisted, however, the EPA decided that it would be ill-advised to require pH control. Because many other factors protect against harmful effects from metals, such as the soil-plant barrier and other elements present in biosolids that bind pollutants. In addition, in soils with a pH below 5.5, high levels of biosolids pollutants are potentially to toxic to plants, and so are naturally occurring soil metals, such as aluminum and manganese. Given the potential toxicity from these widespread soil metals, most agronomic plants do not grow well at very low pH. Under these conditions, farmers and home gardeners must lime soils to obtain a reasonable yield of edible food, regardless of whether biosolids are used for their soil conditioning and fertilizing value.

"Time Bomb" Theory

What is the so-called time bomb theory? The time bomb theory involves the belief that the organic matter present in biosolids is the primary component that binds metals and thus reduces their bioavailability. The basic premise of the theory is that, as soon as the organic matter degrades, the metals become more bioavailable.

Do pollutants in biosolids become more bioavailable after having been added to soil and after the organic matter in biosolids has decayed? Evidence does not support this claim. Biosolids are typically about 50% organic and 50% inorganic. The experts who assisted the EPA in the risk assessments cited evidence that much of the binding is attributable to the inorganic part of biosolids, namely oxides of iron, aluminum, and manganese, and also phosphate compounds. This binding effect is so strong that it persists after the biosolids have been applied to soils, except at very low pH as described in the soil pH Question and Answer section above. Examination of field data, gathered as many as 60 to 100 years after the use of irrigation wastewater and/or biosolids on soils, supports the concept of binding by the inorganic fraction of the biosolids and indicates that metal binding persists after biosolid organic matter has degraded.

A few scientists question this belief, but experimental data support this inorganic binding concept, and experimental data do not refute it. A leading proponent (34) of the time bomb theory who attempted to prove it, dropped advocacy of the theory after conducting a series of experiments that failed to support it (35).

Is there a direct relationship between the amount of biosolids metals applied to soil and the amount of metals absorbed by plants? No, metals are bound by the biosolids matrix, which reduces their phytoavailability. As an example, assume that there is a constant amount of metal pollutants in biosolids. As more and more of the biosolids are added to soils, the total amount of the metal pollutants in the soil/biosolids mixture increases. However, the metal phytoavailability does not proportionately increase because the inorganic part of the biosolids matrix in the soil/biosolids mixture simultaneously increases. This increasing inorganic matrix strongly binds the metal and competes with and limits the ability of a plant to absorb the metal.

Does the Part 503 rule take into account that reduced bioavailability is associated with the use of biosolids? No, because the EPA did not adjust Part 503 pollutant limits based on bioavailability. Yes, because the Agency did, however, use biosolids field data on plant uptake of pollutants to the extent possible, which invariably showed that there is less uptake (ie, a reduced uptake slope) that if only metal salts were added to soils. Nonetheless, in the Part 503 risk assessments, the Agency assumed that the plant uptake slope was linear. Because the uptake slope is actually less than linear, the final rule overestimates the phytoavailability of biosolids metals.

Why weren't more plant uptake data used from experiments in which metals salts were added to soils in the risk assessments for the final rule? Experts determined that salt data are not relevant to biosolids because metals are bound by the biosolids matrix during generation and processing of the biosolids. This binding does not occur when metal salts are added to soils. Data from metal salt studies were used only when no other data were available.

Phytotoxicity

What is phytotoxicity as it relates to the Part 503 biosolids rule? Phytotoxicity refers to the retardation in plant growth caused by plant toxicity from metal pollutants in biosolids. The Part 503 pollutant limits were set to preclude phytotoxicity.

Is it true that the risk assessments assume that phytotoxicity has not occurred unless there is a 50% reduction in plant growth? No, the EPA used several procedures to determine the concentration of the potentially phytotoxic metals (zinc, copper, nickel, and chromium) in plants that cause phytotoxicity. A 50% retardation in growth of young corn and bean seedlings was involved in only one of the alternative approaches used to establish phytotoxic limits. Even in this approach, other levels of growth retardation were evaluated (ie, 8, 10, and 25% plant growth retardation), although the 50% level was used. In another approach, data on plant tissue concentrations associated with reduced yield were taken from the available literature to define phytotoxic effects for sensitive crops, such as lettuce. These sensitive plant species are more susceptible than corn to metal-induced inhibition of growth (phytotoxicity). These data were used to develop plant tissue levels of metals associated with first detectable yield reductions, which were identified as phytotoxic thresholds. These data, in turn, were used, in conjunction with data on plant uptake of metals, to identify metals application rates that exceed the phytotoxic threshold. The more restrictive of the values determined by these approaches was chosen as the pollutant limit for phytotoxicity in the risk assessment.

Why is it difficult to set a phytotoxic limit? The problem facing the experts who assisted the EPA with the phytotoxicity risk assessment was that many things cause phytotoxicity, as well as apparent phytotoxicity, during the growth of seedlings. Furthermore, the retardation in early vegetative growth, may or may not be associated with reduced harvestable crop yield. Factors that cause phytotoxicity or apparent phytotoxicity include cold weather; insoluble salts, low nutrients, and high nutrients and metals in soils; pesticides and herbicides; and ozone and other impurities in the air. In carefully conducted field tests, yields commonly vary by as much as 15–25% with good fertility and management. An ultimate yield reduction of at least that much must be attained to support a determination that the reduction was significant, especially over several seasons and with various crops grown.

Synergistic Effects of Biosolids Metals

Is there evidence of any synergistic (additive or more than additive) negative effects associated with metals in soils amended with biosolids? The only evidence of a synergy is observed in soils freshly amended with metal salts (not biosolids). The Agency is not aware of any evidence suggesting synergy even in pot studies where metal-rich biosolids were used as to amend soil.

Is there any evidence of positive interactive effects from biosolids metals? Yes, when biosolids are used as a source of fertilizer, there is a built-in protection for people who eat crops that accumulate metals, including cadmium because invariably biosolids also contain iron, calcium, and zinc, which are absorbed into the edible portion of the plant. The presence of these other three metals in the crop consumed reduces the potential for cadmium absorption into the person's intestines and body, and hence reduces the health risk from cadmium (2,12,32,33).

Use of Data With Zero or Negative Plant Uptake Slopes

Were data used from experiments that had a zero or negative plant uptake slope? Yes, however such data were given a protective minimum value, that is, when the slope was negative or zero, a minimum, slightly positive value of 0.001 was used. This procedure allowed using such data in determining plant uptake slopes. This minimum value, however, overestimates uptake to some degree.

Lead

How was risk from exposure to lead in land-applied biosolids determined? The critical exposure pathway for lead was Pathway 3, children ingesting biosolids containing lead. Experts assisting the EPA with the Part 503 rule initially recommended a lead limit of 300 mg/kg. This level was based on observations of absorbed and retained lead in the bodies of cows, sheep, pigs, and chickens whose diets consisted of up to 10% biosolids. In these studies, body burdens of lead (ie, the content of lead in blood and bone) did not increase unless the lead concentration of biosolids fed as part of the animals' diet exceeded 300 mg/kg. It should be pointed out that if there is no increase in the blood and bone tissue, there can be no increase in any meat or milk from the ingesting animal. Hence, the ingesting animals are protected, and individuals who might consume the meat or milk from these animals are also protected.

Prior to promulgating the proposed Part 503 rule, there was an extended period during which an internal Agency review took place. EPA reviewers argued that the Agency should be using the Integrated Exposure Uptake Biokinetic (IEUBK) model to estimate soil/biosolids lead concentration limits that protect children who ingest biosolids-amended soils from potential risks. The IEUBK method is the procedure used by the EPA's Office of Research and Development (ORD) to calculate protective limits against lead risks. The IEUBK model, used for this calculation, assumes that

- the lead blood level does not exceed 7.0 μg of lead per deciliter of blood (10 μg of lead per deciliter is the current critical level that should not be exceeded)
- the portion of bioavailable lead is 60% as high as lead absorbed by children if they were to ingest lead from soluble lead salt sources
- the percentage of the population that could exceed the designated blood level was 5%

Using these IEUBK values, the EPA calculated an allowable lead concentration in biosolids of 500 ppm. The EPA made a conservative policy decision to use the lower of the two sets of lead data, 300 ppm, as the pollutant concentration limit in the final Part 503 regulation, thus providing an additional margin of safety for growing children. Studies on rats fed biosolids containing up to 300 ppm lead per kilogram of biosolids as part of their diet (about 10%) show the bioavailability of the biosolids-bound lead is only 5%, compared with 60% bioavailability assumed in the IEUBK model calculations. This 12-fold overestimate of actual bioavailability adds even more conservativeness to the calculated pollutant limit.

Management of Nutrients to Enhance Agronomic Use and Protect Groundwater

Were nutrients managed to enhance agronomic use and protect groundwater? Groundwater is protected because the use of biosolids as a fertilizer must be at the agronomic rate (ie, the rate of nitrogen application must not exceed the amount of nitrogen needed to supply estimated crop needs). Ground water is also protected by Part 503's requirement that nitrate-nitrogen be monitored at biosolids surface disposal sites. However, although the EPA also considered using the Part 503 rule as part of an overall nutrient management model for regulating the application of a number of nutrients from various sources, the Agency decided not to address this complex issue in the Part 503 rule. Many other sources of nutrients are involved in a nutrient management program (eg, chemical fertilizers, animal manures, other wastes), which the EPA does not regulate under the Clean Water Act. Moreover, the EPA believes that other agencies and knowledgeable parties should be involved in developing such a program. In addition, the EPA felt that biosolids should not be singled out from other nutrient sources, particularly because biosolids pose less of a public health and environment risk from lower nutrient levels in biosolids than many other sources and because no EPA nutrient requirements currently address these other sources of nutrients. Guidance and rules for managing nutrients from all sources are being developed on a state-by-state basis. Nutrient management may be regulated nationally at some future time.

Sensitive Soil Microorganisms

Why didn't the Part 503 rule base its regulation on the observation that a strain of Rhizobium lost its ability to fix nitrogen on one strain of white clover in one study on the Woburn experimental plots in England? This loss in ability to fix nitrogen was noted after a 19-year period of biosolids application with moderately high concentrations of metals (eg, 100 mg cadmium/kg biosolids and 3,000 mg zinc/kg biosolids) to a sandy soil on which vegetable crops were being grown. (Nitrogen-fixing microbes are important in agriculture and the environment. They have the unique capability, while in symbiosis in modules on the plant root, of converting nitrogen gas from the air into organic nitrogen, rather than requiring the plant to absorb fertilizer nitrogen from the soil. The organisms live on the root in irregular, rounded, lump-shaped growths with mutual benefit to both the microbes and plant.) Most studies involving the land application of biosolids have shown no adverse effects on soil microbial activity (including nitrification and mineralization of nitrogen, as well as normal development and functioning of nitrogen-fixing bacteria for legumes.)

The circumstances at the Woburn experimental plots where biosolids were applied from 1942 to 1961 were uniquely atypical. No other researchers have been able to repeat this observation with *Rhizobium* on other sites rich in biosolids-applied metals. Furthermore, several studies have found effective strains of white clover *Rhizobium* in farm fields rich in metals. One such study involved soils near a zinc smelter in Pennsylvania, where zinc and cadmium levels in the soil were much higher than in the Woburn study (36,37). Another similar study was reported (38). Other research on mine spoils with high levels of metals, analogous to free salt in soil, has shown that nitrogen fixation was inhibited in free-living bacteria (39), but not by white clover *Rhizobium* until metals levels were so high that phytotoxicity to white clover plants was observed. Finally, a new study (40) provides very strong evidence that biosolids were not the cause of *Rhizobium* becoming ineffective on the Woburn plots. Instead, researchers determined that soil pH below 6.0 caused selection of ineffective strains of *Rhizobium* on experimental controls (soils without biosolids added) and biosolids-amended soils.

Pathogens

Are the pathogen operational standards risk-based? Risk assessment methodologies had not been developed sufficiently to make such calculations. Instead, the pathogen operational standards, which are technology-based, require that pathogens in biosolids be reduced below detectable levels or to levels that, when coupled with crop harvesting and site access restrictions, protect public health and the environment.

Acceptable Concentrations of Biosolids Pollutants in Soils

The biosolids risk assessments were designed to determine pollutant application rates or pollutant concentrations in biosolids. Based on the risk assessment results and the Part 503 pollutant limits, what are the acceptable concentrations of biosolids pollutants in soils? How are these soil concentrations derived? The following equation shows how soil concentrations (RLC) can be derived from biosolids risk assessment pollutant limits (RPs), which are equivalent to the Part 503 cumulative pollutant loading rate (CPLR) limits:

$$\frac{\text{RP}}{\text{MS} \times 10^{-9}} = \text{RLC(in } \mu\text{g/g)} \times 10 = \text{RLC(in mg/kg)}$$

where RP = cumulative application rate of pollutant in biosolids (kg/ha), MS = 2×10^9 g/ha (assumed mass of soil in upper 15 cm), and RLC = allowed soil concentration of pollutant from biosolids (μg/g or mg/kg).

For copper, the soil concentration RLC is

$$\frac{\text{RP}}{\text{MS} \times 10^{-9}} = \frac{1500}{20} = 75(\text{RLC in } \mu\text{g/g}) \times 10$$
$$= 750(\text{RLC in mg/kg})$$

The copper pollutant concentration in soil from biosolids (RLC) calculated from the above equation is further adjusted by adding in the background median (50th percentile) soil concentration for the pollutant in question, in this case for copper to determine acceptable concentration for biosolids pollutants in soils (2):

RLC copper = 750 mg/kg in biosolids
+ median background
× soild concentration RLC copper
= 19 mg/kg
= acceptable concentration of 769 mg/kg in soil
–biosolids mixture

Should soil cleanup standards be compared with Part 503 CPLR or Part 503 concentration pollutant limits? No, instead soil cleanup standards should be compared with acceptable soil concentration values, as derived in the biosolids risk assessments.

How do these acceptable soil concentrations compare with state and other EPA cleanup standards for soils? In most cases, the acceptable soil concentrations calculated from the Part 503 risk assessments are greater than those for state and other federal EPA programs; however, some of the state and other federal acceptable soil concentrations are greater. Almost no set of soil concentrations agree. Furthermore, most of the other sets of numbers are only preliminary, have not been completed, and have been calculated, for other purposes (eg, in connection with efforts to clean up soils contaminated by hazardous wastes). Some of the concentration levels have been calculated based on best available technology and others are based on risk assessments using different data sets, approaches, assumptions, models, and/or pathways than used in the Part 503 risk assessments.

Dioxins and Furans

Were dioxins and furans formally evaluated during the development of the Part 503 rule? The EPA is in the process of reevaluating the potential toxicity and harmfulness of dioxin and furans. Therefore, it would not have been desirable to set limits with such an uncertain toxic endpoint and with the limited data that exists. However, measurements of dioxin concentrations in biosolids during the National Sewage Sludge Survey (19) and informal evaluations during the development of the Part 503 rule were made but did not estimate significant risk. The same pathways were evaluated as for PCBs during via this informal Part 503 risk assessment evaluation. More recently, a simple pathway assessment procedure was devised that focuses on transfers into milk and meat from pasture land based on total equivalents of dioxins and furans applied (14). Worst cases were compared in urban settings vs less risky rural settings. The worst-case scenario projected an average exposure that was still nearly a factor of three below the tolerable daily intake.

What If?

Does everyone agree that the Part 503 rule is scientifically correct and protective of human health and the environment? Although a majority of the scientific community supports the scientific validity of the requirements of the Part 503 rule, a small minority takes issue with parts of the rule. Also a small part of the public does not believe that the rule is protective enough. These persons often cite the no net degradation approach as evidence that the Part 503 limits are not protective enough. They also state that studies have been continued only for 60 to 100 years and that is not long enough.

How can the EPA determine that it is all right to use or dispose of biosolids? What if some unanticipated problem is discovered? The use of biosolids has been one of the most extensively studied waste management practices in the United States. Some public uses have occurred in the United States for 70 years. Throughout this long history of use, biosolids have been a valuable soil conditioning and fertilizing product. While there are no absolute guarantees, the past use of biosolids has been very reassuring when they have been used in accordance with acceptable practices. In the few instances in the past where problems occurred from biosolids use, the implementation of various management practices has rectified most situations, as is the case with any farming practice where stewardship of the land is management-based (ie, managing soil pH, insect pests and plant disease, weeds, water, levels of macro- and micronutrient, crops, microclimate, and harvesting methods).

As discussed in the following section, biosolids are also valuable where lands are mismanaged. It is commonly known that lands disturbed by mining are reclaimed through effective use of biosolids. More recently, it has been determined that arid lands devastated by overgrazing are recovered considerably with the use of biosolids. Also, studies now underway suggest that lead in soils from paint and automotive exhausts are bound by applying biosolids to high lead-containing soils, making the soil lead less available to children who eat soil.

Science continues to show new uses for waste resources, such as biosolids. All field research to date leads to the conclusion that the agronomic use of high quality biosolids is sustainable and safe. A study was undertaken by the National Research Council of the National Academy of Science to provide an independent assessment of risks associated with land application of biosolids. Their report (19) concludes that, when properly treated and used in an agricultural setting, treated biosolids are safe and effective in food crop production. Thus, it seems prudent to use biosolids as a highly recyclable resource.

BENEFITS OF USE

Biosolids are a natural resource. Practices that beneficially use biosolids include land application as a soil amendment or fertilizer supplement and various procedures that derive energy from them or convert them to useful products (1). Practices which recycle and use biosolids beneficially reduce the volume of biosolids requiring disposal by either landfilling or incineration. Other benefits derived from recycling biosolids include improved soil fertility and tilth, reduced need for and enhanced response to inorganic fertilizers, better growth and quality of crops, and decreased energy consumption.

Biosolids are a Natural Fertilizer. Biosolids induce a major emotional response in many individuals. Parents teach children that human waste is unclean and should be avoided. Compare this with the lifelong experience of most persons familiar with animal wastes as a material to be managed and used. Like animal waste, biosolids are part of the natural cycle of life.

Agricultural Use of Biosolids. EPA policy promotes the beneficial use of municipal biosolids (23). This policy is based on years of extensive study and experience. Hundreds of studies have been conducted as a basis for the safe use of biosolids. Moreover, thousands of publicly owned treatment works (POTWs) are currently using their biosolids as an organic fertilizer and soil conditioner on land throughout the United States. For example, more than 55 and 90%, respectively, of all biosolids produced in Ohio and Maryland are used on land.

Examples of communities recycling their biosolids include Hannibal, Missouri (19,000 population), Madison, Wisconsin (250,000 population), and Seattle, Washington (1.1 million population). Each of these communities charges farmers for using their biosolids. Hannibal recovers 100% of the costs of hauling and spreading biosolids from its sales to farmers, and Madison receives $12.00 per acre for applying its biosolids. Madison fertilizes 3,000 to 4,000 acres of farmland with biosolids each year and has farmers waiting with a total of 22,000 acres of farmland available for application. Seattle also receives payment for biosolids it applies to forest and agricultural land.

Since 1974, nearly all the biosolids from Metropolitan Washington, D.C. (3 million population) have been used on land. In 1995 about 75% (90,000 dry tons) of the dewatered biosolids produced was used on agricultural land in Maryland (4000 acres) and Virginia (4000 acres). The remaining 25% was composted for use by landscapers, horticulturists, and the general public. The dewatered biosolids were applied to private farmland by private contractors at no charge to the farmers. Each application of dewatered biosolids applied contained $100 to $135 worth of needed nitrogen, phosphorus, trace nutrients, lime, and organic matter per acre assuming about application of 5 dry tons per acre. Collectively, in 1995 the farmers received approximately $2 million worth of nitrogen, phosphorus, lime, trace elements, and organic matter from metropolitan Washington biosolids that improved soil tilth and crop yield.

Additional agricultural benefits of biosolids include the enrichment of soil microorganisms and improved retention of soil moisture, which increases the capability of plants to absorb nutrients. Biosolids use also suppresses pathogenic soil organisms, such as nematodes, that damage plant roots and also suppress specific plant root diseases that otherwise cause damage in commercially grown potted plants.

Nonagricultural Use of Biosolids. The beneficial uses of biosolids are not limited to farmland application. Biosolids are used in silviculture to increase forest productivity and to revegetate and stabilize lands harvested or disturbed by mining, construction, fires, land slides, or other natural disasters. The application of biosolids to forest land shortens pulp wood and lumber production cycles by accelerating tree growth, especially on marginally productive soils. Studies by the University of Washington, and the U.S. Forest Service on the use of biosolids as a fertilizer in silviculture have shown as much as a threefold increase in tree growth compared to controls for certain tree species.

Biosolids are used productively to stabilize and revegetate areas destroyed by mining, dredging, and construction activities. Alkaline, stabilized, digested, air-dried, and composted biosolids are frequently used to help revegetate mine spoil, highway embankments, median strips, and other construction sites. Alkaline, stabilized biosolids are also used as a soil substitute for intermediate and final landfill cover. The use of biosolids in land reclamation efforts is very successful and comparable in cost to other commercial methods in both large- and small-scale projects. For example, in a stripmined area in Fulton County, Illinois, reclamation with municipal biosolids cost about $3700 per acre, as compared with a range of $3400 to $6300 per acre using commercial methods. In another example, biosolids have been used in Palmerton, Pennsylvania, to reclaim more than 3000 acres of lands devastated by mining and smelting activities. This superfund hazardous waste site is being revegetated with a blend of biosolids and fly ash. The Palmerton site was so highly contaminated from 90 years of smelting zinc that all vegetation in the surrounding area was destroyed. The procedure was developed by a team of practitioners and researchers from Allentown and the Pennsylvania State University, respectively.

Studies in New Mexico have shown sustained improved growth and nutritional quality of desirable native vegetation on rangeland and reduced runoff of rain water from a one-time, 10–20 dry tons per acre surface application of biosolids. Studies in Colorado, with 1–15 dry tons per acre of biosolids applied, are being conducted to determine optimum rates for improving range quality and minimizing public health and environmental risks. Early results from these studies show similar improvements in range quality and reduced water runoff proportional to the rate of biosolids application.

Biosolids Recycling: Practices and Benefits

Biosolids are used separately or in conjunction with chemical fertilizers. Particularly in soils low in organic matter, biosolids provide benefits unavailable from chemical fertilization. The biosolids organic matter enhances the soil rooting media thus providing better water retention, improved air exchange around plant roots, and increasing the soil's ability to hold nutrients available for plants (increased cation exchange capacity). In highly leachable soils, the release of biosolids organic nitrogen at a rate consistent with plant uptake, mitigates the loss of excess nitrogen into groundwater.

Biosolids organic matter had other impacts on the same Yuma, Arizona farm that initially might have seemed undesirable. Herbicides became less effective because of their interaction with the changing soil and the organic biosolids matrix. Those fields, previously weed free, now contained more weeds. On the other hand, the plants became more vigorous and better able to compete with weeds and withstand damage from insect pests. These changes occurred because of biosolids usage that allowed the farmer to decrease his costs for fertilizer, herbicides, and pesticides by approximately \$170 on each acre of his 12,000 acre farm. In some instances the total yield decreased compared to weed-free fields. However, the farmer's net return per acre increased (more dollars per acre profit). The same Yuma farmer decided, because of his enhanced yield and lowered costs from use of biosolids, to dedicate 10% of his land each year to producing grains for wildlife. Because of the farming changes that left more cover from weeds on all 12,000 acres and the 1,200 acres left each year with unharvested grain, the dove and wildlife population increased so substantially in six years that the Yuma region began to realize an unexpected \$3.5 million increased annual benefit from hunting.

Other Uses for Biosolids

The sale of biosolids products to the public for many kinds of garden, nursery, household, and lawn uses is increasing. Treatment, such as heat-drying, composting, and treatment with alkaline materials converts biosolids into useful products that can be considered of "exceptional quality" if pollutant concentrations in the biosolids do not exceed the minimum levels specified in the Part 503 rule. These products are safe for unrestricted use by the general public. Generators of these products must have a monitoring program to ensure that the biosolids continually meet the "exceptional quality" requirements. Examples of these stabilized products include Milwaukee's heat-dried biosolids product, Milorganite which it has been producing and selling throughout the United States since the 1920s. Products of this type have sold in bulk for as much as \$190 per ton if high in nitrogen content and aesthetically pleasing. Kellogg Supply Company (a private firm in California) has been producing and marketing composted biosolids products (eg, Nitrohumus, Topper, Gro-Mulch) mostly in California, Arizona, and Nevada for a similar period of time. Their products include composted biosolids that come predominantly from Los Angeles County wastewater treatment facilities. Both Milorganite and Nitrohumus have been used to establish and maintain grass playing fields in sports stadiums across the country, including the Rose Bowl in California. A composted biosolids product from Philadelphia called Earthgro has been used with great success for growing container plants and chrysanthemums.

Even the White House has used composted biosolids to reestablish its lawns. Several years ago, 825 t composted biosolids (Compro) were used in this highly successful project. Similarly, the lawns at Mount Vernon, the Washington Monument grounds, and the Governor's Mansion in Annapolis, Maryland, were renewed with Compro. The first use of composted biosolids on the Washington, D.C. Mall (nearly 6000 t) was in 1976 to establish the Constitution Gardens in time for the U.S. Bicentennial celebration. Compro is being sold for \$13–\$65/m^3 in bulk depending on quantity and delivery. The cost for the bagged product is \$175–\$210/m^3 or \$5–\$6 for 28 liters.

Research on the use of biosolids to remediate soils containing high levels of lead shows promise (26). The research indicates that appropriately produced and applied biosolids help protect child health because the biosolids matrix reacts with the lead in contaminated soils reducing the bioavailability of the lead. The research involved feeding laboratory animals an otherwise completely balanced diet containing 9% of either a low or high-lead containing urban soil mixed with 1% of different biosolids products. The preliminary results from these animal feeding studies, depicted in Figure 9, show up to 50% reduced bioavailability of ingested lead, (ie, reduced absorption of ingested soil lead into the blood and body tissues reflected by bone lead levels). Such data suggest that children ingesting biosolids-treated soil and dust have decreased absorption of lead into the blood stream, thus lessening the potential for lead-induced nerve and brain damage. Additional research is needed with laboratory animals to determine the best form of biosolids for use and the extent to which reduced bioavailability is possible.

One of the most economical and agriculturally beneficial methods for stabilizing and using biosolids is anaerobic digestion. This stabilization process generally yields a Class B biosolids product that has been spread for years on agricultural land in liquid form and as a dewatered product. The methane gas generated is another benefit of the anaerobic digestion of biosolids. As an example, the Tampa, Florida, treatment works recover about \$700,000 worth of electricity each year from methane produced during anaerobic digestion. This is equivalent to approximately \$65 worth of net electricity produced per dry ton of volatile biosolids removed from the digester. Tampa also uses the heat removed from the electrical generators to provide more than 95% of the warmth needed for the digesters. All but 10–15% of Tampa's anaerobically digested biosolids are heat-dried and marketed for between \$85 and \$120 per dry ton. The balance is land applied in dewatered form.

Future Opportunities

Individuals need to join the effort to creatively turn the environmental problems now caused by improper use or disposal of inorganic and organic by-products and wastes into benefits to society. The view should now be that biosolids, animal manures, food processing wastes, and the recyclable organic fraction of municipal solid wastes are part of the natural cycle of life and are feedstocks for a continuum of agronomic and nonagronomic uses (24). Greater attention must be given to the properties and the opportunities for altering and tailor-making these by-products so that they are even better feedstocks. In fact, the metal bioavailability of these by-products can be managed. The bioavailability of metal pollutants in a waste material, such as biosolids is reduced due to matrix components such as iron, aluminum and manganese oxides, humic acids from the biosolids organic matter, and phosphates (12,13,25). Dewatering biosolids with an iron salt, such as ferric chloride, enhances its ability to reduce bioavailability of metal pollutants in biosolids and in soils to which biosolids have been added, even when added at an agronomic rate. Moreover, the processing of a waste affects a by-product's appearance, its odor potential, and its public acceptability. Processing also determines stability and nutrient content and also determines the extent of reduction of any pathogens and attractiveness of the by-products to vectors. The processing also affects the product's storage and use. And, because of all these impacts, the form of processing leads to varying degrees of restriction on use (16). Some agronomic opportunities for enhanced useability are mentioned below.

After research, the potting industry has adopted specially prepared compost to suppress plant root disease (41). Research with municipal solid waste compost as a soil amendment has resulted in increased yield and quality of field and fruit crops (42). The use of composted biosolids alone and in combination with other feedstocks aid in leaf retention and hence more vigorous roots, enhanced plant growth, and in the case of cantaloupes, more yield and sugar content, because of longer leaf retention, so that newly forming fruits could mature. Research (43) has shown how to use composted biosolids in potting media as a substitute for peat and at the same time reduce the need for chemical macro- and micronutrient fertilization. Recent cooperative studies at the Rodale Institute involve comparative root and plant growth responses to inorganic fertilization and organic composts. This research is being conducted to examine how plant root structure and effectiveness are affected in soils by the stimulation of fungi and other microorganisms, better physical structure, and other related factors. Blending biosolids with kiln dust provides a more balanced fertilizer material that is also useful as a lime substitute (44,45). In addition, certain blends of biosolids with other feed stocks make high quality mulches that reduce soil erosion.

Other unique opportunities exist for tailor making by-products, such as alterations prior to feeding animals. For example, the breeding of low phytate corn seed results in corn grain whose phosphorus is much more bioavailable than phosphorus in normal high phytate containing corn grain. Thus, the litter from poultry that has consumed low phytate corn grain contains reduced levels of phosphorus. Another example is the development of phytase-overexpressing bacterial strains to enhance phosphorus availability to animals and thereby reduce phosphorus in manures. Still another example is the design of feed rations which more closely match nutrient contents to animal nutritional needs.

As can be seen, greater knowledge of the beneficial properties of biosolids leads directly to improved methods of benefiting society including remediating other environmental problems. The proper use of these waste products helps achieve the goal of sustainable farming. The production of these tailor-made by-products creates opportunities for new jobs in rural and urban areas.

BIBLIOGRAPHY

1. U.S. EPA, *Biosolids Recycling: Beneficial Technology of a Better Environment,* EPA 832-R-94-009, Office of Water, Washington, D.C., 1994.
2. U.S. EPA, *A Guide to the Biosolids Risk Assessments for the EPA Part 503 rule,* EPA/832-B-93-005, Office of Wastewater Management, Washington, D.C., 1995.
3. U.S. EPA, *Environmental Regulation and Technology: Use and Disposal of Municipal Wastewater Sludge,* EPA 625/10-84-003, Office of Water, Washington, D.C., 1989.
4. J. M. Walker, R. E. Lee, and R. K. Bastian, "The Impact of the Part 503 and State Biosolids Rules on Distribution and Marketing Practices and Costs," *Proc. of the 9th Annual Biosolids and Residuals Conference of the Water Environ. Fed.,* Kansas City, Mo, 1995.
5. U.S. EPA, *A Plain English Guide to the EPA Part 503 Biosolids Rule,* EPA/832/R-93/003, Office of Wastewater Management, Washington, D.C., 1994.
6. C. Lue-Hing, D. R. Zenz, and R. Kuchenrither, Water Quality Management Library, Vol. 4, *Municipal Sewage Sludge Management: Processing, Utilization and Disposal,* Technomic Publishing Co. Inc., Lancaster, Pa., 1992.
7. U.S. EPA, *Environmental Regulation and Technology: Control of Pathogens and Vector Attraction in Sewage Sludge,* EPA/625/R-92/013, Office of Research and Development, Cincinnati, Ohio, 1992.
8. Council for Agricultural Science and Technology, *Application of Sewage Sludge to Cropland: Appraisal of Potential Hazards of the Heavy Metals to Plants and Animals,* EPA 430/90-76-013, Washington, D.C., 1976.
9. T. J. Logan, and R. L. Chaney, "Utilization of Municipal Wastewater and Sludges on Land–Metals," in A. Page, T. Gleason, J. Smith, I. Iskander, and L. Sommers, eds., *Proc. Workshop on Utilization of Municipal Wastewater and Sludge on Land.* University of California, Riverside, 1983.
10. A. L. Page, T. L. Gleason III, J. E. Smith, Jr., I. K. Iskander, and L. E. Sommers, *Proceedings of Utilization of Municipal Wastewater and Sludge on Land,* University of California, Riverside, 1983.
11. A. L. Page, T. G. Logan, and J. A. Ryan, *Land Application of Sludge: Food Chain Implications,* Lewis Publishers, Chelsea, Mich., 1987.
12. R. L. Chaney, and J. A. Ryan, *Risk-Based Standards for Arsenic, Lead, and Cadmium in Urban Soils,* Deutsche Gesellschaft für Chemisches Apparatewesen (DECNMA), 1995.
13. R. Chaney, and J. Ryan, "Heavy Metals and Toxic Organic Pollutants in MSW-Composts: Research Results on Phytoavailability, Bioavailability, Fate, Etc", in *Science and Engineering of Composting,* Renaissance Publications, Worthington, Ohio, 1993.
14. K. C. Jones, and A. P. Sewart, submitted to *Crit. Rev. Environ. Sci. Technol.* **27**(1), 1–85 (1997).
15. J. A. Ryan, R. M. Bell, J. M. Davidson, and G. A. O'Connor. *Chemosphere* **17,** 2299–2323 (1989).
16. J. M. Walker, "Production, Use, and Creative Design of Sewage Sludge Biosolids," in *Land Utilization and the Environment,* Soil Science Society of the American Proc. Miscel. Pub. Madison, Wisc., 1994, pp. 67–74.
17. T. Hellstrom, and A. Goran–Dahlberg, "Swedish Experience in Gaining Acceptance for the Use of Biosolids in Agriculture," in *Proceedings of the WEF Specialty Conf., The management of water and wastewater solids for the 21st century: A global perspective,* Water Environmental Federation, Alexandria, Va., June 1994, pp. 9–25 to 9–36.
18. S.P. McGrath, A. C. Chang, A. L. Page, and E. Witter, *Environ. Rev.* **2,** 108–118 (1994).
19. National Research Council, *Use of Reclaimed Water and Sludge in Food Crop Production,* National Academy of Sciences, National Academy Press, Washington, D.C., 1996.
20. J.M. Walker, U.S. EPA's roles in waste utilization: The need for partnership, in *Agricultural Utilization of Urban and Industrial By-Products,* Soil Science Society of America, Special No. 59, Washington, D.C., 1995, pp. 11–18.
21. Powell Tate, Inc., *Communications Plan on Biosolids,* Water Environmental Federation, Alexandria, Va., July 1993.
22. P. Machno, R. O'Dette, J. Forste, D. Garvey, G. Gruber, M Lang, L. Naylor, and P. Matthews, *Water Environ. Technol.* **8**(5), 41–44 (1996).
23. U.S. EPA, *Fed. Reg.* **49**(114) 24358–59 (June 12, 1984).
24. D. F. Galloway, and J. M. Walker. "An entrepreneurial view of the future for the use of wastes and by-products," in J. E. Rechergl and H. C. Mackinnon, eds., *Agricultural Uses of By-Products and Wastes,* American Chemical Society, Washington, D.C., 1997, pp. 22–27.

25. R. Corey, L. King, C. Lue-Hing, S. Fanning, J. Street, and J. Walker, "Effects of Sludge Properties on Accumulation of Trace Elements by Crops," in *Land Application of Sludge: Food Chain Implications,* Lewis Publishers, Chelsea, Mich., 1987, pp. 25–51.
26. J. B. Heneghan, D. M. Smith, Jr., H. W. Mielke, R. L. Chaney, and J. M. Walker, "Composts/Biosolids Reduced Soil Lead Bioavailability in Rat Feeding Studies" (Poster Abstract), *National Conference of the Composting Council,* Composting Council, Alexandria, Va., Nov. 1994.
27. D. Rossi, "The New Jersey-New York Information Sharing Group Approach to Overcoming Factors Limiting the Beneficial Use of Biosolids," in *Proceedings of the Water Environment Federation Specialty Conf.,* Water Environmental Federation, Alexandria, Va., pp. 9–37 to 9–43, June 1994.
28. U.S. EPA, *Nomination Guidance: 1997 Beneficial Use of Biosolids Awards Program,* EPA 832-B-97-002, Office of Wastewater Management, Washington, D.C., 1997.
29. J. Ryan and R. Chaney, "Issues of Risk Assessment and its Utility in Development of Soil Standards: The 503 Methodology an Example," in *Proceedings of the Third International Symposium on Biogeochemistry of Trace Elements,* Paris, France, May 15, 1995.
30. K. Nogawa, R. Honda, T. Kido, I. Tsuritani, and Y. Yamada, *Trace Subst. Environ. Health* **21,** 431–439 (1987).
31. K. Nogawa, A. Ishizaki, and S. Kawano. *Environ. Res.* **18,** 397–409 (1978).
32. H. Morgan, and D. Simms, eds., *Sci. Total Environ.* **75,** 1–143.
33. D. Strehlow, and D. Barltrop. 1988. *Health Studies Sci. Total Environ.* **75,** 101–133 (1988).
34. P. Beckett, R. Davis, and P. Brindley, *Water Pollut. Control* **78,** 419–445 (1979).
35. N. B. Johnson, P. H. T. Beckett, and C. J. Waters, "Limits of Zinc and Copper Toxicity from Digested Sludge Applied to Agricultural Land," in R. D. Davis, G. Hucker, and P. L'Hermite, eds., *Environmental Effects of Organic and Inorganic Contaminants in Sewage Sludge,* D. Reidel Pub. Dordrecht, the Netherlands, 1983.
36. J. Angle and R. Chaney. *Trace Subst. Environ. Health* **22,** 321–336 (1988).
37. J. Angle, S. McGrath, A. Chaundri, R. Chaney, and K. Giller, *Soil Biol. Biochem.* **25:**575–580 (1993).
38. J. Obbard and K. Jones, *Environ. Pollut.* **A79,** 105–112 (1993).
39. J. Rother, J. Millbank, and I. Thornton, *J. Soil Sci.* **33,** 101–113 (1982).
40. A. Ibekwe, J. Angle, P. van Berkum, and R. L. Chaney, *Agron. Abstr.*, 235 (1995).
41. H. A. J. Hoitink, M. J. Boehm, and Y. Hadar, "Mechanisms of suppression of soilborne plant pathogens in compost-amended substances," in H. A. J. Hoitink, et. al., eds., *Science and Engineering of Composting. Design, Environmental, Microbiological and Utilization Aspects,* Renaissance Publications, Worthington, Ohio, 1992, pp. 601–621.
42. J. H. Peverly and P. B. Gates. "Utilization of MSW and Sludge Composts in Crop Production Systems," in *Sewage Sludge: Land Utilization and the Environment,* Soil Sci. Soc. Amer. Miscel. Pub., 1994, Chapter 26, pp. 193–204.
43. A. F. Falahi–Ardakani, R. Gouin, J. C. Bouwkamp, and R. L. Chaney, *J. Environ. Hort.* **5,** 112–116 (1987).
44. T. J. Logan, and J. C. Burnham, in D. L. Karlen, ed., *Agricultural Utilization of Urban and Industrial By-Products,* American Society of Agronomists, Special Publication 58, Madison, Wis., 1995, pp. 209–224.
45. T. J. Logan, B. J. Lindsey, and S. Titko, "Characteristics and Standards for Processed Biosolids in the Manufacture and Marketing of Horticultural Fertilizers and Soil Blends," in J. E. Rechergl and H. C. Mackinnon, eds., *Agricultural Uses of By-Products and Wastes,* American Chemical Society, Washington, D.C., 1997, pp. 63–71.

BIOTA, HERBICIDES IN.

See HERBICIDE RESIDUES IN BIOTA, DETERMINATION.

BIOVENTING.

See SOIL REMEDIATION BY AIR SPARGING.

BRINE-IMPACTED SOILS, REMEDIATION OF

THOMAS M. HARRIS
University of Tulsa
Tulsa, Oklahoma

Petroleum reservoirs typically contain a mixture of hydrocarbons and salt water (brine). During production of the oil, this brine is also brought to the surface. Although oilfield brine is typically disposed of by reinjection into a geological formation, it is still common for some to be spilled at the surface during handling. As a result, brine-impacted soil remains the most common environmental problem associated with onshore oil and gas wells.

Soil that comes in contact with concentrated salt water is converted to a saline condition. High soil salinity results in outright death for established plants; it is impossible for the roots to extract water by osmosis. Soil that is unable to support plant life is eventually susceptible to erosion. Oilfield brines also convert soil to a sodic state, in which most of the cations associated with the surfaces of clay minerals have been displaced (through ion exchange) by Na^+. Sodic soils exhibit poor structure, characterized by poor hydraulic conductivity (reduced permeability to water) and the formation of a hard crust on the surface.

The characterization of brine-impacted soil begins with the preparation of a saturated paste. This mixture is then vacuum filtered, yielding a small volume of fluid known as the saturated paste extract. The electrical conductivity (EC) of this extract provides a measure of the overall salinity of the soil. The soil extract may also be analyzed by atomic absorption spectrophotometry for the concentrations of individual metal ions. The concentrations of Na^+, Ca^{2+}, and Mg^{2+} are used in calculating the sodium adsorption ratio (SAR), a parameter used in assessing the sodicity of a soil.

The remediation of brine-impacted soils can be accomplished by leaching with rainwater or irrigation water. Periods of dryness are likely to be encountered in the former case. Such conditions slow the downward movement of salt and actually reverse some of the results achieved when the surface was covered with water. The basin-furrow levee design eliminates this problem by forcing the rainwater to meander across the surface of the soil, solubilizing salt as it passes. In this case, the direction of salt migration is always upward, toward the surface.

When irrigation water is available, it may be used to speed up the leaching process. In the most commonly used irrigation treatment, known as ponding, the water is added rapidly until a prescribed depth covers the entire treatment area. Ponding is effective for sodicity and salinity; even though the hydraulic conductivity of the soil may be poor, the column of liquid at the surface forces water through the limited number of pores available. Sprinkling the water onto the surface is more efficient than ponding if the surface is covered with mulch. The mulch serves to limit evaporation of water from the surface, thus maintaining the downward movement of salt.

The soil may be treated with an amendment to achieve an enhanced rate of leaching. The application of some source of calcium ion, which helps to minimize the swelling and dispersion of clay particles, is the most cost-effective treatment. The mineral gypsum, composed primarily of sparingly soluble calcium sulfate, is the most common amendment. Soluble calcium salts (eg, calcium chloride) provide more rapid results. For soils that are both sodic and alkaline, sulfuric acid is added to solubilize the calcium already present in the soil. Synthetic polymers, such as polyacrylamide, are effective only after most of the soluble salts have been leached from the soil, and then only in conjunction with gypsum.

The fate of salt-laden leachate is of concern in some cases. The leachate is collected by standard agricultural drainage technology. The Texon Scar remediation project utilized 40,000 meters of underground drainage and eight sumps, from which the leachate was pumped to an injection well. Over 2 million m^3 (13 million bbl) of leachate was handled by the system during the first five years of the project. Primarily as a result of this drainage system, the capital cost of this remediation project was approximately $440/ha ($200/acre).

Finally, it should be noted that halophyte species, which can tolerate soils of moderate-to-high salinity, have been successfully employed in revegetating brine-impacted soils. Of the more than 1600 halophyte species identified to date, nearly 90% are perennials; thus, the replanting of ground cover need not be performed annually. Also, nearly 50% of the halophytes are classified as trees or shrubs, which assist in the general downward movement of salt; deep root systems maintain a low water content in the subsoil, thus enhancing permeation downward through the soil profile.

PETROLEUM PRODUCTION AND BRINE-IMPACTED SOIL

Petroleum reservoirs typically contain a mixture of hydrocarbons and salt water (brine). During production of the oil, this "produced water" is also brought to the surface. At one time it was common practice to discharge the brine at the surface, resulting in extensive contamination of soil with salt. For example, in the Big Lake Field in Texas approximately 32 million m^3 (200 million barrels) of brine were discharged during forty years of production, resulting in approximately 2,800 ha (11 mi^2) of brine-impacted soil (1). Such "historical spills" account for nearly 174,000 ha (683 mi^2) of brine-impacted soil in the State of Texas alone (2).

In general, it is no longer accepted practice to discharge produced water at the surface. Instead, the brine is disposed of by reinjection into a geological formation (sometimes that from which it was withdrawn). However, it is still common for some brine to be spilled during its handling at the surface. Such spills typically occur in the vicinity of the wellhead or tank battery (where the crude oil and brine are stored temporarily), or in the pipelines that connect them. In fact, brine-impacted soil remains the most common environmental problem associated with onshore wells.

Although production practices continue to be refined to minimize brine release, current drilling practices still result in brine-impacted soil (2). During the drilling of the well, excess drilling fluid is discharged to an earthen basin known as the reserve pit. The drilling fluid typically contains dissolved salts, as a result of intentional addition and inflow from brine-containing geological formations encountered by the wellbore. When drilling of the well has been completed, the reserve pit is closed by pushing in the earthen dikes. This results in a small but nonetheless significant volume of brine-impacted soil at virtually every contemporary onshore well site.

Soil that comes in contact with salt is converted to the saline condition. High soil salinity results in death for established plants and reduced rates of seed germination and seedling growth (3). Such effects occur because soluble salts in the soil make it more difficult for plant roots to extract water by osmosis. Soil unable to support plant life is eventually susceptible to erosion. Thus, a significant scar often develops some time after the contamination of soil with oilfield brine. For example, at the Texon Scar there are mesquite tree stumps with up to one meter of exposed roots, clearly indicating the extent of topsoil loss (1).

Oilfield brines typically exhibit high concentrations of sodium ions. Such brines also produce sodic soils, where most of the cations associated with the surfaces of clay minerals have been displaced (through ion exchange) by sodium. In addition to possessing reduced concentrations of nutrient cations (eg, calcium, magnesium, zinc), sodic soils exhibit poor structure, characterized by poor hydraulic conductivity (reduced permeability to water). The structure of a soil is dominated by the extent to which clay particles aggregate (4). Aggregation is, in turn, controlled by the total concentration of ions in the soil water, the types of ions exchanged onto the surface of the clays, and the amount of organic matter in the soil. The primary consequences of elevated soil sodicity are that water has a difficult time penetrating the hard crust formed on the surface of the soil and, once infiltration occurs, water percolation occurs slowly.

The chloride and sodium ions of brine-impacted soils are typically quite mobile. Thus, the salt that leaches from brine-impacted soil eventually finds its way to nearby surface waters or groundwaters. In regions in which rainfall is not plentiful, this secondary contamination of water resources is of greater consequence than the primary soil contamination. For example, it has been observed (5) that the electrical conductivity of water in a stock pond approximately 700 m downhill from a recent oilfield brine spill was nearly 10 times greater than that of stock ponds located in adjacent watersheds.

REVEGETATION OF BRINE-IMPACTED SOIL

As noted above, contamination of soil with oilfield brine typically denudes the surface of vegetation. This, in turn, sets the stage for erosion, which results in the loss of valuable topsoil. Thus, it is important that vegetation be reestablished as soon as possible on the contaminated soil.

It is not necessary to reduce the soil salinity to zero before attempting revegetation. Many plant species, including some crops, can survive in slightly saline soils. Other species, referred to as halophytes, tolerate soils of moderate-to-high salinity. Of the more than 1600 halophyte species identified to date, nearly 90% are perennials (6). Thus, the replanting of ground cover for brine-impacted soil would not need to be performed annually. Also, nearly 50% of the halophytes are classified as trees or shrubs. The primary disadvantage of trees and shrubs is that they are not fast growing and must, therefore, be introduced by transplanting seedlings.

One revegetation field test considered the treatment of covered reserve pits (2). With surface mulching, standing crops of a variety of grasses were established. The fourwing saltbush (*Atriplex canescens*) was unique in that it survived without mulch, presumably because its roots penetrated deeper into the soil. When transplanted into the reserve pits, this species exhibited a 100% survival rate in moderately contaminated soil (EC < 15 dS/m). Even when the EC was 115 dS/m, the survival rate was still 25%.

It was pointed out (6) that certain halophyte species also assist in the general downward movement of salt in a saline soil. Halophytes characterized by particularly deep root systems take up water deep within the soil profile. This water is then given off to the atmosphere through transpiration from the leaves of the plants. In addition to promoting the downward movement of water, maintaining the surface in a dry condition employs some of the crack stabilization techniques discussed above in enhancing the downward movement of salt in the soil.

The surface application of the gypsum to saline–sodic soils is made in conjunction with revegetation. The application of gypsum at a rate of 56 mg/ha to the surface of bentonitic mine spoils (composed primarily of dispersible clay minerals) enhanced the survival rate of a variety of native and introduced grass species that germinated following the application of mulch (7).

BIBLIOGRAPHY

1. M. L. Weathers, K. R. Moore, D. L. Ford, and C. K. Curlee, *Trans. Gulf Coast Assoc. Geol. Soc.* **44,** 737 (1994).
2. M. L. McFarland, D. N. Ueckert, and S. Hartmann, *J. Range Manage.* **40,** 122 (1987).
3. H. L. Bohn, B. L. McNeal, and G. A. O'Connor, *Soil Chemistry,* John Wiley & Sons, Inc., New York, 1985, pp. 217–246.
4. S. L. Barbour and N. Yang, *Can. Geotechnical J.* **30,** 920 (1993).
5. R. Yates, private communication, 1996.
6. J. Aronson and E. Le Floc'h, in R. Choukra-Allah, C. V. Malcolm, and A. Hamdy, eds., *Halophytes and Biosaline Agriculture,* Marcel Dekker, New York, 1996, pp. 55–71.
7. G. E. Schuman and J. L. Meining, *Soil Sci. Soc. Am. J.* **57,** 1083 (1993).

BTEX QUANTITATION IN OILS BY GC/MS

ZHENDI WANG
MERV FINGAS
Environment Canada
Ottawa, Ontario, Canada

BTEX is the collective name of benzene, toluene, ethylbenzene, and the three xylene isomers (*o*-, *m*-, and *p*- xylene). BTEX compounds have many direct chemical applications and are widely used in the manufacture of paints, synthetic rubber, agricultural chemicals, and chemical intermediates. BTEX levels in crude oil, weathered oil, petroleum products, and oil-spill-related environmental samples are very important parameters for environmental and toxicity concerns. The concentration of BTEX in petroleum products directly affects the physical and chemical properties of petroleum, such as density, viscosity, flash point, dispersibility, emulsion stability, solubility, and weathering. On the other hand, BTEX compounds frequently enter soil, sediments, and ground water, and volatilize into the atmosphere because of accidental oil spills, leakage of gasoline and other petroleum fuels from underground storage tanks and pipelines, or improper oil-related waste disposal practices. Petroleum hydrocarbons including BTEX, *n*-alkanes, and polynuclear aromatic hydrocarbons (PAHs) are some of the most common pollutants found in water in the United States, Canada, and other industrialized countries where petroleum products are used heavily. For example, a U.S. Environmental Protection Agency (EPA) report indicated that there were approximately two million underground tanks storing gasoline in the United States in 1990, and there have been 90,000 confirmed releases reported in the last two years (1). In these cases, BTEX compounds are also used as markers in identifying the presence and determining the composition of weathered fuels originating from leaking underground fuel tanks or pipelines. BTEX are hazardous, carcinogenic, and neurotoxic compounds and are classified as priority pollutants regulated by Environment Canada and the U.S. EPA, especially benzene, which has a very low tolerance standard in drinking water of 5 μg/L. Therefore, the rapid, accurate, and reliable analysis of BTEX in various oil and petroleum product samples is extremely important in view of their wide distribution and frequent pollution in the environment.

Overview of Physical and Chemical Properties

Benzene (C_6H_6) and toluene (C_7H_8) are volatile, colorless, and flammable liquid aromatic hydrocarbons with distinct and characteristic odors. The odour of toluene is somewhat milder than that of benzene. Xylenes and ethylbenzene are liquid C_2-benzene homologues with the molecular formula C_8H_{10}. The three xylene isomers are *o*-xylene, *m*-xylene, and *p*-xylene, which differ in the positions of the two methyl groups on the benzene ring.

Table 1 summarizes some of the important physical and thermodynamic data for BTEX compounds (2,3).

Spacing between the neighbouring carbon atoms of benzene and toluene is constant (0.139 nm), which is less than typical carbon–carbon single-bond spacing (0.154 nm) and more than typical carbon–carbon double-bond spacing (0.134 nm). Each hydrogen atom is 0.108 nm from the carbon atom to which it is bonded. All 12 atoms lie in a single plane, and all bond angles in benzene are exactly 120°.

Benzene has great thermal stability. It has lower heat of formation from the elements than the corresponding structure of three, fixed, ethylene-type double bonds. Similarly, when benzene is decomposed into carbon and hydrogen, it absorbs more energy than predicted by the Kekule formula. Benzene does not react as a compound with double bonds. However, it is more chemically reactive than single-bond hydrocarbons. Benzene undergoes reactions of substitution, addition, and cleavage of the ring, and thus serves as a source for producing many petrochemicals and hydrocarbon materials. Toluene behaves as a base in forming charge-transfer π-complexes and complexes with super acids because of the high electron density in the

Table 1. Physical and Thermodynamic Properties of BTEX Compounds[a]

Property	Benzene [*71-43-2*]	Toluene [*108-88-3*]	*p*-Xylene [*106-42-3*]	*m*-Xylene [*108-38-3*]	*o*-Xylene [*95-47-6*]	Ethylbenzene [*100-41-4*]
Molecular weight	78.11	92.13	106.16	106.16	106.16	106.16
Density at 25°C, g/cm^3	0.8736	0.8623	0.8610	0.8642	0.8802	0.8671
Boiling point, °C	80.094	110.626	138.37	139.12	144.41	136.19
Freezing point, °C	5.530	−94.965	13.263	−47.872	−25.182	−94.975
Refractive index at 25°C	1.4979	1.4969	1.4958	1.4971	1.5054	1.4959
Surface tension, mN/m(= dyn/cm)	28.20	27.92	28.27	31.23	32.50	31.50
Dielectric constant at 25°C	2.28	–	2.27	2.367	2.568	2.412
Flash point, °C	−11	4.0	25	23.2	17	15
Ignition temp, °C	595	552	500	562	502	460
Critical properties						
Critical density, g/cm^3	0.309	0.291	0.286	0.286	0.288	0.286
Critical volume, cm^3/mol	259.0	316.0	379.0	376.0	369.0	374.0
Critical pressure, MPa[b]	4.898	4.109	3.511	3.535	3.730	3.701
Critical temperature, °C	289.01	318.64	343.05	343.90	357.15	343.05
Heat of combustion, kJ/mol[c]	−3275.3	−3911.3	−4559.8	−4556.9	−4570.7	−4568.6
Gross heating value, J/g[c]	41932	42454	42977	42949	43079	43060
Heats of transition, kJ/(mol · K)[c]						
Vaporization at 25°C	33.899	37.990	42.036	42.036	43.413	42.226
Fusion	9.866	6.615	17.112	11.569	13.598	9.164
Formation at 25°C (liquid)	49.08	12.00	−24.43	−25.418	−24.439	−12.456
Entropy of formation	–	219.6	247.4	252.2	246.5	255.2
Solubility in water, g/100 mL	0.18 (25°C)	0.05 (20°C)	0.0191 (22°C)	0.0187 (22°C)	0.0228 (22°C)	0.015 (20°C)

[a] Refs. 2 and 3.
[b] To convert MPa to psi, multiply by 145.
[c] To convert J to cal, multiply by 4.184.

aromatic ring. In this regard, toluene is intermediate between benzene and xylenes in relative basicity and reactivity toward electrophiles. Chemical derivatives of toluene are formed by substituting the hydrogen atoms of the methyl group, by substituting the hydrogen atoms of the ring, and by adding to the double bonds of the ring.

Environmental Concern

Generally speaking, wide industrial and commercial applications of volatile organic compounds (VOC) including BTEX have created many problems involving human health and contamination of air, soil, and various aqueous environments including groundwater, surface water, and wastewater.

Benzene is toxic to aquatic life and hazardous to waterfowl and animals. At room temperature and atmospheric pressure, benzene is sufficiently vaporized to pose an inhalation hazard. Benzene produces acute and chronic adverse health effects. Poisoning occurs by inhalation, ingestion, or by rapid absorption through the skin. Exposure to approximately 4000 ppm benzene results in complete loss of consciousness. Insomnia, agitation, headache, nausea, and drowsiness may persist for weeks after exposure. Prolonged or repeated exposure to benzene results in serious damage to human health, such as blood dyscrasia and bone marrow damage. The toxicity of benzene has been the subject of numerous studies.

Because of the hazards associated with benzene, exposure to benzene in the workplace has been heavily regulated in Canada, the United States, and many other countries. Benzene is one of approximately 40 known human carcinogens. In the United States, the earliest limit on benzene exposure, recommended in 1927 was 100 ppm. Over the decades the upper allowable limits were reduced to 50, 35, then 25 ppm. In 1971, the U.S. Department of Labor, Occupational Safety and Health Administration (OSHA) adopted a permissible exposure limit of 10 ppm benzene measured as an 8-h time-weighted average (TWA). Effective December 10, 1987, the existing standard for benzene was amended by OSHA. The revised standard reduced the permissible exposure limit from 10 ppm (32 mg/m^3) to 1 ppm (3.2 mg/m^3) in an 8-h TWA. The short term exposure limit (STEL) of 5 ppm was set over a 15 min period. The standard also established action level requirements for exposures of over 0.5 ppm of benzene.

Benzene is classified as a hazardous waste by the U.S. EPA under subtitle C of the Resource and Recovery Act (RCRA). Effective September 25, 1990, solid wastes containing more than 0.5 mg/mL benzene must be treated in accordance with applicable RCRA regulations. Benzene emissions and effluent streams from petroleum refineries or benzene processing plants are subject to strict federal regulations of both Canadian and the U.S. Governments. Any industrial process involving manufacture, transport, treatment, or disposal of benzene must comply with federal waste management procedures.

Toluene generally resembles benzene closely in its toxicological properties. However, it is devoid of benzene's chronic negative effects on blood formation. The odor threshold for toluene is approximately 2.5 ppm. In the human system, toluene is oxidized to benzoic acid which in turn reacts with glycine to form hippuric acid (*N*-benzoyl glycine) which is excreted in urine.

The xylenes are not very toxic. They are mild irritants, and skin protection and the air-purifying respirators are recommended. The oral LD_{50} value for rats is 4000 ppm. Prolonged (8-h) exposure of humans should be limited to 200 ppm. Xylenes are mildly toxic to aquatic life, and the threshold limit for crop damage is 800–2400 ppm. Xylenes biodegrade slowly.

Review of Analytical Methods

Analytical techniques for VOCs including BTEX have advanced significantly in recent years, and the development continues. Modern technologies include gas chromatography (GC), mass spectrometry (MS), infrared spectroscopy (IR), ultraviolet (UV) and fluorescence spectroscopy, and ^{1}H- and ^{13}C-nuclear magnetic resonance (NMR); combined techniques, such as GC/MS, GC/Fourier transform infrared (FTIR), laser two-photon ionization with a conductivity measurement, a microwave-assisted process combined with GC, and supercritical fluid chromatography/GC (SFC/GC) are also used. Among these techniques, high-performance capillary GC equipped with flame ionization detector (FID) and capillary GC/MS are the most important and most widely used techniques for analyzing VOCs including BTEX compounds. In recent years, GC/MS systems, which combine chemical separation by GC with spectral resolution by MS, are popular for analyzing lower molecular weight volatile organics. For quantitating target compounds, the selected-ion monitoring (SIM) mode of GC/MS is most selected. GC/MS operating in SIM mode has higher accuracy and precision than produced by conventional full-scan GC-MS methods. The method detection limit for the target analytes are generally lower by almost an order of magnitude. Furthermore, the use of GC/MS in SIM mode often increases the linear range of the instruments for trace analysis. With multichanneled, double-focusing GC-MS, determination of benzene at concentrations as low as 0.1 ppb in air or breath samples has been reported (4). Recently, direct determination of VOCs, including BTEX compounds, at the parts per trillion level in complex aqueous matrices (including untreated river water, seawater, and matrices containing strong acids and base) using membrane-introduction mass spectrometry has been demonstrated (5).

The U.S. EPA has developed a series of matrix-specific methods for VOC analysis, EPA Methods 602, 624, and 8240, based on original work in the early 1970s (6). Although there are minor differences among the methods, all employ a dynamic headspace or purge and trap technique to extract the analytes from the matrix and introduce them into a gas chromatograph. GC separation was originally accomplished with a nonpolar packed column, and, although this is still specified in most of the methods, substituting a 0.53 mm ID capillary column is widely accepted. As mentioned above, the most common method of detection for BTEX and other volatile compounds involves a series combination of a photoionization detector and a flame ionization detector or a mass spectrometer. Although there has been considerable discussion within the EPA about moving toward performance-based methodologies, this has not yet resulted in significant changes in the requirements faced by analytical laboratories.

In most European countries, requirements to use specific analytical methods for environmental analyses of VOC compounds are uncommon. Rather, the analyst is allowed to use professional judgment to specify performance criteria. Several different sample extraction and introduction techniques are used there, including solvent extraction and injection, static headspace analysis, dynamic headspace analysis, and closed-loop stripping. Solvent extraction is the most common, but static headspace is used more frequently and is firmly established in Europe, especially in central Europe and the United Kingdom (7). There are substantial differences between the techniques, and each has its strengths and weaknesses. Several studies have demonstrated that the accuracy and precision of analytical data for VOCs in aqueous samples by the use of the static headspace technique is comparable to that by the dynamic purge and trap methods.

The American Society for Testing and Materials (ASTM) developed a series of test methods especially for determining BTEX in gasolines; Method D4420 for determining BTEX content in finished gasoline by GC, Method D3606 for determining benzene and toluene in finished motor and aviation gasoline by GC, and Methods D3797 and D3798 for determining benzene, toluene, and ethylbenzene impurities in xylene by GC. All ASTM standard test methods employ packed or capillary columns and a gas chromatograph coupled with a thermal conductivity or hydrogen flame ionization detector.

Table 2 presents a brief summary of EPA, API, and ASTM-approved methods for petroleum hydrocarbon analysis (in particular, for analysis of BTEX compounds).

To obtain real-time data to guide field investigations at sites contaminated with VOC constituents, such as BTEX from gasoline, the static headspace method with portable gas chromatography is increasingly used. A method was recently reported which combines the sequential purging technique with static headspace-portable GC analysis to remove coeluting compounds and to improve accuracy in quantitating methyl-*tert*-butyl ether and BTEX compounds in gasoline-contaminated groundwater samples (8).

In 1990, an alternative solvent-free extraction procedure employing solid-phase microextraction (SPME) of organic compounds from aqueous samples was reported. This new extraction method integrates sampling, extraction, concentration, and sample introduction into a single step (9,10) (SOLID-PHASE MICROEXTRACTION). In the SPME method, a polydimethylsiloxane-coated fused-silica fiber is used and is immersed directly into a rapidly stirred aqueous solution with the analytes partitioning between water and the hydrophobic stationary phase on the fibre. The fiber with its holding assembly then is withdrawn, and the analytes are thermally desorbed in the GC injection port onto a capillary GC column for analysis. So far, the method has been successfully used for analyzing a variety of organic compounds, including BTEX and other alkyl-substituted benzenes in water; chlorinated hydrocarbons in water and air; and PAHs, polychlorinated biphenyl, and phenols in water. In addition, the VOCs defined by EPA Method 624 have been analyzed using SPME followed by GC ion trap mass spectrometry. Very recently, quantitative analysis of fuel-related hydrocarbons in surface water and

Table 2. Summary of the EPA, API, and ASTM-Approved Analytical Methods for Oil and Petroleum Products

Method	Technique	Analyte	Description
EPA 8015	GC/FID	TPH	Applicable to nonhalogenated volatile organics and modified for VOC and TPH in soil samples
EPA 8020 and 8020A[a]	GC/PID	BTEX	Applicable to aromatic volatiles
EPA 602	Purge and trap GC/MS	VOC	Analyzing a limited number of purgeable aromatics in petroleum products
EPA 624 and 8240A and 8240B[a]	Purge and trap GC/MS	VOC	For identifying and determining a number of purgeable volatile hydrocarbons
EPA 610	GC/FID	PAH	Determining priority PAHs in municipal and industrial wastes
API-PHC[b]	GC/FID	PHC	Determining PHC
API-GRO[b]	GC/FID	GRO	Determining GRO compounds
API-DRO[b]	GC/FID	DRO	Determining DRO compounds
ASTM 2306	GC	C_8 aromatics	For analysis of ethylbenzene and xylenes
ASTM 3054	GC	Benzene	Determining benzene content in cyclohexane
ASTM 3606	GC	Benzene and toluene	Determining benzene and toluene in finished aviation gasoline
ASTM 3797	GC	BTEX	Determining BTEX impurities in *o*-xylene
ASTM 3798	GC	BTEX	Determining BTEX impurities in *p*-xylene
ASTM 4053	IR	Benzene	Determining the percent benzene in full-range gasoline
ASTM 4420	GC	BTEX	Determining BTEX and heavier aromatics in finished motor gasoline
ASTM 4534	GC	Benzene	Determining benzene content of cyclic products

[a] These methods can be found in the proposed SW-846 Update II, Nov. 1992, Revision 2.
[b] API = American Petroleum Institute; PHC, GRO, and DRO represent petroleum hydrocarbons, gasoline range organics, and diesel range organics, respectively.

wastewater samples by SPME with GC/FID was reported (11). The experimentally determined FID detection limits ranged from ~30 ppt for high molecular weight PAHs to ~1 ppb for low molecular weight BTEX compounds.

BTEX and Alkylated Benzenes in Oils and Petroleum Products

Oils and petroleum products have exceedingly complex compositions. BTEX compounds are the most common aromatic compounds in oil and petroleum products (the amount of BTEX can be up to a few percent of the total weight for some crude oils) and occur in amounts greater than other aromatic compounds. BTEX constitute the most soluble, most mobile fraction of crude oil and many petroleum products. There is no simple and direct method to determine BTEX and other alkyl-substituted benzene constituents in crude oil, weathered oil, and oil-spill-related sediment or soil samples. Obviously, there is a gap for determining BTEX between petroleum and the light petroleum product, gasoline. Because of the low boiling points and very volatile nature of BTEX compounds, analytical methods for determining BTEX in crude oil and oil-spill-related samples therefore must be (*1*) specific, ie, the method must be able to provide information about specific analytes of crude oil and spilled oil samples (BTEX and other target alkyl benzenes for this work). In addition, the analytical procedures must be specific, preferably without a solvent evaporation and concentration step which loses BTEX compounds. (*2*) sensitive, which is especially important for determining BTEX and other substituted benzene constituents in weathered and biodegraded oil samples.

To achieve these objectives, a rigorous but rapid and reliable approach to identifying and directly determining BTEX and other alkyl-substituted benzene components in oils and oil-related environmental samples has been developed using GC/MS. The efficiency and reproducibility of the method were demonstrated by separating and determining a standard aromatic alkylbenzene mixture whose the alkyl groups ranged from C_1 to C_7. The BTEX and other alkyl-substituted benzene components were identified by mass spectral data, comparison of GC retention data with reference standards, and calculation of retention index values and comparison with those reported in the literature. Quantitation of BTEX and (BTEX + C_3-benzenes) was accomplished with an internal standard method. The extensive qualitative and quantitative information obtained about alkylbenzenes is useful for evaluating weathering behaviour, toxicity, composition, and concentration changes of oil in the short term following a spill. The data are also used to evaluate the fate and transport of alkylated benzene compounds and other petroleum hydrocarbons in the environment and to assess the biological effects and damage of the spill to the environment and natural resources.

GC/MS METHODS FOR DETERMINING BTEX AND OTHER ALKYL SUBSTITUTED BENZENES IN OILS

The Emergencies Science Division (ESD) hydrocarbon fingerprinting program uses GC/FID analysis to evaluate total petroleum hydrocarbons and to categorize the product type in fresh to moderately degraded hydrocarbon samples. The GC/MS analysis provides data on BTEX and alkylbenzenes, PAHs, and biomarkers of the sample, which are used to support GC/FID results, to characterize and quantify the composition changes in weathered/degraded oil products, and to provide additional sensitivity (parts per trillion in water samples) for low-level hydrocarbon source-tracking studies. Figure 1 summarizes the method for oil hydrocarbon analysis.

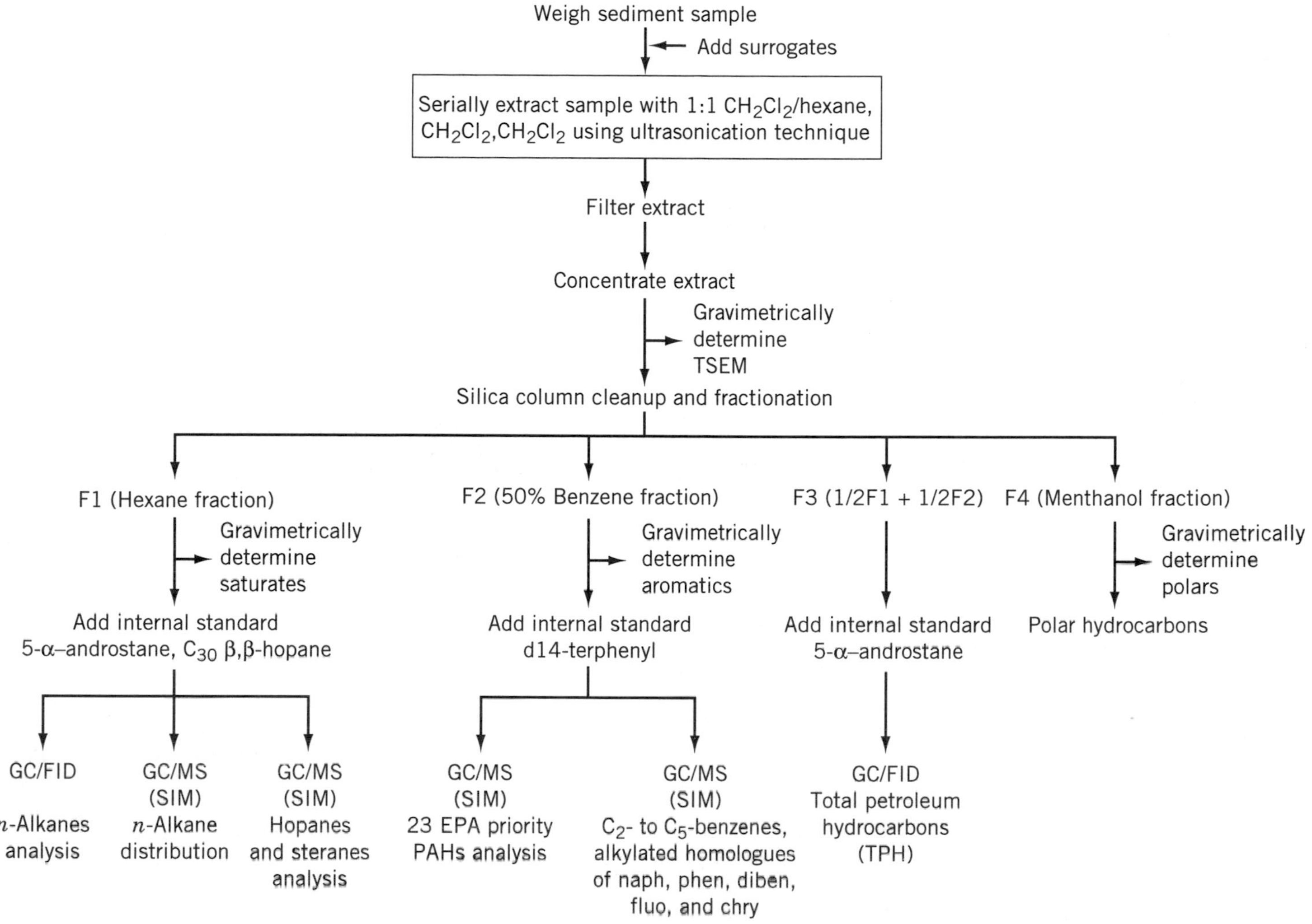

Figure 1. Diagram of analytical protocol for oil.

FINGERPRINTING OF ALKYLBENZENE CONSTITUENTS IN OILS AND PETROLEUM PRODUCTS

The type and concentrations of specific oil constituents in environmental samples are dictated by the origin and nature of the spilled oil. Each oil has a different "fingerprint" and distribution of hydrocarbons including *n*-alkanes, aliphatics, alkylbenzenes and alkylated PAHs, and biomarker terpanes and steranes (12). For crude oil, the hydrocarbon distribution depends greatly on its geological source; for weathered oil, the hydrocarbon distribution depends on the weathering conditions and also on the weathering time (short-term or long-term). The low molecular mass hydrocarbons in weathered oil samples are lost and some degradation-resistant compounds at the same time increase in relative concentration because of weathering effects. This results in significant changes in chemical composition and concentration of the oil. BTEX and alkyl benzene compounds are important constituents of oil and probably are the most volatile aromatic compounds in oil. The composition and concentration of alkylbenzenes directly affect the physical and chemical properties of petroleum. Although the fingerprints of alkylbenzenes are generally not used to differentiate and source-track oil, they provide valuable information about the presence of spilled oil, the fate of spilled oil in the environment, and changes in chemical composition caused by weathering and biodegradation. In addition, the immediate toxic effects of this group of compounds following a spill emphasize the usefulness of analytical data for alkylbenzenes in biological assessment studies.

Figure 2 presents the alkylated PAH homologue distributions in representative light ASMB oil (density = 0.839), medium Maya oil (density = 0.923), heavy Bunker C oil (density = 0.974), and Shell diesel oil. For comparison, the same concentration scale (0–5000 μg/g oil) was applied to the *y*-axis for the alkylated PAH distribution graphics. These three oils and one diesel oil obviously exhibited different PAH signatures. The unusually high abundances of alkylnaphthalenes and alkylphenanthrenes in the Shell diesel oil and Bunker C oil, relative to other alkylated PAH homologous series, are very pronounced. Another significant feature of the alkylated PAH distribution observed from Figure 2 is that, compared to the three crude oils, the diesel oil has extremely low concentrations of the multiring alkylated chrysene and sulfur-containing alkylated dibenzothiophene series, which were determined to be only 2.4 and 125 μg/g oil, respectively. The distinctive character of each oil, as evidenced by the alkylated PAH homologue distribution, is apparent.

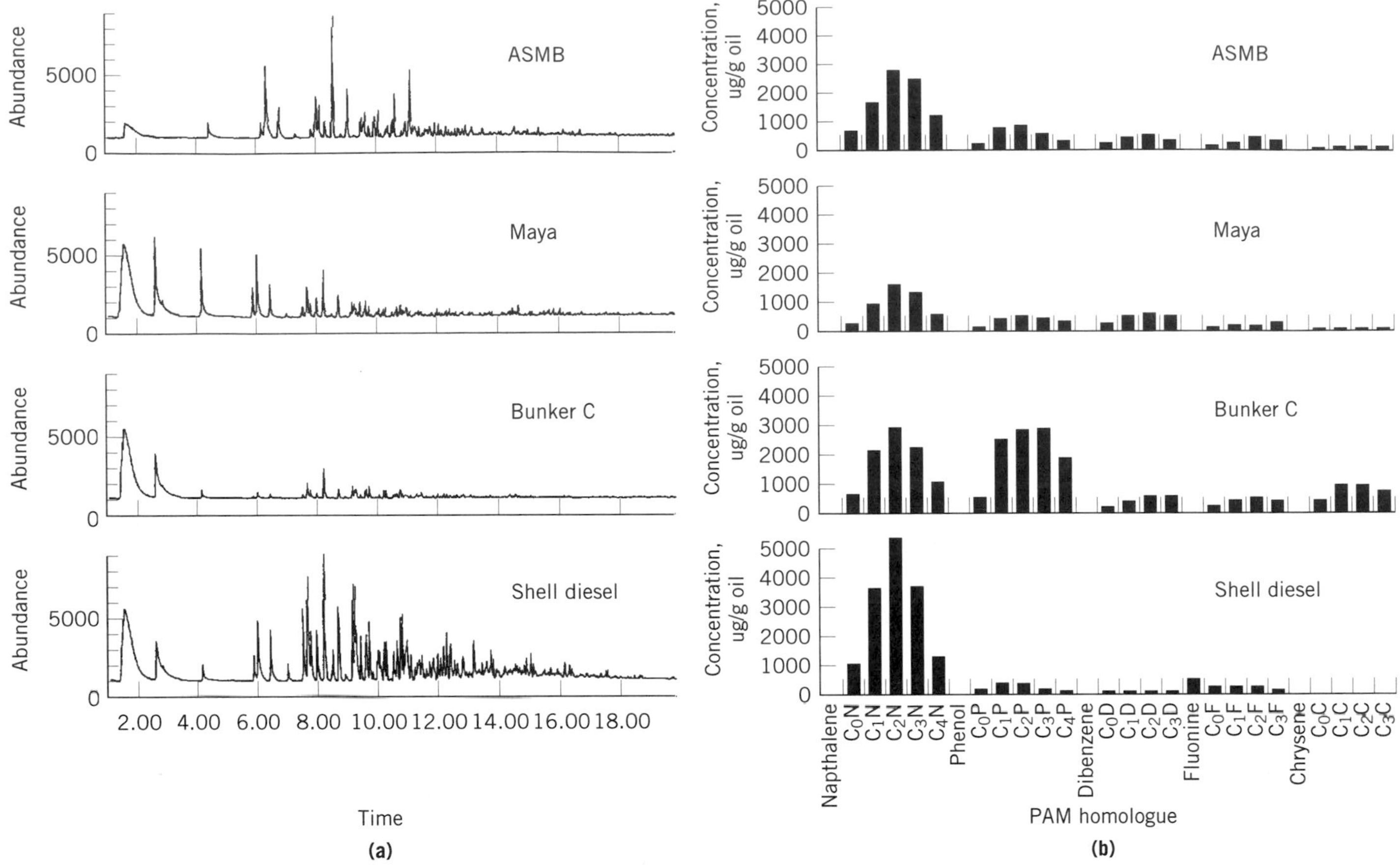

Figure 2. Alkylated PAH homologue distributions (with time in minutes) (**a**) and corresponding GC/MS total ion chromatograms of alkylbenzene compounds (**b**) in representative light ASMB oil (density = 0.839), medium Maya oil (density = 0.923), and heavy Bunker C oil (density = 0.974), and a petroleum product, the Shell diesel. N, P, D, F, and C represent naphthalene, phenanthrene, dibenzothiphene, fluorene, and chrysene, respectively; 0, 1, 2, 3, and 4 represent carbon numbers of alkyl groups in alkylated PAH homologues.

The ability to distinguish between different crude oils on the basis of alkylated PAH distributions is further enhanced when the distribution fingerprints of BTEX and alkylbenzenes are used. Figure 2**a** presents the corresponding total ion chromatograms of the alkylbenzene distribution for the three oils representative and the diesel oil. Figure 2 shows that the distribution of alkyl benzene compounds in these three oils and the Shell diesel differ significantly, because of their different distribution patterns and profiles of alkylated benzenes and also because of the strikingly higher abundances of C_3- and C_4-benzenes relative to BTEX in the Shell diesel oil.

BIODEGRADATION OF BTEX AND LIGHT AROMATIC HYDROCARBONS

Bioremediation is of interest worldwide as an oil spill cleanup option, especially for inaccessible or sensitive environments. Petroleum hydrocarbons are divided into four classes: the aliphatics, the aromatics, the resins (including pyridines, quinolines, carbazoles, sulfoxides, and amides), and the asphaltenes. Petroleum hydrocarbons differ in their susceptibility to microbial attack and have generally been ranked in the following order of decreasing susceptibility: *n*-alkanes > branched alkanes > low molecular weight aromatics > high molecular weight PAHs > multicyclic biomarker terpane and sterane compounds. In many cases, biodegradation rates are highest for the aliphatics, followed by the light aromatics including BTEX and low molecular weight alkylbenzenes. Once BTEX components have entered an aquifer, biological transformation and volatilization are the major removal mechanism for BTEX contamination of soils and groundwater.

The biodegradation of oils and petroleum products in the environment is a complex process, whose qualitative and quantitative aspects depend on many factors, such as the nature and amount of the oil or hydrocarbons present, the ambient and seasonal environmental conditions (temperature, oxygen, nutrients, salinity, pressure, water activity, and pH), and the composition of the autochthonous microbial community. Bacteria are the key agents of degradation and play the dominant role in marine ecosystems and in freshwater and terrestrial environments (13).

Remediation of BTEX-contaminated soils focused historically on either excavation and disposal or treatment of contaminated soils or pumping of groundwater to remove dissolved contaminants from the aquifer. Both approaches are extremely costly. Recent emphasis has focused on developing *in situ* remediation processes to reduce remediation cost and site disruption. Aerobic and anaerobic biodegradation are the two basic approaches (14).

In recent years, more and more laboratory studies on biodegradation of oil hydrocarbons including BTEX compounds have been carried out under various conditions. Interim guidelines for evaluating the efficacy of oil spill bioremediation agents (OSBA) have been produced by ESD of Environment Canada (15) and a standard method has also been published by the National Environmental Technology Applications Centre (NETAC) for the U.S. EPA (16). Both contain simple laboratory shaker flask tests. The Environment Canada test uses a defined microbial inoculum under freshwater conditions, and the U.S. EPA test currently uses a natural seawater inoculum under warm marine conditions. Protocols for evaluating oil biodegradation potential were later to develop, and work has been done by two groups to date (17), the Emergencies Science Division and the U.S. National Oceanographic and Atmospheric Administration (NOAA).

CONCLUSIONS

Analyzing the aromatic hydrocarbons benzene, toluene, ethylbenzene, xylene isomers, and alkylbenzenes is important to research scientists and chemists in the petrochemical, polymer, fine chemical, and environmental industries. A method using high performance gas chromatography mass spectrometry operating mainly in the selected ion monitoring mode can characterize alkylated benzene compounds and directly quantify BTEX and BTEX + C_3-benzenes in oils and petroleum products. Analytical data are used to differentiate oils and to monitor the oil-weathering and biodegradation processes under variable environmental conditions.

For analysis of volatile hydrocarbons including BTEX and alkylbenzenes in complex matrices, such as crude oils and oil-spill-related environmental samples, two criteria must be met to provide both qualitative and quantitative information on the composition of monoring aromatic components: (*1*) simplicity and (*2*) reliability. The simplicity and directness of the method described in this article is apparent. The reliability of the method has been demonstrated by separating and determining standard aromatic alkyl benzene mixtures with alkyl groups ranging from C_1 to C_7, guaranteed by rigorous quality control measures. The distribution fingerprints of alkyl-benzene homologues vary from oil to oil, and are readily affected by sample treatment because the compounds are volatile. However, this method needs neither fractionation nor concentration. Also, by using GC/MS technique in SIM mode, the interference from light aliphatics compounds in determining of target aromatic compounds (each compound has its own characteristic target and qualifying ions at different retention times) is largely minimized. Thus, the required analytical accuracy and precision are obtained.

Using this method, the concentrations of BTEX and C_3-benzenes in over 200 oils were quantified. In addition, the effect of weathering on compositional changes of alkylbenzenes were studied, and the plots of the concentrations of BTEX and (BTEX + C_3-benzenes) versus weathering percentages for several weathered oil series were graphically depicted. Graphs are used to estimate the weathering extent, especially for those oils weathered for short-periods when the loss of BTEX and C_3-benzenes is significant.

BIBLIOGRAPHY

1. *OUST, LUST Trust Fund Monthly Progress Report,* Office of Underground Storage Tanks, U.S. EPA, Washington, D.C. 1990.
2. *Kirk-Othmer Encyclopedia of Chemical Technology,* 4th ed., Vol. 4, 1993; 3rd ed., Vols. 13 and 24, 1984; John Wiley & Sons, Inc., New York.
3. *Ullmann's Encyclopedia of Industrial Chemistry,* 5th ed. Vol. 3, VCH, Weinheim, Germany, 1985.
4. L. D. Gruenke and co-workers, *J. Anal. Toxicol.* **10,** 225–233 (1986).
5. S. Bauer and D. Solyom, *Anal. Chem.* **66**(24), 4422–4431 (1994).
6. T. A. Bellar and J. J. Lichtenberg, *J. Am. Water Works Assoc.* **66,** 703 (1974).
7. T. C. Voice and B. Kolb, *J. Chromatogr. Sci.* **32,** 306–311 (1994).
8. M. J. Lacy and co-worker, *J. Hazard. Mater.* **43,** 31–44 (1995).
9. C. L. Authur and J. Pawliszyn, *Anal. Chem.* **62,** 2145–2148 (1990).
10. Z. Zhang, M. Yang, and J. Pawliszyn, *Anal. Chem.* **66,** 844A–853A (1994).
11. J. J. Langenfeld, S. B. Hawthorne, and D. J. Miller, *Anal. Chem.* **68,** 144–155 (1996).
12. Z. D. Wang, and M. Fingas, *J. Chromatogr.* **712,** 321–343 (1995).
13. J. G. Leahy and R. R. Colwell, *Microbiolog. Rev.* **54**(3), 305–315 (1990).
14. G. F. Bowlen and D. S. Kosson, in L. Y. Young and C. E. Cerniglia, eds., *Microbial Transformation and Degradation of Toxic Organic Chemicals,* Wiley–Liss, New York, 1995, Chapt. 14, pp. 515–542.
15. S. Blenkinsopp and co-workers, in *Proceedings of the 1995 International Oil Spill Conference,* American Petroleum Institute, Washington, D.C., 1995, pp. 91–96.
16. National Environmental Technology Applications Centre (NETAC), *Evaluation Methods Manual: Oil Spill Response Bioremediation Agents,* University of Pittsburgh Applied Research Center, Pittsburgh, Pa., 1993.
17. R. Z. Hoff and co-workers, in *Proceeding of the 18th Arctic and Marine Oil Spill Program (AMOP) Technical Seminar,* June 14–16, 1995, Edmonton, Canada, Environment Canada, Ottawa, 1995, pp. 1233–1241.

BURIED RADIOACTIVE WASTE

JAFFER MOHIUDDIN
U.S. Department of Energy
Germantown, Maryland

GEORGE SCHNEIDER
U.S. Department of Energy
Idaho Falls, Idaho

ANDREW BOUIE
BDM International
Gaithersburg, Maryland

BACKGROUND

During World War II and the Cold War, the United States developed a massive industrial complex to research, produce, and test nuclear weapons. This nuclear weapons complex included nuclear reactors, chemical processing buildings, metal machining plants, laboratories and maintenance facilities that manufactured tens of thousands of nuclear warheads and conducted more than one thousand nuclear explosion tests.

Weapons production stopped in the late 1980s, initially to correct widespread environmental and safety problems and was later ended indefinitely because of the end of the Cold War. The work remaining and the subject of this analysis is the legacy of thousands of contaminated areas and buildings and large volumes of "backlog" waste and special nuclear materials requiring treatment, stabilization, and disposal. Approximately one-half million cubic meters of radioactive high level, mixed, and low level waste must be stabilized and safeguarded, including enough plutonium to fabricate thousands of nuclear weapons.

In 1989, the U.S. Department of Energy (DOE) established the Environmental Restoration and Waste Management program, now called the Environmental Management program, to consolidate activities and accelerate efforts to deal with the inactive production facilities and the accumulated waste, contamination, and materials. This program is responsible for the maintenance stabilization, and the environmental restoration and waste management work at virtually the entire nuclear weapons complex not used for continued weapons activities. The Environmental Management program is one the largest environmental stewardship programs in the world with 150 sites in over 30 states and Puerto Rico.

ENVIRONMENTAL MANAGEMENT PROGRAM

The Environmental Management program (EM) encompasses six major functional areas: (*1*) environmental restoration, (*2*) waste management, (*3*) nuclear material and facility stabilization, (*4*) science and technology development (*5*) landlord, and (*6*) national program planning and management. These six areas are all interrelated. Waste management involves the safe treatment, storage, and disposal of existing waste and waste yet to be generated. Environmental restoration activities address remediation of contaminated soil and water and decommissioning of contaminated surplus facilities. Nuclear material and facility stabilization involves stabilizing and consolidating special nuclear materials, such as plutonium and highly enriched uranium and deactivating surplus facilities to a safe, low maintenance condition while awaiting final decommissioning. Science and technology development refers to a variety of basic and applied research activities that explore more effective and less expensive remedies to address the environmental and safety problems of the EM. Landlord functions represent cross-cutting site support activities, such as road maintenance and fire and ambulance services. National program planning and management encompasses headquarters functions.

Environmental Restoration Program

The Environmental Restoration Program (ER) is the primary user of technologies being developed by the DOE for the remediation of buried waste and, hence, warrants expanded discussion. The ER's overall mission is to protect human health and the environment from risks posed by inactive, surplus facilities and contaminated areas. The program is accomplishing this mission by remediating sites and facilities in the most cost-efficient and responsible manner possible to provide for future use while complying with applicable environmental regulations. Environmental restoration activities are ranked by various factors, including the goals of reducing risks at all sites and complying with existing laws, regulations, and agreements. Most actions are designed to clean up or contain contamination in the environment (including reactors and chemical processing buildings). Related activities to support these actions include immediate treatment of contaminated soils or ground water, packaging of waste for commercial treatment and/or disposal, and on-site disposal of consolidated contaminated media, such as soils and building rubble. Cleanup goals and remedies for each contaminated area are decided through processes established by federal and state laws and other legal agreements. These processes involve decision makers outside of the DOE, such as the states, the US Environmental Protection Agency (EPA) and the U.S. Nuclear Regulatory Commission. The environmental restoration process described below is a generic approach based primarily on requirements of the Comprehensive Environmental Response, Compensations and Liability Act (CERCLA). Other statutes that influence the process include Resource Conservation Recovery Act (RCRA) and National Environmental Policy Act (NEPA).

The Remediation Process

Initially, the DOE characterizes a contaminated area to identify contaminants, determine the extent of contamination, and assesses potential threats to public health and the environment. If there is a significant contamination problem and a fast and limited cleanup or containment action can mitigate this problem, the DOE conducts an expedited response or interim action. To date, the DOE has completed over 500 such limited actions, avoiding larger contamination problems that could have resulted from delay.

Upon completing characterization, the DOE performs a detailed analysis to quantify existing risks and evaluate remedial alternatives. The analysis is followed by a formal decision process, including public meetings and a formal comment period. If the results of the analysis indicate that a contaminated area is not a threat to public health or the environment or that a previously completed limited action adequately remediated the contamination, the DOE decides to take "No Further Action" in conjunction with the regulators, the EPA, and the host state. If, however, there is a threat, the Department identifies and implements the appropriate remedial action.

The DOE reviews potential activities to determine how much waste will be generated in the cleanup and provides for storage, treatment, and/or disposal. If actual cleanup (for example, removal) is not practical or not required because of decisions regarding future land use, the DOE may take steps to stop or slow the spread of contamination by containment technologies. Actions depend on the contaminants and the medium (for example, soil and groundwater) in which they are found. Contaminants, such as hazardous organic chemicals or fuel oil, are highly mobile but are effectively removed from the media and destroyed. Heavy metals and radioactive materials are often less mobile but cannot be destroyed, even when it is possible to remove them from the media.

Radioactivity decays naturally over time, but it can take from a few days to hundreds of thousands of years to become less harmful. During this time, heavy metal contaminated soils and radioactive waste that pose threats to public health and the environment must be contained, stabilized, or moved to a safer place. Contaminated containment structures that have not been fully remediated or stabilized must be continuously monitored and maintained.

To date, the DOE has completed 119 remedial projects. Another 111 projects are under way. These projects have included cleanup of contaminated soils, construction of groundwater treatment facilities, and retrieval of buried waste. The Department is positioned to accomplish even more cleanup in the near term as many characterization activities are complete or nearly complete, and many formal cleanup decisions will be made over the next few years.

Figures 1 and 2 are pictures of old waste sites and means by which waste was buried. These picture provide a sense of the complex nature of the DOE environmental restoration and waste management problem.

Applicable Statutes

The DOE must manage its sites and facilities in compliance with applicable federal, state, and local regulations. The major applicable federal statutes are the following:

- Resource Conservation and Recovery Act (RCRA), as amended
- Comprehensive Environmental Response, Compensation, and Liability Act (CERCLA), as amended
- National Environmental Policy Act (NEPA)
- Federal Facility Compliance Act (FFCA)
- Clean Air Act (CAA), as amended
- Clear Water Act (CWA), as amended
- Safe Drinking Water Act (SDWA), as amended
- Toxic Substances Control Act (TSCA), as amended
- Atomic Energy Act (AEA), as amended
- Uranium Mill Tailings Radiation Control Act (UMTRA)
- Low-Level Waste Policy Act (LLWPA), as amended

BURIED WASTE CHARACTERISTICS

The following subsections address the waste volumes, characteristics, and basic assumptions on which the information presented is based.

Types of Waste

The waste at DOE facilities is divided into categories by regulatory definitions and include high level, transuranic, mixed transuranic, low level mixed, uranium mill tailings, hazardous, sanitary, and special case wastes.

High level wastes are highly radioactive material resulting from reprocessing spent nuclear fuel and irradiated targets, including liquid waste produced directly in reprocessing and any solid material derived from such liquid waste that contains fission products in sufficient concentrations. Most of the Department's high level waste came from the production of plutonium. A smaller fraction is related to recovering enriched uranium from naval reactor fuel and typically contains hazardous chemicals and heavy metals. It must be isolated from the environment for thousands of years. Liquid high-level waste is typically stored in large tanks, whereas powdered waste is stored in bins. All high level waste is managed as mixed waste.

Transuranic (TRU) and mixed transuranic wastes are generated during nuclear weapons production, fuel reprocessing, and other activities involving long-lived transuranic elements. They contain plutonium, americium, and other elements with atomic numbers higher than uranium. Some of these isotopes have half-lives of hundreds of thousands of years and, therefore, require long-term isolation. Since 1970, transuranic waste has been stored temporarily in drums at sites throughout the complex. Mixed transuranic waste contains both radioactive and hazardous waste.

Low level wastes (LLW) are any radioactive waste not classified as high level waste, transuranic waste, spent nuclear fuel or by-product tailings containing uranium or thorium from processed ore. They are produced by every process involving radioactive materials. Low level waste has a wide range of characteristics, but most of it contains small amounts of radioactivity in large volumes of materials. Some waste in this category (for example, irradiated metal parts from reactors) has more radioactivity per unit volume than the average high level waste from nuclear weapons production. Most low level waste has been buried in shallow trenches. A limited inventory remains stored in boxes and drums.

Low level mixed wastes (LLMW) are low level radioactive wastes that also contain hazardous waste. A significant portion of the Department's mixed waste is low level mixed waste.

Uranium mill tailings are large volumes of material left from uranium mining and milling. Although this material is not categorized as waste, tailings are of concern because they emit radon and because they are usually contaminated with toxic heavy metals, including lead, vanadium, and molybdenum.

Hazardous wastes are regulated under Subtitle C of the RCRA. They contain hazardous constituents but no radionuclides. Hazardous waste are generated at most Department of Energy installations in a variety of quantities and forms (for example, laboratory solutions, acids, bases and degreasing agents).

Sanitary wastes include solid sanitary waste (for example, garbage, rubble, or debris) regulated under Subtitle D of the RCRA and liquid sanitary waste regulated under the CWA.

Special case wastes are not high level or transuranic, but require greater confinement than shallow land burial.

Overall Complex Summary

Through 1990 approximately 2.1 million cubic meters of radioactive solid wastes have been buried or stored at the six sites included in this assessment.

Prior to 1970, TRU and LLW were commonly commingled in the same burial structure at many DOE facilities. During this time, TRU was defined as material with a radionuclide content greater than 10 nCi/g.

Beginning in 1970, TRU-bearing wastes were packaged and sealed in containers and segregated from LLW in retrievable storage trenches. In 1982, the lower radionuclide content limit for TRU was revised to 100 nCi/g. Consequently, certain assumptions must be made to quantify the volume of waste likely to be LLW and suspected to be TRU. These assumptions vary by facility.

The RCRA established the mixed waste classification to account for solid waste contaminated with regulated hazardous waste. Volumes of buried and/or stored low level mixed waste (LLMW) and transuranic mixed waste (TRUM) are included in this assessment; no assumptions are made to estimate the volume of mixed wastes buried prior to 1986.

Buried waste is further delineated by identifying its various physical characteristics. Among the categories included in this assessment for TRU are absorbed liquids or sludge, combustibles, noncombustibles, particulates, filters, and other physical characteristics. The volumes in this table were derived from the Integrated Data Base Report (IDB), 1993: U.S. Spent Nuclear Fuel and Radioactive Waste Inventories, Projections, and Characteristics, December 1994.

Among the categories included in this assessment for LLW and LLMW are biological materials, contaminated equipment, debris, dry solids, sludge, and other physical characteristics.

Hazardous constituents of the buried waste are also important to quantify for this assessment. However, since mixed wastes were not regulated until 1986, only limited characterization is available from each of the DOE facilities.

INTEGRATED APPROACH

The U.S. Department of Energy (DOE) Office of Environmental Management has developed a new approach in using science and technology resources to solve its environmental problems. This process has been outlined in *A New Approach to Environmental Research and Technology Development at the US Department of Energy Action Plan* (1). According to this plan, the Office of Environmental Management is faced with five major high priority problems. Therefore, the Office of Environmental Management established five focus areas, as follows: Mixed Waste Characterization, Treatment, and Disposal; High Level Waste Tank Remediation; Landfill Stabilization; Contaminant Plume Containment and Remediation; and Facility Transitioning, Decommissioning, and Final Disposition. More focus areas may be established as needed. The Office of Environmental Management in cooperation with other DOE research organizations will use the best science and technology available to solve the most challenging set of environmental problems in the world. Integral to this new approach is an awareness of program needs obtained from the Focus Area's customers (DOE Office of Environmental Management and the DOE Office of Waste Management), other potential users, regulators, and stakeholders. Then these needs are matched to the suppliers of technological solutions. The goal of this new approach is to conduct a research and technology development program focused on overcoming the major obstacles and taking advantage of the best talent in DOE, other national science organizations, industry, and universities to significantly reduce the cost and to advance the cleanup of the DOE sites.

To manage and implement the *New Approach* to environmental research and technology development, DOE has assembled an integrated team. This structure is depicted in Figure 4. Elements of this structure are briefly discussed below:

Steering Committee: provides top-level oversight and ensures that the technology development program is responsive to the Office of Environmental Management's priorities. Membership includes senior-level representatives from within DOE and other government agencies, such as the U.S. Environmental Protection Agency, National Institutes of Standards and Technology, and the Department of Defense.

Management Team (one for each focus area): translates the Steering Committee's policy guidance and direction,

Figure 1. Buried waste at the Hanford reservation.

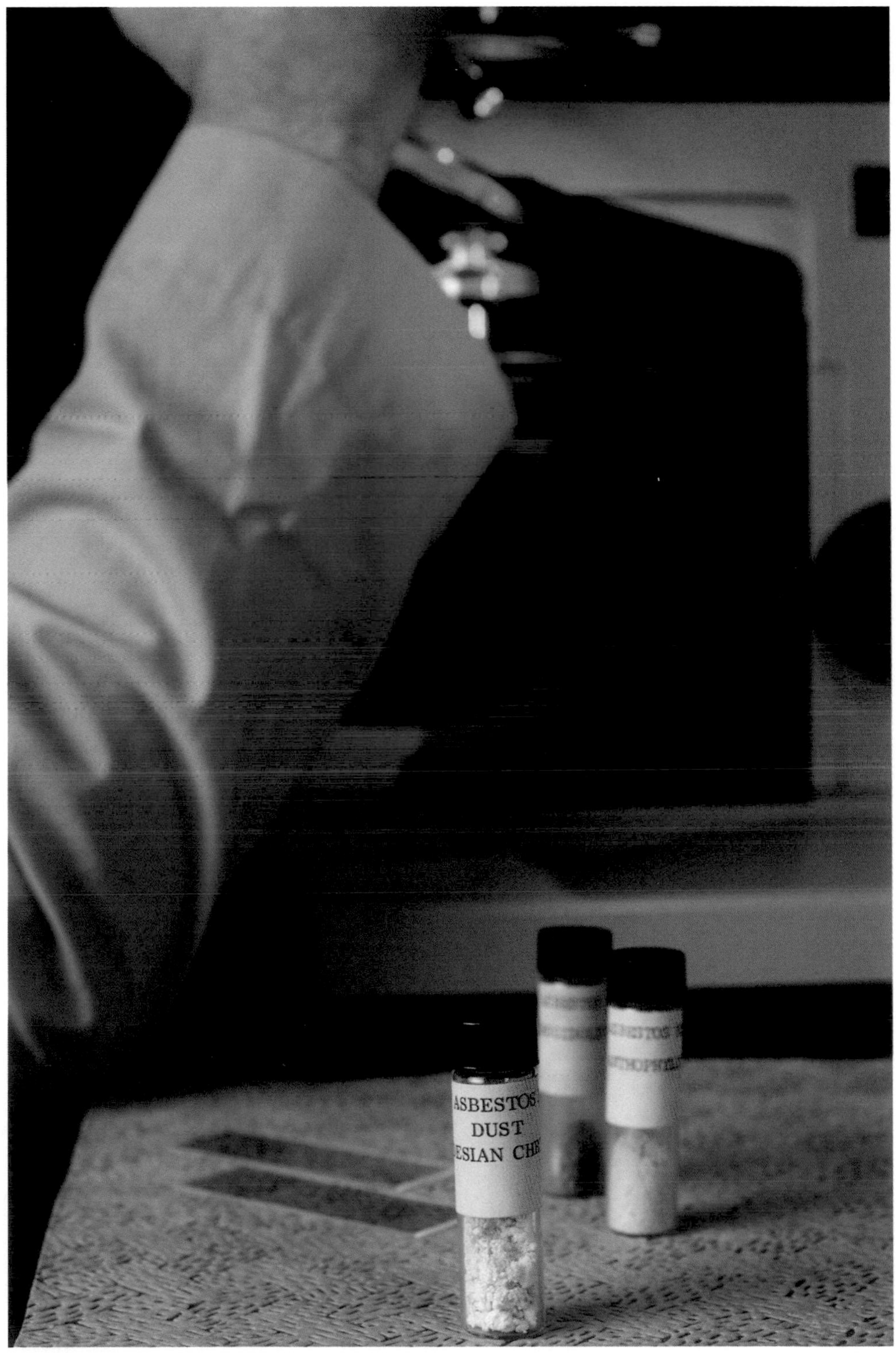

Plate 1 Asbestos Dust Sample. See entry "Asbestos, Analysis of"

Plate 2 Smoke from Exhaust of Car. See entry "Automobile Emissions, On Road"

Plate 3 Hospital Waste. See entry "Biohazardous Waste"

Figure 2. Buried waste at Oak Ridge National Laboratory.

validates and prioritizes needs within the focus area, issues program guidance for technology development, commits resources, coordinates and integrates with the other focus areas, solicits and receives customer needs from the Site Coordination Groups and appropriate Headquarters elements. In addition, it manages and is accountable for project cost and schedule, integrates systems, including technical, cost and schedule components within the focus area, develops a life-cycle approach to solve problems and satisfy customer needs, maintains awareness of available and developing technologies, integrates national technology transfer and commercialization by involving the private sector.

Focus Area Review Groups (one for each focus area): conducts independent reviews to recommend priorities and evaluate the technical merit of proposed and ongoing programs and technologies.

Site Technology Coordination Groups (one for each DOE site): established at each DOE site to rank site needs and required implementation dates. Involvement at the local level provides early stakeholder input which is coupled with technology needs to improve technology selection and development.

Cross-Cutting Technology Development Areas: develops technology common to all or several of the focus areas to avoid duplication and to coordinate development activities. These technology areas are characterization (site characterization, waste assay, and monitoring technologies), robotic technology (remote controlled, telerobotic, and automated technologies), and efficient separation technologies (technologies for separating heterogeneous wastes and for selectively separating a waste component from a mixture before or after treatment).

The overall objectives of this *New Approach* are to (*1*) accelerate development of graded risk-based technology solutions to meet the Office of Environmental Management's highest priorities; (*2*) ensure that applicable technologies are identified and/or developed and delivered ready for use; (*3*) facilitate decision making, reduce uncertainty, and minimize/reduce net remediation costs through better understanding of site problems; (*4*) ensure that basic research necessary to develop remediation technologies is addressed; and (*5*) maintain a program that is under managerial and financial control.

Systems Approach

The systems approach looks at the entire remediation solution from cradle to grave to enhance all or part of a system to remediate a site by using a configurational option scheme illustrating the technology development options available and necessary. The options begin with site assessment and branch into other configurations as follows: (*1*) retrieve the waste; (*2*) treat the waste *in situ* (stabilization); (*3*) contain the waste; and (*4*) monitor the waste. The resultant technology groupings are assessment technology for site and waste assessment; containment technology (which we define as a process that confines the waste to an area); stabilization technology (which we define as a process which does something to the waste); *ex situ* treatment technology; and disposal technology.

CHARACTERIZATION TECHNOLOGIES

Characterization technologies have traditionally relied on environmental media sampling and laboratory analysis to detect contamination or to determine soil properties. The major weaknesses in assessment methods are generally their inability to resolve spatial values in heterogeneous soil to detect contamination and its location. Monitoring is the study of the progress of or final success in a remedial action or closure method. These methods are generally an extension of environmental media sampling and laboratory analysis for characterization. These methods also are limited by near real-time feedback to remediation operations and assessments of long-term verification of stabilized or contained buried waste after closure.

CONTAINMENT TECHNOLOGIES

Containment technologies are based on the installation of surface or subsurface barriers to isolate the waste or soil contamination. This type of approach eliminates near-term threats to human health and the environment. However, natural geologic forces could eventually allow some wastes to escape. Containment technologies for in surface covers currently use clay-based hydraulic barriers to limit precipitation infiltration into buried waste zones. Current trends show that in arid environments traditional clay-based covers desiccate, crack, and fail. Subsurface barriers have been limited to groundwater containment or diversion with little application as unsaturated soil barriers.

STABILIZATION TECHNOLOGIES

Waste stabilization technologies temporarily or permanently reduce landfill contaminant mobility or toxicity. This is often accomplished through *in situ* treatment or encapsulation. *In situ* stabilization avoids removal costs or reduces removal risks by controlling the contaminant mobility during waste handling. Existing stabilization technology is primarily cement-based grout injection into the waste or soil. Cement type grouts do not have the necessary long-term durability or widespread applicability across the DOE complex.

RETRIEVAL

Removal of contaminated soils and/or buried waste is achieved today but at a high cost and significant risk to on-site workers. For radioactive or mixed wastes, costs are driven by contamination control issues, waste segregation and characterization, the limited capacity for off-site disposal, and liability for transfer. Unless the restoration site is very small and has a high demand for future use, excavation as a remedial measure is often far too costly and is not selected in favor of containment.

TREATMENT TECHNOLOGIES

Treatment (*ex situ*) technologies must be versatile to handle many waste forms and waste mixtures and provide a long-lived (greater than 100 years) final waste form. Incineration has been the technology chosen to treat a large number of hazardous wastes. However, incineration has come under much scrutiny in the last few years because of questions about stack emissions. Non-thermal processes generally are capable of handling only limited waste forms and mixtures.

ECONOMICS OF TECHNOLOGY DEVELOPMENT

For the DOE to evaluate the economics of technology development, there are three major factors to consider: life-cycle costs; mortgage reduction potential; mission enabling ability. These factors provide the critical information necessary for decision making relative to the technologies selected for development, go no-go decision making during the developmental process, and technology implementation/commercialization.

Life-Cycle Costs

Life-cycle costing has two components. The first is the life-cycle cost for the development. The second is the life-cycle cost for implementing the technology in full-scale operation. During the developmental process go no-go decisions are made if there is a clear understanding of the relative cost of continuing development vs the incremental increase in the capability of the technology. These two pieces of life-cycle costing, however, cannot be truly separated. The complete set of information is necessary to calculate the mortgage reducing potential.

Mortgage Reduction

The mortgage reducing potential is the factor that tells the DOE what it can save if a particular technology or technology system is developed and implemented to solve a like set of problems. The difference between the life-cycle cost and the mortgage reducing potential is the basis for making decisions about which technologies to develop. The larger the difference, the more compelling is the economic reason to develop the technology.

Mission Enabling

The other main factor for consideration within the DOE complex is whether a particular technology or technology system speeds up or enables the completion of a mission. In the DOE, as stated earlier, there are several environmental problems for which there are no current technical solutions. The life-cycle cost for technology development must be compared with the life-cycle cost and the potential environmental impact of no action at these sites to form an economic basis for decision making.

DOE Orders

DOE orders provide the structure and/or guidance for the management and operations of DOE. These Orders prescribe the methods and procedures for procurement, operational safety, management, and technology development for all DOE operations. The technology development gates (stage of development) are defined in DOE Orders. These gates provide the structure/methods by which developing technologies are moved through the developmental process, from basic science at gate one to engineering demonstration and commercialization at gate six.

BIBLIOGRAPHY

1. *A New Approach to Environmental Research and Technology Development at the US Department of Energy Action Plan, U.S. DOE, Washington, D.C., Jan. 25, 1994.*

C

CARBONACEOUS BOD AND SUSPENDED SOLIDS: REMOVAL FROM MUNICIPAL WASTEWATER

RAO Y. SURAMPALLI
U.S. Environmental Protection Agency
Kansas City, Kansas

SHANKHA K. BANERJI
University of Missouri
Columbia, Missouri

Since the 1960s, the implementation of the Clean Water Act (CWA) tripled the number of treatment plants serving communities and municipalities. According to the U.S. Environmental Protection Agency's (EPA) recent needs survey report (1) there are approximately 15,500 wastewater treatment facilities operating in the United States. Nearly 80% of the plants are designed for less than 3785 m^3/d (1 mgd). Approximately 4% of the plants are larger than 37,850 m^3/d (10 mgd) and provide approximately 65% of the nation's wastewater treatment capacity.

The purpose of municipal wastewater treatment is to prevent pollution of receiving waters. Wastewater collected from households, industries, and other sources in a town or municipality are collected and transported to a wastewater treatment plant for treatment and final effluent disposal to streams, rivers, lakes, or estuaries. Other means of effluent disposal include irrigation, infiltration, evaporation from nondischarging lagoons, water reuse, and ocean disposal.

Wastewater Treatment Requirements

The respective state, local, and the EPA water-quality standards or criteria that have been established for receiving waters define the degree of wastewater treatment required for effluent disposal. Stream classification standards, published by each state as required by EPA, categorize surface waters according to their most beneficial use, such as for body-contact recreation, drinking-water supplies, etc. Many states have adopted more stringent standards than those required by EPA.

Wastewater effluent standards are established under the National Pollutant Discharge Elimination System (NPDES) for regulatory purposes to achieve compliance with their stream standards. In general, the minimum required is secondary treatment. However, many communities and municipalities located on major rivers still discharge lesser quality effluent. At the same time, some communities or municipalities have been asked to install advanced waste treatment systems to remove nitrogen and phosphorus to protect aquatic life, retard eutrophication in receiving waters, or to facilitate wastewater reuse. Wastewater reuse is expected to increase in the future, particularly in arid and semiarid regions where fresh water is in short supply. According to EPA's needs survey (1), approximately 11% of the treatment facilities have less than secondary treatment, 76% have secondary treatment or greater, and 12% have no discharge. In the future, significantly greater number of facilities will be upgraded to higher levels of treatment because of stringent and changing water quality standards.

Sources, Quantities, and Characteristics of Wastewater

Quality, quantity, and characteristics of municipal wastewater depend to a great extent on the nature of contributing area uses, the type of sewer collection system, and the industrial/other wastes entering the sewer collection system. Generally, infiltration/inflow and industrial discharges significantly alter the quality, quantity, and characteristics of the domestic wastewater.

Sources. Typically, four components determine the wastewater flow from a community or municipality depending on the type of collection system. These are domestic wastewater, industrial wastewater, infiltration/inflow (I/I), and stormwater. Where only one sewer system (combined sewer) is used, wastewater flow consist of all these components. Infiltration is groundwater entering sewers and building connections through defective joints and cracks in pipes and manholes. Inflow is water discharged that enters the sewer system from storm drain connections, roof, foundation and basement drains, or through manhole covers. Storm water is runoff resulting from rainfall and snow melt.

Quantities. Wastewater flow rates are commonly determined from existing records or by direct field measurements for areas now served with sewers. For new developments, wastewater flow rates are derived from an analysis of population data and corresponding projected unit rates of water consumption or from estimates of per capita wastewater flow rates of similar situations or communities. The Ten States Standards (2) recommend the use of an average daily flow value of 378 L (100 gallons) per capita per day for new domestic service in undeveloped areas unless water use data or better flow estimates are available. This suggested per capita flow value includes an allowance for moderate sewer infiltration and commercial/institutional use. EPA construction grant guidelines (3) consider infiltration to be excessive if the plant flow for nonrunoff conditions and adjusted for nondomestic use is greater than 454 L (120 gal) per day per capita. The guidance suggests that inflow for stormwater-induced flow conditions would be excessive if plant flow were greater than 1040 L (275 gal) per day per capita.

Characteristics. Understanding the characteristics of wastewater is important in the design and operation of wastewater treatment facilities. Wastewater is characterized in terms of its physical, chemical, and biological composition. The important contaminants of concern in municipal wastewater treatment are biodegradable organics, suspended solids, nutrients, pathogens, heavy metals, and priority pollutants (organic and inorganic compounds). Secondary treatment standards are mainly concerned with the removal of biodegradable organics, suspended solids, and pathogens, whereas advanced treatment facilities are concerned with the additional removal of nutrients and priority pollutants.

Typical pollutant composition ranges of domestic wastewater are 5-day biological oxygen demand (BOD) of 110–220 mg/L, suspended solids 100–220 mg/L, nitrogen (total)

20–35 mg/L, phosphorus (total) 4–8 mg/L, and total coliforms (per 100 mL) 10^6–10^8 (4). Generally, wastewater characteristics are determined through wastewater characterization by means of sampling and analysis. However, the Ten States Standards (2) recommend that new treatment plants be designed for a domestic load contribution of at least 0.08 kg (0.17 lb) of BOD per capita per day and 0.09 kg (0.20 lb) of suspended solids per capita per day unless available information justifies other design criteria.

Table 1. Treatment Level and Number of Wastewater Facilities operating in the United States[a]

Level of Treatment	Number of Facilities
No discharge	1,981
Less than secondary	868
Secondary	9,086
Greater than secondary	3,678
Total facilities	*15,613*

[a] Ref. 1.

Aerobic Wastewater Treatment Options

Conventional wastewater treatment generally consists of preliminary, primary, and secondary treatment. Advanced treatment is provided when nutrient or priority pollutant removal is required. The preliminary processes that can be used in a municipal wastewater treatment are screening, comminution, flow measuring, pumping, grit removal, preaeration, and chemical treatment. Figure 1 shows a schematic diagram of conventional wastewater treatment. Chemical treatment is not common in municipal wastewater treatment. Primary treatment is mainly for removing heavy solids and floatable materials. Primary settling is used in most of the large municipal plants and generally precedes secondary biological process. Primary settling removes 30–50% of the suspended solids from incoming raw municipal wastewater. Remaining organic matter is removed in secondary biological treatment process. There are mainly three biological wastewater treatment systems: suspended growth, attached growth, and natural treatment systems. Generally a secondary biological process consists of either a suspended growth or attached growth system. Advanced wastewater treatment is typically made up of second-stage biological processes and may include chemical precipitation and filtration.

Suspended Growth Systems. Suspended growth systems are those aerobic processes that achieve relatively high microorganism (biomass) concentrations through the recycle of biological sludge solids. These microorganisms convert biodegradable organic wastewater constituents and certain inorganic fractions into new cell mass and by-products, both of which can subsequently be removed from the system by gaseous stripping, settling, and other physical means. Suspended growth systems (activated sludge, oxidation ditches, sequential batch reactors, etc) are already the most common of the secondary treatment processes in use today and will remain so as new treatment facilities are constructed.

Fixed-Film Systems. It was observed in aquatic environments that the growth of slime-producing organisms occurred with the self-cleaning of water in streams. Simply stated, the concept is to provide a surface on which the microbial layer can grow and to expose this surface repeatedly to the wastewater for adsorption of organic material and to the atmosphere for oxygen. Trickling filters and the rotating biological contactors are the most commonly used fixed-film biological processes. Trickling filters use rock, wood, or plastic media and presettled wastewater is sprayed through rotary distributors onto the media surface. Rotating biological contactors use a fixed-film biomass on rotating media for biological treatment. The rotating medium, typically made from sheets of high density plastic, provides a surface on which organisms grow and contact organic wastewater constituents and oxygen from the air.

Natural Treatment Systems. In natural wastewater treatment vegetation, soil, microorganisms, and to a limited extent higher animal life provide the desired treatment. Natural systems for wastewater treatment include soil absorption, wastewater ponds or lagoons, land treatment, floating aquatic plants, and constructed wetlands. Natural treatment systems are better suited for small communities and rural areas because of the need for and availability of large land areas. Table 1 shows treatment level and number of wastewater treatment systems operating in the United States.

Fundamentals of Biological Wastewater Treatment

An empirical relationship depicting the stabilization of biodegradable organic matter in aerobic suspended growth

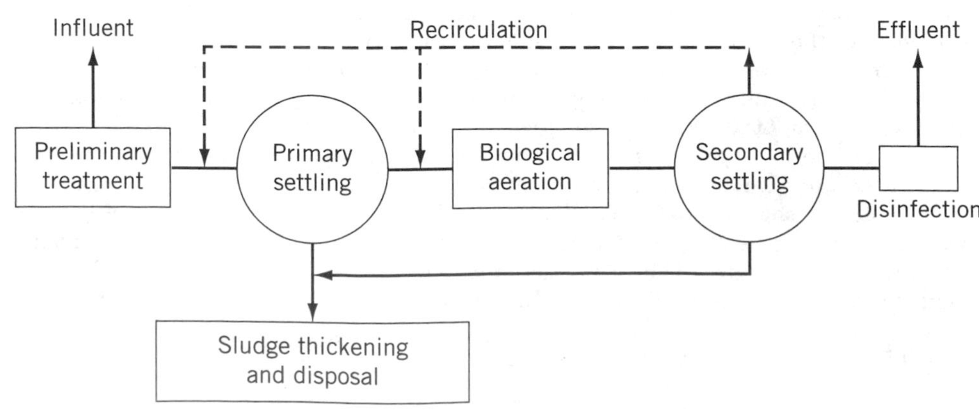

Figure 1. Schematic diagram of conventional wastewater treatment.

systems can be expressed by the following equation:

$$\text{Organic matter} + O_2 + \text{nutrients} + \text{microbes} \rightarrow \text{new microbes} + CO_2 + H_2O$$

The organic matter is consumed by the microorganisms with a portion of the substrate (waste material) being oxidized for energy and a portion for synthesis of new cell mass.

Bacterial Growth. The classical growth curve for a batch-fed reactor is shown in Figure 2. As illustrated, the rate of biomass growth increases logarithmically during the early stages when food is unlimited (microbe population is limiting).

In a short time, insufficient food is available to sustain log growth and the rate of biomass increase begins to decline. At the end of this declining growth phase the substrate remaining is essentially zero and the organisms enter the endogenous phase.

Based on experimental studies, a relationship was developed intended to describe growth kinetics throughout both the log and declining growth phases (5,6):

$$\mu = \mu_{\max} s / k_s + s$$

where μ = specific substrate utilization rate per unit mass of organisms, 1/time; $\mu_{\max}$ = maximum substrate utilization rate per unit mass of organisms, , 1/time; s = growth limiting substrate concentration, mass/volume; and k_s = substrate concentration at which the rate of substrate utilization is one-half of the maximum rate, mass/volume.

Environmental Conditions. Environmental conditions of dissolved oxygen, temperature, pH, and nutrients or trace elements have significant effect on the survival and growth of bacteria. Particularly, the temperature, dissolved oxygen, and pH have important effects on the survival and growth of bacteria. In general, optimal growth occurs within a fairly narrow range of temperature and pH, although the bacteria may be able to survive within much broader limits. Temperatures below the optimum typically have a more significant effect on growth rate than temperatures above the optimum. It has been observed that the growth rates double with approximately every 10°C increase in temperature until the optimum temperature is reached. According to the temperature range in which they function best, bacteria may be classified as psychrophilic (10–30°C) mesophilic (20–50°C), and thermophilic (35–75°C). The pH of the environment is also a key factor in the growth of organisms. Most bacteria cannot tolerate pH levels above 9.5 or below 4.0. The optimum pH for bacterial growth lies between 6.5 and 7.5.

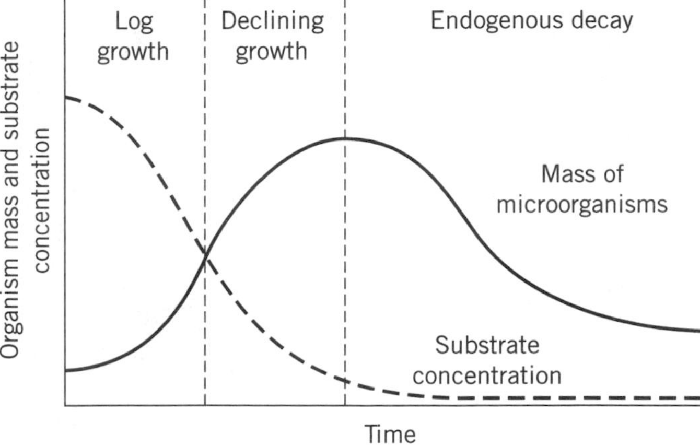

Figure 2. Growth curve for a batch-fed reactor.

FIXED-FILM PROCESSES

As defined earlier, the fixed-film processes use a surface media on which microbial growth takes place which exposes the surface repeatedly to wastewater for adsorption of organic material and atmosphere for oxygen. The most common fixed-film processes are trickling filters and rotating biological contactors. They are used for BOD removal as well as to achieve ammonia nitrification.

Trickling Filters

Earlier installations of trickling filters used rock media, but use of plastic and wood media has become a standard practice to provide more surface area per unit volume and improved ventilation. It is suggested that adequate hydraulic flushing and forced ventilation are needed to provide maximum efficiency, control of nuisance organisms, and to reduce odors. Trickling filters need recirculation for optimum performance and if covers are not provided, high temperature loss through trickling filters can be expected in colder climates.

Trickling filters are referred to as low rate, intermediate rate, high rate, and super rate according to applied hydraulic and organic loadings. Hydraulic loading is the total volume of liquid, including recirculation, per unit of time per unit for filter cross-sectional area. The units for hydraulic loading are $1/m^2 \cdot s$ or $m^3/m^2 \cdot h$ (gpm/sq ft) and organic loading is expressed as kg $BOD/m^3 \cdot d$ ($lb/d/1000\ ft^3$) or $kg/m^2 \cdot d$ ($lb/d/1000\ ft^2$). Typical hydraulic and organic loadings for various filter classifications are shown in Table 2. Roughing filters are used to provide intermediate treatment or as the first step of multistage biological treatment.

Factors Affecting Performance and Design

Trickling filter design and subsequently the performance of trickling filters is affected by numerous variables such as wastewater composition and treatability, NPDES requirements, primary treatment, media type and depth, hydraulic and organic loadings, recirculation, ventilation, temperature, and configuration.

Wastewater Composition and Treatability. Trickling filter wastewater characteristics should be carefully analyzed and documented. Both the volume and strength of influent wastewater may exhibit hourly, daily, and seasonal variations. However, trickling filters can handle the peak conditions without upsetting the system. Influent wastewater with strong industrial contributions should be thoroughly analyzed for treatability. The treatability of a waste is defined by a rate coefficient. Simple organics such as sugars are easily and quickly removed, so they have higher rate coefficients. Complex and inhibitory compounds, as well as nonflocculant suspended solids, degrade slowly resulting in lower rate coefficients.

ROTATING BIOLOGICAL CONTACTORS

Rotating biological contactors (RBCs) were first introduced in Europe in the 1960s and later were installed in the United States. More than 600 RBC plants are presently operating in the United States. Most of the U.S. plants are designed and used for carbonaceous BOD removal, some for combined organic oxidation and nitrification, and a few for nitrification of secondary effluent.

In the past, lower process performance, structural problems with shafts and media, excessive biomass buildup on media, uneven shaft rotation in air drive units, and other process problems have been serious concerns with many RBC installations. These problems have resulted in the process falling out of favor with designers in recent years. Modifications to the equipment have been made to mitigate many of these problems and simplify maintenance.

If designed properly, RBCs performance can be superior to other fixed-film processes because of lower organic loading per mass of biological solids, longer detention time in stage, and better short-circuiting control. Design considerations must include organic and hydraulic loading, staging of the RBC units, and operational flexibility.

BIBLIOGRAPHY

1. *1992 Needs Survey Report to Congress,* U.S. EPA, Office of Municipal Pollution Control, Washington, D.C., 1992.
2. *Recommended Standard for Sewage Works,* Great Lakes–Upper Mississippi River Board of State Sanitary Engineers, Health Education Services, Inc., Albany, N.Y., 1986.
3. *Construction Grants 1985,* U.S. EPA, Office of Water Program Operations, Washington, D.C., 1985.
4. Metcalf & Eddy, Inc., *Wastewater Engineering*, McGraw-Hill, Inc., New York, 1991.
5. J. Monod, *Ann. Rev. Microbiol.* **3** (1949).
6. J. Monod, *Recherches Sur la Croissance des Cultures Bacteriennes,* Herman et cie, Paris, 1942.

CARBON DIOXIDE RECOVERY FROM FOSSIL-FUELED POWER PLANTS

EDWARD J. DANIELS
BASSAM J. JODY
NORMAN F. BROCKMEIER
ALAN M. WOLSKY
Argonne National Laboratory
Argonne, Illinois

BACKGROUND

Industry is focusing more attention on carbon dioxide (CO_2) emissions because they are expected to contribute to the Greenhouse Effect and, therefore, to future climate change. In this article, several technologies are reviewed that are capable of capturing CO_2 from the flue gas of fossil-fuel-fired utility boilers: chemical solvent technologies, cryogenic technologies, membrane technologies, physical absorption methods, and physical adsorption methods. Many of these technologies were developed for reasons unrelated to climate change, eg, to clean up acid gases, to recover CO_2 for enhanced oil recovery, and to recover CO_2 for use in the beverage industry. The logical evolution of these technologies is to investigate whether they could be used to reduce CO_2 emissions. Likewise, can combustion strategies be modified to increase the efficiency and/or cost-effectiveness of CO_2 removal? These strategies include (*1*) combustion of fuels in an oxygen-enriched stream, yielding a mixture that is primarily H_2O and CO_2, from which water is easily separated, and (*2*) partial (or total) elimination of the carbon from the fuels before combustion. To further that inquiry, first the amount of CO_2 emitted by power plants is estimated and then various methods to remove CO_2 from conventional power plants are described. Methods to remove CO_2 from possible future power plant configurations are also discussed.

CO_2 Emissions by Electric Power Utilities

The Energy Information Administration (1) compiles data on the electric power generated by fossil-fueled power plants in the United States. Using the following assumptions and approximations, EIA's data was adopted to estimate (*1*) the thermal efficiencies at which electricity is generated from various fossil fuels and (*2*) the quantities of CO_2 generated by fossil-fueled power plants:

1. The chemical composition of coal is estimated at 60% organic and 40% inorganic. The organic portion is estimated to be $CH_{0.8}$.
2. The chemical composition of petroleum is assumed to be $CH_{1.55}$, and its density is taken as 53 lb/ft^3.
3. The heating values of the different fossil fuels used for power generation were taken as the averages reported by the EIA for July 1991 (coal 20,525,135 Btu/ton; petroleum 6,313,470 Btu/barrel; and natural gas 1,025,250 Btu/10^3 ft^3).

First the total electric energy produced in the United States and that produced from coal, petroleum, and natural gas was determined (Fig. 1). The data indicate that coal was used to generate over 50% of the fossil-fuel-fired electricity. Although

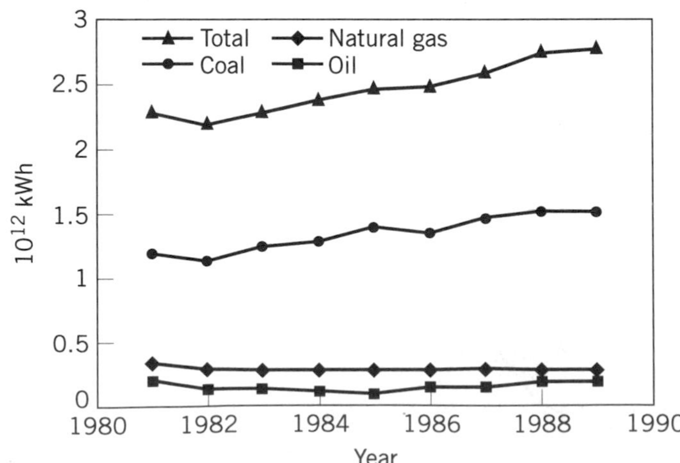

Figure 1. Power production in the United States from different fossil fuels.

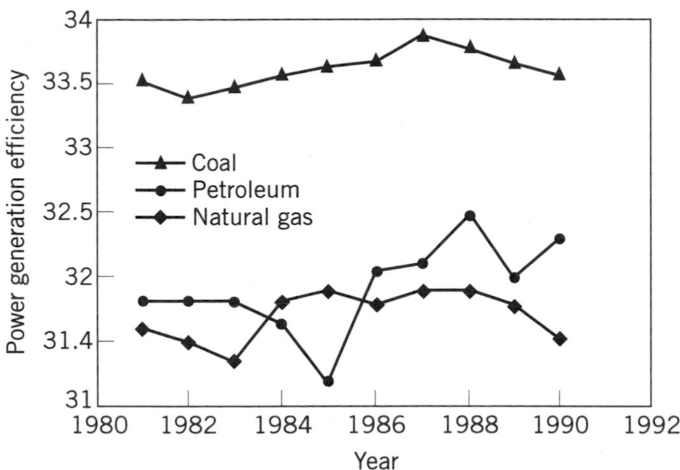

Figure 2. Power generation efficiency using different fuels.

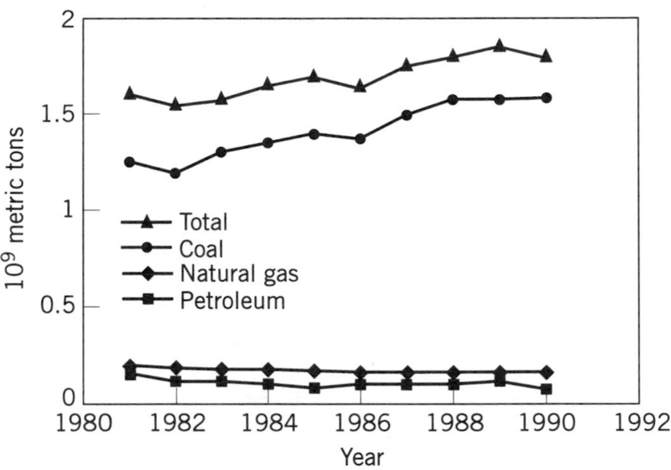

Figure 3. CO_2 production by U.S. fossil fuel-fired power generation plants.

the generation of electricity with natural gas and petroleum fuels did not increase in the 1980s, the use of coal has increased by over 20% during the same period. This trend is significant because of the three fossil fuels considered, coal generates the most CO_2 per unit of fuel energy. Figure 2 shows the power generation efficiencies attained by the power industry. Coal-fired plants are more efficient than either petroleum- or natural-gas-fired facilities, primarily because many of the oil- and natural-gas-fired plants are older and less efficient units that are used only for meeting peak demands. However, recent designs of gas-fired combined-cycle power plants promise very high efficiency. On the basis of the information shown in Figures 1 and 2 and the assumptions stated above, the total amount of CO_2 produced by fossil-fueled power plants was calculated and the data shown in Figure 3. In 1990, the amount of CO_2 produced approached 1.85 billion tons, an increase of 20% over 1982 levels. As expected, coal-fired facilities account for more than 90% of the CO_2 produced by power plants. Consequently, much of the discussion of technology for CO_2 recovery focuses on coal-fired plants. Nonetheless, the technologies also apply to other fossil-fuel-fired plants. Coal-fired power plants, their individual components, and the different process streams are discussed in detail in the literature (2) (see COAL).

Demand for CO_2

CO_2 is consumed in the United States for the following applications: refrigeration (food industry), carbonation (beverage industry), as a chemical feedstock (chemical industry), shielding for welding operations (metals working), inerting and pressurization (aerosol packaging and sprays), and oil and gas well operations (petroleum industry). In 1989, the total consumption of solid and liquid CO_2 in the United States was a little more than 5 million tons (SRI International 1989) (Fig. 4). The total consumption of gaseous CO_2 in 1989 was about 38 million tons (3) (Fig. 5).

CO_2 Supply

Most of the CO_2 used in the United States is recovered as a by-product of the following chemical manufacturing operations:

hydrogen production by steam reforming of fossil fuels;

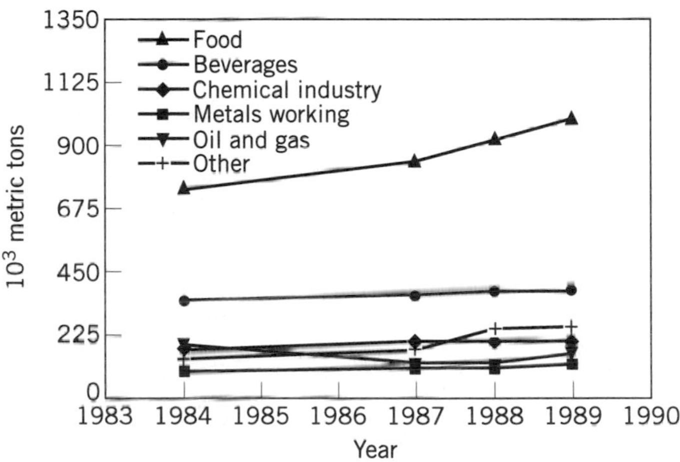

Figure 4. Consumption of liquid and solid CO_2 by U.S. industries.

natural gas operations;

production of ethyl alcohol by fermentation of sugars and starch; and

ethylene oxide production from ethylene.

In addition, substantial quantities of CO_2 are recovered from naturally occurring reservoirs (3). In 1989, the United States was a net importer of about 11,000 tons of CO_2. Total imports were 55,500 tons: 52,600 tons from Canada, 2,600 tons from Japan, and the balance from other countries. Total exports in the same year were 44,500 tons: 17,900 tons to Australia, 5,600 tons to Canada, 17,200 tons to Mexico, and the balance to other countries.

Although there may be a limited opportunity to recover CO_2 from power plants for such industrial markets as enhanced oil recovery for oil and gas operations, the total market for CO_2 is tiny relative to the quantities of CO_2 recoverable from fossil-fuel-fired power plants. Consequently, recovery of CO_2 is only part of the issue: cost-effective sequestration is also required.

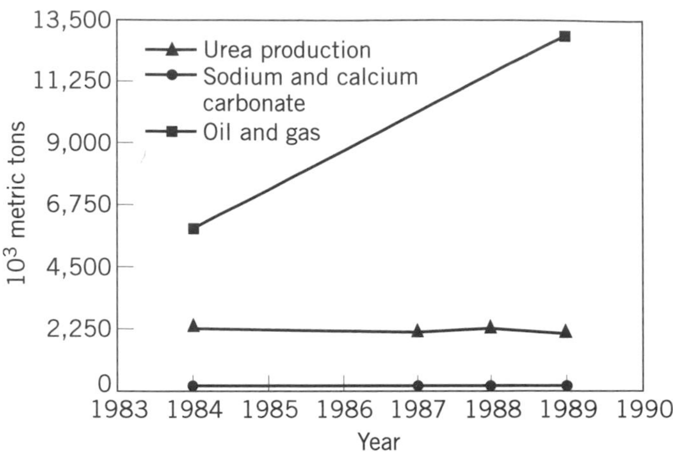

Figure 5. Consumption of gaseous CO_2 by U.S. industries.

TECHNOLOGIES FOR RECOVERING CO_2 FROM FOSSIL-FUEL-FIRED POWER PLANTS

State of the Art

Technologies applicable to CO_2 recovery from flue gas are summarized in Figure 6. The separation of CO_2 from N_2, H_2O, SO_x, NO_x, and excess O_2 in flue gas is accomplished in a one-step operation, such as separating the CO_2 by using a CO_2-selective membrane, which would leave everything else behind (such a membrane is not commercially available), or in a multistep operation, such as compression and cooling to separate H_2O first, followed by separation of CO_2.

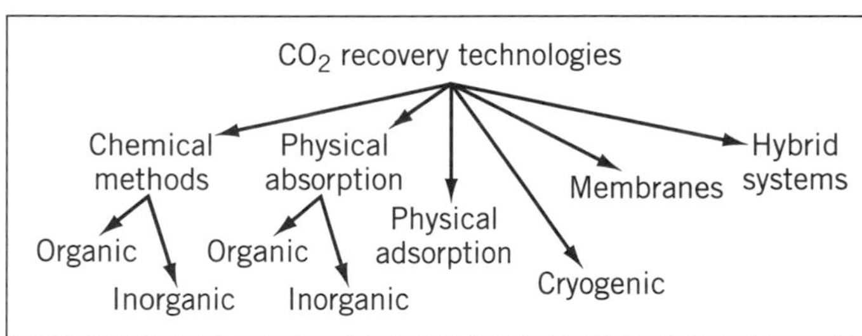

Figure 6. Summary of technologies for removing CO_2 from flue gas streams.

However, recovering CO_2 from flue gases (direct recovery) is energy-intensive and costly. Figure 7 shows the minimum theoretical energy required to separate CO_2 from the flue gas of coal-fired boilers. The minimum theoretical energy required for CO_2 separation from the flue gas (excluding CO_2 compression) represents about 2.5% of the electricity generated by the power plant. The actual separation process requires between five and ten times the theoretical minimum because of the numerous thermodynamic irreversibilities involved and, therefore, represents 12 to 25% of the electricity generated by the power plant. The energy required to compress the CO_2 constitutes another 10–15%. The energy requirement and the cost of separating the CO_2 from flue gas are high for the following reasons:

1. CO_2 is present in the flue gas at low concentrations and thus low partial pressures. For example, when coal is burned with 110% theoretical air, the CO_2 mole concentration in the flue gas is about 15%. It is even lower for oil (~13%) and natural gas (~9%). An obvious way to increase the CO_2 partial pressure is to reduce the partial pressure of other species, primarily N_2. N_2 concentration is decreased by separating some of it before or after combustion. In either case, the separation process is costly and energy-intensive. Many membrane systems, however, operate to produce relatively pure N_2 from air for industrial use. These membrane systems reject a stream that is about 30% O_2. Using such a stream increases the concentration of CO_2 in flue gas by about 50%. This higher concentration enables more efficient recovery of the CO_2, regardless of which technology is applied.
2. Both N_2 and CO_2 require very low temperatures and/or high pressures to condense (CO_2 condenses at −78.6°C and N_2 condenses at −195.8°C, both at atmospheric pressure). Formation of solid CO_2, ice, and H_2O/CO_2 clathrates at low temperatures is another complication for low temperature separation of flue gas.

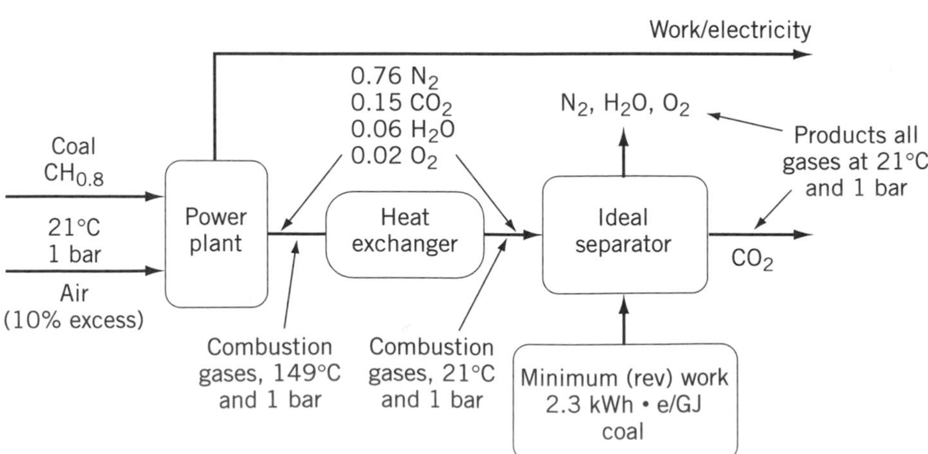

Figure 7. Minimum reversible (rev) work required to separate CO_2 from coal combustion products.

3. The large number of species present (N_2, CO_2, H_2O, H_2S, COS, CO, O_2) and the corrosive nature of some of these species complicates all applicable separation processes. For example, even though sulfur-bearing compounds and O_2 are present in small quantities, they result in irreversible chemical reactions when chemical solvent methods are used. For membranes, the selectivity decreases as the number of species present increases.
4. The presence of water vapor in flue gas increases the cooling load on the separation process. Condensing water also contains appreciable quantities of dissolved acid gases including CO_2, which makes it acidic and corrosive. However, water separation is a problem more manageable than separating CO_2 or N_2.

The technologies presented in Figure 6 are discussed in greater detail in the following section.

Chemical Absorption of CO_2

Chemical absorption of CO_2 involves one or more reversible chemical reactions between CO_2 and another material, such as mono-, di-, or triethanolamine; diisopropanol amine; sodium hydroxide; sodium carbonate; and potassium carbonate to produce liquid or solid species that break down upon heating to liberate CO_2 and regenerate the material used to react with CO_2. For example, the monoethanolamine (MEA) process is represented by the following reversible reaction (A more detailed discussion of the process is available (4)):

$$C_2H_4OHNH_2 + CO_2 + H_2O \underset{\text{Hot}}{\overset{\text{Cold}}{\rightleftarrows}} C_2H_4OHNH_3^+ + HCO_3^-$$

Table 1 lists many of the processes, the solvents they employ, and the conditions under which the forward (absorption) and backward (regeneration) reactions proceed. Many of these processes are in full-scale operation for recovering acid gases (including CO_2 from natural gas reservoirs) and other applications. Table 1 also summarizes the CO_2 absorption capabilities of the various methods and their approximate energy requirements. The data presented here are not directly comparable because they are for streams at different conditions, including different compositions. They are intended to illustrate the magnitude of the effectiveness and energy required to drive these processes and not for quantitative comparison. The data show that these processes, in general, are energy-intensive. Most of the energy consumed is in the form of steam at pressures around 4–5 atm. The steam is used essentially in the solvent regeneration (CO_2 boiling) step. The energy intensity of these processes is related partly to their low capacity (CO_2 absorbed per unit mass). Therefore, a large body of liquid has to be heated to release a small amount of gas. These systems also require electric energy and cooling water. The electric energy is mainly for pumping the solvent and compressing the gas, and the cooling water is for cooling the CO_2 lean solution after regeneration.

Generally, the pretreated flue gas stream is brought in contact with an aqueous (12–30%) scrubbing solution. As a result, the acid gases are removed in the solution. Typical loadings of these solutions are from less than 15 m^3 of acid gas per cubic meter of solution for the 25% hot carbonate solution to about 60 m^3 of acid gas per cubic meter of solution for the ECONAMINE FG solvent (5). Oxygen, normally present in flue gas (~2 mol %) along with sulfur-containing gases, contributes to solvent degradation by forming undesirable species via irreversible side reactions. Particulates in flue gas must be recovered before the gas comes in contact with chemical solvents used in desulfurization and CO_2 recovery methods.

Since the early 1980s, these systems have been tested on commercial and pilot-plant scale for CO_2 recovery from power plant flue gases (5,6,7). Dow Chemical developed its Gas/Spec FT-1 amine-based process in the late 1970s. Fluor Daniel Inc. acquired it later, and it is now known as the ECONAMINE Process (5). In 1978, an 800-ton/day (TPD) amine-based plant (based on a process developed by Kerr McGee and later licensed to Lummus Crest) was started up in Trona, California to recover commercial-grade CO_2 from flue gases. A 200-TPD unit was started up in January 1991 in Oklahoma, and another 300-TPD plant (Soda Ash Botswana Ltd.) was started up in March 1991 (6). An amine-based pilot plant was constructed and tested at the Sundance power plant in Alberta, Canada (7). Another pilot plant was constructed and tested in Saskatchewan, Canada (7). Experience with these and other plants identified two major problems: solvent degradation from irreversible reactions with O_2, SO_x and COS and corrosion in the absorber and in the regenerator. Corrosion inhibitors and additives brought these problems under acceptable control.

The steam used in regeneration is produced by a boiler or it is extracted from a turbine after partial expansion. Extraction of the steam from the turbine reduces the net power generation because of incomplete expansion of the steam. Yet it is a more economical alternative than installing a boiler dedicated to providing the steam. In both cases, the amount of CO_2 generated per kilowatt-hour (kWh) of electricity increases.

Analysis of the data for the commercially available MEA process (8) shows that the actual energy requirement is about 10 times the theoretical minimum value required for separating CO_2 from flue gas. Data reported indicate that retrofitting a coal-fired power plant with an ECONAMINE FG_{sm} system for 90% CO_2 removal and with a Thiosorbic® magnesium-enhanced, wet lime (mag-lime) SO_x removal process resulted in a 35% reduction in net power production (a nominal heat rate of 15,000 Btu/kWh vs ~10,000 Btu/kWh for a power plant without CO_2 removal) (8). The data also indicate that the cost of retrofitting is greater than the initial capital cost of the power plant itself. The cost includes transporting the CO_2 to and disposing of it in large, depleted natural gas fields or in the ocean. In this design, the amine is regenerated with steam extracted from the turbine.

When comparing the costs of CO_2 recovery from fossil-fuel-fired power plants, the costs should be normalized relative to net power production. For example, the MEA process consumes about 35% of the energy generated by a coal-fired plant. Therefore, the plant with an MEA CO_2 recovery system needs a power production design capacity of 1.54 times that of a plant without an MEA CO_2 recovery system to supply an equivalent quantity of electricity to the grid. The essential question therefore is not "What is the cost of CO_2 removal?" Rather, it is "What is the cost of reducing CO_2 emissions while maintaining the original electrical output?" Consequently, the most appropriate comparison of alternative CO_2 recovery technologies is a side-by-side comparison of power plants that provide the same quantity of electricity to the grid. The objective in this article is to provide the background for making

Table 1. Comparison of CO_2 Chemical Stripping Processes

Process	Solvent	Adsorption Conditions T(°C)/ P(Bar Gauge)	Regeneration Conditions T(°C)/P(Bar Gauge)	CO_2 Removal Levels	Other Information
Adip	30–40 wt % Aqueous (DIPA) diisopropanol amine	40–60/0–70	120–135/0	90% removal	Noncorrosive
Alkazid M	Sodium alanine	25/7–75		98% removal	Degrades in presence of O_2; noncorrosive
Alkazid DIK	K salt of diethyl or dimethyl glycine				
Benfield	Potassium carbonate aqueous solution	15–140/7 → 140	–/0.1–0.7	10 ppm in effluent	Corrosive, stainless steel required for some parts
Catacarb	Potassium carbonate and amine borates	15–140/7→140	–/0.1–0.7	10 ppm in effluent	
Fluor Econamine	Aqueous solution ~65 wt % diglycolamine	43/61		99% removal	More corrosive than MEA
Giamarco-Vetrocoke	Mixture of sodium carbonate, arsenites, and arsenates	<150/independent of P		Nearly complete	Inherently stable solution
Fluor ECONAMINE FG	Monoethanolamine (30 wt %)	Near ambient/near ambient	~120/0	90%	Corrosion inhibitors are normally added
MDEA	Methyl diethanolamine	27–52/0 → 70	115–120/0	30–65% removal	Greater selectivity for H_2S over CO_2; more expensive than MEA and DEA; reacts Irreversibly with HCN and organic acids
Seabord	Sodium carbonate dilute (3–3.5 wt %)				Reacts irreversibly with O_2
SNAPA-DEA	Diethanolamine 20–30 wt % solution			500 ppm range	More stable than MEA; reacts irreversibly with O_2; can use carbon steel
Sulfiban	Monoethanolamine 12–20 wt % solution	15–40/0–70			No selectivity between CO_2 and H_2S
Sulfinol	Diisopropylamine carbon steel	40–50/0–70		ppm range	Noncorrosive
Vacuum Carbonate	Sodium carbonate dilute (3–3.5 wt %)	60/0.4–0.7 (abs)			Reacts irreversibly with O_2

such comparisons, recognizing that the actual comparison is subject to site-specific conditions and a host of other plant design considerations.

The economics of the commercially available amine process developed by Kerr McGee and licensed by ABB Lummus Crest, Inc. have been discussed (6). This process uses a 15–20% monoethanolamine (MEA) solution and tolerates oxygen and limited amounts of COS (<100 ppm). The operating costs were determined by assuming that electricity costs \$0.05/kWh, and chilled water costs \$0.10/M gallons. These costs are for a ton of recovered CO_2 and not based on avoided CO_2.

Energy savings are realized if a solvent with high absorption capacity at moderate temperatures and pressures is developed. The higher the absorption capacity, the less solvent required to handle a given gas stream. Thus, less energy is required for pumping and cooling, thereby resulting in smaller and less expensive equipment. Further, a solvent that is regenerated at lower temperatures stands a better chance of using waste heat instead of extracted steam from the turbine, which results in lower turbine power output.

Another way to save energy is to develop a solvent that is regenerated at higher pressures. In this case, the rich solution is pumped to a high pressure and then heated to release its CO_2 at the high pressure, thereby avoiding the need to compress the CO_2 before final disposal. If the solvent is chemically stable upon thermal recycling and can be used after regeneration in an organic Rankine power cycle, the regenerated solution can be externally heated to produce a superheated vapor that expands in a turbine to produce mechanical work. This concept is particularly attractive when used to recover CO_2 from the exhaust of a gas turbine, because the solvent bottoming cycle could be driven by heat available in the turbine exhaust. A simplified schematic of such a concept is shown in Figure 8.

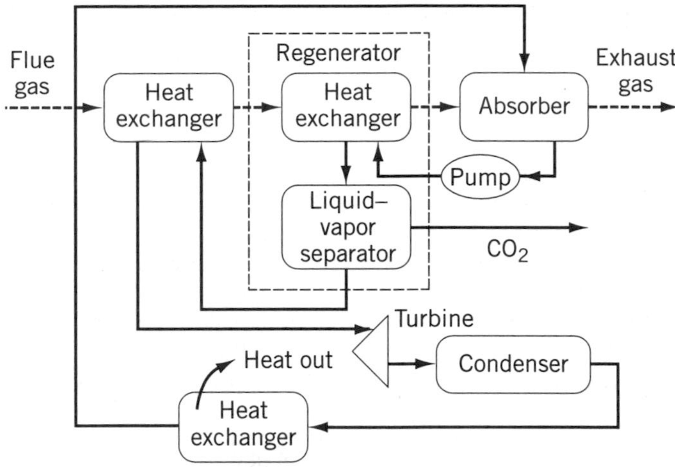

Figure 8. An integrated solvent regenerator/bottoming cycle.

We also made scoping estimates of the capital investments and operating costs for carbon dioxide (CO_2) removal from the flue gas of coal-fired electric power plants in the range of 513 to 600 MW gross power and compared the results with the economics for a 600-MW base power plant having no CO_2 removal facility as a base case. It has been estimated that for a 513-MW (gross) power plant, the combined capital cost of the CO_2 absorber and recovery facilities would be nearly equal to that of the base power plant (8).

Low Temperature (Cryogenic) Separation of CO_2

Cryogenic separation of gas mixtures was first demonstrated by Linde (9) in the early 1900s. Since then, the process has been improved significantly. For example, whereas the energy required to liquefy air by the early Linde process was 4400 Btu/lb, the Claude process uses about 1370 Btu/lb (10). (The theoretical minimum for a fully reversible process is 324 Btu/lb.) Separation of CO_2 from combustion gas streams is more involved than separating air, but the principle is the same. It involves compression and cooling of the gas stream to low temperatures to induce a phase change in CO_2 and thus separate it from the mixture. Before CO_2 recovery, the water vapor in flue gas has to be recovered down to a dew point that is less than the lowest temperature encountered in the separation process. If the water vapor is not recovered, solid ice crystals form, and CO_2 clathrates may form. Both plug process piping. After the bulk of the water is recovered by condensation, a desiccant is used to absorb residual water. The desiccant is regenerated by heating it to higher temperatures to drive off its moisture content. The heat content of the flue-gas stream contributes significantly to the regeneration process.

CO_2 separation through the conventional cryogenic procedure is accomplished by multistage compression of the dried flue-gas followed by cooling of the compressed gas after each stage. Figure 9 illustrates the concept. The high pressure gas leaving the cooler downstream of the last compression stage is expanded to cool it to very low temperatures. As a result of this cooling, the CO_2 is liquefied and automatically separated from the gas phase. A distillation column (or a double column) is needed to purify the liquefied product by distilling off the dissolved impurities, including some dissolved N_2. Then the

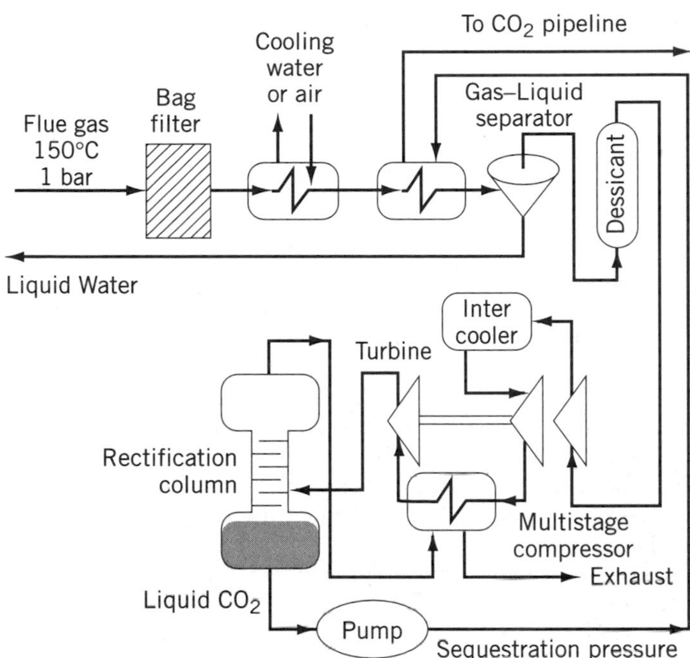

Figure 9. Schematic for a low temperature (cryogenic) process to separate CO_2 from flue gas.

residual very cold gas is cascaded through the gas coolers of the lateral compression stages. The CO_2 liquid product is pumped to the desired sequestration pressure and evaporated by recovering heat from other process streams, as indicated in the second-stage heat exchanger before the gas/liquid separator.

A conventional vapor compression cycle supplements the gas compression/cooling process to provide additional cooling, which enables the condensation of CO_2 with less gas compression and at higher pressures. Then the collected, liquefied CO_2 is pumped to the desired sequestering pressure, if necessary. This process also eliminates the risk of forming solid CO_2. In this case, the residual N_2 gas (which is at high pressure) is heated to high temperatures by external heat sources and expanded in a multistage gas turbine with reheat to provide at least part of the power consumed in the compression process. The capital cost of the process increases, but the efficiency of CO_2 recovery also increases, thereby decreasing the reduction in net power production that otherwise occurs.

As discussed earlier, the low partial pressure of CO_2 in the flue gas and the possibility of forming solids are major impediments to efficient separation of CO_2 by cryogenic techniques. Compressing the flue gas stream to higher pressures increases the partial pressures of all of the combustion products and therefore allows separating them from the combustion products at higher temperatures. A refrigerator's operating efficiency is generally expressed as its Coefficient of Performance (COP). The COP is the dimensionless ratio of the useful refrigeration divided by the net energy required to operate the cooling system. The closer to ambient the required refrigeration temperature, the higher the COP of the refrigeration cycle, and thus less work is required per ton refrigeration. At higher pressures (>50 atm), the formation of the CO_2 clathrates is also suppressed. Compression requires mechanical work and is expensive. The optimum total stream pressure/refrigeration temperature may be estimated analytically. When determining the optimum pressure, one should consider the solubility of the N_2 and the acid gases (SO_x, COS, NO_x) in liquefied CO_2. Even though the solubilities have a minimal impact on the mass and volume of the CO_2 product, they reduce the usefulness of the CO_2 as a commercial product. In the presence of water, acid gases also exacerbate the corrosion problem in transportation pipelines. The presence of SO_x, COS, and NO_x also limits the options available for sequestering the CO_2 after it is removed. The corrosive effects of the acid gases in the presence of moisture or high temperatures also adversely affect the components of the refrigeration system and increase maintenance costs.

Particulates, such as those generated by coal combustion, must be removed from the flue gas stream before CO_2 recovery to prevent erosion of the equipment and to prevent plugging and fouling of the heat exchanger.

Several researchers have published information about processes employing cryogenic principles for separating CO_2 from gas mixtures (11–15). The controlled Freezing Zone process (15) employs distillation, solidification, and melting in a single column to separate CO_2 from hydrocarbon gas mixtures. First the feed stream passes over a desiccant to dehydrate it to very low dew points, then it is precooled through heat exchange with the residual stream from which CO_2 has been separated. Then the gas stream is fed to the bottom of the column, maintained in the case of natural gas feed, at about −1.1°C (30°F) and 38 atm. As the gas moves up the column and encounters colder temperatures (185 K at the top), solid CO_2 forms and drops to the bottom of the column.

Koch Process Systems patented a hybrid process, called Ryan–Holmes, that combines a cryogenic system with an absorption system (12,13). Organic additives, such as butanes and aromatics, are added to suppress clathrate formation. In the Ryan–Holmes process, the feed gas is cleaned by recovering the particulates, desulfurized, then dried. The dry gas is compressed to about 25–35 bars and cooled to allow about one-half of the CO_2 content to condense. The condensed CO_2 cools the inlet compressed gas stream. The condensed CO_2 boils via heat exchange with the inlet gas stream, then the gas is compressed to the desired delivery pressure (11). Normally, it is less energy-intensive to pump the liquid to higher pressure than to compress the gas. (Pumping the liquid to higher pressure increases the liquid temperature slightly and increases enthalpy but barely reduces its cooling capacity, because its cooling capacity is related mainly to its latent heat of vaporization). The overhead gas (flue gas with the balance of CO_2) is brought in contact with a CO_2-absorption column to recover more of the CO_2. Then the depleted stream is expanded to recover some of its energy and is used to cool the incoming compressed flue gas, along with the liquid CO_2 product. The absorption solution is thermally regenerated to release the absorbed CO_2, which is also recovered.

In general, refrigeration is accomplished by compressing a gas, cooling it close to ambient temperature, then expanding it so that its temperature falls. In practice, the above single stage of refrigeration is augmented by having several stages and by exchanging warm inlet streams against cool exit streams. The cost of refrigeration equipment is a function of refrigeration temperature and system size (16).

Cryogenic Compression Process (CCP) for CO_2 Recovery

Argonne National Laboratory (ANL) has developed a conceptual process for CO_2 recovery from flue gas of fossil-fuel-fired electric power plants, using a combination of cryogenic temperature and elevated pressure to separate solid CO_2 (dry ice). This concept has been studied and developed using a sophisticated process simulator (Aspen Plus) to perform the required thermodynamic calculations, including the CO_2 phase changes.

At pressures below its triple point (216.6 K and 5.18 bar), CO_2 forms a nearly pure solid on cooling from a vapor mixture that contains gaseous impurities. These conditions are achievable with available industrial compression and refrigeration equipment. This behavior forms the basis for a conceptual process that separates CO_2 ice, by gravity, from liquid or vapor mixtures that include such other components as nitrogen, oxygen, hydrogen sulfide, and carbonyl sulfide. The recovered CO_2 is melted and pumped to storage. The separation equipment must permit significant temperature differences between the CO_2 ice and the melted liquid. For instance, typical flue gas, which contains only 16% CO_2 after moisture recovery, must be cooled to 165–175 K to effect approximately 90% CO_2 recovery. If the melter is well into the liquid region, its temperature is 235 K; this kind of separation, therefore, must maintain a 60–70 K difference.

An integrated cryogenic compression process (CCP) for CO_2 recovery is used to compare cost and performance relative to the conventional MEA CO_2 recovery process. The following design considerations were used:

- all moisture is recovered from the flue-gas feed by refrigeration and desiccation, before it enters the cryogenic section;
- heat-exchanger temperature approaches are ≥10 K;
- CO_2 ice is melted with a 15 K approach;
- delta-*P*s (pressure drops within a single stream) for heat exchangers are 5% of the average pressure;
- product CO_2 liquid is pumped to 110 bar pressure;
- energy integration is used as follows:
  - unrecovered cold gas is turboexpanded back to one atmosphere pressure;
  - selected cold streams are used to precool warm feed streams; and
  - refrigerant waste heat is used to melt separated CO_2 ice for recovery.

Membranes

Membranes are generally solid porous, semiporous, or nonporous structures through which some of the species in a mixture permeate at a higher rate than others. Another type of membrane, called facilitated transport or liquid membrane or gas absorption membranes (17,18), impregnates the solid porous structures with a liquid capable of selectively dissolving the species, whose permeation is desired, to assist them in permeating through the membrane. Many membrane systems are operating in the field in various applications, and new generations of improved membranes are under development (17,20,19,22). The primary applications of membranes for separating gaseous mixtures are as follows (21,22):

1. selective separation of gaseous mixtures

- separation of H_2 from CO, N_2, and hydrocarbons
- separation of acid gases (CO_2, H_2S) and H_2O from natural gas
- N_2 separation from air

2. Filtration of micrometer and submicrometer suspended solid and dispersed liquid particulates from liquid streams (microfiltration)
3. Removal of macromolecules from liquids containing ionic species (ultrafiltration)
4. Selective transport of ionic species (hyperfiltration)
5. Removal of macromolecules from liquids containing ionic species (ultrafiltration)
6. Selective transport of ionic species (hyperfiltration)
7. Removal of macromolecules from liquids containing ionic species

Microfiltration has the largest market share and gas separation has the smallest (22). There are five basic membrane designs: hollow fibers, spiral wound, plate and frame, tubular, and capillary (22). The hollow-fiber and spiral-wound modules are the most common for gas separation because they provide a high surface area per unit volume. The hollow fibers consist of bundles of very fine membrane fibers, such as polyamide, cellulose triacetate, and sulfonated polysulfone, housed in a cylindrical shell. The exterior diameter of the fibers is about 80–200 μm, and wall thickness depends on the size and operating conditions of the unit (22).

Two criteria for a good membrane separator are selectivity and high permeability or mass flux (the rate at which desired species permeate or cross from one side of a unit area of the membrane to the other). Unfortunately, the two criteria are inversely related for the common polymeric membranes. The permeability of the membrane depends on many factors (19,21,22,28), including the chemical potential gradient of the species being separated. This gradient develops from gradients in one or more of the following parameters: partial pressure, concentration, electrical potential, or a combination of these. The mass flux also depends on the nature and design of the membrane itself. The overall mechanism involves diffusion of the targeted species from the bulk of the gas mixture to the surface of the membrane, into which it will be adsorbed or dissolved. (Some molecules, such as H_2, also diffuse through the membrane without being dissolved in the membrane.) The dissolved molecules diffuse through the membrane to the other side, driven by the concentration gradient across the membrane. The diffusing molecules desorb from the surface and diffuse back into the bulk of the gas downstream from the membrane. Therefore, the net effect (or permeability and selectivity of a membrane system) depends on both the solution and diffusion steps, which are normally the rate-limiting steps. In some cases, one of the mixture constituents has favorable solubility but unfavorable diffusion characteristics, whereas another has opposite characteristics. These differences lead to reduced selectivity of the membrane, as is suspected when H_2 and CO_2 are present.

The dissolution process is governed by Henry's law: the concentration of the species separated in the membrane is equal to the product of its partial pressure in the gas phase times its solubility coefficient or Henry's constant. It is reported that this coefficient increases as the condensibility of the gas increases, but it does not depend on temperature (23). Diffusion through the membrane also depends on the nature and structure of the membrane and on the size or diameter of the diffusing gas molecules (24). The larger the molecules, the more difficult it is for them to diffuse. The diffusion coefficient increases as the temperature increases, perhaps because the kinetic energy of the molecules increases and so do the pores or membrane voids through which the molecules diffuse.

Many membrane systems are operating in the field in various applications, and new generations of improved membranes are under development. Table 2 lists some common and successful applications of membranes for gas stream separations.

Flue gas streams generated by coal combustion contain many species in different concentrations. Each has its own solubility in and diffusivity through a membrane. For example, whereas CO_2, H_2S, and H_2O are reasonably highly soluble in membrane materials at moderately low partial pressures, O_2 and N_2 have higher diffusivities. Because the overall permeability and selectivity depend on both of these parameters, the development of a CO_2-selective membrane system is not an easy task. Several polymer materials are used to achieve substantial separation of the CO_2 from the N_2. It is reported that cellulose acetate polymers have a CO_2/N_2 selectivity (permeability of CO_2/permeability of N_2) of about 66.5, which makes it technically feasible to achieve substantial separation of these two gases (19). Selectivity over 20 is needed for an adequate single-stage separation.

Very fine particulates and acidic conditions decrease performance and increase maintenance requirements of a membrane system. Further, membranes are more energy-efficient when excessive compression for proper operation is not required. The relatively low partial pressure of CO_2 in conventional power plant flue gas dictates that some compression (>5 atm) is necessary to increase the partial pressure of the gas constituents and to compensate for the pressure loss across the membrane. Another concern is the ability to maintain a very low partial pressure for the separated species on the downstream side of the membrane. Water vapor is a potential sweeping gas for the separated CO_2 (which could be easily separated by condensation). Evacuation of the downstream chamber or absorption of the CO_2 into a chemical or physical absorption material are alternative methods. These alternatives are energy-intensive. The evacuation alternative could be integrated with the product compression step, provided that the product is low in non-CO_2 constituents.

Most membranes are commercially available only in small modules (<20 m^2). However, it is reported that inorganic membranes capable of handling one million pounds of air per hour do exist for special applications and that such membranes could be adaptable, in the future, for flue gas separation (25). In a recent study, it is reported that gas separation is one of the most active membrane research areas and is likely to yield breakthroughs in both membrane and module technology (26). At present, membrane technology is not readily adaptable to the recovery of CO_2 from the large volume of flue gas produced by a typical power plant.

Physical Adsorption Systems

Physical adsorption of gases on solid adsorbents (such as activated carbon) is a surface phenomenon in which a few

Table 2. Commercial Applications of Membrane Separation of Gas Mixtures

Application	Removal of	Recovery
Natural gas purification	CO and H_2S	98% of CH_4
Landfill gas and biogas	Nonhydrocarbons	81% of hydrocarbons
Synthesis gas preparation	H_2 to adjust H_2/CO ratio	96% pure H_2 and 1.3/1 H_2/CO ratio in retentate
High purity H_2	H_2 (selectively)	Pure H_2
Synthesis gas separation	H_2	Very pure CO
N_2 production from air	N_2	>95% pure N_2 and an O_2-enriched air

layers of the adsorbed gas are held on the adsorbent surface by weak surface forces, rather than by chemical bonding. Because only a few layers of adsorbed gas are physically adsorbed on the surface, the key issue for commercial application of these systems is the surface area per unit mass or volume of the adsorbent. The capacity of the adsorbent is also a function of temperature and pressure. The higher the partial pressure and/or the lower the temperature, the greater the adsorption capacity of the system. Physical adsorbents are used in many applications, including the recovery of hydrocarbon contaminants from air and water streams. They are also used, on a small scale, for such air-separation applications as the UCIG VPSA (vapor pressure swing adsorption) process (27).

The main advantage of physical adsorbents is their simple and energy-efficient operation and regeneration, which is achieved via temperature swing or pressure swing cycles. In the temperature swing cycle, adsorption takes place at a low temperature. Heating of the adsorbent results in desorption of the adsorbed gases. In the pressure swing cycle, the adsorption step occurs at an elevated pressure. Desorption occurs upon depressurization of the system. Because the adsorption process is highly sensitive to temperature, the flue gas stream must be cold. Competition from other flue gas constituents, especially H_2O, for the active adsorption sites reduces the adsorption of CO_2. For example, data in Table 3 show that water vapor is adsorbed preferentially over CO_2. Particulates in the flue gas also result in the gradual deactivation of the adsorption sites because of the difficulty in removing them by heating or depressurization. A common method to activate carbon is thermal treatment of the carbon by steam and/or CO_2 at a temperature of about 750–1000°C.

The energy consumed by these systems is used in compressing and cooling the feed gas and regenerating the adsorbent. The gas-compression and cooling steps are generally required by most flue-gas separation technologies. Regenerating the solid adsorbent is achieved, as mentioned earlier, via temperature swing or pressure swing cycles. Thermal regeneration of the adsorbent requires heating the material to a temperature high enough to desorb the adsorbed species. The energy required in the process consists of the energy required to volatilize the adsorbed species, the energy required to heat the adsorbent to the required temperature, and the energy lost during the regeneration process. The total amount of energy required per unit mass of the adsorbed material depends on the loading of the bed and on the type of species present. Removal of adsorbed water requires about 1,000 Btu/lb of water desorbed. Thermal regeneration of activated carbon, for instance, involves heating the adsorbent material to about 100°C to dry it and desorb the adsorbed species. Steam, hot gas, or infrared heating is used as a heat source for the *in situ* regeneration of the adsorbent. In addition, because of the low partial pressure of CO_2 in flue, gas compression is required. In general, physical absorption systems are more likely to be cost-effective relative to amine-based systems in applications with high CO_2 partial pressures.

The use of solid adsorbents for gas separation applications is generally limited to small and medium size applications. For instance, it is estimated that a carbon/zeolite system for the separation of air is more economical than its cryogenic counterpart at capacities less than about 100 tons of O_2 per day (27). Because these systems are somewhat modular, scale economics are minor. Therefore, these systems are probably not competitive with the chemical and cryogenic methods for CO_2 separation from flue-gas streams. The economics could be more favorable if the combustion process occurred at elevated pressures and if adsorbent materials that are specific for CO_2 are developed. To the best of our knowledge, no effort is underway to develop such materials for CO_2 separation from flue gases.

Physical Absorption Systems

Physical absorption systems differ from physical adsorption systems in that they are a bulk phenomenon instead of a surface phenomenon. Inorganic and organic liquids are used to preferentially absorb some gaseous species, separating them from the gas mixture. The absorption process depends on the operating temperature and pressure, and on the nature of the gases and the nature of the absorber. Absorption takes place at high pressures and/or low temperatures, and regeneration occurs at high temperatures and/or low pressures. Because these processes are temperature-sensitive, the flue gas stream must be cooled. Dissolution of other flue gas constituents (mainly H_2O) adversely affects CO_2 absorption. Particulates in the flue gas cause the absorptive material to degrade rapidly, unless it is filtered frequently to remove trapped particulates. Table 4 lists some of the physical absorption processes that use organic solvents and their characteristics. Each of the processes listed in Table 4 is implemented on a commercial scale in applications other than flue gas separation.

The energy consumed in such dissolution processes is used essentially for compressing and cooling the feed gas, pumping the liquid absorber, and regenerating and cooling the absorber. High boiling solvents are generally preferred to keep solvent losses to a minimum and to prevent contamination of the released gas with solvent vapors. Solvent regeneration is achieved via temperature swing or pressure swing cycles.

The cost of these processes (when applied to the flue gas of a conventional power plant) is over \$61/ton of recovered CO_2 primarily because the gas must be pressurized for effective operation.

However, the solubility of CO_2 in water is limited (Fig. 10) (28). Seawater may be used as a solvent for CO_2. The use of

Table 3. Characteristics of Some Physical Adsorbents

Adsorbent	Surface, m^3/g	Capacity for CO_2, adsorption, wt %[a]	Capacity for H_2O, adsorption, wt %[b]
Molecular sieve, zeolites	[c]	(6–16)	(9–25)
Activated alumina	300–350	2	7
Activated carbon	300–2500	(5–7)	1
Silica gel	750–830	3	11

[a] At 250 mm Hg and 25°C.
[b] At 4.6 mm Hg and 25°C.
[c] Adsorption is a matter of pore filling, and the surface area concept is not applicable. Pore size = 0.3−0.8 nm.

Table 4. Physical Absorption Processes for CO_2 Separation

Process	Solvent	Absorption Temperature, °C	Pressure, bars	CO_2 Removal, %
Fluor	Propylene carbonates	<Ambient	>4 CO_2	95–99
Purisol	*N*-Methyl-2-pyrrolidone	25–40	8→70	97–99
Rectisol	Methanol	>37	22–138	>99
Selexol	Dimethyl ether of polyethylene glycol	−6–37	35–70	>95
Sulfinol	Sulfolane	37–52	0–70	>95

seawater offers some advantages over organic solvents: water is essentially free, unlike organic solvents, thus regeneration is not necessary; water is environmentally more acceptable than organic solvents; and sea water can be used to return the CO_2 to the ocean, thereby acting as a capturing mechanism and as a sequestering vehicle. However, because the solubility of the CO_2 in water is low compared with organic solvents, larger quantities of water and very large process equipment are required. As Figure 10 shows, the solubility of CO_2 in water is a function of temperature and pressure. These data show that high pressures, low temperatures, and large quantities of water are required to significantly remove CO_2 with a water scrubber. Pressures shown in Figure 10 are the partial pressures of the CO_2 and not the total pressures of the flue gas stream. The cost of compressing the gas stream, the cost of cooling the gas stream and the water, and the cost of pumping the water are likely to make this process uneconomical. Figure 10 shows that a CO_2 partial pressure greater than 25 atm is probably required. To obtain a CO_2 partial pressure of 25 atm requires compression of the flue gas to more than 200 atm, which is extremely energy-intensive and cost-prohibitive.

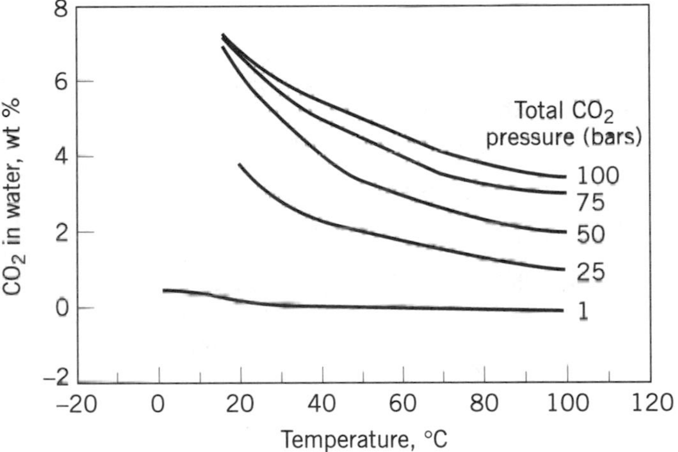

Figure 10. Solubility of CO_2 in water (28).

CO_2 RECOVERY USING ALTERNATIVE POWER PLANT CONFIGURATIONS

In general, the cost of recovering CO_2 from the flue gas of conventional fossil-fueled power plants is excessive because of the low partial pressure of CO_2 in the flue gas. In this section, alternative plant configurations are discussed that use pure O_2, rather than air, as a combustion oxidant to yield a flue gas with a high CO_2 partial pressure. First O_2 production technology is discussed.

Combustion with O_2 increases the partial pressure of CO_2 in flue gas from about 15% to about 66% because N_2 is avoided. The remainder of the flue gas is predominantly water (26%), which can be separated readily from the CO_2. Beyond the advantage of capturing CO_2, combustion with O_2 offers other advantages: (*1*) energy losses associated with heating N_2 from ambient temperature to the flue gas exhaust temperature (of about 300°F) is reduced (this corresponds to a loss of about 6% of the coal's heating value (over two percentage points loss in the overall thermal efficiency of a conventional coal-fired power plant)) and (*2*) burning a fuel in the absence of N_2 reduces the formation of NO_x. The major drawback of combustion with O_2 is that O_2 production is energy-intensive and expensive. About 25–30% of the electric power output of a conventional fossil-fueled power plant will be consumed by the O_2 production process, and the capital cost of the O_2 generation system will be about $520/kW, or about 25–30% that of a conventional power plant's capital cost.

Oxygen Production

A number of technologies produce O_2 from air or water. However, the most widely practiced technology for the large-scale production of oxygen from air is cryogenic air separation, because it is the most cost-effective for large-scale applications. The state-of-the-art equipment for O_2 generation consumes about 220–300 kWh of electricity per ton of gaseous O_2. The cryogenic method is relatively mature, so efforts to improve its performance are minimal.

Splitting of water by electrolysis for O_2 production is simply too energy-intensive for integration into a large-scale power plant. This technology would consume more energy for O_2 production than the net power plant output.

Over the last two decades, researchers have actively investigated thermochemical cycles for producing H_2 from water as an alternative to steam-reforming of methane. The by-product in steam reforming is CO_2. The overall reaction in these cycles is $H_2O \rightarrow H_2 + 1/2O_2$. Unlike electrolysis of water, which consumes electricity, these cycles consume high temperature thermal energy to split water. Thus, thermochemical cycles for splitting water could be energy-effective when integrated with a power plant. None of these multistep thermochemical cycles have yet reached the commercial stage. Therefore, reliable cost data for these systems do not exist. Corrosion problems at the operating temperatures of these cycles may require exotic materials of construction. The potential for a significant reduction in the inefficiency could, however, decrease the cost of CO_2 removal. Little effort, if any, is directed toward improving this technology because it is considerably more expensive than methane reforming to produce H_2. However, it could be a more energy-efficient way to produce O_2 when integrated with a power generation plant.

Two thermochemical cycles for separating O_2 from air (the Moltox process and the barium oxide/peroxide process) were also evaluated. Figure 11 illustrates a possible way for integrating these cycles with a power plant. The barium-oxide/peroxide cycle was used commercially in the late eighteenth century for making O_2. It was displaced by cryogenic distillation. Both of these processes are energy-intensive as stand-alone O_2 production processes. However, when integrated with power generation systems, most of the thermal energy consumed in the air-separation process can be returned to the power generation system. Integration of these high temperature chemical cycles with high temperature fuel-cell systems is another potentially very energy-efficient approach. With steam recovery, it is claimed that the Moltox process consumes about 200–250 kWh of electricity per ton of O_2 recovered. This process competes favorably with state-of-the-art cryogenic technology. The impact of integrating such processes with a power generation system on the overall economics is not readily obvious and requires additional research. However, the potential for efficiency improvement could contribute to reduced cost of CO_2 recovered.

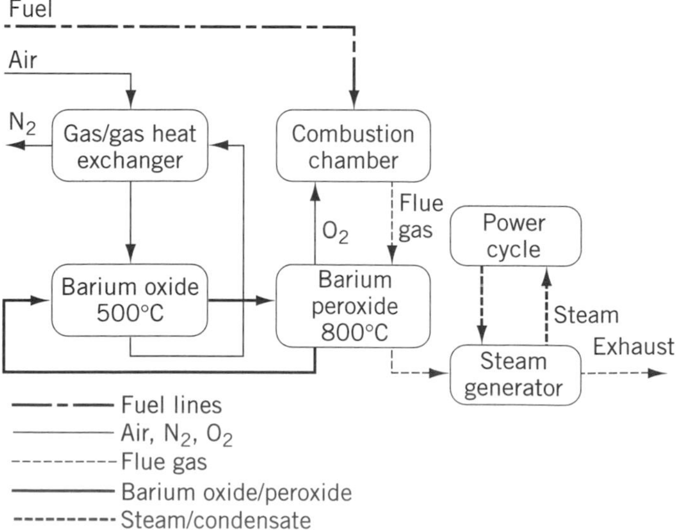

Figure 11. Integration of a chemical O_2 production cycle with a power plant.

Researchers are also investigating the use of membranes for the small-scale production of N_2 from air, which would yield an O_2-rich by-product. At the large scales required for power plant operations, the process is uneconomical compared with the cryogenic method.

Recent advances in physical adsorption suggest that it is more economical than even the cryogenic methods at design capacities below about 150 tons of O_2/day. However, this advantage is lost at the higher capacities required for large-scale power plants.

Retrofitting Conventional Power Plants for Combustion of Fossil Fuels Using O_2

O_2 combustion cannot be substituted for air combustion in existing plants because of the differences in combustion characteristics. A thermal diluent is required to replace N_2. Argonne National Laboratory (ANL) developed and tested a process that recycled a split stream of the flue gas back to the combustor to achieve the same heat-transfer characteristics as experienced when combustion is conducted with air (29,30). The purpose of this step is to minimize the number of changes necessary to the existing hardware. To investigate the technical feasibility of this concept, three experiments were conducted which are described below.

1. *Battelle Columbus Experiment.* The Battelle Columbus Division conducted tests with a research furnace capable of burning 25–65 lb/h of coal. Eight tests were conducted for four oxidant mixtures: air and three mixtures of O_2 and CO_2. The tests showed that coal could be burned with CO_2/O_2 as the oxidant at CO_2-to-O_2 mole ratios of 3.65 to 2.23 and achieve high combustion efficiencies.
2. *Black Hills Power and Light Company Tests.* Tests were conducted at the Black Hills Power and Light Company's Customer Service Center Boiler No. 2 in Rapid City, South Dakota. The boiler is a stoker-fed heating boiler rated at 2.2×10^6 Btu/h. The boiler was retrofitted with a CO_2 recycle system and operated in three phases: normal, air-fired; transition to O_2/flue-gas-fired; and O_2/flue-gas-fired. The tests showed no noticeable effects on the performance of the boiler. The tests also showed that transition from air-fired to oxygen-fired operation did not present any difficulties. Because no attempt was made to seal the coal bunker from the atmosphere, the maximum CO_2 concentration achieved was 48.5%.
3. *Energy and Environmental Research Corporation Tests.* Energy and Environmental Research Corporation ran a series of tests using its pulverized-coal

Table 5. Attributes of O_2 Combustion Technologies

Attribute	Technology Characteristics	
	ANL Process-PC	FBC
Purity of separated CO_2, %	>93	>93
Net CO_2 reduction, %	~100	~100
Cost, \$/t of CO_2 recovered	32	32
Energy consumption, % of net power production	35–40	35–40

experimental facility. The objectives of these tests were (*1*) to establish thep recycled gas/oxygen ratio necessary to achieve heat transfer performance similar to that with coal/air combustion, (*2*) to quantify changes in important parameters (such as burn-out and flame stability) that might affect system performance, and (*3*) to provide a basis for scaling the experimental results to commercial-scale equipment. This work confirmed that satisfactory combustion of pulverized coal is achieved with mixtures of O_2 and recycled flue gases under conditions representative of utility boilers. The CO_2 concentration in the flue gas exceeded 90% on a dry basis.

Table 5 summarizes the attributes of the O_2 combustion processes.

A reference case for evaluating the cost of CO_2 recovery using the CO_2 recycle concept is shown in Figure 12 (30). The capacity of such a plant would be about 20–25 MW, depending on its heat rate. Oxygen would be supplied for the plant boiler/steam generator by an on-site oxygen plant. A split-stream of the flue gas would be recycled to the combustor to maintain the thermal stability of the operation. The remainder of the flue gas stream would be dehydrated and compressed downstream of recycle piping. The research and development effort described above showed that the optimal recycle ratio, expressed as a mole ratio $[(CO_2 + H_2O)/O_2]$ in the oxidant, was about 3.25. The energy required for the oxygen plant varies from about 250 to 300 kWh/ton of O_2, depending on plant size, altitude, and plant design (300 kWh is conservative). The actual cost of electricity charged to the oxygen plant depends on a number of factors, such as mode of operation (ie, peak vs off-peak), the financial structure of the project, and the local utility's and regulatory commission's willingness to accommodate favorable pricing mechanisms.

To estimate the design capacity of the oxygen plant, the following simplifying assumptions were made: (*1*) coal composition is $CH_{0.8}$ and (*2*) combustion occurs stoichiometrically. On the basis of the data generated in the experimental effort discussed previously, these two assumptions underestimate the actual oxygen consumed per ton of flue gas produced by about 2% by weight for a commercial process. Given these assumptions, the oxygen required to produce 455 t/day of CO_2 is 398 t/day. To compensate for the error introduced by the two assumptions, the oxygen plant is sized at 410 t/day. The oxygen plant would consume from 112,500 to 135,000 kWh of electricity per operating day.

The total cost of CO_2 removed is about \$41/t. These costs are based on a 25-MW plant. As the scale of the plant increases, these costs decrease to about \$32/t of CO_2 recovered at a plant output of about 350 MW. The most significant component of the cost is that the oxygen plant, which accounts for almost 60% of the cost of the CO_2 recovered for small power plant design capacities (about 25 MW). At 350 MW, the capital cost of the oxygen plant accounts for about 40% of the \$30/t of CO_2 recovered.

The tests conducted at Black Hills showed that the boiler could be operated intermittently with air or O_2 and CO_2. The implication is that the oxygen plant could be run during off-peak periods and maintained in cold standby condition during peak periods. In this case, the price of electricity decreases to the cost of fuel used to produce the electricity, which is on the order of \$0.01/kWh for coal. If the cost of electricity charged to

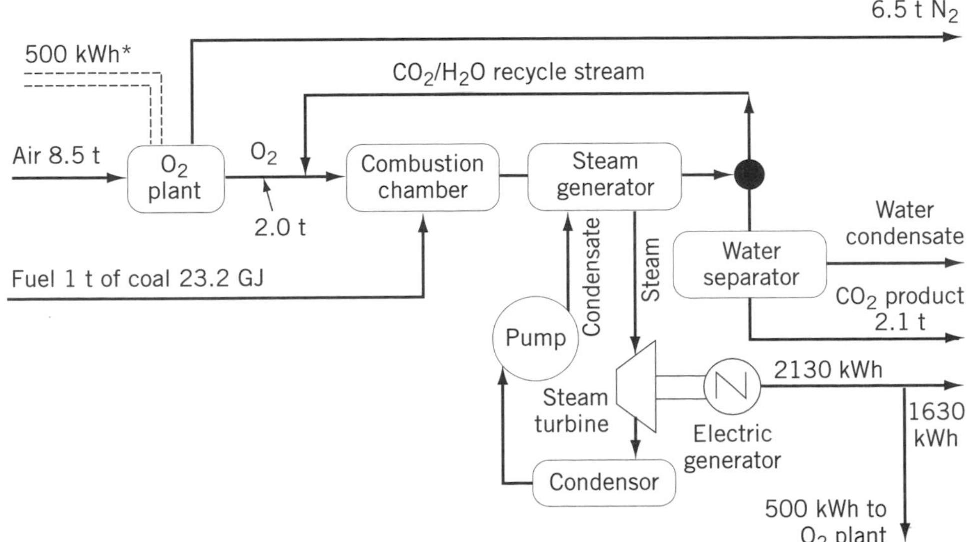

Figure 12. Reference plant configuration for an O_2 plant and recycle piping retrofitted to a coal-fired power plant, where * indicates 250 kWh/t of O_2.

the oxygen plant is \$0.01/kWh, the cost of CO_2 decreases by more than \$11/t (0.21/scm [standard cubic meter]). Of course, the quantity of CO_2 recovered in this mode of operation is significantly lower than that recovered in a continuous operation.

Although the experimental effort for the ANL concept is based on coal as a fuel, the process should also be adaptable to other fossil fuels. Figure 13 shows approximate material and energy flows for coal and natural-gas-fired plants using O_2 in the combustion process.

Note that a smaller O_2 plant is required for a gas-fired plant relative to a coal-fired plant, which results in a lower cost of CO_2 recovery from the gas-fired plant.

CO_2 Recovery from the Effluent of Future Power Plant Designs

A number of advanced technologies have been examined to evaluate the cost and feasibility of CO_2 recovery: integrated coal gasification-combined cycle (GCC), CO_2 turbines, and high temperature fuel cells.

CO_2 Recovery from the Effluent of Integrated Coal GCC Power Plants. This topic was studied by Fluor Daniel, Inc., for the Electric Power Research Institute (EPRI) (8). Detailed engineering and cost information was developed for this study.

Coal GCC plants consist of three main building blocks: a coal gasification plant, a gas cleaning system, and the power generation facility (2,31,32). The gasification plant produces a gas stream rich in CO and H_2. Of particular interest to this discussion is the O_2-blown gasifier, which produces a gas stream containing very little N_2 (31). The main components of these streams are CO, H_2, CO_2, H_2O, and gaseous hydrocarbons. The gas cleanup system of the plant involves particulate removal and acid gas scrubbing. In a typical GCC plant, the crude syngas is first fed to a scrubbing facility to remove the particulates from the gas stream. Then the gas leaving the particulates scrubber is cooled and dewatered. At this point, it consists mainly of CO, H_2, and CO_2 and acid gases, which are removed in a desulfurization system. The acid-gas scrubbing process is generally designed to remove only the sulfur-bearing compounds. Very little CO_2 is removed in the process. However, removal of CO_2 at this stage is possible by using physical absorption technology because the gas is at high pressure, as we discuss below. The power generation facility consists of a gas-turbine system followed by a steam-turbine bottoming cycle. The desulfurized gas is first injected with steam and partially expanded in gas expanders to recover some work. Then the partially expanded gas, which is rich in CO and H_2, is burned with air and expanded in a gas turbine to recover more work. The residual thermal energy in the gas turbine exhaust is used to produce steam for the steam turbine bottoming cycle. A typical plant has been discussed in detail (8).

Although the oxygen plant accounts for about 15–20% of the total plant cost, it is cost-effective vis-a-vis air-blown gasification because the cost savings in the gasification section and gas cleanup systems more than offset the cost of the O_2 plant for large facilities (33). The use of an oxygen-blown gasification system (as opposed to air) simplifies the O_2 recovery process. To provide for CO_2 recovery, a shift reactor and a physical absorption (eg, Selexol) acid gas separation system are incorporated. In the shift reaction, the CO in the syngas reacts with H_2O to convert the CO to H_2 and CO_2. This reaction maintains the fuel value of the CO in the form of H_2 and further increases

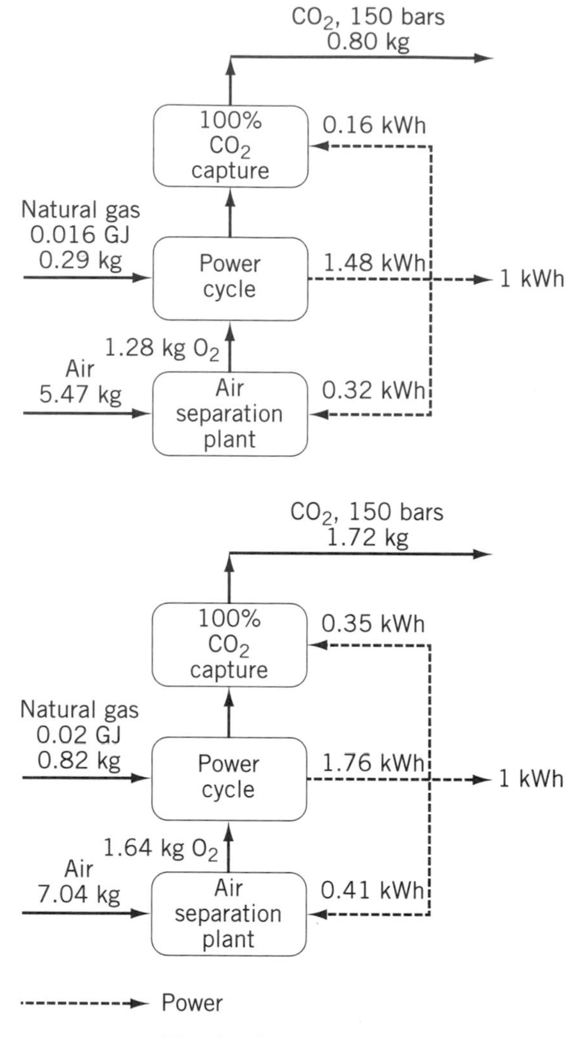

Figure 13. Simplified material and energy flow streams in natural gas and coal-fired power plants using O_2 for combustion. The energy requirement for CO_2 dehydration and compression from 1–150 bar is assumed to be 0.097 and 0.103 kWh/kg of CO_2 from coal and natural gas, respectively, because the flue gas from natural gas contains more water vapor. The energy requirement for O_2 production is assumed to be 250 kWh/t of CO_2.

the partial pressure of CO_2 in the synthesis gas stream. Then recovery of the CO_2 is accomplished in the acid-gas separation system. An alternative to physical absorption is an H_2-selective membrane system for CO_2 recovery because the syngas, after shifting, is predominantly H_2 and CO_2, and the gas is still at high pressure. The Netherlands Energy Research Foundation is working on the development of such a system (34).

The EPRI-sponsored study (8) reported that a 90% recovery of CO_2 from the gasification products, as described above, using a Selexol process, results in 20% reduced power output relative to a plant in which no CO_2 recovery is practiced. However, only about 20% of the 20% reduction in net power is directly attributable to the recovery of the CO_2; the balance is primarily attributable to CO_2 compression. The total capital cost of the plant with CO_2 recovery is about \$310/kWe. Compared with the ECONAMINE FG system, which yields an estimated cost of about \$61/t of CO_2 avoided, the incremental cost of CO_2 recovery in a gasification combined cycle system is estimated at about \$20/t of CO_2 recovered.

CO_2 Turbines. Installing a membrane system downstream of the gas cleanup system to separate the cleaned gas into H_2-rich and CO-rich streams was proposed (35). The CO-rich stream then is burned with O_2 mixed with recycled CO_2, and the product CO_2 stream expands in a CO_2 turbine to recover some of its available energy. Then the CO_2 turbine exhaust is used to produce steam to operate a steam turbine. The cooled CO_2 stream then is split into a recycled stream and a recovered stream. The H_2-rich stream is burned with air, and the products expand in a separate turbine. The exhaust of that turbine is used to produce steam, which is combined with the steam generated from the exhaust of the CO_2 turbine.

At the heart of this concept is the membrane separator. It was calculated that a commercially available, hollow-fiber polyamide membrane developed by Ube Industries can separate the products of Texaco or Shell gasifiers to produce two streams (35). The data are on a dry basis because the H_2O is not reported. Although the membrane system does not achieve clean separation of the gases, it indeed concentrates the carbon-containing species, making it easier to recover the CO_2.

It was pointed out that CO_2 turbines are not commercially available and require development for this application (35). Specially designed CO_2 turbo-machinery is in use in CO_2-cooled nuclear power plants (36). The H_2 is burned and expanded in a conventional turbine, because the working fluid going through the turbine is essentially water vapor. The need for more selective membranes and the importance of high gasification pressures for more effective CO_2 recovery is also pointed out (35).

Another approach presented at the First International Conference on CO_2 removal (36) involves a combined CO_2/evaporative steam cycle, which eliminates the need for two separate turbine systems, which then reduces plant cost.

It was estimated that the cost of the GCC plant will increase by about 22% over the cost of a reference plant (35). The cost of electricity will increase by about 36%, assuming a discount rate of 10%. Table 5 summarizes the key attributes of the GCC technology as it pertains to CO_2 removal from flue gas streams when a CO_2 turbine is used.

CO_2 Recovery from High Temperature Fuel Cells. CO_2 recovery from the anode exhaust of the molten carbonate fuel cell (MCFC) when integrated with a coal gasifier has been discussed (37). The MCFC, originated in Holland in the late 1950s, uses a mixture of molten potassium and lithium carbonates (~40% and 60%, respectively) immobilized in a porous ceramic matrix as the electrolyte material. It is currently under development in the United States and in Japan. This cell uses nickel and nickel oxide electrodes. The anode is a porous, sintered nickel structure stabilized against pore growth and loss of surface area by chromium and/or cobalt additives. Its forecasted efficiency is about 60%. The present state of this technology enables the fabrication of 1-m^2 planar cells. A 250-kW module is under fabrication at MC Power Inc.

This cell operates at high temperatures (~650°C). Therefore, it internally reforms a hydrocarbon fuel (such as methane) to produce H_2. In addition, the high grade heat available from the cell is also used to produce steam for a bottoming cycle.

The key reactions involved in the MCFC are as follows:

Anode reaction $$H_2 + CO_3^{2-} \rightarrow H_2O + CO_2 + 2e^-$$

Cathode reaction $$CO_2 + 1/2O_2 + 2e^- \rightarrow CO_3^{2-}$$

Overall reaction $$H_2 + 1/2O_2 \rightarrow H_2O$$

The fuel is normally a coal gasification product ($CO + H_2$) or a gaseous hydrocarbon. When the fuel is a gaseous hydrocarbon, it is normally preheated and then mixed with steam that is raised by using heat generated within the cell. The fuel steam/mixture is passed on to the anode side of the fuel cell. This process results in reforming the fuel into H_2, CO, and CO_2. The H_2-rich fuel (from reforming or gasification) reacts at the anode as described above, resulting in some shifting of CO to produce more H_2 and CO_2. The anode effluent, which is about 40–75% CO_2, 20–40% H_2O, 2–5% H_2, and 5–10% CO (38,39), is cooled by heating the anode feed stream, then it is burned, and the products are sent to the cathode as a source of CO_2. At the cathode, the CO_2 is consumed, and thus CO_2 must be continuously supplied to the cathode. Because CO_2 participates in the cell reactions and because CO does not poison the cell, *in situ* reforming of hydrocarbon fuels is possible. The cathode exhaust is split into two streams. One part is recycled back to the cathode, and the other is cooled by heat recovery and exhausted.

In the configuration investigated, the anode exhaust is first cooled and then fed to a shift reactor (37). A simplified flow diagram is presented in Figure 14. The effluent of the shift reactor is separated by a membrane separator. The H_2 that leaves the membrane is recycled back to the fuel cell anode, and the residual H_2O and CO_2 stream is split. One part is recycled to the cathode, and the other part is processed for CO_2 recovery.

The recovery of the CO_2 anode stream requires the following additional equipment: a heat exchanger to cool the anode exhaust and produce steam from it; a shift reactor; a membrane system; another heat exchanger to cool the separated CO_2/H_2O stream to separate the water, followed by a desiccant drier before compressing the CO_2; and a CO_2 compressor. A H_2 compressor is also required to compress the separated H_2 for recycling back to the anode. The cost of these components is likely to be about 25–50% of the initial capital cost of the gasification/MCFC plant. We estimate that the process costs \$10/t of CO_2 recovered.

The research needs should focus on the membrane system and on the fuel cell itself. Considerable research on MCFC is under way in the United States and in Europe. The most significant R&D efforts on this system are related to materials and are targeting the following items (40,41):

- controlling corrosion promoted by the molten carbonate salt
- replacing the NiO cathode, which slowly dissolves in the molten carbonate salt and is reduced at the anode to produce nickel, which short-circuits the cell (leading candidates to replace nickel oxide, which have been studied extensively, are lithium ferrite ($LiFeO_2$) and lithium cobaltate ($LiCoO_2$))

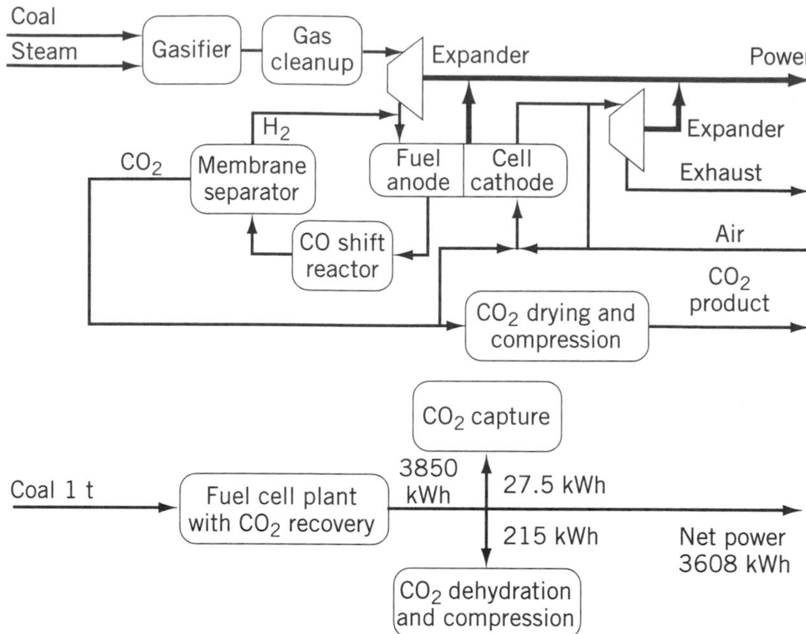

Figure 14. Simplified diagram of an MCFC sysem with CO_2 recovery.

CONCLUSIONS

Several technologies for recovering CO_2 from the flue gas of conventional fossil-fueled-fired power generation systems and from advanced power systems are commercially available. In general, technologies for capturing CO_2 are expensive and energy-intensive. Therefore, they substantially increase the overall cost of power generation. Cryogenic separation of solid CO_2 and chemical absorption of CO_2 (commercially available and proven technologies) are the most economical alternatives for separating CO_2 from the flue gas of conventional fossil-fuel-fired power plants. These technologies also result in the least reduction in net power generation and achieve very high CO_2 removal levels. The estimated costs for the cryogenic separation of CO_2 in the solid form and for the chemical absorption of CO_2 in amines are $27 and $37/t of CO_2 removed, respectively. The estimated costs for all other technologies are greater than $37/t of CO_2 removed.

Recovering CO_2 from the flue gas of future power plants, from current designs, when retrofitted for combustion using O_2 instead of air, and from advanced designs that may be adopted during the next 20 years (such as IGCCs) is simplified because all of the technologies described involve air separation in one form or another. Combustion with O_2 increases the partial pressure of CO_2 in flue gas from about 15% to about 66% because N_2 is avoided. The remainder of the flue gas is predominantly water (26%), which is readily separated from CO_2. About 25–30% of the electric power output of a conventional fossil-fueled power plant is consumed by the O_2 production process, and the capital cost of the O_2 generation system is about $520/kW, or about 25–30% that of a conventional power plant. In advanced plants, such as those using combined-cycle gasification, oxygen is generally more cost-effective than air in the gasification of coal. However, the cost and the inefficiency of using oxygen are only justified in the fuel-conversion process. Because oxygen is potentially cost-effective in the fuel-conversion process, separating the carbon content of the gasified product before combustion could result in the cost-effective removal of CO_2.

BIBLIOGRAPHY

1. EIA, U.S. Department of Energy, Energy Information Administration, Office of Coal, Nuclear, Electric, and Alternate Fuels, report number DOE/EIA-0226 (91/11), Washington D.C., Nov. 1991.
2. M. M. El-Wakil, *Power Plant Technology,* McGraw-Hill Book Co., New York, 1984.
3. SRI International, *Chemical Economics Handbook,* Menlo Park, Calif., 1989.
4. A. Kohl and F. Reisenfeld, *Gas Purification,* 4th ed., Gulf Publishing Co., Houston, Tex., 1985, p. 33 ff.
5. M. T. Sanders, and C. L. Mariz, *The Fluor Daniel ECONAMINE FG Process: Past Experience and Present Day Focus,* First International CO_2 Conference, Amsterdam, Mar. 1992.
6. R. Barchas and R. Davis, The Kerr-McGee/ABB Lummus Crest Technology for the Recovery of CO_2 from Stack Gases, First International CO_2 Conference, Amsterdam, Mar. 1992.
7. M. A. Wilson, R. M. Wrubleski, and L. Yarborough, *Recovery of CO_2 from Power Plant Gases Using Amines*, First Int. CO_2 Conference, Amsterdam, the Netherlands, Mar. 1992.
8. S.C. Smelser and co-workers, *Engineering and Economic Evaluation of CO_2 Removal from Fossil-Fuel-Fired Power Plants,* Electric Power Research Institute report EPRI IE-7365, Vol. 1, June 1991.
9. J. L. Threlkeld, *Thermal Environmental Engineering,* 2nd ed., Prentice–Hall Inc., Englewood Cliffs, N.J., 1970.
10. R. B. Scott, *Cryogenic Engineering,* D. Van Nostrand Co., Inc., Princeton, N.J., 1959.
11. D. Colomb, H. Herzog, J. Tester, D. White, and S. Zemba, *Feasibility, Modeling and Economics of Sequestering Power Plant CO_2 Emissions in Deep Ocean,* MIT-E1 89-003, Dec. 1989.
12. A. S. Holmes, 1982, *Process Improves Acid Gas Separation,* Hydrocarbon Processing, May, 1982, p. 131.
13. A. S. Holmes, "Recovery of CO_2 from Man-Made Sources Using Cryogenic Distillation Techniques", in *Recovering Carbon Diox-*

ide from Man-Made Sources: Proceedings of a Workshop Held in Pacific Grove, California, Feb. 11–13, 1985, Argonne National Laboratory Report, ANL/CNSV/TM-166, Argonne, Ill., 1985, pp. 63–75.

14. G. C. Schianni, *IChemE Symp. Soc.* (44), 50–55 b(1976).
15. D. J. Victory and J. A. Valencia, *Hydrocarbon Proc.,* 44–46 (May 1987).
16. Hagler, Baily and Co., *Magnetic Heat Pump—Assessment of Potential Industrial Cooling Applications,* prepared for U.S. Dept. of Energy Office of Industrial Programs by Hagler, Baily & Co., Contract No. DE-ACO1-87CE40762 Subcontract No. 3090-1, Dec. 1987.
17. P. H. M. Feron, A. E. Jansen, and R. Klaassen, *Membrane Technology in Carbon Dioxide Removal,* TNO Institute of Environmental and Energy Technology, the Netherlands, Jan. 1992.
18. S. Saha and A. Chakma, *Separation of CO_2 from Gas Mixtures with Liquid Membranes,* First International CO_2 Conference, Amsterdam, Mar. 1992.
19. J. Sluijs, C. Hendriks, and K. Blok, *Feasibility of Polymer Membranes for Carbon Dioxide Recovery from Flue Gases,* Department of Science, Technology and Society, University of Utrecht, the Netherlands, NW&S No. 92006, Dec. 1991.
20. D. J. Stookey, and W. M. Pope, "Application of Membranes in Separation of Carbon Dioxide from Gases," in *Recovering Carbon Dioxide from Man-Made Sources: Proceedings of a Workshop Held in Pacific Grove, California, Feb. 11–13, 1985,* Argonne National Laboratory report ANL/CNSV/TM-166, Oct. 1985, pp. 53–62.
21. W. J. Koros, and R. T. Chen, "Separation of Gaseous Mixtures Using Polymer Membranes," in R. W. Rousseau, ed., *Handbook of Separation Process Technology,* John Wiley & Sons, Inc., New York, 1987, Chapt. 20.
22. K. Scott, *Membrane Separation Technology—Industrial Applications and Targets,* Scientific and Technical Information, Oxford, London, 1990.
23. D. R. Paul and G. Morel, "Membrane Technology", *Kirk-Othmer Concise Encyclopedia of Chemical Technology,* John Wiley & Sons, Inc., New York, 1987, pp. 741–743.
24. M. D. Dinello, R. S. Narayan, and C. J. Patton, *Bulk CO_2 Removal Achieved through Membrane Separation,* Society of Petroleum Engineers-Production Engineering, Feb. 1989.
25. D. E. Fain, G. E. Roettger, and R. R. Judkins, *Potential for CO_2 Removal with Inorganic Membranes,* paper number 95d, AIChE 1992 Summer National Meeting, Minneapolis, Minn, Aug. 9–12, 1992.
26. R. W. Baker, E. L. Cussler, W. Eykamp, W. J. Koros, R. L. Riley, and H. Strathmann, *Membrane Separation Systems—A Research Needs Assessment, Final Report,* DOE/ER/30133-H1, Vol. 2, Apr. 1990.
27. CPI Purchasing, 1989, *New Technology Provides Low-Cost Supplies,* Oct. 1989, p. 69.
28. R. H. Perry and D. Green, *Perry's Chemical Engineers Handbook, 6th ed.,* McGraw–Hill, New York, 1984.
29. N. Brockmeier and co-workers, ANL Report ANL/ESD/TM-75, Argonne, Ill., Oct. 1994.
30. A. M. Wolsky, E. J. Daniels, and B. J. Jody, *J. Air Waste Manage. Assoc.* **41,** 449–454 (1991).
31. IGT, *Clean Fuels from Coal—Symposium II,* sponsored by Institute of Gas Technology, IIT Center, Chicago, Ill., June 23–27, 1975.
32. IGT, "Hydrogen for Energy Distribution," symposium papers presented at *Hydrogen for Energy Storage,* a conference sponsored by Institute of Gas Technology, IIT Center, Chicago, Ill., July 24–28, 1978.
33. T. A. Matchak, and C. H. Lawrence, *Economic Assessment of the Impact of Plant Size on Coal Gasification-Combined-Cycle Plants,* prepared by Fluor Engineers, Inc., for Electric Power Research Institute, EPRI AP-3084, May 1983.
34. D. Jansen, A. B. J. Oudhuis, and V. Veen, *CO_2 Reduction Potential of Future Coal Gasification-Based Power Generation Technologies,* First International CO_2 Conference, Amsterdam, Mar. 1992.
35. C. A. Hendriks and K. Blok, *Carbon Dioxide Recovery Using a Dual Gas Turbine IGCC Plant,* First International CO_2 Conference, Amsterdam, Mar. 1992.
36. J. D. Ruyck, *Efficient CO_2 Capture through a Combined Steam and CO_2 Gas Turbine Cycle,* First International CO_2 Conference, Amsterdam, Mar. 1992.
37. A. B. J. Oudhuis, D. Jansen, and P. C. Van Der Laag, *Concept for Coal-Fueled Fuel-Cell Power Plant with CO_2 Removal,* First International CO_2 Conference, Amsterdam, Mar. 1992.
38. T. G. Benjamin, E. H. Camara, and L. G. Marianowski, *Handbook of Fuel Cell Performance,* prepared by Institute of Gas Technology for the U.S. Dept. of Energy under contract No. EC-77-C-03-1545, Jan. 1980.
39. E. J. Daniels, C. B. Dennis, and M. Krumpelt, 1986, *Evaluation of Fuel Cells for Cogeneration Applications,* prepared by Argonne National Laboratory for the U.S. Dept. of Energy, Morgantown Energy Technology Center, Nov. 1986.
40. N. Q. Minh, *High-Temperature Fuel Cells, Part 1: How the Molten Carbonate Cell Works, and the Materials that Make it Possible,* ChemTech (American Chemical Society) 0009-2703/91, Jan. 1991.
41. B. C. H. Steele, 1989, *Materials Engineering and Fuel Cell Development,* MRS Bulletin, June 1989.

CARCINOGENS: IDENTIFICATION AND RISK ASSESSMENT OF

WILLIAM E. PEPELKO
LAWRENCE VALCOVIC
U.S. Environmental Protection Agency
Washington, D.C.

Risk assessment is an organized approach to evaluating scientific information that involves considerable uncertainties. Risk assessment for carcinogenicity attempts to answer two questions. First, what is the likelihood that the agent in question is a carcinogen? Second, if it is a carcinogen, how potent is it? The approach to answering these questions differs somewhat from that used to evaluate noncancer toxicity. Noncancer toxic endpoints occur with varying degrees of severity, are often reversible, and generally have a threshold. A common means of estimating safe dose levels involves dividing the administered dose at the highest no-observable-adverse-effect level or the lowest observable effect level by an uncertainty factor. Another approach involves deriving a dose-response curve, then determining the lower 95% bounds of the dose at which a predetermined percentage of animals respond, for example, 10%. Cancer, on the other hand, is usually irreversible and ultimately is fatal unless death occurs first from other causes. Most carcinogens also have no threshold. Therefore, dose-response modeling is based on the premise that a finite risk occurs with any dose, unless adequate data exist to indicate otherwise. If it is necessary to assess the degree of risk in populations exposed either in the ambient environment or in the workplace, then two additional steps are necessary—exposure assessment and risk characterization. In characterizing risk, hazard identification, dose-response, and exposure assessment information is integrated to estimate the likelihood and the severity of the effects in exposed individuals. This chapter focuses on assessing hazard and estimating dose-response for potential carcinogens (see RISK ASSESSMENT).

HISTORY OF CANCER IDENTIFICATION AND RISK ASSESSMENT

Identification of Carcinogens

In the late eighteenth century, the British surgeon Percivall Pott was the first to note that cancer might relate to specific causes. He observed that there was a likely association between scrotal cancer and occupation in chimney sweeps. Moreover, his recommended means of prevention, bathing at the end of the workday, proved effective. Early in the present century, experimental studies with combustion products, coal tars, were shown to cause cancer following dermal application to rabbits. In the 1920s, Kennaway and co-workers (United Kingdom) fractionated coal tar and discovered the carcinogenicity of polynuclear aromatic compounds dibenz[*a,h*] anthracene and benzo[*a*]pyrene. Subsequently, it was found that these compounds and other polynuclear aromatics, occur commonly during combustion of organic substances. In the late 1800s, clusters of bladder cancer cases in humans were associated with exposure in the dye industry, thus establishing a link between aromatic amines and cancer. Shortly after the turn of the century, Ehrlich in Germany reported that exposure to bioazo dyes, scarlet red or C.I. Solvent Red 24, led to a reversible proliferation of liver cells, providing suggestive experimental evidence for carcinogenicity. However, the carcinogenicity of these azo dyes, was not confirmed experimentally until 1937 in studies with dogs.

In following years, it was found that many classes and types of chemicals are carcinogenic in animals. At present, new carcinogens are being discovered, largely in studies designed to establish structure-activity relationships and with chemicals found genotoxic. Routine safety testing undertaken for potential commercial products and national testing programs, such as those at the National Cancer Institute/National Toxicology Program and the Japanese Cooperative Program on Long-Term Assays for Carcinogenicity, are also yielding evidence for new cancer-causing agents. For further details, see (1).

The most recent class of agents shown experimentally to be carcinogenic are very fine, insoluble, biochemically inert particles administered via inhalation. Most of the investigations of this class of agents have been carried out since 1980. Among those agents that tested positive are diverse particle types, such as carbon black and titanium dioxide. The carcinogenic mechanisms are thought to differ from chemical agents in that they do not act directly on target cells. Rather, they are ingested by phagocytic cells in the lungs, which then are induced to secrete a variety of factors toxic and carcinogenic to lung epithelial cells (2).

Development of Risk Assessment Methodology

Before 1950 few agents, either chemical or physical, were regulated in the United States on the basis of their carcinogenic activity. One exception was ionizing radiation. Initially permissible exposures to radiation were set similarly to those of noncarcinogenic agents, by applying safety factors specific exposures. However, during the 1950s, evidence emerged to support a nonthreshold theory for cancer induction. Probably the most visible expression of this was the Delaney Amendment to the Federal Food, Drug, and Cosmetic Act in 1958, which stated that any substance that induced cancer in humans or animals could not be registered as a food additive. Thus, at that time, dose-response determination was not considered necessary for carcinogens.

A variety of factors contributed to the later development and use of dose-response assessment for carcinogens. During the 1960s and 1970s, large numbers of newly identified carcinogens were found among commercial products. There was also the growing perception during this period that many kinds of risks could not be eliminated completely without unacceptable social and economic consequences. Thus, requirements for quantitatively assessing carcinogenic risk arose. Over a period of two decades, various expert committees sponsored by government agencies and other organizations published numerous reports dealing with carcinogenic evaluation.

In 1983, the National Research Council (NRC) (3) was asked to issue recommendations regarding the scientific basis of risk assessment and the institutional arrangements under which it was being conducted and used. The risk assessment framework and specific definitions of risk assessment and its component steps from the 1983 NRC report have been widely adopted. Many of the recommendations from that report have been implemented by EPA and other regulatory agencies. They include two especially important recommendations. First, that a clear distinction between risk assessment and risk management should be maintained. Second, that regulatory agencies should develop and use inference guidelines to detail the scientific basis for the conduct of risk assessment and to set forth the default options. The NRC committee did not specify any particular methodologic approach to risk assessment, nor did it address the issue of which default options should be used by regulatory agencies. However, it did note that provisions should be made for continuing review of the science underlying the guidelines and of the basis of the default options incorporated in them.

After the publication of the 1983 NRC report, the Office of Science and Technology Policy (OSTP) brought together scientists from regulatory agencies, the National Institutes of Health, and other federal agencies to review the status of risk assessment and make recommendations. The result was a comprehensive review of the scientific basis of risk assessment of chemical carcinogens (4). The OSTP review provided the framework for risk assessment proposed by the NRC committee and provided the individual regulatory agencies with a basis for developing the type of guidelines recommended by the committee. In 1986, EPA (5) finalized a revision of its 1976 interim guidelines (6) based on the NRC and OSTP documents. These guidelines are being updated with a planned publication date of 1997. Increasing activity at the state level was first indicated in 1985 by California's publication of *Guidelines for Chemical Carcinogen Risk Assessments and Their Scientific Rationale* (7).

The major international organization responsible for cancer risk assessment is the International Agency for Research on Cancer (IARC). IARC was first requested to provide information on environmental chemical carcinogens in 1968, and the project was initiated in 1971. By 1977, 16 volumes of monographs had been published. Since then, hundreds of chemicals have been evaluated. These assessments were limited to qualitative evaluation, that is, hazard identification. The IARC guidelines for evaluating carcinogenic risk were first published in 1987 (8).

BIOLOGY OF CANCER

Definition of Cancer

Cancer is a generic term covering a variety of diseases for which a multiplicity of biological, physical, and chemical causes are known or suspected. It is a malignant and invasive growth, recurring after excision and metastasizing to other sites. Such growths are not induced by inflammatory processes from bacterial or fungal infections. Cancer may be of physical, chemical, or viral origin. The important difference between cancerous and inflammatory lesions is that cells in the latter mature and proliferate only to a very limited extent. Cells in cancer tissue, on the other hand, are formed from abnormal cells showing various degrees of dedifferentiation, for example, loss of specialized cellular form, and proliferate unrestrainedly.

Other terms commonly used for abnormal growths include tumors and neoplasms. These are broader terms that include both malignant and benign growths. A distinction is usually made between benign and malignant tumors. Generally the latter are called cancers. Malignant tumors invade adjacent tissue and metastasize to other regions of the body. With no medical intervention, they ultimately kill the host. On the other hand, benign tumors do not invade surrounding tissue and are not fatal unless they interfere with the function of a vital organ.

The nomenclature and biological classification of tumors are based on the criterion of site of origin or resemblance to a particular type of tumor. The suffix *-oma* attached to the root of the name of the tissue generally denotes benign tumors originating in or resembling that tissue. For example, lipoma arises from adipose tissue, fibroma from connective tissue, chondroma from cartilaginous tissue, osteoma from bone tissue, rhabdomyoma from striated muscle and leiomyoma from smooth muscle tissue, angioma from blood or lymph vessels, cholangioma from bile duct cells, papilloma from skin or mucous surfaces, adenoma from glandular tissue, and hepatoma from the liver. However, the -oma suffix does not always provide a distinction between benign and malignant tumors. Carcinoma, for example, designates malignant tumors that arise from epithelial tissue regardless of the germ layer. Sarcoma designates malignant tumors of mesenchymal origin, such as connective tissue, muscle, or bone. Adenocarcinoma denotes a malignant tumor of glandular origin. Melanoma is usually a highly malignant tumor arising from pigment-producing cells in the skin. Malignant tumors of the liver are designated as either hepatocellular carcinomas or cholangiocarcinomas, but the term hepatoma is used to designate both benign and malignant liver tumors. In addition, a variety of lung cancer types have been named, including small cell carcinoma and oat cell carcinoma (9). However, it is beyond the scope of this chapter to characterize all details of the terminology.

Mechanisms of Carcinogenesis

Carcinogenesis is a very large topic that can be treated only superficially here. It is also an area in which there is much to be learned (1,10,11). Carcinogenesis in animals and humans is a complex multistage process involving one or more genetic mutations. The genes most often involved in the carcinogenic process are operationally called oncogenes and tumor suppressor genes. Although the names refer to functions associated with cancer, both types represent genes with important roles in the development and differentiation of normal tissues. Normal oncogenes are often called protooncogenes. In carcinogenesis, protooncogenes are inappropriately activated as oncogenes and cause deregulation of growth. Inactivation of tumor suppressor genes also results in deregulation of growth and enhances the probability of neoplastic transformation. More than 30 protooncogenes and 6 or more tumor suppressor genes have been identified thus far. Other genes, including those involved in cell recognition, also may play a role in the development of cancers. These protooncogenes and tumor suppressor genes are associated with transcribing various cyclins and cyclin-dependent kinases responsible for mediating the central control of cell cycling. With the onset of tumorigenesis, proteins that drive the cell cycle forward, such as cyclins, are frequently overexpressed, whereas proteins that restrain cell proliferation are inactivated (12).

The process of carcinogenesis is often thought of as consisting of three stages called initiation, promotion, and progression. Initiation is the process of the normal cell acquiring a changed *phenotype* by which the cell has a changed response to stimuli for growth and differentiation. Generally initiation is assumed to be a stable, irreversible genetic change occurring either spontaneously or from damage by a carcinogen. During promotion, the initiated cell grows selectively in response to internal or external factors, forming a clone of initiated cells. Promoting influences vary. Some act by directly influencing gene activity, whereas others first interact with specific receptors resulting in an ensuing cascade of events. These are called epigenetic events. An epigenetic change is any change in a phenotype that does not result from an alteration in a DNA sequence (a mutation). It may include alteration in DNA methylation, transcriptional activation, translational control, or post-translational control. Epigenetic events are often induced by hormones and steroids. The progression stage involves the acquisition of the malignant cancer properties by some of the cells in the clone of initiated cells. Cells with increasingly abnormal properties appear, probably as a result of increased genetic instability. The deletion or inactivation of a tumor suppressor gene is probably a critical event in progression.

Although the biological processes associated with each of these stages are complex and not well defined, it is commonly assumed that initiation and progression involve DNA alterations and promotion does not. Initiation of a cell, or even initiation followed by promotion, does not always lead to a cancer. Production of cancer requires the linked occurrence of a number of events over a period of time. Mutations in more than one gene involved in the cell cycling are usually necessary. Other cellular changes or even cell death could interrupt the process. Indeed, therapeutic strategies for new cancer treatment drugs are based on blocking one or more of the processes. Testing conditions may assist in identifying whether a chemical is involved in one or more of the processes. Treatment with a chemical that has only promotional properties will not increase the incidence of tumors when given alone but will enhance the tumor production when animals are first treated with an initiating agent (see below).

Definition of Carcinogens

Chemical carcinogens are defined operationally by their ability to increase the occurrence of neoplasms (1). Four types of increased response generally are accepted as evidence: (*1*) the development of types of neoplasms not seen in controls, (*2*) an increased incidence of the types of neoplasms seen in controls, (*3*) the occurrence of neoplasms earlier than in controls, and (*4*) an increased multiplicity of neoplasms in individual animals. Under certain conditions, the production of benign neoplasms is accepted as evidence of carcinogenicity. This practice is justified by two considerations. First, the distinction between benign and malignant neoplasms is often made entirely on the microscopic appearance of the neoplasms, and the interpretation varies with the expertise and biases of the pathologist. Second, agents that induce benign neoplasms usually yield malignancies under severe conditions. Substances capable of producing an increase in neoplasms and, thereby classified as carcinogens, are an extremely diverse group. They include both organic and inorganic agents. Carcinogens may be organic chemicals, metals, hormones, immunosuppressants, viruses, particles, fibers, etc.

Chemical carcinogens are frequently classified as either mutagenic carcinogens or nonmutagenic carcinogens. Because of the universality of DNA, mutagenicity is detected in almost any biological system; the popular Ames test uses various strains of the bacteria *Salmonella typhimurium*. More than 100 different tests are available, measuring different genetic endpoints in a variety of phylogenetic systems. Conversely, promotion may involve a wide variety of biological events, some of which may be unique or at least specific to mammals. Currently, no short-term assays identify promoting agents; hence, they may be a part of the group of nonmutagenic carcinogens.

DESCRIPTION OF STUDIES USED IN ASSESSING CARCINOGENIC RISK

Epidemiologic Studies

Historically, epidemiology was the study of epidemics. One of the earliest studies was British physician John Snow's identification of contaminated water from the Thames river as the source of cholera during 1853–1854. This led to the eradication of cholera epidemics in London (13). During the latter half of this century, epidemiology has moved beyond epidemics and infectious disease. Its techniques have been adapted to identifying a variety of sources of diseases, including chemically induced cancer. Presently epidemiology is defined as the study of the distribution and determinants of health-related states and events in specified populations and the application of this study to the control of health problems.

The preferred source for assessing human health risks from environmental hazards is epidemiologic data. Using human data avoids the uncertainties resulting from species extrapolation. Large numbers of exposed individuals are often available, allowing detection of effects at lower concentrations than animal studies and thus partially overcoming uncertainties from low dose extrapolation. However, unlike animal studies, humans are generally exposed to a variety of environmental insults, often making it difficult to assess the effects of the agent in question. Moreover, in many cases, lack of accurate exposure data precludes quantitation of risk.

Animal Bioassays

Animal bioassay data are used widely in occupational and environmental risk assessment to predict toxicity and health effects of chemical substances in human populations. Such data also provide a scientific base for quantitative risk assessment for standard setting and regulation. There are a number of reasons for resorting to animal bioassay data. Two primary reasons are lack of sufficient human data and severe ethical limitations to obtaining such data. Other reasons include the ability to test over a wide range of exposures, the ability to test combinations of agents and conditions, the availability of a wide range of endpoints, and the ability to assess mechanisms of toxicity. Although a variety of uncertainties related to use of chronic animal bioassay data exist, no method available today is better in predicting carcinogenic effects before they occur, at unacceptably high levels, in the human population. For a more detailed discussion, see Reference 14.

Short-Term and *In Vitro* Assays

A large body of data supports the notion that direct or indirect alterations of DNA structure or function are critical events in the carcinogenic process. A variety of DNA lesions, formed by reaction of electrophilic compounds with nucleophilic centers in DNA, have been identified in DNA recovered from cells taken from people exposed to carcinogens, from cultured cell and intact organisms treated with carcinogens, and from reactions performed with carcinogens *in vitro*. Changes in gene structure in tumors induced by certain chemicals are sometimes similar (or even identical) to mutational changes caused by the same chemicals in short-term tests. For example, mouse mammary tumors induced by *N*-methyl-*N*′-nitro-*N*-nitrosoguanidine (MNNG) had an activated *c-Ha-ras* gene containing a guanine to adenine transition; this transition is a mutation induced by MNNG. Various genes that involve the control of cell growth, proliferation, and differentiation are implicated in the cancer process. Activated cellular oncogenes and/or tumor suppressor genes are detected in tumors from humans and laboratory animals, suggesting that cellular oncogenes are critical targets of chemical carcinogens.

Two decades of research and testing in genetic toxicology have established a positive correlation between carcinogenicity and mutagenicity. Based on the data available in the early 1980s, it was expected that *in vitro* tests for mutagenicity would accurately predict the outcome of all *in vivo* carcinogenicity testing. As more data accumulated, especially under the auspices of the NTP, it became apparent that this expectation would not be fulfilled (15). Increasingly, there are carcinogens (as detected in rodent bioassays) that are negative in a variety of mutagenicity tests, leading to the categories of "mutagenic carcinogens" and "nonmutagenic carcinogens."

Structure–Activity Relationships

Analyses and models of structure-activity relationships are used to predict molecular properties, surrogate biological endpoints, and carcinogenicity. Overall, these analyses provide valuable initial information about agents that strengthen or weaken concern for an agent's carcinogenic potential (16).

Relationships among chemical structure, mutagenicity, and carcinogenesis were determined for 301 chemicals (17). The chemicals were classified as structurally alerting or nonalerting. Alerting chemicals were those that were considered, based on structure, electrophilic and likely to react with DNA. From these studies, it became apparent that a group of structurally alerting, mutagenic chemicals exist that are predominantly carcinogenic in more than one species and at multiple sites. Another group exists that are nonalerting, nonmutagenic, and noncarcinogens. Between these two groups is a diffusely spread assortment of species/sex/tissue-specific carcinogens, only some of which are structurally alerting. Thus, for some chemicals, structure provides reasonable predictions of carcinogenicity or noncarcinogenicity. It is among the last group of chemicals that further research is required to understand their mechanism of action and their significance to humans.

The following parameters are useful in comparing an agent with its structural analogues and congeners that produce tumors and affect related biological processes, such as receptor binding, activation, mutagenicity, and general toxicity (18):

- nature and reactivity of the electrophilic moiety(ies) present
- potential to form eletrophilic reactive intermediate(s) through chemical, photochemical, or metabolic activation
- contribution of the carrier molecule to which the electrophilic moiety(ies) is attached
- physiochemical properties (eg, physical state, solubility, octanol-water partition coefficient, half-life in aqueous solution)
- structural and substructural features (eg, electronic, steric, molecular, geometric)
- metabolic pattern (eg, metabolic pathways and activation and detoxification ratio)
- possible exposure route(s) of the agent

Suitable analyses of non-DNA-reactive chemicals and of DNA-reactive chemicals that do not bind covalently to DNA require knowledge or postulation of the probable mode or modes of action of closely related carcinogenic structural analogs (eg, receptor-mediated, cytotoxicity-related). Then examination of the physiochemical and biochemical properties of the agent may provide the rest of the information needed to assess the likelihood of the agent's activity by the mode of action.

Biomarkers

A biomarker is a biological measurable endpoint signaling a critical pretoxicological or precarcinogenic event. The most commonly investigated biomarkers for carcinogenesis are protein or DNA adducts. These adducts are chemicals bound to the protein or DNA molecule, usually covalently. Unless repaired before cell division occurs, DNA adducts cause errors in transcription. As a result, the normal activity of the transcribed protein is often changed or inactivated. Historically, methods for adduct measurement were developed to determine target organ doses that, combined with measurement of genotoxic potencies, are used for risk assessment.

Epidemiologic studies generally lack sufficient sensitivity to detect and properly assess risk factors that do not occur in large excess in specifically exposed populations. Although much higher exposure levels are achieved in animal experiments, sensitivity is limited by group size. A consequence of relying on data for disease or death from cancer may be that nonacceptable risks are liable to escape detection because of this low resolving power. The insensitivity gap is bridged by the introducing macromolecule adduct measurement as an auxiliary tool. With mass spectrometric methods, adducts are quantified at levels thousands of times lower than those in which the cancer incidence associated with exposure is detectable in disease-epidemiologic studies. Another advantage of this approach is its ability to identify the chemical structure of the adducts, with clues to the nature of the causative factor. This is particularly important in the case of mixed exposures. Adduct monitoring is also a way to overcome the long latency times from exposure to full disease expression (19).

At present, biomarkers have been used only to a limited extent in cancer risk assessment. The evidence linking these biomarkers to cancer induction is circumstantial. For example, subjects who harbored high levels of polycyclic aromatic hydrocarbon (PAH)-DNA adducts had other increased types of genetic damage. Because such changes are common in malignant cells, however, results are consistent with the notion that increased quantities of adducts reflect increased likelihood of cancer. Blood samples from patients with lung cancer also contained increased amounts of PAH-DNA adducts. Although the linkage between biomarkers and cancer still requires definitive proof, as knowledge increases it should become possible to predict the carcinogenicity of environmental pollutants at lower concentrations and/or with shorter exposure durations than required to detect the onset of tumors. Establishing links between adduct levels and tumor incidence also may allow the use of biomarkers in quantitating risk (20) (see BIOMARKERS).

EVALUATION OF CARCINOGENESIS STUDIES TO DETERMINE CAUSALITY

The goal of epidemiologic studies or animal tumorigenicity experiments is to determine whether a test substance is associated with an increase in tumor development. A secondary objective may be to estimate responses at very low dose levels to predict human risk.

QUALITATIVE EVALUATION OF CARCINOGENIC RISK

Most classification systems are based on a weight-of-evidence assessment using information from a variety of sources. To be considered a known human carcinogen, sufficient evidence in human studies usually is required. In some circumstances, positive bioassay data from more than one species, supported by evidence from short-term tests and structure–activity relationships, are adequate to consider an agent a known human carcinogen. Generally agents are classified as probably carcinogenic based on nearly sufficient human evidence or sufficient evidence in chronic animal studies. An unclassifiable as to carcinogenicity category, based on inadequate information, is included in most systems. Some systems also include categories for possibly carcinogenic, unlikely to be carcinogenic, or known to be noncarcinogenic.

The most widely used classification systems are those developed by IARC (21) and EPA (5). The approaches used by these agencies for evaluating the weight of evidence for carcinogenicity are described in the following paragraphs. While differing in some details, the two methods are quite similar. Similar guidelines have also been published by agencies in several European countries (22).

DOSE–RESPONSE ASSESSMENT OF CARCINOGENIC RISK

Dose–response assessment addresses the relationship of dose to the degree of response observed in an animal experiment or a human study. In assessing human risk at environmental exposure levels, as many as three extrapolations may be necessary. These include extrapolations from high to low doses, from animal to human responses, and from one route of exposure to another. The data are first assessed in the range of empirical observation. This is followed by extrapolations either by modeling, if there are sufficient data to support a model, or by a default procedure based as much as possible on information about the agent's mode of action.

CONCLUSIONS

Great strides have been made in recent years in evaluating carcinogenic hazards. A large number of chemicals have been tested in a variety of experiments. Development of quality assurance guidelines has resulted in improved design, conduct, and reporting of studies. Mechanistic data have resulted in greatly increased knowledge of the carcinogenic process. However, much work still remains to be done.

At the present, relatively few agents in the environment or workplace are human carcinogens. Even fewer agents can be determined confidently to be noncarcinogenic in humans. The great majority are classified as probable human carcinogens, possible human carcinogens, probably noncarcinogenic in humans, or having insufficient data to determine their carcinogenicity. Ideally, every agent would be known to be carcinogenic or noncarcinogenic in humans. With rare exceptions, human epidemiologic studies have provided the definitive basis for determining that an agent is a known human carcinogen. Epidemiologic studies, however, are unlikely to provide adequate information for determining carcinogenicity of many agents in the future. Few populations are sufficiently large, are exposed to great enough concentrations, and/or are not concomitantly exposed to other agents that may be carcinogenic. Because long-term animal bioassays are quite costly, only a small percentage of potential carcinogenic agents are likely to be tested by this method in the future. Moreover, some uncertainty still remains regarding extrapolation to humans. It is hoped that further research will provide sufficient mechanistic knowledge so that short-term test, biomarkers, structure-activity data, and so on will be adequate to provide the basis for establishing the carcinogenicity or noncarcinogenicity of an agent.

Further knowledge of mechanisms is also necessary to establish responses at ambient exposure concentrations. Because of limitations on exposure numbers, animal studies lack sufficient sensitivity to directly determine responses at low exposure levels. Although effects of chemicals that directly damage DNA may be linear at low doses, for less direct acting agents, such as those that influence endocrine secretions, the shapes of the dose-response curves are much less certain. It is hoped that biomarkers will become useful in determining low-dose responses. Thus, although much is known, much remains to be learned.

BIBLIOGRAPHY

1. H.C. Pitot III and Y.P. Dragan, in C.D. Klaassen, M.O. Amdur, and J. Doull, eds., *Casarett and Doull's Toxicology,* 5th ed., McGraw–Hill, New York, 1996, pp. 201–267.
2. U. Heinrich, in U. Mohr, D.L. Dungworth, J.L. Mauderly, and G. Oberdörster, eds., *Toxic and Carcinogenic Effects of Solid Particles in the Respiratory Tract,* ILSI Press, Washington, D.C., 1994, pp. 57–73.
3. National Research Council, *Risk Assessment in the Federal Government: Managing the Process,* National Academy Press, Washington, D.C., 1983, pp. 1–191.
4. Office of Science and Technology Policy, *Fed. Reg.* **49,** 21594 (1984).
5. U.S. Environmental Protection Agency, *Fed. Reg.* **51,** 33992–43003 (1986).
6. U.S. Environmental Protection Agency, *Fed. Reg.* **41,** 21402 (1976).
7. California Department of Health Services, *Guidelines for Chemical Carcinogen Risk Assessments and Their Scientific Rationale,* California EPA, Sacramento, Calif., 1985.
8. International Agency for Research on Cancer, Preamble, *IARC Monographs on the Evaluation of Carcinogenic Risks to Humans,* IARC Internal Technical Report No. 87/001, Lyon, France, 1987, pp. 1–39.
9. J.C. Arcos, M.F. Argus, and G. Wolf, *Chemical Induction of Cancer,* Vol. I, Academic Press, New York, 1968, pp. 303–315.
10. International Agency for Research on Cancer, *Mechanisms of Carcinogenesis in Risk Assessment: A Consensus Report,* IARC Technical Report No. 91/002, Lyon, France, 1991, pp. 1–60.
11. H. Pitot, *Mutat. Res.* **333,** 3–14 (1995).
12. B.E. Clurman and J.M. Roberts, *J. Natl. Cancer Inst.* **87,** 14–17 (1995).
13. R.E. Gots, *Toxic Risks, Science, Regulation and Perception,* Lewis Publishers, Boca Raton, Fla., 1993, pp. 1–275.
14. D.P. Rall, in M.C. Smith, D.C. Christiani, and K.T. Kelsey, eds., *Chemical Risk Assessment and Occupational Health,* Auburn House, Westport, Conn. 1993, pp. 105–111.
15. E. Zeiger, J.K. Haseman, M.D. Shelby, B.H. Margolin, and R.W. Tennant, *Environ. Mol. Mutagen.* **16**(Suppl. 18), 1–14 (1990).
16. U.S. Environmental Protection Agency, *Proposed Guidelines for Carcinogen Risk Assessment,* EPA/600/P-92/003C, U.S. EPA, Cincinnati, Ohio, 1996, pp. 1–170.
17. J. Ashby and R.W. Tennant, *Mutat. Res.* **257,** 229–306 (1991).
18. Y.T. Woo and J.C. Arcos, in N.N. Ragsdale and R.E. Menzer, eds., *Carcinogenicity and Pesticides: Principles, Issues and Relationship,* ACS Symposium Series No. 414, Academic Press, San Diego, Calif., 1989, pp. 175–200.

19. L. Eherenberg, F. Granath, and M. Törnqvist, *Environ. Health Perspect.* **104,** 423–428 (1996).
20. F. Perera, *Sci. Am.* **274**(5), 54–62 (1996).
21. International Agency for Research on Cancer, Occupational Exposure to Mists and Vapours from Strong Inorganic Acids and Other Chemicals, Preamble, *IARC Monographs on the Evaluation of Carcinogenic Risks to Humans,* Vol. 54, Lyon, France, 1992, pp. 13–32.
22. Laboratory for Environmental and Occupational Cancer, *Classification of Carcinogens by Different Countries and Organisations,* Institute for Cancer Research, Oslo, Norway, 1993, pp. 1–15.

CARCINOGENS: MONITORING OF INDOOR AIR

MAX R. PETERSON
R. K. M. JAYANTY
Research Triangle Institute
Research Triangle Park, North Carolina

Indoor air contains a complex mixture of chemical species and particles. The number of chemical species present at one time can range into the thousands, especially if cigarette smoke is present. Concentrations of chemical species and particles, while usually quite low from an analytical perspective, may be significant, and the concentrations vary with location and with time within a given indoor air space. Some of the pollutants within this complex mixture appear to be harmless; but some are irritants, some are toxic, and some are known or suspected human carcinogens (see INDOOR AIR RISK ASSESSMENT AND MANAGEMENT).

CARCINOGENS IN INDOOR AIR

Pollutants, including known or suspected carcinogens, found in indoor air come from a variety of sources. Some pollutants come from building materials, carpets, fabrics, adhesives, sealants, paints, stains, and other materials used to construct or maintain a home or other building; some pollutants are brought in or tracked in from outdoors; and some pollutants are generated by human activities such as smoking tobacco, cooking, or using personal care and cleaning products. Even seemingly innocuous activities like sweeping or vacuuming floors and furnishings can fill the air in an indoor environment with respirable particles containing heavy metals tracked in from outdoors and carcinogenic species formed during heating and cooking.

Categories

For monitoring purposes, chemical species, including carcinogens and other pollutants, are often grouped on the basis of chemical class, chemical properties, or physical state. Species that can be measured by a single method are treated as a group, but particularly hazardous members of a group may also be treated separately. The categories of carcinogens in the following sections have been arranged in the general order of simplest to most complex, although there is considerable overlap. Major categories include radon, organic compounds, inorganic species, asbestos, and particles. Radon and asbestos could be considered with inorganic species and particles, respectively, but methods to measure these two carcinogens are quite different from methods for measuring other carcinogens.

Physical States

Carcinogens in indoor air may exist in the gaseous state, the liquid (or solution) state, the solid state, or in any combination of states. Some, like radon and formaldehyde, exist primarily in the gaseous state, while others, like asbestos and some heavy metals, exist only in the solid state as suspended particles or fibers. A third group of carcinogens, which includes low volatility compounds like some pesticides and some components of environmental tobacco smoke, may exist simultaneously in any combination of states.

Complex Mixtures

Indoor air is a complex mixture of gaseous species, suspended liquid and solid particles, and fibers. Chemical species, including carcinogens, can be found in any or all of those states. Some individual chemical species may exist in the gaseous phase; be dissolved in suspended liquid particles; and be adsorbed onto solid particles at the same time, with molecules constantly moving from one state to another. The distribution of a particular species among the various physical states is determined by its vapor pressure (or volatility), its solubility in whatever liquid aerosol droplets are present, and its affinity for active sites on the surface of whatever solid particles are present; and its concentrations in the gaseous, dissolved, and particle-bound states are affected by the temperature of the air. In addition, the sources of a species and the so-called sinks that remove the species from air are not evenly distributed, nor are source contribution rates and sink removal rates constant and perfectly balanced. Thus the concentration of a species also varies with time and with location in a given indoor environment.

Sampling and Analysis

Measurement methods generally involve sampling indoor air to collect the species of interest and analyzing the collected sample to measure those species. The two parts, sampling and analysis, may be done on-site and essentially simultaneously to obtain more or less real-time measurements. Alternatively, collected samples may be transported to a laboratory for analysis.

Carcinogens are typically present in indoor air at extremely low concentrations (1). The volume of air sampled must yield an amount of the species of interest that is measurable by the chosen analysis method. Both the practical quantitation limit of the analysis and an approximate or a maximum allowable concentration (eg, threshold limit value) of the species of interest must be known to determine the volume of air that must be sampled.

Fundamental Objectives of Sampling. The fundamental objective of sampling is to obtain samples that contain representative concentrations of the species of interest for the purposes of the measurement. Samples are generally collected in a fashion that either mimics human exposure or provides data that can be related to human exposure. Whole-air samples can be collected in suitable containers, or the species of interest may be removed by some physical or chemical process from the air sample during collection.

Particles and fibers are typically collected on a filter, whereas gaseous species may be collected in a stainless-steel canister, a Tedlar, or an aluminized bag; bubbled through a liquid-filled impinger; or concentrated on a sorbent. Filters and sorbents remove and concentrate species of interest from a known volume of sampled air. Impingers are often used for collecting reactive or unstable species and generally are filled with a solution containing a derivatizing agent, which converts reactive species to stable derivatives. Species existing in both gaseous and particle-bound states are typically sampled using both a filter and a sorbent. Thus collected samples may be whole-air samples collected in canisters or bags, particles or fibers collected on filters, or chemical species collected on sorbents or in a liquid-filled impinger.

The volume of air sampled (sample volume) must be large enough to provide a measurable amount of the species of interest. On the other hand, the volume of air sampled is limited by the capacity of the sampling device or medium. Canisters, for example, generally have volumes $\leq$6L. Thus using negative pressure to drive sampling into an evacuated canister limits the volume of air that can be collected to something less than the volume of the canister.

The volume of air that can be sampled through a filter is limited by increasing pressure drop across the filter as filter pores become clogged with particles or fibers. Samples are invalid if the pressure drop across the filter exceeds the maximum drop specified in the chosen sampling method.

The volume of air that can be sampled through a sorbent is limited by the weakness of the attachment of the species of interest to the sorbent. A sorbent is generally selected on the basis of its ability both to retain a species during sampling and to release the species for analysis. Thus molecules of the target species that are trapped from the air moving over the sorbent subsequently move back into the gas phase, flow with the sampled air for a short distance, and then reattach to the sorbent. This process is repeated over and over again so that molecules of the target species move farther and farther into the sorbent bed. Breakthrough is said to have occurred if a measurable amount of the target species actually passes beyond the downstream end of the sorbent bed. The total volume of air sampled when breakthrough occurs is called the breakthrough volume. The volume of air sampled through a sorbent must be less than the breakthrough volume of the target species for that sorbent bed.

Whatever the method of sampling, the temperature, pressure, and volume of air sampled must be known for later calculations of species concentrations. In some cases, the humidity of the sampled air is also a factor.

Fundamental Objectives of Analysis. The fundamental objective of analysis is to provide reliable, accurate, and precise measurements of the species of interest. The analysis must provide an unambiguous measurement of the species without loss (eg, through decomposition) or enhancement (eg, through contamination) of the species. Often the species of interest must be separated from other species present in collected samples before measurement. In this situation, the analytical system may be used both to isolate and to measure species of interest.

An appropriate analytical system is one that responds in a characteristic and quantifiable way to species of interest. In modern instruments, the response of an analytical system to a species is often converted into an electrical signal with a magnitude that is directly related to the amount of the species present in the detection device of the instrument. In such systems, the response of the instrument to the species in a collected sample is compared to the response of the instrument to standards containing known amounts of the species. Analysis of standards containing different amounts of a species and the subsequent plotting of instrument response versus amount of species analyzed is referred to as calibration. The range of the calibration (or range of amounts of species in the standards analyzed for the calibration) must bracket the amounts of species measured in collected samples.

Basic Monitoring Approaches

A monitoring approach must be chosen based on the goals of the desired measurement, the sampling and analysis methods available for the species of interest, and the concentrations of the species of interest in the indoor environment to be measured.

RADON

Radon is considered a carcinogen because of its radioactivity. The actual harm to human health comes from the radioactivity of the decay products of radon. Radon is a gas and, if inhaled, is not retained well by the lungs. Radon decay products are not gases, and they become attached to airborne particles, which, if inhaled, can be retained in the lungs indefinitely. Thus it is the particle-bound radon decay products that are actually the cause of the increased risk of lung cancer attributed to radon.

Concentrations of radon and radon decay products (or progeny) in indoor air vary with location and with time. The primary isotope of interest, radon-222 ($^{222}_{86}$Rn), and its decay products are part of the radioactive decay series leading from a naturally occurring, long-lived radioactive isotope (uranium-238) to a stable isotope (lead-206). A portion of this series, showing the formation of radon-222 and its decay products of interest (polonium-218, lead-214, bismuth-214, and polonium-214), is shown in Figure 1. The primary decay mode of an isotope is denoted by the type of radiation associated with fission of that isotope. An alpha (α) particle is a helium-4 nucleus ($^{4}_{2}He^{2+}$) ejected from the nucleus of an isotope during radioactive decay, and a beta (β) particle is a high energy electron ejected in a similar fashion. Loss of an α-particle reduces the mass number (number of protons plus number of neutrons) of an isotope by four and reduces the atomic number (number of protons) by two. A β-particle is formed when a neutron in the nucleus of an atom is converted to a proton, which remains in the nucleus, and a high energy electron (the β particle), which is ejected from the nucleus. Loss of a β-particle does not affect the mass number but increases the atomic number by one. In both types of radioactive decay, an atom of one element is changed into an atom of another element because the atomic number changes. Emission of gamma (γ) rays, which are high energy photons, during radioactive decay does not affect the mass number or the atomic number of an isotope. The half-life of an isotope is the time required for half of the atoms of that isotope in a sample to undergo radioactive decay.

Isotope	$^{226}_{88}$Ra →	$^{222}_{86}$Rn →	$^{218}_{84}$Po →	$^{214}_{82}$Pb →	$^{214}_{83}$Bi →	$^{214}_{84}$Po
Decay mode	α	α	α	β	β	
Half-life	1600 yr	3.82 days	3.04 min	26.9 min	19.7 min	164 μsec

Figure 1. Formation of radon-222 and its decay products.

Isotope	$^{224}_{88}$Ra →	$^{220}_{86}$Rn →	$^{216}_{84}$Po →	$^{212}_{82}$Pb →	$^{212}_{83}$Bi →	$^{212}_{84}$Po
Decay mode		α	α	α	β	β
Half-life	3.66 d	55.6 s	0.15 s	10.6 h	60.6 min	298 ns

Figure 2. Formation of radon-220 and its decay products.

A second radon isotope, radon-220, sometimes called thoron ($^{220}_{86}$Rn), and its decay products are part of the radioactive decay series leading from another naturally occurring, long-lived radioactive isotope (thorium-232) to a different stable isotope of lead (lead-208). A portion of this series, showing the formation of radon-220 and its decay products of interest (polonium-216, lead-212, bismuth-212, and polonium-212), is shown in Figure 2. Radon-220 and its decay products, while just as deadly as isotopes associated with radon-222, pose a lower risk to human health because of the short (55.6-s) half-life of radon-220, which limits its time in the gas phase and, therefore, its migration distance.

Of all of the isotopes in the two decay series, only the isotopes of radon are gases at ambient conditions. Existence as a gas, which allows radon to migrate from soil or rock into indoor air in the first place, provides a distinguishing characteristic that can be used to separate radon from other radioactive species.

When atoms of radon gas undergo radioactive decay, the newly formed polonium atoms (and any subsequently formed polonium decay products) immediately condense or attach onto any surface with which they come into contact. In indoor air, that surface is usually associated with airborne particulate matter.

Monitoring of radon and radon decay products in indoor air is complicated by the nuclear instability of the target species. With half-lives ranging from almost four days for radon-222 to a fraction of a second for polonium-216, measurements based on emitted radiation must be made shortly after collection.

Indoor air may be sampled for radon or radon progeny measurements by grab sampling, integrated sampling, or continuous sampling. Separation of radon from progeny may be accomplished by allowing the radon to diffuse through a passive barrier (eg, foam rubber) that does not allow the passage of particulate material containing the radon progeny (2).

Methods for measuring radon and radon progeny actually measure emitted radiation. In the absence of other radioactive species, emitted radiation is directly proportional to the concentration of the radioactive species in the sample. The emitted radiation may be in the form of α- or β-particles or γ-rays, all of which can be measured with the appropriate instrumentation (4). Table 1 provides an overview of applications and sensitivities of devices used to measure radon and its decay products.

ORGANIC COMPOUNDS

Organic compounds are simply compounds containing the element carbon. The vast majority of pollutants that may be present in indoor air are organic compounds.

Classification of Organic Pollutants

In addition to the usual classification of organic compounds by functional group (eg, alcohols, ketones, aromatic compounds, halogenated compounds, etc), organic pollutants in indoor air are often classified on the basis of boiling points. The boiling point of a compound is indicative of the compound's volatility, which to a large extent determines the distribution of the compound between gaseous and condensed phases. The boiling point ranges assigned to various classes are often disputed, and compounds with boiling points outside the range for a particular class may be included with that class in a particular study or for analysis by a particular analytical method (4).

Very volatile organic compounds (VVOCs) exist almost completely in the gaseous state in indoor air. Some VVOCs actually have boiling points below room temperature. Volatile organic compounds (VOCs) can exist in the gaseous state in air at relatively high concentrations. Semivolatile organic compounds (SVOCs) are much less volatile and tend to have low gas-phase concentrations, but SVOCs can exist at high concentrations in liquid aerosol droplets and on solid particles. Some organic compounds, often called particulate organic matter (POM), have such low volatilities that they exist in air almost exclusively as or on particles.

Table 1. Applications and Sensitivities of Some Radon Measurement Devices[a]

Instrument Type	Sampling	Sensitivity, Bq/m^3[b]	Purpose[c]
Direct measurement			
Scintillation cell	Grab or continuous	3.7	Screening, diagnostic
Ionization chamber	Grab or continuous	3.7	Screening, diagnostic
Passive barrier method[d]			
Scintillator	Continuous	3.7	Screening, diagnostic
Thermoluminescent detector chip	Integrating	0.08–8.1	Screening, large-scale survey
Two-filter method	Grab or continuous	3.7	Diagnostic
Passive sampling devices			
Activated charcoal	Integrating	7.4 for 100-h exposure	Screening, large-scale survey
α-Track	Integrating	18.5 for 30-d exposure	Screening, large-scale survey

[a] Adapted from Ref. 3.
[b] A sensitivity less than the value shown is generally achieveable, depending on the specific instrument used. To convert Bq/m^3 to pCi/L, multiply by 37.
[c] *Screening,* rapid and inexpensive detection; *diagnostic,* measurement of specific parameters; *large-scale survey,* collection of many time-averaged measurements.
[d] With radon progeny collected on or close to a scintillator or thermoluminescent detector chip.

Complexity of Organic Mixtures

Measuring concentrations of specific organic compounds in indoor air can be a challenge because of the complexity and variability of the indoor air matrix. Literally thousands of organic compounds are present in such complex mixtures as environmental tobacco smoke. If suspended liquid and solid particulate matter are present, the VOCs and SVOCs exist in a complex, constantly changing distribution among gaseous, condensed, and dissolved states.

General Monitoring Methods for Organic Compounds

Monitoring of organic compounds is generally accomplished in three steps: collection, separation, and measurement.

Organic Carcinogens of Particular Concern

Organic compounds that are classified as known or suspected human carcinogens and that may be present in indoor air include benzene, *p*-dichlorobenzene, carbon tetrachloride, chloroform, methylene chloride, formaldehyde, polycyclic aromatic hydrocarbons (PAHs), and some pesticides.

INORGANIC SPECIES

Inorganic carcinogens include some of the heavy metals and their compounds. The heavy metals of primary concern are arsenic (salts, arsenates, and arsenites), beryllium, cadmium (oxide, bromide, and chloride), chromium (hexavalent), nickel (carbonyl and subsulfide), and selenium (sulfide). Most, but not all, of these inorganic carcinogens exist as salts. Because of the omnidirectional nature of ionic chemical bonds, salts tend to exist as high melting solids. Such salts are found in air only as a component of suspended particulate matter.

Oxidation States

The oxidation state of an element in a compound is assigned by a set of rules that allot valence (or outermost) electrons, which are involved in chemical bonding, to the more electronegative (or electron-attracting) element and leave the less electronegative element electron deficient. Thus the more electronegative element is assigned a negative oxidation state (more electrons assigned than protons in the nucleus), and the less electronegative element is assigned a positive oxidation state (fewer electrons assigned than protons in the nucleus). Metal atoms in compounds are usually assigned positive oxidation states. Oxidation state is a factor in the carcinogenicity of a metal. Hexavalent chromium has an oxidation number of +6 and is classified as a carcinogen, whereas chromium in other oxidation states (eg, +3) is not.

General Monitoring Methods for Inorganic Species

Standard methods have been developed for measuring individual metals in workplace atmospheres. Those methods should also be appropriate for indoor air. In some cases, more accurate methods of collection and analysis have been reported, but they have not yet been accepted as standard methods. Inorganic salts, which exist in air as particulate matter, are collected on filters. Volatile inorganic compounds are usually collected in liquid-filled impingers.

Measurement of metals in nonvolatile species is usually accomplished by some type of atomic absorption or atomic emission spectroscopy. More sophisticated instruments, including inductively coupled argon plasma with atomic emission spectroscopy or mass spectrometry, can measure many different metals in a single analysis. This type of measurement can often be carried out on an aliquot of the sample with no separation of species. Unfortunately, because of chemical changes caused by the rather severe analytical conditions, such approaches measure only the total amount of a metal present and cannot distinguish between different oxidation states of the same metal.

Measurement of a metal in a particular oxidation state, eg, chromium(VI), requires that the species be separated from the matrix by ion chromatography (or some other means) or the species must exhibit some chemical property that is unique among the components of the matrix.

Inorganic Carcinogens of Particular Concern

An overview of some published methods for collecting and measuring carcinogenic metals in air and the analyte actually measured is available in the literature (6,7). A general method for metals in air has been developed. The U.S. National Institute for Occupational Safety and Health (NIOSH) has published a single method (method 7300) that can be used to measure 29 elements, including all of those addressed here except nickel (6). Particulates, including the metal salts, are collected on a cellulose ester membrane filter and analyzed by inductively coupled argon plasma with atomic emission spectroscopy.

Arsenic. NIOSH 7900 measures arsenic and its compounds as total arsenic (6). The sample is collected on a cellulose ester membrane and analyzed by atomic absorption, flame arsine generation. The estimated limit of detection for this method is 0.02 μg/sample.

Beryllium. Beryllium also occurs in air as a component of suspended particulate matter, but beryllium compounds are sufficiently volatile to be analyzed by gas chromatography, although a pretreatment step offsets some of the advantage of using GC. Beryllium is collected on a membrane filter and can be measured by inductively coupled plasma emission or by graphite furnace atomic absorption. The limit of detection for the latter, which is used in NIOSH 7102, is 0.005 μg/sample (6).

Cadmium (Oxide, Bromide, and Chloride). According to NIOSH 7048, cadmium compounds in air can be collected on a cellulose ester membrane filter and analyzed by atomic absorption with flame. The estimated limit of detection is 0.05 μg/sample (6).

Hexavalent Chromium. Chromium(VI) is usually collected from air on filters of cellulose, polyethylene, polystyrene, poly(vinylchloride) (PVC), or glass and may be separated from matrix materials by ion chromatography. In NIOSH 7600, samples are collected on a PVC membrane filter and chromium is measured by visible absorption spectrophotometry. The estimated limit of detection by this method is 0.05 μg/sample (6) (see CHROMIUM IN SOILS).

Nickel (Carbonyl and Subsulfide). Most compounds of nickel exist in ambient air as a component of suspended particulate matter and may be collected on a filter. Unfortunately, nickel carbonyl is quite volatile and cannot be collected efficiently on a filter. Volatile compounds of nickel are usually collected in liquid-filled impingers. Analytical methods for nickel include atomic absorption spectrophotometry with and without flame, x-ray fluorescence spectrometry, inductively coupled argon plasma spectroscopy, colorimetry, spark source mass spectrometry, neutron activation analysis, and flame emission spectrophotometry.

Selenium (Sulfide). Selenium may be measured by NIOSH 7300, as described above. The estimated limit of detection for selenium is 1 μg/sample (6).

ASBESTOS

Asbestos is a group of naturally occurring fibrous minerals that are nonflammable and chemically unreactive. Unfortunately, the inertness and chemical stability that made asbestos popular as a component of insulation materials and floor and ceiling tiles are the properties that keep inhaled asbestos fibers in lung tissue intact for decades (see ASBESTOS, ANALYSIS OF).

Sampling Airborne Fibers

Asbestos fibers are typically collected on a cellulose ester or polycarbonate filter, transferred to a slide, and measured visually under relatively high magnification. The volume of air sampled and the surface area of the filter are used in calculating the concentration of fibers.

Laboratory Analysis Methods

Asbestos fibers in collected samples are magnified for counting by either phase contrast microscopy (PCM), scanning electron microscopy (SEM), or transmission electron microscopy (TEM). The extremely small size of airborne asbestos fibers makes them difficult to distinguish from other natural and synthetic fibers. Only TEM offers definitive identification of asbestos fibers.

PARTICLES

Respirable particles (ie, particles with aerodynamic diameters $\leq$10 μm) represent not only a source of respiratory irritation and allergic response but a pathway for delivery of significant amounts of chemical species, including nonvolatiles, directly to the lungs. Virtually all known or suspected carcinogens can be delivered to the lungs and ultimately to the bloodstream through inhalation of particles containing those species. It is often the chemical species attached to particles, not the particles themselves, that pose the greater threat to human health. An overview of sampling devices and analytical methods that are used to collect and measure particles in industrial workplaces is available in the literature (8,9).

Particles of Particular Concern

Particles that deliver extremely hazardous or carcinogenic species directly to the lungs over extended periods of time represent a serious threat to human health. Several sources of particles that are found in indoor air and that carry known or suspected carcinogens are discussed below.

Environmental Tobacco Smoke. Environmental tobacco smoke (ETS) is a complex mixture of volatile, semivolatile, nonvolatile, and particulate matter that may contain more than 3800 chemical species, including heavy metals. Many of the organic and inorganic species discussed above are present in tobacco smoke. The composition and distribution of the components of ETS change continuously as compounds are adsorbed onto or desorbed from particles and as the more reactive compounds are converted into other species.

Because of the complexity of the mixture and the fact that some of the compounds in ETS also come from other sources, exposure to ETS is quite difficult to measure directly. Typically, respirable suspended particle concentration or the concentration of nicotine (5), an organic compound that is specific to ETS, is measured by procedures described in previous sections for particles and organic compounds.

Yet another way to assess exposure to ETS is through analysis of physiological fluids of exposed persons. In this case, exposed persons have become the collection medium. The presence of nicotine or its major metabolite, cotinine, in saliva, blood, or urine is entirely due to exposure to tobacco, tobacco smoke, or environmental tobacco smoke. The measurement of nicotine and cotinine in physiological fluids of active smokers or persons exposed to ETS (sometimes called passive smokers because they are exposed to ETS but do not smoke themselves) is generally accomplished by GC with a nitrogen-sensitive detector. The GC method can be used to measure nicotine concentrations as low as 1 ng/mL fluid and concentrations of cotinine as low as 6 ng/mL.

Particles from Cooking and Heating Activities. Soot particles from poorly vented heaters and oily mist and smoke from cooking can contain PAHs and other carcinogens. As described earlier, PAHs are present in both particle-bound and gaseous states. The particles are collected as described above, and the PAHs are solvent desorbed from the collection media and are typically measured by GC with an appropriate detector.

Other Respirable Particles. Any kind of respirable particle, whatever the source, has the potential to introduce carcinogens or other harmful species directly to the lungs and bloodstream of exposed persons.

CONCLUSION

Monitoring of carcinogens in indoor air is obviously not a simple task. The methods described here have, for the most part, been thoroughly evaluated by many different laboratories. A steady stream of newer, more sensitive techniques and instruments is appearing in the environmental literature and in the marketplace. As an example, a recent journal article reported the measurement, by an automated analysis, of 1 pg of a VOC, which corresponds to less than 1 part per trillion (ppt) of the compound in a 1-L air sample (10). Such a feat would have been impossible only a few years ago. Unfortunately, so-called allowable concentrations of some toxic and carcinogenic species are still defined by the sensitivity limitations of the methods available to measure them rather than by the risk those species pose to human health.

BIBLIOGRAPHY

1. K. W. Gold, D. F. Naugle, and M. A. Berry, in B. Seifert, H. J. van de Wiel, B. Dodet, and I. K. O'Neill, eds., *Environmental Carcino-*

gens Methods of Analysis and Exposure Measurement, vol. 12, International Agency for Research on Cancer, Lyon, France, 1993, pp. 41–71.

2. W. W. Nazaroff, in W. W. Nazaroff and A. V. Nero Jr., eds., *Radon and Its Decay Products in Indoor Air,* John Wiley & Sons, Inc., New York, 1988, pp. 491–504.
3. C. R. Cothern and J. E. Smith Jr., eds., *Environmental Radon,* Plenum Press, New York, 1987.
4. M. R. Peterson, D. F. Naugle, and M. A. Berry in Ref. 1, pp. 126–135.
5. W. T. Winberry Jr., L. Forehand, N. T. Murphy, A. Ceroli, B. Phinney, and A. Evans, "Compendium of Methods for the Determination of Air Pollutants in Indoor Air," *PB90-200288, EPA/600/4-90/010,* U.S. Environmental Protection Agency, Research Triangle Park, N.C., 1990.
6. P. M. Eller, ed., *NIOSH Manual of Analytical Methods,* 4th ed., U.S. Department of Health and Human Services, Public Health Service, Centers for Disease Control and Prevention, National Institute for Occupational Safety and Health, Division of Physical Sciences and Engineering, Cincinnati, Ohio, 1994.
7. I. K. O'Neill, P. Schuller, and L. Fishbein, *Environmental Carcinogens Selected Methods of Analysis,* Vol. 8: *Some Metals: As, Be, Cd, Cr, Ni, Pb, Se, Zn,* International Agency for Research on Cancer, Lyon, France, 1986.
8. P. J. Lioy and M. J. Y. Lioy, eds., *Air Sampling Instruments for the Evaluation of Atmospheric Contaminants,* 6th ed., American Conference of Governmental Industrial Hygienists, Cincinnati, Ohio, 1983.
9. R. K. M. Jayanty, M. R. Peterson, D. F. Naugle, and M. A. Berry, *Risk Anal.,* **10**(4), 587–595 (1990).
10. E. Woolfenden, *J. Air Waste Manage. Assoc.,* **47**(1) 20–36 (1997).

CFCS. See under CHLOROFLUOROCARBON.

CHEMICAL WARFARE AGENTS

JOSEPH R. DONNELLY
Lockheed Martin Environmental Services
Las Vegas, Nevada

Several chemical warfare agents and their degradation products are target analytes in soils at Superfund sites related to military activities. There exists the potential that the agents themselves might be encountered even after environmental weathering. Workers with these agents, and those potentially exposed, must develop and follow strict safety procedures to prevent exposure, and to obtain proper treatment in the event of accidental exposure. All workers must understand the physical and toxic properties of the materials potentially present, and must take appropriate measures of protection.

ABBREVIATED HISTORY OF ORGANOSULFUR AND ORGANOPHOSPHORUS AGENTS

Probably the earliest modern usage of effective chemical warfare agents occurred near Ypres, Belgium, in 1915 by the Germans. In 1914–1915, the French had used ethyl bromoacetate to little effect. Haber introduced chlorine gas to the German military. The term *gas* has since been used for a variety of chemical weapons, such as *mustard gas* and *nerve gas,* even though these later-developed agents were liquids of various vapor pressures. Mustard, also known as HD ($ClCH_2CH_2SCH_2CH_2Cl$), was first used later in World War I, again at Ypres (1917). Mustard was named after its odor which is reportedly like garlic, onions, or mustard (1,2). Its preparation from thiodiglycol was described by Meyer in 1886 (3).

In 1932, Lange and von Krueger reported (4) the synthesis of some organophosphorus esters (dialkylfluorophosphates), in a search for new pesticides. They noted that "the physiological effect of these esters is quite strong. Their odor is pleasant and aromatic, but a few minutes after inhalation, headache and suffocation occur, then vision troubles; the effects last a few hours and are caused by minute quantities."

During the latter half of the 1930s and throughout World War II, Schrader, employed by the German chemical conglomerate IG Farben, worked on toxic organophosphorus compounds, with an emphasis on activity as pesticides. In England during World War II, Saunders prepared diisopropyl fluorophosphate (DFP) (**1**). Schrader continued work, based on his finding (5) in 1937, that certain compounds of the following general formulas had contact insecticidal activity:

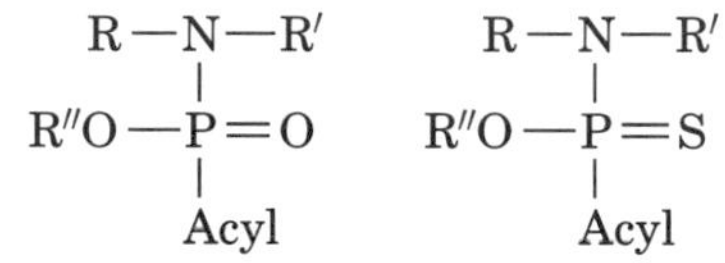

```
    R—N—R'          R—N—R'
      |               |
R''O—P=O        R''O—P=S
      |               |
     Acyl            Acyl
```

where R, R′, and R″ are alkyl groups such as methyl (CH_3), ethyl (C_2H_5), or isopropyl (i-C_3H_7), and acyl is an organic or inorganic acid group such as Cl, F, SCN, or (CH_3COO).

Saunders reported that the high toxicity of the fluorophosphonates and fluorophosphates was not found in compounds where fluorine was replaced with a variety of other substituents such as H, C_2H_5, OH, OC_2H_5, Cl, $NHCH_3$, CN, etc. As an example, Bladan was first marketed in Germany about 1943–1944 as an insecticide. It contained tetraethyl pyrophosphate (TEPP) (**2**), a compound that had been originally synthesized about a century earlier by researchers who did not recognize its insecticidal and toxic potential. Schrader discovered its insecticidal value in 1938. Schrader discovered parathion (**3**) in 1944, a pesticide that is also quite toxic to mammals. However, numerous related compounds have been developed that show less mammalian toxicity, including (**4**, Ar = 3-chloro-4-nitrophenyl) chlorthion, (**4**, Ar = 3-methyl-4-thiomethylphenyl) fenthion, (**4**, Ar = 3-methyl-4-nitrophenyl) fenitrothion, and malathion (**5**).

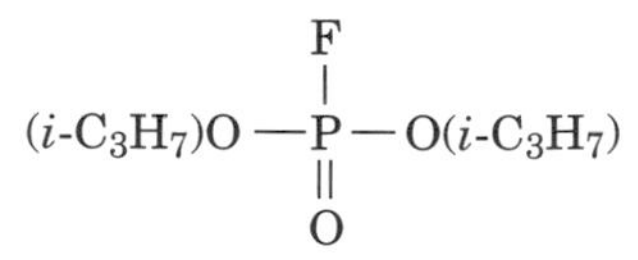

```
                 F
                 |
(i-C3H7)O — P — O(i-C3H7)
                 ||
                 O
```

(**1**)

```
    C2H5O         OC2H5
        |             |
C2H5O — P — O — P — OC2H5
        ||            ||
        O             O
```

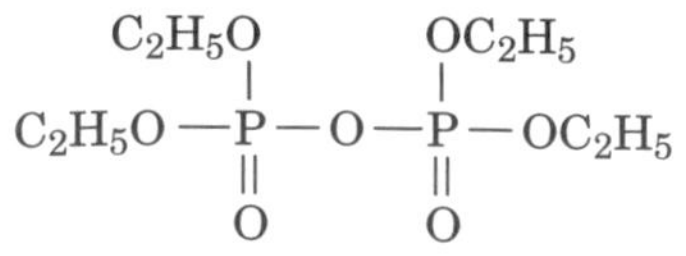

(**2**)

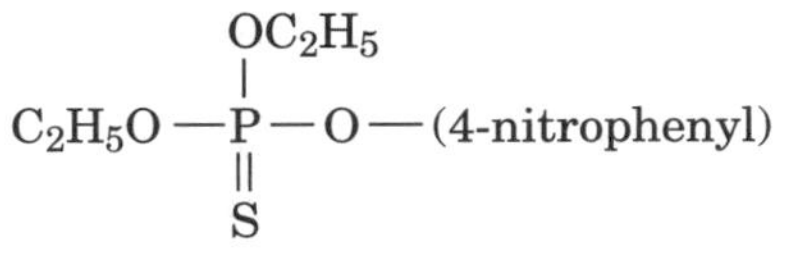

$$C_2H_5O{-}\underset{\underset{\displaystyle S}{\|}}{\overset{\overset{\displaystyle OC_2H_5}{|}}{P}}{-}O{-}(\text{4-nitrophenyl})$$

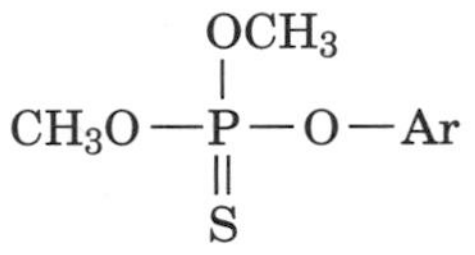

(**3**)

$$CH_3O{-}\underset{\underset{\displaystyle S}{\|}}{\overset{\overset{\displaystyle OCH_3}{|}}{P}}{-}O{-}Ar$$

(**4**)

$$CH_3O{-}\underset{\underset{\displaystyle S}{\|}}{\overset{\overset{\displaystyle OCH_3}{|}}{P}}{-}S{-}\underset{\underset{\displaystyle CH_2COOC_2H_5}{|}}{CHCOOC_2H_5}$$

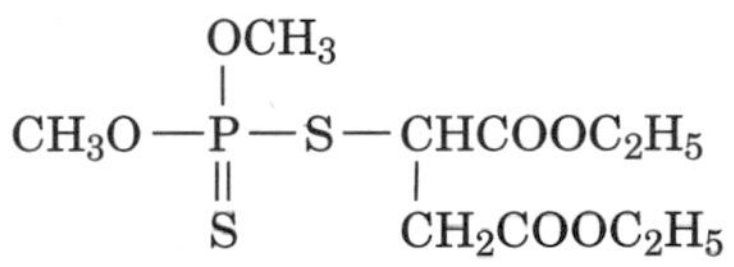

(**5**)

Schrader understood the effectiveness of the phosphorus–fluorine (P-F) moiety in the toxicity of the organophosphorus compounds at least as early as Saunders, the preparer of DFP (6,7). The German chemical warfare agents Sarin (**6**) and Soman (**7**) had this feature, whereas Schrader's earlier compound, Tabun (GA) (**8**) did not. Under the American designation, Sarin is GB: G for German, B for second agent. Tabun was synthesized in December 1936. Its great toxicity was discovered in January 1937 after a very small amount was spilled in the laboratory. It is reported that one drop of Tabun, which has about 1/36 the vapor hazard of Sarin, was spilled on a lab bench in Schrader's laboratory in January 1937. The staff needed three weeks to recover (8). Sarin was developed in 1938, and Soman (GD) in 1944. The pinacolyl substituent reduced volatility of the agent. Parenthetically, there was no German agent designated GC by the American military, because American military medical nomenclature already used GC for gonorrhea. Agent GF (**9**) was not a German development, but is structurally related. Instead of isopropyl or pinacolyl, a cyclohexyl substituent is used.

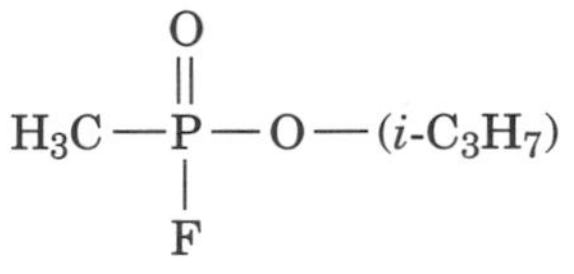

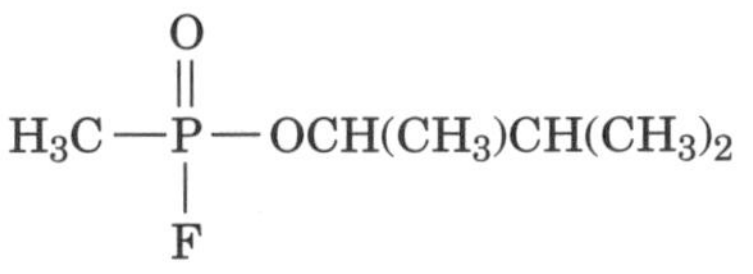

$$H_3C{-}\underset{\underset{\displaystyle F}{|}}{\overset{\overset{\displaystyle O}{\|}}{P}}{-}O{-}(i\text{-}C_3H_7)$$

(**6**)

$$H_3C{-}\underset{\underset{\displaystyle F}{|}}{\overset{\overset{\displaystyle O}{\|}}{P}}{-}OCH(CH_3)CH(CH_3)_2$$

(**7**)

$$C_2H_5O{-}\underset{\underset{\displaystyle CN}{|}}{\overset{\overset{\displaystyle O}{\|}}{P}}{-}N(CH_3)_2$$

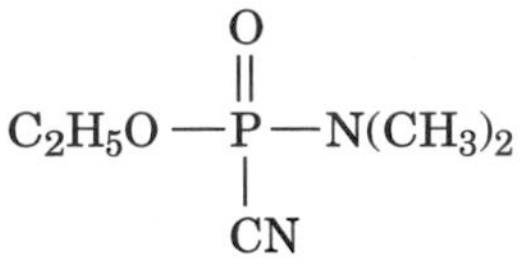

(**8**)

Sarin has the more reactive P-F group compared to the P-CN of Tabun, and more easily oxidized isopropyl groups as opposed to ethyl. Sarin is about 10 times as fast-acting as Tabun, and has about 36 times the vapor hazard (vapor spread) due to its volatility. Soman is less volatile than Sarin, and is more easily absorbed by the skin. Soman apparently never reached production status in World War II. Whereas tons of agents were produced and stored by Germany, there are no reports of any military use by them in World War II.

After World War II, Ghosh of Imperial Chemical Industries (England) developed VX (**10**) with one objective being lower volatility. As with the early German research, the goal was to prepare new pesticides. Its consistency is about that of 30-weight motor oil, compared to Soman which has the consistency of light machine oil (8).

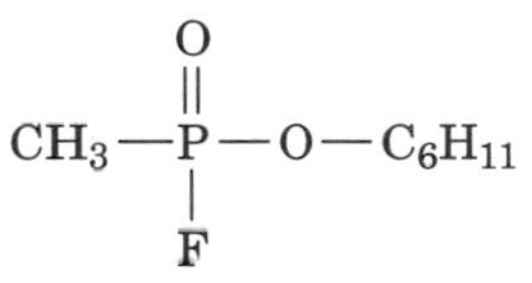

$$CH_3{-}\underset{\underset{\displaystyle F}{|}}{\overset{\overset{\displaystyle O}{\|}}{P}}{-}O{-}C_6H_{11}$$

(**9**)

$$C_2H_5{-}\underset{\underset{\displaystyle CH_3}{|}}{\overset{\overset{\displaystyle O}{\|}}{P}}{-}S{-}CH_2CH_2N(i\text{-}C_3H_7)_2$$

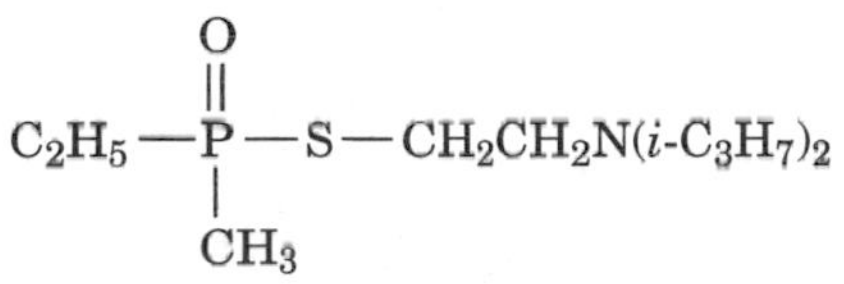

(**10**)

PHYSICAL AND CHEMICAL PROPERTIES

Mustard is an oily liquid with a boiling point 215–217°C; on cooling, it forms prisms (crystals) of melting point 13–14°C. It is very sparingly soluble in water, but very soluble in most organic solvents, and is hydrolyzed by alkali. The recommended neutralizing and deactivating agents are chlorinated lime (bleaching powder) or sodium hypochlorite (bleach). Mustard is reportedly corrosive to canisters. A conformational analysis has been reported (9).

Tabun is a liquid of mp −50°C, bp 240°C. It is soluble in organic solvents, and miscible with water, but quickly hydrolyzed. Bleaching powder decomposes Tabun but forms cyanogen chloride (ClCN).

Sarin (**6**) is a liquid of mp −57°C, bp 147°C. It is soluble in organic solvents, and miscible with and hydrolyzed by water. Dilute aqueous sodium hydroxide or sodium carbonate hydrolyzes Sarin to relatively nontoxic products (6). Water alone removes the fluorine, producing the nontoxic acid $CH_3PO[OCH(CH_3)_2]OH$. Its half-life in water at 30°C and pH 9 is 15 minutes. Sarin has moderate vapor pressure, forms an aerosol cloud if sprayed, and can be readily absorbed through the skin. Constriction of the pupils, difficulty in breathing, and convulsions result, followed by death. Atropine and oximes such as 2-PAM chloride may be effective as antidotes.

Soman (**7**) and VX (**10**) are soluble in water to about 1–1.5%. The half-life of Soman in water at 30°C and pH 7.6 is 577 minutes. At 25°C, the hydrolysis rates of VX are 200 days at pH 2–3; 32 minutes at pH 13, and 2.6 minutes at pH 14. Soman, and especially VX, have low vapor pressure. They are readily absorbed through the skin, and can form persistent aerosol clouds if sprayed.

For comparison, parathion (**3**) is a pale yellow liquid of mp 6°C, bp 375°C. It is soluble in organic solvents such as alcohols, ethers, esters, ketones, and aromatic hydrocarbons. It is practically insoluble in water (soluble to 20 ppm). It degrades at pH above 7.5.

The odors of the organophosphorus chemical agents have been reported as fruity and like bitter almonds (no odor if pure) for Tabun (**8**); fruity, almost odorless if pure, for Sarin; like overripe peaches or camphor-like for Soman; and odorless or like rotten fish for VX. These odors apparently result more from impurities remaining in the formulations than from the agents themselves.

Mustard is a deadly vesicant, producing delayed effects including conjunctivitis and blindness, and damage to the respiratory tract and exposed skin. The blisters are slow to heal. Eye and respiratory tract damage can be permanent.

The organophosphorus agents and related pesticides such as parathion are cholinesterase inhibitors, with LD_{50} in rats or mice reported to be 3.6–13 mg/kg for parathion, 0.6 mg/kg for Tabun, 0.42 mg/kg for Sarin, 0.62 mg/kg for Soman, and 0.0154 mg/kg (rabbits) for VX. Human toxicities may be about 50 times greater for some of these compounds. Soman has two chiral centers: carbon and phosphorus. It has been reported that the C^+, P^+ stereoisomer of Soman is least toxic, with an LD_{50} greater than 5 mg/kg. The C^-, P^+ isomer has a toxicity of about 2, whereas the toxicities of C^+, P^- and C^-, P^- are 0.099 and 0.038 mg/kg, respectively. The (−)-enantiomer of Sarin has been reported to be more toxic than the (+)-antipode. The use of chiral gas chromatography (GC) columns to separate these enantiomers has been reported (10).

SYNTHESIS

One synthesis of mustard involves treating ethylene with sulfur chloride. In another process, 2,2′-dihydroxyethyl sulfide is reacted with gaseous HCl.

Numerous pathways to organophosphorus pesticides and chemical warfare agents have been devised since Lange and Krueger reacted the silver salt of H_2PO_3F with alkyl iodides in a sealed tube at 50 to 100°C (or higher temperatures for higher alkyl homologues) for two hours.

Tabun has been prepared from dimethylamidophosphoryl dichloride and sodium cyanide in ethanol. Saunders prepared Tabun by another pathway as follows:

$$(C_2H_5O)_2PCl + HN(CH_3)_2 \longrightarrow (CH_3O)_2PN(CH_3)_2 \xrightarrow{(+\,ICN)} \quad (\mathbf{8})$$

Compounds such as chlorthion, fenthion, and fenitrothion may be prepared from suitably substituted phenols (ArOH) and dimethyl phosphorochloridothionate in methyl isobutyl ketone, in the presence of potassium carbonate, as follows:

$$CH_3O{-}P(=S)(OCH_3)(Cl) + ArOH \longrightarrow CH_3O{-}P(=S)(OCH_3)(OAr)$$

Malathion may be prepared by the addition of *O,O*-dimethyl phosphorodithioic acid to diethyl maleate, with catalysis by a base:

$$(CH_3O)P(=S)(OCH_3)(SH) + C_2H_5OOC{-}CH{=}CH{-}COOC_2H_5 \longrightarrow H_3CO{-}P(=S)(OCH_3){-}S{-}CH(COOC_2H_5)CH_2COOC_2H_5$$

Derivatives of fluorophosphoric acid include the diamido-substituted insecticides which were used agriculturally in the early stage of development of organophosphorus insecticides. The fluorophosphoric acid derivatives of the fluorophosphono- or fluorophosphoroester types were developed for chemical warfare.

An example of the diamido-substituted insecticidal class is dimefox (**11**):

$$Cl{-}P(=O)(N(CH_3)_2){-}N(CH_3)_2 + NaF \longrightarrow F{-}P(=O)(N(CH_3)_2){-}N(CH_3)_2 \quad (\mathbf{11})$$

Sarin may be prepared as follows:

$$H_3CP(=O)Cl_2 + NaF + i\text{-}C_3H_7{-}OH \longrightarrow \quad (\mathbf{6})$$

DFP (**1**) was prepared from the chloridate analogue:

$$PCl_3 + (i\text{-}C_3H_7)OH + (i\text{-}C_3H_7)Cl + HCl \longrightarrow Cl{-}P(=O)(O{-}(i\text{-}C_3H_7)){-}O{-}(i\text{-}C_3H_7) \longrightarrow \quad (\mathbf{1})$$

Pyrophosphate esters may be prepared as shown below for TEPP:

$$Cl{-}P(=O)(OC_2H_5)_2 \longrightarrow HO{-}P(=O)(OC_2H_5)_2 + Cl{-}P(=O)(OC_2H_5)_2 \longrightarrow \quad (\mathbf{2})$$

Another preparation for TEPP is as follows:

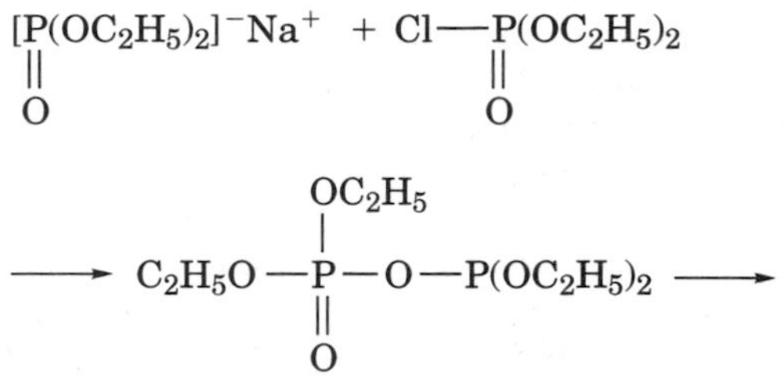

$$[\mathrm{P(OC_2H_5)_2}]^-\mathrm{Na^+} + \mathrm{Cl{-}P(OC_2H_5)_2} \quad (\text{each P}{=}\mathrm{O})$$

$$\longrightarrow \mathrm{C_2H_5O{-}P(=O)(OC_2H_5){-}O{-}P(OC_2H_5)_2} \longrightarrow \quad \mathbf{(2)}$$

The methyl homologue of parathion is 4–7 times less toxic to mammals. Adding a methyl group to the aromatic ring of methyl parathion results in fenitrothion (**4**), which has low mammalian toxicity.

DESTRUCTION PROCESSES

At relatively low temperatures and pressures, on small-scale amounts, it has been found that Sarin can be hydrolyzed and detoxified by aqueous sodium hydroxide. VX was successfully treated with sodium hydroxide and hydrogen peroxide in water. Some data exist to suggest that an alcohol or ethanolamine can be substituted for water, providing better solubility for the agents (6,8). Relative rates of hydrolysis for various functionalities X on $O = PX(OC_2H_5)_2$ are as follows: $OC_2H_5 = 1$; $OC_6H_5 = 100$; $SC_6H_5 = 1.3 \times 10^4$; $F = 5.6 \times 10^5$.

The hydrolysis of mustard and related compounds has been extensively studied (11–13). Mustard can be decomposed with aqueous calcium hydroxide at 90–100°C. A review of decontamination and destruction processes has been published (14).

Oxidation processes, including supercritical water, may be suitable. High temperature kilns, plasma arcs, molten metal pyrolysis, steam gasification, molten salt, fluidized-bed, and catalytic fixed-bed oxidation processes have potential for large-scale destruction of agents. Additionally, incineration tests since the early 1980s on ton-quantities of mustard, VX, and Sarin at Tooele Army Depot (Utah) and on Johnston Island were successful, yielding destruction efficiency ratings that were >99.9999% for these relatively easy-to-incinerate (Class 4 or lower) agents (15). The Retech Plasma Arc Centrifugal Treatment (PACT) process uses 1650° and a 30-second residence time to ensure destruction of even the hardest-to-incinerate (Class 1) compounds that were tested such as chlorinated benzenes.

SELECTED REACTIONS

The oxidation of mustard to the sulfoxide and sulfone was reported in 1920 by Helfrich and Reid (16). The major hydrolysis product of mustard is thiodiglycol. Methyl sulfonium derivatives were described in 1990 (17).

Phosphorus acids are strong acids, so their anions can serve as good leaving groups. As an example, consider the general case below:

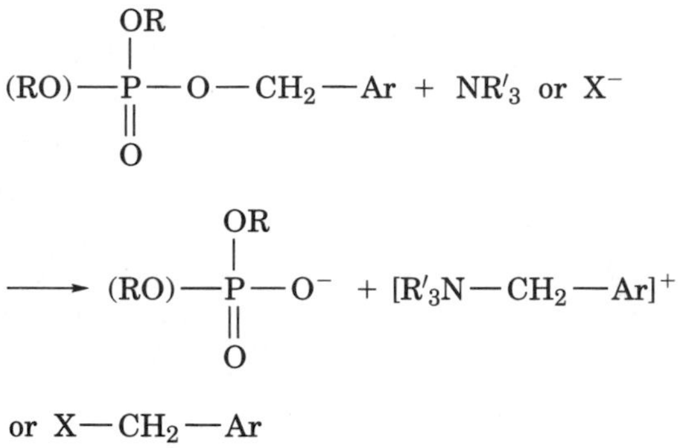

$$\mathrm{(RO){-}P(=O)(OR){-}O{-}CH_2{-}Ar} + \mathrm{NR'_3} \text{ or } \mathrm{X^-}$$

$$\longrightarrow \mathrm{(RO){-}P(=O)(OR){-}O^-} + [\mathrm{R'_3N{-}CH_2{-}Ar}]^+$$

$$\text{or } \mathrm{X{-}CH_2{-}Ar}$$

Phosphorothioates can isomerize upon heating. For example, parathion is 90% converted by heating at 150°C for 24 hours, as shown below:

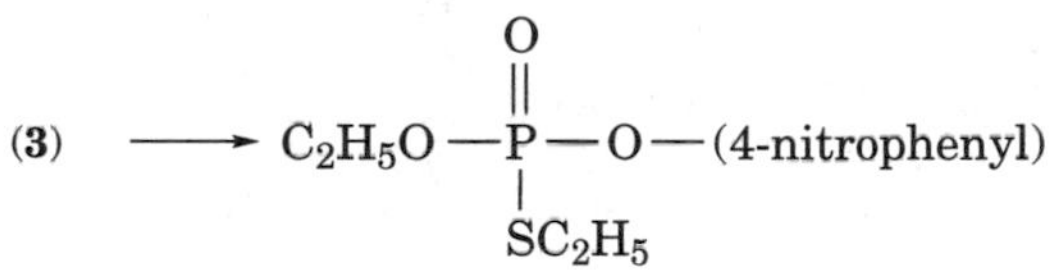

$$\mathbf{(3)} \longrightarrow \mathrm{C_2H_5O{-}P(=O)(SC_2H_5){-}O{-}}(\text{4-nitrophenyl})$$

This class of compound can also oxidize from the P=S thiophosphoryl group to P=O phosphoryl upon treatment with an oxidizer such as 3-chloroperbenzoic acid. The phosphoryl group is generally a much stronger cholinesterase inhibitor. In this way, fonofos (**12**) is converted to its oxo-analogue (**13**):

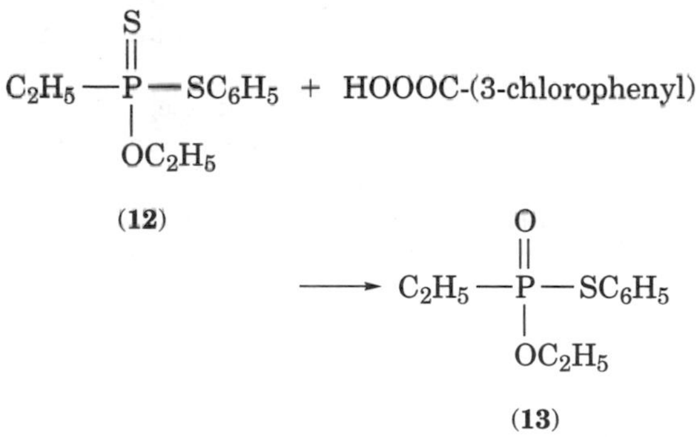

$$\underset{\mathbf{(12)}}{\mathrm{C_2H_5{-}P(=S)(OC_2H_5){-}SC_6H_5}} + \text{HOOOC-(3-chlorophenyl)}$$

$$\longrightarrow \underset{\mathbf{(13)}}{\mathrm{C_2H_5{-}P(=O)(OC_2H_5){-}SC_6H_5}}$$

The thioesters themselves, such as parathion and malathion, are poor cholinesterase inhibitors. However, they are metabolized to the oxo analogues, which are much stronger inhibitors. The process, involving a mixed function oxidase enzyme, is analogous to that shown for a peracid:

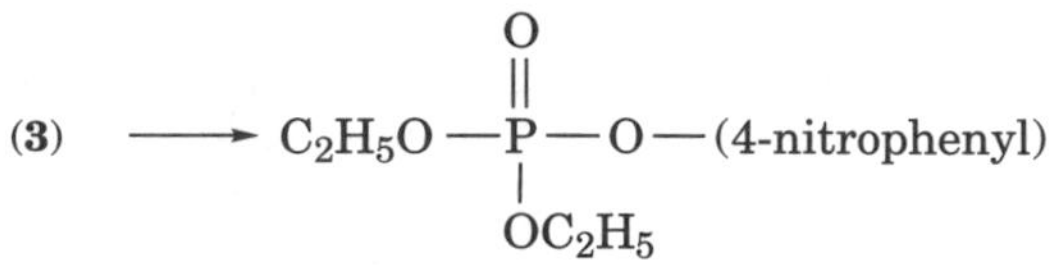

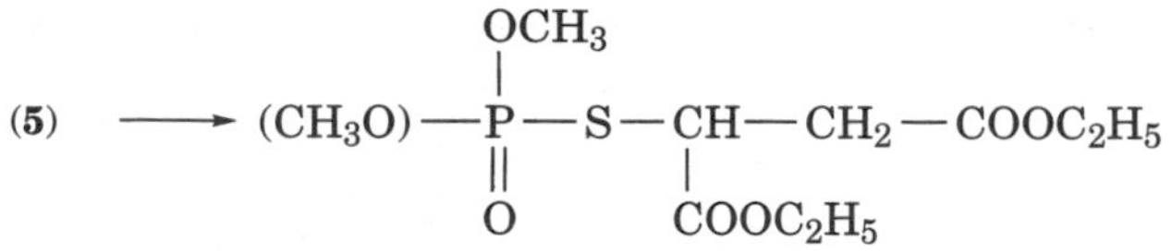

$$\mathbf{(3)} \longrightarrow \mathrm{C_2H_5O{-}P(=O)(OC_2H_5){-}O{-}}(\text{4-nitrophenyl})$$

$$\mathbf{(5)} \longrightarrow \mathrm{(CH_3O){-}P(=O)(OCH_3){-}S{-}CH(COOC_2H_5){-}CH_2{-}COOC_2H_5}$$

Compounds with a thioether group may undergo metabolic oxidation to the sulfoxide or sulfone, for example:

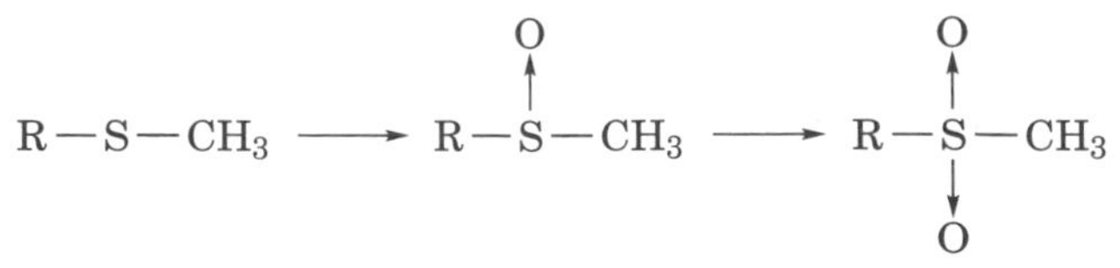

$$R-S-CH_3 \longrightarrow R-\overset{\overset{O}{\uparrow}}{S}-CH_3 \longrightarrow R-\overset{\overset{O}{\uparrow}}{\underset{\underset{O}{\downarrow}}{S}}-CH_3$$

Thioether Sulfoxide Sulfone

Malathion decomposes in base by one of two processes, depending on whether the medium is an organic solvent or water:

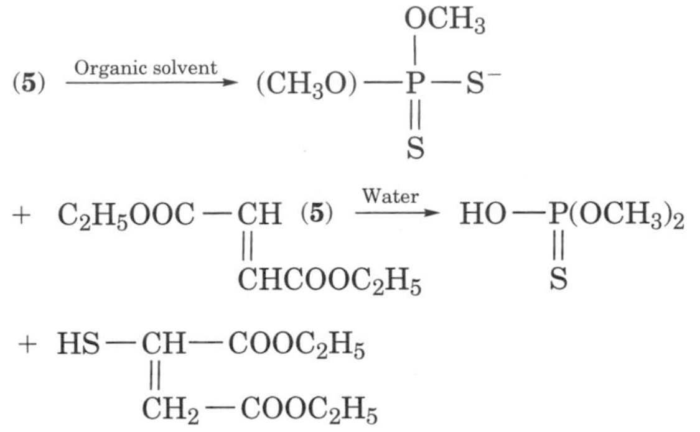

$$(\mathbf{5}) \xrightarrow{\text{Organic solvent}} (CH_3O)-\overset{\overset{OCH_3}{|}}{\underset{\underset{S}{\|}}{P}}-S^-$$

$$+\ C_2H_5OOC-\underset{\underset{CHCOOC_2H_5}{\|}}{CH} \quad (\mathbf{5}) \xrightarrow{\text{Water}} HO-\underset{\underset{S}{\|}}{P}(OCH_3)_2$$

$$+\ HS-\underset{\underset{CH_2-COOC_2H_5}{\|}}{CH}-COOC_2H_5$$

BIOCHEMISTRY OF ORGANOPHOSPHORUS NERVE AGENTS

The target site of organophosphate and organophosphonate esters is acetylcholinesterase (18). This enzyme is one of the hydrolysis enzymes for acetylcholine. Inhibiting this enzyme disrupts the nervous system of animals, including insects and other pests, and mammals including humans. Different organophosphates and organophosphonates have differing inhibiting capabilities toward different species.

The nervous system works through electrical nerve impulses, triggered by transmitters such as acetylcholine and noradrenaline (norepinephrine). Acetylcholine is synthesized in the nerve ending by the action of cholinacetylase on choline. After the electrical impulse, the acetylcholine is rapidly hydrolyzed by acetylcholinesterase to acetic acid and choline. Inhibition of this hydrolytic process prevents the nerve synapse from reverting to its original state.

The reaction of the organophosphorus ester occurs at the hydroxyl group of serine in the enzyme. This reaction of Sarin with the enzyme, enzyme-OH, can be depicted as follows:

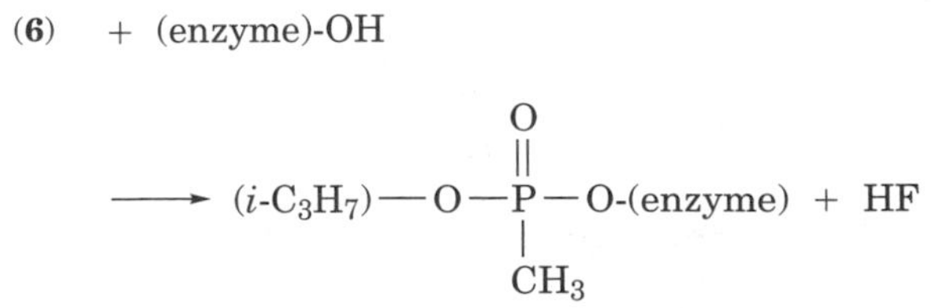

$$(\mathbf{6}) \quad + \text{(enzyme)-OH}$$

$$\longrightarrow (i\text{-}C_3H_7)-O-\overset{\overset{O}{\|}}{\underset{\underset{CH_3}{|}}{P}}-O\text{-(enzyme)} + HF$$

This reaction may be reversible, by the action of therapeutic chemicals (antidotes) such as hydroxylamine and its derivatives (oximes and hydroxamic acids) or atropine, or by choline. The reaction may be essentially irreversible in some cases. Tabun produces a phosphoramidate that is relatively unreactive to this reverse reaction. In general, phosphorus amides, such as dimefox, are not very active inhibitors, but the polarizable cyano group of Tabun makes it an exception. This reverse reaction is the mechanism of recovery by the victim.

It is very difficult to reverse the reaction of Soman with the enzyme, apparently due to a second reaction. This second reaction is the dealkylation of the phosphonate ester. It renders the resultant phosphorylated enzyme unreactive to the reactivating reverse reaction that can be caused by a therapeutic agent (antidote) or by biological recovery by the victim's body. It has been found that ethoxy (OC_2H_5) groups undergo this dealkylation slowly, whereas O-CH_3, O-i-C_3H_7, and other secondary alkyl groups (like the pinacolyl group, $CH(CH_3)\ CH(CH_3)_2$, on Soman) dealkylate many times faster. While the half-life of O-C_2H_5 is 40 hours, O-CH_3 is fully reacted in 24 hours. The i-C_3H_7-O group has a half-life of about 2.5 hours, and the pinacolyl group of Soman has a half-life of only six minutes. Once this dealkylation occurs, antidotes would presumably be ineffective.

The stereochemistry of the phosphorylating agent is important, as might be expected since the enzymes are themselves geometrically complex and optically active. The dealkylation proceeds 1000 to 7000 times faster with the most active stereoisomer compared to the others.

Certain esterases, called A-esterases, arylesterases, or phosphotriesterases, can hydrolyze organophosphorus esters. These enzymes are found in mammals and similar enzymes are found in insects. Such enzymes can hydrolyze DFP to $(i-C_3H_7)_2POOH + HF$, for example. The oxygen analogue of parathion (**3**), paraoxon (**14**), is similarly hydrolyzed:

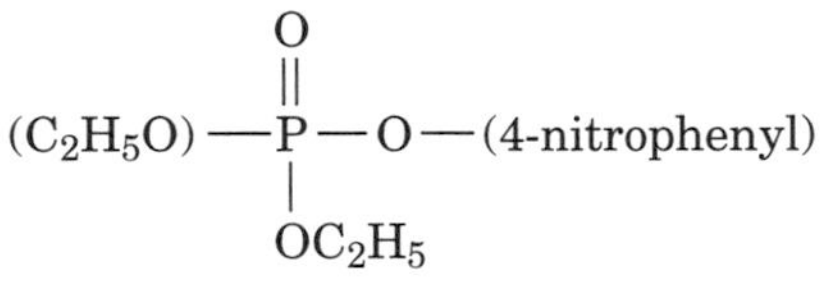

$$(C_2H_5O)-\overset{\overset{O}{\|}}{\underset{\underset{OC_2H_5}{|}}{P}}-O-\text{(4-nitrophenyl)}$$

(**14**)

$$\longrightarrow (C_2H_5O)-\overset{\overset{O}{\|}}{\underset{\underset{OC_2H_5}{|}}{P}}-OH + HO-\text{(4-nitrophenyl)}$$

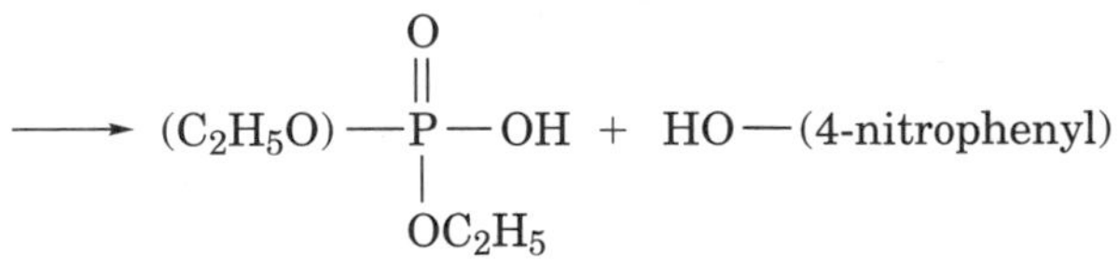

Diethyl phosphate

ANALYTICAL METHODS POTENTIALLY SUITABLE FOR ENVIRONMENTAL ANALYSIS

The analysis of mustard, bis(2-chloroethyl)sulfide, and of the related 2-chloroethyl ethylsulfide (CEES) have been reported using gas chromatographic or liquid chromatographic systems. The use of HPLC with UV detection to analyze water samples for CEES and two degradation products, 2-hydroxyethyl ethyl sulfide (HEES) and vinyl ethyl sulfide (VES) has been reported (19). A method has been developed to detect mustard collected from air using GC/EC (20), and in 1970 mustard was analyzed with GC/FID (21). The extraction of mustard-contaminated soil in a vial with dichloromethane by "tumbling," followed by GC/MS analysis on a 15-m J& W DB-5 column has been reported (22).

GC retention indices have been reported (23) for 37 sulfur compounds associated with mustard on DB-1, DB-5, and DB-1701 columns. Among the analytes were 1,4-oxathiane, 1,4-dithiane, and thiodiglycol. Soil was extracted with chloroform by sonication. They reported retention indices for agents including Sarin, Soman, Tabun, VX, and mustard, as well as degradation products.

The UV absorption spectra were reported (24) for compounds including Tabun, Sarin, Soman, VX, mustard, diisopropyl methylphosphonate (DIMP), and 2-Chloroethyl ethyl sulfide. Knowledge of the UV spectral properties of these compounds facilitates the selection of an appropriate wavelength for LC/UV analysis.

The development of acoustic resonance spectroscopy to identify chemical warfare agents and other compounds in sealed containers was recently reported (25).

Gas chromatographic analysis of organophosphorus agents or decomposition products was reported in 1982 with packed-column GC (26). More recent work has generally used capillary GC columns.

Gas chromatography/mass spectrometry (GC/MS) has been used to analyze both mustard and related compounds, and the organophosphorus compounds. Capillary column GC coupled with MS was used in the selected ion monitoring mode to analyze CEES and the two decomposition products, HEES and VES (27). GC/MS was used in EI and CI ionization modes (28,29). Mass spectral data was published for a number of compounds, along with proposed fragmentation pathways. Soil samples were analyzed (23) for agents and degradation products, using trimethylsilyl derivatization as necessary. A sequential hexane and methylene chloride extraction procedure was found to be effective.

Some related phosphorus-containing pesticide metabolites were benzylated from the acid form to phosphate and phosphonate esters, and analyzed with GC/FPD (30). GC/MS has been used to analyze methylphosphonic acid decomposition products (31). The acids were formed from Sarin, Soman, and VX by forming ion pairs with trimethylphenylammonium hydroxide followed by methylation in the GC injection port. Injection port temperatures above 260°C were used to facilitate the methylation process *in situ*. The analysis of *tert*-butyldimethylsilyl derivatives of phosphonic acids by GC with flame ionization, flame photometric, and mass spectrometric detection has been reported (32). This scheme provides easily-prepared, stable derivatives having good sensitivity to MS detection. DIMP has been analyzed directly, and derivatives of phosphonic acid degradation products have been analyzed using DB-1 and DB-5 capillary columns. Acetonitrile and chloroform solvents are also used.

Liquid chromatography has also been used to analyze for organophosphorus agents and decomposition products. LC/fluorescence has been used to analyze alkylphosphonic acids derivatized with *p*-(9-anthroyloxy)phenacyl bromide to provide a fluorophore (33). This technique requires the extra derivitizing step, but provides improved detection limits compared to UV detection. Thermospray LC/MS was used to analyze agents and hydrolysis products (34), which avoided the need to form fluorescent or UV-absorbing derivatives that would be required by such LC detectors.

Ion chromatography (35,36) and capillary electrophoresis (37,38) have been used to separate and analyze the organophosphorus nerve agent hydrolysis products in soil and water. These techniques have been limited by the difficulty of achieving sensitive confirmatory detection.

HEALTH CONSIDERATIONS FOR MUSTARD AND RELATED COMPOUNDS

Symptoms of Exposure to Mustard

Exposure to the eye usually causes no immediate symptoms, but lachrymation (tearing) and irritation may occur after 2–3 hours; severe conjunctivitis (pink-eye), nasal drip, and sneezing are other symptoms that may occur at a similar time after exposure. In 4–6 hours, burning pain, irritation, lacrimation, and swelling of the eyelids occurs. In some cases, symptoms may appear within minutes of exposure. Photophobia follows, the cornea becomes hazy, and visual impairment can be permanent in the more severe cases. The eye is the most vulnerable organ to mustard.

Mustard penetrates the skin rapidly, without any immediate symptoms. Redness of the skin is followed by blistering, usually 4–24 hours after exposure, and generally associated with itching, burning, and pain. Symptoms are especially intense in armpit and genital regions. Regions of thin, moist skin are more easily affected (eg, face, inside elbows and knees, neck). Penetration of vesicant vapors into skin was studied in 1943–1944 and reported in 1946 (39).

Respiratory tract susceptibility is second only to that of the eyes. After 48 hours, lung and respiratory tract damage is seen. Exposure can be limited to the upper tract (light exposures), or involve the lower tract as well (heavy exposure). The first symptom is hoarseness, with a cough soon after. Other symptoms include mouth and nostril pain, rhinorrhea, sneezing, hoarseness, coughing, and choking. Pulmonary function may be impaired. Respiratory infection is common, and was the major cause of death in World War I cases.

Nausea, retching, vomiting, and abdominal pain occurs later. Mustard also affects the central nervous system, causing apathy, depression, intellectual dullness, and languor.

In summary, the symptoms are delayed for minutes to hours from the time of exposure. Eye, throat, and skin irritation, nausea, and hoarseness are among the early symptoms of exposure (8,18).

Medical Treatment for Mustard Exposure

Immediate irrigation of the eye is needed to prevent injury. A 2% sodium bicarbonate solution could be used. Mydriatic eye drops (1% cyclopentolate) and an antiseptic (20% sulfacetamide) have been recommended for use four times daily.

One report (18) states that exposed skin areas should be washed with powdered calcium chloride and magnesium oxide, followed by soap and water, followed with an antibiotic and sterile dressing. Charcoal and talcum powder have also been used to absorb mustard. In another case, washing with 2% sodium bicarbonate solution, followed by soap and water, was recommended. Treatment should be performed within 10 minutes of exposure. Guinea pig skin was found to be protected by an iodine/povidone–iodine preparation, even when applied 20–60 minutes after exposure (18).

Respiratory tract treatment includes a vaporizer, moist oxygen, bromohexine, codeine (for coughing), and antibiotics (not sulfonamides, because mustard suppresses bone marrow) (18).

Short- and Long-Term Health Effects

Short-term effects include burns, fever, nausea, vomiting, hoarseness, difficulty in breathing, tearing, and irritation to eyes. Long-term effects include difficult healing of blisters, eye and respiratory tract damage, and cancer (18,40).

Precautions for Safe Handling of Samples

Mustard can penetrate clothing and some types of rubber. A protective nonabsorbent suit with gloves, face-shield, and chemical respirator should be used at a minimum if exposure to vapors or liquid is possible (41). In a laboratory, a fume hood or glove box should also be used. After potential exposure, decontaminate even in the absence of symptoms.

Decontamination

Early decontamination is the only effective way to prevent injury. Because mustard binds to tissues within minutes, but symptoms often do not appear for 4–24 hours, the potential for exposure cannot be assessed soon enough by observation of symptoms. Mustard has a very short lifetime in the body before chemically binding; a few minutes in blood, for example.

Strong aqueous bleach (sodium hypochlorite), bleaching powder (chlorinated lime), and sodium hydroxide solutions are reported to be effective decontaminating agents. Live steam and fire have been effective in military situations (1,8). Water solubility is low (0.7 g/L), but mustard is very soluble in organic solvents. Mustard hydrolyzes in water and in basic aqueous solutions.

Most reported analytical methods have utilized gas chromatography (GC), although high performance liquid chromatography (HPLC) and thin-layer chromatography (TLC) have also been used. For quick, facile verification of decontamination, the method of Reference 42 may be useful.

A review of decontamination and destruction processes was recently published (14).

HEALTH CONSIDERATIONS FOR ORGANOPHOSPHORUS COMPOUNDS

Comparison of Different OP Compounds

Organophosphorus (OP) compounds vary widely in their ability to inhibit cholinesterase activity. A number of early pesticides, such as parathion, were very toxic to mammals, including humans. As empirical evidence was gathered about the features that affected toxicity, other pesticides were synthesized (from about the 1950s to present) to minimize mammalian toxicity. Most OP pesticides in current use are much safer to use than parathion and the nerve agents: fonofos is around 0.1–0.01 as toxic as parathion, which itself is about 0.05–0.01 as toxic as Sarin. Cases of death from parathion and other OP compounds have been reported. Except for untreated cases of heavy exposure to nerve agents, these deaths were associated with accidental ingestion or heavy dermal contact, combined with late or no treatment. For example, a young field worker died after severe exposure to parathion that was applied to the field on the preceding day. The problem was not diagnosed, therefore the worker was neither bathed nor was a change a clothes provided, and prompt medical treatment was not obtained (43). In contrast, there are numerous reported cases of accidental or deliberate exposure of people to the extremely toxic OP nerve agents. Proper post-exposure decontamination procedures and medical treatment have enabled the victims to reach full recovery. Available information indicates that di-isopropyl methylphosphonate (DIMP) is minimally active with respect to cholinesterase inhibition, and is also very slowly hydrolyzed (44).

Symptoms of Exposure

In 1932, it was reported that the odor of the fluorophosphate esters is pleasant and aromatic, but a few minutes after inhalation, headaches and suffocation, followed by vision troubles were experienced. The effects lasted a few hours and were caused by minute quantities (4). Based on the past sixty years of experience with OP pesticides and chemical warfare agents, the "classical" symptoms for OP poisoning include the following (6,8,18):

- pinpoint pupils (myosis) and dim vision
- difficulty in breathing
- drooling and excessive sweating
- nausea, vomiting, cramps
- involuntary defecation and urination
- twitching, jerking, and staggering
- headache, confusion, drowsiness
- convulsions, coma

The effective toxicity may be increased by the carrier (solvent) which can increase the rate of absorption.

The order in which Sarin poisoning symptoms appear varies with the route of exposure. Respiratory problems precede gastrointestinal tract symptoms due to inhalation exposure. In a study of Sarin exposure (45), dilute solutions were given orally over three days. In the case of 0.025 of the LD_{50} dose (0.28 mg/kg), no symptoms were observed. At 0.16 of the lethal dose, mild symptoms were observed, including anorexia, abdominal cramps, sweating, and dreaming. At 0.33 of the lethal dose, the above symptoms and tightness of the chest were more pronounced. Bradycardia, frequent urination, subjective dyspnea, fatigue, muscle twitching, giddiness, tension, anxiety, depression, restlessness, and tremors were also seen in some subjects.

It has been reported (6,7) that diisopropyl fluorophosphate (DFP) was tested on human subjects in the early 1940s shortly after its development. Volunteers were subjected to exposure at the 1 ppm level (0.0082 mg/L) in a 10 cubic meter glass chamber for 5 minutes. Intense miosis occurred after 5 minutes and lasted 7 days. Additionally, powers of accommodation were affected and photophobia, headaches, and tightness of the chest was experienced. Even under carefully controlled laboratory conditions, minute traces sometimes adhered to clothing, vaporized later, and caused miosis and headaches. Higher concentrations caused miosis, salivation, muscular weakness, loss of muscular coordination, gasping, diarrhea, and finally respiratory failure. Intense constriction of the bronchioles caused asphyxiation. Extremely small quantities of DFP killed flies in the room (6,7).

The effects can be grouped as muscarine-like, nicotine-like, and central nervous system effects. Muscarine-like effects include gastrointestinal (cramps, nausea, vomiting, diarrhea), respiratory (bronchoconstriction: tight chest, wheezing), increased bronchial secretion, nasal discharge), and ocular (miosis, conjunctival congestion, ciliary spasms, pain in or around the eye). The nicotine-like effects include muscular cramps, twitching, and weakness. Central nervous system effects include excessive dreaming, insomnia, nightmares, headaches, slurred speech, loss of reflexes, paralysis, and coma. Additionally, changes in heart rate and blood pressure are often found. Lingering effects include forgetfulness, irritability, impaired judgment, decreased comprehension, tenseness, depression, insomnia, and nightmares.

Dim vision has been reported in the absence of miosis, and causes the patient trouble (eg, driving) in dim light. Severe rhinorrhea (runny nose) has been reported after exposure to only a small amount of OP agent. The tight chest or shortness of breath (dyspnea) has been reported from small exposures. This symptom is related to respiratory failure, the cause of death from OP poisoning. Respiratory failure has been found in animal studies to occur before circulatory failure (18). Fasciculation of muscles is the visible contraction of small numbers of fibers, and looks like ripples under the skin. This effect can be a localized response to a drop of agent on the skin. Convulsions and seizures have been reported from heavy exposures, and have been correlated to increased likelihood of brain damage. After a few minutes of muscular hyperactivity, flaccid paralysis occurs. Diazepam has been used as an anticonvulsant. Eye effects are often the first to be noted, because the eye is very sensitive to these OP agents. With severe exposure, the patient may become unconscious, have muscular twitching or flaccid paralysis, involuntarily void the bladder or bowels, and have labored and gasping, irregular respiration (apnea) (18).

Summary of Symptoms from Exposure to OP Agents

The effects of exposure to OP vapors can be summarized as follows:

- Exposure to small amounts
  - Miosis
  - Rhinorrhea
  - Slight bronchoconstriction
- Exposure to moderate amounts
  - Miosis
  - Rhinorrhea
  - Bronchoconstriction, secretions
- Exposure to large amounts
  - Miosis
  - Rhinorrhea
  - Bronchoconstriction, secretions
  - Loss of consciousness
  - Convulsions
  - Fasciculations
  - Flaccid paralysis
  - Apnea
  - Involuntary micturition/defecation

The effects of dermal exposure to OP agents can be summarized as follows:

- Minimal exposure
  - Sweating at site of exposure
  - Muscular fasciculations at site
- Moderate exposure
  - Sweating at site of exposure
  - Muscular fasciculations at site
  - Nausea, vomiting, diarrhea
  - General weakness
  - Symptoms may occur 4–18 hours after exposure
- Severe exposure
  - Sweating at site of exposure
  - Muscular fasciculations at site
  - Nausea, vomiting, diarrhea
  - General weakness
  - Symptoms may occur 10–30 minutes or more after exposure
  - Loss of consciousness
  - Convulsions, seizures
  - Generalized fasciculations
  - Flaccid paralysis
  - Apnea
  - Involuntary micturition/defecation

Medical Treatment

There are three widely used, basic treatments for OP poisoning:

1. Artificial respiration
2. Atropine to block the acetylcholine that builds up
3. reactivation of the enzyme with 2-PAM or another oxime

Additionally, tubocurarine was reported in a few references for relieving muscle tension (18).

Atropine is the treatment of choice for patients exposed to cholinesterase inhibitors such as the OP agents (18). Atropine blocks the effects of excess acetylcholine and protects the receptor site from further stimulation. It is effective for muscarine-type effects and sites. The military use 2 mg of atropine as a compromise between the desirable dose and the ease of dealing with side effects in the field.

The side effects of a 2-mg dose of atropine in a normal, young person who has not been exposed to nerve agents include an increase of heart rate (about 35 additional beats per minute), drying of secretions (eg, saliva), decreased sweating, mydriasis, and paralysis of accommodation. Most effects dissipate in 4–6 hours, except for visual blurring, which may take 24 hours to subside. In the military tests, the suppression of sweating was a potentially serious problem if work or strenuous hiking were to be performed, especially in hot weather. Under medical care, the initial dose of atropine could be 6 mg or more. For severe OP poisoning, an intravenous drip could be used. Muscular abnormalities (fasciculations or twitching) continue after the patient is conscious, and are not a reliable indicator of need for more atropine (18).

Certain oximes greatly enhance the usefulness of atropine, but they are not effective when administered by themselves. 2-PAM (pralidoxime, Protopam chloride) is generally the drug of choice. About 15–25 mg/kg is given (or 1–2 g) initially at a rate not exceeding 0.5 g/min. The dosage may be repeated in 1–2 hours, and then at 10–12 h intervals afterward. The oxime removes the OP agent from the inhibited enzyme. It is ineffective if dealkylation (aging) of the OP agent has occurred, and therefore, 2-PAM is of little use in Soman poisoning, where the dealkylation reaction is fast (half-life of about 6 minutes).

Most of 2-PAM (80–90%) is excreted in the urine intact within 3 hours; its half-life in blood is about one hour. Note that oximes are not beneficial for cholinesterase poisoning by the other class of pesticides that exhibits this activity, carbamates (eg, carbaryl) (18).

Diazepam, an anticonvulsant, should be administered (2–10 mg) in the preconvulsive stage to a patient with seizure activity, or to those known to have moderate to severe exposure. Prolonged seizures have been associated with brain damage. Miosis can be reversed by topical application of atropine, but produces blurred vision for a day or longer (18).

Short- and Long-Term Health Effects

Several hours or days are required for full muscular strength and coordination to return, along with cognitive abilities. Regression is not likely. Fine visual discrimination and critical mental judgments should not be expected for one week to several months, depending on job circumstances. For example, air traffic controllers might need more than a month for full recovery, but soldiers have performed adequately even with miosis, rhinorrhea, and mild dyspnea (18).

Precautions for Safe Handling of Samples

Ordinary clothing provides little protection from OP agents. The common OP pesticides are generally safe to handle with moderate precautions: avoid respirating the agent into the lungs; avoid contact with skin and especially with eyes, mucous membranes, and sweat glandular areas. Use of a fume hood and appropriate gloves is desirable for laboratory operations. After any potential exposure, wash thoroughly and change clothes. Wash potentially contaminated clothing thoroughly at least twice with soap and hot water, separately from other clothing. For more powerful cholinesterase-inhibiting OP agents (eg, parathion and especially the nerve agents), a face shield or respirator and even full body suiting such as for hazardous spill responses are wise precautions if spilling or generation of aerosols (eg, high temperatures or winds) is at all likely (41).

Decontamination

In general, sodium bicarbonate, sodium carbonate, and sodium hydroxide solutions are effective. Hot soapy water works well in the field, and even hot water is useful, particularly for easily hydrolyzed compounds such as Sarin (1,8).

Visible droplets of agent can be blotted from the skin, while being careful not to abrade the skin (which would enhance penetration by the agent). Then, flush with water and treat with an alkaline substance, such as soap water or calcium carbonate solution in water, or with a chlorine releasing chemical such as bleach solution (concentrated bleach should be diluted; the military dilute by a factor of 10, but less dilution may be preferred if further rinsing can be done afterward). Decontamination is not immediate, but may take from minutes to an hour or longer. The military M258A1 kit reportedly contains two towels, one with a hydroxide and phenol, and the other with chloramine (1,14).

Personal decontamination can be assisted by absorbent powder such as Fuller's earth and by showering. However, it has been found that simple showering tends to spread the chemical agent rather than cleanly remove it. Following an initial shower, chemical treatment and washing with a lateral shower spray using pulsing water jets is more effective, but must be performed within 10 minutes of exposure. The military agencies have developed decontamination kits (1,6,14,18).

CONCLUSIONS

The organosulfur and organophosphorus chemical warfare agents are historically connected to basic organic chemistry research and to research for the discovery of new pesticides. These agents can cause severe, adverse health effects. Superfund and other environmental investigators need to be aware of these effects and to take appropriate safety measures when investigating sites where they may be present.

Environmental monitoring methods are designed for the analysis of pesticides and other hazardous organic compounds in soils, water, and other matrices. The standardized environmental GC and GC/MS methods for organics and pesticides are similar to the research methods that have been reported for the analysis of the chemically related chemical warfare agents and their degradation products. The chemical literature contains results from the considerable research that has been performed on the agents and their degradation products. These research studies can provide a basis for the modification and optimization of environmental methods and the associated quality assurance/quality control practices. Additional safety precautions need to be observed, utilizing the experience of researchers with chemical warfare agents to incorporate safe sample collection and handling procedures, and adequate medical surveillance.

ACKNOWLEDGMENTS

The author would like to thank R. Haun and Retech Division of Lockheed Martin for partial funding for this project.The U.S. EPA, through its Office of Research and Development (ORD) and the Environmental Sciences Division Technology Support Center partially funded this study and the preparation of this article. It has been subjected to the Agency's peer review and has been approved as an EPA publication. Mention of trade names or commercial products does not constitute endorsement or recommendation for use.

BIBLIOGRAPHY

1. E. M. Spiers, *Chemical Warfare,* University of Illinois Press, Chicago, 1986.
2. J. Cookson and J. Nottingham, *A Survey of Chemical and Biological Warfare,* Monthly Review Press, New York, 1971.
3. V. Meyer, *Ber.* **19,** 3259–3266 (1886).
4. W. Lange and G. V. Krueger, *Ber.* **65B,** 1598–1601 (1932).
5. G. Schrader, *Die Entwicklung neuer insektizider Phosphorsaureester,* Verlag, Weinheim/Bergstrasse, 1963.
6. B. C. Saunders, *Some Aspects of the Chemistry and Toxic Action of Organic Compounds Containing Phosphorus and Fluorine,* Cambridge University Press, London, 1957.
7. B. C. Saunders, "Physiological Action of Compounds Containing Fluorine," in *Advances in Fluorine Chemistry,* M. Stacey and coworkers, eds., Vol. 2, Butterworths, Washington, D.C., 1961.
8. J. A. F. Compton, *Military Chemical and Biological Agents,* Telford Press, Caldwell, N.J., 1987.
9. W. H. Donovan and G. R. Famini, *J. Phys. Chem.* **98,** 3669–3674 (1994).
10. J. Smith and J. Schlager, *J. High Resol. Chromatogr.* **19,** 151–154 (1996).

11. W. Stein, S. Moore, and M. Bergmann, *J. Org. Chem.* **11,** 664–674 (1946).
12. S. P. McManus, N. Neamati-Mazraeh, B. A. Hovanes, M. S. Paley, and J. M. Harris, *J. Am. Chem. Soc.* **107,** 3393–3395 (1985).
13. Y.-C. Yang, J. R. Ward, and T. Luteran, *J. Org. Chem.* **51,** 2756–2759 (1986).
14. Y.-C. Yang, J. A. Baker, and J. R. Ward, *Chem. Rev.* **92,** 1729–1743 (1992).
15. National Research Council, Committee on Alternative Chemical Demilitarization Technologies, *Alternative Technologies for the Destruction of Chemical Agents and Munitions,* National Academy Press, Washington, D.C., 1993; U.S. Environmental Protection Agency, *Handbook: Guidance on Setting Permit Conditions and Reporting Trial Burn Results,* EPA/625/6–89/019, Washington, D.C., Jan. 1989.
16. O. B. Helfrich and E. Emmet Reid, *J. Am. Chem. Soc.* **42,** 1208 (1920).
17. N. M. Mozier and J. L. Hoffman, *FASEB J.* **4,** 3329–3333 (1990).
18. S. M. Somani, *Chemical Warfare Agents,* Academic Press, New York, 1992.
19. P. C. Bossle, J. J. Martin, E. W. Sarver, and H. Z. Sommer, *J. Chromatogr.* **283,** 412–416 (1984).
20. A. Casselman, N. Gibson, and R. Bannard, *J. Chromatogr.* **78,** 317–322 (1973).
21. P. Albro and L. Fishbein, *J. Chromatogr.* **46,** 202–203 (1970).
22. R. M. Black, R. J. Clarke, D. B. Cooper, R. W. Read, and D. Utley, *J. Chromatogr.* **637,** 71–80 (1993).
23. P. A. D'Agostino and L. R. Provost, *J. Chromatogr.* **331,** 47–54 (1985); **436,** 399–411 (1988); **589,** 287–294 (1992).
24. R. T. Rewick, M. L. Schumacher, and D. L. Haynes, *Applied Spectros.* **40,** 152–156 (1986).
25. P. Lewis and O. Vela, *Dateline: Los Alamos,* Los Alamos National Laboratory, Los Alamos, N.M., Aug. 1995.
26. S. Sass, T. L. Fisher, R. J. Steger, and G. A. Parker, *J. Chromatogr.* **238,** 445–456 (1982).
27. D. A. Cataldo, M. W. Ligotke, B. D. McVeety, H. Bolton, Jr., and co-workers, *Acute Environmental Toxicity and Persistence of a Chemical Agent Simulant: 2-Chloroethyl Ethyl Sulfide (CEES),* Chemical Research, Development & Engineering Center Report CRDEC-CR-007, U.S. Army Armament Munitions Chemical Command, Aberdeen Proving Ground, Md., Nov. 1988.
28. P. A. D'Agostino and L. R. Provost, *J. Chromatogr.* **598,** 89–95 (1992); **645,** 283–292 (1993); *Biomed. Environ. Mass Spectrom.* **15,** 553–564 (1988); P. A. D'Agostino, L. R. Provost, A. S. Hansen, and G. A. Luoma, *Biomed. Environ. Mass Spectrom.* **18,** 484–491 (1989); P. A. D'Agostino, A. S. Hansen, P. A. Lockwood, and L. R. Provost, J. Chromatogr. **347,** 257–266 (1985); P. A. D'Agostino, L. R. Provost, and P. W. Brooks, *J. Chromatogr.* **541,** 121–130 (1991); P. A. D'Agostino, L. R. Provost, and K. M. Looye, *J. Chromatogr.* **465,** 271–283 (1989).
29. E. R. J. Wils, *Fresenius J. Anal. Chem.* **338,** 22–27 (1990).
30. C. G. Daughton, A. M. Cook, and M. Alexander, *Anal. Chem.* **51,** 1949–1953 (1979).
31. J. A. Tornes and B. A. Johnsen, *J. Chromatogr.* **467,** 129–138 (1989).
32. J. G. Purdon, J. G. Pagotto, and R. K. Miller, *J. Chromatogr.* **475,** 261–272 (1989).
33. B. H. Reitsma and E. S. Yeung, *Anal. Chem.* **59,** 1056–1059 (1987).
34. E. R. J. Wils and A. G. Hulst, *J. Chromatogr.* **454,** 261–272 (1988); *Fresenius J. Anal. Chem.* **342,** 749–758 (1992).
35. P. C. Bossle, D. J. Reutter, and E. W. Sarver, *J. Chromatogr.* **407,** 399–404 (1987).
36. A. Kingery and H. Allen, *Anal. Chem.* **66,** 155–159 (1994).
37. S. A. Oehrle and P. C. Bossle, *J. Chromatogr.* **692,** 247–252 (1995).
38. J.-P. Mercier, Ph. Morin, M. Creux, and A. Tambute, *J. Chromatogr.* **741,** 279–285 (1996).
39. S. M. Nagy, C. Golumbic, W. H. Stein, J. S. Fruton, and M. Bergmann, *J. Gen. Physiol.* **29,** 441–469 (1946).
40. *Toxicological Profile for Mustard 'Gas',* TP-91/22, U.S. Dept. of Health and Human Services, Washington, D.C., Sept. 1992.
41. *Safety Data Sheets,* MDL Information Systems, Inc., San Leandro, Calif. 1996; MSDS, Zeneca Ag Products, Inc., Wilmington, Del., 1993.
42. S. Sass and M. H. Stutz, *J. Chromatogr.* **213,** 173–176 (1981).
43. E. M. Lores, D. E. Bradway, and R. F. Moseman, *Arch. Environ. Health,* **33,** 270–276 (1978).
44. P. Van Voris, D. A. Cataldo, M. W. Ligotke, R. J. Fellos, B. D. McVeety, H. Bolton, S. W. Li, and R. S. Wentsel, *Acute Environmental Toxicity and Persistence of Selected Chemical Agent Simulants,* Chemical Research, Development & Engineering Center Report CRDEC-SP-88013, U.S. Army Armament Munitions Chemical Command, Aberdeen Proving Ground, Md., Nov. 1988.
45. D. F. Heath, *Organophosphorus Poisons,* Pergamon Press, London, 1961.

CHEMICAL WEAPONS CONVENTION: NMR ANALYSIS

MARKKU MESILAAKSO
University of Helsinki
Helsinki, Finland

Nuclear magnetic resonance (NMR) spectroscopy is a powerful analytical technique for determining the molecular structures and dynamics of pure chemicals, even proteins (1–4). In basic form, NMR data are presented as a spectrum. A one-dimensional (1D) NMR spectrum gives information in the form of chemical shifts of resonances and their multiplicities, coupling constants, and resonance areas/intensities, which can be interpreted for determining molecular structure or used like a fingerprint to identify a chemical. Two-dimensional (2D) homonuclear and heteronuclear correlation NMR spectra give information about through-bond interactions between nuclei and can be interpreted to obtain the molecular structure (3,4). Compared with gas chromatography/mass spectrometry (GC/MS), NMR spectroscopy is an insensitive technique, and resonances from other chemicals and background in environmental samples may overlap the resonances of interest. However, in the international interlaboratory comparison (round-robin) tests for verification of chemical disarmament (5–7), and later in the trial and the official international interlaboratory proficiency tests (8), NMR spectroscopy proved itself a useful technique complementary to mass spectrometry (MS) and infrared spectroscopy (IR) in analyzing samples of chemical warfare (CW) agents and related chemicals spiked at trace level. The low sensitivity of NMR was compensated for in part by proper sample preparation and long-term experiments. Today, MS, IR, and NMR, with their numerous technical applications, are the main instrumental techniques for analyzing CW agents. NMR spectroscopy conventionally stands alone, whereas the other techniques usually are combined with gas chromatography (GC) to provide the superior analytical power achieved by combining selectivity with sensitivity. NMR spectroscopy in environmental analysis is particularly useful for analyzing alkylphosphonates (eg, precursors or degradation products of nerve agents), thioalcohols, aminoalcohols, and their salts, but is not limited to these.

CHEMICAL WEAPONS CONVENTION

General Obligations

As a reaction to the horrors of World War I, an international convention was signed on June 17, 1925, for the Prohibition of the Use in War of Asphyxiating, Poisonous or Other Gases, and of Bacteriological Methods of Warfare (the Geneva Protocol of 1925). This Convention, which prohibited the use, but not the owning and developing of chemical weapons, was unfortunately not respected and was openly violated. The time for a new convention came only seven decades later. The Convention on the Prohibition of the Development, Production, Stockpiling and Use of Chemical Weapons and on their Destruction (the Chemical Weapons Convention, CWC) was signed on January 13, 1993 (9) and entered in force on April 29, 1997. "Each State Party to the Convention undertakes under any circumstances never to use, develop, produce, otherwise acquire, stockpile or retain chemical weapons, or transfer, directly or indirectly, chemical weapons to anyone." Further, "Each State Party undertakes to destroy chemical weapons it owns or possesses" and "shall submit declarations" as required by the CWC (15). The State Parties also recognize that "the Convention reaffirms principles and objectives of and obligations assumed under the Geneva Protocol of 1925."

Organization and Verification

The Organisation for the Prohibition of Chemical Weapons (OPCW) has been established in the Hague, the Netherlands, to "achieve the objectives and purposes of the CWC, to ensure the implementation of its provisions including those for international verification of compliance with it, and to provide a forum for consultation and cooperation among States Parties." Declarations of existing chemical weapons are to be confirmed through on-site inspection. It is the responsibility of the inspectors, inter alia, to "verify the quantity and identity of chemicals and the types and number of munitions, devices, and other equipment." Where possible, samples should be analyzed on-site, but if necessary, they can be transferred for analysis off-site at designated laboratories.

The Director-General of the Technical Secretariat of the OPCW is to "establish a stringent regime governing the collection, handling, transport, and analysis of samples in verification, to certify the laboratories designated to perform different types of analysis, to oversee the standardization of equipment and procedures at these designated laboratories and the mobile analytical equipment and procedures, and to monitor quality control and overall standards relative to certifying the designated laboratories, mobile equipment, and procedures. Selection will be made among the designated laboratories to perform analytical or other functions relative to specific investigations" (9).

VERIFICATION OF CHEMICAL WEAPONS

Definition of Chemical Weapon

Chemical weapons are defined as "(*1*) toxic chemicals and their precursors, except where intended for purposes not prohibited under the CWC, as long as the types and quantities are consistent with such purposes; (*2*) munitions and devices, specifically designed to cause death or other harm through the toxic properties of those toxic chemicals specified in (*1*), which would be released as a result of the employment of such munitions and devices; (*3*) any equipment specifically designed for use directly in connection with the employment of munitions and devices specified in (*2*)" (9). A toxic chemical means "any chemical which, through its chemical action on life processes, can cause death, temporary incapacitation, or permanent harm to humans or animals." CW agent is a synonym for a chemical that acts as a chemical weapon. *Warning:* in view of their toxicity, the chemicals classified as chemical weapons should be handled only in specialized laboratories.

Scheduled Chemicals

The toxic chemicals identified for the application of verification measures are listed in Schedules 1–3 below contained in the Annex on Chemicals of the CWC (9). 3–Quinuclidinyl benzilate (Schedule 2.A.3) is subject to special thresholds for declaration and verification. By way of example, Schedule 1 includes chemicals developed, produced, stockpiled, or used as chemical weapons as defined above. Altogether the Schedules list 43 specified chemicals or chemical groups of toxic chemicals and their precursors. An idea of the total number of chemicals to be verified can be obtained by considering how many different chemicals are included in group A.1 of Schedule 1. The Schedules contain mainly organic chemicals with different chemical and physical properties, which are neutral chemicals, acids, bases, volatiles, and nonvolatiles, where heteroatoms P, F, S, Cl, N, and O occur frequently (see eg, ref. 10 for information about CW agents). Riot control agents are not included in the Schedules, but their use as weapons of warfare is prohibited.

SCHEDULE 1

	A. TOXIC CHEMICALS:	[CAS Reg. No]
(*1*)	*O*-Alkyl ($\leq C_{10}$, incl. cycloalkyl) alkyl (Me, Et, *n*-Pr or *i*-Pr)-phosphonofluoridates, eg, Sarin: *O*-Isopropyl methylphosphonofluoridates	*[107-44-8]*
	Soman: *O*-Pinacolyl methylphosphonofluoridate	*[96-64-0]*
(*2*)	*O*-Alkyl ($\leq$C10, incl. cycloalkyl) *N,N*-dialkyl (Me, Et, *n*-Pr or *i*-Pr) phosphoramidocyanidates, eg, Tabun: *O*-Ethyl *N,N*-dimethyl phosphoramidocyanidate	*[77-81-6]*
(*3*)	*O*-Alkyl (H or $\leq$C10, incl. cycloalkyl) *S*-2-dialkyl (Me, Et, *n*-Pr or *i*-Pr)-aminoethyl alkyl (Me, Et, *n*-Pr or *i*-Pr) phosphonothiolates and corresponding alkylated or protonated salts, eg, VX: *O*-Ethyl *S*-2-diisopropylaminoethyl methyl phosphonothiolate	*[50782-69-9]*
(*4*)	Sulfur mustards:	
	2-Chloroethylchloromethylsulfide	*[2625-76-5]*
	Mustard gas: bis(2-chloroethyl)sulfide	*[505-60-2]*
	bis(2-chloroethylthio)methane	*[63869-13-6]*

Sesquimustard: 1,2-bis(2-chloroethylthio)ethane [3563-36-8]
1,3-bis(2-chloroethylthio)-*n*-propane [63905-10-2]
1,4-bis(2-chloroethylthio)-*n*-butane [142868-93-7]
1,5-bis(2-chloroethylthio)-*n*-pentane [142868-94-8]
bis(2-chloroethylthiomethyl)ether [63918-90-1]
O-Mustard: bis(2-chloroethylthioethyl)ether [63918-89-8]

(*5*) Lewisites:
Lewisite 1: 2-chlorovinyldichloroarsine [541-25-3]
Lewisite 2: bis(2-chlorovinyl)chloroarsine [40334-69-8]
Lewisite 3: tris(2-chlorovinyl)arsine [40334-70-1]

(*6*) Nitrogen mustards:
HN1: Bis(2-chloroethyl)ethylamine [538-07-8]
HN2: Bis(2-chloroethyl)methylamine [51-75-2]
HN3:Tris(2-chloroethyl)amine [555-77-1]

(*7*) Saxitoxin [35523-89-8]

(*8*) Ricin [9009-86-3]

B. PRECURSORS:

(*9*) Alkyl (Me, Et, *n*-Pr or i-Pr) phosphonyl-difluorides
eg, DF: Methylphosphonyldifluoride [676-99-3]

(*10*) *O*-Alkyl (H or ≤C10, incl. cycloalkyl) *O*-2-dialkyl (Me, Et, *n*-Pr or i-Pr)-aminoethyl alkyl (Me, Et, *n*-Pr or i-Pr) phosphonites and corresponding alkylated or protonated salts eg, QL: *O*-Ethyl *O*-2-diisopropylaminoethyl methylphosphonite [57856-11-8]

(*11*) Chlorosarin: *O*-isopropyl methylphosphonochloridate [1445-76-7]

(*12*) Chlorosoman: *O*-Pinacolyl methylphosphonochloridate [7040-57-5]

SCHEDULE 2

A. TOXIC CHEMICALS: [CAS Reg. No]

(*1*) Amiton: *O,O*-Diethyl *S*-[2-(diethylamino)-ethyl] phosphorothiolate and corresponding alkylated or protonated salts [78-53-5]

(*2*) PFIB: 1,1,3,3,3-pentafluoro-2-(trifluoromethyl)-1-propene [382-21-8]

(*3*) BZ: 3-quinuclidinyl benzilate (*) [6581-06-2]

(*4*) Chemicals, except for those listed in Schedule 1, containing a phosphorus atom to which is bonded one methyl, ethyl or propyl (normal or iso) group but not further carbon atoms, eg, Methylphosphonyl dichloride [676-97-1]
Dimethyl methylphosphonate [756-79-6]
Exemption: Fonofos: *O*-Ethyl *S*-phenylethyl-phosphonothiolothionate [944-22-9]

(*5*) *N,N*-Dialkyl (Me, Et, *n*-Pr or i-Pr) phosphoramidic dihalides

(*6*) Dialkyl (Me, Et, *n*-Pr or i-Pr) *N,N*-dialkyl (Me, Et, *n*-Pr or i-Pr)-phosphoramidates

(*7*) Arsenic trichloride [7784-34-1]

(*8*) 2,2-Diphenyl-2-hydroxyacetic acid [76-93-7]

(*9*) Quinuclidin-3-ol [1619-34-7]

(*10*) *N,N*-Dialkyl (Me, Et, *n*-Pr or i-Pr) aminoethyl-2-chlorides and corresponding protonated salts

(*11*) *N,N*-Dialkyl (Me, Et, *n*-Pr or i-Pr) amino-ethane-2-ols and corresponding protonated salts
Exemptions: [108-01-0]
N,N-Dimethylaminoethanol and corresponding protonated salts
N,N-Diethylaminoethanol and corresponding protonated salts [100-37-8]

(*12*) *N,N*-Dialkyl (Me, Et, *n*-Pr or i-Pr) aminoethane-2-thiols and corresponding protonated salts

(*13*) Thiodiglycol: Bis(2-hydroxyethyl)sulfide [111-48-8]

(*14*) Pinacolyl alcohol: 3,3-Dimethylbutan-2-ol [464-07-3]

SCHEDULE 3

A. TOXIC CHEMICALS:

(*1*) Phosgene: Carbonyl dichloride [75-44-5]
(*2*) Cyanogen chloride [506-77-4]
(*3*) Hydrogen cyanide [74-90-8]
(*4*) Chloropicrin: Trichloronitromethane [76-06-2]

B. PRECURSORS:

(*5*) Phosphorus oxychloride [10025-87-3]
(*6*) Phosphorus trichloride [7719-12-2]
(*7*) Phosphorus pentachloride [10026-13-8]
(*8*) Trimethyl phosphite [121-45-9]
(*9*) Triethyl phosphite [122-52-1]
(*10*) Dimethyl phosphite [868-85-9]
(*11*) Diethyl phosphite [762-04-9]
(*12*) Sulfur monochloride [10025-67-9]
(*13*) Sulfur dichloride [10545-99-0]
(*14*) Thionyl chloride [7719-09-7]
(*15*) Ethyldiethanolamine [139-87-7]
(*16*) Methyldiethanolamine [105-59-9]
(*17*) Triethanolamine [102-71-6]

Recommended Operating Procedures (ROPs)

After more than ten years of methods development for identifying CW agents (11–13), the Finnish Research Project on the Verification of Chemical Disarmament (hereafter, the Institute) published Standard Operating Procedures (SOPs) in 1988 and 1989 for the Verification of Chemical Disarmament (14). In 1989 the first international interlaboratory comparison (round-robin) test for verification of chemical disarmament took place (5). At the meeting of 10 participating laboratories (countries), the reports and results were discussed, and the SOPs were renamed ROPs (Recommended Operating Procedures). This test and three subsequent tests (5) were aimed at testing and developing the existing procedures, the ROPs. Through sustained international collaboration, the Recommended Operating Procedures for Sampling and Analysis in the Verification of Chemical Disarmament were upgraded annually (5,14). The OPCW/PTS coordinated its first international interlaboratory comparison test in 1994 (7), three trial interlaboratory proficiency tests in 1995, and two official interlaboratory proficiency tests in 1996. The purpose of the trial proficiency tests was to train and test laboratory evaluation criteria. In the official interlaboratory proficiency tests, the aim is selection of the designated laboratories (see previous section). In all these tests the ROPs were widely and successfully applied. The test participants were also encouraged to use and report alternative procedures if these were considered superior.

The ROPs (14) give instructions for sampling, sample preparation, and instrumental analysis. In the case of specific investigations defined by the OPCW, only the sample preparation and subsequent analysis of the samples, not the sample collection, is carried out by the off-site laboratory. Rather than being the best choice for a single analyte or matrix type, ROPs provide an optimal approach for all in an unknown situation.

Analytical Strategy in Off-Site Laboratory

In verifying the absence or presence of chemicals related to the CWC, the analysis of a specified sample in an off-site laboratory includes preparing and analyzing samples by different chromatographic and spectrometric techniques. In general, the chromatographic techniques are used in screening whereas the spectrometric techniques are used in a complementary manner to obtain unambiguous identification of scheduled chemicals. Analyses are mostly qualitative (identification). Work is conducted according to the ROPs or other documented procedures of demonstrated performance. Laboratories worldwide have very different instrumentation, which means that each laboratory requires its own analytical procedure. The procedure employed at the Institute is schematically shown in Figure 1 and described in the following.

After arrival, check-up, and coding, sample and blank (if available, represents sample background) are divided into portions, and sample preparation is begun. Quality control (QC) samples may be prepared. Samples are prepared for the analytical techniques relying on GC. Water and aqueous extracts are prepared straightforwardly for analysis by liquid chromatography/mass spectrometry (LC/MS) and capillary zone electrophoresis (CZE). Samples for NMR are normally prepared to obtain 5–10-fold concentration. In the next stage screening of the samples for analytes is done, if considered necessary, by different analytical techniques.

The samples are always screened by GC. Screening by GC is done with element (nitrogen, phosphorus, sulfur)-specific detectors (NPD and FPD) and retention index monitoring. This is particularly advantageous, because almost all of the scheduled chemicals are then detectable without interference from hydrocarbons. Results from screening by GC may be useful in analyses carried out by techniques relying on GC. Other screening techniques include ^{19}F and/or ^{31}P{^{1}H} NMR spectroscopy (^{31}P{^{1}H} denotes ^{31}P observation with simultaneous broadband decoupling of protons ^{1}H) and capillary zone electrophoresis (CZE). Screening by NMR reveals phosphorus- and fluorine-containing chemicals, and screening by CZE, for instance, shows the presence of nonvolatile alkylphosphonic acids and their monoesters. One advantage of CZE over conventional GC is the ability to separate and detect polar analytes without sample pretreatment and derivatization. The boundary between screening and the main instrumental analysis carried out by NMR is not always clear-cut.

Analysis for identifying known, volatile chemicals, which are fully resolved from the matrix is carried out with low-resolution gas chromatography/mass spectrometry: electron impact ionization (GC/EI-MS) provides a spectrum characteristic for chemical, whereas chemical ionization (GC/CI-MS) gives the molecular weight (Fig. 1). Strong matrix interference is solved by using gas chromatography/tandem mass spectrometry (GC/MS/MS) or gas chromatography/high-resolution mass spectrometry (GC/HRMS). Polar and nonvolatile chemicals in aqueous solutions or extracts are analyzed by liquid chromatography/mass spectrometry (LC/MS) for confirmatory information. Gas chromatography/Fourier transform-infrared spectroscopy (GC/FTIR) offers pure and characteristic spectra useful for identification. ^{1}H, ^{13}C{^{1}H}, ^{19}F, ^{31}P{^{1}H}, and ^{31}P NMR

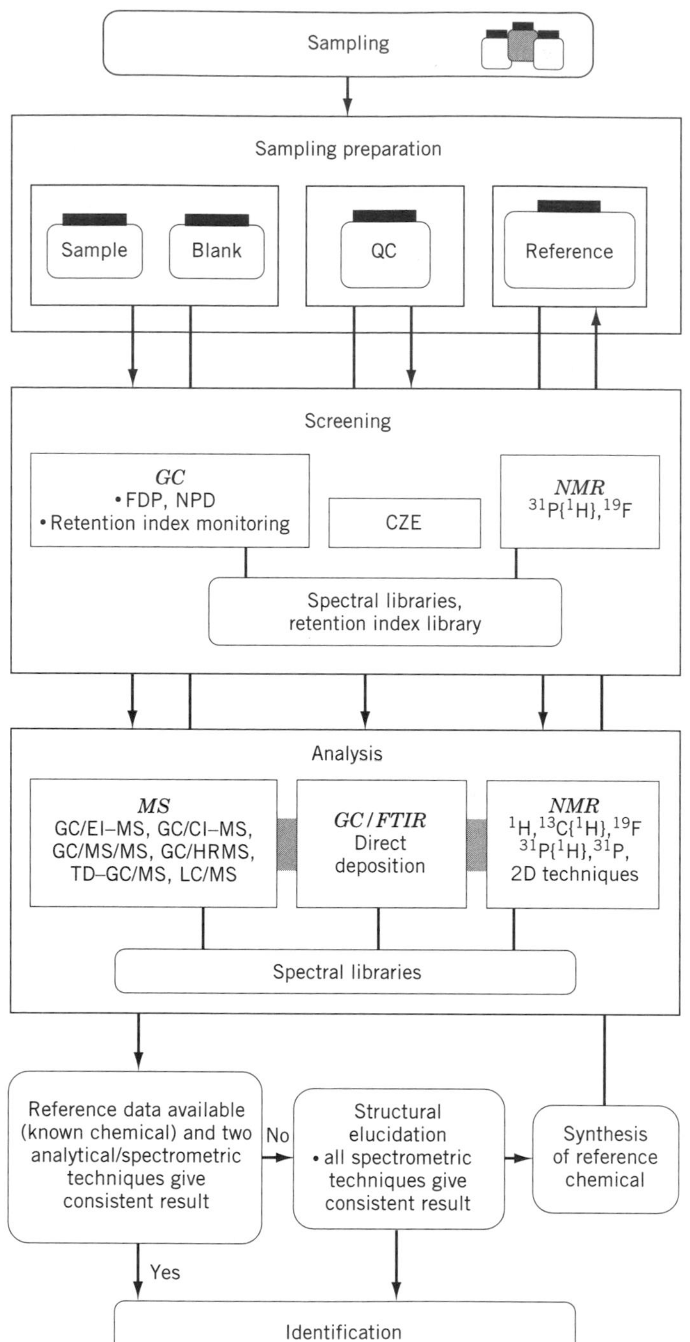

Figure 1. A general procedure for identifying chemical warfare agents in a test sample.

spectra are also characteristic of a chemical, but in ^{1}H NMR, sample background may cause problems of resonance overlapping. In the case of MS, IR, and NMR, the spectrum recorded from a sample is compared with library spectra, and, upon fulfillment of certain requirements, an identification is obtained. In-house MS, IR, and NMR spectral libraries and retention index libraries for GC of chemicals related to the Chemical Weapons Convention have been constructed as follows. Mass spectrometry: GC/EI–MS, GC/CI–MS(methane), GC/CI–MS(isobutane), GC/CI–MS(ammonia), GC/MS/MS(isobutane), and GC/MS/MS(ammonia); GC/FTIR spectroscopy: condensed phase (Tracer); NMR spectroscopy: ^{1}H, $^{13}C\{^{1}H\}$, ^{19}F, $^{31}P\{^{1}H\}$, ^{31}P; and gas chromatography: RI against M-series; RI against C-series.

According to the rules of OPCW/PTS, Criteria for Acceptable Performance of Laboratories in Proficiency Testing, unambiguous identification of a chemical is obtained if two analytical, preferably spectrometric techniques, give consistent results. Similar identifications obtained by GC/EI-MS and GC/CI-MS constitute an unambiguous result.

If preliminary results indicate a scheduled chemical which is not in the spectral libraries, it is structurally elucidated. The GC/EI–MS and GC/CI–MS spectra are interpreted. GC/MS/MS assists in interpretation by solving the fragmentation of ions of interest. Elemental composition of the molecular and fragment ions is determined with GC/HRMS. Interpretation of the GC/FTIR spectrum provides information about functional groups, the different types of 1D and 2D NMR spectral information about type and number of nuclei, and their relations in the molecule. A consistent result from all spectrometric techniques is needed for unambiguous identification. The most reliable confirmation of all is obtained if the suspected chemical is synthesized as a reference chemical and the spectra of the sample and the synthesized chemical are the same.

The report of the analysis should include all analytical data (chromatographic and spectrometric) supporting the identifications made, describe the sample preparation and analytical methods (or make reference to ROPs, SOPs, or other procedures) in detail, and give information about the chemical identified.

NMR SPECTRUM

1D NMR spectra are used to identify chemicals and in structural elucidation. Because, the resonance area under certain conditions is directly proportional to the number of nuclei, NMR spectroscopy is suitable for quantitative work (7).

The 1D ^{1}H NMR spectrum provides information on the number and type of protons (number of resonances, area of resonance, chemical shift δ_H[ppm]) in the molecule, and in the form of couplings (J[Hz]), reveals other magnetic atoms (eg, ^{1}H, ^{31}P, ^{19}F) one to four bonds from a hydrogen atom (here and below, a maximum distance of four bonds between the coupling nuclei should be considered approximate). The $^{13}C\{^{1}H\}$ NMR spectrum provides information on the number and type of carbons (number of resonances, δ_C) in the molecule and, in the form of couplings, may reveal other magnetic atoms (eg, ^{31}P, ^{19}F) one to four bonds from a carbon atom. The notation $^{13}C\{^{1}H\}$ means that, when ^{13}C is observed, protons are irradiated so that the interaction of protons and carbons disappears; however, interaction with other magnetic nuclei remains. The ^{19}F NMR spectrum provides information on the number and type of fluorine atoms (number of resonances, δ_F) in the molecule and, in the form of couplings, reveals other magnetic atoms (eg, ^{1}H, ^{31}P) one to four bonds from a fluorine atom. The $^{31}P\{^{1}H\}$ NMR spectrum provides information on the number and type of phosphorus atoms (number of resonances, δ_P) in the molecule and, in the form of couplings, reveals other magnetic atoms (eg, ^{31}P, ^{19}F) one to four bonds from a phosphorus atom. The ^{31}P NMR spectrum provides information on the number and type of phosphorus atoms (number of resonances, δ_P) in the molecule and, in the form of couplings, reveals other magnetic atoms (eg, ^{1}H, ^{31}P, ^{19}F) one to four bonds from a phosphorus atom.

Identifying Chemicals. Identification by NMR of a chemical relevant to the CWC is made by referring to spectra (^{1}H, $^{13}C\{^{1}H\}$, ^{19}F, $^{31}P\{^{1}H\}$, and/or ^{31}P) included in a spectral library, or comparing with a spectrum of the authentic reference chemical or spiking the test sample with an authentic reference chemical (14). A sample spectrum acceptable for identification should be a pure spectrum, ie, resonances of other chemicals do not overlap with resonances of the identified chemical; partial overlapping is acceptable if the resonances of the identified chemical are after all assigned. For the identification, the NMR spectral parameters for a chemical must also agree within certain limits with those of the library or the authentic reference chemical. If the sample of the chemical to be identified has been prepared so as to correspond in solvent, concentration, pH, and chemical shift references with the chemical in the library (or authentic chemical), the values of δ_H, δ_C, and δ_F should agree with the corresponding values in the library (or authentic chemical) within limits of ± 0.1 ppm, and those of δ_P within ± 1 (δ is a symbol for chemical shift in ppm). The values of coupling constants $^{n}J_{H,H}$, $^{n}J_{H,F}$, $^{n}J_{H,P}$, $^{n}J_{C,P}$, $^{n}J_{C,F}$, and $^{n}J_{P,F}$ should agree within limits of ± 0.5 Hz (in $^{n}J_{H,H}$, n denotes the number of chemical bonds between coupled nuclei, which are given in the subscript; J is a symbol for the coupling constant always expressed in Hz). The resonance patterns (multiplicity) and their intensities must also match. Normally, when the condition of a test sample and reference are comparable, the differences in the NMR parameters (sample vs reference) are much smaller than suggested above; sometimes, within reasonable limits, even greater differences are acceptable. Identification by spiking a sample with the suspected chemical requires exact overlapping of the resonances. Where insufficient data are obtained, eg, because of overlapping, the resonances revealed are useful to support identifications based on other analytical techniques.

NMR Spectral Parameters of Selected CWC-Related Chemicals. The NMR spectral library the Institute of chemicals related to the Chemical Weapons Convention consists of the 400 MHz ^{1}H, 376 MHz ^{19}F, 162 MHz $^{31}P\{^{1}H\}$ and ^{31}P, and 100 MHz $^{13}C\{^{1}H\}$ NMR spectra in paper format. Each library page contains the spectrum of the whole spectral range (eg, ^{1}H NMR spectrum from -0.5 to 10.0 ppm), sufficient expansions to see the spectral details (see Fig. 3), the name, the Chemical Abstracts Service (CAS) registry number, the molecular structure of the chemical, the NMR spectral parameters (chemical shifts and coupling constants), the sample conditions, the experimental conditions, the data processing conditions, and information on the spectrum quality and traceability. Many of the chemicals have been recorded in $CDCl_3$, CD_2Cl_2, and $(CD_3)_2CO$, or at different pHs in D_2O.

Table 1. NMR Spectral Parameters of Selected CWC-Related Chemicals

Trivial Name; NATO Code[a]; IUPAC Name [CAS][b] (Schedule)[c]

Chemical shift δ[ppm][d](*multiplicity*),[e] coupling constant J[Hz];[f] concentration and solvent, pH; chemical shift reference(s); remarks.

Sarin; GB; Isopropyl methylphosphonofluoridate [107-44-8] (1.A.1)

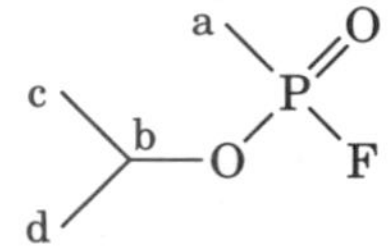

$\delta_{Ha} = 1.623(dd)$, $\delta_{Hb} = 4.908(dspd)$, $\delta_{Hc} = 1.375(d)$, $\delta_{Hd} = 1.382(d)$, $^2J_{Ha,P} = 18.7$, $^3J_{Ha,F} = 5.7$, $^3J_{Hb,(Hc,Hd)} = 6.2$, $^3J_{Hb,P} = 7.7$, $^4J_{Hb,F} = 0.4$, $\delta_{Ca} = 10.47(dd)$, $\delta_{Cb} = 72.73(d)$, $\delta_{Cc} = 23.840(dd)$, $\delta_{Cd} = 23.845(d)$, $^1J_{Ca,P} = 150.1$, $^2J_{Cb,F} = 27.5$, $^2J_{Cb,P} = 7.0$, $^3J_{Cc,P} = 5.9$, $^4J_{Cd,F} = 1.7$, $^3J_{Cd,P} = 3.5$, $\delta_F = -57.43(dq)$, $^1J_{F,P} = 1046.3$, $\delta_P = 29.15(ddq)$; 14.1 mg/1.1 mL $CDCl_3$; TMS, $CFCl_3$, ext. H_3PO_4.

Soman; GD; 1,2,2-Trimethylpropyl methylphosphonofluoridate [96-64-0] (1.A.1)

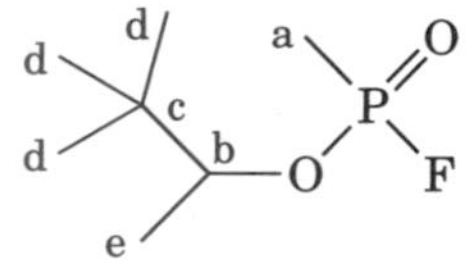

More abundant diastereomer: $\delta_{Ha} = 1.633(dd)$, $\delta_{Hb} = 4.487(ddq)$, $\delta_{Hd} = 0.936(s)$, $\delta_{He} = 1.317(d)$, $^2J_{Ha,P} = 18.7$, $^3J_{Ha,F} = 5.7$, $^3J_{Hb,P} = 7.9$, $^4J_{Hb,F} = 0.5$, $^3J_{Hb,He} = 6.4$, $\delta_{Ca} = 10.05(dd)$, $\delta_{Cb} = 82.23(d)$, $\delta_{Cc} = 34.89(dd)$, $\delta_{Cd} = 25.43(s)$, $\delta_{Ce} = 16.93(s)$, $^1J_{Ca,P} = 151.3$, $^2J_{Cb,F} = 28.2$, $^2J_{Cb,P} = 8.1$, $^3J_{Cc,P} = 7.2$, $^4J_{Cc,F} = 1.5$, $^3J_{Ce,P} = 0.9$, $\delta_F = -55.48(dq)$, $^1J_{F,P} = 1047.2$, $\delta_P = 29.75(ddq)$.

Less abundant diastereomer: $\delta_{Ha} = 1.636(dd)$, $\delta_{Hb} = 4.438(ddq)$, $\delta_{Hd} = 0.936(s)$, $\delta_{He} = 1.328(d)$, $^2J_{Ha,P} = 18.7$, $^3J_{Ha,F} = 5.7$, $^3J_{Hb,P} = 8.9$, $^3J_{Hb,He} = 6.4$, $\delta_{Ca} = 10.63(dd)$, $\delta_{Cb} = 83.73(d)$, $\delta_{Cc} = 34.80(dd)$, $\delta_{Cd} = 25.43(s)$, $\delta_{Ce} = 17.06(s)$, $^1J_{Ca,P} = 150.7$, $^2J_{Cb,F} = 28.5$, $^2J_{Cb,P} = 7.9$, $^3J_{Cc,P} = 5.5$, $^3J_{Ce,P} = 2.0$, $^4J_{Ce,F} = 2.0$, $\delta_F = -58.33(dq)$, $^1J_{F,P} = 1046.4$, $\delta_P = 28.73(ddq)$; 75.5 mg/1.1 mL $CDCl_3$; TMS, $CFCl_3$, ext. H_3PO_4; molar ratio of diastereomers: approx. 55/45.

Tabun; GA; *O*-Ethyl *N,N*-dimethyl phosphoramidocyanidate [77-81-6] (1.A.2)

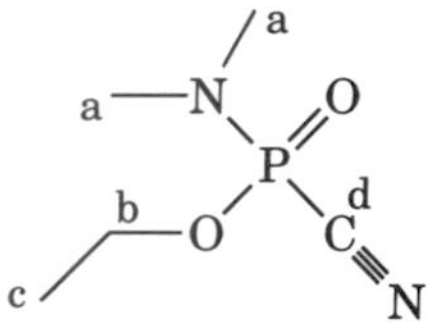

$\delta_{Ha} = 2.779(d)$, $\delta_{Hb} = 4.261(d)$, $\delta_{Hb'} = 4.269(d)$, $\delta_{Hc} = 1.433(dt)$, $^3J_{Ha,P} = 11.3$, $^2J_{Hb,Hb'} = -10.2$, $^3J_{(Hb,Hb'),Hc} = 7.1$, $^3J_{Hb,P} = 8.6$, $^3J_{Hb',P} = 9.8$, $^4J_{Hc,P} = 0.7$, $\delta_{Ca} = 35.55$, $\delta_{Cb} = 63.93$, $\delta_{Cc} = 15.96$, $\delta_{Cd} = 115.01$, $^2J_{Ca,P} = 4.9$, $^2J_{Cb,P} = 5.5$, $^3J_{Cc,P} = 6.8$, $\delta_P = -9.02$; 16.4 mg/1.1 mL $CDCl_3$; TMS, ext. H_3PO_4.

VX; *O*-Ethyl *S*-2-diisopropylaminoethyl methylphosphonothioate [50782-69-9] (1.A.3)

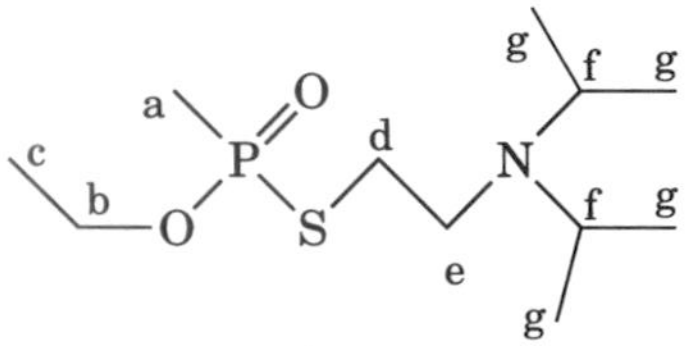

$\delta_{Ha} = 1.783(d)$, $\delta_{Hb} = 4.094(ddq)$, $\delta_{Hb'} = 4.187(ddq)$, $\delta_{Hc} = 1.347(t)$, $\delta_{Hd} = 2.799(m)$, $\delta_{Hd'} = 2.834(m)$, $\delta_{He} = 2.684(m)$, $\delta_{He'} = 2.703(m)$, $\delta_{Hf} = 3.011(sp)$, $\delta_{Hg} = 1.018(d)$, $^2J_{Ha,P} = 15.6$, $^3J_{Hb,P} = 9.3$, $^2J_{Hb,Hb'} = -10.2$, $^3J_{(Hb,Hb'),Hc} = 7.1$, $^3J_{Hb'P} = 8.4$, $^3J_{Hd,P} = 12.4$, $^2J_{Hd,Hd'} = -12.5$, $^3J_{Hd,He} = 10.0$, $^3J_{Hd,He'} = 5.6$, $^3J_{Hd',P} = 12.6$, $^3J_{Hd',He} = 5.7$, $^3J_{Hd',He} = 10.0$, $^2J_{He,He'} = -14.1$, $^3J_{Hf,Hg} = 6.6$, $\delta_{Ca} = 20.06(d)$, $\delta_{Cb} = 61.15(d)$, $\delta_{Cc} = 16.26(d)$, $\delta_{Cd} = 31.62(d)$, $\delta_{Ce} = 46.62(d)$, $\delta_{Cf} = 48.92(s)$, $\delta_{Cg} = 20.94(s)$, $20.99(s)$, $^1J_{Ca,P} = 110.4$, $^2J_{Cb,P} = 6.9$, $^3J_{Cc,P} = 7.1$, $^2J_{Cd,P} = 2.8$, $^4J_{Ce,P} = 4.6$, $\delta_P = 54.80(m)$; 50 μL/0.8 mL $CDCl_3$; TMS, ext. H_3PO_4; 2 μL of tributylamine was added.

Mustard gas; HD; Bis(2-chloroethyl)sulfide [505-60-2] (1.A.4)

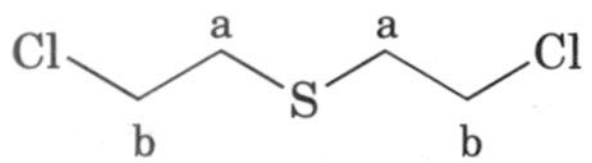

$\delta_{Ha} = 2.923(XX')$, $\delta_{Hb} = 3.645(AA')$, $^2J_{Ha,Ha'} = -10.9$, $^2J_{Hb,Hb'} = -13.9$, $^3J_{Ha,Hb} = 9.5$, $^3J_{Ha,Hb'} = 5.9$, $\delta_{Ca} = 34.61$, $\delta_{Cb} = 43.04$; 17.0 mg/1.0 mL $CDCl_3$; TMS.

O-Mustard; T; Bis(2-chloroethylthioethyl)ether [63918-89-8] (1.A.4)

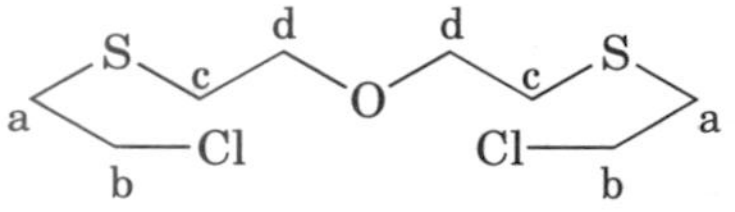

$\delta_{Ha} = 2.924(XX')$, $\delta_{Hb} = 3.653(AA')$, $\delta_{Hc} = 2.762(t)$, $\delta_{Hd} = 3.653(t)$, $^2J_{Ha,Ha'} = -10.8$, $^2J_{Hb,Hb'} = -13.8$, $^3J_{Ha,Hb} = 9.8$, $^3J_{Ha,Hb'} = 5.8$, $^3J_{Hc,Hd} = 6.4$, $\delta_{Ca} = 34.82$, $\delta_{Cb} = 43.16$, $\delta_{Cc} = 32.04$, $\delta_{Cd} = 71.08$; 26.2 mg/1.0 mL $CDCl_3$; TMS.

Lewisite-1; L; 2-Chlorovinyldichloroarsine [541-25-3] (1.A.5)

Table 1. NMR Spectral Parameters of Selected CWC-Related Chemicals *(continued)*

Trivial Name; NATO Code[a]; IUPAC Name [CAS][b] (Schedule)[c]
$\delta_{Ha} = 7.146(d)$, $\delta_{Hb} = 6.936(d)$, $^3J_{Ha,Hb} = 14.8$, $\delta_{Ca} = 139.27$, $\delta_{Cb} = 133.17$; 22.3 mg/1.1 mL $CDCl_3$; TMS.
Nitrogen mustard one; HN-1; Bis(2-chloroethyl)ethylamine [538-07-8] (1.A.6)
$\delta_{Ha} = 2.865(XX')$, $\delta_{Hb} = 3.507(XX')$, $\delta_{Hc} = 2.654(q)$, $\delta_{Hd} = 1.051(t)$, $^3J_{Hc,Hd} = 7.1$, $\delta_{Ca} = 55.97$, $\delta_{Cb} = 42.08$, $\delta_{Cc} = 48.52$, $\delta_{Cd} = 12.35$; 30 μL/0.8 mL $CDCl_3$; TMS.
DF; Methylphosphonyldifluoride [676-99-3] (1.B.9)
$\delta_{Ha} = 1.886(dt)$, $^2J_{Ha,P} = 19.4$, $^3J_{Ha,F} = 5.8$, $\delta_{Ca} = 8.49(dt)$, $^1J_{Ca,P} = 148.4$, $^2J_{Cb,F} = 22.5$, $\delta_F = -59.57(dq)$, $^1J_{F,P} = 1109.8$, $\delta_P = 24.47(tq)$; 20μL/1.0 mL $CDCl_3$; TMS, $CFCl_3$, ext. H_3PO_4.
Chlorosarin; Isopropyl methylphosphonochloridate [1445-76-7] (1.B.11)
$\delta_{Ha} = 1.955(d)$, $\delta_{Hb} = 4.942(dsp)$, $\delta_{Hc/Hd} = 1.399(d)$, $\delta_{Hd/He} = 1.413(d)$, $^2J_{Ha,P} = 17.6$, $^3J_{Hb,(Hc,Hd)} = 6,2$, $^3J_{Hb,P} = 10.5$, $\delta_{Ca} = 20.71(d)$, $\delta_{Cb} = 73.35(d)$, $\delta_{Cc/Cd} = 23.39(d)$, $\delta_{Cd/Cc} = 24.01(d)$, $^1J_{Ca,P} = 130.8$, $^2J_{Cb,P} = 8.0$, $^3J_{Cc/Cd,P} = 5.0$, $^3J_{Cd/Cc,P} = 4.7$, $\delta_P = 39.21(dq)$; 50μL/0.8 mL $CDCl_3$; TMS, ext. H_3PO_4.
Isopropyl methylphosphonate [1832-54-8] (2.B.4)
$\delta_{Ha} = 1.480(d)$, $\delta_{Hb} = 4.671(dsp)$, $\delta_{Hc} = \delta_{Hd} = 1.334(d)$, $^2J_{Ha,P} = 17.9$, $^3J_{Hb,(Hc,Hd)} = 6.2$, $^3J_{Hb,P} = 8.3$, $\delta_{Ca} = 12.37(d)$, $\delta_{Cb} = 70.08(d)$, $\delta_{Cc} = \delta_{Cd} = 23.98(d)$, $^1J_{Ca,P} = 148.2$, $^2J_{Cb,P} = 6.5$, $^3J_{(Cc,Cd),P} = 4.4$, $\delta_P = 33.40(dq)$; 5 μL/0.8 mL $CDCl_3$; TMS, ext. H_3PO_4.
$\delta_{Ha} = 1.555(d)$, $\delta_{Hb} = 4.661(dsp)$, $\delta_{Hc} = \delta_{Hd} = 1.326(d)$, $^2J_{Ha,P} = 17.4$, $^3J_{Hb,(Hc,Hd)} = 6.2$, $^3J_{Hb,P} = 8.2$, $\delta_{Ca} = 13.80(d)$, $\delta_{Cb} = 74.46(d)$, $\delta_{Cc} = \delta_{Cd} = 25.84(d)$, $^1J_{Ca,P} = 139.3$, $^2J_{Cb,P} = 6.1$, $^3J_{(Cc,Cd),P} = 4.2$, $\delta_P = 32.21(dq)$; 10 μL/0.8 mL D_2O, pH 0.0; TSPSA, ext. H_3PO_4.
$\delta_{Ha} = 1.275(d)$, $\delta_{Hb} = 4.444(dsp)$, $\delta_{Hc} = \delta_{Hd} = 1.253(d)$, $^2J_{Ha,P} = 16.4$, $^3J_{Hb,(Hc,Hd)} = 6.2$, $^3J_{Hb,P} = 8.6$, $\delta_{Ca} = 14.98(d)$, $\delta_{Cb} = 71.26(d)$, $\delta_{Cc} = \delta_{Cd} = 26.16(d)$, $^1J_{Ca,P} = 137.1$, $^2J_{Cb,P} = 5.5$, $^3J_{(Cc,Cd),P} = 3.7$, $\delta_P = 26.22(dq)$; 10 μL/0.8 mL D_2O, pH 7-8; TSPSA, ext. H_3PO_4.
$\delta_{Ha} = 1.275(d)$, $\delta_{Hb} = 4.442(dsp)$, $\delta_{Hc} = \delta_{Hd} = 1.253(d)$, $^2J_{Ha,P} = 16.4$, $^3J_{Hb,(Hc,Hd)} = 6.2$, $^3J_{Hb,P} = 8.6$, $\delta_{Ca} = 14.98(d)$, $\delta_{Cb} = 71.26(d)$, $\delta_{Cc} = \delta_{Cd} = 26.15(d)$, $^1J_{Ca,P} = 137.1$, $^2J_{Cb,P} = 5.4$, $^3J_{(Cc,Cd),P} = 3.8$, $\delta_P = 26.20(dq)$; 10 μL/0.8 mL D_2O, pH 13.8; TSPSA, ext. H_3PO_4.
Pinacolyl methylphosphonate; 1,2,2-Trimethylpropyl methylphosphonate [616-52-4] (2.B.4)
$\delta_{Ha} = 1.481(d)$, $\delta_{Hb} = 4.199(dq)$, $\delta_{Hd} = 0.919(s)$, $\delta_{He} = 1.286(d)$, $^2J_{Ha,P} = 17.9$, $^3J_{Hb,P} = 8.9$, $^3J_{Hb,He} = 6.4$, $^4J_{He,P} = 0.3$, $\delta_{Ca} = 12.36(d)$, $\delta_{Cb} = 80.89(d)$, $\delta_{Cc} = 34.88(d)$, $\delta_{Cd} = 25.58(s)$, $\delta_{Ce} = 16.88(s)$, $^1J_{Ca,P} = 149.6$, $^2J_{Cb,P} = 7.7$, $^3J_{Cc,P} = 6.3$, $^3J_{Ce,P} = 1.1$, $\delta_p = 33.52(dq)$; 5μL/0.8 mL $CDCl_3$; TMS, ext. H_3PO_4.
$\delta_{Ha} = 1.277(d)$, $\delta_{Hb} = 3.959(dq)$, $\delta_{Hd} = 0.895(s)$, $\delta_{He} = 1.199(d)$, $^2J_{Ha,P} = 16.3$, $^3J_{Hb,P} = 9.1$, $^3J_{Hb,He} = 6.4$, $\delta_{Ca} = 14.03(d)$, $\delta_{Cb} = 84.65(d)$, $\delta_{Cc} = 36.97(d)$, $\delta_{Cd} = 27.51(s)$, $\delta_{Ce} = 19.05(d)$, $^1J_{Ca,P} = 140.2$, $^2J_{Cb,P} = 7.4$, $^3J_{Cc,P} = 5.9$, $^3J_{Ce,P} = 1.1$, $\delta_p = 31.50(dq)$; 21 mg/0.7 mL D_2O, pH 1.3; TSPSA, ext. H_3PO_4.
$\delta_{Ha} = 1.277(d)$, $\delta_{Hb} = 3.960(dq)$, $\delta_{Hd} = 0.895(s)$, $\delta_{He} = 1.200(d)$, $^2J_{Ha,P} = 16.3$, $^3J_{Hb,P} = 9.1$, $^3J_{Hb,He} = 6.4$, $\delta_{Ca} = 15.29(d)$, $\delta_{Cb} = 82.17(d)$, $\delta_{Cc} = 36.97(d)$, $\delta_{Cd} = 27.82(s)$, $\delta_{Ce} = 19.21(d)$, $^1J_{Ca,P} = 138.0$, $^2J_{Cb,P} = 6.6$, $^3J_{Cc,P} = 5.3$, $^3J_{Ce,P} = 0.9$, $\delta_P = 25.97(dq)$; 7.6 mg/0.7 mL D_2O, pH 6.3; TSPSA, ext. H_3PO_4.
$\delta_{Ha} = 1.274(d)$, $\delta_{Hb} = 3.958(dq)$, $\delta_{Hd} = 0.895(s)$, $\delta_{He} = 1.198(d)$, $^2J_{Ha,P} = 16.3$, $^3J_{Hb,P} = 9.1$, $^3J_{Hb,He} = 6.4$, $\delta_{Ca} = 15.29(d)$, $\delta_{Cb} = 82.11(d)$, $\delta_{Cc} = 36.94(d)$, $\delta_{Cd} = 27.81(s)$, $\delta_{Ce} = 19.20(d)$, $^1J_{Ca,P} = 137.9$, $^2J_{Cb,P} = 6.8$, $^3J_{Cc,P} = 5.6$, $^3J_{Ce,P} = 1.0$, $\delta_P = 25.94(dq)$; 10.9 mg/0.8 mL D_2O, pH 13.7; TSPSA, ext. H_3PO_4.
Methylphosphonic acid [993-13-5] (2.B.4)

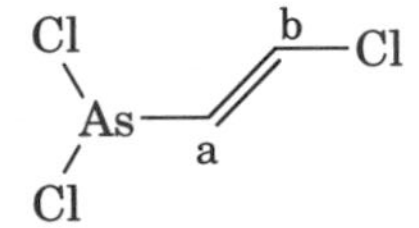

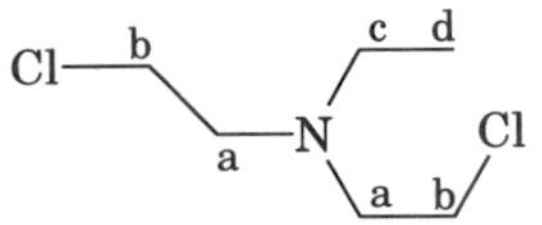

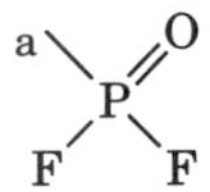

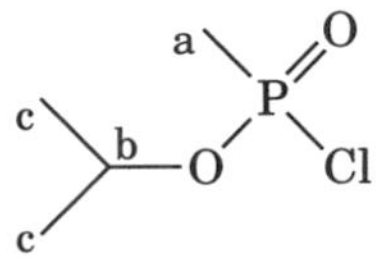

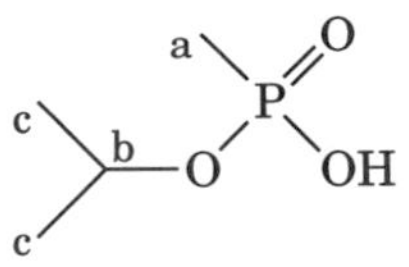

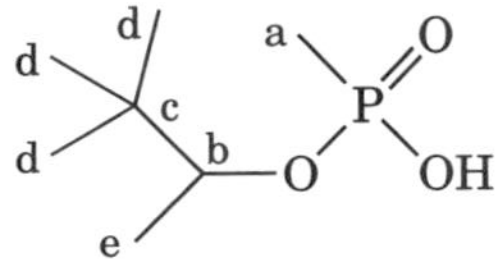

Table 1. NMR Spectral Parameters of Selected CWC-Related Chemicals *(continued)*

Trivial Name; NATO Code[a]; IUPAC Name [CAS][b] (Schedule)[c]
$\delta_{Ha} = 2.633(XX')$, $\delta_{Hb} = 3.460(AA')$, $\delta_{Hc} = 3.044(sp)$, $\delta_{Hd} = 1.029(d)$, $\delta_{OH} = 3.14(b)$, ${}^3J_{Hc,Hd} = 6.6$, $\delta_{Ca} = 45.26$, $\delta_{Cb} = 58.36$, $\delta_{Cc} = 47.61$, $\delta_{Cd} = 20.92$; 40 μL/0.8 mL $CDCl_3$; TMS.
$\delta_{Ha} = 3.288(XX')$, $\delta_{Hb} = 3.886(AA')$, $\delta_{Hc} = 3.776(sp)$, $\delta_{Hd} = 1.351, 1.364(d)$, ${}^3J_{Hc,Hd} = 6.6$, $\delta_{Ca} = 51.40$, $\delta_{Cb} = 60.22$, $\delta_{Cc} = 58.54$, $\delta_{Cd} = 19.28, 20.99$; 20 μL/0.8 mL D_2O, pH 0.5; TSP-d_4.
$\delta_{Ha} = 3.297(XX')$, $\delta_{Hb} = 3.899(AA')$, $\delta_{Hc} = 3.785(sp)$, $\delta_{Hd} = 1.371(d)$, ${}^3J_{Hc,Hd} = 6.6$, $\delta_{Ca} = 51.28$, $\delta_{Cb} = 60.15$, $\delta_{Cc} = 58.37$, $\delta_{Cd} = 20.02$; 23 μL/0.8 mL D_2O, pH 8.4; TSP-d_4.
$\delta_{Ha} = 2.590(XX')$, $\delta_{Hb} = 3.564(AA')$, $\delta_{Hc} = 3.020(sp)$, $\delta_{Hd} = 1.036(d)$, ${}^3J_{Hc,Hd} = 6.5$, $\delta_{Ca} = 50.13$, $\delta_{Cb} = 64.78$, $\delta_{Cc} = 52.71$, $\delta_{Cd} = 21.66$; 23 μL/0.8 mL D_2O, pH 12.8; TSP-d_4.
Thiodiglycol; Bis(2-hydroxyethyl)sulfide [111-48-8] (2.B.13)
$\delta_{Ha} = 2.757(t)$, $\delta_{Hb} = 3.755(t)$, ${}^3J_{Ha,Hb} = 6.3$, $\delta_{Ca} = 36.10$, $\delta_{Cb} = 63.09$; 9.7 mg/0.6 mL D_2O, pH 6.5; TSPSA.
Ethyldiethanolamine [139-87-7] (3.B.15)
$\delta_{Ha} = 2.646(XX')$, $\delta_{Hb} = 3.613(AA')$, $\delta_{Hc} = 2.633(q)$, $\delta_{Hd} = 1.046(t)$, $\delta_{OH} = 3.29(b)$, ${}^3J_{Hc,Hd} = 7.1$, $\delta_{Ca} = 55.53^{g}$, $\delta_{Cb} = 59.66^{g}$, $\delta_{Cc} = 48.22$, $\delta_{Cd} = 11.62$; 40 μL/0.8 mL $CDCl_3$; TMS.
$\delta_{Ha} = 3.38(XX')$, $\delta_{Hb} = 3.934(AA')$, $\delta_{Hc} = 3.370(q)$, $\delta_{Hd} = 1.327(t)$, ${}^3J_{Hc,Hd} = 7.3$, $\delta_{Ca} = 57.04^{g}$, $\delta_{Cb} = 58.20^{g}$, $\delta_{Cc} = 51.95$, $\delta_{Cd} = 10.89$; 40 μL/0.8 mL D_2O, pH 0.5; TSP-d_4.
$\delta_{Ha} = 3.378(XX')$, $\delta_{Hb} = 3.942(AA')$, $\delta_{Hc} = 3.362(q)$, $\delta_{Hd} = 1.334(t)$, ${}^3J_{Hc,Hd} = 7.3$, $\delta_{Ca} = 56.94^{g}$, $\delta_{Cb} = 58.21^{g}$, $\delta_{Cc} = 51.81$, $\delta_{Cd} = 10.84$; 40 μL/0.8 mL D_2O, pH 7.7; TSP-d_4.
$\delta_{Ha} = 2.678(XX')$, $\delta_{Hb} = 3.685(AA')$, $\delta_{Hc} = 2.617(q)$, $\delta_{Hd} = 1.039(t)$, ${}^3J_{Hc,Hd} = 7.2$, $\delta_{Ca} = 57.51^{g}$, $\delta_{Cb} = 61.54^{g}$, $\delta_{Cc} = 50.88$, $\delta_{Cd} = 13.05$; 40 μL/0.8 mL D_2O, pH 12.5; TSP-d_4.
$\delta_{Ha} = 1.547(d)$, ${}^2J_{Ha,P} = 17.5$, $\delta_{Ca} = 14.40(d)$, ${}^1J_{Ca,P} = 136.5$, $\delta_P = 31.47(q)$; 9.6 mg/0.9 mL D_2O, pH 0.0; TSPSA, ext. H_3PO_4.
$\delta_{Ha} = 1.272(d)$, ${}^2J_{Ha,P} = 16.4$, $\delta_{Ca} = 15.88(d)$, ${}^1J_{Ca,P} = 134.1$, $\delta_P = 24.49(q)$; 10.4 mg/0.9 mL D_2O, pH 6.9; TSPSA, ext. H_3PO_4.
$\delta_{Ha} = 1.076(d)$, ${}^2J_{Ha,P} = 15.5$, $\delta_{Ca} = 16.88(d)$, ${}^1J_{Ca,P} = 130.5$, $\delta_P = 20.93(q)$; 11.6 mg/0.9 mL D_2O, pH 13.8; TSPSA, ext. H_3PO_4.
N,N-Diisopropylaminoethane-2-ol [96-80-0] (2.B.11)

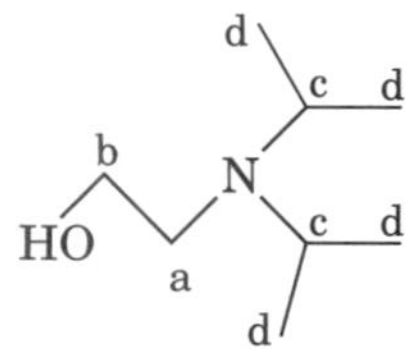

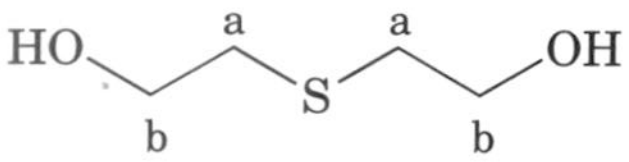

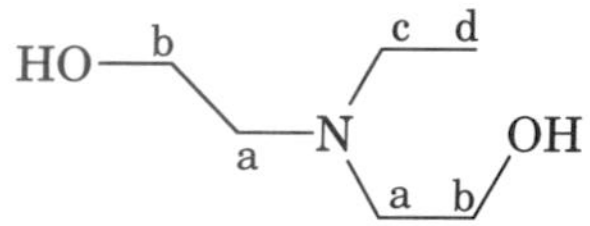

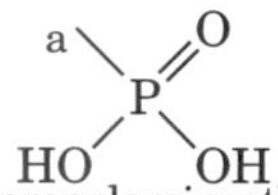

[a] Standard NATO agreement (STANAG) code.
[b] Chemical Abstracts Service registry number.
[c] Schedule number.
[d] Chemical shifts are given relative to $\delta_H(TMS) = \delta_H(TSP - d_4) = 0.00$, $\delta_H(TSPSA) = 0.015$, $\delta_C(TMS) = \delta_C(TSP - d_4) = \delta_C(TSPSA) = 0.00$, $\delta_F(CFCl_3) = 0.00$, and $\delta_P(H_3PO_4) = 0.00$ ppm.
[e] Multiplicity: b = broad, d = doublet, dd = doublet of doublets, ddq = doublet of doublet of quartets, dq = doublet of quartets, $dspd$ = doublet of septet of doublets, dt = doublet of triplets, m = multiplet, q = quartet, s = singlet, sp = septet, t = triplet, $AA' = AA'$ part of an $AA'XX'$ spinsystem, $XX' = XX'$ part of an $AA'XX'$ spin system. If no multiplicity is given, the resonance appears as a singlet.
[f] Only the couplings ${}^2J_{HH}$ are marked negative, although the correct sign of some other couplings also may be negative.
[g] Tentative assignment.

Molecular structures, NMR spectral parameters, and sample conditions of selected chemical warfare agents and some of their precursors and hydrolytic products are presented in Table 1. The NMR spectral parameters of Sarin, Soman, VX, Tabun, Mustard gas, *O*-mustard, Lewisite-1, HN-1, MPDF, and chlorosarin are given only in $CDCl_3$; spectral parameters of the precursors and hydrolytic products are given in $CDCl_3$ and/or in D_2O at different pH.

Structural Elucidation. The 1D NMR spectra are interpreted (see above) to reveal indicative structural details reveal, eg, in the ^1H NMR spectrum methyl, ethyl, propyl, and isopropyl linked directly with phosphorus: all show different resonance patterns. In the next stage, to confirm the heteronuclear coupling, experiments like $^1H\{^{19}F\}$ and/or $^1H\{^{31}P\}$ are run. After these, and in particular where a strong background is present, different 2D NMR spectra are run. By way of example, homonuclear experiments like COSY and TOCSY (or some of their variants) provide information on coupled protons, even networks of protons (3,4); inversely detected heteronuclear correlation experiments like HMQC and HMQC–TOCSY provide similar information, but for coupled protons and heteronuclei, eg, ^1H and ^{31}P or ^1H and ^{13}C. These data are interpreted for the molecular structure. In addition, different techniques of MS and GC/FTIR are used to complement and/or confirm the structure. Unambiguous identification of a chemical is based on consistent results from all spectrometric techniques used.

ACKNOWLEDGMENTS

Thanks to H. Björk, V. Häkkinen, M–L. Kuitunen, O. Kostiainen, M. Rautio, M. Söderström, and P. Vanninen of the Institute and E. Rahkamaa of the University of Oulu for their helpful comments during preparation of the manuscript. Thanks as well to E–L. Tolppa for expert assistance during the proficiency tests.

BIBLIOGRAPHY

1. R. R. Ernst, *Pure Appl. Chem.* **66,** 1583–1588 (1994).
2. G. C. K. Roberts, ed., *NMR of Macromolecules. A Practical Approach,* Oxford University Press, Oxford, U.K., 1993.
3. W. R. Croasmun and R. M. K. Carlson, eds., *Two-Dimensional NMR Spectroscopy. Applications for Chemists and Biochemists,* VCH Publishers, Inc., New York, 1994.
4. G. E. Martin and A. S. Zektzer, in *Two-Dimensional NMR Methods for Establishing Molecular Connectivity. A Chemist's Guide to Experiment Selection, Performance, and Interpretation,* VCH Publishers, Inc., New York, 1988.
5. M. Rautio, ed., F.1 Testing of Existing Procedures; F.2 Testing of Procedures on Simulated Industry Samples; F.3 Testing of Procedures on Simulated Military Facility Samples; F.4 Validating of Procedures for Water and Soil Samples. F. International Interlaboratory Comparison (Round-Robin) Test for the Verification of Chemical Disarmament, *Methodology and Instrumentation for Sampling and Analysis in the Verification of Chemical Disarmament,* The Ministry for Foreign Affairs of Finland, Helsinki, 1990–1993.
6. *Finnish Research Project on the Verification of Chemical Disarmament,* 1990–1993; see Ref. 5.
7. M. Rautio, ed., H.1 First Interlaboratory Comparison Test H. Interlaboratory Comparison Test Coordinated by the Provisional Technical Secretariat for the Preparatory Commission for the Organisation for the Prohibition of Chemical Weapons, *Methodology and Instrumentation for Sampling and Analysis in the Verification of Chemical Disarmament,* The Ministry for Foreign Affairs of Finland, Helsinki, 1994.
8. *Provisional Technical Secretariat for the Preparatory Commission for the Organization for the Prohibition of Chemical Weapons* (OPCW/PTS), 1994; see Ref. 7.
9. *Convention on the Prohibition of the Development, Production, Stockpiling and Use of Chemical Weapons and on their Destruction;* Jan. 1993; Printed and distributed by the OPCW/PTS; the Depositary of this Convention is the Secretary-General of the United Nations from whom a certified true copy can be obtained.
10. J. A. F. Compton, *Military Chemical and Biological Agents. Chemical and Toxicological Properties,* The Telford Press, Caldwell, U.K., 1987; J. Cookson and J. Nottingham, *A Survey of Chemical and Biological Warfare,* Sheed and Ward, London, 1969.
11. A.1 Chemical and Instrumental Verification of Organophosphorus Warfare Agents; A.2 Technical Evaluation of Selected Scientific Methods for the Verification of Chemical Disarmament, A. General Reports, *Methodology and Instrumentation for Sampling and Analysis in the Verification of Chemical Disarmament,* The Ministry for Foreign Affairs of Finland, Helsinki, 1977, 1984.
12. B.1 Identification of Organophosphorus Warfare Agents. An Approach for the Standardization of Techniques and Reference Data; B.2 Identification of Degradation Products of Organophosphorus Warfare Agents; B.3 Identification of Non-phosphorus Warfare Agents; B.4 Identification of Precursors of Warfare Agents, Degradation Products of Non-phosphorus Agents, and Some Potential Agents; B.5 Identification of Selected Mycotoxins: Trichothecenes, Aflatoxins, and Related Mycotoxins. B. Systematic Identification of Chemical Warfare Agents, *Methodology and Instrumentation for Sampling and Analysis in the Verification of Chemical Disarmament,* The Ministry for Foreign Affairs of Finland, Helsinki, 1979, 1980, 1982, 1983, and 1986.
13. C.1 An Approach to the Environmental Monitoring of Nerve Agents; C.2 Air Monitoring as a Means for the Verification of Chemical Disarmament, Part I. Development and Evaluation of Basic Techniques; C.3 Air Monitoring as a Means for the Verification of Chemical Disarmament, Part II. Field Tests; C.4 Air Monitoring as a Means for the Verification of Chemical Disarmament, Part III. Further Development and Testing of Methods. C. Trace Analysis of Chemical Warfare Agents, *Methodology and Instrumentation for Sampling and Analysis in the Verification of Chemical Disarmament,* The Ministry for Foreign Affairs of Finland, Helsinki, 1981, 1985–1987.
14. M. Rautio, ed., *Recommended Operating Procedures for Sampling and Analysis in the Verification of Chemical Disarmament,* The Ministry for Foreign Affairs of Finland, Helsinki, 1993 and 1994.
15. M. Batley and J. W. Redmond, *J. Magn. Reson.* **49,** 172–174 (1982).

CHEMICAL WEAPONS, DESTRUCTION OF.

See MILITARY TOXIC MATERIALS, DESTRUCTION OF.

CHLORINATED ETHENES: ANAEROBIC BIOREMEDIATION

HUGH H. RUSSELL
CHR_2 Environmental Services
Oilton, Oklahoma

GUY W. SEWELL
U.S. EPA National Risk Management Research Laboratory
Tulsa, Oklahoma

The failure of what have been termed conventional methods to reclaim contaminated environmental media has over the years led to the interest and development of alternative technologies. One of these technologies is bioremediation, or the utilization of microbial agents to reduce the overall mass or toxicity of contamination within soil and groundwater systems. Bioremediation has its origins in the SunTech process which was patented in 1974 (1). The process relied on the addition of oxygen and nutrients to contaminated portions of an aquifer to stimulate the reduction of contaminants through microbial processes. The SunTech process used an in-well air sparging apparatus to supply oxygen to the microflora within contaminated systems. For years, variants of this technology have been used in remediation efforts with varying degrees of success. All utilized the more thermodynamically favorable aerobic process.

As with all technologies, limitations were soon discovered for aerobic biological processes. Two major limitations of aerobic bioremediation important in the development of anaerobic process were noted: the solubility limit of oxygen in water and the recalcitrant nature of some oxidized chemicals to aerobic degradation.

The solubility limit of oxygen in water is approximately 40 ppm. Depending on the concentration of contaminants within a contaminated aquifer, it may not be logistically possible to deliver the required concentration of oxygen, given the aforementioned solubility constraints. As a result researchers in some instances have turned to hydrogen peroxide (which decomposes to liberate oxygen) or other infinitely soluble alternate electron acceptors such as nitrate. The efficacy of each process has been debated and is the subject of numerous papers.

The recalcitrant nature of certain contaminants such as tetrachloroethene (also known as perchloroethylene or tetrachloroethylene) under aerobic conditions has also led to the development of anaerobic systems. Even though tetrachloroethene is recalcitrant to aerobic processes, it does lend itself to anaerobic degradation. Tetrachloroethene (and other chlorinated ethenes) under anaerobic conditions may be transformed through a stepwise removal of halogens and can be degraded to ethene, an environmentally benign chemical (2).

To harness this potential, reductive anaerobic biological *in-situ* treatment technologies (RABITT) are being developed and tested. Although RABITT is applicable to the degradation of all chlorinated ethenes, it may be most useful in the treatment of aquifer systems contaminated with tetra- and trichloroethene.

CHLORINATED ETHENES

Of all anthropogenic chemicals or contaminants within soil and groundwater systems throughout the world, some of the more ubiquitous are the chlorinated solvents. Framed within this large group are the chlorinated ethenes: notably, tetrachloroethene (PCE), trichloroethene (TCE), the dichloroethenes (*cis*-1,2-dichloroethene or *cis*-1,2-DCE; *trans*-1,2-dichloroethene or *trans*-1,2-DCE; and 1,1-dichloroethene or 1,1-DCE), and vinyl chloride (VC). Chlorinated ethenes are widely used in industry as degreasing agents, in dry-cleaning processes, and chemical manufacturing. The major uses of chlorinated ethenes are given in Table 1.

Because of their widespread use, and the fact that two (trichloroethylene and vinyl chloride) were listed in the 1995 Agency for Toxic Substances and Disease Registry (ASTDR) top-20 hazardous substances list, chlorinated ethenes are a major environmental liability. Numerous resources have been mobilized to effect the removal of chlorinated ethenes from the environment in the most cost-effective manner.

Physicochemical Properties

Evaluation of the physicochemical properties of contaminants is a necessary first step in any site assessment as well as implementation of any remedial alternative. Physicochemical properties are important factors in determining both fate and transport of contaminants.

As a group, the chlorinated ethenes are heavier than water. Their weight per milliliter or density is greater than 1.0 g/mL, the density of pure water. In the liquid form (phase separate from water) chlorinated ethenes are termed dense nonaqueous phase liquids or DNAPLs. DNAPLs as a rule, if the spill is of sufficient magnitude, move downward through the subsurface under the influence of gravity, until low permeability fea-

Table 1. Major Uses of Chlorinated Ethenes

Chlorinated Ethene	Major Uses
Tetrachloroethylene	Textile industry: dry-cleaning and for processing and finishing Metals: cold cleaning and vapor degreasing Chemical: as an intermediate in manufacture
Trichlorethylene	Textile industry: dry-cleaning Metals: cold cleaning and vapor degreasing Chemical: as an intermediate in manufacture and solvent Also used as a refrigerant and heat exchange liquid
trans-1,2-Dichloroethylene	Chemical: as a solvent and intermediate in manufacture; pharmaceutical manufacturing
cis-1,2-Dichloroethylene	Chemical: as a solvent and chemical intermediate
1,1-Dichloroethylene	Chemical: as a monomer and chemical intermediate
Vinyl chloride	Chemical: as a solvent and intermediate Manufacturing: automotive industry, building industry, rubber, paper, and glass industries Also used as a refrigerant and formerly as a propellant

tures impede their progress. As a result, DNAPLs tend to pool on top of these lower permeability features, whether they are clay lenses within a sandy aquifer or a competent aquitard (defined as an impermeable layer without cracks or fissures into which DNAPLs may flow) which determines the aquifer base. Porosity, permeability, relative permeability, and mean pore throat diameter are the properties which govern the mobility of DNAPL (3). DNAPLs may move into cracks or discontinuities in clayey aquitards. Whether penetration into fractures occurs depends on the DNAPL driving gradient, DNAPL–water interfacial tension, and aperture of the fissure (4).

As DNAPLs move through the subsurface, they force water from the pore spaces. The pore spaces then move from a water saturated condition (water-wet) to an oil saturated condition (oil-wet). The mobility of water and DNAPL within the pore spaces changes as the oily phase moves in and through the voids. Initially, the pore space is saturated with water which is the mobile phase. As the concentration of DNAPL within pore spaces increases, more and more water is forced out and the water left within the pore spaces becomes less mobile until an irreducible saturation point for water is reached. At this point (irreducible water saturation) water will not flow from the pore spaces. The DNAPL within the pore spaces now becomes the mobile liquid. As DNAPL moves out of the pore spaces, water fills the voids. At some point, capillary forces impede and finally immobilize the DNAPL remaining within the pore spaces. Water once again becomes the mobile fraction, and an irreducible DNAPL saturation limit is reached. Depending on the size of the pore throats and characteristics of the DNAPL, the percentage of the pore spaces occupied by immobile DNAPL may be as high as 50%. This irreducible fraction is termed residual saturation.

Pools result when both vertical and horizontal movement of DNAPL from the pore spaces is limited such as by depressions within low permeability zones. Here, forces acting on the free-phase liquid such as gravity or head are not sufficient to force further movement of the liquid. In the case of DNAPL pools, the pore spaces are filled with nonaqueous phase liquid.

Residual saturation and DNAPL pools serve as long-term sources for groundwater contamination. Equilibrium concentrations are pulled into the free-flowing aqueous phase as groundwater moves through or around them. The time frame for pools to naturally dissolve into groundwater can be measured in centuries.

Once dissolved in the free-flowing aqueous phase, contaminants tend to follow the movement of groundwater and are dependent on vertical and horizontal gradients. Density difference effects may be apparent at low solute concentrations (5). As an example, TCE dissolved in groundwater may still tend to move vertically within a system, especially if there is no upward vertical gradient to counteract density effects. DNAPLs within an aquifer system provide technical challenges in both detection and remediation.

Figure 1 is a depiction of a spill of chlorinated ethenes within an hypothetical aquifer. The oily-phase moved through the vadose zone, impacted the aquifer, and was of sufficient mass not only to overcome capillarity but to move vertically through the aquifer under the force of gravity, until an impermeable feature in the middle of the aquifer was encountered. Vertical migration of DNAPLs can be quite rapid. Pure PCE and TCE have been shown to move at 1–4 cm/s in clean-water-saturated sand (6,7). Because vertical movement

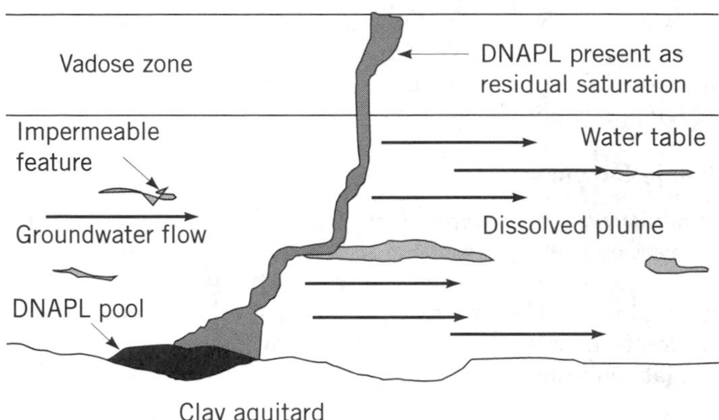

Figure 1. DNAPL movement within the subsurface.

and horizontal movement (in the direction of groundwater flow) was impeded by the structure of the lower permeability feature, the DNAPL moved horizontally, against the local groundwater gradient, until vertical movement was again possible. The DNAPL continued to move downward until the aquitard was encountered. Here, a pool of DNAPL formed within a depression. This pool is immobile, because lower permeability features present barriers in the horizontal and vertical directions which cannot be overcome by forces such as gravity or head. Water moving across the top of the DNAPL pool will become contaminated as soluble DNAPL diffuses from the oily phase into the free-flowing aqueous phase. Even though a large separate phase of solvent may be in contact with groundwater, the observed groundwater concentration may only reflect a small percentage of the aqueous solubility limit for the solvent. The reasons for this can include monitoring well construction, mass-transfer limitations, and the small fraction of groundwater in actual contact with the DNAPL. It is a misconception that observed solvent concentrations of less than 1–10% of saturation *do not* represent the presence of DNAPL. DNAPL pools represent a significant source of contamination within the subsurface that are extremely difficult to locate and remediate.

In addition, the residual saturation that exists from the surface to the top of the DNAPL pool may represent pore spaces filled with as much as 50% DNAPL. This DNAPL, held in place by capillary forces, is immobile but represents a significant source area for the dissolved plume.

If the spill is of sufficient magnitude, chlorinated ethene contamination as with other DNAPLs can span the entire thickness of an aquifer. This provides problems in locating free-phase concentrations, because either pools or residual saturation increases the mass of aquifer containing contamination.

In a discussion of transport and fate, the water solubility of a chemical is important. The more water soluble a contaminant, the more unit mass that can be removed per volume of extracted water. It can also be assumed that the more soluble a compound is in water, the more available it is for biological processes. Of the chlorinated ethenes, PCE is the least soluble.

In general, as chlorines are removed, mobility is increased. The breakpoint occurs between TCE and the DCEs. As the number of chlorines decrease, both the organic carbon and oily phase partitioning is expected to decrease. Within the vadose or saturated zone, the mobility of chlorinated ethenes within the aqueous phase from least mobile to most would be PCE, TCE, and DCEs, and VC. This will have a significant impact on the technology chosen to remediate an aquifer contaminated with a particular chlorinated ethene.

Biotransformation

In addition to physicochemical properties, the potential for the interactions with microorganisms is an important factor in determining the environmental fate of both organic and inorganic contaminants. The rate as well as extent of microbial transformation processes will be determined by how favorable the immediate environment is to supporting microbial life.

MICROBIAL ECOLOGY

Bacteria are generally divided into three categories in regard to their oxygen requirements: (*1*) strict aerobes which require oxygen for growth and utilize same as a terminal electron acceptor or oxidant, (*2*) facultative aerobes/anaerobes which prefer growth in the presence of oxygen (it is more thermodynamically favorable) but switch to alternative electron acceptors such as nitrate, manganese, or even iron when oxygen tensions are low, and (*3*) anaerobes, bacteria which cannot survive in the presence of even low concentrations of oxygen because of its toxicity.

Until recently, it was thought that anaerobes were restricted to areas in which oxygen had been completely depleted by aerobic microorganisms at the expense of oxidizable carbon. The environmental niches to which anaerobes were restricted included the rumen, alimentary canal, slow moving bodies of water, and sludges. It is now known that in a consortium with aerobic bacteria, in biofilms or other suitable environments, anaerobes can thrive in any environment, in concert with an aerobic population. The function of the aerobic population is to prevent toxic concentrations of oxygen from reaching the anaerobes and possibly to produce necessary substrates.

As with all living creatures, the main drive in bacteria is to replicate. To accomplish this task, energy in the form of adenosinetriphosphate (ATP) is required. Aerobic bacteria gain energy through the use of an electron transport chain, in which electrons or reducing equivalents are passed from a series of oxidants to oxygen to form water. Anaerobic bacteria acquire ATP through two basic mechanisms, substrate-level phosphorylation and electron transport chains. Substrate-level phosphorylation can also be termed fermentation as electrons or reducing equivalents are shunted from oxidized products of bacterial metabolism to reduced. In some instances, especially in the case of facultative anaerobes, the lines between substrate level phosphorylation and electron transport may not be distinct because some fermentation products can be linked to a respiratory chain.

Degradation of Organic Carbon

In any favorable environment, once oxidizable or degradable carbon is introduced, aerobic bacteria will utilize that carbon at the expense of oxygen. Once oxygen is depleted or nearly depleted, if sufficient carbon is still present, bacteria may switch from oxygen as a terminal electron acceptor to the utilization of nitrate. We currently accept that many bacteria which utilize nitrate as an electron acceptor are facultative aerobes who utilize nitrate when oxygen tensions are low. Oxygen inhibits enzymes responsible for nitrate reduction and the inducer for enzyme synthesis is nitrate. Some anaerobes may reduce nitrate to nitrite or assimilate nitrate. There are then two known biological mechanisms for the removal of nitrate from a system: dissimilatory reduction of nitrate and assimilatory reduction. In dissimilatory nitrate (NO_3) reduction, nitrate can be reduced to nitrogen (N_2), or reduced to nitrite (NO_2). Nitrate can either be excreted or further reduced to ammonia (NH_3). In assimilatory nitrate reduction, nitrate/nitrite is reduced to ammonia. The ammonia is not excreted but utilized for the biosynthesis of cellular material.

There is no strict requirement that oxygen be completely absent for nitrate reduction to occur in the case of facultative aerobes. We have noted that in an activate sludge digestor with unlimited carbon and excess nitrate, nitrate reduction occurred with as much as 2 ppm oxygen present. It is possible that the molecular oxygen in limited supply is utilized by oxygenases, to initiate oxidation of carbon, while nitrate replaces oxygen as the terminal electron acceptor in respiration processes.

When nitrate concentrations reach a finite level, manganese oxide becomes the dominant electron acceptor. Manganese within the environment will continue to be utilized as a terminal electron acceptor until either the degradable carbon is exhausted, manganese oxides are exhausted, or another more favorable electron acceptor such as oxygen is reintroduced into the system. Once the manganese oxides are removed, bacteria may switch to iron reduction. As with manganese oxides, iron reduction will continue until one of the previously discussed three conditions are met.

When iron within the immediate environment is exhausted, the rather ubiquitous sulfate reducing bacteria (SRB) begin growth. The SRBs are generally represented by bacteria of the genus *Desulfovibrio, Desulfomonas,* and *Desulfotomaculan.* At the level of sulfate reduction, the complex carbon introduced initially has been degraded by aerobic, facultative, and fermentative bacteria to less complex and oxidized endproducts, such as volatile fatty acids (VFAs). These are the preferred substrates of SRBs, although recent work has proven SRBs can utilize more complex substrates.

Once sulfate is depleted, the methanogenic bacteria are capable of growth within the environment. Traditionally, methanogenic bacteria have been known to grow only on simple one-carbon compounds and produce methane. Originally, these compounds were thought to be acetate, carbon dioxide, and hydrogen or the endproducts of sulfate reducing bacteria. It is now known that under methanogenic conditions a number of more complex organic compounds such as aromatic hydrocarbons are subject to degradation.

In any particular environment, any of these electron acceptors may be used in concert. This will be especially true in an environment in which there is little competition for either substrate or electron acceptor. Although the potential of the various reactions is limited to a specific range, the activity of the dominant population is not. A thin film of aerobic bacteria protecting an active population of SRBs in an oxygen-limited, sulfate, and carbon-rich environment will allow both populations to coexist. Aerobic bacteria prevent toxic concentrations of oxygen from reaching the SRBs and degrade complex carbon compounds to those which can be utilized by the SRBs. In an environment such as an aquifer, aerobic and anaerobic bacteria may coexist within biofilms or exist in close but spatially distinct areas.

This discussion would not be complete without some mention of fermentation. Fermenters utilize substrates as both electron donor and acceptor. In effect, an organic compound is metabolized with a portion of the molecule becoming reduced end-products and another becoming oxidized end-products. The best-known example of fermentation is the production of alcohol from the fermentation of starch. Carbon dioxide (CO_2) is the oxidized product and ethanol is the reduced product. A number of fermenters regenerate reducing power, transfer reducing equivalents, or remove hydrogen through reduction of intermediates. Chlorinated ethenes might replace these intermediates and become the reduced end-product.

The electrode potential (E_h) is described by the redox (or oxidation–reduction) status of the system, or the potential for the transfer of electrons to reducible material. An organic chemical would be expected to be reduced if the electrode potential of the specific soil or groundwater system in which the chemical is present is less than that of the organic chemical (8).

Aquifer Microbial Ecology

Microbial communities under ambient (pristine) conditions within an aquifer are defined ecologically by the flux of available donors (carbon). The function of microbial communities with an aquifer system are that of a living filter to remove exogenous carbon, and serendipitously in some instances, inorganic contamination such as nitrate. This living filter in addition to abiotic methods determines the assimilative capacity of a system. The assimilative capacity is merely the capacity of environmental media to adsorb anthropogenic or natural electron donors (carbon) or acceptors.

If small amounts of carbon are introduced into the system, either from natural or human-made sources, the ambient conditions may remain aerobic. In most instances natural sources are not sufficient to drive an aquifer from aerobic to anaerobic conditions. Natural carbon might be expected in an aquifer system that has dissolved carbon from soil organic matter that has leached into the groundwater. The formation of anaerobic conditions usually results from the introduction of anthropogenic carbon sources in significant amounts.

Once bio-oxidizable organic contaminants are introduced into an aquifer system, the local environment progresses from an oxic environment to an anoxic environment as discussed in the previous section. The time required to reach anoxic conditions will be defined by ambient conditions within the aquifer, concentration and degradability of introduced carbon and the availability of appropriate electron acceptors. The microbial communities are for all intents and purposes ecologically defined by the flux of available electron donors (carbon) and acceptors.

Once anaerobic conditions are met and maintained, most organic compounds are degraded by groups of interacting microorganisms referred to as a consortium. In the consortium, individual types of organisms carry out different specialized reactions which when combined can lead to the complete mineralization of a particular compound. The tightly linked metabolic interactions between members of a stable consortium have in some cases led to consortia being identified as a single organism.

There appear to be several advantages to the consortial system:

1. The system allows for the creation of microenvironments where certain types of organisms can survive in otherwise hostile environments.
2. Thermodynamically unfavorable reactions can be driven by favorable reactions when they are metabolically linked within the consortium.
3. This system takes advantage of the diverse metabolic capabilities of microorganisms by allowing for the formation and enrichment of associations that can utilize an introduced substrate faster than a single species could evolve a novel complex degradation pathway for the same compound.

Anaerobic communities thrive within aquifer systems where conditions allow, especially when ambient conditions favor their development or within microsites when conditions are not as favorable. We have studied aquifer sediments from a number of sites impacted by chlorinated ethenes for the presence of anaerobic communities capable of degrading chlorinated ethenes, notably PCE. In every instance, anaerobic bacteria capable of sequential removal of chlorine from the ethene moiety were recovered from the sediments when excess carbon was added, or in some instances in the presence of natural dissolved organic carbon (9–11).

DEGRADATION OF CHLORINATED ETHENES

Under anaerobic conditions, chlorinated ethenes are not utilized as sole sources of carbon and energy. There is a strict requirement for readily degradable or oxidizable carbon as well as the requisite electron acceptor within the system. This readily degradable or oxidizable carbon can be referred to as the electron donor. Electrons or reducing equivalents from the electron donor are transferred to the electron acceptor, in this case chlorinated ethenes. As a result, a chlorine atom is removed and replaced with a hydrogen atom.

What occurs under ideal conditions (excess oxidizable carbon, lack of competing electron acceptor, appropriate levels of chlorinated ethenes, and favorable environmental conditions) is a stepwise removal of one chlorine at a time such that there is a sequential degradation from PCE to TCE, TCE to DCEs, DCEs to VC, and VC to ethene (12–16). This scheme is outlined in Figure 2. With each reaction, it is necessary that the product be more reduced than the reactant; thus, there is a progression from the highly oxidized PCE to the reduced ethene. This process is referred to as reductive dechlorination or biodehalogenation, the linking of oxidation of molecular hydrogen or organic compounds to the reduction of chloroethenes. In return, the anaerobic bacteria gain useable metabolic energy. It can be termed anaerobic respiration, even though the exact mechanism is currently unknown. However, it is clear that in some instances chlorinated ethenes are able to substitute for other terminal electron acceptors. With each stepwise chlorine removal from PCE to VC, the aerobic degradation potential also increases. To our knowledge, no reports of PCE degradation under aerobic conditions exist within the peer-reviewed literature; given its oxidized nature this is not surprising.

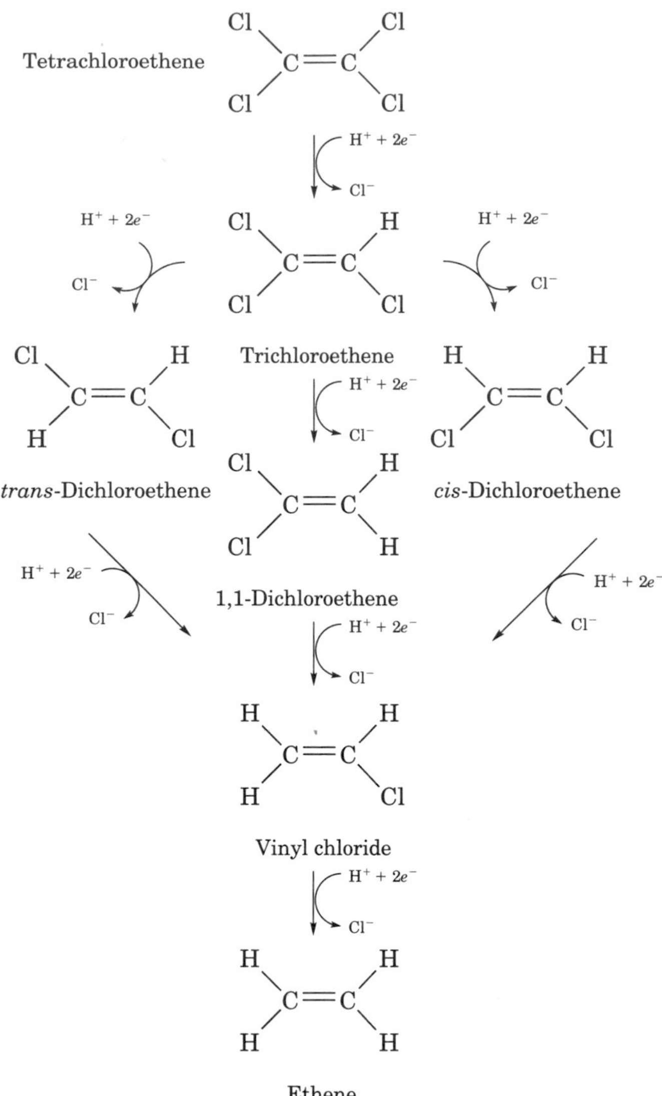

Figure 2. Stepwise reduction of chlorinated ethenes.

From TCE to VC, the aerobic degradation potential begins and increases.

Reductive dechlorination is nothing more than an oxidation–reduction reaction in which a reductant or reducing agent is through the loss of electrons oxidized and an oxidant which gains these electrons is reduced. Reductive dehalogenation is then in fact the transfer of electrons from a reductant to an oxidant. There is no oxidation without a concomitant reduction.

Microcoms have been constructed using aquifer sediments from an aquifer impacted by both chlorinated ethenes and alkyl benzenes (9). Reduction of PCE to TCE and DCE was found in toluene-amended microcosms, versus no reduction or biotic loss of toluene in autoclaved control microcosms.

From 90 to 97% of the reducing equivalents did not participate in reduction of chlorinated ethenes. Reduction of PCE to TCE (initial step) could not be related to degradation of toluene, although it could possibly have been linked to the formation of acetate.

This minor fate of reducing equivalents (reduction of chlorinated ethenes) appears to be common to all reductive dechlorination reactions (chlorinated ethenes, ethanes, chlorinated aromatics) in the case of all donors studied to date. It has been estimated (17) that in most studies of anaerobic biotransformation of chlorinated solvents only about 0.005–1.6% of the electrons available from the donor are used for dechlorination. Even though researchers have managed to shuttle larger percentages of reducing equivalents to reductive dechlorination, the efficiency of the process is still lacking, especially in the field.

DESIGN OF BIOLOGICAL TREATMENT SYSTEMS

Bioremediation systems for anaerobic degradation of chlorinated ethenes may be either *in-situ* or *ex-situ*. The basic premise of all bioremediation systems is that environmental, biological, and chemical barriers to growth must be alleviated. One definition of bioremediation is the growth of microorganisms within a controlled environment to bring about desired changes. The controlled environment is a reactor or merely any system in which we wish to perform biological processes. This controlled environment or reactor can be activated sludge tanks, a prepared bed reactor, or even the contaminated portion of an aquifer. The desired changes are the immobilization, detoxification, or mineralization of toxic chemicals to protect human health and the environment or reduction in the overall mass of contamination within a system. In many instances it may not be possible to meet site-specific cleanup goals if they are maximum concentration levels (MCLs). However, ACLs (alternate concentration limits) may be achieved through an overall reduction in the mass of contamination.

NATURAL ATTENUATION

An alternative to active remedial processes which seems to have gained favor in recent years is natural attenuation or the utilization of the natural processes of biodegradation, dispersion, dilution, sorption, volatilization, and immobilization to reduce the mass, toxicity, mobility, volume, or concentration of contaminants. In effect, this alternative affords protection of human health and the environment.

The current challenge with the implementation of natural attenuation at solvent sites is not whether natural processes can be protective of human health and the environment, but do we currently have the tools and expertise to determine if they are protective, under site-specific conditions? It is evident that in most cases, the dominant natural processes of destruction will be microbially mediated and there is no doubt that with almost all plumes that contain biodegradable contaminants, indigenous microflora are removing these contaminants at some site-specific rate. The question is not usually are biological processes working. The question is merely that of time and distance or the retention time needed under ambient conditions. Of course, this may be directly related to the size of the initial spill or the size of the source area remaining once the plume is discovered. Plumes that result from massive contamination events in permeable media, with high seepage velocities and low sorption potential, must necessarily be larger than smaller events in the same media or similar events in less permeable media with low seepage velocities. In most every instance of environmental contamination the size of the event is suspect and our ability to accurately determine the current mass is limited.

Implicit in the natural attenuation alternative is long-term monitoring of the plume to assure that natural processes continue to provide protection of human health and the environment. In the case of chlorinated ethenes, long-term monitoring is extremely important. Loss of the required concentration of either donor or acceptor will result in the cessation of anaerobic processes. Although not quite as significant in regard to the lessor chlorinated ethenes (which degrade under aerobic conditions), for PCE and TCE reductive dechlorination is the major destructive process and requires exogenous carbon.

CONCLUSIONS

Reductive dechlorination is not a panacea for the treatment of aquifers contaminated with chlorinated ethenes. Implementation may depend on site-specific definitions of success (MCLs, ACLs, reduction in toxicity, mobility). A limitation that has been previously addressed (18) is the propensity, under ambient conditions in the field and in laboratory microcosm for the reductive dechlorination process to stop at the level of DCE or VC. Chlorinated solvent plumes appear to fall into three categories, without regard to age of the original event.

1. Plumes in which the parent chloroethene has not been reduced at any observable level. In some instances this has been simply because during analysis of groundwater samples no one looked for daughter products. In other instances there simply is no (observable) reduction.
2. Plumes that dechlorinate, but seem to never get past the product DCE or VC.
3. Plumes that reductively dechlorinate to ethene.

Incomplete dechlorination is believed to occur under electron donor (degradable carbon) limitations and not as the result of physiological limitations of the dechlorinating population. Others (13) suggest that this may not be the case. The production and buildup of VC should then be alleviated in the presence of the appropriate concentration of requisite donor.

BIBLIOGRAPHY

1. U.S. Pat. 3,846,280, (Nov. 5, 1974), R. Raymond.
2. J. Autien, in *Cosarett and Doull's Toxicology,* J. Doull, C.D. Klaasser, and M.O. Andur, eds., MacMillan, New York, pp. 351–556.
3. D.G. Grubb and N. Starr, *Evaluation of Technologies for In-Situ Cleanup of DNAPL Contaminated Sites,* EPA-600/R-94/120.
4. B. Kueper, and D. McWhorter, *Groundwater* **29,** 716–728 (1991).
5. H.H. Russell, J.E. Matthews, and G.W. Sewell, *TCE Removal from Contaminated Soil and Groundwater,* EPA/540/S-92/002, 1992.
6. F. Schwille, *Dense Chlorinated Solvents in Porous and Fractured Media: Model Experiments,* Lewis Publishers, Chelsea, Mich., 1988, pp. 146.
7. B. Kueper and E. Frind, *Water Resources Res.* **27,** 1049–1057 (1991).
8. J. Dragun, *The Soil Chemistry of Hazardous Materials,* Hazardous Materials Control Research Institute, Silver Springs, Md.
9. G.W. Sewell and S.A. Gibson, *Environ. Sci. Tech.* **25,** 982–984 (1991).
10. S.A. Gibson and G.W. Sewell, *Appl. Environ. Micro.* **58,** 1392–1393 (1992).
11. S.A. Gibson, D.S. Robertson, H.H. Russell, and G.W. Sewell, *Environ. Tox. Chem.* **13,** 453–460 (1994).
12. G. Barrio-Lange, F.Z. Parsons, R.S. Nassar, and P.H. Conanzo, *Environ. Tox. Chem.* **6,** 571–578 (1987).
13. J.M. Odom, J. Tabinowski, M.D. Lee, and B.Z. Fathepure, in *Bioremediation of Chlorinated Solvents,* Battelle Press, Columbus, Ohio, 1996, pp. 17–24.
14. R.A. Kleopfer, D.M. Easley, B.B. Haas, Jr., T.G. Deike, D.E. Jackson, and C.J. Worney, *Environ. Sci. Tech.* **19,** 277–280 (1985).
15. T.M. Vogel and P.L. McCarty, *Appl. Env. Micro.* **49,** 1080–1083 (1985).
16. D.L. Freedman and J.M. Gossett, *Appl. Env. Micro.* **55,** 2144–2157 (1989).
17. E.J. Bouwer, "Bioremediation of Chlorinated Solvents Using Alternate Electron Acceptors" in *Handbook of Bioremediation,* Lewis Publishers, Chelsea, Mich., 1994, pp. 149–175.
18. G.W. Sewell, S.A. Gibson, and H.H. Russell, *Anaerobic In-Situ Treatment of Chlorinated Ethenes,* WPCF Conference, Miami, Fla., Oct. 7–11, 1990.

CHLORINATED HYDROCARBONS: DESTRUCTION OF

BAOSHU CHEN
RONALD L. COOK
JOHN D. WRIGHT
TDA Research, Inc.
Wheat Ridge, Colorado

Today, about 15,000 types of chlorinated compounds are widely used in industry because of their many advantages, eg, low melting point, high solubility or detergency, and noncombustibility. However, due to the toxicity or harmfulness of chlorinated hydrocarbons (CHCs) to human health and the environment (1), emissions are currently regulated under the 1990 Clean Air Act Amendment (CAAA), resulting in a need for advanced emission control technologies. Several types of control methods, such as activated carbon adsorption, thermal incineration, and catalytic incineration, are widely used in industries while new technologies are continuously being developed or investigated. The new technologies being studied include hydrogenolysis, absorption, photo-oxidation, microwave discharge plasma degradation, and biological destruction.

Sources of CHC Emissions

Chlorinated hydrocarbons are chlorine-substituted hydrocarbons that contain carbon, hydrogen, and chlorine atoms. These compounds are formed when the hydrogen atoms of the hydrocarbons are partially or fully replaced by chlorine atoms. Among the widely used CHCs are partially chlorinated hydrocarbons such as chloroform ($CHCl_3$), methylene chloride (CH_2Cl_2), methyl chloroform (CH_3CCl_3), dichloroethane (CH_3CHCl_2), and vinyl chloride (CH_2CHCl), and fully chlorinated carbons such as carbon tetrachloride (CCl_4), tetrachloroethylene (CCl_2CCl_2), and hexachlorobenzene (C_6Cl_6).

Chlorinated hydrocarbons are widely utilized as solvents and chemical intermediates. Since CHCs have high solvent capacities for greases, oils, resins, and waxes, and are nonflammable, some CHCs, eg, CH_2Cl_2, CH_3CCl_3, C_2HCl_3 (trichloroethylene), and C_2Cl_4 (perchloroethylene), are used as solvents in adhesives, aerosols, extraction solvents, industrial cleaning solvents, paint and coating solvents, and pharmaceuticals. They are also used in dry cleaning, vapor degreasing, metal cleaning, and textile processing. CHCs are also intermediates in the manufacture of numerous chemicals. For example, dichloroethane is converted to vinyl chloride for the production of poly(vinyl chloride) (PVC). Methyl chloride (CH_3Cl) is used almost exclusively for the manufacture of silicone. Vinylidene chloride is chiefly used to produce polyvinylidene chloride copolymers used in household food wrap. Chlorobenzenes are important chemical intermediates in the production of disinfectants, thermoplastics, and room deodorants.

The manufacture and use of these compounds has often resulted in the contamination of subsoil and surface waters or air pollution. Many of the waste streams contain significant amounts of chlorinated organics, such as the polychlorinated ethanes and ethylenes vented in the manufacture of vinyl chloride monomer or released in their use as industrial and military solvents. It was estimated that over 90.9×10^6 kg/yr (2×10^9 lbs/yr) of CHC by-products were produced by vinyl chloride monomer plants in the United States (2). Some waste-gas streams may come from area vents, eg, reactor building vents, and process vents including reactor stripper, condenser, drier, and storage vessel vents. Besides the pollution from the use or manufacture of CHCs, chlorinated hydrocarbons are frequently found in hazardous waste streams burned in incinerators (3,4). Many CHCs are resistant to thermal degradation and may form toxic reaction by-products such as dioxin (5). In addition, the vents from some air stripping units used to clean groundwater or contaminated soil contain CHCs. Trichloroethylene (TCE) has been found throughout the United States in groundwater and in aquifers used for potable water.

Hazardous Properties of CHCs

Chlorinated organic compounds have been found to be either toxic or harmful to the health and the environment. Although CHCs, in general, exhibit low acute toxicity, inhalation toxicity is great for some gaseous or volatile liquid compounds, eg, phosgene (6). Permitted exposure levels (PEL) for a selection of chlorocarbons typically used as solvents and cleaners, except for vinyl chloride which is the precursor to poly(vinyl chloride) (PVC), are given in Table 1. The level of concentration permitted depends on the toxicity of each compound. Some CHCs, including 2,3,7,8-tetrachlorodibenzo-*p*-dioxin, polychlorinated biphenyls, polychlorinated dioxins, and chlorinated dibenzofurans, are highly toxic (7,8). Since the late 1970s, chlorinated C_1 to C_3 alkanes have been shown to be mutagens and carcinogens (9). These compounds include trichloroethylene (C_2HCl_3), a common degreasing solvent, perchloroethylene, dry-cleaning fluid, and other commonly used chemicals and solvents such as methylene chloride and chloroform. All of the CHC solvents are skin and eye irritants and may produce adverse effects on the central nervous systems of workers (10). In addition, some CHCs such as carbon tetrachloride and methylchloroform are harmful to the ozone layer (11,12).

Table 1. U.S. Occupational Safety and Health Administration Permissible Exposure Limit–Time Weighted Average (PEL–TWA) Levels

Halocarbon	PEL–TWA, ppm
Phosgene	0.1 (TLV[a])
Vinyl chloride	1
1,1-Dichloroethylene	1
1,2 Dichloroethane	1
1,1,2,2-Tetrachloroethane	1
Carbon tetrachloride	2
Chloroform	2
1,1,2-Trichloroethane	10
Tetrachloroethylene	25
Methyl chloride	50
Trichloroethylene	50
Chlorobenzene	75
1,2-Dichloropropane	75
Methylene chloride	100
1,1,1-Trichloroethane	350

[a] Threshold limiting value.

CHC Emission Regulations

Environmental and health concerns about CHCs have led to stricter emission control regulations. In 1990, ten Titles of the U.S. CAAA set emission standards, compliance dates, and enforcement provisions. Of these Titles, Title III of the CAAA focuses on those industrial processes that typically emit significant quantities of the 189 chemicals designated as hazardous air pollutants (HAPs), including more than 40 CHCs (13). The U.S. EPA is required to issue control standards, based on maximum achievable control technology (MACT) for major and area sources. Major sources are defined as units that emit or have the potential to emit at least 9091 kg/yr (20,000 lbs/yr) of any one of the listed toxics, or 22,727 kg/yr (50,000 lbs/yr) of any combination of them. An area source is a facility that routinely emits HAPs but is not classified as a major source. EPA is required to ensure that 90% of the emissions from the 30 most serious area-source pollutants are regulated by the year 2000 (13). In addition, some CHC solvents, such as CCl_4, are ozone depleters and must be phased out under the Montreal Protocol.

In accordance with U.S. EPA guidelines, chlorinated pollution reduction can be accomplished by material substitution, production reformulation, recycling, process modification, or destruction (14). Thus, the strategies to control CHC emissions can be divided into three categories: reducing the use and manufacture of CHCs by altering the design, operation, maintenance, or manufacturing strategy to reduce the quantity or toxicity of air emissions produced; recovering used CHCs that have high value or high concentration; and destroying the CHCs in the waste stream prior to emission.

The use of chlorinated solvents is predicted to decline by the year 2000 and effective substitutes need to be found. For example, $CHCl_3$ use has declined as a result of environmental concerns, and the use of CH_2Cl_2 will decline in aerosol and paint stripping applications. The Montreal Protocol calls for emissive uses of CCl_4 to be phased out by the year 2000. The production of CH_3CCl_3 is frozen at 1989 levels and will end by 2002. Several large chemical firms in the United States such as DuPont, Monsanto, and Dow Chemical are developing long-range strategic plans to cut back on their use and production of chlorinated organics. Vehicle manufacturers in both Europe and the United States are now trying to move away from the use of chlorinated solvents in their paints (10).

Although much effort has been devoted to finding effective substitutes for chlorinated solvents, the use of these compounds will continue for years and their disposal or destruction will remain an important ecological problem. Thus, waste streams containing CHCs will need to be treated to control the emission of CHCs. Valuable chlorinated compounds or high concentration of CHCs can be recovered by various technologies including activated carbon adsorption and condensation (15). Carbon adsorption systems can remove and recover CHCs by adsorbing them in a packed bed of activated carbon. However, carbon adsorption may not be effective at low CHC concentrations and normally is less effective at high humidity levels (16). Thus, for low CHC concentrations, destruction is more economically attractive.

Technologies for CHC Emission Control

Depending on the reaction regent, destruction technology can be divided into several processes, including pyrolysis, incineration (both thermal and catalytic), and hydrogenolysis. In pyrolysis, organic compounds are decomposed to smaller molecules at high temperatures in the absence of oxygen. In comparison, hydrogenolysis converts CHCs to nonchlorinated hydrocarbons and hydrogen chloride by reacting CHCs with hydrogen. Nonchlorinated hydrocarbons can be burnt and used as a fuel, and hydrogen chloride can be scrubbed with a basic solution. This process requires high pressure and usually is used for the destruction of concentrated or liquid CHCs.

The most commonly used destruction techniques are thermal and catalytic incineration in the presence of oxygen. Normally, the thermal incineration process operates at higher temperatures than catalytic incineration. In catalytic or thermal incineration of CHCs, the CHC-laden waste gas is preheated by heat exchange with the hot exit gases, is further heated by the direct combustion of natural gas or oil, passes through a mixing zone, and enters the catalyst bed or combustion chamber. As the CHC-laden air passes over the reaction zone, the CHCs are oxidized, releasing heat. The cleaned air exits the bed, passes through a heat exchanger, and enters an absorption column where it is contacted with a dilute stream of aqueous caustic that removes the acid (HCl) formed from the combustion of the chlorine.

ACTIVATED CARBON ADSORPTION

There are two main applications of carbon-adsorption technologies: the recovery of valuable industrial solvents for reuse, and the removal of organic compounds for air pollution control and purification. Since it captures rather than destroys pollutants, adsorption is the preferred method for streams containing valuable and recoverable chemicals. Several innovative VOC recovery systems have been commercialized, including open-loop VOC recovery, closed-loop benzene recovery, and rotor concentrator (17). CHCs are a subset of VOCs, and can also be controlled by commercially available VOC-adsorption technologies. For example, carbon adsorption is used in the cleaning of landfill gases, which contain mainly methane and are burnt to produce thermal and electrical energy. Since the industrial utilization of such gases is restricted because of the presence of numerous impurities, including CHCs, they are removed by activated carbon (18).

A carbon adsorption process normally consists of three steps: adsorption (organic compound concentrating), desorption (carbon regeneration), and condensation (organic compound recovery). Typically, the vapor-laden gas is passed through a packed bed of activated carbon. Organic vapors are adsorbed onto the surface and condensed in the pores of the carbon. These adsorbed compounds are periodically desorbed from the carbon through regeneration by hot air or steam, and are ultimately recovered by condensation (15). To operate this adsorption/regeneration cycle continuously, two carbon absorber beds are used (Fig. 1). After a fixed amount of time, or when breakthrough occurs, the stream or air is switched to another bed, so the first bed can be regenerated (see ACTIVATED CARBON ADSORPTION FUNDAMENTALS).

THERMAL DESTRUCTION OF CHCs

Thermal destruction of CHCs is a process by which CHCs decompose at high temperature (normally ≥900°C or 1652°F) and are completely converted to products such as CO_2 and HCl or to other hydrocarbons depending on the inlet composition. The destruction of organic compounds in high temperature environments involves the pyrolytic process (pyrolysis) and the oxidative process (combustion or incineration). The combustion and pyrolysis of chlorinated compounds have been studied extensively in relation to fire suppression (19,20). The mechanisms are generally free radical in nature.

Thermal Incineration of CHCs

Thermal incineration (thermal oxidation) is one of the best known methods for industrial waste disposal and is an effective technology for the total oxidation of the chlorinated organic wastes (21). Incineration employs thermal decomposition in the presence of oxygen at high temperatures (usually ≥900°C (1652°F)) to destroy organic hazardous materials with an efficiency of more than 99%. The oxidation process is highly exothermic in nature. When excess oxygen is present, the breakdown process is enhanced, carbon atoms are bonded to oxygen, and the formation of many of the unsaturated struc-

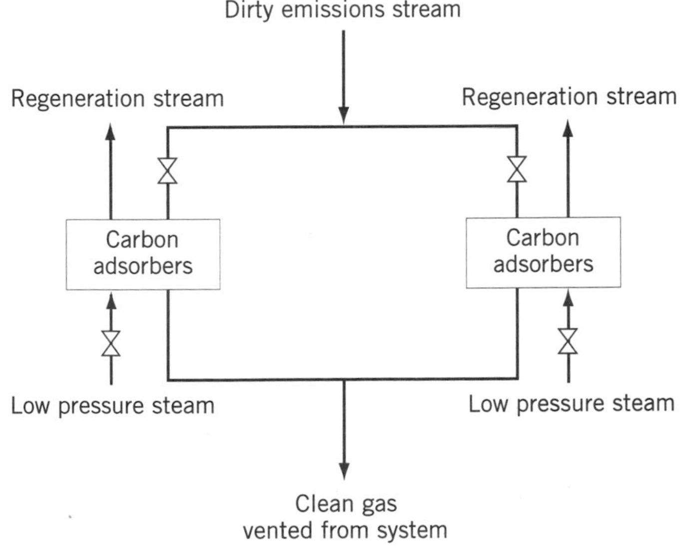

Figure 1. A simplified carbon adsorption process for organic compound removal.

tures is eliminated. Therefore, along with the chain-induced decomposition, there are sharp increases in temperatures, more radicals are created, and the reactions are driven to completion. Generally, wastes with significant organic content are the most appropriate feedstocks for incineration. Technically, however, any waste (solid, liquid, or gas) with a hazardous organic fraction, no matter how small, is at least a functional candidate for incineration (4). As a rule of thumb for the thermal incineration of VOC, thermal incineration is used when the VOC concentration in exhaust is greater than 15% of the lower explosive limit (LEL), or compounds are present that could contaminate a catalyst, or there is a large variability in VOC concentrations (22).

Basically, the hazardous waste stream is first mixed with O_2 and auxiliary fuels and then heated (the combustion process) to the reaction temperature at which the organics are to be decomposed or oxidized. In principle, a completely efficient incinerator converts organic hazardous material to carbon dioxide, water, and a small amount of inert ash. During the incineration of chlorinated hazardous waste streams or when CHCs are present in the waste streams, significant quantities of chlorine acids are also formed. After the incineration, the effluent passes through some type of heat exchanger to recover heat. The gas stream is then passed through a base solution to remove acid gases such as HCl. Carbon dioxide and water vapor can be vented safely and legally to the atmosphere.

The destruction efficiency of the incineration process is determined by several factors, including the gas stream temperature, residence time, turbulence and mixing, and availability of oxygen. The reaction time and temperature depend on the specific compound being destroyed. The gas stream needs to be heated to the temperature at which the oxidation reaction is initiated, and the oxidation temperature must be maintained long enough for the reaction to proceed to a high degree of completion. For destruction of general industrial hydrocarbons that are not extraordinarily toxic, a reaction temperature between 700°C (1292°F) and 815°C (1499°F) and a reaction time between 0.5 and 2.0 s are required (23). These parameters allow the oxidizer to destroy between 95 and 99.9% of the incoming hydrocarbons. Chlorinated compounds can be destroyed in an incinerator's normal operating temperature range (830–1230°C (1526–2246°F)) for liquid injection incinerators, and 630–1330°C (1166–2426°F) for rotary-kiln incinerators (3). The exhaust gas stream must contain a sufficient amount of oxygen and be well mixed to complete the oxidation reaction and to eliminate the formation of byproducts by pyrolysis.

Summary

Thermal incineration is an effective and mature technology for the destruction of CHCs, and the destruction efficiency is normally higher than 99%. However, a high operation temperature (>1000°C (1832°F)) is typically needed for the incineration of CHCs. Thus, the costs of capital (such as building the incinerator) and operations (such as fuel consumption) are usually higher than the cost of competing technologies. Several factors can affect the destruction efficiency of the process, including the operating temperature, residence time, turbulence and mixing, oxygen concentration, and the type of incinerator. Higher efficiency can be achieved with higher temperature, longer residence time, and better mixing. Although excess oxygen is present in the oxidative process, the localized oxygen-starved stoichiometries (pyrolysis condition) result in the formation of PICs, which may be more toxic than the feed materials. A properly designed incinerator can increase heat recovery efficiency, reduce operation cost, improve fuel/air mixing, and therefore decrease the formation of PICs. The nature and yield of PICs and their formation mechanisms depend on the oxygen concentration, temperature, molecular structure of the waste feed components, and the ratio of H:Cl in the feed stream. In general, the amount of PICs significantly decreases with higher oxygen concentration, higher temperature, and higher H:Cl ratio in the feed stream.

CATALYTIC OXIDATION OF CHCs

The catalytic oxidation (incineration) of CHCs is receiving increased attention because a properly designed catalytic process can completely destroy all CHCs at lower temperatures and does not produce the toxic by-products that can result from incomplete combustion of parent CHCs (24,25). Destruction efficiency depends on gas inlet composition and concentration, operating temperature, oxygen concentration, catalyst characteristics, residence time, and space velocity (the volume of incoming gas divided by the volume of catalyst bed). Catalytic oxidizers must be designed to accommodate a specific ratio of incoming air and catalyst volume. If the design ratio is inadequate, or if the ratio changes because large amounts of air must be processed, the systems's destruction efficiency is likely to drop significantly.

Oxidation catalysts offer a degree of freedom not inherent in other control technology strategies. Catalysts can be used to improve the performance of existing thermal oxidation equipment. Use of oxidation catalysts can, in some cases, enhance process operations, reduce operating costs, and be successfully incorporated into existing processes, thus minimizing capital costs. Catalytic incineration is not applicable to all streams, since some airborne contaminants can poison or mask the active surface of the catalyst, rendering it ineffective. Prior knowledge of exhaust stream composition and proper selection of the catalyst can prevent this problem.

A catalytic oxidizer (incinerator) normally contains three major sections: a preheating section, a catalyst bed, and a heat-exchanger section. Three types of systems are used for the catalytic oxidation of VOCs for industrial exhaust control (26). The simplest uses a direct contact open flame to preheat the gas stream upstream of the catalyst. The second involves only a catalyst bed over which the gas stream passes, usually after some indirect preheating. The third involves more extensive indirect preheating and heat exchange. The difference among the three configurations is the method for preheating the gas.

During catalytic oxidation, waste gas is heated to the reaction temperature (typically <600°C or 1112°F) in the preheating chamber (usually by a burner). It is then passed through the catalyst bed where oxygen and CHCs diffuse to the catalyst surface and are adsorbed in the catalyst pores. Oxidation occurs at the active sites on the catalyst surface. Reaction products are ultimately desorbed from the active sites and diffused back into the gas phase. The heat energy of the effluent is then recovered by a heat-exchanger and used either to preheat the process gas before it enters into the preheating section or to generate steam or high temperature air. After heat exchange, the gas stream passes through a wet scrubber that removes acid gases such as HCl. The typical application range for catalytic incineration is 56.6–425 m^3/min (2000–15,000 ft^3/min) and inlet contaminant concentrations that are <15% of the LEL (22).

Catalytic oxidation systems have been used in industries for the destruction of CHCs because a properly designed catalytic process can completely destroy CHCs to form CO_2 and HCl at lower operation temperatures (normally <600°C (1112°F)) without forming PICs and by-products. Catalytic incineration of CHCs have several advantages over thermal incineration, including lower construction cost, lower operating cost, less NO_x formation, less space requirement, and onsite or mobile use. However, the biggest disadvantages of catalytic incineration are the deactivation of the catalyst by fouling or masking, poisoning, aging, or sintering, and the loss of catalyst material through the formation of volatile metal compounds, thereby increasing capital and maintenance costs.

A large effort has been put on the investigation and development of the catalyst for the oxidation of CHCs, and several catalyst systems have been commercialized and used in fixed-bed and fluidized-bed reactors. In general, CHC oxidation catalysts can be divided into three categories: precious metal catalysts (such as Pt, Pd, and Rh), transition-metal oxide catalysts (such as Co_3O_4, NiO, and Cr_2O_3), and acidic oxide catalysts (such as zeolites and TiO_2–Al_2O_3). Precious metal catalysts normally have higher activities for the oxidation of hydrocarbons, but are less active for the destruction of CHCs than the transition-metal oxide catalysts. Higher metal loading and higher operation temperature are usually needed for precious metal catalysts to increase the reaction rate and to overcome chlorine poisoning. Transition-metal oxide catalysts are normally poisoned by water vapor, which is present in virtually all applications. Similarly, acidic oxide catalysts are deactivated in the presence of water vapor. Also, some partially oxidized products (such as CO) and by-products (such as Cl_2) tend to form during the oxidation of CHCs over acidic oxide catalysts. Further investigation is needed to develop a catalyst with high oxidative activity at lower temperatures, high selectivity to HCl and CO_2, long lifetime, and lower cost (see AIR POLLUTION CONTROL, CATALYTIC).

HYDROGENOLYSIS OF CHCs

Detoxification by catalytic hydrogenolysis is a promising alternative to the thermal incineration of CHCs (27). In catalytic hydrogenolysis, the chlorinated hydrocarbons are converted to hydrocarbons and HCl. For example, C_2HCl_3 can react with hydrogen to form CH_4 and HCl, as shown in the following formula:

$$HClC{=}CCl_2 + 5\,H_2 \rightleftarrows 2\,CH_4 + 3\,HCl$$

The resulting hydrocarbons are far less toxic than the chlorinated compounds, and can either be used as a chemical feedstock or burned. The temperatures and residence times needed to destroy hydrocarbons are far lower than those needed for CHCs, and because hydrocarbon incineration is much less likely to pruduce partial oxidation products, hydrocarbon incineration is much cleaner and less expensive than CHC incineration. The HCl product can be readily removed from the reaction stream using an aqueous basic scrubbing bath and converted into an innocuous product such as NaCl (table salt).

The catalytic hydrogenolysis of CHCs has a number of other attractive features. The hydrogenolysis of CHC waste streams can be carried out in a closed reactor system, resulting in a CHC destruction technology with no environmental emissions. Catalytic hydrogenolysis of a CHC waste stream can also produce a hydrocarbon, which can be recycled and reused as a solvent or fuel. Furthermore, a recent analysis of PCB waste treatments found that hydrogenolysis is both more efficient and less expensive than incineration (28).

The catalytic hydrogenolysis of hydrocarbons is a widely used process that is usually carried out at high hydrogen pressures over precious metal catalysts such as platinum and palladium (29). Catalytic hydrogenolysis as a process for the exhaustive dechlorination and detoxification of CHCs is a relatively new idea. There are three main hydrogenation processing methods for CHC detoxification, the primary difference being the temperature at which the hydrogenation is carried out. At high temperatures (>650°C or 1202°F), direct thermal hydrogenolysis is effective for CHC destruction. At more moderate temperatures, 200–600°C (392–1112°F), catalytic hydrogenolysis of CHC vapors can be carried out. At the lowest temperature range (<200°C or 392°F), liquid phase catalytic hydrogenolysis has been successfully carried out. The limitations and advantages of each of these approaches and a brief economic discussion of the use of hydrogenation for CHC detoxification are discussed in the following sections.

Thermal Hydrogenolysis

Thermal hydrogenolysis of CHCs is carried out at high temperatures (650–1000°C or 1202–1832°F) at moderate residence times (5–10 seconds) in the presence of excess hydrogen (5–20 times the CHC concentration). Reactions are generally carried out in quartz reactors or at lower temperatures in quartz-lined steel reactors. At the lower temperatures and shorter residence times, conversions of 10–50% can be achieved. At the highest temperatures and longer residence times, conversions approaching 100% can be achieved. Aliphatic, unsaturated, and aromatic CHCs can be hydrogenated to the respective hydrocarbons and HCl by thermal hydrogenolysis.

An evaluation of the kinetics and thermodynamics of the thermal hydrogenolysis of CHCs has been carried out (30). With few exceptions, the activation enthalpy for hydrogenation of C–Cl bonds ranges from 147 to 255 kJ/mol (35 to 61 kcal/mol). The range of C–Cl bond formation enthalpies is much more restricted and ranges from 314–356 kJ/mol (75–85 kcal/mol) and the enthalpy of formation of HCl is 435 kJ/mol (104 kcal/mol). A direct consequence of these thermodynamic values (ie, the activation energy for hydrogenation is approximately one-half to one-third that of the formation enthalpies of the reactant bonds) is that hydrogenolysis of C–Cl bonds takes place rapidly, with low specificity and similar reaction rates found for different CHCs. Polychlorinated hydrocarbons undergo spontaneously successive dechlorination steps making the overall reaction self-sustained. The overall exothermic nature of the CHC hydrogenolysis reaction makes the treatment of CHC waste streams economically attractive.

Studies of CHC hydrogenolysis have also been carried out using chloroform (31) as a model reactant. Reaction of hydrogen with chloroform leads to the formation of less chlorinated species (ie, CH_2Cl_2, CH_3Cl, $CH_2{=}CHCl$) that are subsequently destroyed at higher temperatures. Formation of CH_3Cl also increases with temperature, with maximum formation rates occurring near 850°C (1562°F). These results suggest that the less chlorinated methanes are more stable with regard to hydrogenolysis, consistent with the bond strengths of the C–Cl bonds in chloromethanes, which increase with decreasing chlorine content. At temperatures above 700°C (1292°F), methane and benzene formation are observed and the formation of acetylene occurs at temperatures above 800°C (1472°F). Significant amounts of carbon (ie, graphite) also occur during hydrogenation of $CHCl_3$ at temperatures between 550°C (1022°F) and 900°C (1652°F). The formation of graphite is thought to occur via free radical chain branching reactions of $:CCl_2$ or $\cdot\ CHCl_2$ radicals that are the principle initial decomposition products of chloroform.

Catalytic Hydrogenolysis

Catalytic hydroprocessing (hydrogenolysis) has long been used in industry for the hydrorefining of petroleum. Large-scale industrial catalytic hydrotreating processes include hydrodenitrogenation (HDN, N elimination), hydrodesulfurization (HDS, S elimination) and hydrodeoxygenation (HDO, O elimination). These processes are highly reliable and the technology continues to advance due to the intensive research performed by oil companies throughout the world. Catalytic hydrogenolysis (ie, hydrotreating) is also a well-known method in synthetic organic chemistry, generally carried out at high hydrogen pressures using carbon-supported precious metals such as platinum and palladium. Unlike the HDN, HDS, and HDO processes, the catalytic hydrodehalogenation of aliphatic and aromatic compounds has not been extensively studied, and, as a method of exhaustive dechlorination for detoxification of chlorinated hydrocarbons, is a relatively recent phenomena. Most of the methods reported for catalytic hydrogenolysis of CHCs or for CHC detoxification use high hydrogen pressures and precious metal catalysts to carry out the dechlorination process (32). However, typical hydrotreating catalysts (ie, sulfided Ni–Mo–γ-Al_2O_3), metal-exchanged zeolites, and transition metal carbides that possess precious metal-like catalytic properties have also been studied. These studies are reviewed below.

Outlook

The outlook for the future of CHC hydrogenolysis techniques is as yet cloudy. Precious metal catalysts and metal carbides deactivate with time during the hydrogenolysis reaction. The Ni–ZSM-5–γ-Al_2O_3 catalysts were also found to deactivate with time. For liquid-phase hydrogenations over Pd–C or Ni–Mo–γ-Al_2O_3 catalysts, little has been done to characterize the lifetimes of these materials. Therefore, the outlook for catalytic hydrogenolysis as a technique for CHC destruction still awaits identification of catalyst compositions that provide high rates and resist chloride deactivation. From an economic standpoint, the advantages of catalytic hydrogenation of chlorinated organics include lower capital and operating costs, relatively low temperature operation, recovery of the dehalogenated hydrocarbon product and HCl as value-added materials, and the lack of highly toxic products of incomplete combustion.

EMERGING TECHNOLOGIES

Several unconventional technologies have recently been reported or patented for the CHC emission control. These technologies include the absorption of CHCs by a solution containing an aprotic polar compound, destruction and absorption by alkali metal oxides, catalytically stabilized thermal (CST) combustion, photo-oxidation of CHCs, microwave discharge plasma degradation of CHCs, and biological destruction of CHCs.

Absorption of CHCs

An absorption process for treating halogenated hydrocarbon-containing gas was recently patented (33). This process comprises two steps: (*1*) the gas-containing halogenated hydrocarbon is contacted with a solution, which contains an aprotic polar compound having a five- or six-membered ring and a nitrogen at an alpha-position of a carbonyl (examples are shown in Fig. 2), to absorb and recover a portion of the halogenated hydrocarbon, and (*2*) the gas effluent exhausted from the first step contacts, with heating, with a mixture solution of an aprotic polar compound and a caustic alkali to decompose the halogenated hydrocarbon.

It has been found that the aprotic polar compounds are good absorption solvents for the halogenated hydrocarbons, such as hydrocarbon chloride or Freon gas. In the first step, a conventional gas–liquid contact process is employed. The absorbed halogenated hydrocarbon may be recovered or desorbed in a recovery step, which is carried out by distillation, evaporation, or introduction of heated air or nitrogen at a high recovery or a high purity. The recovered halogenated hydrocarbon and aprotic polar compound can be used again as an absorbent. The second step is also conducted using the same gas–liquid contacting apparatus as is explained in the first step. The halogenated hydrocarbon is reacted with the caustic alkali to decompose it to lower molecular weight compounds. In case of 1,1,1-trichloroethane, the following reaction may occur (in a post-combustion chamber with a slight excess of air):

$$CH_3CCl_3 + 4\,NaOH \rightleftarrows 3\,NaCl + CH_3COONa + 2\,H_2O$$

According to the invention, the halogenated hydrocarbon is effectively absorbed in the absorbing process, and then is decomposed at a high rate, thus reducing the content of the halogenated hydrocarbon to a degree at which it is not substantially present. The absorbed halogenated hydrocarbon is recovered (33).

Catalytic/Absorbing Process

A process for the chemical thermodecomposition of higher halogenated hydrocarbons by reaction with calcium oxide or calcium hydroxide was developed (34). In this process, excess

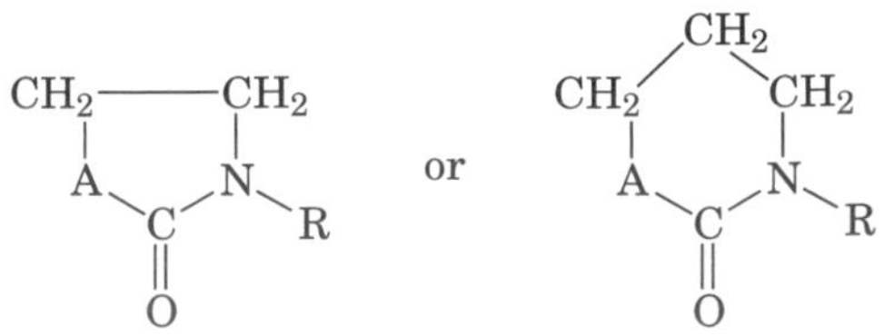

Figure 2. Chemical formula for aprotic polar compounds.

amounts of CaO or $Ca(OH)_2$ with a portion of iron oxide was used at 600–800°C (1112–1472°F) to react with halogenated hydrocarbons to form $CaCl_2$ and nonhalogenated hydrocarbons in a stirred-bed reactor. The waste gases escaping from the reactor during the process are free from halogenated hydrocarbons and are post-burned

An interesting catalytic system based on oxides of copper and manganese supported on alkali or alkaline-earth carbonates, which combines the oxidation and removal steps, has been reported (35–38). In this process, CHCs are first completely oxidized over the $CuO-MnO_2$ catalyst to CO_2 and metal chlorides such as $CuCl_2$. The chlorides then react with supports such as Na_2CO_3 to form NaCl and CuO (38). Once the sodium carbonate support is used up, the catalyst/reactant can be reclaimed by dissolving the salt and reusing the metals. The advantages of this process are inexpensive catalyst and equipment, the possibility of small units due to the simple nature of the process, and the ease of the solid's disposal. The catalytic metals and support are environmentally friendly: typical support materials include limestone, MgO, and potassium carbonate.

Laboratory experiments have shown that 99% conversion of the VOC can be achieved in the 270 to 400°C (518–752°F) temperature range and 6,700 h^{-1} space velocity. For example, with 31 ppm trichloroethylene, a 99% conversion level is reached at 270°C (518°F). A chlorine balance shows that more than 98% of the chlorine from converted TCE remains in the catalyst sorbent. Under the conditions of 7.5 ppm TCE; GHSV 12,000 h^{-1}; and 400°C (752°F), this catalyst would be expected to last approximately one month (38).

Catalytically Stabilized Thermal Combustion

By combining the catalytic oxidation and thermal incineration processes, the catalytically stabilized thermal (CST) combustor was designed to use catalytic surface oxidation reactions to help ignite and sustain homogenous gas-phase reaction (39–41). The CST combustor is different from catalytic oxidation reactors. Catalytic oxidation reactors are low-throughput, moderate-temperature, and high–residence time devices, whereas CST combustors are high throughput, high temperature (>927–1277°C (1700–2330°F), and low residence time devices. Catalytic oxidation reactors (eg, the catalytic convertor in a car) primarily utilize heterogeneous catalytic surface reactions and lower operating temperatures than CST combustors. Because the CST combustor utilizes gas-phase combustion, it is capable of achieving high rates of destruction (per unit volume) not achievable with only catalytic surface reactions. The CST combustor also has advantages over the conventional thermal incinerators. With a CST combustor in a tubular flow configuration, the size of an effective incinerator may be reduced significantly by reducing the required average residence time. This results from catalytic stabilization overcoming halogen flame inhibition sufficiently to permit plug flow combustion (ie, without significant back-mixing). With combustor size requirements significantly reduced, capital and operating costs could also be brought down and on-site or mobile incineration made more economic.

The CST combustor has been demonstrated to be an effective burner for the thermal destruction of CH_3Cl and CH_2Cl_2 (109). For all cases where stable combustion was achieved, destruction effectiveness of total hydrocarbons was beyond detector limits (corresponding to >99.994% destruction efficiency) using a 5-ms catalytic section residence time. Both Pt and a binary $Cr_2O_3-Co_3O_4$ catalyst were tested, with the $Cr_2O_3-Co_3O_4$ catalyst observed to be the more effective in the presence of the chlorinated fuels. Lean stability limits for $CH_3Cl-C_3H_8-$air mixtures and $CH_2Cl_2-C_3H_8-$air mixtures each appear to be the same function of the total Cl:H ratio for the conditions studied (42).

Photo-Oxidation of CHCs

Volatile organics can be photo-oxidized in air using an ultraviolet light produced by a xenon flash lamp (43). In this investigation, a high-intensity six-inch xenon flash lamp was inserted in the middle of the 208-L (7.3 ft^3) steel cylindrical reactor, and the photo-oxidation of trichloroethane, perchloroethane, 1,1,-dichloroethane, chloroform, and methylene chloride were demonstrated at the temperature ranged from 23 to 63°C (73–145°F). Preliminary products are dichloroacetyl chloride and phosgene for TCE oxidation. It is reported that species with higher uv absorbances have higher destruction rates. The destruction rates in some mixtures may exhibit favorable synergistic interactions if one compound with higher absorbance induces a chain carried on by a compound of low uv absorbance. A full-scale air emission control system (>95% removal) for TCE has been constructed, capable of continuous treatment of at least 85 m^3/min (300 standard ft^3/min) with a residence time of about two seconds. It is concluded that the application of this technology to groundwater treatment via air stripping is possible.

Microwave Discharge Plasma Degradation of CHCs

A microwave discharge plasma contained in a tubular flow reactor is reported to effectively detoxify diluted trichloroethylene and 1,1,1-trichloroethane with water vapor or molecular oxygen in an argon gas stream at atmospheric pressures (44). Destruction of removal efficiencies ranging from 95 to 99% are observed. Both the extent of reaction and the product distribution depend upon the power input to the reactor; the primary products are CO, CO_2, HCl, Cl_2 and H_2.

Biological Destruction of CHCs

Another method for promoting the destruction of toxic vapors was reported to use fixed-bed bioreactors in which suitable microorganisms are immobilized on an inert support (45). The degradation of methane, trichloroethylene, and *p*-xylene using methanotrophic bacteria was investigated, and gas-phase bioreactors appeared to offer significant potential for the cleanup of selected gaseous effluent. It is suggested that this technology is in competition with conventional aqueous-phase bioreactors, and that solutes with higher aqueous solubilities may be treated more effectively as liquids.

In summary, new technologies for CHC destruction are continuously being investigated and developed. Technologies such as the catalytic/absorbing process are encouraging because they are competitive with commercially available technologies in destruction efficiency and catalyst cost. Further investigation is needed to narrow the gap between laboratory and industrial-scale applications. To prove that the process is economically feasible for the destruction of CHCs in some cases, scale-up and economic analysis need to be further investigated before these new technologies can be commercialized.

PRODUCT ANALYSES FOR CHC DESTRUCTION

At the present time there are no simple methods for detecting CHCs and their destruction products, especially when the concentration of CHCs is low. Depending on the destruction processes (eg, pyrolysis, incineration, and hydrogenolysis), the products of CHC destruction are different. Commonly existing compounds in the outlet of the reactor can be broadly divided into three groups: nonchlorinated organics (eg, hydrocarbons and aldehydes), chlorinated hydrocarbons (including perchlorinated and partially chlorinated carbons), and inorganic products such as CO, CO_2, HCl, Cl_2, and $COCl_2$. The analysis of all of these compounds require the use of several different detection techniques.

Gas chromatography (GC) is a common technique for the analysis of CHCs and their destruction products. The electron capture detector (ECD) (46) and Hall electrolytic conductivity detector (ELCD) (47,53) are normally used for the detection of CHCs. The ELCD is highly sensitive and selective to chlorine-containing compounds. The detection limit of the ELCD is normally at the ppb level, and nonchlorinated hydrocarbons do not interfere with the signals of CHCs. Although the ECD is sensitive, too, it is not convenient for analyzing low molecular weight gaseous halocarbons. GC using a flame ionization detector (FID) is applicable for ppm-level detection of both CHCs and nonchlorinated organics. CO and CO_2 can be analyzed using GC with an FID. After separating the fixed gases, they are individually hydrogenated over a nickel catalyst to methane. Thus, CO and CO_2 are quantitatively measured as CH_4 which is detected by FID (48). For the detection of high concentration of CHCs, GC equipped with a thermal conductivity detector (TCD) can be used (49). Several GC columns (both packed and capillary column) can be used to separate CHCs and nonchlorinated organics; examples of these columns include Porapak, Chromosorb, Carbograph, and Tenax.

Halocarbons that are EPA-listed priority pollutants may be analyzed by EPA Methods 601 (for waste waters), 502 (for potable waters), and 8010 (for soils, sediments, and hazardous wastes). These are all purge-and-trap GC methods, in which the aqueous samples or an aqueous solution of a soil sample is purged with an inert gas, trapped over adsorbent columns, desorbed with a carrier gas onto a capillary or packed column, and analyzed by GC using a Hall detector. Gaseous substances may be analyzed either at low temperatures under cryogenic conditions or by using capillary column 60 m or greater in length. Chlorinated hydrocarbons in air may also be analyzed by various NIOSH (National Institute for Occupational Safety and Health) methods (6,50). A known volume of air is passed over coconut shell charcoal. The adsorbed chlorohydrocarbons are desorbed with a suitable solvent such as carbon disulfide, propanol, benzene, or toluene, and analyzed by GC using mostly FID or, for certain compounds, ECD as detectors. Alternatively, chlorinated hydrocarbons may be analyzed by GC/MS (gas chromatography/mass spectrometry) as using a purge-and-trap technique (EPA Methods 624, 524, and 8240) or a thermal desorption method. GC/MS analysis is a confirmatory test in which the primary and secondary ions characterizing the compounds can be identified and quantified.

High concentrations of Cl_2 and HCl may be detected by GC-TCD techniques (49). However, low concentrations of HCl or Cl_2 react with or be strongly adsorbed on GC-column (either metal tubing or packing materials) or fittings, and thus it is difficult to qualitatively detect these gases. In addition, even though both HCl and Cl_2 can pass through some columns (eg, capillary column), the separation of HCl and Cl_2 may be difficult. Usually, chemical titration can be used for the detection of HCl, and detector tubes or papers can be used for the analysis of HCl, Cl_2, $COCl_2$, etc. In the chemical titration method, the HCl must first be collected by trapping it into either a water or base solution and then titrated by $AgNO_3$. This method can be used during the steady-state operation, but it is difficult to quickly indicate the product distribution during the change of operating conditions because of the time required for the collecting step. The Dräger gas color detector tubes are easy to use, and usually can selectively indicate the concentrations of different compounds such as Cl_2, HCl, and $COCl_2$ (51,52,53). The Dräger tube can be easily used by directly attaching it to the outlet of the reactor. After a known volume of gas has passed through the tube, the concentration of the detecting compound can be calculated by the scale shown on the tube. However, these detectors do not have a high degree of precision; according to the manufacturer, concentrations determined using Dräger tubes are $\pm 10-15\%$ accurate.

The quantitative measurement of products of destruction of CHCs may be possible using Fourier transform infrared (FTIR) spectrometry equipped with a long-path gas cell (54,55). Infrared analysis is attractive because the C–Cl stretching region is from 500 to 750 cm^{-1}, a region relatively free from water and carbon dioxide interferences, and the spectra of HCl and other simple chlorine-containing species are well known (56). The spectrometer has a large dynamic range, allowing quantification from the sub-ppm range to the percent range, especially when used with adjustable path length cells. A high-resolution interferogram (0.25 cm^{-1}) of a sample may be taken in a few minutes, much faster than conventional GC analysis. The technique, however, cannot be used without careful consideration of spectral conditions for each species to be measured. Several calibration spectra must be examined to ensure correct quantitative analysis, and large extrapolations should be avoided (55).

SUMMARY

The wide use of chlorinated hydrocarbons (CHCs) in industries have resulted in the regulation of CHC emissions by the EPA under the 1990 CAAA because of their toxicity or harmfulness to health and the environment. Thus, emission control technologies are needed to meet the stringent regulations. Several types of control methods, such as activated carbon adsorption, thermal incineration, and catalytic incineration, are extensively investigated and widely used in industry while new technologies are continuously being developed or investigated. The new technologies that are being studied include hydrogenolysis, absorption, photo-oxidation, microwave discharge plasma degradation, and biological destruction.

Similar to the nonhalogenated VOCs, chlorinated hydrocarbons can be removed or recovered by activated carbon adsorption from waste air streams. Activated carbon first adsorbs and concentrates CHCs in the adsorption step, and then releases them in a concentrated form during the regeneration step. The adsorption of CHCs is usually quite costly because the adsorption rate of CHCs on carbon is normally poor, especially in the presence of water vapor, and the regeneration of carbon is expensive and energy intensive. Carbon adsorption technology is not a destruction process; it merely concentrates CHCs. If CHCs are removed from the activated carbon, the removal process produces water or air streams, which are polluted with these compounds. The stream then causes a disposal problem. On the other hand, if the chlorinated hydrocarbons are not removed from the carbon, the spent carbon becomes a hazardous solid waste disposal problem.

Thermal incineration is an effective technology for the destruction of CHCs, and is widely used in industry because of its maturity. The CHC-contaminated gas streams are mixed with air and/or fuel, and burnt at high temperatures (normally >1000°C (1832°F)) in an incinerator. The destruction efficiency is normally higher than 99%, and the higher efficiency can be achieved with higher temperature, longer residence time, and better mixing. The capital cost (such as building the incinerator) and operation cost (such as fuel consumption) are usually higher than those of competing technologies. Although excess oxygen is present in the oxidative process, the localized oxygen-starved stoichiometries (pyrolysis condition) result in the formation of PICs, which may be more toxic than the feed materials. A properly designed incinerator can increase the heat-recovery efficiency, reduce the operation cost, improve the fuel–air mixing, and therefore decrease the formation of PICs. The nature and yield of PICs and their formation mechanisms depend on the oxygen concentration, temperature, molecular structure of the waste feed components, and the ratio of H:Cl in the feed stream. In general, the amount of PICs significantly decreases with higher oxygen concentration, higher temperature, and higher H:Cl ratio in the feed stream.

Catalytic oxidation systems have been widely used in industries for the destruction of CHCs because the properly designed catalytic process can completely destroy CHCs to form CO_2 and HCl at lower operation temperatures (normally <600°C (1112°F)) without the formation of PICs and by-products. Because of the use of low operating temperatures, the catalytic incineration of CHCs has several advantages over thermal incineration, including lower construction cost, lower operating cost, less NO_x formation, less space requirement, and on-site or mobile use. However, the biggest disadvantage of the catalytic incineration is the deactivation of the catalyst. In general, the CHC oxidation catalysts can be divided into three categories: precious metal catalysts, transition-metal oxide catalysts, and acidic oxide catalysts. Each group of catalysts has its own advantages and disadvantages; thus, further investigation is needed to develop a catalyst with high activity at low temperatures, high selectivity to HCl and CO_2, long lifetime, and lower cost.

The technologies for the hydrogenolysis of CHCs are still under development. From an economic standpoint, the advantages of catalytic hydrogenation of chlorinated organics include lower capital and operating costs, relatively low temperature operation, recovery of the dehalogenated hydrocarbon product and HCl as value-added materials, and the lack of highly toxic PICs. The major problem for this process is catalyst deactivation. Therefore, the outlook for catalytic hydrogenolysis as a technique for CHC destruction still awaits identification of catalyst compositions that provide high rates and resist chloride deactivation. Some emerging technologies for CHC destruction are continuously being developed. However, further investigation is needed to narrow the gap between the laboratory scale and the industrial scale and reduce the cost of the destruction of CHCs.

BIBLIOGRAPHY

1. D. Hanson, *Chem. Eng. News,* 4 Mar. 12, (1990).
2. J. S. Benson, *Hydrocarbon Processing,* **58 (10),** 107 (1979).
3. E. T. Oppelt, *J. Air Pollut. Control. Assoc.* **37,** 558 (1987).
4. C. R. Dempsey and E. T. Oppelt, *Air & Waste,* **43,** 25 (1993).
5. P.H. Taylor and B. Dellinger, *Environ. Sci. Tech.* **22,** 438 (1988).
6. P. Patnaik, *A Comprehensive Guide to the Hazardous Properties of Chemical Substances,* NY, Van Nostrand Reinhold, 1992, pp. 333.
7. B. Hileman, *Chem. Eng. News,* 11 Apr. 19, (1993).
8. D. Hanson, *Chem. Eng. News,* 22, Jan. 1, (1996).
9. E. Eder, *Chemosphere,* **23,** 1783 (1991).
10. B. Hileman, *Chem. Eng. News,* 5 Sept. 6, (1993).
11. F. M. Church and F. E. Shephard, *Gas Eng. Managem.* **29,** 282 (1989).
12. M. McFarland and J. Kaye, *Photochem. Photobiol,* **55,** 911 (1992).
13. J. C. Summers, J. E. Sawyer, and A. C. Frost, R. G. Silver and co-workers, eds., *Catalytic Control of Air Pollution: Mobile and Stationary Sources,* American Chemical Society, Washington, D.C., 1992, p. 98.
14. L. R. Ember, *Chem. Eng. News,* 7 (July 8, 1991).
15. K. S. Kumar, R. L. Pennington, and J. T. Zmuda, *Chem. Eng.,* 12 (June 1993).
16. M. S. Jennings and co-workers, *Catalytic Incineration for Control of Volatile Organic Compound Emissions,* Noyes Publications, Park Ridge, N.J., 1985.
17. J. R. Graham and M. Ramaratnam, *Chem. Eng.,* 6 (Feb. 1993).
18. M. Schäfer, H. Schröter, and G. Perschel, *Chem. Eng. Tech.,* **14,** 59 (1991).
19. M. M. Ibiricu and A. G. Gaydon, *Combust. Flame,* **8,** 51 (1964).
20. W. E. Wilson, J. T. O'Donovan, and R. M. Friston, *12th Symposium (International) on Combustion, The Combustion Institute,* Pittsburgh, Pa, 1969, p. 929.
21. J. Josephson, *Environ. Sci. Technol.* **18,** 222A (1984).
22. D. Drohan, *Pollution Engineering,* 30 (Sept. 15, 1992).
23. J. M. Klobucar, *Chem. Eng. Prog.,* 57 (Apr. 1995).
24. WO 90/13352 (1990), G. R. Lester.
25. W. Chu and H. Windawi, *Chem. Eng. Prog.,* 37 (March, 1996).
26. R. L. Berglund, in J. I. Kroschwitz and M. Howe-Grant, eds., *Kirk-Othmer Encylopedia of Chemical Technology,* 4th ed., Vol. 9, John Wiley & Sons, Inc., New York, 1994, p. 1022.
27. F. Gioia, E. J. Gallagher, and V. Famiglietti, *J. Hazard. Materials,* **38,** 277–291 (1993).
28. T. N. Kalnes and R. B. James, *Environ. Progr.* **7**(3) 185 (1988).
29. M. Freifelder, *Catalytic Hydrogenation in Organic Synthesis,* John Wiley & Sons, Inc., New York, 1978.
30. A. Converti, M. Zilli, D. M. De Faveri, and G. Ferraiolo, *J. Hazard. Mater.* **27,** 127 (1991).
31. S. C. Chuang and J. W. Bozzelli, *Environ. Sci. and Technol,* **20,** 568–574 (1986).
32. E. N. Balko, E. Przybylski, and F. V. Trentrini, *Appl. Catal., B: Environ.* **2,** 1 (1993).
33. U.S. Pat. 5,340,555, (Aug. 23, 1994), F. Mashio and N. Itoya (to Nippon Paint Co., Ltd., Japan).
34. U.S. Pat. 4,654,203, (Mar. 31, 1987), P. G. Maurer and D. Neupert (to Nukem GmbH, Germany).
35. U.S. Pat. 5,021,383 (June 4, 1991), J. M. Berty (to Berty Reaction Engineers, Ltd.).
36. U.S. Pat. 5,114,692 (May 19, 1992), J. M. Berty (to Berty Reaction Engineers, Ltd.).

37. J. M. Berty and co-workers, in L. Guczi and co-workers, eds., *Proceedings of the 10th International Congress on Catalysis,* July 19–24, Elsevier Science Publishers, Budapest, Hungary, 1993, p. 1571.
38. H. G. Stenger, Jr., G. E. Buzan, and J. M. Berty, *Apply. Catal. B: Environ.* **2,** 117 (1993).
39. W. C. Pfefferle and L. D. Pfefferle, *Prog. Energy Combust.* **12,** 25 (1986).
40. W. C. Pfefferle and L. D. Pfefferle, *Catal. Rev.-Sci. Eng.* **29,** 219 (1987).
41. D. L. Trimm, *Appl. Catal.* **7,** 249 (1983).
42. S. L. Hung and L. D. Pfefferle, *Environ. Sci. Technol.* **23,** 1085 (1989).
43. P. G. Blystone, and co-workers, *Emerging Technologies in Hazardous Waste Management III,* in D. W. Tedder and F. G. Pohland, eds., ACS Symposium Series 518, 1993, p. 380.
44. T. R. Krause and J. E. Helt, in D. W. Tedder and F. G. Pholand, eds., *Emerging Technologies Hazardous Waste Management III,* ACS Symposium Series 518, 1993, p. 393.
45. W. A. Apel and co-workers, in D. W. Tedder and F. G. Pohland, eds., *Emerging Technologies in Hazardous Waste Management III,* ACS Symposium Series 518, 1993, p. 411.
46. D. Miller and E. Grimsrud, *Anal. Chem.* **51,** 851 (1979).
47. C. P. Nulton, C. L. Haile, and D. P. Redford, *Anal. Chem.* **56,** 598 (1984).
48. T. C. Yu, H. Shaw, and R.J. Farrauto, *Catalytic Control of Air Pollution: Mobile and Stationary Sources,* R. J. Silver and co-workers, eds.,American Chemical Society, Washington, D.C., 1992, p. 141.
49. P. Manning, *Hazard. Waste,* **1,** 41 (1984).
50. *Manual of Analytical Methods,* 3rd ed., National Institute for Occupational Safety and Health, Cincinnati, Ohio, 1984.
51. K. Ramanathan and J. J. Spivey, *Combust. Sci. Tech.,* **63,** 247 (1989).
52. D. M. Papenmeier and J. A. Rossin, *Ind. Eng. Chem. Res.* **33,** 3094 (1994).
53. B. Chen, C. Bai, R. Cook, J. Wright, and C. Wang, *Catal. Today,* **30,** 15 (1996).
54. R. Seeker and C. P. Koshland, *Combust. Sci. Tech.* **74,** i (1990).
55. M. J. Hall, D. Lucas, and C. P. Koshland, *Environ. Sci. Technol.* **25,** 260 (1991).
56. K. Nakamoto, *Infrared Spectra of Inorganic and Coordination Compounds,* John Wiley & Sons, Inc., New York, 1963.

CHLOROFLUOROCARBON REPLACEMENTS: APPLICATIONS OF

RAJAT S. BASU
AlliedSignal Inc.
Buffalo, New York

CHEN-HWA CHIU
Bechtel Corporation
Houston, Texas

Starting in the 1940s, chlorofluorocarbons (CFCs), a class of chemicals which combined low toxicity, nonflammability, and good stability, had revolutionized the refrigeration industry along with other industries such as foam insulation, aerosol, and specialty cleaning. Over a billion kg of CFCs were used globally before 1986, as a refrigerant in refrigeration and air conditioning systems, an insulating gas in foam insulation, a specialty solvent in cleaning printed circuit boards and aerospace components, and in aerosol spray cans for various uses. In 1974, the depletion of the stratospheric ozone layer by a series of human-made chemicals, namely CFCs, was discovered. Molina and Rowland were awarded the Nobel Prize in chemistry in 1995 for this monumental discovery (1). The stratospheric ozone layer protects the earth from the sun's harmful ultraviolet rays and its depletion will have numerous severe and harmful effects on the earth.

This discovery led to enormous global research activity to understand the phenomena and its effect on the ozone layer, and to find alternatives to CFCs. As a result of this research activity, the depletion of the ozone layer hypothesis was firmly established, and was considered the biggest environmental problem facing the earth.

In 1986, the United Nations Environment Programme (UNEP) proposed the Montreal Protocol to ban the use of CFCs by the year 2000 in developed countries, and later in lesser developed countries. This treaty and its amendments were ratified by the UN member countries to have an even earlier ban of CFCs by the year 1995 (2). This resulted in the complete phase-out of CFCs and other ozone-depletion chemicals by the end of 1995.

Besides ozone depletion, another harmful effect of CFCs is the so-called greenhouse effect or global warming effect. Global warming occurs because of the entrapment of thermal radiation by water vapor, carbon dioxide, and other gases in the atmosphere, thus preventing its re-emission as long-wave radiation. This leads to an increase in the planet's surface temperature and induces climate changes. Global warming due to CFCs is minimal compared to the effects of carbon dioxide. However, when developing alternatives, compounds having a minimal contribution to global warming should be chosen.

HISTORICAL PERSPECTIVE

The first CFC, dichlorodifluoromethane (CFC-12), was invented by Midgley (3), an engineer at Frigidaire. CFC-12 had low boiling points and appropriate physical properties to be used as a refrigerant. CFC-12, with its extremely inert properties of nonflammability, low toxicity, and chemical stability, caused changes in the refrigeration, air conditioning, aerosols, and later in polyurethane and specialty cleaning industries. The use of CFCs at home, in industrial locations, and manufacturing facilities made these areas safer and more energy efficient. Later, other CFCs were invented and used in various applications. Table 1 below shows CFC properties, lifetimes, ozone depletion potentials, and uses. Worldwide CFC production peaked at about one billion kg in 1973.

After the discovery of ozone depletion, extensive atmospheric measurements were done to determine the extent of the damage, if any, to the stratospheric ozone layer. To determine the thickness of the ozone layer, atmospheric scientists measured back-scattered ultraviolet light from the ozone layer using remote sensing satellites. Ozone profile measurement from ground-based stations were also recorded. It was confirmed that the ozone hole appeared to be largest above Antarctica; polar stratospheric clouds of ice particles catalyzed the destruction of ozone molecules in the stratosphere (4).

Table 1. Chlorofluorocarbon Properties and Uses

Compound	Code Number	Normal bp, °C	Refrigerant Application	Azeotrope and bp, °C	Other Applications	Ozone Depletion Potentials, CFC − 11 = 1	Greenhouse Warming Potentials, $CO_2 = 1.0$[a]
CF_3H	R-23	−81	Biomedical freezers	With R-13, −88		0	12,100
CF_3Cl	R-13	−82	Biomedical freezers	With R-23, −88		1.4	11,700
CF_3Br	R-13Bl	−57	Biomedical freezers		Fire extinguisher		
$CHClF_2$	R-22	−41	Room air conditioning, supermarket freezers, industrial refrigeration, Japanese domestic refrigerators	With R-115 (R-502), −46	Foam blowing agent, feedstock for polymers	0.055	1600
CF_3CF_2Cl	R-115	−38	Supermarket freezers	With R-22 (R502), −46		0.5	7000
$CClF_2$	R-12	−30	Domestic refrigeration, automobile air conditioning, supermarket fresh food stores, industrial refrigeration, hot climate air conditioning		Aerosols, foam blowing agents, sterilant gas	1.0	7100
CHF_2CH_3	R-152a	−24		Azeotrope with R12 (R-500)		0	150
$CClF_2CClF_2$	R-114	3	Low pressure air conditioning			0.8	4500
CCl_3F	R-11	24	Water chiller air conditioning		Foam blowing agent, solvent, medical aerosols	1.0	3400
$CF_2ClCFCl_2$	R-113	48	Train switch gear coolant		Solvent, dry-cleaning agent, contact cleaner, carrier agent	0.8	1600

[a] 100-yr time horizon.

Because of their exceptionally high chemical stability, CFCs emitted at the earth's surface are totally stable in the troposphere and diffuse unchanged to the stratosphere, where they undergo photolytic dissociation under the intense UV radiation of the sun to produce highly reactive Cl atoms:

$$CCl_3F + h\nu \rightarrow CCl_2F + Cl^{\cdot}$$

Chlorine atoms thus liberated react with ozone to form chlorine monoxide and oxygen. Chlorine monoxide reacts with molecular oxygen to liberate atomic chlorine along with oxygen:

$$Cl^{\cdot} + O_3 \rightarrow ClO + O_2$$

$$ClO + O \rightarrow Cl^{\cdot} + O_2$$

This chlorine atom continues the chain reaction to destroy ozone molecules. Approximately one chlorine atom can destroy a million ozone molecules before reacting to form some stable and inactive molecule. Water molecules in the form of ice particles act as an essential catalyst in this continuing reaction. Similar but much less pronounced ozone depletion has been found at the North Pole. The effect is less because of the absence of polar stratospheric clouds in the North Polar region. However, extensive atmospheric studies of the ozone trends have shown that ozone depletion has occurred over the entire earth. Depletion is less near the equator.

In 1985, the Vienna Convention outlined the nations' responsibility to protect human health and environment against the adverse effects of CFCs. Later, the United Nations Environment Programme proposed the Montreal Protocol to phase out production and consumption of CFCs and other ozone-depleting compounds.

In its final form, the Montreal Protocol stipulates the phase-out of CFCs, some chlorocarbons (eg, CCl_4), and Halons by the year 1995; lesser developed countries (LDCs) are allowed 10 more years to phase these out. HCFCs as a class were included later in the phase-out by the UNEP committee because they also have a smaller nonzero effect on the depletion of the ozone layer. Their phase-out dates are stretched out further in the future.

Environmental Acceptability Study

To test alternatives to CFCs for their environmental, health, and safety characteristics, 17 of world's largest CFC producers joined to form the Alternative Fluorocarbon Acceptability Study (AFEAS) and the Programme for Alternative Fluorocarbon Toxicity Testing (PAFT). These two programs use laboratories and resources worldwide to assess environmental and toxicity studies of alternatives.

In 1988, AFEAS began analyzing available data on CFCs and proposed alternatives. These assessments indicated that the potential alternatives have little or no ozone-depletion potential and have far less effect on global warming. The environmental impact resulting from the breakdown of alternatives such as HCFCs and HFCs in lower atmosphere was also studied. An important component of the AFEAS program is the global warming and energy efficiency study co-funded by U.S. DOE and UNEP.

The PAFT program started in 1987 to evaluate the toxicological study of HFC-32a, HFC-134a and HCFC-123. Later other proposed alternatives, such as HCFC-124, HFC-125, HCFC-225 (ca,cb), and HFC-32, were included. These studies are done in the foremost toxicological laboratories worldwide. The AFEAS and PAFT programs are independent efforts, whereas the chemical industry in general has joined together to provide valuable, scientific, and credible information for developing alternatives to CFCs and ensuring global phase-out in the shortest amount of time.

REFRIGERATION AND AIR CONDITIONING

The largest use of CFCs is in refrigeration and air conditioning (a/c) as a working fluid. This area can be divided into three major segments: (*1*) commercial industrial refrigeration and commercial air conditioning, (*2*) domestic refrigeration and air conditioning, and (*3*) mobile air conditioning. Possible candidates and various alternatives in refrigeration and air-conditioning applications are discussed herein.

Refrigerant Replacements

There are three basic alternatives for CFC refrigerant replacement: hydrocarbons, HCFCs, and HFCs (5). Many blends of these alternatives are used (6).

The refrigerants used for the cascades and simple refrigeration systems can benefit from the use of mixtures. These mixtures may combine hydrocarbons, HCFCs, and HFCs to match existing refrigerant performance and to deal with other issues such as cost, flammability, and compatibility.

In the automobile industry, both in the United States and abroad, the refrigerant HFC-134a is the most popular substitute for CFC-12, despite its greenhouse warming potential. HFC-134a is incompatible with mineral oils used in automotive air conditioning, therefore synthetic lubricants, eg, polyethylene glycol-based or polyester-based, have been developed for use in these systems.

In Germany the refrigerator industry has switched from CFC-12 to isobutane, accepting its risk of flammability (7) over the greenhouse warming potential of HFC-134a. Heat pump manufacturers have also used propane as a substitute for HCFC-22.

For industrial applications, R-502 and HCFC-22 are the most popular refrigerants in the temperature range of −20 to −60°C evaporation temperature. There has been no drop in replacement for HCFC-22 in the market, although the chemical industry is developing many binary and ternary mixtures of several HFCs to minimize the ozone-depletion potential (ODP), and also the Greenhouse Warming Potential (GWP). The most favorable solution is an azeotropic mixture, which behaves like a single component in practice.

Historically, many refrigerants such as CO_2, NH_3, SO_2, and later, CFC-12 and HCFC-22 were used. In 1974, when Molina and Rowland (1) brought up the theory of the relation between halocarbons and the breakdown of ozone, the dominant refrigerants in the United States were CFC-11, CFC-12, HCFC-22, and R-502 (an azeotrope of HCFC-22 and CFC-115). CFCs have already been phased out and the HCFCs will be phased out in the early part of the twenty-first century.

BLOWING AGENTS IN INSULATION FOAMS

CFC-blown low density polymeric foams have been the choice in high end insulation where polymeric foam thickness is important. Because of the higher insulation value of CFCs, indicated by R-value, higher insulation can be obtained with thinner foams. Examples are refrigeration appliances, efficient thermal insulation in building and construction (especially for roofing), and insulation in the transportation industries. Blowing agents with a very low thermal conductivity are encapsulated in a closed-cell structure and provide excellent thermal insulation properties. Nonflammability and nontoxicity of CFCs has made their use widespread in the insulation industry.

CFC-11 was most used as a blowing agent in rigid polyurethane and polyisocyanurate foams. CFC-12 is also used in this application as a blowing agent in polystyrene boardstock insulation foams.

The primary alternative used today in place of CFC-11 as a blowing agent for polyurethane and polyisocyanurate foam is HCFC-141b, and for CFC-12 in extruded polystyrene foam is HCFC-142b (8). Following the recent changes in the Montreal Protocol and United States Clean Air Act Amendments, HCFC-141b will be phased out as a blowing agent by the year 2003 in the United States and most of the developed countries. Several alternatives, including a hydrofluorocarbon compound, HFC-245fa ($CF_3CH_2CF_2H$) (9,10) HFC-365mfc ($CF_3CH_2CH_2$ CHF_2) (11), and cyclopentanes are being considered for replacing HCFC-141b, scheduled to be phased out by 2020. Various compounds including HFC-134a are considered alternatives to HCFC-142b.

INDUSTRIAL SPECIALTY SOLVENTS

A wide range of chemicals are used for various solvent applications. Most industrial products need solvents for cleaning. Other uses for solvents in industry include carrier solvents, dry-cleaning solvents, and aerosol formulations. In the early twentieth century, hydrocarbon-type solvents, such as hexane and mineral oils; ketones, such as acetone and methyl ethyl ketone (MEK); esters such as methyl, ethyl, or butyl acetate; ethers; and aromatic compounds, such as benzenes or toluenes, were very popular in cleaning. The primary uses were in cleaning oils and greases, and the drawbacks included high flammability characteristics. Most of these solvents had a flash point that made them hazardous in use.

In the mid-twentieth century, chlorinated hydrocarbons solvents were introduced and cleaning using vapor degreasers went through a period of large growth. Chlorinated hydrocarbons are nonflammable solvents with good solvency characteristics for a wide variety of soils. In a vapor degreaser, a solvent is boiled, the vapor is condensed using coolants, and then returned to the boiling chamber. Soiled machine parts are immersed in the vapor where the solvent condenses on them and washes off the soils. These vapor degreasers became popular because of their ease of use and individual parts did not have to touch the soiled solvents. Solvents are boiled in chamber A, then condense on the condenser coils and accumulate into chamber B. When chamber B is full, solvents cascade from chamber B into chamber A for reuse in boiling. Material to be cleaned is generally lowered into the vapor zone where the solvent condenses on the cold substrates and rinses the soil. Many sophisticated vapor degreasers have been designed with spray, in-line operations, etc.

In the 1960s, as workplace safety became a greater issue, the toxicity of chlorinated solvents was a deterrent to their use in the vapor degreasing industry. Chlorinated solvents were also found to be too aggressive toward plastics. As more plastics are used in medical, optical, and other highly technological products, solvents having suitable properties toward these applications became necessary. CFC-based solvents were introduced and replaced chlorinated hydrocarbons in many of these applications.

CFCs used as solvents are 1,1,2-trichloro-1,2,2-trifluoroethane (CFC-113) and trichlorofluoromethane (CFC-11). CFC-11 has been used in limited niche applications. Methyl chloroform (1,1,1-trichloroethane), another chlorinated hydrocarbon, was used as a replacement for trichloroethylene and perchloroethylene because it was less toxic. Methyl chloroform was also found to have nonzero ozone-depletion potential and was phased-out with the CFCs. Growth in printed circuit board production spurred growth in CFC-113 usage. CFC-113 boils at about 48°C, is stable, nonflammable, and nontoxic, having good solvency characteristics and miscibility with a large number of chemicals. CFC-11 has properties similar to CFC-113 except its boiling point (25°C) is lower, restricting its use in vapor degreasing.

The principal application of CFC-113 as a solvent has been the post-solder defluxing of printed circuit boards. It has also been used in precision cleaning of optical and medical instruments as well as instruments such as gyroscopes, accelerometers, and other aerospace components, in metal degreasing, in spot-free drying of parts after aqueous cleaning, and in dry-cleaning of fine fabrics and leather. Smaller uses include as carrier solvents for deposition of fluorinated lubricants in disk drives, deposition of silicone fluids in medical devices, as dispersing agents for mold-release agents, as a carrier solvent in various aerosol applications, and in the preservation of rare books.

As CFCs are phased out under the Montreal Protocol many new solvents are introduced. In some applications the chemical industry has reverted to older solvents such as petroleum distillates, alcohols, or chlorinated solvents.

Hydrochlorofluorocarbons: In this category HCFC-141b and azeotropic blends containing HCFC-141b and methanol with stabilizer are considered primary components (12) and are sold under the trade names Genesolv 2000 and Genesolv 2004, respectively. The U.S. EPA limited the use of HCFC-141b to an earlier phase-out (the same as CFC-113 for uses where an adequate alternative existed) due to a nonzero ozone-depletion potential. This date was later extended to the end of 1996 in the United States, however, HCFC-141b may be used in many applications until the year 2003, considered the ultimate phase-out date of HCFC-141b for all nonregulated application. Similar blends are marketed by Elf Atochem of France in limited applications. In vapor degreasers, the cleaning procedure wath HCFC-141b-based blends remained the same as in CFC-113-based degreasers. Some modifications were made to the degreasers to control emissions because of the lower boiling point of HCFC-141b.

The next HCFC blend to be introduced was HCFC-225 ($C_3HCl_2F_5$) and its blends containing alcohol (ethanol) (13). There are two isomers of HCFC-225: HCFC-225ca and HCFC-225cb. Of these, HCFC-225ca is more toxic. A blend of these two isomers is currently used, and although there are no U.S. manufacturers, the blend is marketed by Asahi Glass (Japan) and is considered a drop-in replacement for CFC-113. Due to its higher toxicity, tighter emission standards must be met for vapor degreasers using this solvent, which will be phased out by the year 2020. HCFC-123 has also been proposed as a solvent and is currently used in limited applications.

Hydrofluorocarbons and Hydrofluoroethers: Although conceptually there are a number of possibilities in this solvent category because all of these compounds are nonozone depleting, only a small number of solvents and their blends have been introduced in the marketplace. These are HFC-43-10 ($C_5H_2F_{10}$), a hydrofluorocarbon marketed by DuPont (14) and HFE-7100 ($C_4F_9OCH_3$), a hydrofluoroether, marketed by 3M Company (15,16). Their azeotropic blends are also being marketed. These compounds are not comparable to CFC-113 or HCFC-141b in their solvency characteristics. However, unlike HCFCs these solvents are nonozone-depleting although they do have some greenhouse warming potential.

These solvents are also used in a new generation of degreasers called Advanced Vapor Degreasers (AVD) (17). In these degreasers high boiling, light hydrocarbon-based mineral oil, terpenes, similar solvents are used to wash the soils, then these HFC or HFE solvents rinse away the remaining organics. In this manner the low solvency characteristics of highly fluorinated solvents can be used in cleaning. Disadvantages of the method include distilling the wash solvent from the rinse solvents and handling the flammable wash solvents.

Perfluorocarbon: Perfluorinated fluorocarbons as a class were proposed as CFC-113 alternatives, the principal ones being perfluorohexanes (C_6F_{14}) and perfluoroheptanes (C_7F_{16}) (18). Perfluoromethylmorpholine and perfluoroethylmorpholines were also introduced as alternatives. These molecules are relatively nontoxic, nonflammable, and have properties similar to CFC-113. However, their solvency for most soils is not as good as CFC-113, HCFC-141b, or HFCs. However, these have been used commercially in other applications and show good cleaning ability for fluorinated oils and greases. These solvents also replaced CFC-113 as a carrier fluid for fluorinated oils and greases. Perfluorocarbons are very inert and stay in the atmosphere for thousands of years, thereby contributing to the greenhouse warming potential. The use of these molecules is also very strongly regulated by U.S. EPA. These solvents can also be used in AVD machines.

Hydrocarbons: Hydrocarbon solvents, despite their flammability characteristics, have re-appeared in the industry. These hydrocarbons, such as hexanes, heptanes, are primarily from petroleum fractions. Aromatic hydrocarbons such as toluenes are also used in some applications.

These solvents are generally inexpensive, but must be handled carefully in the workplace and preferably used in explosion-proof degreasers because of their flammability hazards. Hydrocarbons are volatile organic compounds (VOCs) under the U.S. Clean Air Act because they break down in the lower atmosphere and create photochemical smog.

Chlorinated Hydrocarbons: In this class, the four major solvents used are perchloroethylene, trichloroethylene, methyl chloroform (1,1,1-trichloroethane), and methylene chloride. Of these, methyl chloroform is ozone-depleting and was phased out in 1995. The other three are considered toxic and are also in OSHAs list of suspected carcinogens. In the 1970s, CFC-113 replaced some of these solvents because of their toxicity, but in the 1990s the cleaning industry is reverting to these chlorinated hydrocarbons. Chlorinated hydrocarbons are good solvents for most soils except they are in general more aggressive toward plastics. The solvents are used in degreasers with very tight emission controls and workers are monitored for exposure. Over half a billion kg of these chlorinated hydrocarbons are used worldwide.

Aqueous Solvents: Aqueous solvents use water as the main ingredient. They are composed of three components: a builder, which is a blend of two or more alkali metal orthophosphates; an additive, such as glycol or glycol ether; and a surfactant, which provides detergency, emulsification, and wetting characteristics of a cleaner (19,20).

Aqueous cleaning is nonozone-depleting and nonpolluting. These solvents are good cleaners specially for polar and water-soluble soils. Aqueous chemistry has multiple degrees of freedom in cleaning process design and formulation (21,22). At the same time, aqueous cleaning has many disadvantages. Because of the high surface tension of water it is difficult to clean in smaller spaces. Surfactant residues are sometimes difficult to remove and cannot be used in cleaning agents where trace residues are critical. Water advances metal corrosion and is incompatible with various substrates. In addition, to comply with the Clean Water Act, wastewater must be treated before discharge which adds to the cleaning cost.

Semiaqueous Solvents: A significant innovation in cleaning came with the use of semiaqueous blends of solvent. This method uses a combination of wash in an organic solvent and rinse in water to clean the assemblies. The wash is done with a petroleum distillate, a terpene, or a dibasic ester that is then rinsed in water to be removed from the surface (23). This has been more effective than an alkaline water cleaner in many cases.

Semiaqueous solvent methods have similar advantages and disadvantages as aqueous cleaning, but better solvency characteristics for organic soils. The flammability characteristics of the wash solvent are of some concern especially if spray equipment is used. Terpenes also may have an objectionable odor. Separation of these organics from water is also an extra step in the method.

No-Clean Technology: No-clean technology is also another option in cleaning assemblies and has taken over a large share of the industry. Specially designed low solids fluxes or pastes do not require cleaning and the activators are so low they do not create dendritic growth in circuit board assemblies and provide reliable solderability (24). An advantage of this no-clean technology has been that it minimizes waste generated by solvents or aqueous detergents.

NIK Technology: Nitrogeneous or reactive wave soldering is another method developed by the printed circuit board industry (25). In this method, wave soldering is done under a blanket of nitrogen, the oxygen level is kept at a low level, and the circuits are not oxidized to improve solderability. Before the boards enter the wave soldering machine they pass through a reactive atmosphere containing adipic or stearic acid to remove the oxide layer from exposed circuitry.

Another technology which has found some use is supercritical cleaning fluid (SCF) technology. Cleaning is done using a high pressure supercritical gas such as CO_2. At high temperatures and pressures, SCFs have high diffusivity and low density and viscosity compared to liquid solvents that help to dissolve various soils. This technology, although unique, has limited potential because of its high pressure applications.

Plasma cleaning is used to produce a very high degree of cleanliness. A plasma is an electrically charged gas containing highly ionized particles. These gases react with organic contaminants to form volatile gases that are removed by the flow of process gases. Plasma cleaning is a very expensive technology but has found use in the aerospace and electronics industries for high precision cleaning.

Other Solvents: A wide range of aliphatic hydrocarbons falls in this category. These are obtained from petroleum fractions, commonly known as mineral spirits or kerosene, and are widely used in cleaning. Alcohols, such as methanol, ethanol, isopropanol, and *n*-propanol, are also used extensively in cleaning. These are fairly volatile, relatively nontoxic, and good cleaners. Disadvantages include flammability characteristics and their classification as VOCs. Similarly, ketones such as acetone, methyl–ethyl ketone (MEK), and esters such as methyl and ethyl acetate are also used as solvents and have similar disadvantages.

Among other compounds, *n*-methyl pyrrolidone (NMP) (C_5H_9NO) has also found many uses as a solvent replacement for CFC-113 (26). NMP is miscible with water and many other organic solvents including esters, ethers, alcohols, ketones, aromatic and chlorinated hydrocarbons, and vegetable oils.

Volatile methyl siloxanes marketed by Dow Corning have found uses in many applications (27). The two principal uses are dimethyl and hexamethyl siloxanes sold under the trade name OS-10 and OS-20. Although these are flammable, they are good solvents for many silicone-based soils and do not have any significant greenhouse warming potential.

Because there are many possible alternatives for CFC-113 and CFC-11 solvents, tremendous market fragmentation has occurred in the use of replacements.

STERILANT GASES

CFCs have been used in the sterilization of medical equipment, eg, CFC-12 has been used as a carrier gas for ethylene oxide in sterilization. Medical devices such as hypodermic needles, artificial organs, and catheters need to be sterilized, and ethylene oxide is a preferred sterilizing gas. However, ethylene oxide is highly explosive and handling of it is dangerous. CFC-12 forms a nonflammable mixture with ethylene oxide and can be used relatively safely in sterilization.

The industry has chosen HCFC-124 ethylene oxide as an alternative nonflammable, azeotropic blend. However, because HCFC-124 will be phased out by the year 2020, some in the industry have decided to utilize pure ethylene oxide, and must, therefore, install special explosion-proof chambers.

HFC-227ea (CF_3CHFCF_3) is also a proposed alternative for use in combination with ethylene oxide for sterilization to replace HCFC-124.

AEROSOL PROPELLANTS

The use of CFCs in aerosol containers was begun in 1941 by the U.S. Department of Agriculture to provide improved insecticides, eg, DDT. In 1947 the preferred blend for propellant for insecticide delivery was a blend of CFC-12/CFC-11(60/40 by wt) (28,29).

Household products sold as aerosols in the 1950s included air fresheners, glass cleaners, shoe and leather dressings, hard surface cleaners, personal care products (eg, hair sprays, shaving creams, perfumes, and pharmaceutical products), and paints and coatings. In current industrial aerosol applications, HCFC-141b and HFC-134a are used as replacements.

OTHER APPLICATIONS

Other applications of CFC-113 include (*1*) carrier agents for disk drive lubricants and in medical devices, (*2*) spot-free dewatering agents, and (*3*) mold-release agents.

As carrier agents, the two principal uses are in deposition of fluorolubricants onto disk drives and application of silicone-based lubricants to medical devices such as hypodermic needles. In these applications, the lubricants are dissolved in CFC-113 and applied to the disk drive or hypodermic needles. The solvent evaporates and leaves a uniform film of lubricants. Perfluorocarbons, eg, perfluoropentanes and perfluoroheptanes, are presently used as alternatives to CFC-113 in this application. However, perfluorocompounds have high greenhouse warming potential and will be replaced when suitable replacements are developed. HFC-43-10 or HFE-7100 have been tested as alternatives to replace these highly fluorinated compounds. In the application of silicone lubricants volatile methyl siloxanes are considered as alternatives. Siloxanes are marketed by Dow Corning Corporation.

As dewatering agents, CFC-113 mixed with a surfactant has been used to displace water from devices that are cleaned or processed in aqueous media. Surfactant-based drying is used because the typical substances, eg, jewellery, disk drives, and plated metals, are heat labile. These materials are immersed in a typical degreaser in solutions containing CFC-113 and surfactants. The surfactant is then rinsed in CFC-113 which provides spot-free drying of the materials. No suitable solvent-based alternatives have been marketed as of 1997. Some processes use blow drying or low energy infrared drying. Surfactant-based HFE blends are in development.

As mold release agents, CFC-113 is used as a dispersant. Mold release agents are generally fluorinated polymers and may be dispersed in CFC-113. Perfluorinated compounds are used to replace CFC-113 in this application.

CONCLUSIONS

After Rowland and Molina's discovery of stratospheric ozone depletion (1) caused by CFCs, it became clear that this was one of the major environmental problems facing the earth. CFCs have since been phased out and more environmentally benign chemicals have been developed. Before 1973, half a billion kg of CFCs were used globally in refrigeration, air conditioning, foam insulation, and industrial solvents. Replacements have been found for most applications and by the end of 1995, CFCs had been phased out of production in the developed countries, except for essential uses. In some cases, HCFCs are used as interim replacements. HCFCs do have a nonzero ozone-depleting effect, but it is far lower than that of CFCs.

HFCs, which are nonozone-depleting, are used and proposed as long-term replacements in most refrigeration and air-conditioning applications. Foam blowing still uses HCFCs. Solvent use has become very diversified where HCFCs, HFCs, HFEs, PFCs, aqueous and semiaqueous technology, various other solvents, and many not-in-kind technologies are used.

The stratospheric ozone layer continues to be monitored. More alternatives for CFCs are being sought to help the industrialized world and help solve this environmental problem.

ACKNOWLEDGMENTS

The authors would like to acknowledge M.W. Spatz, G.M. Knopeck, H.M. Hughes, R.G. Richard, I.R. Shankland, D.J. Williams, R.L. Bishop, R.R. Singh and H. Magid, all of AlliedSignal Inc. for providing valuable information used in preparing the manuscript and also to M.W. Spatz, G.M. Knopeck, I.R. Shankland, and R.L. Bishop for a careful reading of relevant parts of the manuscript.

BIBLIOGRAPHY

1. M. J. Molina and F. S. Rowland, *Nature,* **810,** 474 (1974).
2. R. L. Powell and J. H. Steven, "Refrigerant Alternatives," *Encyclopedia of Energy Technology,* Vol. 4, John Wiley & Sons, Inc., New York, 1995.
3. Brit. Patent 357,263 (1930), T. Midgley; U.S. Pat. 1,833,847 (1932).
4. M. J. Kurylo, "The Chemistry of Stratospheric Ozone: Its Response to Natural and Anthropogenic Influences," in *CFCs: Today's Options—Tomorrow's Solutions,* Proceedings of ASHRAE's 1989 CFC Technology Conference, ASHRAE, New York, 1989.
5. J. D. Oakey, *Tenth Intersociety Cryogenic Symposium, AIChE Spring National Meeting,* Houston, Tex., Mar. 20–23, 1995, pp. 63–85.
6. J. M. Calm, *Refrigerant Database,* Report DOE/CE/23810-59C (June 1995), Air Conditioning and Refrigerant Technology (ARTI), Arlington, Va., 1996.
7. M. Kuver and U. Hesse, *Gastech 94, The 16th International LNG/LPG Conference & Exhibition Proceedings,* Kuala Lumpur, Malaysia, Oct. 25–28, 1994, pp. 357–363.
8. I. R. Shankland, in Ref. 12, pp. 87–95.
9. G. M. Knopeck, R. C. Parker, R. G. Richard, and M. C. Bogdan, *Proceedings of the Polyurethanes 1995,* Chicago, Ill., p. 2.
10. D. J. Williams, R. C. Parker, and M. C. Bogdan, *Proceedings of the Polyurethanes Expo '96,* Las Vegas, Nev., 1996, p. 394.
11. J. D. Roux, A. Albouy, and D. Mouton, *UTECH Asia '97,* Suntec City, Singapore, Paper 24, 1997, pp. 1–5.
12. R. S. Basu, P. B. Logsdon, and E. M. Kenny-McDermott, *Proceedings of the International CFC and Halon Alternatives Conference,* Washington, D.C., 1991, p. 188–198.
13. K. Kitamura, S. Morikawa and M. Yamabe, HCFC-225 as a Drop-in Alternative to CFC-113: Update on its Application *Proc. of Int'l Conf. On CFC and Halon Alternatives,* Washington, D.C., 1994, pp. 58.
14. A. Merchant, "New HFC Replacement Products for CFC-113," *Proceedings of International Conference on CFC and Halon Alternatives,* Washington, D.C., 1992.
15. M. W. Grenfell, F. W. Klenk, and J. G. Owens, "New Fluorinated Solvent Alternatives," *Precision Cleaning '95,* Chicago, Ill., 1995.
16. T. Koenig and J. G. Owens, "The Role of Hydrofluoroethers in Stratospheric Ozone Protection," *Proceedings of the International Conference on Ozone Protection Technologies,* Washington, D.C., 1996.
17. M. Hayes, "The AVD Process: Vapor Degreasing without Chlorinated Solvents," paper presented at *The International Conference on CFC and Halon Alternatives,* Washington, D.C., 1992.
18. R. M. Minday and E. A. Reiner, *Proceedings of the International CFC and Halon Alternatives Conference,* Washington, D.C., 1992, p. 516.
19. F. Cala, R. Burke, and C. Lowell, *Circuits Assem.* **3**(6), 36–39 (1992).
20. F. Cala and A. Winston, *Proceedings of NEPCON East 94,* Boston, Mass., 1994, pp. 99–106.
21. F. Cala and R. Reynolds, *Surface Mount Technol.* **9**(8), 40–42 (1995).
22. F. A. Cala, *Precision Cleaning,* 13–20 (Jan. 1997).
23. M. Hayes, *Proceedings of International Conference on CFC and Halon Alternatives,* Washington, D.C., 1993, p. 440.
24. J. Linton, *Proceedings of the International CFC and Halon Alternatives Conference,* Vol. I, Washington, D.C., 1993, p. 385.
25. J. Baker, "Inert Gas Wave Soldering," paper presented at the *International CFC and Halon Alternatives Conference,* Washington, D.C., 1992.
26. M. W. Waldrop and W. C. Walsh, in Ref. 11, p. 586.
27. R. F. Burrow, *Proceedings of the International CFC and Halon Alternatives Conference,* Washington, D.C., 1993, p. 654.
28. M. A. Johnson, *Aerosol Age,* 27–32 (Jan. 1991).
29. M. A. Johnson, *Spray Technol.* 20 (July 1996).

General References

F. Cala and A. Winston, *Handbook of Aqueous Cleaning Technology for Electronics Assemblies,* Electrochemical Publications Limited, Asahi House, U.K., 1996.

G. Beisman, R. Devos, and I. D. Rosenbluth, *Proceedings of the Polyurethane World Congress 1993,* p. 498.

ICOLP Technical Committee, *Eliminating CFC-113 and Methyl Chloroform in Precision Cleaning Operations,* U.S. EPA, Air and Radiation (ANR-445), EPA/400/1-91/018, June 1991.

K. W. Cooper and K. E. Hickman, "Refrigeration," *Kirk-Othmer Encyclopedia of Chemical Technology,* 3rd ed., Vol. 20, John Wiley & Sons, Inc., New York, 1982, pp. 78–107.

M. Bogdan, D. Orlowski, J. Bennington, and J. Welch, *Proceedings of the Polyurethanes 1995,* Chicago, Ill., p. 34.

M. Hayes, *Proceedings of the International Conference on CFC and Halon Alternatives,* Washington, D.C., 1993, p. 497.

M. Hayes, E. Boomer, and A. Pullen, *Proceedings of the International Conference on CFC and Halon Alternatives,* Washington, D.C., 1994, p. 600.

N. Kattan, J. R. Thome, and D. Favrat, *ASHRAE Tech. Data Bull.* **11**(4), 52–69 (1995).

N. Kattan, J. R. Thome, and D. Favrat, *ASHRAE Tech. Data Bull.* **11**(4), 70–80 (1995).

P. S. Lermond, R. B. Ramsey, and J. Yellow, "New HCFC Blend for Metal Cleaning," *Proceedings of the International CFC and Halon Alternatives Conference,* Baltimore, Md., 1990, Vol. I.

R. A. Corbett, *Oil & Gas J.* 32–38 (May 29, 1995).

R. S. Basu and K. P. Murphy, *Precision Cleaning,* 25–29 (Oct. 1994).

S. M. Sami, *ASHRAE Trans.* **97**(2), 41–47 (1991).

S. S. Seelig, *Today's Chemist at Work,* June 1996, pp. 48–53.

W. K. Snelson, J. W. Linton, A. R. Triebe, and P. F. Hearty, *ASHRAE Tech. Data Bull.* **11**(4), 44–51 (1995).

World Meteorological Organization (WMO), *Scientific Assessment of Ozone Depletion: 1994,* WMO Ozone Rep. 37, Geneva, 1994.

ASHRAE 1969, ASHRAE Thermodynamic Properties of Refrigerants, ASHRAE, New York.

CHLOROFLUOROCARBONS: DESTRUCTION OF

WILLIAM R. WILLIAMS
DuPont Fluoroproducts
Deepwater, New Jersey

Concern around the ozone depletion potential and global warming contributions of chlorofluorocarbons (CFCs) has led to the cessation of CFC production in industrialized nations in the 1990s. Emissions of existing CFCs are strictly controlled and, in general, prohibited. Although CFCs are no longer produced, a substantial inventory of CFCs is still in use as working fluids (primarily air conditioning and refrigeration), as solvents, or as material trapped in foams (blowing agents). Further, CFCs continue to be used as intermediates in chemical processes and can be produced as by-products. Until the 1990s, these materials were often vented to the atmosphere after their useful life, but now they must be recovered and either recycled or destroyed. While recycling delays the need to destroy these materials, society ultimately requires processes to destroy CFCs.

CFC DESIGNATIONS AND NOMENCLATURE

CFC designations are often confusing. The CFC numbering system is actually a code that was originally designed to protect trade secrets; it therefore has an element of confusion built into the system. It has been adopted by the American Society of Heating and Refrigeration Engineers (ASHRE) as ASHRE Standard 34, for designating refrigerants (1).

The first digit on the right in CFC nomenclature is the number of fluorine atoms on the molecule. For example, CFC-12 contains two fluorine atoms. The second digit from the right is one greater than the number of hydrogen atoms. For example, CFC-12 contains 0 ($1 - 1 = 0$) hydrogen atoms. The third digit from the right is one less than number of carbon atoms; if it is 0, it is omitted from the number. CFC-12 therefore is actually CFC-012, and the number of carbons on the atom is 1 ($0 + 1 = 0$). For molecules containing a double bond, the numeral one is used as the fourth digit from the right. All compounds contained in this report are saturated, so none of these materials uses the fourth digit.

Because chlorine is designated by difference, any remaining bonds to the carbon atom are assumed to be chlorine. Therefore, CFC-12 defines only two fluorine atoms and no hydrogen atoms connected to the lone carbon. The remaining two atoms must be chlorine, so the molecule is CF_2Cl_2.

For molecules with two carbons and multiple isomers, the most symmetric isomer is given a name with no extension according to the rules above. In order of decreasing symmetry, the remaining isomers are designated a, b, c, etc. Symmetry is determined by summing the molecular weights of the atoms on the two carbons. For example, HFC-134 is CHF_2CHF_2, whereas HFC-134a is CF_3CFH_2.

BACKGROUND

Since the discovery (1) that chlorofluorocarbons could be used as safe alternatives to NH_3 and SO_2 as refrigerants, CFCs have become widely used. This is because CFCs have unique properties that make them excellent working fluids for compressors, propellants, solvents, and, if we include the brominated counterparts, fire extinguishants. They are also remarkably stable. While this makes CFCs nontoxic and therefore safe for the above applications, their stability allows them to accumulate in the atmosphere (2).

This atmospheric stability, coupled with the chlorine content in the molecules, is what causes CFCs to be so effective in depleting the ozone. In the stratosphere, the molecules can liberate chlorine radicals and set off a chain reaction that depletes ozone molecules via a mechanism similar to that depicted below for CFC-11 (CCl_3F):

Initiation $$CCl_3F + uv \rightarrow Cl + CCl_2F \quad (1)$$

Propagation $$Cl + O_3 \rightarrow ClO + O_2 \quad (2)$$

Propagation $$ClO + O \rightarrow Cl + O_2 \quad (3)$$

The chain reaction has an efficiency such that up to 10^5 molecules of ozone are depleted per molecule of Cl liberated from a CFC (3). Additionally, CFCs are significant global warmers. It is estimated that total CFC emissions contribute up to 40% of the amount of global warming that CO_2 emissions contribute, with CFC-11 and CFC-12 (CF_2Cl_2) alone contributing almost 30 out of the 40% (4).

By the late 1970s, it was widely recognized that the depletion of the ozone from CFC emissions was a potential environmental problem, but aside from the banning of CFCs as aerosol propellants in some countries, CFCs continued to be produced and demand continued to rise. Society had become dependent on these materials. Thus began an unprecedented effort to discover and commercially synthesize CFC alternatives (5). The urgency of this effort was punctuated by the discovery of the ozone hole during springtime over the Antarctic (6), which ultimately led to the Montreal Protocol of 1987. This agreement called for a 50% reduction in CFC production in industrialized nations by 1998. This phaseout was accelerated shortly thereafter, and by the new modified agreement, 50% reduction was due in 1990 and 100% by the year 2000.

Thanks to the discovery and commercial availability of the CFC alternatives such as the replacement refrigerant for CFC-12, hydrofluorocarbon HFC-134a (CF_3CFH_2), society is well on its way to achieving the goals of the modified 1987 Montreal Protocol. However, there are remarkable amounts of CFCs in use as working fluids or solvents. It is difficult to estimate the full amount of CFCs currently in inventory. Shortly after the Montreal Protocol, the amount of CFCs were estimated for the purpose of establishing a system of credits for CFC destruction (Table 1) (7). Initially, the idea was to give production credits for any CFCs destroyed, thus maintaining the CFC bank at a constant level. However, with the more accelerated phaseout time table, production credits cannot be used to justify new CFC production. The 1988 report estimated more than one million metric tons of CFCs in the U.S. bank of CFCs and halons.

Although there has been remarkable success in recovering and reusing these materials, they must ultimately be destroyed. Primary places where CFCs reside are in the atmosphere, as trace components in water, trapped in foams, or as captured materials either recovered or as working fluids. Methods have been proposed for handling CFCs in all of these environments.

METHODS OF DESTRUCTION OF ATMOSPHERIC CFCS

This is work that seeks to solve the ozone depletion problem by destroying the CFCs already in the atmosphere. There is much appeal to this. First, until recently, there were not many alternatives to CFCs. The initial idea was, rather than give up the conveniences of refrigeration and air conditioning, to try to reduce the amount of CFCs in the atmosphere to an acceptable level. Second, and more importantly, much of the

Table 1. Estimated U.S. Bank of CFCs in 1988

CFC	Composition	Primary Use	Amount, t
CFC-12	CF_2Cl_2	Mobile air conditioners, refrigeration	387,260
CFC-11	$CFCl_3$	polyurethane foam	654,200
CFC-113	$CF_2ClCFCl_2$	solvent	50,200
CFC-114	CF_2ClCF_2Cl	refrigeration	1,140
CFC-115	CF_3CF_2Cl	refrigeration	16,740

CFCs in the atmosphere have not yet made it to the stratosphere where they wreak their damage. The lifetime of CFCs in the troposphere is greater than 75 years (8), so despite great gains in preserving the ozone layer, continued depletion of the ozone layer is expected until well after the last CFC is produced under the current plan. Thus destroying CFCs in the atmosphere appears attractive from this standpoint. These methods, in general, rely on destroying CFCs in the troposphere, thus preventing the CFCs from entering the stratosphere. The target often is to run the same decomposition chemistry that is carried out by uv light in the troposphere for liberation of chlorine radicals. In the troposphere, however, enough hydrogen-containing molecules are present to react with chlorine to make HCl, which washes out in rain (9). Presumably, this process would not significantly increase the severity of acid rain.

Infrared Laser Processing of the Atmosphere

The practicality of using infrared lasers to selectively excite and dissociate CFCs in the troposphere has been evaluated (10). It is pointed out that destroying atmospheric CFCs is a daunting task because the earth's atmosphere contains about 5×10^{15} metric tons of air. Further, to process this amount of material at sea level would take a device having 500-km^2 area 10 years, with 80 km/h of air passing through the device.

The advantage of using lasers to process the atmosphere is that the laser can be tuned to excite vibrational modes of CFCs to dissociate them. The key is to choose wavelengths where atmospheric absorption is relatively weak but CFC absorption is quite strong. In this way, the energy could be devoted only to the destruction of CFCs, and by operating at such wavelengths, greater than 90% transmission is expected through 50 km of mid-latitude summer atmospheres at 4300-m altitude. Large arrays of such powerful infrared lasers could be used to dissociate large amounts of CFCs in the troposphere. Natural wind currents would be used to circulate air through the array (11).

The challenges to such a method are predominantly sources of energy loss. Stimulated rotational Raman scattering effectively limits the path length to 5 km. To compensate for this, it is proposed to place selective mirrors every 5 km, which would absorb the sideband frequencies and reflect the target frequency. The implication is that the energy from the sideband frequencies could be recovered with high efficiency. A practical way to accomplish this was not discussed.

Additionally, collisional quenching is a competing process that prevents the CFCs from dissociating by allowing an alternate path to liberate the energy of the excited molecule. In practice, this would mean that the required fluence (intensity $\times$ time) would have to be delivered in less than 60 ns. This effectively fixes the intensity of such a beam at 130 MW/cm^2. Further, it is estimated that this would have to be tripled to compensate for losses to collisions during excitation.

Based on this analysis, it is concluded that to remove CFCs from the atmosphere at a rate of one million metric tons per year using this method would cost at least \$6/lb of CFC removed, a figure that seems impractical. However, it is pointed out that there are many clever ways of reducing the fluence requirement. These rely on using a series of lower energy resonant frequencies to achieve dissociation through a series of transitions (12). This may also reduce losses resulting from Raman scattering. Through this and possibly other advances, the cost of processing CFCs in this manner may become practical.

Microwave Processing of the Atmosphere

Another method proposed for processing CFCs in the atmosphere is to utilize microwave radiation to heat up plasma electrons in the ionosphere (13). This weakens electron recombination and shifts the charge equilibrium to increase the number of chloride ions. Because chloride ions do not participate in ozone-depleting reactions (eqs. 1–3), this would reduce the amount of ozone depletion.

It has been claimed (9) that this basic method cannot be applied at levels of the atmosphere where either chloride ions or ozone are concentrated enough to make a significant impact. In a modified method, a freely localized microwave discharge is used to produce a relatively high concentration of electrons (9). If done with relatively high concentrations of CFC present, the electrons would react with the CFCs to form chloride ions directly. For CFC-12, the reaction would be as follows:

$$CF_2Cl_2 + e_\nu^- \rightarrow CF_2Cl + Cl^- \quad (4)$$

It is proposed that pulsed microwave beams would be converged at a given altitude somewhere in the troposphere (<10-km altitude). The altitude would be chosen such that the radicals and chloride ions would be consumed by natural atmospheric processes. Chloride ions would react with hydrogen-containing molecules to produce HCl, which would be washed out with rain (3). At higher elevations, the microwaves could actually lead to the accumulation of ozone-depleting materials such as NO_x and chlorine atoms, thus defeating the purpose of the technology (14). While this method seems more versatile than the other microwave method, it does not appear any more cost-effective than the laser method.

Although there are compelling theoretical and laboratory-scale experimental results demonstrating these methods, none have actually been tested in the atmosphere. Additionally, significant technological advancements must be made before such methods become feasible. Therefore, with current technology, destroying CFCs in the atmosphere is impractical, to say nothing of the political and cost factors involved in implementation. The general opinion in the 1990s is that humankind will have to deal with the ozone depletion from CFCs already released, and be vigilant in keeping additional CFCs from escaping to the atmosphere. This makes the methods discussed in the remainder of this article even more critical because of the urgent need to develop efficient, selective, and safe methods of destroying all CFCs currently being stockpiled as the world de-inventories their CFCs for replacement chemicals.

METHODS OF DESTRUCTION OF TRACE CFCS

Most work in this area is involved with destroying CFC-113 ($CF_2ClCFCl_2$) in water or CFC-11 ($CFCl_3$) in landfills. CFC-113 is an excellent cleaning agent for circuit boards and other electronic devices because of its low conductivity, low toxicity, and nonflammability. As a solvent, it often ends up as a trace component in water supplies. CFC-11 is a foam blowing agent often trapped in foam products. These products upon disposal liberate CFC-11 into landfills or foam recycling facilities.

Destruction of trace CFCs poses challenges which are not unlike those encountered in the atmospheric destruction of CFCs. Because the systems are dilute, the method must selectively target the trace component to avoid excessive waste of energy or a valuable reagent.

Destruction of CFCs in Water

The major challenge of removing CFC-113 and CFC-11 in water results from their stability and relative volatility compared to water. Like most CFCs, CFC-113 and CFC-11 are very stable and generally require high temperatures to break the C–F bonds. However, they are also volatile, so attempts to heat the solution by conventional means result in volatilization of CFC-113 and CFC-11.

One unique method that shows promise is the use of the sonochemical effect to react the CFCs with water (15). Using a solution initially consisting of approximately 50-mg/L CFC-113 or CFC-11 in water at 5–10°C and under moderate pressure (5 psig), it has been demonstrated that less than 5% of CFCs volatilize during this process, with most of the remaining material destroyed in solutions. Results using either CFC were similar. Destruction is fairly rapid; most of the CFCs are destroyed within 5 to 10 minutes in a batch system. Circulating systems take a bit longer (30–40 min).

It is unclear exactly what reaction is occurring, but it has been suggested that hydrogen and hydroxyl radicals are attacking the CFCs to make HCl and HF in solution (15). The reduction in pH is used as proof of acid generation.

The key to the success of this method is that sonochemical destruction allows localized heating from cavitation while the bulk solution temperature remains essentially unchanged. In this way, radicals can be formed in the local hot spots without volatilizing the CFCs. However, the researchers acknowledge that advances would have to be made in the development of ultrasonic energy delivery systems for this to be practical on an industrial scale. As of this writing (1996), this would be an extremely expensive way to destroy low levels of CFCs in water.

Another proposed method for destroying trace CFCs in water is electrified microheterogeneous catalysis (EMC) (16). This involves applying a d-c voltage gradient to a medium containing micrometer-size catalyst particles in a low ionic strength fluid. In this particular study Na–Y zeolite was used in powder form in a water-based dispersion. Experimental results demonstrated that CFC-113 could be destroyed at 0°C. The reaction pathway for CFC-113 destruction is not clear, but it is suggested that a strong oxidant is formed from species such as OH^-, H_2O, H_2O_2, or O_2. The problem with this method is that it is estimated that 50% of CFC-113 is volatilized before reaction, and only qualitative evidence is offered for reaction.

Although more development needs to be done, both of these techniques offer much promise. Moreover, they are clever techniques to address the core challenge of removing trace CFCs, namely, to target the CFCs without affecting the great excess of material in the surrounding media, in this case water.

Destruction of CFC-11 in Foams

CFC-11 destruction in foams often does not present the problems of CFC-113 destruction in water because the foam is frequently also being incinerated. In these cases, the technical issues of CFC-11 destruction in foams is simply a matter of making sure that the efficiency and impurity levels of CFC-11 destruction during coincineration in foams is acceptable.

This does not appear to be a problem. A study of cocombusting CFC-11 with polyurethane foams showed CFC-11 destruction to be greater than 99.998% (17). Any products of incomplete combustion (PICs) of CFC-11 are insignificant compared to the PICs from the incineration of urethane foams.

METHODS OF DESTRUCTION OF RECOVERED CFCS

Because the best option for destroying CFCs with current technology requires the recovery of CFCs, great strides have been taken in this area. Environmental regulations require recovery of all CFCs when deinventorying systems with CFCs as working fluids; until the 1990s CFCs were often bled to the atmosphere from these systems.

Much of the recovered materials can be reused simply by separating out impurities (predominantly compressor oils and lubricants). However, some of the material cannot be practically refined; furthermore, as the CFC phaseout continues, there will be less demand for recycled CFCs. Methods are needed for the destruction of these CFCs, which represent by far the largest class of materials that need to be destroyed.

Destruction of CFCs by Thermal Incineration

Incineration is the predominant method of destroying CFCs because it is remarkably efficient and simple. An often omitted or underemphasized fact of CFC incineration is that CFCs cannot be oxidized by themselves. This is because the chlorine and fluorine cannot be readily oxidized by oxygen and the liberation of chlorine or fluorine gas is thermodynamically unfavorable.

Incineration can be done either catalytically or thermally. Thermal incineration is by far the more widely practiced primarily because thermal incinerators are readily available, low cost, and simple. Many commercial facilities have thermal incinerators to comply with environmental regulations in other areas, so incorporating CFCs into the incinerator is of only incremental cost, although the presence of aqueous HF and HCl in the stack gas may require exotic materials of construction for the incinerator. However, new technologies are making catalytic processes more attractive.

Process Studies of the Thermal Incineration of CFCs. For many of the above reasons, thermal destruction of CFCs is probably the most mature field of all the methods of CFC destruction. The attractiveness of thermal incineration is usually the ability to practice it in an incinerator which is already required for other environmental reasons. The key to this is often the assurance that the addition of CFCs does not lead to any unintended hazardous by-products.

A specific example of this which frequently appears in the literature is the possibility of forming hazardous materials in municipal waste incinerators where CFCs are being burned, either as they occur in the waste such as CFC-11 (foams) or CFC-113 (solvent), or intentionally added. The concern is not a direct reaction of the CFCs with other organics, but the possibility that HCl formed during CFC incineration might react with other organics to create new and potentially hazardous impurities.

Specifically, the concern is that the higher acid concentration may lead to higher halogenated hydrocarbons such as chlorobenzene and chlorophenol. These are believed to be precursors for polychlorinated-*p*-dioxins (PCDD) and dibenzofurans (PCDF) (18). However, in studies of CFC-113 and CFC-11 coincineration with polyurethane foams (17) and municipal waste (18), no evidence for PCDDs, PCDFs, or their precursors were observed. Additionally, CFC-113 incineration was studied with *n*-hexane (19). In these studies, the only PIC formed was CFC-12 at lower temperature (760°C). However, by operating at high temperatures (820°C), CFC-12 could also be destroyed. An extensive study was carried out where CFC-11 and CFC-12 were burned in pilot-plant incinerators, and PCDD/PCDF emissions were less than 140 ng/m^3 (20). Based on these studies, no additional hazardous materials are expected to be produced from the addition of CFCs to an incineration process.

Studies of the incineration of CFCs have all shown fairly high destruction efficiencies. If enough fuel is provided, conversions greater than 99.9% are fairly easily obtained. In the studies discussed above, CFC-11 and CFC-113 incineration efficiencies were >99.9% when the systems were optimized for CFC/HC ratio and incineration temperature. This has also been demonstrated for CFC-12 (21). In an extensive study of burner flames and the effect of various parameters on performance, it was shown that CFC-12 could be destroyed with an efficiency of >99.2% over a wide range of conditions.

What is particularly interesting is the effect produced as the minimum CFC/HC ratio required for combustion is approached. At this ratio, very little water is formed. One of the problems with all incineration methods for CFCs is the formation of HCl and HF, particularly when water is present. Anhydrous forms of these materials are easier to handle than aqueous forms. Aqueous HF and HCl have little value and are often neutralized, although the fluorine content can be recovered by precipitating the aqueous F^- ion with Ca as CaF_2 (22). The CaF_2 can then be converted to anhydrous HF (23). This is an involved, multistep process, so it would be advantageous if it could be operated in such a way that no water was formed. In this way, HF could be recovered in anhydrous form directly.

In some of the above studies the operation was attempted near the minimum CFC/HC ratio. Extensive studies (21) were carried out for the coincineration of CFC-12 and methane at varying CFC-12/methane ratios. As the CFC-12/methane ratio was increased toward the maximum allowable value of 1, the flame became unstable and the flame velocity decreased. Additionally, above about 0.2 CFC-12/methane ratio, combustion efficiency of CFC-12 started to decrease until at around 0.6 CFC-12/methane, the CFC-12 destruction efficiency had dropped to around 90% for stoichiometric fuel–air mixtures. Although it seems that better burner design would increase flame stability, it is not clear whether better burner methods would improve destruction efficiency for CFC-12/methane ratios greater than 0.2.

In the work on CFC-113 incineration (19) similar results were obtained. When operating at or near the maximum CFC-113/HC ratio, efficiency dropped. In fact, the reported CFC-12 production occurred to a much greater extent (10–50X) when they ran close to the maximum ratio. This could be caused by chlorine possibly liberated in small amounts in lieu of HCl when the system is hydrogen-starved. This would add to C_1 fragments to make CFC-12.

Another issue is the amount of excess oxygen provided to the system. In this study, CO formation increases in systems that have less than the stoichiometric amount of oxygen. However, this problem is not unique to CFC destruction. Incinerator design and operation must always take into account the possibility of CO formation, and operate in regions where this is not favored. Another problem which is present in any incineration is the potential to create NO_x. This arises because of a frozen equilibrium which exists with N_2 and O_2. At flame temperatures the equilibrium favors NO_x, but rapid cooling after the flame zone slows the reaction so that the NO_x cannot react back to N_2 and O_2, although this is thermodynamically favored. As a partial oxidation product, CO formation is typically favored in fuel-rich systems, whereas as a deep oxidation product, NO_x is typically formed in fuel-lean systems. Therefore these materials present the same challenges for CFC incineration as incineration in general; often the goal is to operate in a narrow region very close to the stoichiometric ratio for combustion to CO_2 and H_2O (and HX for CFC incineration) where the fuel/air ratio favors formation of neither CO nor NO_x.

Global Warming Issues. Although the most frequently discussed impact of CFCs on the environment is their ozone depletion potential, CFCs also possess high global warming potentials (GWP). Therefore, although CFC combustion yields the global warming gas CO_2, CO_2 has a considerably lower GWP (about 1/10,000) of any CFC (24). Despite their lower concentration in the atmosphere, CFCs are a significant contributor to the greenhouse effect, as Table 2 illustrates. Therefore, the incineration of CFCs actually significantly

Table 2. Global Warming Potentials and Impact for CFCs vs CO_2

Material	GWP, °C/ppb[a]	Estimated contribution to global warming[b]
CO_2	4×10^{-6}	49%
CFC-11	0.07	[c]
CFC-12	0.08	[c]

[a] Ref. 25.
[b] Ref. 26.
[c] 14% for CFC-11 and CFC-12 combined.

Table 3. Summary of Technologies for CFC Disposition

Disposition Method	Best CFC Conversion Demonstrated	Scale[a]	CFC Source	Advantages	Disadvantages
Laser destruction	[b]	L	Atmosphere	Destroy CFCs that have already been released to the atmosphere; only possible way to prevent further depletion of ozone layer	Large energy requirements; requires processing of entire atmosphere
Microwave destruction	[b]	L	Atmosphere	Destroy CFCs that have already been released to the atmosphere; only possible way to prevent further depletion of ozone layer	Large energy requirements; requires processing of entire atmosphere
Sonochemical destruction	95%	L	Water	Excellent control of CFC volatilization by using localized heating	Sonochemical sources require further development for application on commercial scale
Electrified micro-heterogeneous catalysis	50%	L	Water	High dispersion of catalyst gives high activity	Much of the CFCs volatilize before reacting
Incineration of urethane foams	>99%	C	Foams	Very efficient; utilizes existing incinerators	Requires destruction of foam; cannot be used in foam recycling
Thermal incineration	>99%	C	Captured or recovered	Efficient, demonstrated technology; particularly advantageous to use with existing incinerator	At higher CFC concentration, may require exotic materials of construction; aqueous HF and HCl recovery required
Catalytic oxidation, CFC + O_2	>99%	L	Captured or recovered	Nonaqueous, less corrosion	Catalyst is really a reagent, or requires a guard bed
Catalytic oxidation, CFC + H_2O + O_2	>99%	P	Captured or recovered	Longer catalyst life than without using water cofeed	Aqueous; catalyst and reactor must be corrosion-resistant
Catalytic reaction with H_2O, CFC + H_2O	96%	L	Captured or recovered	Catalyst does not need to have activity for O_2 oxidation, only requires activity toward H_2O	Avoiding CO formation, limited to C_1 CFCs; allowing CO formation, limited to C_1 and C_2 CFCs; aqueous and corrosive environment

Table 3. *(continued)*

Disposition Method	Best CFC Conversion Demonstrated	Scale[a]	CFC Source	Advantages	Disadvantages
Reaction of CFCs with wet air or supercritical H_2O	>90%	L	Captured or recovered	Same principle as catalytic CFC + H_2O, but without catalyst stability issues	High temperatures and aqueous acid environment may necessitate exotic materials of construction
Catalytic reaction of CFCs with alcohols or ethers	63%	L	Captured or recovered	Large number of catalysts are active and stable for these reactions	Requires CFC recycle because of low conversion
Reaction of CFCs with ammonia to form HCN	80%	L	Captured or recovered	Appears to be generally applicable; nonaqueous	Many by-products; highly hazardous materials
Hydrodechlorination	varies	L	Captured or recovered	Convert CFC waste into valuable product	Only applicable if product of reaction is useful
Chlorination	50% while controlling by-products	L	Captured or recovered	Widely applicable; easily incorporated into existing processes with CFC by-products	Unless incorporated into an existing process, requires separation; disproportionation by-products include CFCs
Plasma	65%	L	Captured or recovered	Nonaqueous; higher conversions possible	Requires development in the area of plasma sources before commercial application
Sodium naphthalenide	>99%	L	Captured or recovered	Widely applicable; products are nonhazardous solids; can be used on HCFCs and HFCs	F and Cl value lost as sodium salts; recovery of naphthalenide products
Sodium oxalate	>99%	L	Captured or recovered	Widely applicable; products are nonhazardous solids	F and Cl value lost as sodium salts; cannot be used on HCFCs or HFCs
Solar	[b]	W	Captured or recovered	Takes advantage of heating rates, temperatures, and range of power levels associated with solar thermal energy	Not demonstrated as of 1996

[a] Largest scale for which technology has been demonstrated: W, white paper or proposal; L, laboratory experiments; P, pilot-plant or demonstration reactor; and C, practiced commercially.
[b] Because these methods are designed to process the entire atmosphere, per pass conversion is not a relevant parameter.

decreases the global warming potential because a material with a GWP on the order of 10^{-1} can be replaced with a material with a GWP on the order of 10^{-6}.

Disposition of CFCs via Catalytic Reactions

Often the solution to CO and NO_x formation in thermal incinerators is to use catalytic oxidation as an alternative. The proper catalyst offers many advantages over thermal incineration. First, the proper catalyst provides excellent selectivity to the desired products, which in this case are CO_2, H_2O, and HX. The selectivity is usually not strongly dependent on fuel/air ratios so that one does not need to tune fuel/air ratios to avoid the unwanted inorganic by-products in thermal incineration. Because catalytic oxidation typically occurs at much lower temperatures than thermal incineration, NO_x formation is thermodynamically unfavored. Operationally, this means that because one does not have to minimize NO_x formation, one can operate in great excess of air to minimize the possibility of creating an oxygen-starved situation and the resulting CO.

Aside from the oxidative methods, the remaining technologies are not oxidative and, for the most part, not destructive. Rather, these approaches to CFC disposition aim to recover the C and F value of the material often as a useful organic material.

Plasma Destruction of CFCs

Plasma destruction provides the opportunity to react CFCs at high temperatures encountered in plasmas where Cl_2 and F_2 are readily formed. In this way CFCs can be destroyed without requiring a fuel to be cofed as with thermal incineration, or water co-fed as with catalytic incineration. As part of a larger study of the destruction of volatile organic chemicals (VOCs), CFC-113 decomposition has been studied in the presence of dry air in a pulsed corona plasma reactor (27), and a maximum efficiency of 65% CFC-113 destruction was observed. However, the authors believed that higher efficiencies could easily be obtained in this system. First, the reactor used in these experiments was limited by pulse frequency, and their results showed that CFC-113 destruction efficiency increased with increasing pulse frequency. Second, CFC-113 destruction efficiency was also increasing with reactor residence time. Therefore, the results indicated that improvement in CFC-113 destruction efficiency could be obtained by increasing pulse frequency and/or reactor residence time.

CFC Destruction by Mineralization

This method uses electron migration to remove halogen atoms from the CFC. The negative charge on the anion is transferred to the halogen (X) on the CFC. The C–X bond is then broken, forming NaX and a very active CFC radical or an olefinic CFC material.

Other Methods for CFC Destruction

Other processes have also been proposed for CFC destruction, but most are proposals or paper studies. One such proposal is the use of solar technology for the destruction of CFC waste (28). Herein it is claimed that thermal solar energy has the capability to provide the necessary heating rates, temperatures, and range of power levels to run a variety of reactions such as pyrolysis, incineration, or reaction with metal oxides. Interestingly, they do not believe solar energy is appropriate for direct photochemical reactions because of the potential for formation of other CFCs via radical recombination.

Another method uses supercritical water or wet air to react with CFCs (29). Because of the additional reactivity of supercritical water, catalyst is not required. Thus, many of the materials issues around finding a catalyst that can decompose water and does not corrode in aqueous acid environments can be avoided. However, for both wet air oxidation and supercritical water reaction with CFCs, there is a high probability that exotic materials of construction would be required for process materials on account of the corrosive environment of supercritical water and/or aqueous acids.

SUMMARY

The rapid phaseout of CFCs is one of the more remarkable technological achievements of the twentieth century. Since 1987, society has progressed from having virtually no in-kind alternatives to CFCs, to having enough alternatives to allow a total phaseout of CFCs in developed nations. However, more challenges are ahead. During the development of CFC alternatives, recovered CFCs were stockpiled because the efforts were focused on developing CFC alternatives. Now, as the need to dispose of these materials is recognized, the technologies outlined in this article will become increasingly important.

In Table 3, the methods for CFC destruction are summarized. Along with the destruction efficiency, the stage of development is estimated and advantages and disadvantages are listed.

The intent of this table is not to give a critique of the methods, but rather to illustrate the interconnectedness of the technology. Many of these technologies work remarkably well for narrow applications, whereas others work fairly well on a wide variety of applications. For instance many techniques that seek to react CFCs to useful materials, such as chlorination of CFCs, create by-products that are difficult to reprocess. In these cases, the use of an incinerator along with the primary recovery process is prudent. Because the bulk of the CFCs would be processed in the primary unit, the duty on the incinerator would be greatly diminished. In this way, many of the methods in Table 3 are complementary. It is clear that what is needed is an arsenal of technologies that can be chosen alone, or in concert with other technologies, to address the wide variety of challenges and applications of CFC destruction. Much progress has been made in developing this arsenal, but as of the mid-1990s, only thermal incineration is practiced.

Although the benefits of the CFC phaseout can already be seen, continued CFC emissions will delay the recovery of the ozone layer. Further, even after the last recovered CFC is destroyed, CFCs will still be encountered as intermediates and by-products in processes. Therefore there is a great need for the development of the promising technologies described in this article. Efficient disposition of CFCs plays, and will continue to play, a crucial role in preventing further emissions of CFCs and damage to the ozone.

BIBLIOGRAPHY

1. *Refrig. Eng.* **65,** 49 (1957); T. Midgely and A. L. Henne, *Ind. Eng. Chem.* **22,** 542 (1930).
2. J. E. Lovelock, *Nature (London),* **230,** 379 (1971).
3. M. J. Molina and F. S. Rowland, *Nature (London),* **249,** 810 (1974).
4. J. Hansen, A. Lacis, and M. Prather, *J. Geophys. Res.* **94,** 16417 (1989).
5. L. E. Manzer and V. N. M. Rao, *Adv. Catalysis,* **39,** 329 (1993).
6. J. C. Farman, B. G. Gardiner, and J. D. Shanklin, *Nature,* **315,** 207 (1985).
7. J. C. Dickerman, T. E. Emmel, G. E. Harris, and K. E. Hummel, *EPA Report No. /600/7-89/011* (1989).
8. F. S. Rowland and M. J. Molina, *Rev. Geophys. Space Phys.* **13,** 1 (1975).
9. G. A. Askaryan and co-workers, in G. Bonizzoni, W. Hooke, E. Sindoni, eds., *Industrial Applications of Plasma Physics,* Ed. Compos., Bologna, Italy, 1993, p. 463.
10. T. H. Stix, "Controlled Active Global Experiments (CAGE)," in E. Sindoni and A. Y. Wong, eds., *Proc. Int. Workshop,* 1991, p. 143.
11. T. H. Stix, *J. Appl. Phys.* **66,** 5622 (1989).
12. W. S. Warren and A. H. Zewail, *J. Chem. Phys.* **78,** 3583 (1983).
13. K. T. Tzang and co-workers, in Ref. 10, p. 143.
14. G. A. Askaryan and co-workers, *Dokl. Akad. Nauk. SSSR,* **302,** 566 (1988).
15. H. M. Cheung and S. Kurup, *Environ. Sci. Technol.* **28,** 1619 (1994).
16. E. A. Hayes, J. Z. Stemple, and D. R. Rolison, *Proc. Electrochem. Soc.* **94,** 120 (1994).
17. C. Rittmeyer and co-workers, *Chemosphere,* **28,** 1455 (1994).
18. C. Rittmeyer and J. Vehlow, *Chemosphere,* **26,** 2129 (1993).
19. M. Takeuchi and co-workers, *Kogai,* **27,** 117 (1992).
20. J. V. Ryan, C. W. Lee, and S. Korn, *EPA Report No. /600/A-94/008* (1994).
21. K. Tokuhashi and co-workers, *Combust. Sci. Tech.* **72,** 117 (1990).
22. Fr. Pat. 2,137,790 (Feb. 2, 1973), N. E. West (to E. I. Du Pont de Nemours and Co., Inc.).
23. U.S. Pat. 4,120,939 (Oct. 17, 1978), W. H. Ehlig (to E. I. Du Pont de Nemours and Co., Inc.).
24. H. Ouchi, *Kogai to Taisuku,* **25,** 1489 (1989).
25. V. Ramanathan and co-workers, *J. Geophys. Research,* **90,** 5547 (1985).
26. J. Hanse and co-workers, *J. Geophys. Research,* **93,** 9341 (1988).
27. T. Yamamoto and co-workers, *IEEE Trans. Ind. Appl.* **28,** 528 (1992).
28. D. M. Blake, *Int. J. Refrig.* **11,** 239 (1988).
29. T. Sugeta and M. Sato, *Kaguku Kogyo,* **42,** 471 (1991).

CHROMATOGRAPHY. See TWO-DIMENSIONAL GAS CHROMATOGRAPHY; THIN-LAYER CHROMATOGRAPHY; CHROMATOGRAPHY/PLASMA MASS SPECTROMETRY; COUNTERCURRENT CHROMATOGRAPHY; ION CHROMATOGRAPHY.

CHROMATOGRAPHY/PLASMA MASS SPECTROMETRY

DOUGLAS T. HEITKEMPER
BRENDA S. ZIMMER
KAREN A. WOLNIK
U.S. Food and Drug Administration
Cincinnati, Ohio

JOSEPH A. CARUSO
University of Cincinnati
Cincinnati, Ohio

Many trace elements are of toxicological concern to living organisms, including humans. Toxic trace elements are found throughout the environment in soils, sediments, air, and water with both natural and anthropogenic sources. Their effects on the environment and humans can be harmful at much lower levels than previously thought. In addition, the chemical form or species of an element can greatly affect its toxicity, bioavailability, mobility in the environment, and environmental impact. Specific oxidation states of inorganic molecules, organometallics, complexes, and biomolecules play significant roles in the environment. For these reasons, the determination of total elemental content in environmental samples is inadequate to evaluate environmental risks. The identification and quantitation of the chemical form of an element is commonly referred to as elemental speciation. The complex nature of environmental samples mandates a technique with a high degree of specificity. Additionally, it is becoming increasingly evident that even ultratrace levels of some chemical forms of an element may pose a serious risk. Thus, detection of a particular chemical species at trace and ultratrace levels in environmental samples is necessary. The combination of plasma mass spectrometry (1,2) with various separation techniques can, in many cases, meet these challenges.

INDUCTIVELY COUPLED PLASMA—MASS SPECTROMETRY

Inductively coupled plasma—mass spectrometry (ICP/MS) is a widely recognized technique for trace and ultratrace elemental analysis (1,2) (see INDUCTIVELY COUPLED PLASMA–MASS SPECTROMETRY). The ICP consists of a partially ionized argon gas at a temperature of approximately 6000–10,000 K. The plasma is maintained in a quartz torch made up of three concentric tubes. Three separate flows of argon are directed through the torch. The outermost flow of argon, called the support or cooling flow, is added tangentially resulting in the formation of a toroidal- or doughnut-shaped plasma. The central flow of argon forms the central channel in the plasma and transports the sample aerosol or gas to the plasma. The intermediate or auxiliary flow of argon is used to keep the plasma away from the glass torch. Radio-frequency (r-f) power is coupled to the plasma through a water-cooled induction coil. The radio-frequency generator operates at either 27.12 or 40.68 MHz with output levels typically in the 0.5–1.5 kW range. Once argon is flowing through the torch and power is applied, the plasma can be ignited using a spark to seed electrons. The plasma is self-sustaining as collisions of gaseous atoms, ions, and electrons result in further ionization and ohmic heating.

Typically, liquid samples are introduced to the plasma by means of a nebulizer and spray chamber arrangement.

Pneumatic nebulization is most often used to form an aerosol which is directed through the spray chamber and into the inner tube of the ICP torch. The spray chamber separates larger droplets and directs them to waste, and water-cooling reduces the amount of solvent vapor reaching the plasma. The aerosol is then injected into the central channel of the plasma where it undergoes the processes of desolvation, vaporization, atomization, and ionization. A portion of the ions formed in the atmospheric-pressure plasma are extracted through a differentially pumped interface into the mass spectrometer where they are separated and analyzed based on their mass-to charge (m/z) ratios. Ions are sampled from the plasma through the orifice of a water-cooled metal sampling cone. The gas expands as it passes the sampler into a region of reduced pressure and ions are again transmitted through a second metal skimmer cone while the majority of argon is pumped away. The ions are passed through a series of ion lenses which direct the ion beam into the low pressure ((Pa) $< 10^{-5}$ mbar) region of the quadrupole mass analyzer where the ions are separated and finally detected with an electron multiplier. A schematic diagram of a quadrupole ICP/MS system is shown in Figure 1.

MICROWAVE-INDUCED PLASMA MASS SPECTROMETRY

The argon ICP is not an efficient ion source for nonmetals such as the halogens. One alternative to the argon ICP which has been investigated for use with chromatographic sample introduction is the helium microwave-induced plasma (MIP) (3). Helium has a higher ionization potential than argon (24.5 vs 15.8 eV) potentially making it a better ionization source for high ionization potential elements such as the halogens and many metalloids. In addition, the use of helium plasmas reduces or eliminates spectral interferences arising from the three isotopes of argon and polyatomic species containing argon. In argon ICP/MS, $^{40}Ar^+$, $^{40}Ar^{56}O^+$, and $^{40}Ar_2^+$ interfere with the major isotopes of Ca, Fe, and Se at $m/z = 40$, 56, and 80, respectively.

The He MIP is formed in a single quartz discharge tube placed inside a resonant cavity. Microwave power, supplied at a frequency of 2450 MHz, sets up a standing electromagnetic wave inside the cavity, and powers of less than 400 W are generally used. The He MIP source can be adapted to existing ICP/MS instruments by replacing the ICP torch box and matching network with the microwave cavity and increasing the pumping capacity of the ICP/MS interface region.

COUPLED CHROMATOGRAPHY/PLASMA MASS SPECTROMETRY

The on-line coupling of various chromatographic techniques with plasma mass spectrometric detection provides a powerful tool for environmental analysis (4–7). The most studied combination of liquid chromatography (LC) and ICP/MS has been used for the elemental speciation of a variety of elements including As, Se, Sn, Pb, and Cr. Additionally, LC/ICP/MS has been used to circumvent some of the limitations of quadrupole ICP/MS. Preconcentration and matrix elimination via LC have been useful in the analysis of, for example, rare-earth elements. Although less studied, the high resolving power of gas chromatography (GC) in combination with plasma mass spectrometric detection by either ICP/MS or MIP/MS has been shown to be an extremely sensitive and selective technique for elemental speciation.

Specific advantages of elemental mass spectrometric detection for chromatographic analysis include excellent sensitivity, the ability to determine the elemental composition of chromatographically resolved peaks, and to differentiate unresolved peaks based on differences in elemental composition. When chromatographic peaks are unresolved, the mass spectrometer can be set to detect only analytes at a specified mass. For example, if a chromium compound and a tin compound coelute, the plasma mass spectrometer could be used to determine these species based on differences in their mass-to-charge ratio (m/z). This feature is especially helpful for separating interferences which are often present at much greater levels than the analyte of interest from unresolved peaks. The mass spectrometer can be used in the single ion monitoring mode in which only a selected isotope is monitored (ie, lead at m/z 208) or in some sort of time-resolved acquisition mode in which several isotopes of the same element can be monitored and/or several different elements can be monitored. The ability to monitor specific masses is very useful when full chromatographic separation of species with differing elemental composition cannot be achieved.

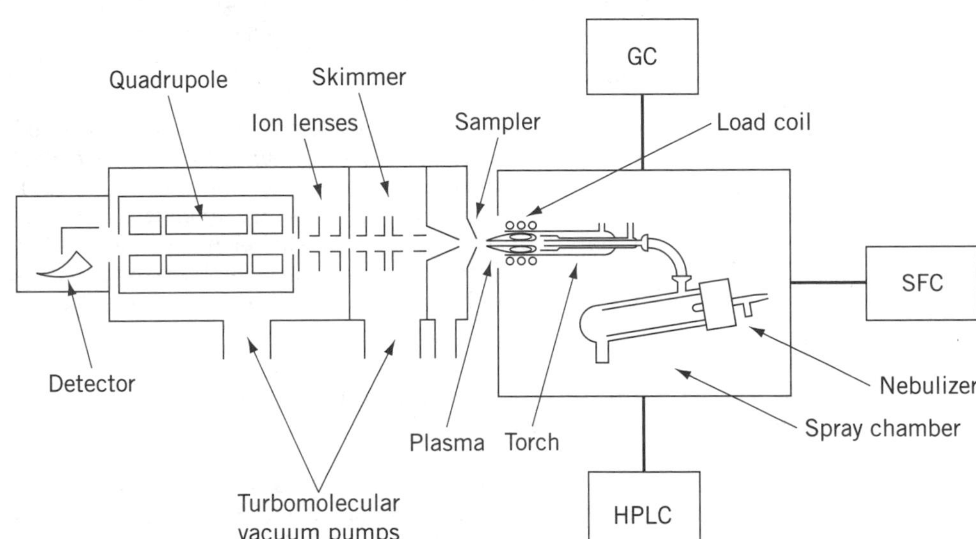

Figure 1. Schematic diagram of ICP/MS with chromatographic sample introduction.

LIQUID CHROMATOGRAPHY/PLASMA MASS SPECTROMETRY

Liquid chromatography (LC) provides a means of separation for nonvolatile and thermally labile compounds. The combination of high performance liquid chromatography (HPLC) with element-specific detection using ICP/MS detection was first reported in 1986 (9). LC/ICP/MS has been widely studied since because of its applicability to elemental speciation, on-line preconcentration, and the elimination of spectral interferences (4–7).

The interface requirements for LC/ICP/MS are dominated by the choice of eluent composition and flow rate. Typically, liquid samples are introduced to the ICP/MS via pneumatic nebulization at flow rates comparable to those used with standard bore, ie, 4–4.6 mm ID, HPLC and IC columns (0.5–2.0 mL/min). Thus, the physical interface between HPLC and ICP/MS is relatively straightforward. The simplest arrangement is to connect the outlet of the HPLC separation column to the inlet of either a cross-flow or concentric-type nebulizer by means of a length of inert tubing (ie, narrow-bore polyetherether ketone (PEEK) or polytetrofluoroethylene (PTFE)). The length of tubing used in the interface is minimized to ensure minimal extra-column band broadening of chromatographic peaks. A number of variations to this interface have been utilized including the use of a UV detector in series with the plasma MS detector. Post-column addition of an internal standard via flow injection or continuous nebulization has been used to correct for ICP/MS signal drift.

The choice of eluent composition affects both the chromatographic separation and performance of the plasma MS detector. The introduction of organic solvents to the ICP causes increased solvent vapor pressure which destabilizes the plasma. Mobile phases containing high concentrations of organic solvents can also result in decreased sensitivity, increased numbers of polyatomic carbon-containing species in the mass spectrum, and carbon deposition on the sampling orifice. The use of higher r-f power and water-cooled spray chambers can help to alleviate some of these problems. In addition, a small amount of oxygen is often added to the nebulizer gas to reduce carbon deposition on the sampling cones. Mobile phases containing high concentrations of buffer salts should also be avoided because of the potential for nebulizer and sampling orifice blockages. Obviously, compromise conditions may be required to ensure both adequate chromatographic resolution and ICP/MS performance.

Gradient elution has been utilized in HPLC ICP/MS on a limited basis. The majority of separations using ICP/MS detection have been under isocratic conditions. Although gradient elution can significantly improve a chromatographic separation, the changes in mobile-phase composition which result tend to destabilize the plasma and cause significant changes in the background intensity.

A number of alternatives to the LC/ICP/MS interface described above have been investigated. The major drawbacks of the above approach are the inefficient nature of conventional pneumatic nebulization and the incompatibility of the plasma with various LC mobile phases. Efficiency for pneumatic nebulizers ranges from 2–5% with the remainder of sample going to waste. More efficient nebulizers such as the ultrasonic nebulizer (USN) or hydraulic high pressure nebulizer (HHPN) have been used as alternatives to pneumatic nebulization (8,9). The increased aerosol transport efficiency of these systems generally results in detection limit improvements of roughly an order of magnitude. The use of a desolvation system is required with the USN and HHPN, especially if an appreciable amount of organic solvent is in the mobile phase. The desolvation system increases the dead volume and can lead to increases in peak width. Direct injection nebulization (DIN) provides very low dead volume but is limited to an eluent flow rate of less than 100 μL/min and is thus generally limited to use with microbore LC separations (10).

GAS CHROMATOGRAPHY/PLASMA MASS SPECTROMETRY

Gas Chromatography

Gas chromatography (GC) is used in the separation of compounds which are volatile and thermally stable at elevated temperatures. Volatility can be increased through the formation of derivatives. The GC technique is commonly used to separate organic compounds containing nonmetallic elements. Separations are based on the partitioning of the sample into and out of a liquid stationary phase. The liquid phase is either coated onto particles in a packed column or bonded in a thin film to the walls of a capillary column. An inert gas such as helium is generally used as the carrier gas to move the sample through the column and into the detector. Other than column selection, temperature programming is the primary means of controlling retention.

GC/ICP/MS

The coupling of GC with ICP/MS is attractive because gaseous sample introduction to the plasma is theoretically 100% efficient and results in pg to sub-pg absolute detection limits (5). In addition, GC provides very high resolving power for complex environmental samples. Despite these advantages, there have been significantly fewer numbers of GC than LC studies reported using plasma mass spectrometric detection. There are several reasons for the slower development of GC/ICP/MS. GC is not readily applicable to nonvolatile compounds. Many compounds of interest in elemental speciation studies are nonvolatile ionic species which would require a derivitization step in order to increase their volatility. In addition, the argon ICP is not an efficient ionization source for high ionization potential nonmetals such as C, H, N, O, Cl, and Br. Argon ICP/MS is not as sensitive for these elements as it is for metals.

GC/MIP/MS

Microwave-induced plasmas are more commonly associated with GC than inductively coupled plasmas (3) and a He MIP emission-based GC detector is commercially available (11). Helium microwave-induced plasmas are able to ionize the higher ionization potential nonmetallic elements such as the halogens, P, and S and thus have found application to the determination of pesticides, PCBs, and dioxins in environmental samples.

SUPERCRITICAL FLUID CHROMATOGRAPHY/PLASMA MASS SPECTROMETRY

Supercritical fluid chromatography (SFC) provides an important alternative to conventional LC and GC separation techniques. A supercritical fluid is defined as a substance above its critical temperature and pressure. Under these conditions, the diffusion constants of the supercritical fluid are increased in comparison to liquids and resistance to mass transfer is lower thereby increasing the speed of the separation in comparison to LC. In comparison to gas chromatography, SFC can be used for thermally labile, nonvolatile, and high molecular weight compounds. In SFC, separations can be controlled through the adjustment of a number of variables including the use of polar mobile-phase modifiers, temperature gradients, and density or pressure programming. Carbon dioxide and nitrous oxide have been the most studied mobile phases used in SFC, and flame ionization detection (FID) is the most commonly used detector.

The coupling of capillary SFC with plasma mass spectrometric detection shows considerable promise for the determination of organometallic compounds in the environment (12,13). To date, studies have focused on interface design, characterization of background spectral features, effects of varying mobile-phase parameters such as pressure programming or the use of alternatives to carbon dioxide, and the demonstration of detection capabilities with a few representative organometallic compounds. Organometallic compounds of tin, lead, chromium, mercury, and arsenic have been used to characterize this technique. Studies have compared ICP/MS to flame ionization detection (13). Typically, ICP/MS detection has been approximately one order of magnitude more sensitive than FID detection with sub-picogram to picogram absolute detection limits obtained.

CONCLUSIONS

The coupling of chromatographic techniques such as high performance liquid chromatography (HPLC), gas chromatography (GC), and supercritical fluid chromatography (SFC) (5,8) with plasma mass spectrometry provides a powerful analytical tool with wide applicability in environmental measurement and elemental speciation. The primary advantages of plasma mass spectrometric detection include excellent sensitivity and selectivity, with the capability of monitoring multiple elements and/or isotopes. The ability to measure isotope ratios on eluting peaks is extremely valuable in providing confirmation of the elemental content of chromatographic peaks and in the application of isotope dilution mass spectrometry (14).

The majority of chromatography/plasma mass spectrometry studies have focused on the coupling of two techniques and demonstration of the hyphenated technique's capabilities. However, a number of important issues remain regarding the application of these methods to routine environmental analysis. As with all elemental speciation methods, the accuracy of the method can be difficult to assess. A limited number of species-specific certified reference materials are currently available. In addition, the analyst must keep in mind the potential for the interconversion of species during the preparation and storage of samples prior to analysis (see MASS SPECTROMETRY).

BIBLIOGRAPHY

1. R. S. Houk, *Anal. Chem.* **58,** 97A (1986).
2. G. Horlick and Y. Shao, in A. Montaser and D. W. Golightly, eds., *Inductively Coupled Plasmas in Analytical Atomic Spectrometry,* 2nd ed., VCH Publishers, Inc., New York, 1992, pp 551–612.
3. L. K. Olsen and J. A. Caruso, *Spectrochim. Acta B,* **49B,** 7–30 (1994).
4. N. P. Vela, L. K. Olson, and J. A. Caruso, *Anal. Chem.* **65,** 585A–597A (1993).
5. S. J. Hill, M. J. Bloxham, and P.J. Worsfold, *J. Anal. At. Spectrom.* **8,** 499–515 (1993).
6. F. A. Byrdy and J. A. Caruso, *Environ. Sci. Technol.* **28,** 528A–534A (1994).
7. A. Seubert, *Fresenius J. Anal. Chem.* **350,** 210–220 (1994).
8. J. J. Thompson and R. S. Houk, *Anal. Chem.* **58,** 2541–2548 (1986).
9. N. Jakubowski, B. Jepkens, D. Stuewer, and H. Berndt, *J. Anal. At. Spectrom.* **9,** 193–198 (1994).
10. S. C. K. Shum, R. Neddersen, and R. S. Houk, *Analyst,* **117,** 577–582 (1992).
11. R. L. Fior, *Am. Lab.* **21,** 40 (1989).
12. W.-L. Shen, N. P. Vela, B. S. Sheppard, and J. A. Caruso, *Anal. Chem.* **63,** 1491–1496 (1991).
13. N. P. Vela and J. A. Caruso, *J. Anal. At. Spectrom.* **7,** 971–977 (1992).
14. A. Brown, L. Ebdon, and S. J. Hill, *Anal. Chim. Acta,* **286,** 391–399 (1994).

CHROMIUM: ENVIRONMENTAL ANALYTICAL CHEMISTRY

SIDNEY A. KATZ
Rutgers University
Camden, New Jersey

Much of the new information about the environmental chemistry of chromium results from continuing progress in trace element analytical chemistry research. Sometimes driven by the pressures of regulatory compliance monitoring, many of the problems associated with the collection, preservation, and storage of samples, with sample preparation and pretreatment, and with the determination of chromium have been addressed. The enhancement of old methodologies and the development of new ones have been applied to the solving these problems. The current analytical methodologies have adequate selectivity and sensitivity for determining chromium in a wide variety of environmental matrices, but only a few of them provide usable information about the chemical speciation of chromium in environmental systems. Much of the current effort in trace element analytical research is directed to identifying both the qualitative and the quantitative parameters associated with the environmental chemistry of chromium.

COLLECTION AND PRESERVATION OF SAMPLES FOR CHROMIUM ANALYSIS

The New Jersey Department of Environmental Protection and Energy (1) has developed an excellent manual for the theory and practice of environmental sampling. Samples collected to determine chromium, however, demand special consideration. In addition to the traditional problems associated with loss and/or contamination, the analytical chemistry of chromium must also preserve the oxidation state during the collection and storage of environmental samples.

Air Samples

The adverse health effects associated with inhaling some hexavalent chromium compounds necessitates ambient air monitoring and the monitoring of air in the workplace to determine chromium in airborne particulates. Airborne particulates are collected by absorption in liquid impingers or by filtration with glass fiber or cellulose ester-based membranes. Although the glass fiber membranes are less likely than the cellulose ester-based membranes to reduce hexavalent chromium, the levels of impurities in the former are significantly higher than those in the latter. Two studies have tabulated the concentrations of impurities in samples of these filter materials (2,3). NIOSH has specified 0.8 μm cellulose ester membranes to collect airborne particulates for determining total chromium (4) and 5.0 μm PVC membranes to collect airborne particulates for determining hexavalent chromium (5). The latter recommendation should be reevaluated considering possible losses of respirable $<5\ \mu$ particles. ASTM procedure D 4185 recommends 0.8 μm cellulose ester or cellulose nitrate membrane filters for collecting airborne particulates prior to determining metals in workplace atmospheres (6). It has been reported that some cellulose ester filters (Millipore AA and Gelman GN 4) are contaminated with from 1 to 7 μg of chromium (7). NIOSH (8) confirmed that samples collected on PVC membrane filters for hexavalent chromium determinations were stable for two weeks at 25°C, but provided no specific directions for their storage or preservation aside from suggesting expeditious shipment to the laboratory with dry ice or with a sealed liquid-refrigerant.

Although one study (9) cited several reasons for not using solution-filled impingers to collect airborne emissions from a municipal sewage treatment plant (STP) sludge incinerator to determine their hexavalent chromium content, another study (10) found that Greenberg–Smith impingers filled with a dilute sodium bicarbonate buffer are suitable for sampling airborne particulates from work place and environmental settings before determining their chromium content. It was reported (11) that both hexavalent and trivalent chromium are stable in the sodium bicarbonate collection medium for at least 100 days. Losses were found of approximately 40% from 5 μg spikes of hexavalent chromium on glass fiber filters during the first month of postsampling storage (12); Subsequent losses during the next three months of storage were negligible. Stable isotopic tracers, $K^{50}{}_2Cr_2O_7$ and $K^{53}{}_2Cr_2O_7$, were used (13) to study the rate of reduction for hexavalent chromium under pseudo-first-order conditions on the quartz fiber filters (Pallflex, type 2500QAT-UP) used to collect particulates from a STP sludge incinerator. A rapid reduction of hexavalent chromium halftime of approximately 60 minutes occurred during active sampling, and a much slower reduction of hexavalent chromium halftime of approximately 21 days, occurred during the postsampling storage. Clearly, further studies are needed to establish the stability of hexavalent chromium in samples of airborne particulates collected by filtration.

TECHNIQUES FOR QUANTIFYING CHROMIUM

In considering the techniques for determining trace elements in biological and environmental materials, sensitivity and selectivity are usually of major significance. Of importance are speed, simplicity, cost, and multielement capabilities. No single technique satisfies all of these criteria for determining chromium in every biological or environmental sample. Molecular, atomic, and nuclear spectrometry and the electroanalytical and chromatographic techniques are frequently applied to analyze biological and environmental samples for chromium.

Spectrophotometry (SPEC) of chromophores formed with organic reagents, such as diphenylcarbazide, continues to be employed as a selective and sensitive technique for quantifying chromium in biological and environmental samples. Atomic absorption spectrometry (AAS) with electrothermal atomization (ETA-AAS) is now the most commonly used technique for measuring ambient levels of chromium in environmental samples. Inductively coupled plasma atomic emission spectrometry (ICP-AES), X-ray fluorescence (XRF) spectrometry, charged particle, usually proton-induced X-ray emission (PIXE) spectrometry, spark source mass spectrometry (SSMS), and instrumental and radiochemical neutron activation analysis (INAA and RNAA) are powerful techniques for determining chromium in a wide variety of sample matrices. These latter techniques have multielement capabilities. They frequently allow the simultaneous collection of data for studying synergistic and/or antagonistic actions of trace elements in biological and environmental systems. Neutron activation analysis has the added advantage of requiring little if any chemical pretreatment of the sample. The atomic and nuclear spectrometric techniques, AAS, ICP-AES, XRF, PIXE, SSMS, and NAA, are techniques for elemental analysis independent of the chemical form. Without selective chemical pretreatments, the results obtained with these techniques do not provide information about the chemical speciation of chromium.

Electroanalytical techniques are species dependent. A cathodic stripping voltammetric (CSV) technique is reported (14) for determining hexavalent chromium in the presence of trivalent chromium. A recent development with the combined technique of ion chromatography–inductively coupled plasma mass spectrometry (IC/ICP MS) (15) determined hexavalent chromium in airborne particulates from the incineration of STP sludge.

Analytical methodologies are reviewed (16) for determining the concentrations of trace elements beneficial to and harmful to human biochemistry. The review includes a tabulation of the sensitivities (S) and the detection limits (DL) of the most frequently used analytical techniques. The tabulated values for chromium were AAS (S) 0.06 ppm; ICP-AES (DL) 1 ppm; XRF (DL) 0.7 ppm; SSMS (DL) 0.0005 ppm; and NAA (S) 0.3 ppm. Among the many reviews on techniques for quantifying trace elements in environmental samples, two (17,18) provide particularly practical comparative information. The U.S. EPA (19) has tabulated analytical methods for determining chromium in a wide variety of biological and environmental samples.

CHROMIUM SPECIATION

Numerous chemical forms of chromium exist in biological and environmental systems, and determining their speciation is important to understanding the biological and environmental chemistry of chromium. As a first step in establishing the chemical speciation of chromium, significant progress has been made in differentiating between its more toxic hexavalent forms and its less toxic trivalent forms. Liquid–liquid extraction and liquid–solid extraction procedures have been developed for separating these two oxidation states of chromium in water and soil samples. In addition, several chromatographic procedures have been developed for determining hexavalent chromium in water and air samples. The speciation of several elements, including chromium, by tandem chromatography-mass spectrometry techniques was recently reviewed (20), and application of the electron probe microanalyzer to elemental X-ray mapping with valence state information was described (21). The synchrotron X-ray microprobe was used to determine of hexavalent chromium in oxides, silicates, glasses, and cementitous waste forms using X-ray absorption near-edge structure (22). Some of the needs are discussed for differentiating between the hexavalent and the trivalent forms of chromium in environmental materials, and some of the procedures by which this has been accomplished are presented (23). These and some of the more-recently developed procedures are described in the following sections.

QUALITY ASSURANCE FOR CHROMIUM DETERMINATIONS

To interpret analytical results with some defined degree of confidence and apply them to further understanding the biological and environmental chemistry of chromium, it is necessary that the measurements be adequately conceived, documented, and executed. A formalized quality assurance program insures control and documentation of data quality.

Data Validation in the Analysis for Chromium

Quality assurance is the overall system of procedures which insure that measurement data meet defined standards of quality within established criteria for dependability, adequacy, and economy. Quality assurance consists of two separate but related activities: quality control and quality assessment. Quality control is the system of procedures which reduce random and systematic errors and maintain them within tolerable limits so that the analytical results meet the needs of those interpreting and applying them. The system of procedures includes qualified personnel, reliable equipment, appropriate methodology, and adherence to standard operating procedures and good laboratory practices. Quality assessment is the system of procedures which monitors quality control by continuously evaluating the performance of the system and the quality of the data. Internal quality assessment evaluates precision and bias by repetitive measurements, interchange of operators and/or equipment, determinations by alternate methods, and the use of control charts. External quality assessment evaluates bias by participating in interlaboratory comparisons, splitting samples with other laboratories, and measuring standard reference materials (SRMs) and/or certified reference materials (CRMs). Quality assurance procedures are combined with a work plan in a guidance document for efficient and effective administration. Such a guidance document includes

1. a description of the project;
2. the organization chart for the project;
3. specification of tasks and responsibilities;
4. data quality requirements and assessments;
5. sampling procedures;
6. chain of custody procedures;
7. instrument calibration and maintenance procedures;
8. documentation, data reduction, and reporting;
9. data validation;
10. performance audits;
11. corrective actions; and
12. records and reports.

These aspects of the guidance document have been elaborated upon (24,25), and guidelines for quality statements have been developed by the ASTM (26). The sources of error in determining chromium in environmental matricies by flame atomization and by electrothermal atomization atomic absorption spectrometry were exhaustively studied (27).

The Standard Operating Procedures Manual prepared by the New Jersey State Department of Health (28) is an excellent document for practical quality assurance in public health and environmental laboratories. Adoption of the quality assurance standards described in this manual enhances the scientific validity of laboratory data related to the biological and environmental chemistry of chromium.

Standard/Certified Reference Materials for Chromium

Subsequent to the preparation of kale powder in the early 1960s (29), Standard Reference Materials (SRMs) and Certified Reference Materials (CRMs) for determining trace elements in biological and environmental samples have been developed by a variety of national (30) and international agencies (31). In addition, individual researchers (32,33,34,35) have reported on the preparation and calibration of reference materials developed in their laboratories. The chromium reference values for some of these materials are tabulated below, and the suppliers of these materials are listed at the end of Table 1. Also included in Table 1 are the 2 σ relative errors associated with the reference values. The reference values cited in Table 1 refer to total chromium. The preparation of reference materials for hexavalent and trivalent chromium has been hindered by problems of long-term stability.

PRACTICAL CONSIDERATIONS FOR DETERMINING CHROMIUM

Among the benefits of developing a guidance document is the need to address questions regarding the chemical form and expected concentrations of chromium to be determined, the accuracy demanded by the use to which the data will be put, the physical form of the sample, the number of samples to be evaluated, and the presence or absence of interferences to determining chromium. The sensitivity and selectivity of ETA-AAS for quantitating chromium at normal environmental concentrations are well documented, and some of the recently described IC and HPLC techniques show great promise for separating trivalent from hexavalent chromium. Hexavalent chromium has been successfully speciated and quantified in air, water, and soil samples. The chemical pretreatment of biological tissues for speciation has not yet been developed to a point where the oxidation state of chromium remains unaltered. The chromium concentrations in environmental

Table 1. Some Environmental Reference Materials for Chromium

Reference Material[a]	Reference Value	2 σ error
NRCC-NASS-1 seawater	0.000184 mg/L	±8.7%
NRCC-NASS-2 seawater	0.000175 mg/L	±5.7%
NRCC-NASS-3 seawater	0.000175 mg/L	±13%
NRCC-CASS-1 seawater	0.000118 mg/L	±18%
NRCC-CASS-2 seawater	0.000121 mg/L	±13%
NRCC-SLEW-1 estuarine water	0.000139 mg/L	±12%
NRCC-SLRS-1 river water	0.00036 mg/L	±11%
NRCC-SLRS-2 river water	0.00045 mg/L	±16%
NIST-SRM-1643b water	0.0186 mg/L	±2.2%
IAEA-W-4 freshwater	0.0099 mg/L	
NIST-SRM-1645 river sediment	29600 mg/kg	±9.5%
IAEA-SL-1 lake sediment	104 mg/kg	±8.7%
NIES-CRM-2 pond sediment	75 mg/kg	±6.7%
IAEA-SD-N-1/2 marine sediment	149 mg/kg	±10%
NRCC-BCSS-1 marine sediment	123 mg/kg	±11%
NRCC-MESS-1 marine sediment	71 mg/kg	±15%
NRCC-PACS-1 marine sediment	113 mg/kg	±7.1%
NIST-SRM-1646 estuarian sediment	76 mg/kg	±3.9%
BCR-CRM-146 sewage sludge	784 mg/kg	±9.4%
BCR-CRM-144 sewage sludge	485.4 mg/kg	±18%
BCR-CRM-145 sewage sludge	105.4 mg/kg	±27%
US EPA sewage sludge	193 mg/kg	±25%
BCR-CRM-143 soil	228 mg/kg	
BCR-CRM-141 soil	75 mg/kg	
BCR-CRM-142 soil	74.9 mg/kg	
IAEA-SOIL-7 soil	60 mg/kg	±21%
IAEA-AIR-3/1 air filter	0.0050 mg	±6.0%
NIST-SRM-1648 urban particulate	403 mg/kg	±3.0%
BCR-CRM-176 incinerator ash	863 mg/kg	±3.5%
NIST-SRM-1633a coal fly ash	196 mg/kg	±3.1%
IRANT-ECH coal fly ash	183 mg/kg	±5.2%
IRANT-EOP coal fly ash	183 mg/kg	±6.3%
IRANT-ENO coal fly ash	96.1 mg/kg	±8.3%
BCR-CRM-038 fly ash	178 mg/kg	
SABS-SARM-20 coal	67 mg/kg	
SABS-SARM-19 coal	50 mg/kg	±11%
SABS-SARM-18 coal	16 mg/kg	±13%
NIST-SRM-1635 coal	2.5 mg/kg	±12%
BCR-CRM-040 coal	31.3 mg/kg	±6.4%
NIST-SRM-1634a fuel oil	0.7 mg/kg	
IAEA-v-10 hay powder	6.5 mg/kg	±12%
NIST-SRM-1573 tomato leaves	4.5 mg/kg	±11%
NIST-SRM-1572 citrus leaves	0.8 mg/kg	±25%
NIST-SRM-1575 pine needles	2.6 mg/kg	±7.7%
Bowen's kale	0.369 mg/kg	±27%
NIES-CRM-1 pepper bush	1.3 mg/kg	
NIST-SRM-1569 brewer's yeast	2.12 mg/kg	±2.4%
IAEA-MA-A-2 fish flesh	1.3 mg/kg	±7.7%
NRCC-DORM-1 fish muscle	3.60 mg/kg	±69%
NRCC-DOLT-1 fish liver	0.40 mg/kg	±18%
IAEA-MA-M-2 mussel tissue	1.25 mg/kg	±27%
NIES-CRM-6 mussel	0.63 mg/kg	±11%
NIST-SRM-1566 oyster tissue	0.69 mg/kg	±39%
NRCC-TORT-1 lobster	2.4 mg/kg	±25%
NRCC-LUTS-1 lobster hepatopancreas	0.53 mg/kg	±15%
IAEA-MA-A-1 copepoda	1.1 mg/kg	±18%
IAEA-A-11 milk powder	0.0177 mg/kg	
NIST-SRM-1549 milk powder	0.0026 mg/kg	±27%
IAEA-V-9 cotton cellulose	0.11 mg/kg	
NIST-SRM-2670 spiked urine	0.085 mg/L	±7.1%
NIST-SRM-2670 normal urine	0.013 mg/L	
BI-CUM-3 urine	0.0556 mg/L	±14%
BI-CUM-2 urine	0.0290 mg/L	±17%
BI-CUM-1 urine	0.0028 mg/kg	±25%
BI-CBM-1 blood	0.0099 mg/L	±12%
NIST-RM-8419 bovine serum	0.00030 mg/L	±17%
NIST-SRM-909 reference serum	0.108 mg/L proc A	
NIST-SRM-909 reference serum	0.0913 mg/L proc B	
IAEA-H-4 animal muscle	0.0091 mg/kg	
NIES-CRM-5 human hair	1.4 mg/kg	±14%

[a] BCR = Community Bureau of Reference, Commission of the European Communities, 200 Rue de la Loi, B - 1049 Brussels, Belgium; BI = Behring Institute, Postfach 1140, D - 3550 Marburg 1, Germany; IAEA = International Atomic Energy Agency, Postfach 100, A - 1400 Wien, Austria; IRANT = Institute of Radioecology and Applied Nuclear Techniques, Post Office Box A - 41, 040 61 Kosice, Czechoslovakia; NIES = National Institute of Environmental Studies, Japan Environment Agency, Post Office Yatabe, Tsukuba Ibaraki 300-21, Japan; NIST = Office of Standard Reference Materials, National Institute of Standards and Technology, Gaithersburg, Md. 20899; NRCC = Division of Chemistry, National Research Council of Canada, Ottawa, Ontario K1A OR6, Canada; SABS = South African Bureau of Standards, Private Bag X191, Pretoria, Republic of South Africa; and USEPA = United States Environmental Protection Agency, Quality Assurance Branch, EMSL, Cincinnati, Ohio 45268.

systems are not constant in the time-space continuum. The sampling protocol must recognize this inconsistency, and analytical data based on a single determination must be interpreted with extreme caution. Normal environmental concentrations of chromium are quite low. Contamination or loss causes significant errors. Blanks, splits, and spikes must be included in the quality assessment protocol. The current lack of suitable reference materials for hexavalent chromium is an impediment to a comprehensive quality assurance program. Continued research and cooperation by scientists in the academic, governmental, industrial, and commercial sectors is needed to overcome some of the current deficiencies in the environmental analytical chemistry of chromium.

BIBLIOGRAPHY

1. N.J. DEPE, *Field Sampling Procedures Manual*, New Jersey Department of Environmental Protection and Energy, Trenton, 1992.
2. S.A. Katz and S.W. Jenniss, *Regulatory Compliance Monitoring by Atomic Absorption Spectroscopy,* Verlag Chemie International, Deerfield Beach, Fla., 1983, p. 29.
3. J.C. Van Loon, *Selected Methods of Trace Metal Analysis; Biological and Environmental Samples,* John Wiley & Sons, Inc., New York, 1985, p. 262.
4. P.M. Eller, ed., NIOSH Manual of Analytical Methods, 4th. ed., DHHS (NIOSH) Publication No. 94-113, U.S. Department of Health and Human Services, Cincinnati, Ohio, 1994, Method 7024.
5. Ref. 4, Method 7600.
6. ASTM, 1991 Book of ASTM Standards, Vol. 11.03, *Atmospheric Analysis; Occupational Health and Safety, Designation D 4185-90, Standard Practice for Measurement of Metals in Workplace*

Atmospheres by Atomic Absorption Spectrometry, American Society for Testing and Materials, Philadelphia, Pa., 1991, p. 263–270.

7. M.L. Shaw, D.K. Verma, and K.L. Ferguson, *Appl. Occupat. Environ. Hyg.* **6** 1027–1031 (1991).
8. Reference 4, Method 7604.
9. J.E. Arar, S.E. Long, T.D. Martin and S. Gold, *Environ. Sci. Technol.* **26** 1944–1950 (1992).
10. P. Sheehan, R. Ricks, S. Ripple, and D. Paustenbach, *Am. Indust. Hyg. Assoc. J.* **53,** 57–68 (1992).
11. Reference 10, p. 58.
12. D.L. Ehman, V.C. Anselmo, and J.M. Jenks, *Spectroscopy* **3** 32–35 (1987).
13. Ref. 9, p. 1946.
14. J. Wang, *Stripping Analysis: Principles, Instrumentation and Applications,* VCH Publishers, Deerfield Beach, Fla., 1985, p. 123.
15. Ref. 9, p. 1946.
16. K.M. Stika and G.H. Morrison, *Fed. Proc.* **40,** 2115–2120 (1981).
17. M. Stoeppler and H. Nuernberg, *Critical Review of Analytical Methods for the Determination of Trace Elements in Biological Materials,* International Workshop on Biological Specimen Collection, CEC-WHO-USEPA, Luxembourg, 18–22 Apr. 1977.
18. S.A. Katz, *Am. Biotechnol. Lab.* **3**(2), 10–17 (1985).
19. U.S. EPA, *Health Assessment Document for Chromium,* EPA-600/8-83-014F, U.S. Environmental Protection Agency, Research Triangle Park, N.C., 1984, pp. 4:10–4:12.
20. N.P. Vela, L.K. Olson, and J.A. Caruso, *Anal. Chem.* **65,** 585A–597A (1993).
21. F.C–Y. Wang, *Spectroscopy* **8,** 32–36 (1993).
22. S. Bajt, S. Clark, S. Sutton, M. Rivers, and J. Smith, *Anal. Chem.* **65,** 1800–1804 (1993).
23. S.A. Katz, *Environ. Health Prospect.* **92,** 13–16 (1991).
24. M.W. Brossman, T.J. Hoogheem, and R.C. Splinter, in J.K. Taylor and T.W. Stanley, eds., *Quality Assurance Project Plans—A Key to Effective Monitoring Programs in Quality Assurance for Environmental Measurements,* ASTM SP 867, American Society for Testing and Materials, Philadelphia, Pa., 1985, p. 53–61.
25. *Guidance for the Preparation of Combined Work / Quality Assurance Project Plans for Environmental Monitoring,* OWRS QA-1, Office of Water Regulations and Standards, U.S. Environmental Protection Agency, Washington, D.C., 1984.
26. Standard Guide for ASTM Standard Specification Quality Statements, E 1267-88, *1992 Annual Book of ASTM Standards,* Vol. 14.02, American Society for Testing and Materials, Philadelphia, Pa., 1992, pp. 799–803.
27. A. Sahuquillo, R. Rubino, and G. Rauret, Detection of Sources of Error in the Determination of Cr in Environmental Matricies by FAAS and ETAAS, in E.A. Maier and B. Griepink, eds., *Quality Assurance for Environmental Analysis, P. Quevauviller,* Elsevier, Amsterdam, 1995; pp. 39–62.
28. *Standard Operating Procedure Manual for the State of New Jersey Department of Health Division of Public Health and Environmental Laboratories,* New Jersey State Dept. of Health, Trenton, N.J., 1987.
29. H.J.M. Bowen and P.A. Cawse, in J.M.A. Lenihan and S.J. Thompson, eds., *The Production of a Homogeneous Biological Material from Intercomparison of Elementary Analysis by Different Laboratories in Activation Analysis: Principles and Practice,* Academic Press, London, 1965, pp. 149–151.
30. E.S. Gladney, C.E. Burnes, D.R. Perrin, I. Roelandts, and T.E. Gills, *1982 Compilation of Elemental Concentration Data for NBS Biological, Geological and Environmental Standard Reference Materials,* NBS SP 260-88, U.S. Government Printing Office, Washington, D.C., 1984.
31. Y. Marumatsu and R.M. Parr, *Survey of Currently Available Reference Materials for Use in Connection with the Determination of Trace Elements in Biological and Environmental Materials,* IAEA/RL/128, International Atomic Energy Agency, Vienna, Austria, 1985.
32. Y. Dokiya, M. Taguchi, S. Toda, and K. Fuwa, *Anal. Chem.* **50,** 533–537 (1978).
33. I. Torelm, L–B. Croon, K. Kolar, and T. Schroder, *Fresenius J. Anal. Chem.* **338,** 435–437 (1990).
34. J. Kumpulainen and R. Tahvonen, *Fresenius J. Anal. Chem.* **338,** 461–465 (1990).
35. T. Anglov, J.M. Christensen, and E. Holst, *Fresenius J. Anal. Chem.* **338,** 530–533 (1990).

CHROMIUM(VI) IN SOILS

JOHN C. PETURA
Applied Environmental Management, Inc.
Malvern, Pennsylvania

BRUCE R. JAMES
University of Maryland
College Park, Maryland

ROCK J. VITALE
GEORGE R. MUSSOLINE
Environmental Standards, Inc.
Valley Forge, Pennsylvania

NECESSITY TO QUANTIFY CR(VI) IN SOILS AND SEDIMENTS

Chromium (Cr) is an economically important element found in the earth's crust that finds its way into many materials and products used in the developing and industrialized world. Major categories of chromium use include the following:

- Manufacture of stainless steels (10–18% Cr) ≈ 75% of consumption
- Hardening of steel
- Ingredient in alloys
- Plating hard, shiny surfaces
- Bricks and cement linings for industrial furnaces
- Chromium chemicals

Important uses of chromium chemicals and compounds (eg, chromates of potassium and sodium, dichromates, and chromic acid) include the following:

Wood preservation (CCA), 38%	Oxidizing agents
Metal finishing/corrosion protection, 15%	Granules in asphalt roofing
Leather tanning, 10%	Catalysts
Pigments (inks, dyes, paints), 8%	Fire retardants
Oil drilling mud, 4%	Ceramic coatings
Textile moridant, 3%	Analytical reagents

Magnetic tape, 2%	Abrasives and refractories
Etchant for plastics	Explosives and pyrotechnics
Glues and adhesives	Safety matches
Synthetic perfumes	
Colored glass	Waterproofing

It is among several unique metals from an environmental analysis and remediation perspective. It is an oxidation–reduction (redox) active metal with two commonly encountered valence states, trivalent [Cr(III)] and hexavalent [Cr(VI)], that are the most stable forms found in the environment (1). However, these two valences exhibit markedly different physical/chemical behavior and toxicological properties, which is manifested in significantly different environmental mobility and health concerns. Accordingly, the accurate differentiated quantitation of Cr in each species is important when performing investigatory activities to characterize the level of Cr in soils, sediments, and like waste materials (2–10).

Cr(VI) is designated a human carcinogen by inhalation, and is also considered hepatotoxic and nephrotoxic (2,7,11). Additionally, most forms of Cr(VI) are relatively soluble in aqueous environments (12,13). Cr(III) is considered an essential trace nutrient for human health, and most common forms found in the environment are sparingly soluble and immobile in soils (14,15).

For these reasons it is essential to be able to selectively extract and quantify Cr(III) and Cr(VI) separately in soils and sediment samples when the intended use of such data is to determine the potential need for remediation of site soils. This chapter addresses the selective extraction and quantitation of Cr(VI) from soils and like media. It is intended to serve as a guide for environmental scientists and engineers, hydrogeologists, chemists, and other practitioners who may be confronted with the need to speciate chromium in soils or waste-amended soils for waste classification, remediation assessment, determination of cleanup standards attainment, and other related reasons.

The analysis of airborne dusts potentially ladened with Cr(VI), which can be derived from disturbed soils, is not specifically addressed in this chapter. For further information on quantification of Cr(VI) in airborne dust samples, relevant references are provided (16–18), all of which relate to the basic methodology embodied in ASTM Method D5281-92, *Standard Test Method for Collection and Analysis of Hexavalent Chromium in Ambient, Workplace, and Indoor Atmospheres.*

CHROMIUM VALENCE OCCURRENCE AND TOXICOLOGICAL SIGNIFICANCE

Chromium Valence Occurrence

The existence of Cr was discovered by Vauquelin in 1789 and its physical/chemical properties have subsequently been determined (Table 1) (19). Cr(III) occurs naturally in soils and mineral deposits, while Cr(VI) is rarely encountered in natural, unpolluted soils. Cr(III) is relatively immobile in soils or sediment. The most common forms of Cr(VI) are readily solubilized from most soils and transported through water that contacts the soils. Because chromium chemicals are so widely distributed and used in both developing and industrialized nations for myriad industrial and commercial products (Table 2), chromium, and more particularly Cr(VI), have become prevalent environmental contaminants. Chromate is ranked among the top 20 contaminants of concern found at Superfund sites within the United States. CERCLA's ranking criteria includes (*1*) frequency of occurrence, (*2*) toxicity, and (*3*) potential for human exposure. A partial list of CERCLA Superfund Priority Hazardous Substances follows (21,22):

Table 1. Properties of Chromium

Property	Value
Atomic weight	52
Specific gravity	7.1
Melting point, °C	1890
Boiling point, °C	2482
Naturally occurring	Silver-white, lustrous, hard metal
Valence states	2^- to 6^{+a}

[a] Predominantly found as 3^+ and 6^+ in the environment; other valences are much less stable in the environment.

1. Lead	*60. Chromium*
2. Arsenic	73. Asbestos (magnesium silicates)
3. Metallic mercury	87. Copper
6. Cadmium	96. Barium
18. Chromium, hexavalent	157. Tin
45. Nickel	158. Aluminum
58. Zinc	*203. Chromic acid*

Natural Soils. Chromium is the 10th most abundant metal and is ranked 21st among all elements in the earth's crust (23). It is naturally present in typical U.S. soils from 1 to 2000 mg/kg, apart from chromium-bearing domestic ore deposits which can range up to 27% Cr (24,25). Like many metallic elements, it can exist in multiple valence states from 2^- to 6^+, although only the 3^+ and 6^+ valence forms are typically encountered unless specifically produced (1,12).

Geology and Mineralogy. Chromium in natural solids varies widely with the type and nature of the rock or sediment deposit. Among different natural solids, shales, lithosphere, sandstone, and river suspended matter typically exhibit relatively high concentrations of Cr, while carbonate, granite, and sandy sediments generally contain low Cr contents (12). When found in geologic deposits, chromium is principally identified as $FeCr_2O_4$ (chromite). Until the early 1930s, high grade ores from New Caledonia, Zimbabwe, India, and Turkey were the primary sources of chromite ores (up to 38% Cr) used to manufacture chromium chemicals (see Table 2) (19). Since the mid-1950s, the high grade ores (20–26% Cr) of Russia and South Africa have become the primary sources for more than 50% of the world's production. South Africa and Zimbabwe have 84 and 11%, respectively, of the world's known reserves (19). Ores with similar quality have been identified in Montana, Oregon, and California, but these resources are small and economically less attractive than those found elsewhere (23).

Most forms of Cr(III) present in or added to soils are low in solubility and reactivity, indicative of its general kinetic inertness (3,26,27). In most soils, the predominant forms of Cr(III) are chromic oxide (Cr_2O_3), $Cr(OH)_3$, or mixed oxides of Fe(III)

Table 2. Characteristics and Uses of Selected Chromium Chemicals[a]

Chromium Compound	Chemical Formula	Cr Valence	Description	Water Solubility[b]	Uses
Chromic acid	CrO_3	6^+	Dark, purplish red crystals	S	Chemicals (chromates, oxidizing agents, catalyst), chromium-plating intermediates, anodizing, ceramic glazes, textile mordant, etchant for plastics, metal cleaning, inks, dyes, paints, colored glass
Chromic chloride	$CrCl_3$	3^+	Violet crystals	I	Chromium plating, olefins polymerization catalyst, waterproofing, textile mordant
	$CrCl_3$ 6H_2O	3^+	Greenish black or violet deliquescent crystals	S	
Chromic hydroxide	$Cr(OH)_3$	3^+	Green, gelatinous precipitate	I	Guignet's green, catalyst, tanning agent, textile mordant
Chromic oxide	Cr_2O_3	3^+	Bright green, extremely hard crystals	I	Metallurgy, green paint pigment, ceramics, organic synthesis catalyst, green granules in asphalt roofing, abrasive, refractory, brick component
Chromic sulfate	$Cr_2(SO_4)_3$	3^+	Violet or red powder	I	Chromium plating, chromium alloys, catalyst, green paint, varnishes and ink, ceramic glazes, leather tanning, textile mordant
	$Cr_2(SO_4)_3 15H_2O$	3^+	Dark, green amorphous scales	S	
	$Cr_2(SO_4)_3 18H_2O$	3^+	Violet cubes	S	
Chromium	Cr	O	Hard, brittle, semigray metal	I	Alloying and plating element on metal and plastic, substrate for corrosion resistance, nuclear and high temperature research, inorganic pigment constituent
Chromium ammonium sulfate	$CrNH_4(SO_4)_2 12H_2O$	3^+	Green powder or deep violet crystals	S	Leather tanning, textile mordant
Chromous chloride	$CrCl_2$	2^+	White, deliquescent needles	S	Reducing agent, catalyst, reagent
Chromous sulfate	$CrSO_4 5H_2O$	2^+	Blue crysals	S	Oxygen scavenger, reducing agent, analytical reagent
Lead chromate	$PbCrO_4$	6^+	Yellow crystals	I	Pigments in industrial paints, rubber, plastics, ceramic coatings, organic analysis
Potassium chromate	K_2CrO_4	6^+	Yellow crystals	S	Analytical reagent, aniline black, textile mordant, enamels, pigment in paints, inks
Potassium dichromate	$K_2Cr_2O_7$	6^+	Bright, yellowish red transparent crystals	S	Oxidizing agent, analytical reagent, brass pickling compositions, electroplating, pyrotechnics, explosives, safety matches, textiles, dyes and inks, glues and adhesives, leather tanning, wood stains, paint pigments, synthetic perfumes, ceramics, bleaching fats and waxes

Table 2. *(continued)*[a]

Chromium Compound	Chemical Formula	Cr Valence	Description	Water Solubility[b]	Uses
Sodium chromate	$Na_2CrO_4 10H_2O$	6^+	Yellow, translucent, efflorescent crystals	S	Inks, dyes, paint pigment, leather tanning, iron corrosion protection, wood preservatives
Sodium dichromate	$Na_2Cr_2O_7 2H_2O$	6^+	Red or orange deliquescent crystals	S	Colorimetry (copper determination), complexing agent, oxidation inhibitor in ethyl ether

[a] Ref. 20.
[b] I = insoluble, S = soluble.

or Cr(III) (28,29). Under moderately oxidizing to reducing conditions, Cr(III) minerals are significant controls for dissolved chromium. Eskolaite, Cr_2O_3(s), even if scare, is a possible solubility control and could maintain low concentrations of aqueous Cr when the sum of the PE (−log of the electron activity) and pH yields values lower than 16 (30,31). Amorphous (Fe(III, Cr(III)) hydroxide is likely the principal Cr solubility-controlling phase in the natural environment. For typical natural conditions, the fraction of Cr in water is low (<1%) (12).

Wherever Cr(III) is found in nature, it is in close association with Al(III) and Fe(III), principally because of the ubiquitous occurrence of Al and Fe in the earth's crust (3). The atomic radii of Fe, Cr, and Al (54, 64, and 67 pm, respectively), and their ionic potentials are sufficiently similar that substitution of one for another is a common occurrence in many mineral and organic compound structures. Although Al(III) has a greater ionic potential, Fe(III) and Cr(III) are stronger Lewis acids due to their preference to complete their noble gas electron configurations, while the loss of three electrons for Al allows it to acquire the stable neon configuration (12). Thus, Cr(III) has a tendency to substitute for small amounts of octahedral aluminum in clay minerals, and in chromite ore, it echoes the common magnetite structure by substituting for two atoms of Fe(III), forming $FeCr_2O_4$ (3). Both Al(III) and Cr(III) may become anionic at pH $>$ 7 and both form strong organic complexes, as well as complexes with phosphate and fluorine (32–34).

Apart from the previously mentioned similarities, Al(III) and Cr(III) are fundamentally very different in that Cr(III) has three unpaired electrons to donate when it is oxidized to Cr(VI), whereas Al(III) has no available electrons to donate, and therefore does not oxidize. Cr(III) behaves chemically as a typical hard Lewis acid and readily forms complexes with a variety of ligands: ammonia, sulfate, hydroxide, cyanide, fluoride, and chloride, as well as natural and synthetic ligands (30,34–37). The plethora of Cr(III) complexes, both inorganic and organic, exhibit a variety of colors. The red color in rubies and the green in emeralds are examples of natural concentration of chromium in mineral deposits (3). However, because Cr complexation occurs slower than aluminum complexes, stronger bonds are formed, which is a major reason why, historically, chromium has replaced aluminum for tanning leather (3).

Due to their high solubility and the requirement to maintain oxidizing and alkaline pH conditions, natural Cr(VI) minerals are rare in nature. However, in environments rich in Ba or Pb, hydrated minerals $XCrO_4 \cdot nH_2O$ can maintain low CrO_4^{2-} concentrations (38,39). Important mineral phases governing the solubility and transport of Cr(VI) include $PbCrO_4$ (crocoite), $PbCrO_4 \cdot H_2O$ (iranite), $K_2Cr_2O_4$ (tarapacaite), and $BaCrO_4$ (hashemite) (40,41).

Man-Made Fill Materials. Because chromium and its compounds have so many diverse uses (16,20,22,42), it is common to find chromium as a component in many waste materials, including discarded consumer products, construction debris, combustion process ash, and industrial wastes. Such wide diversity is based on the use of chromium in stainless steels and metal alloys (single largest consumption) and as an ingredient in bricks (eg, refractories) and cement linings for industrial furnaces. The presence of chromium in man-made fill materials, ranging from construction debris containing discarded stainless steel to off-specification waste materials, is a routine occurrence. However, the fraction of Cr(VI) as a percentage of the total chromium in such materials is generally quite small or negligible. Although stainless steel is the largest single application of Cr, the Cr in stainless steel is not readily available to leaching and migration into the environment unless severe corrosion were to occur. Chromium chemicals, however, many of which are highly soluble, are widely used for industrial and commercial purposes. Wastes or discarded off-specification products are typically generated in each industrial sector that participates in the manufacture or use of chromium chemicals, including wood preservatives, metal finishing/corrosion protection, leather tanning, pigments, oil drilling muds, textile mordants, magnetic tapes, etchants for plastics, oxidizing agents, granules for asphalt roofing, catalyst and fire retardants, ceramic coatings, analytical reagents, photosensitive materials, and abrasives (20,42).

A particularly prevalent form of chromium waste, which is generally associated with chromate chemical manufacturing, is chromite ore processing residue (COPR), frequently referred to as chrome slag. It is generated in a high temperature, alkaline roasting process of the chromite ore that leads to the production of multiple chromate chemicals (43). Visually, COPR has soil-like properties, typically with reddish brown or grayish colored spherical particles varying widely in size. It was commonly used as a construction fill material for many decades before the potential environmental hazards associated with its Cr(VI) content were recognized during the late 1970s and early 1980s.

At sites where COPR has been deposited, crusts containing chromate salts have periodically appeared on the surface of poorly drained soils. Here, shallow groundwater exists through capillary upward movement and evaporation of soil water containing soluble CrO_4^{2-} salts (29). These crusts, or blushes, have been observed as yellow crystals and/or yellowish water in contact with soils that contain elevated concentrations of Cr(VI). This phenomenon appears during dry periods when evaporation of soil water has occurred at the soil surface. Oxidation of Cr(III) by O_2 has been theorized to explain the blush phenomenon; however, it has been demonstrated that oxidation of Cr(III) by O_2 is not a plausible hypothesis to explain blush formation under field conditions, and that some threshold level of soluble Cr(VI) must be exceeded before blush is visible, likely in the range of about 200 to 650 mg Cr(VI)/L in COPR-enriched soil solutions (29).

Chromium Valence Toxicology

Trivalent Chromium. Trivalent chromium has low acute and chronic toxicity to humans at high doses, is considered an essential trace nutrient for human health as an insulin activator, and has historically been regulated only in the workplace (44). The inability of Cr(III) to penetrate cell membranes severely limits or precludes the possibility of carcinogenic activity (45–48). Compounds containing Cr(III) are generally considered to be inactive in most test systems, except those where they can directly interact with DNA (2). A panel of experts concluded there is no reliable evidence that Cr(III) could cause cancer in humans or animals (5). The latest toxicological information on Cr(III) is available from the U.S. EPA Integrated Risk Information System (IRIS) (7), a continuously updated electronic database, and the USEPA Health Effects Assessment Summary Tables (HEAST) (8).

Concerns for the other potential adverse effects of Cr(III) in humans and animals by the various exposure routes have been studied extensively and summarized in a report prepared for the World Health Organization (2), which includes the citation of more than 500 relevant references. The potential hazards of chromium to fish, wildlife, and invertebrates has been extensively documented (49), in a study commissioned by the U.S. Fish and Wildlife Service, that contains over 160 references relevant to these animals. The potential toxicological effects of chromium on plants has been reported (3,50–60). The effects of chromium on microorganisms have also been scrutinized and documented (61–67). Overall, Cr(III) compounds are considered to be much less toxic than like concentrations of Cr(VI) compounds exposed to the same organisms.

Hexavalent Chromium. The hexavalent form of chromium is designated and widely recognized to be a human inhalation carcinogen (7). Epidemiological studies of chromate production workers indicate that Cr(VI) is a human respiratory carcinogen, typically affecting the lung, and in a few cases, the nasal/pharyngeal area (68). This is consistent with the designation of Cr(VI) as a respiratory carcinogen in the U.S. EPA IRIS database maintained by the National Library of Medicine in Bethesda, Maryland (7). Studies of chromate workers that addressed cigarette smoking suggest that Cr(VI) may be a less potent carcinogen than predicted earlier, and that a threshold for Cr(VI) carcinogenesis may exist (69–71).

The noncarcinogenic toxicity of Cr(VI) varies greatly among a wide variety of very different chromium compounds, and it is generally considered to be 10–100 times more toxic than the trivalent form when both are administered orally (72). High doses of Cr(VI) compounds are also associated with nephrotoxicity (5,8), with accidents involving hot chromic acid (CrO_3) burns reported to be the cause of most kidney damage in the workplace (73). At sufficient doses, Cr(VI) can adversely affect excretion and filtration by the kidney (74). Acute exposure to high levels of Cr(VI) can produce nervous system and liver damage (7,8). Numerous reports indicate that Cr(VI) can adversely affect nasal mucous membranes and that lung function can be impaired following occupational exposure (75–77).

A temporary irritant and/or allergic dermatitis reaction can result from direct skin contact with high concentrations of chromium, particularly Cr(VI) compounds. Clinical studies indicate, however, that there is only a small percentage of the population, less than 0.5%, who exhibit dermal sensitization as allergic contact dermatitis to solubilized forms of Cr(VI) (5). A panel of experts indicated that a dose greater than 1750 mg/L of potassium dichromate is generally required to sensitize persons to Cr(VI) (5). Based on more recent studies to assess the potential for elicitation of allergic contact dermatitis in Cr(VI)-sensitive subjects following prolonged contact with Cr(VI) in solution, ChemRisk (78) determined that none of the 21 individuals exposed repetitiously to soluble chromate with an equivalent Cr(VI) concentration of 25 mg/L developed allergic contact dermatitis.

Significance of Toxicological Differences. Due to the marked toxicological differences between Cr(III) and Cr(VI), the guidance levels for Cr-bearing soil remediation are among the last to be established for the more than 20 metals commonly encountered at hazardous waste sites. Many states and various EPA regions continue to use guidelines rather than promulgated cleanup standards based on different interpretations of toxicological data and assessment of risk by the various pathways of exposure for Cr(III) and Cr(VI). Based on telephone survey information, the range of cleanup guidelines for Cr(III) and Cr(VI) in soil being used across the country varies dramatically (79). In most jurisdictions, a promulgated standard has yet to be established. The most restrictive guidance criteria in use was 0.2 mg/kg of Cr(VI) for residential soils, while another state considered 2000 mg Cr(VI)/kg to be an acceptable criterion for a residential soils. Similarly for Cr(III), the approximate range of guidance criteria was 300–600,000 mg/kg, a difference of more than three orders of magnitude. The middle range criteria were about 100–400 mg/kg for Cr(VI) and about 80,000 mg/kg for Cr(III). Apart from these wide disparities, some of which stem from the use of different exposure pathways among the jurisdictional authorities, particularly for Cr(III), there is agreement in the need for distinguishing between the two valences in addressing remediation of soil containing elevated levels of chromium.

In the midst of such variability, the EPA Region III (9) risk-based criteria (RBCs) for Cr(III) and Cr(VI) for ingestion exposures are 78,000 mg/kg for Cr(III) and 390 mg/kg for Cr(VI). Additionally, the EPA has developed soil screening levels to be used as guidance at Superfund and Brownfields sites. The generic soil concentrations are 270 mg Cr(VI)/kg to protect from exposure to Cr(VI) by humans via inhalation and 38 mg Cr(VI)/kg for the protection of ground water (10). The maximum concentration level (MCL) for Cr in potable water under the Safe Drinking Water Act is 0.1 mg/L of total Cr with no designation of valence state. There are no comparable levels for Cr(III) for these two exposure routes. According to the EPA, the soil screening levels are to be used as default guidance values. If conditions related to Cr-contaminated soils at a site are considered to be substantially different than the exposure scenarios identified by EPA in establishing these guidance levels, a site-specific risk assessment may be performed to determine appropriate Cr(III) and Cr(VI) soil cleanup levels.

BIBLIOGRAPHY

1. J. O. Nriagu and E. Nieboer, eds. *Chromium in Natural and Human Environments,* Wiley-Interscience, New York, 1988.
2. WHO, *Chromium,* Environmental Health Criteria 61, World Health Organization, Geneva, Switzerland, 1988.
3. R. J. Bartlett and B. R. James, "Mobility and Bioavailability of Chromium in Soils," in, Ref. 1, pp. 267–304.
4. P. J. Sheehan, D. M. Meyer, M. M. Sauer, and D. J. Paustenbach, "Assessment of the Human Health Risks Posed by Exposure to Chromium-Contaminated Soils," *J. Toxic. and Environ. Health,* **32,** 161–201 (1991).
5. D. J. Paustenbach, W. E. Rinehart, and P. J. Sheehan, "The Health Hazards Posed by Chromium-Contaminated Soils in Residential and Industrial Areas: Conclusions of an Expert Panel," *Reg. Toxicol. Pharmacol.* **13,** 195–222 (1991).
6. U.S. HHS, *Toxicological Profile for Chromium,* U.S. Dept. of Health and Human Services, Agency for Toxic Substances and Disease Registry, Washington, D.C., 1993.
7. U.S. EPA, IRIS (Integrated Risk Information System), A continuously updated electronic database maintained by U.S. EPA, Bethesda, Md., 1997.
8. U.S. EPA, *Health Effects Assessment Summary Tables: FY-1995 Annual,* EPA540/R-95-036, NTIS PB95-921199, Office of Research and Development, Office of Emergency and Remedial Response, Washington, D.C., 1995.
9. U.S. EPA, *Risk Based Concentration Table,* Region III, Philadelphia, Pa., 1997.
10. U.S. EPA, *Soil Screening Guidance: Technical Background Document,* EPA/540/R-95/128, PB96-963502, U.S. EPA Office of Solid Waste and Emergency Response, Washington, D.C., 1996.
11. A. Yassi and E. Nieboer, "Carcinogenicity of Chromium Compounds," in Ref. 1, pp. 443–496.
12. F. C. Richard and A. C. M. Bourg, "Aqueous Geochemistry of Chromium: A Review," *Wat. Res.* **25,** 807–816 (1991).
13. D. B. Kent, J. A. Davis, L. C. D. Anderson, and B. A. Rea, "Transport of Chromium and Selenium in the Suboxic Zone of a Shallow Aquifer: Influence of Redox and Adsorption Reactions," *Water Resourc. Res.* **30**(4), 1099–1114 (1994).
14. R. A. Anderson, "Essentiality of Chromium in Humans," *Sci. Total Environ.* **86,** 75–81 (1989).
15. S. E. Fendorf, "Surface Reactions of Chromium in Soils and Waters," *Geoderma.* **67,** 55–71 (1995).
16. M. Falerios, K. Schild, P. Sheehan, and D. J. Paustenbach, "Airborne Concentrations of Trivalent and Hexavalent Chromium from Contaminated Soils at Unpaved and Partially Paved Commercial/Industrial Sites," *J. Air Waste Manage. Assoc.* **42,** 40–48.
17. P. Sheehan, R. Ricks, S. Ripple, and D. J. Paustenbach, "Field Evaluation of a Sampling and Analytical Method for Environmental Levels of Airborne Hexavalent Chromium," *Am. Ind. Hyg. Assoc. J.* **53**(1), 57–68 (1992).
18. ASTM, *Standard Test Method for Collection and Analysis of Hexavalent Chromium in Ambient, Workplace, or Indoor Atmospheres,* Desig. D5281-92, American Society for Testing Materials, Philadelphia, Pa., 1992.
19. J. H. Westbrook, "Chromium and Chromium Alloys," in J. I. Kroschwitz and M. Howe-Grant eds., *Kirk-Othmer Encyclopedia of Chemical Technology,* 4th ed., Wiley-Interscience, New York, 1991, pp. 228–263.
20. R. J. Lewis, Sr., ed., *Hawley's Condensed Chemical Dictionary,* 13th ed., John Wiley & Sons, Inc., New York, 1997.
21. B. J. Page and G. W. Loar, "Chromium Compounds," in Ref. 19, pp. 263–311.
22. "EPA Updates CERCLA Priority List of Hazardous Substances", *The Hazardous Waste Consultant,* 10(5), McCoy and Associates, Inc., Lakewood, Colo., 1993, pp. 2.26–2.30.
23. R. C. Weast, ed., *CRC Handbook of Chemistry and Physics,* 59th ed., CRC Press, West Palm Beach, Fla., 1978.
24. R. L., Copson, "Production of Chromium Chemicals," in M. J. Udy, ed., *Chromium I: Chemistry of Chromium and Its Compounds,* Reinhold Publishing Corp., New York, 1956.
25. J. Dragun and A. Chiasson, *Elements in North American Soils,* Hazardous Materials Control Resources Institute, Greenbelt, Md., 1991.
26. C. F. Baes, Jr. and R. E. Mesmer, *The Hydrolysis of Cations,* Wiley-Interscience, New York, 1976, pp. 211–220.
27. F. A. Cotton and G. Wilkinson, *Advanced Inorganic Chemistry,* John Wiley & Sons, Inc., New York, 1980, pp. 719–749.
28. D. Rai and R. W. Szelmeczka, "Aqueous Behavior of Chromium in Coal Fly Ash," *J. Environ. Qual.* **19,** 378–382 (1990).
29. B. R. James, "Hexavalent Chromium Solubility and Reduction in Alkaline Soils Enriched with Chromite Ore Processing Residue," *J. Environ. Qual.* **23,** 227–233 (1994).
30. E. Deltombe, N. De Zoubov, and M. Pourbaix, in M. Pourbaix ed., *Chromium, Atlas of Electrochemical Equilibria in Aqueous Solutions,* Pergamon Press, Oxford, U.K., 1966, pp. 256–271.
31. D. Rai, L. E. Eary, and J. M. Zachara, "Environmental Chemistry of Chromium," *Sci. Total Environ.* **86,** 15–23 (1989).
32. W. Mertz, "Chromium Occurrence and Function in Biological Systems," *Physiol. Rev.* **49,** 163–239 (1969).
33. H. Stunzi and W. Marty, "Early Stages of Hydrolysis of Chromium(III) in Aqueous Solution: I. Characterization of a Tetrametric Species," *Inorganic Chem.* **22,** 2145–2150.
34. R. J. Bartlett and J.M. Kimble, "Behavior of Chromium in Soils: I. Trivalent Forms," *J. Environ. Qual.* **5,** 379–383 (1976).
35. E. Nakayama, T. Kuwamoto, S. Tsurubo, H. Tokoro, and T. Rujinaga, "Chemical Speciation of Chromium in Sea Water: 1. Effect of Naturally Occurring Organic Materials on the Complex Formation of Chromium(III)," *Analyt. Chim. Acta* **130,** 289–294 (1981).
36. B. R. James and R. J. Bartlett, "Behavior of Chromium in Soils: V. Fate of Organically Complexed Cr(III) Added to Soil," *J. Environ. Qual.* **12,** 169–172 (1983).
37. A. B. Payà Pérez, L. Gotz, S. D. Kephalopoulos and G. Bignoli, "Sorption of Chromium Species on Soil," in M. Astruc and J. N. Lester, eds., *Heavy Metals in the Hydrocycle,* Selper, London, 1988, pp. 59–66.
38. B. R. James and R. J. Bartlett, "Behavior of Chromium in Soils: VII. Adsorption and Reduction of Hexavalent Forms," *J. Environ. Qual* **12,** 177–181 (1983).
39. D. Rai, B. M. Sass, and D. A. Moore, "Chromium(III) Hydrolysis Constants and Solubility of Chromium(III) Hydroxide," *Inorg. Chem.* **26,** 345–349 (1987).
40. R. W. Puls, D. A. Clark, R. S. Kerr, and C. J. Paul, *Distribution, Speciation, and Transformation of Chromium in Contaminated Soils and Aquifer Sediments,* NTIS PB92-179092, 1992.
41. R. W. Puls, D. A. Clark, and J. Vardy, "Transport and Transformation of Hexavalent Chromium Through Soils and into Groundwater," *J. Soil Contam.* **3**(2), 203–224 (1994).

42. M. J. Udy, ed., *Chromium* I, *Chemistry of Chromium and Its Compounds,* Reinhold Publishing Corp., New York, 1956.

43. G. T. Austin, *Shreve's Chemical Process Industries,* 5th ed. McGraw-Hill, New York, 1984.

44. *Documentation of Threshold Limit Values,* 4th ed., American Conference of Governmental Industrial Hygienists, Cincinnati, Ohio, 1989.

45. J. M. Davies, "Lung Cancer Mortality Among Workers Making Lead Chromate and Zinc Chromate Pigments at Three English Factories," *Br. J. Ind. Med.* **41**(2), 158–169 (1984).

46. F. Bianchi and A. G. Levis, "Mechanisms of Chromium Genotoxicity," in E. Merian, R. W. Fre, W. Haradi, and C. Schlatter, eds., *Carcinogenic and Mutagenic Metal Compounds Environmental and Analytical Chemistry and Biological Effect,* Gordon and Breach, London, 1985, pp. 269–294.

47. L. Polak, "Immunology of Chromium," in D. Burrows, ed., *Chromium: Metabolism and Toxicity,* CRC Press, Boca Raton, Fla., 1983, pp. 69–72.

48. *Final Rule Air Contaminants—Permissible Exposure Limits,* Title 29, Code of Federal Regulations, Part 1910.1000, Occupational Safety and Health Administration, Washington, D.C., 1989.

49. R. Eisler, *Chromium Hazards to Fish, Wildlife, and Invertebrates: A Synoptic Review,* Biol Rep. 85(1.6), U.S. Dept. of Interior, Fish and Wildlife Service, Laurel, Md., 1986.

50. E. E. Cary, W. H. Allaway, and O. E. Olsen, *J. Agric. Food Chem,* **25,** 300–304 (1977).

51. E. E. Cary, W. H. Allaway, and O. E. Olsen, "Control of Chromium in Food Plants. II. Chemistry of Chromium in Soils and Its Availability to Plants," *J. Agric. Food Chem,* **25,** 305–309 (1977).

52. A. Wallace, S. M., Soufi, J. W. Cha, and E.M. Romney, "Some Effects of Chromium Toxicity on Bush Bean Plants Grown in Soil," *Plant Soil* **44,** 471–473 (1976).

53. R. J. Bartlett and B. R. James, "Behavior of Chromium in Soils: III Oxidation," *J. Environ. Qual.* **8,** 31–35 (1979).

54. M. Lahauti and P. J. Peterson, "Chromium Accumulation and Distribution in Crop Plants," *J. Sci. Food Agr.* **30,** 136–142 (1979).

55. S. Silva and B. Beghi, "The Use of Chromium-Containing Organic Manures in Rice Fields," *Riso* **28,** 105–113 (1979); *Abstr. Soils Fert.* **43,** 4100 (1979).

56. P. D. Parr and F. G. Taylor, Jr., "Incorporation of Chromium in Vegetation Through Root Uptake and Foliar Absorption Pathways," *Environ. Exp. Bot.* **20,** 157–160 (1980).

57. R. L. Sykes and N. J. Earl, "The Effect of Soil-Chromium (III) on the Growth and Chromium Absorption of Various Plants," *J. Am. Leather Chem. Assoc.* **76,** 102–125 (1981).

58. M. I. Sheppard, S. C. Sheppard, and D. H. Thibault, "Uptake by Plants and Migration of Uranium and Chromium in Field Lysimeters," *J. Environ. Qual.* **13,** 357–361 (1984).

59. A. L. Stomberg, D. D. Hemphill, Jr., and V. V. Volk, "Yield and Elemental Concentration of Sweet Corn Grown on Tannery Waste-Amended Soil," *J. Environ. Qual.* **13,** 162–165 (1984).

60. E. Nieboer and A. A. Jusys, "Biologic Chemistry of Chromium," in Ref. 1.

61. F. L. Petrilli and S. deFlora, "Toxicity and Mutagenicity of Hexavalent Chromium on *Salmonella typhimurium,*" *Appl. Environ. Microbiol.* **33,** 805–809 (1977).

62. F. L. Petrilli and S. deFlora, "Oxidation of Inactive Trivalent Chromium to the Mutagenic Hexavalent Form," *Mutation Res.* **58,** 167–174 (1978).

63. D. S. Ross, R. E. Sjögren, and R. J. Bartlett, "Behavior of Chromium in Soils: IV. Toxicity to Microorganisms," *J. Environ. Qual.* **10,** 145–148 (1981).

64. H. Babich, M. Schiffenbauer, and G. S. Stotzky, "Comparative Toxicity of Trivalent and Hexavalent Chromium to Fungi," *Bull. Environ. Contam. Toxicol.* **28,** 452–459 (1982).

65. H. Babich, M. Schiffenbauer, and G. S. Stotzky, "Effect of Sterilization Method on Cr^{3+} and Cr^{6+} to Fungi," *Microbios Lett.* **20,** 55–64 (1982).

66. B. R. James and R. J. Bartlett, "Nitrification in Soil Suspensions Treated with Chromium(III, VI) Salts or Tannery Wastes," *Soil Biol. Biochem.* **16,** 293–295 (1984).

67. B. R. James and R. J. Bartlett, "Plant-Soil Interactions of Chromium," *J. Environ. Qual.* **13,** 67–70 (1984).

68. A. Hamilton and H. L. Hardy, *Industrial Toxicology,* Publishing Sciences Group, Acron, Mass., 1974, pp. 71–76.

69. P. L. Bidstrup, "Perspective on Safety: Personal Opinions," *Am. Ind. Hyg. Assoc. J.* **50**(10), 505–509 (1989).

70. J.A. Hathaway, "Role of Epidemiologic Studies in Evaluating the Carcinogenicity of Chromium Compounds," *Sci. Total Environ.* **86,** 169–176 (1989).

71. R. E. Jones, "Hexavalent Chromium: Threshold Concept for Carcinogenicity," *Biomedical Environ. Sci.* **3,** 20–34 (1990).

72. S. A. Katz and H. Salem, "The Toxicology of Chromium with Respect to its Chemical Speciation: A Review," *J. Applied Tox.* **13** (3), 217–224 (1993).

73. *Occupational Disease: A Guide to Their Recognition,* rev. ed., National Institute for Occupational Safety and Health, U.S. Dept. of Health, Education, and Welfare, Public Health Service, Center for Disease Control, Cincinnati, Ohio, 1977.

74. S. Langård and T. Norseth, in L. Friberg, G. F. Norbert, and V. B. Vouk, eds., *Handbook on the Toxicology of Metals, II,* Elsevier Science, Amsterdam, the Netherlands, 1986, pp. 185–210.

75. D. Burrows, in Ref. 47, pp. 138–158.

76. R. M. Adams, *Occupational Skin Diseases,* 2nd ed., W. B. Saunders, Philadelphia, Pa., 1990.

77. U. Korallus, K. Ulm, and W. Steinnann-Steiner-Haldenstaeff, "Bronchial Carcinoma Mortality in the German Chromate Producing Industry: The Effects of Process Modification," *J. Occup. Environ. Health* **65,** 171–178 (1993).

78. ChemRisk, *Assessing Potential for Elicitation of Allergic Contact Dermatitis in Cr(VI)-Sensitized Subjects Over Prolonged Period,* submitted to N.J. Dept. of Environ. Protection, ChemRisk Div. of McLaren/Hart Environmental Engr. Corp., Wexford, Pa., 1996.

79. ChemRisk, *Communication re Unpublished Survey Results,* ChemRisk Division of McLaren/Hart Environmental Engineering Corp., Wexford, Pa., 1996.

CLEAN AIR ACT: REGULATIONS AND IMPLEMENTATION

EDWARD C. MORETTI
Baker Environmental, Inc.
Coraopolis, Pennsylvania

Hazardous air pollutants (HAPs), volatile organic compounds (VOCs), and air toxics all describe several classes of airborne pollution. Adverse health and environmental effects attributable to these pollutants have prompted the worldwide regulation of airborne emissions of some HAPs and VOCs. In the United States, the Environmental Protection Agency

Table 1. EPA List of Nonphotochemically Reactive Compounds[a]

Chemical Name	CAS Registry Number
Methane	[74-82-8]
Ethane	[74-84-0]
Methylene chloride (dichloromethane)	[1665-00-5]
1,1,1-Trichloroethane (methyl chloroform)	[71-55-6]
1,1,2-Trichloro-1,2,2-trifluoroethane (CFC-113)	[76-13-1]
Trichlorofluoromethane (CFC-11)	[75-69-4]
Dichlorodifluoromethane (CFC-12)	[75-71-8]
Chlorodifluoromethane (HCFC-22)	[75-45-6]
Trifluoromethane (HFC-23)	[75-46-7]
1,2-Dichloro-1,1,2,2-tetrafluoroethane (CFC-114)	[76-14-2]
Chloropentafluoroethane (CFC-115)	[76-15-3]
1,1,1-Trifluoro-2,2-dichloroethane (HCFC-123)	[306-83-2]
1,1,1,2-Tetrafluoroethane (HFC-134a)	[811-97-2]
1,1-Dichloro-1-fluoroethane (HCFC-141b)	[1717-00-6]
1-Chloro-1,1-difluoroethane (HCFC-142b)	[75-68-3]
2-Chloro-1,1,1,2-tetrafluoroethane (HCFC-124)	[2837-89-0]
Pentafluoroethane (HFC-125)	[354-33-6]
1,1,2,2-Tetrafluoroethane (HFC-134)	[359-35-3]
1,1,1-Trifluoroethane (HFC-143a)	[420-46-2]
1,1-Difluoroethane (HFC-152a)	[75-37-6]
Parachlorobenzotrifluoride (PCBTF)	[98-56-6]
Cyclic, branched, or linear completely methylated siloxanes	
Acetone	[67-64-1]
Perchloroethylene (tetrachloroethylene)	[127-18-4]
3,3-Dichloro-1,1,1,2,2-pentafluoropropane (HCFC-225ca)	[422-56-0]
1,3-Dichloro-1,1,2,2,3-pentafluoropropane (HCFC-225cb)	[507-55-1]
1,1,1,2,3,4,4,5,5,5-Decafluoropentane (HFC 43-10mee)	[138495-42-8]
Cyclic, branched, or linear completely fluorinated alkanes	
Cyclic, branched, or linear completely fluorinated ethers with no unsaturations	
Cyclic, branched, or linear completely fluorinated tertiary amines with no unsaturations	
Sulfur-containing perfluorocarbons with no unsaturations and with sulfur bonds only to carbon and fluorine	

[a] From Ref. 1.

(EPA), acting in response to statutory mandates in the Clean Air Act (42 USC 7401–7671), has focused on four strategies to regulate airborne emissions of HAPs and VOCs: prevention, recovery, environmentally safe treatment, and controlled release to the atmosphere. As the EPA develops more national programs and participates in international initiatives to protect human health and the environment from the effects of HAPs and VOCs, and as public interest in the environment and public participation in EPA rule making increase, it is important to understand the legal and scientific bases behind existing VOC and HAP regulations in the United States. To develop fair, responsible, and cost-effective regulations in the future, all stakeholders in VOC and HAP rule makings can use the information in this article to leverage the benefits of existing programs with new programs to maximize emission reductions, to identify research opportunities aimed at better understanding of health and environmental risks, and to promote innovative solutions for controlling VOCs and HAPs.

POLLUTANTS

Volatile Organic Compounds

The EPA defines a volatile organic compound as "any compound of carbon, excluding carbon monoxide, carbon dioxide, carbonic acid, metallic carbides or carbonates, and ammonium carbonate, which participates in atmospheric photochemical reactions" (1). Nonphotochemically reactive compounds of carbon are not VOCs according to EPA's definition. In addition to the exclusions listed above in the main definition of VOC, the EPA attaches a list of nonphotochemically reactive compounds to its definition of VOC; this list appears as Table 1. At its own discretion or in response to the petition of another party, EPA may add or remove compounds from the list on the basis of sound scientific data.

Although the compounds listed in Table 1 are not considered VOCs and, therefore, are exempt from regulatory programs for VOCs, the EPA regulates some of these compounds under other regulatory programs. For example, methyl chloroform (1,1,1-trichloroethane) and the chlorofluorocarbons (CFCs) and hydrochlorofluorocarbons (HCFCs) listed in Table 1 are regulated as stratospheric ozone-depleting compounds; also, methyl chloroform (1,1,1-trichloroethane), methylene chloride (dichloromethane), and perchloroethylene (tetrachloroethylene) are regulated as HAPs.

The photochemical reactivity of VOCs creates photochemical oxidants in the earth's troposphere (ie, the ground-level atmosphere). The main photochemical oxidants that are formed in the troposphere from VOCs include ozone (O_3) and peroxyacyl nitrates (PANs). Tropospheric ozone is regulated by the

EPA because of its adverse health and environmental effects. Measurable human health effects of ozone exposure include increased susceptibility to respiratory infection; aggravation of pre-existing respiratory diseases such as asthma; reduced lung function; respiratory distress (eg, chest pain, coughing, sneezing, and pulmonary congestion); lung inflammation; impairment of lung defense mechanisms; and irreversible changes in lung structure that can lead to chronic respiratory illnesses such as emphysema, chronic bronchitis, and premature aging of the lungs (2). Environmental effects of ozone exposure include reduction in agricultural and commercial forest yields; reduced growth and decreased survivability of tree seedlings; increased tree and plant susceptibility to disease, pests, and other environmental stresses; and potential long-term damage to forests and ecosystems (2).

Ozone and other photochemical oxidants are the main components of photochemical smog, the characteristic haze that hangs over urban or industrialized areas mainly during hot, sunny weather. The human health effects of photochemical smog include the same effects attributable to tropospheric ozone mentioned above as well as the less severe effects of PANs and other oxidants (eg, eye, nose, and throat irritation).

The chemistry of ground-level ozone formation and smog formation is complex because of the influence of numerous factors, including ambient temperature, intensity of sunlight, atmospheric mixing, concentrations of VOC and other ozone precursors (eg, nitrogen oxides) in the ambient air, and the reactivity of the ozone precursors. An understanding of the chemistry of ground-level ozone formation begins with a naturally occurring phenomenon known as the photolytic cycle, shown below:

$$NO_2 + \text{sunlight} \rightarrow NO + O \quad (1)$$

$$O + O_2 + M \rightleftarrows O_3 + M \quad (2)$$

$$O_3 + NO \rightleftarrows NO_2 + O_2 \quad (3)$$

where $M = N_2$, O_2, or H_2O. The photolytic cycle occurs because of the presence of nitrogen oxides (NO_x) in the lower atmosphere (the troposphere); NO_x occurs naturally in the troposphere through several mechanisms, including downward transport from the upper atmosphere (ie, the stratosphere), emissions from soils, and lightning strikes. In the presence of sunlight, NO_x in the form of NO_2 dissociates to NO and atomic oxygen (eq. 1). The atomic oxygen reacts with abundant molecular oxygen (eq. 2) to form ozone (O_3), which oxidizes NO to NO_2 (eq. 3), thus completing the photolytic cycle.

In the absence of any anthropogenic (human-made) influences, there is no net change to the atmosphere in the photolytic cycle. However, adding VOCs upsets the balance of the photolytic cycle as shown:

$$RH + OH + O_2 \rightarrow RO_2 + H_2O \quad (4)$$

$$RO_2 + NO \rightleftarrows RO + NO_2 \quad (5)$$

$$NO_2 + O_2 \rightarrow NO + O_3 \quad (3a)$$

$$RH + OH + 2O_2 \rightarrow RO + O_3 + H_2O \quad (6)$$

where R = volatile organic compounds. Equations 3–6 present an oversimplification of the complex chemistry occurring in the troposphere. More information on this topic is available (2,3).

Equation 4 represents the atmospheric oxidative degradation of VOCs. In equation 4, hydroxyl radicals present in the troposphere (mainly from the photodissociation of ozone by solar radiation) chemically degrade VOCs to organic peroxy radicals (RO_2), which react with NO to form NO_2 (eq. 5). While hydroxyl radical-initiated reactions are the primary mechanism responsible for the atmospheric degradation of VOCs, other atmospheric constituents such as nitrate radicals (NO_3) and ozone also contribute to VOC degradation, although the chemical reactions are not shown here; more information on these VOC formation mechanisms is available (2,3).

The NO_2 formed in equation 5 upsets the balance of the photolytic cycle by driving equation 3 to the left, resulting in elevated levels of tropospheric ozone (eq. 3a). The net reaction (eq. 6) is a chemical conversion of VOCs (in the presence of NO_x) to ozone.

The soup of chemicals formed by the combined effect of equations 1–6, consisting mainly of ozone and other photochemical oxidants (eg, PANs), VOCs, and nitrogen oxides, is photochemical smog. The contribution of nitrogen oxides to tropospheric ozone formation and photochemical smog formation is complex. As shown above, naturally occurring NO_x is responsible for maintaining the balance of the photolytic cycle. Human-made NO_x emissions are formed mainly as a product of incomplete combustion of fuels by motor vehicle engines and fuel-fired boilers, furnaces, and stoves used to produce electricity or heat. Nitric oxide (NO) is the predominant product of incomplete combustion; however, small amounts of nitrogen dioxide (NO_2) also are formed. High levels of anthropogenic NO_x can increase tropospheric ozone levels by driving equation 5 to the right, resulting in elevated levels of NO_2 and, as represented by equation 3a, elevated levels of tropospheric ozone. However, reducing NO_x levels too much can disrupt the photolytic cycle, driving equation 3 toward the production of more ground-level ozone.

A combination of meteorological, geographical, anthropogenic, and biogenic (naturally occurring) factors influence the formation of tropospheric ozone and photochemical smog. Urban areas, with high concentrations of anthropogenic NO_x from motor vehicles and fuel combustion, typically experience the worst episodes of photochemical smog on hot, sunny days when the air is still or is trapped by natural terrain (eg, mountains). Thus cities such as Los Angeles, where the combination of smog-producing factors prevails, are infamous for their smoggy skies.

Hazardous Air Pollutants

The EPA defines a hazardous air pollutant as "any air pollutant listed in or pursuant to Section 112 (b) of the Clean Air Act" (4,5). Table 2 identifies the EPA's current list of HAPs. At its own discretion or the petition of another party, the EPA may add compounds to the list or remove compounds from the list based on sound scientific data (eg, in June 1996, caprolactam was removed from the list). The list of HAPs includes VOCs and fine particulate matter.

Generalizing the health and environmental effects of HAPs is difficult because of the diverse chemical properties of the HAPs, the limited health effects data for some HAPs, and the reactive nature of some HAPs after they become airborne. Like VOCs, HAPs affect human health and the environment either directly through emissions to the air or indirectly as the HAPs

Table 2. Hazardous Air Pollutants[a]

Chemical Name	CAS Registry Number
Acetaldehyde	[*75-07-0*]
Acetamide	[*60-35-5*]
Acetonitrile	[*75-05-8*]
Acetophenone	[*98-86-2*]
2-Acetylaminofluorene	[*53-96-3*]
Acrolein	[*107-02-8*]
Acrylamide	[*79-06-1*]
Acrylic Acid	[*79-10-7*]
Acrylonitrile	[*107-13-1*]
Allyl Chloride	[*107-05-1*]
4-Aminobiphenyl	[*92-67-1*]
Aniline	[*62-53-3*]
o-Anisidine	[*90-04-0*]
Asbestos	[*1332-21-4*]
Benzene (including benzene from gasoline)	[*71-43-2*]
Benzidine	[*92-87-5*]
Benzotrichloride	[*98-07-7*]
Benzyl chloride	[*100-44-7*]
Biphenyl	[*92-52-4*]
Bis(2-ethylhexyl)phthalate (DEHP)	[*117-81-7*]
Bis(chloromethyl)ether	[*542-88-1*]
Bromoform	[*75-25-2*]
1,3-Butadiene	[*106-99-0*]
Calcium cyanamide	[*156-62-7*]
Captan	[*133-06-2*]
Carbaryl	[*63-25-2*]
Carbon disulfide	[*75-15-0*]
Carbon tetrachloride	[*56-23-5*]
Carbonyl sulfide	[*463-58-1*]
Catechol	[*120-80-9*]
Chloramben	[*133-90-4*]
Chlordane	[*57-74-9*]
Chlorine	[*7782-50-5*]
Chloroacetic acid	[*79-11-8*]
2-Chloroacetophenone	[*532-27-4*]
Chlorobenzene	[*108-90-7*]
Chlorobenzilate	[*510-15-6*]
Chloroform	[*67-66-3*]
Chloromethyl methyl ether	[*107-30-2*]
Chloroprene	[*126-99-8*]
Cresols and cresylic acid (isomers and mixture)	[*1319-77-3*]
o-Cresol	[*95-48-7*]
m-Cresol	[*108-39-4*]
p-Cresol	[*106-44-5*]
Cumene	[*98-82-8*]
2,4-D, Salts and esters	[*94-75-7*]
1,1-Dichloro-2,2-bis(*p*-chlorophenyl)ethylene) (DDE)	[*72-55-9*]
Diazomethane	[*334-88-3*]
Dibenzofurans	[*132-64-9*]
1,2-Dibromo-3-chloropropane	[*96-12-8*]
Dibutylphthalate	[*84-74-2*]
1,4-Dichlorobenzene(*p*)	[*106-46-7*]
3,3-Dichlorobenzidene	[*91-94-1*]
Dichloroethyl ether (bis(2-chloroethyl)ether)	[*111-44-4*]
1,3-Dichloropropene	[*542-75-6*]
Dichlorvos	[*62-73-7*]
Diethanolamine	[*111-42-2*]
N,N-Diethyl aniline (*N,N*-dimethylaniline)	[*121-69-7*]
Diethyl sulfate	[*64-67-5*]
3,3-Dimethoxy benzidine	[*119-90-4*]
Dimethyl aminoazobenzene	[*60-11-7*]
3,3′-Dimethyl benzidine	[*119-93-7*]
Dimethyl carbamoyl chloride	[*79-44-7*]

Table 2. [a] *(continued)*

Chemical Name	CAS Registry Number
Dimethyl formamide	[*68-12-2*]
1,1-Dimethyl hydrazine	[*57-14-7*]
Dimethyl phthalate	[*131-11-3*]
Dimethyl sulfate	[*77-78-1*]
4,6-Dinitro-*o*-cresol and salts	[*534-52-1*]
2,4-Dinitrophenol	[*51-28-5*]
2,4-Dinitrotoluene	[*121-14-2*]
1,4-Dioxane (1,4-diethyleneoxide)	[*123-91-1*]
1,2-Diphenylhydrazine	[*122-66-7*]
Epichlorohydrin (1-chloro-2,3-epoxypropane)	[*106-89-8*]
1,2-Epoxybutane	[*106-88-7*]
Ethyl acrylate	[*140-88-5*]
Ethyl benzene	[*100-41-4*]
Ethyl carbamate (urethane)	[*51-79-6*]
Ethyl chloride (chloroethane)	[*75-00-3*]
Ethylene dibromide (dibromoethane)	[*106-93-4*]
Ethylene dichloride (1,2-dichloroethane)	[*107-06-2*]
Ethylene glycol	[*107-21-1*]
Ethylene imine (aziridine)	[*151-56-4*]
Ethylene oxide	[*75-21-8*]
Ethylene thiourea	[*96-45-7*]
Ethylidene dichloride (1,1-dichloroethane)	[*75-34-3*]
Formaldehyde	[*50-00-0*]
Heptachlor	[*76-44-8*]
Hexachlorobenzene	[*118-74-1*]
Hexachlorobutadiene	[*87-68-3*]
Hexachlorocyclopentadiene	[*77-47-4*]
Hexachloroethane	[*67-72-1*]
Hexamethylene-1,6-diisocyanate	[*822-06-0*]
Hexamethylphosphoramide	[*680-31-9*]
Hexane	[*110-54-3*]
Hydrazine	[*302-01-2*]
Hydrochloric acid	[*7647-01-0*]
Hydrogen fluoride (hydrofluoric acid)	[*7664-39-3*]
Hydroquinone	[*123-31-9*]
Isophorone	[*78-59-1*]
Lindane (all isomers)	[*58-89-9*]
Maleic anhydride	[*108-31-6*]
Methanol	[*67-56-1*]
Methoxychlor	[*72-43-5*]
Methyl bromide (bromomethane)	[*74-83-9*]
Methyl chloride (chloromethane)	[*74-87-3*]
Methyl chloroform (1,1,1-trichloroethane)	[*71-55-6*]
Methyl ethyl ketone (2-butanone)	[*78-93-3*]
Methyl hydrazine	[*60-34-4*]
Methyl iodide (iodomethane)	[*74-88-4*]
Methyl isobutyl ketone (hexone)	[*108-10-1*]
Methyl isocyanate	[*624-83-9*]
Methyl methacrylate	[*80-62-6*]
Methyl-*tert*-butyl ether	[*1634-04-4*]
4,4-Methylene bis(2-chloroaniline)	[*101-14-4*]
Methylene chloride (dichloromethane)	[*75-09-2*]
Methylene diphenyl diisocyanate (MDI)	[*101-68-8*]
4,4′-Methylenedianiline	[*101-77-9*]
Naphthalene	[*91-20-3*]
Nitrobenzene	[*98-95-3*]
4-Nitrobiphenyl	[*92-93-3*]
4-Nitrophenol	[*100-02-7*]
2-Nitropropane	[*79-46-9*]
N-Nitroso-*N*-methylurea	[*684-93-5*]
N-Nitrosodimethylamine	[*62-75-9*]
N-Nitrosomorpholine	[*59-89-2*]
Parathion	[*56-38-2*]
Pentachloronitrobenzene (quintobenzene)	[*82-68-8*]

Table 2. Hazardous Air Pollutants [a] *(continued)*

Chemical Name	CAS Registry Number
Pentachlorophenol	*[87-86-5]*
Phenol	*[108-95-2]*
p-Phenylenediamine	*[106-50-3]*
Phosgene	*[75-44-5]*
Phosphine	*[7803-51-2]*
Phosphorus	*[7723-14-0]*
Phthalic anhydride	*[85-44-9]*
Polychlorinated biphenyls (aroclors)	*[1336-36-3]*
1,3-Propane sultone	*[1120-71-4]*
β-Propiolactone	*[57-57-8]*
Propionaldehyde	*[123-38-6]*
Propoxur (baygon)	*[114-26-1]*
Propylene dichloride (1,2-dichloropropane)	*[78-87-5]*
Propylene oxide	*[75-56-9]*
1,2-Propylenimine (2-methyl aziridine)	*[75-55-8]*
Quinoline	*[91-22-5]*
Quinone	*[106-51-4]*
Styrene	*[100-42-5]*
Styrene oxide	*[96-09-3]*
2,3,7,8-Tetrachlorodibenzo-*p*-dioxin	*[1746-01-6]*
1,1,2,2-Tetrachloroethane	*[79-34-5]*
Tetrachloroethylene (perchloroethylene)	*[127-18-4]*
Titanium tetrachloride	*[7550-45-0]*
Toluene	*[108-88-3]*
2,4-Toluene diamine	*[95-80-7]*
2,4-Toluene diisocyanate	*[584-84-9]*
o-Toluidine	*[95-53-4]*
Toxaphene (chlorinated camphene)	*[8001-35-2]*
1,2,4-Trichlorobenzene	*[120-82-1]*
1,1,2-Trichloroethane	*[79-00-5]*
Trichloroethylene	*[79-01-6]*
2,4,5-Trichlorophenol	*[95-95-4]*
2,4,6-Trichlorophenol	*[88-06-2]*
Triethylamine	*[121-44-8]*

transform to other HAPs (predominantly formaldehyde and acetaldehyde). The chemistry associated with the atmospheric transformation of HAPs to other HAPs occurs through ozone-, nitrate radical-, and hydroxyl radical-initiated reactions similar to the complex chemistry associated with tropospheric ozone and photochemical smog formation. More information on this topic is available (6).

Scientific studies have linked HAP exposure to a variety of sinister health effects, including increased cancer risk, reproductive and developmental defects, upper respiratory tract irritation, neurotoxic effects (visual blurring, tremors, delirium, unconsciousness, coma, convulsions), and chronic health disorders (aplastic anemia, pernicious anemia, pulmonary structural changes) (7,8).

Emission Sources

Emissions of VOCs and HAPs into the earth's atmosphere occur from biogenic sources as well as anthropogenic sources. The EPA categorizes anthropogenic emission sources of VOCs and HAPs into three groups: point sources, area sources, and mobile sources.

Point sources are large, discrete sources such as chemical plants; petroleum refineries; steel mills; and fuel-fired boilers, furnaces, and stoves. Emission units at point sources can be generally classified as process sources and fugitive sources. Process sources feature some type of capture device (eg, hood, vent, closed system) and, possibly, a control device (see VOLATILE ORGANIC COMPOUNDS, CONTROL AT INDUSTRIAL PLANTS) that direct the air emissions to a stack for atmospheric release. Fugitive sources are uncaptured and are released to the atmosphere by means other than a stack.

Figure 1 illustrates several types of point source emission units (process and fugitive) that typically are regulated by the EPA. The five types of emission units shown in Figure 1 are storage tanks, process vents (eg, reactors, recovery equipment, purification equipment), wastewater operations, material transfer operations, and equipment leaks. The EPA regulates both routine and accidental releases (eg, explosion, fire, operations error, equipment failure, runaway chemical reaction) of VOC and HAP from the point sources shown in Figure 1.

Area sources are small, widespread sources. Area sources include parts cleaning tanks, dry cleaning machines, residential heaters, and gas stations. Some states even regulate backyard grills, lawn mowers, and home fireplaces as area sources. Area sources include both process and fugitive emission units. Mobile sources include automobiles, trucks, buses, motorcycles, aircraft, locomotives, and nonroad vehicles (eg, lawn mowers, construction and farm equipment).

Estimated Emissions

Figure 2 summarizes the estimated annual emissions of VOC from biogenic and anthropogenic sources in the United States

Table 2. [a] *(continued)*

Chemical Name	CAS Registry Number
Trifluralin	[*1582-09-8*]
2,2,4-Trimethylpentane	[*540-84-1*]
Vinyl acetate	[*108-05-4*]
Vinyl bromide	[*593-60-2*]
Vinyl chloride	[*75-01-4*]
Vinylidene chloride (1,1-dichloroethylene)	[*75-35-4*]
Xylenes (isomers and mixture)	[*1330-20-7*]
o-Xylenes	[*95-47-6*]
m-Xylenes	[*108-38-3*]
p-Xylenes	[*106-42-3*]
Arsenic compounds[b] (inorganic including arsine)	
Antimony compounds[b]	
Beryllium compounds[b]	
Cadmium compounds[b]	
Chromium compounds[b]	
Cobalt compounds[b]	
Coke oven emissions[b]	
Cyanide compounds[b,c]	
Glycol ethers[b,d]	
Lead compounds[b]	
Manganese compounds[b]	
Mercury compounds[b]	
Fine mineral fibers[e]	
Nickel compounds[b]	
Polycyclic organic matter[f]	
Radionuclides (including radon)[g]	
Selenium compounds[b]	

[a] From Ref. 5.
[b] Unless otherwise specified, listings defined as including any unique chemical substance that contains named chemical (ie, antimony, arsenic, etc) as part of that chemical's infrastructure.
[c] $X'CN$ where $X = H'$ or any other group where a formal dissociation may occur, eg, KCN or $Ca(CN)_2$.
[d] Includes monoethers and diethers of ethylene glycol, diethylene glycol, and triethylene glycol $R{-}(OCH2CH2)_n{-}OR'$ where $n = 1, 2$, or 3; R = alkyl or aryl groups; R′ = R, H, or groups which when removed yield glycol ethers with the structure $R{-}(OCH2CH)_n{-}OH$. Polymers are excluded from the glycol category.
[e] Includes mineral fiber emissions from facilities manufacturing or processing glass, rock, or slag fibers (or other mineral derived fibers) of average diameter of 1 μ or less.
[f] Includes organic compounds with more than one benzene ring that have a boiling point ≥100°C.
[g] A type of atom that spontaneously undergoes radioactive decay.

(9). This figure reflects the best available information on biogenic and anthropogenic VOC emissions; however, it may not accurately represent the true contribution from those categories. Biogenic VOC emission inventories are based on algorithms that relate VOC emissions from a particular vegetation class to ambient temperature, land use, and light intensity. Uncertainties in these input parameters can affect the estimated emissions. Furthermore, VOC emissions for anthropogenic sources are only as accurate as the emission factors or engineering equations used to estimate them. The National Research Council has concluded that anthropogenic VOC emissions may be underestimated by 60–80% (3).

The EPA has estimated that annual VOC emissions in the United States total 50 Tg (55 million t). Approximately 29 Tg (32 million t) are attributable to biogenic sources, mainly vegetation, that naturally emit VOCs in the presence of sunlight. Approximately 21 Tg (23 million t) are attributable to anthropogenic sources, including motor vehicles; fuel combustion to produce electricity or heat (eg, coal, oil, gas, wood); industrial processes such as chemical manufacturing, petroleum refining, solvent use, and organic liquid storage and transport; and waste disposal and recycling. As Figure 2 shows, biogenic sources account for 58% of the total VOC emissions, mobile sources account for 16%, solvent use (eg, surface coating, graphic arts, degreasing, dry cleaning) accounts for 12%, and industrial processes account for 5%.

Figure 3 summarizes the estimated annual HAP emissions in the United States (9). The contribution of HAP emissions from biogenic sources is negligible. The EPA has estimated that annual HAP emissions in the United States total 4 Tg (4.4 million t). Approximately 1.6 Tg (1.8 million t) are attributable to motor vehicles; and approximately 1.5 Tg (1.7 million t) are attributable to industrial processes such as chemical manufacturing, petroleum refining, solvent use, and organic liquid storage and transport. As Figure 3 shows, motor vehicles account for 40% of the total HAP emissions, industrial processes account for approximately 38%, and fuel combustion accounts for 15%.

CLEAN AIR ACT

Clean air mandates date back to 1273 when King Edward I banned the use of sea coal in London to control smoke (ie,

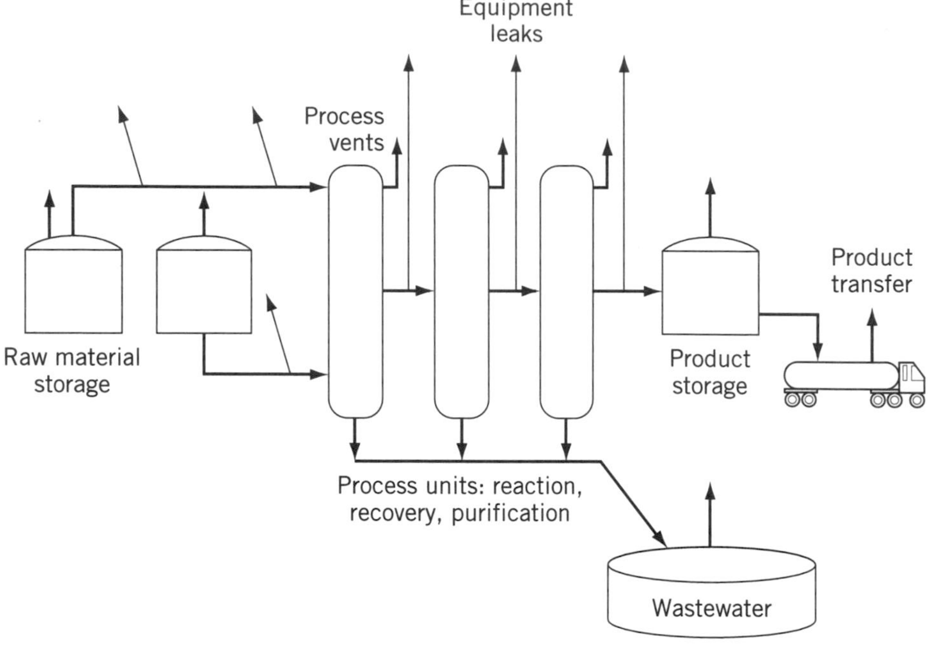

Figure 1. Typical point source emissions.

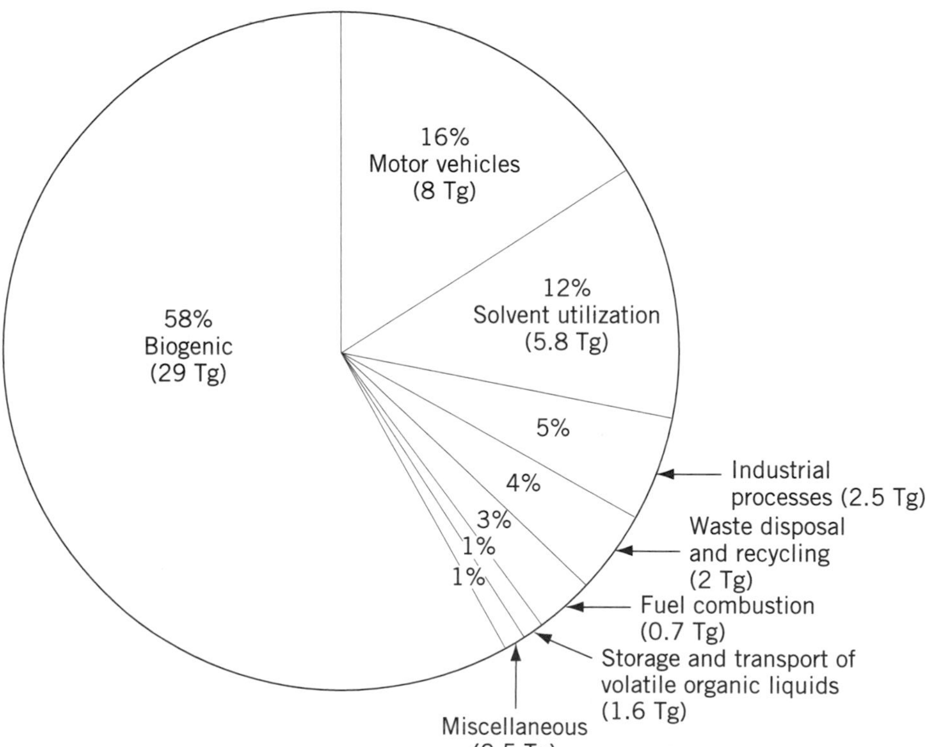

Figure 2. Estimated annual VOC emissions (total = 50 Tg/yr), where Tg = tera-grams. The largest contributors to the industrial processes sector include chemical and allied products manufacturing, petroleum and related industries, metals processing, and agriculture, food, and kindred products; in the miscellaneous sector, the largest contributors include structural fires, agricultural fires, prescribed burning, and forest wildfires (9).

particulate matter). The first smoke control laws in the United States were passed in Chicago and Cincinnati in 1881. Other cities adopted similar smoke control laws throughout the 1900s, including Pittsburgh and Los Angeles.

Following the post–World War II industrial boom in the United States, the need for national standards protecting environmental resources became evident. In 1956, Public Law 84-159 was passed to provide funding for research into the effects of air pollution on health and the environment; it also established a Division of Air Pollution in the U.S. Public Health Service. This legislation marked the first major involvement of the U.S. federal government in air pollution control. In 1960, Public Law 86-493 was passed to direct the surgeon general to study motor vehicle air pollution. In 1963, responding to demands for greater federal assistance to states and local governments in evaluating air pollution problems and in developing effective air pollution control programs, the U.S. Congress passed the first Clean Air Act. This law provided federal funding to state air pollution control agencies to develop, establish, and improve programs for the prevention

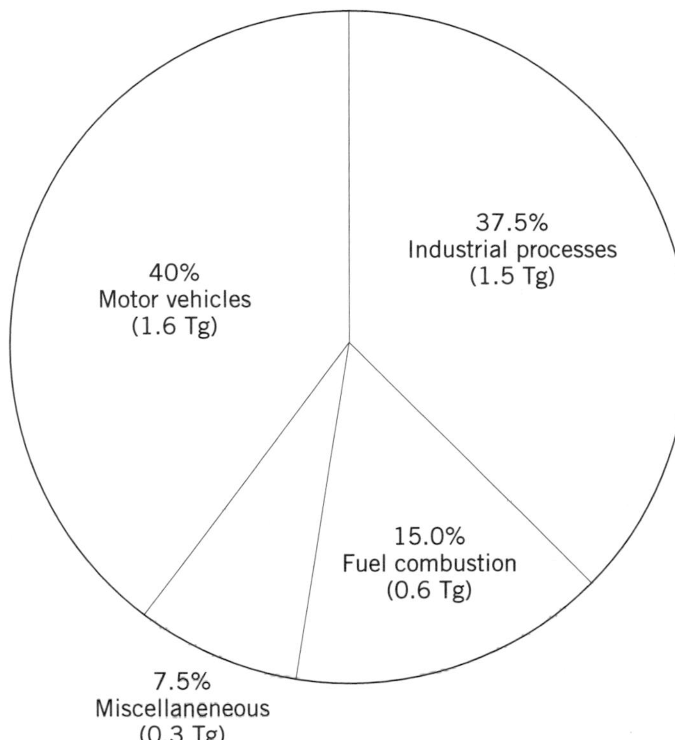

Figure 3. Estimated annual HAP emissions (total = 4 Tg/yr), where Tg = teragrams. The largest contributors to the industrial processes sector include solvent use (eg, surface coating, graphic arts, degreasing, dry cleaning), storage and transport of volatile organic liquids, chemical and allied products manufacturing, and petroleum and related industries; in the miscellaneous sector, the largest contributors include structural fires, agricultural fires, prescribed burning, and forest wildfires (9).

and control of air pollution. Amendments were added to the Clean Air Act in 1966 and 1967. The 1967 amendments, which provided the framework for today's Clean Air Act, established air quality control regions—generally one or more adjacent counties within a state—to identify areas with serious air pollution. It represented the first acknowledgment of the interstate effects of air pollution and attempted to provide federal oversight of the issue.

On December 2, 1970, the EPA was created by executive order of President Richard Nixon. Its mission, according to William D. Ruckelshaus, the EPA's first administrator, was to "gather together in a single agency the responsibility for research, standard setting, monitoring, and enforcement ... to protect and improve our environment" (10). The EPA was designated the authority to implement and enforce federal environmental legislation, including the Clean Air Act.

Major legislative amendments to the Clean Air Act were made in 1970, 1977, and 1990. Following passage of the Clean Air Act Amendments of 1990, the Clean Air Act now contains seven titles (Table 3) containing numerous parts and subparts. Titles I and II of the Clean Air Act contain requirements for VOCs and HAPs. Table 4 identifies the programs within the Clean Air Act that contain requirements for VOCs and HAPs. Also shown in Table 4 are the Clean Air Act sections that mandate each program and the citation where each program can be found in the *Code of Federal Regulations,* the collection of rules and regulations (published annually) developed by the EPA and all other executive departments and agencies of the U.S. government.

Table 3. U.S. Clean Air Act

Title	Name	Citation
I	Air Pollution Prevention and Control	42 USC 7401–7415
II	Emission Standards for Moving Sources	42 USC 7521–7590
III	General Provisions	42 USC 7601–7627
IV	Noise Pollution	42 USC 7641–7642
IV-A	Acid Deposition Control	42 USC 7651
V	Permits	42 USC 7661
VI	Stratospheric Ozone Protection	42 USC 7671

VOC REQUIREMENTS

National Ambient Air Quality Standards

National Ambient Air Quality Standards (NAAQS) have been developed by the EPA in response to a statutory mandate in the 1970 Clean Air Act Amendments requiring the identification of pollutants that "may reasonably be anticipated to endanger public health and welfare" (11) and the issuance of air quality criteria for them. These air quality criteria are to reflect the latest scientific knowledge useful in indicating the kind and extent of all identifiable effects on public health or welfare that may be expected from the presence of a specific pollutant in the ambient air. The act directs the EPA to develop primary and secondary NAAQSs. The act defines a primary standard as one "the attainment and maintenance of which, in the judgment of the Administrator [of the USEPA], based on the criteria and allowing an adequate margin of safety, is requisite to protect the public health" (12). A secondary standard must "specify a level of air quality the attainment and maintenance of which, in the judgment of the Administrator, based on the criteria, is requisite to protect the public welfare from any known or anticipated adverse effects associated with the presence of the pollutant in the ambient air" (13). Welfare effects include, but are not limited to, "effects on soils, water, crops, vegetation, manmade materials, animals, wildlife, weather, visibility and climate, damage to and deterioration of property, and hazards to transportation, as well as effects on economic values and on personal comfort and well-being" (14).

NAAQS are federal limits that must be met by each state. Each state, in turn, must estimate the amount of emission reduction required to meet NAAQS limits. Each state is divided into air quality control regions, which generally consist of one or more adjacent counties. Air quality control regions that meet NAAQS are referred to as "attainment" areas, and air quality control regions that do not meet NAAQS are referred to as "nonattainment" areas. Attainment status is pollutant specific; therefore, an area may be in attainment for some criteria pollutants but nonattainment for others.

The EPA has established NAAQS for six criteria pollutants: ozone, nitrogen dioxide, sulfur dioxide, fine particulate matter, carbon monoxide, and lead. The NAAQS for ozone was established on February 8, 1979, to minimize the health and environmental effects of tropospheric ozone; it replaced a NAAQS

Table 4. VOC and HAP Requirements in the Clean Air Act

Regulatory Program	Clean Air Act Citation	*Code of Federal Regulations* Citation
	VOC	
National Ambient Air Quality Standards	§§ 105–110 (42 USC 7405–7410); § 160–193 (42 USC 7470–7515)	40 CFR 50–53, 58, 65, 81
Ozone Nonattainment Program	§ 110 (42 USC 7410); §§ 181–185 (42 USC 7511)	40 CFR 51, 52
New Source Performance Standards	§ 111 (42 USC 7411)	40 CFR 60
Mobile Source Standards	§§ 202–250 (42 USC 7521–7590)	40 CFR 80, 85–87
Air Permitting Rules	§§ 160–169B (42 USC 7470–7492); §§ 171–173 (42 USC 7501–7503)	40 CFR 51, 52
	HAP	
National Emission Standards for Hazardous Air Pollutants	§ 112(q) (42 USC 7412)	40 CFR 61
Maximum Achievable Control Technology Standards	§ 112 (42 USC 7412)	40 CFR 63
Accidental Releases	§ 112(r) (42 USC 7412)	40 CFR 68
Mobile Source Standards	§§ 202–250 (42 USC 7521–7590)	40 CFR 80, 85–87
Air Permitting Rules	§ 112(g) (42 USC 7412)	40 CFR 63

for photochemical oxidants that was established in April 1971. The primary and secondary NAAQSs for ozone both are maximum hourly average concentrations of 0.12 parts per million (ppm) (235 $\mu g/m^3$). The standard is attained when the expected number of days per calendar year with a maximum hourly average concentration above 0.12 ppm (235 $\mu g/m^3$) is ≤ 1. The number of days exceeding the NAAQS each year may be averaged over the past three calendar years. An area is in compliance as long as this average remains ≤ 1.

The Clean Air Act requires that EPA periodically review all NAAQS. On March 9, 1993, the EPA announced that it had completed such a review of the ozone NAAQS and had concluded that revision of the NAAQS was not appropriate. However, based on a large number of recent scientific papers and ongoing research on the health and welfare effects of ozone, EPA announced that it would proceed as rapidly as possible with the next review of the ozone NAAQS. On July 18, 1997, the EPA replaced the 1-h primary ozone NAAQS with an 8-h standard at a level of 0.08 ppm to protect against longer exposure periods. The EPA also replaced the secondary NAAQS with a standard identical to the new primary NAAQS.

Ozone Nonattainment Program

Each state is required to prepare and submit a State Implementation Plan (SIP) to the EPA that describes how the state will attain and maintain the NAAQS. The SIP is unique to each state, since each state has unique air pollution problems based on geography, climate, population, and types of industries located within the state. Some SIP provisions apply to specific facilities within the state, while other SIP provisions apply to categories of industrial and consumer sources. The SIPs describe the emission limits, mass reduction limits, concentration limits, control equipment specifications, work practices, and schedules that affected facilities must obey for the state to bring nonattainment areas into compliance with the ozone NAAQS.

While ozone is the criteria pollutant for which a NAAQS has been established, little ozone is actually emitted from industrial or other anthropogenic processes. As shown earlier, VOCs and NO_x are the precursors to tropospheric ozone formation. Indirectly, VOC and NO_x create the harmful health effects of tropospheric ozone and photochemical smog. Because most of the sources of photochemical smog formation (eg, industrial plants, motor vehicles, fuel combustion equipment) emit VOC and NO_x but not ozone, the precursors are regulated in lieu of the criteria pollutant. Historically, most SIPs established VOC regulations to meet the NAAQS for ozone. Acknowledgment of the contribution of NO_x to photochemical smog formation was reflected in the 1990 Clean Air Act Amendments, when the EPA was granted authority to consider NO_x an equivalent contributor to photochemical smog formation and states were required to update their SIPs to include NO_x regulations designed to attain and maintain the ozone NAAQS.

The ozone nonattainment program was substantially modified in the Clean Air Act Amendments of 1990. Acknowledging that dozens of air quality control regions still had not attained the ozone NAAQS 11 years after it was established and that millions of people still were not adequately protected from the harmful effects of tropospheric ozone, lawmakers upgraded the ozone nonattainment area program requirements.

First, ozone nonattainment areas were divided into five categories, acknowledging that the extent of the ozone nonattainment problem varies by air quality control region and establishing realistic schedules for attaining the ozone NAAQS. Table 5 lists the categories of ozone nonattainment areas, the ozone concentration ranges of each category, and the statutory deadline by which each air quality control region within a category must attain the ozone NAAQS. The EPA assigned a category of ozone nonattainment to each nonattainment air quality control region based on air quality monitoring data from January 1, 1987, to December 31, 1989.

Second, the number of major stationary sources required to meet state VOC regulations was increased, acknowledging that previous VOC control requirements did not capture enough stationary sources in some air quality control regions for states to attain the ozone NAAQS. Before the 1990 Clean Air Act Amendments, major stationary sources were those emission units or facilities that emit or could emit under maximum design and operating conditions $\geq$90.8 Mg/yr (100 t/yr)

Table 5. Ozone Nonattainment Area Categories[a]

Ozone Nonattainment Area Category	Ozone Concentration, ppm	Deadline for Attaining Ozone NAAQS
Marginal	0.121–0.138	November 15, 1993
Moderate	0.138–0.160	November 15, 1996
Serious	0.160–0.180	November 15, 1999
Severe-15	0.180–0.190	November 15, 2005
Severe-17	0.190–0.280	November 15, 2007
Extreme	>0.280	November 15, 2010

[a] From Ref. (5), § 181(a), 42 USC 7511.

Table 6. Major VOC/NO_x Emission Units[a]

Ozone Nonattainment Area Category	*de minimis* Major Source Emission Limit,[b] Mg/year (t/yr)	
	VOC	NO_x
Marginal	90.8 (100)	90.8 (100)
Moderate	90.8 (100)	90.8 (100)
Serious	45.4 (50)	45.4 (50)
Severe	22.7 (25)	22.7 (25)
Extreme	9.08 (10)	9.08 (10)
Ozone Transport Region	45.4 (50)	90.8 (100)

[a] From Ref. 5, § 182, 42 USC 7511.
[b] A major source is any stationary source or group of sources located within a contiguous area and under common control that emits, or has the potential to emit, VOC or NO_x equal to or above the emission levels shown in the table. The EPA has published numerous guidance documents that define *stationary source* and *potential to emit.*

of VOCs. The amendments expanded the definition of major stationary sources to include sources emitting <90.8 Mg/year (100 t/yr), as shown in Table 6.

Third, in addition to increasing the number of major stationary sources subject to VOC rules, the 1990 Clean Air Act Amendments strengthened the rule requirements themselves. States were required to upgrade their SIPs to strengthen VOC control requirements for existing stationary sources. Before the 1990 Clean Air Act Amendments, certain existing stationary sources were required to install Reasonably Available Control Technology (RACT) to reduce VOC emissions, thereby helping the state attain the ozone NAAQS. RACT is defined as the "lowest emission limit that a particular source is capable of meeting by the application of control technology that is reasonably available considering technological and economic feasibility" (15). Typically, RACT is an emission limit, mass reduction limit (eg, 85% mass reduction from a stack), concentration limit, equipment standard (ie, use of a specific control technology meeting specific operating specifications), or work practice (eg, leak detection and timely repair). States may choose the industries to be regulated by the RACT program, and the EPA issued several Control Technique Guideline (CTG) documents to assist states in establishing "presumptive RACT" for certain industry source categories. For a variety of reasons (eg, states' failure to implement RACT for some sources, states' overestimates of the contribution of RACT to ozone NAAQS attainment), these RACT provisions were insufficient to attain the ozone NAAQS in some air quality control regions. Therefore, the 1990 Clean Air Act Amendments required states to fix up their SIPs to implement existing RACT requirements. Furthermore, the 1990 Clean Air Act Amendments required states with moderate and above ozone nonattainment areas to catch up to attainment areas by implementing RACT rules for VOC sources for which EPA issued CTG documents and for all other major stationary sources of VOC located in the nonattainment area.

Fourth, federal preconstruction permitting (known as New Source Review, NSR) of new or modified VOC sources was tightened to ensure that new emissions in an ozone nonattainment area would be offset by emission reductions from other sources that impact the same nonattainment area. Before the 1990 Amendments, NSR required new or modified air emission sources locating in ozone nonattainment areas and emitting greater than significant amounts of VOC (defined before 1990 as 36.3 Mg/year (40 t/yr)) to install Lowest Achievable Emission Rate (LAER) technology and to offset every new ton of VOC emissions with a permanent removal of 1 t of existing VOC emissions from another source in the same nonattainment area. LAER is defined as "the most stringent emission limitation derived from either the most stringent emission limitation contained in a State's SIP for such class or category of source or the most stringent emission limitation achieved in practice by such class or category of source." The 1990 Clean Air Act Amendments preserved the LAER requirement in NSR preconstruction permitting. The amendments tightened the offset requirements, requiring greater than one-to-one offsets depending on a region's level of ozone nonattainment, as shown.

Ozone Nonattainment Category	VOC Emission Offset Ratio
Marginal	1.1:1
Moderate	1.15:1
Serious	1.2:1
Severe	1.3:1
Extreme	1.5:1
Ozone transport region	1.15:1

Thus for example, a new VOC source locating in a marginal nonattainment area would be required to offset every ton of new VOC emissions by 1.1 t (2200 lbs) of permanent VOC emission decreases.

The 1990 Clean Air Act Amendments also lowered the significant threshold at which NSR preconstruction permitting is triggered to 22.7 Mg/year (25 t/yr) of VOC for new and modified sources locating in serious and above nonattainment areas.

Fifth, continuous improvement toward meeting the ozone NAAQS was emphasized. The amendments established Reasonable Further Progress requirements to check periodically on a state's progress in attaining the ozone NAAQS. Within moderate ozone nonattainment areas and above, states must demonstrate a 15% reduction in VOC emissions by 1996 over 1990 baseline estimates. Furthermore, within serious ozone nonattainment areas and above, states must demonstrate VOC reductions of 3%/year until affected air quality control regions attain the ozone NAAQS.

Sixth, as mentioned earlier, NO_x equivalency was established. While NO_x is itself a criteria pollutant, it also is a precursor to photochemical smog formation and thus contributes to the adverse health and environmental effects attributable to tropospheric ozone. The Clean Air Act Amendments of 1990 acknowledge the contribution of NO_x to ozone NAAQS nonattainment by requiring RACT controls on major stationary sources of NO_x. Table 6 shows the minimum level of NO_x emissions that constitute major stationary source status (and, therefore, require RACT controls) for each ozone nonattainment area. The EPA has assisted states in identifying "presumptive RACT" for NO_x emission sources through the issuance of several Alternative Control Techniques (ACT) documents. States that are able to demonstrate to the EPA that any NO_x reductions or excess NO_x reductions above certain predicted levels do not contribute to ozone NAAQS attainment in certain air quality control regions can limit the application of the NO_x equivalency rules.

Seventh, the 1990 amendments established the concept of "ozone transport regions," acknowledging that wind and other meteorological effects can contribute to interstate ozone problems. Currently, the only ozone transport region in the United States consists of the states of Connecticut, Delaware, Maine, Maryland, Massachusetts, New Hampshire, New Jersey, New York, Pennsylvania, Rhode Island, and Vermont and the District of Columbia.

Eighth, regulations on mobile VOC sources were expanded. Before the 1990 Clean Air Act Amendments, the EPA established transportation control measures including emission limitations on new motor vehicles and motor vehicle inspection and maintenance programs. Like the pre-1990 RACT program, these measures were insufficient to attain the ozone NAAQS in some air quality control regions. Therefore, the 1990 amendments required states to upgrade their SIPs to incorporate increasingly stringent transportation control measures, depending on an area's level of ozone nonattainment. The mobile source requirements are described elsewhere in this article.

Finally, the contribution of VOC emissions from nonconventional stationary and mobile sources to the tropospheric ozone problem was acknowledged. Section 183(e) of the Clean Air Act requires the EPA to study VOC emissions into the ambient air from the use, consumption, storage, disposal, destruction, or decomposition of consumer and commercial products. Based on the study results, the EPA is required to list and schedule for regulation those categories of products that account for at least 80% of the total VOC emissions in ozone nonattainment areas from consumer and commercial products.

The EPA has concluded that annual VOC emissions from consumer and commercial products total 5 Tg (6 million t), accounting for approximately 10% of the biogenic and anthropogenic VOC emissions shown in Figure 1 (16). VOC emissions may be reduced from consumer and commercial products through product reformulation, raw material substitution, and repackaging (17).

New Source Performance Standards

New Source Performance Standards (NSPS) evolved from a statutory mandate in the 1970 Clean Air Act that required the EPA to identify industry source categories that emit criteria or other pollutants in sufficient quantities to "cause or contribute significantly to air pollution which may reasonably be anticipated to endanger public health or welfare" (18). The intended effect of NSPS is to require new, modified, and reconstructed sources to control emissions to the level "achievable through the application of the best system of emission reduction which (taking into account the cost of achieving such reduction and any nonair quality health and environmental impact and energy requirements) has been adequately demonstrated" (19).

NSPS are not intended to attain a prescribed level of emission reduction. Instead, they are designed to establish the best demonstrated technology (BDT) for controlling criteria or other pollutants from industry source categories. Air pollutants that currently are regulated under the NSPS program include VOC, particulate matter, sulfur dioxide, carbon monoxide, nitrogen oxides, lead, fluorides, total reduced sulfur, sulfuric acid mist, hydrogen sulfide, total suspended particulates, and three pollutants from municipal waste combustion (20).

Mobile Source Standards

Figure 2 shows that mobile sources contribute 16% of all VOC emissions in the United States and 38% of anthropogenic VOC emissions. It is difficult for any state containing ozone nonattainment areas to attain the ozone NAAQS without establishing and implementing programs for the control of VOC and NO_x emissions from mobile sources; failure to control emissions from mobile sources forces greater control of stationary sources to attain the ozone NAAQS. Both mobile and stationary sources must share the ownership of ozone NAAQS nonattainment and the legal responsibility of ozone NAAQS attainment.

Controlling VOC and NO_x emissions from mobile sources involves the difficult challenge of convincing individual consumers of the need for controlling automobile engine exhausts, producing and using cleaner-burning fuels, mandating automobile inspection and maintenance programs, and advocating transportation control measures such as carpooling. In the United States, with its historic legacy of personal independence, its current shift toward less government intervention, and its heavy dependence on automobiles, many individual consumers resist any program to restrict in any way their rights to own and operate a motor vehicle. The challenge to federal, state, and local regulators, therefore, is to implement mobile source standards that attempt to balance consumers' rights with their responsibilities as stakeholders in clean air attainment.

Mobile source standards can be classified into the following categories.

Motor vehicle control measures.
Fuel control measures.
Transportation control measures.
Refueling control measures.
Motor vehicle inspection and maintenance (I&M) programs.
Clean fuel vehicle programs.

Except for the clean fuel vehicle programs, mobile source standards were in effect for these categories before the passage of the 1990 Clean Air Act Amendments. However, some of the standards were implemented only by certain states, and some of the standards were insufficient to attain the ozone NAAQS. Therefore, the 1990 Clean Air Act Amendments strengthened the mobile source standards through federal oversight and more stringent requirements.

Motor Vehicle Control Measures. Motor vehicle control measures partially consist of requirements for motor vehicle manufacturers to develop cleaner burning engines on automobiles, trucks, buses, motorcycles, aircraft, locomotives, and nonroad vehicles (eg, lawn mowers, construction and farm equipment). These standards are based either on vehicle weight or engine size. Vehicle weight standards generally impose tailpipe emission limits on VOC, carbon monoxide, and NO_x emissions in units of grams of pollutant per vehicle mile. Engine size standards generally impose tailpipe emission limits on VOC, carbon monoxide, and NO_x emissions in units of grams of pollutant per brake horsepower hour. The EPA has established a durability program to test new motor vehicle engines and to issue certificates of conformity to engines that EPA believes will conform to the applicable emission limits for the useful life of the engine.

Federal engine test standards have been in effect since the 1970s. They apply to gasoline- and diesel-fueled engines. Per the 1990 Clean Air Act Amendments, the EPA established more stringent tailpipe emission limits for automobiles and light-duty trucks (ie, less than 3860 kg (8500 pounds) gross vehicle weight rate) and a two-tier schedule for meeting the new limits. Tier I standards apply to model years 1994 to 1998, and tier II standards, which will be implemented only if the EPA deems them necessary to attain or maintain the ozone NAAQS, will begin with model year 2004.

Motor vehicle control measures also consist of requirements for gasoline-powered motor vehicle manufacturers to reduce evaporative emissions from motor vehicle fuel tanks and carburetors. Evaporative emissions include breathing losses in the fuel tank caused by the expansion of gasoline vapors in a fuel tank due to daily temperature changes, working losses from the fuel tank as the motor vehicle is running, and residual losses in the carburetor and fuel lines after the motor vehicle is shut off. Evaporative emission control systems consisting of small carbon canisters have been installed on gasoline-powered motor vehicles. The EPA has established test procedures and evaporative emission limits to evaluate the performance of evaporative emission control systems. The test procedures, which originally were developed to test breathing and residual losses, were modified per the 1990 Clean Air Act Amendments to include working losses and to include methanol-fueled as well as gasoline engines.

Fuel Control Measures. Fuel control measures consist of requirements on fuel providers to develop cleaner burning fuels for motor vehicles. Federal fuel control measures have been in effect since 1973, when the EPA began a phase-out of leaded fuels. The 1990 Clean Air Act Amendments prohibit the sale of leaded gasoline after December 31, 1995. The 1990 amendments also mandated greater particulate control of diesel fuel; the EPA responded by issuing sulfur content limits on diesel fuel. The 1990 amendments also mandated the use of oxygenated gasoline to effect better fuel combustion and reduce carbon monoxide (a product of incomplete combustion) in certain carbon monoxide nonattainment areas.

Fuel control measures for controlling VOCs partially consist of gasoline volatility rules. Gasoline volatility rules establish maximum vapor pressures for gasoline sold in summer months. (Higher vapor pressures result in higher evaporative losses.) The 1990 Clean Air Act Amendments allow the EPA to establish more stringent vapor pressure limits in ozone nonattainment areas. Fuel control measures for controlling VOCs also consist of reformulated gasoline requirements. Reformulated gasoline must meet a minimum oxygen content (to improve fuel combustion), a maximum benzene content (to reduce benzene emissions), and a NO_x emissions level less than or equal to conventional gasoline. The EPA requires reformulated gasoline in certain ozone nonattainment areas, although any other area of the country may voluntarily join the program.

Transportation Control Measures. Transportation control measures, which include mass transit improvements (eg, bus lanes, car pools), remote parking facilities, and motor vehicle travel restrictions, have been in effect in some cities before a regulatory mandate to establish such measures. The 1990 Clean Air Act Amendments required the EPA and the Department of Transportation to quantify the emissions reduction potential of numerous transportation control measures, including bus lanes, high occupancy vehicle (HOV) lanes, remote parking facilities, vehicle use restrictions in urban areas during peak ozone-forming times (generally morning rush hour), bicycle lanes, mass transit, employer-sponsored plans such as car pools and flexible work schedules, and voluntary retirement of older, less fuel-efficient motor vehicles. States with serious, severe, and extreme ozone nonattainment areas are required to use this quantitative information to estimate VOC emission reductions and establish transportation control measures to attain and maintain the ozone NAAQS.

Transportation control measures also include employee trip reduction programs. Per the 1990 Clean Air Act Amendments, these plans were to be developed by employers with 100 or more employees located in severe and extreme ozone nonattainment areas to establish methods to reduce the number of vehicle miles traveled (21). The EPA has been only minimally successful in implementing this program due to powerful resistance by affected facilities.

Transportation control measures also include transportation conformity rules. Per the 1990 Clean Air Act Amendments, conformity ensures that new highway or other roadway construction projects do not produce new air quality violations, worsen existing violations, or delay timely attainment of the ozone NAAQS.

Refueling Control Measures. VOC emissions associated with motor vehicle refueling include vapor displacement losses associated with transferring fuel from tank trucks to fuel storage tanks, vapor displacement losses associated with motor vehicle refueling, and spillage losses associated with motor vehicle refueling. VOC emissions associated with fuel transfer from tank trucks to storage tanks arise from the vapor displacement within both the truck and the storage tank and also from vapor displacement associated with "splash-filling" a storage tank (ie, introducing the fuel into the tank above the tank's current liquid level). Control of VOC emissions associated with fuel transfer from tank trucks to storage tanks is commonly referred to as stage I vapor control. Stage I vapor control mandates the use of a submerged fill fuel loading pipe to prevent splashing of the fuel as it is loaded into a storage tank and a vapor balance system that is designed and operated to route the displaced vapors from the storage tank to the tank truck.

Control of VOC emissions associated with vapor displacement during motor vehicle refueling is commonly referred to as stage II vapor control. Stage II vapor control mandates the use of a nozzle boot to capture emissions during refueling and a vapor balance system that is designed and operated to route the displaced vapors from the motor vehicle fuel tank to the storage tank.

The EPA has attempted to control VOC emissions associated with spillage during motor vehicle refueling by limiting the dispensing rates for gasoline and methanol fuel pumps to less than or equal to 38 L/min (10 gal/min) for most refueling facilities.

Motor Vehicle Inspection and Maintenance Programs. In addition to the tailpipe emission standards to regulate emissions from new motor vehicles, the EPA has established motor vehicle inspection and maintenance (I&M) programs to ensure proper maintenance of in-use motor vehicles as a means of reducing VOC emissions. Basic I&M programs, which were first required in 1977 for some states' ozone nonattainment areas, consist of emission tests during engine loading (ie, running) and idling. Per the 1990 Clean Air Act Amendments, all ozone nonattainment areas are required to implement basic I&M programs. Furthermore, ozone nonattainment areas classified as serious, severe, and extreme must implement enhanced I&M programs based on EPA guidance.

I&M programs may be conducted by state or municipal authorities at centralized inspection stations or by licensed service stations at decentralized inspection stations. The 1990 Clean Air Act Amendments require centralized testing unless a state can prove that decentralized testing is equally efficient.

Clean Fuel Vehicle Programs. The 1990 Clean Air Act Amendments established a new program for controlling VOC and NO_x emissions from motor vehicles. Referred to as the clean fuel vehicles program, this program establishes tailpipe emission limits for automobiles and trucks that are more stringent than the limits for gasoline- and diesel-fueled motor vehicles. These emission limits apply to motor vehicles fueled by clean alternative fuels, including "methanol, ethanol, or other alcohols (including fuel mixtures comprising at least 85 percent alcohol), reformulated gasoline, diesel, natural gas, liquefied petroleum gas, hydrogen, and electricity" (22).

The clean fuel vehicle program currently applies to certain centrally fueled fleets and to the state of California or any other state that voluntarily opts into the program. Fleets of 10 or more automobiles and trucks that can be centrally fueled and that are located in serious, severe, and extreme ozone nonattainment areas with populations >250,000 (based on the U.S. 1980 census) must meet specific schedules for phasing in the use of clean-fuel vehicles. In California, at least 150,000 clean-fuel vehicles must be sold in the state each year starting with model year 1996, and at least 300,000 clean-fuel vehicles must be sold in the state each year starting with model year 1999.

Air Permitting Rules

Federal and state air quality management programs generally can be classified into emission standards and permitting requirements. Emission standards, such as the NSPS and mobile source standards described above as well as pollutant-specific (eg, VOC) regulations developed to attain and maintain the NAAQS, establish mass emission rate limits, concentration limits, percent reduction requirements, equipment requirements, or work practices to control air pollution. Permitting requirements generally are procedural requirements that may require installation of pollution controls and adequate compliance assurance methods to ensure that the construction of new air emitting units or the modification of existing air emitting units does not adversely affect existing air quality.

Most states and the EPA have both preconstruction and operating permit programs. Preconstruction permits are required before constructing or modifying air emitting units for the permitting authorities to establish the level of pollution control required to attain or maintain the NAAQS. Operating permits are required after construction or modification is complete and before operation of the air emitting unit to establish compliance with the preconstruction permit terms and conditions and to ensure substantiated compliance after the preconstruction permit expires.

Most states have preconstruction and operating permit programs for VOC emitting sources. These programs assume presumptive compliance with existing VOC emission standards and also may require the installation of BDT.

The EPA also has established preconstruction and operating permit programs for VOC emitting sources. As mentioned earlier, EPA's preconstruction permit program, known as the NSR program, requires the installation of LAER on major VOC sources locating in ozone nonattainment areas. The NSR program also requires the installation of best available control technology (BACT) as well as a detailed air quality impacts analysis before the construction or modification of major VOC sources located in ozone attainment areas. This portion of the NSR program affecting attainment areas is known as the Prevention of Significant Deterioration (PSD) program because of its purpose to prevent significant deterioration of air quality in areas that already meet the NAAQS.

The EPA's operating permit program evolved from Title V of the 1990 Clean Air Act Amendments. It is a largely procedural program to identify and consolidate into a single document all of the existing air regulations affecting a facility and to ensure compliance with these existing requirements.

HAP REQUIREMENTS

National Emission Standards for Hazardous Air Pollutants

National Emission Standards for Hazardous Air Pollutants (NESHAP) evolved from a statutory mandate in the 1970

Clean Air Act that required the EPA to (*1*) identify HAPs, (*2*) identify stationary source categories emitting HAPs, and (*3*) develop risk-based emission standards based on an acceptable level of risk and an ample margin of safety to limit HAP emissions from new and existing sources.

The EPA struggled to develop NESHAPs based on the legislative mandate. While it conducted and examined numerous scientific studies to identify HAPs, the EPA encountered strong opposition each time it identified a compound as a HAP. Even greater resistance was encountered every time a NESHAP was developed; many NESHAPs were delayed in court proceedings as stakeholders (eg, regulated industries, environmental advocacy groups, health-based organizations) challenged the EPA's interpretation of acceptable risk levels and ample safety margins. Before the passage of the 1990 amendments, NESHAPs were developed for seven HAPs: asbestos, benzene, beryllium, inorganic arsenic, mercury, radionuclides, and vinyl chloride (23).

Maximum Achievable Control Technology NESHAPs

Litigation stifled NESHAP rule making and prompted a fundamental revision to the NESHAP program in the 1990 Clean Air Act Amendments. The traditional three-step approach of identifying HAPs, identifying source categories emitting HAPs, and developing risk-based rules to control HAPs was revised with the primary purpose of reducing litigation. Under the revised NESHAP program, the EPA was not required to identify HAPs; the list of HAPs was written into the act itself, although the EPA and the public have the opportunity to revise the list (see Table 2). The NESHAP program was converted from a risk-based program to a technology-based program similar to the NSPS program. Instead of developing HAP rules based on controversial risk levels and safety margins, the EPA was authorized to develop HAP rules based on the controls and practices used by the best-performing HAP-emitting sources.

The intent of the HAP program is to regulate HAP emissions from new and existing facilities that emit HAPs. Affected facilities may include both point and area sources. Affected point sources are major sources of HAPs within identified source categories; major HAP sources are stationary sources with potential HAP emissions (based on maximum design and operating parameters) of 9.08 Mg/year (10 t/yr) or more of a single HAP or 22.7 Mg/year (25 t/yr) or more of two or more HAPs. Affected area sources generally are nonmajor HAP emitters (eg, dry cleaning machines), but their combined accumulation within localized areas produces high HAP emissions.

The technology-based emission limit that EPA must establish for stationary point sources is known as maximum achievable control technology (MACT). MACT may be a performance standard (eg, a prescribed HAP emission limit, concentration limit, or percent reduction mandate), an equipment standard (eg, required use of a thermal or catalytic oxidizer, a flare, or a carbon adsorber), or a work practice standard (eg, periodic leak detection and repair).

The Clean Air Act Amendments of 1990 established a dual definition of MACT for new and existing point sources. MACT for new and reconstructed point sources corresponds to the level of HAP control achieved in practice by the single best controlled similar source. MACT for existing point sources corresponds to the level of HAP control achieved in practice by the best performing 12% of the existing similar sources for which the EPA has emissions information.

The 1990 Amendments allow the EPA to establish a technology-based control level for area HAP sources that is less stringent than MACT. This control level is known as generally available control technology (GACT). Like MACT, GACT may be a performance standard, an equipment standard, or a work practice standard.

Under the MACT–GACT program, the EPA still is required to identify source categories that emit HAPs and for which MACT and/or GACT rules will be developed. The EPA also is required to develop a schedule for issuing the rules. The EPA is obligated to review and revise the list of source categories periodically (24).

The risk-based approach to HAP standards development was not completely abandoned in the 1990 Clean Air Act Amendments. Within 8 years after issuing each MACT rule for point sources, the EPA must perform a health-based residual risk evaluation to determine if MACT successfully reduced the public's risk of cancer to <1 in 1 million. If the public's cancer risk still exceeds 1 in 1 million, the EPA is required to issue a residual risk standard more stringent than the MACT standard to reduce the cancer risk.

Accidental Release Standards

Numerous chemical accidents over the past 20 years that resulted in mass casualties and injuries and were reported to the public in graphic detail prompted two U.S. government agencies to develop rules designed to lessen the number and severity of serious chemical accidents. These programs—known as Risk Management Programs (RMPs)—were developed by the U.S. Occupational Safety and Health Administration (OSHA) to mitigate on-site impacts and by the EPA to mitigate off-site impacts of chemical accidental releases. The EPA's program applies to facilities that use, store, manufacture, handle, or internally move more than a threshold quantity of one or more of the regulated 77 HAPs or 63 flammable substances. Table 7 lists the regulated substances and their threshold quantities. The EPA developed this list based on existing lists of regulated hazardous substances.

Facilities subject to the RMP are required to register with the EPA as an affected facility and to prepare a written Risk Management Plan for dealing with accidental chemical releases. The risk management plan must contain a hazard assessment estimating the potential off-site consequences to the public and the environment caused by accidental releases of HAPs, flammables, and explosives. It also must contain a prevention program consisting of management systems, training programs, and written procedures for preventing accidental releases and for conducting accident investigations to identify root causes of accidental releases. Finally, the risk management plan must contain an emergency response program identifying specific actions to be taken by each employee and emergency responder in the event of an accidental release.

Mobile Source Standards

In addition to regulating HAP emissions from stationary sources through MACT and from accidental releases of HAPs, part of the EPA's strategic plan to regulate HAPs is the control

Table 7. Accidental Release Chemicals List[a]

Chemical Name	Threshold Quantity, kg (lb)
Acetaldehyde	4,540 (10,000)
Acetylene [ethyne]	4,540 (10,000)
Acrolein [2-propenal]	2,270 (5,000)
Acrylonitrile [2-propenenitrile]	9,070 (20,000)
Acrylyl chloride [2-propenoyl chloride]	2,270 (5,000)
Allyl alcohol [2-propen-1-ol]	6,800 (15,000)
Allylamine [2-propen-1-amine]	4,540 (10,000)
Ammonia (anhydrous)	4,540 (10,000)
Ammonia (concentration 20% or greater)	9,070 (20,000)
Arsenous trichloride	6,800 (15,000)
Arsine	450 (1,000)
Boron trichloride	2,270 (5,000)
Boron trifluoride	2,270 (5,000)
Boron trifluoride compound with methyl ether	6,800 (15,000)
Bromine	4,540 (10,000)
Bromotrifluorethylene [ethene, bromotrifluoro-]	4,540 (10,000)
1,3-Butadiene	4,540 (10,000)
Butane	4,540 (10,000)
Butene	4,540 (10,000)
1-Butene	4,540 (10,000)
2-Butene	4,540 (10,000)
2-Butene-*cis*	4,540 (10,000)
2-Butene-*trans* [2-butene, (*E*)]	4,540 (10,000)
Carbon disulfide	9,070 (20,000)
Carbon oxysulfide	4,540 (10,000)
Chlorine	1,130 (2,500)
Chlorine dioxide	450 (1,000)
Chlorine monoxide	4,540 (10,000)
Chloroform [methane, trichloro-]	9,070 (20,000)
Chloromethyl ether [methane, oxybis[chloro]]	450 (1,000)
Chloromethyl methyl ether [methane, chloromethoxy-]	2,270 (5,000)
1-Chloropropylene [1-propene, 1-chloro-]	4,540 (10,000)
2-Chloropropylene [1-propene, 2-chloro-]	4,540 (10,000)
Crotonaldehyde [2-butenal]	9,070 (20,000)
Crotonaldehyde, (E)- [2-butenal, (E)-]	9,070 (20,000)
Cyanogen [ethanedinitrile]	4,540 (10,000)
Cyanogen chloride	4,540 (10,000)
Cyclohexylamine [cyclohexanamine]	6,800 (15,000)
Cyclopropane	4,540 (10,000)
Diborane	1,130 (2,500)
Dichlorosilane	4,540 (10,000)
Difluoroethane [ethane, 1,1-difluoro-]	4,540 (10,000)
Dimethylamine [methanamine, *N*-methyl-]	4,540 (10,000)
Dimethyldichlorosilane [silane, dichlorodimethyl-]	2,270 (5,000)
1,1-Dimethylhydrazine [hydrazine, 1,1-dimethyl-]	6,800 (15,000)
2,2-Dimethylpropane [propane, 2,2-dimethyl-]	4,540 (10,000)
Epichlorohydrin [oxirane, (chloromethyl)-] (1-chloro-2,3-epoxypropane)	9,070 (20,000)
Ethane	4,540 (10,000)
Ethyl acetylene [1-butyne]	4,540 (10,000)
Ethylamine [ethanamine]	4,540 (10,000)
Ethyl chloride (chloroethane) [ethane, chloro-]	4,540 (10,000)
Ethylene [ethene]	4,540 (10,000)
Ethylenediamine [1,2-ethanediamine]	9,070 (20,000)
Ethyleneimine (aziridine)	4,540 (10,000)
Ethylene oxide [oxirane]	4,540 (10,000)
Ethyl ether [ethane, 1,1′-oxybis-]	4,540 (10,000)
Ethyl mercaptan [ethanethiol]	4,540 (10,000)
Ethyl nitrite [nitrous acid, ethyl ester]	4,540 (10,000)
Fluorine	450 (1,000)
Formaldehyde	6,800 (15,000)
Furan (total tetra through octachlorinated dibenzofurans)	2,270 (5,000)
Hydrazine	6,800 (15,000)
Hydrochloric acid (concentration 30% or greater)	6,800 (15,000)
Hydrocyanic acid	1,130 (2,500)
Hydrogen	4,540 (10,000)
Hydrogen chloride (anhydrous) [hydrochloric acid]	2,270 (5,000)
Hydrogen fluoride [hydrofluoric acid]	450 (1,000)
Hydrogen selenide	230 (500)
Hydrogen sulfide	4,540 (10,000)
Iron, pentacarbonyl-	1,130 (2,500)
Isobutane [propane, 2-methyl]	4,540 (10,000)
Isobutyronitrile [propanenitrile, 2-methyl-]	9,070 (20,000)
Isopentane [butane, 2-methyl-]	4,540 (10,000)
Isoprene [1,3-butadiene, 2-methyl-]	4,540 (10,000)
Isopropylamine [2-propanamine]	4,540 (10,000)
Isopropyl chloride [propane, 2-chloro-]	4,540 (10,000)
Isopropyl chloroformate [carbonochloridic acid, 1-methylethyl ester]	6,800 (15,000)
Methacrylonitrile [2-propenenitrile, 2-methyl-]	4,540 (10,000)
Methane	4,540 (10,000)
Methylamine [methanamine]	4,540 (10,000)
2-Methyl-1-butene	4,540 (10,000)
3-Methyl-1-butene	4,540 (10,000)
Methyl chloride (chloromethane)	4,540 (10,000)
Methyl chloroformate [carbonochloridic acid, methylester]	2,270 (5,000)
Methyl ether [methane, oxybis-]	4,540 (10,000)
Methyl formate [formic acid, methyl ester]	4,540 (10,000)
Methylhydrazine [hydrazine, methyl-]	6,800 (15,000)
Methyl isocyanate [methane, isocyanato-]	4,540 (10,000)
Methyl mercaptan [methanethiol]	4,540 (10,000)
2-Methylpropene [1-propene, 2-methyl-]	4,540 (10,000)
Methyl thiocyanate [thiocyanic acid, methyl ester]	9,070 (20,000)
Methyltrichlorosilane [silane, trichloromethyl-]	2,270 (5,000)
Nickel carbonyl	450 (1,000)
Nitric acid (concentration 80% or greater)	6,800 (15,000)
Nitric oxide	4,540 (10,000)
Oleum (fuming sulfuric acid)	4,540 (10,000)
1,3-Pentadiene	4,540 (10,000)
Pentane	4,540 (10,000)
1-Pentene	4,540 (10,000)
2-Pentene, (*E*)-	4,540 (10,000)
2-Pentene, (*Z*)-	4,540 (10,000)
Peracetic acid [ethaneperoxoic acid]	4,540 (10,000)
Perchloromethylmercaptan [methanesulfenyl chloride, trichloro-]	4,540 (10,000)
Phosgene [carbonic dichloride]	230 (500)
Phosphine	2,270 (5,000)
Phosphorus oxychloride	2,270 (5,000)
Phosphorus trichloride	6,800 (15,000)
Piperidine	6,800 (15,000)
Propadiene [1,2-propadiene]	4,540 (10,000)

Table 7. Accidental Release Chemicals List[a] *(continued)*

Chemical Name	Threshold Quantity, kg (lb)
Propane	4,540 (10,000)
Propionitrile [propanenitrile]	4,540 (10,000)
Propyl chloroformate [Carbonochloridic acid, propylester]	6,800 (15,000)
Propylene [1-propene]	4,540 (10,000)
Propylene oxide [oxirane, methyl-]	4,540 (10,000)
Propylenimine [aziridine, 2-methyl-]	4,540 (10,000)
Propyne [1-propyne]	4,540 (10,000)
Silane	4,540 (10,000)
Sulfur dioxide (anhydrous)	2,270 (5,000)
Sulfur tetrafluoride	1,130 (2,500)
Sulfur trioxide	4,540 (10,000)
Tetrafluoroethylene	4,540 (10,000)
Tetramethyllead [plumbane, tetramethyl-]	4,540 (10,000)
Tetramethylsilane	4,540 (10,000)
Tetranitromethane [methane, tetranitro-]	4,540 (10,000)
Titanium tetrachloride [titanium chloride (TiCl4) (T-4)-]	1,130 (2,500)
Toluene 2,4-diisocyanate [benzene, 2,4-diisocyanato-1-methyl-]	4,540 (10,000)
Toluene 2,6-diisocyanate [benzene, 1,3-diisocyanato-2-methyl-]	4,540 (10,000)
Toluene diisocyanate (unspecified isomer) [benzene,1,3-diisocyanatomethyl-]	4,540 (10,000)
Trichlorosilane	4,540 (10,000)
Trifluorochloroethylene [ethene, chlorotrifluoro-]	4,540 (10,000)
Trimethylamine [methanamine, *N,N*-dimethyl-]	4,540 (10,000)
Trimethylchlorosilane [silane, chlorotrimethyl-]	4,540 (10,000)
Vinyl acetate [acetic acid ethenyl ester]	6,800 (15,000)
Vinyl acetylene [1-buten-3-yne]	4,540 (10,000)
Vinyl chloride [ethene, chloro-]	4,540 (10,000)
Vinyl ethyl ether [ethene, ethoxy-]	4,540 (10,000)
Vinyl fluoride [ethene, fluoro-]	4,540 (10,000)
Vinylidene chloride [ethene, 1,1-dichloro-] (1,1-dichloroethylene)	4,540 (10,000)
Vinylidene fluoride [ethene, 1,1-difluoro-]	4,540 (10,000)
Vinyl methyl ether [ethene, methoxy-]	4,540 (10,000)

[a] Ref. 25.

of HAP emissions from mobile sources. Figure 3 shows that mobile sources contribute 40% of all HAP emissions in the United States. Since many of the VOC constituents of motor vehicle fuels are HAPs, the EPA expects that the motor vehicle regulations described above—motor vehicle control measures, fuel control measures, transportation control measures, refueling control measures, motor vehicle I&M programs, and clean fuel vehicle programs—will successfully reduce HAP as well as VOC emissions.

Air Permitting Rules

Before the 1990 Clean Air Act Amendments, the seven HAPs for which NESHAPs had been developed were subject to new source review. Thus new equipment installations or modifications were subject to BACT or LAER evaluations for HAP emissions control if the air emissions associated with the installation or modification exceeded *de minimis* emission levels. However, the 1990 Clean Air Act Amendments exempted HAPs from new source review requirements. Instead, the amendments established a separate preconstruction permit program for HAP emissions. This new program, known as the 112(g) program (referring to the section of the Clean Air Act where the statutory mandate for the program resides), requires new source MACT-level control of new and reconstructed HAP emission units (26). Where categorical MACT standards have not been developed, a new installation or reconstruction that triggers 112(g) permitting is subject to a case-by-case MACT determination; ie, a MACT evaluation must be performed (following the same general guidelines as BACT and LAER evaluations for new source review) and MACT must be proposed at the time of submittal of a preconstruction permit application even if MACT has not been established for the planned emission unit's industry source category.

Research Provisions

While the primary function of the EPA is to develop regulations to protect human health and the environment, an important secondary function is to perform and promote research to improve understanding of the health and ecological risks of pollution. The 1990 Clean Air Act Amendments include several required studies of HAP emissions. The EPA is required to perform studies on the following subjects:

HAP deposition into selected bodies of water (ie, the Great Lakes and the coastal waters).

HAP emissions from electric utilities, coke ovens, publicly operated treatment works (POTWs), oil and gas well and pipeline facilities, and hazardous waste facilities.

The urban HAPs problem (ie, area sources).

The EPA's risk assessment procedures and methods for estimating the health effects of HAPs.

Hydrofluoric acid hazards.

INNOVATIONS IN EPA RULE MAKING

Since the EPA's creation in 1970, substantial progress has been made in the development, implementation, and enforcement of VOC and HAP abatement programs to comply with legislative clean air mandates. The EPA strives to maintain its successful management of air quality programs by responding to constructive suggestions by regulated entities to improve on the current regulatory system. On March 16, 1995, President Clinton and Vice President Gore issued a report titled "Reinventing Environmental Regulation" that identified 25 initiatives to improve the federal government's environmental management systems. In response to this report and to other suggestions it has received, the EPA has developed or is in the process of developing the following innovative approaches to VOC and HAP rule making: integrated rules, consolidated rules, partnerships, emissions trading rules, compliance assurance monitoring rules, flexible compliance plans, and voluntary programs.

Integrated Rules

Integrated rule making refers to the EPA's initiative to streamline its rule making for a particular industry by combining information gathering tasks or by developing a multimedia rule. For example, the EPA intends to develop a MACT rule for the paper, film, and foil coatings industry (17,24). Although separate rule makings are required to meet the statutory bases for MACT (a HAPs program) and consumer and commercial products (a VOC program), the development of the individual rules will be integrated to maximize EPA resources and avoid duplication of data gathering efforts. As another example, the EPA intends to develop a multimedia rule for the pulp and paper industry under the MACT program of the U.S. Clean Air Act and the U.S. Clean Water Act. The regulations are being developed jointly to provide greater protection to human health and the environment, to promote the concept of pollution prevention, and to enable the pulp and paper industry to plan compliance more effectively.

The EPA also intends to integrate MACT rules for 21 miscellaneous organic chemical processes scheduled for finalization by November 15, 2000. Information collected by the EPA reveals that these organic processes involve similar process equipment and similar emission points and control equipment; also, many of these processes are co-located within individual facilities. Facilities with co-located organic chemical processes could more easily comply with a single set of emission standards than with individual standards for each of the chemical processes.

The EPA also intends to integrate MACT rules for fuel combustion sources such as boilers, process heaters, solid waste incinerators, stationary gas turbines, and stationary internal combustion engines (24). The EPA has established the Industrial Combustion Coordinated Rule making Advisory Committee, made up of EPA and industry representatives, to develop consistent MACT regulations with greater environmental benefits at a lower cost than regulating each source category individually. This integrated rule making will avoid the potential for conflicting or duplicative regulations.

Consolidated Rules

Consolidated rule making refers to the EPA's initiative to combine and simplify regulatory requirements in existing air rules that apply to a specific industry. Consolidated rule making is intended to eliminate duplicative, near duplicative, and overlapping federal air pollution requirements, resulting in a regulation that will be simpler to implement and comply with than the underlying individual rules. The EPA currently is working with the synthetic organic chemical manufacturing industry (SOCMI) and state regulators in Texas and Louisiana to develop a consolidated rule for the SOCMI, which currently is subject to VOC and HAP rules per the NSPS, NESHAP, MACT, and accidental releases programs.

Partnerships

Partnering is an initiative that evolved in response to the EPA's challenge to meet statutory rule-making deadlines with fewer resources (ie, labor and funding). By working with stakeholders (eg, regulated industries, state regulatory agencies, environmental advocacy groups) to develop air rules, the EPA hopes to fulfill its mandate to issue timely regulations, acceptable to all stakeholders, that provide adequate protection to human health and the environment. Two examples of partnering include the EPA's Common Sense Initiative and MACT partnerships.

In October 1994, the EPA established the Common Sense Initiative Council—made up of representatives from the EPA and the computer and electronics, printing, automobile manufacturing, iron and steel, and metals finishing industries—to find ways to improve rule making through increased coordination and public participation; to promote pollution prevention; to find easy ways to provide, use, and publicly disseminate relevant environmental information; to encourage innovation; and to provide incentives for regulated entities to develop technologies that exceed environmental standards cost-effectively.

MACT partnerships involve two phases for each MACT rule making. The first phase is the development of a presumptive MACT. This phase begins with a meeting among EPA and state and local regulatory agencies to establish presumptive MACT based on available information. The draft presumptive MACT then undergoes a review by industry and environmental groups. After this review, the EPA and the state and local regulatory agencies determine a final presumptive MACT. The second phase is the formal development of a MACT rule. The EPA may work with other regulatory agencies or regulated industries during this phase to develop the underlying data and analyses from which the EPA would determine appropriate MACT levels, or the EPA may opt for a more traditional rule making without outside support. The EPA has announced >25 MACT partnerships since this program was established in March 1995.

Emissions Trading Rules

Since the late 1970s, the EPA has developed several emissions trading approaches to provide industry with innovative and cost-effective compliance options for meeting their Clean Air Act requirements for VOC abatement. Three examples of long-established emission trading programs include bubbles, netting, and offsets. Bubbles are plantwide emission caps. Companies can decide which emission sources can be controlled most cost-effectively to meet a plant's emissions cap. Netting is an approach used in preconstruction permitting to bypass some costly and time-consuming steps in the permitting process by including net emission increases and decreases from contemporaneous process changes in determining the environmental significance of a new installation or modification. Offsets are emission reductions that are required before a new installation or process modification may be permitted in an ozone nonattainment area.

The EPA recently proposed additional emissions trading approaches designed to provide emission reduction benefits using market-based incentives. One approach, referred to as a cap and trade approach, establishes plantwide emission caps for groups of similar industries (eg, power plants, chemical plants) within an ozone nonattainment area. Plants that overcontrol emissions below their emissions cap can sell the excess emission credits to other plants that are not able to meet their cap.

The second approach, referred to as an open market trading rule (OMTR), was the number one initiative in the Clinton–Gore report for reinventing environmental regulations. The OMTR allows all types of sources to trade emissions of ozone precursors and will expedite the process for state and EPA approval of emission trades.

Compliance Assurance Monitoring Rules

Compliance assurance monitoring (CAM) rules were developed to address concerns of the EPA and state and local regulatory agencies that some sources of air pollution are noncompliant with emission control regulations because those regulations do not impose adequate requirements (eg, monitoring, testing, recordkeeping, reporting) to demonstrate compliance. CAM rules, which will be finalized in October 1997, will define what monitoring, testing, recordkeeping, or reporting procedures should be conducted by a facility to provide a reasonable assurance of compliance with air emission standards. The EPA intends to incorporate CAM procedures in all of its subsequent air rules.

Flexible Compliance Plans

The EPA has attempted to respond to concerns about the cost of complying with environmental regulations by designing flexible compliance plans. These plans, referred to in various rules as emissions averaging, or flexible compliance plans, allow facilities to undercontrol some emission points that are too costly to control to mandated levels as long as these units are balanced by overcontrolling other emission units that are more cost-effective to control. The EPA has permitted emissions averaging in many of the MACT standards, and it intends to include flexible compliance planning in the rules it develops for consumer and commercial products.

Voluntary Programs

The EPA promotes HAP emissions reduction not only through mandatory regulations but also through voluntary outreach programs. Three such programs are known as the 33/50 Program, the Early Reduction Program, and Project XL. The 33/50 Program was created by the EPA with the goal of achieving a nationwide 33% reduction (from 1988 levels) of 17 specific HAPs by 1992 and a 50% reduction by 1995. The response from industry was positive, and both goals have been met.

The Early Reduction Program was mandated in the 1990 Clean Air Act Amendments. As developed by the EPA, the program requires facilities subject to a MACT regulation to reduce HAP emissions by 90% (from 1990 levels) before the development of the MACT regulation by the EPA. In return for these early reductions in HAP emissions, a facility is granted a 6-year extension from compliance with the MACT standard after it is issued. This extension offers at least three benefits. First, it delays implementation of recordkeeping, reporting, monitoring, and testing requirements of a MACT regulation. Second, for facilities that achieve the early reductions through pollution prevention instead of installation of control equipment, it delays (and may permanently avoid) costly capital improvements to meet MACT. Third, it requires less than MACT-level HAP emission reductions (90% versus typical MACT levels of 98% HAP reduction), which allows more time to consider alternatives to expensive control equipment for meeting MACT.

The EPA's Project XL (eXcellence and Leadership) was announced on March 16, 1995 as part of the EPA's effort to reinvent environmental protection. The XL program allows a limited number of facilities to develop pilot projects to provide regulatory flexibility that results in environmental protection superior to current and reasonably anticipated future regulations. This program will enable the EPA to test new regulatory strategies that reduce regulatory burden and promote economic growth while achieving better environmental and public health protection. To participate in Project XL, applicants must develop alternative pollution reduction strategies pursuant to eight criteria: superior environmental performance; cost savings and paperwork reduction; local stakeholder involvement and support; test of an innovative strategy; transferability; feasibility; identification of monitoring, reporting, and evaluation methods; and avoidance of shifting risk burden. The EPA evaluates all results to see if any specific elements can be broadly applied to other regulated sources.

OTHER VOC AND HAP REQUIREMENTS

The U.S. Clean Air Act is not the only legislation that seeks to improve air quality. Other laws that contain provisions for controlling airborne emissions of VOC, HAP, and other pollutants include the Pollution Prevention Act (42 USC 13101–13109), the Comprehensive Environmental Response, Compensation, and Liability Act (CERCLA) (42 USC 9601–9675), the Emergency Planning and Community Right-to-Know Act (EPCRA) (42 USC 11001–11050), the Radon Gas and Indoor Air Quality Research Act (Public Law 99–499, Title IV), and the Resource Conservation and Recovery Act (RCRA) (42 USC 6901–6992k).

Under the Pollution Prevention Act, the EPA has established the Pollution Prevention Office to supplement the development, implementation, and enforcement of environmental regulations using waste reduction strategies. Under this act, facilities are required to provide pollution prevention information annually to the EPA when they file annual toxic chemical release reports under § 313 of EPCRA. Pollution prevention strategies may include equipment, process, or procedure modifications; reformulation or redesign of products; substitution of raw materials; and improved management, training, inventory control, or materials handling. This act declares a national policy of preventing pollution whenever feasible. Heeding this policy, the EPA incorporates pollution prevention as MACT for several regulated source categories (eg, industrial process cooling towers).

CERCLA is commonly referred to as Superfund; CERCLA was amended in 1986 by the Superfund Amendments and Reauthorization Act (SARA). The act established federal authority to respond to releases or threats of releases of listed hazardous substances from vessels and facilities. The list of hazardous substances in this act include VOCs and all of the HAPs listed in Table 2. Any federally unpermitted releases (eg, pressure relief system releases, malfunctions) of any of the substances above a listed threshold quantity must be reported to the federal government's National Response Center and potential injured parties. (Although not required, notification to state and local regulatory agencies is recommended.) Reports must include the name and estimated amount of the hazardous substance released, the location of the release, and corrective actions taken to minimize the release's effects.

EPCRA was passed as part of the Superfund amendments to CERCLA, although it constitutes a free-standing act. It was developed to assist state and local governments in chemical emergency planning and to provide the public with information on the type, amount, location, use, disposal, and release of chemicals at certain facilities. The act requires facilities that store or use listed hazardous substances exceeding threshold quantities to notify state emergency response commissions and participate in local emergency planning activities. Many of the listed substances are VOCs or HAPs. Releases of any of the substances in excess of listed quantities must be reported to the emergency planning commissions. Furthermore, EPCRA requires annual reporting of the maximum amount of hazardous chemicals present at a facility during the preceding year and the location of the chemicals at the facility. Finally, EPCRA requires annual reporting of routine emissions (including air emissions) of toxic chemicals from a second list of chemicals, many of which are VOCs or HAPs. As required by the Pollution Prevention Act, this report also must contain pollution prevention information requested by the EPA. The EPA uses all of the information it collects from these annual reports to maintain a Toxics Release Inventory (TRI) database that is accessible to the general public.

The Radon Gas and Indoor Air Quality Research Act also was passed as part of the Superfund amendments to CERCLA, although it constitutes a free-standing act (see RADON IN BUILDINGS). The act recognizes the public health risks attributable to radon and indoor air pollutants. Indoor air quality issues are being addressed by both the EPA and OSHA, which focuses on workplace indoor air quality issues. Indoor air pollutants, many of which are VOCs or HAPs, may be created by many factors, including microbial growth in old or poorly maintained building ventilation systems, natural gas fumes, and the use of certain consumer and commercial products (17). OSHA has been directed to study the occupational health hazards of poor indoor air quality and environmental tobacco smoke; indoor air pollutants have been linked to a variety of health hazards, including heart disease, lung cancer, respiratory disease, Legionnaire's disease, and asthma (27). The EPA has established a research program to understand the health effects of indoor air pollution and radon and has issued public information brochures about indoor air pollution.

RCRA authorizes the EPA to issue regulations to control hazardous wastes from their generation to ultimate disposal (cradle to grave). The EPA is required to determine which hazardous wastes are subject to regulation by identifying hazardous waste characteristics (eg, toxicity, bioaccumulation, flammability, corrosiveness) and by listing specific substances as hazardous wastes. The act regulates air emissions from hazardous waste incinerators; boilers and industrial furnaces (BIFS) that burn hazardous wastes; hazardous waste treatment, storage, and disposal facilities (TSDF); and nonhazardous solid waste landfills. RCRA also requires the EPA to develop regulations for underground storage tank leak detection, prevention, and cleanup.

In addition to the laws mentioned in this section, numerous other U.S. environmental laws include clean air provisions among their broader mandates. A few of these laws include the Toxic Substances Control Act (TSCA) (15 USC 2601–2692), the National Environmental Policy Act (NEPA) (42 USC 4321–4370d), and the Environmental Quality Improvement Act (42 USC 4371–4375).

STRATOSPHERIC OZONE DEPLETING COMPOUNDS

Many of the nonphotochemically reactive compounds listed in Table 1 are stratospheric ozone depleting compounds. They do not contribute significantly to ground-level ozone problems, but they do impact ozone levels in the earth's stratosphere, which begins approximately 15 km (9 miles) above the earth's surface and extends to approximately 50 km (30 miles) above the earth's surface. While tropospheric ozone is undesirable because it contributes to photochemical smog, stratospheric ozone is desirable because of its protective effect against harmful ultraviolet radiation from the sun. Exposure to excessive levels of ultraviolet radiation is known to cause skin cancer, retinal damage, and crop yield reductions (28,29).

Stratospheric ozone depletion occurs through the following series of chemical reactions:

$$Cf_xCl_{4-x} + \text{Energy} \rightarrow CF_xCl_{3-x} + Cl \quad (7)$$

$$Cl + O_3 \rightarrow ClO + O_2 \quad (8)$$

$$ClO + O \rightarrow Cl + O_2 \quad (9)$$

$$O_3 + O \rightarrow 2O_2 \quad (10)$$

Chlorofluorocarbons (CFCs) play a significant role in stratospheric ozone depletion. CFCs have been widely used as propellant gases in aerosol spray cans and as refrigerant gases. Although they are emitted into the troposphere, CFCs are able to migrate into the stratosphere unreacted because of their chemical inertness. In the stratosphere, CFCs are bombarded with high energy photons from the sun (eq. 7). These high energy collisions cause chemical bonds within the CFCs to rupture, creating highly reactive atomic chlorine. The atomic chlorine reacts rapidly with ozone (eq. 8) to form chlorine oxide and molecular oxygen. Chlorine oxide reacts with atomic oxygen to reform atomic chlorine (eq. 9). The net reaction—Equation 10—is a chemical conversion of ozone to oxygen, thus depleting the supply of stratospheric ozone.

Hydrochlorofluorocarbons (HCFCs), which are used as CFC substitutes, also play a role in stratospheric ozone depletion. However, the ozone-depleting potential of HCFCs is lower than for CFCs. Numerous studies of stratospheric ozone levels have failed to establish a quantitative effect of CFCs on stratospheric ozone depletion (30). Additional research on this issue is ongoing.

Requirements for stratospheric ozone depleting chemicals began on an international scale in September 1987, when 150 nations and other interested parties gathered in Montreal, Canada, to sign the Protocol on Substances that Deplete the Ozone Layer. Known as the Montreal Protocol, this international treaty established a schedule for phasing out specific stratospheric ozone-depleting substances. The protocol was modified in London in 1990 and Copenhagen in 1992 to add more stratospheric ozone-depleting substances and to accelerate the phase-out schedule.

In the United States, the 1990 Clean Air Act Amendments (Title VI) provide the legislative mandate to implement the provisions of the Montreal Protocol. The EPA's strategy to regulate stratospheric ozone-depleting substances consists of the following programs (31):

- Production and consumption controls on stratospheric ozone-depleting substances. The EPA developed two classes of ozone-depleting substances based on each substance's ozone depletion potential. Class I substances consist mainly of chlorofluorocarbons and other substances with high ozone-depleting potential; a current

Table 8. Class I Stratospheric Ozone-Depleting Substances[a]

Group[b]	Chemical Name
I	Chlorofluorocarbon-11 (CFC-11)
	Chlorofluorocarbon-12 (CFC-12)
	Chlorofluorocarbon-113 (CFC-113)
	Chlorofluorocarbon-114 (CFC-114)
	Chlorofluorocarbon-115 (CFC-115)
II	Halon-1211
	Halon-1301
	Halon-2402
III	Chlorofluorocarbon-13 (CFC-13)
	Chlorofluorocarbon-111 (CFC-111)
	Chlorofluorocarbon-112 (CFC-112)
	Chlorofluorocarbon-211 (CFC-211)
	Chlorofluorocarbon-212 (CFC-212)
	Chlorofluorocarbon-213 (CFC-213)
	Chlorofluorocarbon-214 (CFC-214)
	Chlorofluorocarbon-215 (CFC-215)
	Chlorofluorocarbon-216 (CFC-216)
	Chlorofluorocarbon-217 (CFC-217)
IV	Carbon tetrachloride
V	Methyl chloroform (1,1,1-trichloroethane)
VI	Methyl bromide (bromomethane)
VII	Hydrobromofluorocarbons

[a] Ref. 31.
[b] Producers of these chemicals are given allowances that authorize production of a given amount of these chemicals. Companies are permitted to trade allowances, but the trades must occur within the groups listed.

list of class I substances is shown in Table 8. Class II substances consist of hydrochlorofluorocarbons; a current list of class II substances is shown in Table 9. The production and consumption of class I and II substances will be phased out starting in 1996. Producers of these substances are given allowances that authorize production of a given amount of the substances. Consistent with EPA's innovative market-based strategies, companies are permitted to trade allowances under certain constraints.

- Servicing of motor vehicle air conditioners. The EPA requires the use of approved CFC recycling equipment and technician training and certification for repairing motor vehicle air-conditioners.
- Ban on nonessential products containing or manufactured with stratospheric ozone-depleting substances. The EPA prohibits the use of CFCs and HCFCs in nonessential products, which include plastic party streamers and noise horns, cleaning fluids for electronic and photographic equipment (some exceptions apply), plastic flexible or packaging foam products (some exceptions apply), and aerosol dispensers (some exceptions apply).
- Federal procurement provisions. The EPA requires federal departments and agencies to adopt procurement regulations that conform to the policies and requirements of the stratospheric ozone-depleting substances program.
- Product labeling requirements. The EPA requires warning labels on containers of, and products containing or manufactured with, certain ozone-depleting substances. The required warning label must state that the product contains a substance that harms public health and environment by destroying ozone in the upper atmosphere.
- Recycling and emissions reduction. The EPA requires safety practices, technician training and certification programs, and leak detection and repair programs by any person servicing, maintaining, or repairing refrigerated appliances and air-conditioners. The EPA requires the removal of ozone-depleting substances from refrigerated appliances and air-conditioners before their disposal. The EPA also requires that manufacturers of refrigerated appliances and air conditioners equip the appliances with a servicing aperture to facilitate the recovery of refrigerants.
- Significant new alternatives policy. The EPA established a program "to identify substitutes for ozone-depleting compounds, to evaluate the acceptability of those substitutes, to promote the use of those substitutes believed to present lower overall risks to human health and the environment, relative to the Class I and Class II compounds being replaced, as well as to other substitutes for the same end use, and to prohibit the use of those substitutes found, based on the same comparisons, to increase overall risks" (31).

Table 9. Class II Stratospheric Ozone-Depleting Substances[a]

Substance
Hydrochlorofluorocarbon-21 (HCFC-21)
Hydrochlorofluorocarbon-22 (HCFC-22)
Hydrochlorofluorocarbon-31 (HCFC-31)
Hydrochlorofluorocarbon-121 (HCFC-121)
Hydrochlorofluorocarbon-122 (HCFC-122)
Hydrochlorofluorocarbon-123 (HCFC-123)
Hydrochlorofluorocarbon-124 (HCFC-124)
Hydrochlorofluorocarbon-131 (HCFC-131)
Hydrochlorofluorocarbon-132b (HCFC-132b)
Hydrochlorofluorocarbon-133a (HCFC-133a)
Hydrochlorofluorocarbon-141b (HCFC-141b)
Hydrochlorofluorocarbon-142b (HCFC-142b)
Hydrochlorofluorocarbon-221 (HCFC-221)
Hydrochlorofluorocarbon-222 (HCFC-222)
Hydrochlorofluorocarbon-223 (HCFC-223)
Hydrochlorofluorocarbon-224 (HCFC-224)
Hydrochlorofluorocarbon-225ca (HCFC-225ca)
Hydrochlorofluorocarbon-225cb (HCFC-225cb)
Hydrochlorofluorocarbon-226 (HCFC-226)
Hydrochlorofluorocarbon-231 (HCFC-231)
Hydrochlorofluorocarbon-232 (HCFC-232)
Hydrochlorofluorocarbon-233 (HCFC-233)
Hydrochlorofluorocarbon-234 (HCFC-234)
Hydrochlorofluorocarbon-235 (HCFC-235)
Hydrochlorofluorocarbon-241 (HCFC-241)
Hydrochlorofluorocarbon-242 (HCFC-242)
Hydrochlorofluorocarbon-243 (HCFC-243)
Hydrochlorofluorocarbon-244 (HCFC-244)
Hydrochlorofluorocarbon-251 (HCFC-251)
Hydrochlorofluorocarbon-252 (HCFC-252)
Hydrochlorofluorocarbon-253 (HCFC-253)
Hydrochlorofluorocarbon-261 (HCFC-261)
Hydrochlorofluorocarbon-262 (HCFC-262)
Hydrochlorofluorocarbon-271 (HCFC-271)

[a] Ref. 31.

BIBLIOGRAPHY

1. *Code of Federal Regulations,* Title 40, Part 51, § 100(s).
2. U.S. Environmental Protection Agency, "Air Quality Criteria for Ozone and Related Photochemical Oxidants," *EPA/600/P-93/004aF,* Washington, D.C., July 1996.

3. National Research Council, *Rethinking the Ozone Problem in Urban and Regional Air Pollution,* National Academy Press, Washington, D.C., 1991.
4. In Ref. 1, Part 63, § 2, 1996.
5. Clean Air Act, §112b, 42 USC 7412b.
6. A. S. Kao, *J Air Waste Manage. Assoc.,* **44,** 683–696 (1994).
7. U.S. Environmental Protection Agency, "Technical Background Document to Support Rulemaking Pursuant to the Clean Air Act Section 112(g). Ranking of Pollutants with Respect to Hazard to Human Health," *EPA-450/3-92-010.*
8. U.S. Environmental Protection Agency, "Hazardous Air Pollutants: Profiles of Noncancer Toxicity from Inhalation Exposures," *EPA/600/R-93/142,* Sept. 1993.
9. U.S. Environmental Protection Agency, "National Air Pollutant Emission Trends, 1900–1995," *EPA-454/R-96-007,* RTP, N.C., Oct. 1996.
10. W. D. Ruckelshaus, Keynote address presented at the Air Pollution Control Association's Second International Clean Air Congress, Dec. 1970.
11. Ref. 5, Title I, §108(a)(1)(A).
12. Ref. 5, Title I, § 109 (b)(1).
13. Ref. 5, Title I, § 109 (b)(2).
14. Ref. 5, Title III, § (h).
15. *Federal Register,* **44,** 53762 (Sept. 17, 1979).
16. *Federal Register,* **61,** 4531–14543 (Apr. 2, 1996).
17. *Federal Register,* **60,** 15264–15268 (Mar. 23, 1995).
18. Ref. 5, Title I, § 111 (b)(1)(A).
19. Title I, § 111 (a)(1).
20. Ref. 1, Part 60, 1996.
21. U.S. Environmental Protection Agency, "Transportation Control Measure Information Document," *EPA/400-R-92-006,* Ann Arbor, Mich., Mar. 1992.
22. Ref. 5, Title II, § 241 (2).
23. Ref. 1, Part 61, 1996.
24. Ref. 1, Part 63, 1996.
25. Ref. 1, Part 68. 130, 1996.
26. *Federal Register,* **61,** 68384–68404 (Dec. 27, 1996).
27. *Federal Register,* **59,** 15968–16039 (Apr. 5, 1994).
28. A. Jeevan and M. L. Kripke, *World Resource Rev.,* **5** 141–155 (1993).
29. R. M. Adams and R. D. Rowe, *J. Environ. Manage.,* **30,** 321–335 (1990).
30. F. S. Rowland, *Ambio,* **19,** 281–292 (1990).
31. Ref. 1, Part 82, 1996.

CLIMATE–BIOSPHERE INTERACTIONS

JOHN E. OLIVER
Indiana State University
Terre Haute, Indiana

YUK YEE YAN
Hong Kong Baptist University
Hong Kong

ENERGY FOR THE SYSTEM

Solar Radiation

All bodies above absolute zero (degrees K) radiate energy in the form of electromagnetic waves. Using wavelength as a criterion, radiant energy exists along a spectrum from very short to very long waves (Fig. 1). The characteristics of radiation emitted by an object are determined by its temperature. Hotter objects radiate more energy per unit area than colder objects. This relationship is expressed by the Stefan-Boltzmann law, which is given as follows:

$$F = \sigma T^4$$

where F is flux of radiation in langleys per minute or W/m^2, T is absolute temperature in K, and σ is a constant (0.813×10^{-10} cal/cm^2/min or 5.67×10^{-8} W/m^2 Ks).

The wavelength of maximum emission for an object at a particular temperature is inversely proportional to its temperature, ie, the hotter the object, the shorter the wavelength of its maximum radiation. It is expressed by Wien's displacement law:

$$\lambda_{\max} = \alpha / T$$

where $\lambda_{\max}$ is the wavelength of maximum emission, α is a constant that equals 2897, and T is temperature in K.

The surface temperature of the sun is about 6000 °K (11,000°F). The sun radiates at a peak wavelength of 5 μm which is primarily in the visible portion of the spectrum (see Fig. 1). The earth, on the other hand, with the surface temperature at about 288 °K (59°F), emits radiation in wavelengths longer than 4 μm, predominately in the infrared spectrum with a peak near 10 μm. For descriptive purposes, solar radiation is referred to as shortwave and terrestrial radiation as longwave.

The Solar Constant

The sun radiates about 100,000 cal/cm^2/min. Because the distance between the sun and earth is 149.5 million km (93 million miles), the earth receives only a minute fraction of the sun's energy (1/2,000,000,000). The amount of solar radiation received on a surface held perpendicular to the sun's rays at the outer edge of the atmosphere is called the solar constant. Although observations of the solar constant have been made, its value is not precisely known. The current estimate is 1370 W/m^{-2} (1.96 cal/cm^2/min), a value that fluctuates by $\pm 1.5\%$ about its mean because of varying earth–sun distances (1). Variations of the solar constant over longer time periods have been related to climatic change (2).

Figure 2 summarizes the astronomical features responsible for the annual variation of the solar constant. Revolving around the sun, the earth is at its closest position to the sun on January 3 (perihelion) and at its farthest position on July 4 (aphelion). This annual distance, varying by 1.7% leads to variations in solar constant of 45 W/m^{-2} with a maximum at perihelion and a minimum at aphelion (4).

The tilt of the earth's axis causes variations in the length of day and night. At extreme positions, the noontime sun is over the Tropic of Cancer and Tropic of Capricorn, creating the summer and winter solstices, respectively. At the former, there is 24-h daylight in areas north of the Arctic Circle and 24-h darkness in areas south of the Antarctic Circle. Long hours of darkness decrease the amount of radiation. As a result, seasonal variations in the length of day and night over the earth's surface are important in determining the radiation received.

The altitude of the sun in the sky and latitude also determine the amount of solar radiation received. The intensity of

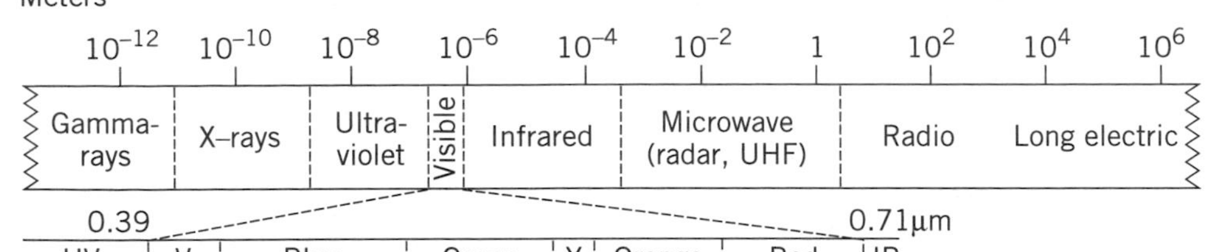

Figure 1. The electromagnetic spectrum, where UV is ultraviolet, V is violet, Y is yellow, and IR is infrared.

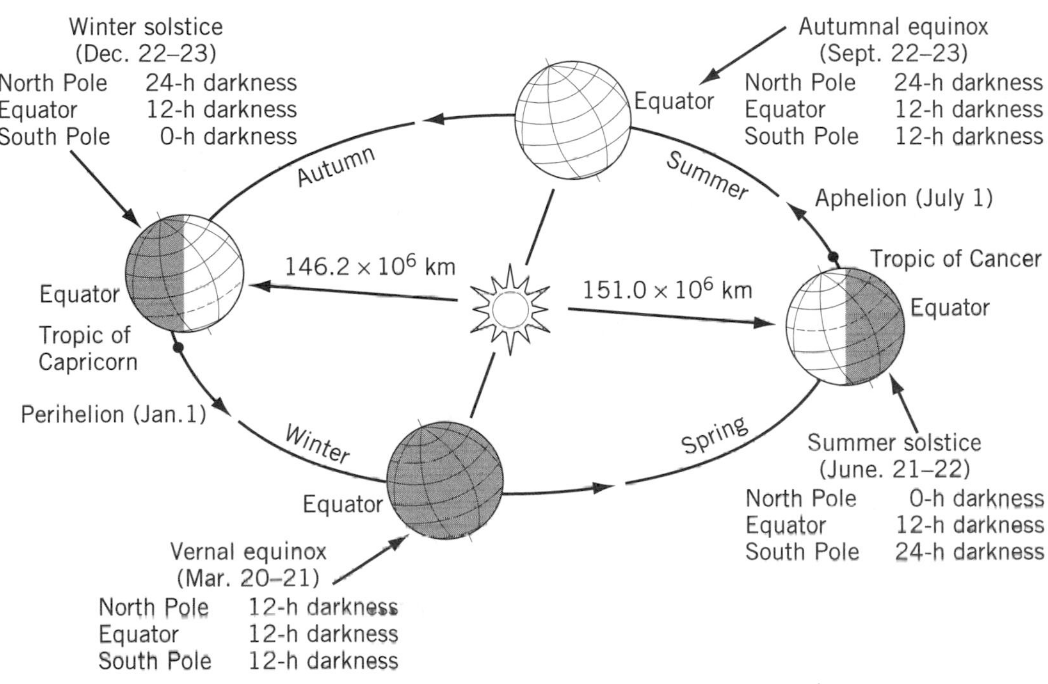

Figure 2. Orbit of the earth (1 orbit-revolution = 365.5 d; 1 rotation = 24 h) around the sun, the reasons, and changes in length of day and night (3).

incoming solar energy received on a horizontal surface at the top of the atmosphere, I_h, is related to the incident radiation, I_0, when the sun is at an elevation of 90°. The relationship is given by the following:

$$I_h = I_0 \sin a$$

where a is the sun's altitude. If the sun is directly overhead, ie, $a = 90°$, sine 90° is 1, and thus I_h equals I_0. The sun's altitude varies with latitude, declination of the sun (sun's angular distance north or south of the equator), and hour angle. In relation to the latter, the hour angle is zero when the sun is at solar noon (when the sun is directly north or south of the observation point) and increases by 15° for every hour before or after solar noon. Accordingly, $\sin a$ can be computed from the following equation:

$$\sin a = \sin\phi \sin\delta + \cos\phi \cos\delta \cos h$$

where ϕ is the latitude, δ is declination, and h is hour angle.

The result of the interplay of these factors is shown in Figure 3, which gives the latitudinal distribution of solar radiation at the top of the atmosphere. Least annual variation is found in the equatorial regions; values are most extreme at the poles. In winter, no radiation is received, whereas in summer the values are quite large despite the low angle of the sun.

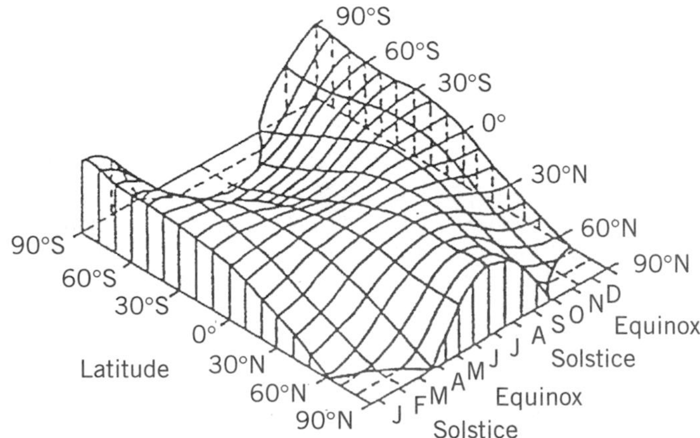

Figure 3. Radiation at the top of the atmosphere (5).

Energy Flow Through the Atmosphere

As solar radiation passes through the atmosphere, some of the solar beam is modified by reflection, scattering, and absorption so that atmospheric transmission of energy varies over time and space. That part of the solar radiation which is reflected back to space by clouds and the earth's surface is called the albedo, expressed as the percentage of incident radiation reflected. The albedo of clouds ranges from 40 to 90%, depending on types and thickness. The earth's surface reflectivity depends on the nature of the surface cover.

A small portion of solar radiation is scattered back to space by atmospheric particles. A greater amount is scattered downward and reaches the earth's surface as diffuse shortwave solar radiation. The amount and direction of scatter depends on the ratio of scattering particles to the wavelength of the energy. Minute particles produce Rayleigh scattering, in that the amount of scatter is inversely proportional to the fourth power of the wavelength. It is the primary cause of the blue sky and is the dominant scattering process in the upper atmosphere. Larger particles with sizes roughly equal to the wavelength of the radiation, such as dust and water droplets, cause Mie scattering, which is greatest in the lower atmosphere (4).

Ozone absorbs ultraviolet radiation effectively whereas carbon dioxide absorbs the mid- and far-infrared range, with the strongest absorption at 13–17.5 μm in the far-infrared. Water vapor absorbs the bands between 5.5 and 7.0 μm, and above 27 μm. Because of differential absorption, the atmosphere is not transparent to all radiation. Wavelengths that are easily transmitted through the atmosphere are called atmospheric windows.

Gases in the atmosphere, notably carbon dioxide and water vapor, absorb longwave terrestrial radiation across broad bands. The gases then reradiate the absorbed energy, some passing to space and some (counterradiation) returning to the earth's surface. This process is known as the greenhouse effect and the gases as greenhouse gases.

Transmission of radiation occurs when radiation passes through the atmosphere. Transmissivity depends on the distance the sun's rays pass through the atmosphere and the state of the atmosphere. The relative distance the sun's rays travel through the atmosphere is optical air mass, which equals 1 when the sun is overhead, or 90° above the horizon. Optical air mass increases as the angle of the sun above the horizon decreases. Solar beams travel a longer distance through the atmosphere at higher latitudes than that at lower latitudes. The larger the optical air mass, the less intense the solar radiation received at the surface. A polluted atmosphere transmits solar radiation less effectively than a clear one. It is found that atmospheric transmittance is about 0.8 in clear and dry skies but only 0.4 in a polluted atmosphere (7). The presence of clouds also influences transmission. Transmission of solar radiation through nimbostratus can be less than 10%, whereas transmission through cirrus can be 60% (8).

Earth's Energy Balance

Averaged over the earth's surface, the energy balance for the net exchange between incoming and outgoing radiation can be represented by the following equation:

$$Q^* = (K\downarrow - K\uparrow) + (L\downarrow - L\uparrow)$$

where Q^* is net radiation, $(K\downarrow - K\uparrow)$ is the sum of direct and diffuse solar radiation less outgoing shortwave radiation, and $(L\downarrow - L\uparrow)$ is incoming longwave radiation less outgoing longwave.

The energy balance of the earth is presented in Figure 4. The percentages of total annual energy inflow reflected, scattered, absorbed, and transmitted are shown. Approximately 20% of the solar energy is absorbed by the atmosphere, whereas the average albedo of the earth–atmosphere system is ca 30%.

Although global radiation is balanced over the year, it is not balanced over both time and space. The highest input values are found in low latitudes while the lowest values occur in the polar regions. Because there is spatial and temporal imbalance of the energy budget, energy transfers are required to maintain a balanced budget at the earth's surface. They are represented in the heat budget equation:

$$Q^* = H + LE + \Delta F + G$$

where Q^* is the net radiation, H is the net-sensible heat transfer between surface and atmosphere by conduction, LE is the latent heat of vaporization, ΔF is the horizontal heat transfer, and G is the flux of heat between the surface and soil or water. Over land, horizontal heat transfer, ΔF, is negligible and can be omitted. It is only important over water bodies, where currents can transport heat energy. In treating the equation on an annual basis, the flux changes in G can be excluded, and thus the equation becomes the following:

$$Q^* = H + LE$$

It is found that LE is greatest at the equator, and LE values are twice as high over oceans as over land. Sensible heat, H, is highest over land and decreases with latitude. Over the ocean, values increase with latitude (9).

CLIMATIC VARIABLES AND PLANTS

Climate exerts a significant influence on plant life. This intimate climate–plant relationship has been recognized and examined since Menestor in the fifth century BC (10). The association between climate and plant can be assessed by evaluating the role of various climatic elements in plant life.

Energy and Sunlight

Solar energy is converted to chemical energy by plants through photosynthesis. This conversion represents that part of solar radiation that enters the biosphere. Photosynthesis is most efficient within the 0.4 to 0.7-μm spectral bands known as photosynthetically active radiation (PAR). In the photosynthesis process, carbon dioxide diffuses through stomata into the leaves and with water in the cell walls to form carbohydrates or dry matter. The overall reaction can be represented by the following:

$$6\ CO_2 + 6\ H_2O + \text{solar energy} \rightleftarrows C_6H_{12}O_6 + 6\ O_2$$

The work of reaction between light and plants is termed quantum efficiency or quantum yield. Light is transmitted in discrete units called quanta and the energy of a quantum is a photon, which depends on the frequency of electromagnetic wave. Because frequency of spectrum is inversely related to wavelength, longer wavelengths have less energy per quantum than shorter wavelengths. Quantum energy is absorbed and utilized by the pigments of plant cells to initiate photosynthesis. Quantum efficiency measures the ability of photons to produce chemical change, and is also the number of O_2 molecules evolved (or CO_2 fixed) per quantum of light absorbed. This work is measured as photon flux in moles

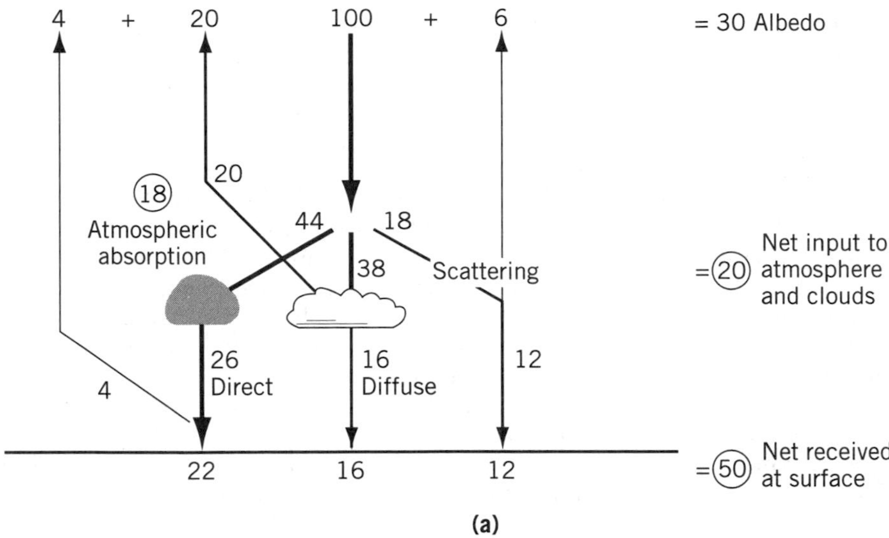

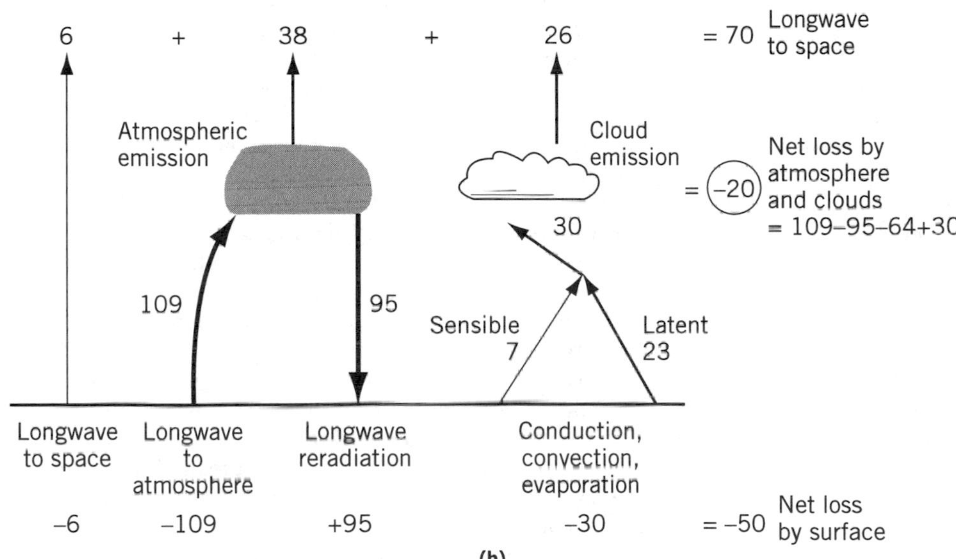

Figure 4. Energy balance of earth where (**a**) is the shortwave (visible) and (**b**), the language (infrared). The solar irradiance is shown as 100 units (5).

of photons (mol m^{-2} s^{-1}), which is the number corresponding to Avogadro's number of particles (6.023×10^{23}) and can be considered as the number of photons required to convert a mole of a matter to another form with 100% efficiency. A mole of photon is also called an einstein which is not compatible with SI units.

It is evident that sunlight is a basic requirement for photosynthesis. Its rate depends on the amount of light available. There is a lower limit of light intensity, or a compensation point, at which, assuming attainment of required temperature and moisture conditions, photosynthesis begins. There is also an optimum intensity or saturation point at which photosynthesis reaches a maximum rate.

Plants can be classified according to their relative requirement of direct sunlight. Those that grow best in high light intensity are heliophytes, and those that tolerate or perform best at low light intensity are sciophytes. Most trees, grasses, and the majority of arable crops are heliophytes. Mosses and ferns are sciophytes. However, there are facultative sciophytes that can grow in dim light but attain maximum development under high intensity, whereas obligate sciophytes grow best in shade.

Seasonal and diurnal changes in the duration of light also affect plant growth. Photoperiodism, the response of plants to length of daylight, is an important stimulus in plants' physiological rhythms. There are four basic daylight plant groups: short-day plants flower only when daily light period is <12 h; long-day plants flower when daylight is >12 h; daylength-intermediate plants flower only within a narrow range of daylight, usually between 10 and 14 h; and day-neutral plants continue growth regardless of daylight period.

Temperature and Moisture

Temperature and moisture are closely studied aspects of the climate–plant relationship. The interrelationship between these two factors and other environmental elements greatly affect plant growth (11).

Temperature exerts a determining effect on the distribution and physiological processes of plants. Plants can be categorized according to their temperature requirements. Temperature controls the rate of biological processes in plants. Both respiration and photosynthesis increase as temperature rises. High night temperature increases respiration rate whereas high day temperature raises photosynthesis rate. However, photosynthesis rate decreases when temperature exceeds a threshold that varies with plant species (generally 30–37°C (86–98°F)). This deceleration of the photosynthesis rate at a higher temperature is caused by the limiting effect of light reactions. A further rise in temperature leads to a decline in photosynthesis, which results from thermal inactivation of enzymes.

All plants have a minimum, optimum, and maximum temperature for growth (10). When the minimum or maximum threshold temperature is exceeded, plant injury occurs. Excessive heat can result in leaf burn, destruction of protoplasm at 54°C (130°F), and can also lead to rapid moisture loss and then wilting. Under such conditions, respiration exceeds photosynthesis, and a metabolic imbalance results. The plant will die if such an imbalance is sustained.

Low temperature can also result in destruction of plant life. Rapid temperature reduction $<0°C$ freezes plant cells. Slow freezing results in the formation of intercellular ice that removes liquid water from the leaf. Because vapor pressure over ice is lower than that of water at the same temperature, additional water is removed from the cell at the site of cellular ice, leading to cell dehydration. The proportion of leaf water remaining unfrozen decreases hyperbolically below the temperature at which extracellular ice is formed (12). Frost damage can also occur in plants when intercellular water freezes resulting in cell dehydration. A sudden warm spell during a cold period can result in transpiration, but water used is not replenished and as temperature falls again, winter burn may occur.

Plants react and tolerate thermal conditions differently. In response to low temperatures, plants undergo some fundamental biological changes including increases in unsaturated lipids, accumulation of water-soluble solutes (eg, sugar), and accumulation of membrane proteins. It is believed that plants can survive low temperatures by undergoing energy-requiring processes related to the protection of cell membranes (12). Most plants cease activities when soil temperature falls below 5°C (42°F). If soil temperature is low, the rate of moisture intake through soil decreases, and plants cannot restore water lost by transpiration. High temperatures can limit plant growth but seldom directly induces death. Increased evapotranspiration caused by heat leads to dehydration of plant cells.

Moisture is essential for plant growth and development because it is the medium by which chemicals and nutrients are conveyed through plants. Moisture is required for efficient metabolism so that plants are kept turgid. It is also a reagent in photosynthesis and a major product of respiration. Few plants can survive if their cellular water drops below 10% of the saturation level, and most plants die if cellular water falls below 50% of the saturation level.

The immediate source of moisture is soil, although mosses and lichens can absorb moisture directly from the atmosphere. The amount and availability of soil moisture are associated with precipitation, surface drainage conditions, ability of soil to retain water, and losses due to evaporation. Insufficient moisture limits plant growth whereas excessive amounts can result in waterlogging and oxygen deficiency. The rate at which plants absorb water is determined by the temperature and chemical composition of the soil solution. Low temperatures, high acidity, alkalinity, and salinity all can reduce the rate of absorption. High evapotranspiration can cause water stress, which results in temporary wilting or reduced growth. Plants die with sustained water stress.

Because moisture is not always available in the same amount at various locations, plants have adapted in a various ways to water availability. Some plants thrive in the condition of moisture surplus; others can survive extreme drought. There are four different groups of plants classified according to moisture requirement: hydrophytes grow in watery environments; hygrophytes exist in wet, marshy conditions; xerophytes grow in arid areas; and mesophytes are found in regions with neither excessive nor deficient water.

Other Climatic Variables

Wind influences plants directly by causing physical damage. Trees are blown over, leaves shredded, or stems twisted and broken. Abrasion by windblown sand, gravel, or ice particles can also damage plants. Wind can have further significant indirect effects on plants. For example, wind affects plant temperature through convective heat transfer influencing plant leaf temperatures. Wind also plays a role in controlling transpiration as transpiration rates increase with increasing wind speed. Carbon dioxide intake of plants is also a function of wind speed. Additionally, wind can dry out the soil, leading to a lack of available soil moisture. Desiccation of plants is another result of the influence of wind and with rapid loss of water, some trees lack turgidity and are unable to grow to normal size. Extreme effects can cause morphological changes in plants, such as deformation of canopy in the direction of a prevailing wind or lack of branch development on the windward side. Wind can, however, act positively by assisting the spread of plant species and is a prominent agent of seed dispersal. The transport of atmospheric pollutants, however, to which plants are sensitive, is also wind-dependent. Pollutants such as ozone and oxidants may cause yellowing of leaves or severe leaf burn.

Humidity affects transpiration rate, latent heat transfer, and certain diseases. Most plants grow at relative humidities between 20 and 80% if they have a well-developed root system. Transpiration is controlled by a plant's stomatal openings, air velocity, and vapor pressure difference between the leaf and air. Low vapor pressure induces an increase in transpiration rate. Very low humidity can cause wilting because evaporation losses are higher than the plant can replenish. The effect of relative humidity on leaf temperature depends on air temperature as well. Relative humidity between 40 and 80% has little effect on leaf temperature at 20°C (68°F) but has significant effect at −12°C (10°F). High humidity provides a good environment for pathogenic organisms. Many pathogenic spores germinate when relative humidity is 96% or higher. Other pathogenic organisms may grow at a somewhat lower relative humidity.

Humidity has variable significance depending on the type of plant and the soil moisture available to it. Many plants can withstand low humidities as long as their roots are supplied with adequate moisture. Xerophytes of arid regions adapt to limited moisture in several ways. Some have thick waxy bark and leaves that inhibit transpiration loss, whereas others have

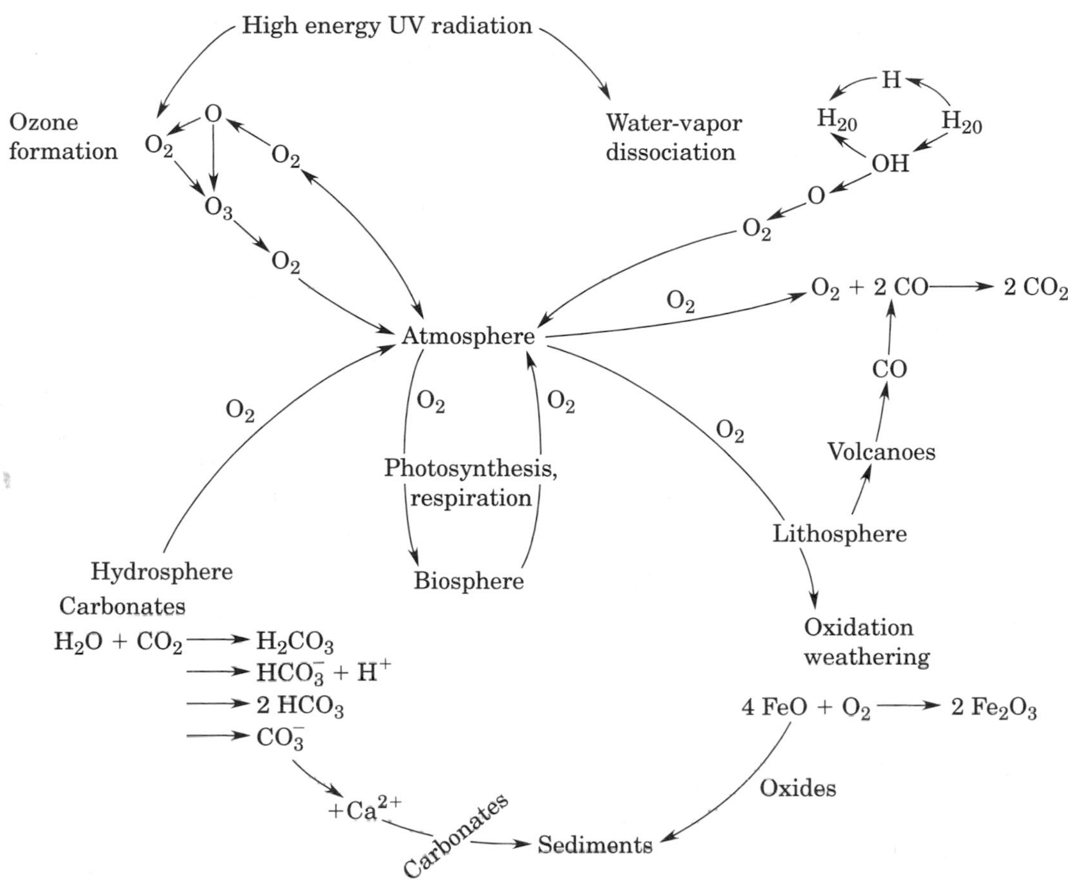

Figure 5. The oxygen cycle. Complexities occur because oxygen appears in so many forms and is interrelated to other cycles (14).

far-reaching root systems situated deep in the soil to gather moisture or, like cacti, store water during wet periods for utilization in a subsequent dry season (13).

CLIMATE AND BIOGEOCHEMICAL CYCLES

Plants do not respond only to climatic elements, but also various environmental factors. To examine the relationship between plants and the physical environment it is essential to understand all the processes, including those derived through the analysis of biogeochemical cycles. Biogeochemical cycles describe the global transfer and transformation of chemical substances. Although climate plays a substantial part in many of these cycles, clearly its role is most evident in those cycles with a gaseous phase.

Oxygen makes up 21% of the atmosphere. It occurs as dissociated atomic oxygen, O; molecular oxygen, O_2; and ozone, O_3, in the atmosphere. Figure 5 indicates the global oxygen cycle. The amount of atmospheric oxygen from the breakdown of water during photosynthesis is $\sim 1.9 \times 10^{17}$ g O/yr.

Oxygen can also be produced through the carbon burial process, which is the physical removal of fixed carbon to anaerobic sediments, where the reaction with atmospheric oxygen does not occur until the sediments are exposed to the surface by tectonic processes. Carbon burial also indicates the difference between photosynthesis and respiration and decomposition. The rate of carbon fixation and the probability that fixed carbon escapes oxidation determine the rate of carbon burial. The rate of carbon fixation is independent of the atmospheric oxygen quantity, whereas the probability that fixed carbon escapes oxidation reduces as oxygen amount rises. The rate of carbon burial decreases as the oxygen pressure increases (15).

Oxidation of reduced materials, such as reduced carbon, sulfides, and ferrous iron, brought to the surface by tectonic activities, can remove atmospheric oxygen. The relationship between oxygen and other elements is expressed as redox potential, O:R, which is the ratio of oxidizing to reducing atoms per oxygen atom used. The lower the ratio, the less oxygen can be used. Increased use of oxygen in the burning of fossils fuels and the elevated demand for oxygen due to increased sewage and organic growth in water bodies could lower atmospheric oxygen content. This potential environmental problem has yet to be fully explored.

The global carbon cycle (Fig. 6) is complex because it includes all life forms, inorganic carbon reservoirs, and the links among them. Carbon exists in the atmosphere mainly as carbon dioxide, CO_2, with small amounts of methane, CH_4, carbon monoxide, CO, and other gases. Oceanic carbon exists primarily as dissolved inorganic carbon (DIC), dissolved organic carbon (DOC), particulate organic carbon (POC), and the marine biota. The basic input of organic carbon in the oceans is considered the primary production in the photic zone (16).

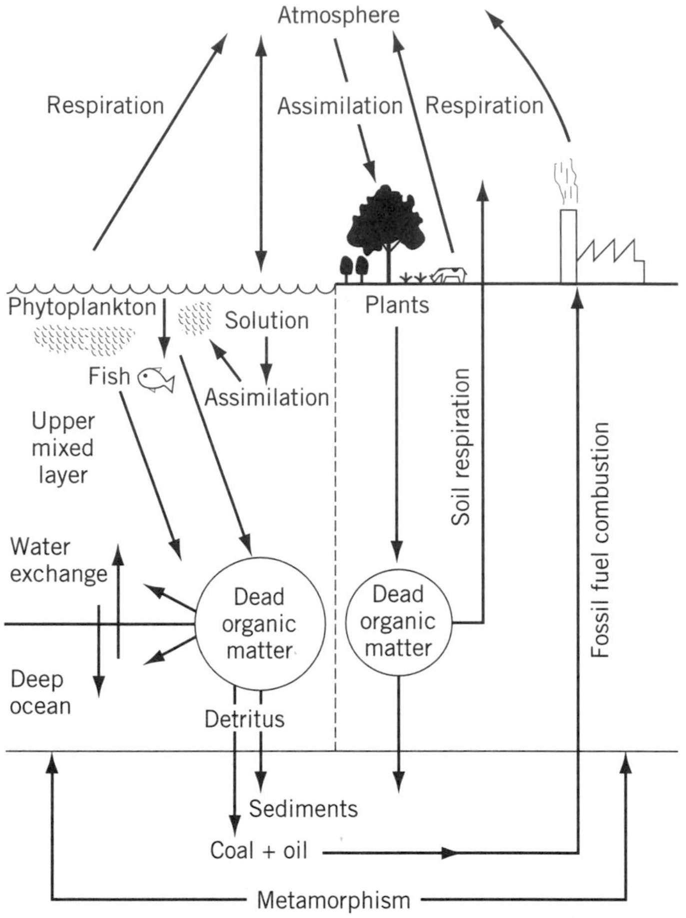

Figure 6. Schematic of the carbon cycle (5).

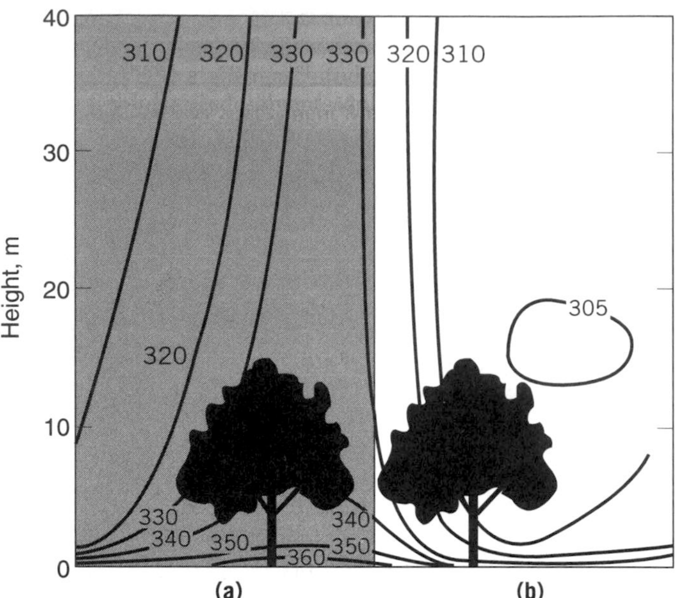

Figure 7. Vertical distribution of carbon dioxide, ppm, in a forested area (amount varies over time). At night (**a**), without ongoing respiration, levels are higher than during the day (b), whereas the relatively high values near the ground are related to soil respiration (14).

There are large reservoirs of carbon in the lithosphere. However, the exchange of carbon within its global cycle is small. One means of transferring carbon from the lithosphere to the atmosphere is the burning of fossil fuels. Erosion also can release carbon from the lithosphere. Chemical weathering of the crust can add and remove carbon from the atmosphere. Oxidation of carbon releases CO_2 to the atmosphere whereas dissolution of carbonates is associated with CO_2 uptake.

There is a rapid and continuous exchange of carbon between the atmosphere and the biosphere. An important source of carbon is the oxidation of methane, produced through fermentation by animals and anaerobic production in paddy fields or marshy areas. Oxidation of terpenes and isoprenes emitted by the forest is another source (17). Carbon dioxide is returned to the atmosphere through respiration. Most plants take up CO_2 during the day and emit it at night. This diurnal cycle (Fig. 7) causes spatial and temporal variations in CO_2 amount. Emissions of carbon monoxide from the biosphere are small. Despite increased CO emission from anthropogenic sources, there is no noticeable atmospheric increase because soil serves as a sink for this gas.

The exchange of carbon between the atmosphere and the oceans takes place most rapidly in the upper 100 meters of the water bodies. Organic and inorganic carbons, particularly carbonates and bicarbonates leached from the land surface, are added to the ocean water. Phytoplankton absorbs inorganic carbon, some of which is recycled in decomposition, whereas much is lost from the surface water as dead organic matter (DOM) which sinks to the sea floor. Organic carbon, with carbonate and bicarbonate derived from marine shells and terrestrial leaching, make sea floor sediments a major sink for carbon. This oceanic carbon may not be brought back into contact with the atmosphere for years, thus oceans function both as a reservoir and a sink for carbon.

Because the CO_2 content of the atmosphere is the best known quantity of the global cycle, increased atmospheric CO_2 resulting from anthropogenic activities, such as the use of fossil fuels and increasing deforestation, is of great concern. The important role of gas in the enhanced greenhouse effect, which potentially leads to global warming, makes understanding the carbon cycle a high scientific priority.

The global nitrogen cycle is more complex than that of oxygen and carbon cycles because it involves a variety of abiotic and biological processes. The major form of nitrogen in the lithosphere is dead organic matter (DOM), with ~4% biomass and 6.5% inorganic nitrogen on the global average (18). The principal form of nitrogen in the oceans is dissolved gaseous nitrogen, N_2, which accounts for ~95% of the total oceanic nitrogen. The remainder is primarily nitrates and DOM. N_2 is the principal nitrogen component in the atmosphere, and the remaining nitrogen is found as nitrous oxide, N_2O.

Although nitrogen accounts for ~78% of the atmosphere, it is an inert gas which only a few organisms can directly utilize. Most available nitrogen is fixed either by bacteria or algae. The symbiotic organisms are considered to be more significant. Because nitrogen fixation is sensitive to the presence of O_2, only those organisms in anaerobic environment fix nitrogen. The most significant terrestrial symbiotic nitrogen fixers are those bacteria associated with leguminous plants. Once nitrogen is fixed as ammonia, NH_3, or ammonium it is either oxidized to nitrate or nitrite by nitrification, or assimilated by an organism as part of its biomass by ammonia assimilation. Nitrification is the oxidation of ammonia to nitrate or nitrite by an organism as a means of energy production. Nitrification is responsible for the production and release of large amounts of nitrous oxide, N_2O, to the atmosphere.

Organic matter is another important pool of available nitrogen in the biosphere. Ammonification is the process by which reduced nitrogen is produced through the decomposition of organic matter. A small portion of reduced nitrogen is volatilized as a result of the breakdown of animal excreta or microbial decomposition of organic matter. Denitrification is the process of reducing nitrate to gaseous forms. Gaseous nitrogen, N_2, is the end product of denitrification, although nitrogen monoxide, NO, and nitrous oxide, N_2O, are also common. Denitrification is believed to be a significant natural source of both N_2 and N_2O to the atmosphere. Some nitrogen is taken out of this exchange in the form of organic matter which sinks to deep sea floor sediments. Denitrification in the ocean is not fully understood. Most data indicate that the ocean is supersaturated with nitrous oxide, N_2O, which is positively related to nitrate and negatively correlated with O_2 (19). However, there are different views on whether the ocean is the source or sink of nitrous oxide. One view is that nitrous oxide is produced during oxidative production of nitrate; another suggests that nitrous oxide could be used by denitrification in anaerobic environments, and this could mean a missing sink for nitrous oxide.

The nitrogen cycle also involves abiotic processes. Photochemistry produces some reactive gases in the atmosphere, such as OH radicals, which are responsible for the oxidation of nitrogen gases. Nitrogen dioxide, NO_2, reacts with OH radicals to produce nitric acid, HNO_3, that can be removed in precipitation. Nitrogen dioxide, NO_2, may also photolyze and produce ozone. Ammonia is also oxidized by OH radicals. In the atmosphere, large amounts of nitrogen compounds, such as the very soluble nitrate and ammonia, are removed by precipitation. These reactions represent a prominent source of atmospheric aerosols. The ocean is the net sink of ammonia, NH_3. Volatilization of gaseous nitrogen compounds from soils is another pathway into the atmosphere. Anthropogenic activities have added inputs to the nitrogen cycle through the use of inorganic nitrogen fertilizers, the emission of nitrogen compounds from agriculture, and industrial fixation of nitrogen.

PRODUCTIVITY

Plant productivity measures energy accumulation over time and space and is a function of photosynthesis. Plants use about one-sixth of the absorbed energy in the process of photosynthesis, and the rest is converted into chemical energy. The amount of energy fixed over a given period of time is gross primary production (GPP), much of which is further utilized in metabolic processes and dissipated in the process of respiration. The remainder is stored in plant tissue as net primary production (NPP).

A plant's capacity to utilize and intercept available light energy, temperature, supply of carbon dioxide, and the availability of water and mineral nutrients is a significant determinant of the rate of photosynthesis and NPP. Only 50% of light falling on the leaves is photosynthetically active radiation (PAR). The rate of carbon assimilation increases with increasing PAR. Both light and thermal conditions affect the rate of photosynthesis. Maximum photosynthesis is attained at optimum light intensity and temperature, which vary with species. Temperature also influences the availability of water to plants through its effect on relative humidity and evaporation. As temperature increases, the ratio of water loss to carbon dioxide uptake increases. This relationship is expressed as water-use efficiency, ie, the amount of water loss through transpiration per unit of fixed carbon. In arid regions, high water-use efficiency and the method of reducing the rate of water loss are crucial to plant growth and survival.

The photosynthetic capacity of plants also depends on the ability to intercept solar radiation. This relates to the total surface area of foliage. Leaf area index (LAI) is the ratio of leaf area to ground area. The increase in leaf area inevitably raises photosynthesis, and hence productivity. However, great leaf area is accompanied by shading of the lower leaves by upper leaves, which have a higher rate of photosynthesis. Disposition of leaves also influences the amount of light intercepted, eg, horizontal leaves tend to intercept more light than vertical ones. There are plants called solar trackers (20) whose leaves change orientation relative to the angle of light.

For effective photosynthesis, plants require various nutrients from the environment. Deficient nutrient supply may depress the photosynthetic and physiological processes. Nitrogen deficiency lowers the content of protein and chlorophyll in leaves, leading to decreased photosynthesis. Deficiency in phosphate induces a decrease in the number and size of leaves, as well as in their protein and pigment contents.

The rate of photosynthesis also varies with different biochemical processes or carbon pathways employed by different plants. The difference between C_3 and C_4 plants lies in the initial procedure of carbon assimilation. The initial steps are the production of a three-carbon compound and four-carbon carboxylic acid, respectively, for C_3 and C_4 plants. C_4 plants can use carbon dioxide and water more efficiently than C_3 plants. Some succulent plants with crassulacean acid metabolism (CAM) have lower photosynthetic rates. The four-carbon acids are formed at night and are photosynthesized during the day when stomata are closed to minimize water loss.

There are various measurements of plant or primary production. Assessment of photosynthetic efficiency is a fundamental calculation of plant productivity. This measures the efficiency with which a plant can convert light energy to chemical energy. It is expressed as a ratio between the energy of organic matter produced and that of either light absorption or availability. Although estimates of photosynthetic efficiency vary, the mean annual net photosynthetic efficiency for the earth's surface is about 0.1–0.2%, which represents an annual conversion of over 200×10^9 t of carbon to sugar. The unit of time used, ie, annual or monthly, for the calculation of photosynthetic efficiency also affects the final estimation. Because of these limitations, GPP or NPP is preferred when studying the conversion of solar energy to plant material.

A different measure of productivity is the calculation of biomass, expressed as volume or weight measured. Theoretically, biomass is the total organic matter at a locale. Because the assessment of the below-ground biomass or the root systems is difficult, frequently only the above-ground biomass is measured. Leaf area index (LAI) can also be considered as another measure of productivity. The higher the LAI, the greater the potential for productivity.

CLIMATE AND NATURAL VEGETATION

So far, consideration of climate–plant relationships has been in terms of the response of plants to individual climatic elements; however, just as the climatic elements are interrelated, so are plants, for they do not exist in isolation but rather as part of a larger plant community. Plants thus respond not only to external physical elements, but also to experience limitations placed on them by other, often competitive species. The term vegetation is used to describe the plant cover of an area, irrespective of the species making up the group. The concept of vegetation has been described (21):

> The term vegetation ... refers to the general aspects of the plants of an area taken collectively and regardless of the kinds of plants which produce that aspect. It is based upon the impression which the plants make through our eyesight, not individually but en masse. ... Why do a forest and a meadow look unalike? Because each is characterized by the prevalence of plants of a certain vegetative form. ... The great bulk of plants in a forest consist of trees ... under the trees are shrubs and herbs and even grasses, but the chief impression on the mind is made by the trees. In a meadow the bulk of plant life consists of grasses; various other plants of different growth form are also in the meadow, but it is the grasses which give the impression which we call meadow.

It is this collective concept of plant life that has most occupied plant geographers and ecologists.

Natural Vegetation and Climatic Classification

The early development of climatic classification is related closely to vegetation studies. Of significance in these early studies of vegetation was the approach used. Systematic botanists arranged plants of the world into classes, genera, etc, using what might be termed a floristic approach. Plant geographers used a physiognomy of vegetation in which plant communities expressed in such terms as forests, meadows, moors, etc, were identified.

The earliest comprehensive floristic classifications were introduced in 1860s and included a moisture factor in which the ratio of precipitation to monthly annual temperature was used to identify five major zones. In 1874, De Candolle developed a classification that was designed for tracing the development of plants through geologic time (*Géographie Botanique Raisonnée*). Six subdivisions were initially used, but the megistotherms (mean annual temperature >30°C), although prevalent in earlier geologic times, are found today only in the vicinity of hot springs. The remaining five, megatherms, mesotherms, microtherms, hekistotherms, and xerophiles, were designated with letters A through E, with B referring to xerophiles. Because the groups were identified zonally, the insertion of a moisture-based unit, B, in a temperature-based grouping is logical.

Vegetation was first classified as an aggregate of various species in a community of mutually dependent plants in 1898 (25). This physiognomy of vegetation identified three principal formations: woodland, grassland, and deserts, each of which was appropriately subdivided. The floristic approach of the botanist is quite secondary in such an approach. The equatorial forest, the savanna, and the tundra are identified, not by the individual plants of which they are comprised, but by the collective.

From the pioneer work on vegetation classification came climate organizational schemes based on vegetation distribution. This realm of classification is dominated by the work of Vladimir Köppen (1846–1940). Köppen's work in classification may be viewed as a climatic determinism of vegetation types and his greatest contribution was to stress the fundamental unity of pattern in the location of climatic regions throughout the world. The 1918 Köppen classification provides the basis for that used today and is reproduced in many sources (22).

Divisions of the Biosphere

To gain some insight into the role played by climate in determining major vegetation groups, it is useful to assess the impact of climate on the grouping found in an orderly vegetation classification scheme. A useful system has been devised (25) using the approach illustrated in Figure 8.

Examination of this system shows different environmental variables operating at various levels. The biosphere is, of course, the total earth system, and its subdivision into biocycles represents the dominant physical media that comprise its near-surface characteristics, the land and water in both sea and land. Climate has its first impact in the derivation of the biochores. It is clear that the four groups named, ie, forest, savanna, grassland, and desert, are differentiated based on moisture availability. Within these biochores, however, vegetation associations are recognized that, under conditions of environmental equilibrium, might produce climax communities called biomes (23). The basis for differentiating among these vegetation associations depends on thermal characteristics. Within each group there is a spectrum of vegetation association that depends on prevailing temperatures. For example, within the desert biochore, the hot dry environment of the tropical desert is at one extreme, whereas the Arctic fell field of the cold dry climate is at the other.

Passing to the lower orders of the scheme, topographic, geomorphic, and pedogenic factors become of prime concern. Thus, although a climax community of any biochore might exist, variations will be found within it as a result of such differences. It is in these lower orders that microclimatology becomes significant, with modifications resulting from slope and surface moisture playing an important role.

World Biomes

It should be emphasized that in each of the biomes identified (see Fig. 8), the stress is on the dominant natural vegetation and, for reasons discussed earlier, there are variations in the type of ground cover that occurs.

Figure 9 provides a general guide to major world biomes. The distribution is keyed to the various biomes that occur. The discussion proceeds through the four biochores, treating first the forest biochore and then the savanna, grassland, and desert biochores.

Equatorial and Tropical Rain Forest. The main characteristics of this biome is the rich diversity of the flora. The layered trees, often buttressed and bearing epiphytes and lianas, form a continuous canopy limiting penetration of sunlight to the forest floor. The humus of the forest floor is superficial because the hot, humid conditions lead to rapid breakdown of organic materials. The difference between these two forest types is essentially locational, with the tropical rain forest extending from

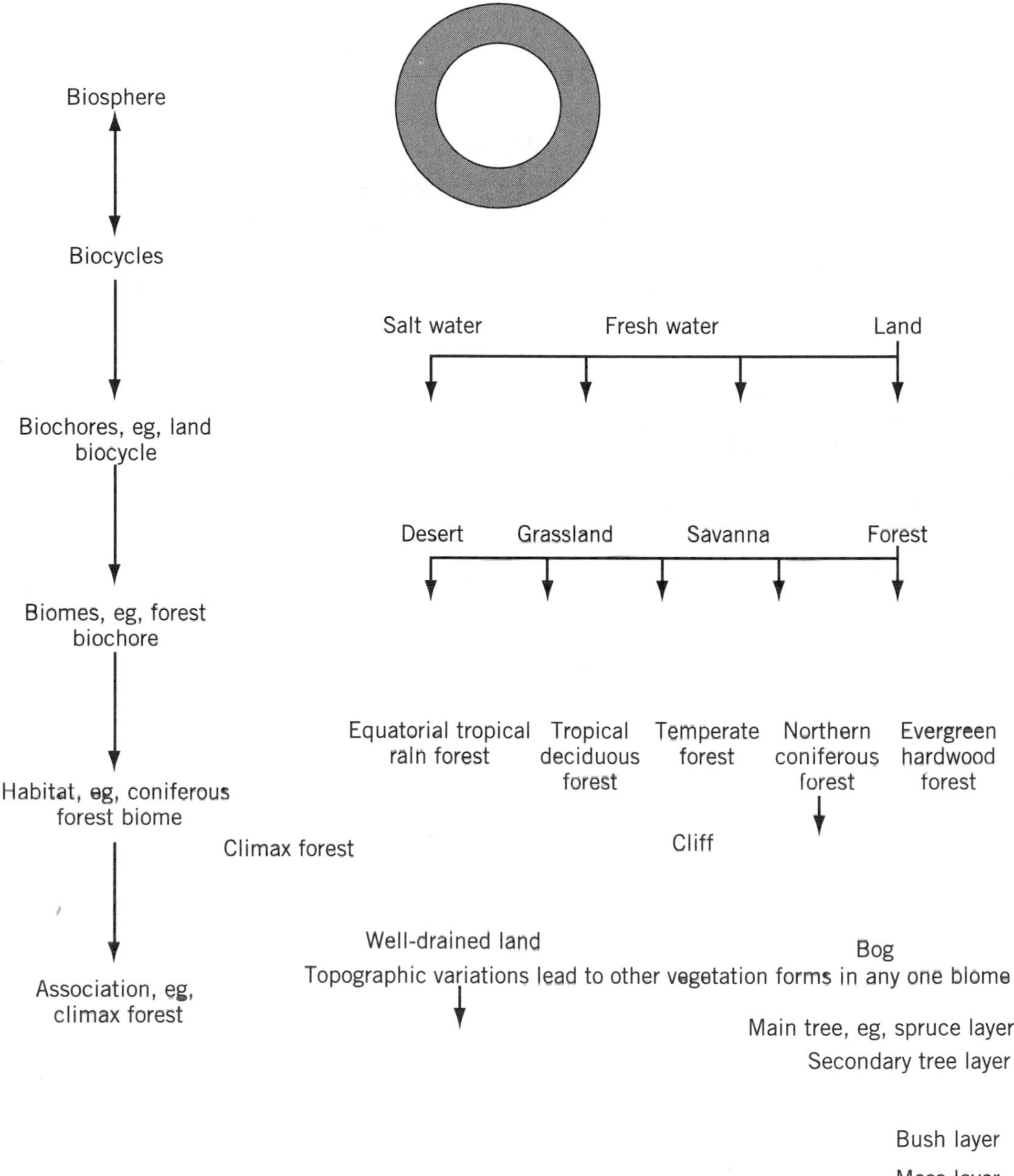

Figure 8. Divisions of the biosphere (22).

10°N and 10°S to the respective tropics on the windward coasts influenced by the trade winds.

Tropical Deciduous Forest. This type of forest is largely associated with tropical areas having monsoon climates. The shedding of leaves is a response to the stress imposed by a long, dry season rather than cold temperatures. This type of forest contains many species of trees and is characterized by thick undergrowth. Clumps of bamboo are frequently found, and some important commercial woods, such as teak, also occur.

Temperate Deciduous and Temperate Rain Forests. Temperate rain forests are marked by relatively few tree species and large populations of a single type. In areas such as the southeastern United States and southern Japan, stands of laurel and evergreen oak trees are common. Temperate rain forests are also found on the west coasts of middle latitudes, between about 35° and 50°, where abundant rainfall and moderate temperatures permit growth. The temperate deciduous forest contains tall broadleaf trees that form an extensive canopy in summer. In winter, the trees shed their leaves. Undergrowth develops best in the spring, before the canopy of leaves prevents sunlight from passing through. Typical trees include oak, beech, elm, and maple.

Northern Coniferous Forests. This is the most extensive of the forest biomes. It extends across the continents of the Northern Hemisphere, almost encircling the earth. Although coniferous trees are found throughout, the actual dominant species of trees varies from North America to Europe and Asia (Eurasia). Spruce and fir are most abundant in Canada, whereas in

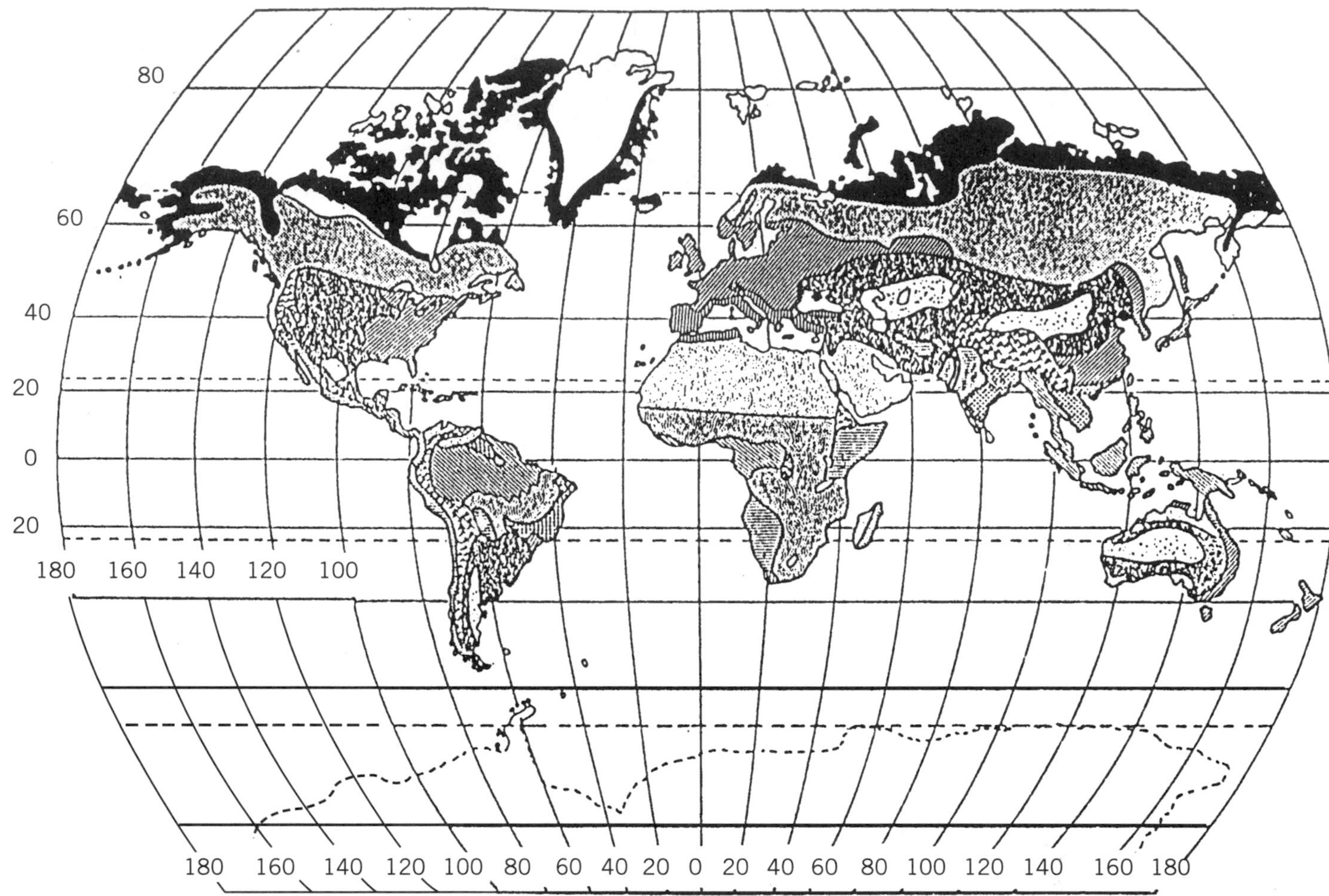

Figure 9. A generalized map showing major world biomes where ■ is grassy tundra; ▩, northern coniferous forest; ▨, temperate deciduous/rain forest; ▧, temperate grasslands; ▥, evergreen hardwood forest; ▦, desert; ▧, equitorial/tropical rain forest; ▩, tropical deciduous forest; ▤, tropical scrub; ▢, savanna/savanna woodland; □, ice desert; and ▢ mountains (complex zonation) (14).

Eurasia pine, especially Scotch and stone pine, is found more widely.

The conifers of this biome are not usually large trees, varying in height from 10 to 15 meters (30–50 ft). Often, though, the trees are fairly dense, and foliage intercepts most of the sunlight. This factor, combined with the generally low temperatures, produces a dark, humid forest floor populated with only fungi and small plants that require little light.

Like other biomes, coniferous forests are interrupted by other vegetation types. A common feature in this biome is the taiga bog. Regions occupied by this biome were glaciated, and poor drainage exists over wide areas. The slow infilling of small lakes and bogs scattered throughout give rise to many bogs, where mosses, cranberries, and small shrubs are found.

Evergreen Hardwood Forest. These forests are closely associated with the Mediterranean type of climate, ie, hot, dry summers and warm, wet winters. Trees are adapted in many ways to the drought conditions. Generally they are low branched with more wood than foliage, whereas their leaves are often thick and waxy to retain moisture. Cork oak, olive, and Aleppo pine are examples. These forests have been greatly disturbed by human activities, and in many places a spiny, shrub-like ground cover exists. This low evergreen shrub cover is known by different names in various parts of the world. In the area bordering the Mediterranean Sea it is called maquis or garrigue, the former applying to shrub growth that is predominantly from 1 to 6 m (3–20 ft) tall, and the latter to widespread shrubs less than 1 m (3 ft) tall. In the United States, the word chaparral is commonly used to describe the scrubby vegetation.

Tropical Scrub. This biome occurs between desert and savanna and contains vestiges of each. Low woody shrubs adapted to dry conditions occur in clumps that are separated by bare ground. Frequently consisting of thorny buses, these stands of scrub are called thorn forests in regions where vegetation covers the ground fairly completely.

Savanna and Savanna Woodlands. These tropical grasslands differ appreciably from temperate grasslands such as the prairies. One noticeable item is the discontinuous grassy covering with bare ground appearing between bunched grass clumps. Another difference is that trees, although widely spaced, are always present.

Today the savanna is found almost entirely in south and central America, Africa, and Australia. The general appearance of the biome varies in and within these locations. In some, the grass grows to heights of 3.6 m (12 f); in others, it is only 3 meters tall. The umbrella-like acacia tree dominates the tree population in some areas, giving way to palms, such as the Palmyra palm, in others.

Trees in the savanna are most dominant along river courses. They often form a corridor across the predominant grasses and are called galeria, a name derived from the dense tangle of trees and canes (or reeds) that occurs on either side of a stream to form a tunnel-like gallery over the stream.

Temperate Grasslands. A number of terms are used to describe the temperate grasslands of the world, and the varied use of these designations has led to some confusion. The word steppe, for example, is used to mean a semidesert in the Köppen climatic classification (22). Russians, however, use the word to designate any grassland of large extent, hence the steppes of Russia. Herein steppe is used to designate temperate grasslands that consist mainly of short grasses. This definition differentiates the steppe from the tall grass cover that can be designated prairie. Trees are absent in both. Grasses of the prairie are deep rooted and make up a continuous ground cover. In contrast, the steppe, which occurs in drier regions, has grasses that tend to bunch, leaving bare ground between their clumps. This short-grass vegetation eventually gives way to semidesert. Today, little remains of the original temperate grassland biome because the land it had occupied has proved ideal for the production of cereal grains. In fact, in some locations in the United States, it has been necessary to identify areas where the original grasslands occur and set them aside as preservation sites.

Grassy Tundra. The tundra is associated with Arctic regions, where vegetation development is limited by the short growing season and abundant moisture resulting from poor drainage.

Deserts. At first view, the world's deserts seem to have little to interest the plant geographer, but plant life in the hot deserts is interesting and more varied than might be expected, especially in the deserts of North America. There, stands of drought-resistant plants such as cacti are found in scattered locations. In deserts of the Old World, appreciably less vegetation exists. There are many excellent sources that describe the desert and other world biomes in more detail (24,25).

Biome Productivity and Vulnerability

Vegetation response in a given area reflects the utilization of energy in that realm and, specifically, the rate at which radiant energy is stored by photosynthetic activity of producer organisms, chiefly the green plants, in the form of organic substances used as food materials. As already noted, the gross primary production of a system represents the total rate of photosynthesis, including organic matter used in respiration.

Global primary production and biomass are illustrated in Table 1. Primary productivity on land is believed to be about 20% higher than that in the ocean. However, it has been estimated that average marine and land biomass production were at 50 g/m^{-2} and 50 kg/m^{-2}, respectively (26). This discrepancy is attributed by greater energy loss through respiration by plant plankton. Global terrestrial primary production correlates with climate, the highest in warm, humid areas and the lowest in cold, arid regions. Most of the biomes represented experience vulnerabilities, often as a result of climate. In the tundra, for example, the slow regenerative growth of vegetation poses great problems for utilization by humans. The slow growth of vegetation allows accelerated erosion to occur and large disruptions of the surface may take many decades to recover.

The vulnerability of both hot desert and equatorial forest biomes are also related to climate. In deserts, rapid evaporation rates may well result in the upward movement of water with eventual salinization of the surface. This fragility has been well demonstrated in relationships of the decline of Mesopotamian agriculture to increased soil salinity (27). A similar problem has been shown to exist in irrigated areas of dry regions today. Just as climate can cause excessive soil salinity in arid regions, so can it cause rapid depletion of soil nutrients in the equatorial rain forest regions. High temperatures and rapid bacterial decay, together with copious water supply, cause excessive leaching in such climates. In agricultural development of the tropics, this is one aspect of applied climatology that needs much consideration. All of the other biomes have similar climatic problems: the desiccation of plants and resulting forest fires in the Mediterranean realm; the variability of precipitation in monsoon regions; and the periodic droughts that may occur even in well-watered biomes of the world. To explain and possibly mitigate such problems requires much groundwork in climatology.

INFLUENCE OF VEGETATION ON CLIMATE

From all that has been studied, it is evident that climate has a marked influence on vegetation. This is not a one-way relationship, however, and it is equally evident that vegetation cover must influence climate. Such a reciprocal effect is evident at various levels of study. It fact, the very evolution of the atmosphere and of vegetation are intimately connected (28). It can be shown, too, that at a more local level, vegetation has a marked impact on the climatic regime on both a meso- and microscale level.

One of the most visual dynamic relationships between climate and vegetation is the seasonal green-up or onset of vegetation leafing in middle latitudes in spring. Because leafing generally occurs in lower middle latitudes first and advances northward over time, it is referred to as the green wave. The green-up represents a vegetation response to atmosphere conditions that are conducive to vegetation growth. Although vegetation responds to climate, once the greening occurs the energy exchanges that occur at the surface are changed and this, in turn, modifies the climate of that surface. A complex feedback loop occurs. The spring green wave has been extensively examined (29).

Another striking example of how climate can be modified by vegetation is provided by forests. Because of their vertical extent, forests modify the energy budget well above the earth–atmosphere interface. The results of such modification are well known to anyone who has walked into a forest and experienced the modification that occurs in relation to the surrounding nonforested areas. The way in which forests modify climate has intrigued workers for many years. One early controversy

Table 1. Primary Productivity of Main Ecosystem Types

Ecosystem Type	Area, 10^6 km^2	LAI, m^2/m^{-2}	Net primary productivity/unit area, g/m^{-2}/yr Normal range	Mean	Biomass, kg/m^{-2} Normal range	Mean
Tropical rain forest	17.0	6–16.6	1000–3500	2200	6–80	45
Tropical seasonal forest	7.5	6–10	1000–2500	1600	6–60	35
Temperate evergreen forest	5.0	5–14	600–2500	1300	6–200	35
Temperate deciduous forest	7.0	3–12	600–2500	1200	6–60	30
Boreal forest	12.0	7–15	400–2000	800	6–40	20
Woodland and shrubland	8.5	4.2	250–1200	700	2–20	6
Savanna	15.0	1–5	200–2000	900	0.2–15	4
Temperate grassland	9.0	5–16	200–1500	600	0.2–5	1.6
Tundra and alpine	8.0	0.5–1.3	10–400	140	0.1–3	0.6
Desert and semidesert scrub	18.0		10–250	90	0.1–4	0.7
Extreme desert rock, sand, ice	24.0		0–10	3	0.1–0.2	0.02
Cultivated land	14.0		100–3500	650	0.4–12	1
Swamp and marsh	2.0	11–23	800–3500	2000	3–50	15
Lake and stream	2.0		100–1500	250	0–0.1	0.02
Total continent	*149*			*773*		*12.3*
Open ocean	332.0		2–400	125	0–0.005	0.008
Upwelling zone	0.4		400–1000	500	0.005–0.1	0.02
Continental shelf	26.6		200–600	360	0.001–0.04	0.01
Algal beds and reefs	0.6		500–4000	2500	0.04–4	2
Estuaries	1.4		200–3500	1500	0.01–6	1
Total marine	*361*			*152*		*0.01*
Full total	*510*			*333*		*3.6*

centered on the problem of whether the forest caused the climate or the climate caused the forest. The reasoning behind this controversy relates to the modified moisture balance of a forest, with graphic evidence of moisture contribution to the air being shown by smoking forests. It was assumed that the large amount of moisture transpired by forests would provide the moisture necessary to be returned as rain. Although there is little probability that forests do, in effect, cause themselves, there is still some argument about whether or not forests do induce higher precipitation.

The boundary layer of the forest is its canopy, for it is at this level that energy exchanges occur. Some insolation is returned directly to space, the amount depending on the albedo or reflectivity of the canopy layer. In some forests this varies enormously from season to season. Some energy is trapped within the canopy layer, while some penetrates to the forest floor. Compared to the flow of air over open areas, wind inside the forest is slight. The amount of decrease depends on the type and structure of the forest concerned. Modification of wind speed in this way has been put to good use in constructing shelter belts in areas that lack a natural tree cover. The forest environment also modifies local moisture conditions. Evaporation from the forest floor is relatively low because of reduced insulation. This is counterbalanced by the fact that, with the profuse vegetation, high transpiration occurs. As such, the humidity within a forest depends on the density of the forest and rates of transpiration. The essential result of the interaction of these factors is that temperatures inside a forest are moderated, the maximum is lower, and the minimum higher than those in nonforested areas experiencing a similar climatic regime. The amount of variation is seasonal, the mean difference in summer being as much as 2.8°C (5°F) within a low-altitude, middle-latitude forest. Exceptions to such moderating influences do occur.

Although the forest, with its extensive vertical cover, provides the most marked influence of vegetation on climate, almost all other vegetation types result in modified climatic regimes. Clearly, many such modifications will occur at the microlevel for, as has been shown, the energy exchange that occurs at any given location depends, in part, on the nature of that surface. Differences caused by varying albedos, transpiration rates, soil characteristics, etc, are aspects of plant growth being actively investigated.

THE ROLE OF CHANGING CLIMATE

Current climate–biosphere interactions are the result of the climate that prevails on earth at this time. But climate is not a static entity, and as Table 2 shows, it has changed over time, and will continue to change in the future. Of considerable importance at present is the potential that exists on earth for global warming. Should this occur, extensive changes in the global system will follow, changes that are modeled in scenarios of the future. If it is assumed that global temperatures are rising and will continue to rise over the twenty-first century, then there are a number of potential impacts. Some of the scenarios created for selected impacts are considered herein.

Of the many impact studies completed and published, of considerable interest is that contained in a report to Congress by the U.S. Environmental Protection Agency (EPA) in 1989 (30). Findings contained in this study were derived from the global circulation models produced by various research agencies on computers and supplemented by analog studies from actual data.

Biological diversity (biodiversity) can be defined as the variety of species in ecosystems, and the genetic variability

Table 2. Scale of Geological Time and Biospheric Evolution

Periods	Epochs	Apparent ages (years before present)	Notable Events in Evolution of Organisms
		Cenozoic	
Quatemary	Holocene (Recent) Pleistocene	2×10^6	Humans appear
Tertiary	Pliocene	10×10^6	Elephants, horses, large carnivores become dominant
	Miocene	25×10^6	Mammals diversify
	Oligocene	36×10^6	Grasses become abundant, grazing animals spread
	Eocene	58×10^6	Primitive horses appear
	Palaeocene	63×10^6	Mammals develop rapidly, dinosaurs become extinct, flowering plants appear
		Mesozoic	
	Cretaceous	135×10^6	
	Jurassic	180×10^6	Dinosaurs reach climax, birds appear
	Triassic	230×10^6	Primitive mammals appear; conifers and cycads become abundant, dinodaurs appear
		Palaeozoic	
	Permian	280×10^6	Reptiles spread, conifers develop
	Carboniferous[a]	340×10^6	Primitive reptiles appear, insects become abundant, coal-forming forests widespread, fishes diversify
	Devonian	400×10^6	Amphibians (first known land vertebrates) appear, forests appear
	Silurian	440×10^6	Land plants and animals first recorded
	Ordovician	500×10^6	Primitive fishes (first known vertebrates) appear
	Cambrian	570×10^6	Marine invertebrate faunas become abundant
		Precambrian	
	Proterozoic	2.5×10^9	Life forms abundant
	Archaean		Primitive life forms, eg, blue-green algae

[a] In North America the Carboniferous is divided into two periods (Pennsylvania in and Mississippian) at about 310×10^6 years.

within each species and the variety of ecosystems around the world. Over 400 species of mammals, 460 species of reptiles, 660 species of freshwater fishes, and tens of thousands of invertebrate species can be found in the United States, in addition to some 22,000 plant species. About 650 species of birds reside in or pass through the United States annually. Biological diversity is needed to provide food, medicine, shelter, and other important products. The EPA report examined the impacts of climatic change on specific plants and animals by using climate change scenarios and models of particular species or systems within a region. Of particular note was the observation that extinction of species could increase.

Historic climatic changes, such as the Ice Ages, have led to the extinction of many species. More recently, human activities, such as deforestation, have greatly accelerated the rate of species extinction. The faster rate of climate warming resulting from the an enhanced greenhouse effect would most likely lead to even greater loss of species. The uncertainties surrounding the rate of warming, the response of individual species, and interspecies dynamics make it difficult to assess the probable impacts, although natural ecosystems are likely to be destabilized in unpredictable ways.

Forests occupy one-third of the land area of the United States with temperature and precipitation ranges of major importance as determinants of forest distributions. The EPA scenario assumes that climate change could move the southern boundary of such species as hemlock and sugar maple northward by 600–700 km (ca 400 miles), whereas the northern boundary would move only as fast as the rate of migration of forests. Assuming a migration rate of 100 km (60 mi) per century, or double the known historic rate, the inhabited ranges of forests could be significantly reduced because the southern boundary may advance more quickly than the northern boundary. Even if the climate stabilized, it could take centuries for migration to reverse this effect. If the climate continued to warm, migration would continue to lag behind shifts in climate zones. If elevated CO_2 concentrations increase the water use efficiency of tree species and pest infestations do not worsen, the declines of the southern ranges could be partly alleviated. Reforestation could help speed the migration of forests into new areas.

As with trees, other plants and animals may have difficulty migrating at the same rate as a rapidly changing climate, and many species may face extinction or reduced populations. The presence of urban areas, agricultural lands, and roads would restrict habitats and block many migratory pathways. These obstacles may make it harder for plants and wildlife to survive future climatic changes. On the other hand, some species may benefit from climatic change as a result of an increase in habitat size or a reduction in the population of competitors. The extent to which society can mitigate negative impacts through such efforts as habitat restoration is not clear (31). To aid in such developments, data banks for both vegetation (32) and climate are available for research, application, and teaching.

BIBLIOGRAPHY

1. W. D. Sellers, *Physical Climatology,* University of Chicago Press, Chicago, Ill., 1965.

2. D. Schrove, in J. E. Oliver and R. W. Fairbridge, eds., *Encyclopedia of Climatology,* Van Nostrand Reinhold, New York, 1986.
3. J. E. Oliver and J. J. Hidore, *Climatolgy: An Introduction,* Merrell, Columbus, Ohio, 1984.
4. T. R. Oke, *Boundary Layer Climates,* Methuen, New York, 1987.
5. A. Henderson-Sellers and P. J. Robinson, *Contemporary Climatology,* Longman Scientific and Technical, Essex, U.K., 1986.
6. R. S. Fraser and R. J. Curran, *Effect of the Atmosphere in Remote Sensing,* Addison-Wesley, Reading, Mass., 1976, p. 35.
7. D. M. Gates, *Biophysical Ecology,* Springer-Verlag, New York, 1980.
8. J. L. Monteith, *Principles of Environmental Physics,* Edward Arnold, London, 1973.
9. J. J. Hidore and J. E. Oliver, *Climatology, An Atmospheric Science,* Macmillan, New York, 1993.
10. A. G. Morton, *History of Botanical Science,* Academic Press, Ltd., London, 1981.
11. J. F. Griffiths, ed., *Handbook of Agricultural Meteorology,* Oxford University Press, New York, 1994.
12. F. I. Woodward, *Climate and Plant Distribution,* Cambridge University Press, London, 1987.
13. D. D. Evans and J. L. Thames, eds., *Water in Desert Ecosystems,* Academic Press, Inc., New York, 1981.
14. J. E. Oliver, *Climate and Man's Environment,* John Wiley & Sons, Inc., New York, 1973.
15. S. S. Butcher, ed., in *Global Biogeochemical Cycles,* Academic Press, Inc., San Diego, Calif., 1992, pp. 73–92.
16. K. Mopper and E. T. Degens, in B. Bolin, ed., *The Global Carbon Cycle,* John Wiley & Sons, Inc., New York, 1979, pp. 293–316.
17. P. I. Crutzen, in B. Bolin, ed., *Major Biogeochemical Cycles and Their Interactions,* John Wiley & Sons, Inc., New York, 1983, pp. 67–112.
18. R. Södderlund and B. H. Svensson, in B. H. Svensson and R. Södderlund, eds., *Nitrogen, Phosphorus and Sulphur Global Cycles,* Swedish National Science Research Council, Örsundsbro, Sweden, 1976, pp. 23–73.
19. J. Hahn, in C. C. Delwiche, ed., *Denitrification, Nitrification and Atmospheric Nitrous Oxide,* John Wiley & Sons, Inc., New York, 1981, pp. 191–240.
20. H. A. Mooney, in M. Crawley, ed., *Plant Ecology,* Blackwell Scientific Publications, Oxford, U.K., 1986, pp. 345–375.
21. H. A. Gleason and A. Cronquist, *The Natural Geography of Plants,* Columbia University Press, New York, 1974.
22. J. E. Oliver, *Phys. Geog.* **13,** 231–251 (1991).
23. P. Dansereau, *Biogeography: An Ecological Perspective,* Ronald Press, New York, 1957.
24. J. Tivy, *Biogeography: A Study of Plants in the Ecosphere,* Longman Scientific and Technical, Essex, U.K., 1993.
25. W. E. Akin, *Global Patterns: Climate, Vegetation and Soils,* University of Oklahoma Press, Norman, 1990.
26. C. J. Kreb, *Ecology: the Experimental Analysis of Population Distribution and Abundance,* Harper and Row, New York, 1985.
27. T. Jacobson and R. M. Adams, *Science* **128,** 1251–1258 (1958).
28. J. E. Oliver and R. W. Fairbridge, *Encyclopedia of Climatology,* Van Nostrand Reinhold, New York, 1986.
29. M. D. Schwartz, *Monthly Weather Rev.* **120,** 2570–2578 (1992).
30. J. B. Smith and D. Tirpak, eds., *The Potential Effects of Global Climate Change on the United States: Report to Congress,* Environmental Protection Agency, Washington, D.C., 1989.
31. R. W. Kates, J. H. Ausubel, and M. Berberian, *Climate Impact Assessment,* Scope 27, John Wiley & Sons, Inc., New York, 1985.
32. EPA Global Climate Research Program, *NOAA/NGDC Global Change Database Program,* Disk Version 1.0, NOAA, Boulder, Colo., 1992.

General References

D. A. Jaffe, *The Nitrogen Cycle,* Academic Press, Inc., San Diego, Calif., 1992.

K. Holmen, *The Global Carbon Cycle,* Academic Press, Inc., San Diego, Calif., 1992.

CLIMATE SYSTEM MODELS: LAND–ATMOSPHERE INTERACTIONS

GORDON B. BONAN
National Center for Atmospheric Research
Boulder, Colorado

The definition of land–atmosphere interactions depends to a large extent on the background and interests of an individual. For many people, land–atmosphere interactions are the societal and ecological consequences of climate change. Human life depends on the processes that occur at the land–atmosphere interface. The climate of this interface, typically described in terms of temperature, precipitation, relative humidity, and wind speed, determines the most fundamental aspects of our lives: the food we eat; the water we drink; the clothes we wear; and the buildings in which we live. The land surface climate also determines the geographic distribution of natural vegetation, soils, agriculture, and glaciers. To many other people, however, land–atmosphere interactions are the ecological and hydrological processes by which land affects climate. Long-term (thousands of years) changes in the geographic distribution of vegetation and glaciers in response to natural climate changes feed back to enhance or mitigate these climate changes. Land use practices such as deforestation and agriculture have significant impacts on regional climates. Even simply watering a lawn, when multiplied by the thousands of lawns in a city and the tens of cities in a region, can have an impact on local and regional climate. Land–atmosphere interactions is, therefore, a rapidly evolving, interdisciplinary field of research whose goal is to develop a common framework to study the processes by which land affects climate and weather and the impacts of climate and weather on terrestrial ecosystems and water resources. It combines elements of atmospheric sciences (eg, micrometeorology, boundary layer meteorology, atmospheric chemistry), ecology (eg, plant physiology, biogeography, ecosystem dynamics), hydrology (eg, soil water, runoff, channel flow), and soil science (eg, soil physics, soil chemistry).

HOW LAND AFFECTS CLIMATE

Land surfaces influence climate through vertical exchanges of energy, water, momentum, and greenhouse gases with the atmosphere. The meteorological characteristics of the boundary layer, usually to a height of one kilometer, are strongly determined by the surface. As the properties of the surface (eg, albedo, roughness, stomatal resistance, hydrology, Bowen ratio) vary, so too will the temperature, humidity, wind, and chemistry of the boundary layer. The following discussion is

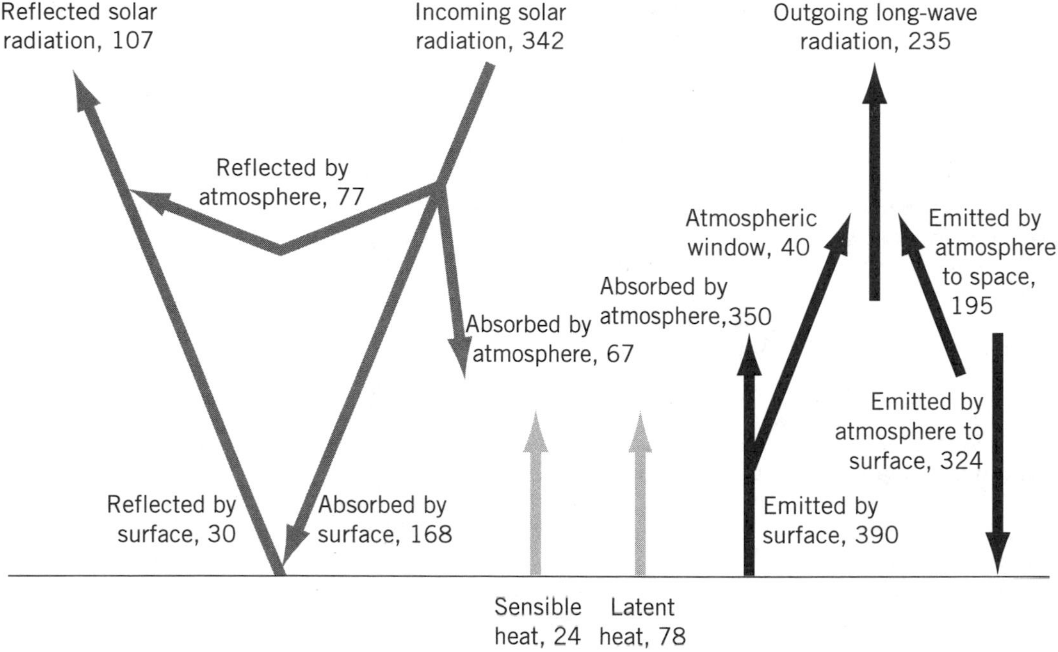

Figure 1. The earth's annual radiation and energy balance in W/m^2. Solar radiation fluxes are shown in dark gray; long-wave radiation fluxes in black; and surface energy fluxes (sensible and latent heat) in light gray.

only a brief introduction to these energetic, hydrological, and biogeochemical land-atmosphere interactions. A more complete discussion can be found in ecological (1), hydrological (2), climatological (3,4), boundary layer (5), and plant biophysics references (6,7).

Energy

The climate of the earth surface can, to first order, be understood in terms of its energy budget. As the sun's radiation passes through the atmosphere, some of it is absorbed by the atmosphere and some is reflected back to space (Fig. 1). Of the remainder that reaches the ground, some is absorbed by the surface and the rest is reflected back to space. The fraction of the incident radiation that is reflected, the albedo, depends on surface conditions. Fresh snow has a very high albedo, reflecting up to 90% of the incoming radiation. Vegetation, on the other hand, has a much lower albedo, generally reflecting 10 to 20% of the incoming radiation. Under certain conditions, lakes are very "dark," reflecting as little as 5% of the solar radiation. Annually, and when averaged over the earth's surface, about 30% of the solar radiation available at the top of the atmosphere is reflected back to space. About 20% is absorbed by the atmosphere, with most of this absorbed by water vapor and clouds in the troposphere, ie, the lowest 10 to 15 km of the atmosphere. The remaining 50% is absorbed by the surface, warming the surface and lower atmosphere.

The warm surface emits infrared (long-wave) radiation, most of which is absorbed by water vapor, clouds, carbon dioxide, and other atmospheric gases in the troposphere and reemitted both up to space and back down toward the surface (see Fig. 1). Globally, about 390 W/m^2 are emitted by the surface. However, only about 40 W/m^2 are lost to space; the remainder is absorbed and reemitted by the atmosphere. The troposphere emits 324 W/m^2 downward on to the surface, so that the net loss of long-wave radiation from the surface is only 66 W/m^2. The large downward reemission of surface long-wave radiation from the atmosphere is essential for maintaining the warm surface and contributes to the greenhouse effect, in which the climate system warms because the atmosphere is relatively transparent to solar radiation and opaque to surface long-wave radiation. Without this downward reemission, the surface would be 33°C cooler. The importance of the greenhouse effect can be seen locally in the diurnal temperature range. Semiarid and desert regions become hot during the day, when there are few clouds and little water vapor to reflect solar radiation. At night, they cool quickly because the dry air does not reemit the surface long-wave radiation. Even in more humid regions, clear nights tend to be colder than cloudy nights.

The net radiation at the surface drives fluxes, ie, exchanges of matter or energy between the land and the atmosphere, that evaporate water and warm or cool the surface. Objects that absorb radiation become warmer than their surroundings and may lose some of that energy by convection, ie, mass motions of air that result in transport and mixing of the properties of the air. This convective heat flux affects air temperature and is known as sensible heat because it can be felt. For example, sensible heat is normally lost from a warm surface to the cooler air above it during the day. Latent heat (evapotranspiration) is also an important means of dissipating heat because of the large quantities of energy required to change water from a liquid to vapor.

Sensible and latent heat fluxes occur because small-scale atmospheric motions transport heat and moisture from the surface into the atmospheric boundary layer and mix them throughout the troposphere. When vertical gradients in temperature or water vapor exist, turbulent fluctuations in wind (eddies) will cause fluctuations in these properties. Sensible and latent heat fluxes are the eddy fluxes of heat and moisture.

Wind flow and turbulence in the boundary layer are controlled by the frictional drag imposed on the flow by the rigid underlying surface (eg, vegetation, soil) and by buoyancy arising from surface heating. Surface drag slows the flow close to the surface, causing a sharp reduction in horizontal wind speed near the surface and a transfer of momentum to the surface. The momentum flux, as well as the sensible and latent heat fluxes, depends on the surface roughness, increasing with increased roughness. Forests are rougher than grasslands; calm lakes are exceptionally smooth.

The momentum, sensible heat, and latent heat fluxes also depend on atmospheric stability. In the daytime, when the surface is typically warmer than the atmosphere, strong solar heating of the land provides a source of buoyant energy. During these unstable conditions, there is a lot of turbulence and the surface fluxes increase. At night, long-wave emission cools the surface more rapidly than the air above. The lowest levels of atmosphere become very stable, with cold, dense air trapped near the surface. Under these conditions, turbulence and vertical mixing are greatly suppressed.

Sensible and latent heat fluxes also vary geographically and seasonally, depending on the net radiation at the surface, the vegetation type, and surface wetness. Globally, about three times as much energy is returned to the atmosphere as latent heat than as sensible heat. Locally, these fluxes are important determinants of temperature, humidity, and wind near the surface.

Some of the net radiation at the surface can also be stored in the soil. Heat storage in the soil is very important for the diurnal cycle of temperature over land. Soils have a much smaller heat capacity, ie, the amount of energy needed to raise temperature by one degree, than water. Consequently, land heats and cools much faster than oceans.

At the surface, therefore, net radiation determines the amount of energy available to evaporate water or to warm or cool the surface. More formally, this is expressed as

$$R_n = H + \lambda E + G$$

where R_n is the net radiation (solar plus long wave) absorbed at the surface, H the sensible heat flux, λE the latent heat flux, and G the soil heat flux. A large net gain of radiation at the surface warms the surface. A transfer of sensible or latent heat from the surface to the atmosphere cools the surface, as does a large flux of heat into the soil. Over the course of a day or year, the heat into the soil is closely balanced by the heat loss from the soil so that G is negligible. At these time scales, the net radiation absorbed by the surface must be transfered back to the atmosphere via sensible and latent heating:

$$R_n = H + \lambda E$$

Hydrologic Cycle

As with the energy budget, the hydrologic cycle is a fundamental feature of the climate system. Water is stored in land as the difference between water input from precipitation (P) and water loss from evapotranspiration (E) and runoff (R):

$$\Delta W = P - E - R$$

This hydrologic balance varies greatly with location and time of year. For example, in temperate regions water is stored as snow in the winter. In these regions, spring snow melt is an important source of water that recharges the soil for the summer months when evapotranspiration exceeds precipitation. In tropical, seasonally dry regions, water is stored in the soil during the rainy season, when precipitation exceeds evapotranspiration, but is lost during the dry season, when precipitation is low. Over long periods of time (eg, annually) and over large geographic regions, storage is small so that

$$P - E = R$$

Total precipitation over land is about 746 mm of water per year, of which 480 mm evaporates back to the atmosphere and the remaining 266 mm flows to oceans as runoff (4). In other words, 36% of the annual precipitation is lost as runoff. Although oceans comprise 70% of the earth's surface they contribute about 85% of the water evaporated from the earth's surface. Much of the land surface is relatively dry, causing small latent heat fluxes. Although certain surface types can be locally important, most of the water in the atmosphere comes from oceanic evaporation. Over land, however, about 60% of the precipitation comes from land, which indicates strong local recycling of water.

Evapotranspiration losses of water from the surface are the source of water vapor to the atmosphere and influence the climate through energy budget considerations, ie, the partitioning of net radiation into sensible and latent heat. Globally, three times as much energy is transferred from the surface to the atmosphere as latent heat than as sensible heat. Locally, sensible heat can be much more important depending on surface conditions. For example, dry surfaces have high sensible heat flux and low latent heat flux. Because more energy goes into sensible rather than latent heat, the lower atmosphere is likely to be warm. Wet surfaces, on the other hand, have high latent heat flux and low sensible heat flux. As a result, the air is likely to be cool and moist. Typical values of the Bowen ratio ($\beta = H/\lambda E$) are 0.1 to 0.3 for tropical rain forests, where high annual rainfall keeps the soil wet year round; 0.4 to 0.8 for temperate forests and grasslands, where less rainfall causes drier soils; 2.0 to 6.0 for semiarid regions with extremely dry soils; and greater than 10.0 for deserts (5).

Plant physiology is also an important determinant of evapotranspiration. Transpiration occurs when stomata, ie, microscopic pores in foliage, open to allow the plant to absorb CO_2 during photosynthesis. At the same time, however, water diffuses out of the leaves as transpiration. The air inside these pores is saturated with water vapor. If the air surrounding the foliage is drier, water will diffuse out of the foliage to the air. This creates a water potential gradient internal to the plant, causing water to move from the soil into the plant to replenish the lost water. Plants have evolved compromises between the need to open stomata to absorb carbon dioxide and the need to close stomata to prevent desiccation. Some plants are better at this than others and have a higher water use efficiency, which is defined as the ratio of CO_2 uptake during photosynthesis to water loss during evapotranspiration.

Runoff is an important component of the climate system because it is a source of fresh water to oceans. Oceans store large amounts of heat in tropical latitudes, where the net radiation into the surface is large. As with the atmosphere, large-scale ocean circulations transport this energy poleward.

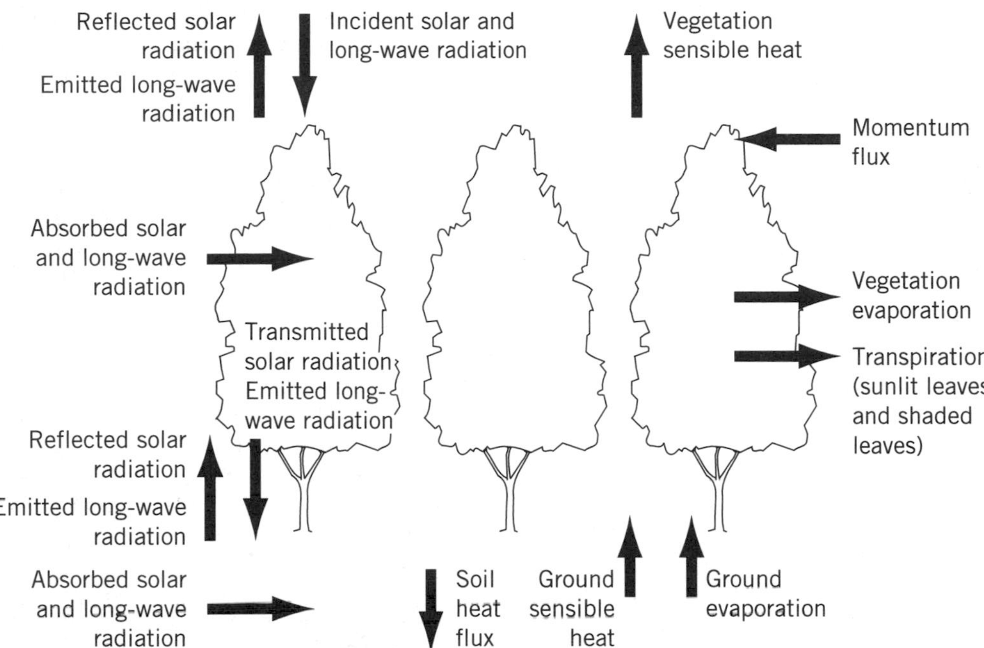

Figure 2. Surface fluxes simulated by land-surface process models. In this diagram, the surface energy budget has two components: the vegetation canopy, where net radiation is balanced by sensible and latent heat; and the ground, where the soil heat flux is an additional term. Solar radiation incident on the vegetation is absorbed, reflected, and transmitted by the vegetation. Some of the transmitted radiation is absorbed by the ground; the rest is reflected. Long-wave radiation has similar fluxes: some of the incident radiation is absorbed by the vegetation; the vegetation emits radiation back to the atmosphere and down onto the ground; the ground absorbs some of this radiation and emits radiation back to the atmosphere. These solar and long-wave radiative fluxes determine the net radiative heating of the atmosphere (Q_{rad}). Sensible and latent heat are exchanged between the atmosphere, the vegetation, and the ground. Sensible heat determines atmospheric temperature (T); latent heat determines atmospheric specific humidity (q). Latent heat is generally partitioned into ground evaporation, evaporation of water intercepted by the vegetation, and transpiration. Transpiration may be further divided into that arising from sunlit and that from shaded leaves. The momentum flux determines atmospheric winds (u, v). The soil heat flux determines soil temperatures.

These circulations, called thermohaline circulations, arise because of density differences from variations in the salinity and temperature of water. In the Atlantic, warm saline surface water flows northward to the North Atlantic, where it cools, becomes denser, sinks, and returns southward at deep depths. Changes in freshwater input to the North Atlantic could affect this poleward transport of heat by decreasing the salinity of the surface water, making it less dense so that it no longer sinks. Runoff of water from land surfaces is a large source of freshwater to oceans. It has been hypothesized that long-term changes in runoff, especially at high latitudes, arising from changes in precipitation, evaporation, or glacial melt could be an important climate feedback by altering the North Atlantic thermohaline circulation.

Biogeochemistry

Terrestrial ecosystems are important sources and sinks of many gases that influence atmospheric radiation, eg, carbon dioxide (CO_2), methane (CH_4), and nitrous oxide (N_2O). The concentration of these gases in the atmosphere is increasing, but the exact causes of this increase and the global cycles of these gases are poorly understood (8). The concentration of CO_2 in the atmosphere is increasing primarily as a result of the combustion of fossil fuels. However, the rate of increase is only about 50% of that of the total emission. The other 50% is taken up by oceans and terrestrial ecosystems. The mechanisms by which this uptake occurs, the magnitudes and geographic locations, and the relative importance of ocean and land are subjects of considerable debate. The major uncertainties for terrestrial carbon storage are related to quantifying: (*1*) the release or uptake of CO_2 from changes in land use practices such as deforestation and agriculture; (*2*) photosynthetic uptake of CO_2 by plants; (*3*) long-term storage of carbon in biomass; (*4*) the response of these ecological processes to increasing atmospheric CO_2 concentration and climate change; and (*5*) possible climate system feedbacks such as release of CO_2 to the atmosphere with a warmer, drier climate. Methane is increasing in the atmosphere because of agricultural activities (rice paddies, animal husbandry), landfills, and biomass burning, but the magnitudes of these fluxes

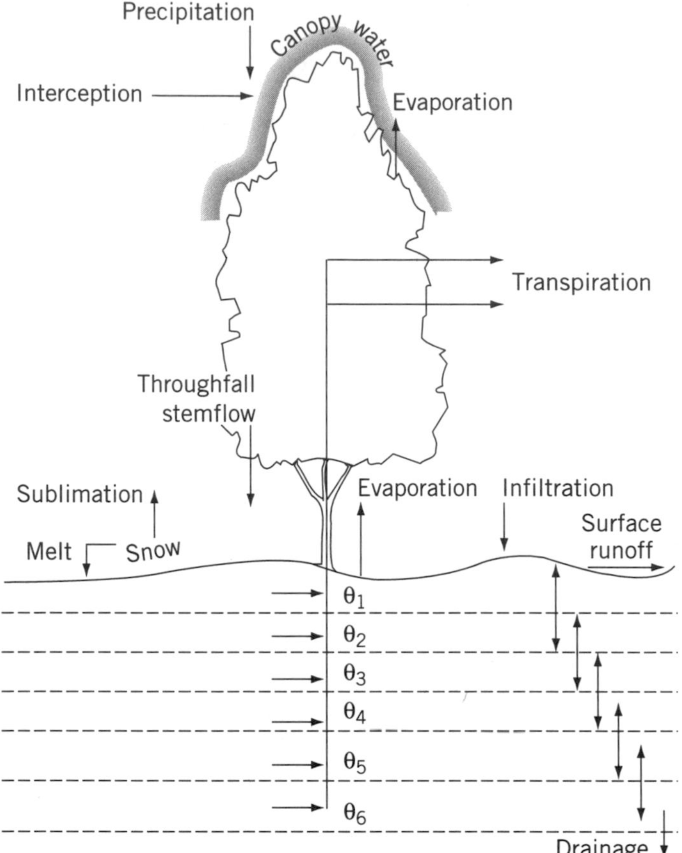

Figure 3. Generalized diagram of the hydrologic cycle within land-surface process models. Some of the precipitation is intercepted by leaves and stems. This canopy water evaporates to the atmosphere. The water that is not intercepted falls as throughfall and stemflow to the ground. If it is cold enough, this water accumulates as snow on the ground. The water in snow can be transferred back to the atmosphere via sublimation or is lost as snow melt when temperatures warm. Some of the flux of liquid water at the surface, ie, throughfall, stemflow, and snow melt, infiltrates into the soil. The rest is lost as surface runoff. Once in the soil, water evaporates, is taken up by plants during transpiration, or moves vertically within the soil. Evaporation, sublimation, and transpiration determine the specific humidity of the atmosphere. The water lost as surface runoff and subsurface drainage eventually reaches the oceans, where it affects salinity.

are not well known. Natural emission of CH_4 from cold, waterlogged boreal forest and tundra ecosystems is highly sensitive to temperature and soil water. It has been speculated that release of methane under a warmer, wetter climate is an important climate feedback. The main anthropogenic sources of N_2O are agriculture, eg, the development of pastures in tropical regions, and biomass burning. As with CO_2 and CH_4, the global budget of N_2O is poorly understood.

MODELING THE EFFECTS OF LAND ON CLIMATE

Land surface process models are a class of models developed for atmospheric numerical models, eg, numerical weather prediction models and global climate models, that simulate the relevant land–atmosphere interactions at the appropriate spatial and temporal scales. Fundamentally, all models solve for a surface temperature, T_s, that balances the energy budget (Fig. 2):

$$R_n[T_s] = H[T_s] + \lambda E[T_s] + G[T_s]$$

In this equation,the components of the energy budget are written as a function of T_s and numerical techniques are used to solve for T_s. The sensible heat and latent heat fluxes are used in the atmospheric model to calculate the temperature and water vapor of the lowest levels of the atmosphere. The difference in net radiation at the top of the atmosphere and at the surface is the net radiative flux into the atmosphere. This, together with H and λE, is the source of energy that drives atmospheric motions. Two other variables, the zonal and meridional surface stresses (τ_x and τ_y), are also calculated because these determine the zonal and meridional winds in the lowest levels of the atmosphere. In combination with precipitation, the evapotranspiration flux E is used to update the hydrologic state of the surface, which feeds back to determine the latent heat flux (Fig. 3). The soil heat flux is used to update the thermal state of the soil.

In calculating these processes, land surface models account for the ecological effects of different vegetation types and the hydrologic and thermal effects of different soil types on the surface fluxes. In more complex models, separate energy budgets are used for the vegetation and the ground surface and solutions for vegetation and ground temperatures that balance these energy budgets are obtained. Processes simulated by many land surface models include:

- vegetation phenology, ie, seasonal changes in leaf area
- absorption, reflection, and transmittance of solar radiation
- absorption and emission of long-wave radiation
- sensible and latent heat fluxes, partitioning latent heat into canopy evaporation, soil evaporation, and transpiration
- momentum fluxes
- heat transfer in a multilayer soil
- stomatal physiology
- interception, throughfall, and stemflow
- snow accumulation and melt
- infiltration and runoff
- soil hydrology, including water transfer in a multilayer soil.

Some land surface models are beginning to simulate biogeochemical processes such as CO_2 uptake during photosynthesis and CO_2 loss during respiration.

Modeling land surface processes is an inexact science. The equations needed to model surface fluxes are well understood from micrometeorological and boundary layer theory (6,7,9,10). Likewise, the soil physics equations are also well understood (2). However, these equations are derived for idealized conditions in which the surface or the soil is relatively homogeneous. In reality, the surface is quite heterogeneous, showing considerable variability in ecological and hydrologic properties at spatial scales much smaller (eg, 1 km^2) than those resolved by atmospheric models (eg, 10,000 km^2). Even a single tree can have considerable variability from leaf-to-leaf in important physiological properties. The inexactness of land

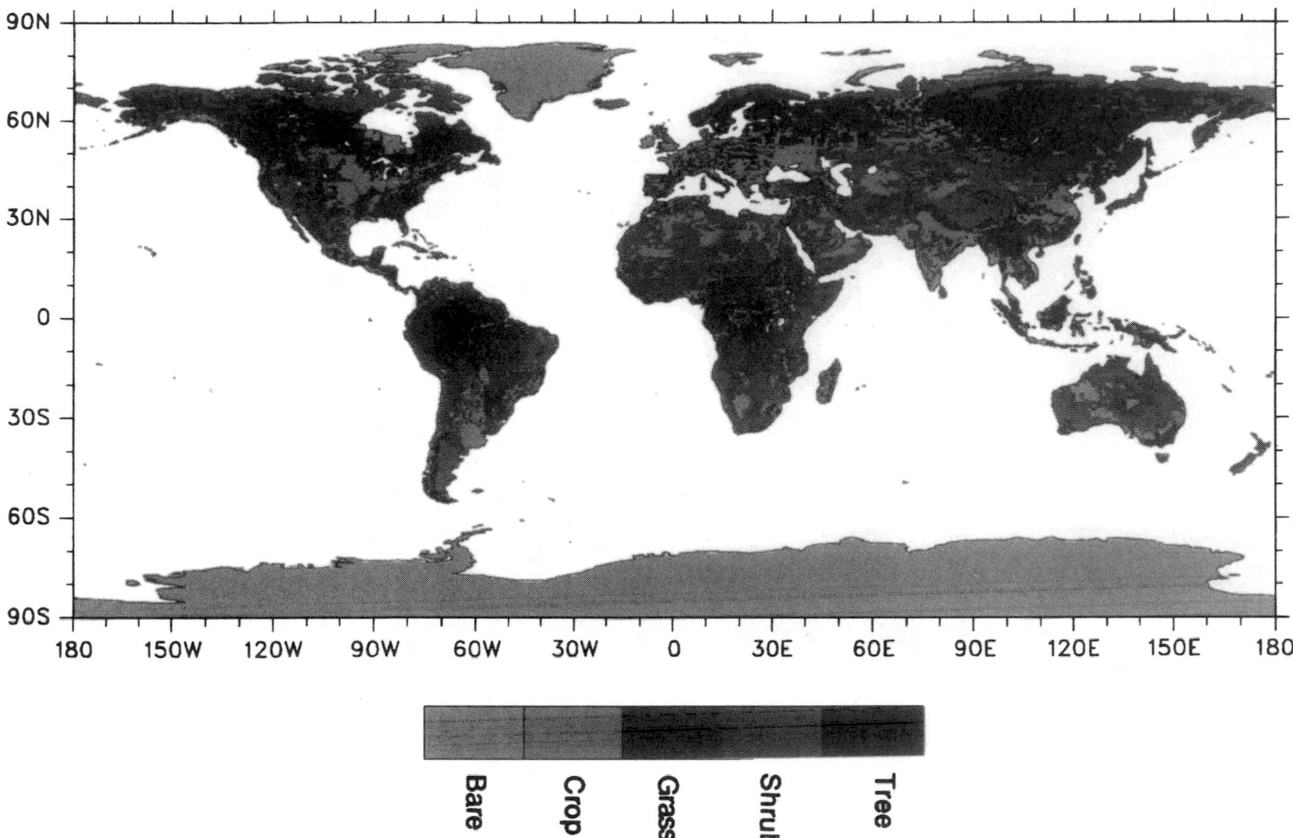

Figure 4. Global distribution of bare ground, crop, grass, shrub, and tree plant types.

surface models arises from attempts to account for this heterogeneity.

IMPORTANT FEATURES OF LAND

Land covers about 150 million km^2, or 30% of the earth's surface. About 70% of this land area is in the Northern Hemisphere. Because of the differences in albedo, surface temperature, sensible heat, and latent heat between land and ocean, this asymmetry causes important differences in climate between the Northern and the Southern Hemispheres.

The ecological characteristics of the surface vary greatly among plant species: some plants absorb more solar radiation; some have higher rates of photosynthesis. It is impossible in a global model to account for all the thousands of different plant species. Rather, plants are grouped into functional types, eg, grass, crops, shrubs, and trees, in which the ecological variability within a type is less than the variability between types. These functional types can be further divided into, for example, C_3 and C_4 grasses (which differ markedly in photosynthesis and stomatal physiology) and needleleaf evergreen, needleleaf deciduous, broadleaf evergreen, and broadleaf deciduous trees (which differ in phenology, physiology, and optical properties). Figure 4 shows the distribution of bareground, crop, grass, shrub, and tree vegetation. Greenland and Antarctica, which are dominated by glaciers, and deserts have no vegetation. Crops are scattered throughout North America, Europe, and tropical regions. Extensive grasslands are found in central North America, northern and central South America, central and southern Africa, central Asia, and Australia. Shrubs occur in high latitude tundra regions and in dry, semidesert areas. Tropical forests occur near the equator; temperate forests occur in mid-latitudes; boreal forests occur at high latitudes.

There is considerable heterogeneity in vegetation structure, eg, leaf area index, biomass, and roughness, within these surface types. Land surface models do not currently account for this type of heterogeneity, in part because it is difficult to estimate these parameters. However, remote-sensing techniques are being developed to determine important land surface characteristics, including vegetation type, leaf area index, and biomass, over large regions and may provide a means to estimate these parameters for global models (11). Accurate classification of vegetation type is particularly important because albedo, roughness, stomatal physiology, and other ecological characteristics that affect surface energy fluxes vary greatly among vegetation types. Leaf area index is an important source of variation within a vegetation type, eg, a young forest with a low leaf area index has different surface fluxes than a mature forest with a high leaf area index.

Thermal and hydraulic properties vary with soil type, eg, loam and clay, and even within a soil type. As with vegetation, soils are generally classified into different texture types based on the sand and clay content of the soil. Figure 5 shows the global distribution of soil types based on sand and clay contents. Sandy soils are found in desert and semidesert regions and at high latitudes. Clay soils are found in the tropics and mid-latitudes.

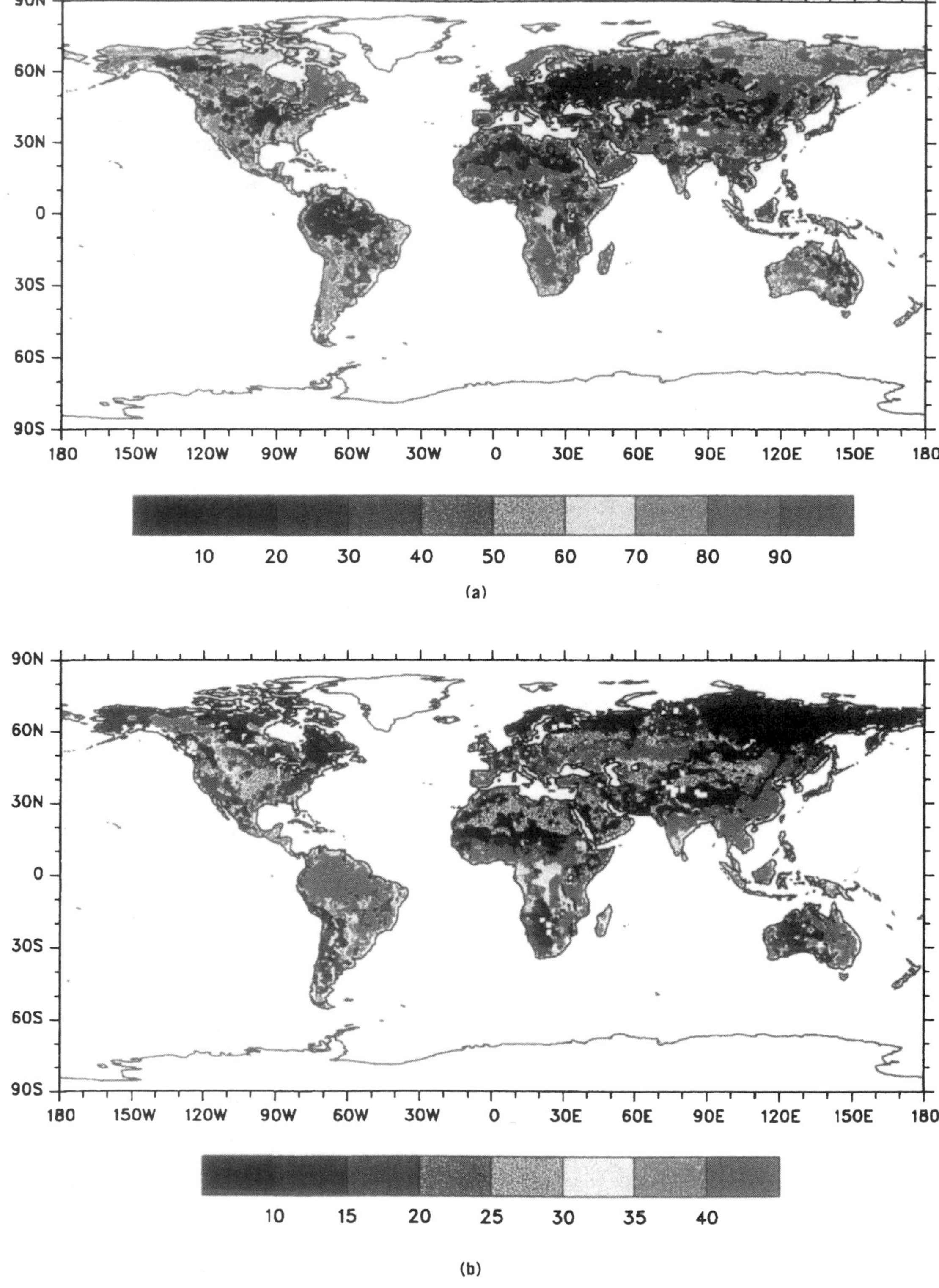

Figure 5. Global distribution of the sand and clay content of soil: (**a**) percentage of sand; (**b**) percentage of clay.

BIBLIOGRAPHY

1. W. H. Schlesinger, *Biogeochemistry: An Analysis of Global Change,* Academic Press, Inc., San Diego, Calif., 1991.
2. R. L. Bras, *Hydrology: An Introduction to Hydrologic Science,* Addison-Wesley Publishing Co., Inc., Reading, Mass., 1990.
3. K. E. Trenberth, *Climate System Modeling,* Cambridge University Press, Cambridge, U.K., 1992.

4. D. L. Hartmann, *Global Physical Climatology,* Academic Press, Inc., San Diego, Calif., 1994.
5. T. R. Oke, *Boundary Layer Climates,* 2nd ed., Routledge, London, 1987.
6. J. L. Monteith and M. H. Unsworth, *Principles of Environmental Physics,* 2nd ed., Edward Arnold, London, 1990.
7. H. G. Jones, *Plants and Microclimate: A Quantitative Approach to Environmental Plant Physiology,* 2nd ed., Cambridge University Press, Cambridge, U.K., 1992.
8. J. T. Houghton and Co-workers, *Climate Change 1995: The Science of Climate Change,* Cambridge University Press, Cambridge, U.K., 1996.
9. W. Brutsaert, *Evaporation into the Atmosphere: Theory, History, and Applications,* D. Reidel Publishing Company, Dordrecht, Holland, 1982.
10. S. P. Arya, *Introduction to Micrometeorology,* Academic Press, Inc., San Diego, Calif., 1988.
11. Special issue on remote sensing of land surface for studies of global change, *Remote Sensing Environ.* **51**(1) (Jan. 1995).

COAL

KARL S. VORRES
Argonne National Laboratory
Argonne, Illinois

The use of coal, known as the rock that burns, was recorded in China, Greece, and Italy over 2000 years ago. Coal mining began in Germany around the tenth century AD and enough coal was mined in England for export in the thirteenth century. Coal mining began in the United States in about 1700.

Coal is usually a dark black color, although geologically younger deposits of brown coal have a brownish red color. The color, luster, texture, and fracture vary with rank, type, and grade. Coal is the result of combined biological, chemical, and physical degradation of accumulated plant matter over geological ages. The relative amounts of different plant parts lead to different types of coal, which are sometimes termed banded, splint, nonbanded (cannel and boghead); hard or soft; or lignite, subbituminous, bituminous, or anthracite. In Europe the banded and splint types are generally referred to as ulmic or humic coals. Other terms refer to the origins of the plant parts through maceral names such as vitrinite, liptinite, and inertinite. The degree of conversion of plant matter or coalification is referred to as rank. Brown coal and lignite, subbituminous coal, bituminous, and anthracite make up the rank series with increasing carbon content. The impurities in these coals cause differences in grade.

Coal consists primarily of carbon, hydrogen, and oxygen and contains lesser amounts of nitrogen and sulfur, and varying amounts of moisture and mineral matter. The mode of formation of coal, the variation in plant composition, the microstructure, and the variety of mineral matter indicate that there is a mixture of materials in coal. The nature of the organic species present depends on the degree of biochemical change of the original plant material, on the historic pressures and temperatures after initial biochemical degradation, and on the finely divided mineral matter deposited either at the same time as the plant material or later. The principal types of organic compounds have resulted from the formation and condensation of polynuclear and heterocyclic ring compounds containing carbon, hydrogen, nitrogen, oxygen, and sulfur. The fraction of carbon in aromatic ring structures increases with rank.

Nearly all coal is used in combustion and coking. At least 80% is burned directly in boilers for generation of electricity or steam for industrial purposes. Small amounts are used for transportation, space heating, firing of ceramic products, etc. The rest is essentially pyrolyzed to produce coke, coal gas, ammonia, coal tar, and light oil products from which many chemicals are produced. Combustible gases and chemical intermediates are also produced by the gasification of coal, and different carbon products are produced by various heat treatments. A small amount of coal is used in miscellaneous applications such as fillers, pigments, foundry material, and water filtration.

In 1995 the annual coal production averaged ca 900×10^6 metric tons in the United States and 4.7×10^9 metric tons for the whole world (3.8×10^9 t coal and 0.9×10^9 t brown coal and lignite) (1,2). Proved recoverable reserves of subbituminous, bituminous, and anthracite coal are ca 7.9×10^{11} tons of coal equivalent, ie, 29.3 GJ/t (12.6×103 Btu/lb) and lignite are 3.5×10^{12} tons of coal equivalent (1). For economic and environmental reasons coal consumption has been cyclic.

ORIGIN OF COAL

Coal evolved from partially decomposed plants in a shallow-water environment. Various chemical and physical changes have occurred in two distinct stages: biochemical and physicochemical (geochemical) (3–7). Because some parts of plant material are more resistant to biochemical degradation than others, optical variations in petrologically distinguishable coals resulted. The terms *vitrain* and *clarain* refer to bright coals, *durain* is a dull coal, and *fusain* is structured fossil charcoal. Exposure to pressure and heat during the geochemical stage caused the differences in degree of coalification or rank that are observable in the continuous series: peat, brown coal and lignite, subbituminous coal, bituminous coal, and anthracite. Carbon-containing deposits in which inorganic material predominates, such as in oil shale and bituminous shale, are not classified as coal.

Complete decay of plant material by oxidation and oxygen-based bacteria and fungi is prevented only in water-logged environments (swamps) in regions where there is rapid and plentiful plant growth. Peat is formed in such swamps from plant debris such as branches and twigs, bark, leaves, spores and pollen, and even tree trunks which are rapidly submerged in the swamp water. A series of coal seams have been formed from peat swamps growing in an area that has undergone repeated subsidence followed by deposition of lacustrine or marine intrusion material. Periods during which vegetation flourished and peat accumulated were followed by rapid subsidence, resulting in submergence of the peat swamp and covering of the deposit with silt and sand. It has been suggested that in the United States the Dismal Swamp of Virginia and North Carolina, which is gradually being flooded by Lake Drummond, is an area undergoing active subsidence (8).

According to the autochthonous (*in situ*) theory of coal formation, peat beds, and subsequently coal, were formed from the accumulation of plants and plant debris in place. According to the allochthonous theory, the coal-producing peat bogs or swamps were formed from plant debris that had been transported, usually by streams or coastal currents, to the observed burial sites.

Biochemical Stage

The initial biochemical decomposition of plant matter depends on two factors: the ability of the different plant parts to resist attack and the existing conditions of the swamp water. Fungi and bacteria can cause complete decay of the plant matter if it is exposed to aerated water or to the atmosphere. The decay is less complete if the vegetation is immersed in water containing anaerobic bacteria. Under these latter conditions the plant protoplasm, proteins, and starches, and to a lesser extent the cellulose, are easily digested. Lignin is more resistant. The most decay-resistant plant parts for both anaerobic and aerobic decomposition are the waxy protective layers, ie, cuticles, spore, and pollen walls, as well as resins. Vitrain results from the partial decay of lignin and cellulose in stagnant water. The original cell structure of the parent plant tissue can be recognized in many samples.

The clarain (9) and bright attritus (8) are finely banded bright parts of coal that evolved from the residues of fine woody material such as branches, twigs, leaves, spores, bark, and pollen. In aerated waters the plant parts were more decomposed and show a higher concentration of resins, spores, and cuticles. Dull coal, called durain (9), was formed under these conditions, and occurs commonly in Europe. It is not as widely found in the United States where it is known as splint or block coal. More selective chemical and biochemical activity, probably in a drier environment, led to the formation of soft, charcoal-like fusain from woody plant material. The conversion was rapid and probably complete by the end of the peat formation stage. Cannel coal is believed to have formed in aerated water which decomposed all but the spores and pollen. The name is derived from its quality of burning in splints with a candle-like flame. Boghead coal closely resembles cannel coal, but was derived from algae instead of plant spores.

Geochemical Stage

The conversion of peat to bituminous coal is the result of the cumulative effect of temperature and pressure over a long time (millions of years). The sediment covering the peat provides pressure and insulation so that the earth's internal heat can be applied to the conversion (temperature increases about 4–8°C for each 100 m of depth). Changes in plant matter are termed normal coalification.

Moisture is lost and the chemical composition changes during coalification. Oxygen and hydrogen decrease, and carbon increases. These compositional changes are accompanied by decreases in volatile matter and increases in calorific value. The volatile matter and calorific content are the main criteria used for commercial classification in the United States and for the International Classification (see below).

The change in rank from bituminous coal to anthracite involves the application of significantly higher pressures, ie, as with mountain building activity, and temperatures, ie, as with volcanic activity. The more distant the coal from the disruption, the less proportionate the alteration. Tectonic plate movements involved in mountain building provide pressure for some changes to anthracite. As a general rule, the older the coal deposit, the more complete the coalification, and the higher the rank of the coal. Most commercial bituminous coal fields were deposited during the Pennsylvanian (ca 285–320 million years ago), Upper Cretaceous (ca 65–100 million years ago), and early Tertiary (ca 20–65 million years ago) ages. The lower rank coals came primarily from the Tertiary and Upper Cretaceous ages, and peat deposits are relatively recent, less than one million years old. However, age alone does not determine rank. The brown coal of the Moscow basin is not buried deeply, and although it was deposited during the Lower Carboniferous age (Mississippian, ca 320–360 million years ago), there was not enough heat and pressure to convert it further.

COAL PETROGRAPHY

Careful examination of a piece of coal shows that it is usually made up of layers or bands of different materials which upon microscopic examination are distinct entities distinguishable by optical characteristics (10–20). The study of the origin, composition, and technological application of these materials is called coal petrology. Whereas coal petrography involves systematic quantification of the amounts and characteristics by microscopic study, petrology of coal may involve a macroscopic or microscopic scale.

On the macroscopic scale two classifications have been used: humic (banded) coals and sapropelic (nonbanded) coals. Stratification in the banded coals, which results from plant parts, is quite obvious; nonbanded coals, which derive from algal material and spores, are much more uniform. The physical and chemical properties of the different layers in a piece of coal or a seam can vary significantly. Therefore relative amounts of layers are important in determining the overall characteristics of the mined product. Coal petrography has been widely applied in coke making and is important in coal liquefaction programs.

If mineral matter in the coal exceeds about 40%, then the material is referred to as a coaly or carbonaceous shale. If the mineral matter is a finely divided clay, well dispersed in the coal, then the material may be described as a stony coal or bone coal.

Macerals

Coal parts derived from different plant parts are referred to as *macerals* (13). Maceral names end in-*inite,* as do the mineral forms of rocks. The most abundant (about 85%) maceral in U.S. coal is vitrinite, derived from the woody tissues of plants. Another maceral, called liptinite, is derived from the waxy parts of spores and pollen, or algal remains. The liptinite macerals fluoresce under blue light permitting a subdivision based on fluorescence. A third maceral, inertinite, is thought to be derived from oxidized material or fossilized charcoal remnants of early forest fires.

A number of subdivisions of the maceral groups have been developed and documented by the International Commission on Coal Petrology (14). Table 1 lists the Stopes-Heerlen classification of higher rank soils. Periodic revisions include descriptions of the macerals, submacerals, morphology, physical properties, and chemical characteristics. Theories on the mode of formation of the macerals and their significance in commercial applications are available (14).

The macerals in lower rank coals, eg, lignite and subbituminous coal, are more complex and have been given a special clas-

Table 1. Stopes-Heerlen Classification of Maceral Groups, Macerals, and Submacerals of Higher Rank Coals[a]

Maceral Group	Maceral	Submaceral
Vitrinite	Telinite	Telocollinite
	Collinite	Gelocollinite
		Desmocollinte
		Corpocollinite
Liptinite	Sporinite	
	Cutinite	
	Suberinite	
	Resinite	
	Alginite	
	Liptodetrinite	
	Fluorinite	
	Bituminite	
	Exudatinite	
Inertinite	Fusinite	
	Semifusinite	
	Macrinite	
	Micrinite	
	Sclerotinite	
	Inertodetrinite	

[a] Ref. 9.

sification. The term *huminite* has been applied to the macerals derived from the humification of lignocellulosic tissues. Huminite is the precursor to the vitrinite observed in higher rank coals.

The elemental composition of the three maceral groups varies. The vitrinite, which frequently is about 85% of the sample in the United States, is similar to the parent coal. The liptinites are richer in hydrogen, whereas the inertinites are relatively deficient in hydrogen or richer in carbon. The liptinites also contain more aliphatic materials; the inertinites are richer in aromatics. The term *inertinite* refers to the relative chemical inertness of this material, making it especially undesirable for liquefaction processes because it tends to accumulate in recycled feedstock streams.

Vitrinite Reflectance

The amount of light reflected from a polished plane surface of a coal particle under specified illumination conditions increases with the aromaticity of the sample and the rank of the coal or maceral. Precise measurements of reflectance, usually expressed as a percentage, are used as an indication of coal rank.

CLASSIFICATION SYSTEMS

Prior to the nineteenth-century, coal was classified according to appearance, eg, bright coal, black coal, or brown coal. A number of classification systems have since been developed. These may be divided into two complementary types: scientific and commercial. Both are used in research, whereas the commercial classification is essential industrially. In the scientific category, the Seyler chart has considerable value.

Seyler's Classification

The Seyler chart (Fig. 1) is based on the carbon and hydrogen content of coals determined on a dry mineral-matter-free basis (17). Classification is based on the observation that the amounts of two elements (C and H) present in coals are useful

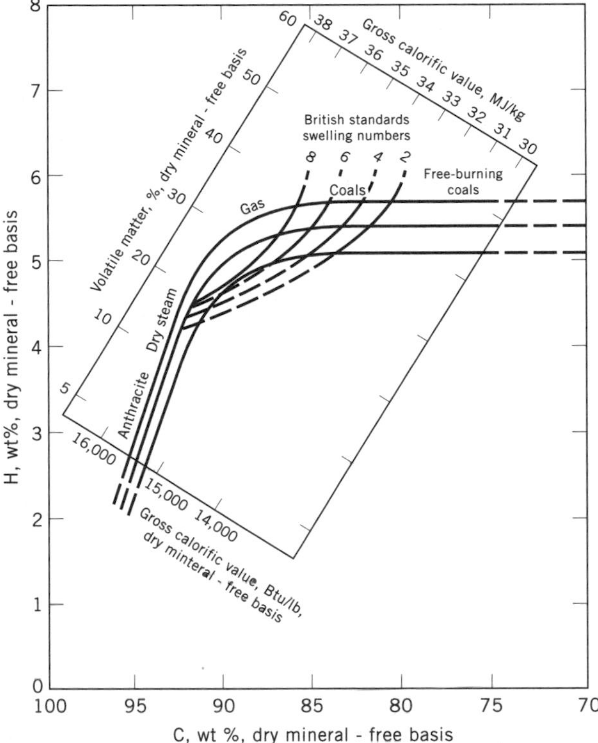

Figure 1. Simplified form of Seyler's coal classification chart (17). An updated version of Seyler's coal classification has been described (6). ASTM uses the free-swelling index (18).

in correlating properties (18). A plot of wt % H vs wt % C, on a dry mineral-matter-free basis, is used. Points representing different coal samples lie along a broad band. The center band on the chart shows the properties of coals rich in vitrinite. The location of the band indicates the range and interrelationship of the properties. Coals above the band are richer in hydrogen, eg, cannel and boghead coals, and the liptinite macerals in the usual coals. Coals below the band are represented by the maceral inertinite.

Other properties, such as moisture and swelling indexes, also fit into specific areas on this chart. The curve in the solid band represents a composition range where the properties of the coal change rapidly. Swelling indexes, coking power, and calorific values are maximized, and moisture is minimized. The lowest rank coals lie on the right side and the highest rank coals on the left at the lower part of the band.

ASTM Classification

The ASTM classification system was adopted in 1938 as a standard means of specification. This system is used in the United States and many other parts of the world, and designated is D388 in the ASTM Standards (18). The higher rank coals are specified by fixed carbon $>$69%, or for volatile matter $\leq$31%, on a dry mineral-matter-free basis. Lower rank coals are classified by calorific value on the moist mineral-matter-free basis. These parameters are given in Table 2. Calorific value depends on two properties: moisture absorbing capacity and the calorific value of the pure coal matter. When some overlap between bituminous and subbituminous coals occurs, it is resolved on the basis of the agglomerating properties.

Table 2. Classification of Coals by Rank

Coals	Fixed Carbon, %[a]		Volatile Matter, %[a]		Gross Calorific Value, kJ/kg[b]		Agglomerating Character
	≥	<	>	≤	≥	<	
				Anthracitic			
meta-Anthracite	98			2			Nonagglomerating
Anthracite	92	98	2	8			
Semianthracite[c]	86	92	8	14			
				Bituminous			
Low volatile	78	86	14	22			
Medium volatile	69	78	22	31			
High volatile							Commonly agglomerating[d]
A		69	31		32,500[e]		
B					30,200[e]	32,500	
C					26,700	30,200	
					24,400	26,700	Agglomerating
				Subbituminous			
A					24,400	26,700	
B					22,100	24,400	
C					19,300	22,100	
				Lignitic			
A					14,600	19,300	Nonagglomerating
B						14,600	

[a] Dry, mineral-matter-free basis.
[b] To convert from kJ/kg to Btu/lb, multiply by 0.4302; moist mineral-matter-free basis, ie, contains inherent moisture but not water visible on the surface.
[c] If agglomerating, classify in low volatile group of the bituminous class.
[d] There may be nonagglomerating varieties in the groups of the bituminous class, and there are notable exceptions in high volatile C bituminous group.
[e] Coals having 69% or more fixed carbon on the dry, mineral-matter-free basis are classified according to fixed carbon, regardless of gross calorific value.

National Coal Board Classification for British Coals

The classification proposed in 1946 by the U.K. Department of Scientific and Industrial Research led to the system used by the National Coal Board for coals in the United Kingdom. There are two parameters: the quantity of volatile matter on a dry mineral-matter-free basis and the Gray-King coke-type assay, a measure of coking power as designated in the *British Standards* (18). This latter assay is used as a primary means of classification for lower rank coals. The classification applies to coals with less than 10% ash. High ash coals are cleaned before analysis by a float-sink separation to reduce the ash content below 10%.

International Classification

Hard Coal. The amount of coal in international commerce since ca 1945 necessitated an international system of coal classification and in 1956 the Coal Committee of the European Economic Community agreed on a system designated the International Classification of Hard Coal by Type (3). Volatile matter and gross calorific value on a moist ash-free basis are among the parameters considered. Table 3 shows the various classes of the international system and gives the corresponding national names used for these coals.

A three-digit classification is employed in the International System (10) where the first digit indicates the class or rank, such that higher digits correspond to lower ranks; the second digit indicates the group indicated by caking properties, such as the free-swelling index or Roga index; and the third digit defines a subgroup based on coking properties as measured with a dilatometer or the Gray-King assay. Coals with volatile matter up to 33% are divided into classes 1–5; coals having volatile matter up to 33% are divided into classes 6–9. The calorific values are given for a moisture content obtained after equilibrating at 30°C and 96% rh. The nine classes are divided into four groups determined by caking properties as measured through either the free-swelling index (17) or the Roga index. These tests indicate properties observed when the coal is heated rapidly.

Brown Coal and Lignite. The brown coals and lignites, defined as those coals having heating values less than 23,860 kJ/kg (10,260 Btu/lb, 5700 kcal/kg), are classified separately. A four-digit code is used for classification. The first two digits (class parameter) are defined by total moisture content of freshly mined coal on an ash-free basis. The third and fourth digits are defined by the tar yield on a dry ash-free basis.

COMPOSITION AND CONSTITUTION

The constitution of a coal involves both the elemental composition and the functional groups that are derived therefrom. The structure of the coal solid depends to a significant extent on the arrangement of the functional groups within the material.

Coal Composition

The functional groups within coal contain the elements C, H, O, N, or S (3,4,5,19). The significant oxygen-containing groups

Table 3. The International and Corresponding National Systems of Coal Classes[a]

International System			National Classifications							
Class No.	Volatile Matter, %	Calorific Value, kJ/g[b,c]	Belgium	Germany	France	Italy	The Netherlands	Poland	United Kingdom	United States
0	0–3					Antraciti speciali		*meta*-Antracyt		*meta*-Anthracite
1A	3–6.5		Maigre	Anthrazit	Anthracite	Antraciti communi	Anthraciet	Antracyt	Anthracite	Anthracite
1B	6.5–10							Polantracyt		
2	10–14		$\frac{1}{4}$ gras	Magerkohle	Maigre	Carboni magri	Mager	Chudy	Dry steam	Semianthracite
3	14–20		$\frac{1}{2}$ gras	Esskohle	Demigras	Carboni semigrassi	Esskool	Polkoksowy	Coking steam	Low volatile bituminous
			$\frac{3}{4}$ gras					*meta*-Koksowy		
4	20–28			Fettkohle	Gras à courte flamme	Carboni grassi corta fiamma	Vetkool	*orto*-Koksowy	Medium volatile coking	Medium volatile bituminous
5	28–33		gras	Gaskohle	Gras proprement dit	Carboni grassi media fiamma		Gazowo koksowy		High volatile bituminous A
6	>33 (33–40)	32.4–35.4				Carboni da gas	Gaskool			
7	>33 (32–44)	30.1–32.4			Flambant gras	Carboni grassi da vapore	Gasvlamkool	Gazowy	High volatile	High volatile bituminous B
8	>33 (34–46)	25.6–30.1		Gas flammkohle	Flambant sec	Carboni secchi	Vlamkool	Gazowoplomienny		High volatile bituminous C
9	>33 (36–48)	<25.6						Plomienny		Subbituminous

[a] Ref. 3.
[b] Calculated to standard moisture content.
[c] To convert kJ/g to Btu/lb, multiply by 430.2.

found in coals are carbonyl, hydroxyl, carboxylic acid, and methoxy. The nitrogen-containing groups include aromatic nitriles, pyridines, carbazoles, quinolines, and pyrroles (20). Sulfur is primarily found in thiols, dialkyl and aryl–alkyl thioethers, thiophene groups, and disulfides. Elemental sulfur is observed in oxidized coal (20).

The relative and absolute amounts of the various groups vary with coal rank and maceral type. The principal oxygen-containing functional groups in vitrinites of mature coals are phenolic hydroxyl and conjugated carbonyls as in quinones. Spectroscopic evidence exists for hydrogen bonding of hydroxyl and carbonyl groups. There are unconjugated carbonyl groups such as ketones in exinites. The infrared absorption bands are displaced from the normal range for simple ketones by the conjugation in vitrinites. Interactions between the carbonyl and hydroxyl groups affect normal reactions.

A range of quantitative organic analytical techniques may be used to determine the functional group concentrations. Acetylation and *O*-alkylation are used to determine hydroxyl groups, whereas carbonyl groups are difficult to quantify with simple procedures. A variety of instrumental techniques has also been used to aid in the understanding of coal structure and constitution. Magnetic resonance techniques have been particularly helpful in determining relative amounts of different carbon species within a coal.

Aromaticity of coal molecules increases with coal rank. Calculations based on several models indicate that the number of aromatic carbons per cluster varies from nine for the lignite to 20 for the low volatile bituminous, and the number of attachments per cluster varies from three for the lignite to five for subbituminous through medium volatile bituminous. The value is four for low volatile bituminous (21).

Reactions of coals and mild selective oxidizing agents like benzoquinone (20,21) causes the coals to lose much of their hydrogen content. Similarly, a palladium catalyst can cause the evolution of molecular hydrogen (22,23). These methods may give an indication of the minimum amount of hydrogen in the coal that is involved in hydroaromatic rings. This amount is close to the total nonaromatic hydrogen determined for lower rank coals. Other hydrogen-determining methods involve dehydrogenation using sulfur (24) and halogens (25). The values obtained by these last methods are somewhat lower than that of benzoquinone.

Hydrogen can be added to the aromatic structures converting them to hydroaromatic rings. Hydrogen addition and removal is generally but not entirely reversible (24).

High resolution mass spectrometry has been used with extracts of a series of coals to indicate the association of different heteroatoms (26). Various types of chromatography have been used to identify the smaller species that can be extracted from coal.

Coal Constitution

Chemical composition studies (27,28) indicate that brown coals have a relatively high oxygen content. About two-thirds of the oxygen is bonded carboxyl, acetylatable hydroxyl, and methoxy groups. Additionally, unlike in bituminous coals, some alcoholic hydroxyl groups are believed to exist.

The anthracites, which approach graphite in composition, are classified higher in rank than bituminous coals, have less oxygen and hydrogen, and are less reactive and insoluble in organic solvents. These characteristics become more pronounced as rank increases within the anthracite group. The aromatic carbon fraction of anthracites is at least 0.9, and the number of aromatic rings per cluster is greater than that for the low volatile bituminous coals, with a value of about 10 for anthracite with 95 wt % C. There is X-ray diffraction evidence (29) to indicate that the aromatic rings are more loosely and variably assembled than those in bituminous coal clusters. The anthracites have greater optical and mechanical anisotropy than lower rank coals, and the internal pore volume and surface increase with rank after the minimum below about 90 wt % C.

Mineral Matter in Coal

The mineral matter (7,30) in coal results from several separate processes. Some comes from the material inherent in all living matter; some from the detrital minerals that were deposited during the time of peat formation; and a third type from secondary minerals that have crystallized from water which has percolated through the coal seams.

A variety of instrumental techniques may be used to determine the mineral content. Typically the coal sample is prepared by low temperature ashing to remove the organic material. Then one or more of the techniques of X-ray diffraction, infrared spectroscopy, differential thermal analysis, electron microscopy, and petrographic analysis may be employed (7).

The various clay minerals are the most common detrital mineral; however, other common ones include quartz, feldspar, garnet, apatite, zircon, muscovite, epidote, biotite, augite, kyanite, rutile, staurolite, topaz, and tourmaline. The secondary minerals are generally kaolinite, calcite, and pyrite. Analyses have shown the presence of almost all elements in at least trace quantities in the mineral matter (31). Certain elements, ie, germanium, beryllium, boron, and antimony, are found primarily with the organic matter in coal, whereas zinc, cadmium, manganese, arsenic, molybdenum, and iron are found with the inorganic material. The primary elemental constituents of mineral matter in coal are aluminum, silicon, iron, calcium, magnesium, sodium, and sulfur. The relative concentrations depend primarily on the geographical location of the coal seam, and vary from place to place within a given field. In the eastern United States the most abundant mineral elements are silicon, aluminum, and iron with much lower amounts of alkali and alkaline-earth elements. West of the Mississippi River the relative amounts of the silicon, aluminum, and iron are much less and the alkaline and alkali elements are much greater.

PROPERTIES

Pieces of coal consist of a mixture of materials somewhat randomly distributed in differing amounts. The mineral matter can be readily distinguished from the organic part. The organic material is a mixture of constituents that have undergone a range of changes under coalification conditions. The properties of the coal reflect the individual constituents and their relative proportions. The emphasis of this section is on the properties of the organic matter, with attention to the macerals. By analogy with geology, the macerals are the constituents that correspond to minerals that make up individual rocks. For coals, these macerals represent particular classes of plant parts that have been transformed into their present state in coal (31). Macerals tend to be consistent in their properties. For this reason most detailed chemical and physical studies have been made on macerals or samples rich in a particular maceral, since separation of macerals is so time consuming. The most predominant maceral group in U.S. coals is vitrinite. The other important maceral groups include inertinite consisting of micrinite, a dull black amorphous material; fusinite, a dull fibrous material similar to charcoal; and the liptinite group, including sporinite which is relatively fusible and volatile. The differences in macerals are evident over the range of coals from brown coal or lignite to anthracite. These coals cover the full range of coalification or rank. The definition of rank is that generally accepted as the wt % C, on a dry mineral-free basis, in the vitrinite associated with the given coal in the seam. The range of ranks in which differences between macerals are most significant is 75–92 wt % C content of the vitrinite. These coals are bituminous.

The U.S. commercial classification of coals is based on the fixed carbon (or volatile matter) content and the moist heating value. One scientific correlation is made by plotting the hydrogen content, on a mineral-free basis, against the corresponding carbon content. A similar plot, made using the commercial criteria of volatile matter and heating placed on axes at an appropriate angle to the C–H axes, forms the Seyler coal classification chart (18) (see Fig. 1).

Physical Properties

Most of the physical properties discussed herein depend on the direction of measurement compared to the bedding plane of the coal. Additionally, the properties vary according to the history of the piece of coal. They also vary between pieces because of the brittle nature of coal and of the crack and pore structure. One example is electrical conductivity. Absolute values of specific conductivity of coal samples are not easy to determine. A more characteristic value is the energy gap for transfer of electrons between molecules. This is determined by a series of measurements over a range of temperatures and is unaffected by the presence of cracks. The velocity of sound is also dependent on continuity in the coal.

The specific electrical conductivity of dry coals is very low (specific resistance 10^{10}–10^{14} ohm · cm), although it increases with rank. Coal has semiconducting properties. The conductivity tends to increase exponentially with increasing temperatures (4,6). As coals are heated to above ca 600°C the conductivity rises especially rapidly owing to rearrangements in the carbon structure, although thermal decomposition contributes somewhat below this temperature. Moisture increases conductivity of coal samples through the water film.

The dielectric constant is also affected by structural changes on strong heating. Also it is very rank dependent, exhibiting a minimum at about 88 wt % C, and rises rapidly for carbon contents over 90 wt % (4,6,32). The presence of polar functional groups is responsible for this with lower ranks, and for higher ranks it is caused by the increase in electrical conductivity. Information on the freedom of motion of the different water molecules in the particles can be obtained from dielectric constant studies (32).

Magnetic susceptibility measurements indicate that the organic part of the coal is diamagnetic, with traces of paramagnetic behavior due to free radicals or unpaired electrons (6).

Density values (4,6) of coals differ considerably, even after correcting for the mineral matter, depending on the method of determination. The true density of coal matter is most accurately obtained from measurement of the displacement of helium after the absorbed gases have been removed from the coal sample. Density values increase with carbon content or rank for vitrinites. They are 1.4–1.6 g/cm^3 above 85 wt % carbon where there is a shallow minimum. A plot of density versus hydrogen content gives almost a straight-line relationship. If the reciprocal of density is plotted, the relationship is improved. Values for different macerals as well as for a given maceral of different ranks are almost on the same straight line.

The thermal conductivity and thermal diffusivity are also dependent on the pore and crack structure. Thermal conductivities for coals of different ranks at room temperature are in the range of 0.23–0.35 W/(m · K). The range includes the spread owing to crack variations and thermal diffusivities of (1–2) × 10^{-3} cm^2/s. At 800°C these ranges increase to 1–2 W/(m · K) and (1–5) × 10^{-2} cm^2/s. The increase is mainly caused by radiation across pores and cracks.

The specific heat of coal can be determined by direct measurement or from the ratio of separate measurements of thermal conductivity and thermal diffusivity. The latter method gives values decreasing from 1.25 J/(g · K) (0.3 cal/(g · K)) at 20°C to 0.4 J/(g · K) (0.1 cal/(g · K)) at 800°C. The specific heat is affected by oxidation of the coal (33).

Ultrafine Structure

Coal contains an extensive network of ultrafine capillaries (3,4,6,34) that pass in all directions through any particle. The smallest and most extensive passages are caused by the voids from imperfect packing of the large organic molecules. Vapors pass through these passages during adsorption, chemical reaction, or thermal decomposition. The rates of these processes depend on the diameters of the capillaries and any restrictions in them. Most of the inherent moisture in the coal is contained in these capillaries. The porous structure of the coal and products derived from it have a significant effect on the absorptive properties of these materials.

A range of approaches has been developed for studying the pore structure. For example, heat of wetting by organic liquids is one measure of the accessible surface. The use of liquids with different molecular sizes gives information about restrictions in the pores. Measurements of the apparent density in these liquids give corresponding information about the volume of capillaries. Measurement of the adsorption of gases and vapors provide information about internal volume and surface area. Pores have been classified into three size ranges: (*1*) micropores (<0.4–2.0 nm) measured by CO_2 adsorption at 298 K; (*2*) mesopores (2.0–50 nm) from N_2 adsorption at 77 K; and (*3*) macropores (>50 nm) from mercury porosimetry. For coals with less than 75 wt % C, macropores primarily determine porosity. For 76–84 wt % C about 80% of the pore volume is primarily due to micro- and transitional pores. For higher rank coals porosity is caused primarily by micropores (35).

Bituminous coals have specific internal surfaces in the range of 30–100 m^2/g, almost entirely from ultrafine capillaries with diameters less than 4 nm. The surface area of the very fine capillaries can be measured accurately by using methods not too far below room temperature, depending on the gas or vapor used, since diffusion into the particle is very slow at low temperatures (36). Therefore, measurements at liquid air temperature relate to the external surfaces and macro- as well as mesopores, and may yield areas that are lower than ambient temperature measurements by factors of 100. Sorption by neon or krypton near room temperature, and heat of wetting in methanol have been favored methods of obtaining surface area values. The methanol method is affected (factor up to 4) by polar groups but is faster. Pore characteristics may also be determined by NMR measurements, as with xenon (37). Total porosity volumes of bituminous and anthracite coal particles are about 10–20% with about 3–10% in the microrange. There are shallow minima in plots of internal area or internal porosity against the rank in the range of vitrinite carbon content of 86–90% (6).

It is possible to use low angle scattering of X-rays to obtain a value of internal surface, but this does not distinguish between accessible capillaries and closed pores (38).

The electron microscope has been applied to different coals. Two pore-size ranges appear to have been observed, one of >20 nm and the other <10 nm (39). Fine pores 1–10 nm across have been observed with a lead impregnation procedure (40). Effectiveness of coal conversion processes depends on rapid contact of gases with the surface. Large internal surfaces are required for satisfactory rates (41).

Mechanical Properties

Mechanical properties (4,6,42) are important for a number of steps in coal preparation from mining through handling, crushing, and grinding. The properties include elasticity and strength as measured by standard laboratory tests and empirical tests for grindability and friability; and indirect measurements based on particle size distributions.

Properties Involving Utilization

Coal rank is the most important single property for almost every application of coal. The rank sets limits on many properties such as volatile matter, calorific value, and swelling and coking characteristics. Other properties of significance include grindability, ash content and composition, and sulfur content.

Combustion. Most of the mined coal is burned in boilers to produce steam for electric power generation. The calorific value determines the amount of steam that can be generated. The design and operation of a boiler requires consideration of a number of other properties.

In general, high rank coals are more difficult to ignite, requiring supplemental oil firing and slower burning with large furnaces to complete combustion. The greater reactivity of lower rank coals makes them better suited for cyclone burners which carry out rapid, intense combustion to maximize carbon utilization and minimize smoke emission. The "burning profile", a derivative thermogravimetric analysis of oxidation, is used to characterize different coals for oxidation or combustion behavior.

Volatile matter is important for ease of ignition of coal. High rank coals have low volatile matter contents. They burn more slowly with a short flame and are used for domestic heating, where heat is transferred primarily from the fuel bed. For kilns, long hot flames are needed and the coal should have medium to high volatile matter.

The swelling and caking properties of coal are not important for most boiler firing, such as pulverized coal-fired use. Some units, however, such as retort stokers, form coke in their normal operation. The smaller domestic heating units require noncaking coal for satisfactory operation.

For pulverized coal firing, a high Hardgrove index or grindability index is desired. This implies a relatively low energy cost for pulverizing since the coal is easier to grind. The abrasiveness of the coal is also important since this determines the wear rate on pulverizer elements.

Moisture content affects handling characteristics. It is most important for fine particles <0.5 mm. The lower rank coals have higher moisture contents that affect the freight costs, especially when the coal fields are considerable distances from the consumption locations. The moisture acts as a diluent, lowering flame temperatures and carrying sensible heat out with the flue gases. For pulverized coal firing the moisture content must be low to maximize grindability and avoid clogging. For this reason it may be more desirable to buy dry run-of-mine coal with up to 30% ash than to clean and dry the coal. Moisture is sometimes desirable. About 8% is necessary for prevention of combustible loss from a chain-grate stoker.

The moisture content of peat or brown coal that is briquetted for fuel must be reduced to about 15% for satisfactory briquetting. Mechanical or natural means are used because of the cost of thermal drying.

Ash content is also important. Ash discharge at high temperature, as molten ash from a slagging boiler, involves substantial amounts of sensible heat. However, the higher cost of washed coal of lower ash content (about 10%) does not always merit its use. Ash disposal and extra freight costs for high ash coals enter the selection of coal. The current use of continuous mining equipment produces coal with about 25% ash content. The average ash content of steam coal burned is about 15% in the United States. For some applications, such as chain-grate stokers, a minimum ash content of about 7–10% is needed to protect the metal parts.

Ash fusion characteristics are important in ash deposition in boilers. Ash deposition occurring on the furnace walls is termed slagging, whereas accumulation on the superheater and other tubes is termed fouling. A variety of empirical indices have been developed (43,44) to relate fouling and slagging to the ash chemical composition through parameters including acidic and basic oxides content, sodium, calcium and magnesium, and sulfur.

A related property is viscosity of coal ash, which affects the rate at which ash deposits may flow from the walls, and affects the requirements for ash removal equipment such as wall blowers and soot blowers. The preferred coal ash has a narrow temperature range through which it passes the plastic range (ca 25–1000 Pa · s or 250–10,000 P) (45).

Some minor constituents can cause trouble in firing. High chlorine (>0.6%) is associated with high sodium and complex sulfate deposits that appear to be required to initiate deposition on superheater tubes, as well as initiate stress corrosion cracking of superheater tubes. Phosphorus (>0.03% of the coal) contributes to phosphate deposits where high firing temperatures are used. Sulfur is also involved in the complex sulfates; however, its most damaging effect is corrosion of the boiler's coolest parts through condensation of sulfur oxides as sulfuric acid. Control is achieved by setting flue gas temperatures above the acid dew point in the boiler areas of concern.

The sulfur content is important in meeting air quality standards. In the United States, the EPA has set a limit of 516 g S02 emission per million kJ (490 g SO_2 10^6 Btu) of coal burned. To meet this, steam coals have to contain less than 1% sulfur. New regulations, involved in the Clean Air Act of 1991, call for reduction of the total amount of sulfur oxide emissions by 8–9 million tons per year. Half of the reduction is required by 1995, with the remainder by 2000. A cap on the total emissions is given. Reductions in NO_x and particulate emissions are also mandated. Credits will be given for reductions of emissions beyond the amount indicated for a given plant. The credits may be sold to other facilities unable to meet their requirements. This requirement will probably force the addition of SO_2, NO_x, and particulate removal equipment to all boilers. New technology is being developed to control SO_x and NO_x through a combination of sorbent injection into the furnace and scrubbing and/or baghouse treatment to neutralize the acid gases and catalytically convert the NO_x to nitrogen (46).

The advent of fluidized-bed boilers has seen the size of units go to 150 megawatts for commercial power generation and cogeneration since the mid-1980s. This technology is displacing some stoker-fired units at the low capacity applications and smaller pulverized coal units at the large size of the fluidized-bed range. Bubbling and circulating bed designs are used, with operating temperatures in the 816–899°C (1500–1650°F) range. Sulfur oxides are controlled with dolomite or limestone injection in the bed. Higher Ca:S ratios (2:1–5:1) are needed for fluid-bed units than wet scrubbing (1:1) or spray dryers (1.2–1.5:1). Nitrous oxide emissions are higher than with other methods, possibly due to formation and oxidation of hydrogen cyanide (47,48).

Coke Production. Coking coals are mainly selected on the basis of the quality and amount of coke that they produce. Gas yield is also considered. About 65–70% of the coal charged is produced as coke. The gas quality depends on the coal rank and is a maximum (in energy in gas per mass of coal) for coals of about 89% carbon (dry mineral-matter-free) or 30% volatile matter.

Coals with 18–32% volatile matter are used to produce hard metallurgical coke. Methods have been developed to blend coals with properties outside this range to produce coke. Several coals are frequently blended to improve the quality of the coke (6,49). Blending also affects the shrinkage required to remove the coke from the ovens after initial swelling.

Lower rank coals with up to 40% volatile matter may be used alone or in blends at a gas-making plant. This coke need not be as strong as metallurgical coke, is more reactive, and is used in the domestic market.

Coking coal is cleaned so that the coke ash content is not over 10%. An upper limit of 1–2 wt % sulfur is recommended for blast furnace coke. A high sulfur content causes steel to be brittle and difficult to roll. There are seams whose coking properties are suitable for metallurgical coke, but whose sulfur content prevents that application. Small amounts of phosphorus also make steel brittle. Low phosphorus coals are needed for coke production, especially if the iron ore contains phosphorus.

Solvent Extraction. The preparation of synthetic liquid fuels by direct liquefaction uses this property in the beginning of the process. Coal partially dissolves in a number of solvents. This property has been used in characterization of coal material, because the composition of extracts is sometimes similar to the coal.

A wide range of organic solvents can dissolve part of coal samples (6,50) but dissolution is never complete and usually requires heating to temperatures sufficient for some thermal degradation or reaction with the solvent to take place (ca 400°C). At room temperature the best solvents are primary aliphatic amines, pyridine, and some higher ketones, especially when used with dimethylformamide. Above 300°C large amounts can be dissolved with phenanthrene, 1-naphthol, and some coal-derived high boiling fractions. Dissolution of up to 40 wt % can be achieved near room temperature, and up to 90% near 400°C. Coals with 80–85% carbon in the vitrinite give the largest yields of extract. Very little coal above 90 wt % C dissolves. Recent work indicates ultrasonic enhancement of extraction will increase the yield of product by about 2.5 times the nonirradiated material (51). The increase occurs only in solvent mixtures which significantly swell coals ranging in rank from lignite to high volatile bituminous coal.

When the concentration of dissolved coal exceeds about 5% of the solution by weight, the extracted material resembles the parent coal in composition and some properties. The extract consists of the smaller molecules within the range of the parent coal. The recovered extract is relatively nonvolatile and high melting.

A kinetic study of coal dissolution indicated increasing heats of activation for increasing amounts of dissolved coal. These results indicate that weaker bonds were broken initially, and the stronger bonds later (52).

Gasification. Many of the coal selection criteria for combustion apply to gasification, which is typically a form of partial oxidation. Gasifiers are primarily described as fixed bed, fluidized bed, entrained or rotating bed (53). The fixed bed involves an upward flow of reaction gas through a relatively stationary bed of hot coal. The gas velocity is slow enough to avoid blowing the coal out of the bed. The fluidized bed operates at higher gas velocities than the fixed bed, and utilizes somewhat smaller particles. The entrained bed operates with parallel flows of reaction gas and finely pulverized coal particles to minimize reaction time and maximize throughput of product. The rotating bed is similar to a kiln which operates with the coal entering at the upper part of the inclined kiln. Rotation avoids clinkering and exposes fresh surfaces to enhance completion of the reaction. The range of coals that may be used vary from one gasifier type to another with entrained flow gasifiers able to handle the widest range of raw coals. Fixed bed gasifiers require mildly caking or noncaking feedstocks for normal operation.

The Lurgi fixed-bed gasifier operates with lump coal of a noncaking type with an ash composition chosen to avoid a sticky, partly fused ash in the reactor. A slagging version of this gasifier has been tested in Westfield, Scotland. Other fixed-bed gasifiers have similar coal requirements.

The Shell-Koppers-Totzek gasifier is an entrained-bed type. It can gasify lignite and subbituminous or bituminous coal. The coal is fed as a pulverized fuel, usually ground to 70% <74 m(−200 mesh) as used for pulverized coal-fired boilers. Residence times are only a few seconds, therefore coal reactivity is important. The gasifier operates at >1650°C so that coal ash flows out of the gasifier as a molten slag. Coal ash composition must permit continuing molten ash flow.

Table 4. Estimated Total Original Coal Resources of the World by Continents[a]

Continent	Identified Resources,[a] 10^9 t	Hypothetical[b] Resources,[c] 10^9 t	Estimated Total Resources,[d] 10^9 t
Asia[c]	3,635[d]	6,362	9,997[e]
North America	1,727	2,272	3,999
Europe[f]	273	454	727
Africa	82	145	227
Oceania[g]	64	55	118
Central and South America	27	9	36
Totals	*5,808*	*9,297*	*15,104*

[a] Ref. 69.
[b] Original resources in the ground in beds ≥30 cm thick and generally <1299 m below surface but includes small amount between 1200 m and 1800 m.
[c] Includes European Russia.
[d] Includes about 2090 × 10^9 metric tons in Russia.
[e] Includes about 8600 × 10^9 metric tons in Russia.
[f] Includes Turkey.
[g] Australia, New Zealand, and New Caledonia.

Fluidized-bed gasifiers typically require a coal feed of particles near 2–3 mm dia. Caking coals are to be avoided since they usually agglomerate in the bed. This can be avoided with a pretreatment consisting of a surface oxidation with air in a fluidized bed. A useful fuel gas is produced. Examples of this type include the commercially available Winkler, and the U-Gas technology developed at the Institute of Gas Technology in Chicago.

CHEMISTRY

Reactions

The reactions of coal follow much of what is expected from organic chemistry for the different functional groups that are found in this material. These reactions are complicated by the structure of coal. The reactions on heating are important for the production of coke and synthetic fuels.

Chemical Reactions. Mature coals (>75% C) are built of assemblages of polynuclear ring systems connected by a variety of functional groups and hydrogen-bonded cross-links (3,4,7,54). The ring systems themselves contain many functional groups. These so-called molecules differ from each other to some extent in the coal matter. For bituminous coal, a tar-like material occupies some of the interstices between the molecules. Generally coals are nonvolatile except for some moisture, light hydrocarbons, and carbon dioxide. The volatile matter produced on carbonization reflects decomposition of parts of the molecule and the release of moisture. Rate of heating affects the volatile matter content. Faster rates give higher volatile matter yields. The composition of a coal primarily depends on the rank. Rank increases with the carbon content and decreases with increasing oxygen content. Table 4 gives a listing of the empirical formulas in terms of carbon, hydrogen, oxygen, nitrogen, and organic sulfur per 100 carbon atoms for a set of eight premium coal samples.

The surface of coal particles undergoes air oxidation. This process may initiate spontaneous combustion in storage piles or weathering with a loss of heating value and coking value during storage. Combustion produces carbon dioxide and water vapor as well as oxides of sulfur and nitrogen. The SO_x results from oxidation of both organic sulfur and inorganic forms such as pyrite. Nitrogen oxides are formed primarily from the nitrogen in the coal during high temperature combustion, rather than from the air used for combustion.

Partial oxidation as carried out in gasification produces carbon monoxide, hydrogen, carbon dioxide, and water vapor. The carbon dioxide reacts with hot carbon from the coal to produce carbon monoxide, and steam reacts with carbon to produce carbon monoxide and hydrogen. The hydrogen can react with carbon through direct hydrogen gasification:

$$C + 2\,H_2 \rightarrow CH_4$$

at high hydrogen pressure, frequently 6.9 MPa (1000 psi) and moderate temperatures of 650–700°C. Methane may also be produced:

$$CO + 3\,H_2 \rightarrow CH_4 + H_2O$$

in a nickel-catalyzed reactor. The latter reaction is highly exothermic and is used to provide steam for the process. The correct 3:1 ratio of hydrogen and carbon monoxide is achieved using the water gas shift reaction:

$$CO + H_2O \rightarrow H_2 + CO_2$$

A mixture of CO and H_2, called synthesis gas, may also be used in other catalytic reactors to make methyl alcohol or hydrocarbons:

$$CO + 2\,H_2 \rightarrow CH_3OH$$

or

$$n\mathrm{CO} + 2n\mathrm{H_2} \rightarrow (\mathrm{CH_2})_n + n\mathrm{H_2O}$$

Surface oxidation short of combustion, or with nitric acid or potassium permanganate solutions, produces regenerated humic acids similar to those extracted from peat or soil. Further oxidation produces aromatic and oxalic acids, but at least half of the carbon forms carbon dioxide.

Treatment with hydrogen at 400°C and 12.4 MPa (1800 psi) increases the coking power of some coal and produces a change that resembles an increase in rank. Hydrogenation with an appropriate solvent liquefies coal. Noncatalyzed processes primarily produce a tar-like solvent refined coal used as a boiler fuel. Catalysts and additional hydrogen were used in the H-Coal process developed by Hydrocarbon Research, Inc. to produce a higher quality liquid product. A 500 t/d plant was built in Catlettsburg, Kentucky, to demonstrate this process by making a coal-derived refinery feedstock. The reactor used a catalyst suspended in a process derived liquid (ebulated bed). Hydrogen reactions over short times (0.1–2 s) with very rapid heating produce a range of liquids (benzene, toluene, xylene, phenol). A less rapid heating with lower maximum temperatures permits removal of some sulfur and nitrogen from the coal (55). These effects have not been commercialized.

Treatment of coal with chlorine or bromine results in addition and substitution of the halogens on the coal. At temperatures up to 600°C chlorinolysis produces carbon tetrachloride, phosgene, and thionyl chloride (56). Treatment with fluorine or chlorine trifluoride at atmospheric pressure and 300°C can produce large yields of liquid products.

Hydrolysis with aqueous alkali has been found to remove ash material including pyrite. A small pilot plant for studying this process was built at Battelle Memorial Institute in Columbus, Ohio (57) and subsequently discontinued. Other studies have produced a variety of gases and organic compounds such as phenols, nitrogen bases, liquid hydrocarbons, and fatty acids totaling as much as 13 wt % of the coal. The products indicate that oxidation and other reactions as well as hydrolysis take place.

The pyritic sulfur in coal can undergo reaction with sulfate solutions to release elemental sulfur. Processes to reduce the sulfur content have been sought (58,59). The reaction of coal and sulfuric acid has been used to produce cation exchangers, but it was not very efficient and is no longer employed. Efforts have turned to the use of hot concentrated alkali in a process called Gravimelt or molten caustic leaching (60). The process uses one part coal and one to two parts sodium hydroxide heated to 350–400°C for up to two hours. Up to 90% of the sulfur and 98% of the mineral matter is removed. Plans for a pilot plant to begin production in 1998 in Czechoslovakia have been described (60).

Many of the products made by hydrogenation, oxidation, hydrolysis, or fluorination are of industrial importance. Concern about stable, low cost petroleum and natural gas supplies is increasing the interest in some of the coal products as upgraded fuels to meet air pollution control requirements as well as to take advantage of the greater ease of handling of the liquid or gaseous material and to utilize existing facilities such as pipelines and furnaces. A demonstration plant was built in North Dakota for conversion of coal to methane, also known as substitute natural gas (SNG) production. This plant, operated by Great Plains Gasification Associates, and later the Dakota Gasification Company. The plant may be converted to produce methanol instead of methane. A chemistry based on synthesis gas conversion to other materials has been developed and applied extensively in South Africa to the production of liquid fuels and many other products. A small-scale production is used in the manufacture of photographic film materials from coal-derived synthesis gas in the Eastman Kodak plant (Kingsport, Tennessee). At present the major production of chemicals from coal involves the by-products of coke manufacturing.

Reactions of Coal Ash. Mineral matter impurities have an important effect on the utilization of a coal. One of the constituents of great concern is pyrite because of the potential for sulfur oxide generation on combustion. The highest concentrations of pyrite are associated with coal deposition under marine environments, as typified by the Illinois Basin, including parts of Illinois, Indiana, and Kentucky. Other concerns deal with the tendency of the mineral matter to form sticky deposits in a boiler. This is most pronounced with mixtures that are rich in water-soluble alkalies such as are found in the Western Plains states (North Dakota, South Dakota, Wyoming, and Montana). These coals are typically low in the sulfur bearing constituents and therefore otherwise desirable as fuels.

Coal deposits from east of the Mississippi river generally have mineral constituents richer in acid constituents (silica, alumina) and tend to make higher melting ash mixtures. These materials do not soften until very high temperatures (above 1000°C) and tend to have limited problems with deposition on the inside walls of the boiler (slagging) or on the superheater tubes inside the boiler (fouling).

The coal ash passes through many reactors without significant chemical change. High temperature reactions that exceed the ash-softening temperature for the coal permit reactions of the simpler ash constituents to form more complex species. The molten ash behavior affects the slagging or ash removal of these reactors. Correlations of viscosity have been made with a variety of chemical parameters. Iron may be interconverted between the Fe(II) and Fe(III) states with significant reduction in viscosity occurring with increases in ferrous concentrations. Utilization of acid–base concepts appears feasible to correlate the observed effects (57).

Corrosion effects on boiler tubes appear to be initiated in some cases with the formation of a "white layer" of general composition $(Na,K)_3Al(SO_4)_3$. Conditions for initiation of the deposit are favored by coals with high alkali and sulfur contents. The white layer bonds to the tubes and permits growth of ash deposits that insulate the layer and permit further corrosion.

Plasticity of Heated Coals. ***Cause and Nature.*** Coals with a certain range of composition associated with the bend in the Seyler diagram (see Fig. 1) and having 88–90 wt % carbon soften to a liquid condition when heated (4,6). These are known as prime coking coals. The soft condition is somewhat reversible for a time, but does not persist for many hours at 400°C, and will not be observed above ca 550°C if the sample is continuously heated as in a coking process. This is caused by degradation of coal matter, which releases vapors and results in polymerization of the remaining material. The coal does not behave like a Newtonian fluid and only permits empirical measurements of plasticity to be made. Studies indicate that about 10–30% of the coal becomes liquid with a melting point below 200°C. The molten material plasticizes the remaining solid matrix. The molten part of a vitrinite is similar to the gross maceral. A part of the maceral is converted to a form that can be melted after heating to 300–400°C. The molten material is unstable and forms a solid product (coke) above 350°C with rates increasing with temperature. The decomposition of the liquid phase is rapid with lower rank noncoking coals, and less rapid with prime coking coals. The material that melts resembles coal rather than tar and, depending on rank, only a slight or moderate amount is volatile.

The fluidity of coal increases, then decreases at a given temperature. This has been interpreted in terms of the sequence of reactions: coal > fluid coal > semicoke. In the initial step, a part of the coal is decomposed to add to that which normally becomes fluid. In the second step, the fluid phase decomposes to volatile matter and a solid semicoke. The semicoke later fuses with evolution of additional volatile matter to form a high temperature coke.

Formation of a true coke requires that the fluid phase persist long enough during heating for the coal pieces to form a compact mass before solidification occurs from the decomposition. Too much fluidity leads to an expanded froth owing to formation of dispersed bubbles from gas evolution in the fluid coal. Excess bubble formation results in a weak coke. The porous nature of true coke is caused by the bubble formation during the fluid phase. The strength of semicoke is set by the degree of fusion during the fluid stage and the thickness of the bubble walls formed during the frothing. In the final conversion to a hard high temperature coke, additional gas evolution occurs while the solid shrinks and is subjected to thermal stresses. The strength of the resultant coke and the size of the coke pieces are strongly affected by the crack structure produced as a result of the thermal stresses. Strong large pieces of coke are desired to support the ore burden in blast furnaces.

Pyrolysis of Coal

Most coals decompose below temperatures of about 400°C (5,6), characteristic of the onset of plasticity. Moisture is released near 100°C, and traces of oil and gases appear between 100–400°C, depending on the coal rank. As the temperature is raised in an inert atmosphere at a rate of 1–2°C/min, the evolution of decomposition products reaches a maximum rate near 450°C, and most of the tar is produced in the range of 400–500°C. Gas evolution begins in the same range but most evolves above 500°C. If the coal temperature in a single reactor exceeds 900°C, the tars can be cracked, the yields are reduced, and the products are more aromatic. Heating beyond 900°C results in minor additional weight losses but the solid matter changes its structure. The tests for volatile matter indicate the loss in weight at a specified temperature in the range of 875–1050°C from a covered crucible. The weight loss represents the loss of volatile decomposition products rather than volatile components.

A predictive macromolecular network decomposition model for coal conversion based on results of analytical measurements has been developed (61). Data are obtained on weight loss on heating (thermogravimetry) and analysis of the evolved species by Fourier transform infrared spectrometry. Separate experimental data on solvent swelling, solvent extraction, and Gieseler plastometer data are also used in the model. Six concepts which are the foundation of the Functional Group, Depolymerization, Vaporization, Cross-linking (FG-DVC) model are considered. (*1*) The decomposition of functional group sources in the coal yield the light gas species in thermal decomposition. The amount and evolution kinetics can be measured by thermogravimetry/Fourier transform infrared spectrometry (TG-FT/IR), the functional group changes by FT/IR and NMR. (*2*) The decomposition of a macromolecular network yields tar and metaplast. The amount and kinetics of the tar evolution can be measured by TG-FT/IR and the molecular weight by field ionization mass spectrometry (FIMS). The kinetics of metaplast formation and destruction can be measured by solvent extraction, by Gieseler plastometry, and by proton magnetic resonance thermal analysis (PMRTA). (*3*) The molecular weight distribution of the metaplast depends on the network coordination number (average number of attachments on aromatic ring clusters). The coordination number can be determined by solvent swelling and NMR. (*4*) The network decomposition is controlled by bridge breaking. The number of bridges broken is limited by the available donatable hydrogen. (*5*) The network solidification is controlled by cross-linking. The changing cross-link density can be measured by solvent swelling and NMR. Cross-linking appears to occur with evolution of both CO_2 (prior to bridge breaking) and CH_4 after bridge breaking. Thus, low rank coals (which form a lot of CO_2) cross-link prior to bridge breaking and are thus thermosetting. High volatile bituminous coals (which form little CO_2) undergo significant bridge breaking prior to cross-linking and become highly fluid. Weathering, which increases the CO_2 yield, causes increased cross-linking and lowers fluidity. (*6*) The evolution of tar is controlled by mass transport in which the tar molecules evaporate into the light gas species and are carried out of the coal at rates proportional to their vapor pressure and the volume of light gases. High pressures reduce the volume of light gases and hence reduces the yield of heavy molecules with low vapor pressures. These changes can be studied with field ionization mass spectrometry.

Nature and Origin of Products. Volatile matter yields decrease with increasing coal rank. For slow heating the final weight loss depends on the maximum temperature. A variety of reactions take place. Increasing temperatures provide thermal energy to break the stronger chemical bonds. Much of the decomposition takes place in a short time (<1 s) but is limited by the rate of diffusion of the volatile products through the solid. The liquids result from initial decomposition and gases result from decomposition of liquid material. Very rapid heating rates produce weight losses as high as 72% at 1900°C, suggesting that the intrinsic volatile matter is limited only by the vapor pressure of the initial pyrolysis fragments, and would be expected to increase with temperature and decreasing coal rank (62).

The residual solid or char heated to 500°C contains 3–3.5 wt % H and up to 5 wt % 0. On further heating to 900°C the solid contains only 0.8 wt % H and up to 0.3 wt % 0. An aqueous liquor is produced that comes from the moisture in the coal as well as hydroxyl and possibly other oxygen-containing groups. Phenols in the tar are probably derived from hydroxyl aromatic groups in the coal. The total tar yield appears to be proportional to the fraction of aromatic carbon in the coal. Coke oven gas is obtained from a variety of reactions that include cracking some of the tar. The hydrogen in the gas is generated after the char is heated to 400°C but most is evolved in the conversion of the fluid coal to semicoke or coke at 550–900°C. The steam in the ovens can also produce hydrogen on reaction with hot coke.

Pyrolysis Reaction Mechanisms. An overall picture of the process is generally accepted but the detailed mechanism is still controversial. Information has been obtained from the sequence of volatile material appearing in a coke plant as determined by gas chromatography; laboratory work simulating coking and minimizing secondary reactions by working in vacuum or sweeping with inert gas (63); laboratory studies with model organic compounds to determine the mechanism by which they are converted to coke, liquid, and gaseous products; and laboratory work with more complex materials including specially synthesized polymers to better provide a model of coal (4,64). Radioactive tracers have been used in the last two studies to follow the transformation to known materials in the products (4). In the last study, gas-generating materials were added to aid in simulating the swelling process. Further observations such as the dehydrogenation of coal, which can alter the distribution of products, provides additional information for the formation mechanism (65). The mechanism of formation of metallurgical coke and its effect on coke properties has been described (66).

The understanding of the mechanism of coal pyrolysis continues to grow (61,63,66). The early stages involve formation of a fluid through depolymerization and decomposition of coal organic matter containing hydrogen. Around 400–550°C aromatic and nonaromatic groups may condense after releasing hydroxyl groups. The highest yields of methane and hydrogen come from coals with 89–92 wt % C. Light hydrocarbons other than methane are released most readily below 500°C; methane is released at 500°C. The highest rate for hydrogen occurs above 700°C (61,63).

RESOURCES

World Reserves

Amounts of coal of some specified minimum deposit thickness and some specified maximum overburden thickness existing in the ground are termed resources. There is no economic consideration for resources, but reserves represent the portion of the resources that may be recovered economically using conventional mining equipment. The first inventory of world coal resources was made during the Twelfth International Geological Congress in Toronto in 1913. An example of the changes since 1913 can be seen from an examination of the coal resources for Canada. These were estimated to have been 1217×10^9 metric tons in 1913, based on a few observations and statistical allowance for all possible coalbeds to a minimum thickness of 0.3 m and to a maximum depth of 1220 m below the surface. In 1974, however, the estimate of solid fossil fuel resources (excluding peat) from the World Energy Conference gave the total resources as only 109×10^9 metric tons, and in 1986 the proven recoverable resources and estimated additional amount in place was given as 50×10^9 metric tons (67), less than 5% of the earliest figures. In 1994 the proved recoverable reserves was 8.6×10^9 t (1).

Comprehensive reviews of energy sources are published by the World Energy Conference, formerly the World Power Conference, at six-year intervals (67). The 1986 survey included reserves and also gave total resources. In 1986 the total proven reserves of recoverable solid fuels was given as 6×10^{11} metric tons. One metric ton is defined as 29.2×10^3 MJ (27.7 million Btu) to provide for the variation of calorific value in different coals. The survey includes reserves as either recoverable resource or total reserve, and also gives total resources. Surveys are also provided in intermediate years and are quoted in other sources (1). In 1995 the total proven reserves of recoverable solid fuels was 1.1×10^{12} metric tons. In 1986 the total estimated additional reserves recoverable and total estimated additional amount in place are 2.2×10^{12} and 7.7×10^{12} t, respectively. These figures are nearly double the 1913 estimates, primarily because significantly increased reserves have been indicated for Russia.

The part of the resource that is economically recoverable varies by country. The estimates made in the survey show that the 1994 proven recoverable reserves of anthracite and bituminous coal (ca 710×10^9 t) would last about 180 years at the 1994 annual rate of production (3.8×10^9 t).

In Table 4 (67,68) a somewhat different basis is used. The estimated total original coal resources of the world include beds 30 cm thick, and generally <1220 m below surface but also include small amounts between 1220 m and 1830 m. The data from column 1 are from earlier World Power Conference Surveys, whereas the figures for hypothetical resources (column 2) and total estimated resources (column 3) may be less reliable. This estimate represents about one-third more than the World Energy Conference Survey.

Reserves in the United States

Coal is widely distributed and abundant in the United States as indicated in Figure 2. A large portion of the coal fields contain lignite and subbituminous coal, however, and another portion of the coal in these fields is contained either in thin or deep

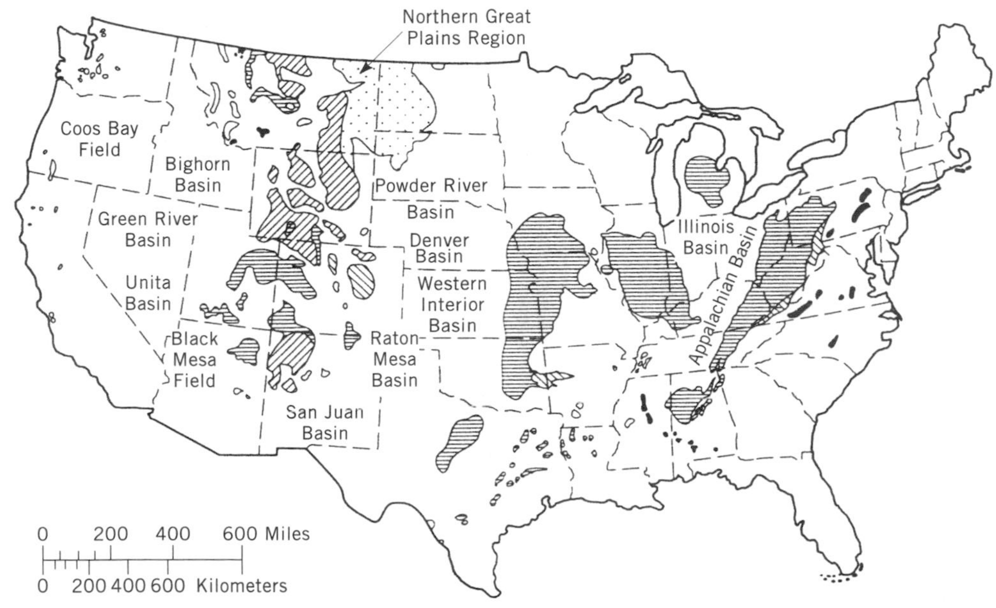

Figure 2. Coal fields of the counterminous United States where ■ represents anthracite and semianthracite; ▨, low volatile bituminous coal; ▤, medium and high volatile bituminous coal; ▧, subbituminous coal; and ▩, lignite (67).

beds that can be mined only with difficulty or great cost. Reserve estimates for the United States as of 1974 were on the order of 1.6×10^{12} t (67). Little mapping and prospecting has been done since the report to change the 1974 estimate. This information is useful for showing the quantitative distribution of reserves, selecting appropriate areas for further exploration of development, and in planning coal-based industrial activity.

The reserves of 21 states have been classified by overburden thickness, reliability of estimates, and bed thickness. This coal represents about 60% of the total identified tonnage. Of this, 91% is less than 305 meters from the surface, 43% is bituminous, and 58% is in beds thick enough to be mined economically.

On a uniform calorific value basis, coal constitutes 69% of the total estimated recoverable resources of fossil fuel in the United States. Petroleum and natural gas are about 7% and oil in oil shale, which is not as of this writing used as a fuel, is about 23%. The 1994 total recoverable reserves of coal are about 180 times the 1994 annual production, whereas the smaller reserves of oil and gas are emphasized by a combined production and consumption rate of oil and gas in the United States that is three times that of coal (2).

Coal Production

In 1860 world coal production was 122×10^6 t/yr. Production increased to 1140×10^6 t in 1913, giving a 4.2 %/yr average rate of increase. The rate slowed and has been erratic since that time. World coal and lignite production rose to about 4310×10^6 t in 1995 (1).

United States Coal Production. Bituminous coal production in the United States, which dates back to 1702, started in earnest about 1820. It has increased with fluctuations to 617×10^6 t in 1976, 755×10^6 t in 1982, 878×10^6 t in 1989, and 935×10^6 t in 1995 (2). In 1995 the United States produced 25.0% of the world's bituminous coal supply (1). U.S. coal production by state, 1984–1995, is given in Table 5 (2). Bituminous coal exports in 1995 totaled 79.6×10^6 t (2). Domestic consumption was 854×10^6 t. Of that, electric utilities received 753×10^6 t (88%), coke and gas plants 30×10^6 t (3.5%), and industrial, commercial, retail, and transportation 71×10^6 t (8.3%) (2).

Anthracite production, which takes place in Pennsylvania, fell from 90×10^6 t in 1918 to 42×10^6 t in 1938; it increased in 1944 to 58×10^6 t but declined steadily to 23×10^6 t in 1957, 15×10^6 t in 1961, 6×10^6 t in 1974, and 3.9×10^6 in 1994. Of the 6 million tons produced in 1974, half were mined and the other half recovered from culm banks (mine waste piles).

The demand for energy is continually increasing and the highest energy consumption in the world occurs in the United States. In 1992 it was 8.7×10^{13} MJ(82.2×10^{15} Btu) or about 11.7 metric tons of coal-equivalent per capita (1). World recoverable reserves are about 180 times annual coal production in 1994, and about 10 times that for the additional reserves believed to be in place (1). Estimated coal consumption will reduce the present recoverable reserves at about 1–2%/yr. Whereas the use of bituminous coal will continue to increase in terms of tonnage, the percentage of coal used in the United States is slowly decreasing, as shown in Table 6.

In the absence of a clear U.S. energy policy, the market forces will determine the mix of energy sources used to meet the demand. Construction of nuclear power plants has stopped in the United States. Coal-fired electric power generation facilities will add substantial amounts of air pollution control equipment to meet more stringent requirements. Concern about global warming due to the greenhouse effect will lead to efforts to increase power plant efficiency, in order that the release of carbon dioxide can be reduced for a given amount of electric power generation.

MINING AND PREPARATION

Mining

Coal is obtained by either surface mining of outcrops or seams near the surface or by underground mining depending on the geological conditions, which may vary from thick, flat seams to thin, inclined seams that are folded and need special methods of mining. Coal mining has changed from a labor intensive activity to one that has become highly mechanized. However, after the 1969 Coal Mine Health and Safety Act, underground production decreased from 15.6 metric tons per day to 8.5 metric tons per day in 1976. In 1994 the average output per person per day in underground mines in the United States was 24.7 metric tons. For surface mines the output was 62 metric tons per person per day (2).

Strip or open-pit mining involves removal of overburden from shallow seams, breaking of coal by blasting or mechanical means, and loading of the coal. The development of very large excavating equipment, including draglines, bulldozers, front-end loaders, and trucks, has been responsible for significantly increased production from strip mining.

The two methods of underground mining commonly used are room-and-pillar and longwall. In room-and-pillar mining the coal is removed from two sets of corridors that advance through the mine at right angles to each other. Regularly spaced pillars, constituting about half of the coal seam, are left behind to support the overhead layers in the mined areas. This method is used extensively in the United States and other nations with large reserves. The pillars may later be removed, leading to probable subsidence of the surface. Longwall mining is used to permit recovery of as much of the coal as possible (70). Two parallel headings are made 100–200 m apart and at right angles to the main heading. The longwall between the two headings is then mined away from the main heading. The equipment provides a movable roof support system that advances as the coal is mined and allows the roof to collapse in a controlled manner behind it. This method also leads to subsidence of the overhead layers. The mechanical equipment used in room-and-pillar underground mining usually involves a series of specific operations with continuous mining equipment. Continuous miners use rotating heads equipped with bits to pick or cut through the coal without blasting and load it into a shuttle car for conveying to a belt system.

Another method used in Europe for steeply inclined seams is called horizontal mining. Horizontal shafts are cut through rock below the coal seams. Vertical connections are made to the seam to permit coal removal.

Preparation

Coal preparation is of significant importance to the coal industry and consumers (6,42,71). Preparation normally involves

Table 5. U.S. Coal Production by State, t × 10³, 1984–1995[a,b]

State	1984	1986	1988	1990	1991	1992	1993	1994	1995[c]
					East				
Alabama	24,596	23,450	24,078	26,359	24,760	23,423	22,489	21,126	23,029
Georgia	114	0	0	0	0	0	0	0	0
Illinois	57,902	56,174	53,203	54,837	54,714	54,350	37,317	47,940	43,861
Indiana	34,100	29,830	28,394	32,604	28,573	27,663	26,600	28,082	23,356
Kentucky									
Eastern	106,485	102,361	106,712	116,584	106,436	108,399	109,133	112,998	106,752
Western	38,378	37,411	36,618	40,793	37,918	37,851	32,786	33,773	30,462
Maryland	3,726	3,547	2,944	3,166	3,426	3,034	3,046	3,298	3,367
Ohio	35,644	33,088	30,911	32,009	27,757	27,606	26,165	27,146	22,591
Pennsylvania									
Bituminous	66,585	61,159	60,919	60,843	56,238	59,472	50,298	52,315	52,350
Anthracite	3,779	3,897	3,228	3,183	3,128	3,163	3,910	4,196	3,737
Tennessee	6,640	6,238	5,911	5,623	3,895	3,156	2,767	2,712	2,410
Virginia	36,654	37,390	41,664	42,601	38,094	39,066	35,700	33,713	31,422
West Virginia	118,955	117,956	131,665	153,638	151,956	147,245	118,517	146,893	149,014
Total East	*533,559*	*512,500*	*526,246*	*572,240*	*536,895*	*534,427*	*468,728*	*514,190*	*492,350*
					West				
Alaska	780	1,426	1,584	1,549	1,304	1,393	1,454	1,423	1,486
Arizona	10,462	10,493	11,257	10,264	11,988	11,361	11,053	11,855	10,813
Arkansas	74	152	251	54	47	53	40	46	12
California	0	0	49	55	52	94	0	0	0
Colorado	16,314	13,835	14,448	17,170	16,193	17,457	19,872	22,976	22,709
Iowa	479	439	310	346	312	262	159	42	0
Kansas	1,206	1,349	669	655	378	330	310	258	265
Louisiana	0	2,047	2,623	2,893	2,861	2,942	2,846	3,144	3,107
Missouri	6,114	4,256	3,785	3,166	2,092	2,620	593	761	421
Montana	29,964	30,852	35,304	34,155	35,311	35,311	32,613	37,809	35,903
New Mexico	19,321	19,518	19,797	22,057	19,538	22,290	25,667	25,461	24,263
North Dakota	20,078	23,281	26,996	26,525	26,813	28,824	29,031	29,316	27,334
Oklahoma	4,213	2,768	1,939	1,542	1,672	1,581	1,596	1,735	1,958
Texas	37,360	44,120	47,471	50,626	48,873	50,004	49,547	47,530	47,458
Utah	11,189	12,956	16,492	20,029	19,926	19,376	19,837	22,154	22,733
Washington	3,516	4,178	4,694	4,541	4,670	4,768	4,303	4,443	4,410
Wyoming	118,870	124,238	148,925	167,298	176,019	172,676	190,797	215,280	239,689
Total West	*279,939*	*295,907*	*336,596*	*362,163*	*367,459*	*371,342*	*389,718*	*424,233*	*442,651*
Total	*813,498*	*808,407*	*862,842*	*934,401*	*904,353*	*905,769*	*858,446*	*938,422*	*935,001*

[a] Ref. 2; U.S. Dept. of Energy/Energy Information Administration.
[b] Data includes mines producing less than 9000 t/y.
[c] Data is preliminary.

Table 6. Energy Source Usage in the United States, % Energy Basis[a]

Source	1942	1952	1962	1972	1988	1995[b]
Coal	68.7	45.2	32.5	25.6	23.5	22.5
Petroleum	17.1	25.4	23.6	22.8	42.7	39.7
Natural gas	9.0	23.3	38.3	44.4	23.1	25.5
Hydroelectric	5.2	6.1	5.5	6.0	3.3	3.9
Nuclear			0.1	1.2	7.1	8.2

[a] Ref. 2; U.S. DOE/EIA, *Monthly Energy Rev.* (Mar. 1996).
[b] Data is preliminary.

some size reduction of the mined coal and the systematic removal of some ash forming material and very fine coal. The percentage of mined coal that is mechanically cleaned in major coal producing countries has risen since the 1960s. There are a number of reasons for this. The most important in the United States is the increased use of continuous mining equipment. The nature of this operation tends to include inorganic foreign matter from the floor and ceiling of the seams, and run-of-mine coal currently includes about 25% mineral matter.

The size consist of the mined coal is smaller when produced with continuous mining equipment. The depletion of the better quality seams, which are low in ash and sulfur, in many coal fields necessitates cleaning of the remainder.

The economic need to recover the maximum amount of salable coal has led to cleaning of the finer sizes which had previously been discarded. Stringent customer demands for coal meeting definite specifications, regulations requiring the removal of pyrites to control air pollution, increased freight rates, and ash disposal costs all contribute to the upsurge in coal cleaning.

Earlier, the coal was hand-picked for removal of larger pieces of refuse, but higher labor costs have made this method uneconomical. Mechanical separation methods are used, most of which depend on the difference in density between the coal and refuse.

The washability characteristics of a coal determine the extent to which the refuse may be removed, and other problems in cleaning. The laboratory float-and-sink analysis gives information on the percentages and quality of the coal material occupying different density ranges. From this information, graphs are constructed showing the composite quantity and coal quality that will be obtained by cleaning at different specific gravities. This information is considered with the economic factors involved in the sale of the washed coal to choose an optimum method of cleaning. Cleaning plants are usually designed to handle the output of specific mines and to clean for a specific market. The plant includes various types of cleaning methods designed to move the different fractions through several cleaning circuits to maximize optimum recovery.

In some areas, run-of-mine coal is separated into three products: a low gravity, premium priced coal for metallurgical or other special use, a "middlings" product for possible boiler firing, and a high ash refuse. The complete preparation of coal usually requires several processes (see COAL PREPARATION).

Storage

Storage of coal may be necessary at any of the major steps in production or consumption, ie, at the mine, preparation plant, or consumer location. Electric utilities have the largest amounts of coal in storage, with stockpiles frequently able to meet 60–100 day normal demand to protect against delays, shortages, price changes, or seasonal demands. In recent years the stockpiles have tended to be smaller, 65 days supply reported for 1989, and 56 days for 1995 (2).

For utilities, two types of storage are used. A small amount of coal in storage meets daily needs and is continually turned over. This coal is loaded into storage bins or bunkers. Long-term reserves are carefully piled and left undisturbed except as necessary to sustain production.

Coal storage results in some deterioration of the fuel owing to air oxidation. If inadequate care is taken, spontaneous heating and combustion result. As the rank of coal decreases it oxidizes more easily and must be piled more carefully. Anthracite does not usually present a problem.

The surface of the coal particles oxidizes or weathers resulting in cracks, finer particles, and reduced agglomeration which may destroy coking properties. If spontaneous heating takes place, the calorific value of the coal is reduced. Hot spots must be carefully dug out and used as quickly as possible. Without spontaneous heating and with good compaction, calorific value losses below 1%/yr have been recorded.

Coal piles are carefully constructed to exclude air, or allow adequate ventilation. The latter requires larger sizes, graded as 4 cm+ without fines, for avoiding heating by ventilation. For exclusion of air, mixed sizes provide fines to fill the gaps between larger pieces. Large pockets must not be allowed because they provide access for air. The coal should be compacted to maximize the bulk density of the coal pile.

Several approaches have been effective for storage:

Large compacted layered piles with sides and top sealed with an oil or asphalt emulsion (4 L oil seals 1 m^2).

Large compacted layered piles with sides and top covered with fines to seal the pile and coarse coal to protect fines from wind and weather (sides may slope at angles ≥30°).

Piles of compacted layers in open pits with tight sides so that the air has access only at the top.

Sealed bins or bunkers (airtight storage can be provided for smaller amounts of coal for longer times).

Underwater storage in concrete pits (expensive and rarely used, but effectively prevents deterioration although introduces other problems related to handling wet coal).

Large compacted storage piles should be located on hard surfaces, not subject to flooding. A layer of fines may be put down first to facilitate recovery. Each layer of coal should be compacted after it is deposited. The top of the pile should have a slight crown to avoid water accumulation. Excessive heights should be avoided to prevent air infiltration caused by wind. Coal removal should be done in layers followed by compacting and smoothing the surface. Piles should be limited to the same rank of coal depending on the intended use.

Transportation

The usual means of transporting coal are railroad, barge, truck, conveyer belt from mine to plant, and slurry pipelines (2,4). In 1994 846 × 10^6 t of coal was transported to U.S. destinations; of this, 62.7% was shipped by railroad, 13.3% by barge, 11.1% by truck, and 12.9% by ship, conveyer, slurry pipeline, and other methods (2). Electric utilities consume a growing share (81.5% in 1981 to 88.1% in 1995) of the coal transported in the United States (2).

The unit train handles about half of rail transportation (72). Most unit trains consist of about 100 rail cars, and are dedicated to coal haulage from the mine to the consuming plant. Almost no time is wasted at either the loading or unloading site due to efficient loading and unloading equipment. Diesel and diesel-electric trucks with capacities up to 350 tons (2) handle off-highway transport of coal.

A 439 km (273-mi) slurry, 45 cm (18-in.) diameter pipeline started in 1970 to move coal from Arizona to southern Nevada. The coal is crushed and ground to the fineness needed for proper viscosity and settling. About 18–20% is <44 μm (−325 mesh), 35–45% is <149 μm (−100 mesh), and 0–2% is <1070 μm (+14 mesh). The solids content of the slurries has approached 70% with additives used to stabilize the mixture. The slurry is dewatered with centrifuges before combustion of the coal.

For shipment in cold climates, a freeze-proofing treatment with inorganic chemicals or oil spray is used. An oil spray on the inside of the coal cars is also effective at 3.7–5.6 L (1–1.5 gal)/car for four round trips of 1287 km (800 mi) each. Oil treatment has also been used for dust-proofing; wind loss can be prevented by use of an asphalt emulsion on the top of rail cars (6).

Coal pipelines have been built in other countries, eg, France (8.8 km) and Russia (61 km), and pipelines are used for transporting limestone, copper concentrates, magnetite, and gilsonite in other parts of the world.

The first coal pipeline, built in Ohio, led to freight rate reductions. The pipeline stopped operation after introduction of the unit train, used exclusively to transport coal from the mine to an electric power generation station.

Hydraulic transport is used in mines and for lifting of coals to the surface in Russia, Poland, and France. Pneumatic transport of coal is used over short distances in power plants and steel mills. The longest (14.6 km) single flight conveyer belt in the world near Uniontown, Kentucky, has a capacity of 1360 t/h.

HEALTH AND SAFETY FACTORS

Coal mining has been a relatively dangerous occupation (2,70,71). During the period of 1961–1967 the average fatality rate in the United States for each million person hours worked was 1.05. In the seven years after the passage of the Federal Coal Mine Health and Safety Act of 1969, the average fatality rate decreased to 0.58, and by 1995 the rate was 0.20 (2).

The rates of occupational injuries are reported per 200,000 employee-hours (about 100 employee-years). In 1994, the total for all mines was 9.07 or 9.07%. Over the three-year periods before and after passage of the act the rates for underground mining were 48.60 and 40.07%, respectively. The major causes of fatalities are falling rock from mine roofs and faces, haulage, surface accidents, machinery, and explosions. For disabling injuries the major causes are slips and falls, handling of materials, use of hand tools, lifting and pulling, falls of roof rock, and haulage and machinery (2).

Gases

Methane is of greatest concern, although other gases including carbon monoxide and hydrogen sulfide may be found in some mines. Methane must be detected and controlled because mixtures of air with 5–15% of methane are explosive.

The U.S. Mine Health and Safety Act of 1969 provides that a mine will be closed if there is 1.5% or more methane in the air. The use of an electrical methane detection device is required. High capacity ventilation systems are designed to sweep gases from the cutting face and out of the mine. These systems remove all gases before they become harmful.

Coal Dust Explosions

The explosion from methane tends to be localized, but may start coal dust explosions with more widespread injury and loss of life. All coal-breaking operations result in formation of fine coal particles; some are controlled with water during the mining operation. Breakage associated with hauling disperses dust. Dust accumulations can be made safe by rockdusting. Powdered limestone is spread over the mine surfaces to cover the dust.

Drainage

Some mines are located beneath subsurface streams, or the coal seams may be aquifers. These mines may become flooded if they are not continually pumped. In Pennsylvania anthracite mines as much as 30 t of water may be pumped for each ton of coal mined (73).

Air or biological oxidation of pyrite leads to sulfate formation and dilute sulfuric acid in the mine drainage. This pollutes streams and water supplies into which the mine water is drained. The U.S. Bureau of Mines and other organizations are studying means of controlling this problem.

Other Hazards

Rocks falling from the roofs of mines used to cause the largest number of accidents. Roof bolts are placed in holes drilled into the roofs of working areas to tie the layers of rock together to prevent rock falls. A disease called pneumoconiosis, also called black lung, results from breathing coal dust over prolonged periods of time. The coal particles coat the lungs and prevent proper breathing.

Government Regulations

The U.S. Bureau of Mines (Mining Enforcement and Safety Administration, MESA) studies hazards and advises on accident prevention. MESA also administers laws dealing with safety in mines. Individual states may also have departments of mines to administer their own standards.

The Federal Coal Mine Health and Safety Act set standards for mine ventilation, roof support, coal dust concentration levels, mine inspections, and equipment. As a part of this comprehensive act, miners must receive medical examinations at the employers expense, and payments are made from the U.S. government to miners who cannot work because of black lung disease.

USES

Coal As Fuel

Coal is used as a fuel for electric power generation, industrial heating and steam generation, domestic heating, railroads, and coal processing. About 87% of the world's coal production is burned to produce heat and derived forms of energy. The balance is practically all processed thermally to make coke, fuel gas, and liquid by-products. Other uses of coke and fuel gas also contribute to coal consumption for heat. In the United States coal use for power generation has increased to 88.1% in 1995, whereas coking coal use has dropped to 3.5% and the industrial/retail market declined to 8.4% (2).

Electric Power Generation. Coal is the primary fuel for thermal electric power generation. Since 1940 the quantity of bituminous coal consumed by electric utilities has grown substantially in each succeeding decade, and growth is expected to continue for many years. Coal consumed by electric utilities increased from about 536×10^6 t in 1981 to 753 t $\times 10^6$ in 1995 (2). The reasons for increased coal demand include availability, relative stability of decreasing coal prices, and lack of problems with spent fuel disposal as experienced in nuclear power plants (see NUCLEAR POWER FACILITIES SAFETY).

The overall efficiency of coal-fired electric power plants has plateaued. The overall efficiency of power plants consisting of coal-fired boilers and steam turbines reached about 39% and is not expected to exceed this with the present configuration of boilers. The addition of pollutant control equipment has increased the internal power use on the stations and lowered the effective efficiency of the plant. The increased efficiencies have been achieved through use of larger units (up to 1500 MW) and higher pressures to 24.1 MPa (3500 psi) and reheat, but concerns about reliability and ability to match power generation and demand have kept plant sizes below these values. Maximum temperatures have not been increased due to difficulties with corrosion owing to coal ash constituents, materials properties, and costs of better alloys. Future increases in efficiency will depend on development of new systems of power generation including fluid-bed boilers, gasification of coal to power a gas turbine with hot exhaust directed to a waste heat boiler in a combined cycle (gas turbine and steam turbine), and magnetohydrodynamics.

Almost all modern large coal-fired boilers for electric power generation use pulverized coal. The cyclone furnace, built mainly for use in Germany and the United States, uses coarser pulverized coal. The ash is removed primarily as a molten slag from the combustor. This design is no longer offered in the United States. This method of firing has not been accepted in the United Kingdom because of the higher softening temperature of the ash of the British coals. Stoker firing is generally limited to the smaller obsolete standby utility plants and generation plants used by industrial companies.

One significant advantage of pulverized coal boilers is their ability to use any kind of coal, including run-of-mine or uncleaned coals. With the advent of continuous mining equipment, the ash content frequently is ca 25%, and some preparation is frequently practiced. There were 716 coal preparation plants in the United States in 1994, mainly in Kentucky, West Virginia, and Pennsylvania (2).

The advent of fluidized-bed boilers has seen the size of units go to 150 megawatts for commercial power generation and cogeneration since the mid-1980s. This technology is displacing some stoker-fired units at the low capacity applications and smaller pulverized coal units at the large size of the fluidized-bed range. Bubbling and circulating bed designs are used, with operating temperatures in the 815–898°C (1500–1650°F) range. Sulfur oxides are controlled with dolomite or limestone injection in the bed. Higher Ca:S ratios (2:1–5:1) are needed for fluid-bed units than wet scrubbing (1.0:–1) or spray dryers (1.2–1.5:1). Nitrous oxide emissions are higher than with other methods, possibly due to formation and oxidation of hydrogen cyanide (47). Several processes are being developed to reduce nitrogen oxides emissions. Three pressurized fluid-bed combustors are being demonstrated under the U.S. Department of Energy's Clean Coal Technology Program (48).

Integration of coal gasification with a combination of a gas turbine for power generation and a waste heat boiler for power generation is termed integrated gasification combined cycle (IGCC). Efficiencies are currently about 42% and promise to be higher as gas turbine technology improves (46). Five plants using this technology have been announced in the United States. The technology uses sulfur gas removal techniques resulting in higher removal rates than conventional scrubbers, in part due to the improved efficiency of scrubbing more concentrated gases.

A primary concern in coal-fired power generation is release of air pollutants. Limits on SO_2 output, 0.52 g/MJ of coal input to a new plant, have been established. For a bituminous coal of 27.9 MJ/kg (12,000 Btu/lb) there is thus an upper limit of 0.72% sulfur content. Relatively few coals can meet this requirement. The U.S. Department of Energy indicated recoverable reserves of 52×10^9 t in 1992 (2). These were categorized by sulfur content (g/MJ coal); 37% had 0.52 or less, 18% had between 0.52 and 0.72, 16% had between 0.72 and 1.45, 12% had between 1.45 and 2.17, and 16% had more than 2.17. The lowest sulfur coal, ca 85%, is found west of the Mississippi River (mainly in Montana and Wyoming), quite distant from the electric power demand centers in the eastern United States. A trend to utilization of western coals has developed.

Industrial Heating and Steam Generation. The primary users of coal in this category include the iron and steel industry and the food, chemicals, paper, engineering, bricks and other clay products, and cement industries, and a group of miscellaneous consumers (mainly for space heating) including federal and local government installations, armed services, and small industrial concerns. Most of the coal is burned directly for process heat, ie, for drying and firing kilns and furnaces, or indirectly for steam generation for process needs or space heating, and for a small amount of electric power generation.

The use of coal by industry in the United States has diminished significantly in past decades, especially among small users, because of the greater convenience in storing and handling gas and liquid fuels and the higher initial cost of coal-fired equipment.

Several developments are being pursued to utilize coal directly, ie, automation of controls, coal and ash handling equipment for smaller stoker and pulverized coal-fired units, design of packaged boiler units, and pollution control equipment. In the cement industry coal firing has been used, since the sulfur oxides react with some of the lime to make calcium sulfate in an acceptable amount.

Coal Processing to Synthetic Fuels and Other Products

The principal approaches to coal processing or coal conversion are thermal decomposition, including pyrolysis or carbonization (5,6), gasification (6), and liquefaction by hydrogenation (6). The hydrogenation of coal is not currently practiced commercially.

The U.S. Office of Coal Research and portions of the successor ERDA and DOE have been authorized to contract for coal research, use it more efficiently, and convert it to more desirable energy forms. The Clean Coal Technology program was created to develop and demonstrate the technology needed to use coal in a more environmentally acceptable manner. This was implemented in several phases. Activities range from basic research and establishing integrated operation of new processes in pilot plants through demonstration with commercial-scale equipment.

Gasification

Gasification of coal is used to provide gaseous fuels by surface and underground applications, liquid fuels by indirect liquefaction (catalytic conversion of synthesis gas), and chemicals from conversion of synthesis gas. There are also applications in steelmaking.

Liquefaction

Liquefaction of coal to oil was first accomplished by Bergius in 1914. Hydrogen was used with a paste of coal, heavy oil, and a small amount of iron oxide catalyst at 450°C and 20 MPa (200 atm) in stirred autoclaves. The process was developed by the I.G. Farbenindustrie AG to give commercial quality gasoline as the major product. Twelve hydrogenation plants were operated during World War II to make liquid fuels.

Imperial Chemical Industries in Great Britain hydrogenated coal to produce gasoline until the start of World War II. The process then operated on creosote middle oil until 1958. None of these plants is currently being used to make liquid fuels for economic reasons. The present prices of coal and hydrogen from coal have not made synthetic liquid fuels competitive. In those cases, as in South Africa where availability of cheap coal and fuel liquids is very important, there is incentive for development of processes.

The Pott-Broche process (74) was best known as an early industrial use of solvent extraction of coal, but was ended owing to war damage. The coal was extracted at about 400°C for 1–1.5 h under a hydrogen pressure of 10–15 MPa (100–150 atm) with a coal-derived solvent. Plant capacity was only 5 t/h with an 80% yield of extract. The product contained less than 0.05% mineral matter and had limited use, mainly in electrodes.

Solvent extraction work was carried out by a number of organizations in the United States. Pilot plants for producing solvent refined coal (SRC) were built and initially sponsored by the Southern Company Services and Electric Power Research Institute (Wilsonville, Alabama) in 1973 and built with Department of Energy sponsorship near Tacoma, Washington in 1974 with capacities of 5 and 45 t/d of coal input, respectively. The Wilsonville plant is expected to close in 1992 after many modifications from the initial design, but the Tacoma plant closed much earlier.

In SRC work, coal was slurried with a process-derived anthracene oil and heated to 400–455°C at 12.4–13.8 MPa (1800–2000 psi) of hydrogen for 0.0–1 h. A viscous liquid was extracted. The product stream contains some hydrocarbon gases and H_2S. The residue is gasified to generate hydrogen for the process. The remaining filtrate is separated into solvent, which is recycled, and SRC, a low ash, tar-like boiler fuel.

Heating value of the product (SRC) is ca 37 MJ/kg (ca 16,000 Btu/lb). Sulfur contents have been reduced from 2–7% initially to 0.9% and possibly less. Ash contents have been reduced from 8–20% to 0.17% (75). These properties permit compliance with EPA requirements for SO_2 and particulate emissions. The SRC is primarily intended to be used as a boiler fuel in either a solid or molten form (heated to ca 315°C). The solid has a Hardgrove index of 150 (76). Boiler tests have been successfully carried out with a utility boiler.

A series of process improvements have been developed at Wilsonville to produce high quality transportation fuels. Two integrated stages of liquefaction separated the initial coal dissolution from the hydrogenation to upgrade the product. This was known as SRC-II. An intermediate step, critical solvent deashing, was added to remove mineral matter to extend the life of catalysts used in hydrogenation. Later efforts involved the use of an ebulated bed developed by Hydrocarbon Research, Inc. (HRI), and eliminated the mineral matter removal between stages. Temperatures were lowered to reduce contamination of catalysts which were also added to the first stage. This approach has been called the integrated catalytic two-stage liquefaction process (77).

Several processes progressed to demonstration scales but have not been commercialized, primarily because of economic inability to compete with available petroleum products. The H-Coal process developed by Hydrocarbon Research, Inc. was demonstrated at Catlettsburg, Kentucky, with a 545 t/d plant and DOE support. The Exxon Donor Solvent Liquefaction process has not been commercialized.

Processes for hydrogen gasification, hydrogen pyrolysis, or coking of coal produce liquid coproducts. The Hygas process described above produces about 6% liquids as benzene, toluene, and xylene. Substitution of petroleum residua for the coal-derived process oil has been used in studies of coal liquefaction and offers promise as a lower cost technology (77).

Bioprocessing and Biotreatment of Coal

The use of biotechnology to process coal to make gaseous and liquid fuels is a new and emerging field (78). Reactions of bacteria and enzymes have been studied to establish the technical feasibility of conversion and obtain some engineering data for possible commercial application. The earliest work was done by Fakoussa in Germany (79) on microbial decomposition of German hard coals.

Reactors have been designed to use a variety of bacteria and fungi to react with coal to both break down the large molecular structure into smaller units which may be useful as intermediates (solubilization) or as liquid and gaseous fuels (conversion). The primary efforts have focused on lower rank coals (lignite or brown coal and subbituminous) because of their greater reactivity.

The conversion processes frequently introduce chemically combined oxygen through hydrolysis or related reactions to make the solid soluble in the reaction mixture as an initial step. Further reaction involves biological degradation of the resulting material to form gases or liquids.

The large-scale processing of coal is expected to involve plants similar to sewage treatment facilities in the handling of liquid and solid materials. The reaction rates are substantially lower than those achieved in high temperature gasifiers and liquefaction reactors. Consequently, much larger systems must be built to achieve comparable coal throughput.

Biological processes are also being studied to establish the ability to remove sulfur species in order to remove potential contributors to acid rain (qv). These species include benzothiophene-type materials, which are the most difficult to remove chemically, as well as pyritic material. The pyrite may be treated to enhance the ability of flotation processes to separate the mineral from the combustible parts of the coal. Genetic engineering as well as modification of known species is being used to develop more effective species.

Other Uses

The quantity of coal used for purposes other than combustion or processing described above is relatively insignificant (2,6).

Coal, especially anthracite, has established markets for purifying and filtering agents, in either the natural form or converted to activated carbon. The latter can be prepared from bituminous coal or coke, and is used in sewage treatment, water purification, respirator absorbers, solvent recovery, and in the food industry. Some of these small markets are quite profitable and new uses are continually being sought for this material.

Carbon black from oil is the main competition for the product from coal, which is used in filters. Carbon for electrodes is primarily made from petroleum coke, although pitch coke is used in Germany for this product. The pitch binder used for electrodes and other carbon products is almost always a selected coal-tar pitch.

The use of coal in metallurgical operations has already been indicated. The preparation of pelletized iron ore represents a substantial market for coke and anthracite for sintering. Direct injection into the blast furnace of an auxiliary fuel, coal, or oil is practiced to provide heat for the reduction and some of the reducing agent in place of the more expensive coke that serves these purposes.

Some minor uses that may grow in the future include the use of fly ash, cinders, or even coal as a building material; soil conditioners from coal by oxidation to humates; and a variety of carbon and graphite products for the electrical industry, and possibly the nuclear energy program.

The growth of synthetic fuels from coal will also provide substantial quantities of by-products including elemental sulfur, fertilizer as ammonia or its salts, and a range of liquid products. The availability of ammonia and straight-chain paraffins may permit production of food from fossil fuels.

BIBLIOGRAPHY

1. *International Coal,* 1996 ed., National Mining Association, Washington, D.C.
2. *Coal Data,* 1996 ed., National Mining Association, Washington, D.C.
3. D. W. van Krevelen, *Coal,* Elsevier Scientific Publishing Co., Amsterdam, the Netherlands, 1961.
4. H. H. Lowry, ed., *Chemistry of Coal Utilization,* Vols. 1 and 2, John Wiley & Sons, Inc., New York, 1945.
5. *Ibid.,* Suppl. Vol. 1963.
6. M. Elliott, in Ref. 4, 2nd Suppl. Vol., John Wiley & Sons, Inc., New York, 1981; an exceptionally extensive source.
7. H. J. Gluskoter, N. F. Shimp, and R. R. Ruch, "Coal Analyses, Trace Elements, and Mineral Matter," in Ref. 6, Chapt. 7.
8. R. C. Neavel, in Ref. 6, "Origin, Petrography and Classification of Coal," Chapt. 3.
9. R. Thiessen, *U.S. Bur. Mines Inform. Circ.,* 7397 (1947).
10. M. C. Stopes, *Proc. R. Soc. London Ser. B.* **90** 470 (1919); *Fuel* **14,** 4 (1935).
11. C. R. Ward, ed., *Coal Geology and Coal Technology,* Blackwell Scientific Publications, London, 1984; an excellent text on this area.
12. A. Davis, in Ref. 11, Chapt. 3.
13. M. C. Stopes, *Fuel* **14,** 4–13 (1935).
14. R. D. Harvey and R. R. Ruch, in K. S. Vorres, ed., "Mineral Matter and Ash in Coal", *ACS Symposium Series,* No. 301, American Chemical Society, Washington, D.C., 1986, Chapt. 2.
15. H. J. Gluskoter and co-workers, *Ill. State Geol. Survey Circ.,* 499 (1977).
16. A. Hood, C. C. M. Gutjahr, and R. L. Heacock, *Bull. Am. Assn. Petrol. Geol.* **59,** 986–996 (1975).
17. R. J. Gray, A. H. Rhoades, and D. T. King, *Trans. Soc. Min. Engrs. A.I.M.E.* **260,** 334–341 (1976).
18. C. A. Seyler, *Fuel,* **3**(15), 41, 79 (1924); *Proc. S. Wales Inst. Eng.* **53,** 254, 396 (1938).
19. *Annual Book of ASTM Standards,* Vol. 5.05, *Gaseous Fuels, Coal and Coke,* American Society for Testing and Materials, Philadelphia, Pa., published annually.
20. I. Wender and co-workers, in Ref. 6, "Chemical Reactions and Constitution of Coal", Chapt. 8.
21. L. M. Stock, R. Wolny, and B. Bal, *Energy Fuels,* **3,** 651 (1989).
22. M. E. Peover, *J. Chem. Soc.,* 5020 (1960).
23. R. Raymond, I. Wender, and L. Reggel, *Science,* **137,** 681 (1962).
24. L. Reggel, I. Wender, and R. Raymond Fuel, **43,** 75 (1964).
25. B. K. Mazumdar and co-workers, *Fuel,* **41,** 121 (1962).
26. B. K. Mazumdar, S. S. Choudhury, and A. Lahiri, *Fuel,* **39,** 179 (1960).
27. T. Green, J. Kovac, D. Brenner, and J. W. Larsen, in R. A. Meyers, ed., *The Macro-molecular Structure of Coals in Coal Structure,* Academic Press, Inc., New York, 1982, p. 199.
28. P. R. Solomon, P. E. Best, Z. Z. Yu, and S. Charpenay, *Energy Fuels,* **6,** 143 (1992); P. R. Solomon, D. G. Hamblen, R. M. Carangelo, M. A. Serio, and G. V. Deshpande, *ibid.* **2,** 405 (1988).
29. D. M. Grant, R. J. Pugmire, T. H. Fletcher, and A. R. Kerstein, *Energy Fuels,* **3,** 175 (1989).
30. L. Cartz and P. B. Hirsch, *Phil. Trans. R. Soc. London Ser. A,* **252,** 557 (1960).
31. W. Spackman, "What Is Coal?", *Short Course on Coal Characteristics and Coal Conversion Processes,* Pennsylvania State University, University Park, Pa., Oct. 1973.
32. I. Chatterjee and M. Misra, *J. Microwave Power Electromag. Energy* **25**(4), 224–229 (1990).
33. R. A. MacDonald, J. E. Callanan, and K. M. McDermott, *Energy Fuels,* **1**(6), 535 (1987).
34. O. P. Mahajan, *Coal Porosity in Coal Structure,* Academic Press, Inc., New York, 1982, p. 51.
35. H. Gan, S. P. Nandi, and P. L. Walker, *Fuel,* **51,** 272 (1972).
36. J. W. Larsen and P. Wernett, *Energy Fuels,* **2**(5), 719 (1988).
37. P. C. Wernett, J. W. Larsen, O. Yamada, and H. J. Yue, *Energy Fuels* **4**(4), 412 (1990).
38. Z. Spitzer and L. Ulicky, *Fuel,* **55,** 212 (1976).
39. J. T. McCartney, H. J. O'Donnell, and S. Ergun, *Coal Science, Advances in Chemistry Series,* Vol. 55, American Chemical Society, Washington, D.C., 1966, p. 261.
40. G. H. Taylor, in Ref. 6, p. 274.
41. W. H. Wiser, "Some Chemical Aspects of Coal Liquefaction", in Ref. 39.
42. H. F. Yancey and M. R. Geer, in J. W. Leonard and D. R. Mitchell, eds., *Coal Preparation,* 3rd ed., American Institute of Mining, Metallurgical and Petroleum Engineers, Inc., New York, 1968, pp. 3–56.
43. R. C. Attig and A. F. Duzy, *Coal Ash Deposition Studies and Application to Boiler Design,* American Power Conference, Chicago, Ill., 1969.
44. E. C. Winegartner and B. T. Rhodes, *J. Eng. Power,* **97,** 395 (1975).
45. *Steam, Its Generation and Use,* The Babcock & Wilcox Co., New York, 1972, p. 15–4.
46. R. Smock, *Power Eng.* **95**(2), 32 (1991).
47. J. Makansi, *Power,* **135**(3), 15 (1991).
48. G. A. Nelkin and R. J. Dellefield, *Mech. Eng.* **112**(9), 58 (1990).
49. J. A. Harrison, H. W. Jackman, and J. A. Simon, *Ill. State Geol. Survey Circ.,* 366 (1964).
50. T. Takanohashi and M. Iino, *Energy Fuels* **4**(5), 452–455 (1990).
51. M. G. Matturro, R. Liotta, and R. P. Reynolds, *Energy Fuels,* **4**(4), 346 (1990).
52. G. R. Hill and co-workers, in *Advances in Chemistry Series,* Vol. 55, American Chemical Society, Washington, D.C., 1966, p. 427.
53. Ref. 6, Chapt. 24.
54. M. S. Solum, R. J. Pugmire, and D. M. Grant, *Energy Fuels,* **3**(2), 187–193 (1989).
55. D. K. Fleming, R. D. Smith, and M. R. Y. Aquino, *Preprints, Fuel Chem. Div., Am. Chem. Soc.* **22**(2), 45 (1977).
56. S. C. Spalding, Jr., J. O. Burckle and W. L. Teiser, in Ref. 68, p. 677.
57. E. P. Stambaugh, in *Coal Desulfurization, Chemical and Physical Methods,* ACS Symposium Series 64, American Chemical Society, Washington, D.C., 1977, p. 198.
58. J. W. Hamersma, M. L. Kraft, and R. A. Meyers, in Ref. 70, pp. 73, 84; K. S. Vorres, in Ref. 70, **22**(4), 118 (1977).

59. H. Hofmann and K. Hoehne, *Brennstoff Chemie,* **35,** 202, 236, 269, 298 (1954).
60. *PETC Rev.* **7,** 12 (Winter 1992–1993); F. Titl, R. A. Meyers, and M. S. Storsul, *Gravimelt Coal Refinery Project in Northern Bohemia: Conference on Alternatives for Pollution Control from Coal-Fired Low Emission Sources,* Apr. 26–28, 1994, Pizen, Czech Republic.
61. P. R. Solomon and co-workers, *Preprints, Fuel Chem. Div., Am. Chem. Soc.* **36**(1), 267 (1991).
62. M. D. Kimber and M. D. Gray, *Combust. Flame* **11,** 360 (1967).
63. D. Fitzgerald and D. W. van Krevelen, *Fuel,* **38,** 17 (1959).
64. K. Ouchi and H. Honda, *Fuel,* **38,** 429 (1959).
65. B. K. Mazumdar, S. K. Chakrabartty, and A. Lahiri, *Proceedings of the Symposium on the Nature of Coal,* Central Fuel Resource Institute, Jealgora, India, 1959, p. 253; S. C. Biswas and co-workers, *ibid.,* p. 261.
66. H. Marsh and D. E. Clark, *Erdol und Kohle,* **39,** 113 (1986); *Proc. Iron Steel Soc.* (Apr. 1992); *Iron Steel Soc. AIME* (Aug. 1992).
67. *Surveys of Energy Resources 1986,* World Energy Conference, Central Office, London, 1986.
68. Ref. 4, p. 472.
69. A. Lahiri and co-workers, *J. Mines Metals Fuels,* **7,** 13 (1959).
70. *Coal Age,* **82,** 59 (1977).
71. *Coal Age,* **68,** 226 (1963).
72. K. S. Vorres, *Energy Fuels* **4**(5), 420–426 (1990).
73. *World Book Encyclopedia,* Field Enterprises Educational Corp., Chicago, Ill., 1975, p. 566.
74. A. Pott and co-workers, *Fuel,* **13,** 91, 125, 154 (1934).
75. *Environ. Sci. Technol.* **8,** 510. (1974).
76. W. Downs, C. L. Wagoner, and R. C. Carr, *Preparation and Burning of Solvent Refined Coal,* presented at American Power Conference, Chicago, Ill., Apr. 1969.
77. *PETC Rev.,* **3** (Mar. 1991).
78. D. L. Wise, *Bioprocessing and Biotreatment of Coal,* Marcel Dekker, Inc., New York, 1990, 744 pp.
79. R. M. Fakoussa, *Investigations of the Microbial Decomposition of Untreated Hard Coals: Coal as a Substrate for Microorganisms,* doctoral dissertation, Bonn, Germany, 1981; prepared for U.S. Dept. of Energy, Pittsburgh Energy Technology Center; translated by the Language Center Pittsburgh under Burns and Roe Services Corp., Pittsburgh, Pa., June 1987.

General References

D. L. Crawford, ed., *Biotransformations of Low Rank Coals,* CRC Press Inc., Boca Raton, Fla., 1992.

D. L. Wise, ed., *Bioprocessing and Biotreatment of Coal,* Marcel Dekker, New York, 1990, 744 pp.

Fuel **70**(3) (1991) contains a series of papers presented at the *International Conference on Coal Structure and Reactivity: Chemical, Physical, and Petrographic Aspects,* Sept. 5–7, 1990, Cambridge, U.K.

Fuel **70**(5) contains a series of papers presented at *Biotechnology for the Production of Clean Fuels,* Aug. 27–28, 1990, Washington, D.C., pp. 569–620.

W. Francis, *Coal,* 2nd ed., Edward Arnold & Co., London, 1961.

E. Stach, M. T. Mackowsky, M. Teichmuller, G. H. Taylor, D. Chandra, and R. Teichmuller, *Stach's Textbook of Coal Petrology,* 3rd ed., 1982, Gebruder Borntraeger, Berlin, 535 pp; excellent text on coal petrography.

K. S. Vorres, "Coal" in J. I. Kroschwitz and M. Howe-Grant, eds., *Encyclopedia of Chemical Technology,* 4th ed., Vol. 6, John Wiley & Sons, Inc., New York, 1993, pp. 423–489.

COAL PREPARATION

JANUSZ S. LASKOWSKI
University of British Columbia
Vancouver, B.C., Canada

MARK S. KLIMA
The Pennsylvania State University
University Park, Pennsylvania

Objectives of Coal Preparation

Coal preparation is the conversion of as-mined coal into marketable coal, whose composition meets the specifications required for its utilization (eg, combustion, coke production, coal conversion, etc). The as-mined coal, referred to as run-of-mine (ROM) material, consists of particles, differing in size and mineralogical composition, made into a clean, graded, and consistent product suitable for the market by physical processes that upgrade the quality of coal by regulating its size and reducing the impurities (1). The impurities include mine rock, clays, fine shales, pyritic sulfur, organic sulfur, and even trace elements (2,3).

Coal preparation ranges from a simple crushing/sizing operation to a complex operation that involves cleaning the entire size range of feed coal to reject the majority of the impurities associated with ROM material. The major unit operations are crushing, screening, cleaning (washing, beneficiation), mechanical dewatering, and thermal drying. By manipulating the arrangement of the unit operations, a process flow sheet is developed, which is used to establish process flows, plant size, product coal yield and quality, and capital and operating costs.

Flow sheet development requires knowledge of raw coal properties related to particle behavior during processing and equipment operation to generate the circuitry that will do the job most economically. While doing this, the two principal objectives of coal preparation, ie, the liberation of the mineral matter from the coal and the separation of the coal from the mineral matter, must be accomplished to maximize the recovery of available clean coal and maximize the rejection of the mineral constituents in the clean coal. The third objective, uniformity of quality, automatically follows.

Levels of Cleaning

Conventional coal preparation is broadly classified into four levels of cleaning related approximately to the sizes of coal being cleaned. Level one preparation involves no mechanical cleaning but only crushing the ROM coal to control the maximum (top) size of the material, often followed by sizing to produce several product size fractions. However, undesirable components inadvertently included during mining, such as mine timbers, metal from mining equipment, and mine rock are typically removed. This level of preparation is commonly employed in many low rank coal producing areas. The product from a level one operation is referred to as raw coal and is utilized directly in many cases.

The raw coal is sized into several streams before being fed to various concentrating operations to reduce the naturally occurring mineral constituents, usually at the expense of an increase in the moisture content. The degree of process complexity depends on the sizes of raw coal processed.

Table 1. Coal Inorganic Impurities[a]

Type	Origin	Examples	Physical Separation
Strongly chemically bonded elements	From coal-forming organic tissue material	Organic sulfur, nitrogen	NO
Adsorbed and weakly bonded groups	Ash-forming components in pure water, adsorbed on coal surface	Various salts	Very limited
Mineral matter			
Epiclastic	Minerals washed or blown into peat during its formation	Clays, quartz	Partly separable by physical methods
Syngenetic	Incorporated into coal from earliest peat-accumulation stage	Pyrite, siderite, some clay minerals	Intimately intergrown with coal macerals
Epigenetic	Stage subsequent to syngenetic; migration of mineral-forming solutions through coal fractures	Carbonates, pyrite, kaolinite	Vein-type mineralization; epigenetic minerals concentrated along cleats, preferentially exposed during breakage; separable by physical methods

[a] Ref. 5.

Level two preparation involves cleaning the coarser sizes of coal (eg, greater than ~10 mm) utilizing some type of mechanical cleaning process, which will be described in a later section. The finer material less than 10 mm (ie, −10 mm) is generally combined with the clean +10 mm coal as part of the product.

Level three preparation extends the bottom size that is treated to include the intermediate coal fraction (eg, 10 × 0.6 mm). The −0.6 mm material is combined with the cleaned +0.6 mm coal or, in some instances, disposed of as part of the refuse.

In many cases it is necessary to process the fines to recover additional coal, while rejecting additional impurities. This scenario is referred to as level four preparation or total cleaning. However, even with this level of cleaning, some portion of the fines (eg, −0.1 mm) is discarded because of inefficiencies in the processes treating this fraction, particularly in regard to moisture reduction. Total cleaning has become more prevalent because it is simply no longer cheaper to mine more coal. Metallurgical coke feed stock or compliance steam generating fuels usually require total cleaning of raw coal.

Product yields for most cleaning facilities range from 60 to 90%, with thermal recoveries from 85 to 98%.

Generic Areas of Coal Preparation

To develop the proper coal preparation circuits required for the more sophisticated levels of cleaning, four generic areas must be considered: characterization, liberation, separation, and disposition (4). The characterization procedure, based on particle size and density fractionation, is called a washability study. This technique determines the maximum yield of coal expected with a particular attribute (ash, sulfur, etc) content. Liberation involves reducing the size of the raw coal (typically by crushing) to free the inorganic constituents that are locked to the coal particles.

Once liberated, the particles must be separated from one another. Because the densities of the mineral impurities are higher than those of coal, the common concentrating processes that separate the inorganic constituents from coal exploit these density differences, while the particles move through water. However, for fine particles, separations are also based on surface property differences.

After separation, the clean coal and refuse fractions must be treated to provide a product that can be handled by the customer and a refuse that can be disposed of in an environmentally acceptable manner. This is referred to as product disposition and includes dewatering, thickening, reconstitution (eg, pelletizing), refuse disposal, and materials handling.

COAL CHARACTERISTICS RELATED TO COAL PREPARATION

Coal is an organic sedimentary rock, which contains combustible organic matter in the form of macerals and inorganic matter mostly as minerals. Coal preparation upgrades raw coal by reducing its impurities (mineral matter). The most common criterion of processing quality is ash, which is not removed from coal during beneficiation, but particles with a lower inorganic matter content are separated from those with a higher inorganic matter content. The constituents of ash do not occur in coal but are formed by chemical changes that take place in mineral matter during combustion. Ash is defined as all elements in coal except carbon, hydrogen, nitrogen, oxygen, and sulfur.

Coal is heterogeneous at a number of levels (5). At the simplest level, it is a mixture of organic and inorganic phases, but because the mineral matter of coal originated from the inorganic constituents of the precursor plant, from other organic materials, and from the inorganic components transported to the coal bed, its textures and liberation characteristics differ (Table 1). The level of heterogeneity is classified as follows:

1. At the seam level, a large portion of mineral matter in coal arises from inclusion during mining of roof or floor rock.
2. At the lithotype level, the mineral matter occurs as deposits in cracks and cleats or as veins.
3. At the maceral level, the mineral matter is present as very finely disseminated, discrete, mineral particles.
4. At the submicroscopic level, the mineral matter is present as strongly, chemically bonded elements. Carboxylic (−COOH), phenolic (−OH), mercapto (−SH), and imino ([=NH]) groups are part of organic coal.

Many elements and most trace elements (especially in brown coal and lignites) are ion exchangeable and are associated with these groups.

Even in ROM coal, a large portion of coal and shale is already liberated to permit immediate concentration. This is so with heterogeneity level one and to some extent with level two; at heterogeneity level three, only crushing and very fine grinding can liberate mineral matter, whereas at level four, which includes chemically bonded elements and probably syngenetic mineral matter, separation is possible only by chemical methods.

Recent findings indicate that most of the mineral matter in coal down to the micrometer particle-size range is indeed a distinct separable phase that is liberated by fine crushing and grinding.

The terms *extraneous mineral matter* and *inherent mineral matter* were usually used to describe an ash-forming mineral, separable and nonseparable from coal by physical methods. Traditionally in coal preparation processes, only the mineral matter at the first level and, to some extent, at the second level of heterogeneity was liberated; the rest remained unliberated and, remaining with the cleaned coal, contributed to the inherent mineral matter. The use of very fine grinding, which also liberates mineral matter at the third level of heterogeneity, has changed the old meanings of the terms *inherent and extraneous.* The content of the "true" inherent ash-forming material (ie, the fraction left in coal after liberating and removing the mineral matter at the first, second, and third levels of heterogeneity) is usually less than 1%.

Coal combustion is the largest single source of sulfur oxide pollution. Therefore, in recent years, the emphasis in coal preparation has been placed on the recovery of combustible material, but, primarily, on reducing sulfur content in coal (6,7).

Sulfur in coal is present in both organic and inorganic forms. The dominant form of inorganic sulfur is pyrite, but marcasite has also been reported in many coals. Pyrite occurs as discrete particles, often of microscopic size. It comprises 30 to 70% of the total sulfur in most coals. Other forms of inorganic sulfur present are gypsum and iron sulfates. The sulfate level in fresh unoxidized coals is generally less than 0.2%.

Organic sulfur in coal is contained in groups such as thiophene, organic sulfides, mercaptans, and organic sulfides:

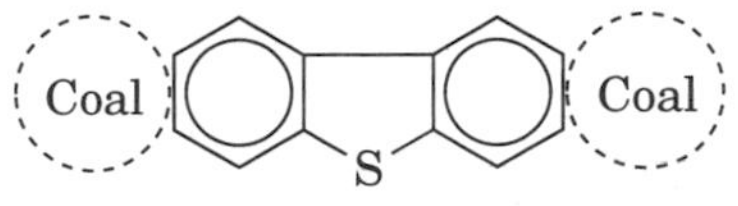

Coal — S — Coal

Thiophene

R—S—R′ R—SH R—SS—R′

Organic sulfide **Mercaptan** **Organic disulfide**

The organic sulfur content in some coals ranges from 0.5 to 2%.

Physical cleaning methods remove inorganic sulfates (gypsum) and most of the coarse pyrite; the finely disseminated microcrystalline pyrite and organic sulfur are usually not separable by such processes. In the case of coal containing 70% of sulfur in pyritic form and 30% as organic sulfur, physical cleaning reduces the sulfur content by about 50%.

Chemical cleaning methods remove pyritic and also organic sulfur from coal. Two chemical desulfurization processes have been tested under pilot-plant conditions: the ferric ion leaching process and the Gravimelt process. The former removes up to 90% of pyritic sulfur, and the latter removes over 90% of both pyritic and organic sulfur.

The removal of pyrites from coal by chemical means is not easy because pyrites are insoluble in all known liquids. Hydrochloric, hydrofluoric, and sulfuric acids have little or no effect on iron pyrites. However, pyrites may be oxidatively converted to sulfates, which are soluble in strong acids, by strong oxidizing agents, such as nitric acid or hydrogen peroxide. These reagents have long been used to analyze the pyritic sulfur content in coal. These reagents, however, have never been advanced as a method for lowering the sulfur content of coal, because they also oxidize the coal matrix (nitric acid also nitrates the coal). In the search for a proper oxidizing reagent, it has been pointed out (8) that such a reagent has to be (*1*) highly selective for pyrite without attacking the coal matrix, (*2*) regenerable, (*3*) highly soluble in both oxidized and reduced forms, and (*4*) inexpensive. It was discovered that aqueous solutions of ferric salts meet this combination of requirements and these reagents are the basis for the ferric ion leaching process.

The Gravimelt process, developed by TRW Inc., California, under contract with the U.S. Department of Energy during the past 10 years, utilizes molten, caustic NaOH (or mixtures of NaOH and KOH), as the primary reactant. The process is effective in reducing levels of ash-forming minerals, pyritic sulfur, and organic sulfur (9–11).

FLOAT-AND-SINK ANALYSIS

Most coal-cleaning processes used to remove inorganic impurities from coal are based on the gravity separation of the coal from its associated refuse. ROM coal consists of some proportion of both coal and shale particles, already sufficiently liberated, together with coal particles with inclusions of refuse (ie, bands of shale). Commercially cleaned coal contains only very disseminated impurities and has a density ranging from 1.2 to 1.6. Carbonaceous shale density ranges from 2.0 to 2.6, and pure shale, clay, and sandstone have a density of about 2.6. The density of pyrite is about 5.0. The difference in density between pure coal and these impurities, in a liberated state, is sufficient to achieve almost complete separation fairly easily. However, the inorganic impurity content, and hence, the ash content, ranges from pure coal containing only microscopic impurities to shale which is entirely free from carbonaceous matter. Generally speaking, the mineral matter content of any coal particle is proportional to its density and inversely proportional to its calorific value.

Before the proper coal preparation circuits can be developed, the coal must be sufficiently characterized to determine the quantity and quality of clean coal that can be produced. This procedure is called a washability analysis. The technique involves systematically partitioning (by size and density) the ROM or raw coal into smaller and smaller fractions to produce samples with similar physical properties. The resulting samples are analyzed to determine the appropriate attribute content (ash, total sulfur, calorific value, etc). These data are used to generate the theoretical washability curves, which represent graphically the experimental separation data obtained under ideal conditions in the float–sink tests. The curves show the maximum amount of clean coal that can be produced at a specified attribute content.

For a new plant, a washability study is used to develop the plant flow sheet. Float-and-sink tests are also used to determine the Tromp curve, which measures the practical results of a density separation. The practical results of separation are compared with the ideal and a measure of efficiency calculated. These data are also used to determine set points in the appropriate process equipment.

The principle of the float–sink testing procedure is as follows. A weighed amount of a given size fraction is gradually introduced into a heavy liquid of low density. The floating fraction is separated from the fraction that sinks and is dried and weighed. The sink material is tested in the next higher density liquid. The procedure is repeated successively with liquids extending over the desired range of densities. The fraction that sinks in the liquid of highest density is also obtained. The weight and ash contents of each density fraction are determined. Additional attributes, such as the calorific value and sulfur content, are also determined.

SIZE REDUCTION

Size reduction controls the top size of the coal or liberates to create coal particles sufficiently free of mineral matter to meet the product specifications after separation. The objective of size reduction varies with the stream under consideration. For example, crushing of the feed controls the top size of the coal entering the plant and improves liberation. During the beneficiation of coal, the problem of treating middlings sometimes arises. This material is not of sufficient quality to be included with the high quality clean coal product, yet it contains potentially recoverable coal. If this material is simply recirculated through the cleaning circuit, little or no quality upgrading can be achieved. Liberation is accomplished if the nominal top size of the material is reduced, which permits recovery of the coal in the cleaning unit. Normally, the nominal top size is reduced to the range of 20 to 6 mm. Product crushing is done for top size control. For example, most power generating plants are not equipped to receive and handle coal much coarser than about 40 mm.

Size reduction actually begins in the mining operation, and the top size and particle size distribution of the ROM material are directly affected by the type of equipment used to mine the coal. Size reduction is categorized as primary breakage, crushing, and grinding. These relate to an approximate size range, which roughly corresponds to the size reduction device employed, although there is often an overlap of the classes.

Primary breakage is the term applied to size reduction of large material (eg, +75 mm) and *crushing* to particle size reduction below 75 mm; the term grinding covers the size reduction of material to below about 6 mm (12). However, these terms are loosely employed. A general term for all equipment is *size reduction equipment,* and because the term *comminution* means size reduction, another general term for the equipment is *comminution equipment.*

Primary Breakage

Primary breakage involves reducing the ROM coal to top sizes ranging from about 200 mm to 50 mm, using one of several devices, such as a rotary breaker, roll crusher, or hammer mill (13). The rotary breaker serves two functions, reduction in the top size of ROM material and rejection of oversize rock. It is an autogenous size-reduction device in which the feed material acts as a crushing medium. It consists of a rotating cylindrical shell with openings in its perimeter to match the desired top size of the product (Fig. 1). The coal is fed into one end where it is lifted and then dropped onto the revolving shell causing breakage as it moves along the length of the device. The broken coal exits through the openings, and the material that did not break (eg, mine rock) is rejected from the other end. Thus this device is capable of reducing the top size and providing some level of cleaning. Rotary breakers are up to 3.7 m in diameter and over 6.1 m long with capacities up to 1800 metric tons per hour (mtph).

In a roll crusher, the coal is fed between a revolving roll and a fixed breaker plate (single-roll crusher) where it is compressed and subsequently crushed. Single-roll crushers reduce ROM material to a product top size in the range of 200 to 18 mm in a single pass, depending on the top size of the feed coal. Double-roll crushers consist of two rolls rotating in opposite directions. Normally, one roll is fixed, and the other roll is movable against spring pressure. This permits the passage of uncrushable material (eg, metal, mine timbers) without damage to the unit. The drive units are normally equipped with shear pins for overload protection. The gap setting between the rolls can be adjusted to control the product size. Triple- or four-roll models are also available. For double-roll crushers, roll sizes up to 76 cm in diameter by 122 cm long are used to crush up to 650 mtph depending on the coal top size and gap setting.

In a hammer mill, the coal is dropped onto swinging hammers or bars, which are attached to a rotating shaft, and is subsequently thrown against breaker plates. The impact with the hammers and plates causes breakage, and a steel grid within the device helps to regulate the product size discharged from the unit. The discharge openings range from less than 10 mm to more than 50 mm. Hammer mills are available to handle up

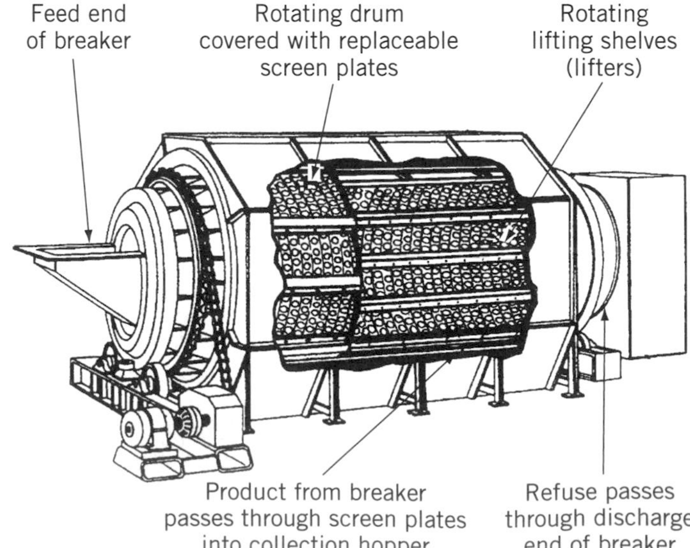

Figure 1. Cutaway view of a rotary breaker.

to 800 mtph, depending on the type of coal and the product size required. An impactor operates on a principle similar to a hammer mill, although no grate is used to control the product size.

Another impact reduction device is a cage mill. It involves multiple (eg, four) stages of rotating bars or cages. The coal is fed inside the innermost cage and is broken as it passes through the cages, which rotate in opposite directions at the same speed. A 2.4 m diameter cage mill processes over 3000 mtph.

SIZING

Sizing is one of the most important unit operations in coal preparation, defined as the separation of a heterogeneous mixture of particle sizes into fractions in which all particles range between a certain maximum and minimum size. Size separations are done primarily with screens, although other sizing devices, such as hydrocyclones, are also used. Sizing with hydrocyclones is called size classification.

SEPARATION

To produce a desirable product, ash-forming minerals must be rejected from clean coal. This processing step is referred to as separation, which is the third generic area of coal preparation. The bulk of the coal is cleaned by gravity concentration, primarily the higher rank coals. Gravity concentration separates particles based primarily on the density differences between the coal and refuse particles. Because certain devices work best on a certain size range of particles, it is necessary to divide the feed coal into several fractions based on particle size prior to processing. The approximate size ranges can be classified as: coarse, +10 mm; intermediate, 20 mm to 0.6 mm; and fine, −0.6 mm. Furthermore, gravity concentration devices fall into two general classifications, dense-medium separators and water-only devices. Fine coal is typically processed by froth flotation.

DISPOSITION

After the coal is separated from the refuse, it must be in such a form that it can be handled in downstream operations and, eventually, by the customer. Likewise, the refuse must be suitable for disposal at the designated waste site. The final area can be designated as disposition and includes dewatering, reconstitution, and refuse disposal, along with the necessary materials handling for both in-plant and out-of-plant operations.

PROCESS EVALUATION

The yield and quality of the clean-coal product from an industrial coal preparation plant and the theoretical yield and quality determined from washability curves are known to be different. In the ideal cleaning process, all coal particles lower in density than the density of separation would be recovered in the clean product, and all material of greater density would be rejected as refuse. Under these conditions the product yield and quality from the actual concentration process and the yield and quality expected from the washability curves would be identical.

Separator performance, however, is never ideal. As a result, some coal particles with densities lower than the separation density end up as rejects, and some high ash particles with densities higher than the separation density end up in the clean coal. These particles are called misplaced material.

In general, particles with densities well below or well above the density of separation are carried to the proper products, ie, the clean coal and refuse, respectively. However, for particles with densities approaching the density of separation, the proportion of misplaced material increases rapidly.

COAL PREPARATION FLOW SHEETS

In run-of-mine coal, a large portion of both coal and shale is already liberated sufficiently to permit immediate concentration. Because dewatering of coarse coal is much more efficient and cheaper than that of fines and because many users prefer larger size fractions for their particular processes, the preparation of feed before cleaning has traditionally consisted of as little crushing as possible. Therefore, coal preparation involves mostly gravity concentration and, to some extent, flotation, complemented by screening and solid/liquid separation processes.

The precrushing needed to reduce the top size of the material is widely done by rotary breakers. This process, as already pointed out, is based on the selective breakage of fragile coal and hard inorganic rock and combines two operations: size reduction and preconcentration (because it rejects larger pieces of hard rock). A further development in coal preparation has been the application of the three-product separators to treat coarse and intermediate size fractions. The recrushed middlings produced in such a cleaning are reprocessed, together with the fine part of the feed, increasing the total coal recovery. This complete concentration sequence is typical for processing a metallurgical coal or a compliance thermal coal for power generation.

Coal preparation plants are classified according to the extent to which coal is cleaned, as follows (Fig. 2): Types I–IV correspond to cleaning levels 1–4, respectively.

- Type I: only the large size coal is reduced and screened into various size fractions.
- Type II: plant cleans coarse coal by gravity methods; refuse content in coal is reduced in the form of coarse reject from the rotary breaker and high gravity rejects (yield of saleable product is 75 to 85%).
- Type III: plant cleans coarse and intermediate size coal, leaving fines untreated.
- Type IV: coal preparation plants clean full size range of raw coal.
- Type V: plant produces different saleable products that vary in quality, including deep-cleaned coal low in sulfur and ash.
- Type VI: plant incorporates fine grinding to achieve a high degree of liberation, using froth flotation (or oil agglomeration) and water-only cyclones (or spirals).

Type VI represents an important stage in the development of coal preparation strategy (1). Types I–V are all predominantly based on coarse coal cleaning, with the amount of fines reduced to a minimum, but in a Type VI plant all coarse coal cleaning operations are aimed at eliminating coarse, hard

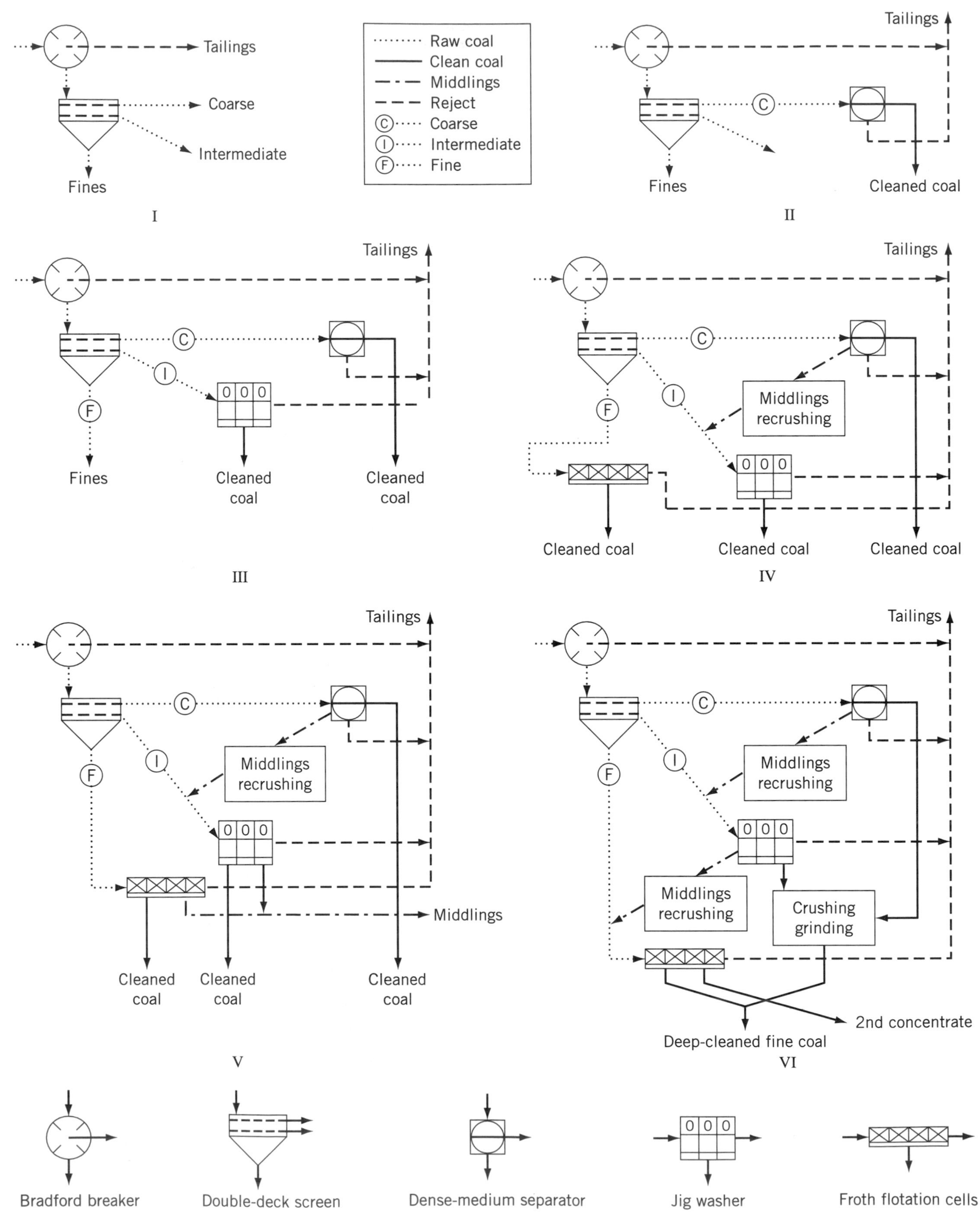

Figure 2. Classification of coal preparation circuits, Types I–VI.

refuse and constitute a pretreatment stage. The final cleaning follows after fine crushing and grinding. This future development in coal preparation technology is aimed at cleaning and utilizing thermal coal and mainly results from the new environmental protection restrictions.

In the future, Type VI coal preparation plants will be combined with coal–water slurry technology, and such slurries, to some extent, will replace liquid fuels.

ON-LINE ANALYSIS

On-line measurements of parameters, such as ash, moisture, and sulfur use an inferential technique in which some property can be measured that in turn relates to the parameter in the material (14). Most on-line techniques use some form of electromagnetic radiation. When radiation passes through matter, there is a loss of intensity by absorption and scattering.

RESEARCH INTO NEW BENEFICIATION PROCESSES

The processes discussed represent the bulk of the coal preparation unit operations utilized in conventional coal-cleaning facilities. However, other technologies have been investigated or are being considered, particularly for fine-coal cleaning.

BIBLIOGRAPHY

1. J. S. Laskowski and A. D. Walters, "Coal Preparation", in *Encyclopedia of Physical Science and Technology,* Vol. 3, Academic Press, Inc., New York, 1992, pp. 409–435.
2. K. S. Vorres, ed., *Mineral Matter and Ash in Coal,* ACS Symposium Series 301, American Chemical Society, Washington, D.C., 1986.
3. D. J. Swaine, *Trace Elements in Coal,* Butterworths, London, 1990.
4. W. J. Brumbaugh and P. T. Luckie, *BTU Special Issue,* McNally Coal, Pittsburg, Kans., 1982, pp. 4–6.
5. A. C. Cook, *Sep. Sci. Technol.* **16,** 1545–1570 (1981).
6. T. D. Wheelock, ed., *Coal Desulfurization,* ACS Symposium Series 64, American Chemical Society, Washington, D.C., 1977.
7. R. A. Meyers, *Coal Desulfurization,* Marcel Dekker, New York, 1977.
8. R. A. Meyers, J. W. Hamersma, J. S. Land, and M. L. Kraft, *Science,* **177,** 1187–1188 (1972).
9. R. A. Meyers and W. D. Hart, "Chemical Removal of Organic Sulfur from Coal," *Symposium on the Removal of Heteroatoms from Fuel,* ACS Meeting, Houston, Tex., Mar. 23–28, 1980.
10. C. D. Chriswell, N. D. Shah, and R. Markuszewski, *Proceedings of the 5th Annual International Pittsburgh Coal Conference,* Pittsburgh, Pa., Sept. 12–16, 1988, pp. 446–459.
11. C. D. Christwell, R. Markuszewski, and G. A. Norton, in P. R. Dugan, D. R. Quigley and Y. A. Attia, eds., *Processing and Utilization of High-Sulfur Coals IV,* Elsevier, Amsterdam, the Netherlands, 1991, pp. 385–397.
12. D. W. Horsfall, *Coal Preparation for Plant Operators,* South African Coal Processing Society, Cape Town, South Africa, 1980.
13. L. G. Austin, in J. W. Leonard, ed., *Coal Preparation,* 5th ed., SME, Littleton, Colo., 1991, pp. 187–219.
14. J. T. Canwell, in Ref. 13, pp. 717–782.

CONE-PENETROMETER DEPLOYED CHEMICAL SENSORS

PAUL V. DOSKEY
Argonne National Laboratory
Argonne, Illinois

ERNESTO R. CESPEDES
U.S. Army Corps of Engineers
Vicksburg, Mississippi

The traditional methods for determining the location and levels of subsurface chemical contamination in widely varying types of soil and groundwater are time consuming and costly and often result in less than adequate site characterization. Conventional environmental site investigations require installation of many monitoring wells and subsequent analysis of discrete soil and groundwater samples by using standard laboratory methods. Samples are often collected without any *a priori* knowledge about the exact location and extent of contaminant plumes. Zones or plumes of contamination can be completely missed or over- or underestimated. Effective site characterization is often hampered by the inability to select optimal locations for monitoring wells (see SITE CHARACTERIZATION, EXPEDITED). The locations of the monitoring wells are usually based on information gleaned from site historical data, groundwater hydrology, or indirect chemical screening by measuring volatile contaminants in soil gas or some combination of the three. Because of uncertainties in the available information, placement of monitoring wells is at best an inexact science; historical data are often incomplete or inaccurate; knowledge of groundwater hydrology at the site may not provide the level of detail required to understand site characteristics; and interpretations of soil gas measurements may be complicated by erratic movement of vapor in the soil due to impervious layers and changes in atmospheric temperature and pressure. Consequently, many wells are not properly positioned and, therefore, yield information of marginal utility. Furthermore, the ability to resolve horizontal and vertical features in the distribution of chemical contaminants is limited by the spacing between wells and the vertical spacing between samples. Accurately delineating the boundaries of contaminant plumes and defining small-scale vertical structure in the distribution of contamination have important implications with respect to site remediation. The more precisely the area of contamination is defined, the less likely it is that clean material will be unnecessarily removed or subjected to costly remediation procedures. For detailed spatial mapping of contaminated areas, they must often be sampled and analyzed in an interactive manner. Such an approach can be prohibitively costly and labor-intensive when conventional techniques are used. In addition, in a number of subsurface environments, the delay from sample collection to interpretation of laboratory analyses can severely hamper the response time and hinder contaminant containment, possibly resulting in a much larger extent and expense of cleanup (1).

As a result of these shortcomings, much current research has focused on the development of rapid, cost-effective *in situ* techniques to monitor both organic and inorganic pollutants. These new methods provide a thorough mapping of contaminated areas by being able to sample the subsurface with high spatial resolution. This mapping makes it possible to rapidly interpret the extent and location of contaminants and efficiently plan and direct cleanup operations. Furthermore, *in situ* methods minimize the risk of sampling artifacts and can be performed in hostile environments without exposing personnel to toxic contaminants.

One technology that is gaining widespread acceptance for rapid environmental site characterization involves the use of cone-penetrometer systems. The electronic cone-penetrometer test (CPT) was developed to provide detailed geotechnical information on the subsurface. By combining the results of the CPT with a detailed chemical characterization of the groundwater, soil, and soil gas, the migration of subsurface contaminants can be investigated (2–8). Many sampling and sensing technologies have been interfaced with the cone penetrometer to provide a chemical characterization of the subsurface. These sampling and sensing technologies provide near-real-time quantitative or qualitative data or both of subsurface chemical contamination with high spatial resolution. The devices include samplers that collect water, soil, and soil gas from the subsurface. The medium is typically transferred to an inert container and returned to the laboratory for analysis. For some analytes, particularly volatile organic compounds (VOCs), these transfer steps can lead to incomplete recoveries of the chemical substances and erroneous results. On-line analytic techniques that include an *in situ* isolation of the chemical substance from the subsurface media in the cone penetrometer and a direct transfer of the analyte to an instrument or chemical sensor at the surface are being developed to eliminate these transfer steps. In a more direct approach, chemical sensors have been interfaced with the cone penetrometer to perform an *in situ* analysis of the subsurface media, thereby eliminating the isolation and transfer steps of the on-line analytic techniques.

The Site Characterization and Analysis Penetrometer System (SCAPS) was initially developed by the U.S. Army during the late 1980s and at that time was the first cone-penetrometer system specifically designed for rapid environmental site characterization. The SCAPS (Fig. 1) is based on a custom-engineered 20-ton truck capable of hydraulically pushing an instrumented probe to a maximum depth of 50 m. The truck houses two separate, protected work spaces to allow access to contaminated sites with minimal risk to the work crew. One of the work spaces contains the penetrometer tool. The other work space houses optical systems, chemical analysis equipment, and the equipment for digital data acquisition, processing, and display. The penetrometer unit can be equipped with a number of sensors and samplers designed to detect a variety of classes of subsurface contaminants, including petroleum, oils, and lubricants (POLs), VOCs, metals, explosives, and radioactive wastes (9). Incorporated into the SCAPS rig, as well as into a number of other cone-penetrometer units, is equipment for decontaminating the push rods as they are being retracted that consists of rubber wipers and steam cleaning equipment mounted under the push room. As the push rods are retracted following a cone-penetrometer push, the rods are scrubbed and steam cleaned, and the wastewater is collected for off-site disposal. Also incorporated into a number of cone-penetrometer probes is equipment that allows for grouting of the hole during probe retraction to prevent any migration of contaminants. The SCAPS trucks are currently being operated by the U.S. DOD, DOE, EPA, and private industry to conduct site characterization tasks (10). The various sensors and samplers used on SCAPS and similar cone-penetrometer systems are described in subsequent sections.

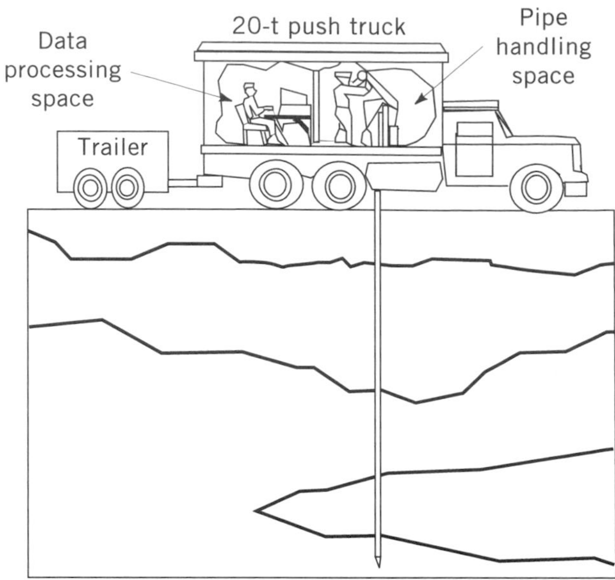

Figure 1. Typical cone penetrometer test (CPT) system for environmental site characterization and monitoring. Courtesy of the U.S. Army Corps of Engineers Waterways Experiment Station.

CONE-PENETROMETER TESTING

The CPT is used to gather geotechnical information on the subsurface. Mechanical cone penetrometers were developed in the early 1900s to identify strong soil layers for the design of pilings and building foundations. The mechanical CPT provides a discontinuous record of geotechnical properties at shallow depths. The electronic cone penetrometer was developed in the 1960s to accurately examine soil stratigraphy at depths approaching 90 m (2,6,8,11–13). The electronic cone penetrometer, which consists of an instrumented conical probe (cone) and 1-m sections of metal pipe (rods), is advanced with hydraulic jacks from a skid- or truck-mounted rig that can weigh between 20 and 40 tons. The maximum thrust the rods can withstand is 45,000 pounds force (200 kilo-Newtons (kN)). This thrust advances the cone to 30 m in stiff material and as much as 80 m in looser soils. The probe is advanced in 1-m intervals at 1–2 cm/s and is interrupted at each interval to attach additional rods to the top of the string of rods in the cone-penetrometer truck. The electronic CPT represents a continuous record of the resistance to penetration.

The cone consists of a hollow steel rod that is 3.0–4.1 cm (OD) (2.5-cm ID) with a 60° conical point and a friction sleeve 13.3 cm long. Several sensors located in the cone are connected to an automatic data recording system in the cone-penetrometer truck by an electrical cable that runs through all of the rod sections. Strain gauges located behind the tip and friction sleeve measure the displacement associated with the resistance to penetration. The tip resistance is a measure of the total stress or intergranular pressure plus water pressure, and the friction resistance is the sum of friction and adhesion. The ratio of these two resistances is used to determine soil type and derive a stratigraphic log of the CPT. Soil classification by using the CPT is empirically based (14–16). The classification differentiates gravel, sand, silt, clay, some intermediate soil types, and some overconsolidated soils. The American Society for Testing and Materials (ASTM) standard D3441-86 is used to ensure uniformity with regard to the cone characteristics, calibration, and testing and reporting procedures (17).

The cone can contain several other sensors for measuring parameters that are useful in interpreting the migration of subsurface contaminants. Dynamic and equilibrium pore water pressures are measured by using a porous stone filter in the cone that is connected to a pressure transducer (18,19). A probe to determine soil moisture has also been developed (20). Sensors to measure the inclination are used for correcting depth to the lateral drift of the cone. Electrical conductivity sensors and nuclear and seismic probes have also been used in the cone penetrometer. Soil resistivity can be used as a rough measure of the contamination of the soil (19,21). The hydraulic conductivity, hydrostatic head, independent soil classifications, and soil permeability can be derived from measurements obtained from these additional sensors.

SAMPLERS

Many samplers have been interfaced with the cone penetrometer to perform an *in situ* collection of groundwater, soil, and soil gas. Following sample collection, the soil and groundwater probes are usually brought to the surface, and the sample is transferred to a container and analyzed on-site or transported off-site to a laboratory for analysis. Soil gas probes directly transfer the gas from the probe through tubing to a container or gas analyzer at the surface. Careful attention must be paid to the materials that are used in the device and in the sample container so that the media is not contaminated or the analytes are not sorbed to the device or container. Groundwater and soil gas are typically sampled through a porous element in the probe. Soil cores are collected in liners located in the rods of the penetrometer. The soil cores can be retrieved and subsampled in the field, or the ends of the liner can be capped and the entire core returned to the laboratory.

ON-LINE ANALYTIC TECHNIQUES

On-line analytic techniques have been interfaced with the cone penetrometer to quantify VOC and semivolatile organic compound (SVOC) contamination of the subsurface. These techniques include isolation, preconcentration, transfer, and chemical detection steps. This section focuses on the first three steps of the method, and chemical detection techniques are discussed in a later section. *In situ* isolation techniques have been developed to eliminate the transfer of groundwater, soil, and soil gas from *in situ* sampling probes to a container prior to chemical analysis. Erroneous results have been found, particularly for VOCs, when this transfer step is included in the sampling technique (22). Following the isolation step, the organic contaminants can be concentrated at depth, analyzed *in situ* by a chemical sensor or instrument, or transferred through tubing to a chemical sensor or instrument in the cone-penetrometer truck. Laboratory methods have been developed to isolate organic analytes from groundwater and soil by water purging and thermal desorption techniques, respectively. Cryogenic preconcentration, solid sorbent preconcentration methods, solid-phase microextraction, and membrane extraction techniques have all been used to preconcentrate the analyte before analysis. The analytes may also be transferred directly from the isolation probe to an instrument or chemical sensor at the surface. This technique allows for multiple sample collection from a single penetrometer push provided the porous element can be cleaned between sample collections; however, a critical aspect of the on-line techniques is the quantitative transfer of analytes from the sample probe to the instrument or chemical sensor at the surface through tubings that can approach 100 m in length.

CHEMICAL DETECTION TECHNIQUES

A variety of chemical detection technologies have been used to measure subsurface contaminants (23,24). These technologies include sensors to provide *in situ* chemical detection and instrumentation such as gas chromatographs and mass spectrometers that are used in on-line analytic techniques with the cone penetrometer.

Chemical Sensors

A chemical sensor is a transducer consisting of a physical layer and a chemically selective layer that is capable of providing direct information about the chemical composition of its environment (25). Chemical detection can be based on the mass of the analyte or its optical, electrochemical, or radiochemical properties.

Gas Chromatography

A gas chromatographic system consists of an injection system, an analytical column, and a detector. The smallest field-portable instruments, with the exception of the Cone-GC (26), have not been miniaturized for incorporation into the cone penetrometer; however, field-portable and transportable gas chromatographs (GCs) have been used to analyze samples collected from the subsurface (27) and have been interfaced with cone-penetrometer isolation devices (28,29). Injection systems typically include a module for analyte preconcentration and water management. Capillary columns allow the complete resolution of complex mixtures in about 30 min. Detectors typically include electrolytic conductivity, electron-capture, flame ionization, photoionization, and thermal conductivity types. Sensitivities of these detectors range from 10 ppt to 1 ppm. Comparison of the retention times of analytes and standard materials under identical chromatographic conditions provides a tentative identification of the contaminant; however, a more positive identification can be made by analyzing the sample on columns of widely different polarities. Detector specificity also permits some tentative identifications to be made. Consequently, GC systems that have multicolumn and multidetector capabilities are very useful in subsurface contaminant applications.

Mass Spectrometry

The mass spectrometer (MS) has been used extensively to monitor subsurface contamination because of its ability to accurately quantify organic compounds and simultaneously provide positive identification of the analytes (23,30). Field-portable and transportable systems can be installed in the cone-penetrometer truck for ease of analysis in the field. Mass spectrometer systems have been directly interfaced to sampling devices in the cone penetrometer and also used in the field for the analysis of samples collected by conventional sampling devices in the cone penetrometer. The mass analyzers that are most commonly used in field instruments are the linear quadrupole and the quadrupole ion trap (30). The linear quadrupole has moderate to low resolution capability. The ion trap is 10–100 times more sensitive than the linear quadrupole; however, the instrument may suffer from air and water ion molecule reactions that complicate the mass spectra.

Direct-sampling ion-trap mass spectrometry (DSITMS) has been developed to provide faster and less costly methods for the analysis of organic contaminants in environmental samples (31,32). The analytes can be introduced directly into the MS with no sample preconcentration and little or no sample preparation. The analytes are monitored in real-time or near-real-time at low ppb levels. Analysis times range from instantaneous to approximately 3 min. Several sample introduction systems have been developed for the DSITMS system that include a soil/water purge module, a thermal desorption module for sorbent tubes, and a direct air sampling module for real-time monitoring of VOCs. Detection limits for VOCs in water and soils are about 1–5 ppb and 10–30 ppb, respectively. Air samples can be quantified down to 1 ppb. Sample preconcentration can extend the detection limits down to 50–100 pg.

A field-transportable thermal desorption (TD) MS has been used to characterize PCBs, PAHs, and VOCs in subsurface samples (33,34). Analysis of VOCs in groundwater, PCBs in soil, and PAHs in soil required about 10, 10–15, and 20–30 min per sample using a short GC column sampling probe as a fast GC.

TECHNOLOGY DEMONSTRATION AND VALIDATION

As with any new technology, before the novel techniques described in this article can gain widespread acceptance and begin to replace conventional methods, the following questions must be addressed:

1. Are there significant and quantifiable improvements in the speed, cost, sensitivity, and accuracy of the technique over conventional methods?
2. What is the quality of the data in widely varying soil and groundwater types?
3. How rugged, operable, and user friendly are the technologies under actual field conditions?
4. Will the innovative methods be accepted by the appropriate regulatory agencies?

A large number of technology demonstrations have been and are currently being conducted to transfer the innovative cone penetrometer technologies from the laboratory development and testing phases to their implementation in the field. For example, the DOE began the Rapid Commercialization Initiative (RCI) to assist inventors of environmental technologies, including cone penetrometer devices, in bringing their devices to the commercialization phase (35). The RCI assists the private sector and state and federal agencies in (*1*) finding appropriate sites for the demonstration and testing of mature environmental technologies, (*2*) verifying the performance and cost of the technologies, and (*3*) facilitating and expediting the issuance of permits. The following discussion contains a brief synopsis of several field demonstrations and validations of cone-penetrometer chemical technologies. The data from these demonstrations were made available but do not include the entirety of the data available from field demonstrations of cone penetrometer devices. Other devices previously described are also in the laboratory and field validation phases, and the reader is encouraged to contact the investigators directly for additional information on their validation, regulatory acceptance, and commercialization.

CONCLUSIONS

The interfacing of chemical technologies with the cone penetrometer has made it possible to perform characterizations of many contaminated sites in a less intrusive, more cost-effective manner than performing the same investigation by the conventional method of drilling, sampling, and chemical analysis in the laboratory. The cone penetrometer is evolving from a device used only to gather geotechnical information about the subsurface to a tool that can also collect samples of groundwater, soil, and soil gas for chemical analysis and can provide *in situ* chemical detection. The most recent technological developments include the incorporation of on-line analytic and *in situ* chemical detection techniques into the cone penetrometer. It is apparent that a large number of probes have been invented that are capable of measuring inorganic and organic contamination in screening and quantitative modes over a wide range of concentrations; however, it is also apparent that many of these probes have not undergone the rigorous field demonstrations, validations, and implementation in site characterizations that will be necessary to obtain regulatory acceptance of the methodologies. To gain regulatory acceptance, data from the probes must be validated against data from verification samples that are collected and analyzed by accepted methods. The field validation should be performed at a variety of sites containing different soil and groundwater types to evaluate the effects of the sample matrix on the mechanical and chemical performances of the probe. Results from site characterizations performed by conventional methods of drilling, sampling, and laboratory analysis should then be compared at the same site with the results of characterizations performed with the cone penetrometer technologies to quantify improvements in the accuracy, speed, and cost of the site characterization.

ACKNOWLEDGMENTS

Thanks to our colleagues for providing valuable information on the cone-penetrometer devices. The technical reviews by Dr. N. A. Marley and Ms. C. M. Rose of Argonne National Laboratory and by Dr. W. M. Davis of the U.S. Army Corps of Engineers Waterways Experiment Station are greatly appreciated. The assistance of Ms. E. Brazelton of Argonne National Laboratory with editing is also greatly appreciated. The writing of this manuscript was supported by the U.S. Dept. of Energy under contract W-31-109-Eng-38 as part of the Characterization, Monitoring, and Sensor Technology Integrated Program of the Office of Technology Development, Assistant Secretary for Environmental Management (PVD), and by the Strategic Environmental Research and Development Program (SERDP), Washington, D.C. (ERC). The authors also wish to acknowledge SERDP, the Environmental Security and Technology Certification Program (ESTCP), the Army Environmental Center, and the Army's Installation Restoration Research Program (IRRP) for funding the SCAPS sensor and sampler development and demonstration projects described in this article.

BIBLIOGRAPHY

1. S. H. Lieberman, G. A. Theriault, S. S. Cooper, P. G. Malone, R. S. Olsen, and P. W. Lurk, *Proceedings of the 2nd International Symposium on Field Screening Methods for Hazardous Wastes and Toxic Chemicals,* Air and Waste Management Association, Pittsburgh, Pa., 1991, pp. 57–63.
2. R. A. Klopp, *Proceedings of the Third National Outdoor Action Conference on Aquifer Restoration, Ground Water Monitoring, and Geophysical Methods,* National Water Well Association, Dublin, Ohio, 1989, pp. 329–343.
3. M. Smolley and J. C. Kappmeyer, *Proceedings of the Conference on New Field Techniques for Quantifying the Physical and Chemical Properties of Heterogeneous Aquifers,* National Water Well Association, Dublin, Ohio, 1989, pp. 717–724.
4. M. Smolley and J. C. Kappmeyer, *Ground Water Mon. Rev.* **11,** 101–106 (1991).
5. N. A. Berzins, *Ground Water Manag.* **11,** 327–340 (1992).
6. C. Y. Chiang, K. R. Loos, and R. A. Klopp, *Ground Water* **30,** 428–436 (1992).
7. P. D. Kuhlmeier and T. E. Sturdivant, in D. M. Nielsen and M. N. Sara, eds., *Current Practices in Ground Water and Vadose Zone Investigations,* American Society for Testing and Materials, Philadelphia, Pa., 1992, pp. 183–198.
8. B. Manchon, "Introduction to Cone Penetrometer Testing and Groundwater Samplers," presented at a *Workshop of the Fifth National Outdoor Action Conference on Aquifer Restoration, Ground Water Monitoring, and Geophysical Methods,* National Water Well Association, Dublin, Ohio, 1991, 26 pp.
9. E. R. Cespedes, B. H. Miles, and S. H. Lieberman, *Optical Sensing for Environmental Monitoring,* SP-89, Air and Waste Management Association, Pittsburgh, Pa., 1994, pp. 621–632.
10. Advanced Sciences, Inc., *Review of Instrumentation and Sensors: Cone Penetrometer Applications,* Task 6 Report, Contract No. DE-AC05-87OR21706, U.S. Dept. of Energy, Martin Marietta Energy Systems, Inc., Oak Ridge, Tenn., 1994, 100 pp.
11. S. S. Cooper, P. G. Malone, R. S. Olsen, and G. B. Mohrman, *Proceedings of the 5th National Conference on Hazardous Wastes and Hazardous Materials,* Hazardous Materials Control Research Institute, Silver Spring, Md., 1988, pp. 424–427.
12. C. Y. Chiang, C. C. Stanley, L. Hekma, and G. F. Boehm, *Proceedings of the Conference on Petroleum Hydrocarbons and Organic Chemicals in Ground Water: Prevention, Detection, and Restoration,* National Water Well Association, Dublin, Ohio, 1989, pp. 141–157.
13. A. R. Holguin and S. H. Edelman, *Proceedings of the 1994 Federal Environmental Restoration III & Waste Minimization II Conference and Exhibition,* Vol. I, Hazardous Materials Control Resources Institute, Rockville, Md., 1994, pp. 135–144.
14. P. K. Robertson and R. G. Campanella, *Can. Geotech. J.* **20,** 718–733 (1983).
15. *Ibid.,* pp. 734–745.
16. P. K. Robertson, *Can. Geotech. J.* **27,** 151–158 (1990).
17. ASTM, *Soil and Rock, Building Stones: Geotextiles,* D3441-86, Vol. 4.08, American Society for Testing and Materials, Philadelphia, Pa., 1988, pp. 409–414.
18. P. K. Robertson, J. P. Sully, D. J. Woeller, T. Lunne, J. J. M. Powell, and D. G. Gillespie, *Can. Geotech. J.* **29,** 539–550 (1992).
19. A. I. Strutynsky, R. E. Sandiford, and D. Cavaliere, in Ref. 7, pp. 169–182.
20. R. Knowlton, *Proceedings of the 2nd Annual Site Characterization and Analysis Penetrometer System (SCAPS) Sensor Development Workshop,* SERDP-96-2, U.S. Army Corps of Engineers Waterways Experiment Station, Vicksburg, Miss., 1996, pp. 243–258.
21. R. G. Campanella and I. Weemees, *Can. Geotech. J.* **27,** 557–567 (1990).
22. R. L. Siegrist and P. D. Jenssen, *Environ. Sci. Technol.* **24,** 1387–1392 (1990).
23. S. E. Carpenter, P. V. Doskey, M. D. Erickson, and P. C. Lindahl, *Performance Specifications for Technology Development: Application for Characterization of Volatile Organic Compounds in the Environment,* Technical Report ANL/ER/TM-3, Argonne National Laboratory, Argonne, Il., 1994, 96 pp.
24. E. N. Koglin, E. J. Poziomek, and M. L. Kram, in L. G. Wilson, L. G. Everett, and S. J. Cullen, eds., *Handbook of Vadose Zone Characterization & Monitoring,* Lewis Publishers, Ann Arbor, Mich., 1995, pp. 657–700.
25. J. Janata and A. Bezegh, *Anal. Chem.* **60,** 62R–74R (1988).
26. N. L. Jarvis and H. Wohltgen, *Miniature GC for In-Situ Monitoring of VOCs within a Cone Penetrometer,* Final Technical Report, Contract No. DE-AR21-94MC31187, U.S. Dept. of Energy, Morgantown Energy Technology Center, Morgantown, W.V., 1996, 65 pp.
27. K. F. Myers, J. M. Brannon, R. A. Karn, C. B. Price, D. Y. Eng, A. B. Strong, and S. S. Cooper, *Laboratory Evaluation of a Volatile Organic Compound Analysis System for the Site Characterization and Analysis Penetrometer System,* Technical Report IRRP-95-3, U.S. Army Corps of Engineers, Waterways Experiment Station, Vicksburg, Miss., 1995, 51 pp.
28. P. V. Doskey, J. H. Aldstadt, J. M. Kuo, M. S. Costanza, and M. D. Erickson, in Ref. 42, pp. 151–155.
29. J. H. Aldstadt, C. H. Batson, A. F. Martin, M. D. Erickson, M. S. Costanza, M. B. J. Foster, T. R. James, and P. V. Doskey, *Field Anal. Chem. Technol.* **1,** 239–247 (1997).
30. W. C. McDonald, M. D. Erickson, B. M. Abraham, and A. Robbat, Jr., *Environ. Sci. Technol.* **28,** 336A–343A (1994).
31. M. B. Wise, C. V. Thompson, M. V. Buchanan, R. Merriweather, and M. R. Guerin, *Spectroscopy* **8,** 14–22 (1993).
32. M. B. Wise, C. V. Thompson, M. R. Guerin, and R. A. Jenkins, in Ref. 89, pp. 543–556.
33. A. Robbat, Jr., T-Y. Liu, and B. Abraham, *Anal. Chem.* **64,** 358–364 (1992).
34. *Ibid.,* pp. 1477–1483.
35. E. Lightner, U.S. DOE, Germantown, Md., 1997.

COUNTERCURRENT CHROMATOGRAPHY

ALAIN BERTHOD
Université Claude Bernard
Villeurbanne, France

It is important to realize that in countercurrent chromatography (CCC) there is no countercurrent circulation in most apparatuses. CCC can be defined as any chromatographic process that uses a liquid stationary phase without a solid supporting matrix (1,2). The term *countercurrent chromatography* was coined by Yoishiro Ito, its inventor who also established the widely used CCC acronym (3,4).

It is important to point out the differences between the extraction and chromatographic processes. In a liquid–liquid extraction, a mixture of solutes is introduced in a biphasic liquid system. Each solute distributes or partitions between the two liquid phases according to its distribution constant's commonly denominated partition coefficient. Next the two phases are separated. A second extraction can be performed with the separated phases and fresh solvents. The chromatographic process can be viewed as a continuous back and forth extraction process between a stationary phase and a mobile phase. The solutes are separated and sorted according to their distribution constants. They enter and leave the chromatographic apparatus, the column, in the mobile phase. The separated solutes are located in bands of mobile phase. The band width is correlated to a number of plates. A narrow band corresponds to a high number of plates and vice versa. The plate number can be viewed as the number of extraction steps that were necessary to obtain the band of solute. A CCC apparatus having an efficiency of 1000 plates is able to separate solutes as well as 1000 successively performed liquid–liquid extractions would do.

Descriptions of the CCC devices show that it is technically difficult to maintain a liquid phase while another immiscible liquid phase is percolated through the stationary liquid phase (3,4). The benefits of a liquid stationary phase without any solid support are as follows:

- The absence of a solid support precludes any solute adsorption.
- With a biphasic liquid system, the upper, lighter phase can be used as the stationary phase as well as the lower, denser phase. It is possible to switch the phase function during a run, the mobile phase becoming the stationary phase; this is the dual-mode operation (8).
- There are no pH limitations.
- The retention mechanism of solutes is simple and is determined solely by their liquid–liquid partition coefficient.
- The solutes can access the whole volume of the stationary phase, not only the interface between the solid phase and the liquid, or gas (mobile) phase.

The latter point means that the CCC technique is able to deal with large amounts of sample. CCC is a preparative tool; and used as a classical chromatograph, CCC is able to purify milligram to gram quantities of crude samples. The solvent polarity ranges from the most polar solvent, water, to the less polar solvents, perfluoroalkanes, with any intermediate polarity possible by mixing various solvents.

Three uses of CCC in environmental analysis and remediation are for: (*1*) the determination of liquid–liquid partition coefficients, emphasizing octanol–water partition coefficients; (*2*) the use of a CCC machine to concentrate trace amounts of a wastewater–contained organic pollutant in a small solvent volume; and (*3*) the extraction, the concentration, and separation of metallic ions from industrial sewage. Complete descriptions of devices are available in the literature (1,2,5,6,7).

CCC AND ENVIRONMENTAL ANALYSIS

In environmental analysis, CCC can be used for (*1*) liquid–liquid partition coefficient determination, (*2*) concentration in a small solvent volume of trace amounts of pollutants contained in a large amount of aqueous phase, and (*3*) concentration and separation of heavy atom ions in wastewater. These three uses of CCC are described as nonexhaustive examples.

BIBLIOGRAPHY

1. N. B. Mandava and Y. Ito, *Countercurrent Chromatography,* Chromatographic Science Series no. 44, Marcel Dekker, Inc., New York, 1988.
2. A. Berthod, *Instrument. Sci. Technol.* **23**(2), 75–89 (1995).
3. Y. Ito, *J. Chromatogr.* **214,** 122–133 (1981).
4. Y. Ito, in J. C. Giddings, E. Grushka, and J. Cazes, eds., *Advances in Chromatography,* Vol. 24, Marcel Dekker, Inc., New York, 1984, Chapt. 6, pp. 181–234.
5. W. D. Conway, *Countercurrent Chromatography, Theory and Applications,* VCH Publishers, Weinheim, Germany, 1990.
6. W. D. Conway and R. Petrowski, *Modern Countercurrent Chromatography,* ACS Symposium Series no. 593, ACS, Washington, D.C., 1995.
7. A. P. Foucault, *Centrifugal Partition Chromatography,* Chromatographic Science Series no. 68, Marcel Dekker, Inc. New York, 1995.
8. A. Berthod in A. P. Foucault, ed., *Centrifugal Partition Chromatography,* Chromatographic Science Series, Vol. 68, Marcel Dekker, Inc. New York, 1995, Chapt. 1, pp. 1–24; Chapt. 7, pp. 167–198.

CRUDE OIL BIODEGRADATION

ROGER C. PRINCE
Exxon Research and Engineering Company
Annandale, New Jersey

Crude oils fuel our modern industrial economy, and they are used on an enormous scale; global use is of the order of 3.8×10^{12} L/year (10^{12} gal/year). Crude oils are complex mixtures of molecules, principally hydrocarbons; and while the major classes are well defined, there is no inventory of all the individual species. Nevertheless, it is known that the majority of molecules in most crude oils are biodegradable. This poses a problem while the oil is under human control, since biodegradation can degrade the product, and microbial biomass can accumulate in critical locations. On the other hand, stimulating the natural process of biodegradation (bioremediation) offers an opportunity for reducing the environmental impact when crude oil escapes into the environment.

Origins of Crude Oil

The great organic carbon cycle of our planet is driven principally by photosynthesis, as atmospheric carbon dioxide is fixed to form biomass at the expense of solar energy. When the photosynthetic organisms die, or are eaten, their biomass is oxidized back to CO_2, and the process is repeated. A small fraction of the primary productivity, however, is preserved from biodegradation by entombment in sedimentary material where oxygen supply is limited. If this occurs on a large enough scale, such as on the continental margins or in large, landlocked basins, it can give rise to substantial amounts of buried organic matter. Over the years this material is subjected to a range of biological, chemical, and geophysical processes; and under some circumstances it can be converted to fossil fuels (1).

The process of conversion of buried biomass to fossil fuels is divisible into diagenesis and catagenesis. Diagenesis, which typically occurs at temperatures at which microbes partially degrade the biomass, results in dehydration, condensation, cyclization, and polymerization. Subsequent catagenesis at higher temperatures and pressures completes the transformation of the biomass to fossil fuel by thermal cracking and decarboxylation. While there is a general conservation of the elements in the maturing biomass, known as kerogen, there is significant rearrangement of the molecules, and a redistribution of hydrogen so that some carbon atoms gain hydrogen while others lose it (1,2).

It is generally accepted that higher plants gave rise to the great coal reserves of the world, typically in nonmarine environments, while aquatic algae, albeit sometimes with some terrigenous material, gave rise to petroleum. Nevertheless, some coal reserves are associated with commercially valuable oil and gas. The generation of oil and gas invariably occur at significant depth, and under great pressure, and this pressure forcibly expels the oil and gas from the initial source rock. Commercially valuable reservoirs are formed if this migrating hydrocarbon is trapped in a geological structure so that the oil and gas can accumulate. The majority of petroleum accumulations are found in sandstones, siltstones, and carbonates, with porosities that typically range from 5 to 30%. Traps are never completely full of oil, and there is always some water, usually containing substantial amounts of inorganic salts (brine) (1).

The complete maturation of fossil fuels is usually a rather slow process, and the majority of commercial oils today are from biomass that was produced of the order of 100 million years ago (71% between 180 and 85 million years ago). The oldest commercially valuable oils are from biomass generated in the Ordovician (486 million years ago), while others are as young as the late Tertiary (a few million years ago). Unusual circumstances, such as those that prevail at the Guaymas hydrothermal vent site, can even result in the formation of petroleum from biomass that is only approximately a thousand years old. This marine vent site, in the Gulf of California, lies in an area of high primary productivity and rather low current. The dying algae thus fall to the bottom (some 2 km) where there are hydrothermal vents. The hydrothermal activity heats the organic-rich sediment to high temperatures, and the processes of diagenesis and catagenesis are speeded up dramatically (1,3). It is generally thought that hydrothermal oil generation is unusual and does not give rise to commercially valuable reservoirs.

The migration of petroleum from its source rock can give rise to surface seeps, either because there is no trap or because the trap has become full. Petroleum (literally rock oil) seeps have been known and used since the beginning of recorded human history. Bitumen, the least volatile fraction of crude oil, was used as a hafting material for gluing stone tools to wooden handles by our Paleolithic ancestors, reputedly provided the mortar for building the Tower of Babel, and was used by the Egyptians in the mummification of their dead. Terrestrial seeps were also the first locations to be drilled when oil production began in earnest in the nineteenth century. For example the first well drilled in the United States, the 1860 Drake well in Pennsylvania, was drilled on a seep site. Most of the known terrestrial seep sites had been drilled by the 1920s, but marine seeps continue to be explored today. It is estimated that approximately 260×10^6 L (68×10^6 gal) of crude oil enters the world's oceans each year from marine seeps, accounting for approximately 8% of the total input. These marine seeps are the basis of some production, and in the early 1980s ARCO installed steel pyramids above some of the seeps in the Santa Barbara Channel, California, to collect some of this seepage. Nevertheless, most marine seepage escapes into the surface environment (4,5); fortunately, the majority of the molecules in crude oil are biodegradable.

Components of Crude Oils

The range of elemental compositions of typical crude oils is shown in Table 1. The inorganic elements are important in refinery operations and in refinery wastewater treatment, but they are present at such low levels that they are generally ignored when oil enters the environment, and they will not be considered further here.

Crude oils are principally hydrocarbons, with a hydrogen to carbon ratio of between 1.5 and 2.0; the organic molecules are thus principally saturated molecules (ie, the predominant form of carbon is —CH_2—). The convention in the oil industry is to call linear alkanes *paraffins,* and cyclic alkanes *naphthenes.* There are also significant amounts of aromatic carbon in all crude oils and polar molecules containing the heteroatoms oxygen, nitrogen, and sulfur. These latter molecules can be fractionated by column chromatography, and the fractions have a variety of names, including resins, asphaltenes and NSOs (compounds containing nitrogen, sulfur, and/or oxygen). The average composition of 527 crude oil samples has been quoted as 58.2% saturates, 28.6% aromatics, and 14.2% polar compounds, although the absolute values vary widely in different oils (1). On average, there is rough parity between paraffins, naphthenes, and aromatics.

The paraffins include the linear alkanes from methane up to waxes with at least 40 carbons. The largest alkane reported in petroleum is $C_{79}H_{160}$. Linear alkanes typically make up 15–20% of a crude oil, although their content can be essentially undetectable, or as high as 35%, depending on source and reservoir conditions. There are also branched alkanes; the most abundant are found in the C_6 to C_8 range, but since these are rather volatile they are not often found in samples collected from crude oil spills. Pristane and phytane, molecular relics of the phytol chains of chlorophylls and perhaps other biomolecules, are usually the most abundant individual branched alkanes. Pristane ($C_{19}H_{40}$) is thought to be the result of initial partial degradation of phytol in the presence of oxygen, while phytane ($C_{20}H_{42}$) is thought to be the result of initial diagenesis in the absence of oxygen. The ratio of pristane to phytane can thus be interpreted to give some information about the initial depositional environment of the biomass that gave rise to the petroleum (1,6). Other

Table 1. Elemental Composition of Typical Crude Oils[a]

Element	Weight
C	84–87%
H	11–14%
S	<0.1–8%
O	<0.1–1.8%
N	<0.1–1.6%
Ni	Trace to 1000 ppm
V	Trace to 1000 ppm
Se	Trace to 510 ppb

[a] Refs. 1 and 2.

branched molecules are individually much less abundant than the linear alkanes, pristane and phytane, but together they are the bulk of the paraffins. They are largely responsible for the "unresolved complex mixture" seen in gas chromatograms of the paraffin fraction of crude oils (1,2,7). Unsaturated alkanes, alkenes, are present at only trace levels in crude oils.

The naphthenes include parent compounds, such as cyclopentane, cyclohexane, and decalin, together with their alkylated congeners. The average composition of the naphthene fraction of 299 crude oils is 54.9% one- and two-ring naphthenes, 20.4% tricyclic naphthenes, and 24.0% tetracyclic and pentacyclic naphthenes (1). These latter molecules are among the better understood molecular biomarkers in crude oils, and they are used extensively in correlating reservoirs and source rocks and in assigning the depositional environment of source rocks (8).

Because of the separation procedures used in the characterization of crude oils, any molecule containing at least one aromatic ring is included in the aromatic fraction, regardless of the presence of saturated rings and alkyl substituents. Crude oils typically contain aromatic molecules with up to four aromatic rings; one series contains just six-membered rings and their alkylated derivatives; benzene (one ring), naphthalene (two rings), phenanthrene and anthracene (three rings), chrysene (four rings), etc. Another includes one five-membered ring in addition to the six-membered ones; fluorene (three rings), fluoranthene (four rings), etc. Sulfur aromatic heterocycles, such as thiophenes, benzothiophenes, and dibenzothiophenes, are included in the aromatic category, and their ratio to phenanthrenes can provide important insights into the depositional environment of an oil (6). Indoles and carbazoles, usually the most abundant nitrogen-containing species, and the less abundant basic nitrogen species, such as quinolines, are also included in the aromatic category. Alkylated aromatic species are usually more abundant than their parent compounds, with monomethyl, dimethyl, and trimethyl derivatives usually being most abundant. Nevertheless, the median aromatic structure probably has one or two methyl substituents, together with a long-chain alkyl substituent (1,2).

The polar molecules are the most difficult to characterize because they are often unamenable to gas chromatography, the usual method of choice for the molecular characterization of petroleum. All are thought to contain heteroatoms such as nitrogen, oxygen, and/or sulfur; and the category includes the porphyrins, usually nickel or vanadium species, and naphthenic acids. Some of these molecules have molecular weights into the thousands and even higher, and many are suspended in the oil rather than dissolved in it. Others are dissolved in the whole oil, but are precipitated if more alkanes (eg, pentane) are added (1,2,9,10). The polar fraction of the oil contains the majority of the color centers in crude oil, and in isolation these materials are difficult to distinguish from more recent biological residues, such as humic and fulvic acids.

Physical Properties of Crude Oils

Oils are classified by several criteria, but among the most important is the specific gravity. The oil industry uses a unit known as the American Petroleum Institute (API) gravity, which is defined as (142.5/specific gravity) −131.5, and expressed as degrees. Thus water has an API gravity of 10°, and denser fluids will have lower API gravities. Less dense fluids, eg, most hydrocarbons, will have API gravities $>10°$. For convenience, oils with API gravities $>40°$ are said to be light oils, and those with API gravities of $<16°$ are said to be heavy. Light oils have higher proportions of small molecules; heavy oils are rich in larger molecules. Viscosity is inversely proportional to API gravity, but it also depends on the physical state of the polar compounds and longer alkanes in the oil and strongly depends on the temperature (1).

Fates of Oil in the Environment

When crude oils are released into the environment the smallest molecules, those containing up to about 15 carbons, are likely to volatilize. This process can remove one- to two-thirds of surface spills (5). Some terrestrial seeps produce so much volatile material that they may periodically ignite, but in the absence of ignition, it is thought that almost all the volatilized material is photochemically oxidized in the atmosphere. The residuum is much more viscous, and some seeps (eg, in Azerbaijan, Los Angeles, and Trinidad) give rise to large lakes of pitch (4). Some molecules are sufficiently soluble that they are washed out of exposed seeps or dissolve from slicks at sea (11), and this process also increases viscosity. Floating slicks will absorb large amounts of water (20–80%), and stable water-in-oil emulsions, known as mousses, often form (5,12). If enough energy is available, such as in a major storm, a floating oil spill will disperse as tiny drops of oil in water. This is what happened in the *Braer* spill of Shetland in 1993 (13). This process can be aided by the addition of dispersants (14).

Regardless of the initial fate of a crude oil in the surface environment, if it is not burned or photochemically oxidized, it will persist until it is biodegraded. Fortunately, biodegradation is a potent natural phenomenon, and the rest of this section describes our current view of the process.

Biodegradation in Oil Reservoirs

Oil reserves that have been invaded by aerobic surface waters have often been subject to substantial biodegradation. This is recognized by the absence of linear alkanes, the smaller aromatic hydrocarbons, and even some of the naphthenes. The Athabasca deposits of northern Alberta are one example of this phenomenon. Biodegradation typically increases the specific gravity and viscosity of an oil, and indeed the Athabasca deposits are viscous (1).

While early (or primary) production of oil from an oil field may occur due to the inherent high pressure in the reservoir, later (secondary) production is usually aided by pumping water into injection wells. Often this is accompanied by a "souring" of the production, an industry term for an increase in H_2S in the produced oil. While some of this may be the result of subterranean geochemistry, it is likely that the majority is the result of the activity of sulfate-reducing bacteria. These organisms typically grow by oxidizing small organic acids at the expense of reducing sulfate to sulfide (15), and they are widely implicated in metal corrosion. Thermophilic and mesophilic sulfate-reducing bacteria have been isolated from many wells (16–18), but it has usually been assumed that they have entered during drilling and production. More recently, it is becoming accepted that oil reservoirs may well contain microorganisms, even in the absence of substantial water incursions. Thus it has been shown that anaerobic hyperthermophilic microorganisms are abundant, even in previously untapped oil reservoirs, and may well be abundant throughout the deep crust of the earth (19). Whether these bacteria play any role in the maturation of crude oils remains to be seen.

Biodegradation of Oil and Refined Products during Use and Storage

Oil and refined products are frequently stored for some time before use, and several nations have strategic oil reserves in underground caverns. These caverns usually become contaminated with water, and oil-degrading microorganism are readily isolated from such environments (20). Diesel, kerosene, and jet fuel storage facilities, and even airplanes, can have similar problems; this is a particular concern because some of the organisms can block fuel filters, with potentially disastrous results (21). Other problems associated with fuel biodegradation include an increase in tank and fuel line corrosion, and production of surfactants that encourage water in oil emulsions. Proprietary biocides are often added to slow biodeterioration in such situations (22).

Biodegradation of Oil and Refined Products in Surface Environments

Hydrocarbons are a natural part of the surface environment. Plants and algae produce large amounts of hydrocarbons and liberate them into the biosphere (5,23,24). Furthermore, as discussed above, large amounts of crude oil get into the surface environment by entirely natural causes, usually as seeps, but also by erosion of oil sands and shales. These processes have been going on for millennia, so it is no surprise that microbes take advantage of this potential carbon source. In 1985, the National Research Council (5) estimated that the total annual input of petroleum into the world's oceans was about 3200×10^6 L(850×10^6 gal), with the majority coming from municipal sources and routine shipping. Fortunately, most of this is biodegraded; and the residuum lacks many of the physical properties of the original oil and is biologically inert.

The extent of biodegradability depends on many factors. On the one hand it depends on the chemical composition of the oil, but it also depends on the environment in which the biodegradation is to occur. The short-term biodegradability of 17 crude oils in soil microcosms has been examined as a function of 78 different parameters that might determine the extent of biodegradation (25). These included 67 individual chemical species, percent sulfur, etc. It was found that the API gravity was the most useful predictor of biodegradability, at least for the most degradable fraction of the oils. At loadings of 0.5 wt % oil in a loam soil with appropriate nutrients, moisture, and aeration, >61% of the most degradable oil (API = 46°) was lost in 4 weeks, while only 10% of the least degradable oil (API = 15°) was consumed under the same conditions. Further degradation occurred on a longer time scale, and the literature reports biodegradation potentials as high as 97% for particularly light oils (26).

Many factors can conspire to hinder biodegradation in the environment. Temperature extremes are likely to inhibit degradation, especially low temperatures such as those found in Arctic environments (27). This effect is probably attributable to the increase in oil viscosity at low temperatures; and some degradation, at least of the smaller molecules, occurs even under ice (27). On land, excess salinity, such as that found in some produced water brines, can have a strong inhibitory effect on oil biodegradation (25) until the salt is washed out by precipitation.

Perhaps the most acute potential limiting factor is oxygen, which will be treated in a separate section. In the presence of adequate oxygen, the most limiting nutrients are likely to be nitrogen, phosphorus, and iron (28). The relief of these limitations has been the underlying strategy in the successful use of bioremediation in treating crude oil spills. Thus the addition of oleophilic and slow-release nitrogen- and phosphorous-containing fertilizers was successful in substantially increasing the rate of biodegradation of oil after the *Exxon Valdez* spill in Alaska (29,30), and nutrient addition has been successful in several terrestrial applications as well (31).

ANAEROBIC DEGRADATION OF CRUDE OIL

For many years it was assumed that oil biodegradation could only occur under aerobic conditions. This was a logical view, since any degradation must involve oxidation, and the very existence of oil reservoirs indicated that anaerobic degradative processes in such environments must be slow. Indeed it remains a truism that the ability to insert oxygen atoms into hydrocarbons is the hallmark of oil-degrading organisms. Once a hydrocarbon possesses a carboxylate or alcohol functionality, it is almost invariably a readily degradable compound. A simple example is the difference between oleic acid, a high calorie food for mammals, and octadecane, present in mineral oil, which is so inert in mammals that it serves as an intestinal lubricant.

In recent years it has become clear that at least some hydrocarbons are oxidized by bacteria under completely anaerobic conditions, in which the oxygen is probably coming from water. Hydrocarbon biodegradation has now been shown under sulfate-, nitrate-, carbon dioxide- and ferric iron-reducing conditions (Table 2), and all seem to be important in the biodegradation of smaller hydrocarbons in subsurface contaminated aquifers (39). The phenomenon is still poorly understood, however, and at present the largest molecules demonstrated to undergo biodegradation under these conditions are hexadecane, heptadecene, and naphthalene. Rather larger molecules predominate in most crude oils. It is also not known to what extent such processes might be occurring in oil reservoirs.

Organisms Involved in Anaerobic Hydrocarbon Degradation of Hydrocarbons

Because anaerobic hydrocarbon degradation is a recently discovered phenomena, relatively few microorganism capable of this process have been isolated and characterized. The few that

Table 2. Papers on Anaerobic Degradation of Hydrocarbons

Electron Acceptor	Substrate	Reference
Sulfate (to sulfide)	Hexadecane, alkylbenzenes	32
	Benzene	33
Nitrate (to nitrogen)	Heptadecene	34
	Toluene, ethylbenzene, xylene	35
	Naphthalene	36
CO_2 (to methane)	Toluene, xylene	37
Iron(III) (to iron(II))	Toluene	38

have been are all bacteria: *Azoarcus*, *Desulfobacterium*, *Desulfobacula*, *Geobacter*, *Pseudomonas*, and *Thauera*. No archaea or eukarya have been isolated to date.

Biochemical Pathways of Anaerobic Degradation

The anaerobic degradation of toluene with nitrate as the terminal electron acceptor seems to occur by at least three disparate routes. One proceeds by attack on the methyl group, and benzaldehyde and benzoate have been identified as intermediates of toluene degradation (40). Another involves hydroxylation of the ring at the para position to generate *para*-cresol, whereas a third involves oxidative addition of acetyl coenzyme A to generate β-phenylpropionyl coenzyme A (41). This is an area of active current research, and significant advances can be expected in the near future.

Hydrocarbons with heteroatoms such as oxygen and nitrogen are readily degraded anaerobically. Benzoate, the cresols, and similar compounds are readily degraded with all the electron acceptors listed in Table 2 (42), and pyridine has been shown to be completely degraded, with nitrate as the terminal electron acceptor, by an *Alcaligenes* (43).

AEROBIC DEGRADATION OF CRUDE OIL

Aerobic biodegradation of oil has been a focus of research for over 100 years and has been recently reviewed (26,28,44). Highly volatile compounds with more than six carbon atoms are biodegradable (45); but since they are so volatile, they are rarely present in significant concentrations after spills. Methane is a particularly special case, with all known methane oxidizers depending on methane and methanol as their sole source of energy (46). These will not be discussed in detail here.

Organisms Involved in Aerobic Degradation of Hydrocarbons

Aerobic microorganisms able to degrade hydrocarbons have been isolated from almost all environments where they have been diligently searched for. They have been isolated from the bottom of the sea, including the Guaymas deep sea site described above, the tops of mountains, soils and shores everywhere (including the Arctic and Antarctic), and from deserts. Extreme halophiles have been isolated as have thermotolerant organisms. No extreme theremophiles have been isolated to date, but there is every reason to expect that they will be found.

This diversity of habitats is reflected in the taxonomic diversity of aerobic hydrocarbon-degrading microorganisms; they are scattered throughout the kingdoms of living organisms, and there is no clear taxonomic distinction between hydrocarbon degraders and their nondegrading kin. The principal reaction that distinguishes the hydrocarbon-degrading organisms is their ability to insert the first oxygen into hydrocarbons, and the enzymes that catalyze these reactions are often (although certainly not always) plasmid-borne. It is thus quite likely that hydrocarbon degradation sometimes represents horizontal gene transfer between species, and perhaps even between kingdoms, rather than descent from specific hydrocarbon-degrading ancestors.

Biochemical Pathways of Aerobic Hydrocarbon Degradation

Aerobic pathways of hydrocarbon degradation have been studied in far greater detail than have the anaerobic ones described above; and pathways are beginning to be clearly defined, at least for relatively simple hydrocarbons. It is the insertion of the initial oxygen atoms into hydrocarbons that initiates hydrocarbon degradation, and the discussion that follows will focus on these reactions. Subsequent oxidation is often through pathways with a much broader specificity; these pathways are usually ubiquitous.

BIOREMEDIATION

With the foregoing discussion in mind, questions of what limits biodegradation in the environment, and whether one can relieve these limitations in a safe and effective way to speed the natural removal of petroleum and its refined products from the environment can be asked.

One generalization that can be made is that microorganisms that can degrade at least some hydrocarbons are ubiquitous, and their absence is unlikely to be a limitation, at least for the majority of molecules in crude oil. Pristine environments have relatively low populations of hydrocarbon-degrading microorganisms, but any spill is likely to cause a bloom of these organisms within a few days or weeks (26). To date, the addition of hydrocarbon-degrading microorganisms has not met with any statistically validated success (47).

Another generalization is that while anaerobic bioremediation of benzene, toluene, and xylene is now well documented (32), the rapid degradation of larger molecules requires oxygen, which can be limited in some soils and some aquatic sediments. Protocols for adding oxygen by the addition of various peroxides are being considered (48).

The most likely limiting nutrient in marine systems is available nitrogen and phosphorus, and the addition of these nutrients was successful in stimulating the biodegradation of oil spilled in Prince William Sound, Alaska, by the *Exxon Valdez* (29,30). The addition of these nutrients has also proved successful in stimulating the biodegradation of oil in soil (31,25,49).

Nitrogen-containing fertilizers are toxic at inappropriate levels, and even nontoxic levels of nutrients may have environmentally undesired effects if they significantly stimulate algal growth in an affected water. Nevertheless, fertilizers can be applied safely and effectively when appropriate levels are used (29).

Other limitations to degradation, such as adsorption of hydrocarbons onto soil and sediment particles, have also been studied (50), but as discussed above, attempts to modify this by the addition of surfactants have met with mixed results (51).

Another generalization is that while individual bacterial strains may excrete metabolic dead-end products, microbial consortia usually metabolize those hydrocarbons that they are able to attack to completion, ie, to CO_2, H_2O, and biomass. This biomass forms the basis of a new food web, and it has been suggested that natural hydrocarbon seepage, and its subsequent microbial degradation, contributes significantly to benthic productivity and the fisheries of Atlantic Canada (52). The new food web that develops after an oil spill has received only cursory attention. It has been known for some time that predation on oil-degrading bacteria stimulates the overall rate of biodegradation of a contaminating crude oil (53), but there have been few attempts to exploit this phenomenon.

Smaller linear alkanes and one- and two-ring aromatic hydrocarbons are usually the first crude oil components to be degraded in the environment, followed by the three-ring aromatics and the simply branched and longer alkanes, and then the four ring aromatics. Within each family of aromatics, the parent compound is degraded more rapidly than its alkylated forms, and the greater the alkylation, the slower the rate (54).

Despite the tremendous diversity of molecules in crude oil, and the concomitant difficulties in analyzing many of them, the judicious choice of analytes can provide important insights into the source of a spilled oil and progress in its biodegradation. Ratios of trialkyl-dibenzothiophenes to trialkyl-phenanthrenes provide a reliable fingerprint for an individual oil, while the ratio of trialkyl-phenanthrenes to trialkyl-chrysenes provides an estimate of the extent of biodegradation if a sample of the initially spilled oil is available. A simpler assay for the early onset of biodegradation is the ratio of *n*-heptadecane to pristane or *n*-octadecane to phytane, but this becomes unreliable when the pristane and phytane become significantly degraded. Hopanes have proved to be a useful conserved internal standard for quantifying oil biodegradation in the environment.

In sum, the majority of the molecules in crude oils and refined products are biodegradable under appropriate conditions. While these hydrocarbons are in commerce, this biodegradation can be prevented by careful storage and by the addition of biocides. If the hydrocarbons get into the environment they will be biodegraded, and this biodegradation can be stimulated by the alleviation of limiting factors. There has been well-documented success in this enterprise with the addition of fertilizers and oxygen.

BIBLIOGRAPHY

1. B. P. Tissot and D. H. Welte, *Petroleum Formation and Occurrence,* Springer-Verlag, Berlin, 1984.
2. H. H. Schobert, *The Chemistry of Hydrocarbon Fuels,* Butterworths, London, 1990.
3. B. R. T. Simoneit and P. F. Lonsdale, *Nature,* **295,** 198–202 (1982).
4. S. A. Tedesco, *Surface Geochemistry in Petroleum Exploration,* Chapman & Hall, New York, 1995.
5. National Research Council, *Oil in the Sea: Inputs, Fates and Effects,* National Academy Press, Washington, D.C., 1985.
6. W. B. Hughes, A. G. Holden, and L. I. P. Dzou, *Geochim. Cosmochim. Acta,* **59,** 3581–3598 (1995).
7. M. A. Gough and S. J. Rowland, *Nature,* **344,** 648–650 (1990).
8. K. Peters and J. M. Moldowan, *The Biomarker Guide; Interpreting Molecular Fossils in Petroleum and Ancient Sediments,* Prentice-Hall, Englewood Cliffs, N.J., 1993.
9. P. R. Waller, A. Williams, and K. D. Bartle, *Fuel,* **68,** 520–526 (1989).
10. V. Calemma, P. Iwanski, M. Nali, R. Scotti, and L. Montanari, *Energy Fuels,* **9,** 225–230 (1995).
11. C. L. Yaws, X. Pan, and X. Lin, *Chem. Eng.,* **100,** 108–111 (1993).
12. A. Bharwaj and S. Hartland, *Ind. Eng. Chem. Res.,* **33,** 1271–1279 (1994).
13. D. Thomas, *Petrol. Rev.,* **47,** 263-265 (1993).
14. J. R. Clayton, J. R. Payne, and J. S. Farlow, *Oil Spill Dispersants: Mechanisms of Action and Laboratory Tests,* Lewis Publishers, Boca Raton, Fla., 1992.
15. F. Widdel, in A. J. B. Zehnder, ed., *Biology of Anaerobic Organisms,* Wiley-Interscience, New York, 1988, pp. 469–586.
16. M. Magot, and co-workers, *Int. J. Sys. Bacteriol.,* **42,** 398–403 (1992).
17. K. O. Stetter and co-workers, *Nature,* **365,** 743–745 (1993).
18. I. Vance and D. E. Brink, *Appl. Microbiol. Biotechnol.,* **40,** 920–925 (1994).
19. S. L.'Haridon, A.-L. Reysenbach, P. Glenat, D. Prieur, and C. Jeanthon, *Nature,* **377,** 223–224 (1995).
20. M. Bock, P. Kampfer, K. Bosecker, and W. Dott, *Appl. Microbiol. Biotechnol.,* **42,** 463–468 (1994).
21. M. P. Reynders, P. Stegmann, R. W. Hiley, R. E. Penfold, and J. F. Pedley, paper presented at the Third International Conference on Stability and Handling of Liquid Fuels, 1988.
22. J. M. Sharpley, *Elementary Petroleum Microbiology,* Gulf Publishing, Houston, Tex., 1966.
23. R. L. Geneve, *Am. Horticult.,* **74**(6), 21–23 (1995).
24. T. D. Sharkey and E. L. Singsaas, *Nature,* **374,** 769 (1995).
25. S. J. McMillen and co-workers, in R. E. Hinchee, C. M. Vogel, and F. J. Brockman, eds., *Microbial Processes for Bioremediation,* Battelle Press, Columbus, Ohio, 1995, pp. 91–99.
26. R. C. Prince, *Crit. Rev. Microbiol.,* **19,** 217–242 (1993).
27. R. Siron, D. Delille, E. Pelletier, and S. Roy, *Mar. Env. Res.,* **35,** 273–302 (1993).
28. R. M. Atlas and R. Bartha, *Adv. Microbial Ecol.,* **12,** 287–338 (1992).
29. R. C. Prince and co-workers in R. E. Hinchee, B. C. Alleman, R. E. Hoeppel, and R. N. Miller, eds., *Hydrocarbon Remediation* Lewis Publishers, Boca Raton, Fla., 1994, pp. 107–124.
30. J. R. Bragg, R. C. Prince, E. J. Harner, and R. M. Atlas, *Nature,* **368,** 413–418 (1994).
31. National Research Council, *In Situ Bioremediation. When Does It Work?* National Academy Press, Washington, D.C., 1993.
32. P. Rueter and co-workers, *Nature,* **372,** 455–458 (1994).
33. D. R. Lovley, J. D. Coates, J. C. Woodward, and E. J. P. Phillips, *Appl. Environ. Microbiol.,* **61,** 953–958 (1995).
34. M. Gilewicz, M. Acquavira, and J. C. Bertrand, *Appl. Microbiol. Biotechnol.,* **36,** 252–256 (1991).
35. P. J. J. Alvarez and T. M. Vogel, *Water Sci. Technol.,* **31,** 15–28 (1995).
36. N. D. Durant, L. P. Wilson, and E. J. Bouwer, *J. Contam. Hydrol.,* **17,** 213–237 (1995).
37. E. A. Edwards and D. Grbic-Galic, *Appl. Environ. Microbiol.,* **60,** 313–322 (1994).
38. D. R. Lovley and co-workers, *Arch. Microbiol.,* **159,** 336–344 (1993).
39. R. C. Borden, C. A. Gomez, and M. T. Becker, *Ground Water,* **33,** 180–189 (1995).
40. B. Seyfried, G. Glod, R. Schocher, A. Tschech, and J. Zeyer, *Appl. Environ. Microbiol.,* **60,** 4047–4052 (1994).
41. P. J. Evans, W. Ling, B. Goldschmidt, E. R. Ritter, and L. Y. Young, *Appl. Environ. Microbiol.,* **58,** 496–501 (1992).
42. W. C. Evans and D. Fuchs, *Ann. Rev. Microbiol.,* **42,** 289–317 (1988).
43. Z. Ronen and J.-M. Bollag, *Can. J. Microbiol.,* **37,** 725–729 (1991).
44. J. G. Leahy and R. R. Colwell, *Microbiol Revs.,* **54,** 305–315 (1990).
45. S. Hartmans, J. A. M. de Bont, and W. Harder, *FEMS Microbiol. Rev.,* **63,** 235–264 (1989).

46. J. D. Lipscomb, *Ann. Rev. Microbiol.,* **48,** 371–399 (1994).
47. P. H. Pritchard, *Curr. Opin. Biotechnol.* **3,** 232–243 (1992).
48. D. L. Pardieck, E. J. Bouwer, and A. T. Stone, *J. Contam. Hydrol.* **9,** 221–242 (1992).
49. E. N. Drake and co-workers in R. E. Hinchee, G. S. Douglas, and S. K. Ong, eds., In *Monitoring and Verification of Bioremediation,* Battelle Press, Columbus, Ohio, 1995, pp. 19–28.
50. W. D. Weissenfels, H. -J. Klewer, and J. Langhoff, *Appl. Microbiol. Biotechnol.* **36,** 689–696 (1992).
51. J. D. Rouse, D. A. Sabatini, J. M. Suflita, and J. H. Harwel, *Crit. Rev. Environ. Sci. Technol.* **24**, 325–370 (1994).
52. E. M. Levy and K. Lee, *Can. J. Fish. Aquat. Sci.* **45,** 349–352 (1988).
53. A. Rogerson and J. Berger, *J. Gen. Appl. Microbiol.* **29,** 41–50 (1983).
54. D. L. Elmendorf, C. E. Haith, G. S. Douglas, and R. C. Prince in R. E. Hinchee, A. Leeson, L. Semprini, and S. K. Ong, eds., *Bioremediation of Chlorinated and Polycyclic Aromatic Hydrocarbon Compounds.* Lewis Publisher, Boca Raton, Fla., 1994, pp. 188–202.

D

DETECTION AND QUANTIFICATION OF ENVIRONMENTAL POLLUTANTS

ROBERT D. GIBBONS
University of Illinois at Chicago
Chicago, Illinois

The earliest possible detection of hazardous environmental constituents in air, surface water, groundwater, and drinking water is of critical concern in protecting human health and the environment. The very presence of anthropogenic compounds in environmental media leads to assessment and potential remediation of industrial activity. As an example, the presence of vinyl chloride, a known carcinogenic initiator, in a groundwater sample obtained from a monitoring well located downgradient of a hazardous or municipal waste disposal facility, results in an expensive site assessment and corrective action or closure of the faculty. In some cases, these consequences result from a single detection decision on a single environmental sample. In other cases the measured concentration of a constituent is compared to a concentration limit, possibly health-based, and if the measured concentration exceeds the limit, remedial action is taken. Unfortunately, in many cases the limit of detection for the constituent in a particular media (eg, surface water) is reported to be considerably larger than the health-based concentration limit. There is public concern regarding protection of human health and the environment such that potentially harmful levels of a chemical may exist in the environment, but, because they are undetectable, no regulatory impact decision can be made.

In response to these concerns, there has been tremendous motivation to develop new analytical methods capable of detecting hazardous compounds at lower levels. Furthermore, these concerns have led to a proliferation of new definitions and statistical approaches to estimating limits of detection. These new definitions and statistical estimators often lead to further confusion and suffer from statistical and conceptual errors (1–3). New methods proposed for setting limits on detection may result in lower limits regardless of the actual analytical capabilities of the method.

Unfortunately, virtually all practical applications of environmental detection decisions in the United States are based on the most severely limited of the definitions (4,5). Not only are the statistical problems with U.S. EPA's estimator, termed the method detection limit (MDL), insurmountable, but the use of this estimator has led to the routine practice of setting detection limits at lower and lower levels regardless of the true analytical capabilities of the method and/or laboratory. Public policy in these areas has not been adequately examined in either the statistical or chemometric literatures.

There are myriad terms for limits of detection (eg, method detection limit (MDL), detection limit (DL), instrument detection limit (IDL), practical method detection limit (PMDL) etc). However, their conceptual foundations and implementation algorithms are often quite diverse leading to tremendous inconsistency in practice. The pioneering definitions of Currie (6) are used, to provide a framework to compare and contrast these various methodologies because they form the basis of most approaches. Currie defined two levels, the decision limit and the detection limit. The decision limit is the measured concentration "at which one may decide whether or not the result of an analysis indicates detection" (6). The detection limit is the true concentration "at which a given analytical procedure may be relied upon to lead to a detection" (6). From a statistical perspective, the decision limit is the critical value for a test of the null hypothesis "analyte absent" versus the alternative hypothesis "analyte present." The detection limit is a direct function of the selection of the critical value.

Perhaps the most significant contribution that this new approach to environmental testing makes is to demonstrate that the most critical source of bias in estimating limits of detection and quantification is the effect of nonconstant variance. Certain findings have revealed that myriad detection limit estimates are obtained from exactly the same data and analytic method solely as a function of the concentration at which samples are spiked. The only solution in this case is to base detection limits on multiple concentration calibration data and directly model the association between variability and concentration or use a variance-stabilizing transformation.

Despite the guidance provided by the developing literature on analytical detection decisions, U.S. EPA continues to base regulations on the MDL or, for cases in which quantifiable measurements are required, a simple multiple of the MDL (ie, the minimum level (ML)). This practice can lead to bias, most notably an extreme dependence on the selection of the spiking concentration. Despite some guidance in selecting the spiking concentration, actual practice is guided by demonstrating compliance with published standards rather than using the true analytical capabilities of the laboratory. For example, MDLs are routinely determined in laboratories by spiking as many as 70 constituents at a fixed concentration in each of seven replicate samples. In this way it is impossible to iteratively adjust each constituent's concentration in the way suggested by U.S. EPA. In contrast, if calibration-based methods were used, U.S. EPA could designate the range of concentrations in advance to ensure that the computed MDL is not adversely affected by a poor selection of the initial spiking concentration. In the future, the U.S. EPA should consider calibration-based approaches for determining limits of detection and quantification.

BIBLIOGRAPHY

1. C. A. Clayton, J. W. Hines, and P. D. Elkins, *Anal. Chem.* **59,** 2506–2514 (1987).
2. C. B. Davis, "Environmental Regulatory Statistics," in G. P. Patil and C. R. Rao, eds., *Handbook of Statistics: Environmental Statistics* **12,** Chapter 26, Elsevier, Amsterdam, the Netherlands, 1994.
3. R. D. Gibbons, *J. Environ. Ecol. Stat.* **2,** 125–167 (1995).
4. *Fed. Reg.* **49**(209), 43430 (Oct. 26, 1984).
5. J. A. Glaser, D. L. Foerst, G. D. McKee, S. A. Quane, and W. L. Budde, *Environ. Sci. Technol.* **15,** 1426–1435 (1981).
6. L. A. Currie, *Anal. Chem.* **40,** 586–593 (1968).

DIOXIN EMISSIONS INVENTORY

V. M. THOMAS
T. G. SPIRO
Princeton University
Princeton, New Jersey

Inventories of dioxin emissions provide quantitative estimates of emissions from various dioxin source categories. Dioxin emissions are highly variable, and emissions inventories for some source categories are often based on only a few measurements, with the result that emission estimates remain highly uncertain. Emissions inventories have identified municipal and hospital waste incineration as major sources of dioxin, although it is possible that other sources, unidentified or not yet well characterized, may also prove to be significant dioxin sources (see DIOXIN-LIKE COMPOUNDS, SCREENING ASSAYS).

CONSTRUCTION OF AN EMISSIONS INVENTORY

A dioxin emissions inventory should include all of the known sources of dioxins. Most of the source categories are some kind of combustion process: virtually all combustion processes have been found to emit some dioxin. In addition, dioxins are found as contaminants in the herbicide 2,4,5-T and other chlorophenols, and they are emitted to wastewater as a by-product of the chlorine bleaching of wood pulp.

Dioxin emissions inventories typically are inventories of emissions of the tetra- through octachlorinated dioxins and furans. These emission levels are reported either as the sum of the masses of the tetra- through octachlorinated dioxins and furans, which is denoted here as PCDD/F, or as a toxicity-weighted sum. The standard, international toxicity weighting, I-TEQ, is used here, and is denoted as TEQ (1). The ratio of TEQ mass to PCDD/F mass is determined by the congener distribution, and is roughly 1:60 for combustion processes.

A dioxin emissions inventory is constructed by multiplying the emission factor for each source category, ie, the mass of dioxin emitted per mass of material combusted, by the total combustion in each category. For most combustion categories, the total amount combusted can be fairly well determined, usually to better than 50%. The emission factor is much less certain, however, because of the difficulty in obtaining a meaningful average when the actual dioxin output is highly variable.

Dioxin emissions from combustion processes, responsible for the bulk of known dioxin emissions, depend on the completeness of combustion, the combustion system and air pollution control (qv) equipment, and the material being burned. These factors differ from site to site, and at any given site they change over time. Moreover, dioxin emissions are measured only infrequently. Municipal waste incinerators typically are tested for dioxins once a year at best. For many other source categories, emission estimates are based on measurements from only a small number of facilities. For these reasons, dioxin emissions inventory estimates should be interpreted on an order of magnitude basis only.

RESULTS OF EMISSIONS INVENTORIES

Dioxin emissions inventories have been constructed for a number of countries, including Canada (2), the United States (3,4), Sweden, Switzerland, West Germany (5), Austria, the Netherlands (4), and the United Kingdom (6,7), and have been estimated for the world as a whole (8). A few of these are discussed below.

One of the first dioxin emissions inventories was a 1985 study for Canada (2). It concluded that forest fires were the largest source of dioxin, followed by municipal waste incineration, wood waste combustion, and sewage sludge incineration. Total emissions from Canada were estimated to be in the range of 25 to 110 kg PCDD/F (2).

An emissions inventory for West Germany found that, as of about 1990, municipal waste incinerators are one of the largest sources, but also reported that nonferrous metal plants had dioxin emissions of a similar magnitude. Hazardous waste incineration, steel mills, hospital waste incinerators, and combustion of leaded gasoline all ranked as significant sources. Total emissions for West Germany were estimated to be 68–929 g TEQ, (or roughly 4–60 kg PCDD/F) (5).

A 1992 emissions inventory from the United Kingdom found that municipal waste incinerators were the largest source, followed by industrial and domestic coal combustion, and hospital waste incinerators. Pentachlorophenol was also suggested as a major source. Total emissions were estimated to be about 30 kg PCDD/F (6). However, another U.K. emissions inventory, published in early 1995 by the U.K. Department of the Environment, found that iron ore sintering might be the largest source of U.K. dioxin emissions, with emissions estimated at 90–2220 g TEQ (roughly 5–130 kg PCDD/F). Municipal waste incineration emissions were estimated to be in the range of 400 to 700 g TEQ (roughly 24–40 kg PCDD/F); total emissions were estimated to range from 590 to 3700 g TEQ (roughly 35–220 kg PCDD/F) (7).

A global inventory for 1990 estimated that municipal waste incineration was the major source, with biomass combustion, ferrous metals production, and cement kilns burning hazardous and nonhazardous wastes as additional major sources. Total emissions were estimated to be 3000 $\pm$ 600 kg PCDD/F (8).

Figure 1 shows the results of two U.S. emissions inventories, one for 1989 (3) and one for the early 1990s (4). Total emissions were estimated to be on the order of 400 kg PCDD/F, and 600 kg PCDD/F, respectively. Municipal and hospital waste incineration were estimated to be the largest sources. The next largest sources were estimated to be cement kilns and industrial boilers burning hazardous wastes and other materials; biomass combustion of all kinds, including industrial wood burning, residential wood burning, and forest fires; and secondary copper (recycling) smelters (which may have chlorinated compounds as contaminants in the feedstock). Much of the difference between the two studies was due to differing estimates of emissions from hospital waste incinerators. Details of the 1989 emissions inventory are shown in Table 1, which lists the emission factor, quantity of the material burned annually, and emissions for each source.

In summary, dioxin emissions inventories typically include similar sources, and most report waste incineration as the largest source of dioxin. However, results diverge on the relative importance of metallurgical processes, coal combustion, biomass combustion, and other sources of dioxin. Some of this divergence can be explained by differing quantities of combustion in different countries. For example, the Canadian study (2) found that forest fires were the largest source of dioxin emissions, but most other studies found that municipal

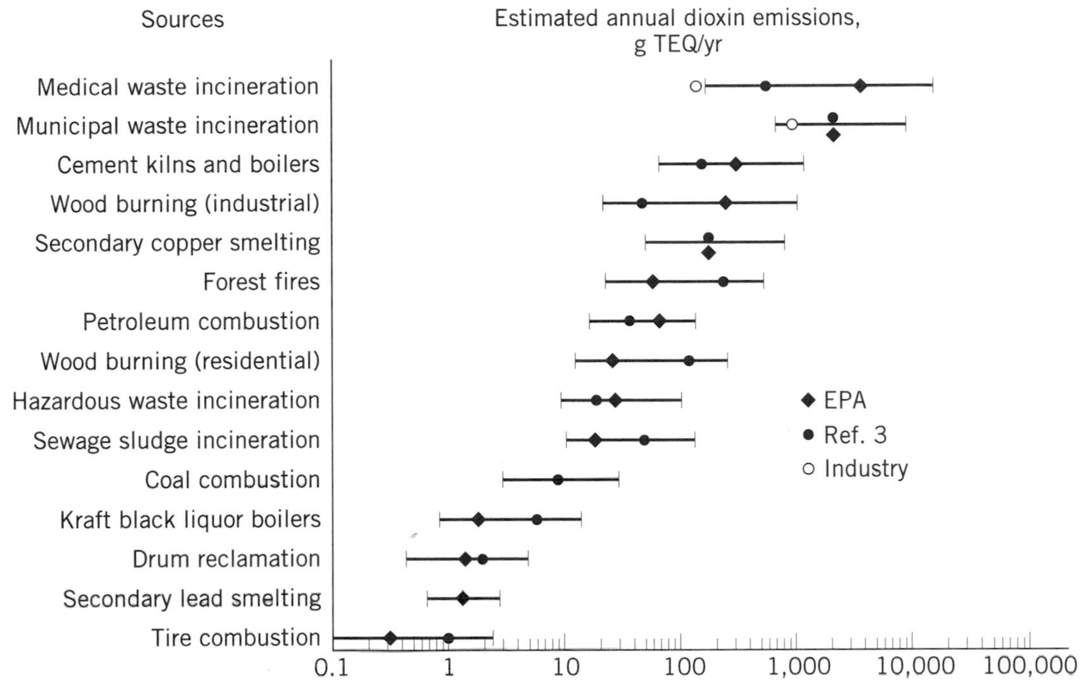

Figure 1. Estimated U.S. annual dioxin emissions from combustion sources (TEQ) (9).

waste incineration was the largest source and that forest fires were relatively minor sources. Because Canada has less waste combustion than, for example, the United States, but a large forested area, it is not inconsistent that forest fire combustion is estimated to be one of Canada's major dioxin sources (10).

Another reason to expect differing results between countries is that emissions from incinerators and other combustion sources may be more tightly regulated in some countries than in others. The date of the study is important, since tighter regulations are being implemented in many countries to reduce dioxin emissions.

Much of the difference between emissions inventories can be attributed to uncertainty in the dioxin emission factors. For example, the U.S. emissions inventories have not included estimates of emissions from most types of metallurgical plants, due to lack of data, but these were estimated to be the largest source in the most recent U.K. study, which was extrapolated from German steel mill data (11).

DISCUSSION OF SELECTED DIOXIN SOURCES

Municipal Waste Incineration

Emissions inventories typically classify dioxin emissions by type of activity, as shown in Table 1, with little accounting for different types of combustion and pollution control systems. However, the data for municipal waste incinerators are sufficiently detailed to permit such categorization (12).

Hospital and Medical Waste Incineration

Dioxin emission factors for hospital and medical waste incineration are estimated to be higher than those for municipal waste incineration. These incinerators tend to be smaller, less regulated, and less well operated than typical municipal waste incinerators. In addition, hospital waste has a higher chlorine concentration than municipal waste, due to the intensive use of disposable plastics, and this may be a factor in their high dioxin emissions (3).

As shown in Figure 1, even within the United States there has been a wide divergence between the estimates of hospital waste incineration emissions. Estimates differ both as to the amount of hospital waste combusted, as well as to the average quality of pollution control. An emission factor of 20 μg/kg PCDD/F was used in the 1989 study (3); the U.S. EPA used an emission factor of 40 μg/kg PCDD/F (4); and the American Hospital Association used a weighted emission factor of 0.25 μg/kg TEQ (roughly 15 μg/kg PCDD/F) (13). For West Germany, an emission factor of 70 ng/kg TEQ (about 4 μg/kg PCDD/F) was used (5).

Hazardous Waste Incineration

Although the incineration of hazardous waste may have the potential for high dioxin emissions, measurements show that dioxin emission factors for hazardous waste incinerators are lower than for typical municipal waste incinerators. This can be attributed to the emphasis on low emissions, through incinerator design, operation, and pollution control equipment, that is required of hazardous waste incinerators.

Dioxin emissions from dedicated hazardous waste incinerators, cement kilns burning hazardous wastes, and boilers and industrial furnaces burning hazardous wastes have been cited for the United States (3,4,9,12). It has been shown that the median emission factor in each category of incineration is much lower than the average emission factor, which means that the average is dominated by a small number of high emissions facilities. As of 1989 in the United States, about 1.6 million metric tons of hazardous waste were burned at commercial and private hazardous waste incinerators, and about 2 million metric tons at cement and aggregate kilns, boilers, and other furnaces. These data were used to derive

Table 1. U.S. Dioxin Emissions Inventory, 1989 [a]

Emission Source (Number of facilities tested)	Emission Factor, μg/kg feed		Material Consumed, kg/yr	PCDD/F Emissions, kg/yr	
	Total	TEQ		Total	TEQ
	Consumer Waste				
MSW incinerators (14)	10	0.2	2×10^{10}	200	3
Hospital incinerators (6)	20	0.35	2×10^{9}	40	0.7
Apartment incinerators[b]	60	1	1×10^{9}	60	1
Open garbage burning[b]	60	1	$\sim 2 \times 10^{8}$	10	0.2
Sewage sludge incinerators (3)	1	0.02	3×10^{9}	4	0.07
	Industrial Waste				
Hazardous waste incineration (35)	3	0.06	4×10^{9}	10	0.2
Copper recycling (2)	20	0.4	7×10^{8}	10	0.2
Steel recycling (9)	0.1	0.004	3×10^{10}	3	0.1
Steel drum reconditioning furnaces (3)	30/drum	0.5/drum	4×10^{6} drum/yr	0.1	0.002
Used motor oil burners (2)	0.04	0.001	3×10^{9}	0.1	0.002
Bleached pulp production (3)	0.01	0.0002	3×10^{10}	0.4	0.007
Tire fires (uncontrolled)[b]	0.09	0.004	3×10^{8}	0.03	0.001
Carbon regeneration (1)	0.06	0.001	5×10^{7}	0.003	0.00005
Tire incineration (controlled) (1)	0.009	0.0004	3×10^{8}	0.003	0.0001
	Biomass Combustion, etc				
Forest and agricultural burning[b]	0.4	0.004	8×10^{10}	30	0.3
Residential wood burning (3)	0.4	0.004	5×10^{10}	20	0.2
Industrial wood combustion (4)	0.05	0.001	6×10^{10}	3	0.06
Structural fires[b]	0.4	0.004	5×10^{9}	2	0.02
PCP-treated wood combustion (1)	8	0.1	$\sim 1 \times 10^{8}$	1	0.02
PCB fires[b]	1000	20	7×10^{5}	0.7	0.01
Cigarettes (1)	0.1	0.002	5×10^{8}	0.05	0.0008
	Fossil Fuels				
Oil combustion (except gasoline)[b]	0.003	0.00005	5×10^{11}	1	0.02
Leaded gasoline (5)	0.03	0.0005	3×10^{10}	1	0.02
Unleaded gasoline (3)	0.003	0.00005	3×10^{11}	0.8	0.01
Coal combustion (1)	0.001	0.00002	6×10^{11}	0.6	0.01
	Dioxin-Contaminated Chemicals				
PCP wood preservative (to air) (1)	2×10^{6}	2×10^{4}		10	0.1
2,4-D herbicide (to soil) (1)	200	0.2	23×10^{7}	5	0.005
Tetrachloroethylene (to air) (4)	10	0.1	3×10^{8}	3	0.03
Total			2×10^{12}	*400*	*6.3*

[a] Ref. 3.
[b] Direct emissions data are unavailable; emission factor estimate is based on data from combustion of similar materials.

a weighted average emission factor of 3 μg/kg for the United States (3). An estimated emission factor of 0.6–90 ng/kg TEQ (roughly 0.036–5 μg/kg PCDD/F) was used for West Germany (5) (see HAZARDOUS WASTE INCINERATION).

Biomass Combustion

There are no direct data on dioxin emissions from forest fires. All of the emission factor estimates are based on measurements of wood combustion or other biomass combustion, and estimates vary widely. Forest fires and other uncontrolled biomass combustion have an estimated emission factor of 0.4 μg/kg PCDD/F (see Table 1), whereas the U.S. EPA estimated an emission factor of about 0.06 μg/kg PCDD/F (4). In an estimate of global dioxin emissions, an emission factor of 0.04 μg/kg PCDD/F was used, with the result that biomass was estimated to be a minor source of global dioxin emissions (8). Other emissions inventories, such as for the U.K. (6) or for West Germany (5), do not include biomass combustion as a dioxin source.

There has been some argument that forest fires could be the major source of dioxin emissions (14). In regions with few other sources of dioxin, such as some regions of Canada, forest fires might indeed be the major dioxin source (2). However, as shown in Table 1, dioxin emission factor estimates for biomass combustion are more than an order of magnitude lower than those for municipal or hospital waste incineration. Thus in industrialized areas, biomass combustion is unlikely to be the major source of dioxin emissions. In the United States, for example, it has been estimated that forest fires, agricultural burning, and residential wood combustion combined account for about 10–15% of total dioxin emissions as of 1989 (see Table 1).

Fossil Fuel Combustion

Detection of dioxin emissions from fossil fuel combustion is difficult because the emission factors are below the detection limits of many experiments. Nevertheless, because of the great quantities combusted, fossil fuels have the potential to be one of the significant sources of dioxin.

Gasoline combustion, especially leaded gasoline combustion, has been reported to release some dioxin. Whereas unleaded gasoline may contain some trace quantities of chlorine (on the order of 10 ppm), leaded gasoline has chlorinated and brominated compounds added as scavengers to prevent lead buildup in the engine. For motor gasolines, the standard "motor mix" has a lead to chlorine weight ratio of 0.3. Emissions of dioxin from leaded gasoline combustion have been reported to be larger for gasoline with higher lead (and chlorine) concentrations, with emission factors of 0.1 μg/kg PCDD/F for higher lead gasolines (~2 g/L lead), 0.03 μg/kg PCDD/F for low lead gasoline (0.15 g/L lead), and <0.005 μg/kg PCDD/F for unleaded gasoline (15,16).

Combustion of diesel fuel has also been suggested as a potential source of dioxin emissions (17). There has been one direct measurement study of dioxin emissions from diesel fuel vehicles, which found the average emissions of a diesel car, bus, and truck to be 0.2 μg/kg PCDD/F (18). However, these measurements were below the limit of resolution of the experimental technique, and the researchers characterized the results as inconclusive.

Coal (qv) combustion has also been found to emit dioxins, but estimated emission factors vary widely. An emission factor of 0.6 μg/kg PCDD/F was used for domestic coal combustion in the U.K. (6), while 1 ng/kg TEQ (roughly 0.06 μg/kg PCDD/F) was used for domestic coal combustion in West Germany (5). An estimate of 0.02 μg/kg for poorly controlled combustion, and 0.001 μg/kg for coal-fired power plants with particulate control equipment were used in one study (3).

Dioxin from Chlorophenols

Production of some chlorinated chemicals results in the inadvertent production of small quantities of dioxins as contaminants. Chlorophenols, which include the herbicides 2,4,5-T and 2,4-D, as well as the wood preservative pentachlorophenol (PCP), are similar in structure to dioxin, and are especially prone to dioxin contamination. Dioxins have also been found in other chlorinated chemicals, including polychlorinated biphenyls (PCBs).

Pentachlorophenols, now used primarily to preserve utility poles, currently account for almost all chlorophenol production in the United States. PCP manufactured in the United States contains about 2000 ppm of PCDD/F. Unlike PCDD/F emissions from combustion, which typically have a fairly even distribution of dioxins and furans with different numbers of chlorine atoms, most of the dioxins in PCP are octachlorinated dioxins and furans, which are thought to be less toxic than the tetrachlorinated dioxins (19).

Until the mid-1980s, PCP was used as a home and garden herbicide. Although this use of PCP accounted for only about 4% of total PCP manufacture, it accounted for almost all of the estimated emissions of PCPs to air and dispersed in soils (20). When these uses were phased out in the United States, estimated emissions of PCP to air and soils decreased by more than an order of magnitude. The major remaining source of PCP air emissions in the United States is from manufacture and application of PCP. Dioxin from PCP amounts to an estimated 2.5% of total dioxin air emissions in the United States as of 1989; before the 1980s it is estimated to have been the largest single source (3) (see PENTACHLOROPHENOL-CONTAMINATED SOILS, BIOREMEDIATION).

2,4,5-T, a trichlorophenol, was used as a herbicide in the United States from the mid-1940s until it was banned in 1984 (21). The 2,3,7,8-TCDD content is estimated to have averaged about 5 ppm (22), and the average PCDD/F concentration has been estimated to be about 80 ppm (3). 2,4,5-T, in combination with another herbicide, 2,4-D (a dichlorophenol), was used as the defoliant Agent Orange during the Vietnam War. An estimated 1200 kg PCDD/F were in the Agent Orange used in Vietnam between 1965 and 1971 (23).

Pulp and Paper Manufacture

Dioxin can also be formed as a by-product of chlorine bleaching of pulp and paper. Typically, most of this dioxin is emitted with wastewater into local streams, which can result in high dioxin exposure for fish and other aquatic organisms. Use of chlorine dioxide, rather than chlorine gas, has been found to produce less dioxin (24), and a number of U.S. pulp and paper manufacturers have switched to chlorine dioxide bleaching. Other manufacturers, particularly in Europe, use chlorine-free bleaching processes, which do not have dioxin as a by-product (25).

VERIFICATION OF INVENTORIES

In principle, dioxin emissions inventories can be verified through measurements of dioxin deposition, or dioxin concentrations in soil or sediments (3,8,26).

In the future, more data on dioxin emissions and deposition is expected to become available. Estimates of dioxin emissions might rise as sources are better understood, whereas implementation of regulations should reduce dioxin emissions from some sources. More and better data on deposition will be especially important to better assess the reliability of dioxin emissions inventories.

BIBLIOGRAPHY

1. *Interim Procedures for Estimating Risks Associated with Exposures to Mixtures of Chlorinated Dibenzo-p-dioxins and Dibenzofurans (CDDs and CDFs) and 1989 Update,* PB90-145756, U.S. EPA, Washington, D.C., 1989.
2. A. Sheffield, *Chemosphere* **14**(6/7), 811–814 (1985).
3. V. M. Thomas and T. G. Spiro, *Toxicol. Environ. Chem.* **50,** 1–37 (1995).
4. *Estimating Exposures to Dioxin-Like Compounds,* EPA/6006/6-88/005Cb, U.S. EPA, Washington, D.C., 1994.
5. H. Fielder and O. Hutzinger, *Chemosphere* **25**(7), 1487–1491 (1992).
6. S. J. Harrad and K. C. Jones, *Sci. Total Environ.* **126,** 89–107 (1992).
7. R. Abel, *United Kingdom Comments on the United States Environmental Protection Agency's External Review Draft Reassessment of Dioxins,* CDEP/2/37/34, U.K. Dept. of the Environment, Toxic Substances Division, London, 1995.
8. L. P. Brzuzy and R. A. Hites, *Environ. Sci. Technol.* **30**(6), 1797–1804 (1996).
9. V. M. Thomas and T. G. Spiro, *Environ. Sci. Technol.* **30**(2), 82A–85A (1996).
10. T. G. Spiro and V. M. Thomas. *Science* **266,** 349 (Oct. 21, 1994).
11. U. Lahl, *Organohalogen Compounds* **11,** 331–314 (1993).

12. *Locating and Estimating Air Emission Sources of Dioxins and Furans,* Draft Final Report, Office of Air Quality Planning and Standards, U.S. EPA, Research Triangle Park, N.C., 1993.
13. *Comments on the U.S. EPA Dioxin Exposure and Health Documents,* American Hospital Association, Chicago, Ill., Jan. 13, 1995.
14. G. Gribble, *Environ. Sci Technol.* **28,** 310A–314A (1994).
15. S. Marklund, *Dioxin Emissions and Imissions: A Study of Polychlorinated Dibenzodioxins and Dibenzofurans in Combustion Processes,* Institute of Environmental Chemistry, University of Umea, Sweden, 1990.
16. S. Marklund, C. Rappe, M. Tysking, and K. Egeback, *Chemosphere* **16**(1), 29–36 (1987).
17. K. H. Jones, *Risk Anal.* **13**(3), 245–252 (1993).
18. *Determination of PCDD and PCDF Emissions from Motor Vehicles,* Draft Report, Monitoring and Laboratory Division, Engineering and Evaluation Branch, Report No. C-86-029, California Air Resources Board, Sacramento, Calif., Oct. 1987.
19. C. Rappe. *Environ. Sci. Technol.* **18**(3), 78A–90A (1984).
20. *Materials Balance for Chlorophenols,* EPA 560/13-80-004, U.S. EPA, Washington, D.C., Feb. 1980.
21. M. Gough, *Dioxin, Agent Orange: The Facts,* Plenum Press, New York, 1986.
22. Executive Office of the President, *Report on 2,4,5-T: A Report of the Panel on Herbicides of the President's Science Advisory Council,* Office of Science and Technology, U.S. Government Printing Office, Washington, D.C., Mar. 1971.
23. P. H. Schuck, *Agent Orange on Trial,* Harvard University Press, Cambridge, Mass., 1986.
24. U.S. Congress, Office of Technology Assessment, *Technologies for Reducing Dioxin in the Manufacture of Bleached Wood Pulp,* OTA-BP-O-54, PB89-223291, U.S. Government Printing Office, Washington D.C., May 1989.
25. D. G. Meadows, *Tappi J.* **77**(11), 3, 51–58; G. E. Maples and co-workers, *Tappi J.* **77**(11), 71–80; T. R. Brooks and co-workers, *Tappi J.* **77**(11), 83–92; A. Mokfienski and B. J. Demuner, *Tappi J.* **77**(11), 95–103; J. Brown and co-workers, *Tappi J.* **77**(11), 105–109; A. Suurnakki and co-workers, *Tappi J.* **77**(11), 111–116; Y.-S. Perng and co-workers, *Tappi J.* **77**(11), 119–125; O. H. Samuelson and U. O. Ojteg, *Tappi J.* **77**(11), 127–133; M. Macleod, *Tappi J.* **77**(11), 227–230 (Nov. 1994).
26. C. Rappe, "Sources of and Human Exposure to PCDDs and PCDFs," in *Banbury Report 35: Biological Basis for Risk Assessment of Dioxins and Related Compounds,* Cold Spring Harbor Laboratory Press, Cold Spring Harbor, N. Y., 1991.

DIOXIN-LIKE COMPOUNDS: SCREENING ASSAYS

NIGEL J. BUNCE
JOHN R. PETRULIS
Toxicology Program, University of Guelph
Guelph, Ontario, Canada

Dioxin-like compounds is a broad term applied to compounds with toxicological properties similar to the reference toxicant 2,3,7,8-tetrachlorodibenzo-*p*-dioxin (TCDD). Typically, dioxin-like compounds include other polychlorinated dibenzo-*p*-dioxin congeners, especially those chlorine-substituted in the 2, 3, 7, and 8 positions (Fig. 1), the corresponding chlorinated dibenzofurans, bromo- and bromochloro-substituted dibenzo-*p*-dioxins and dibenzofurans, and the "coplanar" members of the polychlorinated biphenyl family. The term is also extended to include appropriately halogen-substituted naphthalenes, anthracenes, diphenyl ethers, and azobenzenes. The mechanism of action is an important criterion for identifying dioxin-like compounds. The term dioxin-like compounds (DLCs) comprises those halogenated aromatic compounds that bind to the Ah (aryl hydrocarbon) receptor protein or that cause overt dioxin-like endpoints in living systems.

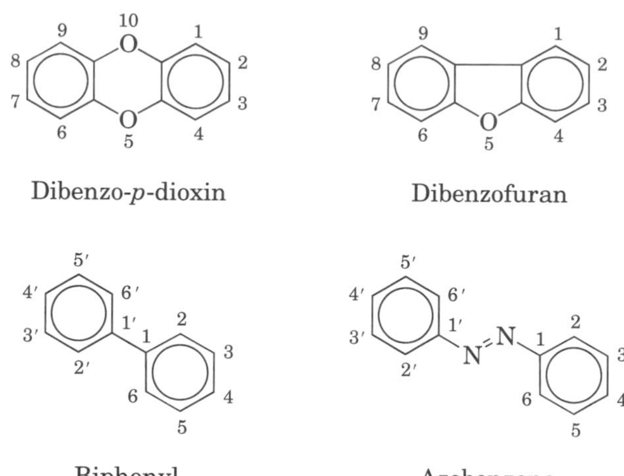

Figure 1. General chemical structures and numbering system for several dioxin-like compounds (DLCs).

The standard chemical method of analyzing DLCs is experimentally demanding, tedious, and therefore expensive. After spiking with an isotopically labelled dioxin surrogate, such as $[^{13}C_{12}]$-TCDD as a recovery standard, conventional analysis involves solvent extraction, multiple chromatographic procedures to isolate the polychlorinated dibenzodioxin/polychlorinated dibenzofuran (PCDD/PCDF) fraction, and quantitation of the individual PCDD and PCDF congeners by capillary gas chromatography (GC)/high resolution mass spectrometry (MS) (1,2). This yields the concentration of each congener in units such as ng congener per gram sample.

Usually in the context of environmental remediation, however, an estimation of the toxic potency of the whole sample is desired, rather than the concentrations of the sample's components. One such measure, the toxic equivalence (TEQ), is obtained from the analytical data by using Toxic Equivalency Factors (TEFs) relating the potency of DLC "X" to the potency of TCDD, taken as a reference toxicant (TEF = 1). The TEQ of the sample is calculated as the sum of terms (concentration of X) × (TEF of X). In the commonly used I-TEF scheme (3), agreed-upon toxic equivalency factors are available for the 17 most toxic PCDD and PCDF congeners, chlorinated in the 2, 3, 7 and 8 positions, and provisional TEFs have been proposed for the coplanar polychlorinated biphenyl (PCB) congeners (4). The potencies of different DLC congeners vary greatly, as do their relative concentrations in environmental samples. For example, the concentrations of coplanar PCBs in many animal tissues are orders of magnitude higher than those of PCDD/PCDFs (5,6), making the PCBs the major contributors to the TEQ, despite their much smaller TEF values.

There are several important drawbacks to the TEQ approach, because a unique TEF must be assigned to each substance with dioxin-like properties, so that its contribution to the TEQ can be included. First, it is inherently cumbersome. Each addition to the list of DLCs adds to the complexity of analytical separation and to the cost of analysis; yet the TEQ is underestimated if the sample contains DLCs for which no TEF has been established. A second problem is assigning the TEF value for a given DLC. This done by examining reports on the toxicity of the candidate substance and trying to reach a consensus value that is "typical" across many endpoints. It is difficult to reach consensus on an appropriate TEF because potency rankings among DLC congeners differ according to the endpoint examined, because not all DLCs lead to all endpoints, and because not all endpoints are seen with the same intensity in all species. Therefore, the relevance of the consensus TEF with regard to a specified toxic endpoint is questionable. Worse, some of the data used to support a TEF are derived from *in vitro* bioassays such as those described in this chapter, rather than from toxicity assays. A third problem is that the TEF approach explicitly takes the toxic potency of a complex sample as the sum of the contributions of its components, ignoring the possibilities of synergism or antagonism among different DLCs. Finally, the underlying assumption of additivity in the TEF approach is questioned by arguing that, when dose-response curves lie parallel (as is common for DLCs), the apparent TEF of a given DLC relative to TCDD is not constant, but depends on the position on the dose-response curve (7). Considerable inaccuracies result when TEFs derived from high-dose laboratory experiments are used in extrapolations to low dose environmental situations.

Thus simple and inexpensive assays are needed that numerically estimate the TEQ of the sample as a whole. The purpose of an assay is to describe (graphically or algebraically) the intensity of "response" as a function of the dose of pure reference toxicant. Then the same assay is applied to the environmental sample, its response intensity determined, and the potency of the sample is expressed as the dose of toxicant that would cause the same response intensity. For DLCs, the reference toxicant is normally TCDD, and the "TCDD equivalent" is a measure of the sample's TEQ.

Simple as this concept appears, it is nevertheless fraught with difficulties because the substances responsible for the response are not identified chemically. This makes it unlikely that regulatory decisions could ever be based on the results of such assays alone. Some of these regulatory issues are identified in the context of using immunoassays (8), but they apply equally to other bioassays. The concerns include (a) extraction efficiency and comparison of assay results with those from previously validated procedures; (b) the occurrence of false negatives and false positives; and (c) cross-reactivity with similar and dissimilar analytes. These will be discussed in turn.

Extraction efficiency is difficult to monitor even when a surrogate is added, if the added surrogate is more easily removed from the matrix than endogenous DLCs. In addition, the practice of adding a surrogate, such as [$^{13}C_{12}$]-TCDD, in chemical analysis by GC/MS is impractical in bioassays, because such a surrogate is itself a DLC and thus contributes to the assay (in principle, the surrogate could be used in the "standard addition" mode (9), but we are not aware that this has been tried in practice). As for validation against other procedures, few proposed assays for DLCs have been rigorously tested in this way (10). Ideally, the assay should require the minimum of extraction and other work-up to save on analyst's time (for discussion and example, see (11)).

False negatives and false positives: The chief use of bioassays involves screening samples for activity and determining which are below and above an agreed cut-off TEQ, so that only those samples assaying above the cut-off require the conventional, expensive analysis to identify the specific chemicals present. False positives occur when a substance is present that produces the endpoint being monitored without causing toxicity, and to this extent false positives make the bioassay "conservative" because the sample's contamination is overestimated. In consequence, false positives increase the number of samples that must be subjected to confirmatory analysis; a high rate of false positives negates the benefit of a low-cost bioassay. False negatives, on the other hand, suggest that contaminated samples are clean; from a regulatory viewpoint, any significant occurrence of false negatives is unacceptable. False negatives occur if a relatively nontoxic congener masks the activity of a more potent congener in the assay under consideration. Such masking effects between DLCs are known and will be addressed later in this article.

Cross-reactivity in bioassays is both an advantage and a disadvantage. It is always a problem if one wishes to monitor the concentration of a single compound, say, a certain pesticide or the specific congener TCDD. However, in the case of DLCs, the same toxic endpoints are common to many congeners, and the toxic equivalence concept assumes explicitly that the total toxic potency of a sample is the sum of the potencies of its DLC components. When estimating the total TEQ of a sample, cross-reactivity is advantageous if each DLC behaves additively in the assay according to its individual toxic potency, if one DLC does not antagonize or otherwise mask the activity of another, if non-DLCs do not act as analytes in the assay, and if the assay potencies of different DLCs match their TEFs reasonably well.

One must ask whether the TEQ determined by a particular assay reflects the sample's toxic potency. For example, if radioimmunoassay is used to rank the DLC activity of a series of fish samples, does the ranking reflect their relative toxic potencies for humans eating the fish? If an enzyme assay is used to rank the activities of a series of sediment samples, does this ranking reflect the probable cancer incidence in bottom-feeding fish found in those waters? The relative potencies of the different DLCs in the assay compared with their potencies toward the endpoint of interest help to answer this question (similar to the point already discussed with respect to obtaining TEQs from TEFs in the chemical analysis). Another contributing factor is the "biological plausibility" of the assay, which addresses the more fundamental issue whether there is any biological connection between the assay endpoint and the toxic endpoint of interest. Is there any biological connection between an enzyme assay and cancer in feral fish? Although specific examples will be discussed later, we will state here our prejudice in favor of *mechanism-based bioassays,* which use a point along or minimally divergent from the established metabolic pathway leading to toxicity, as the indicator of potency.

In principle, three kinds of bioassays are possible. First is the replication of the actual endpoint, using live organisms as the test system. For example, the acute toxicity of industrial effluents to aquatic organisms is assessed by mortality tests, such as the 96-h rainbow trout test or the 24-h *Daphnia magna* test, both of which employ standard organisms as surrogates for aquatic species in general. Other acute effects, such as developmental toxicity, are also conveniently studied in this way (see Early Life-Stage Bioassays). When humans are the organism of concern, laboratory mammals must be used as surrogates. Extrapolation of results obtained with laboratory animals to humans is difficult because of the varying susceptibilities of different species and strains towards intoxication by DLCs (12). Tests involving live animals are very time-consuming and expensive when chronic toxic effects are involved, for example, in carcinogenicity testing. In the second type of bioassay, live organisms or cell cultures are used to monitor an overt endpoint that is different from the endpoint of interest to environmental or human health. In the past, acute toxicity tests, such as the LD_{50}, have often been used incorrectly as a measure of the toxic potencies of substances toward chronic intoxications. This is particularly inappropriate for DLCs, which cause lethality by mechanisms different from those involved with most other endpoints (13). Finally, the assay might have a chemical or biochemical endpoint as in the case of an immunoassay or an enzyme assay; then the relevance and intensity of this endpoint must be correlated with that of the toxic endpoint of interest. This last type of test has advantages of cost, speed, and convenience, but may or may not be directly relevant to the biochemical mechanism of toxicity, depending on the degree of biological plausibility linking the assay to the endpoint of interest.

Important parameters for the success of any analytical method are reproducibility, sensitivity, and detection limit. Reproducibility affects the standard deviation associated with the analytical result. Sensitivity (usually calibration sensitivity) is the slope of the graph of Response intensity vs Concentration of analyte; in chemical analysis a linear relationship between the two is desirable, but this is rarely possible in bioassays, because most toxic responses are saturable. Consequently, there is often only a limited concentration range over which the response intensity is meaningful. Detection limit, in the analysts' sense that the concentration corresponds to the response intensity three standard deviations away from that of a blank assay, is rarely determined. The consequence of inattention to these analytical considerations is that bioassay results are very imprecise.

Environmental remediation requires screening programs for large numbers of samples to determine which sites, for example, require remediation, and ongoing analysis at sites where remediation is ongoing to monitor its progress. These samples usually involve difficult matrices, such as soil, sediment, and biota. The extent of pretreating the sample (clean-up) needed before using the assay is a major consideration. As noted earlier, much of the cost of conventional chemical analysis for DLCs is associated with the lengthy clean-up of the sample prior to chemical analysis by GC/MS. Therefore, monitoring or screening assays are of only limited use unless a major part of the effort of clean-up is avoided (11,14,15). This question has not been routinely addressed by the advocates of new assay techniques.

TCDD is a carcinogen in laboratory animals, such as rats (16), and a disputed human carcinogen (17). This adds to the difficulties of developing DLC assays because the linear models used to extrapolate cancer incidence to the origin from experimentally observable doses assume no threshold to toxicity (18). "Acceptable" human exposure is set by calculating the dose from this linear relationship which corresponds to an arbitrary risk such as one excess cancer in 10^6 lifetimes (19). Such exposures typically correspond to very low concentrations of DLCs, making it increasingly difficult to devise an inexpensive screening assay that allows classifying environmental samples as either contaminated (needing further analysis) or uncontaminated (not needing further study). Yet this is precisely the need, because the high cost of chemical analysis of DLCs places practical limits on society's ability to mount screening programs for these analytes in matrices such as food and the environment.

TOXICITY AND MECHANISM OF ACTION OF DLCs

Most of our knowledge of the toxicity of DLCs is based on that of TCDD, which elicits a wide range of toxic responses. Besides the widely reported acute lethality, which occurs at very low doses and proceeds via a wasting syndrome, other effects include reduced immunocompetence; hyperplasia and carcinogenicity; developmental toxicity (including teratogenicity and errors in sexual differentiation); and altered porphyrin metabolism. TCDD does not induce all effects in all species, and evidence is lacking that the same effects are exhibited by all DLCs (for reviews, see 20,21,22).

The complexity of the toxic responses to TCDD, limited knowledge on whether all other DLCs behave similarly, and almost complete lack of knowledge about interactive effects among DLCs make it difficult to assess the toxic potential of environmental samples, which usually contain complex mixtures of DLCs. Combustion byproducts, such as fly ash, contain a wide range of PCDDs and PCDFs (23), whereas biota, such as fish, meat, and milk, contain a large number of organochlorines, of which the PCDDs and PCDFs are usually minor constituents compared with PCBs, at least some of which are dioxin-like, and the older types of organochlorine pesticides such as DDT, that in general are not. Among dioxin-like compounds, PCDDs, PCDFs, and coplanar PCBs combine toxicity with environmental persistence, and their lack of chemical and metabolic reactivity allows them to bioaccumulate through food webs. Other compounds, such as partly methylated PCDDs and PCDFs (24) and the chlorinated azobenzenes are of less concern because they are metabolized more rapidly.

A mechanism-based bioassay utilizes current knowledge about the mechanism of action of the toxicant. Such a bioassay exploits some biochemical event that takes place in the sequence leading to toxicity to estimate the TEQ. The closer the relationship between the biochemical event being monitored and the toxic endpoint of interest, the greater the biological plausibility of the assay, provided that concerns about cross-reactivity, false negatives and positives, and so on, can be met.

The currently accepted mechanism of action of TCDD is summarized in Figure 2. As already noted, the behaviour of other DLCs is generally assumed to be similar (25,20,22).

Step 1: It is believed that TCDD enters cells by passive diffusion, because it has a high affinity for the interior of the lipid bilayers constituting animal cell walls, and for adipose tissue, through its great lipophilicity (26). The initial biochemical target of TCDD in the cytoplasm of vertebrates is a protein called the Ah receptor (aryl hydrocarbon receptor, AhR), so named because it was first detected as a binding protein for certain aromatic hydrocarbons. Most, if not all, biochemical effects of TCDD are considered AhR-mediated, meaning that association of TCDD with the receptor is a necessary precondition for toxicity. The AhR is an oligomeric 280–kDa protein complex found

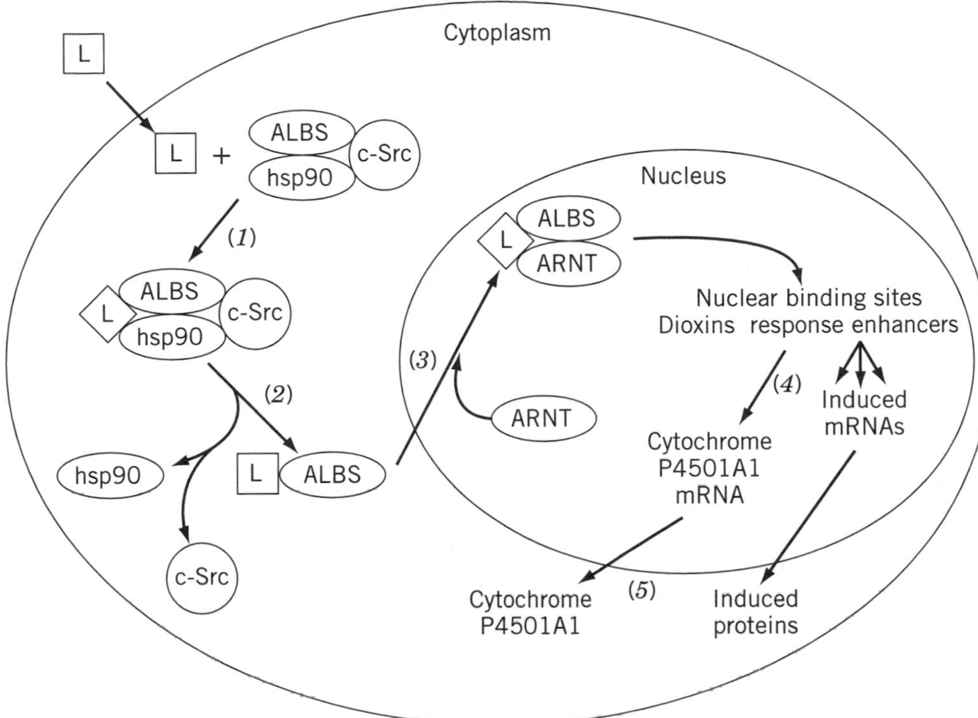

Figure 2. Schematic representation of the mechanism of the action of DLCs. A ligand (L) enters the cell and binds to the ligand binding subunit (ALBS) of the cytosolic AhR. The receptor undergoes transformation, nuclear translocation, and dimerization with the Ah receptor nuclear translocator protein (ARNT), following which it binds to DREs, resulting in the induction of cytochrome P4501A1 and a variety of other proteins. The numbers indicate the stages in this mechanism on which bioassays can be based: (*1*) receptor-ligand binding, (*2*) receptor transformation, (*3*) nuclear translocation, (*4*) DNA binding, and (*5*) enzyme induction.

in the cytosolic fraction of typical tissue preparations. It contains a ligand binding subunit (ALBS) with a molar mass in the range of 90–120 kDa, depending on the species (27), and a molecule of the 90-kDa heat-shock protein (hsp90), which is thought to mask the DNA-binding site (28) and to keep the receptor in a conformation which can bind the ligand (29). Recent evidence suggests that a molecule of the tyrosine kinase c-Src (30) accounts for the remaining mass of the the oligomeric AhR.

In susceptible animals, such as the C57BL/6 mouse, TCDD has high affinity for the AhR. Estimates of its association constant range to $>10^{11}\ M^{-1}$ (31,32). Other strains, such as the DBA/2 mouse, are less susceptible to intoxication by TCDD and have a lower affinity form of the AhR (33,34). The human form of the AhR also has low affinity for TCDD (35,36,37).

Lipophilic molecules of the size and shape of TCDD (planar, molecular dimensions ca 1.0 nm × 0.3 nm (38)) bind preferentially to the AhR; the poorer the fit between the DLC and the receptor binding site, the lower the affinity. The suggestion that toxic responses of TCDD are AhR-mediated (25,39) comes from QSAR studies, which show that AhR binding affinity broadly correlates with toxic potency. Much thought has been given to the "natural" role of the AhR and to finding endogenous ligands for it; one possible role is binding to foreign substances, such as polycyclic aromatic hydrocarbons and indole-3-carbinol (present in cruciferous vegetables) (40), leading to induction of elevated levels of mixed-function oxidase enzymes which metabolize the foreign substances. According to this view, many of the toxic responses of DLCs result from altering the intracellular levels of these powerful oxidants. Alternatively, the AhR may be a regulatory protein that controls the expression of many unrelated genes (22).

Step 2: Following association with a ligand, such as TCDD, the AhR-ligand complex undergoes a poorly understood process termed transformation. Immunoprecipitation experiments have shown that hsp90 and c-Src are released at this stage (30,41), leaving behind free ALBS in the form of its TCDD complex. The rate of transformation depends on temperature and species. Although the initially formed AhR-TCDD complex releases its ligand, TCDD is bound almost irreversibly after transformation has occurred; this suggests that a significant conformational change in ALBS is associated with transformation and that hsp90 acts as a chaperone, maintaining ALBS in a conformation that binds ligand. An ambiguity exists in many literature reports of ligand binding affinity to the AhR, because it is unclear whether the untransformed or transformed AhR-ligand complex is the species present after incubation of the ligand and the receptor preparation.

Step 3: Then the transformed ALBS-TCDD complex penetrates the nuclear membrane, after which it heterodimerizes with the Ah receptor nuclear translocator protein (ARNT), which is localized in the nucleus (42,43), through helix–loop–helix motifs. Phosphorylation may occur at this stage, after which the heterodimer binds to recognition elements on DNA known as dioxin response enhancers (DREs) (44). The term enhancer reflects the *increase* in the expression of various gene products following exposure to substances such as TCDD. DRE binding with TCDD occurs with high affinity (45,46). Several DREs have been identified and are characterized by similarity of their nucleotide sequences. *In vitro* binding experiments carried out with oligonucleotides containing the consensus sequence suggest that more than one DRE is occupied, when toxicity is seen, and present the possibility that different toxic responses to TCDD result from binding to different suites of DREs.

Steps 4 and 5: The DREs act as genomic switches, initiating the transcription of the corresponding mRNAs, which in turn yield the appropriate protein gene products (21). Transcription of *CYP1A1* (cytochrome P-450 1A1 m-RNA) begins shortly after exposure to TCDD, and all necessary components are present constitutively, provided that the cells possess functional AhR and ARNT.

In the context of mechanism, one of the uses of mechanism-based bioassays is to follow a particular DLC stepwise through its mechanism of action to determine whether it acts as an agonist or as an antagonist at each step. An agonist is defined as a substance that elicits the toxic or biochemical endpoint under study, whereas an antagonist is a substance that blocks or masks the biological activity of an agonist. Consider the case of two substances that both bind the AhR. An agonist binds the AhR and successfully surmounts all subsequent hurdles leading to the toxic response. An antagonist binds the AhR, thereby reducing the stock of AhR molecules available to bind an agonist, but does not necessarily give a positive result at each subsequent step. The use of a series of assays, step by step in the mechanism, allows the investigator to probe the exact mechanism of action of such compounds. For example, in the case of AhR binding assays, not all substances that bind to the AhR are necessarily toxic, as clearly demonstrated by indole-3-carbinol, present in cruciferous vegetables (47,48). This compound would show up as "active" in an AhR binding assay or in an assay based on the induction of mixed-function oxidases. Although it acts as a xenobiotic, activating the defense mechanisms designed to remove foreign compounds, it does not cause dioxin-like toxicity.

RECEPTOR–LIGAND BINDING ASSAYS

Binding assays, including ligand–receptor association, DNA binding assays, and immunoassays are carried out either as saturation assays or as competitive assays. These two types are illustrated by AhR-ligand binding assays. An experimental complication is that the concentration of AhR in cells is so small (about 10^5 receptors per cell) (25) compared with other proteins that it is not possible to monitor the concentrations of the liganded and unliganded protein directly. This requires a radiolabeled ligand (such as [^{3}H]-TCDD). The parameter actually measured is the concentration of radiolabelled receptor-ligand complex following incubation of the receptor preparation and the ligand(s).

POST-TRANSFORMATION ASSAYS

This group of assays includes tests for producing the transformed AhR-ligand complex and the nuclear AhR-ligand complex, which bind to both native DNA and synthetic oligonucleotides. Much less work has been done in this area than with receptor binding assays.

ENZYME INDUCTION ASSAYS

Several assays have been developed in which the induction of a particular gene product has been monitored *in vivo* or in cell culture following exposure to TCDD. The most common endpoints monitored are the induction of monooxygenase activities, such as aryl hydrocarbon hydroxylase (AHH) and 7-ethoxyresorufin-*O*-deethylase (EROD), which are markers for the gene product CYP 1A1 (cytochrome P450 1A1). This is the form of cytochrome P450 induced by TCDD and polycyclic aromatic hydrocarbons, such as 3-methylcholanthrene. The isozyme CYP 1A2 is induced by phenobarbital and nonplanar PCB congeners (49); its activity is monitored by following the marker enzyme acetanilide-4-hydroxylase (50). In addition, 7-pentoxyresorufin *O*-deethylase (PROD) is a marker for the phenobarbital-inducible CYP2B family of cytochrome P450s (51), and 7-ethoxycoumarin *O*-deethylase (ECOD) is a marker for both phenobarbital- and TCDD-type inducers (52). The AHH induction assay (53) was refined (54) as a rapid screening tool for detecting planar halogenated aromatic compounds.

MARKERS OF ABNORMAL METABOLISM

A thoroughly examined endpoint is the hyperaccumulation of hepatic uroporphyrin III (UP-III). As in all cases of abnormal metabolism, the question arises as to whether the endpoint observed can be more than a marker of exposure. Does it really parallel toxicity?

The overproduction of uroporphyrin UP-III in the liver and the related human disease porphyria cutanea tarda are observed in animals after exposure to many halogenated aromatic compounds. The link to an Ah receptor-mediated pathway was established by showing that inhibitors of cytochrome P450 1A1 suppress the overproduction of UP-III (reviewed (55)) and that UP-III accumulation parallels AHH activity in AhR-responsive and AhR-nonresponsive mice (56).

IMMUNOASSAYS

Immunoassays depend on the recognition of an antigen (or foreign molecule) by an antibody, which is a soluble protein produced by a live animal or cell culture. Exposure to a particular antigen causes the proliferation of cells that produce the antibody complementary to that antigen, so that an immune response results from subsequent exposure to the antigen. Immunoassays depend on collecting the antibodies and using them to bind selectively to the antigen, usually by immunoprecipitation of an insoluble antigen–antibody complex. The antibodies are removed from the spleen of small laboratory mammals or from the blood of large animals. Then preparations are developed for recognizing the analyte molecule *in vitro*. The actual assay requires both the recognition and quantitation of free analyte in a test solution. The latter is usually accomplished by a competitive assay, although a wide variety of assay methods have been reviewed (57).

Immunoassays are used to detect and quantify large biomolecules, such as specific proteins or parts of proteins, polysaccharides, or nucleic acids at low concentrations, eg, <10 fmol of the N-terminal region of the AhR itself (58). Immunoassays are standard analytical methods in clinical laboratories, where they are adapted for the rapid, automated analysis of large numbers of samples.

Antibodies are normally produced in response to molecules with molecular mass >1000. Small molecules, such as TCDD, therefore do not produce an immune response on their own, and so the antibodies must be raised by covalently binding the small molecule (termed a hapten) to a carrier protein, which produces a response, and then injecting the conjugate into a live animal. For TCDD analysis, an extensively used hapten is 1-(N-adipylamino)-3,7,8-trichlorodibenzo-p-dioxin, which is a TCDD molecule, one of whose chlorine substituents is replaced by a linking moiety that connects it covalently to the carrier protein (59,60). If such a protein-hapten conjugate is injected into a live animal, some of the antibodies produced may subsequently recognize the hapten-protein conjugate whereas others recognize only the unattached hapten.

Antibodies produced as just described are termed polyclonal antibodies, because they are produced by multiple clones of cells, each clone contributing one type of antibody. The antibodies bind to the antigen by way of many different specific recognition elements or epitopes that are functions of the size, shape, and hydrophobicity of different regions of the protein-hapten conjugate. Most of the polyclonal antibodies interact with epitopes on the carrier protein itself, rather than recognizing the hapten. Even if the hapten is part of the recognition site, it is unlikely that an antibody recognizes the hapten alone, rather than the hapten plus part of the carrier, unless the hapten has a molecular mass above 1000.

Polyclonal antibody assays for DLCs were reported around 1980 (for TCDD, 60; for TCDF, 61). In these assays, the DLC molecule was bound covalently to the carrier protein by an N-adipylamido chain, and the antibodies were produced in rabbit. The assay was carried out by a radioimmunological method in which the unlabeled DLC and the appropriate N-adipylamino-trichlorodibenzofuran or -dibenzo-p-dioxin, labeled with ^{125}I in the adipyl chain, competed for the antibody. The complexed antibodies were precipitated with goat antirabbit γ-globulin and counted for ^{125}I. The greater the concentration of unlabeled DLC, or, in the case of cross-reactivity experiments, the higher its affinity for the antibody, the less radioactivity was detected in the pellet. For pure TCDD, as little as 0.08 pmol could be detected by this method. Such competitive radioimmunoassays suffer from the same disadvantages as competitive ligand-binding assays: it is the labeled competitor rather than the analyte that is actually detected, the response is saturable, and the "signal" (intensity of radioactivity) declines to zero as the concentration of analyte increases.

A general problem with polyclonal antibodies is poor recognition of the hapten and interference by those antibodies that recognize different structural motifs. In addition, the particular range of antibodies formed varies with time in a given preparation and from animal to animal among different preparations, making it impossible to standardize methods between different laboratories. These limitations are avoided by using a monoclonal antibody, which is a unique protein derived from a single cloned cell line and which recognizes the analyte specifically. Rather than attempt to separate a single protein from the polyclonal antibody mixture, a more successful approach is separating the individual cells that produce the specific monoclonal antibody. This is done by fusing cells from an immortal cell line to spleen cells that have been previously exposed to the protein-linked hapten. Then the resulting hybridomas are screened to select a particular cell that secretes an antibody which recognizes the hapten. With luck, a hybridoma can be selected which forms antibodies specifically recognizing the free ligand. Then this cell line is grown indefinitely in culture.

The monoclonal antibody is uniform and is distributed to different laboratories, allowing the development of standardized assay procedures. Among the advantages of an optimally selected monoclonal antibody are ease of use, minimal sample preparation, low cost per sample, reproducibility, high selectivity for the analyte, and potential for automation (57,62). Against this must be factored the long development time and high cost of obtaining the monoclonal antibody, requiring the testing and selection of thousands of hybridomas. In the case of TCDD analysis, monoclonal antibodies were first developed by Kennel (63) and by Stanker (59). Kennel (63) used the same assay approach as Albro (60), namely, radioimmunoassay in which the DLC competes with a ^{125}I-labeled reference ligand. The hybridomas were obtained by fusing murine spleen cells with murine myeloma cells, and the hybridomas were screened for their ability to recognize TCDD by identifying those that bound to bovine serum albumin that was covalently coupled to trichlorodibenzo-p-dioxin. As stated by Gilman (64), "(The) ideal patterns for PCDD analysis are either high selectivity for only one specific PCDD congener or a pattern of crossreactivity for a wide range of PCDD-like compounds that mimics (their) relative toxicity." These authors have reported the development of immunoassays, based on haptens structurally closer to TCDD than the original adipylaminotrichlorodibenzo-p-dioxin used by Stanker (59), to improve specificity.

An ELISA (enzyme-linked immunosorbent assay) for PCDDs was developed (59). In an ELISA, an enzyme is covalently linked to the monoclonal antibody, and the final assay method involves detecting the product of the enzyme-substrate reaction by some convenient chemical or physical method of analysis. A standard aliquot of the hapten or a simple hapten derivative is initially adsorbed to a plastic surface, such as those on a 96-well plate. The antibody and the solution to be analyzed are placed in the well, where the antibody complexes competitively to the plastic-bound hapten and the free analyte in solution. After equilibration, the solution phase is removed. The plastic-bound antibody complex is quantitated by adding the substrate for the enzyme that had been linked to the antibody; the enzyme product is detected by a standard method, such as fluorescence or colorimetry, and so the concentration of enzyme product is a measure of the adsorbed antibody-hapten complex. In this type of competition assay, a 100% reading is obtained when the solution tested contains zero hapten; the greater the concentration of hapten in the test solution, the smaller the concentration of enzyme product detected.

An alternative protocol involves immobilizing the antibody on plastic, and a series of incubation steps allows competition for a limited amount of antibody between DLCs in the test sample and a fixed aliquot of a competitor-enzyme conjugate. After removing the solution phase, the enzyme substrate is added, and the amount of enzyme product is determined by the amount of competitor-enzyme conjugate bound to the antibody (again, the less DLC in the sample, the more enzyme product is formed). Detection limits in the range of 0.1 ng of TCDD are reported (64), using a mouse monoclonal antibody specific for tetra- and penta-chlorinated dibenzo-*p*-dioxins that roughly parallels their I-TEFs. A similar approach was used to prepare monoclonal antibodies that recognized coplanar PCBs in the presence of larger amounts of the non-planar congeners (65), using as the hapten a 4-alkoxy-3,3′,4′-trichlorobiphenyl linked to the protein through a terminal carboxyl on the 4-alkoxy substituent. However, various "PCB-fragments" (ie, substituted dichlorobenzenes) are more reactive with the antibody than a "toxic" coplanar PCB (3,3′,4,4′-tetrachlorobiphenyl), which severely limits the usefulness of this approach.

In principle, monoclonal antibodies are developed as probes to recognize individual DLCs by screening appropriate hybridomas. In practice, this has not been very successful because of cross-reactivity (59). The recognition sites on monoclonal antibodies are fairly large (~2–3 nm in diameter) compared with the size of the analyte of interest (0.3 nm × 1.0 nm in the case of TCDD), and so other molecules of similar size and hydrophobicity are accommodated almost equally well (66). This makes the monoclonal antibody assay more applicable to DLCs as a class than to the quantitation of individual DLC congeners. A standard curve is constructed using test solutions containing known concentrations of TCDD, in which case the assay measures the total concentration of DLCs as TCDD-equivalents.

In some ways an assay for DLCs based on a monoclonal antibody resembles the competitive Ah receptor, ligand-binding assay. In each case a protein complexes the DLC, and in each case a reference ligand (in this case the plastic-bound hapten) and the DLCs in the test solution compete for a limited amount of protein. Because the monoclonal antibody recognizes the ligand, its binding site is a model for the binding site of the AhR. Cross-reactivity between DLCs is observed because each DLC in the test solution binds to the antibody according to its relative affinity and its concentration. In a sense, the binding site of the antibody serves as a surrogate for the AhR: how close a surrogate depends on the similarities between the binding sites, which is a matter of chance. However, the TCDD-equivalents obtained by immunoassay differ from those of the receptor binding assay, because of different relative affinities of DLCs for the two proteins. The immunoassay is not mechanism-based because the DLCs do not interact with the antibody as part of the mechanism of toxic expression.

Like the AhR binding assay, immunoassay tends to be conservative in recognizing more analytes than are toxicologically relevant. This has been noted in the case of recognition of "PCB fragments" (65), and was also observed when comparing immunoassays of several environmental matrices to analyses by GC/MS (59,66). Immunoassay consistently shows more of the "toxic" PCDDs and PCDFs than GC/MS, and in some cases the discrepancy is substantial.

EARLY LIFE STAGE BIOASSAYS

TCDD shows dose-dependent AhR-mediated teratogenicity in mice. The easiest effect to identify is cleft palate (67–70).

Recent interest in environmental estrogens and the more general phenomenon of endocrine modulation (71) has focused attention on the role of TCDD (and potentially other DLCs) in developmental and reproductive toxicology. Although the mechanisms of action are not yet clear, it is already apparent that TCDD has strong interactive effects with steroid hormones (although there is no cross-reactivity between TCDD and steroid hormone receptors). In one recent paper, TCDD blocked the binding of the estradiol-estrogen receptor complex to estrogen response elements on DNA, whereas, conversely, estradiol blocked the association of the TCDD–AhR complex to DREs (72). TCDD and other DLCs, notably 6-methyl-3,7,8-trichlorodibenzofuran, inhibit the growth of the human breast cancer cell line MCF-7 (73,74). TCDD also causes developmental toxicity in laboratory rats, causing demasculinization of male pups when administered to the pregnant dams at sublethal doses (75). These effects, which persist into maturity, may be caused by TCDD-induced elevation of the levels of MFOs, thereby enhancing the metabolism of steroid hormones, but this point is not yet settled. These recent observations have led to interest in developing laboratory test systems to test for adverse effects of DLCs in early stages of life.

This interest in developmental toxicity is not really new, because some of the earliest concerns about TCDD were "toxic fat syndrome" in young chickens in the 1960s, and the debate in the 1970s over whether forest spraying programs cause human birth defects and miscarriages. Both these situations involved the use of the herbicide 2,4,5-trichlorophenoxyacetic acid (2,4,5-T), which contained parts per million of TCDD as a contaminant introduced during manufacturing (76).

In the case of toxic fat syndrome, a characteristic finding was the development of pericardial edema in newborn chicks, whose early death caused substantial economic loss in the broiler chicken industry. This was suggested as a bioassay endpoint (77). Related work showed a dose-dependent reduction in hatchability when hens' eggs were treated with TCDD (78). Histopathological changes including hepatic edema in chick embryos were correlated with the induction of AHH and EROD activity (79), and hatchability was correlated with EROD induction, in each case after dosing the eggs with various PCB congeners (80).

The toxicity of TCDD toward fish and their embryos was noted (81,82), and an assay was developed based on the lethality of DLCs toward yolk sac fry of *Salmo gairdneri* (83). Similar assays have been introduced for rainbow trout, fathead minnow, and Japanese medaka. In more recent studies, the toxicants have been injected into the developing fish, rather than exposure through the water, to specify dose more precisely (for technique, see 84, 85). It was found that pairs of PCDD and PCB congeners behave nearly additively (no deviations greater than 2–3-fold) in an early life-stage assay in rainbow trout (86). The same group assessed that a complex PCDD/PCDF/PCB mixture in lake trout has close to the additive lethality of its components (87). LD_{50}s were in the range of 200 pg of TEQ per egg. A disadvantage of these protocols is the relatively long assay time of several weeks.

The Japanese medaka (*Oryzias latipes*) is a small fish convenient for laboratory assays because it is easily reared in captivity; it can be induced to lay eggs in the laboratory year round; and its larval envelope is transparent, permitting easy evaluation of developmental toxicity in addition to mortality. The survival of Japanese medaka embryos to three days post-hatch was studied, after exposing the embryos to the test chemicals within 1–2 h after fertilization. The LC_{50} for TCDD was 13 ng/L (40 p*M*) nominal concentration, and, as in mammalian systems, the congeners fully chlorinated in the 2, 3, 7, and 8 positions are substantially the most toxic. An AhR-mediated mechanism of toxicity is apparent because neither lethality nor morphological change is seen until after the liver forms (88), and lethality occurs in the same concentration range as AHH induction (89). In addition to synthetic DLC mixtures, it was demonstrated that lipophilic extracts from rainbow trout taken from Lake Ontario are embryotoxic to the Japanese medaka (90).

OTHER ASSAYS

An assay system was developed for TCDD based on culturing the XB keratinocyte cell line, which is derived from a mouse teratoma (91). When cultured at low cell density in the presence of lethally irradiated 3T3 fibroblasts, the XB cells replicate and differentiate to give characteristic flattened, keratinized cells identifiable by their shape and because they stain red with the dye Rhodanile Blue. At high cell density, the XB cells replicate but do not keratinize. However, low concentrations of TCDD cause the high-density cultures to keratinize in a dose-dependent fashion. Therefore the extent of keratinization is correlated with the intensity of color development when the cells are stained. Keratinization requires a minimum concentration of TCDD of 50 p*M*. Additionally, the efficacy of keratinization by a PCDD congener roughly parallels its binding affinity for the AhR, thus suggesting that this is a mechanism-based assay. It was shown that this assay is applicable to other DLCs, such as 3,3′,4,4′-tetrachloroazobenzene. A drawback to the assay is the relatively long incubation time (nearly two weeks) required; this mitigates against the use of the assay for following the progress of a remediation "on the fly".

A variant of the keratinization assay, employs a strain of XB cells denoted XBF (92). These do not keratinize in the presence of TCDD, but, during a 14-day incubation, form "cobblestone-like" cells (flat cells) that are visualized by staining. The detection limit in this assay is 10 p*M* TCDD in the culture medium. PAHs have $10^3 - 10^4 \times$ less response than TCDD. The flat cell assay correlates linearly with chemical analysis by GC/MS for a series of soot samples contaminated with PCDFs (92), it agrees semiquantitatively with chemical analysis for the DLC content of a range of fish tissue and sediment samples (93) and of several transformer fluids (94).

As a related endpoint, one of the most characteristic and well documented (95) responses of halogenated aromatic compounds in humans is chloracne, caused by hyperkeratosis of the squamous skin cells. This endpoint is also seen in experimental animals, such as rabbits and hairless mice. Indeed, an early example of a bioassay for DLCs is the production of chloracne in the rabbit pinna (96).

One of the characteristic responses of DLCs is an increase in cell proliferation, which is linked mechanistically to a change in the binding between the epidermal growth factor and its receptor (97,98). Cancers induced by TCDD also are principally those of epithelial cells (91,97).

In addition to the above responses, acute exposure of laboratory animals to lethal doses of DLCs cause death via a wasting syndrome in which animals waste away over a period of several days to weeks (13,20).

The effect of DLCs on the immune systems of intact animals has been known for many years; a dose-dependent reduction in thymic weight in rodents is observed. This was shown to be an Ah receptor-mediated response by correlating the immune activities of different DLCs with their ability to induce mixed-function oxidase enzymes (99,100). Numerous interactive effects on these endpoints involving TCDD and PCB congeners are reported (100,101,102,103,104), yielding evidence that certain PCB congeners mask the activity of TCDD. Mice were treated with TCDD and PCB 153 (2,2′,4,4′,5,5′-hexachlorobiphenyl), alone or in combination, and both thymus weights and the ability of their spleen cells to form plaques upon treatment with sheep red blood cells were examined (105). PCB 153 alone has little effect on either endpoint, but, depending on its dose, it enhances or suppresses the responses of TCDD. Such complex interactive effects make it unlikely that a broadly useful bioassay for DLCs can be developed, using these endpoints.

CONCLUSIONS

Present regulations establish allowable concentrations of dioxin-like compounds either on the concentration of TCDD or the total TEQ of named congeners, in each case based on congener-specific analysis. Under these circumstances, surrogate assays can never replace an absolute analytical method, such as GC/MS, because they do not provide chemical identification. However, an inexpensive and rapid surrogate assay usefully acts as a screening tool based on a "cut-off" TEQ, to eliminate negative samples from further analysis. This greatly reduces costs, especially if most samples are expected to have low levels of contamination and the assay has a low rate of false positives.

We have noted that AhR-binding assays and immunoassays have high false positive rates (a conservative feature) because they respond to substances that bind to the receptor (or antibody) without causing toxicity. A corresponding benefit is that different DLCs behave additively and do not mask each other's response. With their inherent conservatism, this means that the analyst may be confident that a sample assaying below the cut-off level is genuinely free of contamination.

A critical limitation occurs when screening assays have a significant rate of false negatives, which imply incorrectly that a contaminated sample is "clean". Enzyme induction and porphyrin accumulation are examples of assays in which one DLC masks (antagonizes) the dioxin-like activity of another, and the analyst cannot be sure that the TEQ obtained from the assay truly reflects the TEQ of the sample. Such assays are of limited value in screening programs unless masking analytes are independently known to be absent, although they may remain viable options to monitor the progress of a remediation project.

The question of biological relevance always arises with an assay based on a biological principle. Whereas human health concerns might be addressed by an enzyme induction assay, preferably using a human cell culture, protection of fish in an aquatic environment suggests an early life-stage assay to monitor the potential for reproductive or developmental abnormalities. It is unrealistic to imagine that a single procedure will apply to all situations.

BIBLIOGRAPHY

1. US EPA Office of Water, Method 1613, Revision B, EPA 821-B-94-005, Washington, D.C., Oct. 1994.
2. D. Firestone, *J. Assoc. Off. Anal. Chem.* **74,** 375–384 (1991).

3. NATO-CCMS Report No. 176, *International Toxicity Equivalency Factor (I-TEF) Method of Risk Assessment for Complex Mixtures of Dioxins and Related Compounds,* 1988.
4. S. Safe, *CRC Crit. Rev. Toxicol.* **21,** 51–88 (1990).
5. S. Tanabe, N. Kannan, An. Subramanian, S. Watanabe, and R. Tatsukawa, *Environ. Pollut.* **47,** 147–163 (1987).
6. D. L. Stalling, R. J. Norstrom, L. M. Smith, and M. Simon, *Chemosphere* **14,** 627–643 (1985).
7. R. M. Putzrath, *Organohalog. Comp.* **25,** 111–115 (1995).
8. B. Lesnik, *Organohalog. Comp.* **23,** 203–208 (1995).
9. M. Bader, *J. Chem. Educ.* **57,** 703 (1980).
10. T. Zacharewski, L. Safe, and S. Safe, *Environ. Sci. Technol.* **23,** 730–735 (1989).
11. K. Hu, N. J. Bunce, B. G. Chittim, C. H. M. Tashiro, B. R. Yeo, B. J. Sharrat, F. J. Campbell, and D. W. Potter, *Environ. Sci. Technol.* **29,** 2603–2609 (1995).
12. E. K. Silbergeld, T. A. Gasiewicz, *Am. J. Indust. Med.* **16,** 455–474 (1989).
13. K. Rozman, *Exp. Toxic. Pathol.* **44,** 473–480 (1992).
14. J. Höckel, L. Dusterhoft, W. Korner, and H. Hagenmaier, *Organohalog. Comp.* **23,** 135–138 (1995).
15. S. M. G. Schwirzer, F. J. Wiebel and K.–W. Schramm, *Organohalog. Comp.* **23,** 85–88 (1995).
16. R. J. Kociba, D. G. Keyes, J. E. Beyer, R. M. Carreon, C. E. Wade, D. A. Dittenber, R. P. Kalnins, L. E. Frauson, C. N. Park, S. D. Barnard, R. A. Hummel, and C. G. Humiston, *Toxicol. Appl. Pharmacol.* **46,** 279–303 (1978).
17. E. S. Johnson, *Environ. Health Perspect.* **99,** 383–390 (1993).
18. D. J. Paustenbach, M. W. Layard, R. J. Wenning, and R. E. Keenan, *J. Toxicol. Environ. Health* **34,** 11–26 (1991).
19. R. L. Sielken, Jr., *Food Chem. Toxicol.* **25,** 257–267 (1987).
20. J. P. Landers and N. J. Bunce, *Biochem. J.* **276,** 273–287 (1991).
21. J. P. Whitlock, Jr., *Chem. Res. Toxicol.* **6,** 754–763 (1993).
22. A. B. Okey, D. S. Riddick, and P. A. Harper, *Toxicol. Lett.* **70,** 1–22 (1994).
23. H. Fiedler, O. Hutzinger, and C. W. Timms, *Toxicol. Environ. Chem.* **29,** 157–234 (1990).
24. S. D. Gilman, J. R. Sanborn, S. J. Gee, M. S. Denison, L. H. Stanker, A. D. Jones, and B. D. Hammock, *Organohalog. Comp.* **23,** 231–235 (1995).
25. A. Poland, and J. C. Knutson, *Ann. Rev. Pharmacol. Toxicol.* **22,** 517–554 (1982).
26. L. Marple, R. Brunck, and L. Throop, *Environ. Sci. Technol.* **20,** 180–182 (1988).
27. A. Poland, and E. Glover, *Biochem. Biophys. Res. Commun.* **146,** 1439–1449 (1987).
28. A. Wilhelmsson, S. Cuthill, M. Denis, A.–C. Wikström, J.–Å. Gustafsson, and L. Poellinger, *EMBO J.* **9,** 69–76 (1990).
29. I. Pongratz, G. G. F. Mason, and L. Poellinger, *J. Biol. Chem.* **267,** 13728–13734 (1992).
30. E. Enan and F. Matsumura, *Biochem. Pharmacol.* **52,** 1599–1612 (1996).
31. C. A. Bradfield, A. S. Kende, and A. Poland, *Mol. Pharmacol.* **34,** 229–237 (1988).
32. N. J. Bunce, J. P. Landers, and S. H. Safe, *Arch. Biochem. Biophys.* **267,** 384–397 (1988).
33. A. B. Okey, L. M. Vella, and P. A. Harper, *Mol. Pharmacol.* **35,** 823–830 (1989).
34. A. Poland, D. Palen, and E. Glover, *Mol. Pharmacol.* **46,** 915–921 (1994).
35. D. K. Manchester, S. K. Gordon, C. L. Golas, E. A. Roberts, and A. B. Okey, *Cancer Res.* **47,** 4861–4868 (1987).
36. P. A. Harper, R. D. Prokipcak, L. E. Bush, C. L. Golas, and A. B. Okey, *Arch. Biochem. Biophys.* **290,** 27–36 (1991).
37. A. Lorenzen, and A. B. Okey, *Toxicol. Appl. Pharmacol.* **107,** 203–214 (1991).
38. A. Poland, W. F. Greenlee, and A. S. Kende, *Annals of the New York Academy of Sciences* **320,** 214–230 (1979).
39. S. H. Safe, *Ann. Rev. Pharmacol. Toxicol.* **26,** 371–399 (1986).
40. L. F. Bjeldanes, J.–Y. Kim, K. R. Grose, J. C. Bartholomew, and C. A. Bradfield, *Proc. Natl. Acad. Sci. USA* **88,** 9543–9547 (1991).
41. D. S. Riddick, P. A. Harper, and A. B. Okey, *FASEB J.* **6,** A1201 (abstr.) (1992).
42. E. C. Hoffman, H. Reyes, F.–F. Chu, F. Sander, L. H. Conley, B. A. Brooks, and O. Hankinson, *Science* **252,** 954–958 (1991).
43. R. S. Pollenz, C. A. Sattler, and A. Poland, *Mol. Pharmacol.* **45,** 428–438 (1994).
44. H. Reyes, S. Reisz–Porszasz, and O. Hankinson, *Science* **256,** 1193–1195 (1992).
45. E. F. Yao and M. S. Denison, *Biochemistry* **31,** 5060–5067 (1992).
46. L. Wu and J. P. Whitlock, Jr., *Nucl. Acid Res.* **21,** 119–125 (1993).
47. S. Safe, *Organohalog. Comp.* **26,** 7–13 (1995).
48. I. Chen, N. Harper, and S. Safe, *Organohalog. Comp.* **25,** 57–60 (1995).
49. M. A. Denomme, S. Bandiera, I. Lambert, L. Copp, L. Safe, and S. Safe, *Biochem. Pharmacol.* **32,** 2955–2963 (1983).
50. M. J. DeVito, W. E. Maier, J. J. Diliberto, and L. S. Birnbaum, *Fundam. Appl. Toxicol.* **20,** 125–130 (1993).
51. R. A. Lubet, R. T. Mayer, J. W. Cameron, R. W. Nims, M. D. Burke, T. Wolff, and F. P. Guengerich, *Arch. Biochem. Biophys.* **238,** 43–48 (1985).
52. F. P. Guengerich, G. A. Dannan, S. T. Wright, M. V. Martin, and L. S. Kaminsky, *Biochemistry* **21,** 6019–6030 (1982).
53. A. Poland, and A. Kende, *In Origins of Human Cancer (Book B): Mechanisms of Carcinogenesis;* Hiatt, H. H., Watson, J. D., Winsten, J. A., eds., Cold Spring Harbor Laboratory, Cold Spring Harbor, N.Y., 1977, pp 847–867.
54. T. W. Sawyer, and S. Safe, *Chemosphere* **14,** 79–84 (1985).
55. H. L. Bonkovsky, P. R. Sinclair, W. J. Bement, R. W. Lambrecht, and J. F. Sinclair, *Annals of the New York Academy of Sciences* **514,** 96–112 (1987).
56. K. G. Jones, and G. D. Sweeney, *Toxicol. Appl. Pharmacol.* **53,** 42–49 (1980).
57. J. P. Sherry, *Crit. Rev. Anal. Chem.* **23,** 217–300 (1992).
58. G. H. Perdew, B. Abbott, and L. H. Stanker, *Hybridoma* **14,** 279–283 (1995).
59. L. H. Stanker, B. Watkins, N. Rogers, and M. Vanderlaan, *Toxicol.* **45,** 229–243 (1987).
60. P. W. Albro, M. I. Luster, S. K. Chaudhary, G. Clark, L. D. Lawson, J. T. Corbett, and J. D. McKinney, *Toxicol. Appl. Pharmacol.* **50,** 137–146 (1979).
61. M. I. Luster, P. W. Albro, K. Chae, L. D. Lawson, J. T. Corbett, and J. D. McKinney, *Anal. Chem.* **52,** 1497–1500 (1980).
62. J. Sherry, *Organohalog. Comp.* **23,** 181–186 (1995).
63. S. J. Kennel, C. Jason, P. W. Albro, G. Mason, and S. H. Safe, *Toxicol. Appl. Pharmacol.* **82,** 256–263 (1986).
64. R. O. Harrison, R. E. Carlson, and H. Shirkhan, *Organohalog. Comp.* **23,** 187–192 (1995).
65. R. E. Carlson, Y.–W. Chiu, K. L. Marcus, and A. E. Karu, *Organohalog. Comp.* **23,** 193–198 (1995).

66. M. Vanderlaan, L. H. Stanker, B. E. Watkins, P. Petrovic, and S. Gorbach, *Environ. Toxicol. Chem.* **7,** 859–870 (1988).
67. L. S. Birnbaum, H. Weber, M. W. Harris, J. C. Lamb, IV, and J. D. McKinney, *Toxicol. Appl. Pharmacol.* **77,** 292–302 (1985).
68. R. Bannister, and S. Safe, *Toxicol.* **44,** 159–169 (1987).
69. J. M. Haake, S. Safe, K. Mayura, and T. D. Phillips, *Toxicol. Lett.* **38,** 299–306 (1987).
70. R. E. Morrissey, M. W. Harris, J. J. Diliberto, and L. S. Birnbaum, *Toxicol. Lett.* **60,** 19–25 (1992).
71. B. Hileman, *C&EN* **January 31,** 1994 pp. 19–23.
72. I. Kharat, and F. Saatcioglu, *J. Biol. Chem.* **271,** 10533–10537 (1996).
73. J. F. Gierthy, J. A. Bennett, L. M. Bradley, and D. S. Cutler, *Cancer Res.* **53,** 3149–3153 (1993).
74. S. Safe, B. Astroff, M. Harris, T. Zacharewski, R. Dickerson, M. Romkes, and L. Biegel, *Pharmacol. Toxicol.* **69,** 400–409 (1991).
75. D. L. Bjerke, T. J. Brown, N. J. MacLusky, R. B. Hochberg, and R. E. Peterson, *Toxicol. Appl. Pharmacol.* **127,** 258–267 (1994).
76. J. R. Long and D. J. Hanson, *C&EN* **61(23),** 23–36 (1983).
77. G. R. Higginbotham, A. Huang, D. Firestone, J. Verrett, J. Ress, and A. D. Campbell, *Nature* **220,** 702–703 (1968).
78. D. F. Flick, D. Firestone, J. Ress, and J. R. Allen, *Poultry Science* **52,** 1637–1641 (1973).
79. A. B. Rifkind, A. Firpo, Jr., and D. R. Alonso, *Toxicol. Appl. Pharmacol.* **72,** 343–354 (1984).
80. B. Brunstrom, *Arch. Toxicol.* **64,** 188–192 (1990).
81. T. Helder, *Sci. Total Environ.* **14,** 255–264 (1980).
82. T. Helder, *Toxicology* **19,** 101–112 (1981).
83. T. Helder, and W. Seinen, *Chemosphere* **14,** 183–193 (1985).
84. C. D. Metcalfe, V. W. Cairns, and J. D. Fitzsimons, *Aquat. Toxicol.* **13,** 347–356 (1988).
85. M. K. Walker, L. C. Hufnagle, M. K. Clayton, and R. E. Peterson, *Aquat. Toxicol.* **22,** 15–38 (1992).
86. E. W. Zabel, M. K. Walker, M. W. Hornung, M. K. Clayton, and R. E. Peterson, *Toxicol. Appl. Pharmacol.* **134,** 204–213 (1995).
87. M. K. Walker, P. M. Cook, B. C. Butterworth, E. W. Zabel, and R. E. Peterson, *Fundam. Appl. Toxicol.* **30,** 178–186 (1996).
88. J. D. Wisk, and K. R. Cooper, *Environ. Toxicol. Chem.* **9,** 1159–1169 (1990).
89. J. D. Wisk, and K. R. Cooper, *Arch. Toxicol.* **66,** 245–249 (1992).
90. G. E. Harris, T. L. Metcalfe, C. D. Metcalfe, and S. Y. Huestis, *Environ. Toxicol. Chem.* **13,** 1393–1403 (1994).
91. J. C. Knutson, and A. Poland, *Cell* **22,** 27–36 (1980).
92. J. F. Gierthy, and D. Crane, *Fundam. Appl. Toxicol.* **5,** 754–759 (1985).
93. J. F. Gierthy, D. W. Lincoln, P. O'Keefe, R. Smith, C. Meyer, D. Hilker, and K. Aldous, *Chemosphere* **18,** 793–800 (1989).
94. J. F. Gierthy, K. Swami, R. Narang, A. Narang, and G. Eadon, *Chemosphere* **26**(6), 1225–1235 (1996).
95. S. Safe, *CRC Crit. Rev. Toxicol.* **13,** 319–395 (1984).
96. E. Jones, and H. Krizek, *J. Invest. Dermatol.* **39,** 511–517 (1962).
97. R. Osborne, and W. F. Greenlee, *Toxicol. Appl. Pharmacol.* **77,** 434–443 (1985).
98. B. V. Madhukar, D. W. Brewster, and F. Matsumura, *Proc. Natl. Acad. Sci. USA* **81,** 7407–7411 (1984).
99. J. B. Silkworth, L. Antrim, and L. S. Kaminsky, *Toxicol. Appl. Pharmacol.* **75,** 156–165 (1984).
100. D. Davis and S. Safe, *Toxicology* **63,** 97–111 (1990).
101. L. Biegel, M. Harris, D. Davis, R. Rosengren, L. Safe, and S. Safe, *Toxicol. Appl. Pharmacol.* **97,** 561–571 (1989).
102. N. Harper, K. Connor, and S. Safe, *Toxicology* **80,** 217–227 (1993).
103. R. Bannister, D. Davis, T. Zacharewski, I. Tizard, and S. Safe, *Toxicology* **46,** 29–42 (1987).
104. D. Davis and S. Safe, *Toxicol. Lett.* **48,** 35–43 (1989).
105. R. J. Smialowicz, M. J. DeVito, M. M. Riddle, W. C. Williams, and L. S. Birnbaum, *Organohalog. Comp.* **25,** 299–302 (1995).

DISSOLVED ORGANIC CARBON. See ORGANIC CARBON, ANALYSIS METHODS FOR AQUEOUS SAMPLES.

DRILLING WASTE MANAGEMENT

WILLIAM A. PIPER
Amoco Corporation
Houston, Texas

Waste management, the planning and implementation of a prudent disposal plan, has been standard in most industries for decades. There are many reasons for this: regulatory pressures have forced tighter controls; public outcries about waste disposal practices have resulted in environmental pollution and public health risks; the increasing costs of handling and disposal; and the enormous costs associated with liability for wastes and remediation of prior disposal sites. Whatever reasons provide the most impetus, the bottom line is that good waste management is a common and accepted part of good business management. Despite this widespread recognition of waste management as a business factor in most industries, the drilling industry has not embraced it as an important or critical issue for many operations. Often, management of waste from a drilling project is not a planned and programmed activity, but simply a reaction when problems occur. This reactive mode of management can increase costs and problems while leading to less than ideal handling and disposal actions.

Although waste management is standard in most industries, it is not standard in the drilling industry. In order to understand this issue, one must understand the drilling industry and the waste that is generated. The drilling industry is increasingly working in remote and sensitive areas, where very little or no infrastructure exists. Furthermore, there may be inadequate or poorly designed regulation. On top of that, there is a pervasive feeling within the industry that the wastes being generated are simply dirt and water.

A comprehensive financial plan can save considerable costs while ensuring the use of best available options for waste management in an existing waste management infrastructure (commercial disposal site). With the commercial disposal of drilling wastes costing about $10.00 per barrel (plus storage, handling, and transportation), waste disposal for a well generating 25,000 barrels of wastes can easily cost in excess of $300,000. Comprehensive planning for waste management affords the opportunity to address these costs and look at alternatives (1). In most cases, significant reductions in costs (direct from operator's viewpoint) can be realized. The commercial disposal site has the same options as the operator; it must operate prudently, otherwise it tends to concentrate and compound the problem. Stockpiling waste in a giant dump has never proven to be an effective solution.

WASTE GENERATION

Quantification

Monitoring waste generated during the drilling process is a challenge (see WASTE GENERATION). The following exercise helps illustrate this. Estimate the amount of candy contained in an irregularly shaped jar. Discussions on how to do this usually yield the possibilities of counting the number of pieces, estimating weight, or estimating volume. With more thought, the amount of calories, fat content, or value of the candy may be estimated. This exercise demonstrates that estimating waste is also an inexact art subject to inaccuracy and interpretation. Generally, all estimates of the amount are in weight, volume, or number of pieces. Volume estimates are generally low, with an accurate estimate being around 80% of the inventoried value. Number of weight estimates are generally high. An accurate weight estimate may be 175% of inventoried weight. Even in a class of experienced drilling personnel, unless they have had specific training, the range of estimates will be very high, despite the value estimated.

Extending this example into the reality of dealing with drilling waste estimation is valuable. When faced with estimating the amount of waste stored in an earthen pit, is it more valuable to estimate based on weight, volume, or some other value? How should the various components be separated into reasonable components, such as oily cuttings, water-based mud, wash water, and other components that may be in a single large pit? There is no single answer to these questions.

Sources and Examples

Waste generation on a drilling rig comes from a variety of sources. Because these sources are directly related to the process, it is important to understand the process involved. The following can serve as a guide for categorizing wastes.

Hole-Related Wastes. Drilling is the act of building an oil and gas well into the ground. This is done by drilling concentric holes, getting progressively smaller, deep into the earth. Each hole section size is terminated when a technical condition is met. These conditions are determined by pressure limitations, hole wall stability, or some other technical criteria. When this predetermined criteria is met, drilling is stopped and a steel casing is inserted into the hole and cemented in place. This protects the previously drilled hole, but reduces the next hole size. Ultimately, the last hole size is determined by the objectives of the well. It is preferable to finish the well with a large enough hole to install the production equipment. In some cases, the last hole section must be evaluated with electric logging tools. Because this equipment has certain hole size requirements, each hole size above it must be large enough to accommodate the required internal hole size. At this time, the most common final hole size is ~21.5 cm (8.5 in.), which accommodates a 17 cm (7 in.) production casing and the internal associated production equipment. It also accommodates full evaluation by electric logging tools. Because a hole created in the earth is determined by hole size, the volume of earth that should have been removed from the hole is easily calculated.

The hole is drilled into the earth using a drilling bit connected to and driven by a powerful motor at the surface. The connecting tubes are hollow to allow fluid to be pumped down to the bit, internally flushing the earth to be removed externally to the surface. The connecting tubes (drill string), the space between them and the wall of the hole (annulus), the fluid removing the earth (mud), and the earth being removed (cuttings) are important drilling terms. When the mud, laden with cuttings, reaches the surface, an attempt to remove the cuttings is made by the solids control system. The circulating system starts with mud stored in large tanks at the surface, the mud is pumped down-hole (by a large pump) inside the drill string, returns up-hole in the annulus, and is returned to the large tanks at the surface. It is desirable to remove all of the cuttings from the mud before recovering all of the mud to the tanks, but this is never possible. This is because of two factors: (*1*) mud adheres to the cuttings and is removed with the cuttings; (*2*) some cuttings are too fine to be removed by the equipment that makes up the solids control system. This is a simplistic description of a complex and technical subject. Both mud and solids control are major areas of expertise within the drilling specialty.

This means that there are two waste streams generated from the drilling process: *solids* are the cuttings and residual mud associated with them; *liquids* comprise the whole dirty mud discarded as a result of the solids control system being unable to remove a sufficient amount of cuttings. The solids waste stream is related to the volume of dirt removed from the hole, ie, the amount can usually be estimated as twice the gauge (or calculated) hole volume. This can be explained in the following manner. In hard rock drilling, cuttings are generally larger when they get to the surface and can be removed more easily by the solids control system. This means that the absolute volume of rock removed approaches the calculated hole volume. Likewise with larger cuttings, the surface area is smaller than with the same amount of smaller cuttings, which means that the amount of residual mud clinging to the cuttings is minimized. It is generally accepted that the amount of residual mud clinging to the cuttings is about equal in volume, or a ratio of 1:1 by volume. The additive effect of these volumes means that the total volume is about twice the calculated hole volume. This is verified in cases where cuttings have been collected in containers and removed from the drilling site. In softer rock drilling, the efficiency of the solids control system goes down due to the smaller size of the cuttings arriving at the surface. However, the surface area of the cuttings is larger creating a larger amount of residual mud. The net effect is that the solids portion of the drilling process-generated waste can be estimated as about twice the calculated hole volume with some degree of accuracy. This solids portion is generally handled as a solid despite the large amount of associated liquid (ie, it is handled with a scooping device).

The liquids volume generated with the drilling process is much more complex to estimate. There is a relationship to hole volume and hardness of drilling conditions. Hard and soft, as applied to drilling, are really related terms affected by many different factors including the mud. In general terms, in hard rock drilling conditions, the efficiency of the solids control system is high enough that little dirty mud has to be discarded. Hence, the liquids waste generation is low. On the other hand, in soft rock drilling conditions, the efficiency is fairly low, so liquids generation rate is high. Experience factors indicate that for soft rock drilling conditions the liquids generation rate is from six to eight times hole volume, whereas for hard rock conditions the liquids generation rate is from one to three times hole volume. When drilling with an extremely inhibitive fluid, such as oil-based mud, soft rock drilling conditions may generate liquids waste as in hard rock drilling conditions. In reality, liquids waste generation is difficult to estimate without an experience factor based on the area, but the above numbers can be used for planning purposes.

Impact of Collection Method. Because large volumes of liquids are generated on-shore, they are typically stored in an earthen pit. In fact, in remote areas, due to the difficulty of transporting large volumes to appropriate disposal sites, earthen pits are generally required and preferred. Because of the convenience of the pit, most times other process liquids are stored there as well. All of these sources of liquid may be referred to as drilling process liquid wastes. Other sources include cooling water, wash water, and storm water, which are potential major sources of liquid waste. These waste sources are related to time (days) operating on the well site.

Storm water volume can be estimated by the amount of time on the site, the amount of area enclosed, and the amount of expected rainfall. Generally, the entire drilling site is enclosed by ditches or berms in order to prevent potential runoff from the location. In addition, the area encompassed by the pits must be included in the estimate. The amount of wash water and cooling water can be estimated by the amount of time operating and daily average usage. On a rig without any controls for the amount of water used, this amount may be $\geq$160 m^3 per day; with reasonable control, the amount can be reduced to 50 m^3 (it has been reduced to about 10 m^3). Knowledge of drilling crew experience and their attention to this waste stream is required for a better estimate.

Earthen pits cannot be used off-shore. To answer the question of how wash water and storm water runoff is controlled, some knowledge of the off-shore drilling vessel is required. As with all marine vessels, drilling vessels have internal bilge tanks. Internal compartments and external areas not subject to heavy mud spillage can be drained to the vessel's bilge tanks. When drainage control is necessary, areas around the rig that are subject to heavy mud spillage must be drained, or pumped to a separate tank. Catch pans must be placed under areas normally open to the sea. All holes must be welded shut. Under normal circumstances, this is not done and drainage is directly to the sea. When drainage is directed to the sea, it is not possible to estimate the amount. When the rig is sealed, most drilling personnel are surprised at how much water is collected. It seems appropriate to collect the wash water and storm water runoff when using toxic or oil muds. It is, however, probably correct to allow wash water and storm water runoff to go directly to the sea when mud and cuttings are being disposed there. Other reasons for collecting drainage from the rig include possible damage to reefs or other sensitive environments. Other conditions that make using earthen reserve pits undesirable include arctic environments or very salty muds. In these cases special precautions must be taken to safely collect wash water and storm water runoff that might be contaminated.

Trash. Industrial trash is generated on a drilling rig as with any other industrial process. In 1995, a large land rig in Romania generated an average of 636 kg of trash per day. A jack-up rig off-shore in the North Sea generated 300 kg per day. Neither rig used compactors, but the land rig in Romania had more significant problems dealing with the large amount of trash.

Air Emissions. Air emissions are a good example of where a secondary source of monitoring can be used to estimate emissions. Fuel is tracked on every drilling rig. Although it is difficult to monitor the exhaust emissions directly, it is routine to monitor fuel usage. Correlations between fuel usage and exhaust emissions can be readily found in the literature (2). It is interesting to note that fuel usage and thus air emissions are basically dependent on two factors: rig size and number of operating days. The largest daily fuel usage is from a large rig operating in a remote location. Fuel usage is probably high because a large number of support vehicles (mainly trucks and earth-moving equipment) were being fueled from the same storage source as the drilling rig.

Well Testing. After spending millions of dollars drilling a well, significant testing is desired to give information to help a company decide whether or not further exploration/exploitation should occur. In remote locations, this often means flaring the gas, condensate, or oil produced during the test, since transporting large amounts of live oil poses serious safety risks. With environmental groups demanding less and less impact on the surrounding terrain, the drilling location size is often minimized. More often there is a conflict between minimizing the location size and flaring large amounts of hydrocarbons. The basic problem is how to flare a well with a minimum amount of area affected by the radiated heat. This same process applies to flare pit location. The flare pit, while only to be used during emergency flaring operations, must be preplanned and placed a safe distance from the rig. As an example, consider an extended well test in the middle of an isolated jungle. The purpose of the extended test is to evaluate the quality of the reservoir. The produced fluids cannot be stored or transported safely, since there is no infrastructure in the area. Thus, flaring is required.

Cleanup fluids are returned to the test pit through a 17-cm open-ended casing. The initial cleanup fluids anticipated are mud, water, and sand. Cleanup is expected to last about 24 hours with a gradual increase to the maximum 397 m^3/d (2500 bbl/d) of condensate and 70 million standard ft^3/d (MMSCFD; = 2 million m^3) of gas. As cleanup changes from the initial conditions to maximum production, the flow is gradually moved from the test pit to a flow separator. From the separator, the gas and condensate are directed to a burner (or combination of burners) and the condensate is atomized to facilitate complete combustion. Significant radiant heat is generated by this flame. Radiated heat is fairly well understood and can be modeled with accuracy (3). Radiated heat is described in Btu/h/ft^2 (MJ/h/m^2) and is often depicted on isopleth charts.

These values indicate that the "safe" area is outside the 500 Btu/h/ft^2 isopleth. Workers are only able to work intermittently inside the 1500 Btu/h/ft^2 isopleth. For all practical purposes, everything inside the 3000 Btu/h/ft^2 isopleth would be burned or charred. For the test described, the safe area is almost 122 m (400 ft) in front of the flare, over 60 m (200 ft) behind the flare, and about 91 m (300 ft) to the side of the flare. Equipment should be located 60 m (200 ft) away to the side, 76 m (250 ft) away to the front, and about 30 m (100 ft) away behind the flare in order to prevent damage.

After addressing the safety concerns, the emissions can be examined. Alternate flow and shut-in periods may last a week or two. The concentration of combustion products will probably not be a significant concern because the area is remote. The main concern is with incomplete combustion products. Onshore, the burner head is usually directed at a pit or bermed area that catches unburned material. As it collects, it generally burns on the ground as the heat rises. Ultimately, other than the large area required to control the material and the large amount of material burned, there is no long-term effect. Off-shore, however, incomplete combustion products cannot be collected. The burner head is extended away from the rig, high above the water. The dispersion area for the unburned material is probably quite large. Nearer shore, on smaller rigs that are much closer to the water, there may also be some localized heat effect on the water. These areas currently do not have adequate solutions.

Sewage. Sewage can also be estimated from water usage, as long as the usage for the camp is separate from that used in the drilling process. Usually, the camp uses potable water for all living quarter functions. The drilling process uses good quality (drill) water, but not necessarily potable water. When using this technique, the term sewage is applied to a combined stream of "black" and "gray" water. Occasionally, barges or off-shore vessels use seawater (or bay water) for toilet flushing, but this is not a large quantity in comparison to the rest. There is a concern with using seawater in the sewage system. To prevent rapid corrosion of the sewage system, the contractor must flush large quantities of corrosion inhibitor through the system, which is obviously not beneficial to the environment. The desire, in this case, is not seawater usage, but the amount of chemicals used. Many contractors have therefore decided that the problems with using seawater outweigh the benefits. Usage rates are given in volumetric terms per day, whereas usage is dependent on the number of people using the water, which varies daily. Typically, the number of people on a rig is around 50 people, but can be as high as 100, which means that daily usage may be as high as 20 m^3 per day. Average usage is much lower. Water usage rate data from land-based U.S. rigs cannot be used because no camps are associated with them.

Special Considerations. Some items do not fit into a discussion of typical drilling wastes. These include thread compounds and completion brines, both of which need to be monitored along with other wastes that are generated.

MINIMIZATION TECHNIQUES

Minimization efforts, in order to avoid waste generation, are of major concern on a drilling location. As with any industrial plant, drilling generates certain waste streams common to all processes. These waste streams, like sewage, trash, etc, require minimization efforts, but the main thrust of this article is on drilling-specific waste streams. These drilling-related waste streams are simplified by grouping them into three major categories. (*1*) The mud/hole related category is the cuttings, mud, cement, etc, associated with circulating down the hole. This category is further subdivided into two parts: solid and liquid. Solids are the cuttings and associated mud that are dry enough to handle as a solid (with a scoop); liquids are the muds dumped for dilution, cement, etc, that are handled with a pump. This is a breakdown by handling characteristic only. The liquids may have a large amount of solids in them and the solids may have a large amount of liquid in them. (*2*) Storm water runoff is rainfall that is collected on the drilling location. For a land job, this water is usually collected in ditches inside a perimeter berm where all of the equipment is inside the berm. In coastal waters or swamps, this may include several flat-deck barges used for storing equipment or chemicals. (*3*) Rig use water is all of the water used for washing, cooling, and lubricating around the rig.

A good approach to waste minimization is to involve the entire drilling team, ie, the contractor's crews may need to get involved. Frequently, this means providing training, but it is helpful to have the same crews from well to well. A monitoring function be performed to establish goals and measure performance. Field environmental supervision is highly recommended. Often, the intended rewards become disincentives if they are not linked to realistic results or do not recognize the appropriate party.

Some techniques that have been used successfully for minimizing waste in the rig-use water category are:

- Recycle all cooling water used on the rig
- Reuse wash water until it becomes too dirty
- Use rain water for washing or cooling
- Use "pistol grips" on hoses for washing
- Use vacuums for cleanup rather than washing

Most of these are generally accepted practices (4). Minimization is not an excuse for a dirty rig; cleanliness is not discouraged and safety is not sacrificed. Although it is generally true that a clean rig is a safe rig, these are ways that the rig can be cleaned without generating excessive amounts of waste. In numeric terms, the amount of waste generation by rig-use water can be estimated. A rig with no concern for water conservation might use 160 m^3/d (1000 bbls/d), which can fairly easily be reduced to 48 m^3 (300 bbls/d) and, with strict control, be further reduced to $\sim$12 m^3 (75 bbls/d). A scale based on these numbers to reward rig crews has been used. Regarding rainfall, trapped storm water is linked to location size, and although the amount of rainfall cannot be controlled, the location size can be minimized. Another way to minimize waste from storm water is to reuse the water elsewhere, such as for washing the rig, cooling the pump rods, or for mud makeup. In Louisiana, an area with heavy rainfall, some water wells and fresh water pits have been eliminated by collecting and reusing rain water. On a two hectare location, a half meter of rain during the drilling operation translates to over 10,000 m^3 (60,000 bbls) of water. This should be viewed as a resource rather than a disposal problem. Storm water, falling on an area external to the location, should be diverted around the location. This is a particularly important point in hilly or mountainous areas.

Waste generation from mud and cuttings is related to hole volume, and less importantly, to time. Inhibitive muds, such as oil-based mud, can be used to significantly reduce the amount of dumping of whole dirty mud. This is the most significant and viable alternative available to the drilling engineer. The volume of waste generation from mud and cuttings (mud/hole related) can be minimized by:

- Minimizing hole size
- Selecting inhibitive muds where needed
- Using adequate solids control
- Minimizing excess waste from related activities, such as cementing

An interesting side effect to "no discharge rules" is that companies have reverted to oil-based mud. Because they can no longer discharge cuttings from any type of mud system, it was decided to use the most inhibitive system available (this also works well with polycrystalline diamond cutter (PDC) bit to provide fast drilling benefits). This procedure minimized the amount of waste, but increased the potential contamination level and may not have been the desired effect of the regulations. What then are the results of applying these waste minimization techniques?

One concept used to determine whether or not a good job of conservation has been done is to compare waste generation to hole volume, although it is more sophisticated to segregate the sources in order to compare those related to hole volume and those related to time, and the data generated can be used to forecast where effort is needed in the future. Even from a simplistic view, however, it is useful to compare waste-to-hole volume.

DISPOSAL TECHNIQUES

Air emissions, trash, and sewage are waste streams common to all industrial operations. Air emissions from the diesel engines of drilling rigs are generally released. Due to the intermittent and short duration operations involved in drilling, these are not generally considered to be major contributors to air pollution. A drilling rig operates under peak load only occasionally. Further, it only operates in one location for a short period of time and then is moved. Most rigs are fairly small; a large rig is rated at 2.2–2.6 megawatts (3000–3500 HP). Thus, by comparison to large factories or small towns, the loading in the air with pollutants is fairly minimal. Where there is concern, alternatives seem to involve switching the fuel source (eg, to natural gas) or utilizing existing electrical power, which transfers the source elsewhere. Trash is usually collected in containers at the rig site and taken for disposal in nearby municipal landfills (qv). The trash is handled in the same way as is the trash from towns or other industrial sites. As in municipalities, segregation of trash, minimization, recycling, and reuse are being explored. Used motor oil is recycled, as is scrap metal and plastic. Larger containers are being used to reduce the amount of waste. However, this is still a challenge in very remote areas where, compactors should be used to minimize the bulk, then the materials should be returned to the staging area. Sewage may be handled by a small portable sewage unit. These units generally dilute the waste, biodegrade it via aeration, then disinfect it with chlorine prior to releasing it. In most areas, however, the sewage is simply hauled to a municipal treatment facility. Again, the link into existing infrastructure is important.

Unlike the above wastes, drilling process-generated wastes, both solids and liquids, are not generally handled by municipalities or industrial processes. When dealing with the large amounts of relatively benign solids and liquids generated during the drilling process, it must be understood that all disposal ultimately is placed on the land, in the water, or is put into the earth. The trick to responsible disposal is to have negligible impact on the surroundings, ie, criteria must be developed that allow a distinction to be made between what is "good" and what is "bad". This requires a knowledge of the character and amount of waste, as well as a knowledge of the environmental receptor and the interaction between the two. The disposal technique should also have some kind of control placed on it so that quality is assured and this control should be documented. It is also recommended that the disposal process be permitted so that regulatory agencies have some recourse against violators or sloppy companies rather than banning the disposal practice for all companies. Five major disposal methods are currently being used and each has certain criteria that can be used to ensure limited impact by the disposal: (*1*) evaporation and burial, (*2*) treatment and discharge, (*3*) injection, (*4*) land farming, and (*5*) direct discharge (off-shore). In addition there are exotic techniques being used where some form of treatment aimed at removing a given contaminant is employed prior to disposal.

LOCATION PREPARATION CONSIDERATIONS

There has recently been a recognition that location preparation has an impact on the environment. Issues like total area impacted, how to manage pits or drill without pits, how to drill wells in swampy or submerged areas, and how to build locations in hilly, jungle terrains are topics concerning authorities who must issue permits. Drilling is the relatively short-term, temporary act of constructing a production well. In the exploratory mode, establishing a well is not guaranteed. This is an important concept and has direct bearing on two aspects of the environmental design: time and area. The time to drill a well is generally short, measured in months. If successful, the resulting production well is designed to last for decades. Likewise, the area required to drill a well (including all of the necessary safety considerations) is fairly large, whereas the area required for the operation of the production well is quite small. In other words, the drilling location should be designed to be as strong as required to construct the production well, however, it should be temporary, so that the pad can be reduced to the size required to operate the well. Overall, these location construction issues may be just as important to the environment as waste handling for the drilling industry.

The connection between waste management in drilling and the location preparation issues is vital. Consider that a typical drilling location may be 2 hectares in size, a size that may be required for four to six months. In addition, 3.2 km (2 mi) of road accessing the location may be required. If the well is successful, the road will be required throughout the life of the production well. Multiply this by the 20 or 40 wells required to develop a field. Add the area required for pipelines and production facilities. The 1000 hectares will be dotted with locations, criss-crossed with roads and pipelines, and impacted by the continuous activity of humans. It is, after all, the purpose of the oil and gas industry to develop natural resources found within the earth. However, if this development is done in a manner eliminating any other use of that land, then the land has been wasted. Any land should be capable of sustaining multiple uses. Thus, from a broad perspective, just the use of the land could be considered a waste issue. There are more direct links between site construction and waste. In order to access that land, trees and forest may have been removed. Timber or scrub disposal is directly linked with waste disposal. While the reference has been to land operations, many operations occur in the water, either sea or shallow water. Water operations are usually performed by special drilling rigs that can float to the location, however in very shallow waters, the vessels may not be able to float. Thus, access and location in shallow waters may need to be dredged in order to position a rig over the well site. Dredge spoils are directly linked to waste management. An issue relating directly to waste management is the final site clearance. When the well is no longer needed for production of the field, it is permanently abandoned. At this time, the well-site should be returned to either its native state or some other desired use. In doing this, waste will be generated and this waste will require disposal.

Decommissioning (sometimes referred to as abandonment, site clearance, or restoration) occurs years after the drilling of the well. However, if the drilling site is to be constructed in such a way as to minimize waste from site clearance, then the process of site restoration must be understood. The goal of site restoration is to restore the site to its original condition or prepare it to be suitable for another purpose. For instance, if the site was originally forested land, but is now surrounded by farmland, then the restoration could include either reforestation or creation of farmland to be acceptable. If the land is not owned by the oil company, then the private land owner will probably make that choice. The general process of restoration is to remove all production equipment, plug the well, and to remove all vestiges that made the pad a firm, level pad for drilling. The E&P Forum has guidelines on how to plan and execute a program of restoring a site (5) (see ACID MINE DRAINAGE).

BIBLIOGRAPHY

1. *Developing Area-Specific Waste Management Plans for Exploration and Production Operations Participants Manual,* American Petroleum Institute, June 15, 1991.
2. EPA Ap-42 "Supplement E to Compilation of Air Pollutant Emission Factors," Vol. 1, *Stationary Point and Area Sources*, U.S. EPA, Washington, D.C., Jan. 1995.
3. D. Drysdale, *An Introduction to Fire Dynamics,* John Wiley & Sons, Ltd., Chichester, U.K., 1985.
4. J. T. Cline and W. A. Piper, "Drilling Waste Controls," SPE paper 27162, presented at the *Second International Conference on Health, Safety, & Environment,* Jakarta, Indonesia, Jan. 1994.
5. *Decommissioning, Remediation and Reclamation Guidelines for Onshore Exploration and Production Sites,* No. 2.70/242, E&P Forum, London, Oct. 1996.

DRINKING WATER, BACTERIAL PATHOGENS IN.

See PATHOGENS IN DRINKING WATER.

DRINKING WATER DISINFECTION BY-PRODUCTS

SUSAN D. RICHARDSON
U.S. Environmental Protection Agency
Athens, Georgia

Although the importance of purifying water has been recognized since the time of the ancient Egyptians, thousands of years ago, waterborne diseases, such as cholera and typhoid, continued to cause illness and death on a significant scale in industrialized nations until the early twentieth century. Microorganisms visible only under the microscope were the cause. Contaminated drinking water was first linked to human health in 1855, when an outbreak of cholera in London was linked to the consumption of polluted water. When chlorine was introduced combined with filtration for disinfecting drinking water in the early 1900s, the incidence of waterborne diseases dramatically declined. Although chlorine has its drawbacks the benefits gained from its ability to kill harmful microorganisms cannot be overemphasized.

Over the last hundred years, many lives have been saved, and the quality of life has improved through the use of chlorine for maintaining the safety of our drinking water. A recent outbreak in Milwaukee (April 1993) was a dramatic reminder of what can happen when our water is not properly treated. In that outbreak, approximately 400,000 people became very ill, and more than 100 people died when a microorganism called *Cryptosporidium* (which is resistant to chlorine) was not properly removed from the drinking water by filtration. Therefore, removing or inactivating infectious microorganisms remains a primary concern. At the same time, however, the chemical by-products produced from chlorine and other disinfectants are not completely understood, and the risk of these chemicals to human health remains a question.

ALTERNATIVES TO CHLORINE

Although much benefit has been gained over the last hundred years by using chlorine to disinfect drinking water, there are alternatives. The most popular of these alternative disinfectants are ozone, chlorine dioxide, and chloramine. Their biocidal efficiencies (ability to kill microorganisms) compare to chlorine in the following manner: ozone > chlorine dioxide > chlorine > chloramine. Thus, ozone is the most effective biocide, with chlorine dioxide better than chlorine, and chloramine less effective than chlorine. Although these alternative disinfectants are relatively new to water treatment plants in the United States, they have been used more extensively in Europe for decades. Ozone was first used in Oudshoorn, the Netherlands, in 1893 (1). Later, ozonation plants were constructed in Paris, France (1898), Wiesbaden, Germany (1901), Paderborn, Germany (1902), St. Petersburg, Russia (1905), Nice, France (1906), and Madrid, Spain (1910) (1). Ozone was tried in the early part of the century in Niagara Falls and later in Philadelphia and then discontinued. It was not until the 1970s that ozone usage began to increase significantly in the United States (1).

The motivations for using these alternative disinfectants are very different, however. In the United States, the primary reason for using them is to limit the production of the chlorinated organic by-products formed when chlorine is used to treat drinking water. The primary concern is the trihalomethanes (THMs) produced when chlorine is used to treat drinking water. Three of them, chloroform, bromodichloromethane, and bromoform, cause cancer in laboratory animals (2). Another recent cause for concern is haloacetic acids, which also cause adverse health effects in test animals (2). Because of these concerns, the U.S. EPA established a maximum contaminant level (MCL) of 100 μg/

L for total THMs (3), which is planned to be lowered to 80 μg/L, and is establishing a new MCL for some of the haloacetic acids (4). Because some treatment plants in the United States have found it difficult to meet the MCL for THMs, they have changed their treatment method from chlorine to alternative disinfectants.

On the other hand, Europe began using alternative disinfectants long before concerns about by-products were raised. The primary reason is for taste and odor control. Ozone and chlorine dioxide eliminate compounds that produce bad tastes and odors in drinking water. Bad-tasting water is caused by pollutants and metals and also by microorganisms, such as blue-green algae, which produce tastes and odors by excreting chemicals as metabolites. Ozone and chlorine dioxide eliminate these compounds and typically do not produce malodorous compounds when used in drinking water treatment. In contrast, chlorine produces compounds (chlorinated phenols, THMs, and others) that contribute to the "medicine" or "chlorine-like" tastes sometimes associated with chlorinated water. Thus, another motivation for using alternative disinfectants is for taste and odor control.

LISTS OF BY-PRODUCTS

The following section covers many of the specific by-products produced when humic material is reacted with chlorine, ozone, chlorine dioxide, and chloramine or when these disinfectants are used in actual drinking water treatment plants. These lists of disinfection by-products (DBPs) include only those compounds believed to be actual by-products formed by reaction with the disinfectants. Therefore, the list does not include compounds identified in drinking water that were pollutants in the raw water prior to chlorination. Pesticides, such as atrazine, are examples of pollutants identified in drinking water, but not true disinfection by-products. However, some of the by-products listed are formed by the reaction of the disinfectant with a pollutant in the water. For example, it has been suggested that some of the halophenols listed as chlorine DBPs (Table 1) are formed by the reaction of chlorine with phenols already present as pollutants in raw water (5).

Identifications given in Tables 1–9 are listed as they appear in the original references, except for derivatized DBPs. When derivatives were reported, the original DBP in the treated water that gives rise to the derivative is listed in the table. These identifications were performed by many different researchers who used varying amounts of spectroscopic information; therefore, the quality of the data varies. In some cases, precise isomeric identifications were possible; in other cases, they were not. As a result, some identifications are listed without specifying the exact position of the substituent (eg, trichloro-4-oxopentenoic acid is listed without specifying the exact location of the chlorine atoms), and some identifications are listed as isomers of another compound identified (eg, isomer of 5-hydroxy-5-trichloromethyl-2-furanone). Although many of the DBPs listed in Tables 1–9 were identified by library mass spectrum matches and further spectral analyses, their identifications remain tentative until they have been confirmed by standards. Numerous chlorinated and unchlorinated compounds of unknown structure have also been reported in the literature; therefore, the lists of DBPs cannot be considered entirely complete. These compounds of unknown structure are commonly listed as "unknown" or "chlorinated compound" in the original references. As discussed previously, compounds not present in the library databases are difficult to identify, and many compounds are impossible to identify without standards, due to a lack of spectral information provided by the unknown compounds. Because no structural identifications were made for these unknown DBPs, they are not included in the tables presented here.

When known, toxicological information is given for the specific by-products identified. It must be stressed, however, that for many of the by-products (particularly those formed by the newer alternative disinfectants), much of the toxicity information is unknown. Animal toxicological experiments are generally very expensive (typically \$1.5 million per chemical tested in a cancer study); as a result, these experiments often lag behind the identification of the chemical by-products. For some chemicals, mutagenicity information (from Ames assays), is available. Although this assay does not accurately measure carcinogenicity for every chemical, it is a predictor of carcinogenicity for some classes of compounds.

CHLORINE

Chlorine was one of the first chemical disinfectants used to disinfect drinking water. It was used initially in drinking water treatment in 1884 in Germany, and first was used in the United States in the early 1900s. As mentioned earlier, the use of chlorine (combined with filtration) for disinfecting drinking water significantly decreased the incidence of waterborne diseases. Chlorine is a strong oxidant effective for killing most microorganisms. Chlorine is currently used in hundreds of treatment plants in the United States. An estimated 150 million people drink chlorinated water. Chlorine is still, by far, the most common disinfectant being used. The main reasons for this are that it is the cheapest of all chemical disinfectants, is relatively easy to use, and is highly effective.

Chlorine is applied to drinking water in a gaseous form (Cl_2), called free chlorine, or as sodium hypochlorite (NaOCl). When gaseous chlorine is added to water, it undergoes rapid and almost complete hydrolysis to form hypochlorous acid and chloride ion.

$$Cl_2 + H_2O \rightleftharpoons HOCl + H^+ + Cl^-$$

Hypochlorous acid is a weak acid, whose equilibrium is shown below.

$$HOCl \rightleftharpoons H^+ + OCl^-$$

Because the pK_a of hypochlorous acid (HOCl) is 7.5, the species present in water (which generally has a pH range of 6–8) are HOCl and the hypochlorite ion (OCl^-). HOCl is much more potent as a disinfectant than OCl^-, so the pH of the water significantly affects the disinfecting efficiency of chlorine and the distribution of by-products.

Concerns over the use of chlorine for treating drinking water surfaced in the 1970s when chloroform was discovered in most chlorinated tap water. In 1974, it was first found that the chlorination of water containing natural humic substances and inorganic bromide leads to the formation of chloroform and other chlorobrominated methanes (6). The next year, EPA published the results of a national survey, which showed that chloroform is ubiquitous in chlorinated drinking water. In 1976, the National Cancer Institute (7)

Table 1. Chlorine Disinfection By-Products[a,b]

Class of By-Products	Specific DBPs
	Halogenated DBPs
Trihalomethanes	Chloroform[c,d], bromodichloromethane[c,d], dibromochloromethane[c,d], bromoform[c,d], dichloroiodomethane[d], chlorodiiodomethane[c], iodoform[c], dibromoiodomethane[c], bromochloroiodomethane[c] (iodo-methanes not included in total THM regulation)
Other haloalkanes	Chloromethane, dichloromethane, bromochloromethane, dibromomethane[c], carbon tetrachloride[c], hexachloroethane[c,d], trichloropropane[e], 2-bromobutane
Haloalkenes	Pentachloropropene[c], tetrachlorocyclopropene, hexachlorocyclopentadiene[c,d]
Haloacetic acids	Chloroacetic acid[c,d], dichloroacetic acid[c,d], trichloroacetic acid[c,d], bromoacetic acid[d], dibromoacetic acid[d], bromochloroacetic acid[c]
Haloaromatic acids	2-Chlorobenzoic acid[c,d], 4-chlorobenzoic acid[d,e], 5-chloro-2-hydroxybenzoic acid[c], 3,5-dichlorobenzoic acid[c], isomer of 3,5-dichlorobenzoic acid[c], dichlorohydroxybenzoic acid[c], 2-chlorophenylacetic acid, 4-chlorophenylacetic acid[d,e], 2,6-dichlorophenylacetic acid[e], 2,4-dichlorophenylacetic acid[e], chlorohydroxybenzene dicarboxylic acid (4 isomers)[c],dichlorohydroxybenzene dicarboxylic acid (2 isomers)[c], chlorohydroxybenzene tricarboxylic acid (2 isomers)[c]
Other halomonocarboxylic acids	2-Chloropropanoic acid[c], 2,2-dichloropropanoic acid[c,d], 3,3,3-trichloropropanoic acid[e], isomer of 3,3,3-trichloropropanoic acid[e], 2,2-dichloro-3-hydroxypropanoic acid[e], trichloropropanoic acid (2 isomers)[c], 3,3,3-trichloro-2-hydroxypropanoic acid[c], 3,3,3-trichloro-2-hydroxy-2-methylpropanoic acid[c], 2-chlorobutanoic acid[e], 2,2-dichlorobutanoic acid[c], isomer of 2,2-dichlorobutanoic acid[c], pentachlorohydroxybutanoic acid[c], trichlorooxobutanoic acid[c], 2,4-dichloro-3-hydroxybutanoic acid[c], 4,4,4-trichlorohydoxybutanoic acid[c], 2-chloropentanoic acid[e], 2,2-dichloropentanoic acid[e], isomer of 2,2-dichloropentanoic acid[e], dichloro-4-oxopentanoic acid[c], trichloropentanoic acid[c], tetrachlorohydroxypentanoic acid[c], 3,3,5,5,5-pentachloro-4-hydroxypentanoic acid[c], 2,3,3,5,5,5-hexachloro-4-hydroxypentanoic acid[c], 5,5,5-trichloro-4-oxopentanoic acid[c,d], 2-chlorohexanoic acid[e], 2,2-dichlorohexanoic acid[c], isomer of 2,2-dichlorohexanoic acid (2 isomers)[c], tetrachlorohydroxyoxohexanoic acid[c], isomer of tetrachlorohydroxyoxohexanoic acid[c], tetrachlorooxohexanoic acid[c], pentachlorooxo-hexanoic acid, 2-chloroheptanoic acid[e]
Unsaturated halocarboxylic acids	3,3-Dichloropropenoic acid[c], *cis*-2,3-dichloropropenoic acid[c], *trans*-2,3-dichloropropenoic acid[c], 2,3,3-trichloropropenoic acid[c,d], trichloro-4-oxo-methylbutenoic acid[c], chlorobutenoic acid[c], isomer of chlorobutenoic acid[c], 2,3-dichlorobutenoic acid[c], trichlorobutenoic acid[c], tetra-chlorobutenoic acid[e], 2,5,5-trichloro-4-oxopentanoic acid[c], 2,3-dichloro-4-oxopentenoic acid[c], trichloro-4-oxo-pentenoic acid (2 isomers)[c], 2-chloro-3-(dichloromethyl)-4-oxopentenoic acid[c], isomer of 2-chloro-3-(dichloromethyl)-4-oxopentenoic acid[c], 3-(dichloromethyl)-2,5,5,5-tetrachloro-4-oxopentanoic acid[c], 2,5,5,5-tetrachloro-4-oxopentenoic acid[c], 3,5,5,5-tetrachloro-4-oxo-2-methylpentenoic acid[c], isomer of 3,5,5,5-tetrachloro-4-oxo-2-methylpentenoic acid[c], pentachlorooxopentenoic acid[c], 2,5,5-trichloro-4-oxopentenoic acid[c], 3,5,5-trichloro-4-oxopentenoic acid[c], 2,3,5,5-tetrachloro-4-oxopentenoic acid[c], 2,5,5,5-tetrachloro-4-oxopentenoic acid[c], 3,5,5,5-tetrachloro-4-oxopentenoic acid[c], pen-tachloropentenoic acid[c], tetrachlorooxohexenoic acid[c], heptachloro-5-oxo-hexenoic acid[e], trichlorocyclohexadienone carboxylic acid[c]
Halodicarboxylic acids	2,2-dichloropropanedioic acid[c], chlorobutanedioic acid (2 isomers)[c], 2,2-dichlorobutanedioic acid[c], isomer of 2,2-dichlorobutanedioic acid[c], trichlorobutanedioic acid[e], 2-chloro-3-(trichloroacetyl)butenedioic acid[c], *cis*-2-chlorobutenedioic acid (chloromaleic acid)[c,d], *trans*-2-chlorobutenedioic acid[c], 3-chloro-2-(dichloromethyl)-butenedioic acid[c], *cis*-dichlorobutenedioic acid (also called dichloromaleic acid)[c,d], *trans*-dichlorobutenedioic acid (also called dichlorofumaric acid)[c,d], 2,2-dichloropentanedioic acid[c], isomer of 2,2-dichloropentanedioic acid[c], dichloro-oxopentanedioic acid[e], trichlorohydroxypentanedioic acid[c], tetrachlorohydroxypentanedioic acid[c], dichloromethyl isomer of tetrachlorohydrox-ypentanedioic acid[c], chloropentenedioic acid (2 isomers)[c], dichloropentenedioic acid[c], trichloropentenedioic acid[c], tetrachloropentenedioic acid[e], 2-carboxy-3,5,5,5-tetrachloro-4-oxopentenedioic acid[c], chlorohexanedioic acid[e], 2,2-dichlorohexanedioic acid[e], tetrachlorohexanedioic acid[e], 2-chloroheptanedioic acid[e], 2-chlorooctanedioic acid[c], 2,2-dichlorooctanedioic acid[e], dichlorooctenedioic acid[c], 2-chlorononanedioic acid[e]
Halotricarboxylic acids	Chloroethenetricarboxylic acid[e], 2-chloro-3-dicarboxy-2-butenoic acid[c]
MX and analogues	3-Chloro-4-(dichloromethyl)-5-hydroxy-2(5*H*)-furanone (MX)[c,d], (*E*)-2-chloro-3-(dichloromethyl)-4-oxobutenoic acid (EMX)[c,d], 3-chloro-4-(dichloromethyl)-2-(5*H*-furanone (red-MX, for reduced MX)[c,d], 2-chloro-3-(dichloromethyl)-butenoic acid (ox-MX, for oxidized MX)[c,d], (*E*)-2-chloro-3-(dichloromethyl)-butenedioic acid (ox-EMX, for oxidized EMX)[c,d], 2,3-dichloro-4-oxobutenoic acid (mucochloroic acid)[c,d]

Table 1 . Chlorine Disinfection By-Products[a,b] *(continued)*

Other halofuranones	5-Hydroxy-5-trichloromethyl-2-furanone[c,d], isomer of 5-hydroxy-5-trichlomethyl-2-furanone
Haloketones	Chloropropanone[c], 1,1-dichloropropanone[c,d], 1,3-dichloropropanone[c,d], 1,1,1-trichloropropanone[c,d], 1,1,3-trichloropropanone[c,d], 1,1,1,3-tetrachloropropanone[c,d], 1,1,3,3-tetrachloropropanone[c,d], 1,1,1,3,3-pentachloropropanone[c,d], hexachloropropanone[e], 3-chloro-2-butanone[c], 1,1-dichloro-2-butanone[c,d], 1,3-dichloro-2-butanone[c], 3,3-dichloro-2-butanone[c,d], 1,1,1-trichloro-2-butanone[c,d], 1,1,3-trichloro-2-butanone[c], 1-chloro-3-buten-2-one[c], 3-chloro-3-buten-2-one[c], tetrachloro-3-buten-2-one[e], pentachloro-3-buten-2-one[c], 2,2-dichloro-3-pentanone, trichloropentanone[e], trichlorocyclopentenedione[c]
Haloacetonitriles	Chloroacetonitrile[c,d], dichloroacetonitrile[c,d], bromochloroacetonitrile[c,d], dibromoacetonitrile[c,d], trichloroacetonitrile[c,d], bromodichloroacetonitrile
Other halonitriles	Cyanogen chloride[d], dichloropropanenitrile[c], dichloropropenenitrile[c], trichloropropenenitrile, 2,3-dichloro-3-bromopropanenitrile, 3,4-dichlorobutanenitrile, *cis*-2,3,4-trichloro-2-butenenitrile, *trans*-2,3,4-trichloro-2-butenenitrile
Haloaldehydes	Dichloroacetaldehyde, trichloroacetaldehyde (chloral hydrate or chloral)[c,d], dichloropropanal[c], trichloropropanal[c], 2-chloropropenal[c,d], 2,3-dichloropropenal[c], 3,3-dichloropropenal[c,d], 2,2,3-trichloropropenal[c,d], trichlorobutanal[c], dichlorobutenal[c], 2-phenyl-2,2-dichloroethanal[f], Chloro-4-hydroxybenzaldehyde[f], 2-(chloro-4-hydroxyphenyl)-ethanal[f]
Haloalcohols	Chloroisobutanol[e], 3-Chloro-2-methylbutanol[e], tetrachloroisobutanol[e], chlorodimethylbutanol[e], trichloropentanol[e], pentachloroisopentanol[e], chlorodinaphthol[e]
Haloamides	2,2-Dichloroacetamide[c]
Haloesters	1-Chloroethanolacetate, 2-chloroethanolacetate
Halophenols	2-Chlorophenol[d], 2,4-dichlorophenol[d], 2,4,6-trichlorophenol[d], 3-bromophenol[c], bromodichlorophenol[c], tribromophenol[c], pentachlorophenol[c], dichlorodihydroxybenzene[c], dibromodihydroxybenzene[c], trichlorodihydroxybenzene[c]
Other haloaromatics	Chlorobenzene, 1,3-dichlorobenzene, 1,4-dichlorobenzene, 1,2-dichlorobenzene, chlorotoluene, (2-chloroethenyl)-benzene, dichloroethylbenzene[e], dichloromethoxytoluene (2 isomers)[e]
Halonitromethanes	Chloropicrin (trichloronitromethane)[d]
Halothiophenes	Tetrachlorothiophene[c], bromotrichlorothiophene[c], tetrabromothiophene[c]
	Nonhalogenated DBPs
Monocarboxylic acids	2-Methylpropanoic acid[e], 2,2-dimethyl-1-(2-hydroxy)-2-methyl propanoic acid[c], butanoic acid[c], 2-methylbutanoic acid[c], pentanoic acid[c], 2-methylpentanoic acid[e], pentenoic acid[c], hexanoic acid[c], heptanoic acid[c], octanoic acid[e], nonanoic acid[c], decanoic acid[c], undecanoic acid[e], dodecanoic acid[d], tridecanoic acid[d,e], tetradecanoic acid[d,e], pentadecanoic acid[d,e], hexadecanoic acid[d], heptadecanoic acid[d], octadecanoic acid[d], nonadecanoic acid[d], isomer of nonadecanoic acid[c], eicosanoic acid[c], heneicosanoic acid[c], docosanoic acid[c], tricosanoic acid[c], tetracosanoic acid[e], pentacosanoic acid[c], hexacosanoic acid[e], heptacosanoic acid[e], octacosanoic acid[c]
Dicarboxylic acids	Ethanedioic acid[d], propanedioic acid[c], butanedioic acid[d], *cis*-butenedioic acid[c], *trans*-butenedioic acid[e], 2,2-dimethyl butanedioic acid[c], pentanedioic acid[d], 2-methylpentanedioic acid[c], 2,2-dimethylpentanedioic acid[c], hexanedioic acid[d] (+2 isomers), heptanedioic acid[d], octanedioic acid[d], nonanedioic acid[d], isomer of nonanedioic acid[c], decanedioic acid[d,e], tridecanedioic acid[c]
Aromatic carboxylic acids	Benzoic acid[d], 3-hydroxybenzoic acid[c], 3-methylbenzoic acid[c], 4-methylbenzoic acid[c], 3,4-dihydroxybenzoic acid[c], trihydroxybenzoic acid[c], phenylacetic acid[d], 1,2-benzenedicarboxylic acid[d], 1,3-benzenedicarboxylic acid[d], 1,4-benzenedicarboxylic acid[d], methylbenzenedicarboxylic acid (2 isomers)[c], dimethylbenzenedicarboxylic acid (2 isomers)[c], 1,2,3-benzenetricarboxylic acid[d], 1,2,4-benzenetricarboxylic acid[d], 1,3,5-benzenetricarboxylic acid[d], methylbenzenetricarboxylic acid[c], (methylcarboxy)-benzenetricarboxylic acid[c], benzenetetracarboxylic acid[c], benzenepentacarboxylic acid[c]
Heterocyclic carboxylic acids	5-Methyl-2-furancarboxylic acid[c], methylfurandicarboxylic acid[c]
Cyanocarboxylic acids	3-Cyanopropanoic acid[e], 4-cyanobutanoic acid[e]
Aromatic nitriles	Benzonitrile[c], benzeneacetonitrile[c]
Phenols	Methylphenol[c], trihydroxybenzene[c]
Other aromatic compounds	Benzene, toluene, C3-benzene (2 isomers)[g]
Aliphatic aldehydes	Formaldehyde[d], acetaldehyde[d]
Aromatic aldehydes	Benzaldehyde[c,d], phenylacetaldehyde[f], 2-phenyl-2-hydroxybenzaldehyde[f]
Ketones	3-Methyl-2-pentanone[c], 3-methyl-1,2,4-cyclopentanetrione[c]
Alcohols	1-Ethoxy-1-hydroxymethane[c], 1-hydroxy-3-methyl-2-hexene[c]
Other heterocyclic compounds	1,4-Dioxane[c], 1,4-benzodioxin[c], indole[e]

[a] Refs. 2, 7–29.
[b] Underlined DBPs were found in actual drinking water samples.
[c] DBPs identified in laboratory-scale chlorination of fulvic or humic acids.
[d] Identifications confirmed by analysis of authentic standards.
[e] DBPs formed only at high chlorine doses (with fulvic or humic acids).
[f] DBPs formed by chlorination of commonly occurring amino acids (phenylalanine or tyrosine).
[g] Designation C3-benzene indicates there are three carbons (with hydrogens) attached to benzene molecule in an unknown manner (true structure could be trimethylbenzene or methylethylbenzene).

Table 2. Ozone Disinfection By-Products[a,b]

Class of By-Products	Specific DBPs
Inorganic compounds	Bromate[c,d], hypobromite[c,d], hydrogen peroxide[c,d]
Aliphatic aldehydes	Formaldehyde[c,d], cyanoformaldehyde[c,d], acetaldehyde[c,d], propanal[c,d], methylpropanal, butanal[c,d], methylbutanal, pentanal[d], methylpentanal, hexanal[c,d], methylhexanal, 2-hexenal[c,d], heptanal[d], octanal[d], nonanal[c,d], decanal[c,d], 2-methyldecanal, undecanal[d], dodecanal[d], tridecanal[d], tetradecanal[d]
Dialdehydes	Glyoxal[d], butanedial
Ketoaldehydes	Methyl glyoxal[d], ethyl glyoxal[c]
Aromatic aldehydes	Benzaldehyde[c,d], ethyl benzaldehyde[c,d], 2-(methylphenyl)-acetaldehyde[c]
Aliphatic ketones	Acetone[d], butanone[d], dioxobutane, dioxopentane, 2-methyl-3-pentanone[c], 4-methyl-3-penten-2-one[d], 4-methyl-3-penten-2-one[c], 4-hydroxy-4-methyl-2-pentanone[d], 3-methylcyclopentanone, 3-hexanone[c], 1,5,5-trimethylcyclohexane-3-one[c], 3,3,6-trimethylbicyclohexane-2-one[c], 2,6-bis-(1,1-dimethylethyl)-2,5-cyclohexadiene-1,4-dione, 4-heptanone[c], 6-methyl-2-heptenone, 6-methyl-5-hepten-2-one, 2,6-dimethyl-2,5-heptadiene-4-one[c], 6,10-dimethyl-5,9-undecadiene-2-one[d], heptadecadienone
Aromatic ketones	Acetophenone[c], 4-hydroxy-3-methoxyacetophenone[c], 4-phenyl-2-butanone[c]
Diketones	Dimethylglyoxal[d], dioxopentane, 2,5-hexanedione[c]
Monocarboxylic acids	Acetic acid[d], 2-methylpropanoic acid, 2-methoxypropenoic acid[c], 2-ethylpropenoic acid[c], pentanoic acid, 2-methylpentanoic acid, hexanoic acid[c], 2-ethylhexanoic acid[c], heptanoic acid[d], octanoic acid, nonanoic acid[d], decanoic acid[d], tridecanoic acid[d], tetradecanoic acid[d], hexadecanoic acid[d], isomer of hexadecanoic acid[c], 9-hexadecenoic acid, dimethylheptadecanoic acid[c], octadecanoic acid[c], heneicosanoic acid[d], tetracosanoic acid[d]
Dicarboxylic acids	Ethanedioic acid (oxalic acid)[d], propanedioic acid (malonic acid)[d], methylpropanedioic acid[d], butanedioic acid (succinic acid)[d], 2-methylbutanedioic acid[c], hydroxybutanedioic acid[c], *tert*-butyl-*cis*-butenedioic acid (*tert*-butylmaleic acid)[d], oxobutanedioic acid, pentanedioic acid[c], hydroxypentanedioic acid[c], oxopentanedioic acid, hexanedioic acid[c], heptanedioic acid[c], octadecanedioic acid[c], nonanedioic acid[c], ethylnonanedioic acid[c], decanedioic acid[c], tetrahydrofurandicarboxylic acid[c]
Tricarboxylic acids	3-Carboxypentanedioic acid[c]
Aromatic acids	Benzoic acid[c], 4-methylbenzoic acid[c], 3,5-dimethylbenzoic acid[c], ethylbenzoic acid[c], phthalic acid[c]
Aldoacids	Glyoxylic acid[c,d], dioxopropanoic acid, 4-oxo-2-butenoic acid[c]
Ketoacids	Pyruvic acid (2-oxopropanoic acid)[c,d], 1,2-dioxopropanoic acid, ketomalonic acid (oxopropanedioic acid)[d], oxobutanoic acid[d], dioxobutanoic acid, 2-oxopentanoic acid[c,d], 4-oxopentanoic acid[c,d], 2-oxobutanedioic acid (ketosuccinic acid)[c], oxopentanedioic acid (ketoglutaric acid)[c], 5-methyl-3-oxohexanoic acid[c]
Hydroxy acids	2-Hydroxy-2-methylpropanoic acid[c,d], hydroxypropanedioic acid (hydroxymalonic acid)[c], hydroxybutanedioic acid, hydroxypentanedioic acid
Phenols	Methylphenol[c], 4-methoxy-*tert*-butylphenol
Other aromatic compounds	Toluene[c], *o*-xylene, acetophenone[c], 1-methoxyethylbenzene[c], naphthalene[d]
Alcohols	4-Ethyl-1,3-benzenediol[c]
Nitriles	Benzeneacetonitrile, cyanopyridine[c]
Esters	Dibutylbutene-1,4-dioate[c,d], methylbenzoate[c]
Alkanes	Methylcyclohexane[c], 1,1-dimethoxycyclohexane[c], undecane[d], cyclododecane, hexadecane[c,d], heptadecane[c,d], nonadecane[c,d], eicosane[c,d], docosane[c,d], tricosane[c,d], tetracosane[c,d], pentacosane[c,d], hexacosane[c,d], heptacosane[c,d], octacosane[c,d], nonacosane[c,d], triacontane[c,d]
Heterocyclic compounds	3,4-Dimethyltetrahydrofuran[c], 1-valerylpyrrolidine[c], 4-methoxy-6-methyl-2*H*-pyran-2-one[c], isobenzofurandione[c]

[a] Refs. 17, 32–44.
[b] Underlined DBPs were found in actual drinking water samples.
[c] DBPs identified in laboratory-scale ozonations of fulvic acid, humic acid, or natural waters.
[d] Identifications are confirmed by analysis of authentic standards.

published results linking chloroform to cancer in laboratory animals. Following five years of extensive research on the formation of trihalomethanes in laboratory reactions and at pilot and full-scale drinking water treatment plants, the U.S. EPA published federal regulations in 1979 limiting levels of THMs in drinking water. THMs comprise a portion of the organic halogen compounds generated from water chlorination. One of the principal bases of the regulation was their use as indicators of the other unidentified by-products.

Table 1 lists compounds reported to date as by-products of chlorine (either from reaction of chlorine with humic material or chlorination at an actual drinking water treatment plant). A few by-products were formed only at high chlorine doses (higher than commonly applied at treatment plants). Because these may or may not be formed under typical conditions, they are labeled differently from others in the table. Also, DBPs, whose structures have been confirmed by analyses of authentic standards, are also noted in the table, in addition to some by-products formed by the reaction of chlorine with commonly occurring amino acids.

Concentrations of individual chlorine DBPs typically range from ppt (ng/L) to approximately 100 ppb (μg/L).

Table 3. Disinfection By-Products Formed by Ozone in the Presence of Elevated Bromide[a,b]

Class of By-Products	Specific DBPs
Inorganic compounds	Bromate[c,d], hypobromite[c,d]
Trihalomethanes	Bromoform[c,d]
Bromoacetic acids	Bromoacetic acid[c,d], dibromoacetic acid[d], tribromoacetic acid[c,d]
Bromoketones	Bromoacetone[c], 1,1-dibromoacetone
Bromonitriles	Cyanogen bromide[c,d], dibromoacetonitrile[c,d]
Bromonitromethanes	Bromopicrin (tribromonitromethane)[d]
Brominated alcohols	1-Bromo-2-methyl-2-propanol[c,d], 3-bromo-2-methyl-2-butanol[c,d], 1-bromo-2-methyl-2-butanol[c,d], *threo*-2-bromo-3-pentanol[c,d], *threo*-3-bromo-2-pentanol[c,d], *erythro*-2-bromo-3-pentanol[c,d], *erythro*-3-bromo-2-pentanol[c,d], 2-bromocyclopentanol[c,d], 2-bromo-3-ethoxy-propanol[c], *threo*-1,2-dibromo-3-pentanol[c], *erythro*-1,2-dibromo-3-pentanol[c,d], *threo*-1,4-dibromo-3-pentanol[c,d], *erythro*-1,4-dibromo-3-pentanol[c,d], *threo*-3,5-dibromo-2-pentanol[c], *erythro*-3,5-dibromo-2-pentanol[c]
Bromoalkanes	1,3-Dibromopentane[c], *erythro*-2,3-dibromopentane[d], *threo*-2,3-dibromopentane[c], 3,4-dibromo-1-pentene[c], *erythro*-1,2,4-tribromopentane[c], *threo*-1,2,4-tribromopentane[c]

[a] Refs. 32,33,45,46.
[b] Underlined DBPs were found in actual drinking water samples.
[c] DBPs identified in laboratory-scale ozonations of fulvic acid, humic acid, or natural waters.
[d] Identifications are confirmed by analysis of authentic standards.

Table 4. Chlorine Dioxide[a,b]

Class of By-products	Specific DBPs
Inorganic compounds	Chlorite[c,d], chlorate[c,d]
Aliphatic monocarboxylic acids	Butanoic acid, pentanoic acid, 2-methylpentanoic acid,[c] hexanoic acid, 2-ethylhexanoic acid, heptanoic acid[c], octanoic acid, nonanoic acid, decanoic acid, undecanoic acid, tridecanoic acid, tetradecanoic acid[c]
Dicarboxylic acids	Butanedioic acid (succinic acid)[c], 2,2-dimethylbutanedioic acid (+2 isomers)[c], hexanedioic acid (adipic acid) (+5 isomers)[c], *tert*-butylmaleic acid (*tert*-butyl-*cis*-butenedioic acid)[d], 2-ethyl-3-methyl maleic acid (2-ethyl-2-methyl-*cis*-butenedioic acid), heptanedioic acid (+3 isomers)[c], octanedioic acid (+1 isomer)[c], decanedioic acid, undecanedioic acid[c]
Heterocyclic carboxylic acids	Methylfurancarboxylic acid[c]
Aromatic acids	Benzoic acid[c], benzylic acid[c], 1,2,4,5-benzenetetracarboxylic acid[c], methylbenzenedicarboxylic acid[c], methylbenzenetricarboxylic acid[c], methylbenzenetetracarboxylic acid[c], carboxyphenylglyoxylic acid[c], phthalonic acid[c], dicarboxyphenylgloxylic acid[c], tricarboxyphenylglyoxylic acid[c]
Haloacids	Dichloroacetic acid[c], chloromalonic acid (chloropropanedioic acid)[c], chlorosuccinic acid (chlorobutanedioic acid)[c]
Aldehydes	Propanal, methylpropanal, 3-methylbutanal, pentanal, hexanal, heptanal, octanal, ethylbenzaldehyde
Halogenated ketones	1,1,3,3-Tetrachloropropanone
Unsaturated ketones	2,3,4-Trimethylcyclopent-2-en-1-one, 2,6,6-trimethyl-2-cyclohexene-1,4-dione
Halogenated aromatics	(1-Chloroethyl)dimethylbenzene
Other aromatic compounds	3-Ethylstyrene, 4-ethylstyrene, naphthalene, 1-methylnaphthalene, 2-methylnaphthalene
Esters	Hexanedioic acid, dioctyl ester
MX and EMX	3-Chloro-4-(dichloromethyl)-5-hydroxy-2(5*H*)-furanone (MX)[c], (*E*)-2-chloro-3-(dichloromethyl)-4-oxobutenoic acid (EMX)[c]

[a] Refs. 50,53,54.
[b] Underlined DBPs found in actual drinking water samples.
[c] DBPs identified in laboratory-scale chlorine dioxide reactions of fulvic or humic acids.
[d] Identifications are confirmed by analysis of authentic standards.

Trihalomethanes and haloacetic acids are usually the highest concentrated of the by-products. As a result, some treatment plants have found it difficult to meet EPA's regulatory limit of 100 ppb for total trihalomethanes. When bromide is present in the source water, THMs, haloacetic acids, and other chlorinated DBPs shift toward the more brominated species (16). For example, when bromide is present, bromoform and dibromoacetonitrile concentrations are greater than those of chloroform or dichloroacetic acid, which generally predominate in the absence of bromide.

OZONE

Ozone, the most powerful oxidant used for treating drinking water, is effective for killing most microorganisms. Ozone exists as an unstable gas with resonance structures. Because of its instability, ozone must be produced on-site. It is produced by an ozone generator, which converts oxygen to ozone through an electrical discharge in the generator. Ozone is currently used in more than 100 treatment plants in the United States and several thousand in Europe. As mentioned previously,

Table 5. Chloramine Disinfection By-Products[a,b]

Class of By-Products	Specific DBPs
Trihalomethanes	Chloroform[c], bromodichloromethane[c], dibromochloromethane[c], bromoform
Haloacetic acids	Chloroacetic acid, dichloroacetic acid[c], trichloroacetic acid[d], bromoacetic acid[d], dibromoacetic acid[d]
Other halomonocarboxylic acids	2,2-Dichloropropanoic acid[c], 2-chloro-4-hydroxybutanoic acid[c], 4,4,4-trichloro-3-hydroxybutanoic acid[c], 2,5-dichloro-2-methylpentanoic acid[c], 2-methyl-5,5-dichloropentanoic acid[c], 6,6-dichlorohexanoic acid[c]
Unsaturated halocarboxylic acids	3,3-Dichloropropenoic acid[c], 3-chloro-2-methyl-4-hydroxy-2-butenoic acid[c], 2,3,4-trichloro-2-butenoic acid[c], 4-chloro-5-hydroxy-3-pentenoic acid[c], 9,9-dichloro-8-oxo-3-nonenoic acid[c], 3,4-di(chloromethyl)-5-hydroxy-3-pentenoic acid[c], 3-(hydroxychloromethyl)-4-methyl-5-chloro-3-pentenoic acid[c], 3-methyl-4-chloromethyl-5-chloro-5-hydroxy-3-pentenoic acid[c], 3-methyl-4-hydroxy-5,6-dichloro-4-hexenoic acid[c]
Halodicarboxylic acids	2,2-Dichloropropanedioic acid[c], 2,3-dichlorobutanedioic acid[d], 2,3-dichlorobutenedioic acid[c], 2-chloro-3-methyl-2-butenedioic acid[c], 2,3,3-Trichloropentenedioic acid[d]
MX and analogues	3-Chloro-4-(dichloromethyl)-5-hydroxy-2(5*H*)-furanone (MX)[d], (*E*)-2-chloro-3-(dichloromethyl)-4-oxobutenoic acid (EMX)[c,d], (*E*)-2-chloro-3-(dichloromethyl)butenedioic acid (ox-EMX, for oxidized EMX)[c,d], 2,3-dichloro-4-oxobutenoic acid (mucochloric acid)[c,d]
Haloketones	1,1-Dichloropropanone[d], 1,1,1-trichloropropanone[d]
Halonitriles	Cyanogen chloride[d], dichloroacetonitrile[d], bromochloroacetonitrile[d], dibromoacetonitrile[d], trichloroacetonitrile[d]
Halonitromethanes	Chloropicrin (trichloronitromethane)[d]
Haloaldehydes	Trichloroacetaldehyde (chloral hydrate)[d]
Haloethers	Chloromethyl-(1,1-dimethyl-2-hydroxy-ethyl)-ether[c], chloromethyl-(2-chloro-3-hydroxy-propyl)-ether[c], chloromethyl-(1-methyl-2-chloromethoxy-ethyl)-ether[c], 2-hydroxyethyl-(2,3-dichloro-3-butenyl)-ether[c]
Haloalcohols	1-Chloro-2,2-dihydroxypropane[c], 1,2-dichloro-2-hydroxypropane[c], 1,2-dichloro-2,3-dihydroxypropane[c], 3-chloro-3-methyl-2,2-dihydroxybutane[c], 2,3-dichloro-2-hydroxypentane[c], 3-chloro-3-hydroxy-5-methylhexane[c]

[a] Refs. 7,9,51,61.
[b] Underlined DBPs were found in actual drinking water samples.
[c] DBPs identified in laboratory-scale monochloramination of fulvic or humic acids.
[d] Identifications are confirmed by analysis of authentic standards.

Table 6. Chloramine Disinfection By-Products: Methyl Derivatives of Diols or Aldehydes[a,b]

Methyl Derivative	Original By-Product
$(COOCH_3)CHClCCl(OCH_3)_2$	$(COOH)CHClCCl(OH)_2$ or $(COOH)CHClCOCl$
$(COOCH_3)CH(CH_2Cl)CH_2CCl(OCH_3)_2$	$(COOH)CH(CH_2Cl)CH_2CCl(OH)_2$ or $(COOH)CH(CH_2Cl)CH_2COCl$
$(COOCH_3)(CH_2)_2\ C(C_2H_5){=}C(CCl_3)CH(OCH_3)_2$	$(COOH)(CH_2)_2\ C(C_2H_5){=}C(CCl_3)CH(OH)_2$ or $(COOH)(CH_2)_2\ C(C_2H_5){=}C(CCl_3)COH$
$CH(OCH_3)_2\ CCl{=}CCl(CH_2)_4\ COOCH_3$	$CH(OH)_2\ CCl{=}CCl(CH_2)_4\ COOH$ or $COHCCl{=}CCl(CH_2)_4\ COOH$
$CH_2\ ClCH(OCH_3)_2$	$CH_2\ ClCH(OH)_2$ or $CH_2\ ClCOH$
$CH_2\ ClCH_2\ CH(OCH_3)_2$	$CH_2\ ClCH_2\ CH(OH)_2$ or $CH_2\ ClCH_2\ COH$
$CH_3\ CHClCH_2\ CH(OCH_3)_2$	$CH_3\ CHClCH_2\ CH(OH)_2$ or $CH_3\ CHClCH_2\ COH$
$CHCl_2\ CH(OCH_3)_2$	$CHCl_2\ CH(OH)_2$ or $CHCl_2\ COH$
$CH_2\ ClCH(CH_3)CH(OCH_3)_2$	$CH_2\ ClCH(CH_3)CH(OH)_2$ or $CH_2\ ClCH(CH_3)COH$
$CH_3\ CHClCH_2CH(OCH_3)_2$	$CH_3\ CHClCH_2\ CH(OH)_2$ or $CH_3\ CHClCH_2\ COH$
$CHCl_2\ CH_2\ CH(OCH_3)_2$	$CHCl_2\ CH_2\ CH(OH)_2$ or $CHCl_2\ CH_2\ COH$
$CHCl_2\ C(CH_3)_2\ CHClCH(OCH_3)_2$	$CHCl_2C(CH_3)_2CHClCH(OH)_2$ or $CHCl_2\ C(CH_3)_2\ CHClCOH$
$CH_2(OCH_3)CH_2\ CCl{=}CClCH(OCH_3)_2$	$CH_2(OH)CH_2\ CCl{=}CClCH(OH)_2$ or $CH_2(OH)CH_2\ CCl{=}CClCOH$
$CH_2\ ClCH_2\ CH(CH_3)CH(OCH_3)CH(OCH_3)_2$	$CH_2\ ClCH_2\ CH(CH_3)CH(OCH_3)CH(OH)_2$ or $CH_2\ ClCH_2\ CH(CH_3)CH(OCH_3)COH$

[a] Ref. 9.
[b] All DBPs were produced by laboratory-scale monochloramination of fulvic acid.

the first reported use of ozone for drinking water treatment occurred in Oudshoorn, the Netherlands, in 1893, whereas the first reported use of ozone in the United States occurred in Niagara Falls, New York, in 1903. Although ozone is not used in as many U.S. treatment plants as the other disinfectants discussed in this article, it is used in a few large cities, representing a significant portion of t'e population. An example is an ozonation plant that serves the City of Los Angeles and 3 million people. In addition, two other plants in the Los Angeles area are planned to be retrofitted to ozone; if implemented, these plants would affect approximately 16 million people in Southern California. Many other new ozonation

Table 7. Disinfection By-Products Formed by Primary Ozone Treatment Followed by Secondary Treatment with Chlorine[a,b]

Class of By-Products	Specific DBPs
Trihalomethanes	Chloroform[c,d], bromodichloromethane[c,d], dibromochloromethane[d], bromoform[d]
Other halomethanes	Carbon tetrachloride
Haloalkanes	Bromopentane, chlorohexane, 2,3-dichlorobutane, 1,2-dichloro-2-methyl butane, 1,1,1-tribromo-2-bromo-2-chloroethane, 2-chlorododecane, 1,2,3,4,5,5-hexachloro-1,3-cyclopentadiene[c]
Haloacids	Chloroacetic acid[d], bromoacetic acid[d], dichloroacetic acid[c,d], dibromoacetic acid[d], bromochloroacetic acid, bromodichloroacetic acid, dibromochloroacetic acid, trichloroacetic acid[c,d], 2-chloro-3-methyl maleic acid (2-chloro-3-methyl-*cis*-butenedioic acid), 2,2-dichloropropanoic acid[c], 3,3,3-trichlorolactic acid[c], 2-chlorobutenedioic acid[c], 2,3-dichlorobutenedioic acid[c], 2,2-dichlorobutanedioic acid[c], 5,5,5-trichloro-4-oxopentanoic acid[c], bromopropanoic acid, bromochloropropanoic acid, dibromopropanoic acid, bromochloropropenoic acid
MX and analogues	3-Chloro-4-(dichloromethyl)-5-hydroxy-2(5*H*)-furanone (MX)[c,d], (*E*)-2-chloro-3-(dichloromethyl)-4-oxobutenoic acid (EMX)[c,d]
Haloaldehydes	Chloroacetaldehyde, dichloroacetaldehyde, trichloroacetaldehyde (chloral hydrate)[c,d], 4-chloro-3-keto-1-butanal, bromodichloroacetaldehyde, tribromoacetaldehyde
Haloketones	1-Chlorodimethylglyoxal, chloropropanone, 1,1,-dichloropropanone[c,d], 1,3-dichloropropanone, 1,1,1-trichloropropanone[c,d], 1-bromo-1,1-dichloropropanone, 1,1,3-trichloropropanone, 1,1,3,3-tetrachloropropanone, 1,1,1,3-tetrachloropropanone, 1,1,1,3,3-pentachloropropanone, trichloromethyl ethyl ketone, 3,3-dichloro-2-butanone[c], 2,2,4-trichloro-1,3-cyclopentenedione
Halonitro compounds	Chloropicrin (trichloronitromethane)[d], bromopicrin (tribromonitromethane)[d], 1-nitro-1,1-dichloroethane, bromodichloronitromethane, nitrodibromomethane
Halonitriles	Cyanogen chloride[d], cyanogen bromide[d], dichloroacetonitrile[c], trichloroacetonitrile[d], bromochloroacetonitrile[d], dibromoacetonitrile[d], bromodichloroacetonitrile, dibromochloroacetonitrile
Halophenols	2,4,6-Trichlorophenol[d]
Haloalcohols	4,5-Dichloro-2-pentanol
Haloamides	2,2,2-Trichloroacetamide
Haloesters	2-Chloroethanol acetate, 1,2-dichloroethanol acetate, 3-chloro-2-butanol acetate
Halosulfones	Dichloromethyl methylsulfone[c], trichloromethyl methylsulfone[c], methane sulfonyl chloride
Haloaromatics	Benzyl chloride, 4,6-dichloro-1,3-benzenediol, 2-bromobenzothiazole
Other halogenated compounds	Hexachlorocyclopentadiene[c]
Aliphatic aldehydes	Formaldehyde[d], acetaldehyde[d], hydroxyacetaldehyde, propanal, butanal, hexanal, 2-hexenal, heptanal, nonanal
Dialdehydes	Glyoxal[d]
Ketoaldehydes	Methylglyoxal[d]
Carboxylic acids	2-Methyl propanoic acid, butanoic acid, 2-methyl pentanoic acid, hexanoic acid, *tert*-butyl maleic acid (*tert*-butyl-*cis*-butenedioic acid), benzoic acid, dodecanoic acid, tetradecanoic acid, pentadecanoic acid, hexadecanoic acid, octadecanoic acid
Aldoacids	Glyoxylic acid
Ketoacids	Oxopentanoic acid
Nitriles	3-Methylbutanenitrile, benzeneacetonitrile
Alcohols	2-(2-Butoxyethoxy)-ethanol, 1-[2-(2-methoxy-1-methylethoxy)-1-methylethoxy]-2-propanol, 2-[2-(2-butoxyethoxy)ethoxy]-ethanol
Ketones	Acetone[c], dimethylglyoxal[d], hexanedione
Heterocyclic compounds	4-Dodecyl-5-ethyl-2(5*H*)furanone

[a] Refs. 17,32,38,51,61,62.
[b] Underlined DBPs were found in actual drinking water samples.
[c] DBPs identified in laboratory-scale treatments of fulvic or humic acids.
[d] Identifications are confirmed by analysis of authentic standards.

plants are scheduled to be completed soon. Ozone is one of the fastest growing disinfectants in the United States.

Ozone does not produce significant levels of THMs, which more easily allows treatment plants to meet EPA's current regulatory requirements, and has other desirable features. First, ozone produces much smaller amounts of mutagens than chlorine or chlorine dioxide. Ozone is also excellent for controlling turbidity in water. Turbidity often indicates the presence of microorganisms and can be difficult for some treatment plants to control, based on the quality of the incoming raw water supply. Turbidity control was the primary reason for constructing an ozonation plant serving Los Angeles. By using ozone as a coagulant aid and as the disinfectant to kill microorganisms, ozone serves a dual purpose by reducing the amount of coagulant needed to control turbidity.

In addition, ozone is much more effective for killing harmful microorganisms than chlorine. For example, ozone is much more effective against *Cryptosporidium* and *Giardia* than

Table 8. Disinfection By-Products Formed by Primary Ozone Treatment Followed by Secondary Treatment with Chloramine[a,b]

Class of By-products	Specific DBPs
Trihalomethanes	Chloroform[d], bromodichloromethane[d], dibromochloromethane[d], bromoform[d]
Haloalkanes	2,3-Dichlorobutane, 1-chlorooctane
Haloacetic acids	Chloroacetic acid[d], bromoacetic acid[d], dichloroacetic acid[d], bromochloroacetic acid[d], dibromoacetic acid[d], trichloroacetic acid[d]
MX and analogues	3-Chloro-4-(dichloromethyl)-5-hydroxy-2(5*H*)-furanone (MX)[c,d], (*E*)-2-chloro-3-(dichloromethyl)-4-oxobutenoic acid (EMX)[c,d]
Haloaldehydes	1-Chlorodimethylglyoxal, trichloroacetaldehyde (chloral hydrate)[d], 4-chloro-3-keto-1-butanal
Haloketones	1,1-Dichloropropanone[d], 1,1,1-trichloropropanone[d], 1,1-bromochloropropanone, 1-bromo-1,1-dichloropropanone, 1,1,3,3-tetrachloropropanone, 1,1,1,3,3-pentachloropropanone
Halonitriles	Cyanogen chloride[d], cyanogen bromide[d], dichloroacetonitrile[d], dibromoacetonitrile[d], bromochloroacetonitrile[d], trichloroacetonitrile[d]
Halonitromethanes	Chloropicrin (trichloronitromethane)[d], nitrodibromomethane
Haloalcohols	3-Chloro-2-butanol, 4,5-dichloro-2-pentanol
Haloesters	2-Chloroethanol acetate, 1,2-dichloroethanol acetate, 3-chloro-2-butanol acetate
Aliphatic aldehydes	Formaldehyde[d], acetaldehyde[d], butanal[d], pentanal[d], heptanal[d], octanal[d]
Dialdehydes	Glyoxal[d]
Ketoaldehydes	Methylglyoxal[d]
Carboxylic acids	Butanoic acid, hexanoic acid
Alcohols	4-(1-Methylethyl)-benzene methanol
Ketones	4-Methyl-2-pentanone
Aromatic compounds	1,2-Bis(1-methylethenyl)-benzene, 1-[4-(1-methylethenyl)phenyl]-ethanone, isomer of 1-[4-(1- methylethenyl)phenyl]-ethanone, 1,1-(1,4-phenylene)bis-ethanone, 1-[4-(1-hydroxy-1-methylethyl)phenyl]-ethanone, isomer of 1-[4-(1-hydroxy-1-methylethyl)phenyl]-ethanone
Nitriles	Heptanenitrile, benzeneacetonitrile
Other nitrogen-containing compounds	2,6-Di-*tert*-butyl-4-nitrophenol, 5-methyl-3-isoxazolamine

[a] Refs. 17,32,51,61,63,64.
[b] Underlined DBPs were found in actual drinking water samples.
[c] DBPs identified in laboratory-scale treatments of fulvic or humic acids.
[d] Identifications are confirmed by analysis of authentic standards.

Table 9. Disinfection By-Products Formed by Primary Chlorine Dioxide Treatment Followed by Secondary Treatment with Chlorine[a,b]

Class of By-Products	Specific DBPs
Trihalomethanes	Bromodichloromethane, dibromochloromethane, bromoform
Other haloalkanes	Carbon tetrachloride, chlorotribromomethane, 3,3,3-trichloro-2-methyl-1-propene
Haloketones	1,1,1-Trichloropropanone, 1-bromo-1,1-dichloropropanone, 1,1,3,3-tetrachloropropanone, 1,1,1,3,3-pentachloropropanone, 2-chlorocyclohexanone
Haloaldehydes	Dichlorobutanal
Haloacetates	1-Chloroethanol acetate, 2-chloroethanol acetate
Haloethers	3-Bromopropyl chloromethyl ether, 1-chloro-2-ethoxy-2-methoxy ethane
Halonitriles	Dibromoacetonitrile, dibromochloroacetonitrile
Carboxylic acids	Butanoic acid, pentanoic acid, hexanoic acid, 2-ethylhexanoic acid, heptanoic acid, octanoic acid, nonanoic acid, decanoic acid, undecanoic acid, tridecanoic acid, benzoic acid
Ketones	Cyclohexanone
Furanones	4-Dodecyl-5-ethyl-2(5*H*)furanone
Alcohols	2-Ethylhexanol
Esters	Hexanedioic acid, dioctyl ester
Phenols	*tert*-Butylphenol
Other aromatic compounds	Benzyl cyanide

[a] Ref. 50.
[b] Underlined DBPs were found in actual drinking water samples.

chlorine. *Cryptosporidium* and *Giardia* are very resistant microorganisms that have been linked to many outbreaks of illness when present in the drinking water supply. As mentioned earlier, ozone is also effective for taste and odor control and for controlling algae, oxidizing iron and manganese, oxidizing many pesticides and other pollutants in the raw water, and bleaching color in water. Controlling iron and manganese is important because these elements discolor water and also cause the growth of iron- and manganese-oxidizing bacteria, which leads to a general deterioration in water quality. The oxidation of pesticides and other pollutants by ozone is important because some exhibit toxicity, and they can be oxidized to form less harmful compounds, such as CO_2 and NaCl. Interestingly enough, although there are generally minimal health concerns associated with color in water, colored water, along with taste and odor, is responsible for the majority of complaints from drinking water customers. Cities in South Florida had problems with colored water, because much of the raw water comes from the Everglades, which is high in dissolved organic material. As a result, ozonation plants were constructed to control the color in drinking water.

Ozone reacts with organic matter and forms disinfection by-products by two different mechanisms. At low pH, ozone reacts primarily through molecular ozone O_3. At high pH, ozone reacts by direct reaction of molecular ozone and also through the formation of hydroxyl radicals ($OH^{\cdot}$), one of the most powerful oxidants known.

Although ozone is used in many U.S. cities, there have been relatively few studies conducted to determine the identity of ozonation by-products. Some studies have been carried out on a laboratory scale, by reacting ozone with humic material or natural waters, and some studies have been carried out at actual ozonation drinking water treatment plants. Table 2 lists the by-products identified. Ozone does not produce any chlorinated by-products, and compared with chlorine, ozone appears to produce fewer by-products. Studies have shown that the total mutagenicity of water treated with ozone is much lower than with chlorine (30,31).

OZONE TREATMENT WITH ELEVATED BROMIDE LEVELS

Recently, studies have been carried out to identify ozonation by-products produced when elevated levels of bromide ion (Br^-) are in the raw water. Bromide is important because it is in the raw water of coastal cities, because of salt water intrusion, and is also in the ground water of cities in the Western United States with high mineral content in their soils. Also, bromide has been introduced in the ground water of many areas of Texas by the use of brine (which contains natural levels of sodium bromide) in extracting crude oil. The highest bromide levels reported in the United States have approached 2 ppm (mg/L). When bromide is present in raw water, it is oxidized by ozone to form hypobromous acid (HOBr), which reacts with natural humic material to form many brominated compounds. And because brominated compounds are generally more toxic than their chlorinated analogs, there is concern about the toxicities of these by-products.

Table 3 lists the by-products found when ozone is used to treat bromide-containing waters. Inorganic bromate and hypobromite ions and several organic by-products, including bromoform, dibromoacetonitrile, bromoacetone, cyanogen-bromide, and bromoacetic acids have been identified (33,45). A series of 15 brominated alcohols, called bromohydrins, in addition to bromoacetone, bromoform, and a few bromoalkanes, including 1,3-dibromopentane, 2,3-dibromopentane, 3,4-dibromo-1-pentene, and 1,2,4-tribromopentane were also identified (46). This study was the most comprehensive research performed to identify by-products of ozone in the presence of elevated bromide; all observed by-products were identified, not just targeted compounds.

CHLORINE DIOXIDE

Chlorine dioxide is a powerful oxidant, more effective than chlorine for killing most microorganisms. Chlorine dioxide exists as a gas, is yellow-green when dissolved in water and must be generated on-site, because it explodes in the presence of heat (or an electrical discharge) if stored at concentrations above 10% in air (47). For drinking water treatment, it is generated from sodium chlorite. Most generating techniques use the oxidative process in which chlorine (either as a gas or in solution) is mixed with a sodium chlorite solution (48). Chlorine dioxide is currently used in approximately 400 drinking water treatment plants in the United States and in several thousand plants in Europe. The first reported use of chlorine dioxide in drinking water treatment occurred in 1944 at the Niagara Falls, N.Y., water treatment plant (49). Although chlorine dioxide is a widely used disinfectant, relatively few studies have been carried out to determine the DBPs of chlorine dioxide in drinking water. One well-documented aspect about chlorine dioxide is that it does not produce significant levels of THMs when used to treat drinking water; often no detectable levels of THMs are found. Hence, the use of chlorine dioxide more easily allows drinking water treatment plants in the United States to meet EPA's regulatory limits on THMs.

Like ozone, chlorine dioxide is also beneficial for controlling taste and odor problems, for oxidizing iron and manganese, for oxidizing pollutants in the raw water, and for bleaching color in water. From a practical perspective, chlorine dioxide is also five times more soluble in water than chlorine and is effective over a wide pH range. Chlorine dioxide also produces fewer chlorinated by-products than chlorine, as evidenced by lower total organic chlorine levels produced.

Table 4 lists the by-products identified when chlorine dioxide was reacted with humic material and when chlorine dioxide was used at a drinking water treatment plant. Compared to chlorine, chlorine dioxide produces relatively few chlorinated by-products, making it an attractive alternative to chlorine for many treatment plants. The levels of the by-products are lower, as well. Semiquantitative concentrations of organic by-products identified in a pilot plant study (50) were approximated at 1 to 10 ppt (ng/L). These levels are about 1000× lower than levels found for chlorination by-products. Studies have shown that the total mutagenicity observed for water treated with chlorine dioxide is much lower (less than 15%) than with chlorine (51,52).

Their identification involves analysis by ion chromatography. Chlorite is the predominant by-product from chlorine dioxide. There have been some concerns raised about its formation, because chlorite causes anemia in rats at high exposure levels. As a result, the U.S. EPA is preparing to regulate chlorite at an MCL of 1 mg/L (4). Although chlorite is formed as a major by-product of chlorine dioxide, its levels can be controlled at treatment plants by the addition of a reducing agent, such as ferrous chloride ($FeCl_2$) (55). Granular activated carbon (GAC) filtration is also effective in removing chlorite ions from coagulated and filtered drinking water (56). There is insufficient toxicological data to determine whether there are significant health risks associated with chlorate, but chlorate formation also is minimized by optimizing chlorine dioxide generation.

CHLORAMINE

Chloramine has the lowest oxidizing potential of the disinfectants discussed herein, but it is effective for killing many microorganisms. Because of its low oxidizing power, chloramine is used more as a secondary disinfectant than a primary disinfectant. Chloramine is formed by mixing ammonia with free chlorine. Ammonia is added in the form of ammonium chloride, ammonium sulfate, or aqueous ammonia (57). As discussed earlier, free chlorine forms hypochlorous acid (HOCl) when dissolved in water. When mixed with dissolved chlorine, ammonia reacts with HOCl to form the three chloramine species shown below (57):

$$HOCl + NH_3 \rightleftarrows NH_2Cl(\text{monochloramine}) + H_2O$$

$$HOCl + NH_2Cl \rightleftarrows NHCl_2(\text{dichloramine}) + H_2O$$

$$HOCl + NHCl_2 \rightleftarrows NCl_3(\text{nitrogen trichloride}) + H_2O$$

These reactions are governed primarily by pH and chlorine-to-nitrogen weight ratios used to form the chloramine (expressed as $Cl_2{:}N$) (57). As the pH decreases and the $Cl_2{:}N$ increases, the chloramine species becomes successively more chlorinated, forming more dichloramine and nitrogen trichloride. At pH levels commonly present at drinking water treatment plants (7.0–9.5), monochloramine is the predominant species. Many studies performed at very controlled pH and $Cl_2{:}N$ ratios produce almost entirely monochloramine, and therefore, these studies refer to it as such. In actual treatment plants, low levels of dichloramine and sometimes nitrogen trichloride are formed; as a result, those studies frequently use the term chloramines or chloramine to describe the disinfectant, because in reality, there is usually more than one species present, even if only at small levels. This treatment process is also referred to as combined chlorine. For this paper, the term chloramine is used to refer to the three species of chloramine, although monochloramine is the principal species present.

Chloramine is currently used in approximately 300 drinking water treatment plants in the United States. The first reported use of chloramine in drinking water treatment occurred in February 1917 in Ottawa, Canada, and the first reported use in the United States was later in 1917 at a Denver, Colo., treatment plant. In practice, chloramine is the most inexpensive of the alternative disinfectants discussed in this paper, and it is effective for controlling taste and odor. Another advantage of using chloramine as a disinfectant is that the lower reactivity of chloramine allows it to remain in the distribution system longer, a desirable property when it is used as a secondary disinfectant. A disadvantage of chloramine is that it attacks rubber gaskets and valves in a distribution system or in a toilet. Chloramine causes more cracking and a more rapid loss of elasticity in elastomeric materials than chlorine.

Like the other alternative disinfectants, chloramine produces lower levels of trihalomethanes (less than 3%) than chlorine (58). These THM levels depend on how the chloramine is produced. When ammonia is applied prior to addition of chlorine, levels of 3–5 μg/L of total THMs are observed (much lower than the regulated level of 100 μg/L) (59). However, when ammonia is applied after chlorine, total THM levels increase can to 46–72 μg/L (59). This is attributed to reaction of natural organic material with free chlorine prior to the formation of chloramine. In addition to producing lower THM levels, chloramine also produces lower levels of chlorinated by-products, as evidenced by total organic halide (TOX) levels that are 9–49% as much as from chlorine alone (60). There is also a lower total mutagenicity observed for water treated with chloramine than that treated with chlorine (51,52).

Although chloramine is now widely applied in many cities across the United States, there are only a few published studies that have investigated the formation of by-products from sources other than model compounds. The first study (60) investigated the by-products of the reaction of monochloramine with natural, isolated fulvic acid. No by-products were detected, although there was a low, but measurable total organic halide (TOX), which indicated that low levels of by-products are produced. The investigators suggested that monochloramine reacted slowly and incompletely with fulvic acid. In another study using natural water (51) MX and EMX were identified as by-products of chloramination. A study was carried out at an actual drinking water treatment plant of the formation of trihalomethanes, haloacetic acids, haloketones, haloacetonitriles, aldehydes, and four other compounds, chloral hydrate, chloropicrin, cyanogen chloride, and 2,4,6-trichlorophenol (61). Although chloramine produces some traditional chlorine by-products, namely trihalomethanes, haloacetic acids, haloketones, haloacetonitriles, and chloral hydrate, they form at much lower levels (3–20%) compared to chlorine treatment. Chloropicrin is produced at approximately the same levels found with chlorination, and cyanogen chloride forms at higher levels with chloramination. The most recent study (9) identifies a number of by-products formed by reaction of natural, isolated fulvic acid with monochloramine. In addition to identifying MX and its structural analogs, several chlorinated carboxylic acids, ethers, alcohols, and aldehydes are identified. Nonchlorinated compounds are not identified as by-products in the treated samples.

Table 5 lists the by-products identified when chloramine reacts with humic material (9) and with natural waters (51). Methylation derivatizations aided the identifications of acids, alcohols, and aldehydes (9). The types of derivatives are shown below.

$$RCOOH \rightarrow RCOOCH_3 \quad (1)$$

$$ROH \rightarrow ROCH_3 \quad (2)$$

$$R(OH)_2 \rightarrow R(OCH_3)_2 \quad (3)$$

$$RCOH \rightarrow RCH(OCH_3)_2 \quad (4)$$

However, because both terminal diols and aldehydes (eqs. 3 and 4) produce the same type of derivative, it is not possible to determine whether the original by-product is a diol or an aldehyde. These $R(OCH_3)_2$ derivatives are listed in Table 6, but are not included in Table 5, because their identities are uncertain.

COMBINATION TREATMENTS OF DISINFECTANTS

Drinking water treatment plants sometimes use combinations of disinfectants for treating drinking water. A combination of ozone and chlorine or chloramine is an important example of combining disinfectants. In this case, ozone is the primary disinfectant (at the beginning of the treatment process) and chlorine or chloramine is a secondary disinfectant at the end of the process (just before the water leaves the treatment plant to enter the distribution system). The primary reason that a secondary disinfectant is often used with ozone is because ozone does not persist long in water, and a secondary disinfectant is necessary to maintain disinfection in the distribution system (the pipes between the treatment plant and the tap). Chlorine, chloramine, and chlorine dioxide are all used for this purpose, as they persist long enough to provide disinfection throughout the distribution system. If a secondary disinfectant is not used with ozone, bacteria and other microorganisms regrow in the distribution system, causing illness in people who drink the water. Although chlorine is commonly a secondary disinfectant with alternative disinfectants, its use does not usually defeat the purpose of using alternative disinfectants to minimize THM levels. This is because lower levels of chlorine (or another secondary disinfectant) are required as a secondary disinfectant, as opposed to the primary disinfectant which must kill the higher numbers of microorganisms that enter with the raw water (at the beginning of the treatment process). And, when lower levels of chlorine are applied, THMs and other chlorinated DBPs are usually found at lower levels than those with chlorine as the primary disinfectant. Although chlorine is widely used as a secondary disinfectant, chloramine and chlorine dioxide are better choices when THM control is difficult because these disinfectants do not produce significant levels of THMs. When chloramine is a secondary disinfectant, however, care must be exercised to ensure that microorganisms are killed effectively, because chloramine is a weaker oxidant. Many drinking water treatment plants, however, successfully use chloramine as a secondary disinfectant.

Because a combination of disinfectants can create by-products different from those for each individual disinfectant, studies have been carried out to identify by-products from these combinations. A discussion of the by-products formed from some of the more commonly used disinfectant combinations follows.

Ozone Followed by Chlorine

Relatively few studies have been carried out to identify by-products of treatments involving the combination of primary ozonation with secondary chlorination. A few studies analyzed for targeted compounds, and others have analyzed more broadly for any by-products that were formed. One study (62) analyzed for chloroform, dichloroacetic acid, trichloroacetic acid, dichloroacetonitrile and 1,1,1-trichloropropanone after treating natural fulvic acid with ozone, followed by chlorine. All of these compounds are formed by this combination treatment (and all are known to be chlorination by-products). Lower levels of chloroform, trichloroacetic acid, and dichloroacetonitrile are produced relative to treatment with just chlorine. Interestingly, there was a net increase in the formation of 1,1,1-trichloropropanone with the combination of disinfectants, relative to chlorine treatment.

The formation of trihalomethanes, haloacetic acids, haloketones, haloacetonitriles, aldehydes, and four other compounds, chloral hydrate, chloropicrin, cyanogen chloride, and 2,4,6-trichlorophenol was also studied (62). All of these traditional chlorination by-products are observed when ozone and secondary chlorination are used, but the levels of most of the by-products are lower than when chlorine is the sole disinfectant. However, levels of the haloketones actually increase, but only by small amounts of 1.0 μg/L or less. Another effect observed is a shift to more brominated species for THMs and haloacetic acids when chlorine is a secondary disinfectant with ozone. In other studies (51,52), a lower total mutagenicity (50–85%) was observed for water treated with ozone-chlorine than with chlorine alone.

Other studies (17,38) focused on identifying all by-products produced by ozone-chlorine treatment (not just a few targeted compounds). The first study was carried out by analyzing samples resulting from applying ozone with secondary chlorination to humic material in the laboratory and also to finished water at a pilot drinking water treatment plant. The second study was carried out at a pilot drinking water treatment plant. Table 7 lists the by-products identified. For the most part, traditional chlorination by-products (eg, THMs, haloacetic acids, etc) were identified along with a few ozonation by-products (eg, aldehydes) and a few ozone-chlorine combination by-products (eg, 1-chlorodimethylglyoxal, 4-chloro-3-keto-1-butanal). By-products identified include trihalomethanes, halo-alkanes, haloacids, haloaldehydes, haloketones, halonitromethanes, halonitriles, halosulfones, aldehydes, ketoaldehydes, carboxylic acids, nitriles, alcohols, and ketones. The appearance of by-products, such as 1-chlorodimethylglyoxal and 4-chloro-3-keto-1-butanal, provides evidence of reactions between ozonation by-products (eg, aldehydes) and secondary chlorine. As with the other studies mentioned, ozone in conjunction with chlorine produces lower levels of chlorinated by-products than chlorine as the sole disinfectant. This is believed to be caused by the destruction by ozone of much of the precursor material responsible for by-product formation. MX was also found in studies using ozone-chlorine (38,51), but it was identified at a level two to three times lower than in the corresponding chlorinated sample. Likewise, EMX was identified in a study using ozone-chlorine, and its level was two times lower than in a corresponding chlorinated sample.

Ozone Followed by Chloramine

As with ozone-chlorine, there have been relatively few studies carried out to identify the by-products from the combination of primary ozonation with secondary chloramination. Table 8 lists the by-products identified with ozone-chloramine treatment. Among the compounds identified are trihalomethanes, haloacetic acids, haloaldehydes, haloketones, halonitriles, halonitromethanes, haloalcohols, aldehydes, ketoaldehydes, carboxylic acids, and nitriles. MX and EMX were also identified in ozone-chloramine samples (51). It is interesting that their levels were twice the levels found in water treated with chloramine as the sole disinfectant, but their levels were half of the levels observed with chlorination. The formation of trihalomethanes, haloacetic acids, haloketones, haloacetonitriles, aldehydes, and four other compounds, chloral hydrate, chloropicrin, cyanogen chloride, and 2,4,6-trichlorophenol was studied, when ozonation was followed by secondary chloramination (61). The levels of trihalomethanes, haloacetic acids, haloacetonitriles, and chloral hydrate, identified as chloramination by-products, are lower than when chloramine is the sole disinfectant. Therefore, the use of ozone as the primary disinfectant helps to minimize the formation of these compounds. However, levels of aldehydes and chloropicrin increase in the ozone-chloramine samples, compared with chlorine. A lower total mutagenicity (about 50%) was observed for water treated with ozone-chloramine than for water treated with either chlorine or chloramine alone (51,52).

Chlorine Dioxide Followed by Chlorine

A study was conducted of primary disinfection with chlorine dioxide, followed by secondary treatment with chlorine, at a pilot drinking water treatment plant (50). Table 9 lists the by-products identified, among which are trihalomethanes, haloketones, haloacetates, haloethers, halonitriles, carboxylic acids, and a haloaldehyde. Derivatizations to analyze for haloacetic acids (such as dichloroacetic acid) and MX were not performed in this study; therefore, these compounds were not detected. However, it is very likely that they are present, because both chlorine dioxide and chlorine produce these compounds. The majority of the by-products identified in this study are traditional chlorination by-products. It is not apparent that there are any true chlorine dioxide-chlorine by-products, where chlorine dioxide by-products have reacted with chlorine. Although many chlorinated by-products form, they appeared to be at low concentration (1–10 ppt, ng/L) for the semivolatile DBPs. The total mutagenicity of the chlorine dioxide–chlorine treated water is approximately half of the mutagenicity observed for water treated with chlorine as the sole disinfectant.

REMOVAL OF CHEMICAL DBPs

The knowledge of which chemical by-products are produced from the various disinfectants has enabled methods to be developed for minimizing these by-products in treated drinking water. DBP precursor material has been removed by coagulation with alum, prior to treatment with a disinfectant (62), and biological filtration has been applied to remove the DBPs themselves after treatment with a disinfectant. Biological filtration is accomplished with sand, anthracite coal, or granular activated carbon (GAC) (65). Using biological filtration with sand, levels of halogenated by-products (which can be produced from chlorine and to a lesser extent by chloramine and chlorine dioxide) are reduced by as much as 77% (66). Biologically active GAC also reduces the levels of aldehydes, produced by ozonation (65). In one study, formaldehyde is completely removed by GAC filtration (65). In another study, GAC filtration lowers formaldehyde levels by >95%.

All chemical compounds are not removed equally, however. The dialdehyde glyoxal is more difficult to remove. A 41% removal is reported in one study (65) and a >86% removal is reported in another (67). However, to achieve these removal efficiencies, the filtration process must be optimized. Much of the removal efficiency depends on the filtration rate, the contact time, and the type of media (GAC, sand, or anthracite coal) used. A slower filtration rate increases the percentage of compounds removed. Filtration is also more effective when the filters are not backwashed with a secondary disinfectant (eg, chlorine) or when a secondary disinfectant is applied before biological filtration because backwashing with a secondary disinfectant kills the biological organisms that help remove the organic compounds. Backwashing of filters is often used to remove algal growth, which builds up and inhibits water flow through the filter. Of the different filtration media studied, GAC is the most effective for removing chemical by-products.

CONCLUSIONS

Many studies have been conducted to identify DBPs from the different disinfectants used today, and many by-products have been identified. However, it is suspected that many by-products are still unidentified, particularly for the alternative disinfectants that have not been used very long in the United States. In fact, even for chlorine, much remains to be identified. Total organic halide (TOX) is a measure of how much organic material has been chlorinated. To date, all of the known DBPs of chlorine account for less than half of the TOX present in chlorinated humic material or chlorinated drinking water. Therefore, 50% remains to be identified. In addition, only about 25% of the total mutagenicity of chlorinated drinking water has been identified. Some by-products are unidentified because spectral information is not sufficient to allow their identification. Others are unidentified because their high molecular weight and lack of volatility prevent them from being analyzed by gas chromatography/mass spectrometry (GC/MS). Still others are unidentified because their polarity makes these compounds difficult to extract from water and to analyze. As analytical methods improve, more DBPs should be identified. For example, a relatively new analytical method that analytes for polar and high molecular weight compounds is liquid chromatography/mass spectrometry (LC/MS). With LC/MS, treated water samples can be analyzed directly. There is no need for extraction into an organic solvent. LC/MS also permits analyzing high molecular weight compounds that can be ionized in the mass spectrometer. High molecular weight proteins and peptides have been successfully analyzed recently. In addition to improvements in analytical instrumentation, it is likely that extraction procedures will improve and that new derivatization methods will be discovered to allow the identification of new by-products.

For a proper assessment of the possible health implications of drinking water treated with various disinfectants, it is important to identify as many by-products as possible, so as to know which chemicals people are exposed to in their drinking water. Drinking water is a major route of exposure to chemicals. An estimated 226 million people in the United States drink disinfected water daily (68). Once most disinfection by-products are identified, better informed decisions can be made about which disinfectant poses the lowest risk to human health (by exposure to chemical by-products), and, at the same time, kills harmful microorganisms in drinking water. In addition, it is possible that improved methods for removing harmful chemical by-products can be developed, once the identities of by-products are known.

BIBLIOGRAPHY

1. B. Langlais, D. A. Reckhow, and D. R. Brink, *Ozone in Water Treatment: Application and Engineering,* Lewis Publishers, Chelsea, Mich., 1991, pp. 2–3.
2. R. J. Bull and F. C. Kopfler, *Health Effects of Disinfectants and Disinfection By-products,* AWWA Research Foundation, Denver, Colo., 1991.

3. *National Interim Primary Drinking Water Regulations. Fed. Reg.* **44,** 68624–68707, 1979; J. A. Cotruvo, *Environ. Sci. Technol.* **15,** 268–274 (1981).
4. *Fed. Reg.* 59:145:38668–38829, July 1994.
5. A. A. Stevens, L. A. Moore, C. J. Slocum, B. L. Smith, D. R. Seeger, and J. C. Ireland, in *Water Chlorination: Chemistry, Environmental Impact and Health Effects,* R. L. Jolley, L. W. Condie, J. D. Johnson, S. Katz, R. A. Minear, J. S. Mattice, and A. Jacobs, eds., Lewis Publishers, Chelsea, Mich., 1990, Vol. 6, pp. 579–604.
6. J. J. Rook, *Water Treat. Exam.* **23(2),** 234–243 (1974).
7. *National Cancer Institute Report on Carcinogenesis Bioassay of Chloroform.* Carcinogenesis Program, Division of Cancer Cause and Prevention, Bethesda, Md., Mar. 1976.
8. L. Kronberg, R. F. Christman, R. Singh, and L. M. Ball, *Environ. Sci. Technol.* **25,** 99–104 (1991).
9. R. Kanniganti, J. D. Johnson, L. M. Ball, and M. J. Charles, *Environ. Sci. Technol.* **26,** 1998–2004 (1992).
10. E. W. B. de Leer, J. S. S. Damste, C. Erkelens, and L. de Galan, *Environ. Sci. Technol.* **19,** 512–522 (1985).
11. W. E. Coleman, J. W. Munch, W. H. Kaylor, R. P. Streicher, H. P. Ringhand, and J. R. Meier, *Environ. Sci. Technol.* **18,** 674–681 (1984).
12. R. F. Christman, D. L. Norwood, D. S. Millington, and J. D. Johnson, *Environ. Sci. Technol.* **17,** 625–268 (1983).
13. R. F. Christman, J. D. Johnson, F. K. Pfaender, D. L. Norwood, M. R. Webb, J. R. Hass, and M. J. Bobenrieth, in *Water Chlorination: Environmental Impact and Health Effects,* R. L. Jolley, ed.; Ann Arbor Science: Ann Arbor, Mich., 1980, Vol. 3, pp. 75–83.
14. F. C. Kopfler, H.P. Ringhand, W. E. Coleman, and J. R. Meier, in *Water Chlorination: Chemistry, Environmental Impact and Health Effects,* R. L. Jolley, R .J. Bull, W. P. Davis, S. Katz, M. H. Roberts, Jr., and V. A. Jacobs, eds., Lewis Publishers, Chelsea, Mich., 1985, Vol. 5, pp. 161–173.
15. J. R. Meier, H. P. Ringhand, W. E. Coleman, J. W. Munch, R. P. Streicher, W. H. Kaylor, and K. M. Schenck, *Mutat. Res.* **157,** 111–122 (1985).
16. S. W. Krasner, M. J. McGuire, J. G. Jacangelo, N. L. Patania, K. M. Reagan, and E. M. Aieta, *J. Am. Water Works Assoc.* **81,** 41–53.
17. a. S. D. Richardson, A. D. Thruston, Jr., T. W. Collette, T. V. Sullins, K. S. Patterson, and B. W. Lykins, Jr., *Alternative Disinfectants for Drinking Water,* in *Chlorine and Chlorine Compounds in the Paper Industry,* 210th American Chemical Society National Meeting, August 1995, Ann Arbor Press: Ann Arbor, Mich., in press; b. S. D. Richardson, A. D. Thruston, Jr., T. W. Collette, T. V. Sullins, K. S. Patterson, and B. W. Lykins, Jr., unreported data.
18. J. J. Rook, *Environ. Sci. Technol.* **11,** 478–482 (1977).
19. B. D. Quimby, M. F. Delaney, P. C. Uden, and R. M. Barnes, *Anal. Chem.* **52,** 259–263 (1980).
20. W. E. Coleman, R. D. Lingg, R. G. Melton, and F. C. Kopfler, in *Identification and Analysis of Organic Pollutants in Water,* L. H. Keith, ed., Ann Arbor Science, Ann Arbor, Mich., 1976, pp. 305–327.
21. F. C. Kopfler, R. G. Melton, R. D. Lingg, and W. E. Coleman, in *Identification and Analysis of Organic Pollutants in Water,* L. H. Keith, ed., Ann Arbor Science, Ann Arbor, Mich., 1976, pp. 87–104.
22. R. D. Kleopfer, in *Identification and Analysis of Organic Pollutants in Water,* L. H. Keith, ed., Ann Arbor Science, Ann Arbor, Mich., 1976, pp. 399–416.
23. W. E. Coleman, J. W. Munch, P. A. Hodakievic, F. C. Kopfler, J. R. Meier, R. P. Streicher, and H. Zimmer, in *Biohazards of Drinking Water Treatment,* R. A. Larson, ed., Lewis Publishers, Chelsea, Mich., 1989, pp. 107–121.
24. W. E. Coleman, J. W. Munch, P. A. Hodakievic, F. C. Kopfler, J. R. Meier, R. P. Streicher, and H. Zimmer, *GC/MS Identification of Mutagens in Aqueous Chlorinated Humic Acid and Drinking Waters Following HPLC Fractionation of Strong Acid Extracts.* Presented at the 193rd American Chemical Society National Meeting, New Orleans, La., August 1987.
25. R. F. Thompson, R. F. Christman, and J. D. Johnson, in *Water Chlorination: Chemistry, Environmental Impact and Health Effects,* R. L. Jolley, L. W. Condie, J. D. Johnson, S. Katz, R. A. Minear, J. S. Mattice, and A. Jacobs, eds., Lewis Publishers, Chelsea, Mich., 1990, Vol. 6, pp. 171–178.
26. M. Fielding, and H. Horth, *Water Supply* **4,** 103–126 (1986).
27. D. R. Seeger, L. A. Moore, and A. A. Stevens, in *Water Chlorination: Chemistry, Environmental Impact and Health Effects,* R. L. Jolley, R. J. Bull, W. P. Davis, S. Katz, M. H. Roberts, Jr., and V. A. Jacobs, eds., Lewis Publishers, Chelsea, Mich., 1985, Vol. 5, pp. 859–873.
28. J. D. Johnson, R. F. Christman, D. L. Norwood, and D. J. Millington *Environ. Health Perspect.* **46,** 63–72 (1982).
29. S. D. Richardson, A. D. Thruston, Jr., T. W. Collette, K. S. Patterson, B. W. Lykins, Jr., and J. C. Ireland, *Environ. Sci. Technol.* **30,** 3327–3334 (1996).
30. D. K. Noot, W. B. Anderson, S. A. Daignault, D. T. Williams, and P. M. Huck, *J. Am. Water Works Assoc.* **81,** 87–102 (1989).
31. J. A. Cotruvo, V. F. Simmon, and R. J. Spanggord, *Ann. New York Academy of Sciences* **298,** 124–140 (1977).
32. W. H. Glaze, and H. S. Weinberg, *Identification and Occurrence of Ozonation By-Products in Drinking Water,* American Water Works Association Research Foundation, Denver, Colo., 1993.
33. W. H. Glaze, H. S. Weinberg, and J. E. Cavanagh, *J. Am. Water Works Assoc.* **85,** 96–103 (1993).
34. W. H. Glaze, *Environ. Health Perspect.* **69,** 151–157 (1986).
35. W. H. Glaze, M. Koga, D. Cancilla, K. Wang, M. J. McGuire, S. Liang, M. K. Davis, C. H. Tate, and E. M. Aieta, *Environ. Sci. Technol.* **81,** 66–73 (1989).
36. W. R. Haag, and J. Hoigne, *Environ. Sci. Technol.* **17,** 261–267 (1983).
37. S. D. Killops, *Water Res.* **20,** 153–165 (1986).
38. W. E. Coleman, J. W. Munch, H. P. Ringhand, W. H. Kaylor, and D. E. Mitchell, *Ozone Sci. Eng.* **14,** 51–69 (1992).
39. J. Lawrence, H. Tosine, F. I. Onuska, and M. E. Comba, *Ozone Sci. Eng.* **2,** 55–64 (1980).
40. L. J. Anderson, J. D. Johnson, and R. F. Christman, *Org. Geochem.* **8,** 65–69 (1985).
41. M. J. Sclimenti, S. W. Krasner, W. H. Glaze, and H. S. Weinberg, in *Proc. of the American Water Works Association Water Quality Technology Conference,* Denver, Colo., AWWA, 1990.
42. R. M. LeLacheur, L. B. Sonnenberg, P. C. Singer, R. F. Christman, and M. J. Charles, *Environ. Sci. Technol.* **27,** 2745–2753 (1993).
43. F. Xiong, Contribution à l'Etude de l'Ozonation des Acides Fulviques Aquatiques. Ph.D. Diss., Université de Potiers, France, 1990.
44. D. A. Reckhow, *Overview of Alternative Disinfectants,* in *Disinfection By-products in Drinking Water: Critical Issues in Health Effects Research,* Workshop Report, International Life Sciences Institute, Washington, D.C., 1996.
45. I. N. Najm, S. W. Krasner, *J. Am. Water Works Assoc.* **87,** 106–115 (1995); S. W. Krasner, J. T. Gramith, E. G. Means, N. L. Patania, I. N. Najm, and E. M. Aieta, in *Proc. of the American Water Works Association Water Quality Technology Conference,* Denver, Colo., AWWA, 1991.
46. T. W. Collette, S. D. Richardson, A. D. Thruston, Jr., *Appl. Spectrosc.* **48,** 1181–1192 (1994).

47. W. J. Masschelein, *Chlorine Dioxide: Chemistry and Environmental Impact of Oxychlorine Compounds,* Ann Arbor Science, Ann Arbor, Mich., 1979, p. 156.
48. E. M. Aieta and J. D. Berg, *J. Am. Water Works Assoc.* **78,** 62–72 (1986).
49. J. F. Synan, J. D. MacMahon, and G. P. Vincent, *Water Works & Sewer* **91,** 423 (1944).
50. S. D. Richardson, A. D. Thruston, Jr., T. W. Collette, K. S. Patterson, B. W. Lykins, Jr., G. Majetich, and Y. Zhang, *Environ. Sci. Technol.* **28,** 592–599 (1994).
51. P. Backlund, L. Kronberg, and L. Tikkanen, *Chemosphere* **17,** 1329–1336 (1988).
52. K. S. Patterson, B. W. Lykins, Jr., and S. D. Richardson, *J. Water SRT—Aqua* **44,** 1–9 (1995).
53. C. A. Colclough, J. D. Johnson, R. F. Christman, and D. S. Millington, in *Water Chlorination: Environmental Impact and Health Effects,* R. L. Jolley, ed., Ann Arbor Science, Ann Arbor, Mich., 1983, Vol. 4, pp. 219–229.
54. A. A. Stevens, *Environ. Health Perspect.* **46,** 101–110 (1982).
55. M. H. Griese, K. Hauser, M. Berkemeier, and G. Gordon, *J. Am. Water Works Assoc.* **83,** 56–61 (1992).
56. E. W. Howe, E. M. Aieta, S. Liang, and M. J. McGuire, in *Chlorine Dioxide: Scientific, Regulatory, and Application Issues* (Workbook for the International Chlorine Dioxide Symposium). Chemical Manufacturers Association, Washington, D.C., 1989, pp. 126–133.
57. R. L. Wolfe, N. R. Ward, and B. H. Olson, *J. Am. Water Works Assoc.* **76,** 74–88 (1984).
58. A. A. Stevens, C. J. Slocum, D. R. Seeger, and G. G. Robeck, *J. Am. Water Works Assoc.* **68,** 615–620 (1976).
59. S. W. Krasner, S. E. Barrett, M. S. Dale, and C. J. Hwang, *J. Am. Water Works Assoc.* **81,** 86–93 (1989).
60. J. N. Jenson, J. D. Johnson, J. J. St. Aubin, and R. F. Christman, *Org. Geochem.* **8,** 71–76 (1983).
61. J. G. Jacangelo, N. L. Patania, K. M. Reagan, E. M. Aieta, S. W. Krasner, and M. J. McGuire, *J. Am. Water Works Assoc.* **81,** 74–84 (1989).
62. D. A. Reckhow and P. C. Singer, *J. Am. Water Works Assoc.* **76(4),** 151–157 (1984).
63. R. G. Rice, *By-products of Ozonation Formed During Treatment of Drinking Water,* Presented at the 200th American Chemical Society National Meeting, Division of Environmental Chemistry, Disinfection By-Products Symposium, Washington, D.C., August 1990.
64. W. H. Glaze, M. Koga, and D. Cancilla, *Environ. Sci. Technol.* **23,** 838–847 (1989).
65. H. S. Weinberg, W. H. Glaze, S. W. Krasner, and M. J. Sclimenti, *J. Am. Water Works Assoc.* **85,** 72–85 (1993).
66. W. E. Koffskey, *Disinfection By-Product Formation by Alternative Disinfectants and Removal by Granular Activated Carbon,* U.S. Environmental Protection Agency. Cincinnati, Ohio, Sept. 1993, EPA-600/SR-93/136.
67. S. W. Krasner, M. J. Sclimenti, and B. M. Coffey, *J. Am. Water Works Assoc.* **85,** 62–71 (1993).
68. *Water Industry Database: Utility Profiles,* American Water Works Association, Denver, Colo., 1992.

DRINKING WATER, MICROBIAL QUALITY. See MICROBIAL QUALITY OF DRINKING WATER.

DRINKING WATER, OZONATION. See OZONATION IN TREATMENT OF DRINKING WATER AND WASTEWATER.

DUST: MEASUREMENT OF TRACE ELEMENTS IN

ROD BARRATT
YAPING FENG
The Open University
Milton Keynes, United Kingdom

There is a significant body of literature describing the use of deposited dusts as indicators of environmental pollution, especially by trace metals. There are a number of approaches for sampling and analysis of these samples, and care is needed to select the appropriate method applicable for meeting the objectives of the investigation. There are no standard methods. There is similarly a variety of options for presenting the results of the investigation. Only by the application of good quality control, the use of reference materials, reliable techniques, and an understanding of the significance of the results can confidence be placed in a useful approach to environmental quality assessment.

A good understanding of the quality of the environment is a precursor to rational decision-making for environmental management. A variety of indicators serve in monitoring environmental quality, and in recent years dusts have been popular environmental samples for study. As a first impression, the concept of analyzing dusts may appear naive, but for many years investigators have used this medium as a convenient indicator of pollution levels. Dust analysis continues to be attractive to environmental scientists, influenced no doubt by its ubiquity and hence apparent ease of sampling. Dust is, indeed a common material throughout the environment, although the degree of accumulation and its composition depend on many variables, including weather conditions, building construction materials, traffic density, and industrial activity.

The role of dust as a potential vehicle for the intake of trace metals by children, especially in their home environment, has prompted many studies on dust composition. Its relevance is identified, *inter alia,* in a U.S. statute (1) following recognition by the U.S. Congress that:

1. Low level lead poisoning is widespread among American children, afflicting as many as 3,000,000 children under age 6, with minority and low income communities disproportionately affected.
2. At low levels, lead poisoning in children causes intelligence quotient deficiencies, reading and learning disabilities, impaired hearing, reduced attention span, hyperactivity, and behavior problems.
3. Pre-1980 American housing stock contains more than 3,000,000 tons of lead in the form of lead-based paint, with the vast majority of homes built before 1950 containing substantial amounts of lead-based paint.
4. The ingestion of household dust containing lead from deteriorating or abraded lead-based paint is the most common cause of lead poisoning in children.
5. The health and development of children living in as many as 3,800,000 American homes is endangered by chipping or peeling lead paint, or excessive amounts of lead-contaminated dust in their homes.
6. The danger posed by lead-based paint hazards can be reduced by abating lead-based paint or by taking in-

terim measures to prevent paint deterioration and limit children's exposure to lead dust and chips.

Similar observations on the role of paint have been made elsewhere (2) (see LEAD-BASED PAINT: TESTING).

Probably most investigations on dust have concerned lead, with emphasis on sources from industry, traffic, or paint. However cadmium has also featured in such work, especially in relation to incineration of municipal waste, its major source in urban areas. Some studies have included a broad spectrum of metals.

Careful sampling and analysis of dust are fundamental to any of these investigations, but the inhomogeneous nature of dust presents many problems to the investigator. Several of these aspects are dealt with in this article, with particular emphasis on obtaining data that are comparable with those from other studies and guidelines, such as exist.

WHAT IS DUST?

According to the International Union of Pure and Applied Chemistry (IUPAC), dust is (3):

> *Small, dry, solid particles projected into the air by natural forces, such as wind, volcanic eruption, and by mechanical or man-made processes such as crushing, grinding, milling, drilling, demolition, shoveling, conveying, screening, bagging, and sweeping. Dust particles are usually in the size range from about 1 to 100 μm in diameter, and they settle slowly under the influence of gravity.*

In the context of air pollution there is no definition of dust in the legal provisions of the U.K. Clean Air Acts. "Grit" is defined in regulations (4) as "particles exceeding 76 μm in diameter," whereas "fume" is defined in the Clean Air Act, 1993, as "solid particulate matter smaller than dust." Hence a legal interpretation may be that dust is particulate matter smaller than 76 μm but with a lower limit of 1 μm.

These categories are supported by other guidelines:

1. *Grit,* which is defined as "particles of solid matter which are retained in a sieve of 75 μm nominal aperture in conformity with BS 410: 1986 of the British Standards Institution".
2. *Dust,* particles of solid matter which will pass through a sieve of 75 μm nominal aperture in conformity with BS 410: 1986;
3. *Fume,* usually regarded as solid particulate matter smaller than dust, ie, <1 μm in particle diameter.

Such a size limitation may be applied by those involved in air pollution control, but more general environmental investigations tend to follow a definition such as that in the *Shorter Oxford English Dictionary:*

> *Earth or other solid matter so contaminated as to be easily raised and carried in a cloud by the wind; any substance pulverized; powder. Often extended to include ashes etc. from a house.*

The terms "dirt" or "sediment" may be used as synonyms for dust, but all refer to the general heterogeneous mixture of particulate matter deposited on surfaces. Clearly the sources contributing to deposited dust are legion, but it has been suggested that approximately 60% of house dust is nonorganic in nature, and this probably contains most of the lead (5). Of the inorganic material in house dust, 30–40% comes from soil, 25–30% from street dust, and 1–2% from aerosol which has settled in the house.

It has been suggested that soil contributes more to larger than to smaller dust particles, and hence nonsoil particles probably contribute more to smaller dust particle (6). As around 50% of dust occurs as large particles (963–213 μm) soil is a major component of street dust. This was so for all cities studies, although was greater for some (75–90% for Halifax, Christchurch, and Kingston) than others (60% in London and 57% in New York). More recently, it has been claimed that there is no reliable experimental method by which to quantify the soil component of house dust (7).

Clearly soil is one of the major components of dust, whether from within or outside buildings, but the distinction between whether an external sample is dust or soil is problematic. It may be argued that size is not the critical factor in definition of both types of particulate matter. To some extent, the term dust depends on the source of the sample, so that samples from the roadside or inside buildings are called dust. Further, soil is particulate matter which contains nutrients for plant growth; dust does not have this quality. Studies on the composition of soil in dust do not use this criterion. This article is concerned with deposited dust collected from hard external surfaces and from inside buildings.

DEFINITION OF OBJECTIVES

A variety of objectives have driven investigations involving dusts. These have included health risk assessments in domestic, workplace, and general environmental settings. Other studies have focused on citywide surveys and comparisons between cities. It is by no means certain that the design of the research programs has been properly based on achieving these objectives. A clear, precise definition of objectives makes the subsequent stages in an investigation capable of being designed efficiently, and ensures no wasted effort or wasted resources.

Three broad objectives summarize the general reasons for carrying out dust analyses:

1. Investigations of the chemical and physical characteristics of dust, to determine factors such as its composition, content, solubility to certain solvents, particulate size.
2. General surveys of the distribution of trace elements in dust in certain areas, to elucidate the relationships between the trace element levels and factors, such as traffic, road surface, meteorology. In this category the concerns are usually to evaluate the quality of the environment and to identify sources of pollution.
3. To evaluate the quality of dust in relation to possible effects on health. Such studies often focus on the content of certain toxic metals in dust in areas where exposure of humans, and especially children, may occur.

The profile of an environmental issue which has often been claimed to be relatively neglected was raised following the publication of the House of Commons Environment Committee Sixth Report (1990). Indoor exposure to many air pollutants can often exceed outdoor exposures. The report notes that the relative neglect is because indoor pollution is often chronic and not directly or immediately life-threatening. This is true for pollution by metals in dusts, and there have been many studies looking at dusts collected within buildings, where there is greater potential for exposure of more sensitive subjects, namely children.

Relatively few studies have been concerned with organic compounds in dust, but this is equally important and practicable as the measurement of trace metals.

A sequential consideration of the factors involved in the design of a dust measurement program may follow the scheme in Figure 1.

Sampling

The aim of sampling is to collect samples whose quality represents the quality of the whole. To achieve this it is first necessary for the sample to truly represent the temporal and spatial variation of the quality of the whole, and this requires consideration of the sampling locations and the time and frequency of sampling. A second prerequisite is that the analyte(s) of interest have the same values as in the original. Hence it is essential to consider methods of sample collection, transportation, and storage, with implications for representativeness again, as well as for avoiding contamination or analyte loss. Ultimately the results of analysis must be expressed and interpreted to relate to the objectives, as well as being comparable with other work. Relevance of environmental measures to the actual exposure of subjects is probably one of the most difficult aspects of environmental research (8). While some compromise is usually necessary, some of the sampling procedures are utterly irrelevant. In particular, it has been cited that dust samples taken from a roadside gutter rather than from pavement, or from under a carpet are of unlikely relevance to human exposure (8). The need for rigor increases as a field is progressively explored. This rigor starts by clarifying the objectives and following a sequence such as that outlined in Figure 1.

Selection of Analytical Methods

Measurements of toxic metals like lead and cadmium in dust are necessary for their hazard evaluation as well as for identification of their natural and man-made sources. Reliable, convenient, and economic, sensitive analytical methods play

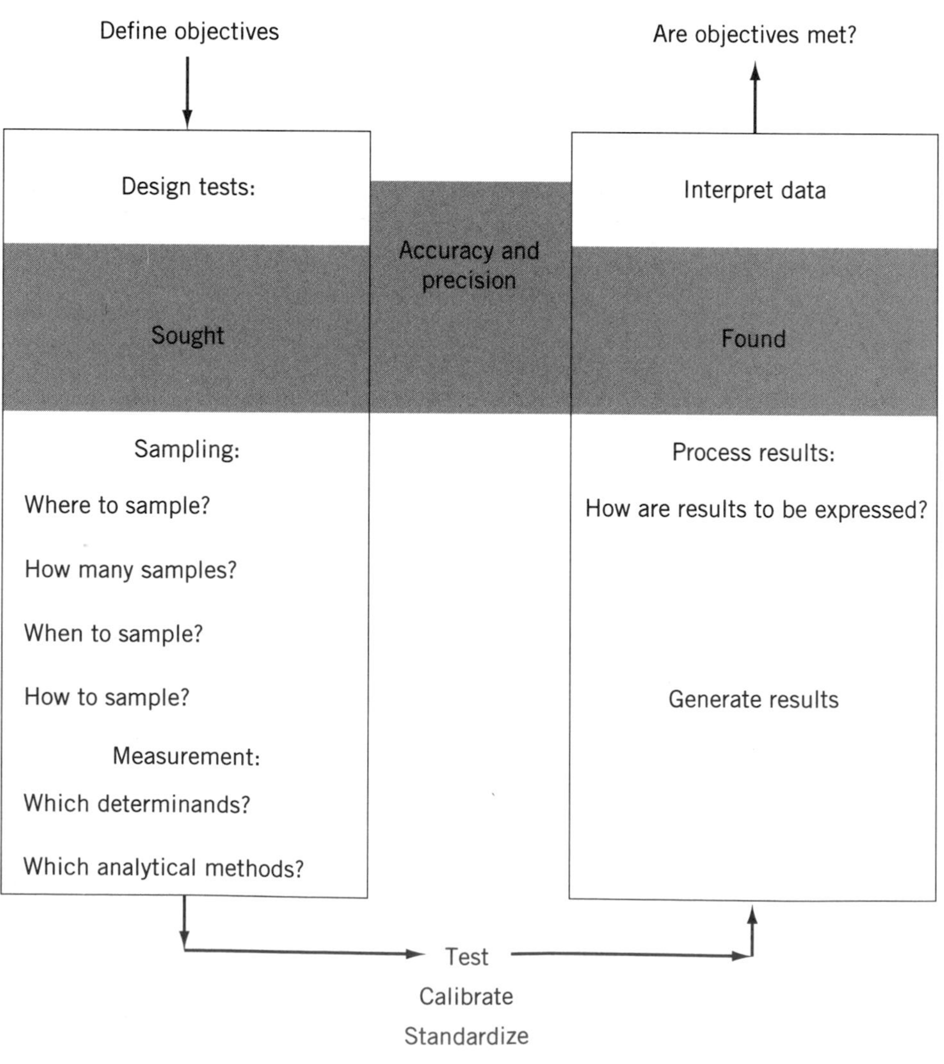

Figure 1. A dust measurement program for assessing environmental quality.

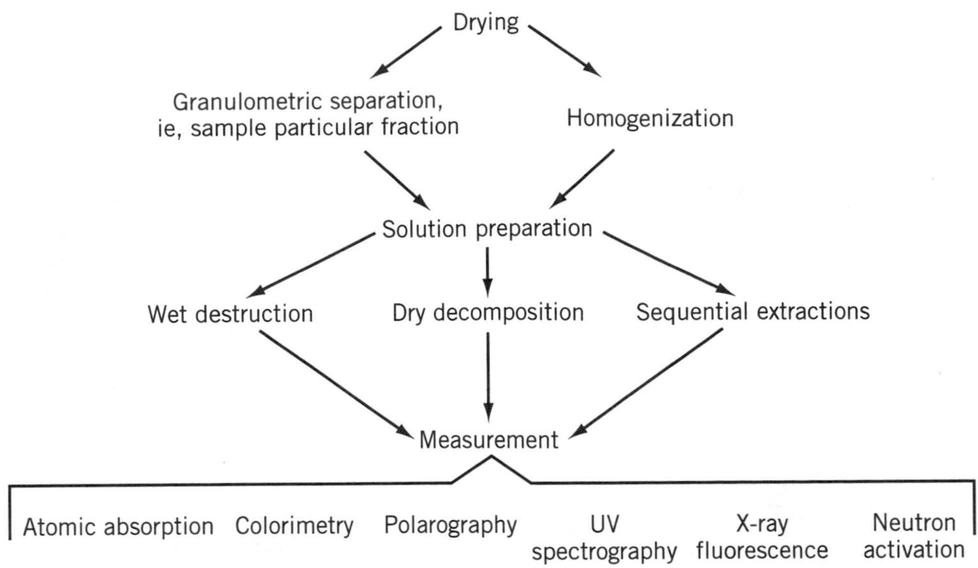

Figure 2. Analytical options for dusts.

Table 1. Common Analytical Methods for Trace Metal Determinations, %

Method Pollutant	Atomic Absorbance	Colorimetry	Polarography	UV Spectrometry	XRF	Neutron Activation
Pb	93	0.5	4	2.5	<0.1	2.5
Cd	88	2.5	6	0.7		

an extremely important role in determining content of toxic metals in dust. The options for sample analysis are varied and published descriptions of methods are sometimes full of detail or too brief to repeat. Some of these procedures are presented in Figure 2.

There are many possibilities in each step of the sequence, and several steps may be interchanged or associated. However, the first stage invariably involves preparing the sample in a form suitable for subsequent analysis. This preparation stage depends to some degree on the analytical method to be used later.

Metal Determination. A general review of analytical methods used throughout Europe (9) for principal pollutants demonstrated the dominance of atomic absorption methods for trace metals such as lead and cadmium, as Table 1 indicates. This distribution of methods is generally found in reports on subsequent work on metals in dusts, with flame atomic absorption predominating for Pb, Cd, Zn, Ni, and Cu. Polarography was used for 10% of samples in studies on lead, and agreement was found with atomic absorption analysis better than ±3% (6). The same two techniques, with polarography in the form of differential pulse anodic stripping voltammetry, were used by others (10). Other techniques have included neutron activation and inductively coupled plasma atomic emission spectroscopy.

The literature on the use of differential pulse anodic stripping voltammetry (DPASV) for the determination of Pb and Cd in dust is limited; the method has been used primarily for the determination of metals in airborne particles and in soil. By comparison with AAS, DPASV offers rapid and simultaneous determination of Pb and Cd, as well as Zn and Cu, if desired. A simpler treatment of dust samples and less expensive equipment are further advantages.

Conditions found effective in the determination of trace metals in dusts. Organic substances can also interfere in these determinations, but a technique for overcoming such problems by the addition of potassium iodide has been reported. This gives the added benefit of enhancing the limit of detection by DPASV (11).

How to Relate Pollutants in Dusts to Sources. When applying dust data to environmental management there is clearly a need to identify the source of the particulate matter. Simply identifying the presence of a contaminant does not associate it with the source. Typical procedures involve the following:

- Determination of the topographic distribution to establish concentration profiles which relate sources and exposure points. This is similar to the approach around the incinerator.
- Determining the relationships between elements, one of which may be a tracer characteristic of a specific source.
- Determination of isotopic ratios to identify contributions from sources with different isotopic ratios.

One scheme for identifying further distinguishing features involves a sequence of steps (Fig. 3) (12). Fractionation of each dust sample according to particle size, density, and ferromagnetic characteristics is followed by scanning electron microscopic examination and energy dispersive X-ray analysis of individual particles in each subsample. This enables the morphological and chemical characteristics of each sample to

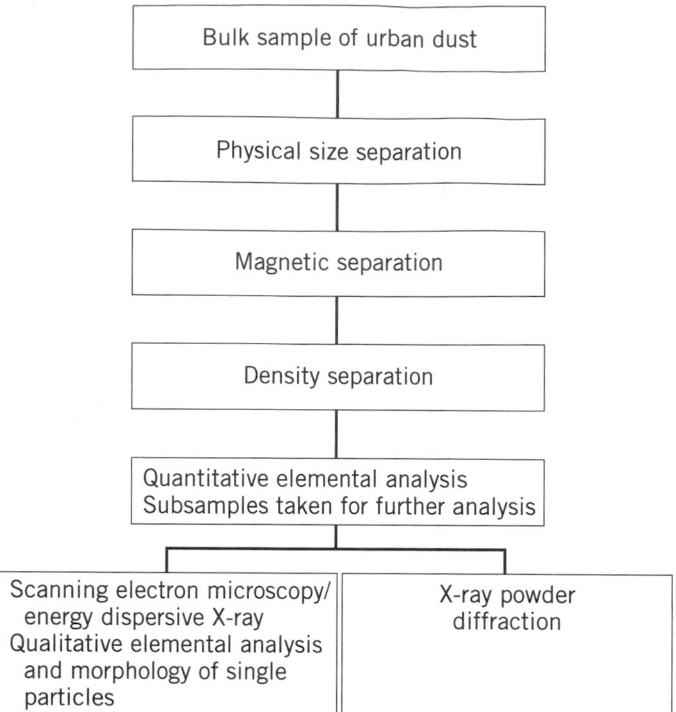

Figure 3. Sequence for fractionation of dust samples to identify sources.

be determined and related to those observed in the particulate matter from suspect sources. Associations of elements in specific particles may suggest the origin of the particle, for example, particles containing lead and bromine linked to car exhaust emissions (13), and lead-to-silicon ratios linked to soil, and lead to potassium linked to refuse burning (14). Other techniques for source characterisation include the use of the stable isotope ratio (15).

Measurements of Organic Substances in Dust. Organic material may be present in surface dust from a variety of sources, many of which have been listed in detail in the context of potential contamination of water through surface runoff (16). However, relatively few studies have examined the organic composition of surface dusts. Nevertheless, fine dust resuspended from the road has been linked to ambient levels of polycyclic aromatic hydrocarbons in air (17), although there are suggestions that there are significant particle size implications.

Is the Environmental Quality of Dust Acceptable? In terms of cause and effect, an incident involving the release of a potentially harmful substance into the environment can be represented as follows:

Event → impact on receptor → consequences or effects

The system may be further developed as a simple model for describing any incident scenario as a result of a release from processes generating dust.

In land contamination, harm and loss of human life represent potential consequences at the site. The cross-boundary dimension is represented by dispersion of particulate matter into the atmosphere, and subsequent deposition. The effects of an incident will depend on many factors including primarily on the nature of the substance(s) concerned, its toxicity or other harmful effects, its pathways through the environment, its fate in the human body or other receptor. Clearly biological responses depend on the duration of exposure and the ability of the substance(s) to reach target organs of the human body, for example, and all of these considerations are outside the scope of this review, which concerns itself primarily with analytical methods for dusts. However, some routes and their effects are shown in Figure 4, while the following shows mechanisms of entry for a variety of species.

Target	Route
Humans and animals in the terrestrial environment	Exposure of skin and outer surface
	Exposure of eyes, mucous membrane of nose, mouth, trachea
	Exposure of pulmonary tissue
	Entry through skin
	Entry through lungs
	Entry through gastrointestinal tract
	Entry by ingestion of food and water
	Entry by ingestion of contaminated material
Plants in the terrestrial environment	Exposure of parts above ground to air and to rain
	Entry through parts above ground (eg, via stomata)
	Entry through roots
Organisms in the aquatic environment	Exposure of external surface
	Entry through external surface
	Entry through gills
	Entry through gastrointestinal tract
	Entry through ingestion of other organisms via the food chain
	Entry through ingestion of suspended matter
	Entry through ingestion of benthic detritus (ie, residues at bottom)

Clearly, in dealing with deposited dusts, this review is concerned with only a limited number of these routes of entry, and further confines itself to human health effects. However, it would be remiss if awareness of the wider implications of dust contamination were not pointed out. Inhalation is an obvious route of entry as dust is transported through the atmosphere on its way to deposition. For airborne dusts there are well-established standards for health protection.

The lack of nationally agreed reference levels for judging the acceptability of trace metal concentrations in dust has been noted (18). Standards on pollution by trace metals in dust do not exist, although in their absence, tentative guidelines have been used by organizations such as the former Greater London Council. A guideline concentration for lead was adopted at 500 g/g

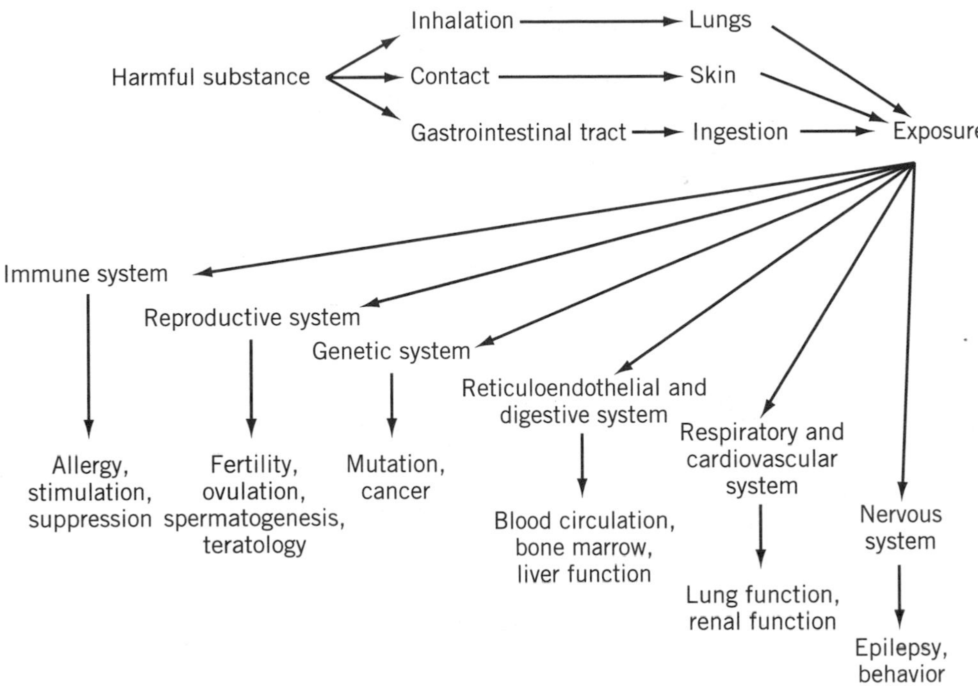

Figure 4. The potential of chemical substances in dust to cause harm.

to indicate the need for further investigation, while a level of 5000 g/g indicated the need for control action. The derivation and use of these guidelines merits further discussion.

ASSESSING SIGNIFICANCE OF TOXIC METALS IN DUSTS

The potential for the toxic materials entering the food chain through the soil–plant–human, soil–plant–animal–human, or even direct soil–human pathway is considered. This may involve estimating the dose of intake according to conventional principles, which assume that the reference dose (RfD) for lead is 1.4×10^{-3} and 1.0×10^{-3} for cadmium with units of mg/kg/d. Another consideration for cadmium, a potential carcinogen, is the potency factor of 6.1/(mg/kg/d). Further assumptions are that there is 40% bioavailability of the toxic metals from the gut, that a person is subjected to an accumulation of dust over a lifetime amounting to 16,786 g, and the estimate of dust ingestion by children is around 100 mg/d.

The loadings of cadmium closely match the locations for the highest loadings of dust, whereas the highest lead loading is found at only one of these locations. The highest Cd levels were found to the northeast of the incinerator in the line of the prevailing wind direction. However, the sites showing the highest Cd loadings were located to the southwest of the incinerator.

The limits for domestic gardens as set out in Table 2 are 3 mg/kg for Cd and 500 mg/kg for Pb, while for parks and public open space, the limits are 15 and 2000, respectively. On the basis of the more stringent criteria applied to the finest dust fraction, cadmium levels failed at one-quarter of the sites sampled, and most severely at the site showing the highest cadmium level at all particle sizes. Nowhere was the public open space limit breached for cadmium. No guidelines are available for loadings of the pollutants, although by combining the level with the amount of dust it may be argued that this offers a more meaningful representation of environmental quality than level alone. Both the numerical values and the contour maps indicate that in terms of the levels of trace metals, the incinerator appears to have little impact on lead levels, but the highest cadmium level is located near to the site at which dispersion modeling predicts an impact.

The potential exposure pathways for an individual exposed to the releases from the incinerator are as follow:

1. Inhalation of the air
2. Ingestion of foodstuffs contaminated by chemicals deposited from the atmosphere
3. Ingestion of drinking water contaminated likewise
4. Dermal contact with contaminated soil
5. Ingestion of contaminated soil
6. Contact with contaminated water

The significance of these pathways vary according to the situation. In this case pathways 3 and 6 can be eliminated from further consideration, whereas a judgment on the inhalation route or the route via foodstuffs is not possible from this work, which only concerns deposited material. Only dermal contact and ingestion of contaminated soil can be dealt with here.

Using the information available, the dermal intake I can be estimated as:

$$I_{\text{dermal}} = \frac{ABC}{WL}$$

where A = accumulated dust over a lifetime (16,786 g) (20); B = bioavailability, assumed to be 40% (more relevant to oral

Table 2. Tentative Trigger Concentrations for Selected Inorganic Contaminants[a]

Contaminant	Planned Uses	Trigger Concentration, mg/kg Air-Dried Soil
Contaminants which may pose health hazard		
As		10
		40
Cd		3
		15
Cr (hexavalent)		25
Cr (total)	Domestic gardens, allotments, parks, playing fields, open space	600
		1000
Pb		500
		2000
Hg		1
		20
Se		3
		6
Phytotoxic contaminants, not normally hazardous to health[b]		
B (water soluble)		3
Cu	Any uses where plants are to be grown	130
Ni		70
Zn		300

[a] Ref. 19.

[b] A pH of about 6.5 is assumed; lower values may increase the uptake of toxic metals. Grass may be more resistant than other plants.

intake, but introduces a safety factor into the estimation by considering a worse-case scenario); W = body weight, (70 kg); and L = lifetime in days(70 yr × 365 d/yr = 25,550 d).

Using these assumptions, the lifetime average daily dose (LADD) from outdoor dermal contact becomes for lead:

$$I_{\text{dermal}} = \left(\frac{565.6\ \mu\text{g}}{\text{g}}\right)(16{,}786\ \text{g})\,(0.4)\left(\frac{1}{25{,}550\ \text{d}}\right)\left(\frac{1}{70\ \text{kg}}\right) = 2.1\ \mu\text{g/kg/d}$$

Similarly cadmium is calculated at 0.03 μg/kg/d.

An ingestion uptake can be calculated using:

$$I_{\text{oral}} = \frac{RDBC}{WL}$$

where R = the rate for dust ingestion and D = exposure duration.

Clearly the risk from soil ingestion is predominantly one for young children, typically under five years, and so the daily exposure over that five-year period must be averaged over a lifetime of 70 years. By the same argument, it is the weight (17 kg) that is considered. Hence, for lead:

$$I_{\text{oral}} = \left(\frac{100\ \text{mg}}{\text{d}}\right)\left(\frac{565.6\ \mu\text{g}}{\text{g}}\right) \times \left(\frac{1\ \text{g}}{1000\ \text{mg}}\right)\left(\frac{365\ \text{d} \times 5\ \text{yr}}{\text{yr}}\right) \times \left(\frac{1}{25{,}550\ \text{d}}\right)\left(\frac{1}{17\ \text{kg}}\right)(0.4) = 0.095\ \mu\text{g/kg/d}$$

The conversion factor is introduced to normalize the units of mass. Similarly, cadmium is calculated at 0.0015 μg/kg/d. The LADD is therefore the sum of all routes of exposure, so for Pb = 2.1 + 0.095 = 2.195 μg/kg/d and for Cd the total becomes 0.03 + 0.0015 = 0.0315 μg/kg/d.

The final step is to relate the LADD to the RfD. For lead, the ratio RfD/LADD equates to:

$$\frac{1.4 \times 10^{-3}\ \text{mg/kg/d}}{2.195\ \mu\text{g/kg/d}}\left(\frac{1000\ \mu\text{g}}{1\ \text{mg}}\right) = 0.64$$

Similarly for Cd the safety factor becomes 32. The key criterion should be that the ratio of reference dose to daily intake should not be less than unity. This is so for cadmium, but not so for lead.

Sometimes the ratio is inverted, with the criterion for acceptability being that the predicted intake divided by the standard does not exceed 1. This latter expression is sometimes called the hazard index (HI), which is the computed maximum daily dose received by an individual, divided by a computed acceptable daily intake, ie:

$$\text{HI} = \frac{\text{Maximum daily dose}}{\text{Acceptable daily intake}}$$

In the scenario of the worst case lead and cadmium values, the HI values computed in this manner are 1.6 and 3×10^{-2}, respectively. The conclusions remain the same, with lead being questionable, and cadmium exposure suggesting no problems. However, it should be recognized that the source of contamination is not the incinerator alone; a motorway network was nearby, representing a significant potential source of lead aerosol for ultimate deposition.

For cadmium, there is the additional aspect of its potential carcinogenic properties. The risk is determined by multiplying the dose by the potency factor, which in this case gives a value of 1.9×10^{-4}. This equates to a lifetime risk of 1 in 5200 or an annual risk of 1 in 3.6×10^{5}, which is greater than the normal acceptable value of 1 in 10^{6}. However, the worst-case scenario for this calculation required that several assumptions be made. Moreover, there is the caveat about the use of microwave digestion as compared with biological digestion to release metals from the dust matrix. Nevertheless, the example illustrates an approach to environmental risk assessment using dust analyses. Standards would help in the generation of reliable data and their interpretation, but in the absence of such standards, it is essential to take a critical view of all uses of deposited dusts as indicators of environmental quality.

Deposited dusts will always present a potential route of contamination by trace metals and organic substances, and the need for future work in this area will not diminish. Only by the application of good quality control, reference materials, and appropriate techniques for decomposition of the matrix to match the objectives of the study can data be viewed with confidence.

BIBLIOGRAPHY

1. *The Residential Lead-Based Paint Hazard Reduction Act of 1992,* U.S. Congress, Washington, D.C.
2. L. Johnson, P. J. Barlow, and R. S. Barratt, *J. Roy. Soc. Health,* **104,** 96 (1984).
3. *Glossary of Atmospheric Chemistry Terms,* IUPAC, Oxford, U.K., 1989.
4. *Clean Air (Emission of Grit and Dust from Furnaces) Regulations,* SI 1971, HMSO, London, 1971.
5. J. E. Fergusson and R. J. Schroeder, *Sci. Tot. Environ.* **46,** 61 (1985).
6. J. E. Fergusson and D. E. Ryan, *Sci. Tot. Environ.* **34,** 101 (1984).
7. I. Thornton, *Sci. Tot. Environ.* **75,** 21 (1988).
8. P. C. Elwood, in M. L. Richardson, ed., *Roy. Soc. Chem. London* (1986).
9. G. M. Raab, D. P. H. Laxen, and M. Fulton, *Env. Geochem. Health,* **9,** 80–85 (1987).
10. R. S. Hamilton, D. M. Revitt, R. S. Warren, and M. J. Duggan, *Sci. Tot. Environ.* **59,** 365 (1987).
11. Y. Feng and R. S. Barratt, *Analyst,* **119,** 2805–2808 (1994).
12. R. W. Linton, D. F. S. Natusch, R. L. Solomon, and C. A. Evans, J., *Env. Sci Technol.* **14,** 159–164 (1980).
13. J. M. Watt & D. L. Johnson, in I. Thornton and E. Culbard, eds., *Lead in the Home Environment,* Science Reviews Ltd., Northwood, England, 1987, pp. 85–96.
14. W. Van Borm, L. Wouters, R. Van Grieken, and F. Adams, *Sci. Tot. Env.* **90,** 55–66 (1990).
15. Y. Yaffe and co-workers, *Arch. Env. Health.* **38,** 237–245 (1983).
16. R. E. Pitt and G. Amy, *Toxic Materials Analysis of Street Surface Contaminants,* Report EPA-R2-73-283, U.S. EPA, Washington, D.C., 1973.
17. R. M. Harrison and co-workers, *Env. Sci. Technol.* **30,** 825 (1996).
18. R. S. Barratt, *Intern. J. Environ. Analyt. Chem.* **40,** 77–97 (1990).
19. *Interdepartmental Committee on the Development of Contaminated Land,* 2nd ed., ICRCL 59/83, Dept. of the Environment, London, July 1987.
20. D. J. Paustenbach, ed., *The Risk Assessment of Environmental Hazards,* John Wiley & Sons, Inc., New York, 1989, p. 954.

DYES AND PIGMENTS: ENVIRONMENTAL CHEMISTRY

ABRAHAM REIFE
Consultant
Toms River, New Jersey

DON BETOWSKI
U.S. Environmental Protection Agency
Las Vegas, Nevada

HAROLD S. FREEMAN
North Carolina State University
Raleigh, North Carolina

Although synthetic organic dyes and pigments are essential for coloring a wide variety of industrial products, some of them present hazards and environmental problems. Since 1970, the dyestuffs and allied industries have become subject to increasingly stringent international, federal, and state regulations designed to protect human health and the environment.

WASTEWATER TREATMENT METHODS

Overview

Methods employed to treat wastewater containing organic dyes and pigments are classifiable into three main categories: physical, chemical, and biological (1). Physical methods involve adsorption, sedimentation, flotation, flocculation, coagulation, fractionation, reverse osmosis/ultrafiltration, ionizing radiation, and incineration. Chemical methods include neutralization, reduction, oxidation, electrolysis, ion exchange, and wet-air oxidation, whereas biological methods utilize stabilization ponds, aerated lagoons, trickling filters, activated sludge, anaerobic digestion, bioaugmentation, and bioremediation. In addition, the various wastewater treatment methods involve preliminary, primary, secondary, and tertiary treatments which differ mainly in the number of operations performed on the waste streams (2,3). Commonly used preliminary treatment processes include equalization, neutralization, and disinfection. The primary treatment processes are mainly physical and include sedimentation, flotation, and flocculation, where the objective is to remove undissolved chemicals and particulate matter. The secondary treatments reduce the organic load and are a combination of physical/chemical separation and biological oxidation. Tertiary treatment is a polishing step for effluent treatment. This final stage often involves adsorption, ion exchange, chemical oxidation, hyperfiltration (reverse osmosis), and electrochemical processes.

Heavy Metals

Metals, such as copper, chromium, mercury, nickel, and zinc, are used as catalysts and complexing agents to synthesize dyes and dye intermediates and are now considered priority pollutants (4). Acid Blue 25 is one dye which uses copper as a catalyst in its synthesis. In this case, 1-amino-2-sulfonic-4-bromoanthraquinone is condensed with aniline in the presence of copper salts, a process known as the Ullmann reaction (5). Another example involves the formation of the triazole ring of Direct Yellow 106 (6,7).

Metal ions are used to make dyes, such as copper-complexed direct and reactive dyes for cotton and chromium-complexed acid dyes for nylon and wool. Examples are Direct Blue 218 (8), Reactive Violet 2 (9), and Acid Black 52 (10). Similarly, zinc is used to prepare the double salt of basic dyes, an example of which is Basic Red 22 (11).

A number of research studies and patents have appeared dealing with the removal of heavy metals from the effluents of dyestuff and textile plants. The methods used involve coagulation (12–15), polymer adsorption (16), ultrafiltration (17,18,19), carbon adsorption (20,21), electrochemistry (22,23), incineration and land disposal (24), PACT (25,26), neutralization (26), iron reduction followed by neutralization (27), and bisulfite-catalyzed borohydride reduction (28).

Of interest is the possible removal of these heavy metals by other methods, such as ion exchange (29), sulfide precipitation (29), chelation with trimercaptotriazine (30), mercaptobenzothiazole (31), diethyldithiocarbonate (32), carbonotrithioic acid (33), adsorptive filtration (34), and carbon adsorption followed by stripping the metals with an acid regenerant and recovering them by electrolysis (35).

Pollution Prevention

The United States Environmental Protection Agency (U.S. EPA) has placed a high priority on reducing or minimizing hazardous chemical waste by limiting the amount of pollutants produced, ie, reducing pollutants at their source rather than controlling them at the end of the manufacturing process with water treatment facilities (36). This initiative led to collaborative efforts involving industry, government, academia, and private environmental groups to implement pollution prevention (37–46). Manufacturing and textile plants now give high priority to waste minimization and source reduction (47). For instance, the U.S. EPA and the Ecological and Toxicological Association of the Dyestuffs Manufacturing Industry (ETAD) jointly established a program for preventing pollution in the dyestuff industry, published a pollution prevention guidance manual, and conducted a baseline survey of industry prevention practices in dye manufacture (48).

INFORM, an environmental research group, reported the waste minimization efforts of two dye manufacturers (49). Also, two papers were published describing the reduction of toxic wastewater in the manufacture of disperse azo dyes (50,51). There are several research papers and patents pertaining to recycling dye manufacturing and textile industry wastewater for reuse of dye, textile auxiliaries, and water (52,53,54,55–60). Today, recycling is considered a key part of pollution prevention. Studies involving waste minimization and reduction of textile wastewater in dyehouses using computer modeling and automatic process control have been published (61,62).

A dye standardization and packaging operation has established a waste minimization process to lower overall production costs by reducing the flow of wastewater from the dye processing operations to the plant's wastewater treatment plant. The key features of the process are dye separation from each wastewater stream using ultrafiltration (UF) or reverse osmosis (RO) technology and recirculating the water produced. The recovered dye concentrate is reused in mixes and the water permeate is clean enough for recycling to dye mixes and in kettle and equipment cleaning. This effort resulted in a 98% reduction of wastewater discharged to the pretreatment system and a 68% reduction of treatment costs (17, 63–65).

The dyestuff and pigment manufacturing industries have long practiced pollution prevention/waste minimization/source reduction, especially striving to obtain high yields (as close to 100% as possible) in preparing intermediates, dyes, and pigments. This has been achieved in synthesizing disperse azo dyes, such as Disperse Red 167 (66), Disperse Red 177 (67), Pigment Red 3 (68), Pigment Red 48 (69), Pigment Red 21 (70), Pigment Yellow 14 (71), and Pigment Yellow 1 (72).

One of the most serious environmental problems in the dyestuff, textile, and leather industries results from the manufacture and use of metallized azo dyes complexed with chromium and/or cobalt to obtain desirable stability to light, rubbing, and wet conditions (73). This is particularly important because the superior light stability required for nylon carpet used in automobile interiors is achieved only through the use of metallized azo dyes in which the azo linkage is protected from UV degradation. Examples of two of these important dyes are Acid Black 172 and Acid Blue 171.

The methods used to date for removing metallized dyes from wastewater effluents involve chemical, physical, and biological pretreatment which are expensive and, in many cases, result in sludges which must be disposed of by incineration or land filling. In keeping with U.S. EPA's top priority of pollution prevention, researchers at North Carolina State University investigated the merits of substituting Fe for Cr and Co atoms in a group of commercially important dyes for nylon and wool (74). Although iron is innocuous, little had previously been published about the suitability of iron complexes as environmentally friendly alternatives to currently used chromium and cobalt metal complex acid dyes. In fact, only one commercially important iron-complexed dye is known, Acid Brown 98.

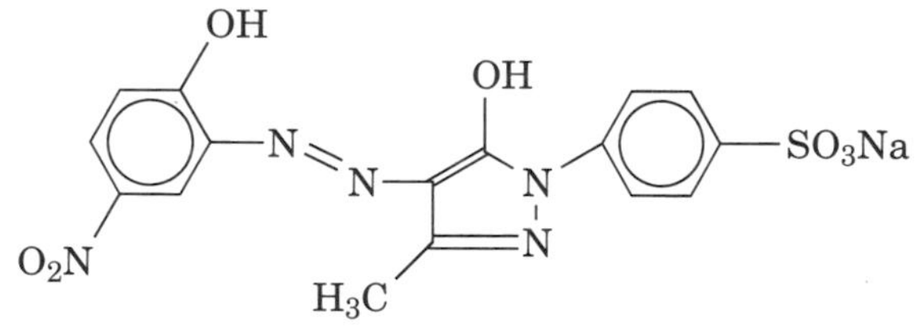

The iron dye research project led to the synthesis of environmentally friendly, metal-complexed azo dyes for nylon and wool with high stability to UV light and washing. Examples are the Fe analogues of azo dyes, such as C.I. Acid Red 182 and C.I. Acid Blue 171 (75–78), and formazan dyes, such as C.I. Acid Black 180 (79,80). Interestingly, in all three cases, the introduction of Fe in the commercial dyes produces nonmutagenic photostable black dyes. In the case of Fe-complexed formazan dyes, it was also possible to produce violet and blue colors, colors not reported for Fe complexes prior to these studies.

One of the most drastic pollution prevention measures implemented to date is terminating the use of toxic and carcinogenic compounds produced by metabolic breakdown of dyes and pigments. Examples are β-naphthylamine and especially benzidine. In the case of benzidine, this decision has curtailed the manufacture and sale of 67 dyes, including large volume products such as Direct Black 38, Direct Blue 6, and Direct Brown 95 (81–83). The most drastic pollution prevention measure is one imposed by Germany. Its new Consumer Goods Ordinance bars the distribution of fabric containing azo

Table 1. List of the Specified Amines

Name of Amine	CAS Number
o-Aminoazotoluene	[*97-56-3*]
4-Aminobiphenyl	[*92-67-1*]
Benzidine	[*92-87-5*]
p-Chloroaniline	[*106-47-8*]
4-Chloro-*o*-toluidine	[*95-69-2*]
3,3′-Dichlorobenzidine	[*91-94-1*]
3,3′-Dimethoxybenzidine	[*119-90-4*]
3,3′-Dimethylbenzidine	[*119-93-7*]
4-Methoxy-*m*-phenylenediamine	[*615-05-4*]
6-Methoxy-*m*-toluidine	[*120-71-8*]
4-4′-Methylenebis (2-chloroaniline)	[*101-14-4*]
4,4′-Methylenedianiline	[*101-77-9*]
4,4′-Methylenedi-o-toluidine	[*838-88-0*]
4-Methyl-*m*-phenylenediamine	[*95-80-7*]
2-Naphthylamine	[*91-59-8*]
5-Nitro-*o*-toluidine	[*95-55-8*]
4,4′-Oxydianiline	[*101-80-4*]
4,4′-Thiodianiline	[*139-65-1*]
o-Toluidine	[*95-53-4*]
2,4,5-Trimethylaniline	[*137-17-7*]

dyes capable of forming any of the 20 aromatic amines, listed in Table 1, by reductive cleavage of azo bonds. The presence of benzidine congeners, such as 3,3′-dimethoxybenzidine (*o*-dianisidine), 3,3′-dimethylbenzidine (*o*-tolidine), and 3,3′-dichlorobenzidine on this list affect the manufacture and use of important dyes, such as Direct Blue 218, Direct Blue 281, and large-volume diarylide yellow pigments (84).

In the future, a significant factor in implementing pollution prevention/waste minimization/source reduction in the dyestuff, textile, and pigment industries worldwide will be the International Organization for Standardization (ISO) and its ISO 14000 standards. More than 60 countries have been involved in this organization and its voluntary standards. These international standards cover environmental auditing, environmental labeling, environmental aspects in products standards, life-cycle assessment, and emphasize a strong commitment to pollution prevention (85–88).

ANALYTICAL METHODS

The analysis of environmental samples for organic colorants presents interesting challenges to the analytical chemist, because numerous dyes and most pigments are not volatile, and many decompose at elevated temperatures. Dyes and pigments also encompass a wide range of chemical functionalities, leading to wide variations in solubility, ionization efficiency, and reactivity. In addition, the fact that certain types of azo dyes are classified as cancer suspect agents, eg, benzidine-based dyes, and other dyes undergo reduction to aromatic amines regarded as carcinogenic, the environmental monitoring of colored effluents requires sophisticated, specific, sensitive, accurate, and cost-effective methods for detecting potentially genotoxic compounds. Because the standards needed for qualitative and quantitative analysis of commercial dyes are not readily available, purification of commercial formulations is an important step in generating the required standards.

Although organic components in colored wastewater are a major concern, many dye wastewater samples contain significant levels of metals that also pose environmental problems. Besides metallized dyes, the sources of these metals in dye wastewater are residual catalysts, corrosion of equipment, raw water, and raw materials (89). The metals most likely to be present in dye wastewater are As, Cd, Co, Cu, Pb, Hg, and Zn (90). Recently, inductively coupled plasma mass spectrometry (ICP/MS) has been identified as the method of choice for determining trace metals present, replacing atomic absorption spectrophotometry. A good review on ICP/MS has been published (91).

Extractions

Because of the complex composition of environmental waste samples, separation techniques must be employed before the characterization step begins. In fact, before separation methods are employed, complex waste samples must undergo various extractions to remove the compounds of interest from the environmental matrix. The goal of extractions is to separate analytes of interest from the matrix material, thus avoiding interference during characterization. A concentration step is often included as part of the extraction procedure to increase the detectability of the compounds of interest. The extraction method employed depends on the sample matrix and nature of the compounds to be extracted. The most common extraction techniques are solid-phase extraction, liquid–liquid partition, and extraction disk technology. There are three principal methods used to extract analytes of interest from a solid sample. Soxhlet extraction, sonication, and supercritical fluid extraction (SFE). Any one of these extraction techniques or a combination of them can be used for a particular sample. The choice depends on the composition of the sample. In addition, in samples containing water-soluble dyes, methods involving rotary evaporation or lyophilization (freeze-drying) may be required.

Liquid samples are extracted by liquid–liquid partition or extraction disk technology. In liquid–liquid partition procedures, the pH of the sample is adjusted to a level appropriate for the dye class undergoing analysis, and then extraction proceeds with an organic solvent, such as methylene chloride. Extraction is often carried out overnight in a continuous extractor. The extract is then evaporated to near dryness using a nitrogen stream, a little heat, or both. The residue is dissolved in a solvent, such as acetonitrile, to a predetermined volume and analyzed. The actual solvent or the solvent mixture is usually chosen for compatibility with subsequent chromatographic analysis. Extraction disk technology has been used only recently for dye analysis (92,93). Empore disks are used to extract various dyes from liquid samples. The conditioned disks are mounted on a vacuum filtration apparatus, and the liquid sample is added. The sorbed dyes are then eluted with methanol or methanol/water, depending on the nature of the dye. For instance, acid dyes are eluted with the latter solvent combination.

Solid samples are extracted to remove dyes by Soxhlet extraction or sonication. Soxhlet extraction employs an apparatus into which the solid sample is placed at an appropriate pH. Methylene chloride is a common extraction solvent. Sonication employs an apparatus which uses a disruption horn to disperse sonic energy through solvents, such as methylene chloride. The extracts from both methods are dried and concentrated, and the residues are redissolved in an appropriate solvent to a predetermined volume. SFE is a newer technique for solid-phase extractions that commonly uses carbon dioxide (CO_2) at supercritical pressure. Supercritical fluids have the beneficial properties of gases and liquids and have been employed for solid wastes with good results. In this method, only a small amount of solvent needs to be added to a collection vial, thereby minimizing solvent use.

Sludge samples are processed by separating them into their liquid and solid phases followed by the extractions outlined above. Liquid extracts and solid samples are separated using extraction disks, as described above, if too many interferences are carried through the original extraction. In addition, particle filtration techniques are required to remove suspended particles from the sample extracts.

Spectrophotometric Methods

Spectrophotometric methods for analyzing dyes, their precursors, or their degradation products detect absorption at a certain wavelength of the visible spectrum. Precursors or degradation products, such as aromatic amines, are frequently subjected to a chemical reaction, such as oxidation or diazotization and coupling, and the colored product is then amenable to spectrophotometric analysis. Although such methods have been used to identify and quantify organic dyes, the products of such reactions are often unstable and light sensitive, making rapid analysis critical. Positive interferences are also a problem with this approach, as absorption maxima of the target compounds are sometimes very close to the maxima of interferences. This method is used when monitoring a single dye in fairly clean samples. Because structurally related dyes have similar absorption maxima, monitoring samples before matrix removal turns up many interferences. Consequently, the identification of dyes in waste samples must include a separation step before spectrometric analysis.

Chromatographic Methods

A variety of chromatographic methods have been developed for separating the components in complex mixtures including paper chromatography (PC), thin-layer chromatography (TLC), gas chromatography (GC), and high performance liquid chromatography (HPLC). A related separation method for ionic dyes is capillary electrophoresis. Although these techniques are used alone, the utility of some of them for characterizing organic compounds is enhanced by interfacing them with techniques, such as mass spectrometry.

Mass Analysis

Most of the separation techniques mentioned in the previous section have been interfaced with mass spectrometry (MS) or have been used in tandem with mass spectrometry. Mass analysis of an analyte often provides the molecular weight of the compound and also structural information about the molecule. These two types of data, in turn, depend on the ionization technique used in the mass spectrometer and any ancillary techniques that fall under the general category of tandem mass spectrometry or MS/MS. The ionization technique varies with the sample introduction interface. The major advantages of mass spectrometry are sensitivity and specificity. Most of the mass spectrometers currently sold are based on the quadrupole mass filter design, in which ions are filtered according to their mass-to-charge ratio in a electrostatic field. Quadrupole ion traps and time-of-flight instruments are currently becoming a popular alternative to the quadrupole instrument. The availability of many different interfaces is another advantage for MS.

Three other techniques that merit mentioning here, because of their effectiveness in generating mass spectra of nonvolatile compounds, are field desorption (FD), laser desorption (including matrix-assisted laser desorption ionization [MALDI]), and ^{252}Cf particle bombardment. All improve the direct probe technique for analyzing nonvolatile and thermally unstable compounds. FD requires a high electric field to desorb the sample from an electrode. MALDI relies on rapid energy transfer from photons to the sample. As in fast-atom bombardment (FAB), MALDI also relies on a sample matrix to assist in the ionization process. ^{252}Cf particle bombardment uses the ^{252}Cf fission fragments to initiate ionization in a sample matrix.

SUMMARY

Although the analysis of environmental samples for dyes, synthetic precursors, and degradation products remains a challenging problem, many innovative tools have been developed in the past 15 years to meet this challenge. Among the more effective tools is liquid chromatography/mass spectrometry (LC/MS). Improvements in chromatography, coupled with creative interfaces to mass spectrometers, make it possible to characterize complex samples. The analysis of water soluble anionic by MS remains difficult, though a number of recent developments have permitted progress in this area.

Notice

The U.S. Environmental Protection Agency (EPA), through its Office of Research and Development (ORD), partially funded and collaborated in the research described in this article. It has been subjected to the Agency's peer review and has been approved as an EPA publication.

BIBLIOGRAPHY

1. J. Park and J. Shore, *J. Soc. Dyers Colour* **100,** 383 (1984).
2. G. McKay, *Am. Dyestuff Rep.* **68**(4), 29 (1979).
3. E. A. Clarke and R. Anliker, *Environmental Chemistry, Anthropogenic Compounds,* Springer-Verlag, New York, Vol. 3, Part A, 1980, p. 181.
4. E. L. Barnhardt, *Symposium Proceedings of the Textile Industry Technology,* Williamsburg, Va., Dec. 1978, p. 17.
5. *Colour Index,* Vol. 3, CI No. 62055, Society of Dyers and Colorists, Bradford, Yorkshire, U.K., 1971, p. 3502.
6. *Colour Index,* Vol. 6, CI No. 40300, Society of Dyers and Colorists, Bradford, Yorkshire, U.K., 1975, p. 6400.
7. U.S. Pat. 2 029 591 (Feb. 4, 1936), H. Schindhelm and C. T. Schults (to General Aniline).
8. *Colour Index,* Vol. 4, CI No. 24401, Society of Dyers and Colorists, Bradford, Yorkshire, U.K., 1971, p. 4209.
9. *Colour Index,* Vol. 4, CI No. 18157, Society of Dyers and Colorists, Bradford, Yorkshire, U.K., 1971, p. 4120.
10. *Colour Index,* Vol. 3, CI No. 15711, Society of Dyers and Colorists, Bradford, Yorkshire, U.K., 1971, p. 3076.

11. A. A. Mamontova, *Khim. Tekhnol. Vody.* **12**(8), 738 (1990).
12. J. He and B. Wang, *Huanjing Wuran Yu Fangzhi* **12**(5), 18 (1990).
13. E. Sindelarova, J. Piskor, J. Vesely, and R. Kocian, CS 196 814 (Mar. 31, 1982).
14. J. Urbas, *Przegl. Skorzany,* **34**(8), 267 (1979).
15. H. A. Fiegenbaum, *Am. Dyestuff Rep.* **67**(3), 43, 36 (1978); *Ind. Wastes* **23**(2), 32 (1977).
16. *Am. Dyestuff Rep.,* **61**(8), 57 (1972).
17. J. Elliot, "Membrane Filtration Techniques in Dyestuff Recovery," in *Environmental Chemistry of Dyes and Pigments,* John Wiley & Sons, Inc., New York, 1996.
18. Fr. Demande 2 619 727 (Mar. 3, 1989), D. Lemordant, P. Letellier, M. Rumeau, and C. Soma (to Universiti Pierre et Marie Curie).
19. J. L. Gaddes and H. G. Spenser, Report EPA/600/2-79/118, Order No. PB80-113889, USEPA, Washington, DC, 1979.
20. U.S. Pat. 4 005 011 (Jan. 25, 1977), C. D. Sweeney.
21. C. A. Pitkat and C. L. Berndt. *Proc. Ind. Waste Conf.* **35,** 178 (1981).
22. K. D. Uhrich and T. R. Demmin, Book Papers, *AATCC International Conference Exhibition,* Nashville, Tenn., 1988, p. 97.
23. T. M. Yohe and J. E. Rich, *Ind. Wastewater* 30,31,33,35 (May/June 1995).
24. F. Petrini and D. Grechi, *Ing. Ambientale* **13**(6), 308 (1984).
25. D. G. Hutton, J. A. Meidl, and G. J. O'Brien, "The PACT System for Wastewater Treatment," in A. Reife and H.S. Freeman, eds., *Environmental Chemistry of Dyes and Pigments,* John Wiley & Sons, Inc., New York, 1996.
26. D. C. McIntyre, "Operating Experience with the PACT System," in A. Reife and H.S. Freeman, eds., *Environmental Chemistry of Dyes and Pigments,* John Wiley & Sons, Inc., New York, 1996.
27. U.S. Pat. 4,194,973 (Mar. 25, 1980) R. E. Smith (to American Color and Chemical).
28. M. M. Cook and co-workers, "Sodium Borohydride Reductions-Novel Approaches to Decolorization and Metals Removal in Dye Manufacturing and Textile Effluent Applications," *203rd National Meeting of the American Chemical Society,* San Francisco, April 5–10, 1992.
29. W. J. Eilbeck and G. Mattock, *Chemical Processes In Waste Water Treatment,* John Wiley & Sons, Inc., New York, 1987.
30. U.S. Pat. 3 778 368 (Dec. 11, 1973), Y. Nakamura, A. Morioka, and Y. Itsuyo (to Sankyokasel).
31. Jpn. Pat. No. 7 600 764 (1976), M. Ichiki, N. Ogawa, and M. I. Shu.
32. Jpn. Pat. No. 7 671 290 (1976), Y. Horie, M. Idohara, and T. Gomyo.
33. Drew Industrial Division, Ashland Chemical Co., Boonton, N.J.
34. M. M. Benjamin and R. S. Sletten, EPA Reports 540/R-93/515 (Sept. 1993); EPA Report 600/S2-91/041 (Sept. 1991).
35. T. Lewis, *Ind. Wastewater,* 16 (July/Aug. 1994).
36. "Pollution Prevention Strategy," *Fed. Regist.* (Feb. 26, 1991).
37. *Chem. Week* 22 (Aug. 19, 1987).
38. C. V. Cox, *Hazardous Waste: Selection, Control, Treatment,* Elsevier Science Publications, Amsterdam, the Netherlands, 1988, p. 353.
39. E. Doughterty, *R&D Magazine,* 62 (Apr. 1990).
40. R. L. Berglund and C. T. Lawson, *Chem. Eng.,* 120 (Sept. 1991).
41. A. J. Goldner, *R&D Magazine,* 48 (Sept. 1991).
42. R. Gager, *Hazmat. World,* 42 (Sept. 1991).
43. L. R. Ember, Strategies for Reducing Pollution at the Source are Gaining Ground, *C&E News,* 7 (July 8, 1991).
44. L. R. Ember, *C&E News,* 10 (May 29, 1995).
45. S. J. Ainsworth, *C&E News,* 9 (Aug. 9, 1993).
46. A. M. Thayer, *C&E News,* 8 (July 26, 1993).
47. F. C. Cook, *Text. World* **141,** 84 (May 1991).
48. Pollution Prevention News, Office of Pollution Prevention, USEPA, Washington, D.C., Apr. 1990, p. 2; Pollution Prevention Guidance Manual for the Dye Manufacturing Industry, USEPA/ETAD, Washington, D.C., 1992.
49. *Cutting Chemical Wastes - What 29 Organic Chemical Plants are Doing to Reduce Hazardous Wastes,* INFORM, Inc., New York, 1985.
50. A. Reife, *Reduction of Toxic Wastewaters in Disperse Azo Dye Manufacture, Colour Chemistry,* Elsevier Applied Science, London and New York, 1991.
51. A. Reife, "Reduction of Toxic Components and Wastewater of Disperse Blue 79," *41st Southeast Regional American Chemical Society Meeting,* Raleigh, N.C., Oct. 9–11, 1989.
52. C. A. Brandon, *Ind. Water Eng.* **12**(6), 14 (1975).
53. C. A. Brandon, J. S. Johnson, R. E. Minturn, and J.J. Porter, *Text. Chem. Color.* **5,** 134 (1973).
54. J. J. Porter, *Text. Chem. Color.* **22**(6), 21 (1990).
55. G. M. Elgal, *Text. Chem. Color.* **18**(5), 15 (1986).
56. U.S. Pat. 4,200,526 (Apr. 29, 1980), R. Anewalt, E. W. Johnson, H. J. King, and M. J. Reider (to Gerber Prod.).
57. B. Lieberherr and W. Beck, *Melliand Textilberichte* **69**(8), 572 (1988).
58. *Chemiefsern Textilindustrie,* **36/88** (11), 888 (1986).
59. V. I. Popova, N. Kravchenko, E. V. Muravera, and G. E. Krichevskii, *Tekh. Texstil. Prom.* **4**(166), 67 (1985).
60. D. Fiebig and G. Schulz, *Lenzinger Berichte* **58,** 109 (1985).
61. B. Glover and L. Hill, *Text. Chem. Color.* **25**(6), 15 (1993).
62. J. J. Porter, "Reduction of Textile Wastewater Using Automatic Process Control, Recycle, and Filtration," in *Environmental Chemistry of Dyes and Pigments,* John Wiley & Sons, Inc., New York, 1996.
63. R. Ballina and P. Douvres, Reducing Wastewater Effluent - The SWEAT Project, ACS Symposium, Div. Environ. Chem., Chicago, Ill., Aug. 22–27, 1993, p. 270; *Effluent Reduction in Dye Processing,* Tech Application, Electric Power Research Institute (EPRI) Process Industry Publication, **4**(5), 1 (1992).
64. D. Rotman and A. Wood, *Chem. Week,* (Dec. 9, 1992), p. 66.
65. B. L. Steelman and J. D. Spencer, *Ind. Wastewater* **4**(1), 29 (1996).
66. A. Reife, "Reduction of Toxic Wastewaters in Disperse Azo Dye Manufacture," in *Colour Chemistry, The Design and Synthesis of Organic Dyes and Pigments,* Elsevier, London, 1991.
67. J. Sokolowska-Gajda, H. S. Freeman, and A. Reife, "Chemical Removal of Phosphate Ions From Disperse Dye Filtrates," in *Environmental Chemistry of Dyes and Pigments,* John Wiley & Sons, Inc., New York, 1996.
68. FIAT 764, *Hansarot B,* PB 74123, p. 5843.
69. FIAT 764, *Permanentrot BB,* PB85593, p. 212.
70. FIAT 764, *Permanentrot FR,* PB 91693, p. 126.
71. FIAT 764, *Vulcanechtgelb G,* FIAT Report No. 1313, Vol. III, p. 475.
72. FIAT 764, *Hansagelb G,* PB 91693m p. 51.
73. M. S. Reisch, *C&EN,* 9 (July 5, 1993).
74. A Newman, *Environ. Sci. Tech.* **28**(11), 463A, 1994.
75. J. Sokolowska-Gajda, H. S. Freeman, and A. Reife, *Textile Res. J.* **64**(7), 388 (1994).
76. H. S. Freeman, J. Sokolowska-Gajda, A. Reife, L. D. Claxton, and V. S. Hook, *Text. Chem. Color.* **27**(2), 13, (1995).

77. U.S. Pat. 5 376 151 (Dec. 27, 1994), H. S. Freeman, A. Reife, and J. Sokolowska-Gajda (to North Carolina State University).

78. A New Look at Iron-Complexed Dyes, H. S. Freeman, L. G. Cleveland, J. Sokolowska–Gajda and A. Reife, *AATCC Int. Conf. Exh.,* Atlanta, Ga., Oct. 8–11, 1995.

79. U.S. Pat. 3 663 528 (May 18, 1972), V. G. Witterholt and J. W. Dolgain (to DuPont).

80. J. Sokolowska, H. S. Freeman, and A. Reife, *Dyes Pigments,* **30**(1), 1 (1996).

81. A. Reife, "Dyes, Environmental Chemistry," in J. I. Kroschwitz and M. Houe-Grant, eds., *Kirk-Othmer Encyclopedia of Chemical Technology,* 4th ed., John Wiley & Sons, Inc., New York, 1993, Vol. 8, p. 768.

82. H. M. Smith, "U. S. Safety, Health, and Environmental Regulatory Affairs for Dyes and Pigments," in A. Reife and H.S. Freeman, eds., *Environmental Chemistry of Dyes and Pigments,* John Wiley & Sons, Inc., New York, 1996.

83. E. A. Clarke, "Regulatory Affairs (International Perspective)," in A. Reife and H.S. Freeman, eds., *Environmental Chemistry of Dyes and Pigments,* John Wiley & Sons, Inc., New York, 1996.

84. "German Ban of Use of Certain Azo Compounds in Some Consumer Goods," ETAD Information Notice No. 6, *Tex. Chem. Color.* **28**(4), 11 (1996).

85. A. M. Thayer, *C&E News,* 11 (Apr. 1, 1996).

86. G. Crognali and M. Water-Slepicka, *Environ. Solutions,* 14 (Mar. 1996).

87. E. F. Greenberg, *Packaging Digest,* 26 (Mar. 1996).

88. L. Bridgewater, *Ind. Wastewater,* 7 (Nov./Dec. 1995).

89. M. S. Yllo, "Analytical Techniques for Ecological and Toxicological Monitoring," in K. Vent Katavaman, ed., *The Analytical Chemistry of Synthetic Dyes,* John Wiley & Sons, Inc., New York, 1977, p. 555–580.

90. *Dyes and the Environment,* American Dye Manufacturers Institute, Inc., 1973, 1974, Vol. I and II.

91. D. W. Koppenaal, *Anal. Chem.* **62,** 303R (1990).

92. A.J. Borgerding and R.A. Hites, *Environ. Sci. Technol.* **28,** 1278 (1994).

93. C. Pace and M. Roby, EPA 600/X-91/147 (Oct. 1991).

E

EARTH SURFACE MONITORING FROM SPACE

DAVID J. BELSON
Consultant
Nottingham, England

In application to the environmental sciences, remote sensing (RS) is defined as the use of sensors of electromagnetic radiation to record images of the terrestrial environment which can be interpreted to yield useful information. Sensors borne on earth-orbiting satellites have become of rapidly increasing importance since the launch of Landsat 1 in 1972, particularly in the production of nonphotographic images. Satellites like those of the Landsat series orbit at a height of several hundred kilometers but can resolve objects as small as 10 m (33 ft). There is a number of systems whose status can be regarded as operational. They are operated, and the images processed and distributed, by commercial companies, but development of equipment continues to improve production and processing of images. The potential for application of remotely sensed images is enormous but, despite rapid growth worldwide, the true commercial potential has yet to be realized. Emphasis is no longer on improving technology but on meeting customer needs. A large stockpile of data exists, going back to the 1970s, and images can also be supplied "tailor-made" to customer requirements. Taking advantage of the opportunities for application requires a more widely spread understanding of RS, the sort of problems it can be used to solve, and its costs and efficiency. The chief advantage of satellite RS as opposed to other types (principally aerial photography) is in the provision of regular monitoring in detail of all parts of a large area over periods of weeks, months, or years. Frequent repeats of the ground track allow comparison of scenes of the same area, under the same lighting conditions, to be obtained at different times. Satellite imagery is ideal for monitoring all changes in the environment, not just the very slow; satellite data are the only kind that can keep up with rapid changes. The availability of a readily accessible store of data maximizes diversity of information and area of coverage at minimal cost. Several problems can be addressed more rapidly by satellite RS than by any other method.

REMOTE SENSING SATELLITE SYSTEMS

Meteorological Satellites

Many of these satellites are in geostationary orbit, at a height of around 36,000 km (19,000 nautical mi). The time taken to complete one such orbit is 24 hours, so the satellite always looks down on the same part of the earth's surface (Fig. 1). Satellites such as Meteosat (EUMETSAT), Geostationary Operational Environmental Satellites (GOES) from the United States, and Geostationary Meteorological Satellites (GMS) from Japan are primarily designed to produce the images of cloud cover over large parts of the globe that are familiar from television weather forecasts. When there are no clouds, pictures of large areas of the earth's surface are obtained, but these lack the high resolution of images produced by spacecraft in lower orbits.

Other meteorological satellites, such as the NOAA (U.S.), Defence Meteorological Satellite Programme (DMSP, U.S.), and Meteor (Russia) series, occupy polar orbits at 800–1200 km (420–630 nautical mi) similar to those of dedicated RS satellites. However, because of the purpose for which they were designed, these satellites, although giving better resolution than geostationary spacecraft, do not match RS systems in this respect. Data for current NOAA satellites are included in Table 1 (orbital data) and 2 (sensor characteristics). It is possible to receive images from NOAA and Meteosat satellites directly with equipment readily available in many educational and industrial establishments.

Manned Spacecraft

Orbital heights of Space Shuttle orbiters, Soyuz, and the MIR space station are rarely greater than 300 km (160 nautical mi). Their orbits have an upper latitude limit of ca 50°N or S. Thus they cannot give global coverage, neither can their orbits be sun-synchronous. Thus far it has not proved practical to launch polar-orbiting manned spacecraft and U.S. plans to launch the Space Shuttle into polar orbits from the Vandenberg launch site on the West Coast (California) have not come to fruition. Mission durations are also generally short and data acquisition has been limited to the use of experimental techniques, some of which have been, or are expected to be, deployed on polar-orbiting unmanned satellites. These include hand-held cameras, the metric and large-format cameras, Shuttle imaging radar, modular opto-electronic multispectral scanner, thermal infrared multispectral scanner, and high resolution imaging spectrometer. Images from some of these instruments are available commercially. Initially, plans for the International Space Station included a sun-synchronous polar orbit, but the latest version (Alpha) is intended for a 440 km (230 nautical mi) orbit. Thus global coverage of the earth will not be possible from the Station itself, but ancillary, unmanned platforms are planned for environmental RS.

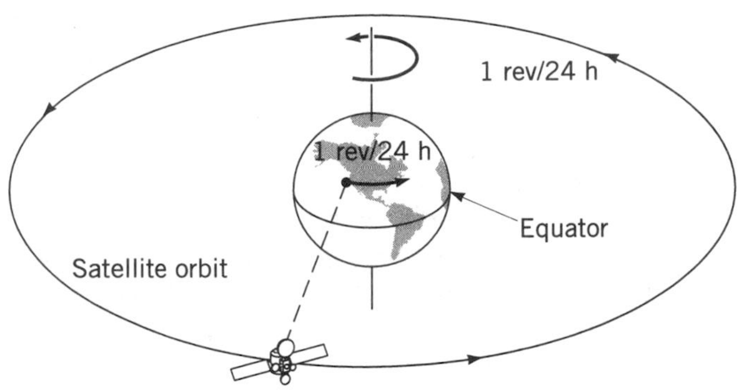

Figure 1. Geostationary orbit.

Table 1. Operational Earth RS Satellite Systems

Satellite	Operational Dates	Orbital Data	
		Property	Value
Landsat		Altitude, km	919
1	1972–1978	Inclination, deg	99.1
2	1975–1982	Period, min	103
3	1978–1983	ECT[a], am	9:30
Landsat		Altitude, km	715
4	1982–1997+	Inclination, deg	98.2
5	1982–1997+	Period, min	99
		ECT[a], am	9:45
SPOT		Altitude, km	830
1	1986–1997+	Inclination, deg	98.7
2	1990–1997+	Period, min	101
3	1993–1997+	ECT[a], am	10:30
ERS		Altitude, km	785
1	1991–1997+	Inclination, deg	98.5
2	1995–1997+	Period, min	101
		ECT[a], am	10:30
Radarsat		Altitude, km	791
1	1995–1997+	Inclination, deg	98.6
		Period, min	101
NOAA		Altitude, km	850
9	1984–1988 and 1993–1995	Inclination, deg	99
		Period, min	102
10	1986–1991	ECT[a], am	7:30, descending
11	1988–1993	ECT[a], pm	2:30, ascending
12	1991–1997		
14	1995–1997		

[a] ECT = Equator crossing time. All are given as local times. For NOAA, these are at launch.

Dedicated RS Satellites

Since 1972, several spacecraft have been launched to provide continuous monitoring of the earth's surface. They occupy near-polar orbits (Fig. 2) at heights between 700 and 1000 km (370–525 nautical mi). The advantages of this type of orbit follow:

- a small area of the earth's surface is covered by each image at remarkably high resolution
- the swathe of surface viewed on a particular orbit is repeated every 2–4 weeks which means that a particular small area can be imaged repeatedly with that frequency
- each image of a given small area is obtained under the same light conditions (the orbit is sun-synchronous)

There are currently five operational systems in polar orbit, for which orbital details are given in Table 1.

1. *Landsat* (U.S.): There have been four more spacecraft since Landsat 1 in 1972, but Landsat 6 was unfortunately lost after launch in 1993.
2. *Satellites Pour l'Observation de la Terre* (France): SPOT 1, SPOT 2, and SPOT 3 are currently operational.
3. *European Remote Sensing System* (European Space Agency, ESA): ERS 1 and ERS 2 are operational.
4. *Radarsat* (Canada): Radarsat 1 was launched in November 1995.
5. *Indian Remote Sensing satellites:* IRS-1B, IRS-1C, and IRS-P2 are currently operational. During the hiatus in the Landsat program, EOSAT (the U.S.-based company which receives and markets data from Landsats) signed a contract to market data from IRS satellites worldwide.

All five systems are operated by commercial organizations that also market "off-the-peg" images from their archives and "made-to-measure" images to suit customer requirements.

Other countries have, or are developing, RS spacecraft, including Japan (ADEOS, MOS, JERS) and Russia (Resurs). China has launched several short-lifetime satellites for earth RS, but is collaborating with Brazil in the development of monitoring RS satellites (CBERS) and with Canada in the field of earth observation by radar (1).

Orbital Coverage

Landsat orbits are typical of sun-synchronous satellites. It is the high inclination of the orbit to the equator that ensures that the satellite crosses the same latitude on successive orbits at the same local time on each successive orbit (the north to south motion of the satellite compensates for the rotation of the earth). Local times of equator crossing on each orbit for various satellites are given in Table 1.

On a Mercator projection map, the orbital tracks appear as a series of sigmoid curves. Figure 3 shows one day's descending orbits of Landsat. Ascending orbits pass on the night side of the earth. Each successive ground track is shifted 2100 km (1100 nautical mi) to the west, but each orbit is only 120 km (63 nautical mi) further west of the corresponding orbit the previous day. Because the swathe of surface imaged on each

Table 2. Sensor Characteristics of Operational Vis/IR Earth RS Satellites

Satellite(Sensor)[a]	Band	Waveband, μm	Resolution, m	Main RS Application Field
Landsat (MSS)	4	0.50–0.60	80	Mapping sediments, reefs, etc, in shallow water
1, 2, 3	5	0.60–0.70	80	Human-made features, ie, towns, roads, etc
	6	0.70–0.80	80	Vegetation studies, boundaries between landforms, land, and water
	7	0.80–1.10	80	As Band 6, best penetration of atmospheric haze
	8	10.40–12.50	120	Thermal imaging and mapping
Landsat (MSS)	1			
4, 5	2	Same as Bands 4, 5, 6, and 7, respectively, in Landsats 1, 2, and 3		
	3			
	4			
(TM)	1	0.45–0.52	30	Coastal waters, differentiation of soil/vegetation, tree types
	2	0.52–0.60	30	Vegetation mapping
	3	0.63–0.69	30	Differentiation of plant species
	4	0.76–0.90	30	Biomass surveys
	5	1.55–1.75	30	Snow/cloud differentiation
	6	10.40–12.55	120	Thermal mapping
	7	2.08–2.35	30	Vegetation moisture, hydrated minerals
SPOT (HRV)	XS	0.50–0.59	20	Vegetation studies, these 3 bands give improved spectral response to 3 important absorptions of chlorophyll
1, 2		0.61–0.68	20	
		0.79–0.89	20	
	PAN	0.51–0.73	10	Applications requiring fine geometric detail, eg, cartography
NOAA (AVHRR) 9, 10, 11, 12, 14	1	0.58–0.68	1100 (LAC) 4000 (GAC)	Snow and ice monitoring
	2	0.725–1.10	1100 (LAC) 4000 (GAC)	Location of water bodies, ice and snow melt, vegetation and agricultural assessment
	3	3.55–3.93	1100 (LAC) 4000 (GAC)	Sea surface temperature, land/water delineation, volcanic activity, forest fire monitoring, straw burning
	4	10.30–11.30	1100 (LAC) 4000 (GAC)	Sea surface temperature, soil moisture
	5	11.30–12.50	1100 (LAC) 4000 (GAC)	Same as Band 4

[a] MSS = multispectral scanner, HRV = high resolution visible imaging system, PAN = panchromatic (visible) mode, LAC = local area coverage, TM = thematic mapper, XS = multispectral mode, AVHRR = advanced very high resolution radiometer, and GAC = global area coverage.

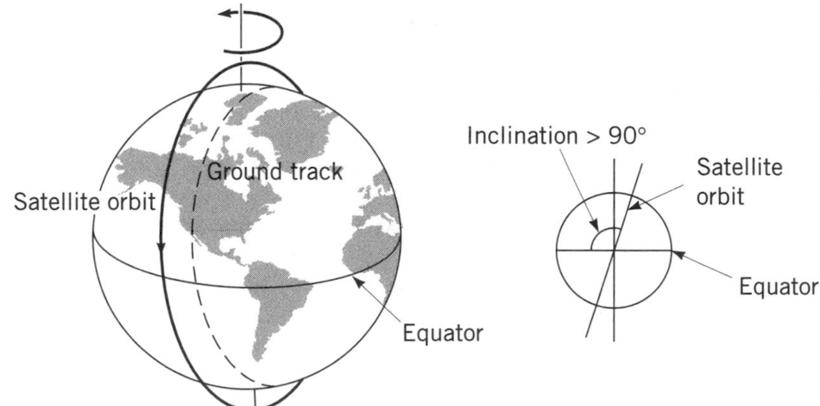

Figure 2. Near-polar orbit.

orbit is 185 km (97 nautical mi) wide, images taken from a particular orbit on successive days overlap (Fig. 4). The degree of overlap is greater at higher latitudes. This geometry means that swathes imaged coincide exactly at intervals shown in Table 1, every 16 days for Landsat 4 and 5. Put another way, the same scene is revisited by Landsat, at the same time of day and the same sun angle, every 16 days.

The NOAA system operates two satellites, one crossing the equator in the morning and the other in the afternoon. It is extremely important to remember that the Equator Crossing Times (ECTs) given for NOAA satellites in Table 1 apply only directly after launch. Because of the 102-minute nodal period of NOAA satellites, slightly more than 14 orbits are completed each 24 hours. This results in a nine-day period between

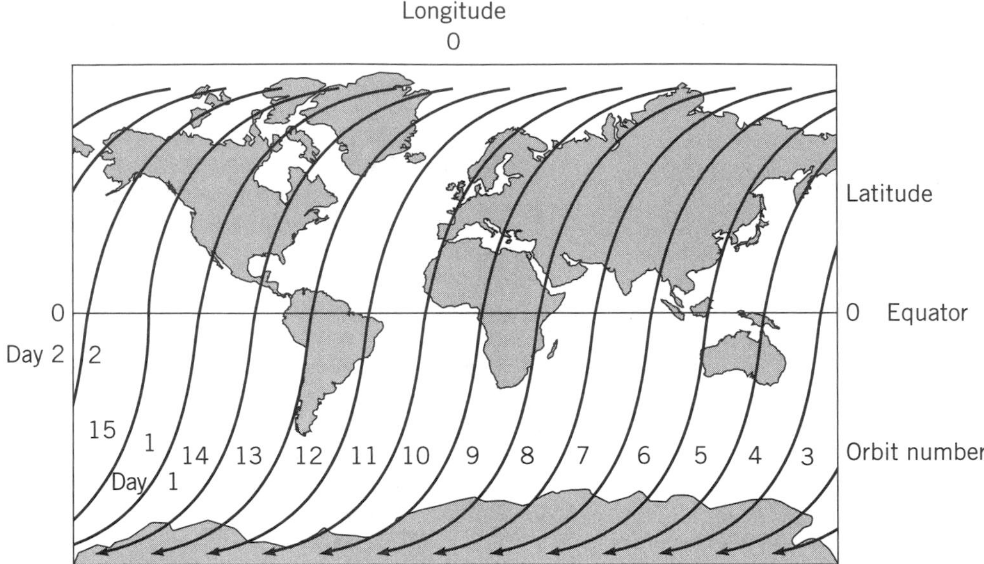

Figure 3. Landsat orbital tracks. Orbit 1 repeats at the Day 1 position every 18 days.

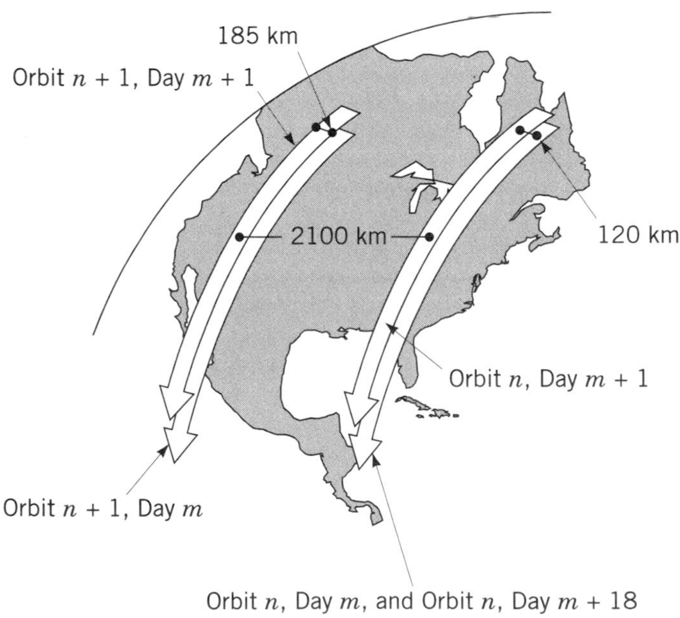

Figure 4. Overlap of Landsat orbital swathes.

revisiting of the same scene (cf 16 days for Landsat), and means that the geometry of a particular orbit is repeated 127 orbits later, although not exactly (2). This drift in the orbit results in a gradually changing ECT. For example, the ECT for NOAA-11 had become 17.00 by 1994. Respective night-time passes are 12 hours later than day-time passes. Each orbit moves 821 km (430 nautical mi) west each day, but this spacecraft images an extremely wide swathe, around 2800 km (1450 nautical miles). The high degree of overlap means that any part of the globe (except the poles) is revisited twice a day in daylight and can be imaged four times a day in the infrared, which can be used at night as well.

Future Satellites

Following the launch failure of Landsat 6, no projected date for the launch of Landsat 7 is available at the time of writing. Both satellites were intended to have modified payloads compared to Landsat 5, giving better resolution, but Landsat 7 will probably have a pointable "pushbroom" instrument (see below), in complete contrast to the rest of the Landsat series (3).

SPOT 4 is in the course of production and scheduled for launch at the end of 1997. It will have the new SPOT Mk. 2 platform, which is strengthened, more powerful, and will have an increased lifetime of five years. The high resolution visible and infrared (HRVIR) instruments, derived from the current HRVs (see Table 2), will have a new channel in the medium-IR giving applications to the analysis of vegetation and harvest forecasting. In addition, there will be the vegetation monitoring instrument, which will give daily global coverage to a resolution of 1 km (ca 0.5 nautical mi) in the same bands as HRVIR. Giving a larger field of view, it will be complementary to the higher resolution instruments. The PASTEL optical link (part of ESA's SILEX program) will enable experiments to be done via high speed interorbit laser communications. SPOT 5A and SPOT 5B are planned to be launched around 2002 and 2007, respectively. Their payload will consist of three of the new, improved high resolution geometry (HRG) instruments. Each will give resolution down to 5 m, one looking forward, one downward, and one backward.

The lifetime of Radarsat 1 was originally projected as five years but, because of the accuracy of the orbit achieved, this has been extended by two and a half years. The satellite became operational on March 21, 1996, and is producing good data (4). The launch of Radarsat 2, an identical satellite, is planned for 2000, but will probably be set back because of the extension to Radarsat 1's lifetime. A third, improved satellite is in the planning stage and may be a joint effort between the Canadian Space Agency and SPOT Image Corporation.

ENVISAT 1 (ESA) is the heaviest and most complex spacecraft yet designed for RS. It will use the SPOT Mk. 2 platform to carry two metric tons (4400 lbs) of optical and microwave instruments in a near circular, polar orbit of about 800 km (420 nautical mi) altitude. Many different instruments are to be carried, giving a wide variety of applications and launch is scheduled for December 1998.

The Earth Observing System (EOS) is planned to operate in conjunction with the International Space Station. It will initially comprise of three unmanned platforms carrying a wide variety of earth RS instruments in sun-synchronous, polar orbits at around 700–800 km (370–420 nautical mi) altitude. One is to be provided by ESA (*Columbus*) and will have a morning equator crossing, the other two will be provided by the United States and will have afternoon equator crossings. Japan may construct a fourth platform at a later date. As mentioned earlier, this altitude is presently too high for a permanent manned station and the EOS platforms will periodically descend to lower altitudes for the servicing necessary to maintain the indefinite monitoring lifetime expected of a permanently manned facility. This servicing will be carried out by the Space Station, a Space Shuttle Orbiter, or Hermes (ESA). The first U.S. launch (EOS-AM1) is planned for June 1998. Its 1155-kg (2541-lb) payload will include, *inter alia,* a Japanese very near-IR (VNIR) radiometer giving 15 m (50 ft) resolution and a Canadian instrument for measurement of pollution in the troposphere.

IRS-1D, to be launched in 1997, will be similar to IRS-1C, with an improved camera operating in three bands in the visible and near-IR giving a resolution of 70 m (230 ft) and a panchromatic camera giving better than 10 m (30 ft) resolution. A wide-field VNIR sensor has a resolution of 188 m (610 ft) for a swathe width of 774 km (410 nautical mi). The IRS-P series use advanced technologies which reduce weight and enable the two linear imaging self-scanners to have identical bandwidths. Future satellites in this series will carry payloads intended for specific earth observation applications (5).

The Japanese Advanced Earth Observing Satellites are intended to observe global environmental changes. ADEOS-1 was launched into a 798 km near-polar orbit in August 1996.

The first of the cooperative China–Brazil Earth Resources Satellites, CBERS-1, has a planned launch in late 1997 or early 1998. Brazil has completed construction of the equipment (including a wide-field camera) and qualification tests are being carried out by China, which will launch the spacecraft.

BIBLIOGRAPHY

1. "China Space News", *Spaceflight* **38**(12), 398 (Dec. 1996).
2. G. Gutman, *J. App. Meteorol.* **27**(9), 973 (Sept. 1988).
3. T. Pirard, *Spaceflight,* **37**(1), 20 (Jan. 1995).
4. B. Shanko, *Spaceflight,* **39**(1), 6 (Jan. 1997).
5. H.P. Mama, *Spaceflight* **38**(6), 209 (Jun. 1996).

General References

A.P. Cracknell and L.W.B. Hayes, *Introduction to Remote Sensing,* Taylor and Francis, London, 1991.

P.J. Curran, *Principles of Remote Sensing,* Longman Scientific and Technical, Harlow, U.K., 1985.

S.A. Drury, *A Guide to Remote Sensing,* Oxford University Press, Oxford, U.K., 1990.

T.M. Lillesand and R.W. Kiefer, *Remote Sensing and Image Interpretation,* 2nd ed., John Wiley & Sons, Inc., New York, 1987.

K-H. Szekielda, *Satellite Monitoring of the Earth,* Wiley-Interscience, New York, 1988.

A Guide to Earth Observing Satellites, National Remote Sensing Centre Ltd., Farnborough, U.K., 1990.

Remote Sensing, Course Book 1, Open Universiteit, Heerlen, the Netherlands, 1989.

ECOLOGICAL TOXICITY TESTING

JOHN CAIRNS, JR.
B. R. NIEDERLEHNER
J. R. BIDWELL
Virginia Polytechnic Institute and State University
Blacksburg, Virginia

Ecotoxicology is the study of the adverse effects of materials on ecological systems. Ecotoxicologists assess existing environmental condition, monitor trends in environmental condition over time, predict environmental effects of proposed actions, diagnose causative agents for existing environmental degradation, and guide the selection of remediation practices to maximize ecological benefits. All of these environmental management activities require sound scientific tools for evaluating the relative risks that various materials and actions pose to the environment. One of these tools is the ecological toxicity test. An ecological toxicity test characterizes the nature and magnitude of an ecological response to a chemical. In an ecological toxicity test, a manageable piece of an ecosystem is exposed to a material or action and responses are observed. Obviously, the most convincing evidence for the success of environmental management practices is obtained by observing the ecological health of the natural systems being protected. However, field surveys assessing existing environmental condition cannot provide all the information managers need to integrate ecological concerns into day-to-day planning and decision making. Ecological toxicity tests are essential if information about a proposed action is to be evaluated before the fact, if multiple causative agents and confounding factors occur in the field, or if a definitive cause and effect between a chemical and an adverse ecological effect must be established.

A BRIEF HISTORY

Humankind has employed poisons throughout recorded history, but a more formal discipline of experimental toxicology arose in the early 1800s in concert with developments in organic chemistry. The field was primarily concerned with the beneficial uses and adverse effects of the ever increasing number of chemicals isolated and synthesized as chemical science progressed. Human health and well being were the primary focus, but studies often employed any convenient mammal. Although some ecological toxicity tests are similar to toxicity tests for assessing effects on human health, there are important differences. The object of protection in tests to assess effects on health is one single species, *Homo sapiens.* In contrast, the object of protection in ecotoxicology is the thousands of species and all their interactions that characterize any ecosystem.

Ecological toxicity testing also has practical roots. Gross environmental damage from human use was judged unacceptable. To repair and prevent such damage in the future, increased attention to the environmental consequences of human actions was necessary. Tests with nonhuman organisms were conducted in the early 1800s to answer questions about adaptation (2). However, tests arising from a concern about the effects of human wastes on other species arose later (3). The most developed ecological toxicity test methods are those for aquatic systems. Rivers have been used as convenient and effective waste disposal systems since the beginning of civilization. But, increasing population made the adverse effects of sewage on river ecosystems readily apparent; rivers could no longer be used in all the ways they previously had. The problem was already well understood at the turn of the century as seen in the saprobic system developed (4,5). In addition, the relationship between mining wastes and the destruction of river fisheries was investigated with early toxicity tests (6–8). By the 1940s, standard methods for controlled ecological toxicity tests with fish were available (9). Methods to assess materials in soil, sediment, and air followed. Although ecological toxicity test methods continue to develop and expand, the basic approach remains unchanged. Ecological toxicity tests expose a piece of an ecosystem to a material and observe the response.

PURPOSES OF ECOLOGICAL TOXICITY TESTS

The general goal of any ecological toxicity test is to characterize the ecological response to a chemical. However, at least five distinct purposes can be given for any particular ecological toxicity test: (*1*) prediction, (*2*) diagnosis, (*3*) ranking, (*4*) regulation, or (*5*) contaminant mapping. Distinguishing among these purposes is important because a test designed for one purpose is unlikely to produce information optimal for another purpose. Although ecological toxicity tests are also occasionally used to assess existing conditions in natural systems, field surveys are often better tools (10).

Prediction

To prevent damage, one must predict the ecological consequences of an action before it is taken. The ecological toxicity test tries to answer the question, What will happen if this chemical is released? Predictions consider environmental factors early in the planning and design process, before there is a discharge and damage. For example, in the United States, the Toxic Substances Control Act requires that chemical manufacturers provide data on the environmental effects of a chemical before it is manufactured on a large scale. If a chemical has severe environmental consequences, alternative chemicals are preferentially developed. Other predictive applications include pesticide registration and evaluation of toxicity reduction.

One formal approach to predicting environmental effects is ecological risk assessment (11). Ideally, an ecological risk assessment produces a probabilistic estimate of the outcome of a human action in a form similar to a weather report, ie, instead of a 30% chance of showers, a 30% chance of a 50% decrease in catch per unit effort for walleye. Ecological risk assessments have three basic elements: problem formulation, analysis, and risk characterization. Ecological toxicity testing is an essential component of the analytical stage in which likely environmental exposures and likely ecological effects at various exposure levels are determined. After this information is combined into a characterization of risk, a risk management stage incorporates political and social considerations into a comparison of the relative risks of alternative actions. An example of an ecological risk assessment and its use of data from ecological toxicity tests is available (12).

Another predictive application of ecological toxicity tests is characterizing a progression of impact from early, subtle damage to severe damage with continued exposure. After ecological toxicity tests identify subtle responses that precede severe damage, field-monitoring programs use these responses for an early warning of environmental degradation before damage proceeds to unacceptable levels (13). The end point early in the progression of impact is used to anticipate other end points associated with unacceptable damage.

Diagnosis

In the case of obvious environmental damage, the causative agent must be identified before the problem is remediated. In natural systems, multiple factors often contribute to the observed damage. Controlled tests separate factors that co-occur in the real world to establish which factors or combination of factors cause the observed effects. In this case, the purpose of the ecological toxicity test is to answer the question, Can this chemical cause this type of damage?

The simple co-occurrence of a damage and a chemical does not establish that one causes the other, ie, correlation is not causation. For example, it was once believed that breathing swamp air caused malaria because nothing was known about mosquitoes and their parasites. The process of building a case for cause and effect between an agent and environmental degradation is derived from the example suggested by Robert Koch, a German microbiologist investigating tuberculosis (14,15). The ecotoxicological equivalent of Koch's postulates include the following:

1. The damage is regularly associated with the toxicants; they occur in the same space and time, but damage must follow exposure. The evidence for cause and effect is stronger if the association is observed at multiple locations that vary in many other environmental factors such as habitat, soil or water chemistry, community structure, or climate. The cause-and-effect relationship is supported further if responses and exposure change together along a gradient from mild exposure and response to severe exposure and response.
2. An indicator of exposure to the toxicant is found in damaged organisms (eg, body burdens or changes in enzyme activity related to the toxicant, such as D-aminolevulinic acid dehydratase for lead exposure or metallothionein for metal exposure).
3. Damage is seen when normal organisms are exposed to the toxicant; this is ecological toxicity testing.
4. The same damage and the same indicator of exposure are seen in both the field observations and the ecological toxicity test. Few ecological problems have satisfied these rigorous requirements for establishing definitive cause and effect (15). The relationship between the decline of the perigrine falcon and the effect of DDT on egg

viability is one case study where these data have been collected.

In most cases, the pressures of environmental decision making necessitate decisions based on available information. In many cases, the preponderance of evidence is clear without definitively establishing cause and effect. So, although ecological toxicity tests are necessary to assign definite causation, it is possible to make reasonable judgments about causation without meeting all of the criteria.

Ranking

Comparisons are necessary to choose between courses of action based on environmental considerations. There are 65,000 chemicals in daily use (16). Which chemicals should be of concern? On which chemicals should limited time and resources be spent for cleanup? Ecological toxicity tests provide an objective basis for ranking risks. The results of standardized ecological toxicity tests provide basic and essential comparative information about the nature of a chemical. This basic information about the biological activity of a chemical is comparable to simple characterizations of its water solubility or vapor pressure (17). Such information is used to screen for ecological effects at the earliest stages in planning an action. For example, at a concentration of 13 ppt ethanol, rainbow trout die; however, a concentration of 0.005 ppm DDT causes similar effects. If a particular task is accomplished by using either ethanol or DDT in similar concentrations, ethanol is the better choice based on relative toxicity. Similar rankings allow assessing the relative benefits of different waste treatment processes based on biological benefit.

Regulation

Some regulatory applications of toxicity tests call for absolutely objective, equitable, and decisive standards for permissible discharges (17). Either a waste passes the biological response-based standard and is discharged, or it fails and requires additional treatment. In this case, fairness is as important as realism, and large safety factors mitigate the lack of realism.

Contaminant Mapping

The importance of contamination at hazardous waste sites is sometimes mapped by ecological toxicity tests in addition to chemical analyses (18). Chemicals can have adverse biological effects at very low concentrations, particularly in complex mixtures. In the case of some complex mixtures, complete chemical characterization is extremely expensive and still does not adequately predict biological hazard because of uncharacterized interactions between component chemicals in producing biological effects.

END POINTS, SCALE, AND COMPLEXITY

End Points

The number of ecosystem attributes that might be measured in response to stress is virtually unlimited. Obviously, not all can be monitored. End points are those few responses chosen for monitoring in ecological toxicity tests.

Various desirable qualities in an end point have been suggested (Table 1). Any scientifically defensible end point must be sensitive, measurable, and interpretable. However, some of the remaining desirable characteristics of end points are mutually exclusive. For example, a single end point cannot be both broadly applicable and diagnostic. Similarly, a timely end point may not be the most socially relevant because it must precede obvious and important damage to be timely. The tradeoffs between these characteristics depend on the specific purpose of the test as previously described and on scale.

The biological units affected by a chemical stress vary from organelles or enzymes within individual cells to landscapes. Each level of this scale of biological organization has its own characteristic structure and behavior (Table 2). Many appropriate end points for ecological toxicity tests have been suggested for each of these biological units (19,22,23). Table 3, drawn from these various sources, provides a list of end points appropriate for ecological toxicity tests. Both structural and functional end points can be measured. Structural end points measure the numbers and kinds of components at one time, and functional end points describe performance.

Scale

The fourteenth desirable property of an end point listed in Table 1 suggests that the end point must have scale appropriate to the management problem addressed. Many scales are important: spatial scale, temporal scale, scale of biological organization, and progression of impact (24).

Environmental problems have their own inherent scales based largely on the spatial and temporal distribution of the contaminant but also on the size, range of movement, and life span of the biological units affected. Spatial and temporal scales co-vary. A point source discharge of a degradable material into water or soil is a local problem whose effects are restricted in time and space. In contrast, air pollution and pesticide usage are problems spanning entire regions. Chernobyl affected a continent, and carbon dioxide accumulations in the atmosphere affect the entire globe. Effects of chemicals also vary across a scale of impact from subtle to gross. In this progression of impact, given equal exposure, some effects precede others because they occur earlier in a causal chain. For example, physiological changes in sensitive species precede community level structural responses, such as loss of species, because they occur earlier in a causal chain. Physiological changes are the mechanism by which competitive dynamics change. A species suffering from chemically induced physiological inefficiencies may undergo local extinction. Not only that species, but all species functionally similar in the community and the entire pool of immigrating species that can functionally replace the extinct species, must be adversely affected before a loss of taxa occurs. As such, end points early in a progression of impact are affected at lower concentrations than end points of biological units with compensatory mechanisms. Structural and functional changes alternate as chemical effects are evaluated along a causal chain upward in a hierarchical system of biological units (see Table 2). Often, a functional end point or mechanism is used to anticipate a structural change at a greater hierarchical level and spatiotemporal scale.

Scale is also an important covariable when extrapolating from smaller scale ecological toxicity test systems to ecological outcomes in larger natural systems. The spatial scale of the

Table 1. Ideal Properties of End Points Selected for Ecological Toxicity Tests[a]

Number	Description
(1)	*Biologically relevant,* ie, important in maintaining a balanced community.
(2)	*Socially relevant,* ie, of obvious value to and observable by shareholders or predictive of a measure that is.
(3)	*Sensitive* to stressors without an all-or-none response or extreme natural variability.
(4)	*Broadly applicable* to many stressors and sites.
(5)	*Diagnostic* of the particular stressor causing the problem.
(6)	*Measurable,* ie, capable of being operationally defined and measured by a standard procedure with documented performance and low measurement error.
(7)	*Interpretable,* ie, capable of distinguishing acceptable from unacceptable conditions in a scientifically and legally defensible way.
(8)	*Cost effective,* ie, inexpensive to measure, providing the maximum amount of information per unit effort.
(9)	*Integrative,* ie, summarizing information from many unmeasured indicators.
(10)	One for which *historical data is available* to define nominative variability, trends, and possibly acceptable and unacceptable conditions in a variety of natural systems.
(11)	*Anticipatory,* ie, capable of providing an indication of degradation before serious harm occurs, early warning.
(12)	*Nondestructive* of any ecosystem.
(13)	One with potential for *continuity* in measurement *over time.*
(14)	Of an *appropriate scale* to the management problem being addressed.
(15)	*Not redundant* with other measured indicators, ie, providing unique information.
(16)	*Timely,* ie, providing information quickly enough to initiate effective management action before unacceptable damage occurs.

[a] Ref. 20 with permission.

physical model for ecological toxicity tests predictably affects the richness of species and the relative importance of "edge effects" (25). Extrapolative models must account for these distortions.

Complexity

As the scale of a problem increases, the end points characteristic of that scale change and the complexity of the biological system affected increases, ie, more components are interacting and contributing to and modifying any biological response. Ideally, all relevant factors, ie, all those behaviors and interactions that materially affect outcome, are incorporated into the ecological toxicity test or into the model used to extrapolate ecological toxicity test results into outcome predictions in natural systems. Simple tests require the most complex extrapolative models to translate measurement end points into predictions of relevant adverse effects in natural systems. Conversely, some site-specific complex ecological toxicity tests require simpler extrapolative models.

Without a substantial amount of background information about an ecosystem, it is very difficult to know which factors are critical to include in either physical or extrapolative models. Many notable failures in predicting environmental effects have been errors of omission (26). For example, if sediment had been included in early tests on the toxicity of mercury, its subsequent methylation, biological uptake, and adverse effects would have been no surprise. This example is a powerful argument for conducting more inclusive tests as at least one part of an assessment (26–29).

Although some community level tests are quite practical, there are severe practical limits to the size of an ecological unit for ecological toxicity tests. It is expensive and unethical to pollute experimentally many fields, lakes, streams, or watersheds. So when an environmental problem is inherently large and complex, many smaller tests that provide models for the various forms of interactions are the only practical approach (30). From these pieces of information, an extrapolative model is constructed.

Before the information from ecological toxicity tests and their accompanying extrapolative models are useful for environmental decisions, some index of their quality is needed. Validation compares predictions to observed effects in natural systems and characterizes the margins of error. Ideally, validations compare predictions and real-world outcome on the same scales. However at times, validations for large-scale models must also proceed component by component for practical reasons (30).

Tradeoffs in Approaches

In summary, ecological toxicity tests on small, simple pieces of ecosystems have the advantages of being quick,

Table 2. Levels of Biological Organization and Their Characteristic Structural and Functional Attributes[a]

Organization	Attributes
Individuals	One whole organism. Individuals have size, shape, health or condition; they grow, reproduce, and die over time.
Populations	A group of individuals of the same species occupying the same area at a given time. Populations have abundance, biomass, and size and age class structures; they compete, exploit prey, and produce biomass.
Assemblages	A set of coexisting populations defined by phylogeny, location, or lifestyle. Assemblages are intermediate to populations and communities and have some of the properties of both groups depending on how they are defined.
Communities	Collections of all organisms that live in a specific region at a given time. Communities have biomass, diversity, evenness, richness, connectedness, and a trophic structure; they produce biomass, process materials, and change by succession over time.
Ecosystem	All organisms in an area together with the physical environment with which they interact. Ecosystems have biological, physical, and chemical structure; they move energy, materials, and nutrients.
Region	An area greater than 1 km^2 that includes heterogeneous patches. Regions are usually defined in terms of ecological characteristics (ie, ecoregions), watersheds, or airsheds, but occasionally political or management boundaries are used. Regional structure is characterized by geology, soils, hydrology, climate, land use, and organisms. Functional attributes of regions include hydrologic cycling, nutrient cycling, exchanges between heterogeneous units, and rates of change in structural characteristics.

[a] Ref. 21 with permission.

Table 3. End Points at Various Levels of Biological Organization

Hierarchical Level	Structural End Points	Functional End Points
Individual	Condition	Growth
	Overt symptomatology (eg, histopathology, skeletal abnormalities)	Fecundity
	Tissue concentration	Neural function
	DNA breakage	Enzyme function
	Death	Avoidance
		Prey-capture rate
Population	Occurrence	Reproduction
	Abundance	Yield/production
	Age structure	Rate of gross morbidity
	Age specific morality	
Community/ecosystem	Number of species	Connectedness
	Species evenness	Production (P)
	Species diversity	Respiration (R)
	Trophic structure	P/R
	Biomass (B)	P/B
	Tolerance (eg, saprobian index)	R/B
	Dominance	Material cycling/export
	Proportion exotics/hybrids	Extinction rate
Region/landscape	Habitat proportions	Material cycling/export
	Patch size	Rate of pest outbreaks, floods, fires, low flows
	Perimeter-to-area ratio	Regional production
	Contagion	Resistance to stress
	Connectivity	Sustainability
	Net embodied energy	

inexpensive, and easy to replicate. These tests are repeatable, they are excellent ranking tools, and provide a basic index of the biological activity of a chemical. However, the ability to predict environmental outcome in complex natural systems from these tests often depends on unvalidated models or large safety factors. Although the cause of observed changes is known, the nature and magnitude of the environmental outcome in the real world remain undetermined. Ecological toxicity tests with large, complex pieces of ecosystems (ie, artificial ponds) allow measuring the same end points of concern in natural systems, but they are difficult to replicate and expensive to test. Including a number of interactions allows for many indirect effects, so the ecological response may be realistic but not repeatable or generalizable. So, there is an inescapable tradeoff between precision and accuracy in selecting the scale of an ecological toxicity test.

BARRIERS TO PROGRESS AND FUTURE TRENDS

Ecological toxicity testing remains unsurpassed for screening potential toxicants before they reach the environment. By having an evidence-based estimate of the concentrations likely to cause harm to ecosystems for certain periods of exposure under certain conditions, a dramatic reduction is possible in the number of mistakes in using pesticides, industrial chemicals, and a variety of wastes.

Despite the often stated uniqueness of each ecosystem, ecosystems share much in common: energy flow, nutrient cycling, community structure, numbers of organisms in various taxonomic categories, and a large number of other attributes. In the same way, humans are not identical, but they share much in common. Therefore, ecological toxicity testing should focus on the attributes and generalities that ecosystems share while always bearing in mind their unique attributes. Several studies (31–33) discuss some of these commonalities in the responses of different ecosystems to stress.

Some uncertainty about the ecological effects of toxicants is unavoidable. However, uncertainty is dramatically reduced by validating or calibrating predictive models, that is, if a prediction is made that a particular pesticide acts in certain ways when applied under certain conditions, this prediction is confirmed by testing under natural conditions in test plots or, ultimately, in real ecosystems or agricultural systems. If the ecological response is as predicted, this constitutes confirmation of the predictive model, and, if not, the experiment usually indicates how the predictive model can be improved.

It is of interest to human society to carry out ecological toxicity tests because compounds harmful to other species are very frequently, although not invariably, harmful to humans. In a very real sense, these organisms provide an early warning system to alert humans to the dangers of various chemical compounds. In the early days, monarchs had wine tasters to be sure that the wine was not poisoned before they drank it. It was hard on the wine taster, but protected the king. In the early days before sophisticated air analysis was available, coal miners in deep mines used canaries to be sure that the atmosphere was breathable and not dangerous to them. When the canary died, the coal miners quickly left the mine until the situation was rectified. Some situations exist where toxicity tests are not, of themselves, accurate predictors of environmental risk. Chlorofluorocarbons (CFCs) proved harmless to humans and other organisms following direct exposure. In fact, metered dose sprays for people with asthma use CFCs to carry the active ingredient. It is an indirect effect that may cause unacceptable harm. The CFCs damage the ozone layer (as the well-publicized holes demonstrate) that protects humans and other species from ultraviolet radiation. It is these indirect effects that are unlikely to be identified if only simple toxicity tests are conducted.

Human society's life support system is both technological and ecological (34). The capacity of natural systems to continue providing ecosystem services is not infinite. The natural systems providing these services are being destroyed at an unprecedented rate at the same time that the human population is growing at an unprecedented rate. Therefore, balancing the benefits of technological services upon which human society now greatly depends with those ecological services on which it is even more dependent is necessary. Ecological toxicity testing provides a crucial set of data to make this possible, although it is only one of the many components of an effective sustainable use management policy. Ecological toxicity testing will be needed more in the future than it is today.

BIBLIOGRAPHY

1. J.A. Zapp, Jr., in J.G. Eaton, P.R. Parrish, and A.C. Hendricks, eds., *Aquatic Toxicology,* STP 707, American Society for Testing and Materials, Philadelphia, Pa., 1980, pp. 2–10.
2. B.G. Anderson, in A.L. Buikema, Jr. and J. Cairns, Jr., eds., *Aquatic Invertebrate Bioassays,* STP 715, American Society for Testing and Materials, Philadelphia, Pa., 1980, pp. 3–35.
3. W.J. Adams, in D.J. Hoffman, B.A. Rattner, G.A. Burton, Jr., and J. Cairns, Jr., eds., *Handbook of Ecotoxicology,* Lewis Publishers, Boca Raton, Fla., 1995, pp. 25–46; C. Penny and C. Adams, *R. Comm. Pollut. Scotland, Lond.* **2,** 377 (1863).
4. R. Kolkwitz and K. Marsson, *Ber. dt. bot. Ges.* **26A,** 505–519 (1908).
5. R. Kolkwitz and K. Marsson, *Int. Revue ges. Hydrobiol. Hydrogr.* **2,** 126–152 (1909).
6. K.E. Carpenter, *Ann. Appl. Biol.* **11,** 1–23 (1924).
7. K.E. Carpenter, *Ann. Appl. Biol.* **12,** 1–13 (1925).
8. K.E. Carpenter, *British J. Exp. Biol.* **4,** 378–390 (1927).
9. W.B. Hart, P. Doudoroff, and J. Greenbank, *The Evaluation of the Toxicity of Industrial Wastes, Chemicals and Other Substances to Fresh Water Fishes,* Waste Control Laboratory, Atlantic Refining Co., Philadelphia, Pa., 1945.
10. J.R. Karr, *Ecol. Appl.* **1,** 66–84 (1991).
11. S.B. Norton, D.J. Rodier, J.H. Gentile, M.E. Troyer, R.B. Landy, and W. van der Schalie, in D.J. Hoffman, B.A. Rattner, G.A. Burton, Jr., and J. Cairns, Jr., eds., *Handbook of Ecotoxicology,* Lewis, Boca Raton, Fla., 1995, pp. 703–716.
12. K.R. Solomon, D.B. Baker, R.P. Richards, K.R. Dixon, S.J. Klaine, T.W. LaPoint, R.J. Kendall, J.M. Giddings, J.P. Giesy, L.W. Hall, Jr., and W.M. Williams, *Environ. Toxicol. Chem.* **15,** 31–76 (1996).
13. J.F. McCarthy and L.R. Shugart, *Biomarkers of Environmental Contamination,* Lewis, Boca Raton, Fla., 1990.
14. A.S. Evans, *Yale J. Biol. Med.* **49,** 175–195 (1976).
15. G.W. Suter, II, ed., *Ecological Risk Assessment,* Lewis, Boca Raton, Fla., 1993.
16. T.H. Maugh, III, *Science* **220,** 293 (1983).
17. D.I. Mount, "Development and Current Use of Single Species Aquatic Toxicity Tests," in J. Cairns, Jr. and B.R. Neiderlehner, eds., *Ecological Toxicity Testing; Scale, Complexity, and Relevance,* Lewis, Boca Raton, Fla., 1995, pp. 97–104.
18. M. Simini, R.S. Wentsel, R.T. Checkai, C.T. Phillips, N.A. Chester, M.A. Major, and J.C. Amos, *Environ. Toxicol. Chem.* **14,** 623–630 (1995).
19. G.W. Suter, II, *Environ. Manage.* **14,** 9–23 (1990).
20. J. Cairns, Jr., P.V. McCormick, and B.R. Niederlehner, *Hydrobiologia* **263,** 1–44 (1993).
21. J. Cairns, Jr., B.R. Niederlehner, and E.P. Smith, in G.M. Rand, ed., *Fundamentals of Aquatic Toxicology: Effects, Environmental Fate, and Risk Assessment,* 2nd ed., Taylor and Francis, Washington, D.C., 1995, pp. 589–608.
22. J.R. Kelly and M.A. Harwell, in S.A. Levin, M.A. Harwell, J.R. Kelly, and K.D. Kimball, eds., *Ecotoxicology: Problems and Approaches,* Springer-Verlag, New York, 1989, pp. 9–35.
23. K. Macek, W. Birge, F.L. Mayer, A.L. Buikema, Jr., and A.W. Maki, in J. Cairns, Jr., K.L. Dickson, and A.W. Maki, eds., *Estimating the Hazard of Chemical Substances to Aquatic Life,* STP

657, American Society for Testing and Materials, Philadelphia, Pa., 1978, pp. 27–32.

24. J. Cairns, Jr. and D.S. Cherry, in P. Calow, ed., *Handbook of Ecotoxicology,* Blackwell Scientific Publications, London, 1993, Vol. I, pp. 101–116.
25. K.T. Perez, in J. Cairns, Jr. and B.R. Niederlehner, eds., *Ecological Toxicity Testing: Scale, Complexity, and Relevance,* Lewis, Boca Raton, Fla., 1995, pp. 49–72.
26. K.D. Kimball and S.A. Levin, *Bioscience* **35,** 165–171 (1985).
27. National Research Council, *Testing for Effects of Chemicals on Ecosystems,* National Academy Press, Washington, D.C., 1981.
28. E.P. Odum, *BioScience* **34**(9), 558–562 (1984).
29. R.V. O'Neill and J.B. Waide, in B.W. Cornaby, ed., *Management of Toxic Substances in our Environment,* Ann Arbor Science Publishers, Ann Arbor, Mich., 1981, pp. 43–73.
30. J.D. Aber, C. Driscoll, D.A. Federer, R. Lathrop, G. Lovett, J.M. Melillo, P. Steudler, and J. Vogelmann, *Ecol. Model.* **67,** 37–47 (1993).
31. E.P. Odum, *BioScience* **35,** 419–422 (1985).
32. D.L. Rapport, H.A. Regier and T.C. Hutchinson, *Am. Nat.* **125,** 617–640 (1985).
33. D.W. Schindler, *Oikos* **57,** 25–41 (1990).
34. J. Cairns, Jr., in P.C. Schulze, ed., *Engineering Within Ecological Constraints,* National Academy Press, Washington, D.C., 1996, pp. 13–30.

ECOTOXICOLOGY. See BIOMARKERS.

ECOZONES, GLOBAL

JÜRGEN SCHULTZ
Geographisches Institut
Aachen, Germany

Ecozones are global divisions that have their own characteristic interplay of climatic factors, morphodynamics, soil-forming processes, living conditions for plants and animals, and production potentials for agriculture and forestry. Correspondingly, they are distinguished by different climates, landforms, soil units, plant formations and biomes, and land use systems. The major aim, by delimitating and characterizing the ecozones, is to aid in establishing a kind of *global classification system* (orientation guide) which allows an immediate listing of several important characteristics of any area on earth, and is suitable as a basis for further detailed investigation (starting with the question: In what way does a certain area in an ecozone differ from the overall characteristics of the ecozone in which it is found?).

METHODOLOGY OF DIVIDING THE WORLD INTO ECOZONES

Anyone who travels, for example, from a place in New England to places in northern Alaska, central Ontario, Kansas, or coastal regions of California, realizes immediately that the scenery changes by many aspects, including climate, vegetation cover, soil types, crops, and animal husbandry. Most likely this is what a traveler expects (and wants) to find. If this traveler goes abroad, eg, to various places in Europe and Asia, it might, however, be surprising to come across landscapes that are quite similar to those in North America; eg, in central Europe summer-deciduous forests are similar to those in the northern United States; in northern Europe and Siberia there is tundra similar to that in Barrow, Alaska; farther south, boreal forests may be found that are like those in Ontario, Canada; in the Ukraine and further east, there are steppes like in the Great Plains of Kansas; and around the Mediterranean Sea evergreen shrub formations are similar to the chaparral in California. All these vegetational affinities go together with similarities in climate, soils, and land use, despite the great distances between the continents and despite the different features in land–sea relation, geology, and history, respectively. In fact, there are many more conformities between the tundras, boreal forest regions, steppe-lands, etc, of North America and Eurasia than between these land types within North America or Eurasia.

In principle, the same applies to all tropical regions, eg, to the tropical rainforests, savannas, and deserts found in South America, Africa, southern Asia, and Australia. Anyone who travels from the forests of the Amazonas Basin to those of the Congo Basin (or Malaysia), or from the open grasslands of Llanos del Orinoco to those of Central African Republic or of Northern Territory in Australia will have, at first, the impression of being in the same (or very similar) environment (surrounding).

It is these far-reaching, conspicuous conformities, found in regions sometimes far apart from each other, that justify a differentiation of earth, independent of orographical, geological, cultural, political, and economical peculiarities, into basic zonal land units. With respect to the comprehensive nature of these land units (they are, up to a certain degree, equally climatic zones, vegetational zones, soils zones, agricultural zones, etc) it is most adequate to call them *ecozones.*

Ecozones are global divisions which have their own characteristic interplay of climatic factors, morphodynamics, soil-forming processes, living conditions for plants and animals, and production potentials for agriculture and forestry. Correspondingly, they are distinguished by different climates, landforms, soil units, plant formations and biomes, and land use systems. The term *ecozone,* as defined in this way, was introduced in 1988 (1). In the (hierarchical) system of regional classification, in which the basic unit is the *ecotope,* this new term represents the highest class heading (or the first subdivision of the *ecosphere*). If necessary, further subdivisions can be included between, for which, more recently, terms such as *ecoregions, ecoprovinces,* and *ecodistricts* have been proposed (2,3).

Existing studies on major land ecosystems that have been published since the 1970s in relatively high numbers and great detail suffer from the fact that individual aspects (mainly botanical–ecological ones) are given the utmost consideration, whereas other aspects (often pedological and geomorphological ones) are almost entirely neglected. The same applies to the some publications on ecological divisions of the geobiosphere (4,5).

Furthermore, various studies have often failed to establish consistent terminology, units of measurement, and organization of the data for all global ecosystems, thus making it difficult to make comparisons. The individual nature of each ecozone therefore remains unclear, since it is precisely these differences that allow those characteristics peculiar to a specific ecozone to be unmistakably recognized. Moreover, many studies deal only with a selection of the major geographical regions, and do not attempt to give a comprehensive introduction to the great ecozonal divisions of the geosphere.

Figure 1 gives an overview of the ecozonal divisions of the world (6). The boundaries have been selected from a map on seasonal climates of the earth (7). Table 1 lists the approximate surface area of each of the ecozones with respect to Figure 1, and Table 2 shows where the individual ecozones are found in relation to countries and/or states.

As regards delimitation, ecozonal boundaries (*1*) must to a certain extent be drawn arbitrarily (for example, along climatic thresholds), (*2*) are applicable only for a certain number of geographic characteristics, and (*3*) encircle a high variation in conditions. As a result the following premises and concessions are accepted:

1. *Variations within the ecozones cannot be viewed as inconsistent with their boundaries.* The decisive point is that at least some highly significant common elements do exist within each of the ecozones. Significance is measured, in the case of characteristic factors, according to their scale and functional dominance, and in the case of characteristic forms according to their distribution and conspicuousness. Significant factor and form characteristics in this sense are, for example, *inhibited decomposition of organic matter* and *thick raw humus layers on the ground* in the Boreal Zone, or *winter rain* and *sclerophyllous (hard-leaved) vegetation* in the Mediterranean-type subtropics. To recognize common elements within zones, an adequate (global) yardstick is the most important prerequisite. When this tool is applied, many discrepancies disappear on their own accord. A figurative comparison may clarify this point: when a section of a small-scale world map is compared to a detailed, large-scale map of the same area, the former is found to be inexact and incomplete, and generalized to the point of being erroneous; however, this map is still of value.
2. *Ecozones can only be characterized by average conditions that are predominant in them.* Average conditions are found in areas which have no excessive runoff and denudation, exhibit neither considerable lateral water inflow and sedimentation nor waterlogging, are at sea level or slightly above, and have neither a distinctly continental nor a distinctly oceanic climate.
3. *Special cases in an orographic or edaphic sense can be included* in the characterization if they are typical for a zone; this applies, for example, to vertisols, halomorphic soils, or histosols, which are typical final links in relief-related soil sequences (catenas) of the Seasonal tropics, the Arid mid-latitudes, and the Boreal zone, respectively.
4. *The drawing of boundaries between the ecozones is of secondary importance.* The main emphasis must be placed on determining the (average conditions of the) core areas.
5. *All quantitative data* mentioned herein as characteristic for a certain ecozone, *can only be viewed as guidelines* (even when ranges are indicated, these do not reflect the actual extremes but rather the ends of the spectrum within which most data lie). The data are intended to make the global differences between the ecozones clearer and can serve as a measure for determining local deviations within each zone.

The primary aim, by listing and characterizing the ecozones in the manner described, is to aid in establishing a *global classification system* (orientation guide) which

- allows immediate listing of several important characteristics of any area on earth
- is suitable as a basis for further detailed investigation (starting with the question: In what way does a certain area in an ecozone differ from the overall characteristics of the ecozone in which it is found?).

MAIN FEATURES OF ECOZONES

Qualitative Characteristics

Table 3 may be useful as an introduction to major qualitative characteristics of the ecozones. The rigid structuring chosen is intended to assist in finding and comparing information on the various types of features, and in verifying the differences between the ecozones. Figures 2 and 3 should give further assistance. A comparison of the maps on ecozones and soil zones reveals the following correspondences between the distribution of both:

Gelic leptosol–gelic gleysol zone	Tundra and frost debris zone of Polar/subpolar zone
Podzol–cambisol–histosol zone	Boreal zone
Haplic luvisol zone	Humid mid-latitudes
Kastanozem–haplic phaeozem–chernozem zone	Grass steppes (moist) of Arid mid-latitudes
Xerosol zone	Grass steppes (dry) of Arid mid-latitudes, as well as thorn steppes of Tropical/subtropical arid lands
Yermosol zone	Deserts and semideserts of Arid mid-latitudes and Tropical/subtropical arid lands
Chromic luvisol–calcaric cambisol zone	Mediterranean-type subtropics
Acrisol–lixisol–nitisol zone	Seasonal tropics and humid parts of Central America
Acrisol zone	Humid subtropics, Humid tropics in southeast Asia, and (moist savannas of) Seasonal tropics in South America
Ferralsol zone	Humid tropics (except in southeast Asia and Central America)

New names are needed for the *xerosol zone* and the *yermosol zone,* since both name-giving soil units have been deleted from the revised (1988) FAO–UNESCO classification. The problems in both cases are, however, that the new soil units replacing xerosols and yermosols within the respective zones will not be known exactly until a revised edition of the World Soil Map has been published. Thus, the former zonal names referring to the original legend and the only existing Soil Map must be retained. Most likely, they represent some

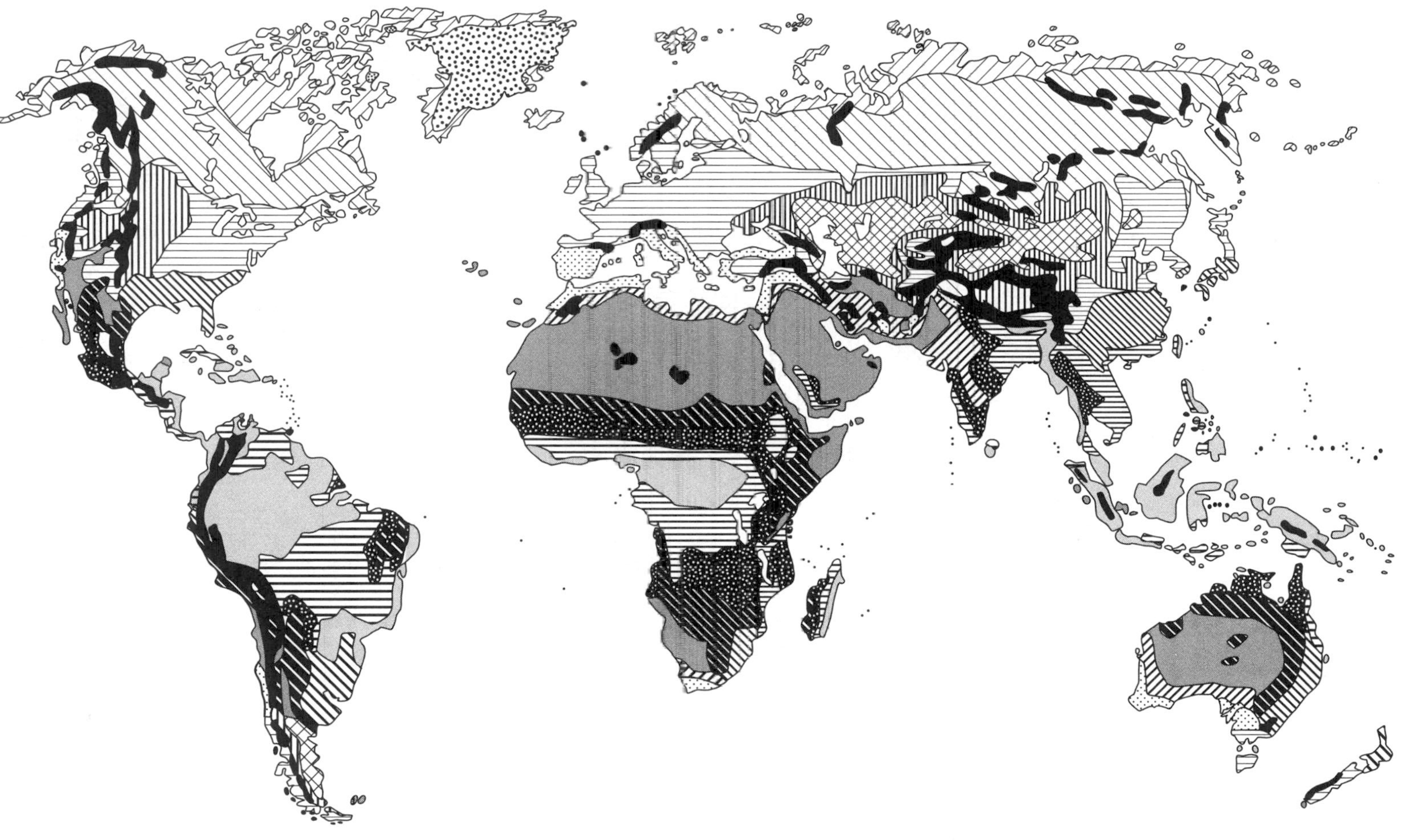

Figure 1. Ecozonal divisions of the world: Polar/subpolar zone; Ice deserts including Tundras and frost debris zone; Boreal zone; Humid mid-latitudes; Arid mid-latitudes; Grass steppes including Deserts and semideserts; Tropical/subtropical arid lands; Deserts and semideserts including Subtropical grass and scrub steppes with winter rainfall (transitional to Mediterranean-type subtropics); Tropical (Sahel-type) thorn savannas and subtropical thorn steppes, both with summer rainfall (transitional to Seasonal tropics and Humid subtropics, respectively); Mediterranean-type subtropics; Seasonal tropics including Dry savannas; Moist savannas; Humid subtropics; Humid tropics; and Mountanous areas (altitudinal belts).

of the following units. (*1*) Xerosol zone: cambic arenosols, vertisols, calcic luvisols, calcisols, and solonetz; (*2*) yermosol zone: regosols, leptosols, calcisols, gypsisols, and solonchaks; unconsolidated sands (dunes).

The order in which the FAO soil units are listed reflects the geographical and evolutionary background (8,9). The following groups can be distinguished (on right: equivalent orders of the U.S. Soil Taxonomy):

Gleysols, regosols, leptosols, arenosols, and *vertisols:* soil formation conditioned by parent material or other edaphic factors; except vertisols, all are weakly developed and not bound to specific climatic conditions.	Entisols, Vertisols
Cambisols: stand alone as soils of initial soil formation (cambic B horizon); all climatic zones.	Inceptisols
Calcisols, gypsisols, solonchaks, and *solonetz:* accumulation of salt, frequently low humus content (ochric A horizon); aridic regions.	Aridisols
Kastanozems, chernozems, phaeozems, and *greysemz:* marked accumulation of base-saturated organic matter in upper soil layer (mollic A horizon); steppes.	Mollisols
Luvisols, planosols, podzoluvisols, and *podzols:* soils with illuvial clay and, in case of podzols, organic matter accumulation in subsurface horizons (argic B and spodic B horizon respectively); high and mid-latitudes.	Alfisols, Spodosols
Lixisols, nitisols, acrisols, and *alisols:* soils with low activity clays (mainly kaolinite) and illuvial clay accumulation (argic B horizon); tropical and subtropical regions.	Alfisols, Ultisols
Ferralsols and *plinthosols:* soils with ferralic B horizon; humid tropics.	Oxisols
Histosols: organic soils in contrast to all the mineral soils listed above; mostly high latitudes.	Histosols

The pole–equator profile of Figure 3 shows how abiotic features such as air temperature, precipitation, evaporation, types of rock weathering and clay formation, translocation processes in soils, decomposition rates of soil organic matter, etc, change with latitude, and how these changes are related to (zonal plant formations of) ecozones.

Quantifiable Characteristics

For quantifiable characteristics, Figure 4 shows how the ecozones differ from each other. Despite the fact that data are available from numerous measurements/surveys all over the world, it is not yet possible to give typical zonal quantities or performances for all of them with certainty. Therefore, only relative, comparative values are shown.

Primary Production. One result that becomes evident from Figure 4 is the difference regarding the amounts of phytomass present and their productiveness. In fact, the differences in the *potential for (natural and agriculturally promoted) plant growth* from one ecozone to another are of such magnitude that they could be used as a major criteria for differentiation; that is, most zones represent regions of the earth having their own particular potential for primary production.

Table 1. Approximate Surface Areas of Ecozones

Ecozones and Subdivisions	Surface areas Mio km^2	%
Polar/subpolar zone	22.0	14.8
Ice deserts	16.0	
Tundras and frost debris zone	6.0	
Boreal zone	19.5	13.1
Humid mid-latitudes	14.5	9.7
Arid mid-latitudes	16.5	11.1
Grass steppes	12.0	
Deserts and semideserts	4.5	
Tropical/subtropical arid lands	31.0	20.8
Deserts and semideserts	18.0	
Subtropical grass and scrub steppes with winter rains	3.5	
Tropical thorn savannas and subtropical thorn steppes, with summer rains	9.5	
Mediterranean-type subtropics	2.5	1.7
Seasonal tropics	24.5	16.4
Dry savannas	10.5	
Moist savannas	14.0	
Humid subtropics	6.0	4.0
Humid tropics	12.5	8.4
Total	*149.0*	*100.0*

The decisive causes for the major regional differences in production are found in the varying *length of the growing season* and the *radiation energy available during this season.* The latter forms the solar growth potential.

With respect to the incoming (short-wave) radiation, data of sufficient accuracy are available for all parts of the world. Uncertainties arise, however, when defining the growing season, although it is generally agreed that in most places on earth plant growth is either limited by the lack of heat or water, or both, for a longer or shorter period during each year. For instance, in all ecozones of mid- and high latitudes, the air temperatures during the winter months drop to levels that cause an interruption of plant growth (except in a few coastal areas and on some islands with exceptionally mild winters); in continental regions, limitations might be dictated by lack of water as well. In the tropical and subtropical ecozones, water deficiencies only (or almost only) may restrict plant growth. Besides this general consent, however, different arguments exist for defining temperatures and moisture supplies for use as thresholds.

As far as *heat requirement* is concerned, the assumption is made that all months with an average temperature of at least +5°C form the growing season. The monthly average of +5°C has been shown through experience in many plant communities to represent the minimum requirement for any notable amount of plant production (growth), although in individual cases, certain species and growth stages of plants can lead to considerable differences in requirements and tolerances.

The monthly temperature +5°C was selected as the threshold value for practical reasons as well: it makes it possible to more quickly determine growing seasons worldwide from climatic data available than would be the case using daily

Table 2. Distribution of Ecozones in Relation to States/Countries

Ecozones	North America	Eurasia	South and Central America	Africa	Australia (with New Zealand)
Polar/subpolar zone	Canada: arctic islands to south end of Hudson Bay (Barren Grounds, Labrador); northern and western Alaska	Arctic islands and Iceland; mainland: continuous coastal belt north of Arctic Circle (northernmost Scandinavia and Russia, mainly Siberia)			
Boreal zone	Central Alaska to Newfoundland (including much of Canada south of tundra)	Scandinavia to Kamchatka (including much of Siberia south of tundra)			
Humid mid-latitudes	West Coast regions from California to Canada, northeast U.S.; (~east of 95° W; north of 36° N), neighboring parts of Canada	Parts of western, central, and eastern Europe; northeast China (Manchuria, Beijing); Korea (except southernmost), and northern Japan	Southern Chile (western Patagonia)		Southeastern Australia (Melbourne) and Tasmania; South Island (New Zealand)
Arid mid-latitudes	Great Plains from Saskatchewan and Alberta (Canada) to Texas (U.S.), and Great Basin	Ukraine to Gobi Desert (Mongolia)	Southern Argentina (eastern Patagonia)		New Zealand: small area on eastern South Island
Tropical/ subtropical arid lands	Southwestern U.S.	Most of Near East, Pakistan, and northwestern India	Central America: Sonoran and Chihuahuan Deserts (from northern Mexico to southwestern U.S.); South America: Caatinga in northeast Brazil, Gran Chaco from southern Bolivia to northern Argentina; Atacama (Chile and Peru)	Northern Africa: Sahara, Sahel; eastern Africa: Somalia to northern Tanzania; southern Africa: Namib, Kalahari, and Karroo; southwestern Madagascar (Toliara)	Most of Australia, except coastal regions in north, east, and south
Mediterranean-type subtropics	California	Coastal areas of Mediterranean basin	Central Chile	Northern Africa: coastal Morocco, Algeria, and Tunisia; Cyrenaica (Lybia); southern Africa: Cape region	Southwestern Australia (Perth), southeastern Australia (Adelaide)
Seasonal tropics		Eastern and southern India, Thailand, Cambodia, Vietnam, and southern China	Central America: most of southern Mexico; South America: Llanos, Mato Grosso (most of Brazil south of Amazonas Basin)	Sudan Zone (south of Sahel); most of eastern Africa, Central Africa south of Congo Basin; central Madagascar	Northernmost Australia (Arnhem Land, York Peninsula)
Humid subtropics	Southeastern U.S., ~35°N to tall-grass prairie of Texas	Central China (south of ~32°N), southern coastal Korea lowlands, southern Japan (north ~37°N)	Southern Brazil (São Paulo, Paraná, Santa Catarina, and Rio Grande do Sul); eastern pampa (Argentina and Uruguay)	Eastern of South Africa: Natal, Transkei, eastern Cape Province	Coastal belt and Great Dividing Range of eastern Australia, ~23° and 37° S; North Island (New Zealand)

Table 2. (*continued*)

Ecozones	North America	Eurasia	South and Central America	Africa	Australia (with New Zealand)
Humid tropics		Parts of Sri Lanka, Burma, Malaysia, Indonesia, Philippines, and New Guinea	Central America: Chiapas (southern Mexico) to Panama; most Caribbean islands; South America: Amazonas Basin, coastal areas in Brazil (Rio de Janeiro)	Guinea Zone south of Sudan Zone (west Africa); Congo Basin (central Africa); eastern Madagascar	

temperature data (of the same or similar level), as is usually done in regional studies.

With respect to a *hygric threshold,* the (empirically developed) Walter-Lieth method has been followed. In principle, other methods of calculation could also satisfy these purposes. On a global scale they all provide similar guidelines for the length of humid periods; the differences in any given region are for the most part less than half a month and at the most one month, and therefore remain within the range which normally applies to the differences in humidity within the individual ecozones. The Walter-Lieth method is preferable because it is most widely applied and, the climates publication contains thousands of climatic diagrams which provide the user with a rapid overview of significant differences in humidity, in terms of magnitude, between the earth's various regions.

The Walter-Lieth climatic diagrams show, for each month, the mean precipitation and air temperature. Their peculiarity is that they keep a strict relation between the ordinate scales for both climatic parameters in such a way that differences of 2 mm precipitation and 1 K air temperature are equally spaced; this means, for instance, that the ordinate position of 40 mm precipitation is the same as of 20°C air temperature. At this ratio, times during which the precipitation curve is above the temperature curve are considered humid; the remaining periods are arid.

The growing season, then, is the sum of those months within a year whose mean temperatures are at least +5°C and whose precipitation (in millimeters) surpasses twice the air temperatures (in centigrades). The period during which these conditions prevail can be quickly and accurately determined using the climatic diagrams. The data on growing seasons given herein for the various ecozones (see Table 3, column 1) are derived from measurements of many such diagrams.

The utilization of the solar potential, ie, the radiation energy available during the growing season, for the production of organic matter depends on the *exploitation of radiation by (apparent) photosynthesis,* ie, on its efficiency in transforming radiation energy into chemical (biologically useful) energy. This transformation, which is variously called the energy yield or useful effect of photosynthesis, or the (net) photosynthetic efficiency, can be calculated in different ways. In this context it is expressed as the *relation between the annual primary production (or, more exactly, its energy content) and the incoming radiation during the growing season.*

Annual production values should be viewed as *relative magnitudes of ecozonal vegetation production potentials.* They can also serve as a relative measure of (climate-determined) *agricultural and forest yield potentials* in the individual ecozones. As a rule, crop growth is promoted by a longer growing season, higher solar radiation, and higher air temperatures, as is the case with natural vegetation. This means that the growth performance of suitable crops should be at least as high as the primary production found in the ecozone as a whole (in the case of annuals, however, this is the case only if the growing season is fully exploited). The yields can be raised above this level if the growing season is artificially extended (in arid regions through irrigation; in regions with cold winters through the use of greenhouses) or when new varieties of crop exhibiting higher photosynthetic efficiency are developed. Both of these methods have been applied extensively.

Nutrient Balance. Ecozones differ a great deal with respect to the relative importance of their nutrient stores, nutrient pathways, nutrient-use efficiencies, and nutrient turnover rates. Figure 5 provides an overview of the primary stores and flows of mineral nutrients generally found in any one ecosystem. Nutrient balances of plant stands or ecosystems are usually calculated in kilograms per hectare and year.

INDIVIDUAL ECOZONES

In contrast to other planetary divisions, the ecozones are more distinctly understood as *geozonal ecosystems.* This means that the conventional (mostly qualitative) description of single-feature complexes such as soils, vegetation, and relief, are replaced or at least supplemented by a more *quantitative* and *integrative analysis* of matter and energy reserves and flows (turnovers) within and between different ecosystem compartments. The matter *reserves* to be dealt with include, for instance, the biomass of plants and animals, dead soil organic matter, and mineral nutrients in vegetation and soil; the matter *turnovers* to be dealt with are those of primary production, animal consumption, secondary production, litter fall, decomposition of organic wastes, and mineral and water cycling; with regard to organic substances and transfers, *energetic* aspects must also be considered. This way of representing ecozonal characteristics has become possible because modern ecological research has provided rich data on a large variety of regional ecosystems.

In this article, some examples will be given for this way of analyzing and describing ecological divisions in the planetary

Table 3. Qualitative Characteristics of Ecozones

Ecozones	Hygrothermic Growth Conditions	Weathering, Denudation, and Landforms	Soil Units	Plant Formations	Land Use
Polar/subpolar zone[a]	Brief summer warming allows 3 months plant growth; air temperatures remain below optimum for life processes; annual precipitation <250 mm, mainly as snow (snow cover lasts ~9 months)	Frost splitting produces frost debris; freeze–thaw cycles in upper (active) permafrost create landforms[b]; on slopes: gelifluction; runoff that occurs during snow melt, leads to hill wash and fluvial erosion	Gelic gleysols and gelic leptosols prevail, followed by gelic cambisols and gelic histosols; podzolization moderate; leptosols (formerly lithosols or regosols) on slopes and recent deposits, and in polar deserts	Species-poor dwarf shrub, meadow, moss, or lichen tundra; toward pole, plant cover becomes more discontinuous until found at only a few advantageous sites; polar desert plant cover <10%	Predominantly settlement free, reindeer herding, fishing, and seal hunting; no cultivation
Boreal zone	Winters long and cold (6 months snow cover); summer growing season between 4–6 months; air temperatures (except continental climate areas) below optimum; annual precipitation 250–500 mm (more rain than snow)	Permafrost widespread; frost action creates palsas, hummocks, string bogs, and thermokarst landforms (eg, alases); gelifluction and hill wash less effective than in tundras and frost debris zones, due to dense and deep-rooting vegetation cover	Podzols predominate; cambisols (gelic or dystric) and leptosols in mountainous; gleysols (gelic and dystric) and histosols (gelic and fibric) in poor drainage areas; dystric podzoluvisols or albic luvisols in transition zone to humid mid-latitudes	Species-poor coniferous forests and mires (mainly peat bogs); forest tundra between polar forest boundary and timber line	Moderately low population density; logging of natural and commercial forests; peat exploitation; except in extreme north, cropping of barley, oats, rye, potatoes, and fodder grass (of little economic importance)
Humid mid-latitudes	Compared to Boreal zone: shorter and less severe drop in temperature during winter limits growing season to 7–11 months; summer temperatures below optimum; annual precipitation 500–1000 mm, some as snow; changeable weather conditions; dry spells in summer may hamper crops	Morphological effects of linear downcutting, slope degradation, and weathering moderate; Pleistocene landforms still exist; soil–water balance different between winter (buildup of water reserves) and summer, when water intake drops below evapotranspiration	Mostly young (post-Ice Age) soils with lithomorphic properties; most prevalent zonal soils haplic luvisols and dystric and eutric cambisols; in transitional area to Boreal zone, eutric podzoluvisols; in transitional area to the Arid mid-latitudes, greyzems	Moderately species-rich winter-deciduous broadleaf (or mixed) forests; four distinct, phenological seasons; temperate broadleaf rain forests in winter-mild coastal regions; temperate coniferous forests in coastal area between northern California and southern Alaska	In Northern Hemisphere: most densely populated areas on earth, greatest economic activity; intensive farming and forestry; small and medium-sized farms prevail, exhibiting high yields per hectare; major crops wheat, maize (from subtropics), potatoes, sugar beets, fodder beets, rape, apples, pears, cherries, plums; dairy and beef farming
Arid mid-latitudes	Low temperatures in winter and aridity in summer limit growing season to ≤5 months; summer air temperatures very high	Salt and temperature weathering more important than chemical weathering, result in shattered rocks, granulate disintegration, desert varnish, tafoni, etc;	In extreme arid regions: weakly developed, humus-poor soils like regosols, leptosols, calcisols, and gypsisols, and unconsolidated sands; in basins, solonchaks; in	In extreme arid regions: deserts, semideserts, or desert steppes; patchy or wide-spaced stand of ligneous or semiligneous plants with stunted growth;	In extreme arid regions: sparse population except with irrigation agriculture; nomadic herding (camels, goats, sheep); with increasing moisture:

Table 3. Qualitative Characteristics of Ecozones (*continued*)

Ecozones	Hygrothermic Growth Conditions	Weathering, Denudation, and Landforms	Soil Units	Plant Formations	Land Use
		typical relief catenas: steep mountains with bare rock and coarse debris; gently sloping pediments (zones of degradation), less inclined bajadas (zones of aggradation, mostly sands), flat playas[c]; river flow episodic, ends in playas; Eolian land-forms[d]conspicuous but of second rank behind fluvially sculptured forms[e]	semi-arid regions: humus-rich kastonozems, chernozems, and phaeozems, with high cation-exchange capacities and base saturations; in basins, solonetz	increasing moisture: short-grass, mixed-grass, tall-grass, and forested steppes, with continuous herbaceous layer (mainly grasses)	extensive ranching,large-scale cereal (wheat, sorghum) farming; dry farming systems widely practiced (with fallows to accumulate available soil water for succeeding crop)
Tropical/subtropical arid lands	Growth conditions usually limited by aridity (<5 mildly humid months); summer temperatures high, winter temperatures remain above freezing (tropics) or fall slightly below (subtropics)		Arenosols, regosols, leptosols, calcisols, gypsisols; in basins, solonchaks and solonetz; unconsolidated sands	Change from deserts to semideserts with decreasing aridity; in tropics, to (sahel-type) thorn savannas; in subtropics, thorn steppes (with summer rains) or grasslands/shrub steppes (with winter rains) replace thorn savannas; in semi-arid vegetation types: short, discontinuous grass cover, stunted shrubs/ trees often with thorns	Similar to Arid mid-latitudes, except for large-scale wheat cultivation, preferably practiced during dry season with irrigation
Mediterranean-type subtropics	Summer dryness limits growing season to 5–9 (winter) months, or restricts plant growth during summer; air temperatures during winter growing season below optimum for life processes (light frosts occur yearly); during dry summer, in most regions ≥4 months at ≥18°C	Intensive slope wash and strong fluvial processes widespread in winter, especially at start of rainy season, when plant cover is patchy; small rivers may become raging torrents within short time	Bright red or red-brown colored chromic luvisols or chromic cambisols (Terra rossa, Terra fusca), or brown eutric or calcaric cambisols; in level or depressed relief planosols, sometimes with sodic properties	Evergreen sclerophyllous forests were predominant in many areas; most have been degraded to sclerophyllous shrub formations (maquis, garrigue); all plant formations rich in plant species	Medium population density; lowlands used for vines, olives, almonds, figs, peaches, apricots, and citrus, or irrigated crops (rice, sugar cane, cotton, maize), or wheat cultivation in winter; mountainous areas largely used for grazing (sheep and goats); transhumance widespread
Seasonal tropics	Aridity in winter limits growing	Deep chemical weath-ering leads	Low activity clay (LAC; eg,	Continuous grass cover,	Population density greater than other

Table 3. *(continued)*

Ecozones	Hygrothermic Growth Conditions	Weathering, Denudation, and Landforms	Soil Units	Plant Formations	Land Use
	season to 5–9 months or, reduces plant growth; during rainy season, precipitation between 500–1500 mm, with air temperatures at least as high as in Humid tropics; winter significantly cooler than summer, although at sea level, mean temperatures ≥18°C	to thick surface layers of loose material (regolith) and rotted rock (saprolite); discharge of rivers periodic; denudation mainly by overland flows (sheet erosion) from rain storms; has led (on old land surfaces) to peneplains; sometimes with inselbergs	kaolinite) soils predominate: CEC (at pH 7) <24 cmol(+)/kg clay, low base saturation; important zonal soil types: lixisols and nitisols; ferralsols and acrisols widespread; vertisols in poorly drained depressed or level relief	discontinuous or missing tree/ shrub layer; During dry season, grass dries and woody plants loose foliage; with decreasing aridity, change from dry (short-grass) to moist (tall grass) savannas, or from dry woodlands/ forests to moist forests, respectively	tropical regions (except in SE Asia); predominant land use: small-scale rain-fed cropping (land rotation, frequently combined with cattle husbandry); production mainly for subsistence; in Latin America and Australia: ranching widespread
Humid subtropics	Temperature drop in winter severe enough (light frosts regular) to limit plant growth for 1–2 months (monthly averages remain above +5°C); in summer, ≥4 months averages >18°C; inland, winter rainfalls decline so that subhumid or semi-arid conditions can arise	Deep chemical weathering; soils usually more susceptible to erosion than in Humid tropics; tropical cyclones with high wind speeds and heavy rain result in considerable erosion, flooding, and storm damage	Acrisols are zonal soil unit; Bt horizon of soil is acid, with a base saturation of <50%; LAC clays predominate; high percentages of Al found on exchangers; then alisols	In wettest parts (close to coasts), subtropical rain forests; in other areas, semi-evergreen or (moderately sclerophyllous, mostly broadleaved) laurel forests	Most areas with dense settlement and advanced economic development; predominant crops: sorghum, groundnuts, rice, soybeans, sesame, sweet potatoes, cotton, and tobacco; in moist regions with mild winters, permanent crops (tea, citrus fruit) grown
Humid tropics	Unlimited growth conditions year-round, slight water stress for 1–2 months; mean temperatures ~25–27°C; annual precipitation frequently 2000–3000 mm; no distinct seasonality	Extremely intensive chemical weathering created regoliths many meters thick; fluvial processes important in shaping relief: river density very high; yearly ≥1000 mm rain water runs off; slope erosion insignificant due to dense vegetation cover and good permeability of soils; landslides may be important	Ferralsols: high clay content mainly of kaolinite, and Fe and Al oxides; few weatherable silicates; CEC (pH 7) <16 cmol(+)/kg clay, base saturation low; tendency to P fixation and Al toxicity; high Fe oxide contents in B horizon: plinthosols; other soil units: ferralic cambisols, ferralic arenosols	Evergreen hardwood forests (tropical rain forests): extremely species-rich and dense; trees of uppermost canopy ~30–40 m; conspicuous physiognomic and floristic differences depending on soil fertility and drainage (eg, Terra firme forests, Swamp forests, heath forests)	Low population density: traditional farming systems: shifting cultivation (slash and burn, for few years cropping) with cassava, taro, yam, etc, or (in parts of SE Asia only) paddy rice farming; hunting and gathering; modern land use types: plantations (large-scale) of oil palms, rubber, cacao, etc; logging, foresty

[a] Ice-free regions only.
[b] For example, ice wedge polygons, pingos, and patterned ground.
[c] Basins with clay-rich sediments, frequently with salt crusts, seasonally water covered.
[d] For example, dunes, mushroom rocks, yardangs, desert pavements, and deflation basements.
[e] Refers to both Arid mid-latitudes and Tropical/subtropical arid lands.

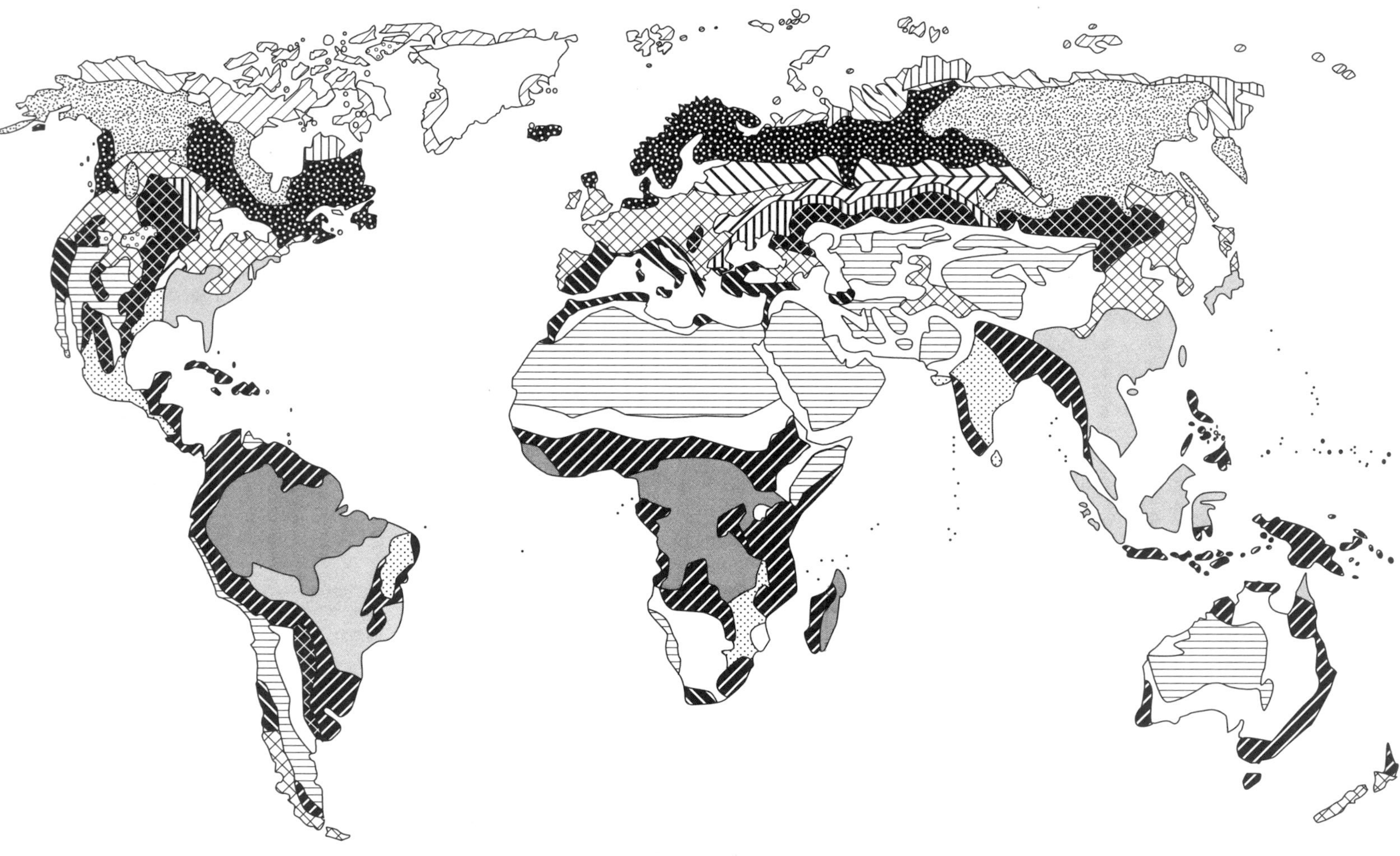

Figure 2. Soil zones of the world (based on FAO–UNESCO soil classification): ▥ Gelic leptosol–gelic gleysol zone; ▨ Gelic leptosols predominate; ▧ Gelic gleysols predominate; ▩ Podzol–cambisol–histosol zone (dystric and gleyic podzoluvisols, gelic cambisols, fibric histosols); ▩ Podzols and dystric and gleyic podzoluvisols; ▩ Haplic luvisol zone (eutric podzoluvisols, luvic phaeozems, luvic chernozems, greyzems, dystric and humic cambisols); ▨ Luvic chernozems, luvic phaeozems, greyzems; ▧ Eutric podzoluvisols; ▩ Kastanozem–haplic phaeozem–chernozem zone; ▥ Chernozems; □ Xerosol zone (cambic arenosols, vertisols, calcic luvisols, calcisols, solonetz); ▤ Yermosol zone (regosols, leptosols, calcisols, gypsisols, solonchaks, dune areas); ▧ Chromic luvisol–calcaric cambisol zone (eutric cambisols, rendzinas); ▨ Acrisol–lixisol–nitisol zone (ferralic arenosols); ▩ Vertisols; □ Acrisol zone (alisols); ▩ Ferralsol zone (plinthosols); and ■ Ice caps.

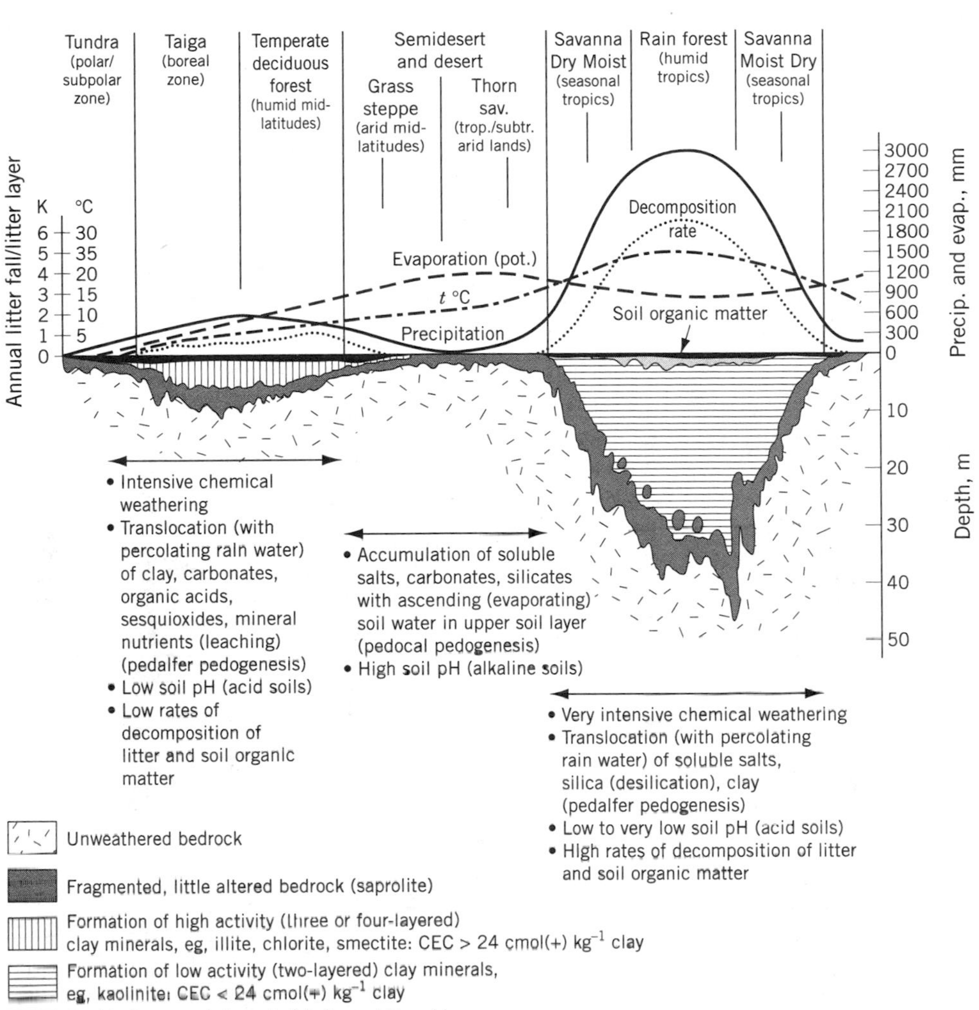

Figure 3. Schematic depth and structures of the weathered mantle (regolith and saprolite) in different climates. The curves for precipitation, evaporation, temperature, and decomposition are based on mean annual values.

dimension. For the sake of better comparison, all have been selected from the Boreal zone and the Humid mid-latitudes.

Ecosystem Models for the Boreal Zone and the Humid Mid-Latitudes

This example presents typical ecosystems of both ecozones in the form of a unified model (Fig. 6). This assumes *steady-state* conditions, ie, gains (inputs) and losses (outputs) of individual stand reserves balance each other out. For example, plant wastes (above and below ground) and animal consumption are equal to the net primary production (PP_N), and the rate of decomposition of dead organic matter is equal to the litter fall and below-ground waste input. As a consequence of this *dynamic equilibrium,* all material and energy stores are constant.

Such steady states are seldom, if ever, found in nature (they are most likely to occur, averaged over the year, in grass steppes) and are instead an imaginary average condition of, at the most, mature plant stands (juvenile and old-age phases of the same stands are always fundamentally different), around which the actual stores and rates of transfer fluctuate (sometimes closely, sometimes not; sometimes in the very long run only). One has further to admit that there are changes in ecosystems which are not cyclical at all, but directional, such as in successions.

Average mature condition can only be roughly estimated because, among other reasons, it never occurs at the same time in the different compartments (when, for example, the PP_N and loss of plant tissue are in balance, and the amount of phytomass thus remains constant, the rate of decomposition can lag behind the litter input and the litter reserve would continue to grow). To use this as the basis for rejecting the *steady-state concept* is not advisable, at least as long as there is no other method of describing representative temporal and spatial conditions more accurately. As regards the existing and in some cases highly detailed descriptions of ecosystems, it should, however, be noted that their accuracy applies only to the state of a particular plant community for a brief moment under very specific environmental conditions which will, most likely, never occur again in exactly the same constellation. If the investigations were repeated at the same location, different results would inevitably be found.

In the two ecosystem models presented here, biomes that are characteristic for the Boreal zone and the Humid mid-latitudes are described (not necessarily average conditions in these ecozones). Numerical values given should be considered rough estimates.

A characteristic feature of the boreal coniferous forest is that the stocks of dead organic matter on and in the soil (litter and humus) achieve the same magnitude as the stocks

Feature \ Ecozone	Polar/subpolar zone (Ice desert / Tundra and frost debris zone)	Boreal zone	Humid mid-latitudes	Arid mid-latitudes (Grass steppes / Deserts and semideserts)	Tropical/subtropical Arid lands (Trop. thorn savannas and subtrop. steppes / Deserts and semideserts)	Mediterranean-type subtropics	Seasonal tropics	Humid subtropics	Humid tropics
Annual precipitation (P)	Very small	Small	Medium	Small / Very small	Small / Very small	Medium	Small	High	Very high
Mean annual temperature	Very small	Small	Medium	Medium	High	High	Very high	High	Very high
Potential annual evapotranspiration	Very small	Small	Medium	High	Very high	High	High	Medium	High
Run-off — Amount (Q)	Small	Medium	Medium	Very small	Very small	Small	Medium	High	Very high
Run-off — Coefficient (Q/P)	Very high	High	Medium	Very small	Very small	Small	Small	High	Very high
Annual global radiation	Very small	Small	Medium	High	Very high	High	Very high	High	High
Length of growing season	Very small / Small	Medium	High	Small / Very small	Small / Very small	High	High	Very high	Very high
Incoming short-wave radiation during growing season	Very small	Small	Medium	Small	Small	Small	High	Very high	Very high
Mean temperature during growing season	Very small	Small	Small	Medium	High*	Small	Very high	High	Very high
Phytomass — Total	Small	Medium	High	Small / Very small	Small / Very small	Small	Medium	Very high	Very high
Phytomass — Root/shoot ratio	High	Small	Small	High / Very high	Medium / Very high	Medium	Medium	Very small	Very small
Leaf area index	Small	High	Medium	Medium / Very small	Small / Very small	Small	Medium	High	Very high
Primary production (PP_N)	Very small	Small	Medium	Medium / Very small	Small / Very small	Small	High	High	Very high
Litter accumulation	High	Very high	Medium	Small / Very small	Small / Very small	Medium	Small	Medium	Small
Dead soil organic matter	Very high	Small	High	Very high / Very small	Very small	Medium	Small	Medium	Small
Decay period of organic wastes	Very high	High	Medium	Small	Small	High	Small	Small	Very small

● = Very high (value) ◕ = High (value) ◑ = Medium (value) ◔ = Small (value) ○ = Very small (value)

No symbol = Inapplicable * =For thorn savannas only

Figure 4. Comparison of ecozones according to selected (quantifiable) features, where * indicates for thorn savannas only.

of living matter (phytomass). It is particularly significant that two-thirds of the dead organic matter belong to the litter layer (in the slightly decomposed form of raw humus). This is because of the extraordinarily low rate of decomposition of the litter (barely 3% per annum). Observations suggest that there is, on average, a continuous increase of carbon contents in boreal soils to an extent that may be of significance for the global carbon budget (27–30). This is likely for all excessively moist forests and peatlands which, together, cover several million square kilometres in the Boreal zone (31,32). This would mean that the Boreal zone is one of the few places on earth where there is a net carbon sink under natural conditions; the only other one is most likely the tundra in the Polar/subpolar zone (33). As a consequence of low mineralization rates, the amounts of soil nutrients available to plants are very small. The broadleaf winter deciduous forest ecosystems are better off, regarding all these aspects in that they have higher primary production, much faster litter decomposition, soil organic matter consisting of more fertile moder or mull, and a higher soil nutrient status.

Organic and Mineral Stocks and Turnovers in Forest Ecosystems

Figure 7 illustrates the various stocks of organic matter and of some selected mineral nutrients as well as their turnovers found in typical mature broadleaf winter deciduous forests of the Humid mid-latitudes. The main characteristics follow (see Fig. 7).

The phytomass of the leaves per unit area is much smaller (~half) than that of the needles in coniferous forests, although stands of broadleaf trees generally have slightly higher wood masses than those of coniferous trees. On the other hand, at 4.3%, the nutrient content of the leaves is more than twice as high as that of needles; thus, both contain about the same

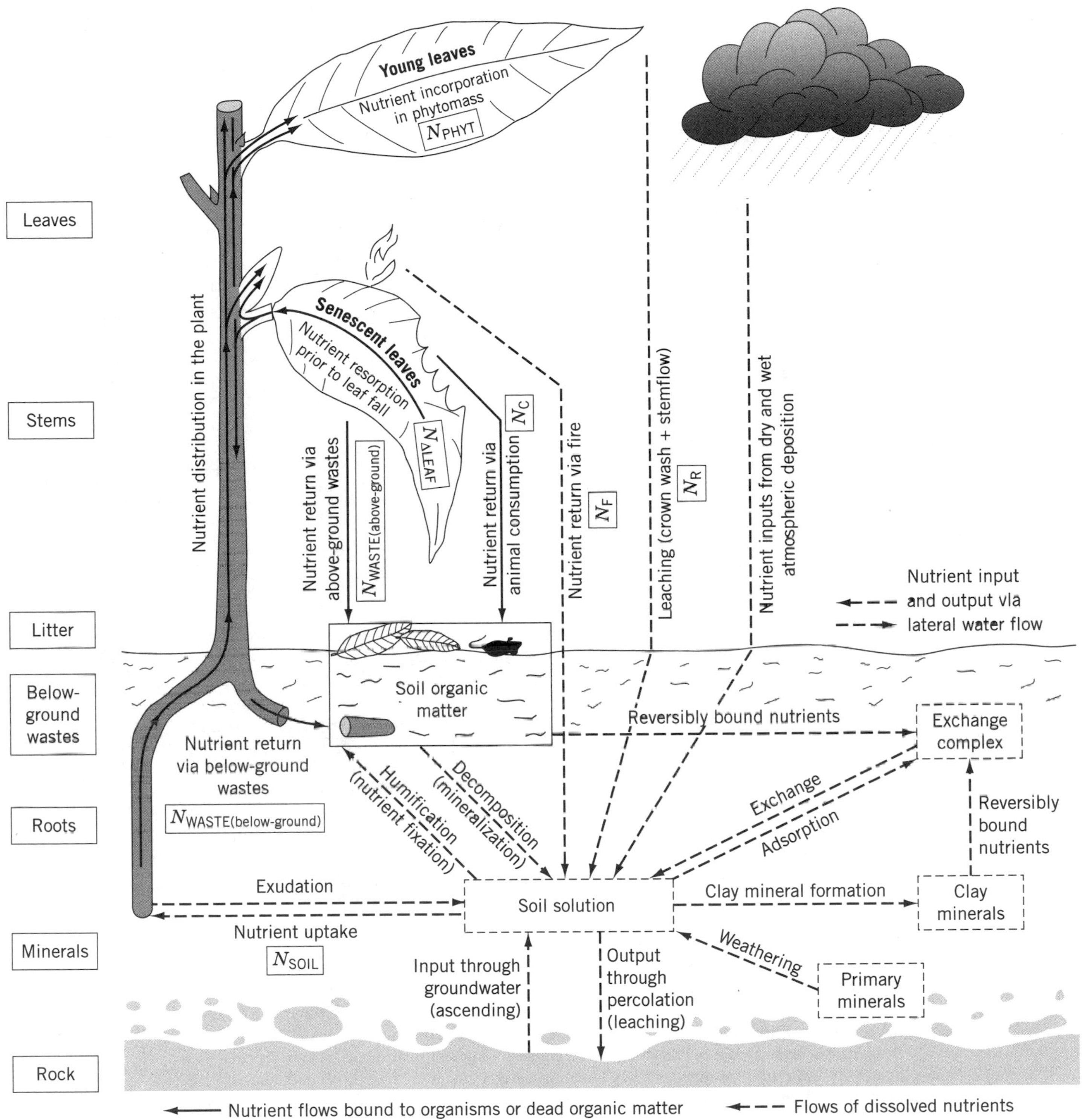

Figure 5. Principal stores and flows of mineral nutrients in ecosystems (biological N_2-fixation and mycorrhiza not included). Nutrient requirements for primary production are fed from two sources: uptake from the soil (N_{SOIL}) and return (resorption, retranslocation, internal recycling) from the leaves before they fall ($N_{\Delta LEAF}$). Release of nutrients from their organically bound form in litter, animal excreta, and carcasses, and dead roots on or in the soil occurs by means of biological–chemical decomposition processes which lead finally to mineralization. The time it takes for this to occur can last less than a year or several centuries. Nutrients returned to the soil by crown wash or stemflow (N_R), on the other hand, are readily available for reuse by the plants (this also applies to nutrients involved in internal recycling). Thus, a differentiation can be made between medium- to long-term (indirect) recycling ($N_{WASTE} + N_C$) and short-term (direct) recycling ($N_R + N_{\Delta LEAF}$). The former can be shortened by fire (N_F).

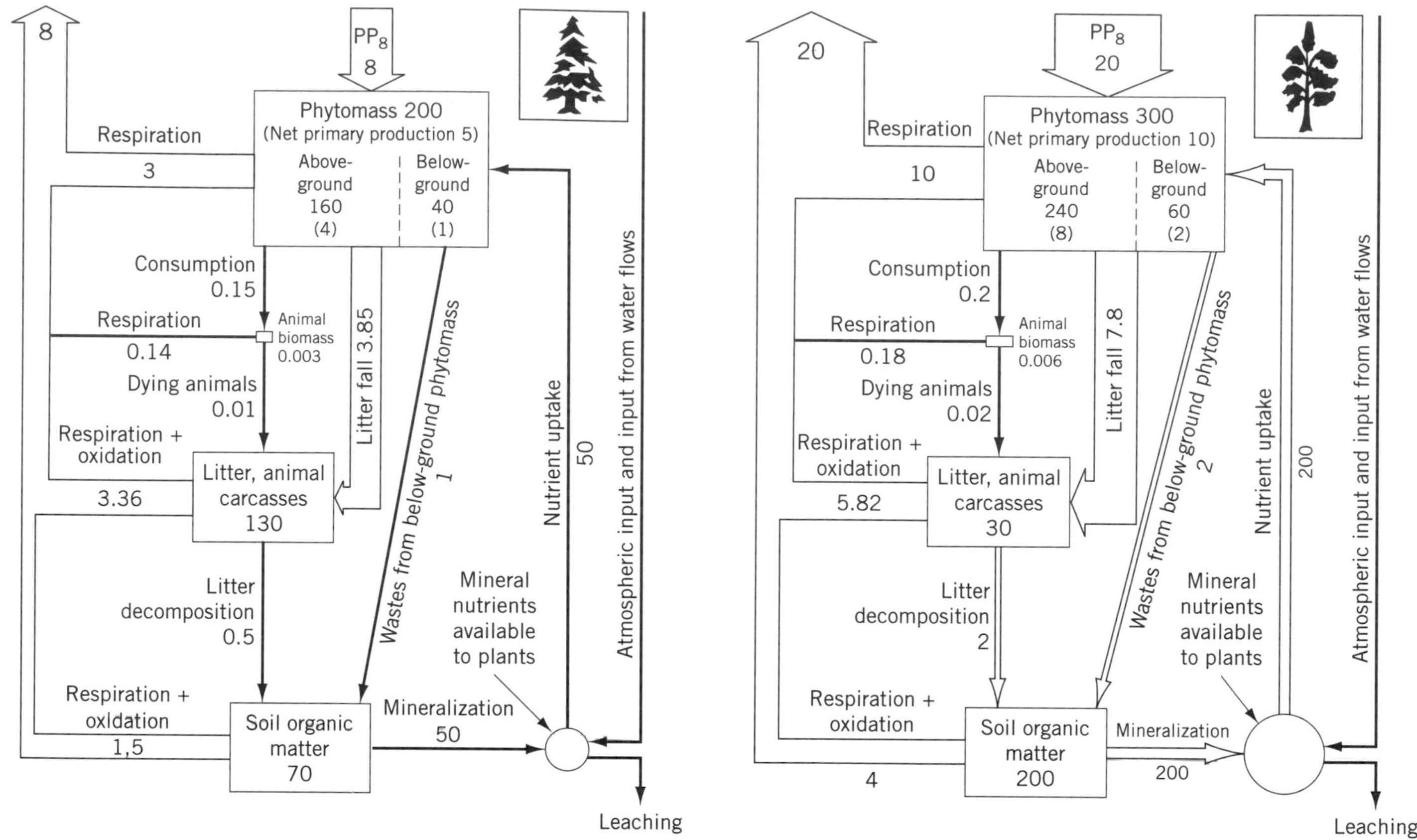

Figure 6. Simplified ecosystem models of a coniferous forest (10–16) in the Boreal zone and of a broadleaf winter deciduous forest (11,17–26) in the Humid mid-latitudes. Areas of boxes and widths of arrows correspond to quantities of material stores and turnovers (models drawn on same scale). Circular areas represent mineral nutrients available to plants (mainly nutrients adsorbed on exchangers) reflecting the approximate magnitudes in comparison to those in the other ecozonal system. The unit of measure for organic pools and turnovers is t/ha or t/(ha·a), respectively, and for nutrient turnovers, kg/(ha·a). The width of the arrows for inorganic nutrient turnovers are given on a 10 × larger scale (eg, 0.1 t nutrient width = 1 t organic material). Animal biomass includes above-ground animals (mainly herbivores). Litter and soil fauna, which belong to the decomposer compartment, are not included; the organic substance from these creatures is contained in litter and soil organic matter. Unless more detailed data are provided, transfer of energy/matter from animal biomass (through death) was calculated as a flat 1/10 of ingested food (animal consumption) and the rest was attributed to losses caused by respiration. The difference in residence time between the elements contained in the organic matter was ignored. Instead, the model is based on the assumption that all nutrients taken up are released only when mineralization takes place. Ecosystem imports (from surroundings) and exports (to surroundings) depend on local conditions and are thus almost impossible to describe on a representative basis for ecozones (their presence is simply indicated by arrows). It should be noted, however, that both can have an important influence, for instance in tropical rain forest ecosystems, after trees have been felled (net exports), and with regard to forest regeneration (net imports).

total amount of nutrients per unit area. Since roughly the same applies to the wood masses, both forest types contain about the same (in absolute terms) amount of mineral nutrients. However, broadleaf forests do contain slightly higher amounts of nutrients when they have more phytomass.

More distinct differences are found with regard to all mineral and organic turnovers. In deciduous forests each year about 40% of the $PP_{N(\text{tree layers})}$ goes into *leaf production;* ie, it is used only to provide seasonal assimilation organs, not long-term stand increments. Each spring, deciduous trees must create new organs in order to produce again. Due to the high nutrient content of the leaves (4.3% as compared to only 0.6% for wood of stems), the nutrient requirement for leaf production represents a full 80% of the total required for annual production in the tree layer.

Considerably more nutrients per unit of production are therefore required for the PP_N of temperate deciduous trees, ie, they have a much lower *nutrient use efficiency* (in particular,

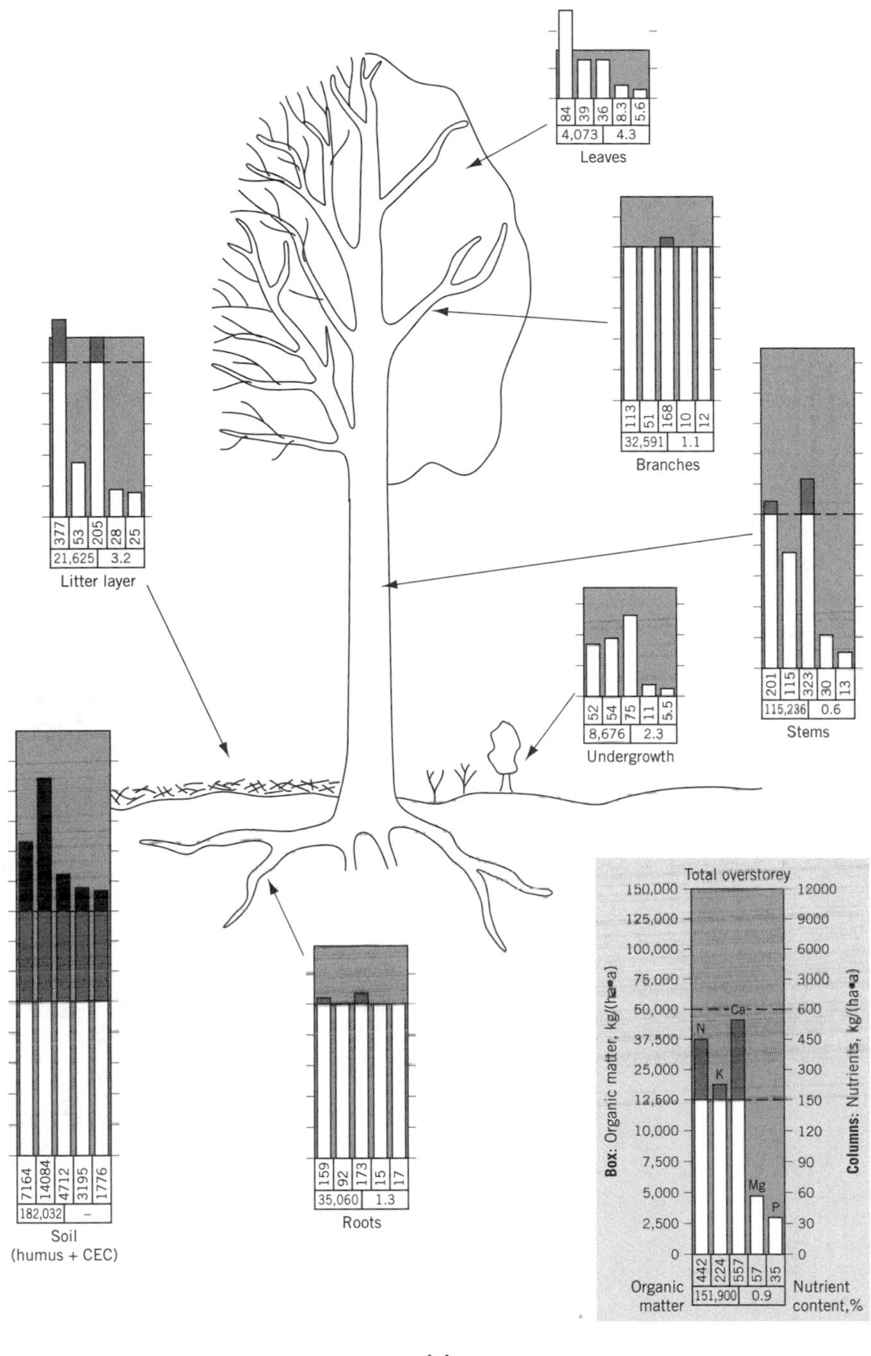

Figure 7. Matter reserves (**a**) and turnovers (**b**) in broadleaf winter deciduous forests in the Humid mid-latitudes, mean values computed from up to 14 separate stands in Europe and North America (34). Organic matter is represented by shaded boxes, and nutrients (N, K, Ca, Mg and P, from left to right) by individual columns. (**a**) Stand reserves: scales reduced to 1:5, starting at 150 kg (nutrients) and 12,500 kg (organic matter) and to 1:100 starting at 600 and 50,000 kg, respectively. Stand reserves for the tree layer (overstory) are not equal to the sum of the individual reserves because only 12 values were available for averaging of branches and stems. Nutrient data on Soil (humus + CEC) refer to values presented (34) under the heading "soil-rooting zone." It is unclear whether these values always refer to nutrients bound in the soil organic matter and the exchangeable fraction. (**b**) Turnovers: scales reduced to 1:12.5, starting at 6 kg (nutrients) and 500 kg (organic matter). (Minor) differences between the values for PP_N for Tree Layer and Return, on the one hand, and the sums of the individual values for each, on the other, can be attributed to the fact that the averages taken were based on different samples sizes (depending on data available).

a lower nitrogen use efficiency). On average they produce only 103 kg organic matter per kg of nutrients, less than half compared with boreal coniferous trees.

Less obvious differences are found in the *nutrient uptake* from the soil, since the deciduous trees meet their higher nutrient requirements, especially for nitrogen, partly from those nutrients that were returned to the branches from the leaves prior to leaf fall the previous autumn (see below). Uptake is reduced accordingly, but still remains pronounced.

Because broadleaf trees also exhibit a higher PP_N, the differences in nutrient requirements and nutrient uptake expressed *per unit area* are even greater. If the PP_N is assumed to be twice that of the boreal forests, both the nutrient requirement and the nutrient uptake will be almost four times as high.

Temperate deciduous forests thus place much higher demands on soil fertility than do boreal coniferous forests. Retention of needles over many years (more than 25 in the case of Alaskan black spruce) allows conifers to flourish even on sites of marginal quality. This is confirmed from woodlands data sets produced by the International Biological Programme (34). The mean nutrient contents of humus and exchangers under broadleaf forests were found at 30,931 kg/ha compared with only 5,027 kg/ha in the podzols under boreal forests.

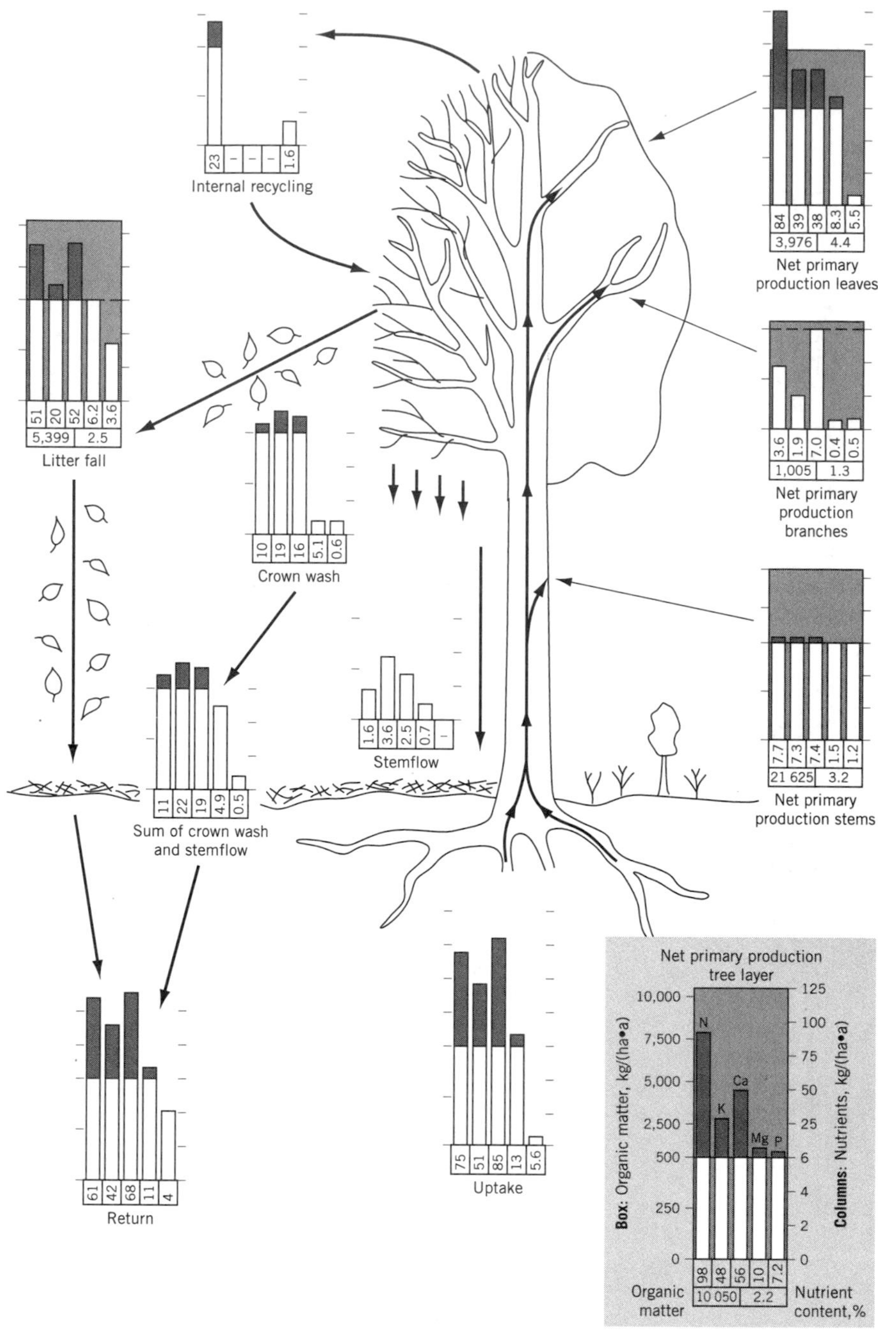

Figure 7. (*continued*) **(b)**

Vast differences also exist with regard to *nutrient return.* Total litter fall (leaves, twigs, branches, fallen trees) found on average in the 14 forests studied is 5.4 t ha/a. Due to the large proportion of leaves present (frequently 3–5 t ha/a in temperate deciduous forests), the average amount of nutrients contained in the organic matter is a respectable 2.5%. In addition, crown wash and stem flow contribute notable amounts of nutrients in mineral form. This explains why over 80% of the nutrient uptake is returned to the soil annually, although litter fall accounts for only 54% of the PP_N during this period. These return processes take place on a smaller scale in boreal coniferous forests.

When *absolute* amounts of nutrients returned each year by the two types of forest are compared, it is found that the values for temperate deciduous forests may be 20+ times higher than those for boreal coniferous forests.

Despite the considerable amount of wastes they produce, only thin layers of *litter* (in our example, 21.6 t/ha) are encountered in temperate deciduous forests because most wastes are rapidly decomposed. The ecosystem of the boreal coniferous forest behaves in the opposite manner. Here, the much smaller amounts of litter produced go with litter layers many times as thick because decomposition takes place at a much slower rate. However, the respective absolute nutrient contents of the two

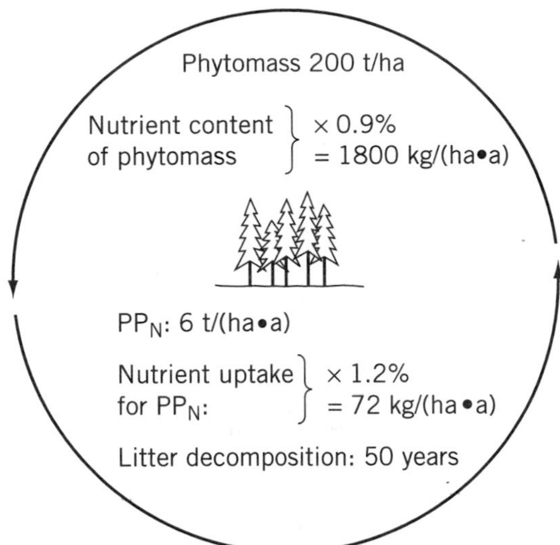

Figure 8. A comparison of nutrient cycles in broadleaf deciduous forests of the Humid mid-latitudes and in evergreen coniferous forests of the Boreal zone. The percentage nutrient content of the phytomass is similar in both forest types, although the larger amounts of phytomass in deciduous forests lead to higher absolute nutrient quantities. Much more distinctive differences are seen in the fact that nutrient uptake, requirement, and return are much higher, and turnover processes (only litter decomposition is shown here) are much shorter in temperate deciduous forests than in boreal coniferous forests. Figure 8 assumes steady-state conditions under which the PP$_N$ is equivalent to the litter input, and nutrient uptake equivalent to nutrient return.

litter layers are in a ratio of only 1 to 2 because the leaf-rich litter of the Humid mid-latitudes is much richer in nutrients (3.2% vs slightly more than 1%). If steady states are assumed (ie, litter is decomposed at the same rate at which it accumulates), the turnover time for litter in deciduous forests is four years, as compared to perhaps hundreds of years for that in coniferous forests. The residence time for individual nutrients may exceed or fall short of these values by varying amounts. In both deciduous and coniferous forests, K is released more quickly than other nutrients.

Nutrients bound in the leaves, which are seasonally produced and, after falling, undergo rapid decomposition, thus become available for (reuse) much more quickly than those contained in the needles, which have a longer life span and break down less easily.

In summary (Fig. 8) it can be said that:

- The temperate deciduous forests in the Humid mid-latitudes go through a brief but intense nutrient cycle. Nutrient uptake in spring and summer is high, but most of it is returned to the soil along with the leaf fall in the following autumn, then released from the litter (including woody parts) on average within four years.
- The coniferous forests in the Boreal zone, on the other hand, have a lengthy, less intense nutrient cycle. Requirement for PP$_N$ is low because the relatively nutrient-rich needles are perennial, therefore annual nutrient losses are low. At the same time, however, the release of nutrients from the organic matter takes considerably longer; bottlenecks in nutrient supply occur more frequently in the Boreal zone than in the more demanding deciduous forests of the Humid mid-latitudes (see FOREST ECOSYSTEMS IN A CHANGING GLOBAL CONTEXT).

BIBLIOGRAPHY

1. J. Schultz, *Die Ökozonen der Erde,* 1st ed., Ulmer, Stuttgart, Germany, 1988, 488 pp.; *ibid.,* 2nd ed., 1995, 535 pp.
2. R. G. Bailey, *Ecosystem Geography,* Springer, Berlin, 1996, 204 pp.
3. F. Klijn, ed., *Ecosystem Classification for Environmental Management,* Dordrecht, the Netherlands, 1994, pp. 85–116.
4. H. Walter and S. W. Breckle, *Ecological Systems of the Geobiosphere,* Springer-Verlag Berlin, 1983–1991, 3 Vols.; ibid., vol. 4, Fischer, Stuttgart, Germany, 1991, in German.
5. O. W. Archibold, *Ecology of World Vegetation,* Chapman and Hall, London, 1995, 510 pp.
6. J. Schultz, *The Ecozones of the World,* Springer-Verlag, Berlin, 1995, 449 pp.
7. C. Troll and K. H. Paffen, *Karte der Jahreszeiten-Klimate der Erde,* Erdkunde 18, Bonn, 1964, pp. 5–28; English transl.: "Seasonal Climates of the Earth", in Landsberg and co-workers, eds., *Weltkarten zur Klimakunde,* Springer-Verlag, Berlin, 1963.
8. FAO–UNESCO, *Soil Map of the World,* Report 60, Vols. I–X and 18 maps 1:5, Mio, Paris, 1974–1981.
9. FAO–UNESCO, *Revised Legend of the Fao–Unesco Soil Map of the World,* World Soil, Rome, 1988, 119 pp.
10. J. A. Larsen, *The Boreal Ecosystems,* Academic Press, Inc., New York, 1980, 500 pp.

11. D. E. Reichle, ed., *Dynamic Properties of Forest Ecosystems*, International Biological Programme 23, Cambridge University Press, Cambridge, U.K., 1981, 683 pp.
12. H. H. Shugart, R. Leemans, and G. B. Bonan, eds., *A Systems Analysis of the Global Boreal Forest,* Cambridge University Press, Cambridge, U.K., 1992; 565 pp.
13. H. Walter and S. W. Breckle, *Temperate and Polar Zonobiomes of Northern Eurasia, Ecological Systems of the Geobiosphere,* Vol. 3, Springer, Heidelberg, 1989.
14. H. Walter and S. W. Breckle, *Spezielle Ökologie der Gemäßigten und Arktischen Zonen außerhalb Eurasiens, Ökologie der Erde,* Vol. 4, Fischer, Stuttgart, Germany, 1991, 536 pp.
15. R. H. Whittaker and G. E. Likens, in H. Lieth and R. Whittaker, eds., *Primary Productivity of the Biosphere,* Ecology Studies 14, Springer, Berlin, 1975, pp. 305–328.
16. K. Van Cleve, F. S. Chapin III, and P. W. Flanagan, eds., *Forest Ecosystems in the Alaskan Taiga,* Ecological Studies 57, Springer, Berlin, 1986, 230 pp.
17. F. H. Bormann and G. E. Likens, *Pattern and Process in a Forested Ecosystem,* Springer, New York, 1979, 253 pp.
18. P. Duvigneaud, ed., *Productivity of Forest Ecosystems,* UNESCO, Paris, 1971, 707 pp.
19. H. Ellenberg, R. Mayer, and J. Schauermann, *Ökosystemforschung, Ergebnisse des Sollingprojekts 1966–1986,* Ulmer, Stuttgart, Germany, 1986, 507 pp.
20. J. B. Falinski, *Vegetation Dynamics in Temperate Lowland Primeval Forests, Ecological Studies in Bialowieza Forest,* Geobotany 8, Dr. W. Junk, Dordrecht, the Netherlands, 1986, 537 pp.
21. W. F. Harris, P. Sollins, N. T. Edwards, B. E. Binger, and H. H. Shugart, in National Academy of Sciences, ed., *Productivity of World Ecosystems,* Seattle, Wash., 1975, pp. 116–122.
22. P. Jakucs, ed., *Ecology of an Oak Forest in Hungary,* Akadkiadó, Budapest, 1985, 545 pp.
23. G. E. Likens and F. H. Bormann, *Biogeochemistry of a Forested Ecosystem,* Springer, New York, 1995, 2nd ed., 159 pp.
24. D. E. Reichle, *Temperate Forest Ecosystems,* Ecological Studies 1, Springer, Berlin, 1970, 304 pp.
25. E. Röhrig and B. Ulrich, eds., *Temperate Deciduous Forests,* Ecosystems of the World 7, Elsevier, Amsterdam, the Netherlands, 1991, 635 pp.
26. J. E. Satchell, in P. Duvigneaud, ed., *Productivity of Forest Ecosystems,* UNESCO, Paris, 1971, pp. 619–630.
27. S. M. Fan and co-workers, *Oecologia* **102** 443–452 (1995).
28. W. A. Kurz and M. J. Apps, *Environ. Poll.* **83** 55–61 (1994).
29. T. P. Kolchugina and T. S. Vinson, *Ambio* **24**(5), 258–264 (1995).
30. E. S. Kasischke, N. L. Christensen, Jr., and B. J. Stocks, *Ecolo. Appl.* **5**(2), 437–451 (1995).
31. E. Paavilainen and J. Päivänen, *Peatland Forestry: Ecology and Principles,* Ecological Studies 111, Springer, Berlin, 1995, pp. 248.
32. E. Gorham, *Ecolo.* Appl. **1**(2) 182–195 (1991).
33. M. J. Apps, W. A. Kurz, R. J. Luxmoore, L. O. Nilsson, R. A. Sedjo, R. Schmidt, L. G. Simpson, and T. S. Vinson, *Water, Air, Soil Poll.* **70** 39–53 (1993).
34. D. W. Cole and M. Rapp, in D. E. Reichle, ed., *Dynamic Properties of Forest Ecosystems,* International Biological Programme 23, Cambridge University Press, Cambridge, U.K., 1981, pp. 341–409.

ELECTRIC AND MAGNETIC FIELDS

MARKUS ZAHN
JAMES C. WEAVER
Massachusetts Institute of Technology
Cambridge, Massachusetts

We live in natural electromagnetic fields due to atmospheric electricity and the earth's magnetism. In addition, living organisms have internal electric fields due to electrolytic processes for cell and nerve functions and magnetic fields from the magnetic moments of molecules. Superposed onto these natural fields are man-made fields from electric power lines, appliances and wiring. Some recent studies have linked man-made fields to adverse health effects but these results are often disputed because the statistical correlation is not strong and no biophysical mechanism has been identified for biological interaction with such low level nonionizing fields. To better understand how electromagnetic fields interact with living systems, this article reviews the fundamentals of electric and magnetic field interactions with media and basic issues for possibly causing biological effects. It remains for future health research to ultimately decide on how we can safely live in our electromagnetic field environment.

SOURCES OF ELECTRIC AND MAGNETIC FIELDS

All matter is composed of atoms which are themselves composed of negatively charged particles called electrons and positively charged particles called protons. Throughout this article, SI units (Systéme International d'Unités) are generally used for electrical quantities for which the base units are taken from the rationalized MKSA system of units where distances are measured in meters (m), mass in kilograms (kg), time in seconds (s), and electric current in amperes (A). The unit of charge is a coulomb = 1 ampere-second. The charge magnitude on an electron and proton $e = 1.6 \times 10^{-19}$ coulombs so that one coulomb of electricity contains about 6.24×10^{18} elementary charges. Each of these charges has associated with it an electric field **E** which has SI units of volts/meter (V/m). The electric field emanates radially from a point charge and is proportional to the force exerted by the charge on other charges as given by Coulomb's inverse square law with distance and where opposite charges attract and like charges repel. Usual matter is charge neutral with an equal amount of protons and electrons so that there is no external electric field. An electric field arises from an excess of protons or electrons on a material so that it is not charge neutral.

Humans often experience nuisance frictional electrification due to charge separation when we walk across a carpet or pull clothes out of a dryer. When we comb our hair with a plastic comb, our hair often becomes charged and stands up as the like charged hairs repel one another. In a dry environment, where large amounts of charge accumulate, these effects often result in small sparks because the electric forces from large amounts of charge actually pull electrons from air molecules. These sparks occur when the electric fields are larger than the electrical breakdown strength of air $\sim 3 \times 10^6$ V/m.

When charges move, they constitute a current, and they give rise to the magnetic flux density **B** which has SI units of teslas (T). Because a 1 T (tesla) magnetic field is uncommonly large, a more common unit for **B** is gauss (G) (1 tesla = 10, 000 gauss). A related quantity, the magnetic field intensity **H** with units of amperes/meter (A/m) is also often used to describe magnetic fields where $\mathbf{B} = \mu_0 \mathbf{H}$ in free space. The quantity $\mu_0 = 4\pi \times 10^{-7}$ henry/meter is called the magnetic permeability of free space. The magnetic field from a small element of current is given by the Biot–Savart law, is in the direction perpendicular to the current element, and is proportional in magnitude to the force exerted by the current element on other current elements as given by the Lorentz force law where opposite flowing currents repel and like flowing currents attract. The magnetic field decreases inversely with the square of the distance from the current element. The magnetic Lorentz force on a current element is in the direction perpendicular to both the direction of current flow and the direction of the magnetic field acting on the current element. Currents generally flow in conductors, such as metallic wires, where electrons easily flow. Insulators do not allow easy charge flow.

Net positive charge accumulates on clouds with a negatively charged earth giving rise to a "fair weather" vertical dc electric field at ground level typically of the order of 130 volts/meter (V/m). However, in thunderstorms, these charges increase and, when the electric field in air exceeds $\sim 3 \times 10^6$ V/m, air breaks down with luminous spark discharges. As the charge passes from the cloud to another cloud or toward the ground, we see it as lightning and hear the pressure wave of heated air as thunder. A typical lightning strike passes 10,000–100,000 amperes of current for about 100 μs, a total charge of 1–10 Coulombs.

The earth's core consists of iron. Molecules of iron behave as if its nucleus of protons is spinning. This moving charge creates a dc magnetic field distributed around the earth, and which has values of about 0.25–0.6 gauss at ground level. Because motion through a magnetic field creates an electric field, walking through the earth's magnetic field induces body currents that add to normal biological body currents.

Our bodies are composed of electrolytic materials, with bodily functions which depend on ion concentrations and membrane potentials of the order of the thermal voltage kT/q, where $k = 1.38 \times 10^{-23}$ joule/K is Boltzmann's constant, T is the absolute temperature in degrees Kelvin and $q \approx 1.6 \times 10^{-19}$ coulombs is the charge on an electron. At room temperature $T \approx 300$K, the thermal voltage is about 25 millivolts (mV) (1000 millivolts = 1 volt). The brain communicates with the rest of the body by electrical impulses. A typical nerve cell fires at ~50–100 millivolts. We monitor the health of the body by measuring these electrical signals using electrocardiograms (ECG) of the heart and electroencephalograms (EEG) of the brain. Heart pacemaker electrodes in contact with heart muscle provides current pulses of 0.1–10 mA (1000 milliamperes(mA) = 1 ampere(A)) a few milliseconds (1000 milliseconds(ms) = 1 second(s)) long to synchronize the firing of heart cells. Magnetic resonance imaging (MRI) images internal organs from the magnetic moments of molecules in the body using dc magnetic fields up to 2T (20,000 G).

Man-made electricity adds additional electric and magnetic fields to our environment which differ from natural electromagnetic fields in amplitude, direction, and frequency (1). In the United States, electric power uses 60-Hertz (Hz) alternating current whereas Europe and some other parts of the world use 50 Hz. To get some perspective on the relative levels of man-made electromagnetic fields, Table 1 lists representative values. More extensive tabulations are given (2–4).

The typical household background at 60 Hz is 1–10 V/m for the electric field and 0.1–10 milligauss for the magnetic field (7). People generally sense 60-Hz electric fields above 20 kV/m (1000 volts(V) = 1 kilovolt(kV)) through hair and skin sensations although there is great variability among individuals (7,8). Some individuals sense electric fields as low as 0.35 kV/m (8). People cannot generally sense magnetic fields even up to 20,000 gauss (2T), although low frequency (~20 Hz) magnetic fields greater than 100 gauss produce luminous images known as "phosphenes" apparently caused by electric currents induced in the retina (7,8).

In the United States, maximum exposure standards for 60-Hz electromagnetic fields at transmission line rights of way are typically over the range of 1–10 kV/m rms for the electric field and up to 250 mG (milligauss) in Florida and 200 mG in New York for the magnetic field (7,9,10). Elsewhere there are no exposure standards for 50/60-Hz magnetic fields. Because these representative environmental fields do not differ greatly from natural electromagnetic fields, it was generally thought that there were no health hazards.

BRIEF OVERVIEW OF EVIDENCE OF ADVERSE HEALTH EFFECTS

A number of epidemiological *in vitro* and *in vivo* studies (8–14) hypothesize that long-term exposure to low level electric and magnetic fields in the home and workplace leads to an increased incidence of cancer after an initial epidemiological study noted an association between childhood cancer and homes near high current distribution lines (15). Studies, which find adverse health effects, link effects to magnetic field exposure and not to electric field exposure. They find that the incidence of cancer in populations exposed to magnetic fields are 2–3 times higher than in unexposed populations. Children are more prone than adults. In contrast, cigarette smoking increases the risk of lung cancer by 20–60 times (7). Some believe that there is a severe health problem due to electromagnetic field exposure (16–18) but subsequent studies in other communities and other countries have led to ambiguous and varying results (8–14,19–21). Some agreed with the first study (15), finding a slight statistical correlation of cancer rate to the location of high current power lines. The data often have large error bars which include the case of no effect whereas other studies did not find such a correlation. There are also some suspicions that any disease correlation is caused by other confounding causes, such as vehicle exhaust, pesticides, water and air pollution (qv), radon, ozone, tobacco, diet, alcohol, and drugs.

Laboratory work has shown that exposure to electromagnetic fields for *in vitro* experiments with animal and human tissues and cell cultures results in complex interactions that depend on the intensity, frequency, time dependence, and direction of the fields (11,12,19–21). Whether or not such effects are harmful is not yet known. *In vivo* experiments with live animals and people are similarly unclear. Electromagnetic fields have some effect on bodily functions, but it is unknown if there are adverse health effects. As discussed subsequently, there are presently no proven interaction mechanisms between electric and magnetic fields and living organisms that are

Table 1. Typical Electric and Magnetic Power Frequency Fields[a]

	E,V/m rms[b]	B, milligauss rms
Home wiring	1–10	0.5–10
At electrical appliances	30–300	5–3000
Under distribution lines serving homes	5–60	1–20
Inside railroad cars on electrified lines (13 kV)	350–700	35–300
Under high voltage transmission lines	1000–7000	20–100

[a] Refs. 5 and 6.
[b] rms = root-mean-square. It is equal to the square root of the average of the square of periodic voltage or current over a period. For a sine wave with peak amplitude A, the rms value $A/\sqrt{2} \approx 0.707\ A$.

understood to cause or promote cancer or disease. The normal dose-response relationship of other toxins, "if exposure is harmful, a greater exposure is more harmful," is not true for electromagnetic fields because an increase in exposure did not increase the incidence of disease. Higher intensity fields are not necessarily worse, nor are lower intensity fields necessarily better.

Many experiments report biological effects, but, even after prolonged exposure to 100-gauss magnetic fields and 100-kV/m electric fields, no pathological effects have been confirmed. In the absence of a biophysical mechanism and with unclear experimental results, there is no way to decide what level of electromagnetic field is safe. Many feel that the evidence is too weak to conclude that the present level of electromagnetic fields is harmful (6,22,23), but advise "prudent avoidance" by minimizing exposure to electromagnetic fields, especially during long intervals, such as sleeping (7,10,20,21). Some ways to practice "prudent avoidance" are to place beds and other furniture at locations with a minimum magnetic field, move alarm clocks, clock radios and other electric devices that are on all the time as far away as possible, avoid standing in front of major appliances while in use, stay at least 2–3 feet away from computer terminals and televisions, and do not use electric blankets or water beds with electric heaters or at least turn them on before bed and then turn them off while sleeping (10).

Because electric and magnetic fields are everywhere in our environment, it is difficult to quantify exposure levels. Other studies examined the health histories of workers in electrical occupations, assuming that their exposure to electromagnetic fields is greater than the average population (10–12,19–21). Such workers, however, are often also exposed to other potential industrial toxins, such as chemicals, and sample sizes were often small so that definitive conclusions could not be reached.

REPRESENTATIVE STANDARDS

The American National Standards Institute's (ANSI) first electromagnetic field standard established in 1966 limited exposure to 100 W/m^2 over the frequency range of 10 MHz to 100 GHz corresponding to an electric field limit of ~200 V/m rms and magnetic field limit of ~6.5 mG rms (24). In 1982 the ANSI standard was revised to make absorbed power depend on frequency with a goal of limiting absorbed body average power density to 0.4 watt/kg in the frequency range of 300 kHz–100 GHz.

The International Radiation Protection Association (IRPA) formed the International Non-Ionizing Radiation Committee (INIRC) to formulate limits of exposure of 50/60-Hz electric and magnetic fields as given in Table 2 (25,26). These values were chosen so that the induced body current density is less than 10 mA/m^2, because this value does not exceed normal biological values. Australia has adopted these limits. The United Kingdom has approved a 1988 standard for a 50/60-Hz magnetic field at 20 gauss for public and occupational exposure. Sweden has adopted only a magnetic field standard for computer monitors at less than 2.5 milligauss at 50 cm from the screen (10).

Table 2. IRPA/INIRC Limits of Exposure to 50/60-Hz Electric and Magnetic Fields[a]

	E, kV/m rms	B, Gauss rms
Occupational Workers		
Whole working day	10	5
Short-term	30	50
General public		
Whole day	5	1
Few hours per day	10	10

[a] Refs. 25 and 26.

IRPA/INIRC guidelines on radiofrequency and microwaves given in Table 3 assume no adverse health effects for energy deposition rates below 4 W/kg. They further use a safety factor of 10 for long-term exposure and a further safety factor of 5 for general public limits.

PERSONAL MEASUREMENTS

In 1994, Amy E. Zahn, as part of a high school (9th grade) science project, was assisted in obtaining first-hand experience in measuring environmental electric and magnetic fields in the home and community in Lexington, Mass., a suburb of Boston. A Holaday Industries, Inc. HI-3600-02 Power Frequency Field Strength Meter was used to assist in evaluating electric and magnetic fields associated with 50/60-Hz electric power transmission and distribution lines and electrically operated equipment and appliances (27). A Field Star 1000 magnetic field meter with memory was also used to record magnetic fields as a function of time.

Electric fields are detected with a capacitive current sensor which consists of two thinly separated conductive disks electrically connected. When placed in an alternating electric field, charge is redistributed on the two parallel disks so that the electric field between the two disks at the same potential is zero. In a sinusoidally varying electric field, the redistribution of charge also changes sinusoidally with time giving a measured sinusoidal current in the connecting wire between the

Table 3. IRPA/INIRC Guidelines for Exposure Limits at Radio and Microwave Frequencies[a,b]

	Occupational Exposures		General Public Exposures	
f(MHz)	E, V/m rms	B, mG rms	E, V/m rms	B, mG rms
0.1–1	614	$20/f$	87	$2.9/f^{1/2}$
1–10	$614/f$	$20/f$	$87/f^{1/2}$	$2.9/f^{1/2}$
10–400	61	2	27.5	0.9
400–2000	$3f^{1/2}$	0.1	$1.375f^{1/2}$	$0.0465f^{1/2}$
2000–300,000	137	4.5	61	2

[a] Frequency f in megahertz.
[b] Refs. 1, 25 and 26.

disks whose measured amplitude is proportional to the local electric field.

The magnetic field in the Holaday instrument is measured by a coil consisting of several hundred turns of fine gage wire. When placed in an alternating magnetic field, a current is induced in the coil whose amplitude is proportional to the magnetic field strength perpendicular to the coil. The Field Star 1000 instrument has three mutually perpendicular coils to measure the three vector components of a magnetic field. The magnetic field magnitude is also computed by the instrument by taking the square root of the sum of squares of the three components.

The measured ambient rms electric field in the Zahn house varied over the range of 3–1600 V/m. The highest value was in front of an operating television. Far from an appliance, the ambient electric field was of the order of 10 V/m. The rms magnetic field far from appliances was about $H = 90$ milliamperes/m (mA/m) ($B \sim 1$ mG). The highest values were 3330 mA/m ($B \sim 42$ mG) at a stove heating element at a high setting. During a walk of about 1/2 mile from their home to Lexington Center, the outdoor electric field varied over the range of 0.35–17.2 V/m rms, and the magnetic field varied from $H = 5.2–620$ mA/m rms($B = 0.065–7.8$ mG rms). Within 10 feet of a substation transformer in Lexington Center, the electric field was 0.8 V/m and magnetic field was $H = 1083$ mA/m($B = 13.6$ mG). For a typical high school student, the Field Star 1000 recorded a school day magnetic field exposure of 0–2 milligauss which increased to about 16 milligauss in the school lunchroom (28).

BIOPHYSICAL MECHANISMS

The main conceptual problem in understanding reported "weak" electric and magnetic field effects has to do with coupling fields to biological systems and understanding how weak field exposures cause changes that compete with naturally occurring changes. A number of biophysical mechanism theories have been considered. Some theories, such as the ion cyclotron resonance models, are fundamentally flawed and should not be taken seriously. For example, although cyclotron resonance is a well-established phenomenon, it can take place only in a near vacuum, ie where there are very few collisions between ions and other (usually neutral) atoms and molecules. Unlike a vacuum, biological systems are based on solids and liquids, with aqueous electrolytes (eg, salt water) the predominant liquid. In water the average distance between collisions is extremely small, just a few molecular diameters. For this reason the spiral orbital motion of cyclotron resonance cannot become established, and the concept of cyclotron resonance is fundamentally irrelevant (29). Quantum mechanical versions of ion resonance involving caged ions also fail (30).

Other theories have focused on the cell membrane, the site of voltage-gated channels and voltage-sensitive enzymes. The electrically insulating lipid bilayer portion of the cell membrane plays an important role: application a of 50–60-Hz magnetic field to the human body results in an induced 50–60-Hz electric field within tissues, and this electric field changes a cell's transmembrane voltage U. The cell membrane is very thin (about 7 μm) compared with a typical mammalian cell (about 10 to 100 μm). The ratio of cell size to membrane thickness provides a kind of geometric amplification, such that the 50–60-Hz component of the transmembrane voltage is about 1,000 to 10,000-fold larger than the magnetically induced electric field outside the cell. The cell membrane, therefore, has been identified as a likely "target" of any 50–60-Hz effects.

Initial theoretical estimates of the plausibility of biological effects considered a purely physical competition. The field-induced 50–60-Hz change in the transmembrane voltage was compared to the spontaneous random fluctuations in the transmembrane voltage, the so-called thermal noise. This purely physical approach showed that, in principle, typical individual mammalian cells might respond to small magnetically induced electric fields (31,32). A simple extension to cell clusters with gap junction electrical connections between cells suggested that still smaller fields might win this physical competition for a few of the cells in the cluster (33).

Purely physical analysis has also been applied to the provocative case of biological magnetite particles (magnetosomes), which are biologically synthesized, microscopic permanent magnets encapsulated in lipid bilayer membranes. Biological magnetite exists in many tissues, including human brain (34,35). Moreover, other human tissues contain contaminant magnetic particles (36–38). Except for these magnetic particles, human tissue is "magnetically transparent", because low frequency magnetic fields penetrate the body with negligible interaction. In the case of biologically synthesized magnetite, the magnetic moment of a magnetite particle interacts with 50–60-Hz fields and attempts to rotate (translational movement turns out to be negligible). But thermal fluctuations impart a jittery, changing rotation that competes with the 50–60-Hz field. Several physical theoretical analyses have treated this problem (39–41), and, although their estimates of the minimum field magnitude required vary somewhat, it appears that very weak fields cannot compete successfully with this form of thermal noise. Moreover, although it has been suggested that biological magnetite might be coupled to mechanogated membrane channels (42), it is otherwise not clear how small particle rotations could alter a cell.

This emphasizes a general shortcoming of purely physical analyses: there is no recognition of the way physical changes translate into biochemical changes. This is not surprising, because purely physical estimates do not address the coupling of electric and magnetic fields to biochemical processes. Molecular change due to environmental fields is really the critical issue. Even if physical competition is satisfied, this alone does not imply that a molecular change capable of causing a biological effect has occurred. The controversy thus involves how weak electric and magnetic fields alter biochemistry to generate causal molecular changes.

In contrast, there is no conceptual controversy surrounding the idea that a small number of molecules cause significant changes in a biological system. For example, some olfactory sensory systems detect a few individual molecules. Further, about 400 bleomycin molecules introduced into the cytoplasm of a cell kill that cell (43). For this reason, analyzing the biophysical mechanisms which couple electric and magnetic fields to biochemical reactions and to biochemical transport has emerged recently as a critical topic.

Specifically, threshold field exposures to cause biological effects have lower bounds that result from quantitative competition: a field-induced molecular change must equal or exceed molecular changes due to other sources. The field-induced molecular change can be estimated theoretically. To be conservative, this involves using a most field-sensitive model that exaggerates the molecular change due to a field exposure. Fundamental noise is present, because of the stochastic nature of biochemical reaction and transport processes. By quantitatively comparing the field-induced molecular change to these fluctuations and all other changes (eg, natural physiological molecular changes and those due to natural temperature variations), a lower bound for threshold exposures is estimated. Actual thresholds should be larger, because of the conservative assumptions and the ability of biological systems to repair externally imposed molecular changes.

Voltage-gated cell membrane channels and enzymes comprise the most well studied biophysical mechanism (44,45). Essentially all of electrophysiology, including the exquisite electrosensory systems of sharks and rays (elasmobranch fish), are based on this biophysical mechanism. Within a second, these animals sense and respond to electric fields of 5×10^{-7} V/m (46,47). The basis of their extraordinary sensitivity is about 100 specialized organs (the ampullae of Lorenzeni) that concentrate the electric field in the seawater so that a voltage change occurs across a layer of cells. Yet, a quantitative theoretical analysis of field-induced molecular change for large isolated cells exposed to a 50–60-Hz field yields an extracellular electric field threshold of about 0.1 V/m (48), which corresponds to about 10^{-3} T (10 G) for 50–60-Hz magnetic fields that induce electric fields within the human body. Unless some human tissues act somewhat like ampullae of Lorenzeni, it appears that voltage-gated membrane proteins cannot be a biophysical mechanism for any biological effect due to weak, 50–60-Hz environmental magnetic fields in humans.

Transients and local sources of magnetic fields are, in principle, more plausible candidates for causing effects than time weighted average (TWA) 50–60-Hz fields. Even here, molecular change analysis shows that there are severe constraints. For example, if magnetic particles, such as biological magnetite, were attached to a cell membrane, then even a short but large magnetic field transient might create metastable pores (49). Even if the transient is of short duration (eg, 1 ms), the metastable pores could transport molecules into a cell over many seconds, in principle, yielding a molecular change which exceeds competing molecular changes due to other sources. However, transients, and local residential and industrial sources, are very different from the weak, continuous fields that have led to much of the concern about environmental electric and magnetic fields. Research is continuing, some now using molecular change analysis based on biophysical mechanisms that quantitatively predict the magnitude of the molecular change, which can then be quantitatively compared to all other sources of molecular change, such as natural physiological variations and temperature variations. Until such analyses are completed, the minimum exposures to cause biological effects will not be fully established. Presently, however, known biophysical mechanisms seem incapable of yielding causal molecular changes for the weak 50–60-Hz fields implicated by epidemiology.

CONCLUSIONS

Despite epidemiological, laboratory, and human and animal studies that show possible biological effects from nonionizing electromagnetic fields, there are no identifiable biophysical mechanisms or conclusive proof of adverse health effects. Nevertheless, it remains prudent that when possible, for people to minimize their exposure to electromagnetic fields. This paper has shown that typical exposures from power frequency and radiofrequency fields result in voltages, currents, fields, and heating comparable or less in magnitude to those naturally occurring in the environment or from normal metabolic body processes.

Because the body is a good conductor of electric fields at frequencies below ≈10 MHz but a poor conductor for magnetic fields, the presence of a body in an ambient electric and magnetic field perturbs the electric field distribution but has no effect on the magnetic field distribution. The electric field inside the body is greatly reduced from the electric field just outside the body, and the external electric field must terminate essentially perpendicularly to the body resulting in a surface charge distribution and a surface current distribution if the external electric field is time-varying. Those studies that show adverse health effects linked to the magnetic field but not the electric field may perhaps be due to the body conductivity that almost completely shields the body interior from the electric field but not the magnetic field.

Charges and voltages are the source of electric fields and currents are the source of magnetic fields. Appliances are charge neutral with current flow in a closed loop and thus have electric fields approximately like that of a point electric dipole and magnetic fields like that of a point magnetic dipole, fields that decrease with distance as $1/r^3$. Home wiring and balanced power lines have electric and magnetic fields like line dipoles that decrease with distance as $1/r^2$ but lines with unbalanced currents have larger magnetic fields due to the net current flow on the lines that decreases with distance as $1/r$.

Because normal residential power in the U.S. is at 60 Hz, 120 volts rms whereas in Europe and elsewhere it is 50 Hz, 240 volts, normal household current in the U.S. is about double that in Europe. Thus in Europe the typical residential electric field is twice that in the United States, whereas the European magnetic field is about half that in the United States Health studies around the world need to recognize this difference in voltage and electric field magnitudes and in current and magnetic field magnitudes and perhaps frequency, although interpretation is difficult because health studies thus far have not shown the normal dose-response relationship that greater exposure is more harmful. While continuing health research tries to sort out biophysical mechanisms of possible health hazards, it is prudent for everyone to minimize unnecessary exposure to electric and magnetic fields.

ACKNOWLEDGMENT

Thanks to D. L. Haes, Jr., Assistant Radiation Protection Officer in the MIT Environmental Medical Service for providing constructive comments on the manuscript and for lending the Holaday Industries Power Frequency Field Strength Meter and the Field Star 1000.

BIBLIOGRAPHY

1. O.P. Gandhi, ed., *Biological Effects and Medical Applications of Electromagnetic Energy,* Prentice Hall, Englewood Cliffs, N.J., 1990.
2. W.R. Bennett, Jr., *Health and Low-Frequency Electromagnetic Fields,* Yale University Press, New Haven, Conn., 1994.
3. D.L. Mader and S.B. Peralta, *Bioelectromagnetics* **13,** 287–301 (1992).
4. National Research Council (U.S.) Committee on the Possible Effects of Electromagnetic Fields on Biological Systems, *Possible Health Effects of Exposure to Residential Electric and Magnetic Fields,* National Academy Press, Washington, D.C., 1997.
5. Oak Ridge Associated Universities Panel for the Committee on Interagency Radiation Research and Policy Coordination, *Health Effects of Low Frequency Electric and Magnetic Fields,* Executive Summary, Oak Ridge, Tenn., June 1992, p. 4.
6. W.R. Bennett, Jr., *Physics Today,* 23–29 (Apr. 1994).
7. M.G. Morgan, *"Electric and Magnetic Fields from 60 Hertz Power: What Do We Know about Possible Health Risks?"* Dept. of Engineering and Public Policy, Carnegie-Mellon University, Pittsburgh, Pa., 1989.
8. L.E. Anderson, "Interaction of ELF Electric and Magnetic Fields with Neural and Neuroendocrine Systems," in B.W. Wilson, R.G. Stevens, and L.E. Anderson, eds. *Extremely Low Frequency Electromagnetic Fields: The Question of Cancer,* Battelle Press, Columbus, Ohio, 1990, Chap. 7.
9. C. Polk and E. Postow, eds., *CRC Handbook of Biological Effects of Electromagnetic Fields,* 2 Edition, CRC Press, Boca Raton, Fla., 1996, Appendix 2.
10. W.F. Horton and S. Goldberg, *Power Frequency Magnetic Fields and Public Health,* CRC Press, Boca Raton, Fla., 1995.
11. M. Blank, ed., Electricity and Magnetism in Biology and Medicine, Review and research presented at The First World Congress for Electricity and Magnetism in Biology and Medicine, Orlando, Fla., 1992, San Francisco Press, 1993.
12. E.L. Carstensen, *Biological Effects of Transmission Line Fields,* Elsevier, Amsterdam, the Netherlands, 1987.
13. W.T. Kaune and L.E. Anderson, "Physical Aspects of ELF Electric and Magnetic Fields: Measurements and Dosimetry," in B.W. Wilson, R.G. Stevens, and L.E. Anderson, eds., *Extremely Low Frequency Electromagnetic Fields: The Question of Cancer,* Battelle Press, Columbus, Ohio, 1990, Chap. 3.
14. T.S. Tenforde, "Biological Interactions and Human Health Effects of Extremely Low Frequency Magnetic Fields," in B.W. Wilson, R.G. Stevens, and L.E. Anderson, eds., *Extremely Low Frequency Electromagnetic Fields: The Question of Cancer,* Battelle Press, Columbus, Ohio, 1990, Chapt. 12.
15. N. Wertheimer and E. Leeper, *Am. J. Epidemiol.* **109,** 273–284 (1979).
16. P. Brodeur, "Annals of Radiation, the Hazards of Electromagnetic Fields," in *The New Yorker,* "I-Power Lines," 51–88 (June 12, 1989); II-"Something is Happening," 47–73 (June 19, 1989); III-"Video-Display Terminals", 39–68 (June 26, 1989); "The Cancer at the Slater School," 86–119 (Dec. 7, 1992).
17. P. Brodeur, *Currents of Death: Power Lines, Computer Terminals, and the Attempt to Cover Up Their Threat to Your Health,* Simon and Schuster,New York, 1989; review by K.R. Foster, *Proc. IEEE,* **78**(11) 1763–1764 (1990); review by M.G. Morgan, *Sci. Am.* 118–123 (Apr. 1990).
18. P. Brodeur, *The Great Power-Line Cover Up,* Little Brown and Company, Boston, Mass., 1993.
19. Electric Power Research Institute (EPRI) Resource Papers, *"EMF Health Effects Research: A Selected Bibliography,"* Mar. 1992; *Exposure Assessment Fundamentals,* Nov. 1992; *Fundamentals of Epidemiology: Parts I and II,* Palo Alto, Calif., Oct. 1993.
20. I. Nair, M.G. Morgan, and H.K. Florig, "Biological Effects of Power Frequency Electrical Fields," Prepared for U.S. Congress Office of Technology Assessment, NTIS PB89-209985, Washington, D.C., May 1989.
21. K. Fitzgerald, I. Nair, and M.G. Morgan, *IEEE Spectrum,* 22–35 (Aug. 1990).
22. E.R. Adair, *Currents of Death Rectified: A Paper Commissioned by the IEEE-USA Committee on Man and Radiation in Response to the Book by Paul Brodeur,* IEEE-USA, New York, 1991.
23. G. Taubes, *The Atlantic Monthly,* 94–108 (Nov. 1994).
24. K.R. Foster and A.W. Guy, *Scientific American,* **255**(3), 32–39 1986.
25. A.S. Duchene, J.R.A. Lakey, and M.H. Repacholi, eds., IRPA (International Radiation Protection Association) *Guidelines on Protection against Non-Ionizing Radiation,* Pergamon Press, New York, 1991.
26. M. Grandolfo and M.H. Repacholi, "Risk Assessment and IRPA/INIRC Guidelines on Protection Against Electromagnetic Fields and Waves," in M. Blank, ed., *Electricity and Magnetism in Biology and Medicine,* San Francisco Press, San Francisco, Calif., 1993, p. 77–80.
27. Holaday Industries, Inc., *User Manual, HI-3600-2 Power Frequency Field Strength Meter,* #600040, Sept. 1989; *User Manual, HI-3600 VDT Radiation Survey Meter,* #600031 B, England, Jan. 1988.
28. A.E. Zahn, *Environmental Electric and Magnetic Fields in the Lexington, MA Community,* 9th Grade Science project, Lexington High School, Lexington, Mass., Apr. 1994.
29. J. Sandweiss, *Bioelectromagnetics* **11,** 203–205 (1990).
30. R.K. Adair, *Bioelectromagnetics* **13,** 231–235 (1992).
31. J.C. Weaver, "Transient Aqueous Pores: A Mechanism for Coupling Electric Fields to Bilayer and Cell Membranes" in M. Blank and E. Findl, eds., *Mechanistic Approaches to Interactions of Electromagnetic Fields with Living Systems,* Plenum, New York, 1987, p. 249–270.
32. J.C. Weaver and R.D. Astumian, *Science* **247,** 459–462 (1990).
33. J.C. Weaver and R.D. Astumian, *Bioelectromagnetics*, Suppl. 1, 119–138 (1992).

34. J.L. Kirschvink, A.K. Kirschvink, and B.J. Woodford *Proc. Nat. Acad. Sci.* **89,** 7683–7687 (1992).
35. J.R. Dunn, M. Fuller, J. Zoeger, J. Dobson, F. Heller, J. Hammann, E. Caine, and B.M. Moskowitz *Brain Res. Bull.* **36,** 149–153 (1995).
36. D. Cohen, *Science* 180, 745–748 (1973).
37. P. Gehr, J.D. Brain, S.B. Bloom, and P.A. Valberg *Nature* **302,** 336–338 (1983).
38. P.A. Valberg, *Science* **224,** 513–516 (1984).
39. R.K. Adair, *Phys. Rev.* **A43,** 1039–1048 (1991).
40. J.L. Kirschvink, *Phys. Rev.* **A46,** 2178–2184 (1992).
41. C. Polk, *Bioelectromagnetics* **15,** 261–270 (1994).
42. J.L. Kirschvink, A. Kobayashi-Kirschvink, J.C. Diaz-Ricci, and S.J. Kirschvink *Bioelectromagnetics* Suppl. 1, 101–113 (1992).
43. B. Poddevin, S. Orlowski, L. Belehradek, Jr., and L.M. Mir, *Biochem. Pharmacol.* **Suppl. 42,** S67–S75, (1991).
44. B. Hille, B, *Ionic Channels of Excitable Membranes,* 2nd ed., Sinauer Associates, Sunderland, Mass., 1992.
45. B. Robertson and R.D. Astumian, *Biochemistry* **31,** 138–141 (1992).
46. A.J. Kalmijn, *Nature* **212,** 1232–1233 (1996).
47. A.J. Kalmijn, *Science* **218,** 916–918 (1982).
48. R.D. Astumian, J.C. Weaver, and R.K. Adair, *Proc. Nat. Acad. Sci.* **92,** 3740–3743 (1995).
49. T.E. Vaughan and J.C. Weaver, *Biophysical J.* **71,** 616–622 (1996).

ELECTRIC CARS: ENVIRONMENTAL EFFECTS

VLADISLAV BEVC
California Public Utilities Commission
San Francisco, California

The development of electric cars is driven by the desire to substitute them for the existing conventional internal combustion engine powered cars because electric cars do not emit pollutants in the atmosphere. Examination of the extent to which electric cars can be expected to mitigate the pollution caused by conventional cars requires a clear idea of the number of the conventional vehicles that are expected or may have to be replaced by the electric cars. The total world car population in 1995 was 480 million. The world production of new cars in 1995 was 36 million. New car world production has increased by an average of 2.6% per year in the past 7 years (1). If this trend continues, the annual new car world production will be 54 million in the year 2010. In the United States in 1993, there were a total of 194 million motor vehicles, 146 million of which were automobiles (2), practically all of them powered by internal combustion engines. Estimates of environmental pollution caused by these vehicles are well known. A replacement program will clearly be most effective in areas where there are large concentrations of conventional cars. In general it is impractical to speak of global replacement as opposed to replacement in a given area with known specific characteristics and pollution levels.

Consequences of the extrapolations of the proposed local programs to the overall picture have to be carefully examined. Conventional cars are responsible for two-thirds of all air pollution. If the goal should be that all emissions attributed to the existing conventional automobiles are to be removed, 146 million of cars would have to be replaced, any lesser number would remove proportionally less pollution. The currently mandated programs of various states require some 900,000 electric cars in operation by the year 2003. At this level of electric vehicle penetration, about 0.6%, it is questionable if their contribution to the reduction of the overall car-related environmental pollution would be measurable at all. Only in urban areas with high concentration of conventional cars, the substitution of electric cars may be expected to have a locally measurable and beneficial effect such as reduction of ozone-forming pollutants.

Electric cars, although not directly emitting significant amounts of pollutants to the atmosphere, are nevertheless contributing to environmental pollution. Two principal areas of concern exist. First, any car, whether it is powered by an internal combustion engine or an electric motor, requires about the same amount of energy for propulsion. There may be differences in the efficiency of the engines or motors but the order of magnitude is the same. In case of the electric car, the electrical energy has to be generated somewhere, and the generation process is associated with a degree of environmental pollution, although not always at the place where the transportation activity takes place. Second, storage batteries that are available at present as well those most likely to be produced in the foreseeable future have as one of their principal components metals which, in the quantities in which they may be required, pose substantial environmental problems because of their toxicity.

Accordingly, electric cars are not zero-emission vehicles. Although they do reduce emissions in congested urban areas, additional emissions are generated at the electric power plants which supply the required energy. In some areas of the country, stationary power plants have an important effect on local air quality. For global warming emissions the location of the source is irrelevant.

The environmental impact of the electric cars will become apparent when the amount of the electric power plant pollution attributed to the electric cars is compared with the pollution caused by conventional vehicles and, in addition, estimates of the total amounts of pollution and toxic material discharges introduced in the environment at the different levels of electric-car penetration on the environmental scene are made. Only then a conclusion can be drawn as to whether, on balance, the electric car possesses a decisive advantage.

EMISSIONS CAUSING ENVIRONMENTAL POLLUTION

The principal pollutants emitted by internal combustion engines are carbon dioxide (CO_2), nitrogen oxides (NO_x), nonmethane volatile organic compounds (NVOC), sulfur oxides (SO_x), suspended particulate matter (SPM), carbon monoxide (CO), and lead (Pb).

About 98.5% of carbon dioxide attributable to human activities in the United States is caused by the combustion of fossil fuels. The causes of changing carbon dioxide emissions can be found in energy consumption trends and changes in the composition of fossil fuels burned to provide energy services. Carbon dioxide is a greenhouse gas, 1,406 million metric tons of it have been emitted in 1993, which is 85.5% of the total weight of greenhouse gases. The other greenhouse gases and their respective annual emission levels in 1993 are methane, 178 million metric tons; nitrous oxide, 40 million metric tons; and hydrofluorocarbons (HFCs) and perfluorocarbons (PFCs), 20 million metric tons (3).

Most carbon monoxide is produced from incomplete combustion of fuels. Transportation sources account for most of the carbon monoxide: consistently 75–80% of total emissions. In the atmosphere, carbon monoxide is relatively short-lived, lasting only two to three months before being oxidized into carbon dioxide. Carbon monoxide also functions as the primary sink for atmospheric hydroxyl radicals (OH^-). Through the process of oxidation, hydroxyl radicals chemically break apart methane molecules. By reducing the concentration of hydroxyl radicals, carbon monoxide plays a role in the production of tropospheric ozone. In 1993, 88 million metric tons of carbon monoxide were emitted to the atmosphere, representing an overall decline of 16% from 1980 levels, a reduction that can be attributed to pollution controls and replacement of older, less fuel efficient vehicles. Carbon monoxide is a *criteria pollutant;* that is, pollutants that have adverse effects on public health or welfare for which the Clean Air Act of 1970 requires that they be limited by established standards (4).

Nitrogen oxides commonly denote the molecules NO and NO_2. These gases are important for their role as catalysts in the chemical reactions that create tropospheric ozone; they, too, are criteria pollutants. Nitrogen oxides may also reduce stratospheric ozone if they are emitted directly from aircraft at high altitudes. Fossil fuel combustion is the main origin for nitrogen oxide emissions, particularly from electric power generation and transportation sources. Minor amount of noncombustion emissions are produces by the chemical industry. Nitrogen oxide emissions in 1993 totaled 21 million metric tons. This figure has been relatively stable since 1980.

Nonmethane volatile organic compounds are compounds that play an active role in atmospheric photochemical reactions. These include propane, butane, and ethane, as well as numerous other complex chemicals. Volatile organic compounds affect climate change in two ways: first, as with carbon monoxide, by indirectly increasing atmospheric methane concentrations through reduction of hydroxyl; and second, by causing production of tropospheric ozone. In 1993, emissions of nonmethane volatile organic compounds were estimated to be 21 million metric tons. This is an overall reduction of 10% since 1980 levels. Industrial processes continue to be the primary source of nonmethane volatile organic compounds emissions due in part to solvent use. Emissions from highway vehicles have declined by approximately 32% since 1980 as a result of tailpipe emission controls mandated by the Clean Air Act. Nonmethane volatile organic compounds also are criteria pollutants.

Sulfur oxides are compounds containing sulfur and oxygen such as sulfur dioxide SO_2 and sulfur trioxide SO_3. Emissions of sulfur dioxide, as well as nitrogen oxides, are byproducts of fossil fuel combustion, the most significant emissions of SO_2 result from the combustion of coal oil that contains sulfur. Oil and coal burning power plants are a primary contributor to sulfur oxide emissions. In the atmosphere sulfur dioxide reacts with oxygen and water and produces rain containing sulfuric acid. Acid rain deposition changes the pH factor on the surfaces and in waterways, damages some plants and building materials, and may cause respiratory damage.

Particulates are tailpipe emissions from the internal combustion engine powered vehicles, tire and brake particles, and wind-blown dust. Particulates are generated by dust disturbance by traffic, farming, construction, and combustion, including power plants burning gas and oil.

Prior to 1972, leaded gasoline sold in the United States contained tetraethyl lead in the amount of 2.1 g of lead per gallon. Lead contents discharges from leaded gasoline sold prior to the 1970 Clean Air Act were 320 kilotons per year. Pollutant emissions in the United States for the year 1993 are shown on Table 1 (5–7).

AIR POLLUTION REDUCTIONS EXPECTED FROM ELECTRIC CARS

Electric cars would eliminate pollution emanating from the tailpipe of the conventional cars but they need electrical energy which must be generated somewhere, a process that causes a degree of environmental pollution. Moreover, electric cars must store their energy in batteries which contain heavy metals, all of which have toxic effect in large concentrations or in cumulative deposits. Assessment of the magnitude of these deleterious effects associated with the electric cars must be made and compared to those of conventional cars to see if the benefits will prevail on balance.

With the substitution of electric cars local carbon monoxide emissions would be reduced by 99%, reactive organic gases and volatile organic compounds by 96–99%, and nitrogen oxides emissions by 67–91%. Most studies show that sulfur compound emissions and particulates would increase because of increased generation of electric power, the increase of sulfur oxides would be from 96 to 407% and particulates from 10 to 165%, although these two pollutants do not occur at a very high level to begin with (8). There is a degree of uncertainty over the effect on the nitrogen oxide emissions because of the assumptions about fuel formulations and emission controls (9). These reductions refer to emissions from the vehicles but not to those in electric power generation. Some overall reduction may occur in some areas with cleaner fuel generation mix such as in California, where natural gas is used for electric power generation in preference to oil. This reduction will occur only if electric cars are charged at the electric off-peak periods but not during the peak time when old oil-fired power plants are cut in to supply the peak capacity. The electric rate tariffs would make it unattractive to charge electric vehicles at peak time (10).

In general, comparisons of electric cars to conventional vehicles show that substitution of the former with the latter might decrease emissions of most pollutants, including carbon monoxide, hydrocarbons, and volatile organic compounds but would not remove them to a significant extent.

NUMBER OF INTERNAL COMBUSTION ENGINE POWERED VEHICLES IN THE UNITED STATES

In considering the effect of introducing the electric car as an alternative to the conventional cars it is important to be aware of the levels at which the replacements would have to be made in order that an environmental remediation may be noticeable. As already mentioned, there were a total of 194 million automobiles, trucks, and buses registered in the United States in 1993. Of these, 146 million were automobiles. These vehicles traveled a total of 2,300 billion miles (11). In California 17.1 million automobiles, of which 1.67 million were new vehicles,

Table 1. Annual Pollutant Levels in the United States in 1993 (Millions of Metric Tons Except Lead)[a]

Pollutant	Total (All Sources)	Energy Consumption	Electric Utilities	Transportation
Carbon dioxide	1,406.2	1,372.5	494.9	446.3
Carbon monoxide	88.12	73.14	0.29	68.26
Nitrogen oxides	21.22	20.05	7.06	9.45
Volatile organic compounds	21.14	8.11	0.03	7.53
Sulfur Oxides	21.89		19.27	0.44
Particulates	45.49		0.27	0.20
Lead, tons	4,885		62	1,383

[a] Ref. 5.

were registered in 1993 (12), with 4.8 million automobiles in the Los Angeles County. Vehicle registration trends show that these numbers remained within 1% of the totals since 1991, and the California Department of Motor Vehicles forecasts predict essentially no substantial change in the year 1996. Using the current registration numbers in estimating the electric car penetration levels is reasonable because by the end of the year 2003 the totals will most likely not change by more than 10%.

EXPECTED PENETRATION LEVELS OF ELECTRIC CARS

The goals set for the number of the electric cars to be introduced in California were set by legislation (13) and the California Air Resources Board (14). The regulations require that of all new cars sold in California, zero-emission vehicles will have to constitute 2% of them by the year 1998, 5% by 2001, and 10% by 2003. For all practical purposes *zero-emission-vehicles* means electric cars although, as shown elsewhere in this article, they also are associated with emissions occurring at the power plants supplying the required energy for electric cars and other toxic discharges.

If the mandated goals should be met there would be 434,000 low emission vehicles in California by the year 2003. This figure is in agreement with the expectations quoted by the California Electric Vehicle Task Force (15). However, early in 1996, the California Air Resources Board modified its mandate to the effect that automobile manufacturers are now only required to have available a production capacity adequate for meeting specified production levels rather than actually having to sell a prescribed number of vehicles (16).

Following California's lead, 11 other states (17) enacted similar legislation. Fulfillment of these mandates would result in a total of 919,000 electric vehicles in operation by the year 2003 in the United States (18); that is, only 0.63% of all automobiles in operation today. The legislative bodies are aware that such small percentage of electric cars cannot be expected to result in a significant improvement in the air quality; their expectation appears to be that forcing the technology might trigger a technological breakthrough and as a consequence significantly higher electric car penetration levels.

Early in 1991 the 1990 Clean Air Act became effective (19). It provides that states may either adopt the federal emission standards or California's low emission vehicle program including the zero-emission vehicle requirements.

In a study on utility emissions associated with the electric car charging commissioned by the United States Department of Energy, two different scenarios for the year 2010 envision 12 million electric and hybrid vehicles and 6 million electric cars, respectively (20). Such penetration levels are believed to be required to achieve compliance with the Clean Vehicles Act. Implications of the demand on the supply of battery materials for so many electric cars are discussed later in this article.

COMPARISON OF EMISSIONS ATTRIBUTED TO ELECTRIC CARS AND EMISSIONS OD INTERNAL COMBUSTION ENGINE POWERED VEHICLES

Very roughly, one could make an estimate about how much of a given pollutant per mile traveled can be attributed to the conventional motor vehicles. Although such estimates would provide an order of magnitude type of information, it would be problematic whether all the internal engine powered vehicles included in this statistic could in fact be replaced by electric cars. In actual studies that are used for comparison of the pollutant emission per mile attributed to the conventional vehicles and electric cars, respectively, emissions attributable to electric vehicles are compared to low emission internal combustion engine powered vehicles (21). Low Emission Vehicles are classified in four categories: transitional low emission vehicles (TLEV); low emission vehicles proper, ultralow emission vehicles (ULEV); and zero-emission vehicles (ZEV). EPA also proposes a category of inherently low emission vehicles (ILEV). These categories are divided in tiers: tier 0 includes vehicles with emission standards in effect up to and including 1993, whereas tier I refers to the standards applicable since 1994. These tiers are further subdivided by the type of maintenance performed on the vehicles: for instance, enhanced inspection and maintenance.

Studies comparing emissions of power plants due to incremental power generated to recharge electric car batteries must necessarily be specific to a given area and depend on the type of fuel mix that is used by the power plants in the area. The U.S. Environmental Protection Agency (EPA) initiated a study of the possible increase in power-plant emission that could arise from charging a large number of electric vehicles. In areas where power plants are fueled by coal and oil, such as is the case in the northeastern part of the United States, the agency's studies suggest that the energy demands of electric vehicles could impair the air quality. Increased nitrogen dioxide emissions, for example, which are common in many power-plants, are an ingredient of smog.

The method used consists essentially of taking the additional energy and associated pollution expected for a given number of electric vehicles in an area and calculating the average amount that can be attributed to a typical electric vehicle.

This figure is then compared to the tailpipe emissions of a conventional automobile meeting the federal standards for tailpipe exhaust (22). In its preliminary assessment of November 1993, EPA compared the emissions from internal combustion engine powered vehicles ranging from tier I, enhanced inspection and maintenance to inherently low emission vehicles with maximum inspection and maintenance with electric cars powered by lead–acid batteries and sulfur–sodium batteries, the two variants representing a bracketing of the likely options.

The EPA made a comparison between representative low emission conventional vehicles (23) and electric cars powered by lead–acid and sodium–sulfur (24) batteries. These vehicles are designated by type as tier I, Enh I/M; CA.LEV, Enh I/M; ILEV, Enh I/M; CA. LEV, MAX I/M; ILEV, MAX I/M. 40 CFR 88.101-94 and 40 CFR 88.302-93 define the types of vehicles as follows: low emission vehicle, transitional low emission vehicle, ultra low emission vehicle, inherently low emission vehicle and zero-emission vehicle mean any light-duty vehicle conforming to the applicable low emission vehicle standard, or any heavy-duty vehicle with an engine conforming to the applicable low emission vehicle standard, the transitional low emission vehicle standard, ultra-low emission vehicle standard, inherently low emission vehicle standard, and zero-emission vehicle standard, respectively. Lead–acid batteries are available today; sodium–sulfur batteries are still considered prohibitively expensive by some manufacturers. These two types of batteries were used because they bracket the range of costs and energy densities available. Gasoline vehicles were evaluated for the assumed in-use fuel economy of 23.1 miles per gallon (mpg) which corresponds to a nominal fuel economy of 27.5 mpg and available energy of 114,000 btu/gallon (12.3 kWh/kg) gasoline (SAE 930376).

The following national mix of stationary source plant types was assumed: coal, 56.3%; oil, 3.2%; gas, 9.4%; other fuel, 31.1%. Other fuel sources consist of nuclear, hydroelectric, solar, wind, geothermal, waste, and co-generation. These sources were assumed to produce no emissions. Stationary source power generation emissions in gram per kilowatthour (g/kWh) for the various pollutants were taken from the EPA Special Report (25) and multiplied by the stationary source plant type mix. EPA Special Report estimates average emissions from power plants on a regional basis for the three fossil fuels (coal, oil, natural gas) based on regional fuel type and power plant technology.

The outcome of the EPA preliminary assessment attributed 393 g of carbon dioxide to lead–acid battery operated cars, which is about 10% less than the federal standard of 354 g per mile (26–28). Lead–acid batteries are available today (27). By 1998, lead–acid and nickel–cadmium batteries will be available commercially for electric vehicles. Sodium–sulfur (Na–S) batteries are still considered to be prohibitively expensive by some manufacturers. These two battery types were used because they bracket the range of costs and energy density of most available battery types.

As is apparent the comparison of this particular assessment is not very favorable for the electric cars and the study has drawn considerable criticism primarily from the agencies promoting the introduction of electric cars (29). These were represented mainly by the Northeast States for Coordinated Air Use Management (NESCAUM) whose members are Connecticut Bureau of Air Management, Maine Bureau of Air Quality Control, Massachusetts Division of Air Quality Control, New Hampshire Air Resources Division, New Jersey Office of Energy, New York Division of Air Resources, Rhode Island Division of Air and Hazardous Materials, and Vermont Air Pollution Control Division. These agencies contend that (*1*) the electric vehicle energy consumption was overestimated by 50–130% by using a calculation of the required power rather than the figures provided by automobile manufacturers; (*2*) nitrogen oxide emissions rates for power generating plants were overestimated by as much as 300% because geographical differences of power generating mix and future compliance with the Clean Air Act were not taken in account; and (*3*) benefits of the electric cars were understated by a factor of two for volatile organic compounds because the comparison was made with internal combustion engine powered vehicles which included inspection and maintenance and reformatted gasoline programs that would not be needed for electric cars. Although there is some merit to these observations they only point out the sensitivity of this type of analyses to the initial assumptions. They do not necessarily vitiate the conclusion that the power requirement for electric cars may well be associated with pollution levels of the same order as those for the internal combustion engine powered vehicles, although the emissions may occur at a different place than the location where the electric cars are operated. The criticism apparently was not directed at the emission levels of carbon dioxide for which it does not matter where it occurs so far as the global warming is concerned.

A study made by the Electric Power Research Institute (EPRI) for the year 2011, considering electrical energy generation from power plants within the urban air quality districts served by California's largest utilities, found that electric car related emissions of reactive organic gases, nitrogen oxides, and carbon monoxide would be 93–99% lower than those of gasoline-powered ultralow emission vehicles, whereas the particulate matter would be 72–82% lower (30). The model assumed strict emission controls in effect for the California power generating districts. However, in 1997, the California Public Utilities Commission deregulated electric power generation so that local electric power distribution companies in California will be able to purchase power from the cheapest available sources; for example, from coal- or oil-burning power plants outside of California who might purchase pollution credits to avoid cleaning up their emissions. This would tend to vitiate analyses based on the exclusive use of natural gas fired power generating plants.

Examination of the environmental implications of having 500,000 battery-powered vehicles in Southern California and another 500,000 in the New York City area (31) under present conditions shows that peak ozone levels would decrease approximately one part per billion in Southern California and less than one part per billion in New York City. In contrast to the small ozone-related benefits from lead–acid battery powered cars, the potential environmental problems are considerable. As discussed subsequently, recycling more than 500 kg of lead batteries per vehicle would result in environmental lead releases 80 times greater than those of a gasoline-powered vehicle with a simple starter battery. Half a million of these vehicles would increase national lead discharges by about 20%. It is interesting to note in this connection that even if every conventional vehicle were to disappear from the streets of Los Angeles, the city would still not meet the allowed health-based air quality standards (32).

ENERGY REQUIRED BY ELECTRIC CARS

Determination of the energy requirements for the electric cars is essentially no different than that for the conventional automobiles. The energy required to obtain the specified performance can be readily estimated on the basis of the well-known general principles of mechanical engineering and this, in turn, provides the lower bound of the required energy capacity of the battery needed to power the vehicle. The typical conservative requirement on an electric car within the constraints imposed by the size and capacity of its battery is that it should have a range of some 80 miles and a top speed of 60 mph. We shall use these in the following discussion because it would not be realistic at this time to adopt a range and performance equal to those of conventional cars.

To move any car over the desired range it is necessary to overcome the air and rolling resistance which are of about the same order of magnitude (33). For a typical American car weighing 2,000 kg (4,410 lb) the combined resistances amount to about 100 kg (220 lb) at 60 mph (26.8 m/s) which translates to a traction power of 26.8 kW or 35 HP. This power sustained over a range of 80 miles at the specified speed requires an energy of 36 kWh. The capacity of the battery must be at least that much. In practice it is also necessary to consider the charging losses and the battery efficieny; that is, the amount of the energy available from a fully charged battery is always less than it takes to charge it up. This result is compared with the specifications for two commercially available electric vehicles.

Consider first the G-Van manufactured by the General Motors (34). Although with its weight of 3,900 kg (8,600 lb) the G-Van is not a personal automobile, it is useful to compare its battery capacity with the previous estimate. According to the specifications, the 36 lead–acid batteries constituting the power storage pack have a capacity of 44.28 kWh. On the other hand, General Motors Impact, weighing 1,000 kg (2,200 lb) is much lighter than the average American car. Its specifications give the capacity of a power pack consisting of 32 batteries as 13.6 kWh (35). These two battery capacities can be considered as bracketing values for the energy storage requirements.

DEMAND ON THE ELECTRIC UTILITY SYSTEMS

The effect of electric cars on utility systems is to increase demand for both peak capacity and total energy. Once the energy requirements of the electric cars are known it is of interest to estimate the magnitude of the demand the charging of the electric cars will represent for the serving electric utilities. Estimates of this kind have to be made for specific locations and geographic areas taking in the account the number of electric cars in operation, the amount of driving, and the characteristics of the electric utilities serving the area. A comprehensive study for the Los Angeles Basin, using values of the electric car penetration expected if the original California Air Resources Board mandate is fulfilled, including also the service territory of the Los Angeles Department of Water and Power, has been prepared by the California Energy Commission. The study sets forth the method used in making such studies and specific projections for the case considered (36).

An example of such estimates is a study prepared by the Southern California Edison Company which serves the greater part of the Los Angeles Basin (37). Edison proceeds from the assumption that by the year 2010 there will be at least as many electric cars in the Los Angeles Basin as required by the California Air Resources Board mandate. These projections mean that there will be 1.7 million electric cars in California of which 1.3 million will be in the Los Angeles Basin and 910,000 in Edison's service territory. It is noteworthy that the utility did not use projections of additional market penetration of the electric cars in excess of what has been mandated. According to Edison, 8 million of internal combustion engine powered vehicles operate in the Los Angeles Basin every day. As already mentioned, 4.8 million automobiles were registered in the Los Angeles County in 1993. The study assumes that each electric car will on average be driven 10,000 miles per year with an energy consumption of 0.6 kWh/mile or a total annual consumption of 6,000 kWh—a light passenger gasoline powered automobile with a mileage of 27.5 mpg uses approximately 0.8 kWh/mile. Assuming further that these vehicles will be driven on the weekdays their weekday energy requirement per vehicle will be 25 kWh. With a charging time of 5 hours, each of these electric cars will represent a demand of 5 kW on the utility system, a total of 4,500 MW for 900,000 cars if the charging is coincidental. By the year 2010 Edison expects its peak load to reach 25,000 MW; the electric cars would thus represent up to 18% of the system's total demand. The utility points out, however, that with staggered charging at the off-peak time, it will not be required to add a full 4,500 MW of power generating capacity to its system, an addition of 195 MW would be adequate to meet the additional demand presented by the electric cars.

Pacific Gas and Electric Company, the largest electric utilities in California with a service territory of 94,000 square miles, made detailed estimates for three different penetration scenarios that might occur as a result of different technological and political events. The first scenario represents an environment of low penetration levels such as might be expected with a softening regulatory approach implying only 126,000 electric vehicles in the utility's service area by the year 2100. The second scenario represents an environment in which regulatory mandate is maintained and results in a moderate penetration but still less than what is mandated by the California Air Resources Board, ie, 547,000 electric vehicles in the service area by 2010. The third scenario represents an environment in which a significant technological breakthrough occurred and created a market driven industry with penetration levels reaching 1,147,000 electric vehicles in 2010.

Scenarios for three different penetration levels by the utility's various service area divisions were used. It was assumed that a typical commuter traveling less than 50 miles per day five days in a week. The total energy delivery requirement per division resulted in an average of 30 kWh per vehicle per day, equivalent to 44 mpg for regular internal combustion engine powered cars. Charging scenarios also were specified to project the time varying demand added to the system by the electric cars.

Two elements have to be determined in developing possible charging scenarios. The first is a uniform charge cycle, or a load profile of a single charger that can represent all other chargers on the distribution system. The second is the accumulated load profile of all chargers on the system over a 24-h period. The accumulation describes the electric vehicle charging demand which the utility must deliver, Identifying electrical load characteristics for future electric vehicle battery charging involves the uncertainty in the behavior of the electric vehicle users and future charger characteristics (38). It has been shown that in the highest potential impact areas, new peak load in excess of 10% of the total demand could occur for the high penetration levels for the year 2011, whereas for the low penetration levels it would have significant impacts on the system.

Merely meeting the California Air Resources Board mandate by presumably substituting 910,000 electric cars for internal combustion engine powered automobiles out of a total of 5.6 million cars means removing only 16% of the locally polluting vehicles that are responsible for only 10.6% of the total local air pollution. Suppose, on the other hand, that one wanted to do better and replace all 5.6 million cars in Edison's service territory by electric cars. The required demand on the utility system for charging these cars would be 28,000 MW, more than Edison projects for its system-wide peak in the year 2010. Distributing this load over three shifts would flatten the additional demand to roughly one-third of the above. Still, a substantial addition of the power generating plant or purchased energy and a substantial restructuring of the local power distribution system would be needed to meet this load.

It is of interest to consider how much energy would be required if all conventional cars were replaced by electric cars. The energy needed for propulsion would be the same. In 1994 all cars in the United States used 73.8×10^9 gallons of gasoline (39). The energy content of this gasoline is $2,760 \times 10^6$ MWh; assuming an efficiency of 25% for the internal combustion engine, 690×10^6 MWh of energy was needed for propulsion. If the electric motor has an efficiency of 90%, electric cars would require 766×10^6 MWh for propulsion. In 1994 the total electrical energy generated in the United States was $2,910 \times 10^6$ MWh or about the same as the total energy content (heating value) of gasoline burned in automobiles (40). To generate the amount of energy needed for propulsion 1.277×10^9 barrels (bbl) of fuel oil would be required, assuming an electric power plant heating value of 600 kWh/bbl. On the other hand, to produce 73.8×10^9 gallons of gasoline at least $2,510 \times 10^6$ bbl of petroleum is needed (41).

It has been suggested that using fuel cells powered by hydrogen would avoid the problems associated with storage batteries (42). Fuel cells with an efficiency of 50% would require $1,532 \times 10^6$ MWh of energy contained in 16.365×10^{12} cu ft of hydrogen (43). To produce this amount of hydrogen by electrolysis, with 150 kWh required to produce 1,000 cu ft (44), $2,455 \times 10^6$ MWh is required, which is nearly 85% of the total electric energy generation in the United States in 1994. It has been further suggested that, rather than doubling the nation's thermal power plant capacity, the required energy could be supplied by solar power (45). Assuming an average 8 h of operation per day, the required solar power plant capacity would have to be 840,000 MW, approximately 20% more than the total United States capacity in 1994 (46). The largest solar power generating plant in operation today is the Carizo Solar Corporation's photovoltaic plant in California with a capacity of 5 MW (47). The worldwide goal for the solar capacity by the year 2000 is 1,400 MW (48). These values apply, of course, only in the case where all internal combustion cars would be replaced by electric cars; the values would be proportionally lower at lower values of electric car penetration.

STORAGE BATTERIES

Batteries are a major and indispensable component of the electric car as they store the energy necessary for the propulsion. Their capacity or rather specific energy; that is, energy that can be stored by a given amount of mass of active materials is limited by the electrochemical properties of the materials of which they are made. The chemical reactions utilized for energy storage and recovery determine the amount of the electricity a given amount of mass can store, representing the theoretical limit of the battery's specific energy (49). Typically these are of the order of a few hundred watthours per kilogram, considerably less than the energy stored in gasoline, which contains about 13 kWh/kg. There is no way in which the theoretical upper bound of the specific energy can be exceeded because it is determined by the electronic configuration of the elements involved in the electrochemical processes.

In practice, the theoretical capacity is considerably lower because a battery consists of other materials in addition to those active in the electrochemical process. A number of electrochemical pairs would appear to be suitable for energy storage and some of these have quite respectable theoretical specific energies, but unfortunately the technology that would make use of them has not yet been developed. Among the various storage batteries that have been considered for use are (*1*) lead–acid, (*2*) sodium–sulfur, (*3*) sodium–nickel chloride, (*4*) nickel–iron alkaline cell, (*5*) nickel–cadmium, (*6*) nickel–metal hydrides, (*7*) nickel–zinc, (*8*) zinc–bromine, (*9*) zinc–chlorine, (*10*) zinc–air, (*11*) silver–zinc, (*12*) lithium–iron sulfides, (*13*) lithium–vanadium oxides, and (*14*) conductive polymer batteries (50–58).

ACCUMULATION OF TOXIC METALS IN THE ENVIRONMENT

Large-scale use of metals and other materials contained in the electric storage batteries raise environmental concerns because of their toxic effects. Continued contamination of the air, soil, and water with nondegradable toxic metals has to be viewed with concern; the problem has become global. At the current pollution rates the abilities of many ecosystems to cope with toxic metal pollution are being rapidly depreciated, and the margin of safety for many organisms is disappearing rapidly. From 1900 to 1980 there was an exponential increase in industrial emissions of toxic metals paralleling the rates of population growth; during that period the rates of emissions increased as shown. A decline from 1980 to 1990 reflects the effects of the pollution control programs in developed countries.

In 1993 the annual world production of lead was 2.93 million metric tons, a 13% decrease from the 1990 level, of which 12% or 352,000 metric tons were produced in the United States (50). More than half of the lead produced is used in batteries. Other major uses include semifinished sheet metal and pipes, alloys, cable sheathing, additives in gasoline, and ammunition. Between 1960 and 1990 lead consumption decreased in all product categories except batteries and ammunition. In the United States many lead products have been effectively phased out; however, because lead use in batteries has increased by more than a factor of two worldwide, overall lead use has increased about 40% in the same period.

Lead is a neurotoxin: it causes reduced cognitive function and behavioral problems even when present at very low levels in the blood and as such it has major detrimental health effects both on human population and many bird populations. The major known health and environmental effects are due to leaded gasoline, lead-based paint, and lead shot. For these reasons tetraethyl lead has been eliminated from gasoline in the United States with the benefit that the blood levels of lead in children has been greatly reduced (51). Although leaded gasoline has been banned in the United States it is still used in most countries. In the United States, lead poisoning is one of the most common and preventable pediatric health problems (52). In a typical child, absorption of about 10 μg of lead per day results in a blood level of about 5 μg/dL. To have a blood lead level of 15 μg/dL, a child must typically absorb about 40 μg of lead per day. In the presence of lead-based paint, lead levels of indoor dust can reach 2×10^3 ppm. At this level, ingestion of 0.1 g per day of dust would result in absorption of as much as 100 μg of lead daily which is well above the threshold for health effects. Wild waterfowl and other birds, predominantly in popular hunting areas, as well as domestic animals grazing near lead smelters also are victims of lead poisoning. Concentrations of lead in soil above 10^3 parts per million have been found to decrease soil microorganism activity (53). Lead materials flow balances for the United States and the world are generally incomplete (54,55).

Human health and ecosystem damage from cadmium is found only in areas of high localized contamination. Currently, risks appear to be present in trace cadmium contamination in phosphate fertilizers and in coal. Cadmium is cumulatively deposited in livers and kidneys causing damage in kidney functions, and inhaled cadmium causes lung cancer.

In ecosystems, damage to soil organisms has been observed in the vicinity of smelters with cadmium concentrations of about 100 ppm. In laboratory studies cadmium has been shown to be toxic at concentrations only 10 times those typically found in soils. The main concern about long-term cadmium exposure has been that a buildup of cadmium in soil might lead to dangerous exposures in the future. Clarification of its effects and projected future quantities are needed before it can be determined which cadmium products might be leading to increased exposure (56).

ENVIRONMENTAL PROBLEMS ASSOCIATED WITH STORAGE BATTERIES.

In addition to the pollution associated with the electric power production required to charge electric vehicles the batteries required to store the energy necessary for the propulsion of electric cars also represent a serious environmental concern. Most batteries, and certainly those that are commercially available for electric vehicle manufacture contain wastes that are toxic or hazardous. It is also important to consider environmental issues associated with battery systems, such as end-of-life recycling or reclamation of electric vehicle batteries, identifying important environmental health and safety issues, and assessing where further analysis of these issues is required. The amount of pollution and exposure problems may be exacerbated when the industry has to process large quantities of the materials used in the manufacture of batteries.

The environmental consequences of producing and reprocessing large quantities of lead–acid batteries have been the subject of considerable controversy. Smelting and recycling of lead for the batteries will result in substantial releases of lead to the environment.

If a large number of electric cars are produced, the demand for lead batteries will surge, requiring that more lead be mined. It is instructive to examine how much lead would be required for the lead–acid batteries at the contemplated electric car penetration levels. The specific energy of currently available batteries ranges from 20 to 40 Wh/kg (57). At the highest available specific energy figure of 40 Wh/kg, the required battery weight for the battery capacities of the two representative electric car models discussed previously, Impact and G-Van, would be 340 kg (750 lb) and 1,000 kg (2,200 lb), respectively. Lead constitutes approximately 65% of the weight of a battery (58), accordingly, from 220 to 650 kg of lead would be required for each car. With 6 million cars this would amount to something between 1.32 to 3.90 million metric tons of lead. This is a sizable order in view of the fact that in the total 1993 world production of lead, 2.93 million metric tons, of which 12% or 352,000 metric tons, were produced in the United States (59). It would be necessary to evaluate the impact on the production, processing, and recycling of quantities of lead approaching the world production figures. Prior to the enactment of the Clean Air Act of 1970, approximately 320 kilotons of lead per year were emitted from leaded gasoline. This source of pollution has now been removed—electric cars would reintroduce this pollutant at very high levels.

The problem associated with the recycling of large quantities of lead becomes apparent by looking at how the lead recycling is handled. Over 70 million batteries are replaced in the United States every year, of which up to 23 million are improperly discarded either in the incinerators, landfills, or along the side of the road. These 23 million batteries contain over 180 million kg of lead. Introduction of 6 million electric cars with battery packs of 30 batteries each would add an additional 180 million batteries to be replaced every four years. If it is assumed that the battery pack would be good for 500 cycles at 80 miles each and 10,000 miles driven by an average electric car every year, that is 45 million more batteries. American smelters conforming to the strict environmental and health regulations report that they keep lead discharges down to 36.7 ppm of processed lead. They, however find it expensive to continue operating at ever-increasing expenses required by the pollution control measures and over half of them have shut down their plants because they were no longer profitable (60).

In industrial countries, the environmental and occupational health regulatory cost of operating lead battery recycling is continuously increasing, whereas the prices offered for secondary lead are low. In many industrial countries it is simply not profitable to operate secondary lead smelters. Battery brokers are finding more profitable markets in places workers' wages are low and environmental regulations weak or nonexistent. Lead battery smelters have been transferring out of industrial countries in recent years as environmental regulations have tightened and domestic lead prices have dropped. The secondary lead industry has already made a massive exodus out of North America. Over half of the secondary lead smelters have closed their operations because of the inability to economically install emission controls and purchase liability insurance because waste disposal is becoming a very significant expense and often a difficult task to perform. Foreign smelters, on the other hand, can pay a higher price for scrap because their capital, labor, and environmental costs are lower than those of the American producers. American smelters are facing a similar situation as their counterparts in Great Britain, ie, continually dropping prices because of low demand and excessive inventories.

Without a global dumping ground, the lead–acid battery manufacturing industry would be forced to become clean by eliminating the use of lead in batteries. The demise of lead smelting companies in industrial countries reflect the desire of those societies to avoid further lead contamination. However, the flourishing international trade of lead–acid battery waste is providing the manufacturers with cheap and easy outlets for their toxic waste.

The battery industry's efforts to persuade legislators and consumers that lead battery recycling was remarkably successful notwithstanding the continual decline of the lead recycling industry in industrial countries. Model laws sponsored by the battery industry have been adopted in many parts of the United States requiring retailers to accept used car batteries when consumers purchase new ones (61). According to Greenpeace, however, the returned batteries do not necessarily go to recycling but are instead channeled into the international lead–acid waste trade and result in substantial harm to the environment and people of the Third World. The recipient countries include Indonesia, Mexico, the Philippines, South Africa, Taiwan, and Thailand.

Although the operators of the lead–acid battery recycling processes are able to recycle 90% or more of lead we have to be concerned about the amounts that would not be recycled which would find their way into the environment and, of course, lead that would never be processed by properly operated recycling facilities. Substantial and extensive changes in the processing of the volumes that have hitherto not been handled would have to be made. The implication that introduction of electric cars on a massive scale would increase the discharges to the environment of lead, a toxic substance which humans should avoid, several times over its present amount is inescapable.

Sodium–sulfur batteries and lithium polymer batteries are not likely to be available soon, sodium–sulfur batteries are expensive and thus not attractive to the manufacturers. Nickel–cadmium and nickel–metal hydride are likewise toxic substances. For vehicles that could be mass produced in 1997, lead–acid batteries are the only practical technology. Environmental discharges of lead are a major concern. The United States Department of Energy has commissioned comprehensive studies to assess important environmental, health, and safety issues, pertaining to the cell and battery materials, and workplace, processing, shipping, and vehicle safety.

Lithium solid polymer electrolyte batteries with metallic lithium or nonmetallic lithium compounds in the negative electrode have been evaluated with respect to these issues (62). For the time being it is not known how reactive the various nonmetallic lithium solid polymer electrodes materials would be if the cell was ruptured or if the electrolyte failed and the cell was short-circuited. There is potential for human exposure to lithium solid polymer electrolyte materials during use and during the manufacture and recycling of batteries. Toxicity and carcinogenic issues exist for some potential cell materials, including lithium hexafluoroarsenate, nickel dioxide, vanadium oxides, and titanium dioxide. Facilities that dispose of lithium batteries (nonsolid polymer electrolyte type) use hydrolysis to deactivate the metallic lithium. It is not economical for these facilities to recycle the lithium, and it may likewise not be economical to recycle the lithium from solid polymer electrolyte batteries. The bulk of a lithium solid polymer battery, however, will be composed of metals such a stainless steel, and plastics which may be recycled. In many types of lithium solid polymer batteries, such as for instance batteries using negative electrodes with intercalated carbon, lithium in its ionic form rather than metallic lithium would be used. Nonmetallic lithium compounds are less reactive than metallic lithium, the level of reactivity would affect the feasibility of using hydrolysis to deactivate lithium solid polymer cells in case of a rupture.

Health and safety hazards associated with nickel metal-hydride electric vehicle cell are generally small because of the nontoxic nature of the electrode materials (63). Internal pressure buildup and accumulation of hydrogen gas is a potential hazard which should be handled by a pressure release valve and appropriate design features. Nickel and cadmium compounds are probably carcinogens; therefore, compliance with the OSHA standards for exposure to these compounds at the manufacture and in the workplace should be assured. Shipping nickel metal-hydride batteries should take place under the provisions for spillable substances until testing can confirm either a dry or nonspillable classification. Thus far no major in-vehicle safety problems have been identified except the hazards associated with hydrogen gas. Recycling is the main environmental issue for nickel metal-hydride batteries. Test results show that nickel metal-hydride AB_2-type electric vehicle batteries would not be considered hazardous under the existing EPA regulations, although they would be considered hazardous under California and European Community environmental regulations (64). Other nickel-based electric vehicle batteries share many common environmental health and safety characteristics of the nickel battery family. Nickel–cadmium batteries, for example, are considered hazardous waste under EPA regulations.

In the evaluation of sodium–sulfur batteries, failure and life testing of cells have shown that the chemical and thermal risks associated with the presence of sodium, sulfur, and sodium polysulfides can be satisfactorily mitigated by engineering design at the cell level but not, as yet, on the battery pack level. The safety of sodium–sulfur batteries depends on the electrical and thermal functioning of thousands of individual cells and the electrical, thermal, and mechanical systems that control these cells. A comprehensive, long-term testing program of sodium–sulfur batteries in vehicles under actual driving conditions is needed to demonstrate that the technology has evolved to a point where commercialization is possible. The results available so far are not sufficient for acceptance under the United States Federal Motor Vehicle Standards (65). Recycling and disposal processes of spent sodium–sulfur batteries are in the beginning stages of development. Long-term sodium–sulfur battery disposal and recycling will have to be capable of handling large quantities of spent batteries in a cost-effective and environmentally acceptable way in accordance with the provisions of the Resource Conservation and Recovery Act (66). Transport of sodium–sulfur beta batteries must comply with shipping regulations for hazardous materials because they contain large quantities of elemental sodium (67).

The typical sodium–sulfur battery for traction, like all other electric vehicle batteries, operates at a voltage between 200 and 300 V which poses a potentially lethal shock hazard particularly during charging, maintenance, and in the event of a severe collision. Fire resulting from short circuiting, overheating, or cell rupture is a related chemical hazard. Overheating might result from overcharging, a failure of the thermal management system, or cell failure. Failure tests under overcharging conducted by one major sodium–sulfur battery manufacturer show that through proper safety design this problem can be avoided. The thermal management system of a sodium–sulfur battery is designed to control the initial heating of the battery, dissipate waste heat, and insulate the system typically at temperatures between 300 and 400°C; however, excessive temperatures resulting from a rapid rate of discharge could cause cell failure and further heating if the cooling system does not respond adequately (68).

BIBLIOGRAPHY

1. *Motor Vehicle Facts and Figures,* American Automobile Manufacturers Association, Detroit, Mich., 1995.
2. *U.S. Bureau of the Census, Statistical Abstracts of the United States 1995,* Washington DC, 1996, p. 632.
3. *Emissions of Greenhouse Gases in the United States 1997–1994,* Energy Information Administration, Office of Integrated Analysis and Forecasting, US Department of Energy, Washington, D.C., 1995, p. xi.
4. Public Law 91-604, Sections 2 to 11a, 12 to 15(a), (c), Dec. 31, 1970; 84 Stat 1676 to 1713; *United States Code, Title 42, Public Health and Welfare, Sections 7403 to 7418, 7521 to 7525, 7541 to 7546, 7550, 7571 to 7574, 7601 to 7616, 7641, 7642–7671.*
5. *Emissions of Greenhouse Gases in the United States 1987–1994,* Energy Information Administration, Office of Integrated Analysis and Forecasting, United States Department of Energy, Washington, D.C., 1995, pp. 11, 12, 63–65.
6. "Sulfur Dioxide; Air Pollutant Emissions, by Pollutant and Source," U.S. Bureau of the Census, *Statistical Abstracts of the United States 1995,* Washington, D.C., 1995, p. 233, Table No. 375.
7. "Both PM-10 and PM-10 Fugitive Dust: Air Pollutant Emissions, by Pollutant and Source," U.S. Bureau of the Census, *Statistical Abstracts of the United States 1995,* Washington, D.C., 1995, p. 233, Table No. 375.
8. T. S. Yau, H. W. Zaininger, M. J. Bernard III, K. Heitner, M. K. Singh, and C. L. Saricks, *Utility Emissions Associated With Electric and Hybrid Vehicle (EHV) Charging, Interim Report,* U.S. Dept. of Energy, Apr. 1993; D. Sperling: *Future Drive,* Island Press, Washington, D.C., 1995, p. 45.
9. T. P. Henderson and M. Rusin, *Electric Vehicles,* American Petroleum Institute, Washington, D.C., Jan. 1994, Appendix IV, p. 66.
10. *Southern California Edison Company's Tariffs: Original Cal. PUC Sheet 19864-E, Schedule TOU-EV-1, Effective Mar. 29, 1996; Southern California Edison Company's Tariffs: Revised Cal. PUC Sheet 19069-E, Schedule D, Effective Jan. 1, 1995,* rates for electric cars are heavily discounted and subsidized by all other ratepayers.
11. U.S. Bureau of the Census, *Statistical Abstracts of the United States 1995,* Washington, D.C., 1996, *loc. cit.*
12. *Estimated Fee-Paid Registration by County; as of December 1993, Annual Report,* State of California Department of Motor Vehicles, Sacramento, Calif., Revised Mar. 1994.
13. *California Public Utilities Code,* Section 740.2 (Deering 1994).
14. *Title 13, California Code of Regulations,* Section 1960.1(g) (12), note (9), Barclay Law Publishers, South San Francisco, 1994, 214.3.
15. California Electric Vehicle Task Force, *A California Plan for the Commercialization of Electric Vehicles I,* Sacramento, Calif., 1989, p. 7.
16. California Air Resources Board, *News Release 96-4,* Sacramento, Calif., February 8, 1996.
17. Delaware, District of Columbia, Maine, Maryland, Massachusetts, New Hampshire, New Jersey, New York, Pennsylvania, Rhode Island, and Vermont.
18. L.G. O'Connell, "Establishing a National Electric Vehicle Recharging Structure," *Proceedings: International Electric Research Exchange Workshop on Electric Vehicles, Advanced Batteries, and Impacts on Electric Utilities,* Report EPRI TR-101138, Electric Power Research Institute, Palo Alto, Calif., Dec. 1992, 4–263.
19. Public Law 101-549, Nov. 15, 1990; 104 Stat 2399; *United States Code, Title 42, Public Health and Welfare, Sections 7401–7671.*
20. T. S. Yau and co-workers, U.S. Dept. of Energy, *Utility Emissions Associated With Electric and Hybrid Vehicle (EHV) Charging, Interim Report,* Apr. 1993, pp. 1–3 to 1–5.
21. *U.S. Code of Federal Regulations,* Title 40, Sections 88.102-94 and 88.302-93.
22. *Preliminary Electric Vehicle Emission Assessment,* Table 1, U.S. EPA National Vehicle and Fuel Emission Laboratory, Ann Arbor, Mich., 1993, p. 4.
23. *U.S. Code of Federal Regulations, Title 40,* Sections 88.102-94 and 88.302-93.
24. *Preliminary Electric Vehicle Emission Assessment,* Table 1, U.S. EPA, National Vehicle and Fuel Emission Laboratory, Ann Arbor, Mich., 1993, p. 4.
25. M. Safoutin, *Fuel Efficiency of Electric Vehicles as Affected by Range, Technology, and Battery Type,* Special Rept., U.S. EPA, Ann Arbor, Mich., 1992; *Methodology for Analyzing the Environmental and Economic Effects of Electric Vehicles: An Illustrative Study,* prepared for U.S. EPA by ICF, Inc., Ann Arbor, Mich., Sept. 1991; Ref. 24, Table 9, p. 13.
26. *Preliminary Electric Vehicle Emissions Assessment,* Table 1, United States Environmental Protection Agency, Regulation Development and Support Division, National Vehicle and Fuel Emissions Laboratory, Ann Arbor, Mich., Nov. 3, 1993, p. 4.
27. *California Air Resources Board Battery Technology Audit Panel Preliminary Findings,* Update, U.S. Advanced Battery Consortium, Dearborn, Mich., Fall 1995, p. 5.
28. Ref. 24, Table 8, p. 12; S. Hoffman, *The Transportation Program, Electric Vehicles,* Technical Rept. 102880, Electric Power Research Institute, Palo Alto, Calif., 1993, p. 3; J. Carey, D. Woodruff, and P. Coy, *Bus. Week,* 110 (May 30, 1994).
29. M. J. Brady, *Memorandum Re: EPA Draft Rept. on Electric Vehicles* (see Ref. 24), Northeast States for Coordinated Air Use Management, Boston, Mass., Mar. 21, 1994; see Ref. 9.
30. *EV Emissions in California,* Technical Brief RP2882, Electric Power Research Institute, Palo Alto, Calif., 1994.
31. L.B. Leave, A.G. Russell, C.T. Hendrickson, and F.C. McMichael, *Environ. Sci. Technol.,* **39**(9), 402A–407A (1996).
32. D. Sperling, *Scientific American,* **275,** 56 (1996).
33. T. Baumeister, ed., *Marks' Standard Handbook for Mechanical Engineers,* 7th ed., McGraw-Hill Book Co., New York, 1967, p. 11–4.
34. *Electric Vehicle Highlights,* Southern California Edison Co., Rosemead, Calif., 1990.
35. L.G. O'Connell, *EPRI J.* 13 (Apr./May 1991).
36. A. Tanghetti and P. McAuliffe, *Electric Vehicle and Powerplant Emissions,* California Energy Commission, Sacramento, Calif., 1995.

37. M. Merlo, *Proceedings of the International Electric Research Exchange (IERE) Workshop on Electric Vehicles, Advanced Batteries, and Impacts on Electric Utilities,* Electric Power Research Institute (EPRI), Palo Alto, Calif., 1992, pp. 4–161 to 4–169.
38. S. Rahman and G. B. Shresta, *An Investigation into the Impact of Electric Vehicle Load on the Electric Utility Distribution System,* IEEE/PES Winter Meeting, New York, Jan. 1992.
39. U.S. Bureau of the Census, *Statistical Abstracts of the United States:* 1996, 116th ed., Washington, D.C., 1996, Table No. 1019.
40. U.S. Bureau of the Census, *Statistical Abstracts of the United States:* 1996, 116th ed., Washington, D.C., 1996. Table No. 942.
41. *Marks' Standard Handbook for Mechanical Engineers,* 7th ed., McGraw-Hill Book Co., New York, 1967, pp. 17–60.
42. *See,* for instance, D. Sperling, *Scientific American,* **275,** 56 (1996).
43. *Marks' Standard Handbook for Mechanical Engineers,* 7th ed., McGraw-Hill Book Co., New York, 1967, p. 4–72, Table 43.
44. *Collier's Encyclopedia,* P.F. Collier, Inc., New York, 1993, p. 431.
45. D. Sperling, *Sci. Amer.* **275,** 56 (1996).
46. U.S. Bureau of the Census, *Statistical Abstracts of the United States:* 1996, 116th ed., Washington, D.C., 1996, p. 592, Table No. 942.
47. Pacific Gas and Electric Co., *Resource,* 2nd ed., San Francisco, Calif., 1992, p. 422.
48. J.E. Rannels and J.E. Bradley, *Progress in Photovoltaic Power Production,* Power Generation Technology, Sterling Publishing Group, London, 1994, p. 99.
49. D. Linden, "Batteries Handbook", in *Handbook of Batteries and Fuel Cells,* McGraw-Hill Book Co., New York, 1984, p. 1–7.
50. U.S. Bureau of the Census, *Statistical Abstracts of the United States: 1995,* 115th ed., Washington, D.C. 1995, Table No. 1182.
51. W. Thomas and T. Spiro in R. Socolow, C. Andrews, F. Berkhout, and V. Thomas, eds, *Industrial Ecology and Global Change,* Cambridge University Press, Cambridge, U.K., 1994, Chapt. 21.
52. Committee on Measuring Lead in Critical Populations, *Measuring Lead Exposure in Infants, Children and Other Sensitive Populations,* National Academy Press, Washington, D.C., 1993.
53. International Lead and Zinc Group, *Principal Uses of Lead and Zinc,* 1960–1990, London, 1992.
54. D.T. Allen and N. Behmanesh, *Wastes as Raw Materials,* National Academy of Engineering, Washington, D.C., 1994, pp. 69–89.
55. R.A. Frosh, *Physics Today,* **47,** 63 (1994).
56. R. Socolow, C. Andrews, F. Berkhout, and W. Thomas: *Industrial Ecology and Global Change,* Cambridge University Press, Cambridge, U.K., 1994, p. 297.
57. *Horizon Advanced Battery Technology, Technical Summary,* Electrosource Inc., Austin, Tex., 1993.
58. D. Linder, *Handbook of Batteries and Fuel Cells,* McGraw-Hill, Inc., New York, 1984, p. 14–15.
59. U.S. Bureau of the Census, *Statistical Abstracts of the United States: 1995,* Table No. 1182, 115th ed., Washington, D.C., 1995.
60. M. Cobbing and S. Divecha, *The Myth of Automobile Battery Recycling,* Toxic trade Update, 4th quarter, 1993. Contains extensive references to documentation.
61. A bill called *The Lead Battery Recycling Act of 1991* was considered but apparently not adopted by Congress.
62. D. Corbus and C.J. Hammel, *Current Status of Environmental, Health and Safety Issues of Lithium Polymer Electric Vehicle Batteries,* NREL/TP-463-6831, National Renewable Energy Laboratory, Golden, Colo., 1995.
63. D. Corbus, C.J. Hammel, and J. Mark, *Current Status of Environmental, Health and Safety Issues of Nickel Metal-Hydride Batteries for Electric Vehicle Batteries,* NREL/TP-463-5474, National Renewable Energy Laboratory, Golden, Colo., 1993.
64. J.C. Sabatini, E.L. Field, I-C. Wu, M. R. Cox, B.M. Barnett, and J.T. Coleman, *Feasibility Study for the Recycling of Nickel Metal-Hydride Electric Vehicle Batteries,* NREL/TP-463-6153, National Renewable Energy Laboratory, Golden, Colo., 1994.
65. J.M. Ohi, *Environmental, Health and Safety Issues of Sodium–Sulfur Batteries for Electric and Hybrid Vehicles, Volume I: Cell and Battery Safety,* NREL/TP-262-4678, National Renewable Energy Laboratory, Golden, Colo., 1992.
66. D. Corbus, *Environmental, Health and Safety Issues of Sodium–Sulfur Batteries for Electric and Hybrid Vehicles, Volume II: Battery Recycling and Disposal,* NREL/TP-262-4680, National Renewable Energy Laboratory, Golden, Colo., 1992.
67. C.J. Hammel, *Environmental, Health and Safety Issues of Sodium–Sulfur Batteries for Electric and Hybrid Vehicles, Volume III: Transport of Sodium–Sulfur and Sodium Metal Chloride Batteries,* NREL/TP-463-4951, National Renewable Energy Laboratory, Golden, Colo., 1992.
68. J. Mark, *Environmental, Health and Safety Issues of Sodium–Sulfur Batteries for Electric and Hybrid Vehicles, Volume IV: In-Vehicle Safety,* NREL/TP-463-4952, National Renewable Energy Laboratory, Golden, Colo., 1992.

ELECTROANALYSIS IN ENVIRONMENTAL ANALYSIS

LARRY TAYLOR
Nova Tech
Westtown, Pennsylvania

This is the information age and the importance of information is clearly evident in environmental analysis, assessment, and remediation. Information from many sources is necessary to determine if there is environmental pollution, how serious the pollution is, if the pollutants can cause health problems, which methods are effective in remediation efforts, and when remediation is complete. Electroanalysis, the topic of this article, is one of the tools that can be employed to collect the information necessary to perform environmental analysis, assessment, and remediation. The purpose of this article is to describe how electroanalysis can be used and to show the advantages of it for environmental studies.

ROLE OF ELECTROANALYSIS IN ENVIRONMENTAL ANALYSIS

The role of electroanalysis is primarily in the analysis and assessment areas of the environmental field. Electroanalytical techniques are important in the environmental arena because they provide quantitative and qualitative information about substances present in the environment. Both kinds of knowledge are necessary to assess the extent of an environmental problem before, during, and after remediation.

As an analytical tool, electroanalysis has broad applicability because it can determine organic, inorganic, and organometallic substances. It can also be used to differentiate more subtle differences within a single substance, such as the oxidation state of inorganic substances (Cr^{6+} vs Cr^{3+}) or between structurally similar organic substances (maleic acid vs fumaric acid vs maleic anhydride). These capabilities are rarely found within a single analytical technique and are important because the toxicity of substances in the environment is often dependent on their structure.

Electroanalytical techniques also combine sensitivity with this selectivity. One of the unique capabilities of these techniques is their ability to do *in situ* preconcentration. This group of specialized electroanalytical techniques goes by the name stripping analysis and has been instrumental in making electroanalysis useful for environmental studies. This inherent sensitivity, along with the ability to interface easily with liquid and ion chromatography, makes electroanalytical techniques ideal for complex and variable environmental matrices. Electroanalytical techniques are also among the most cost-effective methods to do environmental analysis. The capital cost of these techniques is generally one-half to one-third the cost of chromatographic techniques and one-fourth to one-tenth the cost of spectroscopic and x-ray techniques. Operating costs are also much lower for electroanalytical techniques.

Electroanalysis is not a single technique. It is actually a broad spectrum of techniques that have utility for performing analytical analyses, researching chemical reaction pathways, manufacturing chemicals, and electroplating metals. Electroanalysis also has several branches that are distinguished by the variable that is controlled: voltage or current. Figure 1 shows a family tree for electrochemistry and some of the more important techniques within each branch.

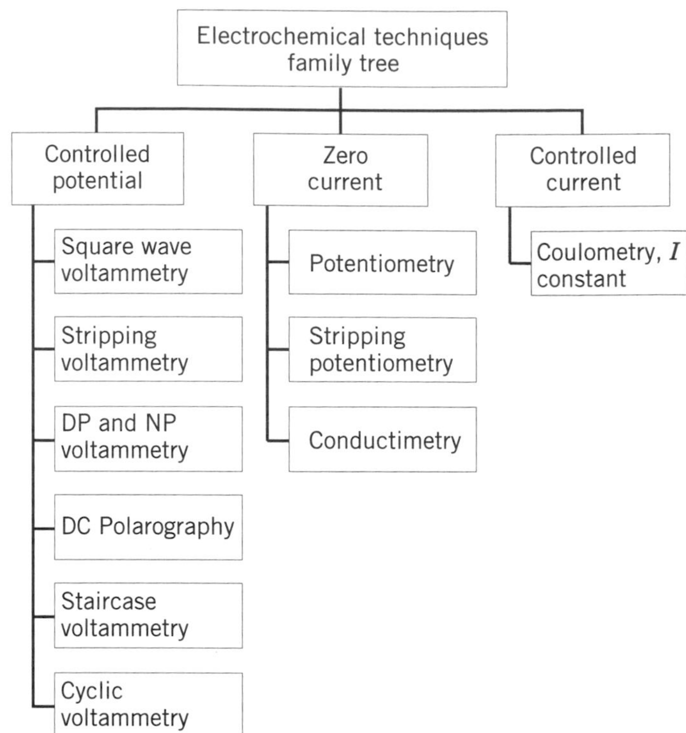

Figure 1. The family tree of electrochemical techniques based on the controlled variable where DP = differential pulse, NP = normal pulse, and DC is direct-current.

Theory and Instrumentation

Electrochemistry, in the broadest sense, involves the use of potential, current, or conductance to make qualitative and quantitative measurements on species. The usual practice is to apply one of these variables, for example potential, to a solution containing the analyte species and measure one of the other variables, for example current. From a plot of the measured variable versus the applied variable, information regarding the concentration and identity of electroactive species in solution is determined. Of the many electrochemical techniques available, only a few are routinely used for environmental analysis and assessment. The electrochemical tree in Figure 1 is organized to show the relative importance of each technique. The controlled potential and zero current branches of Figure 1 are the most relevant to the environmental field.

Voltammetry is the name usually given to the family of techniques in which current is measured as a function of applied potential. This rather large family of techniques includes DC and pulse polarography, stripping voltammetry, alternating current voltammetry, square wave voltammetry, amperometry, and others. The usual way to differentiate these techniques is by the potential waveform applied to the voltammetric cell (Fig. 2). A novice to the field of electrochemistry might easily become overwhelmed by the number of voltammetric techniques available and confused as to which are best suited for environmental determinations. Each of these techniques has advantages, but only a few are extensively used for routine analyses.

ELECTROANALYSIS BASICS

An electroanalysis method can be broken down into four basic steps: sample preparation (including collection), technique selection, electrochemical analysis, and data analysis.

ENVIRONMENTAL APPLICATIONS

Overview

The major application of electrochemistry in the environmental field is to analyze samples for their elemental, organic, and/or organometallic content. Quantitative analysis of these environmental samples occurs much more frequently than qualitative analysis because the strength of electrochemical techniques is in the measurement of a known species rather than in the identification of an unknown. Looking at the various types of environmental matrices also shows that it is the ability to determine species that brings electrochemistry to the forefront.

The quantitative and qualitative information obtained with electrochemistry is used in many different ways. Some of the more important follow:

- Measurement of pollutants in water, solids, and air to assess compliance with government regulations, track pollutant levels in the environment, and monitor industrial processes
- Follow environmental species through geochemical and biochemical pathways
- Determine lability and inertness of metal ions and correlate these with toxicity, bioavailability, and geochemical availability
- Characterize groundwater before, during, and after remediation
- Determine amount and type of toxic species present to do risk assessments

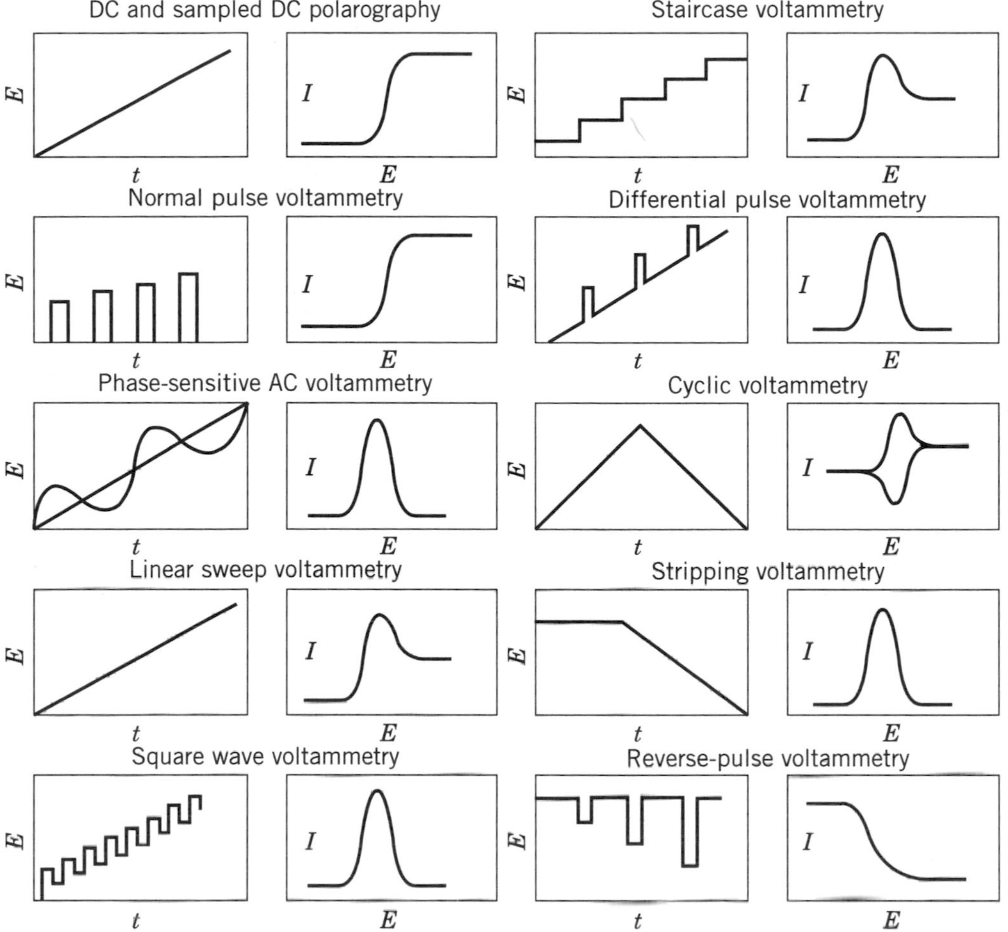

Figure 2. Commonly used voltammetric waveforms and their current–potential outputs where k = potential, t = time, and I = current.

The applications of electrochemistry are examined by looking at the types of environmental samples in which determinations are made, ie, water (sea, natural, ground, potable, and waste), solids, and air. The analysis of water samples is, by far, the type of environmental sample analyzed most frequently because this is the medium in which electrochemistry is normally done. Solids must be dissolved or digested and air scrubbed through a solvent before electrochemistry can be used on these types of samples.

Water

Seawater. Seawater, as a matrix, causes problems for many analytical techniques because of its high salt content and corrosivity. However, these are not drawbacks for electrochemical techniques and in some cases become advantages. Electrochemical techniques are useful for the analysis of seawater because they are sensitive enough to determine substances at their natural concentrations (typically <20 ppb in unpolluted seawater), selective to speciation and oxidation states of substances (especially the elements), suitable for making direct determinations in seawater, and can be used to determine substances at remote sites or laboratories. These advantages make the use of electrochemical techniques ideal for monitoring pollutants in seawater and help determine the distribution of substances found naturally in seawater. Electrochemical techniques are also excellent for determining the bioavailability of elements and lability of metal complexes with natural ligands. Because of these advantages, seawater is analyzed more than other environmental matrix by electrochemistry.

Any substance that can be determined using electrochemical techniques can potentially be determined in seawater as well. In actual practice, metal ions are the substances most often determined because they are more easily determined by highly sensitive electrochemical stripping techniques than organic or organometallic species. Table 1 lists some of the species that have been determined in seawater and differentiates those that are determined as total versus species. Although this is not a complete list, it illustrates the breadth of electrochemistry for this matrix. There are relatively few organic species on this list, but the use of electrochemistry in this area is expected to grow.

The determination of metal ions in seawater is done almost exclusively with stripping techniques because of their sensitivity. Natural concentrations of metal ions are typically $<20\mu g/L$ and the *in situ* preconcentration (deposition step) of stripping techniques makes these techniques fast, selective, sensitive, inexpensive, and portable. Anodic stripping voltam-

Table 1. Species Determined in Seawater and Electrochemical Techniques Used

Species Determined	Electrochemical Technique Used[a]	Reference
Al^{3+}	AdSV	1
β-Carotene	DPV	2
Cd^{2+}	PSA	3
Co^{2+}	AdSP	4
Cr^{6+}	FIA	5
Cu^{2+}	AdSV	6
F^-	ISE	7
Fe_{total}, Fe^{3+}, Fe^{2+}	AdSV	8
Folic acid	AdSV	9
Hg^{2+}	ASV, Voltammetry at chemically modified GCE	10
I_{total}, I^-	SW AdSV	11
IO_4^-	DPP	11
Mn^{2+}	AdSV	12
Mo^{6+}	AdSV	13
NH_3	AdSV	14
Ni^{2+}	AdSP	4
NO_3^-	IC using conductivity detection	15
O_2	Tetrimetry	16
Pb^{2+}	PSA	3
Phenols	HPLC with amperometric detection	17
PO_4^{3-}	IC using conductivity detection	15
Se^{4+}, Se^{6+}	DP AdSV	18
SO_3^{2-}	Enzyme ISE	19
Ti^{4+}	AdSV	20
Tributyl tin	A-c polarography	21
U^{6+}	SW ASV	22
V^{5+}	AdSV	23
Zn^{2+}	A-c ASV	24

[a] AdSV = adsorptive stripping voltammetry, DPV = differential pulse voltammetry, PSA = potentiometric stripping analysis, AdSP = adsorptive stripping potentiometry, FIA = flow injection analysis, ISE = ion-selective electrode, ASV = anodic stripping voltammetry, SW = square wave, DPP = differential pulse polarography, IC = ion chromatography, and HPLC = high performance liquid chromatography.

metry was the first stripping technique used extensively to analyze seawater for metal ions such as Cd, Cu, Pb, and Zn. With the development of adsorptive stripping voltammetry (ADSV), it becomes possible to determine many more metal ions by stripping voltammetry. Most of the remaining elements can be determined using either ion chromatography (IC) or ion-selective electrodes (ISEs). The periodic chart in Figure 3 graphically shows those elements that can be determined using voltammetric electrochemical techniques.

Without any sample pretreatment, anodic stripping voltammetry (ASV) measures the concentration of free metal ion, Cd^{2+} for example, plus the concentration of easily dissociable (labile) metal ion. This later category is difficult to define, but is a function of the strength of the ligand–metal ion complex. If a ligand–metal ion complex dissociates in the diffusion layer during deposition, it will be counted as free metal ion. This dissociation is deposition potential-dependent (25). This has been demonstrated for Cu^{2+} and has shown that AdSV gives 50–60% higher ligand concentrations than ASV. Thus, without sample pretreatment, stripping voltammetry measures free metal ion plus some fraction of metal ion complexes. If the total metal concentration in seawater is desired, pretreatment to release metal ions from ligand complexes is necessary.

Pretreatment of seawater may be needed to destroy organics, convert all of an element to a single species, or filter out particulates. The most common pretreatment, by far, is to irradiate seawater with UV light to destroy organics, which can be done in a batch or flow process. The length of time required to pretreat a sample with uv light depends on the concentration of organics, UV lamp intensity and wavelength, seawater pH, and type of organics present. Ultraviolet irradiation is preferred over other pretreatment techniques for environmental analyses, ie, acid digestion, because it minimizes the amount of chemicals added and reduces the number of operator steps necessary to do the pretreatment.

Ultraviolet irradiation is less likely to change an element's oxidation state, therefore addition of chemical oxidants or reductants is required when an element is present in several oxidation states and the total concentration is desired. The determination of iodate, IO_3^-, in seawater is an example of how oxidants or reductants may be used. Acidification and addition of sulfite converts IO_3^- to I^-. I^- is first determined using AdSV, then total I^- is determined after conversion from IO_3^-. The IO_3^- concentration is then measured by subtraction (11). Other chemical oxidants or reductants used to convert elements to one oxidation state include H_2O_2, O_2, NaClO, and concentrated acids.

The ability to provide speciation information in seawater is an important area in which electrochemistry is used. Speciation encompasses three different types of studies: bioavailability, geochemical cycling, and oxidation state determination. As indicated earlier, it is not just the total concentration of a toxic species that is important in environmental analysis, but how much of it is available to living organisms (bioavailability), which forms are stable in seawater (geocycling), and how much is in a species (oxidation state) that is toxic to living organisms. Electrochemical techniques, especially stripping techniques, are very good at determining this information because they are capable of determining the concentration of bioavailable species directly. Application of extraction, chromatography, and other sample pretreatments significantly disturb the equilibrium in the natural environment and may lead to false conclusions regarding bioavailability of a species. Studies that have examined the applicability of electrochemical techniques to measure bioavailability generally do so by classifying metal ions by how strongly they complex with organic compounds or solids. Typical class nomenclature used is labile, quasi-labile, inert, and particulate.

Labile refers to the metal ion fraction that can be measured directly by ASV or AdSV, quasi-labile and inert refer to the metal ion fractions measured after separation by another technique, such as ion exchange, and the particulate fraction is that which is retained by filtration. Several investigators have developed systematic classification schemes by which a metal ion in seawater can be analyzed to determine the percent in each class. This system has proved effective and shown a positive correlation to toxicity. Table 2 lists some toxic species vs their electrochemically determined lability (26). Two classification systems based on stripping voltammetry are illustrated in Figures 4 and 5 (27,28). Although there are other ways to perform this classification, stripping voltammetry provides the simplest and most direct method.

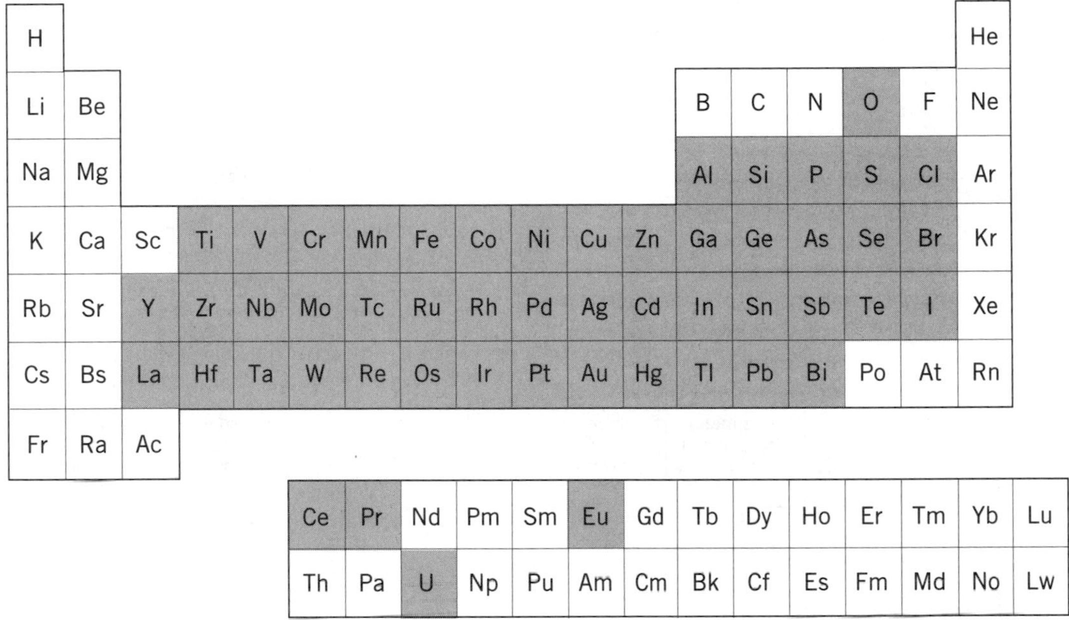

Figure 3. Periodic chart of the elements. Those most frequently determined by voltammetry are highlighted.

Table 2. Comparison of Toxicity and Electrochemical Lability for Species in Natural Waters[a]

Species	Toxicity	Electrochemical Lability
As(III)	High	High
As(VI)	Low	Low
Cr(III)	Low	Low
Cr(VI)	High	High
Tl(I)	High	High
Tl(III)	Low	Low
Cu^{2+}	High	High
$CuCl_2$	High	High
$CuCO_3$	Low	High
Cu^{2+}fulvic acid	Low	Low
Cu^{2+}humic acidFe_2O_3	Medium	Medium
Cu^{2+}2,9-dimethyl-1,10-phenanthroline	High	Low

[a] Ref. 26.

Two other ways that electrochemical techniques are used to study bioavailability are to measure the amount of metal ion that can be complexed by natural ligands and to measure the strength of metal ion–ligand complexes. Measurement of the complexing capacity of natural or model ligands is done by titration. Aliquots of a metal ion are added to a solution, either a natural seawater stripped of metal ions or a synthetic seawater with model ligands, and the free metal ion concentration is determined by ISE or stripping voltammetry. The result is a complexing capacity titration curve which measures how much metal ion the natural or artificial seawater can retain as metal ion–organic complexes. Figure 6 shows a typical titration curve for Cu (29). These curves also provide the means to measure the complexing strength which is a function of the metal ion–ligand equilibrium constant. Examples of the use of these techniques include the study of transfer mechanisms of Se species between water–soil–plants–animals (30), mobility of metals in fly ash (31), and speciation as a function of depth in seawater for a variety of species including Fe, I, and Cd (8,32,33).

Qualitative determination of the ligands in seawater is generally not done using electrochemical techniques. However, use of gel permeation chromatography (GPC) with ASV detection has been reported to determine the molecular weight of Cd-organic complexes (34) and this combination of techniques may find greater utility in the future because of the selectivity of ASV.

Although stripping techniques are the most important electrochemical techniques for analyzing seawater, others are used. IC, high performance liquid chromatography/electrochemical (HPLC/EC), ISE, and voltammetry may be used where stripping techniques are not applicable. IC is used to analyze seawater for ions such as NO_3^- and PO_4^{3-}, HPLC/EC for organics such as phenols, ISEs for ions such as F^-, Cu^{2+}, and SO_3^{2-}, and voltammetry for species such as NH_3, I_2, and tributyl tin. Usually these techniques require a separate pretreatment or preconcentration step because they are inherently less sensitive than stripping techniques.

Portability is rapidly becoming one of the most important features of an environmental analytical instrument and electrochemistry is well endowed in this respect. Three examples to illustrate portability are referenced here in which adsorptive stripping voltammetry is used on board a ship to determine Co, Ni, and Al (1,35,36). For on-ship use an instrument was adapted consisting of a potentiostat, static mercury drop electrode, personal computer, digital I/O board, and plotter. It was found that using a high-frequency staircase stripping waveform reduced the interference from dissolved oxygen and engine vibrations, simultaneously reducing the analysis time. Another method of

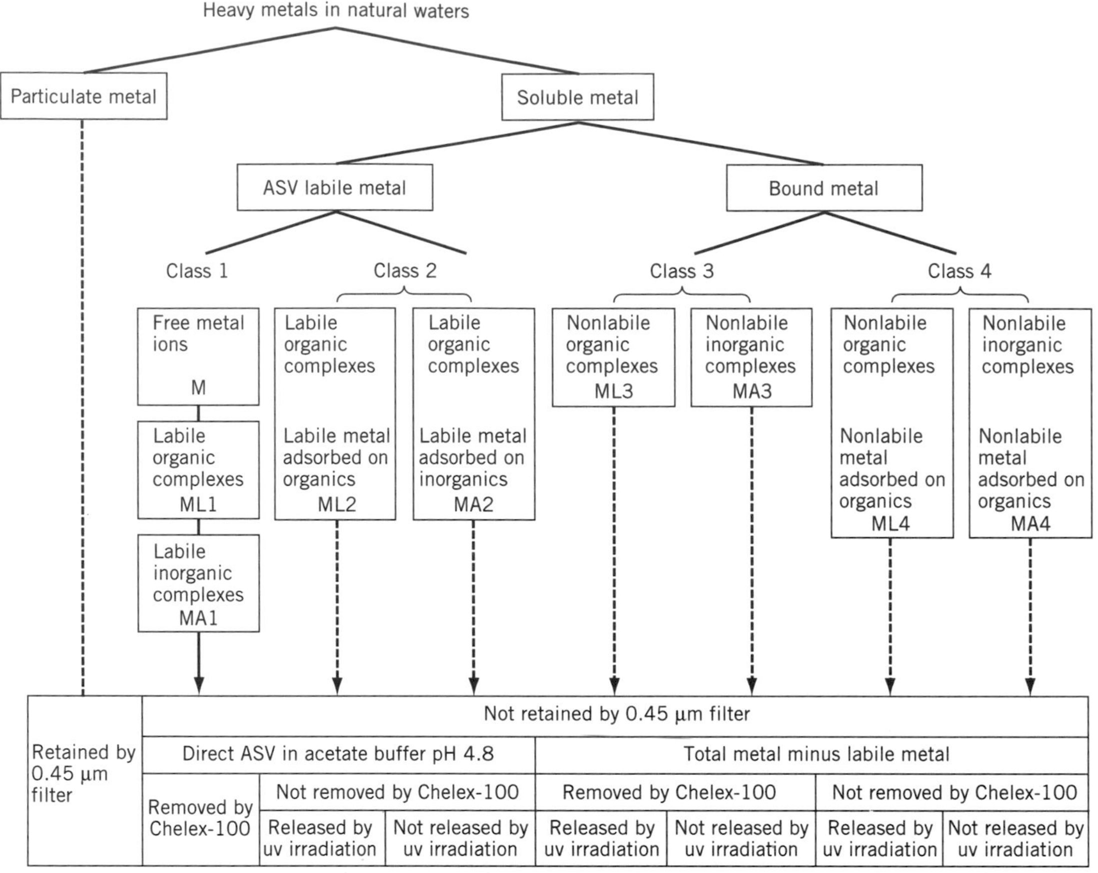

Figure 4. Flow scheme classifying metal ion–organic complexes based on lability, where TM = total metal. Courtesy of CRC Press, Inc.

doing remote analysis also eliminates collection of the sample. Submersible probes have been developed that can be immersed directly in seawater and is connected to the potentiostat using a long shielded cable (37). Their use has been demonstrated for Cd, Cu, Hg, Mn, Pb, Se, and Zn in seawater.

Natural and Potable Water. Natural waters (rivers, lakes, ponds, rain, ground water) are good candidates for analysis using electrochemical techniques because the concentrations of pollutants may range from sub-ppb to ppm, the ability to determine species is often more important than the total concentration, and the ability to store samples for long time periods is difficult. The natural water matrix, although more variable than that of seawater, is lower in salt content but has a much wider concentration range of natural and/or synthetic organic substances, inorganic complexants, and particulates. Potable water, on the other hand, has few interferents but the analyte concentrations of interest are often very low and may require large preconcentration factors.

The major use of electrochemistry for natural and potable waters is to do quantitative analysis of metal ions; stripping techniques are the ones most often used for this purpose. *In situ* preconcentration and the ability to do speciation directly in samples are the reasons why electrochemical techniques are used in these matrices. Natural and potable waters also need to be analyzed for various organic pollutants, eg, pesticides, herbicides, and chlorinated organics, and electrochemical techniques may be used to determine them. Rapid throughput and selectivity are important for analyzing these types of organics and this has fueled the development of biosensors and HPLC/EC for the determination of organic environmental pollutants.

The concentration of metal ions in natural and potable waters is normally very low making detection difficult for most analytical techniques without some form of preconcentration. Stripping techniques, with their built-in deposition step, provide a rapid, simple, and portable analytical technique for metal ions. Direct analysis of natural waters, after addition of an electrolyte, is used to determine the free and labile forms of a metal ion. Metal ions that are strongly complexed with ligands can also be determined, but only after freeing them from the organic ligands present in most natural water systems. ASV, AdSV, and potentiometric stripping analysis (PSA) are the preferred stripping techniques. The metal ions determined in natural and potable waters are virtually identical to the ones determined in seawater (see Table 1).

Techniques used to release metal ions from organics and particulates include UV irradiation, UV irradiation + H_2O_2, and microwave acid digestion. The use of UV irradiation is preferred when organic concentrations are not too high because it reduces the chance of contaminating the sample. When organic levels are high and UV irradiation alone is insufficient to free the metal ions, the addition of H_2O_2 may be used to enhance the efficiency of UV irradiation. Samples that have high concentrations

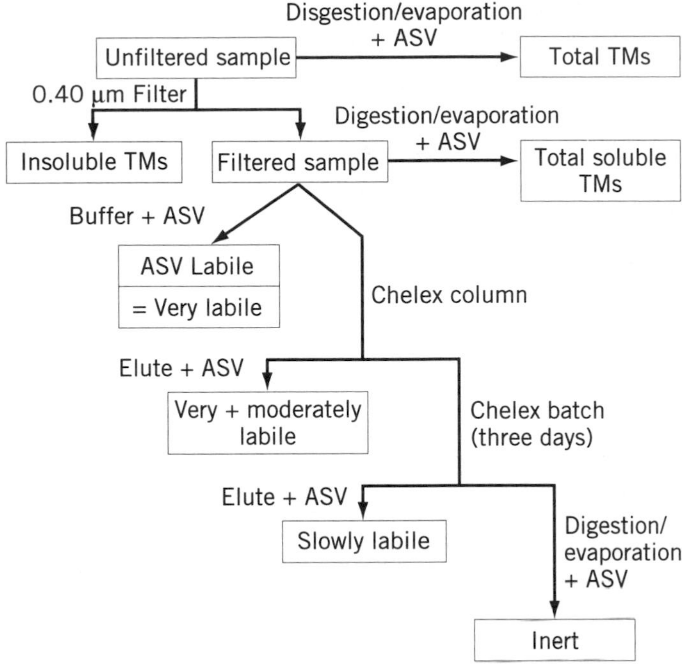

Figure 5. Alternative flow scheme classifying metal ion–organic complexes based on lability. Courtesy of the American Chemical Society.

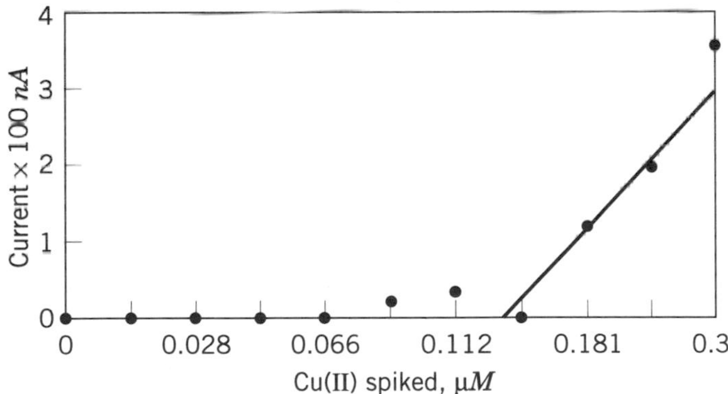

Figure 6. Complexation capacity curve for Cu^{2+} in river water.

of organics must be digested with strong acids using microwave or hot plate heating.

The interest in analyzing samples with little or no sample pretreatment has resulted in the development of coated electrodes for environmental scientists to use. Coated electrodes are prepared by polymerizing, onto an electrode, a polymer that has a functional group that interacts with an analyte. The polymer may enhance reduction or oxidation of the desired analyte while preventing reduction or oxidation of other electroactive species, it may inhibit adsorption of organics that block the electroactivity of the analyte, or it may selectively extract and concentrate the analyte. Two examples of coated electrodes are the use of a polypyrrole-coated electrode to amperometrically determine anions, eg, Cl^-, Br^-, NO_3^-, NO_2^-, ClO_4^-, CO_3^{2-}, SO_4^{2-}, and PO_4^{3-}, by FIA or IC (38), and use of a poly(3,3′-diaminobenzidine) film on a Au electrode to concentrate and determine Se^{4+} (39).

Speciation of metal ions is as important in natural and potable waters as it was in seawater. Stripping techniques, voltammetry, and ISEs are used to study speciation and measure the effect of ionic species and oxidation states on bioaccumulation and geochemical cycling. The same procedure used in seawater is applicable to natural and potable waters: divide the metal ion species into particulate and soluble fractions and classify the soluble fraction by the strength of the metal ion complexes into labile, quasi-labile, and inert classes. Using these procedures it has been possible to use DPP, differential pulse anodic stripping voltammetry (DPASV), and ASV to track soluble and particulate forms of Pb^{2+} as they are transported to the sea (40). Investigation of a geocycle involving dissolved and particulate forms of Mn, Fe, Co, Ni, Cu, Zn, Cd, and Pb has been done by DPP and DPASV in anoxic sulfidic waters (41).

Many nonnatural organic species are found in natural and potable waters and several different electrochemical techniques are used in their analysis: HPLC/EC, IC, voltammetric-based biosensors, and AdSV. HPLC/EC is used when the greater selectivity available with HPLC is required, and has been used for the determination of chlorinated phenols, phenylurea herbicides, nitrophenols, and many other electroactive pollutants. To meet regulatory requirements, methods with very low LODs are needed and preconcentration prior to analysis by HPLC is one technique to do so. SPE is one of the preferred preconcentration techniques and detection limits as low as 1 ng/mL have been reported using a 100 mL sample, SPE, and HPLC/EC (42).

Chromatography techniques are more suited for laboratory than field use, but this is not true for voltammetric-based biosensors. This electrochemical technique, based on biosensor technology, is highly selective if the enzyme–analyte reaction is selective and is easily adapted to field use. A biosensor is constructed by attaching an enzyme to an electrode. The analyte reacts with the enzyme to generate an electroactive species which is then oxidized or reduced producing a current proportional to the analyte concentration. The excellent selectivity of voltammetric-based biosensors has been demonstrated for phenolic analytes, using a tyrosinase electrode, which have been determined in untreated river and ground water (43). The pesticides carbaryl and paraxon are determined by biosensor technology without sample preparation using a different mechanism. These pesticides inhibit the acetyl cholinesterase-catalyzed hydrolysis of 4-aminophenyl acetate and are thus detected (44).

Stripping voltammetry, usually thought of as a technique for metal ions, has also been applied to the determination of pesticides. This is possible because organics can be adsorbed at Hg electrodes and stripped off similarly to metal ion–organic complexes. The determination of Fenchlorazide-ethyl and Terbutryn are examples of the use of AdSV for organics. LODs for organics are excellent using AdSV: Fenchlorazide-ethyl can be detected down to 0.2 μg/L (45) and Terbutryn to $2 \times 10^{-9} M$ (46).

The last point to be made is that natural and potable waters may be analyzed at remote sites. Battery-powered instruments able to do DPASV have been developed for Pb and Cu and used to analyze drinking water (47) and disposable amperometric electrodes that have been shown to be suitable for analyzing hydrazines (48).

Wastewater. Wastewaters differ from sea, natural, and potable waters in their proximity to the pollution source and complexity of their matrix. The identity of the pollutant being determined in wastewater is generally known because they are reactants, products, or by-products of the process. The purpose for analyzing wastewaters may be to report the concentration of a regulated pollutant being discharged, to monitor for breakthrough of a pollutant being removed by a cleanup process, or to develop a pollutant cleanup process. Wastewaters are usually more chemically complex than sea, natural, or potable water placing an added selectivity burden on the electrochemical technique.

The determination of metal ions in wastewater with electrochemical techniques is not done as frequently as it is in other types of water. There are two reasons for this. First, the concentrations of metal ions are often not as low as they are in natural waters, and second, total metal ion concentrations are more important than speciation which means that electrochemical techniques are not as advantageous. However, there are still times when electrochemical techniques are the most desirable. The simplicity of electrochemical techniques, with their excellent precision and accuracy, make them ideal for on-line monitoring of waste streams. For example, total As determination in the treated effluent from Zn and Cd refining plants (49) is normally done using atomic absorption spectroscopy (AAS)/hydride methodology in laboratories, but chemical treatment with L-cysteine or SO_3^{2-} to reduce As^{5+} to As^{3+} makes on-line voltammetric determination of total As possible. In the nuclear technology industry, on-line determinations are very desirable for safety reasons and AdSV of uranium in raffinates can be done this way (50).

When electrochemical techniques are used for wastewater determinations it is generally because they have a selectivity advantage that makes them easier and less expensive to operate. Selectivity can be achieved using voltammetry directly by modifying carbon paste electrodes (CPE) with complexants or biosensors. A CPE modified with 2,9-dichloro-1,10-phenanthroline was used to selectively determine Ag^+ (51). A biosensor incorporating tyrosinase on a wax impregnated graphite (WIG) electrode illustrates this technology for the determination of CN^- (52). The chemistry used to achieve selectivity is often well known, but its application with voltammetry makes it fast, easy, and portable.

When very low LODs are not needed, the use of ISEs should be considered. Commercial ISEs are available for many anions and some metal ions and may be capable of direct analysis of wastewaters. ISEs are most useful when the analyte being determined is one that the ISE has been designed to determine, as in the determination of Cr^{6+} in electroplating industrial waste with a solid-state ISE (53). Of course, some wastewaters may have interferences that prevent direct determination by ISE. When this occurs, modification of the ISE may be made so that a direct determination is feasible. The ease with which this can be done for the analysis of S^{2-} produced by sulfate-reducing bacteria has been demonstrated (54). A commercial S^{2-} ISE, found to work for samples below pH 5, was modified and used with a pH electrode to analyze samples above pH 7 as well.

The complexity of wastewaters, however, often requires the greater selectivity provided by chromatography with electrochemical detection. Simple ions, such as F^-, HCO_3^-, CL^-, NO_2^-, NO_3^-, $H_2PO_4^-$, and SO_4^{2-}, are readily determined using IC with conductivity detectors in wastewaters. Phenols and chlorinated phenols are easily determined in wastewater with HPLC and amperometric detection at a GCE. Cellodextrins, which are degradation products of the reaction between cellulose and ozone, are more difficult to determine because they adsorb onto and foul Pt electrodes when normal amperometric detection is used. Use of HPLC/EC with pulse amperometric detection eliminates electrode fouling (55).

Air

Electrochemical determinations are not possible directly in air because the conductivity of this medium is too low. Despite this limitation, environmental applications have been developed for environmentally important species. All involve passing the air sample through a liquid, which absorbs the analyte, or through a solid filter that collects particulates containing the analyte. The most important step in the analysis of air is this collection step which must absorb the analyte in a liquid, most often aqueous, that contains an electrolyte.

The first consideration, therefore, is scrubbing technology. Typically, the air sample is passed through one or more scrubbers designed to efficiently remove the analyte in as short a time as possible. An efficient scrubber breaks the air into tiny bubbles which are passed through the collection liquid. Efficiency depends on the physical design of the scrubber, solubility of the analyte in the collection liquid, and flow rate of air through the scrubber towers. LOD is influenced primarily by the length of time air flows through the scrubber towers, their efficiency, and the sensitivity of the electrochemical technique.

Unless the scrubbing technology used has been tested and validated previously, breakthrough of the analyte should be tested. Breakthrough of analyte may be the result of inefficient scrubber design, collecting too long, or exceeding analyte solubility in the collection solvent.

Electrochemical techniques that have been utilized for analysis of air samples include voltammetry, IC/conductivity detection, and HPLC/EC. Examples of voltammetric techniques include differential pulse polarography (DPP) to determine SO_2 in air after collection in dimethyl sulfoxide (DMSO) or dilute HCl/SeO_3^{2-} (56) and differential pulse voltammetry (DPV) to determine pyrocatechol in indoor air after collection in dilute HCl (57). An automated analysis of 10 anions and 10 cations in atmospheric aerosols was developed using IC/conductivity detection to avoid keeping samples in storage longer than 24 hours (58). Finally, airborne cocaine (LOD = 2 ng) and heroin (LOD = 0.5 ng) was collected with a high throughput sample prior to their determination by HPLC/EC (59).

Metals in aerosols or dust can also be determined by electrochemical techniques. As with the analysis of water, collected particles can be analyzed and classified by lability. Also, by fractionating the particles collected, the amount of metals, such as Pb and Cd, can be associated with different sized particles (60).

Solid Waste

Solid wastes cover a range of matrices including soil, sewage, sediment, dust, ash, and many others. To analyze these types of environmental samples with electrochemical methods it is first necessary to get the analyte into solution, which may be accomplished by dissolution into a solvent, extraction, or digestion. Rarely, except for some ISE analyses, can determinations be done in solid waste without any sample preparation. Solid wastes are the most difficult type of matrix to analyze for electrochemical as well as most other analytical techniques. Some of the advantages inherent in electrochemical methods, such as the ability to determine oxidation states and species, are not as useful for solid wastes because of the harsh methods needed to get analytes into solution.

Digestion using strong acids/heat, strong acids/microwave, and UV irradiation/hydrogen peroxide are the most common sample preparation techniques for solid wastes when metals are to be determined. UV irradiation/hydrogen peroxide is the weakest of these sample digestions. Electrochemical techniques, especially stripping techniques, are adversely affected by organics in solid wastes, thus potent dissolution techniques are needed to remove them. After digestion, total metal analysis can be done by any of the usual electrochemical techniques, ie, voltammetry, stripping techniques, and ISE. If the metal analyte is weakly held in the solid matrix it may not require digestion, as illustrated by the DPASV determination of Tl^+ in soil (61), thus preserving speciation and oxidation state information.

Stripping techniques, because they are so sensitive and selective, are the most frequently used techniques for analyzing metals in solid wastes. The choice of stripping technique used should be based on the one that gives the best selectivity, sensitivity, linear working range, and LOD for the desired analyte. Anodic, adsorptive, and potentiometric stripping techniques have all been used successfully. Polarography and voltammetry are used less frequently because they are less sensitive, but are generally faster if high sample throughput is needed.

Digestion of the solid waste matrix is not always desirable because it may be necessary to determine organic species or a particular oxidation state. For these types of determinations extraction techniques may be appropriate if the analyte can be freed from the solid waste matrix. Soxhlet or conventional extraction methods may be utilized to remove analytes from solid wastes, particularly soil. The types of extractants employed vary with the analyte. Water, acid–base solutions, or solvent–complexing agents work best for extracting anions, metal ions, and water-soluble organics. Organic solvents are better for extracting organic species from solid wastes. Examples of extraction procedures for selected analytes include the following:

- Dilute HCl used to extract Fe ions from sediments to preserve the oxidation state of Fe ions (Fe^{2+}/Fe^{3+}) for detection using AC or DC polarography (62)
- NaOH used to extract phenols from soil, prior to HPLC/EC, because the base ionizes the phenols enhancing their extractability (63)
- Solution of $HBr-Br_2$ in diethyl ether found to efficiently extract Tl^+ from cement dust and sediment (64)
- Acetone and CH_2Cl_2 able to extract organophosphorus pesticides, eg, Fenitrothin, Diazinon, Parathion Et, Mevinphos, and Heptenophos, from soil for determination with a butyryl cholinesterose biosensor (65).

Despite the high organic content of solid wastes, the ability to determine analytes without sample preparation is possible. To do such a determination, the analyte is not complexed with any organic or inorganic species or one must be looking for the free analyte. ISEs are the technique most often used for determinations without sample preparation. Nitrate in soil, for example, can be determined by ISE without any sample preparation using conventional electrodes (66) or coated wire electrodes (67). The latter has been shown to work for unattended, remote site analysis of nitrate in which the ISE is placed directly in soil.

CONCLUSION

As more environmental scientists learn about the advantages of electrochemistry, its use in the area of environmental assessment and remediation will grow. These techniques give scientists fast, accurate, and precise information about the environment which is critical for assessment and remediation. Electrochemical techniques are also portable, inexpensive, and easy to use which are important for on-site utilization. They also have capabilities other techniques lack or can only do with great difficulty, such as the ability to provide speciation and oxidation state information.

The desire to do environmental analyses in the field continues to grow and electrochemical technology is ideally suited for on-site work. The use of microchip technology to make selective sensors based on voltammetry is being pursued and is likely to spawn the development of dedicated instruments for many inorganic and organic species of importance to environmental scientists. One example combines cyclic voltammetry, microsensor technology, ceramic metallurgy, and neural networks to make a versatile and selective gas detector (68).

The use of electrochemistry is also being expanded to include remediation by developments in the field of electrocatalysis. Electrochemical destruction of hazardous wastes has been pursued for many years, but has not been economical on a large scale because of cost issues. The development of soluble catalysts that enhance charge-transfer reactions at the interface of an electrode and electrolyte may change these economics. Electrochemical oxidation processes using catalysts are demonstrating their ability to detoxify wastes at lower temperatures and without the formation of undesirable side products, such as dioxins (69).

BIBLIOGRAPHY

1. J. J. Hernandez-Brito, M. D. Gelado-Caballero, J. Perez-Pena, and J. A. Herrera-Melian, *Analyst,* **119**(7), 1593–1597 (1994).
2. B. V. Pfund, A. M. Bond, and T. C. Hughes, *Analyst,* **117**(5), 857–861 (1992).
3. D. Jagner, E. Sahlin, and L. Renman, *Talanta,* **41**(4), 515–522 (1994).
4. C. G. Nan and T. J. Cardwell, *Electroanalysis,* **7**(11), 1068–1074 (1995).
5. H. Ohura, T. Imato, S. Tamasaki, and N. Ishibashi, *Bunseki Kagaku,* **43**(1), 31–37 (1994).
6. F. Quentel, C. Elleouet, and C. Madec, *Electroanalysis,* **6**(8), 683–688 (1994).
7. C. J. Rix, A. M. Bond, and J. D. Smith, *Anal. Chem.* **48**(8), 1236–1239 (1976).
8. M. Gledhill and C. M. G. van den Berg, *Mar. Chem.* **50**(1–4), 51–61 (1995).

9. A-C. Le Gall and C. M. G. van den Berg, *Anal. Chim. Acta,* **282**(3), 459–470 (1993).
10. I. Turyan and D. Mandler, *Electroanalysis,* **6**(10), 838–843 (1994).
11. G. T. F. Wong and L. S. Zhang, *Mar. Chem.* **38**(1–2), 109–116 (1992).
12. G. R. Scollary, G. N. Chen, T. J. Cardwell, and V. A. Vicente-Beckett, *Electroanalysis,* **7**(4), 386–389 (1995).
13. P. A. M. Farias, A. K. Ohara, A. W. Nobrega, and J. S. Gold, *Electroanalysis,* **6**(4), 333–339 (1994).
14. A. M. Harbin and C. M. G. van den Berg, *Anal. Chem.* **65**(23), 3411–3416 (1993).
15. S. Carrozzino and F. Righini, *J. Chromatog. A* **706**(1,2), 277–280 (1995).
16. K. Furuya and K. Harada, *J. Oceanog.* **51**(3), 375–383 (1995).
17. M. T. Galceran and O. Jauregui, *Anal. Chim. Acta,* **304**(1), 75–84 (1995).
18. F. Seby, M. Potin-Gautier, and A. Castetbon, *Fr. Hydrol.* **24**(1), 81–90 (1993).
19. L. Campanella, P. Cipriani, T. M. Martini, M. P. Sammartino, and M. Tomassetti, *Anal. Chim. Acta,* **305**(1–3), 32–41 (1995).
20. H. Li and C. M. G. van den Berg, *Anal. Chim. Acta,* **221,** 269–277 (1989).
21. M. Ochsenkuhn-Petropulu, K-M. Ochsenkuhn, G. Parissakis, A. N. Giannakis, and H. D. Smith, *Can. J. Appl. Spectrosc.* **40**(3), 66–71 (1995).
22. A. Economou, P. R. Fielden, and A. J. Packham, *Analyst,* **119**(2), 279–285 (1994).
23. M. Vega and C. M. G. van den Berg, *Anal. Chim. Acta,* **293**(1–2), 19–28 (1994).
24. C. Locatelli, T. Garai, and F. Fagioli, *Mikrochim. Acta,* **107**(1–2), 41–48 (1992).
25. C. M. G. van den Berg, *Analyst,* **117**(3), 589–593 (1992).
26. T. M. Florence, *Analyst,* **111,** 489–505 (1986).
27. T. M. Florence, “Electrochemical Techniques for Trace Metal Speciation,” in G. E. Batley, ed., *Trace Element Speciation: Analytical Methods and Problems,* CRC Press, Inc., Boca Raton, Fla., 1989.
28. P. Figura and B. McDuffie, *Anal. Chem.* **52,** 1433–1439 (1980).
29. V. N. Iyer and R. Sarin, *Anal. Lett.* **25**(10), 1915–1927 (1992).
30. M. Potin-Gautier, F. Seby, and M. Astruc, *Fresenius' J. Anal. Chem.* **351**(4–5), 443–448, (1995).
31. J. Fernandez-Turiel, W. Carvalho, M. Cabanas, X. Querol, and A. Lopez-Soler, *Environ. Geol. (Berlin),* **23**(4), 264–270 (1994).
32. R. C. Tian and E. Nicolas, *Mar. Chem.* **48**(2), 151–156 (1995).
33. E. Helmers, *Fresenius' J. Anal. Chem.* **350**(1–2), 62–67 (1994).
34. P. M. Linnik and I. V. Iskra, *Microchem. J.* **50**(2), 184–190 (1994).
35. J. Perez-Pena, J. J. Hernandez-Brito, J. A. Herrera-Melian, C. Collado-Sanchez, and C. M. G. van den Berg, *Electroanalysis,* **6**(11–12), 1069–1076 (1994).
36. J. J. Hernandez-Brito, P. Cardona-Castellano, V. Siruela-Matos, and J. Perez-Pena, *Electroanalysis,* **6**(11–12), 1141–1146 (1994).
37. J. Wang, N. Foster, S. Armalis, D. Larson, A. Zirino, and K. Olsen, *Anal. Chim. Acta,* **310**(2), 223–231 (1995).
38. P. Ward and M. R. Smyth, *Talanta,* **40**(7), 1131–1137 (1993).
39. Q. Cai and S. B. Khoo, *Anal. Chem.* **66**(24), 4543–4550 (1994).
40. M. S. Goncalves, L. Sigg, and W. Stumm, *Environ. Sci. Technol.* **19,** 141–146 (1985).
41. M. Oeztuerk, *Mar. Chem.* **48**(3–4), 329–342 (1995).
42. D. Puig and D. Barcelo, *Anal. Chim. Acta,* **311**(1), 63–69 (1995).
43. J. Wang and Q. Chen, *Anal. Chim. Acta,* **312**(1), 39–44 (1995).
44. C. La Rosa, F. Pariente, L. Hernandez, and E. Lorenzo, *Anal. Chim. Acta,* **308**(1–3), 129–136 (1995).
45. M. Vega, R. Pardo, E. Barrado, M. A. de la Fuente, and J. L. del Valle, *Fresenius' J. Anal. Chem.* **350**(3), 139–144 (1994).
46. M. Pedrero, M. R. Alonso, F. J. M. de Villena, and J. M. Pingarron, *Electroanalysis,* **7**(7), 644–648 (1995).
47. V. A. Argent, J. M. Southall, and E. D'Costa, *Proc. Annu. Conf. Water Works Assoc.,* 43–54 (1994).
48. J. Wang and P. V. A. Pamidi, *Anal. Chim. Acta,* **42**(3), 463–467 (1995).
49. R. I. Mrzljak, A. M. Bond, T. J. Cardwell, R. W. Cattrall, O. M. J. Newman, B. R. Champion, and J. Hey, *Analyst,* **119**(5), 1051–1055 (1994).
50. J. T. van Elteren, C. M. G. van den Berg, H. Zhang, T. D. Martin, and E. Achterberg, *Anal. Chem.* **67**(21), 3903–3010 (1995).
51. S-S. Huang, Z-G. Chen, B-F. Li, H-G. Lin, and R-Q. Yu, *Analyst,* **119**(8), 1859–1862 (1994).
52. X. Hu and Z. Leng, *Analyst,* **120,** 1555–1557 (1995).
53. A. K. Jain, L. P. Singh, and P. K. Jain, *Sens. Actuators,* **B25**(1–3), 729–732 (1995).
54. E. Schmidt, A. Marton, and J. Hlavay, *Talanta,* **41**(7), 1219–1224 (1994).
55. L. Van Nifterik, J. Xu, J. L. Laurent, J. Mathieu, and C. Rakoto, *J. Chromatogr.* **640**(1–2), 335–343 (1993).
56. G. Somer and A. Kocak, *Analyst,* **118**(6), 657–659 (1993).
57. I. Eskinja, Z. Grabaric, and B. S. Grabaric, *Atmos. Environ.* **29**(10), 1165–1170 (1995).
58. E. Dabek-Zlotorzynska and J. F. Dlouhy, *J. Chromatog.* **640**(1–2), 217–226 (1993).
59. S. Zaromb, J. Alcaraz, D. Lawson, and C. S. Woo, *J. Chromatog.* **643**(1–2), 107–115 (1993).
60. Y. Feng and R. S. Barratt, *Sci. Total Environ.* **152**(3), 261–267 (1994).
61. C. von Laar, R. Reinke, and J. Simon, *Fresenius' J. Anal. Chem.* **349**(8–9), 620–624 (1994).
62. K. Wallmann, K. Hennies, I. Koenig, W. Petersen, and H-D. Knauth, *Limnol. Oceanog.* **38**(8), 1803–1812 (1993).
63. C. Webster, M. Smith, P. Wilson, and M. Cooke, *J. High Resolut. Chromatog.* **16**(9), 549–551 (1993).
64. M. A. Allus and R. G. Brereton, *Analyst (London),* **117**(7), 1075–1084 (1992).
65. S. Kumaran and M. Morita, *Talanta,* **42**(4), 649–655 (1995).
66. J. Thottan, J. F. Adsett, K. J. Sibley, and C. M. MacLeod, *Commun. Soil Sci. Plant Anal.* **25**(17–18), 3025–3034 (1994).
67. M. M. Wander, L. M. Shuman, and R. B. Pitts, *Soil Sci. Plant Anal.* **26**(15–16), 2391–2401 (1995).
68. V. Comello, *R&D Mag.,* 53–54 (Jan. 1996).
69. *R&D Mag.,* 24 (May 1996).

ELECTROCHEMICAL REMEDIATION OF SOILS

JOHN P. ZELINA
JAMES F. RUSLING
University of Connecticut
Storrs, Connecticut

The use of electrochemistry to help clean up contaminated soils appears quite promising. A variety of fundamental electrochemical processes are applicable. Electrokinetic phenomena provide powerful bulk separation technologies for pollutant ions and molecules. Electrolytic processes can render pollutants harmless while operating at close to ambient temperature. They are also compatible with separation and recovery of product materials. Oxidations and reductions of pollutants can be accomplished cleanly with electricity, rather than by using stoichiometric amounts of chemical oxidizing or reducing agents, whose products may create additional disposal problems. Electrochemical processes are applicable to the treatment of mixed wastes containing organic and inorganic pollutants. Electrochemical methods can be applied to soils in place or to slurries of soils in water or electrolyte solutions.

Electrochemical processes typically employ an electrochemical cell or reactor with two or more electrodes dipping into an electrolyte solution. An electrical power supply is used to apply a current or voltage across this cell. The resulting applied electric field drives electrochemical reactions at the electrodes as well as electrokinetic mass transport within the cell. These electrode reactions can be designed to destroy pollutants or to convert them to easily recoverable forms. Electrokinetic processes refer specifically to movement of materials by using electric fields. They can be used to move pollutants in soils to collection points for physical removal.

Thus contaminated soil seems quite amenable to treatment by electrochemical technology. Processes for electrochemical remediation of soil are actively being developed for the removal of inorganic and organic pollutants (1–20). Specific applications to soils include removal and recovery of heavy metals, removal of organic materials, and decomposition of organic and inorganic pollutants. In some instances, electrochemical processes are combined with other types of operations to provide complete cleanup technology.

Metals such as lead, chromium, cadmium, copper, uranium, mercury, and zinc in soils are possible candidates for electrochemical remediation and recovery. Metal concentrations at uncontrolled sites can be larger than at uncontaminated sites. For example, more than 50,000 mg of lead per kilogram of soil was reported at the site of a former battery recycling plant (1) compared to about 700 mg/kg in natural soils. A study of six U.S. Army installations found metal concentrations ranging from 0.05–453 mg/kg for cadmium; 1,200–3,000 mg/kg for chromium; and 1.2–4940 mg/kg for lead (2). Organic pollutants such as polychlorinated biphenyls (PCBs), polychlorinated benzenes, phenols, chlorophenols, toluene, trichloroethane, chloroacetic acids, and acetic acid should also be possible to remove.

Electrochemical remediation has generated considerable recent interest from the scientific, business, and military communities because it is one of the few technologies that may work *in situ* in low permeability clay and silt-laden soils. European and U.S. research and development has resulted in numerous patents on electrochemical soil remediation processes. European accomplishments include the removal of toxic metals (lead, copper) from Rhine River sludge in Rotterdam, Germany (3), a severely environmentally damaged area. Russians in collaboration with a private U.S. company have successfully remediated sites in Uzbekistan and Kazakhstan, where groundwater contamination was caused by acid leachates from mines (4). Extraction efficiencies have been reported at 97% for uranium and 69% for sulfates (5).

Presently much *in situ* electrochemical remediation methodology is in the developmental stage, with large-scale fieldwork beginning to be conducted. This article reviews the basic experimental methodology being considered, then summarizes research into possible applications. The fundamentals of electrochemical processes are reviewed first to provide a basis for understanding the details of the applications to follow. The basic electrochemical processes covered include electroosmosis, electrokinetic phenomena, and electrolytic and voltaic (galvanic) reactions.

Applications are illustrated by specific examples taken mainly from the recent literature; many more fine examples of studies on electrochemical remediation can be found in the literature.

The success of electrochemical remediation depends on specific conditions encountered in the field, including types and amounts of contaminants present, soil type, pH, and organic content. Electrode materials and process conditions must be chosen carefully for goals to be achieved. Every pollution problem has its own set of solutions, and electrochemical methods may not be applicable in all cases.

In some situations, combinations of electrochemical and other technologies may lead to viable cleanup processes where one or the other technique cannot do the job alone. Electrochemical methods have been used *in situ* in conjunction with methods such as biodegradation, chemical catalytic dechlorination, absorption, and the use of chemical additives to improve system performance. For example, subsurface injection of surfactant solutions has been found to aid the electrokinetic removal of insoluble organic pollutants absorbed to soils (15,16). Chelating agents such as ethylene–diaminetetraacetic acid (EDTA) have been explored as a means to remove metals from soil and solid sediments (17,18). EDTA can be recycled in such processes.

Electrochemical methods to destroy mixed wastes (19) and to remove radioactive contamination (20) are also being developed. Electrochemical catalysis or mediation, in which electrons are shuttled between pollutants and electrodes by a chemical mediator, may be included in the system (14). For example, in a reduction a mediator can deliver electrons from the electrode to the pollutant. This approach can lower reduction potentials required for a specific reaction and enhance electrolysis efficiency for decomposing pollutants. Both of these factors decrease power requirements for a given process.

FUNDAMENTAL ELECTROCHEMISTRY

A variety of electrochemically driven processes can be employed in remediation of contaminated soils, including electromigration and electroosmosis, which occur in electrokinetic extraction, as well as chemical oxidation and reduction reactions at the electrodes.

Electrokinetic Phenomena

When direct-current electric fields are applied to contaminated soil by electrodes placed into the ground, migration of charged species (eg, ions) occurs. Positive ions are attracted to the negatively charged cathode, and negative ions move to the positively charged anode. Electromigration is the term used to describe this movement of ions in an electric field from one electrode to another (Fig. 1). It is the basis of a highly successful analytical separation method called electrophoresis.

Most organic molecules found in soils are not ionized. Therefore, electromigration is primarily for highly soluble ionized inorganic species, including metal cations, chlorides,

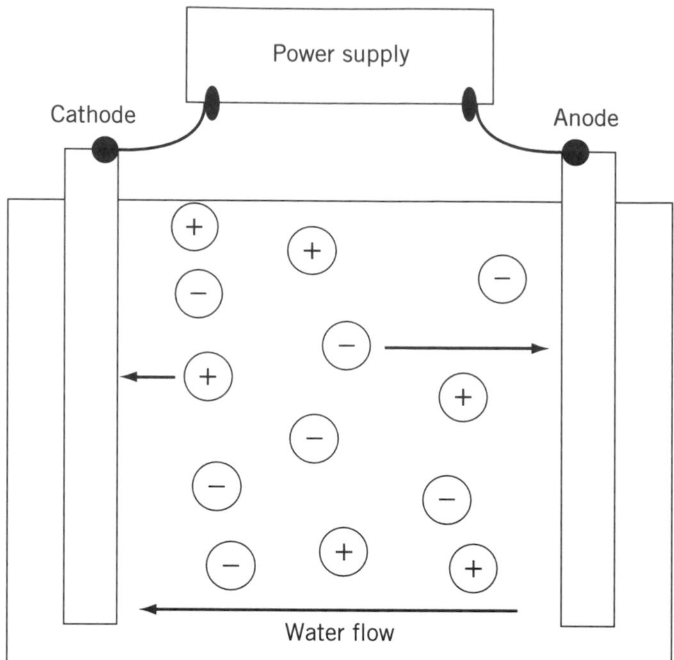

Figure 1. Simplified conceptual diagram illustrating movement of ions (electromigration) and water (electroosmosis) under the influence of an electric field applied with electrodes placed into wet soil.

nitrates, and phosphates in most soil environments (5). Efficient electromigration usually requires low permeability water-saturated soils. However, removal of contaminants such as chromium from unsaturated sandy soils is also being investigated (5).

Electroosmosis is another electrochemically driven transport process that can be employed for soil remediation. Electroosmosis in water-saturated soil is the movement of water relative to the soil under the influence of an imposed electric field gradient. This water flow is a consequence of ion movement past charged surfaces and usually accompanies electromigration in soils (see Fig. 1).

A classic demonstration of electroosmosis results from an experiment in which a voltage is imposed across a glass capillary filled with an electrolyte solution. The inside surface of the glass capillary is charged due to adsorbed anionic sites, which are actually part of the glass surface. There is a diffuse space charge, consisting of a double layer of ions, in which positive ions predominate in a layer of solution close to the negatively charged capillary walls (21). Near these walls, the electric field tends to cause the positive ions in the space charge region to flow to the cathode, dragging the solution with them. Viscous drag exerted by solution movement near the walls pulls the solution in the center of the capillary along with it. Thus a net electroosmotic flow of solution occurs to the cathode.

Electroosmosis works well in wet (eg, water-saturated), fine-grained soils and can be used to remove soluble pollutants even if they are not ionic (5). The dissolved neutral molecules simply go with the flow. As in the capillary experiment described above, flow occurs because mobile cations in the soil–fluid system are attracted to the cathode, whereas many of the anions are not mobile. An excess negative surface charge exists in all kinds of soils. For example, many clays, such as montmorillonite, are anionic colloidal polyelectrolytes. Thus a significant proportion of anionic sites are bound to the solid surfaces in the soil and are stationary. The cations near the soil surface migrate and carry water with them, as in the capillary experiment. This is illustrated conceptually in Figure 2, showing ion and water movement between two polyanionic soil particles.

The surface charge density on soil particles increases as the surface area per unit weight of the soil mineral increases. The surface charge density increases in the following order: sand < silt < kaolinite < illite < montmorillonite (6). The result of increasing surface area at fixed charge density is more rapid cationic ion movement than anion movement, leading to more efficient water flow to the cathode.

Injection of clean fluid at the anode can improve the efficiency of pollutant removal. Contaminated fluid can be removed at the cathode. For example, such a flushing technique using electroosmosis has been developed for the removal of benzene, toluene, trichloroethane, and *m*-xylene from saturated clay (7). These neutral molecules travel with the electroosmotically mobilized fluid.

Electrokinetic extraction is a technique that uses electroosmosis and electromigration, resulting in the flow of colloidal particles and dissolved and undissolved contaminants. This method can be employed with injection of a flushing solution at the anode, and collection of contaminated solution at the cathode. Electrokinetic extraction has been used successfully to remove chromium from a Superfund site at Corvallis, Oregon (8) (see HEAVY METAL IONS, REMOVAL FROM WASTEWATER). It has also been employed to remove metals such as arsenic, cadmium, cobalt, chromium, calcium, mercury, nickel, manganese, molybdenum, lead, antimony, and zinc from clay, peat, and sand (9). This technique is effective for the removal of dissolved organic pollutants such as acetic acid, phenol, and trichloroethylene from contaminated soils (10,11). Again, the molecules travel with the electroosmotic flow.

Superimposing the removal mechanisms of electroosmosis and ionic migration induced by an imposed electrical field provides an effective means for the removal of contaminants from fine-grained soils. Cationic contaminant removal by electromigration is enhanced by electroosmosis. Both mechanisms cause cations and other materials to move to the cathode. Experimental results indicate that ionic migration often predominates. This situation can be used for the removal of anionic contaminants at the anode, despite electroosmosis, which tends to move materials in the opposite direction (12).

Electrode Reactions

When electrical energy is supplied to an electrolytic cell, it is possible to drive nonspontaneous oxidation–reduction reactions (21–23). An external power supply injects electrons into the cathode and removes them at the anode. To maintain electrical neutrality in the cell, oxidation–reduction reactions involving these electrons take place at the electrodes in the cell. Oxidation reactions occur at the anode, whereas reductions occur at the cathode. These reactions result in chemical transformations of the participating reactants and can be designed to destroy pollutants.

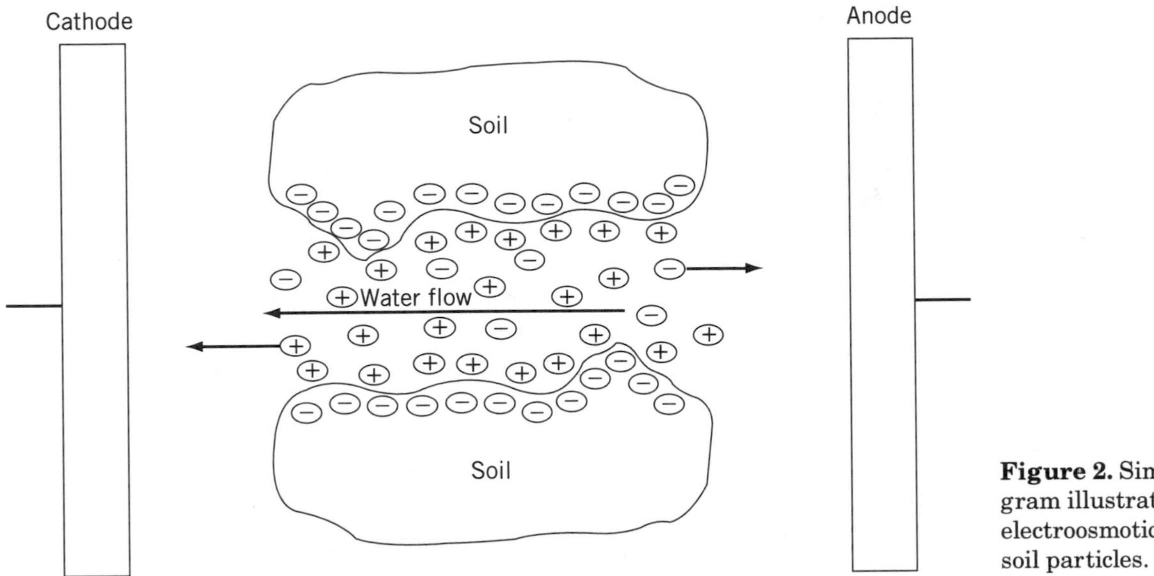

Figure 2. Simplified conceptual diagram illustrating the details of electroosmotic flow between charged soil particles.

An electrolytic cell consists of two or more electrodes connected to a power supply with a conducting medium between the electrodes. The conducting medium is often a solution of an electrolyte in water or another solvent. By convention, the cathode is assumed to be negative and the anode positive. Electrons in the external electrical circuit flow from the anode (+) to the cathode (−). The oxidation and reduction reactions that occur at the electrodes within the electrolytic cell result in an effective electron flow from cathode to anode within the cell. The overall process is called electrolysis. The oxidations and reductions are called Faradaic reactions, after Michael Faraday, who discovered the quantitative relation between amount of electricity passed and the amount of reactant converted to products by electrolysis.

In a reduction, a reactant accepts electrons. Ions or molecules in the electrolyte solution accepting electrons from the cathode are reduced. In an oxidation, a reactant loses electrons. An anode accepts electrons from ions or molecules in the electrolyte solution. These electron acceptors are thereby oxidized.

APPLICATIONS

Electrokinetic Soil Remediation: Fundamental Studies

Electrokinetic-based soil treatment methods have been labeled with various names, including electrochemical soil processing, electromigration, electrokinetic decontamination, and electroreclamation. Electrokinetic soil processing relies on the combination of several interacting mechanisms, all of which are connected to the application of an electric field to the soil. These mechanisms include advection, which is generated by electroosmotic flow and externally applied hydraulic gradients; diffusion of the acid front to the cathode; and migration of cations and anions toward respective electrodes (24).

Advection is a term used to describe the movement of fluid and dissolved and microparticulate matter from one electrode to the other and includes electroosmosis. It can be enhanced by injecting a fluid at the anode and removing it at the cathode.

The dominant electron transfer (Faradaic) reactions that occur at anodes and cathodes during electrokinetic process usually involve the electrolysis of water. Although secondary reactions at the cathode may be favored at lower electrochemical potentials, electrolysis of water predominates at early stages of most processes. Secondary reactions are expected at later stages due to contaminants being carried to the cathode by the acid front (19).

The acid front is carried toward the cathode by electrical migration, diffusion, and advection. The production of hydrogen ions at the anode decreases the pH near this electrode. At the same time, the increase in hydroxide ion concentration causes an increase in the pH near the cathode. The mobility of protons under the influence of an electrical field is about twice that of the hydroxyl ions (25) and is enhanced by electroosmotic flow. Therefore, the acid front eventually flushes across the cell and neutralizes the base generated at the cathode.

The injection of acids into soils to solubilize basic metal hydroxides, carbonates or adsorbed species, and protonate electron-rich organic functional groups has been discussed (26). This procedure gives some of the pollutant molecules and ions a more cationic character and promotes their migration to the cathode, facilitating their removal.

It has been suggested that hydrogen ion is the master variable in electrokinetic pollutant removal from soils (6,15), because hydrogen ions play key roles in many chemical and physical events throughout the cell, including electrolysis, water ionization, and ion exchange at surfaces. Figure 3 illustrates the diffuse charge distribution that exists near surfaces of clay or other anionic soil particles throughout the porous media to which an electric field is applied. This diffuse charge or space charge distribution is called a diffuse double layer and is common to any charged surface in contact with an electrolytic solution. The interaction between the cations in the soil fluid and the negatively charged soil surfaces results in this diffuse double layer (see Fig. 2). When an electrical field is applied, it is the excess cations from the double layer that migrate to the cathode.

The movement of cations in the diffuse double layer near soil particles brings along water and establishes viscous drag on the pore fluid, resulting in bulk flow to the cathode.

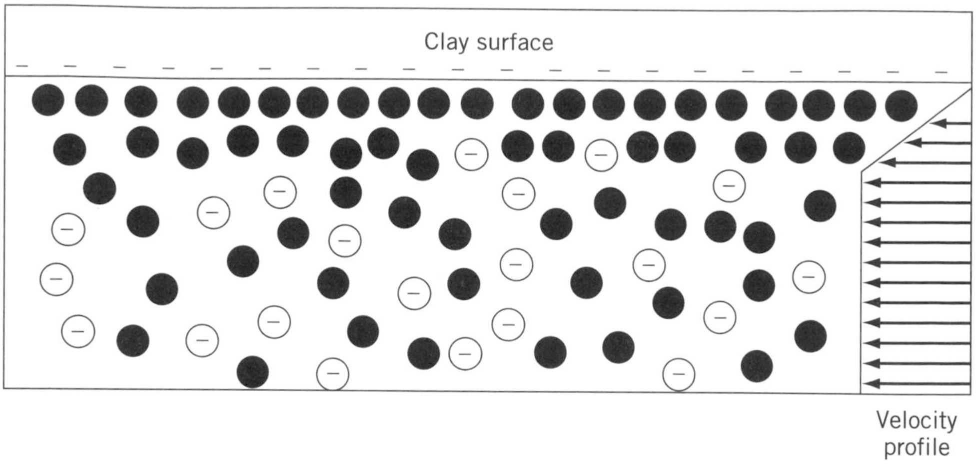

Figure 3. Simplified schematic diagram of the diffuse double layer of ionic species near a negatively charged clay surface in an electric field, and the pore fluid velocity profile at the electrolytic solution–soil interface where (●) represents cations. Adapted with permission from Ref. 6.

The thicker the diffuse double layer, the stronger the electroosmotic flow. The thickness of the diffuse double layer depends on the surface charge density of the soil, the concentration of ions in the pore fluid, the charge on the individual cations, and the dielectric properties of the pore fluid. As the net concentration of ions in the system increases, the thickness of the diffuse double layer decreases and the extent of electroosmosis will decrease. A maximum electroosmotic flux of about 10^{-4} cm^3s^{-1}/cm^{-2} can be obtained by using an electric gradient of 1V cm^{-1} in low activity clays at high water content, low electrolyte concentrations, and low soil charge densities (6). Under these conditions, the thickness of the diffuse double layer will be maximum, and the conductivity of the water in the pores minimized.

Electrokinetic cell design generally consists of an open flow arrangement at the electrodes, which allows for injection of processing fluid into the porous medium with later removal of the contaminated fluid. Imposition of an electrical gradient by placing inert electrodes into a wet contaminated soil mass results in electroosmotic flow to the cathode. Thus clean water or a cleaning agent can be injected continuously at the anode and carried to the cathode. Contaminated water can be removed at the cathode. Extraction and removal of contaminants are accomplished by electrodeposition, precipitation, or ion exchange either at the electrodes or in an external extraction system (26,27,28).

Electrode materials are generally chosen to avoid introduction of secondary corrosion products. The nature of electrode reactions during electrokinetic treatment of soils depends on the electrode material and the type and concentration of chemical species in the fluid around the electrodes. The electrolyzed products of some electrodes introduce other ions, thus changing the chemistry generated at the electrodes; therefore, it is often necessary to use electrodes that are inert to anodic dissolution. Common electrodes used include graphite, platinum, and gold, although the later noble metals may add considerable expense to a process. These electrodes, however, minimize introduction of harmful products from the electrode itself into the medium.

The conductivity of the soil and the pore fluid affects the electroosmotic flow rate. Conductivity is inversely related to the resistance offered to current flow. In soil, this resistance changes due to variation in pore size (porosity), tortuosity in the porous media, and variations in pore volume and electrolyte concentration (24). The conductivity of the soil depends on the concentration and the mobility of the ions present. As the conductivity increases, the electrical potential difference between the electrodes decreases, resulting in a decrease of electroosmotic flow (25,29). Electroosmotic flow has been minimized by the addition of electrolyte (29).

Studies have shown that hydrogen ions in the pore fluid significantly affect the conductivity of a system once H^+ concentrations are of the order of or greater than the initial electrolyte concentration (30). During long testing periods, the conductivity profiles at the anode demonstrate that low pH pore fluid increases the conductivity, diminishing the electroosmotic flow to the cathode (30).

pH and electroosmotic flow can also be influenced by ion exchange. It has been found that the removal efficiencies decrease as the concentration of contaminant is reduced (31). Hydrogen ions exchange with cationic contaminants on the soil surface, freeing the contaminants. As the contaminant is removed, the hydrogen ion concentration in the pore fluid increases, resulting in an increasing fraction of the current being carried by the hydrogen ion rather than by the cationic contaminant. The removal efficiencies decrease as the contaminant concentration decrease.

A better understanding of a process can be developed when the conductivity across the cell is evaluated (30). Electroosmotic experiments usually employ constant electrical current. As the experiment proceeds, the potential across the sample cell increases and conductivity decreases. The degree of saturation decreases across the cell, resulting in an increase in sample tortuosity (ie, an increased compaction). When contaminant ions are removed from the soil near the electrodes, the electrical conductivity at these locations decreases more rapidly than that of the central region, leading to several electrical gradients near the electrodes. Cations are retarded by lower pH and anions are accelerated (32). Complex formation and precipitation of chemical species also depend on pH. In summary, the following variables were found to influence the efficiency of removing contaminants from soils: chemical processes at electrodes, water content of the soil, saturation of the soil, pH and pH gradients, type and concentration of chemicals in the soils, applied current density, and sample conditioning.

Another factor that decreases electroosmotic flow occurs when the permeability of the soil is low. Low permeability results in an insufficient supply of pore fluid to the cathode zone from the anode. A suction of pore fluid to the cathode occurs, causing soil consolidation in the anode region.

Several studies have documented suction pressures and soil consolidation (33,34). It has been suggested that the developing negative pore pressure (suction) offsets the electroosmotic flow and decreases the net water flow (34). The chemistry taking place across the cell varies with time, electrical gradient, electrolyte concentration, and pH. It influences flow rates throughout the system, inducing a nonlinear flow velocity.

Insufficient pore fluid supply can induce precipitation of other chemicals present throughout the soil chemistry matrix. An example is the precipitation of calcium in the form of bicarbonates and hydroxides. These precipitates have been observed to clog pores and prevent contaminant transport (35).

A pH buffer is a chemical system that resists changes in acidity (22) and usually consists of a weak acid and its conjugate base. Several occur naturally in soils. Acidic front generation can be limited by the buffering capacity of a soil (15). Electrokinetic enhancement with the aid of added chemical constituents may alleviate some of these problems.

Electroosmotic Removal of Pollutants

Electroosmosis is an electrokinetic process (25,37–41) involving the flow of an electrolyte solution under the action of an applied electric field relative to a charged surface, and it can be induced through a porous material such as clay or soil, whose internal surfaces are charged (26,37). The rate of flow is relatively insensitive to the size of the pores. It allows a uniform flow distribution and a high degree of control of flow direction.

Decontamination can be achieved by flowing liquid through the soil matrix as a means to transport dissolved pollutants, even if they are not ions. Such applications, which are termed electroosmosis. This method is similar, if not identical, to electrokinetic remediation, since the electrokinetic processes of electromigration and electroosmosis usually occur simultaneously.

Galvanic Electrochemical Oxidation and Reduction

The thermodynamics of metal oxidations can drive reductions of certain pollutants without the input of external energy (21,22,36,42–49). These electrochemical reactions are called galvanic reactions.

A promising technique for subsurface remediation involves coupling the corrosion of iron with pollutant reduction, using a permeable subsurface barrier. This barrier prevents migration of contaminants by adsorption, precipitation, or biotransformation as the contaminants encounter the barrier and are exposed to altered pH, redox activity, adsorptive substrates, or biological activity (46–49). This type of barrier has been evaluated for reductive dehalogenation of organic compounds such as trichloroethylene (TCE) and tetrachloroethylene.

Electrolytic Reductions

Unlike galvanic processes, electrolysis requires the input of electrical energy. Electrolytic transformation offers promising approaches for the prevention and remediation of pollution. Attractive features include versatility, avoidance of large quantities of chemical redox agents, and energy efficiency (26). Oxidations and reductions can be done directly at electrodes or indirectly by using a chemical mediator to shuttle electrons between electrodes and reactants. Phase separations, precipitations, and other processes can be effected. Volume scales from microliters to many thousands of liters are possible.

Electrolytic Oxidations

Electrochemical oxidation is capable of degrading many pollutants to harmless materials. Research in this area is beginning to intensify. A review of this topic has been published (26).

FUTURE OUTLOOK

This review has shown that there is considerable promise for electrochemical methods in the cleanup of polluted soil. Most of these methods are in the early stages in regard to practical application. Among them, electrokinetic soil remediation appears to be the farthest along in development. Experiments under controlled conditions have contributed to improved understanding of the relevant process in soils. Mathematical modeling can assist in understanding and using the complex chemistry involved. The success or failure of an electrokinetic method depends on careful site selection and characterization (50).

At present, there is a need for additional pilot-scale and field studies to evaluate electrokinetic technology (6). Furthermore, methods are required at the end of the processes to permanently destroy organic pollutants or convert them to useful materials. Otherwise, it is a matter of simply trading one pollution problem for another.

There are indications that innovative use of chemistry and electrode materials could lead to improved technologies for electrokinetic soil remediation. For example, a recent study employed an iron anode and a carbon cathode for *in situ* decontamination of Cr(VI) in sand (51,52). Cr(VI) is reduced by Fe(II) formed by dissolution of the anode. The acidic front is stabilized near the anode by participating in iron corrosion, and an alkaline front generated at the anode flushes across the cell. The process was modeled mathematically (51). This novel method combines electrokinetics with galvanic reduction of Cr(VI) by iron.

Galvanic and electrolytic techniques have significant capabilities for pollutant destruction, but applications to real polluted sites lag well behind demonstrations of laboratory effectiveness. New pilot and field studies are needed to assess practical effectiveness and to evaluate problems due to site differences.

A possible future role for electrolysis could be in destroying pollutants as they are collected in a fluid medium from contaminated soils, eg, by electrokinetic remediation or pump-and-treat technology. On-site electrolytic generators for pollutant destroying electrochemical mediators, eg, hydroxyl radicals or hydrogen peroxide, would also be valuable in future cleanup efforts. However, it should be stressed that oxidations of organics must be complete, so that toxic products do not form, eg, chlorodioxins from oxidation of polychlorinated biphenyls (PCBs). With this in mind, electrolytic reductions may play a useful role, since they can selectively yield less toxic hydrocarbons from halogenated organics.

For most electrochemical methods, economic factors that have an impact on practical pollutant remediation have yet to be fully considered in a general fashion. However, they have the potential to be cost effective. Most methods can employ inexpensive electrodes and operate at ambient temperature and pressure. A major cost will be the electricity required. For electrokinetic methods, the soil itself is the reactor. For electrolytic methods, simple batch or flow through reactors can be employed. Only the future, however, can reveal the economic competitiveness of electrochemical methods in soil remediation.

Many different problems are being faced in cleaning up and maintaining a healthy environment for ourselves and future generations. No single technology will be a panacea; many different methods will be needed (chemical, microbiological, catalytic, electrochemical, and others) in the toolkit of environmental remediation. Technologies and conditions required for cleanup are highly site dependent. The correct tool must be chosen to fit each specific job. Within this scenario, the future appears bright for a significant role of electrochemical processes in pollutant cleanup technology.

ACKNOWLEDGMENT

The author's research described in this article was supported by U.S. PHS grant No. ES03154 from the National Institute of Environmental Health Sciences (NIEHS).

BIBLIOGRAPHY

1. H. Allen, R. A. Evangelista, and A. P. Zownir, paper presented at the *Fifth National RCRA/Superfund Conference,* Las Vegas, Nev., Apr. 1988.
2. J. Kesari, P. S. Puglionesi, S. Poop, and M. H. Corbin, *Heavy Metal Contaminated Soil Treatment: Conceptual Development,* U.S. Army Toxic and Hazardous Material Agency Report, AMXTH-TE-CR-86101, Weston, Inc., West Chester, Pa., 1987.
3. R. Lageman, W. Pool, and G. Seffinga, *Chem. Ind.* 585 (Sept. 1989).
4. H. L. Lomasney, in *Proceedings: EPRI Workshop on In Situ Electrochemical Soil and Water Remediation, TR-104170,* EPRI, Palo Alto, Calif., 1994, pp. 159, 347.
5. J. Trombly, *Environ. Sci. Technol.* **28,** 289A (1994).
6. Y. B. Acar, R. J. Gale, A. N. Alshawabkeh, R. E. Marks, S. Puppala, M. Bricka, and R. Parker, *J. Haz. Mater.* **40,** 117 (1995).
7. C. J. Bruell, B. A. Segall, and M. T. Walsh, *J. Environ. Eng.* **118,** 68 (1992).
8. S. Banerjee, *Land Disposal, Remedial Action, Incineration and Treatment of Hazardous Waste, EPA/600/9-87/015,* U.S. Environmental Protection Agency, Cinncinnati, Ohio, 1987, p. 193.
9. R. Lageman, W. Pool, G. A. Seffinga, *Forum on Innovative Hazardous Waste Treatment Technology, EPA/540/2-89/056,* U.S. Environmental Protection Agency, Atlanta, Ga., 1989, p. 57.
10. A. P. Shapiro, P. C. Renaud, and R. F. Probstein, *Physiochem. Hydrodynam.* **11,** 785 (1989).
11. A. P. Shapiro, P. C. Renaud, and R. F. Probstein, *In-Situ Extraction of Contaminants from Hazardous Waste Sites,* Battelle Press, Columbus, Ohio, 1989, p. 346.
12. A. T. Yeung and J. K. Mitchell, *Geotechnique* **43,** 121 (1993).
13. Y. Liu, J. Schwartz, and C. L. Cavallaro, *Environ. Sci. Technol.* **29,** 836 (1995).
14. J. F. Rusling, *Accounts Chem. Res.* **24,** 75 (1991).
15. Y. B. Acar, A. N. Alshawabkeh, and R. J. Gale, *Waste Manage.* **13,** 141 (1993).
16. Y. B. Acar, R. J. Gale, A. Ugaz, S. Puppala, and C. Leonard, "Phase I Project Report," *USEPA-RREL Cooperative Agreement No. CR816828-01-1,* U.S. E.P.A., Washington, D.C., 1992, p. 243.
17. H. E. Allen and P. H. Chen, *Environ. Prog.* **12,** 285 (1993).
18. S. B. Martin and H. E. Allen, *Chemtech,* 23 (Apr. 1996).
19. D. E. Wedman, H. E. Martinez, and T. O. Nelson, *Chemtech,* 26 (Apr. 1996).
20. J. Dziewinski, S. Marczak, and W. Smith, *Chemtech,* 30 (Apr. 1996).
21. P. H. Reiger, *Electrochemistry,* Prentice Hall, Englewood Cliffs, N.J., 1987, pp. 86–100.
22. W. L. Masterton and E. J. Slowinski, *Chemical Principles,* 4th ed., W. B. Saunders, Philadelphia, Pa., 1977.
23. D. Pletcher and F. C. Walsh, *Industrial Electrochemistry,* 2nd ed., Blackie Academic, London, 1993.
24. Y. B. Acar, H. Li, R. J. Gale, *J. Geotechnical Eng.* **118,** 1837 (1992).
25. N. C. Lockhart, *Coll. Surf.* **6,** 253 (1983).
26. K. Rajeshwar, J. G. Ibanez, and G. M. Swain, *J. Appl. Electrochem.* **24,** 1077 (1994).
27. U.S. Pat. 5,137,608 (Aug. 11, 1992), Y. B. Acar and R. J. Gale.
28. Eur. Pat. 0,312,174 A1 (Apr. 19, 1989), W. Pool.
29. N. C. Lockhart, *Colloids Surfaces* **6,** 239 (1983).
30. J. Hamed, Y. B. Acar, and R. J. Gale, *J. Geotech. Eng.* **117,** 241 (1991).
31. D. D. Runnels and J. L. Larson, *Ground Water Monitor. Rev.* **6,** 85 (1986).
32. A. T. Yeung and S. Dalta, *Can. Geotech. J.* **32,** 569 (1995).
33. M. I. Esrig and J. P. Gemeinhardt, *J. Soil Mech. Found Eng. Div.* **93,** 109 (1968).
34. Y. B. Acar, R. J. Gale, G. A. Putnam, J. Hamed, and R. L. Wang, *J. Environ. Sci. Health, Part A,* **A25,** 687 (1990).
35. Y. B. Acar and A. N. Alshawabkeh, *Environ. Sci. Technol.* **27,** 2638 (1993).
36. A. J. Bard and L. R. Faulkner, *Electrochemical Methods,* John Wiley & Sons, Inc., New York, 1980.
37. A. P. Shapiro and R. F. Probstein, *Environ. Sci. Technol.* **27,** 283 (1993).
38. B. A. Segall and C. J. Bruell, *J. Environ. Eng.* **118,** 84 (1992).
39. C. J. Casagrande, *Boston Soc. Civ. Eng.* **69,** 255 (1983).
40. L. Bjerrum and co-workers, *Geotechnique* **17,** 214 (1967).
41. M. I. Esrig, *J. Soil Mech. Found. Div. Am. Soc. Civ. Eng.* **94,** 899 (1968).
42. J. C. Sculley, *The Fundamentals of Corrosion,* Permagon Press, New York, 1975.
43. U. R. Evans, *The Corrosion and Oxidation of Metals,* Edward Arnold Ltd., London, 1960, p. 1094.
44. U. L. Snoeyink and D. Jenkins, *Water Chemistry,* John Wiley & Sons, Inc., New York, 1980, p. 463.
45. R. M. Powell, R. W. Puls, S. K. Hightower, and D. A. Sabatini, *Environ. Sci. Technol.* **29,** 1913 (1995).
46. R. R. Spangler and S. J. Morrison, *Environmental Restoration,* U.S. Department of Energy, Pasco, Wash., 1991.
47. S. J. Morrison and R. J. Spangler, *Environ. Sci. Technol.* **26,** 1922 (1992).
48. J. S. Fruchter, "In-Situ Redox Manipulation," *PNL-SA-21731,* Pacific Northwest Laboratory, Richland, Wash., 1993.
49. S. J. Morrison and R. R. Spangler, *Environ. Prog.* **12,** 175 (1993).
50. F. G. Will, *Chem. Ind.* **15,** 377 (1995).
51. B. S. Haran, B. N. Popov, G. Zheng, and R. E. White, *J. Haz. Mater.,* in press.
52. B. S. Haran, B. N. Popov, and G. Zheng, *Environ. Progr.* **15,** 166 (1997).

ELECTROTHERMAL ATOMIC ABSORPTION SPECTROMETRY

DIMITER L. TSALEV
University of Sofia
Sofia, Bulgaria

Atomic absorption spectrometry (AAS) is one of the most valuable techniques in the vast application area of environmental analysis. It is now a very well established, reliable, and cost-effective analytical tool in thousands of laboratories throughout the world for quantitative determinations of chemical elements (up to 60–70 analytes) at trace levels down to nanograms (ng, 10^{-9} g) and picograms (pg, 10^{-12} g). This chapter is devoted to the electrothermal atomic absorption spectrometry (ETAAS), which is the AAS technique offering the lowest absolute limits of detection (LOD), applicability to solid or slurried microsamples, direct analysis of intact or simply diluted microsamples, and a high degree of automation.

THEORETICAL BACKGROUND AND INSTRUMENTATION OF AAS

The principle of AAS is simple: the analyte element is transformed into atoms in an atomizer; most of these atoms are in their electronic ground state with energy, E_o, and are able to absorb energy ($\Delta E = E_1 - E_O = h\nu = hc/\lambda$) from a special radiation source (lamp), where E_1 is the energy of the (first) excited state, h is Planck's constant, c is the velocity of light, and ν and λ are the frequency and the wavelength of the atomic spectral line. The lamp emits the atomic spectrum of the analyte element, ie, just the energy that can be absorbed in a resonance manner by the analyte atoms in the atomizer. The measuring beam from the lamp passes through the observation volume of the atomizer wherein the radiation is proportionally absorbed, according to the Beer's law, while the analyte atoms are excited to one of their permitted upper energy levels (E_1, E_2, etc). The ratio of the incident radiant power, I_o, and the power transmitted by the analyte vapor cloud, I, is then processed by the electronics of spectrometer to yield the analyte signal, absorbance, A:

$$A = \log_{10} I_o/I = -\log_{10} T$$

$$A = abc (\text{Beer's law})$$

where $T = I/I_o$ is transmittance, a is absorptivity, b is the sample path length (the length of atomic vapor cloud), and c is the concentration of the absorbing analyte atoms. Although concentration can be calculated under certain conditions (absolute analysis) (1), the common approach in analytical AAS is to relate concentration and absorbance by plotting a calibration graph or curve: $c = f(A)$.

Radiation sources in AAS are hollow cathode lamps (HCL) or electrodeless discharge lamps (EDL) for each individual element. Multielement HCLs are available but cannot be recommended since measurements are prone to spectral interferences. The lamp emits a number of intensive, narrow ($\leq$0.01 nm) spectral lines of the analyte element. Most lines are atomic ($\lambda_1 = hc/E_1 - E_0, \lambda_2 = hc/E_2 - E_0$), but there are also some ionic lines. The atomic lines that originate from the ground state, E_o, are referred to as resonance lines. Usually the most sensitive resonance line, often λ_1, is isolated from the adjacent lines in the monochromator (Fig. 1) by employing a narrow bandpass setting (within 0.1 and 2 nm). Both emission and absorption lines are not indefinitely narrow because of effects of natural, Doppler, and Lorentzian broadening. The emission line of the lamp must be narrower than the absorption line of the analyte atoms in the atomizer by a factor of $\geq$2–3, so as to ensure good sensitivity and linearity in AAS measurements. This condition is generally fulfilled in analytical AAS, since both the Doppler and Lorentz broadening of spectral lines are less pronounced for the radiation source than for the atomizer, owing to lower temperature and pressure in the spectral lamp versus those in the atomizer. Radiation of the lamp is modulated electronically or mechanically (chopped) with a certain frequency of 100–300 Hz in order to be recognized by the electronics and be separated from other emission signals in the atomizer, originating from the excited atoms of the analyte, concomitant molecules, radicals, and incandescent walls of the atomizer, etc.

Several AAS techniques can be distinguished, depending on the mode of sample introduction and atomization.

Flame AAS (FAAS). The liquid sample is pneumatically nebulized, the aerosol is mixed with acetylene in a spray chamber, and then is introduced in a flame atomizer, ie, burner with an air–C_2H_2 or N_2O–C_2H_2 flame. A continuous, steady-state signal is obtained and measured after a few seconds, unless small sample aliquots are introduced by flow injection (FI) or pulse-nebulization mode, wherever a peak signal is produced. FAAS is applicable for ca 70 elements (2,3).

Hydride Generation AAS (HGAAS). The analyte is reduced to its volatile hydride (AsH_3, H_2Se, etc) in a hydride generator of a batch, continuous flow or FI type; the hy-

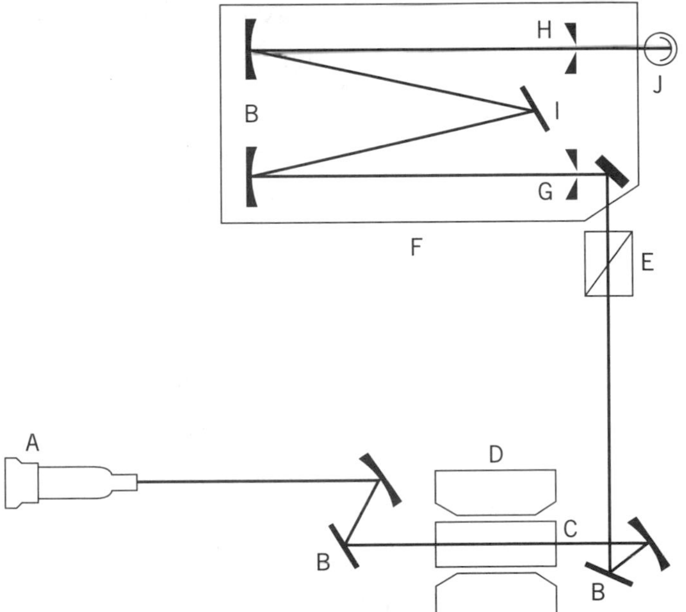

Figure 1. Zeeman optical system of the SpectrAA-30/40/300/400 Atomic Absorption Spectrometers with A, radiation source; B, mirrors; C, graphite tube atomizer; D, enclosed by a magnet; E, prism polarizer; F, monochromator; with, G, entrance and, H, exit slits; I, holographic grating; and J, detector. Reproduced by permission of Varian Australia.

dride is stripped-out from solution by an inert purge gas, Ar, and is atomized in an externally heated quartz tube (QTA), flame-in-tube, flame, etc (4). This technique is applicable for only 10 elements which yield volatile covalent hydrides (4).

Cold Vapor AAS (CVAAS). The analyte element, mercury, is reduced to its elemental state in a vapor (mercury/hydride) generator in a manner analogous to the HGAAS technique: the quartz cell may or may not be heated, thus, called the cold vapor technique (CVT) (2,4–7).

Electrothermal AAS (ETAAS). The analyte element is atomized in an electrothermal atomizer, a device that is heated to the required temperature by the passage of electrical current through its body (8). This is applicable for ca 60 elements (Table 1); see Refs. (2,5–7,9–17).

Hyphenated (Coupled) AAS Techniques. Some techniques in this category are HG–ETAAS (4,18), gas chromatography–AAS (GC–AAS) (19–22), high performance liquid chromatography–AAS (HPLC–AAS) (19–22), and FI–AAS (23–27).

GENERAL ASSESSMENT OF THE AAS TECHNIQUES

As with every analytical method, the AAS techniques (FAAS, ETAAS, HGAAS, and CVAAS) have their own scope, advantages, and limitations, which are compared with a view to environmental applications.

1. *Selectivity,* ie, an absence of interference from other concomitants or their reliable control, is undoubtedly one of the most attractive features in analyses of complex matrices. In this respect the AAS techniques can be ranked in the following decreasing order:

   FAAS > CVAAS > ETAAS > HGAAS

2. *Sensitivity,* ie, the slope of the calibration graph, dA/dc or dA/dm, where m is the analyte mass:

   ETAAS > CVAAS > HGAAS

   > FAAS(absolute sensitivity, dA/dm)

CVAAS, HGAAS > ETAAS > FAAS(relative sensitivity, dA/dc)

3. Relatively *simple sample pretreatment:*

   FAAS, ETAAS > CVAAS > HGAAS

4. *Automation,* all techniques are presently automated, except for the batch-type vapor generation AAS:
5. High *sample throughput rates:*

   FAAS > HGAAS/CVAAS > ETAAS

   Examples: >150, 30–60, and 30–40 samples/h, respectively.

6. *Large elemental coverage:*

   FAAS > ETAAS > HGAAS > CVAAS

Table 1. Representative Ranges for Characteristic Masses, m_o, and Instrumental Limits of Detection, LOD, of the ETAAS Technique for Individual Elements[a]

Analyte	m_o, pg[k]	LOD, pg[k]	Comments
Ag	0.7–1.7	0.1–0.5	b
Al	5–12	2–5	b,c,d
As	7.5–40	9–30	b,e,f
Au	4.4–18	10	b
B	500–1000	500–2,000	d,e,g
Ba	6.5–15	5–10	d,h,i
Be	0.5–1	0.5–1	b
Bi	9–28	10	b,e,f
Ca	0.5–1	5	c,h,j
Cd	0.2–1.3	0.3	b,e,f
Ce	80,000		d,g,h,i
Co	2.9–17	1–3	b,i
Cr	0.8–3.5	0.3–1	b,d
Cs	4–18	5	f,h
Cu	4–17	1–2	b,c
Dy	20–140		d,h,i,k
Er	70–280		d,i,k
Eu	18–34		d,h,i,k
Fe	1.2–12	1–2	c,i,j
Ga	4.5–42	10	e
Gd	3,500–14,000		d,g,h,i,k
Ge	8–34	13–20	d,e,f,i
Hg	20–220	25–100	e,k
Ho	61–280		d,h,i
In	7–15	5	b,e
Ir	135–275	200	d,i
K	0.4–1.1	2	c,f,h,j,k
La	7,400–26,400		d,g,h,i,k
Li	1–4	2–5	h
Lu	52–3,500		d,g,i,k
Mg	0.2–0.35	0.4–0.5	c,j
Mn	0.6–6.3	1	b,c
Mo	2–12	4–9	b,d,g,k
Na	0.4–1.2	5	c,h,j,k
Nd	900–1,800		d,g,h,i,k
Ni	4.8–20	10	b,i
Os	1,400	270	d,g,i,k
P	1,200–3,000	3,000	d,e,f
Pb	5.5–12	5	b,c,e,f
Pd	8.6–52	5–25	
Pr	4,400–10,000		d,g,h,i
Pt	54–220	33–50	d
Rb	1–10	3–5	f,h
Rh	8–24	10	d
Ru	15–45		d
Sb	10–19	15–20	b,e,k
Sc	30–130		d,g,h,i
Se	20–45	10–30	b,e,f
Si	15–120	40	b,c,d,i
Sm	230–800		d,g,h,i,k
Sn	10–90	2–20	b,d,e,f
Sr	1–4	2	h
Tb	3.5–4,100		d,g,h,i
Tc	60		d
Te	9–17	10	e,f
Ti	32–70	50–100	d,f,g,i
Tl	10–19	10	b,e,f
Tm	13–18		d,i,k
U	12,000		d,g,i,k
V	13–50	3.5–48	d,g
Y	1,300–4,000		d,g,h,i,k

Table 1. *(continued)*

Analyte	m_o, pg^k	LOD, pg^k	Comments
Yb	2.5–3.0		d,i,k
Zn	0.04–1	0.1–1	c,e,f,j

[a] Based on data from different instrument manufacturers, as compiled in Refs. 2, 6, 7, 9, 28, and 29.
[b] ETAAS is the technique of choice.
[c] Drastic problems with airborne contamination and blanks.
[d] High heating rates and temperature-controlled heating during the atomization step particularly important for the indicated elements.
[e] Chemical modification essential for the most volatile analytes.
[f] EDL radiation sources available for these analytes.
[g] Poor sensitivity and carryover because of strong carbide formation.
[h] Deuterium background correction not applicable at all.
[i] Narrow bandpass settings mandatory.
[j] FAAS is adequately sensitive and preferable for most applications.
[k] Except for m_p based on peak-height measurements, A_p, for most refractory analytes.

Examples: ca 70, 50–60, 8–9 and 1–2 (Hg and Cd), respectively. AAS is not at all applicable to the following elements: Ar, Br, C, Cl, F, H, He, I, Kr, N, Ne, O, Rn, S, Xe, and actinoids. Several elements Hf, Nb, Ra, Re, Ta, Th, Zr and W, cannot be determined in graphite atomizers because of strong carbide formation. Sensitivity for several other refractory elements is extremely poor and their ETAAS determinations are practically ruled out. These elements are Ce, Gd, La, Nd, Os, Pr, Sm, Tb, U, and Y (see Table 1).

7. Small *sample size requirements:*

$$\text{ETAAS} < \text{FI}-\text{FAAS} < \text{FI}-\text{HGAAS/FI}$$

$$-\text{CVAAS} < \text{FAAS} < \text{HGAAS/CVAAS}$$

Examples: ca 0.01–0.05, 0.1–0.5, 0.1–2, 0.5–2 and 5–20 mL, respectively.

8. *Adequate precision,* eg, instrumental within-run repeatability of measurements characterized by relative standard deviations (RSD) within 0.2–3% for FAAS and 0.5–5% for ETAAS and vapor-generation techniques.
9. Moderately priced, *affordable* equipment.
10. *Well-established methodology* and detailed documentation.

Some limitations and drawbacks of AAS are as follows:

1. There is a limited potential for *simultaneous* multielement analysis. Although several manufacturers offer instrumentation with multielement capabilities (two to six elements simultaneously) and there has been remarkable progress in recent years (30), the AAS method is seriously rivaled by other modern, truly multielement techniques, such as inductively coupled plasma–optical emission spectrometry (ICP–OES) and inductively coupled plasma–mass spectrometry (ICP–MS).
2. The range of rectilinear calibration is fairly narrow. The *linear range* is typically about two orders of magnitude. This is a serious disadvantage since analyte concentrations in environmental samples vary within broader intervals, hence part of the overrange samples have to be reanalyzed after dilution or sensitivity attenuation.
3. Samples brought into a *liquid state* are generally preferred, since they are more conveniently homogenized, diluted, handled by autosamplers, and introduced into atomizers than are solids, powders, suspensions, slurries, and emulsions. Calibration is also facilitated with liquid samples. ETAAS is more suitable for solid and slurry sampling than the other AAS techniques:

$$\text{ETAAS} \gg \text{FAAS} > \text{CVAAS} > \text{HGAAS}$$

4. As far as *speciation* analysis is concerned, the FAAS technique often lacks sensitivity, whereas ETAAS is not a particularly suitable detector for GC, HPLC, and other chromatographic techniques. Therefore AAS techniques can be ranked in the following decreasing order of their potentialities for speciation:

$$\text{HGAAS} > \text{CVAAS} \gg \text{ETAAS} > \text{FAAS}$$

5. The large variety of instrumentation offered by different manufacturers is far from standardized. Therefore, each newly adopted procedure from the literature needs modification and adaptation to the particular available apparatus. The ETAAS technique is the most dependent on the manufacture of hardware (and software):

$$\text{ETAAS} > \text{HGAAS} > \text{CVAAS} > \text{FAAS}$$

6. The *apparent simplicity* of AAS instrumental operation may well be a misleading and discouraging factor in understanding that rather *complex* physical–chemical processes are responsible for the accuracy and precision of results. Therefore these techniques, ETAAS and HGAAS in particular, can produce not only highly sensitive, accurate, and reproducible results but also artefacts (15).

In conclusion, several generalizations about the choice of AAS technique and analytical strategy can be made. Whenever possible, preference is given to the flame AAS technique owing to its inherent selectivity, high sample-throughput rates, low cost, simple operation, and easy maintenance. Installing FAAS apparatus is less expensive by a factor of 2–4, 5–10, and 10–15 as compared with ETAAS, ICP–OES, and ICP–MS, respectively, whereas the running costs for FAAS are lower by more than one order of magnitude. FAAS should be considered first (or second, if simultaneous ICP–OES is available) in the determinations of Al, Ba, Ca, Cd, Cr, Cu, Fe, K, Li, Na, Mg, Mn, Ni, Pb, and Zn. Alkaline metals, Li, Na, K, Rb, and Cs, are often determined by flame atomic emission spectrometry (AES), using the AES mode of AAS spectrometers or simple, dedicated flame photometers. Sensitivity for volatile elements can be improved by several fold using such simple accessories as slotted tubes, which are adapters for flame gases and atom traps (3,7,31). Sensitivity down to an impressive characteristic concentration of 0.1 and 1 μg/L for Cd and Pb, respectively can be achieved. FAAS is presently scaled down to sub-mL sample volumes by means of FI (25–27) or pulse nebulization (3,5–7) and is efficiently combined with preconcentration techniques in both off-line, eg, with simultaneous multielement-chelate extraction (3,5–7,33,34), as well as with automated on-line approaches such as FI preconcentration (23,25–27,34).

ETAAS is the current technique-of-choice at ng and pg levels for about 20 elements, as indicated by footnote [b] in Table 1. Alternative techniques at these levels are the HGAAS (4) and HG–ETAAS (4,18,24) for As, Bi, Ge, Sb, Se, Sn, and Te. The generation of the lead hydride, plumbane, is not straightforward, however, (4). ICP–MS is also very attractive because of its simultaneous multielement capabilities. No generalization is possible for the selection between HGAAS and ETAAS for the hydride-forming elements. The choice would be guided by considerations such as particular matrix, analyte levels, availability, experience, etc. Thus, biological fluids, waters with low total dissolved solids (TDS) contents, and slurries are preferably analyzed by direct ETAAS. The HGAAS technique may be preferred for digests–leachates from soil, plant, and solid wastes.

CVAAS is the technique-of-choice for mercury, with LODs down to 0.02–0.2 and 0.001–0.01 μg/L in the direct and amalgamation mode, respectively (2,5–7,24). Determinations of Hg by ETAAS are prone to serious problems caused by the extreme volatility of this analyte (2,5–7,13); nevertheless some interesting ETAAS applications have been documented, particularly with solid and slurried samples (2,6,7,9,12–16).

Switching between AAS techniques is often practiced in small laboratories. Wherever affordable, at least one dedicated instrument for each of three techniques would be a good solution: one modern, high performance ETAAS spectrometer and two more simple instruments for FAAS and vapor generation (mercury/hydride) techniques. Depending on the sample workload, an ICP–AES instrument could be a good alternative to FAAS. It is a common that the analytical strategy often depends on technical and economic considerations rather than on the actual metrological and methodological requirements. Thus, each new analytical task can be solved with an approach that is either more instrumental or more chemical (15). The first relies on modern versatile apparatus wherein direct analysis of intact, simply diluted, or slurried samples is expedited, whereas sample treatment is performed *in-situ* and on-line by means of elaborated temperature programs, alternate gases for facilitating ashing, etc. Such an approach is preferred in busy laboratories with high sample turnover, if laboratory is equipped with versatile, expensive ETAAS apparatus, and with ultratrace analyte levels where exogenous contamination becomes drastic. Thus, sample-pretreatment errors such as contamination and losses are confined or eliminated, reagent consumption and personnel expenses are reduced, and a high degree of automation is provided. However, the instrumental time with certain intact samples may happen to be longer and also a more complicated calibration may be involved. On the other hand, the philosophy of the more chemical approach is viable enough. The use of simple and reliable procedures for preinstrumental treatment of solid samples in a batch mode such as acid leaching, pressurized microwave decompositions, and complete digestions is well known and documented. The use of more chemistry and even preconcentration is indispensable in some cases such as inadequate instrumental LODs and/or very difficult matrices.

ETAAS INSTRUMENTATION

The common atomizer material in ETAAS is graphite and the recent International Union of Pure and Applied Chemistry (IUPAC) recommendations refer to a graphite atomizer (8) rather than to the older, yet very popular terms, graphite furnace (GF) and GFAAS. The terms electrothermal atomizer and graphite atomizer are used practically as synonyms, since materials other than graphite such as glassy carbon tube, tungsten strip, tantalum ribbon, and molybdenum microtube, have found very limited application in ETAAS. The actual physical form of most atomizers is a graphite tube (GT); hence terms such as graphite-tube atomizer (GTA) (Fig. 2), heated-graphite atomizer (HGA) and transversely heated graphite atomizer (THGA) (28) are popular names for some commercial electrothermal atomizers. Other forms of atomizers are much less popular: cup, boat, and probe (eg, for solid sampling), rod, filament, braid, strip, loop, and wire. The ordinary graphite material (polycrystalline electrographite, EG) is rather reactive, porous, and permeable. Therefore, graphite components are often coated with a thin layer of ca 50 μm of pyrolytic graphite. Pyrolytic graphite coatings (pyrocoatings) are produced by thermal decomposition of hydrocarbons at temperature above 2000°C. Graphite components made of total pyrolytic graphite are more expensive. Those coated with refractory carbides (see the Section, Carbide Coatings) are not available commercially.

The graphite atomizer (Fig. 2) is isolated from the atmosphere by means of protective gas (argon), usually at atmospheric pressure, and is heated to different pre-set temperatures by the passage of electrical current through its body. The protective gas is typically divided into two flows: outer flow (usually kept constant) and inner flow through the observation volume of the atomizer, entering from the tube ends and leaving through the introduction hole. The inner argon flow may be stopped during the atomization stage, ie, stopped-flow operation, thus, reducing the forced removal of analyte atoms and increasing sensitivity for volatile elements. However, background absorption is also increased under stop-flow conditions. Argon can be alternated by another gas or gas mixture, *alternate gas,* during a pre-set stage of the temperature program.

During the atomization stage, the analyte is atomized and a transient peak signal, absorbance profile, is produced as a

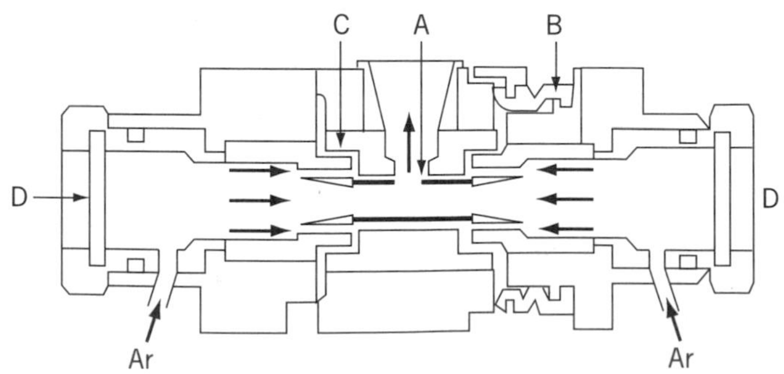

Figure 2. Workhead assembly of the Graphite Tube Atomizer GTA-100 with, A, a graphite tube; B, graphite contacts; C, graphite shroud; D, argon protective gas inlets and quartz windows. Argon, Ar, flows indicated by arrows. Reproduced by permission of Varian Australia.

result of the combined effect of the atom supply and removal processes–functions (Fig. 3). Some fundamental analytical characteristics of the AA peak are shown in Figure 3: peak height absorbance, A_p; integrated absorbance, peak area, Q_A; appearance time, τ_{app}, which can be related to the appearance temperature, T_{app}, for the analyte; and peak time, τ_p. Although both Q_A and A_p can be used for calibration (Fig. 4), the integrated absorbance measurements are strongly preferred and recommended in almost all analytical cases (see the Section, Calibration). In Figure 4 are also defined the characteristic mass for integrated absorption, m_o, and characteristic mass for peak absorption, m_p, ie, the mass of analyte that produces a signal equal to $Q_A = 0.0044$ s and $A_p = 0.0044$, respectively (8). These characteristic masses can be easily converted to their corresponding characteristic concentrations, C_o, if the injected sample volume and concentration are known. Representative data for the ranges of characteristic masses and LODs for 64 chemical elements are compiled in Table 1. It is not surprising that these values vary broadly between different electrothermal atomizers–manufacturers, thus reflecting the effects of various fundamental and instrumental parameters on characteristic mass (1): eg, atomizer size, sample path length, and efficiency of atomization. For a given instrument, however, the m_o values exhibit remarkable stability.

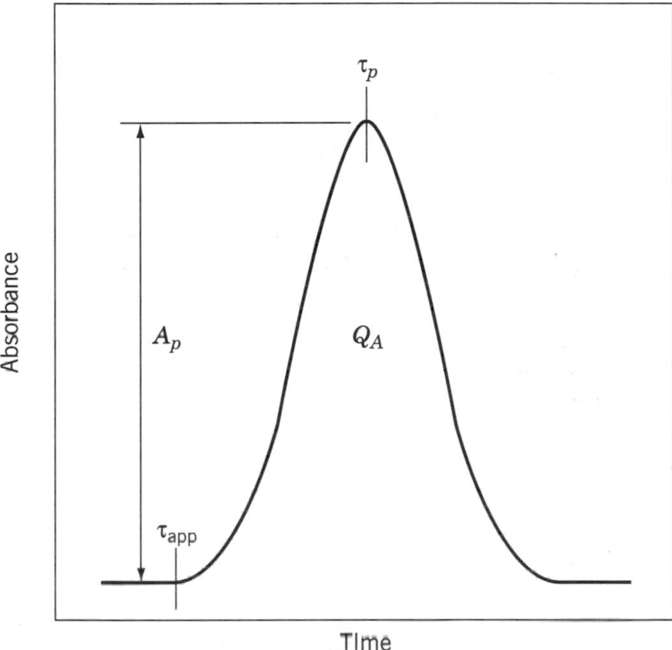

Figure 3. Atomization peak signal with indicated appearance, τ_{app}, and peak times, τ_p, peak height absorbance, A_p, and integrated absorbance, peak area, Q_A.

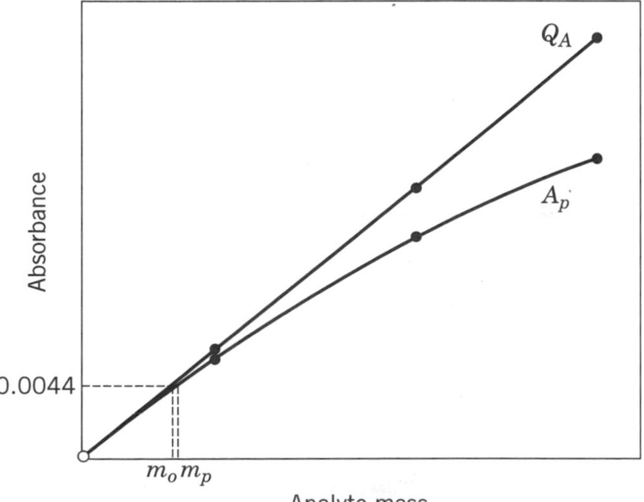

Figure 4. Calibration graph: peak height absorbance, A_p, or integrated absorbance, Q_A, versus analyte mass–concentration. Characteristic mass for Q_A measurements, m_o, and for A_p measurements, m_p, shown on the abscissa.

ETAAS METHODOLOGY

Environmental analysis is a rather complex application field featured by (*1*) a large variability of sample matrices and their composition; (*2*) low analyte levels, prone to contamination, and losses; (*3*) broad and nonpredictable ranges of analyte concentrations; (*4*) various forms (species) of certain analytes (oxidation and binding states, etc) which eventually may exhibit drastic differences in their behavior during sample preparation and instrumental quantitation; (*5*) sporadic presence of unexpected concomitants as potential interferents; and (*6*) critical effect of sample pretreatment stage(s) on sample throughput rates and accuracy of results (analyte losses and contamination). Complexity of problems in ETAAS analyses generally increases in the following order: water (low TDS) < water (high TDS) < biological fluids < air < fly ash < sludge–sediment < soil < plant tissues < biological tissues.

The scope of ETAAS and availability of analytical procedures for various environmental materials are summarized in Table 2. Three groups of sample pretreatment approaches are distinguished: (*1*) direct analysis of liquids, digests, leachates, extracts, etc; (*2*) slurry or solid sampling; and (*3*) procedures with preconcentration for enrichment or separation from the matrix. For protocols of analytical procedures, see the recent monographs (2,4,5–7,9–11), reviews, and original publications. Approved United States EPA methods (32) are available for the ETAAS determination for the following (Method number given in parens): Ag (7761), As (7060A), Ba (7081), Be (7091), Cd (7131A), Cr (7191), Co (7201), Cu (7211), Fe (7381), Mn (7461), Mo (7481), Pb (7421), Sb (7041), Se (7740), Tl (7841), V (7911), and Zn (7951).

Important methodological considerations in ETAAS are as follows.

- Selection of an appropriate *sample preparation* method (see the Section, Rational Sample Pretreatment).
- *Interference control* (see the Section, Interference Control).
- *Optimization* of instrumental and chemical parameters for achieving the required LODs, acceptable accuracy, precision, and speed of analysis (see the Section, Optimization of Instrumental Parameters).
- Selection of proper *calibration* mode and preparation of adequate standards (see the Section, Calibration).
- Data processing, quality control, and documentation.

Table 2. Availability and Classification of the ETAAS Procedures for Analysis of Environmental and Related Materials[a]

Sample Type	Element and Analytical Approach[b] Liquid, Digest or Extract	Slurry or Solid Sampling	Preconcentration
Air (particulate matter)	Ag, **Al, As,** Ba, **Be, Cd,** Co, **Cr, Cu,** Fe, Ga, **Mn,** *Mo,* **Ni, Pb, Sb, Se,** *Sn,* Te, **Tl,** V, **Zn**	Ag, As, Ba, Be, Bi, **Cd,** Co, **Cr, Cu, Fe, Hg,** Li, **Mn, Ni, Pb,** Se, Si, Sn, V, Zn	As, Cr, Hg, Sb, Se, Te, Tl
Air (vapors or gases)			**As, Hg,** Mn, Ni, **Pb, Sb,** Se, Te
Biological fluids	*Ag,* **Al, As,** *Au, Be, Bi,* **Cd,** *Co,* **Cr, Cu, Fe,** Hg, *Li,* **Mn,** *Mo,* **Ni, Pb,** *Pt,* Rb, **Sb, Se, Si,** Sr, *Te, Ti,* **Tl, V, Zn**		Ag, As, Au, **Be, Bi, Cd, Co, Cr, Cu,** Gd, Ge, **Hg,** Mn, Mo, **Ni, Pb,** Pt, **Sb, Se, Sn,** Te, **Ti, V**
Biological tissues	*Ag,* **Al, As,** *Au,* Ba, *Be, Bi,* **Cd, Co, Cr,** Cs, **Cu, Fe, Hg,** *Li,* **Mn, Ni,** P, **Pb,** *Pr, Pt,* **Rb,** *Ru, Sb,* **Se, Si,** *Sn,* **Sr,** *Tb, Te, Ti, Tl, V,* **Zn**	*Ag,* **Al, As,** Ca, **Cd,** Co, Cr, Cs, **Cu, Fe, Hg, Mn, Ni, Pb,** *Pt,* Rb, *Sb,* **Se,** *Tl,* **Zn**	**As,** Au, **Be,** Cd, **Co,** Cu, Gd, Ge, Mo, **Ni,** Pb, Pd, Pt, Rh, **Se, Sn,** Te, **Tl,** V
Fly ash[c]	Ag, **Al,** As, **Be, Cd,** Co, **Cr, Cu** Fe, **Ga, Ge, Hg, Mn, Ni, Pb,** *Sb,* **Sc,** Se, *Sn,* Ti, **Tl, V,** Zn	As, **Cd,** Cr, Cu, Ga, Ni, **Pb,** *Se,* Tl, V	Er, In, Pb, Tl
Food, feed, including beverages	*Ag,* **Al, As, Cd,** Co, **Cr, Cu, Fe, Hg,** *La,* **Mn,** *Mo,* **Ni,** P, **Pb,** Rb, Se, Si, **Sn,** *Te, V,* Zn	*Ag,* Al, As, **Cd, Cr,** Cs, **Cu, Fe, Hg, Mn,** *Mo,* **Ni, Pb,** Rb, **Se,** *Ti, Tl,* **Zn**	**As,** Be, **Cd, Co,** Cr, **Cu, Hg, Mo, Ni, Pb,** Pt, Rh, **Sb, Se, Sn,** Tl, **V**
Food/tissues (marine)	*Ag,* Al, **As,** Ba, **Cd,** Co, Cr, Cs, **Cu, Fe, Hg, Mn, Pb,** Rb, **Se,** *Sn,* Sr, *Ti,* Zn	Al, **As, Cd,** Co, Cr, **Cu, Fe, Hg, Mn, Ni, Pb, Se,** *V,* Zn	**As,** Ba, Cd, Cu, Ni, Pb, Pt, **Sb, Se, Sn,** Tl
Plant tissues	*Ag,* **Al,** *B,* Ba, **Cd,** Co, **Cr,** Cs, **Cu,** Fe, *Ge, Hg, La,* Li, **Mn,** *Mo,* **Ni, P, Pb,** Rb, *Sb, Se, Sn,* Sr, *Te, Tl, V,* Zn	*Ag,* **Al,** *B,* **Cd,** Co, **Cr, Cu, Fe, Hg, Mn,** Ni, **Pb,** Rb, *Ti, Tl,* **Zn**	Ag, **As, Au, Be,** Bi, **Cd, Co,** Cr, **Cu,** Er, Ga, Ge, **Hg,** Ir, **Mo, Ni, Pb,** Pd, Pt, Rh, **Sb, Se,** Sn, Ti, Tl, V
Sediments, sludges	**As,** Ba, *Be, Bi,* **Cd, Co, Cr, Cu,** Fe, **Hg, Mn,** *Mo,* **Ni, Pb,** Se, **Sn,** Ti, *Tl,* V	Ag, **Al, As,** *Bi,* Ca, **Cd, Co, Cr, Cu,** Fe, **Hg,** Mg, **Mn,** *Mo,* **Ni, Pb,** Sb, Si, *Sn, Te, Tl,* Zn	**As,** Bi, Co, Cu, Hg, In, Ir, Ni, Pt, **Sb, Se, Sn,** Te, **Tl**
Soil	Ag, Al, **As,** Ba, *Be,* **Cd, Co, Cr,** Cs, **Cu,** *Hg,* Li, Mn, *Mo,* **Ni, Pb,** *Sb, Se,* Sn, Sr, Ti, *Tl,* **V, Zn**	Ag, **As, Cd,** Co, **Cr, Cu,** *Hg,* **Ni, Pb,** *Tl,*	**As, Be,** Bi, **Cd, Co, Cu, Hg,** In, **Mo, Ni, Pb, Sb, Se,** Sn, Tl
Miscellaneous samples from the lithosphere[d]	**Ag,** As, *Au,* Ba, *Be,* **Cd, Co, Cr,** Cs, **Cu,** *Ga, Hg,* **Li,** Mn, **Ni,** P, **Pb,** Rb, *Se, Sn,* Sr, Ti, *Tl,* V, Zn	**Ag,** As, *Au,* Be, **Cd,** Co, **Cr,** Cs, **Cu,** *Ga, Ge,* **Hg, Mn, Ni, Pb,** Rb, *Se, Sn, Tl,* V, **Zn,**	**Ag, As, Au, Be, Bi, Cd, Co, Cu,** Er, Ga, **Ge, Hg,** In, Ir, **Mo, Ni, Pb, Pd, Pt,** Rh, **Sb,** Sc, **Se, Sn, Te, Tl,** Y
Water (with low TDS)[e]	Ag, **Al, As,** *B,* Ba, *Be, Bi,* **Cd, Co, Cr,** Cs, **Cu, Fe,** *Hg,* Li, **Mn,** *Mo,* **Ni,** P, **Pb,** Rb, *Sb, Se,* **Si,** *Sn,* Sr, *Te,* Ti, Tl, V, **Zn**		Ag, Al, **As, Au,** Ba, **Be, Bi, Cd, Co, Cr, Cu,** Ga, **Ge, Hg,** In, Ir, **Mn, Mo, Ni,** P, **Pb,** Pt, Rh, **Sb, Se, Sn,** Tc, **Te, Tl,** U, **V,** Yb, **Zn**
Water (with high TDS)[f]	*Ag,* Al, *As,* Ba, **Cd, Cr,** Cs, **Cu,** Li, **Mn,** Mo, Ni, **Pb,** Rb, *Se,* Sr, V, **Zn**		**Ag,** Al, **As, Au,** B, Ba, **Bi, Cd, Co, Cr, Cu,** Fe, Ga, Ge, **Hg,** In, Ir, Li, **Mn, Mo, Ni,** P, **Pb,** Pd, Pt, **Sb, Se, Sn, Te, Tl,** U, V, Yb, **Zn**

[a] For references and details, see recent books, reviews, and bibliographies as listed in General References.
[b] Symbols in bold used for routine, straightforward applications; symbols in italics denote lack of sensitivity at normal analyte concentrations yet good performance at elevated levels of the indicated analyte.
[c] Coal or oil fly ash, incinerator ash, etc.
[d] Examples are coal, rocks, silicates, etc, except for ores and minerals.
[e] Samples with low total dissolved solids (TDS) such as drinking and surface waters.
[f] Samples with high TDS such as marine and estuarine waters, brines, etc.

The role of instrumental parameters cannot be overemphasized. Some twenty variables and decisions about alternative hardware and/or software options are to be made by the operator, currently helped by an interactive software, with these primary aims of: (*1*) minimizing interferences, and particularly, keeping background at correctable levels; (*2*) achieving good sensitivity, the m_o figure claimed by the manufacturer is the target value; and (*3*) best signal-to-noise (S/N) ratios, which are prerequisite for low LODs and good reproducibility. Several more secondary considerations can be noted; (*4*) facilitating calibration, ideally against simple aqueous standard solutions; (*5*) ensuring broader dynamic range; (*6*) duration of temperature program that affects the sample throughput rates and costs; and (7) lifetime of atomizer. The recommendations in manufacturers' manuals, "cookbooks", and standard conditions and default options, as well as previous experience, are a good starting point. Some general guidelines are given below and more information can be found in original protocols and specialized monographs (2,5–7,9,11).

CONCLUSIONS

ETAAS is a modern and well-established analytical technique for the determination of over thirty chemical elements in large variety of environmental materials. Its main advantages are low limits of detection, good selectivity, small sample-size requirements, simple preinstrumental treatment, and automation. Confronting the challenge by more expensive multielement techniques such as inductively coupled plasma–optical emission spectrometry (ICP–OES) and inductively coupled plasma–mass spectrometry (ICP–MS), the ETAAS has achieved several important technical and methodological developments in recent years. These developments include automated sequential and simultaneous analysis, thus, increasing the elemental throughput rates by a factor of 4–6; automated slurry and solid sampling; automated flow-injection preconcentration down to ng/L levels; *in situ* treatment of samples by means of more universal chemical modifiers, permanent modification, and alternate gases; more reliable interference control by approaching stabilized temperature atomization conditions in platform-equipped atomizers with Zeeman background correction; introduction of fast temperature programs (40–50 s); simplified calibration; and compliance with the stringent requirements for quality control/quality assurance (QC/QA) and good (automated) laboratory practice.

Anticipated future progress could result in improved coupling with other techniques for on-line speciation analysis; better understanding of mechanisms controlling interferences, thermal stabilization and atomization; use of more chemometric approaches for facilitating optimization, method development, diagnostics, intelligent autosampling and calibration; and interactive on-line guidance to the vast literature information and alternative solutions.

BIBLIOGRAPHY

1. B.V. L'vov, V.G. Nikolaev, E.A. Norman, L.K. Polzik, and M. Mojica, *Spectrochim. Acta Part B* **41,** 1043–1053 (1986).
2. B. Welz, *Atomic Absorption Spectrometry,* 2nd ed., VCH Verlagsgesellschaft mbH, Weinheim, Germany, 1985, pp. 1–506; B. Welz and M. Sperling, *Atomic Absorption Spectrometry,* 3rd, completely revised edition, 1997, in press.
3. M.S. Cresser, *Flame Spectrometry in Environmental Chemical Analysis: A Practical Guide,* Royal Society of Chemistry, Cambridge, UK, 1994, pp. 1–108.
4. J. Dědina and D.L. Tsalev, *Hydride Generation Atomic Absorption Spectrometry,* John Wiley & Sons, Chichester, UK, 1995, pp. 1–526.
5. D.L. Tsalev and Z.K. Zaprianov, *Atomic Absorption Spectrometry in Occupational and Environmental Health Practice,* vol. I, *Analytical Aspects and Health Significance,* CRC Press, Boca Raton, Fla., 1983, pp. 1–252.
6. D.L. Tsalev, *Atomic Absorption Spectrometry in Occupational and Environmental Health Practice,* vol. II, *Determination of Individual Elements,* CRC Press, Boca Raton, Fla., 1984, pp. 1–295.
7. D.L. Tsalev, *Atomic Absorption Spectrometry in Occupational and Environmental Health Practice,* vol. III, *Progress in Analytical Methodology,* CRC Press, Boca Raton, Fla., 1995, pp. 1–349.
8. T.A.M. Ure, L.R.P. Butler, B.V. L'vov, I. Rubeska, and R. Sturgeon, *Pure Appl. Chem.* **64,** 253–259 (1992).
9. W. Slavin, *Graphite Furnace AAS. A Source Book,* Perkin-Elmer Corp., Ridgefield, Conn., 1984, pp. 1–229.
10. A. Varma, *Handbook of Furnace AA Spectroscopy,* CRC Press, Boca Raton, Fla., 1990, pp. 1–428.
11. C. Minoia and S. Caroli, eds., *Applications of Zeeman GFAAS in the Chemical Laboratory and in Toxicology,* Pergamon Press, Oxford, UK, 1992, pp. 1–675.
12. Z.-m. Ni and X.-q. Shan, *Spectrochim. Acta, Part B* **42,** 937–949 (1987).
13. D.L. Tsalev, V.I. Slaveykova, and P.B. Mandjukov, *Spectrochim. Acta Rev.* **13,** 225–274 (1990).
14. D.L. Tsalev, *At. Spectrosc.* **12,** 169–198 (1991).
15. D.L. Tsalev, *J. Anal. At. Spectrom.* **9,** 405–414 (1994).
16. C. Bendicho and M.T.C. de Loos-Vollebregt, *J. Anal. At. Spectrom.* **6,** 353–374 (1991).
17. K.S. Subramanian, *Spectrochim. Acta, Part B* **51,** 291–319 (1996).
18. H. Matusiewicz and R.E. Sturgeon, *Spectrochim. Acta, Part B* **51,** 377–397 (1996).
19. R.M. Harrison and S. Rapsomanikis, eds., *Environmental Analysis Using Chromatography Interfaced with Atomic Spectroscopy,* Ellis Horwood Ltd., Chichester, UK, 1989, pp. 1–370.
20. G.E. Batley, ed., *Trace Element Speciation: Analytical Methods and Problems,* CRC Press, Boca Raton, Fla., 1989, pp. 1–350.
21. A.M. Ure and C.M. Davidson, eds., *Chemical Speciation in the Environment,* Blackie, London, 1995, pp. 1–408.
22. S. Caroli, ed., *Element Speciation in Bioinorganic Chemistry,* John Wiley & Sons, Chichester, UK, 1996, pp. 1–474.
23. J.L. Burguera, ed., *Flow Injection Atomic Spectroscopy,* Marcel Dekker, New York, 1989, pp. 1–359.
24. *The FIAS-Furnace Technique: Recommended Analytical Conditions and General Information,* Publication B3212.20, Part No. 0993-5204, Bodenseewerk Perkin-Elmer GmbH, Überlingen, Germany, 1993.
25. Z.-l. Fang, *Flow Injection Separation and Preconcentration,* VCH Verlagsgesellschaft mbH, Weinheim, Germany, 1993, pp. 1–259.
26. Z.-l. Fang, *Flow Injection Atomic Absorption Spectrometry,* John Wiley & Sons, Chichester, UK, 1995, pp. 1–306.
27. Z.-l. Fang, S.-k. Xu, and G.-h. Tao, *J. Anal. At. Spectrom.* **11,** 1–24 (1996).
28. *The THGA Graphite Furnace: Techniques and Recommended Conditions,* Publication B3210.20, Part No. B050-5538, Bodenseewerk Perkin-Elmer GmbH, Überlingen, Germany, 1992.
29. G. Tyler, "AA or ICP - Which do you choose?", *ICP Instruments at Work,* ICP-3, Varian, Sunnyvale, Calif., Sept. 1991, pp. 1–6.
30. B. Radziuk, G. Rödel, H. Stenz, H. Stenz, H. Becker-Ross, and S. Florek, *J. Anal. At. Spectrom.* **10,** 127–136 (1995).
31. D.L. Tsalev, E.I. Tserovski, A.I. Raitcheva, A.I. Barzev, R.G. Georgieva, and Z.K. Zaprianov, *Spectrosc. Lett.* **26,** 331–346 (1993).
32. A.D. Turner and D.J. Roberts, *J. Anal. At. Spectrom.* **11,** 231–234 (1996).
33. I.I. Petrov, D.L. Tsalev, and A.I. Barzev, *At. Spectrosc.* **1,** 47–51 (1980).

34. R.A. Nickson, S.J. Hill, and P.J. Worsfold, *Anal. Proc.* **32,** 387–395 (1995).
35. United States Environmental Protection Agency, *Test Methods for Evaluating Solid Waste,* Volume IA of *Laboratory Manual Physical/Chemical Methods, (SW-846),* 3rd ed., Office of Solid Waste and Emergency Response, Washington, D.C., Nov. 1986; Update IIB, Jan. 1995.

General References

References 2–7, 9–11, 19–26 (books), 12,13,15–18,27,34 (reviews), 10,14 (bibliographies) and 35 (official texts) are good general references for a fast start-up.

Manufacturers application information pertinent to their particular instruments.

Annual reviews (Atomic Spectrometry Updates, ASU) on "Environmental Analysis" (Feb. issue), "Clinical and Biological Materials, Foods and Beverages" (April issue), and "Advances in Atomic Absorption and Fluorescence Spectrometry and Related Techniques" (Aug. issue) are published in *Journal of Analytical Atomic Spectrometry (JAAS).*

Biannual Application Reviews on "Air Pollution", "Environmental Analysis", "Water Analysis", etc, are published in June, odd years, by *Analytical Chemistry.*

A Fundamental Review on "Atomic Absorption, Atomic Emission and Flame Emission Spectrometry" appears in June, even years, in *Analytical Chemistry.*

Bibliography on "Atomic Spectroscopy," *Atomic Spectroscopy* (1) (Jan./Feb.) and (4) (July/Aug.) (bibliography terminated since Feb. 1996).

Databases on analytical chemistry and applied spectroscopy are available on diskettes, compact discs and on-line: *Chemical Abstracts, Analytical Abstracts, Current Contents, JAAS Base.*

ELEMENTAL POLLUTION IN AIR

DAVID E. KIMBROUGH
Castaic Lake Water Agency
Santa Clarita, California

Some air pollutants derive their hazardous characteristics from the synergistic effects of the different elements that make up their molecular structure. N_2O and NO_2 have the same elemental composition but have very different environmental impacts. The hazards of others are largely derived from their elemental composition, with the molecular structure either attenuating or augmenting the hazard or having no effect at all. Lead, mercury, cadmium, beryllium, and arsenic are examples of elemental air pollutants. These are often referred to as "metals" or "heavy metals" even though they may not be transition metals (eg, arsenic is metalloid) in a narrow sense, or in a metallic form (eg, lead oxides are powders) in the broader sense. Elemental air pollutants are of great public health concern because of their highly toxic nature, their readily assimiable form, and the broad exposure that is possible.

INDUSTRIAL PROCESSES

Elemental pollutants share a number of characteristics: they have similar industrial use streams; all elemental pollutants are mined from the earth and purified by thermal or electrochemical means; the purified material, often a metal in either sense of the word, is then sold to various industries and reprocessed by either thermal, electrochemical, precipitative, or abrasive processes; and the final product can often be recycled when the price of the material makes it worthwhile, eg, gold, silver, platinum, palladium, aluminum, copper, mercury, lead, antimony, etc, repeating many, if not all, of these processes.

Whereas most of these materials can exist in one part of their use-stream as a purified metallic form, at most points they exist as mixtures with other elements, as ores, alloys, compounds, components, etc. For example, chromium ore, chromite, is a mixture of chromium, iron, and oxygen. Chromium is alloyed with iron, precipitated with zinc, plated onto nickel plating, and oxidized with lime (1). Lead is often mined in conjunction with antimony, silver, gold, and copper (2). Lead is used in the majority of its applications alloyed with antimony, arsenic, cadmium, silver, selenium, and zinc, to name a few.

During the industrial life-cycle of any elemental pollutant, the ratio of itself to other elements changes. The industrial use-stream of lead for lead–acid batteries (over 25% of all lead is used for these batteries) is a good example (3). Lead ore from a particular mine can have characteristic quantities of silver, cadmium, antimony, and gold. When the ore is milled, the elemental ratios remain the same, but when the milled ore is smelted, the ratios change. Lower boiling point elements volatilize, other elements being less dense than lead raise to form a lead-poor dross, and molten elemental lead settles to the bottom of the smelter. The lead can be oxidized to produce lead oxide powders (Pb_2O, PbO, and PbO_2). These oxides are then mixed with elemental lead which may be alloyed with antimony, arsenic, cadmium, and other elements (cadmium may be an impurity or added intentionally). When lead–acid batteries are recycled, they are ground (an abrasive process), resmelted, drossed, alloyed, and cast in ingot of various alloys of characteristic elemental ratios.

For any elemental material to be an air pollutant, it must be capable of becoming airborne. Most elemental air pollutants have very high vapor pressures and are not gases under normal conditions; therefore they usually enter the atmosphere as particles and aerosols. There are, however, some important exceptions, mainly in the hydride and alkyl forms. There are four main industrial processes that produce airborne elemental pollutants: abrasion, bubbling, precipitation, and volatilization. Each process generates a different type of particle and the dispersion of the elemental air pollutants is controlled by the aerodynamic properties of the particles of which they are a part.

Abrasive processes are those where a larger body of material is broken down into smaller bodies by mechanical means, such as milling, grinding, or polishing. For nonfriable materials such as metals and ores, the particles produced by abrasion tend to be very large ($>10\ \mu m$ diameter) in mean mass aerodynamic diameter (MMAD) and are of environmental interest. (MMAD is a statistical measure of the apparent behavior of particles, irrespective of their shape) (4). There are inherent physical limitations on abrasive processes that prevent large numbers of particles smaller than 10 μm from forming. Friable materials, on the other hand, are often composed of very fine particles which may be held together by some binding agent. When such a material is ground or milled, the fine particles can then be released. These fine particles can be made airborne by the abrasive process itself or, if left exposed, become suspended by winds. Because this process does not alter the elemental composition of the material, the elemental ratio present in the parent material will be the same in the abraded products.

A common industrial process is thermal treatment, eg, smelting, melting, roasting, welding, soldering, brazing, etc. These processes may be used to chemically reduce oxidized materials (eg, smelting lead ores to get elemental lead), oxidize reduced materials (air roasting of chromite to get chromates), add or remove alloys (alloying or drossing), or physically joint two pieces of metal. However, in all of these processes the materials are heated sufficiently to volatilize some portion of the original material. The particles formed by these processes are extremely small (0.001–10 μm). Thermal processes alter the elemental composition of the parent material. In smelting, different elements volatilize at different temperatures, so the elemental content of the air is different from the heated materials. Similarly, oxidation or reduction of a given element gives rise to materials of volatilities different than the untreated material and hence ratios of elements found in air and parent material are different. The elemental composition of the volatilized particles, the final product, and any by-products (such as dross) will all have different elemental compositions from each other and from the parent material.

There are a few species of elemental pollutants that form stable room temperature gases. Elemental mercury, alkylated mercury, alkylated lead, alkylated tin, and the hydrides of arsenic (AsH_2), selenium (SeH_2), lead (PbH_2), tin (SnH_2), mercury (HgH_2), boron (BH_4), and silicon (SiH_4), are examples of gaseous elemental pollutants. All of these species are extremely toxic, both because they are gases and because they are far less polar than most elemental pollutants and can thus pass through cellular membranes more rapidly. Most of these materials are either no longer used (such as alkylated lead) or only used in limited applications. However, it is possible to accidentally form the hydride species when the toxic element is mixed with a strong reducing agent, such a mercury and stannous chloride or arsenic and sodium borohydrate.

Precipitation is another industrial process that can generate respirible particles. Chemicals are solubilized, then combined with other components to form an insoluble compound. The resulting precipitate, when the solute is removed, can become airborne when exposed to shear winds. Many toxic elements are also used in plating. Here the elements are solubilized in an acidic water solution, then an electric current is applied drawing the ions toward one of the electrodes, the material to be plated. At each electrode, gases, eg, hydrogen or oxygen, escape the bath leading to the formation of aqueous acidic chromate mists. The droplets formed during electroplating have an MMAD of around 100 μm (5).

FATE AND TRANSPORT

Once airborne, gaseous and particulate toxic elements are subject to a variety of physical and chemical forces. The smallest particles act like gases, but unlike gases they are not stable at room temperature. Because they are created by very high temperatures, once they begin to cool they rapidly begin to interact with other particles. They may attach themselves to larger particles (scavenge) or coalesce with other particles the same size (coagulate) to form larger particles. Irrespective of the initial size distribution of particles of material, the final distribution after coagulation is always they same. Most of the mass of an aerosol ends up in the 0.2–10 μm range, although the number (count) of particles of smaller size is a great deal larger. This process can also occur with chemically unstable gases and is called the gas-to-particle conversion. Particles of greater than 10 μm MMAD do not generally coagulate and are not broken down into smaller particles. Conversely, particles smaller than 10 μm MMAD rarely coagulate into larger particles. Thus, for most practical purposes, there are two populations of particles, those larger than and those less than 10 μm (this is generally referred to as particulate matter less than 10 μm or PM_{10}) (4).

Generally, only PM_{10} is of environmental interest. One reason for this is that larger particles that behave like traditional Newtonian projectiles are not carried far from their source. PM_{10} acts less like particles and more like gases and can thus be carried much farther from the source. Another reason is that particles larger than 10 μm MMAD are not respirible. When they are inhaled they are trapped in the nose or on the mucus of the upper airways and either expelled or ingested. Particles that contain materials that are directly irritating to the mucus lining of the nose, such as chromium and antimony, are a concern even at larger MMADs. PM_{10} is of great human health concern as this can penetrate deep into the lungs where absorption into the blood system is most likely to occur. Although ingestion can also result in absorption, it is usually far less than absorption through the lungs. Finally, smaller particles have a much larger surface-to-volume ratio and thus the likelihood of solubilization and intermedia transfer of toxic elements is increased.

Most particles released by industrial activities possess some initial momentum (or they would not have become airborne) but rarely enough to carry them far. Most of the momentum for large-scale transport is provided by bulk motion of air. This may initially be motion with heated air leaving the industrial process, such as flue gases from a power plant or smelter. Once released into the atmosphere, bulk motion is provided by winds.

Toxic elements entrained in aerosols may be removed from the atmosphere by either gravitational fallout (dry deposition), washout, or rainout (wet deposition), which are all particle size-dependent processes. In dry deposition the particles settle and are captured by the soil or surface waters via gravitational sedimentation, impaction, and interception. The overall deposition velocity of atmospheric chromium is a function of the distribution of chromium within the particle phase and the deposition velocities of chromium-containing particles, which are in turn a function of the diameter of the individual particle (6).

Wet deposition is the process where aerosol particles are actively entrained or scavenged by atmospheric moisture, such as rain, snow, fog, or dew. Removal of atmospheric chromium by wet deposition is dependent on the size of the chromium-containing particles and the size of the scavenging droplet. In both dry and wet deposition the relationship between removal rates and particle size are all direct: the larger the particle size the faster it is removed to the atmosphere. Smaller particles have sufficiently long atmospheric half-lives that enable them to travel significant distances from their source, especially those with an MMAD of less than 10 μm. Elemental pollutants can also be introduced or re-introduced into the atmosphere via the suspension of contaminated soil particles. For particles with a diameter of <50 μm MMAD, the resuspension process is induced by both mechanical and wind disturbances which provide sufficient energy to overcome gravitational forces and allow particles to be dispersed by the wind (7).

Elemental pollutants can undergo a variety of chemical transformations after emission including reduction, oxidation, sorption, desorpton, solubilization, and precipitation which can occur in the gas phase, in liquids, or on the surface of solids. All of these reactions and media can occur on the surface of aerosols.

Oxidation/reduction chemistry of elemental pollutants is controlled by several factors: the presence and concentrations of the toxic species and oxidizing or reducing agents, the electrochemical potentials of the oxidation and reduction reactions, ambient temperature, light, sorbents, acid–base reactions, complexing agents, and precipitation reactions. Concentration of reactants is also an important factor. For example, ozone can react with chromium(III) to form chromate, a reaction which has a redox potential of 0.87 volts. However, the concentration of ozone in most environmental media is rarely high enough to accomplish this oxidation. Conversely, the reduction of chromium(VI) by iron(II) has a smaller redox potential (0.56 volts), but the concentration of iron in many media is high enough to actually achieve this reduction (8). Elemental mercury, on the other hand, is readily oxidized to Hg(I) by ozone (9).

Elemental pollutants can also undergo precipitation–dissolution reactions. Most elemental pollutants exist in many forms that have a wide range of solubilities. Solubilization reactions are governed by the solubility of the dissolved species, the kinetics of the dissolution, and the presence precipitants. For example, many common forms of silver are very water soluble, such as silver nitrate or sulfate. However, these species react rapidly with free chloride to form water-insoluble silver chloride, a common environmental component. Thus, in many environmental media, irrespective of the initial form of silver, it ends up as silver chloride. However, the water solubility of silver chloride is proportional to the free chloride content of the aqueous phase and so it is soluble in salt water. The sorption and desorption of an elemental pollutant refers to the removal of a solute from the aqueous phase of an environmental matrix to the surface of a solid. Chemical transformation is especially important to gaseous elemental forms such as tributyl tin. Although physically stable, most of these materials are not chemically stable. Most of these materials either react rapidly with water, are degraded by hydroxyl radicals (a common air pollutant), or are photochemically reactive. For example, tributyl tin is a stable room temperature gas; however, when exposed to atmospheric hydroxy radicals, it is oxidized to dibutyl, monobutyl, or a nonalkyl tin species. These species can then undergo gas to particle conversions.

SAMPLING

The principal mechanism of dispersion is wind and thus only by understanding wind patterns can the fate of air pollutants be estimated. Elemental air pollutants carried by the wind are deposited on the surfaces upon which they come in contact, for example, plants, animal, walls, and soils. Animals can inhale, ingest, and absorb these pollutants directly or indirectly through the consumption of these plants or soils. Thus any of these media can accumulate the elements of interest and be used for sampling and analysis. Once associated with these media the elements can become mobile, either by being removed from the media or if the media is itself mobile. For example, rain can wash deposits off of the surface of plants or walls, soil can be resuspended by wind or washed away by rain, elements deposited in soils can be mobilized by ground water or absorbed into the roots of plants, plants can be consumed by animals or removed by people, or animals and people can move from a given site. The trail left by airborne pollutants can extend into a variety of environmental media which may or may not be stable, thus any of these media may be used to trace the fate and transport of airborne pollutants. The wider the variety of media used, the stronger the analysis (10).

The most common type of sampling for toxic air emissions is air sampling. There are two types of air samplers for elemental air pollutants: those that collect aerosols and those that collect gases. Both types of air samplers use pumps to pull air through sampling media where the elements are trapped (a fixed volume of air can be sampled in a Tedlar bag, glass vial, or can, but this is rarely used for elemental pollutants). In both cases, the larger the volume of air sampled, the more sensitive the sampling. Gaseous elemental pollutants are usually trapped in a liquid, where they are solubilized or react with soluble reagents, and the liquid can then be analyzed. These impinger samplers can be used for personal sampling or for the sampling of stacks for flue gases. The chemistry of the target analytes and the analytical procedure determines the type of trapping liquid used. Stack gases are often very hot and therefore many elements, even those that do not form stable room temperature gases, can be hot enough that coagulation into particles does not occur (11) (see AIR SAMPLING INSTRUMENTS).

Because the majority of elemental air pollutants are associated with aerosols, this is the most common type of sampling. As with gases, air is pumped through a trap; however, in aerosol sampling the particles are trapped on a filter. Fiber filter materials include glass and quartz, plastic, and cellulose. Filters do not sieve the particles out of air by forcing particles through holes too small for particles to pass, rather the particles adhere to filter fibers after collision, interception, or diffusion depending on particle size and chemistry. Filter materials should be selected based on sampling and analytical need. Filter materials must be able to be analyzed without interfering with the analytical procedures. For example, if the filter is going to be weighed before and after sampling, then filter materials should not absorb moisture from the air or be capable of being dried. The filter material should also be efficient at collecting particles in the size of interest. Different filters have different collection efficiencies at different particle sizes. Finally, filter materials should be able to collect a target mass of sample without excessive pressure drop. Pressure drop is change in air pressure across the filter as air is being forced through it. Excessive pressure drop can bias the sampling process. Pressure can be caused by the filter material itself, if it is too resistant, or by the accumulation of sample material which clogs the filter (4).

Air samplers come in many sizes, from personal samplers, which sample a few liters per minute, to large high volume (HiVol) samples that can sample thousands of liters per minute. They can hold a single filter for a range of particle sizes or contain impactors which segregate particles according to particle size. In environmental testing the concentrations of elements are often very low, in the ng/m^3 range. Therefore, very large volumes must be sampled and large masses of particles collected. Personal samplers can be attached to mobile individuals or remain stationary for sampling air in a fixed area.

Wipes are used to sample surfaces exposed to deposition. A wipe is simply a sheet of cellulose-based material (similar to a facial tissue) that is free of contamination. A selected area of surface is measured and that area is wiped. The elemental composition of materials collected on the wipe can then be determined. The wipe material should be free of contamination and not interfere with the chemical analysis (blank wipes are important for this reason). A template can be prepared with a pre-measured area so that the same area can be wiped with each sample. Wipes can be collected from undisturbed surfaces at several distances both upwind and downwind from sites to measure historic deposition.

Soils and solids can be more difficult to collect and sample in a useful fashion than is immediately obvious. The natural tendency is to simply shovel dirt or solids into a glass jar, a technique which can be useful in some situations. However, deposited particles are mostly found on the top few centimeters of the surface of soils so sampling only this top layer will likely yield the largest concentration of target elements. Sampling deeper layers of soil and mixing with shallower dilutes the concentration. Sampling several different layers of the same soil column can show if toxic elements have been leached from the surface and are being mobilized. It is also useful to remove extraneous materials such as rocks, wood, and debris. Solids, such as industrial materials and wastes, can also be misleadingly easy to sample. Different industrial processes not only produce different types of airborne particles, they also produce by-products of different properties. A mixture of different sized solids will, if agitated, separate with the finer particles at the bottom and courser particles at the top. These different sizes may represent different products or by-products of different industrial processes with different elemental compositions. If a goal of the investigation is to correlate different industrial processes with air emissions, such differences will bias the results.

Living organisms can absorb and store toxic elements. Thus the sampling of plant and animal tissues can be used to measure long-term air emissions. However, sampling animal tissues and excrement for elemental pollutants requires a great deal of expertise, both to design and to physically collect the samples. Without a thorough knowledge of the toxicology of the particular elements, including route of exposure and metabolism, the wrong tissues may be sampled. For example, all industrial forms of antimony and chromium can cause damage to the stomach but are poorly absorbed by the gastrointestinal tract; however, both can also damage the respiratory tract and be absorbed by lung tissue, as well as passing through the lung tissue into the blood system.

Chromium is excreted rapidly from the blood into the urine, whereas antimony is excreted much more slowly. Thus, ingested chromium and antimony are most likely to be found in the feces within a short period after ingestion and do not accumulate in any tissue. Inhaled chromium and antimony accumulate in the lung tissue, and antimony, but not chromium, remains in the blood for a significant amount of time after inhalation. In contrast, lead and arsenic accumulate in the blood, bone, and hair of animals. Thus, even if elements are released and absorbed together, they will end up in different parts of the body. Additionally, these elements can be deposited on the hair of mammals as opposed to (or in addition to) being incorporated into the hair. The sampling of animal tissues also raises health and safety issues for the sampler and analyst. The sampling of human tissues and excrement is even more complex than animals as there are additional legal and ethical considerations. This type of sampling should only be undertaken with thorough preparation and the guidance of experienced investigators (12).

Plants are stationary and can have large surface areas for collection of deposited particles. The sampling of leaf surfaces is a common procedure similar to wipe sampling. Leaves can be sampled at various distances from a source, both up and downwind. Accumulation can be measured over a period of time by sampling the same plant or area repeatedly over a period of time. Elements deposited in the soil and that are water soluble can be absorbed by the roots of plants and sometimes transported into other parts of the plant. As with mammals, elemental pollutants can be deposited on leaves as well as absorbed into leaves through the roots. Thus the sampling of the actual tissues can both reveal long-term emissions and transport (13–15).

Multielement Sampling

Elemental air pollutants move through and collect in a variety of media, therefore sampling for these elements can be performed in those same media. It has been a common practice to study air emissions in a single medium, typically by sampling the air for a single element. This is, in part, an artifact of analytical technology and regulations. Until the 1970s, the most common analytical procedures could only analyze a single element so that analyzing several elements was costly and time consuming. Further, environmental and occupational regulations have been single medium-focused, so sampling and analytical methods specified in either regulations or promulgated rules have been single medium, single element in nature (10).

Researchers have often followed this lead, often for the sake data comparability. Using this approach, high volume PM_{10} air samplers (see below) are placed up and downwind of a facility (or in several directions if wind patterns are unstable). If, for example, the site is a brass foundry, then, if the copper concentration is higher downwind than up, the source and emission rates can be determined. There are many situations where this is adequate, such as where only a single toxic element is released in an area isolated from other possible sources. However, as has been noted, toxic elements are rarely processed, or released, alone. In urban industrial settings there are a variety of possible sources of air emission of a given element. For example, copper can be emitted by both a brass foundry and a copper recycling facility. If two such facilities are near one another, determining which is emitting airborne copper or apportioning emissions if both are emitting is difficult using only a single element and medium. Further, as has been noted, airborne elemental pollutants eventually end up in other media and can pose different risks to human health and the environment. All of these reasons, combined with the development of multielement analytical techniques, have encouraged the approach of multimedia and multielement analysis. In the example above, air, soil, structure walls, plant surfaces, plant tissues, ground water, and storm drains upwind and downwind of the facilities could be sampled and analyzed for a wide range of elements.

Brass is made up of copper (45–90%) and zinc and sometimes alloyed with lead, nickel, aluminum, iron, or tin. There are industry standards (eg, for Admiralty Brass, White Brass, etc) that determine the ratios of these elements in the brass. Brass is made by melting copper (bp 2567°C) and adding solid zinc (bp 907°C) in a foundry (when brass is worked into a final product, it may be abraded or brazed, but cannot be precipitated or electroplated). Because zinc is far more volatile than copper, it may be a larger component of the air emissions than would be indicated by the nature of the product, which is mostly copper. Likewise lead (bp 1740°C) and tin (bp 2270°C) may be over-represented in air emissions, whereas nickel, iron, and aluminum, which all have comparable boiling points to copper, would not be. Depending on what type of brass is being manufactured, air emissions of copper, zinc, lead, and tin could be more revealing than copper alone.

Samples collected from around these sites would also allow for different measurements of copper emissions. For example, PM_{10} and personal air samplers would give a snapshot of emissions on a particular day (this day may or may not be representative of normal wind patterns or emission rates). Wipe samples of walls and samples of plant surfaces would give a more intermediate sample of total deposition rates depending on how often it rains, when plants lose their leaves, or when other processes that remove deposition occur. Soil samples reveal a longer time frame of deposition patterns as elemental pollutants are bound more tightly to soil particles than plant and wall surfaces. However, elements can be removed from soil by leaching from rain and groundwater and absorption by plants. The rates of migration and absorption depend on the chemical form of the element and on the chemistry of the soils and waters. For example, chromium(VI) oxides are far more water soluble and phytoavailable than commonly occurring chromium(III) compounds (see CHROMIUM(VI) IN SOILS).

ANALYSIS

There are generally two parts to chemical analysis: sample preparation and instrumental analysis. Today there is a wide variety of analytical techniques available, both for sample preparation and analysis. Techniques can be grouped as either destructive or nondestructive. Destructive techniques separate the elements of interest from the sample matrix, thus preventing further analysis of that particular sample. This is usually done to accommodate the needs of the analytical instrumentation. Nondestructive techniques leave the elements of interest in the sample matrix for analysis.

There are a number of factors that must be considered when selecting analytical methods: sensitivity, limit of detection, selectivity, range of analytes, linear range, and cost. It must be emphasized that these factors apply to both the sample preparation procedures and instrumentation. The results obtained from an instrument are only as good as the sample preparation procedures allow them to be. Sensitivity is a measure of the ability of a technique to discriminate between two concentrations. One procedure may be able to distinguish 40 from 50 mg/kg, but not 45 from 46 mg/kg. A closely related concept is that of limit of detection, a measure of the smallest quantity of analyte that can be reliably identified or quantified. Generally, the more sensitive a procedure is, the lower the limit of detection. There is a wide variety of definitions and procedures for determining the limit of detection and although some have been given official recognition the subject remains highly controversial.

Selectivity is the ability of a procedure to discriminate between a signal from the target analyte and a signal from an interfering species. Although there is no procedure that is interference-free, some are more interference-prone than others. Selectivity impacts sensitivity (and hence the limit of detection) as the presence of interfering species prevents the identification and quantitation of the target element. There are a variety of interference-correcting procedures available, but there are no universal solutions. The greater the selectivity of procedure, the fewer other analytes can necessarily be analyzed. This is not always a drawback, but as has been noted, it is often useful to analyze a number of elements in the same sample. If a number of separate analytical procedures are needed, the cost of analysis will increase.

The linear range of a procedure is the range of quantities having a linear correlation between the actual quantity in the sample and the measured quantity. Each analytical procedure has upper and lower limits beyond which accurate and reproducible results cannot be obtained. The wider this range, the less additional manipulation (such as dilutions) laboratory personnel have to make to the sample, reducing the likelihood of error. Finally, few investigators have unlimited budgets, and there is generally a need for getting the most analytical product per dollar. Hence, the use of a selective and sensitive procedure is likely to limit the number of samples that can be analyzed.

As noted, it is possible to separate particles into different particle-size classes through the use of impactors. Having separated the particles according to particle size the chemical makeup of each particle-size population can be determined. This can reveal important information, as different industrial processes release particles of different sizes and chemical composition. Further, particle size is important for determining how far air emissions travel before deposition.

Physical Preparation

Samples containing wide ranges of moisture, such as soils, sediments, and animal and plant tissues, must be dried prior to analysis. This is done for a variety of reasons, including so that comparisons between samples are more realistic and not skewed by differences in moisture content. Drying can also make other sample preparation procedures easier, such as milling and sieving. Finally, some analytical procedures are sensitive to moisture content and so by removing moisture the samples are more similar to calibration standards. Some samples, such as soils, sediments, and solids, can be dried simply by placing them in drying ovens at 105°C for 24 hours. Others sample types, such as animal tissues, are often freeze-dried. Glass and quartz fiber filters are generally conditioned prior to sampling in a controlled environment of fixed relative humidity (4).

Milling solid samples is often performed to achieve a more uniform particle size. Some analytical techniques are very sensitive to particle-size effects and milling minimizes this problem. Particle-size reduction also increases the surface-to-volume ratio which is important for effective acid digestion procedures (see below). Milling also make sieving easier. Sieving is used to remove portions of solid samples that are extraneous such as rocks, wood, and debris. It also complements milling by removing larger particles in need of further milling.

Destructive Procedures

These procedures remove the elements of interest from the sample matrix, transfer them to liquid matrix, then analyze the liquid extract by instruments and compare them to other liquid standards. By far the most common approach to this is the use of hot acids to destroy chemical and physical bonds between the elements of interest and the matrix and to convert those elements into water-soluble forms, often leaving the bulk of nontarget elements, especially silicacious materials, as solids that can be filtered out.

In selecting an acid digestion procedure, two important factors must be considered: (*1*) is the acid (or acid mixture) aggressive enough to break the bonds between the analyte and the matrix? and (*2*) will the element be soluble in the acid? For example, a number of acid digestion procedures use hot nitric acid. If copper oxide is bound to the surface of a carbonaceous particle, a significant portion of the carbon on the surface of the particle is mineralized to CO_2 when exposed to hot nitric acid, thus freeing the copper atoms. The copper is converted to copper nitrate which is soluble in nitric acid and water. In contrast, consider antimony trioxide in a blood sample. Nitric acid is not strong enough to break up or mineralize the fatty and oily materials in blood, therefore the acid cannot react with the antimony. Further, even if the nitric acid is strong enough to break up the blood, antimony reacts with nitric acid to form Sb_2O_5, which is very insoluble in either nitric acid or water (16).

For glass and quartz fiber filters, this process means separating toxic elements (*1*) attached directly to the fibers and/or (*2*) from particulate matter (significant quantities of silicon can be solubilized during this procedure). Cellulose acetate filters and cellulose paper wipes can be completely or partially combusted (converted to CO_2) leaving only particulate matter. Soils and sediments generally consist of a silica–aluminum oxide core surrounded by a layer of organic matter. Anthropogenic toxic elements are generally found in this outer layer which is as highly susceptible to acid digestion, as the core is not. The sample preparation of soils and sediments is complicated by the presence of extraneous materials, such as rocks, wood, and debris, which are less likely to contain deposited particles. These can be removed by drying and sieving the sample through a U.S. Standard #10 sieve (although moisture content should be determined as part of the drying process).

Plant tissues are amenable to acid leaching and generally dried prior to acid treatment, but as with soils and sediments, the moisture content should be determined. The hair, nails, and urine of animals are also quite amenable to acid treatment. All of these materials are also generally free of major interfering elements.

Instrumentation

One of the most commonly used instruments is inductively coupled plasma (ICP) (see INDUCTIVELY COUPLED PLASMA). Liquid samples are made into a fine aerosol by a nebulizer which are injected into an argon gas that has been heated to 10,000 K, ie, has been ionized (ie, the outer-most electron has been ejected from the atom) into a plasma by radio-frequency (r-f) energy. As the elements dissolved in the aerosols enter the plasma they are vaporized, atomized, and then ionized as they absorb the energy in the plasma. At this point there are two techniques for measuring the number of atoms of given element in the plasma: mass spectroscopy (ICP/MS) and atomic (or optical) emission spectroscopy (ICP/AES or ICP/OES) (17).

ICP/MS samples the plasma, takes a portion into a vacuum, and accelerates the ions along a flight path via a high voltage differential. A magnetic field is then applied to separate ions from one another by difference in their mass-to-charge ratios (m/z). For example, a silicon atom has a mass of 28 atomic mass units (AMUs) which, if singly charged, would have a m/z of 28, which can be distinguished from aluminum at 27 m/z or phosphorus at 31 m/z. Interferences can be caused by double-charged atoms, such as iron which has mass of 56 AMU, but with two electrons removed has a m/z of 28. Likewise, there are different mass isotopes for each element, such as magnesium 26 which interferes with silicon 26. There are also molecular interferences caused by high temperature dimers such as argon–oxygen with a m/z of 56. These interferences are relatively few and predictable and to some extent correctable. ICP/MS is a powerful method as it has a wide linear dynamic range (six orders of magnitude), can measure 90% of elements in the Periodic Table, is extremely sensitive, and fairly selective. ICP/MS is, however, an expensive instrument and requires significant analyst skill. Further, it is limited by the fact that it can only analyze aqueous samples low in total dissolved solids (TDS) which means that samples high is TDS, such as acid digestates, must be diluted significantly, reducing sensitivity.

ICP/MS has the additional capability of distinguishing between different sources of lead. Lead has a number stable isotopes (204, 206, 207, and 208 AMU) and different natural sources of lead have different ratios of these elements. Thus, lead mined from these sources, and by extension materials manufactured from them, will also have distinct patterns of isotopes. This can be applied to air emissions: if an emission source is using lead from a single geological source, it can be both identified and distinguished from another emission source that uses lead from a different geological source.

ICP/AES is based on the fact that when the elements leave the plasma they cool, and electrons drop back to their ground state, releasing the absorbed energy in the form of ultraviolet and visible radiation. The emission spectra can be separated by diffraction gratings and prisms and the individual wavelengths measured by either photomultiplier tubes (PMTS) or, more recently, by charge-coupled or charge-injection (solid-state) devices. Each element has a unique combination of electron transitions giving rise to a distinct analytical spectrum. The strongest emission wavelengths (lines) can be measured and quantified. There are two types of ICP/AES: sequential and simultaneous. Sequential ICP/AES measures one line at a time by a single PMT, whereas sequential ICP/AES has a fixed array of PMTs or solid-state detectors that measure a fixed set of wavelengths. ICP/AES, like ICP/MS, has a very wide linear dynamic range and can analyze a large number of elements. It is, however, subject to a large number of interferences from overlapping wavelengths from other elements (but not from molecules) and is therefore not very selective. It is less expensive than ICP/MS and far more robust, being far less sensitive to differences in analyst skill. It is also more tolerant of TDS in samples, so that fewer dilutions are needed. Simultaneous ICPs are very sensitive, in some cases comparable to ICP/MS.

Flame atomic absorption spectroscopy (FAAS) is based on the same principle as ICP/AES. A liquid sample is made into an aerosol, injected into an acetylene flame (1000–4000°C), the aerosol is vaporized, then atomized (there is far less ionization than in ICP techniques). A hollow cathode lamp (HCL) or electron discharge lamp (EDL) containing a fixed amount of the target element is mounted perpendicular to the flame. An electric field is applied to the lamp and a cool plasma is created containing the target element. The target element emits radiation characteristic of the element just as with ICP/AES; however no other elements are present so there are no other emission spectra. The emissions are passed through the flame and the target element present in the flame absorbs the emitted radiation in direct proportion to the number of atoms present. FAAS is extremely selective and can be very sensitive (this varies greatly from element to element). However, FAAS has a narrow linear dynamic range (1–3 orders of magnitude) and is generally limited to a single element at a time (one per HCL or EDL, although some newer instruments can mount several lamps simultaneously). FAAS is very robust and does not require great analyst skill, so whereas the FAAS instrument is the least expensive technology, if multiple elements are measured, labor can drive up the costs.

Nondestructive Procedures

Destructive procedures attempt to make all sample types into one sample by extracting the target elements into aqueous liquids, thus the calibration standards are automatically matrix-matched to the samples. Nondestructive procedures make no changes in the elemental composition of the sample, so it is generally necessary to find calibration standards that are of a very similar matrix to that of the sample. The nondestructive procedures all use emission spectrographic techniques wherein the sample is bombarded by high energy particles which result in the emission of X-ray or gamma-ray radiation of wavelengths characteristic of particular elements.

In neutron activation analysis (NAA) the sample is exposed to a fixed flux of thermal neutrons for a fixed time. The neutrons impart their energy to the nucleus of the atom, which releases this excess energy as gamma-radiation. Each different element releases a characteristic wavelength that can be quantified. NAA is extremely sensitive, has a robust linear range, and can simultaneously measure a large number of elements. However, NAA is subject to interelement interferences that can be minimized by carefully matching the matrix of the calibration standards to the matrix of the samples. Matrix-matching involves matching nontarget elemental concentrations as well as uniformity of particle sizes. NAA requires specialized facilities and expensive equipment. It is more useful for glass and quartz fiber filters as the majority of the matrix (the silicon oxide portion) is exactly the same for all samples and thus is easy to matrix-match samples and standards. NAA can be used for any sample type, but more extensive sample preparation, such as milling control for particle size and drying to control for moisture (to which the method is sensitive), may be required. NAA is not recognized in any standardized environmental or occupational methods books.

X-ray emission spectroscopy is similar to NAA in that the sample is irradiated, but instead of thermal neutrons, either nuclear particles (particle induced X-ray emission (PIXE)) or X-rays (X-ray fluorescence (XRF)) are used. Also, instead of acting on the nucleus, the target is the innermost electrons. When one of these electrons is hit by a particle or X-ray, it is ejected from the atom. This causes a cascade as higher orbiting electrons fall to fill the vacancy. As the electrons drop in energy, they emit X-rays characteristic of the electronic structure of the element. These X-rays are measured and quantified. Like NAA, X-ray emission is subject to interelement interferences which can be minimized by matching the matrix of the calibration standards to the matrix of the samples. X-ray emission is also particle-size and moisture sensitive. PIXE is very sensitive and can measure many elements simultaneously over a range of concentrations, but it requires specialized facilities and equipment and is expensive. XRF is commonly used for glass and quartz fiber filters because it requires almost no sample preparation and easily matrix-matches samples and standards. It is sensitive enough for many common applications such as routine monitoring of a characterized site and, although less sensitive than PIXE and NAA, is far less expensive. Very small-field portable XRF units are finding extensive application for preliminary field investigation of soils and other solids and surfaces. Field-portable XRF units allow for semiquantitative analysis in identifying gross differences in elemental concentrations, which can be used to more carefully select sample locations. None of these nondestructive techniques has received official recognition.

BIBLIOGRAPHY

1. R.G. Reese, *Minerals Yearbook,* Vol. 1, *Metals and Minerals,* U.S. Dept. of the Interior, Bureau of Mines, Washington, D.C., 1991, pp. 1389–1412; *Knight-Ridder CRB Commodity Yearbook 1993, Silver,* Commodity Research Bureau, Knight-Ridder Financial Publishing, John Wiley & Sons, Inc., New York, pp. 231–235.
2. M. Grayson, ed., *Kirk-Othmer Encyclopedia of Chemical Technology,* 3rd ed., Vol. 6, John Wiley & Sons, Inc., New York, 1978, pp. 54–82.
3. Ref. 2, Vol. 3, pp. 569–669.
4. W.C. Hinds, *Aerosol Technology,* John Wiley & Sons, Inc., New York, 1982.
5. R.M. Stern, in S. Langard, ed., *Topics Environ. Health, Biolog. Aspects Chrom.* **5,** 16–47 (1982).
6. Montieth and Unsworth, *Principles of Environmental Physics,* Edward Arnold, 1990.
7. D.E. Kimbrough, Y. Cohen, A.M. Winer, L. Creelman, and C. Mabuni, "A Critical Assessment of Chromium in the Environment," *Crit. Rev. Environ. Sci. Technol.* (1997) (in press).
8. P.M. Grohse, W.F. Gutknecht, L. Hodson, and B.M. Wilson, *The Fate of Hexavalent Chromium in the Atmosphere,* Research Triangle Institute, prepared for the California Air Resources Board, Sacramento, Calif., 1988.
9. E.D. Stein, Y. Cohen, and A.M. Winer, "Environmental Distribution and Transportation of Mercury Compounds," *Crit. Rev. Environ. Sci. Technol.* (1995).
10. D.E. Kimbrough and I. Suffet, *Environ. Sci. Technol.* **29,** 2217–2221 (1995).
11. "Air Sampling Instruments for Evaluation of Atmospheric Contaminants," 5th ed., *American Conference of Government Industrial Hygienists,* Washington, D.C., 1990.
12. N. Ainsworth, J.A. Cooke, and M.S. Johnson, *Ecotoxicol. Environ. Safety,* **21,** 165–170 (1991).
13. N. Ainsworth, J.A. Cooke, and M.S. Johnson, *Environ. Poll.* **65,** 65–77 (1990).
14. *Ibid.,* pp. 79–87.
15. P.M. Haygarth, A.F. Harrison, and K.C. Jones, *Environ. Sci. Technol.* **27,** 2878–2884 (1993).

16. D.E. Kimbrough and J.R. Wakakuwa, *Proceedings of the Sixth Annual USEPA Symposium on Solid Waste Testing and Quality Assurance,* U.S. Government Printing Office, Washington, D.C., July 1990, pp II-214; reprinted in *Environ. Sci. Technol.* **26,** 173–178 (1992).

17. P.W.J.M. Boumans, ed., *Inductively Coupled Plasma Emission Spectroscopy—Part 1: Methodology, Instrumentation, and Performance,* John Wiley & Sons, Inc., New York, 1987.

ENVIRONMENTAL IMPACT STATEMENTS

J. I. BREGMAN
Bregman & Company, Inc.
Bethesda, Maryland

The National Environmental Policy Act (NEPA) is probably the most comprehensive piece of legislation protecting the environment that has been passed by the U.S. Congress. It introduces the requirement for the preparation of environmental impact statements (EISs) before a project approach is finalized. These EISs, if properly carried out, make certain that possible environmental impacts of the proposed action are noted as well as those of alternative actions. Mitigating measures are considered where feasible.

NEPA covers the gamut of environmental factors and includes both the natural and socioeconomic environments. It requires federal agencies to consider the possible effects of proposed major federal actions on both the ecology of the area and the welfare of the nearby human inhabitants before an action is implemented. Major federal actions range from the proposal to construct and operate a facility to the granting of federal permits, leases, etc, for industrial and commercial activities.

In 1969, NEPA changed the national philosophy concerning negative environmental impacts of major projects. Until then, the general attitude concerning those adverse effects was to ignore them during the planning stages of the project. After the work was completed and the negative environmental effects were apparent, the attitude was generally one of "Too bad, but it could not be avoided." On a rare occasion, minor mitigative measures would be installed after the fact.

Recognizing that this attitude of "build now and worry later" was pervasive among federal agencies, The U.S. Congress enacted NEPA. This Act, in effect, said that environmental impacts should be considered *before* the plans for a project were finalized (1).

Almost anyone can think of actions taken before NEPA was passed that could have been modified or changed if environmental impact statements had been required. For example, thousands of acres of wetlands that were destroyed by highways built with federal funds, would have been preserved.

The NEPA approach is a common sense one. It requires that federal decision makers think through the environmental consequences before taking action. If those consequences include undesirable effects, then NEPA requires that consideration be given to mitigating measures that can be built into the action and/or that alternatives to the action be considered that would produce a similar end result but be less damaging to the environment.

The passage of NEPA has caused federal agencies to incorporate environmental values in their decision making. NEPA is now an integral part of program planning. The NEPA process has led to alterations in project design, location, or operation and agency consideration of a greater range of alternatives and implementation of mitigation measures. The NEPA review process also has enabled agencies to address compliance with other environmental laws as part of a single review process rather than separate reviews under each law, thereby reducing the amount of paperwork, staff time, and effort.

LEGAL REQUIREMENTS

NEPA

Title I, § 101, Declaration of National Environmental Policy states:

> *The Congress, recognizing the profound impact of man's activity on the interrelations of all components of the natural environment, particularly the profound influences of population growth, high density urbanization, industrial expansion, resource exploitation, and new and expanding technological advances and recognizing further the critical importance of restoring and maintaining environmental quality to the overall welfare and development of man, declares that it is the continuing policy of the Federal Government, in cooperation with state and local governments, and other concerned public and private organizations, to use all practicable means and measures including financial and technical assistance, in a manner calculated to foster and promote the general welfare, to create and maintain conditions under which man and nature can exist in productive harmony, and fulfill the social and economic, and other requirements of present and future generations of Americans.*

NEPA then goes on to state the results it expects as follows:

1. *Fulfill the responsibilities of each generation as trustee of the environment for succeeding generations*
2. *Assure for all Americans safe, healthful, productive, and aesthetically and culturally pleasing surroundings*
3. *Attain the widest range of beneficial uses of the environment without degradation, risk to health or safety, or other undesirable and unintended consequences*
4. *Preserve important historic, cultural, and natural aspects of our national heritage, and maintain, whenever possible, an environment which supports diversity, and variety of individual choice*
5. *Achieve a balance between population and resource use which will permit high standards of living and a wide sharing of life's amenities and*
6. *Enhance the quality of renewable resources and approach the maximum attainable recycling of depletable resources.*

The Act then develops requirements for evaluating environmental effects before finalizing a project by stating in § 102 that

> *all agencies of the Federal Government shall—*
>
> *(A) Utilize a systematic, interdisciplinary approach which will insure the integrated use of the natural and social sciences and the environmental design arts in planning and in decision-making which may have an impact on man's environment*
>
> *(B) Identify and develop methods and procedures which will insure that presently unquantified environmental amenities and values may be given appropriate consideration in decision-making along with economic and technical considerations*

(C) Include in every recommendation or report on proposals for legislation and the major Federal actions significantly affecting the quality of the human environment, a detailed statement by the responsible official on—

(i) The environmental impact of the proposed action

(ii) Any adverse environmental effects which cannot be avoided should the proposal be implemented

(iii) Alternatives to the proposed action

(iv) The relationship between local short-term uses of man's environment and the maintenance and enhancement of long-term productivity, and

(v) Any irreversible and irretrievable commitments of resources which would be involved in the proposed action should it be implemented.

A requirement for consultation with all interested federal agencies before issuing an EIS is imposed as follows:

Prior to making any detailed statement, the responsible Federal official shall consult with and obtain the comments of any Federal agency which has jurisdiction by law or special expertise with respect to any environmental impact involved. Copies of such statement and the comments and views of the appropriate Federal, state, and local agencies, which are authorized to develop and enforce environmental standards, shall be made available and shall accompany the proposal through the existing agency review processes.

Section 102 of NEPA goes on to make the following requirements:

A study of appropriate alternatives to any project.

International cooperation on the prevention of degradation of the world environment.

Accessibility of information on environment upgrading to state and local agencies.

Use of ecological information in resource oriented projects.

Section 103 of NEPA requires that all federal agencies make certain that their statutory authority allows them to comply with NEPA.

Title II of NEPA creates the Council on Environmental Quality (CEQ; three members to be named by the president) in the executive office of the president, giving it the following responsibilities:

Assist in the preparation of the annual environmental quality report.

Report on environmental quality trends that might interfere with NEPA objectives.

Regulate federal environmental programs and recommend new programs.

Conduct appropriate studies on ecological systems and environmental quality.

Document and analyze environmental changes.

CEQ Regulations

NEPA and two subsequent Executive Orders (11514 and 11991) required that the CEQ issue regulations that would tell all federal agencies how to carry out NEPA requirements. The regulations state their purposes as follows:

NEPA procedures must insure that environmental information is available to public officials and citizens before decisions are made and before actions are taken. The information must be of high quality. Accurate scientific analysis, expert agency comments, and public scrutiny are essential to implementing NEPA. Most important, NEPA documents must concentrate on the issues that are truly significant to the action in question, rather than amassing needless detail. The NEPA process is intended to help public officials make decisions that are based on understanding of environmental consequences and take actions that protect, restore, and enhance the environment. These regulations provide the direction to achieve this purpose.

These CEQ regulations, in turn, may be modified by each federal agency to meet its own needs. Under no circumstances, however, can the modified regulations be less stringent than the CEQ regulations or subvert CEQ requirements.

The primary purpose of the CEQ regulations is to define the procedures required to carry out the intent of NEPA. The process begins with a definition of what a major federal action is, since this is the activity that is subject to NEPA.

Major federal actions are defined as including the following areas:

1. *Adoption of official policy, such as rules, regulations, and interpretations; treaties and international conventions or agreements; and formal documents establishing an agency's policies which will result in or substantially alter agency programs*
2. *Adoption of formal plans, such as official documents which guide or prescribe alternative uses of Federal resources, upon which future agency actions will be based*
3. *Adoption of programs, such as a group of concerted actions to implement a specific policy or plan*
4. *Approval of specific projects, such as construction or management activities located in a defined geographic area. Projects include actions approved by permit or other regulatory decision as well as Federal and federally assisted activities.*

Categorical Exclusions. The first step in the NEPA process for most agencies is to develop a list of categorical exclusions, which are actions that have been predetermined by the agency as those that will not individually or cumulatively have a significant effect on the human environment. Therefore, preparation of an environmental assessment or an environmental impact statement is not required. Thus routine actions are not subject to NEPA review and can be implemented immediately. A few examples of categorical exclusions of the Department of Defense (DOD) are as follows:

Normal personnel, fiscal, and administrative activities involving military and civilian personnel.

Recreation and welfare activities not involving off-road recreational vehicle management.

Commissary and postexchange (PX) operations, except where hazardous material is stored or disposed.

Routine repair and maintenance of buildings, roads, airfields, grounds, equipment, and other facilities, except when requiring application or disposal of hazardous or contaminated materials.

Routine procurement of goods and services.

Simulated war games and other tactical and logistical exercises without troops.

Training entirely of an administrative or classroom nature.

Routine movement of personnel; routine handling and distribution of nonhazardous materials in conformance with Department of the Army (DA), Environmental Protection Agency (EPA), Department of Transportation (DOT), and state regulations.

Conversion of commercial activities to contract performance of services.

Preparation of regulations, procedures, manuals, and other guidance documents that implement, without substantive change, the applicable agency regulations, procedures, manuals, and other guidance documents that have been environmentally evaluated.

Environmental Assessments. If the proposed major federal action cannot be categorically excluded and the decision maker is unsure if there is a potential for significant environmental impact, an environmental assessment (EA) may be prepared. CEQ regulations define an EA as follows:

(a) Means a concise public document for which a Federal agency is responsible that serves to:

1. *Briefly provide sufficient evidence and analysis for determining whether to prepare an environmental impact statement or a finding of no significant impact*
2. *Aid an agency's compliance with the Act when no environmental impact statement is necessary*
3. *Facilitate preparation of a statement when one is necessary*

(b) Shall include brief discussions of the need for the proposal, of alternatives as required by sec. 102(2)(E), of the environmental impacts of the proposed action and alternatives, and a listing of agencies and persons consulted.

In content, an EA is similar to an EIS. Most often, an EA is less detailed than an EIS, resulting in lower preparation costs and shorter periods of analysis. If it is apparent that there will be major negative impacts, then the agency has the option of going directly to an EIS and bypassing the EA.

A typical EA will contain the purpose of and need for the project, alternatives to be considered, a description of the existing environment, and the environmental effects of the alternatives. Based on the EA, the decision maker recommends that one of the two following courses be pursued: (*1*) because of the potential for significant negative impacts, an EIS should be prepared or (*2*) because there is no likelihood for significant adverse impacts, a Finding Of No Significant Impact (FONSI) should be issued. Other differences between the EA and EIS processes include the following: no formal scoping meeting is held; almost all of the data collected are already available, rather than being new material; notification of the public of the availability of the draft EA is not announced in the *Federal Register;* and the comment period is shorter for the EA, if comments are requested.

FONSI. If the EA analysis supports a conclusion that significant adverse environmental impacts are unlikely, then a FONSI is issued. The CEQ regulations define a FONSI as follows:

A document by a Federal agency briefly presenting the reasons why an action, not otherwise excluded, will not have a significant effect on the human environment and for which an environmental impact statement therefore will not be prepared. It shall include the environmental assessment or a summary of it and shall note any other environmental documents related to it. If the assessment is included, the finding need not repeat any of the discussion in the assessment but may incorporate it by reference.

If the project is controversial, or of national interest, a FONSI may be published in the *Federal Register.* Otherwise, the announcement of a FONSI and the public availability of the final EA are published in the local media.

If a *Federal Register* notice is made, public comments may be received for a specific time period, although this is not required except under limited circumstances (1501.4(8)(2)). The agency publishing the FONSI usually does not wait for comments before it starts action. However, if legitimate comments are received that question the validity of the FONSI, the agency may reevaluate its decision to proceed with the proposed project.

The EIS. An agency determines that an EIS is necessary based on the potential for adverse effects or by such conclusions from an EA. When only one federal agency is involved in a proposal, it becomes the lead agency for the EIS preparation. If more than one agency is involved or interested in a proposed action, then the lead agency must be designated. The CEQ regulations give the following factors (in order of descending importance) in determining the lead agency designation:

1. *Magnitude of agency's involvement.*
2. *Project approval / disapproval authority*
3. *Expertise concerning the action's environmental effects*
4. *Duration of agency's involvement*
5. *Sequence of agency's involvement*

After a lead agency has been designated, cooperating agencies then are established. The CEQ regulations state that:

Upon request of the lead agency, any Federal agency which has jurisdiction by law shall be a cooperating agency. In addition any other Federal agency which has special expertise with respect to any environmental issue, which should be addressed in the statement may be a cooperating agency upon request of the lead agency. An agency may request the lead agency to designate it a cooperating agency.

The duties of cooperating agencies are listed as follows:

Each cooperating agency shall:

1. *Participate in the NEPA process at the earliest possible time*
2. *Participate in the scoping process*
3. *Assume on request of the lead agency responsibility for developing information and preparing environmental analyses including portions of the environmental impact statement concerning which the cooperating agency has special expertise*
4. *Make available staff support at the lead agency's request to enhance the latter's interdisciplinary capability*
5. *Normally use its own funds.*

EIS Preparation. To begin the EIS process, the lead agency inserts a Notice Of Intent (NOI) to prepare an EIS in the *Federal Register.* The NOI announces the date and location of scoping meetings to determine the issues to be addressed in the EIS. Invitations to attend are sent to "affected Federal, state and local agencies, any affected Indian tribe, the proponent of the action, and other interested persons (including those who might not be in accord with the action on environmental grounds)."

The CEQ regulations require that agencies shall prepare environmental impact statements in the following manner:

> *(a) Environmental impact statements shall be analytic rather than encyclopedic*
>
> *(b) Impacts shall be discussed in proportion to their significance. There shall be only brief discussion of other than significant issues. As in a Finding of No Significant Impact, there should only be enough discussion to show why more study is not warranted*
>
> *(c) Environmental impact statements shall be kept concise and shall be no longer than absolutely necessary to comply with NEPA and with these regulations. Length should vary first with potential environmental problems and then with project size*
>
> *(d) Environmental impact statements shall state how alternatives considered in it and decisions based on it will or will not achieve the requirements of sections 101 and 102(1) of the Act and other environmental laws and policies*
>
> *(e) The range of alternatives discussed in environmental impact statements shall encompass those to be considered by the ultimate agency decision maker*
>
> *(f) Agencies shall not commit resources prejudicing selection of alternatives before making a final decision*
>
> *(g) Environmental impact statements shall serve as the means of assessing the environmental impact of proposed agency reactions, rather than justifying decisions already made.*

It is recommended by CEQ that "the text of final environmental impact statements shall normally be less than 150 pages and for proposals of unusual scope or complexity shall normally be less than 300 pages." Further, "environmental impact statements shall be written in plain language and may use appropriate graphics so that decision makers and the public can readily understand them."

The CEQ regulations state that

> *except for proposals for legislation, environmental impact statements shall be prepared in two stages and may be supplemented.*
>
> *(a) Draft environmental impact statements shall be prepared in accordance with the scope decided upon in the scoping process. The agency shall make every effort to disclose and discuss at appropriate points in the draft statement all major points of view on the environmental impacts of the alternatives including the proposed action*
>
> *(b) Final environmental impact statements shall respond to comments. The agency shall discuss any responsible opposing view which was not adequately discussed in the draft statement and shall indicate the agency's response to the issues raised*
>
> *(c) Agencies:*
>
> **1.** *Shall prepare supplements to either draft or final environmental impact statements if:*
>
> *(i) The agency makes substantial changes in the proposed action that are relevant to environmental concerns; or*
>
> *(ii) There are significant new circumstances or information relevant to environmental concerns and bearing on the proposed action or its impacts.*

The format for EISs that is recommended by CEQ is as follows: cover sheet, summary, Table of Contents, purpose of and need for action, alternatives including the proposed action, affected environment, environmental consequences, list of preparers, list of agencies, organizations, and persons to whom copies of the statement are sent, index, and appendices (if any). Details of some of the material to be included in the above items are described as follows.

> *COVER SHEET*
> *The cover sheet shall not exceed one page. It shall include:*
>
> *(a) A list of the responsible agencies*
>
> *(b) The title of the proposed action*
>
> *(c) The name, address and telephone number of the person at the agency who can supply further information*
>
> *(d) A designation of the statement as a draft, final or draft or final supplement*
>
> *(e) A one paragraph abstract of the statement*
>
> *(f) The date by which comments must be received...*

SUMMARY

The summary shall stress the major conclusions, areas of controversy and the issues to be resolved. The summary will normally not exceed 15 pages...

PURPOSE AND NEED

The statement shall briefly specify the underlying purpose and need to which the agency is responding...

ALTERNATIVES

This section should present the environmental impacts of the proposal and the alternatives in comparative form, providing a clear basis for choice among options by decision maker and the public...

AFFECTED ENVIRONMENT

...The EIS shall succinctly describe the environment of the area(s) to be affected or created by the alternatives under consideration...

ENVIRONMENTAL CONSEQUENCES

The discussion will include the environmental impacts of the alternatives including the proposed action, any adverse environmental effects which cannot be avoided should the proposal be implemented, the relationship between short-term uses of man's environment and the maintenance and enhancement of long-term productivity, and any irretrievable commitments of resources which would be involved in the proposal should it be implemented...

LIST OF PREPARERS

The EIS shall list the names, together with their qualifications, of the persons who were primarily responsible for preparing the environmental impact statement or significant background papers.

The CEQ regulations call for wide distribution of the draft and final EIS. They go to all the agencies concerned: federal, state, and local; the applicant (if any); and all interested members of the public or organizations.

Relevant cost–benefit analyses may be discussed in the EIS and included in detail in the appendix. The draft EIS (DEIS) should list all federal permits, licenses, and other entitlements that must be obtained while implementing the proposal.

Commenting on the Draft EIS. *Inviting Comments.* When the DEIS is made available to all interested parties and an announcement is placed in the *Federal Register,* comments are solicited from any federal agency that has jurisdiction by law or special expertise with respect to any potential environmental impacts or is authorized to develop and enforce environmental standards. Comments also are requested from appropriate state and local agencies, American Indian groups (when reservations may be affected), any agency that has requested that it receive statements on actions of the kind proposed, the applicant (when a permit is involved), and the public or organizations that may be interested or affected.

Response to Comments The final EIS includes the agency's assessment and consideration of all comments. The Final EIS may include any of the following changes:

1. *Modify alternatives including the proposed action*
2. *Develop and evaluate alternatives not previously given serious consideration by the agency*
3. *Supplement, improve, or modify its analyses*
4. *Make factual corrections*
5. *Explain why the comments do not warrant further agency response, citing the sources, authorities, or reasons which support the agency's position and, if appropriate, indicate those circumstances which would trigger agency reappraisal or further response*
6. *All substantive comments received on the draft statement (or summaries thereof where the response has been exceptionally voluminous), are to be attached to the final statement in the Appendix*
7. *If changes in response to comments are minor and are confined to the responses described above, agencies may write them on errata sheets and attach them to the statement instead of rewriting the draft statement.*

Agency Disagreement Resolution. The CEQ regulations establish a procedure whereby interagency disagreements concerning proposed major federal actions that might cause unsatisfactory environmental effects are referred to CEQ for resolution. This is especially the case in situations in which the EPA, using its authority under § 309 of the Clean Air Act to comment publicly on federal EISs, is in disagreement with the conclusions of another federal agency. If an agency cannot resolve its disagreements with the lead agency, a procedure is established for referring the matter to CEQ for resolution. CEQ then makes a determination as to the resolution within 60 days.

Records of Decision. At the time of its decision or, if appropriate, its recommendation to Congress, each agency shall prepare a concise public record of decision (ROD). The ROD must include the following items:

(a) State the decision

(b) Identify all alternatives considered, specifying the alternative or alternatives considered to be environmentally preferable. Discuss all such factors including any essential considerations of national policy which were balanced by the agency in making its decision and state how those considerations entered into its decision

(c) State whether all practicable means to avoid or minimize environmental harm from the alternative selected have been adopted, and if not, why they were not. A monitoring and enforcement program shall be adopted and summarized where applicable for any mitigation.

Implementing the Decision. The lead agency shall include appropriate conditions in grants, permits or other approvals, as well as condition funding of actions on mitigation.

Other Key Items. Other key requirements of the CEQ regulations include the following:

No actions are to be taken on proposals that will have negative impacts or limit alternatives until after a ROD is issued.

Duplication with state and local procedures is to be avoided.

Draft and final EISs prepared by another agency may be adopted if they fit a particular situation.

Environmental documents may be combined.

Intense public involvement in NEPA procedures is required, including notices of hearing, insertion in the *Federal Register,* state clearinghouses, notices to American Indian groups, notices in local papers and other media, community organizations, business groups, newsletters, mailings to nearby residents, and posting of notices.

Public hearings on EISs are to be held if the proposed action is controversial or if they are requested by another agency. Details are presented for NEPA studies for legislation that affects the human environment. In practice, Congress does not pay much attention to this requirement. All EISs, together with comments and responses are to be filed with EPA, which delivers one copy to the CEQ. EPA publishes a notice in the *Federal Register* each week of the environmental impact statements filed during the preceding week. No decision on the action may be made by a federal agency until at least 90 days after the *Federal Register* notice for a draft EIS or 30 days after the notice for a final EIS. All federal agencies must comply with these CEQ regulations and must adopt procedures to supplement and implement them. Agency procedures may include the criteria for identification of classes of actions that do not require EISs or environmental assessments (categorical exclusions).

Clean Air Act

Section 309 of the Clean Air Act has the following provisions

(a) The Administrator shall review and comment in writing on the environmental impact of any matter relating to duties and responsibilities granted pursuant to this chapter or other provisions of the authority of the Administrator, contained in any (1) legislation proposed by any federal department or agency, (2) newly authorized federal projects for construction and any major federal agency action (other than a project for construction) to which section 4332(2)(c) of this title applies, and (3) proposed regulations published by any department or agency of the Federal Government. Such written comment shall be made public at the conclusion of any such review.

(b) In the event the Administrator determines that any such legislation, action, or regulation is unsatisfactory from the standpoint of public health or welfare or environmental quality, he shall publish his determination and the matter shall be referred to the Council on Environmental Quality.

This section, passed 1 year after NEPA, requires the EPA to review and comment on the environmental impact of all federal EISs. As a result, the EPA's comments on draft EISs, which usually are done by the appropriate regional office, tend to be the most comprehensive and worthwhile comments received (see CLEAN AIR ACT, REGULATIONS AND IMPLEMENTATION).

Executive Order 11514

Executive Order 11514—titled Protection and Enhancement of Environmental Quality—was promulgated on March 5, 1970, and was amended by Executive Order 11991 on May 24, 1977.

It requires federal agencies to conform with NEPA under the guidance of the Council on Environmental Quality.

Section 3 gives the CEQ overview responsibility for federal policies and activities directed to pollution control and environmental quality. The CEQ is to

(e) Promote the development and use of indices and monitoring systems (1) to assess environmental conditions and trends, (2) to predict the environmental impact of proposed public and private actions, and (3) to determine the effectiveness of programs for protecting and enhancing environmental quality

(f) Coordinate Federal programs related to environmental quality

(h) Issue regulations to Federal agencies for the implementation of the procedural provisions of the Act.

State and Local Requirements

Currently, 16 states, the District of Columbia and Puerto Rico have passed acts that are essentially state NEPAs. Furthermore, 18 states and the District of Columbia have limited environmental review requirements established by statute, executive order, or other administrative directives. State and local requirements tend to emphasize facilities not subject to NEPA. Requirements tend to vary from state to state and from municipality to municipality as a function of the environmental concerns that are of greatest importance to the local authorities.

In an increasing number of states, delegation of the NEPA process by the EPA has been accomplished. The EPA maintains oversight authority on these environmental documents and may overrule the state if it believes the documents are inadequate. It is generally assumed that the state will have a more relaxed attitude toward NEPA than will EPA, because of the state's proximity to the project and the economic benefits deriving therefrom. This turns out to be the case in only some situations.

THE EIS PROCESS

The Decision to Prepare an EIS

Preparatory Activity. Before deciding whether an EIS is required, certain steps are followed, as described earlier in the discussion on CEQ regulations.

Categorical Exclusion. The first step each agency takes in the process is to determine whether the proposed action falls within the categorical exclusion (CX) area. These are the routine actions that have no significant impact on the human or natural environment. For example, 29 activities were listed under a categorical exclusion by the Department of the Army in Appendix A of AR200-2, which governs army NEPA functions. In a similar vein, the following types of municipal wastewater treatment plant projects may be eligible for categorical exclusions (1): minor rehabilitation of existing facilities, functional replacement of equipment, construction of new ancillary facilities, and minor upgrading and minor expansion of existing treatment works in unsewered communities of $<$10, 000 people.

Environmental Assessment. If the proposed action is not covered by a categorical exclusion, then the agency should move toward an environmental assessment. The first step in that direction is to obtain all of the information that already exists on the proposed action. If the applicant is an industrial firm that requires a permit or lease, it probably has done a considerable amount of preliminary environmental work on the matter. If the proposed action is an agency activity, it likewise should have its own pre-NEPA study. In any event, all available information should be gathered and made ready to turn over to the person who will draft the EA.

The EA is a concise public document designed to provide sufficient evidence and analysis for an agency to determine whether to prepare an EIS or a FONSI. It has the same basic contents as an EIS but is completed in a much shorter time, eg, 6–12 weeks.

Major differences between the EA and EIS include no scoping meeting is held, almost all of the data collected are already available (new material is not sought), and publication of the availability of the draft EA is not placed in the *Federal Register*. The EA concludes with a recommendation that one of the following courses be pursued.

Because of possible significant negative impacts, an EIS should be prepared.

Because there will not be any significant impacts, a FONSI should be prepared, which allows the project to proceed without further environmental studies but does not preclude the project proponent from being responsible for adopting mitigating measures to lessen any possible negative impact (1).

Finding of No Significant Impact. As indicated earlier, if the agency finds, based on the EA, that its proposal will have no significant impact on the environment, it prepares a FONSI, which is a document explaining why the proposal will have no impacts. The FONSI must include the EA or a summary of it and must note any other environmental documents related to the EA.

Steps in the EIS Preparation

The preparation of an EIS is accomplished through the following series of steps. These steps start with the assumption that a lead agency has been selected.

Notice of Intent. The first item in the EIS process is the preparation of the NOI. The Notice should briefly describe the proposed action and possible alternatives, describe the agency's proposed scoping process, and state the name and address of a person within the agency who can answer questions about the proposed action and the environmental impact statement. The NOI then is submitted for publication in the *Federal Register*.

Scoping Meetings. The purpose of the scoping meeting(s) is to determine the scope of the draft EIS and to identify the major project-related issues to be addressed and emphasized in it. Comments to this effect by the attending agencies and public are solicited. Invitees include all of the federal agencies that may have an interest in the project. State and local agencies invited include all of the pollution prevention, natural resource, historical, and archaeological agencies and any others who express an interest. Public participation includes groups or individuals. The arrangements are made for meeting areas and required newspaper advertisements. An agenda is prepared. The EIS scoping meetings are usually held within 30 days of project start. The specific places of the scoping meetings are usually located close to the alternative areas proposed for the project. The scoping meetings are held in accordance with the CEQ's EIS scoping requirements (40 CFR 1.501.7) and minutes or transcripts of the discussion are taken.

The product of the scoping meeting is a brief paper (called the scoping report) that summarizes the results of the meeting. The paper reflects issues and extent of coverage to be contained in the draft EIS. A brief discussion of the scoping process and the comments received from the public are included in the report. This material will be included in the appendix to the draft EIS.

Data Collection of the Existing Environment. The environmental data already on hand are evaluated for use in the descriptions of existing environmental conditions. Work then focuses on the gathering and analysis of additional data, when necessary; but every attempt should be made to avoid duplication of already available data. Environmental information includes local climatology, topography, geomorphology (including geology and soils), biology, water quality (including surface waters, storm water and groundwater), noise, air quality, land use, historic preservation and archaeology, demography, housing, local economy and other socioeconomic aspects, hazards and nuisances, aesthetics and urban design, community services, and transportation. A detailed discussion of each of the items is presented later.

Assessment of Possible Environmental Impacts. The possible impacts of each proposed project alternative are assessed, as well as of the "do nothing" alternative, for each of the environmental and social components delineated. The primary level of impact analysis is site specific. Emphasis is placed on the key issues identified and discussed during the scoping phase of the project.

Secondary impacts also are projected. They are primarily socioeconomic in nature and are concerned with effects of the induced growth (or alternately, population loss) in the area surrounding the project as a result of the project. Items such as demography, taxes, property values, community infrastructure, police and fire protection, recreation, aesthetics, and transportation are considered. However, environmental factors such as water, air, noise, and ecology must not be overlooked.

In both cases, identification is made of the potential short- and long-term impacts associated with the project. Short-term impacts resulting from the proposed project may be those associated with the construction phase, including such disturbances as noise, dust, erosion, and wildlife displacement. Long-term, post-construction impacts may include such factors as pollution from storm-water runoff, air or surface water pollution, noise, consumption of energy, depletion and contamination of groundwater sources, overloading of roadways and other infrastructures, and placing of heavy demands on community services such as sewage treatment and the disposal of solid wastes. These impacts are further characterized as "avoidable," "unavoidable," and "capable of being mitigated."

Irreversible and irretrievable resource commitments resulting from the implementation of the proposed action are evaluated. The consumption of resources is categorized in terms of environmental and human effects.

Mitigation actions should be reviewed here. Where possible, their costs and benefits should be quantified. These facts will be critical in the final decisions of which alternative to select.

Preparation of the Draft Environmental Impact Statement. The format of the draft EIS should be in accordance with the CEQ regulations and also should be consistent with the procedures and requirements of the sponsoring agency. The appendices are bound separately from the rest of the EIS and contain materials such as scoping meeting details, detailed data from which information in the EIS is drawn, and letters from agencies and the public. In addition, space is left for the chapter that will be inserted in the final EIS that will address comments received on the draft EIS.

Distribution of Draft EIS. Care should be taken to see that the draft EIS is distributed to all interested parties. This includes all attendees of the scoping meeting, all appropriate governmental agencies, special interest groups, and concerned members of the public. Copies also should be sent to public officials. A notice of the availability of the draft EIS and the time period for comments (usually 45 days) should be placed in the *Federal Register* through the EPA.

Public Hearing. Public hearings may be held at the discretion of the lead agency if it believes that the proposal is controversial. Practically every proposal that requires an EIS has some controversial elements in it, so the agency is well advised to hold one or more public hearings on it. Furthermore, the lack of adequate public participation is a major cause of court rulings that have declared EISs inadequate.

During the time when the draft EIS is out for comments, public hearings on the draft EIS are held. The purpose of the public hearings is to review the draft EIS and to solicit comments on it from the attendees. The hearings should be held at locations and times that are convenient for all interested parties. Adequate notice in the form of letters, newspaper advertisements, etc should be given to all potential attendees. The minutes and attendance at the hearings should be kept and inserted in the appendix to the final EIS.

The hearings themselves should start with a review of the draft EIS presented in plain, easy-to-understand English, since much of the audience will have had no technical training. Questions on the draft EIS should be answered in the same manner.

Response to Comments. The preparation of the final EIS commences with consideration of comments received during the draft EIS public comment period as well as a review of comments made at public hearings. Administrative or policy questions are answered by the agency and given to the EIS preparer, who develops answers to the technical comments.

The draft EIS is changed where necessary in response to comments. The chapter that was left blank in the draft EIS for response to comments is used in the final EIS for that purpose. This chapter breaks the comments down into (*1*) comments that have been addressed in the draft EIS that are inappropriate or with which the EIS authors disagree, (*2*) comments that are general and deserve a response but do not require an EIS change and (*3*) comments that have resulted in a change in the EIS. All of the comments received, either in writing or at the public hearings, are included in full in the appendix to the FEIS. The first item is discussed briefly in the final EIS. Item 2 is discussed at somewhat greater length, and explanations of why no changes have been made are given. The third item is broken down by commentor. The name of the commentor and an abstract of the comments are presented, followed by a discussion of the reason for accepting the change and a description of where the EIS has been changed and how. With this action, the final EIS has been completed and is ready for distribution.

Final EIS Distribution. The final EIS is distributed to the same listing as the draft EIS plus all people and groupings who have requested it since that time. A notice of the availability of the final EIS also is inserted (via the EPA) in the *Federal Register*.

Record of Decision. The ROD is signed by the highest-ranking person in the agency or department that prepared the EIS. It is that person's rationale for approving a proposed action and it should be written accordingly. The ROD should be 5–10 pages long and should state the need for the action; review the EIS findings; and describe any environmental impacts, planned mitigation measures, and future monitoring. The ROD must be a document that is complete and makes the case for the action. After the ROD is signed, it should be published in the *Federal Register* and distribution is made to the same people and groups that received the final EIS. The appropriate comment period should be allowed before the action proposed in the ROD is taken. If no comments of substance are received during that period, the agency may proceed with the project. If comments of substance are received, responses should be made. It is conceivable that a supplemental EIS may have to be completed if the comments, even at that late stage, point out a weakness in the NEPA activity.

Public Participation

Intensive public participation in the NEPA process is essential for the following reasons: The act requires public involvement, and the lack of adequate public participation is one of the major reasons for litigation resulting in a judicial decision that throws out an EIS and requires a new one. The public becomes a partner to the process. Rumors are laid to rest, and the public has the facts about the proposed project. Inevitably, this lessens the public tensions and hostility to the project. The public often has good suggestions for items to be incorporated or stressed in the EIS, resulting in a better product. NEPA requires the following actions of federal lead agencies:

- A diligent effort to involve the public in NEPA procedures.
- Public notice of meetings and available documents to specific requestors, the *Federal Register* for actions of national concern, statewide and areawide clearinghouses, American Indian groups on reservations, local newspapers or other local media, community organizations, newsletters, and individuals by direct mailing (for local actions).
- Public hearings and meetings in which there is substantial environmental controversy concerning the proposed action.
- Solicitation of information from the public.
- Making the EIS and supporting information readily available in conveniently located public places (1).

Every federal agency has adopted procedures to comply with the CEQ regulations for public participation in the EIS process. These procedures incorporate the actions listed above.

In an agency development of a public participation program for an EIS, the following questions should be answered.

- What parts of the public are affected by the proposed action?
- Where do we find them?
- What do we want from them?
- How can mechanisms be provided for input?
- When in the study is input needed?

The U.S. EPA has developed an extensive NEPA public-participation program. Since the EPA is involved in more NEPA activities than any other agency or group, this program has become a model. EPA regulations identify the key components of public participation such as publication of Notices Of Intent, conduct of public meetings or hearings, public review of Findings Of No Significant Impact, and dissemination of the Record Of Decision. Requirements also address how copies of these documents are to be made available to the public. The regulations present requirements for how public hearings, public meetings, and advisory groups are to be used as mechanisms of a public-participation program. The regulations describe the scheduling and conduct of hearings. Public involvement is required to achieve balanced participation of all interest groups. The EPA requirements call for responsiveness summaries to demonstrate the efforts made at key decision points to facilitate public participation. The agency also must prepare evaluations of effectiveness of public participation.

For approved state programs, the EPA is required to monitor state compliance with public participation requirements and is empowered to withdraw approval for noncompliant programs. For example, a federal court in Chicago ordered the EPA to withdraw approval of the Illinois National Pollutant Discharge Elimination System (NPDES) permit program because of the failure to establish guidelines for citizen participation as required by § 101(e) of the Clean Water Act.

The Superfund program managed by the EPA has been granted an exemption from participation in NEPA. This is because it has been judged to be "functionally equivalent" to NEPA. It is stated to have all of the activities of NEPA and to carry them out intensively. This is certainly true of public participation.

At each Superfund site where a remedial action is undertaken, a community relations plan is completed, which presents a brief case history of the contamination problem and recommendations for public-participation activities that the EPA should undertake to address residents' concerns. The underlying rationale for this approach is that if an agency first analyzes the nature and intensity of community concerns for a site-specific project, it will be able to develop a more effective public-participation package.

The EPA also has recognized that even when a comprehensive, sensitive public-participation program has been implemented at a hazardous waste site, disputes may still arise between concerned community residents and agency officials. For this reason, the agency has been exploring alternative means for resolving these disputes through mediators.

The EPA public participation requirements use these mechanisms: public notices, consultation, and hearings; advising groups; and responsiveness summaries.

It is desirable at this point to emphasize the activities with respect to public participation after the draft EIS has been written. The agency sponsoring it must request comments from the following: federal agencies, state and local agencies, American Indian groups (when affected), any other interested agency, the applicant (if any), and the interested public (by soliciting comments from persons and organizations). Comments to the draft EIS may be responded to in any of the following methods:

Modification of alternatives, including the proposed action.

Development of new alternatives.

Modification of analyses in the EIS.

Factual corrections.

Explaining why the comments do not apply.

A summary of the comments along with the responses should be incorporated into the final EIS. In addition, the appendix must contain all of the comments in their totality.

TOPICS CONSIDERED IN AN EIS

The topics most frequently found in EISs include (*1*) earth resources, (*2*) biology, (*3*) water, (*4*) air, (*5*) noise, (*6*) hazards and nuisances, (*7*) historic and cultural resources, (*8*) transportation, and (*9*) socioeconomics. These topics are not all-inclusive. They are the ones that are most frequently found in EISs; but others may be required, depending on the nature of the EIS. A brief description of each topic follows.

Earth Resources. Earth resources include soils (characteristics, bearing strength, susceptibility to erosion), geology (bedrock, surficial, seismicity), and physiography and geomorphology (topography).

Biology. Biological factors include both animal and plant components in all possible locations—water, air, and land. A broad and incomplete listing follows. Aquatic forms include all species from phytoplankton to whales. Birds, bats, and flying insects inhabit the air. Terrestial organisms include vegetation of all types, reptiles, animals of all types, and insects.

Water. When describing water environments, include surface water, groundwater, and storm water. Each item has specific laws and regulations that must be followed (see WATER POLLUTION TOXICOLOGY).

Air. The control of air quality is dictated almost entirely by federal and state laws and regulations. Most of these legal requirements apply to the following pollutants, which are known as the "criteria pollutants" because of criteria that the EPA has developed concerning their health effects: particular matter 10 μ (PM_{10}), sulfur oxides (SO_x) carbon monoxide (CO), lead (Pb), ozone (O_3) (levels are set for the chemicals—precursors—that react to create it, and nitrogen oxides (NO_x).

The 1970 version of the Clean Air Act required that the EPA establish primary and secondary air quality standards for each of the criteria pollutants. Human health-related standards were designated as primary ambient air quality standards, and welfare-related standards were designated as secondary ambient air quality standards. Welfare may be defined as effects on human activities that are nonhealth related, eg, corrosion of automobile bodies and growth of crops.

For each of the pollutants, the EPA must set a maximum concentration level, specify an averaging time over which the concentration is to be measured, and identify how frequently the time-averaged concentration may be violated per year. The primary standards are uniform across the country (though the states may impose stricter standards if they wish) and are set without regard to the costs or technical feasibility of attainment. The deadlines for their attainment are long past. No deadlines have been set for attaining the secondary standards, but the Clean Air Act calls for their attainment as promptly as practicable.

Noise. Highway traffic noise is the largest single source of noise that is considered in an EIS. This noise is not constant. It changes with the number, type, and speed of the vehicles that produce it. Traffic noise variations can be plotted as a function of time. However, it is usually inconvenient and cumbersome to use such a graph to represent traffic noise. A more practical method is to convert the noise data to a single representative number.

The two most common statistical descriptors used for traffic noise are L_{10} and L_{eq}: L_{10} is the sound level that is exceeded 10% of the time, and L_{eq} is the constant average sound level, which over a period of time, contains the same amount of sound energy as the varying levels of the traffic noise. L_{eq}, for typical traffic conditions is usually about 3 dBA (ie, decibels averaged the way the normal person hears sounds) less than the L_{10} for the same conditions.

Aviation has its own peculiar set of noise problems. The noise appears to consist primarily of short-term, high intensity bursts at specific time periods in specified flight paths. The effect of the noise is specific to the sensitive receptor source.

The federal law controlling noise is the Noise Control Act of 1972. Under the act, the EPA published information on what was known about the levels of noise necessary to protect public health. The noise control part of the EPA was dissolved in the early 1980s and practically all of the noise control activities now reside in the states and municipalities.

Hazards and Nuisances. The material in the hazards and nuisances portion of an EIS covers two areas: conventional hazards and nuisances, such as high voltage power lines, buried oil, and gas lines; and toxic and hazardous waste sites.

Conventional hazards and nuisances include items such as major natural gas and petroleum pipelines, odors, high wind velocities, and high voltage transmission lines. These types of hazards and nuisances are site-specific.

Toxic and hazardous materials are covered by major federal laws—RCRA and Superfund—including hazardous and solid waste activities that are of the most concern to people doing a NEPA study. RCRA is intended to provide cradle-to-grave management of hazardous wastes, management of solid wastes, and regulation of underground storage tanks containing chemical and petroleum products. Wastes excluded from regulation as hazardous wastes are household wastes, crop or animal wastes, mining overburden, wastes from processing and beneficiation of ores and minerals, flyash, bottom ash, slag waste, flue gas emission control waste, and drilling fluids from energy development.

Historic and Cultural Resources. The term *historic and cultural resources* covers a variety of features, including archaeologic sites, historic sites (recent history, ie, up to 200 years old), architecturally important buildings, locations and facilities that have a cultural significance to the local communities (including those of ethnic, Indian, or racial significance), and possibly unique geological locations. The term *historic* is broad enough to include any site at which history may have been made.

The National Historic Preservation Act is the principal legislation that establishes the protection of significant historical properties as a national policy. Recent amendments to that Act have pulled together aspects of earlier requirements—such as the Antiquities Act of 1906, the Historic Sites Act of 1935, the Reservoir Salvage Act of 1960, and Executive Order 11593—and incorporated them into the National Historic Preservation Act of 1966.

The term *cultural resources* is a general concept and does not appear in the major historic preservation laws. It is understood by professionals working in the field to refer to particular places (sites), buildings, and objects that symbolize either important historical events and people or long-term trends and continuities that have helped form our common experience as a nation and society. Unique and unusual places, groups of people, structures, and objects also may be included. While these resources may be considered significant in particular situations or contexts or to particular individuals or groups, they will not all necessarily meet the legal and regulatory definition of historic properties.

Transportation. Transportation factors always should be considered in the preparation of a NEPA statement. In some cases, the existing transportation and the impacts on it may be major factors in the study, eg, large buildings and facilities in urban areas. In other cases, transportation issues may be relatively minor. The basic points to be considered include items such as local traffic considerations, mass transit, private vehicles, parking, road and highway adequacy, and pedestrian access.

Socioeconomics. Socioeconomics covers a wide range of topics, including demography, local economic base, local government finances, land use, housing, community services, recreation, and aesthetics. *Demography* is considered to be the social makeup of a site. Factors such as population estimates, the labor force, employment level, rate and nature of population changes, median age, median household income, age structure, and the proportion of households below the poverty level are included.

The *local economic base* is an economic profile of the project area and its surrounding region. It includes a description of the labor force, employment and unemployment characteristics, economic activity, and trends in each major economic sector.

Local government finances includes information about the municipal budget, sources of revenue, categories of expenditures, and tax structure. The overall population increase and induced economic growth directly affect the account and level of community services provided by the municipality to the area.

The existing *land use* of the areas in the sites and their vicinities may include commercial and retail establishments, parking lots, housing, etc. Land use plans for the study area and zoning designation are documented. Farmland also is included in this category.

Housing is defined as the types and availability of housing in the study area. *Community services* are multifold and include such things as water supply systems, stormwater drainage, wastewater collection and treatment systems, solid waste disposal and energy utilities, schools, police, fire, recreation facilities, health care, and shopping areas. *Recreational resources* are parklands, ball fields, and the like. The *aesthetics* of an area are particularly important where tourism is a major industry.

BIBLIOGRAPHY

1. J. I. Bregman and K. M. Mackenthun, *Environmental Statements*, Lewis Publishers, Inc., Chelsea, Mich., 1992.

ENVIRONMENTAL RISK ASSESSMENT. See RISK ASSESSMENT.

ETHANOL MOTOR FUEL. See ALCOHOL FUELS.

EXPEDITED SITE CHARACTERIZATION. See SITE CHARACTERIZATION, EXPEDITED.

EXPLOSIVES - CONTAMINATED GROUNDWATER. See GROUNDWATER TREATMENT, EXPLOSIVES CONTAMINATION.

EXPLOSIVES CONTAMINATION: DETERMINATION IN WATER AND SOIL

JEHUDA YINON
Weizmann Institute of Science
Rehovot, Israel

SHMUEL ZITRIN
Israel Police Headquarters
Jerusalem, Israel

It has been known for many years that most explosives are toxic (1). 2,4,6-Trinitrotoluene (TNT) poisoning can lead to severe diseases such as aplastic anemia or toxic jaundice. In aplastic anemia, which in many cases is fatal, the blood-forming organs fail to function, resulting in a progressive loss of the blood elements. Toxic jaundice is an indication of severe liver damage. The most severe symptoms resulting from 1,3,5-trinitro-1,3,5-triazacyclohexane (RDX) poisoning are convulsions followed by loss of consciousness. Symptoms of glycerol trinitrate (NG) poisoning, eg, headache, throbbing in the head, and palpitation of the heart, are mainly the result of reduction of blood pressure. In severe cases, the heart muscle is affected directly and the heart beats are weakened. There may also be delirium and convulsions or sudden collapse. It is therefore evident that explosives present an occupational health hazard to workers employed in the explosives and ammunition industries and to military personnel who handle such materials.

However, beyond the occupational health hazard, explosives constitute a general environmental contamination problem, the result of the disposal of ammunition and explosives waste into the environment. For many years obsolete explosives and munitions were disposed of by being buried in the ground or dumped in the sea. Explosives and ammunition manufacturing plants produce large amounts of wastewater containing nitro explosives and other nitro compounds. The discharge of this wastewater into rivers and streams has contributed to environmental contamination.

In order to assess the extent to which an area is contaminated by explosives, it is necessary to analyze trace amounts of explosives and their degradation products in ground and surface water and in soil. Such analyses would reveal the degree of contamination of areas suspected to be polluted by improper disposal of explosives or by wastewater from ammunition plants.

ANALYTICAL TECHNIQUES

Analytical techniques for the detection and identification of explosives in water and soil have been extensively reviewed (2,3).

Types of Explosives

The main group of explosives causing environmental pollution are the military explosives. They include 2,4,6-trinitrotoluene (TNT); 1,3,5-trinitro-1,3,5-triazacyclohexane (RDX), also known as hexogen or cyclonite; 1,3,5,7-tetranitro-1,3,5,7-tetrazacyclooctane (HMX), also known as octogen; pentaerythritol tetranitrate (PETN), also known as penta or ten and smokeless powders, containing mainly cellulose nitrate (nitrocellulose) (NC), glycerol trinitrate (nitroglycerin) (NG), and nitroguanidine. NG is also used as an industrial explosive. In addition, the degradation products of explosives may also contribute to environmental contamination. The degradation products of some of the above-mentioned compounds are well known, while others are still under investigation. Most military explosives contain additives, such as plasticizers, binders, and stabilizers. Some are accompanied by impurities which are by-products of their production, such as 2,4-dinitrotoluene (2,4-DNT) in the synthesis of TNT.

Thin-Layer Chromatography

Thin-layer chromatography (TLC) (4,5) has the advantage over gas chromatography (GC) and high performance liquid chromatography (HPLC) in that it is simple, rapid, and inexpensive. For explosives that are in a complex matrix (such as humic acid), the results are relatively independent of the matrix.

Automated multiple-development (AMD) high performance thin-layer chromatography (HPTLC) has been applied to the analysis of explosives in water and soil extracts (6,7). The method enables identification and quantitation of explosives as well as their degradation products and by-products. Up to 20 samples can be chromatographed simultaneously. The technique is well suited for on-site screening of explosives-contaminated areas. A schematic diagram of the apparatus (6) is shown in Figure 1. Samples are applied as spots on the plate which is placed in the developing chamber. A mobile-phase gradient is generated by mixing the solvents in the mixing chamber. A fixed volume of the mobile phase is forced into the developing chamber and the separated spot is developed along a preselected distance (1–5 mm). The mobile phase is then sucked from the developing chamber into a waste bottle. The development is successively repeated in the same direction for 10–30 cycles. A stepwise gradient is generated by changing the composition of the mobile phase for each successive cycle. Densitometric evaluation is done by UV absorbance at various wavelengths. Identification of compounds is done by comparison of the flow rate (R_f) with cochromatographed standards.

This method has been used to detect and identify explosives in groundwater near former ammunition plants (6). Automated TLC separation was done on precoated Merck HPTLC silica gel 60 plates with fluorescence indicator F254, using a gradient starting from 100% ethyl acetate to an 85:15 mixture of petroleum ether and ethyl acetate, or a gradient consisting of mixtures of methanol, ethyl acetate, and petroleum ether. Figure 2 shows the chromatogram of the extract of a groundwater sample. Detection limit was found to be 10–20 ng. Although the analysis time is relatively long (about 4–5 h), this method allows the simultaneous analysis of up to 20 samples, which include unattended development. Additional explosives detected by this method included picric acid and HMX.

Gas Chromatography

GC is a well-established analytical technique (4), widely used in environmental analysis of pollutants in air, water, and soil. The use of capillary columns allows the separation of complex mixtures. The improvements in injection techniques and the incorporation of new detectors with high sensitivity and selectivity have greatly contributed to the extensive use of GC.

The GC detectors (2) used for the analysis of explosives are the flame-ionization detector (FID), the electron-capture

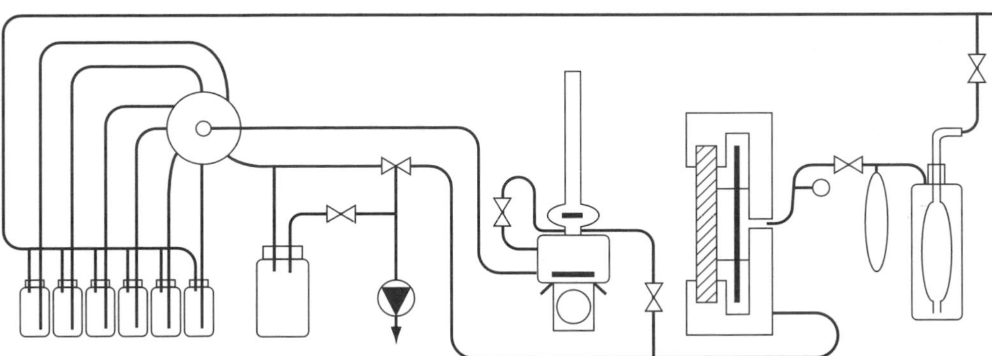

Figure 1. Schematic diagram of apparatus for automated TLC multiple development (6). Copyright American Chemical Society.

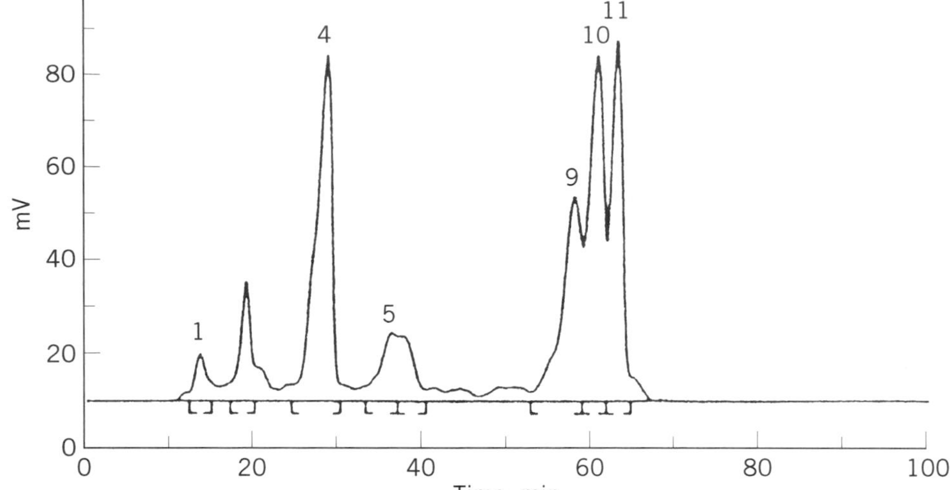

Figure 2. Thin-layer chromatogram of a groundwater sample from a former ammunition plant (6). Peak identification: 1 = hexyl; 4 = RDX; 5 = 2-amino-4,6-dinitrotoluene; 9 = 2,4-DNT; 10 = 2,6-DNT; and 11 = TNT. Copyright American Chemical Society.

detector (ECD), the nitrogen–phosphorus detector (NPD), and the pyrolysis–chemiluminescence detector, mainly known as the thermal energy analyzer (TEA) detector. The incorporation of the TEA detector into GC has increased the specificity of the latter for the identification of explosives containing nitro groups. The FID has been useful in the analysis of nitroaromatic explosives, but much less for nitrate esters and nitramines, where the C:O ratio is much smaller (4).

Trinitro-, dinitro-, and mononitrotoluenes in groundwater and in soil were determined in the ppb-range by injecting chloroform extracts into a gas chromatograph (GC) with an FID (8). The column used was a glass capillary ((12 m × 0.2 mm inside diameter (ID)) coated with deactivated carbowax OV101, programmed from 60 to 150°C, at a rate of 15°C/min. The samples were injected into a splitless injector heated to 230°C. The FID was kept at 250°C.

GC–ECD has been used to determine contamination by explosives of water wells in the neighborhood of a former munition-filling plant (9). The GC analysis of extracts (by solid-phase extraction) from water samples was carried out on a DB-5/30N capillary column, maintained at 180°C. Nitrogen was used as carrier gas. The limit of detection for RDX was 0.3 ng/μL. Other contaminants found in the water samples included DNT isomers, TNT, and TNT degradation products such as 2-amino-4,6-dinitrotoluene and 4-amino-2,6-dinitrotoluene, which were detected at the μg/L level.

Liquid–liquid extracts of nitroaromatic compounds from contaminated water have been determined by GC with ECD and TEA detection (10). The columns used were DB-17 and OV-225 (30 m × 0.32 mm ID). The column temperature was 70°C for 1 min, then raised to 250°C at 3°C/min, using nitrogen (ECD) or helium (TEA) as carrier gas.

An ECD, an NPD, and a TEA were compared for the analysis of nitroaromatic compounds in ammunition wastewater and in aqueous samples from former ammunition plants (11). A DB-5 column (30 m × 0.32 mm ID) was used with nitrogen as carrier gas. The column temperature was 100°C for 1 min, then raised to 200°C at 3°C/min. Both on-column and split/splitless injections were used for sample introduction. Sylilation of the injector insert was recommended in order to avoid adsorption of TNT on the walls of the insert (11).

Although ECD was found to be the most sensitive detector, exhibiting relatively high selectivity, the NPD detector can be more advantageous for the analysis of ammunition wastewater because of its higher selectivity. As expected, the TEA detector has the highest specificity, because it detects only compounds with a nitro or nitroso group, but it was found to be the least sensitive of all three detectors. Detection limits for TNT were found (11) to be 9.4 pg with the NPD and 0.04 pg with the ECD; for 2,4-DNT, 15 pg with the NPD, 0.08 pg with the ECD, and 85 pg with the TEA. In another study (12), detection limits of GC–TEA were found to be between 1 to 10 ng, depending on the type of explosive.

Aqueous samples can be enriched by a factor of 1000 after liquid–liquid or solid-phase extraction. This should result in an increase of 10^3 in the detection limit (assuming 100% recovery and 1 μL injection). In practice, this detection limit can be reached using the selective TEA and in some cases with the NPD, but not with the ECD, because of interferences from solvent impurities.

Figure 3 shows a gas chromatogram (with an NPD) of a water sample from a drained field near a former ammunition plant. The main components found were 2,6-DNT (593 μg/L), 2,4-DNT (103 μg/L), and TNT (592 μg/L). Using supercritical fluid extraction–gas chromatography with a TEA detector (13), the minimal detectable quantity for the analysis of 2,4-DNT in soil samples was found to be 2.6 ppm.

High Performance Liquid Chromatography

The application of HPLC to the analysis of explosives has been widely reviewed (2,4,14). Two detectors have been used with HPLC in the analysis of explosives and their degradation products in soil and in water: UV detectors and, to a smaller extent, electrochemical detectors. The U.S. Environmental Protection Agency (EPA) has issued an HPLC method, Method 8830, for the analysis of nitroaromatic compounds and nitramines in water and soil at ppb levels (15). This method is based on solvent extraction of analytes from soil using sonication and salting-out extraction of aqueous samples, followed by an isocratic HPLC separation and UV detection. The method has been recommended for the analysis of the following com-

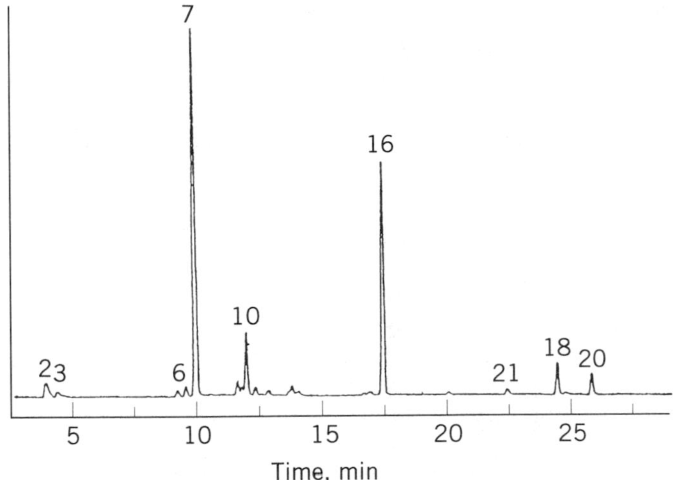

Figure 3. Gas chromatogram (with NPD) of a water sample from a drained field near a former ammunition plant (11). Peak identification: 2 = 2-nitrotoluene; 3 = 4-nitrotoluene; 6 = 1,3-dinitrobenzene; 7, = 2,6-DNT; 10 = 2,4-DNT; 16 = TNT; 21 = RDX; 18 = 4-amino-2,6-dinitrotoluene; and 20 = 2-amino-4,6-dinitrotoluene. Copyright VCH Verlagsgesellschaft.

pounds: RDX; HMX; TNT; 4-amino-2,6-dinitrotoluene and 2-amino-4,6-dinitrotoluene, which are degradation products of TNT; tetryl, 2,4- and 2,6-DNT; 2-, 3-, and 4-nitrotoluene; 1,3,5-trinitrobenzene (TNB); 1,3-dinitrobenzene (1,3,-DNB); and nitrobenzene. The compounds are separated on a C-18 reversed-phase column (25 cm × 4.6 mm ID; 5-μm particle size) and confirmed on a CN reversed-phase column (25 cm × 4.6 mm ID; 5-μm particle size). UV detection wavelength is 254 nm.

Electrochemical detectors (EC) are based on the electrochemical characteristics of certain groups of compounds, which are expressed by the transfer of electrons (16). A potential of sufficient magnitude is applied to a working electrode, over which the analyte is passing, forcing an oxidation or reduction reaction to take place. The flow of electrons to or from the working electrode can be measured as a current, the magnitude of which is proportional to the amount of analyte passing through the detector. Because the occurrence of the oxidation or reduction reactions is dependent on the type of compound being analyzed, HPLC/EC has better selectivity than HPLC/UV detection.

HPLC/EC has been used for the analysis of munition compounds, including RDX, HMX, PETN, NG, TNT, and 2,4- and 2,6-DNT in surface and groundwater (17). The method is based on Porapak resin adsorption of the explosives from the aqueous samples. Following desorption with acetone, the compounds containing nitro groups are measured by HPLC/EC at a gold–mercury electrode, maintained at −1.0 V, vs a silver–silver chloride reference electrode. The column was a Zorbax octadecyl silica (ODS) column (25 cm × 4.6 mm ID; 5-μm particle size). Mobile phase was 1-propanol–0.025 *M* sodium acetate:0.025 *M* monochloroacetic acid (30:70).

Detection limits were found to be about 1 μg/L for most of the analyzed components. The major limitations of the EC detector are the inherent presence of oxygen in the system (even though the system is fitted with an oxygen removal apparatus and the samples are deoxygenated) and the deterioration of the electrode with time, causing an increase in the background current.

Gas Chromatography/Mass Spectrometry

Gas chromatography/mass spectrometry (GC/MS) is a well-established analytical method for the separation and identification of pollutants in various matrices. The scope and importance of its use for the analysis of explosives has been described (2). Nitroaromatic compounds in water have been determined using GC/MS (10). Three types of ionization were used: electron ionization (EI), chemical ionization (CI), and negative-ion chemical ionization (NCI) (2,4).

Liquid Chromatography/Mass Spectrometry

Liquid chromatography/mass spectrometry (LC/MS) combines the separation capabilities of the high performance liquid chromatograph (HPLC) with the identification power of the mass spectrometer. As the HPLC operates at room temperature, it is well suited for thermally labile or low volatile compounds, such as explosives. LC/MS has been used mainly in forensic applications of explosives analysis (18,19), and only to a limited extent in environmental applications. An example of the latter is the analysis of the generated wastewater associated with the manufacturing process of TNT. The main component of this wastewater, called redwater, contains a large variety of nitroaromatic compounds, such as dinitrotoluenesulfonic acid isomers, TNT isomers (other than 2,4,6-TNT), di- and trinitrobenzene, and di- and trinitrobenzoic acids (2,20). Particle beam (PB)-LC/MS has been used for the analysis of redwater components (20). Best results were obtained in the negative-ion chemical ionization (NCI) mode, using methane as moderating gas. HPLC consisted of a C-18 (15 cm × 2 mm ID) column, using acetonitrile–water with 1% acetic acid as mobile phase, at a flow rate of 0.3 mL/min. Mobile-phase solvent ratio was CH_3CN-H_2O (15:85) for two minutes, then changing with a linear gradient to 100:0 within 15 min. Detection limits for the nitroaromatic compounds in redwater ranged from 50 to 500 ng. It seems likely that LC/MS will find its way as a major method for the analysis of explosives in environmental matrices.

On-Site Field Tests

Although a variety of laboratory methods are available for detection and analysis of explosives in water and soil in areas suspected to be contaminated by explosives, the use of field methods enables fast screening of such contaminated areas. This is especially important for the determination of explosives in soil before they decompose and leach through the water-unsaturated zone into groundwater.

Some of the analytical methods described above (TLC, GC, HPLC, and GC/MS) have been adapted for field tests on the contaminated sites. In addition, several other on-site field tests have been developed. A color field test was developed for the screening of TNT in soil (21). The test is based on the well-documented (4) reaction between polynitroaromatic compounds and alkalies, producing colored complexes known as Meisenheimer complexes.

Immunoassays are immunochemical detection methods based on a reaction between a target analyte and a specific antibody (22) (see IMMUNOASSAY, ENVIRONMENTAL). Quantitation is performed by monitoring a color change or by measuring fluorescence. In the enzyme-linked immunosorbant assay (ELISA), the specificity of the antibody for the analyte, producing the resulting immune complex, is the basis for the specificity of immunoassays. A sample is spiked with a known amount of a labeled analyte. The label is typically an enzyme. A chromogenic substrate is added to serve as an indicator of the analyte concentration in the sample. Several enzyme-linked immunosorbent assays have been developed for TNT and other nitroaromatic compounds (23).

Groundwater samples were analyzed without preparation. Soil samples were extracted with methanol, followed by filtration with a glass-fiber filter. Quantitative determinations were made by reflectance measurements with a portable reflectometer. A quantitative color response to concentrations of TNT in the range of 1–30 μg/L in water and 50–1000 μg/kg in soil was obtained. Similar polynitroaromatic compounds, such as TNB and tetryl, showed high cross-reactivities, while the heterocyclic nitramines RDX and HMX were totally inactive.

A competitive, fluorescence immunoassay has been developed for the detection of TNT (24). TNB–SO_3H was labeled by coupling with a fluorophore (an amine-containing derivative of sulfoindocyanine–succinimidyl ester) and used as the analyte competitor. A solution containing 7.5 ng/mL of the resulting fluorescently labeled derivative was exposed to an antibody-coated optical fiber. When these antibodies bind the fluorescently labeled TNT analogue, laser light excites the fluorophore, generating a signal that corresponds to the reference (100%) signal. When TNT is present in the sample, the binding is prevented, thereby decreasing the signal. Inhibition of the reference signal is proportional to the TNT concentration in the sample. Detection sensitivities of 10 ng/mL TNT in water samples were obtained.

PREPARATION OF SAMPLES

Soil Samples

The common method for the recovery of explosive contaminants from soil is the extraction of the explosives and their metabolites with an organic solvent. This is because the problem of environmental contamination by explosives has been usually associated with military explosives, such as TNT and RDX. Munitions containing military explosives have contaminated soils in sites where those munitions were manufactured, tested, distributed, or disposed of by burying in the ground. As most of the military explosives are organic compounds, which dissolve in organic solvents (with various yields), it is not surprising that extraction by organic solvents has been the normal method for the recovery of explosive residues from soil.

The Soxhlet extraction, using a variety of organic solvents, such as 1,2-dichloromethane, benzene, toluene, acetone, and many other solvents, is probably the most widely used method for the extraction of organic residues from soil (25). The efficiencies of four extraction techniques, ie, Soxhlet, ultrasonic bath, mechanical shaker, and homogenizer-sonicator, were compared for the recovery from soil of 2,4-DNT, TNT, TNB, RDX, and HMX, using two different solvents, acetonitrile and methanol (25).

Water Samples

Explosives have been extracted from water by two principal methods: liquid–liquid extraction, where extraction from water is carried out by an organic solvent, and solid-phase extraction (SPE), in which the analytes are adsorbed on a solid phase.

DISPOSAL AND REMEDIATION

The disposal of large quantities of explosives in an environmentally acceptable manner is a serious problem which has not yet been solved. The practice of open burning or detonation of obsolete explosives and ammunition is not acceptable. These methods contribute to air pollution through production of particulates, nitrogen oxides, and other toxic compounds (26). Some methods of incineration, which overcome part of these difficulties, are available, but are very costly.

Microbial processes could be used for the remediation of explosives-contaminated soils and wastewater because a variety of different microorganisms are able to metabolize certain explosives. In some cases a complete mineralization has been found, whereas in others only biotransformation reactions took place (27,28).

Nitroaromatic compounds are biodegradable under aerobic conditions to a certain extent. However, with an increasing number of nitro groups on the ring, these molecules become more and more persistent. The aerobic biodegradation of RDX and HMX has had so far very limited success. Satisfactory results were reported with the white rot fungus *Phanerochaete chrysosporium* (29). The same fungul species was able to degrade 2,4-DNT and nitroglycerin and their corresponding intermediates in the presence of glucose as a primary carbon source (30). The explosives were mineralized into carbon dioxide and water.

Under anaerobic conditions, TNT and RDX formed amino-substituted compounds; this makes further degradation easier (28). The development of genetic engineering offers new possibilities of constructing bacterial strains capable of mineralizing the desirable xenobiotics. Recombination techniques were reported to have designed microorganisms that were able to mineralize nitrotoluenes completely (31).

A method that is being investigated (32,33) is supercritical water oxidation. The first stage of this two-step method involves hydrolysis of the explosive compound. This breaks the compound down into water-soluble, nonexplosive products. These are then introduced into a flow reactor containing supercritical water. The combustible materials are destroyed using an oxidant in water at temperatures of 350–600°C and pressures of (17,240 kPa) (≥2500 psi). The products formed are water, CO_2, nitrogen, metal oxides, and inorganic salts. Several hydrolysis reagents were tested, including sodium hydroxide, ammonium hydroxide, and concentrated aqueous nitric acid.

Ammonium hydroxide was found (33) to be the safest and most controllable hydrolysis reagent, as well as the most compatible with the supercritical water oxidation method. Hydrolysis products include glycerol, ammonium acetate, ammonium formate, ammonium oxalate, ammonium nitrate, ammonium nitrite, ammonium perchlorate, and hydrated alumina (solid).

NOMENCLATURE

Symbol	Definition
AMD	automoted multiple development
CI	chemical ionization
1,3-DNB	1,3-dinitrobezene
2,4-DNT	2,4-dinitrotoluene
2,6-DNT	2,6-dinitrotoluene
3,5-DNT	3,5-dinitrotoluene
EC	electrochemical detector
ECD	electron-capture detector
EGDN	ethylene glycol dinitrate
EI	electron ionization
ELISA	enzyme-linked immunosorbant assay
FID	flame-ionization detector
FTIR	Fourier transform-infrared
GC	gas chromatography
GC/MS	gas chromatography/mass spectrometry
HMX	1,3,5,7-tetranitro-1,3,5,7-tetrazacyclooctane (octogen)
HPLC	high-performance liquid chromatography
HPTLC	high-performance thin-layer chromatography
ID	inside diameter
LC/MS	liquid chromatography/mass spectrometry
MIBK	methyl isobutyl ketone
MS/MS	mass spectrometry/mass spectrometry (tandem mass spectrometry)
NB	nitrobenzene
NC	cellulose nitrate (nitrocellulose)
NCI	negative-ion chemical ionization
NG	glycerol trinitrate (nitroglycerin)
NPD	nitrogen–phosphorus detector
ODS	octadecyl silica
PB-LC/MS	particle beam-liquid chromatography/mass spectrometry
PETN	pentaerythritol tetranitrate
RDX	1,3,5-trinitro-1,3,5-triazacyclohexane (hexogen, cyclonite)
RSD	relative standard deviation
SEX	1-aceto-3,5,7-trinitro-1,3,5,7-tetrazacyclooctane
SFC	supercritical-fluid chromatography
SFE	supercritical-fluid extraction
SIM	selected-ion monitoring
SPE	solid-phase extraction
SPI	septum-equipped programmable injector
SPME	solid-phase microextraction
TAX	1-aceto-3,5-dinitro-1,3,5-triazacyclohexane
TDM	thermal-desorption modulator
TEA	thermal-energy analyzer
TLC	thin-layer chromatography
TNB	1,3,5-trinitrobenzene
TNB-SO_3H	2,4,6-trinitrobenzenesulfonic acid
TNT	2,4,6-trinitrotoluene

BIBLIOGRAPHY

1. J. Yinon, *Toxicity and Metabolism of Explosives,* CRC Press, Inc., Boca Raton, Fla., 1990.
2. J. Yinon and S. Zitrin, *Modern Methods and Applications in Analysis of Explosives,* John Wiley & Sons, Inc., New York, 1993.
3. M. E. Walsh, T. F. Jenkins, and P. G. Thorne, *J. Energ. Mater.* **13,** 357–383 (1995).
4. J. Yinon and S. Zitrin, *The Analysis of Explosives,* Pergamon Press, Oxford, U.K., 1981.
5. C. F. Poole and S. K. Poole, *Anal. Chem.* **61,** 1257A–1269A (1989).
6. C. Steuckart, E. Berger-Preiss, and K. Levsen, *Anal. Chem.* **66,** 2570–2577 (1994).
7. J. Sohr, W. Janes, and A. Bongartz, *Analusis Mag.* **23,** M25–M26 (1995).
8. H. Kohler, *Proceedings of the 3rd International Symposium on Analysis and Detection of Explosives,* Mannheim-Neuostheim, Germany, 1989, pp. 6.1–6.20.
9. R. Haas and co-workers, *Fres. Z. Anal. Chem.* **338,** 41–45 (1990).
10. J. Feltes and co-workers, *J. Chromatogr.* **518,** 21–40 (1990).
11. K. Levsen and co-workers, *Acta Hydrochim. Hydrobiol.* **21,** 153–166 (1993).
12. J. Feltes and J. Koll, *LC-GC Int.* **7,** 698–701 (1994).
13. E. S. Francis and co-workers, *J. Microcol. Sep.* **7,** 23–28 (1995).
14. J. B. F. Lloyd, *Adv. Chromatogr.* **32,** 173–261 (1992).
15. U.S. Environmental Protection Agency, Office of Solid Waste, SW-846 Method 8330 (Revision 0), Washington, D.C., 1994.
16. P. T. Kissinger, *J. Chem. Educ.* **60,** 308–311 (1983).
17. M. P. Maskarinec and co-workers, *J. Chromatogr.* **302,** 51–63 (1984).
18. R. D. Voyksner and J. Yinon, *J. Chromatogr.* **354,** 393–405 (1986).
19. D. W. Berberich, R. A. Yost, and D. D. Fetterolf, *J. Forensic Sci.* **33,** 946–959 (1988).
20. T. Chow, *Proceedings of the 205th ACS National Meeting,* Denver, Colo., 1993.
21. R. T. Medary, *Anal. Chim. Acta,* **258,** 341–346 (1992).
22. J. M. Van Emon and V. Lopez-Avila, *Anal. Chem.* **64,** 78A–88A (1992).
23. D. L. Eck, M. J. Kurth and C. Macmillan, in J. M. Van Emon and R. O. Mumma, eds., *Immunochemical Methods for Environmental Analysis,* American Chemical Society, Washington, D.C., 1990.
24. L. C. Shriver-Lake and co-workers, *Anal. Chem.* **67,** 2431–2435 (1995).
25. T. F. Jenkins and C. L. Grant, *Anal. Chem.* **59,** 1326–1331 (1987).
26. J. Yinon, R. A. Yost, and S. Bulusu, *J. Chromatogr. A* **688,** 231–242 (1994).
27. W. C. Roberts and W. R. Hartley, eds., *Drinking Water Health Advisory: Munitions,* Lewis Publishers, Boca Raton, Fla., 1992.
28. T. Gorontzy and co-workers, *Crit. Rev. Microbio.* **20,** 265–284 (1994).
29. T. Fernando and S. D. Aust, in D. W. Tedder and F. G. Pohland, eds., *Emerging Technologies in Hazardous Waste Management II,* American Chemical Society, Washington, D.C., 1991.
30. N. Pal and C. Christodoulatos, *J. Energ. Mater.* **13,** 259–282 (1995).
31. E. Duque and co-workers, *J. Bacteriol.* **175,** 2278–2283 (1993).
32. I. Stambler, *R&D Mag.* 14 (Dec. 1993).
33. M. H. Spritzer, D. A. Hazlebeck, and K. W. Downey, *J. Energ. Mater.* **13,** 185–212 (1995).

F

FENTON'S CHEMISTRY IN WASTEWATER TREATMENT

RICHARD J. BIGDA
The Technotreat Corporation
Sedona, Arizona

Fenton's Chemistry is an old technology, but an environmental chemist's new tool for the treatment of toxic organic chemicals in wastewater and remediating contaminated soils. Simply, Fenton's reagent is the combination of hydrogen peroxide and common ferrous sulfate to form exceedingly powerful hydroxyl radicals that readily react with a wide number of organic chemicals. At ambient temperatures and pressures, Fenton's reagent is applied to solutions containing high chemical concentrations, mixtures of compounds, dirt, colored solutions, and oily wastes. Carbon dioxide, water, nitrogen, and smaller, less toxic molecules are the usual end products. Industrialization of the process has allowed the cleanup of waste waters created by paint stripping, engine cleaning, explosives, plastic production, wood treating, oil refining, and many other processes. The principle of Fenton's Chemistry and its applications are presented to guide waste water treatment engineers and chemists in developing new techniques for protecting and remediating the environment.

Considerable laboratory effort has been expended to elucidate the chemistry of catalyzed hydrogen peroxide. This research work has been directed to explaining the formation and effect of the powerful hydroxyl radical and the interaction of the iron catalyst. Usually, systems containing pure reactants are necessarily used to study these mechanisms. Unfortunately, waste water streams are not pure or consistent, nor do they ideally react. Because of the complex nature of waste water streams, this entry is presented as a guide to the chemical environmental engineers in using this impressive oxidation tool. Various references are presented to allow deeper insight into the chemistry, but there is little published data on wastewater streams. However, successful projects are reviewed with suggestions for treating other industrial streams.

Commercial Fenton reaction design and use is presented to provide a prudent approach to toxic wastewater treatment. The industrial process scheme and safety concerns are explained. Lists of chemicals known to react with the hydroxyl radical are included, as well as unreactive organics. These guidelines will aid the engineer in developing an operating procedure for a specific wastewater stream. The economics of the process are considered and mainly depend on the quantity of hydrogen peroxide required to meet discharge regulations.

FENTON'S CHEMISTRY IN WASTEWATER TREATMENT

The term Fenton's Chemistry originated in 1876, when Dr. H. J. H. Fenton sent a letter to the editor of the *Chemical News* in London, entitled "On A New Reaction Of Tartaric Acid", describing the reactions of tartaric acid with hydrogen peroxide catalyzed by ferrous sulfate. However, a more significant event occurred in 1894 with his publication of "The Oxidation of Tartaric Acid In The Presence Of Iron", published in the *Chemical Society Journal* of London (1). The oxidation of malic acid was also described at that time. This research was the genesis of "Fenton's Chemistry." Following this description of a new oxidative chemistry came "New Formation of Glycolic Aldehyde", also in the *Chemical Society Journal,* 1895. In 1899, the *Chemical Society Journal* published "The Oxidation of Polyhydric Alcohol In The Presence Of Iron." A paper appeared in the *Chemical Society Journal* in 1900 entitled "The Oxidation Of Organic Acids In The Presence of Iron". These basic research publications presented chemical researchers with a new tool for oxidizing organic chemicals. During the dawn of the age of synthetic chemistry, most reactions were explored as tools for developing new organic compounds. Fenton's Chemistry became relegated to obscurity for a period, because it was not ideal in preparing novel organic products. *The Merck Index* (9th edition) lists Fenton's reaction as a reaction for the oxidation of a hydroxy acid using hydrogen peroxide and ferrous salt to synthesize keto acids and convert 1,2-glycols to hydroxy aldehydes. Fenton's reagent was said to be widely used for the hydroxylation of many organic compounds. It is doubtful that chemists of the late nineteenth century had any thoughts at all of using oxidative chemistry to destroy organic compounds. They were intent upon synthesizing new chemicals, while having few environmental concerns. Wastes and failed experiments were dumped into the sewer or in waste pits.

The first chemical mechanisms for the oxidative properties of Fenton's reagent was presented by E. Able in 1920 in Germany. W. C. Bray and R. S. Livingston proposed the Fenton reaction mechanisms in the *Journal of American Chemical Society* in 1923. The next important step in developing the novel Fenton chemistry occurred with a publication of in 1934 proposing that the hydroxyl radical is the active oxidant formed by the catalytic decomposition of hydrogen peroxide by iron salts (2). It was not until 1940 that Merz and Waters developed the currently recognized stoichiometric relationships. In 1947, they also published "The Mechanism of Oxidation of Alcohol With Fenton's Reagent" in "Discussions of the Faraday Society" (3). Renewed interest was shown in Fenton's chemistry in the late 1940s and increasing research was performed in the 1970s, which was fundamental to understanding the oxidative power of the hydroxyl radical. Recently, in the 1980s and 1990s, Fenton's chemistry has been rediscovered by environmental chemists as a unique and powerful process to eliminate a wide variety of toxic organic chemicals from wastewater by completely destroying the molecules or so altering them that they are not toxic.

REACTIVITY OF FENTON'S REAGENT WITH VARIOUS ORGANIC CHEMICALS

The industrial use of Fenton's reagent, along with research reports, shows that a wide variety of organic chemicals react with hydrogen peroxide. Nearly all chemical groups react to some extent, but not all chemicals in the group react at the same rate. Literature gives experimental reaction rates for a number

Table 1. Relative Reactions of Selected Compounds

Alcohols	K_3/K_2
Methyl	4.03
Ethyl	7.00
n-Propyl	11.3
Isopropyl	6.65
Isobutyl	14.0
tert butyl	1.9
Neopentyl	12.
Ether	
Tetrahydrofuran	6.72
Carbonyl compounds	
Acetone	0.322
Acetic acid	0.089
Amide	
Dimethylformamide	6.0

Table 2. Peroxide Reactivity

Compound	Reactivity
Phenol	High
Cresols	High
2,4-Dimethylphenol	High
2,4-Dichlorophenol	High
Toluene	Moderate
o-Xylene	Moderate
Ethylene glycol	Low
Aromatic distillates	Low
Methyl ethyl ketone	Low
Total organic halides	Little to no reaction
Chloroform	Little to no reaction
Methylene chloride	Little to no reaction
Trichloroethane	Little to no reaction
Tetrachloroethylene	Little to no reaction
Carbon tetrachloride	Little to no reaction

of pure chemicals in water. Because of the different activities exhibited by compounds or the tendency for some compounds to be oxidized, individual chemicals in complex mixtures do not show the same rates of disappearance as in the pure form. A number of functional groups are reactive: acids, alcohols, aldehydes, aromatics, amines, dyes, ethers, and ketones. Within the groups, reactivities vary considerably. Table 1 compares the reactivities of the hydroxyl radical with alcohols.

The relatively high reactivities of isobutyl alcohol and neopentyl alcohol compared with methyl and *tert*-butyl alcohol illustrate variations in one group. Acetone with a K_3/K_2 of 0.322 and acetic acid with 0.089 demonstrate how difficult it might be to oxidize a chemical mixture completely. The data in Table 1 show the reactivity relative to the reaction of Fe^{2+} and hydrogen peroxide (K_2) and the reaction of RH with H_2O_2 (K_3). These studies were conducted at a very low pH, less than 1. It was also observed that hydroxyl radicals add rapidly to double bonds. The literature on the reactivities of various specific organic chemicals with Fenton's reagent is somewhat mixed, in that different values are reported for the reactions. This may be due to variations in the pH at which the reaction occurred, H_2O_2/substrate concentrations, etc. Thus, the rates are not completely comparable. One set of data gives a list of compounds, beginning with phenol as the most reactive, to organic halides with very low reactivity. Table 2 is included as a guide to the reactivity of different substrates with Fenton's reagent.

Data have been developed for the batch kinetics of nonchlorinated and chlorinated aromatics (4). This test data was based on a pH of 3.5 an average ratio of hydrogen peroxide to the substrate of about 14 to 1, and iron at a ratio of H_2O_2 to iron of about 700 to 1 on a mole basis.

More literature is being published on Fenton's reactions in the last decade of the twentieth century, and this new research will provide a rich source of relative reactivities and perhaps even more elucidation of the kinetics of the Fenton's reaction with various substrates. It is wise to review as much of the late literature as possible and to conduct bench scale tests on wastewater prior to treating large commercial batches.

Along with this survey of reactivities, it is important to consider those compounds which are final reaction products or compounds which do not seem to react at all with Fenton's reagent. To aid in developing treatment procedures, a list of compounds known to react follows:

Chemicals Susceptible to Fenton's Oxidation

	Glycerol	1,4-Butanediol	1,2-Propenediol
	t-Butyl	1,3-Propenediol	2,2-Trifluorethyl
Acids	Formic	Propionic	Lactic
	Hydracrylic	Pyruvic	Glyoxylic
	Gluconic	Glycollic	Tartaric
	Phenylacetic	Malic	Dihydroxytartartic
Alcohols	Methanol	Ethanol	Isopropanol

Chemicals Susceptible to Fenton's Oxidation

	Benzyl	Ethylene glycol	2,2,2-Trichloroethyl
	Neopentyl		
Aldehydes	Formaldehyde	Acetaldehyde	Glycolaldehyde
	Glyoxal	Isobutyraldehyde	Glyceraldehyde
	Benzaldehyde	Chloral (trichloroacetaldehyde)	
Aromatics	Benzene	Phenol	Toluene
	Xylene	Cresol	*p*-Nitrophenol
	Quinone	Hydroquinone	Chlorophenol
	Creosote	Dichlorophenol	Trichlorophenol
	Chlorobenzene	Phenylacetic acid	Benzyl alcohol

	Benzoquinone	Catechol (1,2-benzenediol)	
	Naphthol	Nitrophenol	TNT
			RDX
Amines	Aniline	Dimethylformamide	Quaternary ammonium
	Primary	Secondary cyclic amines	Tertiary n-propylamine
	Diamine		
	1,3-Propane diamine		Diethylamine
	Ethylenediaminetetraaceticacid (EDTA)		
Dyes	Monoazo	Diazo	Anthraquinone
	Drimarene	Drimarene Navy	Remazol Yellow GL
	Scarlet K26	V-RBL	
Ether	Tetrahydrofuran		
Ketones	Dihydroxyacetone	Methyl ethyl ketone	
Organic Halides	Chlorophenol	Dichlorophenol	Trichlorophenol
	Pentachlorophenol	Chloral	Chlorobenzene
	Perchloroethylene	Dichloroacetyl	Chloride (DCAC)
	Polychlorinated biphenyls (PCBs)		
	2,3,7,8-Tetrachlorodibenzo-p-dioxin (TCDD)		
	Trichlorobenzene		
Pesticides			
	Dieldrin		
	Trifloralin		
	Pentachlorophenol		
	n-hexadecane		
	Phenoxyacetic herbicides		
PAH			
	Acenaphthylene		
	Naphthalene		
	2-Methylnapthalene		
	Benzo(a)pyrene		
	Dibenzoforan		
	Chrysene		
	Benzo(b)fluoranthene		
	Benzo(a)anthracene		
	Pyrene		
	Fluorane		
	Acenaphthene		
	Fluoranthene		
	Phenanthrene		
	Anthracene		

Compounds known to be unreactive are listed below.

Acetic acid	Acetone
Chloroform	Fumaric acid
Maleic acid	Malonic acid
Methylene chloride	Muconic acid
Oxalic acid	n-Paraffins
Tetrachloroethane	Trichloroethane
Carbon tetrachloride	

These lists are not complete and it may be that, with different pH or catalyst concentration, oxidation occurs. Even though some organics are not completely destroyed, some reaction products of the reactive compounds are classifiable as nonpriority pollutants, reducing the disposal problem.

APPLICATIONS FOR FENTON'S CHEMISTRY IN WASTEWATER TREATMENT

The treatability of any industrial wastewater, of course, depends upon its composition. A wastewater, containing any of the compounds susceptible to Fenton's oxidation, is a candidate for treatment. Industries produce wastewaters which may contain some toxic organics susceptible to oxidation. For example, the petroleum industry produces waste streams containing phenols, benzenes, toluenes, and xylenes. Plywood and adhesive manufactures frequently have phenol, formaldehyde, and resorcinol wastes. The aerospace industry uses a combination of chemicals including phenol, formic acid and benzyl alcohol in stripping paint from aircraft. Industries generating wastewater with Fenton's treatment possibilities are listed below.

Civilian	Military
Aviation paint stripper wastewater	Air Force, Navy, Coast Guard, paint stripper wastewater
Aircraft and engine cleaning wash water treatment	Engine cleaning wastewater
Electroplating	Electroplating
Circuit board wastewater	Circuit board wastewater
Plastics industry	Rocket fuel wastewater
Chemical process wastewater	Explosives wastewater
Refinery phenol and BTX removal	Remediation of contaminated soil

Industrial boiler heat-exchanger cleaning solution treatment
Paint manufacture and clean-up rinse
Process plant wastewater neutralization, heavy metal and oil removal
Explosives wastewater
Pipeline and fuel terminal tank bottom wastewater and dissolved BTX
Tank Car, tank truck, and marine wash water
Dye wastewater
Textiles wastewater
Pesticide production and application
Drugs and pharmaceuticals
Wood treating
Coal-tar refining
Remediation of contaminated soils

Initially, to determine treatability, examine the components of the waste mix and separate the components into categories which are oxidized by Fenton's reagent. The waste is characterized by the chemical known to be used in the operation and by complete analysis of the wastewater. Determine the chemicals which cannot be oxidized and those inorganic volatiles and oily insolubles which are removable by other means. By source segregation, some of the untreatables can be eliminated from the process or be prevented from contaminating the stream. Other unit operations or separation processes can be applied to the wastewater to improve treatability. For example, methylene chloride and other volatile organic chemicals are more cheaply removed by air sparge or steam stripping. Oils and grease are removed by skimming or by adsorbing the oils on clay or other media. Obviously, highly acidic or highly basic solutions are neutralized to the optimum pH before Fenton's reaction. After reaction, the pH is usually adjusted to 7 or above. If a batch reactor is employed, heavy metals are precipitated as the pH is raised and hexavalent chromium is reduced to trivalent chromium by Fenton's reagent and precipitated as the insoluble chromous hydroxide. Occasionally these separations must be performed before Fenton's treatment because of economics, process control, maintenance of equipment or potential interference with the oxidative process. Each case requires careful process planning and probably some bench tests to determine the best procedure for treating the complex mixture. Stepwise processing is not difficult and is generally performed successfully in a properly designed batch treatment system. Continuous treatment of wastewater is complicated, usually because the mixture is too complex or because of rapidly changing compositions. In the batch, the procedure is accomplished move easily, because it is not discharged until all the operations have been completed.

SUMMARY

The treatment of toxic waste waters and remediation of soils is greatly facilitated by oxidizing the chemicals with Fenton's reagent. Many different chemicals are destroyed or degraded to less hazardous products by altering the chemical structure of the molecule through reaction with the powerful hydroxyl radical. This simple to use oxidant is easy to prepare, is inexpensive and is used at ambient temperatures and pressures. Industrial sized Fenton reactors with customized automatic controls are available as skid-mounted units. The treatment process is generally not inhibited by dirt, extraneous materials, oils, or light absorbing colors. During the oxidation of high concentrations of the substrate, caution must be taken to avoid the rapid evolution of heat. A dangerous exothermic reaction occurs if excess quantities of the peroxide coexist with reactive chemicals. Fenton's reagent is a combination of hydrogen peroxide and ferrous sulfate, both readily available and cost effective. This advanced oxidation process is the subject of intensive research and development after nearly one hundred years of neglect. Many new applications are being investigated and publications on the use of Fenton's chemistry are increasing.

BIBLIOGRAPHY

1. H. J. H. Fenton, *Chem. Soc. J. (London),* **65,** 899 (1894).
2. F. Haber and J. J. Weiss, *Proceedings of Royal Society of London (Ser. A)* **14,** 332 (1934).
3. J. H. Merz and W. A. Waters, *Discussions of the Faraday Society* **2,** 179 (1947).
4. F. J. Potter and co-Workers, "Oxidation of Aromatics in Wastewater with Hydrogen Peroxide," *1989 AIChE Spring Meeting,* Houston, Tex.

FIELD ANALYTICS: DYNAMIC WORK PLANS

ALBERT ROBBAT, JR.
Tufts University
Medford, Masschusetts

The ability to rapidly assess the disposition of environmental contaminants at purported or existing hazardous waste sites is an essential component of the nation's environmental restoration program. Each site, whether owned by the public or private sector, must be evaluated to determine whether risk to human health or the environment exists. If the data obtained support the notion that no risk or an acceptable level of risk exists for the intended land usage then no further action may be required. If, on the other hand, sufficient risk has been determined to warrant a full site characterization, the site investigation effort must delineate the nature, extent, direction, concentration, and rate of movement of the contamination along with the physical and chemical site attributes.

Despite the best efforts of the U.S. Environmental Protection Agency (EPA) and other federal agencies, including the Departments of Defense and Energy, to validate field analytical technologies, field analytics have not played a significant role in either hazardous waste site assessments or cleanup (1). In 1995, the EPA issued a request for proposals in support of President Clinton's efforts to promote application of innovative environmental technologies and to address the many factors that might pose barriers toward their commercialization. The president's Environmental Technology Initiative (ETI) is focused on accelerating environmental protection, strengthening

America's industrial base, and increasing exports of U.S. technologies and expertise. The Tufts project was directed at two key objectives identified in the FY95 strategic plan; namely, strengthening the capacity of technology developers and users to succeed in environmental innovation and strategically investing EPA funds in the development and commercialization of promising new environmental monitoring, control, and remediation technologies.

The dynamic work plan guidance document represents one aspect of these objectives. The document is aimed at helping federal and state regulators, site owners and their consulting engineers, and remediation companies understand what is involved in constructing and carrying out a dynamic work plan. The purpose of the document is to illustrate the many factors that should be considered in incorporating field analytical instrumentation and methods into an adaptive sampling and analysis program for expediting the site investigation process. This dynamic process should result in a faster, better, and (it is hoped) cheaper site characterization and cleanup. With these goals in mind, field analytical technologies developed by the Tufts Center for Field Analytical Studies and Technologies and with in-kind support from several commercial companies were demonstrated in the context of a dynamic work plan and adaptive sampling and analysis strategy. The ETI project, in part, supported an ongoing soil investigation study at Hanscom Air Force Base in Bedford, Massachusetts (2). With the assistance of the EPA Region 1 and the Air Force and its contractor (CH2MHill), a videotape was produced illustrating the dynamic site investigation process.

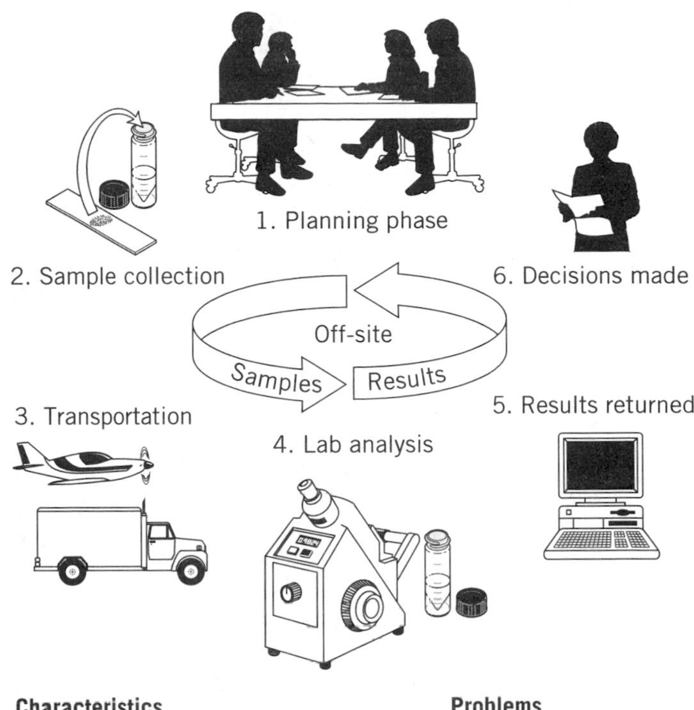

Figure 1. Traditional site investigation.

DYNAMIC WORK PLANS

Successful hazardous waste site investigations should be focused, with goals and objectives being clearly defined. This does not mean, as has been practice, that the site investigation process should result in work plans that are etched in stone. Figure 1 depicts a traditional sampling and analysis program. The work plan relies on prespecified sampling locations, numbers of samples collected, and the types of analysis to be performed. The traditional site investigation is static in its application. It does not provide a framework for changes in direction based on what is learned in the field. Samples are collected, packaged, and typically sent off-site for analysis. Because data turnaround times range from several weeks to several months, analytical results are unavailable during the field investigation phase to address surprises or concerns while the sampling team is still on site. Experience has shown that multiple field investigations within the same or subsequent seasons are required to fill data gaps. The traditional process results in several trips to the field by the sample collection and survey teams before the site investigation can be completed. This static process typically occurs during hazardous waste site cleanups as well.

Dynamic work plans (Fig. 2) provide an alternative to the traditional approach. Dynamic work plans rely, in part, on an adaptive sampling and analysis strategy. Rather than dictate the details of the sample analysis to be performed and the location and number of samples to be collected, dynamic work plans specify the decision-making logic that will be used in the field to determine which chemical compounds require analysis, where to collect the samples, and when to stop sampling. Adaptive sampling and analysis programs change as the conceptual model for the site is refined based on the analytical results produced in the field. A successful adaptive sampling and analysis program requires analytical methods and instrumentation that are field practical and can produce data fast enough to support the dynamic work plan process.

Factors to Be Considered

Several factors should be considered before embarking on a dynamic work plan and adaptive sampling and analysis program for projects consisting of complex chemical and physical site conditions, environmental contamination, and long duration. The following questions should be asked:

Is it possible to assemble a well-rounded core technical team, including analytical chemists, engineers, geologists, geochemists, geophysicists, hydrogeologists, risk assessors, and regulators?

Will the core technical team be in the field for the duration of the field investigation? Is the decision-making process well defined and is the authority vested in an appropriate technical team member?

Has the action level for field decisions, which rely on developing an understanding of the scientific and engineering questions under investigation, been established as part of the data quality objectives?

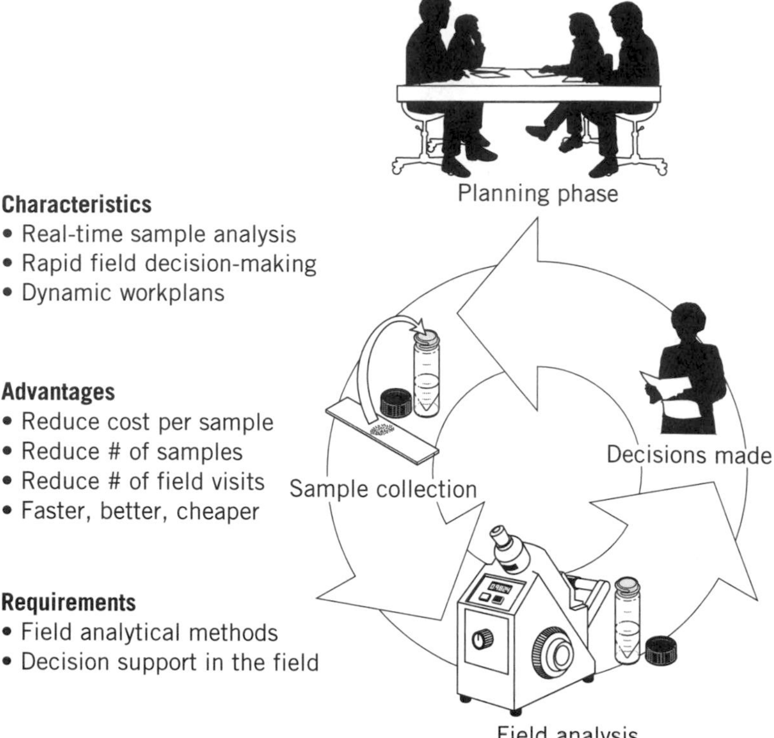

Figure 2. Dynamic work plan approach.

Will the project objectives permit screening and semiquantitative data or will only quantitative data be required to meet data quality objectives?

Will more than 10% of the samples analyzed in the field be sent off-site for laboratory confirmation analysis? Has the methodology for determining field and laboratory data comparisons been addressed?

When selecting the field instrument or method, have measurement selectivity, sensitivity, precision, accuracy, representativeness, and action levels been addressed?

When selecting the field instrument or method, have the measurement attributes listed above been addressed in terms of sample throughput rates and cost? (Note that the number of sample cleanup steps and the time needed to prepare samples for analysis to meet the site-specific data quality objectives may limit throughput rates and increase sample costs.)

Can standard operating procedures and method detection limit studies be completed before mobilization to evaluate matrix interferences that might be associated with a particular field technology?

Will data management tools and geostatistical sampling tools be integrated into the field investigation?

Is the site accessible for field analytic deployment, including mobile laboratories, electrical power (line voltage versus a generator), and water, if necessary?

Has sufficient space been provided to house analytical instruments and staff, sample preparation, and data management in the field laboratory? Has proper ventilation been incorporated into the field laboratory?

Does the length of the project and the potential overall cost savings warrant this approach?

DYNAMIC WORK PLAN GUIDELINES

Dynamic work plan investigations are site dependent. They incorporate field-based technologies and methods that produce chemical, physical, geological, and hydrogeological information about the site. The data generated must be of sufficient quality, with respect to measurement precision, accuracy, sensitivity, and completeness, to support the objectives of the site investigation or cleanup. The dynamic work plan guide described here is not intended to be all inclusive. It does not address subsurface sampling tools; methods for collecting soil, water, or air samples; remote sensing and geophysical surveys; and mathematical or computer modeling. Nor will it discuss computer-based statistical sampling or the various site visualization tools. Depending on project objectives, a successful dynamic hazardous waste site investigation or cleanup will require one or more of these tools.

The guidance document is aimed at integrating field analytics into the dynamic work plan and adaptive sampling and analysis process. It is intended to lay the foundation for incorporating an iterative process into the static but widely used data quality objectives (DQOs) framework for decision-making and planning. The guideline outlines field analytical instrument implementation, an adaptive sampling and analysis strategy, and site requirements.

THE DYNAMIC WORK PLAN PROCESS

In the traditional approach, major decisions concerning the direction of the site investigation or cleanup are generally made by the project manager after the fieldwork has been completed. A report is prepared presenting the findings to the appropriate regulatory body. Discussions begin about whether sufficient information has been obtained to address the scientific and engineering questions of concern. Typically, several field mobilizations occur, reports are written, with many meetings held among the site owner and its environmental consulting company and the site owner and federal and/or state regulatory agencies. In contrast, these same decisions are made in the field in an adaptive sampling and analysis program. In constructing the dynamic work plan, it is important to determine before mobilization what decisions will be made, how these decisions will be made, and who will make them in the field.

INTRODUCTION TO FIELD ANALYTICS

The selection of field analytical methods critically depends on the need to make decisions in the field rapidly. Field analytical techniques should be capable of providing data from minutes to tens of minutes. They should have documented measurement sensitivity, precision, and accuracy to meet site investigation and cleanup DQOs. The simpler the technique the more likely it will be used in the field. Field instruments must be transportable, operate under adverse conditions, and provide improved costs and benefits over laboratory analysis. For projects of short duration and low sample volume, staff and equipment mobilization expenses may make field analytics a cost-prohibitive option. In addition, if quantitative measurements are required for all samples, field analytics may not provide a cost-effective means for obtaining site data. Rarely is this the case. Almost all projects will require screening or semiquantitative data during the field screening phase of the site investigation. Even short projects of 1–3-days, in which 6–12 samples per day may be collected, will benefit from field measurements. For example, head space gas chromatography (GC) can be simple and fast for the analysis of volatile organic compounds (VOCs) in soil and water samples during underground storage tank removal or well installation and monitoring. Immunoassay kits can provide rapid detection of polychlorinated biphenyls (PCBs) or explosives during site characterization or remediation. Field instrumentation, such as *in situ* fiberoptics and electrochemical sensors or portable GCs can be used to provide a security system to monitor underground subsurface contamination migration, process control, or fugitive emissions during site cleanups or long-term monitoring operations.

Field analytics can be routinely used to monitor worker and community health and safety during site investigations and cleanups. For example, the protection of workers from exposure to hazardous substances during sampling is of primary concern. In this case, sampling speed and limited sample handling are important aspects of the measurement process. The sampling and measurement methods must be suitable to meet guidelines set forth by the National Institute for Occupational Safety and Health.

DYNAMIC VERSUS TRADITIONAL INVESTIGATION AND CLEANUP COSTS

Dynamic work plans provide the framework for collecting chemical, physical, geological, and hydrological data in one or two field efforts compared to the phased engineering approach of collecting data, evaluating them; collecting more data, evaluating them until sufficient information is obtained to meet the study objectives. Fixed-based (commercial) laboratories should be able to generate data of comparable (either screening or quantitative data) quality at lower per sample costs than field or mobile laboratories. Economies of scale should be more easily achieved by fixed-based laboratories, since they are designed for mass production. However, steep sample surcharges (100–200%) are generally added to the base price if samples are moved up in the queue to obtain 1- to 3-day data turnaround times. Moreover, fixed-base laboratory sample analysis costs vary greatly between regional (typically local non-CLP) and national laboratories.

Comparing the selection of field instruments as a function of cost is difficult. Field instruments and methods should be chosen first to meet the data quality requirements and second based on their ability to match the rate at which samples are collected.

The second point is not a trivial or obvious statement. If sample analysis lags behind sample collection, sample collectors and decision support staff sit idle waiting for data to be produced. On the other hand, if sample collection is operating below capacity, analytical instruments and field laboratory personnel sit idle. In both cases, site investigation efficiency and cost is lost. Therefore, it is essential that the analytical team member play an integral role in designing the sample collection program. Moreover, combining field screening and on-site quantitative analysis into the program should increase the total number of samples analyzed while decreasing the number of samples sent off-site for traditional laboratory analysis.

Table 1 summarizes the field and laboratory sample charges and data turnaround times for the analysis of VOCs, PCBs, polyaromatic hydrocarbons (PAHs), explosives, and semi-VOCs. Commercial laboratory charges vary widely, depending on the size and revenue amount of the laboratory and the number of national programs the laboratory participates in (eg, contract laboratory program (CLP), U.S. Army Corps of Engineers, Hazardous Waste Remedial Actions Program (HAZWRAP), and state certified programs). Field analytical technologies can provide analyses comparable in cost to regional or local laboratories employing EPA standardized methods with same or next day data turnaround times compared with 14–35 days by commercial laboratories. Field analytics compete best when total project cost is considered and when it is incorporated into the dynamic work plan and adaptive sampling and analysis program. Cost savings can be realized when:

Sample selection and locations are optimized. Increased sampling efficiencies result in more targeted sample collection efforts, minimizing the handling of samples that provide little value toward answering site-specific DQOs—faster site characterizations and verification of cleanup.

The identity of the contaminants becomes known. Increased field analytical productivity is obtained when the type of analysis performed is more targeted, resulting in more samples analyzed per day—*faster* site characterizations and verification of cleanup.

More data are produced in less time. More informative decisions are made that improve the site delineation process,

Table 1. Field and Laboratory Cost and Data Turnaround Time Comparison

Analyte	Regional Laboratory, Data Turnaround = 14 Days	National Laboratory Contract Laboratory Program, Data Turnaround = 35 Days	Field TDGC/MS, Data Turnaround = 1 Day	Field GC/PID or GC/ECD, Data Turnaround = 1 Day	Strategic Diagnostic Enzyme Kits, Data Turnaround = Same Day
VOCs	*$125/sample*; SW 846 Method 8240/8260; 25-min/sample analysis	*$165/sample*; SW 846 Method 8240/8260; 25-min/sample analysis	*$100/sample*; modified Method 8260; 20-min/sample analysis	*$88/sample*; modified Method 8021/8015 or headspace analysis, 25 min/sample analysis	Not applicable
PCBs	*$100/sample*; SW 846 Method 8080 20 min/sample analysis; sample preparation, 2 h/batch for 20 samples	*$150/sample*; SW 846 Method 8080; 20 min/sample analysis; sample preparation, 2 h/batch for 20 samples	*$100/sample*; field method; 10 min/sample analysis; sample preparation, 1-h/batch for 20 samples	*$88/sample*; field method; 20 min/sample analysis; sample preparation, 1-h/batch for 20 samples	*$102/kit*; field method; 10 min/sample analysis; sample preparation, 1-h/batch for 20 samples
PAHs	*$145/sample*; SW 846 Method 8100/8310; 20 min/sample analysis; sample preparation, 2 h/batch for 20 samples	*$255/sample*; SW 846 Method 8100/8310; 20 min sample analysis; sample preparation, 2 h/batch for 20 samples		Not applicable	*$102/kit*; field method; 10 min/sample analysis; sample preparation, 2 h/batch for 20 samples
Explosives	*$180/sample*; SW 846 8330/USAED 30; 20 min/sample analysis; sample preparation, 18 h/batch for 20 samples	*$220/sample*; SW 846 8330/USAED 30; 20 min/sample analysis; sample preparation, 18 h/batch for 20 samples	*$100/sample*; field method; 10 min/sample analysis; sample preparation, 1-h/batch for 20 samples	Not applicable	*$102/kit*; field method; TNT and RDX kits require 20 min/sample analysis; sample preparation, 1 h/batch for 20 samples
Semi-VOCs	*$400/sample*; SW 846 Method 8270; 40 min/sample analysis; sample preparation, 4 h/batch for 20 samples	*$450/sample*; SW 846 Method 8270; 40 min/sample analysis; sample preparation, 4 h/batch for 20 samples	*$150/sample*; field method, 20 min/sample analysis; sample preparation, 1-h/batch for 20 samples	Not applicable	Not applicable

ie, the separation of highly contaminated areas from noncontaminated areas—*better* site characterizations and verification of cleanup.

A more detailed picture of the site is obtained. Increased confidence in evaluating the risk to human health and the environment results—*better* site characterizations and verification of cleanup.

More efficient use of human and financial resources is obtained. Increased project efficiencies lead to more data obtained at lower total project costs—*cheaper* site characterizations and verification of cleanup.

The rationale for selecting an adaptive sampling and analysis program should be based on the inherent efficiencies obtained when decisions are made in the field and the overall total project cost savings that can accrue.

ACKNOWLEDGMENT

The many helpful suggestions offered by J. Smaldone and N. Conlon from the U.S. Environmental Protection Agency, Region I, in preparing this guideline are greatly appreciated. Financial support was provided through assistance agreement No. X99163–01, by the U.S. EPA, Environmental Technology Initiative.

BIBLIOGRAPHY

1. E. Koglin and L. R. Williams, *Trends Anal. Chem.*, **13**, 294–299 (1994).
2. A. Robbat Jr., *Tufts University Case Study: Dynamic Workplans and Field Analytics: The Keys to Cost-Effective Site Investigations*, Tufts University, Medford, Mass., 1997.

FIELD-PORTABLE INSTRUMENTATION

MICHEL G. GOEDERT
Hewlett-Packard Laboratories
Palo Alto, California

In recent years, the area of environmental laws governing emissions has been one of the fastest-growing areas of legislation. Its primary aim has been to protect the environment by exercising control over emissions from industrial processes. From country to country, different agencies monitor compliance, and failure to meet the standards can result in prosecution and heavy penalties.

Sources of emissions can be broadly split into two areas: mobile (vehicles) and stationary (plants). The volume of legislation governing automotive exhaust emissions has been on the increase for some time. Most of this legislation involves the measurement of oxygen, carbon monoxide, carbon dioxide, unburned hydrocarbons, and to some extent, nitrogen oxides. An instrument used to monitor mobile sources for field measurements has some requirements: it must be small and light; use a minimum of consumable fuel (electrical power and gases); and operate for hours, even days, without need of recharging.

Of primary interest among stationary sources of emissions are the emissions from flues or stacks. The composition of flue gas depends on the fuel or chemical process causing the emission, but the main components are emitted as a result of fossil fuel burning. The gases to be analyzed consist mostly of oxygen, unburned hydrocarbons, oxides of carbon, sulfur, and nitrogen in varying concentrations. The majority of legislation governing emissions is concerned with carbon monoxide, carbon dioxide, sulfur dioxide, nitric oxide, and nitrogen dioxide.

In many instruments designed specifically to monitor a single chemical, interferences are minimized, and concentrations within the healthy range—permissible exposure limit (PEL) and threshold limit value (TLV)—can be reliably and accurately measured. When used correctly and appropriately, measurements made with these dedicated instruments can often be used in lieu of traditional instruments, such as gas chromatographs.

Generally, instruments dedicated to a single chemical are less expensive and easier to use in the field than are instruments that detect many different chemicals, such as gas chromatograph (GC) and broad-spectrum infrared (IR) units. The instruments with broader measurement capabilities can also be dedicated to the measurement of a single chemical, but this is generally not cost-effective. IR instruments dedicated to the measurement of a single chemical are constructed more simply than those with broader monitoring capabilities. One available variation is a single-instrument housing that can be adapted to monitoring different chemicals by replacement of the sensors. The choice of instrument depends on the situation to be monitored; in general, for atmospheres for which the types of contaminants are well characterized, the more specific the monitor the better. On the other hand, with a number of unknowns the more appropriate instrument will be a nonspecific general-survey instrument.

The intent of this article is to provide a guide to nonoptical techniques used for the analysis of gaseous samples. The most commonly available technologies are described here, along with descriptions of commercial products using those technologies in practical applications, if available. Most of these techniques are mature, and their theoretical bases well established. On the other hand, the hardware used for field measurements is under continuous development, so information is valid only up to the time of writing. It is important to check the current state-of-the-art information in specialized journals and contact vendors to get updated information about products.

Measurements can be performed off-line by collecting the samples and analyzing them later, generally in the laboratory. Several approaches are described in detail. Sample collection can be discrete (grab sampling, use of canister) or done over time (solid sorbent, impingers).

When the decision is made to take a field measurement, the proper analytical technique must be selected. Although many approaches are available, the final decision is based on what is already known about the sample. If the components of the gaseous sample are known and the user wants to determine only the presence and the concentration level of the analytes, the use of cartridge tubes or any other technique based on colorimetry is perfectly acceptable. These tubes are inexpensive, sensitive, and give visual confirmation of the level of the gases present. However, they are irreversible and cannot be used as permanent monitoring devices. Electrochemical sensors are used when an unknown element of the gaseous sample must be determined. These sensors, described below, are relatively selective and sensitive enough to detect gas or combinations of gases at the sub–parts per million level. They can be operated in an alarm (monitoring) condition and are reversible.

Portable monitors, such as flame ionization detectors (FID) or photoionization detectors (PID), are also described; they give some useful information regarding the family of compounds of the analytes. They are sensitive and reliable but provide only a little information about the nature of the sample itself. They are easy to use and give the go/no-go information necessary for a quick decision. Several commercial units are described.

The next category of measurements uses instruments that separate the components of the mixture, at least partially, before detecting them and thus allows quantification of the sample components. Gas chromatographs fall into this category. Gas chromatography is a mature technology that separates the various constituents of a gaseous mixture before detecting them. GC is not an absolute method of identification. Since GC identification is based only on the comparison of retention times and not on the physical properties of the analytes, multiple compounds may be eluting at the same time and not identified. Therefore, it is not possible to detect with 100% certainty all the components of a totally unknown sample. However, GC provides a reasonable amount of information and is so convenient that it is commonly used. It is fast; inexpensive; easy to use; and when combined with other analytical techniques, such as mass spectrometry, provides an accurate method of identification. The components of a GC, from the injection to the separation to the detection of the sample, are described. Portable units are becoming smaller, more sophisticated, and more versatile. However, the quality of data provided with portable units is still significantly less than the quality of measurements obtained with laboratory instruments. Some of the portable units available will be described, and the pros and cons of those techniques will be identified.

The next category of measurement is mass dispersion methods. Two of these methods are discussed: mass spectrometry

(MS) and ion mobility spectrometry (IMS). IMS was developed to detect toxic gases in chemical warfare applications. IMS is more qualitative than quantitative. IMS is sensitive (parts per billion level), fast (within seconds), and ideal for detecting suspected gases, but it is not as good at identifying totally unknown mixtures. IMS instruments are portable and easy to use, even by unskilled operators. Since interest in this technique is growing for civil applications, it will be described here.

Mass spectrometry, especially combined with GC separation, is the most powerful but also the most expensive and difficult to use analytical tool available. MS provides positive identification, even in complex situations. It is fast (less than 1 s) and accurate. Its dynamic range is large (in decades) and sensitive. This technique is now becoming portable, thanks to the technological improvements of portable vacuum technics (better material, turbomolecular pumps). It is usually coupled with GC to simplify the interpretation of spectra.

TECHNIQUES OF COLLECTION FOR VAPORS AND GASEOUS SAMPLES

Regulations and Government Agencies

The analysis of air, soil, and water samples is still, and will continue to be, an important component of the overall effort to resolve the environmental problems associated with undesirable and unhealthy chemicals. Environmental laws and regulations and government agencies are obviously major participants in the activities related to the protection of the environment (1). Any analytical work performed in conjunction with a federal or state regulation or government agency must conform to the guidelines that it provides. A working knowledge of the guidelines and of the capabilities of the various analytical techniques available is essential for any analyst involved in environmental analysis (2).

At the federal level, the U.S. Environmental Protection Agency (EPA) is the most recognized regulatory organization. From the regulatory perspective, the primary function of the EPA is to interpret laws passed by Congress and to formulate operating practices. Other prominent federal agencies, such as the Department of Energy (DOE), the U.S. Army Corps of Engineers (COE), the Department of the Interior (DOI), the Department of Labor (DOL), the Occupational Safety and Health Administration (OSHA) and the National Institute of Occupational Safety and Health (NIOSH) are also charged with the protection of the environment. Detailed information on a particular government agency or regulation may be obtained directly from the agency.

Ambient Air Quality Standards

At the federal level, the quality of the air is measured by comparing pollutant concentrations with the National Ambient Air Quality Standards (NAAQS). At the state and local level, it is measured by comparing the pollutants with NAAQS, as well as with state ambient air quality standards (SAAQS) and local ambient air quality standards, if any. Recent legislation (the 1977 and 1990 federal Clean Air Act amendments) has added other parameters, such as aesthetics (visibility) and risk (air toxicity), to the evaluation of air quality.

Atmospheric Dispersion Modeling

Modeling is used to determine what happens to pollutants emitted into the atmosphere. When a pollutant is emitted, it mixes with the existing air, and its concentration decreases because of atmospheric turbulence. This process is termed *atmospheric dispersion.* Most of the point-source atmospheric models that are used in the regulatory process are termed *Gaussian dispersion* models. Non-point-source models are used to access regional pollutants (eg, urban air shed modeling) or vehicle emissions (eg, freeway or traffic impact modeling). Summaries of preferred air quality models such as SCREEN, SCREEN3, and ISC have been described (3). More information can be found in the EPA guidelines and model handbooks.

Classification of Contaminants

Contaminants may be divided, depending on physical characteristics, into gases, vapors, and particulate matter. The terms *gases* and *vapors* are often used interchangeably, although they are not identical. At ordinary temperature and pressure, both gases and vapors will diffuse rapidly. Gases are generally noncondensable at room temperature; whereas vapors are considered to be condensable, since they are derived from volatile liquids. Therefore, under ordinary conditions, gases remain in the gaseous state even when present at high concentrations. Vapors, on the other hand, may condense at high concentrations and coexist in both gas and aerosol forms.

Particulate matter can be further differentiated as dust, fumes, smoke and soot, and liquid particles. Dusts are solid particles formed from inorganic or organic materials reduced in size by mechanical processes. These particles range in size from the visible to the submicroscopic, but the principal concern of industrial hygienists is with those under 10 μm in size, because such particles stay suspended in the atmosphere for a long period of time and can reach the deepest parts of the lungs.

Fumes are fine particles formed from solid materials by evaporation, condensation, and gas phase molecular reactions. These particles range in size from 0.1 to <0.01 mm. Smoke and soot are products of incomplete combustion of organic materials. The size of smoke particles is usually <0.5 μm. Liquid particles are produced by condensation and atomization from the gaseous state. Droplets formed from atomization are usually >5 μm in diameter.

There are two methods for collecting gaseous samples. In grab sampling, an actual sample of air is taken in a flask, bottle, bag, or other suitable container; in continuous or integrated sampling, gases or vapors are removed from the air and concentrated by passage through an absorbing or adsorbing medium.

Grab sampling usually involves the collection of instantaneous or short-term samples, normally within a few seconds or minutes. This type of sampling is acceptable when peak concentrations are sought or when concentrations are relatively constant. An important feature of grab samples is that their collection efficiency is normally 100%. However, sample decay does occur for various reasons, such as reaction or adsorption on the inner surfaces of the collector.

Grab sampling is of questionable value when the concentration of the contaminant varies with time or is at low concentration, when the measurement device is barely sensitive enough to measure a sample, or when a time-weighted average exposure is desired.

Integrated sampling consists of extracting the gas or vapor and concentrating it by solution in an absorbing liquid, reaction with an absorbing solution, or collection onto a solid adsorbent. Collection efficiency is frequently <100%; therefore, individual efficiency percentages must be determined for each case.

Guidelines for the selection of sampling devices and analytical procedures are available (4–6). Secondary sources are also available in books (7) or in journals such as the *American Industrial Hygiene Association Journal* and *Applied Occupational and Environmental Hygiene.*

Generally speaking, nonreactive and nonabsorbing gaseous substances may be collected as grab samples. Gases and vapors that are water soluble or that react rapidly with absorbing solutions can be collected in simple gas-washing bottles. Gaseous substances that are volatile and less soluble or that react slowly with absorbing solutions require more liquid contact and may require more elaborate sampling devices. Insoluble and nonreactive gases and vapors are collected by adsorption onto activated charcoal, silica gel, or another suitable adsorbent.

Grab Sampling

The preparation steps required for trace analysis of most air or gaseous samples differ from those required for solid or aqueous samples in that they are performed partially in the field; the gaseous samples themselves are never really transported to the laboratory. The organic contaminants contained in these samples are usually extracted or removed from the gaseous matrix and concentrated during the sampling procedure.

In its simplest design, a measured volume of the sample is drawn into the sampling system or directly through the measuring instrument by some type of pump. The simplest sampling system consists of only a collection medium. The analytes are retained on or in the collection medium, which is transported to the laboratory for preparation and analysis.

Gaseous samples can be collected in rigid glass or metal flasks, in soft plastic bags, or in combinations of these with aluminum foil in sizes up to 120 L. For plastic bag sampling, the sample is introduced into the bag by a hand- or battery-operated pump or a squeeze bulb. Pyrex gas-collecting tubes of 300 mL capacity with a capillary standard taper stopcock at each end may be used when the atmosphere sampled contains components incompatible with plastic bag materials. The flask may be evacuated in the laboratory and then opened in the environment to be sampled, or the flask may be evacuated in the field with a pump. Specially passivated stainless-steel pressure canisters, are used for short-period sampling of reactive gases and vapors.

Sample Bags. When low detection limits are not required, sample bags are used to collect gaseous samples and are transported to the laboratory for analysis without intermediate preparation steps. Transportation of gaseous samples in this manner is useful only when the analytes of interest are stable and when other methods of collection are not applicable. For instance, bags are used when sorbents have a low breakthrough volume or when collection of time-weighted average (TWA) samples are required (8).

These bags can be made of plastics but are more often made of materials such as Tedlar, Mylar, Scotchpak, and Teflon. Figure 1 shows a plastic bag equipped with a septum injector.

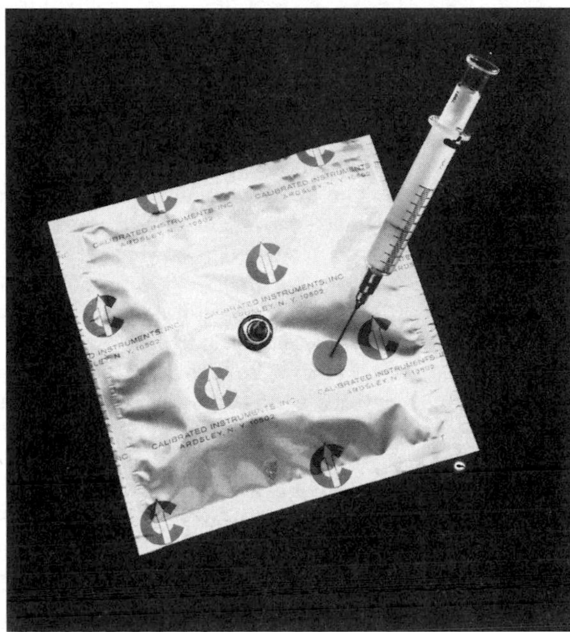

Figure 1. Plastic bag used for the collection of gaseous samples. This bag can also be used to prepare calibration samples by injecting a known quantity of chemicals with a syringe. (Courtesy of Calibrated Instruments Inc., Hawthorne, N.Y.)

The sampling apparatus consists of a 20 L (5-gal) open-top container with a sealable lid, a Teflon stopcock, Teflon bulkheads, Teflon tubing, and a sampling pump (8). The bag is attached to a low-flow exhaust outlet and to an adjustable pump. The sampling period determines the size of the sample bag and the flow rate of the pump. Most bags have a septum for injection using a syringe.

Personal bags are often used to collect gases and vapors for high-resolution analysis of environmental samples and occupational sampling. This method is commonly used for collecting air samples in open fields and from vapor wells. Bulk air samples are collected in small bags carried by an operator. Personal sampling requires that the collection device be small and light. A pump and a battery are provided with the device and must be carried away by the operator. Sampling bags are used when other methods are not applicable, for instance, when sorbent tubes have a low breakthrough volume or when there is no alternative method. Sampling bags are also used when a mixture of gases is incompatible. Sampling bags come in various materials and shapes but most have a valve to allow for injection filling (with a syringe) using a pump and a septum. Sample losses from bags are caused by leaks, chemical reaction of the sample with the bag walls, adsorption of the sample into the bag material, and permeation of the sample through the walls of the bag.

Evacuated Vessels. In many cases, the sampler simply wants to collect a quick sample of air from a given area. Evacuated flasks and cans can be used for this type of collection. Evacuation-type devices are sealed vessels in which a vacuum has been created, with only one inlet. These devices can be used to collect gross gas contamination. The concentration of contaminants must be high enough to be detected in a small-sample volume.

Evacuated cans, similar to those that hold shaving cream, are available commercially. By pushing in the top of the can, the seal is broken, allowing air to enter. Vacu-Sampler (MDA Scientific, Lincolnshire, Illinois) is an example of a container completely evacuated in the factory. When the sample is collected, a volume of 120 cm^3 vapor or gas can be obtained.

Before use, bulk air sampler bags should be cleaned, tested for leaks, and preconditioned by flushing with the gas to be sampled (if known) or zero air.

Stainless-Steel Canisters. Collecting ambient air samples in canisters provides a convenient method of integrating ambient samples over a period of time, such as 24 h. Collection can be unattended and provides enough samples to allow detection of small quantities of pollutants. Two sampling modes are used: passive and pressurized. In passive sampling, an evacuated canister is opened to the atmosphere through a flow restrictor adjusted so that the differential pressure causes the canister to fill over a designated period of time. For pressurized sampling, an initially evacuated canister is filled by the action of a flow-controlled pump from a near atmospheric vacuum to a positive pressure that should not exceed 1.2×10^{-5} N/m^2 (25 psi). Pressurized sampling is used when long-term integrated samples or higher volume samples are required. Commercial canister units are sold with a filter that traps particulate material. Figure 2 shows a canister used for sampling ambient gas and vapor. Canisters are generally not used for the collection of polar compounds, such as methanol, ammonia, and hydrogen chloride, because these adsorb to the walls of the vessel. The list of volatile organic compounds (VOCS) that can be collected is defined in EPA canister method TO-14 (9).

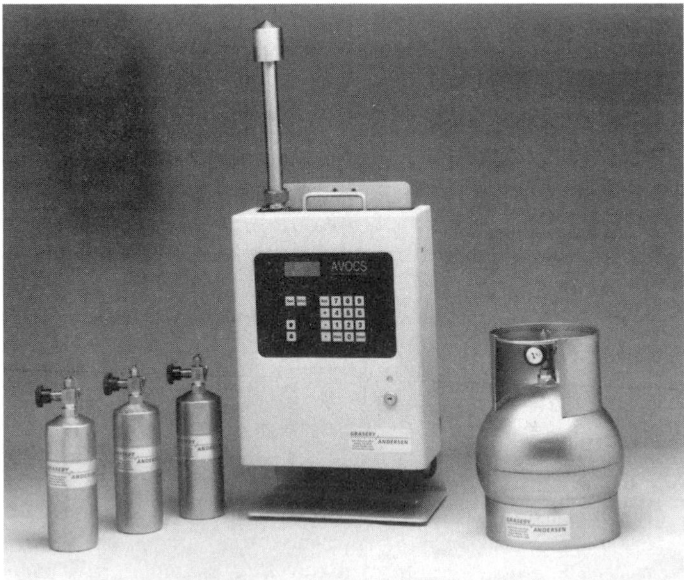

Figure 2. Canister for sampling ambient gas and vapor for VOCs. Samples can be collected in two modes: passive (the canister is under vacuum) and pressurized (the canister is filled with a pump). (Courtesy of Graceby Andersen Instruments, Inc., Smyma, Ga.)

Integrated Sampling

Gas and vapor samples may be collected in a solvent with wash bottles, impingers, and adsorbers; on adsorbents (eg, activated charcoal); by condensation; and in large plastic bags that are filled using a battery-operated pump. After sampling, the contents may be analyzed in the field or laboratory by nondispersive infrared (carbon monoxide), gas chromatography (hydrocarbons, chlorinated solvents), and other methods.

Absorber characteristics vary, depending on the gas or vapor to be collected. Simple bubbling devices, such as impingers and Drechsel bottles, are adequate for readily soluble gases such as hydrogen chloride, hydrogen fluoride, and sulfur dioxide. For less easily absorbable materials, such as chloride and nitrogen dioxide, multiple contact washing is required (as with fretted glass absorbers). Sometimes it is desirable to burn the gas or vapors in a furnace and sample the oxidation products, eg, for chlorinated hydrocarbons.

Adsorption tubes are the method of choice for insoluble or nonreactive vapors. Commonly used adsorbents include activated charcoal, silica gel, and molecular sieves. Gas adsorption traps are sometimes preceded in the device by one or two water vapor adsorption stages containing calcium chloride, calcium sulfate, or silica gel, all of which have excellent water vapor adsorption characteristics and poor adsorption capacity for most organic molecules. In the laboratory, the collected vapors may be desorbed thermally or stripped from the adsorbent with carbon disulfide and the recovered vapors quantified by gas chromatography using a suitable detector.

Adsorption tubes used for personal integrated sampling of many organic gases and vapors contain two interconnected chambers in series filled with gas adsorption charcoal. The first chamber, containing 100 mg charcoal, is separated from a backup section, containing 50 mg carbon, by a plastic foam plug. Sampling can be conducted for as long as 8 h without saturating the first chamber when occupational exposures are at or below the TLV. The contents of the two chambers are analyzed separately to determine whether the first-stage adsorbent has become saturated and has lost an excessive amount of the sample to the second stage.

Gas chromatographic column packing materials (eg, Tenax) are also used in adsorbent traps for field sampling of organic vapors. They are particularly useful for sampling high boiling compounds. The resealed trap is returned to the laboratory for analysis. Absorption of inorganic atmospheric constituents by liquid coatings on solid supports has been used to collect nitrogen dioxide on triethanolamine-coated molecular sieves.

Chemically active compounds may react with each other or with oxygen in the air after adsorption and make it difficult or impossible to recover and quantify the adsorbed gases and vapors. In such instances, the best collection method may be to react the contaminants chemically to a stable derivative or to condense them at low temperatures using a mixture of dry ice and acetone or liquid nitrogen as the coolant. Direct reading instruments combine sampling and analytical functions and usually display results rapidly. Many in this category are also capable of storing continuous readings and displaying averages for selected time intervals on command.

Organic compounds and other pollutants present in air can exist either as gases or as particulate matter. These compounds can originate from various manufacturing and mechanical processes as a result of the incomplete combustion

Table 1. Factors Affecting the Collection Behavior of Solid Sorbents

Factor	Effects
Temperature	Reduced adsorption at high temperature; breakthrough volume is reduced
Humidity	Water vapor is adsorbed by polar sorbents and reduces breakthrough volumes
Flow rate	Varies with type of sorbents; at higher flow, reduces breakthrough volumes
Concentration	Breakthrough volumes decrease at highest concentrations
Mixture	When several compounds are present, the compound most strongly held will displace the other compounds, down the length of the tube
Nature of sorbent	Decreases in sorbent particle size are proportional to increases in sampling efficiency and drops in pressure
Size of tube	Breakthrough volumes are proportional to the sorbent volume

of the fuel material or as a result of the spontaneous volatilization of material from a chemical spill, uncontained waste, or uncontained feedstocks. Vegetation sources also release a significant volume of organic compounds into the atmosphere.

Pollutants that exist primarily in the gaseous state are described as volatile organic compounds (VOCs). These are compounds with boiling points typically around or below 100°C and vapor pressures of $>10^{-1}$ torr. The semivolatile organic compounds (SVOCs) have vapor pressures in the range of $10-10^{-5}$ N/m^2 ($10^{-1}-10^{-7}$ torr) and can exist in the gaseous phase or in the particle-bound phase, either as an adsorbed material or as a liquid droplet or mist. These adsorbed compounds can remain attached to the particles as they are dispersed through the atmosphere, or they can undergo chemical reaction in the atmosphere. Alcohols, aldehydes, ketones, organic acids, esters, and organic nitrates have been identified as components of organic smog aerosols, which are essentially liquids.

Solid Sorbent Sampling. Generally charcoal, silica gel, and chromosorbs are used for occupational air sampling; and Tenax and carbon molecular sieves are preferred for environmental samples. Since the levels expected are much lower in environmental samples, the collection period is much longer. Typically, environmental samples are collected on a 24-h average. Sampling time and volume must be determined experimentally, based on the overall concentration and the number of components.

Solid sorbents are specific for groups of compounds, and one sorbent will not work with all compounds (8). Most solid sorbents do not differentiate among compounds during collection, so unwanted compounds may be collected along with the target compounds. On the other hand, because of the wide range of vapor pressures and breakthrough volumes for organic gases of interest, there is no one solid sorbent that can collect all gases and vapors simultaneously. Sometimes sorbents are coated with reagents to enhance collection of a specific compound. For example, acrolein is collected on 2-hydroxymethylpiperide-coated XAD-2.

The biggest concern in collecting material on a sorbent tube is whether breakthrough can occur. Breakthrough occurs when the front section of a tube is saturated and enough compound accumulates in the backup section that it begins to exit the tube with the airstream (10).

When selecting sorbents, the potential for water absorption must be considered. Both the compound being sampled and the water compete for the absorbent's surface, resulting in concentrations that appear lower than they actually are. Table 1 lists factors affecting the collection behavior of solid sorbents (11).

Charcoal Sorbent. Charcoal is one of the most commonly used sorbents, since it is useful for sampling a wide variety of organic gases and vapors, including several different compounds at a time. Ordinary charcoal is activated by steam at 800–900°C, causing it to form a porous structure. Charcoal tubes are generally analyzed using the solvent carbon disulfide to collect adsorbed contaminants. The adsorbing capability of activated charcoal varies from batch to batch. Charcoal tube collection efficiency for various hydrocarbons may be affected by such variables as sampling rate, vapor concentration and totally adsorbed hydrocarbon mass. Nonpolar compounds, preferentially sampled on charcoal, displace polar compounds in charcoal media. Competitive adsorption also occurs among polar compounds. Factors that may contribute to the affinity of the molecules for charcoal include hydrogen bonding, molecular size, volatility, and the dipole movement.

Breakthrough volumes are variable and a function of the carbon, temperature, humidity, storage times, and pollutant (11). Since charcoal tubes have been shown to be affected by high humidity, sample volume should be reduced when high humidity is present.

Silica Gel Sorbent. Silica gel is considered a more selective sorbent than activated charcoal, and gases and vapors are more easily desorbed from it. It is the adsorbent recommended for collecting organic amines, both alkyl and aromatic, such as aniline and *o*-toluidine.

Factors that affect the dynamic adsorption of materials onto silica gel include the size range of the gel particles, tube diameter and length, temperature during sampling, concentration of contaminants being sampled, air humidity, and duration of sampling. Since the polarity of the adsorbed compounds determines the binding strength of silica gel, compounds of higher polarity will displace compounds of lower polarity. Therefore, when attempting to collect relatively nonpolar compounds, the presence of coexisting polar compounds may interfere with collection on silica gel.

Silica gel shows a decrease in breakthrough capacity with increasing humidity. Under high humidity conditions, the sample is lost because of saturation with water vapor (12).

Molecular Sieves. The carbon molecular sieve is a spherical, macroporous structure, the result of pyrolysis of the synthetic polymeric or petroleum pitch precursors. The diameter of the micropores and their number are responsible for the differences in tube retention volume, adsorption coefficient, and equilibrium sorption capacity for a given compound (13). The limiting factor with molecular sieves is humidity. Carbon molecular sieves are most commonly used to collect environmental samples of highly volatile nonpolar organic compounds.

Porous Polymeric Sorbents. Porous polymers are another class of sorbent used for air sampling and include Tenax GC, Porapaks, Chromosorb 101–108, and XAD tubes. Their wide variety offers a high degree of selectivity for specific applications. Limitations include displacement of less volatile compounds, especially by carbon dioxide; irreversible absorption of some compounds, such as amines and glycols; oxidation, hydrolysis, and polymerization reactions of the sample; chemical changes in the contaminant in the presence of reactive gases and vapors, such as nitrogen oxides, sulfur dioxide and inorganic acids; artifacts arising from reaction and thermal desorption; limited retention capacity; thermal instability; and limitations of sampling volume, flow, and time (12).

The Porapaks are a group of porous polymers that exhibit a wide range of polarity. The least polar member, Porapaks P, is used in gas chromatography columns, and the most polar, Porapaks T, separates water and formaldehyde. The Chromosorbs 101–108 are similar to the Porapaks. Chromosorb 101 is the least polar and Chromosorb 104, the most polar. XAD resins include a number of different porous polymer types. XAD-2, the most commonly used, is equivalent to Chromosorb 102 and is used to collect anisidine and tetraethyl lead.

Tenax. Tenax is one of the most widely used porous polymers, especially for environmental sampling. It has been used for studies of VOC levels in indoor air. Tenax is a polymer of 2,6-diphenyl-*p*-phenylene oxide and can be used to collect organic bases, neutral compounds, and high boiling compounds. Tenax is used mostly for sampling low concentrations of volatile compounds.

Tenax GC has a high thermal stability and can withstand temperatures of up to 350°C, which permits it to be used for thermal desorption. Thermal desorption is desirable because the entire sample is introduced into the analytical system, whereas solvent extraction dilutes the sample and allows only a portion of it to be injected. One limitation of thermal desorption is that the sample can be injected only once.

Tenax has other properties besides its temperature stability that make it useful for concentrating compounds of medium volatility. It is relatively inert and has low, but not zero, affinity for water vapor. Because of its extensive use, its advantages and limitations have been well characterized. It is commonly used for multicomponent qualitative analysis by gas chromatograph/mass spectrometry (GC/MS) and quantitative analysis with standards (14).

However, retention or breakthrough volumes for a variety of highly volatile compounds on Tenax are low (15). Other limitations of Tenax include a laborious cleanup procedure to control blank problems, a short useful half-life after sample collection, and a tendency to decompose during sampling to produce acetophenone and benzaldehyde; the humidity in the air being sampled may affect sample retention; high concentrations of or a high number of compounds may exceed the retention capacity of the Tenax bed; and artifact formation may occur due to chemical reactions during sampling and/or thermal desorption. Tenax is not effective for low molecular weight hydrocarbons (C_4 and below) and midrange (C_5–C_{12}) highly polar compounds.

Tenax also reacts with strong oxidizing agents, such as chlorine, ozone, nitrogen oxides, and sulfur oxides, to form benzaldehyde, acetophenone, and phenol. Another documented chemical transformation with Tenax is that oxidizers react with organics, such as styrene, to produce compounds such as benzaldehyde and chlorostyrene.

Other Sorbents. Other sorbents occasionally used for integrated gas and vapor sampling include alumina gel and florisil. Alumina gel, a form of aluminum oxide, is rarely used as an air-sampling sorbent except for special applications such as those using formaldehyde. Alumina gel permits a several thousandfold concentration of pollutants. It selectively absorbs polar and higher molecular weight compounds, with the degree of polarity determining the binding strength of a compound on this gel. Alumina gel has been used to collect polar compounds (such as alcohol, glycols, and ketones) and aldehydes, such as formaldehyde. Like all polar sorbents, it has a high affinity for water and thus high humidity samples are difficult to handle.

Florisil, based on silicic acid, is used for polychlorinated biphenyl (PCB) collection as well as for some pesticides. Table 2 lists types and uses of solid sorbents (12).

Impingers for Solvent Extraction. Impingers are suitable for collecting nonreactive gases and vapors that are highly soluble in the absorbing liquid as well as those that react rapidly with a reagent in the solution, such as occurs in the neutralization of

Table 2. Types of Solid Sorbents

Solid Sorbent	Characteristics
Activated charcoal	Very large surface area; reactive surface; high absorptive capacity; breakthrough capacity is a function of the source of the charcoal, its particle size, and packing configuration in the sorbent bed
Silica gel	Less reactive than charcoal; polar; hygroscopic
Porous polymers	Less surface area and less reactive surface than charcoal; adsorptive capacity is lower; reactivity is lower
Molecular sieves	Zeolites and carbon molecular sieves retain adsorbed species according to their molecular size; water may displace organics
Coated sorbents	These are sorbents on which a layer of reagent has been deposited; adsorptive capacity is determined by the capacity of the reagent to react with a particular analyte

strong acids and bases. These devices are usually made of glass with an inlet tube connected to a stopper fitted into a graduated vial. The inlet tube rests slightly above the vial bottom. A measured volume of absorber liquid is placed into the vial, and the unit is connected to a pump by flexible tubing. The function of these absorbers is to provide sufficient contact between the sampled air and the liquid surface to provide complete absorption of the gas or vapor. Figure 3 shows how impingers are set up to collect personal samples.

Problems with the use of impingers include condensation of material in the sampling lines and losses by adsorption or volatilization from the equipment. Impingers are bulky and are susceptible to breakage. For these reasons, occupational sampling with impingers is being phased out and replaced by sorbent methods. However, there are still applications, such as in high humidity or when only certain samples (eg, formaldehyde) are collected, for which the impinger approach is necessary (16).

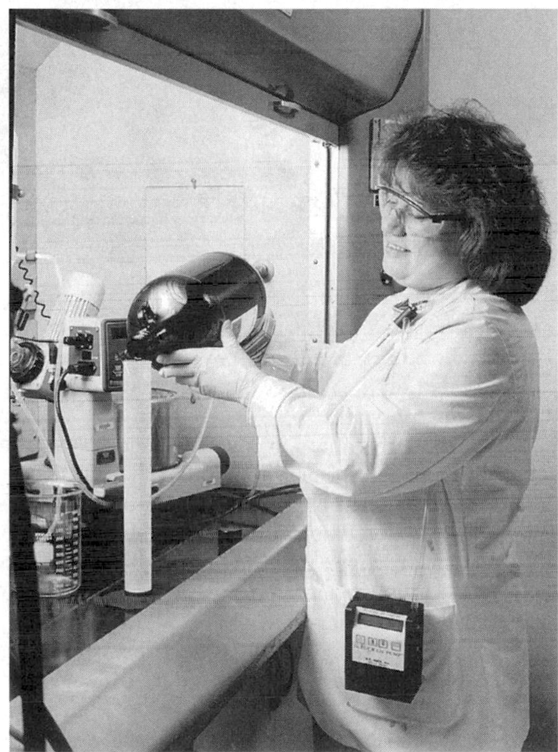

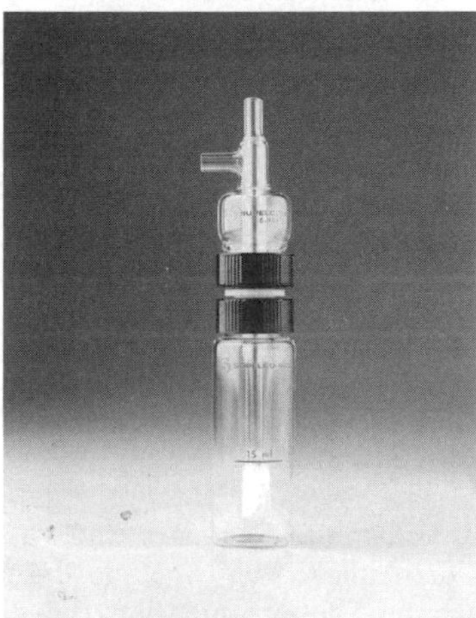

Figure 3. Personal sampling impingers and bubblers. These devices are usually made of glass and filled with the absorbing liquid. (Courtesy of Gilian Environmental.)

Badges. Another passive colorimetric sampling device is the spot plates, or badges. These have indicators strips or buttons designed to change color when a critical accumulation of a given gas is reached. Badges should not be used to warn employees of levels in excess of the PEL. Real-time instruments with audible alarms should be used for these situations.

An example of a badge indicator is the ChromAir (Gilian Environmental, West Caldwell, New Jersey) (Fig. 4). This passive monitor badge is for indoor air quality and OSHA compliance screening. The passive monitor relies on the principle of diffusion. The contaminant gas or vapor being monitored diffuses through a series of six cells on the front of the badge, each having an increased resistance to diffusion. After diffusing through the cells, the gas or vapor reacts with an indicator layer containing a specific reagent for the contaminant, and the cells change color. When sampling is completed, the exposure dose is clearly indicated on the opposite side of the badge.

ChromAir badges are available for ammonia, acetone, chlorine, carbon monoxide, sulfur dioxide, nitrogen dioxide, hydrogen sulfide, carbon disulfide, methanol, ethanol, hydrazine, and formaldehyde. The sensitivity of these badges depends on the application and is typically in parts per million. To obtain the average concentration in parts per million, the exposure dosage (parts per million) is divided by the exposure time in hours.

Passive Collection of Gases and Vapors. Passive monitors are lightweight badge assemblies that rely on natural wind currents rather than pumps to move contaminated air to the collection surface. Most of these units are dosimeters that accumulate an average dose and give an integrated measurement. Some units are specific for a given gas, such as chlorine, ammonia, sulfur dioxide, hydrogen sulfide, nitrogen dioxide, carbon monoxide, and mercury. Others can sample multiple gases such as organic vapors. The obvious advantage of passive monitors is their ease and simplicity of use. When used for personal monitoring, they do not interfere with worker activity and are unlikely to affect the behavior pattern of the wearer, whereas wearing a pump with its tubing and media might. The most commonly used passive monitors contain a solid sorbent similar to that used in sorbent tubes. There are also devices that use liquid absorption solutions and passive devices that change color proportional to the concentration present.

Passive devices rely on two basic collection principles: diffusion and permeation. With diffusion-controlled monitors, the mass uptake of the monitor is controlled by the length and diameter of the badge cavity and the physicochemical properties of the contaminants. When permeation is used for sampling, the mass uptake of the monitor is controlled by the physicochemical characteristics of the membrane and the contaminants. The mass uptake is a direct function of the badge permeation sampling rate, the ambient concentration, and the sampling time.

In diffusion devices, the most common passive devices, molecules pass through a barrier or draft shield that minimizes the effect of air currents on a stagnant air layer; they

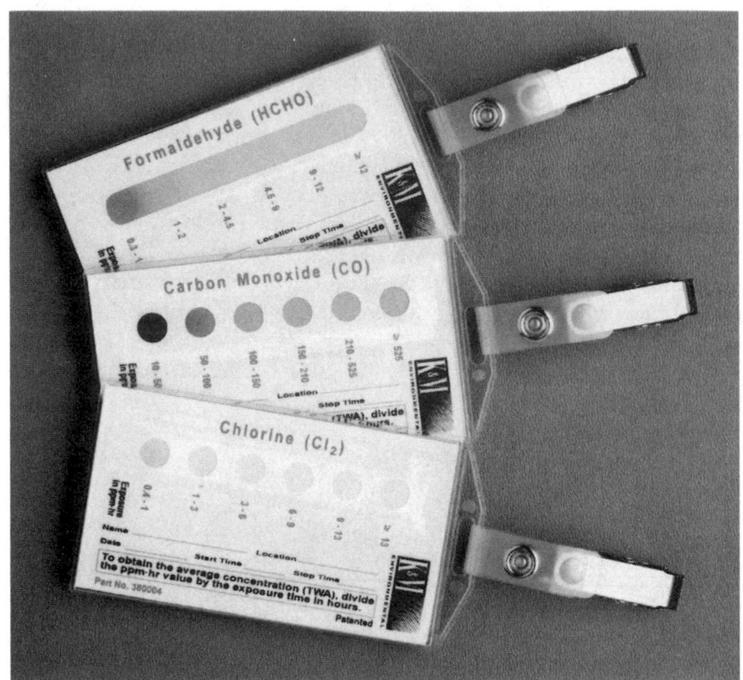

Figure 4. ChromAir passive monitoring badges from Gilian Environmental.

are then collected on an adsorbent material such as charcoal. A concentration gradient is created within the cavity of stagnant air, and the amount of gas or vapor transferred is proportional to the ambient concentration. Each compound has a unique diffusion coefficient for each type of badge. The diffusion coefficient necessary to calculate the final concentration after analysis in the laboratory limits sampled materials to those for which this value is established.

On permeation dosimeters, the gaseous contaminants dissolve in a polymeric membrane and are then transferred to a collection medium, such as a solution. Permeation across the membrane is controlled by the solubility of the gas or vapor in the membrane material and by the rate of its diffusion across the membrane under a concentration gradient. Factors influencing permeation include thickness and uniformity of the membrane, affinity of the membrane, affinity of the membrane for the contaminant, swelling or shrinking of the membrane, and possible etching by corrosive chemicals. The efficiency of these devices depends on finding a membrane that is easily permeated by the contaminant of interest and not by the others.

As for diffusion monitors, sampling rates must be determined for each analyte and type of permeation monitor. One difficulty in making such determinations is the need to use thin and fragile membranes to obtain practical sampling rates. The long response times required to reach equilibrium, the sampling rate, and the degree of permeation may be affected by changes in temperature and ambient humidity.

The accuracy and precision of passive monitors depend on sampling time, air currents, and temperature and humidity effects. In the presence of mixtures, passive monitors may be affected by competition among compounds for adsorptive sites, resulting in displacement of one compound by another or preferential adsorption of one compound over another. Passive monitors must be handled with care, since a penetration in the membrane would cause variations in the amount of contaminant collected. Sampling in a dusty environment may also be a problem if the gas being collected is capable of adsorbing onto the dust.

Collection of organic vapors is one of the most common uses for passive monitors. Organic vapor badges contain twice as much charcoal as a typical charcoal tube. The charcoal is usually impregnated in some sort of a matrix. Regardless of the design of the organic vapor monitor, the analytical procedure is generally the same: desorption with carbon disulfide and analysis by gas chromatography.

The affinity of various compounds for the activated carbon in a badge varies; and because diffusion coefficients exist only for certain gases and vapors, the use of badges is limited to these compounds regardless of the wide applicability of the badges for sampling organic vapors. The accuracy of an organic vapor monitor scan varies, depending on the concentrations being sampled as well as on the exposure period.

The difference in the design of the individual monitors, type of charcoal used, quality control, type of adsorption pad, particle size of the charcoal, amount of charcoal present, surface area of the adsorption material, and membrane or draft shield selection create each monitor's unique performance with respect to the compounds collected.

DIRECT-READING INSTRUMENTS FOR ENVIRONMENTAL MEASUREMENTS

Direct-reading instruments for gases and vapors operate on many different principles. They include detector tubes and instruments such as flame ionization detectors (FID) and photoionization detectors (PID). Chemical sensors are part of another family of direct-reading sensors that is rapidly displacing some of these techniques. Chemical sensors are currently undergoing intense research and development. These detectors are generally not specific but can be calibrated for a single vapor or known mixture. Gas chromatographs, mass spectrometers, and ion mobility spectrometers represent a distinct class of instruments in that they specifically address the issue of separation as well as detection, and as such they will be discussed in later sections. This section reviews direct-reading instruments. A comprehensive review of direct-reading gas and vapor sensors, electrochemical devices, ionization detectors, optic sensors, and instruments is available (17).

Introduction to Direct-Reading Instruments

Chemical sensors are a fast-growing part of the direct-reading instrument area. They are found almost everywhere when a single-point measurement such as oxygen, carbon monoxide, or carbon dioxide must be performed. This type of sensor often operates in multisensor combinations. For instance, oxygen meters are often paired with combustible gas detectors. Oxygen measurements are needed not only to evaluate the presence of the concentrations necessary to sustain life but also to evaluate correctly the readings of combustible gas indicators.

Chemical sensor applications are found almost everywhere: in environmental measurements of air and water, industrial applications for food and pharmaceuticals, in clinical applications, and in many other applications. This great interest in chemical sensors is due to their small size, low cost, low power requirements, and capability of being integrated into portable instruments. This discussion is limited to gas detection.

Indicator (or detector) tubes are direct-reading industrial hygiene air analysis instruments. They are small, light, hand operated, safe in all atmospheres, and generate immediate readouts. In addition, an indicator tube is the simplest and most economical air analysis method available for many common air contaminants. Tube detectors for more than 160 chemicals are currently available, some in more than one concentration range. The most popular types are offered by commercial sources. Most of the chemicals measured by indicator tubes are included in the current tabulation of American Conference of Governmental Industrial Hygienists (ACGIH) threshold limit values. Color-developing reagents give point measurements that are almost instantaneous and integrated and that are collected for a long time. The approach chosen is usually dictated by the regulator and by the sensitivity of the measurement device. The instant repartition of specific contaminants may be lost when integrated measurements are chosen over a long period of time, but this approach is better at identifying contaminants that are harmful to the exposed community.

One of the most important developments in air-sampling technology has been the commercial appearance of passive dosimeters that measure the exposure to low concentrations of a broad list of airborne volatile substances. Many of these dosimeters use the principle of diffusion to a nonspecific adsorbent with subsequent laboratory analysis. These direct-reading devices use permeation through a plastic film barrier to a compound-specific chemical that bonds to color-developing reagents.

Instruments such as photoionization and flame ionization detectors respond to a broad class of organic and inorganic gases and vapors and are used widely for survey measurements. The only specificity of these instruments is determined by the class of molecules they detect. For instance, FID meters detect only organic compounds and do not respond to constituents commonly found in air and water. The threshold of specificity of the PID meter is controlled partially by the selectable level of energy emitted by the lamp.

Chemical Gas Sensors

Chemical sensors transform chemical information into electrical signals. These sensors are characterized by small dimensions and relatively low costs and lead naturally to portable instruments for survey and personal measurements. These sensors are generally classified according to the mode of transduction, eg, electrochemical, optical, or acoustic. Some well-established chemical sensors include electrolyte potentiometric sensors for oxygen, catalytic sensors for combustible gases, and semiconductor oxide sensors. These find use in the detection of both the combustible gases and the toxic gases associated with chemical systems.

New techniques for chemical sensors are emerging, such as field effect transistors, surface acoustic wave devices, fiberoptic gas sensors, and pattern recognition methods used with arrays of sensors in so-called electronic noses that evaluate flavors (18). Solid electrolyte sensors for oxygen and semiconductor gas sensors, based on the response characteristics of a wide range of inorganic oxides and of metallo-organic materials, are also part of this new generation of sensors. Some of the designs are described below.

Electrochemical sensors with aqueous electrolytes, operating on fuel cell principles, have been adopted widely to meet the growing demand for reliable low cost gas-measuring devices in safety and process control applications. They are available for the detection of organic and inorganic gases such as ammonia, sulfur dioxide, hydrogen sulfide, chlorine, carbon monoxide, isocyanates, nitrogen compounds, oxygen, phosgene, formaldehyde, ozone, and hydrogen cyanide. General information about chemical sensors and electroanalytical methods is available (19–22). Applications of real time measurements using electrochemical sensors have been published (8).

The greatest advantage of electrochemical sensors is that they provide a direct relationship between the concentration of gas being measured and the electrical signal at the output (current or voltage, depending on the type of sensor). Also, they can be constructed using well-established engineering methods to produce small, compact devices with a high degree of robustness and reliability, freedom from orientation effects, intrinsically low cost, and suitability for volume production. Such sensors can operate within an ambient temperature range of −20 to 55°C without the need for external heating. Their power requirements are, therefore, low; some designs are available that are completely self-powered, so that additional power is required only for extrasensor functions such as alarm monitoring, recording, and transmitting data.

A wide range of gas concentrations may be measured using electrochemical gas sensors. Thus commercially available systems to monitor oxygen levels range, eg, from nearly 100% in medical and gas-purity applications to around 20% for general air quality systems, to just a few percent in flue gas analysis, down to a few parts per million in some gas purity and process control applications. Toxic gases can be measured from around 20% down to a few parts per billion.

Electrochemical sensors do not suffer from variations in ambient humidity, a problem for many semiconductor and solid state devices. Even through cross-interferences can pose problems, as with any chemical sensor, electrochemical sensors can achieve high specificity when suitably designed and operated.

Principle of Operation of Electrochemical Sensors. The following discussion is limited to potentiometric and amperometric sensors, since they are the most commonly used with portable instruments.

Potentiometric Sensors. If a cell signal reflects a change in the potential of the sensor, it is classified as potentiometric. Potentiometric devices have been well studied and can serve as sensors for a variety of gases such as ammonia, hydrogen cyanide, carbon dioxide, hydrogen sulfide, and sulfur dioxide. Potentiometric devices include ion-selective electrodes (of which the pH electrode is a well-known example). These operate on the basis of the Nernst equation, for which electrode potential is a function of concentration:

$$E = E^* + \frac{RT}{z} \log C \tag{1}$$

where E is the potential of the sensing electrode, referred to as the standard hydrogen electrode (SHE), in volts; E^* is the standard electrode potential of the sensing electrode (SE), in volts; R is the ideal gas constant (8.3143 J/ K/mol); T is the absolute temperature, in degrees Kelvin; C is the quotient of the product of the activities of the product species of the electrochemical reaction, written as an oxidation reaction, over the product of the reactant species of that electrochemical reaction (in dilute solutions, activities may be approximated by concentrations); and z is the number of electrons transferred in the electrochemical reaction.

Potentiometric sensors for gases monitor the direct chemical reaction of the gas with the electrolyte, thereby changing the potential of the sensing electrode from the initial potential to produce an electrochemical couple with one of the electrolyte ions. The potentiometric sensor observes the potential difference between the sensing electrode and another electrode. This difference occurs when the chemical species of choice is detected. The output logarithmically depends on the concentration of the species being detected. The recovery time of the sensor is a function of the concentration of the sample. The higher the concentration, the longer the recovery time. At a high concentration the sensor may never recuperate.

Amperometric Sensors. Amperometric sensors are important in portable instrument design because they are relatively small, inexpensive, and lightweight, and use little power to generate significant signals. These sensors exhibit fast responses, detect parts per million levels of electrochemically active gases and vapors, can be engineered to have significant selectivity, and operate over a wide range of temperatures. Their accuracy is also considered better than that of other methods such as potentiometry. Amperometric sensors are further subdivided into two classes: potentiostatic and coulometric.

In potentiostatic amperometry, an indicating electrode is polarized against a reference electrode to a constant potential by passing a polarizing current. The sample flows through the cell, and the electric current proportional to the concentration of the species to be measured is recorded. A potentiostatic sensor uses a constant-voltage circuit, ie, the sensing electrode is held at a fixed voltage.

Electrochemical sensors can be categorized as galvanic or electrolytic. A galvanic cell consists of two electrodes and an electrolyte undergoing a spontaneous chemical reaction, which develops an electrical potential difference between the electrodes. The term *galvanic* refers to any device that can convert chemical energy to electrical energy, be it in a gas sensor, battery, fuel cell, or any other device. *Electrolytic* is a term applied to the conversion of electrical charges into chemical energy, as in electroplating or charging a battery. Potentiostatic amperometry involves galvanic devices, but requires an external variable current supply to maintain a fixed electrode potential; coulometric amperometry involves electrolytic devices. The process that occurs in these sensors is electrooxidation or electroreduction of the species to be analyzed, resulting in a current. Some require oxygen to function and others do not.

A typical potentiostatic amperometric sensor consists of several major parts: filter, membrane, working or sensing electrode, electrolyte, counterelectrode, and reference electrode. The gaseous species of interest is transported (by pump or diffusion) across the membrane to the sensing electrode of the cell; it then migrates to the electrolyte boundary, dissolves in the electrolyte, diffuses to the electrode surface, and reacts electrochemically. The products of the reaction diffuse away from the electrocatalytic surface.

As a way of understanding this method of detection, a typical potentiostatic curve is shown in Figure 5. Oxidation and reduction current are represented on the $+y$ and $-y$ axes, respectively, while voltage is plotted along the x-axis. Below the reaction potential of a given electrooxidizable gas there is no reaction; therefore, the current is zero (except for a small charging effect). As the reaction potential is approached and exceeded, the current rises sharply until it reaches a maximum limited only by the diffusion of the gas to the reaction site. The magnitude of this current is directly proportional to the number of electrons per mole, the Faraday constant, the diffusion coefficient, the surface area of the reaction site, and the concentration of the gas; it is inversely proportional to the diffusion gradient resulting from the varying concentrations

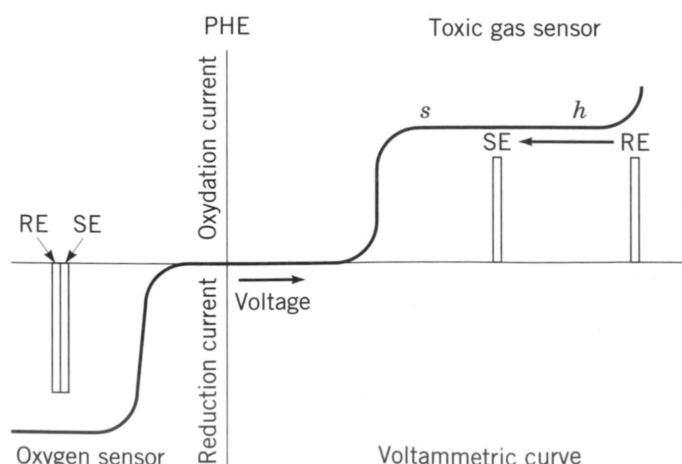

Figure 5. Basic principles of operation of a potentiostatic sensor (Courtesy of Interscan Corp.)

of reacting gas that range from the highest concentration found in the electrolyte, to zero concentration at the electrode surface. There is a linear relationship between the current generated and the concentration of the gas. An even further increase in voltage results in a high current because of electrooxidation of water to oxygen gas, and no detection takes place.

For the device to work, the sensing (working) electrode must be held at a potential within the region of the limited diffusion current. This is represented by SE in Figure 5. A reference electrode (RE) with a lower or higher oxidation potential than that of SE is used to maintain the sensing electrode in the region of the limited diffusion current. The potential of the reference electrode must not change with the passage of current. Depending on whether the potential of RE is higher or lower than SE, it will have a voltage that is biased in either the negative or positive direction. The height of the diffusion plateau from the voltage axis is linear with the concentration of the sample present. This is an example of an externally biased sensor. Typical features of potentiostatic sensors include fast response, linear response in a broad concentration range, nonlinear dependence on the flow rate, and temperature dependence of the signal.

Commercial coulometric monitors are constant-current devices. Coulometric monitors measure the quantity of electricity (in coulombs) that passes through a solution during the occurrence of an electrochemical reaction. Therefore, measurement of the amount of electrical energy transferred across an electrode-solution interface (number of electrons) in terms of the coulombs required to carry the reaction of a specific substance to completion is called coulometry. During coulometric sensing, the gas being monitored is consumed by electrolysis during passage through the sensor. The current is controlled by the feed rate of the sample and corresponds to the charge passed in a given time unit:

$$\text{Current} = \frac{\text{Charge}}{\text{Time}} \tag{2}$$

It is generally a wet-chemical titration method in which one of the reactants is generated in the test cell by electrolysis of the solution. These detectors can be made specific by adjusting the concentration, pH, and composition of the electrolyte. An important feature of the coulometric sensor is that the current is, within certain limits, independent of changes in the working electrode; thus the electrode characteristics are unimportant. Electrodes may be constructed from platinum foil, wire, or mesh. The accuracy of the sample flow rate controls the accuracy of measurement and must be kept as constant as possible; the signal is, however, independent of the temperature. Coulometric cells are reusable after being cleaned and recharged with a fresh electrolyte solution.

Basic Description of a Fuel Cell Electrochemical Sensor. In its simplest form, a fuel cell electrochemical sensor consists of two similar gas diffusion fuel cell electrodes, a concentrated ionically conducting aqueous electrolyte, a low impedance external electrical circuit, and a diffusion barrier (Fig. 6) (19). In clean air, when reactant gases are absent, both electrodes assume the same potential and no current flows in the external circuit. If an electrochemically oxidizable gas is present, it diffuses to the sensing electrode first and causes its potential to shift in the negative direction. The resulting potential difference between the sensing electrode and the other electrode, the so-called counterelectrode then causes a current to flow in the external circuit, which is sustained by electrochemical oxidation of the reactant gas at the sensing electrode and matched by an equivalent amount of oxygen reduction at the counterelectode.

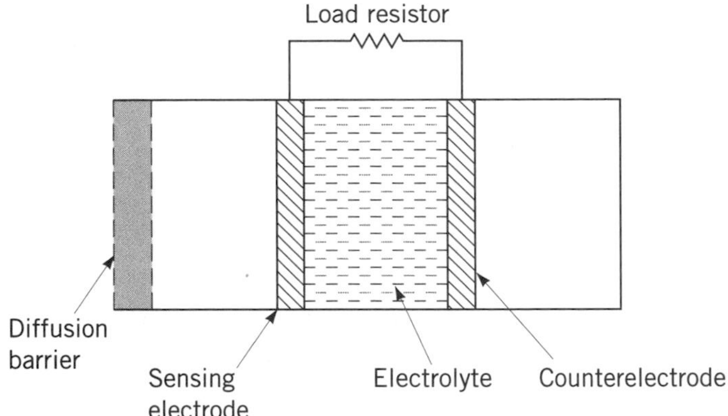

Figure 6. Schematic diagram of the basic elements of a fuel cell gas sensor (19)

A carbon monoxide sensor with an acid electrolyte is represented by the following equations.

Sensing electrode:

$$2\ CO + 2\ H_2O \rightarrow 2\ CO_2 + 4\ H^+ + 4\ e^- \tag{3}$$

Counterelectrode:

$$O_2 + 4\ H^+ + 4\ e^- \rightarrow 2\ H_2O \tag{4}$$

Cell reaction:

$$2\ CO + O_2 \rightarrow 2\ CO_2 \tag{5}$$

Carbon monoxide diffuses via the diffusion barrier to the sensing electrode, where it is oxidized to carbon dioxide and protons, consuming water and releasing electrons to the external circuit. The carbon dioxide product is rejected by the acidic electrolyte and diffuses out of the cell via the diffusion barrier. Protons produced at the sensing electrode migrate to the counterelectrode and recombine with electrons and oxygen to form water.

This electrochemical power source is converted to a sensor by the inclusion of the diffusion barrier at the sensing electrode of the cell. This barrier is fundamentally important to the operation of the sensor and is designed to restrict access of the reactant gas so severely that it becomes completely oxidized at the sensing electrode.

A reactant molecule present in ambient air passes through the following stages in undergoing detection by an electrochemical sensor:

1. Mass transport in the gas phase to the diffusion barrier.
2. Diffusion, under a concentration gradient, through the sensor barrier.
3. Gas phase diffusion from the barrier to the electrode, through any porous electrode support and into the electrolyte structure itself.
4. Diffusion through an aqueous electrolyte film to the catalyst surface.

5. Adsorption from the liquid phase onto the catalyst surface, followed by surface migration to the catalyst active sites.
6. Electron transfer between catalyst surface and adsorbate, at the active center, to form products.
7. Desorption of products and their subsequent transport back into the gas phase and ambient air, and/or into the bulk electrolyte

Diffusion Barrier. The nature of the diffusion barrier exerts a profound influence on the characteristics of the sensor. Two broad categories of diffusion barrier have been used, identified according to the diffusion process through the barrier.

In the first type, the rate-controlling process is one of diffusion in solution, sometimes designated activated diffusion. It is characterized by a high exponential response to temperature. The two variants of activated diffusion barriers involve diffusion in solution through an electrolyte film and through a solid polymer membrane.

In the second type, the gas remains in the gas phase as it diffuses through a porous barrier. It is characterized by a relatively low temperature coefficient and also exists in two variants involving either macropores or micropores. The most common barrier employed for macropore diffusion uses a capillary or pore size that is much greater than the mean free path of the gas molecules undergoing diffusion. Here, intermolecular collisions dominate the diffusion process, resulting in a sensor output that has a considerably lower temperature coefficient than do solid-state membranes. It should be mentioned that the sensor will have a $T^{1/2}$ temperature effect.

For barriers in which the pore diameter is much smaller than the mean free path of the reactant gas molecules that occur in the micropore barriers, the molecules pass through the pore, ricocheting off the pore walls with substantially no intermolecular collisions (23). This diffusion process has fundamentally different properties from the macropore process and leads to different sensor properties. The diffusion rate is determined by the rate of capture of molecules and the molecule velocity. The rate of capture varies as T^{-1} and the velocity as $T^{1/2}$, the combination resulting in proportionality to $T^{-1/2}$. The capture rate and hence the diffusion rate, vary as d^{-1} (where d is the pore diameter), in contrast to the macropore case, for which diffusion rates are independent of pore size.

Of practical significance is that macropore membrane barriers can be easily fabricated and do not suffer the problems associated with solid membranes. By combining both categories of porous diffusion barriers, the negative and positive temperature coefficients can be offset to produce a sensor with negligible temperature coefficient (24).

Electrodes. The electrochemical reactions occur at the electrode–liquid interface, where the solid catalyst surface is in contact with the gaseous reactant dissolved in the liquid electrolyte. To maximize the area of contact, gas diffusion electrodes are employed (25,26). These consist of finely divided catalyst powders with large surface areas, mixed with a polytetrafluoethylene (PTFE) and supported on a highly porous PTFE backing membrane. This design controls catalyst wetting and provides gas channels to the wet electrode interior.

Electrolyte. The electrodes of an electrochemical sensor are separated by a high ionic conductivity, negligible electronic conductivity medium. All commercial liquid electrolyte fuel cell sensors employ aqueous solutions of strong mineral acids, alkaline hydroxides, or neutral salts that are absorbed in a separator material between the electrodes, which consists of plastic or glass mats as used in conventional electrochemical power source systems. The electrolyte should have a high boiling point and a low freezing point (typically $<-20°C$).

The choice of electrolyte can affect the electrocatalytic activity of the electrode and can provide a means of controlling selectivity. Some gases are completely inactive, electrochemically, in alkaline media, but react readily in acidic media, while other gases react conversely.

The work in the field of solid ion conductors in the 1950s led to increasing interest in all solid electrolyte cells. A solid state cell in which the electrolyte was a zirconia-based ceramic was developed in the 1960s for the measurement of oxygen partial pressures. Zirconia sensors have been used in industry ever since, and the technology is now considered mature. A major application for zirconia sensors operating on the electrochemical principle is the monitoring of the air:fuel ratio in fossil fuel combustion systems; this monitoring, combined with closed-loop control, enables maximum efficiency and minimal pollution. Zirconia sensors are versatile and can be tailored to the application; the response time to a change in oxygen gas concentration can be rapid (eg, 50 ms for oxygen partial pressures >100 Pa at 700°C); the sensor can be highly selective to oxygen. Potentiometric devices can measure oxygen pressure levels over an extremely wide range (10^6-10^{-25} Pa); alternatively, amperometric devices deliver a linear response, which is preferred in some applications. It is likely that zirconia sensors will continue to dominate the oxygen-sensing market for applications requiring operations at moderate and elevated temperatures (27).

Two-Electrode and Three-Electrode Designs. In its simplest form, a fuel cell sensor comprises two electrodes connected by a load resistor. Such a device is self-powered, and the voltage developed across the load resistor R_L by the system constitutes the signal from the sensor. The response time of the cell increases proportionally with the current. The lowest possible value is desirable, and an alternate approach consists of using a potentiostatic circuit that runs the cell at nearly zero current. The equivalent circuit is shown in Figure 7(**a**).

A larger problem with the two-electrode approach derives from polarization at the counterelectrode, particularly when an oxygen reduction counterelectrode is used. Such a counterelectrode has the advantage of unlimited capacity for ambient air monitoring but has the disadvantage of significant polarization. The sensing-electrode potential must follow the counterelectrode potential fairly closely. Changes in potential of the sensing electrode can cause interfering currents from side reactions caused by other electrochemical couples.

A practical consequence is that the useful concentration range of such two-electrode sensors is limited. However, they are simple and self-powered and, therefore, are used successfully commercially.

The effect of polarization at the counterelectrode can be completely eliminated electronically using three-electrode sensors (Fig. 7**b**).

In a three-electrode sensor, all three electrodes may be constructed of the same material. The way they are connected to the circuit determines whether they are sensing, counter, or reference electrodes. The sensing electrode interacts with the gas to be monitored. The counterelectrode acts as an electron

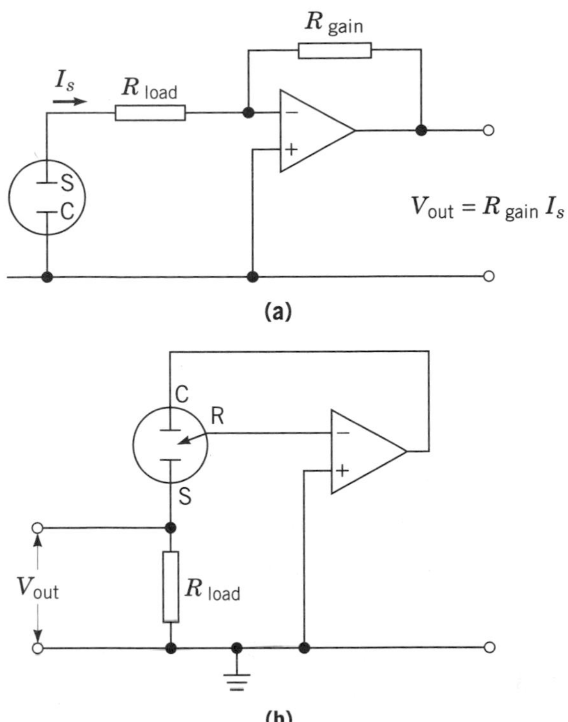

Figure 7. (**a**) A two-electrode potentiostatic circuit. (**b**) A three-electrode potentiostatic circuit. *S,* sensing electrode; *C,* counterelectrode; *R,* reference electrode.

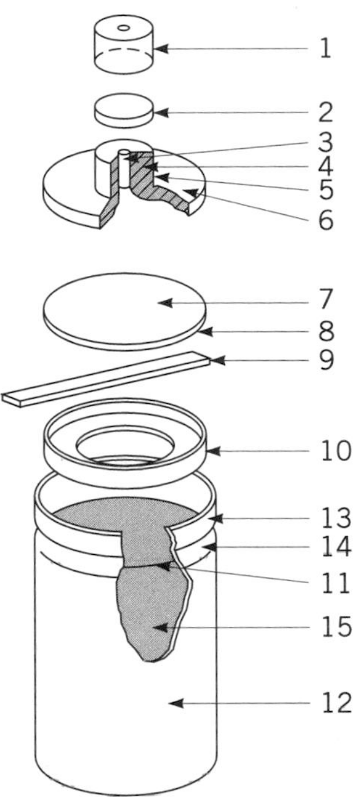

Figure 8. Schematic diagram of the construction of an oxygen–lead, alkaline–electrolyte electrochemical sensor. *1,* end cap; *2,* antibulk flow membrane; *3,* capillary; *4–6,* capillary plates; *7* and *8,* sensing electrode; *9,* sensing electrode contact; *10,* sealing grommet; *11,* separator; *12–14,* container, anode contact; *15,* anode. Courtesy of City Technology, Ltd.

sink for anodic sensing reactions by completing the circuit for the sensor. The third electrode acts as a reference. No appreciable current passes through the reference electrode, because it is connected to the high impedance input of an amplifier. The reference electrode provides a bias reference for either an applied or an internal potential, so that the sensing electrode is at a potential within the region of the limited-diffusion current. Using a potentiometric feedback between the reference and the sensing electrodes maintains the latter at a constant potential.

As with a two-electrode circuit, the signal is determined by the voltage generated across R_L, except that an amplifier with negative feedback is used to maintain a constant potential relative to the reference electrode at the ground side of the load resistor. This potential is quite independent of the counterelectrode potential. With this arrangement, the change in potential of the sensing electrode is due solely to the current generated at the sensing electrode by the reactant gas; this current is used as the signal.

Three-electrode circuits also allow the sensing-electrode potential to be biased with respect to its rest potential, which can be beneficial in promoting desired reactions that do not occur at the reference electrode potential.

Construction of a Typical Gas Electrochemical Sensor. There are two basic types of sensor designs available commercially, depending on the electrolyte (alkaline or acid) and the operating principle (metal-air battery or fuel cell).

The construction of an alkaline metal-air, capillary-type oxygen sensor is show in Figure 8. Conventional nickel–cadmium battery hardware is used for the containment can, which is filled with a lead anode material. This containment can also acts as the counterelectrode electrical output. A sealing grommet holds the oxygen-reducing sensing-electrode assembly, which inserts into the top of the anode can after electrolyte charging. Cell closure and sealing are effected by crimping the anode can over the grommet, thus avoiding electrical contact with the cathode (28).

Various modifications of sensors are possible. For example, one manufacturer may choose to slightly acidify the hydrogen peroxide electrolyte in a sulfur dioxide conductimetric sensor to decrease the effect of certain interferents. The gas exposure path can be modified to obtain differing gas flow rates that reach the electrode, thus affecting the magnitude of the signal. In units with active sampling, high flow rates are used to pull gases to the sampling electrode to obtain the highest sensitivity. Control of the electrode exposure area is important, and in practical designs as much catalyst as possible is packed into this area (by using material with high surface area) to ensure adequate reacting capability and high sensitivity. Each chemical species has its own unique interaction at an electrode, but for almost all the electrochemical cells, the primary variables are working electrode potential, analyte concentration, current, and time.

The construction materials of the sensor also influence its operating characteristics. Choosing construction materials and sensor geometry is critical and has a profound influence on the accuracy, precision, response time, sensitivity, background signal, noise, stability, lifetime, and selectivity of the resulting sensor. For example, selection of a gold rather than a platinum electrocatalyst for the sensing electrode allows for selective determination of hydrogen sulfide in the presence of carbon monoxide.

Membranes are usually chosen for their ability to protect the sensing electrode. However, if the membrane has a low permeability to air, the sensor will have a slow response time. Materials used for the membrane construction are typically Teflon and high density plastics like polypropylene, because such materials must be compatible with reactive gases and corrosive electrolytes.

It is important to select the proper electrolyte for each sensor, since the electrolyte composition can affect the solubility and the rate of diffusion of the reactant gas to the electrode (catalyst) surface. Electrolyte composition can also alter the chemical being monitored before it reaches the electrode surface. For example, the use of an acidic electrolyte to detect ammonia causes the formation of an ammonium ion that may not be as electrochemically active as ammonia under the conditions of the cell. The electrolyte profoundly influences the response characteristics observed for sensors with strongly acidic and basic electrolytes.

Another means of chemically controlling the sensor properties is by altering the composition of the electrocatalyst. Each catalyst formulation has unique properties. The reactivity of platinum with carbon monoxide oxidation has been found to be 10^3-10^6 times better than that of gold. Although both reactions occur on metal, one is orders of magnitude faster than the other.

The selectivity of the sensor can also be improved by controlling the electrochemical potential of the working electrode. For instance, proper selection of a gold electrode potential allows the determination of nitrogen dioxide in the presence of nitric oxide. Response time, linearity, zero drift, repeatability, sensor stability, and even sensor life depend on sensor design and methodology.

As noted, gas sensors are often covered with a membrane that is selectively permeable to a given contaminant. This membrane minimizes the likelihood of poisoning the electrodes by an electroactive and surface-active species. The resolution of these systems is also enhanced if other contaminants (interferents) that also undergo electron transfer at the electrode can be excluded by tuning the circuitry. The appropriate choice of electrocatalyst can help in achieving selectivity in some applications. However, although choosing a membrane or an electrolyte for its selectivity is a good idea, it does not eliminate all the possible interferents, since virtually any compound for which a similar type of detection method is used is a potential interferent.

Multiple Gas Instruments Based on Gas Electrochemical Sensors. Multiple gas instruments that are able to monitor several gases at the same time are available. Advantages to this approach are almost simultaneous monitoring of various gases at a low concentration level without need for a separation column, the low cost associated with the use of electrochemical sensors, relatively fast access to the information, and portability caused by the small dimensions of the sensors. The principal goals of such a system are to control the sampling conditions and to limit cross-sensitivities to other compounds (29).

Figure 9 shows a typical arrangement for a multisensor system equipped with a water trap, particle filter, pump, and in-line SO_x-NO_x filter. Moisture condensing from the gas stream can cause blockages and flooding, and absorbs sulfur dioxide and nitrogen dioxide from the sample. Any water, therefore, must be removed as soon as it condenses. The gas stream must also be cooled to the ambient temperature. The particle filter should trap solid particles but not remove any gas the user wants to measure. The choice of the materials in contact with sampled gas should have low absorption properties.

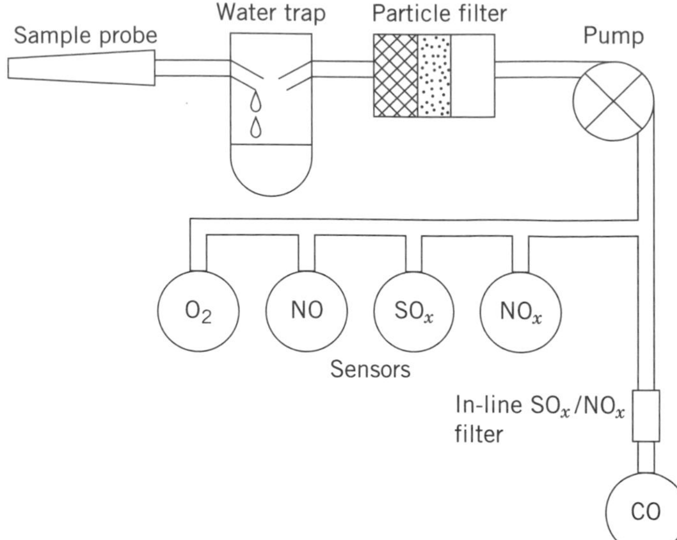

Figure 9. Typical arrangement of a multisensor instrument equipped with sensors for oxygen, nitrous oxide, sulfur dioxide, nitrogen dioxide, and carbon monoxide. In addition the instrument is equipped with a water trap, particle filter, pump, and selective trap to capture SO_x and NO_x compounds and increase the selectivity of the carbon monoxide sensor Courtesy of City Technology Ltd.

Figure 10 shows the Minigas Multigas monitor from Neotronics of North America Inc.

The Minigas is a confined space gas detector that can display up to four gases:

- Oxygen (0 up to 35%)
- Flammable (percent lower explosive level (LEL) (0–99%)
- Up to two toxic: carbon monoxide, hydrogen sulfide, sulfur dioxide, or chlorine

The sensitivity of the Minigas for toxic gases varies with the nature of the gas. The unit can detect 0.1 ppm of chlorine and 1 ppm for the other gases. The sampling is one either by diffusion or with an optional internal pump.

The Minigas can operate on nonrechargeable batteries (40 h minimum with four AA batteries) or with a rechargeable battery (nicad: 12–15 h, NiMH: 21 h minimum). The dimensions of this instrument follow: depth 1.75-(4.6 cm) × width 2.75 in. (7.0 cm) × length 7.12 in. (18.5 cm). It weighs less than 31 oz. (831 g) with the zinc rechargeable battery version. The Minigas is equipped with an alpha-numeric liquid display which simultaneously gives the reading of gas concentrations and is programmed to give an audible alarm.

Typical Performance of Gas Electrochemical Sensors. Signaling the presence of a hazardous atmosphere

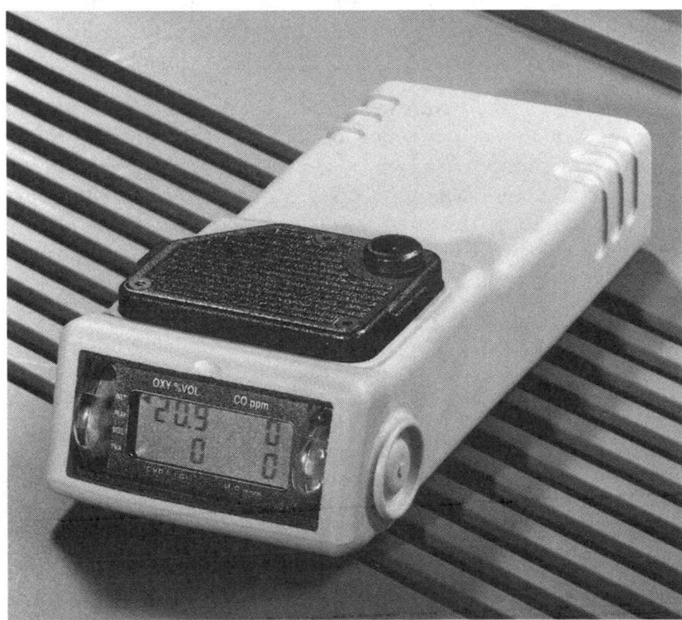

Figure 10. View of the Minigas Multigas monitor from Neotronics of North America. The Minigas multigas personal monitor simultaneously monitors and digitally displays up to four gases. Courtesy of Neotronics of North America, Inc. (Flowery Branch, Ga.).

is often more important than measuring its precise content, at least for portable monitoring instruments. Safety sensors do not normally need to have a range capability far in excess of the legal exposure limits.

Typical Linearity Curves. For low concentrations, ie, below a few percent of reactant gas, the sensor outputs are linear with respect to reactant concentration. At concentrations above a few percent, sensors become increasingly nonlinear and require some partial electronic compensation.

Design Parameters Affecting the Selectivity of an Electrochemical Sensor. Two key features of sensors are that they be highly specific and exhibit minimal effects from cross-interfering gases. Three possible approaches to reaching these goals are to develop specific electrode catalyst and electrolyte systems, control the operating potential of the sensing electrode, and use chemical filters to remove interfering gases selectively. Each method has limitations in practice. For instance, electocatalysts can be developed that exclude some potential cross-interfering compounds, but it is difficult to develop specific catalysts that are sufficiently reactive toward the less reactive gases and do not respond to the more reactive cross-interfering substances.

The selectivity of the sensor can also be improved by controlling the electrochemical potential of the working electrode. For instance, proper selection of a gold electrode potential enables the determination of nitrogen dioxide in the presence of nitric oxide. Response time, linearity, zero drift, repeatability, sensor stability, and even sensor life depend on sensor design and methodology.

Chemical filters placed in the main gas stream of the input sensor remove cross-interfering gases by selective adsorption, chemical oxidation, or reduction. These reactions can be specific. The main disadvantage of this technique is its short filter life; as the filter is exposed to the full gas stream and reaches exhaustion, it must be replaced or regenerated periodically. An alternate and much more efficient method is to place the filter between the electrode and the diffusion barrier (28). In this design, the filter material is exposed to the relatively much smaller amount of interferent gas diffusing to the electrode.

Most applications involve ambient air monitoring, for which the background gas is essentially nitrogen plus oxygen and remains constant. However, if the composition changes, recalibration of the sensor to the new environment becomes necessary. This is called background effect.

Limitations of Electrochemical Sensors. Operating temperature of these sensors is typically from −20 to 50°C. This is not a problem for most applications, but the user should be aware that lower temperatures tend to result in longer response times. When the sensors are used outdoors in cold climates, ambient temperatures should be monitored to ensure that they do not fall below the lower limits.

Electrochemical sensors designed to measure specific toxic gases may have cross-sensitivities to other compounds. Response specificity is determined by the semipermeable membrane selected, the electrode material, and the retarding potential (the potential used to retard the reaction of gases other than those being sampled). The primary gases that interfere are those that other electrochemical sensors are designed to monitor, because they are easily electrooxidizable or electroreductible. Gases that cannot be oxidized or reduced, such as methane, do not interfere.

Interference can be eliminated with selective chemical filters or by prescrubbing the sampled atmosphere. Both of thesemethods have been used by manufacturers for some applications. However, not all interfering gases can be eliminated, and the sampling professional must be familiar with those that may pose a problem.

Poor stability of a potentiostatic sensor is primarily due to the effect of temperature on the inherent background current present in every sensor. This current is due to internal electrooxidation or electroreduction. Most passive electrochemical sensors should not be exposed to the air when not in use, as the water loss from evaporation through the porous membrane will shorten sensor life.

Sensor longevity is also a function of the gases being tested. Since most chemical sensors depend on a specific chemical reaction to generate a signal, the reactions involved in these sensing mechanisms are often not sufficiently reversible, so that the sensitive material (electrolyte) is used up during the life of the sensor. Sensor lifetime has many determinants, including heat, humidity, dirt, and cumulative gas exposure; the importance of maintenance should not be underestimated. Sensors need periodic recalibration; costs associated with calibration represent the majority of expenses associated with these sensors.

Thermochemical Sensors. Gases and vapors have certain thermal properties that can be exploited during analysis. For the instruments used in industrial hygiene applications, one of two thermal properties, conductivity or heat of combustion, is measured.

Thermal conductivity detectors are simple devices that operate on the principle that a hot body loses heat at a rate that depends on the composition of the surrounding gas (8), ie, the ability of the surrounding gas to conduct heat away from the hot body can be used as a measure of the composition of the gas. In usual practice, a thermal conductivity detector consists of an electrically heated element or sensing device whose temperature at constant electrical power depends on the thermal conductivity of the surrounding gas. The resistance of the sensing device is used as a measure of its temperature. Thermal conductivity detectors are universal detectors, responding to all compounds that have a different thermal conductivity than the ambient air. They have large dynamic ranges, on the order of 10^5, and limits of detection on the order of 10^{-8} solute per milliliter of carrier gas (10–100 ppm for most analytes). Thermal conductivity detectors require good temperature and flow control.

The heat-of-combustion detectors, which make up the largest single class of direct-reading instruments for analyzing airborne gases and vapors, measure the heat released during combustion or reaction of the contaminant gas of interest (8). The released heat is a particular characteristic of combustible gases and may be used for quantitative detection.

The heat mechanism in combustion instruments employs the use of catalysts—catalytically heated filaments or oxidation catalysts. This mechanism may measure either resistance change or temperature change via thermocouples or thermistors. The sensor contains pairs of elements, each containing a platinum wire coil embedded within a catalytic bead bed. An active detector element oxidizes combustible gases, while an inert reference element compensates for changes in ambient conditions. Both elements are in a Wheatstone Bridge circuit and the out-of-balance signal produced is proportional to the concentration of combustible gases.

Each heat-of-combustion detector is specific to some combustible gases. Some specificity can be introduced by manipulating the temperature to combust interfering gases or by carefully selecting the oxidation catalyst. Combustible gas instruments also use a semiconductor as the sensitive element. Semiconducting properties can be altered by adding doping impurities. These oxide coatings can be designed to selectively accelerate the oxidation of certain compounds, thus providing more specificity than does a conventional catalytic combustion sensor. Many instruments can detect both combustible gases and toxic gases. When specificities are needed, sensors with the appropriate selective characteristics are installed. The combustible gas sensor has a broad response to various combustible gases and is left unmodified. Selectivity is made possible through different mixtures of oxides and selected operating temperatures.

Limitations of Thermochemical Sensors. Thermochemical sensors for combustible gas measurements are used primarily for occupational measurements near sources of combustible material. Combustible gas indicators (CGI) were the first direct-reading instruments to be developed. Their initial use was to detect explosive methane in mines. A variety of CGI designs is on the market, ranging from simple sensors to sophisticated microprocessor-based units with multiple detectors.

Prolonged exposure to certain substances (eg, silicones, chlorinated hydrocarbons, and lead- or sulfur-containing compounds) results in an irreversible decrease in the sensitivity of thermochemical sensors. Some compounds (eg, halogenated hydrocarbons) can temporarily inhibit performance, but sensors normally recover after a period of operation in clean air.

Other Types of Chemical Sensors. ***Semiconductor-Type Gas Sensors.*** Among other types of chemical sensors (30), the first type of sensor based on adsorption of gases on metal oxide semiconductors was developed by Figaro TGS >30 years ago. The gas to be detected is adsorbed on the surface or reacts with the reactive surface oxygen (adsorbed oxygen) of the oxide semiconductor. This induces a change in its surface space charge layer, which is then converted into a change in electrical resistance of the polycrystalline elements. The gas recognition is carried out through the surface chemical processes; more specifically, the adsorption or reaction sites play the role of receptor for the gases. On the other hand, the conversion of such an interaction to an electrical resistance change is carried out through the physical processes that cause the transport of electrons. This is determined not only by the semiconducting properties of the oxide used but also by the microstructure of the element, such as the grain size of semiconductor particles and the geometry of the connection between particles.

Metal oxide semiconductors (MOS) are another category of gas sensors. They are devices in which the conduction between the source and drain is modified by a charge or potential on the gate electrode. Because they are able to detect changes in charge or potential, MOS transistors have been investigated as possible chemical sensors. For example, if ions are absorbed or adsorbed onto the gate, the charge on the gate is modified, which leads to a detectable signal. This can occur by a number of different mechanisms. Either the gas can chemically interact with the gate material and change the carrier concentration on the gate, and hence its Fermi energy, or the gas can induce dipoles in the gate that alter the potential gradients at the gate–oxide interface, or a change in a chemical cell can occur between the gate and the oxide.

Semiconductor gas sensors are based on devices that are manufactured using silicon technologies (30). Various types of sensors have been developed using either this or MOS transistor technology, in which the material used to fabricate the base is palladium instead of polysilicon (31) and thus has a high selectivity for hydrogen. Sensor technology is based on ion-sensitive field effect transistors (ISFET), for which an ion-sensitive membrane provides selectivity (32).

These types of solid state sensors detect a variety of gases: hydrogen (33), ammonia (34), unsaturated hydrocarbons (35), saturated hydrocarbons (36), carbon monoxide (37), hydrogen sulfide (38), alcohol (39), arsine (40), and oxygen (41). Many reviews of these technologies are available (42,43).

Surface Acoustic Wave Sensors for Gas Sensing. Devices that exploit the transmission of surface acoustic waves (SAW) have been used in electronic applications for more than two decades. The principle of operation of a SAW for gases is conceptually quite simple (44). A SAW—a periodic deformation to the normal surface—is transmitted across the surface of a solid, which is presented to the atmosphere to be analyzed. If the solid is a piezoelectrical crystal (PZX), then the wave can be induced electrically. The transmission of the SAW is sensitive to the mass of material adsorbed on the surface and may be damped by the environment. It is altered by the interaction of gaseous analyte molecules with the surface, usually via an intermediate (coating) layer; the process is monitored by measuring changes in the frequency, amplitude, or phase of the transmitted wave.

There are many reports of SAWs being proposed as gas detectors, and these reports identify the problems associated with the choice of coating, the specificity, the coating thickness, and the obtaining of reproducible amounts of polymer coating on the surface of the sensing element. Applications of SAW devices for the detection of nitrogen dioxide (45), hydrogen (46), sulfur dioxide, and hydrogen sulfide (47) have been investigated. Sensitivity is sample dependent and is typically at the level of a few parts per million. Strong effects on the performance of the detectors at ambient temperature have been reported.

SAWs have been used to sense a variety of organic gaseous species such as nitromethane, methanol, propanol, benzene, tetrachloromethane, *N,N*-dimethylacetimide, 1-butanol, 2-butanol, diethyl sulfide, 1,2-dichloroethane, iso-octane, methanesulfonyl fluoride, and tributylphosphate (48).

Optical Fiber Gas Sensing. The development of inexpensive, high quality optical fibers for the communications industry has provided the essential component for the development of optical fiber sensors (20). The physical principle underlying optical fibers is that of total internal reflection. If a material with a higher refractive index (the core) is surrounded by a material with a lower refractive index (the cladding), then repeated total internal reflections result in optical energy travelling along the core. This structure can be planar for waveguides or cylindrical for optical fibers. The evanescent wave is important for sensing applications, because it can interact with analytes outside the light-guiding core by absorption, luminescence, or scattering, thereby modifying the characteristics of the guided light.

There are several advantages for sensor applications, such as small size and flexibility (typically a 200-μm plastic-clad silica fiber can be bent around a 1-cm-radius mandrel). The current development of polyimide, amorphous carbon, and metallic coatings means that fibers can operate at temperatures of up to 400°C. These fibers operate remotely up to several kilometers without significant degradation of performance. In addition, fibers can be multiplexed, meaning that expensive optical equipment can be shared. They are intrinsically safer when flammable or explosive reagents are present.

Most of the disadvantages of optical-fiber gas sensors are associated with the multiphase chemical-sensing technique. For instance, the response time of the sensor, which is determined by the mass transport in the reagent phase, is directly related to the thickness of the immobilized reagents; response times on the order of minutes are not unusual. Another common problem is that the sensitivity of the sensor can change when the chemical properties of the immobilized reagents change.

Fiber optic sensors can be subdivided into two types: extrinsic and intrinsic. For extrinsic sensors, the optical fiber merely acts as a light-guiding link between the measurement point and a remote spectrometer. With this approach, the optical properties of the analyte are modified by an added reagent and measured directly. For intrinsic sensors, the fiber, probably in some modified form, is the sensing element. The refractive index or the effect of temperature on the optical transmission can be used for this type of measurement. The examples that follow are illustrative of the capabilities of optical fiber sensing.

Methane has been detected at a level of 700 ppm by an extrinsic sensor and a remote spectrometer at a distance of 1 km from the source. Measurements over even longer distances (up to 5 km) have been demonstrated, showing a decrease in sensitivity but still giving useful data (49).

Irreversible sensors using immobilized reagents have detected hydrogen cyanide in air within 1 min at parts per million levels (50). The irreversibility limits its potential applications.

One example of a sensor that is based on luminescence quenching is that for oxygen under partial pressure in the range of 0–20% oxygen with an accuracy of 0.4% (51). Another gas that has been detected by fluorescence quenching is sulfur dioxide in the range of 0.01–6% (52). Nitrogen dioxide, hydrogen chloride, and chlorine have been detected in the 1–10 ppm range using luminescence quenching of a tetraphenylporphine Langmuir-Blodgett film (53). One important limitation is again the irreversibility of the sensor.

For intrinsic sensors, interaction with the analyte or an immobilized analyte-sensitive reagent can occur only within the waveguide or in its vicinity (by an evanescent wave interaction). Sensors based on the principle of direct propagation of light through the reagent phase have been shown to be reversible up to around 10 ppm for ammonia and 5 ppm for hydrogen chloride gases (54).

Evanescent wave sensors using planar and cylindrical optical waveguides have been described. Ammonia concentrations as low as 60 ppm were detectable within 1 min (55). In this application, a coated capillary tube was used instead of a solid glass rod to increase the number of reflections with the outer surface and hence the sensitivity. Methane has also been detected without the addition of any further reagent at a concentration level of 5% in air (56). Another interesting evanescent wave sensor is an integrated optical device designed to sense hydrogen at a level of 20 ppm but with a response time of several minutes (57).

Surface Plasmon Resonance Sensor. Surface plasmon resonance can also provide the basic transduction mechanism for optical gas sensors. A surface plasmon is a particular form of electromagnetic wave that propagates along the surface of a metal. It can be excited optically by light undergoing total internal reflection at the surface of a glass substrate onto which the metal film has been deposited. With the proper choice of metal, usually silver or gold with a thickness of a few nanometers, excitation occurs at a particular angle of incidence, leading to a sharp dip in the intensity of the reflected beam at that particular angle. This angle depends on the surface plasmon's resonant frequency and is sensitive to variations in the refractive index of the medium immediately adjacent to the metal surface. The sensitivity of the resonant frequency to variations in the refractive index of the neighboring medium, falls exponentially as a function of distance from the metal surface, having the same form as and a similar decay constant to the evanescent field interaction discussed previously. This principle of analysis has been applied to anesthetic gas concentrations with a detectable limit of 10 ppm and a response time of 1 s (58).

Detector Tubes and Other Colorimetric Direct-Reading Indicators

Detector tubes and other direct-reading colorimetric indicators are sampling media that change color when exposed to contaminated gases or vapors. A typical sampler is a glass tube filled with a solid granular material that has been coated with a chemical substance that reacts to change when contaminated air moves through the tube. The primary use of these devices is for occupational sampling, since they are generally not sensitive enough to detect the low levels of contaminants measured in environmental detection.

Three types of colorimetric indicators are used for determining contaminant concentration in air: liquid reagents, chemically treated papers, and glass detector tubes that contain solid indicating chemicals (59).

Liquids reagents determine alkaline gases by measuring the volume of air required to produce a color change. This method is somewhat inconvenient and requires skill but is capable of good accuracy. Convenient laboratory procedures using liquid reagents have been simplified and packaged for field use. Reagents are supplied in sealed ampoules or tubes and are diluted at the time of use. Unstable mixtures are freshly prepared by mixing the ingredients at the time of the measurement.

The TDI/MDI Analyzer Kit (Sensidyne, Inc., Clearwater, Florida) is based on liquid reagent technology. It provides a rapid method for field determination of toluene diisocyanate and methylene bis (4-phenylisocyanate) in air. A sample is drawn through a special absorbing solution using the BDX 55 pump and a midget impinger at 2.8 L/min (0.1 cf/min) for 10 min. The solution is transferred to a test tube, and a series of reagents are added to produce a blue-red color. The color is compared to a color reference card graduated in parts per million by volume. Results within 0.01 to 0.35 ppm limits can be obtained in about 30 min. This approach is labor intensive and not geared for fast analysis but should be mentioned since it is the basic of all colorimetric measurements.

Chemically treated papers are used to detect and determine the nature of toxic gases. The method is simple, convenient, and compact. Papers may be used wet, dry, or freshly prepared, as needed. Special chemical chalks or crayons are also used to sensitize ordinary papers to gases. Semiquantitative determinations may be made by hanging the paper in contaminated air. The accuracy of such a procedure is limited by the fact that the volume of air sample is rather indefinite and the degree of color change in the paper is influenced by air currents and temperature. More accurate quantitative results may be obtained by using a sampling device capable of passing a measured volume of air. Visual evaluation of the stains on the paper may be made by comparison with color charts or by photoelectric instruments.

Tape-based instruments make up the next level of sophistication of chemically treated papers that can provide quantitative results. These instruments can be purchased as dedicated systems or as instruments in which different cassettes can be inserted depending on the nature of the gases to be monitored. In either case, the instrument will sample only one chemical at a time and has poor selectivity. An example of a toxic gas monitor using a gas detector cassette tape is shown in Figure 11.

During exposure to a contaminant the reaction produces a continuous characteristic stain on the paper tape. After exposure, the tape is moved incrementally to an optical reader. The reflected optical density of the stain, which is proportional to the concentration of the contaminant in the air, is compared to the reference reflectance and displayed on a meter. The advantage of the optical reader is that the system is calibrated optically and does not require gases for the calibration of the instrument. The disadvantages of this approach are the intermittent nature of this type of detection and measurement as used with paper tapes and the relatively long response time of the system (typically 60 s).

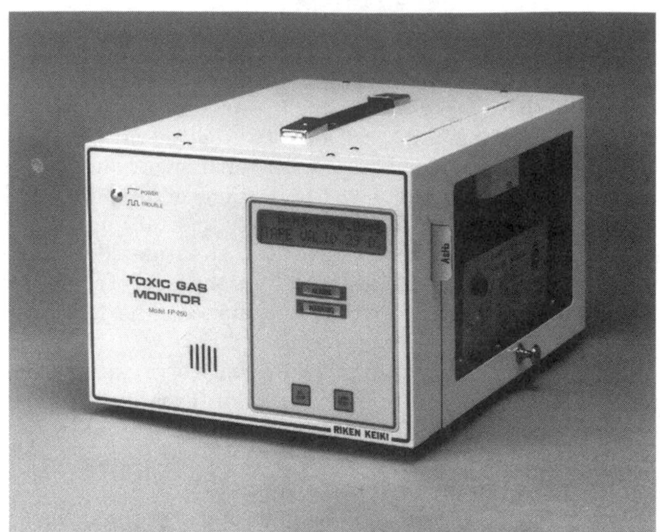

Figure 11. Gas monitor that uses a gas detection cassette (Model FP-250A from RKI Instruments, Inc.).

The model FP-250A from RKI Instruments, Inc. (Hayward, California) uses cassette tapes impregnated with various reagents to obtain a selective detection. Gases such as hydrogen sulfide and phosphine arsine can be identified, but potential interferences will not be detected if both gases are present at the same time. Consequently, this type of instrument is best adapted to monitor the level of toxic gases of a known nature such as phosphine. The minimum detection level also depends on the gas monitored. Arsine, for instance, is detected at the 2.5 ppb level in 20 s, whereas it takes 60 s to detect silane at a 250 ppb level with the same tape.

Detector tubes are available for short-term measurements (grab sample measurements) or long-term measurements (dosimeters). Short-term measurements take typically 1–2 min, whereas long-term collection can take 4–8 h, using either active or passive methods. Active methods involve the use of a portable battery-powered pump or hand-powered pump. Passive methods, such as the badges described earlier, rely on diffusion.

Applications of Detector Tubes. There has been a great expansion in the development and use of detector tubes, and more than 400 different types are now available. Detector tubes are convenient for quantitative and qualitative evaluations of toxic hazards in industrial atmospheres, process control, and for rapid evaluation of spills of hazardous materials (60,61). Detector tubes are also used for detecting explosive hazards, checking compressed breathing air, and confirming carbon monoxide in exhaled breath or in gas released from a sample of blood. They are used for law enforcement to measure alcohol in the breath and signs of arson. Several manuals provide comprehensive descriptions and listings (62–64).

Detector tubes are simple, fast, inexpensive and easy to use, but are also subject to many limitations and potential errors. The results may be dangerously misleading, so the sampling procedure should always be supervised and the results interpreted by an adequately trained operator.

Operating Procedure for Detector Tubes. The use of detector tubes is extremely simple. After its two sealed ends are broken open, the glass tube is exposed to a known quantity of air, using either a calibrated squeeze bellows or a piston pump. The observer then reads the concentration in the air by examining the exposed tube. More recently developed types of tubes interpret the length of the stain produced on the indicator gel rather than color changes. The stain length is measured against a calibration scale that can be printed either directly on the tube or on the provided chart. Selection of the detector tubes depends on the chemical for which monitoring will be done and on the concentration range of interest.

Specificity and Sensitivity of Detector Tubes. Many tubes exhibit cross-sensitivity to other gases, and the manufacturers' literature should be consulted before use. Most tubes are not specific. In the presence of mixtures, the tube reading should be interpreted by a trained operator. The lack of specificity of some tubes may be an advantage for detection of substances not indicated by the manufacturer, but such use requires specific knowledge of the identity of the reagent and of the proper correction to the calibration scales. Most of the tubes are generally designed for detection of relatively high gas concentrations (a few parts per million).

Equipment. Figure 12 shows two models of pump sold by National Dräger. The Accuro bellows pump and short-term detector tubes form a portable sampling unit for use in measuring concentrations of various gases and vapors. Dräger detector tubes are available for measuring approximately 350 air contaminants and for gas analysis. The pump delivers 100 mL sample air with each pump stroke. After a prescribed number of pump strokes, the stain length or the discoloration of the tubes gives a direct measure of the gas or vapor concentration. Calibration scales are printed directly on most types of tubes. The Accuro is a modular system. For large-volume measurements, the Accuro slides into the electronically programmable Accuro 2000. The Quantimeter 1000 is a programmable, battery-operated bellows pump with the same flow characteristics as the hand-operated pump and is intrinsically safe. The complete Dräger Accuro Deluxe Pump Kit with spare parts and tube opener weighs approximately 1.5 kg (3.3 lbs). The detector tubes are usually specific for particular gases and vapors. This specificity is achieved not only by the use of specific and stable reagents but also by the use of precleansing layers placed in front of the actual reactive layer to selectively absorb interfering components that may be contained in the gas or vapor sample. The reading deviations for many of the detector tubes are not more than ±25% from the true value. The long-term diffusion detector tubes operate on the principle of gaseous diffusion to give a long-term TWA measurement without a pump. The contaminant gas diffuses into the tube by means of the concentration gradient between the ambient atmosphere and the interior of the tube.

Dräger recently introduced a chemical measuring system (CMS) based on a chip, which is in effect the gas detection sensor and allows multimeasurements on one plastic carrier (Fig. 13). The reactive preparation necessary for detection is kept in a hermetically sealed glass capillary until needed. The housing of the chip also protects the capillaries from potential external mechanical influences. Prelayers filter out potential interferences from other gases to ensure that the result of the measurement is substance specific.

Figure 13 shows the analyzer and the chip containing 10 microchannels filled with small quantities of reagents.

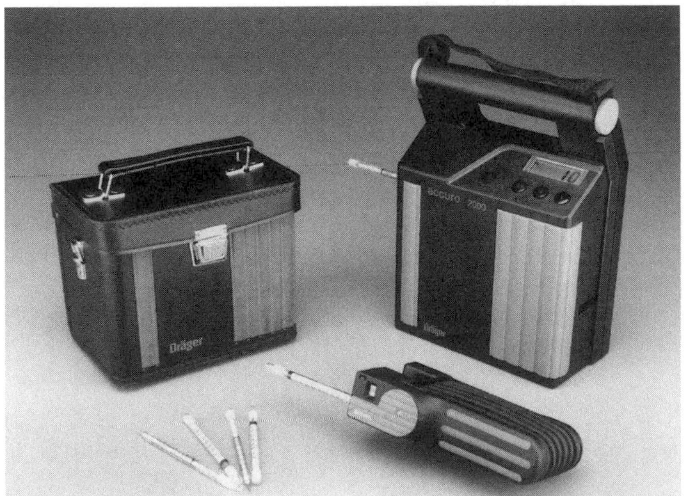

Figure 12. Dräger Accuro, Accuro 2000, and Quantimeter 1000. The figure shows the tubes operating in a manually and electronically programmable pump. Courtesy of National Dräger, Inc.

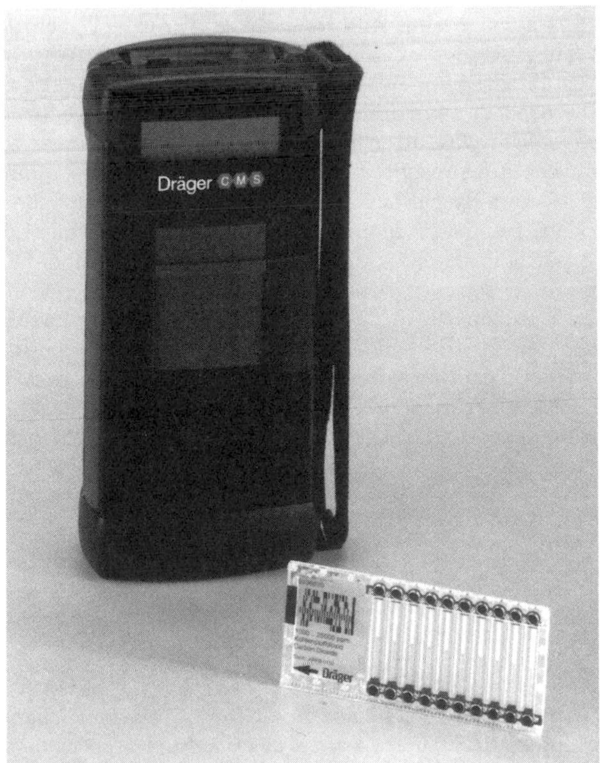

Figure 13. The Dräger CMS. The optical reader is shown with a chip containing 10 measurement capillaries filled with a substance-specific reagent system. Courtesy of National Dräger, Inc.

Low quantities of reagent ensure a quick and reproducible reaction with the contaminant gas or vapor (typically 1 or 2 min). The analyzer records the reaction products that occur optoelectronically. The measurement principle of the CMS is based on a dose measurement that depends on concentration, ie, a dynamic dose measurement. The measurement time is not constant but adapts itself automatically to the concentration, ie, the higher the concentration, the shorter the measuring time, thereby providing information about possible gas dangers more quickly. For instance, the standard measuring time of nitrogen dioxide at a concentration of 5 ppm is 30–35 s; for a fivefold value this drops to about 10–12 s.

A pump system with mass flow sensor, which is part of the analyzer, draws a constant mass of air; the final results are, therefore, totally independent of external ambient pressure conditions. The analyzer operates on batteries and has an autonomy of 7 h.

The Dräger CMS unit can analyze ammonia, benzene, carbon dioxide, carbon monoxide, chlorine, hydrochloric acid, hydrogen cyanide, hydrogen sulfide, nitrogen dioxide, nitrous fumes, perchloroethylene, and sulfur dioxide. Chips with other specificities are in development.

Ionization Detectors

General-survey instruments are used to monitor ambient conditions whenever immediate results are needed, even though the identity of the compound present may be unknown or a mixture of compounds may be present. The primary detectors used are the PIDs and FIDs. General descriptions of the fundamental theory behind their operation are available (65–67).

These instruments are used for general surveying on hazardous waste sites and for industrial hygiene monitoring (68). They screen air, soil, water, and drum bulk samples and establish priorities for laboratory analysis. They can also be used to determine if decontamination procedures are effective. They are often used in a qualitative manner to see if volatiles are present, but they also can be used quantitatively if properly calibrated. Other applications include leak detection, perimeter monitoring, and continuous monitoring using periodic checks. Another application is boundary line sampling for total hydrocarbons to detect airborne releases to the environment from an industrial operation or hazardous waste site. All these instruments are calibrated for known gases and respond to different compounds with different levels of sensitivity. A response factor is provided with these units, but it requires that the identity of the gas to be analyzed is known.

Flame Ionization Detector. In a conventional FID used in gas chromatography (FIDGC), a gaseous sample is pyrolyzed in a hydrogen flame, producing ions and electrons. When hydrocarbons in the sample are introduced to the detection zone, ions are produced by the following reaction:

$$RH + O \rightarrow RHO^+ + e^- \rightarrow H_2O + CO_2 \quad (6)$$

where R is the carbon compound. Ionized carbon fragments move under the influence of an electric field present between the electrodes surrounding the flame. As the positive ions are collected, a current corresponding to the collection rate is generated on the input electrode. The current is directly related to the hydrocarbon number R present in the sample. The FIDGC response is proportional to the number of carbon atoms present in the sample. Typically, the response of the FIDGC to a C_6 hydrocarbon is six times the response to methane, which has only one carbon atom.

The FID used for general survey (FIDGS) is different from the FIDGC in design and performance. Instead of being mixed with the fuel and carried away to the jet, the sample is fed into the air sheath around the flame, and thus the sample reaches the flame only by a process of diffusion. This FIDGS is not selective. All hydrocarbons produce almost the same response. The basic design of the FIDGS is demonstrated by the Foxboro (East Bridgewater, Massachusetts) model shown in Figure 14.

Instrument characteristics, such as sensitivity, are generally given as methane equivalents with a correcting factor (65). Water and carbon dioxide molecules are also present and result from the ionization process and from the recombination of oxygen, hydrogen, and carbon molecules, but they do not affect the response of the FIDGS. This detector is insensitive to water, inert gases, and inorganic compounds and has a negligible response to carbon monoxide and carbon dioxide. It is only when the oxygen level is displaced by a high concentration of other gases that the response varies and possibly extinguishes the flame, ie, when there is $<14\%$ oxygen.

Typically, the FIDGS has a wide dynamic range of four to five orders of magnitude, which can be expanded even further by diluting the sample. The dilution kit can also be used to enrich oxygen-deficient samples by adding ambient air that is rich in oxygen, as is modeled by Foxboro's design.

The fuel (H_2) is provided in a small tank under pressure (typically several atmospheres), and oxygen is supplied from ambient air. To keep the background noise low, purified air must be used; ambient air is cleaned by drawing it through a charcoal trap and is then pumped into the detector. The sample is introduced into the ionization chamber by means of a pump.

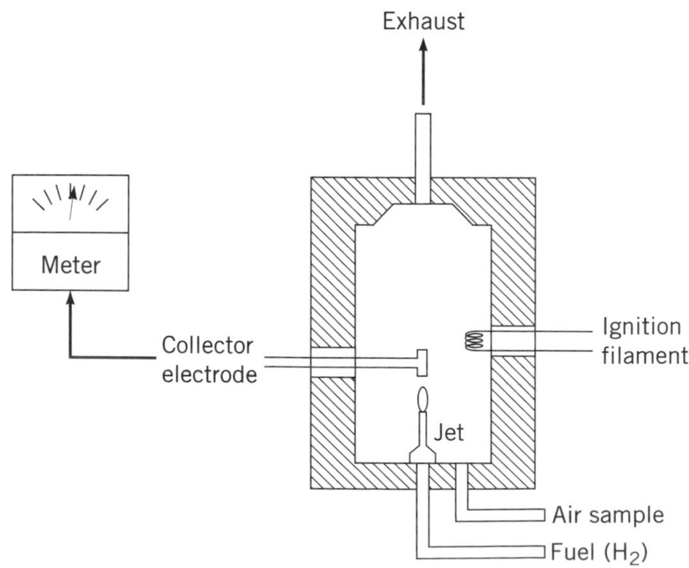

Figure 14. Basic schematic of a flame ionization detector used for general survey. The sample is fed into the air sheath and gives a response that is less dependent on the number of carbons than is the conventional detector used in gas chromatography FIDGC. Courtesy Foxboro.

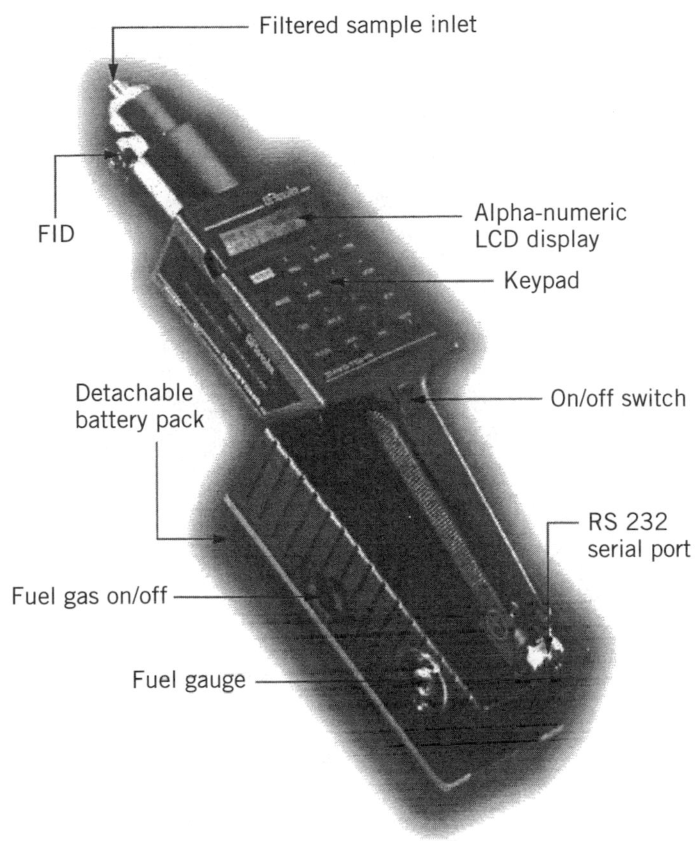

Figure 15. MicroFID from PE Photovac. The unit has a built-in hydrogen cylinder. Courtesy of PE Photovac.

Figure 15 shows a commercial FIDGS package from PE Photovac (Markham, Ontario, Canada). This model is hand carried and has a built-in hydrogen cylinder. The unit displays the measurement results digitally at concentrations ranging from 0.1 to 50,000 ppm. The readings are displayed in parts per million of methane and require a correcting factor for other gases. This instrument has an autonomy of >10 h on the same batteries and hydrogen tank. When used in combination with a PC, the MicroFID becomes a powerful sensor and complies with EPA Method 21 (fugitive emissions monitoring).

Common applications for a FIDGS include, but are not limited to:

- Industrial hygiene surveys and short-term health and safety monitoring.
- Leak detection and fugitive emissions monitoring.
- Detection of soil gases and headspace of water samples.
- Detection of arson.
- Solvent vapor extraction (SVE) systems in groundwater remediation surveillance.

Photoionization Detector. Photoionization is a flameless ionization technique. A PID consists of a UV lamp of a specific energy and an ionization chamber (Fig. 16). Compounds passing through the chamber are excited by photons of UV energy from the lamp and ionized. The resulting ions are collected by an electric field; the current measured gives an accurate representation of the amount of compound present (69).

Whether a compound can be detected by a PID depends on the lamp energy and the energy required to ionize the sample (the ionization potential is the energy required to remove an electron from the sample). If the lamp energy is greater than the compound's ionization potential, the PID will detect it. Lamps of 9.6, 10.0, 10.6, and 11.8 eV are usually available. The 11.8-eV lamp permits detection of many compounds; whereas the lower energy lamps allow more selectivity, theoretically, by not responding to undesired compounds with a higher ionization potential. However, since the radiation from the UV source is not entirely monochromatic, compounds with higher ionization potentials will still be detected to some degree with lower energy specified PID lamps as evidenced by PID/GC analysis of equal concentrations of compounds with both high and low ionization potentials.

Figure 17 shows a hand-held PID unit, Model 2020 from PE Photovac. It does require compressed calibration gases. The only maintenance operations are the replacement of the water/particle inlet filter and the occasional cleaning of the UV lamp.

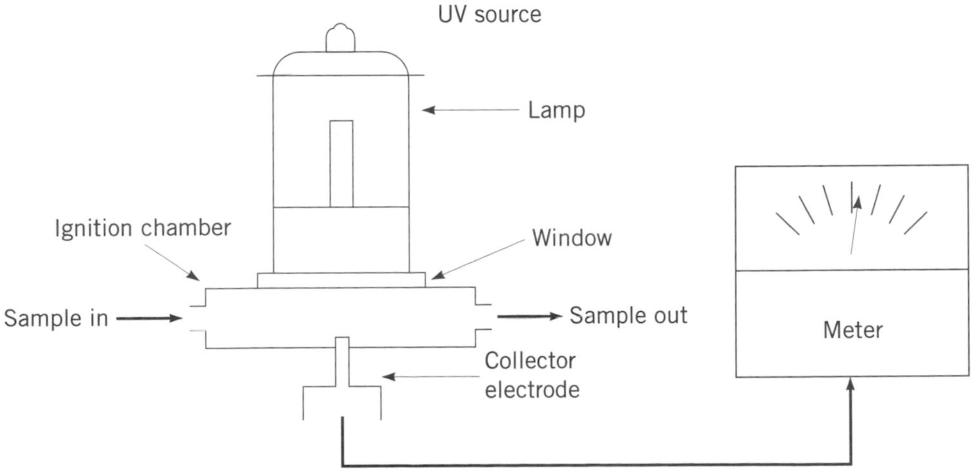

Figure 16. Typical photoionization detector. Ions are produced by photoionization initiated by the absorption of a high energy photon by a molecule or atom. Courtesy Foxboro.

Figure 17. Hand-held PID Model 2020 from PE Photovac. Courtesy of PE Photovac.

Figure 18. The Portable toxic vapor analyzer Model TVA-1000, from Foxboro, combines a FID and a PID in the same package and can operate both detectors simultaneously. Courtesy of Foxboro.

PID Model 2020 is characterized by its small size and low weight. The dynamic range of the PID is limited to 2000 ppm. The PID is used for the detection of VOCs, hazardous wastes, fugitive emissions, and all operations related to emergency response. Another application is industrial hygiene, for which it is used to monitor toxic gases and vapors in a workplace environment. The unit can operate automatically for 10 h. The internal data logger retains a complete record of the day's activity.

The Foxboro Portable Toxic Organic Vapor Analyzer. The TVA-1000 is a compact portable instrument for the simultaneous analysis of organic and inorganic compounds contained in the air (Fig. 18). The instrument uses individual or concurrent PID and FID in the same package. The combination of detectors with fundamentally different responses provides unique response ratios for additional qualitative information about the analytes (70).

Each unit is factory-calibrated with methane (in the case of the FID) or isobutylene (in the case of the PID). However, both detectors respond to many different compounds with different levels of sensitivity. To adjust the analyzer reading to the compound of interest, a correction factor must be entered into the analyzer. The result is compound-specific (corrected) readings.

Comparison of FID and PID. The FID and the PID have both advantages and disadvantages and complement rather than compete with each other (71). The FID has a wider dynamic range than does the PID (0–50,000 as compared to 0–2,000 ppm) and is, therefore, well suited for measuring high concentrations. The FID is also less susceptible to water vapor interference than is the PID.

The PID does not require hydrogen to operate and is the detector of choice if fuel gas is limited or not available. Thus the overall PID package is typically smaller than one with an FID. It can also operate as a selective detector, depending on the ionization lamp selected. The PID is also sensitive to aromatic and chlorinated compounds and can measure some inorganic compounds that the FID does not detect at all (eg, ammonia, carbon disulfide, arsine, phosphine, carbon tetrachloride, chloroform, ethylamine, formaldehyde, and hydrogen sulfide).

The two detectors operating simultaneously in a single instrument offer a net advantage for survey monitoring by providing information quickly about organic and inorganic vapors. The responses of the two detectors relative to each other help identity the compounds being analyzed. For instance, the PID does not respond to methane at all, but the FID responds well. A high FID reading with virtually no PID response may indicate the presence of a naturally occurring methane, commonly found at excavation sites. The PID responds well to some inorganic gases that the FID does not detect. A high PID reading with no FID reading may suggest the presence of an inorganic compound.

GAS CHROMATOGRAPHY FOR ENVIRONMENTAL MEASUREMENTS

The introduction of gas–liquid chromatography in the 1950s and its subsequent enhancements coincided roughly with the increased demand for quick, reliable, and sensitive analytical methods for the determination of the many chemical compounds that contaminate the environment. Gas chromatography proved invaluable for many of these applications and continues to play a major role. Gas chromatography is a physical method of separation that relies on differences in solubility of components distributed between a mobile and a stationary phase. In the GC procedure of interest herein, a sample is vaporized and injected into a mobile phase such as hydrogen, helium, or nitrogen and passed into a column containing the stationary (solid or liquid) phase. The components of the sample demonstrate varying affinities for the stationary phase, thus slowing (to varying degrees) their passage through the column. In the perfect case, each sample component has a different retention time in the column and is thereby separated from all others. The magnitude of the signal that each component yields as it emerges from the column and enters the detector is a measure of its concentration.

Two objectives are paramount for any analysis using gas chromatography: the data must be accurate and precise and they must be produced in a minimum of time. Sampling and separation/measurement are important in some cases than in others. In some instances, it is more difficult to obtain a representative sample than to analyze it. Gas chromatography can reduce the difficulty in obtaining a representative sample because the instrument is small enough to be brought to the field, thus eliminating errors caused by the degradation of samples during transport.

This section details some practical hints about chromatographic hardware, choice of columns, selection of injection systems, and choice and performance of detectors. Several portable gas chromatographs are described. Compound techniques such as GC/MS are described later.

Basic Components of a Gas Chromatograph

A basic chromatograph may be extremely simple (Fig. 19) (66). It consists of a supply of carrier gas (commonly helium, nitrogen, or hydrogen) with appropriate flow regulation. In this example, the sample is introduced by means of a hypodermic syringe through a rubber septum cap into the carrier gas stream, where it is rapidly vaporized at the head of the column before passing through the column, where it is separated. The column, in turn, is housed in a suitable temperature-controlled oven. Finally, a detector is provided to indicate when the components are eluting from the column. This detector signal, when recorded as a function of time, produces a chromatogram.

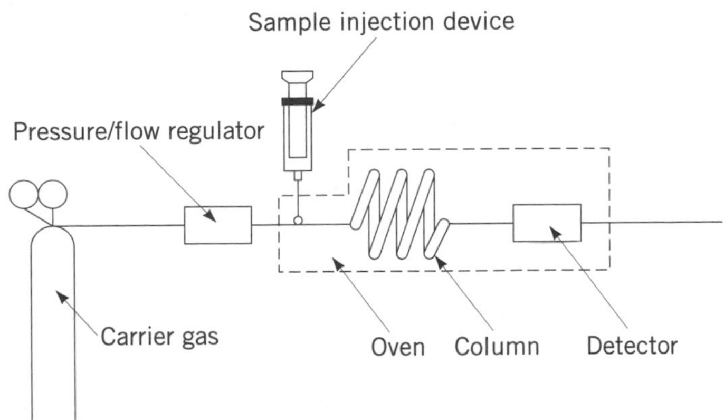

Figure 19. Schematic of a basic gas chromatograph (66).

The time required for each of the components to elute is called the retention time, and the magnitude of the detector signal (peak height and peak area) depends on the amount of component present in the sample. In this figure, the unretained peak (usually air) is defined as t_M, and all other retention times are corrected from t_M since this time is a characteristic of only the void volume of the column and has nothing to do with chromatography.

$$t'_R = t_R - t_M \tag{7}$$

Classification of Gas Chromatographs Used for Field Measurements

The gas chromatographic field measurement is only one part of the overall analytical procedure, which includes sample collection, sample extraction, sample concentration, sample derivatization (conversion of a chemical to another chemical), sample separation (in the separation column), and the data collection and reduction necessary to interpret the chemical information. The selection of the field instrument to be used must be based on the understanding of the complexity of the application. The instrument can be hand-portable, portable or transportable. Each type has its own features.

Portability implies that the total weight and size of the instrument are within some reasonable limits, ie, easily carried by the average person. Versatility roughly depends on the instrument's weight and size; the smaller the instrument, the more specialized it will be. Thus the difference between portable and nonportable gas chromatographs depends on the manufacturer's chosen trade-offs between size and analytical capability. Analytical capability depends on the features available—type of columns, maximum temperature of the overall analytical system, type of detectors, and type of injectors. Therefore, when comparing portable gas chromatographs, it is important to understand the trade-offs that the manufacturer makes between analytical capability and portability.

The analysis of gaseous samples is the simplest of field measurements, since most of the time sample preparation is not required. Hand-portable gas chromatographs fit this category and are oriented exclusively to the analysis of gases. They are light, have limited analytical capabilities (but can be sensitive), and are tailored to give a quick answer to a specific application. A typical hand-portable gas chromatograph is a self-contained unit that includes its own gas supply and a battery pack for autonomous operation. Figure 20 shows an example of field measurement using a hand-portable instrument during an emergency situation. In this application the instrument must identify the presence of a volatile chemical without disturbing the site. No front-end sample preparation is necessary and only the vapors coming from the suspected source of contamination are analyzed. A typical analysis with a sub-parts per million sensitivity can be accomplished in a matter of minutes, but this requires some knowledge about the nature of the sample to be analyzed. There is a high probability of a miss, if the hand-portable instrument is not tuned to the compound present.

The field analysis of medium temperature boiling volatile samples by gas chromatography is the next step of difficulty. This analysis requires that the pneumatic lines of the instrument be heated to a temperature of up to 200°C to avoid con-

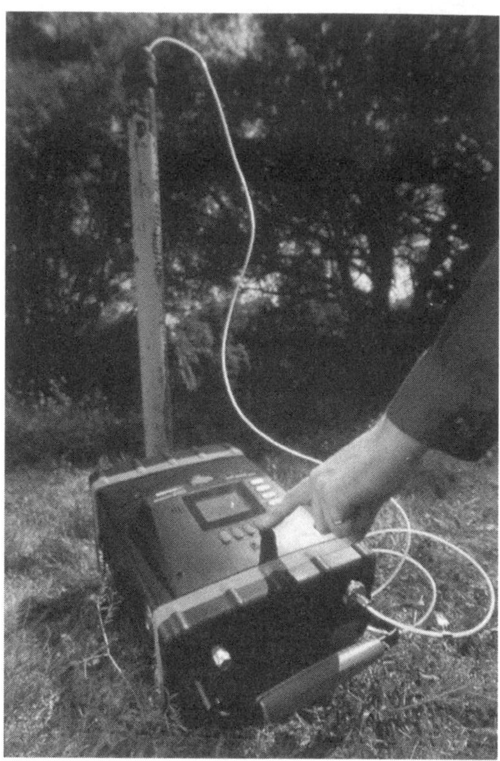

Figure 20. Field intervention using a hand-portable instrument. Courtesy of PE Photovac.

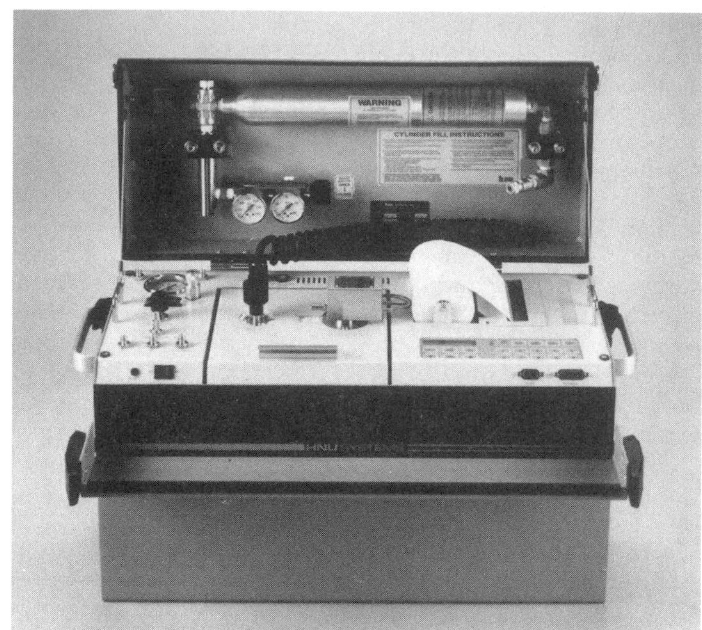

Figure 21. Portable instrument used for field intervention. Courtesy of HNU Systems, Inc.

densation of samples. The instrument has more features than a hand-portable instrument, but fewer than a laboratory instrument. For instance, it can accept liquid samples, and the column can be temperature programmed to reduce the analysis time (but at the expense of the size and weight of the instrument). Some basic sample extraction or sample preparation is also possible before analysis. The instrument can also be run directly at the field site, provided electrical power is available (AC line or portable power generator). The portable GC carries its own gas supply. Figure 21 shows the field operation of a portable gas chromatograph from HNU Systems, Inc. (Newton Highlands, Massachusetts). This instrument is smaller than a laboratory instrument and can be carried by an operator in the back of a station wagon. This portable instrument provides a broader knowledge about the sample to be measured than does a hand-portable GC, but it is bulkier and, therefore, more difficult to use for spot checks in the field.

The most elaborate and more expensive approach to quality measurements in the field is to use conventional laboratory instruments operating in a mobile laboratory. A mobile laboratory can be equipped with most of the analytical tools found in a conventional laboratory. Laboratory analytical instruments, such as gas chromatographs and mass spectrometers, are housed in commercial vehicles that have been modified to accept these instruments and to allow sample preparation. Depending on the vehicle, it can be modified to accommodate wet chemistry, analytical evaluation, and data handling at the same time. Such vehicles also provide a shelter and climate control for the operator and the equipment, and can be designed to operate in extreme weather conditions. Figure 26 shows a mobile laboratory, modified by E-N-G (Concord, California), that offers more than 120 square feet of laboratory space. The vehicle is air-conditioned and can provide up to 14 kVA of electrical power. Gas cylinder storage is provided for the transportation of compressed gases. The unit shown in Figure 22 can accommodate a sample preparation unit, several analytical instruments, and a computing facility, including wire or wireless data transmission. A larger version can handle gaseous, liquid, and solid samples and provide valuable field measurements with the highest quality results. It can also communicate to the outside world via wire or wireless technology.

Categories of Analysis

Analysis can be categorized by the range of concentration of the component being analyzed—major components (5–10%), minor components (100 ppm to 5%), and trace analysis (1 ppb to 100 ppm). Only the last category will be considered.

Trace analysis for which the sample is <100 ppm is certainly the most difficult to perform and is routinely required for environmental analysis. Methods of trace analysis require the highest level of instrumentation sensitivity and sample preparation to enhance the concentration level of the analyte. Analytical procedures for air or gaseous sample matrices are required for three types of analysis: air emission methods for stationary or point sources like incinerators, power plants, and various industrial processes; ambient air monitoring methods; and methods for monitoring indoor air contamination in the workplace. Air emissions and ambient air monitoring fall within the jurisdiction of state agencies and several offices of the EPA. Indoor air monitoring falls into the realm of industrial hygiene monitored by OSHA. The basic principles for the sample preparation and analysis of industrial hygiene samples are the same as those applied to outdoor air pollution (72).

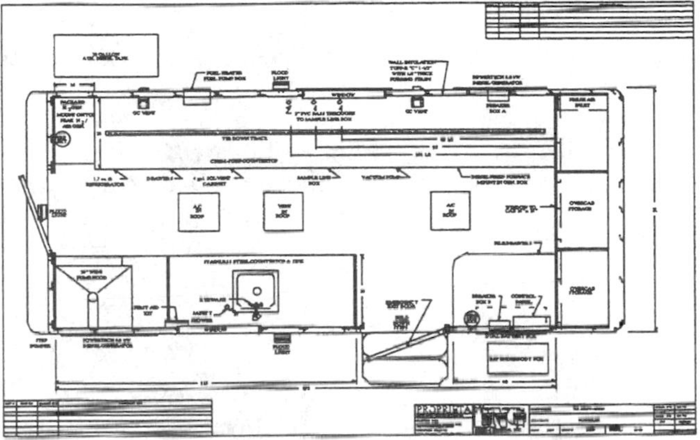

Figure 22. The Mobile laboratory unit Mobilab Model 216 from E-N-G Mobile Systems Inc. can be equipped with the same type of analytical equipment used in a conventional laboratory. *Top* the inside of the vehicle with some analytical equipment. *Bottom* a floor plan of the unit. Courtesy of E-N-G Mobile Systems Inc.

Trace Analysis of Gaseous Samples

With the sophisticated analytical instruments available, it is easy to lose sight of the importance of sample preparation techniques. Even the most powerful of the analytical instruments will be useless with a sample that is improperly collected or prepared.

Gases and vapors diffuse rapidly, mix freely with the general atmosphere, and can reach equilibrium in a short time. They thus offer fewer difficulties in air sampling than do aerosols. For air sampling purposes, contaminants can be grouped with respect to solubility and vapor pressure. Many gases and vapors of hygienic significance are water-soluble and can be collected in aqueous media with or without a dissolved reacting chemical to suppress the vapor pressure of the solute. Gases and vapors that are not water-soluble but are soluble or reactive in other agents can be adsorbed in a suitable solvent. Gases and vapors that are neither soluble nor reactive may be collected on adsorbent (eg, activated charcoal, silica gel, or molecular sieves) in either active or diffusive (passive) samplers. Adsorbents have become the sample collection medium of choice for all gases and vapors because of their convenience and generally high collection efficiency.

As indicated previously, analysis is the final step of a complex process that depends on the nature of the sample, the matrix in which the sample is found, the level of concentration of the sample, and the availability of a field instrument. In the most general case, the steps preceding analysis are identified as sample collection, sample concentration, sample extraction, and sample derivatization. It is only after all or some of these steps are completed that the sample can be analyzed.

Sample collection can be accomplished just before the analysis by drawing the sample through a sampling line connected to the gas chromatograph. The probe itself can be the gas chromatograph (73). Tedlar bags or appropriate containers (glass or inert metal) can also be used to collect samples that are then transported to a laboratory for analysis. The use of evacuated stainless-steel canisters to sample VOCs is a standard procedure in the industry (74). These approaches are applicable to clean samples, such as hydrocarbons in air. The drawback of sample storage is the risk of sample degradation, either by adsorption on the walls of the container or by selective adsorption in the trap, creating a change in the relative concentration of the compounds present in the sample.

When the sample is part of a complex matrix, whether gaseous, liquid, or solid, the extraction stage consists of isolating the vapor sample from its matrix. This extraction stage is usually accomplished thermally (headspace, purge and trap, thermal desorber, pyrolyzer). It is seldom done directly at the collection point, since extraction equipment is not readily transportable to the field. Care should be taken in evaluating the analysis as the extraction process can degrade the sample and introduce a bias.

Solid adsorbents are useful for trapping a gaseous sample before analysis. This trapping approach also offers the advantage of concentrating the sample at the time it is stored, by passing a known quantity of air through the adsorbent and retaining only the sample of interest. The analyte collected on a solid adsorbent can then be desorbed with an organic solvent. This sample-rich solvent can then subsequently be injected into the gas chromatograph. The analyte may also be concentrated by partitioning it into a liquid phase coated onto a support material. Impingers can be used to concentrate analytes in a gaseous matrix by passing the sample through an appropriate solvent that retains the analytes but allows the gas to escape. The solvent can be further concentrated or directly injected into the GC. The adsorption is usually sample- and volume-dependent. All these devices must be carefully prepared for the samples of interest when quantitative measurements are performed. Sorbents and impingers were reviewed earlier; solid adsorbents have been described in the literature (75).

Derivatization is an additional preparation step that is used less frequently. It consists of chemically converting an analyte into another chemical species to improve stability, sensitivity, or instrument performance. One popular application of derivatization techniques is to convert a compound that does not respond to an electron capture detector (ECD) into a product with electronegative properties and a favorable ECD response (76). Another example of derivatization is the conversion of carbon monoxide or carbon dioxide to methane before introduction to the FID.

The following general discussion considers aspects of sample preparation procedures so the analyst can evaluate practical applications and identify the logic behind the various modifications to the procedures.

Solid Extraction Thermal Desorber Unit

The use of solid sorbent sampling is illustrated with an application of a field-operable unit for VOC analysis, developed by Louisiana State University (77). The unit includes two traps and provides an enrichment factor up to 1000:1. The first trap provides the concentration of analytes on a large break-through trap (the concentration trap), desorption of analytes on a low volume desorption plug trap (the focusing trap), desorption of analytes into the analyzer instrument in a narrow plug, and back-flushing of the system between analyses to prevent cross-contamination.

Figure 23 illustrates the VOC sample processing device. The concentrating trap can be exposed to large volumes of gases, up to 2 L, on a low thermal mass concentrating trap packed with hydrophobic adsorbents. Break-through volumes are, of course, compound dependent; but with dual-bed series, packed traps are typically in the 1–2 L range for common VOC analytes when using Tenax and carbon molecular sieve adsorbents.

Water in the sample is not significantly retained by these sorbents, and thus its concentration is reduced in the extracts. Since the desorption plug from the concentrating trap is several milliliters in volume (typically 5–10 mL), the analytes in this plug are reconcentrated on a small and low thermal mass focusing trap into a final volume of approximately 300 μL. This small concentrated plug of analytes, typically 1 μL and generally devoid of large quantities of water, is then passed to an appropriate analyzer for qualitative and quantitative determinations.

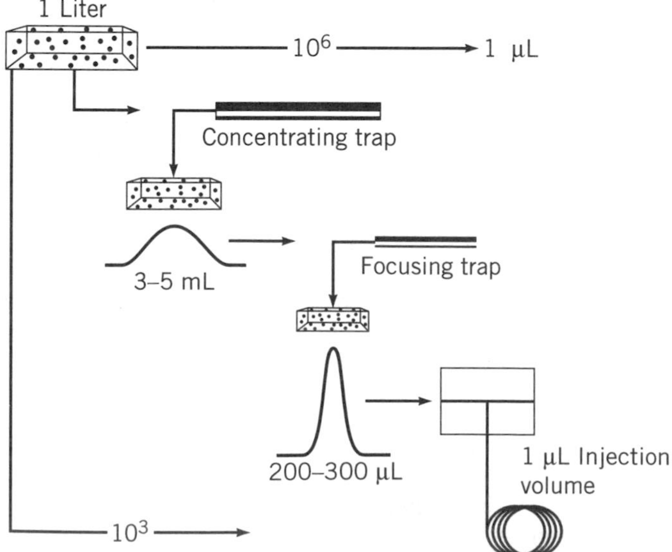

Figure 23. Pneumatic configuration of a two-stage preconcentration unit. The first trap is used for concentrating several liters of air and the second trap for focusing the sample in a plug of a few hundred microliters. Courtesy of Analytical Specialists Inc.

The unit manufactured by Analytical Specialists Inc. (Baton Rouge, Louisianna) is microprocessor controlled and can be operated remotely. The total amount of power necessary to operate the unit and desorb the samples from the traps is approximately 50 W. Furthermore, since it is a dual-trap system, selective adsorbent chemistry can be used to enhance performance in special applications.

Transfer Lines

The use of fluidic (no moving mechanical parts) sampling and column-switching valves in gas chromatography has been reviewed (78). Enviroprobe, the air probe developed by the University of Utah and commercialized by FemtoScan (Salt Lake City, Utah) is an example of this approach (Fig. 24) (79). The Enviroprobe operates with a small vacuum pump and provides the functions of collection, injection (from 20 ms), and separation (column of 100 μm inner diameter, 1 m long). The unit output is connected to another analytical instrument, here a portable mass spectrometer.

Airborne Pollutants

Ash or fly ash, the solid debris that remains after incineration processes, is another class of solid sample material that is of environmental concern. These airborne and potentially carbonaceous particles contribute to visibility reduction and have complex chemical compositions that include carcinogenic and

Figure 24. View of a transfer line (Enviroprobe) connected to a portable mass spectrometer. The unit integrates several functions of sample collection, sample injection, and sample separation. Courtesy of FemtoScan Corp.

mutagenic organic compounds. The chemical composition and structure of particle organic matter can provide important information regarding origin, distribution, and fate of respirable aerosols of small dimensions (typically 10 μm). Although current regulations for these types of small particles matter (PM_{10}) are concerned only with total particle densities over a period of time (typically 24 h), future regulations must address the obvious link between their chemical composition (eg, heavy metal, carcinogen, or allergen content) and associated health risks (80). Permits for PM_{10} emissions for specific particles must be obtained from the EPA (81). Another subset of particle matter is $PM_{2.5}$, particles with mean aerodynamic diameters of less than 2.5 μm. The EPA is considering adding a $PM_{2.5}$ standard, primarily to protect vision.

Numerous organic compounds may be adsorbed on particles as they enter the atmosphere and contribute to air, soil, and water pollution. These organic compounds are difficult to extract from particulate airborne pollutants.

Many of the same compounds that must be determined in soil and water samples must also be determined in airborne pollutants. The major procedural difference required for the analysis of air samples is concerned with the collection of the sample itself. The gas chromatographic portion of the analysis for a given class or group of compounds is essentially the same as that described for the analysis of extracts prepared from soil and water samples.

Collection and Analysis of Airborne Pollutants

Methods for particle collection and separation are still empirical. A field procedure will be used as an example of a fieldable system that encompasses sample collection, sample transfer, and sample analysis units, including a gas chromatograph and a mass spectrometer operating in a mobile laboratory unit (82).

Atmospheric particles with an aerodynamic diameter <10 μm (PM_{10}) are obtained by drawing an aerosol through an impaction jet and depositing larger particles on a plate immediately in front of the nozzle. The impactor is contained in a stainless-steel delivery tube to channel the air flow to a filter holder and a particle counter. Particles are collected in a clean filter. The sampling port is raised approximately 1.5 m (5 ft.) above the roof of a transportable laboratory module (Fig. 25).

Aerosol concentration and particle-size distribution measurements (parallel to aerosol collection on quartz fiber filters) are performed at 10-min intervals over eight size-resolved channels covering the 0.3–10-μm-diameter range using a multichannel particle counter.

Samples are positioned inside a special glass reaction tube lined with a ferromagnetic foil. Volatile and semivolatile organic compounds are desorbed by flash desorption at a temperature of 315°C and nonvolatile organic matter is pyrolyzed with a Curie-point desorption–pyrolysis reactor at a temperature of 650°C.

Volatile analytes from the reactor are transferred into the mass analyzer via a short capillary column (transfer line) that can be temperature programmed up to 20°C/s. The capillary is inserted into the ion trap mass spectrometer.

Theory of Packed and Fused-Silica Capillary Column Operation

The introduction of inert fused-silica columns markedly changed the practice of gas chromatography, enabling high resolution chromatography in most laboratories. A further decrease in the use of packed columns occurred after the introduction of megabore capillary columns of 0.53 mm inner diameter that, with their increased sample capacity, served as direct replacements for the packed column. These developments, in conjunction with the emergence of immobilized or cross-linked stationary phases especially tailored for fused-silica capillary columns, have been responsible for the greater acceptance of capillary GC.

Figure 25. Mobile laboratory module on a three-quarter-ton pick-up truck used for the collection of PM_{10} particles. Courtesy of the Center for Micro Analysis and Reaction Chemistry (Salt Lake City, Utah).

The use of packed columns represents 20% of gas chromatographic analysis for laboratory applications. This number is even smaller for field applications. Packed columns are employed primarily for preparative applications, for fixed gas analysis, and in those separations for which high resolution is not always desirable. Packed columns will continue to be used for GC methods that were previously validated on packed columns and for which the time and cost of revalidation on capillary columns would be prohibitive.

The introduction of commercially available fused-silica capillary columns in late 1979 followed an investigation of the applicability of fused silica in fiber-optic technology to chromatographic columns (83). The inherent strength and flexibility of fused silica make it easier to use and less fragile than silicate glass capillary columns. In addition, fused silica provides a more inert surface for improved capillary performance and less adsorption of active components.

Uncoated fused-silica capillary tubing is susceptible to moisture in the air. Water molecules attack the silicon-oxygen bonds, forming silanol groups and opening cracks that propagate into fractures. Protecting the silica from moisture and scratches is the main reason for the outer coating of either polyimide or some other equally resistant material.

A good understanding of the use of capillary columns begins with an understanding of the terminology. Perhaps the most important terms to become familiar with are selectivity, resolution, and efficiency. Further information on the theory of capillary gas chromatography and its practical implications can be found in the literature (66,67,84).

Column Selectivity. The critical step in developing a gas chromatograph method is the selection of a stationary phase for which each sample component has a different affinity, thus ensuring a separation in the column. Analysis on a large number of phases has been reported through the years and has been reviewed in the literature (85,86). By far, the best method for classifying stationary phases was introduced in 1966 (87) and later modified (88). In this procedure a group of compounds (benzene, butanol, 2-pentanone, nitropropane, and pyridine) containing different functional groups are chromatographed on the phase to be classified, and their Kovats retention indices I are recorded. The differences in the indices (ΔI) for the compounds from those recorded on the squalane stationary phases are the Rohrschneider indices for that phase and are a measure of the phase's selectivity to the particular functional group contained in the probes (66). Some of the more popular phases, along with their modified Rohrschneider indices are shown in the table. It should be noted that a casual inspection quickly reveals that there is no significant difference in the selectivities of the older SE-30 phase and the newer highly purified methyl siloxane, OV-1.

A uniform expression for retention data that is relatively free of instrumental contribution is given by the Kovats retention index system (89).

Parameters Influencing a Separation

The major factors that influence column efficiency h, and thus the required column length, are a function of several interrelated column, extracolumn, and operational parameters. Major factors that influence column efficiency are column internal diameter, column length, the type of stationary phase and its film thickness, the type of carrier gas, the carrier gas velocity, and the column temperature, as well as other instrumental limitations, such as the sample injection width, dead volumes between injector and column and detector, detector volume, data acquisition time constants, and data rates.

Fortunately, a computer model is now available that not only computes L_{reqd} from the imputed values of the column and instrument variables, the identity of the stationary phase, and the Kovats indices of the sample components, but also generates the predicted chromatogram. Use of such a program saves invaluable laboratory method development time (90). Examples of computer-aided column selection and optimization are described later.

Computer Optimization of the Separation

Chromatographic separation optimization can be broken into two major areas: stationary phase selection and optimization of the analysis conditions. Several tools are now available to aid the chromatographer in making a decision with few or no laboratory experiments. Pro ezGC (Analytical Innovations, Inc., Kettering Ohio) and Dry Lab (LC Resources, Walnut Creek, California) are models that use the temperature and pressure programming approach to GC analysis. OnLine's (Duxbury Massachusetts) GC Lab is an isothermal model that uses the appropriate lengths of serially coupled columns containing the stationary phase of widely different selectivities. One example using Pro ezGC to optimize the run time on a given liquid phase is given below.

In the following computer-aided optimization example, Pro ezGC is used to optimize the temperature separation of the mixture of 20 phenols and pesticides on cross-linked 5% phenylmethyl siloxane DB-5. The list of the components is given in Table 3. An arbitrary temperature of 100°C, and a flow velocity of 45 cm/s hydrogen were used. This run was isothermal with no attempt to optimize the separation.

Obviously, it is not necessary to be a chromatographic expert to see that the total run time can be shortened considerably by programming the column temperature, by shortening the column, and by selecting another flow velocity. Program simulation gives the user the capability to study the separate effect of each of these parameters and to optimize all the conditions simultaneously with a minimum of experimental chromatographic work. The operator defines the simulated runs with parameters similar to those used with a gas chromatograph, but yielding the tremendous advantage of obtaining the results in minutes instead of days.

The boundary conditions of the simulation are the following:

Length of column: 7.5, 15, 30, and 60 m

Liquid phase: DB-5

Flow velocity: 45–120 cm/s, in increments of 15 cm/s

Initial temperature: 100–200°C, in increments of 25°C

Temperature programming conditions: 5–45°C/min, in increments of 5°C/min

Final temperature: 300°C

Table 3. Computer Simulation of the Separation of Phenols and Pesticides on DB-5

Component Number	Component Name
1	Phenol
2	2-Chlorophenol
3	2,4-Dimethylphenol
4	2-Nitrophenol
5	2,4-Dichlorophenol
6	Dichlorvos
7	4-Chloro-3-methylphenol
8	2,4,6-Trichlorophenol
9	2,4-Dinitrophenol
10	4-Nitrophenol
11	2-Methyl-4,6-dinitrophen
12	Ethroprop
13	Chloroproham
14	Trifluralin
15	Atrazine
16	Pentachlorophenol
17	Diazinon
18	Disulfoton
19	Alachlor
20	Oxyflurofen

This range of conditions requires 1080 simulations, which are performed in just a few seconds on a laptop computer. The optimized conditions of the run are given below:

Resolution: 1.5
Column: 15 m × 0.25 mm × 0.25 μm, DB-5
Initial temperature: 150°C
Temperature rate: 30°C/min up to 300°C
Inlet pressure: 54.6 kPa (8 psi)

General Comments. The computer tools that are now available, although currently limited to gas–liquid applications, offer the tremendous potential of operating gas chromatographs much more efficiently using less skilled personnel. A program such as Pro ezGC aids the analyst in minimizing run time by optimizing the programming conditions of flow and temperature.

The GC Lab simulation tool gives the user the capability of optimizing the overall selectivity of the column (probably the most important step in method development) by optimizing the selectivity of two independent segments of column individually coated with different liquid phases. Also, optimization may be focused either on designing for minimal time analysis or minimum detection limits. Most of the simulation programs allow for the addition of compounds to the library, either through in-house laboratory work or by purchasing commercial libraries containing many compounds in a variety of stationary phases.

The simulation tools allow the analyst to experiment with various methods, in essence, developing a GC method at the computer keyboard. The parameters of the proposed method are obtained quickly and can then be used to design the proper laboratory experiments for its validation. All of this saves the valuable laboratory time normally required in the initial method development stage. The experimental results, however, are only as good as the chromatographic model that is used to predict them. In the case of fast chromatography, for instance, the dead volumes upstream and downstream of the column contribute to peak broadening, and accurate quantitative analysis is sacrificed as peak widths decrease to below the minimum of 20 points/peak needed for accurate and precise quantitative determination. One of the simulation tools (GC Lab) allows for user inputs of these extra column contributions. It has been verified that when this is done, the predicted chromatograms generated with the model match those obtained in the laboratory (91,92).

As with all automated tools, the user should be conscious of their limitations, regarding them as somewhat imperfect mathematical tools. It is only by iteratively correcting to the model using results obtained in the laboratory that accurate predictions of experimental results are generated by the computer.

Sample Injection Devices

The main function of a gas chromatographic inlet is to provide the accurate, reproducible, and predictable introduction of a sample into the column. The sample can be gaseous, liquid, or solid, and several injection modes can be used. There is no single inlet that can satisfy all analytical requirements. Inlets are usually divided into packed-column inlets and capillary-column inlets. Sample introduction has been discussed at length in the literature (93–95).

Inlets must be selected carefully for each analysis and used to optimize chromatographic efficiency, analytical accuracy, and the reproducibility of results. Because capillary columns have higher efficiency and lower sample capacities than do packed columns, inlet performance is much more vital to obtaining accurate results with capillary rather than with packed-column systems. Once an inlet has been selected based on sample, column type, and analysis goals, all inlet variables must be set appropriately to achieve optimal results.

The type of analysis and the composition of the sample itself are the primary factors that must be considered when setting or determining the interrelated variables related to the injection itself (type of injector, injected volume, injection temperature) and to the chromatographic analysis (column selection and column temperature).

Effect of Inlet on Peak Broadening. A basic functional requirement of GC inlets is that they introduce the sample into the column as a narrow band with a composition that is identical to the original sample. The inlet should produce peaks (width-of-injection function) that are much narrower than the expected peak width resulting from the column band-broadening process. More efficient columns require narrower initial peak widths. This can be accomplished by using inlets that generate narrow peaks initially, or by using subsequent focusing techniques.

Initial bandwidths are broadened by band broadening in time or by band broadening in space. Band broadening in time is caused by the slow transfer of sample vapor from the inlet to the column. The initial peak width is equal to the time it takes for the sample to be transferred to the column. Band broadening in space is a direct consequence of the migration and spreading of a liquid sample within the column, either after the cool on-column injection of the sample into the column or after recondensation of the sample in the column. The condensed liquid, which starts by occupying only a few centimeters of column, becomes too thick to be stable and spreads over a longer length of column. The carrier gas pushes the plug farther into the column, creating a flooded zone. The solute material is spread over the full length of the flooded zone, creating an initial peak width that equals the length of the flooded zone.

Sample Focusing as a Way to Reduce Peak Broadening. Several focusing techniques are used to narrow peak broadening in time and space (96): stationary phase focusing, solvent focusing, thermal focusing, and retention gap.

Stationary Phase Focusing. Stationary phase focusing is the most frequently used focusing technique and is possible only in temperature-programmed analysis. Retention times of solutes are an exponential function of temperature. As the initial temperature of the column is lowered, the speed at which solutes travel down the column slows down drastically. As a vaporized sample moves from the inlet to the column, it comes in contact with the stationary phase and is trapped in a narrow zone. The lower the temperature, the more effective the focusing.

Solvent Focusing. As the condensed solvent starts to evaporate, solutes with volatility similar to that of the solvent tend to concentrate and focus on the solvent tail. This solvent focusing, or the "solvent effect," yields narrow peaks for these early eluting compounds (96).

Thermal Focusing. Thermal focusing relies on the condensation of gases in a tube or at the head of the column. Peaks narrow as the solute volume is reduced during condensation.

Solutes do not migrate into the chomatographic system until the temperature is raised, then they are vaporized again. Sometimes cryogenic temperatures (cryogenic focusing) are used to focus peaks from inlets or auxiliary sampling devices that generate peaks broadened in time. Thermal focusing narrows bandwidths effectively only when the column temperature is approximately 150°C below the boiling points of the solutes. In this sense, thermal focusing does not rely on chromatographic processes. It requires only a surface on which vapors can condense.

Retention Gap. A retention gap is an empty piece of column that accommodates the condensed sample but does not retain solvent or solutes once they have been vaporized. The primary function of the retention gap is to reduce the length of the flooded zone created whenever solvent is condensed in the column. An equally important function is to protect the column from nonvolatile sample components, especially when doing cool on-column injection (96).

Types of Capillary-Column Inlets Using Syringes

Capillary-column inlets that use hypodermic syringes to introduce the samples fall into two categories: vaporizing and nonvaporizing (cold) injections (94).

Vaporizing. Capillary direct, or vaporizing, injections are used with some capillary columns of ID >0.5 mm and are made by substituting a special insert inside a packed-column inlet to protect the column from nonvolatile sample components.

Split mode injection uses a vaporizing inlet that vents most of the sample in the split mode and transfers most of it to the column in the splitless mode. It is also column protecting but can cause discrimination and decomposition of the sample. Split injection is used for general analysis, whereas splitless injection is most frequently used for trace analysis.

The programmed temperature vaporizer (PTV) inlet offers a mixture of injection possibilities, including cool sample introduction, split or splitless modes, and sample concentration (solvent elimination mode); it is column protecting. Because of this flexibility, PTV inlets are useful for both general and trace analyses.

Nonvaporizing. Cool on-column inlets give high accuracy and reproducibility, do not lead to sample decomposition, have the least solute discrimination among all the inlets, and work by depositing the sample directly into the column. Unlike vaporizing sample introduction techniques, here the sample is not exposed to high temperatures during injection or transfer to the column. Cool on-column inlets are used for the analysis of samples with a wide boiling-point range or those that are thermally sensitive for trace analysis.

Types of Capillary-Column Inlets Using Other Mechanisms

Other sample introduction devices include gas and liquid sampling valves, headspace autosamplers, thermal desorbers, purge and trap samplers, pyrolyzers, and transfer lines. Depending on the sampling device and the GC column used, auxiliary sampling devices can be connected directly to the column or to an existing inlet.

Valves. Valves can consistently introduce fixed volumes of gas or liquid samples and are simple to automate. Valves are versatile, can operate at elevated temperatures (several hundreds of degrees), have a fast response time (down to fractions of a second), and can be small enough to be mounted into a portable instrument. Valves are frequently used for sampling gases and liquids in moving streams (process and on-line monitoring).

Sample Introduction Based on Gas–Liquid Extraction. The principle of sample introduction based on a gas–liquid extraction is rather simple. Once an equilibrium has been achieved, an analyte distributes itself between two immiscible phases (here the gas and the liquid sample phases) at some constant ratio. The two modes of gas–liquid extraction are static and dynamic. In the static mode, the analysis begins with the addition of an accurate volume of sample into the vial. The sample volume must be selected to allow for the appropriate volume of gas phase or headspace in the remaining volume of the vial.

In the dynamic or purge-and-trap mode, extraction of the sample is effected by passing a constant flow of gas through the aqueous sample. The analytes are continuously purged from the sample and carried by the gas flow to a medium that can collect and concentrate the analytes.

Microextraction. Microextraction has been shown to be an efficient method for the introduction of samples into a high-speed GC and is applicable to liquid–liquid extraction (97).

Headspace Autosamplers (Static Gas–Liquid Extraction). Headspace analyzers perform static gas–liquid extractions and are used to determine volatiles in liquids, solids, or complex matrixes. Once the sample has been added, the vial is sealed with a septum cap, placed in a temperature-controlled environment, and allowed to equilibrate for a certain period of time. After the equilibrium has been achieved, a sample of the gaseous phase in the vial is removed by a gas-sampling syringe and injected into the gas chromatograph. Headspace autosamplers inject a portion of the gas that is in equilibrium with a sample into a thermostated sealed vial (98). Headspace analysis is used for the analysis of residual solvents, fragrances, and volatile pollutants in soil and water. Reproducible headspace analysis requires excellent control of all the parameters affecting the vapor pressure.

Purge-and-Trap Automatic Samples (Dynamic Gas–Liquid Extraction). Purge-and-trap combines dynamic headspace trapping with thermal desorption. Volatile components are continuously purged out of a water sample (dynamic headspace), trapped on an adsorbent, and then desorbed quickly for introduction into GC (thermal desorption). Purge-and-trap is used mainly for analysis of environmental pollutants in water and for analysis of volatiles in beverages.

Thermal Desorbers. Thermal desorbers are used in environmental sampling and are complementary to headspace analysis and analytical pyrolysis. Volatile sample components that are contained in a solid sample or have been absorbed onto a solid adsorbent are thermally liberated in the sample in a stream of carrier gas and carried to the column. The advantages and disadvantages of sorbent sampling, active and passive techniques, solvent and thermal desorption, grab-sampling containers, cryogenic traps, and cold-trap concentration for the analysis of C_1-C_4 hydrocarbons can be found in the literature (99). This technique is used for monitoring hazardous gases in the workplace and for environmental air analysis.

Pyrolyzers. Pyrolyzers are used to thermally cleave nonvolatile samples into volatile fragments, which are then analyzed by GC. Temperatures in the 500–1000°C range are normally used for the analysis of polymers, fibers, microorganisms, and geological samples.

Transfer Lines (Ambient Vapor Sampling Inlet Module). The ambient vapor sampling (AVS) inlet module (79), mentioned earlier, combines injection of gas with crude gas chromatographic separation. Its main application is for field mass spectrometry and for IMS measurements.

Figure 26 shows a schematic view of the Enviroprobe. The AVS is a valveless approach (100) to injecting variable quantities of gases by modulating the time during which the carrier gas exchanges the carrier gas flow in tube two for the sample flow in the opposite direction. This approach enables resolution in milliseconds for the injection pulse width (73). The GC separation is performed in the transfer line between the injector and the detector. This approach minimizes the upstream dead volumes and at the same time reduces the sample adsorption sites.

Cryofocusing Trapping and Fast Desorption. Cryofocusing of samples, followed by fast thermal desorption, at a rate of several thousands degrees Celsius per second, is an approach used to generate narrow injection pulses. One successful approach, originated at the University of Michigan (101), uses a cryofocusing capillary dimension trap for sample collection and capacitive discharge for heating the trap rapidly (102). This approach is now commercialized by Chromatofast (Ann Arbor, Michigan).

Figure 27 shows an application of fast injection to the determination of solvents from paint. A total of 15 solvents were separated in less than 1 min on a DB-1 column (103). A conventional GC analysis for these compounds required a 1–2-h cycle time. This application analyzes the air in paint booths at automotive plants.

Gas Chromatographic Detectors—Basic Terminology

Whatever detector is used, there are three major response characteristics of detectors: sensitivity, selectivity, and dynamic range. To be useful in a practical sense, stability and repeatability must also be considered.

Sensitivity. This is the response per amount of sample, ie, the slope of the response/amount curve. The minimum amount on the curve is defined as the minimum detectable quantity (MDQ). The MDQ is practically defined as the amount of sample for which the peak height is two times the noise height (signal:noise ratio = 2).

Selectivity. Selectivity identifies which categories of compounds give a detector response. Some detectors, such as the thermal conductivity detector (TCD), sense almost everything and are considered universal. Others, such as the ECD, respond only to certain types of components and are quite

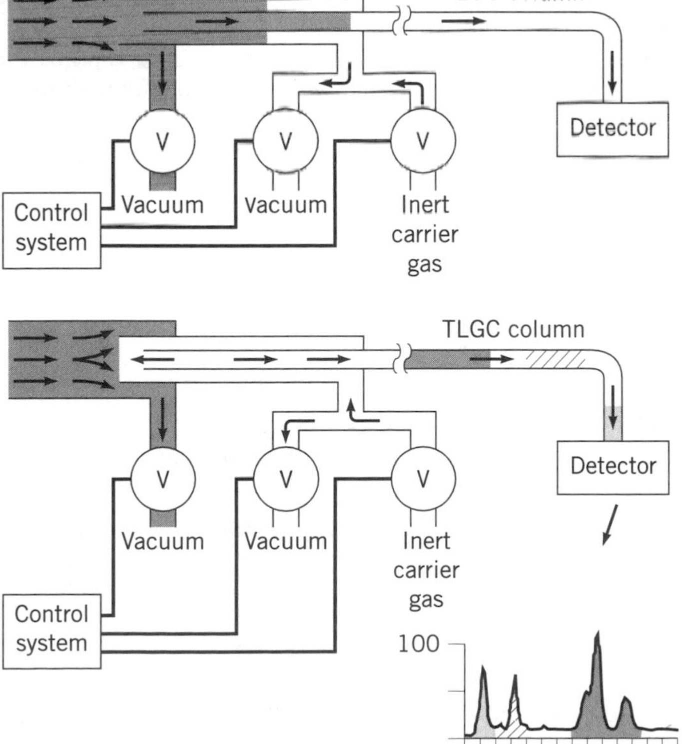

Figure 26. Schematic view of the Enviroprobe (AVS) from FemtoScan. *Top,* In the injection mode, the two vacuum lines are opened and the sample gas is injected into the column. *Bottom,* By switching the second vacuum line (middle connection), the inert carrier gas is forces its way through the column and creates a barrier. Courtesy of FemtoScan Corp.

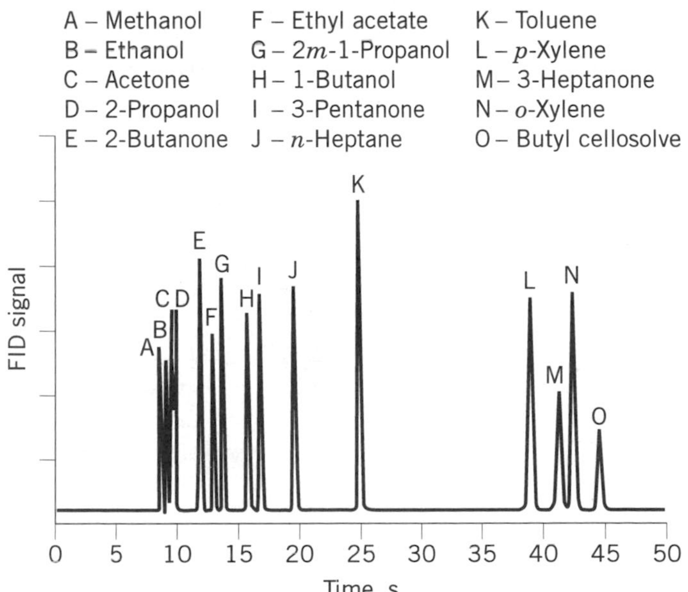

Figure 27. Example of fast GC injection and separation, showing the separation of paint booth solvents. Experimental conditions: column DB-1, 8 m, 0.25 mm ID, 0.25 μm film; Temperature program: 35°C for 0.1°C/min, 50°C/min to end of analysis: Column flow: 4 mL/min. Courtesy of Chromato-FAST, Inc.

useful when the detected component is present in a complex, nondetected matrix. Selectivity is usually expressed as the ratio of the injected quantities of two compounds that give the same output response. For instance, the selectivity of the ECD for anhydrides relative to hydrocarbons is 10,000:1.

Dynamic Range. Dynamic range is the range of sample concentrations over which a relative incremental change in sample concentration produces a similar relative change in the signal. This range can be quite large for certain detectors (seven orders of magnitude for the flame ionization detector). For others, such as the flame photometric detector (FPD), it can be quite limited.

Gas Chromatographic Detector Selection

The selection of the GC detector used is determined by the information desired. The compound structure is identified by a mass spectrometric detector (MSD) or an infrared detector (IRD). High sensitivity is achieved by the ECD, the nitrogen phosphorus detector (NPD), and the MSD. High selectivity is provided by the ECD, NPD, FPD, and the MSD in single-ion monitoring mode. The electrolytic conductivity detector (ELCD) and the PID also offer high selectivity for particular species. A general description of these detectors follows. Gas chromatography detectors are covered in detail in the literature (65,67,104).

Flame Ionization Detector. The FID consists of an oxygen-rich hydrogen flame that burns organic molecules, producing ionized molecular fragments. The resulting ions are collected by a polarized collector (Fig. 28). The FID is selective for materials that ionize in an air/hydrogen flame. Few ions are formed until an organic compound elutes into the flame. The resulting large increase in ions is collected by a polarized collector. The FID is characterized by a large dynamic range (10^7), high sensitivity (15 mC/g), low detection limit (10^{-12} g/s), and small dead volume (nL). For these reasons, the FID is the most popular GC detector today. With optimized operating parameters and a good analytical column, it is possible to determine components in as little as 20 pg or about 5 ppb of sample gas.

The response of the FID depends on the flow rate of all three gases—carrier gas, hydrogen, and air—with hydrogen being the most critical. The flame is usually optimized by successive injections until the signal reaches its maximum.

Photoionization Detector. The PID consists of a UV lamp and an ionization chamber. The lamp is filled with a gas (usually argon or hydrogen) and produces emission lines characteristic of the gas when excited. The radiation passes through a metal fluorite window and into the ionization chamber or cell. There the sample absorbs the radiation and ionizes. Electrodes collect the ions; the current measured is proportional to the sample concentration (Fig. 29).

Most of the energy is emitted at one frequency; the lamps are characterized according to the energy of this major emission line. The lamp is selected based on the ionization potential of the compound being measured and on the selectivity and sensitivity required. Lamps of 8.3, 9.5, 10.2, and 11.7 eV are available. Using a 10.2-eV lamp, the PID does not respond to C_1 though C_4 hydrocarbons (ionization potential of 10.5–12.98 eV) but does respond to C_5 hydrocarbons and above. Likewise, it does not respond to permanent gases (IP $>$ 10.5 eV). It does respond, however, to aromatics, olefins, ketones, amines, etc. Thus for complex samples such as gasoline, PID chromatograms are much easier to interpret than the ones obtained with the FID.

The PID is a concentration detector with a dynamic range of 10^7 and an MDQ of 2–3 pg benzene. Detector cell volume is an important specification for the capillary column. The available cell volume today is $<$30 μL, which may be acceptable for normal speed capillary chromatography but not for high speed variations.

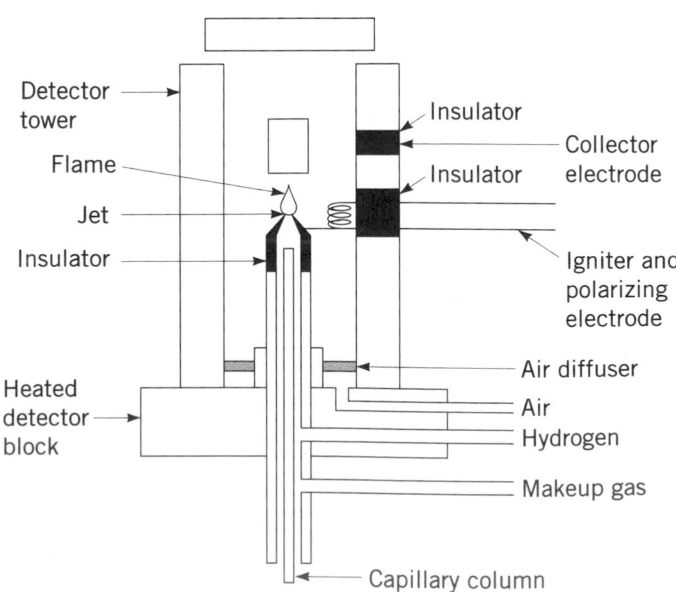

Figure 28. Schematic of an FID.

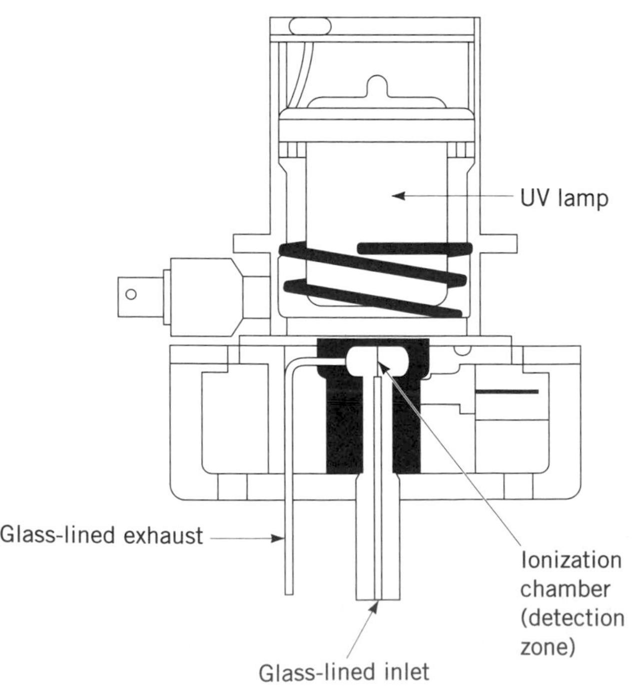

Figure 29. Principal elements of a PID. Courtesy of HNU Systems, Inc.

Electron Capture Detector. The cavity of the ECD contains two electrodes with the effluents of the column passing between them. The cathode is coated with β-emitting radioactive material. The emitted high energy electrons produce large amounts of positive ions and low energy (thermal) electrons in the carrier gas. These electrons are normally collected by the positively charged electrode, thus producing a standing current of typically $10^{-8}-10^{-9}$ amps. However, when a material with an affinity for thermal electrons is present in the carrier gas, it captures some of these electrons; and consequently, the standing current that is observed is reduced. The essential elements of the ECD are summarized in Figure 30.

Instead of applying a steady DC polarization voltage, a pulse of 1–3 μs at 50 V is applied to the electrodes every 150 μs. This pulse collects only free electrons, since the heavier negative ions cannot respond in this small interval and are swept out of detection by the carrier gas. The most common carrier gas selections for ECD are nitrogen and 5% methane in argon.

The ECD is highly selective. It responds to electrophilic species such as halogenated compounds. Pesticides give excellent response with MDQs at the picogram level and over a dynamic range of 10^4.

For successful ECD operation, it is important that the carrier and purge gases are clean and dry (99.9995%). Moisture, oxygen (leaks), or other contaminants must be avoided for stable operation of the detector. The column must be preconditioned before use. Finally, the use of a radioactive source is regulated by many governments and usually requires a license to be portable.

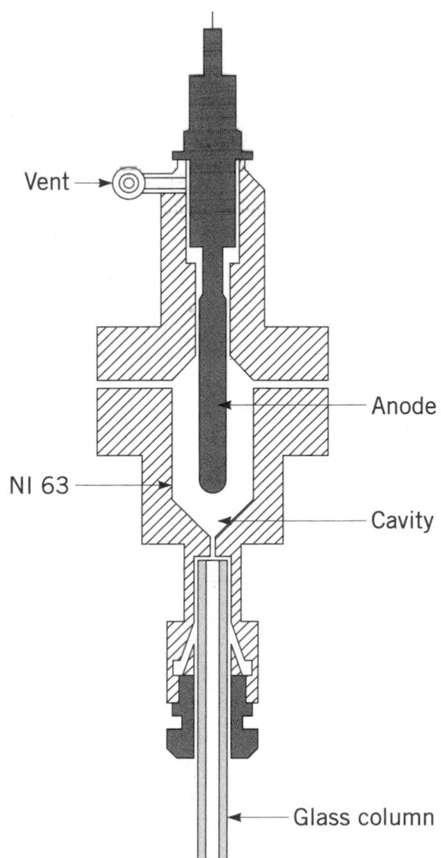

Figure 30. Cross-section of an ECD. Copyright 1997, Hewlett-Packard Co. Reproduced with permission.

Nitrogen Phosphorus Detector. The NPD (also called a thermoionic detector) uses a jet and collector similar in appearance to the FID. However, this detector is made selective for nitrogen- and phosphorus-containing compounds by a source of alkali salt positioned above the jet. This thermoionic source efficiently ionizes nitrogen- and phosphorus-containing organic molecules. Ions are collected, and the resulting current measured for the chromatogram.

The NPD has proven valuable in pesticide analysis, where its selectivity replaces costly extraction or concentration steps. It can detect a few picograms of nitrogen- and phosphorus-containing organic compounds. The NPD has a dynamic range of 10^4 and has a selectivity of 10,000:1 to 100,000:1 by weight of nitrogen and phosphorus to weight of carbon. It does not detect inorganic nitrogen, such as N_2 and ammonia.

Flame Photometric Detector. An FPD is essentially a flame emission spectrometer with its design optimized as a GC detector. This is an element-specific detector that responds to a particular molecular species. It is used primarily for the determination of sulfur- or phosphorus-containing compounds.

The sample contained in the column is fed to a flame, where the individual atoms are excited to higher electronic states by the energy of the flame. These excited atoms and molecular fragments subsequently return to the ground state with the emission of characteristic atomic or molecular band spectra. In the case of sulfur, the band emission from the S_2 species at 394 nm is used, while the 526 nm emission is used for phosphorus.

A narrow bandpass filter is used to isolate the appropriate analytical wavelength, and its intensity is measured by a photomultiplier (Fig. 31). The FPD response to sulfur and phosphorus is a factor of 10^4 greater than its response to hydrocarbons,

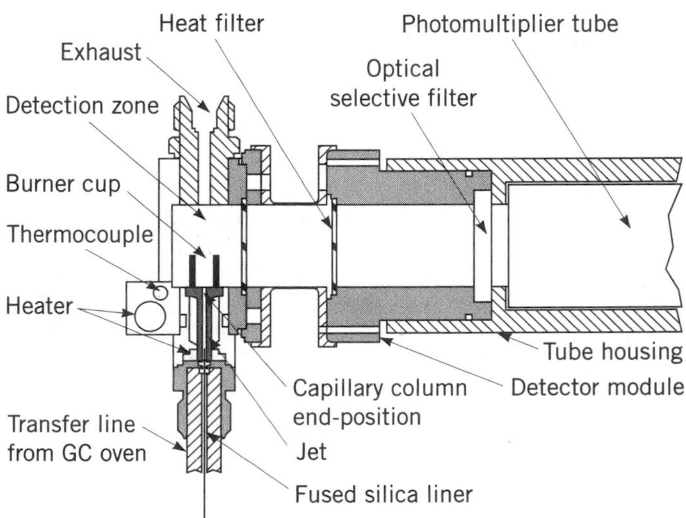

Figure 31. Diagram of a commercially available FPD. The detection zone (chemiluminescent zone) is above the flame; since the flame is shielded by a burner cup, the photomultiplier can see only the area above the flame (115). Copyright 1997, Hewlett-Packard Co. Reproduced with permission.

and the MDQ is at the subnanogram level. The response to sulfur concentrations is not linear. It should be proportional to the square of the sulfur concentration; in practice, a factor of 1.8 provides a better fit to the calibration curve (104,105).

One potentially serious problem with the FPD is the undesired light absorption that can occur in the flame. This can happen by hydrocarbon quenching from collision when a high concentration of carbon dioxide occurs in the flame at the same time as the sulfur atom species. Also, self-quenching can occur at high concentrations of the heteroatom species. In this case, collisional energy absorption or reabsorption of the photon by an inactivated species could effectively prevent the photon from reaching the photomultiplier.

The flame conditions are critical to successful operation. Gas flows are also critical to optimizing selectivity and sensitivity. Temperature can have a significant effect on sensitivity, too. Finally, condensation of water, especially with halogenated solvents or samples, will cause corrosion in the detection zone or fogging of the window leading to the photomultiplier.

In summary, the FPD has a number of limitations. These include its response dependency on the O_2:H_2 ratio, the H_2 flow rate, the type of sulfur compound, and even the length of time the flame has been burning. In addition, the best results are obtained by generating a calibration curve for each compound in the mixture.

Thermal Conductivity Detector The TCD responds to any compounds with a thermal conductivity that is different than that of the carrier gas. Hydrogen and helium are recommended carrier gases because of their exceptionally high thermal conductivity differences from all compounds. The principle of operation of a TCD is quite simple: when the analyte is present in the carrier gas, the thermal conductivity drops and less heat is lost to the cavity wall. Under constant temperature operation the voltage supply to the filament drops. This change is recorded and measured.

In practice, the TCD is extremely temperature sensitive. One can show that a difference of 10^{-6}°C is detectable in a single-filament detector; for this reason, a TCD is usually composed of two identical cells physically closed from each other and part of a Wheatstone bridge. One of the cells sees only pure carrier gas and the other is connected at the output of the column. Although this procedure reduces the variations in signal caused by changes in block temperature and any other common mode signal, such as variations in carrier gas flow rate, it does not eliminate their effect on the final signal. Careful thermal design is of critical importance.

An alternative solution to the stability problem has been offered (106,107). It has been shown that it is possible to produce a stable and fairly rapidly responding TCD by using only one TC filament and alternatively switching (at 10 Hz) the column and reference flow of carrier gas and effluents from the column. This pneumatic modulation has the advantage of almost completely eliminating the common mode signal and producing a signal with much lower baseline noise, drift, and wander. Assuming that a minimum of 10–20 data samples are required to accurately define a peak, pneumatic modulation restricts the minimum peak width to approximately 1 s and is not suited for fast chromatography unless makeup gas is used.

The dynamic range of the TCD is greater than 10^4, and its detection limit is less than 1 ppm. Samples that react with the filament can reduce the TCD's sensitivity and create instability in the baseline. Nonmodulated designs are particularly sensitive to common mode signals such as flow and temperature variations, and require long stabilization times. Generally, it is difficult to operate a nonmodulated design for high-sensitivity field operations.

Electrolytic Conductivity Detector. The ELCD can be used to detect halogen-, sulfur-, or nitrogen-containing compounds selectively. This is done by mixing the effluents with a reaction gas (oxiding or reducing, depending on the analysis) in a reaction tube (usually nickel). The resulting products are then mixed with ionizing solvents, producing a conductive solution. This conductivity is measured and recorded.

In the case of halogen analysis, the column effluent is reduced with H_2 at 850°C in a nickel reaction tube, producing strong acids (eg, hydrogen chloride from chlorinated components). This gaseous product is then dissolved in n-propanol, and the change in solvent conductivity is measured. Other reduction products besides the hydrogen halides do not ionize and, therefore, do not cause an increase in conductivity in the slightly acidic n-propanol solvent.

The sensitivity for chlorinated compounds is <1 pg and the dynamic range is 1×10^6. For sulfur, the sensitivity is 2 pg and the dynamic range is 1×10^4. For nitrogen, the sensitivity is 4 pg and the dynamic range is 1×10^4. Selectivity is 100,000:1 by weight of halogen relative to carbon and 10,000:1–100,000:1 for sulfur and nitrogen, depending on the compound and element.

Higher reactor temperatures (950°C) are necessary for compounds that are more difficult to reduce, such as PCBs. Using different solvents and gases, the ELCD has been made to detect heteroatoms different from halogens. Good results have been reported for sulfur and nitrogen selectivity.

Summary. Figure 32 shows how common detectors compare with respect to sensitivity and dynamic range. A minimum detectable quantity below the nanogram level (in parts per million) can routinely be detected. A MDQ at the picogram level (in parts per trillion) is at the limit of detectability of most detectors operating without preconcentration. The dynamic range of GC detectors is in excess of 4–5 decades.

Table 4 summarizes the main characteristics of GC detectors in terms of minimum detectable quantity, dynamic range,

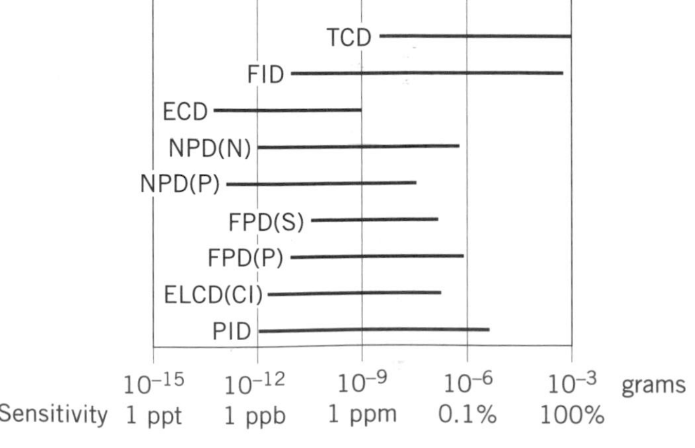

Figure 32. Comparison of various common GC detectors (115). Courtesy of Hewlett-Packard Co.

Table 4. Summary of Main Characteristics of GC Detectors

Name	Type	Selective for	Typical Minimum Detection Level ($S/N = 2$)	Linear Dynamic Range
FID	Selective	Material that ionizes in air–H_2 flame	5 pg C/s	10^7
TCD	Universal	Anything with thermal conductivity different from carrier	400 pg/mL carrier	$>10^4$
ECD	Selective	Gas phase electrophores	0.1 pg Cl/s (varies with structure)	10^4
PID	Selective	Compounds ionized by UV	2 pg C/s	10^7
Thermoionic	Selective	N, P, heteroatoms	0.4 pg N/s 0.2 pg P/s	10^4
ELCD	Selective	Halogens, N, S	0.5 pg Cl/s 2 pg S/s 4 pg N/s	10^6 10^4 10^4
FPD	Specific	P, S	20 pg S/s 0.9 pg P/s	10^3 10^4
FTIR	Universal	Molecular vibration	1000 pg strong absorber	10^3
MSD	Universal	Tunable for any species	10 pg to 10 ng (depending on SIM vs. scan)	10^5
AED	Universal	Tunable for any element	0.1–20 pg/s (depending on element)	10^4

and selectivity. All these detectors can be adapted to field operations but some of them will require more attention than others when used to maximum sensitivity. The main difficulties for the user are the time it takes to stabilize the detector and its susceptibility to environmental parameter variations, such as ambient temperature.

Other detectors such as mass spectrometers and ion mobility spectrometers are used for field operation and will be discussed below. The atomic emission detector (AED) has field applications for which unequivocal analytical information is needed (108). It has been successfully tested in various quantitative applications, such as the evaluation of lead, sulfur, and oxygenates in gasoline. The AED has also been used to identify base neutral pollutants, herbicides and PCBs. For instance, it has been used successfully in unfriendly territory for treaty verification.

The appropriate detector is determined by the class of compounds to be analyzed. Some of the applications of the various detectors are summarized in Table 5.

The FID and the PID have long been used in hand-portable units, and their ruggedness has been demonstrated under difficult conditions. The NPD is similar in design to the FID; it can be operated easily and provides valuable information.

The TCD, at least in the flow modulation mode of operation (one filament), can also provide reliable data and can stabilize in a matter of minutes. When operated in a differential mode at the maximum sensitivity, the dual filament TCD detector has a tendency to drift endlessly. Since in this mode the TCD relies on the equality of resistance between the reference and detection filaments, it is also susceptible to aging and becomes more sensitive to common mode signal variation after the resistance of one filament changes irreversibly. This may happen when the detection cells see a corrosive sample. Again, the single-filament TCD is insensitive to common mode signal variations.

Table 5. Choice of GC Detectors Based on Class of Compounds to be Analyzed[a]

Compounds	Detectors
Fixed gases	TCD, PID
Hydrocarbons	FID, PID
Halogen compounds	ECD, ELCD
Nitrogen compounds	NPD, ELCD
Sulfur compounds	FPD, ELCD
Phosphorus compounds	FPD, NPD
Oxygen compounds	IRD
Isomers	IRD

[a] The FID detects the organic portion of the compounds; the MSD is universal, and gives structural information.

The ECD has many potential field applications but may not be practical, since it can be easily contaminated; for this reason it requires a lot of attention when operating in the low range of detectability and may not be practical. The FPD is difficult to operate quantitatively in the field because it requires a calibration for each compound; also, the sensitivity of the FPD varies with the time the flame has been lighted. Finally, the ELCD has some potential applications for field measurements but is bulky and not designed for field operation.

Portable and Transportable Gas Chromatographs

This section gives general information about field gas chromatographs and describes some of the instruments available in 1996. The intent is not to promote or criticize the models selected but to provide the reader with a tool for selecting an instrument by understanding the pros and cons of each.

The market for portable gas chromatographs is growing steadily, and manufacturers are developing innovative ways to reduce the size and weight of GC instruments. Each new design represents a trade-off between functionality and portability.

High temperature boiling compounds require the GC column to operate above ambient temperature and, therefore, necessitate higher amounts of electrical power than can be realistically supplied by batteries. The most common trade-off for portability is to provide analysis of only VOCs and to run the column and the detector just above ambient temperature. This approach allows the instrument to be operated from battery packs for several hours.

Portable instruments usually eliminate interactive interfaces, such as keyboards and displays, and provide only computer interfaces to reduce weight and size. However, this compromise places the burden on the user to provide a computer device or network.

One method of reducing the chromatograph to its simplest functionality is to limit its analytical capabilities. The column used is a narrow-bore column of >1 m, which is long enough to give a crude separation but no baseline separation of complex mixtures. The column is heated to several hundred degrees in a crude oven and has enough separation ability to serve as the front end of a mass spectrometer or an ion mobility spectrometer for which the peaks can be identified without baseline separation (73,109–111). In this type of trade-off, the burden of analytical separation shifts from the gas chromatograph column to the sophisticated detector and data interpretation.

PE Photovac. PE Photovac has two notable portable instruments: the SnapShot GC, which has a low weight 4.5 kg (<10 lb), and the Voyager, which is a combined GC and total VOC analyzer with built-in computation and data-logging capabilities. The Voyager is also low in weight 6.8 kg (15 lb). The column used in both these instruments operates isothermally <100 °C and is, therefore, used primarily for VOCs. Raw data are stored during analysis and are available later for data reduction. Each of these instruments can be connected to a network for either remote control or data retrieval. Both have a small canister for the supply of carrier gas and a battery pack that allows typically 4–9 h of field operation; both have internal sampling pumps for automatic headspace analysis of water or soil.

An exploded view of the SnapShot instrument is shown in Figure 33. This instrument is application-driven and operates with interchangeable modules. Factory-programmed application modules include *Compound Library, Analyzer Set-Up,* and *Analytical Method Software for Automatic Compound Identification and Measurement.*

The Voyager portable GC is shown in Figure 34. This instrument has a number of unique features, including a built-in three-column set for analysis of VOCs with a wide range of boiling points and molecular weights. A dual PID-ECD system allows determination of both chloalkane and chloroalkene solvents of importance in environmental monitoring. Inorganic gases such as hydrogen sulfide, carbon disulfide, nitric oxide, arsine, and phosphine can also be detected. Confirmation analysis can also be performed by injection of the sample on each of the three columns, each of a different polarity. The Voyager GC, like the SnapShot GC, has Intrinsic Safety Classification, which makes it useful for hazardous confined space entries to refineries and chemical plants.

MTI Analytical Instruments. MTI Analytical Instruments (Fremont, California) uses silicon wafer fabrication technology to micromachine parts of its portable gas chromatograph. This technology combines precision etching (micromachining) and insulator-to-silicon (anodic) or silicon-to-silicon (fusion) multilayer bonding with standard integrated circuit processing techniques, such as photolithography, diffusion, ion implementation, and thin-film deposition.

MTI Analytical Instruments combines the varied aspects of silicon micromachining technologies to fabricate a gas injector and a thermal conductivity detector (Fig. 35) (112). The gas injector is a hybrid combination of micromachining technology and conventional machining (113). The small physical dimensions of the valves and restrictor channels result in the sample reaching the column as a sharp plug.

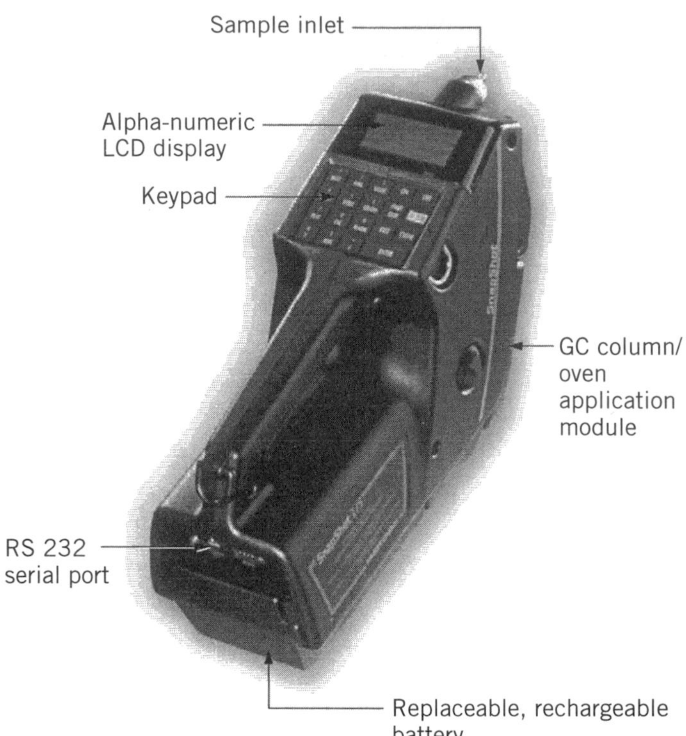

Figure 33. The SnapShot hand-portable gas chromatograph. Courtesy of PE Photovac.

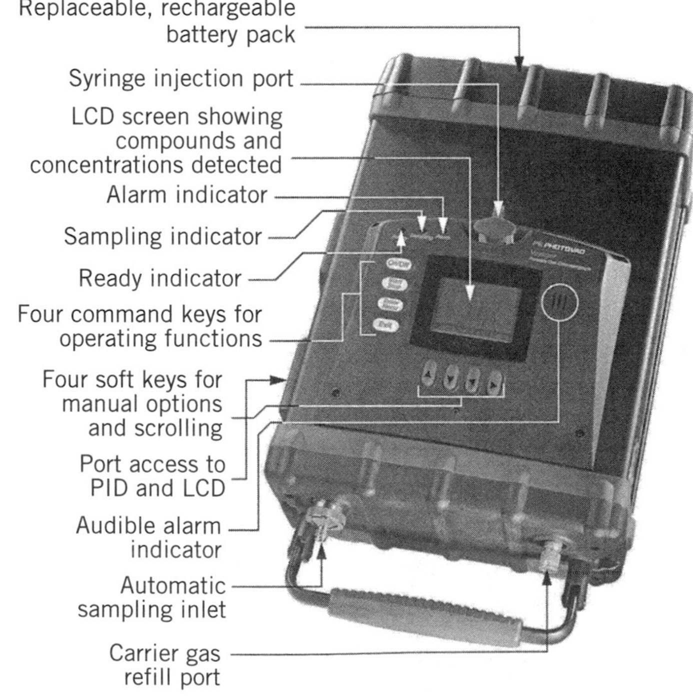

Figure 34. Portable gas chromatograph, model Voyager GC. Courtesy of PE Photovac.

Figure 35. Micromachined thermal conductivity detector. Courtesy of MTI Analytical Instruments.

The small internal volume of these micromachined parts has allowed MTI to design a portable GC that has little extra column dead volumes. The instruments can, therefore, use short, narrow-bore capillary columns and perform a typical gas analysis in 20–30 s (114), whereas several minutes are needed for conventional GCs.

The low and high temperature versions of the P200 have battery packs and gas supplies that make field operations possible for several hours (depending, of course, on the amount of power drawn from the battery pack). The instruments accept gaseous samples only and can operate in isothermal conditions. The two modules differ in stationary phase (column material), operating temperature of the column, and injected volume. The sample transfer line and the injector operate at lower temperatures than the column and the detector. This design may create cold spots and condensate some high temperature boiling compounds. Therefore, it should be considered primarily for VOCs. It cannot be used for the injection of a liquid sample, but vapors with high boiling points up to 220°C can be analyzed.

Figure 36 shows a graphic description of the inside of the P200 Micro GC. The sample inlet, at the front of the instrument, is heated to 110°C to prevent water condensation. A gaseous sample is pumped into the instrument with a built-in pump and analyzed in up to four GC modules simultaneously. A rechargeable carrier gas tank provides up to 40 h of operation. The rechargeable battery pack can operate a maximum of 8 h before needing to be recharged.

Model 311D from HNU Systems Inc. HNU offers a full VOC/semi-VOC gas chromatograph for operation in the field or in a laboratory. This GC includes a column oven with multiramp temperature programming or isothermal operation at up to 200°C, heated syringe injector and detectors (one or two), sample pump with automatic gas sample valve, or a Tenax sample concentrator/injector. Detector options include PID, far UV, ECD, TCD, FPD, and FID. Detectors can be configurated singularly or in series, with single or multiple columns in the

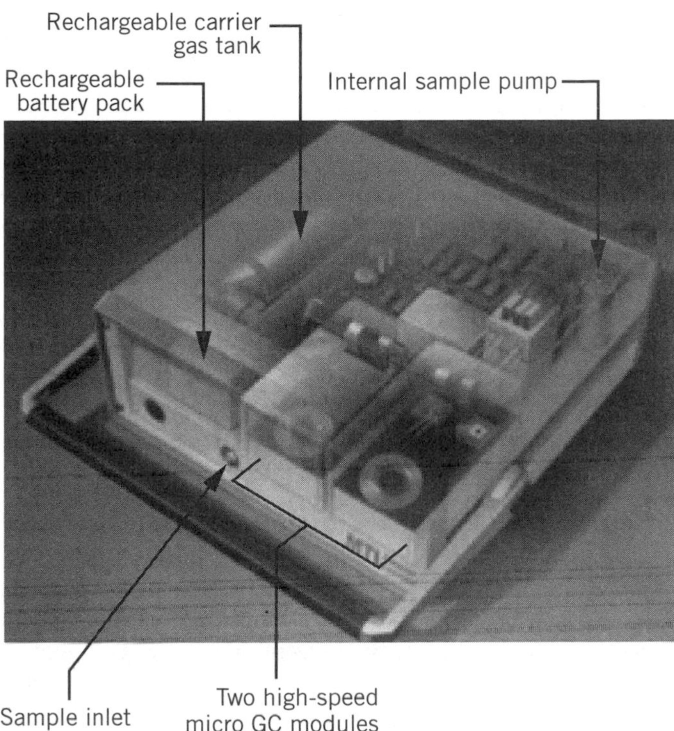

Figure 36. Portable instrument, model P200, from MTI. Courtesy of MTI Analytical Instruments.

oven. The GC includes a built-in carrier gas supply and requires external AC to operate.

The illustration of the 311D is shown in Figure 37. In the figure, the GC is controlled from an external PC. The carrier gas supply is in the detachable instrument cover. The heated injector is used for syringe injections of liquids or gases. Built into the GC is a sample pump and either a gas sample valve or Tenax gas sample concentrator/injector for single or continuous analysis. The

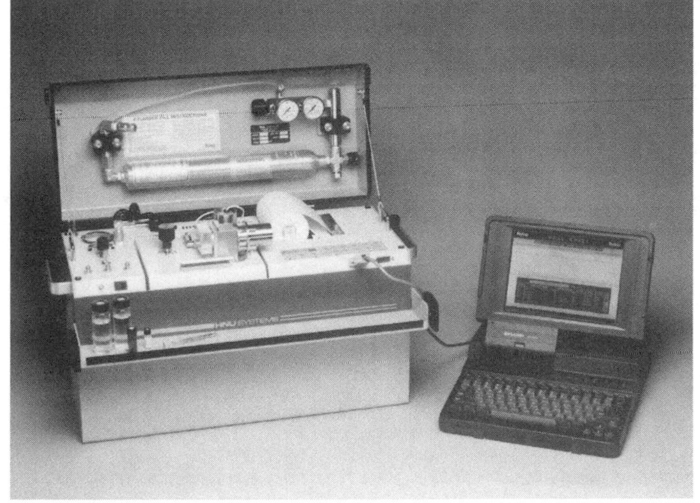

Figure 37. Transportable GC, Model 311D, shown with a laptop computer used for data handling and control. Courtesy HNU Systems, Inc.

concentrator improves the lower detection levels for most compounds.

The performance of this gas chromatograph has not been degraded for portability; the 311D performance in the field is similar to a laboratory GC. One way to check the performance of an instrument is to periodically inject a test mixture and compare the resulting chromatogram to a standard (115,116). The measured retention times and relative peak heights should not vary from the standard by more than some agreed-on value. For example, according to HNU Systems, the repeatability of retention times (which is a measure of temperature and flow stability) is better than 0.5%.

Hewlett Packard 6890 GC. Mobile laboratories are equipped with conventional nonportable instruments or combinations of instruments, such as GC/MS. Field applications requiring versatile, high quality chromatographic data rely on laboratory instruments operating from mobile shelters. The 6890 GC from Hewlett-Packard Co. (Palo Alto, California) is one such instrument used in the field.

The 6890 GC is at the extreme end of the spectrum of field instruments (Fig. 38). It is not hand-portable, and requires a bench and several kilowatts of electrical power. This instrument is not designed to operate in extreme temperature and dust conditions; thus one could easily question its choice for field applications. However, it is versatile, given the range of detectors and the nature of samples it can analyze.

A major advantage of the 6890 GC is that it operates with specific detectors, including MSD and AED, and thus provides positive identifications that cannot be achieved with conventional GC detectors for gas, liquid, and solid samples. Since this instrument must operate inside a shelter, it can be connected to sophisticated data reduction and communication tools, providing an integrated solution to high quality analysis.

General Comments. Gas chromatography is a simple technology that is inexpensive and powerful. This technique has been used for several decades and still has an attractive future.

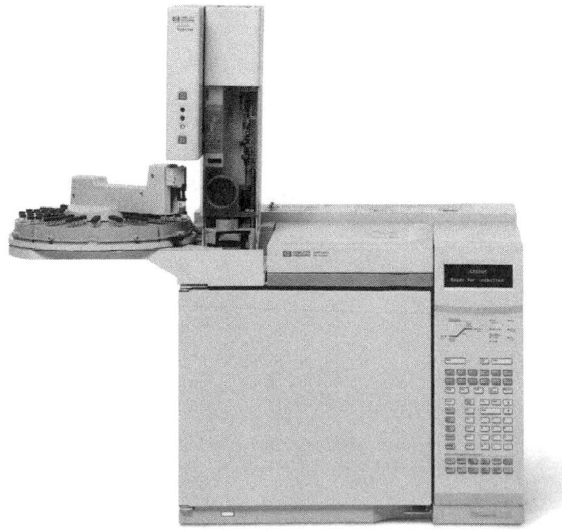

Figure 38. A Model 6890 gas chromatograph. The instrument is seen equipped with an autosampler. In a more elaborate version, the GC is equipped with a mass spectrometer or an elemental analyzer. Copyright 1997, Hewlett-Packard Co. Reproduced with persmission.

The development of portable GCs is relatively recent and shows that the field is growing, although the performance limitations of these instruments still requires the field operation of conventional instruments. Portable GCs available today do not yet offer the same versatility as laboratory instruments. However, by operating directly at the site, they can give valuable on-the-spot qualitative analytical information.

The major advantage of portable GCs lies in their ability to make measurements directly in the field, thus eliminating the transport and manipulation of samples. If used properly, they can provide limited but immediate information about the nature and the concentration of compounds. The more specialized portable instruments provide unskilled users with tools adapted to their applications. The Photovac SnapShot offers modules for specific applications and is a good example of a clever compromise between portability and versatility.

Just enough GC is an interesting approach applied to the design of portable instruments. This concept, combined with the wireless networking of analytical instruments in the field, opens the market for totally new types of portable instruments. The portable GC is a sensor under interrogation by a remote unit.

Portable instruments are limited by the range of samples that they can handle. For instance, it is not possible today to analyze semivolatile compounds with a portable unit operating on batteries, since the amount of power necessary to heat the oven and the column cannot be supplied by battery packs. Only alternative thermal designs to the conventional heating of the column by a convection oven may provide a solution to this limitation.

The GC hardware is also a limiting factor. When conventional GC hardware is used with portable instruments it increases the power demand and size of the instrument to unacceptable levels. Most of the injectors and detectors (with only a few exceptions, such as the microthermal conductivity hardware developed by MTI Analytical Instruments) have been designed for laboratory instruments. By microfabricating its detector with technologies borrowed from the semiconductor industry, MTI has demonstrated that small-volume detectors can be developed that are well adapted to small-bore columns. The result is faster analysis of field samples.

The major challenge faced by the manufacturer of field gas chromatographs is to determine the maximum range of compounds that can be analyzed by a given instrument. HNU, with the Model 311D, can operate at up to 200°C and accepts a broad range of columns, including packed columns. HNU also offers a range of specific detectors for analysis of gases, hydrocarbons, and halogen compounds. The result is versatile instruments that can analyze PCBs, perform static headspace analysis or monitor landfill gas. The price to pay for this versatility includes their larger size and weight and the fact that they need greater electrical power.

On the upper end of price and size, the 6890 GC from Hewlett Packard is by far the least portable and field-operable instrument, but it offers several valuable advantages when high quality data are needed. Since this instrument must operate in a mobile laboratory, it can operate with a broad range of nonconventional detectors such as mass selective and atomic emission spectrometers. The analytical tools provided give the user the capability to get the same quality and versatility of measurement for gas, liquid, and solid samples as in the laboratory. Although the instrument operates in the field, there is still a need to transport samples to the mobile laboratory, which can be remote from the measurement site with the risk of sample degradation that this implies.

MASS DISPERSION METHODS FOR ENVIRONMENTAL MEASUREMENTS

Two popular techniques illustrate analysis based on analyte mass or ion size: mass spectrometry and ion mobility spectrometry. The first method is widely accepted by government agencies such as the EPA. It is applicable to most of the compounds found in air, soil, and water at a detection level of a few parts per billion; differentiation of molecules is based on their mass. IMS uses ion size to differentiate between analytes. This is a truly hand-portable technique in which the chemistry has been adapted to a specific application. Both techniques are used with gas chromatography to get a separation before analysis.

Mass Spectrometry

Mass spectrometry is a microanalytical technique that requires only a few nanomoles of sample to obtain characteristic information regarding the structure and molecular weight of the analyte. It is a destructive technique in which the sample is consumed during analysis. In all cases, some form of energy is transferred to the analyte molecules to effect ionization. In most cases the nascent molecular ions of the analyte "explode" in a predictable fashion into a variety of fragment ions; the resulting fragmentation pattern constitutes the mass spectrum (117). In principle, the mass spectrum of each compound is unique and can be used as a chemical "fingerprint" to characterize the sample.

Figure 39 gives a synopsis of the entire process of analysis by mass spectrometry, culminating in the bar graph mass spectrum often published the literature (117).

For analysis by a conventional mass spectrometer, the sample must have a vapor pressure 1.3 N/m^2 ($>10^{-2}$ mm Hg), because molecules of the sample migrate by diffusion from the inlet system into the ionization chamber. The inlet may be either a direct probe or a gas chromatographic inlet. As the neutral molecules randomly diffuse throughout the ion source to the ionization chamber, only a few hundredths of a percent of them are ionized. This fraction is referred to as the ionization efficiency.

The most common ionization process, electron-impact ionization (EI), is achieved by bombarding the sample with electrons at 70 eV of energy. The ionization process in general is nothing more than the transfer of energy to the neutral molecule in the vapor stage, giving the neutral molecule sufficient energy to eject one of its own electrons and, therefore, become charged with a residual positive charge. This process produces a molecular ion with a positive charge (Fig. 39). This molecular ion still has considerable energy, and much of that energy can be dissipated by fragmentation of its chemical bonds. This decomposition of various chemical bonds leads to the production of fragment ions whose mass is equal to the sum of the atomic masses of the group of atoms retaining the positive charge during the decomposition process. It is important to realize that not all of the molecular ions decompose into fragment ions. Molecules producing a molecular ion that is stable tend to survive or not to fragment, and an intense molecular-ion peak is recorded. During the analysis of a compound having a molecular ion that is unstable, nearly all of the molecular ions decompose into fragment ions; and in these cases the mass spectrum contains only a small peak for the molecular ion.

The task now is to analyze all the ions created in the ionization chamber according to mass. Ions have an electrical charge that permits them to be focused by various ion optics potentials to form a well-collimated ion beam. This beam of ions is accelerated out of the ionization chamber and into a mass analyzer.

Mass Analyzers

The mass analyzer can be any one of several types: quadrupole (which includes quadrupole filter and ion trap), Fourier transform (FT/MS), ion cyclotron resonance (ICR), time-of-flight (TOF), and magnetic sector. Regardless of which is used, the result is the same. The ions are analyzed according to their abundance along a mass scale. During the process of data recording and reduction, the data can be arranged in tabular form or in bar graph format. Additional details concerning the design and operating principles of mass spectrometers in general can be found in the literature (117–119).

Time-of-Flight. In time-of-flight mass spectrometers, ions are given a kinetic energy with a high-voltage pulse. Another

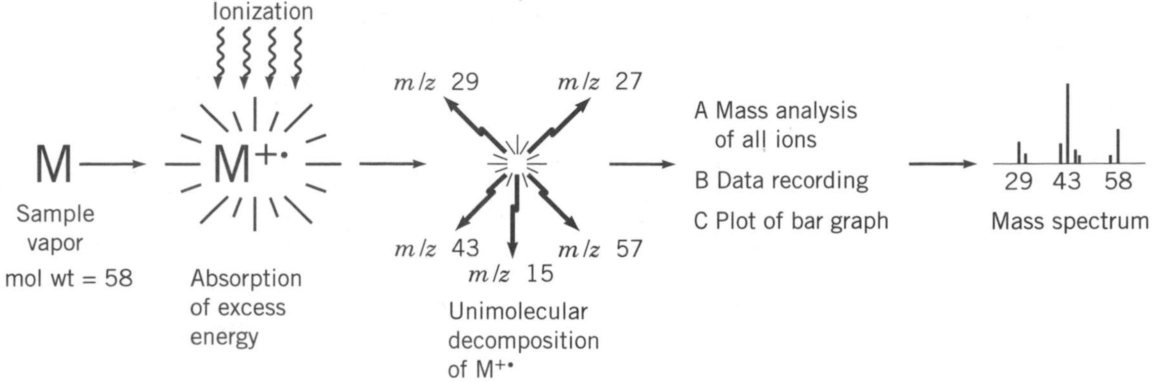

Figure 39. Synopsis of analysis by mass spectrometry. This example illustrates the fragmentation of the original vapor sample into fragment ions (128). *M*, molecules of the sample in the vapor state.

important parameter is the duty cycle of the instrument defined as the ratio of the time when the TOF is operating to the period between pulses. For instance, in orthogonal acceleration TOF, the duty cycle is the ratio of fill time of the sample in the deflector to the period of the repetition pulses. The duty cycle is a measure of the efficiency of measurement of the instrument. The higher the duty cycle number the more efficient the TOF.

Ions then drift through a tube to a detector. The length of the tube (between a fraction of a meter to several meters) determines the resolution of the TOF. The longer the drift tube, the better the resolution. Although all ions leave the source together, ions with smaller mass:charge ratios accelerate to a higher velocity than heavy ions of the same charge and will reach the detector first. Identification is done by measuring the transit time of the ion. Since the typical drift time for the largest ions is <50 μs, the accelerating pulse could be repeated at approximately 20 kHz.

In a time-of-flight system, a burst of ions (with a distribution of masses) that is accelerated to a constant energy will eventually be resolved in space (ions of similar mass will be grouped together), because each ion is traveling at a different velocity. The $w = z\text{eV} = 1/2\ mv^2$ equation predicts an inverse quadratic relationship between v and m/z.

In this instrument some ions reach the detector at different times, and the m/z value of an ion is determined by its time of arrival at the detector. Ions of low mass reach the detector before those of high mass. Voltages within the system, except for the "pulser," are constant (typically a few hundred to a thousand volts DC). This contrasts with the other mass filter types, in which nonselected ions are filtered out by changing magnetic or electric fields.

There are two principal modes of operation in TOF/MS: pulsed operation and continuous operation. In the continuous mode, ions are produced nearly all the time, thereby increasing the sensitivity of the TOF by at least one order of magnitude. The ionization process takes place for at least 70% of the TOF cycle (typically 100 μs for a mass range of 1000 Da) and is shut off only during the period of ion acceleration (typically a few microseconds). In the pulsed mode, the cycle is initiated by ionizing the molecules for typically 1 μs and then by pulsing the ions into the acceleration region for 1–2 μs, where mass separation is effected. The larger bundle of ions formed in the continuous mode leads to increased sensitivity but decreased resolution compared to the pulsed mode of operation.

TOF/MS requires a data acquisition system capable of digitizing at a rate of about 100 MHz to maintain a timing resolution of 10 ns in the instrument. The data must be processed further at similar rates. Although TOF/MS is well suited to acquiring full spectra, it is also well suited for selected ion monitoring by integrating the ion current at each given m/z value for several TOF cycles.

Manufacturers are highly motivated to reduce the size of drift tubes to make their products more portable. With the use of V-shaped and linear reflectors, which are well suited for field operations, Kore Technology (Cambridge, UK) has been able to reduce the length of its Model T-CAT drift tube to 0.2 m (120).

Quadrupole Filter. The quadrupole mass filter consists of four conductive rods that are used to shape an electrodynamic field. This nonmagnetic MS employs a combination of direct current and radio frequency potential as a mass filter. Ideally, the four rods arranged symmetrically, should have the shape of a hyperbola in cross-section (121,122). This can be accomplished, as in the case of the Hewlett Packard Model HP5973A MSD, by using a single piece of 20-cm-long quartz, with a portion of the inner surface made conductive. The magnitude of the field determines which ion will pass through the quadrupole to the detector.

The entire mass spectrum is scanned as voltages are swept from a preestablished minimum to a maximum value (approximately 400 VDC and 2000 VRF), but at a constant DC:RF ratio. This ratio defines the operating stability, sensitivity, and resolution of the instrument (117).

The quadrupole features reasonable sensitivity (picogram range) and scanning rate (several thousand Daltons per second), which are adequate for use with small-bore GC columns that produce peaks in the hundreds of millisecond range. Another advantage of the quadrupole is that it produces mass spectra with a constant mass resolution, ie, the separation of the peaks at m/z 400 and 401 is the same as for m/z 30 and 31.

A quadrupole can be operated in one of two modes: scanning or selected ion monitoring (SIM). In the scanning mode, voltages are swept from a preestablished minimum to a maximum value at a constant DC:RF ratio, therefore scanning the sample over the overall mass range of the analyzer. In the SIM mode, the DC:RF ratio is set constant to a value that sets a measurement to a predetermined m/z ratio. In the SIM mode, the analyzer acts as a narrow-band filter and rejects all the other masses. To summarize, the scanning mode leads to more versatility (no knowledge about the exact masses to be monitored is required) but less sensitivity than the SIM mode.

The quadrupole mass spectrometer is more capable of maintaining quality output at higher pressures than other types of mass spectrometers; this may be an advantage in GC/MS where a constant flow of carrier gas is supplied to the analyzer. The only harmful effect of higher pressures (10^{-4} torr) is that once the mean free path is less than the tube length, molecular collisions result in loss of ions to the poles and thus loss of sensitivity linearly proportional to the increase in pressure.

The quadrupole is well suited for selected ion monitoring because selected ions from any region of the mass spectrum can be monitored without altering the optimum conditions in the ion source or mass analyzer. Furthermore, the parameters (superimposed RF and DC fields) that control the mass scale can be changed rapidly with good response and are well stabilized through the mass range. These features are particularly desirable for computer-controlled operation of the MS for ion monitoring.

Quadrupole MS systems produce EI spectra that can be matched with accepted classic library spectra for positive identification of either target compounds or unknowns. A major advantage of this feature is the predictability of data. Even if the spectrum of an unknown is not in a library, the spectrum can be interpreted by using isotope ratio calculations and well-established fragmentation patterns.

One limitation is that the sweeping time cannot be reasonably reduced to below 10 ms, thereby limiting the rate of acquisition of chromatographic signals to a peak width of approximately 100 ms. When the ratio peak width to sweep time is too small (ie, 10:1), the spectra can be distorted by the slope of the GC peak, for which heavier masses are augmented relative to lighter ones on the GC peak's upslope, and diminished on the downslope.

Ion Trap. In the ion trap, ionization and mass filtering occur in the same place (117). A conceptual schematic of the ion trap detector (ITD) is shown in Figure 40 (123). The top end cap, a ring electrode, and bottom end cap are electrically insulated from one another. Sample molecules enter the central cavity of the ion trap from the inlet. A filament produces a beam of electrons that penetrates into the cavity and interacts with the molecules in the cavity. An electrical field is generated by an AC voltage applied to the ring electrode. This electrode is between caps on each end of the cavity. Ions of selected mass:charge ratios are trapped within the electrical field. Ions over a specified range of m/z values are trapped by appropriately selected RF and DC voltages (124).

A mass spectrum is produced by ramping the electrical field, which ejects ions from the cavity according to their mass:charge ratio, thus producing a mass spectrum. The ion trap is scanned from low to high mass by increasing the voltage on the ring electrode.

The ion trap has some advantages over most other forms of MS in terms of size, weight, and pumping requirements. These advantages make the ion trap attractive for field applications. Under favorable conditions, almost 100% of the parent ions can be converted to product ions, although a 10–50% conversion is more typical. The analogous conversion in most beam-type MS experiments is one to three orders of magnitude lower, implying significant reductions in detection limits.

The major drawbacks of the ion trap as a mass analyzer arise from the much greater ion densities that can be encountered in an ion trap instrument relative to those encountered in ion beams in most analytical mass spectrometers. High densities of ions of like charge can adversely affect the ion trap experiment in several ways, such as imposing a limit on the number of ions that can be stored, thereby limiting considerably the dynamic range.

Ion-Cyclotron Resonance Mass Spectrometer. Although a product based on ICR technology is not yet available, the technology is attractive and thus will be discussed. One prototype of such an analyzer was developed by Lawrence Livermore National Laboratory (LLNL) (Livermore, California).

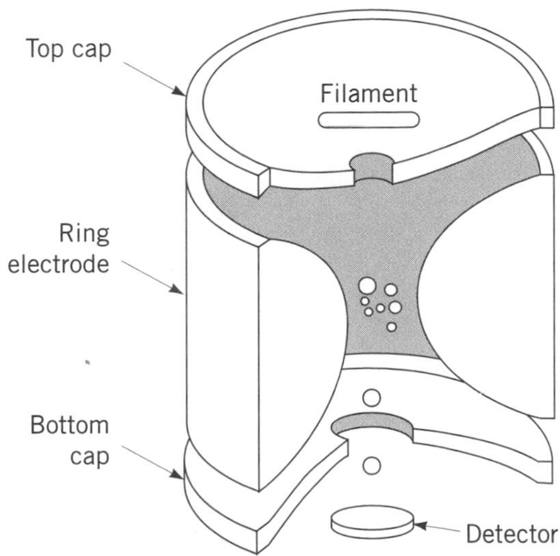

Figure 40. Exploded view of an ion trap. Courtesy of Hewlett-Packard Co.

ICR is the operating principle of the Penning ion trap. Ions are created inside the trap, and analysis and detection are done there as well. Both features reduce ion loss and allow the ions to be sampled many times. A traditional Penning ion trap is a rectangular box that sits inside a powerful electromagnet (Fig. 41**a**). A gas is introduced into the box and ionized by an electron beam. Two opposing ends of the box receive a positive electrical charge; the four sides, forming a square tube along the magnetic field direction, receive a less positive charge. The positively charged ions are repelled by the positively charged ends; thus the electrical voltage prevents the ions from leaving the box along the direction of the magnetic field. If the positive ions try to leave the box across the direction of the magnetic field, ie, toward any of the less positive sides, they are deflected by the magnetic field into cyclical orbits, called cyclotron motion.

The frequency of this cyclical motion is determined by the magnetic field and the ratio of the ions' electrical charge to their mass. Because the value of the magnetic field and the ions' electrical charge are known, the mass of a trapped ion can be determined from the frequency of its orbit. This method of determining an ion's molecular weight and thus its identity is known as ion cyclotron resonance.

Fourier transform algorithms have been applied to this technique, thereby improving its performance. For FT/MS, the ICR instrument is operated in a pulsed mode. In Figure 41**a,** a burst of electrons forms an assortment of ions in the cell. In the presence of a constant and uniform magnetic field (Fig. 41**b**), all ions move in circular orbits with a characteristic cyclotron frequency ω_c, which depends only on the ions' m/z value:

$$\omega_c = (ze \times B)/(2\pi \times m) \tag{8}$$

In Figure 41**b** a frequency-swept chirp signal is applied to the cell. Ions that experience oscillating fields equal in frequency to their cyclotron frequency absorb energy from the circuit generating the chirps. During this process, these ions move to orbits of larger radii and become in phase with the exciting field. In Figure 41**c** the translationally excited ions move coherently between the receiver plates and establish an image current in the external conducting network attached to these plates. Because the chirp contains all the excitation frequencies, the ions transmit a complex RF signal that contains the frequency components characteristic of all of the ions present. The resulting total time-dependent image current is subjected to Fourier transformation, which resolves these components and produces a database for the mass spectrum. For low resolution GC/MS, the pressure requirements are comparable to those of other common MS types, but to reach levels of performance identical to quadrupole or ion trap, high vacuums are likely to be needed.

FT/MS is unique in that increased measurement times increase both sensitivity and resolution. This advantage derives from the fact that in FT/MS the ions are not consumed during the detection process. Furthermore, all the ions at all masses are detected simultaneously in a process that could be viewed as frequency-array detection. The FT/MS procedure provides an integrating effect that optimizes S/N and minimizes the problems due to a changing partial pressure of the analyte in the ion source, as in GC/MS applications.

At the present time there are not enough data to conclude that this technique is going to be a potential competitor for field applications. LLNL developed an award-winning prototype of

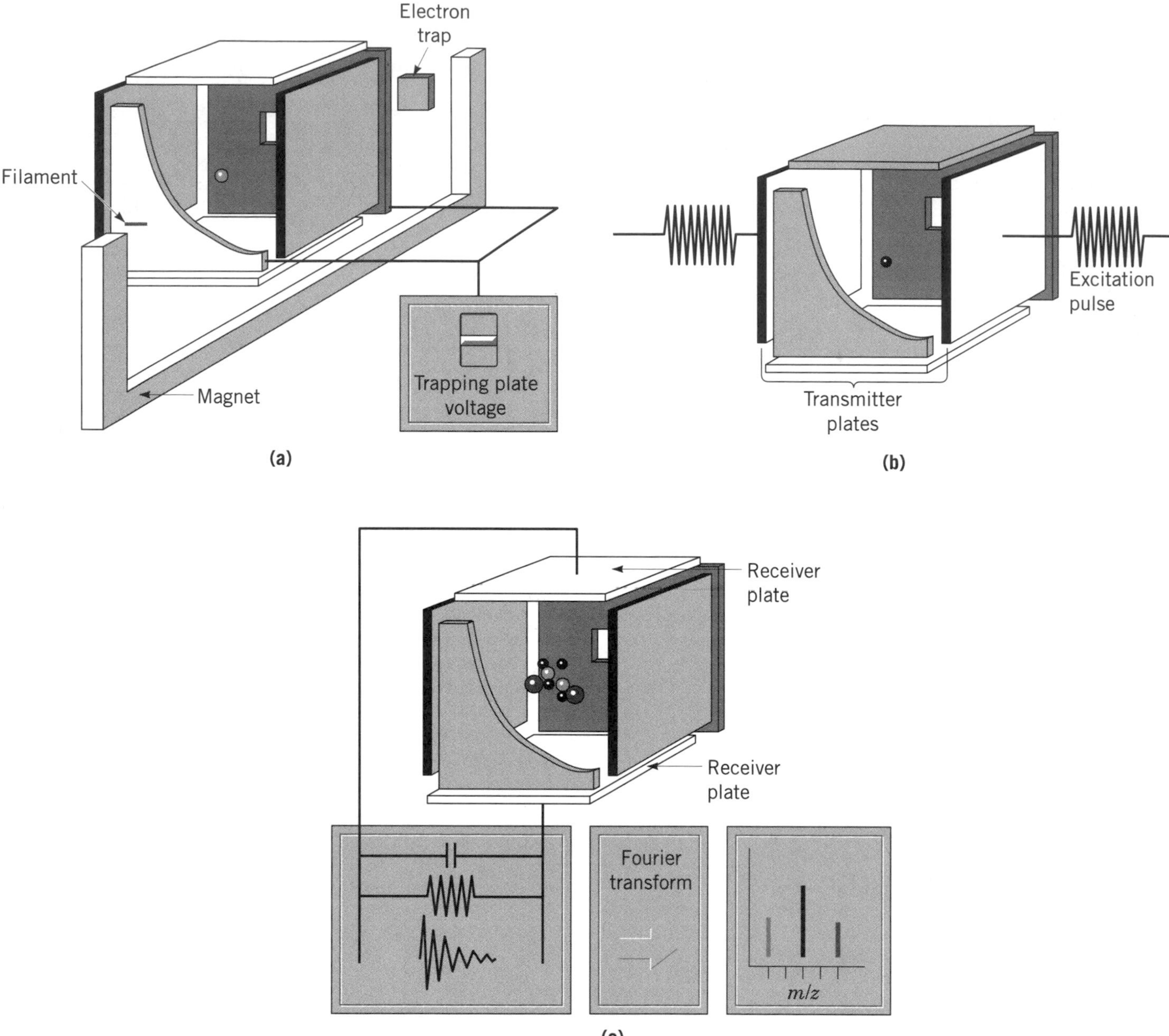

Figure 41. Schematic diagram of an ICR/MS. (**a**) Basic configuration of the instrument, showing the trapping plates, the direction of the magnetic field, and the introduction of the ions into the cubic cell. (**b**) A frequency sweep is applied to the transmission plate (horizontal plates) forcing the ions into oscillation. (**c**) Ions move coherently between the receiver plates (vertical plates) where they are collected. Fourier transform is then applied to the signal. Courtesy of Hewlett-Packard Co.

a FT/MS based on ICR. The prototype weight <12.3 kg (33 lb), uses a battery, fits into a small attaché case, and has an accuracy of 1:10^3. The permanent magnet is designed to optimize the magnetic field. It is cylindrical, 7.6 cm (3 in.) across and 5.7 cm (2.3 in.) high, with a center bore of slightly less than 2.5 cm (1 in.). A thin-walled vacuum chamber containing the ion trap fits inside the hole. A small ion pump, along with a miniature cryogenic pump, maintains the tube's vacuum pressure. A miniature Piezo electric inlet valve admits gas to be sampled into the tube. There is also a pulse electron beam that ionizes the gas in the trap (125). This unit is claimed to have a mass range of 3–1000 Da and a detection limit of 20–100 atoms, but it requires a high vacuum, in the range of 10^{-8} torr, to operate with this kind of sensitivity, making it less attractive for a portable instrument coupled with a GC.

Magnetic Sector. The magnetic sector mass spectrometer uses a magnetic field to separate ions. Sample ions are accelerated out of the ion source and through a magnetic field. The mass analyzer of this type of instrument is a sector-shaped magnetic field that disperses the total ion beam from the ion source into discrete ion beams of individual m/z values through a process of momentum dispersion and direction focusing.

The m/ze of any ion traversing a fixed radius (fixed magnet geometry) can be related to the magnetic field and accelerating potential by the following equation:

$$\frac{m}{ze} = r^2\frac{B^2}{2V} \tag{9}$$

It is common practice not to deal with the absolute charge on the electron, and for convenience it is assumed to be unity. Thus equation 9 is often found in the simplified form:

$$\frac{m}{z} = r^2\frac{B^2}{2V} \tag{10}$$

The radius of curvature of an ion trajectory is proportional to its momentum (mv). Therefore, the total ion beam is dispersed into many individual ion beams, each having its characteristic radius of curvature in the region of the magnetic field. The imposed accelerating potential V and magnetic field B permit those ions of a specific mass m_1 to follow a trajectory to the detector, whereas the heavier ions of mass m_2 ($m_2 > m_1$) follow a trajectory that leads to collisions with the walls of the analyzer. Because the instrument has a fixed radius of curvature, either the magnetic field or the accelerating potential must be scanned to collect the mass spectrum.

Magnetic scanning is advantageous in that the entire mass range may be scanned while the instrument maintains optimum accelerating potential. A minor disadvantage is related to the quadratic dependence of m/z on the value of magnetic field B. As the mass spectrum is scanned, the peaks at the high mass end appear closer together than do those at the low mass end.

The combination of magnetic field and the forward velocity of the ions exerts a lateral force on the ion, causing it to follow a curved path. The radius of the curve depends on the m/z ratio of the ion, the strength of the magnetic field, and the charge and velocity of the ion (the latter two values depend on ion source behavior). For a given strength, only a single m/z ion will curve at the radius needed to reach the detector opening. All others will miss and be pumped away by the vacuum system.

Advances in gas chromatography continue to impose demands for faster scan rates on the mass spectrometer in GC/MS applications. The scan rate of a magnetic instrument is limited by the reluctance of the electromagnet. Laminated magnets permit more rapid changes in the magnetic field so that the mass range can be scanned at rates approaching 0.1 s/decade. Two major disadvantages of magnetic MS for field applications are the weight associated with the magnet and the potential effect of the magnetic field surrounding the instrument. If scan speeds exceed a certain rate, the magnetic field can change significantly during the ion transit period (the time required for an ion to traverse the magnet), causing ions to follow a distorted path to the detector. This results in losses in both resolution and sensitivity.

Cycloidal Mass Spectrometry. Cycloidal (trochoidal) mass spectrometry is another emerging technology that has applications with residual gas analyzers. This technique could also potentially be used as an analytical tool, although no data are yet available to back this theory up. The technique is briefly discussed.

When a charged particle is moving in a plane that is perpendicular to a uniform magnetic field that is in turn perpendicular to a uniform electrical field, the particle follows a cycloidal trajectory. The ions' path has a repetition distance, or cycloidal pitch, that is independent of ion energy and the angle of injection into the crossed fields. Given a fixed electrical field E and fixed magnetic field B, this pitch depends only on the mass:charge ratio of the ion. If p represents the pitch, this may be expressed as:

$$P = 2\pi[E(m/e)]/B^2 \tag{11}$$

If an ion source produces ions that are forced to enter such a crossed set of fields through a narrow slit, ions of a given mass:charge ratio move with a trajectory that comes back to the initial position after moving a horizontal distance equal to the pitch. Again, this is independent of the initial entrance angle of the ion into the field and the initial energy of the ion. The cycloidal mass spectrometer uses fixed electric fields and a collector array to collect a wide band of mass-selected ions simultaneously. The dimensions of the analyzer are potentially small (centimeters).

Small trochoidal mass spectrometers such as the MG2100, from Industrial Scientific Corporation (Oakdale, Pennsylvania) are suitable for the measurement of residual gas composition in ultrahigh vacuum systems and could potentially be used in applications for environmental analysis. A potential drawback of the FT/MS technique is the need for powerful magnets and the interaction of high magnetic fields with instrumentation. There is no published information on the effect of a downgraded vacuum (due to the coupling of a GC) on the performance of this instrument and it is, therefore, too early to speculate on potential GC/MS applications.

Portable GC/MS Detectors

Over the past few years, GC/MS has emerged from the laboratory and into the field. In spite of its complexity and size limitations, the sheer analytical power of this hybrid separation and analysis technique has helped it earn its place in the field. A number of workers have demonstrated that transportable GC/MS systems can be moved to a field site and operated by bringing samples to the instrument either manually or via long, heated analytical transfer lines. The next step is to develop field-portable equipment with analytical performances similar to those achieved with laboratory instrumentation.

There is a broad range of commercially available benchtop instruments, including quadrupoles and ion traps. Several specialized MS instruments operating in a mass range up to 1000 have been developed for applications for which transportability is a prime requirement. Well-known examples include the Bruker Franzen (Bremam, Germany) MEM and EM650 systems, originally developed for military applications involving chemical agent detection, and the Viking SpectraTrak system, primarily designed for environmental applications. All the instruments used for field analytical work require only unit resolution.

Portable GC/MS Based on a Quadrupole Mass Analyzer. Figure 42 shows a view of the Viking SpectraTrak 572 (Viking

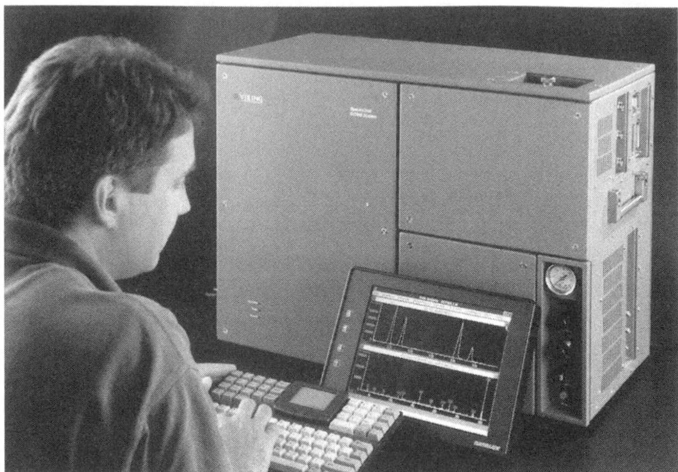

Figure 42. Spectra Trak 572 portable GC/MS. Courtesy of Viking Instrument Corp.

Instruments Corporation, Chantilly, Virginia). This instrument is self-contained and requires only the power from an AC line. This transportable GC/MS is configured for GC injection with a capillary-direct transfer line. The miniaturized GC oven accommodates standard capillary columns and provides temperature programming and a split or splitless injector. Other options are available and should be checked with Viking Instruments Corporation.

The Hapsite. The Hapsite instrument is designed to cover mass ranges from 1 to 300 Da and, therefore, is useful only for the analysis of VOCs. The mixture is first separated by a GC column before being analyzed by a quadrupole filter. One of the particularities of this instrument is that it is self-contained and operates on batteries. The instrument includes the vacuum pump, the carrier gas supply, an internal gas canister for calibration, rechargeable batteries, and all the electronics to control the instrument and acquire and transfer the data to a foreign site.

The vacuum system is the part of the instrument that typically consumes the most power. This subsystem is also the heaviest portion of the instrument. The Hapsite portable GC/MS manufactured by INFICON (East Syracuse, New York) uses a commercial bulk getter pumping system. In this vacuum subsystem vanadium–iron–zirconium–alloy sintered pellets are heated to 200°C to chemically pump most gases and vapors. The lack of pumping capacity for noble gases represents a problem because of argon accumulation due to small air leaks in the vacuum system. A small secondary ion getter system is added to provide sufficient noble gas pumping. This instrument requires a separate service unit with an AC supply, vacuum pump, and carrier gas supply to prepare the portable unit before field operations.

Time-of-Flight Mass Analyzer. The performance of the TOF/MS analyzer is a direct function of flight-tube length. The resolving power of a commercially available instrument with a 2-m flight tube is approximately 500. Various improvements have been developed to reduce the length of the instrument without limiting the flight path. Kore Technology has patented and developed a spectrometer geometry that starts with a ring-shaped ion source with a dimension of 20 cm (8 in.) and a mass range of 1000 Da (120).

The ions are extracted perpendicular to the ring and deflected inward so that, as they proceed through the spectrometer, each annular bunch of ions converges to a relatively small size by the time it reaches the detector. This allows a relatively large source to be used with a small detector giving parts per billion ppb detection limits. In addition, an aperture can be placed at the crossover to allow mass or energy selection and enhance mass resolution.

General Comments. Miniaturized spectrometry is a powerful analytical technique that has found wide use in laboratory and applied monitoring areas. Applications have included planetary orbiting mission (Mars *Viking* Lander GCMS), earth orbiting satellites (*Atmospheric Explorer* satellites and others), manned spacecraft respiratory experiments (*Skylab*), submarine atmosphere monitors, and commercial, medical, and industrial process monitors. Research efforts in the late 1990s have demonstrated that miniaturized systems can be built with today's technology.

To miniaturize a mass spectrometer, some trade-offs must be accepted. For instance, it is generally accepted that sensitivity falls off rapidly with increasing mass range and decreasing size and that it is difficult to get it back by increasing gas load, implying a correspondent higher pumping speed and larger pump.

In addition to understanding the parametric relationships for each type of instrument, a practical knowledge of its performance is also essential. Some general comments regarding the various analyzers can be made. For instance, the ion trap is a proven sensitive analyzer with a dynamic range that is more limited than some other systems. The ion trap requires a collision gas to achieve good operation, which places a high gas load on the instrument, affecting the size and type of vacuum pump. The quadruple MS does not have these limitations and operates at lower pressure than the ion trap. It has a larger dynamic range, but is somehow less sensitive and is not as fast as the ion trap.

Although the TOF may have appeal, there are duty cycle issues that must be understood before its net performance can be determined. The TOF also puts a lot of burden on the electronics and vacuum system to operate at its optimum. The drift tube also must be folded to reduce the overall dimensions of the instrument. This design may degrade the performance of the instrument.

The ICR must be operated in a Fourier transform mode, which requires highly sophisticated electronics to achieve superior performance. Like other magnetic sector instruments the ICR's also have electronics problems due to the size of the magnet and to the effect of the magnetic field on surrounding instruments.

Magnetic sectors and trochoidal MS also have a duty cycle limitation, but using an array detector (focal plane detector) could, at least in theory, overcome this limitation. The overall mass range of the instrument could easily be monitored simultaneously. The major disadvantage for field applications is the weight of the magnet.

Ion Mobility Spectrometry

Ion mobility spectrometry, like mass spectrometry, falls in the category of mass dispersion methods. IMS was first described as an analytical method for trace analysis in 1974 (126); since then, its potential for field analysis has received some recognition, especially in the military, for the detection of chemical-warfare agents such as nerve gases and blister agents. In addition to these applications, IMS has also been recognized as useful in a number of narrow technical niches for chemical analysis of vapors. Several broad categories in the development of IMS have been identified:

- A selective sensor in which the chemistry of ionization can be tailored or manipulated for the detection of a specific compound or classes of compounds.
- A stand-alone analyzer for samples of predictable composition.
- A potential catalyst for improving the performance of gas chromatography for crude separation.

When experimental conditions are proper (low moisture and elevated temperature, ionization mechanism), all chemicals introduced into an IMS can be ionized, which produces a spectrum of responses. In short, the conversion of molecules to ions at an ambient pressure followed by ion characterization is nearly as comprehensive as MS.

The IMS detector has some unique advantages over gas chromatographic detectors, including MS, but also has some limitations, the most serious being its limited dynamic range. In addition, IMS should not be compared to time-of-flight mass spectrometry. While certain functional similarities exist between time-of-flight mass spectrometry and IMS, comparisons can create unrealistic expectations for IMS performance. IMS peaks are typically rather broad compared to the range of possible drift times, and the amount of qualitative information available from IMS is less than that from MS and IR spectrometry but more than that from FID and ECD. On the positive side, IMS detectors have been engineered to be truly hand-portable stand-alone units, with a chemistry adapted to specific applications in order to enhance their sensitivity to some compounds. A brief introduction to IMS follows, with an emphasis on field applications and equipment. An excellent source of information for this technology is available (127).

Description. Ion mobility spectrometry has received a fresh examination in the last few years (128–130). The interest in this technique is due to its intrinsic response features (excellent detection limits) when experimental conditions are proper (low moisture and elevated temperature) and to practical considerations (size, weight, and power advantages). Interest has also grown, following successful, proven application of IMS technology to significant challenges, including chemical-warfare agent sensing, contraband screening, and toxic vapor monitoring.

Figure 43 shows a chemical agent monitor CAM unit developed by Graseby Ionics, Ltd. (Watford, Herts, U.K.). Its simplicity of design, atmospheric pressure operation, universality of response, and ion-resolving capability have enticed both scientists and instrument manufacturers to miniaturize and modify IMS detectors to provide analytical determination of targeted compounds in nonlaboratory environments.

The IMS technique is illustrated schematically in Figure 44 (131). The sample is introduced by means of a carrier gas into a reaction region, where it is ionized. Although positive and negative gas-phase ions for IMS have been produced by a variety of methods, including photoionization, laser multiphoton ionization, thermoionic emission, and corona spray, the most common ionization source is radioactive ^{63}Ni foil (130). In the presence of sample molecules these ions undergo ion/molecule reactions to generate product ions. The nature of the product ions depends on the type of ion/molecule reaction that occurs, and early models for their formation can be found (132,133). Once formed, product ions are extracted from the ion source region under the influence of an electrical field, and ions are drawn toward an ion shutter. The ion shutter is periodically pulsed, and ions are injected into the drift region.

Figure 43. Warfare application of IMS. The operator is using the CAM instrument for the decontamination of the military vehicle. Courtesy of Graseby Ionics Ltd.

In the drift region, the ions travel toward the collector still under the influence of the electric field imposed on the entire drift tube. Because ions with different cross-sections travel with different velocities, the ions are separated and arrive at the collector at different times. When ions of different types and mobilities exist in the drift tube, the ions can be separated through differences in their mobilities.

Drift Velocity. The ion mobility is characteristic of a particular ion and can provide a means for detecting and identifying vapors. Ion mobilities are determined from ion velocities, which are measured in a drift tube with supporting electronics. In practice, a vapor sample is introduced into the reaction region of a drift tube where neutral molecules of the vapor undergo ionization, and the resulting ions (product ions) are injected into the drift region for mobility analysis. Mobility K is determined from the drift velocity V_d attained by ions in a weak electric field E of the drift region at atmospheric pressure:

$$V_d = KE \tag{12}$$

Ion velocities are inversely dependent on the effective collisional cross-section of an ion, making IMS a kind of molecular size analyzer. Commonly, electrical fields of 200 V/cm and mobilities of 1–2 cm^2/Vs result in ion speeds of 200 V/cm to 1000 cm/s or drift times of 5–20 ms in 4- to 20-cm-long drift regions at ambient pressure. As the reactant ions and product ions are drawn toward and collide with the detector plate, current is registered and a signal is generated as a mobility spectrum. When ions of different types and mobilities exist in

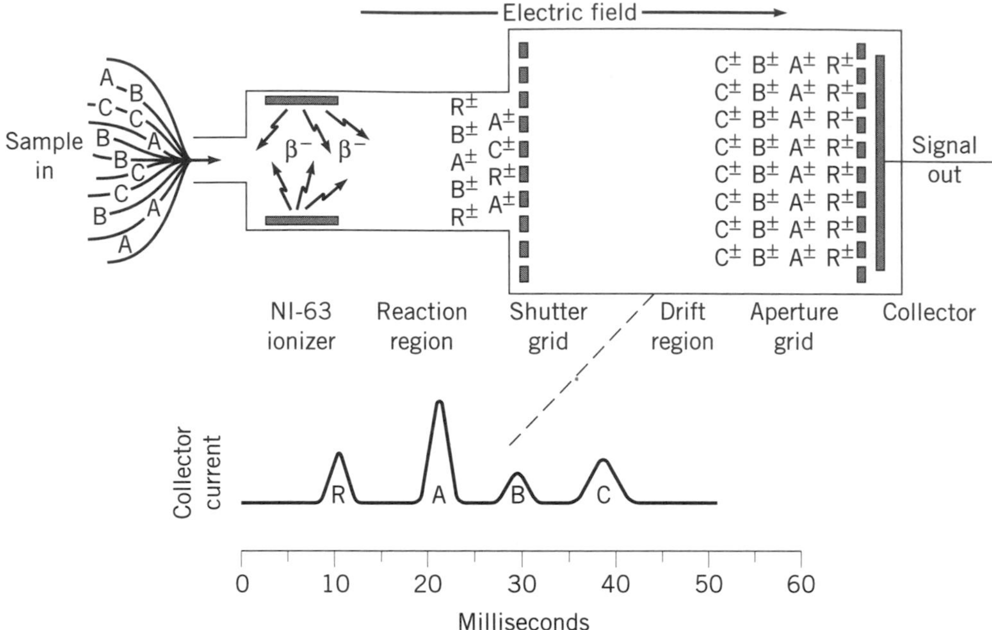

Figure 44. Basic principle of an IMS. The various ions resulting from the ionization process are introduced into the drift region, where they will be separated according to their drift time (131).

the drift tube, the ions can be separated through differences in their mobilities (127).

Ionization Process. Vapors from nearly all chemical classes or functional groups can be ionized through atmospheric pressure chemical ionization (APCI) processes. However, a given substance can be converted into various and different product ions, depending on the ionization mechanism. In addition, quite a few product ions can be formed from a substance at a given concentration and temperature, but most of these are fragile clusters that do not survive transit even from the source region to the shutter. Thus in an IMS spectrum, only the stable ions survive and the bulk of all ions decomposes, sometimes into stable ions. In practice, higher clusters are formed in the source and are rapidly depopulated to protonated monomers and proton-bound dimers in the drift region.

A fortuitous consequence of proton or electron transfers is that molecules form ions that retain much of the original shape and size of the molecules. In addition, product ions rarely dissociate or fragment, since the energetics of the APCI process are weak; mobility spectra thus often comprise only one ion or a few ions with a direct structural association to the original molecular neutral. The extensive fragmentation of molecules in electron-impact ion sources for mass spectrometers does not occur as a general rule in IMS.

Ions are generated in a purified nitrogen or air carrier gas containing a trace of water vapor, by the action of β particles emitted from a ^{63}Ni foil. The primary N_2^+ ion resulting from the ionization of nitrogen gas is too short lived to appear in the IMS spectrum but begins a series of ion/molecule reactions with nitrogen and trace amounts of water principally.

In the absence of a sample, the resulting stable ion clusters have been identified as $(H_2O)_nNH_4^+$, $(H_2O)_nNO^+$ and $(H_2O)_nH^+$ (134,135). These background ions are called reactant ions because they undergo further ion/molecule reactions with neutral gas phase analytes to produce analyte product ions. The formation of product ions occurs in the ionization region predominantly by collisions between the reactant ions and the sample molecules. In addition, product ions may be formed through association reactions in which an adduct is made by a weak attachment of a reactant ion to a neutral vapor. A trace organic molecule injected into the carrier gas undergoes reactions with these ions and electrons, forming reaction product ions that show both quasi-molecular and simple dissociated ions in their mobility spectra. This method is extremely sensitive and reasonably quantitative. It is capable of detecting 10^{-12} g or less of a compound and can provide identification through the characteristic positive and negative mobility spectra.

Positive and negative ions can be independently observed merely by the choice of the electrical field polarity. Positive product ions are formed mainly through a proton-transfer reaction. Early models for the formation of product ions of positive polarity have been summarized (132). By monitoring positive reactant ions, one can gain essentially the same information from IMS that is obtained from an FID (136).

The formation of negative ions in IMS is closely related to those chemical events that occur in an ECD. Early reviews of the ECD form the foundation for the current understanding of negative ion chemistry in IMS (137–139). Two general observations have evolved: the data contained in positive mobility spectra bear a close similarity to those obtained in chemical ionization mass spectrometry, and the data contained in negative mobility spectra correspond to those obtained with the GC ECD. Interpretation of the data based on this model is available (126).

Ions formed in the reaction region are injected as a short pulse (200–300 μm), into a drift tube, where they acquire constant velocities under the influence of a weak electrical field and collisions with a counterflowing neutral drift gas (air or nitrogen). Various ions injected into the IMS drift region will arrive at the Faraday cup detector at different times, as determined by their mass, charge, and collisional cross-section with a drift gas. Consequently, different ion species, even if characterized by the same mass, may arrive at the detector at different times (140).

An important variation of IMS ionization that can be used to enhance the sensitivity or selectivity of the technique for particular classes of compounds involves the modification of reactant ion populations by adding carefully controlled concentrations of dopants to the drift gas (141). For example, Cl reactant ions have been used to increase sensitivity to explosives; NH_4^+ reactant ions enhance selectivity to amines; and $(CH_3COCH_3)_2H^+$ reactant ions are used to selectively ionize organophosphonates (142,143).

In summary, ion mobility spectrometry comprises and is governed by two distinct, independent or separate events: gas phase ionization in air at atmospheric pressure through collisional charge exchanges or ion/molecule reactions and ion characterization using mobilities of gas phase ions in a weak electrical field.

Instrumentation for Ion Mobility Spectrometry. It is convenient to divide the IMS system into three major sections: the sample introduction system, which is the interface of the instrument with the ambient atmosphere; the drift tube, where ionization and the separation and detection of the ions take place; and the detection device, where the signal is acquired, analyzed, and displayed.

Sample Introduction. The use of a nonporous dimethylsilicone membrane barrier to the interface between the IMS and the ambient atmosphere is the most common way to introduce samples into the IMS, but in addition to membranes, other sample to drift tube connections have been made by GC inlets, electrospray, thermal desorption, and pyrolysis. The membrane retards the flux of water and ammonia molecules present in the sampled ambient air, thus limiting the number of polar molecules available for cluster formation. Excessive clustering reduces the resolution of the IMS and may lead to erratic behavior of the ion signals. Membrane interfaces are especially important in a portable instrument when the drift tube operates near ambient temperature, as clustering presents a serious problem under this condition. The main advantage of employing a membrane interface is that it makes ambient-temperature operation possible and can lead to considerable reduction in power consumption and thus in size and weight, which is required for handling instruments. The disadvantages of using a membrane are diminished sensitivity and an increase in the response time of the instrument. Other ways of increasing the selectivity and sensitivity of the IMS rely on preseparation by gas chromatography and will be discussed later.

Reagent gases can also be selected as a means of increasing the sensitivity and specificity of the IMS by preferentially ionizing the target compound or by enhancing the separation of the target ions from the reactant ions without affecting the response time of the instrument. For instance, adding a reagent gas that has a proton affinity just slightly below that of the target eliminates most of the interfering ion signals, and the presence of the target compound is easily detected. For example, monitoring *i*-propylamine (proton affinity = 218.6 kcal/mol) using ethylamine (proton affinity = 217.0 kcal/mol) as a reagent gas should exclude interferences from all compounds that have proton affinities (PA) below that of ethylamine.

The Drift Tube. The first patented commercial IMS analyzers were created by Frankin-GNO, Inc. (144,145). The drift tube was made of circular elements biased at different electrical potentials to maintain a uniform electric field gradient of 200–300 V/cm along the central axis of the tube. These elements were alternating conducting and insulating guard rings. Molecules were ionized by a polished β-emitting radioactive ^{63}Ni source inserted into the first guard ring. The shutter grid was of the Bradury-Nielson type, in which a potential difference is placed between a set of interdigitated thin wires with a voltage difference applied to each of the independent wires, thus creating a strong electrical field (of about 600 V/cm) perpendicular to the axis of the field gradient across the drift tube (146). A collector plate, somewhat like a Faraday cup, is placed close to the collector to increase detection efficiency and to filter out the noise caused by the pulse on the gate, thus improving the signal:noise ratio.

The basic design has been modified in several ways (147). The drift tube is now segmented with a unidirectional flow. With this modification, the sample is introduced into the reaction region just downfield of the ionization source and is swept by the drift gas through the source. The unidirectional flow drastically reduces both memory effects and residence or response time and clarifies ionization chemistry, thus improving the performance of the cell drastically. The ions are moved down the length of the tube by the electrical field, while the neutral sample molecules that are not ionized spend only a short time in the reaction region before being swept out.

Detection Devices and Signal Processing. The ion mobility spectrum is a plot of the signal intensity (ion current) as a function of the drift time, taking zero as the time the shutter-grid gate is opened to allow ions to enter the drift region. The most common and simple detection method used to measure the ion current in the IMS drift tube is an electrode of flat plate detector that serves as a Faraday cup. The output signal from the collector is, therefore, an analog signal that is digitized and displayed as a digital signal.

There are several basic modes of operation: signal averaging (up to hundreds of spectra are averaged to increase the signal:noise ratio), hardware monitoring, and Fourier transform (148). Another method of registrating a signal is through the use of two shutters with boxcar integration. In this technique, a ion shutter near the source region is operated at a fixed frequency of approximately 30 Hz for a conventionally sized drift tube, and the shutter pulse width may be 100–300 μs. After the first shutter is opened, a delay is measured and the second shutter, near the detector, is opened for a short interval and then closed. Only ions with a drift time corresponding to the delay are allowed to pass to the detector. This delay is gradually increased to a maximum, and the spectrum is generated during the sweep of the delay time.

Quantitative Limitations of the IMS

IMS has been shown to be a selective and extremely sensitive detection technique for monitoring vapors or air pollutants. However, systems with membrane-type inlets, such as

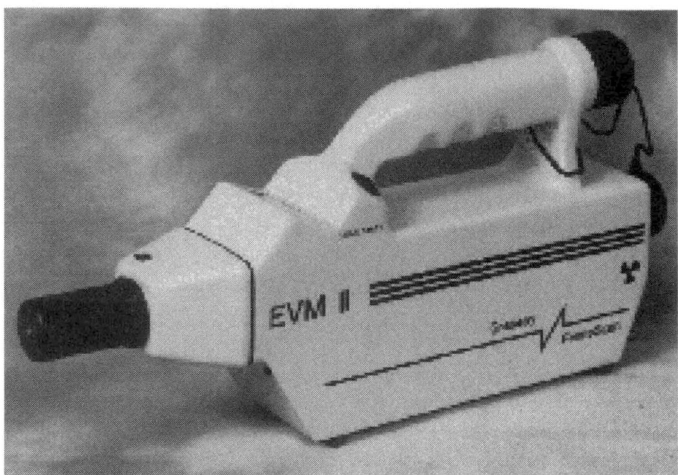

Figure 45. Civilian GC/IMS hand-held unit model EVM II. This environmental vapor monitor is used for the detection of mixtures of toxic vapors and gases. (Courtesy of FemtoScan Corp.)

the chemical agent monitor (CAM), are subject to serious limitations in regard to both qualitative and quantitative performance. Moreover, an ion mobility spectrometer, as with any secondary ionization detector, exhibits a nonlinear response with respect to the concentration of analytes introduced into the ionization chamber. Such a response increases exponentially up to the depletion of the reactant ions that is observed as the disappearance of the reactant ion peak (RIP) peak in IMS spectra.

Hand-Held IMS Instruments

The widest application of hand-held IMS instruments is the detection of chemical warfare agents. The chemical agent monitor made by Graseby Ionics, Ltd. and manufactured under license in the U.S. by Environmental Technologies Group, Inc. (Baltimore, Maryland), is slightly larger than a cordless telephone. Weighing less than 1.8 kg, with 38 cm as its longest dimension, the CAM represents an exceptional work of engineering (127). The drift tube, sampling and flow systems, electronics, signal processing, and a simple bar display have been compressed into a hand-held module. A civilian spin-off of the CAM, called the airborne vapor monitor (AVM), is also produced by Graseby Ionics, Ltd. A recent development in hand-held IMS instruments is the inclusion of a gas chromatograph column as a preconcentrator that is interfaced with the AVM commercialized by FemtoScan (Fig. 45).

The production of commercial applications is, to a large extent, a spin-off of the military chemical agent detector and monitor program. When discussing industrial and environmental applications of IMS, it is convenient to differentiate between fixed-point monitors and hand-held instruments. The former are used to detect, measure, and monitor a specific compound or a number of chemicals at a fixed location on a continuous basis. The hand-held instruments are generally used as monitors for a short duration at a given spot and are easily transportable from point to point. Thus the fixed-point monitors are useful for obtaining the concentration versus time profile at a given point, whereas the hand-held instruments are useful for obtaining concentration versus location profiles. Hand-held instruments are ideally suited for detecting leaks or hot spots at which the chemical of interest has accumulated.

The major environmental and industrial applications of IMS instruments include monitoring toxic chemicals, stack gases, and chemicals that are considered hazardous to humans or to the environment. To date, few detailed accounts of the performance of IMS have been made public. Applications of IMS can be found in reviews (149,150,151) (Table 6). A comprehensive list of chemicals that can be monitored by IMS is available (152); it includes compounds such as toluidine, pyridine, diaminotoluene, DMSO, glutaraldehyde, methyl salicylate, acetic acid, acetic anhydride, halothane, enfluarane, isoflurane, and nicotine.

Partial Review of Published IMS Applications. The use of IMS for field applications has been described in a number of symposia related to waste monitoring. Some key environmental and industrial applications of IMS have been reviewed (127). Nickel carbonyl, which is highly toxic with a TLV of 50 ppb, has been targeted (153). Another application of IMS is monitoring the toxic chemicals hydrazine and monomethylhydrazine (MMH) at levels below TLVs above 10 ppb (154). An advanced environmental application for commercial IMS is the monitoring of hydrogen fluoride levels at parts per million (155). Another application is the monitoring of highly toxic (TLV of 5 ppb) toluene diisocyanate (TDI) (156).

Unlike industrial and environmental applications, use of IMS-based technology to detect contraband explosives or drugs does not require quantitative measurements. Because of the high sensitivity of the IMS, its specificity, and the fact that it operates at atmospheric pressure, it is apparently a leading candidate for detection of hidden contraband materials (157). The major problem is getting enough molecules of the target compound into the detector. This problem is most severe with chemicals that have low vapor pressures.

IMS is currently being successfully applied to the problem of on-line trace detection of plastic and other explosives in airports and other facilities (158). The methods of sample retrieval consist primarily of batch sampling for particle residue on a filter card for introduction into the IMS. The sample is desorbed using air as the carrier into the IMS, where negative ions of the explosives can be detected, some as an adduct with a reagent ion such as Cl. Based on studies and tests conducted by different airport authorities, this method seems to work well for low vaporpressure explosives, such as RDX and PETN as well as TNT. These are highly adsorptive and can be found in nanogram quantities on contaminated surfaces.

An ion mobility spectrometer is also capable of detecting minute amounts of narcotics with few false alarms and a minimum of operator interventions. Sampling rates of 10 suitcases or passengers per minute have been reported (159).

Recent Developments in GC/IMS Techniques. In gas chromatography, constituents of a mixture are physically separated by the column and identified by their elution time. With the exception of the mass spectrometer, GC detectors do not provide any information about the chemical identities of the compounds. The most important feature that gas chromatography brings to IMS is the individual delivery of compounds to the IMS, thus reducing the problem of interferences and providing near-ideal ionization conditions. The GC part of the system is usually unsophisticated, compared to a laboratory

Table 6. Compounds of Environmental Interest Suitable for IMS with Typical Detectable Limits[a]

Compound	Minimum Detectable Limits (ppm)	Compound	Minimum Detectable Limits, ppm
Acid and stack vapors			
HCN	0.1	HCL	0.1
NH_3	0.1	NO_2	0.1
Hl	0.1	HF	0.1
Phosgene $COCl_2$	0.1	Nitric acid (HNO_3)	0.1
Other chemicals			
Acetaldehyde CH_3CHO	0.1	Aliphatic amines	0.005
Aromatic amines	0.005	Aniline	0.005
Methylene dianiline	0.01	Chlorine Cl_2	0.1
TDI	0.005	H12MDI	0.1
1,6-Diisocyatohexane	0.1	TXMDI	0.1
IPDI	0.1	TDA	0.005
NDI	0.005	Nitrobenzene	0.005
PCBs	0.1	Vinyl acetate	0.1
Solvents			
Furans	0.1	Alcohols	0.1
Cyclohexanone	0.01	Acrylonitrile	0.01
Phenols	0.1	Toluidine	0.1
Chlorinated solvents	1	Acetonitrile	0.01
Ketones	0.01	Esters	0.01
Ethers	0.1		
Semiconductor gases			
PCl_3	0.1	NF_3	0.1
Dry etchant gases	0.1		

[a] From Environmental Technologies Group, Inc.

gas chromatograph, and provides only crude separation of the components.

The addition of a short capillary column between a specialized valveless inlet and the ion mobility spectrometer allows for the separation of vapors in less than a few seconds (160,161).

The pumps, used originally in hand-held IMS to control drift gas and sampling flows, are used to lower the pressure in the drift tube. This results in a pressure drop across the column, and the air is drawn through the column as the carrier gas.

The ability to produce an effective hand-portable instrument depends on achieving several goals: limited sample preparation and preseparation (short chromatographic column, for instance), lightweight design (simplified design, use of air instead of nitrogen), and high speed response (just enough chromatography, no baseline separation). The efficiency of the GC column in this example is only 2000–5000 theoretical plates, yet it is suitable for delivering the phosphonate esters individually to the ion source and reducing interference problems.

Figure 46 illustrates some advantages of operating an IMS in the AVS transfer line gas chromatograph (TLGC) mode compared to the membrane inlet for air analysis of an equimolar mixture of model alkylphosphonates (162). The IMS spectrum of the ternary mixture of alkylphosphonates without GC preseparation (ie, in the case of a membrane-type inlet) reveals several mixed ion peaks (eg, DMMP + diethylmethyl phosphonate (DEMP) and DEMP + diisopropylmethyl phosphonate (DIMP), etc) in addition to protonated monomer, proton-bound dimer and trimer formation by particular components (Fig. 46, A). In Figure 46, B, the IMS spectrum of the same mixture is presented, but it is obtained by an average of 32 scans recorded after AVS injection in a 2-m-long capillary column. Preseparation by TLGC introduces each component separately into the ion source. Therefore, the relative intensities of particular component signals reflect the composition of the original mixture, since no mixed-ion peaks or competitive ionization processes are observed. All individual IMS spectra recorded after sampling the analyte mixture are presented in Figure 46, C in the form of a pseudo-three-dimensional waterfall graph. Furthermore, a reconstructed transfer line gas chromatograph of this ternary mixture, obtained by summing all ion current intensities at drift times between 5 and 7 ms is shown in Figure 46, D. It is clear that a 2-m-long capillary column provides sufficient resolving power to produce just enough separation of the ternary mixture components within 30 s, simplifying ionization processes and dramatically improving qualitative and quantitative capabilities of the CAM for analysis of mixtures.

The use of short, fast separation columns reduces many problems associated with analysis of complex mixtures. However, under real-world conditions, coeluting compounds is inevitable in the IMS and can complicate the interpretation of spectra. Therefore, part of the optimization should include the chromatographic separation.

Recent Progress in IMS Technology. The limited linear response of the IMS to one or two orders of magnitude is one of the limitations to using IMS as a gas chromatographic detector. To extend this dynamic range there has been some attempt

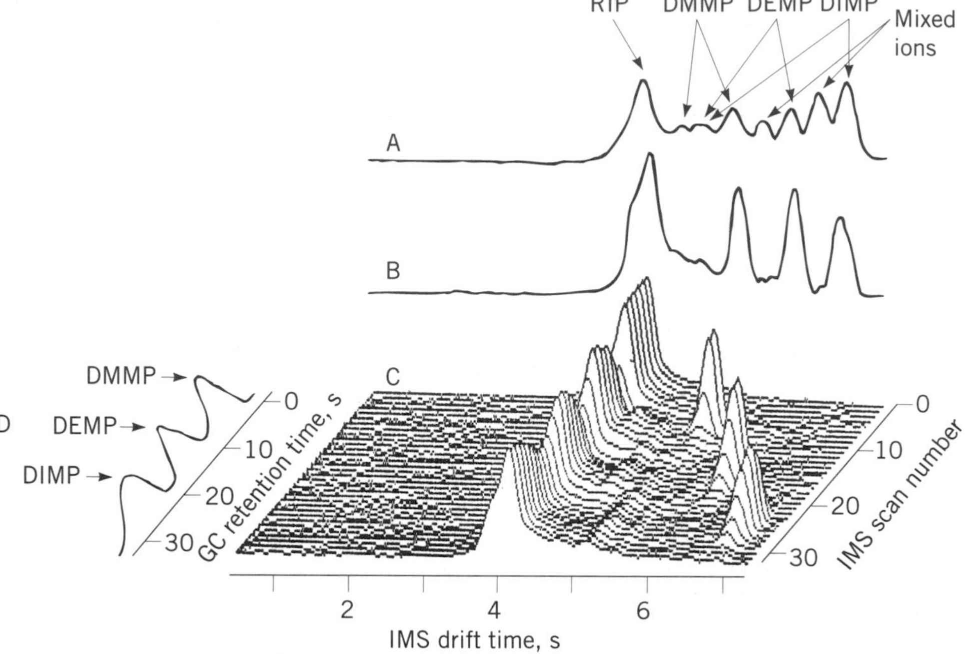

Figure 46. Comparison of a direct and an AVS-GC/IMS inlet mode. A, Direct inlet IMS spectrum of three methylphosphonates (without TLGC preseparation). B, sum of the 32-ion mobility scans from the GC/IMS spectra of the methylphosphonates shown in C; C, quasi three-dimensional waterfall display of a GC/IMS profile. D, chromatogram of the GC run shown in C (162).

to operate the IMS inlet on a dynamic dilution sampler (163). In this concept, the sample intake is pulsed with a fast valve (down to 20 μs), into the inlet against a flow of clean air or nitrogen. The amount of sample entering the inlet is governed by the pulse rate or the pulse width of the valve. In an effective and practical inlet, either of these would be controlled by the reactant ion peak intensity from the mobility spectrum. The valve is fully open until the reactant ion peak intensity falls below a preset value, which then triggers pulsing of the valve.

General Comments about Industrial Applications of IMS. IMS is a sensitive detector with a broad response that has found favor in certain applications, owing to the advantages of some IMS drift tube configurations. The industrial products based on IMS technologies are still limited, probably because the use of these devices requires a meticulous understanding of the measurement process. Both the target chemical and the matrix must be evaluated when considering IMS for an application.

IMS limitations and possible solutions to some of its identified problems have been published (150) and are summarized below:

- The chemical compound to be detected must be in the vapor state and must be ionized in the reaction tube of the IMS; some materials, including small hydrocarbons, are not easily ionized with a radioactive source when using water-reactant ions, but may be seen using other ionization techniques, such as charge exchange ionization.
- IMS response varies with gas pressure in the drift tube; small atmospheric pressure variations can disrupt measurements and do require frequent calibration when the IMS is used at high sensitivity.
- Ion clouds are formed by stealing charge from the reactant ion, which may significantly change the concentration; this problem is usually corrected by doping the carrier gas stream with a compound that has a proton or electron affinity slightly below that of the compound to be analyzed.
- IMS peaks, and particularly the reactant ion peak, change position with varying humidities because of differences in water clustering; in a portable IMS this can be controlled by using a recycling loop of air that is continuously dried by scrubbers.
- The limited dynamic range of one to two orders of magnitude is a serious limitation; in addition, the signature of a compound varies with the concentration. Working at a constant concentration (dynamic dilution) close to the one used for calibration is an attractive but not yet practical approach to reaching quantization results over a broad dynamic range.
- Water and ammonia, even at small concentrations, can create a change in the performance of the IMS, which can take a long time to recover.

Although IMS was begun over 25 years ago as an analytical technique, the technology is still in a nascent stage of development. The current implementations of IMS have shown technology "fences" rather than fundamental limitations. For example, high resolution IMS drift tubes of a small size have been demonstrated in Germany, and, recently, injection pulse width has been recognized as critical, opening up a broad area of research and potentially new products. In conclusion, significant technical improvements can be expected in the near future and, as the price of IMS decreases, a broader use of this technology for field industrial applications will also be seen.

BIBLIOGRAPHY

1. R. Erbes, *A Practical Guide to Air Quality Compliance,* 2nd ed., John Wiley & Sons, Inc., New York, 1996.
2. N. Stoloff, *Regulating the Environment: An Overview of Federal Environmental Laws,* Oceana, Dobbs Ferry, New York, 1991.

3. M. Katz, *J. Air Pollut. Control Assoc.* **30,** 528 1980.
4. National Institute for Occupational Safety and Health, *NIOSH Manual of Analytical Methods,* 2nd ed., DHEW (NIOSH) Pub. No. 75–121, 1975; 3rd ed., DHEW (NIOSH) Pub. No. 84–100, 1984 (revised 1990); 4th ed., DHHS (NIOSH) Pub. No. 94–113, 1994.
5. Occupational Safety and Health Administration, *OSHA Analytical Methods Manual,* OSHA Analytical Laboratories, Salt Lake City, Utah, 1985; (Suppl. 1991, 1993).
6. Health and Safety Executive, *Methods for the Determination of Hazardous Substances,* HSE Occupational Medicine and Hygiene Laboratory, Sheffield, U.K. 1981–1993.
7. Intersociety Commitee's Methods for Air Sampling and Analysis, *Methods of Air Sampling and Analysis,* 3rd ed., Lewis Publishers, Inc., Chelsea, Mich., 1988.
8. S. Ness, *Air Monitoring for Toxic Exposures,* Van Nostrand Reinhold, New York, 1991.
9. U.S. EPA, *Compendium Method T0141: The Determination of Volatile Organic Compounds (VOCs) in Ambient Air Using Summa® Passivated Canister Sampling and Gas Chromatographic Analysis,* U.S. EPA, Washington, D.C., May 1988.
10. L. Pagnotto, *Air Sampling Instruments,* 6th ed., American Conference of Governmental Industrial Hygienists, Cincinnati, Ohio, 1983.
11. A.T. Saalwaechter and co-workers, *Fundamentals of Analytical Procedures in IH, AIHA,* Akron, Ohio, 1987, pp. 363–373.
12. S. Crisp, *Ann. Occup. Hyg.* **23,** 47–76 (1980).
13. W.R. Betz and co-workers, *AIHA J.* **50**(4), 181–187 (1989).
14. P.M. Eller, ed., National Institute for Occupational Safety and Health, *NIOSH Manual of Analytical Methods,* NIOSH, Cincinnati, Ohio, Feb. 1984.
15. R.H. Brown and C.J. Purnell, *J. Chromatogr.* **178,** 79–90 (1979).
16. R.R.A. Miksch, *Personal Monitoring Employing Knudsen Diffusion: Development and Application to the Measurement of Formaldehyde in Indoor Air,* Air Technology Labs, Inc., Fresno, Calif.
17. M.L. Woebkenberg and C.S. McCammon, *Air Sampling Instruments Direct-Reading Gas and Vapor Instruments,* 8th ed., ACGIH, Cincinnati, Ohio, 1995, pp. 430–510.
18. E. Kress-Rogers, ed., *Handbook of Biosensors and Electronic Noses,* CRC Press, New York, 1997.
19. P. Moseley, J. Morris, and D.E. Williams, eds., *Techniques and Mechanisms in Gas Sensing,* Adam Hilger, New York, 1991.
20. S. Yamauchi, ed., *Chemical Sensor Technology,* Vol. 4, Elsevier, New York, 1992.
21. H. Strobel and W. Heineman, *Chemical Instrumentation: Systematic Approach to Instrumental Analysis,* 3rd ed., John Wiley & Sons, Inc., New York, 1989, Chapt. 27–32.
22. J. Wang, *Analytical Electrochemistry,* VCH Publishers, New York, 1994.
23. A. Foust, *Principles of Unit Operations,* John Wiley & Sons, Inc., New York, 1960.
24. U.S. Pat. 4,324,632 (Apr. 4, 1982), A. Tantram and Y. Chan.
25. A. Tseung and A. Tantram, *Nature,* **221,** 167–168 (1969).
26. J. Giner and C. Hunter, *J. Electrochem. Soc.* **116,** 1124–1130 (1969).
27. Steele and co-workers, *Adv. Ceram.* **3,** 286–309 (1981).
28. U.S. Pat. 4,132,616 (Jan. 2, 1979), A. Tantram.
29. U.S. Pat. 4,633,704 (Jan. 6, 1987), A. Tantram and Y. Chan.
30. Modwa, *Chemical Sensor Technology,* Shigeru Yamauchi Editing, Vol. 4, Elsevier, New York, 1992.
31. Lundström and co-workers, *Appl. Phys. Lett.* **26,** 55–57 (1975).
32. J. Janata, *Sens. Actuators,* **12,** 121–128 (1987).
33. K. Lundström, M. Shirvaman, and C. Svensson, *J. Appl. Phys.* **46** 3876–3881 (1975).
34. T. Fare, A. Spetz, M. Armgarth, I. Lundstrom, *Sens. Actuators,* **14** 369–386 (1988).
35. F. Winquist and I. Lundstrom, *Sens. Actuators,* **12** 255–261 (1987).
36. T. Poteat and B. Lalevic, *IEEE Electron. Devices Lett. EDL,* **2** 82–84 (1981).
37. Maclay and co-workers, *Sens. Actuators,* **14** 331–348 (1988).
38. Z. Weixin and Z. Yibing, *Sens. Actuators,* **15** 85–93 (1988).
39. Ackelid and co-workers, *IEEE Electron Device Lett. EDL,* **7** 353–355 (1986).
40. Mokwa and co-workers, *Sens. Actuators,* **12,** 333–339 (1987).
41. Kang and co-workers, *Sens. Actuators,* **12** 349–366 (1987).
42. J. Janata and co-workers, *Anal. Chem.* **66** 207R–228R (1994).
43. D. Johnson, M. Ryan, and G. Wilson, *Anal. Chem.* **60** 147R–162R (1988).
44. C.G. Fox, and J.F. Alder, in P. Moseley, J. Norris, and Dewilliams, eds., *Techniques and Mechanisms in Gas Sensing,* Adam Hilger, New York, 1991, (Chapt. 13).
45. M. Nieuwenhuizen and A. Nederlof, *Anal. Chem.* **60** 236–240 (1988).
46. D. Amico and co-workers, *Appl. Phys. Lett.* **41** 300–301 (1982).
47. A. Bryant and co-workers, *Sens. Actuators,* **4** 105–111 (1983).
48. Martin and co-workers, *Transducer '85: Proc. Int. Conf. on Solid State Sensors and Actuators, New York,* IEEE, Philadelphia, Pa., 1985, pp. 71–73.
49. K. Chan and co-workers, *Appl. Phys. B,* **38** 11–15 (1985).
50. A. Bentley and J. Alder, *Anal. Chim. Acta,* **222,** 63–73 (1989).
51. Lippitsch and co-workers, *Anal. Chim. Acta,* **205** 1–6 (1988).
52. O. Wolfbeis and A. Sharma, *Anal. Chim. Acta,* **208** 53–58 (1988).
53. R. Beswick and C. Pitt, *J. Colloid Interface Sci.* **124** 146–155 (1988).
54. Muto and co-workers, *J. Appl. Phys.* **28** 125–127 (1989).
55. J. Giuliani, H. Wohltjen, and N. Jarvis, *Opt. Lett.* **8** 54–56 (1983).
56. Tai and co-workers, *Opt. Lett.* **12** 437–439 (1987).
57. Nishizawa and co-workers, *Proc. 4th Int. Conf. on Optical Fiber Sensors, Tokyo, 1986,* p. 131.
58. C. Nylander, B. Liedberg, and T. Lind, *Sens. Actuators,* **3** 79–88 (1982).
59. E. Campbell and H. Miller, *Chemical Detectors, A Bibliography for the Industrial Hygienist with Abstracts and Annotations,* LAMS-2378, Los Alamos Scientific Laboratory, N.M., Vol I, 1961; Vol II, 1964.
60. A. Grote, W. Kim, and R. Kupel, *Am. Ind. Hyg. Assoc.* **39** 880 (1978).
61. V. Brown, *Proceedings of the HazTech International Conference, Aug. 1–16, 1986, Denver, Colo.,* Colorado Ground Water Association, 1986, pp. 125–136.
62. *Dräger-tube Handbook,* National Dräger, Inc., Pittsburgh, Pa., 1997.
63. *Precision Gas Detector System Manual,* Sensidyne/Gastec, Largo, Fla., 1997.
64. *Direct Reading Colorimetric Tubes—A Manual of Recommended Practices,* American Industrial *Hygiene Association,* Fairfax, Va., 1997.
65. M. Dressler, *J. Chromatog. Lib. Elsevier* **36** (1986).
66. R. Annino and R. Villalobos, *Process Gas Chromatography Fundamentals and Applications,* Instrument Society of America, Research Triangle Park, N.C., 1992.
67. R.L. Grob, *Modern Practice of Gas Chromatography,* 3rd ed., A Wiley-Interscience Publication, New York, 1995.

68. J.N. Driscoll and J.H. Becker, *Amer. Lab.* 69–76 (1979).
69. R.T. Smyth and D. Bingham, "The Use of Nonspecific PID Monitors for Remedial Investigations and Field Studies", *HazTec International Conference Proceedings,* Denver, Colo., Aug. 11–15, 1986.
70. Butler, *Amer. Lab.* **26**(19), 20 (1994).
71. I.N. Lee and co-workers, *AIHA J.* **48**(5), 437–441 (1987).
72. C.W. Bayer, *J. Chromatogr. Sci.* **32,** 312–316 (1994).
73. N. Arnold, W. McClennen, and H.L.C. Meuzelaar, *Anal. Chem.* **63** 299–304 (1991).
74. G.F. Evans, T.A. Lumpkin, D.L. Smith, and M.C. Somerville, *Air Waste Manage. Assoc.* **42**(10) 1319–1323.
75. M. Harper, *Stud. Surf. Sci. Catal.* **80** 267–274 (1993).
76. D.W. Lehmpuhl and J.W. Birks, *J. Chromatogr. A,* **740** 71–81 (1996).
77. R.W. Sherman, M.K. Solecki, E.S. Collard, T.H. McKinney, L.H. Grande, and E.B. Overton, "Development of A Field Portable Concentrator/Purge and Trap Device for Analysis of VOC in Ambient Air and Water Samples," *Proceedings of the First International Symposium for Field Screening Methods for Hazardous Waste Site Investigations,* Las Vegas, Nev., Oct. 1988.
78. R. Annino, in J. C. Giddings, E. Grushka, and P. Brown, eds., *Advances in Chromatography,* Vol. 26, Marcel Dekker, New York, 1987, p. 67.
79. U.S. Pat. 4,970,905 (Nov. 20, 1990), W.H. McClennen and co-workers.
80. U.S. Environmental Protection Agency, Technical Assistance Document, *Recommended Recordkeeping Systems for Air Pollution Control Equipment, Part 1: Particle Matter Controls,* EPA-340/1-86-021, Stationary Source Compliance Division, Washington, D.C., Aug. 1987.
81. R. Erbes, "Draft Model Title V Permits for PM_{10} Emissions Form Lime Manufacturing Plants", *A Practical Guide to Air Quality Compliance,* John Wiley & Sons, Inc., New York, 1996.
82. J. Dworzanski, H. Meuzelaar, W. Maswadeh, X. Nie, P. Cole, and N. Arnold, *Proceedings of the 1993 U.S. EPA/A&WMA,* 517–541 (1993).
83. R.D. Dandeneau and E.H. Zerenner, *HRC & CC,* **2,** 351–356 (1979).
84. F.I. Onuska, F.W. Karasek, "*Open Tubular Column Gas Chromatography in Environmental Sciences*", Plenum Press (1984).
85. L. Blomberg, *HRC & CC,* **5** 520–533 (1982).
86. J.K. Haken, *J. Chromatogr.* **300** 1–77 (1984).
87. I. Rohrschneider, *J. Chromatog.* **6** 22 (1966).
88. W.O. McReynolds, *J. Chromatogr. Sci.* **8** 685–692 (1970).
89. E. Kovats, in J.C. Giddings and R.A. Keller, eds., *Advances in Chromatography,* Vol. 1, Marcel Dekker, New York, 1966, pp. 229–247.
90. R. Annino and R. Villalobos, *J. Testing and Eval.* **29**(3) 143–148 (1996).
91. R. Villalobos and R. Annino, *JHRC,* **12** 149–160 (1989).
92. R. Villalobos and R. Annino, *JHRC,* **13** 764–773 (1990).
93. K. Grob, *On-column Injection in Capillary Gas Chromatography,* Huethig, New York 1987.
94. P. Sandra, ed., *Sample Introduction in Capillary Gas Chromatography,* Vol. 1, Huethig, New York, 1985.
95. K. Grob, ed., *Classical Split and Splitless Injection in Capillary Gas Chromatography,* 2nd ed., Huethig, New York, 1988.
96. M.S. Klee, *Gas Chromatography Inlet: An Introduction,* Ref. 5958-9468, Hewlett Packard, Calif., 1990.
97. T. Gorecki and J. Pawlazyn, *Anal. Chem.* **67**(18), 3265–3274 (1995).
98. B.V. Ioffe and A.G. Vitenberg, *Head-Space Analysis and Related Methods in Gas Chromatography,* John Wiley & Sons Interscience, New York, 1984.
99. K. Des Tombe, D.K. Verma, L. Stewart, E.B. Reczek, *Am. Ind. Hyg. Assoc. J.,* **52** 136–44 (1991).
100. D.R. Deans, *J. Chromatog.* **43,** 43 (1984).
101. M. Klemp, L. Akard, and R. Sacks, *Anal. Chem.* **65** 2516–2521 (1993).
102. L. Lanning, R. Sacks, R. Mouradian, S. Levine, Foulke, *J. Anal. Chem.* **60** 1994–1996 (1988).
103. M. Klemp, Chromab FAST, Inc., private communication, 1997.
104. *Detectors for Gas Chromatography, A Practical Primer,* Ref. 5958–9433, Hewlett Packard, Calif., 1987.
105. Hewlett-Packard Appl., *Publication No. 43-5954-7615,* Hewlett-Packard Calif., Nov. 1986.
106. U.S. Pat. 4,254,654 (Mar. 10, 1981), D.E. Clouser and J.S. Craven (to Hewlett-Packard Corp.)
107. J.S. Craven and D.E. Clouser, "A Fresh Design for Thermal Conductivity Detector," *30th Pittsburgh Conference on Analytical Chemistry and Applied Spectroscopy,* Cleveland, Ohio, Mar. 7,1979.
108. P.C. Uden, *Chromatogr. Forum* **17** (Nov.–Dec. 1986).
109. E.U. Ehrmann, H.P. Dharmasena, K. Carney, and E.B. Overton, *J. Chromatogr. Sci.* **34** 533–539 (1996).
110. R. Arenas, K. Carney, and E. Overton, *Am. Lab. (Shelton, Conn.),* **25**(11), 25–31 (1993).
111. Snyder and co-workers, *Anal. Chem.* **65** 299–306 (1993).
112. M. Bruns, *IECON '92,* San Diego, Calif., 1992, pp. 1640–1644.
113. S.J. Santy, "Portable Gas Chromatograph", paper presented at the *Gas Quality Measurement Symposium,* Institute of Gas Technology, Chicago, Ill., June 10–12, 1991.
114. R. Siemers, D. Heigel, and A. Spilkin, *Amer. Lab.* **23** 44L–44R (Mar. 1991).
115. M. Goedert and G. Guichon, *Anal. Chem.* **42** 962–968 (1970).
116. M. Goedert and G. Guichon, *Anal. Chem.* **45** 1188–1196 (1973).
117. J.T. Watson, *Introduction to Mass Spectrometry,* 3rd ed., Raven Press, New York, 1997.
118. J.B. Farmer, in C.A. McDowell, ed., *Mass Spectrometry,* McGraw-Hill, New York, 1963 pp. 7–44.
119. J. Roboz, *Introduction to Mass Spectrometry,* Wiley Interscience, New York, 1968.
120. Eur. Pat. 6,873,381 (Dec. 20 1995), S.J. Mullok.
121. P.H. Dawson and B. Yu, *Int. J. Mass Spectrom. Ion Proc.* **56** 25–50 (1984).
122. W. Paul and M. Raether, *Das elektrishe Massenfilter, Z. Physik,* **140** 262–271 (1955).
123. *Mass Spectrometry Fundamentals,* Hewlett-Packard manual part number G1626-90002, Hewlett-Packard Corp., Palo Alto, Calif.
124. J. Todd, G. Lawson, and R. Bonner, in P. H. Dawson, ed., *Quadrupole Mass Spectrometry and its Applications,* Elsevier, New York, 1976, pp. 181–224.
125. *Sci. Tech. Rev.,* 30–31 (Nov./Dec. 1995).
126. F. Karasek, *Anal. Chem.* **46**(8), 710A–720A (1974).
127. G.A. Eiceman and Z. Karpas, *Ion Mobility Spectrometry,* CRC Press, Boca Raton, Fla., 1994.
128. G.A. Eiceman, *CRC Crit. Rev. Anal. Chem.* **22** 17–36 (1991).
129. R.H. St. Louis and H.H. Hill, Jr., *CRC Crit. Rev. Anal. Chem.* **21** 321–355 (1990).
130. H.H. Hill, Jr., W.F. Siems, R.H. St. Louis, *Anal. Chem.* **62** 1201A–1209A (1990).
131. Spangler and co-workers., *ISA Trans.* **23**(1), 17–28 (1984).

132. P. Keberle, *Ann. Rev. Phys. Chem.* **28** 445–476 (1977).
133. G. Nicol, J. Sunner, and P. Kebarle, *Int. J. Mass Spectrom. Ion Proc.* **84** 135–155 (1988).
134. A. Good, D.A. Durden, and P. Kebarle, *J. Chem. Phys.* **52** 212–221 (1970).
135. M. Shahin, *J. Chem. Phys.* **45** 2600–2605 (1965).
136. F.W. Karasek, H.H. Hill, Jr., S.H. Kim, S. Rokushika, *J. Chromatogr.* **135** 329–339 (1977).
137. F.W. Karasek and G.E. Spangler, in A. Zlatkis and C.F. Poole, eds., *Electron Capture—Theory and Practice in Chromatography,* Elsevier, Amsterdam, 1981, pp. 377–406.
138. E.P. Grimsrud, *Mass Spectrom. Rev.* **10** 457–489 (1991).
139. E.P. Pellizari, *J. Chromatogr.* **98** 323 (1974).
140. S.H. Kim and G.E. Spangler, *Anal. Chem.* **57** 567–569 (1985).
141. U.S. Pat. 4,551,624 (Nov. 5, 1985), G.E. Spangler and J.H. Cox.
142. G.E. Spangler, J.P. Carrico, and D.N. Campbell, *J. Test. Eval.,* **13**(3), 234–40 (1985).
143. G.E. Spangler, D.E. Campbell, and J.P. Carrico, Presented at the 1983 *Pittsburgh Conference and Exposition on Analytical Chemistry,* Atlantic City, NJ, March 1983.
144. F.W. Karasek, *Res. Dev.* **21** 34–37 (1970).
145. M.J. Cohen and F.W. Karasek, *J. Chromatogr. Sci.* **8** 330–37 (1970).
146. N.E. Bradury and R.A. Nielsen, *Phys. Rev.,* **49** 388–93 (1936).
147. H. Hill and D. McMinn, *Detectors for Capillary Chromatog.,* **121** 297 (1992).
148. F.J. Knorr, R.L Eatherton, W.F. Siems, and H.H. Hill., Jr., *Jr. Anal. Chem.,* **57** 402–406 (1985).
149. H. Hill, and G. Simpson, *Field Anal. Chem. Technol.* **VI**(N3) 119–134 (1997).
150. J. Roehl, *Appl. Spectros. Rev.* **26** 1–57 (1991).
151. H. Hill and co-workers, *Anal. Chem.,* **21,** 321–355 (1990).
152. Technical brochures, *TDI/AVM Monitor, PUMA, Airbone Vapor Monitor (AVM), AVM material list and E.IMS,* Graseby Ionics, Ltd., Watford, United Kingdom, 1996.
153. W.M. Watson and C.F. Kohler, *Environ. Sci. Technol.* **13** 1241–1243 (1979).
154. R.C. Young and J.C. Travis, *A Feasibility Study of an Ion Mobility Spectrometer (IMS) for Hydrazine Vapor Detection,* NASA Technical Memo 102150, Resource Technology, 1989 Ann. Re., JFK Space Center, Fla., 1989.
155. T.A. Bacon, R. Getz, and J. Reafegui, *Chem. Eng. Prog.* **87** 61–66 (June 1991).
156. J.L. Brokenshire, V. Dharmarajan, L.B. Coyne, and J. Keller, *J. Cell. Plast.* **26** 123–142 (1990).
157. A. Lawrence, ed., *Cargo Inspection Technologies,* San Diego Calif., July 1994, pp. 25–27.
158. L.L. Danylewych-May and C. Cumming, *Adv. Anal. Detect. Explos., Proc. Int. Sym., 4th,* 1992, pp. 385–401.
159. J. Schultz *ICAO J.* **45** 19–20 (June 1990).
160. H. Meuzelaar, D.T. Urban, and N. Arnold, *Proceedings of the International Symposium on Field Screening Methods for Hazardous Waste Site Investigations,* 1991 p. 289.
161. N. Arnold and co-workers, *Proceedings of the Field Screening Methods for Hazardous Wastes and Toxic Chemicals,* Las Vegas, Nev., Vol. 2, 1995, p. 1038.
162. J. Dworzanski, M.G. Kim, N. Arnold, and H. Meuzelaar, *Anal. Chim. Acta,* **293** 219–235 (1994).
163. U.S. Pat. 4,797,554 (Oct. 1, 1989), W.C. Blanchard and A.T. Bacon.

FLUE GAS DESULFURIZATION. See AIR POLLUTION CONTROL FOR STATIONARY SOURCES.

FLUORESCENCE SPECTROSCOPY

MITCHELL E. JOHNSON
Duquesne University
Pittsburgh, Pennsylvania

Fluorescence spectroscopy is used for the trace and ultratrace analysis of organic and inorganic species. An analyte that is present at low concentration in environmental samples is usually contained in a highly complex matrix, which entails a need for not only sensitivity but also high selectivity. All successful techniques in use today for environmental trace analysis have varying degrees of both attributes. Fluorescence spectroscopy has the potential for extremely high sensitivity. This property arises from the fact that it is essentially a zero-background technique: in contrast to absorption spectroscopy, no signal is generated in the absence of analyte. In addition, fluorescence intensity depends on the intensity of the excitation source, and emission is of a different wavelength than the excitation light. However, fluorescence is in practice a blank-limited technique. Scattering from the solvent and fluorescence from impurities (either in the matrix or solvent) limits the ability to distinguish the fluorescence of the analyte from the blank signal. Therefore, selectivity is of ultimate importance for maintaining high sensitivity. Furthermore, as sensitivity increases, the need for selectivity likewise increases. A fluorescing molecule has a high degree of information intrinsically contained in its signal, and that information is relatively easy to obtain, thus giving fluorescence an inherent degree of selectivity. In addition, most molecules do not fluoresce to any appreciable extent, giving an additional advantage. Unfortunately, neither advantage is always sufficient, and other techniques sometimes must be employed, such as chromatographic separation.

BASIC THEORY

In order to understand how luminescence (fluorescence and phosphorescence) spectroscopy is used to advantage in chemical analysis, and particularly in environmental analysis, it is necessary to understand a certain basic level of theory. For example, the nature of the excitation and emission processes and, from a practical standpoint, the factors affecting luminescence. Luminescence efficiency is highly dependent on the local environment (ie, matrix) of the molecule: solvent, pH, temperature, concomitant species, etc. Since the analyst is often at the mercy of either the analytical matrix or conditions imposed by separation techniques, it is vital that the analyst understand these effects. In addition, there are several methods for employing luminescence spectroscopy. In order to understand how these techniques operate, and to appreciate compromises in performance required for their success, the analyst must have some familiarity with basic processes.

These subjects are covered in several excellent books. A good comprehensive spectrochemistry textbook is available (1). A readable introduction to fluorescence spectroscopy has been provided (2). The classic book by Guilbault has recently been revised for the second edition and includes an excellent chapter on environmental applications of fluorescence (3). An excellent introductory book is available (4), as well as an ongoing series of volumes that cover theory, instrumentation, and numerous applications in fluorescence (5,6). There are a number of recent reviews covering fluorescence detection in chromatography that contain good background material on luminescence (7,8). The basics of lasers and laser-induced fluorescence are covered well in several books (9–12). In addition, the interested reader is strongly encouraged to consult the biannual Fundamental Reviews in *Analytical Chemistry* and the alternating biannual series of Application Reviews on a variety of subjects.

Fluorescence can be an extremely sensitive technique. There is an old saw, though, that says only about 10% of all molecules fluoresce: the 10% you do not want. Fluorescent impurities and quenchers in the matrix are the biggest hindrance to low detection limits in real samples. It is not uncommon to leach fluorophores from plastic pipet tips (colored tips are notorious), or to get a fluorescence background from using cheap glass cuvettes. One of the main reasons for using fluorescence is its extraordinary sensitivity, so excessive care must be taken to ensure cleanliness in sample preparation. Spectroscopic grade solvents (or, better fluorescence grade or redistilled solvents) should always be used, and great care must be taken with sample cells. Fingerprints fluoresce nicely. Note that techniques used by atomic spectroscopists are not always applicable. For example, potassium dichromate fluoresces and is the worst possible cleaning medium. Potassium hydroxide/ethanol or the newer surfactant-based cleaners are generally better. Likewise, sample cleanup is important. Although fluorescence is relatively selective, the broad bands encountered in the condensed phase limit the spectral selectivity, and some form of sample cleanup is often necessary.

THE USE OF LUMINESCENCE IN ENVIRONMENTAL ANALYSIS

There are two inherent advantages to using fluorescence spectroscopy for environmental analysis: sensitivity and selectivity. Sensitivity arises from the fact that it is a zero-background technique, with an excitation intensity-dependent signal. Selectivity arises from the lack of widespread natural fluorophores and from the fact that, spectroscopically, fluorescence is information-rich. Two spectroscopic channels as well as lifetime information are available. Fluorescence is not as information-rich, except under special conditions, as mass spectrometry, but it is distinctly more compatible with liquid chromatography. In terms of atomic spectroscopy, it is one of the most sensitive techniques available, although it lacks the multielement capability of atomic emission or mass spectrometry. In addition, it is relatively inexpensive and simple to implement. These advantages, taken together, make fluorescence (and phosphorescence) useful for environmental analysis. Three major analysis methods have emerged as being particularly useful: screening, sensing, and detection following separation. The moderate selectivity and simple instrumentation are excellent qualifications for field screening methods. Remote sensing is quite compatible with fluorescence, mainly due to its sensitivity and the ability to use long optical fibers as probes. The major application, at least in terms of number of samples, is detection following liquid chromatography.

ANALYTICAL METHODS

Detection Following Chemical Separation

In analytical matrices that are as complicated as most environmental samples, the moderate selectivity of fluorescence spectroscopy is not sufficient to distinguish one sample from another. However, a program of sample cleanup, for example, by solid-phase extraction, followed by some form of chromatography, provides a reasonably good means of obtaining the needed selectivity. Several different techniques are available for separation. The majority of environmentally important samples, of course, are still analyzed by gas chromatography (GC), but there are a large number of polar analytes that are more amenable to liquid chromatography (LC).

Fiber Optic Sensing

A persistent problem in environmental analysis is the maintenance of sample integrity during the sampling process. Sampling a site *in situ* has great potential for minimizing contamination and disturbance of the site, as well as an advantage in time and cost savings. Furthermore, extremely hazardous sites are best studied remotely. Fiber optic probes have been investigated as one means of performing such analyses. The concepts and instrumentation of fiber optic sensors based on fluorescence have been reviewed (13), as have waveguide chemical sensors (14), and environmental applications of fiber optics (14,15). The basic idea of a fiber optic probe is simple: use a fiber to deliver the excitation light and another to collect the emission and deliver it to a spectrometer. There are a few alternatives: a bifurcated fiber is often used, so the probe (distal) end is the same for excitation and emission, or a single fiber can be used, with a mirror and dichroic filter to separate excitation light from the fluorescence. Most commercial spectrofluorometers provide fiber sampling attachments based on the former. Both methods have advantages and disadvantages. One limitation with fiber optic sensors is the fact that transmission of fibers in the UV is poor and depends strongly on wavelength, which limits useful fiber length to under 25 m below approximately 300 nm, and requires frequent replacement of the fibers. This does not necessarily obviate their use, however; water table monitoring, for example, does not require greater depth than 25 m, and many soil contamination applications are within a few tens of meters of the surface. Another problem is the lack of UV laser sources that are compatible with fibers. The HeCd laser is an exception, having been available for many years, with continuous-wave (CW) output and near-UV and blue wavelengths. Other exceptions are the recent introduction of intracavity frequency-doubled or UV-optimized argon ion lasers. Most UV lasers (excimer, frequency-quadrupled Nd:YAG), however, are pulsed, and the high peak powers can easily damage the face of the fiber. Also, the beam quality of such lasers is generally of poor enough quality to make coupling into the fibers difficult. Other blue and UV CW sources are becoming available, however, and may help the situation. The standard Xe lamp may be coupled easily into larger multimode fibers or fiber bundles.

Total Luminescence Methods

The fact that the excitation and emission processes are decoupled from one another gives rise to the following observed phenomena: The emission spectrum is independent of the excitation wavelength, and vice versa. Any differences must be attributable to additional species in solution (if we assume the excited state does not undergo any wavelength-dependent reactions). This property can be exploited to enhance selectivity in luminescence measurements. Adding additional parameters, such as lifetime, will further enhance selectivity, although at the expense of instrumental complexity.

High Resolution Techniques

There are several techniques that increase the selectivity of fluorescence by eliminating inhomogeneous broadening. High resolution Shpol'skii spectroscopy (HRSS) (16), matrix isolation (MI), and fluorescence line narrowing (FLN) (17) do so by incorporating the molecule in a frozen matrix and supersonic jet spectroscopy (SJS) entrains the sample in a kinetically cooled gas. All these techniques are rather specialized, and none are broadly applicable. For maximum resolution, a narrow-band laser must be used as the source, a high resolution spectrometer should be used for measuring spectra, and all require special sample-handling equipment. Thus, although these techniques are not at present applicable to routine environmental analysis, they do offer excellent capabilities that should not be overlooked for research applications. When used properly, these techniques offer the possibility of very simple sample preparation.

Complexometric Methods for Inorganic Species

Few metal ions fluoresce naturally in aqueous solution. A number of the trivalent lanthanides (Sm, Eu, Gd, Tb, Dy) are exceptions. For those few metals that can be determined directly, selectivity is inherently high and sensitivity can approach parts per quadrillion (18). Lanthanides have very narrow absorption and emission bands and long lifetimes, which makes it very easy to filter out stray light and interferences. They may also be sensitized by binding to a ligand that is more efficiently excited; energy is transferred to vacant orbitals on the metal via the ligand triplet state. A number of uranium species fluoresce in solution (ASTM STM D2907-91), most notably the uranyl ion (UO_2^-); commercial instruments exist simply for measuring U fluorescence. Uranium chemistry in aqueous solution can be extremely complex; researchers have shown that the use of time-resolved fluorescence spectroscopy can provide a tremendous amount of information for sorting out the contributions of various species to overall equilibrium (19). However, the vast majority of metal ions must be complexed with a fluorescent ligand to provide reasonable quantum yields in solution. As with UV-Vis spectrophotometry, complexation with (an) organic ligand(s) provides a means to quantitate metal ions using conventional instrumentation. Ligands exist for nearly every metal ion in the periodic table and most simple inorganic anions. Dean gives a comprehensive summary of the more common reagents (20). The major advantages to analyzing metal ions this way are simplicity and better detection limits than UV-Vis absorption spectrophotometry. Since most complexation methods involve a dynamic equilibrium between the analyte and ligand, sensitivity is highly dependent on the binding constant (in addition to fluorescence considerations). Binding constants in many cases may be high enough to provide subnanomolar (sub-parts-per-billion) detection limits. Dynamic range is limited for the same reason to within a maximum of two orders of magnitude above and below the binding constant; precision will degrade at more than one order of magnitude away from the binding constant due to the inherent nonlinearity of the concentration response. The disadvantage is that most reagents lack specificity for a given metal ion (but see below), requiring masking or removal of interferents, so in that sense, fluorescence cannot compete with atomic absorption (AA) and emission (AE) techniques. Furthermore, most AA and AE techniques offer multielement capability (modern instruments for AA offer increasingly better multielement capabilities) and excellent detection limits. Therefore, any discussion of the merits of fluorescence spectroscopy using complex formation must be understood in that context.

The real advantages are simplicity and a much better prospect for field portability than AA or AE techniques. The use of flow injection analysis offers better precision and higher throughput than traditional "beaker" methods and has been shown to easily transportable to the field (21,22). For example, zinc can be determined by complexation with *p*-tosyl-8-aminoquinoline (*p*-TAQ) at subnanomolar concentrations in seawater following preconcentration on an 8-hydroxyquinoline (8-HOQ) column (22). The 8-HOQ column removes most of the potential interferents except for cadmium, which also forms a fluorescent complex with *p*-TAQ. Interestingly, the authors claim, although they do not demonstrate, that the fluorescence lifetime can be used to distinguish between zinc and cadmium complexes, which again demonstrates the superior selectivity of fluorescence methods over absorption methods. Detection limits can be quite respectable. Butyltin species were determined by ion-exchange chromatography followed by either inductively coupled plasma mass spectrometry (ICP/MS) or fluorescence after complexation with morin (23). The fluorescence method is only worse by an order of magnitude or so compared to ICP-MS in the single-ion monitoring mode, which in many cases may be a perfectly reasonable tradeoff for a much cheaper and easier method. That work also demonstrates the fact that alkylated metal species can be determined using a complexation method. In the case of tin species, the trialkyl species require hydrolysis before complexation (24).

In some cases, selectivity can be achieved by using the (often) inherently high sensitivity of biological systems. For example, pyoverdin was immobilized on controlled-pore glass (CPG) and the resulting biosensor used to determine Fe(III) with a moderately high degree of selectivity over many other cations and anions. The selectivity is based on the fact that pyoverdin, found as an iron transport compound in *Pseudomonas fluorescens,* is naturally selective for Fe(III). Such approaches are becoming more common for both inorganic and organic species as researchers learn how to immobilize or isolate the specific agent from its cellular environment without destroying its function. A new generation of reagents for ionic species is based on the rational design of binding structures in order to provide high specificity as well as favorable fluorescence properties (25). The probes are generally highly conjugated systems with a binding portion based on various types of chelators, including simple ligands, coronands (such as crown polyethers), and cryptands (bicyclic structures). The higher the order in the binding structure, the greater the binding specificity. Spectroscopic selectivity comes about by some form of intramolecular charge transfer. For example, the ligand itself may be higher in energy than the HOMO of the fluorophore, so upon excitation the ligand transfers an electron to the HOMO, filling it and quenching fluorescence. Upon metal binding, however, the energy of the ligand is lowered enough to prevent the charge transfer, and the quantum yield of the fluorophore is enhanced significantly. Even if several metals can bind to the ligand, they may have spectral or lifetime differences that can be exploited to differentiate among species. Such probes are ideally suited to fiber optic sensors. Such approaches are routinely employed in biological imaging applications, but have yet to be fully exploited for environmental monitoring or screening.

One means of achieving better detection limits and lower interference is to incorporate the ligand into a solid-phase matrix such as an ion-exchange gel. The sample (extract or aqueous sample) is passed through a column containing the gel, the analyte is adsorbed, the interferents are passed, it is hoped, and the gel is packed into a solid-phase accessory for a conventional spectrofluorometer. In fact, a flow-through cell would be even better. This method has been successfully applied to molybdenum, for example, using complexation with carminic acid followed by adsorption on strong anion-exchange beads (26). The beads are collected by filtration, packed into a solid-phase cell, and analyzed in a standard spectrofluorometer. Detection limits are sub-parts-per-billion, but with a rather small (order of magnitude) linear dynamic range. Preconcentration on the beads enhances the detection limit.

Atomic Fluorescence

The atom cells used for atomic absorption spectrometry (flame, electrothermal atomizer) are eminently suited for atomic fluorescence spectroscopy (AFS), since they produce gas-phase atoms in the ground state. There is a high degree of specificity, because lineshapes are very narrow and both excitation and emission can be specific to the element of interest. The sources of band-broadening are different in AFS than in the solution-phase spectroscopy of molecules. Because there are no vibrational or rotational lines and there are no solute–solvent interactions, the transitions are discrete. Major sources of band-broadening are pressure and Doppler broadening. Pressure broadening arises from collision with other gas-phase atoms, and Doppler broadening results from the Doppler shifting of emitted radiation. In a typical flame, total broadening is much less than 1 Å (1). If a high intensity laser is used for excitation, saturation of the transition will cause much worse broadening than either of these sources. The net result, in the absence of high-intensity excitation, is a fairly simple spectrum. However, since many more transitions are accessible in the UV and visible, spectral crowding may result, particularly with lanthanides and actinides. The selectivity of AFS is then a real advantage compared to high-temperature AE spectroscopic sources such as the inductively coupled plasma (ICP).

Other types of cells are used besides the flame and electrothermal atomizer. The ICP can be used, and indeed it is the atom cell of the only commercially available general-purpose AFS spectrometer (Baird Corporation). Excitation is accomplished well above the load coil, where atoms have cooled and returned to the ground state. Specialized instruments for hydride-generating elements have been used, and several commercial instruments based on cold vapor generation are available for the AFS of mercury. Sources are typically narrow-band (usually dye) lasers (tunability is required for general-purpose instruments), hollow cathode lamps (HCLs), and electrodeless discharge lamps. Mercury analyzers usually use low pressure mercury arc lamps. Broadband sources are also useful, but eliminate most of the advantage to AFS, that is, selective excitation. The Baird instrument used a circular array of paired HCLs and PMTs to provide multielement capability.

The advantage to AFS, besides selectivity, is very high sensitivity. For example, researchers describe a copper-vapor laser-pumped, frequency-doubled dye laser system for the excitation of lead vaporized in an electrothermal atomizer (27,28). Detection limits for lead without any form of preconcentration are parts per trillion or femtograms absolute, due in great part to "negligible" background levels (27). Such sensitivity allows the direct determination of lead in Great Lakes water and seawater, in good agreement with other techniques. The ability to perform direct determination is a distinct advantage. The major drawback, of course, is the lack of multielement capability. Cost is much higher than for a typical atomic absorption or inductively coupled plasma spectrometer, but compares favorably to ICP-MS or multielement electrothermal atomization atomic absorption with Zeeman background correction. Many schemes for AFS use resonance excitation (ie, excite and detect at the same wavelength), which places extra demands on the system for eliminating stray light. A two-step excitation scheme for mercury that gives detection limits in the low parts-per-trillion range (90 fg absolute) has been reported (29).

Instead of using electrothermal atomization or a flame, a number of applications use conventional techniques for generating hydride vapors (30,31), or mercury vapor can be analyzed directly, as noted above. The primary advantage of vapor generation is to remove the element from its matrix. Mercury analyzers have an additional advantage in the preconcentration step on a gold amalgamator. A gold foil or gauze of some sort preconcentrates the mercury vapor, which is then thermally desorbed and swept into the fluorescence cell by a carrier gas. Several commercial mercury analyzers are available for AFS, and this technique has been shown to provide very reliable quantitation with low detection limits (subpicogram) (32). In addition, cold vapor AFS has been used as a detection scheme for mercury speciation following gas chromatography (33,34). One of the few comparisons has shown that detection limits for mercury are low enough using this technique that the blank limits detection, and AFS is therefore comparable in utility to CV/AAS or atomic emission detection with a microwave-induced helium plasma (32).

ANALYTES AND MATRICES

Generally speaking, fluorescence is not the primary means of standard analysis for any but a few compounds, such as polycyclic aromatic hydrocarbons (PAHs). However, the point of this section is to show what types of compounds have been or can be analyzed by fluorescence, because fluorescence spectroscopy, as has been pointed out in several places above, offers some advantages to more sophisticated techniques in terms of screening or field methods. This discussion is by no means comprehensive. The reader is referred to a recent review (18), the biannual application reviews in *Analytical Chemistry,* and other reviews mentioned in this article.

Pesticides (Including Insecticides, Herbicides, and Fungicides)

By far the most widely used methods for the analysis of pesticides, herbicides, etc, are based on GC with various detection methods. However, there are a number of very polar or thermally unstable pesticides unsuitable for GC, although derivatization may sometimes help. For these compounds, high-performance liquid chromatography HPLC is the most widely used alternative for multiresidue analyses. Detection is mainly performed by UV absorption, but a large number of pesticides fluoresce or can be derivatized to fluoresce. Fluorescence is therefore well-suited for a number of screening applications.

The fluorescence and LC properties of a number of pesticides has been surveyed (35,36). Naphthalene derivatives are typically very highly fluorescent. The hydroxycoumarins are part of a large class of coumarin compounds, many of which are sold commercially as laser dyes and, as such, are strongly fluorescent and photostable.

Although HPLC with fluorescence detection is the method generally used for the analysis of these pesticides, and others, there are other means that take advantage of the selectivity of fluorescence. In any case, the selectivity advantage over UV absorbance often leads to reduced sample cleanup. Thiabendazole (36) and morestan and *o*-phenylphenol (OPP) (37,38) have been determined by solid-phase spectrofluorometry. The sample was stirred with ion-exchange gel beads, which were filtered out of solution and packed into a fluorescence cell in a standard spectrofluorometer. By choosing the gel and operating conditions properly, interferences could be minimized. Little interference was found from dichlone, morestan, or OPP. Carbendazim was determined in natural waters by a flow-injection analysis method that employed micelles for a 12-fold enhancement of sensitivity and 13-n*M* detection limit (39). Thiabendazole and carbendazim were resolved using first-derivative synchronous scanning fluorescence spectroscopy (40).

Biphenyls are reasonably fluorescent, but increasing the number of chlorine substituents has a detrimental effect on quantum yield due to the internal heavy atom effect (41). Polybrominated biphenyls fluoresce poorly if at all. The use of micellar solutions increases the fluorescence intensity of chlorobiphenyls; the effect is reasonably selective for tri- and tetra-chlorobiphenyls (42). The use of solid-phase techniques has shown a great deal of potential for screening applications, rather than identification and quantitation. Both photoactivated luminescence and solid-surface, room-temperature phosphorescence have been demonstrated as screening methods for polychorinated biphenyls (PCBs) (43,44). The use of phosphorescence is a reasonable approach, given the enhanced intersystem crossing. Parts-per-billion detection limits or below are possible. Additional heavy atom salts, such as thallium acetate, are often used for the further enhancement of phosphorescence (41).

A number of pesticides that do not fluoresce naturally can be derivatized to form fluorophores. The *N*-methylcarbamates (NMCs) are probably the most common in this group (see above), although the NMCs are not themselves derivatized (EPA 531.1, RCRA 8318). The determination of NMCs and their metabolites has been recently reviewed (45,46). One distinct advantage to fluorescence detection is the fact that cleanup requirements are eased by the inherent selectivity of fluorescence and the derivatization reaction. Glyphosate likewise is first oxidized and the product primary amine derivatized with various amine derivatizing agents, including OPA (47). Glyphosate (EPA 547), gluphosinate, and aminomethylphosphonic acid are examples of compounds that are best analyzed by HPLC rather than GC. Anilines can be derivatized also. Although aniline itself is moderately fluorescent, many of its derivatives, especially the chlorinated derivatives, are not. Derivatization with dansyl chloride was used to detect phenylurea herbicides and their aniline derivative metabolites by thin-layer chromatography (TLC) or HPLC (48). TLC has an advantage in that derivatization can be performed *in situ*. Phenoxyacid herbicides can be readily derivatized through the carboxylic acid group (49); several fluorescent derivatization agents exist that are specific for carboxylic acids. Finally, there are a number of nonfluorescent nitrogen-containing pesticides (eg, Alachlor, butachlor, paraquat) whose quantum yield is significantly enhanced upon UV photolysis (50).

Organic Pollutants

Polycyclic aromatic hydrocarbons are by far the most important group of pollutants that are determined regularly by fluorescence spectroscopy (EPA 550, 550.1, 610, TO-8, RCRA 8310, ASTM STM D4657-92, D5412-93). The reason is that most polycyclic aromatic hydrocarbons (PAHs) and their derivatives are naturally fluorescent. In addition, most PAHs have spectra with partial vibrational resolution, making identification and selectivity easier than for most compounds. These spectral band intensities are extremely sensitive to their environment, so care must be taken during spectral identification and calibration. The most common situation in the determination of PAHs is to use fluorescence as a detector for liquid chromatography. For an in-depth discussion of HPLC in determining PAHs, the reader is referred to reviews (51). The use of HPLC has been compared to other methods (52,53). In most cases, LC compares favorably to gas chromatography/mass spectrometry (GC/MS) in terms of cost and competes quite well with respect to the ease of sample preparation, selectivity, and detection limits. If field portability is an issue, LC/fluorescence is clearly the preferred choice. However, in terms of resolution and peak capacity, capillary GC is still unmatched, although electrochromatography (CEC) and capillary LC may be considered potential competitors (54,55). It must be noted that, as discussed above, the use of multiple-wavelength pairs or diode array/spectrographic detection dramatically increases selectivity and identification capabilities.

Since PAHs are fluorescent, spectrofluorometry is often used as a means of screening environmental samples for "hot spots." In some cases, simple spectrofluorometry (or room-temperature phosphorescence) can be used for screening (56,57,58), but it is more common to use HPLC, such as size-exclusion or reversed-phase LC, for additional selectivity (59,60,61). As long as calibration is performed with care, screening can be performed in numerous types of environmental samples. A very common application is screening for oil pollution or contamination following oil spills or pipeline leaks (62,63,64,65–67) (ASTM STM D3650-93). The PAHs in the oil provide the fluorescent signal. Researchers have used a graphite furnace electrothermal atomizer with a slow-temperature ramp to take advantage of different PAH volatility in screening for PAHs with laser-induced fluorescence and wavelength-resolved detection (68). Synchronous scanning methods can also be used for additional selectivity, which may be sufficient for simple samples (see above). Solid-phase preconcentration coupled with synchronous scanning has also been used for the selective step (69). As has been discussed above, PAHs are nicely compatible with Shpol'skii matrices, making this an excellent means of identifying and quantitating PAHs in a variety of matrices (60,70,71,72). In fact, Shpol'skii spectroscopy can be comparable to GC/MS/single-ion monitoring in terms of selectivity but better in detection limits (60).

A number of PAH derivatives have reduced fluorescence, most notably the halogenated and nitrated PAHs. Nitrated PAHs can be reduced to amines, which are fluorescent (73,74). Relatively little information is available on halogenated PAHs. However, quantum yield is still sufficient for the determination of halogenated benz[*a*]anthracenes following HPLC (75). Because of the internal heavy atom effect, enhanced intersystem crossing is expected and indeed observed for brominated naphthalenes. Phosphorescence quantum yield is high enough at room temperature for HPLC detection (76), whereas chlorinated species require a sensitization scheme (77).

Other pollutants that can be determined by fluorescence methods include phenol, cresols, and their derivatives (EPA TO-8) (62,78,79,80); benzenes, toluenes, and xylenes ethylbenzene (BTXEs) (81); and chlorinated hydrocarbons (82). Sample types include sediments, soils, aerosols, particulates, and gaseous samples. Fluorescence has a tremendous advantage in gaseous or aerosol samples in that it may have sufficient sensitivity to allow *in situ* monitoring, thus diminishing the major source of contamination in gaseous or air-borne analysis, that is, sampling (83). ASTM D4763-88(1995)e1 outlines procedures for the use of fluorescence for the quantification of more than 90 chemicals in water.

Toxins, Carcinogens, and Mutagens

The most widely studied group of carcinogens or suspected carcinogens are the PAHs, which were discussed in the previous section.

Liquid chromatography is widely used for the analysis of mycotoxins and phycotoxins, due mainly to the fact that the majority of these compounds are large, polar molecules with ether, ester, and amide ester linkages that have low thermal stability (84). In many cases, fluorescence is the preferred method for detection because it provides selectivity in complex matrices (grains, foodstuffs, dairy products) and has the requisite sensitivity for determining very low concentrations of highly potent toxins. Aflatoxins G_2 and B_2, ochratoxin A, zearalenone, and a few other mycotoxins fluoresce naturally with enough intensity for direct determination. Most commonly, some form of derivatization is required. In many cases, postcolumn oxidation is sufficient for accomplishing this task. Aflatoxins G_1 and B_1 and, especially, the paralytic shellfish poisons (PSPs) are amenable to such a simple procedure. Oxidation is often moderately selective, however, which can lead to problems in multiresidue techniques. The use of precolumn derivatization is preferred in this case because the oxidant is nonfluorescent, thus avoiding most of the drawbacks of precolumn derivatization, but careful attention must be paid to conditions of both the reaction and separation conditions (85). A number of other toxins contain amino groups (fumonisins, domoic acid, tetrodotoxin), carboxylic acid groups (diarrhetic shellfish poisons), or hydroxyl groups (trichothecenes) that are amenable to fluorescent derivatization (84).

Inorganic Pollutants; Metals

The methodology for the determination of metal species by fluorescence spectroscopy was discussed in the sections on complexometric methods and atomic fluorescence. The comments apply in general to other inorganic species, such as the simple anions like nitrate, sulfate, sulfite, and the halides. Nitrate is generally reduced to nitrite and then complexed with various agents, such as tryptophan (86). Halides can be readily determined by their quenching effect on numerous fluorophores (87).

Free Radicals and Peroxides

Peroxides and free radicals are extremely important as intermediates in atmospheric chemistry, but because they are highly reactive, they tend to be transient species found at very low concentration. As a consequence, fluorescence spectroscopy is one of the few methods available for the direct detection of these species. Although chemiluminescence or enzymatic assays are most common for the determination of hydrogen peroxide, fluorescent methods exist that are comparable in terms of sensitivity and lack the disadvantages of the enzymatic methods, such as irreproducibility and high cost. For example, a simple method has been developed based on Fenton's reaction for the determination of hydrogen peroxide, hydroxymethylhydroperoxide, and methyl hydroperoxide in aerosols (88) and gaseous samples (89). The method is quite simple: Hydrogen peroxide and Fe(III) react to form the hydroxyl radical, which is scavenged by benzoic acid, producing fluorescent hydroxybenzoic acid species. Conditions have been optimized to eliminate sulfur dioxide interference, which is most important given the generally accepted reaction pathways of sulfur in the atmosphere (89). It is also possible to determine these species directly in the gas phase or aerosols using laser-induced fluorescence. Major interferences are scattering from particulates and aerosols and fluorescence from other species. For example, the hydroxyl radical, in principle, can be easily determined in the atmosphere, but fluorescence from aerosol and gaseous concomitants and hydroxyl radicals produced by the high energy excitation can be major interferences. Time-resolution and multiphoton excitation schemes help overcome these limitations (90). Some of the schemes envisioned, which were inappropriate because of instrumental limitations (90), appear now to be within reach due to advances in laser technology over the last few years. This is fortunate, because laser remote sensing has significant advantages over sampling methods, and laser-induced fluorescence appears to be one of the few techniques capable of the sensitivity needed for real-time measurements.

Simple Gases

Simple gases, either as dissolved gas or in their native state, can be determined by direct fluorescence, fluorescence quenching, or derivatization following solubilization or reduction/oxidation to a soluble salt. Most of the simple gases fluoresce naturally in the gaseous state and can be determined fairly readily, provided interferences may be avoided. The hydroxyl radical example given above is a case in point. Quenching is perhaps the most common method, as simple gases dissolved in a membrane or solution tend to be fairly efficient quenchers of many fluorophores. In most cases, the fluorophore can be chosen for its optical characteristics (a luxury not often found) (91) and also for lack of interference from other gases. For example, gaseous hydrogen sulfide can be determined by scrubbing followed by a quenching method (92,93). Fluorescence quenching has been used extensively as a means of oxygen sensing. In most cases, the fluorophore is bound to a membrane, usually a plastic such as PVC, ion-exchange beads, sol-gels, or CPG. Phosphorescence is generally much more sensitive to oxygen quenching than fluorescence and is therefore often used instead. Various luminophores, such as metal chelates (91,94,95) and PAHs (often pyrene or its derivatives) (15), are used.

New Problems: Bioengineered Analytes

New approaches to agriculture and new problems in water purification are demanding novel resources from the analytical chemist. Concern over the metabolites of genetically engineered pesticides and herbicides demands analytical methods that until recently have been the purview of molecular biologists and biochemists. Fluorescence spectroscopy has traditionally played a large role is such laboratory work. The major methods for sequencing DNA or the sizing of DNA fragments all rely on fluorescence staining or tagging for quantitation. In addition, the newer methods for detecting specific sequences in DNA or RNA rely on fluorescence microscopy for detection. There are several means of detecting specific sequences of DNA in a sample. The simplest in concept is called fluorescence *in situ* hybridization (FISH). In FISH, illustrated in Figure 1, an oligonucleotide fragment with its sequence complementary to the sought-after sequence is synthesized, and the fragment is labeled with a fluorophore (usually only a single fluorophore, to avoid quenching). The oligonucleotide "finds" and binds to its complement, the excess probe is washed away, and the sample is viewed under a microscope or in a flow cytometer. The longer the oligonucleotide sequence, the lower the probability of a false positive. Multiple probes with different emission characteristics may also be used for higher specificity. This methodology has been incorporated into a fiber optic probe (96) and a micro-machined "gene chip" (Affymetrix Inc., Santa Clara, California). These approaches have high potential for screening for large amounts of genetic material.

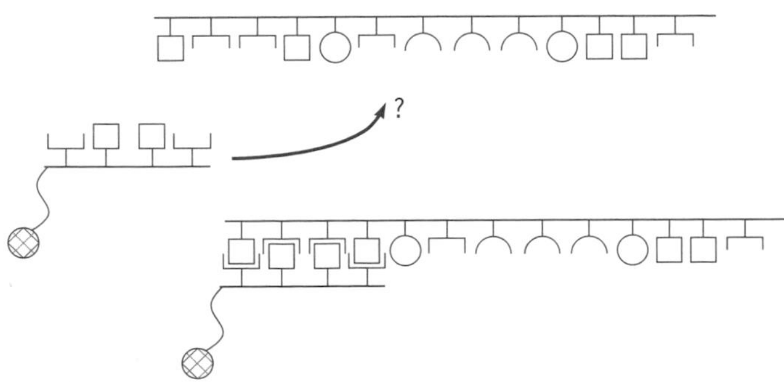

Figure 1. Illustration of fluorescence *in situ* hybridization (FISH) showing binding of the labeled oligonucleotide probe to its complementary sequence on a strand of DNA. The shapes represent bases, the dark line the phosphate backbone, and the cross-checked circle the fluorophore.

Another powerful approach is that based on the polymerase chain reaction (PCR). Its use in environmental analysis has been reviewed recently (97). DNA strands are separated by heating, oligonucleotide probes are bound to a primer sequence on either end of the stretch to be amplified, the PCR enzyme extends the primers by "filling in" the DNA strand, and the cycle is repeated. The number of copies of the target sequence increases exponentially with the number of cycles, so a billion-fold amplification of the target can be achieved in only 30 cycles. The resulting DNA is often quantified using fluorescent intercalating dyes. This is a very powerful technique for detecting extremely small amounts of DNA, say, from a virus or pathogen, but the PCR enzyme is sensitive to its environment, and sequence information must be known beforehand. These methods were suggested, for example, for use in detecting *Cryptosporidium,* for which conventional fluorescent antibody staining was inadequate due to high interferences from algae and other natural fluorophores (98).

It is also possible to use standard enzyme assays for these types of analyses. For example, β-D-galactosidase is an enzyme marker for fecal coliform bacteria, a common bacterial analyte for water purity testing. Numerous substrates are available that can be converted into fluorescent molecules by the enzyme. An example is methylumbelliferyl-β-D-galactoside. A procedure for the enzyme analysis of fecal coliform bacteria based on a simple enzyme assay has been shown to have excellent performance (99). The utility of an enzyme assay is similar to that of PCR: amplification of the analyte, in this case by the accumulation of product fluorophore.

CONCLUSIONS

Although the majority of environmentally interesting compounds do not exhibit enough luminescence for direct detection, there is a tremendous advantage to employing luminescence when they do, or when they can be derivatized or complexed with a fluorophore. The use of high resolution techniques, although experimentally more complex, may yield tremendous advantage in eliminating chromatography or sample cleanup steps. The newer, high resolution capillary separation techniques are particular beneficiaries when fluorescence detection is used due to the requirements of high mass and concentration sensitivity. Given the importance of these techniques for separation, it is likely that environmental applications of fluorescence spectroscopy will grow over the next several years, particularly as performance-based methods are introduced by the EPA and other agencies.

BIBLIOGRAPHY

1. J. D. Ingle, Jr. and S. R. Crouch, *Spectrochemical Analysis,* Prentice-Hall, Englewood Cliffs, N.J., 1988.
2. W. R. Seitz, in P. J. Elving, E. J. Meehan, and I. M. Kolthoff, eds., *Treatise on Analytical Chemistry, Part I: Theory and Practice,* 2nd ed., Wiley-Interscience, New York, 1981, pp. 159–248.
3. G. G. Guilbault, *Practical Fluorescence,* 2nd ed., Marcel Dekker, New York, 1990.
4. J. R. Lakowicz, *Principles of Fluorescence Spectroscopy,* Plenum, New York, 1983.

5. J. R. Lakowicz, ed., *Topics in Fluorescence Spectroscopy,* Vol. 1–4, Plenum, New York, 1991–1994.
6. E. L. Wehry, *Modern Fluorescence Spectroscopy,* Vol. 1–4, Plenum, New York, 1976–1981.
7. M. J. Sepaniak and C. N. Kettler, in E. S. Yeung, ed., *Detectors for Liquid Chromatography,* John Wiley & Sons, Inc., New York, 1986, pp. 148–203.
8. R. J. van de Nesse, N. H. Velthorst, U. A. T. Brinkman, and C. Gooijer, *J. Chromatogr. A* **704,** 1–25 (1995).
9. W. Demtröder, ed., *Laser Spectroscopy: Basic Concepts and Instrumentation,* 2nd ed., Springer-Verlag, Berlin, 1996.
10. D. S. Kliger, ed., *Ultrasensitive Laser Spectroscopy,* Academic Press, New York, 1983.
11. N. Omenetto, ed., *Analytical Laser Spectroscopy,* John Wiley & Sons, Inc., News York, 1979.
12. E. H. Piepmeier, ed., *Analytical Applications of Lasers,* Wiley-Interscience, New York, 1986.
13. R. B. Thompson, in J. R. Lakowicz, ed., *Topics in Fluorescence Spectroscopy, Vol. 2: Principles,* Plenum, New York, 1991, pp. 345–365.
14. R. E. Dessy, *Anal. Chem.* **61**(19), 1079A–1094A (1989).
15. E. L. Wehry, in G. G. Guilbault, ed., *Practical Fluorescence,* 2nd ed., Marcel Dekker, New York, 1990, pp. 367–430.
16. A. P. D'Silva and V. A. Fassel, *Anal. Chem.* **56**(8), 985A–1000A (1984).
17. R. Jankowiak and G. J. Small, *Anal. Chem.* **61**(18), 1023A–1032A (1989).
18. S. M. Steinberg, E. J. Poziomek, and W. H. Engelmann, *Chemosphere* **28**(10), 1819–1857 (1994).
19. C. Moulin, P., Decambox, V. Moulin, and J. G. Decaillon, *Anal. Chem.* **67**(2), 348–353 (1995).
20. J. A. Dean, *Analytical Chemistry Handbook,* McGraw-Hill, New York, 1995, pp. 5.60–5.62.
21. K. N. Andrew, N. J. Blundell, D. Price, and P. J. Worsfold, *Anal. Chem.* **66**(18), 917A–922A (1994).
22. J. L. Nowicki, K. S. Johnson, K. H. Coale, V. A. Elrod, and S. H. Lieberman, *Anal. Chem.* **66**(17), 2732–2738 (1994).
23. J. I. Garcia-Alonso, A. Sanz-Medel, and L. Ebdon, *Anal. Chim. Acta* **283**(1), 261–271 (1993).
24. J. A. Stäb, M. J. M. Rozing, B. van Hattum, W. P. Cofino, and U. A. T., Brinkman, *J. Chromatogr.* **609,** 195–203 (1992).
25. B. Valeur, in J. R. Lakowicz, ed., *Topics in Fluorescence Spectroscopy, Vol. 4: Probe Design and Chemical Sensing,* Plenum, New York, 1994, pp. 21–48.
26. J. L. Vilchez, G. Sánchez-Palencia, R. Blanc, R. Avidad, and A. Navalón, *Anal. Lett.* **27**(12), 2355–2368 (1994).
27. V. Cheam, J. Lechner, I. Sekerka, R. Desrosiers, J. Nriagu, and G. Lawson, *Anal. Chim. Acta* **269**(1), 129–136 (1992).
28. V. Cheam, J. Lechner, I. Sekerka, and R. Desrosiers, *J. Anal. At. Spectrosc.* **9**(3), 315–320 (1994).
29. W. Resto, R. G. Badini, B. W. Smith, C. L. Stevenson, and J. D. Winefordner, *Spectrochim. Acta* **48B**(5), 627–632 (1993).
30. L. Ebdon, P. Goodall, S. J. Hill, P. B. Stockwell, and K. C. Thompson, *J. Anal. At. Spectrosc.* **8**(5), 723–729 (1993).
31. W. T. Corns, P. B. Stockwell, L. Ebdon, and S. J Hill, *J. Anal. At. Spectrosc.* **8**(1), 71–77 (1993).
32. M. Okumura, K. Fukushi, S. N. Willie, and R. E. Sturgeon, *Fresenius J. Anal. Chem.* **345**(8/9), 570–574 (1993).
33. L. Liang, N. S. Bloom, and M. Horvat, *Clin. Chem.* **40**(4), 602–607 (1994).
34. E. Saouter and B. Blattmann, *Anal. Chem.* **66**(13), 2031–2037 (1994).
35. R. T. Krause, *J. Chromatogr.* **255,** 497–510 (1983).
36. F. Capitán, E. Alonso, R. Avidad, L. F. Capitán-Vallvey, and J. L. Vilchez, *Anal. Chem.* **65**(10), 1336–1339 (1993).
37. J. L. Vilchez, R. Avidad, J. Rohand, A. Navalón, and L. F. Capitán-Vallvey, *Anal. Chim. Acta* **282**(2), 445–449 (1993).
38. J. L. Vilchez, A. Navalón, R. Avidad, J. Rohand, and L. F. Capitán-Vallvey, *Fresenius J. Anal. Chem.* **345**(11), 716–719 (1993).
39. J. Sancenón and M. de la Guardia, *Anal. Chim. Acta* **287**(1/2), 49–57 (1994).
40. F. García Sánchez and C. Cruces Blanco, *Anal. Chem.* **60**(4), 323 (1988).
41. J. Hernández Garcia, Z. Sosa Ferrera, A. J. Bermejo Martín-Lazáro, and J. J. Santana Rodríguez, *Anal. Lett.* **27**(7), 1355–1382 (1994).
42. J. Hernández García, J. R. Betancort Rodríguez, A. J. Bermejo Martín-Lázaro, and J. J. Santana Rodríguez, *Anal. Chim. Acta* **290**(1/2), 146–153 (1994).
43. T. Vo-Dinh, A. Pal, and T. Pal, *Anal. Chem.* **66**(8), 1264–1268 (1994).
44. W. Watts, A. Pal, L. Ford, G. H. Miller, T. Vo-Dinh, D. Eastwood, and R. Lidberg, *Appl. Spectrosc.* **46**(8), 1235–1239 (1992).
45. A. De Kok, M. Heimstra, and U. A. T. Brinkman, *J. Chromatogr.* **623,** 265–276 (1992).
46. B. D. McGarvey, *J. Chromatogr.* **642** (1/2), 89–105 (1993).
47. M. E. Oppenhuizen and J. E. Cowell, *J. Assoc. Off. Anal. Chem.* **74,** 317–324 (1991).
48. J. Lantos, U. A. T. Brinkman, and R. W. Frei, *J. Chromatogr.* **292,** 117–127 (1984).
49. Y. Mechref and Z. El Rassi, *Anal. Chem.* **68**(10), 1771–1777 (1996).
50. B. M. Patel, H. A. Moye, and R. Weinberger, *Talanta* **38**(8), 913–922 (1991).
51. S. A. Wise, L. C. Sander, and W. E. May, *J. Chromatogr.* **642**(1/2), 329–349 (1993).
52. K. G. Furton, E. Jolly, and G. Pentzke, *J. Chromatogr.* **642**(1/2), 33–45 (1993).
53. C. Escrivá, E. Viana, J. C. Moltó, Y. Picó, and J. Mañes, *J. Chromatogr. A* **676,** 375–388 (1994).
54. C. Yan, R. Dadoo, H. Zhao, and R. N. Zare, *Anal. Chem.* **67**(13), 2026–2029 (1995).
55. S.-H. Chen, C. E. Evans, and V. L. McGuffin, *Anal. Chim. Acta* **246,** 65 (1991).
56. A. D. Campiglia, J. P. Alarie, and T. Vo-Dinh, *Anal. Chem.* **68**(9), 1599–1604 (1996).
57. M. T. Tena, M. D. Luque de Castro, and M. Valcárcel, *Anal. Chem.* **68**(14), 2386–2391 (1996).
58. C. J. Owen, R. P. Axler, D. R. Nordman, M. Schubauer-Berigan, K. B. Lodge, and J. P. Schubauer-Berigan, *Chemosphere* **31**(5), 3345–3356 (1995).
59. M. M. Krahn, G. M. Ylitalo, J. Buzitis, S.-L. Chan, and U. Varanasi, *J. Chromatogr.* **642,** 15–32 (1993).
60. G. Morel, O. Smhan, P. Literathy, H. Al-Hashash, L. Moulin, T. Saeed, K. Al-Matrouk, M. Martin-Bouyer, A. Saber, L. Paturel, J. Jarosz, M. Vial, E. Combet, C. Fachinger, and J. Suptil, *Fresenius J. Anal. Chem.* **339,** 699–715 (1991).
61. M. M. Krahn, G. M. Ylitalo, J. Buzitis, S.-L. Chan, U. Varanasi, T. L. Wade, T. J. Jackson, J. M. Brooks, D. A. Wolfe, and C.-A. Manen, *Environ. Sci. Technol.* **27**(4), 699–708 (1993).
62. T. A. Taylor, G. B. Jarvis, H. Xu, A. C. Bevilacqua, and J. E. Kenny, *Anal. Instrum.* **21**(3/4), 141–162 (1993).
63. T. A. Taylor and H. H. Patterson, *Anal. Chem.* **59**(17), 2180–2187 (1987).

64. J. M. Shaver and L. B. McGown, *Appl. Spectrosc.* **49**(6), 813–818 (1995).
65. V. A. Lawford, M. Morgan, P. Vessey, and J. E. G. Wheaton, in H. O. Nielson, ed., *Environment and Pollution Measurement Sensors and Systems* SPIE, The International Society for Optical Engineering, Bellingham, Wash., 1990, pp. 158–166.
66. S. D. Alururi, M. Rasas, O. Alamedine, S. Jubian, F. Al-Bahrani, and M. Quinn, *Opt. Eng.* **34**(1), 214–221 (1995).
67. E. M. Filippova, V. V. Chubarov, and V. V. Fadeev, *Can. J. Appl. Spectrosc.* **38**(5), 139–144 (1993).
68. A. Mellone, B. W. Smith, and J. D. Winefordner, *Talanta* **37**(1), 111–118 (1990).
69. L. F. Capitan-Vallvey, M. Del Olmo Irurela, R. Avidad Constandeda, and J. L. Vilchez Quero, *Anal. Lett.* **26**(11), 2443–2454 (1993).
70. F. Ariese, S. J. Kok, M. Verkaik, G. P. Hoornweg, C. Gooijer, N. H. Velthorst, and J. W. Hofstraat, *Anal. Chem.* **65**(8), 1100–1106 (1993).
71. F. Ariese, C. Gooijer, N. H. Velthorst, and J. W. Hofstraat, *Fresenius J. Anal. Chem.* **339,** 722–724 (1991).
72. A. Saber, G. Morel, L. Paturel, J. Jarosz, M. Martin-Bouyer, and M. Vial, *Fresenius J. Anal. Chem.* **339,** 716–721 (1991).
73. S. Matsuzawa, P. Garrigues, H. Budzinski, J. Bellocq, and Y. Simizu, *Anal. Chim. Acta* **312**(2), 165–177 (1995).
74. M. Murayama and P. K. Dasgupta, *Anal. Chem.* **68**(7), 1226–1232 (1996).
75. Y. Mori, S. Goto, S. Onodera, S. Naito, S. Takitani, and H. Matsushita, *Fresenius J. Anal. Chem.* **345**(1), 63–67 (1993).
76. J. J. Donderbroek, C. Elzas, C. Gooijer, R. W. Frei, and N. H. Veltohrost, *Talanta* **28,** 717–723 (1981).
77. J. J. Donkerbroek, N. J. R. Van Eikema Hommes, C. Gooier, N. H. Velthorst, and R. W. Frei, *J. Chromatogr.* **255,** 581 (1983).
78. W. A. Chudyk, M. M. Carrabba, and J. E. Kenny, *Anal. Chem.* **57**(7), 1237–1242 (1985).
79. J. Sherma and S. C. McGinnis, *J. Liq. Chromatogr.* **18**(4), 755–761 (1995).
80. G. Marko-Varga, in D. Barcelo, ed., *Environmental Analysis: Techniques, Applications, and Quality Assurance,* Elsevier, Amsterdam, 1993, pp. 255–271.
81. J. Bublitz, M. Dickenhausen, M. Grätz, S. Todt, and W. Schade, *Appl. Opt.* **34**(18), 3223–3233 (1995).
82. C. S. McEnally, R. F. Sawyer, C. P. Koshland, and D. Lucas, *Appl. Opt.* **33**(18), 3977–3984 (1994).
83. R. Niessner, in D. Barcelo, ed., *Environmental Analysis: Techniques, Applications, and Quality Assurance,* Elsevier, Amsterdam, 1993, pp. 3–22.
84. J. F. Lawrence and P. M. Scott, in D. Barcelo, ed., *Environmental Analysis: Techniques, Applications, and Quality Assurance,* Elsevier, Amsterdam, 1993, pp. 273–309.
85. M. Janecek, M. A. Quilliam, and J. F. Lawrence, *J. Chromatogr.* **644**(2), 321–331 (1993).
86. N. Jie, J. Yang, and F. Meng, *Talanta* **40**(7), 1009–1011 (1993).
87. G. Guoquan, W. Liufang, W. Fenqin, and L. Liqian, *Anal. Lett.* **26**(1), 147–151 (1993).
88. J. H. Lee, I. N. Tang, and J. B. Weinstein-Lloyd, *Anal. Chem.* **62**(21), 2381–2384 (1990).
89. J. H. Lee, I. N. Tang, J. B. Weinstein-Lloyd, and E. B. Halper, *Environ. Sci. Technol.* **28,** 1180–1185 (1994).
90. F. L. Eisele and J. D. Bradshaw, *Anal. Chem.* **65**(21), 927A–939A (1993).
91. W. Xu, K. A. Kneas, J. N. Demas, and B. A. DeGraff, *Anal. Chem.* **68**(15), 2605–2609 (1996).
92. T. W. Andreae, M. O. Andreae, H. G. Bingemer, and C. Leck, *J. Geophys. Res.; [Atmos.]* **98**(D12), 23389–23396 (1993).
93. D. J. Cooper and E. S. Saltzman, *J. Geophys. Res.; [Atmos.]* **98**(D12), 23397–23409 (1993).
94. Y.-M. Liu, R. Pereiro-García, M. J. Valencia-González, M. E. Días-García, and A. Sanz-Medel, *Anal. Chem.* **66**(6), 836–840 (1994).
95. P. Hartmann and W. Trettnak, *Anal. Chem.* **68**(15), 2615–2620 (1996).
96. P. A. E. Piunno, U. J. Krull, R. H. E. Hudson, M. J. Damha, and H. Cohen, *Anal. Chem.* **67**(15), 2635–2643 (1995).
97. B. H. Olson, *Environ. Sci. Technol.* **25**(4), 604–611 (1991).
98. A. Newman, *Anal. Chem.* **67**(23), 731A–734A (1995).
99. S. C. Apte and G. E. Batley, *Sci. Total Environ.* **141,** 175–180 (1994).
100. N. D. Coggeshall and E. M. Lang, *J. Amer. Chem. Soc.* **70**(10), 3283–3292 (1948).

FLUORESCENT SIGNATURES IN ENVIRONMENTAL ANALYSIS

SERGEY BABICHENKO
LARISA PORYVKINA
Institute of Ecology
Tallinn, Estonia

YURI ORLOV
IGOR PERSIANTSEV
Moscow State University
Moscow, Russia

SERGEI REBRIK
University of California
San Francisco, California

The methods of fluorescent diagnostics are widely applied in a variety of environmental studies and analyses. They are known as an effective tool for the analysis of organic compounds in water environments. The principal merits of fluorescent techniques are the high sensitivity and possibility to carry out the proximate analysis in remote mode without time-consuming pretreatment of water samples. Thus, it is possible to build the on-line diagnostic systems based on the method of induced fluorescence. A promising fluorescent method is the registration and analysis of spectral fluorescent signatures (SFS), which include the excitation spectra as well as fluorescence spectra of organic compounds (1). In application to water quality diagnostics the SFS technique allows analysis of major groups of organic substances of water ecosystems, including phytoplankton, dissolved organic matter of natural origin (DOM), and organic pollution. Living phytoplankton is a very sensitive indicator of the ecological status of water ecosystems. It has been shown by numerous studies that organic pollution or enhanced nutrient loading of natural waters may cause exceptional blooms of phytoplankton and changes of its taxonomic composition, including development of toxic species. Oxygen depletion following the die-off of a planktonic bloom can cause water quality problems. DOM in natural waters consist of a large number of organic compounds. In estuaries and coastal waters it may be either of marine origin, derived primarily from activities of planktonic organisms, or it may be of terrestrial origin, transported to the marine environment via rivers and coastal wetlands. The diagnostics of organic pollution is a key monitoring task of natural water quality. The primary objective is to control the influence of human activity on the ecology of the aquatic environment. Direct impact of technogeneous organics into the water are caused by drainage from industrial works, agricultural farms, and urban domestic waters. Risk aquatories are also areas of intensive navigation, shell-oil mining, and processing. Timely detection of the pollution, estimation of its abundance, and identification of the polluter allow adequate reactions to prevent the spreading of pollution.

METHOD OF SPECTRAL FLUORESCENT SIGNATURES

The fluorescence of organic compounds in water is caused by dissolved fraction as well as by emulsion and suspended matter. Excitation and fluorescence spectra, depending on the structure of organic molecules, contain information about complex water organics. Decomposing the conglomerate of substances present in the water into their chemical ingredients for subsequent analysis is a complicated task. The most productive approach to on-line diagnostics is to treat the object as an integral spectroscopic sample, characterized by specific spectral fluorescent signatures (SFS). In general, the SFS is the sum total of the emission spectra of an object at different excitations. It is recorded as a matrix of fluorescent intensity in coordinates of excitation and emission wavelengths in a definite spectral window (two-dimensional spectrum). The correctness of SFS depends on the right combination of spectral bands of excitation and emission wavelengths. The spectral matrix has to include the main fluorescent features of organic compounds. The method of analysis based on SFS includes compiling a systematized library of organic compounds and comparing measured SFS with the reference library of signatures to identify the substances in a water sample (2). Correct identification is the first step in estimating substance quantities. Reliability of this method depends on how representative the reference library of water organics is, as well as how the reference library is organized. The systematization of a reference library is based on preliminary classification and grouping included substances by physical, chemical, biological, or spectral features. The characteristic spectral structure of SFS is used for substance recognition, whereas the intensity of fluorescence serves as a measure of substance quantity. The task of chemical or biological analysis is transformed to the analysis of specific patterns of two-dimensional spectral images. This allows the computer expert systems for the analysis of SFS to be built.

SFS of Phytoplankton

The excitation and emission spectra of various micro-algae differ by their characteristic features, connected primarily with the pigment composition of phytoplankton (chlorophylls, phycobilins, and carotenoids), and light-energy transfer processes from accessory pigments to chlorophyll a. In spite of the complicated dependence of the processes in pigment systems (and, as a consequence, of the fluorescence response), on the diverse changes of external condition the SFS can give the overall characterization of the phytoplankton community. The analysis of spectral features of different taxa makes it possible to systematize the phytoplankton species according to their pigment composition and to arrange them into groups within a reference library (3). Taking into consideration the difficulties which occur when algae have to be strictly divided into established taxonomic groups, the species can be divided into four groups according to their main accessory pigments. The corresponding marker pigments follow: group I, *phycocyanin* (cyanobacteria); group II, *phycoerythrin* (cyanobacteria and red algae); group III, chlorophyll b (green and prasinophyte algae), and group IV, chlorophyll c (diatoms, dinophytes, and others). The spectral features of different phytoplankton pigments reveal themselves in the form of local maxima with fixed coordinates and specific structures of equal fluorescence intensity lines in spectral signatures. For example, very easily detectable spectral structures having the coordinates excitation/emission 620/640 and 550/580 nm correspond to phycocyanin and phycoerythrin. Distinguishing groups III and IV is more complicated, as their accessory pigments do not have their own fluorescence *in vivo*. In the excitation spectra of chlorophyll a the maxima of chlorophylls b and c are located rather close together (480 and 460 nm, respectively). Moreover, their position may vary. There is an additional diagnostic feature of the chlorophyll c-containing algae that is not typical for chlorophyll b-containing algae which is a wide excitation band of chlorophyll a in the region of 500–550 nm caused by the absorption of the carotenoids (4).

SFS of Dissolved Organic Matter

The spectral characteristics of DOM are caused by its highly fluorescent humic substance (HS) portion. It has been demonstrated by numerous investigations that HS composition is subject to spatial and seasonal variability. This leads to changes in fluorescence spectra. Moreover, the external factors such as temperature, acidity (pH), and heavy-metal ions may affect HS fluorescence. SFS demonstrate structural variability in the shape of plot lines due to different compositions of HS. Some classifications of DOM can be done by analysis of the spectral features of different natural waters (5). For definite aquatoria, the seasonal dynamics of HS are expressed mainly by the intensity of fluorescence with possible minor variations of spectral shape due to the conservative nature of DOM. As the water environment carbon balance is heavily influenced by riverine- and salt marsh-derived organic matter, the investigation of DOM itself by using SFS can be important. In pollution detection, SFS of DOM can be considered as a variable fluorescent background and must be taken into account in the analysis of water organics conglomerates.

SFS of Organic Pollution

The fluorescent characteristics of organic pollution, including oil and oil products, phenols and their derivatives, and lignin and their derivatives, depend on polyaromatic hydrocarbons (PAHs) which effectively absorb light in the UV region. Variability of SFSs in general is relatively high, but some types of substances have quite similar spectral signatures, meaning that preliminary classification of substances by physical or chemical parameters may differ from classification by spectral features. When compiling the reference library of a pollutant's SFS, it must be adjusted to the specific diagnostic task. The

SFSs of all potential pollutants must be included, keeping in mind possible changes caused by pollutant degradation. Pollution of natural waters typically consists of a number of organic compounds in a complex mixture. Sometimes it is more important to detect and estimate the abundance of specific conglomerates of substances than to distinguish individual compounds. In this case, the SFS of a mixture must be included in the library, and preliminary classification using SFS is the most effective method for diagnostics.

CONCLUSION

The application of SFS techniques in environmental analysis provides the possibility to carry out the rapid qualitative diagnostics of organic compounds in natural water. Such an approach allows the use of phytoplankton pigments as natural fluoroindicators of the taxonomic groups and allows the monitoring of water quality by analysis of pigment composition in a mixed micro-algae population. The classification system is able to detect and classify organic pollution in both natural and technogeneous environments. The system is insensitive to DOM spectrum variations. Identification of pollutants from a given end-user library can be done by the hierarchical structure of specialized artificial neural networks (ANNs). Within the concept of generalized classes, the system is able to classify up to a quadruple mixture of pollutants from different classes. The concentrations of the pollutants in the mixture are determined by a linear ANN corresponding to the detected combination of classes.

BIBLIOGRAPHY

1. M. Warner, G. Patonay, and M. Thomas, *Anal. Chem.* **57**, 464A–466A (1985).
2. A. Dudelzak, S. Babichenko, L. Poryvkina, and K. Saar, *Appl. Opt.* **30**(4), 453–458 (1991).
3. J. Hilton, E. Rigg, and G. Jaworsci, *J. Plankton Res.* **11**, 65–74 (1989).
4. C.S. Yentsch and C.M. Yentsch, *J. Mar. Res.* **37**, 471–483 (1979).
5. L. Poryvkina, S. Babichenko, and J. Lapimaa, *AMBIO* **21**(7), 465–467 (1992).

FOREST ECOSYSTEMS IN A CHANGING GLOBAL CONTEXT

KATE LAJTHA
Oregon State University
Corvallis, Oregon

Perhaps the most diverse and widespread vegetation classification on the surface of the earth is the *forest ecosystem.* Composed of a dominant canopy cover of trees, forest ecosystems differ in general physiognomy, in the relative diversity of both overstory and understory species, and in the degree of canopy cover. Forests may also differ in the relative composition of deciduous versus evergreen species. While climate, particularly temperature and water regime, determine both the presence or absence of forests as well as the general physiognomy and composition of the tree species, local geology and soil nutrient availability modify this climate control. As will be discussed below, forest ecosystems may also affect the climatic regime of their environment.

Forests are subject to stress from both natural causes, such as climatic fluctuations and pathogens, and anthropogenic causes, such as air pollution, physical disturbance, and acidic precipitation. In many systems there may be multiple stresses on an ecosystem, and these stresses may work synergistically. Unfortunately, the effects of many types of disturbance are similar and the responses of plants are responses to general stress, and are not easily used to diagnose specific stressors. For this reason, cause-and-effect relationships between markers of forest ecosystem stress and specific stressors are difficult to prove conclusively and have confounded much research into causes of, and solutions for, forest damage and decline.

This article will briefly review forest community structure and forest ecosystem function, then discuss the major anthropogenic factors thought to affect forest ecosystem health and forest productivity. For discussion of the division of the earth into broad divisions, see ECOZONES, GLOBAL.

FOREST COMMUNITIES

A forest *community* consists of the particular assemblage of biological organisms that inhabit the forest, including the plants and the animals, as well as soil organisms such as bacteria, fungi, and soil-dwelling invertebrates. An *ecosystem,* on the other hand, consists of the populations of species that inhabit the community taken together with the abiotic environment, such as the soil, atmospheric inputs, and the transfers and circulation of energy and nutrients.

Distinct forest ecosystems intergrade not only with each other over environmental gradients but with other vegetation community types (Fig. 1). For example, temperate deciduous forests change in composition and intergrade with spruce–fir forests with increasing elevation in the mountains of the northeastern United States; with increasing elevation in southwestern states, dry deserts and desert grasslands intergrade with juniper grasslands, pinyon–juniper woodlands, ponderosa pine forests, and finally montane spruce–fir forests. Although these communities are generally continuous with one another and species composition can (but does not always) change gradually over gradients, these communities have been classified by a variety of different criteria into distinct groupings for the sake of convenience and comparison. The various types of forest communities include woodlands and tree savannas; temperate and deciduous forests; tropical rain forests; and coniferous forests.

The two most common and general community classification systems are the *physiognomic* approach and the *dominant species* approach. The physiognomic approach names community types by their overall structural appearance and the growth form of the dominant plant species. The rationale for this approach is based on the general assumption that whenever similar climatic environments occur across continents on the earth, the same growth forms exist, although often with very different species. One example of this is the temperate woodland community type that occurs in western North America, the Mediterranean, and in parts of Australia, characterized by open-growth, usually short trees with a shrub

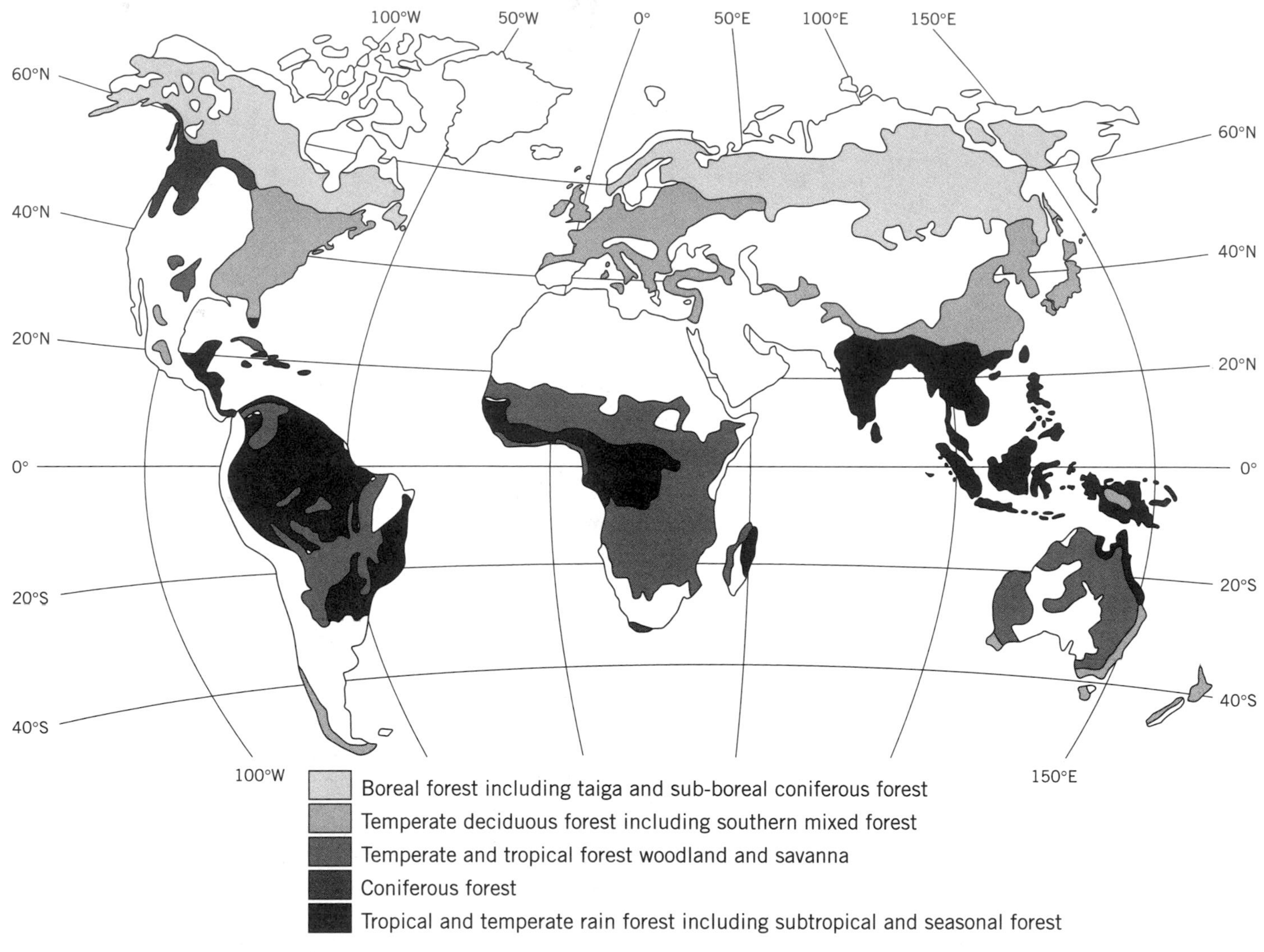

Figure 1. Geographic extent of forest community types.

or grass understory. Although these woodlands have similar physiognomic traits and perhaps similar functional traits, the dominant trees can be needle-leafed, deciduous, sclerophyll, or any combination of these in the different regions. The dominant species approach is perhaps the most direct and natural approach, particularly in communities dominated by one or two major species. However, this approach will not be as effective in communities that are extremely diverse with many overstory dominants, such as in tropical rain forests.

Another more general approach to classifying ecological communities is to use the *biome* or *biome-type* approach. A biome comprises a grouping of plants and their associated fauna on a given continent that are easily recognizable based on general physiognomy, but that might consist of species that differ greatly in actual taxonomic affinity. A biome-type is a grouping of convergent biomes across different continents. The needle-leafed conifer forest biome-type is found across northern Eurasia and North America, for example, yet encompasses the temperate rain forest of the Pacific Coast of North America that is composed of the redwood and sequoia forests of California and the Douglas fir–western hemlock forests of the northwest, the relatively less dense and colder boreal forests of spruce and balsam fir, and the warmer and drier Ponderosa pine forests farther south.

FOREST ECOSYSTEM FUNCTION

Biogeochemical cycles and transformations within forests can be separated into intersystem versus intrasystem transfers (Fig. 2). Intersystem transfers are those processes of chemical gains and losses to other ecosystems on the surface of the earth, including precipitation, dryfall, and gaseous absorption inputs, and streamflow, groundwater, and gaseous emission losses. The uptake of ions, their return to the soil, and the chemical transformations of these ions within the system constitute the intrasystem cycle. In general, the intrasystem circulation of essential nutrient ions through the biota is significantly greater than the annual inputs or outputs to the ecosystem, suggesting that vegetation receives most of

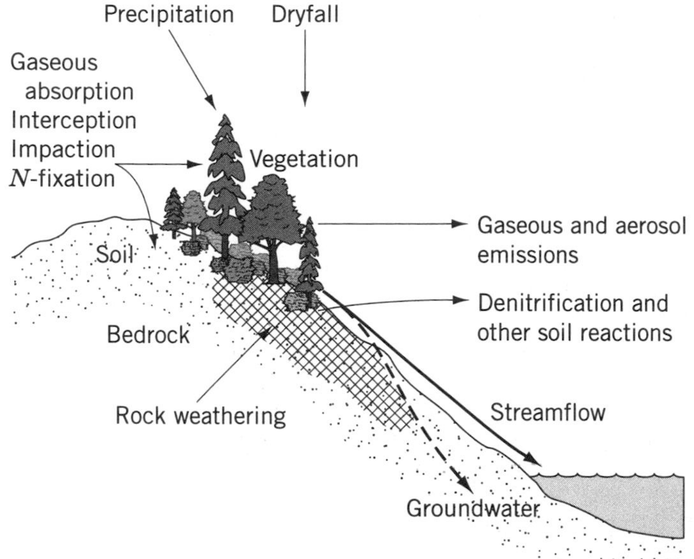

Figure 2. A model of the pathways of nutrient movement through undisturbed forest ecosystems (1).

its annual requirements from recirculation through the soil rather than through new inputs.

With the exception of nitrogen, most nutrient ion inputs to a forest ecosystem come ultimately from the weathering of parent rock material or from dust deposition. Rates of chemical weathering and nutrient release from bedrock depend not only on the hydrologic regime but also on the specific rock type, and thus inputs can vary widely. Inputs of nitrogen, however, come solely from atmospheric deposition and from *in situ* nitrogen fixation by microbes, as there is no fixed, usable form of nitrogen in rocks. Although there is much variation in rates of *N*-fixation, atmospheric deposition, and bedrock weathering among sites, primary production in most temperate forest ecosystems has been found to be limited by the availability of nitrogen. In contrast, many tropical forest ecosystems are phosphorus limited, in part due to the very old, weathered soils that have lost much of the original phosphorus content and that contain minerals with a high phosphorus adsorption capacity. Boreal forest soils have been shown to be N, P, or N and P limited, perhaps due to the slow turnover of organic matter in boreal soils that contain much of the ecosystem's organically bound nutrient capital.

Climatic regime and the chemistry and physiology of the dominant tree species control such ecosystem processes as production of organic matter, decomposition of organic litter, and nutrient remineralization. Consequently, world forests differ greatly in the quantity of organic matter and nutrients stored both aboveground and in the forest floor (Table 1). Although local temperature and moisture regime should theoretically limit maximum net primary production (NPP) of any given forest, correlations between NPP and temperature and moisture variables are generally significant but weak, due to differences in species' rooting depths, length of foliage display, physiological efficiencies, and site nutrient and soil characteristics. Over a wide range of ecosystem types, NPP has been found to be more correlated with the maximum site leaf area than to direct climatic variables.

Temperature and moisture exert strong controls over litter decomposition in forests, with temperature of primary importance. Rates of decomposition of fresh litter can be accurately predicted using site actual evapotranspiration, a variable that combines the effects of temperature and moisture availability. However, the relationship is significantly improved if characteristics of litter chemistry are included, such as nitrogen content and concentrations of resistant, carbon-rich chemicals. Because soil organic matter accumulation reflects the difference between primary production and litter decomposition, soil organic matter content is also strongly controlled by regional differences in temperature and moisture. Thus cold boreal forests, with relatively low biomass in aboveground tree components, accumulate a great deal of carbon in the forest floor, whereas tropical forests accumulate relatively low amounts of carbon and nutrients in the forest floor compared to the aboveground standing biomass (see Table 1).

Because climate and anthropogenic inputs to forest ecosystems affect much of ecosystem functioning, clearly even small-scale alterations of climate can affect such factors as maximum annual production, the balance between production and decomposition, nutrient limitation, and nutrient exports to streams and rivers.

ANTHROPOGENIC ALTERATION OF FOREST STRUCTURE AND FUNCTION

Disturbance and Equilibrium

There is much debate among scientists as to whether forest ecosystems can be considered structurally and functionally equilibrium communities. The recent geologic record clearly shows that the earth's climate at any given location has not remained constant; and paleographic evidence, both from pollen records in preserved sediments and from leaves preserved in fossil structures such as packrat middens, has shown that distributions of forest species change dramatically as climate changes. Pollen records from eastern U.S. forests during the last 10,000 years document the movement northward of individual tree species following the retreat of glaciers, and show that movement of each species is individualistic, rather than as a conglomerate flora moving together (2). For instance, 18,000 years ago spruce grew in association with a variety of sedges in the American midwest, and by 10,000 years ago they formed a closed monospecific canopy in southern Canada. The spruce–birch association of the modern boreal forest is no more than 6,000 years old and is probably as sensitive to variations in climate as its community predecessors. With the potential for accelerated climate change from human activities, forest communities are expected to undergo even more dramatic changes in the future.

Even at smaller spatial scales and over shorter temporal scales there is much heterogeneity and lack of equilibrium in natural communities. Although the term *disturbance* is generally thought of as an event or series of events that are abnormal or unusual to a system, forest ecologists now view disturbance, at least at some scale, as being both commonplace and essential for maintaining the structure of forests and the coexistence of species. Disturbance can be disruptions of ecosystem, population, or community structure that are commonplace or recurring, discrete events defined by environmental fluctuation. Disturbance, however, can also be unusual destructive events

Table 1. Storage of Organic Matter and Nutrients in World Forest Ecosystems[a,b]

	Aboveground Vegetation						Forest Floor					
Ecosystem	Biomass	N	P	K	Ca	Mg	Mass	N	P	K	Ca	Mg
Boreal coniferous	51,300	116	16	44	258	26	113,700	617	115	109	360	140
Temperate												
Coniferous	307,300	479	68	340	480	65	74,881	681	60	70	206	53
Deciduous	151,900	442	35	224	557	57	21,625	377	25	53	205	28
Tropical	292,000	1,404	82	1,079	1,771	290	27,300	214	9	22	179	24

[a] Ref. 1.
[b] Data are in kilograms per hectare.

that occur rarely or perhaps only once, from which ecosystem recovery is only gradual or perhaps incomplete. The death of a canopy tree or several trees, forming a gap, can be considered a commonplace disturbance in most forests, allowing the release of suppressed saplings or germination of new seedlings. Windthrow, hurricane damage, insect outbreaks, and drought occur infrequently but repeatedly in eastern deciduous forests, and affect larger areas at any given time. In general, eastern moist forests tend to be dominated by such small-scale disturbances characterized by the formation of many small gaps. However, larger-scale hurricanes have periodically struck the region, causing larger-scale damage. Moving across the United States to the west, however, and to drier forests, disturbance regimes change. In the Allegheny Plateau of Pennsylvania, for example, sites dominated by dry, sandy soils with a pine overstory are more subject to larger-scale fires and blowdowns than are forest communities on upland moist sites. In the northern hardwood section of Minnesota and Canada, as well as in dry southwestern coniferous forests, widespread fires were prehistorically more important to community structure than local gap formation events.

It is important to remember that anthropogenic disturbances now commonly override and overshadow these natural disturbance regimes. Certainly the presence of humans has changed the nature of fire disturbance in the United States. Before human intervention, fire return intervals in dry, fire-adapted ecosystems were fairly frequent, and thus fuel loads rarely accumulated to a degree where extensive crown burns were common. Fire histories suggest that infrequent burns are generally more severe than frequent burns, and are generally confined to the forest floor. Certainly many extensive burns in the past have been due to natural causes and have been important determinants of community structure (3). With increasing human presence, and increasing human property in forested regions, fire suppression has become more common. Fire suppression has caused the decline of species that depend on fire for regeneration, and has also drastically increased the fuel load in dry forests. This practice has ensured that when such systems do finally burn, fire is much more widespread, intense, and devastating, as the forest fires of 1988 in Yellowstone National Park, and the more recent 1996 fires in California, Oregon, and the southwest, have demonstrated.

Other direct disturbances to the forest community have changed forest structure as well. Forest harvest has changed the face of the western U.S. landscape, as it has for the Amazonian and southeast Asian tropics (see Plate 4). Clearing for agriculture has affected almost all of the world's forest types, and effects include not just forest destruction, but forest fragmentation as well, and the clearing of forests globally is increasing. In the early 1980s the annual rate of forest clearing in Central America and northern South America ranged from 0.9 to 4.6%, and in some African countries rates are over 5%. Net global deforestation between 1980 and 1985 proceeded at about 1% per year, and if this rate were sustained, forest half-life would average 70 years (4). The effects of forest harvest on community structure and ecosystem function is discussed in a later section.

Finally, the introduction of exotic species by humans to a new environment has often profoundly altered species composition or ecosystem function (5). In Hawaii, for instance, invasion of young volcanic sites by an exotic nitrogen-fixing tree species has strongly increased the biological availability of nitrogen, and thus has altered the nature of ecosystem development after volcanic eruptions (6). Many exotic pest species have had a similar effect. The chestnut blight, *Endothia parasitica,* is a fungus that is parasitic on chestnut trees in Eurasia. When introduced to North America and to a chestnut species not accommodated to it, virtually all of the chestnuts (*Castanea dentata*), one of the most important species in the eastern deciduous forest, were killed. As commercial forest crop species are increasingly moved to new plantations in new habitats, it can be predicted that habitat expansion will have profound effects on native vegetation.

Forest Decline

In recent years ecologists have reported declines in the vigor and death of forest trees in many parts of the world. Such decline and dieback has been observed in industrialized areas as well as in areas with relatively low air or land pollution. Decline is usually characterized by a progressive deterioration in the health of older and larger trees as well as stand-level mass mortality. While the ultimate cause of death is often fungal, microbial, or insect disease, initial causes of stress that lead to an increased susceptibility to disease are not always completely understood. Vegetation and forest community composition on any given site is rarely static over long periods of time, and natural occurrences of decline and dieback are known from pollen records of eras unaffected by anthropogenic pollution. For example, there was a widespread decrease in the eastern hemlock in eastern North American forests about 4800 years ago, but not in other co-occurring species with similar habitat requirements; there was a similar decline of elm species in northwestern Europe about 5000 years ago. It is unlikely that a sudden climate change could account for these species-specific mortality events, yet disease agents for ancient decline occurrences are also unknown. Modern plant geographers have also noted sudden decreases in the abundance of specific forest tree species in the twentieth century, with the primary causes of decline unknown.

In the early 1970s, a new type of forest damage was observed in central Europe, particularly in West Germany. Inventories noted a silver fir (*Abies alba* Mill.) dieback in the Black Forest and severe damage to Norway spruce (*Picea abies* Karst.) and Scots pine (*Pinus sylvestris* L.) stands (7). By the early 1980s, large-scale mortality was seen in stands of spruce, beech, and oak in much of industrialized Europe, and by 1986, 54% of the total forest area in West Germany was believed to contain some trees with at least some foliage loss. However, only 19% were marked by distinct damages, defined as foliage losses greater than 25%. Within central Europe similar inventories were carried out, and by 1986 it was clear that Dutch and German forests showed the greatest amount of this new type of forest damage (7).

Different symptoms were described and included in the classification of the new forest damage, which was given the name of *Waldschaeden* (forest damage). As Norway spruce is the most common and most important economic tree species in central Europe, this species received a great deal of attention.

Decline symptoms were noted in deciduous trees in central Europe as well, particularly in beech. Crown thinning, branch dieback, growth reductions, and root damages were recorded.

Simultaneously, quite similar decline symptoms were being observed in high elevation red spruce stands of New England. A 1982 survey of red spruce at sites above 650 meters on Whiteface Mountain in New York, Mount Mansfield in Vermont, and Mount Washington in New Hampshire showed a relationship between elevation and severely declining and dead spruce (8). Reductions of live spruce density and basal area of 50% or more were common above 900 meters between the mid-1960s and early 1980s.

Many natural factors were thought to influence the spruce decline. Age and climate were both suspected; climate is generally more severe at the higher elevations, and thus, perhaps, these sites are more susceptible to variations in temperature and precipitation. Soils are also generally thinner and less fertile at high elevation, and thus ecologists hypothesized that winter damage might initiate and synchronize stress. Patches of high elevation spruce in New Hampshire were thought to have been killed by severe winter conditions in the mid-1960s, as very low temperatures for prolonged periods can cause drought stress and freezing. There was a widespread drought in the 1960s in the northeast, and drought has been shown to be an initiating factor in other tree decline events. It has been observed that a single drought year in the 1990s had a distinct effect on forests across much of the northeastern United States. Although spruce beetles might have contributed to the extent of damage, patterns of spruce beetle outbreaks and spruce decline did not match very well, and thus it was considered unlikely that the immediate cause of the spruce decline was due to a pest outbreak.

Have climate factors alone caused the extensive, and synchronous, tree declines seen at high elevations in Europe and the United States? Many ecologists suspect that air pollution (qv), acid rain (qv), or some combination of the two may be responsible. These factors are discussed below. It is also quite possible that different factors in different regions are responsible for variations in forest productivity; decade-scale climate variations may be responsible in some regions, while acid rain may account for forest damage in other areas, for example.

Air Pollution and Forest Ecosystems

The definition of air pollution is complex, and has changed and broadened in the last few decades in response to resource management concerns. Air pollution can be defined as the alteration of atmospheric chemistry by human activity that has direct or indirect consequences on human health, plant growth, or ecosystem function (9). Thus ozone produced from lightning is not considered a pollutant, although ozone produced from automobile emissions is. CO_2 is a gas that is increasing in the atmosphere due to human activity, and yet scientists are split as to whether it should be included as an air pollutant. Moderately elevated CO_2 appears to have a beneficial effect on plants, and yet global effects of this elevated CO_2 are of serious concern (see below).

Scientists have recognized for decades that certain trace chemicals have increased in the atmosphere due to human activities, and many of these can cause injury to vegetation. Input into an ecosystem can come via wet (direct precipitation or fog and cloud water) or dry (generally particulate or gaseous) deposition. Some of these pollutants act on a local scale (eg, many trace metals are deposited close to a specific source such as a smelter), whereas other have effects at the regional (eg, ozone, acidic precipitation) or global scale (Table 2). The scale at which an individual pollutant acts depends on its residence time in the atmosphere as well as the distribution of sources around the globe. Regions of the globe not immediately adjacent to anthropogenic sources of pollutants are not, however, immune from the effects of pollutants; the geographic extent of air pollution are significantly greater than ecologists once thought. Vast regions once described as pristine, such as remote parts of Alaska, may in fact be affected by long-range atmospheric transport. In addition, most terrestrial ecosystems are impacted by more than one anthropogenic pollutant at a time, and effects of air pollutants may have strong interactions with natural climatic or biotic stresses in each specific ecosystem (11).

Tropospheric ozone (O_3) is a pollutant of increasing concern to ecologists. In the last century concentrations have nearly tripled in Western Europe and North America and are projected to increase an additional 30–40% over the next several decades (10). Acute damage to trees directly caused by high levels of ozone is known from the San Bernadino Mountains of California, as well as from forests near Mexico City. Yet in many areas exposed to elevated ozone such as the San Bernadino Mountains, elevated depositions of H^+, N, and S is co-occurring, and thus all damage observed cannot be attributed to just one single pollutant. However, chamber and greenhouse experiments have demonstrated that ambient ozone concentrations in the San Bernadino Mountains show characteristic signs of ozone damage. Similarly, in the high elevation Appalachian Mountains where red spruce is declining, concentrations of many air pollutants are quite high, and higher than at lower elevations. The 24-hour average ozone concentrations are typically twice those at lower elevations and do not decline substantially at night, as they do at lower elevations, and direct damage to wall membranes of mesophyll cells containing chlorophyll by ozone has been demonstrated. These areas are also subjected to high loads of H^+, N, and S

Table 2. Some Attributes of Atmospheric Pollutants of Interest to Ecologists[a]

Category	Pollutant	Mean Residence Time	Distribution	Source[b]
Carbon	CH_4	Years	Globe	A, N
	VOCs	Days	Region	A, N
	CO_2	Decades	Globe	A, N
	CO	Days	Hemisphere	A, N
Nitrogen	NO	Days	Region	A,N
	NO_2	Days	Region	A, N
	HNO_3	Hours/Days	Region	A, N
	NH_3	Hours/Days	Region	A, N
	N_2O	Years	Globe	A, N
Sulfur	SO_2	Days	Region	A, N
Chlorine–halogen compounds				
Chlorofluorocarbons (CFCs)	CH_3CCl_3	Century	Globe	A
	CF_2Cl_2	Century	Globe	A
Halons	$CBrF_3$	Decades	Globe	A
Miscellaneous	O_3	Days	Region	A, N
	H_2O	Days	Region	N
	Hg^0	Years	Globe	A, N
	CH_3Br	Decades	Globe	A

[a] Ref. 10.
[b] A = anthropogenic and N = natural.

from wet and dry deposition, which can serve as interacting or dominant stressors to the vegetation in addition to the ozone exposure. However, both field fumigation and controlled environment studies have outlined the characteristic injury symptoms from ozone, and thus ozone damage in the field is generally easy to identify. Ozone directly inhibits photosynthesis, and seedling experiments have suggested that ambient concentrations of ozone are stunting the growth of most conifer and hardwood forests in the eastern United States. Experiments with elevated ozone have also noted enzymatic changes and disturbances to plant hormone balance. Early leaf fall, caused by early aging of leaves, is frequently observed. Significant damage to crops, both in terms of visible damage and reduction of photosynthesis and yield, has been reported throughout the United States and in Europe. A recent report suggested that ozone reacts with terpenes and isoprenes from vegetation, especially in the presence of water, to form peroxide, which itself is highly toxic to plants and may be responsible for vegetation decline in the presence of elevated ozone (12). In areas with chronic high ozone, there is evidence that selection for ozone-resistant clones of trees such as *Populus tremuloides* is occurring, and thus alters the genetic structure of the vegetative community (13).

In the case of the new type of forest damage observed since the 1970s throughout Europe and the United States, air pollutants such as ozone, sulfur dioxide, toxic heavy metals, excess nitrogen deposition, and acid rain were implicated, as much of the documented forest mortality was in areas potentially exposed to urban and industrial air pollution. Levels of ozone in West Germany were measured as high as 400 $\mu g/m^3$, 2–3 times as high as concentrations known to cause injury, and thus ozone has been implicated in the European forest decline. However, spatial patterns of decline and high ozone levels do not match perfectly, and decline symptoms do not always match known ozone injury symptoms; thus it is unlikely that ozone alone is responsible for all, or even most, of the new type of decline.

Sulfur dioxide (SO_2) is also known to be toxic to plants, and cases of dramatic SO_2 fumigation damage have been seen in the Ore Mountains of Czechoslovakia and near smelters such as the Sudbury Smelter in Canada. Because mosses and lichens are generally more sensitive to SO_2 pollution than are vascular plants, they are often used as sensitive indicators of SO_2 pollution in the near environment. Although elevated levels of SO_2 were initially implicated in the more widespread European forest decline, the major areas exhibiting forest dieback do not always correspond to patterns of elevated SO_2 in the atmosphere, decline symptoms do not necessarily match known symptoms of SO_2 damage, and lichens and other species sensitive to elevated SO_2 are not always affected in these areas. Due to the fairly rapid transformation of SO_2 to H_2SO_4 in the atmosphere, SO_2 effects tend to be at the landscape, rather than the regional or global level. However, SO_2 certainly can be a contributing agent in certain cases near specific point sources. Although such toxic gases may not be the primary cause of decline even when ecosystems are near sources of these pollutants, it is quite possible that an interaction among factors, or cumulative stresses, might be allowing trees to succumb to disease.

Other toxic gases are often of local concern. For example, an aluminum ore reduction plant's fluoride emissions killed a large stand of trees in Spokane, Washington, in the early 1950s, and significant foliar damage was observed in a 80 km^2 (50 mi^2) area.

In 1989, the U.S. Environmental Protection Agency and the National Research Council, along with several other agencies, organized a committee to examine the biological markers of air pollution damage in trees (14). They had six main conclusions:

1. No pollutant-specific marker for identifying the effects of pollution on trees has been identified, as many other stresses can produce symptoms in plants that mimic or conceal damage caused by air pollution.
2. There are genetic differences within and between plant species in their genetic capacities to respond to air pollutants, and thus exhibit different sensitivities to air pollutants.

3. Most biological markers of responses of trees to air pollution measure changes in aboveground plant parts; markers are needed to measure effects on roots.
4. A better understanding of natural processes, or stresses that affect forest trees is needed in order to be able to detect and measure air pollutant disturbances.
5. Markers of air pollution must be analyzed in conjunction with spatial and temporal data on patterns of air pollutant distributions.
6. The most useful analyses of the effects of air pollutants on forests combine surveys with controlled-exposure studies of cause-and-effect relationships (see AIR POLLUTION).

Acid Rain and Nitrogen Saturation

Acid rain has also been implicated in causing forest decline, both through direct effects on foliage and through indirect effects such as elevated Al levels in soils, which can inhibit Ca uptake by roots and lower mycorrhizal production. Acidity in high elevation mountain fog and cloud water is known to be greatly elevated, and it is in such high elevation forests in the eastern United States that decline symptoms are greatest. As acid mists deposited on needle surfaces dry, pH levels are further reduced, and foliar leaching, damage to stomatal wax plugs, and disruption of cellular structure can occur. In addition, the two species that initially showed decline symptoms, Norway spruce and red spruce, have been shown to be highly sensitive to elevated levels of Al in the root medium in culture experiments.

Recently, a hypothesis of *nitrogen saturation* of forest ecosystems in highly industrialized northern temperate regions has been proposed. Atmospheric deposition of nitrogen compounds is greatly increased in forests near industrial or agricultural centers from as low as 2 kg N/ha/yr to more than 40 kg N/ha/yr in high elevation forests of New England (15,16), with some regions in Europe, particularly in the Netherlands, having maximum deposition levels twice that. Many high elevation forests are subject to high rates of atmospheric nitrogen inputs from wetfall, particulate dryfall, and gas and fog inputs. Sources of this increased nitrogen include emission of NO_x from cars and power plants as well as N_2O and NH_4 losses from fertilized agricultural fields, and volatilized NH_4 from animal manures. Because most of these forests are historically limited by rates of N input and intrasystem turnover, added N deposition should initially have a fertilization effect, increasing forest productivity, and this has indeed been observed. However, preliminary measurements suggest that many forests in Europe and New England have already or may soon reach their limits for biotic uptake and storage of nitrogen, and several negative consequences have been hypothesized to occur. Nitrogen saturation is defined as inputs of N that exceed N requirements of plant and microbial need for maximum production, until growth is limited by some other nutrient. Excess nitrogen fertilization could cause stand-level injury symptoms through several mechanisms:

1. Increased N leads to increased plant growth, primarily of the crown rather than of roots
2. Increased growth leads to an increased demand for water or other nutrient elements as well, such as Ca or Mg, that are not part of the atmospheric fertilization
3. These other elements become limiting to photosynthesis and new tissue growth, and are prematurely withdrawn from old needles to supply new needles, thus causing chlorosis and premature needle-drop
4. This element deficiency is exacerbated by excess NO_3 deposition to the forest floor, because as excess NO_3 is leached from the soil, cations are removed due to exchange processes with the co-deposited H^+

Increased leaching of Ca has been observed in New England watersheds that have high HNO_3 deposition rates or high acidic inputs in general. Mg deficiencies have been shown in several West German forests that show signs of decline, particularly in forests on thin soils; trace nutrient fertilization has led to a reversal of many of the decline symptoms. In addition, excess nitrogen fertilization has been shown to lead to delayed frost hardiness in conifers, as well as a decrease in allocation to defensive chemistry against insects and pathogens, although it has not been shown conclusively that high atmospheric deposition of N has led to this same result. At the extreme, vegetative changes have been predicted to occur. With forest damage and thinning of the forest canopy, high N deposition, low soil pH, and often high Al levels may favor the growth of high nitrogen-requiring grass species.

To date, evidence of widespread stand damage due specifically to excess nitrogen deposition is lacking. What is known is that excess nitrogen deposition may lead to increased leaching of nitrate, and nitrate concentrations have increased in many surface waters draining impacted forests, with varying negative effects both on the ecology of the receiving aquatic ecosystems and on drinking water quality (17–20). Although Mg deficiency has definitely been shown to be a problem in high elevation forests with thin, infertile soils, Mg deficiency has been seen in areas not impacted by excess nitrogen deposition, such as the Kaingaroa Forest of New Zealand, where plantation forests have been developed on base-poor soils similar to the podzols derived from granite in the Black Forest of Germany. Indeed, Mg deficiencies in Norway spruce is only found on acidic base-poor soils that have developed from Mg-poor parent materials (21).

In summary, the relative roles of natural variations in vegetation growth rates and structure due to climatic variations versus anthropogenic pollution effects have not been determined, although they continue to be hotly debated. Although declines of red spruce in the Adirondacks occurred over a century ago near where the modern decline is occurring, this alone is not evidence that the same factors caused both events. Nor is there evidence that all modern-day forest species declines are occurring for the same reason, as species' tolerances to air pollution and toxic element stresses vary considerably. Several researchers have suggested that a mix of elevated ozone and cloud acidity are causing spruce and fir decline through direct impairment of photosynthetic tissues, nutrient leaching, and fine-root growth impairment, whereas ozone alone is causing decline symptoms in beech. In any case, it will be critical to understand the role, if any, of anthropogenic pollution in causing dieback symptoms in world forests (see ACID RAIN).

Deforestation

The greatest loss of forest land from known human disturbance is through the clearing of tropical forests. Although the tropics contain over 40% of the forests of the world and are estimated to contain 64% of the earth's potentially cultivatable land, tropical forest soils are generally characterized as being of marginal quality for repeated cropping, due to low fertility or by being too steep or shallow. However, as population increases in the rural tropics lead to increased demand for fuelwood and new arable land, deforestation is accelerating. Estimates of the rate of deforestation in the tropics vary widely, but current estimates suggest that in 1989 almost eight million hectares were logged or burned in tropical America, primarily in the Amazon, and 1.6 and 4.4 hectares of closed forest were cleared in tropical Africa and Asia, respectively. If open forest types are included in this estimate, nearly 22 million hectares were cleared worldwide, representing nearly a doubling in rate since 1980. Since 1950, forest loss in Africa is estimated at over 200 million hectares; most of the Atlantic coastal forest of Brazil has disappeared, with the exception of small forest reserves.

There are many possible global consequences of this large-scale deforestation that is taking place in the humid and sub-humid tropics. The loss of species is perhaps the most obvious consequence of forest destruction. The magnitude of the effects of tropical deforestation on global species diversity is still being debated, but losses stem not only from direct removal of habitat but also from reduction of landscape heterogeneity after forest clearing. Of the estimated quarter million species of plants inhabiting the earth, roughly two-thirds are found in tropical regions, many of which have not yet been described. The diversity of insect species is equally as great in tropical regions, and yet they are even less well known and described. Of concern to human societies is not only the loss of the aesthetic benefits of a diverse flora and fauna, but also plants potentially useful in medicine, as new fibers, chemicals, or fuels. Less well known is the effect of loss of genetic diversity for ecosystem function.

Other ecological effects include reductions in site fertility, losses of soil through erosion, changes in microclimate and, potentially, regional climate, alterations of the hydrologic cycle, and effects on greenhouse gases. Site fertility has been shown to decrease with short harvest rotation lengths, not only due to direct removal of nutrients in forest products, but also from nutrient runoff in waters draining into forest soils and in the increased sediment erosion from disturbed watersheds. Because vast areas are often felled or burned at any given time, soil stability is greatly reduced. Once the forest canopy is removed and the soil surface is disturbed by felling equipment, raindrop impact on the soil surface is greater, infiltration rate is reduced, and runoff velocity is greater, leading to massive erosion of soils, degrading soil fertility and reducing water quality in local rivers and streams. Often the building of roads through harvested land is the primary cause of increased sediment loss. Such erosion further increases the need to clear additional forest land; it is estimated that almost half of the acreage cleared annually in tropical forests can be attributed to the need to replace degraded agricultural soils.

Another significant impact of deforestation or forest conversion is on local, regional, and potentially global climate. Within the Amazon basin, which contains about half of the earth's tropical forests, it is estimated that 50% of the rainfall is derived from local evaporation, including transpiration from vegetation. When simulations of deforestation and forest conversion to pasture effects have been run using three-dimensional atmospheric global climate models in this area, results predicted a weakened hydrologic cycle, with lower precipitation and evaporation and an increase in surface temperature. Such a disturbance of the regional climate could lead to an irreversible loss of the tropical forest ecosystem. Perhaps the strongest effect on the global climate would be through the increase in albedo from forest conversion. Locally, increases in soil and near-ground air temperature and decreases in relative humidity have been observed in pastures compared to intact forests. On a regional scale, increases in streamflow and total water yield have been observed in areas with high rates of deforestation, again demonstrating strong changes in regional water balance.

The global climate is also potentially affected by deforestation through changes in emissions or uptake of chemically or radiatively active greenhouse gases such as carbon dioxide, water vapor, methane, and nitrous oxide. When forests are cleared or burned, CO_2 is released to the air from decaying or burned wood, leaf litter and the forest floor, and oxidizing soil organic matter. Current estimates suggest that deforestation is now contributing 26–33% of the worldwide annual emissions of CO_2, with the rest being largely from the combustion of fossil fuels. Similarly, CH_4 is released to the atmosphere both as a direct effect of deforestation and fuelwood burning and as an indirect effect of forest conversion, from such activities as cattle ranching and the annual burning of pastures and grasslands. Although total emissions of CH_4 are several orders of magnitude lower than emissions of CO_2, the radiative effects of CH_4 are 25 times greater than that of CO_2, and thus emissions could be a significant source of global warming. Although absolute values of increases in such trace gases as N_2O from forest conversion are less well known, N_2O has a radiative effect 250 times that of CO_2, and preliminary measurements show that new pastures and disturbed soils may contribute significantly to global increases in N_2O concentrations. When differences in radiative forcing of these three gases are taken into account and estimated releases are tallied, tropical deforestation could account for up to 25% of total greenhouse gas emissions globally.

FOREST VEGETATION AND GLOBAL CLIMATE CHANGE

With the buildup of greenhouse gases such as CO_2, the global climate is predicted to change by as much as 1–4°C by the middle of the twenty-first century. Although the earth's climate has alternately cooled and warmed between glacials and interglacials in the past, such a rapid change is unprecedented. It is not known if the flora and fauna of the earth's ecosystems can track this predicted climate change (22–24). Pollen analysis shows that past climate changes caused species to migrate at rates of approximately 10–200 km per century under unusual circumstances, and yet models of predicted climate change suggest that species will be required to move at a rate of up to 500 km per century to maintain life in their same approximate thermal environment (2). This entails very high dispersal rates, rates that are too high for many if not most plant species, if dispersal must be made latitudinally to compensate for the temperature increase. For forested communities in a mountainous region an altitudinal shift can compensate for climatic shifts, yet this leads to a concomitant reduction in total area of habitat, and many species could become extinct. Thus future forest communities may be composed of quite different species assemblages, and slow-dispersing species assemblages and slow-dispersing species with narrow physiological tolerances or large-area requirements may face extinction.

Increases in global temperatures will affect not only the species composition of forests, but ecosystem function as well. Increased mean temperatures can affect the balance between production and decomposition, as decomposition is generally more sensitive to and limited by low temperatures than is NPP. Thus boreal forests, currently net carbon sinks that contain about 13% of the earth's soil carbon pool, could have a net release of CO_2 to the atmosphere and have a positive feedback effect on global temperatures. Carbon stored in temperate forest soils could also face a slow release, although this will depend in part on how natural disturbance regimes are affected by climatic change. If the frequency of extreme disturbance events such as hurricane and drought increase as has been suggested, then forests will be constantly in a cycle of growth and tree loss, decreasing the capacity of the ecosystem to store carbon in reduced organic matter. Changes in disturbance frequently will also affect the rate at which vegetation can respond to climate change, with increased disturbance frequency overcoming lags in vegetation response. The nature of disturbance in specific regions should change with climate change as well; fire may become a more significant factor in many regions where it had not formerly been a frequent disturbance event.

The increase in atmospheric CO_2 can directly affect vegetation as well. Because plants use CO_2 as a building block for growth, there are predictions that a CO_2 increase will serve as fertilizer for the vegetation of the earth, and forest productivity will increase (25). Not all vegetation if any is predicted to respond, however, as other factors such as water and nutrients may overwhelmingly limit plant production. The physiology of certain species may also change in the face of increased CO_2; water-use efficiency, or the ratio of carbon gain in photosynthesis to water loss in transpiration, is predicted to increase, thus perhaps overcoming water limitations in certain areas. Plant allocation to reproduction versus growth may change, thus affecting dispersal ability of individual species. Because many climate factors are expected to change with increased concentrations of greenhouse gases in the atmosphere, such as temperature, CO_2 concentrations, and rainfall regime, it is difficult to make predictive models of direct changes in tree species' physiology with climate change.

Perhaps the greatest difficulty in predicting vegetation response to climate change results from vegetation feedbacks to the global system. Individual species and vegetation type and density can affect such system properties as albedo, surface roughness, snow cover, and ground frost. The models that predict future forest ecosystem composition and function must incorporate not only direct effects of each climate factor on species physiology and dispersal, but also feedbacks of altered forest ecosystems to the global climate system.

BIBLIOGRAPHY

1. R.H. Waring and W.H. Schlesinger, Forest Ecosystems: Concepts and Management, Academic Press, Inc., Orlando, Fla., 1985.
2. M.B. Davis "Invasions of Forest Communities During the Holocene: Beech and Hemlock in the Great Lakes Region," in A.J. Gray, M.J. Crawley, and P.J. Edwards, eds., *Colonization, Succession and Stability,* Blackwell, Oxford, U.K., 1987, pp. 373–393.
3. S.F. Arno, *The Historical Role of Fire on the Bitterroot National Forest,* USDA Forest Service Paper INT INT-187, 1976.
4. B. Freedman, *Environmental Ecology,* Academic Press, Inc., New York, 1995.
5. J. Diamond, "A-Bombs Against Amphibians," *Nature* **383,** 386–387 (1996).
6. P.M. Vitousek, L.R. Walker, L.D. Whiteaker, D. Mueller-Dombois, and P.A. Matson, "Biological Invasion by *Myrica faya* Alters Ecosystem Development in Hawaii," *Science* **238,** 802–804 (1987).
7. R.F. Huettl, "New types of Forest Damages in Central Europe," in J.J. MacKenzie and M.T. El-Ashry, eds., *Air Pollution's Toll on Forests and Crops,* Yale University Press, New Haven, Conn., 1989, pp. 22–74.
8. A.H. Johnson and T.G. Siccama, "Decline of Red Spruce in the High-Elevation Forests of the Northern United States," in Ref. 7, pp. 191–234.
9. W.E. Winner, "Mechanistic Analysis of Plant Responses to Air Pollution," *Ecolog. Appl.* **4:**651–661 (1994).
10. G.E. Taylor, Jr., D.W. Johnson, and C.P. Anderson, "Air Pollution and Forest Ecosystems: A Regional to Global Perspective," *Ecolog. Appl.* **4,** 662–689 (1994).
11. S.B. McLaughlin and D.J. Downing, "Interactive Effects of Ambient Ozone and Climate Measured on Growth of Mature Forest Trees," *Nature* **374,** 252–254 (1995).
12. K.H. Becker, K.J. Brockmann, and J. Bechara. 1990. Production of hydrogen peroxide in forest air by reaction of ozone with terpenes. Nature **346,** 256–258.
13. P. Berrang, D.F. Karnosky, and J.P. Bennett. "Natural Selection for Ozone Tolerance in *Populus tremuloides:* Field Verification," *Can. J. Forest Res.* **19,** 519–522 (1989).
14. Committee on Biological Markers of Air-Pollution Damage in Forests, *Biological Markers of Air-Pollution Damage in Forests,* National Academy Press, Washington, D.C., 1989.
15. J.D. Aber, K.J. Nadelhoffer, P. Steudler, and J.M. Melillo, "Nitrogen Saturation in Northern Forest Ecosystems," *BioScience* **39,** 378–386 (1989).
16. J.D. Aber, A. Magill, S.G. McNulty, R.D. Boone, K.J. Nadelhoffer, M. Downs, and R. Hallett, "Forest Biogeochemistry and Primary Production Altered by Nitrogen Saturation," *Water Air Soil Pollut.* **85,** 1665–1670 (1995).
17. S.C. Nodvin, H. Van Miegroet, S.E. Lindberg, N.S. Nicholas, and D.W. Johnson, "Acidic Deposition, Ecosystem Processes, and Nitrogen Saturation in a High Elevation Southern Appalachian Watershed," *Water Air Soil Pollut.* **85,** 1647–1652 (1995).
18. M.E. Fenn, M.A. Poth, and D.W. Johnson, "Evidence for Nitrogen Saturation in the San Bernadino Mountains in Southern California," *Forest Ecol. Manag.* **82,** 211–230 (1996).
19. M.W. Williams, J.S. Baron, N. Caine, R. Sommerfeld, and R.J. Sanford, "Nitrogen Saturation in the Rocky Mountains," *Environ. Sci. Technol.* **30,** 640–646 (1996).
20. A. Tietema and co-workers, "NITREX: The Timing of Response of Coniferous Forest Ecosystems to Experimentally-Changed Nitrogen Deposition," *Water Air Soil Pollut.* **85,** 1623–1628 (1995).
21. R.F. Huettl, "Mg Deficiency—A "New" Phenomenon in Declining Forests," in R.F. Huettl and D. Mueller-Dombois, eds., *Forest Decline in the Atlantic and Pacific Region,* Springer-Verlag, Berlin, 1993, pp. 97–114.
22. C. Parmesan, "Climate and Species' Range," *Nature* **382,** 765–766 (1996).

23. J. Harte and R. Shaw, "Shifting Dominance within a Montane Vegetation Community: Results of a Climate Warming Experiment," *Science* **267,** 876–880 (1995).
24. J.P. Barry, C.H. Baxter, R.D. Sagarin, and S.E. Gilman, "Climate-Related, Long-Term Faunal Changes in a California Rocky Intertidal Community," *Science* **267,** 672–612 (1995).
25. F.A. Bazzaz, "The Response of Natural Ecosystems to the Rising Global CO_2 Levels," *Ann. Rev. Ecol. System.* **21,** 167–196 (1990).

FORMALDEHYDE IN THE ENVIRONMENT

WING-HONG CHAN
FRANK S. C. LEE
Hong Kong Baptist University
Kowloon, Hong Kong

Properties

Formaldehyde is a flammable, colorless, and pungent gas. Formaldehyde as an industrial product is produced solely from methanol. In the presence of suitable catalysts (ie, metal oxides), methanol undergoes air oxidation or simple dehydrogenation:

Oxidation $$2\,CH_3OH + O_2 \xrightarrow[\Delta]{\text{catalyst}} HCHO + 2H_2O$$

Dehydrogenation $$CH_3OH \xrightarrow[\Delta]{\text{catalyst}} HCHO + H_2$$

Table 1 gives a summary of its physical properties. Formaldehyde is stable only in water solution. Methanol (3–15%) may be presented as a stabilizer. Aqueous solutions (30–50%) that contain formaldehyde as a hydrate or as a mixture of oligomeric oximethyleneglycols are commercially available. Formaldehyde in its pure gaseous form, however, is unstable and readily trimerizes to trioxane **(1)** or polymerizes to paraformaldehyde **(2)**.

$$HO\!\left(CH_2{-}O\right)_n\!H$$

(1) **(2,** *n* = 8–50)

Sources in Environment

Formaldehyde is the most abundant gaseous carbonyl compound in the ambient atmosphere. It is formed from primary emissions of combustion sources, including natural gas, kerosene, cigarette and automobile exhaust. Tropospheric oxidation of gaseous methane and other volatile organic compounds via free radical reaction is responsible for the generation of airborne formaldehyde. Because formaldehyde is a reactive gas, it has a relatively short half-life in the atmosphere. After an average lifetime of several days, most formaldehyde molecules decompose photolytically by sunlight, a process that contributes to the photochemical formation of ozone and other oxidants (1).

Formaldehyde is one of the widely used chemicals in the plastic industry. In the United States alone, the annual production of formaldehyde in 1995 was 3.62 million metric tons (2). Most of it is used to manufacture urea–, melamine,– and phenol–formaldehyde resins which, in turn, are the feed raw materials that produce adhesive, bonding and laminating agents, foam insulation, fabrics, and coatings. Formaldehyde-containing resins are used also to treat many consumer paper products in order to improve their strength when wet. Furthermore, formaldehyde has also found wide industrial usage as a fungicide, bactericidal agent, and disinfectant. As a result, formaldehyde is released from a variety of household products (Table 2). The primary residential sources are hardwood plywood, medium-density fiberboard, particleboard, and other pressed-wood products. All of these pressed-wood products are manufactured by mixing wood pieces and urea–formaldehyde resins with adhesives. Thus in the United States two subpopulations have been identified as having a particularly high potential for formaldehyde exposure: 2.2 million residents of mobile homes containing particle board and plywood have an average exposure of 0.4 ppm formaldehyde, and 1.7 million people living in conventional homes insulated with urea–formaldehyde foam have a potential average exposure of 0.12 ppm.

Indoor vs Ambient Levels

Formaldehyde is now recognized as one of the most widely characterized indoor and outdoor air pollutants. Background concentrations of formaldehyde range from 0.1 ppbv in oceanic regions to 1 ppbv in uncontaminated regions of the land mass as a result of decomposition of organic matter. Although its concentration in urban outdoor air is normally very small (~0.01 ppm) (4), the level of indoor formaldehyde is often of a greater magnitude, averaging about 0.1 ppm because of various emission sources existing in indoor atmosphere. In certain cases, formaldehyde concentration in indoor air may exceed 1 ppm. Table 3 summarizes some formaldehyde measurements which have been made in both outdoor and indoor environments.

Health Effects

Human exposure to formaldehyde principally is through inhalation and skin absorption. Formaldehyde is a known irritant and sensitizer with a detection threshold in humans at approximately 0.8 ppm and the irritating effects have been shown to increase with concentration and length of exposure. Irritant effects have been reported in concentrations as low as 0.5 ppm. A concentration of 2–5 ppm causes eye, nose, and throat irritation in most people. For most individuals, concentrations of 10 ppm is tolerated, but with difficulty. At much higher levels (10–20 ppm) of formaldehyde, which may be encountered in occupational settings, breathing becomes difficult. Cutaneous sensitization and allergic contact dermatitis, on the other hand, have been well documented in individuals chronically exposed to formaldehyde-containing materials. However, up to 2% solution of formaldehyde does not produce acute skin-irritant effects in humans.

Experimentally, formaldehyde has been shown to produce methylene cross-links between methylol groups by a condensation reaction between nucleotides of DNA in a single-strand configuration. Based on sufficient animal and limited human evidence, formaldehyde was classified as a "probable human carcinogen" by the U.S. EPA in 1987 (10). Due to the toxic and possible carcinogenic effects of formaldehyde, the U.S. Occupational Safety and Health Administration (OSHA) has established limits for the manufacturing sector to observe (*1*)

Table 1. Physical Properties of Formaldehyde

Property	Value
Molecular formula	Formaldehyde
Relative molecular mass, g/mol	30.021
Boiling point, °C	−21
Melting point, °C	−92
Density, at −20°C	0.8153
Vapor density	1.075
Vapor pressure, at −88°C, kPa[a]	1.33
Solubility	Very soluble in water; soluble in alcohol and other polar solvents

[a] To convert kPa to mm Hg, multiply by 7.5.

Table 2. Potential Sources of Indoor Formaldehyde[a]

Pressed-wood products	Hardwood plywood, particle board, medium-density fiber-board (MDF), decorative paneling
Insulation	Urea–formaldehyde foam insulation (UFFI), fiber glass made with formaldehyde binders
Combustion sources	Natural gas, kerosene, tobacco, automobile exhaust
Paper products	Grocery bags, waxed paper, facial tissues, paper towels, disposable sanitary products
Stiffeners, wrinkle resisters, and water repellents	Floor coverings (rugs, linoleum, varnishes, plastics), carpet backings, adhesive binders, fire retardants, permanent press textiles
Other sources	Plastics, cosmetics, deodorants, shampoos, disinfectants, starch-based glues, adhesives, laminates, paints, fabric dyes, inks, fertilizers, fungicides

[a] Ref. 3.

Table 3. Measurements of Formaldehyde Concentrations at Different Locations

Location	Data Point, (n)	Concentration, ppm			References
		Minimum	Maximum	Mean	
Conventional homes with chipboard	23	0.07	1.82	0.51	5
Mobile homes	65	<0.10	3.68	0.10	5
Office buildings	10	0.008	0.053	0.025	6
Urban air at Columbus, Ohio	24	0.002	0.004	0.003	7[a]
Forest atmosphere at Boulder, Colorado	8	0.0004	0.0010	0.0008	8[b]
Exhaust gas from automobile	4	3.40	10.90	6.53	9

[a] One-day variation.
[b] 12-h variation.

an 8-h time-weighted-average (TWA) action level of 0.5 ppm; and (*2*) a permissible exposure limit of 0.75 ppm, (*3*) a 2 ppm, 15 min short-term exposure limit (STEL). Similar threshold-limit values for formaldehyde have been established by the occupational safety authority of many other countries (11).

In order to define the level of human exposure to formaldehyde in different environments, reliable and sensitive means for its quantification are required. This article intends to give a full account on the scope and applications of all available commonly used methods for the determination of formaldehyde in various working environments and in ambient indoor and outdoor atmospheres.

SAMPLING METHODS

Basic Considerations

In environmental analyses, samples are collected and analyzed in order to meet specific objectives. Acquiring representative samples is the first and most critical step in the overall effort toward the achieving of these objectives. The development of a sampling plan requires the consideration of many factors including the objectives of the investigation, the nature of the samples and the surrounding environment, and the techniques of sampling themselves. The three basic factors to consider in devising a sampling plan are program objectives, data quality objectives, and analytical methods chosen.

The first factor to consider is the program objective. It defines the types of samples that are to be collected and the information intended to obtain, eg, an exploratory or screening type of study to uncover potential problems, or a systematic monitoring that aims at air quality assessment. A selected target sampling of a few samples may be sufficient in the former case. However, a comprehensive set of samples with adequate special and temporal resolutions would be needed in the latter.

Data quality objectives (DQOs) are statements that specify the tolerable uncertainties of the data, and thus the inherent confidence of the conclusion derived from the data. The DQOs should be set along with the overall program objectives because they help navigate the program in a more quantitative course toward the program goal. The quality of data should be judged on the basis of their uncertainties in relation to their intended end use. For regulatory-oriented monitoring programs, the settings of DQOs are a must because the end use of the data is decision making in the enforcement process. For exploratory studies, the end use of the data is to derive scientific conclusions. Here the DQOs are also desirable in order to avoid misinterpretation or overextrapolation of the data produced.

The analytical method selected should be considered next. Sampling and analyses are interrelated. The sensitivity and detectability of the analytical method determine the optimum size of the sample to be collected. Conversely, the range of contaminant concentrations in the samples dictates the sensitivity of the analytical methods needed for its detection. Furthermore, the nature of the analytical procedure often places special requirements on the sampling and sample preservation techniques to be used.

Environmental Considerations

Source Measurement. The objective of source measurement is to quantify the concentration of formaldehyde in the emission stream from a particular source. Such information is needed to quantify the source emission rate (source strength) for risk assessment. Environmental formaldehyde is either originated from direct-source emissions, or produced indirectly from atmospheric reactions (see below). Source measurement concerns only the direct emissions. The two major formaldehyde emission sources are combustion processes and formaldehyde-containing polymeric or fabric materials.

Combustion Sources. The samplings of combustion-exhaust emissions from stationary or mobile sources have been documented extensively (12). In exhaust-emission sampling, the sampling point and the number of traverse points taken during the sampling process affect the data quality. Because of the large variation in the concentration gradient, a screening test is often needed prior to any systematic sampling.

Formaldehyde is a volatile organic compound that is highly soluble in water. Its concentration at any given point is dependent on the temperature and relative humidity. Because of the high temperature and humidity typically found in combustion-exhaust streams, proper conditioning of the stream before sample collection is often required in quantitative-monitoring work.

Generally, the formaldehyde concentrations in the combustion-exhaust streams are sufficiently high that during the detection process analytical sensitivity is not a major concern. Instead, careful attention should be given to interferences or possible reaction losses of formaldehyde during sampling. The compositions of these emission streams are generally very complex, and are rich in gaseous oxidants such as ozone, NO_x and SO_x, etc. Under the high temperature conditions typically found in these streams, formaldehyde loss may occur due to oxidation reactions between formaldehyde and the oxidants. Another problem is the high moisture content in the stream. Since formaldehyde is highly soluble in water, it has the tendency to be absorbed on the moist surface of the sampling device or along transfer lines. The effects of these variables on sample recovery should be carefully checked when standard sampling protocols are adapted.

For screening purposes, sometimes direct sampling of the hot exhaust from combustion sources can be used. The sampling probe is inserted directly into the exhaust to collect samples in glass bottles or Tedlar bags for subsequent analysis (9). The most common problems in these types of sampling, besides oxidation loss, is loss due to possible wall reactions in the collection media. The stability of formaldehyde in these containers is usually in the order of hours, and therefore, analysis should be carried out immediately after sampling.

Industrial Materials. Direct emissions of formaldehyde from construction materials and household synthetic polymeric products are another major contributing source to atmospheric formaldehyde, especially in indoor environments. The emissions of formaldehyde from these sources are time-dependent. In general, there is a significant decrease of formaldehyde emission with time. For new and fresh material, emissions are mostly due to the release of residual formaldehyde present in the resin. As the material ages, continuing formaldehyde release includes, primarily, emissions from resin hydrolysis. Thus the age of the materials, the surface area exposed to the environment, and the general conditions of the surrounding air are all factors that need to be recorded in sample collection.

The sampling of formaldehyde from contaminated materials can be carried out by a simple distillation method in which the free formaldehyde and bound formaldehyde can be collected upon mild heating. For more systematic emissions measurement, the closed-chamber headspace sampler is applicable (12). The sampling enclosure is placed on the emission surface and the formaldehyde that has built up in the headspace of the sealed chamber can then be collected for analysis. The environment in the chamber can be varied to simulate real-world conditions. Both static grab sampling or dynamic purging techniques can be used for vapor sampling.

Workplace Monitoring The objective of workplace monitoring is to provide data for assessing the health risk associated with acute or chronic formaldehyde exposure by the workers. The temporal resolution of the method required to meet program objectives should first be considered. For chronic exposure, the objective of sampling is to detect formaldehyde at lower concentration levels over an average time period, such as 24 hours or a year. This process can be handled by regular integrated sampling followed by off-site lab analysis. Acute exposure, on the other hand, it is necessary to evaluate the risk to an individual upon one or many short-term exposures. Here a real-time continuous monitor is usually the method of choice, since action levels are based on short-term average concentrations such as 15 minutes, one hour, or two hours.

Acute exposure in a workplace with a large spacial variation of formaldehyde concentrations should also be considered. During working hours, a person may move around to several different locations that have different pollutant concentrations. For personal protection, a personal monitor is needed. The basic requirements for a personal monitor are for it to be compact, portable, and lightweight. The common approach is to use passive samplers in which the pollutant in the air is collected by the sampler through diffusion, rather than through the use of air pumps. An active surface in a small cartridge is used to adsorb the pollutant in passive samplers. After sampling, the cartridge is then analyzed at an off-site laboratory.

The sensitivity and precision of the diffusion-based passive sampling method is usually poorer than those of active sampling, mainly because of the smaller volume of air sampled. Validating the passive sampler against an active sampler is necessary in order to confirm the monitoring data.

Atmospheric Measurement. The measurement of formaldehyde and other aldehydes in the atmosphere has received increasing attention because of influence on photochemical smog and air quality. The formaldehyde is usually the most abundant, accounting for 70–80% of the total aldehyde. The main emission sources in urban air are motor vehicle exhaust and other combustion processes, such as incinerators. In addition, contributions from photochemical reactions involving the nonmethane hydrocarbon precursors are also significant in urban environments. In rural areas, localized participation of biogenic hydrocarbons such as isoprene from plant emissions may also lead to the formation of formaldehyde (13).

In ambient air, formaldehyde has a typical diurnal smog cycle that peaks in the afternoon. The peak concentration of formaldehyde ranges from 10 to 100 ppbv. The daytime levels fluctuate because of changes in meteorological conditions, intensity of photochemical smog processes, and the transport of formaldehyde from other urban pollution sources. Generally, the direct emission sources contribute much more than photochemical reactions to the total formaldehyde concentrations in ambient air. Thus, the formaldehyde level at a given site is affected mostly by the transport from off-site sources, or the release from on-site ones. The locations of these sources must be accounted for in the designing of sampling strategies.

Because of their presence in low concentrations, the measurement of formaldehyde in ambient outdoor air requires a sensitive method for their detection. A large air sample with hours to days of sampling time is required. The large fluctuation in formaldehyde concentration over a short period of time, usually occurring in an ambient atmosphere, makes atmospheric sampling a challenging task. A statistically designed large sample set, along with long-term monitoring, is needed to obtain the background or "average" formaldehyde concentrations in ambient air.

One of the problems with integrated long-term sampling is the sampling artifact, for example, the occurrence of interfering reactions or formation of interfering products *in situ,* in the sampler during sampling. The common cause of the sampling artifact is oxidation reactions that are initiated by trace atmospheric oxidants such as ozone and NO_x. Special considerations should be given while sampling air with high oxidant concentrations.

Indoor Air Measurement Formaldehyde is one of the most important indoor air pollutants because of the extensive use of formaldehyde-containing constructions and furnishing materials in modern buildings. The outgasing or breakdown of these man-made materials such as insulation, sealane and adhesives, paneling, furniture, carpeting and vinyl wall covering release formaldehyde continuously. Another significant source is from combustion processes such as cooking and smoking. The accumulation of formaldehyde indoors affects air quality, which may impact on the occupants' health.

The accumulation of formaldehyde in indoor environments is influenced by many factors. The most significant, besides the emission sources themselves, is the air ventilation and exchanges rate. In modern buildings, the ventilation air exchange is either done by natural diffusion or by air-conditioning systems. The concentration of formaldehyde varies depending on sampling locations relative to the positions of the output and return of the central air system, the entrances to the building, and windows in the room. Locations near the windows or entrances reflect primarily the outdoor concentration. Locations near air output of the air-conditioning system reflect formaldehyde concentration in the air-conditioning devices, whereas those near the air returns reflect most closely the average concentration inside the building.

In order to truly reflect the level of indoor air, the position of sampling must be considered together with the emission sources and ventilation points. For air quality assessment, a statistical design is needed to obtain the average measurement and the distribution of formaldehyde inside the building. Information regarding the historical and current uses of formaldehyde-containing products such as solvents, cleaning reagents, and pesticide applications should be documented. The smoking habits of the occupants should also be recorded.

For exposure study, an 8-h TWA is commonly applied; short-term 15-min exposure limits (STEL) are also frequently used in industrial hygiene studies. The sampling time is a factor of concern, since long-term integrated sampling would average the shot peaks and thus underestimate the health hazard.

Sampling Techniques

Because of its reactive nature, special techniques are required for the sampling and analysis of formaldehyde. The standard containers used for the sampling of more inert volatile organic compounds (VOCs) include solid sorbent traps, canisters, or air bags. The direct application of these containers to formaldehyde, however, is unsuitable because of the reactions of formaldehyde with the sampling media. The post-sampling analytical method for formaldehyde also requires techniques different from those commonly used for other VOCs. The most important trace analysis techniques for the latter species are chromatography coupled with mass spectrometry (MS), flame ionization detection (FID), or other selective detectors. Again, due to its reactive nature, direct analysis of formaldehyde by the chromatographic procedure is often problematic and thus rarely practiced.

Specific techniques have been developed for environmental formaldehyde collection, taking into consideration both the sampling and analytical factors as described above. Absorption in aqueous solutions in an impinger is a classical method-taking advantage of the high aqueous solubility of formaldehyde. To enhance the trapping efficiency, and also to facilitate subsequent analysis, *in situ* derivatization in the impinger is frequently used. Recently, adsorption techniques combined with *in situ* derivatization on solid sorbent is gaining popularity. The solid adsorption technique has the advantage that it is applicable in both active (by air pumps) and passive (by natural diffusion) sampling modes.

ANALYTICAL METHODS

Spectrophotometric

The most widely used method recommended by the U.S. Intersociety Committee for the determination of formaldehyde content in the atmosphere is spectrophotometry. Two analytical reagents, 4,5-dihydroxy-2,7-naphthalenedisulfonic acid (chromotropic acid) and 3-methyl-2-benzothiazolinone hydrazone hydrochloride (MBTH), have been developed as selective agents to react with formaldehyde in order to form purple and blue chromogen, respectively (14,15).

Chromatographic

Because formaldehyde is a labile compound, it is highly desirable to derivatize it into a stable nonvolatile compound during sample collection. This approach not only enhances the stability of samples but also facilitates subsequent analyses. Several derivatization agents, including 2-(hydroxymethyl)piperidine, 2-aminoethanethiol, and 2,4-dinitrophenylhydrazine (DNPH) have been effectively used for the derivatization of formaldehyde. All of them can react rapidly and quantitatively with formaldehyde to give rise to a stable adduct. Subsequent quantitation of these derivatives can be carried out by gas chromatography (GC) or high-performance liquid chromatography (HPLC). DNPH by far is the most commonly used derivatization reagent for formaldehyde determination.

Electroanalytical

Under strong alkali conditions, an aqueous solution of formaldehyde undergoes electroreduction at a hanging mercury drop electrode.

In an alternative approach, formaldehyde is first derivatized *in situ* with Girard's reagent T in an aqueous solution to produce an electroactive adduct (16).

With an attempt to further improve the sensitivity of the electroanalytical method, two approaches have been undertaken. Both approaches take full advantage of the cationic nature of the formaldehyde–Girard's T adduct (6).

CONTINUOUS MONITORING TECHNIQUES

Continuous monitoring is used to monitor the real-time concentration of a pollutant on site. This is mostly used in workplace environments. There, the possible occurrence of high concentrations of a toxic substance could pose an immediate acute exposure risk to on-site workers or the population nearby. Another application is the monitoring of time-trend data. For instance, the declining trend of the concentration of an indoor pollutant in a new construction can be a function of time.

Spectroscopic methods such as Fourier transform infrared (FT/IR), tunable diode laser adsorption spectroscopy, and differential absorption have been used for continuous monitoring. These spectroscopic methods, however, require expensive, complex, and large instrumentation. Their applications are therefore mostly for references purposes, rather than for routine monitoring in the field or at the workplace.

Small and portable wet-chemical monitors have been developed and are available in the market. For instance, the CEA TGA 555 is a portable wet-chemical analyzer that is capable of continuously monitoring airborne formaldehyde.

Recently, more sensitive and selective continuous formaldehyde monitors based on fluorescence detection of the derivative produced by formaldehyde, ammonia, and 1,3-diketone via the Hantzsch reaction have been developed (7).

Many of the formaldehyde-detection techniques as described earlier can be modified for use as continuous monitors. To meet this need, the techniques, besides being sensitive and specific, must also have a fast response time. The common approach to sample formaldehyde is by membrane or solvent scrubbing in a continuous flow system. The extracted formaldehyde will then react with a derivatization reagent to facilitate detection. Either spectrophotometric absorption or fluorescence techniques can be used to detect formaldehyde.

BIBLIOGRAPHY

1. P. Carlier, H. Hannachi, and G. Mouvier, *Atmos. Environ.* **20,** 2079–2099 (1986).
2. *C. & E. News,* **74** (15), 17 (Apr. 8, 1996).
3. *Introduction to Indoor Air Quality: A Reference Manual,* EPA/400/3-91/003, U.S. EPA, Washington, D.C., 1991.
4. J. J. Shah and H. B. Singh, *Environ. Sci. Technol.* **12,** 1381–1388 (1988).
5. S. M. Hays, R. V. Gobbell, and N. R. Gauick, *Indoor Air Quality: Solutions and Strategies,* McGraw Hill Book Co., New York, 1995, p. 68.
6. W. H. Chan and H. Huang, *Analyst* **121,** 1727 (1996).
7. T. J. Kelly and C. R. Fortune, *Intern. J. Environ. Anal. Chem.* **54,** 249–263 (1994).
8. L. Nondek, D. R. Rodler, and J. W. Birks, *Environ. Sci. Technol.* **26,** 1174–1178 (1992).
9. A. Yasuhara and T. Shibamoto, *J. Chromatogr.* **672,** 261–266 (1994).
10. U.S. Environmental Protection Agency, *Assessment of Health Risks to Garment Workers and Certain Home Residents from Exposure to Formaldehyde,* U.S. EPA, Office of Pesticides and Toxic Substances, Washington, D.C., 1987.
11. R. Allen, *Chemical Safety Data Sheets,* vol. 3, *Corrosives and Irritants,* The Royal Society of Chemistry, Cambridge, U.K., 1990, pp. 109–114.
12. S. Batterman, B. McQnown, and A. McFarland, *84th Annual Meeting of Air and Waste Management Association,* Vancouver, B.C., Canada, 1991.
13. L. Nondek, D. R. Rodler, and J. W. Birks, *Environ. Sci. Technol.* **26,** 1174–1178 (1992).
14. J.P. Lodge, ed. *Methods of Air Sampling and Analysis*, 3rd ed., Lewis Publishers, Inc., Chelsea, Mich., 1989, pp. 274–278.
15. Ref. 14, pp. 279–284.
16. W. H. Chan, W. C. Chung, and P. X. Cai, *Analyst,* **120,** 2233–2236 (1995).

FUEL ANALYSIS. See PETROLEUM FUELS ANALYSIS FOR MEETING REGULATORY LIMITS.

FUNGAL TECHNOLOGIES FOR HAZARDOUS WASTE TREATMENT

RICHARD A. HAIMANN
CH2M HILL
Los Angeles, California

The management and treatment of hazardous wastes pose challenges for engineers and scientists. Wastes regulated as hazardous have stringent management and treatment requirements which substantially increase costs associated with the production of those wastes. More cost-effective methods of treating hazardous wastes are actively sought. This article describes the use of fungal technologies for the treatment of hazardous wastes.

Fungal technologies, as the name implies, involve the use of fungi to degrade, detoxify, and destroy waste materials. These technologies are a subset of biological waste treatment methods used to treat wastes of varying types. Most notably, bacterial-based biological processes have traditionally been used to degrade organic wastes such as sewage, and some hazardous wastes, into carbon dioxide and water (1,2). What has been noted by many researchers and practitioners, however, is that there are many xenobiotic compounds that either cannot be degraded by most bacteria or are degraded at such slow rates that the technology is not cost effective (2–7). These compounds tend to be complex, organic molecules with one or more aromatic rings and/or halogen substitutions (8) (see BIOLOGICAL MATERIALS, ANALYSES OF HAZARDOUS SUBSTANCES).

Work with fungi-based biological processes has shown that certain lignin-degrading fungi are capable of degrading these complex xenobiotic compounds which are recalcitrant to bacterial degradation. In some instances, work has progressed from research to commercial applications (2,9). Additionally, in the pulp and paper industry, the same fungi have been investigated for wastewater treatment, wood-pulping, and bleaching applications to reduce costs and waste volumes (10,11).

TYPES OF HAZARDOUS WASTES

What is a hazardous waste? In the United States the Resource Conservation and Recovery Act (RCRA) defines hazardous wastes as those wastes specifically listed under Title 40 of the *Code of Federal Regulations,* part 261 (40 CFR 261) or those wastes which have particular characteristics as measured with standard analytical methods. Outside the Unites States, the definition of a hazardous waste varies depending on the country in which the waste is produced or managed. In general, a hazardous waste contains one or more anthropogenic chemicals resulting from some manufacturing process which, if released to the environment, are expected to present some hazard to human or ecological receptors.

Anthropogenic chemicals which constitute hazardous wastes fall within two broad categories: organic and inorganic. Biological processes for treating hazardous wastes are typically used to degrade the organic constituents of the wastes. Many inorganic constituents in hazardous wastes are elemental such as mercury, cadmium, and other heavy metals and cannot be broken down to more elemental and less harmful products. However, as discussed later in this chapter, some organic processes using fungi and higher plants have shown that metals can be removed from solutions and concentrated in the fungal or plant material.

BIOLOGICAL PROCESSES FOR TREATING HAZARDOUS WASTES

Biological processes have been used for thousands of years for treating wastes produced by humans. The exact date of the invention of the compost pile is unknown, but such processes have been used to treat wastes since long before the industrial revolution. Toward the end of the nineteenth century, biological processes for treating sewage were implemented on a somewhat broad scale. These processes depended primarily on bacteria to degrade organic matter in sewage.

One goal of many waste treatment systems is to convert organic chemicals into carbon dioxide. Doing this requires oxidation of the organic molecules and consequent reduction of other electron acceptor molecules. Because biological organisms catalyze such oxidations and utilize the energy released for growth and sustenance, they can be very suitable mechanisms to degrade such wastes.

Many redox reactions are thermodynamically favorable and, hence, possible abiotically. However, for the reaction to occur, electron acceptor and electron donor molecules must collide with each other and overcome the reaction's activation energy. Enzymes catalyze these reactions by attracting both the electron donor and electron acceptor molecules to an active site on the enzyme where the reaction takes place. Hence, the presence of enzymes speeds the redox reactions considerably over those that would occur through random collisions of molecules in a solution.

Bacteria are some of nature's simplest life forms and are believed to be some of the earliest life forms to have evolved. They play important roles in global nutrient cycles by breaking down complex organic molecules into carbon dioxide and simpler organic molecules, and releasing mineral nutrients which are bound in the complex organic molecules for use by other life forms such as plants. Bacteria are ubiquitous in nature and are found in extreme environments where no other life forms can survive. As a result, bacteria have evolved many species and enzymatic systems which allow them to degrade organic molecules utilizing a wide variety of electron acceptors and in a variety of conditions.

It has been observed that as organic molecules increase in size, number of branching substitutions, number of aromatic rings, and complexity, they become more recalcitrant to bacterial degradation. As a result, bacterial degradation of hazardous waste has been shown to be successful for relatively simple organic molecules such as gasoline and diesel range fuel hydrocarbons, single aromatic ring compounds with few substitutions, and low molecular weight chlorinated aliphatics. More complex hazardous organic molecules such as certain pesticides, wood preservatives, dioxins, and polychlorinated biphenyls (PCBs), have only been degraded by bacterial cultures on a very limited basis (12).

Fungi are a class of organisms in the domain of *Eukarya* which are significantly evolved from bacteria. Comparatively, fungi cells have organelles and a nuclear membranes which bacteria and *Archaea* lack. Fungi can live as single-celled organisms; yeasts are a type of fungi which are single-celled, but many fungi are multicellular organisms with distinct tissue types. Fungi, similar to bacteria, obtain energy by oxidizing organic molecules. Within nature, fungi have evolved to occupy some unique niches: they are the primary organisms responsible for the decay of bulk organic matter such as wood and forest litter.

The white-rot fungi (WRF) are particularly important. WRF include many genera that are capable of degrading lignin, cellulose, hemicellulose, and other wood components. Of particular interest are the lignin-degrading properties of the WRF. Lignins are highly complex, nonspecific organic chemicals which constitute a portion of vegetative cell walls. Lignins are a highly diverse class of very large molecules composed of nonrepeating phenyl proponoid units that are linked by carbon–carbon and ether bonds. These molecules have chiral carbons of both D- and L-configurations within the same molecule and exhibit extreme stereoirregularity.

Additionally, the molecules are highly oxidized. Hence, these molecules are strongly resistant to most known enzymatic oxidation mechanisms. In response to this, the WRF have evolved a highly oxidizing and nonspecific enzymatic system for oxidizing these lignin molecules (13). WRF are responsible for wood decay in nature which is the rate-limiting step in the global carbon cycle (10,14). As would be suspected with such a powerful lignin-degrading enzyme system, WRF have been observed to also degrade complex anthropogenic organic molecules and hazardous wastes (1,13–18).

USE OF FUNGI TO DEGRADE HAZARDOUS WASTE

Upon discovery of the ability of WRF to degrade anthropogenic organic chemicals, further research was conducted. The first report of WRF degrading such chemicals was in 1985 when it was reported that the white-rot fungus *Phanerochaete chrysosporium* successfully mineralized DDT, 3,4,3′,4′-tetrachlorobiphenyl, 2,4,5,2′,4′,5′-hexachlorobiphenyl, 2,3,7,8-tetrachlorodibenzo-*p*-dioxin, lindane, and benzo[*a*]pyrene to carbon dioxide (4). The study also concluded that the metabolism of those compounds depended on the lignin-degrading enzyme system of that fungus. Further investigations have shown white-rot fungi to mineralize a host of xenobiotics, in addition to those mentioned above, under laboratory conditions (3,7,16,19–35). These chemicals include Aroclor 1254, 4-chloroaniline, 3,4-dichloroaniline, pentachlorophenol, triphenylmethane dyes, 2,4,5-trichlorophenoxyacetic acid, phenanthrene, anthracene, fluoranthene, benzo[*b*]fluoranthene, benzo[*k*]fluoranthene, indeno[*ghi*]pyrene, benzoperylene, and other polycyclic aromatic hydrocarbons (PAHs), azo and heterocyclic dyes, 2,4,6-trinitrotoluene, and RDX.

Munitions wastewaters have been successfully treated at the laboratory scale with WRF. Such wastewaters contain relatively large concentrations of nitroaromatic compounds as well as color compounds which require removal prior to discharge. WRF in laboratory batch reactors or immobilized on rotating biological contactors have successfully mineralized such compounds (5,14,31,32,35).

Investigations have indicated that to enhance expression of the lignin-degrading system in white-rot fungi, it is necessary to create conditions similar to those found in plant woody tissue; this includes a source of carbohydrates for growth, nitrogen-limiting conditions, and an aerobic environment (1,2,16). Lignin alone is not sufficient for growth of the organisms. Therefore, another carbohydrate source is necessary. Generally, the expression of the ligninolytic system by WRF is in response to their need to remove lignin in order to obtain access to cellulose in woody tissues which is the primary substrate for growth of WRF. The enzymes responsible for lignin degradation have been observed to be regulated by the concentrations of usable nitrogen in the surrounding environment with low nitrogen concentrations stimulating the production of the lignin-degrading enzymes and high nitrogen concentrations repressing the production of the lignin-degrading enzymes. Conversely, the growth of the fungal tissues themselves require nitrogen from the surrounding environment. Hence, the ligninolytic system is only produced in response to nutrient starvation conditions which is similar to conditions in woody tissue in which the cellulose and cell proteins are bound within a lignin matrix.

APPLICATIONS OF WHITE-ROT FUNGI

Primary applications of hazardous chemical degradation with WRF that have been pursued include (*1*) treating contaminated soils at hazardous waste release sites and (*2*) treating kraft paper pulp mill effluent. Other applications have been investigated including bleaching paper pulp, delignifying wood (pulping), treating styrene vapors, and removing heavy metals from solutions. Each of these processes has been researched at the laboratory scale. Some of the processes have been tested at the field scale. As of this writing, none have yet been widely implemented at the full scale in their respective industries.

OTHER USES OF WHITE-ROT FUNGI

Two more applications to the paper industry have been investigated: the use of WRF to bleach pulp (biobleaching), and the use of WRF to delignify pulp (biopulping). Both of these applications show promise. WRF have evolved to remove lignins from wood and using these organisms rather than the kraft process for pulping wood (removing lignin) can result in significant reductions in the volumes of hazardous wastes, energy usage, and water usage (10,36,37). However, there are some problems because although the WRF have evolved to degrade lignin, they have also evolved to degrade cellulose which they use for energy and growth. Hence, when applying WRF to a biobleaching or biopulping process, stopping their activity before substantial cellulose has been degraded as well as removing the fungi from the pulp after the process is completed can pose problems. There have been investigations in the use of *in vitro* enzyme systems which can be better controlled than an *in vivo* process with live organisms. Generally, it has been discovered that the *in vitro* enzyme systems can achieve significant lignin degradation, but not to the extent than *in vivo* systems can. Additionally, the production of enzymes is relatively costly at this time (38). There are no reports of pulp mills that have incorporated full-scale continuous processes for biopulping or biobleaching with WRF.

Investigations have been conducted to identify the ability of WRF and other fungi to remove metals from solution. Fungal mycelia have shown a propensity to bioaccumulate certain metals including copper and nickel (39). Microbial biomass has been shown to selectively and rapidly take up metal ions from solution (40,41). A fungus, *Aspergillus oryzae,* was reported to take up heavy metal ions in its mycelial mass (41). In another study, cadmium uptake efficiencies of 97% were reported for an isolated fungus (40). Some WRFs, to protect themselves from antagonist microorganisms and desiccation, form melanized pseudosclerotial plates, referred to as zone lines, at colony boundaries. These plates show rapid metal adsorption kinetics which sequester metals from surrounding soils. Metal concentrations in these zone lines have been found to be 15 times greater than in surrounding soils. Results have shown that these tissues can sequester both free and bound ions from soils. This shows potential for metals removal from soils and waters by creating the environmental conditions necessary for these organisms to form zone lines. There is evidence that accumulation of these metals makes these organisms toxic to predators (42). In addition to metals, one study showed uptake of radionuclides by WRF (43). There are no published reports of full-scale systems or commercial applications of fungi for metals removal from waters or soils. Also, research has been conducted with an air-phase biofilter with styrene-degrading fungi in a laboratory setting in which degradation of styrene was observed at a rate of 70 g styrene per m^3 filter bed per hour (44).

Another noteworthy study is the production of recyclable plastic materials with a lignin component (6,45). This study concludes that grafting of lignin with synthetic side chains such as polystyrene will result in a much more biodegradable material than synthesis of a polymer from pure, petroleum-based products.

Finally, nylon degradation by three WRF has been demonstrated. Complete mineralization has not been demonstrated. It has not been determined if the WRF could utilize nylon as a sole energy/carbon source. It has been demonstrated that Mn enhances degradation-indicating activities of the MnP system. Additionally, LiP activities have been measured, and it was determined that oxidations were similar to thermal oxidation mechanisms in which methyl groups next to an N atom were attacked by nitrogen. Hydrolysis of amide bonds (prevalent in nylon) did not occur (46).

FUTURE WORK AND COMMERCIALIZATION

Research continues, some sponsored by the U.S. Environmental Protection Agency and others, to answer questions regarding how these organisms work and how engineers and scientists can enhance the WRF xenobiotic-degrading properties for waste treatment and site restoration (1). Commercial ventures have been established to apply this technology (2,9,47). Intec One-Eighty Corporation (Logan, Utah) was founded to be the exclusive licensee of Utah State University to commercialize white-rot fungus technology. Mycotech (Butte, Montana) was founded in 1982 to develop and produce fungi as biopesticides as well as for bioremediation. Tienzyme (State College, Pennsylvania) was founded in 1991 to provide white-rot fungi enzymes to the research community. The Lambert Spawn Company (Coatesville, Pennsylvania) has produced fungal inoculum for various purposes since 1904 and has produced white-rot fungus inoculum for environmental applications.

There are a number of issues regarding the use of these organisms which further research will help resolve. Of primary importance is the amount of mineralization which occurs during degradation. It is unclear what happens to the degradation daughter products that are not evolved as carbon dioxide. Further studies are necessary to answer these chemical fate questions as well as to provide greater insight into how a WRF system can be engineered to efficiently degrade and mineralize wastes.

COST CONSIDERATIONS

Biological processes have significant advantages over other processes for treating hazardous wastes. Because biological processes have the potential to degrade waste to nonhazardous products, liabilities can be greatly reduced over other processes which only contain or immobilize hazardous chemicals. Such liability reductions can translate directly into cost savings, particularly when organizations incorporate risk management tools into their decisionmaking process. However, the cost of treatment with WRF may not be the lowest cost treatment alternative.

Costs of treatment with WRF will be dependent on a number of factors, most important of which is the efficacy with which WRF will achieve waste treatment goals. Research with WRF has shown it to potentially be an effective, efficient process when environmental conditions are appropriate. However, it is important to carefully consider the environmental conditions in which the ultimate treatment system will operate and evaluate whether treatment with WRF will be adequately cost effective against other treatment alternatives.

Primary factors to consider when estimating the costs of treatment with WRF include (*1*) chemicals to be degraded, (*2*) waste matrix (soils, water, organic liquids, solids, debris), (*3*) initial concentrations of chemicals, (*4*) desired final concentrations of chemicals, (*5*) potential daughter product fate and toxicity, (*6*) volume of waste, (*7*) ultimate discharge, disposal, or use of the treated waste, (*8*) amount of space available for treatment system and staging areas, (*9*) availability of appropriately trained designers, constructors, and operators, and (*10*) maximum allowable project duration. Each of these factors can significantly affect the total project cost and possibly make WRF less cost effective than another treatment alternative.

As an example, in the case of treating contaminated soils, it is necessary to excavate the soils and to have an area large enough to accommodate treatment of the soils. When comparing WRF to *in situ* technologies, the costs for excavation would make WRF more costly when all else is considered equal. However, in certain cases, *in situ* technologies may not be able to treat the contamination or, because of the difficulty and expense of treating the waste with other *in situ* alternatives, even with excavation, the total costs of WRF may be less than the *in situ* technologies.

When comparing WRF to other excavation and treatment technologies such as incineration, thermal desorption, or soils washing, there is great potential for WRF to be more cost effective. However, it is necessary to evaluate the costs on a site and contamination-specific basis. The primary costs outside of excavation, backfill, and disruption to site operations involved with WRF treatment of soils are inoculum and soils handling during treatment. Hence, the longer treatment lasts, the greater the costs of WRF. Identifying the length of time in which WRF is expected to degrade contaminants to remedial goals could require laboratory and/or pilot testing of contaminated soils from the site. There may be great uncertainties associated with the estimates.

For other processes, such as treating kraft pulp effluent or biopulping wood, it is necessary to understand all the impacts the new process will have to plant operations to assess and compare the costs. The processes currently in place for pulping wood and treating kraft effluent are relatively dependable and well understood. WRF is less understood and there is greater uncertainty regarding the amount of effort required to operate it dependably. Because of the complexity of biological processes, it can be assumed that a biological process may require more effort to operate, maintain, and troubleshoot which would increase costs. However, to better estimate such costs, pilot testing would be very beneficial.

CONCLUSIONS

In conclusion, fungal processes have many potential applications for treatment of waste products, particularly those wastes which are recalcitrant to bacterial degradation and expensive to treat with physical or chemical methods. Research is advancing to understand how to harness these processes for waste treatment and to evaluate the cost effectiveness of fungal treatment against other treatment alternatives. Commercialization has begun for some applications and is likely to continue as more research and field-scale successes are achieved.

BIBLIOGRAPHY

1. J. A. Glaser "Hazardous Waste Degradation by Wood Degrading Fungi," *Biotechnology and Biodegradation, Advances in Applied Biotechnology Series, Vol. 4,* D. Kamely, A. Chakrabarty, and G. S. Omenn, eds., Gulf Publishing Co., Houston, Tex., 1990.
2. R. J. Hicks "Fungi Degrade Complex Organic Compounds in Onsite Bioremediation," *The Hazardous Waste Consultant,* May/June, 1993.
3. S. D. Aust, M. M. Shah, D. P. Barr, and N. Chung, "Degradation of Environmental Pollutants by White-Rot Fungi," *Bioremediation of Chlorinated and PAH Compounds, Proceedings of the 2nd International Symposium of On-Site and In-Situ Bioreclamation, San Diego, Calif.,* Battelle Press, Columbus, Ohio, Apr. 1993.
4. J. A. Bumpus, M. Tien, D. Wright, and S. D. Aust, *Science,* **228,** 1434–1436 (June 21, 1985).
5. A. Majcherczyk, A. Zeddel, and A. Hüttermann, "Biodegradation of TNT (2,4,6-Trinitrotoluene) in Contaminated Soil Samples by White-Rot Fungi," *Applied Biotechnology For Site Remediation,* in Ref. 3.
6. O. Milstein, R. Gersonde, A. Hütterman, M. J. Chen, and J. J. Meister, "Rotting of Thermoplastics Made From Lignin and Styrene by White-Rot Basidiomycetes," *Applied Biotechnology Site,* in Ref. 3.
7. A. Zeddel, A. Majcherczyk, and A. Hüttermann "Degradation and Mineralization of Polychlorinated Biphenyls by White-Rot Fungi in Solid-Phase and Soil Incubation Experiments," *Bioremediation of Chlorinated and PAH Compounds,* in Ref. 3.
8. F. Baud-Grasset, S. I. Safferman, S. Baud-Grasset, and R. T. Lamar, "Demonstration of Soil Bioremediation and Toxicity Reduction By Fungal Treatment," *Bioremediation of Chlorinated and PAH,* in Ref. 3.
9. D. L. Illman, "Hazardous Waste Treatment Using Fungus Enters Marketplace," *Chem. Eng. News,* (July 12, 1993).
10. K. T. Kirk, R. R. Burgess, and J. W. Koning, Jr, "The Use of Fungi in Pulping of Wood: An Overview of Biopulpingb Research," *Frontiers in Industrial Mycology,* G. Leatham, ed., Chapman and Hall, New York, 1992.
11. I. D. Reid, and M. G. Paice, "Biological Bleaching of Kraft Paper Pulp," in Ref. 10; review article about biobleaching of pulp.
12. National Research Council, *In Situ Bioremediation When Does It Work?* National Academy Press, Washington, D.C., 1993.
13. D. P. Barr, and S. D. Aust, *Environ. Sci. Technol.* **28**(2), 78A–87A (1994).
14. K. L. Sublette, P. Cho, S. Schwartz, and D. Pak, "Microbial and Biomimetic Degradation of Nitrobodies," *International Conference on Environmental Pollution Proceedings of International Conference on Environmental Pollution—ICEP, Apr. 1, 1991, Lisbon, Portugal,* Interscience Enterprises Ltd., Geneva, Aeroport 15 Switzerland, Apr. 1991, pp. 756–763.
15. I. D. Reid, *Can. J. Botany Rev. Canadienne De Botanique,* **73,** S1011–S1018 (1995).
16. R. T. Lamar, J. A. Glaser, and K. T. Kirk, "White-Rot Fungi in the Treatment of Hazardous Chemicals and Wastes," in Ref. 10.
17. A. Paszczynski, and R. L. Crawford, *Biotechnol. Prog.* **11**(4), 368–379 (1995).
18. C. A. Reddy, *Current Opinion in Biotechnol.* **6**(3), 320–328 (1995).
19. B. W. Bogan, and R. T. Lamar, *Appl. Environ. Microbiol.* **61**(7), 2631–2635 (1995).
20. B. W. Bogan and R. T. Lamar, *Appl. Environ. Microbiol.* **62**(5), 1597–1603 (1996).
21. P. J. Collins and A. D. W. Dobson, *Biotechnol. Lett.* **18**(7), 801–804 (1996).
22. J. A. Field, E. Heessels, R. Wijngaarde, M. Kotterman, W. DeJong, and JAM DeBont, "The Physiology of Polycyclic Aromatic Hydrocarbon Biodegradation by the White-Rot Fungus, *Bjerkandera* sp Strain BOS55," in Ref. 3.
23. M. H. Gold, D. K. Joshi, K. Valli, and H. Wariishi, "Degradation of Chlorinated Phenols and Chlorinated Dibenzo-*p*-Dioxins by *Phanerochaete Chrysosporium,*" in Ref. 3.
24. K. E. Hammel, B. Kalyanaraman, and T. K. Kirk, *J. Biolog. Chem.* **261**(36), 16948–16952 (1986).
25. C. Johannes, A. Majcherczyk, and A. Huttermann, *Appl. Microbiol. Biotechnol.* **46,** 313–317 (1996).
26. L. Roy-Arcand and F. S. Archibald, *Enzyme Microb. Technol.* **13,** 194–203 (1991).
27. S. W. Kullman and F. Matsumura, *Appl. Environ. Microbiol.* **62**(2), 593–600 (1996).
28. D. R. Thomas, K. S. Carswell, G. Georgiou, *Biotechnol. Bioeng.* **40**(11), 1395–1402 (1992).
29. B. C. Alleman, B. E. Logan, and R. L. Gilbertson, *Water Res.* **29**(1), 61–67 (1995).
30. C. D. Barclay, G. F. Farquhar, and R. L. Legge, *Appl. Microbiol. Biotechnol.* **42,** 958–963 (1995).
31. P. Bayman, S. D. Ritchey, and J. W. Bennet, *J. Indust. Microbiol.* **15**(5), 418–423 (1995).
32. T. Fernando and S. D. Aust, "Biodegradation of Munitions Waste, TNT (2,4,6-Trinitrotoluene) and RDX (Hexahydro-1,3,5-Trinitro-1,3,5-Triazine) by *Phanerochaete chrysosporium,*" *ACS Symposium Series 468, Emerging Technologies in Hazardous Waste Management II,* D. W. Tedder and F. G. Pohland, eds., American Chemical Society, Washington, D.C., 1992, pp. 2214–2232.
33. B. S. Aiken and B. E. Logan, *Biodegradation,* **7**(3), 175–182 (1996).
34. T. Fernando, J. A. Bumpus, and S. D. Aust, *Appl. Environ. Microbiol.* **56**(6), 1666–1671 (June, 1990).
35. B. E. Logan, B. C. Alleman, G. L. Amy, and R. L. Gilbertson, *Water Res.* **28**(7), 1533–1538 (1994).
36. K. A. Onysko, *Biotechnol. Adv.* **11**(2), 179–198 (1993).
37. M. B. WAll, D. C. Cameron, and E. N. Lightfoot, *Biotechnol. Adv.* **11,** 645–662 (1993).
38. M. G. Paice, F. S. Archibald, R. Bourbonnais, L. Jurasek, I.D. Reid, T. Charles, and T. Dumonceaux, *ACS Symp. Ser.* **655,** 151–164 (1996).
39. M. M. Kao, H. S. Shen, and F. Y. Chang, "Removal of Heavy Metals from Waste Water by Fungi," *Hazardous and Industrial Wastes—Proceedings of the Mid-Atlantic Industrial Waste Conference Proceedings of the 25th Mid-Atlantic Industrial Waste Conference, July 7–9, 1993, College Park, Md.,* Technomic Publishers, Lancaster, Pa., 1993.
40. R. Campbell and M. H. Martin, *Water Air Soil Poll.* **50,** 397–408 (1990).

41. J. P. Huang, C. P. Huang, and A. L. Morehart, "Removal of Heavy Metals by Fungal (*Aspergillus Oryzae*) Adsorption," in J.-P. Vernet, ed., *Heavy Metals in the Environment,* Elsevier Publishers, New York, 1991.
42. D. N. McDougall and R. A. Blanchette, *Mycologia,* **88**(1), 98–103 (1996).
43. C. White and G.M. Gadd, *J. Chem. Technol. Biotechnol.* **49**(4), 331–343 (1990).
44. H. Cox, H. J. José, H. M. Houtman, H. J. Doddema, and W. Harder, *Biotechnol. Lett.* **15**(7), 737–742 (1993).
45. O. Milstein, R. Gersonde, A. Huttermann, M. J. Chen, and J. J. Meister, *J. Macromolec. Sci. Pure Appl. Chem.* **A33**(5), 685–702 (1996).
46. T. Deguchi, M. Kakezawa, and T. Nishida, *Appl. Environ. Microbiol.* **63**(1), 329–331 (1997).
47. Product literature, *The Lambert Spawn Interim Report On Fungal Technologies;* L. F. Lambert Spawn Co., Inc., Coatesville, Pa., 1993.

FURNACE AAS. See ELECTROTHERMAL ATOMIC ABSORPTION SPECTROMETRY.

G

GAMMA-RAY ACTIVATION ANALYSIS

CHIEN CHUNG
National Tsing Hua University
Hsinchu, Taiwan,

With the substantial increase in scientific and technological activities, considerable interest exists in the development and application of new environmental analytical techniques. Many severe limitations of conventional delayed neutron activation analysis (NAA) exist with respect to its capability (on which research communities rely heavily) to give an instant, online characterization of a field sample. The need for more and better analytical techniques has resulted in a quite spectacular development and diversification of new methods in environmental analysis. Particularly noteworthy has been the development of nuclear techniques, which have an unusually high degree of rapidity, elemental selectivity, and sensitivity. The penetrating radiations usually involved in nuclear methods enable measurements to be made that are characteristic of the bulk environmental sample instead of just its surface condition. It is often possible to avoid chemical separation. More important, nuclear methods have also shown adaptability to extremely harsh field conditions that might preclude the use of other analytical methods. This has led to the development of nuclear analytical techniques for *in situ* analysis of particular samples, such as pollutants in river water. It has stimulated further search for and application of nuclear phenomena as analytical probes.

Among the nuclear methods that have proved useful in neutron activation analysis is prompt γ-ray activation analysis (PGAA), in addition to conventional NAA. Although the PGAA technique has been developed using various neutron sources and such systems are usually installed around a nuclear facility (eg, a research reactor), many demands create field applications such as *in vivo* PGAA (IVPGAA) medical diagnosis, *in situ* PGAA (ISPGAA) environmental surveys, and online PGAA (OLPGAA) detection of unique specimens. These real-time, online techniques do not rely on induced radioactivity and thus avoid the lengthy period of counting during sample decay long after the irradiation.

Despite the extensive use of environmental analysis, no single method, nuclear or chemical, has been found to satisfy all of the sensitivity requirements for the elements of interest to all scientific and engineering disciplines. Each method is complementary to the others.

Nuclear analytical methods are certainly not limited to taking the sample to a fixed laboratory for measurement. In the last decade, many field applications using the PGAA technique have been developed. Among the latest developments is the on-site inspection of abandoned chemical warfare agents and ammunitions by observing neutron-induced γ-rays following detection with a high purity germanium detector (1). The elemental composition of some typical chemical warfare and high explosive munitions, such as lewisite with its high content of arsenic, can be clearly identified by the unique prompt γ-rays. A ^{252}Cf neutron source and a high resolution germanium detector are preferred for such measurement design, and this system has been proven successful in verifying declared chemical warfare agents and ammunitions (2).

A knowledge of the neutron reaction and the prompt γ-rays emitted from it is prerequisite for the PGAA method, and the neutron source and γ-ray spectrometer are indispensable integral parts of a PGAA facility. After the high resolution γ-ray spectrometer was commercialized in the 1960s, and the prompt γ-rays were well documented in the 1970s, PGAA was then rapidly adopted by the analytical community as a reliable nuclear method in the 1980s. Future expansion of PGAA, in particular for field applications in environmental analysis, is probable, owing to its mobility, convenience, and versatility.

PRINCIPLES

The principles of PGAA consist of three stages involving nuclear phenomena: (*1*) the sample is bombarded with neutrons of a neutral ionizing radiation; (*2*) the nuclear reaction caused by neutron interaction with the sample may eject characteristic prompt γ-rays; and (*3*) the prompt γ-rays are subjected to using a spectrometer quantitative measurement.

APPLICATIONS

In the field of environmental research, there is still a considerable interest in improving the techniques leading to sensitive, nondestructive, multielemental analysis. Conventional NAA, although a powerful tool for the quantitative determination of numerous elements in trace amounts, has some severe limitations, since it depends on induced radioactivities with appropriate half-lives for post-irradiation measurement. An interesting example of this kind is again the toxic element cadmium at trace levels in environmental sample. The enormous thermal neutron cross-section (nearly 20000×10^{-24} cm^2) for the 12.2% abundance isotope ^{113}Cd cannot be used for conventional NAA because the reaction product of ^{113}Cd(n,γ) ^{114}Cd is nonradioactive. A similar problem arises with another toxic element, Hg, for which the capture reaction product of high cross-section (2000×10^{-24} cm^2) for 17% abundance ^{199}Hg$(n,\gamma)^{200}$Hg is again stable. PGAA overcomes this problem by using the capture prompt γ-rays emitted from the excited compound nucleus with energy above that of the stable isotope. As the benefits of conventional methods remain, PGAA is an indispensable supplementary technique for analysis.

In the last decade, the technique has evolved to become a powerful nuclear analytical method. It is generally useful for major constituent analysis, in particular for those elements for which no appreciable amount of induced radioactivity is produced by neutron bombardment. It is also suitable for trace analysis for a limited number of elements with large neutron cross-sections. As the elemental sensitivity in nuclear analytical methods depends directly on the intensity of the bombarding particles, a reactor-based PGAA facility having a high neutron flux is preferred in trace analysis, whereas the PGAA setup using a low flux isotopic

Table 1. Detection Limit for Elements Analyzed by the PGAA Technique with Preselected Irradiation Conditions[a]

Detection Limit	Elements
0.1–1.0 μg	B, Gd, Cd
1.0–10 μg	Sm, Hg
10–100 μg	Er, Ag, Nd, Rh, In, Kr, H
100μg–1 mg	Eu, Ti, Dy, Au, Ta, Cl, Lu, Tm, Hf
1–10 mg	Ho, Co, Xe, Os, Mn, La, Ir, Cr, Ge, Ca, Sr, V, Mo, Ni, Te, Br, I, S, Fe, K, Y, Zn, As, Cs, Cu, Li, Yb, Ar, Sc, N
10–100 mg	Se, Ga, Si, W, Ru, Na, Pr, Zr, Al, Re, Pt, Sn, Mg, P, Tb, Ba, Pb, Ce
100 mg–1 g	Nb, Be, Tl, Sb, Ne, C, F, Bi, Rb
1–10 g	O

[a] Ref. 3.

neutron source is preferred in field applications for determining some environmentally vital elements.

The elements that can be determined at trace level using a reactor-based PGAA facility (see Table 1), are B; the toxic Cd and Hg; and some rare-earth elements, such as Gd, Sm, Er, and Nd.

Some major constituents in the sample matrix, in particular those light *Z* elements, cannot be activated by nuclear methods and, therefore, cannot be determined. For instance, the hydrogen content in most environmental samples is high. However, the reaction $^1H(n,\gamma)^2D$ produces a stable isotope, making conventional NAA impossible for determining hydrogen. On the other hand, PGAA allows convenient determination via the 2.223 MeV prompt γ-ray emitted from the $^1H(n,\gamma)^2D$ reaction, with a detection limit down to the submilligram level. Thus, in many trace analyses by reactor-based PGAA, major element analysis or bulk analysis is an added benefit, supplementing analyses that cannot be performed readily by other analytical methods. A typical example is the rapid determination of the major elements H and C as well as the trace elements B and Cd using a reactor-based PGAA facility.

REACTOR-BASED PGAA APPLICATION

In the reactor-based PGAA facility, irradiation can be conducted internally in the core or externally outside the reactor by extraction of a neutron beam through a collimator. The trade-off between high flux with low detecting efficiency for internal irradiation and low flux with high detecting efficiency in external irradiation approximately cancel in the two configurations; however, external geometry is superior for the following reasons:

- Lower interferences from fission γ-rays.
- No restriction on the material (such as toxic environmental sample) being investigated.
- Larger space for special samples, such as an inert atmospheric chamber.
- Larger space to accommodate bulk samples.
- Minimal sample deterioration caused by heating and radiation damage.
- Flexible design for multisample station.
- Flexible design for multidetector arrangement.
- Little induced radioactivity in the irradiated sample.
- Flexible neutron beam profiles using neutron filters.

Hence most reactor-based PGAA facilities are in external geometry, and the one attached to the Tsing Hua Open-pool Reactor (THOR) in Taiwan—a typical PGAA setup using an external reactor beam—is introduced briefly as an example (4). Neutrons from the 1 MW THOR facility are extracted horizontally by means of the throughport ≈30 cm from the nearest fuel elements. A two-section beam tube is inserted in the throughport to collimate and deliver the neutron beam at targets with a cross-section of 5 cm. The thermal neutron flux measured in the beam using gold foil is $1.3 \times 10^6 n \cdot cm^{-2} \cdot s^{-1}$ at the target with a cadmium ratio of 26.4:1. The beam passing through the target is captured with a beam catcher consisting of 40% weight of BO_2 in polyethylene (PE) matrix blocks surrounded by heavy concrete blocks. The concrete blocks also serve as a biological shield for those who must work around the facility.

The main γ-ray detectors are HPGe detector and an annular NaI(Tℓ) detector shield to perform Compton suppression and to detect pairs of 0.511 MeV γ. The 145 cm^3 *n*-type HPGe detector has 30% relative efficiency and a resolution of 2 keV full width at half maximum (FWHM) at 1332 keV, and a peak-to-Compton ratio of 46:1 at the same photopeak energy. The detector shield surrounding the primary HPGe detector is an annual NaI(Tℓ) detector 22.8 cm in diameter and 25.4 cm in length. A 7.6 cm diameter lateral hole can accommodate the 38-cm-long HPGe detector. Eight 7.6 cm photomultiplier tubes (PMT), four on each optically isolated half, view the NaI(Tℓ) detector. The sample for irradiation can be placed as close as 25 cm from the HPGe detector head. The geometric arrangement of the two detectors, together with the shielding materials associated with them (5).

Lead is chosen as the major detector shielding material, and a 8-cm-thick layer is used to surround the spectrometer. To eliminate the X-rays induced by lead–photon interaction, the annular NaI(ℓ) detector is further covered by absorption layers of 1-cm-thick Lucite and 1.5-cm thick copper. In addition, to prevent the neutrons reaching the detectors, the front end-window of the HPGe detector is covered by a 1-cm layer of ^{6}LiF powder; and on the outside, the lead shielding of the spectrometer is surrounded by an additional 5-cm layer of BO_2-loaded polyethylene blocks as a first-line scattered-neutron absorber. The detector assembly is placed on a trolley so that the 1-ton assembly can easily be moved around the PGAA facility (5).

The electronic block diagram for this spectrometer (6). The basic electronic setup is for the two optional data accumulation modes. In the Compton-suppressed mode, γ-ray events collected in the HPGe detector that have coincident events in the annular NaI(Tℓ) detector within 0.1 μs are rejected at ADC 1 input. For the pair spectrometric mode, γ-ray events of 0.511 MeV signal occurring simultaneously in both halves of the NaI(Tℓ) crystal are treated as an allowable gate for a HPGe pulse at ADC 1. The effect of Compton suppression, judged at 0.662 MeV for the ^{137}Cs photopeak's peak:Compton ratio, is improved by a factor of 5.2; at 2.754 MeV (^{24}Na photopeak) the peak:Compton ratio is further improved by a factor of 7 under anti-Compton operation. In a complex radioenvironmental PGAA test, Compton suppression not only yields a much better ratio (11:1) of the Compton continuum between unsuppressed and suppressed spectra but also eliminates most contaminating and interfering gamma-rays.

The neutron beam flux at the target of the present system is quite low compared to that in many other reactor-based PGAA installations, or two orders of magnitude lower than the most intense external flux. However, improvements can be made by reducing the sample-to-detector distance and using a larger HPGe main detector. The detection sensitivities for some elements are thus sensitive enough for many applications. However, extending the irradiation period is expected to provide better detection limits, especially for low interaction rate elements, but not all elements give good results. Further developments are expected to improve the reactor-based PGAA facility, making it even more useful.

In Situ PGAA Application

On the other hand, many field applications of PGAA use isotopic neutron sources. Among them ISPGAA environmental surveys frequently employ portable neutron sources in rugged conditions. The detailed setup, from mobile unit to permanent facility, depends on the particular use (7).

The frontier of major element analysis lies in field applications. Some critical determinations of major element content must be performed on a large sample matrix, eg, a rock formation around a bore hole, and a facility as complex as a nuclear reactor is difficult to adapt to field applications. Convenient PGAA arrangements using portable detection systems and isotopic neutron sources have been developed in the past decade for field applications such as *in situ* environmental surveys and, more recently, on-site scanning for sensitive objects, such as unidentified chemical warfare agents dumped in the environment.

There is growing concern about the environmental pollution problem, particularly in industrialized countries. Waste discharge from factories, mills, and refineries can cause water pollution; and immediate actions should be taken before the toxic chemicals expand into the ecosystem. For this reason, a nuclear technique based on ISPGAA was developed, it certainly offers potential for *in situ,* rapid, and continuous survey of water pollutants.

The basic concept for designing the ISPGAA probe is similar to that of logging sondes employed in surveys of geological formations (8–11). The submerged probe mainly consists of a ^{252}Cf neutron source, a portable HPGe detector, and neutron and γ-shields in between. Signal and prompt γ-rays collected by the HPGe detector are transmitted and analyzed in the battery-powered, portable MCA with built-in AMP, HVPS, and rechargeable batteries. All are placed above the waterline. For *in situ* measurement in effluent body, water is the only major constituent, which serves as a neutron moderator as well as a neutron shield. Thermalized neutrons are confined to the region immediately surrounding the detecting probe, where capture reactions are dominant, and the prompt γ-rays are used for the determination of pollutants.

The detection limit of various industrial pollutants in water using the ISPGAA technique is given in Table 2. For a 2.7-μg ^{252} Cf and 10% HPGe detector setup with 1000-s survey period, the relative sensitivity factor S_r can be calculated for toxic boron, cadtmium, and mercury with the detection limit in the parts per million order; while chlorine, chromium, manganese, iron, and copper can be determined if greater than 20 ppm. Other pollutants are not suitable for ISPGAA investigation since their detection limits are much too high. Therefore, in highly polluted water, the ISPGAA technique using such design and setup can determine three trace pollutants in parts per million as well as another five minor pollutants in the range of 20 ppm.

Table 2. Detection Limit of Industrial Pollutants Using a 2.7-μg ^{252}Cf–10% HPGe probe in a 1000-s ISPGAA Survey

Pollutant	Atomic Mass, A	Detection Limit, ppm
B	10.8	5.7
S	32.0	315
Cl	35.4	21.4
Cr[a]	52.0	19
Mn	54.9	42.6
Fe[a]	55.8	24.5
Cu	63.5	27.0
Zn	65.4	275
As	74.9	432
Se	79.0	273
Ag	107.9	120
Cd	112.4	0.21
Hg	200.6	1.62
Pb	207.2	426

[a] Construction material of the probe must be free of this element.

COMPARISON WITH OTHER NUCLEAR METHODS

Except for the nuclear methods used only for surface analysis, several nuclear analytical techniques can compete with PGAA. Among the nuclear techniques that have proven useful are conventional NAA, which may be conducted using radiochemical separations (RNAA) or completely by instrumental means (INAA) as well as the instrumental methods of photon activation analysis (IPAA) and charged particle activation analysis (ICPAA). All these off-line techniques rely on the production of radioactive nuclides and the subsequent detection and measurement of the decay of the induced radioactivity. These methods demand a log time of at least 1 month to permit short half-lived nuclei to decay and allow measurement of the longer half-lived species with lower specific activities. With PGAA, in contrast, the data are immediately available during irradiation.

A problem often encountered with nuclear methods is the escape of volatile components from the sample after irradiation. This is especially true for environmentally toxic Hg, which is quite volatile in the elemental state. The hot atom Hg is released from its compounds, owing to the recoil from the emission of its prompt γ-rays following the (n,γ) reaction. The recoil is sufficient to rupture the chemical bonds of the Hg compound, and Hg then diffuses out of the container after irradiation, thus precluding the measurement of its radioactive decay. With PGAA, however, this effect is of little concern, since by the time recoil occurs the prompt γ-measurement has already been completed.

There is a synergism between PGAA and other nuclear methods such as INAA in that these techniques employ the same equipment and technology to determine complementary sets of elements. INAA is best known as a method of trace analysis, insensitive in general to the major and minor elements that make up many materials. On the other hand, some of the strongest lines in PGAA Compton-suppressed and pair spectra often come from the major constituents, especially C, H, N, O, P, S, and Si. If the PGAA is applied to an unknown material before nondestructive INAA, the difficult question of neutron self-shielding by strongly absorbing elements in the sample is directly answerable, using the same sample repeatedly if necessary. The two methods together are capable of nondestructively measuring, in a wide variety of materials of scientific, economic, regulatory/and interest, all the major elements except perhaps oxygen, all minor elements, and many trace elements—up to 78 in all. In oxide matrices such as soil, a partial check is possible on the accuracy and completeness of the analysis by adding the concentrations of the elements as oxides and comparing the sum with 100%. This synergism of prompt and delayed radioactive analysis is a major justification of PGAA with the neutron intensity attainable from presently available neutron sources.

In field applications, the *in situ* PGAA for environmental concerns meet with little competition from other analytical techniques, as mentioned above. Numerous prototype PGAA setups have been installed by many users and clients from various scientific and engineering disciplines. Once the radiation safety and reliability for quantitative determination have been satisfactorily resolved, the rapid expansion of field applications of PGAA can be anticipated in the near future.

BIBLIOGRAPHY

1. A. J. Caffrey and co-workers, *IEEE Nucl. Sci.* **39,** 1422 (1992).
2. J. H. Chao, "On-Line Applications," in Z. B. Alfassi and C. Chung, eds., *Prompt Gamma Neutron Activation Analysis,* 1st ed., CRC Press, Boca Raton, Fla., 1995, pp. 159–176.
3. C. Chung, in Z. B. Alfassi, ed., *Chemical Analysis By Nuclear Methods,* 1st ed., John Wiley & Sons, Ltd., Chichester, U.K., 1994, p. 175.
4. Ref. 3, p. 180.
5. Ref. 3, p. 181.
6. Ref. 3, p. 183.
7. Ref. 2, pp. 131–157.
8. C. Chung and T. C. Tseng, *Nucl. Instr. Meth.* **A267,** 223 (1988).
9. J. H. Chao and C. Chung, *Nucl. Instr. Meth.* **A299,** 651 (1990).
10. J. H. Chao and C. Chung, *Appl. Radiat. Isot.* **A42,** 723 (1991).
11. J. H. Chao and C. Chung, *Appl. Radiat. Isot.* **A42,** 735 (1991).

GAS CLEANUP: ABSORPTION

BYRON Y. HILL
Union Carbide Corporation
South Charleston, West Virginia

ROBERT M. COUNCE
JOSEPH J. PERONA
University of Tennessee
Knoxville, Tennessee

Absorption is a unit operation used to remove one or more components, or solutes, from a gas stream. This is accomplished by contacting the gas with a liquid, or solvent, into which the solutes preferentially dissolve. Its purpose may be removal of environmentally hazardous materials, such as volatile organic compounds (VOCs), from a stream to be discharged to the air, removal of undesirable contaminants from a process gas, or recovery of valuable products from a waste gas.

The solute may dissolve in the solvent by virtue of its physical solubility, but in most cases it reacts chemically with the solvent. Common large-scale absorption applications include SO_2 removal from flue gas and acid gas (CO_2, H_2S, COS, etc), and removal from natural gas or chemical process gases (1,2). Absorption is often used to remove VOCs or other contaminants from vent gases, but the units are usually smaller.

Frequently the solute and solvent must be separated in order to recover the solute, reuse the solvent, or both. This is usually accomplished by the converse of absorption, ie, stripping, an operation in which the solute–solvent mixture contacts a gas and the solute evaporates preferentially. The absorption occurs at a higher pressure and/or lower temperature than the stripping.

Absorption and stripping are usually carried out in vertical countercurrent towers. A typical simplified flow sheet is shown in Figure 1. The feed gas enters the bottom of the absorber. The lean solvent, (ie, containing very little solute), enters the top. After absorbing the solute, the rich solvent leaves the bottom of the absorber and flows to the top of the stripper, also often called a regenerator. As the solvent flows down the stripper, contact with the stripping gas removes the solute, which leaves the top of the stripper and is discharged or treated further. The lean solvent leaves the bottom of the stripper and is pumped back to the top of the absorber.

Figure 1 shows a reboiler on the base of the stripper. When the solvent is an aqueous solution, water vapor is used as the stripping medium with the water vapor generated by adding heat to the reboiler. Figure 1 also shows a heat exchanger to interchange heat between the hot, lean solvent and the cooler, rich solvent. If there is no temperature difference between the absorber and stripper, as in some hot potassium carbonate processes, this heat exchanger is unnecessary. A condenser on the top of the stripper condenses the water vapor and returns it to the process. Whereas many different gas–liquid contacting devices may be used for absorption and stripping, most are carried out in packed or trayed towers.

Throughout the absorber the solute partial pressure in the gas is greater than the partial pressure that would exist if the gas were in equilibrium with the liquid. The difference in partial pressures is the driving force for absorption. The larger the driving force, the more rapidly the solute transfers into the liquid. In the stripper, the opposite situation exists; the partial pressure of solute in the stripping gas is less than would be in equilibrium with the liquid, so the solute desorbs from the liquid. The driving force can be expressed in terms of liquid concentrations rather than partial pressures if desired. The temperature and/or pressure difference between the absorber and stripper helps to magnify the driving force in each tower.

The most important factor in the economics of an absorption–stripping system is the solvent circulation rate. The larger the solvent rate, the larger the stripper, heat exchangers, pumps, etc, in the system. Other important factors include the corrosivity of the solvent, the cost of the solvent, amount of solvent losses, and process reliability.

The amount of a solvent required for a given absorption process is determined by the cyclic capacity of the solvent, ie,

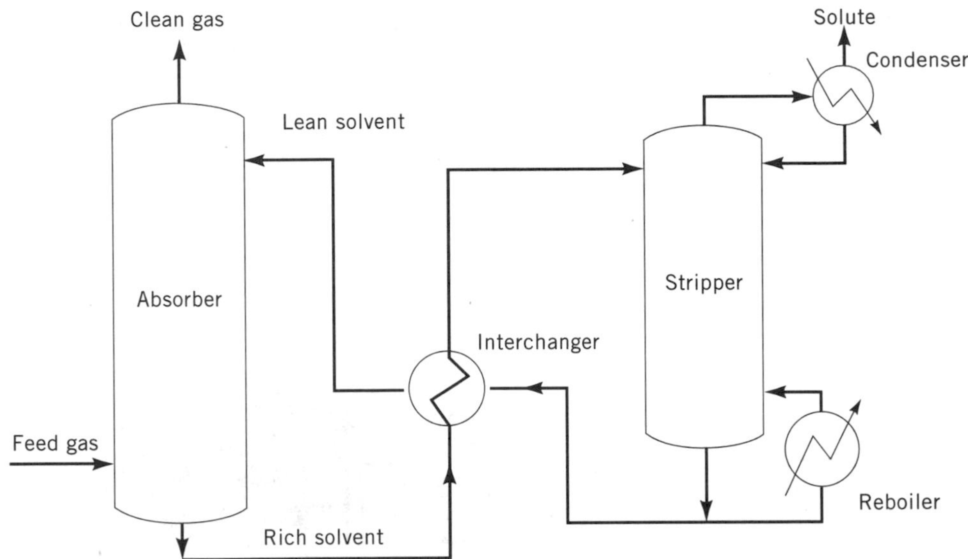

Figure 1. Typical absorber–stripper column arrangement.

the maximum amount of solute that can be absorbed per unit mass or volume of solvent. With a high cyclic capacity less solvent is needed, making the process more economical. When a solute is absorbed the heat of absorption is liberated, causing the temperature of the solvent to increase as it flows down the absorber. The heat of absorption for chemical absorption is generally greater than that for physical absorption, because it includes a heat of reaction in addition to a heat of solution. The increase in solvent temperature at the absorber bottom tends to decrease the driving force at that location; in some cases this effect may limit the capacity of the solvent.

The heat supplied to the stripper reboiler must be sufficient to generate a sufficient amount of stripping vapor, provide the heat of desorption (reverse of the heat of absorption), and provide the necessary sensible heat to raise the solvent to the equilibrium temperature needed to reduce the solute concentration low enough to provide the driving force needed at the top of the absorber. A solvent rate increase causes two of the three heat quantities to increase.

Compared with physical solvents, chemical solvents (those that react with the solute) have the advantages of a high cyclic capacity and the capability of removing a high percentage of the solute. Disadvantages of chemical solvents include a higher heat of absorption, a greater tendency to be corrosive, and additional costs associated with replacing solvent degraded by unwanted side reactions. In general, a physical solvent is preferable when the solute has a high partial pressure in the feed gas and vice versa.

PACKED TOWERS

The packed tower is a shell filled with randomly dumped elements (random or dumped packing) or with corrugated, stacked pieces of wire gauze or sheet metal (structured packing). Packed towers offer relatively high efficiency, and a low pressure drop (3). Random packings have been classified into three generations based on the time period during which they came into common use (4). The Raschig ring, Berl saddle, and Lessing ring comprise the first generation. These packings are rarely used in new installations, but are still common in existing towers. All three packings are made in ceramic and the ring packings are also available in metal.

Second-generation packings include the Intalox saddle and the Pall ring. U.S. Stoneware (now Norton Chemical Process Products Corporation) improved on the Berl saddle design with the Intalox, and later the Super Intalox saddle. Both of these saddles are available in ceramic, and the Super Intalox is also available in plastic.

The Pall ring was a significant breakthrough in packing design. It consists of a metal Raschig-type ring in which slots have been punched, leaving the punched-out material attached to the ring on one end so that fingers protrude into the center of the ring. These fingers provide additional points for creating new surface area, and also provide additional strength so that the metal thickness can be less than for Raschig rings. Later, plastic Pall rings, similar in design, were developed. The Pall ring is the standard of comparison for dumped packings. The Hy-Pak is similar to the Pall ring, but has more internal tongues, is slightly larger than the equivalent Pall ring, and is claimed to have the same efficiency as the Pall ring with greater capacity.

Packings considered equivalent to second-generation packings are marketed by a number of vendors under various trade names. Among the saddles are Flexisaddle, Novalox, and Ballast saddles. Ballast rings and Flexirings are Pall-type rings; K-PAC and Ballast-Plus are similar to Hy-Pak.

Third-generation packings originated in the mid-1970s. There are a large number of different geometries; usually each is marketed by one vendor only. A partial list of these packings follows: Intalox metal tower packing (IMTP), Nutter rings, HcKp, Hiflow ring, Cascade MiniRings (CMR), Jaeger Tri-Packs, Top-Pak, IMPAC, NORPAC (NSW), Chempak or Levapak (LVK), Intalox Snowflake packing, FLEXIMAX, and LANPAC.

A packing's specific surface area a_t is the amount of area per unit packed volume and is indicative of the amount of interfacial area available for mass transfer between gas and liquid. As a general rule, the larger the specific surface area the more efficient the packing.

The packing factor F_p is a proportionality between pressure drop through the packing and flow rates. The higher the packing factor, the higher the pressure drop. Because high pressure drop tends to lower capacity, the packing factor can be considered an inverse indicator of capacity. Smaller packings have greater efficiency, but less capacity.

Because random packings can be made from metals, plastics, and ceramics, a packed tower with a random packing can usually be designed with the most economical choice of materials of construction. When carbon steel is suitable, it is the material of choice because metal packings usually perform better than ceramic or plastic packings. Most plastic packings are limited to temperatures less than 121°C, but some more expensive plastics withstand higher temperatures. Many liquids tend to wet plastics less well than metals or ceramics, and plastics can be chemically attacked by some organic solvents. Ceramic packings tend to be used in highly corrosive services. They are heavy and less convenient to work with and are subject to abrasion and breakage.

In towers containing random packings, the packing should be small enough relative to the tower diameter to prevent excessively large gaps at the tower wall. The ratio of tower diameter to packing size should be at least 8, and 12 is preferred, especially if using a first-generation packing. On the other hand, packings larger than 5 cm (2 in.) seem to have lower efficiency than their specific surface area would suggest, so there is little incentive for choosing packings larger than 5 cm, except for operations requiring high capacity but few separation stages.

Structured packings exhibit extremely good mass-transfer characteristics combined with a very low pressure drop, but they are usually more expensive per volume of packing than dumped packings. Widely used in multistage and vacuum distillations, until recently they have rarely been used for absorption and stripping applications. This situation is changing, however, as costs have become more competitive and appreciation of the performance advantages of structured packing has increased.

In addition to the packing, a packed tower should contain a support plate and a liquid distributor; a bed limiter or hold-down plate is also advisable.

A packed tower is usually preferable to a trayed tower in the following situations: when a low pressure drop is valuable, ie, cycle gas streams, vent gas scrubbers; when use of ceramic or plastic is less costly, ie, in corrosive systems or low temperature aqueous systems; with small diameter towers, $\leq$0.9 m; when low liquid holdup is desirable; and for systems with foaming tendency.

TRAYED TOWERS

The trayed tower is a shell containing a series of flat, usually metal plates or trays. When operating normally, these trays support a pool of frothy liquid while gas rises through the tray and bubbles through the liquid. The froth flows into one or more downcomers, where the liquid degasses and flows to the tray below. Trayed towers offer high efficiency and stable and predictable operation.

The trays are classified by the device used to introduce the gas into contact with the liquid and by the flow path taken by the liquid as it flows down the column. The most common trays based on device classification are the sieve (or perforated) tray and the valve tray. Figure 2 depicts the gas flow through the devices on these trays. Other type trays include the bubble cap tray, the fixed valve tray, and the baffle tray.

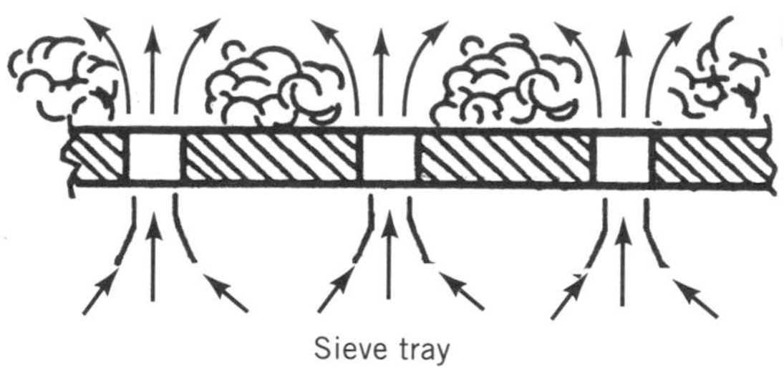

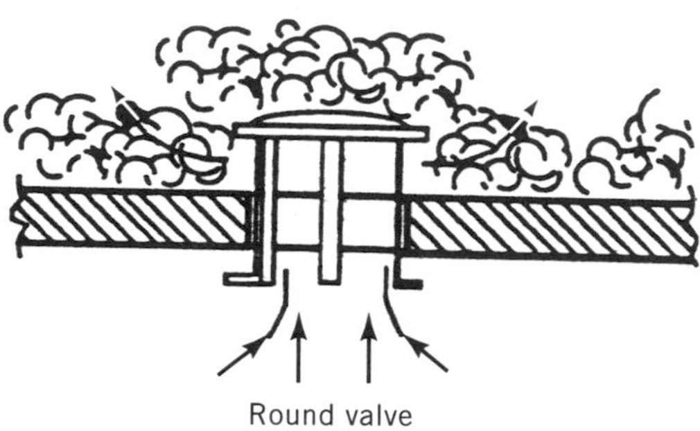

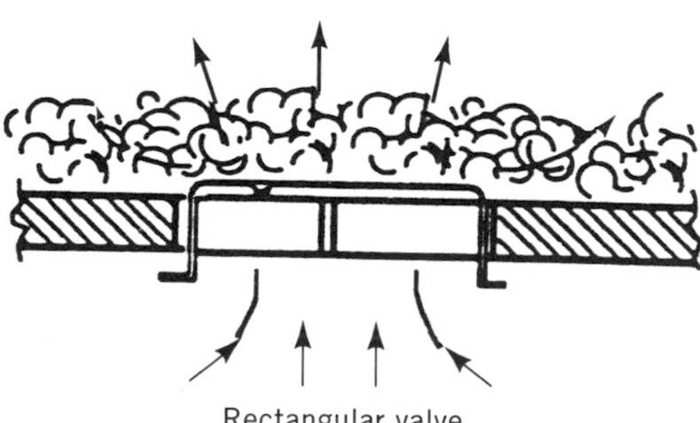

Figure 2. Gas flow through contact devices on trays.

GAS–LIQUID EQUILIBRIA

The most important information for the design of absorbers and strippers is that of the gas–liquid equilibria for the transferring species. This equilibria defines the limitations of the gas–liquid contacting operation for physical absorption. It is also necessary in the calculation of driving force compositions. Securing sound gas–liquid equilibria information is often the most difficult part of the design of gas–liquid contacting equipment.

BIBLIOGRAPHY

1. G. Astarita, D.W. Savage, and A. Bisio, *Gas Treating with Chemical Solvents,* John Wiley & Sons, Inc., New York, 1983.

2. A.L. Kohl and F.C. Riesenfeld, *Gas Purification,* 3rd ed., Gulf Publishing Co., Houston, Tex., 1979.
3. R.F. Strigle, *Random Packings and Packed Towers,* Gulf Publishing Co., Houston, Tex., 1987.
4. H.Z. Kister, *Distillation Design,* McGraw-Hill Book Co., Inc., New York, 1992.

GEOCHEMISTRY, ENVIRONMENTAL

KONRAD B. KRAUSKOPF
Stanford University
Stanford, California

Environmental geochemistry is the study of the chemical composition of rocks, soil, water, air, and organisms near the earth's surface and the chemical reactions among them. These reactions form a system of interrelated processes that constitute a steady state in which concentrations of major substances remain approximately constant for long periods. In environmental geochemistry the concern is the mechanisms by which this steady state is maintained and the influences that cause changes in concentrations and so disturb the steady state. Both natural processes in the environment and the interactions of humans with their environment are considered, especially the possible long-term effects of human activity on reactions in the steady-state system.

The present steady state of the earth's environment is the product of a long history of change from very different conditions on the primitive earth. Slow overall change may be continuing, but reaction rates are now well adjusted so that most materials in the environment are being regenerated about as fast as they are consumed. For example, oxygen continually disappears in the processes of decay and slow oxidation of iron and sulfur minerals, but is constantly replenished by photosynthesis in the green leaves of plants. Reactions in the environment proceed at widely different rates. Some are fast enough that they approach equilibrium. For example, calcium carbonate in contact with water at the earth's surface is commonly in approximate equilibrium with its ions and the carbon dioxide of the adjacent air. Other reactions are so slow that, even in geologic times, equilibrium is not approached. Nitrogen and oxygen in air, for example, are in equilibrium with a dilute solution of nitrate or nitric acid, yet they remain in contact with seawater containing very little nitrate and show no sign of reaction. Most environmental reactions have rates fast enough to make the equilibrium model useful in predicting their general course, but slow enough that kinetic factors are more important determinants of concentrations at a particular time and place.

Humankind is well adjusted to present steady-state conditions and is necessarily concerned with changes in these conditions brought about by natural causes or by human activity. The steady state is reasonably stable, because it has maintained an environment much like the present all through human history. Although our ancestors disturbed the environment in many ways (eg, by setting fires, mining metals, and polluting streams), other reactions in the steady-state system have compensated for these disturbances. In modern times the increased population and greater technical knowledge have enormously increased our ability to modify the environment. Important questions, then, are whether the chemical reactions of the steady-state complex are still adequate to compensate for changes that humankind can produce and whether the system is stable enough to resist changes that come about from predictable natural events.

Such problems loom large in a survey of environmental geochemistry. To understand them, some reactions of the steady-state system must be examined, then how changes in constituents of the reactions are compensated for at present must be determined, and finally consequences of some environmental modifications that can be foreseen may be predicted. Because the concern is mainly with the environment in which humans spend most of their lives, the discussion focuses on the land surface of the continents. Other parts of the earth's environment, for example, the oceans or the high atmosphere, are mentioned only to the extent that they influence the global geochemical system that operates in our immediate surroundings.

SOLID MATERIALS

Surfaces of continents are underlain everywhere by solid rock, and most other solid materials at the surface of the earth are derived from rock. Information obtained from seismic waves indicates that ordinary rock persists downward into the earth for a few tens of kilometers and that it changes rather abruptly into a different kind of rock at a depth averaging about 35 km. The material above this depth is called the earth's crust. The crust also extends under the oceans but has an average thickness of only about 6 km and consists of material different from the kind of rock under most continental areas. Two kinds of substance make up most of the continental crust: compounds of silicon and compounds of carbon. The nature of these compounds and their reactions with air and water at the earth's surface will be discussed.

Silicon, Silica, and Silicates

The many kinds of rock that form the crust of the earth include igneous rocks, such as granite and basalt; metamorphic rocks, such as schist and marble; and layered sedimentary rocks, such as sandstone and shale. Most of them consist chiefly of minerals that are compounds of silicon. Much of the inorganic geochemistry of our surroundings is linked to the properties of this element. Among familiar rock types, only limestone and dolomite are exceptions to the rule that rocks are predominantly aggregates of silicon compounds.

Atoms of silicon, Si, are small and have four valence electrons, so that four strong bonds can be formed with adjacent atoms. In the crystal structures of rock-forming minerals, the adjacent atoms are always oxygen, so that each Si atom is surrounded by four O atoms at the corners of a tetrahedron. Each oxygen in this SiO_4^{4-} group has an unsatisfied valence that can link the group to other atoms, either other atoms of silicon or atoms of metals. If all the bonds are linked with metals, a simple crystal structure can result, with alternating SiO_4^{4-} groups and metal cations. If some of the oxygens are linked to other silicon atoms, however, the SiO_4^{4-} tetrahedra can attach themselves to form many kinds of structures: single chains, double chains, rings, and three-dimensional frameworks, generally with metal ions forming linkages between the structures. It is this unique ability of SiO_4^{4-} groups to join together in many patterns that

Table 1. Average Abundances of Major Elements, wt %

Element	Crust[a]	Granite[b]	Basalt[b]	Shale[b]	Sandstone[c]	Limestone[c]	Soil[c]
O	47.2	48.5	44.1	49.5	49.2	49.4	49.0
Si	28.8	32.3	23.0	23.8	32.7	3.2	33.0
Al	7.96	7.7	8.4	8.8	4.3	0.9	7.1
Fe	4.32	2.7	8.6	4.7	2.9	1.7	4.0
Ca	3.85	1.6	7.2	2.5	3.1	34.0	1.5
Na	2.36	2.8	1.9	0.9	1.0	0.1	0.5
Mg	2.20	0.4	4.5	1.4	1.2	0.6	0.5
K	2.14	3.2	0.8	2.5	1.5	0.3	1.4

[a] Ref. 2.
[b] Ref. 3.
[c] Ref. 4.

gives rise to the extraordinary variety of silicon-oxygen-metal compounds that make up the minerals of ordinary rocks (1).

In the common mineral quartz, the tetrahedra are linked in a three-dimensional structure with no metals. Thus the formula of the mineral is simply SiO_2, or silicon dioxide, a compound called silica. The metal-containing minerals are called collectively silicates. Formulas of many silicates are very complex because metals can substitute extensively for one another in the crystal structures. A few examples, with their formulas somewhat simplified, include the following:

feldspar	$(K,Na)AlSi_3O_8$ and $(Na,Ca)(Si,Al)AlSi_2O_8$
amphibole	$Ca_2(Mg,Fe)_5Si_8O_{22}(OH)_2$
pyroxene	$Ca(Mg,Fe)Si_2O_6$
mica	$KAl_2(AlSi_3O_{10})(OH)_2$ and $K(Mg,Fe)_3AlSi_3O_{10}(OH)_2$
olivine	$(Mg,Fe)_2SiO_4$
kaolinite	$Al_4Si_4O_{10}(OH)_8$
serpentine	$(Mg,Fe)_6Si_4O_{10}(OH)_8$

The first five on this list, along with quartz, are the major constituents of igneous rocks, and the first four are also abundant in metamorphic rocks. Kaolinite is common in sedimentary rocks, and serpentine in some metamorphic rocks. The minerals range in hardness from feldspar, nearly as hard as quartz, to kaolinite, which is soft and chalky. The five silicates that contain iron are dark colored (green to black), and the others are light colored. This sampling gives only a glimpse of the variety of properties found in silicate minerals. They have one property in common: all are resistant to chemical attack and react only slowly to potent reagents.

Composition of the Earth's Crust

The minerals in the list above, plus quartz, make up a major part of the bedrock under all parts of the earth's surface. The mineral formulas include only a small number of elements, and these elements are thus among the most abundant constituents of the earth's crust. In any estimate of the average composition of the continental crust, the eight elements O, Si, Al, Fe, Mg, Ca, Na, and K are more abundant than all others (Tables 1 and 2).

An estimated analysis of average crustal composition, like that in the two tables, is obviously an important item in geochemistry. Much effort has been devoted to the problem of making the figures in such an analysis as accurate as possible. The problem is not easy: it requires assembling analyses of various rock types and then making an estimate of the relative abundance of these rocks in the part of the earth accessible to observation. Inevitably there is much disagreement about relative abundances and a corresponding variety of opinion about the best numbers to use in the estimated overall analysis. Because granite and basalt are the two most common igneous rocks and because the materials of most sedimentary and metamorphic rocks are ultimately derived from these two, a procedure used in calculating the overall figures is simply to combine accurate analyses for granite and basalt assuming roughly equal amounts in the crust. Because of differences of opinion about the proportions and about what rocks to include under the headings granite and basalt, analyses prepared this way by different authors are not identical. A recent particularly careful calculation, based on analyses of a greater variety of rocks and on seismic records as an aid in estimating their relative abundance, is the basis for the numbers in Tables 1 and 2 (2). Comparison of the first three columns in each table shows that the numbers for most elements lie between those for granite and basalt, as they would necessarily in the older analyses, but a few of the minor elements are conspicuous exceptions. Thus the analysis we start with for the overall chemistry of the earth's crust cannot claim complete accuracy or entire agreement among all those who have worked with it, but it still forms a satisfactory basis for our inquiries into the geochemistry of the environment.

Because basalt and granite play such important roles in continental environments, some of their major chemical differences (second and third columns of each Tables 1 and 2) are worth noting. As a rock consisting chiefly of quartz and feldspar, granite has high Si, Na, and K values; basalt, which is made up chiefly of Fe–Mg silicates and Ca–Na feldspar, is rich in Fe, Mg, and Ca. (Because the crust under the oceans is largely basalt, its inclusion in the average analysis would result in larger values for these three elements.) Minor elements in the two rocks tend to follow major elements of similar chemical properties. Thus Ba, Rb, and La are conspicuously more abundant in granite, whereas V, Cr, and Ni are prominent in basalt. Some elements, for example Al, Ga, and Nb, show little difference from one rock to the other and are also the same in the crustal average.

Especially notable in the crustal average is the great abundance of only a few elements. Oxygen and silicon alone make up nearly 75% of the weight of the crust, and, adding the six metals mentioned above, almost 98% is accounted for. Such common and important elements as H, C, S, Cu, and

Table 2. Average Abundances[a] of Minor Elements, ppm

Element	Crust	Granite[b]	Basalt[b]	Shale[b]	Sandstone[c]	Limestone[c]	Soil[c]
Ag	0.07	0.04	0.1	0.1	0.2	0.1	0.05
As	1.7	1.3	1.7	10	1	1	6
Au	0.0025	0.002	0.004	0.003	0.003	0.002	0.003
B	11	15	5	100	30	20	20
Ba	584	700	300	600	300	90	500
Be	2.4	5	0.5	3	<1	<1	0.6
Br	1.0	0.5	0.5	5	1	6	10
C	1900	360	610	14,000	16,000	116,000	20,000
Cd	0.1	0.1	0.2	0.3	0.05	0.03	0.4
Ce	60	90	50	70	80	20	50
Cl	472	200	60	170	500	76	100
Co	24	3	40	20	0.3	0.1	8
Cr	126	20	200	130	35	11	70
Cs	3.4	5	1	7	0.5	0.5	4
Cu	25	12	90	50	30	5.5	30
Dy	3.3	7	4	5	7	2	5
F	525	1300	500	800	180	220	200
Ga	15	19	17	25	6	2.5	20
Hf	4.9	7	4	3	4	0.3	6
Hg	0.04	0.06	0.05	0.1	0.05	0.2	0.06
I	0.8	0.2	0.1	0.1	0.1	0.1	5
La	30	55	20	40	42	10	40
Li	18	30	12	60	38	7.5	25
Mn	716	500	1500	850	460	620	1000
Mo	1.1	1.5	1	2	0.2	0.2	1.2
N	60	20	20	600	120	80	2000
Nb	19	20	20	15	0.05	0.3	10
Nd	27	40	20	30	38	9	35
Ni	56	5	150	80	9	7	50
P	757	700	1400	750	440	700	800
Pb	14.8	24	4	20	10	6	35
Pr	6.7	11	7	9	11	3	7
Rb	78	150	30	160	46	50	150
S	697	300	300	2500	2300	1100	700
Sc	16	8	35	15	1	1	7
Se	0.12	0.05	0.05	0.6	0.05	0.03	0.4
Sm	5.3	9	5	7	8	2	5
Sn	2.3	3	2	6	0.5	0.5	4
Sr	333	300	450	400	20	610	250
Ta	1.1	3	1	2	0.05	0.05	2
Th	8.5	20	2	12	4	2	9
Ti	4010	2100	9000	4500	3500	300	5000
U	1.7	5	0.5	3.5	0.5	2	2
V	98	50	250	130	20	45	90
Y	24	40	30	35	54	23	40
Zn	65	50	100	90	30	20	90
Zr	203	240	140	180	220	20	300

[a] Elements not listed have abundances in crust <1.5 ppm.
[b] Ref. 3.
[c] Ref. 4.

Pb make up only small parts of the remainder. Only because these and other minor elements have been concentrated locally by geochemical processes can we obtain them in the quantities needed for civilized living.

Carbon Chemistry

An all-important minor element, from a human standpoint, is carbon, the chief constituent of materials produced by life processes. Because plants and animals are so conspicuous and omnipresent in the world around us, it is hard to realize that their principal element makes up a bare 0.05% of the weight of the earth's crust. Organic materials, living and dead, form only a thin and discontinuous veneer on the outer surface of the silica-rich materials that make up most ordinary rocks. The rocks themselves are part of a layer that is thin in comparison with the earth's diameter. Thus carbon makes up a very minor portion of the planet's overall chemistry, but it plays a major role in the human environment at the surface.

The properties of organic materials reflect the chemistry of carbon, much as the properties of rocks depend largely on the chemistry of silicon. Like silicon, carbon consists of small atoms with four valence electrons, but the atoms are so small that four oxygen atoms cannot fit comfortably around them. Thus, in contrast to silicon, carbon atoms are not linked with oxygen in tetrahedra, but rather form strong bonds with each other. This property permits them to join into chains and rings of great variety. Attached to the chains and rings are the other elements common in living organisms, chiefly oxygen and hydrogen plus small amounts of nitrogen, phosphorus, sulfur, and many others. Thus the complexity of silicate minerals is traced to the ability of silicon–oxygen tetrahedra to join together in many patterns with metals as linkages, and the complexity of organic compounds is explained by the tendency of carbon atoms to unite with each other in structures whose other components are largely nonmetals.

Organic compounds are usually less stable than silicates in environments at the earth's surface. Their reactions are generally slow, but most of them yield in times that are measurable in human terms to reactions involving oxidation or internal rearrangement of complex molecules to form simpler ones. Common products of oxidation reactions are CO_2 and H_2O; a common ultimate product of nonoxidative decay is methane, CH_4. Some organic compounds are exceptions to the general rule, and, under the right circumstances, persist for geologic times. Coal is a rock formed by the burial of partly decayed vegetation from ancient swamps, and petroleum consists of organic compounds resulting from partial decay of the remains of tiny marine organisms that accumulated in ancient seas. Most sedimentary rocks contain at least ~1% of resistant carbon compounds that have persisted with little change since shortly after the rocks were formed.

One other group of rocks which contain carbon in abundance are carbonate rocks, of which limestone (chiefly $CaCO_3$) and dolomite (chiefly $CaMg(CO_3)_2$) are the principal varieties. Carbonate minerals form in many ways, but in large part they are an indirect product of organic processes. Most limestone consists of material originally deposited as the shells of myriads of marine organisms, and much dolomite is formed by replacement of this material. Some limestone is nearly pure $CaCO_3$, but most contains at least minor amounts of quartz, clay, and other silicate minerals. Mixtures of carbonate and silicate minerals are very common in sedimentary rocks and locally in metamorphic rocks, but carbonate minerals are scarce or absent in common igneous rocks.

Weathering

Rocks and organic materials exposed at the earth's surface are attacked by water and air. The attack involves a set of complex slow reactions, partly physical and partly chemical, collectively called weathering. The chemical part of weathering is a slow approach to a new equilibrium for minerals and compounds that were in very different environments at the time of their origin. The silicate minerals of igneous and metamorphic rocks, for example, were generated under high temperature and often high pressure out of contact with the atmosphere. The carbonate minerals of limestone and dolomite were formed in bodies of water having compositions generally different from those of rain or fresh running water. Organic materials originate in the special chemical environment of living substances. When these materials are exposed to air and water at the earth's surface, they must adjust to temperatures, pressures, and compositions of adjacent materials quite different from the conditions of their origins. The process of adjustment causes the discoloring and softening of most rock surfaces and the disappearance of most organic matter. Because rocks slowly disintegrate by weathering, the ultimate result is the formation of loose debris that remains as a layer of soil over the rock surface or is carried away by wind or running water to be deposited as sediment.

Many kinds of weathering reactions are distinguished (5,6). The simplest is the ordinary dissolving of soluble compounds. When a rock containing salt or gypsum comes in contact with water, the reactions are the following:

$$NaCl \rightarrow Na^+ + Cl^-$$

$$CaSO_4 2H_2O \rightarrow Ca^{2+} + SO_4^{2-} + 2H_2O \qquad (1)$$

Dissolution may be aided by hydrolysis, a reaction with water to form a weak acid or its ion. For example, carbonates dissolve to form bicarbonate ion by the following reaction:

$$CaCO_3 + H_2O \rightarrow Ca^{2+} + HCO_3^- + OH^- \qquad (2)$$

and silicates dissolve to form the even less dissociated silicic acid or its ion:

$$Mg_2SiO_4 + 3\,H_2O \rightarrow 2\,Mg^{2+} + H_3SiO_4^- + 3\,OH^- \qquad (3)$$

These reactions go farther if the water contains acid to combine with the liberated OH^-. The most common acid in water under natural conditions is carbonic acid H_2CO_3; this acid is so readily formed that nearly all water at the earth's surface contains at least small amounts of it. The acid results from the dissolving of atmospheric CO_2 by the following reaction:

$$CO_2 + H_2O \rightarrow H_2CO_3 \qquad (4)$$

The weakness of the acid is shown by its dissociation constants, $K_1 = 10^{-6.4}$ and $K_2 = 10^{-10.3}$. Despite the weakness and the small amount generally present (the solubility of CO_2 in pure water at 25°C is only $\sim 10^{-5}$ M in equilibrium with the $10^{-3.5}$ atm of the gas in air), carbonic acid greatly increases the solubility of carbonates and silicates by the following reactions:

$$CaCO_3 + H_2CO_3 \rightarrow Ca^{2+} + 2\,HCO_3^- \qquad (5)$$

$$Mg_2SiO_4 + 4\,H_2CO_3 \rightarrow 2\,Mg^{2+} + H_4SiO_4 + 4\,HCO_3^- \qquad (6)$$

The effectiveness of the acid markedly increases if weathering takes place (as it commonly does) in contact with soil, because decaying vegetation makes the concentration of CO_2 in the air in soil pores much higher than that in the surrounding atmosphere.

These reactions show simple dissolving of the minerals because their anions form weak acids. The carbonate reaction is rapid enough that a close approach to equilibrium is indicated by the fact that measured concentrations of Ca^{2+} are similar to those calculated from dissociation constants.

Hydrolysis reactions of silicates that have two or more metals in their formulas are more complicated. Such a mineral commonly dissolves incongruently, meaning that at first only part of it dissolves, leaving behind a new mineral and forming a solution with a mole ratio of constituents different from that in the original mineral (7). A good example is the hydrolysis of aluminosilicate minerals like the feldspars, which are especially important because these reactions lead to the formation of clay minerals:

$$\underset{\text{feldspar}}{2\,KAlSi_3O_8} + 2\,H_2CO_3 + 9H_2O \rightarrow \underset{\text{kaolinite}}{H_4Al_2Si_2O_9} + 2\,K^+ + 4\,H_4SiO_4 + 2\,HCO_3^- \quad (7)$$

This process is driven, as in equation 6, by the reaction of an acid (H_2CO_3) with silica to form the undissociated silicic acid (H_4SiO_4). Here, however, only part of the silica reacts; some is left behind in the clay, and the ratio K:Si in solution is 1:2 in contrast to 1:3 in the original feldspar. Aluminosilicate minerals are common constituents of ordinary rocks, and their weathering produces many types of clay. The kind of clay formed depends on the nature of the original mineral and on the weathering conditions. The reactions are often much more complex than equation 7. One or more of the clay minerals (kaolinite, smectite, or illite), is the usual product of aluminosilicate weathering in temperate climates, but, under extreme conditions (particularly in the humid tropics), the clay minerals break down by releasing silica to solution and leaving a hydrate of aluminum oxide (gibbsite $Al(OH)_3$, or boehmite AlOOH) as a solid residue.

Another common reaction of weathering is oxidation, which primarily affects minerals containing iron and/or sulfur:

$$Fe_2SiO_4 + 1/2O_2 + 2\,H_2O \rightarrow Fe_2O_3 + H_4SiO_4 \quad (8)$$

$$ZnS + 2O_2 \rightarrow Zn^{2+} + SO_4^{2-} \quad (9)$$

$$2\,FeS_2 + 15/2O_2 + 4\,H_2O \rightarrow Fe_2O_3 + 4\,SO_4^{2-} + 8\,H^+ \quad (10)$$

The iron oxide product of reactions 8 and 10 is written as Fe_2O_3 (hematite), but it could as well be shown as FeOOH (goethite) or amorphous material with formula $Fe(OH)_3$. These forms of ferric oxide have a variety of colors ranging from red through purple and brown to yellow and orange. The color depends on the particular mineral and the grain size. These are the chief coloring agents in rocks of ordinary environments responsible for the common browns, reds, and yellows of weathered rock surfaces and sediments derived from them. Reaction 10, the oxidation of pyrite, is particularly important, because it shows that upon weathering this most abundant sulfide mineral yields an acid solution and abundant iron oxide, two products generally conspicuous where sulfide ore deposits are exposed to air and water. Oxidation is also a major process in the weathering of organic materials, which leads to carbon dioxide and water as the principal products; for example,

$$C_6H_{12}O_6 + 6\,O_2 \rightarrow 6\,CO_2 + 6\,H_2O \quad (11)$$

If nitrogen is present, it is usually released as N_2; sulfur is converted to sulfate in solution and phosphorus to phosphate.

Long-continued weathering ultimately produces solutions containing most of the original metal ions present in the rock and some of the silica, plus solid products consisting of newly formed compounds (iron oxide and clay minerals) and residual solids little affected by weathering reactions (chiefly quartz and minor amounts of such resistant minerals as zircon and garnet). Because weathering is very slow and its rates differ markedly for different minerals, its products generally also include material that is only partly weathered.

Weathering results in converting igneous and metamorphic rock into sediments and dissolved ions in stream and ocean waters and in the disappearance of most dead organic matter. It is a process of decomposition of complex original compounds into simpler ones and a rough sorting of elements according to their chemistry: silicon mainly into detrital sands because of the resistance of quartz to chemical and physical attack, aluminum into fine-grained sediments because of the stability of clay minerals, and sodium and potassium into the ocean because most of their compounds are soluble. The new substances are nearer to chemical equilibrium with air and water than the original rock minerals, but complete equilibrium is seldom attained.

Formation of Soils

Soils are the ultimate product of rock weathering. As the debris of weathering accumulates, it is subject to continued attack by air and by rainwater percolating through it. Movement of water carries dissolved material and small particles from one layer to another. Organic material from decaying vegetation is mixed with the rock debris by water movement and by the activity of small organisms. Bacteria play an important role in the partial decomposition of organic matter. The result of this complex of reactions is soil.

The chemical changes in soil formation continue the reactions noted in the weathering of the original rock (8). Soluble alkali metals (Na and K) and alkaline-earth metals (Ca and Mg), plus much of the silica in the form of H_4SiO_4, are leached out of the upper layers of the soil and removed in solution. Aluminum goes mostly into clay minerals that are moved mechanically from upper to lower layers. Iron is dissolved from the upper part (where it is kept in the more soluble reduced form by organic matter) and precipitated as ferric oxide as it moves downward. By such processes, soil commonly develops a distinct layered structure, which comprises (*1*) an upper layer (called the A horizon) from which much of the soluble material has been removed and in which the organic matter is concentrated, (*2*) a middle layer (B horizon) in which clay and ferric oxide are prominent constituents, and (*3*) a lower layer (C horizon) made up of little altered rock debris. Chemical change in the organic matter of soil results in the breakdown of complex carbon compounds into simple ones, and the liberation of much CO_2 which aids in further decay of the rock. The decomposing organic matter also provides surfaces that trap some of the alkali and alkaline-earth ions by sorption. Although these ions are in large part leached completely out of the soil, those that are held by sorption on clay particles and organic surfaces are important because plants obtain much of the mineral substances necessary for their growth from these adsorbed ions.

Soils vary greatly in their structure and composition. The differences depend on the rocks from which they are derived and on the prevailing climate. Surprisingly, climate is usually the more important of the two factors. In humid temperate climates, the three major horizons are commonly well developed, with an upper layer impoverished in the more soluble elements and a B horizon enriched in clay and iron oxide, sometimes creating a hardpan. In the tropics, where oxidation is rapid and organic matter decays quickly, ferric oxide may be prominent in all layers, giving rise to the red soils called laterites. Where rainfall is lower, leaching from the A horizon is less, and there is some accumulation of Ca and Mg in the B horizon. Because the more soluble elements are essential for plant growth, these soils are more productive than those of very humid regions. Leaching is less effective with increasing aridity, and the accumulation of alkali and alkaline-earth elements at all levels produces the typical alkaline soils of desert regions. These generalizations apply to soils developed on most kinds of rock; in areas where soil formation has been going on for a long time, the soils found on different varieties of bedrock are often remarkably similar. Where rock compositions are extreme, soils developed on their surfaces depend more on the character of the rock. Soils on limestone, for example, are unusually rich in calcium, and those on serpentine are rich in magnesium.

Another factor influencing the composition of a soil is its age. Any soil represents a steady state between processes of formation and processes of destruction. As new soil is being generated by weathering of bedrock below, some of the surface material is being removed by erosion. A relatively young soil that develops in areas where erosion is rapid and soil-forming processes have little time to operate, as on a steep hillside, has a composition similar to that of the rock beneath it. An older soil, for example one developed over a flat area where the formation of soil horizons has been going on for a long time, has a composition largely determined by climate. Because soil formation is exceedingly slow and because mature soils are better suited for agriculture than raw young soils, the control of erosion of well-developed soils is an important part of environmental management (9).

Trying to specify an average chemical composition for soil is futile, because any one soil differs from top to bottom and because soils differ greatly from place to place. Nevertheless, attempts have been made to give rough estimates of the overall chemistry of soils, and one such effort is included as the last column of Tables 1 and 2. Although the particular numbers may be questionable, they provide a useful summary of the general chemical effect of soil formation.

Ore Deposits

The chemical reactions of weathering and soil formation, together with the more active processes of erosion, lead to enrichment of elements or groups of elements in some deposits and depletion in others. The enrichment of particular elements may become so marked and the amount of accumulation so large that the element or elements can be extracted in commercially important quantities. Such accumulations are called ore deposits. Not all ores, of course, originate in the low temperature processes of weathering and erosion, but some important ones form in this way and have contributed substantially to the materials essential for industrial civilization.

The simplest example is placer deposits, formed when grains of a heavy resistant mineral like gold, originally dispersed in small amounts through bedrock, are released during long periods of weathering of the more easily dissolved part of the rock and then are accumulated by the action of running water or waves. In addition to gold, placers are important sources of tin, thorium, and platinum. Rich hematite deposits, such as the now exhausted high grade ores of the Mesabi Range in Minnesota and similar deposits being exploited in Venezuela and Brazil, were formed from rocks containing abundant iron minerals through processes similar to those that are now observed segregating iron oxides in sediments and soils. Major clay deposits have been formed by weathering of rocks rich in feldspar, and the pure quartz sand needed for making glass comes from accumulations of quartz grains left after the more soluble parts of the original silica-rich rocks were dissolved or winnowed out by erosion. In tropical regions of high temperature and abundant rainfall, much of the clay decomposes to release silica in solution and leave behind a residue of aluminum hydroxide minerals ($Al(OH)_3$ (gibbsite) and AlOOH (boehmite)) called bauxite, the most important aluminum ore.

Many ore deposits, of course, are formed at higher temperatures, especially the temperatures associated with the intrusion of igneous magma into rocks of the crust. Most such deposits consist chiefly of sulfide minerals, which are stable under reducing conditions at depths not in contact with the atmosphere. When the rock material that once covered such deposits is removed by weathering and erosion to directly expose the sulfide ore to air, the metal compounds undergo a variety of reactions. Examples were given in equations 9 and 10; two others are equations 12 and 13:

$$\underset{\text{chalcocite}}{H_2O + Cu_2S} + 5/2O_2 + 2\,HCO_3^- \rightarrow \underset{\text{malachite}}{Cu_2(CO_3)(OH)_2} + SO_4^{2-} + H_2CO_3 \qquad (12)$$

$$17/2O_2 + \underset{\text{chalcopyrite}}{2\,CuFeS_2} + 2\,H_2O \rightarrow 2\,Cu^{2+} + \underset{\text{hematite}}{Fe_2O_3} + 4\,SO_4^{2-} + 4\,H^+ \qquad (13)$$

In all such reactions, the sulfur is oxidized to sulfate, the solution becomes more acidic, and metal ions are released in solution or precipitated as carbonates, oxides, sulfates, or other compounds formed with the common anions of waters near the earth's surface. Because pyrite and other sulfide minerals containing iron are common in sulfide deposits, hematite or a hydrated form of ferric oxide is nearly always present in large amounts at the surface around a weathering deposit of sulfide ore. The resulting patches of brown and yellow staining on surface rocks have long been used by prospectors to indicate the presence of valuable ore beneath. As an ore deposit is opened up for the recovery of its metal(s), sulfide grains are exposed widely over the surface, and the acid water resulting from their weathering carries abundant iron oxide far downstream. The resulting large area of discoloration and stunted vegetation is offensive to environmentalists and requires the expenditure of large sums by mining companies to limit the extent and unsightliness of such areas (10).

Ore deposits are an important part of the environment because their products are indispensable to civilized life. On the other hand, exploitation of an ore deposit can lead all too quickly to degradation of the environment by disfigurement of the landscape, by the spread of acid solutions into large areas of surface and underground material, and by the contamination of water supplies with toxic metals such as lead, cadmium, and arsenic. Striking a reasonable balance between these seemingly contrary effects on the environment has become a problem in the chemistry of ore deposits and in the economics and politics of their exploitation.

Reversal of Weathering reactions

Weathering reactions are part of the huge steady-state system that makes up our environment, which means that somewhere these reactions are reversed. Proof that reversal takes place lies in the simple fact that high mountains still exist, despite continuous weathering of the continental surface for more than four billion years. Were there no compensation for the effects of weathering and erosion, the continents would long ago have become plains mantled with thick weathered debris. Compensation does occur, but just how it occurs and whether it is complete in all details are questions subject to long and lively debate.

Weathering is essentially a destructive process in which complex minerals in igneous and metamorphic rock are broken down by reaction with water and gases from the air into simpler substances deposited as sediments and ultimately converted to sedimentary rocks. The overall process can be summarized in the following symbolic reaction (eq. 14):

$$CaSiO_3 + CO_2 \rightarrow CaCO_3 + SiO_2 \qquad (14)$$

or

$$CaSiO_3 + H_2O + 2\,CO_2 \rightarrow Ca^{2+} + 2\,HCO_3^- + SiO_2 \qquad (15)$$

where $CaSiO_3$ represents all the complex silicate minerals in the original rock; SiO_2, the new solid materials produced by weathering that ultimately become clastic sediments; and $CaCO_3$ (or $Ca^{2+} + 2\,HCO_3^-$), material dissolved during weathering and partly precipitated as chemical sediment. Somewhere among the many processes that shape the earth's surface are the reversals of such reactions.

The most large-scale of these processes are a part of plate tectonics. In subduction zones, great masses of rock and sediment are dragged down into the mantle by crustal plate movement and are heated to form metamorphic equivalents or melted to produce liquid rock (magma) that rises up through the crust in volcanoes or "freezes" under the surface of the earth as granite. The gases that have taken part in weathering are recycled, because the water and carbon dioxide that are added to sedimentary material (according to eqs. 7 and 8) now reappear as the major gases of volcanic eruptions. In effect, high temperature processes in the earth's interior reverse the low temperature weathering reactions symbolized by equations 14 and 15. Silica combines with simple metal compounds to form the complex silicates of metamorphic and igneous rocks, and other elements in the metal compounds are released to maintain the supply of CO_2, H_2O, and other gases in the atmosphere. On this large scale, the destructive work of weathering is adequately compensated for by the synthesis of silicates and liberation of gases brought about by plate tectonics (11).

On a smaller scale, many other processes are involved, and it is less certain that parts of the overall scheme remain in balance or that some of the compensating reactions are fast enough to prevent progressive and continuing changes in the geochemical environment. The formation of limestone, for example, requires that CO_2 be taken from the atmosphere. It is difficult to assess whether all limestone is recycled through subduction zones, in view of the great quantities of this rock that have been preserved on the continents for at least 500 million years. It is possible that CO_2 is being permanently removed from air and locked up in $CaCO_3$ far from any present or likely future subduction zone. A similar possibility exists for oxygen, which is continually reacting with minerals of iron and sulfur. It is uncertain whether these reactions are adequately reversed by subduction and photosynthesis and whether oxygen is being permanently lost to sediments.

The problem of reversibility is particularly important for ore deposits, because both natural and human processes are destroying these concentrations of unusual materials that were formed slowly over long time periods. It would be useful to know to what extent new accumulations of ore are now being formed that can make up for the dispersal of those presently known. Although ore deposits are surely forming today as they have in the past, the rate of formation is so slow that it has no significance in human terms. Here the balance that probably exists in nature between formation and destruction is being upset by human influence, and ore deposits being exhausted now will not be replenished at any times that would interest us or our descendants.

Thus, based on imperfect knowledge, it can be assumed that the earth's chemical environment is being maintained in a generally stable steady state. The obvious reactions of weathering are reversed by less visible processes mostly deep within the earth. Some details of the steady state are uncertain. The dispersal of metals and nonmetals from ores is one example of part of the steady state in which human activity has grossly disturbed the natural balance of compensating reactions.

THE ATMOSPHERE

The earth's atmosphere is a mixture of many gases, some essential to life and some continuously active in weathering rocks. Like the solid materials of the crust, the gases in air are maintained in a steady state. Some of them are being rapidly consumed and as rapidly generated, in a rough state of balance that has persisted without major change for long periods of geologic time. As for rocks and minerals, important geochemical data about the atmosphere relate to the mechanisms by which the balance is maintained, the stability of the balance, and the possible long-term effects of human activities.

Composition and Structure

The part of the atmosphere of direct concern to the environment is the relatively dense lower layer called the *troposphere,* ranging in thickness from about 10 km at the poles to 16 km over the equator. Clouds and storms are limited to this part of the atmosphere. Above the troposphere is the much less dense *stratosphere,* and beyond the stratosphere rarefied gases persist for many tens of kilometers into space.

The troposphere is kept sufficiently stirred so that its composition is constant from place to place, except for variable amounts of water vapor and local additions of gas from volcanic activity and industrial operations. As shown in Table 3, ordinary dry air consists of about 80% nitrogen and 20% oxygen, plus ~1% argon and minor amounts of many other gases. Of the minor gases, carbon dioxide (0.03%) is the most important. Water vapor amounts to 4% in warm humid air and is nearly absent in air over deserts. In addition to gases, air contains variable amounts of tiny solid particles and liquid droplets (aerosols): the water droplets of clouds, the mineral and organic particles of windblown dust, soot particles from combustion, and salt crystals from the evaporation of ocean spray. Such fine-grained material eventually settles to the ground, but may remain suspended for months or years. Especially persistent is the material ejected in the eruption clouds of volcanoes as solids and acid liquids, which sometimes remain in the stratosphere and troposphere in amounts large enough to affect weather patterns for several years.

Oxygen and Carbon Dioxide

From a cosmic perspective, the strangest constituent of the earth's atmosphere is free oxygen O_2. This active gas is absent or never present in more than traces in the atmospheres of other planets and all other known objects in the universe. Yet the earth has oxygen as a major constituent of its gaseous envelope. The chief reason for this is the reaction called *photosynthesis,* the process by which green plants are able to use the energy of sunlight to produce organic compounds from carbon dioxide and water. The overall reaction can be symbolized as in equation 16:

$$6\ CO_2 + 6\ H_2O \rightarrow C_6H_{12}O_6 + 6\ O_2 \qquad (16)$$

but the mechanism is far more complicated than this simple equation suggests. The reaction takes place in green leaves, with CO_2 coming from the surrounding air and H_2O from the soil through the plant's roots. It is catalyzed by the green leaf pigment chlorophyll, which absorbs sunlight and transfers the energy to the reacting molecules. Plants have the further ability to manufacture other organic molecules from those first formed and thus to prepare the foods on which all animal life depends.

Oxygen, a by-product of photosynthesis, is important to animals as a source of energy through respiration. The reaction involved in respiration can be symbolized as the reverse of equation 16, with the understanding that it applies to many other organic molecules in addition to the simple carbohydrate shown. The reverse reaction also shows what happens when organic molecules are burned, as symbolized by equation 11, or when they are oxidized by the slower processes of decay. Thus living things maintain a balance between CO_2 and O_2: plants use CO_2 to produce O_2, and animals accomplish the reverse (Fig. 1). It is no accident that the earth is unique in the abundance of O_2 in its atmosphere and also in the flourishing of life on its surface.

Photosynthesis is not the only source of molecular oxygen in a planetary atmosphere. Tiny amounts are formed in the stratosphere by the action of high energy radiation on water molecules, and circulation of the atmosphere brings it to lower levels. Oxygen thus produced would have only a transitory existence, because it would soon react with other gases or solids, but it may have played a role in the earth's early history before life became abundant. Steps in the development of life from its primitive beginnings in an atmosphere nearly devoid of oxygen to the establishment of photosynthesis as a major reaction can be traced only speculatively. At the present time, however, and for a long while in the geologic past, it is certain that the earth's supply of oxygen is maintained by this one reaction.

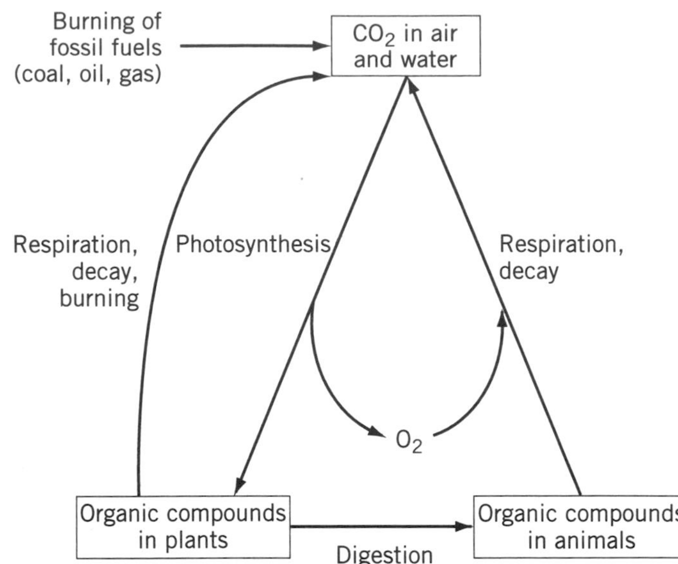

Figure 1. Short-term controls of CO_2 concentrations in air (periods of years).

Carbon dioxide also has sources other than the cycle represented by equation 16 and its reverse (Fig. 2). It is one of the major gases emitted during volcanic eruptions, and it is produced in large amounts by the precipitation of $CaCO_3$ from bicarbonate solutions:

$$Ca^{2+} + 2\ HCO_3^- \rightarrow CaCO_3 + H_2O + CO_2 \qquad (17)$$

This reaction is the reverse of equation 5. Unlike oxygen, CO_2 is a relatively unreactive gas found in the atmospheres of planets other than the earth. It may have been part of the earth's primitive atmosphere and has certainly been added by volcanoes and the precipitation of calcium carbonate sediments throughout earth history. It is presently maintained at a low concentration, probably chiefly by photosynthesis, but it has almost surely been more abundant at times in the earth's past. Because of its low concentration, CO_2 is one gas whose amount can be perceptibly influenced by human activity. The ever-increasing production of CO_2 by the burning of fossil fuels (coal, petroleum, and natural gas) during the twentieth century, plus the large-scale cutting of forests, has raised the concentration in air from about 300 ppm in 1900 to 350 ppm in 1980 (12). This long-term increase of carbon dioxide may be environmentally important because of its influence on the earth's temperature.

Nitrogen

The other principal constituent of air, N_2, is a less active element than O_2. It exists in the atmospheres of some other planets, and most likely was part of the earth's primordial atmosphere before life began. Nitrogen takes part in many organic processes, and these processes play a major role in stabilizing its concentration in air.

Nitrogen is important to all living things as an essential constituent of protein. The nitrogen that plants use to form protein (which is the ultimate source of all animal protein as well) does not come directly from the atmosphere but from nitrogen compounds dissolved in water that enter plants through

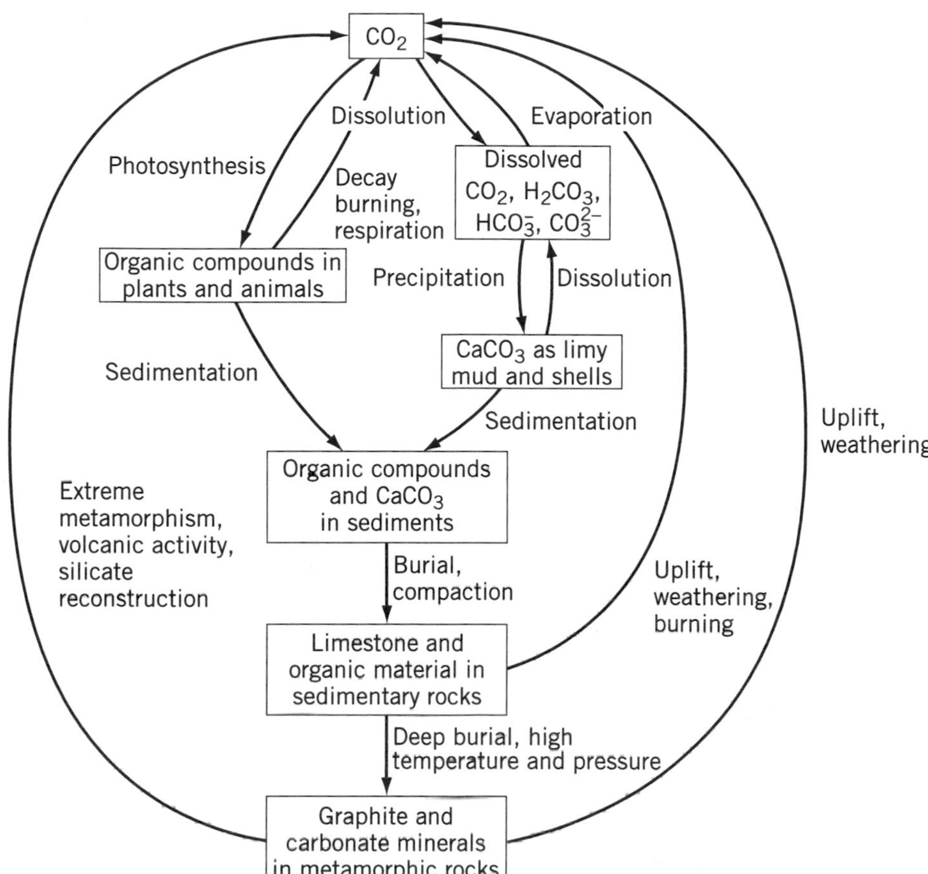

Figure 2. Long-term controls of CO_2 concentrations in air (to millions of years).

their roots. The principal form of nitrogen dissolved in soil water is nitrate, NO_3^-. This is derived from several sources but mainly from oxidation of ammonia and ammonium compounds generated by the decay of organic matter. Not all the nitrogen from the protein of decaying organisms is converted to nitrate. In the steps leading from protein to ammonia and then to nitrate, some of the nitrogen is set free as the stable diatomic molecule N_2, which returns to the atmosphere. Thus the cycle of nitrogen (Fig. 3) from living organisms to dead organisms and back to living is not complete, because some of the element is lost as N_2, which plants cannot use. Without other methods of generating compounds containing nitrogen (so-called fixed nitrogen), the usable nitrogen in soil water would gradually disappear and plant life would be impossible.

Fortunately, nitrogen compounds enter the soil in other ways. One important source is the activity of a particular kind of bacteria that lives in nodules on the roots of legumes. These bacteria have the remarkable ability to convert N_2 from air to compounds (amino acids) that plants can use directly. A few other kinds of bacteria that live in soil and water, plus blue-green algae, change N_2 to ammonia (NH_3) and ammonium compounds. These compounds are oxidized by still other varieties of bacteria to NO_3^-, the form of fixed nitrogen used by most plants. Another natural source of combined nitrogen is atmospheric electricity: lightning discharges lead to reactions between N_2 and O_2 to form nitrogen oxides (chiefly NO and NO_2), which are carried to the soil dissolved in rainwater and further oxidized to NO_3^-.

The reservoir of free N_2 in air is so vast that human activity cannot appreciably alter its amount, and adequate means are at hand for ensuring plentiful supplies of fixed nitrogen for agricultural and industrial purposes. The major environmental question concerning nitrogen at present relates to the role of nitrogen oxides generated by industry and automobile exhaust as contributors to increased acidity of rainfall and to the formation of ozone.

Minor Constituents

Of the minor constituents of air (see Table 3), the most abundant is argon, which makes up nearly 1% of the total. Argon and related elements (He, Ne, Kr, Xe) have little environmental significance because they are chemically inert. Their amounts do not change, except that one isotope of argon (^{40}Ar) is produced by the radioactive decay of potassium (^{40}K) and hence has steadily increased through geologic time. Helium is also a product of radioactive decay, but its concentration does not increase because its atoms are light enough to escape from the earth's gravitational field. The small amount in ordinary air represents a balance between its production from uranium and thorium in rocks and its escape into space from the stratosphere. Other minor constituents of air are produced by a variety of processes, for example, decay of organic matter with or without the aid of bacteria, volcanic eruptions, chemical industry, and automobile exhaust. In addition to the molecules listed in Table 3, trace constituents include free radicals formed by molecular dissociation, generally in photochemical reactions promoted by low-wavelength radiation from the sun. Examples of such reactions are the following:

$$O_2 + \text{radiation} \rightarrow O + O \qquad (18)$$

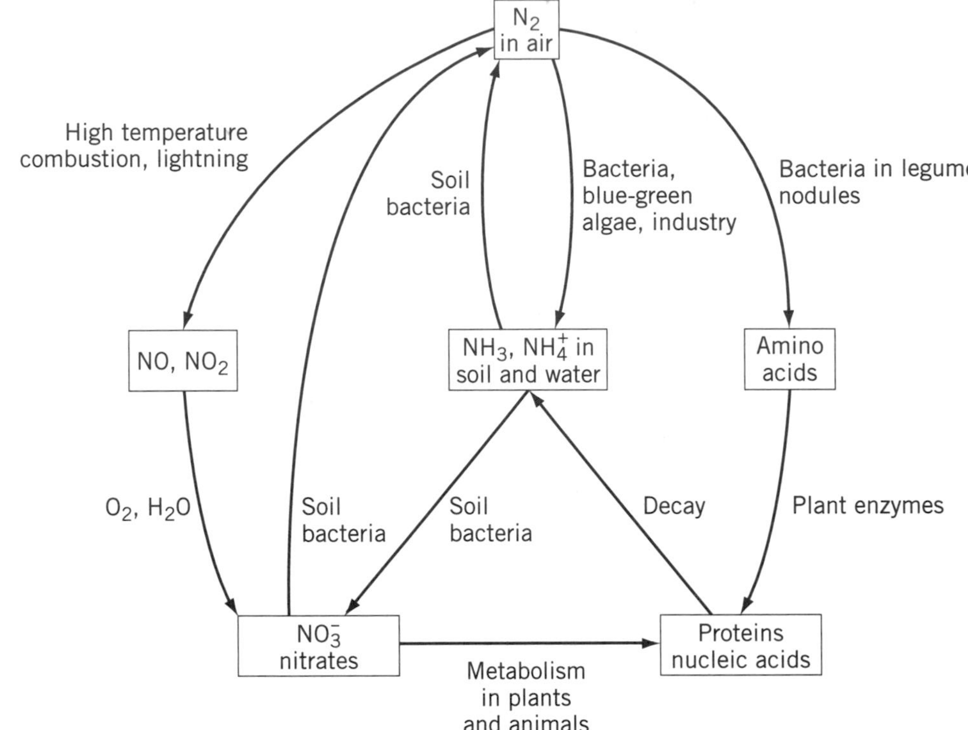

Figure 3. Principal reactions of the nitrogen cycle.

Table 3. Composition of Dry Air[a]

Gas[b]	Vol %
N_2	78.08
O_2	20.95
Ar	0.93
CO_2	0.033
Ne	0.0018
He	0.00052
CH_4	0.00015
Kr	0.00011
H_2	0.00005

[a] Ref. 3.
[b] Other important minor gases <0.0001 vol % : Xe, N_2O, NO, NO_2, NH_3, H_2S, SO_2, CO, O_3, $CClF_3$, and CCl_2F_2.

$$H_2O + \text{radiation} \rightarrow H + OH \tag{19}$$

$$NO_2 + \text{radiation} \rightarrow NO + O \tag{20}$$

These radicals are chemically very reactive, and only exist briefly in the laboratory. At high altitudes in air, however, where molecules are far apart and collisions are infrequent, some of the radicals persist for appreciable times and serve as intermediates in reactions that otherwise would be extremely slow. For example, the formation of ozone from oxygen can be written as follows:

$$3\,O_2 \rightarrow 2\,O_3 \tag{21}$$

but the collision and rearrangement of three oxygen atoms suggested by equation 21 would be immeasurably slow. In air the reaction takes place by a series of steps involving volatile organic compounds and oxygen atoms produced by photochemical dissociation of O_2 or NO_2 (eqs. 18 and 20):

$$O + O_2 \rightarrow O_3 \tag{22}$$

The existence of numerous free radicals makes the chemistry of minor atmospheric gases exceedingly complex.

Other than CO_2, gases containing carbon are principally methane (CH_4) and carbon monoxide (CO), but also include a variety of other volatile organic compounds (VOCs) in very minor amounts. Methane and carbon monoxide are naturally produced by bacterial decay under low oxygen conditions, for example, in swamps, and both are by-products of some industrial processes. CO is also a product of incomplete combustion in gasoline motors. Methane is the major constituent of natural gas and leaks to the surface in small amounts in areas of oil and gas accumulations. It is also known to escape from rocks in other areas and is a minor constituent of some volcanic gases. A much debated question is whether the amount of methane in natural gas is entirely accounted for by the decay of buried organic matter or whether some has a source in the earth's deep interior. Both CH_4 and CO exist only temporarily in air, because they react with oxygen (through free-radical intermediates) to form CO_2. Methane is similar to carbon dioxide in that its concentration in air has markedly increased over the past century, and it may similarly influence earth temperatures.

The two sulfur gases H_2S (hydrogen sulfide) and SO_2 (sulfur dioxide) are also products of both natural and industrial processes and are short-lived in air because of reaction with O_2. The product of oxidation is SO_3, which unites with any available water to form sulfuric acid H_2SO_4. This may appear as a constituent of raindrops, or it may be deposited on the ground surface where it reacts with soil or rock to form sulfates. Where

SO_2 is produced in abundance, as it is in many industrial operations, the resulting H_2SO_4 in the atmosphere has noticeable effects on nearby vegetation, metal surfaces, and human lung tissue. Smelters for sulfide ores are especially efficient generators of SO_2, and are commonly surrounded by wide areas barren of plants if emission of the gas is uncontrolled. On a larger scale, the windblown sulfur gases from industrialized areas are a major source of the acid rain that has become a well-publicized hazard to vegetation and to fish life in the lakes and streams of eastern North America and northern Europe (see ACID RAIN).

Ozone is an unstable and reactive form of oxygen produced in air by the action of ultraviolet rays in sunlight on O_2 molecules (eqs. 18 and 22). The reaction may be accelerated by catalysts, and materials that act catalytically are abundant in the automobile exhaust of large urban areas (13). The smog that sometimes covers such areas is partly a product of the ozone generated. Another source of O_3 is electric discharges; ozone's peculiar odor is often noticeable when lightning strikes nearby. Ozone molecules do not last long, reverting in days or weeks to ordinary O_2, but during their brief existence, they are destructive because they oxidize metals and organic materials more rapidly than the diatomic molecules. The embrittlement and cracking of rubber is a familiar effect of oxidation by ozone.

Although ozone has unpleasant, even poisonous, effects in the surface environment, it plays an important positive role high in the stratosphere (14). The lifetime of an O_3 molecule is longer (averaging a few months) in the attenuated gas of the stratosphere because collisions with other molecules are infrequent; thus the stratosphere has a layer with permanently high concentrations (up to 10^{-4} %) of this active gas. The O_3 absorbs short-wavelength ultraviolet light from the sun, screening the earth from this potent radiation. Without the protection provided by the ozone layer, skin cancer is more prevalent, and some forms of plant and animal life are damaged. Although a nuisance in the troposphere, ozone provides a major benefit to life by its presence in the stratosphere.

Gaseous nitrogen compounds include ammonia (NH_3), nitrous oxide (N_2O), and the higher nitrogen oxides, chiefly NO and NO_2 (symbolized NO_x because proportions of the oxides are variable). All of these compounds are naturally produced by various kinds of bacteria and the higher oxides also by electric discharges. NO_x has an additional source in some industrial processes and the exhaust from gasoline motors. Reaction of the oxides with O_2 (a reaction that also takes place via free radicals produced by photochemical dissociation) generates nitric acid as follows:

$$2\,NO + 3/2O_2 + H_2O \rightarrow 2\,HNO_3$$

$$2\,NO_2 + 1/2O_2 + H_2O \rightarrow 2\,HNO_3 \qquad (23)$$

This reaction is slow enough that the oxides persist for long distances downwind from their source, and the resulting acid (like the acid from sulfur gases) thus becomes part of the rainfall over a wide area. In nature this is beneficial, because fixed nitrogen is thereby added to soil as a plant nutrient. But where the nitrogen oxides are produced in large quantity, as they are in some urban and industrial areas, the acid is a serious contaminant of surface waters. The combination of nitric and sulfuric acids dissolved in rainwater constitutes the acid rain that has become such a menace to big areas of the environment downwind from heavily industrialized regions.

Change in Atmospheric Composition over Time

The nature of the earth's gaseous envelope has unquestionably changed over geologic time. The earliest atmosphere on the primitive earth was presumably similar to atmospheres we now see on some other objects in the universe, made up principally of hydrogen and helium. As these very light gases escaped, the remaining mixture consisted chiefly of compounds of carbon, sulfur, nitrogen, and oxygen. Details of the composition are uncertain. Guesses depend on assumptions about the oxidation state of the early earth: if conditions were oxidizing, the mixture would be chiefly CO_2, SO_2, H_2O, and N_2; if reducing, the major compounds would more likely be CH_4, NH_3, H_2O, and H_2S. With the appearance of life, particularly green plants, profound changes occurred in the oxygen-rich atmosphere (15). Timing and details of of the changes are subjects of lively argument, but the long paleontological record of living forms similar to those of the present suggests that the major changes were completed by early Paleozoic time. For many millions of years, the atmosphere has had close to its present composition, maintained by the balance between processes by which the principal gases are added to air and processes by which they are removed.

This balance is not always perfect. One example of imbalance is argon. This inert gas has increased in amount very slowly through geologic time and is still increasing, because its addition to air by the decay of radioactive potassium is not compensated for by any process of removal. More environmentally interesting are the possible fluctuations in carbon dioxide. If all the carbon now locked up in carbonate rocks and such organic materials as coal and oil came originally from CO_2 in the atmosphere, as seems probable, the amount that would have been required is far greater than the amount now available in air. At some time in its history, the earth had a dense atmosphere of CO_2 that has gradually shrunk to its present small concentration, or more likely, carbon has been gradually added to the atmosphere from the earth's interior as CO_2 or CH_4. If this is a reasonable hypothesis, the addition of carbon need not have been at a uniform rate, but may have varied from one geological period to another. Such variation in the concentration of atmospheric CO_2 is also suggested by other kinds of geological evidence. If CO_2 has changed, concentrations of other gases may have changed as well, especially that of oxygen, which is closely linked to CO_2 through the photosynthesis reaction (eq. 16).

WATER IN THE ENVIRONMENT

Analyses

Waters at and near the earth's surface have a complex and variable composition that is hard to express in a few analyses. Some representative values are given in References 4, 16, 17, 18, but particular samples can deviate widely from the suggested values.

The dissolved elements in rainwater come largely from dust particles and grains of sea salt with which raindrops have come in contact, plus a small amount of CO_2 from the surrounding air that normally makes rainwater slightly acidic. The amount of sea salt carried inland from ocean spray decreases with increasing distance from the coast; thus rain samples can have concentrations of some elements, particularly Na and Cl, that are different from those in the average analysis in (4, 16, 17, 18). On reaching the ground, rainwater picks up additional dissolved material from soil and rocks, and its pH generally increases because of hydrolysis of silicate and carbonate minerals (Eqs. 2 and 7). Water that seeps into the ground has more intimate contact with rock, and its solute content is further increased. The sea is the ultimate repository for all water from the land, and it becomes a concentrated solution of some constituents because of long evaporation. The concentrating effects of evaporation, however, are offset by reactions of the dissolved materials with organisms, sediments on the sea floor, and rock beneath the sediments. Only in restricted basins does evaporation lead to extreme concentrations and the precipitation of all or most of the salts.

The principal dissolved ions in most natural waters are Na^+, K^+, Mg^{2+}, Ca^{2+}, Cl^-, SO_4^{2-}, and HCO_3^-. In addition, most waters contain at least a few parts per million of silica, chiefly in the form of the neutral molecule H_4SiO_4 (commonly reported in analyses as SiO_2). Only rarely is water encountered that contains some other element (eg, F, P, or Sr) in concentrations comparable to those of the eight major constituents. These eight substances are predominant simply because they are fairly abundant in ordinary rock material (see Table 1) and hence accessible to water, and they do not form insoluble compounds with each other or with substances commonly found in rocks and soil.

In addition to the dissolved substances, most natural waters contain at least a little suspended material. This is in the form of clay or silt, which slowly settles out if the water is undisturbed, or finer grained colloidal particles (diameters roughly 1–1000 nanometers) that are invisible to the unaided eye and remain suspended indefinitely. Common colloidal materials are hydrates of ferric oxide and aluminum oxide (4,18). Colloidal silica is often present in addition to dissolved H_4SiO_4. If water contains organic matter, it is commonly present in colloidal form, although some exists in true solution. The distinction between a colloidal suspension and a true solution is not always easy to make, because particle sizes grade imperceptibly from one to the other.

Streams and Lakes

Running water dissolves substances from the atmosphere and the rocks and soil over which it flows. Some of the CO_2 that dissolves combines with water to form carbonic acid (eq. 4). This is a weak acid present only in minor amounts ($\sim 10^{-5}$ *M* in equilibrium with air at usual temperatures), but it is nevertheless an important agent of weathering (Eqs. 5, 6, and 7). It dissociates in the following two steps:

$$H_2CO_3 \rightarrow H^+ + HCO_3^- \qquad K_1 = 10^{-6.4} \qquad (24)$$

and

$$HCO_3^- \rightarrow H^+ + CO_3^{2-} \qquad K_2 = 10^{-10.3} \qquad (25)$$

where K is the equilibrium constant (the product of concentrations of products divided by concentrations of reactants). The resulting H^+, chiefly from equation 24, gives a pH of 5.7 in water with no other dissolved material. The hydrolysis of carbonate or silicate minerals in the stream bed counteracts this slight acidity, so that measured pH values in stream water are commonly between 6 and 8. In unusual kinds of environment, streams show pH values well outside this range; for example, in warm humid areas where weathering is deep, the pH of stream water falls considerably below 6, and in desert climates it often rises to 9 or higher.

Besides CO_2, the other active gas that streams dissolve from air is O_2, giving stream water the capability of oxidizing minerals containing reduced iron or sulfur. As in some of the reactions of weathering (eqs. 8 to 12), the major products of oxidation are ferric oxide or one of its hydrates and sulfate ion in solution. The results of such reactions are often unpleasantly evident in streams that drain areas where sulfide ores are mined: the water is brown from suspended ferric oxide, is markedly acidic, and contains high concentrations of toxic metals from the ore.

Dissolved O_2 also plays an active role in oxidizing the dissolved or suspended organic matter that a stream may carry. This reaction can be effective, within limits, in clarifying stream water and in removing toxic organic chemicals. If organic matter is abundant, however, (eg, where a stream flows sluggishly through a swamp), its dissolved oxygen may be entirely consumed and the water becomes reducing rather than oxidizing.

Water from near the surface of freshwater lakes is similar to that of streams, as expected, because much of it is supplied by streams and because it also is continuously exposed to the CO_2 and O_2 in the atmosphere. Below the surface where the water is out of contact with air, the dissolved O_2 is depleted by reaction with organic matter that drops from the surface. How completely the O_2 is removed depends on how fast the lake water circulates. In a shallow lake, wave action and currents produced by inflowing streams may be sufficient to keep the entire lake aerated. Deeper lakes may have bottom waters with little or no circulation where dissolved O_2 is completely exhausted and where partly decayed organic matter accumulates. In many lakes of temperate regions, the development of stagnant bottoms is prevented by annual or semiannual overturning of the water by temperature changes, which bring bottom water to the surface where its organic content is oxidized. In places where nutrient material favorable to plant growth is supplied to a lake in large amounts (eg, by intensive use of fertilizer on adjacent fields), aquatic plants grow so luxuriantly that the entire lake is choked with organic debris, even if the circulation would be sufficient to keep it oxidized in its natural state. This process, called eutrophication, is common in areas of intensive agriculture and some kinds of chemical industry, and is a menace to the use of lakes for water supplies and recreation (19).

In lakes of arid regions, dissolved materials become concentrated because of long-continued evaporation, often to an extent that plant life is inhibited and the water remains beautifully clear. Extreme evaporation leads to precipitation of some of the dissolved ions as salts. The nature of the salts depends on the extent of evaporation and the composition of the lake waters. Common precipitates are calcite ($CaCO_3$), gypsum ($CaSO_4 \cdot 2H_2O$), and halite (NaCl). Because most arid zone lakes are in isolated basins, their composition varies widely according to the nature of the rock in different basins, and salts different from these familiar ones are often found: Na_2CO_3, Na_2SO_4, $MgSO_4$, and occasionally salts containing less common elements such as B, Sr, Li, and Br. Waters that contain abundant HCO_3^- become very alkaline during evaporation, because of the escape of CO_2:

$$2\,HCO_3^- \rightarrow CO_3^{2-} + H_2O + CO_2 \qquad (26)$$

The liberated carbonate ion is partly precipitated as $CaCO_3$ or Na_2CO_3, and the part remaining in solution hydrolyzes to generate OH^-:

$$CO_3^{2-} + H_2O \rightarrow HCO_3^- + OH^- \qquad (27)$$

Groundwater

Much rainwater does not run off the ground surface immediately to join streams, but seeps into the soil and underlying rock to become part of the great reservoir of underground water that supplies water to wells and feeds the springs and seepages that maintain the flow of rivers through periods of dry weather. Groundwater normally does not form definite streams underground (although it may do so locally in caves), but moves very slowly through the interstices of rock. Measured rates of groundwater motion range from a few centimeters to tens of meters per year, so that the water remains in intimate contact with rock surfaces for long periods. Groundwater is subject to the same sort of chemical controls as water at the surface, except that it soon loses much of its initial dissolved O_2 and CO_2 and more nearly approaches equilibrium with adjacent rock because of its prolonged contact.

Rainwater that moves underground gives up its dissolved CO_2 through reactions with silicate and carbonate minerals (eqs. 5, 6, and 7), and its pH is further raised by hydrolysis. Its dissolved O_2 reacts with organic matter in soil and with minerals containing Fe(II) or reduced sulfur in rock. How fast and how completely these reactions go depends on many variables, particularly the rate of movement and the kind of rock. Thus water moving through pores of clean quartz sandstone may remain dilute, slightly acidic, and oxygen-rich for a long time, whereas water encountering an organic-rich soil overlying an iron-rich rock like basalt quickly becomes alkaline and reducing and acquires a substantial content of dissolved material. As a general rule, most groundwater is alkaline rather than acid and depleted in O_2; its alkalinity and reducing capacity generally increase with depth, but there is much variability. The dissolved cations and anions reflect to some extent the composition of the surrounding rock, although differences in the waters from different rocks are not as great as might be expected.

As groundwater moves from one rock to another, from one depth to another, or from one temperature regime to another, its composition slowly adjusts itself to the new environment. The adjustment is accomplished by dissolving material from the adjacent rock or by precipitating dissolved constituents in the cracks and interstices through which the water moves. An extreme example of dissolution is the formation of limestone caves, which form by water containing H_2CO_3 (and rarely other acids) moving through cracks and dissolving $CaCO_3$ (eq. 5). Examples of precipitation include the cementing of grains of sand and silt to form solid rock and the deposition of material in cracks to form veins. The two substances that are dissolved and precipitated in greatest abundance are SiO_2 and $CaCO_3$, which are the constituents of quartz and calcite veins, respectively, and are also the commonest cementing materials of sedimentary rocks.

One kind of chemical reaction that can be very active in groundwater but less so in surface waters is the process of adsorption mentioned previously, which refers to the ability of some mineral surfaces to temporarily hold ions taken out of solution from water moving past them. Thus groundwater containing Mg^{2+} that is moving through clay can lose most of its Mg^{2+} because the ions are adsorbed onto the surfaces of the clay minerals. The ions are not strongly held, because, when a later flow of groundwater deficient in Mg^{2+} moves through the same clay, some of the adsorbed ions will return to solution. Adsorbed ions are held to mineral surfaces in a variety of ways. The attraction may be electrical, because charges on the ions are partly neutralized by residual charges on the surface, or ions are concentrated on corners or edges of mineral crystals where charges are greater than on flat crystal faces. Ions from solution may replace particular ions in the mineral structure, or the attachment may be the result of a chemical reaction. Generally the attachment of adsorbed ions is less firm than in a normal chemical bond: the ions ordinarily are removed by a simple change in conditions, such as a change in groundwater composition. Adsorption is nevertheless an important part of groundwater chemistry, often greatly delaying the transport of particular ions as groundwater moves through sorbent material. It is less important in the chemistry of waters in streams and lakes because these waters have less exposure to mineral surfaces.

Organic matter, both dissolved and colloidal, is picked up and carried as groundwater moves through soil and persists in water moving through upper layers of the underlying rock. Organic compounds are slowly oxidized (often aided by bacteria) and adsorbed on mineral surfaces, so that at deeper levels organic matter is generally absent or nearly so. The natural organic matter of shallow groundwater is augmented by compounds leaking from underground tanks or from areas where industrial organic wastes have been disposed of near the ground surface. This kind of groundwater contamination has become a serious problem, especially since some of the toxic compounds produced by modern chemical industry are not easily oxidized or attacked by bacteria and persist in the environment almost indefinitely. The problem is particularly difficult compared with the similar problem of contamination in stream water, because the movement of contaminants underground is hard to trace and very expensive to control even where tracing is possible (20,21).

Another sort of groundwater contamination, sometimes accompanying organic contamination, is that produced by toxic metals. Cadmium, mercury, lead, and arsenic are particularly troublesome.

Still another kind of groundwater contamination is that produced by salt, by the incursion of salt water from the sea, when fresh water is pumped too intensively from wells in coastal areas, or by the dissolution of salt in water that is in contact underground with deposits of rock salt. Salt is also troublesome in the groundwater of arid regions, where repeated use of irrigation water leads to slow concentration by evaporation during each use.

Seawater

The oceans serve as gigantic basins for the dissolved and suspended material carried by rivers and for solid debris erupted by volcanoes and scoured from the land surface by winds and glaciers. As might be expected, the major ions in seawater are the same as those in rivers: Na^+, K^+, Ca^{2+}, Mg^{2+}, Cl^-, SO_4^{2-}, HCO_3^-. Evaporation over geologic time has concentrated these ions, but not to the extent expected. This suggests that the ions are steadily being removed from seawater and added. Some of the removal processes are obvious, for example, the use of some elements by the abundant organisms that live in the sea and the carrying of some elements to the bottom as precipitates or adsorbed on solid particles. Less obvious removal processes include adsorption on bottom sediments and reaction with underlying basalt as heated seawater circulates through cracks. Organic processes are particularly effective in controlling Ca^{2+} and SiO_2. Ca^{2+} is controlled by precipitation of $CaCO_3$ in the shells of marine organisms whose remains cover large areas of the sea floor, and SiO_2 by precipitation in the form of opal by other species of small plants and animals. Not all the controlling processes are definitely known, but analyses of samples from many areas and many depths show that chemical controls and distribution by waves and currents are effective in keeping the concentrations of major ions within narrow limits (see 4,16,17,18). The pH of seawater is also kept within a small range (8.0 to 8.3 near the surface, slightly lower at depth), and circulation is rapid enough to bring dissolved CO_2 and O_2 from the surface to great depths in nearly all parts of the ocean.

The sea plays an important role in two ways in the overall system of chemical reactions that maintain a steady state in land environments. One is the addition of salt particles to air as windblown spray evaporates; such salt can be carried far inland and is a major source of the Na^+, K^+, Mg^{2+}, Cl^-, and SO_4^{2-} in stream water. The second, and probably more important, influence of the sea on land chemistry is its contribution in regulating the CO_2 concentration in air. The controls on this concentration are complex. As noted previously, part of the control is a balance between the consumption of CO_2 in photosynthesis (eq. 16) and its release by respiration, combustion, and decay. The life processes of marine plants and animals are included in this balance.

But beyond the strictly organic production and consumption of CO_2, the sea exerts control in other ways. The dissolving of CO_2 at the ocean surface establishes equilibria among CO_2, H_2CO_3, HCO_3^-, and CO_3^{2-}, as represented by equations 4, 24, and 25. Any increase of CO_2 in the atmosphere drives all these reactions to the right, thus increasing concentrations of the ions and making the water more acid, whereas a decrease causes the reactions to reverse and generate CO_2 to make up for the deficiency. Additionally, the equilibrium represented by equation 17 shows that an excess of CO_2 dissolves $CaCO_3$ in sediments and forms HCO_3^-, whereas a deficiency is compensated for by the reaction of this ion with Ca^{2+} to produce more $CaCO_3$. Other equations can be written representing more complicated equilibria involving CO_2 and the constituents of seawater and its sediments, all of them showing the generation of CO_2 if its concentration in air should drop or its depletion if the concentration should rise. Thus the great reservoir of the ocean is a major source and sink for atmospheric CO_2 and a major control on its global concentration.

Long-Term Changes in Natural Waters

Much of the water at the earth's surface probably came initially from the interior, and additions from the interior probably have continued throughout geologic history. How large the later additions have been is a matter of argument, but present opinion is that the great bulk of our present water supply was created early in earth history, shortly after the earth's surface became cool enough for oceans to accumulate. Water comes to the surface today in volcanic eruptions, and presumably has done so in all past ages, but most of this is probably recycled water from sedimentary rocks that have been carried down into the mantle by subducting plates and heated to release their contained water. If volcanic water contains any additions from the deep interior, they are probably minor. The best current guess is that the total amount of surface and near-surface water has been virtually constant for millions of years and will remain constant in the future.

The composition of streams and lakes is necessarily responsive to changes in climate, and many such changes have been recorded in human history as the climate of particular regions has shifted from humid to arid or vice versa. The changes are slow, hardly perceptible in a single lifetime. If the CO_2 content of the atmosphere continues to increase as it has during the past century, an expected result is a slight general increase in the acidity of surface waters. The possible relationship of human activity to such long-term changes is discussed in the next section.

EFFECTS OF HUMAN ACTIVITY

Chemical reactions in soil, air, and water at the earth's surface are part of a huge system that reached an approximate steady state long ago. The system is remarkably stable toward natural events that might disturb it. An excess of any component sets in motion reactions that tend to cut down the excess, and a deficiency automatically leads to reactions that seek to restore the balance. In the operation of natural processes, there are few reactions that might result in permanent long-term changes. The decay of radioactive isotopes, the escape of heat from the earth's interior, and the loss of hydrogen and helium from the upper atmosphere are three examples of processes that produce uncompensated change, but the effects of these on the environment are small. Human beings have modified the chemical environment in obvious ways by clearing forests, cultivating the soil, and extracting metals from ores, but, for most of human history, the effects on the overall steady-state system have been slight and often temporary.

With the huge growth in population over the last century, however, and with the development of ever larger and more efficient machines, the ability of humans to influence the environment has increased enormously. It may well be that some kinds of human activity are producing changes that seem trivial in the short term, but whose cumulative effect over a few generations may modify the environment to an extent not readily compensated for by the reactions that normally maintain the steady-state system. Some of these effects have been noted in preceding sections.

Increase in Carbon Dioxide: The Greenhouse Effect

The extent of the earth's heating by the sun depends on certain characteristics of the atmosphere. Radiation comes to us from the sun in many different wavelengths, which are partly reflected back into space by the surface and partly converted to the long wavelengths of heat radiation (infrared light) responsible for warming the lower atmosphere. If the atmosphere did not exist, the infrared radiation would escape into space; the earth's surface would be intensely hot during the day and bitterly cold at night, just as is the airless surface of the moon. The molecules of some gases in air absorb the radiation, however, and collisions with other molecules distribute the heat through the lower layers of the atmosphere. The moving molecules retain some of the heat overnight and thus keep temperatures from plummeting far below zero as soon as the sun sets. Just as the glass windows of a greenhouse admit many wavelengths of light but prevent the longer wavelengths of heat radiation from escaping, so the atmosphere traps the longer wavelengths and thus warms surface environments (22).

CO_2 is the most efficient of the molecules that absorb infrared rays. Others include some of the variable constituents, particularly H_2O and CH_4. The effectiveness of H_2O is evident in the strong contrast between day and night temperatures in the dry air of deserts as compared to the more even temperatures of humid regions. If the CO_2 content of air increases, surface temperatures would be raised, and a long-continued increase would necessarily affect world climates. Just such an increase is the probable result of many human activities: the burning of wood, coal, and oil, the smelting of metals, and the clearing of forests. These activities have greatly accelerated over the twentieth century, and, during this period, the CO_2 content of air has increased by more than 15%. Further increase seems assured, and the question arises as to how much effect on climate can be expected (23,24).

To calculate the effect seems simple: the rise in temperature because of a given concentration of CO_2 is known, the excess amount of CO_2 produced annually can be estimated, and multiplication of these two figures should give the amount by which the earth's average temperature increases each year. But the calculation turns out to be much less straightforward. One difficulty stems from the fact that a rise in CO_2 concentration in the earth's steady-state system immediately sets in motion the natural reactions that tend to counteract the increase. Some of the excess CO_2 would be used for additional photosynthesis of organic compounds in the green leaves of plants; more of the excess would dissolve in seawater, and some of this would be consumed in the dissolving $CaCO_3$ from sediments. An added complication is that rising temperature would lead to greater evaporation of water and thus an increase in cloud cover, which would shield the surface from the sun's rays and hence moderate the temperature increase. Most of these counterreactions are slow, and questions remain as to whether they will keep pace with the global increase in CO_2 concentration. It is hardly surprising that agreement has not been reached about the probable magnitude of the greenhouse effect. Some have estimated a worldwide temperature rise as high as 5°C by the year 2050, but most predictions are closer to 1°C, a small change, to be sure, but sufficient to alter climates appreciably. One result of such an increase would be the melting of some of the ice in Greenland and Antarctica, with a consequent rise in sea level worldwide. The change in climate will be so gradual as to be hardly noticeable in a single lifetime, but if it continues for several generations, the positions of coastlines and the distribution of deserts and forested lands could be markedly altered (25).

Acid Rain

In recent years the number of fish in lakes and streams of the northeastern United States and the eastern provinces of Canada markedly decreased. The decrease has been correlated with an increase in the acidity of the water, which in turn is thought to result from a change in the character of rain in this area caused by acid emissions from industries concentrated south of the Great Lakes and in the eastern states. Prevailing winds from the west and southwest carry the acidified water droplets into the affected area. This process lowers the pH of surface water during rains and ultimately lowers the pH of the groundwater that maintains stream flow between storms. The decline in fish life and increasing damage to vegetation have been attributed to acid rain. Similar effects have been reported from Scandinavia, where the acid rain comes chiefly from heavily industrialized parts of Germany, England, and the Low Countries. If the correlation of damage to aquatic life and forests with industrial emissions is valid, this is another example of large-scale disturbance of environmental geochemistry by human activities (26,27).

Although the correlation seems to be well established, argument remains as to the culpability of industry relative to acid emissions from automobile exhausts and possible changes in natural processes. The uncertainty stems from the difficulty of obtaining a meaningful average pH value over a large area and from the inadequacy of data on pH values of surface water and groundwater before the expansion of industry in modern times. Because efforts to control the acidity of rain require expenditure of billions of dollars, determining the actual magnitude of damage and assigning responsibility for it are subjects of current intensive research and political controversy.

Whatever their source, the acid emissions consist chiefly of oxides of sulfur and nitrogen. The principal sulfur gas SO_2 oxidizes slowly in the presence of air and water to form sulfuric acid as follows:

$$SO_2 + 1/2\ O_2 + H_2O \rightarrow 2\ H^+ + SO_4^{2-} \qquad (28)$$

Nitrogen forms a number of oxides as previously discussed. To avoid the difficulty of identifying individual oxides, they are symbolized NO_x. Two major oxides NO and NO_2 react with water and oxygen to form nitric acid (eq. 23). These oxidation reactions take place slowly in a series of steps involving free radicals, and the equations give merely a simplified representation of the final result. If present in any quantity, the two strong acids produced cause the kind of damage ascribed to acid rain.

Even in their natural state, surface waters in the humid parts of the north temperate zone are somewhat acidic. In part this results simply from dissolution of CO_2 from the surrounding air (eq. 3), which gives pure water in contact with air a pH of ~5.7. Soils of these regions, however, commonly contain enough decaying organic matter to produce a soil atmosphere much richer in CO_2 than ordinary air (often as great as 0.1 atm partial pressure), and the resulting H_2CO_3 plus small amounts of organic acids cause the pH of groundwater to drop to 4 or even lower. In places where the bedrock is limestone, the pH is not this low, because $CaCO_3$ is effective in neutralizing acid (eq. 4). Thus the pH values in nature are controlled by a complex of reactions and vary a good deal from place to place and from season to season. Addition of acid from rainfall sets in motion the natural reactions that try to neutralize it, especially the dissolution of carbonate and the replacement of adsorbed metal ions by H^+. Because of these buffering reactions, a substantial amount of acid can be added to a soil or lake without greatly affecting its pH. Only when the quantity of excess acid becomes large enough to exceed the capacity of the natural buffers does the pH fall much below its usual value. Measurements of pH values as low as 3 reported from Canadian lakes indicate that rainfall over a long period has indeed supplied quantities of acid sufficient to change the pH markedly.

How serious the problem of acid rain is and how effectively it can be remedied by controlling industrial emissions remain matters of dispute. Certainly the activity of acid rain can be reduced by installing devices to trap or lower the rate of emission of SO_2 and NO_x, but the amount of reduction is hard to estimate. The problem is enormously complicated both technically and politically, and a final solution awaits additional research (see AIR POLLUTION).

Ozone in the Stratosphere

As noted previously, ozone is a very minor constituent of the atmosphere that plays a curious double role in the environment: in the lower atmosphere it is a nuisance, mildly poisonous and responsible for cracking rubber and for generating the unpleasant smog of large cities, yet its presence in the stratosphere helps greatly in keeping the earth's surface safe for living things. Ozone accomplishes its protective mission by absorbing much of the low wavelength, high energy ultraviolet radiation from the sun, thus shielding surface environments from this damaging radiation. Just how harmful the radiation would be if the ozone layer did not exist is a matter of dispute, but it damages some plants and animals and causes skin cancer in humans. Thus preservation of the ozone layer is an important concern, and there was great consternation a few years ago when it appeared that some volatile products of human industry are effectively destroying it. In a natural state, unstable O_3 molecules in the stratosphere are being continually created and destroyed, and the fear was that manmade substances might be disturbing this natural balance by accelerating the reactions destroying O_3 molecules (14,28).

The compounds in question are halogenated hydrocarbons, the so-called CFCs or chlorofluorocarbons, for example, $CFCl_3$ and CF_2Cl_2. These are used widely in ordinary life as refrigerants and propellants for spray cans. The compounds are gases or volatile liquids at ordinary temperatures, so when released they quickly mix with air. In contrast to most organic compounds, they are very stable, resisting oxidation by O_2 for periods of years or decades. Once in the atmosphere, some of the molecules eventually find their way to the stratosphere, where Cl atoms are liberated by photochemical reactions and react with O_3 to destroy it:

$$Cl + O_3 \rightarrow ClO + O_2 \qquad (29)$$

Although any individual use of a device containing one of the CFCs would release only trivial amounts, the sum total of present and future releases by the world's population is sufficient to greatly diminish the amount of ozone in the stratosphere. One indication of the seriousness of the threat was the discovery a few years ago, after a steady increase for several years of chlorofluorocarbon (CFC) releases, of an area in the stratosphere over Antarctica where ozone was practically gone, the ozone hole (24).

Recognition of the danger posed by falling levels of ozone in the stratosphere has led to international governmental action to curtail the free use of halogenated hydrocarbons in spray cans and to substitute other compounds for this purpose. The amount of damage to the ozone layer is thus greatly reduced. Quite obviously, possible effects on the chemical balance of the ozone layer must henceforth be an important consideration in the extensive use of the new organic chemicals that industry is providing in ever-increasing numbers.

Mine and Tailings Drainage

Ores of many metals (eg, Cu, Pb, Zn, Ag, Hg, Mo, and Sb) consist largely of sulfide minerals, which commonly are associated with abundant pyrite (a sulfide of iron FeS_2). On exposure to weathering at the earth's surface, the sulfur of such ores is oxidized to SO_4^{2-} and hydrolysis of the metal ions makes infiltrating water acid (eqs. 11 and 12). Pyrite is particularly susceptible to weathering, and the resulting Fe(III) is strongly hydrolyzed because of the extreme insolubility of ferric oxide (as Fe_2O_3, FeOOH, or $Fe(OH)_3$). In a natural state, much of the Fe(III) oxide remains in place where it forms, accumulating as a red, brown, or yellow capping over the deposit (the iron hat or gossan in the jargon of miners), whereas most of the SO_4^{2-} and ions of other metals are carried off in solution. (In arid climates some remain behind as oxide or carbonate minerals; often conspicuous are the green and blue colors of the copper carbonate minerals malachite and azurite.) Ordinarily the weathering process is slow and the contamination of surface water and groundwater by metal ions and acid is not serious.

When a deposit is opened by a mining operation, however, sulfides are exposed to air and water in large quantity, and weathering is greatly accelerated. Waters draining the deposit, both on the surface and underground, become highly acidic and heavily contaminated with toxic metals. Fe(III) oxide precipitates at the deposit itself and in drainage water as its acidity is gradually neutralized. Exposed pieces of metal are attacked by the acid and their surfaces become coated with iron oxide. Abandoned mines are often conspicuous eyesores in the landscape because of the large area of discolored rock, the heavily rusted machinery, and the brown color of streams that drain them (10). They are also a nuisance to nearby communities by contaminating water supplies and making streams and lakes unsuitable for fish life, even at distances many miles downstream (30,31). Some measure of control is possible, for example, by using limestone to neutralize the acid and precipitate some of the metals, but the best insurance against damage to scenery and water is care in the original mining operation to minimize the dispersal of weathering sulfides and to control the escape of drainage water.

Similar problems arise in coal mining, because most coals contain at least a little iron sulfide. Abandoned coal mines, like abandoned metal mines, are often conspicuous because of iron oxide staining and the acidity of the water that drains them.

Hazards to adjacent areas are also associated with debris from mining and milling operations that do not involve sulfide minerals. A well-publicized example is the huge accumulation of tailings from the milling of uranium ore in many parts of the western United States. The tailings consist of material from which the uranium has been extracted, but in which substantial amounts of the highly radioactive daughter-product elements are left resulting from the long-term decay of uranium in the original ore. These elements, of which radium is the best-known example, pose no danger to the environment as long as the ore remains in place deep underground because little groundwater reaches them. In piles of loose debris at the surface, however, they are

Plate 4 Amazon Deforestation. See entry "Forest Ecosystems in a Changing Global Context"

Plate 5 Tractor at Landfill. See entry "Landfills"

Plate 6 Ship Discharging Waste. See entry "Marine Pollution"

Plate 7 Oil Spill Cleanup. See entry "Oil Spill Containment, Terrestrial"

exposed to infiltrating rainwater and hence make their way into water used for human consumption.

Some of these elements pose an additional hazard, because their radioactive decay produces atoms of the gaseous element radon, which is highly radioactive. Radon produces several radioactive daughters in a series ending with an isotope of lead. Because it is a gas, radon is highly mobile and escapes readily into air from the loose materials of the tailings piles. The radioactive lead produced becomes part of the dust blown into nearby communities. Quantities of the dissolved radium and the escaping radon are generally very small, so that the seriousness of the hazard is open to question, but there is general agreement that some control is needed.

Radon is a possible hazard in places other than the tailings piles, because it escapes from the minute quantities of radioactive elements found in almost any rocks and soils (32). Wherever its concentration is substantially greater than normal, the radon produced builds up to dangerous levels in places where air is confined in a quiet closed space like the interior of a house or a cave, and tests for its presence are recommended.

In summary, the mining of metals and coal, a human activity essential to industrial civilization, presents chemical and even radioactive hazards to the environment that are sure to increase in the future as the demand for resources grows.

Disposal of Chemical Waste

Still another impact of human enterprise on the chemistry of the environment results from efforts to dispose of the huge amounts of waste that modern civilization generates. In preindustrial times, when the world's population was smaller, waste disposal was a relatively minor problem. Garbage accumulated locally and temporarily in unpleasant amounts, but it seldom contained hazardous substances and was quickly rendered harmless by the natural processes of oxidation and bacterial decay. The natural environment has a large capacity for disposing of waste automatically. One can put substantial quantities of waste chemicals and toxic metals on the ground surface or in streams or burial pits and be assured that within a reasonable time most of the material has disappeared by oxidation or is so widely dispersed in surface and underground water that it is no longer a hazard. It is only when the quantity of waste becomes too large for the earth's natural system of compensating reactions to handle it expeditiously or when substances hazardous to health make up a sizable fraction of the waste, that the problem of disposal demands attention.

The magnitude of the current problem in the United States is suggested by recently published figures showing that more than 400 million tons of municipal and industrial solid waste are generated yearly, of which 60 million tons may contain hazardous ingredients. Liquid industrial wastes amount to 10 trillion gallons per year. Much of the waste is harmless and is managed by the conventional techniques of incineration, bacterial digestion, reverse osmosis, shallow land burial, or addition to surface waters. But the waste containing toxic materials, now generated at more than a thousand sites in the United States, cannot be so easily dismissed.

The problem becomes increasingly serious if materials in the waste dissolve in groundwater, as often happens when rain falls on accumulations at the surface or in shallow burial pits and sinks into the ground. In surface waters, contamination is generally apparent at once, and remedial measures are taken. But contamination underground can continue undetected for a long time and is difficult or nearly impossible to correct at a reasonable cost.

The most troublesome contaminants of groundwater and surface water are organic compounds, produced directly or as byproducts in many industrial operations. More than 10,000 such substances are toxic, and their number increases yearly. Many, of course, are valuable materials considered indispensable in modern life, but, as uncontrolled constituents of waste, they do enormous harm. Some of them, probably a majority, are fairly unstable compounds that are decomposed by bacteria (biodegradable) or that oxidize on contact with air into innocuous substances that can be disposed of with only minimal care. Those that cause real concern are the stable compounds that persist unchanged for decades or centuries, that are immune to attack by atmospheric gases or bacteria, and that accumulate to dangerous levels in water that drains disposal sites. Widely published examples are dichlorodiphenyltrichloroethane (DDT) and a group of compounds called polychlorobiphenyls (PCBs). A major difficulty is that many of these compounds have been discovered so recently that their properties are not well known: their degree of toxicity and their long-term behavior in contact with air and water are still subjects of active research. Studies have also shown that some familiar compounds, long considered harmless and widely used in households and industry, are dangerously toxic on long-continued exposure. Prominent examples are benzene, carbon tetrachloride, and formaldehyde.

Several methods are available for reducing the amount of organic chemicals in waste and drainage water. Most such compounds can be broken down by controlled incineration, some can be transformed by enzymatic reactions promoted by bacteria, and some can be adsorbed onto mineral surfaces or particulate organic matter in soils and sediments. But the tailoring of methods for particular kinds of waste is difficult and expensive, especially for wastes containing a variety of chemicals. One of the worst problems is the treatment of old dump sites, places where large amounts of waste were disposed of before the toxicity of its constituents was recognized or before the movement of groundwater in the vicinity had been studied.

Other important contaminants that come from waste disposal sites are toxic metals and their compounds. Arsenic, cadmium, mercury, and lead are well known examples, but several others are damaging if present in unusual amounts. Most of the heavy metals, in fact, are toxic, and even the familiar metal aluminum is harmful to plant life if too much is carried in solution. The metals are toxic only if their concentrations exceed a threshold value, which is different for each metal. For some of the toxic metals, it is not necessary or even desirable to reduce concentrations to zero, because, in very small amounts, they are essential for life processes. Selenium is a good example: it is a well-known poison for cattle if present in large amounts in plants, but small quantities are necessary in the diet of animals, including humans (33). Fortunately the compounds of most potentially toxic metals are nearly insoluble in waters whose pH is close to neutral, as it is in much surface water and groundwater. Exceptions are chromium and selenium, which are soluble in the form of stable anions under oxidizing conditions at near-neutral pH; but their concentrations would be small in groundwater at low redox potentials.

By concentrating large amounts of long-lived organic substances and compounds of toxic metals in waste dumps and by allowing some to be dissolved in groundwater, humans are altering parts of the earth's chemical environment in ways that natural processes cannot correct for a long time in the future.

Disposal of Radioactive Waste

Chemical waste that contains radioactive elements poses particularly troublesome problems of disposal. The difficulties are partly technical and partly political: technicians must find a way to keep the amounts of radioactive constituents that find their way into groundwater and surface water exceedingly small for very long periods, and politicians must heed the general fear of radioactivity by opposing any suggestion to consider disposal of radioactive waste in areas where their constituents live.

Tiny amounts of radioactive elements are widespread in materials of the earth's crust, even in the living tissues of all plants and animals. Radiation similar to that from these elements is produced in the atmosphere by cosmic rays, to which the environment is continually exposed, and the common use of x-rays for medical diagnosis and therapy adds additional radiation exposure to much of the human population. Thus radiation at very low levels is an integral part of our environment and cannot be considered harmful in itself. There are even regions of the earth where natural radioactive minerals are substantially more abundant than normal and radiation is correspondingly higher, yet no harm to the inhabitants has been demonstrated. The hazardous feature of radioactive waste is simply the much greater intensity of the radiation it generates, compared with that in the normal environment.

All such radiation is *ionizing radiation,* meaning that it can displace electrons in atoms through which it travels, thus breaking up compounds and disrupting the activity of cells (34). Small amounts of cell damage cause illness, and large amounts cause death in a short time. Furthermore, the damage from repeated small exposures to radiation is cumulative and may lead to illness many years after initial exposure. Just how much radiation can be tolerated without short-term or long-term effects is uncertain, but, in the absence of this knowledge, it is obviously desirable that radioactive waste is isolated as effectively as possible from the environment.

Radioactive material in waste generally includes a variety of elements. Some of them decay quickly to nonradioactive daughters and thus become harmless, whereas others remain radioactive for long periods, up to hundreds of thousands or millions of years. The problem of isolating waste is difficult chiefly because of these long-lived elements. Certainly barriers can be erected to keep the radioactive elements out of the environment for decades or centuries, but can any manmade structure be completely relied on to endure for tens of thousands of years? Ultimately, whatever is done to dispose of the waste, the possibility exists of breakdown of the artificial barriers and exposure of the waste to natural environmental agents, particularly groundwater, because plans for disposal generally envision placement well below the ground surface. One of the major geochemical problems of waste disposal, therefore, is to find a place where groundwater is so scarce or so slowly moving, where its composition is so noncorrosive, and where the adjacent rock contains so much sorbent material, that no more than minute quantities of radioactive material will dissolve in the groundwater and be carried by it to surface environments, even over very long periods of time.

Many kinds of radioactive waste have been generated, ranging from such low level materials as slightly contaminated laboratory equipment and clothing to the intensely radioactive spent fuel rods of nuclear reactors that have been removed after many months of neutron bombardment of their contained uranium. The different types require different methods of disposal. Low level waste, particularly if most of its radioactive materials are elements having short half-lives, can be safely disposed of in shallow covered trenches near the ground surface. But the problem is much more daunting for high level waste. This is waste consisting of the fuel rods from reactors or material derived directly from them, waste emitting radiation so intense that all contact must be avoided and remote handling is essential. Current plans call for putting such wastes in thick-walled canisters and burying them in mined cavities several hundred meters underground. Seemingly this is a good solution to the disposal problem, but finding a suitable place or places for the underground construction has proved extremely difficult.

Technically, the search for a disposal site means finding an area where the rock at depth is resistant to heat and radiation, where disturbance from earthquakes or volcanoes or deep erosion is unlikely, and where groundwater is so scarce and slow-moving that no movement of dissolved radioactive material to the ground surface in more than minute amounts can be guaranteed for 100,000 years or longer. Politically, the search means seeking an area where the nearby population can be persuaded to allow the construction. To date, no place meeting all these conditions has been found in any of the countries where disposal of high level waste is a problem, despite intensive and persistent exploration. So high level waste continues to accumulate on the earth's surface in the places where it has been or is being generated. This is not a threat to the environment as long as it is scrupulously watched over, but a very serious threat if the care is ever relaxed. Clearly permanent disposal is desirable and should be possible by deep burial, but enough questions remain to prevent its accomplishment at any time in the near future.

In our own immediate future, the overall amount of radiation in surface environments can be kept low only if we are successful in the proper handling of waste, especially in the disposal of high level waste deep underground (35).

Depletion of Metal and Fossil Fuel Resources

One other human activity that significantly affects the environment is the ever-increasing exploitation of ore deposits and accumulations of gas, oil, and coal for the materials and energy on which modern civilization depends. Such concentrations of useful materials, accumulated over geologic ages, are not being renewed by any reactions fast enough to be of human interest. How long the existing accumulations will last is a much-argued question, but it seems certain that within a matter of decades or a few centuries the high grade ores and major oil pools will be gone. Lower grade deposits will serve to maintain supplies for a longer period, but the readily available cheap metals and cheap energy that we have enjoyed will not be available to future generations. It is often said that, if global war reduces humankind to a state of barbarism, civilization will never reappear because the easily obtainable metals on which its early growth would depend will have disappeared. In this respect, human activity has effected a profound and irreversible change in the earth's chemical environment.

CONCLUSION

In the earth's development through geologic time, a complicated system of interlocking reactions has been established among the chemical constituents of the crust, atmosphere, and water bodies that keeps surface environments in a steady state. If one reaction temporarily goes faster than normal, other reactions respond to slow it down. If one substance is added to or subtracted from the atmosphere or oceans, reaction rates change so as to cut down the excess or make up the deficiency. Thus the concentrations of materials in the atmosphere and water bodies are maintained at remarkably constant values. Although some of the concentrations have changed gradually over long periods of geologic time, most of the changes occurring today are so slow as to be nearly imperceptible.

Among known astronomical bodies, the earth is unique in having an abundance of liquid water, abundant free oxygen in its atmosphere, and abundant life on its surface. These three characteristics are interrelated. Oxygen is maintained by a reaction between water and carbon dioxide in the leaves of living plants. Carbon dioxide is kept at a constant level by this reaction and several others, including dissolution in seawater and the dissolution and precipitation of carbonate minerals. Rocks are attacked by water, oxygen, and carbon dioxide, and their complex silicate minerals break down into simpler minerals and other substances in solution, which ultimately become part of sedimentary rocks. The three weathering agents are restored by reactions in the earth's mantle, where gases are released and complex silicate minerals are newly formed. Along with these major reactions are many subsidiary ones that keep all parts of the chemical environment in delicate balance.

Throughout most of history humankind has had only a minor effect on this system of compensating reactions, but in the last two centuries the growth of population and increasing use of energy have produced effects that may mean long-term or even permanent changes in the chemistry of the environment (36,37). Notable among these possible changes are an increase in the carbon dioxide content of air, an increase in the acidity of rainfall, a decrease in the ozone of the stratosphere, the contamination of much surface and underground water, and the depletion of concentrations of metallic ores and fossil fuels (38). The steady-state reactions of the global system will partly compensate for these changes but probably cannot undo them entirely (see CLIMATE–BIOSPHERE INTERACTIONS; CLIMATE SYSTEM MODELS, LAND–ATMOSPHERE INTERACTIONS).

BIBLIOGRAPHY

1. H. D. Holland and U. Petersen, *Living Dangerously,* Princeton University Press, Princeton, N.J., 1995, pp. 108–120.
2. K. H. Wedepohl, *Geochimica et Cosmochimica Acta* **59,** 1217–1232 (1995).
3. K. H. Wedepohl, ed., *Handbook of Geochemistry,* Springer-Verlag, Berlin, 1969–1974; S. R. Taylor, *Geochim. Cosmochim. Acta* **28,** 1273–1286 (1964).
4. H. J. Bowen, *Environmental Chemistry of the Elements,* Academic Press, London, 1979.
5. J. A. Plant and R. Raiswell, in I. Thornton, ed., *Applied Environmental Geochemistry,* Academic Press, London, 1983, pp. 9–14.
6. L. M. François, J. C. G. Walker, and B. N. Opdyke, in E. Takahashi, R. Jeanloz, and D. Rubie, eds., *Evolution of the Earth and Planets,* Geophysical Monograph 74, American Geophysical Union, Washington, D.C., 1993, pp. 143–150.
7. P. W. Birkeland, *Soils and Geomorphology,* Oxford University Press, New York, 1984, pp. 67–74.
8. M. B. McBride, *Environmental Chemistry of Soils.* Oxford University Press, New York, 1994, pp. 207–218.
9. Ref. 1, pp. 123–126.
10. U.S. Geological Survey Bulletin 2220, *Environmental Considerations of Active and Abandoned Mine Lands: Lessons from Summitville, Colorado,* U.S. Government Printing Office, Washington, D.C., 1995, pp. 10–13 and 20–27.
11. W. S. Broecker, *How to Build a Habitable Planet,* Eldigio Press, Lamont-Doherty Geological Observatory, Columbia University, New York, 1985, pp. 211–214.
12. C. D. Keeling and co-workers, in D. H. Peterson, ed., *A Three-Dimensional Model of Atmospheric CO_2 Transport Based on Observed Winds: I. Analysis of Observational Data,* American Geophysics Monograph No. 55, American Geophysical Union, Washington, D.C., 1989, pp. 165–236.
13. J. H. Seinfeld and co-workers, *Rethinking the Ozone Problem in Urban and Regional Air Pollution,* National Academy Press, Washington, D.C., 1991, pp. 19–29.
14. Ref. 1, pp. 373–379.
15. H. D. Holland, in S. Bengtson, ed., *Early Life on Earth,* Columbia University Press, New York, 1994, pp. 237–244.
16. R. M. Garrels and F. T. Mackenzie, *Evolution of Sedimentary Rock,* W. W. Norton, New York, 1971.
17. D. A. Livingston, *Chemical Composition of Rivers and Lakes,* paper No. 440-G, U.S. Geological Society, 1963.
18. W. S. Broecker and T.-H. Peng, *Tracers in the Sea,* Eldigio Press, Lamont-Doherty Geological Observatory, Columbia University, New York, 1982.
19. E. K. Berner and R. A. Berner, *The Global Water Cycle,* Prentice–Hall, Englewood Cliffs, N.J., 1987, pp. 254–262.
20. J. A. Cherry, in M. E. Fleet, ed., *Mineralogical Association of Canada Short Course in Environmental Geochemistry,* Selby Young Printing, London, 1984, pp. 273–277 and 296–301.
21. T. Nystén, in E. Pulkinnen, ed., *Environmental Geochemistry in Northern Europe,* Geologian Tutkimuskeskus, Espoo, 1991, pp. 151–156.
22. Ref. 1, pp. 324–332.
23. P. R. Ehrlich and co-workers, in R. S. DeFries and T. F. Malone, eds., *Global Change and Our Common Future,* National Academy Press, Washington, D.C., 1989, pp. 21–23.
24. C. S. Silver and R. S. DeFries, *One Earth, One Future: Our Changing Global Environment,* National Academy Press, Washington, D.C., 1990, pp. 64–67.
25. K. Clayton, in T. O'Riordan, ed., *Environmental Science for Environmental Management,* Longman Group, London, and John Wiley & Sons, Inc., New York, 1995, pp. 113–120.
26. Ref. 1, pp. 310–323.
27. Ref. 24, pp. 131–141.
28. Ref. 1, pp. 373–379.
29. Ref. 24, pp. 103–110.
30. G. S. Plumlee, K. S. Smith, and W. H. Ficklin, *Geoenvironmental Models of Mineral Deposits and Geology-Based Mineral-Environmental Assessments of Public Lands,* U.S. Geological

Survey Open-File Report 94-203, U.S. Government Printing Office, Washington, D.C., 1994, pp. 1–6.

31. C. M. Ager and co-workers, "Water Transport—a Powerful Way to Move Metals," in T. V. V. King, ed., *Environmental Considerations of Active and Abandoned Mine Lands,* U.S. Geological Survey Bulletin 2220, U.S. Government Printing Office, Washington, D.C., 1995, pp. 20–23.
32. L. C. Gunderson, "Geology of the Reading Prong, New York, and Its Application to Predicting Areas That Have High Radon Potential," in B. R. Doe, ed., *Proceedings of a USGS Workshop on Environmental Geochemistry,* U.S. Geological Survey Circular 1033, U.S. Government Printing Office, Washington, D.C., 1990, pp. 165–168.
33. J. M. McNeal and co-workers, in Ref. 32, pp. 177–178.
34. M. Eisenbud, *Environment* **26,** 6–11 (1984).
35. Board on Radioactive Waste Management, National Research Council, *Rethinking High-Level Radioactive Waste Disposal.* National Academy Press, Washington, D.C., 1990.
36. C. A. Hodges, *Science* **268,** 1305–1312 (1995).
37. P. C. Stern, O. R. Young, and D. Druckman, eds., *Global Environmental Change,* in Committee on the Human Dimensions of Environmental Change, National Research Council, National Academy Press, Washington, D.C., 1992, pp. 44–52.
38. Ref. 1, pp. 269–274.

GLOBAL ECOZONES. See ECOZONES, GLOBAL.

GROUNDWATER SAMPLING. See SOIL WATER AND GROUNDWATER SAMPLING.

GROUNDWATER AND SOIL REMEDIATION

MARVIN H. HYMAN
Bechtel
Richland, Washington

Since the early 1980s, cleanup of contaminated groundwater and soil has become a primary focus of environmental protection throughout the United States. Massive amounts of money and professional/administrative effort have been invested in a campaign to restore the subsurface environment. Prior emphasis for environmental protection was on air and surface water, and the technology challenges have been formidable. In the 1980s it was discovered that whenever an environmental restoration effort is directed below ground, the challenges are doubled.

The main effort in subsurface technology applications in the 1980s was in characterizing the nature and extent of contamination in groundwater and soils. The investment in characterization and in governmental legal action outweighed the investment in applying treatment technologies. Much of the remedial action done was limited to strategies involving containment or disposal, rather than treatment.

With the advent of the 1990s, the emphasis shifted from limited remedial action toward treatment. Although containment, natural attenuation, and disposal are practiced to the greatest extent, application of treatment technologies is expanding at a fast pace. At many complex Comprehensive Environmental Response, Compensation and Liability Act (CERCLA) sites and at some state-controlled sites, interim treatment and emergency actions employing treatment have started without completing characterization. Some interim and emergency actions will become long-term remediation solutions. In some instances, field pilot installations for treatability studies have become interim remediation facilities. Thus, the pace of remediation by treatment is accelerating through pilot, interim, and emergency treatment plants and final remediation solutions.

Another trend in the 1990s that helps jump start remedial action is the growing use of the observational approach in applying treatment technologies. Without waiting to characterize fully the extent of contamination or to find the optimum solution via extensive feasibility analysis and treatability studies, treatment is started and actual results are weighed against cleanup goals. At the same time, characterization of contaminated groundwater or soil is continuing. The process embodied in the observational approach includes greater risks of failure than with complete characterization and feasibility analysis. However, a properly planned project based on the observational approach has contingency actions that are employed in the event of a failure and that accomplish more needed characterization or cleanup. The contingency actions are not employed from the start because they cost more and/or take more time than implementing the trial treatment.

As we look beyond the 1990s, we foresee more development of *in situ* treatment technologies. Remediation costs and sometimes time requirements are reduced if groundwater is remediated without pumping it to the surface or if soil does not have to be excavated.

An inspiring aspect of the field of remediation treatment is the proliferation of new technologies and improvements to proven technologies. In this article on groundwater and soil remediation, emerging treatment processes and improvements are designated "innovative." As used here, this term is not what EPA considers as innovative.

Not all of the emerging technologies are mentioned in this article, and no attempt has been made in this limited space to assemble an exhaustive listing and description. Rather, a number of new processes are described in each section to illustrate the variety of approaches taken to improve remediation applications. References are made to specific suppliers and developers, and these references do not constitute endorsement.

A challenge facing remediation engineers and control agency officials is to choose, among effective treatment processes, the best (or the best grouping of treatment steps) for cleanup of groundwater or soil at a particular site. More detailed information about selecting individual processes for gasoline-contaminated sites is provided (1). The Department of Defense Screening Matrix (2) can help provide selection information on processes for treating all types of contaminants. A scheme is presented for defining groups of treatment steps and selecting among the alternatives (3) based on the Remedial Action Assessment System (4).

REMOVAL OF METALS FROM GROUNDWATER

This section covers the following proven technologies for removing metals from groundwater:

- chemical precipitation
- membrane separation

- ion exchange and adsorption
- evaporation

These technologies are used to remove radionuclides and hazardous metal compounds from groundwater. All but precipitation are also used to remove nitrates.

Innovative technologies covered include multistage reverse osmosis systems; applications of electroprecipitation; biological metals separation systems; membranes and gels for ion exchange and adsorption of selected metals; *in situ* treatment barriers.

Applications

Chemical Precipitation. Precipitation is carried out by adding chemical reagents that form insoluble compounds with metals and adding coagulants that aid in separating the precipitate from the water, followed by clarifying and/or dewatering steps. Precipitation is generally used for these two different applications:

removing toxic heavy metals, eg, removing lead: Pb^{+2}(dissolved) + S^{-2} = PbS (insoluble);

removing calcium, iron, and/or manganese from groundwater fed to air strippers that in turn remove volatile organic compounds.

Cleanup levels approach the solubility of the precipitated metal compound if good mixing and long residence time along with clean separation of the precipitate are achieved.

With sulfide precipitation, the treated effluent contaminant concentration is near the lower detectable limit. With hydroxide and carbonate precipitates, the cleanup levels attained are not as good as with sulfide and depend on the solubility of precipitates at the pH employed. If there is not a clean separation of precipitate, the treated effluent contains contaminants in the suspended solids, which sometimes dissolve if the pH of the discharge is altered downstream.

Problems and limitations with precipitation are as follows: attaining a settleable floc is often difficult; precipitates are in the form of a sludge that is costly to dewater; cleanup levels are limited by solubility; each metal has its ideal pH, so no one pH is optimal for multiple contaminants; sulfide precipitation attains the best cleanup levels, but toxic hydrogen sulfide gas is emitted if conditions are acidic; many metals form anionic complexes difficult to precipitate; mixing of coagulant aids and reagents must be done under carefully controlled conditions (including mixing speed, pH and residence time); precipitation reactions are reversible, so toxics are resolubilized.

Membrane Separation. The membrane separation methods referred to here remove dissolved ions (unlike filtration, which separates undissolved particulates from a fluid). Most membrane applications are carried out by reverse osmosis. Pressure is applied to groundwater on one side of a very fine-pore membrane. Water molecules pass through, (becoming "permeate") afd ions are retained. Electrodialysis reversal competes with reverse osmosis and achieves similar separation at lower pressures. Membrane separation methods require prefiltration and remove 90 to 99% of contaminants.

Problems and limitations with reverse osmosis (RO) include membrane disintegration; high pressure requirements; relatively low removal efficiency for many ions; up to 20% (and sometimes more) of the groundwater is a secondary waste stream. To reduce the volume of the secondary waste stream (retentate), it must be concentrated more, leading to scaling of membranes, which limits the operability of the systems and the permeate fraction. One study states that RO is expensive, not typically used to remove metals, and should be limited to removing nitrates, sulfates, total dissolved solids, and naturally occurring inorganics (5).

Ion Exchange and Adsorption. Ion exchange and adsorption obtain the very highest removal percentages. Most ion exchange systems use synthetic resins to adsorb dissolved ions. The term "medium" will be used here for both ion exchange media and adsorbents that function without ion exchange. The media are used in either of these two modes:

An ion-exchange medium is periodically regenerated with an acid, ammonium ion, or sodium ion solution to remove adsorbed cations and with hydroxide or chloride ion solution to remove adsorbed anions.

Or, the medium is used once and disposed of as solid waste.

For most metals removal problems, there is virtually no limit to the cleanup level attained, especially if regeneration is not practiced. The more the column length and volume of the medium for a given groundwater flow rate, the better the cleanup efficiency. Cleanup efficiencies exceeding 99.9% are attained, with treated effluent contaminant concentrations near or below detectable levels.

Problems and limitations with these processes, include the following: the medium is frequently not ion-specific, so noncontaminant ions compete for adsorption sites; a waste concentrate stream is formed with regeneration; costs for makeup media and solid waste disposal are high without regeneration; early breakthrough of contaminant occurs if there is hydraulic channeling through media beds.

Evaporation. Evaporation is done in a solar basin or with a heated forced-evaporation system. The discussion here is limited to heated forced-evaporation systems. If recovering water from an evaporator is desired, it can be condensed and used as a clean water source, injected into an aquifer, or discharged to a sewer or to surface waters. Evaporation is accomplished with a submerged fuel gas burner, in a vessel with heat from a burner or by steam, with a steam vacuum system, or with a vapor recompression system.

Cleanup levels with evaporation are near or below detectable levels.

Concentrated Wastes. Membrane separation systems, regenerated ion exchange systems, and evaporators all produce concentrated waste streams. With membrane separation and regenerated adsorbent systems, the volume of waste concentrate is up to approximately 20% of the groundwater volume. With evaporation, the concentrated brine volume produced is usually a few percent of the groundwater volume. Concentrated waste streams from using these technologies are often evaporated and/or dried (crystallized) to a semisolid or powdered waste to minimize final disposal volume.

Companion Technologies. If volatile organics are to be air stripped, removal of dissolved calcium, iron, and manganese by precipitation often precedes the stripping operation. Ultrafiltration, a membrane technology used to separate suspended solids and large dissolved organic molecules, is used as a pretreatment step in reverse osmosis. Reverse osmosis is a pretreatment step for ion exchange. Evaporation and drying/crystallization are used to dispose of the concentrated streams produced as secondary wastes from processes, such as reverse osmosis and ion-exchange regeneration (see REVERSE OSMOSIS).

Basic Methods for Metals Removal from Groundwater

Precipitation Methods with Clarification/Filtration. The reagents used to form insoluble metal compounds contain hydroxide, sulfide, or carbonate. Hydroxide reagents are caustic soda (sodium hydroxide) solution and slaked lime (calcium hydroxide) slurry. Sulfide reagents are hydrogen sulfide gas and sodium sulfide or sodium bisulfide solution. Another widely used precipitation agent is sodium carbonate. Magnesium hydroxide is effective for precipitation at a pH of 9. Ferric ion is used to precipitate some anions, such as arsenate.

The reagent and a coagulant aid (eg, organic polymers) are usually mixed in two or more stages. Sometimes an oxidant (eg, ozone, permanganate, hypochlorite, or peroxide) is added. For example, if iron or manganese causes scaling problems in downstream air stripping equipment, an oxidizing agent is also added. This converts ferrous ion to ferric and manganous ion to manganic, resulting in more insoluble precipitates. If chromate is present, precipitation is accompanied by a reduction step changing soluble chromates to chromic ion that can be precipitated as hydroxide or sulfide.

The first mixing step, called flash mixing, is done with a high-speed mixer and a residence time of less than a minute. Various chemicals are used as coagulant aids to hasten settling by forming a floc. Organic polymers are often added in parts per million quantities. Other coagulant aids include inorganic polyelectrolytes, such as alum, ferric chloride, and ferrous sulfate.

Subsequent mixing is done with progressively slower-speed mixers in one or more flocculation chambers. The precipitate formed is allowed to settle as a sludge in a clarifier with a residence time of several hours.

A cone-bottom vessel is used for thickening the precipitated sludge, which is then fed to the filter press. Removal of sludge is periodic. When a substantial volume of sludge accumulates at the bottom of the clarifier, the sludge pumps are temporarily activated.

Clarifiers are very large—over 8500 liters (30 cu ft) per 0.063 L/s (1.0 gpm) of flow. Various schemes are used to reduce the size of clarifying and sludge dewatering equipment. Settlers with slanted plates that form coalescing surfaces for the suspended solids use much less space than clarifiers. Centrifuging dewaters sludge rapidly. Ultrafiltration systems with crossflow through porous membranes, an alternative to clarification, dewaters suspended solids and removes large organic molecules.

If clarified or filtered effluent needs neutralization, the pH is lowered by adding acid or by sparging with carbon dioxide.

Membrane Separation Methods. Reverse osmosis (RO) and electrodialysis (EDR) systems use membranes with pore sizes smaller than ultrafiltration membranes. Whereas ultrafiltration removes metals in precipitated (suspended) solids, RO and EDR remove dissolved ions. Ultrafiltration works with applied pressures well below 100 psig. RO generally works in the 400 to 1000 psig range. EDR works below 100 psig with membrane pore sizes comparable to RO, by applying static charges of opposite polarity across membrane surfaces.

With RO and EDR, the treated effluent (permeate) is generally over 80% of the influent volume. The concentrate (retentate) rejected by the membrane is the remaining percentage. Multiple treatment stages raise the permeate percentage into the middle nineties and, in some instances, into the high nineties. The limiting factor is usually scale formation, which is reduced by injecting acid upstream of the membrane.

The purity of the permeate depends mainly on membrane structure, applied pressure, and the number of treatment stages. Contaminant removal efficiencies are usually above 90%, and typically 95 to 99% removal is achieved. In general, the larger the ion, the better the removal. For even higher removal efficiencies, membrane separation is used in conjunction with downstream ion exchange. If 90% of dissolved ions are first removed with RO, the frequency of ion-exchange resin regeneration (or discarding expensive spent resins if there is no regeneration) is greatly reduced.

An important feature of multistage membrane separation is the fact that percentage removal is generally the same for each stage. Thus, if the removal efficiency is 80% for a particular ionic species, a two-stage system works as follows: the first-stage effluent has 20% of the influent concentration; the final effluent has 20% of 20%, or 4%. The overall efficiency is 96%.

Ion-Exchange and Adsorption Methods. A variety of substances adsorb dissolved ions from groundwater. The most widely used are synthetic demineralization resins that remove virtually all of the cations or all of the anions by ion-exchange. Natural zeolites and synthetic zeolites are also used for ion-exchange. Other substances (eg, bone char, alumina, carbon) adsorb certain metals without ion-exchange taking place. One method of reducing regeneration frequency or spent adsorbent disposal volume is to use media that selectively remove contaminant ions and that let the bulk of the dissolved minerals go through with the effluent. Thus, ion-exchange sites or adsorption sites are not completely taken up by ionic species that are not contaminants. Some examples will illustrate this method:

A synthetic anion-exchange resin is selected for chromate removal. For this groundwater, nitrate concentration is up to 10 times that of chromate but within the drinking water standards, and nitrate removal was not required. In practice, after a period of operating time, the effluent concentration of nitrate and a number of other anions equals the influent concentration. An equilibrium is attained and these anions occupy ion-exchange sites on the resin, but not all the sites. This system continues to reduce groundwater chromate concentrations of hundreds of micrograms/liter to below the target level of 50 ug/L.

A cation removal system is operated to remove radioactive plutonium, strontium, and cesium. Following a particulate filter that removes suspended particulate are three adsorbent beds in series. The first bed contains granular bone char. The calcium phosphate in the bone structure reacts with and adsorbs dissolved plutonium, and essentially 100% removal is achieved. The second bed contains clinoptilolite, a natural zeolite, that removes a high percentage of strontium and virtually all of the cesium. The third bed contains 50% bone char and 50% clinoptilolite and is a guard chamber in the event of contaminant breakthrough from either of the first two beds.

Ion-exchange resin vendors advise on resin selection, expected resin life between regenerations or changeout, optimum pH, backwash velocity, regeneration solutions and volume, and expected pressure drop. Jar tests are used to help choose between competing resins. Such tests determine the amount of contaminant species that an adsorbent retains under equilibrium batch conditions for various amounts of adsorbent and contaminant in a given volume of groundwater. Improved design is achieved if bench-scale flow-through dynamic column tests are performed.

Such column tests are set up to match the velocity of groundwater flow through the bed for the full-scale design. Results determine pressure drop, percent contaminant removal at a given residence time, the volume of groundwater that corresponds to contaminant breakthrough, and the volume of groundwater that corresponds to contaminant saturation of the resin. It is important to distinguish between breakthrough and saturation. Breakthrough occurs when the effluent contains a detectable amount of contaminant or when the effluent concentration is a given percent of the influent.

Systems that discard spent resin use multiple beds in series, an arrangement that allows saturation of the first bed. If the bed lengths are designed adequately, breakthrough has not started out of the last bed, when the first bed just becomes saturated. Effluent quality is totally controlled. Bed replacement costs are minimized without jeopardizing effluent quality by removing only the first (saturated) bed and retaining the last bed and partially spent second bed in service.

While the system is shut down to remove a fully spent first bed, the second bed is placed into first-bed service. The third bed is placed into second-bed service, and the third or last position is filled with fresh adsorbent. Thus, the final effluent is always in contact with the best medium, thereby achieving maximum removal of contaminants. The changing of bed positions is usually accomplished by opening and closing valves in the piping manifold. With small systems, the vessels are physically moved from the last position toward the first.

With staged systems, it is important to note that the percentage removal and bed capacity in a successive stage are not usually as high as in the previous stage that operates with higher influent concentrations.

Evaporation Methods. Most evaporation installations use steam-heated multiple-effect vacuum evaporators or vapor recompression systems. Chemical additions are needed to reduce scaling. Reverse osmosis is used for pretreatment. The groundwater should be deaerated, either by warming first above 100°C or by warming while applying vacuum. An oxygen scavenger is added to reduce corrosion. If water vapor is recovered by condensing it, heat exchangers preheat the groundwater and thereby optimize the energy input.

With vacuum evaporators, the water is boiled at temperatures below 100°C by applying vacuum. The amount of heat input needed is reduced by using heat exchangers between stages that operate at successive vacuum levels (and at corresponding temperature levels).

Vapor recompression systems operate under vacuum with one stage. No external heat or steam is needed. A typical system operates as follows:

Warm deaerated groundwater is fed to the sump of the evaporator vessel where it mixes with the concentrate (or with slurry if some of the mineral content has crystallized). Sump contents are continuously removed as a secondary waste product and some are pumped to the top inlet of vertical heat exchange tubes within the evaporator vessel.

The concentrate falls as a film on the inside of the tubes. Compressed water vapor is condensed on the outside of the tubes. The heat released from condensation is transferred to the falling film, causing a portion of the concentrate to evaporate. The water vapor thus formed is fed to the compressor suction.

The basic energy source is electrical power for the compressor motor. Some of this energy becomes heat of compression, which in turn is used for evaporation in the falling film heat exchanger.

Operating Conditions and Treatability Studies

Precipitation requires selecting reagents, coagulant aids, pH, and, sometimes, temperatures. The apparatus for preliminary testing is simple laboratory equipment, but the combinations of reagents and pH conditions are extensive. The time required for settling precipitates with coagulant aids must be determined for clarifier design.

Most potential reverse osmosis systems must be checked for the permeate fraction that is obtained versus pressure, up to the limitations associated with scaling on the membrane. The efficacy of various antiscalants and pressure selection can be determined on site on a pilot scale with rented equipment or at the vendor's laboratory with a large sample of groundwater. Some vendors have trailer-mounted equipment that operates on-site, complete with prefiltration components and acid (antiscalant) injection for pretreatment.

Potential ion exchange and adsorption systems should be checked for cleanup efficiency and resin capacity or adsorbent capacity in three steps:

Laboratory jar tests or batch equilibrium tests where candidate adsorbents are checked for the partitioning coefficient K_p, the ratio of adsorbed concentration to liquid phase concentration. The tests are done with varying amounts of adsorbent and sometimes at varying pH levels.

Bench-scale tests using adsorbents selected from promising batch equilibrium results using dynamic flow-through conditions in glass columns. The tests are done at varying contact times at the selected pH. Regenerants and regeneration parameters are also tested and optimized.

Field pilot-scale tests using dynamic flow-through conditions with contact time determined from bench-scale tests, where pressure drop, other hydraulic parameters, and bed capacity are monitored.

The bed capacity is expressed as the number of bed volumes of groundwater that pass through the ion exchange or adsorbent column until the medium no longer effectively retains the contaminant. If the bed volume of a column is expressed in liters of medium, then the capacity is determined in a bench-scale test by measuring liters passed through. Most field applications include multiple columns in series, so considerable breakthrough of contaminants is allowed from the lead column to use all the potential capacity of the bed. A good rule for estimating the capacity of the medium is to use the number of bed volumes corresponding to 0.50 breakthrough.

Trends and Innovations in Metals Removal from Groundwater

Precipitation is trending toward more use of ultrafiltration instead of clarifiers. Innovations that use electro-precipitation and co-precipitation of metals (5) have come into use. Research continues to find ways to separate precipitates cleanly. An example is the development of crystallization technology for arsenic removal by forming ferric arsenate at elevated temperatures and gradually raising the pH from below 1.0 (6).

The trend in removing metals from groundwater is toward combination systems employing prefiltration, removal of organic compounds, reverse osmosis, and possibly ion exchange polishing of RO permeate, with evaporation/drying of concentrated streams, such as RO retentate. RO is the main step in these systems. RO does not require regeneration or replacement of spent materials; for most applications, the membranes last for years.

Developments within the membrane separations industry are resulting in a decreasing percentage of retentate volume. The demonstration of improved membrane structures that resist scaling is underway. Such membrane improvements will aid in reaching the goal of maximizing retentate concentrations, with correspondingly reduced secondary waste volume. An example is the membrane crystallization system marketed by Zenon (Burlington, Ontario), where the volume of permeate is over 95%.

Reference 7 summarizes a report in *Science 268,* 700 (1995) on the development of ion-selective metal membranes. This development could lead to innovative membrane separation applications where scale-forming ions that are not contaminants are allowed to pass through.

Supported liquid membranes are used to remove dissolved metals or organics (8).

An application of direct osmosis for removing water from wastewater has been reported (9).

Because sulfide precipitates are more insoluble than hydroxide precipitates, there is a trend toward sulfide precipitation technology. One innovation involves the use of sulfate-reducing bacteria to produce the sulfide for metals precipitation. The precipitation takes place in an anaerobic bioreactor, with sulfate additions if the groundwater does not already naturally contain sufficient sulfate.

Biological systems are used as adsorbents that selectively capture heavy metals. Reference 10 quotes from the June 1994 issue of EPA's *Ground Water Currents* on the use of algae. Microbiological research at Clark Atlanta University (Atlanta, Georgia) has led to the development of algal/bacterial mats that remove heavy metals from water and destroy chlorinated organics. Phytoremediation (plant uptake of contaminants including metals and nitrates) has been described (11).

Selective ion exchange continues to be the main method for removing radionuclides from groundwater. The cleanup goals for radioactive strontium, cesium, technetium, uranium, and plutonium are so stringent that multistage nonregenerative ion exchange systems are sometimes the only effective technology that is not energy intensive. Research continues with the goal of developing more selective, highly efficient resins.

An innovative technique using membranes ("webs") for selective adsorption for analyzing certain radionuclides has been developed by Argonne National Laboratory (Argonne, Illinois) and 3M Company (St. Paul, Minnesota) (12). This analytical technique can be adapted for use as a metals removal treatment system. An ion-exchange or adsorptive medium is held on cartridges that resemble filter cartridges. The medium particulate size is much smaller than resin beads normally used for ion exchange, so the required contact time and the volume of solid waste are greatly reduced.

A gel has been developed at the Department of Energy's Hanford (Washington) site that extracts heavy metals or radionuclides. The gel is recycled to recover valuable constituents.

One of the most important developments for removing metals from groundwater is the use of *in situ* treatment barriers (13). Treatment materials are placed in trenches or in rows of large diameter augered holes within aquifers. In some examples the barrier is a porous reactive wall. Research has been done on sorptive barriers containing a wide variety of materials, including hydrated lime, fly ash, barium chloride, calcium phosphate, titanium oxide, peat, various iron compounds, and zeolite pretreated with surfactant.

Costs for Metals Removal from Groundwater

Precipitation and RO Costs. The investment for units with similar throughput capacity is approximately the same for precipitation or reverse osmosis technology, except that RO units need prefiltration, which adds to their overall cost.

The annual expense for either technology depends mainly on whether a unit is attended. Operating labor cost is the significant item if attended. The cost of chemicals, even with precipitation, is small compared to equipment amortization. RO uses less chemicals, typically only antiscalant, but, at approximately $1 per 3785 L (1000 gallons) treated, uses more pumping energy. A two-person operating crew with attendance 7 days/week for 50 weeks per year expends 5600 labor hours/year. A two-person crew with attendance 24 hours/day plus a technician working 40 hours/week expends 22,600 labor hours per year for a unit running 310 days/year. When these hours are multiplied by the wage rate and worker benefits and overhead expense is added, operating labor expense may eclipse other annual expenses.

The remediation cost estimating "RACER/ENVEST" models developed by Delta Technologies Group (Englewood, Colorado) include chemical precipitation and neutralization. "COMPOSER GOLD" estimating models marketed by Building Systems Design (Atlanta, Georgia) include coagulation/flocculation and neutralization.

Ion-Exchange and Evaporation Costs. Treatment costs of ion exchange or evaporation are highly variable and are best estimated for each situation. Ion exchange costs depend on the adsorbent selected, whether regeneration is applied or the frequency of bed replacement, and the disposal requirements for regeneration solutions or of spent media.

Evaporation costs are highly dependent on the amount of heat recovery equipment included in each design, the amount of energy input required, and the unit cost of energy. With no heat recovery, the energy expended to evaporate water flowing at 1 gpm is approximately 500,000 Btu/h (5 therms/h). If this heat input is derived directly from a fuel gas burner using fuel costing $.70/therm, the fuel cost is $3.50/h per gpm. A 100-gpm unit operating 24 h/day for 310 day/yr has a fuel cost of over $2.5 million/yr. These units, usually attended by operators, have a labor cost comparable to fuel costs. So, the operating costs for evaporation can greatly outweigh equipment amortization costs.

REMOVAL OF ORGANICS FROM GROUNDWATER

This section covers the following proven technologies:

- aqueous phase adsorption
- stripping
- aqueous oxidation
- bioremediation

Innovative technologies described in this section include *in situ* application of zero valent metals for dechlorination/destruction of solvents in groundwater; the use of resins and peat instead of activated carbon; *in situ* stripping with air sparging; in-well stripping; aqueous oxidation with titanium dioxide catalyst or with mixed oxidants; catalyzed electrolytic oxidation; supercritical oxidation; the use of reducing agents and cavitation with ultraviolet oxidation; high energy electron beam destruction of organics; bioremediation with fungi; and bioremediation combined with carbon adsorption.

Applications

Adsorption. The most common aqueous phase adsorption technique employs granular activated carbon in fixed beds within vessels (canisters). Other forms of carbon and certain resins compete with granular activated carbon. Thorough reactivation of spent aqueous phase carbon requires steaming in a high temperature oxygen-deprived furnace. Except for very large systems, it is not economic to use such a furnace for on-site reactivation. For the vast majority of groundwater remediation applications, spent carbon is sent off-site to regional activation facilities.

Activated carbon is made from coal or from cellulosic substances. For remediation, the activated carbons most used are derived from bituminous coal, lignite, or coconut shells.

Activated carbon removes a wide variety of organics. (One drawback is that because carbon is so highly effective and nonselective, it becomes loaded with natural organic matter in groundwater. This deprives an adsorption system of some of its capacity for removing organic contaminants.) Small molecules and some oxygenated organics are poorly adsorbed. Commonly encountered organic contaminants that are poorly adsorbed include vinyl chloride, methylene chloride, and methanol.

Oil and grease rapidly loads activated carbon and should be filtered out of the groundwater. Granular activated carbon is a filter for removing suspended solids, but becomes plugged unless prefiltration is used.

Problems and limitations with aqueous phase granular activated carbon, besides interfering compounds and poorly adsorbed compounds include the following: calcium scale formation calcifies the carbon bed; unwashed fresh carbon causes a pH change; fines cause pressure drop and should be backwashed out of fresh carbon beds; corrosion of steel proceeds rapidly in the presence of groundwater and activated carbon, so canisters must be appropriately lined or made of stainless steel or plastic; air must be removed from the system to avoid maldistribution of water flowing through the bed and shortened residence time; bacteria multiply rapidly on activated carbon; biofouling and other causes of plugging lead to overpressured canisters; maldistribution and shortened residence time occurs with downflow at hydraulic loadings below 1 gpm/sq ft.

Multistage adsorption systems and/or monitoring techniques are applied so that breakthrough of contaminants into the final effluent is avoided. At the same time, economics dictate that adsorbent beds removed from service for reactivation are as close as possible to maximum adsorptive capacity (saturation with adsorbed contaminants).

Stripping. Stripping with air at ambient temperatures is the most common stripping application and is limited to removing only VOC. However, by heating the groundwater or by stripping with steam instead of air, semi-VOC are also removed.

Stripping is sometimes followed by aqueous phase activated carbon for polishing treatment. Stripping serves to greatly reduce the loading of organics on the carbon, with commensurate lowered carbon reactivation costs. For many VOC, stripping typically removes over 98%, and effluent concentrations from carbon treatment are undetectable.

The off-gas from stripping contains the organic contaminants removed from groundwater and often requires vapor treatment (emissions abatement) prior to discharge to the atmosphere. The most common abatement methods are vapor phase activated carbon adsorption and direct thermal or catalytic oxidation.

Problems and limitations with stripping include scaling/fouling of stripper internals when air causes calcium carbonate, ferric oxide, manganic oxide, or bacterial cell mass formation in the groundwater; scaling is promoted at elevated temperatures with heated strippers and steam strippers; the presence of semivolatile and nonvolatile organics that do not strip in unheated systems; scaling of internals if groundwater flow stops and the stripper goes dry; maldistribution of liquid flow causes poor stripping, which often occurs at groundwater flow rates below design in nonrecycled packed strippers; maldistribution of liquid flow in the lower part of tall packed strippers; malfunctions of stripper sump level controls causing backflow of stripped water into the air inlet or loss of effluent pump suction.

Aqueous Oxidation. Proven methods of aqueous oxidation include wet air oxidation (injection of air with heat and pressure), reaction with Fenton's reagent (hydrogen peroxide plus iron), and reaction with hydrogen peroxide and ozone. All of these methods are enhanced by ultraviolet (UV) light. The aim of these methods is the destruction of organic compounds, with the formation of water and of carbon dioxide or relatively nontoxic organic acids. If chlorinated hydrocarbons are present, chloride ion is formed. With wet air oxidation, either suspended or dissolved organics are destroyed. With UV oxidation, only dissolved organics are destroyed.

Hydrogen peroxide alone sterilizes microorganisms, such as bacteria, but only very slowly destroys organic compounds. With Fenton's reagent, nonalkane organics are destroyed with residence times of the order of 30 to 60 minutes (16). The peroxide forms hydroxyl radical and converts ferrous ion to ferric, and oxidation of organics proceeds with hydroxyl radical and ferric ion together.

UV light speeds organic destruction by oxidants, aids in the destruction of a wider variety of organic compounds, and can be applied with wet air oxidation, with Fenton's reagent, or with peroxide and ozone. Hydrogen peroxide is the most used oxidant. Ozone is also used. Some systems use both of these oxidants with or without UV light. UV light promotes the formation of hydroxyl radicals that react rapidly with organics. The destruction percentage of some unsaturated chlorinated compounds (eg, dichloroethylene) approaches the high nineties. Saturated organics (eg, dichloroethane), especially highly chlorinated compounds, are not as readily destroyed.

One study reports that UV-peroxide pilot experiments achieve 99.7% destruction of tetrachloroethylene, an unsaturate, with 0.6 minutes exposure time, and only 78 to 83% destruction of tetrachloroethane with 5.5 minutes exposure time (17).

The organics destruction efficiency depends somewhat on the configuration of each manufacturer's reactor, because the shorter the distance between the UV light source and the dissolved organic molecules, the better the destruction.

Problems and limitations with wet air oxidation systems include erosion of pressure letdown valves; high capital cost per gpm of capacity; high costs for pressurization and temperature elevation; danger of explosion if influent organics concentrations are relatively high. However, wet air systems are excellent for treating organic sludges, whereas UV oxidation systems require prefiltration because turbid water inhibits UV light transmission, as does scaling of UV lamps, which require periodic mechanical scraping to stay clean. If ozone is used, it is too unstable to be stored and must be generated on-site. Undissolved ozone causes an air emissions problem unless abated. Other problems and limitations associated with UV oxidation include sensitivity to adequate residence time, UV dosage, and oxidant dosage; difficulty in dechlorinating certain compounds; limited UV lamp life and need for a lamp replacement program; handling of a hazardous chemical if peroxide is used.

Bioremediation. Organics in groundwater are converted by microorganisms to less toxic organic compounds or are mineralized (forming carbon oxides and water). Both anaerobic and aerobic wastewater biotreatment techniques are applied. However, groundwater generally does not have enough organics concentration to sustain anaerobic bacterial colonies. A notable use of anaerobic biotreatment is denitrification of nitrates, where a dissolved organic compound (eg, methanol, acetic acid) is added to maintain anaerobic colonies.

Most of the remediation that occurs is achieved in the ground by natural aerobic action. Most of the engineered treatment applications employ aerobic bacteria. An aquifer serves as an *in situ* bioreactor either by injecting air or hydrogen peroxide and nutrients or by pumping groundwater to the surface temporarily, aerating and/or injecting additives, and reinjecting the groundwater into the aquifer.

Ex situ applications are done by injecting nutrients and air into completely mixed bioreactors (eg, activated sludge units, fluidized bed units) or into unstirred plug-flow fixed-film bioreactors (eg, submerged fixed film units, rotating disc bioreactors, trickling filters), or with aeration in a lagoon or pond.

Problems and limitations common to all bioremediation systems include the organic contaminants must be biodegradable (not highly "refractory") and not highly toxic to the microorganisms; breakdown intermediate products must not be highly toxic; temperature and pH must be within prescribed ranges; macronutrients (eg, fixed nitrogen compounds and phosphates) and micronutrients concentrations must be adequate. With anaerobic systems, organic loading must be at least approximately 4000 mg/L, so dissolved organic carbon additions are needed. One study points out that microorganisms degrade only very specific classes of organics; minor molecular variations may not be metabolized (18).

With *in situ* bioremediation, potential problems include nonuniform dispersion of injected additives; reactions of additives with noncontaminants; plugging of the aquifer; difficulty in monitoring and controlling all of the elements needed for successful microbial activity. One study cautions that overdosing with hydrogen peroxide destroys beneficial organisms and recommends using magnesium peroxide, which slowly releases oxygen and stops when the groundwater becomes fully saturated with oxygen (19).

Problems and limitations with activated sludge systems include sensitivity to variations in pH, temperature, toxicity, organic loading (substrate or "food"), food/microorganism ratio, and sludge age; amount of air or oxygen that can be injected.

An advantage of completely mixed systems when influent toxicity is marginally high is that the average toxicity through the reactor may be tolerable. Problems and limitations with fixed-film plug-flow systems include potential for killing the entire bacterial colony at marginal influent toxicity; limited organic destruction efficiency.

Problems and limitations with aerobic lagoons include difficulty in attaining good mixing throughout the entire pond area; slowing down of bacterial action in cold seasons and consequent reduced organic destruction; air emissions control not practical; attractive to birds and other wildlife that may be harmed if toxic organics are in the influent; large area basins require expensive liners.

Companion Technologies. Aqueous phase adsorption should always include prefiltration and sometimes is preceded with stripping. Stripping may need pretreatment including sterilization of bacteria; oxidation of ferrous and/or manganous acid and precipitation/removal; removal of calcium; removal of toxic metals; use of an oil/water separator to remove floating undissolved organics. Stripping often requires off gas abatement. Some strippers include acid wash systems that remove scale and biofouling. Groundwater stripping often accompanies soil venting because the organics contamination source is common to the soil and an aquifer. The stripper offgas and the soil gas may have a common vapor treatment system. Either adsorption or stripping can serve as a polishing treatment following aqueous oxidation.

Basic Methods for Groundwater Organics

Adsorption Methods. The vast majority of aqueous phase adsorption installations consist mainly of two or more canisters in series, each containing a fixed bed of granular activated carbon, with downflow through the beds. A residence time of 15 minutes (figured as gross bed volume divided by volumetric water flow rate) results in quantitative removal of adsorbable dissolved organics. Breakthrough of an organic is detected in the effluent from the first adsorption stage when the adsorbent becomes partially spent. More groundwater is still treated in the first stage until the bed is saturated, when the x/m value for a contaminant is attained. The x/m value is the amount of a compound which the adsorbent can hold per unit mass of adsorbent.

The x/m for each type of carbon or other adsorbent is determined by any of these methods:

Freundlich isotherm: a plot of log x/m against log of C, the equilibrium concentration, or a tabulation of x/m values and corresponding concentrations. The log–log plot is a straight line with slope $1/n$; the equation is $\log x/m = (1/n)C + constant$.

Dynamic laboratory experiments done with a sample of groundwater pumped continuously through a column of adsorbent.

Field pilot test conveniently conducted with a pair of 55-gallon drums of adsorbent.

The isotherm data are obtained from published values or preferentially from laboratory experiments done with groundwater samples. The groundwater is in contact with varying quantities of powdered carbon for periods of hours, and the equilibrium concentrations of the organics are recorded. Because a flow-through remediation system does not attain equilibrium conditions, the x/m attained is a fraction of that determined in the laboratory. According to one study the fraction attained for granular activated carbon is usually in the 0.45 to 0.55 range, if the influent organic concentration is used with Freundlich isotherm data (20).

The dynamic laboratory column test is an excellent method of determining a realistic value of x/m, especially because natural organic matter interferes with attaining the pure compound x/m value. To cut down the size of the groundwater sample required and the time it takes to saturate the laboratory test bed, an accelerated test method accompanied by mathematical modeling is used (21).

Data from a field pilot test are the easiest to scale up for full-scale design, but typically many weeks of operation are needed to saturate a carbon bed.

Stripping Methods. Most stripping is done by passing air up through a column with random packing or trays, with groundwater introduced at the top. The VOC volatilize and pass out of the column at the top with the exhaust air. A critical design parameter is the air/water ratio. The higher this ratio, the shallower the packing depth, or fewer trays are needed. For a given depth of packing or number of trays, the organics removal efficiency depends on the air/water ratio and Henry's constant (a function of temperature) for each dissolved organic compound. Henry's law states that a compound's partial vapor pressure or concentration is proportional to its concentration in the liquid, and Henry's constant is the proportionality constant. Henry's constant rises with increases in temperature. Henry's law holds for dilute liquid concentrations, which is the situation considering the very low solubilities of most VOC in water.

Both cylindrical and rectangular cross section stripping towers are used, made of metal, plastic, or concrete. Packings are plastic, stainless steel, or ceramic shapes. Most strippers are packed with plastic shapes ranging in size from 2.5 to 7.6 cm nominal width, the tower being a plastic cylinder with the air supplied under the packing support grid by a forced-draft blower. The cross-sectional area must be large enough to prevent excessive air pressure drop or flooding of water from the top. Usually the tower cross-sectional area is proportional to the water flow rate.

A mist elimination section, usually made of packing or baffles, is usually included within the tower near the top above the groundwater inlet to minimize carryover of water droplets in the off-gas. Instead of using a forced-draft blower, some strippers have an induced-draft blower in the offgas stream. A mist separator is placed at the blower suction to minimize the chance for condensation of water vapor in suction ductwork. This arrangement results in perceptibly improved stripping with a slight vacuum in the tower instead of pressure and in a shorter tower.

At a selected air/water ratio, the depth Z of packing is given by

$$Z = HTU \times NTU$$

where HTU is the height of a transfer unit and NTU is the number of transfer units required for a desired reduction in concentration of a given compound. If multiple organic compounds must be removed, the largest calculated NTU is used for design. HTU is proportional to the liquid loading divided by the specific area of packing surfaces and by the mass transfer coefficient. Packing manufacturers supply data for determining the HTU, or pilot tests are used to derive the HTU. Transfer coefficients are calculated for traditional packings (22), and for modern packings as a function of diffusivities and surface tension (23).

NTU is a complex function of the molal gas/liquid ratio, Henry's constant, and total pressure P. (For most air strippers, the total pressure is close enough to one atmosphere to use P equal to 1.0 if Henry's constant is expressed in atmospheres.) NTU values are available (22). "Air strip" computer programs automatically calculate NTU or packing depth Z and the corresponding effluent concentration for each given organic compound and air/water ratio. Air strip programs also compute the air pressure drop. Such programs are available from some packing manufacturers, the U.S. EPA, or D. Schoeller (Ames, Iowa). Unless the design packing depth is based on pilot data, a safety factor (eg, 1.5) should be applied to calculated depth.

Important concepts to understand with strippers include the following:Increasing the air/water ratio up to a value of 5 for $(H/P)(G/L)$ allows decreasing the depth of packing, where H is Henry's constant (atm), P is total pressure (atm) and G/L is the molal air/water ratio.

A section of packing or group of trays provides a certain factor of concentration reduction for each of the VOC at a given air/water ratio. For example, if the top 3 meters of packing reduce the influent concentration by a factor of 10 (a 90% reduction), the next 3 meters reduce the concentration by a factor of 10 again. The overall reduction in concentration through 6 meters is 99%.

Instead of using conventional packed towers or tray towers, a cooling tower can be used for stripping. For most cooling tower designs, capturing emissions is not practical, so a cooling tower is a good candidate for stripping applications where emissions abatement is not required.

Improved removal of VOC or the ability to remove semi-VOC is attained by heating the groundwater, a costly enhancement of stripping. Heating of water is favored over heating of air because the water has more heat capacity. By applying heat or by using steam as the stripping gas instead of air, Henry's constants and corresponding volatilities are raised. Steam strippers often operate at partial vacuum, which enhances volatilization. With steam, condensation in the stripper reduces the volume of off-gas to a relatively small amount of water vapor and organic vapors that are recovered by condensation.

The steam is injected directly in place of air, or is generated from stripped water by a reboiler. The reboiler is fired, heated heated electrically, or heated with steam. This steam is condensed without directly contacting the process water and returned to the boiler, whereas with direct steam injection clean condensate is not recovered.

Aqueous Oxidation Methods. The higher the pressure, the more air or oxygen water can absorb. The higher the temperature, the more complete and faster is the destruction of organic compounds. Wet air oxidation takes advantage of these phenomena and is carried out with air (or oxygen) injection typically at 123 atmospheres pressure and at temperatures up to 280°C (536°F).

Wet air oxidation and Fenton's reagent systems are not used for groundwater treatment to the extent that UV oxidation systems are used. Wet air oxidation costs more than UV oxidation, but handles sludges and has the potential for much higher organic destruction efficiency. Fenton's reagent is worth considering if alkanes are not included and is especially effective for destroying aromatics, organic acids, aldehydes, ethers, phenol, alcohols, amines, and ketones. Some of these compounds do not adsorb well on carbon.

UV light without added oxidants is used to sterilize fresh water and in some wastewater treatment works. Relatively light UV dosages kill microorganisms, such as bacteria. This operation plays a role in groundwater remediation, especially for sterilizing groundwater fed to a stripper.

UV oxidation systems depend on the formation of hydroxyl radical (OH^*) to destroy chlorinated hydrocarbons (R-Cl) by the following typical reactions:

$$H_2O_2 + UV \rightleftarrows 2\ OH^*$$

$$R\text{-}Cl + OH^* \rightleftarrows H_2O + CO_2 + Cl^-$$

Organic destruction efficiency depends on oxidant dosage, UV light intensity, exposure time, the fraction of organics that are unsaturated (easier to oxidize than saturates), and pH. Raising the pH favors hydroxyl radical formation; lowering the pH removes carbonates that consume oxidants.

Hydrogen peroxide can be purchased from suppliers and stored on-site. Ozone is unstable and must be generated on-site. Ozone is usually generated by passing air or oxygen through an electrical corona discharge.

Bioremediation Methods. The methods reviewed here are as follows:

- denitrification
- *in situ* bioremediation
- semi-*in situ* bioremediation
- activated sludge
- fixed film biotreatment
- treatment in lagoons and wetlands

The reduction of nitrates to nitrogen gas by bacteria is done at certain sewage treatment plants and can be adapted to groundwater treatment. The main difference between these applications is that with groundwater, phosphate nutrient may have to be added. In either case, a dissolved organic carbon compound, usually methanol, must be added to sustain anaerobic activity. Other organic compounds (eg, sugars, organic acids, or sodium salts formed from organic acids) are used that are less toxic than methanol and not flammable, but more expensive. Both conventional and fluidized-bed reactors are used. The destruction of nitrates, which are oxidants (electron acceptors), and the bacterial metabolism of organic carbon results in the formation of carbon dioxide gas, nitrogen gas, and an increased number of bacterial cells, consisting mainly of carbon, oxygen, nitrogen and phosphorus, as a sludge.

Unless excess nitrates or other fixed nitrogen compounds (eg, ammonium ions, urea) are present, fixed nitrogen and phosphate must be added for almost all engineered groundwater bioremediation methods. For some groundwaters, potassium or other trace metal micronutrients may also be needed. The aerobic processes described below require the addition of an electron acceptor, usually oxygen in air or derived from hydrogen peroxide. For some aerobic situations, a cometabolite is added to achieve successful bioactivity. A cometabolite is an organic compound that is not toxic to the microorganisms and is added to assure that adequate substrate is present.

In situ bioremediation is accomplished by injecting nutrients and peroxide solutions in wells. In place of peroxide, air is sparged directly into an aquifer. With hydrogen peroxide, dissolved oxygen contents as high as 500 ppm are attained; with air only 10 ppm is attained. Sometimes, purchased mutant bacterial strains cultured in biological laboratories are added. In situ bioremediation usually is successful only in formations of uniform sand. The requirements for high permeability and homogeneity, and other requirements needed to avoid plugging and adverse reactions are summarized in (14). The chance of success is greater with light hydrocarbons, alcohols, phenols, amines, acids, esters, and amides than with hydrocarbons heavier than C_{11} and especially heavier than C_{20}, chlorinated hydrocarbons, and pesticides.

Semi-*in situ* bioremediation is accomplished by extracting groundwater from wells, aerating the water or injecting peroxide, adding fixed nitrogen and phosphate nutrients, and reinjecting the groundwater into injection wells. With proper spacing of extraction wells and injection wells, adequate time for dispersion of the additives and bioactivity is accomplished, with the aquifer serving as a bioreactor. *Ex situ* aeration is accomplished by passing groundwater through a sparging vessel, a stripping tower, or a cooling tower. With a stripping tower or cooling tower, volatile organics are removed and aeration is accomplished.

Activated sludge is an *ex situ* treatment carried out in a bioreactor where groundwater, nutrients, air, and recycled bacterial sludge are completely mixed. Sludge is settled from treated water in a clarifier. Most of the sludge removed from the clarifier is recycled to the bioreactor; some sludge is removed and is usually dewatered and landfilled or incinerated. The sludge volume is reduced by using anaerobic bacteria in a warmed digester. Good clarification is achieved only with proper sludge age, which depends on recycling enough sludge to achieve a proper food (contaminant substrate)/microorganism ratio. The biokinetics involved have been summarized (15).

With sequencing batch reactors, the bioreactions and clarification occur in the same vessel. The groundwater nutrients are mixed and then aerated and mixed for a period of hours. Then the batch is allowed to settle; the treated water is decanted. All or part of the sludge must be retained from batch to batch to achieve the proper sludge age and to allow time for completing the biological reactions.

Fixed film reactors are not mixed. The submerged types contain plastic baffles that provide a surface for bacterial cell buildup while groundwater flows through horizontally. Air is injected and diffused near the bottom. With rotating biological contactors, plastic discs mounted on a horizontal, slowly rotating shaft are half submerged in groundwater flowing horizontally, where film buildup occurs. Then, as the discs rotate out of submergence into the air, oxygen contacts the film. With trickling filters, water flow is vertical, down through plastic packing in an aerated vessel or over rocks in the open air. With all of these types of fixed film reactors, from time to time the film of bacterial cells becomes so thick that it sloughs off the surfaces and falls to the bottom where it is removed as a sludge, or water and sludge are removed to a clarifier.

Treatment in a lagoon or pond is achieved by mechanical aeration or by natural exposure of the water surface to the air. Very long residence times are attainable at much lower costs than with a bioreactor. Aerobic conditions are achieved with natural exposure at depths down to five feet. High organics destruction efficiencies are attained in warm seasons if good mixing is achieved across the breadth of the pond by baffling or by multiple water inlets and outlets with long residence times.

Constructed wetlands are ponds with aquatic plant growth, such as duckweed. Wetlands are aerobic at the surface, but can be operated anaerobically two feet deep near the base of the plants or in root zones. Thus, nitrates are reduced if dissolved organic carbon is added. If aerobic conditions are to be maintained, the plants should be harvested periodically. With anaerobic operation, the plants are allowed to decay, thereby reducing the amount of dissolved organic carbon additions and nutrients (ie, phosphate and potassium) additions required.

Operating Conditions and Treatability Studies. For adsorption systems design, key parameters are residence time, hydraulic loading (gpm per sq ft of bed cross-sectional area) and pressure drop. The duration of bed life before the adsorbent is spent must be reasonably long. Laboratory column tests or field pilot tests are used to determine cleanup efficiency and to confirm residence time requirements and x/m values. A hydraulic loading is selected in the range of 1 or 2 gpm/sq ft to 10 gpm/sq ft, and bed diameter and pressure drop depend directly on hydraulic loading. For high groundwater flow rates, the required residence time for a given hydraulic loading is attained by using multiple vessels in parallel for each adsorption stage. The bed life duration is proportional to the x/m value.

For stripping systems, the key variables are tower cross-sectional area, air pressure drop, air/water ratio, selection of packing or type of trays, and packing depth or number of trays. If the area corresponding to the liquid rate is inordinately large, parallel towers are used. A select number of air/water ratios are chosen for study. If the corresponding computed pressure drop is high, a different packing or a larger area is considered. Pilot testing is the best method of determining pressure drop and the *HTU,* which is essential for finding the required packing depth.

For aqueous oxidation systems, key design parameters are exposure time, UV light dosage, and oxidant dosage. For wet air oxidation systems, temperature must also be evaluated. Treatability tests are used to evaluate all of these parameters. For systems using Fenton's reagent, the pH should be lowered to the 3 to 4 range for organics destruction; then caustic additions can be used to precipitate the iron for recovery and reuse (16). Laboratory tests help determine acid requirements and optimum pH for organics destruction, caustic requirements, and dosage required for the coagulant aid, which is needed to separate the ferric hydroxide precipitate from the treated groundwater.

Because the use of UV light enhances any aqueous oxidation system, including wet air oxidation and Fenton's reagent systems, treatability tests should generally include various dosages of UV light.

Setting up independent treatability tests using wet air oxidation is economically impractical, even on a laboratory scale. Wet air oxidation laboratory tests are done by Zimpro Environmental (Rothschild, Wisconsin). UV oxidation tests are done by several vendors of UV treatment systems.

Laboratory tests or field pilot tests are strongly recommended for bioremediation projects. Laboratory flask tests with additions of seed bacteria, fixed nitrogen, phosphate, and air within a few days determine whether bioactivity proceeds without adding micronutrients, if the contaminants are biodegradable, and if toxins are not interfering. Pilot tests help confirm residence time requirements, settleability of sludge, carbon dioxide generation rate, destruction of contaminants, and efficacy of purchased bacterial strains. A simple pilot unit is constructed with a cloth bag filter in a plastic bucket. Air is diffused into groundwater through a porous stone purchased from an aquarium supply store with an aquarium air pump. Seed bacteria are added by mixing in native soil, local sewage treatment plant sludge, or cultures purchased from a laboratory. Testing is often be accomplished in a few weeks.

Pilot fluidized-bed bioreactors are constructed with laboratory glassware. Silica sand or granulated activated carbon are fluidizing media often used. If aerobic, air injection is used to fluidize the bed. If anaerobic, recycling of effluent is used to fluidize the bed. Recycling also helps attain the required residence time. Stable bacterial activity may require a few weeks, and then some additional weeks of testing are often involved.

Pilot lagoon tests and wetlands plots are done with children's plastic wading pools or with small lined basins. Lagoon tests take several months, and wetlands tests one to two years.

Trends and Innovations in Groundwater Organics Treatment

A general trend in organics remediation is toward the use of *in situ* techniques, especially for shallow groundwater. The prime example in recent years is the rapid growth of *in situ* stripping, described below under "Stripping" as air sparging. An emerging technology is the use of zero-valent metals (eg, iron filings) in trenches placed across natural groundwater flow paths. Presentations at the September 1995 American Chemical Society Conference on Emerging Technologies in Hazardous Waste Management included over eight papers on this technology. The metals destroy chlorinated solvents upon contact with the groundwater. The University of Arizona has developed a system being marketed by Research Corp. Technologies (Tucson, Arizona) that prevents rust from forming by coating iron particles with palladium (24).

Use of zero-valent metals and other *in situ* reactive applications are used in "funnel and gate" systems for shallow aquifers. The groundwater is guided with vertical barriers (eg, sheet piling) into relatively narrow *in situ* treatment chambers (25). The treatment chambers are filled with a variety of treatment substances:, iron/sand mixes; bioremediation nutrients or bacterial cultures; oxygen-releasing substances; carbon; and oxidants.

One study (26) summarizes trends and innovations in technologies applied at "Leaking Underground Storage Tank" (LUST) sites. In addition to natural attenuation, which is used the most, following are the 1993 rankings of treatment technology applications:

1. pump and treat;
2. air sparging (for *in situ* stripping);
3. slurping (groundwater extraction with soil vapor extraction);
4. *in situ* bioremediation; and
5. air sparging (for bioremediation).

Between 1993 and 1995, thirty states reported decreased use of pump-and-treat and increased use of alternative technologies (26).

Aqueous Phase Adsorption Trends. One trend in adsorption applications is the use of three stages in series instead of two. With only two stages, running the first stage all the way to saturation risks breakthrough in the final effluent unless elaborate monitoring is done. By adding the third stage, the first stage is run well past first stage breakthrough without changing the adsorbent, with only occasional monitoring for breakthrough from the second stage.

The innovative use of synthetic resins that adsorb organics presents an opportunity for on-site regeneration of an adsorbent with steam. With activated carbon, steam regeneration works well for vapor phase adsorption. However, steam regeneration of aqueous phase carbon is usually not adequate for thoroughly removing adsorbed organics, especially heavy organics. With certain resins, steam regeneration is adequate, and the more expensive scheme of off-site reactivation in a regional high-temperature furnace is avoided.

Thermally activated peat absorbs metals and organics in a process developed at the University of Minnesota's Natural Resources Research Institute (Duluth).

Stripping Trends. A trend in recent years is toward the use of low profile tray strippers. These strippers are usually rectangular in cross section and have a long water flow path so that only a few trays are needed. The air/water ratio is generally much higher than with packed towers and accomplishes the same degree of VOC reduction. Such tray strippers should be considered when the large offgas volumes involved do not require emissions abatement.

Another type of low profile stripper is an aeration chamber. Usually rectangular, the air is admitted as fine bubbles through diffusers. By using a large number of small-diameter bubbles, the area per unit volume of air for mass transfer of VOC vapors across the air/water film is maximized. By passing the water in series through multiple chambers (eg, four units), a high degree of VOC removal is achieved.

An innovation applicable to conventional packed strippers and low profile strippers is to apply a relatively high vacuum without heating. This concept avoids accelerated fouling associated with heated stripping, achieves higher percentages of VOC removal than with atmospheric strippers, and is applicable to removal of some semi-VOC.

A very strong trend is underway toward *in situ* stripping, usually accompanied by soil vapor extraction techniques if vapor treatment is required. *In situ* stripping is accomplished with direct air sparging below the water table or with in-well stripping. *In situ* sparging has become widely used, requires less remediation time than conventional pump-and-treat systems, and scrubs saturated soil pores of residual nonaqueous phase liquids. Air sparging also enhances *in situ* bioremediation. The process is enhanced for low permeability soils by hydraulic or pneumatic fracturing (27).

Two types of in-well stripping are used, with wells screened both below the water table and in the vadose zone and with an above-ground vacuum blower. The vacuum system captures VOC released from groundwater in the upper part of the well and discharges them through emissions abatement equipment. The EG&G system used in France sets up a circulation pattern of groundwater out from the upper screen down and into the lower screen with vacuum inducement. The German UVB system uses compressed air similarly, with a well pump (28).

Another form of groundwater stripping is dual extraction, applicable with soil vapor extraction systems where the groundwater table is shallow. The process is patented by Xerox (Stamford, Connecticut). Vapor extraction wells have a single screen interval that traverses the water table. Both soil gas and groundwater are extracted with a liquid-ring vacuum blower. As the groundwater travels up the well, air in the soil gas extracted from the vadose zone strips VOC from the groundwater.

Aqueous Oxidation Trends. A trend in applying chemical oxidation to groundwater is the use of polishing treatment, such as activated carbon adsorption or air stripping, downstream from aqueous oxidation systems. Another trend is in the use of multiple oxidants, the use of catalysts, or the use of UV light with methods that formerly did not employ UV (eg, Fenton's systems; wet air oxidation). By using a variety of oxidation enhancement tools within a single reactor, a wider variety of organic compounds is destroyed with short exposure times. Examples include the injection of peroxide into UV-ozonation systems and the development of the MIOX method by Los Alamos Technical Associates (Albuquerque, New Mexico). MIOX (mixed oxidants) uses a combination of chlorine-based oxidants and hydroxyl radicals formed in a separate cell by low-voltage electrolysis of brine.

A completely different electrolytic method uses diaphragm cells at low voltage with a multivalent cation (eg, silver ion, cobalt ion, or cerium ion) to catalyze oxidation of organics. Known as mediated or catalyzed electrolytic oxidation, these systems produce chlorine gas at the anode, if chlorinated organics are present, and hydrogen gas at the cathode of each cell. A diaphragm separates the region around the anode from the cathode. The metal ions are reduced to a lower valence by the organics. Electrical energy applied at the anode raises the valence. Systems are operated using sulfuric acid as the electrolyte with cobalt at 60°C and using nitric acid with silver. Full-scale units are used in France. Test units are at U.S. Department of Energy sites.

An electrochemical process that uses titanium electrodes generates hydroxyl radicals directly from water, as developed by Sonoma Research Company (Napa, California) (29).

Another innovation that springs from wet air oxidation is supercritical oxidation. Oxygen or air is injected into water at conditions exceeding the critical point of water, 218 atmospheres and 374°C. At such high pressures, oxygen and hydrocarbons are highly soluble in water. Oxidation is quite complete for many organics in 6 to 30 seconds when the temperature is elevated above 500°C.

An innovation in UV peroxidation is to combine it with hydrodynamic cavitation, as developed by Magnum Water Technologies (El Segundo, California). The water being treated is first depressurized from 75 psig to 0.5 atmosphere, causing cavitation bubbles. When the bubbles collapse, some organic compounds solubilize, making UV peroxidation highly effective (30).

Another innovation in UV peroxidation developed by Solarchem (Nashville, Tennessee) is the use of a reducing agent (eg, iodide) to aid in dechlorinating solvent contaminants in groundwater. Solarchem also markets systems using UV and visible light with a catalyst consisting mostly of potassium ferrioxalate.

Much research is under way on applying catalysts with UV oxidation. The catalyst favored by most researchers in this field is titanium dioxide. Presentations at the September 1995 American Chemical Society Conference on Emerging Technologies in Hazardous Waste Management (Atlanta) included nine papers on photocatalytic oxidation, including five on titanium dioxide. One challenge is to develop techniques that employ this dense, white pigment in a manner such that the groundwater contacts the catalyst while UV light penetrates the groundwater. Matrix Photocatalytic (London, Ontario) produces fabric mesh coated with titanium dioxide for this purpose.

A catalytic UV–ozone process has been developed by Eco Purification Systems USA (Baltimore, Maryland).

British Nuclear Fuels Ltd. (Seascale, England) has developed an aqueous oxidation system that uses hydrogen peroxide without UV light. A chromate salt added to the waste stream catalyzes the process (31).

Perhaps the ultimate aqueous organics destruction method is the use of electron beams generated at very high voltages. Full-scale tests have been conducted by the University of Florida and High Voltage Environmental Applications (Miami, Florida) (32).

Bioremediation Trends. Trends in bioremediation are toward ever increasing use of *in situ* methods, and the use of purchased mutant microorganisms cultured in biological laboratories. The ideal use of mutant strains is for initial application with reduction of the additions as production of additional cell mass proceeds. Certain mutant strains metabolize organic contaminants that are toxic to indigenous microorganisms or are generally considered refractory to biodegradation. One study (18) projects that research will stress the search for naturally occurring microorganisms that metabolize a wide variety of contaminants at rapid rates and are tolerant of a broader range of microbial conditions. Another study notes growth in the use of mutant organisms, especially for groundwater and soil contaminated with petroleum products from leaking underground storage tanks (33).

One report (15) sees trending toward more use of methanogenic microorganisms for degrading volatile chlorinated hydrocarbons; sequencing batch reactors; denitrification; and sulfate bioreactors.

A variety of innovations are being developed for *in situ* applications. The Battelle Pacific Northwest National Laboratory (PNNL) (Richland, Washington) is remediating carbon tetrachloride- and nitrate-contaminated groundwater anaerobically at Hanford, Washington. Sodium acetate is injected to supply additional dissolved organic carbon. PNNL has developed a mathematical modeling technique that provides guidance for injection rates of water and nutrients to optimize the location of high bioactivity with respect to extraction and injection wells.

Other innovations include the use of fungi and other nonbacterial microorganisms and the use of activated carbon combined with bioremediation techniques. The advantages over using bioremediation alone are that the bioreactor systems, especially activated sludge and sequential batch reactors, are less subject to upset if conditions change and disturb the bacterial environment; VOCs are captured and then gradually destroyed by the bacteria on the carbon pore surfaces; odors are eliminated; nonbiodegradable contaminants are removed from the groundwater; and less bacterial sludge is produced. The advantages over using granular activated carbon alone are that there is less carbon consumption; groundwater prefiltration is eliminated; no fixed beds of carbon are used that can become plugged; powdered carbon is used that is cheaper than granulated carbon.

An aerator with inclined plates partially exposed to air, developed by the Delft University of Technology (Delft, Netherlands), transfers oxygen into water much faster than rotating disc contractors (34).

Phytoremediation (plant uptake of contaminants including organics) is described (11).

Tracking of progress in bioremediation, and summaries of associated reports are kept up to date in EPA's "Bioremediation in the Field" (35). The computerized data base and additions to the mailing list are available through EPA's Office of Research and Development, 26 West Martin Luther King Drive, Cincinnati, Ohio 45268, phone (513) 569-7562.

Costs for Groundwater Organics

If vapor treatment is not required, then air stripping is often the lowest cost method of remediating VOC-impacted groundwater that has been pumped to the surface for treatment. The price of air cannot be beat, and the power consumed by air blowers for stripping systems is relatively low. Total investment and operating costs are given in the EPA Bulletin EPA/540/2-91/022 (Oct. 1991).

Expenses for operating bioremediation systems are somewhat comparable to air stripping expense. However, bioremediation equipment is generally much larger than stripping equipment for a given throughput capacity, with corresponding higher capital costs.

Aqueous phase carbon adsorption equipment is larger than most UV peroxidation equipment for comparable throughput, because adsorption requires a longer residence time. However, the investment for electrical items is higher for UV systems. So, overall investment is of the same magnitude. Carbon replacement costs for adsorption systems are more than electric power costs for UV oxidation of most organic compounds. However, carbon adsorption removes a higher percentage of organic contaminants.

The RACER/ENVEST computerized system developed by Delta Technologies Group (Englewood, Colorado) evaluates conceptual design parameters and estimates costs of aqueous phase carbon adsorption; air stripping and air sparging; *ex situ* bioremediation with fixed-film reactors; and *in situ* bioremediation.

A comparison of treatment using ozone and hydrogen peroxide with other remediation methods in 1988 was completed for removing 95% of the TCE and PCE in groundwater (36). The other systems considered used aqueous-phase granular activated carbon (GAC), air stripping, and air stripping with vapor-phase GAC for off-gas treatment. The results of the comparison are summarized in Reference 36.

Frequently, cleanup must achieve over 99% VOC removal rather than 95% as in this example, in which event aqueous oxidation would require ultraviolet light treatment and polishing with aqueous phase GAC or air stripping.

Adsorption Costs. Costs for aqueous phase carbon adsorption equipment are given (37), with annual maintenance costs (excluding carbon replacement) at 5 to 10% of capital investment. Costs for installed systems given by (20), include, by way of example, a 400-gpm system with two 20,000-pound canisters in series at $563 per gpm at 1989 costs (approximately 10% below present-day costs).

Carbon replacement costs were approximately $1 per pound, for large quantities, until the early 1990s, and have been coming down in price since.

Stripping Costs. Based on the author's estimating experience, the investment for basic stripping systems ranges from $250 per gpm for capacities of hundreds of gpm to $1000 per gpm for small capacities. The cost of emissions abatement equipment installations can double or triple the capital cost.

Aqueous Oxidation Costs. Cost estimating information is available from vendors for wet air oxidation systems and for ultraviolet light peroxide systems. Wet air oxidation systems were first applied by Zimpro Environmental (Rothschild, Wisconsin) to destruction of sewage treatment sludge and can be applied to groundwater cleanup. Zimpro data indicate that investment costs range from approximately $140,000/gpm down to $100,000/gpm for 50- to 150-gpm units. The main annual expense is for compressing air and pumping water; utilities cost approximately 1 cent/gallon, most of this expense for compressing air.

Equipment investment costs and the main annual expenses for UV peroxide systems are estimated as functions of the electrical power needed for the UV lamps and of organics concentrations. Based on data from Solarchem (Nashville, Tennessee), the power consumption is given by

$$\text{kilowatts} = EE/O \times \log(C_{\text{in}}/C_{\text{eff}}) \times (\text{gal/h})/1000$$

where C_{in} is influent concentration and C_{eff} is effluent concentration. *EE/O* is the electrical energy consumed for each order of magnitude that the concentration of an organic compound is reduced. The *EE/O*s with Solarchem equipment for some groundwater contaminants range up to the following values:

3 Vinyl chloride
4 TCE
5 Monoaromatic hydrocarbons, chlorobenzene, phenol, DCE
8 PCE
15 Chloroform and DCE

For multiple contaminants, the highest value for any one compound is used (the *EE/O* values are not additive).

The equipment cost in 1995 dollars is approximately $1160 \times$ kilowatts $+ 52,000$. The yearly UV electrical power cost is kilowatts $\times$ utility price. The yearly lamp replacement costs typically range up to 50% of the power cost. Hydrogen peroxide costs range up to $0.008 per ppm of peroxide for each 1000 gallons treated. The ppm peroxide concentration required is taken as the greater of 25 ppm or twice the concentration of influent chemical oxidation demand (COD).

One study cites an example where TCE concentrations in well water in the 500 to 1000 μg/L range were reduced to 0.68 to 0.83 ug/L at 60 gpm with 50 ppm hydrogen peroxide, 30 kW of UV power, and 50 seconds residence time. Treatment costs range from $1.01 to $1.35 per 1000 gal (16).

Bioremediation Costs. The Department of Defense Screening Matrix (2) gives these estimates for groundwater bioremediation:

In situ bioremediation with either air sparging or hydrogen peroxide injection: $10 to $20 per 1000 L ($50 to $100 per 1000 gal).

Ex situ bioreactor, one example with a Biotrol (Eden Prarie, Minnesota) reactor: under $0.92 per 1000 L (< $3.50 per 1000 gal).

The Technical Module Formulations Documentation references for the RAAS (4) include these costs from various citations for bioreactors:

activated sludge: $3.42/1000 gal to $90/1000 gal;

capital costs to treat hazardous waste leachate at 25 to 100 gpm: $184,000 to $364,000, with operating/maintenance expense at $18,000 to $47,000/yr.

REMOVAL OF ORGANICS FROM SOIL

This section covers the following proven technologies primarily used for removal of organics from soil:

- soil venting (vapor extraction)
- incineration

- thermal desorption
- bioremediation

The next section covers two other proven technologies, washing and fixation, that remove or immobilize organics and are also frequently used for metal-contaminated soils.

Innovative technologies covered in this section include simultaneous vapor and groundwater extraction within the same well; thermally enhanced vapor extraction; vapor extraction enhanced by soil fracturing; passive vapor extraction; destruction of recovered soil vapors; desorption under vacuum; bioventing; methane augmentation for biodegradation; and use of special microorganisms.

Applications

Soil Venting. Soil venting is the passage of air through soil, usually at negative pressure, most often to extract VOC. Venting is done *ex situ* with piles of excavated soil using imbedded perforated or screened piping or *in situ* using wells or trenches. Air is blown through or pulled through. If the air is pulled through, it is captured with the volatilized organics and treated prior to being emitted to the atmosphere. Most applications are *in situ* with vacuum conditions, the main emphasis here.

A principal objective with *in situ* applications is to have the air flowing horizontally in unsaturated soil through the contaminant plume toward the extraction wells or trenches. Unless there is a confining layer within the soil above the plume, the surface over the contaminated area should be paved or covered with heavy plastic sheeting. Otherwise, the air flow path will short-circuit the plume and flow with a strong vertical component from the surface downward in the vicinity of the extraction point. For the same reason, the upper part of the wells or trenches must be sealed except for the vertical run of air piping.

Except for innovative applications designed to extract groundwater and soil gas simultaneously, the wells or trenches should not be screened below the water table unless the water table is depressed by groundwater extraction. Economics generally favor trenches rather than wells where the depth to the water table is less than 10 feet. Most soil vapor extraction is done with screened wells under vacuum. J.J. Malot of Terra Vac Corporation (Princeton, New Jersey) is a coholder of patents for this technology. The broadest claim of the patents is in dispute in a court battle involving Roy F. Weston, Inc. (West Chester, Pennsylvania) (38).

Depending on the volatility of the organics and the number of months that venting is applied, removal of over 99% of VOC is possible. Organic mixtures, such as gasoline hydrocarbons quickly clean up to 90% removal; after that point the remaining hydrocarbons are relatively much less volatile and more difficult to remove (39).

Problems and limitations with *in situ* vapor extraction include soils with high contents of natural organic matter (eg, peat) which sorb organics strongly and are more difficult to clean up with venting; heterogeneous inclusions of thick, fine-grained formations in sandy soils are not remediated; rising and falling groundwater tables interfere intermittently with vapor extraction; vapor concentrations decrease with time, making costs per kilogram extracted rise progressively; water vapor causes condensation and freezing problems; without heat enhancement the technology generally applies to organics with at least 1.0 mm Hg vapor pressure at 20°C and a Henry's law constant greater than 100 atm/mol fraction (0.01 dimensionless) (40).

Incineration. Organics are destroyed by conversion to carbon dioxide and water vapor (plus hydrogen chloride with chlorinated organics) when subjecting soil to combustion temperatures with adequate oxygen present. Such incineration is most often carried out with any of these three devices:

rotary kiln;
fluidized bed combustor; and
infrared furnace.

Most soil incinerators are rotary kiln types operating at temperatures up to 1000°C (1832°F). To assure essentially complete combustion of all organics, the incinerator exhaust is passed through a secondary chamber ("afterburner") operating at 1000°C, or, depending on the design, over 1200°C (2200°F). The destruction and removal efficiency of organics are over 99.99%. Such efficiencies and correspondingly higher temperatures apply when toxic substances are involved. For polychlorinated biphenyls (PCBs), the TSCA applies, and 99.9999% destruction is attained. The residual PCB concentration is reduced below 1 or 2 mg/kg. Incineration is used for soils, sludges, and sediments contaminated with virtually any form of organics. Incineration, applied to hazardous substances, requires extensive permitting, emissions monitoring, and relatively long residence times at high temperatures, resulting in very high costs. Such incineration systems are not economical for soils contaminated only with petroleum hydrocarbons. Petroleum-contaminated soils are usually treated in simpler incineration systems or thermal desorption systems not permitted for hazardous wastes.

For incineration and thermal desorption, both fixed-base and mobile units are used. Mobile units are competitive with fixed-base units typically at soil quantities in excess of 2000 tons. For such quantities, the equipment transport and setup/demobilization costs associated with a mobile unit are less than costs for hauling soil to a fixed-base unit. With either fixed-base or mobile units, the major portion of the equipment is devoted to emissions control and heat recovery. If wet scrubbing is used for emissions control, wastewater treatment is also involved.

Problems and limitations with incineration, in addition to air emissions control and associated permitting, include volatilization of certain metals; potential presence of explosive materials in the feed; debris and mud that cause plugging; uneven heat content in the feed, leading to intermittent lack of excess oxygen needed for complete combustion; refractory degradation and formation of submicron particulates from soils with high potassium or sodium content; formation of slag that builds up on internal surfaces; difficulty in handling sticky clays and frozen clods in feed handling systems; and release of VOC during feed preparation.

Thermal Desorption. *Ex situ* desorption units have chambers for vaporizing volatile and semivolatile organics. The desorption chamber is frequently a rotary kiln or an aggregate dryer converted from asphaltic concrete service, especially when the burner flame or burner exhaust contacts the soil directly. Some desorbers use electric heating or indirect heating, so that no burner exhaust is mixed with vaporized contaminants. Indirect-fired desorbers thereby have a much smaller emission volume than direct-fired units. This feature greatly reduces the complexity and costs of emissions abatement. Many direct-fired units use a fabric filter baghouse to abate particulate emissions and an afterburner to burn vaporized contaminants. Indirect-fired units are fitted with a much smaller filter and use a condenser or carbon adsorption system to recover vaporized contaminants. In practice, a wide variety of abatement systems is applied, as with incineration.

Indirect-fired adsorbers are used for soils with very high organics content. There is practically no explosion hazard because the gas-phase oxygen content is controlled. With direct-fired units, hydrocarbon contaminant levels should not exceed the 2–3% range (41).

Because only vaporization (not combustion) is needed in the primary chamber, operating temperatures for most desorbers are 149°C (300°F) to the 260–427°C (500–800°F) range. This is adequate for petroleum-contaminated soils and certain hazardous wastes. Most applications are for continuous flow of excavated soil through the desorber. Some desorbers operate on batches of soil. And, *in situ* desorption is possible, as discussed under Trends and Innovations. Reference 41 summarizes the removal efficiencies and cleanup levels with petroleum-contaminated soils.

Soils fed to desorbers containing cobbles and clods are prescreened. Cobbles do not adsorb contaminants and are usually disposed of without treatment. Other oversize material should be crushed and screened. Clays and asphaltic sludges interfere with such pretreatment or form lumps in the desorber. Plugging problems and uneven desorption result. Some treated product is recycled and mixed with fresh clay feed to reduce moisture content and caking. High-moisture soils and sludges up to approximately 70% moisture content are fed.

Moisture has two effects on operational costs. It directly increases fuel costs, because virtually all of it must be vaporized. On the other hand, the steam formed helps strip organics (41). The fraction of organics vaporized is sustained at a lower temperature (and thereby with somewhat less heat input) with steam than without it. Increased moisture reduces the capacity of contaminants to adsorb on soils with high mineral contents (silts and clays) (42).

Problems and limitations with desorption technology, in addition to moisture and caking, include formation of tars and coke at high temperatures, especially with electrically heated units; generation of fugitive dust from fine-grained soils; refractory degradation and formation of submicron particulates from soils with high potassium or sodium content; difficulty in handling frozen clods; and release of VOC during feed preparation.

***Ex Situ* Soil Bioremediation.** Most engineered bioremediation of soil and groundwater is done *ex situ*. (*In situ* bioremediation generally applies only to homogeneous sand formations.) Bioremediation applies to nonvolatile and to volatile organics. A completely degraded organic compound is mineralized, and the final products are carbon dioxide and water. With some contaminants, partial degradation is sufficient for remediation, if the products are nontoxic. Large organic molecules are more difficult to degrade than small ones; heavy fuel degrades only very slowly; gasoline degrades readily. Chlorinated hydrocarbons are more difficult to degrade than hydrocarbons, and the more chlorine atoms per molecule, the more refractory is the compound. Most bioremediation is done aerobically, rather than anaerobically without oxygen additions, and aerobic processes are emphasized in this section. The commonly applied soil bioremediation techniques are the following:

- slurry the soil, and then treat the slurry in one or more aerated, mixed bioreactors;
- spread the soil, for good aeration;
- pile the soil with perforated imbedded aeration piping; and
- apply composting methods.

Cleanup efficiencies depend on the time allowed for the process and the biodegradability or molecular weight of the organics. With conditions that favor bacterial environments, 90% to 99% destruction of biodegradable compounds is achieved within a few months. Higher molecular weight organics take longer to biodegrade.

Problems and limitations with *ex situ* bioremediation include the following: clays and soils with high concentrations of natural organic matter or of oil and grease are more difficult to bioremediate (43); it does not work well with refractory (or non-biodegradable) organics or if biotoxins are either present or generated; spreading and piling techniques require maintenance of high soil moisture contents for optimum performance.

***In Situ* Soil Bioremediation.** The natural attenuation of organic contaminants through *in situ* biodegradation is the most common of all remediation processes. Where biodegradation proceeds aerobically, we help the process by augmenting supplies of oxygen, moisture and nutrients. Oxygen is often applied in two alternative ways: infiltration of the soil with hydrogen peroxide solution; blowing air through or pulling air through the soil (bioventing, as discussed under Trends and Innovations). Maintaining a high moisture content is important, because bacterial action ceases if the soil gets too dry, a condition that occurs naturally in many soils. Adding macronutrients, fixed nitrogen and phosphorus, is important because natural replenishment is the limiting factor when oxygen and moisture needs are satisfied.

Delivery of oxygen and nutrients to the soil pores of all contaminated regions is feasible if the soil is coarse-grained, is homogeneous, has a low content of natural organics that consume peroxide, does not contain inorganic substances that react with oxygen to form acids, and does not adsorb nutrients, such as phosphate, too strongly. One study indicates that soil bioremediation success (whether *in situ* or *ex situ*) requires that the soil have the right pH range (6 to 8 except in occasional cases), limited salinity, low toxicity, and adequate microbial populations (43). Some of these conditions are readily adjusted *ex situ*, such as blending high toxicity soils with low toxicity soils, but are difficult problems for *in situ* applications.

Companion Technologies. Soil venting for groundwater contaminated with organics is often accompanied by pump-and-treat systems, and the air sweeping through the soil promotes aerobic bioremediation. Another companion technology with bioremediation is soil washing. Bioremediation is often applied to washed soil for cleanup of residual organics and is also applied to cleanup of the wash water.

Basic Methods for Soil Organics

Soil Vapor Extraction Methods. Most soil vapor extraction installations are designed for negative-pressure application with abatement of the organic vapor emissions. Devices commonly used to induce negative pressure include centrifugal blowers; positive-displacement blowers; liquid-ring compressors; and internal combustion engines. Where abatement of organic emissions is required, most installations use carbon adsorbers or oxidizers, either direct thermal oxidizers or catalytic units. Condensers are used if the organics concentrations are very high. The soil vapor handled is generally over 95 volume percent air. Treated air is usually vented to the atmosphere, but it can be discharged back into the soil (see SOIL VAPOR EXTRACTION).

The most widely used centrifugal blowers are the regenerative type that produce vacuums of 100 mm (4 in.) Hg to 200 mm (8 in.) Hg of vacuum with one stage. All aluminum or other nonsparking materials of construction should be used.

Internal combustion engines, such as automobile engines, are used in two different modes: (*1*) direct flow of soil vapor through the engine intake manifold; (*2*) flow from a centrifugal blower discharge through the intake manifold. With either mode, combustion within the engine cylinders acts as a thermal oxidation system. A small catalytic converter in the engine exhaust system, as used for automobiles, oxidizes unburned organics and carbon monoxide. Such engines produce very high vacua; use of mode (*1*) with direct injection is a most efficient patented scheme, marketed by VR Systems (Anaheim, California). The engines are applied to extracting vapors from gasoline-contaminated soils. The gasoline vapor is fuel for the engine.

Very often no auxiliary fuel is needed for oxidizers or engines used to abate organics emissions near the beginning of a vapor extraction process. After all the liquid contaminant in the soil is vaporized, the vapor concentration decays with time. Auxiliary fuel (eg, natural gas, propane) must be added when the concentration drops below the level where combustion is self-sustaining.

The concentration decay is proportional to e^{-kt}, where k is a constant and t is time. If extraction is stopped for a few weeks, the concentration rebounds. It pays to stop operations periodically to maximize concentrations and thereby make the most efficient use of vacuum-inducing equipment and of emissions abatement systems.

With direct thermal oxidizers, over 95% destruction of hydrocarbons is achieved at 730°C (1200°F) and 0.3 seconds residence time. Some air pollution control districts require a minimum of 855°C (1400°F) and 0.5 or 1.0 seconds. With these conditions, 99% destruction is achieved if the oxidizer is designed with good mixing throughout the combustion chamber. If chlorinated solvents are extracted, hydrogen chloride gas is emitted. If the concentration exceeds regulatory limits with direct emission, then either an acid gas wet scrubber must be added to the system, or vapor-phase carbon adsorption used instead of oxidation.

With catalytic oxidizers, the soil gas must be preheated to at least 355°C (600°F). Then oxidation proceeds in the catalyst bed without additional auxiliary fuel consumption. The temperature rise in the bed is proportional to hydrocarbon concentration. Monitoring the temperature rise is good practice. If the temperature decreases, either the catalyst activity is falling or the soil vapor concentration is becoming quite low. Most catalyst beds are not designed to operate above approximately 675°C. If the temperature is too high, the catalyst bed will be damaged; the limiting hydrocarbon concentration is approximately 3500 ppmV. To be safe, systems are designed to operate at a maximum of 2500 to 3000 ppmV. When the soil vapor concentration is higher, dilution air is admitted. This concentration range corresponds to about 25% of the lower explosive limit (LEL) for hydrocarbons, which is a good safe design point. LEL values are given in (44). The percent LEL for mixtures of vapors in air is estimated from the vapor densities, concentrations, and molecular formulas (45).

Catalytic oxidizer designs achieve over 95% hydrocarbon destruction efficiencies with 355°C preheat. With more preheat and/or additional catalyst (more residence time) and/or preheat, 98% destruction is achieved. The common catalyst is platinum on an inert substrate, as in most automobile exhaust converters. The platinum catalyst is poisoned by vaporized organic lead compounds present in many gasoline-contaminated soils. The preheat burner converts the organic lead to lead oxide, which does not poison the catalyst, but electric preheaters may let volatile organic lead compounds (eg, tetramethyllead) go through. A simple method of detecting lead at concentrations down to the needed 5 $\mu g/m^3$ level has been developed (46).

Chlorides deactivate platinum catalyst. If the chloride concentration is above the level recommended as the maximum allowable by the catalyst manufacturer, then a nonplatinum catalyst that tolerates chlorides should be used. Table 1 shows field data with a catalyst developed for destroying halogenated hydrocarbons (47).

When organics concentrations are so low that auxiliary fuel is needed, catalytic units use up to 60% less fuel than direct thermal units. Auxiliary fuel consumption is reduced with either catalytic or direct thermal units by including heat recovery in the design. This is usually done by heat exchanging soil vapor with hot exhaust from the oxidizer (recuperative designs). Most such heat exchangers are designed to reduce fuel consumption by 50 to 70%. Over 90% heat recovery is achieved with regenerative designs that use high-density media (eg, ceramic shapes, gravel) heated by hot exhaust to transfer heat to incoming soil vapor (48). An economic analysis of recuperative versus regenerative designs indicates that if the heating value of the organic vapors is enough to raise the temperature of the soil gas by over approximately 165°C (300°F), a recuperative design is favored (49). This approximation corresponds to about 1000 ppmV of gasoline vapor in extracted soil gas.

The heating duty is estimated as follows:

$$\text{Btu/h} = (60\ \text{min/h}) \times \text{scf/min} \times 0.076\#/\text{specific heat} \times (\text{temperature rise, °F}) \quad (1)$$

Table 1. Trichloroethylene Destruction Efficiency with a Halohydrocarbon Destruction Catalyst

	Superficial Catalyst Residence Time		
Preheat temperature, °C (°F)	0.563 s	0.60 s	0.735 s
374 (630)	94.7%	96.7%	98.4%
388 (657)	98.6%	99.0%	99.5%

where scf/min is the air flow rate in standard cubic feet/minute at 1 atmosphere and 60°F and the average specific heat of air in the temperature range of concern is 0.24 Btu/(# °F). If a heat exchanger is used, the Btu/hr calculated is reduced by the factor (1 − percent heat recovery/100). The heating value of the organic vapors is estimated from the next equation, using the ppmV concentration of each flammable vapor and can be subtracted from the heating duty.

$$\text{Btu/h} = (60\text{ min/h}) \times (\text{ppmV}/10^6) \times \text{scf/min} \times \text{density} \times (\text{heat of combustion, Btu/\#}) \quad (2)$$

where the density (#/scf) of a vapor is its molecular weight/379 and the heat of combustion for each flammable compound is found in handbooks giving thermal properties, estimated from equations in (45), or estimated from this approximation from (50): heat of combustion is 184,000 Btu/mol of hydrocarbon for each mole of oxygen needed for stoichiometric combustion.

For oxidizers and internal combustion engines (ICEs), skid-mounted and trailer-mounted assemblies can be purchased, and in some cases rented. The assembly typically includes controls and the main components (eg, water knockout vessel, flame arrester, vacuum blower and oxidizer or ICE, and dilution air system).

Vapor-phase activated carbon adsorption is often used instead of oxidation techniques. Applications are as described for air stripper off-gas control in the previous section. The best flow scheme includes a water knockout vessel, an electric heater or heat exchanger that reduces relative humidity to below 50%, two carbon canisters in series, and a vacuum blower. Placing the carbon ahead of the blower helps in two ways: the heat of compression causes a temperature rise that is beyond the level for effective adsorption if the carbon is placed after the blower; and explosivity of the soil vapor is eliminated as early as practical in the flow scheme.

Unless on-site carbon regeneration is included, carbon replacement costs are higher than oxidizer costs near the beginning of vapor extraction projects. If air pollution control permits can be obtained, the optimum emissions control plan would be as follows: use a direct thermal oxidizer or engine initially; then catalytic oxidation; and end with carbon adsorption when concentrations are so low that carbon replacement is infrequent. Blower power consumption is reduced and most effective use of abatement systems is achieved if extraction is stopped when concentrations become low, and then started a few weeks later when concentrations rebound.

The spacing of vacuum extraction wells should correspond to the well radius of influence for the stratum in which the wells are screened, with at least 40% overlap of radii. The radius of influence, volumetric flow rate, and required vacuum are roughly predicted from the type of soil and well radius and screen length (39). The radius of influence for a given stratum is somewhat proportional to vacuum at the extraction well. Field pilot tests are used to determine the true radius of influence and soil air permeability.

For pilot testing, air is extracted from one well at a time. The change in vacuum with time is logged for a short period upon starting extraction, and contaminant vapor concentration is measured for as long as the test proceeds. After the vacuum becomes steady, the air flow rate and the amount of vacuum at observation wells or probes at various distances from the extraction well are logged. The soil air permeability is calculated from these data and is used to correlate radius of influence, air flow conditions, well diameter, and screen length (39).

Wells should be located so that the air flow path sweeps through the contaminated region of soil and into extraction wells. Air enters the surface of the soil at unpaved or unsealed areas such as planters and gravelled areas. Ventilation wells that convey air into the strata selected for remediation are used to help direct the air flow paths or allow air in where the entire area is completely paved. Such wells are operated passively or with an air injection blower using ambient air or recycled extracted air that has been through an emissions abatement device.

The amount of contaminant mass extracted per unit of time is proportional to the ppmV concentration:

$$\text{Mass rate,\#/min} = (\text{ppmV}/10^6) \times \text{scf/min} \times \text{density} \quad (3)$$

where scf/min is the soil gas flow rate (standard cubic feet per minute) and density is that of the contaminant vapor. For standard conditions of 1 atmosphere and 60°F, density (#/scf) is accurately taken as molecular weight divided by 379. Reference (39) gives a method of estimating the extracted mass during the initial period when liquid contaminant is being vaporized. After a period of steady extraction rate, the decay in mass removal rate is predicted from this equation:

$$M(t) = (M_o - M_f)e^{-kt} + M_f \quad (4)$$

where $M(t)$ is the mass rate at time t, M_o is the initial mass rate, and M_f is the mass rate at the end of remediation. The exponential decay constant k is chosen so that the area of integration under the curve $M(t)$ versus t is equal to the total mass to be extracted after the concentration starts declining. Or, k can be evaluated from the value measured for M_o and a subsequent measurement of $M(t)$ at a known time t by the following relationship:

$$k = \log_e[M_o/M(t)]/t \quad (5)$$

Incineration Methods. The organics destruction efficiency of incineration depends on temperature, residence time, mixing, and availability of oxygen. Temperature is the main independent variable. However, with good mixing, the target destruction efficiency is met with lower temperatures and residence times. Temperature usually has the largest influence on fuel consumption. With sludges, water content has a major influence. If the soil or sludge has a high content of organics, fuel consumption is significantly reduced.

With rotary kilns, the residence time is typically over 20 minutes. The exhaust gases are treated with more excess air at a higher temperature in an afterburner (thermal oxidizer) with a residence time requirement of less than 2 seconds. This is adequate for destroying products of incomplete combustion, considering the high temperatures used in afterburners. Rotation of the kiln cylinder, which is slightly inclined from horizontal, and in some designs flights on the inside cylinder walls provide mixing and convey the solids.

Systems designed for destruction of hazardous wastes include continuous monitoring that assures excess oxygen. If sludges with highly uneven organic content are fed, excessive emissions occur when oxygen is depleted, and the feed is automatically cut off. Air lances or oxygen lances are installed to alleviate this problem (51). Oxygen lances and oxygen burners are used to raise flame temperature, to increase throughput, to lessen nitrogen oxide emissions, and to reduce exhaust volume (with attendant smaller abatement system needs and reduced particulate emissions and slagging) (52).

Heat is recovered by heat exchange between hot exhaust and incoming combustion air. In some fixed-base units where there is a use for steam on-site or nearby, heat is recovered by passing hot exhaust through a boiler (waste heat steam generator).

In many designs much of the heat is wasted and not recovered. To bring afterburner exhaust temperatures down to manageable levels, water sprays are used to quench the hot exhaust. Thus, instead of putting the heat to use, much of it goes into vaporizing quench water to cool the exhaust.

The main equipment used for particulate emissions abatement must capture fine particulate, including sizes near one micron. The types of equipment most used that perform accordingly are Venturi scrubbers, dry scrubbers, fabric filter baghouses, and (to a much lesser extent) electrostatic precipitators. A Venturi scrubber passes the exhaust through a narrow throat, greatly increasing the gas velocity. Scrubbing water is sprayed in just above or at the throat. The high-velocity exhaust shears the spray water droplets and increases contact between water and particulate matter, including some submicron particulate. This is a high energy scrubber with a high gas pressure drop that is reflected in a high power input requirement for the blower motor.

A dry scrubbing system includes a spray dryer where a slurry of scrubbing water and an alkaline agent is evaporated by the hot exhaust, thereby neutralizing most of the acid gases and cooling the exhaust. The dried alkaline agent and particulate are removed in the second part of the system, which is a baghouse or an electrostatic precipitator.

A baghouse is generally the most efficient device for removing all sizes of particulate. It contains rows of filter bags that are intermittently cleaned by air jets, by reversal of air flow, or by mechanical shaking. Baghouses require considerable maintenance and are strictly temperature limited. The highest temperature fabric is woven fiberglass, good up to 260–288°C (500–550°F). A reliable quenching or heat recovery system must be used to ensure that exhaust gas temperature is brought down below the bag fabric limitation for steady service.

The most common device used for abating emissions of acid gases (eg, hydrogen chloride) is a low energy scrubber, a water spray tower with trays or packing. Water, sometimes with a dissolved alkaline agent such as caustic soda, is injected near the top, and exhaust gas flows upward from beneath the bottom tray or packing support, the same arrangement as an air stripper.

Wet scrubbers also remove some volatile metals, polynuclear hydrocarbons, and soluble organic compounds.

Oxides of nitrogen are abated with either of two types of ammonia reduction and with combustion techniques. Thermal deNOx is a process in which ammonia is injected where the exhaust gas is at a relatively high temperature. Selective catalytic reduction employs a catalyst where the ammonia is injected at exhaust gas temperatures as low as 260 to 427°C (500 to 800°F), with over 80% reduction to NO_x to nitrogen gas (50,53,54).

A comparison of a number of abatement devices given in a publication (55) of an EPA predecessor agency has been reprinted by (50).

Carbon monoxide emissions from soils incineration are controlled with combustion techniques and not with abatement devices. Carbon monoxide formation is predicted from data given by the ASME (56). Mathematical models for burners that predict carbon monoxide production are described (57).

With fixed-base incineration systems that have been permitted for handling RCRA wastes or PCBs, the only testing that is required is soil characterization. With mobile units and new fixed-base units, a trial burn with at least 3 tests is also needed. The trial burn determines if the temperature and combustion air rate are adequate and if emissions limitations are met.

Rotary kiln volume typically corresponds to a heat release of approximately 25,000 Btu per cubic foot (50). The volume of the associated afterburner chamber must be equal to the design residence time (converted to minutes) divided by the actual cubic feet/minute flow rate of the exhaust gases. Other design criteria and examples of design calculations are given in references (50,58). Software for rotary kiln, wet scrubber, and heat recovery steam generator computerized design calculations is available from Incinerator Consultants Inc. (Reston, Virginia).

With fluidized-bed combustors, the soil particles are suspended in the combustion air. The mixing is so good with this arrangement that the temperature can be 93.3°C (200°F) less than with a rotary kiln, and the residence time as little as one-tenth as much. The lower temperature results in less NO_x emissions. The good mixing results in less carbon monoxide emissions. Limestone or other solid alkaline agents are mixed with the soils feed to control emissions of acid gases, such as sulfur oxides.

The best mixing and effective residence time with a fluidized bed is achieved with a circulating bed combustor. With this design, there is no distinct dense phase. There is a gradual decrease in particle density as solids move up through the combustor. Except for fine particles, solids that carry over are recovered in a cyclone separator and returned to the combustor. If the cyclone is integral with the combustor, the design is an internal fluidized bed combustor.

Soil usually moves through infrared furnaces on alloy metal conveyors. The soil does not contact the walls, so lightweight insulating materials are used, lending the design to mobile applications. The heat source is either electric resistance elements or gas burners behind radiant metal plates. Exhaust gases flow through an afterburner. Typical residence times vary from minutes to a few hours. In practice, the residence time is readily adjusted by varying the conveyor speed.

Thermal Desorption Methods. Factors that affect *ex situ* desorption include volatility of the contaminants, soil plasticity, and moisture content. The soil must be prescreened to remove debris and cobbles larger than 5 cm (2 in.). Ideally, the desorption temperature should correspond to that at which the vapor pressure of the least volatile contaminant is 1 atmosphere. In practice, the optimum range is 0.5 to 2.0 atmospheres (41). Clayey soils that become plastic when wet interfere with prescreening, with conveying through the desorption chamber, and with heat transfer. The heat input needed to vaporize the moisture in the soil normally vastly outweighs the heat input needed to vaporize the organics.

Except for a few electrically heated units, direct-fired units generally operate at temperatures higher than other designs, with consequent potential for higher cleanup efficiencies. If the cleanup goal is met with indirect firing, then the advantages of dealing with relatively small volumes of emissions and high-organics content soils make indirect-fired units attractive. Some of the indirect-fired designs include double-shell kilns with heating in the annular space; conveyance of the soil with hollow augers that have oil (heat transfer fluid) circulating through the augers from a separate fired heater; infrared metal panels backed up by gas burners. A small amount of burner exhaust gas or an inert gas, such as nitrogen, is used to sweep vaporized organics from indirect-fired and electrically heated desorption chambers.

Soil Bioremediation Methods. Soil bioremediation is an attractive technology from the viewpoint of economics and because a natural process is used to degrade contaminants. Engineering applications that enhance this natural process include assuring oxygen supply for aerobic bacteria; adding macronutrients; applying water so that moisture content is at least about 50% of the saturation value.

The importance of maintaining adequate moisture content for soil bioremediation to succeed cannot be overemphasized. The best way to assure adequate moisture content and the fastest *ex situ* biodegradation is with slurry bioreactors. Slurrying the soil also results in making it dust-free and pumpable in pipelines, which are cleaner and less costly to maintain than conveyors or loaders as with nonslurried soil. Slurry bioreactors are well mixed and are thereby the best soil bioremediation method to ensure thorough cleanup of all the soil treated. Potential VOC emissions and odors are easy to capture or control. Slurry bioreactors consume the least amount of land space of all the *ex situ* bioremediation methods described here. Most bioremediation processes proceed well at temperatures above 10°C (50°F), and temperature is relatively easy to control in a reactor. With slurry bioreactors, air is sparged into the slurry while it is agitated with rotating arms, and nutrients (fixed nitrogen and phosphates) are dissolved in the water.

With slurry bioreactors, pH adjustment and the addition of bacterial cultures or of dispersants is readily accomplished and thorough (59). Required residence time varies from 10 days to 9 weeks. The process has potential for a wide range of contaminants including fuels, pesticides, wood preservatives, PCBs, and other halogenated organics.

The lowest cost *ex situ* treatment consists of spreading the soil in lifts approximately 20 cm (8 in.) thick and spray-irrigating it. Such a thin layer of soil uses much land area, but exposes the soil to air enough that no other action is needed for oxygen supply. Spreading techniques generally require a few months for thorough cleanup but often do not succeed during cold months.

If the soil is spread in lifts 20 cm (8 in.) to 45 cm (18 in.) thick to save land area, it must be aerated by plowing or rototilling every one or two weeks. This process is called land farming. The first lift is spread on top of a layer of clean sand or sawdust so that the underlayment is not damaged. With spreading techniques, nutrients are added with fertilizer pellets (purchased from a nursery) spread on top or liquid fertilizer applied with irrigation water. The leachate is captured and recycled.

If water additions are applied carefully, recontamination of treated lower lifts from newly spread contaminated soil is avoided. Vertical migration of soluble contaminants has little effect on underlying treated soil (60).

Land area required is reduced even further by piling the soil a few meters high and applying mechanical aeration or composting techniques. A common method of aeration is to imbed the piles with perforated piping and use a blower. The air flow rate must be sufficiently slow so that volatiles are not stripped, unless combined soil venting is desired and VOC emissions are controlled. In some strict air pollution control districts, plastic covers must be applied when using spreading or piling techniques for VOC contaminants, and only a portion of the covers are pulled back each day to expose the soil to air.

Composting involves mixing the soil with a bulking agent (eg, straw, wood chips, mushroom farm waste) and turning the piles or windrows every one or two weeks. Or the soil and bulking agent are placed in a vertical reactor with air blown through, and mixing is accomplished with rotating arms. Bulking agents are especially helpful with clay soils, by desorbing the contaminants (43). Nutrients are in the form of synthetic fertilizers applied with the irrigation system or manure. Composting proceeds naturally with considerable heat evolution that results in temperatures over 38°C (100°F) in the interior of a pile. A type of bacteria that thrives at such temperatures metabolizes organic contaminants faster than with other spreading and piling techniques.

Operating Conditions and Treatability Studies. Soil venting involves operating with adequate air flow, vacuum, and soil vapor concentrations. The rate of extraction of contaminants is proportional to air flow and concentrations. The air flow in turn is controlled by the number of blowers or engines used and the vacuum applied. Field pilot tests are used to aid in the design of air flow rates and amount of vacuum. Optional conditions and approaches by opening ventilation wells that are located so that air flows through the contaminant plume toward extraction wells, and by shutting off extraction from wells periodically when concentrations decline. Management of condensate and its removal is vital to maintaining air flow and preventing interference with the operation of emissions abatement equipment.

Soil gas surveys are analyses of air samples taken at various locations and depths. Such surveys help locate the contaminant plume and aid in managing ventilation wells. They are used to select extraction well locations and screened intervals (61). A number of field tests for determining soil air permeability are given by (39,62). A column of soil brought into a laboratory and subjected to air flow with a laboratory vacuum pump yields information about initial soil vapor concentrations.

For incineration and thermal desorption, laboratory pilot units yield information about temperature and residence time requirements. This information is useful for preparing specifications for contract cleanup work; setting parameters for trial burns or initial operating conditions for cleanup; and estimating throughput and corresponding cleanup time and associated costs for labor, equipment, and fuel.

The best laboratory test apparatus for thermal desorption is an oven rigged so that soil batches are conveyed through. Contaminant concentration in the soil is measured at two points, midway and at the exit, for a number of temperatures. Two plots are graphed (the midway concentration and the exit concentration) against temperature. The temperature noted where the midway concentration falls below the target cleanup level is a conservatively designed upper temperature. The temperature noted where the exit concentration falls below that level is the minimum operating temperature.

For bioremediation, treatability tests are essential to determine whether the process proceeds at all and to establish optimum operating conditions. Slurry testing is done in a laboratory flask for *ex situ* or *in situ* bioremediation. One study recommends shaker flask tests under aerobic and anaerobic conditions with both mixed bacterial cultures and isolated bacteria and fungi known to rapidly degrade the contaminants (63). Other methods can be tested in buckets or pans. If contaminants are not all degraded to carbon dioxide and water, it indicates a bacterial toxicity problem, the lack of a micronutrient, or that contaminants are bonded too tightly to soil particles to pass into the water phase where the bacteria act. Leaching tests are carried out with high salinity soils to see if electrical conductivity is reduced to favorable levels.

If the microbial count is too low (below 10^5 cells per gram of soil), purchased cultures are tested. Soils typically contain 10^6 to 10^9 cells per gram (43). One study recommends a series of initial tests followed by slurry or pan tests (43). The initial tests include organics analysis, including leachable fraction; particle size distribution; total organic matter; moisture retention; nitrogen and phosphorus content; conductivity; pH; microbial count; 5-day incubation in aerated stirred flasks with nutrient; and oxygen consumption rate.

To try *in situ* bioremediation, the following tests are recommended in addition to the usual site characterization effort: microorganisms present; biodegradability of contaminants (half-life or rate constant); biodegradation by-products; soil properties including pH, oxygen content, moisture and nutrient contents, organic matter, temperature, texture, and water-holding capacity; and erosion potential and trafficability (63).

Trends and Innovations in Soil Organics Treatment

The trend toward using treatment technologies instead of disposal at landfills skyrocketed from 1993 to 1994 (26). Hauling soil away for landfilling formerly was done for the majority of LUST (leaky underground storage tank) sites. In 1995 treatment was done at over 38% of the sites; natural attenuation at 28%; and disposal at landfills was used for less than 34% of the sites.

Soil Venting Trends. A trend in soil venting is toward more use of heat enhancement, to hasten the remediation process or to volatilize higher molecular weight organics. Demonstrations of various thermal enhancement methods are described in (64).

Electrokinetics (application of direct-current electrodes imbedded in soil) results in heating that helps vaporize organics. Microwave heating (radiofrequency heating) is applied to warm the soil. A mobile radiofrequency heating unit operated by KAI Technologies Inc. (Woburn, Massachusetts) uses dipole antenna applicators. A single applicator is installed horizontally with slant drilling for a small plume. A larger volume of soil is heated with two parallel vertical applicators.

Lawrence Livermore National Laboratory (LLNL) (Livermore, California) has applied both electric resistance heating and steam injection to enhance venting of hydrocarbons. A concern with steam injection is that condensate forms and washes contaminants downward into an aquifer. At LLNL, the soil was heated to above 100°C, and this problem was avoided. The Battelle Pacific Northwest Laboratory (Richland, Washington) has applied 6-phase electric power to heat soil to 100°C at the Department of Energy Savannah River Site where clay was contaminated with chlorinated VOCs. Six electrodes were arrayed, each at 9.1 m (30 ft) from a central vapor extraction well. Removal of 99.7% was achieved in 25 days.

Shallow contamination is removed by applying thermal blankets over the ground surface, using technology marketed by Terratherm Environmental Services (Houston, Texas) and developed by Shell Oil Co. (Houston).

Compressed air heated to temperatures up to 730°C (1200°F) is used in a process patented by Hrubetz Environmental Services (Dallas, Texas), for either *in situ* or *ex situ* venting. For *in situ* applications, the air is injected in wells, and vapor is collected under vacuum in channels near the surface. A trailer-mounted oxidizer designed for 917°C (1500°F) and 0.5 or 0.6 seconds residence time is used for emissions control.

Another trend in soil venting applications is the use of mathematical models. "HyperVentilate" (39), is distributed by EPA's Office of Underground Storage Tanks and is available from the U.S. Government Printing Office, Pittsburgh, Pa. "Venting" is a more sophisticated model (39) and is available from Environmental Systems & Technologies, Blacksburg, Virginia. Models for vapor extraction and air sparging have been developed by Integrity Engineering, West Chester, Pennsylvania. A model developed by Lowney Associates, Mountain View, California optimizes design for hot air injection, ambient air injection, or conventional vapor extraction. A model for vapor extraction with steam injection has been reported (66). Vapor extraction air flow equations have been developed from groundwater hydraulics that are useful if the difference in pressure between any two points in the flow field is less than 0.5 atmospheres (67). Appendix I to Reference 68 describes a model developed by Eckenfelder Inc. (Nashville, Tennessee) that predicts cleanup time and residual concentrations and designs wells or trench placement.

Two trends that utilize *in situ* vapor extraction technology are in the application of air sparging of contaminated groundwater and of simultaneous water and vapor extraction from the same well. Sparging is done in two ways: in-well, with volatilized organics and air traveling from the well to a soil vapor extraction blower; and with sparge nozzles below the water table and conventional vapor extraction wells in the saturated zone. Simultaneous water and vapor extraction (a process termed "Dual-Phase Extraction" patented and licensed by Xerox, Stamford, Connecticut) is applied where the water table is close to the ground surface, by using a liquid-ring compressor to induce vacuum.

An innovative remediation technique that enhances vapor extraction in low permeability soils is pneumatic fracturing. Air is injected at a pressure that overcomes the stresses within the soil matrix. Some of these stresses are matrix tensile strength, overburden pressure, and fluid pore pressure (69). Hydraulic fracturing is accomplished with drilling equipment designed to fill sand into hydraulically induced horizontal fractures in low permeability strata. Pneumatic and hydraulic fracturing are reported (27).

Another innovation that applies near the end of soil venting, when vapor concentrations are low, is passive extraction. Wells that are screened in a contaminated zone breathe vapors to the atmosphere (through a drum of activated carbon if abatement is required) without a vacuum blower. Natural fluctuations in barometric pressure cause intermittent pressure differences between the atmosphere and the soil at the screened interval. Foster Wheeler Environmental Corporation, Richland, Washington has demonstrated this technology. Costs were studied by (70).

A number of innovations are for emissions abatement systems and technologies. A regenerative thermal oxidizer marketed by Eisenmann Corporation (Crystal Lake, Illinois) uses a unique rotating distributor to direct incoming soil vapor to the hottest portion of the ceramic heat sink media. Catalytic oxidizers with fluidized catalysts have better mixing and surface contact than conventional fixed-bed units, resulting in higher conversion at a given preheat temperature. A fixed-bed catalytic unit marketed by North East Environmental Products (West Lebanon, New Hampshire) uses ozone for oxidation at a temperature below 100°C. Carbon adsorption systems now compete with resin adsorption systems that are regenerated on-site under vacuum with heated nitrogen gas. Some of the resins are manufactured by Rohm & Haas (Philadelphia, Pennsylvania) and by Dow Chemical (Midland, Michigan).

New methods of destroying organic vapors are being applied to soil venting systems. Biofilters use beds of soil, compost, or biomass supported on plastic beads to sorb the organics. Soil vapor blown through the beds is biodegraded if the bed is sprayed with a mist of water. One study indicates that costs are 60–94% lower than with carbon adsorption or oxidizers, but destruction efficiencies are not as high (71).

A high voltage approach to organics destruction is beams of electrons in a plasma unit. This technology has been demonstrated at Department of Energy sites by the Plasma Fusion Center of the Massachusetts Institute of Technology (Cambridge, Massachusetts). A system marketed by Zapit Technology (Santa Clara, California) is economical for high air flow rates with VOCs in the 10 to 20 ppm range (24).

Incineration Trends. Because of the high costs and permitting difficulties associated with incineration, there has been a strong trend toward using desorption as much as possible when thermal remediation techniques are involved. Another trend involves destruction of PCBs and other chlorinated compounds. The contaminants are dechlorinated first (in some techniques, with chemical reactants), leaving hydrocarbon compounds that are incinerated at lower temperatures or destroyed by other techniques.

Although the vast majority of soil and hazardous waste incinerators are rotary kilns, there has been somewhat of a trend toward fluidized bed combustors. Among the fluidized bed designs, the trend is toward circulating bed combustors. An innovation in fluidized bed incineration applications is the use of reducing conditions. A circulating bed unit was tested for the destruction of chlorofluorocarbons (72). A reducing atmosphere of methane and hydrogen created from pyrolysis of wood chips destroyed the chlorofluorocarbons, forming hydrogen chloride and hydrogen fluoride gases. These gases were reacted in the fluidized bed with calcium carbonate that was also fed to the unit.

Thermal Desorption Trends. There has been a trend toward the application of higher temperature designs to extend the technology to organics with lower vapor pressures. However, problems with tar and coke formation are encountered that put a damper on this trend. An innovative design that combines combustion with desorption was developed by Soil Tech ATP (Porter, Indiana). The soil organics form cracked gases and coke in the absence of oxygen at approximately 552°C (1025°F). The coked soil is combusted with air in another section.

Another method vaporizes heavy organics at temperatures below where cracking occurs by applying vacuum. The McLaren–Hart (Alameda, California) system has six parallel desorption chambers. Each chamber is operated with batches of soil that are heated while vacuum is drawn on the chamber.

The trend in emissions control for desorption systems is towards complexity applying a variety of abatement techniques. For example, the McLaren–Hart unit treats combustion exhaust with a cyclone separator, baghouse, acid gas scrubber, and carbon adsorption unit, whereas the cracked gases and vapors are treated with a cyclone separator, scrubber, fractionator, and condenser.

In situ desorption is practiced as described above for heat-enhanced soil venting by radiofrequency heating, electric resistance heating, steam injection, or hot air injection.

Soil Bioremediation Trends. Because of their simplicity and low costs, spreading techniques have long been the main choice for *ex situ* remediation and will continue to be so. A strong trend is seen in using laboratory cultures as soil additives to start bacterial action and growth, attack toxins, accelerate degradation, or to maintain bioactivity. Innovations using DNA splicing techniques to form engineered bacteria may result in developing bacterial strains that degrade hitherto refractory contaminants. However, applications using engineered bacteria are inhibited by permitting restrictions. Their use may be limited in most situations to enclosed vessels (18). As with groundwater bioremediation discussed previously, researchers continue to search for naturally occurring microorganisms that metabolize a wide variety of organics at fast rates and tolerate wider ranges of environmental conditions. One study notes growth in the use of cultured microorganisms, especially for groundwater and soil contaminated with petroleum products from leaking underground storage tanks (19).

Treating soils contaminated with polyaromatic hydrocarbons is difficult by any method except incineration. Several innovative techniques, including use of slurry bioreactors with Fenton's reagent (hydrogen peroxide plus iron) for polishing and use of pretreating with Fenton's reagent are described (73).

A successful innovative *in situ* technology with widespread use in recent years is bioventing. Using wells, air is blown into the soil or drawn through as in soil vapor extraction, except at 80 to 95% lower flow rates (73), so that VOCs are not stripped. Just enough air is used so that aerobic bacteria grow. The soil is moistened and nutrients are added, where contamination is shallow, by using an infiltration gallery or surface flooding or by using water injection wells where contamination is deep. As with soil venting, the ground surface is paved or covered with plastic sheeting to induce horizontal air flow. It is reported that

> bioventing is best suited for hydrocarbons that are too heavy for soil venting and is potentially suited for aromatics, ketones, phenols, polyaromatic hydrocarbons, and chlorinated solvents (74).

Treatment time is typically five times longer than with soil venting.

Progress with bioventing is monitored by checking soil gas for oxygen content falling below 21% or carbon dioxide content rising above 0.03% (75). The process is effective for semi-VOC such as polyaromatic hydrocarbons, ketones and other VOC.

Bioslurping is the removal of liquid hydrocarbons from the upper part of an aquifer simultaneously with bioventing (76).

Oxygen is reported to be retained in soil pores longer and with better concentrations in soil moisture if air is injected as colloidal gas aphrons (77). These are air bubbles that are 25 to 50 micrometers diameter, stabilized with surfactant.

An innovative *in situ* approach is reported in which air water is forced by compressed air through the soil and through inserted bioplugs (78). The bioplugs are slotted plastic tubes filled with cultured microorganisms on an inert substrate, with fertilizer pellets added.

Bacterial cell enzymes are instrumental in organics degradation. Innovative bioremediation methods that use cell-free enzymes or fungi that excrete enzymes are being researched. The enzymes, in turn, degrade organics, including large molecules or compounds that are toxic to bacteria. Fungi are mixed into soil piles and composting techniques applied. In some cases the piles do not have to be turned over for aeration (79). White rot fungus shows promise for degrading chlorobenzene derivatives and wood preservatives, such as pentachlorophenol and related polyaromatic hydrocarbons (80).

Injected methane is reported cometabolizing with chlorinated solvents by methanotrophs that produce a bioactive enzyme in a process suitable for both contaminated soils and groundwater (81). A mixture of nutrients, air, and four volume percent methane is injected in pulses through horizontal wells. Methanotrophs are microorganisms that consume methane (80).

Soil Dehalogenation. Dehalogenation techniques break chlorine bonds to aromatics, reducing toxicity and making remediation with contaminants, such as PCBs, easier. In one process, contaminated soil is treated *ex situ* in a reactor with alkaline agents at 100–180°C for 1 to 5 hours. Treated soil is washed with acidified water. The process uses less energy and produces less off-gas than conventional thermal techniques. PCB concentrations are reduced by 99%; residual concentrations of less than one ppm are possible (82).

Costs for Soil Organics

Soil Venting Costs. *In situ* vapor extraction is competitive or less costly than excavation if the contaminated volume exceeds about 500 cu yd, if the contamination is underneath a structure, if the depth is over 20 to 30 ft, or if the area is more than several thousand sq ft (83). A study reports costs for 200 sites at \$15 to \$50 per cu yd, with remediation periods ranging from four months to three years (84). Another study gives a range of \$10 to \$150 per ton of soil, typically \$50/ton (85). The \$10/ton figure applies when no vapor treatment is involved and no wastewater or groundwater need treatment (86). An economic analysis by EPA (87) indicates costs ranging from \$27 to \$66 per ton of soil. Small, shallow sites cost \$30 to \$70 per cu yd (88).

Laboratory column tests cost approximately \$30,000 to \$50,000 and take 3 to 7 weeks. Air permeability tests cost approximately \$1500 to \$2500 per well or about \$10,000 to \$50,000 per site. Engineering design costs range from \$50,000 to \$250,000 (83). Appendix D of that reference includes tables of components costs (83). Field pilot tests and selection studies cost over \$100,000.

Reference 59 reports costs as follows: under \$100,000 capital and startup costs for a small system; capital costs in the \$65,000 to \$135,000 range, plus wells; wells at \$2000 to \$4000 each; operating and maintenance at \$6000 to \$26,000 per year. Costs as follows are outlined (61):

Surface sealing (installed): 40-mil high density polyethylene, \$5/sq yd; 10-mil polyethylene, \$2.25/sq yd; 4 in. bentonite, \$2.22/sq yd; 2 in. asphalt, \$9.24/sq yd.

Water knockout pots (installed): 65 gal, \$2,300; 130 gal, \$3,100; 800 gal, \$15,400.

Centrifugal blowers (installed): 2.5 hp, \$600; 25 hp, \$12,000; 50 hp, \$42,000.

Internal combustion engines: 60–100 cfm, approximately \$50,000 or \$4,000–\$4,800/month rent; 400 cfm, \$7,500/month rent.

Direct thermal oxidizers: 100 cfm, \$11,200; 320 cfm, \$18,400; 570 cfm, \$22,900.

Direct thermal/blower/etc: skid-mounted package: 100 cfm, \$23,000 or \$3,850/month rent; 320 cfm, \$32,000 or \$5,350/month rent; 570 cfm, \$40,000 or \$6,675/month rent.

Catalytic/blower/etc: skid-mounted package: 200 cfm, \$37,500 + installation; 200 cfm, \$63,000 installed or \$6,600/month rent 1 year; 500 cfm, \$50,000 + \$10,000/heat exchanger + installation; 500 cfm, \$78,000 installed; 1000 cfm, \$70,000 + installation; add \$20,000 for automatic air dilution system.

Carbon canisters: 100 cfm, \$650 to \$764; 600 cfm, \$4,000 to \$11,000; 2200 cfm, \$13,750.

Carbon/blower/etc skid-mounted package: 105 cfm, \$18,400 or \$1,540/month rent 1 year; 250 cfm, \$21,900 or \$1,830 rent 1 year; 500 cfm, \$30,000 or \$1,900 rent 1 year. An automated regenerable unit costs \$149,000 or \$7,500/month rent for a 6-month lease.

Electric power: (0.75) × (blower horsepower) × (\$/kWh) × (hours). For example, a 10 hp centrifugal blower (handling 300–500 cfm) with power costing \$0.10/kWh would cost \$18 for 24 hours. (These costs should be multiplied by 1.25 to account for motor inefficiencies and power factor. Also, for catalytic oxidizers operating with rich soil gas, a correction has to applied for adding dilution air at hydrocarbon concentrations above 3000 ppmV.)

Auxiliary fuel: \$800/month for a 200-cfm catalytic oxidizer with zero flammable vapor in the soil gas; (Fuel for a direct thermal unit would cost about double the fuel cost for a catalytic unit with up to 3000 ppmV hydrocarbons.) \$35/day for an internal combustion engine using propane (times 0.75 using natural gas) with zero flammable vapor.

Capital and operating costs are given (49) for direct thermal oxidizers with various heat recovery components for large systems (5,000 to 90,000 cfm). Another study (89) gives costs, dimensions, and fuel consumption for direct thermal oxidizers ranging in size from 3 cfm to 10,000 cfm. Examples of heat recovery economics and a comparison with refrigerated condensing and carbon adsorption are given. Capital costs for direct thermal units with no heat recovery are 20 cfm, $5,890; 100 cfm, $7,700; 200 cfm, $9,400; 400 cfm, $11,100; 800 cfm, $14,100; 1500 cfm, $17,600; 3000 cfm, $21,000. A flame arrester, inlet valve, and second thermocouple add approximately 27% to these costs.

Reference 61 also gives unit costs for wells, piping, instruments, flame arresters, diffuser stacks, and monitoring. The computerized RACER/ENVEST estimating system developed by Delta Technologies Group (Englewood, Colorado) gives installed costs for wells or trenches, piping, blowers, monitoring, maintenance, and condensate collection for *in situ* vapor extraction without emissions abatement. Input information must include area and depth of contamination and type of soil. If not also given, the program computes well or trench dimensions, spacing and number, cfm flow, and blower cfm and horsepower. Separate models give costs for carbon adsorption, overhead electrical service, fencing and signs, site clearing, demolition and hauling of old pavement, and installing new pavement.

Another ENVEST model is for heat-enhanced vapor extraction with steam injection. Costs are computed for the extraction system, plus a trailer-mounted gas-fired boiler with preheating and feedwater treatment systems, steam piping, flowmeters, steam traps, and a condenser or heat exchanger to condense steam from the blower discharge.

The computerized COMPOSER GOLD estimating system marketed by Building Systems Design (Atlanta, Georgia) includes models for *ex situ* and *in situ* vapor extraction, heat-enhanced vapor extraction, and carbon adsorption.

Incineration Costs. The cost for fixed-base incineration of soils with hazardous contaminants is of the order of $1200/ton, plus excavation and hauling costs. This cost includes disposal of the treated soil, which in some cases involves metals fixation and landfill in a RCRA-permitted facility. At some fixed-based units, this cost is somewhat independent of moisture content or organics content. However, oily wastes with a high Btu content cost significantly less.

Capital costs for a fixed-base unit are tens of millions of dollars. A unit rated at 90 million Btu/h costs approximately $60 million (90). A unit of this size typically has a capacity of 20 ton/h.

Costs with a mobile unit are of the order of $500/ton for quantities over 10,000 tons (91). Between 2000 and 6,000–10,000 tons, prices should be obtained from both fixed-base operators and mobile-unit vendors, and a cost analysis made including hauling and final disposal costs. Below 2000 tons, fixed-base incineration is usually more economical than mobile. Mobile unit expenses involve two cost categories: transport, setup, and demobilization, usually in the range of $2 million to $4.5 million, of which one-third to one-half is for permitting and trial burns; and operating, from $100/ton to $400/ton. Operating costs include equipment maintenance and use fees, labor, supplies, utilities, analyses, and site restoration.

The RACER/ENVEST computer software (Delta Technologies Group, Englewood, Colorado) was developed for mobile units using any of these incinerator types: rotary kiln; fluidized circulating bed combustor; liquid injection incinerator; infrared furnace. These ENVEST models are applicable to fixed-base units by the user specifying zero transport distance and using another ENVEST model for soil hauling charges. Estimated costs include installing a slab, transport, screening, shredding, and conveying of soil. Hazardous wastes are distinguished from petroleum-contaminated waste, and the different permitting and analysis requirements are accounted for.

The COMPOSER GOLD computer software marketed by Building Systems Design, (Atlanta, Georgia) includes portable incineration.

Thermal Desorption. The cost (excluding excavation and disposal) for desorption of 10,000 tons of soil with hazardous contaminants ranges from $200 to $300 per ton; with petroleum contaminants from $40 to $70 per ton; with infrared furnaces from $100 to $150 per ton (10). One study (41) tabulates costs that show the effect of moisture for 10,000 tons of petroleum-contaminated soil:

Fixed base units: $57/ton with 10% moisture; $64/ton with 20%; $72/ton with 30%.

Mobile units: $40–$48/ton with 10% moisture; $48–$56/ton with 20%; $56–$68/ton with 30%.

If fixed-base and mobile unit costs are compared, the hauling costs must be added to the fixed-base unit costs. Hauling of soil with hazardous contaminants costs $2 to $4 per ton-mile; with petroleum contaminants $0.08 to $0.15 per ton-mile (59).

The RACER/ENVEST (Delta Technologies Group, Englewood, Colorado) computerized models estimate cost with direct-fired and indirect-fired systems. As with the model for incineration, costs with fixed-base systems are estimated by specifying zero transport distance, and screening/shredding costs are included.

Soil Bioremediation Costs. Costs per cubic yard of soil treated are reported (92) as follows:

bioreactor:	$150–$250
land treatment:	$30–$90
bioventing:	$50–$120

Bioventing costs are much lower than this range, $10 per cubic yard, if an existing soil venting system is adapted or if the site is very large.

RACER/ENVEST computer models (Delta Technologies Group, Englewood, Colorado) estimate the cost of land farming (including tilling, water piping and watering, fertilizing, and use of cultured microorganisms) and of bioventing (including ventilation and air extraction wells, vacuum blower, nutrient additions, and surface water sprinkling). Another ENVEST model is available if the user desires a subsurface infiltration gallery. The bioventing model computes the required well spacing and air flow rate based on the given area, soil type, and length of the contaminated interval.

TREATING SOILS CONTAINING ORGANIC OR METALS CONTAMINANTS

Some remediation technologies are applied to soils contaminated with metals or organic compounds. Proven technologies in this category covered in this section are washing/flushing; and

stabilization/solidification (fixation or immobilization).

Certain innovative technologies are also covered, including vitrification and electrokinetics.

Applications

***Ex Situ* Washing.** Soil washing generally applies to coarse-grained soils and is used in these modes or a combination of them:

using particle size fractionation techniques (such as screening and hydraulic classification or flotation) to reduce the volume that needs final treatment;

using water (usually with additives) to extract contaminants from coarser particles;

using solvents to extract organic compounds.

Contaminants adsorb more strongly on fine-grained soil than on sand and gravel. After application of soil washing to separate the sand and gravel portions from the finer grained fraction, another remediation technique (such as incineration for organics or fixation for metals) is applied to the finer grained fraction. Washing is a relatively low cost technique that reduces the volume needing more expensive treatment by 80% to 90% with sandy or gravelly soils. Reference 93 summarizes contaminant removal efficiencies and residual contamination.

If extraction is done, the washing fluid is usually treated to remove the contaminants and then recycled through the washing process to the maximum extent practical.

Problems and limitations with soil washing include the following: water-based washing is not applicable to fine-grained soils; the extent of cleanup is highly variable and difficult to control; soil may have to be recycled through the system to achieve cleanup goals or multiple extraction stages are used; washing medium needs cleanup; hydrocarbon solvents used for extraction are flammable.

***In Situ* Flushing.** Soil flushing is *in situ* soil washing with water, sometimes with additives. The geologic setting should include an impervious stratum above the aquifer so that the flushing action does not transport soil contaminants downward into an otherwise uncontaminated aquifer.

***Ex Situ* Fixation or Immobilization.** Stabilization is generally the use of chemical reagents to form insoluble metal precipitates that do not migrate. The best example is the use of caustic to stabilize a sludge or soil contaminated with heavy metals. Solidification of soils is the use of microencapsulating agents to immobilize contaminants. Solidification is also used for liquid wastes by applying absorbents. For soils, solidification to trap metals is usually accomplished with Portland cement or lime plus pozzalans (eg, fly-ash or blast furnace slag). Such a mixture in some cases encapsulates organics, but thermoplastic agents (eg, asphalt) work better for many types of organic compounds.

The main goal with fixation techniques is to form a solid matrix from which the contaminants are not leachable. Typically, soil that has undergone fixation must pass a laboratory test, such as the toxicity characterization leaching procedure (TCLP, as given in 40 CFR 261.24). The TCLP calls for mixing the sample with dilute acetic acid and measuring extracted contaminants in the leachate.

If treated soil is to be used as backfill, a granular product is desired rather than a monolith. With cement or lime plus pozzalans, a concrete is formed with a an unconsolidated compressive strength of perhaps only 100 psi. If the product is landfilled, it needs only enough compressive strength to maintain nonleachability when overburden is fully placed on top of it.

Problems and limitations with fixation include interference with binding with some contaminant substances; potential leaching or inadequate strength; large volume and weight increase with many fixation agents; difficulty in determining optimum agent/soil ratio and water content; potential fast setting, leading to a monolithic (instead of granular) structure; failures due to freeze/thaw cycles where product is not placed below the frost line; special apparatus and heating needed with thermoplastics; potential VOC emissions and associated safety concerns.

***In Situ* Fixation.** Fixation is a very common *ex situ* technique, but can be used *in situ* by combining shallow soils or pond sediments with added agents and mixing or by injecting slurried agents through a large, hollow auger at depths down to approximately 120 feet. The auger mixes soil with the agents by rotating and oscillating vertically, thereby forming a round, solidified column. By repeating the process with overlapping columns, soil is fixed areally as well as at depth.

Problems and limitations with *in situ* fixation include inadequate mixing unless large augers are used that are costly to mobilize; depth limits; difficulty in monitoring results; interference from boulders; volume increase must be spread on-site or hauled off and disposed of.

Area Mixing Fixation. Another technique for fixation is area mixing. The soil is spread and agents are spread on top of it and then mixed in. Additional lifts are fixed in the same area. The fixed soil is left in place or removed and disposed of in a landfill.

Problems and limitations with area mixing arise if heavy rainfalls occur; thorough mixing is difficult to achieve.

Companion Technologies. Washing is usually accompanied by two technologies; one involves treatment of the wash water or solvent for recycling, the other remediates the fines fraction.

Fixation is frequently preceded by thermal treatment (desorption or incineration) or by washing for removal of organics. Then fixation traps metal contaminants. It should be noted that, while not companion technologies, enhanced soil vapor extraction and in situ soil biodegradation are accomplished with the same augers used for *in situ* fixation. Steam (for vapor extraction) or nutrients and peroxide solution (for biodegradation) are injected through the auger.

Basic Methods for Soil Organics and Metals

Soil Washing Methods and Flushing with Water Solutions. Surfactants (detergents) are the agents most frequently added to remove insoluble organics by washing with water (94). The same equipment used in the mining/metallurgical industries to separate coarser fractions is applied. Frequently, the finer fractions of ores contain desired metals and minerals, just as finer fractions of soil contain adsorbed contaminants.

Other agents that are used for organics removal with soil washing and soil flushing are bases (eg, caustic or sodium carbonate solutions) and water-soluble polymers. Such agents are used by the petroleum industry for enhanced oil recovery. Caustic is used to extract hydrocarbons from tar sands. Either water or acid solution is used for cationic metals extraction. Reference (6) indicates that basic solutions are used for some phenols and zinc, tin, and lead.

Soil Washing Methods with Solvents (Solvent Extraction). When solvents are used instead of water-based mixes to extract organics, the washing is often effective with finer grained soils as well as with sands and gravels. Solvent extraction effectively treats soils contaminated with VOC (including halogenated solvents), PCBs, and petroleum wastes. A major concern with solvent extraction is the possibility that some residual solvent remains with the treated soil (94). If the percentage of contaminant removal is not high enough with one pass, the washed soil is subjected to repeated solvent applications.

For organics that are insoluble in water, solvents that are not highly toxic are used. A liquid petroleum fraction with a narrow boiling range and a higher vapor pressure than the contaminant organics extracts oils from dry soil. Then the solvent is stripped or distilled from the solvent/oil mix and recycled through the process.

Such systems are not the best for soil contaminants that include hydrophilic compounds. The B.E.S.T. system developed by Resources Conservation Corporation, (Bellevue, Wash.) uses a mixture of triethylamine, sodium hydroxide, and water as a solvent for both hydrophilic and hydrophobic organic contaminants. The amine is highly soluble in water in contact with soil slightly chilled below ambient temperature and insoluble at a somewhat higher temperature. This property is used to separate the amine with some contaminants from the water. The amine is recovered by evaporation or steam stripping and condensing. Dried soil is subjected to warm extraction with amine after cold extraction with the amine/caustic mix.

Most solvent extraction systems operate at atmospheric pressure. However, pressurized systems using liquified petroleum gases (propane or butane) and liquified carbon dioxide have been developed by CF Systems (Woburn, Massachusetts). Propane or butane, liquified under pressure, dissolve nonpolar organics. Liquified carbon dioxide under pressure dissolves polar or water-soluble organics. Either solvent is separated from the contaminants as a vapor by depressurizing the liquids. Then the solvent vapor is compressed and condensed by cooling for recycling in the liquid state. With propane or butane, the soil is given a final wash with warm water to remove any residual hydrocarbon solvent. Such solvent extraction systems are used for a much higher range of contaminant vapor pressures than liquid hydrocarbon systems operating at atmospheric pressure.

Fixation or Immobilization Methods. The most common fixation agent is portland cement. It forms concrete when mixed with soil and water. The soil particles with adsorbed contaminants become microencapsulated. Metals and some oily wastes are fixed with cement. Relatively small amounts of additives are added as binders or to hasten hardening. The most common additives used with Portland cement are sodium and potassium silicates, which are water soluble.

Lime plus flyash or blast furnace slag is used similarly to portland cement. These agents fix soils contaminated with certain organics that interfere with setting (hardening) when cement is used and form a stronger concrete. For most applications, the choice between cement and lime plus pozzalans is based on local availability of the agents.

The ratio of cement to soil is usually in the 1:4 to 1:2.5 range. Lime/soil ratios range from 1:20 to 1:3 (95). It is also possible to form concrete with contaminated soil by using sulfur.

A frequently used *ex situ* method uses a pug mill to mix crushed and screened soil, cement (plus sometimes a small percentage of special additives, such as soluble silicates), and water. Mixing continues for 5 to 20 minutes. The batch is tested for compressive strength and leaching. Batches that fail testing must be crushed and recycled through the process.

Thermoplastics are agents that soften upon warming and are used for fixation more often than thermosetting plastics. Plastics used for fixation of organics include asphalt, polyethylene, urea–formaldehyde, and polyesters. Asphalt is the most common among these. Plastics cost more than Portland cement, need heating to be applied, and are more subject to biodegradation. However, they successfully fix many organic compounds that interfere with binding when silicaceous agents are used. It is reported (96) that thermoplastics are applied in the temperature range 130°C (266°F) to 230°C (446°F).

Other agents used for fixation include clays and proprietary agents, usually containing clays or aluminum silicate. Sorbents, such as silica, clay minerals, feldspars (eg, $KAlSi_2O_4$) and zeolites (eg, $NaAlSi_2O_4 \cdot H_2O$) are used to solidify liquid organic wastes (97).

Operating Conditions and Treatability Studies. A most important parameter in soil washing is the ratio of washing fluid to soil. Allied to this parameter is the fraction of wash fluid that is recycled. The higher the fluid/soil ratio, the better the efficiency of contaminant removal, and the more cost per ton of soil. The costs are higher because equipment is bigger, fluid treatment or recovery processing is more extensive, and pumping energy requirements are higher. Fluid recycling breaks down into two categories: (*1*) the portion that is recycled directly without treatment; (*2*) the portion that must be treated before recyling into the main process. The higher the portion in category (*1*), the lower are treatment costs. The higher either category is, the lower are fluid disposal costs. The treatment required for disposal of spent fluid is often more extensive and costly than treatment required for category (*2*) recycling.

Jar tests and pilot tests are conducted to help determine optimum operating conditions and cleanup capabilities as follows:

- selection of wash fluid and additives;
- ratio of washing fluid to soil;
- fraction of soil that does not require further treatment;
- contaminant removal efficiency;
- fraction of fluid that is recycled; and
- treatment requirements for fluid to be recycled and for fluid to be disposed of.

If solvent extraction is to be applied, tests help determine soil drying requirements, solvent recovery conditions, and percent solvent that adsorbs on washed soil.

The most important operating parameters with fixation techniques are the required ratio of agent to soil and the volume increase. The optimum agent/soil ratio is the minimum amount for a product that passes the leachate test and has sufficient compressive strength. Tests are used to give information on

- selecting the main fixation agent and additives;
- agent/soil ratio;
- durability, strength, and curing conditions;
- leachability;
- moisture requirements;
- heat evolved or cooling requirements with silicaceous agents;
- evolution of hazardous vapors; and
- temperature requirements for thermoplastics.

Trends and Innovations in Soil Organics/Metals Treatment

Trends advancing soil washing techniques include the use of ion exchange resin beads suspended in wash water and increased use of liquified propane. The ion-exchange resins adsorb contaminants released from the soil into the water phase, thereby treating the water without using conventional tanks containing fixed-bed columns of resin. The liquified propane extracts nonpolar organics and is readily recovered and recycled.

One trend in fixation technology, especially for heterogenous soils with toxic contaminants or radionuclides, is toward more use of vitrification. Electric energy at very high voltages causes resistance heating of the soil until it melts. Organic compounds and volatile metals are vaporized and captured as the mass warms up. At higher temperatures, the organics are pyrolyzed. After the mass is melted and cooled, metals are trapped in the glass formed. The glass is durable and has extremely high compressive strength. The technology is applied *in situ* up to depths of 5 meters or *ex situ*. With *in situ* applications, the soil volume shrinks, causing subsidence and the need to import backfill.

An innovation for *in situ* vitrification that has no depth limitation is the use of plasma arcs in narrow boreholes. The Georgia Institute of Technology is developing this technique (98). The arc is started with a plasma torch at the bottom of a borehole, which is withdrawn as a blob of soil surrounding the torch is melted.

An *ex situ* scheme that melts soil at 1316°C (2400°F) with fluxing agents, lime, alumina and ferric oxide has been developed by the Institute for Gas Research (Des Plaines, Illinois). Air is bubbled through the melt to combust organics. Cooled, pulverized product is blended with ash and waste concrete to make a high-strength cement. An *ex situ* fixation scheme that immobilizes metals, including those that volatilize with vitrification, and does not require heat has been developed by Argonne National Laboratory (Argonne, Illinois). A ceramic binder is mixed with soil or sludge to form Caramicrete.

Another trend in fixation technology is toward more use of large augers for *in situ* applications. U.S. technology includes 30–45 cm (12–18 in.) diameter augers. S.W.M. Seiko (Hayward, California) uses Japanese technology with augers up to 10 cm (4 in.) diameter that cut through boulders, and with three interleaved augers at once. The Japanese technology penetrates much deeper than the U.S. technique.

Applications of phytoremediation (uptake by plants of metals, nitrates and organics) are described (11).

The use of water-soluble organic solvents (eg, ethanol) for flushing organic contaminants from soil is reported in (99).

Electrokinetics is an attractive innovation in remediation processes for soils contaminated with metals or acidic organics. An important characteristic of this *in situ* technology is that it is applicable to heterogeneous fine-grained soils including silts and clays. Soil flushing and *in situ* bioremediation, by way of contrast, are applied successfully only to homogenous sand. With electrokinetics, ionic species or organics are transported to electrodes implanted in the soil. The products collected at the electrodes are removed from the subsurface for treatment or disposal. The volumes so recovered are very small compared to the volumes recovered with soil flushing. The soil does not have to be saturated, but does have to contain some moisture.

The main mode of movement toward the electrodes for ions and polar organics is electromigration. Secondary modes of transport include movement of water by electroosmosis and movement of charged particles by electrophoresis. A relatively low direct-current voltage is applied. The aqueous phase moves towards the negative electrode (cathode). Electroosmotic flow of water is relatively rapid, even in clay, and carries nonionic species (eg, nonpolar organics) to a limited extent and ionic species with the water. Ion migration occurs with cations moving toward the cathode and anions toward the anode. An important feature of this electrolytic migration is the flow of hydrogen ions from the anode, where they are generated, toward the cathode, and the flow of hydroxide ions from the cathode, where they are produced, toward the anode. The hydrogen ions move the fastest, resulting in an acid front moving through the soil toward the cathode. Another feature of electrolytic migration is that, when the ionic strength is high, water molecules are dragged to the cathode with the cations.

Movement of the acid front helps extract metals from the soil. However, unless acid is injected at the cathode, precipitates form where the acid front meets hydroxide ions starting to migrate away from the cathode.

Organic acids and organics that exhibit acidic properties (eg, phenol) are removed by electrokinetics. Electrokinetics also aid in situ bioremediation. It warms the soil, which is usually naturally near 12° (55°F), the colder end of the range for most bioactivity. Electromigration helps disperse nutrients. Electrokinetics acts as a source of oxygen or of electron receptors for aerobic bioactivity. However, electrokinetics acidifies soil, which halts bioactivity unless the soil is highly buffered.

Soil mineralogy, porosity, tortuosity, conductivity, chemical composition, and electrochemical properties of the fluids affect the transport of chemical species in an electric field (100). Reference 100 describes the roles of diffusion, ionic mobility, conductivity, permittivity, porosity, viscosity and pH.

The voltage depends on the distance between electrodes. The amperage depends on the vertical cross-sectional area of the soil plot being remediated. Reference 101 suggests a voltage of a few volts per cm and a current density of about 30 to 50×10^{-6} amp/cm^2.

Electroosmosis is used in the "Lasagna" process (layered subsurface treatment zones and granular electrodes) developed by an industry consortium and EPA's Risk Reduction Engineering Laboratory (Cincinnati, Ohio).

Costs for Soil Organics and Metals

Soil Washing/Flushing Costs. Soil washing for physically separating the finer grained fraction typically costs within approximately 30% of $100/ton. Extraction of contaminants, water treatment, and further soil processing adds another 25 to 100% to costs. The ENVEST computerized cost estimating model developed by Delta Technologies Group indicates a cost of $77/ton for 20,000 tons of soil, using a transportable unit rated at 20 ton/hour. This cost includes water treatment and excludes excavation, debris removal, and treatment/disposal of residues generated by treatment. Additional costs apply if a high level of worker protective safety equipment is involved. The estimated cost is $66/ton for 60,000 tons of soil.

Another ENVEST model developed by Delta is for *ex situ* solvent washing of soils (up to 30% moisture) and sludges (31 to 70% moisture). This model assumes that the equipment is assembled at the site, rather than a transportable unit. The minimum volume of soil for economical application is 5,000 cubic yards. The model assumes that the soil has been prescreened so that processing is applied to particle sizes up to one inch. The model includes crushing, conveying, and screening of this material and treatment in one of three different sizes of units: 1,500 cubic yards/month; 6,000 cubic yards/month; or 18,000 cubic yards/month. The user specifies what percentage of treated material is to be recycled through the treatment process. The model includes contractor's markups for overhead and profit as follows: 25% on labor; 20% on equipment; and 18% on materials.

Reference 102 indicates that costs for *in situ* flushing for a 15,400 m^3 (20,000 yd^3) site are $240/$m^3$ ($185/$yd^3$) for capital investment plus $20/$m^3$ ($15/$yd^3$) for expense. A study with surfactant flushing predicts costs in the range of $104/$m^3$ ($79/$yd^3$) to $215/$m^3$ ($164/$yd^3$), including both amortized (7 years) investment cost and expense. These estimates exclude costs for remedial investigation, feasibility studies, permitting, administration, project management, contractor profit, and contingency.

The ENVEST model developed by Delta for flushing estimates costs using water only; alkylbenzene sulfonate detergent (for neutral organics); sulfuric acid solution (for cationic metals and basic organics, such as amines, certain ethers and anilines); or caustic solution (for zinc, tin, lead, and phenols). The model assumes that an infiltration gallery, placed on the surface, or beneath deposits of low permeability, or beneath the frost line applies flushing solutions. The user must specity the area, depth to groundwater or to the lower confining unit, and vertical permeability. The chosen permeability determines the solution application rate over the given area, as follows:

coarse sand/gravel: 12 in./d for permeability of 0.1 cm/s (284 ft/d)

coarse to fine sand: 6 in./d for permeability of 0.01 cm/s (28.4 ft/d)

fine to silty sand: 3 in./d for permeability of 0.001 cm/s (2.84 ft/d)

A soil porosity of 30% percent is assumed, so the pore volume is taken at 0.30 times the soil volume being flushed. Unless the user specifies otherwise, the model assumes that 10 pore volumes of solution will be flushed through. The model uses 1.33 times the pore volume for the initial flush, to account for soil retentivity. The model estimates capital investment for an earth berm, pumps, mixing tank, PVC piping (and later demolition of piping), and operating costs for electric power and additives. Other ENVEST models are available for estimating sampling and analysis costs, water supply, overhead power supply, fencing and signage. No operating labor is estimated (except for sampling labor in the sampling/analysis model), because the model does not include water treatment and unattended operation is feasible.

Stabilization/Solidification Costs. Costs for fixation (in 1990 dollars) using transportable units are estimated for quantities over 5000 cubic yards (95). These costs exclude excavating and stockpiling of soil, mobilization/demobilization, engineering, safety gear, decontamination of equipment, and delivering reagents. The assumed process includes 30% portland cement and 2% sodium silicate (95).

Estimates using CORA software (a program developed for EPA and available until 1994) for pug mill type of applications indicate costs as follows:

10,000 cubic yards: $73/cubic yard
50,000 cubic yards: $58/cubic yard
100,000 cubic yards: $54/cubic yard

If a step-by-step detailed estimate is desired, these items should be accounted for in addition to processing (103):

decontamination of transport, handling, and processing equipment, and in some cases treatment and disposal of the decontamination water;

grading, with a 2 to 3% slope, berming and lining of a cleared area with subdrainage system for storing untreated material;

covering of untreated material;

installing a paved loading ramp;

bins, hoppers, or pallet storage areas that are weatherproofed, for storing reagents; and

back-filling of treated soil in lifts of 8 to 10 inches, adding moisture, and compacting. Moisture is needed to provide lubrication for the soil particles to slide past each other during compaction, and the optimum moisture content is that amount which results in maximum bulk density. Excess moisture lowers the density.

Reference 96 gives detailed examples for estimating fixation costs using 30% portland cement and 2% sodium silicate.

The ENVEST solidification model developed by Delta estimates updated costs for an on-site batch plant with a pug mill using cement for solids, sludges, or incinerator ash. Input from the user must include the total wastes volume and the batch capacity of the pug mill (ie 2 cu yd, 5 cu yd, 10 cu yd, or 15 cu yd). The model assumes that the batch cycle time is 20 minutes, including 5 minutes for loading, 14 minutes of mixing, and 1 minute for unloading. If not specified, the model assumes moisture contents as follows: 15% for solids; 60% for sludges; 20% for ash. If the user does not give the amount of added water and cement, the model computes how much water to add, so that the water/cement ratio is 0.40 and the cement/

waste ratio is 0.15 for solids; 0.40 for sludges; or 0.10 for ash. The model computes the volume increase. Other ENVEST models are used for estimating related costs for excavation, permitting, sampling/analysis, and pilot testing.

BIBLIOGRAPHY

1. G.M. Long, *Chem. Eng. Prog.* 58–67 (May 1993).
2. P.J. Marks, W.J. Wujcik, and A.F. Loncar, "Remediation Technologies Screening Matrix and Reference Guide," 2nd ed., U.S. Dept. of Defense Environmental Technology Transfer Committee, NTIS PB95-104782 (Oct. 1994).
3. M.H. Hyman and L. Bagaasen, *Chem. Eng. Prog.* 22–43 (Aug. 1997).
4. Battelle Memorial Institute, RAAS Version 1.1, Battelle, PNL-8751, Rev. 3, Oct. 1996.
5. E.K. Nyer, *Groundwater Treatment Technology,* 2nd ed., Van Nostrand Reinhold, New York, 1992, p. 226.
6. G. Parkinson, ed., *Chem. Eng.* 23 (Mar. 1996).
7. E. Wilson, *Chem. Eng. News,* 6–7 (May 8, 1995); see also *Science* **268,** 700 (1995).
8. S. Kilambi, *Pollut. Eng.* 46–48 (Aug. 1996).
9. *Pollut. Eng.* 20–21 (Jan. 1997).
10. *Indus. Wastewater,* 29 (Nov./Dec. 1994).
11. I. Kim, *Chem. Eng.,* 39–41 (Dec. 1996).
12. A. Bray and G.N. Brown, "Web Technology in the Separation of Strontium and Cesium from INEL-ICPP Radioactive Acid Waste" (WM-185), Pacific Northwest Laboratory, NTIS DE95006368, Richland, Wash., 1995.
13. *In Situ Remediation Technology Status Report: Treatment Walls,* EPA542-K-94-004, U.S. EPA, Apr. 1995.
14. Anon., *The Hazardous Waste Consultant* 1.16-1.20 (Jan./Feb. 1992).
15. S.K. Bhattacharya, *Remediation* 199–210 (Spring 1992).
16. L. Plant and M. Jeff, *Chemical Engineering* EE-16 to EE-20 (Sept. 1994).
17. R. Sirabian, T. Sanford, and R. Barbour, *Remediation* 189–205 (Spring 1994).
18. M. Atlas, *Chem. Eng. News,* pp. 32–42, Apr. 3, 1995.
19. T.C. Morin, *Pollut. Eng.* 44–47 (Feb. 1997).
20. M.H. Stenzel and W.J. Merz, *Environ. Prog.* **8** (4) 257–264 (1989).
21. J.C. Crittenden and co-workers, *J. Amer. Water Works Assoc.* **83** (1) 77–87 (1991).
22. M.C. Kavenaugh and R.R. Trussell, *J. Amer. Water Works Assoc.* **72** (12) 684–692 (1980).
23. K. Onda, H. Takeuchi, and Y. Okumoto, *J. of Chem. Engineering of Japan* **1** (1) 56–62 (1968).
24. W. Gruber, *Environ. Eng. World* 39 (July–Aug. 1995).
25. D. Smyth, J. Cherry, and R. Jowett, *Soil & Groundwater Cleanup,* 36–43 (Dec. 1995).
26. D. Tremblay, D. Tulls, P. Kostecki, and K. Ewald, *Soil & Groundwater Cleanup* 6–10 (Dec. 1995).
27. *In Situ Remediation Technology Status Report: Hydraulic and Pneumatic Fracturing,* EPA542-K-94-005, U.S. EPA, April 1995.
28. T.W. Schrauf, Sheehan, P.J., and Pennington, L.H., *Remediation* 93–114 (Winter 1993/1994).
29. G. Parkinson, ed., *Chem. Eng.* 17 (Mar. 1996).
30. G. Ondrey, ed., *Chemical Engineering* 17 (Sep. 1995).
31. G. Parkinson, ed., *Chemical Engineering* 19 (Mar. 1996).
32. W.J. Cooper et al., *J. Amer. Water Works Assoc.* **85**(9), 106 (1993).
33. R.A. Jensen, *Environmental Technology* 32–36 (Nov./Dec. 1996).
34. G. Parkinson, ed., *Chemical Engineering* 21 (Mar. 1996).
35. "Bioremediation in the Field," EPA/540/N-94/501, U.S. EPA, (July 1994).
36. E.M. Aeita, K.M. Reagan, J.S. Lang, L. McReynolds, J.W. Kang, and W.H. Glaze, *J. Amer. Water Works Assoc.* **80**(5) (May 1988).
37. M.H. Stenzel, "Adsorption," *Chem. Eng. Prog.* 36–43 (Apr. 1993).
38. *The Hazardous Waste Consultant* Jan./Feb. 1995, pp. 3.2–3.4.
39. P.C. Johnson and co-workers, *Ground Water Mon. Rev.* (Spring 1990).
40. *Soil Vapor Extraction Handbook,* 165, EPA/540/2-91/003 U.S. EPA, (Feb. 1991).
41. W.T. Troxler, J.J. Cudahy, R.P. Zink, J.J. Yezzi, and S.I. Rosenthal, "Treatment of Nonhazardous Petroleum Contaminated Soils by Thermal Desorption Technologies," paper 92-33.05, *Air and Waste Management Association 85th Annual Meeting,* Kansas City, June 21–26, 1992.
42. W.C. Anderson, ed., *Thermal Desorption,* WASTECH/Amer. Academy of Environmental Engineers, Annapolis, Md., 1993.
43. R. Block, H. Stroo, and G.H. Swett, *Chem. Eng. Prog.* 44–50 (Aug. 1993).
44. National Fire Codes NFPA 325, *Guide to Fire Hazard Properties of Flammable Liquids, Gases, and Volatile Solids,* National Fire Protection Association, Quincy, Mass., 1994.
45. H.L. Shelton, *Environ. Eng. World* 22–25 (May–Jun. 1995).
46. C.R. Windmueller and A. Sykes, *Pollut. Eng.* 14–15 (Nov. 1995 Suppl.).
47. F.A.M. Buck and C.W. Hauck, "Vapor Extraction and Catalytic Oxidation of Chlorinated VOC," *Proceedings of 11th Annual Incineration Conference* (sponsored by University of California), Albuquerque, N.M., May 1992.
48. R. Renko, *Pollut. Eng.* 62–63 (Apr. 1994).
49. J.M. Klobucar, *Chem. Eng. Prog.* 57–63 (Apr. 1995).
50. C.R. Brunner, *Handbook of Incineration Systems,* McGraw–Hill, New York, 1991, p. 24.4 (for heat of combustion); p. 23.10 (for ammonia reduction); and pp. 22.36–22.37 (for abatement devices).
51. K. Fouhy, G. Ondrey, *Chem. Eng.* 30–43 (May 1994).
52. *The Hazardous Waste Consultant* 1.12–1.19 (Jan./Feb. 1995).
53. *Environ. Eng. World* 6–15 (Jan.–Feb. 1995).
54. *Environ. Eng.* pp. EE-3–EE-8 (Nov. 1994 suppl. to *Chemical Engineering*).
55. U.S. EPA publication AP-51, *Control Techniques for Particulate Air Pollutants,* Jan. 1969, p. xxiv.
56. American Society of Mechanical Engineers, *Combustion Fundamentals for Waste Incineration,* New York, 1974.
57. V. Haroutunian, *Environ. Eng. World* 26–31 (July–Aug. 1995).
58. J.P. Reynolds, R.R. duPont, and L. Theodore, *Hazardous Waste Incineration Calculations,* John Wiley & Sons, Inc., New York, 1991.
59. *Air Emissions from the Treatment of Soils Contaminated with Petroleum Fuels and Other Substances,* EPA-600/R-92-124, U.S. EPA, July 1992.
60. S.G. Huling and co-workers, *Remediation* 41 (Spring 1995).
61. *Soil Vapor Extraction Handbook,* Appendix C, EPA/540/2-91/003 U.S.EPA, Feb. 1991.
62. Ref. 61, pp. 38–50.
63. G.P. Korfiatis and C. Christodoulatos, *Remediation,* 402 (Autumn 1993).
64. *In Situ Remediation Technology Status Report: Thermal Enhancements,* EPA/542-K-94-009 U.S.EPA, Apr. 1995.

65. *Handbook on In Situ Treatment of Hazardous Waste-Contaminated Soils,* EPA/540/2-90/002, U.S.EPA, Jan. 1990.
66. D.C. Noonan, W.K. Glynn, and M.E. Miller, *Chem. Eng. Prog.* 60 (June 1993).
67. J.W. Massmann, *J. Environ. Eng.* **115**(1), Amer. Soc. of Civil Engr. Paper No. 23208 (Feb. 1989).
68. *Soil Vapor Extraction Handbook,* App. I, EPA/540/2-91/003, U.S. EPA, Feb. 1991.
69. M.E. Dodson, J.A. Pezzullo, and R.J. Piniewski, in G.W. Gee and N.R. Wing, eds., *Thirty-Third Handford Symposium on Health and the Environment, In Situ Remediation,* Nov. 7–11, 1994, Pasco, Wash., Part 2, Batelle Press, Columbus, Ohio, pp. 721–736.
70. M. Cummings and S. Booth, *Remediation* 69–79 (Summer 1996).
71. A. Rozich, *Environ. Eng. World* 32–34 (July–Aug. 1995).
72. *Chem. Eng.,* 19 (Apr. 1995).
73. *Chem. Eng. Prog.,* 13–18 (Nov. 1996).
74. J. Chowdhury and K. Fouhy, *Chem. Eng.,* 33–37 (Feb. 1994).
75. D. Ross and P.L. Sudano, *Indust. Wastewater,* 39–41 (Nov./Dec. 1994).
76. R.S. Baker and J. Bierschenk, *Pollut. Eng.* 38–40 (Mar. 1966).
77. *Handbook on In Situ Treatment of Hazardous Waste-Contaminated Soils,* 38–40, EPA/540/2-90/002 U.S. EPA, 1990.
78. R.J. Portier, K.K. Barton, and J. Koury, *Remediation* 57–75 (Spring 1995).
79. *Chem. Eng.,* 23–25 (May 1994).
80. Ref. 77, pp. 51–52.
81. "Methane Injection for Bioremediation of Soils and Groundwater," *National Environ. J.,* 14 (Nov.–Dec. 1995)
82. Engineering Bulletin, *Chemical Dehalogenation Treatment: APEG Treatment,* EPA/540/2-90/015, U.S.EPA, Sept. 1990.
83. *Guide for Conducting Treatability Studies Under CERCLA: Soil Vapor Extraction,* EPA 540/1-91/019A, U.S.EPA, Sept. 1991.
84. T.A. Pederson, *Chem. Week, Environ. Eng.,* 41 (Feb. 5, 1992).
85. *In Situ Soil Vapor Extraction Treatment,* Engineering Bulletin, EPA/540/2-91/005, U.S.EPA, May 1991.
86. M.K. Stinson, "In Situ Vacuum Extraction Process in Groveland, Massachusetts," *EPA Site Demonstration of the Terra Vac,* Aug. 1989.
87. *SITE Program Demonstration Test, Terra Vac In Situ Vacuum Extraction System, Groveland, Massachusetts,* Technology Evaluation Report, Vol. I, EPA 540/5-89/003a, U.S.EPA, Apr. 1989.
88. R. Ball and co-workers, *Environ. Prog.* (Aug. 1990).
89. J.F. Straitz III, *Environmental Eng. World,* 18–23 (July–Aug. 1995).
90. D.A. Oberacker and co-workers "Remediation (Clean-Up) of Contaminated Uncontrolled Superfund Dumpsites by Incineration and Other Popular Technologies," *1990 Pacific Basin Conference on Hazardous Waste, Honolulu,* Nov. 12–16, 1990.
91. *High Temperature Thermal Treatment for CERCLA Waste: Evaluation and Selection of Onsite and Offsite Systems,* EPA/540/X-88/006, U.S.EPA, Dec. 1988.
92. *Chem. Eng.,* 117 (June 1993).
93. *Soil Washing Treatment,* EPA/540/2-90/017, U.S.EPA, Sept. 1990.
94. *Air Emissions from the Treatment of Soils Contaminated with Petroleum Fuels and Other Substances,* EPA/600/R-92-124, U.S.EPA, July 1992.
95. E.F Arniella and L. Blythe, *Chem. Eng.,* 93–102 (Feb. 1990).
96. *Handbook for Stabilization/Solidification of Hazardous Waste,* EPA/540/2-86/001 U.S.EPA, June 1986.
97. N.J. Sell, *Pollution Eng.* 44–48 (Aug. 1998).
98. L. Circeo, S.L. Camacho, G.K. Jacobs, and J.S. Tixier, "Plasma Remediation of In Situ Materials - The Prism Concept," in G.W. Gee and N.R. Wing, eds., *33rd Hanford Symposium on Health and the Environment, In Situ Remediation, Nov. 7–11, 1994, Pasco, Wash., Part 2,* Batelle Press, Columbus, Ohio.
99. *In Situ Remediation Technology Status Report: Cosolvents,* EPA542-K-94-006, U.S.EPA, Apr. 1995.
100. Y.B. Acar and A.N. Alshawabkeh, *Env. Sci. Tech.* **27**(13) 2638–2647 (1993).
101. Y.B. Acar, A.N. Alshawabkeh, and R.J. Gale, *Waste Manage.* **13**(2), 141–151 (1993).
102. W.C. Anderson, ed., *Soil Washing/Soil Flushing,* WASTECH/American Academy of Environmental Engineers, Annapolis, Md., 1993.
103. Stabilization/Solidification of CERCLA and RCRA Wastes, EPA/625/6-89/022, U.S.EPA, May 1989.

GROUNDWATER MONITORING

CAROL J. MILLER
Wayne State University
Detroit, Michigan

Groundwater monitoring is used to analyze the impact of a variety of surface and subsurface activities, including seawater intrusion, application of agricultural products such as herbicides, pesticides, and fertilizers, residential septic systems, and industrial waste ponds. Another focus of groundwater monitoring has been contamination associated with waste landfills and ruptured underground petroleum storage tanks (1,2).

Groundwater monitoring is a necessary component in any investigation of subsurface contamination. A wide variety of information can be gleaned from the data including groundwater velocity and direction, and contaminant identification and concentration (1,3,4). These data can be combined with other observations to infer various characteristics of the contamination (5). Examples are source and timing of the release, and future location of the contaminant plume.

The design of groundwater monitoring strategy requires a basic understanding of groundwater flow systems. The majority of groundwater flow occurs in formations known as aquifers. At least two types of data can be retrieved using groundwater wells, ie, groundwater pressure and groundwater quality. A monitoring well allows measurement of these properties at a specific point in an aquifer. Monitoring wells come in a variety of sizes and materials, but each is simply a pipe extending from the ground surface to a point in the aquifer at which the pressure or contaminant is to be assessed. Monitoring wells are functional only in the saturated zone of the subsurface. Within the unsaturated soil zone, tensiometers, soil moisture blocks, and psychrometers have been used to assess fluid pressures. Fluid samples are retrieved using suction cup lysimeters for subsequent quality analysis.

AQUIFERS

The term *aquifer* is used to denote an extensive region of saturated material. There are many types of aquifers. The primary distinction between types involves the boundaries that define the aquifer. An unconfined aquifer, also known as a phreatic or water table aquifer, is assumed to have an upper boundary of saturated soil at a pressure of zero gauge, or atmospheric pressure. A confined aquifer has a low permeability upper boundary that maintains the interstitial water within the aquifer at pressures greater than atmospheric. For both types of aquifers, the lower boundary is frequently a low permeability soil or rock formation. Further distinction exist. An artesian aquifer is a confined aquifer for which the interstitial water pressure is sufficient to allow the aquifer water entering the monitoring well to rise above the local ground surface. Figure 1 identifies the primary types of aquifers.

Calculation of the flow in the saturated portion of the subsurface is generally much easier than that in the unsaturated zone. However, calculation of flow in either requires a fundamental understanding of groundwater pressure and energy.

MONITORING WELL DESIGN FOR CONTAMINANT TRANSPORT STUDIES

There are a variety of contaminant problems that may prompt the development of a groundwater monitoring program. The specific details of the program depend on the situation prompting such monitoring. For example, groundwater monitoring may be required in the vicinity of a new or existing landfill, and would serve the purposes of clarifying groundwater flow conditions, identifying background water quality, and leak detection. Groundwater monitoring in the vicinity of known contamination is used to delineate the spatial extent of the contamination, as well as to verify the chemicals present. Groundwater monitoring may be required in association with real estate transactions to verify the existence of a pristine water source for well development.

Monitoring wells are installed by first completing a soil boring to the approximate depth of groundwater measurements. Drilling methods for the borehole include auger, mud rotary, cable tool, jetted wells, and driven wells (6,9). During the drilling, a boring log is prepared that records details of the subsurface materials encountered as the depth progresses. A well casing is installed in the borehole with a well screen at or near the bottom of the borehole. The annular space between the borehole and the casing must be filled properly to allow free passage of groundwater from the monitored zone to the well screen and to preclude passage of moisture from the surface vertically along the sides of the casing. In the vicinity of the well screen, a filter pack of natural, ie, typically sand or pea gravel, or synthetic material is used to preclude clogging of the well screen. The specific design of the filter pack must take into consideration details of the aquifer soil. Often, a secondary filter pack consisting of finer materials is placed above the primary filter. Above this is the virtually impermeable bentonite seal. A neat cement grout above this layer extends to the ground surface. Figure 2 illustrates the primary components of a monitoring well.

A variety of techniques can be used to retrieve the groundwater sample once the well is in place. Pumps, bailers, and syringes are among the devices used to draw the sample to the surface. Typically the well is purged of three to ten casing volumes of fluid prior to retrieval of the sample, to ensure standing water is not being analyzed (3,6,9). Care must be taken during sampling and delivery to the lab. The characteristics of the sample may be altered if protocols are not followed. Volatile gas stripping, oxidation, and pH shifts are examples of modifications that may occur owing to the introduction of oxygen or other gases to the samples (10).

It is often important to quantify the contamination of pore fluid in the unsaturated soil zone, where monitoring wells are ineffective. In this region, suction cup lysimeters are useful (10). These samplers consist of a porous cup, typically ceramic,

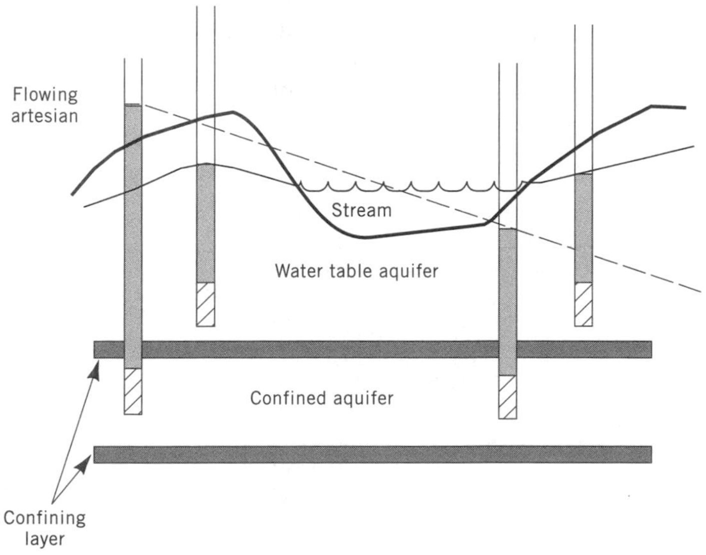

Figure 1. Aquifers and monitoring wells where ▨ denotes the well screen and ▤ the water-filled space in the monitoring well. (—) denotes the water table level, (— —) the potentiometric surface, and (**—**) the ground surface. Terms are discussed in text.

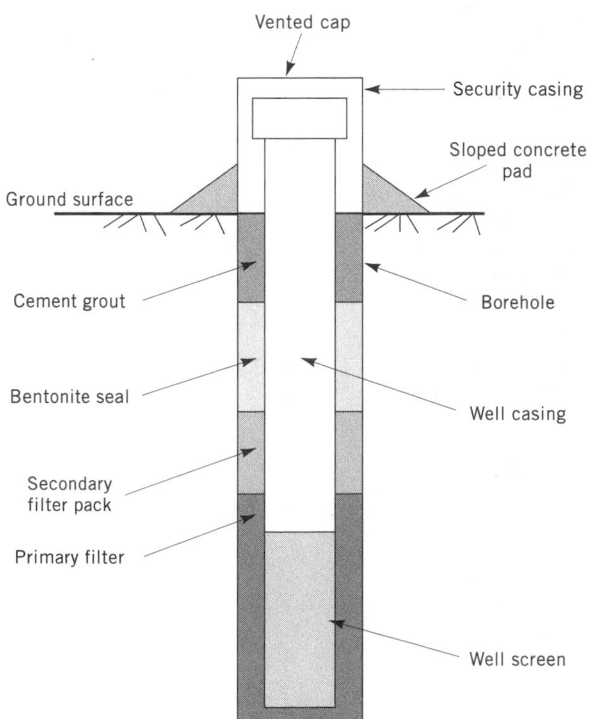

Figure 2. Schematic of a monitoring well.

having two access tubes which are usually Teflon. One access tube provides a pressure vacuum, the other discharges the sampled fluid to the surface. The porous cup, typically between 2 and 5 cm in diameter, is attached to a PVC sample accumulation chamber.

The installation of the probes should ensure good contact between the suction cup portion of the sampler and the surrounding soil, and minimize side leakage of liquid along the hole that has been cored for the sampler and access tube lines. Typically a clay plug of bentonite is used to prevent leakage down the core hole. A silica–sand filter provides good contact with the suction cup and prevents clogging of the cup. To retrieve a sample, the sample tube is clamped and suction is applied to the lysimeter through the air tube, which is then clamped. Moisture enters the accumulation chamber through the porous cup. The suction is released and pressure applied, forcing the sample to the surface through the sample collection tube.

Design of a groundwater monitoring program minimally includes consideration of materials, location, indicator parameters, and timing. Material selection is important for both the well casing and well screen. Materials of construction must be inert to the fluid being tested and to the ambient soil. The material must not release any type of chemical that could be interpreted as present in the groundwater. Typical inert materials include Teflon, polypropylene, PVC, and stainless steel (7,11). Material durability is also an issue, especially because many monitoring systems must be utilized for 50 years or more. The screens should also be evaluated regarding the potential for clogging, either via the porous media or biological activity.

Locational considerations include both surficial location and screened interval, ie, the sampling depth. The surficial location is selected based on whether the sample is to represent background quality or quality at the location of contamination, or potential leak location. In selecting the surficial location, the groundwater flow parameters, velocity and direction, are assumed to be known from other monitoring wells or borings already completed. The sampling depth is selected based on the type of contaminant monitored, ie, light or heavy, aqueous or nonaqueous, and/or the groundwater depth of interest (12,13).

Indicator parameters are those chemicals for which the water sample is analyzed. Often it is a simple matter to select the indicator parameter, if one suspects a discharge of a particular chemical. However, the situation is often much more complex. If monitoring wells are used to assess the occurrence of leachate leaks below a landfill, selection of the indicator parameters should be based on the expected chemical composition of the landfill leachate. In addition, the indicator parameters should be distinct from chemicals known to exist in the background groundwater (9). If the monitoring program is used for leak detection, the indicator parameter should be one that is expected to have an early arrival at the monitoring well, eg, the material having negligible adsorption.

Groundwater monitoring programs typically employ a routine schedule of sampling. Depending on the application, samples may be retrieved for analysis at weekly, monthly, quarterly, or other appropriate intervals. When the monitoring program serves the purpose of leak detection, as around the periphery of a landfill, wells are sampled quarterly. If contamination of an aquifer is known to exist, and monitoring wells are used to track movement of contaminants or the effectiveness of remediation efforts, sampling may occur more frequently. If monitoring wells are used in combination with a tracer test (8,9) to analyze flow characteristics, continuous sampling may be required.

Data analysis is aided by a variety of statistical techniques to assess significance, highlight trends, and form mathematical models of any correlations developed (4,14). It is never possible to design a groundwater monitoring program that samples an aquifer completely. Many pockets of unknown quality remain. A geostatistical technique, such as kriging (4,14), can, however, be used to determine an optimized estimate of groundwater quality at such an unsampled location using observed data from surrounding sampled locations.

BIBLIOGRAPHY

1. S. Karnchanawong, L. Koottatep, and T. Ilkeguchi, *Environ. Int.* **19**(6), 579–587 (1993).
2. S. Hasan, *Geology and Hazardous Waste Management,* Prentice Hall, Upper Saddle River, N. J., 1996.
3. M. J. Barcelona, H. A. Wehrmann, H. Allen, and M. D. Varljen, *Ground Water,* **32**(1), 12–22 (1994).
4. G. L. Grabow, C. R. Mote, W. L. Sander, J. L. Smoot, and D. C. Yoder, *Water Sci. Technol.* **28**(3–5), 327–335 (1993).
5. T. E. Reilly and A. W. Harbaugh, *Ground Water,* **31**(3), 489–494 (1993).
6. D. M. Nielsen, ed., *Practical Handbook of Ground-Water Monitoring,* Lewis Publishers, Inc., Chelsea, Mich., 1991.
7. C. W. Fetter, *Applied Hydrogeology,* 3rd ed., Macmillan Publishers, New York, 1994.
8. P. Domenico and F. Schwarz, *Physical and Chemical Hydrogeology,* John Wiley & Sons, Inc., New York, 1990.
9. M. Barcelona, A. Wehrmann, J. Kelly, and W. Pettyjohn, *Contamination of Ground Water: Prevention, Assessment, Restoration,* Pollution Technology Review No. 184, Noyes Data Corporation, Park Ridge, N.J., 1990.
10. J. Devinny, L. Everett, J. Lu, and R. Stollan, *Subsurface Migration of Hazardous Wastes,* Van Nostrand Reinhold, New York, 1990.
11. E. A. Boettner, G. L. Ball, Z. Hollingsworth, and R. Aquino, *Organic and Organotin Compounds Leached from PVC and CPVC Pipes,* U.S. Environmental Protection Agency Report, EPA-600/1-81-062, Washington, D.C., 1981.
12. J. F. Villaume, *Ground Water Monitor. Rev.* **5**(2), 60–74 (1985).
13. G. B. Wickramanayake, N. Gupta, R. E. Hinchee, and B. J. Nielsen, *J. Environ. Eng.* **117**(5), 686–691 (1991).
14. J. Davis, *Statistics and Data Analysis in Geology,* 2nd ed., John Wiley & Sons, Inc., New York, 1986.

GROUNDWATER, TRANSPORT AND REMEDIATION OF CHEMICALS IN

JAMES W. MERCER
HSI GeoTrans, Inc.
Sterling, Virginia

The ultimate goal at contamination sites is to eliminate or minimize the risks associated with contamination. Understanding chemical transport in groundwater leads to selecting a suitable remedy to manage the contamination. Contamination management varies from site to site and depends on numerous factors, including subsurface conditions, chemical type and distribution, and risks to human health and the environment. Contamination management may be active, such as a pump-and-treat technology, or more passive, such as intrinsic remediation in which monitoring is used to show natural attenuation of contamination. When characterizing a site and attempting to understand controlling transport processes, the ultimate remediation is the guiding factor. For example, if an anticipated remedy takes advantage of a chemical's volatile nature (such as soil vapor extraction), it is important to characterize the vadose zone above the water table. Other remedies will force consideration of other transport mechanisms and appropriate characterization. Considering the ultimate remediation during the investigatory stage ensures that appropriate and sufficient data will be available to make final remedy and design decisions.

DESCRIPTION OF AQUEOUS TRANSPORT PROCESSES

Three factors affect the movement and fate of dissolved chemicals in groundwater: advection of the chemical with water flowing through the subsurface; dispersion of the chemical; and sources and sinks of the chemical, such as chemical reactions, biodegradation, and adsorption onto the soil or rock matrix through which the water is flowing. The description of transport processes in this section includes advection, dispersion, retardation, and biodegradation.

Advection

Advection (or convection) is the movement of solute (eg, a contaminant) caused by the motion or flow of groundwater. Advection is the primary process by which solutes move in the subsurface. Because the chemical is dissolved in groundwater, the chemical moves with the groundwater. Whereas advection carries the solute, other processes such as dispersion and chemical and biological reactions normally tend to reduce the concentration of a solute as it moves with the flowing groundwater. Like most forces of movement, groundwater flows in response to an energy gradient, ie, it moves or flows from areas of higher energy to areas of lower energy. The amount of potential energy possessed by groundwater can be measured by a quantity called hydraulic head (h). Hydraulic head is the elevation to which water will rise in a well. Groundwater moves from areas of higher head to areas of lower head (down gradient) due to gravity. At shallow depths and unconfined conditions, the water table gradient generally conforms to the slope of the land surface. Geological conditions (type and sequence of soils and/or rocks) in the subsurface can have a major control on the direction and rate of groundwater movement. Compared with surface water flow, groundwater moves slowly. Surface water flow usually is measured in centimeters per second (ft/s) while groundwater flow is defined in meters per year (ft/yr). Streams flow freely, whereas groundwater follows tortuous paths through small openings in the soil and rock matrix, a matrix that resists the flow.

Direct measurement of groundwater velocity is difficult. Therefore, groundwater velocities are routinely estimated indirectly using measurements of hydraulic head, hydraulic conductivity K, and Darcy's equation, defined as $q = Ki$. Hydraulic conductivity is a function of both the media and the fluid flowing through the media; it is related to intrinsic permeability k by $K = k\rho g/\mu$, where ρ is density of water, g is the gravitational acceleration constant, and μ is the dynamic viscosity of water. Hydraulic conductivity and intrinsic permeability are measures of a subsurface layer's ability to transmit water. The greater the hydraulic conductivity, the less the resistance to flow, which will potentially yield higher velocities. Hydraulic gradient i is the change in hydraulic head over a given distance. The Darcy velocity q, also called specific discharge, is an average value representative of the entire cross-sectional area for which the hydraulic conductivity is determined. That is, Darcy's equation describes volumetric flow per unit cross-sectional area, which includes the soil and rock matrix. It does not represent the average velocity of the water through the pores. To obtain the velocity through the pores, or the interstitial velocity v (also called average pore velocity, average linear velocity, or seepage velocity), the Darcy velocity must be divided by the effective porosity. Porosity n defines the void space within the soil and rock matrix. When pores between the solid matrix are not connected, dead-end pores exist, and there is no groundwater flow. Similarly, the pore space within the lower conductive parts of the subsurface transmits water less actively than the pores in the higher conductive parts. Effective porosity refers to interconnected pore spaces and is generally less than porosity. Interstitial velocity applies to advective solute transport movement.

Considering advection alone (without other transport processes) is advantageous because only effective porosity and those parameters associated with groundwater flow need to be characterized. In general, these parameters are better understood and easier to obtain in the field than parameters associated with dispersion and chemical and biological reactions. Also, the equation describing groundwater flow is easier to solve than the equations describing dispersion and chemical and biological reactions. Other processes, however, must be considered to account for mass removal and contaminant attenuation.

Dispersion

If advection were the only process operating, nonreactive solutes would be carried at the average rate of water movement (using the seepage velocity). However, contaminants tend to spread rather than to move as a plug, which would be predicted if advection alone was responsible. Contaminant spreading and mixing caused by molecular diffusion and by velocity variations within the groundwater are referred to as dispersion. Mechanical dispersion is the spreading and mixing of a contaminant caused by variations in the water velocity (Fig. 1). Dispersion occurs at different scales. Microscopic dispersion results from variation in water velocity within pores (the velocity is higher in the center of the pore than next to the solid grains) and from the tortuous movement of water around the grains. This dispersion is observed in laboratory experiments. Macroscopic dispersion results from variation in water velocity caused by interbedding or interfingering of materials of differing hydraulic conductivity. In most field situations, macroscopic dispersion is more important than microscopic dispersion, especially in the direction of groundwater flow.

Except where water velocities are low (eg, some clays with low hydraulic conductivity), molecular diffusion (described by Fick's law) is smaller than mechanical dispersion. However, molecular diffusion is important in retarding solute movement

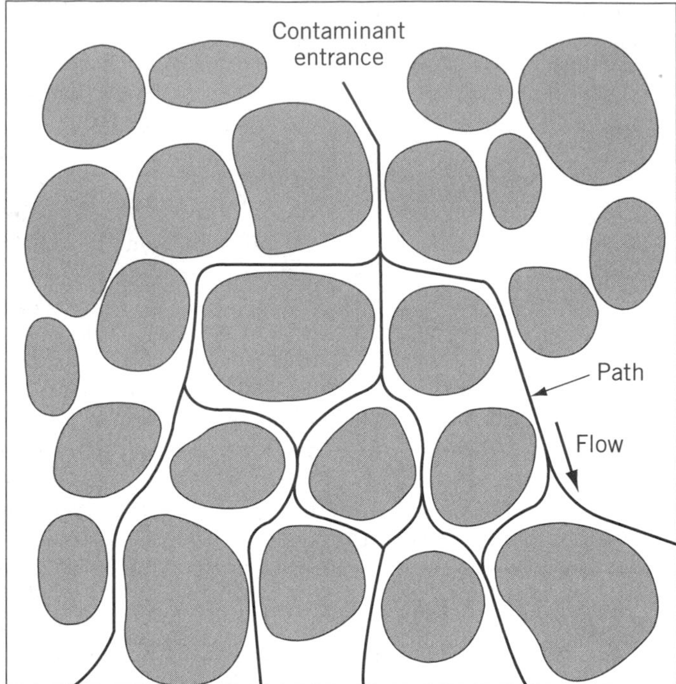

Figure 1. Dispersion, or the spreading of the contaminant plume as a result of mixing in the pores, affects the movement of contaminants. From Ref. 1.

if diffusion occurs from high permeability materials or fractures into materials of low permeability. This is referred to as matrix diffusion. Matrix diffusion causes an apparent retardation of the plume; however, as discussed later, when diffusion is reversed and contaminants move out of low permeability materials, matrix diffusion results in prolonged remediation, making low cleanup goals difficult to achieve. The hydrodynamic dispersion term is complex (2). For isotropic media, the dispersion coefficient is written as a combination of molecular diffusion and the product of dispersivities (longitudinal is in the direction of flow, and transverse is in the direction perpendicular to flow) and the seepage velocity. This form of the dispersion term is the one most commonly used in applications. However, the nature of the dispersion phenomenon and its precise mathematical form are still sources of debate. Values determined by observations from laboratory and field data are scale dependent. Longitudinal dispersivity values vary from a few centimeters for laboratory experiments, through a few meters for tracer studies, to tens or hundreds of meters for regional pollution problems. The commonly accepted explanation for this is that the longitudinal dispersivity coefficient is a measure of the heterogeneity scale not included in the analysis, ie, the larger the area, the larger the heterogeneity.

In general, a calibration process determines the amount of dispersion, adjusting the dispersivities until a match with measurements of concentration is obtained. This requires that concentrations be known over space and time. As opposed to this history matching on an existing contaminant plume to determine dispersion, geostatistical models of hydraulic conductivity may be used to estimate asymptotic dispersivities. The asymptotic longitudinal dispersivity is estimated by multiplying the variance of the log-transformed hydraulic conductivity ($\ln K$, where K is hydraulic conductivity) by the correlation length in the mean direction of flow (3). The correlation length is representative of the length over which $\ln K$ is correlated. This approach requires multiple measurements of hydraulic conductivity over space. It assumes unidirectional mean flow and an exponential covariance (ie, the correlation between any two hydraulic conductivity measurements decays with the separation or lag between measurement points). It also is generally assumed that the medium is stationary, ie, the mean, variance, and correlation length scales of $\ln K$ do not vary in space. The geostatistical approach offers the advantage of allowing longitudinal dispersivity to be estimated without any concentration measurements. The disadvantages include the requirement of multiple hydraulic conductivity measurements, the lack of a transverse dispersivity estimate, and geostatistical assumptions that may not be consistent with known geological conditions at a particular site.

Retardation

Removing solute from the active part of the groundwater system retards the movement of the concentration front relative to groundwater flow. Retardation occurs via several mechanisms (4), including precipitation and/or coprecipitation, sorption, and ion exchange. Although retardation mechanisms generally cause the contaminant to be relatively immobile, under certain conditions, small solid particles (colloids) may be transported with the moving water. This facilitated transport can mobilize contaminants that are thought to be immobile. Equilibrium contaminant partitioning between porous media and groundwater is described by the Langmuir or Freundlich isotherms, among others (5).

Biodegradation

Biodegradation refers to biologically mediated processes that chemically alter a solute or substrate (6). Primary biodegradation is any biologically induced structural transformation in the parent compound. Microorganisms are responsible for biodegradation and are abundant in most aquatic environments. Microbes can initiate chemical transformations because they supply energy for reactions through metabolic processes and catalyze reactions through enzymatic activity. Reactions can take place either in the presence or absence of oxygen. Reactions in the presence of oxygen are aerobic, whereas those in the absence of oxygen are anaerobic. In all biochemical reactions, the number of electrons must be conserved. When one reaction product has carbon atoms in a higher oxidized state through the loss of electrons, then another product must be present in a reduced state by having more electrons per carbon atom. Often the reduced product (electron acceptor) is not part of the original chemical or substrate. In aerobic reactions, oxygen is the electron acceptor and is reduced to water. In anaerobic systems, nitrate (NO_3^-) may be the electron acceptor and be reduced to NO_2^-, N_2O, or N_2. If sulfate (SO_4^{2-}) is present, it will also accept electrons and be reduced to H_2S. Even carbon dioxide (CO_2) is used as an electron acceptor by methanogenic bacteria to form methane (CH_4). The general reduction reaction may be written as follows:

$$\text{Substrate} + \text{electron acceptor} \xrightarrow{\text{biomass}} \text{products} + CO_2 + H_2O + \text{biomass} + \text{reduced electron acceptor}$$

Favorable chemical characteristics that promote biodegradation include small numbers of organic contaminants, nontoxic concentrations, diverse microbial populations, suitable electron-transfer conditions, and pH 6–8. Favorable hydrogeological characteristics include granular porous media, shallow contamination, hydraulic conductivity greater than about 10^{-4} cm/s, uniform mineralogy, homogeneous media, and saturated media.

Before the rate of biodegradation can be quantified and a rate constant calculated, a kinetic expression is needed to describe the pattern of loss over time. Two general rate laws have been proposed to describe biodegradation: the power rate law and the hyperbolic rate law. Both laws express rates of contaminant loss from the groundwater system. The power rate law states that the rate of biodegradation is proportional to some power of the substrate concentration.

The hyperbolic rate law is commonly used to describe the growth of microbial populations. When associated with enzymatic processes, it is known as Michaelis-Menten kinetics; for microbial growth, it is known as Monod kinetics. Based on Monod kinetics, this law expresses the biodegradation rate as a hyperbolic saturation function of the substrate concentration. Although the measured rate generally refers to population growth, it can be converted to a term describing the disappearance of the substrate supporting growth. The biological reaction rate often is modeled using the Monod relationship (6). The Monod equation becomes zero-order at high contaminant concentration and first-order at low contaminant concentration; these simplified equations often are used for modeling microbial kinetics for a well-defined concentration range. A dual-Monod expression that contains a term for the substrate (contaminant) and a term for oxygen often is used to describe aerobic biodegradation. Other more complex kinetic relationships that account for toxic inhibition, competitive inhibition, and noncompetitive inhibition have been used to model biodegradation processes.

Because biodegradation has the potential to remove contaminant mass, it can aid many remedial efforts (see below). There must be a close association between a microorganism and contaminant for biodegradation to occur. The contaminant must be available for uptake and use by the microorganism. Microorganisms and contaminants are distributed among the solid, liquid, and gas phases within the subsurface. Many organic contaminants are hydrophobic and tend to sorb onto soil such that only a small fraction of the compound may be in groundwater. Over long contact time, contaminants slowly diffuse into low permeability material, where they are not available to microorganisms. Most evidence indicates the uptake of compounds by bacteria proceeds via the water phase. Consequently, sorption, matrix diffusion, and volatilization that reduce the solution concentration tend to reduce the biotransformation rate. Cosolvency effects that make chemicals more mobile may also increase the biotransformation rate due to the increased plume spread. Dispersion may also play a role in bioavailability. Sites with high dispersion have plumes that will spread out, making it more accessible to microorganisms and oxygenated groundwater. This tends to increase the biotransformation rate.

Degradation rates are available for many chemicals (7,8). These are determined either through microcosm studies performed in the laboratory or via field studies on existing contamination or controlled pilot tests. A review of data indicated that field measured half-lives tend to be 4–10 times longer than laboratory-determined values, presumably due to scale-dependent rate limitations (9). A rate dependence on initial substrate concentration consistent with Monod kinetics, also was observed as part of the review.

MOVEMENT OF NONAQUEOUS PHASE LIQUIDS

Nonaqueous phase liquids (NAPLs) are associated with many contamination problems and often are sources for dissolved contamination. In addition, contaminants can dissolve into the NAPL and be transported with the moving NAPL. NAPLs are grouped according to whether they are denser than water (DNAPL) or lighter than water (LNAPL). An example of a LNAPL is gasoline in sufficient quantity to form a separate fluid. Examples of a DNAPL are chlorinated solvents, such as trichloroethene (TCE). The density of the NAPL is important, rather than the density of the separate components. For example, a mixture of oil and TCE may be an LNAPL or a DNAPL, depending on the amount of TCE present and the resulting mixture density.

As a separate fluid, a NAPL moves in response to pressure and gravitational gradients. Because more than one fluid occupies the pore space (water and NAPL in the saturated zone; water, air, and NAPL in the vadose or unsaturated zone), multifluid flow concepts are used to characterize NAPL movement. Reference 10 summarizes some of the important properties of the fluid and media. NAPL migration in the subsurface is affected by the following: (*1*) volume of NAPL released, (*2*) area of infiltration, (*3*) time duration of release, (*4*) properties of the NAPL, (*5*) properties of the media, and (*6*) subsurface flow conditions (11). The cross-sectional schematic in Figure **2a** depicts the distribution of organic chemicals in multiple phases resulting from a release of LNAPLs. When introduced into the subsurface, gravity causes the NAPL to migrate downward through the vadose zone as a distinct liquid. To some extent, lateral spreading accompanies this vertical migration due to the effect of capillary forces (12) and media spatial variability (layering) which is not shown in Figure **2a**. As the NAPL progresses downward through the vadose zone, it leaves residual liquid (at the residual saturation) trapped in the pore spaces. This entrapment is due to surface tension effects. In addition to the NAPL migration, some organics may volatilize or evaporate and form a vapor extending beyond the NAPL.

If the NAPL release is sufficiently large, some of the NAPL eventually reaches saturated groundwater, where density influences flow. An LNAPL spreads laterally along the capillary fringe (zone extending up from the water table to a point where soil moisture decreases), forming a complex distribution of water and LNAPL near the top of the water table. This distribution may be a lens that depresses natural groundwater levels. Soluble components may dissolve in the water and migrate because of the movement of the water. DNAPL displaces water and, given sufficient volume, continues its vertical migration until it encounters a layer of low permeability (see Fig. **2b**). Here it moves under pressure and gravity forces along the confining layer. As in the vadose zone, some of the DNAPL will be held in the pore space within the saturated zone. This residual DNAPL is a source of contaminants to a dissolved plume, and depending on the aqueous solubility of the organic compounds, the DNAPL source can last for decades. In the vadose zone,

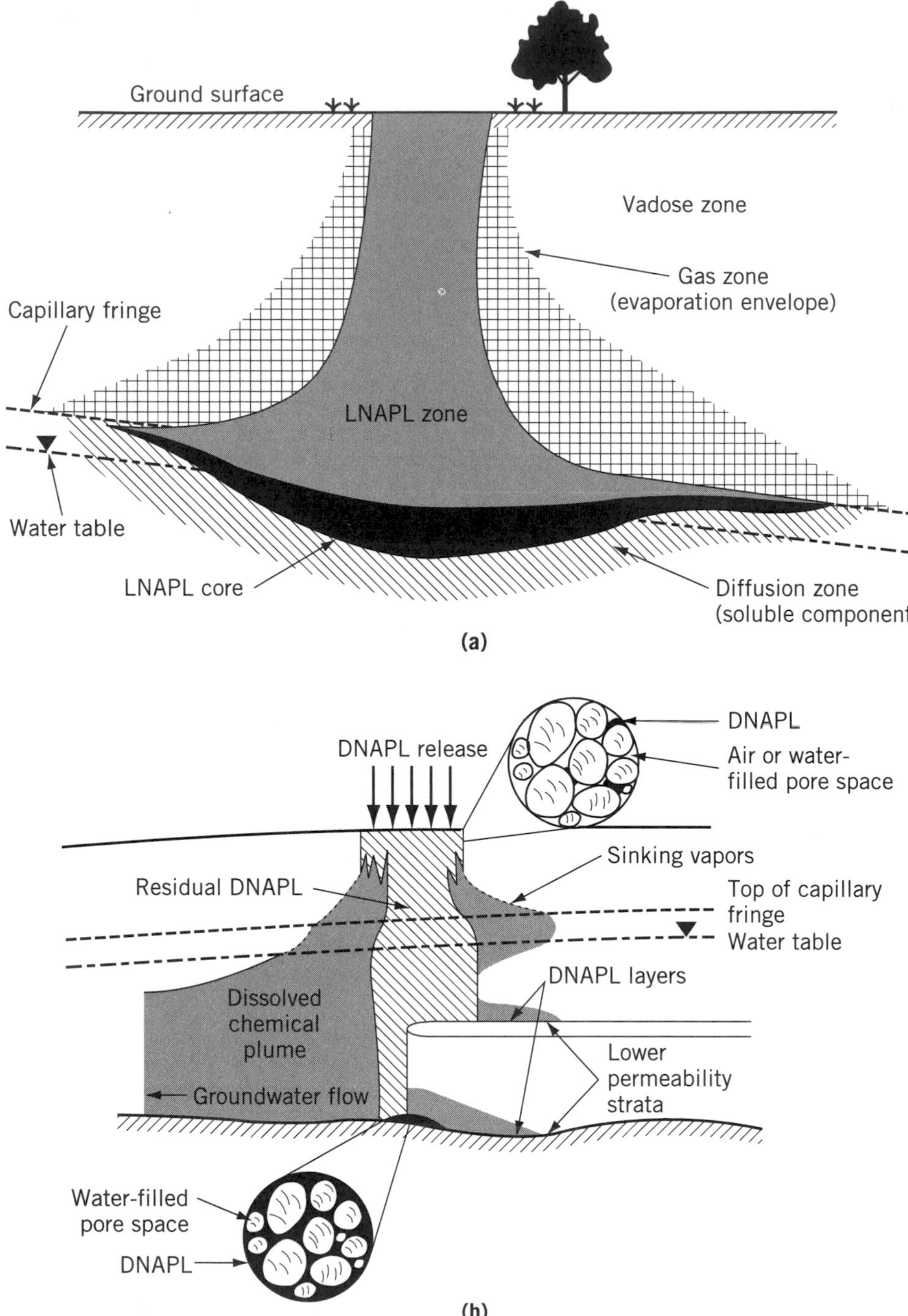

Figure 2. (**a**) LNAPL infiltration schematic. (**b**) Migration to low permeability. From Ref. 11.

infiltrating rainwater may dissolve organic vapors or the residual NAPL and transport these organic components to the saturated region. According to a panel of experts, no proven technologies exist to restore DNAPL zones to drinking water standards (13). More information on NAPLs is available (14–16). Field techniques applied to NAPL sites have been discussed (17), and modeling or simulation approaches for NAPL problems have been reviewed (12).

TRANSPORT PROCESS CHARACTERIZATION

Site characterization, a process following the scientific method, is performed in phases. In the initial phase, a hypothesis about the site or system behavior is made. The hypothesis should identify sources of contamination, whether or not NAPLs may be present, what processes control transport, and the likely distribution of chemicals based on site history and transport mechanisms. Based on the hypothesis, a data collection program is designed in the second phase. Data are collected and an analysis or assessment follows. Results of the analysis are used to perform an iterative step of refining the hypothesis; additional data may be required. As the knowledge of the site increases and complex site conditions are better understood, the working hypothesis may be expanded into the form of either a numerical or analytical site model. Data collection continues until the refined hypothesis is proven sufficiently. The proof

Table 1. Chemical Properties and Information That Influence Fate and Transport in the Subsurface

Property and Information	Influence on Behavior
Aqueous solubility	Determines degree to which chemical dissolves in water; compounds with low solubilities tend to be NAPLs; solubility indicates possible concentrations (max); high solubility indicates low sorption tendencies; therefore, more soluble compounds tend to migrate farther and are more amenable to pump and treat
Chemical speciation	Determines charge and distribution of ions and mobility of inorganic chemicals; ion valence influences whether chemical dissolves and, therefore, is mobile; toxicity also varies according to valence; valence and chemical distribution may also be indicators of anaerobic biodegradation
Henry's law constant	High values may signify volatilization from aqueous phase as an important transport process; used in conjunction with vapor pressure; is directly proportional to vapor pressure of compound and inversely proportional to aquenous solubility of compound; volatile chemicals amenable to soil vapor extraction and air sparging
Vapor pressure	Provides an indication of extent to which compound volatilizes; tendency of compound to volatilize rises proportionately with vapor pressure
Density	For high concentrations, density of contaminated fluid may be greater than density of pure water causing downward vertical movement of contaminants; density of separate-phase chemicals (NAPLs) determines whether they are DNAPLs or LNAPLs
Octanol–water partition coefficient	Indicates chemical's tendency to partition between groundwater and soil; ratio of concentration of compound in octanol to concentration of compound in water; large coefficient signifies a highly hydrophobic compound, indicating strong sorption, and providing similar information as that provided by solubility
Organic carbon partition coefficient	Indicates chemical's tendency to partition between groundwater and soil; ratio of amount of chemical sorbed per unit weight of organic carbon in soil to concentration of chemical in solution at equilibrium; greater organic carbon partition coefficients = greater tendency for compound sorbtion
Half-life and degradation process	Provides information regarding persistence of chemical and which, if any, transformation products might be expected; half-life implies a first-order decay where smaller half-life value – faster chemical degradation

forms the basis for decision making. The ultimate goals of site characterization are to collect sufficient and appropriate data such that informed decisions can be made; to assess the background water quality; to establish the impact of certain facilities, practices, or natural phenomena on water quality; and to predict future groundwater quality trends under various remedial conditions. Whatever the objective, however, assessment of the groundwater flow system and of groundwater chemistry must be made to characterize the site. Since the early 1980s, extensive work has been performed on site characterization tools and techniques (see *General References*).

Flow System Characterization

The groundwater flow system is a key element affecting solute transport. This includes the physical parameters of the hydrologic region influencing the contaminated site (eg, hydraulic conductivity, porosity, and aquifer thickness), system stresses (eg, recharge and pumping rates), and other system characteristics (eg, physical and hydraulic boundaries, groundwater flow directions and rates, and the geology that controls the flow system geometry). Because migrating miscible contaminants travel with moving groundwater, it is important to characterize groundwater flow. Groundwater flows from areas of recharge (commonly via rainfall, surface water bodies, or irrigation) to areas of discharge (surface water or wells). Along the way, subsurface heterogeneities (such as faults) influence its direction. Groundwater velocity, determined from Darcy's equation, is controlled by the porosity and hydraulic conductivity of the media through which it travels and by hydraulic gradients, which are influenced by recharge and discharge. Often it is important to conduct a site characterization quickly; however, groundwater flow systems vary with time. Seasonal variations in water levels, often several feet in magnitude, can affect remediation and should be characterized. Much of this information is obtained by drilling wells to observe the subsurface via cuttings and cores. Water levels are measured to determine flow directions. System geometry can also be inferred from some surface geophysical techniques. Slug tests and aquifer tests are conducted and analyzed to determine hydraulic conductivity.

Contaminant Plume Characterization

To remediate a site, the contaminant distribution must be determined, along with chemical characteristics that affect transport. Contaminant information includes source characterization, concentration distribution of contamination and naturally occurring chemicals, and data associated with the processes that affect plume development. The chemical volume released, the area infiltrated, and the time duration of release constitute source characterization. Often, the release occurred so long ago that source information is difficult to obtain. Chemicals of concern need to be identified based on distribution and associated risk. Determining the chemical properties that affect transport may be done by researching the properties in chemical handbooks. Table 1 presents important chemical properties and information and how they influence fate and transport in the subsurface.

Quantitative characterization of subsurface chemistry can include sampling the vadose and saturated zones to determine the concentration distributions in groundwater, soil, soil water, and soil gas. A network of monitoring wells (also necessary for the hydrogeologic data) is required to

Table 2. Recent Innovations in Site Characterization

Technology	Description[a]	Advantages and Limitations	References
Direct push			
Cone penetrometer	cone penetrometer truck that uses instrumented probe forced into ground; penetration force supplied by hydraulic cylinders; inside probe, 2 load cells independently measure vertical resistance against conical tip and side friction along sleeve; probe can be coupled with LIF device, used to delineate contaminant plume	Penetration, dissipation, resistivity data used to determine site stratigraphy based on penetration pore pressure (function of fluid compressibility and formation permeability); continuous soil sampler can be used to verify penetration pore pressure interpretations; no cuttings brought to ground surface, eliminating need to handle cuttings; process presents a reduced potential for vertical cross-contamination if the openings are sealed with grout from the bottom up upon rod removal; the push capacity may be limited in tight soils or by the presence of hard clays, caliche, or boulders	18–21
Mini wells			
Hydropunch	Sampling device designed to be pushed or driven to desired sample depth, from either ground surface or bottom of drilled borehole; designed to minimize downward transport of contaminants as tool is advanced and uses an air- and watertight sealed intake screen and sample chamber	Can be used to sample both groundwater and free product and provides 1.2-L sample volume; depth to water can also be determined; advantages include lack, thus far, permitting requirements, low visibility with respect to full-size drill rigs, lower overall costs, rapid collection of groundwater samples, and ability to access difficult locations; detractions include data reliability (generally provides one-time sample), inability to provide adequate hydrogeologic information, and constraints on installation with respect to bedrock, areas with caliche or boulders, and clay-rich soils	21–24
Geoprobe	Driven sampling tool; disposable driven tip used, and after reaching desired depth, tubing is threaded through hollow sampling rod and secured to drive point; sampling rod then pulled back to allow groundwater sampling and water depth measurement	Many of the same advantages and limitations as Hydropunch; can be used to sample all subsurface media including soil, groundwater, and soil gas; has capability to insert permanent sampling implants and has been used in sands ≥30m	25,26
Micropurge sampling	Water samples collected at low flow rates, (ie, 100 mL/min) using dedicated sampling pump to prevent disturbing well; samples collected directly from well screen interval are representative of groundwater in surrounding formation	Only sampling tubing and pump are purged (~2 vols to ensure complete removal of stagnant water), thus purging of multiple wellbore volumes associated with conventional sampling eliminated as are large volumes of purge water requiring disposal; where there are a large number of wells, costs associated with storing, testing, and disposing purge water are considerable; does not disturb well introducing colloids and particles into sample, thus sample does not require filtering so representative samples are obtained of total mobile contaminant load of both dissolved and sorbed species	27–31
Chemical sensors			
Probes and electrodes	Various probes available that measure dissolved oxygen, oxidation–reduction potential (*Eh*), and pH in bore-hole fluids, individually, or in combination; ion-selective electrodes developed for ammonia, bromide, calcium, chloride, fluoride, hydrogen sulfide, and nitrate	Advantages: *in situ* measurements less likely to reflect chemical alteration due to pressure changes than measurements taken of water sample brought to surface, combination probes that measure several hydrochemical parameters simultaneously greatly simplify collection, and real-time measurements that are cost-effective; ASTM describes standard terminology for electrodes, measurement technique, and conditions affecting measurements; disadvantages: proper calibration difficult due to interference from different constituents present in many groundwaters, some parameters may inhibit electrode output, and constituents for which specific electrodes have been developed are limited	32–34

Table 2. *(continued)*

Technology	Description[a]	Advantages and Limitations	References
Immunoassay	Based on animal immune systems (produce antibodies that recognize unique molecular structures of foreign proteins, viruses, or bacteria and allow antibodies to destroy invaders) prepared samples and enzyme-labeled analyte placed in antibody-coated tube and allowed to incubate; later, color development reagents added and results measured using colorimeter	Advantages: portability allows on-site analysis; analysis is simple, rapid, and cost-effective; regulatory acceptance for TPH, PCBs, and PCP; disadvantages: each analyte requires different reagents and test procedures, relatively high detection limit, strict quality control may be difficult in field	36
Fiber optics	FOCS also known as remote LIF, RFS-FOCS, immunochemical fiber optic sensors; are made of a reagent phase, which is physically confined or chemically immobilized at end of optical fiber; reagent phase contains chemical or immunochemical indicator that changes optical properties (ie, absorbance or fluorescence) when interacts with analyte; optical fiber is a strand of glass or plastic, ranging from 2–200 πm and acts as conduit to propagate light to and from FOCS; FOCS placed in subsurface using cone penetration rig or into groundwater monitoring well; fiber optic cable attached to spectrophotometer or fluorometer, containing a light source (light bulb or laser) and detector; excitation signal from light source transmitted down cable to FOCS and sensor fluoresces and provides constant-intensity light source that is transmitted back up cable and detected as return signal; if target contaminant present, intensity of return signal reduced and intensity of light recorded by detector is inversely proportional to concentration; fiber optic sensors also used with colorimetric bore-hole dilution techniques	Advantages: provides selective *in situ* real-time measurements in field, eliminates sample handling and chain-of-custody concerns, has potential for specific detection of large number of specific organic compounds, sensors can be placed in small bore holes (1.3 cm dia); reducing drilling and monitoring well installation costs) or used with cone penetration rigs for rapid field screening, field instrumentation potentially portable (small enough to fit in coat pocket), and has potential for greatly reduced costs compared to conventional sampling and analytical methods for organic contaminants; disadvantages: new technique with limited operational and field experience, equipment not yet readily available, field performance of field LIF poorer than laboratory results (perhaps due to temperature fluctuations and affect of increased vibration on optics), numerous separate sensors required for discrimination between specific compounds, and turbidity might interfere with readings	37–46

[a] LIF = laser-induced fluorescence; FOCS = fiber optic chemical sensors; RFS = remote fiber spectroscopy

be installed to collect depth-discrete groundwater samples. Because drilling is invasive, less intrusive techniques, such as surface geophysics and soil gas surveys, are used initially to aid in well siting. Locate wells in areas that supply information on ambient (background) groundwater chemistry and on plume chemistry. At a minimum, soil and groundwater samples are analyzed for the parameters of concern from the waste stream. A full-priority pollutant scan on the first round provides information on plume chemistry and may be useful in differentiating plumes that have originated from several different sources. On subsequent rounds, the parameter list may be refined on the basis of site-specific considerations.

If inorganic contaminants are of concern or interest, analyze a full suite of major and minor ions. Field measurements of pH are especially important. After analyzing the samples, the resulting concentration data should be mapped in three dimensions to determine the spatial distribution of contamination. These plume delineation maps and the results from aquifer tests yield estimates of plume movement and identify locations for extraction wells.

As indicated, traditional sampling involves drilling and purging wells, sampling, and laboratory analysis. This process is expensive and time-consuming; proper disposal of purge water is necessary, and generally, weeks are required to obtain the laboratory analysis. To reduce costs and obtain chemical analysis results in the field, several innovations have been developed. As indicated in Table 2, these innovations include direct push technologies, micropurge sampling, and a variety of chemical sensors. The direct push technologies include cone penetrometer, Hydropunch and Geoprobe. When wells are

Table 3. Biological Sampling and Analysis

Sampling and Analysis	Description	References
Aseptic sampling	Goal: to obtain representative subsurface sample not contaminated by microorganisms from other locations; to do so, (*1*) sampling tool (eg, hollow-stem auger) must be sterilized, (*2*) discard upper 10–12 cm of recovered core obtained, eg, using shelby tube, and (*3*) outside portion of core must be removed or inside of core sampled with sterilized instruments; for microcosm studies, samples below water table should be collected in area with dissolved-phase contamination; care should be taken to prevent volatilization, and sample should be stored at 4°C for transport	47–49
Microbial enumeration	Methods include heterotrophic plate counts, epifluorescence direct counts, hydrocarbon-degrader plate counts, most probable number (MPN) techniques, DNA probes, lipid assays, metabolic indicators, and radioisotopic methods; direct counts include both metabolically active and inactive organisms; heterotrophic plate counts theoretically measure only the fraction capable of growth; MPN techniques and DNA probes indirectly measure microbial density through chemical changes brought about due to metabolite production or extraction and quantitation of cell genetic material (DNA probes, sensitive to 4.3×10^4cells/g soil); lipid assays quantify fatty acid content of cell suspensions and relate this to total microbial population; adenosine triphosphate (ATP) assessment methods use ratio of ATP and adenosine diphosphate (ADP) to total pool of available adenylate as a measure of microbial activity; metabolic indicator methods include tetrazolium dye reduction and measurement of ratio of cell ATP: total adenylate available; radioisotopic methods measure kinetics of cell uptake of tritiated or ^{14}C-labeled compounds (tritiated thymidine or acetate) and translate uptake into cell activity estimates; results generally reported in colony-forming units (CFUs) per gram dry weight soil or CFUs per mL groundwater; cell counts higher in shallow groundwater than in vadose (below root) zone with numbers $> 10^6$CFU/g dry soil observed	9,48,50–57
Microcosm studies	Performed in laboratory to determine transformation rates and products; procedures for conducting anaerobic microcosm studies specified by EPA; collected subsurface material used along with sterile controls; active microcosms and control microcosms sacrificed and analyzed for residual test chemical and formation of degradation intermediates	49,58–60

required, micropurge sampling can replace conventional sampling, which requires the purging of multiple well volumes. Finally, in-field and immediate chemical analyses can be obtained using probes, ion-selective electrodes, immunoassay, and fiber optic chemical sensors.

Biological Characterization

The importance of biodegradation as a process that removes contaminants necessitates more emphasis on characterizing the microbial population in the subsurface. Table 3 discusses biological sampling and analysis methods and includes aseptic sampling, microbial enumeration, and microcosm studies. In addition to this information, other important data include plume behavior over time, distribution of electron acceptors, and appearance of degradation products.

Data Analysis

At many hazardous waste sites, large amounts of data are collected. Tools such as Geographic Information Systems (GISs) and models analyze these data. GISs have grown in importance as tools for solving environmental problems. GIS is a computer software system that stores and links nongraphic attributes or geographically referenced data with graphic map features to allow a wide range of information processing and display operations as well as map production, analysis, and modeling. GIS groundwater applications include data management, analysis, and visualization; integration of diverse data sources; rapid development, visualization, and testing of alternative hydrogeologic and hydrologic system models; incorporation of contouring, attribute tables, and statistical analysis during data analysis and model development; and integration of basic and interpreted databases with numerical modeling for purposes of model design, parameter discretization, statistical evaluation, and sensitivity analysis (71).

Mathematical models play many roles in contaminant hydrology. They serve the important function of codifying knowledge of the physical, chemical, and biological processes that control the transport and fate of contaminants in groundwater. In practical, site-specific applications, mathematical models function as important decisionmaking tools. In this role, models are used to reduce the uncertainty inherent in decision making by providing a rational, logical, self-consistent structure for data collection, site characterization, hypothesis testing, quantification of uncertainty, risk assessment, and the design and evaluation of corrective or remedial actions. During site characterization, a hypothesis or conceptual model is formulated through a preliminary understanding of physical, chemical, and biological processes. The conceptualization often is incorporated into a numerical groundwater model to keep track of all the necessary site information effectively and to ensure that the conceptualization is self-consistent and reflects current knowledge of physical, chemical, and biological processes. In this manner, use of the model can help guide both system conceptualization and data collection activities.

The model subsequently helps identify pathways and potential receptors, which is the exposure assessment portion of a risk assessment. Solute transport models, for example, can be used to compute concentration versus time at a drinking water well. Groundwater models can also be used to evaluate the effectiveness and associated risk reduction of proposed groundwater remediations before they are implemented. This can include the effects of wells, drains, and permeability barriers. In this way, models can aid in the selection and design of remediation techniques. Once the remediation is implemented, models can be used to help evaluate and improve operation of the remedy. Finally, models can be used to help estimate the impact of terminating the remediation. Numerous groundwater flow and transport models exist (62–69). Groundwater modeling standards have also been established by ASTM (D5447-93 and D5490-93).

GROUNDWATER REMEDIATION AND TRANSPORT PROCESS EFFECTS

Contaminant transport processes depend on chemical behavior and subsurface conditions. The contaminant may be present as trapped NAPL, as mobile NAPL, sorbed to the soil or rock matrix, in soil gas, and/or dissolved in groundwater. Contaminant location and contaminant behavior (how long it takes to migrate or degrade) in the subsurface dictate which remediation can be used successfully to attempt cleanup. Contaminant location may necessitate distinguishing between source areas and a dissolved plume. Consequently, some remediations may be more appropriate for the source areas, whereas other remediations may be more appropriate for the dissolved plume. Managing the entire contamination problem may require a combination of remedial actions. This section discusses a variety of remediations. Remedial actions can be subdivided according to whether they are applied to groundwater, soil, or contaminant source areas. This section focuses primarily on groundwater remediation. The remediation technologies are organized in terms of removal, containment, and transformation. Each technology is briefly described, including its advantages and limitations.

Contaminant Removal

Conventional Pump and Treat. A common remediation for contaminated groundwater is to extract the contaminated water via wells or trenches (drains) and treat it at the surface before discharge or reinjection (Fig. 3). This is referred to as conventional pump-and-treat remediation to distinguish it from enhancements to groundwater extraction, such as combining it with a soil vapor extraction (SVE) system. An overview of pump-and-treat groundwater remediation (70) indicates that conventional pump and treat is best suited for homogeneous media having a hydraulic conductivity greater than about 10^{-4} cm/s, ie, it works well for contaminants that are easily advected. This technology, if designed properly, hydraulically contains and controls the movement of contaminated groundwater, which prevents continued expansion of the contaminated zone, and reduces dissolved contaminant concentrations. When cleanup is the goal, concentration targets are in the low micrograms per liter or parts per billion range, which makes achievement difficult.

The advantage of pump and treat is that a large experience base has been developed, rendering it a proven technology with straightforward design. It effectively contains contamination (hydraulic containment) by pumping groundwater from wells or drains (71). Fluid injection (using wells, drains, or surface application) and physical containment options (such as subsurface barrier walls and surface covers) can enhance hydraulic containment systems. However, widespread experience with pump-and-treat systems since the early 1980s indicates that complex hydrogeologic conditions, contaminant chemistry factors, and inadequate system design (71–74) hinder the ability to clean the subsurface in a reasonable time frame (eg, <10–20 yr) at many sites. Levels of contaminants in extracted groundwater generally decline quickly in the early stages of pump and treat, then decrease asymptotically toward some limiting value (usually above the cleanup goal) with greater operational time. This is referred to as tailing. In addition, if

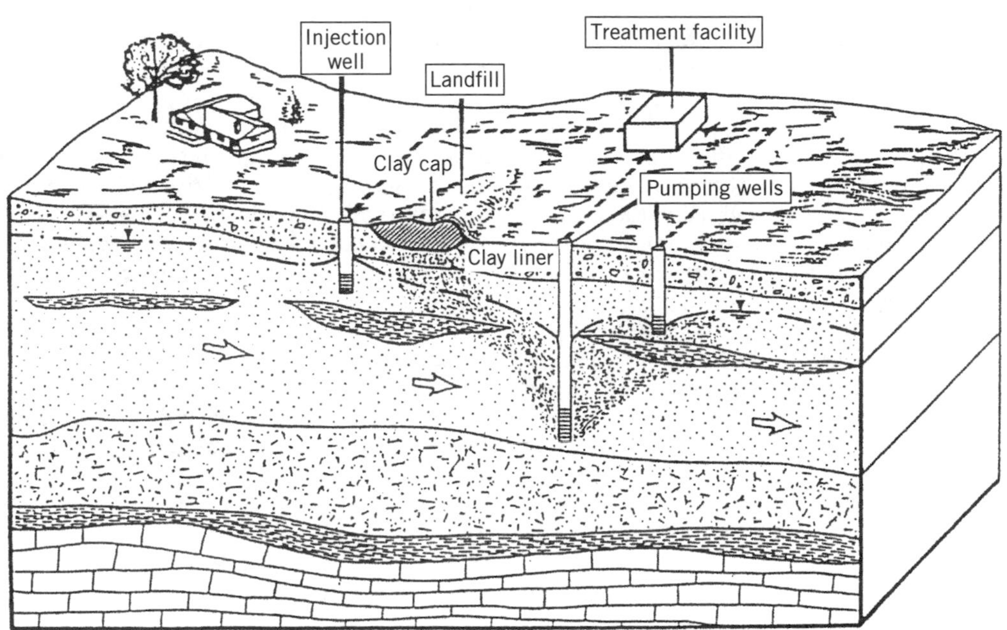

Figure 3. Example of a pump-and-treat system at a landfill. From Ref. 70.

the extraction system is turned off for some time, dissolved concentrations often increase (rebound).

Hydrogeologic conditions that confound groundwater cleanup include complex (heterogeneous) sedimentary deposits, low permeability formations, and fractured bedrock. Chemical processes that cause contaminant concentration tailing and rebound during and after pump-and-treat operation, respectively, and thereby impede aquifer restoration, include the presence and slow dissolution of NAPLs, contaminant partitioning between groundwater and the media through which it is flowing, and contaminant diffusion into and out of low permeability regions that are inaccessible to flowing groundwater. For example, the release of contaminants from areas of relatively low hydraulic conductivity zones is significantly rate limited (75). In addition, nonequilibrium effects on mass transfer can have pronounced effects on cleanup times (76). DNAPLs in particular have exacerbated groundwater restoration efforts. This is due to their prevalence at contamination sites, their complex subsurface migration behavior and distribution, their low aqueous solubility, and limits to DNAPL removal using available technologies (13,15,16). Greater success has been achieved remediating petroleum hydrocarbon LNAPLs using pump-and-treat and enhanced technologies such as SVE, bioremediation, and air sparging.

Inadequate design and implementation of a pump-and-treat system can severely impact groundwater cleanup. Examples of design inadequacies include too few recovery wells, insufficient pumping rates, deficient well locations or completion intervals, and failure to account for contaminant chemistry factors. Similarly, poor system operation, exemplified by excessive downtime and failure to manipulate pumping schemes to limit groundwater stagnation, restricts pump-and-treat effectiveness. Proper design of pump-and-treat systems has been discussed (77). Pulsed pumping, with alternating pumping and resting periods, has been suggested as a means to address tailing, flush stagnation zones by selective well cycling, and increase pump-and-treat efficiency (74,78,79). Dissolved contaminant concentrations increase due to diffusion, desorption, and dissolution in slower-moving groundwater during the resting phase of pulsed pumping. Once pumping is resumed, groundwater with higher concentrations is removed, thus increasing the rate of mass removal during active pumping. Slow mass transfer from immobile phases to flowing groundwater causes contaminant concentrations to decline with continued pumping until the next resting phase begins. Advantages of pulsed pumping are somewhat limited. The disadvantage is that hydraulic containment may be lost during the resting period. Proper operation of a pump-and-treat system, however, likely involves variable pumping of various extraction wells to optimize contaminant recovery.

Soil Vapor Extraction. SVE, also called soil venting, is considered a proven technology that involves the enhanced removal of chemicals from the subsurface (unsaturated or vadose zone) through application of a vacuum (71). Instead of removing dissolved contaminants in groundwater as pump and treat does, SVE removes soil gas containing volatile contaminants (71). The applied vacuum enhances volatilization of compounds from soil and pore water by creating air flow. The technology particularly applies to relatively volatile organic compounds (Henry's law constant $<10^{-3}$ (atmm^3)/mol) residing in the vadose zone (80)). The contaminant's boiling point must be <150 °C or its vapor pressure $>5 \times 10^{-4}$ atm (81,82). The technology also applies to removal of volatile LNAPLs floating on the water table or entrained in the capillary fringe, where the chemicals of concern have high vapor pressures. An advantage of SVE is that flushing air through the subsurface is generally easier than flushing water. This yields higher flow rates and more air volume flushed in a given time frame compared to water. In addition, the air contains oxygen, which may promote aerobic biodegradation of some contaminants (see SOIL VAPOR EXTRACTION).

The SVE process involves installation of vacuum extraction wells or trenches at strategic locations and depths. The SVE system usually consists of these wells as well as vacuum pumps, condensate traps, and a treatment system for the extracted vapors. To avoid short circuiting air flow from the surface, the ground may be covered with a low permeability cap. Air extraction can be combined with air injection. Spacing wells or trenches depends on soil properties, such as permeability and porosity. The technology applies to most soil types; although, just as with pump and treat, removal efficiency generally decreases with decreasing soil permeability and increasing subsurface heterogeneity. Conditions in which SVE is more successful include soil air conductivity $>10^{-4}$ cm/s, soil temperature >20 °C, moisture content <10 % vol, soil surface area <0.1 m^2/g, carbon content <1 % wt, and contaminant water solubility <100 mg/L (78).

During SVE, decreasing soil moisture and low soil organic carbon enhances contaminant removal. As percent moisture decreases, air permeability increases, allowing more air flow. For higher moisture content, volatile chemicals may tend to partition into soil water, rather than remaining in a vapor state, thus contaminant removal efficiency decreases. Increased soil organic carbon content increases sorption to the soil matrix, which decreases removal efficiency. When soil is dry, however, volatile chemicals may also sorb directly onto mineral surfaces, where the magnitude of sorption is increased and, consequently, volatilization is decreased. Henry's law, which describes partitioning between air and contaminants dissolved in water, is not appropriate under these conditions because partitioning is between air and soil phases. When moisture is added to soil, the effect is reversible (84). For contamination by hydrocarbons with multiple components (eg, gasoline), volatilization proceeds such that lower molecular weight chemicals more volatilize preferentially than do higher molecular weight compounds. Through this process of contamination weathering (selective volatilization and dissolution), SVE extraction efficiency decreases. As the lower molecular weight compounds are removed, the boiling point of the mixture increases, requiring either heat input or higher vacuum to maintain extraction efficiency.

Heterogeneous flow conditions also affect the efficiency of contaminant removal, with higher flow zones (preferential flow zones) cleaning up faster than low flow zones (less permeable zones). For low permeability sediments, volatile chemical removal is initially limited by evaporation and advection but is later diffusion limited, as characterized by nonzero asymptotic mass removal rates (85). The effects of permeability heterogeneities on SVE have been evaluated; a rebound of concentrations was observed when extraction wells were temporarily turned off, with an equilibrium concentration determined by Henry's law constant (86). When blowers were turned on, an increase in the concentration of extracted vapor from the soil was observed. As with pump and treat, diffusive release from less permeable zones causes the slow continual release of chemicals into the soil gas (87), prolonging cleanup.

Air Sparging. Air sparging generally uses wells to inject gas (typically air) into the saturated zone into and below (no less than about ~1.5 m) contaminated areas (71). Ideally, dissolved, separate-phase and sorbed contaminants partition into the injected air, effectively creating an *in situ* air-stripping system. This can take place within a single-well system (87) or the stripped contaminants can be transported in the gas phase to the vadose zone and collected by SVE wells (88). The advantage of such a system is that the treatment of groundwater and soil takes place *in situ,* reducing the need for disposal of treated material. The injection wells can be vertical or inclined, ranging to horizontal (89). Although air sparging is a physical and chemical treatment process, the addition of air can potentially promote biodegradation (90,91).

Air sparging is used when volatile contaminants are below the water table. Alternatively, the water table could be lowered and the newly exposed vadose zone treated using SVE. The air sparging technology has been used in Germany since the mid-1980s (92). Air generally is injected outside the well(s) in short screened intervals (0.3–1.5 m). One proposed use of air sparging involves injecting air within a well (93). This causes water in the well to be lifted and stripped, forcing groundwater toward the well, creating a recirculating cleanup zone. The zone of influence of such an in-well method is limited, depending on the size of the well and the flow system. After it was field tested (94), this approach was concluded to efficiently remove VOCs from the test well; however, problems were encountered lifting water out of the borehole and saturating the treated water with dissolved oxygen. High dissolved oxygen can cause iron precipitation and may have corrosion implications.

Design parameters include contaminant type and site hydrogeology and geochemistry. Air flow rates typically range from 3–20 ft^3/min (cfm) per sparge point, with injection pressures 1–2 lbs/in.^{2} (psi) in excess of the hydrostatic head at the top of the well screen (for coarse soils). The compounds most amenable to air sparging are volatile and meet the condition for SVE. As with most other *in situ* methods, air sparging works best for homogeneous conditions and where sufficient permeability allows ease of flow. Heterogeneities allow for preferential flow, bypassing zones of contamination. For example, low permeability lenses contribute to the isolation of zones above the lenses within a contaminated aquifer. Precipitation of dissolved minerals, affecting permeability, is a concern due to changes in oxidation reduction (redox) potential.

The physics of multiphase flow, which requires additional data, partially governs the air flow through the saturated porous media. Part of the physics of multiphase flow concerns capillary pressure effects. Air sparging often is mischaracterized by disconnected bubbles flowing as if through a pot of water. In a porous media with multiple fluids, this situation generally does not occur. The air bubbles must be connected in order to flow. If they become disconnected, the bubbles become trapped by capillary pressure effects. As a result of the connected bubbles, channeling air flow is highly likely. This channeling is influenced by media heterogeneity, leading to an asymmetric sparging zone around a sparge well and to bypassing (discussed above). Pulsing the injected air has been suggested as a way to overcome some of the problems associated with channeling. Air injected into the saturated zone displaces the water table. The movement of this displaced water can lead to a rise in the overlying water table. This is sometimes called water-table mounding and has been used by some as an indicator of the radius-of-influence of the sparge well (88). Mounding is also a design concern because it represents a driving force for lateral movement of groundwater, dissolved contaminants, and any LNAPL mounded on the water table. An interesting aspect of the changes in water table during sparging is that the water table sometimes drops below its initial position after a few hours of continuous sparging. This phenomenon develops during the collapse phase (95) and is thought to be related to the establishment of preferred air flow pathways within the aquifer. Within this region, groundwater is trying to flow back toward the sparge well, following the establishment of a preferred air flow path to the water table.

In summary, air sparging can remove contaminant mass but has several potential problems, including mass transfer tends to be limited by diffusion, preferred pathways develop that result in air stream tubes bypassing the contamination (limited or no treatment concerns), heterogeneities (especially horizontal layering) greatly effect the air movement (for in-well sparging, this could cause a lack of circulation), heterogeneities can result in nonradial flow so that the zone of influence in all directions is difficult to estimate, the zone of influence is difficult to estimate due to heterogeneities and the complexities of two-phase flow (a high gas velocity results from the low gas viscosity and large density difference between water and gas), and it is difficult to monitor the zone of influence to verify it.

Product Recovery. Pumping an immiscible fluid out of the subsurface is known as product recovery. This is performed in a source area containing mobile (pumpable) NAPL. Experience in the recovery of mobile NAPL currently pertains mainly to the recovery of LNAPLs, most notably petroleum products residing at or near the water table (96). Summaries of the special considerations and procedures involved in the recovery of LNAPLs are available (95–104). Where the water table is shallow, open, or permeable trenches equipped with pumps can be used. The trench can be equipped with pumps designed only to skim the LNAPL as it collects in the trench. An alternative approach is to pump groundwater from the trench to reduce water levels and create gradients that will bring the LNAPL into the trench at greater rates and against natural gradients.

Wells also may be used to collect mobile LNAPL. The general approach uses the gradients established in the cone of depression surrounding a pumping well to collect the LNAPL. As the LNAPL collects in the cone of depression, it can be recovered with a pump. This may be accomplished using one well to establish the cone of depression and a second, nearby well to recover the LNAPL as it collects. Alternatively, a single well can be used to both establish the cone of depression and recover the LNAPL. To enhance recovery, injection wells may be used in combination with the recovery wells to flush contaminants from the vadose zone and increase the hydraulic gradient toward the recovery wells. A single pump can be used to collect the groundwater and LNAPL. However, the pump must be positioned near the water–LNAPL interface to ensure that both groundwater and LNAPL are drawn into the well.

Alternatively, separate pumps can be used to draw down the water table and collect the LNAPL. These can be located in the same or separate wells. Screens must be positioned to allow LNAPL intake. Consequently, the screen must extend above the level to which groundwater will be drawn down and into the area in which LNAPL will collect. Otherwise, LNAPL will collect above the screen and not be able to enter the well. Large draw downs increase the flow rate of LNAPL toward the well, thus increasing the speed of recovery. Unfortunately, they also create larger dewatered areas into which the LNAPL collects. These areas will have been previously uncontaminated by LNAPL. Substantial amounts of LNAPL will be retained in the aquifer matrix at residual saturation, even after water levels are allowed to recover (97), referred to as smear zones. Thus recovery systems should maintain recovery while minimizing drawdown.

Bioslurping is a new, innovative approach used by the U.S. Air Force for site remediation at petroleum-release sites that have free-phase light, nonaqueous-phase liquid contamination. By design, bioslurper systems recover free-phase LNAPL, via vacuum-enhanced pumping, while simultaneously initiating the remediation of the vadose zone soils via bioventing. In most applications, a single aboveground vacuum pump can be plumbed to multiple extraction wells to extract LNAPL, groundwater, and soil gas in the same process stream. LNAPL recovery is enhanced by the vacuum-induced gradient, which increases the rate of fluid flow into extraction wells. The system is configured to maximize the removal of LNAPL while minimizing the volume of groundwater that must be extracted. Soil gas is extracted at a low rate to aerate (biovent) the soils and to minimize volatilization.

Procedures for DNAPL cleanup are experimental and are more poorly documented than are procedures for LNAPL cleanup. Removing DNAPL from the subsurface is extremely difficult. Problems associated with DNAPL recovery have been described (10,105). Because of the problems of locating and removing DNAPL, complete cleanup generally is not possible. Instead, the DNAPL source area needs to be managed or controlled via physical or hydraulic containment. The properties that control DNAPL movement are largely responsible for the difficulties associated with their cleanup. Pressure gradients and gravity influence DNAPL movement. Thus, to move DNAPLs toward a well, gradients must be created that will overcome gravity. In cases where DNAPLs are migrating along an inclined confining bed, the required gradients may be large. For this reason, it is best to collect DNAPLs from low points along the bedrock or at some downslope position where the pool is migrating.

Mobile DNAPL may be recovered using the same basic techniques used in other recovery programs (104). If DNAPL is located near the surface, drains may be capable of recovering these contaminants. In this case, recovery is from the bottom of the trench, probably on top of a confining bed. Otherwise, wells are required. In both cases, recovery requires pump intakes placed far into the DNAPL to ensure that as much DNAPL as possible is collected. Pumping rates should be used that discourage mixing and the formation of emulsions at the interface between the DNAPL and groundwater. In most cases, the groundwater above the pool of DNAPL will be contaminated and may require containment. Using separate wells or a single well screen over the entire aquifer equipped with two pumps accomplishes containment.

Once mobile NAPL is collected, immobile NAPL remains at residual saturation. A variety of technologies, many associated with enhanced oil recovery, have been suggested to either mobilize the NAPL by lowering the interfacial tension or increasing the solubility of the NAPL. Lowering the interfacial tension can mobilize DNAPL so that hydraulic control may be lost, resulting in deeper subsurface contamination. These technologies, therefore, can be risky, are generally expensive, and are largely unproven; they include flushing with cosolvents such as alcohol, flushing with surfactants, hot water or steam flood, resistive heating, radio frequency heating, and chemical oxidation. These technologies are discussed elsewhere (15,81).

Electrokinetics. As indicated, the above removal technologies are not effective in low permeability materials. The demand to develop *in situ* remediation technologies for low permeability materials stimulated the effort to employ conduction phenomena in soils using an electric field to remove chemical species from soils (106–111). This technique (also called electrokinetic remediation, electroreclamation, electrokinetic soil processing or remediation, electrochemical decontamination, electrorestoration, or electrochemical soil processing) uses low level direct current between electrodes placed in the ground in an open flow arrangement (112). The process has been called Lasagna due to the layered configuration of electrodes and treatment zones (113).

Electrokinetic extraction technique has gone through the phases of bench-scale testing and pilot-scale testing. Precipitation of species close to the cathode compartment has been a bottleneck for the process. Acetic acid depolarization technique and other depolarization schemes have been developed (108). Heavy metals extraction by bench and pilot-scale tests from soils retrieved from sites across the nation and from Europe demonstrate that the technique may be efficiently and cost-effectively used at selected sites. The technology could be effective for treating contaminated low permeability soils (clayey, silty soils) or heterogeneous soils (clay lens in permeable soils). The technology conceptually can treat organic and inorganic contamination as well as mixed wastes. At this stage, however, it is still an emerging technology.

Contaminant Containment

Pump and treat is used for hydraulic containment. This technology was discussed earlier.

Low Permeability Barriers. Barriers generally are used as a contaminant source control technology to isolate the contaminant source from the environment. Low permeability barriers include vertical barriers (walls), bottom barriers (liners and floors), and surface control barriers (caps) (114). Barriers can be made of a wide variety of materials with a lower permeability than the aquifer. Typical materials include compacted clay, impervious synthetic layers (geomembranes), mixtures of soil and bentonite, mixtures of cement and bentonite, or barriers of engineered materials such as sheet piling (115,116). Construction techniques vary, and for certain technologies (eg, caps) regulatory requirements are specified in detail (117,118).

New developments in the technology for constructing sheet pile cutoff walls have significantly reduced the potential for leakage through these walls (119). Sheet piles can be made of interlocking steel, precast concrete, plastic, or wood sections.

Sheet pile cutoff walls are constructed by driving individual sections into the ground. The interlocks can be grouted to seal the joints between sheets. This configuration can prevent advection of contaminated groundwater through the cutoff wall. Containment systems for groundwater frequently include low flow, pump-and-treat systems to help control groundwater flow and prevent contaminated water from escaping the confined area (114). The pump-and-treat systems can take the form of wells or trenches. The barriers, in effect, reduce the amount of clean water that enters the contaminated zone, thereby reducing pumping and treatment costs. Barriers are considered low maintenance as opposed to no maintenance.

The long-term performance of physical barriers is uncertain. Construction difficulties are common. For example, sheet piling is difficult to install in rocky soils, and bentonite slurry walls are heterogeneous, with relatively high permeability zones occasionally occurring. Questions remain about the ability to key walls into the confining layers below the aquifer without creating leaks in the formation. The layer underlying the aquifer must be reasonably free of flow channels to maintain the integrity of the barrier. In addition, NAPLs may interact with clay barriers, causing shrinkage. Formation of macropores (syneresis cracks) through shrinkage of the clay may lead to rapid propagation of the fluid through clay liners and natural clayey formations (120).

Permeable Reactive Walls. The typical goal of conventional pump and treat is to completely remediate the source and contaminant plume. A low maintenance alternative is to treat the contaminated plume as it passes across some boundary to prevent further contamination downgradient (71). This plume control involves installation of permeable reactive zones or walls within the aquifer. As contaminated water moves under natural hydraulic gradients through the reactive zone, the contaminants are removed or degraded, leaving uncontaminated water to emerge from the downstream side.

Several approaches for installation of the reactive wall or treatment zone are possible (121). A trench can be excavated and backfilled with the reactive material. This method is limited to shallow depths. Slurry wall construction technology can be used to create a deeper and larger permeable curtain. Bentonite is replaced by a biodegradable polymer. The polymer mixed with reactive materials then replaces subsurface materials as excavation proceeds. When excavation is complete, pumping and biodegradation removes the polymer, leaving a permeable wall containing the reactive material. Alternatively, sealable sheet piling can be temporarily installed to allow dewatering and construction of a reactive zone. Costs are likely to be high if a continuous wall is installed across large zones of contaminated water. A promising alternative (122) uses low permeability barriers to funnel the natural groundwater flow to narrow openings in the barrier in which the reactive material has been installed. This method provides greater control of the treatment zone and facilitates replacement or removal of the reactive material.

The reactive zone can use a combination of physical, chemical, and biological processes. One possible physical method uses granular activated carbon or other sorption barriers to adsorb organic contaminants (123). Other possible physical or chemical methods use redox controls to precipitate metals and metal catalysts to degrade halogenated organic compounds (124). For example, in one set of laboratory studies, a calcium apatite barrier system precipitated lead (Pb^{2+}) as lead phosphate via phosphate dissolution from the apatite (125). In another study (126), iron-bearing solids in a laboratory soil column removed chromate (CrO_4^{2-}) by reducing chromium(VI) to chromium(III). An alternative technology uses air sparging as the reactive wall (127) to either remove volatile compounds or promote aerobic degradation.

Another possibility is to add nutrients and/or bacteria to the reaction zone to enhance contaminant biodegradation (128). Researchers are investigating several possible methods for creating biological reaction zones. For example, some researchers have tested biological removal of nitrate from groundwater by periodically injecting a readily biodegradable organic compound to stimulate denitrifying bacteria, which convert nitrate to nitrogen gas (121). When barriers use exhaustible material such as granular activated carbon or chemical reactants, they must be installed as modules to enable periodic replacement. The technology of reactive barriers is in the conceptual and development stages, the principles are being studied on a small scale.

Contaminant Transformation

***In Situ* Bioremediation.** *In situ* bioremediation stimulates the growth of indigenous microorganisms in subsurface regions where degradable contaminants are present. The process typically entails injection of nutrients and one or more electron acceptors (71). Soil and groundwater bioremediation near the land surface (within about 6 m) can be achieved by using infiltration galleys that allow substrates and nutrient-laden water to percolate through the soil. When contamination is located at greater depths, growth-sustaining chemicals can be introduced using injection wells to stimulate microorganisms in the contaminant plume. A dynamic system that includes injection and extraction wells and equipment for the addition and mixing of nutrients can be used to better control flow and movement of electron donor, electron acceptor, nutrients, and contaminants. The objective is to stimulate microorganisms to transform a portion of the desorbed organic contaminant with each pass of water laden with growth-supporting chemicals. Because the microorganisms colonize the soil particle surfaces, contaminants can be biotransformed as they desorb from the aquifer solids. Biotransformation reduces the dissolved contaminant concentration, thus enhancing the rate of desorption or dissolution of an immiscible phase.

Electron donors and acceptors, carbon source, moisture, and nutrients are essential for microbial growth. Besides natural compounds, many contaminants can provide these growth requirements. Organic contaminants are often metabolized as the electron donor. Energy is made available for cell growth when the electron donor transfers its electrons to a terminal electron acceptor. Examples of some contaminants that can serve as primary electron donors or substrates include phenol, chlorophenols, chlorobenzoates, alkanes, benzene, toluene, xylene, and chlorobenzenes. These chemicals tend to biodegrade in natural subsurface environments. Because trace concentrations of organic contaminants are typically present in the subsurface (eg at the low microgram per liter level), they frequently cannot support microbial growth as sole electron donors. In this situation, the organic contaminant can still be transformed by a microbial population that obtains the majority of its energy and carbon from a different compound that serves as the primary substrate. This metabolism is sometimes referred to as secondary use because the organic contaminants available at trace levels become secondary substrates and contribute negligibly to the energy and carbon needed for synthesis of new cell mass (129).

Multiple contaminants or substrates in the subsurface create a complex situation. They compete for the same electron acceptors (eg, dissolved oxygen); therefore, predicting degradation rates is difficult. In addition, some chemicals are not energy sources for microorganisms, yet they degrade in the presence of other chemicals. This process, co-metabolism, has been defined as the transformation of a nongrowth substrate in the presence of a growth substrate or another transformable compound (130). In co-metabolism, enzymes produced by the microorganisms to metabolize the primary substrate can interact with an organic contaminant and bring about its transformation in a fortuitous manner. If biodegradable contaminants are present, then microbial processes can be successful at reducing the mass of contaminants. This mass loss reduces the risk, and it may even be possible to achieve regulatory end points with bioremediation. The literature contains several examples of successful bioremediation efforts (131–136).

Bioventing. Bioventing, an *in situ* bioremediation process for the vadose zone, provides air flow to deliver oxygen to the subsurface to stimulate *in situ* biodegradation of organic contaminants (71). Because air is introduced, care must be taken near structures where gas may accumulate and pose an explosive problem. Bioventing systems operate at lower air flow rates than SVE systems to reduce vapor extraction quantities, and increase biodegradation associated with longer vapor retention. Soil moisture levels necessary for biological activity usually are higher than those recommended for optimum SVE systems. Experience indicates that bioventing is optimized when performed in cycles of oxygenation followed by shut-in. The addition of nutrients may also enhance bioremediation. Numerous field and experimental studies show that bioventing can effectively clean up such relatively nonvolatile hydrocarbons as diesel fuel (137) as well as hydrocarbons ranging in composition from gasoline to heavy fuel oil. The U.S. Air Force is using bioventing at several jet fuel spill sites. As a result, several protocols have been developed for field treatability testing of bioventing (138–140).

Intrinsic Remediation. Intrinsic remediation occurs when naturally occurring attenuation mechanisms, such as biodegradation (aerobic and anaerobic), reduce contaminant mass dissolved in groundwater (49). During intrinsic remediation, contaminants are ultimately transformed to innocuous byproducts (eg, carbon dioxide and water) not just transferred to another phase or location within the environment. Intrinsic remediation results from the integration of several subsurface attenuation mechanisms that are classified as either destructive or nondestructive. Destructive processes include biodegradation, abiotic oxidation, and hydrolysis (reaction of a halogenated organic compound with water or a component ion of water). Nondestructive attenuation mechanisms include sorption, dilution (caused by dispersion and infiltration), and volatilization. Because intrinsic remediation depends on natural attenuation, an effort for this remediation is to provide evidence of intrinsic remediation through monitoring.

Microorganisms preferentially use electron acceptors that provide the maximum free energy during respiration. Of the common electron acceptors, oxygen provides the most free energy during electron transfer (Table 4). Use of nitrate, Mn(IV), Fe(III), sulfate, and carbon dioxide yields less energy during electron transfer according to the order listed in Table 4. In the subsurface, there is limited capacity for mixing, and rates of oxygen replenishment from the atmosphere are extremely slow following consumption by aerobic reactions. The coupling of mass transport and reaction in the subsurface results in spatial gradients of electron acceptor concentrations. Competition between microorganisms deriving energy from oxidation of reduced organic carbon and nitrifiers deriving energy from the oxidation of ammonia consumes oxygen. As oxygen is depleted, anoxic conditions follow with the sequential use of nitrate (denitrification), Mn(IV), and Fe(III), sulfate (sulfate respiration or reduction), and carbon dioxide (methanogenesis) as electron acceptors (140).

In some cases, intrinsic remediation reduces dissolved-phase contaminant concentrations to below maximum contaminant levels (MCLs) before the contaminant plume reaches potential receptors, even if little or no source removal or reduction takes place. When intrinsic remediation will not reduce contaminant concentrations to below regulatory MCLs in an acceptable time frame, less stringent cleanup goals may be implemented. This is especially likely if it can be demonstrated that intrinsic remediation will result in a continual reduction in contaminant concentrations over time such that calculated risk values are reduced. Intrinsic remediation is gaining regulatory acceptance and has been implemented at several sites involving petroleum hydrocarbons over the past few years (49,142,143).

Intrinsic remediation is chemical specific. Light hydrocarbons generally degrade more readily than heavier hydrocarbons. A significant percentage of the specific hydrocarbons found in light to medium range distillates (ie, gasoline, jet fuels, and diesel) are all amenable to biodegradation. The monoaromatic hydrocarbons, ie, benzene, toluene, ethylbenzene, and xylenes (BTEX), are the most soluble and degrade the easiest, followed by the straight-chained alkanes. Compounds resistant to degradation are the isoprenoids (branched-chained alkanes), such as pristane and phytane hydrocarbons, and asphaltenes (144–147). Numerous aerobic and anaerobic mechanisms degrade aromatic hydrocarbons. Under aerobic conditions, benzene and its derivatives are partially oxidized to catechols with subsequent ring cleavage and β-oxidation (148). Under anaerobic conditions, phenols and cresols are partially oxidized to intermediate products. Under aerobic conditions, BTEX compounds have been shown to degrade quickly (1–2 yr). Similar behavior has been observed in anaerobic groundwater systems. Whereas many aromatic hydrocarbons degrade (as evidenced by relatively small plumes associated with gasoline spills), chlorinated aliphatic hydrocarbons (chlorinated solvents) are more resistant to degradation and often are found in plumes extending a mile or more. More recently, evidence has demonstrated that intrinsic remediation of chlorinated solvents occurs at some sites (149). This generally occurs under anaerobic conditions where terminal electron acceptors, other than oxygen, are used.

In summary, evidence of intrinsic remediation includes stable or decreasing contaminant plume with time; disappearance of degradable contaminants relative to a "conservative"

Table 4. Electron Acceptors in Biotransformation Processes[a]

Microbial Process	Electron Acceptor	Reaction	Free Energy Change ($\Delta G°$) at pH 7, kJ/equivalent[b]
Aerobic respiration	O_2	$CH_2O + O_2(g) \rightleftarrows CO_2(g) + H_2O$	−125
Denitrification	NO_3^-	$CH_2O + 0.8NO_3^- + 0.8H^+ \rightleftarrows CO_2(g) + 0.4N_2(g) + 1.4H_2O$	−119
Mn(IV) reduction	Mn(IV)	$CH_2O + 2MnO_2(s) + 2HCO_3^- + 2H^+ \rightleftarrows CO_2(g) + 2MnCO_3(s) + 3H_2O$	−97
Fe(III) reduction	Fe(III)	$CH_2O + 4FeOOH(s) + 4HCO_3^- + 4H^+ \rightleftarrows CO_2(g) + FeCO_3(s) + 7H_2O$	−42
Sulfate reduction	SO_4^{2-}	$CH_2O + 0.5SO_4^{2-} + 0.5H^+ \rightleftarrows CO_2(g) + 0.5HS^- + H_2O$	−25
Methanogenesis	CO_2	$CH_2O + 0.5CO_2(g) \rightleftarrows CO_2(g) + 0.5CH_4(g)$	−23

[a] From Refs. 140, 141.
[b] To convert kJ to kcal, divide by 4.184.

tracer (eg, for gasoline contamination, smaller BTEX plume relative to the oxygenate additive methyl tertiary butyl ether); correlation of low concentrations of electron acceptors with high contaminant concentrations (ie, reduced oxygen, nitrate and sulfate concentrations within the contaminant plume relative to their background concentrations); appearance of degradation products (eg, accumulation of dissolved iron and the production of carbon dioxide, hydrogen sulfide, methane and specific degradation products); ratios of microbial numbers between impacted area and upgradient area that reflect enrichment of the indigenous microbial community by microorganisms with the desired metabolic capabilities; microcosm studies to estimate rates of biodegradation in the lab; and modeling studies to estimate rates of biodegradation in the field.

BIBLIOGRAPHY

1. R. C. Heath, "Basic Ground-Water Hydrology," *U.S.G.S. Water-Supply Paper 2220,* U.S. Geological Survey, Washington, D.C., 1983.
2. P. A. Domenico and F. W. Schwartz, *Physical and Chemical Hydrogeology,* John Wiley & Sons, New York, 1990.
3. L. W. Gelhar and C. L. Axness, *Water Resources Res.* **19**(1), 161–180 (1983).
4. R. A. Freeze and J. A. Cherry, *Groundwater,* Prentice-Hall, Englewood Cliffs, N.J., 1979.
5. C. D. Palmer and W. Fish, "Chemical Enhancements to Pump-and-Treat Remediation," *USEPA Ground Water Issue Paper, EPA/540/S-92-001,* Robert S. Kerr Environmental Research Laboratory, Ada, Okla., 1992.
6. C. W. Fetter, *Contaminant Hydrogeology,* Macmillian Publishing Co., New York, 1993.
7. P. J. J. Alvarez, P. J. Anid, and T. M. Vogel, *Biodegradation* **2,** 43–51 (1991).
8. T. M. Vogel, C. S. Criddle, and P. L. McCarty, *Environ. Sci. Technol.* **21,**(8), 722–735 (1987).
9. P. J. Sturman, P. S. Stewart, A. B. Cunningham, E. J. Bouwer, and J. H. Wolfram, *J. Contam. Hydrol.* **19,** 171–203 (1995).
10. J. W. Mercer and R. K. Waddell, in D. R. Maidment, ed., *Handbook of Hydrology,* McGraw-Hill, New York, 1993.
11. S. Feenstra and J. A. Cherry, paper presented at the International Groundwater Symposium, International Association of Hydrogeologists, Halifax, Nova Scotia, May 1–4, 1988.
12. F. Schwille, *Dense Chlorinated Solvents in Porous and Fractured Media,* Lewis Publishers, Chelsea, Mich., 1988.
13. U.S. Environmental Protection Agency, "Dense Nonaqueous Phase Liquids—A Workshop Summary," *EPA/600/R-92/030,* Robert S. Kerr Environmental Research Laboratory, Ada, Okla., 1992.
14. J. W. Mercer and R. M. Cohen, *J. Contam. Hydrol.* **6**(2), 107–163 (1990).
15. R. M. Cohen and J. W. Mercer, *DNAPL Site Evaluation,* C. K. Smoley, Boca Raton, Fla., 1993.
16. J. F. Pankow and J. A. Cherry, *Dense Chlorinated Solvents and Other DNAPLs in Groundwater,* Waterloo Press, Portland, Oreg., 1996.
17. J. F. Villaume, *Ground Water Monitor. Rev.,* 60–75 (1985).
18. P. K. Robertson and R. G. Campanella, *Guidelines for Use and Interpretation of the Electronic Cone Penetration Test,* 3rd ed., University of British Columbia, Vancouver, 1986.
19. C. Y. Chiang., K. R. Loos, and R. A. Klopp, *Ground Water* **30**(3), 428–436 (1992).
20. P. W. Lurk, S. S. Copper, P. G. Malone, and S. H. Lieberman in *Proceedings of Superfund 1990,* Hazardous Materials Control Research Institute, Silver Spring, Md., 1990, pp. 297–299.
21. M. Smolley and J. C. Kappmeyer *Ground Water Monitor. Rev.* **11**(2), 101–106 (1991).
22. R. W. Edge and K. Cordry, *Ground Water Monitor. Rev.* **9**(3), 177–183 (1989).
23. K. Cordry, paper presented at the *5th National Outdoor Action Conference on Aquifer Restoration, Ground Water Monitoring and Geophysical Methods,* Las Vegas, Nev., 1991.
24. P. D. Kuhlmeier and T. E. Sturdivant in D. M. Nielsen and M. N. Sara, eds., *Current Practices and Vadose Zone Investigations, ASTM STP 1118,* American Society for Testing Materials, Philadelphia, Pa., 1992, pp. 183–198.
25. T. M. Christy and S. C. Spradlin, *The Use of Small Diameter Probing Equipment for Contaminated Site Investigation,* Geoprobe Systems, Salina, Kans., 1992.
26. Geoprobe; *Tools and Equipment Catalog,* Geoprobe Systems, Salina, Kans., 1995–1996.
27. P. M. Kearl, N. E. Korte, and T. A. Cronk, *Ground Water Monitor. Rev.,* 155–161 (1992).
28. D. E. Shanklin, W. C. Sidle, and M. E. Ferguson, *Ground Water Monitor. Remediat.,* **15**(3), 168–176 (1995).
29. R. W. Powell and R. W. Puls, *J. Contam. Hydrol.,* **12,** 51–77 (1993).

30. M. J. L. Robin and R. W. Gillham, *Ground Water Monitor. Rev.,* 85–93 (1987).
31. R. W. Puls and R. W. Powell, *Ground Water Monitor. Rev.,* 167–176 (1992).
32. W. H. Pedlar, M. J. Barvenik, C. F. Tsang, and F. V. Hale, paper presented at the *4th National Outdoor Action Conference on Ground Water Management,* Las Vegas, Nev., 1990.
33. J. L. Newman and G. W. Corbell, paper presented at the *4th National Outdoor Action Conference on Ground Water Management,* Las Vegas, Nev., 1990.
34. J. D. Ritchey, *Ground Water Monitor. Rev.* **6**(2), 108–113 (1986).
35. American Society for Testing and Materials, "Standard Terminology Used with Ion-Selective Electrodes," *D4127-82, vol. 11.01,* ASTM, Philadelphia, Pa., 1982.
36. K. Nesbitt, *Nat. Environ. J.,* 13–15 (Jan./Feb. 1993).
37. L. A. Eccles and S. J. Simon, "In situ Monitoring at Superfund Sites with Fiber Optics, II. Plan for Development," *EPA/600/X-87/415,* U.S. EPA Environmental Monitoring Systems Laboratory, Las Vegas, Nev., 1987.
38. L. A. Eccles, S. J. Simon, and S. M. Kaliner, "In situ Monitoring at Superfund Sites with Fiber Optics, I. Rationale," *EPA/600/X-87/156,* U.S. EPA Environmental Monitoring Systems Laboratory, Las Vegas, Nev. 1987.
39. FSMC System Coordinator, OERR, Analytical Operations Branch (WH-548-A), "Field Screening Methods Catalog: User's Guide," *EPA/540/2-88/005,* U.S. Environmental Agency, Washington, D.C., 1988.
40. U.S. Environmental Protection Agency "First International Symposium, Field Screening Methods for Hazardous Waste Site Investigations," *EPA/600/D-89/189, NTIS PB-132572,* Washington, D.C., 1988.
41. W. F. Arendale and R. Hatcher, paper presented at the *2nd International Symposium of Field Screening Methods for Hazardous Waste and Toxic Chemicals,* Washington, D.C., 1991.
42. S. M. Barnard and D. R. Walt, *Environ. Sci. Technol.* **26,** 1301–1305 (1991).
43. W. Chudyk, K. Pohlig, K. Exarhoulakos, J. Holsinger, and N. Rico in D. M. Nielsen and A. I. Johnson, eds., *Ground Water and Vadose Zone Monitoring, ASTM STP 1053,* American Society for Testing and Materials, Philadelphia, Pa., 1990.
44. W. Chudyk, K. Pohlig, C. Botteron, and R. Najjar, paper presented at the *2nd International Symposium of Field Screening Methods for Hazardous Waste and Toxic Chemicals,* Washington, D.C., 1991.
45. J. W. Griffin and K. B. Olsen, in Ref. 29, pp. 311–328.
46. S. M. Klainer, J. D. Koutsandreas, and L. Eccles in A. G. Collins and A. I. Johnson, eds., *Ground-Water Contamination: Field Methods, ASTM STP 963,* American Society for Testing and Materials, Philadelphia, Pa., 1988.
47. W. J. Dunlap, J. F. McNabb, M. R. Scalf, and R. L. Cosby, "Sampling for Organic Chemicals and Microorganisms in the Subsurface," *EPA-600/2-77-176,* Washington, D.C., 1977.
48. J. T. Wilson, J. F. McNabb, D. L. Balkwill, and W. C. Ghiorse, *Ground Water* **21,** 134–142 (1983).
49. T. H. Wiedemeier, J. T. Wilson, D. H. Kampbell, and R. N. Miller, paper presented at *the 8th National Outdoor Action Conference and Exposition,* Minneapolis, 1994.
50. W. E. Holben, J. K. Jansson, B. K. Cheim, and J. M. Tiedje, *Appl. Environ. Microbiol.* **54,** 703–711 (1988).
51. K. O'Carroll in B. Austin, ed., *Methods in Aquatic Bacteriology,* John Wiley & Sons, New York, 1988, pp. 353–358.
52. J. T. Trevors, *Trends Biotechnol.* **3,** 291–293 (1985).
53. R. C. Wyndham and J. W. Costerton, *Appl. Environ. Microbiol.* **41,** 783–790 (1981).
54. E. S. Pankhurst in D. A. Shapton and R. G. Board, eds., *Isolation of Anaerobes,* Academic Press, Inc., New York, 1971 pp. 223–240.
55. J. G. Jones, B. M. Simon, and S. Gardener, *J. Gen. Microbiol.* **128,** 1–11 (1982).
56. F. E. Clark in C. A. Black, ed., *Methods of Soil Analysis, Part 2, Chemical and Microbiological Properties,* American Society of Agronomy, Inc., Madison, Wisc., 1965.
57. J. J. Molongoski and M. J. Klug, *Appl. Environ. Microbiol.* **31,** 83–90 (1976).
58. H. F. Kaspar and J. M. Tiedje in A. L. Page, ed., *Methods of Soil Analysis, Part 2, Chemical and Microbiological Properties,* American Society of Agronomy, Inc., Madison, Wisc., 1982.
59. M. E. Watwood, J. W. Fitzgerlad, and J. R. Gosz, *Can. J. Forest Resources* **16,** 689–695 (1986).
60. B. H. Wilson, G. B. Smith, and J. F. Rees, *Environ. Sci. Technol.* **20,** 997–1002 (1986).
61. K. E. Kolm, paper presented at the *ASTM International Symposium on Remote Sensing and GIS for Site Characterization: Applications and Standards,* San Francisco, Calif., 1994.
62. P. K. M. van der Heijde and co-workers, *Water Resources Monograph 5,* American Geophysical Union, Washington, D.C., 1985.
63. National Research Council, *Ground Water Models Scientific and Regulatory Applications,* National Academy Press, Washington, D.C., 1990.
64. P. D. Moskowitz, R. Pardi, M. P. DePhillips, and A. F. Meinhold, *Risk Anal.* **12**(4), 591–621 (1992).
65. P. K. M. van der Heijde and O. A. Elnaway, "Compilation of Ground-Water Models," *EPA/600/R-93/118,* U.S. Environmental Protection Agency, Ada, Okla., 1993.
66. J. W. Mercer and C. R. Faust, *Ground Water Modeling,* National Water Well Association, Dublin, Ohio, 1980.
67. M. P. Anderson and W. W. Woessner, *Applied Groundwater Modeling,* Academic Press, Inc., New York, 1992.
68. M. P. Anderson, D. S. Ward, E. G. Lappala, and T. A. Prickett in Ref. 14.
69. J. Bear, M. S. Beljin, and R. R. Ross, "Fundamentals of Ground-Water Modeling, *EPA/540/S-92/005,* U.S. EPA Ground Water Issue, Ada, Okla., 1992.
70. J. W. Mercer, D. C. Skipp, and D. Giffin, "Basics of Pump-and-Treat Groundwater Remediation," *EPA/600/8-90-003,* Robert S. Kerr Environmental Research Laboratory, Ada; Okla., 1990.
71. National Research Council, *Alternatives for Ground Water Cleanup,* National Academy Press, Washington, D.C., 1994.
72. R. M. Cohen, A. H. Vincent, J. W. Mercer, C. R. Faust, and C. P. Spalding, "Methods for Monitoring Pump-and-Treat Performance," *EPA/600/R-94/123,* Robert S. Kerr Environmental Research Laboratory, Ada, Okla., 1994.
73. U.S. Environmental Protection Agency, "Guidance for Evaluating the Technical Impracticability of Ground-Water Restoration," *Directive 9234.2-25, Interim Final,* Office of Solid Waste and Emergency Response, Washington, D.C., 1993.
74. J. F. Keely, "Performance Evaluation of Pump-and-Treat Remediations," *EPA/540/4-89-005,* Robert S. Kerr Environmental Research Laboratory, Ada, Okla., 1989.
75. M. L. Brusseau, *Water Resources Res.* **28**(9), 2485–2497 (1992).
76. S. E. Powers, C. O. Loureiro, L. M. Abriola, and W. J. Weber Jr., *Water Resources Res.* **27**(4), 463–477 (1991).
77. R. M. Cohen, J. W. Mercer, and R. M. Greenwald, "Design Guidelines for Conventional Pump-and-Treat Systems, *EPA Ground Water Issue,"* Ada, Okla., 1995.
78. R. C. Borden and C. M. Kao, *Water Environ. Res.* **64**(1), 28–36 (1992).

79. S. M. Gorelick, R. A. Freeze, D. Donohue, and J. F. Keely, *Groundwater Contamination Optimal Capture and Containment,* Lewis Publishers, Boca Raton, Fla., 1993.
80. U.S. Environmental Protection Agency, "Subsurface Contamination Reference Guide," *EPA/540/2-90/011,* Office of Emergency and Remedial Response, Washington, D.C., 1990.
81. N. Hutzler, B. Murphy, and J. Gierke, "State of Technology Review: Soil Vapor Extraction Systems," *EPA/600/2-89/024,* EPA Risk Reduction Engineering Laboratory, Cincinnati, Ohio, 1989.
82. P. Johnson, C. Stanley, M. Kemblowski, D. Byers, and J. Corthart, *Ground Water Monitor. Rev.,* 159 (1990).
83. P. J. Reidy, W. J. Lyman, and D. C. Noonan, "Assessing UST Corrective Action Technologies: Early Screening of Clean-Up Technologies for the Saturated Zone," *EPA/600/2-90/027,* Risk Reduction Engineering Laboratory, Cincinnati, Ohio, 1990.
84. R. C. Sims, *J. Air Waste Manage. Assoc. Rpt. Ser.* RS-15, 3–30 (1990).
85. T. Buscheck and R. Peargin, paper presented at the *Petroleum Hydrocarbons and Organic Chemicals in Ground Water Conference,* Dublin, Ohio, 1991.
86. C. Gomez-Lahoz, J. M. Rodriguez-Maroto, and D. J. Wilson, *Sep. Sci. Technol.* **26**(2), 133–163 (1991).
87. B. Herrling, W. Buermann, and J. Stamm, paper presented at the *Conference on Subsurface Contamination by Immiscible Fluids,* Calgary, Canada, 1990.
88. M. C. Marley, D. J. Hazebrouck, and M. T. Walsh, *Ground Water Monitor. Rev.* **12**(2), 137–145 (1992).
89. B. B. Looney, D. S. Kaback, and J. C. Corey, 1992, demonstration presented at the *Subsurface Restoration Conference* Dallas, 1992.
90. K. M. Lee and H. D. Stensel, *J. WPCF* **58**(11), 1066–1072 (1986).
91. B. Herrling, J. Stamm, and W. Buermann, paper presented at the *In-Situ and On-Site Bioreclamation International Symposium* San Diego, Calif., 1991.
92. B. Herrling, J. Stamm, E. J. Alesi, P. Brinnel, F. Hirschberger, and M. R. Sick, paper presented at the *3rd Forum on Innovative Hazardous Waste Treatment Technologies: Domestic and International,* Dallas, 1991.
93. H. Gvirtzman, and S. M. Gorelick, *Transport Porous Media* **8** 71–92 (1992).
94. J. A. Coyle, H. J. Borchers Jr., and R. J. Miltner, paper presented at *the American Water Works Association Annual Conference,* Washington, D.C., 1985.
95. P. D. Lundegard and G. Andersen, paper presented at the *Petroleum Hydrocarbon and Organic Chemicals in Ground Water: Prevention, Detection, and Restoration Conference,* National Ground Water Dublin, Ohio, 1993.
96. U.S. Environmental Protection Agency. "How to Effectively Recover Free Product at Leaking Underground Storage Tank Sites," *EPA/510-B-95-013,* Washington, D.C., 1995.
97. S. B. Blake, and R. W. Lewis, *Ground Water Monitor. Rev.* **3**(2), 40–46 (1983).
98. T. L. de Pastrovich, Y. Baradat, R. Barthel, A. Chiarelli, and D. R. Fussell, 1979. Protection of groundwater from oil pollution, CONCAWE (Conservation of Clean Air and Water-Europe), The Hague, 61 pp.
99. American Petroleum Institute (API), 1989. A guide to the assessment and remediation of underground petroleum releases, API Publication No. 1628, 2nd edition, Washington, D.C., 81 pp.
100. American Petroleum Institute (API), 1980. Underground spill cleanup manual, API Publication No. 1628, Washington, D.C., 34 pp.
101. D. R. Fussell, H. Godjen, P. Hayward, R. H. Lilie, A. Marco, and C. Penisi, 1981. Revised inland oil spill clean-up manual, CONCAWE (Conservation of Clean Air and Water-Europe), The Hague, Report no. 7, 150 pp.
102. J. F. Villaume, P. C. Lowe, and G. P. Lennon, paper presented at *the Conference on the Disposal of Solid, Liquid and Hazardous Wastes,* Bethlehem, Pa., 1983.
103. J. F. Villaume, P. C. Lowe, and D. F. Unites, paper presented at *the 3rd National Symposium on Aquifer Restoration and Ground Water Monitoring,* Worthington, Ohio, 1983.
104. J. P. Ferry, P. J. Dougherty, J. B. Moser, and R. M. Schuller, paper presented at the *Petroleum Hydrocarbons and Organic Chemicals in Ground Water: Prevention, Detection and Restoration Conference,* Houston, 1986.
105. D. M. Mackay and J. A. Cherry, *Environ. Sci. Technol.* **23**(6), 620–636 (1989).
106. Y. B. Acar and A. N. Alshawabkeh, *Environ. Sci. Technol.* **27**(13), 2638 (1993).
107. Y. B. Acar, and co-workers, *J. Hazardous Mater.* **40**(3), 117–137 (1995).
108. Electrokinetics, Inc., "An Investigation of Selected Enhancement Techniques in Electrokinetic Remediation," *U.S. Army Waterways Experiment Station Report,* Electrokinetics Inc., Baton Rouge, La., 1994.
109. Electrokinetics, Inc. "Theoretical and Experimental Modeling of Removing Contaminants from Soils by an Electrical Field," *USEPA Report,* Electrokinetics Inc., Baton Rouge, La., 1995.
110. R. Lageman, P. Wieberen, and G. Seffinga, *Electro-Reclamation: Theory and Practice, Chemistry and Industry,* Society of Chemical Industry, London, 1989.
111. A. P. Shapiro, and R. F. Probstein, *Environ. Sci. Technol.* **27**(2), 283–291 (1993).
112. Y. B. Acar and co-workers, paper presented at the *21st Annual RREL Research Symposium* Cincinnati, Ohio, 1995.
113. S. V. Ho and co-workers, *Environ. Sci. Technol.* **29**, 2528–2535 (1995).
114. R. R. Rumer and M. E. Ryan, eds., *Barrier Containment Technologies for Environmental Remediation Applications,* John Wiley & Sons, Inc., New York, 1995.
115. R. Bonaparte, ed., *Waste Containment Systems: Construction, Regulation, and Performance,* American Society of Civil Engineers, New York, 1990.
116. R. K. Rowe, R. M. Quigley, and J. R. Booker, *Clayey Barrier Systems for Waste Disposal Facilities,* E & FN SPON, London, 1995.
117. U.S. Environmental Protection Agency, "Requirements for Hazardous Waste Landfill Design, Construction, and Closure," *EPA/625/4-89/022,* Washington, D.C., 1989.
118. U.S. Environmental Protection Agency, "Technical Guidance Document, Final Caps on Hazardous Waste Landfills and Surface Impoundments," *EPA/530-SW-89-047,* Washington, D.C., 1989.
119. R. C. Starr, and J. A. Cherry, paper presented at the *45th Canadian Geotechnical Conference,* Toronto, 1992.
120. H. I. Inyang, in *Encyclopedia of Environmental Control,* Gulf Publishing Co., New York, 1995, pp. 175–216.
121. R. W. Gilham and D. R. Burris, paper presented at the *Subsurface Restoration Conference,* Dalls, 1992.
122. R. C. Starr and J. A. Cherry, *Ground Water* **32**(3), 465–476 (1994).
123. K. Hatfield, D. Burris, T. B. Stauffer, and J. Ziegler, *ASCE J. Environ. Eng.* **118**(3), 322–337 (1992).

124. R. W. Gilham and S. F. O'Hannesin, *Ground Water* **29**(5) 752 (1991).

125. Y. Xu and F. W. Schwartz, paper presented at the *Subsurface Restoration Conference,* Dallas, 1992.

126. D. W. Blowes, and C. J. Ptacek, paper presented at the Subsurface Restoration Conference, Dallas, 1992.

127. J. F. Pankow, R. L. Johnson, and J. A. Cherry, *Ground Water* **31**(4), 654–663 (1993).

128. J. F. Devlin, and J. F. Barker, *Ground Water* **32**(3), 374–380 (1994).

129. P. L. McCarty, M. Reinhard, and B. E. Rittmann, *Environ. Sci. Technol.* **15,** 40–51 (1981).

130. H. Dalton and D. I. Stirling, *Phil. Trans. R. Soc. London* **97,** 481–496 (1982).

131. M. D. Lee and R. L. Raymond Sr. in Ref. 96, pp. 429–436.

132. L. Semprini, P. V. Roberts, G. D. Hopkins, and P. L. McCarty, *Ground Water* **28**(5), 715–727 (1990).

133. R. E. Hinchee and R. F. Olfenbuttel, eds., *In-Situ Bioreclamation and On-Site Bioreclamation,* Butterworth-Heinemann, Boston, Mass., 1991.

134. J. T. Wilson, L. E. Leach, M. Henson, and J. N. Jones, *groundwater remediation technique, Ground Water Monitor. Rev.* **6**(4), 56–64 (1986).

135. M. D. Lee and co-workers, *CRC Crit. Rev. Environ. Control* **18**(1), 29–89 (1988).

136. J. M. Thomas and C. H. Ward, *Environ. Sci. Technol.* **23**(7), 760–766 (1989).

137. T. L. Bulman, M. Newland, and A. Wester, *Hydrol. Sci. J.* **38**(4), 297–308 (1993).

138. R. E. Hinchee, S. K. Ong, R. N. Miller, D. C. Downey, R. Frendt, *Test Plan and Technical Protocol for a Field Treatability Test for Bioventing,* Air Force Center for Environmental Excellence, Brooks AFB, Texas, 1992.

139. J. A. Kittel, R. E. Hinchee, R. N. Miller, C. Vogel, and R. Hoppel, paper presented at the *GNWA Petroleum Hydrocarbons Conference,* Houston, 1993.

140. E. J. Bouwer, *Environmental Microbiology,* Wiley-Liss, Inc., New York, 1992. pp. 287–318.

141. W. Stumm and J. J. Morgan, *Aquatic Chemistry,* John Wiley & Sons, New York, 1981.

142. D. C. Downey, and M. J. Gier, paper presented at the *USAF Environmental Restoration Technology Symposium* San Antonio, Texas, 1991.

143. T. H. Wiedemeier, P. R. Guest, R. L. Henry, and C. B. Keith, paper presented at the *Petroleum Hydrocarbons and Organic Chemicals in Ground Water: Prevention, Detection, and Restoration Conference,* Houston, Tex., 1993.

144. J. F. Barker, and G. C. Patrick, a paper presented at the *Petroleum Hydrocarbons and Organic Chemicals in Ground Water: Prevention, Detection, and Restoration Conference,* Houston, 1985.

145. J. F. Barker, G. C. Patrick, and D. Major, *Ground Water Monitor. Rev.* **7**(1), 64–71 (1987).

146. K. L. Berry-Spark and co-workers, "The Behavior of Soluble Petroleum Product Derived Hydrocarbons in Groundwater—Phase III, *PACE Report No. 88-2,* Petroleum Association for Conservation of the Canadian Environment, Ottawa, 1988.

147. K. L. Berry-Spark, J. F. Barker, D. Major, and C. I. Mayfield, paper presented at *Focus on Eastern Regional Ground Water Issues: A Conference* (July), Burlington, Vt., 1987.

148. F. H. Chapelle, *Ground-Water Microbiology and Geochemistry,* John Wiley & Sons, Inc., New York, 1992.

149. L. Semprini, P. K. Kitanidid, D. H. Kampbell, and J. T. Wilson, *Water Resources Res.* **31**(4), 1051–1062 (1995).

General References

AASHTO, *Manual on Subsurface Investigations,* American Association of State Highway and Transportation Officials, Washington, D.C., 1988.

L. Aller and co-workers, "Handbook of Suggested Practices for the Design and Installation of Ground-Water Monitoring Wells," *USEPA-600/4-89/034,* National Water Well Association, Dublin, Ohio, 1989.

ASTM, "ASTM Standards on Ground Water and Vadose Zone Investigations," *PCN:03-418192-38,* Philadelphia, Pa., 1992.

M. J. Barcelona, J. P. Gibb, J. A. Helfrich, and E. E. Garske, "Practical Guide for Ground-Water Sampling," *USEPA/600/2-85-104,* Ada, Okla., 1985.

M. J. Barcelona, J. P. Gibb, and R. A. Miller, "A Guide to the Selection of Materials for Monitoring Well Construction and Ground-Water Sampling," *Illinois State Water Survey Contract Report 327 to USEPA R.S. Kerr Environmental Research Laboratory,* Ada, Okla., 1983.

M. J. Barcelona, J. F. Keely, W. A. Pettyjohn, and A. Wehrmann, "Handbook of Ground Water," *USEPA/625/6-87/016,* Ada, Okla., 1987.

E. L. Berg, "Handbook for Sampling and Sample Preservation of Water and Wastewater," *USEPA/600/4-82-029,* Washington, D.C., 1982.

J. R. Boulding, 1995. *Practical Handbook of Soil, Vadose Zone, and Ground-Water Contamination, Assessment, Prevention, and Remediation,* Lewis Publishers, Boca Raton, Fla., 1995.

Canadian Council of Ministers of the Environment, "Subsurface Assessment Handbook for Contaminated Sites," *Report CCME EPC-NCSRP-48E,* Ottawa, 1994.

H. C. Claasen, "Guidelines and Techniques for Obtaining Water Samples That Accurately Represent the Water Chemistry of an Aquifer," *USGS Open-File Report 82-1024,* Washington, D.C., 1982.

D. A. Devitt, R. B. Evans, W. A. Jury, and T. H. Starks, "Soil Gas Sensing for Detection and Mapping of Volatile Organics," *USEPA/600/8-87/036,* Las Vegas, Nev., 1987.

Electric Power Research Institute, "Preliminary Results on Chemical Changes in Groundwater Samples Due to Sampling Devices," *EPRI EA-4118 Interim Report,* Palo Alto, Calif., 1985.

Electric Power Research Institute, "Techniques to Develop Data for Hydrogeochemical Models," *EPRI EN-6637,* Palo Alto, Calif., 1989.

D. Fenn, E. Cocozza, J. Isbister, O. Braids, B. Yard, and P. Roux, "Procedures Manual for Ground Water Monitoring at Solid Waste Disposal Facilities," *USEPA SW-611,* Washington, D.C., 1977.

P. J. Ford, D. J. Turina, and D. E. Seely, "Characterization of Hazardous Waste Sites—A Methods Manual, vol. II, Available Sampling Methods," *USEPA-600/4-84-076 NTIS PB85-521596,* Washington, D.C., 1984.

Gas Research Institute, "Management of Manufactured Gas Plant Sites," *GRI-87/0260,* Chicago, Ill., 1987.

A. Klute, ed., *Methods of Soil Analysis,* Soil Science Society of America, Inc., Madison, Wis., 1986.

D. M. Nielsen, ed., *Practical Handbook of Ground Water Monitoring,* Lewis Publishers, Chelsea, Mich., 1991.

N.J. Department of Environmental Protection, *Field Sampling Procedures Manual, Hazardous Waste Program,* NJDEP, Trenton, N.J., 1988.

B. W. Rehm, T. R. Stolzenburg, and D. G. Nichols, "Field Measurement Methods for Hydrogeologic Investigations: A Critical Review of the

Literature," *Electric Power Research Institute Report EPRI EA-4301,* Palo Alto, Calif., 1985.

M. N. Sara, *Site Assessment Manual,* Waste Management of North America, Inc., Oak Brook, Ill., 1989.

M. R. Scalf, J. F. McNabb, W. J. Dunlap, R. L. Cosby, and J. S. Fryberger, *Manual of Ground-Water Quality Sampling Procedures,* National Water Well Association, Worthington, Ohio, 1981.

M. S. Simmons, *Hazardous Waste Measurements,* Lewis Publishers, Chelsea, Mich., 1991.

S. W. Sisk, "NEIC Manual for Groundwater/Subsurface Investigations at Hazardous Waste Sites," *USEPA-330/9-81-002, NTIS PB82-103755,* Washington, D.C., 1981.

J. Skridulis, "Comparison of Guidelines for Monitoring Well Design, Installation and Sampling Practices," *NUS Corporation Report to USEPA, Contract No. 68-01-6699,* Washington, D.C., 1984.

K. V. Summers and S. A. Gherini, "Sampling Guidelines for Groundwater Quality," *Electric Power Research Institute report EPRI EA-4952,* Palo Alto, Calif., 1987.

U.S. Environmental Protection Agency, "Subsurface Characterization and Monitoring Techniques," *EPA/625/R-93/003a and b,* Washington, D.C., 1993.

U.S. Environmental Protection Agency, "RCRA Ground-Water Monitoring: Draft Technical Guidance," *EPA/530-R-93-001,* Washington, D.C., 1992.

U.S. Environmental Protection Agency, "Seminar Publication—Site Characterization of Subsurface Remediation," *EPA/625/4-91/026,* Washington, D.C., 1991.

U.S. Environmental Protection Agency, "Guidance for Conducting Remedial Investigations and Feasibility Studies under CERCLA," *USEPA/540/G-89/004,* Washington, D.C., 1988.

U.S. Environmental Protection Agency, "Handbook—Ground Water," *EPA/625/6-87/016,* Washington, D.C., 1987.

U.S. Environmental Protection Agency, "A Compendium of Superfund Field Operations Methods," *USEPA/540/P-87/001,* Washington, D.C., 1987.

U.S. Environmental Protection Agency, "RCRA Ground-Water Monitoring Technical Enforcement Guidance Document," *OSWER-9550.1,* Washington, D.C., 1986.

U.S. Geological Survey, *National Handbook of Recommended Methods for Water-Data Acquisition,* Washington, D.C., 1977.

L. G. Wilson, "Monitoring in the Vadose Zone: A Review of Technical Elements and Methods," *USEPA-600/7-80-134,* Ada, Okla., 1980.

GROUNDWATER TREATMENT: EXPLOSIVES CONTAMINATION

MARK E. ZAPPI
Mississippi State University
Mississippi State, Mississippi

Weapons are the tools of warfare. Since early times, humanity has used a wide variety of weapons ranging from clubs and spears to guns and explosive devices. The operative word in any weapon is destruction. Explosives are a class of chemicals that are utilized for destructive purposes because of their reactive properties that occur during rapid oxidation. Rapid oxidation is typically initiated by extreme compression and product combustion which results in the evolution of large volumes of hot gases in a very short time. Explosives are used almost exclusively in weapons production where they either serve as the total weapon or as a component of a weapon system. The three main categories of explosives are propellents (used for powering rockets and missiles), primary explosives (used to initiate the reactions of secondary explosives), and secondary or high explosives (used as the *workhorse* reactants since they are safer to handle and have slower burn rates). High explosives make up the bulk of the explosives production activities because of the much higher volume required within most weapon systems. Subsequently, high explosives represent by far the most prevalent of environmental contamination requiring cleanup (1,2) (see EXPLOSIVES CONTAMINATION: DETERMINATION IN WATER AND SOIL).

High explosives have been a mainstay within the warfare theater since the turn of the twentieth century. The most common high explosives that have been used during modern times are 2,4,6-trinitrotoluene (TNT), hexahydro-1,3,5-trinitro-1,3,5-triazine (RDX), and octahydro-1,3,5,7-tetranitro-1,3,5,7-tetraazocine (HMX). These explosives have been used independently of each other, in combination with each other, or with other explosive chemicals to formulate the vast majority of weapons systems used worldwide by most armies since the early 1900s. Because of these widespread applications and the wide assortment of explosive compounds available, this article focuses on TNT, RDX, and HMX. However, limited data are discussed on other high explosives where applicable.

BACKGROUND

The first key advancement toward the development of modern explosives was the isolation of benzene from coal tar by Rev. Dr. Clayton in England in 1925. He later nitrated benzene to form trinitrobenzene, one of the first nitroaromatic explosives invented. During the 1800s, many important advances in explosives and related weapons components were discovered which included safety fuses, ammonium perchlorate, ammonium picrate, nitroglycerin, nitrostarch, the shaped charge, and dynamite. However, the most important discovery from a military viewpoint was TNT in 1863 by Willibrand, a noted German scientist of the time. Interesting enough, the early use of TNT as a textile dye was relatively benign.

In 1899, Henning of Germany discovered RDX. The United States further developed RDX by great improvements in production techniques. By World War I, the German dye industry accounted for over 85% of the world's dye industrial volume. Dye production facilities were easily modified for explosives manufacturing, thus, giving a large advantage to the Germans for wartime explosives production. Between the two world wars, RDX further developed as an explosive of great importance, especially for use in formulating plastic explosives. The English gave RDX its name because it was used in research laboratories as Research Department Explosive or RDX. TNT became the standard high explosive for bomb filling by the United States during the 1910s. Later, the English used RDX as a secondary filler component after improving its stability by adding beeswax to desensitize it slightly. Finally, during the post-World War II period, HMX was blended with TNT to formulate the octol series of fillers which found widespread usage in many of the U.S. Army's bombs and shells. HMX is reputed to have received its name by the English who labeled it Her Majesty's Explosives or HMX. Other accounts report that HMX is an abbreviation for High Melting Point Explosive (3).

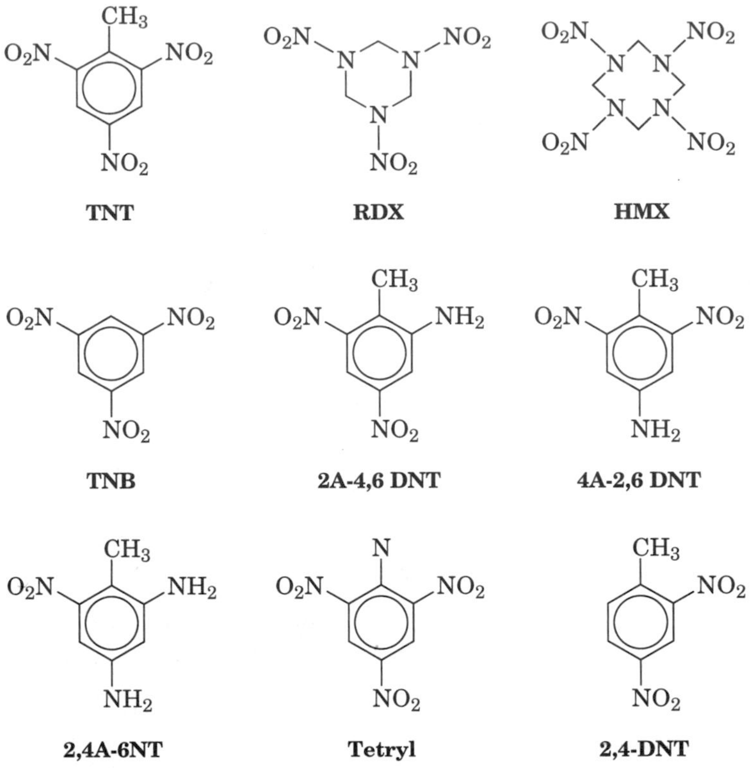

Figure 1. Chemical structures of key explosives and related by-products.

CHARACTERISTICS OF SELECTED SECONDARY EXPLOSIVES

Figure 1 illustrates the chemical structure of TNT, RDX, and HMX and related compounds. TNT is a nitroaromatic compound that is composed of a toluene ring with three nitro-substitutions on the 2, 4, and 6 carbons. Both RDX and HMX are considered nitramines because of the bonding of the nitro groups to the nitrogens within the nitrogen–carbon rings. RDX contains only three nitrogens within the N–C ring (triazine) on which three nitro groups are attached. HMX contains one more nitro-bonded nitrogen within its N–C ring. Table 1 lists the chemical and physical characteristics of all three explosives. None of these explosives are very water soluble nor are they volatile (see Table 1). These physical properties are characteristic of chemicals that adsorb well onto activated carbon (ie, low solubility and volatility).

The current recommended method for analyzing high explosives and their related breakdown products is high performance liquid chromatography (HPLC) as described in the U.S. EPA Analytical Method 8330 (5). Recent work with colorimetric and immunoassay techniques has resulted in the refinement and potential use of these innovative techniques for field analysis of explosives in both water and soils (6–8). Also, the stability of environmental samples containing explosive compounds has been evaluated (6,9). Acidification to pH 2 using sodium bisulfate with a refrigerated holding time of 30 days for water samples is recommended with no adverse effects on Method 8330 noted.

Explosives that have released into the environment pose a serious threat to the biota (1,10,11). Environmental fate studies have found explosives to be relatively mobile, yet reactive with natural soil matrices (4,12). TNT is especially noted for its susceptibility to environmental transformation into reduced by-products (4). The U.S. Environmental Protection Agency (U.S. EPA) has reviewed a variety of epidemiological and toxicological studies on explosives with respect to humans and numerous animals (3,11). Based on these studies, TNT, RDX, and HMX were deemed an EPA Group C, a possible human carcinogen. Using an estimated cancer risk of 10^{-6}, 2.0 μg/L is suggested for TNT and RDX by the U.S. EPA as the current recommended long-term drinking water health advisory (3,6). The HMX drinking water standard as reported (3) is 400 μg/L. The EPA long-term drinking water advisory for TNB, a regulated by-product of TNT oxidation, is reported to be 2 μg/L (13). Reference 14 reports an EPA Health Advisory of 0.11 μg/L for DNT (cancer risk of 10^{-6}).

CONTAMINATION SOURCES

Vast areas of surface and subsurface areas worldwide have become contaminated with explosive compounds owing to past military related activities (4,10,15). Most of these areas are located next to facilities that were involved in one or more of the following activities: (*1*) manufacturing; (*2*) load-and-pack (LAP); (*3*) practice areas; and (*4*) spent weapons disposal.

The source of contamination around manufacturing areas is predominately past improper production and wastewater disposal (10). Contamination levels around these areas are often very high with vast amounts of soil and groundwater contamination present. These sites characteristically contain many types of explosives owing to blending activities often co-located within explosives production areas. Wastewaters generated from manufacturing facilities generally are classed into three types; yellow, red, and pink waters. Only pink water

Table 1. Chemical and Physical Properties of Selected Explosives

Parameter	TNT	RDX	HMX
Synonyms	Trotyl Triton	Hexogen Cyclonite	Octogen
Appearance (STP)	Yellow solid (Monoclinic)	White solid (Orthohombic)	White solid (Monoclinic)
Molecular weight	227.13	222.13	296.17
Density, g/cc	1.654	1.806	1.89
Water solubility, mg/L			
at 20°C	150[a]	50	5[a]
at 40°C	na[b]	250	na[b]
Organics solubility			
Acetonitrile, g/g, °C	na[b]	0.12[c]	0.0198[d]
Acetone, g/g, °C	1.32[d]	0.084[c]	0.0096[d]
Volatility	Very low	Very low	Very low

[a] At 25°C (4).
[b] na = not available.
[c] At 30°C.
[d] At 25°C.

contains TNT in its commonly used isomeric form, 2,4,6-TNT, because this wastewater is produced as a result of production area rinsing (16). Red and yellow waters are produced as actual wastestreams of the production process (17,18). Yellow water results from the washing of crude TNT, a product containing 2,4,6-TNT, and other isomers of TNT. Therefore, yellow water generally contains acids and nitro bodies. Red water is produced by washing sellite. The active agent causing the red color is hexanitrodibenzyl (17). In general, traditional techniques of TNT production results in one pound of TNT produced per approximately ~380 L of wastewaters generated. Load-and-pack facilities involve the loading of the finished explosive products into the weapons systems. The primary source of contamination appears to be from the rinse waters generated during washing of the loading areas. Contamination levels tend to be slightly lower than manufacturing facilities, however, the extent of groundwater contamination tends to be similar to manufacturing areas. Practice areas tend to have relatively lower levels of contamination because of the usually fairly complete oxidation of the explosives. Also, because of usage trends, practice areas generally do contain widespread contamination, but at very much reduced levels. Contamination of the groundwaters within practice ranges is usually rare, however, soil contamination can be present depending on the age and usage rates of the range area. Disposal areas are likely the most widespread explosives contamination problem facing military installations. However, since these areas are usually small in terms of area, the extent of contamination tends to be localized with moderate to low levels of contamination likely. The disposal techniques most widely used in the past has been open burning and open detonation. Both techniques did result in soil and groundwater contamination because of the point sources that form through years of use.

Several compounds of interest, outside of TNT, RDX, and HMX, are also often detected within explosives contaminated groundwaters (see Fig. 1). Among these are the building block compounds of TNT which includes nitrotoluene, 2,4-dinitrotoluene (2,4-DNT), and 2,6-dinitrotoluene (2,6-DNT). TNT environmental breakdown products such as amino-dinitrotoluenes, diaminonitrotoluenes, and trinitrobenzene are also commonly detected in these groundwaters (19–21). Another explosive sometimes detected in soils and groundwaters in the vicinity of explosive handling facilities is 2,4,6-trinitrophenylmethylnitramine or tetryl (4,5).

DEMONSTRATED REMEDIATION TECHNOLOGIES

Remediation technologies may be classified into two main categories in terms of process development; demonstrated and developing. Demonstrated technologies are defined as those processes that have been applied at more than one site on a large-pilot or full-scale basis. Demonstrated processes are developed to such a point that design criteria and process limitations are relatively well-defined. Developing technologies may be in various states of development. In general, the basic scientific mechanisms responsible for contaminant removal are known. Further research may be needed to elucidate these mechanisms completely. The limitations that are inherent in any treatment process are usually not defined with developing processes because of the lack of design and operational experience by the engineering community.

It is important to note that remediation of explosives-contaminated groundwater has only begun to become of engineering significance since the early to mid-1980s. This interest was sparked by increasing regulatory demands by both federal and state agencies on the DOD to remediate their sites and continual base closures. However, it is not correct to state that little work has been done on the remediation of explosives-contaminated waters. In fact, since the late-1970s, much research and technology demonstration efforts have been initiated by the DOD for the treatment of explosives-contaminated waters (19,22–28). To date (ca 1997), only activated carbon adsorption and chemical oxidation can be considered demonstrated processes for explosives-contaminated groundwaters. However, only activated carbon is fully developed to a point where full operational and design criteria for use in explosive-contaminated water treatment have been proven over years of usage (16).

Activated Carbon Adsorption

Basic Concepts. Adsorption is a thermodynamic surface phenomena in which a chemical (adsorbent) is removed from one phase onto that of another (adsorbate). In terms of water treatment, adsorption has been utilized for many years for removal of organic impurities from contaminated water supplies (29,30). The most common adsorbate used in the water purification industry has been activated carbon. In this case, an organic contaminant is the adsorbent, while the activated carbon serves as the adsorbate. The forces that hold an adsorbent onto the adsorption site of the adsorbate include hydrogen bonding, dipole–dipole interaction, and van der Waals forces of electrostatic attraction (29) (see ACTIVATED CARBON ADSORPTION FUNDAMENTALS).

Cost of Activated Carbon Treatment. The final decision on what type of technology is implemented at a particular site is almost always based on economics once a set of candidate technologies that can adequately perform is developed. Process economics are typically broken down into capital and operational costs. Capital costs are essentially all costs associated with design and construction of the system. Capital costs are typically presented as a one-time occurrence. Operational costs are the costs of operating the system over time (usually on an annual basis). These costs include equipment maintenance and bulk-chemical dosing requirements. Site remediation plans and goals can significantly impact on the overall economics of the site. For example, if the site is expected to require only 5–10 years for remediation (usually not the case), then capital costs are of great importance as opposed to operational costs. Also, with some U.S. federal remediation programs, one agency funds the construction and first year costs and a second governmental entity funds the long-term operation of the system. In many cases, the operating agency is a state agency which has very limited resources. Therefore, in this case, a process with a low operational costs, yet with a slightly higher capital cost, may be the better alternative. The training level of the available operations crew may also warrant the use of a slightly more capital or operation cost-intensive option than other options if the system is much easier to operate using lesser trained crews. However, in most cases, a design life evaluation that calculates the present worth value of both capital and operational costs is the best approach (referred to as amortized costing). Design life costing requires either knowledge or strong estimates of the actual design life of the process. Design life should not be taken lightly, because this value strongly dictates process economic viability. Groundwater modeling can provide estimates of the time required for operation of a pump-and-treat system until the site cleanup objectives have been met. Most designs utilize a 20–30 year design life when information is not sufficient to provide better insight into the required system life.

The cost of implementing activated carbon in general terms for drinking water treatment is well documented (29). Reported capital costs for a 75 Mgd system is $9,973,010. Annual operational costs range from $117,920 to $2,097,950 for 1 to 75 Mgd systems, respectively. Amortized total costs are reported to range from $0.52 to $0.12 per thousand gallons treated for systems ranging from 3,785,000–567,750,000 L/d (1 to 150 Mgd), respectively (29). Amortized costs of implementing activated carbon for treatment of a high level explosive wastewater at $7.11 per thousand gallons treated for a 151.4 Lpm (40 gpm) system are reported (25). A total treatment cost of activated carbon of $4.97 per thousand gallons for a similar-sized system treating waters of similar chemical matrix is estimated in Reference 26. Capital and annual operational costs of $669,333 and $86,688, respectively, for a 1 Mgd system treating a low level contaminated groundwater are reported in Reference 3.

Chemical Oxidation

Chemical oxidation processes use powerful chemical oxidizers to destroy organic contaminants in contaminated waters and/or remove pathogens from drinking water sources (disinfection). Typical oxidizers used in chemical oxidation processes include ozone, chlorine, hydrogen peroxide, and potassium permanganate. The technology has historically been used as treatment technology for municipal drinking water (31,32,33), and more recently, industrial water and groundwater treatment (22,34,35).

Process Economics. Several observation can made with regard to the relative cost of the various advanced oxidation processes (AOPs). medium pressure mercury vapor UV lamp (MPUV) systems generally have higher operational cost because of the high energy requirements, whereas low pressure mercury vapor lamp (LPUV) systems tend to have lower energy requirements. Hydrogen peroxide and ozone costs per unit weight of oxidizer are generally about the same for long-term projects (ie, >10 yr). The cost of ozone generation equipment can be quite costly, therefore, short design life projects may not consider ozone because of these costs.

Several case studies have been presented where a commercial UV–hydrogen peroxide system successfully treated a variety of organic contaminants (36). The costs for treatment using MPUV–hydrogen peroxide systems would range from $1.37 to $58.51 per 3785 L (1000 gal) treated (36). This range is within the cost estimates presented for a MPUV–ozone process used for TNT treatment, which was estimated to cost $6.00 per thousand gallons treated (37). A variety of contaminated groundwaters and wastewaters that were successfully treated using a commercially available LPUV–ozone system is available (38). Treatment costs from $0.15 to $86.00 per 3785 L treated are indicated (38). A 378,5000 Lpm (100,000 gpm) LPUV–ozone system would cost $10.71/kgal and $7.41/kgal for using either an air-fed or oxygen-fed ozone generator, respectively, for treatment of a high level explosive contaminated wastewater (39). Based on discussions with commercial industry experts, the treatment of waters with much less explosives contamination (ie, low level contamination), using a LPUV–ozone system would cost within the $1.00–$5.00/kgal (3785 L) treated range. A comparison of the cost of activated carbon treatment and LPUV–hydrogen peroxide for treatment of a high level explosives contaminated wastewater is given in Reference 25. The results indicate that LPUV–hydrogen peroxide costs ranging from $3.08 to $4.69/3785 L treated depending on the influent explosives composition and respective concentrations. This was compared to activated carbon which was estimated at approximately $4.22/3785 L treated. Also it is noted that treatment costs associated with LPUV–hydrogen peroxide increase by over 260% by increasing the hydraulic residence time within the reactor from 30 minutes ($3.08/kgal) to 120 minutes ($8.23/kgal). A capital cost of $510,000 for a 1 Mgd LPUV–ozone system treating low levels of RDX (<5 μg/L) within a groundwater system is reported in Reference 28. This unit had an estimated operational cost of $52,238 per year. Clearly, process economics are very dependent on groundwater chemical quality and targeted treatment levels.

Since peroxone does not require the addition of high concentrations of chemical oxidizers and UV light, it is estimated that reductions in treatment costs as high as an order of magnitude may be realized (40). The results of a recently completed design evaluation for application of peroxone for drinking water treatment in Southern California suggest that potential costs savings of implementing peroxone over ozonation alone could be as high as $150 million (41). Recent cost evaluations between UV-based AOPs and peroxone for remediation of solvents contaminated groundwater indicate that peroxone could be implemented at least half the cost of the UV-based processes (42).

Summary. Treatment of explosives contaminated groundwaters using ozone or hydrogen peroxide alone does not appear promising. These oxidizers do not appear to be aggressive enough to appreciably remove the explosives from contaminated water sources. The UV–ozone systems appears to be quite effective for the removal of explosives from contaminated groundwater. The presence of ozone is almost a necessary to remove 1,3,5-trinitrobenzene (TNB) from the highly contaminated waters. The UV–ozone–hydrogen peroxide systems that utilize MPUV lamps provide the most complete and rapid explosives removal of all the AOP systems. This is not surprising in light of the higher amounts of energy emitted from UV lamps used with this AOP along with the several hydroxyl-radical formation mechanisms afforded within these systems. The combined use of ozone with hydrogen peroxide in a reactor irradiated with light from low pressure mercury vapor UV-lamps also perform well in explosives removal. This system appears to be much more economical than the same system using the medium pressure mercury vapor UV lamps because of reduced operational costs associated with the low pressure mercury-vapor UV lamp. In either system, results reported from a variety of sources all indicate that systems of these types are highly effective for explosives removal.

The UV–hydrogen peroxide systems do not appear to provide the same level of aggressive treatment provided by those UV based AOPs that utilize ozone. However, the UV–hydrogen peroxide systems, especially those that utilize the medium pressure mercury vapor UV lamps, do have potential for effectively treating groundwaters containing relatively low levels of TNT and TNB. The primary limiting factor for these systems is the removal of TNB which can require prohibitively long residence times. Concerns of over-heating of the reactor contents from the high heat output of the medium pressure mercury vapor UV lamps must be taken into account. Based on current designs, if the residence time within a reactor using medium pressure mercury-vapor UV-lamps exceed 10–15 minutes, then some form of water chilling must be employed to prevent boiling of the water within the reactor.

The peroxone and the LPUV–hydrogen peroxide systems appear to be effective for treating low level contaminated groundwaters. The impact of groundwater chemical matrix has a great impact of the effectiveness of these systems for groundwater treatment. The peroxone system is generally much more aggressive than the LPUV–hydrogen peroxide system for TNB removal, whereas the LPUV–hydrogen peroxide systems appear more effective for HMX removal. Both of these AOPs appear to have a significant economic advantage over the other systems discussed. Reported costs of these systems appear to be potentially almost a full order of magnitude cheaper than the two more traditional AOPs.

At this stage of process development, it is suggested that treatability studies be performed to evaluate the various AOP options available before proceeding to pilot studies or full-scale system design. The ozonated UV-based AOPs have had a long history of success, and therefore, may be implemented with very limited bench testing. However, the state of development is still such that bench testing is highly encouraged to ensure that the process can meet the site cleanup demands. In the case of the other AOPs, bench testing is considered an absolute requirement because of the sensitivity of these processes to groundwater chemical matrix. In many cases, it is believed that AOPs may offer an economically attractive alternative to activated carbon adsorption.

DEVELOPING REMEDIATION TECHNIQUES

Several developing treatment processes have recently been reported for treating explosives contaminated groundwaters. A brief overview of each process that appears to have potential for treating explosives contaminated groundwaters at reduced costs and/or improved performance is presented below.

Biological Treatment

A tremendous amount of research and limited field activity has been expended toward the development of biological treatment processes for *ex situ* treatment. Most of this work has been oriented toward remediation of TNT contaminated soils (19,43,44,45). Both aerobic and anaerobic processes appear to have potential for use in groundwater remediation. The aerobic and anaerobic both utilize the same general degradation pathway for TNT, which is the stepwise reduction of the nitro substitution groups into amino groups (21,46,47). Recent work has indicated the potential for the biodegradation of TNT using a novel pathway that involves the stepwise loss of the nitro groups until TNT is converted into toluene (48). In either case, much more process development is required before this promising technology can be effectively fielded. The envisioned biotreatment process that would be used is a fluidized-bed reactor that has been successfully used for treatment of explosives- and methanol-contaminated wastewaters from a production plant (49). This technology is believed to be advantageous over a suspended-bed reactor because of increase stability of the biomass and the additional benefit of the activated carbon for enhancing the attenuation potential of the bioreactor for incoming explosives (50).

Phytoremediation

Recent work by several research groups have indicated that several plant species have nitroreductases that are active toward the removal of explosives from contaminated groundwaters (51,52). The proposed design of these systems is constructed wetlands (51). Preliminary results indicate that constructed wetland systems are effective in removing both TNT and TNB, however, removal of RDX appears to be kinetically very slow (52). The potential cost benefit of this technology based on reported costs for other wetlands-based systems is encouraging plus the design protocols for these systems are fairly well established (53,54).

Photocatalytic Oxidation

Photocatalytic oxidation is an AOP that uses the surface-based reaction of semiconductor transition metals, such as titanium dioxide, with UV photons primarily in the 340–400 nm range (55). Hydroxyl radicals are generated by the formation of electron holes within the surface of the photocatalyst (56). The general theory behind the reaction of the explosives with the hydroxyl radical and UV light is identical to those associated with the more traditional AOPs discussed earlier. However, this innovative AOP has seen very limited commercial application. Some successes are reported for explosives treatment (57,58). The potential for reduction of the explosives in the oxygen-starved electron holes and the potential difficulties with TNB formation–removal requires further investigation.

Sonolytically Enhanced Ozonation

Very limited research has been conducted on the combined use of ultrasound (sonolysis) with ozone. The primary removal mechanism appears to be the generation of the hydroxyl radical during sonolysis of ozone (59). Recent research indicates that this process is very effective for removal of both TNT and TNB (43,60). The benefits of this process is that it appears to be more aggressive than peroxone and is similar to that of LPUV–ozone. However, since it is a dark AOP, then the UV transmissivity of the influent is not an issue.

SUMMARY

Activated carbon and advanced oxidation processes are effective in the removal of explosive compounds from contaminated groundwaters. The economics of both processes appear very competitive with each other. Activated carbon appears slightly more economical at this time. In either case, treatability testing is highly encouraged because of the general lack of performance data for these processes when treating the wide variety of chemical matrices found with explosives contaminated groundwaters. Several innovative alternatives may be available in the near future that do hold some promise for improved technical and economic performance (see CHEMICAL WARFARE AGENTS; CHEMICAL WEAPONS CONVENTION: NMR ANALYSIS).

BIBLIOGRAPHY

1. J. Pennington, *Adsorption and Desorption of 2,4,6-Trinitrotoluene by Soils,* Report No. EL-87-17, USAE Waterways Experiment Station, Vicksburg, Miss., 1987.
2. T. Moore, *Environ. Eng. World,* (May–June 1995).
3. M. Bricka and E. Fleming, *Use of Activated Carbon for the Treatment of Explosives-Contaminated Groundwater at the Picatinny Arsenal,* Report No. TR-EL-95-31, USAE Waterways Experiment Station, Vicksburg, Miss., 1995.
4. C. McGrath, *Review of Formulations for Processes Affecting the Subsurface Transport of Explosives,* Report No. TR-EL-95-2, USAE Waterways Experiment Station, Vicksburg, Miss., 1995.
5. R. Schuster and A. Gratzfeld-Huesgen, *HPLC Analysis of Explosives Constituents in Soil Samples,* Application Note No. 12-5091-7626E, Hawlett-Packard Inc., Wilmington, Dela., 1993.
6. T. Jenkins, P. Thorne and E. McCormick, *Preservation of Water Samples Containing Nitroaromatics and Nitramines,* Special Report 95-16, USAE Cold Regions Research and Engineering Laboratory, Hanover, N.H., 1995.
7. K. Myers, E. McCormick, A. Strong, P. Thorne and T. Jenkins, *Comparison of Commercial Colorimetric and Enzyme Immunoassay Field Screening Methods for TNT in Soil,* Technical Report No. IRRP-94-4, USAE Waterways Experiment Station, Vicksburg, Miss., 1994.
8. P. Marsden, *Environ. Testing Anal.,* **4**(1) (1995).
9. M. Maskarinec, C. Bayne, L. Johnson, S. Holladay, R. Jenkins and B. Tomkins, *Stability of Explosives in Environmental Water and Soil Samples,* Report No. TM-11770, Oak Ridge National Laboratory, Oak Ridge, Tenn., 1991.
10. W. Roberts, *Data Summary for Trinitroluene,* Technical Report No. 8611, U.S. Army Medical Bioengineering Research and Development Laboratory, Fort Detrick, Md., 1986.
11. W. Roberts and W. Hartley, *Drinking Water Health Advisory: Munitions,* Lewis Publishers Inc., Boca Raton, Fla., 1992.
12. B. Greene, D. Kaplan and A. Kaplan *Degradation of Pink Water Compounds in Soil - TNT, RDX, and HMX,* Technical Report No. TR-85/046, U.S. Army Natick Research and Development Center, Natick, Mass., 1985.
13. E. Fleming and K. Christiansen, "Removal of Explosives Using Standard and Innovative Technologies," *The Proceedings of the ACS Emerging Technologies on Hazardous Waste Management VII,* Sept. 17–20, Atlanta, Ga., 1995, abstract.
14. W. Wujcik, W. Lowe and P. Marks, *Use of Activated Carbon for Treatment of Explosives-Contaminated Groundwater at the Badger Army Ammunition Plant (BAAP),* U.S. Army Toxic and Hazardous Materials Agency Report No. CETHA-TE-CR-89216, CETHA, Aberdeen Proving Ground, Md., 1989.
15. R. Turkeltaub and C. Wiehl, *Hazardous Materials Control* (July–Aug 1989).
16. J. Cho, W. Potter and K. Subletter, *Appl. Biochem. Biotechnol.* **45** (1994).
17. G. Schulte, R. Hoehn and C. Randall, "The Treatability of Munitions-Manufacturing Waste with Activated Carbon," *Proceedings of the 28th Industrial Waste Conference,* May 1–3, Purdue University, Lafayette, Ind., 1973.
18. PEI Associates Inc., *Technology Evaluation for Treatment/Disposal of TNT Red Water,* USATHAMA Report No. CETHA-TE-CR-90048, U.S. Army Toxic and Hazardous Materials Agency, Aberdeen Proving Grounds, Md., 1990.
19. M. Zappi, A. Hong and R. Cerar, "Treatment of Groundwater Contaminated With High Levels of Explosives Using Traditional and Traditional Advanced Oxidation Processes," *Proceedings of the 1993 HMCRI Superfund Conference,* Washington D.C., 1993.
20. P. Bradley, F. Chapelle, and J. Landmeyer, "Degradation of 2,4-DNT and 2,4,6-TNT by Indigenous Aquifer Microorganisms," in R. Hinchee, R. Hoeppel, and D. Anderson, eds., *Bioremediation of Recalcitrant Organics,* Battelle Press Inc., Columbus-Richland, Ohio, 1995.
21. P. Reiger and H. Knackmuss, "Basic Knowledge and Perspectives on Biodegradation of 2,4,6-Trinitrotoluene and Related Nitroaromatic Compounds in Contaminated Soils", in J. Spain, ed., *Biodegradation of Nitroaromatic Compounds,* Plenum Press Inc., New York, 1994.
22. M. Roth and J. Murphy, *Ultraviolet-Ozone and Ultraviolet-Oxidant Treatment of Pink Water,* Westgate Research Corp., U.S. Army Report No. ARLCO-TR-78057, 1978.

23. R. Jackson, J. Green, R. Hash, D. Lindsten, and A. Tatyrek, *Nitramine (RDX-HMX) Wastewater Treatment at the Holston Army Ammunition Plant,* Report No. ARLCD-CR-77013, U.S. Army Research and Development Command, Dover, N.J., 1978.
24. C. Andrews, *Photooxidative Treatment of TNT Contaminated Wastewaters,* U.S. Navy Report No. WQEC/C-80-137, U.S. Naval Weapons Support Center, Crane, Ind., 1980.
25. B. Jackson and J. Lachowski, *Alternative Treatment Methods for Pink Wastewater,* U.S. Army Toxic and Hazardous Materials Agency Report No. ARLCO-TR-83009, 1983.
26. D. Burrows and E. Brueggemann, *Tertiary Treatment of Effluent from Holston AAP Industrial Liquid Waste Treatment Facility Report V: Degradation of Nitramines in Holston AAP Wastewaters by Ultraviolet Radiation,* U.S. Army Medical Bioengineering Research and Development Laboratory, Fort Detrick, Frederick, Md., 1986.
27. M. Zappi, "Innovative Oxidation Technologies", *Ground Water Currents,* Document No. EPA/542/N-93/011, U.S. EPA Office of Solid Waste and Emergency Response, Washington, D.C., 1993.
28. E. Fleming, R. Bricka and S. Bailey, *Ultraviolet/Chemical Oxidation Treatment of RDX-Contaminated Waters at Picatinny Arsenal,* Report No. EL-95-7, USAE Waterways Experiment Station, Vicksburg, Miss., 1995.
29. S. Faust and O. Aly, *Adsorption Processes for Water Treatment,* Butterworths Publishing Inc., Boston, Mass., 1987.
30. American Water Works Association, *Water Quality and Treatment: A Handbook of Community Water Supplies,* 4th ed., AWWA, Denver, Colo., 1990.
31. MetCalf and Eddy Inc., *Wastewater Engineering: Treatment, Disposal, and Reuse,* 3rd ed., McGraw-Hill Publishing Inc., New York, 1991.
32. James Montgomery Consulting Engineers Inc., *Water Treatment Principles and Design,* John Wiley & Sons, Inc., New York, 1985.
33. B. Langlais, D. Reckhow and D. Brink, *Ozone in Water Treatment,* Lewis Publishers Inc., Chelsea, Mich., 1991.
34. J. Echols and S. Mayne, *Chem. Eng.* (May 1990).
35. M. Zappi, F. Ragan, D. Guimbellot, N. Francingues, S. Harvey, J. Smith, D. Strang, E. Kaastrop and D. Burrows, *A Laboratory Evaluation of the Feasibility of Chemical Oxidation Processes for Treatment of Contaminated Groundwaters,* Report No. MP-IRRP-95-1, USAE Waterways Experiment Station, Vicksburg, Miss., 1995.
36. D. Hager, C. Loven and C. Giggy, "Chemical Oxidation Destruction of Organic Contaminants in Groundwater," *Proceedings from the 1987 Superfund Conference,* Washington D.C., 1987.
37. D. Reed, "Destruction of TNT in Pink Water from Explosives Manufacturing Using Rayox Enhanced Ozonation", *paper presented at the 18th Environmental Symposium of the American Defense Preparedness Association,* Alexandria, Va., 1992.
38. J. Barich and J. Zeff, "A Review of Ultrox Ultraviolet Oxidation Technology As Applied to Industrial Groundwater, Wastewater, and Superfund Sites," *Proceedings of the 1989 Superfund Conference,* Washington D.C., 1989.
39. W. Layne, R. Nicholson, R. Wahl, P. O'Brian, B. Jackson and J. LaChowski, *Ultraviolet-Ozone Treatment of Pink Wastewater: A Pilot Scale Study,* Report No. ARLCD-CR-82023, U.S. Army Armament Research and Development Command, Dover, N.J., 1982.
40. M. Zappi, *Peroxone Oxidation Treatment of 2,4,6-Trinitrotoluene Contaminated Waters With and Without Sonolytic Catalyzation,* Dissertation Submitted to the Department of Chemical Engineering, Mississippi State University, MSU, Miss., 1995.
41. M. McQuire and M. Davis, *Water Eng. Management* (May 1988).
42. T. Streckfuss, COE-Omaha and J. Talley, COE-Baltimore, personal communication, 1995.
43. M. Zappi, E. Toro, K. Myers, R. Hernandez, P. Kodukula, and R. Gilbertson. *A Laboratory Assessment of the Potential for Treating Groundwater from the Former Nebraska Ordnance Plant Contaminated With Low-Levels of Explosives and Chlorinated Solvents Using Advanced Oxidation Processes,* USAE Waterways Experiment Station Treatability Report, 1995.
44. C. Montemagno and R. Irvine, *Feasibility of Biodegrading TNT-Contaminated Soils in a Slurry Reactor,* CETHAMA Report No. TE-CR-90062, U.S. Toxic and Hazardous Materials Agency, Aberdeen Proving Ground, Md., 1990.
45. S. Eversmeyer, P. Faessler, and C. Bird, *Soil and Groundwater Cleanup* (Dec. 1995).
46. J. Walker and D. Kaplan, *Biodegradation* **3** (1992).
47. F. Marvin-Sikkema and J. de Bont, *Appl. Microbiol. Biotechnol.* **42** (1994).
48. J. Ramos, A. Haidour, A. Delgado, E. Duque, M. Fandila, M. Gil and G. Pinar, "Potential of Toluene-Degrading Systems for the Construction of Hybrid Pathways for Nitrotoluene Metabolism", in J. Spain, ed., *Biodegradation of Nitroaromatic Compounds,* Plenum Press Inc., New York, 1994.
49. S. Berchtold, S. Vanderloop, M. Suidan and S. Maloney, *Water Environ. Res.* **67**(7) (1995).
50. M. Zappi, E. Fleming, D. Thompson, and N. Francingues, "Treatability Study of Four Contaminated Waters at Rocky Mountain Arsenal, Commerce City, Colorado Using Chemical Oxidation With Ultraviolet Radiation Catalyzation," *Proceedings of the 14th Annual Army Environmental R&D Symposium,* Williamsburg, Va., 1989.
51. J. Schnoor, L. Licht, S. McCutcheon, L. Wolf, and L. Carriera, *Environ. Sci. Technol.,* **29**(7), (1995).
52. E. Best, M. Zappi, H. Fredrickson, S. Larson, S. Sprecher and M. Ochman, "Draft-Fate of TNT and RDX in Aquatic and Wetland Plant-Based Systems During Treatment of Contaminated Groundwater," submitted for Publication to the New York Academy of Sciences, 1996.
53. U.S. EPA, *Constructed Wetlands and Aquatic Plant Systems for Municipal Wastewater Treatment,* USEPA Report No. 625/1-88/022, USEPA, Cincinnati, Ohio, 1988.
54. M. Hammer, *Constructed Wetlands for Wastewater Treatment: Municipal, Industrial, and Agricultural,* Lewis Publishers, Chelsea, Mich., 1989.
55. D. Ollis and H. Al-Ekabi, *Photocatalytic Purification and Treatment of Water and Air,* Elsevier Publishers Inc., New York, 1993.
56. H. Gerischer, "Conditions for an Efficient Photocatalytic Activity of TiO_2 Particles", M.D. Ollis and H. Al-Ekabi, eds., *Photocatalytic Purification and Treatment of Water and Air,* Elsevier Publishing Inc., New York, 1993.
57. D. Schmelling and K. Gray, "Feasibility of Photocatalytic Degradation of TNT as a Single or Integrated Treatment Process," in D. Ollis and H. Al-Ekabi, eds., *Photocatalytic Purification and Treatment of Water and Air,* Elsevier Publishers Inc., New York, 1993.
58. E. Selby, *Degradation of Trinitrotoluene Using Photocatalytic Oxidation Processes,* Thesis Submitted to the Civil and Environmental Engineering Department, Howard University, Washington D.C., 1996.
59. C. Huang, C. Dong, and Z. Tang, "Advanced Chemical Oxidation: Its Present Role and Potential Future in Hazardous Waste Treatment," *Proceedings of the 5th Annual Symposium of the Gulf Coast Hazardous Waste Substance Research Center,* Lamar University, Beaumont, Texas, 1993.
60. R. Sierka, *Ozone Science Eng.* **6,** 275–290 (1985).

GROUNDWATER TREATMENT: NITRATE CONTAMINATION

MICHAL GREEN
Israel Institute of Technology
Haifa

Water quality records in the last decades have shown a continuous increase of nitrate concentrations in groundwater, mainly due to intensive farming. Although the best solution to nitrate pollution would be to prevent its input into groundwater, the anticipated results will be achieved only after 10 to 40 years due to the long transportation time through the unsaturated zones. Consequently, to meet the standards for nitrate levels in the near future, the application of advanced treatment technologies is required.

The technologies developed for suitable nitrate removal fall into two categories: biological processes and physicochemical processes. Biological denitrification removes mainly nitrates and oxides of nitrogen, while the physicochemical processes remove nitrate and many other inorganic constituents. Selection of the preferred treatment technology depends on the specific requirements, such as plant size, stability of water demand and nitrate content, regional temperatures, water loss, need for additional water treatment requirements, and waste disposal options.

At present, ion-exchange and heterotrophic denitrification are the most widely used technologies. The technical and economic feasibility of these processes have been confirmed on full-scale operation. Because of the low installation costs and easier control, ion-exchange technology may be an attractive alternative for nitrate removal, especially for small sites. However, because of the problematic disposal of the concentrate produced in this process, in areas where brine discharge to sewers is not allowed, the ion-exchange technique is limited to coastal areas or to sites that are close to other large receiving waters, not susceptible to nitrogen-borne eutrophication. While the costs of the ion-exchange process are unlikely to change in the near future, disposal of the brine is becoming increasingly problematic, involving much higher costs.

Reverse osmosis and electrodialysis can also be used to remove nitrates from drinking water. However, costs for membrane processes have been prohibitive, except for small installations. But recognizing that research and development into higher rejection, lower pressure membranes are currently expanding exponentially, it is likely that in the near future these technologies will have a profound effect on the efficacy of nitrate removal processes.

NITROGEN IN THE ENVIRONMENT

Nitrogen Cycle

Nitrogen is an essential element for all organisms. Elemental nitrogen is a component of amino acids, nucleic acids, coenzyme nucleotides, and other constituents of the cell. Plants (autotrophs) directly and indirectly provide the main source of nitrogen to animals (heterotrophs). Natural sources of nitrogen for the plants are mainly inorganic substances found in animal excreta, animal and plant remains, and in some special cases also in the atmosphere. The earth's atmosphere consists of 78% nitrogen. However, this inert gaseous nitrogen can not be used readily by plants, and it must be combined with oxygen (eg, nitrate, NO_3^-) or hydrogen (ammonia, NH_3) before it is readily available to plants. The principal mechanism by which atmospheric nitrogen is made available to animal and plants is biological nitrogen assimilation (fixation). This is accomplished by some groups of bacteria, both free living and in association with the roots of higher plants, blue green algae, and some fungi. Other atmospheric nitrogen fixation processes are lightning and industrial processes.

In addition, ammonia may volatilize from soil, be released to the atmosphere through transpiration and from combustion processes (eg, from power plants), or be emitted from volcanic eruptions. These gaseous products are eventually washed out of the atmosphere by precipitation and dustfall and transported to the soil.

A direct, natural result of the nitrogen cycle (Fig. 1) is the contamination of water by inorganic nitrogen compounds. In this regard, the nitrate ion, due to its chemical stability, is the major contaminant. Nitrate is produced through biological nitrification, which is the oxidation of ammonium (NH_4^+) to nitrite (NO_2^-) and nitrate (NO_3^-) :

$$NH_4^+ + 1.5\,O_2 \rightarrow 2\,H^+ + H_2O + NO_2^-$$

$$NO_2^- + 0.5\,O_2 \rightarrow NO_3^-$$

Nitrification is carried out mainly by chemoautotrophic bacteria. However, several heterotrophic microorganisms (bacteria and fungi) also contribute to the oxidation of ammonium to nitrate. The two major groups of chemoautotrophic bacteria responsible for the nitrification process are, first, *Nitrosococcus* and *Nitrosomonas* species, which oxidize ammonia to nitrite, and, second, *Nitrobacter* species, which oxidize nitrite to nitrate. In nature, the nitrate formed may be reduced again to either ammonia or nitrogen gas (by the bacterial denitrification process) and released to the atmosphere:

$$NO_3^- + 6\,H^+ + 5\,e^- \rightleftarrows \frac{1}{2}\,N_2 + 3\,H_2O_2$$

$$NO_3^- \rightarrow NO_2^- \rightarrow NO \rightarrow N_2O \rightarrow N_2$$

In an undisturbed natural ecosystem, nitrate concentration in groundwater is usually stable and low, < 3 mg/LNO_3–N

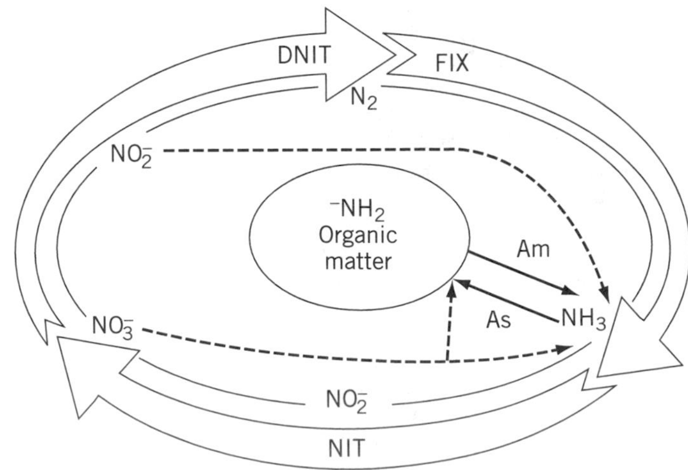

Figure 1. The nitrogen cycle where NIT = nitrification, FIX = fixation, DNIT = denitrification, As = assimilation, and Am = ammonification (decomposition).

due to immobilization of nitrogen compounds by plants and reduction processes (to nitrogen gas), thereby minimizing excess nitrates available for leaching to groundwater.

Nitrate Pollution

Sources and Trends. Nitrate contamination is amplified substantially by human activity. Intensive farming (the application of artificial fertilizers and animal manure and intensive cultivation of nitrogen-fixing legumes), together with municipal and industrial wastes have been identified as the major sources of nitrogen compounds in water (1).

Other anthropogenic sources of nitrate are landfill leachate, urban surface runoff, leaching of nitrate and ammonia from geologic deposits of nitrate salts due to intensive irrigation, mining of minerals of nitrate salts, and industrial processes releasing nitrogen compounds to the atmosphere, which may later be transported to the soil by precipitation and dustfall. Precipitation, farming patterns, and the hydrology of the aquifer are the principal additional factors that determine the extent of groundwater contamination.

Water-quality records have shown a continuous increase in nitrate concentrations in groundwater since the early 1970s, mainly due to intensive farming. In some areas of Denmark and the Netherlands, an increase in the nitrate level of 0.2–1.3 mg/L/yr has been reported (2). In England, the number of groundwater sources containing nitrate concentration >50 mg/L increased from 60 in 1970 to 105 in 1984 and 142 in 1987, and from 154 to 192 between 1989 and 1990 (3).

In Israel, about half of the water wells in the coastal aquifer (supplying about 30% of the total water consumption) exceeds international standards for dissolved nitrate. Nitrate as the major contaminant in groundwater was reported by 75% of the states in the United States (4), where 23% of the violations of primary drinking water standards were the result of excessive nitrate.

Environmental and Health Aspects

Environmental Aspects. Nitrate as a nutrient in surface water might increase eutrophication (ie, cause excessive plant growth and/or algal blooms). Increased eutrophication due to increasing nitrate concentration will occur in water bodies where nitrogen, and not phosphorus, is the limiting nutrient. Eutrophication can result in deterioration in the appearance of water bodies, odor problems, lower dissolved oxygen levels, and fish death. In addition, nitrate in surface water and groundwater flowing to the sea can induce excessive marine blooms.

Health Effects. Sources of nitrate in human body include diet (from water, vegetables, and processed food) and endogenously formed nitrate. The fractions from each source depend mainly on nitrate concentration in the water and on the diet. The U.S. Environmental Protection Agency (EPA) has estimated that in the case of drinking water with maximum contaminant level (MCL) level (ie, 10 mg/L NO_3^-N), approximately 50% of the nitrate intake originates from drinking water (5). Nitrate itself is not toxic; however, nitrate in the body can be reduced to nitrite by the oral and gastrointestinal microflora, and nitrite can be toxic.

Methemoglobinemia in Infants. Transformation of hemoglobin to methemoglobin occurs when nitrite (or some other compounds) oxidizes the iron in the hemoglobin from the ferrous to the ferric state. This transformation causes a reduction in the oxygen-carrying capacity of the blood. The transformation is normally reversible, and levels of 1–2% methemoglobin are normal. At 10%, toxic syndromes appear, and at 50–60% death is possible. Methemoglobinemia is a disease primarily affecting infants because of specific physiologic differences between them and adults: infants have higher stomach pH, which is favorable for the nitrate reducing bacteria; higher water consumption relative to the body weight; higher percentage of fetal hemoglobin, which is easier to oxidize; higher sensitivity to gut infections, which allow for more nitrite absorption; and an immature enzymatic system to effect methemoglobin reduction. Most reported cases of methemoglobinemia arose in areas with a nitrate concentration in the drinking water >100 mg/L, and no cases have been reported in areas with a concentration <50 mg/L.

Nitrate and Cancer. Nitrate and nitrite when digested can form *N*-nitroso compounds, which are potential carcinogens in humans. A direct correlation between nitrosamines and cancer has been proved only in laboratory animals. According to the EPA, the existing studies are not adequate to show that nitrite causes an increased cancer risk; however, data concerning the possible oncongenicity of nitrate or nitrite cannot be entirely dismissed (5).

Nitrogen Control: Regulations and Policy

The problem of nitrate pollution in drinking water has received public and political attention since the 1970s. Standards of 44.3–50 mg/L as NO_3 and 10–11.3 mg/L as N, from WHO, EPA, and Canada, and the EEC, respectively, are presently in effect in most countries. This standard for nitrate is based directly on a nominal observed adverse effect level in infants.

There are several options for controlling nitrate concentrations in drinking water: blending waters containing high nitrate concentrat ions with those containing low concentrations; replacing the water source, and treating the water. The first two options are viable alternatives only if sources of high quality water are available. Increase in costs with distance of transportation (pipes and pumps) eventually surpasses the benefits of these two alternatives.

Although the best solution to nitrate pollution would be to prevent its input into groundwater, the anticipated results will be achieved only after 10–40 years due to the long and variable transportation time through unsaturated soil zones to a water table at variable depth. Therefore, application of advanced technologies for water treatment is a justified alternative of meeting required nitrate levels.

Since the early 1980s thorough research has been conducted, mainly in Europe, to investigate and develop processes for the removal of nitrate from drinking water. The technologies developed for suitable nitrate removal fall into two categories: biological processes and physicochemical processes. Biological denitrification removes mainly nitrates and oxides of nitrogen, while the physicochemical processes remove nitrate and many other inorganic constituents. Selection of the preferred treatment technology depends on the local specific requirements.

PHYSICOCHEMICAL PROCESSES FOR THE SEPARATION OF NITRATES FROM GROUNDWATER

Membrane Separation for Nitrate Removal

Reverse Osmosis. In reverse osmosis (RO) water is forced to flow from a concentrated solution to a dilute solution across a semipermeable membrane by applying a pressure gradient across the membrane that exceeds the osmotic pressure. The osmotic pressure varies with concentration and absolute temperature.

Limitations and Problems. Because of the selectivity of the membranes, salt accumulates at the membrane surface. A boundary layer of high salt concentration is established, which causes the so-called concentration polarization phenomenon. This phenomenon creates several negative effects:

- Decrease in water permeate flux due to higher osmotic pressure close to the membrane
- Increase in salt concentration in the permeate
- Increase in fouling of the membrane via mineral precipitation and accumulation of other ions

Electrodialysis. In electrodialysis an electrical potential is used together with semipermeable ion-selective membranes to separate ions from the bulk solution. These membranes prevent the passage of water and allow the passage of either cations (cationic membrane) or anions (anionic membrane). The separation is accomplished by the application of electrical potential between two electrodes, which causes an electric current (ionic migration) to pass through alternating cationic and anionic selective membranes. The cations pass through a cationic membrane on their way toward the cathode, but their path is halted an anionic membrane. Similarly, anions pass through the anaionic membrane on their way toward the anode, and their path is halted by the cationic membrane. These processes create cells of concentrated and dilute (permeate) solutions and form the basis of electrodialysis.

Usually, an electrodialysis plant consists of 100–250 unit cells. Raw water is fed through a manifold into all compartments. An electromagnetic force of about 1 V per cell is applied, and the current density is set at about 5 mA/cm^2. The reject stream retains about 90% of the total inflow.

Limitations and Problems. In principle, problems arising with the process are those emanating from polarization discussed earlier in conjunction with the reverse osmosis process. The membrane scaling problem can be avoided (or at least minimized) in the electrodialysis process by alternating the polarity of the electrodes. Preferred removal of particular ionic species using ion-selective membranes can reduce the energy consumption and increase the amount of treated water produced per membrane area. Recent developments have produced nitrate-selective membranes. However, at present they can be operated only at low current density, thus a large membrane area is required. Higher current application results in local polarization with respect to NO_3^-, which decreases the selectivity to NO_3^-.

Nitrate Removal by the Membrane Processes. Reverse osmosis and electrodialysis can be used to remove nitrates from drinking water (6). In both processes, the nitrate is not eliminated; it is simply concentrated into a waste (brine) stream that must then be disposed of. The brine stream volume amounts to 5–25% of the total water production, with a corresponding increase in salt concentration (7). For reverse osmosis, a usable water recovery rate of about 75% is optimal (8). This is not acceptable when water resources are at a premium and brine disposal is ecologically objectionable. Under these constraints, the recovery rate should be at least 85%, requiring more expensive systems due to higher operating pressures, increased scaling potential, and reduced membrane lifetime. The situation is exacerbated for hard water containing high concentrations of carbonates species because of the increased potential for precipitation of these species onto the membrane when higher recovery rates are attempted. Prevention of membrane fouling due to carbonate precipitation usually is affected by depressing pH via addition of acid. However, this means addition of chemicals to the environment. Usually, water recovery rate in nitrate removal from well water by reverse osmosis is limited by the saturation concentrations of potential precipitants. In the case of electrodialysis, where higher water recovery rates are possible, zero discharge of salts again is prohibitively expensive.

Reverse osmosis is usually planned for the treatment of only part of the total well capacity (split treatment), with subsequent blending of the permeate with untreated water (6) to give a final quality that meets present standards with minimal cost. This arises because of the near total removal of ions in the process before blending. In contrast, in electrodialysis systems near total removal of ions is not economically justified. Consequently, the total water supply is treated. For both these membrane systems, the feed must be pretreated to remove colloidal matter to prevent membrane fouling. Usually, the pretreatment consists of chemical clarification, filtration, ultrafiltration, and sometimes the removal of iron and manganese to decrease scaling potential.

In the reverse osmosis process, it is not possible to uncouple removal of NO_3^- from the removal of other ions in the solution's ionic matrix. Electrodialysis, on the other hand, offers the potential application of nitrate-specific anaionic membranes (modified cellulose membranes), allowing uncoupling of nitrate removal from other ionic species. Application of such technology significantly reduces process energy consumption and increases volume of permeate water produced per membrane area (9). However, costs have been prohibitive, except for small installations. Nonetheless, recognizing that research and development into higher rejection low pressure membranes are currently expanding exponentially (10), it is likely that in the near future selective membrane technologies will have a profound effect on the efficacy of the nitrate removal processes.

Ion-Exchange Processes for the Removal of Nitrates from Groundwater

General. In the ion-exchange process, ions that are held by electrostatic forces to charged functional groups on the surface of the resin are exchanged with ions originating in the treated solution. Synthetic ion-exchange resins usually consist of hydrocarbon radicals cross-linked in a three-dimensional matrix, to which soluble ionic functional groups are attached. The behavior of the resin (eg, the ion-exchange capacity, equilibrium, and selectivity) depends mainly on the attached ionic group, ie, their number per unit weight and their type, and the internal pore structure of the resin. In cation exchangers, the resins contain functional acid groups derived from either strong or weak acids. In anion exchangers, the resins contain functional groups such as a primary amine, R-NH_2 (when R stands for the organic network of the resin), secondary amine, R-R'NH (when the R' stands for organic radical), tertiary amine, R-R'_2N, or quaternary ammonium group, $R - R'_3N^+OH^-$. Weakly basic resins are defined as those resins containing functional groups derived from weak base amines, whereas in strongly basic resins, the functional group is derived from quaternary ammonium compounds. The charges of the functional groups attached to the resins are balanced by counterions of opposite charge to maintain electroneutrality. These counterions are exchanged during the ion-exchange process. The resins are regenerated with a strong solution of the exchangeable counterion.

When the ion-exchange process is used to treat water containing nitrate, the nitrate ion is exchanged for either chloride or bicarbonate, depending on the resin used. Usually, strong base, macroporous anion resins are used, with either countercurrent or cocurrent regeneration in both fixed beds and continuous systems. Like all physicochemical processes, nitrate is not eliminated, but concentrates into a relatively small volume reject stream. All or part of the water is treated, depending on nitrate concentration in the raw water and on the required concentration in the blend. Usually water containing nitrates also contains significant concentrations of sulfate, chloride, and bicarbonate ions. When the water is passed through an anion-exchange column saturated with chlorides, initially almost all the anions in the water is replaced by chlorides from the resin; thus their concentration in the treated water is close to zero, while the chloride concentration in the treated water increases accordingly. Later in the ion-exchange cycle, attached ions with lower exchange preference (ie, lower affinity of the resin to these ions) are replaced by ions with the higher preference, thus their concentration in the treated water increases (11). A typical order of affinity for strong base anion exchange resins is the following (12):

$$SO_4^{2-} > NO_3^- > Cl^- > HCO_3^-$$

This decreasing affinity order indicates that usually (depending on concentration) HCO_3^- is the first anion to appear again in the treated water (after the chloride) followed by NO_3^-, which is replaced from the resin by SO_4^{2-} if too long a run time is applied. These characteristics of ion-exchange processes are responsible for the reduced nitrate removal efficiency in high sulfate waters and for the increased aggressiveness and corrosiveness in the treated water.

Ion-exchange processes are well established in water treatment technology and are also widely applied to industrial water treatment. Low installation costs and the possibility for automatic control make ion exchange an attractive alternative for nitrate removal, especially for small facilities.

Limitations and Problems. Intensive research at both the laboratory and the pilot levels have led to the full-scale implementation of ion-exchange water treatment plants since 1983 (13–15), especially in Europe. The main drawback of the ion-exchange process is the production of large amounts of waste (up to 2% of the total product water) with high nitrate, sulfate, and chloride or bicarbonates concentrations (4–9 times the equivalent amount of NO_3^- removed), which are difficult to dispose of. Disposal of these concentrates can be problematic financially and environmentally (16).

Environmental considerations regarding the release of undesirable substances originating in the process reject stream (mainly chloride, sodium, and nitrate), prohibit the application of this technology in places that are near the coast or large rivers and where regulations prevent discharge to sewers, due to the potential negative impact on sewage treatment processes.

Moreover, regulations for both chloride discharge to fresh water and nutrient discharge to either fresh or salt water are becoming increasingly stringent. These regulations are aimed at avoiding damage to agricultural crops from irrigation with water containing a high concentration of sodium and chloride and minimizing eutrophication from high concentrations of nitrogen compounds. It is anticipated that in the near future the disposal of chlorides and nitrates to water bodies, either saline or fresh, will be severely restricted.

In the last few years, several approaches have been adapted to ameliorate the problems encountered with the ion-exchange technique.

BIOLOGICAL PROCESSES FOR THE REMOVAL OF NITRATES FROM GROUNDWATER

General

Biological denitrification is the dissimilatory reduction of nitrate or nitrite to nitrogen gas through the reduction of the nitrogen gaseous oxides. Generally, denitrification is considered to be an anoxic process carried out principally by facultative bacteria, ie, bacteria that are able to use either nitrate (where oxygen is limiting) or oxygen as an electron acceptor. Ability to denitrify is widespread among bacteria in water. The reduction of nitrate to nitrogen gas proceeds through four steps: $NO_3 \rightarrow NO_2 \rightarrow NO \rightarrow N_2O \rightarrow N_2$ with many bacterial species able to perform only part of these consecutive steps. Any one of the last three steps can be the final one, depending on the species involved and on the pH (17). The process of denitrification, because of its similarity to oxygen respiration, is termed *nitrate respiration*. The energy yield of nitrate respiration, is slightly lower than that of oxygen respiration and, therefore, enzyme systems of most denitrifying bacteria have evolved to be inhibited by oxygen.

Most of the denitrifiers are heterotrophic bacteria and can use a variety of organic carbon sources, with methanol, ethanol, and acetic acid being the most commonly used in water treatment to effect denitrification. The stoichiometric equations for ethanol as the carbon and electrons source are as follows. Stoichiometric catabolic equation:

$$10\ NO_3^- + 10\ H^+ + 25/(6\ C_2H_5OH) \rightarrow 25/(3\ CO_2) + 5\ N_2 + 35/(2\ H_2O)$$

Stoichiometric anabolic equation:

$$7/(3\ C_2H_5OH) + NO_3^- + H^+ + 1/(3\ CO_2) \rightarrow C_5H_7O_2N + 4\ H_2O$$

Overall stoichiometric metabolic equation:

$$39\ C_2H_5OH + 66\ NO_3^- + 66\ H^+ \rightarrow 6\ C_5H_7O_2N + 48\ CO_2 + 30\ N_2 + 129\ H_2O$$

The most commonly found heterotrophic denitrifying species belong to the bacterial genera *Pseudomonas, Alcaligenes,* and *Flavobacterium.* Denitrification can also be effected by autotrophic bacteria, which cannot use organic carbon as an energy source but use some inorganic reduced species, eg, H_2. Detailed basic information about the biochemistry, microbiology, and factors controlling denitrification has been covered thoroughly (18–20).

The biological technologies for nitrate removal, both heterotrophic and autotrophic, fall into two categories: aboveground and underground. Biological processes can be applied for both groundwater and surface water, with surface water treatment limited to those cases where temperatures are sufficiently high.

In Situ Biological Denitrification

Naturally occurring heterotrophic denitrification and the less commonly found autotrophic denitrification, in both shallow and deep aquifer systems, are well known phenomena (18,21,22). The carbon and energy source used by heterotrophic denitrifying bacteria originates from either surface organic matter or from biodegraded fossilized organic matter present in the aquifer bed. In autotrophic denitrification, the carbon source is dissolved inorganic carbon (CO_2, HCO_3) and the energy source is inorganic substances such as: Fe^{2+}, FeS_2, H_2, and S_x.

The natural *in situ* denitrification process is slow, due to the limited availability of an organic carbon source. Consequently, its engineering application requires the injection of various substrates and nutrients into the aquifer to stimulate and accelerate the natural process. Several investigations have been made in the use of different carbon and energy sources, including methanol, ethanol, sucrose, acetic acid, treated wastewater, methane (18), and hydrogen plus CO_2.

One advantage of *in situ* denitrification is that it inherently includes part of the post-treatment required for all biological denitrification processes. Normally the necessary full post-treatment consists of filtration, degradation of organic residuals, disinfection, and aeration. In most cases, disinfection of the treated water is required, while the necessity for other post-treatment processes is less common.

In situ denitrification processes have been investigated in both the laboratory and field. These experiments used several carbon sources and a variety of arrangements for the wells, including carbon source injection wells, recharging nitrate polluted water wells, and denitrified water pumping wells. Short-term pilot *in situ* denitrification experiments were carried out in the Netherlands, the former Czechoslovakia, Israel, and France (9,19,23,24). Well clogging problems, mainly due to accumulation of product gases, were observed in the Netherlands and Israel, but not in the former Czechoslovakia.

Full-scale denitrification experiments using the DAISY system were conducted in the former Czechoslovakia (25). The system is based on one pumping well surrounded by small-diameter injection wells. To achieve better mixing of the carbon source (95% ethanol + 5% methanol) with the groundwater, part of the pumped water was recirculated to each injection well. Nitrate removal efficiency of approximately 40% was observed. No clogging in this system was observed.

In France, an *in situ* reactor with a flow rate of 30 m^3/h was in operation for five months (26). A total of 30 injection wells were bored in a circle 25 m from the central collection well and were fed with recirculated water containing ethanol as the energy and carbon source. A nitrate removal efficiency of 70% (a reduction from 100 to 30 mg/L as NO_3) was observed at operation conditions of a C:N ratio of 1.25 and hydraulic retention time of 0.95 day. Post-treatment of the denitrified water (oxygenation, filtration, and chlorination) was necessary to obtain drinking water quality. Discontinuous injection pumping regime (1 h rest period) was applied in an attempt to minimize clogging risks. However, this proved to be insufficient and the work was terminated.

A full commercial scale plant (21 5 m^3/h) is being successfully operated in Bisamberg, Austria, using the NITRODEX method (27). A typical NITRODEX plant is composed of a number of injection-pumping wells located on the circumference of two concentric circles and a pumping well at the center, through which the treated water is extracted. The injection wells situated on the outer ring are used for nitrate reduction, while those in the inner ring are used for degassing of the nitrogen gas and oxidation of residual nitrite to nitrate. Dissolved organic carbon (either methanol or ethanol) is injected intermittently through the wells in the outer circle, while degassed aerated water is injected intermittently in the inner circle. At present, the average nitrate reduction in Bisamberg is about 25 mg/L, and nitrate concentration in the treated water never exceeds 30 mg/L as NO_3. No clogging has been reported (28).

Another approach of the *in situ* technique was studied in Germany. Nitrate contaminated groundwater was passed through filter beds composed of soil and aquatic plants (29). Nitrate reduction of about 90% was reported. A demonstration plant (25 m^3/h) has been operating successfully since 1988.

Summarizing, the *in situ* denitrification process has a number of advantages and disadvantages. The advantages include a relatively simple and cheap operation and less intensive post-treatment. The principal disadvantages are a high potential for clogging the soil media, a difficult to control process, and a relatively slow reaction rate. The disadvantages are interlinked via the complex nature of the hydrogeological conditions. As a consequence, no reasonable control of the process is possible, with resultant clogging of the porous media caused mainly by the accumulation of product gases and biomass. These factors limit the successful application of the *in situ* denitrification process to sites with specific geological conditions.

Autotrophic Denitrification

Autotrophic denitrification refers to the dissimilatory reduction of nitrogen oxides by autotrophic bacteria using inorganic compounds such as hydrogen, sulfur, and sulfur derivatives as the electron donor and energy source. Examples of such organisms include *Thiobacillus denitrificants,* which uses sulfur and its derivatives, and *Paracoccus denitrificans,* which uses hydrogen. The main advantage of autotrophic denitrification for water treatment purposes lies in the fact that no organic carbon source residuals remain in the treated water. However, the reaction rate of this process is much lower than the heterotrophic one, due to the slow growth rate of the autotrophic bacteria.

Autotrophic denitrification using either sulfur or reduced sulfur derivatives has been studied by several investigators in laboratory and pilot experiments (19,30,31). The following stoichiometric relationships for sulfur and reduced sulfur compounds have been reported.

Thiosulfate (31):

$$5\ S_2O_3^{-2} + 8\ NO_3^- + H_2O \rightarrow 4\ N_2 + 10\ SO_4^{-2} + 2\ H^+$$

Sulfide (32):

$$5\ S^{-2} + 8\ NO_3^- + 8\ H^+ \rightarrow 5\ SO_4^{-2} + 4\ N_2 + 4\ H_2O$$

Sulfur (33):

$$5\ S + 2\ H_2O + 6\ NO_3^- \rightarrow 5\ SO_4^{2-} + 3\ N_2 + 4\ H^+$$

A process has been developed for nitrate removal using a sulfur–limestone filter (the limestone is used to neutralize the acid formed during the denitrification, whereas sulfur is used as the electron donor and as inorganic carbon source) (19). Oxygen and nitrogen, which interfere with the process, were removed by vacuum deaeration, thus eliminating the need for backwashing. After denitrification, the water was aerated and injected into the aquifer for post-treatment. Based on extensive research with laboratory and pilot units (32) and on five years of operation of a 35 m^3/h demonstration plant, a 100 m^3/h full-scale plant was built in the Netherlands and has operated since 1991. The full-scale plant consists of a vacuum deaerator to remove nitrogen and oxygen gases, an upflow sulfur–limestone denitrification reactor, a cascade aerator, and artificial recharge to the aquifer. Water is extracted after a residence time of 10–12 weeks.

The process has proven to be simple and reliable; however, it results in an effluent with a sulfate content that is probably in excess of the water quality standard. Furthermore, the post-treatment, by artificial recharge, requires a large surface area and high soil permeability, limiting the application of this technique (33).

Hydrogen, which is inherently clean but, unfortunately, only slightly soluble in water (1.6 mg/L at 20°C), is a good reactant for denitrification of drinking water. Autotrophic denitrification using hydrogen has been studied by many investigators (34–36). The reaction can be written as follows:

$$2\ NO_2^- + 5\ H_2 \rightarrow N_2 + 4\ H_2O + 2\ OH^-$$

Stoichiometrically 0.35 mg/L of H_2 are required for complete denitrification of 1.0 mg/L NO_3^-N. The Monod half-saturation constant for hydrogen was found to be less than 1% of saturation (36). In laboratory experiments using a fluidized-bed reactor, a reaction rate of almost 0.13 g N removed per liter per day was reported (36).

One commercial plant using hydrogen has been operating in Monchengladbach since 1986 (35,37). The plant is designed to supply 100 m^3/h and reduce the nitrate concentration from 80 to 40 mg/L NO_3^-. The process used is one of split treatment. Only part of the total water is denitrified (to 5 mg/L NO_3^-) and then mixed with raw water. A series of fixed beds, each filled with three-dimensional corrugated elements (Mellapack) acting as carriers for the biomass are used for the biological denitrification. Groundwater is pumped to surface, saturated with hydrogen gas and then introduced into the biological reactors. Phosphate, inorganic acid, and/or carbon dioxide are added. Reaeration, flocculation, filtration, and uv disinfection are used for post-treatment. Backwashing is required every 2–4 weeks. A denitrification rate of 0.25 kg N/m^3/d was reported. After five years of operation, results indicate that the treated water quality meets the German drinking water standards from all aspects: physicochemical, bacteriological, and hygienic.

Summarizing, the autotrophic denitrification process has some advantages and disadvantages. In regard to advantages, the process does not require the addition of an organic electron donor, which reduces the problems associated with significant dissolved organics in the effluent. Furthermore, a lower biomass production occurs, which reduces both the solid waste treatment and the propensity for clogging the fixed-bed reactor. In regard to the disadvantages, the process is slower than the heterotrophic process. The addition of an inorganic electron donor often proves problematic. In the case of hydrogen, this substrate is only sparingly soluble, requiring high running costs. In the case of reduced sulfur species, the by-product is sulfate, which may be undesirable, especially from a corrosion point of view.

Heterotrophic Denitrification

Heterotrophic denitrification is the most studied and most widely used biological commercial process for groundwater nitrate removal. In this process, an organic carbon source is used as the electron donor and energy source for biogrowth.

Processes for heterotrophic denitrification are characterized, first, by the way in which the bacteria are retained in the system and, second, by the carbon and energy source. Ethanol is the most common carbon source used, followed by methanol and acetic acid. Other, less studied, organic substrates include methane, carbon monoxide, cellulose and poly-β-hydroxy-butyric acid (PHB) (22). Addition of phosphate (together with the organic carbon source), is also required for optimal operation. High biomass concentration in the denitrification reactor is usually obtained by the biological attachment to a solid medium, which is either stationary (fixed bed), moving (rotating discs), or fluidized (fluidized bed). These types of reactors are well suited to conditions of low substrate concentrations and high flow rates, which are typical for denitrification processes of groundwater. Post-treatment is always required, since the effluents from the denitrification reactor contain microorganisms, carbon source residuals, organic debris resulting from bacterial activity, nitrite, and no dissolved oxygen. Conventional post-treatment includes aeration, filtration through sand and activated carbon, and disinfection. A novel alternative for post-treatment is the use of a combined process in which denitrification is performed in an aboveground reactor while the post-treatment and storage occur *in situ*. This is effected by injecting the denitrified water into the aquifer (18,21,32,38). Clogging of the sand aquifer, in this instance, was observed at a few sites.

The main advantages of biological denitrification is that nitrate is selectively removed by its conversion to nitrogen gas and no waste concentrate is produced (as in physicochemical processes), except for small volumes of biomass waste, which are easy to handle. Moreover, the reaction rate in heterotrophic denitrification reactors is much higher than in autotrophic and *in situ* processes. However, the process requires careful control and adequate monitoring to achieve stable operation. In addition, it requires intensive post-treatment process. Also, the process is sensitive to low temperatures.

At present most of the industrial denitrification treatment plants are based on fixed (packed) bed reactors. However, these reactors suffer from clogging and channeling (short circuiting) problems, which can be solved by using upflow fluidized-bed reactors (39). Moreover, the fluidization conditions that prevail in this type of reactor allow for the development of high biomass concentrations. These in turn provide for high treatment capacities (>10 gN(Ld)) and short retention times (<3 min), resulting in a small reactor volume (39,40). Usually, no recirculation is required, since fluidization of the bed is provided by the high flow rates, ie, high upflow velocities. However, fluidized-bed reactors are more difficult to control and require highly skilled labor for proper operation and maintenance.

The costly post-treatment requirement for heterotrophic biological denitrification has been addressed in many research studies. One approach is via entrapment of denitrifying organisms in gel to prevent bacterial contamination of the treated water (40,41). However, formation of a biofilm on the gel particles results in bacterial washouts and further post-treatment. Moreover, the process is slow, due to substrate and metabolite transport limitations, and cost.

A second approach is via the use of membrane reactors. Recent developments of ultrafiltration have given rise to membrane bioreactors in which high biomass concentrations can be retained while bacterial contamination of the effluent water is eliminated (42–44). However, increased diffusion problems caused by the membrane reduce the denitrification rate, and membrane fouling decreases filtration performance. The membrane bioreactor technology seems to be a promising field in denitrification; however, further research is still needed before it can be implemented.

BIBLIOGRAPHY

1. D. C. Bouchard, M. K. Williams, and R. Y. Surampalli, *AWWA* **84**(9), 85–90 (1992).
2. R. F. Packham, *Water Supply* **10**(3), 1–6 (1992).
3. B. McCann, *World Water Env. Eng.* **14:** 29–31 (Sept. 1991).
4. EPA, "National Water Quality Inventory," *Report to Congress* (1986).
5. *Fed. Reg.* **56**(20), 3526 (1991).
6. T. Hall and B. T. Croll in T. P. Burt, A. L. Heathwaite, and S. T. Trudgill, eds., *Nitrate: Processes, Patterns and Management,* John Wiley & Sons, New York, 1993, pp. 369–385.
7. R. Rautenbach, W. Kopp, R. Hellekes, R. Peters, and G.van Opbergen, *Aqua* **5,** 279–282 (1986).
8. J. C. Kruithof and H. M. M. Koppers, *Aqua* **38,** 207–216 (1989).
9. A. Bilyk and R. Szpadt, *Water Sci. Technol.* **24**(7), 165–173 (1991).
10. J. Murrer and G. Filteau, Paper presented at the *Euromembrane Conference,* Bath, U.K., Sept. 1995.
11. B. T. Croll, in A. Dyer, M. J. Hudson, and P. A. Williams, eds., *Special Publication,* vol. 122, The Royal Society of Chemistry, London, 1933.
12. W. S. Midkiff and W. J. Weber Jr., Purdue University Extension Service, Lafayette, Ind., 1970, pp. 137–593.
13. F. Rogalla, D. de Lurminat, J. Coutelle, and H. Godart, *NATO ASI Ser.* **30,** 369–383 (1991).
14. Y. R. Richard, *J. IWEM* **3**(2), 154–167 (1989).
15. I. J. Fletcher, A. J. Woodward, and N. G. Henwood, *J. IWEM* **5,** 566–572 (1991).
16. G. S. Solt, Paper presented at the Second International Conference on Advances in Water and Effluent Treatment, UK, 1993.
17. M. H. Christensen and P. Harremoes, *Prog. Water Technol.* **8**(4–5), 509–555 (1977).
18. W. J. Payne, *Denitrification,* John Wiley & Sons, Inc., New York, 1981; A. Barrenstein, U. Kramer, and P. Obermann, *DVGW Schrfttenr. Wasser* **106,** 99–116 (1986).
19. R. Knowles, *Microbiol. Rev.* **46**(1), 43–70 (1982); J. J. Kruithof, C. A. va Bennekom, H. A. L. Dierx, and W. A. M. Hijnen, *Water Supply* **6,** 207–218 (1988).
20. J. M. Tiedje, in J. B. Zehnder, ed., *Biology of Anaerobic Microorganisms,* John Wiley & Sons, Inc., New York, 1991, pp. 179–244.
21. K. M. Hiscock, J. W. Lloyd, and D. N. Lerner, *Water Res.* **25,** 1099–1111 (1991).
22. B. P. Gayle, G. D. Boardman, J. H. Sherrard, and R. E. Benoit, *J. Environ. Engl.* **115**(5), 930–943 (1989).
23. M. Chalupa, *Water Sci. Tech* **20**(3), 211–213 (1988).
24. A. Mercado, M. Libhaber, and M. I. M. Soares, *Water Sci. Technol.* **20**(3), 197–209 (1988).
25. V. Janda, J. Rudovsky, J. Wanner, and K. Marha, *Water Sci. Technol.* **20**(3), 215–219 (1988).
26. M. Hamon and E. Fustec, *Res. J. Water Pollut. Control Fed.* **63**(7), 942–949 (1991).
27. C. Braester and R. Martinell, *Water Sci. Technol.* **20**(3), 149–163 (1988).
28. G. Jechlinger, G. Schoeller, F. Seidelberger, and F. Zibuschka, *Water Supply* **10**(3), 101–110 (1992).
29. C. J. Soeder, R. Shultz-Hock, and E. Stengel, in R. Schindler and G. Plum, eds., *Berichte aus der Okologischen Forchung,* 1991.
30. Z. Lewandowski, R. Bakke, and W. G. Charaklis, *Water Sci. Technol.* **19,** 175–182 (1987).
31. G. K. Claus and H. J. Kutzner, *Appl. Microb. Biotech.* **22,** 283–296 (1985).
32. W. A. M. Hijnen, D. Koning, J. C. Kruithof, and D. van der Kooij, *Water Supply* **6,** 265–274 (1988).
33. J. P. van der Hoek, J. W. N. M. Kappelhof, and J. C. Sachippers, *J. Water SRT Aqua* **43**(2), 84–94 (1994).
34. D. Dries, J. Liessens, W. Verstraete, P. Stevens, P. de Vos, and J. de Ley, *Water Supply* **6**(3), 181–192 (1988).
35. H. Gros, G. Schnoor, and P. Rutten, *Water Supply* **6**(3), 193–198 (1988).
36. M. Kurt, I. J. Dunn, and J. R. Bourne, *Biotechnol. Bioeng.* **29,** 493–501 (1987).
37. P. Rutten and G. Schnoor, *Water Supply* **10**(3), 183–190 (1992).
38. R. Bockle, U. Rohmann, and A. Wertz, *Aqua* **5,** 286–287 (1986).
39. M. Green, M. Schnitzer, S. Tarre, B. Bogdan, G. Selef, and C. Soeder, *Water Sci. Technol.* **29**(10–11), 509–515 (1994).
40. J. Liessens, R. Germonpre, S. Beernaert, and W. Verstraete, *J. AWWA* **85**(4), 144–154 (1993); D. Lemoine, T. Jouenne, and G. A. Junter, *Appl. Microbiol. Biotechnol.* **36,** 257–264 (1991).
41. S. Nitisoravut and P. Y. Yang, *Water Sci. Technol.* **25,** 923–931 (1992).
42. B. Delanghe, F. Nakamura, H. Myoga, Y. Magara, and E. Guibal, *Water Sci. Technol.* **30**(6), 157–160 (1994).
43. J. Chang, J. Manem, and A. Beaubien, *J. Membr. Sci.* **80,** 223–239 (1993).
44. P. R. McCleaf and E. D. Schroeder, *J. AWWA* **87**(3), 77–86 (1995).

HAZARDOUS MATERIALS TRANSPORTATION

CRAIG WYMAN
Liskow & Lewis
New Orleans, Louisiana

THE SUBJECT MATTER

The U.S. Department of Transportation (DOT) defines hazardous materials as substances or materials capable of posing an unreasonable risk to health, safety, and property when transported in commerce by highway, railcar, and pipeline.

According to DOT estimates, approximately 500,000 hazardous materials shipments (excluding pipeline transportation) occur every day. The amount of hazardous materials moving in commerce by land, sea, and air exceeds 4 billion tons per year. Highway transport of hazardous materials accounts for more than half of all shipments. The remainder consists primarily of shipments by rail or water.

Under DOT regulations (1), hazardous materials are grouped, according to the level and type of risk presented, into several numbered hazard classes, and with corresponding descriptive names such as *corrosive, flammable, combustible,* and *radioactive.* The public probably most frequently encounters DOT regulations in the form of the color-coded shipping placards posted on motor vehicles, freight containers, and rail cars. Those placards are intended to identify the hazard class, and to communicate the type of hazard posed by the shipment.

Although several federal and state agencies are involved in some way in the regulatory scheme, the DOT by far functions as the lead agency. DOT regulations establish packaging, loading, and hauling standards for shippers and carriers of hazardous materials. DOT regulations also set standards for pipeline safety, and for the bulk transportation of materials by vessel.

Key Federal Statutes

Key federal statutes include the Hazardous Materials Transportation Law (Federal Hazmat Law), as recodified by the Hazardous Materials Transportation Authorization Act of 1994 (Public Law 103-311), 49 USC (United States Code) § 5101, *et seq.* superseding the Hazardous Materials Transportation Act (HMTA), and the Natural Gas and Hazardous Liquid Pipeline Safety Law (Federal Pipeline Safety Law), 49 USC § 60101, *et seq.* A new 1994 codification combines the provisions of the Natural Gas Pipeline Safety Act and the Hazardous Materials Pipeline Safety Act.

Regulated Substances

The Federal Hazmat Law. Under the DOT's regulations, 49 CFR § 171.8, a hazardous material is a substance or material which has been determined by the U.S. Secretary of Transportation to be capable of posing an unreasonable risk to health, safety, and property when transported in commerce. These materials especially include "hazardous substances" listed under Section 311 of the Clean Water Act, 33 USC § 1321, and "hazardous wastes" recognized by the U.S. Environmental Protection Agency (EPA) under the Resource Conservation and Recovery Act (RCRA). Moreover, by virtue of the 1986 Superfund amendments, any "hazardous substance" listed under 42 USC § 9601(14) is also listed as a DOT "hazardous material," if the amount transported exceeds the reportable quantity (RQ) of the substance (42 USC § 9656). "Hazardous materials" include petroleum and petroleum products. The DOT continues to amend its list of hazardous substances and reportable quantities in order to reflect additional hazardous substances identified by EPA under CERCLA (2).

The Federal Pipeline Safety Law. These statutes regulate the transportation by pipeline of "gas" and "hazardous liquids." "Gas" is defined as "natural gas, flammable gas, or gas which is toxic or corrosive" (49 CFR § 192.3). "Hazardous liquids" are defined by statute to include petroleum or any petroleum product, and any other substance or material in liquid form, excluding liquefied natural gas, which may pose an unreasonable risk to life or property when transported by pipeline facilities (49 USC § 60101(4)). The DOT regulations define the term to include petroleum, petroleum products, and anhydrous ammonia (49 CFR § 195.2).

DOT ORGANIZATION IN ADMINISTERING MAIN REGULATORY PROGRAMS

With the exception of the U.S. Coast Guard's rulemaking authority under the Port and Tanker Safety Act (ie, covering bulk vessel standards and port safety) and other marine pollution statutes, the Research and Special Programs Administration (RSPA) within DOT issues all departmental hazardous materials transportation regulations.

Originally a part of the DOT, the National Transportation Safety Board (NTSB) was established as an entirely independent agency in 1975. The NTSB has broad investigatory authority over a wide variety of marine, motor carrier, railroad, pipeline, and aviation accidents. The NTSB also is directed to issue recommendations to the DOT Secretary regarding needed regulatory amendments. (Preambles to many DOT rulemakings carry references to previous NTSB recommendations.) Any major transportation disaster is followed by the convening of a NTSB review team. The NTSB, however, has no *enforcement* authority, and NTSB investigation findings are not admissible into evidence in any private litigation arising from a disaster (49 USC § 1154(b)). NTSB policy provides that NTSB employees may neither testify as experts in any private civil action, nor give any opinion testimony regarding accidents investigated by the Board.

THE HAZARDOUS MATERIALS TRANSPORTATION AUTHORIZATION ACT OF 1994

Background

The immediate precursor to the 1994 legislation was the long-established Hazardous Materials Transportation Act of 1994

(HMTA). The HMTA was preceded by the Federal Railroad Safety and Hazardous Materials and Transportation Control Act of 1970. The HMTA amended and expanded on the 1970 legislation. Section 5 of the HMTA expanded the authority of the DOT Secretary to regulate hazardous materials moving through interstate and foreign commerce by any mode of transportation. Regulatory authority over the various modes of transportation, including air, railroad, and highway traffic were consolidated within the DOT. The DOT Secretary was further authorized to issue regulations covering the manufacture, fabrication, marking, maintenance, reconditioning, repair, testing, and distribution of packages and containers used in hazardous material transportation, in addition to regulating the carrier and shipper of such materials. Civil penalties and injunctive relief were added to existing criminal penalties as sanctions for regulatory violations.

The HMTA was reauthorized by the Hazardous Materials Transportation Uniform Safety Act of 1990, Public Law 101-615. One important provision of the reauthorization was the statutory delineation of the "jurisdictional schemes" for federal and state legal authority in this area. The act denoted (*1*) areas of exclusive federal regulation, (*2*) areas of joint federal and state jurisdiction, and (*3*) under certain conditions, matters of state or local regulation. Exclusive federal regulation covers the classification, packaging, handling, marking, and documentation of hazardous materials, and the notification of releases of such materials (unless the state law is substantially the same as the federal regulation.) The areas of joint federal and state jurisdiction primarily concern highway routing of hazardous material shipments.

In addition, the 1990 legislation (*1*) re-established procedures whereby preemption rulings can be issued by DOT, (*2*) specified that any state fees for hazardous materials transportation must be "equitable" and appropriated for hazardous material transportation regulatory purposes, (*3*) established grant programs for "hazmat" planning and training, (*4*) directed DOT to improve its vehicle placarding regulations, (*5*) increased the maximum civil penalty for HMTA violations to $25,000, and (*6*) established a safety permit program for motor carriers which transport five listed commodities, including carriers of liquefied natural gas. The DOT may also recognize additional commodities. Shippers of the listed products will only be able to use permitted carriers.

Finally, as a result of the Hazardous Materials Transportation Authorization Act of 1994, the HMTA has been statutorily replaced by new Chapter 51 of Title 49 of the *United States Code.* The reference to former HMTA provisions is now 49 USC § 5101 *et seq.* The federal Hazmat scheme is generally known as the Federal Hazardous Materials Transportation Law.

DOT Jurisdiction Under Federal Hazmat Law

Federal Hazmat law previously granted the DOT regulatory authority over any hazardous material transportation in interstate and foreign commerce, and further, any transportation that affects interstate commerce, ie, intrastate shipments. 51 USC § 5103(b) now provides that the DOT Secretary "shall prescribe regulations for the safe transportation of hazardous material in intrastate, interstate, and foreign commerce." As a matter of regulatory policy, the DOT had limited its authority over motor vehicle traffic only to *interstate* commerce, except in the case of

hazardous wastes and hazardous substances

flammable cryogenic liquids (refrigerated liquefied gas having a boiling point <54°C)

Presently, the RSPA is operating under a Notice of Proposed Rulemaking which would implement the new statutory authority and begin regulation of all intrastate shippers and carriers (3).

Hazardous Materials Regulations

The Federal Hazmat regulatory program applies to:

- persons who offer hazardous materials for transportation (ie, shippers)
- persons who accept hazardous materials for transportation (ie, carriers)
- persons who manufacture, fabricate, or retest containers of hazardous materials
- persons who are either "hazmat employers" or "hazmat employees" (ie, shippers, carriers, and testers or reconditioners of hazardous materials containers)

The Hazardous Materials Regulations (49 CFR § 171–180) (HMR) generally are divided into three areas of emphasis:

- shipper material designation and "hazard communication" requirements (49 CFR Part 172)
- packaging and container specification requirements (49 CFR Parts 173, 178, 179, and 180)
- shipper and carrier operational rules (49 CFR Parts 171, 173, 174, 175, 176, and 177)

In December 1990, the HMR were comprehensively revised with respect to hazard communication, classification, and packaging requirements (4). Many of the changes are based on the United Nations Recommendations on the Transport of Dangerous Goods. The effective date of the new regulations was October 1, 1991; however, compliance was authorized on and after January 1, 1991. Certain of these compliance dates, set forth in 49 CFR § 171.14, were also revised by RSPA rulemaking (5). Basically, the outside compliance date for the rules was set at October 1, 1996. In addition, however, the RSPA continues to issue modifications to the HMR as appropriate.

The HMR places special emphasis on "shipper" and "carrier" responsibilities.

GENERAL OVERVIEW OF THE REGULATION OF PIPELINE TRANSPORTATION

The Natural Gas Pipeline Safety Act (NGPSA) of 1968 and the Hazardous Liquid Pipeline Safety Act (HLPSA) of 1979

Federal Pipeline Safety law authorizes the DOT to develop and enforce regulations governing pipeline transportation of natural gas and hazardous liquids, such as petroleum and petroleum products. Although federal jurisdiction is exclusive over interstate pipelines, the statutes allow states to impose additional requirements for intrastate lines if compatible with the federal programs. The combined regulated pipeline system is approximately 2.7×10^3 km (1.7×10^3 mi) in length: approximately 2.4×10^3 km (1.5×10^3 mi) of gas lines, and approximately 240,000 km (150,000 mi) of hazardous liquid lines.

DOT Pipeline Safety Regulations

These are divided into four primary areas:

1. Transportation of Natural and Other Gas by Pipeline; Annual Reports, Incident Reports, and Safety-Related Condition Reports (49 CFR Part 191)
2. Minimum Federal Safety Standards for Natural and Other Gas Pipeline Transportation (49 CFR Part 192)
3. Federal Safety Standards for Liquefied Natural Gas Facilities (49 CFR Part 193)
4. Transportation of Hazardous Liquids by Pipeline (49 CFR Part 195)

Minimum Federal Safety Standards for Pipeline Transportation of Natural Gas

These regulations (49 CFR Part 192) prescribe minimum safety requirements for, with certain listed exceptions, pipeline facilities and the transportation of gas, including pipeline facilities and transportation of gas on the OCS. These standards do not apply to the gathering of gas in rural locations outside of certain populated areas (49 CFR § 192.1(b)(2)) 49 USC § 60101(b)(2)(A) directs DOT to clarify its definition of the term "regulated gathering line."

The regulations designate materials for pipeline construction, specifications for pipe design, design of pipeline components, welding of pipelines or the joining of materials other than by welding, general construction requirements for transmission lines and mains, requirements for customer meter service regulations and service lines, corrosion control requirements, testing requirements, upgrading requirements, operational requirements, and maintenance requirements. Certain regulations that could have special "environmental law" application include:

The pipeline testing regulations, 49 CFR § 192.515, contain specific requirements for environmental protection and safety of employees and the general public during testing operations, and further require that the test medium be disposed of in a manner that will minimize damage to the environment.

Subpart L of Part 192 requires the preparation of an operation and maintenance plan. The plan must include instructions to employees on regulatory requirements, including recognition of safety-related conditions, specific programs relating to facilities presenting the greatest hazard to public safety in an emergency, and periodic pipeline inspection procedures.

The regulations include the recognition of different service areas, defined as "Class Locations." The different classes generally recognize the proximity of the pipeline to human population and development. For example, a Class 1 location includes offshore locations, and any location that has ten or less buildings intended for human occupancy; a Class 4 location is a location where buildings with four or more storeys above ground are prevalent (49 CFR § 192.5). Generally, safety precautions increase with the number of the class location.

Any pipeline operations plan must include requirements for continuing surveillance to determine changes in class location, failures, leakage history, corrosion, and other unusual operating or maintenance conditions (49 CFR § 192.613).

49 CFR § 192.614 requires an operator of a buried pipeline to prepare a damage prevention program to prevent damage to buried pipeline by excavation activities. In 1990, the RSPA instituted a requirement for a "one-call" damage prevention program to be adopted by all state pipeline safety programs. This one-call system is intended primarily to facilitate excavations conducted near buried lines (6). In a proposed rulemaking the RSPA announced plans to make mandatory the participation in these programs by pipeline operators (7). In addition, the RSPA has extended existing excavation damage prevention requirements for gas pipelines in urban areas to gas pipelines in rural areas, and now requires (with limited exception) line markers for gas transmission lines in urban areas (8).

49 CFR § 192.615 requires the operator to prepare a written emergency plan to minimize hazards resulting from a gas pipeline emergency. The plan must include notification requirements for local fire, police, or other public officials, response plans for various emergencies such as a detection of gas inside or near a building, fire, explosion, or other natural disasters, measures for emergency system shutdown, if necessary to minimize hazards to life or property, availability of personnel and equipment to respond to an emergency and reduce any hazard, and planning for actions to protect the public first, then property.

Pipeline maintenance requirements, 49 CFR Subpart M, Part 192, include requirements for patrolling lines for indications of leaks, construction activity, and other conditions affecting safety and operations; periodic leakage surveys depending on class locations; line markers for mains and transmission lines (including marker warnings with the telephone number of the operator where the operator can be reached at all times); and, recordkeeping requirements.

Finally, the RSPA has proposed adding a new Subpart N to Part 192 to incorporate specific qualifications for pipeline personnel. Acting on recommendations made by the NTSB, the RSPA believes adding new training requirements for those directly responsible for pipeline inspection maintenance will enhance line safety (a new corresponding Subpart G to part 195 is being proposed) (9).

Federal Safety Standards for LNG Facilities

49 CFR Part 193 of DOT's rules prescribes safety standards for, with certain listed exceptions, LNG facilities used in the transportation of gas by pipeline. An "LNG facility" is defined as a "pipeline facility that is used for liquefying or solidifying natural gas or synthetic gas or transferring, storing, or vaporizing liquefied natural gas." "Liquefied natural gas" (LNG) means "natural gas or synthetic gas having methane (CH_4) as its major constituent which has been changed to a liquid or semisolid" (49 CFR § 193.2007).

Subpart B of Part 193, which contains siting requirements, generally requires that any LNG facility must be located at a site of suitable size, topography, and configuration so that the facility can be designed to minimize the hazards to persons and off-site property resulting from leaks and spills of LNG and other hazardous fluids at the site. Plant design requirements are set forth in Subpart C of Part 193, and include specifications for plant impoundment design and capacity to withstand major spill events. Other strict facility component construction standards are set forth in Subpart D, including the requirement for testing of the integrity of system components, and the retention of construction records (49 CFR § 193.2329).

Transportation of Hazardous Liquids by Pipeline

49 CFR Part 195 prescribes safety standards and accident report requirements for pipeline facilities used in the transportation of hazardous liquids. "Hazardous liquids" include petroleum, petroleum products, or anhydrous ammonia. These regulations, with certain listed exceptions, apply to pipeline facilities, and transportation of hazardous liquids associated with those facilities, in or affecting interstate or foreign commerce, including pipeline facilities on the OCS. In July 1994, the RSPA finalized rulemaking under which the agency has partly discontinued the exemption from federal regulation given hazardous liquid pipelines that are operated at 20% or less of the "specified minimum yield strength" of the pipe. The RSPA's new rules remove the exemption for low stress pipelines which are used for the transportation of highly volatile liquids which traverse a populated area or a navigable waterway (10). The rulemaking was prompted in part by the January 1990 incident in which an otherwise-exempt pipeline spilled 1.8×10^3 L (500,000 gal) of fuel oil into a New York waterway.

Under 49 CFR § 195.4, no person may transport any hazardous liquid unless the hazardous liquid is chemically compatible with both the pipeline, including all components, and any other commodity that it may come in contact with while in the pipeline.

Subpart B of Part 195 requires the reporting of accidents and safety-related conditions:

for accidents, 49 CFR § 195.50 requires the operator to submit a report of any failure of a pipeline system subject to Subpart B where there has been a release of hazardous liquid resulting in either:

- explosion or fire,
- loss of 50 or more barrels of liquid,
- escape to the atmosphere of more than five barrels a day of highly volatile fluids (ie, those with the potential of forming vapor clouds),
- death of any person,
- serious bodily harm, or
- property damage in excess of $5,000.

49 CFR § 195.52 requires telephone notice of any accident involving death or serious personal injury, fire or explosion, significant property damages, or any pollution of any water body at "the earliest practicable time." This notice must include the name and address of the pipeline operator, the name and address of the reporter, the location and time of the line failure, any casualties, and any knowledge of the cause of the failure and the extent of property damages. A written accident report must follow within 30 days (49 CFR § 195.54). Safety-related conditions, which are similar to those defined for gas pipelines under Part 191, must be reported within five working days of when first determined, but not later than 10 working days after a representative or operator discovers the condition (§§ 195.55, 195.56). These reports also are similar to those required under Part 191. Finally, 49 CFR § 195.57 requires the filing of "offshore pipeline condition reports" pursuant to inspections conducted in the Gulf of Mexico under a program instituted under Public Law 101-599.

Subparts C, D, E, and F of Part 195 include design, construction, hydrostatic testing, and operation and maintenance requirements. Subpart F, Operation and Maintenance, includes specific requirements for personnel training, pipeline inspection, and corrosion control measures. 49 CFR § 195.260 requires valves to be located on main lines at locations that will minimize damage or pollution from accidental hazardous liquid discharges. 49 CFR § 195.402 focuses on response planning, and requires the operator to prepare and follow a manual of written procedures for conducting normal operation and maintenance activities and handling abnormal operations and emergencies:

The manual requirements for abnormal operations include a set of written procedures to respond to, investigate, and correct the cause of any failure of the system, and the handling of emergency events, including the notification of local fire and police officials, the containment of accidental releases of hazardous liquids, and the minimization of public exposure to the release.

Planning must include the possible evacuation of nearby residents and the halting of traffic on nearby roads and railroads. In the case of failure by a pipeline system transporting a highly volatile liquid, planning must include the use of appropriate instruments to assess the extent of coverage of a vapor cloud and to determine any hazardous areas.

49 CFR § 195.403 requires employee training in emergency procedures.

49 CFR § 195.408 provides for special communication systems. An operator must have a communication system to provide for transmission of information needed for the safe operation of the pipeline system, including the means to (*1*) monitor operational data; (*2*) receive notices from operator personnel or public authorities regarding abnormal or emergency conditions; (*3*) transmit this information to appropriate personnel or government agencies for corrective action; and (*4*) provide communication with fire, police, and other public facilities during emergency conditions.

49 CFR § 195.410 requires line markers and warnings for buried pipelines.

49 CFR § 195.412 requires regular inspection of right-of-ways and crossings under navigable waters.

49 CFR § 195.440 institutes formalized public education procedures, and requires each operator to establish a continuing educational program to enable the public, appropriate government organizations, and persons engaged in excavation-related activity to recognize and report hazardous liquid pipeline emergencies to the operator or public officials.

49 CFR § 195.442 established a "one-call" system in order to all excavators and the general public to provide notification of any intention to engage in excavation activities that could effect underground lines (RSPA proposal for mandatory operator participation) (7).

In addition, the RSPA has established excavation damage prevention requirements for hazardous liquid and carbon dioxide pipelines by amendments to § 195.442 (8).

Delineation Between Federal and State Jurisdiction: Statement of DOT Policy and Interpretation

The HLPSA expanded the DOT's existing statutory authority for pipeline safety regulation, which was limited to transportation by common carriers in interstate and foreign commerce. The Act also regulated the transportation of facilities "used in or affecting" interstate or foreign commerce. Federal law provides for exclusive federal regulation and enforcement over "interstate pipeline facilities," ie, those used for the pipeline transportation of hazardous liquids in interstate or foreign commerce. For the remainder of pipeline facilities, ie, the "intrastate facilities," federal law will apply unless a state certifies that it will assume those responsibilities. A state so certifying must adopt the federal minimal standards, but may adopt additional stricter standards if compatible with the federal program.

For deciding the division between pipeline systems, the DOT, in Appendix A to 49 CFR Part 195, decided to follow NGPSA procedure and generally recognize the division of "interstate" and "intrastate" pipelines as adopted by the FERC. However, the DOT Appendix points out that, for liquid pipelines, the assertion of first jurisdiction through the filing of a tariff with the FERC only carries a rebuttable presumption of jurisdiction. A state or other person may challenge the filing and jurisdictional status. The Appendix also explains that the DOT will generally follow the FERC and rely on FERC filings unless it is clear that a request for reconsideration by the FERC would be successful. It also sets forth examples of situations in which the DOT will treat a pipeline as an interstate facility despite a lack of a FERC filing, or conversely, treat a facility as a state-regulated intrastate pipeline despite a FERC filing.

Ramifications for Private Tort Law

Brister v. Gulf Central Pipeline Co., 684 F. Supp. 1373 (W.D. La. 1988) involved referencing DOT pipeline regulatory definitions in ruling that the owner of an anhydrous ammonia pipeline owed a duty of "extraordinary care" in the conduct of his operations. In *Williams v. Hill Manufacturing Co.*, 489 F. Supp. 20 (D.C. SC 1980), plaintiff attempted to establish the defendant's *per se* negligence by outlining an alleged hazmat violation. The court found, however, that the plaintiff's injury, which was caused by the handling of an improperly labeled chemical drum which had been transported to plaintiff's workplace 14 months earlier, was outside the scope of protection of the federal hazmat scheme. The *Indiana Harbor Belt Railroad Co. v. American Cyanamid Co.*, 916 F.2d 1174 (7th Cir. 1990) relied in part on the hazmat regulatory program to conclude that the transportation of liquid acrylonitrile by rail car was not an ultrahazardous or abnormally dangerous activity which would subject the shipper to strict liability for the consequences of a spill in transit. The court declined, however, to adopt the shipper's argument that the federal program should preempt common law liability.

BIBLIOGRAPHY

1. *Code of Federal Regulations,* Title 49, U.S. Government Printing Office, Washington, D.C.
2. *Fed. Reg.* **55,** 31822 (June 22, 1994).
3. *Fed. Reg.* **61,** 11484 (Mar. 20, 1996).
4. *Fed. Reg.* **55,** 52,402 (Dec. 21, 1990).
5. *Fed. Reg.* **59,** 67390 (Dec. 29, 1994).
6. *Fed. Reg.* **55,** 38688 (Sept. 20, 1990).
7. *Fed. Reg.* **60,** 14714 (Mar. 20, 1995).
8. Ref. 7, 14646.
9. *Fed. Reg.* **59,** 39506 (Aug. 3, 1994).
10. *Fed. Reg.* **59,** 35465 (July 12, 1994).

HAZARDOUS WASTE

There are various working definitions for *hazardous waste*. One definition relates to the waste having high values of one or more of the following properties: ignitability, corrosivity, reactivity, and toxicity. Another refers to its effects: causing an increase in the human death rate or occurrence of serious illness, or presenting a hazard to the health of organisms or the environment. For a general profile of hazardous waste, see WASTE GENERATION. For the transportation of hazardous waste and other hazardous materials, see HAZARDOUS MATERIALS TRANSPORTATION.

For the environmental planning for dealing with hazardous waste, its cost control, and its disposal by incineration and by solidification and stabilization, see the following four sections: HAZARDOUS WASTE: PLANNING; HAZARDOUS WASTE: COST CONTROL; HAZARDOUS WASTE: INCINERATION; HAZARDOUS WASTE: SOLIDIFICATION AND STABILIZATION.

For substantial coverage of other aspects of hazardous waste see BIOHAZARDOUS WASTE; FUNGAL TECHNOLOGIES FOR HAZARDOUS WASTE TREATMENT; LABORATORY HAZARDOUS WASTE: DESTRUCTION; LABORATORY HAZARDOUS WASTE: MANAGEMENT; SOIL SAMPLING AT HAZARDOUS WASTE SITES.

HAZARDOUS WASTE: COST CONTROL

RICHARD A. SELG
British Nuclear Fuels Limited—Savannah River Corporation
Aiken, South Carolina

RONALD G. STILLMAN
Roy F. Weston, Inc.
West Chester, Pennsylvania

Total cost management techniques are vitally important in our fast-changing world of industry and commercial endeavors. Total cost management is the effective application of professional and technical expertise to plan and control resources, costs, profitability, and risk. Managers, engineers, and educators at commercial industry, governmental facilities, and universities around the world are discovering these benefits of total cost management in the field of hazardous waste management and environmental restoration.

United States industry generates approximately 12 billion metric tons of solid waste annually, of which about 750 million metric tons are hazardous. Large amounts of energy, four to seven quadrillion BTUs (quads), are expended each year to produce and process raw materials that eventually become waste. Two additional quads are used annually in waste treatment and disposal operations. Recent estimates indicate that three leading U.S. government agencies (Department of Energy, Department of Defense, and Environmental Protection Agency) alone spend nearly 16 billion dollars per year. The estimated costs for the environmental cleanup of the United States over the next half-century range from $400 billion to $700 billion (1).

It is no wonder that the cost of environmental cleanup of hazardous waste is a subject that is both timely, relevant, and of interest to so many in the United States and around the world. According to a Harris poll, a large portion (76%) of Americans believe the disposal of toxic chemicals to be a very serious problem. The poll reveals that the public blames industry, the federal government, and state government. The public is asking a number of hard and pointed questions: How are you going to solve the problem? Can we, in fact, handle hazardous wastes? Do we have to have them?

In some measure, hazardous wastes are the inevitable byproducts of the goods that we, as people, demand. Toxic waste products, for an example, are generated by production of the clothes we wear, the food we eat, the cars we drive, and the medicines we take. It is important that the general public understand that the hazardous waste problems can be managed in all its dimensions. Questionable practices are being eliminated. New disposal techniques are being discovered and implemented. New methodologies are being used to manage the cost of hazardous waste through total cost management techniques such as those presented herein.

In managing hazardous wastes, industry and government have their respective roles to play. An ever-increasing part of that role and the key to success are sound cost-control practices and techniques. This article will provide basic (generic) cost-control concepts for managers and engineers in the field of waste management and environmental restoration and an outline for upper-level undergraduate and graduate students in environmental engineering and sciences.

MANAGEMENT OF HAZARDOUS WASTE PROJECTS

Agency Involvement

If a project becomes a state or federal hazardous waste project through the discovery of hazardous waste or contaminated soil or groundwater, there are several steps that must be incorporated into the project plan. The specific steps are dependent on whether the site is being managed by the EPA or by a local state agency; however, the basic steps are the same.

- Collect initial site background information.
- Develop a plan for conducting a site investigation.
- Obtain regulatory agency approval of the investigation plan.
- Conduct a site investigation.
- Compile results and review with regulatory agency.
- Prepare remedial action feasibility study.
- Obtain state and federal agencies approval for cleanup action.
- Conduct site cleanup.
- Conduct sampling and analysis and obtain agency sign off on project.

The number of steps indicates that this process can be lengthy and expensive. A cost of several hundred thousand dollars and a schedule of one to two years is not uncommon; however, it is not correct to assume that this is the case for every hazardous waste investigation and cleanup. Several steps can be combined through an interactive relationship with the governing regulatory agency. Properly handling agency interface and negotiations can save significant costs and reduce the over-project duration.

Typically a regulatory agency will request a minimum of 30 days to review and approve any document submitted during a hazardous waste project. These reviews can have a definite impact on a project schedule with the submittal of both draft and final documents for each step of the process. However, it is possible to reduce the time if a good relationship is established with the agencies and they are brought into the planning process early on in the project. It is also important to remember that all projects and contaminants are not treated the same. If the project involves contaminants that are not exceptionally hazardous and the site is in an area with minimum impact to human health and the environment, it is possible that the project will be placed on a faster track for approval. This can be accomplished by developing an understanding of the regulators' specific areas of concern early on in the project and by developing a project plan to meet those concerns. Quite often it is beneficial to have preliminary meetings with the regulatory agencies to discuss the project and make sure that all parties involved have a clear understanding of the project. This may be more difficult if more than one potentially responsible party (PRP) has been identified and there is an adversarial relationship between the PRPs. It is better for all involved if a good working relationship can be developed with the regulatory agencies.

Project Strategy

One of the key steps to remember is that the primary purpose of the investigation is to gain enough information to prove that a site is not contaminated, or to identify a cleanup process that can be applied to the site. It is important to remember this strategy when developing an investigation plan. It is also important to remember that most hazardous waste projects will also need to be accepted by the public so the investigation that is conducted should be one that the public will believe is representative of the conditions at that site. Gather data that will be needed to design a cleanup strategy; do not just go through the investigation process hoping not to find any contamination. A well-planned and negotiated investigation will result in a shorter total project schedule and will be more cost effective in gathering the field data.

Work with the regulatory agency to determine exactly what their concerns are with the site. If they have only one area of concern, be sure to address that area and perhaps thus avoid the collection of unneeded information. Good relations with the agencies are not always easily attained. In fact, there are often situations when agencies appear to be making unreasonable demands that will greatly increase the cost of the project. It is important to educate the regulator about your concerns and reach a mutual understanding of the project requirements without alienating the regulatory agencies involved. Occasionally the project team will not agree with the regulatory requirements, but remember state and federal agencies win more disagreements than they lose. Choose your battles wisely; select ones that make sense and that you can win and concede the others. Doing so will save time and money even though it may be very hard to accept.

When conducting the cleanup be sure to gain agency approval throughout the process. There is nothing worse than thinking that you have completed a project only to find that the regulators did not understand what you were doing and want the project reworked from the beginning.

Overall there are a growing number of environmental regulations that will impact your existing and upcoming projects. Although a project manager may not be able to predict where the regulations will be next year, it is likely that they will be equal or more restrictive than the current regulations. It is important to become aware of the regulations that currently exist and to incorporate them into a project plan. Ignoring the potential impact of environmental regulations can be very costly and time consuming.

COST RISK DRIVERS IN HAZARDOUS WASTE PROJECTS

There are many sources of cost risk in hazardous projects. Although hazardous waste projects face different types of risk, most of the major cost risk drivers can be categorized into the following four classes: scope definition, contaminant and contaminated media complexity, technical sophistication, and regulatory uncertainty.

Scope Definition

Perhaps the largest source of cost risk is simply the degree to which the project is adequately defined. In fact, according to a Construction Industry Institute study (2), poor scope definition is the leading cause of cost overrun in major construction projects. The problems with scope definition in hazardous waste projects have been noted in several studies (3). In part, adequate project definition is dependent on the outcome of the assessment phase (CERCLA remedial investigations (RI) and RCRA facility investigations (RFI)). Also, project execution factors (laboratory availability, labor availability, equipment availability, etc) are often assumed for hazardous waste projects, which can contribute substantially to cost growth.

Contaminant and Contaminated Media Complexity

Complexity is an ambiguous term. Nevertheless, more complicated, intricate, multifaceted projects offer more opportunities for cost misestimation and therefore tend to engender more cost risk. The relationship between project complexity and cost growth has been documented for hazardous waste projects, and for projects in the oil and chemical industry. In cleanup projects, complexity has a number of dimensions. Generally, projects with multiple classes of contaminated media (eg, volatile organics and heavy metals) and with different types of contaminated media (eg, groundwater contamination and contaminated unsaturated soils) face a greater chance of incurring cost growth due to the unknown amount of underground contamination.

Technical Sophistication

Cleanup projects that employ innovative or first-of-a-kind technologies are more likely to need additional contingency funds in the short term. In the long term, however, cost risk should decrease as utilization of the new technology improves. A study performed at the Oak Ridge National Priorities List (NPL) projects revealed varying degrees of technical complexity and intensity (4).

Regulatory Uncertainty

Regulatory adjustments and new legislation and interpretations often lead to changes in scope of projects. Projects are affected by adjustments to the existing regulations, or may be targets of new policy actions. The direction of cost risk caused by regulatory-driven scope changes is usually upward. In most cases, the need for regulatory-driven contingency is driven by a change in project scope. In some cases, the change in regulatory driver may be expected and contingency can be evaluated deterministically based on this potential change. This new contingency would be determined by costing the scope of the project as driven by the expected change. Differing scenarios contingent on which legal status prevails could be postulated, with risk determined by the probability of occurrence of the scenarios.

One common misconception is that the same factors that drive the cost of a project also contribute to cost risk. This is especially true for quantities of waste to be remediated. In fact, larger projects do not incur greater cost risk (on a percentage basis) than do smaller projects (5).

The federal government has changed regulations over the past several years in regard to remedial activities. For example, the handling and disposal of waste sites has been revised so that approved approaches to work had to be changed or rewritten to reflect these new rules. Projects are delayed so that the work plans can be rewritten, increasing costs and delaying the project.

Hidden Expense of Hazardous Waste Projects

Remediation work has hidden cost impacts associated with the performance of the work such as remediation techniques and bonding restrictions and costs. The estimator has to review these potential impacts and include the cost impacts as necessary. Also, due to the changing remedial marketplace, these cost impacts may be positive or negative.

Remediation Techniques. The technique proposed for the remediation work may be proprietary to a firm or only a few companies can supply the technique. The estimator developing a budget estimate has to be cognizant of the market conditions for the remediation technique and adjust the overhead and profit margins in the estimate accordingly. The unique techniques may limit the completion on many sites.

The selected technique may be difficult to perform. The constructibility of the project has to be reviewed by the estimator. Additional costs may result due to the creative nature of the remediation.

As the environmental market has matured, more competition has entered into the executing of many of these techniques. For example, the thermal treatment marketplace has many more qualified firms as compared to several years ago, when there were only a few, which is driving down costs. Estimates developed in the past may not reflect the current marketplace for thermal work. The estimator should review the market conditions for the selected techniques prior to developing an estimate.

Bonding Restrictions and Costs. Obtaining a performance and payment bond for hazardous waste work can be more costly and more restrictive than encountered in general construction. The bonding limits placed on the company for a remediation project may be lower than for other general construction projects. This bonding restriction may limit the number of competitors for larger projects.

The estimator must be cognizant of the limited competition as well as the higher cost for the bond in the development of a project estimate. The costs of hazardous waste bonds usually run about \$5,000 to \$10,000 dollars higher than for a general construction project (6).

CONCLUSION

Preparing cost estimates and schedules for any hazardous waste project requires time. Due to the critical nature of the work involved with a hazardous waste remediation project and the costs involved in execution, this industry must advance the manner in which cost estimates and schedules are being controlled for these projects. The hazardous waste industry needs to ensure that effective cost and schedule control becomes a normal part of project execution. The lessons learned in this relatively new and advancing field should be considered and continually accounted for, thereby mitigating and avoiding the potential costly impacts of these concerns.

BIBLIOGRAPHY

1. *Eng. News-Record,* **232**(6), 3 (1994).
2. Construction Industry Institute, *Bureau of Engineering Research,* The University of Texas, Austin, Tex., 1986.
3. G. T. Ponce de Leon, T. McManus, and G. Klanac, in R. Selg, ed., *Hazardous Waste Cost Control,* Marcel Dekker, Inc., New York, 1993, p. 29.
4. C. B. Doty, A. G. Crotwell, and C. C. Travis, *Cost Growth for Treatment Technologies at NPL Sites,* ORNL/TM-11849, Oak Ridge National Laboratory, Oak Ridge, Tenn., 1991.
5. *Fed. Reg.* **56**(130), 30996 (July 8, 1991).
6. R. G. Stillman, in Project Management Associates, Inc., *FORM-SPEC Model CPM Specification,* U.S. Domestic Model, Ann Arbor, MI., 1993, p. 89.

HAZARDOUS WASTE: INCINERATION

R. BERTRUM DIEMER, JR.
THOMAS D. ELLIS
E. I. du Pont de Nemours & Co., Inc.

GEOFFREY D. SILCOX
JOANN S. LIGHTY
DAVID W. PERSHING
University of Utah

Municipalities and industries are encouraged to reduce waste generation. Nevertheless, even under maximum use of source reduction and recycling, significant quantities of waste continue to be generated. As of this writing, high temperature incineration is the preferred technology for managing these wastes (1–3). Properly designed incinerators have the capability to destroy nearly 100% of all types of liquid organic wastes and an estimated 60% of solid wastes. However, as shown in Table 1, incineration is extremely limited in the United States. Roughly 60% of total U.S. wastes generated annually are classified as hazardous waste and less than 0.5% is incinerated. About 15% of the municipal solid waste generated is disposed of in incineration systems and effectively all of the medical waste (see WASTE GENERATION).

High (\$300–\$1900/t) capital investment and operating costs help discourage use of incineration for hazardous wastes. These costs for incinerators are well above those for alternative treatment methods (8) such as biological treatment, \$60–\$770/t; landfill in drums, \$260–\$740/t or in bulk, \$90–\$150/t. A primary contributor to the high operating cost of incinerators is the need for auxiliary fuel, particularly for the disposal of liquid wastes having high water content or solid wastes having low heating value. In addition, gas scrubbers can consume large quantities of chemicals, especially if chemical addition is not carefully controlled. Moreover, because incineration systems are typically complex, highly skilled operators are required to ensure efficient and reliable operation.

U.S. REGULATIONS IMPACTING DESIGN AND OPERATION OF INCINERATORS

U.S. regulations governing the design and operation of incinerators include the Resource Conservation and Recovery Act (RCRA), the Toxic Substances Control Act (TSCA), and the Clean Air Act Amendments of 1990 (CAA). Many states are authorized to regulate hazardous waste and incinerator programs, and state regulations are generally more stringent than federal.

Resource Conservation and Recovery Act

The RCRA, Subtitle C, regulates hazardous waste disposal. It identifies wastes as being hazardous if falling into one or more of the following categories: ignitable, eg, having a flash point less than 60°C; corrosive, eg, having a pH less than 2 or greater than 12.5; reactive, eg, reacting violently when mixed with water; and toxic, as determined by the toxicity characteristic leaching procedure (9), or listed in Subtitle C as a hazardous waste from nonspecific sources, industry-specific sources, or as an acute hazardous or toxic waste.

RCRA incinerator regulations include administrative as well as performance standards. Administrative standards include procedures for waste analysis, inspection of equipment, monitoring, and facility security. Steps needed to meet administrative standards are outlined in the permit application; performance standards are demonstrated during a trial burn. Trial burn operating conditions are included in the permit to assure ongoing compliance with performance standards. Performance standards include destruction and removal efficiency (DRE), particulate emissions limits, products of incomplete combustion emission limits, metal emission limits, and HCl and Cl emission limits.

Destruction and Removal Efficiency. In preparation for a trial burn, the owner prepares an analysis of the waste feed stream. Roughly 400 principal organic hazardous constituents (POHCs) are listed in the *Code of Federal Regulations* (10), and

Table 1. Summary of U.S. Waste Generated and Incinerated

Type	Waste, 10^6t/yr Generated	Incinerated	Number of Operating Incenerators
Hazardous[a]	249.3	1.3	171
Municipal[b]	163	26	168
Medical[c]	0.4	0.4	6850[d]

[a] 1987 Estimate (4).
[b] 1988 Estimate (5).
[c] 1990 Estimate (6).
[d] 1991 Estimate (7).

if the waste analysis includes POHCs, these are listed along with their concentrations in the incinerator permit application. A government agency, eg, the U.S. EPA, permit writer selects one or more POHC to be used during the trial burn to demonstrate the incinerator's DRE. The writer bases the selection on incineration difficulty, ie, low heating value or high thermal stability, and concentration. Constituents having concentrations less than about 100 ppm are likely not to be selected because of difficulty in analytical detection of such small quantities at 99.99% or greater DRE. During the trial burn, for each POHC selected, the incinerator must demonstrate a DRE of 99.99% or greater for RCRA wastes, and 99.9999% for polychlorinated biphenyl (PCB) or dioxin wastes.

$$\mathrm{DRE} = \frac{W_{\mathrm{in}} - W_{\mathrm{out}}}{W_{\mathrm{out}}} \times 100$$

where W_{in} is the mass feed rate into an incineration system of a given POHC and W_{out} is the mass emission rate of the same POHC at the stack to go into the atmosphere. If ash from a municipal incinerator contains metals or other contaminants in amounts that would cause the ash to be classified as hazardous, it must be treated as a hazardous waste under RCRA, Subtitle C regulations. Figure 1 shows a schematic of an incinerator system.

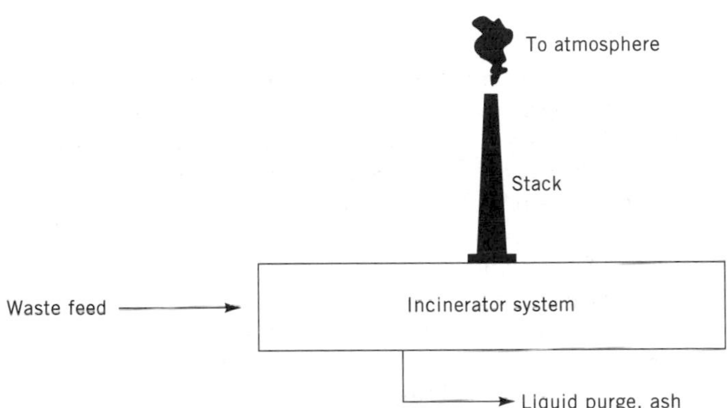

Figure 1. Schematic of an incinerator system.

Particulate Emission Limits. Particulate emissions, including condensables, must be demonstrated during the burn to be on a dry basis less than 0.18 g/m^3 (0.08 grain per dry standard cubic foot (gr/DSCF)) at the incinerator's stack, or lower depending on state regulations.

Products of Incomplete Combustion Emission Limits. Products of incomplete combustion typically are not directly measured during the trial burn. Instead, levels of carbon monoxide emissions are used as an indication of combustion efficiency. High combustion efficiencies are assumed to result in acceptable levels of products of incomplete combustion. If carbon monoxide emissions are measured at less than 100 ppmv dry basis, the standard is met. However, if emissions are greater than 100 ppmv, no more than 20 ppmv of total hydrocarbons are allowed at the incinerator stack during the trial burn.

Metal Emission Limits. Limits for metals, both carcinogenic and noncarcinogenic, are based on an adjusted stack height. Failure to meet these limits requires risk assessments using site-specific factors and modeling to establish limits for each metal. The assessments are based on the probability of developing adverse health effects or cancer, based on an inhalation exposure pathway to maximum exposed individuals located near the incinerator.

HCl and Cl Emission Limits. Hydrochloric acid and chlorine must also meet emission limits for an adjusted stack height.

Other Federal Regulations.

The TSCA regulates the operation of incinerators disposing of waste containing more than 50 ppm PCBs. Such units must demonstrate a 99.9999 destruction and removal efficiency during a trial burn prior to obtaining an operating permit. The CAA governs emissions of ozone, carbon monoxide, particulate matter, sulfur dioxide, nitrogen oxides, and lead from incinerators. It is intended to maintain National Ambient Air Quality Standards (NAAQS) for each of these pollutants in attainment areas, and to improve the quality of air in nonattainment areas. This regulation is generally more restrictive than RCRA because it specifies the type of emission control technology to be used when the potential to emit exceeds specified levels.

SOLID WASTE INCINERATION

Polymeric or carbonaceous solids are degraded by high temperature. In the presence of oxygen, any carbon, hydrogen, and sulfur are oxidized to CO_2, H_2O, and SO_2, respectively. The rate of incineration increases rapidly with temperature. A range of 700–760°C is generally required for combustion, and most general-purpose incinerators operate between 760 and 1100°C. For an incinerator to operate without auxiliary fuel or air preheating, the waste feed or refuse must contain less than 50% moisture or 60% ash, and have more than 25% combustibles.

Atmospheric Conditions

In addition to complete combustion, wastes may be destroyed by treatment at high temperatures either without oxygen (pyrolysis), using limited oxygen (partial combustion), or in reactive atmospheres (gasification), such as those containing steam, hydrogen, or carbon dioxide.

Refuse Benefaction

It is extremely difficult to burn and recover useful energy from unsorted municipal waste because of its heterogeneity in size, shape, chemical composition, and heating value. However, preparation of the waste before thermal treatment facilitates burning. Such pretreatment contributes to the front-end costs but reduces furnace costs. The waste is upgraded by separation of the nonorganic fraction and drying, shredding, and densifying solids. These fuels thus prepared are referred to as refuse-derived fuels (RDF).

Incinerator Types

The various types of incinerators include the moving grate incinerators and the multichamber incinerators. Among the more nonconventional types are the suspension-fired units, starved-air incinerators, vortex incinerators, multihearth furnaces, as well as the method called slagging incineration.

Rotary Kiln Incinerators. The rotary kiln has been used to incinerate a large variety of liquid and solid industrial wastes. Any liquid capable of being atomized by steam or air can be incinerated, as well as heavy tars, sludges, pallets, and filter cakes. This ability to accept diverse feeds is the outstanding feature of the rotary kiln and therefore this type of incinerator is often selected by the chemical and waste treatment industries.

Factors Affecting Destruction of Solid Wastes

The analysis of the evolution and/or destruction of hydrocarbons during the incineration of solid hazardous wastes involves heat transfer, mass transfer, and reaction kinetics. Figure 2 is a generalized flow chart for the processes experienced by solids during incineration. The key phenomena include the flashing of liquid hydrocarbons; the vaporization, desorption, and stripping of hydrocarbons; the pyrolysis and charring of hydrocarbons; and the oxidation of char. To a certain extent these processes occur in parallel and are common to most thermal treatment processes. Emphasis herein is on applications to the thermal treatment of contaminated soils at temperatures less than 800°C. Slagging kilns are not discussed.

The key variables affecting the rate of destruction of solid wastes are temperature, time, and gas–solid contacting. The effect of temperature on hydrocarbon vaporization rates is readily understood in terms of its effect on liquid and adsorbed hydrocarbon vapor pressures.

Pollutant Emissions from Solid Waste Incinerators

NO_x and SO_2 Emissions. Oxides of nitrogen (NO and NO_2) and sulfur (primaily SO_2) are emitted from most combustion systems including hazardous waste incinerators. The two principle mechanisms are summarized in Figure 3 (11). The thermal NO_x pathway is important in any high temperature process containing N_2 and O_2. The fuel NO pathway is also important if the fuel or waste contains nitrogen. NO_x levels are difficult to estimate because the rate of formation of NO by both pathways is determined by complex kinetics and

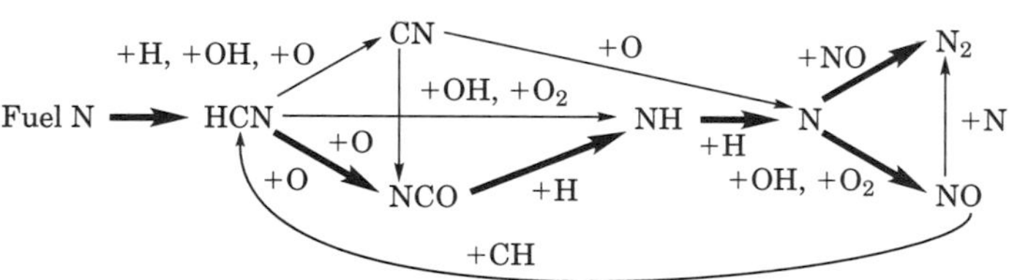

Figure 3. Schematic showing reaction pathways by which fuel nitrogen, N, is converted to NO and N_2. The bold lines indicate the key pathways (11). Thermal NO is formed from N_2: $N_2 + O \rightarrow NO + N$.

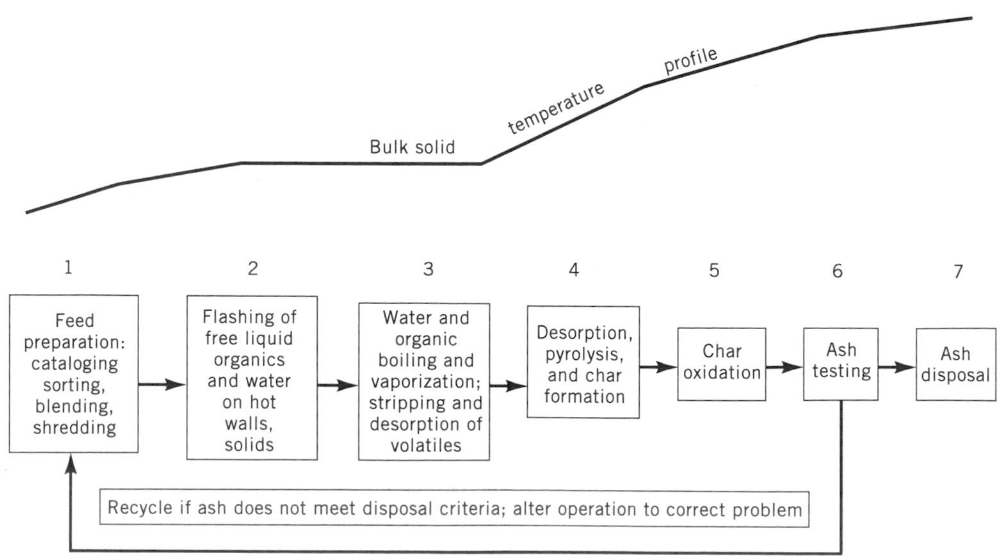

Figure 2. Generalized process flow chart for the thermal treatment of solid wastes. To a certain extent, steps 2, 3, 4, and 5 always proceed in parallel because of mixing limitations, nonhomogeneities in the waste, and unevenness in its heating.

gas-phase mixing. SO_2 production in incinerators is generally easy to predict because nearly all organic sulfur species are completely converted to SO_2 and other products of combustion. The incineration of liquid wastes containing nitrogen are fairly well understood from studies on liquid fuels.

In a rotary kiln, the burner can produce both thermal and fuel NO_x, if the fuel contains nitrogen. Many solid waste streams also contain nitrogen, typically as much as 20 wt %, which contributes to the fuel NO pathway. Key sources of solid waste fuel nitrogen include plastics, nylons, dyes, and other process wastes. Nylon, for example, is 33 wt % nitrogen.

Partitioning of Heavy Metals. Metals entering a solid waste incinerator can leave the system with the bottom ash, the captured fly ash, or the exhaust gases. The fraction leaving with the exhaust gases can include metal vapors such as mercury and submicrometer particles that escape capture in the air pollution control devices. Metals entering the incinerator as liquid streams are usually carried out of the reactor in the fly ash, which is typically enriched with heavy metals relative to the entering solid waste. The level of enrichment increases with increasing incinerator temperature and increasing metal volatility. The enrichment on small particles results from the condensation of vaporized metals. Metal concentrations on particles also typically increase with decreasing particle diameter. The vast literature on the behavior of metals in coal combustion is useful for understanding these phenomena (12). A comprehensive review on metal emissions from incinerators is available (13).

Dioxin and Furan Emissions. The emissions of polychlorinated dibenzo-*p*-dioxins (PCDD) and polychlorinated dibenzo-furans (PCDF) from incinerators (4) are of interest to the public, scientists, and engineers. The U.S. EPA classifies 2,3,7,8-tetrachloridibenzo-*p*-dioxin (2,3,7,8,-TCDD) as the most potent carcinogenic compound it has evaluated. It is also listed as the agency's most potent reproductive toxin (4).

The proposed mechanism by which chlorinated dioxins and furans form has shifted from one of incomplete destruction of the waste to one of low temperature, downstream formation on fly ash particles (14). Two mechanisms are proposed, a *de novo* synthesis, in which PCDD and PCDF are formed from organic carbon sources and Cl in the presence of metal catalysts, and a more direct synthesis from chlorinated organic precursors, again involving heterogeneous catalysis. Bench-scale tests suggest that the optimum temperature for PCDD and PCDF formation in the presence of fly ash is roughly 300°C.

Chlorine may be formed by the Deacon reaction at temperatures below about 900°C,

$$2\ HCl + 1/2\ O_2 \rightleftarrows Cl_2 + H_2O \tag{1}$$

Both Cl_2 and HCl have been shown to chlorinate hydrocarbons on fly ash particles. Pilot-scale data involving the injection of fly ash from municipal waste combustion (14) show that intermediate oxygen concentrations (4–7%) produce the highest levels of PCDD and PCDF. These data also show significant reductions in PCDD and PCDF emissions with the upstream injection of $Ca(OH)_2$ at about 800°C.

LIQUID WASTE INCINERATION

Incinerators

Furnaces. A furnace for combusting both high and low heating value liquid wastes is shown in Figure 4. Vertical furnaces are normally used for wastes containing high salt concentrations. Investment is typically higher than for furnaces of horizontal orientation as burners and controls are located in an elevated position, installation of furnace refractory is more difficult, and additional structural steel to support the furnace is required. For systems having a quench tank downstream of the furnace, the outlet of the furnace is tapered to reduce exposure to the quench and subsequent radiation losses. Furnace outlet velocities are maintained below 24 m/s to minimize erosion of outlet refractory.

For given combustion air, waste, and auxiliary fuel feed rates to the incinerator, furnace residence time decreases as furnace pressure decreases. Often the required pressure drop through the downstream particulate removal device is not established until actual operation, and furnaces are sized assuming little or no pressure drop across the particulate removal device to allow adequate furnace residence time under all operating conditions.

Regulations require that the incinerator furnace be at normal operating conditions, including furnace temperature, before hazardous wastes are injected. This requires auxiliary fuel burners for furnace preheating. In addition, the burners

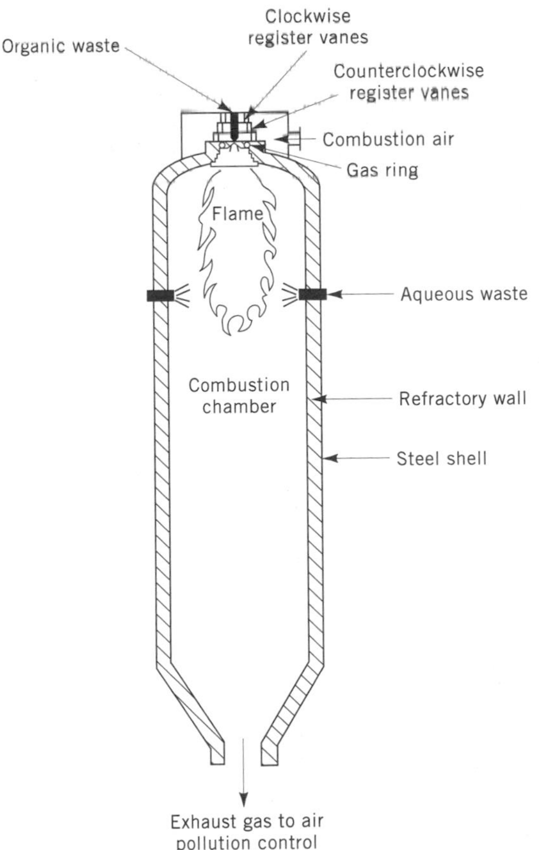

Figure 4. Liquid waste incinerator furnace.

provide heat when the wastes burned are of low heating value. Auxiliary burners are sized for conditions where liquid wastes are injected without the addition of high heating value wastes.

Auxiliary fuels normally include natural gas and No. 2 or No. 6 fuel oils. Natural gas has several advantages: it is relatively clean burning and easily transported and measured. No. 2 fuel oil requires mechanical, steam, or air atomization to ensure combustion. Moreover, No. 2 fuel oil burners, when not properly adjusted, tend to form soot. No. 6 fuel oil is highly viscous, normally in the 400–2000 mPa · s(=cP) range, at ambient temperatures, and as a result requires preheating to 80–120°C to reduce the viscosity to 15–65 mPa · s(=cP) prior to combustion. No. 6 fuel oil piping requires steam or electric tracing to ensure lower viscosities as the fuel is being injected into the incinerator furnace. It normally requires burners capable of steam atomization. In addition, it can contain up to 3% sulfur and 500 ppm vanadium, which adds to the sulfur dioxide and vanadium pentoxide loading, and corrosiveness of flue gases leaving the combustion chamber.

The furnace is constructed with a steel shell lined with high temperature refractory. Refractory type and thickness are determined by the particular need. Where combustion products include corrosive gases such as sulfur dioxide or hydrogen chloride, furnace shell temperatures are maintained above about 150–180°C to prevent condensation and corrosion on the inside carbon steel surfaces. Where corrosive gases are not present, insulation is sized to maintain a shell temperature below 60°C to protect personnel.

Three types of refractory are used. Castable refractory, similar to concrete, is placed in the shell using forms and poured in place or blown in. Plastic refractory is prepared in a stiff consistency and is either hammered or rammed in place. Plastic refractories are typically used for repairs. Fire brick is the most commonly used refractory. It is bonded in place using thin mortar joints. Brick having alumina content of 60–90% is used in areas exposed to hot corrosive gases. To decrease investment, less expensive insulating brick is often placed between the high alumina brick and the furnace shell. The cost of high alumina brick is typically 4–6 times that of insulating brick.

Refractory failures resulting from erosion and corrosion from hot particulate-laden gases can result in incinerator downtime and high maintenance costs. Of particular concern are fluorine, sodium, potassium, and sulfate salts, which penetrate brick surfaces when hot. Upon cooldown, salt hardens and expands, causing the surface which has been penetrated to fail. In addition, organically bound alkali metals in wastes can react chemically with the refractory to form new compounds with lower melt points (eutectics) than furnace operating temperatures. Continued operation at elevated furnace temperatures and close attention to the design and operation of the furnace to keep wastes from impinging on refractory walls, along with controlling the amount of alkali metals fed, help prolong refractory life.

Quench Systems. Quench systems are used to cool hot furnaces gases from 980–1200°C to 120–150°C. This allows less expensive materials of construction such as fiber glass reinforced plastic (FRP) to be used downstream of the quench and in gas cleaning equipment, and reduces the volume of gas flow, resulting in smaller equipment. Water or air quenching systems are typically used. Water quenching systems use the latent heat of evaporation to adiabatically cool gases. Particulate matter collects in quench water requiring that the system be continuously purged. Air quench systems require the addition of large volumes of ambient air resulting in larger downstream gas cleaning equipment than with water quench systems. Air quench chambers are sized to allow particulate matter to be removed manually during incinerator shutdown. Materials of construction and design of spray systems are critical in the design of quench systems.

Control Systems. Control systems are used to regulate the addition of liquid waste feed, auxiliary fuel, and combustion air to the incinerator furnace. In addition, scrubber operation is automated to help ensure meeting emission limits. Flows are measured using differential pressure devices such as flow nozzles and flow orifices when the fluids are relatively nonviscous ($Re > 3000$). Gas and combustion air flows are typically measured using a flow orifice. Highly viscous wastes require measurement with positive displacement systems such as oval gear, target flow, or mass flow meters.

Temperature measurements ranging from 760 to 1760°C are made using iron–constantan or chromel–alumel thermocouples and optical or surface pyrometers. Temperature-measuring devices are placed in multiple locations and protected to allow replacement without incinerator shutdown.

To ensure combustion gases are not present during initial burner light off, the furnace is purged with ambient air. After the purge, a 145–585 kJ/s (0.5–2.0 MBtu/h) pilot establishes a flame in the furnace. A scanner is used to sense flame and is interlocked to shut down all waste and auxiliary fuel flows upon loss of flame. Flame scanners for incinerator furnaces are typically of the ultraviolet type because hot refractory emits in the infrared region. After the pilot has been established, the auxiliary fuel burners are ignited and the furnace is brought up to its normal operating temperature. Once the operating conditions given in the permit are satisfied, including furnace temperture, stack oxygen levels, and combustion air flow rates, wastes are injected into the furnace.

Factors Affecting Destruction of Liquid Wastes

Liquid wastes can be divided into two classes: low and high heating value. The former requires auxiliary fuel. A heating value above 16 MJ/kg (7000 Btu/lb) is generally considered high enough to be burned without auxiliary fuel, but this depends on the specifics of excess air and desired flame temperature. As shown in Figure 4, high heating value wastes are usually injected with auxiliary fuel and oxidant into a swirl burner. Roughly one vessel diameter is allowed in the axial direction for mixing and combustion of the fuels, and then the low heating value waste, which is typically aqueous, is injected radially into the hot gases. Because the high heating value waste is injected with the fuel, more residence time is available for its destruction, so that the destruction of the low heating value waste is generally limiting. Similar mechanisms are involved for both types of waste.

The steps in waste destruction are (*1*) heatup of the waste to its boiling point, (*2*) vaporization of the waste droplets, (*3*) heatup of any waste residue to combustion temperature, and (*4*) destruction by combustion reaction. Destruction of waste species volatilized during the first three steps can begin the moment a species becomes gaseous, because sensible heating of volatilized components is extremely rapid and need not be considered as a separate step.

The factors which govern the efficiency of waste destruction include atomization, ie, mean drop size, and size distribution; temperature; residence time; O_2 concentration; and flow patterns.

The atomization, temperature, and O_2 concentration interact with the various rate processes to set the residence time requirements. A desirable flow pattern is for the fuel, ie, the high heating value waste plus any supplemental fuel, to burn completely in a back-mixed zone, and for the low heating value waste to be injected into a plug flow zone having uniform temperature and O_2 concentration profiles. This pattern generally produces the minimum average residence time requirements. To the extent that part of the incinerator volume is ineffective owing to cold spots, O_2 starved spots, or bypassing of portions of the flow, the average residence time has to be increased to compensate. In cases where the waste contains dissolved or suspended solids, a design allowance should be made for the probability that these solids slowly coat the chamber walls and reduce the effective residence time during the course of operation. Periodic burnouts or shutdowns for cleaning are required to remove this material.

Pollutant Emissions from Liquid Waste Incinerators

Gas Emissions. Wastes considered for incineration are usually organic in nature, so that the vast majority of the waste ends up as CO_2, H_2O, and N_2. Although CO_2 is of increasing concern because it is a greenhouse gas, emission issues have generally been related to the combustion products of sulfur, halogens, and metallic components of the waste. Carbon monoxide, which is not thermodynamically stable under normal incineration conditions, is also regulated. If CO is found in incinerator off-gas it is evidence of poor fuel–oxidant mixing or insufficient effective residence time, possibly owing to cold spots, oxidant starved zones, or bypassing.

Sulfur generally becomes SO_2, although some smaller amounts are possibly converted to SO_3, depending on temperature. Chlorine mostly results in HCl, but some Cl_2 and atomic Cl forms as well. Any atomic Cl recombines to form Cl_2 if quenching is rapid. Low incineration temperatures favor Cl_2, and high temperatures favor atomic Cl. There is an optimal temperature for minimizing the total effective Cl_2, ie, $Cl_2 + Cl/2$.

HCl can be absorbed into water to make a concentrated (usually 20 wt %) HCl solution, whereas SO_2 and Cl_2 must be scrubbed using a basic reagent (usually caustic or lime) to be effectively removed. The scrub liquor pH must be controlled to avoid heavy reagent consumption caused by the removal of CO_2. Even if only HCl is present, it is usually a good idea to follow the HCl absorber with a basic scrubber in order to ensure complete elimination of acids from the exhaust gas. When absorbing chlorine into alkaline solutions, the products are both chloride and hypochlorite. It is necessary to keep the pH above 9 in order to stabilize the hypochlorite and avoid regeneration of chlorine. The blowdown can be treated in one of several ways. Hypochlorite can be decomposed to chloride and oxygen using homogenous transition-metal catalysts (15), reduced to chloride and sulfate with sulfite, or reacted with hydrogen peroxide to form chloride, water, and oxygen.

Oxides of nitrogen, NO_x, can also form. These are generally at low levels and too low an oxidation state to consider water scrubbing. A basic reagent picks up the NO_2, but not lower oxidation states; the principal oxide is usually NO, not NO_2. Generally, control of NO_x is achieved by control of the combustion process to minimize NO_x, ie, avoidance of high temperatures in combination with high oxidant concentrations, and if abatement is required, various approaches specific to NO_x have been employed. Examples are NH_3 injection and catalytic abatement (16).

Engineering Calculations. A good discussion of column design for gas absorption can be found in the literature (17). For staged columns (tray or plate) the classic approach is to estimate the number of equilibrium stages, and translate this to the number of actual stages through models for stage efficiencies. For packed towers, either the height of a theoretical plate is used to translate an equilibrium-stage calculation to a packing height, or the concept of the transfer unit is used to estimate the number or required transfer units, and the height of a transfer unit.

These design methods depend on a knowledge of the vapor–liquid equilibria for estimating the number of equilibrium stages or the number of transfer units. Usually, the scrub liquor is an aqueous electrolyte, possibly containing suspended solids. Techniques for predicting the pertinent equilibria have improved to the point where reliable calculations can be done on solutions of the high ionic strength typical of these scrub liquors (18). The techniques are available in computer models of countercurrent contactors so that the number of required equilibrium stages can be readily estimated (19).

Metal Contaminants and Ash. Alkali metals form basic oxides that are reactive toward acidic species such as the acid gases, silicates, and aluminates. These form stable salts with acid gases if the off-gas contains such gases. Sodium, the most common of these metals, prefers to form chlorides ahead of sulfates. Sodium carbonate only forms in the absence of halides and sulfur oxides, SO_x. There usually is too little NO_x present to form nitrates.

Alkali metal halides can be volatile at incineration temperatures. Rapid quenching of volatile salts results in the formation of a submicrometer aerosol which must be removed or else exhaust stack opacity is likely to exceed allowed limits. Sulfates have low volatility and should end up in the ash. Alkaline earths also form basic oxide. Calcium is the most common and sulfates are formed ahead of halides. Calcium carbonate is not stable at incineration temperatures. Transition metals are more likely to form an oxide ash. Iron, for example, forms ferric oxide in preference to halides, sulfates, or carbonates. Silica and alumina form complexes with the basic oxides, eg, alkali metals, alkaline earths, and some transition-metal oxidation states, in the ash.

Estimates of Composition. The best approach toward estimating the chemistry of most contaminant species is to assume chemical equilibrium. Computer programs and databases for calculating chemical equilibria are widely available (20). Care must be taken that all species of concern are in the database referenced by the program being used, and if necessary, important species must be added in order to get the complete picture. In addition to predicting the exhaust composition of both gases and solids, the ability of these chemical equilibrium programs to do adiabatic calculations makes them useful for computing supplemental fuel requirements and the effect of excess oxidant on temperature.

The equilibrium approach should not be used for species that are highly sensitive to variations in residence time, oxidant concentration, or temperature, or for species which clearly do not reach equilibrium. There are at least three classes of compounds that cannot be estimated well by assuming equilibrium: CO, products of incomplete combustion (PICs), and NO_x. Under most incineration conditions, chemical equilibrium results in virtually no CO or PICs, as required by regulations. Thus success depends on achieving a nearly complete approach to equilibrium. Calculations depend on detailed knowledge of the reaction network, its kinetics, the mixing patterns, and the temperature, oxidant, and velocity profiles.

NO_x formation occurs by a complex reaction network of over 100 free-radical reactions, and is highly dependent on the form of nitrogen in the waste. Nitro-compounds form NO_2 first, and then NO, approaching equilibrium from the oxidized side. Amines form cyano intermediates on their way to NO, approaching equilibrium from the reduced side. Using air as the oxidant, NO_x also forms from N_2 and O_2. This last is known as thermal NO_x.

Through numerical codes which integrate stiff differential equations (21), it is possible to estimate NO_x levels based on kinetic schemes. The reaction network is becoming better defined, both in terms of the important reactions and their kinetic parameters. These estimates only show trends; comparisons with data available for compounds similar to those being incinerated are recommended. Because the reaction trajectory is sensitive to the type of nitrogen-bearing species, chemical similarity is the important factor in making good projections from literature data.

Particulate Pollutant Control Equipment

Venturi scrubbers, electrostatic precipitators, and baghouses are among the various types of equipment used to control particulate pollution.

INCINERATORS FOR VAPORS AND GASES

Undesirable combustible gases and vapors can be destroyed by heating to the autoignition temperature in the presence of sufficient oxygen to ensure complete oxidation to CO_2 and H_2O. Gas incinerators are applied to streams that are high energy, eg, pentane, or are too dilute to support combustion by themselves. The gas composition is limited typically to 25% or less of the lower explosive limit. Gases that are sufficiently concentrated to support combustion are sometimes burned in flares, waste-heat boilers, in conjunction with other fuels in boilers and kilns, or used as process fuel. Occasionally, such gases may be burned in specially designed furnaces incorporating heat or material recovery, eg, chlorine. Protection against flame flashback must be provided between the incinerator and the process or source of the waste gas. Such protection is necessary to guard against unusual gas compositions even when the waste gas stream does not support combustion. Design of the flashback-prevention device depends on stream composition. Parallel-plate or multiple-screen flame arrestors, or a water seal, are usually adequate, but special precautions must be taken when the gas contains appreciable quantities of hydrogen because of high flame speed. A carrier gas such as nitrogen can be used to provide sufficient velocity to prevent flashback. Knockout devices should be considered to prevent carrying slugs of combustible entrained liquid into the incinerator. Safety controls are desirable to protect against overheating caused by erratic flows and sudden composition changes. Noncombustible particles can present problems. These may be melted and collected as a usually corrosive liquid pool. Removal of particulate prior to incineration may be desirable, but if the material is too fine, especially submicrometer in size, efficient removal may not be economical. In such cases, direct-flame incinerators specifically designed for slag removal may be preferred. Generally, catalytic incinerators, direct-flame incinerators, and flames can be used to destroy hazardous vapors or gases.

BIBLIOGRAPHY

1. P. F. Fennelly, *The Role of Incineration in Chemical Waste Management,* American Institute of Chemical Engineers, New York, 1986.
2. *Hazardous Waste Incineration—A Resource Document,* American Society of Mechanical Engineers, New York, 1988, p. ix.
3. U.S. Environmental Protection Agency, *Hazardous Waste Incineration: Questions and Answers,* EPA/530-SW-88-018, Washington, D.C., Apr. 1988, p. 1.
4. C. R. Dempsey and E. T. Oppelt, *J. Air Waste Manage. Assoc.* **43,** 25–73 (1993).
5. J. V. L. Kiser, *Waste Age,* 139 (Nov. 1990).
6. C. C. Lee, *Medical Waste Incineration Handbook,* Government Institutes, Inc., Nov. 1990, pp. 2–10.
7. E. M. Steverson, *Environ. Sci. Technol.* **25**(11), 1810 (1991).
8. K. D. Phillips, *Atlanta J.,* 10A (June 8, 1988).
9. *U.S. Code of Federal Regulations,* Title 40, Part 261, Appendix II, Washington, D.C., July 1991.
10. Ref. 9, Appendix VIII.
11. J. A. Miller and C. T. Bowman, *Prog. Energy Combust. Sci.* **15,** 287 (1989).
12. R. C. Flagen and J. H. Seinfeld, *Fundamentals of Air Pollution Engineering,* Prentice Hall, Inc., Englewood Cliffs, N.J., 1988, pp. 358–390.
13. W. P. Linak and J. O. L. Wendt, *Prog. Energ. Combust. Sci.* **19,** 145–185 (1993).
14. B. K. Gullett, P. M. Lemieux, and J. E. Dunn, *Environ. Sci. Technol.* **28,** 107–118 (1994).
15. M. W. Lister, *Can. J. Chem.* **34,** 479 (1956).
16. C. N. Satterfield, *Heterogeneous Catalysis in Practice,* McGraw-Hill Book Co., Inc., New York, 1980, pp. 229–231.
17. W. M. Edwards, "Mass Transfer and Gas Absorption," in R. H. Perry, D. W. Green, and J. O. Maloney, eds., *Perry's Chemical Engineers' Handbook,* 6th ed., McGraw-Hill Book Co., Inc., New York, 1984.
18. J. F. Zemaitis, D. M. Clark, M. Rafal, and N. C. Scrivner, *Handbook of Aqueous Electrolyte Thermodynamics,* AIChE, New York, 1986.
19. M. Rafal and S. J. Sanders, "The ProChem System for Modeling/Simulation of Aqueous Systems," in *Proceedings of the Second International Airlie House Conference on Aqueous Systems,* Warrenton, Va., May 10–14, 1987.
20. S. Gordon and B. J. McBride, *Computer Program for Calculation of Complex Chemical Equilibrium Composition, Rocket Performance, Incident and Reflected Shocks, and Chapman-Jouget Detonations,* NASA SP-273, Interim Revision, NTIS, Springfield, Va., Mar. 1976.

21. R. J. Kee, J. A. Miller, and T. H. Jefferson, *CHEMKIN: A General-Purpose, Problem-Independent, Transportable, FORTRAN Chemical Kinetics Code Package,* DOE, Sandia Laboratory SAND80-8003, NTIS, Springfield, Va., Mar. 1980.

HAZARDOUS WASTE: PLANNING

J. ANDY SOESILO
Western International University
Phoenix, Arizona

K. DAVID PIJAWKA
Arizona State University
Tempe, Arizona

The economic system can be viewed as a system that transforms energy and natural resources into products and residues. These residues will eventually reach the biophysical environment and may threaten the availability of resources, thereby affecting the sustainability of the economic system. To prevent resource degradation, intense utilization of natural resources must be combined with efficient residuals management. The proper utilization of resources and the efficient management of residuals require good environmental planning.

Environmental planning can be defined as the initiation and operation of activities to manage the acquisition, transformation, distribution, and disposal of resources in a manner capable of sustaining human activities, with a minimum disruption of physical, ecological, and social processes (1). This means that the transformation of energy and natural resources in the economic system must be conducted at minimum environmental, monetary, and social costs. In this context, the goal of environmental planning is to prevent adverse effects of economic activities by ensuring minimum monetary and environmental costs, in a just and equitable manner (2).

Environmental planning encompasses both natural resources planning and residual planning. Residuals can take the form of air emission, wastewater, solid waste, or hazardous waste. Residual planning, therefore, can be broken down further into air quality planning, water quality planning, solid waste, and hazardous waste planning. As far as residual planning is concerned, it not only deals with proper handling of residuals, but also considers any effort to reduce the generation of residuals early in production planning.

Hazardous waste planning focuses on hazardous waste. It does not specifically deal with air, water, or solid waste; however, hazardous waste planning considers those components when addressing integrated multimedia (air, water, waste) environmental management. As an integral part of the economic system, hazardous waste generation is unavoidable. The central question of hazardous waste management is not whether elimination of all hazardous waste is possible, but more realistically, how well that portion that cannot be eliminated is managed.

Prior to the 1990s, the hazardous waste management philosophy in the United States focused primarily on managing hazardous wastes after they were generated. This philosophy is referred to as the "end-of-pipe" approach. Although not desirable, this philosophy typified the U.S. hazardous waste management practice. In 1976, the U.S. Environmental Protection Agency (EPA) stated that the desired approach to managing hazardous waste should be based on a hierarchy with source reduction as the first priority. Next in the hierarchy is recycling, followed by waste treatment, including incineration (see HAZARDOUS WASTE: INCINERATION). Land disposal is considered the fourth and least desirable option because of the continuing threat of contaminant releases (3).

Despite this stated preference, the EPA's effort to promote source reduction has been limited. From 1976 to 1986, the EPA concentrated on developing the regulatory program over the generation, transportation, storage, treatment, and disposal of hazardous waste with a belief that the program would provide sufficient incentive for source reduction efforts (4). The 1984 Hazardous and Solid Waste Amendments (HSWA) formalized the concept of waste reduction. This amendment to the Resource Conservation and Recovery Act (RCRA) law specified that wherever feasible, the generation of hazardous waste is to be reduced or eliminated as expediously as possible. The Pollution Prevention Act of 1990 expanded the commitment to source reduction by establishing the hazardous waste management hierarchy as national policy. The hierarchy represents a governmental preference of available hazardous waste management options and is a strategic instrument to achieve the RCRA goals of improving human health, protecting the environment, and conserving natural resources.

The hierarchy provides an ordering of technologies according to their possible adverse environmental consequences. Source reduction is the most desirable option. This means that industries must be seriously considering *not* producing hazardous waste in the first place. If the manufacturing technology cannot be modified to eliminate hazardous waste, or the input chemicals cannot be substituted with nonhazardous materials, companies ought to give the generated hazardous waste a "second life" by recycling the waste, ie, by reusing it or reclaiming metals and other valuable resources from the waste. When source reduction and recycling options have been exhausted, appropriate treatment of the remaining waste should be explored before considering land disposal.

Some large corporations successfully adopted the source reduction technique as part of the company's business culture long before the shift in philosophy occurred nationally. In 1972, Dow Chemical U.S.A. introduced comprehensive Global Pollution Control guidelines and in 1978 started the Waste Reduction Always Pays (WRAP) program which has since reduced the company's hazardous waste generation significantly (5). The 3M Company's Pollution Prevention Pays (P3) has been implemented since 1975 and in the first 14 years of the program, the pollution prevented has resulted in a savings of $426 million for U.S. operations (6). The departure from the traditional end-of-pipe approach to the source reduction approach for most companies is particularly in response to more regulations and the anticipation of potential future liabilities from improper hazardous waste management practices (7). Some companies are developing better waste reduction methods, but some firms are not strongly motivated to implement the source reduction concept. They are slow to adopt the concept because of unfamiliarity with available reduction technologies, reluctancy in adopting something new, or unwillingness to spend the time needed to examine the program (8,9). It is not surprising that these barriers to implement source reduction persist. When adopting the hierarchy, the industrial community will always consider other factors, such as cost effectiveness of an alternative technology, organizational culture, environmental impacts of the technology, social acceptability, and the existing regulatory framework (10,11).

PLANNING REQUIREMENTS

Regulatory framework is critical to hazardous waste planning. As has been argued, the heart and soul of planning is in the regulations that implement the plans (12). In the planning profession, planners must continue to develop their planning skills and regulatory knowledge to keep pace with the intricate institutional setting and the rapidly expanding rules and regulations. This means that hazardous waste planners should have a good understanding of the legal framework of hazardous waste management. The two federal statutes that are important for hazardous waste management are RCRA and Comprehensive Environmental Response, Compensation, and Liability Act (CERCLA). RCRA regulations outline the procedures necessary to perform the proper handling of hazardous wastes. The provisions specify what materials are regulated as hazardous wastes, what businesses are required to comply with the regulations, and in what way and under what operating conditions compliance should occur.

Although RCRA creates a framework for the proper management of wastes, the Act does not address the problems of hazardous waste found at inactive or abandoned sites or those spills that require emergency response. These problems are taken care of by the CERCLA, which was passed by U.S. Congress in 1980 and was signed into law by President Carter. The statute was created in response to the discovery of a large number of inactive and abandoned hazardous waste dumps in the late 1970s. The CERCLA program contains three objectives: (*1*) to take actions against releases of hazardous chemicals, (*2*) to identify individuals or firms responsible for a contaminated site and require them to take appropriate response action to clean up the site, and (*3*) to clean up a contaminated site where the responsible parties are not found or unable to take cleanup responsibility.

RCRA and CERCLA are separate but interrelated programs. For example, when wastes from a CERCLA site are taken off-site for treatment, storage, or disposal (TSD), the receiving facility must be a RCRA TSD facility, and the waste must be handled in accordance with applicable RCRA requirements including the hazardous waste generator and transporter regulations (see HAZARDOUS MATERIALS TRANSPORTATION). The difference between the two programs is that RCRA handles "hazardous waste" which is narrowly and precisely defined in the hazardous waste regulations, while CERCLA deals with remediation caused by releases of a "hazardous substance". The definition of a hazardous substance is extremely broad and covers most chemicals.

Although knowledge of rules and regulations is vital, hazardous waste data management represents another essential requirement for planners. Many planning activities involve intensive regulatory analysis as well as advanced data management (13). Managing hazardous data needs some understanding of hazardous waste data sources, the nature of the data, and data analysis techniques. Eventually, a data management system should be able to convert the data into meaningful information. It has been argued that if data alone could be used to solve planning problems, there would be little reason to convert them into information (14). Not only must planners understand the information they produce from various data sources, they must also be able to communicate the information precisely to clients, decision-makers, and the public. The process of developing information for planning involves not only locating data, but also evaluating its meaning and its relevance to planning.

The major tasks for developing a basic hazardous waste data management system include (*1*) the identification of hazardous waste data sources, (*2*) the collection and compilation of appropriate data, and (*3*) the analysis of the compiled data and conversion into meaningful hazardous waste information. To perform the first task, it is necessary for planners to first determine whether a facility is indeed handling a hazardous waste. This implies that at the beginning of the data management process, the hazardous waste planner recognizes the spectrum of the wastes under consideration.

The first step in determining if a facility manages hazardous waste is the application of the regulatory definition of solid waste to the waste streams at the facility. Solid waste need not be in a solid physical state; it may be in a liquid, semisolid, or even a contained-gaseous physical state. Once a material is legally classified as a solid waste, the next step is to determine if it is also hazardous. The determination is conducted by using appropriate regulatory hazardous waste determination process.

When a facility handles hazardous waste, a hazardous waste planner must know what types of waste are handled, how much waste is generated on a monthly basis, and how much is shipped off-site from the facility. Hazardous waste regulations establish record-keeping and reporting requirements for hazardous waste handlers and provide the handlers with record-keeping and reporting compliance guidance. For an individual state, the federal requirements represent a mechanism to track the quantities of waste managed and transported by hazardous waste handlers statewide. These record-keeping and reporting provisions consist of the hazardous waste manifest requirement and the submission of the biennial report by hazardous waste handlers.

RCRA establishes a strategy for controlling hazardous waste movement from generation to ultimate disposal. This management scheme is popularly known as the cradle-to-grave system. Generators and transporters of hazardous waste and owners or operators of TSD facilities are the links in the cradle-to-grave chain of hazardous waste management. RCRA requires generators to ensure and fully document that the hazardous waste they generate is properly managed. The regulatory requirements for hazardous waste generators include obtaining an identification number from the U.S. EPA, pre-transport packaging and labeling, manifesting, record-keeping, and reporting.

Per RCRA regulations, any person who has hazardous waste transported off-site must prepare a shipping document called a hazardous waste manifest. The manifest must be signed by the generator of the waste and include the signature of the transporter and the date of acceptance. The generator retains one copy and gives the transporter the remaining copies of the manifest. Copies of each manifest must be kept for three years by the generator after they are signed by the designated TSD facility that received the waste and the manifests returned to the generator.

The manifest requirement is applicable to large quantity generators (LQGs) and small quantity generators (SQGs). An LQG is any facility that generates more than 998 kg (2200 lbs) of hazardous waste (or 0.99 kg (2.2 lbs) of "acutely" hazardous waste as specified in RCRA regulations) in any calendar month. A generator of more than 99 kg (220 lbs) but less than 998 kg (2200 lbs) in any month is defined as an SQG. If a generator produces not more than 99 kg of hazardous waste in any calendar month, it is a conditionally exempt SQG and not subject to the manifest requirement.

A manifest is designed so that generators can track the movement of hazardous waste from the point of generation to the point of ultimate destination. The manifest is part of a controlled tracking system. Each time the waste is transported off-site, the manifest must be signed to acknowledge receipt of the waste. Once the waste is received by the TSD facility, the TSD must send a copy of the manifest back to the generator. This system ensures that the generator has documentation that the hazardous waste has made it to its ultimate destination. Because the purpose of a manifest is to track hazardous waste shipped off-site, it is not required for generators who treat, store, or dispose of their waste on-site.

The hazardous waste manifest serves as both a shipping document and a tracking document. The manifest provides important planning information including:

1. The identity of the generator, the transporter, and the receiving TSD facility,
2. The physical description and the type of the waste being transported,
3. The quantities of the waste being transported.

It is important for the generator to prepare and manage the manifest properly. A good manifest data system is vital to any facility. It is used for regulatory compliance and also represents an essential tool of the facility's hazardous waste management planning process. A good manifest data system enables hazardous waste planners to retrieve the facility's basic hazardous waste information on a monthly, quarterly, or yearly basis. Examples of the types of information that can be retrieved from the manifest data system follow:

1. The amount and type of hazardous waste shipped off-site by the facility,
2. The amount shipped by names of the receiving TSD facility,
3. The amount shipped by names of the transporter.

At the state level, the use of the manifest data in the state's hazardous waste management planning is somewhat restricted. Limitations embodied in the manifest requirements follow:

1. A generator is not required to submit a copy of the manifest to the state.
2. Conditionally exempt SQGs are exempted from the manifesting requirement.
3. The manifest deals with off-site shipments only.

In addition to the manifest requirement, an LQG must also prepare and submit a copy of a biennial report to the U.S. EPA, or the authorized state, by March 1 of each even-numbered year. The report must be submitted on an EPA-approved format and must cover the handler's activities during the previous year. The reporting forms and accompanying instructions are collated in a bound document known as the biennial reporting system (BRS) booklet.

In the BRS, the generators are required to specify the characteristics of the waste generated in the reporting year and activities conducted to manage the waste. The BRS consists of five components:

1. Identity of the generator including certification statement using the identification of certification (IC) form of the BRS
2. Description of waste generation and management provided in the generation of management (GM) form
3. Waste received from off-site, described in the waste received (WR) form
4. Description of waste treated, recycled, and disposed, as specified in the process system (PS) form
5. Identification of off-site TSD facilities receiving the waste listed in the off-site identification (OI) form.

For the regulated industry, correctly completing the BRS forms, hazardous waste manifests, and other record-keeping and reporting documents is critical. Not only is accurate reporting essential for hazardous waste planning, it is also important for regulatory compliance. Failure to do so may result in unnecessary litigation and publicity. RCRA provides for felony penalties for anyone who knowingly omits material information, or makes a material false statement or representation in documents required for purposes of compliance with RCRA (15).

Although the BRS serves as the basis for tracking the generation and movement of hazardous waste in the state, waste flow estimates require more than the mere understanding of the operation of the database management system. At a minimum, accurate estimations require knowledge of federal hazardous waste regulations, state rules, and waste handling practices. When planning a state hazardous waste program, it is imperative that the hazardous waste planner has a good understanding of the patterns of waste flow in the state.

The hazardous waste generator report of the BRS which consists of the IC, OI, and GM forms, provides information on the identity of the generator, the generation level of each type of hazardous waste produced, the destination of the waste, and the transporter used to ship the waste. The generator report can serve as the basis for estimating the flow; however, because of the overlapping functions among hazardous waste generators and TSD facilities, the generator report should be used with caution.

RCRA designations on hazardous waste producers result in a functional grouping of waste generators. The four categories of generators follow:

1. Pure generators who do not perform TSD activities.
2. Generator – TSD, ie, generators who perform TSD activities on-site.
3. Captive TSDs that generate secondary hazardous waste.

4. Commercial TSDs that generate secondary hazardous waste.

Hazardous waste generated from the manufacturing of a product is known as the primary waste. Hazardous waste (re)generated from a treatment process is called the secondary waste. A captive TSD is defined as a TSD facility that manages hazardous waste generated by other generators under the same ownership. If the senders of hazardous waste are not under the same ownership as the receiving TSD facility, the facility is designated as a commercial TSD (16).

Once the classes of hazardous waste generators are identified, the next step is to comprehend how these categories interact in a larger system. In this regard, it is necessary to view hazardous waste flow as a set of interrelated activities performed by all categories of generators (17).

Hazardous waste planning creates plans, policies, or programs. The planning practice can be generalized into planning groups based on the specific planning objectives. The *Hazardous Waste Planning* handbook (18) which presents the principles, regulatory overview, data system, and current practices of hazardous waste planning in the private and public sectors, categorizes the activities into four major types. They include:

1. Hazardous waste management planning
2. Hazardous waste site remediation
3. Emergency response and hazard management planning
4. Citizen participation planning

HAZARDOUS WASTE MANAGEMENT PLANNING

Hazardous waste management planning is an activity that supports the goal of hazardous waste management in assuring the proper management of hazardous waste from the time it is generated to its ultimate disposal. In the context of hazardous waste management, the 1995 handbook indicates that compliance planning represents the first planning category, followed by waste minimization and pollution prevention, capacity assurance planning, facility siting, and state hazardous waste management planning (18).

Regulatory Compliance Program

In carrying out its mission to achieve an effective cradle-to-grave RCRA program, the U.S. EPA and authorized states utilize a combination of enforcement approaches that promote voluntary compliance by the regulated community and that intensify compliance monitoring activities, including inspections to detect violations, and take firm but fair enforcement action against violators to correct violations and create a strong enforcement presence (19).

In the 1990s, one of the most notable propensities in the domain of hazardous waste management is the increase of the government's effort in its environmental crime-prevention enforcement activities, which force more companies to develop sensible hazardous waste compliance programs in order to avoid prosecution. Noncompliance may result in a potentially large monetary fine, and if hazardous waste laws are knowingly violated, criminal prosecution can become a reality.

Regulatory compliance currently represents a hot issue in the industrial community (20). Some companies face obstacles in developing a good compliance program, including limited resources, a lack of information on regulatory requirements, and uninformed and untrained environmental personnel. Because of these obstacles, many small to mid-size companies say they are not achieving satisfactory compliance.

Although larger corporations usually have environmental planners, lawyers, consultants, or established programs to help satisfy compliance needs, smaller and mid-size companies normally do not have that level of expertise or support. Companies that have no compliance program and are recipients of penalties for violations of environmental laws may find that not having a compliance program is more expensive than having one.

Compliance program planning includes a framework for understanding the complex web of regulatory requirements and a guide for waste management decision-making within the regulatory context (21). Overall, the purpose of a regulatory compliance program is to ensure that the facility is in full compliance with all hazardous waste regulations. The most logical place to start with a regulatory compliance program is to determine if the facility generates hazardous waste, and if so, to what extent. This determination can be made by taking an inventory of each waste stream generated from the facility.

The term *waste stream* refers to waste materials generated from a known point of origin. Examples of waste streams are spent solvents from vapor degreasers, paint booth sludges from water wall paint booths, and wastewater treatment sludges from electroplating operations. Once each waste stream is identified, it can then be determined whether or not the waste is hazardous within the regulatory context.

In most cases, a facility that generates hazardous waste also temporarily stores waste, and perhaps treats some waste on-site, before the facility ships it to off-site disposal facilities. Certain on-site treatment is allowable under specified conditions, otherwise a TSD permit is required. With this generate–store–treat scenario, the regulatory inspection protocol can be established with the purpose of (*1*) examining points of waste generation and accumulation, (*2*) analyzing in-plant waste transport from generation points and accumulation points to storage and treatment units, and (*3*) evaluating storage and treatment unit for compliance with applicable standards (22).

The principal objective of in-house inspection is essentially to find regulatory areas that are normally vulnerable to noncompliance. These target areas are (*1*) the current practices of hazardous waste determination, (*2*) the hazardous waste generator status identification, (*3*) the waste accumulation procedures, and (*4*) the recycling exemptions related to the facility. This compliance targeting illustrates a typical compliance planning scenario. The scenario obviously varies from one company to another depending on the size and nature of the company. With those four components as target areas, a detailed schedule of follow-up activities can be developed. The schedule represents a variety of efforts including document reviews, site visits, staff interviews, data analysis, and plan formulation.

The conclusive result of these activities is the development of a final compliance plan. In essence, compliance planning translates the compliance target areas into a set of activities, or tasks, that are necessary to accomplish the ultimate compliance objectives. The tasks include reviewing records, conducting site visits and interviews, documenting the findings of all analyses, and drafting a compliance plan. Once upper management approves the draft, the final compliance plan can be formalized.

The process described above shows that compliance planning represents a retroactive mechanism to compliance. By expanding the scope of a compliance plan, a proactive compliance effort can be accomplished. Proactive compliance takes the form of compliance strategy. It is a strategic plan for achieving an effective compliance program. Its scope is broader than a compliance plan. The basic philosophy of a well-planned compliance strategy is that the best defense to an enforcement action is active and aggressive compliance. A compliance program is truly effective only if self-imposed compliance standards are followed and violations are corrected promptly upon discovery (23).

As part of the compliance strategy, monitoring violations are necessary. These can be performed through an internal environmental audit. As defined by the U.S. EPA, an audit is a systematic, documented, periodic, and objective review by regulated entities of facility operations and practices related to meeting environmental requirements (24). An environmental audit is a preventive measure that can improve facility performance before the state or the EPA inspector arrives.

An audit can identify environmental management deficiencies; however, the audit cannot correct the deficiencies (25). Additionally, an audit report noting a problem that is not later addressed by a facility can be used by prosecutors to prove "knowing" or "negligent" criminal conduct against the facility (26). Companies may take a variety of legal steps to protect audit reports; however, even the best-laid efforts at protection may be overruled by a judge ordering disclosure in specific circumstances (27). Audit reports should always be drafted with the possibility of disclosure in mind.

A well-executed compliance strategy provides a company with an effective hazardous waste compliance program; however, the program does not necessarily result in efficient hazardous waste management. To achieve efficiency and effectiveness in managing its hazardous waste, a company is required, among other things, to minimize the generation and toxicity of its hazardous waste to the extent feasible. This implies that the company should implement some sort of waste minimization and pollution prevention program.

Capacity Assurance Planning

At the state level, waste minimization and pollution prevention programs represent significant components of a Capacity Assurance Plan (CAP). A CAP is a planning document developed by a state to satisfy a federal requirement. The requirement is Section 104(c)(9) of the CERCLA, as amended, which mandates states to assure the EPA that adequate capacity will exist to effectively manage all hazardous wastes that are reasonably expected to be generated within the state during the next 20 years.

Each state must develop a CAP, because no federal Superfund remedial action monies will be provided to a state unless it has provided such an assurance and the EPA deems the assurance adequate. Although the statutory provisions of the CAP is CERCLA, and the sanction of not developing a CAP is tied to a Superfund grant, the CAP is basically a hazardous waste planning document.

Facility Siting

Facility siting has always been a major issue in the field of hazardous waste management. The siting, construction, and operation of hazardous waste management facilities have historically been a private-sector activity. Industries that generate hazardous wastes from their business operations might build storage, treatment, or disposal facilities on their property to accommodate their waste management needs. Alternatively, they might ship their hazardous wastes to commercial TSD facilities because of the cost-effectiveness calculations (see HAZARDOUS WASTE: COST CONTROL). The generators and the TSDs typically have made their own decisions regarding where to dispose of wastes and what waste management technologies to build. State governments did not get involved in those activities, except to oversee that the activities complied with the laws.

With the growing public concern about the potential risks associated with hazardous waste, the "not in my backyard" (NIMBY) syndrome, and most recently the CAP requirements, many state governments currently have an increased role in the siting process. Hazardous waste facility siting is essentially an interplay among the industry, the government, and the public. A successful siting is always characterized by, among other things, involvement of elected officials and good communication with the citizens (28).

As expected, state roles in the siting process vary from state to state. Apogee Research Inc. (29) provides a variety of siting factors that can be used to explain variations in the siting practices among the states.

State Hazardous Waste Management Planning

Facility siting, the CAP, waste minimization, and pollution prevention are all closely related activities in state hazardous waste planning. When the CAP documents were developed and finalized by the states, there was an argument over whether or not the CAP could be called a state hazardous waste management plan. The consensus was that it was not. The CAP is more appropriately categorized as a planning tool. The states, however, have had mixed opinions about whether the CAP is an effective planning tool (30). Most states agreed, however, that the CAP process provides useful forums for discussing waste management needs and plans. Many states thought that the states were developing better hazardous waste management plans through their own initiatives outside the CAP process.

In general, industry views the CAP process as both good and bad (31). The CAP can be destructive insofar as it encourages states to erect barriers to interstate commerce and to balkanize the national markets. It can be constructive if it increases the state's understanding of the market and promotes state cooperation to overcome the NIMBY problem. The states are supportive of the CAP process because the CAP improves data management, promotes waste minimization, improves communication, promotes industrial development, promotes interstate communication, and encourages regional dialogue.

Although the CAP is developed to anticipate a long-term solution to assuring hazardous waste management capacity, it represents only a fraction of a statewide hazardous waste management plan. A more comprehensive state hazardous waste management plan, or state plan, is needed. Currently, the state planning practice is still at its infancy stage. However, a recent trend suggests that the importance of state planning is becoming more apparent in the 1990s, which is particularly due to the increasing awareness of the public about environmental issues.

The purpose of state hazardous waste management planning is to develop strategic recommendations to all state participants in the hazardous waste management field. The recommended actions include plan implementation and plan monitoring, which are designed to protect human health and the environment and prevent unnecessary site remediation activities. The goal of state hazardous waste management planning is formulated by the state's legislature or is based on the state's needs.

Using the rationale planning approach, the state hazardous waste management planning process can be viewed as consisting of the five following interactive stages:

1. Identifying the most important needs, problems, and issues in the current state hazardous waste management practices.
2. Gathering information to characterize the current situation and to project the anticipated future scenarios.
3. Examining the plan's alternatives and specifying action plans to achieve the goals of state planning.
4. Developing a planning report that describes the planning needs, analyses, findings, and recommended strategies for future actions.
5. Implementing the recommended actions and monitoring the results. Monitoring is needed because it is a tool for implementing changes and incorporating new ideas.

The above planning process begins with the identification of the hazardous waste management planning needs. The needs can be identified through a series of surveys, workshops, meetings, and discussions involving stakeholders in the hazardous waste management system, ie, the hazardous waste handlers, the regulatory agencies, and the general public. Identifying planning needs, management problems, and societal issues may result in a broadening of the plan's vision and the refinement of planning goals.

The hazardous waste needs assessment represents a unique element of the planning process, because the formulation of the needs governs the scope and characteristics of the state hazardous waste plan. In other words, the more that is known about the planning needs, the better the construction of the state plan is understood.

An EPA study conducted in 1993 reveals that, in general, the states placed primary emphasis on managing hazardous waste according to the national waste management hierarchy. This reflects the importance of waste minimization and pollution prevention programs as seen by hazardous waste stakeholders. The second most important need is evaluating access to hazardous waste management capacity, which for the most part, is triggered by the CAP requirements. Conducting public education represents the third most important need because of the apparent link between public education and achieving the objectives related to other needs categories. The identified planning needs are basically the same as the planning objectives that the state attempts to achieve.

Most of the state plans contain specific recommendations which many of the states have implemented. The recommendations cover a variety of implementation activities. These activities must be implemented by the state environmental agency, legislature, EPA, local governments, hazardous waste handlers, and the public (32).

Although the implementation activities vary from state to state, the 1993 EPA study found that the activities can be categorized into three major groups: (*1*) follow-up studies, (*2*) legislative changes, and (*3*) administrative changes. The follow-up studies are required as supporting documentation for monitoring and evaluating the plan. The state plan generally requires periodic evaluation and revision to remain as a dynamic and viable document.

CITIZEN PARTICIPATION PLANNING

Under the RCRA statute, the law considers public involvement as extremely important to the success of the RCRA program. The RCRA public participation framework covers a number of major areas. Section 3006, for example, provides requirements that the public be given the opportunity to comment before the issuance of decisions regarding RCRA authorization from EPA to a state to implement RCRA program. Public comment is also required in the permitting of a TSD facility.

RCRA public participation provisions states that public notice must be given for all important permits including the preparation of a draft permit, a 45-day comment period is allowed during which time a person may submit written comments on the draft permit and request a public hearing. When there is a written notice of opposition to the permit and requests for a hearing, or when there is a significant degree of public interest in a draft permit, a public hearing will be held by the regulatory agency.

CERCLA public participation plans are required for removal action, remedy selection, and remedial activities. The major components of public participation under CERCLA include:

1. Public notice requirement
2. Public comments and response to public comments
3. Establishment of information repository
4. Availability of technical assistance
5. Development of community relation plan (CRP)

A community relations program provides interested persons an opportunity to comment on and participate in decisions concerning site actions and ensures that the community is provided with accurate and timely information about site activities.

In many respects the relationships among the business community, government, and the general public have become less confrontational and more cooperative. Public participation is essentially about building and maintaining communication links with all parties involved. Public participation concentrates on communication. The goal of public participation planning is to establish effective and efficient communication among all of the stakeholders in the hazardous waste management system: business, government, and the public for the purpose of building trust and establishing credibility.

Building trust and credibility is an ongoing process. Credibility is a perception of how believable and how trustworthy a project owner is based on the evaluation of its words and actions from the community's perspective. People normally distrust and fear a remediation project or hazardous waste facility on the basis of what they know. When the subject is the long-term operation of a hazardous waste treatment facility and its associated risk, their perception is presumably about something very dangerous and about an uncontrollable situation established in their neighborhood. A project tour may provide the community with the opportunity to examine the project.

Although the cost effectiveness interest of a hazardous waste or remediation project is obvious, the project must make it clear that it also stands on the side of the public interest. The message must be concise, positive, nondefensive, and the community must be convinced that its interests come first. Open dialogue with the community is often beneficial.

As more information flows into the community, the key to public relations today is to involve the public directly in the decision-making process, to communicate clearly, honestly, and constantly about issues that must be faced together. A true partnership based on this fundamental concept is the most viable option for sharing the same environment in a community (33). Working together in a mutually beneficial environment is the goal of a community relations program. An effective program needs hard work and dedication and demands good planning and intensive preparation.

Strategizing a community relations program is done for two basic reasons: the public becomes involved in the process because a project appeals to their interests or because the project threatens their interests (34). Relationships that occur because of mutual interests are usually voluntary. Relationships that grow from a perceived threat to the public's interest are usually involuntary, forced by circumstances, and must be identified early and worked out aggressively.

The term *public participation* entails a broad scope of activities designed to help citizens participate in decisions, convey information, solicit citizen concerns, heighten public awareness, and motivate participation in various programs (35). Successful public participation programs are the result of careful planning. By developing a realistic plan, decision-makers can evaluate the situation, problems, and issues, and know where best to direct their efforts and resources.

A public participation plan can be defined as consisting of the following elements:

1. Identification of issues that need to be addressed
2. Formulation of goals to be reached
3. Description of activities to accomplish each goal
4. Availability of resources for each activity
5. Program scheduling, coordination, and monitoring

There is a broad range of possible activities that can be included in a public participation plan. The activities mostly represent educational undertakings. For this reason, public participation planning is often described as public education and involvement planning. In essence, public participation implies integrating public concerns and values at every stage of the decision-making process. It is a two-way communication that involves both getting information out to the public and getting back from the public ideas, recommendations, issues, and concerns.

PLANNING ISSUES AND TRENDS

Managing hazardous waste in the 1990s and into the 21st century represents a unique challenge for hazardous waste planners and managers. The following five issues will continue to dominate the planning and management of hazardous waste:

1. Escalation of multimedia enforcement
2. Integration of environmental justice into program planning and management
3. Continuing promotion of waste minimization and pollution prevention
4. Continuing effort to establish uniform cleanup standards, including the integration of risk assessment and comparative risk in the hazardous waste management decision-making process
5. Acceleration of the cleanup of contaminated sites.

The integration of a cross-program and multimedia approach (ie, involving components of air, waste, and water) into all stages of the inspection and enforcement program is predicted to be accepted as normal. This multimedia approach will enable the regulatory agency to better pursue its goal of protecting public health and the environment through deterrence while maintaining an efficiency that could not be achieved through the use of traditional single-media approaches alone (36).

The NIMBY syndrome will not go away, and anger over what community activists perceive as a pattern of locating industries near poor, minority residences has awakened the EPA, U.S. Congress, and the industrial community to a new societal issue: environmental justice (environmental equity), which can manifest itself in the form of procedural equity, geographic equity, or social equity (37). Initially a local concern, environmental justice has now garnered broader national support (38).

An awareness of environmental justice will ultimately lead to the reduction or prevention of harmful exposure to hazardous wastes at the community level (39). The mechanism for reducing exposures from future industrial practices will come primarily via increased waste minimization and pollution prevention efforts.

Environmental justice will also become one of many criteria that scientists and the public will be asked to use in assessing, preventing, and remediating environmental hazards. The environmental justice issue will force the hazardous waste industry to make some adjustments in such things as facility siting and cleanup standards (40).

While the cleanup standard for remediating a site will trigger a forum of continuing debate, the number of newly found sites contaminated with hazardous waste will become larger than the number of sites remedied. This situation will stimulate the development of more effective and efficient hazardous waste site remediation.

Parallel to the growing problems and issues in the hazardous waste management field, the role of hazardous waste planners will also continue to grow.

BIBLIOGRAPHY

1. P. Rowe and co-workers, *Principles for Local Environmental Management,* Ballinger, Cambridge, Mass., 1976.
2. J. Baldwin, *Environmental Planning and Management,* Westview Press, Boulder, Colo., 1985.
3. U.S. Environmental Protection Agency, *Waste Minimization Opportunity Assessment,* U.S. EPA, Washington, D.C., 1988.
4. U.S. General Accounting Office, *Hazardous Waste: New Approach Needed to Manage the Resource Conservation and Recovery Act,* U.S. GAO, Washington, D.C., 1988.
5. L. Martin, *Proven Profits from Pollution Prevention,* Institute for Local Self-Reliance, Washington, D.C., 1989.
6. T. Zosel, *Poll. Prev. Rev.* **1,** (1), pp. 67–72 (1990).
7. A. Metry, "Elements of a Balanced Approach Towards Hazardous Waste Management," in S. Maltezou and co-workers, eds., *Hazardous Waste Management,* Tycooly, New York, 1989.
8. F. Hansen, "Pollution Prevention, A New Mandate for Oregon's Environment," *Environ. Forum,* 30–34 (Sept./Oct. 1989).
9. B. Shorthouse, *Poll. Prev. Rev.* **1** (1), 41–53 (1990).
10. S. Maltezou, A. Biswas, and H. Sutter, eds, *Hazardous Waste Management,* Tycooly, New York, 1989.
11. C. Wentz, *Hazardous Waste Management,* McGraw-Hill, New York, 1989.
12. W. Toner, in F. So and J. Getzels, eds., *The Practice of Local Government Planning,* International City Management Association, Washington, D.C., 1988, pp. 117–138.
13. D. Slater, *Management of Local Planning,* International City Management Association, Washington, D.C., 1984.
14. C. Patton, in Ref. 12, pp. 472–502.
15. J. Morgulec and D. Woltte, "Resource Conservation and Recovery Act," in *Environmental Crimes Manual,* U.S. Dept. of Justice, Washington, D.C., 1991, pp. VII.1–VII.78.
16. U.S. Environmental Protection Agency, *Guidance for Capacity Assurance Planning—Capacity Planning Pursuant of CERCLA Section 104(c)(9),* U.S. EPA, Washington, D.C., 1993.
17. J. Soesilo and S. Wilson. *Hazardous Waste Planning,* Lewis Publishers, Boca Raton, Fla., 1995, p. 116.
18. J. Soesilo and S. Wilson, *Hazardous Waste Planning,* Lewis Publishers, Boca Raton, Fla., 1995.
19. U.S. Environmental Protection Agency, *Fundamentals of Environmental Compliance Inspections—Third Printing,* Government Institutes Inc., Rockville, Md., 1993.
20. M. Hopkins, *Waste Tech. News,* **5** (16), 1993.
21. G. Lindgren, *Managing Industrial Hazardous Waste—A Practical Handbook,* Lewis Publishers, Chelsea, Mich., 1989.
22. U.S. Environmental Protection Agency, *Resource Conservation and Recovery Act Inspection Manual—Second Edition,* Government Institutes Inc., Rockville, Md., 1992.
23. C. Green, J. Dunn, and P. Oppenheimer, in K. Kastner, ed., *Current Development in RCRA,* Government Institutes Inc., Rockville, Md., 1993, pp. 75–110.
24. U.S. Environmental Protection Agency, "Environmental Auditing Policy Statement," *Fed. Reg.* 25004 (July 9, 1986).
25. T. Kuusinen, "A Government Perspective," in L. Cahill and R. Kane, eds., *Environmental Audits—Sixth Edition,* Government Institutes Inc., Rockville, Md., 1992, pp. II.1–II.14.
26. B. Shanoff, *World Wastes,* 22 (June 1993).
27. M. Levin, A. Hymes, and S. Mullaney, **24** (36), 1606–1612. (1994).
28. T. Kusterer, *Environ. Decision,* 4–5, 21–22 (Feb. 1991).
29. Apogee Research Inc., *Analysis of Siting Attempts and the Factors that Influence Success,* National Governors' Association, Washington, D.C., 1988.
30. U.S. Environmental Protection Agency, *Guidance for Capacity Assurance Planning—Capacity Planning Pursuant to CERCLA Section 104(c)(9),* U.S. EPA, Washington, D.C., 1993.
31. Environmental Resources Management, *Report on the National Conference on State Capacity Assurances,* Consortium of State Hazardous Waste Siting Authorities, Boston, Mass., 1989.
32. Washington State Department of Ecology, *Washington State Hazardous Waste Plan,* WSDOE, Olympia, Wash., 1992.
33. B. Harrison, *Environmental Communication and Public Relations Handbook—Second Edition,* Government Institutes Inc., Rockville, Md., 1992.
34. J. Lukaszewski and T. Serie, *Waste Age,* 83–94 (Mar. 1993).
35. U.S. Environmental Protection Agency, *Decision-Makers Guide to Solid Waste Management,* U.S. EPA, Washington, D.C., 1989.
36. U.S. Environmental Protection Agency, *Enforcement Accomplishments Report FY 1992,* U.S. EPA, Washington, D.C., 1993.
37. R. Bullard, *Wildlife EnviroAct.,* 6–10 (Nov. 1991).
38. *Environ. Rep.,* 1660–1684 (Jan. 21, 1994).
39. A. Harding and G. Holdren Jr., *Environ. Sci. Technol.* **27** (10), 1990–1995 (1993).
40. Hazardous Waste News, *Hazard. Waste News,* 50 (Feb. 15, 1994).

HAZARDOUS WASTE: SOLIDIFICATION AND STABILIZATION

ARUN S. WAGH
HANN S. HUANG
Argonne National Laboratory
Argonne, Illinois

The ever-increasing world population and associated industrial activities are consuming more and more raw materials. This, in turn, increases the production of waste streams that require proper disposal. As discussed elsewhere (1), hazardous and radioactive waste streams require proper treatment, which includes destruction, recycling, or as an ultimate option, isolation and storage. A search for methods applicable to the third option (isolation and storage) has resulted in the development of solidification and stabilization (S & S) technologies (see WASTE GENERATION).

NEED FOR SOLIDIFICATION AND STABILIZATION

For a waste that cannot be destroyed, its minimization takes precedence over any remediation approaches (2) in waste management. When a waste is produced, its volume reduction by decontamination and recycling is done. If contaminant levels are low in the waste and the waste volume is large, this approach is not economical and it is necessary to treat the waste for ultimate disposal. Often, the waste volume may not be large, but if there is no decontamination technology available to clean up this waste, it is then necessary to treat the waste for disposal.

The treatment approach is to convert the waste stream into a form by which the contaminants are not released into the groundwater, or in air, and thereby into the food cycle or other routes that reach the human body, such as direct contact and ingestion. Waste treated this way is called the *waste form* and the treatment methods are grouped under the name *solidification and stabilization.* Although there is some ambiguity on the exact definition of this title, terminologies with the following definition will be used (3):

Stabilization is the process used for reduction of hazard potential of the waste by converting the contaminants into their least soluble, least immobile, or least toxic form. Other characteristics of the waste may not change in this treatment.
Solidification, on the other hand, physically binds or encapsulates the waste in a monolithic solid of high structural integrity. Thus, solidification may be used for powders, liquids, or gases.

In general, solidification techniques eliminate the free liquid, increase the bearing strength, decrease the surface area of the waste materials, and produce a monolithic solid product of high structural integrity, whereas stabilization techniques typically fix the contaminants chemically into their least soluble, mobile, or toxic form.

The stabilization and solidification of waste are done for two purposes. Transportation of the waste requires that it be packaged according to the regulations established by the U.S. Department of Transportation or another authority in each state so that it is not subject to spillage, leakage, or release to the atmosphere during transportation. In addition, proper disposal of the waste will require a stabilization procedure in which the waste form will meet certain requirements to ensure that there is no release of the contaminants to air and groundwater. These requirements clearly suggest solidification of the sludges, liquids, and powders into a solid form that will not leach out the contaminations during either transportation or long-term storage or disposal.

Both stabilization and solidification may be done by either thermal or nonthermal treatments. Thermal processes are more expensive and hence suitable where criteria are very stringent, such as in radioactive waste management. On the other hand, for hazardous wastes, nonthermal stabilization is the best approach because it leads to the treatment of high volumes at low cost. Thus, most of our discussion here is limited to nonthermal processes, although a brief discussion of thermal *in situ* processes is provided.

CHEMICAL STABILIZATION

If contaminants are water-soluble, it is expected that when they are in contact with groundwater, they will easily leach out. Therefore, it is essential that the contaminants be first converted into their most insoluble form by chemical treatment of the waste. The converted contaminants should have very low solubility in groundwater, which may be slightly acidic in acid rain conditions or basic in limestone areas. The treated waste should also be insoluble in saline water and carbonate water, which represent typical groundwaters. Thus, a superior stabilization process requires that contaminants be immobilized in a wide range of pH.

Table 1 indicates that the conversion of hazardous metal species into their hydroxides, phosphates, silicates, or sulfides may chemically fix them. In nonthermal processes, hazardous contaminants are converted to any one of these four compounds, whereas thermal-treatment processes depend more on silicates.

Table 2 gives the three nonthermal treatment processes that render contaminants in soils and sludges to their most insoluble form. Cement and lime/fly ash processes convert contaminants into hydroxides and this route has been most common in treating superfund sites, but other routes such as the conversion of the contaminants into sulfides and phosphates have great potential, provided they are developed to field-scale applications. Table 3 gives solubility product constants, expressed as pK_{sp}.

Table 1. Solubility of Compounds of Metal Contaminants in Waste Streams

Compound	Solubility
Nitrates	All are soluble.
Chlorides	All, except $AgCl$, Hg_2Cl_2 are soluble.
Hydroxides	None except $Ba(OH)_2$ and $AgOH$ are soluble. These two are moderately soluble.
Sulfates	Sulfates of Ba, Pb, Ag, Hg are slightly soluble, all others are insoluble. Hydrosulfates are more soluble than sulfates.
Carbonates, phosphates, chromates, silicates	All are insoluble except $MgCrO_4$. Carbonates are often not very stable.
Sulfides	All except Cr sulfides are soluble. Cr sulfides are hydrolized.

SOLIDIFICATION

As mentioned before, solidification processes bind or encapsulate waste particles into solids of high integrity. In doing so, they reduce the total surface area by which the possibility of individual particles coming into contact with groundwater is virtually eliminated. Their consolidation removes any chances of dust formation during transport or spillage.

The consolidation of waste particles into solids affects several other properties related to the permanent immobilization of contaminants. Due to the nonpermeable nature of these solids, fluid conductivity is greatly reduced. Because the solid gains in strength, its resistance to physical stress (such as overburden at disposal sites) is enhanced. Individual particles are also not amenable to biological alteration.

As in cement stabilization, solidification also binds water in the matrix. Reduction of free water from the waste eliminates the miscibility of waste with groundwater.

Physical Encapsulation

In addition to the consolidation and binding the particles, solidification also physically encapsulates the contaminants. This physical encapsulation serves an important purpose. When the waste form is disposed, due to their physical encapsulation, the contaminants are isolated from the groundwater environment and are not leached out.

As we shall see later, among the various treatment methods used for chemical fixation that are discussed in "Portland Cement Stabilization" through "Other Inorganic Binders" below the sulfur polymer, Portland cement, and Ceramicrete also provide physical encapsulation. In these methods, while reactions during setting fix the contaminants into their chemically stable form, once the matrix sets, they will also physically encapsulate the reacted contaminants. Thus, both chemical fixation

Table 2. Treatment Processes for Stabilization of Inorganic Contaminants

Treatment	Contaminants	Commercial Process	Comments
Sulfide	All except Be	Lime/sulfide, sulfur polymer cement	Effective in very wide range of pH, most common commercial treatment along with lime.
Hydroxide/ silicate	All except Ba, As, Cr, and Se	Portland cement	Effective only in alkaline environment.
Phosphate	All except Be	MgO and/or phosphoric acid or acid phosphate	Effective in moderate acidic to moderate alkaline range of pH.

Table 3. Solubility Product Constants pK_{sp} of Major Reaction Products in Chemical Fixation[a]

Treatment	Ag	Ba	Be	Cd	Co	Cr III	Cu	Hg	Ni	Pb
Sulfide	50	s^+	d^+	26.1	20.4 24.7	20.4	47.6 35.7	47 52.4 51.8	18.5 24 25.7	27.9
Hydroxide	8	2.3	21.8	13.6	14.8 43.8	30.2	14.8 19.7	23.7 25.5	4.7 1	14.9
Phosphate	16	22.6	s^+	32.6	6.7 (H) 34.7	22.6	36.9	12.4 (H)	30.3	9.9 (H) 42.1

[a] The data are quoted from *Lange's Handbook of Chemistry,* 14th ed., 1992 (Courtesy of McGraw-Hill). Symbols in parentheses indicate state of the product: H = hydrated, s^+ = soluble in water, d^+ = decomposes in water.

and physical encapsulation are achieved in the same process in these methods.

Micro- and Macroencapsulation

Encapsulation of the wastes is achieved in two different ways. It is possible to encapsulate individual contaminants by the matrix material such that, even if there is any intrusion of the groundwater, such microencapsulated contaminant is protected by an impermeable barrier. On the other hand, it is also possible to encapsulate the whole waste form in an impermeable matrix. Such a macroencapsulation is often useful when the treatment process fails to provide the necessary microencapsulation.

Microencapsulation is best done when there is a glassy phase in the matrix material, which coats the contaminant particles.

METHODS OF STABILIZATION AND SOLIDIFICATION

More than 50% of the hazardous waste produced is wastewater treatment sludge that is mostly aqueous-based (1). Because of this, the stabilization technologies that have been developed are aqueous-based and depend on hydration and pozzalinic reactions. Portland cement, the most common binder among structural materials, is an obvious candidate for these sludges because it needs water during setting and hence has been widely studied in the literature. In addition, many pozzalans, such as lime produced from natural limestone or a waste component itself such as ash, have found their way into stabilization technologies as inexpensive binders.

The hydration or pozzalinic reactions stabilize waste in an alkaline environment in which hydroxides of most of the contaminants have their lowest solubility in groundwater. Several other processes have been developed mainly to stabilize low level radioactive wastes that also contain hazardous contaminants. They are based on a different principle, relying on the fact that excellent physical encapsulation alone can reduce the leachability of contaminants very satisfactorily. Organic binders follow this approach. They also eliminate excess water in the waste sludge by evaporation during thermal treatment of the waste. On the other hand, recently developed inorganic binders such as sulfur cement and Ceramicrete are based on the conversion of the contaminants into sulfides and phosphates because they have a lower solubility over a wide pH range than hydroxides and, hence, perform well in extreme ambient environment. The lower solubility of phosphates and sulfides may be seen in Table 2 from their higher values of pK_{sp} than those of the hydroxides. In Figure 1, the solubilities (pC) of Pb compounds are drawn as a function of the pH as an example. The diagram shows that at all pH values, the solubilities of hydrophosphate and sulfide of Pb are lower than

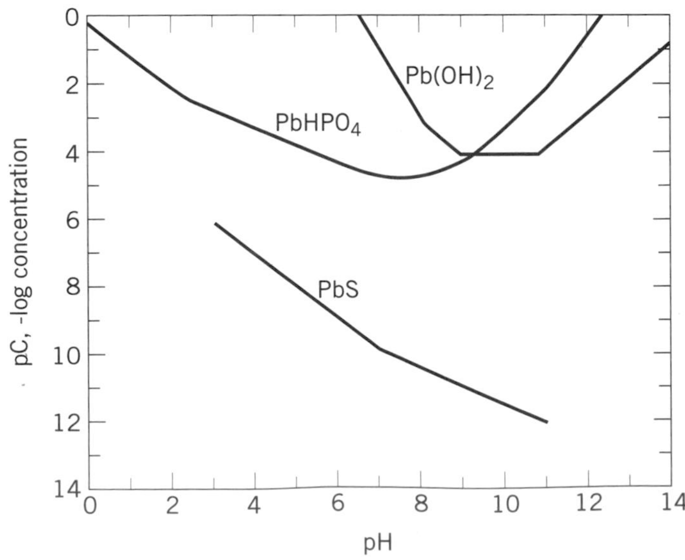

Figure 1. Solubility of hydroxide, hydrophosphate, and sulfide of Pb as a function of pH (3).

that of hydroxide. Sulfur polymer cement and Ceramicrete take advantage of this.

CONCLUSIONS AND DISCUSSION

The stabilization and solidification of hazardous wastes is the ultimate option in waste management. For economic reasons, the prudent use of resources is an essential part of any technology development. This requires that technologies are constantly improved to minimize production of wastes, develop new technologies to reduce and recycle wastes, and, when all these options are exhausted, stabilize and solidify waste so that its environmental effects are minimized.

In spite of the great strides made in solidification and stabilization technologies, there are several limitations in their applications as a whole. For example, gases are not easily amenable to solidification and their permanent fate in the stabilized waste is uncertain. Focus of the technologies has hardly been in converting contaminants into their less toxic forms such as Cr^{+6} to Cr^{+3}. Finally, most stabilization and solidification processes in use do not incorporate both components of immobilization to the same degree. Generally, the objective of the solidification and stabilization seems to be passing regulatory tests at the time of disposal rather than long-term immobilization.

Soils and sludges form the majority of wastes that require stabilization and solidification. Because most of the industrial waste is wastewater sludge, most of the unconfined contamination has occurred in soils. Sludges and soils have drawn considerable attention, resulting in the development of a number of stabilization and solidification technologies.

Most major stabilization and solidification technologies employ aqueous-based stabilization approaches to take advantage of the water fraction that is a constituent of sludges and soils. Cement, lime, and ash are the major candidate stabilizers. They are based on the principle that the contaminants may be chemically stabilized by converting to their hydroxides that are insoluble species. In addition, these processes isolate these precipitates by providing an encapsulating physical matrix. Because of the low cost of these technologies, they are very attractive for the large-scale stabilization of soils and sludges.

Because sulfides and phosphates can provide superior chemical stabilization, technologies are being developed that are also aqueous-based sulfide and phosphate treatments. Chemical stabilization with sulfides is being used along with physical encapsulation provided by cement and lime. Phosphates, on the other hand, are being developed to provide both. These two approaches provide superior stabilization and solidification methods at a penalty of a slightly higher cost than that of lime- and cement-based processes.

Thermal technologies offer an advantage over the aqueous-based stabilization and solidification approach when the wastes contain mainly destructible organics. In these approaches, water is evaporated, organics are destroyed, and considerable waste reduction is achieved. Processes based on vitrification, polymer encapsulation fall into this category. They also provide excellent physical encapsulation of the remaining fraction of the waste. The penalty in these processes is the very high cost of the treatment and the off-gases that are often released during the treatment. Proper controls are needed to monitor and capture these secondary waste streams. In addition, they also require subsequent aqueous-based solidification.

In superfund projects, both aqueous and thermal technologies are used for *in situ* stabilization and soils. The operations have been generally expensive. The high cost is justified, however, on the grounds that contamination in soils is likely to be unconfined and may affect large areas through the water table. On the other hand, the treatment of other industrial sludges is governed by the cost of treatment and aqueous-based technologies are used more commonly. In either case, although considerable gains have been reported, accumulation of waste continues to grow. This clearly indicates that waste management will be one of the prime industries of the twenty-first century in which stabilization and solidification efforts will thrive as major industrial operations.

ACKNOWLEDGMENTS

The authors wish to thank S. Hagamann, J. Wesolowski, S-Y. Jeong, and S. Kersten of Argonne National Laboratory for their help in manuscript preparation. We also acknowledge the efficient technical editing of the manuscript by C. Malefyt of Argonne National Laboratory. Manuscript preparation was partially supported by U.S. Dept. of Energy, Office of Technology Development, as a part of the Mixed Waste Focus Area, under Contract W-31-109-Eng-38.

BIBLIOGRAPHY

1. A. Wagh, "Waste Generation," in *Encyclopedia of Environmental Analysis and Remediation.*, John Wiley & Sons, Inc., New York, 1998.
2. N. Beecher, K. Geiser, and K. Fischer, *J. Haz. Waste Mat.* **5**(2), 177–184 (1988).
3. J. R. Conner, *Chemical Fixation and Solidification of Hazardous Wastes,* Van Nostrand Reinhold, New York, 1990, pp. 322–332.

HEAVY-METAL IONS, REMOVAL FROM WASTEWATER

GREGORY L. RORRER
Oregon State University
Corvallis, Oregon

Many different types of manufacturing operations generate wastewater streams which contain dissolved heavy metals. Several heavy metals, including cadmium, chromium, lead, and mercury, are toxic even at very low concentrations. Consequently, process wastewaters and groundwaters contaminated with heavy-metal ions pose a serious environmental problem. Although source reduction and waste minimization are the long-term solutions for reducing or eliminating the release of toxic heavy metals to the environment, treatment technologies will continue to play a crucial role in controlling heavy-metal ion emissions from current operations and in cleaning up contaminated sites from previous operations. Although organic contaminants in wastewater can be converted to harmless products, dissolved heavy metals must be physically removed from the wastewater and either reclaimed through recycling processes or converted to a solid form suitable for disposal. The purpose of this work is to overview both practiced and emerging process technologies for physically removing heavy-metal ions from wastewater and contaminated groundwater. The recycling or disposal of the separated heavy-metal ions is also overviewed within the context of each process described.

GENERAL STRATEGIES FOR HEAVY-METAL ION REMOVAL

Heavy-Metal Ion Waste Streams

What are heavy metals? In the Periodic Table, the heavy metals constitute the Group III transition metals, the actinide series (uranium, neptunium, plutonium, and americium), the lanthanide series, and three of the Group IV metaloids (arsenic, tellurium, selenium). The terms heavy metal and transition metal are often used interchangeably, although transition metals strictly refer to only Group III elements. The alkali metals (Group I) and alkaline-earth metals (Group II) are not heavy metals. Although some alkali or alkaline-earth metal ions, such as radioactive cesium, are considered highly toxic and pose significant waste disposal problems, they are not considered in this presentation.

Heavy metals can exist in either a water-soluble or water-insoluble form. Oxide, hydroxide, sulfide, and carbonate forms of most heavy metals are not soluble or are marginally soluble in water. Most common salts of heavy-metal ions used in industrial operations are soluble in water, including chloride, sulfate, and nitrate salts. As a dissolved salt, the heavy-metal ion is in cationic form. For example, cadmium chloride ($CdCl_2$) is soluble in aqueous solution and the heavy-metal cation is Cd^{2+}, where the exponent indicates the valence oxidation state of the metal. Most heavy-metal cations are divalent (+2), but some are monovalent (eg, silver Ag^+) or trivalent (eg, aluminum Al^{3+} or chromium Cr^{3+}). Many heavy metals can have multiple oxidation states (eg, iron Fe^{2+} or Fe^{3+}). Heavy metals can also exist in soluble form as oxymetal anions. The most prominent example is chromium, which can also exist as chromate CrO_4^{2-} or dichromate $Cr_2O_7^{2-}$ with a Cr oxidation state +6. Similar oxyanion states can exist for arsenic, including arsenate (AsO_4^{3-}), where As has an oxidation state of +5, arsenite (AsO_{3-}^{3-}), where As has oxidation state of +3, as well as selenium (SeO^{2-3}, SeO_4^{2-})

Sources of wastewater are diverse, but the metals-related industries are the largest contributors. Primary sources of wastewater containing heavy-metal ions include mining operations, metal plating facilities (electroplating, electroless plating), metal cleaning and fabrication operations, electronic device manufacturing facilities, photo processing facilities, battery manufacture and disposal sites, and nuclear weapons manufacturing and storage facilities. This article focuses on the description of process technologies to remove heavy-metal ions from wastewater; detailed information on the sources and extents of heavy-metal ion contamination in wastewaters is available (1,2). Nonmetals-related industries can also generate large volumes of aqueous waste contaminated with very low concentrations of heavy-metal ions below 1 ppm. Prominent examples are coal-fired plants and pulp mills, where heavy metals indigenous to the raw materials (coal, wood, etc) leach out during processing and ultimately find their way into wastewater streams.

Many heavy metals, either in cationic or oxymetal anion form, are highly toxic. The toxicity of common, industrially significant heavy-metal ions has been detailed (3). The most toxic heavy metals and their maximum contaminant drinking water limits are provided in Table 1 (4). Although these heavy metals are toxic, other heavy-metal ions are essential to life at trace levels, such as iron, copper, and zinc.

Increasingly, heavy metals will have to be recycled within manufacturing processes. In fact, five of the top 17 hazardous chemicals in the U.S. EPA's Toxics Release Inventory targeted for source reduction under the EPA 33-50 Voluntary Release Reduction program are heavy metals. The volume of heavy-metal wastes is significant. In 1986, EPA F006-classified waste alone, which represents the heavy-metal sludge generated only from electroplating operations, accounted for 11.2 million tons, or 1.5% of the 747 million tons of waste generated that year (5).

Groundwater and surface waters contaminated with low levels of toxic heavy-metal ions also represent a significant environmental problem. A prominent example is the Berkeley Pit outside of Butte, Montana. According to data collected in May of 1991, the pit was filled with 20 billion gallons (75.7 million m^3) of acidic mine leachate effluent contaminated with heavy-metal ions, principally aluminum (290 ppm), copper (190 ppm), iron (1090 ppm), manganese (182 ppm), and zinc (550 ppm). The water also contained dilute concentrations of toxic heavy-metal ions (see Table 1) which were above the MCL, including arsenic (0.8 ppm), cadmium (1.5 ppm), and nickel (1.1 ppm) (6).

Both private and public sectors are actively involved in the development and implementation of technologies for the removal of heavy-metal ions from dilute wastewater streams. The public sector includes local, state, and federal governments. Federal government agencies within the United States include the Environmental Protection Agency (EPA), the Department of Energy (DOE), the Department of Defense (DOD), and the Department of the Interior. Significant activities within the EPA include the Superfund Innovative Technology Evaluation (SITE) program which promotes development and implementation of new technologies to clean up heavy-metal contaminated soils and groundwater at Superfund sites. Local publicly owned treatment works often work with the U.S. EPA to manage the cleanup of Superfund sites. A significant mission of the DOE is to restore contaminated soils and groundwater from nuclear weapons processing facilities, many of which are contaminated with heavy metals and radioactive heavy metals. All major branches of the Armed Forces within the DOD, including the Army, Air Force, and Navy, also are

Table 1. Maximum Contaminant Levels (MCL) for Heavy-Metal Ions in Drinking Water[a]

Heavy Metal	Common Ionic Forms	MCL, mg/L
Silver, Ag	AG^+	0.05
Arsenic, As	AsO_3^{3-}, AsO_4^{3-}	0.05
Cadmium[b], Cd	Cd^{2+}	0.005[c]
Chromium[b], Cr	Cr^{3+}, CrO_4^{2-}, $Cr_2O_7^{2-}$	0.1[c]
Copper, Cu	Cu^{2+}	1.3[d]
Lead[b], Pb	Pb^{2+}	0[d]
Mercury[b], Hg	Hg^{2+}	0.002[c]
Nickel[b], Ni	Ni^{2+}	0.1[e]
Selenium, Se	SeO_3^{2-}, SeO_4^{2-}	0.05[c]

[a] Under the Safe Water Drinking Act of 1974, as amended in 1977, 1980, and 1986 (4).
[b] Voluntary source reduction under U.S. EPA 33-50 Release Reduction plan.
[c] SOCs and IOCs. Final rule: *Fed. Reg.* **56,** 20:3526 (Jan. 30, 1991).
[d] Lead and copper. Final rule: *Fed. Reg.* **56,** 110:26460 (June 7, 1991).
[e] SOCs and IOCs. Final rule: *Fed. Reg.* **57,** 138:31776 (July 17, 1992).

required to begin cleanup of heavy-metal contaminated soil and groundwater at U.S. military installations, for example, the extensive subsurface contamination of toxic hexavalent chromium at Tinker Air Force Base in Oklahoma City, Oklahoma (7). Prior to its closure in early 1996, the U.S. Bureau of Mines within the U.S. Department of the Interior had very active and successful research and development programs to remove heavy-metal ions from waste streams, particularly those associated with aqueous mining effluents.

This presentation introduces engineers and scientists to the spectrum of practiced and emerging process technologies for removal of heavy-metal ions from wastewaters. Available literature on separation technologies for removal of heavy-metal ions from wastewater is immense; as much as possible, detailed texts and literature reviews are cited for more in-depth treatment and specific processes which illustrate broad classes of heavy-metal ion removal technologies are presented. Both practical examples of established process chemical technologies such as precipitation and electrolytic recovery are provided along with recent literature on emerging biological process technologies for heavy metals remediation.

Chemical Process Technologies

Chemical process technologies for removal of heavy-metal ions from waste are schematically illustrated in Figure 1. There are four major classes of chemical separation technologies: chemical precipitation, electrolytic recovery, adsorption/ion exchange, and solvent extraction/liquid membrane separation. In chemical precipitation, the heavy-metal ions are chemically reacted with a precipitant to form an insoluble metal oxide or metal salt. The heavy-metal precipitate (sludge) is mechanically separated from the treated wastewater stream. In electrolytic recovery, electrochemical reaction processes driven by an externally supplied electric current reduce the heavy-metal ions to solid free metals. Adsorption and ion-exchange processes immobilize the heavy-metal ions onto the surface of a solid adsorbent material which possesses chemical moieties that bind the metal to the surface. The immobilized metals are then eluted to regenerate the adsorbent and provide a concentrated heavy-metal ion stream which can be recycled or hydrometallurgically processed. Solvent extraction and liquid membrane separation processes selectively transfer the heavy-metal ions from the wastewater stream to an extractant phase which is brought into contact with the wastewater stream. In solvent extraction, the contacting extractant phase is immiscible in water. In liquid membrane separation, a physical barrier (the membrane) exists between the two phases which is permeable only to the heavy-metal ions. Comparisons of the major features, advantages, and disadvantages of chemical process technologies for removal of heavy-metal ions from wastewater are provided in Table 2.

There are many comprehensive literature reviews comparing the four major chemical process technologies for removal of heavy-metal ions from wastewater. Detailed compilations of scientific literature (2,8,9) and detailed descriptions of representative chemical process technologies, with emphasis on process design and economics, are provided (10–12).

Biologically Promoted (Bioprocess) Technologies

A summary of emerging biologically promoted process (bioprocess) technologies for removal of heavy-metal ions from wastewater is schematically illustrated in Figure 2. There are four major classes of bioprocess technologies for heavy-metal ion removal: bioprecipitation, bioaccumulation, biosorption, and phytoremediation. A common feature of all bioprocess technologies is the use of biological organisms to carry out the removal of heavy-metal ions from wastewater. Bioprecipitation is similar to chemical precipitation. However, reactions promoted by living microorganisms either produce the chemical precipitant or directly reduce the heavy-metal ion to free

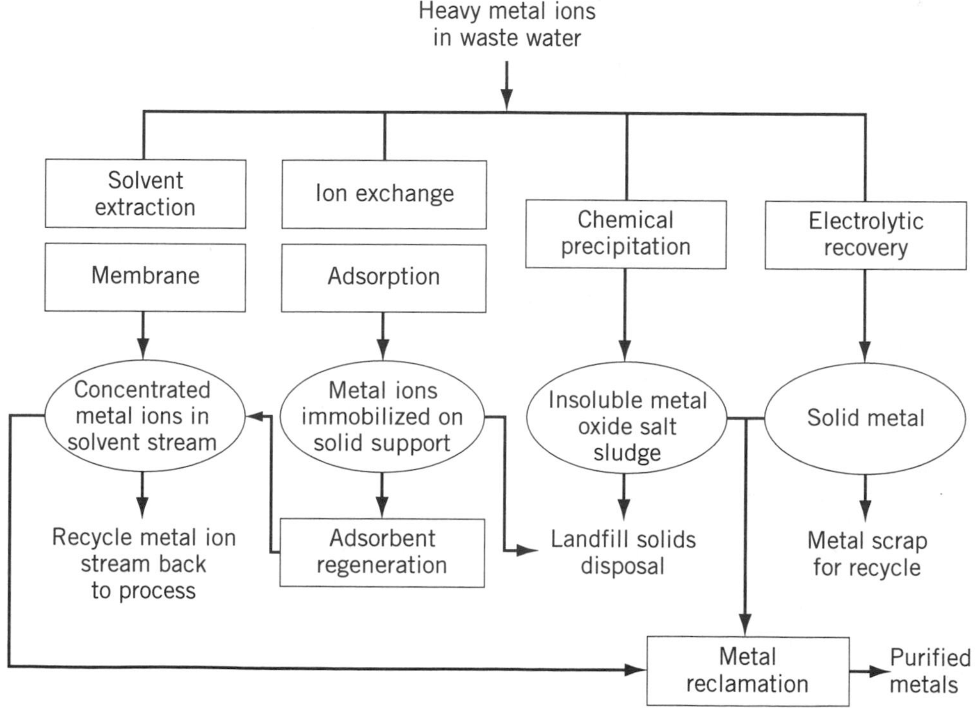

Figure 1. Summary of available chemical process technologies for heavy-metal ion removal from wastewaters.

Table 2. Comparison of Chemical Process Technologies for Heavy-Metal Ion Removal

Process	Chemical/Energy Input	Metal Reclamation	Major Advantages	Major Disadvantages
Chemical precipitation	Precipitant, flocculant, acid, base; mixing and fluid handling	Metal sludge	Well established; low effluent concentrations	High chemical requirements, several process elements
Electrolytic recovery	Electrical power	Solid metal scrap	Well established; direct recovery of solid metal; no chemical consumption	Energy intensive; high capital costs; reduced efficiency at dilute concentrations
Ion-exchange/ regenerable adsorbents	Regenerant solutions; fluid handling	Concentrated soluble metal stream	Highly selective; effective < 100 ppm	Chemical regeneration requirements, adsorbent expense; prone to fouling in mixed waste streams
Disposable adsorbents	Replacement adsorbent; fluid handling	Metal immobilized on solid adsorbent	Simple metal removal process; low adsorbent cost; effective < 100 ppm	Selectivity; recurring cost of new adsorbent, disposal cost of spent adsorbent
Liquid membranes	Extractant for liquid-supported membrane; fluid handling	Concentrated soluble metal stream	Selective; continuous concentrated metal solution recycle	Membrane durability, fouling
Liquid–liquid extraction	Stripping solvent, makeup extraction solvent; fluid handling	Concentrated soluble metal stream	Selective; continuous concentrated metal solution recycle	Capital costs; solvent emissions to air/water; solvent disposal

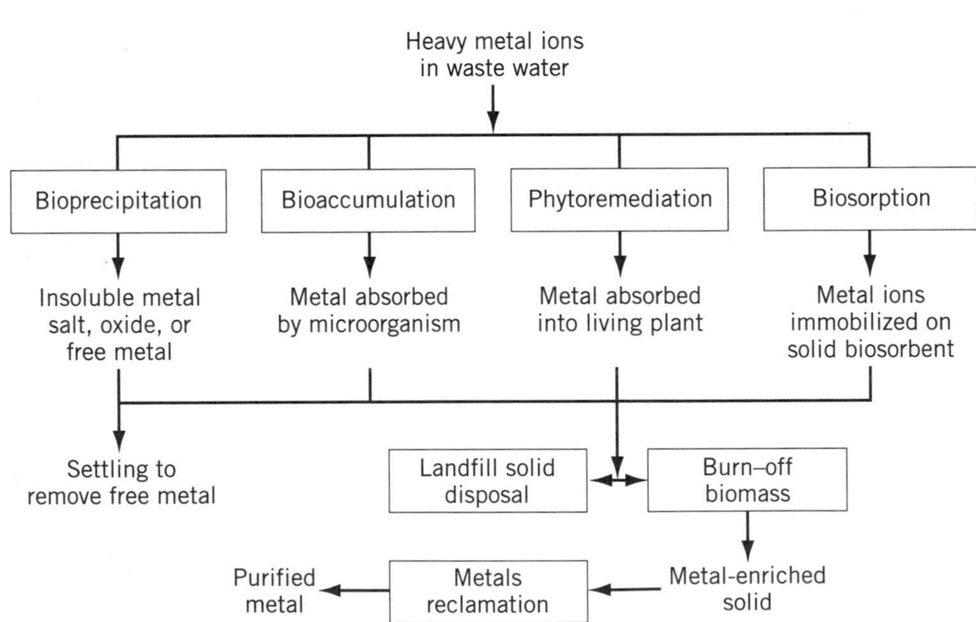

Figure 2. Summary of emerging biologically promoted (bioprocess) technologies for heavy-metal ion removal from wastewaters.

metal. Bioaccumulation is the active uptake of heavy-metal ions by microorganisms and is often accompanied by bioprecipitation. Phytoremediation processes use living plants to actively uptake heavy-metal ions into the plant biomass. In biosorption processes, heavy-metal ions are immobilized on dead biomass. Strictly speaking, biosorption is really an adsorption/ion-exchange process. However, it is classified as a biologically promoted heavy-metal remediation process because the adsorbent material is produced biologically. The bioremediation of metals, focusing primarily on bioprecipitation technologies (ie, chromium, mercury, selenium, manganese, arsenic, and uranium) and biosorbent technologies, has been comprehensively reviewed (13). Bioaccumulation and bioprecipitation of heavy-metal ions using microorganisms has also been reviewed (14).

Choosing a Heavy-Metal Ion-Removal Process

The selection of a specific process for removal of heavy-metal ions from a process water waste stream or contaminated groundwater is governed by many factors. The primary technical considerations are (*1*) characteristics of the wastewater stream to be treated, (*2*) volume of the wastewater stream, and (*3*) recovery, recycle, or ultimate disposal of the separated heavy metals. Of course, process economics underlie these technical considerations. Generally, the specific heavy-metal ion to be treated and the concentration range of the heavy-metal ion in the waste stream has the most important bearing on narrowing process selection. For the purposes of this work, the concentration of a given a heavy-metal ion is considered as low or dilute if it is below 100 mg per liter (100 parts per million (ppm)). Chemical components in the wastewater matrix, including particulates, hydrogen ions (measured as pH or pOH), alkali metal ions, alkaline-earth metal ions, inorganic anions (nitrates, phosphates, sulfates), dissolved organic compounds, and undissolved organic compounds such as oil and grease affect the process choice. Biological components, most often measured as biological oxygen demand (BOD), can also complicate the wastewater matrix.

Often, the advantages of one process over another are not clear cut and many process options may initially be suitable. A few examples of the choices to consider are provided below. Chemical precipitation with sodium hydroxide works superbly with chromium ions at neutral pH, but is effective for lead or nickel ions only at an elevated pH. Ion exchange and chelating resins (solid adsorbents) work well at dilute concentrations for chromium, lead, and nickel, but are prone to fouling by other constituents in the wastewater matrix. Electrolytic recovery does not work well at low concentrations, but is relatively insensitive to other wastewater constituents. Chemical precipitation processes can operate at high flows of wastewater exceeding 100 gpm (387.5 L/min), whereas ion-exchange and electrolytic recovery processes generally work well at wastewater flows below this value.

The recovery, recycle, or ultimate disposal of the separated heavy-metal stream is the second major consideration for selecting a heavy-metal ion removal process or a specific option to pursue within a given process. As shown in Figures 2 and 3 and Table 2, there are several options for disposal or reclamation of the separated metal streams. For example, chemical precipitation processes generate a concentrated heavy-metal sludge which must be landfilled or hydrometallurgically reclaimed. Electrolytic processes recover solid metal. In contrast, solvent extraction, liquid membrane separation, and adsorbent-based processes all generate a concentrated heavy-metal ion stream which can be potentially recycled back to the process. Finally, bioprocesses can generate a sludge-like material concentrated in heavy metals.

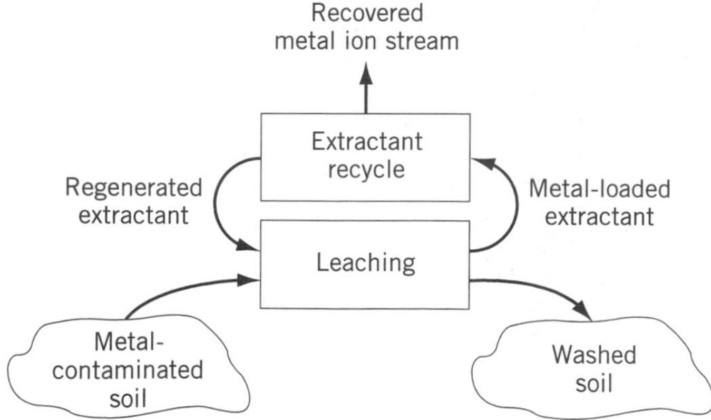

Figure 3. General concept for remediation of heavy-metal contaminated soils (Cognis TerraMet metal remediation process).

The ultimate disposal of the final heavy-metal stream can have a significant bearing on operating costs. For example, if the final heavy-metal waste stream exists as sludge containing leachable toxic heavy metals, then the solid waste must be sent to a secured hazardous landfill, which can be costly. However, if the heavy-metal sludge can be rendered stable to leaching by vitrification or by mixing the sludge with stabilization reagents such as silicates, fly ash, or cement, then the stabilized, nonleachable sludge can be sent to a nonhazardous landfill. General strategies for designing disposal sites to accommodate solid or liquid wastes have been reviewed (15).

Finally, the value, purity, and amount of the recovered metal dictate whether solid metal recovery should be pursued. Common hydrometallurgical options for metal reclamation from concentrated heavy-metal waste streams are not discussed herein, but final disposal of metals will be discussed within the context of a specific process. Pyrometallurgical and hydrometallurgical processes for reclamation of nonferrous heavy metals from scrap, sludge, and concentrated aqueous effluents have been reviewed (16). Large companies are most likely to consider on-site heavy-metal disposal or reclamation processes, whereas small-volume generators are compelled to seek off-site treatment and disposal. Companies specializing in off-site disposal include Chemical Waste Management, Ensco, and SCA Chemical Services.

In addition to metal recovery and recycle, the other major economic considerations are the separation process operating costs and capital costs. For example, chemical precipitation has significant chemical costs but modest energy costs, whereas electrolytic recovery has minimal chemical costs but significant energy costs. Both processes can have significant capital costs; however, in some cases existing equipment within a plant can be retrofitted for use in a chemical precipitation process which lowers capital costs. Bioprocess technologies can have low operating and capital costs if metal disposal costs are not factored in.

Processes for Heavy-Metal Ion Treatment by Concentration of the Wastewater Stream

This presentation focuses on separation technologies that remove heavy-metal ions from wastewater. However, there are several process technologies for processing wastewaters contaminated with heavy-metal ions that fall outside of this definition. Specifically, there are several processes which do not separate the heavy-metal ions from wastewater but instead remove the water as a pure solvent and concentrate the heavy-metal ions into a smaller volume of liquid waste. These processes include evaporation, freeze crystallization, and reverse osmosis. In evaporation (12), water and other volatile components are removed by a combination of heat and reduced system pressure to leave behind a liquid concentrate containing the nonvolatile components, including heavy-metal ions. In freeze crystallization (17) the wastewater stream is subjected to flash evaporation, vapor recompression, and subcooling to remove the water as ice crystals. The heavy-metal ions are enriched in the liquid stream, then separated from the ice crystals. Conventional crystallization of the heavy-metal salt directly from dilute aqueous solution is also theoretically possible, but energy costs for cooling are generally prohibitive. Conventional crystallization is practiced only for the removal of copper sulfate from relatively concentrated aqueous solutions, because its solubility data show that significant precipitation is possible at near-ambient temperatures around 10°C. In reverse osmosis (18), pressure is applied to the wastewater solution to force water through a membrane which is permeable to the solvent but not to the heavy-metal ions. If the applied pressure is higher than the osmotic pressure working in the opposite direction, then water will move against its concentration gradient and flow to the solvent side of the membrane. In this case, the wastewater volume is reduced and concentrated in heavy metals. In 1981, over 100 processes used reverse osmosis to concentrate nickel solutions from electroplating wastewaters at costs ranging from \$0.75 to \$2.00 per 1000 gallons (\$0.188–0.516/m^3) treated (19).

Processes for Removal or Disposal of Heavy Metals from Solid Wastes

Several technologies are being developed for removal or disposal of heavy metals in solid waste streams and soils. Two processes, high gradient magnetic separation (HGMS) and vitrification, deserve mention. HGMS processes can selectively remove heavy-metal containing particulates from other solids, liquids, or gases based on the differential paramagnetic properties of the target materials in the mixture. The waste slurry is passed through a magnetic volume and the paramagnetic materials are captured while the diamagnetic materials in the waste matrix pass through. HGMS processes for selectively removing dilute concentrations of uranium plutonium oxides from contaminated water and soils are being developed at the Los Alamos National Laboratory (20). Vitrification is a process where heavy metals adsorbed on soil matrices are encapsulated by fusing the soil to a glassy solid under high temperatures. This is accomplished by passing an electric current through the soil matrix, either *ex situ* or directly onsite under high voltage. Vitrification is particularly useful for disposal of soils contaminated with radioactive heavy metals. The radioactive heavy metals are encapsulated in an highly inert, nonporous matrix, which can be stored permanently in underground secured radioactive waste storage facilities currently under development. Vitrification has been demonstrated for encapsulation of heavy metals from soils, sludge, and sediments (21).

Processes for remediating heavy metals and radionuclides from soils have been overviewed (22). Several processes are also being developed to mobilize metals from solid waste matrices. Most common are soil washing processes which leach heavy metals adsorbed on soil matrices with water containing chelating agents or weak organic ligands such as organic acids. For example, the TerraMet process for removal of heavy metals from contaminated soils, developed by the Cognis Corporation (Santa Rosa, California), features soil washing with soluble chelation agents followed by recovery of the mobilized metals. The basic concept is illustrated in Figure 3. Bioleaching processes use microorganisms to solubilize heavy metals or heavy-metal oxides from mineral matrices by biologically catalyzed oxidation–reduction processes. In both soil washing or bioleaching processes, the eluted heavy-metal ions are then removed from the aqueous stream by the various chemical or biological process technologies described herein.

Both chemical and biological process separation process technologies for the direct removal of heavy-metal ions from wastewaters, particularly in cases where the heavy-metal ions are present in dilute concentrations, are described herein, as are the basic scientific and engineering principles underlying particular heavy-metal ion removal process classifications.

EMERGING BIOLOGICALLY PROMOTED (BIOPROCESS) TECHNOLOGIES

Common Features of Bioprocess Technologies

Many toxic organic molecules dissolved in wastewaters or physically bound to contaminated soil matrices can be biologically degraded to harmless products such as carbon dioxide and water by microorganisms which consume the organic material through primary or secondary metabolic pathways. However, toxic heavy-metal ions are already in elemental form and so cannot be biologically degraded to harmless products. Heavy-metal ions must be physically removed from the contaminant matrix under an externally supplied driving force. Chemical process technologies that provide this driving force include chemical precipitation, which converts the metal ion to an insoluble form by chemical reaction with a precipitation agent, electrochemical reduction, which reduces the metal ions to free metals by redox processes, and adsorption/ion exchange, which immobilizes heavy-metal ions onto a solid adsorbent material. Bioprocess technologies use living organisms or the dead remains of living organisms to promote many of these basic processes of physicochemical removal.

As shown in Figure 2, there are four major classes of biologically promoted process (bioprocess) technologies for heavy-metal ion removal: bioprecipitation, bioaccumulation, biosorption, and phytoremediation. A common feature of all bioprocess technologies is the use of biological materials to carry out the biologically and/or chemically promoted removal of heavy-metal ions from wastewater. Bioaccumulation processes use living microorganisms, ie, bacteria, fungi, and microalgae, to actively uptake heavy-metal ions into the cells of the living organism. Bioprecipitation processes use living microorganisms to promote chemical precipitation or redox reactions that reduce the heavy-metal ions to an insoluble state. The insoluble forms of the heavy metal (as oxides, sulfides, or free metals) and free cells are separated from the wastewater. Bioprecipitation and bioaccumulation often accompany each other. Phytoremediation is essentially bioaccumulation using living plants. The metal-adsorbed portion of the plant is cropped to remove the metals from the site. Biosorption processes use dead biomass to immobilize heavy-metal ions. The metal-absorbed biomass generated by each of these processes can be disposed of in a secured landfill site which can accommodate heavy-metal wastes. Alternatively, the biomass portion of the metal–biomass complex can be burned off to provide a metal-enriched solid ash or composted to concentrate the metal-adsorbed solid. This concentrated waste can then be refined by conventional hydrometallurgical operations to reclaim the metals.

Microorganisms can also be used to solubilize metals from solid matrices in a process known as biosolubilization or bioleaching. Bioleaching operations for metals reclamation have been reviewed (23,24). Bioleaching can be used to recover toxic heavy metals from ores or contaminated solid matrices. However, since this work focuses only on processes for the removal of heavy-metal ions from wastewater, bioleaching will not be discussed further herein.

There are three principal advantages of bioprocess technologies for heavy-metal ion removal. First, biologically based processes can be carried out *in situ* at the heavy-metal ion contaminated site. Bioprocess systems for heavy-metal ion removal from contaminated groundwater may be ideally suited for use in the subsurface permeable ion barriers described earlier. Specifically, the barrier would be loaded with an inert solid coated with a biofilm of living microorganisms. Alternatively, the barrier could be filled with an environmentally benign biosorbent. As contaminated groundwater flows through the permeable barrier, the metals are immobilized in the biological treatment zone of the contaminant site. Second, bioprocess technologies are usually environmentally benign. However, exogenous organisms must often be introduced to the contaminant site which can potentially cause other problems associated with upsetting the ecological balance of the site and surrounding area. Third, the bioprocess technologies are potentially low cost. The principal operating costs associated with bioprocesses are the generation of the biomass used to remove the heavy-metal ions and the final segregation of the captured heavy metals from the biomass matrix. Energy costs are small and chemical costs are low if the microorganisms can derive organic nutrients from the waste site or the aqueous contaminant stream. Capital costs are modest if on-site heavy-metal ion immobilization strategies are required as opposed to a pump-and-treat process where the heavy metals must be physically removed from the site.

A disadvantage of bioprocess technologies for heavy-metal ion removal is that they are sensitive to process changes and difficult to control. Bioprocess technologies for heavy-metal ion removal carried out by living organisms generally have relatively narrow temperature and pH tolerances. Furthermore, the inherent stochastic performance of biological organisms makes process control a difficult task.

Bioprecipitation and Bioaccumulation

Any microorganism that is resistant to superambient levels of heavy-metal ions is a candidate for bioprecipitation or bioaccumulation. Bacteria, especially genus *Pseudomonas,* are a significant source of candidate microorganisms for bioprecipitation/bioaccumulation. Bioaccumulation and bioprecipitation processes for heavy-metal ion removal have been reviewed (14,25).

Activated sludge from wastewater treatment plants is one often overlooked source of bioaccumulating microorganisms for heavy-metal ions. Heavy metals are the frequently detected trace contaminants in municipal wastewater sewage sludges. Although the microorganism consortia for bioaccumulation of heavy-metal ions in activated sludge is not well defined, it is nonetheless effective in accumulating heavy-metal ions. Process models have been developed to predict how heavy-metal ions from influent wastewater are captured by activated sludge in order to manage heavy-metal ion discharge from wastewater treatment plant effluents (26).

Biosorption

In biosorption, the heavy-metal ions are immobilized onto the outer surface of cell walls of dead or inactive organisms. Comprehensive reviews on biosorption of heavy-metal ions have been provided. Biosorption has been demonstrated for virtually all heavy-metal ions of industrial interest, including arsenic (As), silver (Ag), gold (Au), cadmium (Cd), cobalt (Co), chromium (Cr), copper (Cu), iron (Fe), mercury (Hg), manganese (Mn), nickel (Ni), lead (Pb), platinum (Pt), palladium (Pd), uranium (U), thorium (Th), and zinc (Zn). Saturation adsorption capacities can be as high as 200 mg metal/g biosorbent. Biosorbent materials include dead biomass from microalgae, macroalgae (seaweed), fungi, and bacteria. Biosorbent materials also include polysaccharide-based biopolymers, which are the major constituents of the cell walls of the aforementioned organisms. The heavy-metal ion binding mechanisms of biosorption parallel those of conventional adsorbents, including ion exchange, chelation, and physical adsorption. There are several chemical groups within the biomass that could bind with heavy-metal ions, including acetamide groups of chitin; carboxyl, sulfate, and hydroxyl groups in algal polysaccharides; amino and phosphate groups in amino acids; and amino, sulfhydryl, and carboxyl groups in proteinaceous materials. Pump-and-treat processes using biosorbent materials have been reviewed (27) (see PUMP-AND-TREAT TECHNOLOGY, AQUIFER REMEDIATION).

Phytoremediation

Phytoremediation is an intriguing process for the biologically promoted removal of heavy-metal ions from wastewaters. Phytoremediation is broadly defined as the use of living green plants to remove, contain, or render harmless environmental contaminants (28). Phytoremediation processes are capable of removing heavy metals directly from contaminated soil or from wastewater. Phytoextraction is the use of plants to remove heavy metals from soils (29), whereas rhizofiltration is the use of plants to remove heavy-metal ions from aqueous streams (30).

The basic mechanism of phytoremediation for removal of heavy metals involves two steps: (*1*) active uptake of heavy-metal ions by the root system, and (*2*) translocation of the heavy-metal ions to green shoots. Plants which can sequester high concentrations of heavy-metal ions are called hyperaccumulator metallophytes. For example, the hyperaccumulator *Thalaspi caerulescens* can accumulate several heavy-metal ions, including Ni, Zn, and Cd. In this plant, over 18,000 mg of zinc and 1000 mg of Cd per kg of shoot biomass can be accumulated without inhibition of growth (31). The exact mechanisms for active uptake for heavy-metal ions are under debate, but could involve the use of biologically expressed phytochelatins that complex with heavy-metal ions and thus protect the plant's biochemical machinery from the toxic heavy-metal ions. Genetic engineering offers a means to improve growth rates of hyperaccumulators or introduce heavy-metal remediation activity into plants. For example, the bacterial gene for mercuric reductase, which reduces Hg^{2+} to Hg^0, was recently cloned into Arabidopsis (32).

Engineered reed-bed systems and constructed wetland systems for removal of heavy-metal ions from wastewaters using phytoremediation have been reviewed (33). The root system of the hyperaccumulator plant penetrates a permeable rock bed. The wastewater is introduced into one end of the bed and flows through the permeable rock layer. The rock layer should be inert to heavy-metal ion binding so that it does not unwittingly serve as sink for heavy metals. Heavy-metal ions are sequestered by the root system and translocated to the shoots. The treated effluent is discharged. Periodically, the metal-containing shoots are harvested. The biomass can be burned off or composted to yield a low volume of metal-rich ash.

Phytoremediation processes have two major advantages. First, phytoremediation is an environmentally friendly process for on-site remediation of heavy metals. The only potential drawback is that the species of hyperaccumulator plant introduced to the contaminant site may not be endogenous to the local ecosystem. Second, phytoremediation processes are simple in configuration, require minimal chemical or energy input, and are inexpensive to operate. However, like many biologically based processes, they are difficult to control.

CONCLUSIONS

Although source reduction and waste minimization are the long-term solutions for reducing or eliminating the release of toxic heavy metals to the environment, treatment technologies will continue to play a crucial role in controlling heavy-metal ion emissions from current operations and in cleaning up contaminated sites from previous operations. In this work, both practiced and emerging process technologies for physically removing heavy-metal ions from wastewater and contaminated groundwater have been overviewed.

The broad spectrum of process technologies for physically removing heavy-metal ions from wastewater can be broken down into two basic categories: chemical processes (see Fig. 1) and biologically promoted processes (see Fig. 2). Often, several specific heavy-metal ion-removal processes within a total treatment system are assembled together as unit operations for heavy-metal ion removal, metal recovery/disposal, and final polishing of the treated effluent. With respect to chemical processes, the chemical precipitation of heavy-metal ions to insoluble hydroxides or sulfides is the most widely practiced process for removing heavy-metal ions from wastewaters and contaminated groundwaters. The precipitation process generates a heavy-metal sludge which must be landfilled or hydrometallurgically reprocessed to reclaim the free metals. Electrolytic recovery processes are preferred for removing heavy-metal ions from concentrated wastewater streams. In particular, electrowinning processes plate-out the heavy-metal ions as a solid free metal, which can be easily recycled. However, electrolytic processes do not work well for direct removal of heavy-metal ions from low concentration wastewater streams. Adsorption, ion-exchange, solvent extraction, and liquid membrane processes are useful in more specific heavy-metal ion-treatment applications where selective heavy-metal ion separation at low concentrations is desired. Solid adsorbents, which immobilize and concentrate the heavy-metal ions onto the surface by ion-exchange and chelating resin processes, can be engineered to discriminate between heavy-metal ions and other ions in the wastewater matrix. Ion-exchange and chelating resin adsorbents work particularly well for final polishing of the treated effluent to remove the last traces of heavy-metal ions. All of these extractant mediums, ie, ion exchange/chelating resins, liquid extraction solvents, and supported liquid membranes, are generally expensive and require regeneration. The regeneration process produces an aqueous stream concentrated in heavy-metal ions that can be recycled back to the source process which generated the waste or electrolytically recovered to generate free metals.

Chemical process technologies are well established and continue to evolve. However, chemical process technologies are inherently pump-and-treat processes, meaning that the wastewater stream must be pumped from the contaminant source to the treatment facility. Engineered biological (bioprocess) technologies are relatively new and untested, but offer significant promise as environmentally benign processes for on-site treatment of contaminated ground and surface waters at hazardous waste sites. Biologically promoted processes roughly parallel chemical precipitation and adsorption processes. Bioaccumulation and bioprecipitation processes sequester heavy-metal ions and convert the metal ions to free metals (eg, Hg^{2+} to Hg^0 by mercuric reductase) or precipitate the metal ions to metal hydroxides, phosphates, or sulfides. Biosorbents are inexpensive but still capable of adsorbing heavy-metal ions from extremely dilute solutions with selectivities approaching those of much more expensive ion-exchange/chelating resins. Phytoremediation is a novel process which uses living plants to sequester heavy metals; the root system absorbs heavy-metal ions and transports them to the green shoots, which can be mechanically harvested (mowed) to recover biomass material enriched in heavy metals. In bioprocess systems, the metal-laden biomass can be concentrated significantly by burning at low temperature to obtain a nonvolatile, concentrated heavy-metal ash or by composting. The concentrated solid residue can be landfilled or hydrometallurgically reprocessed to recover the heavy metals.

Four needs are prominent for future development of technologies for the removal of heavy-metal ions from wastewater. First, viable technologies are needed for on-site containment of ground and surface waters contaminated with low concentrations of heavy-metal ions at hazardous waste sites. Engineered permeable barriers loaded with low cost adsorbent materials (eg, iron oxide) is an emerging technology for this purpose. In this regard, the biologically promoted technologies described above need to be integrated into engineered permeable barrier systems for *in situ,* on-site capture and containment of heavy-metal ions from flowing groundwater plumes contaminated with heavy metals. This could include immobilization of microorganisms capable of bioaccumulation and bioprecipitation within the permeable barrier. Alternatively, biosorbent materials could be loaded into the permeable barrier. However, fragile biosorbent materials must be engineered for stability in aggressive chemical and biological environments at low expense while retaining heavy-metal ion selectivity at low concentration.

Second, further research is needed for improving the highly selective capture of specific heavy-metal ions from complicated aqueous waste matrices. A report by the National Research Council (34) targeted selective removal of trace concentrations of heavy-metal ions from complicated waste matrices as a key technology need in separation/purification chemical processing operations. An example of an early success is the Diphonix resin described earlier.

Third, the continued development of specialty precipitants for removal of very low (<1 ppm) concentrations of heavy-metal ions from high volume industrial wastewater streams is needed. Many industries that do not process metals directly still generate significant volumes of wastewater, in some instances near 4000 m^3/d (1×10^6 gal/d), that contain very low levels of heavy-metal ions. Prominent examples are coal-fired plants and pulp mills where heavy metals indigenous to the raw materials (coal, wood, etc) leach out during processing and ultimately find their way into wastewater streams. Because the volumetric flow rates are high, chemical precipitation is the best option. However, commodity chemical precipitants (eg, sodium hydroxide or sodium sulfide) have limited performance in one or more areas. In this context, specialty precipitants need to be effective at very low concentrations (<1ppm) and generate a solid sludge that settles without flocculation and is stable to leaching.

Finally, disposal options for the recovered heavy metals from all heavy-metal ion removal processes require continued development. Disposal options for the separated heavy metals are dependent on the type of process used for removal of the heavy-metal ions from wastewater. For example, combustion combined with vitrification offers a safe and permanent means of final disposal for highly toxic and radioactive heavy metals. However, this disposal process requires the development of low cost, carbon-based "throwaway" adsorbent materials which can adsorb toxic and radioactive heavy-metal ions at low concentrations with selectivities and capacities approaching those of much more expensive ion-exchange/chelating resins.

There is no single ideal process for heavy-metal ion removal. Therefore, process selection must be based on the types of heavy-metal ions to be treated, the concentration of heavy-metal ions in the wastewater, and the final disposal or recycle option desired.

ACKNOWLEDGMENTS

Grateful acknowledgment is given for support from several individuals who helped to make this work possible: John Sharp (Merix Corporation); Dan Hanthorn (City of Corvallis Public Works Department); Gary Hundley (U.S. Bureau of Mines, Albany, Oregon); Mike Gula (Eichrom Industries, Inc.); Ray Maddalone (TRW, Inc.); Scott McCray (Bend Research); and Scott Warner (Geomatrix Consultants, Inc.). The U.S. EPA is acknowledged for supporting research on the development of engineered biopolymer materials for heavy-metal ion separations through Grant #818626-01 provided by Exploratory Research Grants Program.

BIBLIOGRAPHY

1. T. Stephenson, in J. N. Lester, ed., *Heavy Metals in Wastewater and Sludge Treatment Processes*, Vol. I, *Sources, Analysis, and Legislation*, CRC Press, Boca Raton, Fla., 1987, pp. 31–64.
2. C. S. Brooks, *Metal Recovery from Industrial Waste*, Lewis Publishers, Chelsea, Mich., 1991.
3. T. Rudd, Ref. 1, pp. 1–30.
4. F. W. Pontius, *J. Am. Water Works Assoc.* **88**(3), 36–46 (1996).
5. U.S. EPA, *1987 National Report of Hazardous Waste Generators, Treatment, Storage, and Disposal Facilities Regulated Under RCRA*, U.S. Environmental Protection Agency, Report No. EP A530SW4106 Washington, D.C., July 8, 1991.
6. MSE, Inc., *Resource Recovery Project Technology Demonstrations, Series II*, RFP #48255, MSE, Inc., Butte, Mont., 1993.
7. V. M. Reid, K. W. Wyatt, and J. A. Horn, *Civil Eng.* **64**(4), 56–58 (1994).
8. C. S. Brooks, *J. Metals* **38**(7), 50–69 (1986).
9. R. W. Peters, Y. Ku, and D. Bhattacharyya, in R. W. Peters and B. M. Kim, eds., *Separation of Heavy Metals and Other Trace Contaminants*, AIChE Symposium Series 243, Vol. 81, New York, 1985, pp. 165–203.
10. S. Beszedits and N. S. Wei, *Removal of Heavy Metals from Wastewaters*, B & L Information Services, Toronto, Ontario, Canada, 1980.
11. M. Meltzer and co-workers, *Metal-Bearing Waste Streams: Minimizing, Recycling, and Treatment*, Noyes Data Corp., Park Ridge, N.J., 1990, pp. 9–27.
12. E. R. Krishnan and co-workers, *Recovery of Metals from Sludges and Wastewaters*, Noyes Data Corp., Park Ridge, N.J., 1993.
13. P. L. Mattison, *Bioremediation of Metals—Putting it to Work*, Cognis Corp., Santa Rosa, Calif., 1992.
14. J. R. Bennemann and E. W. Wilde, *Literature Review on the Use of Bioaccumulation for Heavy Metal Removal and Recovery*, Vol. 2, Report No. WSRCTR90175, U.S. Dept. of Energy, Washington, D.C., 1991.
15. T. D. Wright, D. E. Ross, and L. Tagawa, in H. M. Freeman, ed., *Standard Handbook of Hazardous Waste Treatment and Disposal*, McGraw Hill Book Co., Inc., New York, 1989, pp. 10.3–10.23.
16. M. Shamsuddin, *J. Metals*, **38**(2), 24–31 (1986).
17. J. P. Tratnyek, in J. B. Erkowitz and co-workers, eds., *Unit Operations for Treatment of Hazardous Industrial Wastes*, Noyes Data Corp., Park Ridge, N.J., 1978, pp. 552–573.
18. P. S. Cartwright, *Plating Surface Finish.* **71**(4), 62–66 (1984).
19. P. S. Cartwright, *Plating Surface Finish.* **68**, 40–45 (1981).
20. A. R. Schake, L. R. Avens, D. D. Hill, D. D. Padilla, and F. C. Prenger, *Magnetic Separation for Environmental Remediation*, Report No. LAUR943373, U.S. Dept. of Energy, Los Alamos National Laboratory, Los Alamos, N.M., 1994.
21. C. J. Dial, J. M. Houthoofd, and E. F. Harris, *J. Air Waste Manage. Assoc.* **41**, 1393–1396 (1991).
22. M. Mishra, R. Mehta, and K. Rajendra, *J. Metals* **47**(9), 45–53 (1995).
23. S. R. Hutchins, M. S. Davidson, J. A. Brieley, and C. L. Brierly, *Ann. Rev. Microbiol.* **40**, 311–336 (1986).
24. D. E. Rawlings and S. Silver, *BIO/TECHNOLOGY* **13**, 773–778 (1995).
25. E. W. Wilde and J. R. Bennemann, *Heavy Metal Removal and Recovery Using Microorganisms*, Vol. 1, *State-of-the-Art and Potential Applications at the SRS*, Report No. WSRCTR90175Vol1, U.S. Dept. of Energy, Washington, D.C., 1991.
26. H. D. Monteith, J. P. Bell, D. J. Thompson, J. Kemp, C. M. Yendt, and H. Melcer, *Water Environ. Res.* **65**, 129–137 (1993).
27. B. Volesky and Z. R. Holan, *Biotechnol. Prog.* **11**, 235–250 (1995).
28. S. D. Cunningham and W. R. Berti, *In Vitro Cell. Dev. Biol. Plant* **29P**, 207–212 (1993).
29. Kumar, V. Dushenkov, H. Motto, and I. Raskin, *Environ. Sci. Technol.* **29**, 1232–1238 (1995).
30. *Ibid.*, p. 1239.
31. S. L. Brown, R. L. Chaney, J. S. Angle, and A. J. M. Baker, *J. Environ. Qual.* **23**, 1151–1157 (1994).
32. S. A. Moffat, *Science* **269**, 302–303 (1995).
33. K. R. Gray and A. J. Biddlestone, *TIBTECH* **13**, 248–252 (1995).
34. C. J. King and co-workers, *Separation and Purification: Critical Needs and Opportunities*, National Academy Press, Washington, D.C., 1987.

HERBICIDE RESIDUES IN BIOTA, DETERMINATION OF

J.V. HEADLEY
L.C. DICKSON
National Hydrology Research Institute
Saskatoon, Saskatchewan, Canada

A.J. CESSNA
Agriculture and Agri-Food Canada, Lethbridge Research Centre
Lethbridge, Alberta, Canada

The determination of herbicide residues in biota is presented from the perspective of the residue analyst. The emerging trends are as follows: (*1*) Herbicides in biota are seldom determined using analyte-specific methods but are generally integrated into multi-residue methods for cost-effective applications to biological and environmental samples. (*2*) There is growing concern to reduce the amount of organic solvent used for sample preparation to minimize analyst exposure and waste generation. Thus, strategies which limit or reduce the amount of solvents have been adopted. (*3*) There is a growing trend to miniaturize apparatus used for sample extraction procedures with smaller sample sizes which demands higher instrumental performance and detection of the analytes at lower detection limits. As well as using less solvent, the resulting micro methods tend to be generally cheaper, faster, and less labor-intensive than macro methods. They also tend to be more amenable to automation using on-line techniques compared to the off-line macro methods. (*4*) For sample extracts with high fat or lipid content where cleanup cannot be avoided, there is a growing trend in the use of automated gel-permeation chromatography and high performance liquid chromatography (HPLC)-based procedures. (*5*) There is widespread acceptance of the use of high resolution gas chromatography (GC) using capillary columns with specific detectors or mass spectrometers. (*6*) Capillary electrophoresis is a new technique providing many advantages over traditional GC and liquid chromatography (LC) separations, including short analysis times and small sample volumes. (*7*) Mass spectrometry (MS) is particularly well-suited to the confirmation of target analytes when operated in selected ion monitoring mode and remains the detector of choice for the tentative identification of unknowns when operated in the full-scan mode. In the last five years, ion-trap MS instruments have been more commonly used in environmental laboratories. These systems can be used for MS-MS applications and are presently less expensive than the conventional quadrupole or sector MS instruments. Consequently, it is anticipated that the use of MS-MS employing ion traps will grow for the determination of herbicides in biota. (*8*) There is still a significant lag in the use of LC methods compared to GC methods for the determination of herbicides in biota. However, as LC-MS systems continue to become more widely available and less expensive, applications in herbicide residue analyses are anticipated to become more common. This area warrants further study to accelerate the applications of LC-MS to polar or thermally labile compounds. (*9*) In recent years noted advances have been made in the development of immunoassays for the rapid screening of herbicide residues. These methods, once optimized, can facilitate high sample throughput at relatively low cost compared to conventional approaches for screening of herbicide residues. Furthermore, the immunoassays kits are easily field portable. To date, however, development has been primarily for aqueous systems and little work has been done for the direct determination of herbicides in biota.

INTRODUCTION

Herbicides are generally synthetic organic compounds used to control unwanted vegetation or weeds. Application may either be directly to plants or soil to control unwanted terrestrial vegetation, or to water bodies to control aquatic weeds. Herbicide use plays an integral role in most crop production systems and agricultural production accounts for the greatest use of herbicides. However, significant amounts of herbicides are also used in the forestry industry and to maintain pipeline, powerline, and highway rights-of-way. Significant use also occurs within the urban setting to control weeds in lawns and gardens, and to maintain parks and golf courses.

The sustainability of agricultural uses of herbicides is most often assessed within an agronomic context and generally implies sufficient efficacy to kill problem weeds or minimize their competitiveness with the crop, good crop tolerance, and an economic return to the producer. It also implies that soil persistence of herbicides will not affect yields of subsequent crops, and that soil quality will not be compromised; that is, there will be no irreversible deleterious effects on soil microbial populations.

However, it is now well-established that herbicides can undergo chemical modification following application and that not all of the herbicide remains within the area to which it was applied (1,2). Herbicides undergo microbial or chemical degradation in soil, metabolism/conjugation in plants, and photodegradation by sunlight. As illustrated in Figure 1, transport mechanisms by which herbicides or their metabolites/degradation products can move off treated areas include drift during application, post-application volatility losses, and wind erosion of treated soil. Such atmospheric inputs are later removed either by dry (particulate) deposition or through rainfall washout, both of which result in wildlife habitat and surface water contamination. Transport into surface waters occurs with snowmelt, rainfall, or irrigation runoff from treated land. Contamination of ground waters can occur via leaching and preferential flow of herbicides under both dry land and irrigated agriculture. Herbicides and their metabolites may also be removed from treated areas as residues in edible portions of crops. Such transport results in human and wildlife exposure to herbicides, and affects environmental quality through wildlife habitat degradation and decreased biodiversity. Herbicides generally express low to moderate mammalian toxicity, as well as toxicity to birds and fish. Thus, it is necessary that sustainability of herbicide use within the agricultural context be considered not only in terms of crop production, but also in terms of human health and the environment. Similarly, human health and environmental quality should also be considered with respect to other herbicide uses (forests, rights-of-way, etc).

In general, there are two types of human exposure to herbicides. One is occupational exposure and generally involves either those manufacturing/formulating herbicides or those applying herbicides (eg, farmers, professional applicators). The other is nonoccupational or environmental exposure and involves the general population. Environmental exposure to

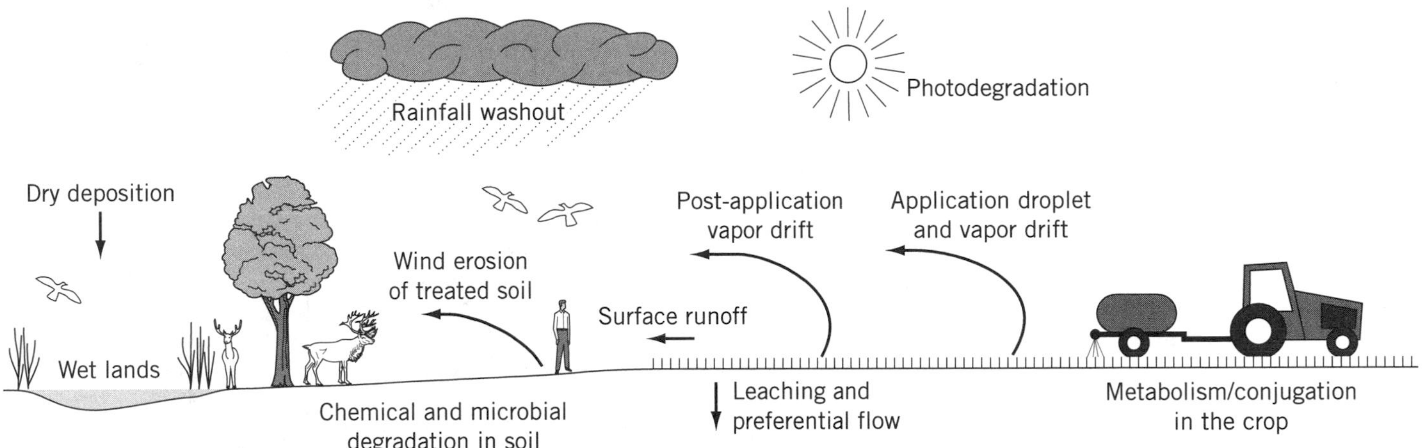

Figure 1. Herbicide degradation processes which occur following application, and transport mechanisms which remove a portion of the herbicide application from the treated area.

herbicides, metabolites, or degradation products results from breathing contaminated air, drinking contaminated water, ingestion of residues in edible portions of crops or animals, atmospheric deposition, and contact with surfaces contaminated by atmospheric deposition. A possible third type of exposure involves urban dwellers who infrequently apply relatively small amounts of herbicides to lawns and gardens.

Terrestrial animals are exposed to herbicides by the same routes by which humans are environmentally exposed. However, animals may be further exposed through contact with treated soils and plants, ingestion of residues on recently treated plants as well as those resulting from biomagnification within the food chain, and by directly intercepting herbicide spray during application. Fish and other aquatic animals are environmentally exposed to herbicides by contact with contaminated waters and sediments, and by ingestion of residues in contaminated plants and animals including those resulting from biomagnification within the food chain.

Environmental contamination of adjacent crops or other plants generally results from droplet and vapor drift during herbicide application or from post-application vapor losses. However, such residues may also result from snowmelt, rainfall, or irrigation runoff or wind erosion of treated soil.

In many countries, current public interest in human health and safety issues with respect to herbicide (and other pesticide) use is high. For example, in Canada, people remain concerned with air quality and the safety of their drinking water and food supply (3). Environmental quality is also important to Canadians as evidenced by the popularity of wildlife- and fish-related activities, and strong commitments to the protection and conservation of abundant and diverse wildlife (4).

Such global public concern regarding the sustainablity of herbicide (and other pesticide) use with respect to human health and environmental quality has contributed to the driving force for development of sensitive analytical methods. However, the initial impetus was derived from the requirement for the agrochemical industry, in order to support the registration of herbicide uses, to provide residue data for herbicides and their metabolites/degradation products in the organs and body fluids of animals used in toxicity testing, edible portions of crops, and soils used to study the persistence of these compounds. Subsequently, the methods also have been used by government regulatory agencies for monitoring purposes and for soil and surface and ground water remediation, and by researchers studying pesticide transport mechanisms, effects of pesticides on human health, and impact of pesticides on wildlife and environmental quality. As a consequence of combined industry needs and public concerns, analysts have developed compound-specific as well as multiresidue methods for determining pesticide residues in a wide variety of matrices including soils, air, surface and ground waters, plant tissues, and human and animal tissues and body fluids.

As suggested by the term, compound-specific analytical methods are developed to determine residues of a single analyte. Currently, such methods are seldom used. Even in studies involving a single pesticide, such as toxicity testing, dissipation in crops, persistence in soils, photolysis, leaching under irrigation, or environmental fate in wetlands, analytical methods are usually developed to include the determination of possible metabolites or degradation products and thus are really multiresidue methods. Government agencies, in their environmental monitoring programs or food basket surveys, characteristically use multiresidue methods, each of which may be used to quantitate residues of more than 100 pesticides and their metabolites and degradation products. It should be recognized that, with a multiresidue method of this nature, not all of compounds analyzed for will be extracted from a given matrix with the same efficiency, nor will they respond with the same sensitivity to a specific detector. Thus, it is not unusual that a range of detection and quantitation limits will result for a given method. It should also be noted that pesticide multiresidue methods are characteristically used to quantitate compounds involving several classes of pesticides (ie, herbicides, insecticides, and fungicides) (5).

Although several classes of pesticides can be analyzed in many matrices, this article will address only the determination of herbicides, and exclusively in biota; that is, terrestrial and aquatic animals and plants. Even though it is acknowledged that body fluids (urine, saliva, and blood) are important matrices in toxicity and human health studies, and that environmental matrices (soil, air, and water) and model exposure systems (semipermeable membranes containing lipophilic materials, simulated plant surfaces, etc), are important to environmental fate studies, analysis of these matrices will not be discussed. The intent of this article is not to review the history of pesticide residue analysis *per se*. Rather, in light of additional options available to the residue analyst arising from the developments in analytical technologies over the past decade, current extraction, derivatization, cleanup, and detection methods will be reviewed as well as emerging technologies. Because herbicides are targeted for uptake by plants, in contrast to animals which are only indirectly exposed, the majority of information regarding the determination of herbicide residues in biota derives from plants and discussion will be weighted accordingly.

It is important to emphasize that in the majority of cases, analysts collaborate with researchers from other disciplines in multidisciplinary studies. For example, crop residues are generally studied in collaboration with an agronomist, whereas transport studies may involve irrigation specialists, hydrologists, soil scientists, and micrometeorologists. Human health studies may involve toxicologists and epidemiologists, as well as social scientists. Environmental impact studies can involve wildlife, freshwater, and marine biologists. A statistician is usually involved in the studies to ensure that a statistically defensible sampling plan is followed and that representative samples are collected for residue analysis. The other researchers, together with the analyst, determine the level of sensitivity and degree of confirmation required from the residue analysis such that the objectives of the study will be achieved.

The article is written from the point of view of the residue analyst and will focus on various strategies used by analysts to develop methods appropriate to the matrices to be analyzed and the ultimate use of the analytical data. Discussion will include considerations for the type of matrix, known or unknown analytes, acidic or base-neutral herbicides, metabolites or conjugates, and the level of sensitivity and degree of confirmation required. Because the reliability of the information to be derived from residue data depends not only on the analytical method but also on the sampling and subsampling procedures, sampling is identified and discussed as an integral component of residue analysis. An exhaustive literature review of methodology currently available in the literature will not be presented; rather, specific examples from the literature will be used which illustrate various strategies or procedures.

Throughout the article, herbicides will be referred to by their common names as adopted by the Weed Science Society of America (WSSA) (6). Acidic herbicides are occasionally formulated as the free acid, but more commonly either as an alkali metal salt, an amine salt, or as an ester. Unless reference is made to a specific salt or ester, acidic herbicides will be referred to by the WSSA common names for the parent acids.

GENERAL CONSIDERATIONS IN ADDRESSING STUDY OBJECTIVES

There are three basic reasons for determining herbicide residues in biota: (*1*) studies in support of registration, (*2*) environmental and toxicological research, and (*3*) regulatory monitoring of residues in food commodities.

Quality Assurance/Quality Control Considerations

The sampling and analysis plan is defined by the hypotheses to be tested and the level of confidence required in drawing conclusions about the hypotheses. The statements about the level of confidence are called the data quality objectives. The QA/QC program ensures that the objectives are met in an efficient and cost-effective manner. QC refers to the technical activities, such as the use of blanks, controls, and spiked blanks, and evaluation of the condition and integrity of analytical standards and stored samples which are used to control and assess the quality of the analytical data being produced. QA is the management strategy than ensures that the QC system is working properly (7). The strategy involves defining the QC activities necessary to meet the objectives, assessing the QC results to ensure that the system is in control, and reporting the QC data along with the analytical results so that the overall quality of the results can be assessed (7). The overall purpose of the QA/QC program is to provide analytical data of a known level of reliability that is acceptable to the end user as defined by the objectives. The importance of QA activities in chemical analyses in general (8) and pesticide analyses in particular (9) has been reviewed.

Type of Matrix

The most basic division of sample types is plant versus animal. Analytical methods can be very different for plants and animals because of differences in tissue, expected levels of contaminants, and the nature of metabolites and degradation products.

Different types of plant and animal tissues are subsampled in different ways. Plant parts, such as fruiting bodies or kernels, leaves, stems, and roots are often separated from each other and analyzed separately. In animals, the focus for research studies is usually on the fatty tissue and the liver where persistent residues are most likely to accumulate; for regulatory studies the focus is on muscle tissue. The choice of extraction and cleanup will be affected by the level of moisture, the level of oils or fat, and the presence of co-extracted compounds and conjugates specific to plant or animal tissues (10). The concerns and approaches to the extraction and cleanup of plant and animal tissues are discussed in subsequent sections.

Acidic or Base-Neutral Herbicides

Although there are exceptions, such as paraquat and diquat, herbicides fall into two basic classes, base-neutrals and acids. Acidic herbicides compose a relatively large group and account for approximately one-third of currently used herbicides. Acidic herbicides include benzoics, oxyphenoxyalkanoics, phenols, phenoxyalkanoics, pyridinecarboxylics, quinolinecarboxylics, substituted alkanoics, and some miscellaneous acidic herbicides (Table 1). Acidic herbicides, whether a salt or an ester, usually hydrolyze in the environment or in the organism to the parent acid. The parent acid herbicide is more readily excreted by animals than the base-neutral herbicides, and readily forms conjugates in plants. Base-neutral herbicides have relatively low water solubilites compared to acidic herbicides and are more likely to accumulate in fatty tissues. These compounds can be extracted from tissues with an aqueous solvent at basic or neutral pH conditions (discussed in subsequent sections). The choice of analyte and the applicable methods will depend

Table 1. Currently Used Acidic Herbicides

	Benzoic Acids	
acifluorfen	imazamethabenz	napthalam
chlorimuron	lactofen	pyrithiobac
dicamba		
	Oxyphenoxyalkanolic acids	
diclofop	haloxyfop	quizalofop
fluazifop		
Phenols		
bromoxynil	ioxynil	
	Phenoxyalkanoic acids	
2,4-D	dichlorprop	2,4-DB
MCPA	mecoprop	MCPB
2,4,5-T		
	Pyridinecarboxylic Acids	
clopyralid	imazethapyr	picloram
imazapyr		
	Quinolinecarboxylic Acids	
imazaquin	quinclorac	
	Sulfonylureas[a]	
bensulfuron	metsulfuron	primisulfuron
sulfometuron		
	Miscellaneous	
benzoylprop	flumetsulam	glufosinate
DCPA	fluroxypyr	glyphosate
endothall		

[a] These herbicides also contain the benzoic acid functional group and alternatively could be classified as benzoic acids.

on whether determinations of the parent compound are sufficient to meet the objectives of the study, or whether the suite of target compounds should include hydrolysis products, metabolites, and conjugates.

Known or Unknown Analytes

The analytical strategies adopted for the determination of herbicide residues in biota fall into two main categories: namely, cases where specific target analytes are required, and where the biota may contain a mixture of unknown herbicides and their transformation products.

When specific target analytes are known, as in the case of residue determinations on sprayed crops, the sample extraction and cleanup options available to the analyst can be optimized to give the highest recoveries for the chosen compounds. The instrumental analysis, usually employing a chromatographic separation, can be optimized to give maximum resolution between closely eluting pairs of analytes. Relatively inexpensive specific detectors can be used in lieu of more expensive mass spectrometric (MS) detectors. With an extraction, cleanup, and detection methodology optimized for a small subset of possible herbicides, the level of effort necessary to confirm that a given analyte, rather than an interfering substance, has actually been detected will be lower than in the case when analytes are not known beforehand.

If the analytes are not known, as in the case of many environmental monitoring studies, compromises must be made to ensure that a wide range of herbicides and their metabolites and degradation products can be extracted and isolated from the matrix components. The recoveries of some compounds will not be as high as in the optimized methodology used for known target analytes but will be sufficiently high to be quantifiable with a given degree of confidence. MS or atomic emission (AE) detection is desirable in this case because these detectors not only provide quantitative data but will give qualitative information which can be used to confirm the presence of an analyte in the sample. Full details of the numerous techniques available to the residue analyst will not be reviewed here but are summarized in Table 2.

Anticipated Levels of Residues

Plants can be either directly exposed to herbicides by spraying, or inadvertently exposed through droplet or vapor drift during application. Animals are usually indirectly exposed by contact with drifting or oversprayed herbicides, or by ingesting contaminated plants, insects, or water. Most herbicides have low toxicities to animals and are usually readily metabolized. Therefore, most herbicides do not bioaccumulate in fatty tissues and are not likely to reach levels that generally prove toxic to animals. With the exception of toxicity studies or accidental ingestion of large amounts of contaminated feeds, animal exposure to herbicides is likely to result in much lower residue levels than those in plants. The expected levels of the herbicide residues will affect the choice of sample sizes and the choice of extraction, cleanup, and detection methods. These concerns are discussed in subsequent sections.

Performance Criteria for Analytical Methods

Residue analysts develop analytical methods to determine residues in a wide variety of matrices. In doing so, analysts generally try to build certain desirable features into the methods. One feature is portability, that is, methods that can be readily and effectively used by different analysts in different laboratories. This requires that the methods use glassware, chemicals, and instrumentation which are readily available in analytical laboratories. Another is reproducibility, which entails low variability or good precision associated with residue recoveries from fortified matrices, and minimization of systematic errors. The method must be rugged or robust, which requires that the reproducibility should not be affected when the method is used in different laboratories by other analysts using slightly different techniques, glassware, equipment, and instrumentation. There must be no matrix effect; that is, the response from an analyte in a spiked matrix must be the same as that from an equivalent amount of analyte from an analytical standard. The method must also have good throughput, and it must have the capability of handling several samples simultaneously. The method must be cost effective through utilization of minimum amounts of solvents, reagents, column packings, etc, and avoiding time-consuming steps. Finally, safety is also a consideration in the development of an analytical method because steps are needed to ensure that a given method does not present the analyst with unreasonable hazards.

Three other attributes by which the performance of an analytical method is often evaluated are detection limit, sensitivity, and specificity or selectivity. There are two types of detection limits in common usage. The instrumental detection limit is the concentration or amount that gives a response two or three times the standard deviation of the background response level of the instrument. This is determined using a solution of an analytical-grade compound dissolved in a suitable solvent and does not contain coextractives usually associated with a real sample. Instrumental detection limits, while useful

Table 2. Summary of Commonly Used Instrumental Techniques for Determination of Herbicides

Physical/Chemical Properties	Herbicide Class	Instrumentation[a]
Volatile and thermally stable	Organo-halogens	GC-ECD, GC-HECD, GC-MS
	Chlorophenoxyalkanoic acids	
	Chlorophenols	
	Chlorobenzoic acids	
	Organo-sulfur	GC-FPD, GC-HECD, GC-MS
	Thiocarbamates	
	Organo-nitrogen	GC-NPD, GC-HECD, GC-MS
	Dinitroanilines	
	Triazines	
Polar and thermally labile	Sulfonylureas	HPLC-UV, HPLC-Fluor
	Ureas	HPLC-MS
	Bipyridylium	
	Phosphonic acids	
	Carbamates	

[a] ECD = electron capture detector; Fluor = fluorescence detector; FPD = flame photometric detector; GC = gas chromatography, HECD = Hall electrolytic conductivity detector; HPLC = high performance liquid chromatography; UV = UV detector; MS = mass spectrometric detector; NPD = nitrogen−phosphorus detector.

to compare the performance of instruments, are not usually achievable in practical situations. The detection limit for the entire analytical system is more properly defined as the concentration or amount corresponding to a measurement level three standard deviation units above the background level, which is determined by analyses of field blanks. In this case, the standard deviation refers to the variation of the entire procedure, including contributions from the nature of the samples, variable recoveries, quality of the reagents, protocols for calibration and reagent blank correction, short- and long-term baseline and calibration drift, the instrument, environmental or location factors, and the analyst (11). The detection limit is also a function of sensitivity. Sensitivity (S) is strictly defined as the slope of the calibration function, $y = Sx + b$, which relates the instrumental response (y) to concentration or amount of analyte (x). The term b is the y-intercept which corresponds to the background level. The term sensitivity is however, sometimes used interchangeably but incorrectly with the term detection limit (12). The analyst can build greater sensitivity into a method by incorporating sufficient matrix extract cleanup to permit concentration of the matrix extract to smaller volumes (for example, 100 μL (or 10^{-6} L)) prior to analysis. When derivatization is necessary, derivatives incorporating functional groups to increase specific detector response can also increase the sensitivity of a method. The selectivity of a method is a measure of its ability to discriminate between the analyte and an interfering compound. Selectivity can be improved by more rigorous cleanup procedures, the use of selective instrumental detectors, or a combination of both.

Maximum Pesticide Residue Limits and Method Detection Limits

Government regulatory agencies will often dictate minimum performance parameters for analytical methods to be used to determine herbicide residues in commodities and food products. The levels of residues in food are monitored for two purposes: to enforce regulations on acceptable levels, and to estimate dietary intakes. These two programs make different demands on the analytical method performance, especially on method detection limits.

Regulatory monitoring programs require analytical methods that can detect pesticide residues at or below a stated maximum residue limit (MRL). This limit, sometimes called a tolerance, is the legal maximum concentration of a particular pesticide residue, metabolite, or degradation product allowed in a particular food. Although national regulatory agencies set their own MRLs, many agencies attempt to set their MRLs at the same levels as does the Codex Alimentarius Commission (13) in order to facilitate international trade (14). MRLs typically range from 0.001 to 110 mg/kg in Canada (15), 0.1 to 50 mg/kg in the United States (16), 0.05 to 70 mg/kg in European countries (17), and 0.02 to 50 mg/kg in Japan (18).

For regulatory monitoring it is desirable that the method detection limit for a particular pesticide be at or below the MRL. Detection limits are variable, and depend on the particular pesticide, commodity, choice of sampling and subsampling methods, and choice of analytical method. Detection limits for methods used for regulatory monitoring in the United States range from 0.005 to 1 mg/kg, although 0.01 mg/kg is a typical value. These values are usually well below the MRL for any particular combination of pesticide and commodity (16). By using Belgium as typical of European countries, the ratio of method detection limits to MRLs ranges from 0.02% to 100% (17). In Japan, residue levels of 0.01 to 0.1 mg/kg are considered to be easily detectable (19).

Total diet studies require detection limits as low as practically possible to give accurate information on dietary residue intakes. Detection limits for analyses used in these studies are 5 to 10 times lower than those used for regulatory monitoring. In general, detection limits are about 0.001 mg/kg (16). In general, detection limits can be lowered by a combination of larger sample size, smaller final extract volume (17), a larger injection volume, or use of a more sensitive detector.

Multiresidue Methods

It is seldom that an analyst is required to analyze only a single analyte within a matrix. Even in simple crop residue studies, there is generally the requirement to determine residues of not only the parent herbicide but also of one or more metabolites or degradation products. Thus, when a method is developed, it is generally multiresidue in nature. In those situations in which analysis of a single analyte is required, rather than develop a method, the analyst will generally choose an appropriate existing multiresidue method (MRM) to use, or adapt if necessary, for the single analyte determination.

Early MRMs were developed primarily by agrochemical companies for the purpose of providing information to support the registration of pesticide uses. They were designed to quantify the parent compound and a small suite of toxicologically significant by-products, metabolites, and degradation products. Similar MRMs involving analysis of a few herbicides and their metabolites or degradation products are currently used for research purposes in crop residue studies, toxicological studies, environmental fate studies, etc.

MRMs have been used by regulatory agencies to monitor a broad spectrum of several pesticides, including insecticides, herbicides, and fungicides (20). Early methods were heavily slanted toward the more stable and low polarity compounds, such as the organochlorines and organophosphates. These early methods also relied largely on GC with selective detectors, such as the nitrogen–phosphorus detector (NPD), the Hall electrolytic conductivity detector (HECD), the flame photometric detector (FPD), and the electron-capture detector (ECD). More recent MRMs incorporate the use of MS detectors which provide much greater specificity and essentially equivalent sensitivity. Because of the increased specificity provided by the MS detectors, larger numbers of analytes are now included in current MRMs. It is not unusual for an MRM utilized by regulatory agencies to monitor greater than 100 analytes. Some MRMs used by the regulatory agencies are optimized for a selected class of compounds. However, in reality a single MRM will not be amenable to the analysis of all pesticides. Instead, MRMs for pesticides are usually based on physicochemical properties of the pesticides or analytical approaches, and do not distinguish between herbicides and other pesticides. Thus, methods commonly used for insecticides and other pesticides can also be used for herbicides.

An advantage of MRMs used in monitoring programs is that they are generally cost effective: that is, the cost per analyte per sample is low. These methods are usually designed to be rugged and handle large volumes of samples with relatively short turnaround times, and to have acceptable levels of performance as defined by the end-use of the data. However, some current MRMs do not include acidic herbicides in the suite of pesticides monitored. Acidic herbicides require pH adjustment of the aqueous homogenate for efficient extraction and derivatization prior to GC analysis.

Sampling Biota

When designing sampling plans for residue analyses in plant and animals in the field, there are many factors that have to be considered that are usually not a concern when sampling water or air. Some of these are large variations within the study population, inhomogenieties in the matrix and in the distribution of the analyte within the matrix, biological degradation processes, bioaccumulation, and the efficiency of collection of individual organisms (7).

A fundamental difference between plants and animals for sampling purposes is that animals are mobile. Sampling protocols for animals have to account for factors such as movement into and out of areas of exposure, migration patterns, feeding, spawning, or other periodic activities that could affect the level of exposure and the concentration and location of residues in the organism (7).

SUMMARY

The determination of herbicides in biota is seldom performed as stand-alone tests and is best integrated into multiresidue methods for cost-effective analyses of herbicides and other analytes. In this context, it is advantageous to utilize strategies that eliminate as much sample handling as possible and minimize the number of steps required in the cleanup of crude extracts. This is desirable because sample cleanup is often the most time-consuming part of the analysis and can lead to significant losses of analyte. Indeed, careful choice of extraction can help reduce or eliminate the need for cleanup steps prior to instrumental detection. For sample extracts with high fat content where cleanup cannot be avoided, the use of automated procedures which utilize gel permeation chromatography (GPC) or HPLC are recommended. For some applications it can also be cost effective to use a detection system which is proven to be rugged and less prone to interferences from the sample matrix. Traditionally, this has entailed the use of selective GC detectors, halogen electron capture detectors (HECD) for Cl and Br, nitrogen–phosphorus detectors (NPD) for N and P, and flame photometric detectors (FPD) for P and S or HPLC with post-column derivatization and fluorescence detection (for carbamates, for example).

There is a growing trend to reduce the amount of solvent used for a given analysis. This is due in part to the relatively high cost associated with disposal of waste solvents and safety considerations in reducing the exposure of laboratory personnel to extraction solvents. Presently, disposal costs for solvents and contaminated solutions (water and solvents) equal or exceed the original purchase price of the solvents. Thus, strategies which limit or reduce the amount of extraction solvents are to be encouraged. An example of this overall trend is the reduction in size of the GPC column reported (21).

On this note, there is a trend to miniaturize the apparatus used for the sample extraction procedure, permitting smaller sample sizes, and to demand higher performance instrumental detection of the analytes at lower levels. A micro method can reduce solvent and reagent use by 1/10 to 1/100 over a conventional method. For example, a micro method may use 1–5 g of sample instead of 20–100 g required for the macro method. Philosophically, it can be argued that residue analytical programs should provide a net benefit to human and environmental health. Thus, it is attractive to use micro methods where appropriate, in view of the much lower requirements for solvents and reagents. Furthermore, the micro methods tend to be generally cheaper, faster, and less labor-intensive than macro methods. They also tend to be more amenable to automation using on-line techniques compared to the off-line macro methods.

For instrumental analyses, there is widespread acceptance of the use of high resolution GC using capillary columns with specific detectors or mass spectrometers. The use of capillary GC columns has virtually replaced the use of packed columns for the determination of herbicides in biota samples. This is largely due to the greater resolving power of the capillary columns with corresponding lower detection limits. These columns are now manufactured with high reproducibility and are relatively rugged, less prone to active sites, and give good reproducibility of retention times. Positive confirmation can thus be based on a match of the combination of the retention time and mass spectrum of the unknown with those of an authentic standard. MS is particularly well suited to the confirmation of target analytes when operated in SIM mode and remains the detector of choice for the tentative identification of unknown herbicides when operated in the full-scan mode.

In the last five years, more ion-trap mass spectrometers were used in environmental laboratories. These systems can be used for MS/MS applications and are presently less expensive than the conventional quadrupole or sector tandem MS instruments. Consequently, it is anticipated that the utility of tandem MS will grow for the determination of herbicides in biota. This is based on the fact that tandem MS (MS-MS) can be a highly sensitive and selective method of detection which is well suited to analysis of crude extracts with little or no cleanup. These systems, however, are not likely to replace the advantages of selective and specific detectors such as atomic emission detectors (AED). Although MS (mass selection detector (MSD) and ion-trap) facilitates analysis of many different classes of pesticides in a single run, complementary studies can be performed, for example, using the atomic emission detector. AED has the advantage of limiting the detection to specific elements, and can thus be more sensitive for some applications.

There is still a lag in the utilization of HPLC methods compared to GC methods for the determination of herbicides in biota. However, as HPLC–MS systems continue to become more widely available and less expensive, applications in herbicide residue analyses are anticipated to become more commonly used. This area warrants further study to accelerate the applications, in general, to polar or thermally labile compounds.

Capillary electrophoresis is a new technique providing many advantages over traditional gas and liquid chromatographic separations, including higher resolution, shorter analysis times, and smaller sample volumes.

In recent years, there have been noted advances in the development of immunoassays for the rapid screening of herbicide residues. These methods, once optimized, can facilitate high sample throughput at relatively low costs compared to conventional approaches for screening of herbicide residues. Furthermore, the immunoassays kits are easily field portable. To date, however, development has been primarily for aqueous systems and little work has been done for the direct determination of herbicides in biota. Despite some attractive features, present applications are limited by problems associated with poor multiresidue performance, and inconsistent results linked to the lot-to-lot variability of such kits.

ACKNOWLEDGMENTS

Thanks to the following people for their contributions to this review: G. Bruns, Envirotest Laboratories; D. Forsyth, Canadian Wildlife Service; V. Martz, Agriculture and Agri-food Canada: J. Purdy, Ciba-Geigy Canada; A. Rowe, Hewlett-Packard; and C. Warfield, Pest Management Regulatory Agency (Canada).

BIBLIOGRAPHY

1. R. Grover, ed., *Environmental Chemistry of Herbicides,* vol. 1, CRC Press, Boca Raton, Fla., 1988.
2. R. Grover and A. J. Cessna, eds., *Environmental Chemistry of Herbicides,* Vol. 2, CRC Press, Boca Raton, Fla., 1991.
3. *National Environmental Strategy for Agriculture and Agri-Food,* Agriculture and Agri-Food Canada, Ottawa, 1995.
4. *The Importance of Wildlife to Canadians: Highlights of the 1991 Survey,* Canadian Wildlife Service, Environment Canada, Ottawa, ON, Cat. no. CW66-103/1993E, 1993.
5. R. A. Baumann, G. F. Ernst, J. T. A. Jansen, A. de Kok, P. D. A. Olthof, L. G. M. T. Tuinstra, W. Verwaal, P. van Zoonen, and F. H. Hernandez, *Fresenius J. Anal. Chem.* **339,** 357–364 (1991).
6. *Herbicide Handbook of the Weed Science Society of America,* 5th ed., Weed Science Society of America, Champaign, Ill., 1983.
7. L. H. Keith, *Environmental Sampling and Analysis: A Practical Guide,* Lewis Publishers, Boca Raton, Fla., 1991.
8. R. J. Mesley, W. D. Pocklington, and R. F. Walker, *Analyst* **116,** 975–990 (1991).
9. H. B. S. Conacher, *J. Assoc. Off. Anal. Chem.* **70,** 941–943 (1987).
10. S. M. Walters, "Clean-up of Samples" in T. Cairns and J. Sherma, eds., *Emerging Strategies for Pesticide Analysis,* CRC Press, Boca Raton, Fla., 1992.
11. Analytical Methods Committee, *Analyst* **112,** 199–204 (1988).
12. J.N. Miller, *Analyst* **116,** 3–14 (1991).
13. Joint FAO/WHO Food Standards Program, *Status of Codex Maximum Residue Limits for Pesticides in Food and Animal Feed,* Codex Committee on Pesticide Residues, 25th Session, Havana, Cuba, 1993.
14. C. L. Trichilo and R. D. Schmitt, *J. Assoc. Off. Anal. Chem.* **72,** 536–538 (1989).
15. Canada Food and Drug Regulations, *Maximum Residue Limits for Agricultural Chemicals,* Division 15, Table II, 1996.
16. *J. AOAC Int.* **77,** 161A–185A (1994)
17. W. Dejonckheere, W. Steurbaut, S. Drieghe, R. Verstraeten, and H. Braeckman, *J. AOAC Int.* **79,** 97–110 (1996).
18. *Japan Pesticide Information No. 54,* Third Update, Ministry of Agriculture, Forestry and Fisheries, Tokyo, Japan, 1996.
19. R. Kakagawa, H. Hirakawa, and T. Hori, *J. AOAC Int.* **78,** 921–929 (1995).
20. J. N. Seiber, "New Trends in Analytical Methods for Pesticide Residues in Foods," in B. G. Tweedy, H. T. Dishburger, L. G. Ballantine, J. McCarthy, and J. Murphy, eds., *Pesticide Residues and Food Safety: A Harvest of Viewpoints,* ACS Symposium Series #446, American Chemical Society, Washington, D.C., 1991.
21. J. A. Van Rijn and L. G. M. Th. Tuinstra, *J. Chromatogr.* **552,** 517–526 (1991).

General References

Analytical Methodology

Official Methods of Analysis, 16th ed., AOAL International, Gaithersburg, Md., 1995.

G. Zweig, ed., *Analytical Methods for Pesticides and Plant Growth Regulators,* vol. X, New and Updated Methods, Academic Press, New York, 1978.

G. Zweig, ed., *Analytical Methods for Pesticides and Plant Growth Regulators,* Vol. XI: Updated General Techniques and Additional Pesticides. Academic Press, New York, 1980.

G. Zweig and J. Sherma, eds., *Analytical Methods for Pesticides and Plant Growth Regulators,* vol. XIV: Modern Analytical Techniques, Academic Press, New York, 1986.

G. Zweig and J. Sherma, eds., *Analytical Methods for Pesticides and Plant Growth Regulators,* Vol. XV: Principles, Statistics, and Applications, Academic Press, New York, 1986.

J. Sherma, ed., *Analytical Methods for Pesticides and Plant Growth Regulators,* vol. XVI: Specific Applications, Academic Press, New York, 1988.

J. Sherma, *Anal. Chem.* **67,** 1R–20R (1995).

J. Sherma, *Anal. Chem.* **65,** 40R–54R (1993).

J. Sherma, *Anal. Chem.* **63,** 118R–130R (1991).

J. AOAC Int. **79,** 239–247 (1996).

J. AOAC Int. **78,** 210–213 (1995).

J. AOAC Int. **77,** 173–182 (1994).

J. E. Conaway, *J. Assoc. Off. Anal. Chem.* **74,** 715–717 (1991).

Herbicide Residue Occurrence in Foods

D. V. Reed, P. Lombardo, J. R. Wessel, J. A. Burke, and B. McMahan, *J. Assoc. Off. Anal. Chem.* **70,** 591–595 (1987).

J. A. T. Pennington and E. L. Gunderson, *J. Assoc. Off. Anal. Chem.* **70,** 772–782 (1987).

J. A. T. Pennington and E. L. Gunderson, *A History of the Food and Drug Administration's Total Diet Study, 1961 to 1987,* Appendices, PB87 151676, National Technical Information Service, Springfield, Va., 1987.

J. A. T. Pennington, S. G. Capar, C. H. Parfitt, and C. W. Edwards, *J. AOAC Int.* **79,** 163–170 (1996).

Anon., *J. AOAC Int.* **78,** 117A–142A (1995).

H. A. McLeod, D. C. Smith, and N. Bluman, *J. Food Safety* **2,** 141–164 (1980).

R. Frank, H. E. Braun, and B. D. Ripley, *J. Assoc. Off. Anal. Chem.* **70,** 1081–1086 (1987).

H. B. S. Conacher, R. A. Graham, W. H. Newsome, G. F. Graham, and P. Verdier, *J. Can. Inst. Food Sci. Technol.* **22,** 322–326 (1989).

E. Neidert, R. B. Trotman, and P. W. Saschenbrecker, *J. AOAC Int.* **77,** 18–24 (1994).

W. Dejonckheere, W. Steurbaut, S. Drieghe, R. Verstraeten, and H. Braeckman, *J. AOAC Int.* **79,** 97–110 (1996).

Properties and Toxicology of Herbicides

D. J. Ecobichon, "Toxic Effects of Pesticides," in C. D. Klaassen, ed., *Casarett and Doull's Toxicology: The Basic Science of Poisons,* McGraw-Hill, New York, 1995.

D. S. Saunders, and C Harper, "Pesticides" in A. W. Hayes, ed., *Principles and Methods of Toxicology,* Raven Press, New York, 1994.

S. A. Briggs, *Basic Guide to Pesticides: Their Characteristics and Hazards,* Hemisphere, Washington, D.C., 1992.

J. H. Montgomery, *Agrochemicals Desk Reference: Environmental Data,* Lewis, Chelsea, Mich., 1993.

C. R. Worthing and R. J. Hance, *The Pesticide Manual, A World Compendium,* 9th ed., British Crop Protection Council, Farnham U.K., 1991.

H. Kidd and D. Hartley, *The Pesticide Index, The Royal Society of Chemistry,* Old Working U.K., 1988.

HOUSEHOLD HAZARDOUS MATERIAL MANAGEMENT

JOAN D. SULZBERG
Science Applications International Corporation
McLean, Virginia

DAVID E. B. NIGHTINGALE
EMCON, Inc.
Olympia, Washington

HISTORY OF HAZARDOUS WASTE MANAGEMENT

Toxics policy in the United States has evolved dramatically from an initial concern with the safer management of hazardous waste to a heightened scrutiny of the pervasive use of toxic products, chemicals, and materials and their possible impact on human health and the environment.

In 1976, Congress passed the Resource Conservation and Recovery Act (RCRA), which directed the Environmental Protection Agency (EPA) to develop and implement a program to protect human health and the environment from improper solid and hazardous waste management practices. Congress designed RCRA to control the management of hazardous waste, from initial generation to ultimate disposal—from cradle to grave. Initially, RCRA's focus was on large-quantity waste generators, because they generated a major portion of the hazardous waste in the United States. In 1980, Congress included smaller waste generators, defined as those producing at least 1000 kg (2200 lbs) of hazardous waste in one month or batch. Then, in the 1984 amendments to RCRA, Congress directed the EPA to lower the regulatory threshold to 100 kg (220 lbs) per month or batch. In 1986, the EPA promulgated the revised regulations, which lowered the generation-rate limits. Small businesses that produced between 100 kg and 1000 kg were no longer exempt from hazardous waste legislation and became known as "small-quantity generators." Businesses that produced less than 100 kg per month of hazardous waste continued to be exempt from most hazardous waste requirements, and became known as conditionally exempt small-quantity generators (CESQGs). The revised regulations retained the unconditional exemption for hazardous waste from households.

Traditional hazardous waste management compliance and enforcement efforts have focused on easily identified point sources of pollution, such as large industrial outputs. Approximately 90% of the nation's hazardous waste is produced by large-quantity generators, but these generators constitute only approximately 2% of the companies now regulated by RCRA (1). The remaining 98% are small-quantity generators. Since the EPA has targeted large-point sources with intense hazardous waste compliance efforts, focus has now turned to nonpoint sources of pollution, such as agricultural runoff, storm water runoff, and household hazardous material (HHM).

CURRENT HOUSEHOLD HAZARDOUS MATERIAL ACTIVITY

Local attention to hazardous materials from homes began in the late 1970s and early 1980s when citizens in selected towns across the country realized that there were potentially negative impacts resulting from the use, storage, and disposal of HHM, and began requesting information. The year 1979 marked the beginning of what has become one of the most mature programs in the United States, the Seattle, Washington metropolitan area (King County) local hazardous waste program. At that time, King County received a waiver for secondary treatment of waste waters in coastal areas. As part of the waiver requirements, the sewage treatment plant was to enhance its industrial pretreatment and other activities to reduce the inflow of toxins, which pass through the system untreated and into receiving waters. In 1980, King County formed a consortium to address citizen concern regarding HHM. At that time, no other HHM efforts existed in the United States. In the mid–1980s, King County's HHM program expanded through numerous HHM collection programs held in a diversity of locations. By the late 1980s, the collection program had grown with such vigor that longer term options were needed. Washington state also required all counties to create and implement a local hazardous waste plan that included, among other components, HHM collection and education activities. This resulted in the development of two permanent HHM collection facilities in Seattle, and the King County Wastemobile, a roving permanent HHM collection center that was setup in various areas of the region for two weeks at a time throughout the year. The focus of HHM management in King County since the 1980s has included not only a collection component, but education and prevention components as well.

Paralleling the King County experience were the efforts in the local HHM collection scene. By the late 1980s, the national HHM collection era was in full bloom. The total number of HHM collection programs nationwide grew from two in 1980, to 31 in 1983, to 273 in 1986, 692 in 1989, and to 1649 in 1994 (2,3).

Today, attention to the management of HHM often stems from vocal citizens concerned about health and environmental impacts related to the use and disposal of HHM. State policies and programs for HHM include labeling laws, retailer permit fees, collection activities and centers, prevention programs, education activities, and deposit-refund mechanisms, among others. However, HHM is much more than a technical and environmental issue. It is a sociopolitical issue with technical and environmental components. As a result, underlying conflicts between science and human values persist when it comes to HHM. Citizens assume negative impacts are associated with the disposal of HHM, whereas scientists do not all agree on the negative impacts associated with the landfill disposal of HHM. In addition, other important issues related to HHM management include sewer and septic system disposal and solid waste handling safety issues. It is much easier to show a cause-and-effect relationship for HHM with the latter issues, but controversy does exist. This discord remains pervasive in the HHM policy arena today. Effective HHM managers must take all considerations into account. This article explores many issues related to HHM, including terminology, definitions, hazards and risks, current management practices, and future practices.

BASICS ISSUES PERTAINING TO HOUSEHOLD HAZARDOUS MATERIAL

Terminology

There is no clear-cut definition for products and materials used in the home that contain hazardous substances. Several terms describe such materials. In the literature and among solid and hazardous waste practitioners, "household hazardous products," "household hazardous substances," "household hazardous materials," "household hazardous wastes," and "toxic household substances" are commonly used terms, although *household hazardous waste* (usually abbreviated as HHW) is by far the most widespread term. Since each term has a certain connotation associated with it, this article will use the term household hazardous material (HHM) which reflects the nature of various issues and refers to hazard-containing products including those (*1*) manufactured by industry, (*2*) sold in retail establishments, (*3*) used in the home, and (*4*) whose residuals are disposed.

Definition

RCRA specifically excludes a number of wastes from being considered hazardous. Subtitle C, Section 261.4(b)(1), unconditionally exempts household wastes from being designated by the EPA as hazardous wastes even when transported, stored, treated, disposed, recovered, reused, or accumulated in quantities that would otherwise be regulated. The EPA interprets this exclusion to include materials that are generated by individuals on the premises of a temporary or permanent residence and composed primarily of materials found in the wastes generated by consumers in their homes, even if the waste exhibits one of the four characteristics of hazardous waste—ignitability, corrosivity, reactivity, or toxicity. A general definition of HMM is any material that

1. contains a substance that is regulated by the EPA under RCRA or any material that fails one of the characteristic hazardous tests,
2. is stored at home, and
3. is typically discarded in the municipal solid waste (MSW) stream.

In addition to this solid waste definition, parallel definitions are used pertaining to these same household materials, which are often disposed to wastewater treatment systems, surface water, the ground, and the air.

Typical Household Hazardous Material Generation

Household hazardous material includes, but is not limited to, household cleaners, personal care products, automotive products, home maintenance and improvement products, household batteries, and yard maintenance products. Although the composition of HHM varies from community to community, from state to state, and from rural to urban areas, a typical breakdown, by weight, is shown in Table 1. Examples of each category of HHM are listed in Table 2. Depending upon whether the product is normally consumed during use, HHM can be further segregated into two categories. Materials that are typically not consumed include motor oil and other lubricants, antifreeze, batteries, light bulbs, and paint thinners/solvents. After the normal life and/or usage of the material, a residual will always exist. This is not the case for HHM which can be used completely, such as household cleaners, paint products, and insecticides.

Environmental Protection Agency Hazard Classifications

The most common system for classifying HHM is the EPA's four hazard characteristics. Although this is the most common method, some states use other characteristics, such as persistence and toxicity shown by bioassay testing. To illustrate what

Table 1. Household Hazardous Material, by Category and Percentage[a]

HHM Category	Percentage (by weight)
Home maintenance	37
Household batteries	19
Personal care	12
Household cleaners	12
Automobile maintenance	11
Yard maintenance	4
Other	5

[a] Ref. 4.

Table 2. Examples of Household Hazardous Material by Category

HHM Category	Examples
Home maintenance	Oil-based paints, latex (water-based) paints, rust proofing, turpentine and paint thinner, paint and varnish remover, wood preservatives, room fresheners, stains and varnishes, roofing tar
Household batteries	General purpose batteries for items such as radios, flashlights, and toys; rechargeable batteries for items such as power tools, cameras, and calculators; and specialized batteries for items such as hearing aids and watches
Personal care	Nail polish, nail polish remover, hair products, cosmetics, cologne, and after-shave
Household cleaners	Toilet, oven, glass, drain, rug and upholstery cleaners, bleach, metal polishes, and ammonia products
Automobile maintenance	Motor oil, brake fluid, transmission fluid, antifreeze, gasoline, windshield-washer solution concentrate, engine degreaser, carburetor cleaner, car wax, chrome polish, and automobile batteries
Yard maintenance	Fungicides, insecticides, herbicides, molluscicides, and rodenticides
Other	Asbestos, flea collars, felt-tip markers, barbecue lighter fluid, mercury from thermometers and thermostats, swimming pool supplies, ammunition, gun cleaning solvent, hobby and art supplies, and photographic chemicals

makes a particular household material hazardous, examples match specific EPA hazard characteristics with specific product types.

Ignitability. Ignitable materials can be easily set on fire. According to Title 40 Code of Federal Regulation (CFR) 261.21, ignitables include (*1*) liquids with a flashpoint of less than 140 degrees Fahrenheit, (*2*) nonliquids liable to cause fire through friction or spontaneous chemical change, (*3*) oxidizers, and (*4*) ignitable compressed gas. In addition to the obvious hazards created from fire, the heat and toxic smoke of burning materials can spread harmful combustion products over a large area. Aerosol propellants such as propane in hair spray are identified as ignitable. Mineral spirits found in paint thinner or varnish are ignitable. Gasoline is denoted as extremely ignitable. Petroleum distillates in floor cleaners and air fresheners, acetone in nail polish remover, lighter fluid, turpentine, and paint stripper or finish remover containing benzene are all defined as highly ignitable.

Corrosivity. Corrosives are characterized as (*1*) aqueous wastes that exhibit a pH less than 3.0 (acids) or greater than 12.5 (bases) or (*2*) liquid substances that dissolve steel and/or burn or destroy living tissue. Drain opener and oven cleaner may contain sodium hydroxide or potassium hydroxide (bases), which are corrosives. Sulfuric acid in automotive batteries is characterized as a corrosive; it is likely to cause severe skin burns and blindness when in direct contact with living tissue or eyes. Other corrosive materials include naval jelly, lye, and photo fixer.

Reactivity. Reactive products are characterized by one or more or the following: (*1*) instability and readiness to undergo violent change; (*2*) violent reactions when mixed with water; (*3*) formation of potentially explosive mixtures when mixed with water; (*4*) generation of toxic fumes when mixed with water; (*5*) ease of detonation or explosive reaction when exposed to pressure or heat; and/or (*6*) ease of detonation or explosive decomposition or reaction at standard temperature and pressure. Examples of a reactive material are ammunition, calcium carbide, ether crystals, picric acid crystals, black powder, fireworks, and TNT.

Toxicity. A toxic or poisonous substance poses a threat to human health, domestic livestock, pets, and/or wildlife through consumption, inhalation, or by contact with and absorption through the skin. For example, many paint strippers and finish removers contain phenols, which cause central nervous system depression, severely affect the circulatory system, and are suspected carcinogens. Toxic metals such as mercury from used batteries or broken fluorescent lamps can damage the brain and nervous system, while cadmium can damage both kidneys and lungs. Some household insecticides contain organophosphates, which affect the nervous system and exposure can cause headaches, dizziness, twitching, and nausea. Significant exposure to naphthalene, contained in mothballs, may result in liver damage, and prolonged exposure may lead to cataract formation. Antifreeze is also toxic and can be fatal if ingested, especially by pets.

Estimated Quantities of Household Hazardous Material in Waste Streams

A common practice to determine the amount of HHM in a municipal solid waste (MSW) stream is to conduct a waste composition study or *garbage sort*. This procedure entails the physical separation of waste stream components to calculate the percentages of each by weight. State environmental protection agencies, local solid waste agencies, universities, and private researchers have conducted waste composition studies to determine HHM content in the MSW stream. Amounts vary from as low as 0.1% to as high as 3.4%. Discrepancies exist due to different methodologies used in calculating percentages, weighing methods, and locations. For instance, some studies may include empty containers and used oil filters, whereas others may not. In addition, these percentages are derived only from MSW and do not take into account the HHM that is disposed of in wastewater systems or directly to the air or ground. King County, Washington estimated that 35 to 40% of HHM is likely to be disposed of outside of the MSW system, which indicates a significant addition for the total HHM residuals generated and discarded.

One percent is typically used to estimate the quantity of HHM residuals in the MSW stream. Nonetheless, the composition of HHM varies by community, season, degree of urbanization, and socioeconomic class. A large proportion of HHM from low income homes are automotive products; a significant amount of HHM from middle income homes are home improvement products such as paints, stains, and varnishes; and the HHM of affluent homes contain more lawn and garden-related products. Also, contrary to what was intended in one instance,

Table 3. Municipal Solid Waste and Household Hazardous Material in the United States[a]

	1960	1986	1993	2000 (projected)
Million tons of MSW	88	160	207	218
HHM in MSW (1%)	0.88	1.60	2.07	2.18
Total HHM generated/disposed	1.47	2.67	3.45	3.63

[a] Data from EPA (5,6) and OTA (4).

the solid waste discarded after a well-publicized HHM collection day contained more than twice as much hazardous material, by weight, than the solid waste that had been discarded before the collection day. Since residents became aware of the hazards of the materials stored in the home but missed the collection day opportunity, they decided to rid themselves and their homes of the hazardous materials in the conventional manner, in the garbage. The amount of HHM in the home also depends on the size of the home, the number of vehicles, income level, and on the presence of a basement and garage. In addition, rural communities tend to generate more household and vehicle-maintenance wastes.

In 1993, 207 million tons, or 1.9 kg (4.4 lb) per person per day of MSW were discarded in the United States. Using the estimate for HHM in the solid waste stream of 1%, HHM totaled 2.07 million tons in 1993 and a projected 2.18 million tons by 2000. Table 3 shows the amounts of MSW and HHM in MSW using a 1% MSW estimate, and the total estimated HHM residuals generated/disposed is shown in the last row, assuming 60% of the total is in MSW. These estimates are projected to the year 2000 in the United States as well as back to 1960.

Household Hazardous Material In Wastewater Systems

In addition to the typical deposit of HHM into the MSW stream, it is also improperly disposed of in wastewater. Households introduce HHM into these systems by everyday use of consumer products and through direct disposal of HHM. While some consumer products are not usually considered HHM, together these products are a significant source of pollutants. Since the 1970s various researchers have investigated these issues. Most have focused on characterizing the quantities and impacts on wastewater systems—both treatment plants and smaller septic systems.

The distribution of sources of HHM that enter the wastewater stream from households can be segregated into seven categories: toilet, garbage disposal and kitchen sink, dishwasher, laundry, bath and shower, utility sink, and bathroom sink (7). The priority pollutants in household wastewater (7) are the organics benzene, phenol, 2,4,6-trichlorophenol, 2-chlorophenol, 1,2-dichlorobenzene, 1,4-dichlorobenzene, 1,1,1-trichloroethane, naphthalene, toluene, diethylphthalate, dimethylphthalate, trichloroethylene, aldrin, and dieldrin and the inorganics (metals) arsenic, cadmium, chromium, copper, lead, mercury, zinc, antimony, and silver.

The most frequent contribution of toxic compounds to household wastewaters are from cleaning products containing solvents and cosmetics (8), and studies indicate that as much as 7.5% of municipal treatment influent is composed of residential organic chemicals, and the residential contribution of heavy metals is approximately 20% of the total metal influent load (9). The Office of Technology Assessment estimates that 15% of regulated toxic pollutants entering treatment plants are from households (8). Specific sources for these pollutants include pesticides, drain cleaners, toilet bowl cleaners, degreasers, detergents, gasoline, and motor oil.

A 1979 National Sciences Foundation study also found significant contributions of heavy metals to wastewater systems from household sources (10). Twenty-eight sources of common household products, from soaps to bleaches and toilet bowl cleaners, contributed seven metals. Similarly, one study (11) found the residential contribution to the wastewater stream of nine metals. Results of these are shown in Table 4. These contribution factors can be compared to systems in other countries by the fact that indoor water use in the United States is typically 40 to 60 gallons per day.

In addition to the metals, studies have found significant quantities of organic chemicals in domestic wastewater. These values, shown in Table 5, are from an EPA study of Cincinnati, St. Louis, Atlanta, and Hartford, and from a parallel Seattle study. The data shown from both studies are mean values.

The concentration of metals and organic compounds in Tables 4 and 5 may seem small, typically in the parts per million to parts per billion range. However, wastewater is 99.9% water and 0.02 to 0.03% suspended solids by weight. Small quantities of contaminants can significantly degrade water quality, making it unsuitable for human or agricultural uses, and/or significantly degrade waters and sediments to which it is discharged. The levels shown in the tables above are also statistical means, which smoothes over the presence of concentrated peaks, or slugs, of certain constituents that are also found at much higher levels in practice.

CONCERNS, HAZARDS, AND RISKS OF HOUSEHOLD HAZARDOUS MATERIAL

Household hazardous material presents hazards to human health and the environment because many chemical components of HHM do not lose their hazardous characteristics when discarded; they are simply transferred to another location or medium. As HHM is not represented by one material, product, or product type, it is difficult to assess the overall risk associated with HHM. Certain products or compounds present hazard and risk and arouse concern in one or more routes of use, storage, transport, and disposal.

Methods of conveying product warnings on certain HHM to consumers include product labels, pamphlets, radio announcements, and television announcements. Long, written messages include more information but are not necessarily more persuasive than shorter messages, because written messages are reviewable by the consumer at his/her convenience. Frequent users of a product tend to pay less attention to warning labels

Table 4. Selected Heavy Metals Contributed to Domestic Wastewater by Households

Heavy Metal	Residential Contribution to the Wastewater Stream, %[a]	Residential Contribution to the Wastewater Stream, μg/capita/d[b]
Arsenic	86	Not available
Cadmium	12	81
Chromium	20	662
Copper	47	510
Lead	14	272
Mercury	50	7.5
Nickel	34	23,449
Silver	9	Not available
Zinc	59	738

[a] Ref. 11.
[b] Ref. 10.

Table 5. Selected Organic Chemicals Contributed to Domestic Wastewater, ppb[a]

Organic Compound	Seattle	EPA
Phenol	14.8	5.8
Naphthalene	3.8	2.1
Phenathrene	0.9	NR[b]
Diethyl phthalate	10.4	9.8
Di-*n*-butyl phthalate	34.5	9.0
Butyl benzyl phthalate	36.4	6.8
Di-octyl phthalate[c]	20.4	6.8
Methylene chloride	77.8	NV[d]
Chloroform	4.2	3.0
Trichloroethylene	4.3	0.4
Tetrachloroethylene	125.0	6.3
Benzene	5.0	0.2
Ethyl benzene	1.0	0.4
Toluene	3.1	2.6

[a] Ref. 12.
[b] Detected, but mean not reported because of low detection frequency.
[c] Includes bis(2-ethylhexyl) phthalate.
[d] No value reported because of unreliability of analyses.

and signs, perceive less danger associated with product hazards, and are more likely to engage in risk-taking behavior with a particular product than infrequent users of the same product (13–16). In addition, a phenomenon known as the *familiarity effect* indicates that as consumers learn more about a product, they (*1*) become less concerned with product safety information and (*2*) feel more comfortable with using the product (17,18).

In research that explored the nature of perceived risk for hazards associated with consumer products containing chemicals, a study of 47 household items using 15 evaluative scales was designed and carried out (19). Table 6 summarizes this information. It was found that the higher the perceived accident danger, the more information and federal labeling the consumer desired.

Overall, implications of the research study indicate that product categories such as baby items (eg, shampoo and laundry soaps) and regularly used personal care items were perceived to be less risky items. High risk, low benefit items, such as mothballs, hair dyes, and hair removers, were labeled as potential *endangered species,* and it was hypothesized (19) that such items were possibly subject to future controversy.

Table 7 summarizes the number of exposures to each type of HHM, by year, and Table 8 summarizes the outcomes of selected exposures of HHM for 1992.

CONCLUSIONS

Management Trends

Life Cycle and Regulatory Control. There continues to be an increase in the collection of HHM residuals across the United States and Canada. Communities often begin with mobile or temporary collection events and then move to building permanent fixed collection facilities. These facilities allow for increased reuse and recycling of HHM and provide increased service levels at a reduced unit cost.

Despite the benefits of increased availability of collection services, there exists a need for increased focus on front-end management methods such as pollution prevention to reduce the amount and/or toxicity and persistence of HHM manufactured, used, and disposed. Similar to the regulation of industrial hazardous waste, the focus of HHM management programs has begun with end-of-the-pipe methods, and has shifted to pollution prevention or source reduction.

According to the EPAs hierarchy of waste management practices—reduce, reuse, recycle, dispose—the collection of HHM for disposal is the least desired practice. The most effective method of controlling HHM is to prevent it from being produced in the first place. This type of activity takes place in preconsumer phase programs. Realizing this, after years of a collection focus, HHM policy and program leaders have begun to focus pre-consumer related practices.

With respect to the preconsumer phase, performance standards and technology-based standards, including product reformulations and product bans, are both widespread and popular techniques for eliminating the hazardous components of consumer products. These approaches remove the hazards of products but offer no incentive to the manufacturers to develop cost-effective approaches to pollution prevention. These control approaches are potentially expensive for the manufacturer and consumer and are counterproductive to usual industrial practices. The costs of controlling effluent and emissions often varies from facility to facility, and, because each production line and manufacturing facility is unique, there exists no one technology that could work across all facilities. Manufacturers' research and development efforts that might lead to attractive

Table 6. Consumer Perceptions of Certain Household Hazardous Material[a]

Product Category	Definition/Consumer Perception	Examples
High accident danger	Products perceived to pose hazards most likely to cause very serious harm or death; accidents are unpreventable, uncontrollable, and unknown to those exposed to them	Oven cleaner, hair dye, ammonia, stain remover, bleach, mothballs
Low accident dangers	Products are perceived to pose hazards that are relatively unlikely to cause very serious harm or death; Accidents are preventable, controllable, and known to those who are exposed to them	Toothpaste, pure soap, baby shampoo, mouthwash, softsheets
High risk	Products perceived to pose relatively high levels of risk to children, adults, the elderly, and the environment; accidents are perceived to be unpreventable, and have serious, uncontrollable, adverse effects	Oven cleaner, ammonia, bleach, adult shampoo, hair dye, mothballs
Low risk	Products are perceived to pose relatively low levels of risk to adults, children, the elderly, and the environment; accidents are perceived to be preventable and to have mild, controllable adverse effects, if any	Pure soap, toothpaste, lotions, dryer softener sheets, mouthwash
High benefit	Products are perceived to be both necessary and beneficial	Aspirin, deodorant, home permanent, laundry powder, toothpaste
Low benefit	Products are perceived to be not very beneficial, luxury items	Hair remover, hair dye, perfume, lemon scent, phosphates

[a] Ref. 19.

Table 7. Household Hazardous Material Exposures in the Home, 1986–1992[a,b]

Type of HHM	1986	1987	1988	1989	1990	1991	1992
Home maintenance	31,468	37,141	42,789	49,296	56,990	61,665	63,469
Household batteries	4,487	4,976	5,664	6,254	7,112	8,334	8,912
Personal care	80,214	94,349	110,546	130,207	140,039	153,424	153,721
Household cleaners	111,839	122,866	147,899	170,463	190,394	203,444	207,557
Auto maintenance	4,525	5,875	7,008	8,583	9,525	10,786	11,566
Yard maintenance	56,215	59,144	65,106	74,658	81,828	89,953	88,480
Other	168,871	175,650	202,820	235,866	254,365	259,878	263,024

[a] Refs. 20–26.
[b] An exposure indicated that a call was received at a Poison Control Center, but did not indicate a significant medical consequence.

Table 8. Outcome of Household Hazardous Material Exposure in the Home, 1992[a]

		Outcome[b]				
HHM	No. of Exposures[c]	None	Minor	Moderate	Major	Death
Home maintenance	63,469	13,587	14,992	1,434	54	8
Household batteries	8,912	2,921	2,447	261	13	1
Personal care	153,721	42,623	28,185	1,044	58	6
Household cleaners	207,557	53,351	63,162	4,484	154	19
Auto maintenance	11,566	2,883	4,263	422	73	13
Yard maintenance	88,480	23,570	17,642	1,423	120	23
Other	263,024	52,415	79,391	10,208	1,089	144

[a] Ref. 26.
[b] Minor = self-limited exposure with no treatment required; moderate = exposure that requires medical treatment and typically lasts < 24 h; and major = life threatening or induced, permanent damage.
[c] Total number of exposures did not equal the total number of outcomes. Only exposures diagnosed as having a significant medical consequence *and* were followed-up by Poison Control Center personnel with respect to consequence were tabulated in the outcome section. The ability to follow-up on an exposure was hindered by refusal, change of phone, change of address, or other reasons.

alternative products can be dampened by the need to adhere to inflexible standards. On the other hand, administrative costs are considerably lower for command-and-control approaches, as compared to market-based approaches.

Market-based approaches to inhibiting HHM purchases in the preconsumer phase fall not only to the manufacturer, but to the transporter, wholesaler, and retailer, and ultimately the consumer. This spreads out the cost burden felt by those depending on household hazardous materials for livelihood. Battery manufacturers have reduced or eliminated mercury levels in their products on their own. They internalized the cost of upgrading their manufacturing processes and now competitively advertise the virtues of their improved products.

Although consumers are the target of HHM collections and education outreach programs, there are usually only internal incentives to participate. And although the burden of hazardous materials in the home are felt heavily by consumers, not enough is typically done to encourage ecologically desirable behavior. Deposit refund mechanisms and product excise taxes are two methods of incorporating incentives into consumer behavior, but these practices are not widespread for most types of HHM.

When devising alternative means to alleviate the effects of HHM, it is necessary to focus on the sector or stage of the product life that can avoid manufacture or use at the lowest cost. After determining what sector this is, incentives should be provided to entice the desired behavior. Since the most costly component of an HHM program is ultimate disposal, it should be avoided through pollution prevention, reuse, and recycling when possible. At the same time, the societal cost of a product and its disposal must be conveyed to the consumer at the point of purchase. This way when the purchase and consumption of HHM occurs, it will reflect the product's true cost, including disposal.

Sociological Implications. Management of HHM is popular with a small portion of the population. Because households are ubiquitous and not easily reached by new rules or regulations, voluntary programs are the primary approach. Household hazardous material managers must use more social and psychological science to study and change the way consumers buy, use, and dispose of products. The difficulty is not only in developing methodologies to accomplish this but also in developing indices to measure success. The first step is to gain baseline information that represents the region prior to program implementation. Then, appropriate measurement of change needs to be assessed after the implementation of the program. Lacking this, a program cannot show what progress has been made or if goals have been achieved toward enhancing environmental or public-health conditions.

Program sponsors need to measure change over time and try to influence the behavior of consumers and manufacturers. This is a formidable challenge. Many manufacturers are reluctant to disclose the actual ingredients in their household products. This makes it more difficult to assess the relative environmental efficacy and levels of health threat between equally effective consumer products. On the other hand, any business with a proprietary formula or ingredient should not be forced to reveal its competitive formulation. Nonetheless, environmentally benign products and advertising have great appeal in certain marketplaces.

Traditional measurements of success with HHM programs has typically begun with the number of households that respond to the media campaign designed to get them to bring in their unwanted or unusable HHM to a collection sits. For HHM, tons collected, percentage of population served, cost per pound, and cost per participant are easy to track and compare over time. But if there is no indication of whether these numbers achieve the objective of the program, then an opportunity is lost and unnecessary costs may be incurred.

To continue forward, program managers need to be aware of how progress can be measured. There must be a demonstration that there is a positive return for the resources spent. Therefore, critical positive self-evaluation, appropriate goal setting, and communication is necessary. This helps focus resources on the most valuable tasks and insures that the program closes the loop between resource allocation and program implementation.

Intersectoral and International Household Hazardous Material Management

The majority of HHM management programs in the United States have been initiated and sponsored by state and local governments. There is, however, an increasing involvement by private and nonprofit enterprises in the management of HHM. This is usually accomplished through a cooperative partnership with the original government sponsors. Some large corporations have sponsored or provided resources for HHM management activities. Many waste management firms and many local and international oil companies, auto parts stores, and automotive service firms have sponsored collection centers for used oil, oil filters, and/or antifreeze. At the same time, local YMCA and Salvation Army organizations, among other public and private organizations, have used, processed, or sold paint or paint-related items. The paint industry and local distributors and formulators also have been integral in finding creative and innovative ways to recycle unused paints.

In Canada there is a strong trend toward close working ties between municipalities and interested businesses. This has included cooperative efforts to develop product-stewardship programs, sponsoring of HHM collection activities, educational efforts, and finding other areas of common interest in HHM programs.

The HHM management system in North America has been influential in other countries. Representatives from all parts of the globe have visited active programs in the United States and Canada, including representatives from Australia, China, Austria, Bermuda, Chile, Belgium, Great Britain, Hong Kong, Germany, and Tunisia. Great Britain, Belgium, Australia, and Austria have begun or are expanding their HHM management programs in certain localities or regions for many of the same reasons that spawned interest in North America. This continuation of cross-pollination of program techniques and information across the oceans has been a feature of environmental management since the dawning of environmental awareness in the 1960s.

Expanded Scope of Programs

Just as HHM is typically not regulated as a hazardous waste, nonhouseholds that generate hazardous wastes below a certain amount are considered conditionally exempt small quantity generators (CESQGs). The nature of the wastes and byproducts from this group of businesses, schools, agencies, and nonhousehold generators is quite similar to many HHM wastes generated by households. Many collection and educational programs for HHM are starting to include CESQG waste management. In 1986, Washington state coined the term *moderate risk waste,* which combined into one term HHM and CESQG wastes. The EPA has recognized this trend toward expanding the scope of these programs. From its inception through the ninth meeting in 1994, the EPA's annual conference on the subject was entitled the National Household Hazardous Waste Conference. At the 1995 conference, the name was changed to the Hazardous Materials Conference on Household, Small Business and Universal Waste. Now, CESQG is a term that is commonly used in affiliation with HHM.

Similarly, as National Pollutant Discharge Elimination System (NPDES) effluent and stormwater permits, biosolids program, wellhead protection programs, coastal zone management plans, and other state and local nonpoint source water-quality-related initiatives and systems continue to develop, HHM management will necessarily become more integrated. Ultimate integration may occur as a watershed-based environmental and public health program.

Research Needs

Despite the emphasis on HHM hazards, risks, and management programs, questions still remain: What is the risk of HHM relative to other environmental and social issues, and how can this risk be measured accurately? Studies that have investigated HHM in the disposal (or post-consumer) phase found that little risk to human health could be *directly* correlated to HHM disposal in landfills. It may, in fact, be a poor use of research efforts to attempt to prove or disprove HHM problems in landfills given all the confounding factors involved. Although it is the nature of science to continuously try to prove or disprove a hypothesis, whether or not HHM poses a risk to human health and the environment is not the issue at the heart of the policy matter. Anecdotal accidents and even deaths due to HHM exposure are in our common consci-entiousness. People are troubled with the nature of chemicals in household products and demand that state, provincial, and local agencies act responsibly to provide information and outlets for alternative management techniques.

There are no large-scale qualitative studies that examine the behavior of consumers, localities, industry, and other interest groups with a stake in the HHM management process. Nor are there large-scale, cross-media risk assessments for HHM. Specific qualitative research should focus on (*1*) local implementors and industry, (*2*) socioeconomic characteristics of consumers concerned with HHM, (*3*) relating short- and long-term success levels of state policies and programs, and (*4*) addressing impacts of specific legislation and regulation (eg, the Superfund) that have implications associated with the management of HHM. Regarding quantitative studies, there is a need for research efforts to focus on (*1*) hazards and risks associated with transportation of HHM by sanitation workers, (*2*) processing accidents that occur when HHM is mixed with MSW, and (*3*) the relative risk level associated with the accumulation of HHM at collection events and facilities versus the no-collection option.

BIBLIOGRAPHY

1. U.S. Environmental Protection Agency, EPA/600/2-84-107, 1984.
2. *Household Hazar. Waste Manage. News,* **5**(19), 7 (1993).
3. *Household Hazar. Waste Manage. News,* **6**(24), 2 (1995).
4. Office of Technology Assessment, *Facing America's Trash: What Next For Municipal Solid Waste?,* Government Printing Office, Washington, D.C., 1989.
5. Environmental Protection Agency, EPA/530-S-92-019, Office of Solid Waste and Emergency Response, Government Printing Office, Washington, D.C., 1992.
6. Environmental Protection Agency, EPA/530-R-94-042, Office of Solid Waste and Emergency Response, Government Printing Office, Washington, D.C., 1993.
7. S. Hathaway, Municipal Environmental Research Laboratory, PB-110942, Cincinnati, 1980.
8. General Accounting Office, GAO/RCED/-92-40, Government Printing Office, Washington, D.C., 1991.
9. Environmental Protection Agency, EPA 430/09-90-016, Office of Water, Washington, D.C., 1990.
10. F. Gurnham, B. Rose, H. Ritchie, W. Fetherston, and A. Smith, *Control of Heavy Metal Content of Municipal Wastewater Sludges,* National Science Foundation, Washington, D.C., 1979.
11. J. Burnam II, H. McDonald, W. Kido, and M. James, *Water Environ. Tech.* **6**(1), 44 (1994).
12. D. Galvin, in *Proceedings of the Third United States Environmental Protection Agency Conference on Household Hazardous Waste Management,* Waste Watch Center, Andover, 1988.
13. S. Godfrey and K. Laughery, *Proc. Human Factors Soc.* (1984).
14. E. Karnes, S. Leonard, and G. Rachwal, *Proc. Human Factors Soc.,* (1986).
15. G. Goldhaber and M. deTurck, *Forensic Rep.* **1**(3), 193 (1988).
16. M. deTurck and G. Goldhaber, *J. Consumer Affairs,* **23**(1), 111 (1989).
17. G. Goldhaber and M. deTurck, *J. Product Liability,* **2,** 29 (1988).
18. G. Goldhaber and M. deTurck, *Forensic Rep.,* **1**(4), 281 (1988).
19. N. Neil, P. Slovic, and P. Hakkinen, *Mapping Consumer Perceptions of Risk,* Chemical Manufacturers Association, 1993.
20. T. Litovitz, T. Martin, and B. Schmitz, *Am. J. Emergency Med.,* 5(5), 405 (1987).
21. T. Litovitz, B. Schmitz, N. Matyunas, and T. Martin, *Am. J. Emergency Med.,* **6**(5), 479 (1988).
22. T. Litovitz, B. Schmitz, and K. Holm, *Am. J. Emergency Med.,* **7**(5), 495 (1989).
23. T. Litovitz, B. Schmitz, and K. Bailey, *Am. J. Emergency Med.,* **8**(5), 394 (1990).
24. T. Litovitz, K. Bailey, B. Schmitz, K. Holm, and W. Klein-Schwartz, *Am. J. Emergency Med.,* **9**(5), 461 (1991).
25. T. Litovitz, K. Holm, K. Bailey, and B. Schmitz, *Am. J. Emergency Med.,* **9**(5), 452 (1992).
26. T. Litovitz, K. Holm, C. Clancey, B. Schmitz, L. Clark, and G. Odera, *Am. J. Emergency Med.,* **10**(5), 494 (1993).

HPTLC. See THIN-LAYER CHROMATOGRAPHY.

HYDROCARBONS IN THE ATMOSPHERE

IAN BARNES
Bergische Universität
Wuppertal, Germany

Hydrocarbons are ubiquitous components of the atmosphere and many have occurred naturally for thousands of years. However, over the last 100 years increasing human activity has progressively lead to the emission of anthropogenic hydrocarbons which are continually adding to the natural background hydrocarbon burden creating a number of serious atmospheric problems. Many of the anthropogenic hydrocarbons emitted are directly hazardous to human health and can damage plants and building material. Others are oxidized, generally in the presence of nitrogen oxides (NO and NO_2), forming new and often more toxic secondary pollutants. These effects can be very local but many of the compounds, due to their volatility and persistence, can be transported over large distances. Apart from toxic effects hydrocarbons are also of great concern in global environmental issues such as stratospheric ozone depletion, photochemical formation of oxidants such as ozone at ground level, and climatic influences due to the greenhouse properties of many of the hydrocarbons or their oxidation products.

The troposphere is that region of the atmosphere which stretches from the earth's surface to the tropopause which constitutes the boundary to the stratosphere (10–12 km above the ground at middle latitudes). Within the troposphere the region from the ground to approximately 1–2 km is called the mixing layer or the atmospheric boundary layer (the height of this layer is highly variable and can sink during the night to 100 m above ground level). The region stretching from the atmospheric boundary layer to the tropopause is termed the free troposphere. The exchange time between the atmospheric boundary layer and the free troposphere is of the order of a few days. Transport of trace gases in the free troposphere is much faster than within the mixing layer, and consequently once trace gases have entered the free troposphere they can be transported very quickly over extremely large distances. Thus, the point of emission of the hydrocarbon is not always the area which will suffer from possible detrimental effects resulting from its photooxidation.

Because of the enormous diversity and number of hydrocarbons emitted to the atmosphere, the coverage of the subject is by no means comprehensive. The reader, if interested, is strongly encouraged to consult textbooks on atmospheric chemistry (1–3), reviews of hydrocarbon chemistry (4), and the various authoritative review articles cited throughout the article for more in-depth treatments of the subject.

DEFINITIONS AND UNITS

Definitions

By strict definition molecules with the empirical formula C_xH_y are *hydrocarbons*. However, this definition is too restrictive and excludes organic compounds containing other elements such as O, N, and S. There are several terms which are frequently used in the scientific literature to refer collectively to the organic compounds found in the atmosphere. The terms most often encountered are nonmethane hydrocarbons (NMHCs), nonmethane volatile organic compounds (NMVOCs), volatile organic compounds (VOCs), and reactive organic gases (ROGs).

The term NMHC embraces all hydrocarbons with the exception of methane; methane is excluded because of the very much larger concentration of this compound compared to other hydrocarbons which are only present at trace levels. NMHCs was the term most frequently encountered in the earlier literature on atmospheric science, but as the subject has developed it has largely been replaced by the term VOCs in recent years. In atmospheric science the term VOC is deliberately used very loosely and is generally interpreted as meaning any carbon-containing compound in the atmosphere, excluding elemental carbon, carbon monoxide, and carbon dioxide. In reality VOC is defined as an organic compound whose vapor pressure at 293 K is less than 101.3 kPa (760 torr) and greater than 0.13 kPa (1 torr); however, this definition if strictly adhered to would exclude many hydrocarbons which play pivotal roles in atmospheric chemistry. The term VOC also includes oxygenated, halogenated, and sulfur-containing hydrocarbons. The term ROG is often encountered in place of VOC. Nearly all hydrocarbons present in the atmosphere can be classified as ROG; exceptions include methane and the fully halogenated organic compounds which are oxidized very slowly in the troposphere and diffuse to the stratosphere. All hydrocarbons with an atmospheric lifetime of 1–2 years will, at least partially, reach the stratosphere. In this article the terms hydrocarbons and VOCs are used interchangeably as collective terms for all hydrocarbons found in the gas phase in the atmosphere, excluding methane.

Units

Several different units are used to express the concentration of chemicals such as hydrocarbons in the atmosphere. The atmospheric concentrations of hydrocarbons are frequently given as mass per unit volume: that is, mg/m^3 (10^{-3}g per cubic meter), μg/m^3 (10^{-6} g per cubic meter). Another common unit is the volume mixing ratio. The units most commonly used are %, percent (volume mixing ratio, 1 in 100); ‰, per thousand (volume mixing ratio, 1 in 1000); 1 ppmv, parts per million (volume mixing ratio, 1 in 10^6); 1 ppbv, parts per billion (volume mixing ratio, 1 in 10^9); and 1 pptv, parts per trillion (volume mixing ratio, 1 in 10^{12}). An atmospheric concentration of say 20 ppbv of a hydrocarbon, for example, means that in every 10^8 volumes of air only 2 volumes are due to the hydrocarbon. Hydrocarbons concentrations are frequently given as ppmC or ppbC, a variant of the volume mixing ratio. These correspond to the volume mixing ratio values multiplied by the number of carbon atoms in the hydrocarbon. For example, 2 ppbV propane is equivalent to (3C-atoms $\times$ 2) = 6 ppbC propane.

For species present at the sub-ppt level, such as radicals, it is customary to give their concentrations as the number of molecules, atoms, or free radicals present in a given volume of air, normally a cubic centimeter (cm^3). Taking again 20 ppbv as an example, from the definition of ppb this would correspond to $(20 \times 2.46 \times 10^{19} \times 10^{-9}) = 4.92 \times 10^{11}$ molecules cm^{-3} at 25°C and one atmosphere total pressure (2.46×10^{19} is the number of molecules per cm^3 in air at one atmosphere pressure and 25°C). Obviously, corrections must be made for different temperatures and pressures. The relationship between the mass concentration and the volume mixing ratio is given by:

$$(\text{mg/m}^3)_{P,T} = [\text{ppm}] \times (P \times MW)/(RT)$$

$$(\mu\text{g/m}^3)_{P,T} = [\text{ppb}] \times (P \times MW)/(RT)$$

$$(\text{ng/m}^3)_{P,T} = [\text{ppt}] \times (P \times MW)/(RT)$$

where P is the pressure in bar, MW is the relative molecular weight of the hydrocarbon, R is the gas constant (0.08314 bar mol^{-1} K^{-1}), and T is the absolute temperature in kelvin (K).

Each of the above units has its own particular merits for certain facets of atmospheric science. Efforts are presently being made to standardize the units but no standard form has yet been agreed on. Reference 1 should be consulted for a detailed discussion and examples of the use of the units.

CLASSES OF ATMOSPHERIC HYDROCARBONS

VOC emissions to the atmosphere are composed of many very different organic substances with very different structures and chemical activity (5,6). The simplest hydrocarbons are composed only of carbon and hydrogen and on account of their chemical behavior can be classified into three groups:

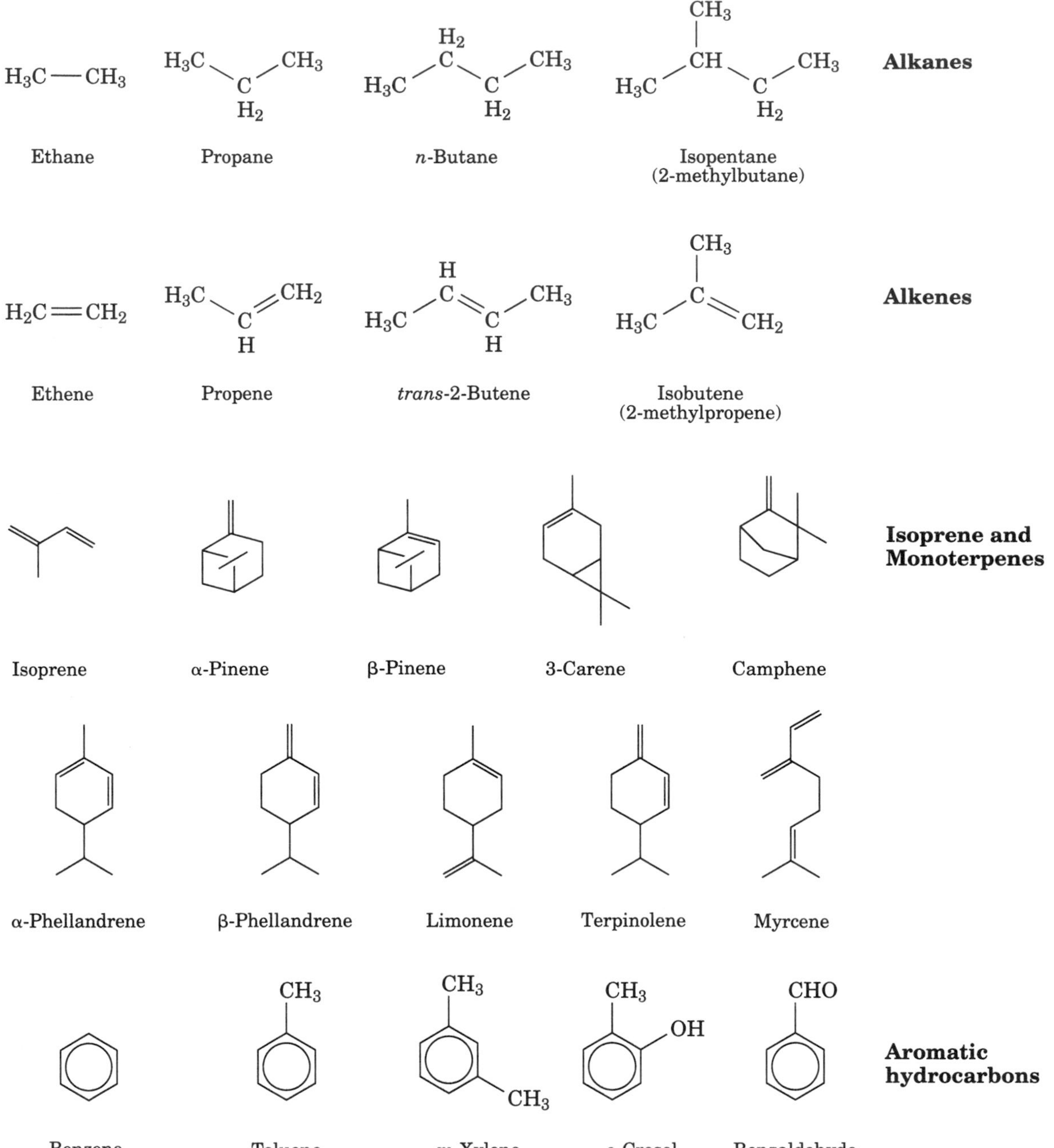

Figure 1. Chemical structures of some hydrocarbons commonly found in the atmosphere.

(*1*) alkanes, molecules containing only single C–C bonds and no functional groups; (*2*) alkenes and alkynes, unsaturated hydrocarbons with double C=C and triple C≡C bonds, respectively; and (*3*) aromatic hydrocarbons, a special case of unsaturated hydrocarbons in which three double bonds form a six-membered ring. Typical representatives of alkanes in the atmosphere are methane, ethane, propane, and the butanes; for the alkenes, ethene, propene, butene, isoprene, and the terpenes are of particular importance; and for the aromatic hydrocarbons, benzene, toluene, and the xylene isomers are important. Figure 1 shows examples of the chemical structures of alkanes, alkenes, oxygenated hydrocarbons (aldehydes, ketones), and some of the most frequently observed monoterpenes which include α- and β-pinene, 3-carene, camphene, α-and β-phellandrene, limonene, terpinolene, and myrcene. Terpenes are either open chain or cyclic hydrocarbons composed of two or more isoprene molecules, whereby the isoprene molecules are usually joined in the head–tail position. Depending on the number of isoprene building blocks the terpenes are classified as monoterpenes, sesquiterpenes, diterpenes, triterpenes, etc (7). Generally, the collective term terpenes is not only taken as meaning compounds with the formula $C_{10}H_{16}$ and molecular weight 136 but also the derivative alcohols and carbonyl compounds.

Each of the hydrogen atoms in a hydrocarbon can be replaced by function groups (or substituents) which contain atoms other than H or C atoms. The functional groups can occur adjacent to one another, singly or multiply; that is, halons contain apart from F- and Cl- also Br-atoms. Some of the more frequently encountered functional groups in atmospheric chemistry are $-OH$, hydroxy (alcohols); $-OOH$, hydroperoxide; $> C{=}O$ ($-HC{=}O, > C{=}O$), keto (aldehydes, ketones); $-C(O)OH$, carboxylic acid; $-C-O-C-$, ether; $-C(O)OR$, ester; $-NO_2$, nitro; $-ONO_2$, nitrate (alkylnitrate); $-C(O)O_2NO_2$, peroxynitrate(alkylperoxynitrates); $-Br$, bromide (haloalkanes); $-F$, fluoride (CFCs, HCFCs); $-Cl$, chloride (CFCs, HCFCs).

As will be seen in the discussion on the different oxidation mechanisms for the hydrocarbons many of the compounds emitted to the atmosphere can result in the formation of much more complicated high oxized substances containing one or more of the above functional groups. Compounds containing two or more functional groups are termed bifunctional and multifunctional compounds, respectively.

SOURCES, AMBIENT CONCENTRATIONS, AND EMISSION INVENTORIES OF VOCs

Sources and Ambient Concentrations

Several hundred different hydrocarbons (VOCs) have already been identified in the atmosphere (5,6,8) and many thousands are probably present. The sources are very varied and scattered. Human activities give rise to emissions of hydrocarbons to the atmosphere which are termed anthropogenic. Major anthropogenic sources include automobile tailpipes, evaporation of gasoline vapor from automobiles, solvent usage, the oil and chemical industries, gasoline storage and distribution, landfilled wastes, food manufacture, agriculture, and biomass burning. For the United States and Western Europe a breakdown of anthropogenic emissions shows that emissions from vehicles are often the most dominant followed by solvent use. Emissions from combustion and the chemical industry sources are composed mainly of alkanes, alkenes and aromatic hydrocarbons; natural gas contributes mainly light alkanes; the use of organic solvents higher alkanes and aromatic hydrocarbons; and biomass burning contributes mainly light alkanes and alkenes. On a global scale, fuel wood combustion and savanna burning are among the largest VOC emission sources, accounting for over 35% of the total global VOC emissions. The production and use of gasoline (16%), solvent use (7%), refuse disposal activities (8%), and organic chemical and rubber manufacturing (7%) were also found to be significant sources of VOC emissions (9).

The overwhelming majority of hydrocarbons emitted naturally to the atmosphere come from plant foliage, emissions from trees in particular contribute significantly to VOC emissions. Other natural sources include marine and fresh water, soil and sediments, microbial decomposition of organic material, and geological hydrocarbon reservoirs. Volatile olefinic compounds such as the hemiterpene (C_5) isoprene and a range of monoterpene (C_{10}) compounds were considered to be the most important natural species emitted to the atmosphere (10); however, it is becoming evident that volatile emissions from vegetation also include alkanes, alkenes, and many oxygenated hydrocarbons such as acetates, alcohols, esters, ethers aldehydes, ketones, sesquiterpenes, and even aromatic hydrocarbons such as *p*-cymene. Characterization of the biogenic VOC is extremely difficult due to the enormous diversity in plant species and the variability in the hydrocarbons emitted from species to species (11).

Methane, CH_4, the simplest and most abundant of the alkanes has both natural and anthropogenic sources. Major sources of methane are the natural wetlands, rice agriculture, and domestic cattle; each of which probably produces 60 Tg yr^{-1} or more. Other minor sources with emission in the 10–60 Tg/yr range include biomass burning, landfills, coalmines, termites, sewage disposal, natural gas leakage, lakes, and oceans. The concentration of methane is presently $\approx$ 1.7 ppmv in the troposphere and $\approx$ 0.5 ppmv in the stratosphere. The CH_4 concentration is increasing at a rate of 1% per year, and the present-day value is approximately 130% higher than that of pre-industrial times due to human activities. However, recent measurements indicate that the rate of increase may be slowing down. It is not clear at present whether emissions of methane are increasing or decreasing; measurements indicate that the growth is not as rapid as observed in previous years. Whether this is due to a change in the strength of the emission sources or an increase in their oxidation rate due to an increase in the atmospheric OH concentration is still open to question.

Table 1 summarizes the concentrations (in ppbC) of some of the most frequently observed hydrocarbons in the atmosphere excluding methane. Examples are given for a remote rural area, the city of London/U.K., and the mean values determined for numerous U.S. cities. Also included in Table 1 are the typical spreads observed in the atmospheric concentrations of the hydrocarbons. The data provide an indication of the diversity of the hydrocarbon classes observed in urban environments and also the variability of their concentrations. Because automobile emissions are a major source of atmospheric hydrocarbons in most urban areas it is of no surprise that measurements have shown that the VOC composition of ambient air usually reflects the type of gasoline being used (15). For example, in the United States the gasoline is comprised of 45–60% alkanes, 10% alkenes, and 30–45% aromatic hydrocarbons which resembles very closely an analysis of the early morning air in Los Angeles, California, which was found to contain 45–50% alkanes, 10–15% alkenes, 20–25% aromatic hydrocarbons, and 3–7% carbonyls. Similar hydrocarbon concentration patterns are also observed in urban areas in Europe.

Isoprene and the monoterpenes are generally the natural hydrocarbon emitted in highest concentration from plant foliage; however, for some species and many agricultural crops isoprene and the monoterpenes constitute less than 20% of the emitted VOCs. Newer investigations also show that isoprene is emitted in the tailpipe emissions from automobiles. The terpenes are among the most reactive hydrocarbons present in the troposphere and can only be measured directly near their source such as a leaf surface where the collective concentrations can reach 10 ppbV. Table 2 shows the concentrations of isoprene and monoterpenes measured at various locations around the globe. At present the global flux of natural hydrocarbons is estimated to be comprised of 44% isoprene, 11% monoterpenes, 22.5% other reactive VOCs, and 22.5% other VOCs.

Table 1. Mean Concentrations (in ppbC) of Frequently Observed VOCs Measured at a Remote Rural Site and in Major Cities

Hydrocarbon	Langenbrugge (Germany)[a] remote rural	London (U.K.)[b] Urban Roadside	U.S. Cities[c]	Typical Urban Concentration Ranges[d]
Ethane	4.4	14.8	33.3	0.2–180
Ethene	1.6	21.7	29.1	20–304
Propane	3.1	10.6	37.6	9–290
Propene	0.6	17.4	17.2	10–210
Isobutane	1.1	25.5	24.8	6–210
Ethyne	2.1	38.6	19.3	20–1000
Butane	1.1	48.9	54.3	20–1100
trans-2-Butene			15.3	
cis-2-Butene			17.4	
Σ-Butenes	1.06	14.4	32.7	5–30
Cyclopentane	0.11		15.1	2–240
Isopentane	1.33	37.3	56.5	5–500
Propyne	0.07	1.0		
Pentane	0.85	12.9	27.9	25–1000
Buta-1,3-diene	0.08	3.2	22.7	
trans-2-Pentene			20.7	
cis-2-Pentene			17.3	
Σ-Pentenes	0.23	4.4	38.0	1.5–400
2-Methylpentane			19.1	10–650
3-Methylpentane			14.0	6–450
Σ-Isohexanes	0.89	5.1	33.1	16–1100
Isoprene	0.10	1.0	8.8	0–14
Hexane	0.63	3.6	15.8	10–750
2-Methylhexane			11.3	7–300
3-Methylhexane			11.8	7–500
Σ-Methylhexanes	0.91		23.1	14–800
Heptane	0.54		13.1	1–700
Benzene	2.11	27.8	17.0	0.4–1200
Toluene	3.09	52.3	45.4	0.7–1100
Ethylbenzene	0.39	9.4	13.7	0.5–320
m- + *p*-Xylene	1.12	30.6	24.6	10–600
o-Xylene	0.50	11.6	11.0	10–300
1,2,3-Trimethylbenzene			30.2	3–160
1,2,4-Trimethylbenzene		2.7	14.8	3–160
1,3,5-Trimethylbenzene		9.7		3–160

[a] Ref. 12.
[b] Ref. 13.
[c] Ref. 14.
[d] Ref. 8; gives typical spread in measured concentrations for urban areas.

Table 2. Ambient Concentrations (ppbv) of Some Frequently Occurring Natural Hydrocarbons During Summer

Hydrocarbon	Georgia Forest[a]	Southeastern U.S.[b]	Niwot Ridge[c], 3 km Rocky Mountains	Brazil[d] Amazon Basin Forest	Southwest Scotland[e]
Isoprene	1.4	0.9	0.63	2.04	0.302
α-Pinene	0.8	0.75	0.05–0.14	0.1	0.025
β-Pinene	0.43	0.4	0.97–0.08	0.03	0.015
3-Carene	0.9		0.05	0.01	
Camphene	0.09	0.062	0.04	0.03	0.006
Sabinene					0.008
D-Limonene	0.08		0.03–0.05		0.015
β-Myrcene	0.068				0.027
p-Cymene		0.15			

[a] Ref. 16.
[b] Ref. 17.
[c] Ref. 18.
[d] Ref. 19.
[e] Ref. 10.

VOC Emission Inventories

There are numerous very important reasons for compiling VOC emission inventories (20). Hydrocarbons contribute to the formation of photochemical ozone and other oxidants on urban and regional scales. They also play a role in acid deposition because they contribute to the generation of radicals which convert NO_x ($NO + NO_2$) and SO_2 to nitric acid and sulfuric acid, respectively. They are also involved in the formation of peroxides which influence acid formation in clouds. Some VOC undergo phase transitions leading to aerosols implicated in visibility degradation. The oxidation of the organic sulfur compound dimethyl sulfide (DMS · CH_3SCH_3) has been suggested to be actively involved in the formation of cloud condensation nuclei (CCN) which regulate the earth's climate.

Because hydrocarbons play a central role in atmospheric chemical processes accurate estimates of VOC emissions are essential for an accurate assessment of chemical impacts on the environment. The emission inventory is a key component of any air pollution control program. It includes many types of emission sources, quantities of emissions, the temporal and spatial characteristics of emissions, and the process and emission control devices used at sources. Air pollution control agencies use emission inventories to identify potential control measures and sources that would be subject to controls, to determine control program effectiveness, and to predict future air quality through air quality simulation models. Therefore, accurate emission inventories are critically essential to provide the least costly basis and most effective pollution control strategies.

Accurate estimates of both natural and anthropogenic VOC emission rates are of importance because they are required for models which provide the basis of global change scenarios and regulatory emission control strategies. Ideally the characterization of VOC emissions requires not just an estimate of total VOCs but also categorization of individual VOC emissions by class and spatial distribution.

Apart for most of North America and Western Europe detailed inventories are available for very few species and very few regions (21). CFCs are an exception because their sources are purely anthropogenic and have been catalogued very extensively because of their threat to the stratospheric ozone layer. Global information also exists for isoprene and terpenes. Even though emission inventories are available for the United States and Europe they are all associated with a high degree of uncertainty due to the difficulties in compiling such inventories. For example, it has been recently recognized in the United States that the emissions of biogenic hydrocarbons have been seriously underestimated; the emission of isoprene, the major component of the biogenic hydrocarbons is currently uncertain by a factor of 3 to 7.

The estimated worldwide emissions of methane are 150 million tons/yr from natural sources and 350 million tons/yr from human activities. Table 3 lists estimates of the emissions of anthropogenic and biogenic VOC for some European countries and the United States. Based on present estimates the United States is the largest emitter (21% of the total global VOC), followed by the former USSR, China, India, and Japan. In the United States, emissions of anthropogenic VOC are currently estimated to be 20 million tons/yr and the natural emissions 29 million tons/yr (22). Present estimations put the European emissions of anthropogenic VOC at 23.8 million tons/yr and biogenic emissions at 5 million tons/yr (23). In Europoe emissions from anthropogenic sources overwhelm natural sources in most countries (Table 4). Although the situation is not entirely clear, in Europe the natural emissions are thought to be comprised largely of isoprene from deciduous trees.

Despite the high degree of uncertainty presently associated with emission inventories it is now reasonably certain

Table 3. Estimation of Anthropogenic VOC Emissions (Excluding Methane, Including Condensable Organics)[a]

Source	Tg product	Tg VOC	Percent
Fuel production / distribution			
Petroleum	3100	8	6
Natural gas	2150	2	1.5
Oil refining	2950	5	3.5
Gasoline distribution	730	2.5	2
Fuel consumption			
Coal	600	3.5	2.5
Wood	1550	25	18
Crop residues	200	14.5	10
Charcoal	20	2.5	2
Dung cakes	400	3	2
Road transport		36	25
Chemical industry	330	2	1.5
Solvent use	20	20	14
Uncontrolled waste burning		8	5
Other		10	7
Total		*142*	*100*

[a] See Ref. 51.

Table 4. Emissions of VOCs from Anthropogenic and Natural Sources, t/yr×10^6, 1989[a]

Country	Anthropogenic	Natural
Albania	33	10
Austria	441	37
Belgium	340	39
Bulgaria	167	203
Czechoslovakia	275	88
Denmark	125	5
Finland	181	72
France	1972	641
German Democratic Republic	110	31
German Federal Republic	2042	118
Greece	358	31
Hungary	358	113
Iceland	8	0
Ireland	124	3
Italy	2793	72
Luxembourg	10	2
Netherlands	473	9
Norway	245	27
Poland	982	77
Portugal	162	46
Romania	386	219
Spain	936	175
Sweden	460	104
Switzerland	304	7
Turkey	263	283
United Kingdom	1777	24
USSR (European part)	9064	2256
Yugoslavia	300	66

[a] Refs. 14 and 29.

that the majority of global VOC emissions are from natural rather than anthropogenic sources. Worldwide it is currently estimated that of the hydrocarbons emitted to the atmosphere 100 million tons/yr can be attributed to anthropogenic sources of VOC and 1000 million tons/yr to natural sources.

TROPOSPHERIC CHEMISTRY OF HYDROCARBONS (VOCs)

Principal Atmospheric Processes Involved in Hydrocarbon Transformation or Removal

The transformation of hydrocarbons to the ultimate end products CO_2 and H_2O, whether in the gas phase, aqueous phase, or on surfaces, involves many intermediate steps. In the atmosphere physical removal processes include wet and dry deposition. Although very little information is available, dry deposition is generally not thought to be important for hydrocarbons. Wet deposition by definition is only important for readily soluble species and alcohols, carboxylic acids and alkyl hydroperoxides are probably the only classes of hydrocarbons for which it is of importance. For the large majority of VOCs photo-induced oxidation by the reactive inorganic species the hydroxy radical (OH), the nitrate radical (NO_3), and ozone (O_3) represents the major primary step in their transformation (4). For VOCs absorbing radiation in the actinic region of the solar spectrum (~ 290–800 nm) photolysis is also important and for some species dominates the removal processes. There is evidence that reaction with halogen atoms or halogen oxides (X and XO, X = Cl or Br) may also play a role in the degradation of hydrocarbons in certain regions of the troposphere. It is, however, presently a matter of considerable debate as to the extent of reactive halogen chemistry in the troposphere. Many research programs are currently directed toward determining the extend of reactive halogen species in the atmospheric chemistry of hydrocarbons which might transpire to be more widespread than previously thought.

Photolysis in the Troposphere

Photochemistry in the atmosphere is generally taken as meaning the light-influenced degradation of substances. This can either be through direct absorption of light resulting in decomposition of the substance or indirectly via chemical reaction of the substance with photochemically produced reactive species. A primary photochemical reaction is a reaction in which the absorption of a photon by an atom or molecules results in the formation of one or more radicals (or other reactive species); in such a case one speaks of photolysis or photodissociation (24,25). The rate of a photochemical reaction depends on the intensity and wavelength of the light and the concentration of the absorbing species and their absorption coefficients:

$$d[X]/dt = J_{\Delta\lambda}[X]$$

where $J_{\Delta\lambda}$ is the photolysis frequency in a wavelength interval $\Delta\lambda$ given by:

$$J_\lambda = \int Q(\lambda)s(\lambda)\phi(\lambda)d\lambda$$

where $Q(\lambda)$, $\sigma(\lambda)$, and $\phi(\lambda)$ are radiation flux, absorption cross-section, and photolysis quantum yield at wavelength λ, respectively.

Figure 2 shows the spectral distribution of the photon flux density for a sun zenith angle of 40° in the spectral region from 200 to 1200 nm, as measured at the top of the earth's atmosphere (curve A) and as measured at the earth's surface (curve B). The reactive species formed in the atmosphere through photolysis of ozone, nitrogen dioxide, aldehydes, etc, react with other components in the air (secondary photoreactions/indirect photochemistry) which often result in the initiation of radical chains.

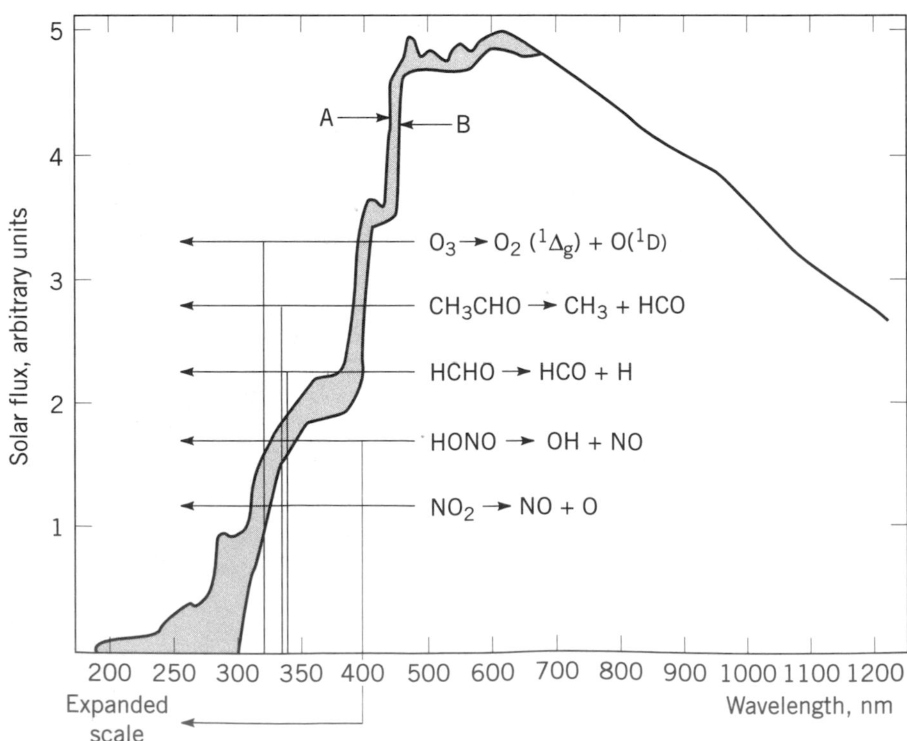

Figure 2. Spectral distribution of sunlight (A) above the stratospheric ozone layer and (B) at the earth's surface and wavelength thresholds for some photochemical reactions.

The basis of chemical transitions in the atmosphere is the direct or indirect photochemical activation. Because of the stratospheric ozone layer photochemistry in the troposphere is limited to wavelengths $\lambda \geq 285$ nm or binding energies of ≤ 435 eV (150 kJ/mol), only molecules capable of absorbing radiation at longer wavelengths (290–800 nm) are able to undergo photodissociation or isomerization. Only a limited number of atmospheric trace gases absorb radiation effectively enough at these long wavelengths in the troposphere to be able to initiate photochemical radical reactions.

Of the photodissociation processes in the troposphere the formation of excited oxygen atoms via the photolysis of ozone below 310 nm is by far the most important because $O(^1D)$ forms OH radicals in a fast secondary reaction with H_2O (J is the photolysis frequency and $O(^1D)$ and $O_2(^1\Delta_g)$ are excited oxygen atoms and molecules, respectively):

$$O_3 + h\nu(\lambda \leq 310 \text{ nm}) \rightarrow O_2(^1\Delta_g) + O(^1D)$$

$$O_3 + h\nu(\lambda \geq 310 \text{ nm}) \rightarrow O_2(^1\Delta_g) + O(^3P)$$

$$O(^1D) + H_2O \rightarrow 2\ OH$$

$$O(^3P) + O_2 + M \rightarrow O_3 + M$$

($J_{O_3} = 1.2 \times 10^{-5}$/s, noon summer, mid-latitudes NH)

The photolysis of ozone at longer wavelengths results in the formation of oxygen atoms in the ground electronic state $O(^3P)$. In the troposphere because of the fast termolecular recombination reaction with O_2 to reform O_3 this reaction is of no chemical significance.

Also important for photochemical processes in the troposphere is the photolysis of NO_2, whose absorption spectrum stretches into the visible and from approximately ≥ 410 nm photodissociates:

$$NO_2 + h\nu(\lambda \leq 410 \text{ nm}) \rightarrow NO + O(^3P)$$

This reaction produces oxygen atoms and is, together with the reaction of NO with peroxy radicals, a vital element of anthropogenic ozone formation.

The photolysis of formaldehyde is an important source for HO_2: radicals when in the presence of NO can produce OH radicals:

$$HCHO + h\upsilon(\lambda \leq 340 \text{ nm}) \rightarrow H + HCO$$

$$H + O_2 + M \rightarrow HO_2 + M$$

$$HCO + O_2 \rightarrow HO_2 + CO$$

$$HO_2 + NO \rightarrow OH + NO_2$$

($J_{HCHO} = 2.3 \times 10^{-5}$/s, noon summer, mid-latitudes)

Under low NO_x conditions HO_2 radicals can also recombine to form H_2O_2 the photolysis of which will also produce OH radicals:

$$HO_2 + HO_2 + M \rightarrow H_2O_2 + O_2 + M$$

$$H_2O_2 + h\nu(\lambda \leq 340 \text{ nm}) \rightarrow 2\ OH$$

($J_{H_2O_2} = 6.7 \times 10^{-6}$/s, noon summer, mid-latitudes)

Other carbonyl compounds (aldehydes and ketones) are, because of their low absorption cross sections, low quantum yields, and low atmospheric concentrations, of little significance as radical sources. Exceptions are α-dicarbonyls, which are formed in the atmosphere from the photooxidation of aromatic hydrocarbons.

Under conditions where the concentration of NO_x is high (eg, see section on Photosmog Formation) the photolysis of HONO (nitrous acid) can be an important source for OH radicals:

$$HONO + h\nu(\lambda \leq 390 \text{ nm}) \rightarrow OH + NO$$

The absorption spectra of other photoactive trace gases (eg, HNO_3) overlap only weakly with the solar spectrum in the region of 300 nm and consequently have low photolysis frequencies and therefore no real importance as radical sources.

OH Radical Sources in the Troposphere

The hydroxyl radical is the most important in atmospheric chemistry and is responsible for the degradation of the majority of VOCs. Without the presence of OH radicals hydrocarbons would accumulate in the atmosphere and reach intolerable levels with disastrous consequences for all forms of life on the earth. The atmospheric concentration of OH radicals is very low, the global 24-h averaged OH–radical concentration is currently estimated to be 8×10^5 radicals cm^{-3} (26). Peak daytime OH radical concentrations can be of the order of 10^6–10^7 radicals cm^{-3}. The most important formation processes for OH radicals in the troposphere is the photolysis of O_3, HCHO, H_2O_2, and HONO as outlined in the section on photolysis, whereby the reaction of electronically excited atoms $O(^1D)$, produed in the photolysis of ozone, with water vapor is by far the major.

NO_3 Radical Formation in the Troposphere

Nitrate radicals, NO_3, are formed virtually exclusively by the reaction of NO_2 with O_3:

$$NO_2 + O_3 \rightarrow NO_3 + O_2$$

An equilibrium is quickly established between NO_3 and NO_2 with N_2O_5:

$$NO_3 + NO_2 + M \rightleftarrows N_2O_5 + M$$

Because of this equilibrium the behaviors of NO_3 and N_2O_5 in the atmosphere are closely connected. During daytime, NO_3 radicals are rapidly destroyed by two mechanisms, either through photolysis or reaction with NO:

$$NO_3 + h\nu(\lambda < 580 \text{ nm}) \rightarrow NO_2 + O \rightarrow NO + O_2$$

$$NO_3 + NO \rightarrow 2\ NO_2$$

The photolysis lifetime of NO_3 radicals by cloudless sky conditions and 90° zenith angle (overhead sun) is ~5 s; also, their rapid reaction with NO ($k = 2.6 \times 10^{-11}$ molecules cm^{-3}/s at 298 K) in the presence, for example, of 0.4 ppbV NO limits its lifetime to 5 s. Because of these loss processes the NO_3 concentration during daytime is very low with values under 0.1 pptV. During the night photolysis is absence and the concentration of NO falls to very low levels due to its rapid reaction with NO which enables NO_3 radicals to reach measurable concentrations. The measured atmospheric concentration of NO_3 radicals ranges from 1 pptV detection limit of the analytical method employed (generally DOAS) to 350 pptV (27). Long-term observations in rural air masses give 24-h averages which are closer to a few pptV.

ATMOSPHERIC LIFETIMES OF VOCs

The residence time or lifetime of a hydrocarbon in the atmosphere is determined by the sum of all the transformation, deposition, and transport processes. The oxidation rate of a hydrocarbon A can be described by $d[A]/dt = -K[A]$, where $[A]$ is the concentration of the species A. The constant K is given by $K = k_{phot} + Sk_i[Y_i]$ whereby k_{phot} is the photolysis frequency of the hydrocarbon, k_i is the rate coefficient for reaction of the hydrocarbon with the reactant i, and $[Y_i]$ is the reactant concentration, so that:

$$K = k_{phot} + S(k_{OH}[OH] + k_{NO_3}[NO_3] + k_{O_3}[O_3] + k_{drydep}[\text{dry dep}] + k_{wetdep}[\text{wet dep}] \ldots \text{etc}$$

The overall lifetime $\tau_{overall}$, is given by $1/K$ and the lifetimes due to individual reactions, for example, with OH radicals, by $\tau_{OH} = (k_{OH}[OH]^{-1})$.

Rate coefficients have been measured for the reaction of OH radicals, NO_3 radicals, and O_3 with a great variety of hydrocarbons (4,28). Where relevant, photolysis frequencies are also given, although these are often not as well established as the kinetic data for OH and NO_3 radicals and O_3 reactions.

Because it is not possible to measure the rate coefficients for every hydrocarbon emitted to the atmosphere, estimation methods have been developed to try and accurately predict their values. Methods exist for OH and NO_3 radicals. The methods for prediction of OH radical reaction rate coefficients have received particular attention because reaction with OH is, for most hydrocarbons, the dominant atmospheric loss process. Such estimations are often applied to calculate the atmospheric lifetime of new chemical substances in order to assess potential environmental risks, such as transport to remote unpolluted regions of the globe or the stratosphere and buildup in the atmosphere to levels which are toxic or could contribute significantly to the greenhouse effect. Many of the estimation methods make use of the chemical properties of the hydrocarbon, such as ionization energy, NMR chemical shifts, bond-dissociation energies, molecular orbitals, and IR absorption frequencies. Correlations between gas- and liquid-phase rate coefficients for the various reactants have also been used. All of these methods suffer, at present, from limited databases which severely constrains their application. Probably the most widely applied method is the structure–reactivity relationship estimation method (30,31), which has been continually updated and improved and can predict rate coefficients for a wide range of hydrocarbons, often to within better than a factor of 2.

PHOTOSMOG: CAUSES AND POSSIBLE REMEDIES

Within the last 100 years human activities have had a measurable influence on the composition of the trace gas composition in the earth's atmosphere, particularly hydrocarbons, NO_x, and SO_2, through emissions from industrialization, power plants, motorized vehicles, etc. Photochemical smog or photosmog was first observed in Los Angeles in the 1940s. Laboratory investigations on the causes of smog in photoreactors established that the irradiation of air enriched solely with NO or NO_2 did not produce photooxidants; this only occurred when hydrocarbons were also present in the reaction mixture. The photosmog formation was observed to consist of several phases and each phase of the changing reaction mixture was associated with particular photooxidants.

Photooxidants are secondary air pollutants which are formed in air plumes by photochemical processes involving reactive precursor emissions often hours but in some cases days after the primary emission. In a board sense the term oxidant embraces the reaction products from the sunlight-induced reactions involving NO_x and VOCs. The most easily measurable and best known component of photosmog, ozone, O_3, processes a natural source in the stratosphere which can be transported to lower levels in the troposphere; however, contributions to the O_3 in the troposphere from this source are approximately 20% or less. In photosmog we are concerned mainly with ozone produced *in situ* at ground level; that is, ozone in the lowest layer of the atmosphere whose concentration is subject to strong and daily variations caused by constantly varying emissions, solar radiation, and meteorological influences. Apart from O_3 other typical photooxidants are also formed and include the radicals (Table 5) OH, HO_2 and RO_2, NO_2, NO_3, N_2O_5, HNO_3, aldehydes and other carbonyl compounds, organic acids, H_2O_2 and other hydroperoxides, organic nitrates, and peroxyacetyl nitrate and other peroxynitrates. Ozone is the component present at the highest concentration and at elevated levels is harmful to humans, plants, and animals and can also erode building material (32).

The processes and conditions leading to the production of harmful levels of ozone and other photooxidants are very complex in nature; the fundamental principals behind the production of ozone in the troposphere are outlined. In urban areas excess ozone production at ground level has been observed in summer, whereas in winter increased destruction of ozone is observed. In the presence of sunlight in summer, photochemical processes occur which result in the formation of excess or surplus ozone compared to the natural present-day background level. It is now firmly established that this is a direct outcome of increased precursor emissions of both VOCs and NO_x ($NO_x = NO + NO_2$). The reverse trend in O_3 in winter and also during the night is due mainly to the NO_x emissions; in winter the photochemical O_3 production is greatly reduced and at night it is nonexistent and O_3 is, therefore, consumed in reactions with NO_x. Model simulations and trends measurements from several European and American observatories indicate that about 100 years ago the mean ozone yearly concentration at ground level was under 30 $\mu g/m^3$ ($\sim$ 15 ppbv). Presently, the ozone concentrations in the free troposphere in the northern hemisphere are approximately 80 $\mu g/m^3$ ($\sim$ 40 ppbv), and at ground level between 30 and 70 $\mu g/m^3$ (15–35 ppbv). These observations support that in the last 100 years the background concentration of ozone in the free troposphere of the Northern Hemisphere has increased by about 200% since industrialization (33), although recently the rate of increase has decreased. An upward trend in the concentration of PAN has also been established.

The phenomenon of photosmog was initially only thought to affect industrialized areas and major cities; however, because of the increasing levels of pollutants and their long-range transport it is now a common occurrence worldwide on urban, suburban, and rural scales (33). Photochemical ozone formation has even been observed in parts of Western and Northern Europe where the solar irradiation was not considered sufficiently intense to produce smog. It is also now clear that the

Table 5. Important Free-Radicals in Atmospheric Chemistry and their Mean Concentrations

Radical	Formula	Main Source(s)	Concentration, Radicals cm^{-3}
Oxygen atoms (ground state)	$O(^3P)$	Photolysis of ozone and NO_2	2.5×10^4
Hydroxy	OH	Reaction of $O(^1D)$ with water vapor	8×10^5
Hydroperoxy	HO_2	Reaction of O_2 with alkoxy radicals	6.5×10^8
Methoxy	CH_3O	Photooxidation of CH_4	1.3×10^6
Methylperoxy	CH_3OO	Photooxidation of CH_4	1.0×10^8
Nitrate	NO_3	Reaction of NO_2 with O_3	3×10^8

contribution from the stratospheric to the O_3 observed in regions of the troposphere considered to be pristine is only ca 20%, it was originally thought that infusions from the stratosphere was responsible for all of the O_3 in these regions.

The back ground O_3 concentration is influenced by both natural and anthropogenic precursor emissions. Excess O_3 formation has also been observed in clean air masses. In these regions natural emissions from vegetation (isoprene and terpenes) form a significant fraction of the precursor substances, whereas No_x is transported from urban areas. Episodes of high concentrations of ozone in the lower troposphere now occur frequently over all parts of Europe and North America every summer. Although regulatory efforts have been made in the United States by reduction in emissions of the anthropogenic precursor compounds, mainly VOC emissions, many regions (termed nonattainment areas) still exceed the current air quality standards. In Europe, short peak ozone concentrations of over 600 $\mu g/m^3$ (300 ppbv) have been observed. During periods of intensive sunshine and stagnant air masses 1-h peak concentrations of 300–400 $\mu m/m^3$ have been frequently observed particularly in the southwest of Germany and east of Vienna. Numerous studies on the conditions leading to smog formation in the United States, Europe, and elsewhere have been published and the reasons for photosmog formation are qualitatively well understood (13).

Ozone Chemistry and Photosmog Formation in the Troposphere

Ozone Formation. The reaction sequences that lead to ozone formation in photosmog situations are very complex and are presented here in a very simplified manner. The only photolysis reaction that results directly in O_3 formation is the photolysis of NO_2 which becomes effective at wavelengths just above 400 mm:

$$NO_2 + h\nu(\lambda \leq 400 \text{ nm}) \rightarrow NO + O(^3P)$$

$$O + O_2 + M \rightarrow O_3 + M(\text{very fast})$$

The O atoms formed in the photolysis recombine with molecular oxygen to form ozone, the formation rate of which is given by:

$$d[O_3]/dt = J_{NO_2}[NO_2]$$

The photolysis frequency, J_{NO_2}, is a measure for the probability of the NO_2 photolysis; for example, during midday in summer in middle Europe J_{NO_2} is of the order of 7×10^{-3}/s. For $[NO_2] = 10$ ppb this would give an ozone formation rate of 250 ppb/h.

The above sequence, however, cannot result in excess O_3 formation because of the fast reaction of O_3 with NO, which reforms NO_2:

$$NO + O_3 \rightarrow NO_2 + O_2$$

In a photostationary equilibrium, that is, when the O_3 formation and O_3 loss rates are the same, the O_3 concentration is given by:

$$[O_3]_s = \frac{J_{NO_2}[NO_2]}{k_{NO+O_3}[NO]} \text{ with } J_{NO_2}/k_{NO+O_3}$$

$$= 16(\text{ppb}) \text{ at midday in summer}$$

In such a system the cycle $NO_2 \rightarrow NO \rightarrow NO_2$ occurs many times during the course of the day, typically around 50 times in winter and can reach 250 times in summer.

Chemical Processes Resulting in Formation of Surplus Ozone and Photosmog. Ozone in excess of the natural tropospheric background level and the associated photosmog is now known to mainly due to efficient VOC oxidation over an OH radical chain propagated in the present of NO_x, whereby NO is oxidized to NO_2 via RO_2 or HO_2 radicals and not via O_3 as outlined by the reactions in Table 6 and shown in Figure 3. This is contrary to earlier opinions when it was thought that transport from the stratosphere was the major source of O_3. Transport from the stratosphere contributes approximately 20%.

The initiating reaction is the photolysis of ozone in the presence of water:

$$O_3 + H_2O + h\nu \rightarrow O_2 + 2\,\mathbf{OH}$$

The chain is then propagated via reaction of VOCs with OH radicals:

$$VOC + \mathbf{OH} + O_2 \rightarrow RO_2 + \ldots$$

The reaction scheme in Table 6 makes clear that during the stepwise degradation of VOC the chain propagation and chain branching steps must dominate over the termination steps in order to form excess ozone. Each propagation step in the radical chain is associated with a new oxidation of NO to NO_2:

$$RO_2 + NO \rightarrow RO + NO_2$$

$$RO + O_2 \rightarrow R'CHO + HO_2$$

$$HO_2 + NO \rightarrow \mathbf{OH} + NO_2$$

The reactions lead in sunlight via NO_2 photolysis to excess ozone formation when the rates of these reactions are greater than the rate of the reaction of NO with O_3. When NO has been nearly completely oxidized to NO_2 the ozone concentration

Table 6. Reactions Involved in O_3 and Photosmog Formation

Reaction	Examples
Initiating	$O_3 + H_2O + h\nu \rightarrow O_2 + 2OH$
Chain propagation (radical interconversion)	$VOC + \mathbf{OH} + O_2 \rightarrow RO_2 + \ldots$
	$RO_2NO_2 \rightarrow RO_2 + NO_2$
	$RO_2 + NO + O_2 \rightarrow R'CHO + NO_2 + HO_2$
	$HO_2 + NO \rightarrow \mathit{OH} + NO_2$
Chain branching	$R'CHO + O_2 + h\nu \rightarrow 2HO_2 + \ldots$
Chain termination (radical loss)	$HO_2 + HO_2 \rightarrow H_2O_2 + O_2$
	$HO_2 + RO_2 \rightarrow ROOH + O_2$
	$RO_2 + R'O_2 \rightarrow$ Alcohols, aldehyde, etc
	$RO_2 + NO_2 \rightarrow RO_2NO_2$
Ozone formation	$NO_2 + O_2 + h\nu \rightarrow NO + O_3$ (Direct O_3 formation)
	$NO + O_3 \rightarrow NO_2 + O_2$ (Titration reaction)

reaches a maximum, after this point chain termination steps such as the following begin to become effective:

$$NO_2 + OH + M \rightarrow HNO_3 + M$$

$$HO_2 + HO_2 + M \rightarrow O_2 + H_2O_2\text{(hydrogen peroxide)} + M$$

$$RO_2 + HO_2 + M \rightarrow O_2 + RO_2H\text{(organic hydroperoxide)} + M$$

$$RO_2 + NO_2 + M \Longleftrightarrow RO_2NO_2\text{(peroxynitrate)} + M$$

The most important chain termination reaction is that of OH radicals with NO_2 to form nitric acid (HNO_3). When the concentrations of NO and NO_2 become low due to conversion to organic nitrates and HNO_3, other removal processes gain in importance and the formation of photooxidants decreases. Reactions which result in the formation of hydroperoxides and carboxylic acids then become dominate. Laboratory investigations have shown that formation of peroxyacetyl radicals, $CH_3C(O)O_2$, are particularly effective in termination of the radical chain.

Besides peroxyacetyl nitrate (PAN, $CH_3C(O)O_2NO_2$), peroxyproprionyl nitrate (PPN) and peroxybenzoyl nitrate (PBZN) have been observed in the atmosphere, albeit at lower concentrations than PAN. PAN together with O_3 is one of the most important indicators of photochemical oxidant activity in air masses. With rapidly decreasing NO_x concentrations, such as in pristine air regions, $RO_2 + HO_2$ and $RO_2 + RO_2'$ reactions gain in significant and control the extend of chain termination. The photolysis of aldehydes and ketones formed in the oxidative decay of the VOCs are the major chain branching steps which stimulate the ozone formation process (see Fig. 3).

$$R'CHO + O_2 + h\nu \rightarrow 2\,HO_2 + \ldots$$

An example of the development of smog formation in a laboratory experiment is shown in Figure 4 for a mixture of air, propene, and NO and is reasonably typical of the chemistry occurring in the polluted atmosphere. During the decay of propene, NO is rapidly converted to NO_2 which photolyses to produce O_3. The formation of PAN removes NO_x from the system which causes the ozone to pass through a maximum. Because of the removal of NO_x the ozone production virtually stops and a slow decay is observed due mainly to heterogeneous loss to the reactor walls. Figure 5 is a very generalized schematic representation of the steps involved in the overproduction of ozone and photosmog. The production of photosmog can be expressed in one very simple equation:

$$\text{Sun} + \text{VOC} + NO_x + \text{air} \rightarrow \text{photosmog}$$

VOC Ozone Creation Potentials (Ranking)

The central role of VOCs in the formation of tropospheric ozone and other photooxidants makes some form of strategy for the reduction of the anthropogenic VOC component an obvious necessity. Because a blanket reduction in all VOCs is not possible its makes sense to reduce the VOCs most actively involved in the production of ozone. This, however, is not an elementary task because the potential for hydrocarbons to form ozone differs not only very greatly between different classes of hydrocarbon but also within the class itself. Further, the ozone forming potential of a particular species may be completely different in a complicated hydrocarbon mix compared to that determined separately for the compound. The reasons for the differences are complex but the difference is essentially due to competing reaction pathways in the complex oxidation process which control the ozone production.

A number of different concepts have been evolved to classify VOCs in terms of their ozone formation potential (OFP) in the presence of NO_x. In order to accurately predict the rates of ozone formation resulting from NO_x and VOC emissions, knowledge of various parameters is necessary. Emission levels of NO_x and VOCs, meteorological conditions, and the VOC oxidation mechanism provide the major input parameters into atmospheric models. These types of models have led to a number of definitions of ozone formation potentials for VOCs. In general, the various methods for determining ozone formation potentials give fairly similar values and provide information concerning the relative rate of reaction of OH radicals with the VOCs, and the role of the products of these reactions in generating ozone and PAN. The calculated OFPs are strongly dependent on the mechanism for the atmospheric oxidation of VOCs, together with the relative concentration of VOCs and NO_x. Despite the differences in the approaches in the formulation of reactivity scales with respect to ozone formation, all of the different methods agree in the general order of reactivity of the VOC species: alkanes and monoalkylbenzenes $<$ 1-alkenes and dialkylbenzenes $<$ trialkylbenzenes and internal alkenes.

The simplest classification which has been made is based purely on the rate constants for the reaction of OH radicals with

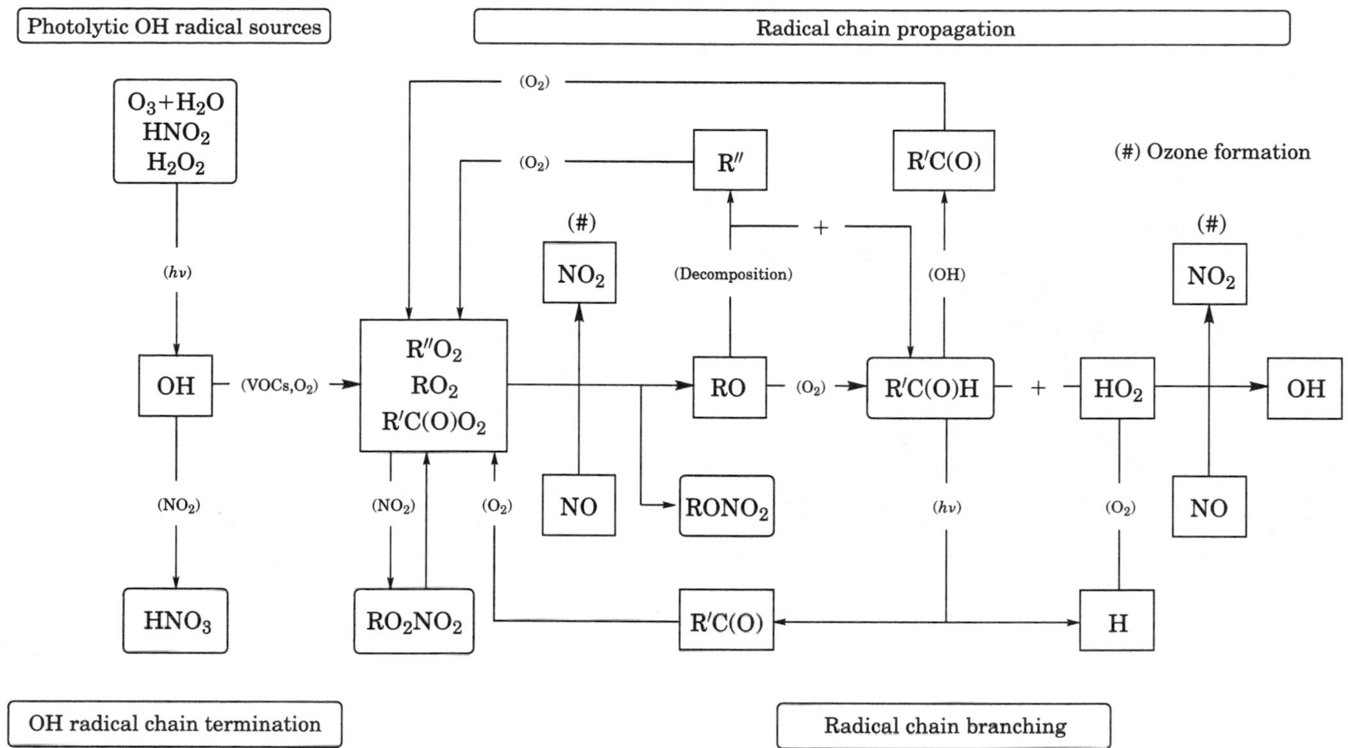

Figure 3. OH-radical reaction chain in the oxidation of VOC.

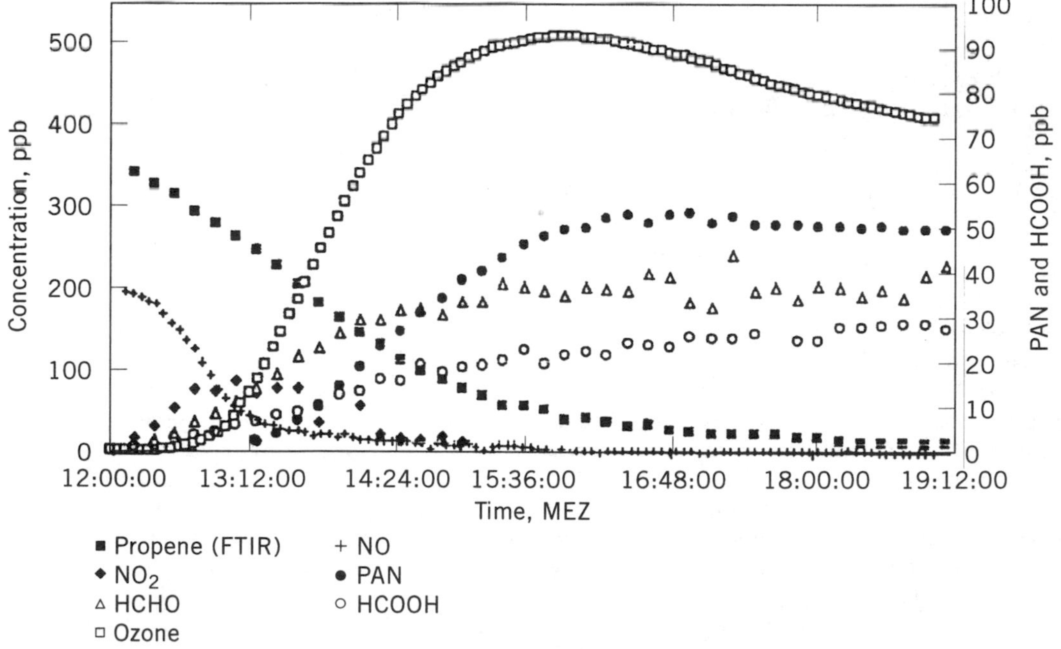

Figure 4. Typical concentration–time profiles of reactants and products in the irradiation of a propene–NO_x–air system.

VOCs. This classification leads to the obvious conclusion that a VOC with a high reactivity with respect to OH radicals, such as aromatic hydrocarbons and alkenes, gives rise to a high rate of ozone formation. This type of classification is, however, a gross over-simplification because it does not take into account the oxidation mechanisms under different atmospheric conditions or the ozone formation potential of the initial oxidation products of the reaction of OH radicals with the VOC. If these products have relatively short lifetimes with respect to photolysis or reaction with OH radicals, such as aldehydes, the ozone formation potential of the VOC, based purely on its reactivity with OH radicals, will be seriously underestimated.

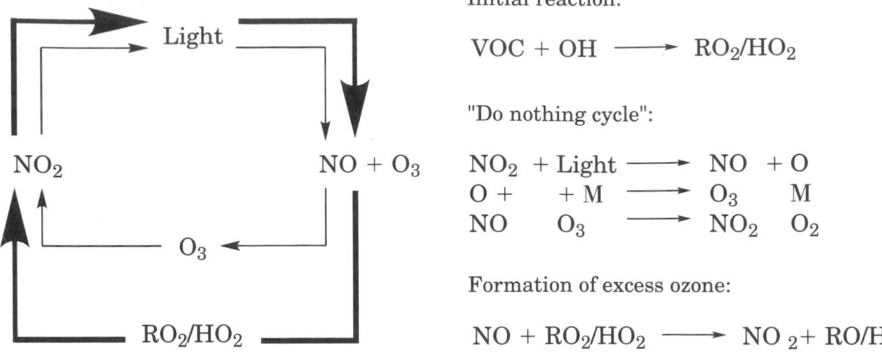

Figure 5. Simplified diagram depicting the formation of excess ozone in the troposphere, summer photosmog: $NO + RO_2/HO_2 \rightarrow NO_2 + RO/HO$.

As discussed in the section on photosmog the conversion of NO to NO_2 other than via the reaction of NO with O_3 in the photochemical oxidation of a hydrocarbon determines the ozone formation rate in such reaction systems. Thus, a potential method of classifying of defining the ozone producing potential of hydrocarbons is a simple counting of the efficiency of the NO to NO_2 conversion in the chemical system. By using known chemical mechanisms of a particular hydrocarbon the number of NO molecules which are converted to NO_2 in the oxidation of the hydrocarbon can be computed by the means of counter species in the model (34). This approach considers not only the NO to NO_2 conversion in the primary reaction steps but also the secondary photochemical oxidation of the products, which can lead to NO to NO_2 conversion and also loss of NO_x further down the oxidation chain. The method, however, is confined to those organics for which explicit chemical oxidation mechanisms are known, which considerably restricts the application of the approach because explicit atmospheric oxidation mechanisms have only been established for a limited number of hydrocarbons.

Another approach is the determination of so-called incremental reactivities. The reactivity of an hydrocarbon is defined as the amount of ozone formed per unit of organic added in a mixture of VOCs being representative of the rural or urban atmosphere (35).

$$\text{Reactivity} = \Delta[O_3]/\Delta[\text{organic}]$$

However, rather than defining reactivities with respect to the amount of ozone formed, in computer modeling and experimental studies the incremental reactivity (*IR*) has been defined as the amount of additional ozone formation resulting from the addition of a small amount of the compound to the system in which ozone is formed, divided by the amount of compound added. This depends not only on the VOC and its atmospheric reactions, but also on the conditions of the system where the VOC is reacting. This definition of reactivity takes into account the most significant processes producing ozone: that is, it is a direct measure of the extent of conversion of NO to NO_2 due to reaction with peroxy radicals and the subsequent photolysis of NO_2, the sole process responsible for O_3 production in the troposphere.

$$IR_t(\Delta[\text{organic}]) = R_t\{\Delta[\text{organic}]\} - R_t(O)/\Delta[\text{organic}]$$

where $R_t(O)$ is the reactivity observed in the base-case mixture with $\Delta[\text{organic}] = 0$. The limiting reactivity is defined as the amount of test compound added approaches zero:

$$IR_t(0) = \lim_{\Delta[\text{organic}]\to 0} [IR_t([\text{organic}])] = dR_t/d[\text{organic}]$$

Reactivity is defined in terms of effects on ($[O_3]/[NO]$) because this quantity has a direct relationship to the fundamental chemical processes responsible for ozone formation in photochemical smog systems.

$$[d(O_3-NO)]_t^{VOC} = \frac{d(O_3-NO)-d(O_3-NO)}{[VOC]}$$

The incremental reactivity of a given VOC is very sensitive to the composition of the initial air mass. Table 7 shows calculated incremental reactivates for CO and a selection of VOCs as a function of the VOC/NO_x ratio for an eight-compound surrogate mixture chosen to be representative of the atmosphere (California). It can seen that the *IR* varies quite considerably with changes in the VOC/NO_x ratio; for acetaldehyde, a precursor of PAN, it varies from 1.34 to 0.24 as the ratio varies from 4 to 20. At higher VOC/NO_x ratios, changes in VOCs have relatively little effect on the peak O_3. However, at low VOC/NO_x ratios, which are characteristic of polluted urban areas, addition of VOCs has a large impact on the peak O_3.

Another approach, variants of which have been applied in Europe, is the concept of the Photochemical Ozone Creation Potential (POCP) (36), which has been used to classify compounds according to their ability to form ozone. In this concept a photochemical trajectory model is used to describe the ozone production from the oxidation of methane and approximately 95 other hydrocarbons in air parcels. By adding small additional amounts of each compound production of additional O_3 was induced. By using the information from the model, a POCP index results, on a mass-emitted basis, and has been generated for the O_3 producing potential for a wide range of hydrocarbons relative to the base-case hydrocarbon, ethene. The studies show that aromatic and olefinic hydrocarbons exhibit the highest POCP and halocarbons the lowest. The results show motor vehicle tailpipe emissions exhibit the highest ozone-forming potential of the hydrocarbon emission source categories evaluated. Table 8 shows a comparison of POCP values and MIR (maximum incremental reactivity) values for a selection of VOCs. The POCP scale shows that only a small number of hydrocarbon species contribute significantly to regional scale ozone formation when taken individually. Certain hydrocarbons, including acetone and methyl acetate, show very low POCPs and have potential as candidates as alternative solvents.

Alternative and Reformulated Fuels and Alternative Solvents

Motor vehicles and solvent use are major sources of VOCs and NO_x to the atmosphere. The Environmental Protection

Table 7. Examples of Calculated Incremental Reactivities (*IR*) as a Function of the VOC/NO_x Ratio for Selected VOCs[a,b]

	IR for VOC/NO_x of							
Compound	4	6	8	10	12	16	20	40
CO	0.011	0.025	0.022	0.018	0.016	0.012	0.010	0.005
Ethane	0.024	0.054	0.041	0.031	0.026	0.018	0.015	0.007
n-Butane	0.10	0.22	0.16	0.12	0.098	0.069	0.052	0.019
Ethene	0.85	1.65	0.90	0.64	0.50	0.33	0.30	0.14
Propene	1.28	2.04	1.03	0.61	0.51	0.39	0.25	0.14
trans-2-Butene	1.42	2.02	0.97	0.62	0.48	0.31	0.23	0.54
Benzene	0.038	0.082	0.033	0.011	0.003	−0.002	−0.004	−0.002
Toluene	0.26	0.52	0.16	0.04	−0.021	−0.036	−0.058	−0.051
Formaldehyde	2.42	3.28	1.20	0.77	0.48	0.32	0.24	0.051
Acetaldehyde	1.34	1.83	0.83	0.55	0.42	0.29	0.24	0.098
Benzaldehyde	−0.011	−0.15	−0.27	−0.34	−0.37	−0.41	−0.41	−0.40
Methanol	0.12	0.27	0.17	0.12	0.091	0.066	0.055	0.029
Ethanol	0.18	0.37	0.22	0.14	0.10	0.06	0.038	0.006

[a] Ref. 35.
[b] Incremental reactivity values molecules O_3/C atom of VOC added.

Table 8. Ranking of Hydrocarbons According to Potential to Form Ozone as Determined by Various Computational Methods

Hydrocarbon	POCPs[a]	IVL[b]	MIR[c]	LOTO[d]	EMEP[e]
		Alkanes			
Methane	3		0		
Ethane	14	17	3		5–24
Propane	41	60	6		
n-Butane	60	55	12		22–85
i-Butane	43	33	15		
n-Pentane	62	61	13		
n-Hexane	65	78	13	40	
		Alkenes			
Ethene	100	100	100	100	100
Isoprene	118	53	145		
		Oxygenates			
Formaldehyde	55	42	127		
Acetaldehyde	65	53	77		
Acetone	18	17	7	20	
Methanol	21	17	7		
Ethanol	45	45	15	20	9–58
Dimethylether	26		11	20	
Methyl-*t*-butyl ether	27		8		
		Aromatics			
Benzene	33	32	5		
Toluene	77	45	35	55	
o-Xylene	83	42	91	65	
m-Xylene	109	58	123		
p-Xylene	95	61	95	90	

[a] POCPs = photochemical ozone creation potentials (36).
[b] IVL is the computer model of the Swedish Environmental Research Institute (Göteborg, Sweden) (37).
[c] MIR = maximum incremental reactivity (35).
[d] LOTO is the computer model of TNO (Delft, the Netherlands) (38).
[e] EMEP is the Cooperative Programme for Monitoring and Evaluation of the Long Range Transmission of Air Pollutants in Europe (39).

Agency (EPA) estimates that, in 1990, on-road motor vehicles accounted for 29% of the total VOC emissions, 33% of the total oxides of nitrogen (NO_x) emissions, and 65% of all carbon monoxide (CO) emissions in the United States nationwide (40). In urban areas, motor vehicle emissions typically account for even larger fractions of total anthropogenic air pollution emissions. It has also been shown that unreacted gasoline contributes a significant fraction (30–50%) of tailpipe hydrocarbon emissions (41). Control and, if possible, reduction of these species is, therefore, an obvious means of obtaining effective control of anthropogenically generated tropospheric ozone and other photooxidants.

Many oxygenated organics are used as solvents and some have also been suggested as alternative fuels. A wide variety of different types of compounds are utilized as solvents and include aromatic hydrocarbons, paraffins, ketones, alcohols, ethers, esters, and halocarbons. Solvent usage results in release of large amounts of these hydrocarbons to the atmosphere often of the same order of magnitude or higher than that from vehicle tailpipe emissions. By comparison of the potential of many of these compounds to produce O_3, as discussed earlier and shown in Tables 7 and 8, it is obvious that a rational selection of solvents should have a marked influence on tropospheric ozone and other photooxidant formation. For example, the POCP values imply that the replacement of all other solvents by acetone, on an equal mass emissions basis, would result in an approximately 70% reduction in the O_3 formation potential from solvent use. Therefore, the consideration of alternative solvent use is also a vital element in control strategies directed toward reduction in tropsopheric photooxidant production. The basis for the introduction of alternative/reformulated fuels and also alternative solvents is to control and reduce the reactivity of the VOCs emitted to the atmosphere. In the case of fuels the manipulation of the fuel composition is also designed to reduce emissions.

A variety of alternative fuels and technologies are or could become available for motorized vehicle use: including natural gas; methanol (and methanol blends); ethanol (and ethanol blends); liquid petroleum gas (LPG), including propane; hydrogen; electricity; and reformulated gasoline. The only alternative fuel that can be introduced on a short-term basis without the requirement of major new technology or distribution systems is reformulated gasoline (RFG). The use of oxygenated additives in motor vehicle fuels is rapidly increasing in the United States, Europe and also Asia. In the United States the change in fuel composition was greatly helped by the 1990 Clean Air Act Amendments. These amendments mandated the use of oxygenated fuels in areas of the country which exceed the National Ambient Air Quality Standard for carbon monoxide during the winter and in the nine worst summer smog areas. The use of reformulated fuels is increasingly required by law in many other countries around the globe even through the potential benefits and their ability to improve air quality, particularly in the long term, are still very uncertain.

The composition of the fuel is altered to make exhaust products less photochemically active and toxic and also to lower total emissions. This is achieved by modifying the refining process and by adding oxygenates. The addition of oxygenated compounds to fuels has three main functions: to increase the fuel oxygen content, to increase the fuel octane value, and to lower the fuel's Reid vapor pressure (RVP). The resulting fuel is lower in olefins and aromatic hydrocarbons. The oxygenated compounds which have been seriously considered as fuel additives include methanol, ethanol, *t*-butyl alcohol, dimethyl ether, diethyl ether, MTBE, ETBE, and *t*-amyl methyl ether (TAME). Other oxygenated compounds, including organic carbonates, have been suggested as potential oxygenated fuel additives because their use can minimize these problems. The present trend is, however, toward the use of ethers. The oxygenate content of reformulated fuels can constitute a few to more than 15 wt % of the fuel. The ethers, because of their low oxygen content, require high blending volumes to meet fuel oxygen requirements. For example, to meet a 2.7 wt % oxygen standard requires 15.1 (% by volume) MTBE in a standard fuel. Other compounds, such as dimethylcarbonate (DMC) and diethylcarbonate (DEC), have much higher oxygen contents (53.3 wt % and 40.6 wt %, respectively) compared to MTBE (18.2 wt %) and also their high boiling points can lead to a reduction in the RVP of the blended fuel.

Possible detrimental impacts on the environment through the use of alternative fuels is a matter that needs to be carefully thought about, preferably before their introduction, and is the subject of considerable controversy. For example, the nation-wide use of ethanol as a fuel in Brazil has resulted in high emissions of acetaldehyde, an important precursor to PAN. Similarly, experience in the North and South America shows that the use of methanol- and ethanol-enriched fuel significantly increases formaldehyde and acetaldehyde atmospheric concentrations, respectively. MTBE has also been found recently in ground water in the United States. It has to be considered whether possible short-term benefits may on the longer term lead to more serve and intangible problems in the future.

The effects of oxygenated gasoline use on vehicle emissions have been studied by several groups (Table 9). The results from measurements on vehicle emissions show that CO emissions are decreased by 16 ± 3% following an increase in gasoline oxygen content from 0 to 2% by weight. The studies indicate that an increase in oxygenates such as MTBE from 0 to 2% or more oxygen by weight results in a reduction of VOC emissions of around 18% with no significant change in NO_x emissions from light-duty vehicles. An increase of 5.1 ± 2.1 in NO_x emissions has been found when fuel ethanol was increased from 0 to 3.5% oxygen by weight.

Addition of oxygenates changes the properties of gasolines, such as the speciation of VOC emissions from motor vehicles. Addition of oxygenates such as MTBE to gasoline affects the speciation as well as the total mass of VOC emissions (Table 10). Other changes required for gasoline are to limit the summertime RVP of gasoline to 7.8 psi, whereas prior to 1992, the summertime RVP limit was 9.0 psi. To reduce RVP, the lightest hydrocarbons such as butanes and pentanes must be reduced or eliminated from gasoline.

Reactive Halogen Chemistry in the Troposphere

It is well established that reactive halogen species play an important role in stratospheric chemistry; in the troposphere,

Table 9. Average Gasoline Properties for Low Oxygenate and High Oxygenate Gasoline as Measured in San Francisco Bay Area, 1994

Fuel Property	Low oxygenate	High oxygenate
Oxygen content, wt %	0.3 ± 0.4	2.0 ± 0.2
Sulfur, ppm by wt	54 ± 47	90 ± 53
Reid vapor pressure, psi	7.2 ± 0.2	7.7 ± 0.3
Paraffins, vol %	47–54	38–46
Aromatics, vol %	34–43	26–35
Olefins, vol %	0.4–7.3	4.3–13.4
Naphthenes, vol %	2.9–10.4	4.1–9.6
Benzene, vol %	1.7–5.1	1.0–3.6

Table 10. Example of Speciation of VOC Emitted from Tailpipes of Light-Duty Vehicles Running On Low and High Oxygenate Fuels

Paraffins	Low oxygenate	High oxygenate	Olefins	Low oxygenate	High oxygenate
Methane	9.9 ± 2.3	11.0 ± 3.5	Ethene	7.8 ± 0.9	7.2 ± 0.5
Ethane	1.3 ± 0.2	1.2 ± 0.3	Propene	3.5 ± 0.4	3.3 ± 0.1
Propane	0.25 ± 0.03	0.23 ± 0.15	Isobutene	1.4 ± 0.2	2.6 ± 0.1
Isobutane	0.49 ± 0.06	0.34 ± 0.06	I-Butene	0.48 ± 0.06	0.43 ± 0.02
n-Butane	1.8 ± 0.2	1.9 ± 0.3	1,3-Butadiene	0.33 ± 0.22	0.36 ± 0.17
Isopentane	8.0 ± 0.6	6.9 ± 0.7	*t*-2-Butene	0.27 ± 0.13	0.36 ± 0.11
n-Pentane	2.1 ± 0.2	2.4 ± 0.3	*c*-2-Butene	0.16 ± 0.07	0.40 ± 0.16
2,2-Dimethylbutane	0.59 ± 0.18	0.38 ± 0.13	1-Pentene	0.16 ± 0.07	0.20 ± 0.05
2,3-Dimethylbutane	1.0 ± 0.2	0.56 ± 0.14	2-Methyl-1-butene	0.23 ± 0.13	0.34 ± 0.05
2-Methylpentane	2.4 ± 0.4	2.3 ± 0.2	*t*-2-Pentene	0.23 ± 0.13	0.36 ± 0.12
3-Methylpentane	1.3 ± 0.2	1.2 ± 0.1	*c*-2-Pentene	0.11 ± 0.03	0.18 ± 0.07
n-Hexane	1.2 ± 0.5	1.3 ± 0.2	2-Methyl-2-butene	0.43 ± 0.06	0.48 ± 0.06
Methylcyclopentane	1.5 ± 0.4	1.4 ± 0.1	Cyclopentene	0.14 ± 0.06	0.16 ± 0.04
2-Methylhexane	1.0 ± 0.2	1.0 ± 0.1	1-Hexane	0.11 ± 0.06	0.16 ± 0.05
2,3-Dimethylpentane	0.42 ± 0.12	0.35 ± 0.03			
3-Methylhexane	0.94 ± 0.17	0.87 ± 0.09			
2,2,4-Trimethylpentane	1.2 ± 0.1	1.2 ± 0.1			
n-Heptane	0.59 ± 0.09	0.72 ± 0.11			
Methylcyclohexane	0.36 ± 0.07	0.63 ± 0.11			
2,3,4-Trimethylpentane	0.34 ± 0.04	0.30 ± 0.11			
2,3-Dimethylhexane	0.24 ± 0.22	0.47 ± 0.15			
2-Methylheptane	0.30 ± 0.08	0.28 ± 0.08			
n-Octane	0.16 ± 0.08	0.12 ± 0.06			
C_{10}+ paraffins	1.4	1.1			

however, it has generally been assumed that OH radicals are the main driving force for tropospheric chemistry during daytime with NO_3 radicals playing a role during nighttime. Over the last few years more and more evidence has been emerging that reactive halogen species, which may include halogen atoms X (X = Cl, Br, I, and F), halogen monoxides XO, halogen acids (HOX), nitrosyl and nitryl halides (XNO, XNO_2), and halogen nitrates ($XONO_2$), may be actively involved in the oxidation of hydrocarbons in some regions of the troposphere (42,43). There is direct and indirect evidence for X/XO activity in the troposphere, particularly in coastal regions which represent major sources of halogen compounds. For example, *in situ* data ranging from 50 to 250 ppt HCl in the marine boundary layer and also up to 0.25 ppb Cl in the form of Cl_2 or HOCl has been reported (44). The time dependence and relative concentrations of hydrocarbons measured during springtime O_3 depletion episodes in the Arctic have been interpreted as being more in line with Cl atom rather than OH radical oxidation chemistry (45). In the same episodes the behavior of acetylene relative to other hydrocarbons has been interpreted as indicating Br radical chemistry. During the episodic destruction of boundary-layer ozone observed in the Arctic (Alert, > 80°N) and also in Ny-Alesund (Spitzbergen) high levels of filterable bromide have been observed. There has recently been positive identification of bromine oxide (BrO) at both Alert and Ny-Alesund using Long-Path DOAS with values up to 20 ppt. The oxides ClO and IO were both below the detection limit of the detection method. The nitrogen-containing species XNO, XNO_2, and $XONO_2$ have never been observed in the troposphere.

Excellent reviews on the formation and sink processes for X/XO and XNO_x in the atmosphere are available (46–48). There are many possible sources for the reactive halogen species and only the most important are mentioned briefly here. Over 2000 partially halogenated organic compounds are emitted naturally to the atmosphere (X) and this natural source is augmented by the release of halogen-containing species from industrial processes. Degradation of these halogenated compounds in the troposphere can lead to the release of halogen atoms. Sea salt aerosol contains (by weight) 55.7% Cl, 0.19% Br, and 0.00002% I and many reactions ultimately result in the release of halogen atoms. Reaction with O_3 in the troposphere will convert any I or Br virtually quantitatively to IO and BrO radicals, respectively. In the case of Cl because of its high reactivity toward hydrocarbons only approximately 50% will be converted to ClO. The reaction of F atoms with water (and also methane) is fast; therefore, because of the high water content of the atmosphere, any F atoms formed in the troposphere will be very quickly and irreversibly converted to HF and thus cannot play any role in tropospheric hydrocarbon chemistry.

Cl, Br, and I Reactions with Hydrocarbons

There is a fairly large kinetic database for the reactions of Cl atoms with hydrocarbons, although not quite as extensive as that for the OH radical. Much of these data are, however, either for room temperature or have only been investigated over a very limited temperature range. The kinetic database for the reactions of Br atoms with hydrocarbons was until recently very limited but has now been considerably extended. Reactions of I atoms with hydrocarbons are considered to be very slow and of no atmospheric importance. As will be explained below, the nature of the reaction conditions is particularly important for the reactions of Br atoms with unsaturated hydrocarbons. The rate coefficients are generally much lower than those for the analogous Cl atom reactions.

The reactions of both Cl and Br atoms with alkanes are straightforward abstraction reactions which produce an alkyl radical and HCl or HBr; the further reactions of alklyl radicals under atmospheric conditions have already been dealt with in the section on alkane chemistry:

$$RH + Cl/Br \rightarrow R + HCl/HBr$$

(R = simple or branched alkyl radical)

The reactions of Br atoms with alkanes are very slow and thus of negligible importance in the atmosphere. The reactions with Cl atoms on the other hand are fast in producing HCl. The fate of HX is important when assessing stationary-state reactive halogen concentrations. In the gas phase it can reform the halogen atom via reaction with OH:

$$HX + OH \rightarrow X + H_2O$$

However, HX is highly water soluble and can be removed irreversibly from the atmosphere by wet or dry deposition. Heterogeneous reactions have also been postulated which could return the reactive halogen atom to the atmosphere.

Both Cl and Br atoms, as for OH radicals, react with alkenes via addition to the double bond of alkenes with the addition to the less substituted carbon atom being favored.

$$C{=}C + X \rightarrow C{-}C(X)$$

For Cl-radicals reactions in many cases H-atom abstraction from alkyl substituents around the double bond will also be important. The further reactions of the halogen radicals will be as described for substituted alkyl radicals; the fate of the haloalkyl radicals will be addition of O_2 to form β-haloalkylperoxy radicals. The further fate of the β-haloalkylperoxy radicals has not been extensively studied. There are very few product and mechanistic studies of the reactions of Cl or Br atoms with alkenes most studies are incomplete because of difficulties in identifying and quantifying the numerous halogen-containing products arising from the secondary reactions of the primary halo-alkylperoxy radicals. Many of the suspected products are not commercially available and their syntheses pose the major hurdle to a quantitative product analysis and thus mechanistic interpretation. Many of the reactions are very similar to the analogue OH reaction. However, in contrast to the OH reaction the major fate of the β-chlorooalkoxy radicals is not decomposition but abstraction of the H atom attached to the alkoxy radical by O_2 to form β-chloroocarbonyl compounds and HO_2 radicals. In studies on the reaction of Br(Cl) with ethene in the absence of NO_x, $BrCH_2CHO$ ($ClCH_2CHO$) and $BrCH_2CH_2OH$ ($ClCH_2CH_2OH$) are the major reaction products.

In cases where halogens or higher alkyl groups are attached to the alkoxy radical, elimination of one of the halogen or alkyl groups occurs. Studies indicate that the bromoketones/aldehydes are much more photoactive than alkylbromides. Some of the most extensive product studies available are those for the reactions of Cl atoms with HCFCs which are being used to mimic the atmospheric OH degradation of these compounds because of the concern over stratospheric O_3 depletion.

Atmospheric Chemistry of Halogenated Carbonyls

Because of the interest in the fate of CFCs and the replacement HCFCs there is a growing body of data on the atmospheric chemistry of halogenated carbonyl compounds: reaction with OH radicals, photolysis and heterogeneous reactions. Many of the studies are concerned with Cl- and F-containing carbonyls and very little data are available on Br-containing carbonyls. Reactions of halogenated organics can release halogenated carbonyl in the lower troposphere; COXH, COX_2, $CH_yX_xCOX(H)$, and CX_3COX (X can be any combination of Cl, Br, F, and I atoms). For many of these compounds dissolution and hydrolysis in clouds and rain is believed to be the major atmospheric sink. Many aspects of the atmospheric chemistry of the halogenated carbonyls are still uncertain, particularly the effect of temperature and also the nature of the release of the halogen atoms.

CURRENT CONTROL STRATEGIES FOR OZONE AND OTHER OXIDANTS

Today, because of the very many serious environmental and human health problems associated with air pollution, there is a very real need in many parts of the world for an optimal control strategy for reduction of oxidant formation in the atmosphere. The development of effective measures requires that the ozone precursor relationship can be reliably described. All hydrocarbons, CO, and the nitrogen oxides are *precursors* for the formation of tropospheric ozone. If organic compounds and CO are greatly in excess over NO_x, then the ozone formation rate is generally determined by the availability of NO_x; in the reverse case the rate is controlled by the availability of VOC. Although there have been many studies on the tropospheric ozone problem these are often of limited value in an analysis of the problem because of limited measurement databases, shaky emission inventories, and inadequate knowledge concerning transported pollutants. The difficulties are compounded by the effects of complex meteorological variations which are often a dominating factor.

The scientific understanding and the advances as well as failures to control air quality with respect to O_3 and other photooxidant formation in the troposphere have been discussed in detail by experts in the United States (20,49) and in Europe the situation is under critical review. One of the first scientific-based approaches to gaining an insight into the intricacies of photooxidant formation was from empirical considerations using O_3 isopleths generated using the EKMA (empirical kinetics modeling approach) model (49). Systematic smog chamber experiments have lead to the development of an isopleth scheme: that is, three-dimensional isopleths generated with help of EKMA which depict the dependence of predicted ozone levels on the initial concentrations of VOCs and NO_x when mixtures are irradiated in air. From these model studies it has been established for US urban areas that close to the emission sources near midday for a [VOC]/[NO_x] ratio of ~ 10, very high ozone concentrations are quickly formed, with [VOC] in ppbC and [NO_x] in ppb. This EKMA strategy of the EPA has many shortcomings and has come under critical review in recent years. Today the preferred tools are urban-scale, three-dimensional Eulerian models, such as Urban Airshed Model often referred to as photochemical grid models or regional-scale Eulerian models (50).

In the United States the Clean Air Act of 1970 and subsequent amendments has led to substantial efforts to identify pollutants, restrict the emission of such pollutants, and to issue air quality criteria for them. On the basis of the experience gained by scientists in the United States, in Europe the European Commission is preparing several directives dealing with air quality and air quality standards. A specific directive is planned to regulate the emissions of volatile organic solvents. VOCs emitted from the use and manufacturing of solvents presently contribute around 34% to the total anthropogenic VOC emission in Europe. Another directive is concerned with Ambient Air Quality—Assessment and Management. Both of these directives, directly or indirectly, are also concerned with ozone and other harmful photooxidants which are produced by the oxidation of VOCs in the troposphere. One emission abatement approach under consideration is integrated pollution control: for example, a 30% reduction of the VOC emissions during the next few years. It has also been suggested that substance-specific characteristics are required to optimize cost-effective measures in controlling air quality. Such a strategy has recently been introduced for automobile emission control in California. In previous EU draft directives the ozone formation potential (OFP) has been identified as one possible characteristic property on which substance specific control could be based. However, as indicated earlier in the article, at present adequate information is not available to enable OFP for organic compounds to be determined with any degree of accuracy.

It has been recognized for some time that O_3 and its precursors are being transported between urban areas and that local controls alone are incapable of attaining a national air quality standards which governments may impose. The situation in the United States shows that the control of the photooxidants in photosmog requires detailed knowledge over transport and the atmospheric chemical behavior of the precursor substances. Success has only been observed after very drastic reductions in emissions. Despite enormous efforts in the United States to reduce ozone concentrations to the National Air Quality Standard of 120 ppb, this standard is often exceeded in many states across the country. It has been proposed by the Ozone Transport Assessment Group that the following five premises should serve as a solid foundation for the development of future ozone control strategies:

1. Urban VOC control may be effective for reducing ozone concentrations in and immediately downwind of urban areas.
2. Regional VOC control may be effective for reducing ozone on the urban scale, but less effective on the regional scale.
3. Regional NO_x control may be effective for reducing ozone on a regional scale.
4. Urban NO_x control may cause an increase in peak ozone concentrations on the local scale, but may be effective in reducing ozone further downwind on the regional scale.
5. Urban VOC control and regional NO_x control may be the most effective strategy.

In order to obtain the maximum benefit from the reductions in VOC emission levels, it is currently considered that an approach based on the OFPs of the VOCs may be preferable to a flat-rate emission reduction. Such an approach requires values of OFPs for the full range of VOCs emitted into the atmosphere. The use of transport and dispersion models to analyze and predict changes in air quality of different spatial scales is absolutely necessary. The reliability of modeling to a great extent depends on the input data employed: meteorology, emission data, and chemical modules. In urban air quality control many model studies and treatments of chemical mechanisms in the United States and in Europe have been published. However, many uncertainties remain in the chemical mechanisms. Our understanding of the chemical atmospheric oxidation mechanisms of hydrocarbons has greatly increased over the past few years and continues to do so. This coupled with the ever-increasing capacity of computers and our improved understanding of the meteorological and other factors involved in developing control strategies gives hope that reasonably effective controls will be possible within the near future.

BIBLIOGRAPHY

1. B.J. Finlayson and J.N. Pitts, Jr., *Atmospheric Chemistry. Fundamentals and Experimental Techniques,* John Wiley & Sons, Inc., New York (1986).
2. J.H. Seinfeld, *Atmospheric Chemistry and Physics of Air Pollution,* John Wiley & Sons, Inc., New York, 1986.
3. P. Warneck, *Chemistry of the Natural Atmosphere,* International Geophysics Series, Vol 41, Academic Press, Inc., London, 1988.
4. R. Atkinson, *J. Phys. Chem. Ref. Data,* Monograph 1 (1989); Monograph 2 (1994).
5. T. E. Graedel, D. T. Hawkins, and L. C. Claxton, *Atmospheric Chemical Compounds. Sources, Occurrence, and Bioassay,* Academic Press, Inc, London, 1986.
6. T. E. Graedel, *Chemical Compounds in the Atmosphere,* Academic Press, New York, 1978.
7. W. F. Erman, "Chemistry of the Monoterpenes. An Encyclopedic Handbook" in P. G. Gassman, ed., *Studies in Organic Chemistry,* Vol. 11, Marcel Dekker, Inc., New York, 1985.
8. V. A. Isidorov, *Organic Chemistry of the Earth's Atmosphere,* Springer-Verlag, Berlin, 1990.
9. J. J. Watson, J. A. Probert, and S. D. Piccot, *Global Inventory of Volatile Organic Compound Emissions from Anthropogenic Sources,* EPA-600/8-91-002 (NTIS PB91-161687), U.S. Environ. Prot. Agency, Air and Energ. Eng. Res. Lab., Research Triangle Park, N.C., 1991.
10. C. N. Hewitt, "Atmospheric VOCs from Natural Sources" in R. E. Hester and R. M. Harrison, eds., *Issues in Environmental Science and Technology 4, Volatle Organic Compounds in the Atmosphere,* Royal Society of Chemistry, London, 1995, pp. 17–36.
11. R. C. McDonald and R. Fall, *Atmos. Environ.* **27A,** 1709–1713 (1993); P. D. Goldan, W. C. Kuster, F. C. Fehsenfeld, and S. A. Montzka, *J. Geophys. Res.* **100,** 25,945–25,963 (1995).
12. S. Solberg, N. Schmidbauer, U. Pedersen, and J. Schaug, *VOC Measurements August 1992–June 1993,* EMEP/CCC Report 6/93, Norwegian Institute for Air Research, Lillesstrom, Norway, 1993.
13. R. G. Derwent, D. R. Middleton, R. A. Field, M. E. Goldstone, J. N. Lester, and R. A. Perry, *Atmos. Environ.* **29,** 923 (1995).
14. R. L. Seila, W. A. Lonneman, and S. A. Meeks, "Determination of C2 to C12 Ambient Air Hydrocarbons in 39 U.S. Cities from 1984 through 1986," *Office Resource Development Report,* 600/3-89/058, Research Triangle Park, N.C., 1989; *Proceedings US/FRG Workshop on photochemical Ozone Problem and its Control-US Experiences and the Situation in Europe,* EPA/600/9-89/058, Research Triangle Park, N.C., 1989.

15. S. K. Hoekman, *Environ. Sci. Technol.* **26,** 1206–1216 (1992).
16. R. W. J. Shaw, A. L. Crittende, R. K. Stevens, D. R. Cronn, and V. S. Titov, *Environ. Sci. Technol.* **17,** 466–475 (1983).
17. F. Fehsenfeld and co-workers, *Global Biogeochem. Cycles* **6,** 389–430 (1992).
18. J. M. Roberts, F. C. Fehsenfeld, D. L. Albritton, and R. E. Sievers, *J. Geophys. Res.* **88,** 10,667–10,678 (1983).
19. P. R. Zimmerman, J. P. Greenberg, and C. Westberg, *J. Geophys. Res.* **93,** 1407–1416 (1988).
20. National Research Council, "Rethinking the Ozone Problem in Urban and Regional Air Pollution", National Academy Press, Washington, D.C., 1991.
21. A. Guenther, C. N. Hewitt, D. Erickson, R. Fall, C. Geron, T. Graedel, P. Harley, L. Klinger, M. Lerdau, W. A. McKay, T. Piece, B. Scholes, R. Steinbrecher, R. Tallamraju, J. Taylor, and P. Zimmerman, *J. Geophys. Res.* **100,** 8873–8892 (1995).
22. S. D. Piccot, J. J. Watson and J. W. Jones, *J. Geophys. Res.* **97,** 9897–9912 (1992).
23. D. Simpson, A. Guenther, C. N. Hewitt, and R. Steinbrecher, *J. Geophys. Res.* **100,** 22,875–22,890 (1995).
24. R. P. Wayne, *J. Geophys. Res.* **98,** 13,119–13,136 (1993).
25. R. P. Wayne, *Principles and Applications of Photochemistry,* Oxford University Press, New York, 1988.
26. R. G. Prinn, R. F. Weiss, B. R. Miller, J. Huang, F. N. Alyea, D. M. Cunnold, P. J. Fraser, D. E. Hartley, and P. G. Simmonds, *Science* **269,** 187–192 (1995).
27. U. Platt and F. Heintz, *Israel J. Chem.* **34,** 289–300 (1994).
28. R. Atkinson, D. L. Baulch, R. A. Cox, R. F. Hmpson, Jr., J. A. Kerr, and J. Troe, *J. Phys. Chem. Ref. Data,* **21,** 1125–1600 (1992).
29. D. Simpson, *Atmos. Environ.* **27A,** 921 (1993).
30. R. Atkinson, *Int. J. Chem. Kinetics,* **29,** 99–111 (1997).
31. E. S. C. Kwok and R. Atkinson, *Atmos. Environ.* **29,** 1685–1695 (1995).
32. World Health Organization, *Air Quality Guidelines for Europe,* WHO Regional Publications, European Series No. 23, Copenhagen, WHO, Regional Office for Europe, 1987.
33. D. Klei, "*Tropospheric ozone in the global, regional and subregional context*", in C. F. Boutron, ed., *Topics in Atmospheric and Interstellar Physics and Chemistry,* Les Editions de Physique Les Ulis, Grenoble, France, 1994, pp 161–184.
34. F. M. Bowman and J. H. Seinfeld, *J. Geophys. Res.* **99 D3,** 5309–5324 (1994).
35. W. P. L. Carter, *J. Air Waste Man. Assoc.* **44,** 881–899 (1994); W. P. L. Carter and R. Atkinson, *Environ. Sci. Technol.* **23,** 864–880 (1989).
36. R. G. Derwent, M. E. Jenkin, and S. M. Saunders, *Atmos. Environ.* **30,** 181–199 (1996).
37. Y. Andersson-Skold, P. Grennfelt, and K. Pleijel, *J. Air Waste. Man. Assoc.* **42,** 1152–1158 (1992).
38. P. J. H. Builtjes, M. G. M. Roemer, A. C. Baart, G. Boersen, and P. Verhoeve, TNO Report IMW R 92/419a, TNO, Delft, the Netherlands.
39. D. Simpson, EMEP MSC-W Note 1/91, The Norwegian Meteorological Institute, Oslo, Norway, 1991.
40. T. W. Kirchstetter, B. C. Singer, R. A. Harley, G. R. Kendall, and W. Chan, *Environ. Sci. Technol.* **30,** 661–670 (1996).
41. W. R. Leppard, L. A. Rapp, V. R. Burns, R. A. Gorse, J. C. Kneper, and W. J. Koehl, *SAE Tech. Pap. Ser.,* No 920329 (1992).
42. P. J. Squillace, J. S. Zogorski, W. G. Wilber and C. V. Price, *Environ. Sci. Technol.* **30,** 1721–1730 (1996)
43. H. B. Singh, "Halogens in the Atmospheric Environment", in H. B. Singh, ed., *Composition, Chemistry and Climate of the Atmosphere,* VNR, New York, 1995, pp. 216–250.
44. A. A. P. Pszenny, W. C. Keene, D. J. Jacob, S. Fan, J. R. Maben, M. P. Zetwo, M. Springer-Young and J. N. Galloway, *Geophys. Res. Lett.* **20,** 699–702 (1993).
45. B. T. Jobson, H. Niki, Y. Yokouchi, J. Bottenheim, F. Hopper and R. Leaitch, *J. Geophys. Res.* **99,** 25,355–25,368 (1994).
46. U. Platt and C. Janssen, *Faraday Discuss.* **100,** 175–198 (1995).
47. R. P. Wayne, G. Poulet, P. Biggs, J. P. Burrows, R. A. Cox, P. J. Crutzen, G. D. Hayman, M. E. Jenkin, G. LeBras, G. K. Moortgat, U. Platt, and R. N. Schindler, *Atmos. Environ.* **29,** 2677–2881 (1995).
48. B. J. Finlayson-Pitts, *Res. Chem. Intermed.* **19,** 235–249 (1993).
49. H. E. Jeffries, "Photochemical Air Pollution", in H. B. Singh, ed., *Composition, Chemistry and Climate of the Atmosphere,* VNR, New York, 1995, pp. 308–347.
50. T. E. Graedel and P. J. Crutzen, *Atmospheric Change, An Earth System Perspective,* W. H. Freeman and Co., New York, 1992.
51. P. Middleton, in Ref. 43, pp. 88–119.

HYDROGEOLOGY

LESLIE SMITH
ROGER BECKIE
University of British Columbia
Vancouver, British Columbia, Canada

Hydrogeology is the study of the movement of water through the subsurface environment, its interaction with the sediment or rock matrix, and the transport of chemical species by groundwater flow. Groundwater within near-surface, permeable sediments or rocks constitutes about 30% of the fresh water within the hydrologic cycle, second only to polar ice as a reservoir of fresh water. Groundwater is a major source of water supply; in 1990, groundwater accounted for about 23% of the daily fresh water requirements of the United States. Locally, groundwater can take on even greater importance. Many communities and cities derive nearly 100% of their water supply from groundwater reservoirs. From a biospheric perspective, groundwater is the main contributor to stream flow in periods between storm or snowmelt runoff events and is often a key component in the hydrologic budget of wetlands. For these reasons, protection of groundwater systems from contamination has emerged as a major issue in environmental management. Problems involving contamination of groundwater resources or issues related to remedial options at contaminated sites cannot be dealt with rationally without adequately characterizing the hydrogeological setting and the hydraulic and geochemical properties of the underlying sediments and/or rock units.

SITE INVESTIGATION

The site investigation is designed to characterize a groundwater flow system to reach a management decision related to a groundwater contamination problem. The effort begins with formulating a conceptual model of the geologic setting and the subsurface flow system. The conceptual model is viewed as a hypothesis describing the main features of the geology, the hydrologic setting, and site-specific relations between geologic structure and patterns of fluid flow. To permit a quantitative analysis, the conceptual model is transformed into a corresponding mathematical model. Mathematical modeling is a process of hypothesis testing, permitting refinement of the conceptual model and its expression in the form of a so-called hydrogeological simulation model.

Site characterization is an iterative process, beginning with the development of a simple model to guide initial collection of field data and proceeding through a sequence of model refinements, data collection, and model evaluation as knowledge about the groundwater system is accumulated. At each stage, model predictions are evaluated in the light of project needs. A decision is made to proceed to project completion or to cycle through an additional round of data collection and hydrogeologic simulation to reduce uncertainties in the analysis. This framework of viewing hydrogeologic analysis as a decision model provides the basis for the presentation of technical material in this article.

HYDRAULIC FRAMEWORK

Geological Setting

Consideration of the geologic setting of a study area is integral to analyzing groundwater flow. A description of the geologic setting includes documenting the sediment and rock types in the study area, characterizing the spatial and temporal relationships among the different geologic units, gaining a sense of the scale of the spatial variability, and a description of the structural features in the area (eg, folds, faults). An understanding of the geological processes forming and modifying the sediments and rock units provides the best basis for extrapolating conditions observed at the ground surface or within boreholes to the much larger volume of the subsurface environment in a study area for which no samples or measurements are available. In this article, geological properties important to hydrogeologic analysis are discussed as needed, rather than isolating the concepts in a separate section of the chapter. A comprehensive review of groundwater conditions in a broad range of geologic settings and geographic areas within North America has been compiled (1).

Hydraulic Potential and Fluid Flux

Groundwater moves through interconnected void spaces in soils, unconsolidated sediments, and rocks. These voids are pore space between grain surfaces, microcracks between grain boundaries, or fractures. Porosity n is defined as the fraction of void space per unit volume of porous medium. It is a dimensionless number less than one, although porosity is frequently reported as a percentage. Hydrogeologists generally distinguish among several types of porosity. Primary porosity refers to the void space between grains. Secondary porosity results from fracturing or chemical dissolution of the mineral grains. Effective porosity includes only the void space that forms interconnected flow paths through the medium and excludes isolated or dead-end pore space. Table 1 lists representative ranges of porosity for a variety of sediment and rock types.

Table 1. Representative Porosity Values of Geologic Materials

Material	Porosity
Clays	0.40–0.60
Silts	0.35–0.50
Fine sands	0.20–0.45
Coarse sands	0.15–0.35
Sandstones	0.05–0.35
Shales (unweathered)	0.01–0.10
Shales (weathered)	0.30–0.50
Igneous rocks (unfractured)	0.0001–0.01
Igneous rocks (fractured)	0.01–0.10

The driving force or hydraulic potential causing groundwater flow is expressed in terms of the mechanical energy of the fluid per unit mass of water (2). For the great majority of groundwater flow systems, the driving force is related simply to differences in fluid pressure and elevation within the flow domain. The driving force is measured in terms of the hydraulic head at a point within the porous medium, where

$$h = z + \frac{p}{\rho g} \qquad (1)$$

Here h is the hydraulic head (units of length), z is the elevation of the point above datum, p is the fluid pressure, ρ is the fluid density, and g the acceleration of gravity. The datum is often conveniently set at mean sea level, but its choice is arbitrary. Hydraulic head is sometimes referred to as the piezometric or potentiometric head. Groundwater always flows from regions of higher hydraulic head toward regions of lower head, independent of the direction in space.

It is conventional to define fluid pressures in terms of values above atmospheric pressure. Furthermore, defining the pressure head (Ψ) as

$$\Psi = \frac{p}{\rho g} \qquad (2)$$

leads to an expression for hydraulic head as the sum of two components, the pressure head Ψ and the elevation head z:

$$h = z + \Psi \qquad (3)$$

Hydraulic head is measured in a piezometer (Fig. **1a**). Standpipe piezometers are constructed by installing PVC or metal casing inside a borehole, with an open screen connected to the bottom of the casing to provide an inflow zone. The diameter of the casing is usually in the range from 4 to 6 cm, with the screen length from 0.5 to 3 m. A clean, coarse-grained sand is placed around the screen, and the remaining annular space is backfilled to the surface with a cement/clay mixture. In this way, an isolated measurement of fluid pressure is obtained, representative of the fluid pressure in the porous medium surrounding the piezometer screen. The hydraulic head is calculated from equation 3, given a measurement of the depth to water in the piezometer. By convention, the midpoint elevation of the piezometer screen defines the elevation head. Multiport piezometer systems have been developed that permit numerous measurements of fluid pressure at different depths within a single borehole (Fig. **1b**). A principal advantage of this technology is the small interval over which a pressure measurement is made, providing a detailed and accurate picture of the variation in hydraulic head with depth.

Contour maps of hydraulic head characterize directions of subsurface flow. Figure 2, for example, shows a map of the hydraulic head and general pattern of groundwater flow in alluvial sediments within the Avra Valley, near Tucson, Arizona.

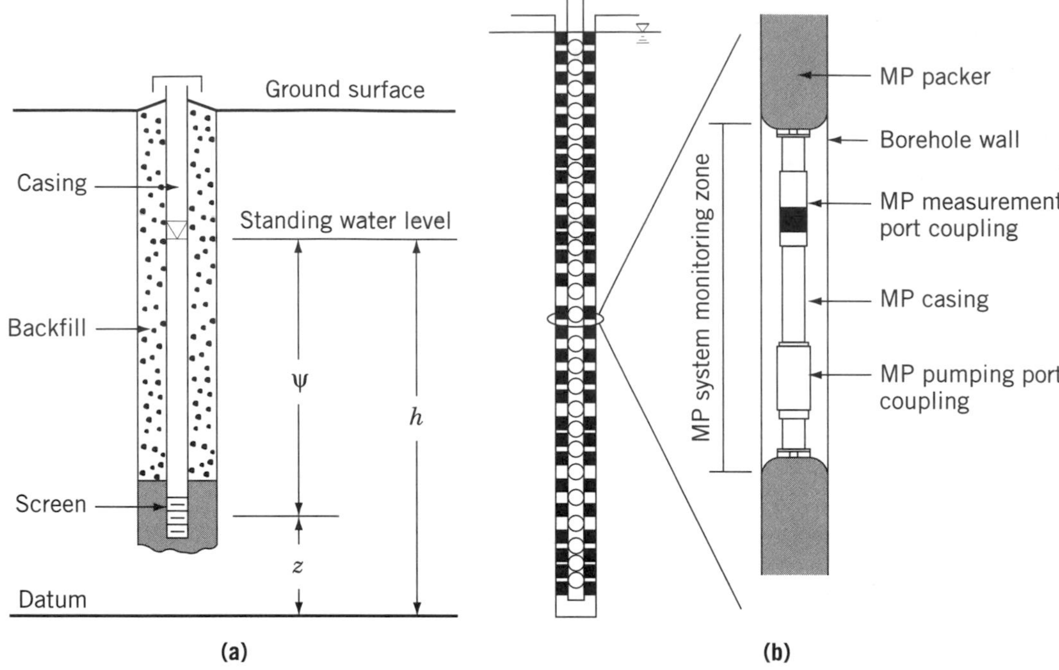

Figure 1. (**a**) Schematic diagram of a piezometer, illustrating the relationships between hydraulic head, pressure head, and elevation head; (**b**) Westbay multiport (MP) piezometer system.

Water below the ground surface occurs in two regimes, the vadose (or unsaturated) zone and the saturated zone. These two zones are separated by the water table. Below the water table in the saturated zone, fluid pressure exceeds atmospheric pressure ($\Psi > 0$). The moisture content, defined as the volume of water contained in a unit volume of porous medium, is equal to the porosity. Above the water table in the vadose zone, not all the void space is filled with water. Air, typically at atmospheric pressure, is also present in the pore space. As a consequence, the moisture content is smaller than the porosity. Water is held in tension so that the fluid pressure is less than atmospheric pressure ($\Psi < 0$). The lower the moisture content, the more negative the fluid pressure. Each soil or sediment type within a study area has a characteristic moisture content–pressure head relationship.

The water table is defined as the surface on which the pressure head is equal to zero. Given a datum of sea level, the hydraulic head at a point on the water table is equal to the elevation of the water table above mean sea level at that point (see eq. 3). A zone of saturation, called the capillary fringe, exists above the water table, where water rises upward from the water table by capillary attraction. Within the capillary fringe, the moisture content is equal to the porosity, but the pressure head is negative because the water is held in tension. The capillary fringe has a greater vertical extent in finer grained sediments than in coarse-grained materials (eg, up to several meters in silts, compared with several tens of centimeters in medium- to coarse-grained sands). The pressure head in the vadose zone and capillary fringe is measured with a tensiometer, a device that measures the suction across a porous ceramic membrane in contact with the water in the pore space. Although the pressure head is negative, equation 3 defines the hydraulic potential above the water table and is used to infer directions of flow and hydraulic gradients within the vadose zone.

The water table is located by noting the depth to water in a borehole that just penetrates the saturated zone. A borehole above the water table does not contain water because water in the surrounding porous medium is held in tension and it cannot flow into the well bore. Water only enters the well bore where the pressure head in the adjacent medium is greater than zero. Boreholes with completion intervals or well screens a considerable distance below the water table poorly indicate the location of the water table, if the borehole is in an area with a significant component of vertical groundwater flow.

The volumetric flow of groundwater in the saturated zone is calculated by Darcy's law. For the component of flow in the x-coordinate direction,

$$Q_x = KA\frac{\delta h}{\delta x} \tag{4}$$

where Q_x is the volumetric flow rate in the x-direction, A is the cross-sectional area of flow, K is the hydraulic conductivity of the porous medium, and $\delta h/\delta x$ is the component of the hydraulic gradient in the x-direction. Similar expressions can be written for the flow components Q_y, Q_z in the other coordinate directions. Darcy's law is valid for laminar flow, a condition met in the great majority of hydrogeologic settings.

Hydraulic conductivity is a measure of the ability of a fluid to move through the interconnected void spaces in a sediment or rock. It has high values for sands and gravels and low values for silts and clays. Hydraulic conductivity depends on properties of the porous medium and the fluid. Hydraulic conductivity has the units of L/T, for example, m/s. To isolate the effects of the medium from those of the water, the permeability k of a

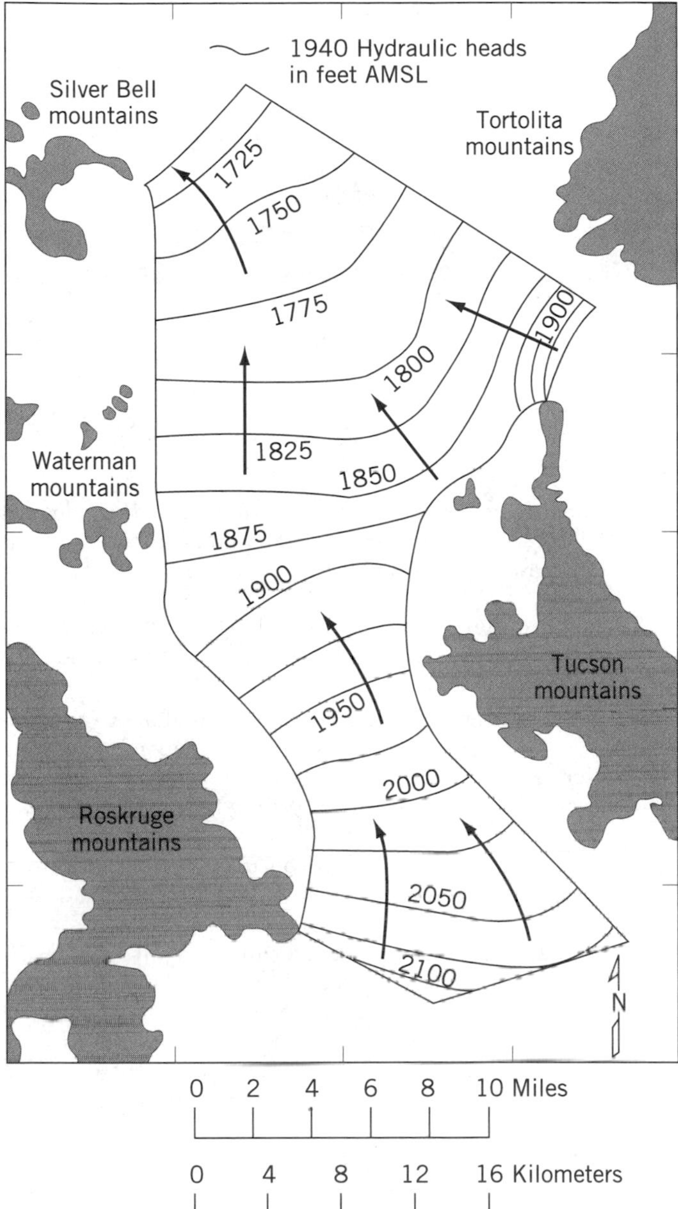

Figure 2. Example of a hydraulic head map and inferred directions of groundwater flow in Avra Valley, Arizona (3).

porous medium is related to the hydraulic conductivity through the expression

$$K = \frac{k\rho g}{\mu} \tag{5}$$

where μ is the dynamic viscosity of the fluid. Permeability has units of L^2, for example, m^2. Table 2 lists representative ranges of hydraulic conductivity for various sediment and rock types. Permeability values are obtained from Table 2 by multiplying the value of hydraulic conductivity by 10^{-7}, for permeability values in m^2.

For the various sediment or rock types listed in Table 2, hydraulic conductivity varies over many orders of magnitude. In contaminant investigations at a site with multiple geologic units, it is not uncommon to see variations in hydraulic conductivity between units of five or six orders of magnitude. For a single sediment or rock type, the hydraulic conductivity can range over two or three orders of magnitude, because of minor changes in the grain size distribution or pore structure.

Table 2. Representative Hydraulic Conductivity Values of Geologic Materials

Material	Hydraulic Conductivity, m/s
Unfractured clay	$10^{-10}-10^{-7}$
Glacial till	$10^{-10}-10^{-4}$
Clean sand	$10^{-3}-10^{-2}$
Silty sand	$10^{-5}-10^{-1}$
Gravel	$10^{-1}-1$
Shales	$10^{-12}-10^{-9}$
Shales (fractured, weathered)	$10^{-9}-10^{-5}$
Igneous rocks (unfractured)	$< 10^{-12}-10^{-8}$
Igneous rocks (fractured)	$10^{-6}-10^{-2}$
Limestone	$10^{-7}-10^{-3}$
Karst limestone	$10^{-4}-1$

It is often convenient to describe groundwater flow in terms of the specific discharge, that is, the flow of water per unit area. In this case,

$$q_x = \frac{Q_x}{A} = K\frac{\delta h}{\delta x} \tag{6}$$

where A is the total cross-sectional area of flow. The specific discharge has units of L/T, for example, m/s. It is sometimes referred to as the Darcy flux, or the Darcy velocity. The pore water velocity v is related to the specific discharge by

$$v_x = \frac{q_x}{n_e} \tag{7}$$

where n_e is the effective porosity. The pore water velocity (also called the seepage velocity) is the average velocity at which dissolved solutes are transported through a porous medium. Under natural flow in near-surface permeable sediments, typical groundwater velocities range from ten to several hundred meters per year. In media with lower hydraulic conductivity, groundwater velocities are correspondingly lower. For example, in compacted, unfractured clays, groundwater velocities are on the order of mm per year, or less.

Hydraulic conductivity in the vadose zone is not a constant, but depends on the moisture content of the porous medium at a point. As the moisture content decreases, there are fewer connected pathways in the water phase through the pore structure, and the hydraulic conductivity of the medium decreases. Darcy's law is rewritten for the vadose zone as

$$Q_x = K(\theta)\frac{\delta h}{\delta x} \tag{8}$$

where θ is the moisture content.

Time-Dependent Changes in Hydraulic Head

If the hydraulic head values in a hydrogeologic system do not change significantly through time, then it is approximated as a steady-state system. One snapshot in time is sufficient to characterize the patterns of subsurface flow. No system, however, is in a true steady state because many processes cause water levels in piezometers to fluctuate on time scales ranging from a few minutes to hundreds of years, depending on the nature of the processes that initiate the fluid pressure variations. Short-term changes in hydraulic head reflect changes in the barometric pressure in the atmosphere, earth tides, or seismic events. Earth tides lead to water level changes of several centimeters, and atmospheric pressure changes cause fluctuations of up to several tens of centimeters. In piezometers completed in the vicinity of the water table, fluctuations in the water level occur in response to time-varying rates in the use of groundwater by plants whose roots penetrate to the water table (phreatophytes). Yearly cycles reflect changing wet and dry seasons and consequent changes in the temporal distribution of groundwater recharge. Multiyear cycles are tied to longer term changes in precipitation patterns.

Groundwater withdrawals by pumping cause large-scale transient changes in hydraulic head as fluid pressure is reduced in the region surrounding the pumping well, creating the hydraulic gradient necessary for groundwater to flow towards the well bore. When fluid pressures are reduced in a porous medium, two responses occur: the fluid volume expands because of the lower fluid pressure, and the pore volume decreases because an additional fraction of the overburden pressure is carried by the solid matrix. The magnitude of fluid volume expansion is controlled by the compressibility of water (β), whereas the magnitude of the reduction in pore volume is controlled by the compressibility of the porous medium (α). Clays and silts have relatively high compressibility values, whereas rocks, such as sandstones or granites, are quite rigid. These two responses are often combined in a single parameter that characterizes the volume of water released from storage with a decline in fluid pressure. Specific storage (S_s) is defined as the volume of water that a unit volume of porous medium releases from storage per unit change in hydraulic head. It is related to the compressibilities of water and the porous medium by the equation

$$S_s = \rho g(\alpha + n\beta) \tag{9}$$

Representative values of specific storage for various geologic media are given in Table 3.

HYDROGEOLOGIC SETTINGS

Hydrogeological models are based on the hydraulic properties of the soil and rock units within the region of interest. Rocks or sediments with similar hydraulic properties are grouped to form hydrostratigrahic units, which define the large-scale heterogeneity in the conceptual model of the flow system. The following terminology provides a simple classification of the major hydrogeologic features at a site (Fig. 3):

Aquifer–a permeable geologic unit that transmits and stores significant quantities of water

Aquitard–a less permeable unit that stores but does not readily transmit water

Unconfined aquifer–a permeable geologic unit with the water table forming its upper boundary

Table 3. Representative Specific Storage Values of Geologic Materials

Material	Specific storage, 1/m
Unconsolidated clays	10^{-4}–10^{-2}
Sands	10^{-5}–10^{-3}
Gravels	10^{-6}–10^{-4}
Compacted sediments	10^{-7}–10^{-5}
Igneous rocks	10^{-7}–10^{-5}

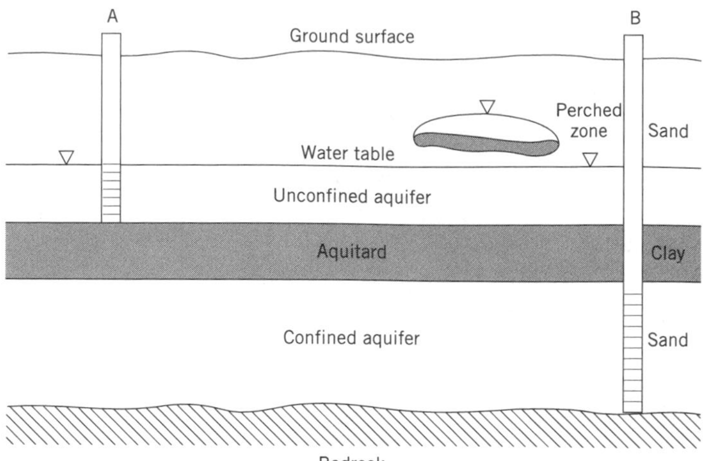

Figure 3. A simple classification of common hydrogeological settings. Well A withdraws water from an unconfined aquifer; well B withdraws water from a confined aquifer.

Confined aquifer–a permeable geologic unit located beneath a saturated, less permeable unit, that is, beneath an aquitard

Perched zone–a zone of limited area, located above the continuous zone of saturation, that occurs when infiltrating water is impeded by a low permeability unit, creating saturated zones above the impeding layer.

If the bedrock beneath the unconsolidated deposits illustrated in Figure 3 is permeable, it too is classified as an aquifer. However, if it has low permeability, it behaves as an aquitard.

It is common to describe groundwater flow in aquifers by a two-dimensional, plan-view model (eg, Fig. 2). In this case, two derived parameters are introduced. The transmissivity of an aquifer is the hydraulic conductivity value multiplied by the aquifer thickness, $T = Kb$. The storage coefficient of a confined aquifer is the specific storage value multiplied by the aquifer thickness, $S = S_s b$.

In a confined aquifer subject to transient flow conditions, the volumes of water released from storage by fluid expansion and aquifer compaction are approximately equal. In a hydrogeologic setting where permeable sands are interbedded with compressible silts and clays, the dominant source of water over the long term is usually associated with compaction of the silt and clay layers. In this setting, large-scale water-supply developments lead to land subsidence (4). In addition to their elastic storage properties, unconfined aquifers yield water by desaturation of the pore space as the water table declines. Water released from storage in an unconfined aquifer greatly exceeds that of a confined aquifer for equal water-level declines. To reflect this difference, storage of an unconfined aquifer is defined in terms of the specific yield, which is equal to the volume of water released from a unit area of the aquifer for a unit decline in the water table. The specific yield is a significant fraction of the effective porosity.

A hydrostratigraphic unit is homogeneous if its hydraulic properties are uniform throughout the unit. However, no unit is truly homogeneous because of spatial and temporal variability in the geologic processes that create and modify porous media. Heterogeneity occurs on a range of spatial scales. The analyst must decide whether or not homogeneity approximations are adequate, and if not, at what scale or level of detail the heterogeneity must be specified. These decisions are incorporated within the conceptual model and then are refined during the development of the hydrogeologic simulation model.

In an isotropic porous medium, hydraulic conductivity is independent of the direction of flow. Groundwater flow is in the same direction as the hydraulic gradient (and therefore, in an unconfined aquifer, groundwater flow is perpendicular to contour lines of equal water table elevation; see Fig. 2). In an anisotropic porous medium, the value of hydraulic conductivity depends on the direction of flow. When considered at the grain scale, anisotropy results from the preferred alignment of elongate sand grains, which creates a greater resistance to flow in a direction across the grain alignment, compared with a direction parallel to the grain alignment. Flow lines are not aligned with the direction of the hydraulic gradient but are rotated toward the coordinate direction with the higher hydraulic conductivity. The direction of the flow line relative to the orientation of the hydraulic gradient (θ_j) and the anisotropic ratio $K_r = K_z/K_x$ is given by the following:

$$\theta_q = \tan^{-1}(K_r \tan\theta_j) \tag{10}$$

where θ_q is the angle of the flow line relative to orientation of the x-axis. In mapping the spatial distribution of contaminants beyond a source area, anisotropy in hydraulic conductivity, if present, must be recognized if boreholes are to be located in an effective pattern to define the contaminant plume.

In sedimentary deposits such as fluvial or deltaic sands or their equivalent lithified forms (sandstone), the hydraulic conductivity in a direction perpendicular to stratification is usually less than that parallel to the stratification. When measured on a core sample, the hydraulic conductivity parallel to the stratification is usually from 2 to 10 times higher than that in a direction perpendicular to the layering. When considered at the scale of tens to hundreds of meters in a geologic system composed of alternating layers of silts and sands, the equivalent hydraulic conductivity for flow parallel to the layering is several orders of magnitude greater than the equivalent hydraulic conductivity for flow across the layering. Anisotropy in the horizontal plane is less common, but it is expected at sites underlain by steeply dipping, layered sedimentary rocks or in fractured rocks.

A surprisingly common issue in contaminant investigations within regions underlain by glacial deposits concerns the continuity of low-permeability layers that protect deeper aquifers from contamination. Glacial deposits are often extremely variable and the possibility always exists that the low-permeability layer is locally absent. These windows in the low-permeability layer provide potential pathways for downward migration of contaminants. If the analyst preceives that this issue is important at a site, this condition should be included in formulating the initial conceptual model.

Carbonate rocks, such as limestone or dolomite, have a diverse hydrologic character; they can be extremely productive aquifers or tight confining beds. Their hydraulic properties depend on three key factors: (*1*) the nature of the original carbonate sediment, (*2*) the degree to which the original porosity of the sediment is lost by post-depositional processes, such as compaction and cementation, and (*3*) the extent to which fresh water flowing through the rock mass creates new porosity and permeability by chemical dissolution. Solution channels and karst features create a highly permeable rock mass. Karst typically consists of a network of open conduits transmitting water through a porous and fissured matrix. The extent of permeability development within each of these two domains and the hydraulic connections between them determine the nature of the flow system. Quantitative analysis of groundwater flow in karst terrain is difficult, and predictions are subject to considerable uncertainty. The most widely adopted investigative techniques are based on the analysis of time-varying discharge rates of karst springs and on the use of chemical tracers to identify recharge zones, estimate groundwater velocities, and to define the conduit geometry (5).

The hydraulic properties of crystalline rocks, such as granite, strongly depend on the extent of fracturing within the rock mass. Regions with intact or sparsely fractured crystalline rock have low permeability. Fractured media at the local scale are extremely heterogeneous. Fracture properties controlling fluid flow include the number of fractures, their spacing, fracture length, orientation of fracture sets, and fracture aperture. Although fracture porosity is on the order of $\leq 0.1\%$, a well-connected fracture network has a large impact on the hydraulic conductivity of the rock mass. Fractures typically have lengths on the order of centimeters to tens of meters and occur in sets that are defined by the orientation of the fracture planes. Because fractures typically occur in sets with preferred orientations, the hydraulic properties of fractured rocks are commonly anisotropic. Field evidence suggests that the majority of the flow within a fractured rock mass occurs in a relatively small percentage of the total population of fractures. The hydraulic conductivity of crystalline rock typically decreases by two to three orders of magnitude in the first several hundred meters below the ground surface as the number of fractures decreases and fractures close under the increasing weight of the overlying rock.

Faults are discontinuities in a rock mass along which shear displacement occurs. They appear as intermixed zones of clay-rich crushed rock (gouge) and highly fractured rock, grading outward into a less fractured rock mass. Permeability of the highly fractured rocks within the fault zone are three to five orders of magnitude higher than in the rock mass surrounding the fault, in which case the geometry of the fault zones controls the patterns of groundwater flow. However, if clay-rich gouge predominates, the permeability of the fault zone is hydrologically similar to that of the intact rock mass, or it may act as a barrier to flow. Offset of geologic strata along faults disrupts the lateral continuity of higher permeability zones or aquifers.

Site characterization in fractured rock terrain requires considerable planning and forethought if cost-effective data are to be obtained. This viewpoint is particularly relevant in groundwater contamination studies because of the important but difficult task of identifying the dominant features controlling the fluid pathways and their hydraulic connections within a fractured rock mass. For detailed investigative approaches useful in fractured rock settings, see References 6,7, and 8.

In regions underlain by lower permeability rocks, a surficial weathered horizon is sometimes identified as a separate hydrostratigraphic unit. Breakdown of the rock matrix by weathering and/or a greater amount of fracturing creates a zone where the hydraulic conductivity is two to three orders of magnitude higher than that of the parent rock. However, if chemical weathering predominates and clay minerals form within this weathered zone, then the hydraulic conductivity of the weathering zone remains similar to that of the parent rock. In this case, a separate hydrostratigraphic unit for the weathered zone is not likely to be incorporated within the conceptual model.

CONCEPT OF A FLOW SYSTEM

The topographic relief within a watershed is a primary factor in determining patterns of subsurface flow. In most instances, the relief on the water table reflects the surface topography; the elevation of the water table is higher beneath topographic highs, and is at or close to the ground surface in the topographically low regions of a basin. Topographically higher areas usually form groundwater recharge areas, whereas the topographic lows form groundwater discharge areas. Piezometer nests located in groundwater recharge areas have lower water levels in the deeper piezometers. The hydraulic head must be lower in the deeper piezometers for water to move away from the water table. In groundwater discharge areas, deeper piezometers have higher water levels. The hydraulic head must be higher in the deeper piezometers for the water to move upward toward the water table. Large regions of a watershed overlie regions of lateral groundwater flow, with little or no net flow across the water table. In these instances, water levels are about the same in piezometers completed at different depths at the same location.

Topographic highs and lows often coincide with groundwater divides which separate one flow system from another. However, without field data it is uncertain to what depth a surface-water divide can be projected downward as a groundwater divide. Flow systems are classified according to scale. A local flow system transmits water from a local high to an adjacent topographic low, whereas a regional flow system transmits water between the topographically higher areas of a basin and the regional topographic low (Fig. 4). Extended discussions of regional flow systems are available (9,10). In assessing the potential impacts of off-site contaminant migration from a waste disposal facility, it is essential to characterize the local groundwater flow patterns and to place the facility in its regional hydrogeological setting.

Discharge areas are not limited only to stream channels or surface bodies of water. A broad region of groundwater discharge may occur in areas with low topographic relief. Groundwater moving up through the water table is removed from the vadose zone by evaporation and transpiration. In this case, salt crusts may develop on the soil surface and provide a simple mapping technique for identifying discharge areas, particularily in semiarid to arid areas.

When flow lines cross from one hydrostratigraphic unit to another with hydraulic conductivity values K_1 and K_2, respectively, the flowlines are refracted according to the tangent law:

$$\frac{K_1}{K_2} = \frac{\tan(\theta_1)}{\tan(\theta_2)} \tag{11}$$

where θ_1 is the deviation of the flowline from the vertical in unit 1 and θ_2 is the deviation in unit 2. As a consequence of flow-line refraction, flow systems reflect a geometry in which the geologic units with higher hydraulic conductivity act as zones of preferential flow. Flow lines tend to cross the zones of lower hydraulic conductivity in the shortest distance possible. This behavior is illustrated in Figure 4, which shows the typical flow pattern where a permeable layer is located beneath an aquitard. Aquifers at depth promote the formation of a regional-scale flow system, whereas substantial local relief on the water table promotes the development of more local flow systems. In upland recharge areas, flow paths that originate in nearby areas may follow entirely different routes through the subsurface if they contribute to flow systems of different scales.

The depth to the water table depends primarily upon climatological factors, the hydraulic properties of surface soils and the subsurface geology, and location within the groundwater flow system. Climate and vegetative cover influence the amount of water available for deep percolation below the root zone. The higher this deep percolation rate, the shallower the water table. The influence of hydraulic conductivity is best understood in terms of the ratio of the deep percolation rate to large-scale effective hydraulic conductivity of the subsurface units within the flow system. The higher the value of this ratio, the greater the hydraulic gradient required to transmit the recharge through the subsurface to its discharge area. A higher water table in the recharge area creates this gradient.

Several examples serve to illustrate these controls on the water table depth. In the desert basins of the southwestern

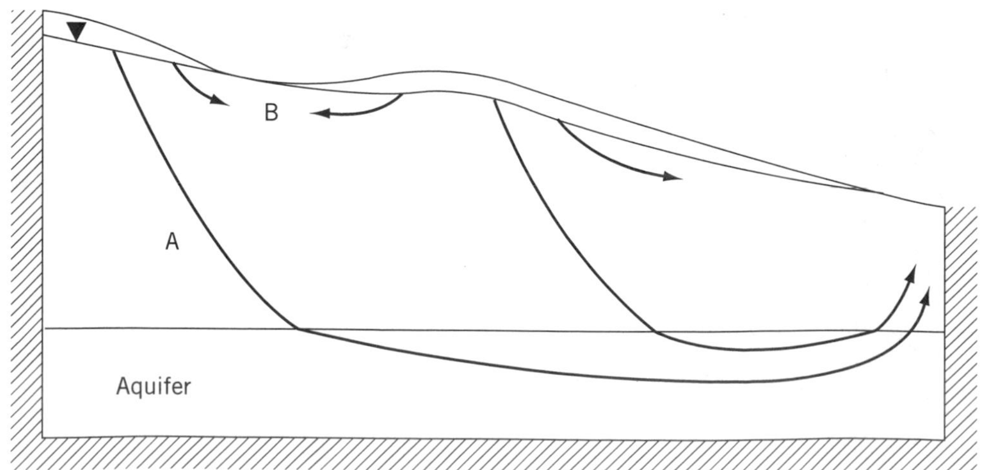

Figure 4. Effects of flow-line refraction on patterns of groundwater flow. Flow line A is part of a regional-scale flow system, whereas those labeled B form a local-scale flow system.

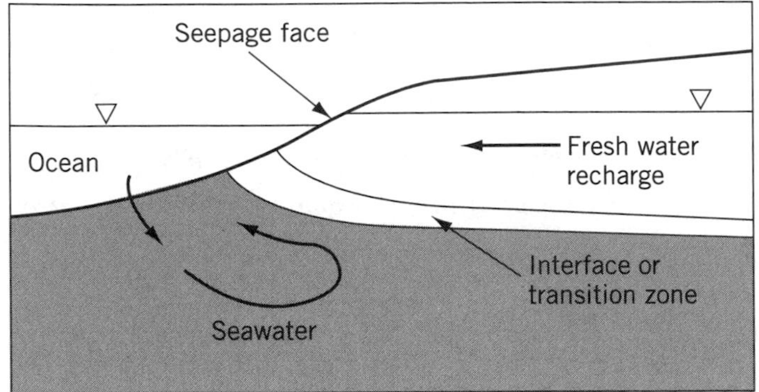

Figure 5. Seawater intrusion in a coastal aquifer. A seepage face forms where the water table intersects the land surface above the mean sea level.

United States away from groundwater discharge zones, the water table is several hundred meters below ground surface, with extremely low moisture contents in the vadose zone. It is thought by some that this thick vadose zone provides suitable siting characteristics for hazardous waste disposal facilities, because of very low rates of recharge in the lateral-flow regions of a basin (11). Recharge is limited primarily to infiltration in the surrounding mountains. At the discharge end of the flow system, the water table locally intersects the ground surface (with groundwater discharging as springs), or it is several meters beneath the surface of a dry lake playa. Groundwater in this latter case is discharged across the surface of the playa by evaporative losses.

In the northeastern United States extensive glacial deposits overlie bedrock. These deposits are quite permeable if they are glaciofluvial sands and gravels, or they are highly impermeable if they are glacial lake clays or clay-rich glacial till. In general, areas with surficial deposits of clay have a shallow water table because of the low fluid fluxes that are transmitted through these types of deposits. Areas underlain by sands and gravels behave more variably. Percolation rates below the root zone are relatively high in these deposits (eg, 20–40% of the annual precipitation) because of the wetter climate and the seasonal losses by evapotranspiration. The water table is a few meters below the ground surface, even though the sediments are permeable, if there is limited topographic relief across the basin. In areas with appreciable local relief, the water table is commonly up to a few tens of meters below the ground surface. These descriptions are, of course, broad generalizations, and the depth to the water table is a site-specific feature that reflects the interaction of a variety of site conditions.

Aquifers in coastal areas and islands are prone to seawater intrusion. Because seawater is denser than fresh water, it invades aquifers that are hydraulically connected to the ocean (Fig. 5). Freshwater recharge forms a lens that floats on top of the intruding seawater. A simple hydrostatic calculation known as the Guyben–Hertzberg relationship predicts that there is about 40 times as much fresh water in the lens beneath sea level as there is freshwater above sea level. The equilibrium condition illustrated in Figure 8 is disturbed by changes in the amount of groundwater recharge and/or by fluid withdrawals from the freshwater lens by pumping. A variety of simulation models have been developed to aid in predicting the response of the interface to groundwater withdrawals and to characterize the extent of the transition zone where fresh water mixes with seawater (12). The freshwater lens is an important source of potable water that must be protected from contamination by encroachments of seawater or by hazardous chemicals released to the environment.

Wells adjacent to tidal bodies of water have water-level fluctuations that follow rising and falling tides. The magnitude of the variation and the time lag relative to the water level in the tidal water depend on the hydraulic properties of the aquifer, the distance of the well from the shoreline, and the extent of the tidal signal in the surface water body. Tidal fluctuations are used to infer hydrogeologic properties of the porous medium (13).

WATER BALANCE FOR A GROUNDWATER BASIN

An evaluation of the water balance of a groundwater basin involves assessing each component of the hydrologic cycle. The water balance equation is

$$P - E_w - T_w - R = \Delta S \qquad (12)$$

where P is the volume of precipitation falling in the basin, E_w is the evaporation, T_w is the transpiration of water from the soil zone by plant activity, R is streamflow, and ΔS is the change in subsurface storage within the vadose zone and the deeper groundwater system. On a seasonal basis, with changing values for precipitation, evaporation, and transpiration, soil moisture is either depleted or replenished. When precipitation exceeds evaporative losses, the consequent increase in soil moisture is available for percolation below the root zone and eventual recharge at the water table. When averaged over a number of years, in the absence of groundwater resource development, the change in groundwater storage is commonly zero ($\Delta S = 0$). Equation 12 applies in the case of a closed groundwater basin. Additional terms are included in this expression if there is subsurface flow into or out of the basin.

It is sometimes necessary to express the water balance relationship in a modified form to estimate the change in groundwater storage on a year-by-year basis. In wet years, higher rates of groundwater recharge augment the groundwater reservoir and are reflected in a sustained rise in the water table. In dry years that follow, groundwater discharge sustains stream flow to a degree, but this is reflected in a decline in the regional water table. In exceptionally dry years, groundwater is the only source of stream flow. Of the incident precipitation, components can be allocated to interception by plants and trees, evaporation from surface water bodies, consumption by vegetation, and storage as soil moisture. The remainder (sometimes termed available water) runs off as stream flow, is consumed by phreatophytic plants adjacent to streams or in wetlands, or percolates as groundwater recharge. Estimates are developed for each of these components, using on-site monitoring data and/or regional data which indicate, for example, potential evapotranspiration rates. In this process, a value for the basinwide groundwater recharge is derived, given the time frame of the calculation (eg, monthly or yearly data). An extended discussion of these approaches is available (14).

GEOCHEMICAL FRAMEWORK

The chemical composition of groundwater is important in determining its suitability as a supply of potable water, in influencing the mobility and persistence of contaminants, and as an aid in interpreting patterns and rates of groundwater flow. Groundwater chemistry is controlled by a complex coupling of chemical and physical processes operating in a heterogeneous environment (15,16,17). The complexity of the geochemical system means that characterizing and predicting groundwater geochemical behavior is usually more difficult than groundwater flow behavior. Indeed, to completely characterize the geochemical state of the subsurface requires much more data than is required to characterize the hydrogeology. The principal processes that control groundwater geochemistry are (*1*) acid–base reactions, (*2*) solid-phase precipitation and dissolution, (*3*) complexation and speciation, (*4*) oxidation–reduction, and (*5*) solid-surface reactions and sorption. Although it is convenient to examine these processes individually, they do not typically operate independently of each other. For example, a change in acid-base conditions affects changes in mineral solubility and sorption chemistry. Dissolution of one mineral causes precipitation of another. One might say that, in groundwater geochemistry, everything depends upon everything.

Samples of groundwater for chemical analysis are obtained by withdrawing water from piezometers like those illustrated in Figure 1. Although simple in concept, strict attention to detail is required to obtain resolute samples that reflect the pore-water chemistry adjacent to the well screen or monitoring port. Sampling protocols have emerged in recent years (18). Water samples from boreholes with long completion intervals or uncased boreholes have little value except as gross characterizations of groundwater chemistry.

Acid–Base Processes

Acid–base reactions involve the transfer of hydrogen ions (protons). pH is the most important chemical characteristic of groundwater because the hydrogen ion participates in many geochemical processes. For example, hydrogen ions affect the solubility of many minerals and metal species. A change in pH causes fouling of pumping wells with precipitates or the dissolution and mobilization of toxic metals at contaminated sites. What controls the pH of groundwater? The most important acid in groundwater is dissolved carbon dioxide gas, which forms carbonic acid in solution. At 20°C, pure water exposed to atmospheric concentrations of carbon dioxide has a pH of 5.6. Biological processes enhance soil–gas carbon dioxide concentrations, increasing the concentration of dissolved carbon dioxide in water infiltrating through the soil horizon. In groundwater systems, the acidic effects of dissolved CO_2 (g) are balanced by the neutralizing effects of mineral dissolution, principally silicates, aluminosilicates, and carbonates. In typical near-surface groundwater systems, the pH of the water is determined by the relative availability of carbon dioxide gas and the solubility and rate of dissolution of the porous matrix.

One is often interested in how the pH of a groundwater changes in response to infiltration of acidic surface waters, such as acid rain or acid rock drainage. The ability of groundwater to neutralize acid is characterized by the alkalinity, or acid neutralizing capacity (ANC). A water with a high alkalinity neutralizes additional acidity without large changes in pH. In alkaline waters, hydrogen ions are neutralized by reacting with bases to form new species. In this fashion, alkaline waters are buffered against changes in pH caused by the addition of acidity. Note that even though a water has a high pH and is basic, it can simultaneously have a low alkalinity, and thus be sensitive to additions of acidity.

Solid-Phase Precipitation/Dissolution and Alkalinity

The alkalinity reported in water-quality analyses characterizes the acid neutralizing capacity of the dissolved constituents in the water. However, the solid phase in the subsurface is also a source of alkalinity. In response to an increase of acidity, minerals dissolve and release bases, such as the carbonate ion. These bases react with hydrogen ions and form new species, thus removing hydrogen ions from solution.

The sensitivity of a groundwater flow system to acid loading is controlled by the mineralogy of the porous medium. Carbonate minerals are the most important source of alkalinity for many systems because of their relative abundance and relatively rapid rate of dissolution. It was shown that the dominant source of alkalinity in North American rivers is carbonate minerals (19). As river water has a large component of groundwater, the study suggests that carbonate minerals are also a significant source of alkalinity in subsurface waters. Although more abundant than carbonates, silicate and aluminosilicate minerals dissolve much more slowly, so that groundwater is often undersaturated by these minerals. The slow dissolution kinetics of silicate minerals translate into lower alkalinity and buffering. For this reason, lakes sensitive to acid rain are almost always found in watersheds with siliceous bedrock, such as granite (20).

The mineralogy of the porous matrix controls the concentration of its dissolved constituents at saturation by precipitation and dissolution. A groundwater at equilibrium with a solid phase must be saturated with respect to that mineral. If it is undersaturated, then the mineral dissolves until saturation is reached or the solid phase disappears. If the water is initially saturated by a mineral phase and one of the mineral's dissolved constituents is added to the solution by some other process, then the mineral precipitates out of solution to reestablish the saturated equilibrium conditions.

It is possible to determine the mineral phases that control the groundwater geochemistry by identifying the minerals that are saturated in solution. In general, basaltic rocks dissolve more quickly and produce basic groundwaters with moderate to high alkalinity. Granitic rocks dissolve slowly and produce neutral or acid groundwaters with low alkalinity. Unfortunately, the analysis of controlling minerals is confounded by the slow dissolution kinetics of many silicates. Furthermore, mineral precipitation is typically a slower process than mineral dissolution. It is not unusual to find groundwaters supersaturated with respect to minerals, such as calcite and quartz (16).

Precipitation reactions are a very important control on metal concentrations in groundwater. For example, iron has low solubility under oxidizing (see below) conditions and precipitates out of solution as iron oxides. Under reducing conditions in groundwater, the concentrations of metals, such as lead, zinc, and copper, are controlled by the low solubility of sulfide minerals. At high pH, metal carbonates exert dominant control on total dissolved metal in solution. In remediation, precipitation immobilizes metals *in situ*.

Although the chemical signature of a groundwater is controlled by local mineral assemblages, six ions typically account for about 90% of the dissolved constituents: sodium, magnesium, calcium, chloride, bicarbonate, and sulfate (21). In general, the total dissolved solids (TDS), defined as the mass of dissolved constituents per mass of water, increases as the residence time of the groundwater in the subsurface increases and as the depth of the groundwater in the subsurface increases. Most potable water has a TDS of 1000 mg/L or less. Some old and deep groundwaters are brines with a TDS that exceeds 300,000 mg/L. However, if soluble minerals, such as gypsum, occur near the ground surface in a recharge area, young groundwater with a relatively high TDS occurs.

Complexation and Speciation

Complexation is defined as a reversible reaction between two dissolved species to form a new aqueous species (17). Complexes are typically formed of a positively charged metal species and a negatively charged anion or ligand. Complexation is an important control on the concentration of major cations, such as calcium and magnesium, secondary cations, such as iron, and trace metals, such as copper, lead, zinc, mercury, and nickel. The most significant complexing anions in groundwater are chloride, sulfate, bicarbonate, hydroxide, and sulfide. The extent of complexation increases as the concentration of these anions increases. Generally, an increase in salinity corresponds to an increase in complexation. Organic species, such as naturally occurring humic acids, are important ligands in shallow, near-surface groundwater systems.

Complexation increases the solubility of minerals and the mobility of trace metals. Mineral saturation is controlled by the concentration of the free, uncomplexed dissolved constituents, not the complexes. Minerals dissolve enough mass to saturate the aqueous solution and in addition to satisfy all the complexation demands. Complexation also mobilizes trace metals by sequestering them from reactions that affect free metal species. For example, many trace metals, which are normally sorbed onto solids and thus are at least temporarily immobile, are not available for sorption when complexed. Ligands which form strong complexes in this fashion reduce the effectiveness of chemical treatments that target the free metal. This condition is particularly true for strong, man-made ligands, such as EDTA, which are sometimes found at contaminated sites (22, 23).

Water quality analyses typically report the total metal concentrations in solution. Total concentrations are the sum of the concentrations of the free species and all their complexes. Often one is not interested in total concentrations but rather the concentration of free species in solution. For example, only the free species of many trace metals are toxic to biota. Complexation reduces the toxicity of these trace metals to organisms. It is difficult to measure the concentrations of individual complexes in solution. In lieu of direct measurements, the equilibrium concentration of complexes is predicted from measured total concentrations by thermodynamic equilibrium relationships. An extended discussion of these methods is available (16,17).

Oxidation–Reduction Reactions

Oxidation–reduction (redox) reactions, involve the transfer of electrons between electron donor species, called reductants, and electron acceptor species, called oxidants. In the process, the electron donor species are oxidized and the electron acceptor species are reduced. A redox reaction occurs only if both electron acceptor and donor species are present.

Groundwater-related redox processes include the oxidation of sulfide minerals to form acid–rock drainage, the oxidation of organic contaminants for remediation, and the fouling of pumps by iron oxidation. Redox chemistry is also fundamental to all biological activity, including that involved in bioremediation (24). To determine an element's electron donating or accepting capability, its oxidation state must be examined. The oxidation state of an element characterizes its electron content. For example, reduced iron has an oxidation state of 2 (indicated as Fe(II) or Fe^{2+}), whereas oxidized iron has an oxidation state of 3 (Fe(III) or Fe^{3+}). Many elements, for example, sulfur, have several oxidation states. The chemistry of an element strongly depends on its oxidation state. For example, sulfide, which is stable under sufficiently reducing conditions, reacts with dissolved metals to form low-solubility minerals which precipitate from solution, whereas sulfate, which is stable under more oxidizing conditions, does not. The overall redox conditions, or the tendency of the system to donate or accept electrons, are measured by pe, which, analogous to pH, is interpreted as the effective concentration of electrons in solution. A low pe indicates reducing conditions and a high pe indicates oxidizing conditions. pe ranges from -10 under reducing conditions to $+20$ under oxidizing conditions. Another commonly used measure of redox conditions is the redox potential *Eh,* which is expressed in units of electrical potential. Redox reactions are also strongly affected by pH. An important feature of redox reactions is that they are slow compared to acid-base reactions. For this reason redox systems are often not in equilibrium. Furthermore, it is difficult to measure a unique pe for a groundwater system which is not in redox equilibrium (25).

In groundwater systems, redox reactions occur in a sequence (26,27). Each reaction in the sequence is associated with a distinct redox zone. The zones range from an oxygen-rich, aerobic zone to a methane-rich, highly reducing zone (Fig. 6). The sequence and zoning is explained by the energetics of redox reactions and the relative abundance of redox species. Consider the addition of dissolved organic matter (DOC), such as gasoline, into a groundwater system. Organic matter is the most highly reduced species, analogous to the strongest of acids. Being highly reduced, organic matter potentially donates electrons to many other more oxidized species. However, the DOC is oxidized at a significant rate only with the electron acceptor that releases the greatest amount of energy during the reaction. It is thus observed that the DOC is first oxidized by oxygen, and then in order by nitrate, manganese, iron, sulfate, and ending with methane producing conditions. In contrast to acid–base and complexation processes, redox reactions proceed until one of the reactants is completely exhausted. A sufficiently large input of organic matter thus strips an aquifer of oxygen, nitrate, oxidized manganese and iron, and then sulfate.

A variety of naturally occurring electron acceptors oxidize reduced contaminants in an aquifer. Near the surface, groundwater recharge from the unsaturated zone is typically aerobic, containing between 8 and 12 mg/L of dissolved oxygen. Reduced substances completely consume the oxygen within a few meters of the water table. Many aquifers are thus anaerobic. Reduced substances are still oxidized if another electron acceptor is available. In aquifers, the most common, naturally oc-

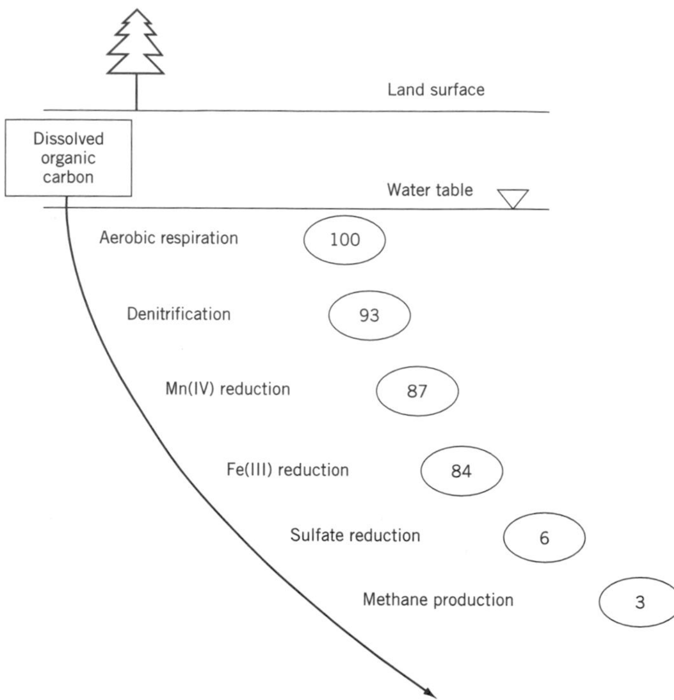

Figure 6. Typical redox zones found in groundwater systems. The figure depicts the redox zones that form as dissolved organic carbon is oxidized by a sequence of redox processes along a flow line. The values in the ellipses indicate the relative energy yield for each redox process (adapted from 27 and 28).

curring electron acceptors are oxidized manganese, iron, and sulfate. Nitrate, also an electron acceptor, does not occur naturally in significant quantities but is associated with agricultural pollution.

It is thought that bacteria mediate most subsurface redox reactions. Bacteria catalyze reactions that normally proceed inorganically at a slow rate. So, for example, the oxidation of organic matter by sulfate proceeds relatively rapidly with bacterial catalysis but is immeasurably slow in sterile systems. In the process of catalysis, bacteria capture some of the energy released for metabolic processes. Classes of subsurface bacteria have developed to extract energy from the dominant reaction in each redox zone (24). For example, aerobes use oxygen as an electron acceptor, denitrifiers use nitrate, etc. Bacteria are particularly damaging to well screens in anoxic aquifers containing significant amounts of reduced iron. This problem is caused by aerobic bacteria adjacent to the borehole that use oxygen to oxidize the iron. As this reaction yields relatively little energy, the bacteria must oxidize great quantities of iron to sustain metabolic processes. Upon oxidation, the relatively insoluble oxidized iron is precipitated out at the well screen, fouling the well. In the absence of bacteria, this reaction proceeds slowly. Bleaching the well provides temporary relief by killing the bacteria and slowing the reaction.

Surface Reactions and Sorption

Surface reactions are a group of processes by which dissolved species are retained at the solid–solution interface (17,29,30). At one extreme, precipitation incorporates species into the solid phase. At the other extreme, a species is loosely held by electrostatic forces. Reactions in which species are retained at the solid surface are often crudely classified as sorption, because these reactions are difficult to distinguish experimentally. In groundwater systems, the most important surface reactions are ion exchange, surface complexation, and the partitioning of organics into solid-phase organic matter.

Mineral surfaces which carry a net surface charge are responsible for ion exchange processes. The net charge of the solid surface exerts an electrostatic force on oppositely charged species in solution. To balance excess surface charge, the solution next to the surface must contain an equal excess of oppositely charged species. These species are bound at the surface by electrostatic forces. In clays, ions are bound at specific sites in the mineral structure. Because the net surface charge must balance the net charge in solution next to the surface, ions are only released from the surface if they are exchanged for ions with the same total charge. In general, smaller, higher charged ions are preferentially held over larger, lower charged ions. Thus aluminum (Al^{3+}) is preferred over sodium (Na^{+}).

The total surface charge of a mineral and hence its capacity to bind ions is a function of mineral type and pH. In general, clays and organics have very large exchange capacities compared to other mineral surfaces. In a porous medium, typically 6 to 10 times more ions are sorbed onto solid surfaces than dissolved in solution (15). As pH decreases, hydrogen ions (protons) are sorbed onto mineral surfaces increasing the positive surface charge. At low pH, surfaces are highly protonated and have a net positive charge, thus attracting anions. At higher pH, surfaces become negatively charged and then are sites of cation exchange. Sorbed cations buffer solutions against changes in cation concentrations. For example, if sodium-rich water infiltrates an aquifer, which initially contains mostly calcium on the exchange sites, the sodium exchanges onto the solid for calcium. Thus, after cation exchange, the infiltrating water is depleted in sodium and enriched in calcium.

Whereas the dominant cations in groundwater systems (calcium, sodium, potassium, and magnesium) are affected by cation exchange, trace metals are more strongly controlled by surface-complexation processes. In surface complexation, ions are held by chemical bonds. Iron oxides and organic matter are particularly active sites of surface complexation. It is often observed at contaminant sites that trace metals are completely sorbed in the near-surface soils. Indeed, soils are rich in both organic matter and iron oxides. It is possible that the sorbed metals are remobilized if the iron oxides are later reduced. Sorption of organic compounds is more appropriately described as a partitioning of the dissolved organic between solid-phase organic matter and the aqueous phase. Dissolved organics preferentially enter into hydrophobic organic matter. The strength of this partitioning is characterized by the octanol–water partitioning coefficient. Measurements of organic carbon content are used with this partitioning coefficient to predict the extent of organic sorption (31).

SIMULATION MODELS

A staged approach to hydrogeologic analysis begins with the formulation of a conceptual model, which is then transformed into a hydrogeologic simulation model. A hydrogeologic simulation model is typically developed as an aid in reaching a management decision. Groundwater models serve at least three roles in this regard: (*1*) to provide insight into the hydrogeologic processes operative in the study area, (*2*) to test alternative hypotheses on system behavior to guide site investigational plans or to increase the confidence level in the management decision, and (*3*) to develop predictions of system behavior under changed conditions.

The conceptual model of the system characterizes the basic elements of the flow system, such as fluid inflows, outflows, and the geometry of the flow domain. It is based initially on a reconnaissance of the area using geologic and soils maps, air or satellite photographs, field observations, and the skill and experience of the analyst. A mathematical statement of the conceptual model includes specifying the governing differential equation describing the variation in hydraulic head in space and/or time, the domain geometry, boundary conditions, and the initial conditions for transient flow. The spatial variation and values of the hydraulic properties of the porous medium are also specified. For fluid flow problems, hydraulic conditions on the domain boundaries are normally assigned as impermeable (no flow orthogonal to the boundary), a specified value of hydraulic head, or a specified fluid flux (eg, a recharge rate at the water table).

If the domain geometry, governing equation, boundary conditions, and initial conditions are simple, an analytical solution of the mathematical model is often possible. Even for cases where the problem is too complex to permit an analytical solution, it is often advisable to simplify the conceptual model in the first instance to the point where an analytic solution is possible. Such a solution is easy to work with and provides valuable insight into the behavior of the groundwater system. As the concept of the flow system is refined, a decision must be made that a more detailed analysis is required or that the existing model is adequate to resolve the management problem. If the decision is to proceed with a more flexible computer-based numerical simulation, it is usually because of complex boundary conditions or spatially variable hydraulic properties that influence the behavior of the flow system in some critical manner.

PARAMETER ESTIMATION FOR SIMULATION MODELS

Field data are used to refine the conceptual model of the site and to estimate parameter values for the simulation model. Specific issues addressed with field data at the conceptual model stage include the geometry of the study area, the location of boundaries, and the identification of important processes and conditions, such as recharge, unsaturated flow, and confining conditions. After a conceptual model is developed and transformed into a mathematical model, observations of the response of the real-world system to known stresses are used to estimate parameter values for the hydrogeological simulation model.

HYDROGEOLOGICAL DECISION ANALYSIS

Predictions from hydrogeological simulation models are uncertain because of uncertainties in system properties, boundary conditions, and perhaps even the geometry of the flow system. In a heterogeneous medium with only limited site characterization data, large uncertainties are the norm. During the site characterization process, the hydrogeologist must consider to what extent the uncertainty in the model prediction influences the management decision for which the model was developed and whether it is worthwhile to collect additional field data to reduce uncertainties in model predictions. This question is commonly addressed informally, but in recent years techniques derived from engineering decision analysis have been developed to aid in making project decisions involving a hydrogeological component (32). Decision analysis provides the link between the economic framework in which decisions are made and the technical analyses on which decisions are based.

The potential advantages of a decision analysis approach are numerous. The approach recognizes the interrelationships and tradeoffs among site characterization activities, facility design, and monitoring activities. It permits a rational approach for ending a remedial investigation or data-collection program. In this context, hydrogeological data have worth only if they aid in making decisions among alternative courses of action. Additional measurements have worth only if the risk reduction they provide exceeds the cost of obtaining those measurements.

BIBLIOGRAPHY

1. W. S. Back, J. S. Rosenshein, and P. R. Seaber, eds., *Hydrogeology, The Geology of North America,* Geological Society of America, Boulder, Colo., 1988, Vol. O-2.
2. R. A. Freeze and J. A. Cherry, *Groundwater,* Prentice–Hall, Englewood Cliffs, N.J., 1979.
3. P. Clifton and S. P. Neuman, *Water Resources Res.* **18**(4), 1215–1234 (1982).
4. J. F. Poland, ed., *Guidebook to Studies of Land Subsidence due to Groundwater Withdrawals,* UNESCO Studies and Reports in Hydrology, No. 40, Paris, France, 1984.
5. D. C. Ford and P. W. Williams, *Karst Geomorphology and Hydrology,* Unwin Hyman, London, 1989.
6. A. Cohen, *Hydrogeologic Characterization of Fractured Rock Formations: A Guide for Groundwater Remediators,* LBL-38142/UC-800, Lawrence Berkeley Laboratory, Berkeley, Calif., 1995.
7. National Research Council, *Rock Fractures and Fluid Flow: Contemporary Understanding and Applications,* National Academy Press, Washington, D.C., 1996.
8. M. Sara, *Standard Handbook for Solid and Hazardous Waste Facility Assessments,* Lewis Publishers, CRC Press, Boca Raton, Fla., 1994.
9. P. A. Domenico and F. W. Schwartz, *Physical and Chemical Hydrogeology,* John Wiley & Sons, Inc. New York, 1990.
10. C. W. Fetter, *Applied Hydrogeology,* 3rd ed., Prentice–Hall, Englewood Cliffs N.J., 1994.
11. National Research Council, *Ward Valley, An Examination of Seven Issues in Earth Sciences and Ecology,* National Academy Press, Washington, D.C., 1995.
12. P. Huyakorn, P. Anderson, J. Mercer, and H. White, *Water Resource Res.* **23**(2), 293–312 (1987).
13. A. D. Erskine, *Ground Water* **29**(4), 556–562 (1991).
14. L. Dingman, *Physical Hydrology,* Prentice–Hall, Englewood Cliffs N.J., 1994.
15. C. A. Appelo and D. Postma, *Geochemistry, Groundwater and Pollution,* Balkema, Rotterdam, the Netherlands, 1993.
16. J. I. Drever, *The Geochemistry of Natural Waters,* 3rd ed., Prentice–Hall, Englewood Cliffs, N.J., 1997.
17. F. M. M. Morel and J. G. Hering, *Principles and Applications of Aquatic Chemistry,* John Wiley & Sons, Inc., New York, 1993.

18. U.S. Environmental Protection Agency, *Ground Water Sampling—A Workshop Summary,* Dallas, Tex., EPA/600/R-94/205 Washington, D.C., 1995.
19. R.M. Garrels and F.T. MacKenzie, *Evolution of Sedimentary Rocks,* W. W. Norton & Co., New York, 1971.
20. J. O. Reuss, B. J. Cosby, and R. F. Wright, *Nature,* **329**(3), 27–32 (1987).
21. J. D. Hem, *Study and Interpretation of the Chemical Characteristics of Natural Water,* USGS Water Supply Paper 2254, USGS, Washington, D.C., 1985.
22. J. M. Cleveland and T. F. Rees, *Science,* **212,** 1506–1509 (1981).
23. F. M. Dunnivant, P. M. Jardine, D. L. Taylor, and J. F. McCarthy, *Environ. Sci. Technol.* **26,** 360–368 (1992).
24. F. H. Chapelle, *Ground-water Microbiology and Geochemistry,* John Wiley, & Sons, Inc., New York, 1993.
25. R. D. Lindberg and D. D. Runnels, *Science* **225,** 925–927 (1984).
26. R. E. Jackson and R. J. Patterson, *Water Resource Res.* **18**(4), 1255–1268 (1982).
27. D. R. Lovely and F. H. Chapelle, *Rev. Geophys.* **33**(3), 365–381 (1995).
28. S. F. Korom, *Water Resource Res.* **28**(6), 1657–1668 (1992).
29. J. A. Davis and D. B. Kent, "Surface Complexation Modeling in Aqueous Geochemistry," in M. F. Hochella and A. F. White, eds., *Reviews of Mineralogy 23,* Mineralogical Society of America, Washington, D.C., 1990.
30. D. A. Dzombak and F. M. M. Morel, *J. Hydraulic Eng.* **113,** 430–475 (1987).
31. C. W. Fetter, *Contaminant Hydrogeology,* MacMillan, New York, 1993.
32. R. A. Freeze, J. Massmann, L. Smith, T. Sperling, and B. James, *Groundwater* **28**(5), 738–766 (1990).

I

IMMUNOASSAY: ENVIRONMENTAL

DIANA S. AGA
Swiss Federal Institute for Environmental Science and Technology
Duebendorf, Switzerland

Immunoassay (IA) is a method of analysis that relies on specific interactions between antibodies and antigens to measure a variety of substances, ranging from complex viruses and microorganisms to simple pesticide molecules and industrial pollutants. The key reagents in IAs are antibodies, which are soluble proteins produced by the immune system in response to infection by foreign substances (called antigen). The use of IA as an analytical tool can be qualitative, semiquantitative, or highly quantitative.

IAs are based on the fundamental concept that antibodies prepared in animals can recognize and bind with exquisite specificity the antigen that stimulated their production. The binding forces involved in specific interactions between antibodies (Ab) and antigens (Ag) are of a noncovalent, purely physicochemical nature. The noncovalent interactions involved in Ab–Ag complex formation include hydrogen bonds, ionic bonds, hydrophobic interactions, and van der Waals interactions (Fig. 1). Each of these interactions is weaker than the covalent bonds. Hence, for an effective Ab–Ag interaction, a large number of such interactions need to be present, and there should be a close fit between the Ab and the Ag for these noncovalent forces to operate (<1 Å or 1×10^{-7} mm) (1). This is reflected in the high degree of specificity characteristic of Ab–Ag interactions.

The strength of the sum total of noncovalent interactions between a single binding site on an Ab and a single epitope (the part of Ag that directly interacts with the Ab binding site, which is also called antigenic determinant) is the affinity (K_{aff}) of the Ab for the epitope. The reaction between the antibody and the antigen can be represented as follows:

$$\text{Ab} + \text{Ag} \underset{K_d}{\overset{K_a}{\rightleftharpoons}} \text{Ab–Ag}$$

where K_a is the association constant and K_d, the dissociation constant. Thus the affinity constant K_{aff} may be expressed as:

$$K_{\text{aff}} = \frac{[\text{Ab–Ag}]}{[\text{Ab}][\text{Ag}]} = \frac{K_a}{K_d}$$

Antibodies may possess affinity constants for individual antigens on the order of 10^4–10^{12} M^{-1}. The high affinity constant of an antibody makes it a suitable tool for environmental analysis because it provides the antibody with the ability to distinguish between the homologous antigen (target compound) and myriad other compounds of widely diverse structure that are found in environmental samples.

The use of antibodies as analytical reagents was first reported in 1959 when the measurement of picogram levels of human insulin in samples of body fluids using radioimmunoassay (RIA) was successfully demonstrated (2). Since then, various IAs for detecting hundreds of molecules of endogenous and exogenous origin have been described. This technique proved to be reliable, fast, and sensitive; hence many other RIAs have been developed for clinical and medical tests since then. Radiolabels were gradually replaced with enzyme labels because of the hazards associated with using radioactive materials. Enzyme-linked immunosorbent assay (ELISA), which was first introduced in 1971 (3), has become perhaps the most popular IA format used in laboratories today.

The use of immunochemical techniques in the environmental field was first proposed in 1971 (4); it was suggested that the use of immunological screening methods could well be used to rapidly detect pesticide residues in food and water resources and that these methods could be used to confirm results of conventional analyses. An RIA for the insecticides aldrin and dieldrin was one of the first reported IAs for environmental contaminants (5). Today, although a few RIAs still exist in the medical field, RIAs are seldom used in environmental and food analyses, because of the need for special handling and disposal of the radioactive materials.

The development of IAs for environmental applications was rather slow. It was only during the early 1990s that the scientific community and regulatory agencies recognized the strength and advantages that IAs can offer. The supportive action given by regulatory agencies and the private sector has helped in the promotion of IA technology in environmental monitoring. Since 1992, the U.S. Environmental Protection Agency (EPA) Characterization Research Division in Las Vegas has sponsored an annual immunochemistry summit meeting to facilitate a dialogue among IA developers, university researchers, instrument manufacturers, and users to discuss current trends and issues concerning regulatory acceptance (6). There is also the Superfund Innovative Technology Evaluation

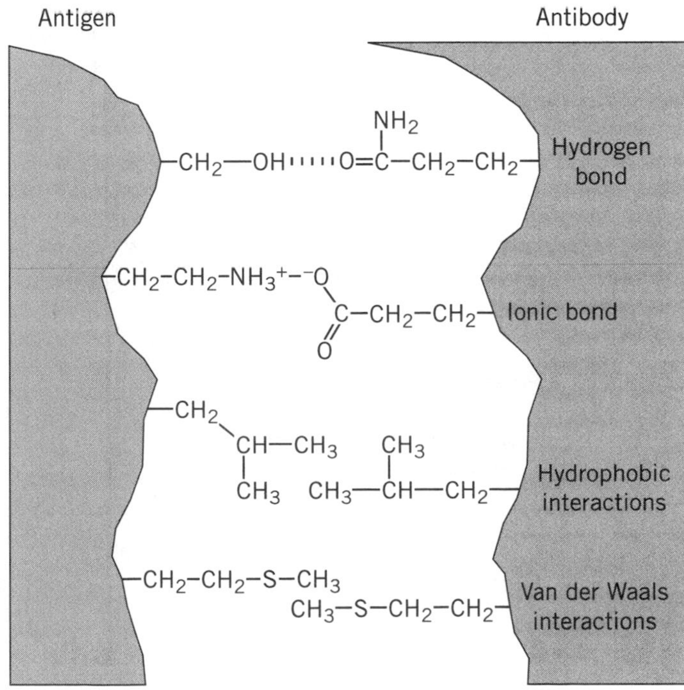

Figure 1. The intermolecular forces involved in antibody–antigen interactions.

(SITE) program, which was developed by the EPA to evaluate remediation and monitoring technologies, such as immunoassays, for use at hazardous waste sites (7). Furthermore, an industry association called the Analytical Environmental Immunochemical Consortium was formed in 1994 to establish guidelines for methods development and performance standards for immunochemical methods in environmental applications. Initial efforts of this group has been directed toward establishing consistent definitions for commonly used terms, developing standardized package insert information, establishing sources of kit calibrators and quality control samples, and providing guidelines for user quality control.

Some commercially available IA kits have been validated by the EPA to be included in SW-846 methods for the analysis of pentachlorophenol (PCP), 2,4-dichlorophenoxyacetic acid (2,4-D), polychlorinated biphenyl (PCB), total petroleum hydrocarbon (TPH), polyaromatic hydrocarbons (PAH), toxaphene, chlordane, dichlorodiphenyltrichloroethane (DDT), 2,4,6-trinitrotoluene (TNT), hexahydro-1,3,5-trinitro-1,3,5-triazine (RDX), and mercury. This list will continue to grow as more kits prove useful and reliable for the analysis of pollutants in several matrices of concern.

The increasing contamination of the environment has prompted regulatory agencies to mandate extensive monitoring of soil, air, water, and food resources. Analysis of environmental samples can be formidable, time-consuming, and costly. As a consequence, immunochemical techniques have become popular for environmental analysis because they are reliable, rapid, and cost effective. During the past 5 years, the use of IAs for environmental monitoring has increased substantially, and their use as an integral analytical tool in many environmental laboratories is now commonplace.

CONCLUSIONS

Despite many clear demonstrations that immunochemical techniques provide high quality data in a fraction of time and price required by common chromatographic methods, the acceptance of this technology in environmental analysis has been slow. However, the rate of acceptance of the technology has increased substantially in the last few years, as manifested by the increase in related literature and participation in many scientific meetings, due to the number of studies demonstrating the reliability of immunoassays in environmental analysis, such as in large water quality surveys, characterization of hazardous waste sites, and postremediation monitoring.

With the acceptance of immunochemical technology, more innovative concepts are now being applied for application to the environmental field. More than simply screening of environmental samples, it is likely that immunochemistry will be among the many technologies in the analytical field. Flow-injection IA and flow immunosensors also show promise in continuous on-site monitoring. Each of these techniques has been developed in response to a particular limitation encountered in the existing methods. As more and more analytical chemists illustrate new applications of IAs, the current limitations of this technology may be addressed and solved accordingly.

The previous limitations of immunochemical assays, such as heterogeneity of polyclonal antibodies, cross-reactivity, or single-assay capability, which hindered their wide acceptance by the scientific and regulatory community, are gradually being resolved. For example, the production of homogeneous monoclonal antibodies against environmentally important compounds has advanced immunosensor research and the development of highly specific assays. In addition, recombinant antibodies and molecularly imprinted polymers are promising and attractive alternatives to conventional antibodies because of their lower cost of production and their potentially better detection capabilities. The problem of cross-reactivity may eventually be eliminated, or even turned into a benefit, through statistical manipulations and neural networks that will deconvolute detection signals to identify constituent analytes. New detection systems have also been employed to increase assay sensitivity and improve detection limits. Finally, as multianalyte immunoassays become available, wider applications in studies on the fate and transport of contaminants in the environment will be realized.

Great efforts by immunochemists are paving the way for the miniaturization and automation of immunoassay techniques. Collaboration among scientists from various disciplines resulted in the realization of usable immunosensors and field-portable assays, automated flow-injection immunoassays, and high performance immunoaffinity chromatography for the analysis of pollutants in many contaminated sites. Monitoring is needed to identify temporal and spatial changes in water-quality attributable to pesticides and industrial chemicals. In addition, characterization of contaminated military and industrial sites is important in ensuring that the decisions made by regulatory agencies and the actions taken achieve their intended objectives effectively and economically. There have been many studies demonstrating that the advantages offered by environmental IAs outweigh the limitations. The applications of this technology in environmental analysis will continue to grow as more sensitive assays are developed and as automated and miniaturized systems become available.

ACKNOWLEDGMENTS

The author would like to thank L. Zimmerman for making many of the scientific illustrations used in the full article. The author also acknowledges the authors of several of the copyrighted illustrations for willingly providing their artwork. Special thanks to the Organic Geochemistry Group of the U.S. Geological Survey, Lawrence, Kansas, headed by M. Thurman, where the author conducted most of her immunoassay works for environmental applications.

BIBLIOGRAPHY

1. K. Kuby, *Immunology,* W. H. Freeman & Co., New York, 1992.
2. S. A. Berson and R. S. Yalow, *J. Clin. Invest.,* **38,** 1196 (1959).
3. E. Engvall and P. Perlman, *Immunochemistry,* **8,** 871 (1971).
4. C. D. Ercegovich, in F. J. Biros, ed., *Pesticide Identification at the Residue Level,* American Chemical Society, Washington, D.C., 1971, pp. 162–177.
5. J. L. Langone and H. Van Vunakis, *Res. Commun. Chem. Pathol. Pharmacol.,* **10,** 163 (1975).
6. J. M. Van Emon and C. L. Gerlach, *Env. Sci. Technol.,* **29,** 312A (1995).
7. U.S. Environmental Protection Agency, "Superfund Innovative Technology Evaluation Program, Technology Profiles," 6th ed., *EPA/540/R-93/526,* EPA, Washington, D.C., 1993.

ICP–MS. See INDUCTIVELY COUPLED PLASMA—MASS SPECTROMETRY.

INDOOR AIR RISK ASSESSMENT AND MANAGEMENT

LIDIA MORAWSKA
Queensland University of Technology
Brisbane, Australia

Inadequate air quality and the risk to human health and well-being resulting from inhalation of airborne pollutants are listed among the top environmental risks in developed countries. The major consequences of air pollution resulting from industrial emissions were realized several decades ago in the early 1950s and 1960s. Pride in urban landscapes marked with smoking factory chimneys, symbolizing development and progress, was replaced by concern, when the extent to which air pollution affects human health became clear. In notable pollution episodes in Donora, Pennsylvania in 1948 and in London, England in 1952, thousands of people died as a result of exposure to dense smog, a mixture of a number of airborne pollutants. Identification of ambient air pollution as a health hazard was followed by steps, undertaken by a broad range of national and international professional and government organizations, aimed at reduction or elimination of the hazard. This process is still in progress.

Indoor air quality (IAQ) emerged as an issue sometime later and was soon recognized as having at least equal importance to outdoor air quality. The U.S. Environmental Protection Agency for example, has identified indoor air pollution as being among the top five environmental risks to public health.

In the modern age humans spend over 90 percent of time indoors, often in environments effectively isolated from outdoor influences. The most susceptible segments of the population—the very young, old, and sick—spend close to 100% of their time indoors. The concentration levels and number of pollutants could be much higher indoors than outdoors due to the presence of indoor-specific sources (1). Efforts to reduce energy consumption and isolate the indoor from the outdoor environment often result in additional deterioration of indoor air quality. The complexity of the indoor environment in comparison with the outdoor environment and the emergence of a range of specific indoor air quality issues required the development of new approaches to indoor air risk assessment and risk management at both technical and management levels (2).

An additional level of complexity in approaching indoor environment issues is related to their interdisciplinary nature. Expert teams to assess and manage the indoor environment should include scientists, government and regulatory professionals, medical practitioners, occupational health professionals, building owners and managers, building, construction and air conditioning engineers, architects, and environmental lawyers.

An important direction in solving indoor environment problems is in developing an integrated approach for health risk assessment and risk management (3). Integration extended to comprehensive characterization of indoor air, to multiple exposure analysis, and to combined health risk assessment (qv). Integration applied to management and control strategies, to policy making, and to risk communication. Until recently the most common approach to monitoring, health risk assessment, and management of indoor air quality was to consider each airborne pollutant individually. However, it has become clear that both the comfort and combined health risk may depend not only on the concentrations of individual pollutants but also on the complexity of the interactions among all the constituents in the air. Any mitigation process aimed at one particular pollutant can affect other pollutants as well, decreasing or increasing their concentrations, or changing interaction mechanisms. Perceived air quality problems can be exacerbated by attempts to improve other indoor factors such as lighting and noise. Integrated strategies should take these processes into consideration and aim at an overall upgrade of the indoor environment.

RISK ASSESSMENT

A well-known definition of the World Health Organization (WHO), states: "Health is a state of complete physical, mental and social well-being, and not merely the absence of disease or infirmity." This definition is particularly applicable to the very broad range of health end points (a common term for ultimate health effects) related to inadequate air quality or more generally to an unhealthy indoor environment.

The main divisions of health responses are (*1*) short-term responses such as discomfort or irritation and (*2*) long-term effects such as cancer. In this classification, discomfort and irritation are considered along with illness because they affect human well-being and over a longer period of time can lead to developing clinical conditions. Acceptable discomfort level is defined quantitatively by the American Society for Heating, Refrigeration, and Air Conditioning Engineers (ASHRAE) (4). According to this definition, air quality is acceptable when there are no known contaminants at harmful concentrations and when a substantial majority (usually 80%) of the people exposed to the air do not express dissatisfaction.

A more detailed list of health end points includes the following (5):

- Allergy and other effects on the immune system
- Cancer and effects on reproduction
- Respiratory effects (other than cancer and allergic effects)
- Irritative effects on skin and mucous membranes of eyes, nose, and throat
- Sensory effects on the nervous system
- Effects on the cardiovascular system

Risk assessment for cancer is different from risk assessment for noncancer illnesses (see RISK ASSESSMENT). Cancer is an unambiguous outcome due to exposures several to many years before, usually to low levels of a specific chemical or radioactive species. The increased lifetime risk of developing cancer is usually relatively small, of the order of 10^{-4} to 10^{-6}, compared with the background lifetime risk of the order of 0.2 to 0.3 (6). The precise mechanisms of developing cancer are not well understood and thus a precise estimate of risk for cancer is difficult to make. Yet there is a need to reduce risks, so a formal method of risk assessment is necessary. The situation is much more complicated with risk assessments for noncarcinogenic agents. The outcomes show gradation and progression over time. There could be many outcomes and some of them could be reversible. In general they are not life threatening, and there is a considerable range of sensitivity in the population (baseline or threshold levels for the general population are difficult to describe). Several outcomes may be associated with a single contaminant, and a single outcome may be affected by several contaminants.

In addition to the specific health end points listed above there are three other terms used to describe syndromes, illnesses, and sensitivities exhibited or acquired as a result of indoor environment exposures: sick building syndrome (SBS), building-related illness (BRI), and multiple chemical sensitivity (MCS).

Sick building syndrome is related to symptoms such as irritated eyes, nose, throat, skin, headache, lack of concentration, or lethargy experienced in a particular building and disappearing after leaving the building. Two types of cases can be distinguished: a relatively clear clinical picture for which a specific etiology can be defined, and nonspecific symptoms related to a particular indoor environment, very commonly an office. Symptoms include mucous membrane and eye irritation, chest tightness, fatigue, headache, and malaise (7) and also odor annoyance, skin irritation, and influenza symptoms.

Building-related illness is an illness directly related to indoor exposures and ranges from lung disease to cancer (8). Rarely, BRI occurs without first experiencing SBS symptoms. One of those rare cases when there are no prior SBS symptoms is exposure to radon in indoor air which could result in the development of lung cancer.

Multiple chemical sensitivities is, "an acquired disorder characterized by recurrent symptoms, referable to multiple organ systems, occurring in response to demonstrable exposure to many chemically unrelated compounds at doses far below those established in the general population to cause harmful effects. No single widely accepted test of physiological function can be shown to correlate with the symptoms" (9).

In addition to measurable physical factors resulting in illness or symptom response, there are psychological factors whose overall role on human health could be comparable with physical factors. Often outcomes normally associated with indoor air quality (IAQ) problems might in fact be wholly or partly psychosomatic in origin. One of the main psychological factors is stress, the role of which as a causal factor in workplace illness has been generally accepted. It has been agreed that stress can cause symptoms that overlap with SBS symptoms; however, it is less clear whether stress-induced illness is masquerading on a large scale as illness induced by the environment (10).

An unhealthy indoor environment can also have behavioral and social effects. An example of a behavioral effect is exposure to carbon monoxide (CO). Levels of 40–50 ppm of CO, often encountered in urban traffic, have been shown to result in impairment of tasks that are complex enough to involve vigilance and reaction time (11). The same conclusion is drawn about exposures to mixed pollutants, including mixtures of volatile organic compounds (VOCs).

Social effects include interpersonal aggression and violent reactions. For example, increased air pollution expressed in terms of ozone readings was related to higher rates of family disturbances, measured by frequency of calls to the police (12). It should be noted here that identification of specific factors affecting social behavior could be very difficult. For example, high ozone readings often occur during extended heatwaves (elevated temperature and humidity). Extreme weather conditions such as high heat and humidity are likely to cause aggressive behavior, thus making it difficult to distinguish what is the prime cause of this behavior: weather conditions or pollution.

In summary, human response, in terms of health end points, to multiple physical and psychological factors in the indoor environment is very individual, complex, and often not well defined. Although there is a general knowledge about the qualitative relation between exposure and end points, in quantitative terms information is often very limited.

General Framework for Risk Assessment

The term indoor environment is very broad and in fact covers a range of different types of environments. General classifications of indoor environments include residential, occupational (nonindustrial), institutional, commercial, public, industrial, and transportation. Although often different factors have to be taken into account in relation to different types of environments, the general framework for risk assessment presented below is applicable across the whole spectrum of environments.

The main aspects of health risk assessment include hazard identification and characterization of pollutant concentrations, exposure, dosimetry factors, dose, response factors, lifetime individual risk and exposed population.

1. Identification of *hazards* in the indoor environment relates to the identification of sources (one or more sources emitting one or more pollutants), or factors (high humidity level, thermal discomfort, draft, stress, etc) adversely affecting human health or well-being.
2. *Pollutant concentration* is the numerical value of the amount of an individual pollutant in the air emitted by one or more sources at a particular point in time or averaged over a period of time.
3. Existence of *conditions* or presence of *factors* that result in the deterioration of air quality or the perception of an unhealthy indoor environment (draft, stuffy air, etc) is very important, but often impossible to quantify.
4. *Exposure* is a product of the concentration of a pollutant and the duration of contact. Two types of exposures are of special importance in the health risk assessment process: (*1*) lifetime exposure which is the summation of times the exposure occurred in different environments—this is particularly important for carcinogenic pollutants; and (*2*) short-term exposures to elevated concentrations
5. *Dosimetry factors* characterize how much of the pollutant present in the air enters the body and is absorbed by different organs. This depends on contact rate, ingestion or inhalation rate, absorption rate, average body weight, average lifetime, regional surface area of the lung, regional dose ratio, breathing pattern, sex, and age.
6. *Dose* is a product of exposure and dosimetry factors, which quantifies the amount of substance available for interference with metabolic processes or biologically significant receptors. This applies to pollutants only but not to environmental conditions or environmental factors. Administered dose is the dose of the contaminant or its metabolites deposited in the body. A biologically effective dose is the dose received at the target cell.

7. *Response factor* is the magnitude of response of an individual to a given dose of a pollutant.
8. *Lifetime individual risk* is a product of lifetime average daily dose and a response factor. This applies to carcinogenic risks only.
9. Risk to *exposed population* is a quantitative assessment that takes into account lifetime individual risk for the population groups affected and specific exposure scenarios of the population.

The main objective of health risk assessment is to link the hazard with the risk to an exposed population in some quantitative way. On this basis, risk management steps could be initiated.

There are obvious differences between health risk assessment for cancer and for noncancer effects, as well as for any irreversible long-term effects and for reversible short-term effects. For noncancer effects the probability of response or the degree to which exposure or dose exceeds a threshold for adverse health effects is usually considered. The Predictive Risk Equation (PRE) was formulated (13) for carcinogenic risk assessment (13).

The PRE is the basis for the Risk Characterisation Framework (RCF) for cancer health effects. The framework that subdivides the four components of the risk assessment process into 10 elements is presented in Table 1 (13). Individual components of the RCF will be discussed in detail later.

For noncarcinogenic risks a modified framework (Table 2) was proposed in which the mathematical operators were dropped (14).

In the above PRE, individual components of the framework are expressed as point estimates. Usually the situation is much more complex due to the variation of concentration in time, complexity of dosimetry factors, and the nonlinear response factor. Point estimates should then be replaced with distribution functions (15).

Risk Assessment Elements

Risk assessment can be conducted as a step-by-step process according to the risk characterization framework, or in certain cases some of the steps can be by-passed. An example of this could be an investigation aimed at directly linking exposure with individual risk. Important elements of the risk assessment process are discussed below.

Exposure Assessment. Exposure assessment consists of the identification of sources of indoor pollution, characterization (through measurements, surveys, questionnaires, etc) of indoor air pollutants and contributing factors such as concentration, duration, frequency, route, distribution, time frame, geography and modeling exposure for different occupancy and lifestyle scenarios.

Sources of indoor pollution include the following:

Occupants
Outdoor air
Soil, water
Cooking
Furnishings
Building materials
Tobacco combustion
Consumer products
Combustion appliances
Maintenance products
Heating, ventilation, and air conditioning (HVAC) systems

Occupants include humans and pets contributing to air pollution through functions natural to life processes or through activities conducted.

Infiltration of outdoor air is a very important factor affecting indoor air. A critical review of indoor–outdoor air quality relationship has been performed (16). For pollutants from predominantly outdoor sources, the outdoor concentration was always higher than indoor concentration. For sulfur dioxide, outdoor concentration was higher than indoor concentration by a factor of two, for ozone by a factor of five and for lead by a factor of one point five. The situation was less clear for pollutants generated by both indoor and outdoor sources. For example in the absence of indoor sources, the carbon monoxide ratio of indoor-to-outdoor concentration was close to one and for nitrogen dioxide, outdoor concentration exceeded indoor concentration by a factor of two, but with indoor sources such as unvented stoves or heaters, indoor levels of both gases could exceed outdoor concentration by a factor of two or more. The most complex was the situation for total particulate matter and for respirable particulate matter (particles smaller than 10 μm), where there was little relationship between indoor and outdoor in terms of concentration and physical or chemical nature. This is due not only to the presence of indoor sources of particulates, but also to different types of interactions of particles and of gases with the indoor environment.

The concentration of pollutants in the indoor environment depends on a number of factors (17):

The type, nature, and number of sources
Source use characteristics
Building characteristics
Infiltration or ventilation rates
Air mixing
Removal rates by surfaces, chemical transformation, or radioactive decay
Existence and effectiveness of air contaminant removal systems
Outdoor concentration of pollutants
Meteorological conditions

The most important indoor pollutants are listed below together with a brief characterization of known or suspected health effects and examples of concentration levels. The examples of concentration levels were chosen from the extensive body of literature as illustrations of the topic and to present a general idea of levels encountered in the indoor environment or range of variation between different environments.

Environmental tobacco smoke (ETS) is one of the most common indoor pollutants. Over 4000 compounds have been identified in ETS including combustion gases: CO, CO_2 SO_x, NO_x, particulates, nicotine and volatile organic compounds. Environmental tobacco smoke is a mixture of two components: mainstream smoke drawn through the tobacco, taken in and

Table 1. Risk Characterization Framework for Cancer Effect[a]

Predictive Risk Equation	Source Factors →	Pollutant Concentration ×	Exposure Duration/ Setting =	Exposure ×	Dosimetry Factors =	Dose ×	Response Factor =	Lifetime Individual Risk ×	Exposed Population[b] =	Risk to Exposed Population
Elements of risk equation	AC Systems	Env Tobacco	Short duration	Integrated by direct measurement	Contact rate	Pollutant mass per body weight per time	Carcinogenic potency	Incidence	Infants	
	Ambient air	Smoke	Long duration		Absorption rate	or	Noncarcinogenic threshold	Endpoint	Children	
	Building materials	Nonionizing	Indoors, work		Avg body weight	Pollutant mass per surface area per time	Noncarcinogenic potency/ severity		Home	
	Copying machines	Radiation	Indoors, home		Avg lifetime (or shorter exposure duration for noncancer endpoints)				School	
	Earth or ground	Organics	Outdoors		Regional surface area of lung				Adults	
	Furnishing Gas stoves Household products	Asbestos Inorganics Biologicals	Other						Male Female Worker	
	Insulation	Radon/progeny							Homemaker	
	Heaters, kerosene								Smoker	
	Tobacco smoking								Nonsmoker	
	Vehicle exhaust Woodstoves								Other	
Qualitative information					Hazard identification					

Table 1. Risk Characterization Framework for Cancer Effect[a] (*continued*)

Predictive Risk Equation	Source Factors →	Pollutant Concentration ×	Exposure Duration/ Setting =	Exposure ×	Dosimetry Factors =	Dose ×	Response Factor =	Lifetime Individual Risk ×	Exposed Population[b] =	Risk to Exposed Population
Quantitative information			Exposure assessment			Dose-response			Risk characterisation	
Qualitative or quantitative analysis					Risk assessment					

[a] Ref. 13.

[b] Ideally, in characterizing an exposed population or subpopulation, a distribution should be obtained that incorporates variability associated with age of the population, exposure levels associated with different activity patterns and microenvironments, and the susceptibility within the population to a specific effect.

Table 2. **Risk Characterisation Framework for Noncancer Health Effects**[a]

Source Factors	Pollutant Concentration	Exposure Duration/ Setting	Exposure	Dosimetry Factors	Dose	Response Factor	Individual Health Effects	Exposed Population	Population Health Effects
AC systems	Organics	Short duration	Integrated by direct measurement	Contact rate	Cumulative and/ or/peak	Noncarcinogenic potency	Adverse effects	Infants	Incidence
Ambient air	Inorganics	Long duration	Calculated from "exposure duration/ setting" and "exposure"	Absorption rate	Pollutant mass per body weight per time	Noncarcinogenic	Multiple organs	Children	
Building materials	Biologicals	Microenvironments		Avg body weight	or	Threshold	Multiple symptoms	Home	
Copying machines	Mixtures	Outdoors		Regional surface area of lung	Pollutant mass per surface area per time	Dose response	Severity of symptom/ effect	School	
Earth or ground	Peak constant variable	Indoors		Other factors		Function	Impairment of function	Adults	
Furnishing	Averaging time	Time activity patterns				Multiple organs		Male	
Gas stoves						Multiple symptoms		Female	
Household products						Mixtures/ synergisms		Worker	
Insulation								Homemaker	
Heaters, kerosene								Smoker	
Tobacco smoking								Nonsmoker	
Vehicle exhaust								Other	
Woodstoves								Susceptibility	
Carpets								Hypersusceptibility	
			Hazard identification						
	Exposure assessment				Dose response			Risk characterisation	
					Risk assessment				

[a] Ref. 14.

exhaled by the smoker, and sidestream smoke, which is emitted by the smouldering cigarette between puffs. Exposure by nonsmokers, called passive smoking, is of great concern as it is an involuntary risk, often incurred by the most susceptible members of society—unborn children, infants, and young children. Infants whose mothers smoked during the pregnancy are noted to have a lower birth weight and increased respiratory problems. The known health effects of exposure to ETS include lung cancer, acute respiratory illness such as bronchitis, chronic diseases such as coronary heart disease, chronic pulmonary disease, chronic bronchitis and emphysema and cardiovascular conditions.

Carbon monoxide (CO) results from the incomplete combustion of substances containing carbon and is most often present in motor vehicle emissions and tobacco smoke. Indoor sources of carbon monoxide include faulty furnaces and hot water heaters, clogged chimneys, and automobile exhaust in houses with attached garages. Carbon monoxide combines with hemoglobin (which transports oxygen in the blood as oxyhemoglobin), forming carboxyhemoglobin (COHB). Carboxyhemoglobin is more stable than oxyhemoglobin, and as a consequence, transport of oxygen to the tissues is reduced.

Chronic health effects may be present for low concentrations of CO and sudden death can occur at high levels of the gas. Carbon monoxide is one of the few contaminants where a direct cause–effect relationship can be established or demonstrated in a short time.

Carbon dioxide (CO_2) is produced by combustion processes in indoor and outdoor environments and is a byproduct of the natural metabolism of living organisms. Sampling for CO_2 is often used as a means for screening areas where potential SBS problems may exist. Use of CO_2 sampling data as a surrogate for SBS is an inexpensive method, providing real-time results. The results are then used to determine if additional, more detailed, monitoring is required.

Nitrogen oxide (NO_2) comes from outdoor (motor vehicles) and indoor combustion sources (unvented combustion appliances such as gas stoves, kerosene heaters, gas-fired water heaters). From controlled human exposure studies, it has been concluded that the health effects are primarily short-term effects of peak concentrations on susceptible individuals (asthmatics), and long-term effects of respiratory disease incidence and prevalence, primarily among children (7,18). A World Health Organization (WHO) evaluation indicated that a 1-h average of 400 $\mu g/m^3$ is a level of concern for susceptible individuals, a recent epidemiological study has suggested that for children, an annual average level of 30 $\mu g/m^3$ could cause adverse health effects (18).

A review of indoor sources and possible contributions to NO_2 has been performed (19). The conclusion from the review was, that for the children, the presence of unvented gas stores is the major determinant of high exposure, whereas for asthma and chronic obstructive pulmonary disease (COPD) patients, kerosene heaters and gas-fired space heaters constitute an important contribution as well. Some aspects related to the risk assessment to NO_2 are further discussed below.

Sulfur oxides (SO_x) are formed in the outdoor air during fuel combustion processes and smelting of nonferrous ores. They are also emitted indoors from kerosene heaters and by certain consumer products. Exposure to sulfur oxides alters the mechanical function of the upper and lower airways.

The short-term direct cause-and-effect relationship between elevated SO_2 readings and health problems is well documented for the Donore, Pennsylvania and London, England episodes.

Volatile organic compounds (VOCs) are generated from such indoor sources as glues, coatings, paints, household chemicals, copiers, and printers but also from outdoor sources usually associated with motor vehicles. The indoor emissions are dependent directly or indirectly on temperature, humidity, place of application, and ventilation rate. Volatile organic compounds are defined as organic chemicals with a boiling point between a lower limit of 50 to 100°C and an upper limit of 240 and 260°C. There are thousands of chemicals that belong to this group. The lists of individual compounds emitted by different indoor sources are extensive (over 60 from copying machine toner powders alone). There have been more than three hundred different VOCs identified in indoor air (20). They differ by orders of magnitude in their toxicity, which has to be reflected by different threshold limit values for different compounds. For example, the threshold limit value is 2350 mg/m^3 for octane and only 0.25 mg/m^3 for acrolein.

Other terms related to this general group of pollutants include total volatile organic compounds (TVOC, for which there is no unique definition), microbiological volatile organic compounds (MVOC, microbiologically formed VOCs), semivolatile organic compounds (SVOC, higher boiling points than VOCs), and particulate organic matter (POM, organic compounds associated with particulates). A comparison between indoor and outdoor concentrations of specific VOCs measured in 52 private homes in the city of Munich, Germany (21). A conclusion from the inspection of the data is that, except for benzene, all values of VOCs are higher indoors than outdoors, which indicates that indoor sources of VOCs are more significant and contribute more to health hazards than outdoor sources.

The negative health effects from exposures to TVOCs are likely to occur if the exposure exceeds 2 mg/m^3 (20). A literature review on associations between SBS and indoor air pollutants was performed (22). The conclusion, with regard to TVOCs, was that despite a large number of studies no consistent associations have been shown between measures of TVOCs and health effects. The inconsistent results from epidemiological studies ranged from reported positive associations between concentrations of TVOCs and symptoms reported, through no associations, to negative associations. The author of the review concluded that TVOCs may not be a valid indicator of indoor air pollution from a health perspective in the low-concentration exposure range, whereas the indoor air chemistry, involving for example, possible transformation products of VOCs, may be of importance.

A similar conclusion comes from a review on VOCs (23). According to this source, the role of VOCs as causative agents in increased IAQ complaints and prevalence of SBS appears to be multifactorial, and it is still not yet clear how VOCs alone may cause SBS. Several hypotheses have been proposed: (*1*) the total exposure approach, where VOCs exhibit a considerable additivity through nonspecific processes; (*2*) the action of specific irritants; (*3*) the multipollutant exposure approach; (*4*) spatial variations, eg, concentration gradients; (*5*) physicochemical properties, such as concentrations expressed as the chemical activity, lipophilicity, or reduction of the surface tension of the eye tear film by surface active substances; (*6*) the possibility of odor triggering certain symptoms; and (7) reactive species formed by indoor air chemistry.

Particulates are a mixture of solid particles and liquid droplets that vary in concentration, nature, and size distribution. The particulates are characterized by the mass concentration as total suspended particles (TSP) with diameters between 0.1 and 100 μm, PM_{10} fraction (particles smaller than 10 μm) or $PM_{2.5}$ fraction (fine particles smaller than 2.5 μm). From the point of view of health hazard, in addition to the nature and concentration in the air the particle size distribution is important. The size of airborne particles determines in which parts of the respiratory tract the particles are deposited and the composition determines in what way the human body is affected. Small airborne pollutant particles have large diffusion coefficients and thus a very high probability of deposition in the respiratory tract. In contrast, larger particles are predominantly deposited in the nasal area from which they are removed by mechanical means. Existing regulations refer to mass concentration of ambient particulates only as TSP and PM_{10}fractions. There are no standards regulating concentration of fine particles, which has currently emerged as a major issue. A comprehensive study (24) for six United States cities, demonstrated an almost linear relationship between mortality rate ratio (adjusted for other risk factors) and concentration of fine particles. By number, the majority of airborne particles are in the size range below 1 μm. Submicrometer particles are generated by most of the indoor sources; the particles could be combustion products, dust, or biocontaminants, and in addition to the their primary effect on health, they can also act as carriers of absorbed chemicals, biocontaminants, radioactive products, or condensed gases.

The size distribution and concentration of environmental tobacco smoke particles in a students' club during a rock music concert is presented in Reference 25. From the ETS distribution presented there, the conclusion can be drawn that a majority of ETS particles are very small, with a count median diameter of approximately 0.1 μm. Similar spectra are generated by other combustion sources.

There have been a number of studies indicating the effects of the presence of particles in the indoor air on air quality deterioration. An example is a cross-sectional study that investigated the correlation between dust quality and the prevalence of general symptoms (26). The conclusion from the study was that a distinct correlation existed between the amount of gram-negative bacteria in the dust, the relative amount of particles in relation to other components of dust, and VOCs absorbed by dust and general symptoms as well as mucosal symptoms in exposed persons.

Radon, mainly ^{222}Rn, is a naturally occurring noble radioactive gas with a half-life of 3.82 days. Its immediate parent, ^{226}Ra, is present in rocks, soil, groundwater, and some building materials. It decays to radionuclides which are chemically active and relatively short lived (^{218}Po, ^{214}Pb, ^{214}Bi, and ^{214}Po). The radiation released, mainly from alpha decay of short-lived decay products within the lungs, imparts the lung dose to which increased risk to lung cancer is attributed. There appears to be a clear relation between other types of cancer and elevated exposures to radon and its decay products (27), for example, cancer of the stomach, liver, salivary glands, esophagus, gall bladder, pancreas, larynx, skin, and bones and leukemia. The average indoor radon concentration in European Economic Community countries is in the range from 20 to 65 Bq/m^3 with concentrations occasionally up to 10,000 Bq/m^3 (28). For example, in Finland according to recent studies, 13% of workplaces in radon-prone areas exceed 400 Bq/m^3 and 5% exceed 800 Bq/m^3 (29). The result is based on 2964 radon measurements in workplaces.

Formaldehyde comes from particleboard, insulation, furnishing, tobacco smoke, gas stoves, and consumer products. It is a colorless gas with a characteristic odor, highly soluble in water. Health effects due to exposure to formaldehyde range from acute (such as eye, throat, or nose irritation, headache, cough, and fatigue) to chronic effects such as cancer, reduction of lung function and asthma. It was demonstrated that certain health effects occur at any concentration level and some only above a threshold level. According to researchers (30), the threshold concentration of 112 μg/m^3 may cause discomfort among some hypersensitive subjects during the first couple of days. On the other hand, there is no threshold in acquiring cancer, and there is also no consensus even as to the order of magnitude of concentration levels as related to levels of lifetime risk. For example, the concentration of formaldehyde associated with a lifetime risk of 10^{-5} for developing cancer ranges from 0.0285 to 36 μg/m^{-3} if calculated from different models (30).

Biocontaminants originate both indoors and outdoors. The long list of biological agents include bacteria, viruses, molds, microbial byproducts, animal dander and saliva, house dust mites, fungi, cockroaches, and pollen antigens. Biological agents produce illness primarily through infection of the respiratory tract and through an immune response. Some of the agents, such as bacteria, originate mainly indoors, whereas others, such as fungi, originate from outdoor sources. An example of a common type of bacteria present in an unhealthy indoor environment is *Legionella.* Cooling water systems in buildings can create conditions suitable for the multiplication of the bacteria, which can then be dispersed into the air. Legionellosis is a building-related illness, which refers to acute bacterial infection with *Legionella pneumophila.* Home humidifiers can have the same effect on *Legionella* and other types of bacteria, as can cooling water systems.

Usually in healthy buildings the air spore population resembles qualitatively the outdoor air, however, the indoor concentrations should be lower. The growth of molds in buildings contributes to the indoor air spore count. Specific fungi in individual cases are involved in allergic respiratory conditions such as asthma, rhinitis, or the rare instances of extrinsic allergic alveolitis. Apart from allergic responses, inhalation of fungi containing various kinds of toxins results in immunological disorders with potential neurological effects. Exposure to fungi in the agricultural context may result in a wide variety of symptoms ranging from respiratory tract bleeding to cancer. A summary of types and effects of fungi in the indoor environment is available (31). Biocontaminants of outdoor origin include pollen from trees, grasses, and other plants that may infiltrate the indoor environment and are believed to cause seasonal allergic responses (hay fever or asthma).

Ozone (O_3) is generated outdoor by photochemical reaction and infiltrates to the indoor environment. Indoor ozone generation is associated with high voltages, the main indoor sources of ozone being copy machines. The health effect related to ozone exposure were reviewed (32) and include both acute and chronic decrease in lung function and also accelerated aging of the human lung.

Mineral fibers may be natural or synthetic (man-made mineral fibers, MMMF). The most common natural mineral fiber is asbestos. The sources of asbestos and other mineral fibers in the indoor environment are fire retardant; acoustic, thermal, or electric insulation, and structural material. The list of health effects resulting from the exposure to asbestos include fibrosis (the excessive deposition of collagen at various thoracic sites), mesothelioma (tumor of the pleura or peritoneum), and lung cancer (33).

Lead in ambient air arises mainly from combustion of leaded gasoline. Indoor sources could include lead glazed ceramic flooring, lead-based paint, and cigarettes. The main risks related to exposure to lead is hemoglobin synthesis and brain function often resulting in incremental mental retardation.

Pollutant mixtures are a natural consequences of the fact that in most cases there are a number of pollutants from different sources present in indoor air. There are various types of physical and chemical interactions between pollutants, including coagulation, condensation, nucleation, gas-to-particle conversion, and gas-phase chemical reactions. The resulting products of interactions often have different physical parameters from the reactants (size distribution, diffusive properties, etc), different chemical reactivity or even different effects on human senses, such as smell. The lung deposition of radon progeny is, for example, higher when the progeny are not attached to the environmental aerosols than when they are attached. Thus the presence of airborne environmental aerosols, to which the progeny can attach, will alter the radiation dose delivered to the lung as well as radiation risk. The pollutants' behavior and interactions also depend on the presence of specific factors or conditions. The most important of them are temperature and humidity, which, for example, can affect the growth of mold or sorption and desorption of VOCs.

It is very difficult to evaluate the health effects of complex mixtures. The combined effects of pollutants may be additive ($C = A + B+$), synergistic ($C > A + B+$) or antagonistic ($C < A + B+$) (20). The area of increasing concern and interest is the human response to multiple exposures. Further progress in indoor environmental health-risk assessment will be difficult without the extension of the knowledge of the risks to multiple exposures.

Several factors also affect the indoor environment, including odor, thermal comfort, humidity, and ventilation. Odor is generated by most chemicals that elicit two nasal sensations: odor and pungency. Odor sensation is detected by the olfactory epithelium in the upper rear portion of the nose and is carried to the central nervous system by the olfactory nerve, whereas Pungency is elicited in the mucosae of the face through chemical stimulation of free nerve endings from the trigeminal nerve (34). Pungency from nonreactive airborne chemicals may rest heavily on a relatively unspecified physical interaction with a susceptible biophase (34). There are significant variations to odor sensation when substances are present individually, as well as when they occur as mixtures.

Thermal comfort in principle relates to temperature. Although temperature is easy to measure, the human comfort aspect of temperature depends also on air movement, radiant heat from the sun, and other sources, and humidity. Air humidity is a very important factor affecting air quality perception, interaction of pollutants in complex mixtures, growth of certain biocontaminants, and interactions of pollutants with the indoor environment. It was concluded (22) that the perception of dry air was associated with SBS symptoms but this perception was not necessarily demonstrated to be associated with physical air humidity. Closely linked with air humidity is dampness which often results in mold growth. Extensive studies performed in Europe and North America have demonstrated the existence of a relationship between self-reported symptoms and home dampness and mold for both children and adults (35).

Ventilation system design and air flow rate are of principal importance to the indoor environment. On the one hand ventilation results in dilution and removal of pollutants from indoor air, on the other hand outdoor air is introduced to the indoor environment. There was, however, an inconsistency reported (22) in the relationship between SBS and ventilation. Although inadequate ventilation is commonly encountered in buildings with SBS, a number of studies reported a lack of relationship between SBS symptoms and the type of ventilation or air flow rate. The author's assessment demonstrated that the type of ventilation system, recirculation of air, and the presence of rotary heat exchangers was not associated with an increase in SBS symptoms, whereas a low outdoor air flow rate in offices was associated with an increased prevalence of SBS symptoms.

A global approach to ventilation should be taken in the areas where outdoor concentration of pollutants could be higher than in indoor concentration. In fact, at times, low/no ventilation may present IAQ problems. The air intake system of a building may be negatively impacted during certain times by peak pollutant concentrations in the outside air. These episodes could include high CO concentrations during rush hours, emissions associated with landscaping (gas powered blowers/mowers operated in the vicinity of air intakes), and downwind proximity to landfills and tidal basins. Proper regulation of intake ventilation, which could include low ventilation during peak episodes, can be beneficial.

Other factors of importance for the indoor environment include lighting, noise, ergonomics, stress, and psychosocial factors.

Methods of Exposure Assessment. The general approach to exposure assessment is presented in Figure 1 (17). In this figure, indirect monitoring or modeling include air monitoring, time budgets, and questionnaires, whereas direct modeling (the term is not used on the diagram), includes biological markers and personal monitoring.

RISK MANAGEMENT

Risk management is a complex process by which, on the basis of risk assessment and taking into account budgetary limitations and available technical, social, or cultural solutions, responsible government bodies make decisions about what level of risk is acceptable (see RISK ASSESSMENT). This decision is translated into standards and guidelines that need to be considered in planning and designing new buildings, introducing new building materials or consumer products, and maintaining new and existing buildings. Risk management strategies should be reviewed frequently and updated as new risk assessment facts come to light, new technologies become available, and socioeconomical situations change.

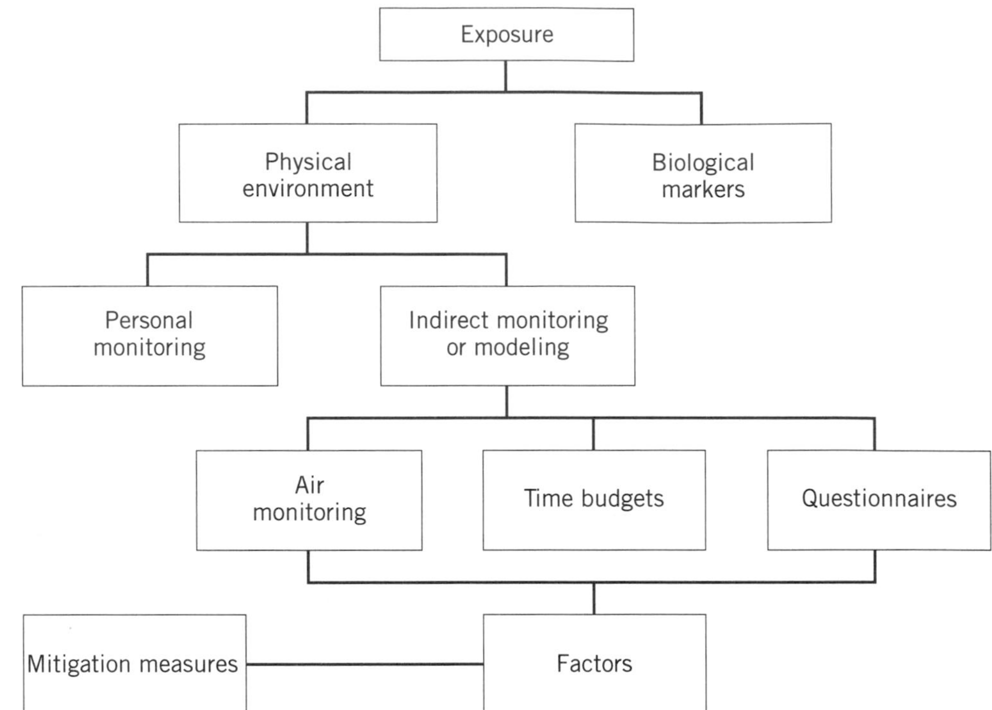

Figure 1. Scheme of approaches for estimating exposure to indoor air contaminants.

RISK MANAGEMENT: STATE OF THE ART

There is a general consensus that indoor exposures could significantly affect human health and well being and thus the exposures should be eliminated or reduced using appropriate management strategies. The main progress in the area of indoor air risk management relates to the establishment of awareness of most decision making bodies at local, national, and international levels of high priority in this area as well as the need for action. The main difficulty in the risk management process arises from insufficient quantitative information available on risk assessment which normally should be the basis for risk management actions. In the years to come, risk assessment techniques will continue to be improved and will provide more quantitative data, a necessary input for developing better management strategies. In the meantime, decision makers will have to continue improving or introducing new strategies based on the best available data, and at the same time actively encouraging and supporting research investigations to obtain more accurate data.

ACKNOWLEDGMENTS

Thanks for constructive comments, suggestions, and editorial support to Drs. N. D. Bofinger, V. Garnys, P. Morgan, and Ms. E. Stein, and Ms. Alina Morawska.

BIBLIOGRAPHY

1. L. A. Wallace, E. D. Pellizzari, T. D. Hartwell, C. M. Sparacino, L. S. Sheldon, and H. Zelon, *Atmos. Environ.* **19**, 1651–1661 (1985).
2. J. D. Spengler, in L. Morawska, N. D. Bofinger, and M. Maroni, eds, *Indoor Air: An Integrated Approach,* 1st ed., Elsevier, Oxford, 1995, pp. 33–41.
3. L. Morawska, in L. Morawska, N.D. Bofinger, and M. Maroni, eds, *Indoor Air: An Integrated Approach,* 1st ed., Elsevier, Oxford, 1995, pp. xix–xxi.
4. American Society of Heating, Refrigerating and Air-Conditioning Engineers, *Ventilation for Acceptable Indoor Air Quality,* (ASHRAE Standard 62-1989), Atlanta, Ga., 1989.
5. M. Maroni, in L. Morawska, N. D. Bofinger, and M. Maroni, eds., *Indoor Air: An Integrated Approach,* 1st ed., Elsevier, Oxford, 1995, pp. 169–174.
6. J. A. J. Stolwijk, in B. Seifert, ed., *NATO/CCMS Pilot Study on Indoor Air Quality and European Collaborative Action. Indoor Air Quality and its Impact on Man,* NATO/CCMS, 1993, pp. 47–53.
7. J. M. Samet, M. C. Marbury, and J. D. Spengler, *Am. Rev. Respir. Dis.* **137,** 221–242 (1988).
8. H. Levin, in L. Morawska, N. D. Bofinger, and M. Maroni, eds., *Indoor Air: An Integrated Approach,* 1st ed., Elsevier, 1995, pp. 253–258.
9. M. R. Cullen, *Occup. Med.* **2,** 4 1987.
10. G. Raw, R. Gammage, in L. Morawska, N. D. Bofinger, and M. Maroni, eds, *Indoor Air: An Integrated Approach,* 1st ed., Elsevier, Oxford, U.K., 1995, pp. 15–17.
11. S. Hygge, in B. Seifert, ed., *NATO/CCMS Pilot Study on Indoor Air Quality and European Collaborative Action. Indoor Air Quality and its Impact on Man,* NATO/CCMS, 1993, pp. 81–91.
12. J. Rotton, J. Frey, T. Barry, M. Milligan, and M. Fitzpatrick, *J. Appl. Soc. Psychol.* **9,** 397–412 (1979).
13. D. F. Naugle and T. K. Pierson, *J. Air Waste Manage. Assoc.* **41**(10), 1298–1307 (1991).
14. T. K. Pierson, R. G. Hetes, and D. F. Naugle, *Environ. Health Persp.* **95,** 121–129, (1991).
15. D. F. Naugle, in B. Seifert, ed., *NATO/CCMS Pilot Study on Indoor Air Quality and European Collaborative Action. Indoor Air Quality and its Impact on Man,* NATO/CCMS, 1993, pp. 1–21.
16. J. E. Yocom, *J. Air Pollution Contr. Assoc.* **32,** 500–520 (1982).
17. B. P. Leaderer, P. J. Lioy, and J. D. Spengler, in B Seifert, ed., *NATO/CCMS Pilot Study on Indoor Air Quality and European Collaborative Action. Indoor Air Quality and its Impact on Man,* NATO/CCMS, 1993, pp. 53–65.

18. L. M. Neas, D. W. Dockery, J. H. Ware, J. D. Spengler, F. E. Speizer, and B. G. Ferris Jr, *Amer. J. Epidemiol.* **134**(2), 204 (1991).
19. B. Brunekreef, in B Seifert, ed., *NATO/CCMS Pilot Study on Indoor Air Quality and European Collaborative Action. Indoor Air Quality and its Impact on Man* NATO/CCMS, 1993, pp. 135–143.
20. H. Rothweiler and C. Schlatter, *Toxicol. Environ. Chem.* **40,** 93–102, (1993).
21. I. L. Gebefuegi, G. Loerinci, and A. Kettrup, in L. Morawska, N. D. Bofinger, and M. Maroni, eds., *Indoor Air: An Integrated Approach,* 1st ed., Elsevier, Oxford, 1995, pp. 51–54.
22. J. Sundell, *Indoor Air Suppl.* (2/94), (May 1994).
23. P. Wolkoff, *Indoor Air Suppl.* (3/95), (1995).
24. D. W. Dockery, C. A. Pope III, X. Xu, J. D. Spengler, J. H. Ware, M. E. Fay, B. G. Ferris, and F. E. Speizer, *New Engl. J. Med.* **329**(24), 1753–1759 (1993).
25. L. Morawska, M. Jamriska, and N. D. Bofinger, *Sci. Total Environ.* **196,** 43–55 (1997).
26. F. Gyntelberg and co-workers, *Indoor Air* **4**(4), 223–238 (1994).
27. F. Steinhäusler, in L. Morawska, N. D. Bofinger, and M. Maroni, eds., *Indoor Air: An Integrated Approach,* 1st ed., Elsevier, Oxford, 1995, pp. 175–181.
28. B. M. R. Green, J. S. Hughes, and P. R. Lomas, *Radiation Atlas—National Sources of Ionizing Radiation in Europe.* Commission of the European Communities, Directorate General for Environment, Nuclear Safety and Civil Protection, Luxembourg, 1991.
29. E. Oksanen, M. Annanmäki, and M. Markkanan, in L. Morawska, N. D. Bofinger, and M. Maroni, eds., *Indoor Air: An Integrated Approach,* 1st eds., Elsevier, Oxford, 1995, pp. 153–156.
30. L. Mølhave, S. Dueholm, and L. K. Jensen, *Indoor Air* **5**(2), 105–119 (1995).
31. B. Flannigan and J. D. Miller, in R. A. Samson, B. Flannigan, M. E. Flannigan, A. P. Verhoeff, O. C. G. Adan, and E. S. Hoekstra, *Health Implications of Fungi in Indoor Environment,* Elsevier, Amsterdam, 1994.
32. C. J. Weschler, D. V. Naik, and H. C. Shields, in J. G. Kay, G. E. Keller, and J. F. Miller, eds., *Indoor Air Pollution,* Lewis, Chelsea, Mich., 1991, pp. 83–99.
33. J. A. Hoskins and R. C. Brown, in M. Maroni, ed., *Healthy Buildings' 95, Proceedings,* University of Milano and International Centre for Pesticide Safety, 1995, pp. 513–518.
34. J. E. Cometto-Muñiz and W. S. Cain, *Indoor Air* **4**(3), 140–145 (1994).
35. J. D. Miller, in L. Morawska, N. D. Bofinger, and M. Maroni, eds., *Indoor Air: An Integrated Approach,* 1st ed., Elsevier, Oxford, U.K., 1995, pp. 159–168.

INDUCTIVELY COUPLED PLASMA—ATOMIC EMISSION SPECTROMETRY

ZOE A. GROSSER
The Perkin-Elmer Corporation
Norwalk, Connecticut

WALTER SLAVIN
Bonaire Technologies
Ridgefield, Connecticut

HISTORY

Atomic emission spectrometry, AES, is one of the most widely used techniques for identifying and quantitating metallic elements. To make this measurement, we must convert the sample from its initial state, solid or liquid, to the vapor phase. This process is vaporization. Then we must break the bonds that bind the element of interest to other elements, which is atomization. Vaporization and atomization are usually accomplished thermally, by heating the sample to a high temperature. If the temperature of the source is high enough, some fraction of the atoms will loose an electron, resulting in ionization. Once the elemental atoms are free, further energy must be provided to excite the atom or ion to a state which will produce emission, and this is excitation. The wavelengths (color) that each metallic element emits are specific to that element and this specificity identifies the element. If we choose an appropriate wavelength emitted by the element of interest, its intensity is generally linearly proportional to the number of atoms of that element in the original sample. This relatively simple sequence is the basis of AES. The practical fact that we are not always successful in the vaporization step, or in the atomization step, or in the excitation step, results in interferences which must be understood and controlled.

There are many books on analytical atomic spectrometry and these should be consulted for greater depth than is possible in this article. Two books devoted entirely to inductively coupled plasma (ICP)/AES are highly recommended (1,2). An older book summarizes the atomic emission methods for spectrochemical analysis before 1970 and the advent of the ICP (3).

In the nineteenth century, flames of methane and other light gases were used to accomplish the tasks required to produce atomic emission, and simple prism devices were used to isolate the spectral line for qualitative or quantitative analysis. The intensities were compared to standard solutions of known concentration of the analyte of interest. The intensities were measured usually by the eye, using filters of varying opacity.

With the advent of photography, early in the twentieth century, the measurement of intensities became more precise. This stimulated the development of large and elaborate spectrometers to separate the often complex spectra of the elements. Flames were not hot enough to provide reliable vaporization, atomization, and excitation for many of the elements, especially in complex or refractory matrices. Electrical sparks or arcs were developed in the first quarter of the twentieth century. The spark was created by storing electricity in large capacitors and discharging the capacitors very quickly across a gap between electrodes on which the sample was placed. Alternatively, an arc was formed by passing a large continuous current, usually at relatively low voltage, between two electrodes on one of which the sample was placed.

By mid-century, several inventions and discoveries provided alternatives to arc and spark spectrometry. Photoelectric devices became available which generated an electrical signal proportional to the intensity of the light falling on them. These were further improved during World War II to incorporate means within the photoelectric device to amplify the signal in a remarkably noise-free manner. These became the photomultiplier, the detector for almost all spectrometric instruments until very recent times.

The early low temperature coal gas or petroleum gas flames were supplemented early in the twentieth century with air–acetylene flames which are very much hotter. A great deal of experimentation in mid-century was devoted to producing flames even hotter than the air–acetylene flame. Oxygen, or oxygen-enriched air, was used to increase the temperature but burners of this type were liable to explosion. A mixture of nitrous oxide and acetylene proved to be controllable and almost as hot as oxy-acetylene. This combination is the basis of modern flame emission and flame absorption instrumentation.

Also in mid-century, the diffraction grating was developed as an alternative to prisms for separating radiation into a spectrum. There are many advantages to diffraction gratings. For one thing, prisms require that the radiation pass through the prism medium and there are no materials that are transparent at all wavelengths. Gratings are typically used in reflection and it is not difficult to find materials that reflect radiation uniformly at all wavelengths. It eventually became possible to replicate gratings in relatively simple vacuum evaporation chambers, thus making the cost of gratings generally much smaller than that of prisms. There are other advantages; thus, gratings are most widely used at this time.

Electric furnaces were shown to be convenient sources for emission spectroscopy in the third quarter of the century, although research use of electrically heated carbon or graphite furnaces is known in the literature from the last century. Simple furnaces are formed by applying a low voltage across the length of a graphite cylinder into which the sample is deposited. This has developed into the device for vaporization and atomization used in graphite furnace atomic absorption spectrometry (GFAAS).

More recently, numerous systems have been used for creating a high temperature arc by passing an electrical current through a support gas, usually argon or helium. In 1961, Reed (4) wrapped an induction coil around a flowing stream of argon, forming an electric arc without the need for electrodes. Thus was born ICP, although Reed developed the device primarily for growing crystals. By 1964–1965, Greenfield et al. (5) in England and Fassel and Wendt at Iowa State University (6) had adapted ICP for emission spectrometry.

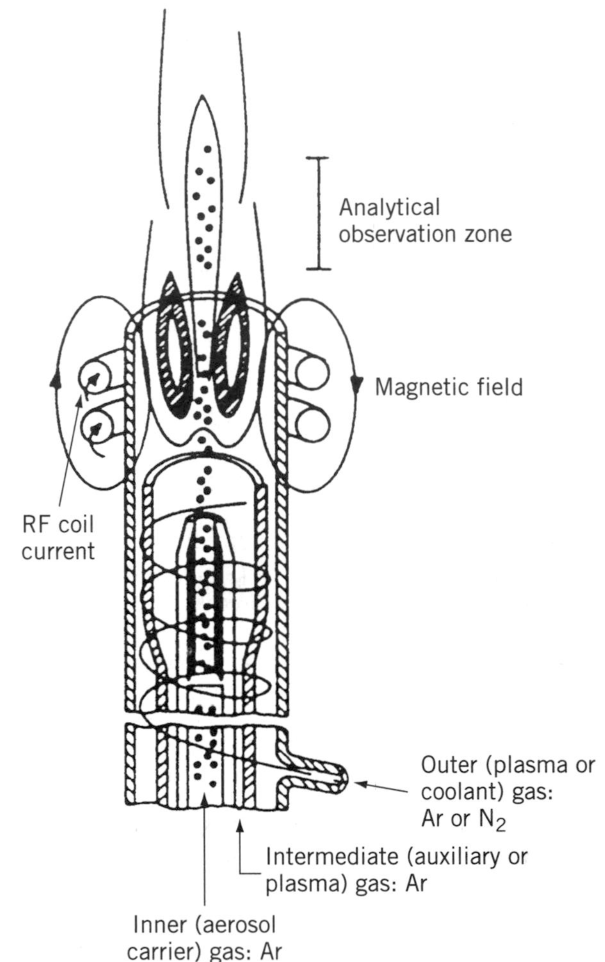

Figure 1. Typical ICP/AES torch. Reprinted with permission (1).

TECHNOLOGY

Torches and Nebulizers

A typical torch design is shown in Figure 1. An outer cylinder, usually quartz, confines several streams of gas. The inner cylinder carries the sample supported as an aerosol from a spray chamber which we will describe. Radio frequency (RF) energy is coupled to the flowing gas using an induction coil around the outer cylinder. The inner cylinder terminates below the induction coil in a narrow orifice to accelerate the gas to high velocity so that it will punch a hole into the center of the plasma which, then, has the shape of a doughnut. The sample is introduced into this inner gas stream where it is enclosed in the plasma. An intermediate cylinder between the inner and outer cylinders provides additional support gas to the extent that it is necessary. Between the intermediate cylinder and the outer cylinder, a stream of coolant gas, either argon or nitrogen, flows to protect the quartz cylinder from the high temperature of the plasma. Radiation from the source is observed just above the induction coil in the analytical observation zone.

Because the sample is usually introduced as an aerosol, it must undergo a sequence of transitions very rapidly as it progresses through the hot zone. First the aerosol droplet must be desolvated, then the small gas-borne particle must be vaporized, atomized, and finally excited to emit the characteristic elemental spectra. It should be no surprise that different elements and different matrices proceed through this environment at different rates, and this process is still being studied in great detail to understand just how and when each step occurs.

In ICP/AES, the sample is typically available as a solution, usually an aqueous solution. This must be converted into a gas-supported aerosol for introduction into the plasma. Most frequently this is done by pneumatic nebulization, passing a gas stream at high velocity over or around a very narrow orifice through which the solution flows or is pumped. There are many different designs of pneumatic nebulizers in use, each having some advantages and disadvantages compared to the others.

In the concentric nebulizer (86), the solution enters a very narrow capillary surrounded by a chamber through which the support gas, usually argon, flows. The gas exits the chamber through a narrow cylindrical capillary surrounding the liquid capillary. The high velocity gas shears the liquid into a fine mist (aerosol).

The cross-flow nebulizer (86), directs the gas stream at the exit of the liquid capillary, usually at right angles to the liquid flow, producing the aerosol. To reduce the wear and chemical attack of the air and liquid orifices, sapphire tips are pressed into the nebulizer body. These tips are machined very accurately, assuring similar performance between different nebulizers of the same design.

In the Babington design (86), the liquid sample flows over a surface which is penetrated by a small hole through which the gas exits. The film of liquid covering the surface is sheared by the gas stream, forming the aerosol. The concept of the Babington nebulizer has been implemented in many commercial designs. In the V-groove nebulizer (86), solution enters through a port in the vee groove above a second very narrow port through which the support gas enters. The sample flows downward under the action of gravity and is sheared into an aerosol as it passes over the gas port. A version of this design is called a fritted nebulizer and in this design the surface over which the thin film of liquid flows is penetrated by numerous small holes. Yet another variant is the cone spray nebulizer (86). Liquid enters at right angles to the axis of the nebulizer and flows into a cone in a sapphire tip. The support gas enters through a narrow opening at the base of the cone.

In general, the concentric and cross-flow nebulizers are convenient in situations where the sample contains only small amounts of matrix, such as drinking waters. One of the Babington designs is preferable when considerable matrix is present. Typically, the precision (and therefore the detection limit) is slightly better with concentric and cross-flow nebulizers for simple samples.

In recent years, reliable ultrasonic nebulizers have become commercially available (86). These are considerably more expensive and they provide five- to tenfold improvement in the sensitivity. The sensitivity is defined as the signal produced by a specific amount of analyte. This can be translated into equivalent improvements in detection limits for samples that have limited matrix. A source of radio frequency (RF) energy is focused on a plate onto which sample flows at a controlled low rate. Some fraction of the sample is converted to a fine aerosol by the ultrasonic energy and the amount of aerosol is greater than can be generated by Babington designs. However, the plasma usually cannot be sustained with the large amount of water vapor, so some of the water must be removed prior to the torch. This is done by heating the aerosol to produce a vapor and condensing the water out of the vapor stream as it flows through a cooled tube.

When the aerosol is formed in the nebulizer, it is carried through a spray chamber which separates out the larger droplets that would be inadequately handled by the plasma torch. Spray chambers come in many designs (1). The argon-borne aerosol enters through the inner tube and passes into a chamber of larger cross-section, therefore at lower linear velocity. In this environment the coarse droplets of aerosol fall to the chamber walls and are carried away through a drain. The fine aerosol exits upward to the plasma.

Spray chambers and torch designs are many and varied. Sample is carried through the inner gas tube which terminates in a very narrow capillary, so that the gas velocity is very high. This punctures the plasma, putting the dried aerosol into the hottest region of the plasma. The aerosol is rapidly desolvated, vaporized, atomized, ionized, and excited. The resulting emission is viewed above the turns of the RF coil, usually at about 15 mm above the coil (86).

In some situations, the sensitivity of ICP/AES is slightly too poor for the requirements of environmental analysis, most especially for the determination of Pb and Se in drinking waters. Therefore, quite recently, the torch has been turned on its side (as it routinely is used in ICP/MS) so that emission is observed along the axis of gas flow, axial observation (86). This provides a longer path, thus greater sensitivity, typically about five times. Some instrumental systems permit the operator to choose between axial and radial viewing (see INDUCTIVELY COUPLED PLASMA—MASS SPECTROMETRY).

The gain in sensitivity afforded by axial viewing is not without drawbacks. The influence of the matrix is much greater in this mode of operation so it is usually limited to solutions of low solids content, such as drinking waters or diluted samples. Thus, both axial torch arrangements and ultrasonic nebulization allow the analyst to reach the Pb and Se detection limits required in drinking waters. But in both cases this is done with some potential for imprecision, especially in samples with a large matrix content.

Optical Systems

The purpose of the spectrometer is to separate the wavelengths emitted by the materials in the plasma. This can be done by a prism but it is very difficult to make prisms that are sufficiently transparent in the far-UV near 200 nm. This is now done almost entirely with gratings. Light reflecting from a very fine grid will be diffracted by an angle that is determined by its wavelength. In one configuration of a grating spectrometer (1), the fine rulings are made directly on a concave mirror. When light of a single wavelength enters the spectrometer through a narrow slit, S1, the concave mirror will form an image at S2. If the light at S1 is composed of numerous spectral lines, the grating rulings on the concave mirror will form a succession of images in the plane of S2, each of a different wavelength. Often, the image at S2 is a narrow slit of about the same size as the entrance slit, S1. To cause the different wavelengths to pass through the S2 slit and fall upon a detector, *D,* either the concave mirror is rotated or the exit slit is scanned.

It is often easier to construct a grating spectrometer using a plane ruled grating and a typical mounting (1). In this mounting, the focussing element is split into two curved mirrors, M1 and M2. To scan the several wavelengths imaged at S2, the plane grating is usually rotated about an axis perpendicular to the page.

Typically, gratings used for ICP spectrometers have 1000–3000 lines/mm. Thus the lines are very closely spaced and must be very straight and very accurately spaced. Very elaborate machines controlled by lasers are used to move a diamond stylus over an aluminum coating on glass (or quartz). Such gratings are extremely expensive. A critical breakthrough occurred when it was found that gratings of very high quality could be replicated. To replicate a plane master grating, a thin layer of parting agent is evaporated on the surface of the master in a high-vacuum chamber. Over this a layer of aluminum is evaporated. A polished blank of glass or quartz is cemented to the evaporated aluminum surface, and the two glass pieces are separated at the parting material, which is washed away. These replicas are relatively inexpensive and of remarkably high optical quality.

Although replicated gratings provide excellent results, small imperfections in the lines or their straightness introduce ghosts, which are lines that are not part of the elemental spectra. Holographic gratings are an important modern advance. An interference pattern is formed by crossing two laser beams and this interference pattern is incident upon a photoresist that has been deposited on a glass surface. Removal of the unexposed photoresist leaves a fine ruling on the surface. Because holographic gratings do not require any mechanical devices for their production, they have far fewer imperfections than ruled gratings and are therefore much freer of ghosts. The technology for producing holographic gratings is still under active development.

The theory that describes exactly how a grating performs is readily available in many books (eg, (1)), and it will not be developed here. In practice, the spectrum from UV to IR is formed and is called an order. If the grating is used at a large angle to the incident radiation, several such spectra (or orders) are formed with a different spacing between the individual spectral lines. Because these will be superimposed one on top of the other, to be useful the orders must be separated by mounting a thin prism within the spectrometer, aligned so that its dispersion is perpendicular to that of the grating. This will separate the orders one above the other.

The detector most frequently used for ICP spectrometers is the vacuum photomultiplier, which has excellent sensitivity spanning the full range of wavelengths required for metal determinations. It is mounted just beyond the exit slit of the spectrometer and the full wavelength is scanned, or the grating is set to the particular wavelength chosen for a specific determination. Such instruments are called sequential scanning ICP spectrometers. When a large number of different elements must be determined, such instruments require time for the scanning and individual measurement to be made at each wavelength setting. Some commercial instruments are available which use individual photomultipliers set permanently at the wavelengths selected for a specific battery of determinations. Such instruments are often called quantometers or direct readers and they collect the data simultaneously from each detector.

Modern ICP spectrometers take advantage of these separated orders, discussed previously, by using a two-dimensional array of diode detectors. Only in the last decade have diode detectors been developed which perform as well as photomultipliers in the far-UV below 200 nm. Echelle grating mounts use gratings specially designed to be used at very large angles to the incident beam and cast numerous orders across the diode detector array. These instruments can access a large number of spectral lines simultaneously. They are thus very fast and efficient.

ENVIRONMENTAL APPLICATIONS

A distinction must often be made between the scientific concepts underlying environmental analyses and those concepts forced on the analyst by governmental regulations. In this section we will discuss applications from a technical point of view and in a later section we will discuss the subject more from the governmental point of view in the United States.

Environmental applications utilizing spectroscopic techniques for metal determination encompass a wide range of materials. These include natural waters and wastewaters, seawater, soils, sediments, biological tissues, and air particulates. ICP emission spectroscopy is widely applied to these problems because most of the metals of interest can be determined at the required levels. With modern instruments, the analyses are rapid and reliable. Major and minor constituents can be determined, often in the same analytical run. An excellent summary of these applications is available (7).

Even though some papers have been published on the direct determination of solid samples, it is very rarely feasible in practice. Solid material must be put into solution and diluted, generally, to 1–3% of total solids before presentation to the ICP. Acid digestions, furnace ashing followed by acid dissolution, and direct fusion are all widely used. Typically, the amount of acid in the samples should be duplicated in the standards. Aside of that, all of the conventional acid dissolution materials are compatible with ICP/AES. Care should be taken to avoid loss of volatile fluorides of As, Se, Sb, Si, Be, and B if HF is used. Fusion methods using sodium metaborate as a flux are often used for soils and sediments.

In many situations in the analysis of solids, the interest is in the determination of a metal that a particular sample treatment will extract from the solid. For example, in the determination of potentially toxic metals in soils, there is little interest in the metal that is so tightly bound that environmental effects will be unable to make those metals available for a toxic action. Therefore, various leaching or extraction methods are often specified in environmental methods. In these situations, care must be exercised if reference materials are subjected to the method, because reference materials are usually certified for the total metal present in the sample. Fusion methods, mentioned previously, are rarely used in U.S. Environmental Protection Agency (EPA) methodology because the fusion is effective in measuring the total amount of metal present in the sample.

An important strength of ICP/AES compared to atomic absorption methods is the wide dynamic range available on the ICP, 5 to 6 orders of magnitude. However, using this range to determine major and minor constituents in the same analytical run requires considerable care in the preparation of standards to cover so wide a range. Also, it must be remembered that at lower values of concentration, the standard deviation of the measurement is relatively constant, whereas at higher concentrations the relative standard deviation (coefficient of variation) is constant. The first two orders of signal, starting at the detection limit, will generally exhibit constant standard deviation, whereas higher orders of signal will exhibit constant relative standard deviation.

The principal sources of potential analytical interferences in ICP/AES are spectroscopic, in contrast to chemical, in nature. That is, the analytical line of the analyte may overlap weak lines of matrix materials that are present at high concentration. Problems of this type are not easily discovered. For example, recovery experiments will not reveal this problem. When a new method is being developed, it is wise to experiment with different concentrations of the major constituents of the sample to observe whether the response is independent of the matrix. Almost all elements have more than one analytical line, so another useful check is to compare the analytical result for the analyte at more than one line. It is very unlikely that coincidence will provide the same incorrect analysis at more than one wavelength.

Spectrometers of higher optical resolution will reduce the likelihood of spectral interferences. The only disadvantage of instrumentation of higher optical resolution is that it is more

Table 1. Summary of EPA ICP/AES Methods

EPA Program	Method Number	Analytes	Matrices	Comments
Water	200.7	Al, Sb, As, Ba, Be, B, Cd, Ca, Ce, Cr, Co, Cu, Fe, Pb, Li, Mg, Mn, Hg, Mo, Ni, P, K, Se, SiO_2, Ag, Na, Sr, Tl, Sn, Ti, V, Zn	Drinking water, wastewater, related matrices such as sludge and sediments	Hotplate and microwave digestion available, none represent total dissolution
RCRA	6010B	Al, Sb, As, Ba, Be, B, Cd, Ca, Cr, Co, Cu, Fe, Pb, Li, Mg, Mn, Hg, Mo, Ni, P, K, Se, SiO_2, Ag, Sr, Tl, Sn, V, Zn	Ground water, wastes, soils, TCLP extracts	Many digestion methods to choose from
CERCLA (CLP)	200.7 CLP-M	Al, Sb, As, Ba, Be, Cd, Ca, Cr, Co, Cu, Fe, Pb, Mg, Mn, Ni, K, Se, Ag, Na, Tl, V, Zn	Water and wastes	Microwave or hotplate hot acid leach sample preparation
Air	Method 29	Sb, As, Ba, Be, Cd, Cr, Co, Cu, Pb, Mn, Ni, P, Se, Ag, Tl, Zn	Metals emissions from stationary sources	(Based on Method 6010) Microwave digestion for samples

expensive. But the expense is rapidly compensated by more rapid method development and fewer concerns about accuracy.

Trace metals in drinking waters are often present at troublesome levels that are below the detection limit of the conventional ICP/AES. These include Pb and Se among others. In this case, ultrasonic nebulizers can often be used to provide about fivefold improvement in the detection limit for these samples which usually have low solids content. Alternatively, the newer ICP instrumentation permits the use of the torch in a horizontal position, thus providing a longer optical path and about fivefold improvement in detection limit.

ENVIRONMENTAL REGULATIONS AND METHODS

Analytical measurements are an integral part of environmental management. They are used to assess health effects, which are important in prioritizing contaminants for regulation. Routine monitoring of regulated contaminants ensures compliance with allowed levels and can indicate a hazardous situation. Cleanups of contaminated sites are driven by measurements indicating the location and extent of contamination. Each EPA program has created analytical methods to specify the way measurements should be performed for different media. Each method is based on a particular analytical technique and lists the target analytes, preparation of standards and samples, instrumental setup, and the quality control required. Table 1 summarizes the methods used in the major monitoring programs. For example, Method 200.7 is an ICP/AES method for drinking water and wastewater and includes an analyte list of 30 elements. The analytes for which the drinking water methods are acceptable are listed in a table in the drinking water section of the U.S. *Code of Federal Regulations* (8). Therefore, Method 200.7 may be used for the determination of nine elements out of a possible 20 primary and secondary metallic drinking water contaminants, based on the regulations. The drinking water program does not take into account possible future changes in technology with this approach. With the advent of axial plasmas and robust ultrasonic nebulizers it may be technically possible to determine additional analytes with Method 200.7. Until the regulations are changed to allow Method 200.7 for additional analytes such as Pb or Cd, new technology may only be allowed on a limited basis if a laboratory requests permission through an Alternate Test Procedure process.

Method 200.7 is also used for the analysis of wastewater. Whether or not Method 200.7 is suitable for a particular determination depends on the data quality objectives of the analysis and the capabilities of Method 200.7. Wastewater discharge is regulated through a system of National Pollution Discharge Elimination System (NPDES) permits. Permits are negotiated based on federal guidelines for the industrial category (9), the receiving waterway, and any state and local guidelines. Recent changes from a best available technology (BAT) approach for discharge cleanup to a Water Quality Criteria (WQC) may require modeling of the watershed to determine the actual effects of the discharge components on the ecosystem. This enables permit levels to be established that may protect the environment while allowing the discharge of contaminants which do not appear to affect a particular ecosystem.

Method 6010 is the ICP/AES method developed for solid and hazardous materials requirements of the RCRA program. It lists 15 analytes and is used as part of a compendium of methods providing analytical tools. The RCRA program provides cradle-to-grave regulation of hazardous materials and, as such, demands the analysis of a variety of matrices although the most common analyses are soil digests and toxicity characteristic leaching procedure (TCLP) matrices.

Method 200.7 CLP-M is Method 200.7 modified for the contract laboratory program (CLP). It is included in a statement of work (SOW) which serves as a contract for analysis. The CLP serves to provide analyses of Superfund sites, designated under Comprehensive Environmental Response, Compensation, and Liability Act (CERCLA) for assessment and cleanup. Twenty-two elements are included and contract required detection limits (CRDLs) must be demonstrated before a particular technology, such as ICP/AES, may be used for analysis. The methods, including rigorous QC must be followed exactly as written and documented in numerous reporting forms. Often the same protocol is followed less rigorously for other samples and this type of analysis is referred to as "CLP-like."

The ICP/AES methods developed for the different programs contain many similarities: technology, similar analyte lists, instrumental setup, and execution. The major difference among the methods is the quality control requirements. The CLP method contains the largest number of quality control tests and was designed to document the quality of data for legal purposes, in the absence of the chemist or laboratory that performed the testing.

The different program requirements have arisen from the perceived different needs of the programs and the media they analyze. However, for laboratories that measure samples for several different programs, the choice of the appropriate method and associated QC can be confusing and implementing several QC schemes can decrease laboratory productivity. Current issues related to these problems include the Environmental Monitoring Management Committee efforts, performance-based measurements, and laboratory accreditation.

The Environmental Monitoring Management Committee was created in 1990 with representatives from several levels and program offices of the EPA. Of their goals, the promotion of uniform methodology and laboratory accreditation have become very visible. They have developed a format for all method write-ups to make it easier to follow across program lines and this has been readily adopted. They have efforts underway to propose several uniform methods for adoption across program lines. Their initial efforts targeted a graphite furnace method, GC/MS volatile organic compound method, and a microwave digestion method. They have also been supportive of national laboratory accreditation efforts.

In the United States most laboratory analyses are conducted by private laboratories, not by government laboratories, as in other countries. To ensure uniform data generation across the variety of laboratories and operator skill levels there is merit to written analytical methods. The disadvantage comes in the situations where current methods do not work properly as written or when method modifications or new technologies offer advantages. Laboratories may have difficulty getting accurate data when performing a method exactly as written, with a difficult matrix. Or new technology may come along that makes a method more rugged or easier or more economical to perform. Or a new technique may be developed that offers advantages in terms of analytical capability or business requirements. EPA methods are very prescriptive and modifications are not encouraged. A different approach to analytical methodology based on a performance measurement system would circumvent some of these difficulties. The data quality objectives would be specified for a particular analysis and any method using any technique that would meet those objectives would be allowed. This is controversial because it would be a major change in the current way of operating and questions about enforcement issues must be addressed. It is moving forward within several EPA program offices and forces in Congress are addressing the issue through legislation such as the Baker Bill proposed in the 1996 session of Congress.

Performance-based measurements are being considered in conjunction with a national laboratory accreditation system. Currently, there is no national system, but laboratories can be accredited in several states in medium-specific programs. It is hoped that a national accreditation system will provide a more economical alternative and assure that laboratories have the capabilities with which to handle a performance-based measurement system.

Application issues of recent interest include field-based methods and speciation. Field-based methods may be semiquantitative or provide full quantitation directly at the field site. Examples of methods that may be field-based include X-ray fluorescence and immunoassay. Field-based methods can quickly define a clean-up area and monitor the on-going clean-up process. They can more economically provide samples to the laboratory, if used for screening purposes, because fewer clean samples will be taken back to the laboratory.

Speciation is an issue of interest to many parties involved in the environmental monitoring process. For inorganic analysis speciation can refer to the organic form of the metallic element, such as ethyl mercury versus methyl mercury. The oxidation state of the metal may instead be referenced, such as the more toxic Cr^{6+} versus the less toxic Cr^{3+}. Speciation research is only now gaining considerable attention in the United States. Sample preparation to maintain the desired form of the element for analysis is a challenge. Hyphenated instrument techniques, such as HPLC/ICP/MS, are required to separate the species of interest and sensitively detect the desired component. Once rugged analytical methodology is developed for speciation it will be possible to evaluate the hazard of various species and perhaps regulate only the toxic forms of an element. This issue is coupled with bioavailability and whether an element is present in a form that may be accessed by the body. In this case the free versus complexed or colloidal form of the element may be the determining factor (see INDUCTIVELY COUPLED PLASMA—MASS SPECTROMETRY).

Literature Review

The literature of the past several years was reviewed for articles that give a flavor of the type of efforts underway to characterize the environment. Speciation was of great interest and a number of different approaches have been taken to separation and measurement of species. Detection limit considerations would favor ICP/MS for trace species measurement, but several researchers have found ICP/AES to be a viable technique for investigating these problems. Wang and co-workers (10) have coupled together a number of techniques such as HPLC with ICP/AES for species separation and simultaneous determinations. Koropchak and co-workers (11) have coupled thermospray sample introduction methods with ICP/AES to separate and measure various oxidation states of Se. Boughriet and co-workers (12) measured the total metal concentrations in marine particles from the English Channel with ICP/AES and used Mossbauer spectroscopy to identify the Fe speciation and oxidation states. Chromium^{+3} and Cr^{+6} have been separated and measured with ion chromatography coupled to ICP/AES (13). Alternatively Cr^{+4} can be measured after extraction from solid waste by using ammonium complexation (14). Arsenic is of continuing interest and six arsenic compounds are separated and measured with a HPLC/UV/HyG/ICP/AES system (15). HPLC coupled with ICP/AES or FI–AAS has been demonstrated to be useful for the measurement of arsenobetaine and arsenocholine (16). Other elements, such as Si (17), Sn (18), and I (19), have been separated into species and detected with ICP/AES. The issue of bioavailability of Al in a variety of water sources is addressed (20).

Sample preparation procedures affect the species and total content available for measurement. Ashing and microwave digestion of wastes were compared (21). Four digestion procedures, using a variety of pressures, acids, and detection techniques showed that HF was required to dissolve river sediment completely (22). Instrumental techniques showed comparable results. Low temperature ashing followed by microwave digestion yielded good recoveries for 16 elements with less time and consumption of reagents than open-vessel digestion procedures (23). The environmental interpretation of soil contamination has been studied by examining metals extracted through a multi-step extraction procedure designed to separate fractions with differing characteristics (24,25). Ultrasonication was used to facilitate the dissolution of biological matrices in a sulfuric acid/peroxide mixture (26). A robotic system to prepare samples using digestion or concentration has been developed (27). Digestion coupled with an on-line aerosol desolvator showed improved spike recoveries compared to open or closed digestions alone without additional acid (28). An automated microdigestion for plant tissues minimized waste disposal and maintained analytical control (29).

Instrumental accessories can make the analysis of difficult environmental samples easier. The analysis may be less time consuming or subject to less contamination. Ultrasonic nebulization can improve detection limits, but may show increased matrix interferences (30,31). Slurry sampling provides the direct insertion of a finely ground sample suspended in a solution. Calibration is generally performed with aqueous solutions (32). Internal standard solutions for ICP/AES can be added on-line, eliminating the need to add the solution to each sample individually (33). Electrothermal vaporization into a plasma can improve detection limits or remove interferences. It may also be useful for solid samples and when small volumes of liquid are available. Researchers have used tungsten boat atomizers (34) or graphite cups (35) for introduction. Alkylation of Cd improved the detection limit in the presence of interferences (36). A generalized standard additions method has been used to assist in calibration (37) and showed good agreement with certified values. Solid samples can be analyzed directly by laser ablation and introduction of the vapor directly into ICP/AES. Tea leaves, milk powder, and sediments were examined (38) showing good agreement with reference values. The technique was used in a field sampling system for screening contaminated soils for trace inorganic elements (39). It was compared with three radioanalytical techniques in a screening demonstration for U (40). A new sample introduction device, fused-silica aperture thermospray, was tested on toxicity characteristic leaching procedure (TCLP) extracts and the detection limits were found to be superior to pneumatic sample introduction, although the interferences were found to be worse (41).

On-line manipulation of samples for chemical reaction, separation from the matrix, or concentration has been an area of study. Hydride and nonhydride forming elements could be sensitively detected using continuous hydride generation with conventional ICP/AES nebulization (42). Continuous flow plumbane generation ICP/AES was applied to the measurement of low levels in soft drinks, sediments, and lichens (43). Several techniques are compared for Se determination and hydride generation ICP/AES was deemed to be useful for routine samples (44). Copper was determined after generation of a vapor species which was detected with an ICP/AES. Efficiency was 50% and interferences were minimal (45). A continuous flow system was used to measure As, Se, and Sb in 10 different environmental matrices simultaneously (46). Two digestion methods were tested in conjunction with the same system as used previously for Sb (47). A review of flow injection applied to on-line preconcentration covers fundamental and practical aspects of the technology (48). On-line preconcentration was applied to seven trace metals in biological and environmental matrices (49). Other separations include anion exchange for the elimination of interferences (50), solvent extraction for concentration with a new crown hydroxamic acid (51), sorption on chelating resins for enrichment (52), and magnesium hydroxide precipitation for more than 20 elements (53).

The factors which contributed to uncertainty in environmental ICP/AES measurements are discussed (54). The role of reference materials in quality control is highlighted in several articles. Mohl and Stoeppler discuss the criteria for choosing the reference material that best matches the sample (55). Three new soil standard reference materials are characterized by a variety of techniques and determined to be useful for geochemical and environmental use (56).

Lead is an element of considerable interest because of toxicity, especially to children, and has been measured in a variety of matrices. Ogunsola characterized Pb in roadside dust in Nigeria (57). Roadside dusts were characterized for Pb and a variety of other elements and compared to results from neutron activation analysis and AAS (58). The variability of Pb in proficiency samples was reported for paints, soils, and dusts (59). A microwave digestion using HNO_3-HF was developed for paints and paint scrapings and showed good performance (60). ICP/AES and GFAAS were explored for Pb determination in soil and water samples (61).

The utility of ICP/AES was explored for environmental samples (62). Sixteen elements were measured in solids, soils, wipes, and glass-fiber air filters over several orders of magnitude concentration range (63). Metals in fly ash were characterized to aid in determining effective stabilization (64). Heavy metals and radioactive elements were measured in coal and fly ash to aid in measuring environmental impact (65). A method to determine Nd and radioactive materials was developed to assess environmental contamination around nuclear facilities (66). Total mercury was determined in a method coupling cold vapor generation with ICP/AES (67). Sulfur measurement by ICP/AES was compared to ion chromatography and significant differences explained (68).

ICP/AES has been used as a routine characterization tool in a variety of environmental studies. Heavy metals were characterized in Spanish bay sediment (69). Bioaccumulation of rare earth elements was studied in carp (70). A marine bivalve was monitored for several elements over the course of a year (71). Barium and Sr were measured as indicators of fresh or marine history of a fish (72). In another study, Ba and Sr were measured to monitor adverse effects on a marine bivalve and Ba was shown to have toxic effects in seawater (73). Mercury was determined in topsoil samples in a historic lead mining and smelting town (74). The mobility of B was studied in peach trees after topical application (75). ICP/AES was used to measure Au or Ag in labeled sediments used to study natural processes (76). Aqueous leaching of heavy metals from municipal compost was compared with total values after acid digestion (77). Leachate from scrap-shredder waste was measured for seven elements (78).

Exposure to metals can be measured in humans or in substances contacting humans, such as dietary items. Several workers have evaluated exposure based on hair analysis (79–81). Elements in rice straw and ash were evaluated to determine the environmental quality around rice paddies (82). Trace elements in cow, goat, and sheep milk and cheese were measured to evaluate potential human exposure (83,84). A crustacean was evaluated as an indicator of trace element contamination (85).

Regulatory Issues

Regulatory issues are often at the heart of environmental analytical decisions. The EPA program to which data are being submitted determines which set of methods may be required. The most suitable technique is then chosen from those allowed in the list of methods based on availability, workload, and analytical capability.

ACKNOWLEDGMENT

Thanks to S. Slavin for her generosity and helpful editorial comments.

BIBLIOGRAPHY

1. P. W. J. M. Boumans, *Inductively Coupled Plasma Spectroscopy, Parts I and II,* Wiley-Interscience, New York, 1987.
2. A Montaser and D. W. Golightly, *Inductively Coupled Plasmas in Analytical Atomic Spectrometry,* VCH Publishers, New York, 1987.
3. M. Slavin, *Emission Spectrochemical Analysis,* Wiley-Interscience, New York, 1971.
4. T. B. Reed, *J. Appl. Phys.* **32,** 821 (1961).
5. S. Greenfield, I. Ll. Jones, and C. T. Berry, *Analyst* **89,** 713 (1964).
6. R. H. Wendt and V. A. Fassel, *Anal. Chem.* **37,** 920 (1965).
7. J. W. McLaren, Environmental Applications in P. W. J. M. Boumans, ed., *Inductively Coupled Plasma Spectroscopy, Part II,* Wiley-Interscience, New York, 1987.
8. *U.S. Code of Federal Regulations 40,* Pt. 141.23, U.S. Government Printing Office, Washington, D.C. (July 1, 1996).
9. *U.S. Code of Federal Regulations 40,* Subchapter N, Pts. 401–471, U.S. Government Printing Office, Washington, D.C., July 1, 1996.
10. X. Wang, Z. Zhuang, P. Yang, B. Huang, *Microchem. J.* **51**(1/2), 88 (1995).
11. J. Yang, T. S. Conver, J. A. Koropchak, *Anal. Chem.* **68**(22), 4064 (1996).
12. A. Boughriet, M. Wartel, C. Cordier, C. Douez, L. Deram, E. Marting, B. Ouddane, H. Chamley, P. Recourt, *Mar. Pollut. Bull.* **28**(9), 541 (1994).
13. J. Prokisch, B. Kovacs, Z. Gyori, J. Loch, *J. Chromatogr.* **683**(1), 253 (1994).
14. J. D. Hwang, W. J. Wang, *Appl. Spectrosc.* **48**(9), 1111 (1994).
15. N. Violante, F. Petrucci, F. La Torre, S. Caroli, *Spectroscopy* **7**(7), 36 (1992).
16. A. J. Murer, A. Abildtrup, O. M. Poulsen, J. M. Christensen, *Analyst* **117**(3), 677 (1992).
17. S. B. Dorn, E. Skelly Frame, *Analyst* **119**(8), 1687 (1994).
18. A. Mazzucotelli, R. Frache, E. Magi, P. Rivaro, T. Gerbino, *Ann. Chim. (Rome)* **82**(7–8), 379 (1992).
19. K. A. Anderson, B. Casey, E. Diaz, P. Markowski, B. Wright, *J. AOAC Int.* **79**(3), 751 (1996).
20. B. Smith, N. Breward, M. B. Crawford, D. Galimaka, S. M. Mushiri, S. Reeder, *Geol. Soc. Spec. Publ.* **113** (Environmental Geochemistry and Health), 141 (1996).
21. R. Moral, J. Pedreno, G. Navarro, J. Mataix, *Commun. Soil Sci. Plant Anal.* **27**(3,4), 753 (1996).
22. P. Krause, B. Erbsloeh, R. Niedergesaess, R. Pepelnik, A. Prange, *Fresenius' J. Anal. Chem.* **353**(1), 3 (1995).
23. M. Achili, R. Barban, B. Zucchi, W. Martinotti, *Water, Air, Soil. Pollut.* **57–58,** 495 (1991).
24. J. Arunachalam, H. Emons, B. Krasnodebska, C. Mohl, *Sci. Total Environ.* **181**(2), 147 (1996).
25. X. Li, B. J. Coles, M. H. Ramsey, I. Thornton, *Chem. Geol.* **124**(1–2), 109 (1995).
26. S. Mamba, B. Kratochvil, *Int. J. Environ. Anal. Chem.* **60**(2–4), 295 (1995).
27. M. J. Zoellner, H. W. Emmel, L. D. Nelson, *Proc. Int. Symp. Lab. Autom. Rob. Meeting Date 1991,* Zymark Corp., Hopkinton, Mass., 1992, 402.
28. L. B. Allen, P. H. Siitonen, H. C. Thompson, Jr., *J. Anal. At. Spectrom.* **11**(7), 529 (1996).
29. K. A. Aderson, *At. Spectrosc.* **17**(1), 30 (1996).
30. E. Magi, M. Grotti, A. Mazzucotelli, *Ann. Chim. (Rome)* **84**(9–10), 417 (1994).
31. P. Galli, N. Oddo, *Microchem. J.* **46**(3), 327 (1992).
32. L. Ebdon, M. Foulkes, K. O'Hanlon, *Anal . Chim. Acta* **311**(2), 123 (1995).
33. J. W. Milburn, *At. Spectrosc.* **17**(1), 9 (1996).
34. Y. Okamoto, H. Kakigi, T. Kumamaru, *Anal. Sci.* **9**(1), 105 (1993).
35. A Golloch, M. Haveresch-Kock, W. G. Fisher, *J. Anal. Appl. Pyrolysis* **25,** 37 (1993).
36. S. Tao, T. Kumamaru, *Anal. Chim. Acta* **310**(2), 369 (1995).
37. S. Boonene, P. Verrept, L. J. Moens, R. F. J. Dams, *J. Anal. At. Spectrom.* **8**(5), 711 (1993).
38. S. F. Durrant, *Analyst* **117**(910), 1585 (1992).
39. E. J. Jaselskis, M. S. Anderson, A. P. D'Silva, D. P. Baldwin, D. S. Zamzow, *J. Environ. Eng. (N.Y.)* **121**(7), 521 (1995).
40. J. C. Cunnane, S. Y. Lee, D. L. Perry, V. C. Tidwell, J. Schwing, K. R. Nuhfer, G. Weigand, *Technol. Programs Radioact. Waste Manage. Environ. Restor.* (**1**), 803 (1993).
41. M. Veber, J. A. Koropchak, T. S. Conver, J. Herries, *Appl. Spectrosc.* **46** (10), 1525 (1992).
42. Zhang Li, S. McIntosh, W. Slavin, *Anal. Proc.* **29**(10), 438 (1992).
43. M. C. Valdes-Hevia y Temprano, M. R. Fernandez de la Campa, A. Sanz-Medel, *J. Anal. At. Spectrom.* **8**(6), 821 (1993).
44. P. M. Haygarth, A. P. Rowland, S. Sturup, K. C. Jones, *Analyst* **118**(10), 1303 (1993).
45. R. E. Sturgeon, J. Liu, V. J. Boyko, V. T. Luong, *Anal. Chem.* **68**(11), 1883 (1996).
46. K. A. Anderson, B. Issacs, *J. AOAC Int.* **78**(4), 1055 (1995).
47. K. A. Anderson, B. Issacs, *J. AOAC Int.* **77**(6), 1562 (1994).
48. Z. Fang, *Spectrochim. Acta Rev.* **14**(3), 235 (1991).
49. P. Schramel, L. Xu, G. Knapp, M. Michaelis, *Mikrochim. Acta* **106**(3–6), 191 (1992).
50. J. Goossens, R. Dams, *J. Anal. At. Spectrom.* **7**(8), 1167 (1992).
51. P. Shrivastav, Y. K. Agrawal, *Analusis* **24**(1), 13 (1996).
52. G. Wuensch, S. Knobloch, J. Luck, W. Bloedorn, *Spectrochim. Acta* **47B**(1), 199 (1992).
53. B. Bostroem, K. Bostroem, *Geol Foeren. Stockholm Foerh.* **113**(2–3), 97 (1991).
54. P. Fodor, B. D. Zs, *Microchem. J.* **51**(1/2), 151 (1995).
55. C. Mohl, M. Stoeppler, *Fresenius' J. Anal. Chem.* **345**(2–4), 164 (1993).
56. S. A. Wilson, P. H. Briggs, J. S. Mee, D. F. Siems, *Geostand. Newsl.* **18**(1), 85 (1994).
57. O. J. Ogunsola, A. F. Oluwole, O. I. Asubiojo, H. B. Olaniyi, F. A. Akeredolu, O. A. Akanle, N. M. Spyrou, N. I. Ward, W. Ruck, *Sci. Total Environ.* **146/147,** 175 (1994).

58. S. S. Ismail, F. Grass, A. E. Arifien, A. G. Mostafa, Sohag *Pure Appl. Sci. Bull.* **7,** 16 (1991).

59. P. C. Schlecht, J. H. Groff, A. Feng, R. Song, *Am. Ind. Hyg. Assoc. J.* **57**(11), 1035 (1996).

60. A. M. Paudyn, R. G. Smith, *Fresenius' J. Anal. Chem.* **345**(11), 695 (1993).

61. S. Azimi, *Environ. Test. Anal.* **3**(1), 54 (1994).

62. P. Capota, G. E. Baiulescu, M. Constatin, *Chem. Anal. (Warsaw)* **41**(3), 419 (1996).

63. D. E. Kimbrough, I. H. Suffet, *Analyst* **121**(3), 309 (1996).

64. M. D. McKinley, G. W. Warren, S. M. Lahoti, K. Sreenivasarao, *J. Hazard. Mater.* **29**(2), 255 (1992).

65. J. Font, M. Casas, R. Forteza, V. Cerda, F. Garcias, *J. Environ. Sci. Health* **A28**(9), 2061 (1993).

66. S. Sumiya, N. Hayashi, H. Katagiri, O. Narita, *Sci. Total Environ.* **130–131,** 305 (1993).

67. K. A. Anderson, B. Issacs, M. Tracy, G. Moeller, *J. AOAC Int.* **77**(2), 473 (1994).

68. A. C. Edwards, R. C. Ferrier, J. D. Miller, *Hydrol. Sci. J.* **37**(3), 277 (1992).

69. M. Rodriguez-Pastor, R. Moral, M. De La Muela, D. Prats, J. Conese, J. Moreno-Caselles, *Fresenius Environ. Bull.* **5**(5/6), 295 (1996).

70. Q. Tu, X. Wang, L. Tian, L. Dai, *Environ. Pollut.* **85**(3), 345 (1994).

71. G. Bordin, J. McCourt, A. Rodriguez, *Sci. Total Environ.* **127**(3), 255 (1992).

72. P. J. Pender, R. K. Griffin, *Trans. Am. Fish. Soc.* **125**(5), 679 (1996).

73. J. V. Spangenberg, G. N. Cherr, *Environ. Toxicol. Chem.* **15**(10), 1769 (1996).

74. J. S. Rieuwerts, M. Farago, *Sci. Total Environ.* **188**(2,3), 167 (1996).

75. Z. H. Shu, G. H. Oberly, E. E. Cary, *J. Plant Nutr.* **17**(7), 1243 (1994).

76. I. Olmez, F. X. Pink, R. A. Wheatcroft, *Environ. Sci. Technol.* **28**(8), 1487 (1994).

77. B. L. Sawhney, G. J. Bugbee, D.E. Stilwell, *ASTM Spec. Tech. Publ. (1993)*, **1162** (Application of Agricultural Analysis in Environmental Studies), 122 (1993).

78. K. M. Attar, B. S. Tawabini, E. R. Pambid, M. A. Abdel-Aal, *Waste Manage. (N.Y.)* **11**(4), 283 (1991).

79. W. Ashraf, M. Jaffar, D. Mohammad, *Sci. Total Environ.* **151**(3), 227 (1994).

80. W. Ashraf, M. Jaffar, K. Anwer, U. Ehsan, *Environ. Pollut.* **87**(1), 61 (1994).

81. J. Kubova, V. Stresko, J. Medved, J. Polakovicova, *Acta Environ. Univ. Comenianae* **2,** 17 (1994).

82. K. Saito, H. Muto, Y. Takizama, M. Kodama, *Toxicol. Environ. Chem.* **41**(1+2), 15 (1994).

83. E. Coni, A. Bocca, D. Ianni, S. Caroli, *Food Chem.* **52**(2), 123 (1994).

84. E. Coni, P. Coppolelli, S. Caroli, C. Cavallucci, M. Marinucci, *Food Chem.* **57**(2), 253 (1996).

85. F. Petrucci, S. Caimi, G. Mura, S. Caroli, *Microchim. J.* **51**(1/2), 181 (1995).

86. C. B. Boss, K. J. Fredeen, *Concepts, Instrumentation, and Techniques in Inductively Coupled Plasma Atomic Emission Spectroscopy,* The Perkin-Elmer Corp., Norwalk, Conn., 1989, expected revision 1997.

INDUCTIVELY COUPLED PLASMA—MASS SPECTROMETRY

ZOE A. GROSSER
The Perkin-Elmer Corporation
Norwalk, Connecticut

WALTER SLAVIN
Bonaire Technologies,
Ridgefield, Connecticut

HISTORY

Mass spectrometry, the sorting of material in a sample according to mass, has long been an important technique, both in organic and inorganic analytical chemistry.

In mass spectrometry, separation is achieved by a combination of the properties of electric charge and mass. The charge on the particle is used to accelerate it in an electric field. A magnetic field is used in classical mass spectrometers to bend the accelerated beam into curves that depend on the mass of the particle. Most modern mass spectrometers for metal determinations use quadrupole spectrometers.

If atomic spectrometry is to be applied to inorganic analysis, it is necessary that the sample be converted to an atomic vapor (vaporization) and that all the bonds between the analyte and everything else are broken (atomization). Vaporization and atomization are usually accomplished thermally, by heating the sample to a high temperature. For atomic emission, it is necessary to supply additional energy to raise the free atom to an excited state (excitation). However, for mass spectrometric detection, the free atom must be ionized by removal of an electron (ionization).

Mass spectrometry (MS) dates back to the beginning of the 20th century when British physicist Francis Aston build the first mass spectrometers (1). He used electric and magnetic fields to bend a beam of charged atoms, separating the atoms according to their ratio of mass to charge. In the United States, the technique was developed to study the abundance of various isotopes of elements and the concentration of elements in various natural materials (2).

The instruments were large and expensive and were gradually used to study complex organic compounds, because of a lack of simpler devices to separate and quantitate such materials. The application to inorganic chemistry continued, using spark and DC arc methods to convert the sample to a charged atomic vapor. Traditionally, spark source mass spectrometry has mostly been applied to solid samples.

In the early 1960s, the inductively coupled plasma (ICP) was developed (3) and adapted as a source for atomic emission spectroscopy (AES) (4,5) (see INDUCTIVELY COUPLED PLASMA—ATOMIC EMISSION SPECTROMETRY). In 1975, it was observed that plasmas were efficient in producing ions (6). The first investigations of the properties of the inductively coupled plasma for producing charged particles for mass spectrometry were in 1981 at Iowa State University (7), and in England (8). The research groups recognized that most of the atoms passing through the plasma were ionized, making the plasma an ideal source for the mass spectrometer and one ideally suited to solutions. These groups have continued to develop ICP/MS for chemical analysis, and their designs provided the foundation for the two major commercial manufacturers. The Perkin-Elmer Sciex instruments stemmed from the Iowa State design through additional important work done at the University of Toronto. It was introduced by Sciex in 1983. The VG instrument was developed in England.

The fraction of each element of the periodic table that is ionized after passage through the argon ICP is close to 100% for all but a few elements. The energy required to doubly ionize an atom is, for almost all elements, much higher than is available in the argon ICP. Therefore, in almost every situation, the argon ICP is extremely efficient at singly ionizing the analyte while relatively little of the analyte is doubly ionized and higher levels of ionization are negligible. It has thus proven to be an attractive ionization device for mass spectrometry.

Since the ICP/MS technology has been developed largely as an outgrowth of ICP emission spectrometry, and by workers in that field, it is more often seen as ICP with mass spectrometry detection, not mass spectrometry with an ICP ion source.

TECHNOLOGY

Introductions to the technology are available (9,10). The plasma is mounted horizontally in front of a quadrupole mass spectrometer that uses a multiplier as a detector of the separated masses. An example of a system is shown in Figure 1. The plasma flame is incident on a cooled sampling cone with a small orifice through which a fraction of the ions pass. A second cone creates a space between the two cones, which is pumped down to a low pressure so that the ions passing through the skimmer can enter the high vacuum chamber of the quadrupole spectrometer. A series of ion lenses focuses the beam into the entrance of the spectrometer.

A major problem in the coupling technology is the fact that the ICP operates at atmospheric pressure while the mass spectrometer requires a high vacuum. Figure 2 shows this part of the system in more detail. The plasma is incident, just above the observation zone, onto a cone with an aperture typically about 1 mm in diameter. A second cone, called a skimmer, has a smaller aperture. The space between is pumped with a high efficiency mechanical vacuum pump that removes most of the gas, down to a pressure of about 1 torr. The pressure in the next chamber is low enough to provide a mean free path of several centimeters. In this chamber, ion cylinder lenses collect and focus the beam into the entrance to the mass spectrometer. A solid metal stop is mounted on the axis of the ion lens to block line-of-sight radiation from the plasma.

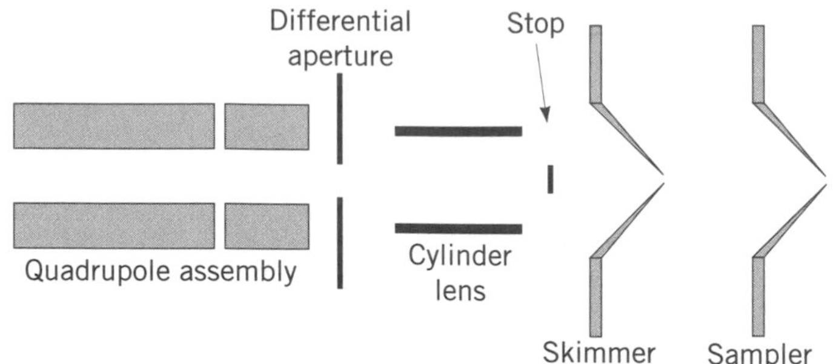

Figure 2. The ICP/MS interface.

The quadrupole mass spectrometer is more accurately a mass filter. Four carefully spaced parallel metal poles are charged to a variable voltage. When atoms of different mass enter the space between the poles, a particular voltage is just correct to keep that mass spiraling between the poles and, therefore, exiting on-axis to the detector. Other masses are out of synchronism and do not exit. When the voltage is changed, a different mass exits successfully. The voltage is scanned at radio frequencies so all masses of interest are collected rapidly. While the ICP/MS is often considered a simultaneous detector, it is really a rapid sequential detector with a mass resolution of about 1 mass unit. If only a few elements are of interest, a peak-hopping mode can be employed, by which the voltage

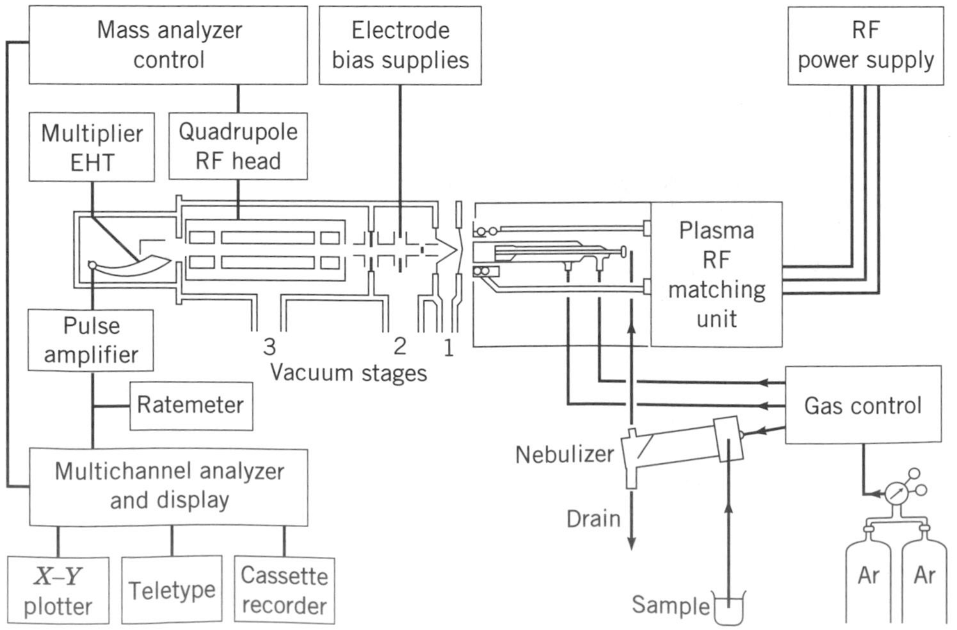

Figure 1. A typical ICP/MS system (9).

switches rapidly between appropriate values, improving precision or reducing analytical time.

ENVIRONMENTAL APPLICATIONS

There are many reviews of ICP/MS that describe the hardware, techniques, potential interferences, and applications (11,12). There are several important advantages of ICP/MS over other spectrometric methods. For solutions containing little besides the analyte itself, ICP/MS is more sensitive than the graphite furnace by more than an order of magnitude. By comparison with ICP/AES, it is more sensitive by almost three orders of magnitude. While sensitivity is defined as the amount of analyte that produces some arbitrary signal, detection limit is a better measure of the difference between different techniques. Detection limit is the smallest amount of analyte that can be measured with some stated level of confidence. Figure 3 compares the detection limit of these techniques. However, some caution is required in interpreting these limits. The operation of ICP/MS permits only small amounts of total solids. Therefore samples with considerable matrix require dilution compared, for example, with furnace AAS, which can handle large amounts of matrix.

If many elements must be determined in a sample, ICP/MS is fast; many times faster than furnace AAS and comparable to ICP/AES. An analytical range of five or six orders of magnitude from the detection limit is available on ICP/MS, comparable to ICP/AES. In contrast, AAS provides no more than two or three orders of magnitude above the detection limit.

The major drawback of ICP/MS is its expense, and that is gradually dropping. It is still the newest of the spectroscopic techniques; and until recently, it was considered unwise to put

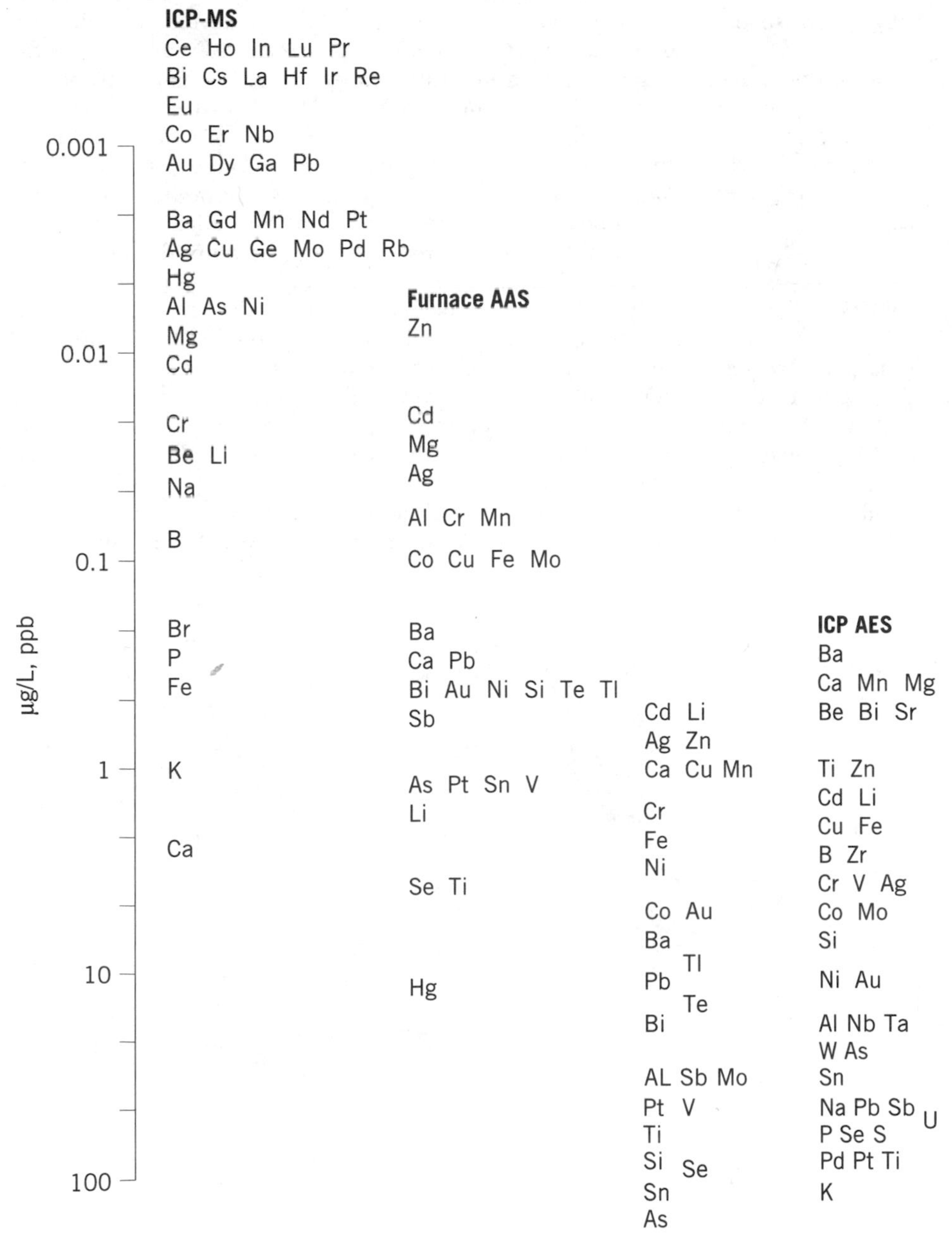

Figure 3. Detection limit comparison for ICP/MS, furnace AAS, flame AAS, and ICP/AES. For furnace AAS, a sample size of 20 μL was used (13).

the equipment into hands that were less than extremely well trained. Now routine methods are handled by technicians.

A major advantage over any other spectrometric technique is the access to isotope determination. Most elements have more than one natural isotope. For some elements, the natural distribution varies geographically, so the isotope ratio can provide information on the source of the metal, for example, in toxicology situations. Radioisotopes are frequently used as tracers in environmental and biochemical situations. But they are dangerous and, in recent times, illegal to administer to humans. Experiments on humans can be conducted with ICP/MS using stable isotopes. Radioisotopes of uranium, technetium, and plutonium, can be sensitively and rapidly determined in environmental samples.

Many accessories that are available for the ICP/AES for sample input provide advantages for ICP/MS also. These include the several nebulizers: concentric, cone, groove, and ultrasonic. In addition, several important accessories have been developed to extend ICP/MS potentialities. These include laser ablation to obtain a representative sample from solids. This will be discussed further below.

The graphite furnace, developed for atomic absorption, has been adapted to ICP/MS and has several advantages. It permits microliter amounts of sample to be analyzed with the sensitivity inherent in ICP/MS. Since the solvent and some of the matrix can be driven off before introduction of the residue into the ICP/MS, potential mass interferences can be greatly reduced.

ENVIRONMENTAL REGULATIONS AND METHODS

In this section the development of the U.S. system of environmental monitoring is discussed. The philosophy of method development and the current methods are tabulated. Issues impacting analytical methodology are discussed. A review of the recent literature completes the survey and expands the scope to techniques under investigation outside the United States.

Environmental regulation in the United States started approximately 30 years ago and is one of the oldest systems in existence. The Public Health Service initially held responsibility for drinking water quality and environmental responsibility increased with the creation of the EPA in December 1969. Since that time, the number of legislative acts passed by Congress for which the EPA must create regulations has increased explosively. Figure 4 is a timeline of legislative acts in their original form and after reauthorization.

Most of the legislative acts, such as the Safe Drinking Water Act and Clean Air Act, are centered on a particular medium. The EPA organizational structure reflects this emphasis and offices have been created for air, water, solid, and hazardous waste emergency responses. The disadvantage of this approach is that regulations for each medium are created separately, with limited communication among the different offices. This makes an overall approach to pollution control at a particular industrial site difficult and somewhat disjointed. The EPA has recognized these limitations and has recently reorganized the major research laboratories in the research and development division to focus on projects that involve multimedia. Figure 5 shows the current organizational structure of the EPA's major offices.

Analytical measurements are an integral part of environmental management. They are used to assess health effects, which are important in prioritizing contaminants for regulation. Routine monitoring of regulated contaminants ensures compliance with allowed levels and can indicate a hazardous situation. Cleanups of contaminated sites are driven by measurements indicating the location and extent of contamination. Each EPA program has created analytical methods to specify the way measurements should be performed for different media. Each method is based on a particular analytical technique and lists the target analytes, preparation of standards and samples, instrumental setup, and the quality control required. ICP/MS is a fairly recent addition to the tools used for environmental monitoring.

Table 1 summarizes the methods used in the major monitoring programs. For example, Method 200.8 is an ICP/MS

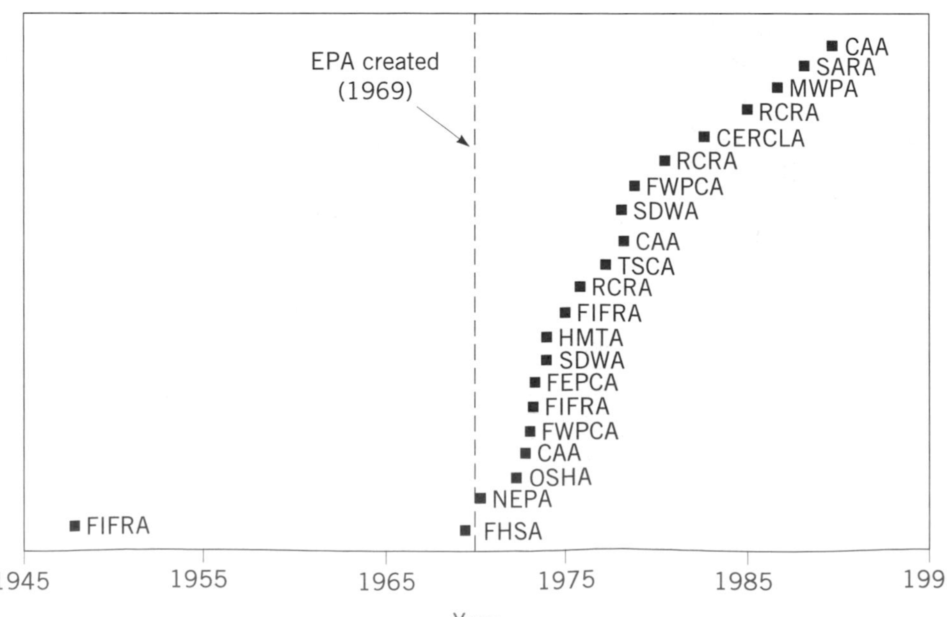

Figure 4. Timeline of legislative mandates involving the EPA.

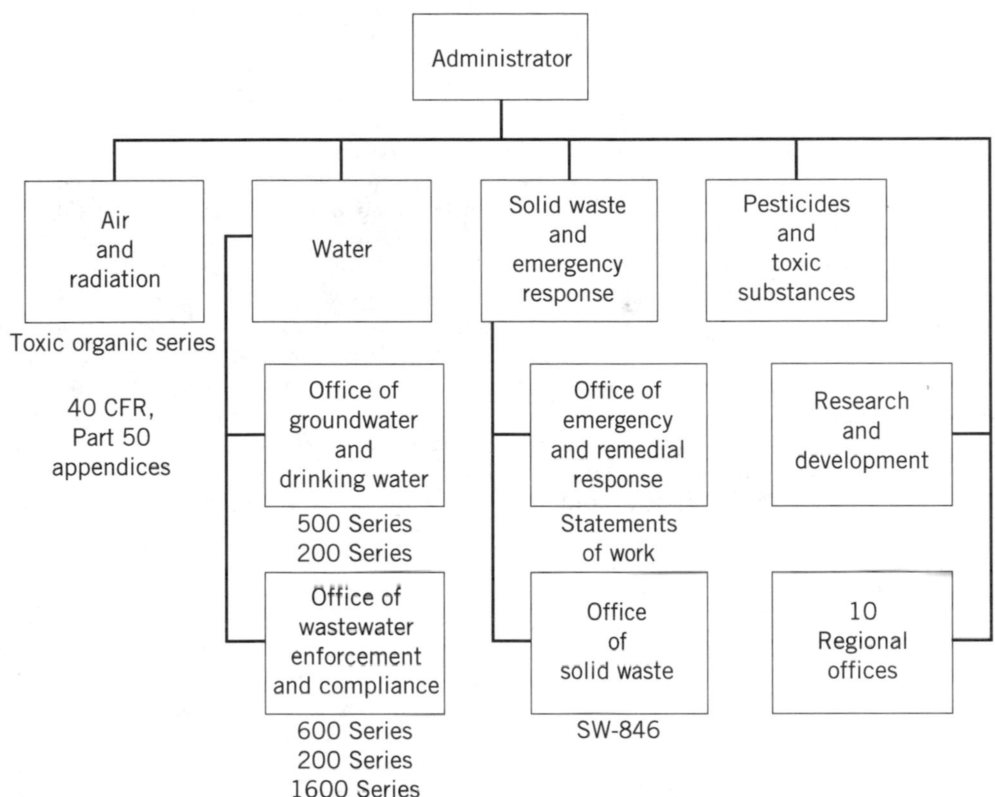

Figure 5. Simplified EPA structure and method sources.

Table 1. Summary of EPA ICP/MS Methods

EPA Program	Method Number	Analytes	Matrices	Comments
Water	200.8	Al, Sb, As, Ba, Be, Cd, Cr, Co, Cu, Pb, Mn, Hg, Mo, Ni, Se, Ag, Tl, Th, U, Sn, V, Zn	Drinking water, wastewater, related matrices such as sludge and sediments	Hotplate and microwave digestion available, none represents total dissolution
	1638	Sb, Cd, Cu, Pb, Ni, Se, Ag, Tl, Zn	Ambient water	Clean sample handling techniques required
	1640	Cd, Cu, Pb, Ni	Ambient water	Clean sample handling techniques required; uses chelation preconcentration
RCRA	6020	Al, Sb, As, Ba, Be, Cd, Cr, Co, Cu, Pb, Mn, Ni, Ag, Tl, Zn	Ground water, wastes, soils, TCLP extracts	Many digestion methods to choose from
CERCLA (CLP)	6020-CLP-M	Al, Sb, As, Ba, Be, Cd, Ca, Cr, Co, Cu, Fe, Pb, Mg, Mn, Ni, K, Se, Ag, Na, Tl, V, Zn	Water and wastes	Microwave, hotplate, or block digestor hot acid leach sample preparation
Air	Method IO-1 (Draft)	Al, Sb, As, Ba, Be, Cd, Cr, Co, Cu, Pb, Mn, Mo, Ni, Se, Ag, Tl, Th, U, V, Zn	Metals emissions captured on glass fiber filters	Based on method 200.8; hotplate digestion for samples

method for drinking water and wastewater and includes an analyte list of 22 elements. The analytes for which the drinking water methods are acceptable are listed in a table in the drinking water section of the *U.S. Code of Federal Regulations* (14). Therefore, Method 200.8 may be used for the determination of 16 elements out of a possible 17 primary and secondary metallic drinking water contaminants, based on the regulations. It is significant that mercury is included in this method. This allows the analysis of drinking water to be streamlined from three techniques (ICP/AES, GFAAS, and mercury cold vapor) to one technique for the analysis of the primary drinking water contaminants. The analysis may become more cost-effective or the frequency of monitoring may be increased for the same cost.

Method 200.8 may also used for the analysis of wastewater. Whether Method 200.8 is suitable for a particular determination depends on the data quality objectives of the analysis and the capabilities of the method. Method 200.8 has not been formally approved for routine wastewater monitoring efforts, although it was proposed for this purpose (15). Laboratories can use the alternate test procedure process to request permission from the EPA to use this method, and many laboratories have taken advantage of this procedure. Wastewater discharge is regulated through National Pollution Discharge Elimination System (NPDES) permits. Permits are negotiated based on federal guidelines for the industrial category (16), the receiving waterway, and any state and local guidelines. Recent changes from a best available technology (BAT) approach for discharge cleanup to a Water Quality Criteria (WQC) may require modeling of the watershed to determine the actual effects of the discharge components on the ecosystem. This enables permit levels to be established that may protect the environment while allowing the discharge of contaminants that do not appear to affect a particular ecosystem.

Methods 1638 and 1640 were developed with the specific purpose of monitoring the ultratrace levels present in ambient water. Examining ultratrace levels makes it possible to more accurately assess bioaccumulation models and chronic health effects from low levels of exposure. The analytical techniques must be coupled with clean sample handling techniques to yield accurate values at such low levels. Although the methods do not specify that the instrument be operated in a clean room environment, sample preparation should be performed in a clean area.

Method 6020 is the ICP/MS method developed for solid and hazardous materials requirements of the RCRA program. It lists 15 analytes and is used as part of a compendium of methods providing analytical tools. The RCRA program provides cradle-to-grave regulation of hazardous materials and, as such, demands the analysis of a variety of matrices, although the most common analyses are soil digests and toxicity characteristic leaching procedure (TCLP) matrices.

Method 6020 CLP-M is Method 6020 modified for the contract laboratory program (CLP). It is available as a Superfund analytical services method. The CLP serves to provide analyses of Superfund sites, designated under CERCLA for assessment and cleanup. A total of 22 elements are included, and contract required detection limits (CRDLS) must be demonstrated before a particular technology, such as ICP/MS, may be used for analysis. The methods, including rigorous quality control (QC) must be followed exactly as written and documented in numerous reporting forms. Often the same protocol is followed less rigorously for other samples, and this type of analysis is referred to as *CLP-like.* ICP/MS has not been used extensively in this program because the method was not included in a routine statement of work, which would receive wide use. It has been available since 1988 for the special analytical services program and has been used on a limited basis by the EPA regions.

The U.S. Departments of Energy and of Defense have cleanup issues related to weapons production and the decommissioning of federal facilities. They have recognized the value of ICP/MS for the determination of long-lived radioisotopes and have developed three methods for this purpose. Method MM100 extends Method 6020 to the analysis of radioactive elements, such as technetium, iodine, radium, thorium, uranium, neptunium, plutonium, and americium. In some cases preconcentration or separation of the matrix to control interferences may be necessary. Method MM210 specifies the ICP/MS determination of ^{99}Tc, ^{230}Th, and ^{234}U using flow injection preconcentration. Method MM800 uses ion chromatography to separate uranium before measurement of total concentration and isotopic abundance with ICP/MS. Additional methods are currently under development.

The different program requirements have arisen from the perceived different needs of the programs and the media they analyze. However, for laboratories that measure samples for several different programs, the choice of the appropriate method and associated QC can be confusing, and implementing several QC schemes can decrease laboratory productivity. Current issues related to these problems include the Environmental Monitoring Management Committee efforts, performance-based measurements, and laboratory accreditation.

The Environmental Monitoring Management Committee was created in 1990 with representatives from several levels and program offices of the EPA. Of their goals, the promotion of uniform methodology and laboratory accreditation has become visible. They have developed a format for all method write-ups to make it easier to follow across program lines, and this has been readily adopted. They have efforts under way to propose several uniform methods for adoption across program lines. Their initial efforts targeted a graphite furnace method, GC/MS volatile organic compound method, and a microwave digestion method. They have also been supportive of national laboratory accreditation efforts.

In the United States most laboratory analyses are conducted by private laboratories, not by government laboratories, as in many other countries. To ensure uniform data generation across the variety of laboratories and operator skill levels, there is merit to written analytical methods. The disadvantage comes in the situations in which current methods do not work properly as written or when method modifications or new technologies offer advantages. Laboratories may have difficulty getting accurate data when performing a method exactly as written, with a difficult matrix. Or new technology may come along that makes a method more rugged or easier or more economical to perform. Or a new technique may be developed that offers advantages in terms of analytical capability or business requirements. EPA methods are prescriptive and modifications are not encouraged. A different approach to analytical methodology based on a performance measurement system would circumvent some of these difficulties. The data quality objectives would be specified for a particular analysis and any method using any technique that would meet those objectives would be allowed. This is controversial, because it would be a major change in the current way of operating and questions about enforcement issues must be addressed. It is moving forward within several EPA program offices, and forces in Congress are addressing the issue through legislation such as the Baker Bill, proposed in 1996.

Performance-based measurements are being considered in conjunction with a national laboratory accreditation system. Currently, there is no national system, but laboratories can be accredited in several states in medium-specific programs. It is hoped that a national accreditation system will provide a more economical alternative and ensure that laboratories have the capabilities with which to handle a performance-based measurement system.

Application issues of recent interest include field-based methods and speciation. Field-based methods may be semi-quantitative or provide full quantitation directly at the field site. Examples of methods that may be field-based include X-ray fluorescence and immunoassay. Field-based methods can quickly define a cleanup area and monitor the on going cleanup process. They can more economically provide samples to the laboratory, if used for screening purposes, because fewer clean samples will be taken back to the laboratory.

QUALITY CONTROL AND TROUBLESHOOTING

Quality data mean many different things to different people. Actively controlling quality is established as an important part of the laboratory process, especially in the United States. However, it should start even earlier in the process to be most successful. The sampling plan is designed to provide the samples that will yield the most pertinent information, and the sampling plan is carried out with procedures compatible with those objectives. For example, if traces of mercury in ambient water are to be determined, it might be necessary to use extensive clean sample handling procedures to collect samples without contamination. Sample transport and storage in the laboratory are continuing parts of the process. Sample preparation must be compatible with both the instrumental technique used and the concentration levels of the analytes desired. The development and application of quality assurance and control differs among countries. It has become highly developed in the United States because of the variety of laboratories and differently skilled analysts providing environmental monitoring services. Quality assurance and control can aid in the process of providing uniform data of documented quality. This section discusses the terminology and ways to measure quality of laboratory operations, instrument performance, and analytical measurements using specific methods.

A brief historical perspective of quality measurement development for regulatory control in the United States may be helpful. In the 1960s, data were generally reported through STORET, a computerized water quality database. Basically, the location and depth at which the sample was taken were reported with the data and results. In the 1970s the development of the National Pollution Discharge Elimination System (NDPES) program required an on-site laboratory inspection and the successful evaluation of performance samples. In the 1980s the CERCLA program was strong and had the unusual requirement that laboratory data must stand on their own in a court of law. If a potentially responsible party was brought to court in an effort to recover some of the cleanup costs of an abandoned hazardous waste site, it might be many years after the laboratory analysis was done. The analyst may have left the laboratory or the laboratory may have merged with another or even ceased to exist. The quality assurance and control requirements of the CLP were extensive and required documentation at each step of the way. In the 1990s the data quality is more often matched to the need of the project. *CLP-like* is a term often used to include selected components from the CLP program, but not all the components.

COMPARISON WITH OTHER SPECTROSCOPIC METHODS

Laboratories may have several inorganic techniques at their disposal to apply to an analytical scenario. How do you know which technique to choose? When the technique is decided, which is the proper analytical method to perform? This section provides a basic overview of AAS, ICP/AES, and ICP/MS. These techniques will be compared in terms of analytical capability. Business issues will be considered. The regulatory issues of choosing methods will be reviewed, and examples of decision making will be presented.

Comparison of Technology

Flame and furnace atomic absorption, ICP/AES, and ICP/MS can all be used to determine part per million to less than part per billion concentrations of elements in a variety of matrices. What then distinguishes them from each other? Figure 6 shows in a simplified form the analytical principles used to make measurements with these techniques. Atomic absorption and ICP/AES are related in that they both measure light, either absorbed or emitted. ICP/MS uses a plasma to generate ions, rather than light, and separates and measures the quantity of each mass:charge ratio. The techniques differ in their suitability for different elements, range of detection limits, linearity, interferences, and multielement capability (13,18).

Atomic absorption can be applied to a wide range of elements, provided suitable light sources are available. Flame atomic absorption is a good general technique, suitable for a wide variety of elements. Some elements, such as aluminum and barium, require a hotter nitrous oxide-acetylene flame to be sensitively determined. Refractory elements have poor detection limits with flame, but can be measured with ICP/AES. Graphite furnace atomic absorption (GFAAS) is most suitable for elements that can be vaporized from a pyrolytic graphite surface. Elements of environmental interest, such as lead and arsenic, can be measured at sub part per billion levels with this technique. Elements that are refractory, such as boron, are difficult to determine by GFAAS because they are not atomized efficiently from the surface into the tube atmosphere. ICP/AES is better at measuring refractory elements because the high temperature argon plasma has more energy to excite elements to emit light. Since light emission is measured, light sources for each element are not required. Boron is easily determined with ICP/AES. ICP/MS can sensitively determine a wide range of elements, isotopes, and some long-lived radioisotopes. Elements such as uranium, thorium, and technetium can be determined faster than with traditional radiochemistry counting methods. Isotopic analysis, such as measuring the various isotopes of lead that vary geographically is a unique capability of this technique. Isotope dilution can also be used to measure elemental concentrations accurately.

Detection limits are a direct measurement of the suitability of a technique for an element. Detection limits for the techniques overlap, but generally become better in the order flame AAS, ICP/AES, GFAAS, and ICP/MS. Detection limits vary on a day-to-day basis and among instruments and methods. The detection limits for ICP/AES vary more than for other techniques because of the wide variety of instrumentation available, using different resolution and different views of the plasma. Method detection limits are one type of detection limit and indicate the ability to measure an element using a particular method and a particular low level sample, which has been carried through any sample preparation procedures. Method detection limits for ICP/MS will vary more among methods than for other instruments because the matrix strongly influences the ability to measure at low concentrations. Instrument detection limits are often mentioned in environmental methods and reflect the performance of the instrument alone (19). They are calculated by

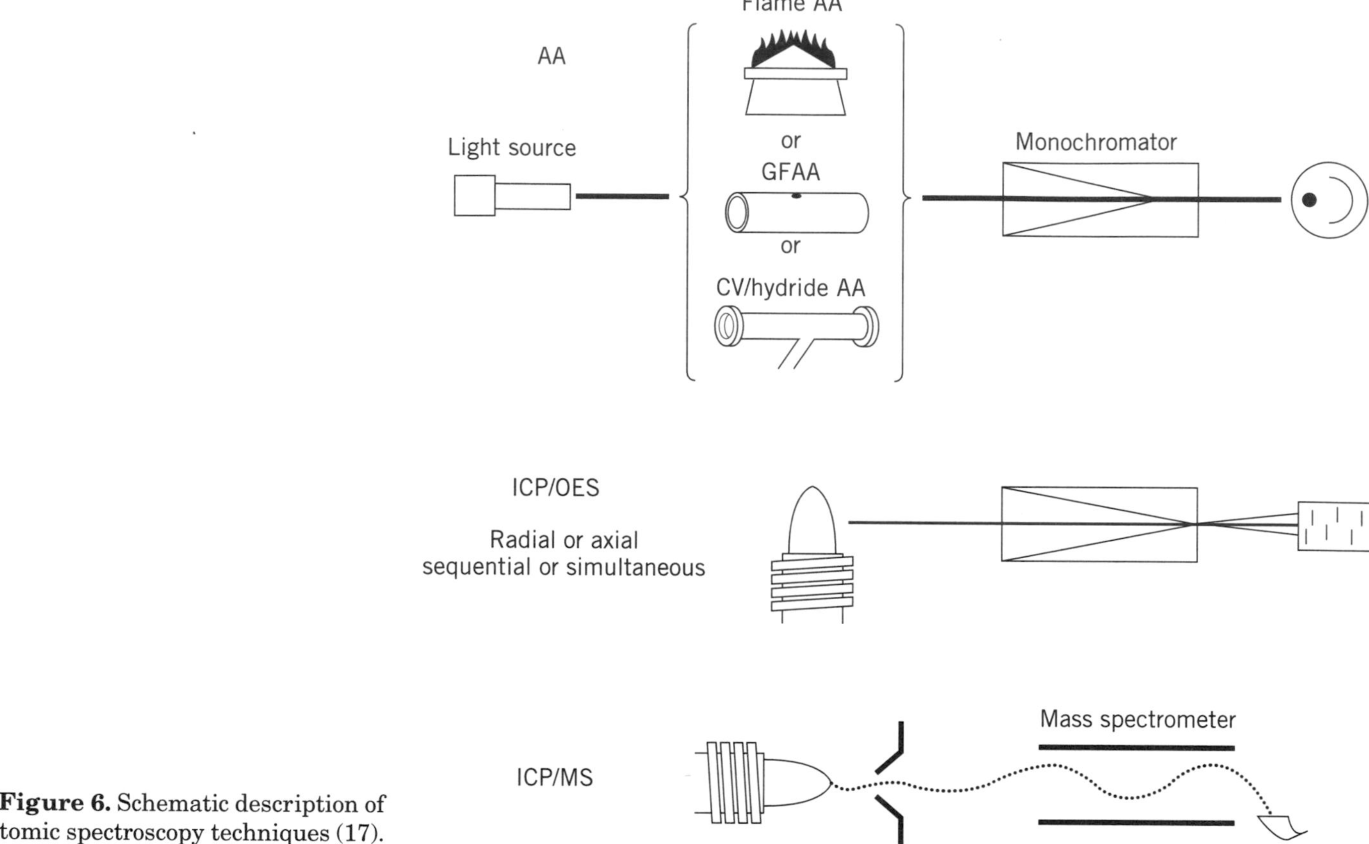

Figure 6. Schematic description of atomic spectroscopy techniques (17).

measuring the low level spike solution directly without any sample preparation. Seven replicates are measured on each of three nonconsecutive days. The average standard deviation is multiplied by three. In Table 2, method detection limits for ICP/AES, GFAAS, and ICP/MS are compared for EPA methods used for the measurement of trace metals in drinking water and wastewater. The method detection limits are quoted from the method write up, except in cases where the information was unavailable, such as for axial ICP/AES. Preliminary experimental results from Perkin-Elmer are reported for axial ICP/AES. The method detection limits vary from day to day, among instrument models, and based on the matrix. The comparison is provided for guidance in comparing the suitability of techniques for environmental analysis using EPA methods.

Table 2. Technique MDL Comparison, μg/L

Element	Radial ICP/ AES, Method 200.7	Axial ICP/ AES, Method 200.7	GFAAS, Method 200.9	ICP/MS, Method 200.8
Al	50	26	7.8	0.03
As	30	5.9	0.5	0.05
Cd	3	3.4	0.05	0.02
Cr	2	1	0.1	0.06
Cu	3	4.7	0.7	0.008
Fe	30	4	—	—
Pb	30	4.2	0.5	0.008
Sb	30	5.5	0.8	0.004
Se	20	2.6	0.6	0.23
Tl	20	7.3	0.5	0.001

No analytical technique is completely free from interferences. Awareness of the potential interference and suitable means of correction reduces the potential for inaccuracy. Flame AAS interferences are well characterized through many years of experience and documentation. Ionization interferences are common when determining sodium and lithium and can be overcome with the addition of an ionization suppressant, such as 1000 mg/L K. GFAAS interferences arise from atomization problems and the absorbance of interfering species. Matrix modifiers and the use of Zeeman background correction can reduce these problems. ICP/AES interferences can arise from spectral overlaps or physical transport effects. ICP/MS interferences consist of physical transport effects and mass overlaps from a few molecular species and dimers in the otherwise relatively simple mass spectrum. Internal standards and interference correction equations can mitigate these problems for ICP/AES and ICP/MS. Interferences are worse for more complex samples. Soil samples can present a variety of difficulties and the most complex samples would likely involve mixed waste, containing both hazardous constituents and radioactive materials.

EPA methods generally specify operation within the linear working range of the technique. This can be defined as the lowest concentration that will be reported, such as the method detection limit, to the highest concentration, where the curve is no longer considered linear (such as $>5-10\%$ deviation from the concentration predicted from a lower calibration standard). Longer linear range will reduce the need for dilutions and reruns. Atomic absorption has a linear range covering several orders of magnitude, which can be extended if nonlinear working curves are permitted. Generally, concentrations four to six times above the top of the linear range can safely be determined. The accuracy and precision of quality control checks in the nonlinear portion of the curve can show whether the DQOs of the analysis will be met with the extended range. ICP/AES and ICP/MS have useful linear ranges of five to eight orders of magnitude of concentration. Different mechanisms cause nonlinearity and these techniques are not used outside the linear range.

Some techniques lend themselves more readily to multielement operation. Flame and GFAAS rely on light sources for each element, and it is difficult to combine the beams from more than four lamps with a satisfactory amount of light passing through the atomization cell. ICP/AES and ICP/MS are generally used as multielement techniques and the elements are measured simultaneously or sequentially in a relatively fast manner.

Business Issues

Profitable operation is the continuing goal of most commercial laboratories. This requires a balance between productivity and initial investment in acquiring new technology and the improvement in productivity expected. Once the technique most suitably matched to the sample load has been identified, the investment can be considered. Flame and GFAAS are generally less expensive to purchase than is ICP/AES, which in turn is less expensive than ICP/MS. However, more detailed analysis, such as the internal rate of return for the instrument over the course of time may show that the investment in a more expensive technique is more beneficial to the laboratory. The payback period and internal rate of return can be favorable in several scenarios. The initial investment may be substantial, but the financial criteria show that the return of the principal is rapid. It is important in any of these analyses to understand the laboratory cost structure and software and financial specialists are available to help with these tasks.

Regulatory Issues

Regulatory issues are often at the heart of environmental analytical decisions. The EPA program to which data are being submitted determines which set of methods may be required. The most suitable technique is then chosen from those allowed in the list of methods based on availability, workload, and analytical capability.

ACKNOWLEDGMENT

The authors thank Dr. R. Wolf of Perkin-Elmer for help with QC examples and method development information.

BIBLIOGRAPHY

1. F. W. Aston, *Phil. Mag.,* **38,** 709 (1919).
2. A. J. Dempster, *Phys. Rev.,* **11,** 316 (1918).
3. T. B. Reed, *J. Appl. Phys.,* **32,** 821 (1961).
4. S. Greenfield, I. L. Jones, and C. T. Berry, *Analyst,* **89,** 713 (1964).
5. R. H. Wendt and V. A. Fassel, *Anal. Chem.,* **37,** 920 (1965).
6. A. L. Gray, *Analyst,* **100,** 289 (1975).
7. R. S. Houk, H. J. Svec, and V. A. Fassel, *Appl. Spectrosc.,* **35,** 380 (1981).
8. A. R. Date and A. L. Gray, *Analyst,* **106,** 1255 (1981).
9. P. W. J. M. Boumans, *Inductively Coupled Plasma Spectroscopy,* 2 parts, Wiley-Interscience, New York, 1987.
10. G. Horlick, S. H. Tan, M. A. Vaughn, and Y. Shao, in A. Montaser and D. W. Golightly, eds., *Inductively Coupled Plasmas in Analytical Atomic Spectrometry,* VCH Publishers, New York, 1987.
11. R. S. Houk, *Anal. Chem.,* **58,** 97A (1986).
12. J. W. McLaren, *At. Spectrosc.,* **13**(2), 81 (1992).
13. W. Slavin, *Spectroscopy,* **6**(8), 16 (1991).
14. 40 *CFR* 141.23.
15. 60 *Federal Register* 53988 (Oct. 18, 1995).
16. 40 *CFR* 401–471.
17. Z. Grosser and C. Schneider, *At. Spectrosc.,* **17**(6), 209 (1996).
18. D. Beauchemin, *Spectroscopy,* **7**(7), 13 (1992).
19. Environmental Protection Agency, "Contract Laboratory Statement of Work for Inorganics Analysis, Multimedia, Multiconcentration," *Document No. ILM03.0,* EPA, Washington, D.C., 1993.

INDUSTRIAL WASTE DUMPS, SAMPLING AND ANALYSIS

W. Rasemann
Freiberg, Germany

B. Markert
Zittau, Germany

GENERAL REMARKS

Environmental sampling for pollution research is a complex subject. Mastering it can be as complicated as the object that has to be evaluated. The importance of this subject has long been ignored compared with improved techniques and methods of chemical analysis. The analytical steps shown in Figure 1 have varying significance relative to their effect on the precision or accuracy of analytical results (1). This is often incorrectly estimated by outsiders who view the physical measurement of the analytical signal as the most important analytical step, often even the only one. This is true to the extent that no quantitative analytical result is obtained unless the signal is measured and correlated with the actual concentration. However, relative to the possibilities for error, physical measurement is one of the most accurate and most easily reproducible analytical steps. The step by far the most susceptible to error is sampling. The first principle of chemical analysis is that the part of the sample to be measured in the analytical device must have exactly the same average composition as the total amount of the sample evaluated. If this is not adhered to, then analytical errors of much greater than 1000% can occur (2). Because the true elemental concentration of a system compartment is unknown and therefore it is not possible to calibrate the sampling procedure, genuine quality control (how accurate or how representative the sample is) is very difficult if not impossible in some cases.

The following three axioms are valid for sampling (3):

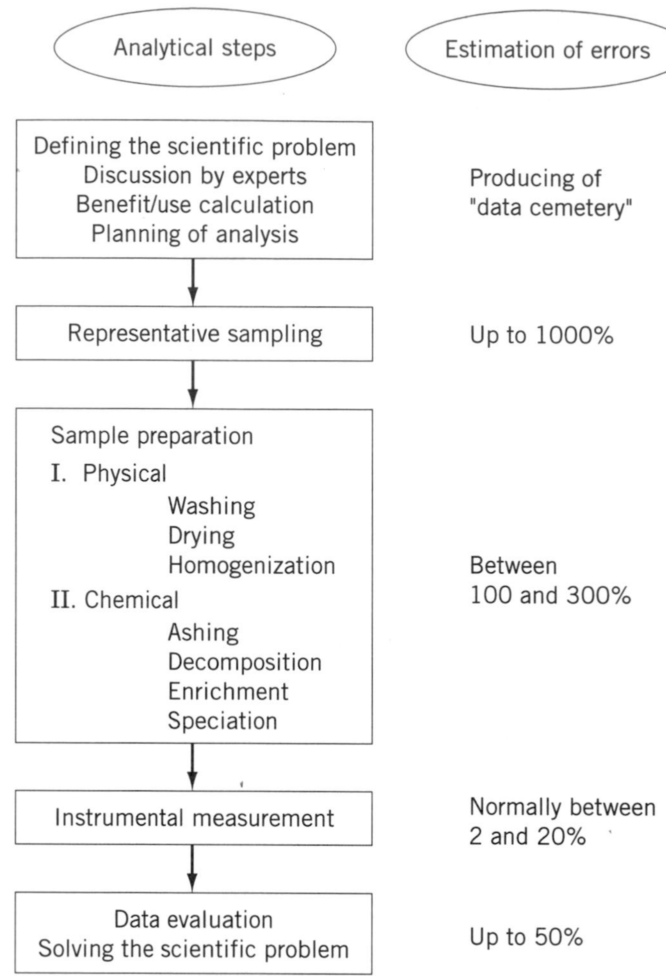

Figure 1. Simplified flow chart for the instrumental chemical analysis of samples.

Axiom 1. The sample taken from the system shall have exactly the same chemical composition as the original sample.

Axiom 2. The probability of being removed from a total population must be the same for each individual.

Axiom 3. The larger the degree of dispersion among individuals and the larger the number of individuals, the larger is the amount of time and effort needed for sampling.

Practical difficulties during sampling prevent satisfying the first two axioms. A sample taken from the field never has exactly the same chemical composition as the original sample. At best it has a similar composition. One reason for this is that often only a very small part of the total sample (eg, 100 mg per 10,000 kg of soil) is analyzed (4,5). The original contents of the source sample can be changed by contamination or volatilization of certain components in the sample when it is removed. Also, it is hardly possible to give each individual the same probability of being removed when they are diffusely distributed in the ecosystem. Rather, one must attempt to approach the first two axioms by a carefully prepared sampling strategy. Some practical principles are helpful here:

Avoid any contamination of the sample by the sampling equipment, the sample container, or the person doing the sampling.

Avoid all volatilization of elements due to microbiological activities, adsorption to the wall of the vessels where the samples are stored, or overheating the samples during transport and storage.

Do not remove too small a sample as long as sufficient material is present in the system and it is not protected by environmental law.

Note seasonal variations in the composition of the total sample and other parameters which affect chemical metabolism, such as temperature, humidity, light, etc.

Carry out random subdivision of the area to be studied.

As Figure 1 demonstrates, every environmental analytical study requires formulating a specific question or goal. In many cases, this formulation is the result of the cooperative effort of several experts from different disciplines (6–10). The adaptation of the analytical plan to the formulation of the question is tied to practical requirements. A checklist of criteria is reproduced below, which can be consulted when planning new experiments. This checklist is not complete; it is merely to support analytical planning (1).

1. Location being studied:

   Which area is best suited for answering the question that has been formulated?

   Will the location be available long enough for the planned purpose, or must one also take into account the possibility that the area could be changed or destroyed by public or private activities? (This is important in long-term studies.)

   Is the area protected by conservation law or does one need authorization from a public office or a private individual to take a sample?

   Is it permissible to drive into the area and do the geographical conditions permit this? This might be important for transporting the sample.

   Does the sampling of the object of investigation, or do any associated studies, change the structure of the region or put its holdings at risk?

2. The object being studied:

   Which individual, which society of individuals, or which system compartments are best suited for working out the question that has been formulated?

   Is there a sufficient quantity of the material to be studied?

3. Schedule and associated studies:

   What period of time is needed to answer the question that has been formulated?

   How urgent are the expected results of the study?

   Do plans need to be made for follow-up studies, or is the formulated question sufficiently addressed?

   In addition to the actual sampling, do more extensive or accompanying studies need to be performed in the test area to gather additional system analytical parameters?

Do additional measuring devices need to be set up and maintained to do so?

Have data relevant to the study already been obtained elsewhere?

4. Chances of success and financing:

Is the sampling program set up so that a solution of the problem can be expected with the aid of the data that will be gathered?

Are the results of the study comparable with results from similar studies done by other research groups?

Does the financial expense justify the effort involved in the scientific statement of the problem?

Have the costs for equipment, personnel, and materials been covered for the entire analytical process from sampling to compiling the final report?

Must the financing be done by soliciting funds from third parties?

Can the expected results possibly be used for problems currently being discussed by the public?

This checklist of general criteria for a decision making makes clear that a specific set of questions must be generated for each study, and the proposal must be discussed with all the participants, scientists, technicians, and sponsors. The list also shows how closely intertwined the statement of the questions and the planning of the analysis are. Planning the study to be conducted is a major component of the analytical process, and if followed carefully, it prevents the accumulation of inexact or incorrect results.

SAMPLING AND GEOSTATISTICAL EVALUATION OF HETEROGENEOUS INDUSTRIAL SITES ON EXAMPLE OF A SAXONIAN SLAG HEAP

Industrial sites where by dumping residuals and wastes, such as slags, ashes, dust and sludges have been dumped, can be suspected of environmental risk or can be a source of valuable materials. The evaluation of any dump of industrial waste materials is quite difficult. A dump generated by changing production history and dumping conditions and by chemical reactions or physical changes after dumping is heterogeneous, as a rule. In contrast to unambiguously defined solids systems, such as ore, coal, bulk materials and salt, there are no regulations for sampling and statistical analysis expressly for such heterogeneous heaps. For this reason, recommendations and methods satisfactory or promising for other solid systems are used. In some cases, solids systems are structurally or materially similar to the particular heap of industrial waste, for example, soils, granular and free running bulk materials, and iron ore. Methods of sampling and evaluation are based on mathematical statistics and geostatistics. Unfortunately, the assumptions underlying geostatistics are not always correctly fulfilled by industrial dumps. A Saxonian slag heap is an example of the application of regulations proven for sampling and known statistical methods for homogeneous bulk solids and uniformly contaminated soils can be applied to evaluate extremely heterogeneous solids systems. Further studies were also presented by (11).

Object Investigated

The slag heap of the smelting plant at Freiberg in Saxonia was studied. The slag was a reject of the thermal processing of zinc blende (ZnS) ore concentrates. The concentrates were roasted in a fluidized-bed furnace which produced calcined zinc oxide (ZnO) and sulfur dioxide (SO_2) in the roaster gas. The roaster gas was used to produce sulfuric acid, and the calcine was leached with sulfuric acid. The leaching was followed by a separation into solid low grade leaching tailings and high grade leachate. The leachate was an electrolyte, from which the zinc was cathodically precipitated. The leaching tailings consisted of remaining portions of zinc and of other valuable components, such as lead, cadmium, and iron; these tailings were washed and dried, and subsequently, they were thermally processed in a rotary furnace to extract the metals. Fine coke was added to reach a high furnace temperature and to reduce the metal oxides. The residue of this process, the rotary furnace slag, was continuously discharged, cooled, granulated and transported to a slag pit. Finally, the slag was moved by truck from the pit to the dump.

The leaching tailings processed were delivered from different factories; they varied in the composition during the entire operating period. Consequently, the rotary furnace slag was composed of widely varying portions of SiO_2 (10–20%), FeO (15–30%), CaO (10–15%), Al_2O_3 (about 6%), C (10–45%), S (about 6%), Zn (1–5%), Pb (0.1–0.5%) and Cu (0.3–0.6%). Cadmium (Cd) and arsenic (As) also occur in low concentrations. The high carbon content resulted from unburnt coke. After the dumping of the slag, the fine coke in the heap was carbonized at low temperatures, parts of the usually grained slags were incinerated, the heap caved in and the slags were sintered to blocks here and there. All in all, the very heterogeneous slag heap consisted of blocks and lumps of slag and of coarse-grained to fine-grained bulk material. The chemical composition was extremely varied. After the first inspection, the heap surface was distinguished by three regions (Fig. 2):

normal grained slag which still partly burns

very incinerated regions

solid blocks and lumps of sintered slag

The heap extended over an area of 1.6 hectares and it was between 4 and 12 m high. The total mass was about 80 kt.

SUMMARY AND CONCLUSIONS

Sampling by a systematic pattern and subsequent geostatistical evaluation are successful in obtaining a realistic picture of the spatial variability of the slag heap, in some cases. Empirical (semi-)variogram functions were calculated using original data in definite directions of the field of investigation and in fixed levels of the heap. The empirical variogram functions were fitted by a spherical model. Nugget effect, sill, and range were easily interpretable parameters of this model. The technique of point kriging was applied to calculate data at additional points on the basis of the fitted variogram functions. Using the original data and the calculated values, three-dimensional presentations show the spatial variability of essential chemical components such as iron, sulfur and lead. Useful results were obtained for the sulfur

Figure 2. View at the rotary furnace slag heap of the smelting plant at Freiberg in Saxonia.

content in all directions and all levels. The geostatistical evaluation of the iron content provided valuable variograms and kriging data only in some directions of the heap. However, the iron content varied so extremely in other directions that geostatistical methods did not work there, either. The cadmium and the arsenic content were untypical of the composition of the slag heap. Semivariograms being difficult to evaluate or useless at all were calculated for which, there were neither preferred directions of variability nor any geostatistical regularity.

To sum up, it can be concluded that the highly expensive sampling, chemical analysis and geostatistical evaluation of extremely heterogeneous heaps of slags and industrial wastes is justified only if for compelling reasons. However, reasonable results are obtained even with the wrong assumptions, because many geostatistical procedures are fairly robust. On the other hand, the sampling regulations based on relationships between grain size and sample mass fail for contaminated sites and wastes where the structure has been considerably changed by physical and chemical processes, such as leaching, carbonizing at low temperatures, precipitation, and other. If a global evaluation of the heap by mean values and variances of certain chemical components has priority, however, the usual methods of sampling and statistical data analysis recommended and developed for bulk materials and soils are promising tools.

The methodical results in evaluating extremely heterogeneous contaminated sites and wastes obtained in the example of the Saxonian slag heap clearly show the possibilities and bounds of known sampling methods and geostatistical models. On the one side, probably the greatest disadvantage in applying geostatistics is the difficulty in knowing when the assumptions, for example, a homogeneous isotropic random field, are acceptable or not in an individual case (12). On the other hand, such a close-meshed sampling and percussion coring pattern as applied in the special case of the Saxonian slag heap cannot be realized under commercial conditions. In practice, a compromise must be made. A promising strategy is to lay a regular grid over the region so that the coordinate origin and axis orientation are random. Then, samples are taken either systematically at each grid point or randomly at each subarea of the pattern. As shown in the example of the slag heap, one advantage of systematic sampling using a randomly placed grid is that the data obtained are very convenient for analysis by random sampling and geostatistical methods.

When the number of samples is too small for parametric geostatistical analysis, such as variogram estimation and point kriging, nonparametric geostatistics and risk assessment (13) should also be applied.

ACKNOWLEDGMENT

The geostatistical calculations reported here were carried out by M. Herbst with the commercial computer program Surfer, which is supported by the computer working surface Windows and is frequently used for geostatistical applications. The authors wish to thank him for providing the calculations and for helpful discussions regarding the interpretation of the results.

BIBLIOGRAPHY

1. B. Markert, *Instrumental Element and Multi-Element Analysis of Plant Samples, Methods and Applications,* John Wiley & Sons, Chichester, U.K., 1996.
2. B. Sansoni and G. V. Iyengar, Sampling and Sample Preparation Methods of Trace Elements, in *Biological Materials,* Forschungszentrum Jülich, Jül. Spez., 13, 1978.
3. B. Markert, ed., *Environmental Sampling for Trace Analysis,* VCH Publisher, Weinheim, Germany, 1994.
4. G. Kraft, *Theoretische Grundlagen der Probenahme, Probenahme-Theorie und Praxis, Schriftenreihe der Gesellschaft Deutscher Metallhütten- und Bergleute,* VCH-Verlagsgesellschaft mbH, Weinheim, Germany, 1980.
5. P. M. Gy, *Sampling of Particulate Materials,* Elsevier Scientific Publishing Co., Amsterdam, the Netherlands, 1982.

6. L. H. Keith, ed., *Principles of Environmental Sampling,* ACS Professional Reference Book, American Chemical Society, Washington, D. C., 1988.
7. L. H. Keith, ed., *Environmental Sampling and Analysis, A Practical Guide,* Lewis Publishers, Chelsea, Mich., 1991.
8. P. Quevauviller, ed., *Quality Assurance in Environmental Monitoring Sampling and Sample Pretreatment,* VCH Publisher, Weinheim, 1995.
9. G. V. Iyengar, *Anal. Chem.* **54,** 554 A–560 A (1982).
10. A. Gomez, R. Leschber, and P. L'Hermite, eds., *Sampling Problems for the Chemical Analysis of Sludge, Soils, and Plants,* Elsevier Applied Science Publisher, London, 1986.
11. W. Rasemann and M. Herbst, *Probenahme und geostatistische Datenanalyse zur Bewertung von Altablagerungen—Möglichkeiten und Grenzen am Beispiel einer Freiberger Schlackenhalde, Preprints der 2,* Tagung des Arbeitskreises "Probenahme", Oct. 25–26, 1996, Freiberg, pp. 25–40.
12. J. W. Einax and U. Soldt, *Möglichkeiten und Grenzen der Anwendung geostatistischer und multivariat-statistischer Methoden zur Bewertung von belasteten Böden,* Preprints der 1, Tagung des Arbeitskreises "Probenahme," Nov. 17–18, 1995, Freiberg, pp. 51–60.
13. A. G. Journel, in L. H. Keith, ed., *Principles of Environmental Sampling,* ACS Professional Reference Book, American Chemical Society, Washington, D.C., 1988, pp. 45–72.

INFRARED SPECTROSCOPY

BARBARA H. STUART
University of Technology
Sydney, Australia

Infrared spectroscopy is certainly one of the most important analytical techniques available today. One of the great advantages of infrared spectroscopy is that virtually any sample in virtually any state can be studied. Liquids, solutions, pastes, powders, films, fibres, gases, and surfaces can all be examined with a judicious choice of sampling technique. Because of improved instrumentation in the field, a variety of new, sensitive techniques have been developed to examine formerly intractable samples. This technique is of great value in environmental applications, such as air pollutants, solid pollutants, and degradation.

THEORY OF INFRARED SPECTROSCOPY

Infrared spectroscopy is a technique based on the vibrations of the atoms of a molecule. An infrared spectrum is obtained by passing infrared radiation through a sample and determining what fraction of the incident radiation is absorbed at a particular energy. The energy at which any peak in an absorption spectrum appears corresponds to the frequency of a vibration of a part of the sample molecule. The basic ideas and definitions associated with infrared spectroscopy are discussed in the first section of this article. The vibrations of molecules are examined in detail, as these are crucial to the interpretation of infrared spectra. The interactions of infrared radiation with matter can be understood in terms of changes in molecular dipoles associated with vibrations and rotations (1).

Normal Modes of Vibration

A molecule can be looked upon as a system of masses joined by bonds with springlike properties. The simple case of a diatomic molecule can be used to illustrate molecular vibrations. Diatomic molecules have three degrees of translational freedom and two degrees of rotational freedom. The atoms in molecules also move relative to one other, that is, bond lengths vary or one atom moves out of its current plane. This describes stretching and bending movements collectively referred to as vibrations. For a diatomic molecule, only one vibration is possible which corresponds to the stretching and compression of the bond. This accounts for one degree of vibrational freedom.

Polyatomic molecules containing many (N) atoms have $3N$ degrees of freedom. Two groups of triatomic molecules, linear and nonlinear, can be distinguished. Two simple examples are CO_2 and H_2O illustrated below. Both CO_2 and H_2O have three degrees of translational freedom. Water has three degrees of rotational freedom but the linear molecule CO_2 has only two, because no detectable energy is involved in rotation around the O=C=O axis. Subtracting these from $3N$ gives $3N - 5$ for CO_2 or any linear molecule and $3N - 6$ for water or any nonlinear molecule. N in both examples is three, so CO_2 has four vibrational modes and water has three. The degrees of freedom for polyatomic molecules are summarized in Table 1.

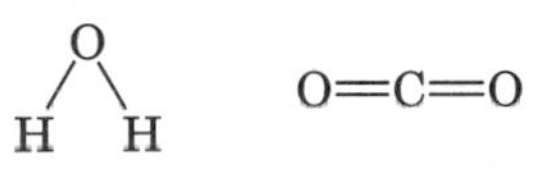

Whereas a diatomic molecule has only one mode of vibration which corresponds to a stretching motion, a nonlinear B–A–B type triatomic molecule has three modes, two of which correspond to stretching motions and the remainder corresponds to a bending motion. A linear type triatomic has four modes, two of which have the same frequency and are said to be degenerate.

Two other concepts are also used to explain the frequency of vibrational modes, the stiffness of the bond and the masses of the atoms at each end of the bond. The stiffness of the bond can be characterized by a proportionality constant termed the force constant k derived from Hooke's law. The reduced mass μ provides a useful way of simplifying our calculations by combining the individual atomic masses and is expressed as follows:

$$1/\mu = 1/m_1 + 1/m_2 \tag{1}$$

where m_1 and m_2 are the masses of the atoms at the ends of the bond. A practical alternative way of expressing the reduced mass is the following:

$$\mu = m_1 m_2/(m_1 + m_2) \tag{2}$$

Table 1. Degrees of Freedom for Polyatomic Molecules

Degrees of Freedom	Linear	Nonlinear
Translational	3	3
Rotational	2	3
Vibrational	$3N - 5$	$3N - 6$
Total	$3N$	$3N$

The equation relating force constant k, reduced mass μ, and the frequency of absorption is the following:

$$\nu = 1/2\pi\sqrt{k/\mu} \tag{3}$$

This equation can be modified as follows so that direct use of the wavenumber values can be made for bond vibrational frequencies:

$$\overline{\nu} = 1/2\pi c\sqrt{k/\mu} \tag{4}$$

where c is the speed of light.

A molecule only absorbs radiation when the incoming infrared radiation has the same frequency as one of the fundamental vibrational modes of the molecule. This means that the vibrational motion of a small part of the molecule increases whereas the rest of the molecule is left unaffected. Vibrations can involve a change in bond length (*stretching*) or bond angle (*bending*) (see below). Some bonds can stretch in phase (*symmetrical* stretching) or out of phase (*asymmetric* stretching) as shown below.

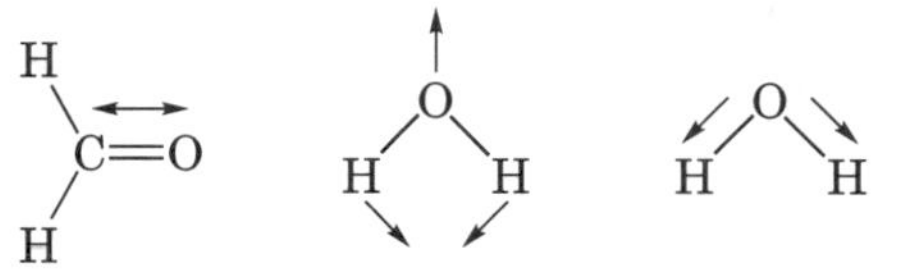

stretching **bending** **stretching**

symmetric stretching **asymmetric stretching**

If a molecule has different terminal atoms, such as HCN, ClCN, or ONCl, then, the two stretching modes are no longer symmetric and asymmetric vibrations of similar bonds, but will have varying proportions of stretching motion in each group. In other words, the amount of *coupling* will vary.

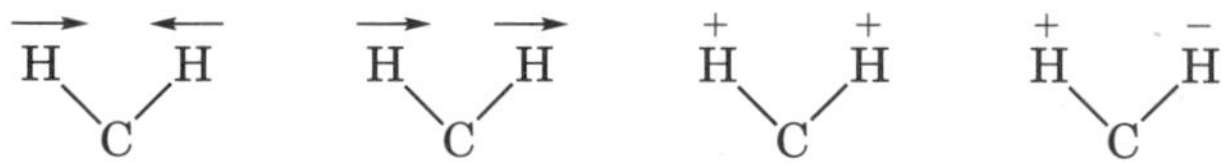

deformation **rocking** **wagging** **twisting**

Bending vibrations also contribute to infrared spectra and these are shown below. It is best to consider the molecule being cut by a plane through the hydrogen atoms and the carbon atom. The hydrogens can move in the same direction or in opposite directions in this plane, here the plane of the page. For more complex molecules the analysis becomes simpler since hydrogen atoms can be considered in isolation as they are usually attached to more massive, and therefore, more rigid parts of the molecule. This results in *in-plane* and *out-of-plane* bending vibrations illustrated below.

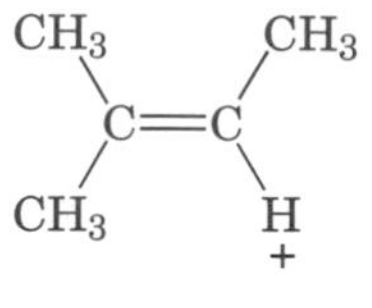

out-of-plane bending 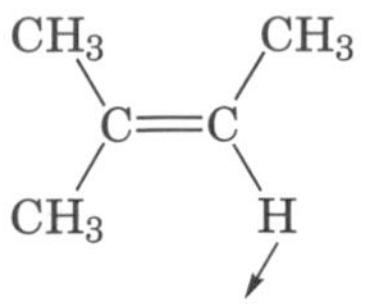 **in-plane bending**

However, there will be many different vibrations for even fairly simple organic molecules. The complexity of an infrared spectrum arises from the coupling of vibrations over a large part of or over the complete molecule. Such vibrations are called *skeletal* vibrations. Bands associated with skeletal vibrations are likely to conform to a pattern or fingerprint of the molecule as a whole, rather than to a specific group within the molecule.

Intensity of Infrared Bands

For a vibration to absorb infrared radiation, it must change the dipole moment of the molecule. The larger this change, the more intense the absorption band. Because of the difference in electronegativity between carbon and oxygen, the carbonyl group is permanently polarized, as shown below. Stretching this bond increases the dipole moment, and hence, C=O stretching is an intense absorption.

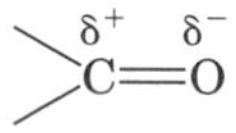

Symmetrical molecules have fewer infrared active vibrations than unsymmetrical molecules. This leads to the conclusion that symmetric vibrations, generally, are weaker than asymmetric vibrations, because symmetric vibrations the former do not change the dipole moment. It follows that bending or stretching of bonds involving atoms in widely separated groups of the periodic table leads to intense bands. Vibrations of bonds, such as C–C or N=N, give weak bands because of the small change in dipole moment associated with their vibrations.

INSTRUMENTATION

Infrared spectrometers have been commercially available since the 1940s. At that time, the instruments relied on prisms as dispersive elements, but, by the mid-1950s, diffraction gratings had been introduced into dispersive machines. The most significant advances in infrared spectroscopy, however, have come about with the introduction of Fourier transform spectrometers. This type of instrument employs an interferometer and exploits the well-established mathematical process of Fourier transformation. Fourier transform infrared (FTIR) spectroscopy has dramatically improved the quality of infrared spectra and minimized the time required to obtain data. Also, with improvements to computers in recent years, infrared spectroscopy has made great strides.

Representation of Spectra

Early infrared instruments recorded percentage transmittance over a linear wavelength range. Now, it is unusual to use wavelength for routine samples and inverse wavelength units are used. This is the wavenumber scale whose units are cm^{-1}. The output from the instrument is referred to as a spectrum. Most commercial instruments present the spectrum with the wave number decreasing from left to right.

The infrared spectrum is divided into three regions: the *far-infrared* (400–0 cm^{-1}), the *mid-infrared* (4,000–400 cm^{-1}) and the *near-infrared* (13,000–4,000 cm^{-1}). Most infrared applications employ the mid-infrared region, but the near- and far-infrared also provide information about certain samples. The majority of instruments scan only the mid-infrared range.

The ordinate scale may be presented in % transmittance with 100% at the top of the chart. It is commonplace to have the choice of absorbance or transmittance as a measure of band intensity. Figures 1 and 2 illustrate the difference in appearance between absorbance and transmittance spectra. Personal preference usually dictates which of the two modes to use, but transmittance is traditionally used for spectral interpretation, whereas absorbance is used for quantitative work.

Fourier Transform Infrared Spectrometry

In recent years, FTIR spectroscopy has found increasing favor in laboratories (2). This more recent method is based on the old idea of the interference of radiation between two beams to yield an interferogram. An interferogram is a signal produced as a function of the change of path length between the two beams. The two domains of distance and frequency are interconvertible by the mathematical method of *Fourier transformation.*

Although the basic optical component of FTIR instruments, the Michelson interferometer, has been known for almost a century, it was not until advances in computing that the technique could be successfully applied. The basic components of a FTIR spectrometer are shown schematically in Figure 3. The radiation emerging from the source passes through an interferometer to the sample before reaching a detector. Upon amplification of the signal, in which high-frequency contributions are eliminated by a filter, the data are converted to a digital form by an analog-to-digital converter and transferred to the computer for Fourier transformation.

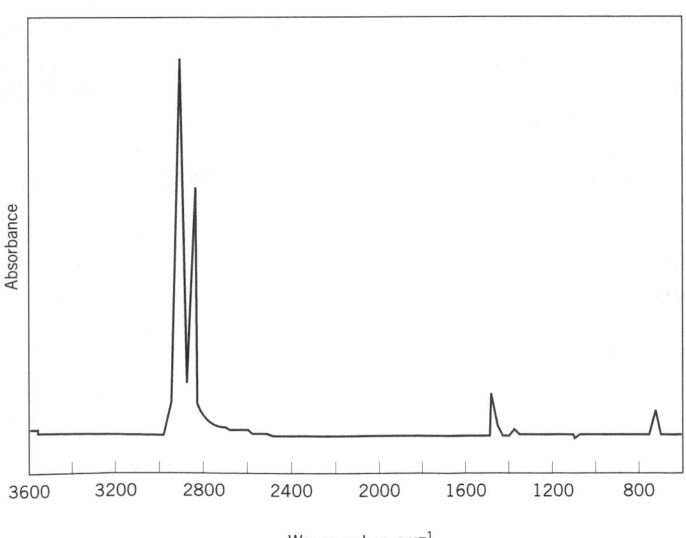

Figure 1. An absorbance spectrum.

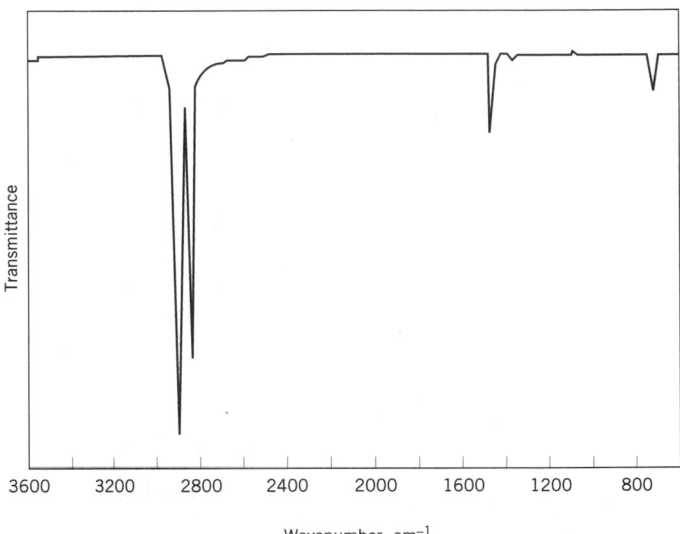

Figure 2. A transmittance spectrum.

The Michelson Interferometer

The most common interferometer is a Michelson interferometer, which consists of two perpendicular plane mirrors, one of which can travel in a direction perpendicular to the plane (Fig. 4). A semireflecting film, the *beam splitter,* bisects the planes of these two mirrors. The beam splitter material has to be chosen according to the region to be examined. Materials, such as germanium or iron oxide, are coated onto an infrared transparent substrate, such as potassium bromide or cesium iodide, to produce beam splitters for the mid- or near-infrared regions. Thin organic films, such as polyethylene terephthalate, are used in the far-infrared region.

If a collimated beam of monochromatic radiation with wavelength λ cm passes into an ideal beam splitter, 50% of the incident radiation is reflected to one of the mirrors and 50% is transmitted to the other mirror. The two beams are reflected from these mirrors, returning to the beam splitter where they recombine and interfere. Fifty percent of the beam reflected from the fixed mirror is transmitted through the beam splitter and 50% is reflected back in the direction of the source. The beam which emerges from the interferometer at 90° to the input beam is called the transmitted beam, and this is the beam detected in FTIR spectrometry.

The moving mirror produces an optical path difference between the two arms of the interferometer. For path differences of $(n + 1/2)\lambda$, the two beams interfere destructively in the case of the transmitted beam and constructively in the case of the reflected beam. The resultant interference pattern is shown in Figure 5 for a source of polychromatic radiation.

Figure 3. The components of a FTIR spectrometer.

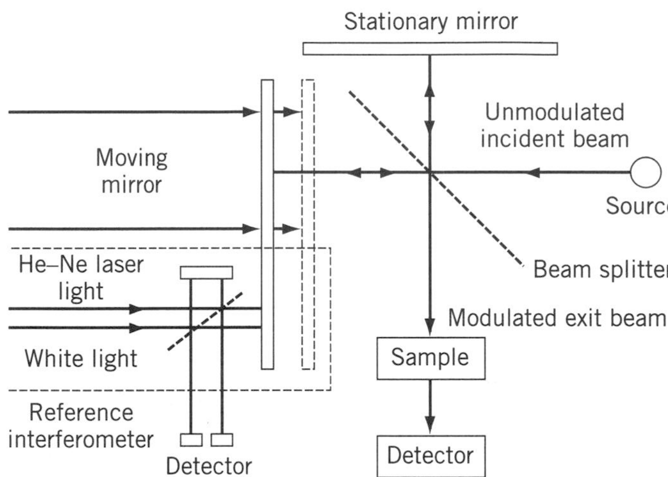

Figure 4. A Michelson interferometer.

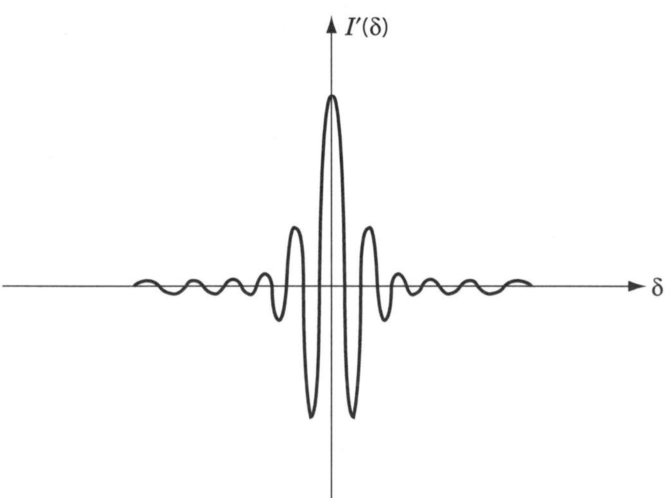

Figure 5. An interferogram for polychromatic radiation.

Fourier Transformation

The essential equations relating the intensity falling on the detector $\mathrm{I}(\delta)$ to the spectral power density at a particular wave number $\overline{v}$, given by $B(\overline{v})$, are as follows:

$$\mathrm{I}(\delta) = \int_0^{+\infty} B(\overline{v}) \cos 2\pi\overline{v}\delta \, d\overline{v} \qquad (5)$$

which is one half of a cosine Fourier transform pair. The other is the following:

$$B(\overline{v}) = \int_{-\infty}^{+\infty} I(\delta) \cos 2\,\pi\overline{v}\delta \, d\,\delta \qquad (6)$$

Equations 5 and 6 are interconvertible and are known as a Fourier transform pair. Equation 5 shows the variation in power density as a function of the difference in path length, which is an interference pattern. Equation 6 shows the variation in intensity as a function of wavenumber. Each can be converted into the other by the mathematical method of Fourier transformation.

The essential experiment to obtain a FTIR spectrum is to produce an interferogram with and without a sample in the beam and transforming the interferograms into spectra of (*1*) the source with sample absorptions and (*2*) the source without sample absorptions. The ratio of the former and the latter corresponds to a double-beam dispersive spectrum.

Sources and Detectors

FTIR spectrometers use a Nernst or Globar source for the mid-infrared region. If the far-infrared region is to be examined, then a high pressure mercury arc lamp can be used. Tungsten lamps are used as sources for the near-infrared.

There are two detectors commonly used in the mid-infrared region. The detector for routine use is a pyroelectric device incorporating deuterium tryglycine sulfate (DTGS) in a temperature-resistant alkali halide window. For more sensitive work, mercury cadmium telluride (MCT) is used, but it has to be cooled to liquid nitrogen temperatures. In the far-infrared, germanium or indium-antimony detectors are employed, and detectors for the near-infrared, generally, are lead sulfide photoconductors.

The Moving Mirror

The moving mirror is the most crucial component of the interferometer. It has to be accurately aligned and must scan two distances so that the path difference corresponds to a known value. A number of factors associated with the moving mirror need to be considered when calculating an infrared spectrum. The interferogram is an analog signal at the detector which has to be digitized to carry out the Fourier transformation into a conventional spectrum. There are two particular sources of error in transforming the digitized information on the interferogram into a spectrum.

First, the transformation carried out in practice involves an integration stage over a finite displacement rather than over an infinite displacement. The mathematical process of Fourier transformation assumes infinite boundaries. The consequence of this necessary approximation is that the apparent line shape of a spectral line may be as shown in Figure 6, where the main band area has a series of negative and positive side lobes (or pods) with diminishing amplitudes. The process of *apodization* is the removal of the side lobes or pods by multiplying the interferogram by a suitable function before the Fourier transformation is carried out. A suitable function must cause the intensity of the interferogram to fall smoothly to zero at its ends. Most FTIR spectrometers offer a choice of apodization options. A good general purpose apodization function is the following cosine function:

$$F(D) = [1 + \cos(\pi D)]/2 \qquad (7)$$

where D is the optical path difference. The cosine function provides an good compromise between reduced oscillations and deteriorated spectral resolution. When accurate band shapes are needed, more sophisticated mathematical functions may be needed.

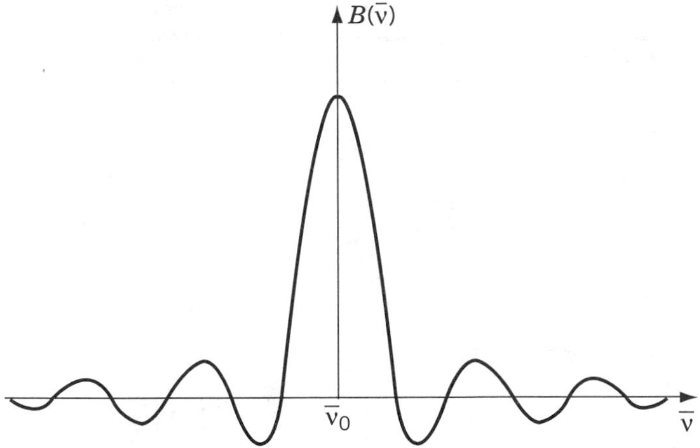

Figure 6. Instrument lineshape without apodisation.

Another source of error arises if the sample intervals are not exactly the same on each side of the maxima corresponding to zero path difference. Phase correction is required, and this correction procedure ensures that the sample intervals are the same on each side of the first interval and correspond to a path difference of zero.

The resolution is limited by the maximum path difference between the two beams. The limiting resolution in wavenumber (cm^{-1}) is the reciprocal of the path-length difference (cm). For example, a path-length difference of 10 cm is required to achieve a limiting resolution of 0.1 cm^{-1}. It may seem from this simple calculation that it is easy to achieve high resolution. However, this is not so because precision of the optics and mirror movement mechanism become more difficult to achieve at longer displacements of path lengths.

Signal Averaging

The main advantages of rapid scanning instruments is the ability to increase the signal-to-noise by signal averaging which leads to an increase of signal-to-noise proportional to the square root of the time. In a rapid-scan interferometer, it follows that the signal-to-noise ratio (SNR) is related to the number of scans by the following relationship:

$$\text{SNR} \propto n^{1/2} \tag{8}$$

There are diminishing returns for signal averaging because it takes an increasingly longer time to achieve greater and greater improvement. The accumulation of a large number of repeat scans makes greater demands on the instrument if it is to reproduce the conditions exactly. It is normal to incorporate a laser monochromatic source in the beam of the continuous source. The laser beam produces standard fringes which line up successive scans accurately and determine and control the displacement of the moving mirror at all times.

Advantages of FTIR Spectroscopy

FTIR instruments have several significant advantages over dispersive instruments. Two of these are the Fellgett (or multiplex) advantage and the Jacquinot (or throughput) advantage (3). The Fellgett advantage derives from an improvement in the SNR per unit time proportional to the square root of the number of resolution elements being monitored. This results from the large number of resolution elements being monitored simultaneously. Also, because FTIR spectrometry does not require a slit or other restricting device, especially at low resolution, the total source output passes through the sample continuously. This results in a substantial gain in energy at the detector, translating to higher signals and improved SNRs and known as Jacquinot's advantage.

Another strength of FTIR spectrometry is its speed advantage. The mirror can move short distances quite rapidly, and this, together with the SNR improvements from Fellgett's and Jacquinot's advantages, makes it possible to obtain spectra on a millisecond timescale. In interferometry, the factor that determines the precision of the position of an infrared band is the precision with which the scanning mirror position is known. By using a helium-neon laser as a reference, the mirror position is known with extremely high precision.

Computers

The computer forms a crucial component of modern infrared instruments. Instruments have a devoted data station incorporated into the machine or interface with personal computers. Software is constantly being updated, and many packages are designed in a convenient format. Computers perform the following functions: (*1*) control the instrument, for example, set scan speeds, scanning limits, start and stop scanning; (*2*) read spectra into the computer memory from the instrument as the spectrum is scanned; this means that the spectrum is digitized; (*3*) plot the spectra on the instrument's plotter; (*4*) manipulate spectra, for example, adding and subtracting spectra, expanding areas of the spectrum of interest; (*5*) scan the spectra continuously and average or add the result in the computer memory; and (*6*) run complex analyses automatically by following a set of preprogrammed commands.

SAMPLING METHODS

An important decision in infrared spectroscopy is the choice of an appropriate sampling technique. The sampling procedure depends very much on the type of sample to be examined, for instance, whether it is a solid, a gas, or a liquid.

APPLICATIONS

There is a wide range of environmental applications for infrared spectroscopy. Some possible approaches to environmental analysis are outlined in the following sections with examples of air pollution, pesticide, and polymer degradation infrared studies.

Air Pollution

The routine monitoring of vapor levels of toxic chemicals in the work environment has led to the development of the infrared technique for automatic quantitative analysis of noxious chemicals in air (4). The technique is based on cell of a long path-length, with a pump to circulate the air sample through the instrument. Instruments are dedicated to a single contaminant in air, employing a filter system for wavelength selection. These instruments are commonly used to analyze for SO_2, HCN, phosgene, NH_3, H_2S, formaldehyde, and HCl. More complex instruments are tunable for a particular analytical band and usually give a direct readout in ppm or % vol/vol of

Table 2. Common Pollutants and Recommended Frequencies

Compound	Analytical Frequency, cm^{-1}	Path Length Recommended, m	Minimum Detectable Concentration, ppm
CH_3CN	1042	20	5.0
NH_3	962	20	0.2
C_6H_6	672	20	0.3
$CH_3CH_2COCH_3$	1176	20	0.15
CCl_4	794	20	0.06
CS_2	2203	20	0.5
CO_2	2353	0.75	0.5
CO	2169	20	0.2
CCl_2F_2 (Freon 12)	1099	0.75	0.02
CBr_2F_2	1087	2.25	0.02
SO_2	1163	20	0.5
HCN	3290	20	0.4
CH_3NH_2	2941	20	0.1
CH_2CHCl	917	20	0.3

the pollutant. The path-length of the cell is also varied from <1 to 20 m.

Typical applications are monitoring of formaldehyde in plastic and resin manufacture, anaesthetics in operating theatres, degreasing solvents in a wide variety of industries, and carbon monoxide in garages. Table 2 shows some common pollutants with the recommended analytical frequencies. The minimum levels quoted are for a 20 m cell.

Pesticides

Infrared spectroscopy has proved effective for investigating solid pollutants. For example, contaminents such as pesticides, organic and inorganic sprays, and dusts on plants may be examined by this technique.

GC/IR has been used to investigate the degradation products of Mirex (dodecachloropentacyclodecane), a pesticide used to control fire ants (5). One of the requirements for pesticides is that they rapidly decompose in the environment, and GC/IR can be used to identify Mirex derivatives. This technique has an advantage over GC/MS, for instance, because it can more readily differentiate the derivatives produced by degradation. During the decomposition of Mirex, some of the chlorine atoms are substituted with hydrogen. The 5,10-dichloro derivative has anti and syn isomers, and these are difficult to separate using GC/MS. However, the infrared spectra of the two isomers show bands at 1100 and 1120 cm^{-1}, respectively. There is some overlap of these bands, but the pure spectrum of each isomer can be obtained from the difference spectra.

Polymer Degradation

Infrared spectroscopy is used to investigate the effects of environmental exposure on samples. In particular, the technique has been applied to the degradation of polymers. The exposure to ultraviolet radiation, chemical attack by free radicals produced by natural processes, and microbial attack are studied by infrared spectroscopy.

The role of infrared spectroscopy in polymer degradation is illustrated by its application to heat and photooxidized polyethylenes (6). During the heat oxidation of polyethylene, carbonyl-containing compounds are formed. These decomposition products create a broad C=O stretching band at about 1725 cm^{-1} consisting of a number of overlapping component bands. When the oxidized samples are treated with an alkali, a shoulder at 1715 cm^{-1} disappears and is replaced by a distinctive peak near 1610 cm^{-1}. This band results from C=O stretching of the COO^- ion of a salt, indicating that the shoulder at 1715 cm^{-1} is characteristic of saturated carboxylic acids. Another shoulder at 1735 cm^{-1} is characteristic of a saturated aldehyde. However, the major contribution to the carbonyl band is from saturated ketones. The broad C=O stretching band is also present in the infrared spectrum of photooxidized polyethylene samples, which show additional bands at 990 and 910 cm^{-1}. The latter bands are characteristic of vinyl groups, and their presence shows that chain-terminating unsaturated groups are being formed, most likely resulting from chain scission.

BIBLIOGRAPHY

1. B. H. Stuart, *Modern Infrared Spectroscopy,* John Wiley & Sons, Ltd., Chichester, U.K., 1996.
2. P. R. Griffiths, *Chemical Infrared Fourier Transform Spectroscopy,* John Wiley & Sons, Inc., New York, 1975.
3. J. C. Case and W. G. Fateley, *Analytical Applications of FTIR to Molecular and Biological Systems,* D. Reidel Publishing, Dordrecht, the Netherlands, 1980.
4. F. W. Fifield and P. J. Haines, eds., *Environmental Analytical Chemistry,* Chapman and Hall, London, 1995.
5. K. S. Kalasinsky, *J. Chromatog. Sci.* **21,** 246–253 (1983).
6. D. I. Bower and W. F. Maddams, *The Vibrational Spectroscopy of Polymers,* Cambridge University Press, Cambridge, U.K. 1989.

INTERNATIONAL ORGANIZATION FOR STANDARDIZATION. See ISO.

ION CHROMATOGRAPHY

S. V. KARMARKAR
Zellweger Analytics, Inc.
Milwaukee, Wisconsin

Ion chromatography (IC), introduced in 1975 by (1), has come into its own in the last 20 years. The technique, a type of high-performance liquid chromatography (HPLC), has gained popularity in laboratories for determining anions and cations in environmental, agricultural, clinical, pharmaceutical, metal plating, power generation, semiconductor fabrication, and industrial samples. Environmental samples could be as simple as drinking water or as complex as persulfate-digested influent and effluent wastewaters.

Several review articles and books provide an elaborate description of the basic principles of IC (2,3,4,5).

BASIC PRINCIPLES OF ION CHROMATOGRAPHY

Ion chromatography falls under the broad category of liquid chromatography (Fig. 1). The typical components of an IC system (Fig. 2) include an optional autosampler, a high-pressure pump, an injection valve with a sample loop of suitable size (typically, 10–250 μL), a guard column (also called a precolumn), an analytical column (also called a separator column), an optional post-column reaction system, a flow-through detector, and a data system ranging in complexity from a chart recorder to a computerized data system. A suitable mobile phase, called the eluent, is constantly flowing through the columns and the detector. Typically, all of the components in contact with the eluent and sample are made from inert materials, such as polyetheretherketone (PEEK). Following suitable sample preparation, usually filtering through a 0.45-μm filter and diluting as required, the sample is introduced into the IC via the injection valve. The valve injects a fixed volume of sample into the guard and analytical columns. The ions of interest, then, are separated by differing affinities for the column packing material as the ions are swept along in the flowing eluent.

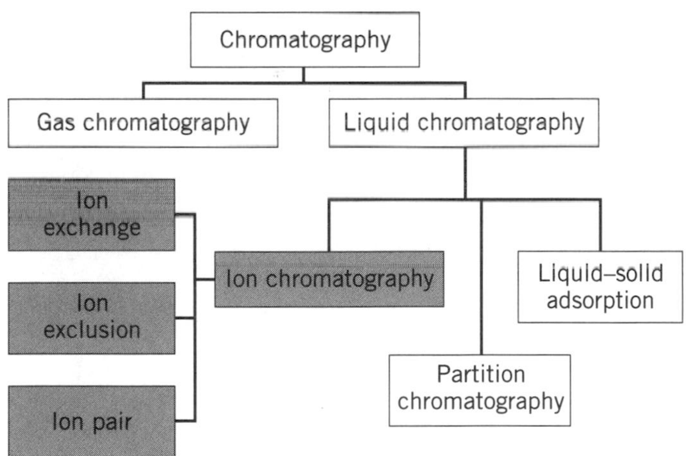

Figure 1. Ion chromatography (IC), a type of liquid chromatography. Ion exchange, ion exclusion, and ion pair are the three separation modes used in IC.

Separation Modes

Ion chromatographic separation takes place by one of three separation modes: ion exchange, ion exclusion, and ion pair. Examples of ion-exchange separation include IC determination of common anions, bromide, chloride, fluoride, nitrate, nitrite, phosphate, and sulfate, as shown in Figure 3, and alkali metals and alkaline-earth cations, ammonium, calcium, lithium, magnesium, potassium, and sodium, as shown in Figure 4. Ion exclusion separation is used in the IC determination of low molecular weight organic acids, adipic, acetic, citric, formic, fumaric, malic, malonic, oxalic, succinic, and tartaric, as shown in Figure 5. Examples of ion-pair separation include IC determination of heavy and transition metals, cadmium, cobalt, copper, iron (II and II), lead, manganese, nickel, and zinc as presented in Figure 6. Details of each of these separation modes are well described in (2). The separated ions are most commonly detected with conductimetric, colorimetric, and electrochemical detectors.

Conductivity Detection

Anions, cations, and organic acids are determined by conductivity detection in which the IC is operated in either a suppressed or nonsuppressed mode. Suppressed IC with conductivity detection was introduced in 1975 (1). Nonsuppressed IC was introduced in 1979 (6).

Suppressed Conductivity. In suppressed IC of anions, a typical eluent is a mixture of Na_2CO_3 and $NaHCO_3$ in a suitable ratio. This eluent is quite versatile and readily provides separation of mono- and divalent anionic species in a short time. The high background conductance of this eluent, however, must be suppressed before any useful data can be obtained.

An excellent signal-to-noise (S/N) ratio is obtained with suppressed IC. The two reactions occurring in a suppression device are background suppression and signal enhancement. These reactions are summarized below using the example of IC for anions. The eluent used was $NaHCO_3 + Na_2CO_3$.

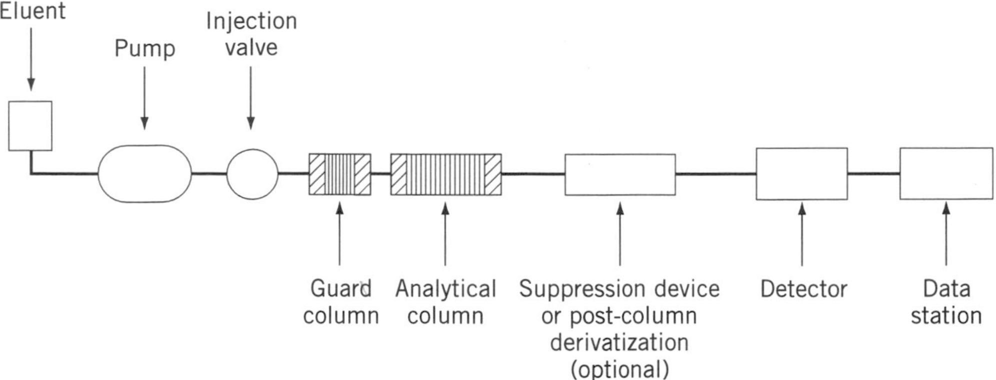

Figure 2. Schematic showing components of a typical IC.

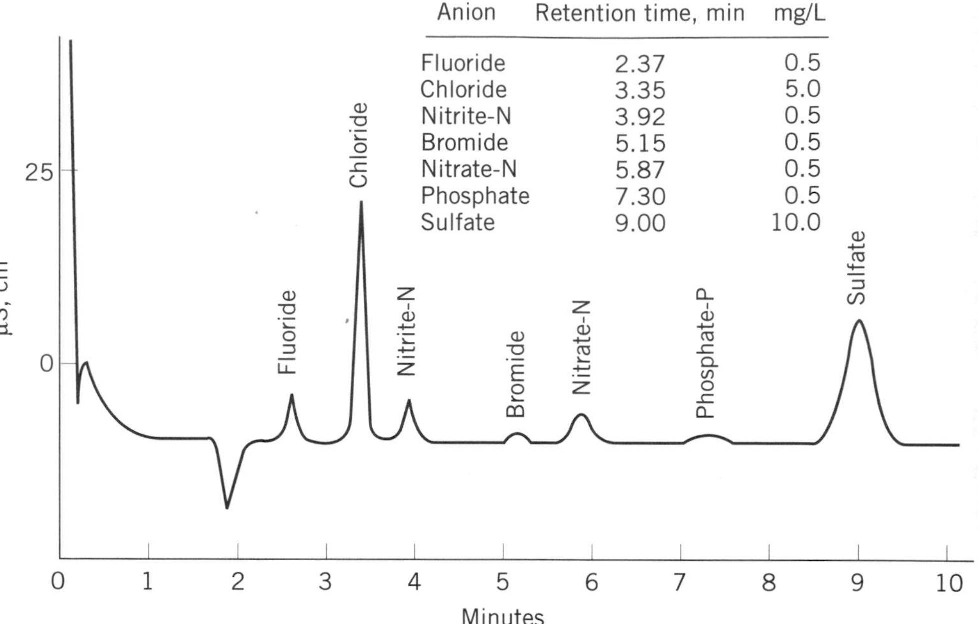

Figure 3. Chromatogram for anions (U.S. EPA method 300.0, part A). Chromatographic conditions: guard column, 5.5 × 50 mm; analytical column, 5.5 × 150 mm; eluent, 2.2 mM $NaHCO_3$ + 2.8 mM Na_2CO_3 at a rate of 2.0 mL/min; QuikChem small suppressor, 4.6 × 20 mm, regenerated with 0.25 M H_2SO_4; and sample loop, 200 μL, conductimetric detection.

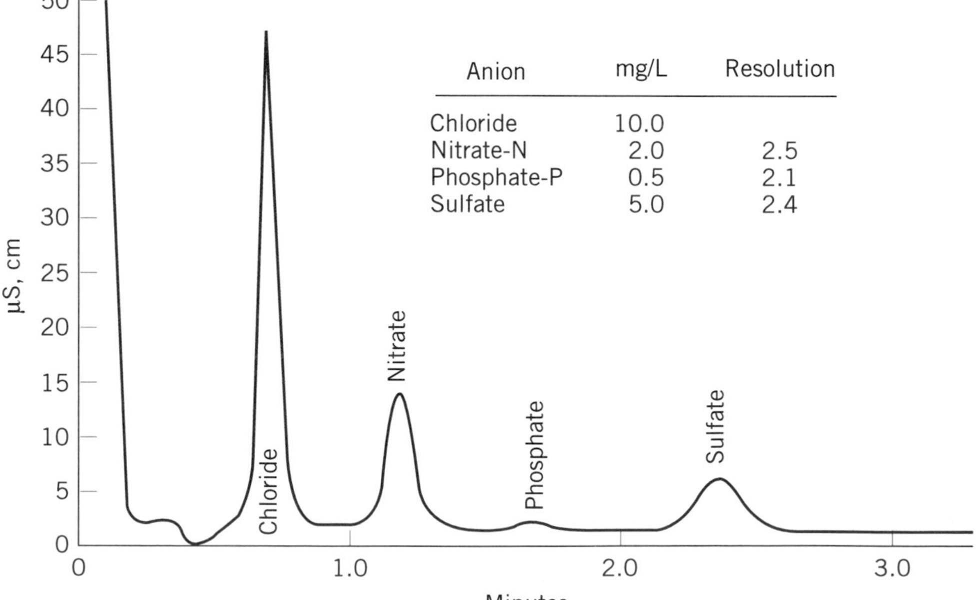

Figure 4. Chromatogram for rapid determination of chloride, nitrate, phosphate, and sulfate. Chromatographic conditions: analytical column, 4.6 × 50 mm , with a guard disk; eluent, 7.0 mM $NaHCO_3$ + 0.5 mM Na_2CO_3 at a rate of 1.6 mL/min; QuikChem small suppressor, 4.6 × 10 mm, regenerated with 0.25 M H_2SO_4; sample loop, 50 μL, conductimetric detection.

Background suppression:

$$R^- - H^+ + Na^+HCO_3^- \rightarrow R^- - Na^+ + H_2CO_3^0$$

$$2\,R^- - H^+ + 2\,Na^+CO_3^{2-} \rightarrow 2\,R^- - Na^+ + H_2CO_3^0$$

Signal enhancement:

$$R^- - H^+ + Na^+A^- \rightarrow R^- - Na^+ + H^+A^-$$

where, R^- is a functional group attached to the resin within the suppressor and A^- is an anionic species in the sample.

The background is suppressed because the end product, $H_2CO_3^0$ has no charge. The signal is enhanced because the sodium counterion Na^+ in the eluent is exchanged for the hydrogen ion H^+. The equivalent conductance value for H^+ is about seven-fold greater than that for Na^+, and, thus, an amplification effect is obtained by the exchange of cations. Because of this effect, ion chromatography systems provide excellent signal-to-noise ratios when equipped with suppressed conductivity detection, usually achieving detection limits two orders of magnitude lower than those obtained in a nonsuppressed mode.

The suppression device introduced by (1) consisted of a suppressor column (dimensions ranging from 9 × 250 mm, 9 × 110 mm, or 2.8 × 300 mm) packed with high-capacity ion exchange material. The eluent and sample steadily exhausted the active sites, $R{-}H^+$, in the suppressor;

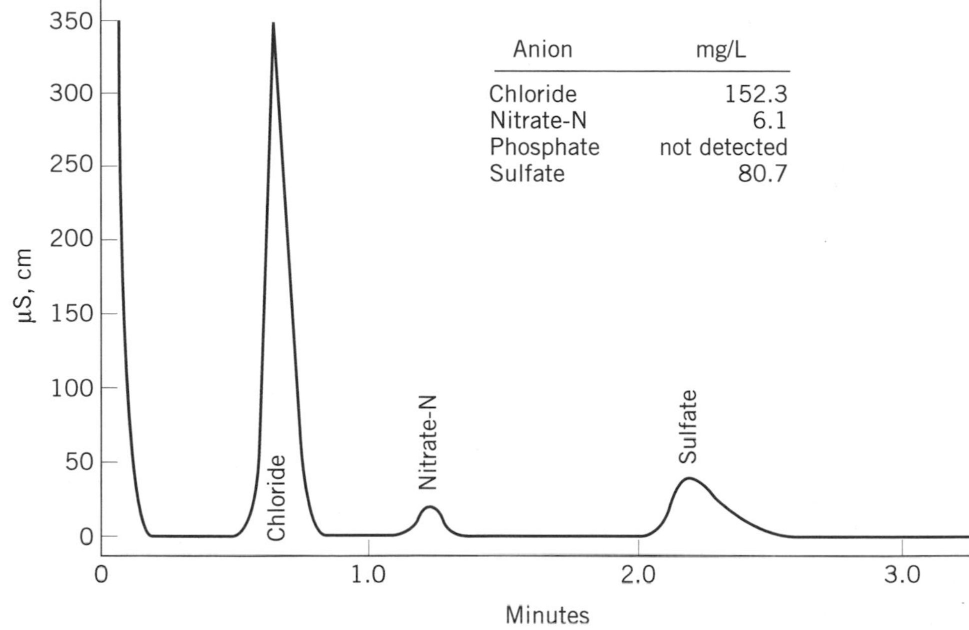

Figure 5. Chromatogram for rapid determination of chloride, nitrate, and phosphate in an effluent sample obtained from a wastewater treatment plant. For chromatographic conditions, see Figure 4.

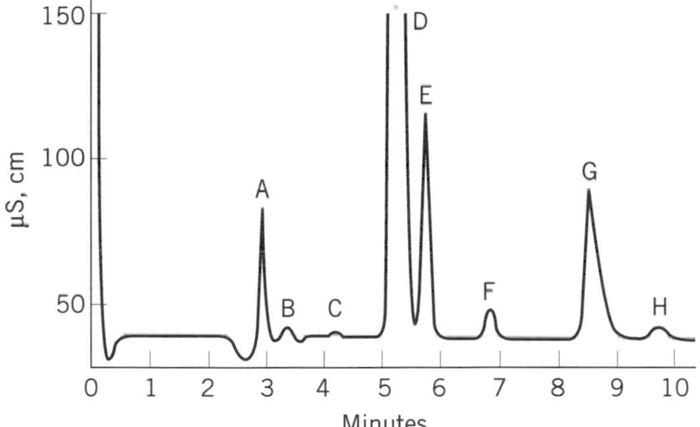

Figure 6. Chromatogram for anions in units of mg/L: A, fluoride, 1; B, chlorite, 0.4; C, bromate, 0.4; D, chloride, 40; E, nitrite-N, 2; F, chlorate, 2; G, nitrate-N, 2; and H, bromide, 1 (U.S. EPA method 300.0, part B). Chromatographic conditions: guard column, 5.5 × 50 mm; analytical column, 5.5 × 250 mm; eluent, 0.24 m*M* $NaHCO_3$ + 2.16 **m***M* Na_2CO_3 at a rate of 1.5 mL/min; QuikChem small suppressor, 4.6 × 20 mm, regenerated with 0.25 *M* H_2SO_4; and sample loop, 250 μL, conductimetric detection.

these suppressor columns had an useful life of 6–24 hours after which the suppressor column had to be regenerated offline. Along with this inconvenience, the device had several disadvantages. First, a relatively large volume of the suppressor column resulted in band broadening and, hence, loss in chromatographic efficiency. Second, the detector response to ions of strong acids or bases decreased, whereas the response for ions of weak acids or bases increased as the active sites in the suppressor were steadily depleted. The lack of a steady state resulted in poor precision for determining ions. Despite these disadvantages, the packed-bed suppressor provided the foundation on which suppressed IC was built. This batch-type device was in use from 1975–1981. In 1981, a continually operated fiber-based device was introduced (7). Currently, the following four suppression devices are commercially available:

1. QuikChem small suppressor that is regenerated after every sample using chemicals (8), commercially available with IC from Zellweger Analytics, Inc., Lachat Instruments Division (Milwaukee, Wisconsin).
2. A set of two parallel, small suppressor columns in which one is regenerated electrochemically while the other is being used (9), commercially available with IC from Alltech Associates, Inc. (Deerfield, Illinois).
3. A device with post-column addition of a colloidal suspension of a high-capacity ion exchange material, also called solid phase reagent (10) available from Sarasep, Inc. (San Jose, California).
4. Membrane-based device that is regenerated electrochemically (7), commercially available with IC from Dionex Corporation (Sunnyvale, California).

Nonsuppressed Conductivity. In a nonsuppressed mode, the analytical column effluent flows directly to a conductivity detector (6). The typical eluents used in nonsuppressed IC are phthalic acid and *p*-hydroxybenzoic acid for determining anions and methanesulfonic acid for determining cations. The equivalent conductance values of chloride, sulfate, and other common anions are appreciably greater than the conductance of the eluent anion, and hence, a positive peak is detected as the anions are carried through the detector. The equivalent conductance values of sodium, potassium, and other common cations are appreciably smaller than the conductance of the eluent cation, and hence, a negative peak is detected as the cations are carried through the detector.

Compared with suppressed IC, nonsuppressed IC is easy to operate. It is also a useful technique for determining ions of weak acid, such as cyanide and sulfide, that do not conduct after chemical suppression. For several reasons, nonsuppressed IC, however, did not gain as much acceptance as suppressed IC, especially in environmental analysis. The first reason was that regulatory methods, such as U.S. EPA method 300.0, are based on suppressed IC. The second reason was that the S/N ratio is much greater with suppressed IC than that with nonsuppressed IC. Lastly, the modern suppression devices developed since 1981, a couple of years after the introduction of nonsuppressed IC, eliminated the drawbacks of the original packed-bed suppressor.

Other Detectors

In colorimetric detection, such as in hexavalent chromium analysis by U.S. EPA method 218.6, the column effluent is merged with a reagent stream containing diphenylcarbazide (Fig. 7). The combined stream flows through a mixing coil where mixing and color reaction occur, followed by colorimetric detection of Cr(VI) at 520 nm. Similarly, heavy and transition metals (Fig. 6) are also detected colorimetrically by post column addition of 4-(2-pyridylazo)resorcinol.

Pulsed amperometric detection is useful for IC analysis of anions of weak acids with pK_a values above 7. These anions cannot be detected by suppressed IC because, after chemical suppression, these anions form undissociated weak acids and, hence, do not conduct. Examples of such anions include cyanide and sulfide. Details on this detection mode are available (3).

APPLICATIONS

Numerous IC applications for environmental analysis reported in the literature are summarized in several books and review articles (2,3,11).

Generally, IC environmental analysis can be divided into the following two major categories: (*1*) analysis performed for compliance monitoring and (*2*) analysis performed to understand the physicochemical fate of polluting ions in soils, plants, sewage sludge, and waters. Regulatory methods, such as those recommended by the United States Environmental Protection Agency (U.S. EPA), the National Institute of Occupational Safety and Hazard (NIOSH), and the Occupational Safety and Health Administration (OSHA), are used in compliance monitoring.

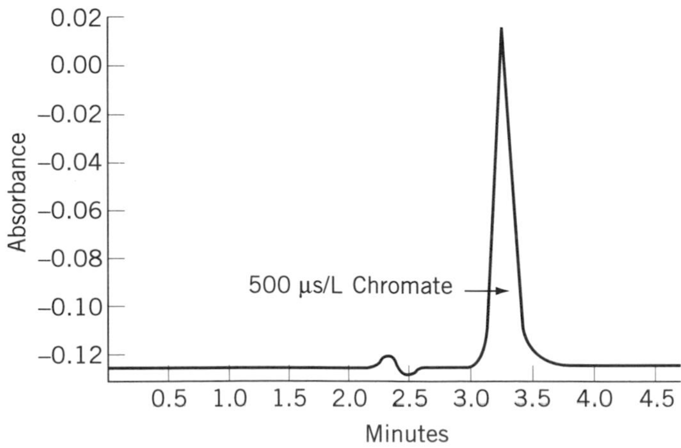

Figure 7. Chromatogram showing Cr(VI) (U.S. EPA method 218.6). Chromatographic conditions: guard column, 5.5 × 50 mm; analytical column, 5.5 × 250 mm; eluent; 55.9 m*M* $(NH_4)_2SO_4$ + 23.1 m*M* NH_4OH at a rate of 1.5 mL/min; post-column reagent, 2 m*M* diphenylcarbazide 10% vol/vol CH_3OH + 1*N* H_2SO_4 at a rate of 0.5 mL/min; and colorimetric detection at 530 nm.

SUMMARY

Since the late 1970s, IC has become a workhorse in routinely determining anions, cations, and metals in environmental samples. Published applications range from disinfection byproducts in drinking water to anions in effluent samples and from toxic gases in ambient air to sulfate in soils and plant tissue. Ion chromatographic methods are accurate, precise, and sensitive. Among the advantages of IC are (*1*) the ability to determine ionic species in low μg/L levels to mg/L levels for compliance and noncompliance work, (*2*) the ability to speciate elements into various ionic forms to understand toxicity and biogeochemical processes, (*3*) minimal need for toxic reagents, (*4*) fully automated system operation, (*5*) the ability to simultaneously determine multiple classes, eg, anions and cations, and (*6*) low operating costs.

Ion chromatographic methods, however, cannot be used for determining all analytes relevant in environmental analysis because of the regulatory constraints placed by the EPA, the harsh nature of a few samples for some analytes, such as total Kjeldahl nitrogen (TKN) and total phosphorus (TP), and a relatively low throughput. A few examples of such parameters include, alkalinity, ammonia, cyanide, hardness, phenol, TKN, and TP. For these parameters, the U.S. EPA-approved methods, instead, are based on manual or automated wet chemistry, such as flow injection analysis (FIA). Recently available commercial instruments allow simultaneous and independent operation of automated wet chemistry by both FIA and IC (12). The FIA methods are EPA-approved for determining these non-IC type parameters, they are rugged for the harsh samples, and they provide throughput ranging from 60–90 samples per hour. Ion chromatographic methods can run simultaneously with FIA methods to provide complete solutions for environmental analysis.

BIBLIOGRAPHY

1. H. Small, T. Stevens, and W. Bauman, *Anal. Chem.* **47**, 1801 (1975).
2. P. Haddad and P. Jackson, *Ion Chromatography—Principles and Applications,* J. Chromatogr. Library, vol. 46, Elsevier, New York, (1990).
3. J. Weiss, *Ion Chromatography,* 2nd ed., VCH Verlagsgesellschaft mbH, Weinheim, Germany, and VCH Publishers, Inc., New York, (1995).
4. *NIOSH Manual of Analytical Methods,* 4th ed., Superintendent of Documents, U.S. Government Printing Office, Washington D.C., (1994).
5. *OSHA Analytical Methods Manual,* 2nd ed., Parts 1–2, U.S. Dept. of Labor, Occupational Safety and Health Administration, Directorate of Technical Support, OSHA Technical Center, Salt Lake City, Utah, (1991).
6. D. Gjerde, J. Fritz, and G. Schmuckler, *J. Chromatogr.* **186,** 509 (1979).

7. A. Henshall, S. Rabin, J. Statler, and J. Stillian, *Am. Lab.,* 20R (Nov. 1992).
8. U.S. Pat. 5,567,307 (Oct. 22, 1996), S. V. Karmarkar (to Lachat Instruments, Inc.)
9. R. Saari–Nordhaus and J. Anderson, Jr., *Am. Lab.,* 329 (1996).
10. U.S. Pat. 5,149,661 (Sept. 22, 1992), D. T. Gjerde and J. V. Benson (to Sarasep, Inc.)
11. W. Frankenberger, Jr., H. Mehra, and D. Gjerde, *J. Chromatogr.* **504,** 211 (1990).
12. S. Stieg and S. Karmarkar, *Determination of Ammonia, Phenol, and Six Anions by Simultaneous Flow Injection Analysis and Ion Chromatography,* Paper No. 1149, The Pittsburgh Conference, New Orleans, La., 1995.

ISO

DOUGLAS J. HILEMAN
Pilko & Associates, Inc.
Woodland Hills, California

ISO AND THE STANDARDS

The International Organization for Standardization (ISO), a federation of more than one hundred national standards organizations, is in the process of developing a series of standards under the designation ISO 14000, which will establish a framework for business enterprises to manage the present and potential effects of their operations and products on the environment. The full set of ISO 14000 standards is not expected to be finalized completely for several years. Nonetheless, ISO 14001, 14004, and the standards in the 14010 series have been published (1,2,3). This chapter addresses the ISO 14000 series of standards on environmental management systems. Introductory concepts are presented first, followed by sections that address each standard in the ISO 14000 series sequentially, with selected quotes from each and discussions of how key points may affect an organization.

The core of the standards series is ISO 14001, which defines a model environmental management system (EMS) (1). ISO 14001 is unique in the series. It is a *specification* of the elements of an EMS that should be in place for an enterprise to be able to manage environmental matters effectively. It is intended that the ISO 14001 specification will provide a basis for an organization's EMS to be reviewed by a third party who can issue a certification that the standard has been met by the enterprise.

The other standards in the series are *guidelines* for the evaluation of an organization and its products. The guidelines provide a basis to assess its environmental management program and the effects of its products in the environment. Neither the specification nor the guidelines set a standard for what an organization's performance should be. They only establish what management tools and processes need to be in place for an organization to consistently meet appropriate performance standards and to improve upon them.

ISO 14000 standards appear deceptively simple; in fact, they create a complex series of commitments, thoughts, actions, and follow-ups for an organization that chooses to commit to them, either in whole or in part. They are all voluntary. It may or may not be advisable for an enterprise to seek 14001 registration or follow the guidance provided by the other standards in the series. Marketplace and other stakeholder concerns about an organization's environmental impact vary from organization to organization. Furthermore, the extent to which the standards become references in the rules and policies of the regional, national, and local jurisdictions in which an organization does business will vary.

Companies whose products, services, and operations have a potentially significant effect on the environment must deal increasingly with the expectations of customers, neighbors, shareholders, and other stakeholders. Customers want products that reduce and minimize their own environmental exposure and costs. Communities do not want facilities nearby that can cause health concerns or environmental or aesthetic damage. Employees want to work for companies that consistently operate in a prudent manner. Shareholders, investors, and lenders understand the effect that environmental problems can have on profits and the value of assets.

In contemplating whether or not to commit to all or part of the ISO 14000 series, an organization should evaluate whether the ISO 14001 Environmental Management System provides a useful, model framework for its business; a third party review and registration is worth the cost and effort involved; and whether using the ISO 14001 EMS specification as the basis for an internal assessment and as a framework for program improvement would be appropriate.

The level of environmental risk posed by an organization's operations and products and the size and simplicity of the organization may not justify the installation of a fully integrated environmental management system. Organizations that have multiple stakeholders, operate in multiple jurisdictions, or are concerned with assuring and demonstrating environmental performance consistent with internal policies may elect to register, certify, or simply follow ISO 14001 and any of the ISO 14000 series of standards, using them in whatever ways contribute to better environmental and business performance.

ISO

The International Organization for Standardization was formed in 1957 to coordinate the development of technical and product standards that would facilitate the trade and use of products on a worldwide basis. The American National Standards Institute (ANSI) represents the United States in ISO.

ISO Standards are developed with input from interested parties (government, industry and others) through a system of committees and open procedures for reviewing and commenting on draft standards and for voting on a final consensus standard. The development process involves a considerable amount of give and take and negotiation at all levels and it may take years to achieve a consensus on a final standard document. All ISO standards are voluntary but are sometimes adopted by reference in the laws and regulations of some jurisdictions and in procurement specifications of both public and private institutions. In fact, many ISO standards become "mandatory" for enterprises, especially those doing business internationally.

The 14000 Series of Standards

"ISO 14000" refers to a series of generic standards that have been or are being developed by ISO to provide businesses with a structure for managing environmental impact. The standards consist of two major types. Organization or process standards

Table 1. ISO 14000 Standards

Number	Title
Organization or Process Standards	
14001[a]	Environmental management systems[b]
14004[a]	Environmental management systems-General guidelines on principles, systems, and supporting techniques
14010[a]	Guidelines for environmental auditing-General principles of environmental auditing
14011[a]	Guidelines for environmental auditing-Audit procedures, auditing of environmental management systems
14012[a]	Guidelines for environmental auditing-Qualification criteria for environmental auditors
14013[a]	Management of environmental audit programs
14014	Initial review
14015	Environmental site assessments
14031	Evaluation of environmental performance
Product-Oriented Standards	
14020	Environmental labeling-General principles
14021	Terms and definitions for self-declaration environmental claims
14022	Environmental labeling-Symbols
14023	Environmental labeling-Testing and verification methodologies
14024	Environmental labeling-Guiding principles, practices, and criteria for multiple criteria-based practitioner programs (Type I)-Guide for certification procedures
14040	Environmental management-Life cycle assessment-Principles and guidelines
14042	Environmental management-Life cycle assessment-Goals and definitions/Scope and inventory analysis
14042	Environmetal management-Life cycle assessment-Life cycle impact assessment
14043	Environmental management-Life cycle assessment-Interpretation
14050	Terms and definitions-Guide on principles for ISO/TC 207/SC6 terminology
14060	Guide for inclusion of environmental aspects in product standards

[a] Adopted as final as of this writing.
[b] The only specification standard in the series.

focus on operations and services, and include environmental management systems (EMS), environmental auditing (EA), environmental performance evaluation (EPE), and environmental aspects in product standards. The other type of standard deals with products, including life cycle assessment, and labeling. The EMS standards have evolved much further than the product standards as of this writing, with five adopted in their final form. The product standards are still in the development process.

ISO standards themselves are of two types: Guidance and specification. Guidance standards are descriptive documents. Specification standards, not surprisingly, *specify;* that is, they are prescriptive in nature; conformance can be verified by an appropriately trained entity, often an independent third party. All the 14000 series of standards are guidance standards except for ISO 14001.

The full series of ISO 14000 standards are listed in Table 1.

The Development Process of the ISO 14000 Series of Standards

ISO established the Strategic Advisory Group on the Environment (SAGE) in 1991 to recommend international standards for environmental management. Shortly thereafter, the United Nations Conference on the Environment and Development (UNCED) was held in Rio de Janeiro in 1992, at which members concluded that there was a need to stimulate an improvement in environmental management worldwide. UNCED asked ISO to develop appropriate environmental management standards for applicability worldwide.

ISO's SAGE studied various national EMS standards, such as the United Kingdom's British Standard 7750 (BS 7750) (4). SAGE evolved into a Technical Committee (TC), designated number 207 (ISO TC 207) in June 1993, to develop a series of environmental management system standards. Since the concept of management systems embodied in the ISO 9000 series of standards on quality management systems (5), which assures consistent quality of products moving in international markets, was already familiar to many organizations worldwide, an approach consistent with the ISO 9000 system for quality management was encouraged. ISO 9000 became the template for writing the ISO 14000 series.

TC 207, following ISO procedures, established an organization of subcommittees and working groups with international representation to develop different portions of the standard. The technical committee created six subcommittees (SCs) and one less formal working group (WG) to deal with different areas, as follows:

SC1 Environmental Management Systems

SC2 Environmental Auditing and Related Environmental Investigations

SC3 Environmental Labeling

SC4 Environmental Performance Evaluation

SC5 Life-Cycle Assessment

SC6 Terms and Definitions

WG1 Environmental Aspects in Product Standards

ANSI, the official U.S. representative to ISO, developed its own structure for American entities to participate in the process of developing the ISO 14000 series of standards. ANSI organized an overall U.S. Technical Advisory Group (US TAG) responsible for input into evolving ISO documents and standards. Within the overall US TAG are several Sub-Technical Advisory Groups (SubTAGs or STs); sub-working groups (SWGs), and task groups (TGs). Although participation in the TAG process is open to all, the TAG administrator has discretion to set reasonable limits on the size of TAGs and SubTAGs. There are over 500 members of the U.S. TAG, with representatives from government organizations (both regulators such as the U.S. EPA and regulated entities such as the

Departments of Defense and Energy), industry, consultants, standardization bodies, and public interest environmental groups. Each of the TAG entities has its own area of interest corresponding to the international TC 207 subcommittees and working groups. The SubTAGs and other groups develop consensus on various issues, and provide recommendations or concurrence only when broad consensus is achieved. As of this writing, six of the ISO 14000 standards have been approved; the remainder are in various states of discussion, comment, or drafting within their SubTAGs, working groups, or appropriate entity.

CONCEPTS AND DEFINITIONS

ISO: Relationship Between 14000 and 9000 Series Standards

Because many businesses have invested heavily in seeking and acquiring certification of quality management systems under ISO 9000, a good first step in understanding ISO 14000 is to understand the similarities and differences between the two related series of standards. Likewise, the process of integrating a comprehensive EMS into an overall organizational culture is similar to the integration of quality management.

Key similarities between ISO 9000 and ISO 14000 include their structure: Both series contain at least one registration standard; ISO 9000 has three and ISO 14000 has one (14001, for Environmental Management Systems) (1). Both also contain standards that are advisory or guidance documents. The two series also share a common emphasis on a strong management style, requiring management commitment to the program, clearly stated policies and objectives, well-defined assignments of responsibility to those in positions to affect program results, and employee training on what is expected and how those expectations are to be met. Additionally, continual improvement is an implicit requirement of ISO 9000 and an explicit requirement of ISO 14000. Both series require measuring activities to gauge performance, management and audit reviews and documentation that all program and system requirements are in place.

Neither the ISO 9000 nor the ISO 14000 standards specifies *what* the performance results should be—only that they be monitored and recorded so that an outside party can observe and report that the system is working.

There are many differences as well as similarities between the two series of standards, with the net result that 14000 may be more difficult to implement than the 9000 series.

First, the subject matter of the two standards is, of course, different. ISO 9000 addresses systems to consistently produce a quality product or service. ISO 14000 focuses on what it takes to make environmental management an integral part of general management.

Second, the nature, goals, and variability of the intended audiences differ. The purpose of implementing ISO 9000 is to demonstrate to customers that the registrant is a quality producer, and to inspire confidence in the customers that the organization is continually focused on improving the quality of their product. The intended audience for ISO 9000 is very focused—customers or potential customers. The objectives of all target audiences is ostensibly the same: the confidence in purchasing a quality product. The audience for an ISO 14000 registration, on the other hand, is much broader and diffuse—and includes entities with different objectives. Interested entities include customers, governments, local communities, employees, shareholders, lenders, and investors; all those who have a concern about the impact of the organization's operations on the environment and whose concerns are important to the organization. Within this spectrum of stakeholders is a wide variety of levels of awareness or sophistication and expectations.

Third, the "boundary" of the entity seeking conformity or accreditation with the two standards is often different. While ISO 9000 registration can be obtained by an entire organization or division, it can also be achieved by an individual facility because, typically, all the relevant functions related to the production of specific goods or services are located at the facility. The ISO 14000 standard has been written to allow similar flexibility in designating the entity's "boundary." Except in the case of a very small organization or a discrete operation, however, a number of the functions involved in making environmental management decisions that could affect the environmental performance of an operating unit and its product are located in other organizational units, often off-site: executive management; sales and marketing; production; purchasing; or research and development. While the ISO 14000 series only applies to "those environmental aspects which the organization can control and over which it can be expected to have an influence," larger and more complex enterprises may involve more thought and documentation for ISO 14000 than for ISO 9000.

Fourth, the applicability and legal interface for the two series of standards differs. The impetus for the development of ISO 9000 was the desire to improve and maintain the consistency of the quality of products within market regions and internationally. The primary driving force was to meet customer expectations. Companies became interested in ISO 9000 because of a desire to solidify, protect, and expand market position. Environmental management, on the other hand, has been driven for the past 25 years by a proliferation of laws and regulations around the world, with widely varying philosophies and issues of concern. Environmental laws and regulations range from strict "command and control" regulations in the United States, which allow for onerous fines and penalties—and even imprisonment, to more cooperative and flexible systems in other countries, to virtually no law or regulations in many developing countries. As companies became more global, some consistent mechanism was desired. ISO 14000 addresses many areas where this proliferation of legal and regulatory requirements still applies; it does not supersede any of them. The issue of environmental management systems is of interest of companies operating facilities and to the public environmental groups interested in monitoring their impact; product standards are of interest to all who purchase products or services.

ISO 14000—Key Questions

An ISO 14000 compliance program makes environmental management an integral part of the strategic thinking of senior management and a part of the day-to-day responsibilities of people in production, marketing, research, and a number of staff functions. Environmental management in the past has been reactive and advisory. A new law has been passed—what does it mean? An accident has occurred—what do we do? We have this waste product—where do we put it?

One key objective of an ISO 14000 program is prevention of pollution, as it is defined most broadly. This objective is met by making changes in the design, production, packaging, sales, and delivery of products and services that result in reduced environmental impact and increased market share. It is met by making changes in operations to reduce environmental impact, often affecting cost and yield. An organization's environmental managers and technicians are not in a position to make these kinds of decisions, nor, typically, able to advise on how they should be carried out. ISO 14000 makes those who have the ability and the authority to make these decisions accountable for making them and for achieving measurable results.

A decision to bring an organization into compliance with ISO 14000 is one to be made by senior management. Its implications need to be understood by virtually the entire organization. To make this decision and to prepare everyone who will be involved in program implementation, answers are needed to the following questions:

Will the organization implement an EMS consistent with ISO 14004 (2)? What is the breadth and scope of such a system?

What will the benefits of an ISO EMS be?

What are the expected or possible effects of the EMS on business or operations?

What are environmental legal requirements? Are there potential conflicts with an ISO EMS?

Are there drawbacks to an EMS? Are there trade secrets within the organization that require special consideration?

What costs can be expected to be associated with the EMS? What are the timing of those costs, and who in the organization will fund them?

What benefits and costs are there associated with demonstrating the EMS via ISO 14001 (1)? If ISO 14001 is to be followed, will the organization register via ISO-trained and certified personnel, or self-declare, or simply follow this pattern for its own EMS?

Are there other elements of ISO 14000 series that provide value to the organization?

What definition of "organization" makes the most sense?

ENVIRONMENTAL MANAGEMENT STANDARDS

ISO 14001—Environmental Management Systems

The requirements of the ISO 14001 environmental management system (EMS) specification are outlined in this section (1). This is the only specification standard in the entire 14000 series; this means that an organization's system can be compared to this one and certified, either by an independent, qualified certification person or entity (see ISO 14012 (6)), or by "self-declaration."

The critical nature of the decision on what constitutes the "organization" cannot be overstated. This definition is entirely up to the entity implementing 14000, and can be any of a number of options, including:

- an individual operating unit at a single location;
- a particular process or product train within a larger unit;
- all operations within a specified footprint (either an entire facility, or a specified portion of a facility);
- a group of operating units with similar operations at a number of locations;
- all operations or services within a particular country;
- a business unit including marketing, engineering, research and development, and staff functions that have some responsibility for the environmental performance of products and facilities;
- the organization as a whole including all the operating units of all the businesses and all the corporate functions that contribute to the organization program.

The ISO 14001 specification accepts any of these definitions—or, indeed, any others—and allows an enterprise to determine the scope of the program that the EMS is going to address.

The EMS specification includes five basic principles, each of which has its own requirements. The requirements are:

1. Commitment and policy: "An organization should define its environmental policy and ensure commitment to its EMS" (7).
2. Planning: "An organization should formulate a plan to fulfill its environmental policy" (8).
3. Implementation: "For effective implementation an organization should develop the capabilities and support mechanisms necessary to achieve its environmental policy, objectives and targets" (9).
4. Checking and corrective action: "An organization should measure, monitor and evaluate its environmental performance" (10).
5. Management review: "An organization should review and continually improve its environmental management system, with the objective of improving its overall environmental performance" (11).

The five principles underlying this standard include the requirement for top level support and the familiar "plan, do, act, check" mantra commonly associated with quality programs.

ISO 14004—Environmental Management Systems—General Guidelines on Principles, Systems, and Supporting Techniques

ISO 14004 is a guidance document for environmental management systems (2). It is a voluntary, internal management tool for organizations that are interested in developing an EMS, or improving an EMS to ISO standards. The guidelines are applicable to any organization, regardless of size, type, or level of maturity. ISO 14004 is not intended to be used as EMS certification or registration criteria, although it provides guidance on many areas subject to certification or registration using ISO 14001.

Definitions. ISO 14004 define many terms, including:

- Environment: The surroundings in which an organization operates, including air, water, land, natural resources, flora, fauna, humans, and their interrelation

- Environmental aspect: The element of an organization's activities, products, or services that can interact with the environment
- Environmental impact: Any change to the environment, whether adverse or beneficial, wholly or partially resulting from an organization's activities, products, or services
- Environmental management system: That part of the overall management system including organizational structure, planning, responsibilities, practices, procedures, processes, and resources for developing, implementing, achieving, reviewing, and maintaining the environmental policy.
- Environmental policy: A statement by the organization of its intentions and principles in relation to its overall environmental performance that provides a framework for action and for the setting of its environmental objectives and targets
- Interested party: An individual or group concerned or affected by the environmental performance of an organization
- Organization: A company, firm, enterprise, authority or institution, or portion thereof, whether incorporated or not, public or private, that has its own functions and administration

Definitions are also provided for continual improvement, environmental objective, environmental performance, environmental target, and prevention of pollution (12).

EMS Principles and Elements. An EMS is a framework by which an entire organization—not just an environmental staff—performs, in terms of how their activities affect the environment. An EMS is a framework that includes "what, who, how, when, where" in relation to environmental management. ISO states that "every individual in an organization should accept responsibility for environmental improvement" (13). To illustrate, consider the analogy between quality and environmental issues. Quality cannot be the function of any one person; it is not any one operation or group; and it is not represented by any one product or service an organization provides. To provide a product or service that is of high quality, every person must be involved. The same holds true for environmental issues. And just as the transition to a quality-oriented organization generally takes several years, implementing a comprehensive EMS is complex and time-consuming.

An EMS has five basic principles and elements:

- Commitment and policy
- Planning
- Implementation
- Measurement and evaluation
- Review and improvement

ENVIRONMENTAL AUDITING STANDARDS

ISO 14010—Environmental Auditing Standards

In its simplest form, an "audit" involves comparing a given entity or situation to a clearly identified "standard," looking for gaps between the two, and communicating the result of the audit to the client. Environmental auditing has been done in the United States since at least the late 1970s. In its infancy, environmental auditing was used to monitor compliance with environmental regulations. As regulations, risks, operations, the nature of the workforce, business arrangements, and public expectations changed, so, too, did environmental auditing. The ISO 14000 standards include three that focus on environmental auditing: general principles (3), audit procedures (14), and qualification criteria for environmental auditors (6). All are guidance documents, not certifiable standards. Each of the three major standards are discussed in this article.

Definitions and Requirements. Definitions provided as part of ISO 14010 (15) include the following:

Audit conclusion: Professional judgment or opinion expressed by an auditor about the subject matter of the audit, based on and limited to reasoning the auditor has applied to audit findings

Audit criteria: Policies, practices, procedures, or requirements against which the auditor compares collected audit evidence about the subject matter.

Audit findings: Results of the evaluation of the collected audit evidence compared with the agreed audit criteria

Client: Organization commissioning the audit

Environmental audit (EA): Systematic, documented verification process of objectively obtaining and evaluating audit evidence to determine whether specified environmental activities, events, conditions, management systems, or information about these matters conform to audit criteria, and communicating the results of this process to the client

Other definitions provided are audit evidence, audit team, environmental auditor, lead environmental auditor, organization, subject matter, and technical expert (16).

ISO 14010 requires that an environmental audit should focus on clearly defined and documented subject matter. Furthermore, the audit should be undertaken only if, after consultation with the client, the lead auditor believes that

- there is sufficient and appropriate information about the subject matter of the audit;
- there are adequate resources to support the audit process; and
- there is adequate cooperation from the facilities.

An audit is an element of an EMS. If these ingredients are not met to the lead auditor's satisfaction, and the lead auditor decides that the EA cannot be conducted to the anticipated standards, the lead auditor is advised to consider the needs of the client before aborting the audit. Environmental audits are often done for third parties, such as prospective customers, vendors, insurance underwriters, prospective purchasers, or financial institutions. There are usually business needs that deny the lead environmental auditor the luxury of time (or authority) to postpone an environmental audit. In these cases, the deficiencies, and their consequences, should be communicated to the client as soon as practicable, with changes agreed to in scope, resources, timing, or other expectations.

Communication is a key element of environmental auditing. With environmental issues sensitive for many persons involved—and with the United States' regulatory scheme including onerous provisions for fines, penalties, and criminal prosecution—the nature, substance, and distribution of communication in written or verbal form must be contemplated and managed appropriately.

General Principles. ISO 14010 includes seven general principles for environmental audits (17):

- Objectives and scope
- Objectivity, independence, and competence
- Due professional care
- Systematic procedures
- Audit criteria, evidence, and findings
- Reliability of audit findings and conclusions
- Audit report

ISO 14011—Guidelines for Environmental Auditing—Audit Procedures—Auditing of Environmental Management Systems

ISO 14011 establishes procedures to plan and conduct an audit of an EMS (14). It defines an EMS (as does 14001), an EMS audit, and EMS audit criteria. An EMS audit is simply a specific type of environmental audit. Much of the material provided in this guideline is more detailed guidance for implementing ISO 14010 for this type of environmental audit.

EMS Audit Objectives, Roles, and Responsibilities. Like any other environmental audit, an EMS audit should have defined objectives. Once the objectives are defined, the scope should be determined, with the appropriate confidence level conveyed to the client. Five entities associated with the EMS audit should delineate their roles and responsibilities in the process: the lead auditor, auditor, audit team, client, and auditee.

ISO 14012-Guidelines for Environmental Auditing—Qualification Criteria for Environmental Auditors

ISO 14012 provides guidance on qualification criteria for environmental auditors, and special criteria for lead auditors (6). Criteria for the selection of an audit team is not provided, but is left to lead auditors for the individual audits. ISO 14012 defines environmental auditor, lead environmental auditor, degree, and secondary education.

Educational and Training. ISO 14012 suggests that auditors should have completed at least secondary education or equivalent. Auditors should have four to five years of work experience, based upon educational level, in at least some of the following areas:

- Environmental science and technology
- Technical and environmental aspects of facility operations
- Relevant requirements of environmental laws, regulations, and related documents
- Environmental management systems and standards against which audits may be conducted
- Audit procedures, processes and techniques

Auditors' on-the-job training should include involvement in the entire audit process under qualified supervision for at least 20 equivalent work days over no more than three years. Competence should be also be demonstrated by "suitable means," which may include internal or external evaluation via interviews, written and/or oral assessment, review of written work, role playing, or peer observation.

Personal Attributes and Skills. ISO 14012 suggests that auditors should possess personal attributes and skills that include:

- competence in oral and written communications;
- interpersonal skills, including listening, diplomacy, and tact;
- ability to maintain independence and objectivity;
- organizational skills;
- ability to exercise logic and reason in developing conclusions;
- ability to adapt to local customs, cultures, and (if necessary) languages.

The auditor, after all, will be finding things that will probably include problems at the auditee's facility or organization. In many instances, the findings will be taken personally.

Lead Auditor. The lead auditor should have relevant experience and should be thoroughly capable of using personal attributes and skills necessary for effective and efficient leadership and management of all aspects of the audit throughout the auditing process. The lead auditor should have 35 equivalent work days of experience, including participation as a lead auditor under qualified supervision.

The keys to a successful environmental audit, whether an EMS audit or any other kind, may be the lead auditor's experience with auditing, experience with operations, understanding of management systems, and organizational skills. When audits are limited by factors beyond the lead auditor's control (time, budget, access to key areas), the lead auditor must act to make the most of the resources or situations available, and to convey the value and limitations of the audit to the client. Often, the key to a successful environmental audit is the proper recognition of non-verbal cues, hidden client or auditee objectives, diplomacy, and trust. When dealing with confidential information, as is the case with most environmental audits, the importance of trust and other personal attributes is critical.

Maintenance of Competence. ISO 14012 suggests that auditors should maintain their competency in the areas of their relevant expertise, including: environmental science and technology; facility operations; environmental laws and regulations; environmental management systems; and auditing techniques. It also suggests that auditors exercise due professional care, and avoid participating unsupported in audits where they do not have proficiency in the language commonly used at the auditee's entity.

OTHER ISO STANDARDS

Labeling and Product Standards

ISO standards for labeling are being developed at the time of this writing. The standards and their intended purpose are shown in Table 2.

Table 2. ISO 14000 Labeling Standards

Number	Title	Purpose
14020	Environmental labeling-General principles	To draft all labeling standards, and to guide development of specific environmental claims.
14021[a]	Terms and definitions for self-declaration environmental claims	General guidelines for environmental claims of goods and services; to harmonize currently conflicting or misleading environmental claims.
14022	Environmental labeling-Symbols	To standarize symbols denoting recyclability and recycled content.
14023	Environnmetal labeling-Testing and verification methodologies	To standarize methods for parity.
14024	Environmental labeling-Guiding principles, practices, and criteria for multiple criteria-based practitioner programs (Type I)-Guide for certification procedures	To provide criteria for practitioners to use in evaluating products and awarding environmental labels to products and services.

[a] Adoption as draft international standard will probably be after adoption of ISO 14022 and ISO 14023.

Table 3. ISO 14000 Standards for Life Cycle Assessment

Number	Title	Purpose
14040	Environmental management-Life cycle assessment-Principles and guidelines	To encourage public policy makers and other organizations to expand the current concept of life-cycle analysis.
14041	Environmental management-Life cycle assessment-Goals and definitions/Scope and inventory analysis	To help formulate the goal and scope of a life-cycle assessment and inventory analysis
14042	Environmental management-Life cycle assessment-Life cycle impact assessment	May propose three categories for consideration in impact assessment component of life-cycle analysis: classification, characterization, and valuation.
14043	Environmental management-Life cycle assessment-Interpretation	Interpret various criteria associated with life cycle assessment.
14060	Guide for inclusion of environmental aspects in product standards	General considerations in developing product standards

Environmental Performance Standards

ISO standards for environmental performance include one to evaluate the performance of the overall EMS management and others that relate to environmental performance of products and services. The ISO 14031 standard is designated as a guideline for environmental performance evaluation (EPE) (18). It is in the process of consensus building, and is not a standard at the time of this writing. Key issues requiring clarification and development of consensus have included

- definitions;
- whether references to life-cycle assessment was appropriate;
- whether EPE should be created as a stand-alone document independent of an EMS, or whether it should be integrated into a organization's EMS;
- how information should flow within an organization;
- how entities within an organization should use the information; and
- the degree to which EPE is an ongoing activity.

The standard include sections addressing scope, references, definitions, overview, planning, evaluation, and improvement of the EPE. Elements of planning include management considerations; environmental aspects; and selection of environmental indicators for management, operations, and the environment. Elements of evaluation include collecting data, analyzing data, evaluating information, and reporting and communicating of information.

At least nine informative "Annexes" (appendices) are included, providing guidance or information on data, environmental indicators, and reference reports.

Intended ISO standards for life cycle assessment and their general intent include the ISO 14040 series and ISO 14060, as shown in Table 3. ISO 14050 (19) provides a list of all definitions in ISO 14000 EMS and auditing standards, with a guide on how they are to be applied.

OTHER SYSTEMS AND GROUPS

The ISO 14000 series is not the only EMS designed for broad application. Organizations may be involved in countries or organizations with other EMSs that may require consideration or implementation. As ISO 14000 is contemplated, these other systems should be contemplated as well, since there are similarities and differences among them.

British Standard 7750, Environmental Management Systems (BS 7750), is the national environmental standard for the United Kingdom (4). It was first adopted as a final standard in March 1992, and was revised in January 1994. The 1994 revision is similar to ISO 14001. BS 7750 requires that a company validate its compliance with its stated environmental policy and demonstrate that compliance to others. Many of the requirements in BS 7750 reflect those in ISO 14001, including development and implementation of an environmental plan, assignment of responsibilities, and continuous improvement. BS7750 also requires personal environmental targets for individual employees.

The Chemical Manufacturers Association (CMA) developed comprehensive guidelines for environmental management systems called Responsible Care (20). Conformance to Responsible Care by individual companies is obligatory for membership in the CMA. Responsible Care, adopted by the CMA in 1988, based on six codes of management practices:

- Product stewardship
- Community awareness and emergency response
- Process safety
- Employee health and safety
- Pollution prevention
- Distribution

Responsible Care, like the ISO 14000 series, focuses on management systems. ISO 14001 is broader because it requires organizations to address all environmental aspects of operations, not just the products they produce. Responsible Care does not require EMS audits and review of audits, nor does it impose the volume of recordkeeping requirements that ISO 14001 does.

The European Union has also worked on broadly applicable EMSs. The Eco-Management and Audit Regulation (EMAR) contains the European Eco-Management and Audit Scheme (EMAS), which establishes specifications for environmental management systems. EMAS is site-specific and voluntary. There are many differences between EMAR/EMAS and ISO. EMAR is a legislative instrument, rather than a voluntary system like ISO. EMAR applies to sites only, not to any definition of an "organization," as ISO allows. EMAS specifically requires public environmental statements and an annual statement including factual data, unlike ISO.

BIBLIOGRAPHY

1. *Environmental Management Systems,* ISO Standard 14001, International Standards Organization, Geneva, Switzerland, 1996.
2. *Environmental Management Systems—General Guidelines on Principles, Systems, and Supporting Techniques,* ISO Standard 14004, International Standards Organization, Geneva, Switzerland, 1996.
3. *Guidelines for Environmental Auditing—General Principles of Environmental Auditing,* ISO Standard 14010, International Standards Organization, Geneva, Switzerland, 1996.
4. *Environmental Management Systems,* BS 7750, British Standards, United Kingdom, 1994.
5. *International Standards on Quality Management,* 6th ed., ISO Standard 9000, International Standards Organization, Geneva, Switzerland, 1996.
6. *Guidelines for Environmental Auditing—Qualification Criteria for Environmental Auditors,* ISO Standard 14012, International Standards Organization, Geneva, Switzerland, 1996.
7. *Environmental Management Systems,* ISO Standard 14001, International Standards Organization, Geneva, Switzerland, 1996, Section 4.1.
8. *Environmental Management Systems,* ISO Standard 14001, International Standards Organization, Geneva, Switzerland, 1996, Section 4.2.
9. *Environmental Management Systems,* ISO Standard 14001, International Standards Organization, Geneva, Switzerland, 1996, Section 4.3.
10. *Environmental Management Systems,* ISO Standard 14001, International Standards Organization, Geneva, Switzerland, 1996, Section 4.4.
11. *Environmental Management Systems,* ISO Standard 14001, International Standards Organization, Geneva, Switzerland, 1996, Section 4.5.
12. ISO Standard 14004, Sections 3.2–3.12, 1996.
13. *Environmental Management Systems—General Guidelines on Principles, Systems, and Supporting Techniques,* ISO Standard 14004, International Standards Organization, Geneva, Switzerland, 1996, Section 4.0.
14. *Guidelines for Environmental Auditing—Audit Procedures, Auditing of Environmental Management Systems,* ISO Standard 14011, International Standards Organization, Geneva, Switzerland, 1996.
15. *Guidelines for Environmental Auditing—General Principles of Environmental Auditing,* ISO Standard 14010, International Standards Organization, Geneva, Switzerland, 1996, Sections 2.1–2.9.
16. *Environmental Management Systems—General Guidelines on Principles, Systems, and Supporting Techniques,* ISO Standard 14004, International Standards Organization, Geneva, Switzerland, 1996, Section 4.3.3.3.
17. *Guidelines for Environmental Auditing—General Principles of Environmental Auditing,* ISO Standard 14010, International Standards Organization, Geneva, Switzerland, 1996, Section 4.
18. *Draft Evaluation of Environmental Performance,* ISO Standard 14031, International Standards Organization, Geneva, Switzerland, 1996.
19. *Draft Terms and Definitions—Guide on Principles for ISO/TC 207/SC6 Terminology,* ISO Standards 14050, International Standards Organization, Geneva, Switzerland, 1996.
20. *Responsible Care,* Chemical Manufacturers Association, Washington, D.C., 1988.

L

LABORATORY HAZARDOUS WASTE DESTRUCTION

GEORGE LUNN
Consultant
Baltimore, Maryland

A wide variety of hazardous chemicals are used in research laboratories, and there is no single process which can be used to degrade all of them. However, a number of validated techniques are available for degrading many of these compounds in the laboratory where they are used. These techniques completely degrade the hazardous compounds without producing hazardous products or mutagenic reaction mixtures and may conveniently be grouped under the following headings: reductive procedures (primarily nickel–aluminum alloy in dilute base or sodium metabisulfite), oxidative procedures (primarily potassium permanganate or sodium hypochlorite), hydrolytic procedures (involving neutral, acidic, or basic solutions), and miscellaneous procedures (eg, photolytic procedures, slow and controlled addition of alkali metals to alcohols, etc). In addition, decontamination procedures, generally involving ion-exchange or polymeric adsorption resins, exist in which the hazardous compounds are not degraded but the volume of waste is greatly decreased, easing disposal problems.

The safe disposal of hazardous chemicals in research laboratories presents a variety of problems to researchers. Unlike the disposal of hazardous biological agents, there is no single process (eg, autoclaving) which can be used to degrade hazardous chemical agents. Laboratory chemical waste presents unique problems. In contrast to industrial plants which have to deal with only a few hazardous compounds with well understood properties, the typical research laboratory contains many different compounds whose properties may be poorly understood. Although these compounds may be present only in small amounts, they may be in a form which makes them particularly difficult to manage. For example, 5 mg of dimethylnitrosamine in 1 liter of water constitutes 1 liter, not 5 mg, of hazardous material. It is also difficult to incinerate large quantities of aqueous solution. Although it will, probably, always be necessary to ship out some waste for disposal, there are a number of steps that researchers can take to mitigate the problem.

Over the last few years a number of papers have been published detailing practical procedures that can be used for degrading hazardous compounds in the laboratory where they are used. Original papers or books (1), containing details of all of these procedures should be consulted.

PROPERTIES OF AN IDEAL DESTRUCTION TECHNIQUE

An ideal destruction technique has a number of features which may be summarized as follows:

The hazardous compound is completely destroyed.

The products of the reaction are innocuous materials and the material balance should be verified. Additionally, the final reaction mixture is not biologically hazardous. Mutagenicity testing (Ames test) is frequently used to ascertain this.

The absence of the starting material and hazardous byproducts are readily verifiable using commonly available analytical techniques.

The equipment and reagents are commonly available, and the procedure is rapid, reliable, and easy to carry out.

The method is of practical use in the laboratory. Thus it can degrade the compound in bulk quantities or dissolved in various solvents. For example, surplus chemicals may come in the form of aged analytical samples, unused contaminated drinking water for animal feeding experiments, or expired pharmaceutical preparations as well as unneeded bulk quantities. Therefore, any procedure must be able to degrade the chemical in a variety of matrices. In addition, the procedure should be able to treat spills.

These criteria may seem obvious but procedures have been proposed in which no effort was made to see if the compound was actually degraded or what the products of the reaction might be. These criteria represent an ideal seldom achieved but toward which one should always strive.

SAFETY CONSIDERATIONS

All reactions involving hazardous chemicals should be carried out with the utmost regard to safety. It is not possible to detail here all the necessary safety precautions, but appropriate protective clothing should always be worn, and all reactions should be carried out in a properly functioning chemical fume hood. Hazardous chemicals should be handled only by workers who have received proper training, and glassware and equipment should be thoroughly decontaminated in the laboratory before being sent out for washing. Numerous texts have been written on the subject of laboratory safety, and these should be consulted before commencing work. It is important to understand the properties of the chemicals being handled. Extensive listings of the known hazards can be found in a number of texts (2,3). Although Material Safety Data Sheets must be made available to laboratory workers, the quality of these documents varies widely, and thus, they are of limited use. Extra hazards may be present when some of these procedures are carried out. For example, reduction with nickel–aluminum alloy involves the generation of flammable hydrogen gas, and oxidations with potassium permanganate involve working with a strong oxidant.

Generally, procedures should not be followed blindly but should be adapted to fit local needs. Due consideration should be paid to the likely products of the reaction and their ultimate disposal. All applicable regulations should be followed. In addition, the various destruction procedures described are not universally applicable. Thus a procedure which works well for one compound may fail or even make matters much worse when applied to another compound. For example, sodium hypochlorite degrades aflatoxins but is unsatisfactory when applied to doxorubicin or polycyclic aromatic hydrocarbons. Destruction procedures should always be thoroughly validated before being applied to a new compound.

If it appears that one of these procedures could be useful, the original references must be consulted before commencing work. Many factors influence the way a degradation procedure is to be carried out. In particular, the matrix in which a hazardous compound is dissolved can have a great influence on how the destruction procedure is carried out. Additionally, different members of the same class of compounds can require different procedures. A number of the procedures are very similar. Nevertheless, small differences do exist and will have important effects on the outcome of the reactions. For these reasons, it cannot be emphasized too strongly that the original literature must be consulted.

VALIDATION PROCEDURES

Almost all of the procedures described have been validated by analytical techniques to ensure that the compounds were completely degraded and by mutagenic testing to ensure that mutagens were not generated in the course of the reaction. In general, the analytical techniques were high-performance liquid chromatography (HPLC) or gas chromatography (GC), although other techniques, such as fluorescence spectroscopy and colorimetric procedures, have been used on occasion. Whenever a new degradation technique is put into routine use, it must be thoroughly validated. Again, the original literature should be consulted for full descriptions of these techniques.

For HPLC a dual-pump, computer-controlled, gradient-capable solvent delivery system was used with ultraviolet (UV) detection using a variable-wavelength, diode array detector. The injection volume was 20 μL and the flow rate was 1 mL/min. A 250 × 4.6-mm id column of Microsorb 5 μm C-8 fitted with a 15 × 4.6-mm guard column of the same material (Rainin Instrument Company) was used, but any equivalent product is suitable. For GC, a 1.8-m × 2-mm id × 0.25 in. od packed, silanized-glass column was used. Various packing materials were used depending on the nature of the analytes. The column was fitted with a guard column packed with the same material. The injector temperature was 200°C, the flame ionization detector, 300°C. The carrier gas was nitrogen flowing at 30 mL/min. Injection was by syringe and sample volumes were in the range of 1–5 μL. Although capillary GC has greater sensitivity and resolution, packed-column GC is much more robust, and we have found that aqueous reaction mixtures can be injected directly. The guard column should be changed periodically (daily or weekly depending on the nature of the reaction mixtures). An electronic integrator was used for each instrument to determine peak areas automatically.

In some cases, it was found that injecting unneutralized reaction mixtures onto the hot GC column degraded the material for which we were analyzing. Thus, degradation during the reaction might have been incomplete, but the peak for the original compound was not observed in the chromatogram because the compound was degraded in the GC column. Spiking experiments can be used to determine if this is a problem. In a spiking experiment, a small amount of the original compound is added to the final reaction mixture and this spiked mixture is analyzed. If an appropriate peak is observed, compound degradation on the GC column is not a problem. If an appropriate peak is not observed, it may be necessary to neutralize the reaction mixture before analysis and/or use a different GC column. Similar problems may be encountered when using HPLC because of the formation of salts or the influence of the sample solvent; again, spiking experiments should be employed. Spiking experiments should be used routinely to test the efficacy of the analytical techniques.

Mutagenicity testing is a very useful technique for assessing the toxicity of final reaction mixtures. In a number of cases, it was found that, although the hazardous compound was completely degraded, the reaction mixture was highly mutagenic. Thus, the destruction procedure could not be recommended. Most of the procedures described below were tested for mutagenic products. Mutagenicity testing is probably not suitable for routine use but testing during the development period is helpful. The residues produced by the destruction methods were tested for mutagenicity using the plate incorporation technique of the *Salmonella*/mammalian microsome assay essentially as recommended (4) with modifications (5). Some or all of the tester strains (TA98, TA100, TA1530, TA1535, TA1537, and TA1538) of *Salmonella typhimurium* were used with and without S9 rat liver microsomal activation. The reaction mixtures were neutralized before testing. In general, basic reaction mixtures were neutralized by adding acetic acid. Acidic reaction mixtures were neutralized by adding solid sodium bicarbonate. Reaction mixtures containing potassium permanganate were decolorized with sodium ascorbate before neutralization. A 100-μL aliquot of the solution (corresponding to varying amounts of undegraded material) was used per plate. Pure compounds were generally tested at a level of 1 mg per plate in dimethyl sulfoxide (DMSO) or aqueous solution. One hundred μL of these solutions were added to each plate. The criterion for significant mutagenicity was set at more than twice the level of the control value. The control value was the average of the cells only and cells plus solvent runs. Only procedures which produce nonmutagenic reaction mixtures can be recommended. The absence of mutagenic activity in the residual solutions, however, does not necessarily imply that they are nontoxic or have no other adverse biological or environmental effects.

More than one type of procedure may be useful for any given compound or group of compounds, but usually one procedure is most generally applicable. The various types of procedures include reductive, oxidative, and hydrolytic, as well as miscellaneous degradative and decontamination procedures.

REDUCTIVE PROCEDURES

Nitrosamines

Nitrosamines are potent carcinogens having a general form (**1**) (eq. 1) where R and R′ are alkyl, aryl, or heteroatoms. Compounds, in which R or R′ is a good leaving group, are called nitrosamides. The degradation chemistry of nitrosamides is quite different and will be dealt with separately. The lower molecular weight nitrosamines are volatile liquids. It has been shown that *N*-nitrosodimethylamine causes cancer at the part per billion level in drinking water. Nitrosamines are widely used in cancer research. Early research showed that the reduction of nitrosamines resulted in two possible products, the corresponding hydrazines (**2**) and amines (**3**) (eq. 1). Because a number of hydrazines are known carcinogens, and may also be reoxidized to nitrosamines, (6), a reduction procedure which only produces the relatively innocuous amines is clearly desirable.

$$\underset{(1)}{R(R')N\text{—}NO} \longrightarrow \underset{(2)}{R(R')N\text{—}NH_2} + \underset{(3)}{R(R')NH} \quad (1)$$

The best reducing agent is nickel–aluminum alloy, Ni–Al, in the presence of a dilute base (7). In the course of the reaction, the aluminum in the alloy reacts with the base to form aluminate ion and hydrogen. The hydrogen reduces the nitrosamine to the corresponding hydrazine. As the aluminum reacts with the base, it leaves behind finely divided nickel with hydrogen adsorbed on the surface (Raney nickel), and this material cleaves the hydrazine to the amine. At the end of the reaction, the nickel is removed from the solution by filtration through Celite, allowed to dry in a fume hood away from flammable solvents, and recycled or discarded with the solid waste. The nickel content of the filtrate is <0.5 ppm.

Hydrazines

Nickel–aluminum alloy also reduces hydrazines, many of which are carcinogenic, to the corresponding amines.

Antineoplastic Drugs

A number of nitrosourea antineoplastic drugs (eg, carmustine, chlorozotocin, lomustine, *N*-(2-chloroethyl)-*N*′-(4-methylcyclohexyl)-*N*-nitrosourea (methyl CCNU), *N*-(2–chloroethyl)-*N*′-(2,6-dioxo-3-piperidinyl)-*N*-nitrosourea (PCNU), but not streptozocin, are also degraded by nickel–aluminum alloy. In all cases, the compounds are completely degraded, and mutagenic products are not produced.

Halogenated Compounds

Nickel–aluminum alloy is a powerful reducing agent and is used to reduce halogenated compounds to the parent hydrocarbons, even when the halogen is fluorine. Halogenated compounds are highly toxic and a number of them are known carcinogens, eg, benzyl chloride, 2-bromoethanol, and 2-chloronitrobenzene. They are widely used industrially and in chemistry laboratories.

Miscellaneous Reductions with Nickel–Aluminum Alloy

Nickel–aluminum alloy is a general reagent for cleaving N–N and N–O bonds (9,10). For example, azoxymethane, a carcinogen in experimental animals used in cancer research, is reduced to methylamine (11).

OXIDATIVE PROCEDURES

Although chromic acid has been widely used to degrade hazardous compounds in the past, the chromium compounds produced in the course of these oxidations are potentially carcinogenic and may not be discharged to the environment. Potassium permanganate ($KMnO_4$) appears to have fewer hazards and has been used to degrade a variety of hazardous compounds. Most reactions involve the use of potassium permanganate in 3 *M* sulfuric acid but degradation reactions are also carried out with potassium permanganate in neutral and basic solutions. At the end of the reaction, excess oxidant is removed by adding sodium metabisulfite. Manganese is precipitated from solution by making it strongly basic (if necessary). The manganese content of the resulting filtrates is 1 ppm or less (12).

Potassium permanganate in sulfuric acid has been successfully used to degrade polycyclic aromatic hydrocarbons (13) (eg, benz[*a*]anthracene, benzo[*a*]pyrene, 7-bromomethylbenz[*a*]anthracene, dibenz[*a,h*]anthracene, 7,12-dimethylbenz[*a*]anthracene, 3-methylcholanthrene), and polycyclic heterocyclic hydrocarbons (14), eg, dibenz[*a,j*]acridine, dibenz[*a,h*]acridine, 7*H*-dibenzo[*c,g*]carbazole, and 13*H*-dibenzo[*a,i*]carbazole. These carcinogenic compounds are used in cancer research laboratories and are found in the environment as products of combustion.

HYDROLYTIC PROCEDURES

Alkylating Agents

When dealing with reactive alkylating agents hydrolysis is frequently an attractive option. For example, dimethyl sulfate, a volatile, highly toxic alkylating agent, is hydrolyzed with dilute base (eg. 1 *M* sodium hydroxide solution) to methanol and methyl hydrogen sulfate (15,16). Methyl hydrogen sulfate, which is nonmutagenic and a poor alkylating agent, subsequently is slowly hydrolyzed to methanol and sulfuric acid. Dimethyl sulfate is used as an alkylating agent in industry and in the laboratory.

DECONTAMINATION PROCEDURES

Unlike degradation procedures, decontamination procedures do not result in destroying the hazardous compound. However, by removing it from solution, the volume of hazardous waste is reduced, and so, the waste disposal problem is diminished.

For example, 200 mL of an aqueous solution containing 1000 ppm lead, a toxic metal, is decontaminated by stirring it with 1 g of Amberlite IR-120(plus) ion-exchange resin for 6 h (1). The final lead concentration is <0.45 ppm. Thus, instead of having to dispose of 200 mL of contaminated solution, it is only necessary to dispose of 1 g of contaminated resin. Of course, the lead can always be recovered from the resin and recycled.

BIBLIOGRAPHY

1. G. Lunn and E. B. Sansone, *Destruction of Hazardous Chemicals in the Laboratory,* 2nd ed., John Wiley & Sons, Inc., New York, 1994.
2. R. J. Lewis, Sr., *Sax's Dangerous Properties of Industrial Materials,* 8th ed., Van Nostrand Reinhold, New York, 1992.
3. L. Bretherick, *Bretherick's Handbook of Reactive Chemical Hazards,* 4th ed., Butterworths, London, 1990.
4. B. N. Ames, J. McCann, and E. Yamasaki, *Mutat. Res.* **31,** 347–364 (1975).
5. A. W. Andrews, L. H. Thibault, and W. Lijinsky, *Mutat. Res.* **51,** 311–318 (1978).
6. G. Lunn, E. B. Sansone, and A. W. Andrews, *Environ. Molec. Mutagen.* **17,** 59–62 (1991).
7. G. Lunn, E. B. Sansone, and L. K. Keefer, *Carcinogenesis* **4,** 315–319 (1983).

8. G. Lunn, E. B. Sansone, A. W. Andrews, and L. C. Hellwig, *J. Pharm. Sci.* **78,** 652–659 (1989).
9. G. Lunn and E. B. Sansone, *Am. Ind. Hyg. Assoc. J.* **52,** 252–257 (1991).
10. G. Lunn, E. B. Sansone, and L. K. Keefer, *Synthesis,* 1104–1108 (1985).
11. G. Lunn and E. B. Sansone, *Appl. Occup. Env. Hyg.* **6,** 1020–1026 (1991).
12. G. Lunn, E. B. Sansone, M. De Méo, M. Laget, and M. Castegnaro, *Am. Ind. Hyg. Assoc. J.* **55,** 167–171 (1994).
13. M. Castegnaro, G. Grimmer, O. Hutzinger, W. Karcher, H. Kunte, M. Lafontaine, E. B. Sansone, G. Telling, and S. P. Tucker, eds., *Laboratory Decontamination and Destruction of Carcinogens in Laboratory Wastes: Some Polycyclic Aromatic Hydrocarbons,* IARC Scientific Publications No. 49, International Agency for Research on Cancer, Lyon, France, 1983.
14. M. Castegnaro, J. Barek, J. Jacob, U. Kirso, U., M. Lafontaine, E. B. Sansone, G. M. Telling, and T. Vu Duc, eds., *Laboratory Decontamination and Destruction of Carcinogens in Laboratory Wastes: Some Polycyclic Heterocyclic Hydrocarbons,* International Agency for Research on Cancer, Lyon, France, 1991.
15. G. Lunn and E. B. Sansone, *J. Chem. Educ.* **67,** A249–A251 (1990).
16. G. Lunn and E. B. Sansone, *Am. Ind. Hyg. Assoc. J.* **46,** 111–114 (1985).

LABORATORY HAZARDOUS WASTE MANAGEMENT

RUSSELL W. PHIFER
Environmental Assets Inc.
West Chester, Pennsylvania

When the U.S. Congress passed the Resource Conservation and Recovery Act (RCRA) in 1976, the primary objective was to address the handling of hazardous wastes generated by industry. Congress charged the Environmental Protection Agency (EPA) with the responsibility of prioritizing numerous issues within the framework of the original act, and again when it was amended in 1984. The deadlines for implementing regulations resulted in enormous pressures to address the most pressing issues first. In many ways, the same conditions exist today, twenty years after passage of RCRA. The regulation of wastes generated by laboratories has never been considered a primary issue by either Congress or the EPA. Because laboratories generate less than 0.1% of all hazardous wastes and an even smaller percentage of other waste categories, there is little or no inclination by either legislators or regulators to consider special rules for laboratories. There are no large Superfund sites with a focus on laboratory wastes. There are few transportation incidents involving laboratory waste materials.

Despite the perceived lack of large problems involving laboratory wastes, laboratory facilities are regulated almost precisely the same as large industrial facilities generating huge volumes of waste. The nature of laboratory wastes, however, is vastly different from any predominant industrial category. This presents a number of ironies.

Number of Laboratory Facilities. No one knows how many laboratory facilities exist. Estimates range from 150,000 to 250,000, depending on the definition of a laboratory. This far exceeds the number of regulated facilities in any other industry group.

Number of Different Wastes Generated. Over six million chemical compounds have been formulated in the laboratory; several hundred thousand of these are used at least occasionally in research. As a result, laboratories have the potential to generate a much wider variety of waste than any other industry group.

Quantities of Individual Wastes Generated. Laboratory processes, whether they are performed routinely or once, utilize extremely small quantities of chemical products. Many processes use less than 100 milligrams of material and, in fact, it is quite possible to purchase specific chemical supply bottles that are legally RCRA-empty at the time of shipment. By contrast, large industrial facilities may generate thousands of gallons of waste a week as by-products of a single manufacturing process.

Cost Per Unit Paid for Disposal. Naturally, the cost for disposal of hazardous wastes varies according to the nature, volume, and location of the materials. When these factors are applied to laboratories, the result is waste management costs that may approach 100 times that of wastes generated by large industries.

DEFINITION AND TYPES OF LABORATORIES

Although the EPA has never felt the need to provide a definition of a laboratory, the Occupational Safety and Health Administration (OSHA) has issued regulations specific to laboratories, and defines a laboratory as follows:

> *A laboratory is a place in which (1) containers used for reactions, transfers, and other handling of substances are designed to be easily and safely manipulated by one person; (2) multiple chemicals or chemical procedures are used; and (3) protective laboratory practices and equipment are available and in common use to minimize the potential for employee exposure to hazardous chemicals. The definition excludes operations (1) in which the procedures involved are part of or in any way simulate a production process or (2) whose function is to produce commercial quantities of materials* (1).

Using this definition it is possible to identify facilities that produce laboratory wastes. Following is a list of the types of facilities that may be defined as laboratories, or may include laboratories in their operation:

1. High school and college chemistry laboratories.
2. Medical and clinical testing facilities.
3. Environmental and industrial hygiene testing laboratories.
4. Industrial research and development operations.
5. Pharmacies.
6. Doctors' and veterinarians' offices.
7. Photographic laboratories.
8. Agricultural research facilities.
9. Utilities.

It would appear improper to include such facilities as pharmacies and doctors' offices as laboratories. However, they meet the OSHA definition and at least a few of the wastes they generate may be considered hazardous under RCRA because it is the *specific hazards* of chemical materials *when discarded* that result in regulation, not *how* they are generated as wastes.

THE NATURE OF LABORATORY WASTES

Some of the previous discussion provides some initial insight into the nature of laboratory wastes, ie, small quantities of a large variety of chemicals. What has not been discussed is the unusually high percentage of chemicals being discarded as wastes which have never been used. Although no serious studies have been performed, no one in the business of managing laboratory wastes would be surprised if this percentage is at least 35%. Many individual disposal projects involve exclusively nonprocess wastes such as surplus, off-specification, and/or out-of-date laboratory chemicals. Out-of-date pharmaceutical products would also fall into this category. Although unused chemicals represent the largest single category of laboratory wastes, there are many other waste streams commonly generated by laboratories.

HOW RCRA APPLIES TO LABORATORIES

Simply put, nearly all hazardous waste regulations that apply to industry also apply to laboratories. There are, however, some intricacies associated with waste determinations, and a few exemptions that apply only to laboratories. In addition, various states and the United States EPA regions have their own hazardous waste controls that may affect laboratory operations relating to waste management.

Waste Determinations

Laboratory waste generators are obligated to evaluate their waste streams to determine if they are regulated under RCRA just as industrial waste generators do. This evaluation can be based on either knowledge of the waste's constituents or on laboratory analysis. Ironic as it may seem, laboratories rarely find it necessary to perform specific waste-classification analysis on their waste materials, whereas industrial facilities frequently have this done on an annual or semiannual basis. It is important to note that most generators can determine fairly easily if their waste is hazardous or nonhazardous through knowledge of the process and a basic knowledge of RCRA. Waste determination analysis is most often performed to meet *disposal facility* permit requirements. The generator has the right to declare a material a hazardous waste based on knowledge of the material. The cost of confirming analysis, if not required for disposal, represents an unnecessary expense. There are a number of reasons that laboratories are rarely required to have wastes analyzed:

1. A high percentage of laboratory waste consists of unused chemical reagents, where the precise constituents are known.
2. A number of common laboratory reagents are shown on the "U" and "P" hazardous waste lists. These designations apply when surplus, out-of-date, or off-specification chemicals are discarded.
3. A significant percentage of nonsolvent laboratory waste is disposed of in labpacks, where a functional grouping or general description of the waste on an inventory sheet is generally all that is required. If a complete waste approval form was required for each chemical in a labpack, appropriate documentation for a single drum could easily run to 200–800 pages. In addition, some facilities require verification of signatures by a notary. Confirmatory analysis for one labpack could easily cost in the range of $10,000–50,000, depending on the cost of specific analytical methods that might be necessary. Complying with such requirements would make the cost of laboratory disposal prohibitive.
4. Disposal facilities usually accept a list of constituents for waste consolidated in bulk, such as mixed solvents.

Thus, the quantities of waste generated and how they are prepared for disposal have a significant impact on how waste determinations are performed. Keeping careful track of what wastes are mixed together and maintaining good labeling practices for all chemicals can result in far less effort or expense in the making of a RCRA waste determination.

Exemptions

Although RCRA, its 1984 amendments, and the resultant regulations have rarely acknowledged the unique situation of laboratories in the development of environmental controls, there are a few situations where exemptions apply. Perhaps the most significant is the laboratory wastewater exclusion (2,3). This exemption states that toxic, corrosive, or ignitable waste contained in a laboratory wastewater stream is exempt from RCRA as long as the total volume of laboratory wastewater does not exceed 1% of the facility's total wastewater volume reaching the facility's headworks. Even though the domestic sewage discharge exemption (4) may allow the discharge of hazardous wastes into a public sewer, the laboratory must still comply with any local publicly owned treatment works (POTW) discharge limits, and discharge is limited by other federal laws. For instance, the Comprehensive Emergency Response and Civil Liabilities Act (CERCLA) limits hazardous discharges to 15 kg/month without meeting release reporting requirements.

Additional RCRA exemptions that may apply to the laboratory include those for analytical samples being held for evaluation and treatability samples. Analytical samples may be stored in the laboratory before testing, transported back to a collector, or stored after testing for a specific reason (such as evidence in a court case), without entering RCRA regulation. These samples are exempt until discarded. Treatability test samples, which are essentially hazardous waste samples undergoing testing to determine how they might be treated, have additional requirements in order to obtain exemption. These include the submittal of a notification to the EPA regional administrator and/or state environmental regulatory agency, limits on the volume treated daily, and volume and time limits on storage. Specific recordkeeping requirements also apply.

Regional and State Differences in Laboratory Waste Regulation

Most states have primacy, or the right to both regulate their own hazardous waste generators and to issue operating permits to Treatment Storage and Disposal Facilities (TSDFs). Although U.S. EPA approval is required for state-authorized programs, virtually the only significant EPA requirement is that the regulations be "at least as strict as" the federal model. As a result, there are huge differences in regulatory structures, permit requirements, and in issues important to the laboratory community.

Regulations and enforcement policies change frequently at the state level; it is therefore difficult, if not possible, to maintain a complete database on all of the differences.

TYPES OF WASTES GENERATED BY LABORATORIES

Surplus Chemicals

There are a number of reasons why laboratory chemicals might be considered surplus. These include chemicals that no longer have a use because of a change in procedure or standard experiment; those that might be remaining from someone's research because of discharge, resignation, retirement or death; those purchased on speculation of a future need; or volumes purchased in excess of the actual need. It is difficult to eliminate surplus chemicals entirely, since these conditions could occur at any facility, anytime.

Off-Specification and Contaminated Reagents

The requirements for quality control in laboratory operations often means that the chemical user needs absolute certainty of the purity of a material. Unless a container is under the strict control of the user from the time it is first opened, there may be concerns over purity. Depending on the application, a user may believe the risk of accidental contamination warrants opening a new container each time a chemical is needed. Materials found to be impure as a result of improper handling or quality control during manufacture, previous use, or storage, are frequently discarded. The decision may be based on either laboratory analysis or appearance change such as physical state or color.

Samples

Samples procured by the laboratory for evaluation may be discarded after the removal of a small amount of material. Many manufacturers send free samples of their commercial preparation upon request; designation only by trade name can make disposal more difficult.

Experimental and Research Waste

Experimental and research laboratory wastes can theoretically include millions of compounds, although it may be possible to consolidate similar materials, particularly by-products of ongoing research efforts.

Academic Laboratory Wastes

Many experiments routinely performed in the teaching of chemical and biological sciences generate by-products that may be regulated as hazardous. Research wastes may also be generated in academic laboratories.

Production Wastes

Clinical Wastes. Wastes generated as a result of medical testing or medical processes are usually regulated as infectious or medical because they may include tissue, blood, or other bodily fluids. There are, however, some instances where medical laboratories generate waste regulated as hazardous. Examples include discarded formaldehyde solution used as a tissue preservative; scintillation fluids containing xylene; various solvents, particularly alcohols; wastewater streams generated as instrument effluent; and certain chemotherapeutic drugs discarded as out-of-specification or beyond their expiration dates. Out-of-date reagent test kits may also contain hazardous wastes, such as trace quantities of mercury compounds, mineral acids, or reactive materials.

Pharmaceutical Wastes. Pharmaceutical operations are among the largest generators of hazardous laboratory waste. Research facilities may generate a great variety of wastes, depending on the nature of the work being performed. Halogenated and nonhalogenated solvents are frequently generated in significant quantities as the result of extractions and other processing. The discarding of surplus, off-specification, and out-of-date samples and reagents is common. If biotechnology laboratories are included in this class, wastes generated could also include certain agriculture-based wastes.

Environmental and Industrial Hygiene Testing Laboratories. Laboratories that perform commercial testing of environmental or industrial hygiene samples generally have less control over the volume or nature of wastes generated. EPA methods (5) allow little or no flexibility in procedures and many routine analyses generate hazardous wastes. Freon, which is specified as an extraction solvent in several petroleum hydrocarbon analytical procedures is an example. In addition, testing laboratories must decide how to handle samples after analysis. Samples are generally either returned to the client or disposed of by the laboratory. The additional handling involved in sample returns adds to overhead expense, and the packaging, consolidation, storage, and related waste handling procedures associated with waste disposal mean that either a surcharge must be made for sample disposal or the costs must be absorbed.

THE FUTURE OF LABORATORY WASTE MANAGEMENT

Regulatory Changes

Long-term efforts have been underway to change how laboratories are regulated since RCRA was first implemented. There are numerous inconsistencies within the regulations that make them difficult to apply to laboratories. One example is that of a university campus with laboratory facilities in a number of buildings. Under current regulations, the university may not be able to consolidate wastes at a central facility without a TSDF permit. If the buildings are not contiguous and public roadways must be crossed to get from one laboratory building to another, each location would be considered as a separate generator. Since waste can only be shipped on a hazardous waste manifest, waste from one location cannot be combined with waste from another for transportation purposes, recycling, reclamation, neutralization, or any other on-site management process. In addition, separate records must be kept for shipments from each location. Although many states have addressed this problem in their own regulations and enforcement policies, it is still part of the federal model.

Efforts within the U.S. EPA and Congress to make changes specific to laboratories have never reached a consensus, but changes to the generator site rule have widespread support. Other regulatory changes that are likely include allowing laboratories to do limited on-site treatment of waste, and easing the restrictions on storage and accumulation times.

Waste Minimization Procedures

Laboratories have continued to pursue waste minimization efforts, and more and more facilities have staff members with specific waste management responsibilities. Environmental controls are an integral part of U.S. industry, and laboratories have made enormous strides in reducing costs and implementing effective waste management programs. It is in the area of source reduction, however, that the greatest advances can be expected. Such relatively new innovations as computer simulation and microglassware for teaching and research have gained wider support in academia and industry (6). There has been much discussion about miniaturization of laboratories, and there are clear advantages, such as:

- A reduction in possible exposure to toxic materials, by a factor of as much as 2000.
- Virtual elimination of fire and explosion risk.
- Substantially reduced air emissions.
- Reduced costs for chemicals, storage, energy, and disposal.
- Significant reduction in volume of hazardous wastes generated (7).

Given these advantages, it is difficult to see why more laboratories do not embrace the concept. It may be that the initial capital outlay for microglassware is a factor; it is more likely that the necessary change in philosophy that miniaturizing requires. It also can mean reevaluating many laboratory procedures. Widespread use of computer simulation will increase as software programs continue to improve, and today's chemistry students are more likely to accept substitution of a computer program for hands-on laboratory experimentation than their predecessors. The use of miniaturization is also likely to increase, based on recent advances in molecular engineering techniques, micro-machining of chemical instruments, and compact chemical separation systems.

Computerization of inventory records allows far greater control over chemical purchasing and waste management planning, and as demand grows, these systems are becoming more sophisticated and more affordable.

Laboratories may always be destined to pay the highest cost per unit for waste disposal, but the quantities of waste generated continue to decrease. As the chemical waste management industry continues to mature, competition for the smaller volumes of waste will result in lower pricing for disposal and better treatment and disposal technologies.

BIBLIOGRAPHY

1. *Code of Federal Regulations,* Title 29, 1910.1450.
2. *Code of Federal Regulations,* Title 40, 261.3(a)(2) (iv)(E).
3. *Code of Federal Regulations,* Title 40, 268.1(e)(5).
4. *Code of Federal Regulations,* Title 40, 261.4(a)(1)(ii).
5. Environmental Protection Agency, *Solid Waste Sampling and Analysis Methods,* Publication #SW-846, EPA, Washington, D.C.
6. "ChEs Seek Big Gains From Process Miniaturization", *Chem. Eng. Prog.* (Apr. 1996).
7. *Laboratory Safety & Environmental Management,* The Target Group Inc., Burbank, Calif., Sept./Oct. 1993.

General References

American Chemical Society Task Force on Laboratory Waste Management, *Laboratory Waste Management A Guidebook,* American Chemical Society, Washington, D.C., 1994.

R.W. Phifer and W.R. McTigue, *Hazardous Waste Management for Small Quantity Generators,* Lewis Publishers/CRC Press, Chelsea, Mich., 1988.

American Chemical Society Task Force on Laboratory Waste Management, *Less is Better. Laboratory Chemical Management for Waste Reduction,* 2nd ed., American Chemical Society, Washington, D.C., 1993.

Draft Report, *Problems Associated with Management of Hazardous Wastes from Educational Institutions,* U.S. Environmental Protection Agency, Office of Solid Waste, Sept., 1987.

P.A. Reinhardt, K.L. Leonard, and P.C. Asbrook, *Pollution Prevention and Waste Minimization in Laboratories,* CRC Press/Lewis Publishers, Boca Raton, Fla., 1995.

Code of Federal Regulations, Title 40, Parts 261–265.

Laboratory Safety & Environmental Management (bimonthly newsletter), The Target Group, Inc., Burbank, Calif.

"Safety Net" (to subscribe, send e-mail to listserv@uvmvm.uvm.edu: SUBSCRIBE SAFETY your name). Maintained by R. Stuart, rstuart@moose.uvm.edu.

LANDFILLS

AMALENDU BAGCHI
Consultant
Madison, Wisconsin

In the U.S., until the late 1960s, waste was disposed of primarily in low lying areas in an uncontrolled fashion. It was generally believed that leachings from waste were completely purified by soil and groundwater, so groundwater contamination from waste disposal was not an issue. Subsequently, many studies have indicated that groundwater contamination can occur from waste disposal practices. Currently, only extremely low radiation nuclear wastes, commonly generated by municipal water treatment systems, are allowed to be disposed of in conventional landfills.

A landfill can be described as a land area where waste is disposed of in a controlled fashion (see Plate 5). Because leachings from waste types have the potential to contaminate groundwater, most modern landfills collect and treat leachate to minimize its impact on groundwater and surface water. In general, there are two types of landfills: natural attenuation (NA) landfills (Fig. 1) and containment landfills (Fig. 2). In NA landfills there is no liner at the base of the landfill, therefore leachate is allowed to percolate through the underlying soil into the groundwater. In containment landfills, a liner is constructed at the base to minimize percolation of leachate into the underlaying groundwater. Depending on the landform where a waste is disposed, three types of landfills have emerged: (*1*) below-grade landfills (see Fig. 2), (*2*) at-grade landfills (1), and (*3*) canyon landfills (1). All of these types are containment landfills.

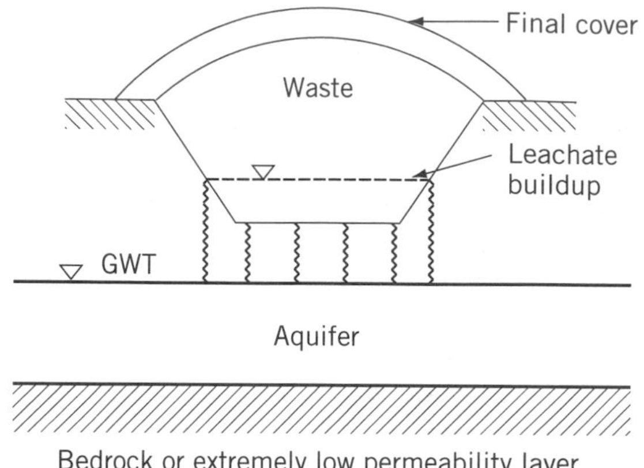

Figure 1. Natural attenuation landfill. Reprinted with permission (1).

An improperly designed and operated landfill can cause the following problems: odor and fire hazard due to gas migration, discharge of leachate to groundwater aquifer, and breeding and harboring of disease vectors. In addition to groundwater, a landfill can impact several elements of the physical environments like surface water, air, forest lands, and wetlands. There are also socioeconomic impacts from a landfill. These impacts are discussed in detail in a later section. The primary impact of landfills on the environment is from leachate, gas, and landfill construction activities.

FORMATION OF LEACHATE AND ITS ESTIMATE

Precipitation that falls on a landfill is the primary cause for the generation of leachate and gas in a landfill. In an active landfill all precipitation falling on the active area percolates through the waste. However, in a landfill which is properly closed with a final cover only a small part of the precipitation percolates through the waste; a significant reduction in leachate volume occurs due to surface runoff and evapotranspiration (1). Leachate is generated primarily due to the percolation of water through the waste. Although percolation of other liquids generated due to chemical reactions within the waste, and liquid squeezed out due to the self-weight of the waste also contribute to leachate generation, the volume of leachate generated due to these two processes is very low. Leachate can therefore be defined as a liquid that is produced when water or another liquid comes in contact with waste. Leachate contains a number of dissolved and suspended materials. The percolating liquid reacts both physically and chemically with the waste and transports the generated chemicals toward the landfill base. In a putrecible waste landfill, microbial activities also contribute toward the formation of leachate (2,3). A detailed discussion of effect of various factors on leachate formation can be found elsewhere (1,4). Characteristics of leachate depend primarily on the waste type disposed of in a landfill. Both quantity and quality of leachate influence landfill design. Discussion on leachate quality from various waste types can be found in a later section.

Quantity of leachate generated in a landfill depends primarily on the volume of precipitation. Although the exact total volume of precipitation and its daily variation in a year is difficult to predict, in most geographic locations it lies within a known range. Thus, approximate total volume of leachate and its monthly variation can be predicted within a reasonable range. The quantity of leachate from an active landfill is significantly more than that from a properly closed landfill. A landfill designer needs to know the quantity of leachate to determine spacing of leachate collection pipes, the size of a leachate collection tank (if one is used), and to design an

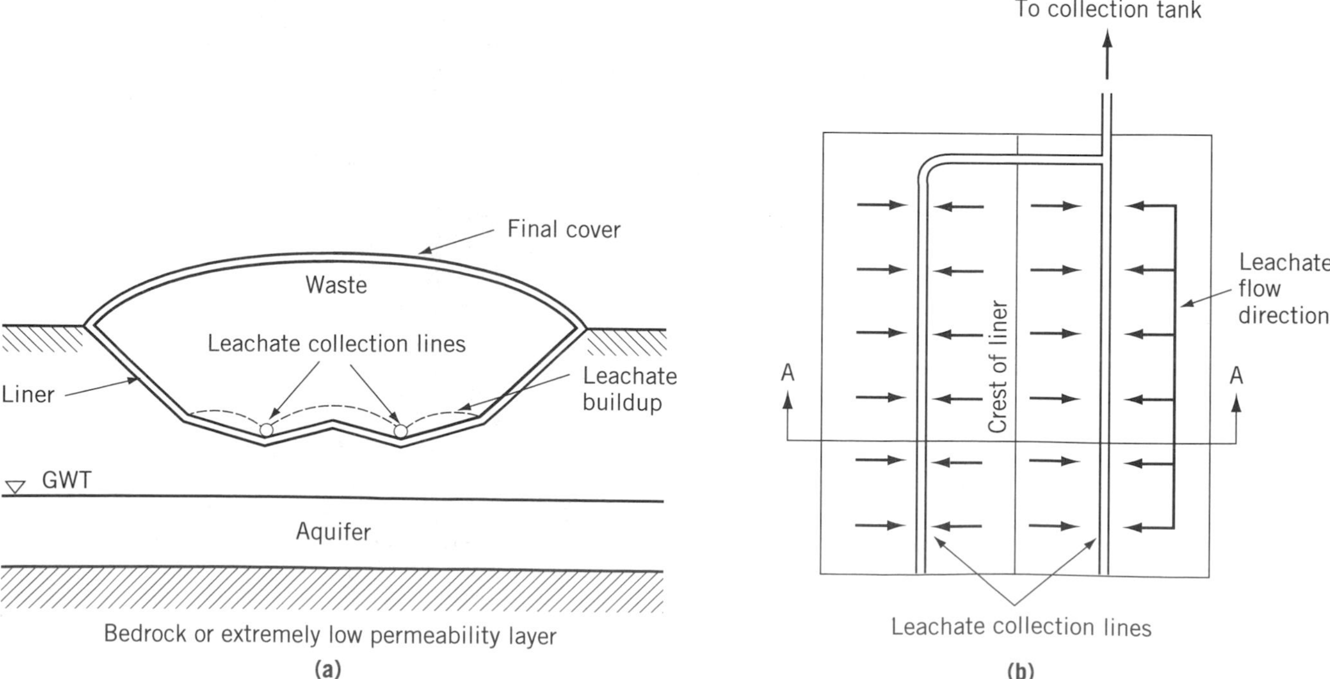

Figure 2. Containment landfill: (**a**) cross section A–A and (**b**) plan view. Reprinted with permission (1).

on-site or off-site leachate treatment plant. The total quantity of leachate from an active municipal waste landfill varies between 15 and 50% of total precipitation (1).

Although a significant quantity of gas is generated in putrescible waste landfill, nonputrescible waste may also generate gas. The quality and quantity of gas depends on waste type. If gas generation is expected then proper arrangements must be made for venting the gas. A detailed discussion regarding gas generation and venting can be found elsewhere (1).

A final cover is constructed over a closed landfill to minimize infiltration. A final cover consists of several layers (1); each layer has a specific function. The thickness of each layer depends on the material used for the construction of the layer and the climatological region in which the landfill is located. The purpose of the grading layer is to provide a stable surface for constructing the final cover. This layer also helps in venting landfill gas if a passive venting system is used. It usually consists of a 15–60 cm (6–24 in.) thick coarse-grained material. A layer of bark or a geotextile may also be used over unstable sludge surfaces (eg, paper mill sludge). The barrier layer provides a barrier against water infiltration. Clay, synthetic clay, synthetic membrane, or a combination of these materials is used for final cover construction. In many landfills a composite layer consisting of a synthetic membrane over a clay or synthetic clay layer is constructed. A low permeability layer (eg, clay) should be constructed below a synthetic membrane layer to ensure low permeability of the barrier layer as a whole. The usual thicknesses of the various materials of the barrier layer are as follows: clay, 45–60 cm (24–30 in.); synthetic clay (prior to swelling), 0.6 cm (0.25 in.); and synthetic membrane, 0.1–0.15 cm (40–60 mils). The drainage layer enhances infiltration reduction. Usually either a 5–15 cm (2–16 in.) thick coarse sand layer or a synthetic net is used as a drainage layer. The protective layer protects the barrier layer from freeze–thaw degradation (in a cold climate) and desiccation cracks. It also serves as a medium for root growth. Ideally a silty loam soil should be used for constructing this layer. The thickness of this layer depends primarily on frost penetration and thus varies between 30–105 cm (12–42 in.). The topsoil layer is for supporting vegetative growth. It is usually 15 cm (6 in.) thick.

The topsoil layer and the protective layer play important roles in evapotranspiration and surface run-off. When properly constructed and maintained, these two layers reduce infiltration by 70–80%. However, the role of the drainage and barrier layers in further reducing infiltration must also be given due consideration. Although the absorption of moisture by soil and waste is high initially, the effectiveness in the long run is negligible.

LANDFILL TYPES

As mentioned earlier there are primarily two types of landfills: NA and containment-type landfill. The approach for designing the final cover is the same in both types. However, in a NA landfill no liner is constructed at the base to reduce infiltration of leachate into the underlying soil and groundwater. In a containment landfill a liner and a leachate collection system are constructed to further reduce leachate infiltration into the groundwater.

TYPES OF WASTE DISPOSED IN A LANDFILL

There are basically two types of wastes that are disposed in a landfill: municipal waste (commonly known as garbage or household waste) and industrial waste. Industrial waste can be subdivided into hazardous and nonhazardous waste. Currently, in the United States and in many other developed countries hazardous waste is not permitted to be disposed of in landfills; these must be treated prior to land disposal and rendered nonhazardous. Waste with the following characteristics should not be disposed of in a landfill (5):

- Waste with high percentages of volatile organic content
- Waste with high percentages of aromatic, halogenated, and nonhalogenated compounds
- Waste with high percentages of metallics, especially arsenic, cadmium, lead, mercury, and selenium
- Waste with high percentages of cyanide and sulfide
- Powdery waste that may cause dust problems in and around the landfill
- Waste with very low shear strength that may preclude construction of a final cover for the landfill
- Waste with high percentages of liquid that may generate too much leachate in the landfill
- Nuclear waste

Table 1 provides components of municipal solid waste (MSW) in the United States. Because of recycling efforts in many communities, the percentages of garden waste, glass, metals, paper, and plastics in municipal waste has been reduced significantly.

Chemical Characteristics of Leachate

From a disposal standpoint, the primary item of interest is the chemical composition of the leachate generated from a waste. Leachate quality from a waste type is not unique. Both the local climate and operational practices heavily influence leachate quality. This also varies with time. The time vs quality curve is somewhat bell-shaped. Although quality of leachate for many nonputrescible waste is known, usually standardized tests may be performed when needed, to assess the quality in most cases. Assessing leachate quality for putrecible waste is somewhat difficult because microbial

Table 1. Percentages of Materials Discarded in Municipal Solid Waste in the United States, 1990[a]

Materials	% of MSW Total
Paper and paperboard	32.3
Plastics	9.8
Yard trimmings	19.0
Ferrous metals	6.4
Rubber and leather	2.7
Textiles	3.3
Wood	7.3
Food wastes	8.1
Other	3.5
Aluminum	1.0
Glass	6.5
Total	*100.0*

[a] Adapted from Ref. 6.

Table 2. Concentration Range of Parameters in Municipal Waste Leachate[a]

Sl Number	Parameter	Range of Concentration, mg/L[b,c]
1	TDS	584–55,000
2	Specific conductance	480–72,000 μmho/cm
3	Total suspended solids	2–140,900
4	BOD	ND–195,000
5	COD	6.6–99,000
6	TOC	ND–40,000
7	pH	3.7–8.9 units
8	Total alkalinity	ND–15,050
9	Hardness	0.1–225,000
10	Chloride	2–11,375
11	Calcium	3.0–2,500
12	Sodium	12–6,010
13	Total Kjeldahl nitrogen	2–3,320
14	Iron	ND–4,000
15	Potassium	ND–3,200
16	Magnesium	4.0–780
17	Ammonia–nitrogen	ND–1,200
18	Sulfate	ND–1,850
19	Aluminum	ND–85
20	Zinc	ND–731
21	Manganese	ND–400
22	Total phosphorus	ND–234
23	Boron	0.87–13
24	Barium	ND–12.5
25	Nickel	ND–7.5
26	Nitrate–nitrogen	ND–250
27	Lead	ND–14.2
28	Chromium	ND–5.6
29	Antimony	ND–3.19
30	Copper	ND–9.0
31	Thallium	ND–0.78
32	Cyanide	ND–6
33	Arsenic	ND–70.2
34	Molybdenum	0.01–1.43
35	Tin	ND–0.16
36	Nitrite–nitrogen	ND–1.46
37	Selenium	ND–1.85
38	Cadmium	ND–0.4
39	Silver	ND–1.96
40	Beryllium	ND–0.36
41	Mercury	ND–3.0
42	Turbidity	40–500 Jackson units

[a] After Ref. 1. Sl represents serial number.
[b] Unless otherwise noted.
[c] ND = nondetectable.

activities heavily influence the results. Tables 2 and 3 provide concentration ranges of various parameters in two putrecible wastes, namely municipal and papermill sludge waste. As mentioned earlier, an industrial waste can be hazardous or nonhazardous. The definition of hazardous waste varies in different countries (7). In the United States hazardous waste is identified by its characteristics or by list (that is, if a listed hazardous chemical is present in the waste, the waste is considered hazardous). A third approach for using toxicological standards for defining hazardous waste is under consideration. Because the approach to define hazardous waste is different in different countries and is subject to change, the appropriate regulatory authority should be contacted to find the definition of hazardous waste applicable to the disposal site.

Physical Properties

Apart from the chemical characteristics, a knowledge about the physical properties of the waste is necessary when designing a landfill. The physical properties of interest for disposal purposes include compacted bulk density, specific gravity, grain size distribution, permeability, static and dynamic strength characteristics, and consolidation characteristics. In most instances standard tests used for soil can also be used to assess these physical properties of waste. However, difficulty arises if the waste is putrescible, and/or heterogeneous.

IMPACTS FROM A LANDFILL

A landfill has the potential to impact both the physical environment and the socio-economic environment.

Table 3. Concentration Range of Parameters in Papermill Sludge Leachate[a]

Sl Number	Parameter	Range of Concentration, mg/L[b,c]
1	pH	5.4–9.0 units
2	TDS	289–9,810
3	TSS	80–320
4	Conductivity	70–14,370 μmho/cm
5	Alkalinity	174–5,500
6	Hardness	682–6,600
7	BOD	36–10,000
8	COD	4–43,000
9	Sulfate	0.9–550
10	Sodium	9–4,500
11	Calcium	5.5–2,400
12	Aluminum	0.008–18
13	Chloride	1–1,200
14	Iron	<0.1–950
15	Zinc	<0.018–0.03
16	Color	1,315–38,300 color units
17	Turbidity	Not reported
18	Phenols	0.0011–4.5
19	Tannin–lignin	13–90
20	Kjeldahl–nitrogen	34.5–385
21	Ammonia–nitrogen	<0.1
22	Nitrate	<0.1–15
23	Nitrite	<0.01–0.018
24	Sulfite	4–64
25	Sulfide	ND
26	Phosphate	0.11–0.58
27	Total volatile solids	211–483
28	Total fixed solids	144–266
29	Barium	0.011–1.1
30	Bromide	ND
31	Cadmium	0.006–0.02
32	Chromium	0–0.15
33	Cobalt	0.005–0.014
34	Copper	<0.01–0.21
35	Lead	0.037–0.1
36	Magnesium	3.8–6,000
37	Manganese	0.1–200
38	Mercury	<0.01–7 g/L
39	Nickel	<0.005–0.024
40	Potassium	140
41	Selenium	75
42	Tin	<0.1
43	Titanium	0.04
44	Vanadium	<0.01
45	TOC	1,350
46	Silicon	<3
47	Phosphorus	0.65
48	Arsenic	0.029
49	Cyanide	0.017

[a] Ref. 1. Sl represents serial number.
[b] Unless otherwise noted.
[c] ND = nondetectable.

Physical Environment

The primary impact from a landfill on the physical environment is on groundwater beneath the landfill. Because of the velocity of groundwater, contaminants can pollute a significant portion of the aquifer in the direction in which the groundwater flows. Precipitation falling on a landfill dissolves the contaminants within the wastes and percolates downward from the base of the landfill. The migration of leachate from a landfill base can be reduced significantly by proper design and maintenance of the landfill. Degradation of groundwater from a natural attenuation landfill is higher than from a containment landfill. A proper groundwater monitoring plan around a landfill should be implemented to ensure early detection of groundwater impact.

A landfill should be designed and operated in such a manner that waste contact liquid does not flow into surface water bodies. Chances of impacting surface water are usually high during landfill construction. Because large areas are stripped, precipitation erodes soil causing an increase in suspended soil loading in nearby flowage. A proper erosion control plan should be implemented during landfill construction. When designed and installed properly, silt fence, hay bales, and a sedimentation pond can reduce discharge of fines into surface water bodies.

Landfills can also pollute surrounding air. Methane, emitted from a putrescible waste landfill, is considered to have a "greenhouse" effect on the atmosphere. In addition, air contaminants such as vinyl chloride may be present in significant quantity in landfill gas. In many instances gas from a closed landfill is burned before discharging it into the atmosphere. Odor from a landfill can have health effects both on landfill workers and on people living around the landfill. Odor can be reduced by minimizing the active phase of the landfill. In some instances, a temporary gas collection system over the active area is installed to minimize an odor problem. Dust from an active landfill area can also have negative health effects on workers and people living around the landfill. Watering of dusty roads and minimizing the active phase of a landfill is undertaken in some cases where dust causes a serious problem.

In many instances landfills are sited on or near a wetland. Although every effort should be made to site a landfill away from a wetland, in some cases it is unavoidable. In many instances liner quality clay is found within or near a wetland. Although wetland mitigation is undertaken when a landfill destroys a wetland partially or entirely, many feel that a wetland ecosystem cannot be regenerated in a newly created wetland area. It is argued that it takes hundreds of years to develop a wetland ecosystem. To minimize impact on a nearby wetland, attempts should be made to minimize any changes in wetland hydrology. In other words, the amount of water flowing in or out of a wetland should remain the same even after the construction of a landfill. There will be some temporary impact from an active landfill because the precipitation falling on an active landfill is not allowed to be discharged into the surface water drainage system. To minimize this impact, the active phase of the landfill should be minimized and whenever possible at least an intermediate cover should be constructed on unused portions of landfills. In a few instances identification of a wetland can be difficult. In such instances a wetland ecologist should be consulted to render an opinion.

In many instances either the landfill or a borrow source for clay and sand are located in forest land. In addition to the actual site of the landfill and or borrow sources, access roads need to be constructed within the forest land. All of these activities impact the forest land. In addition, impact on groundwater by a landfill may also impact nearby forest land. Proper care should be taken to minimize the disturbance of forest land when a landfill or a borrow source is sited within a forest land. The presence of certain contaminants in groundwater may have toxic effects on trees (see FOREST ECOSYSTEMS IN A CHANGING GLOBAL CONTEXT).

A landfill also has a socio-economic impact on the community in which it is located. In general, siting of a landfill within a populated area should be avoided. Opposition of local residents is unavoidable because in most cases landfill sitting reduces property values. There is also the fear of groundwater degradation. A landfill may be opposed because of supposed increases in traffic, noise, and dust problems. Careful planning and design can reduce the impact, and local residents should be made aware of the landfill design features. An open discussion with the neighbors and a genuine attempt to address their concerns should be made to establish a good long-term relationship with the community.

In most cases nonmunicipal landfill siting may increase revenue for the community as a whole, although the property values of the immediate adjoining lands tend to decrease. Monitoring of groundwater and enforcement of permit conditions are essential to earn the trust of the general population, which has both short- and long-term benefits.

NEED FOR A LANDFILL

Prior to undertaking a landfill project, prospective owners (municipal or private) should investigate whether a new landfill is needed and if so, the size of landfill needed for the communities for which it is proposed. Knowledge about the waste generation rate and population projection of the communities over the life of the landfill are essential for estimating the volume of waste in a municipal landfill. For an industrial landfill, estimation of landfill volume is somewhat simpler because both the waste generation rate and necessary site life of the landfill are known.

Every effort must made for reducing, reusing, and recycling both industrial and municipal waste. A proper analysis regarding the percentage of each component of a municipal waste stream should be performed. In the United States the volume of paper in municipal wastes is significantly high. Proper management practice for recycling paper and other recyclables (eg, glass, plastics) should be undertaken to prolong the site life of a municipal landfill. Similarly, an industrial waste stream should be studied to identify recyclables.

Siting a landfill is becoming more and more difficult; it is therefore becoming more important to dispose only those items that can no longer be used. The following factors are considered in choosing a landfill site: haul distance, locational criteria dictated by the regulatory agency, site access, subsoil condition and topography, hydrology and hydrogeology (qv) of the site, and potential use of the landfill area after closure. A modern waste management practice puts emphasis on reduce, reuse, and recycle practices even if the short-term cost for recycling is found to be higher than disposal.

CURRENT DEBATE ON LANDFILL ISSUES

There is a debate regarding the timing of final cover construction on containment-type landfills, especially for putrescible waste landfills. The current practice is to construct a final cover on a landfill as soon as a portion reaches final grade. Opponents of this practice argue that early construction of a final cover prolongs the need for collection and treatment of leachate generated in a landfill. In their opinion the waste should be allowed to be flushed by precipitation water for several years, allowing harmful chemicals to be flushed out of a waste more easily. The number of years for which the landfill needs to be maintained and monitored can be reduced significantly if the final cover construction is delayed by several years. Those in favor of construction of a final cover at an early date argue that the cost of handling and stabilizing the waste is expected to be high. An uncovered putrescible waste also generates odor. A detailed discussion on the issue can be found elsewhere (1).

The cost of long-term care of a landfill is significant. Although it was once thought that long-term care of a landfill was necessary for 20 to 30 years, it is currently estimated that long-term care should last at least 40 years. Further research is needed to address the issues raised in the debate.

ACKNOWLEDGMENT

Thanks to S. Bagchi for administrative assistance.

BIBLIOGRAPHY

1. A. Bagchi, *Design, Construction, and Monitoring of Landfills,* 2nd ed., John Wiley & Sons, Inc., New York, 1994, p. 361.
2. R.P. Caffrey and R.K. Ham, *Compost Sci.* **15**(2), 11–15 (1974).
3. F.A. Roovers and G.J. Farquhars, *J. Environ. Eng. Div., Am. Soc. Civ. Eng.* **99**(EE5), 671–690 (1973).
4. R.K. Ham and M. Barlaz, in D.E. Daniel, ed., *Geotechnical Practice for Waste Disposal* Chapman & Hill, New York, 1993, pp. 113–136.
5. T. F. Stanczyk, *Proc. Purdue Ind. Waste Conf.* **42,** 309–320 (1987).
6. *Characterization of Municipal Solid Waste in the United States,* EPA/530-R-92-013, U.S. EPA, Washington, D.C., 1992.
7. G.W. Dawson and B.W. Mercer, *Hazardous Waste Management,* John Wiley & Sons, Inc., New York, 1986, pp. 43–105.

LASER-INDUCED BREAKDOWN SPECTROSCOPY

DAVID A. CREMERS
ALICE E. PICHAHCHY
Los Alamos National Laboratory
Los Alamos, New Mexico

Laser-induced breakdown spectroscopy (LIBS) is a novel method for determining the elemental composition of a gas, liquid, or solid (1,2). As an analytical method, LIBS has many advantages including rapid analysis, the absence of extensive sample preparation, the ability to analyze a material *in situ,* and simultaneous multielement detection capabilities. For this reason, the method applies to a wide range of environmental analysis problems not suited to conventional analytical techniques. In a LIBS measurement, powerful laser pulses are focused directly on or into a sample to form a microplasma. Because of the high temperature of the plasma, a small amount of sample is atomized and partially ionized, and the resulting atoms and ions are excited to emit light at a unique set of wavelengths which produce an emission spectrum. The spectrum is analyzed to determine the elemental composition of the material. Quantitative and qualitative analysis are possible. Because the laser plasma is formed by focused optical radiation, remote elemental analysis is possible. Because of its unique capabilities, LIBS is being developed for a number of environmental analytical applications including occupational health monitoring, site characterization to determine soil contamination, and monitoring of decontamination procedures.

LASER-INDUCED BREAKDOWN SPECTROSCOPY

Principles of the Technique

Laser-Induced Breakdown Spectroscopy (LIBS) is one form of atomic emission spectroscopy (AES) for determining the elemental composition of a material. In any type of AES-based analysis there are three main steps: (*1*) vaporization of the sample to produce atoms and ions; (*2*) excitation to produce electronically excited atoms and ions that then emit light at their unique characteristic wavelengths; and (*3*) detection and analysis of the emitted light. If the method is calibrated, the concentration of an element of interest, the analyte, is determined. The majority of conventional AES methods use vaporization/excitation produced by a plasma (3). The plasma in these cases is a highly

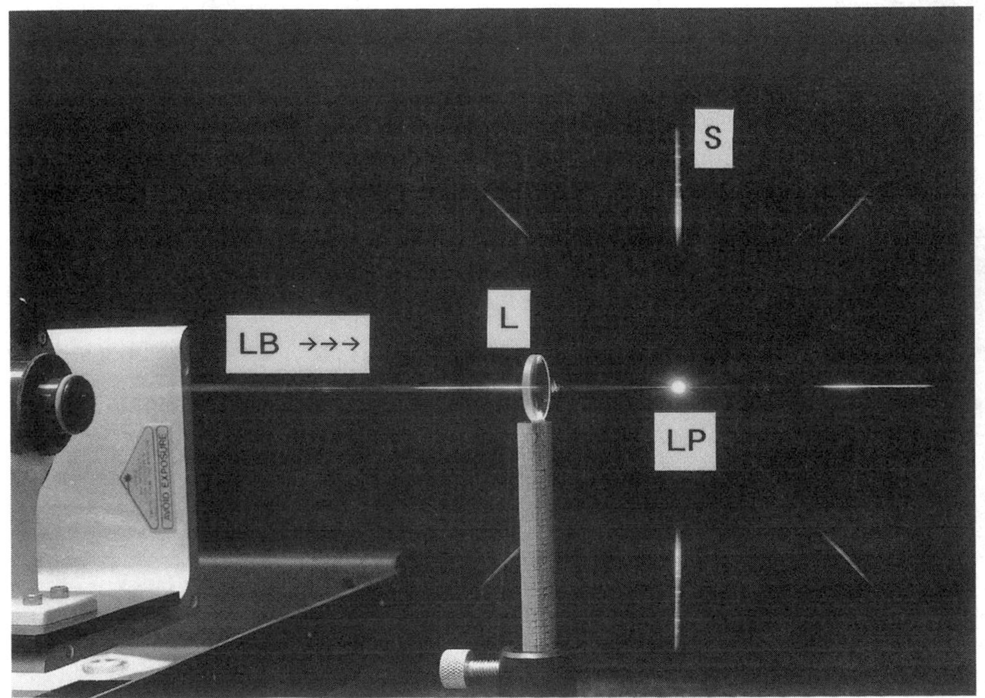

Figure 1. Photograph of the laser plasma (LP) formed in air. The laser beam (LB) is incident from the left and focused by a lens (L). A transmission diffraction grating is positioned between the camera lens and the spark to spectrally disperse the light producing the spectra (S) shown. The emission lines most evident in the top and bottom spectra are caused by nitrogen in air.

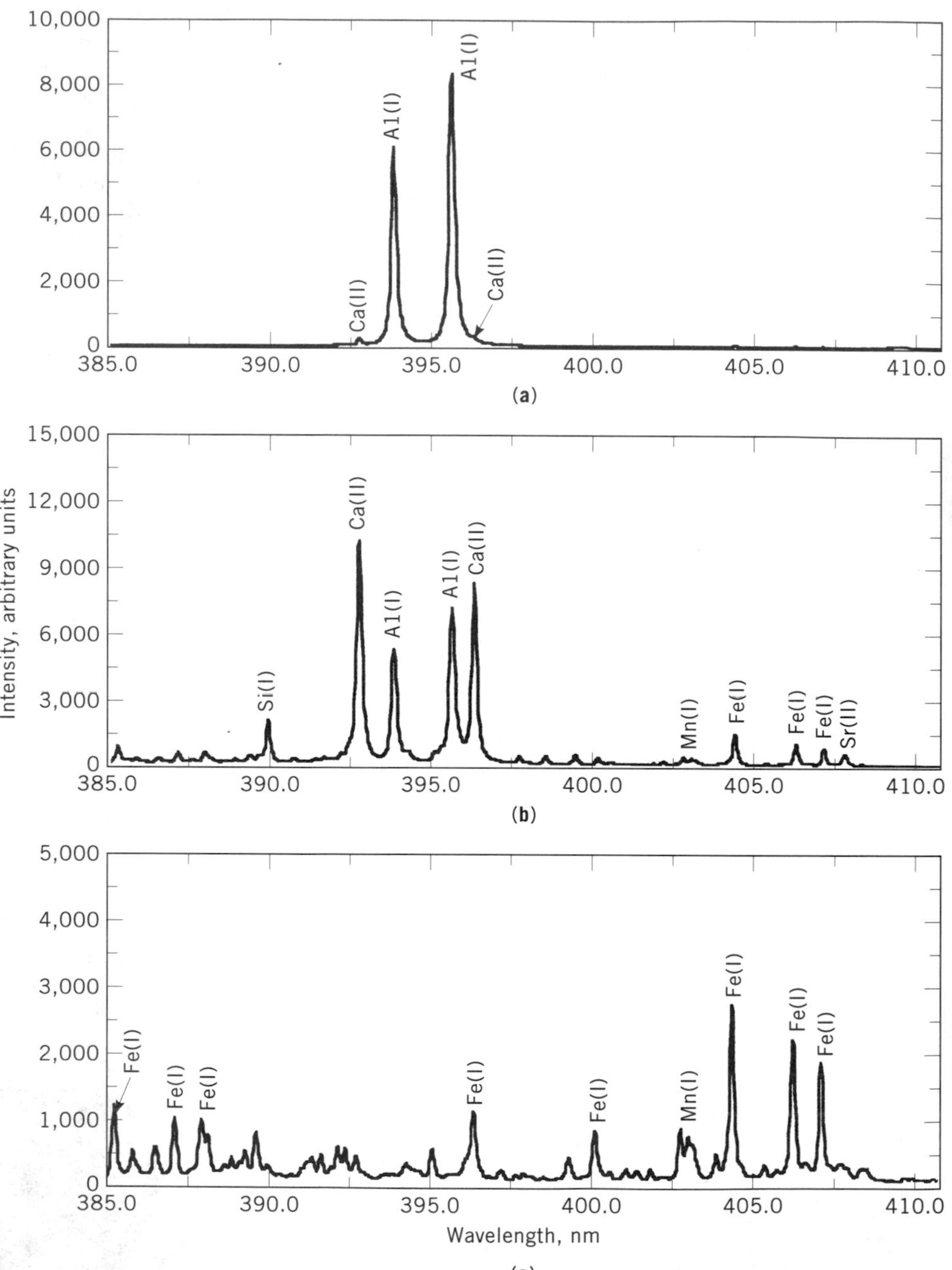

Figure 2. LIBS spectra obtained from three different samples: (**a**) aluminum alloy 6061; (**b**) soil; (**c**) a steel alloy.

ionized gas, typically at atmospheric pressure, formed in air or some inert gas (eg, argon, helium) by introducing strong electromagnetic fields into a small volume via some physical device: conventional arcs and sparks use metal electrodes and the inductively coupled plasma (ICP) is formed by a radiofrequency field applied to a metal coil (see INDUCTIVELY COUPLED PLASMA—ATOMIC EMISSION SPECTROMETRY).

In contrast to these conventional methods of plasma production, a plasma or spark is formed by focusing a high power laser beam in or on a material. Because the laser spark plasma is produced by focused optical radiation, it has certain unique advantages. A photograph of the laser spark produced in air is shown in Figure 1. This spark is produced by focusing, with a lens, a laser pulse of 10 nanoseconds duration containing about 100 mJ (0.0239 cal) of energy directly in air. Visually, the spark appears the same when formed in a gas, liquid, or solid because of the "white light" or spectrally broad continuum formed immediately upon plasma initiation.

The spectra displayed in Figure 1 show the continuous visible spectrum produced by this "white light" along with some line emissions from nitrogen in air. In the hot plasma, molecular nitrogen is atomized, and the nitrogen atoms are ionized and excited to emit light. Spectra recorded from the visible blue spectral region by forming the laser spark on three different solids are shown in Figure 2.

Because the laser spark is a pulsed source, optical emissions from the plasma are time-dependent. Although initially of high intensity, the "white light" from the plasma decays very rapidly during the first microsecond after plasma initiation. This light is from electron-ion recombination in the plasma, which is a prominent process caused by the high density of electrons in the plasma immediately after formation. After about one microsecond, however, there is still sufficient electron density (10^{17} cm^{-3}; 1.6×10^{18} in^{-3}) for the remaining atoms to become electronically excited through energetic collisions with electrons. These collisions result in excitation and subsequent emission of photons characteristic of the elements in the plasma. Because each element has a unique emission spectrum, spectral analysis of the light identifies these atoms. Lists of the emission lines of the elements are compiled in a number of sources and are a guide for determining the stronger analyte emission lines in a laser plasma (4). In addition, because the intensity of the emission is related at least in part to the number density of the element in the plasma, the method is used for quantitative analysis.

The temporal evolution of the laser plasma is divided into three distinct periods of emission:

1. $t < 1$ microsecond-During this early time the spectrum is dominated by the "white light" or spectrally featureless light of strong intensity. Some strong atomic emission features are observed, such as emission from once-ionized nitrogen in air (eg, emission lines in Fig. 1).
2. $0.5 < t < 2$ microseconds-During this period the white light decays rapidly and emission features from once-ionized atoms peak in intensity. Some strong emissions from neutral atoms are also observed.
3. $t > 2$ microseconds-By this time, the white light has decayed significantly and the spectrum is dominated by emissions from neutral atomic species and simple molecules formed as the plasma cools. For example, emissions from AlO molecules are readily observed from the laser plasma produced on aluminum metal in air. Typically, all emissions from a plasma formed at atmospheric pressure decay completely by $t > 100$ microseconds.

Early, the spectrum is generally featureless. At later times, emissions are strongest. Because the spectrum evolves with time, an analytical system capable of time-resolved detection of the plasma light provides optimum detection capabilities. Methods of obtaining time-resolved detection are described later.

A typical LIBS apparatus is shown in schematic form in Figure 3. High power laser pulses are focused into or on the sample to form the laser sparks with a simple lens. The minimum pulse energy required to form a spark on a solid is generally less than that required to generate a spark in a liquid or gas. The light from the spark is collected by a lens or a fiber optic cable. Using a lens, an image of the spark is formed on the entrance slit of the spectrometer as shown in Figure 3. With a fiber optic cable, the simplest approach is to position the fiber against the entrance slit. To obtain optimum light coupling efficiency, a bundle of optical fibers is arranged in a circular configuration at the light collection end and as a linear array at the spectrometer slit. In this way the linear array is positioned along the slit for maximum light transfer into the spectrometer. One advantage of using a fiber optic is that the spectrometer can be located remote from the sample. The spectrometer spectrally disperses the collected light which is then focused

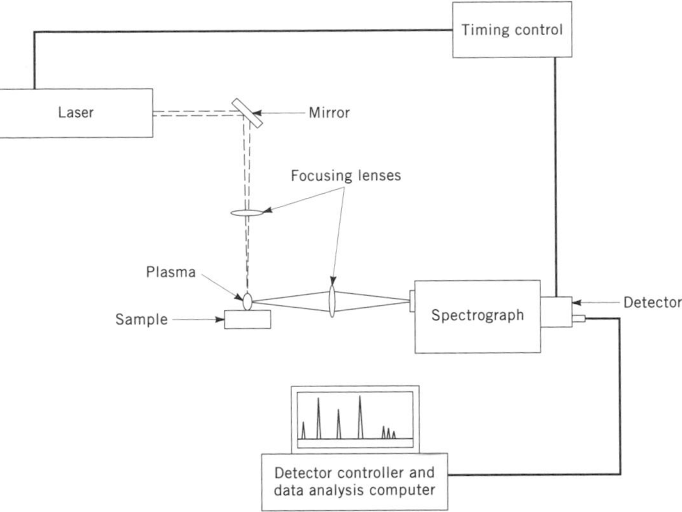

Figure 3. Schematic of a typical LIBS setup. The specifications of the components depend on the analytical requirements for the intended application.

at the exit plane of the spectrometer to produce a spectrum. The light is then detected by an array detector, as shown in Figure 3, which provides spectral coverage over a certain spectral region, or a slit is positioned at the focal plane to monitor only a single wavelength. In this latter configuration, a sensitive light detector, such as a photomultiplier tube, is positioned behind the slit to monitor the light. The spectra shown in Figure 2 were obtained with an array detector.

BIBLIOGRAPHY

1. *Applications of Laser-Induced Plasmas,* L.J. Radziemski and D.A. Cremers, eds., Marcel Dekker, New York, 1989.
2. L. Moenke–Blankenburg, *Laser Microanalysis,* John Wiley & Sons, New York, 1989.
3. To Török, J. Mika, and E. Gegus, *Emission Spectrochemical Analysis,* Adam Hilger, Bristol, 1978.
4. Wavelengths and Transition Probabilities for Atoms and Atomic Ions Part II. Transition Probabilities, NSRDS-NBS 68, U.S. Government Printing Office, Washington, D.C. 1980.

LASER-INDUCED FLUORESCENCE SPECTROSCOPY

WOLFGANG SCHADE
JENS BUBLITZ
Universität Kiel
Kiel, Germany

Recently, fiber optical sensors and laser-induced fluorescence (LIF) spectroscopy have drawn considerable interest for remote sensing of various environmental pollutants in air, water, and soil (1–5). Aromatic hydrocarbons, which are basic components of petroleum products, are of particular interest, because they are mutagenic and carcinogenic in screening tests and animal experiments and because exposure to aromatic hydrocarbons at workplaces has been investigated recently (6). The U.S. Environmental Protection Agency has included 16 polycyclic aromatics on its list of priority pollutants. Concentration limits for aromatic hydrocarbons in industrial sewer systems,

Table 1. Photophysical Properties of Some PAHs[a]

PAH	λ_{abs}, nm[b]	λ_{Fl}, nm[c]	Φ_{Fl}[d]	ϵ, $(M\ cm)^{-1}$[e]	τ_{Fl}, ns[f]
Naphthalene	302	322	0.23	300	96
	319			20	
Acenaphthene	300	347	0.50	4,650	0.9
	320			1,800	
Fluorene	300	310	0.80	10,000	10
Anthracene	356	399	0.36	9,100	4.9
	374			8,900	
Fluoranthene	359	462	0.30	7,840	53
Pyrene	336	383	0.65	55,800	450[g]
	372			140	
Benzo[*k*]fluoranthene	308	402	1.0[h]	68,300	11.3[h]
	402			25,000	
Benzo[*g,h,i*]perylene	300	419	0.29[h]	59,700	54.3[h]
	406			284	
Benzo[*a*]anthracene	300	385	0.23[h]	10,300	49.4[g]
	385			101	32.5[h]
Benzo[*b*]fluoranthene	302	446	0.53[h]	40,600	44.3[h]
	369			7,020	
Benzo[*a*]pyrene	385	403	0.60[h]	30,600	57.5[g]
	404			4,300	42.9[h]

[a] Refs. 8, 25, and 32.
[b] λ_{abs}: wavelength of maximum absorption.
[c] λ_{Fl}: wavelength of maximum fluorescence intensity.
[d] Φ_{Fl}: fluorescence quantum yield.
[e] ϵ: molar extinction coefficient.
[f] τ_{Fl}: fluorescence lifetime.
[g] Ref. 8.
[h] Ref. 25.

drinking water, and soils have been recommended by the World Health Organization and other organizations. In the Dutch list of soil contaminants (7) are reference values for selected polycyclic aromatics. Most of these aromatics are excited in the ultraviolet (UV) and visible spectral range (8) and because of their high fluorescence quantum yields (8) their fluorescence emission is used to detect these species (refer to Table 1).

Among various spectroscopic techniques, LIF is distinguished by its high sensitivity and selectivity. Besides measuring only fluorescence intensities, time-resolved recording of the fluorescence improves the sensitivity and the selectivity of the method considerably (9–11). For laser excitation in the UV and recording fluorescence in the blue spectral range, the LIF-decay spectra of pollutants and surrounding media, eg water or soil, show significant differences (12–18). Also the time analysis of LIF-decay curves offers the possibility of distinguishing different components of aromatics in complex mixtures. However, this is not the case when only stationary LIF spectroscopy is applied for the diagnostics.

In particular, recent investigations show that data accumulation of fluorescence intensities in an "early" and in a "late" time window with respect to the excitation pulse simplifies the time-resolved LIF method so much that it is very attractive for practical applications (19). For example, for oil-contaminated water or soil samples, a characteristic dependence of the ratios of "late" and "early" fluorescence intensities on oil concentration is obtained. Thus, these data are used very efficiently for qualitative and quantitative diagnostics with time-integrated LIF (19,20).

In recent years several groups developed fiber optic LIF sensors for on-line and *in situ* monitoring of oils and aromatic hydrocarbons in water, soil, and air (19–24). These systems are mobile and have several interesting applications, such as process control of industrial oil separator systems (19), field screening of oil contamination in water and soil after environmental accidents (20), localization of subsurface oil plumes at contaminated waste sites (22,23), and detection of aromatics on surface bound particles in air, eg, in smoke stacks (24). For laser excitation in the UV (eg $\lambda = 337$ nm) and recording fluorescence around 400 nm, typical detection limits are 0.5 mg/L for engine oil in water (19), 5 mg/kg for engine oil in soil (18), on the order of ng/L for polycyclic aromatic hydrocarbons(PAHs) in water (24), and on the order of $\mu g/m^3$for PAHs in the air (24).

This more practical research is accompanied by intense laboratory studies in which the photophysical properties are discussed that influence the intensity and the time evolution of LIF signals. In these investigations particular emphasis is given to quenching effects by humic acids, pH-dependence of LIF, and interactions of LIF with complex matrices in natural environments (25,26). Besides this, possibilities for discriminating between mono- (such as benzene, toluene, xylene, and ethylbenzene (BTXE)) and polycyclic aromatic hydrocarbons by UV-laser-excited fluorescence signals are also discussed, as the excitation wavelength is tuned from 240 to 360 nm (19). Whereas BTXE aromatics show only strong absorptions in the spectral range below 280 nm, for the PAHs more red shifted absorption bands (eg $\lambda > 300$ nm) are obtained. Therefore a two-wavelength laser excitation, eg at 246 nm and 355 nm, followed by time-integrated LIF detection discriminates between different components of hydrocarbons, such as BTXE and PAHs (27).

In contrast to LIF applications for monitoring pollutants in water, the heterogeneity and the opacity of soil matrices is still a considerable experimental challenge for the qualitative and quantitative interpretation of LIF data, when this method is applied for measuring PAHs or oil contamination on soil surfaces or in the subsurface (25,28). In this context, high intensity laser pulses are additional possibilities for diagnostics. Because of the high power density, when focusing the laser beam on the surface, a laser-induced plasma is created on the soil surface. If time-integrated detection of the emission follows, the strong background light of the plasma is suppressed, and single spectral lines of species in the sample under investigation are recorded. In recent investigations, laser-induced breakdown spectroscopy (LIBS) successfully measured heavy metal concentrations in soils down to concentrations of parts per million (ppm) (29).

Fiber optics and LIF are also used to measure, very accurately, temporal and spatial distribution of aqueous tracers in different soil samples (30). These data give information on the flow paths of environmental pollutants in the ground. They are used by soil scientists to prove existing transport models and to evaluate the mobility of different soil pollutants.

Since there is a considerable practical interest in the LIF technique because of its capability for on-line and *in situ* measurements, high sensitivity, and short times for data accumulation, miniaturizing existing experimental configurations is of particular value for industrial applications. The possibilities of microchip lasers (31) have been considered for the development of miniaturized LIF probes.

ACKNOWLEDGMENTS

We gratefully acknowledge financial support by the Bundesministerium für Bildung, Wissenschaft, Forschung und Technologie (contract 17F 1852/0), the Technologiestiftung Schleswig-Holstein (contract 93-35), and the Deutsche Forschungsgemeinschaft.

BIBLIOGRAPHY

1. K. Oesgaard, *Trace Anal.* **3,** 163–212 (1984).
2. R. Niessner, W. Roberts, and P. Wilbring, *SPIE* **1172,** 145–156 (1989).
3. S.H. Lieberman, S.M. Inman, G.A. Theriault, S.S. Cooper, P.G. Malone, Y. Shimizu, and P.W. Lurk, *SPIE* **1269,** 175–184 (1990).
4. T. Hengstermann and R. Reuter, *Appl. Opt.* **29,** 3218–3227 (1990).
5. S.E. Apitz, G.A. Theriault, and S.H. Lieberman, *SPIE* **1637,** 241–254 (1992).
6. H. Fiedler and W. Mücke, O. Hutzinger, ed., in *The Handbook of Environmental Chemistry,* Springer, Berlin, 1991, Vol. 3G, pp. 97–138; C.A. Menzie, B.E. Potocki, and J. Santodonato, *Environ. Sci. Technol.* **26,** 1278–1284 (1992).
7. D. Rosenkranz, G. Einsele, and H.-M. Harreβ eds., *Bodenschutz,* E. Schmidt, Berlin, Vol. 2, 1988, Chapt. 8935, pp. 1–27; H. Hein and G. Schwedt, *Richt- und Grenzwerte: Wasser-Boden-Abfall-Chemikalien-Luft,* 3rd ed., Vogel, Würzburg, 1992, Chapt. 4.
8. W. Karcher, *Spectral Atlas of Polycyclic Aromatic Compounds,* Kluwer, Dordrecht, 1985; I.B. Berlman, *Handbook of Fluorescence Spectra of Aromatic Molecules,* Academic Press, New York, 1971; L.K. Patterson, G. Porter, and M.R. Topp, *Chem. Phys. Lett.* **7,** 612 (1979).
9. R.M. Measures, Houston, W. R., and Stephenson, D. G, *Opt. Eng.* **13,** 494–501 (1974).
10. D.M. Rayner and A.G. Szabo, *Appl. Opt.* **17,** 1624–1630 (1978).
11. P. Camagni, A. Colombo, C. Koechler, N. Omnetto, Pan Qi, and G. Rossi, *Appl. Opt.* **30,** 26–35 (1991).
12. S.H. Lieberman, S.M. Inman, and G.A. Theriault, *SPIE* **1172,** 94–98 (1989).
13. R. Niessner, U. Panne, and H. Schröder, *Analytica Chimica Acta* **255,** 231–243 (1991).
14. W. Schade and J. Bublitz, in Spectroscopy, L. Bloomfield, T. Gallagher, and D. Larson, eds., *Amer. Inst. of Phys.* 261–263 (1994).
15. S.M. Inman, P. Thibado, G.A. Theriault, and S.H. Lieberman, *Analytica Chimica Acta* **254,** 45–51 (1990).
16. Y.-B. Zhu and J. Ma, *SPIE* **1637,** 285–290 (1992).
17. W. Schade, J. Bublitz, K.-P. Nick, and V. Helbig, in V. Klein, K. Weber, and C. Werner, Eds., *Laser in Remote Sensing,* Springer-Verlag, Berlin, 1992, pp. 53–61.
18. W. Schade and J. Bublitz, *Laser und Optoelektronik* **24,** 41–48 (1993).
19. W. Schade and J. Bublitz, *Environ. Sci. Technol.* **30,** 1451–1458 (1996).
20. J. Bublitz, M. Dickenhausen, M. Grätz, S. Todt, and W. Schade, *Appl. Opt.* **34,** 3223–3233, (1995).
21. S.H. Lieberman, G.A. Theriault, S.S. Cooper, P.G. Malone, R.S. Olsen, and P.W. Lurk, *Field Screening Methods for Hazardous Wastes and Toxic Chemicals, Second International Symposium,* 1991, pp. 57–63
22. R. Niessner, W. Roberts, and A. Krupp, *Fresenius J. Anal. Chem.* **341,** 207–213 (1991).
23. E. Jäger and H. Lucht, *Labor Praxis* **17,** 872–877 (1993).
24. U. Panne, F. Lewitzka, and R. Niessner, *Analysis* **20,** 533–542 (1992).
25. M.U. Kumke, H.-G. Löhmannsröben, and T. Roch, *J. Fluorescence* **5,** 139–153 (1995); G. Heinrich and H. Güsten, in A. Björseth and A.J. Dennis, eds., *Polynuclear Atomic Hydrocarbons: Chemistry and Biologic Effects,* Battelle, Columbus, Ohio, 1980.
26. M.U. Kumke, H.-G. Löhmannsröben, and T. Roch, *Analyst* **119,** 997 (1994).
27. J. Bublitz, A. Christophersen, and W. Schade, *Fresenius J. Anal. Chem.* **355,** 684–686 (1996).
28. H.–G. Löhmannsröben and T. Roch, *Analytiker Taschenbuch* Bd. 15, Springer, Berlin, 1997, pp. 218–251.
29. R. Wisbrun, I. Schechter, R. Niessner, and H. Schröder, *SPIE* **1716,** 12 (1992).
30. J. Bublitz, A. Knaack, M. Dickenhausen, M. Grätz, W. Schade, T. Baumgartl, and R. Horn, *EARLSeL Advances in Remote Sensing* **3,** 179–187 (1995).
31. J.J. Zayhowski, *Opt. Lett.* **21,** 588 (1996).
32. I.B. Berlman, *Handbook of Fluorescence Spectra of Aromatic Molecules,* Academic Press, New York, 1971.

LC/MS IN ENVIRONMENTAL MEASUREMENT. See OSHA STANDARDS, SAMPLING AND MEASUREMENTS.

LEAD-BASED PAINT: HAZARD MITIGATION

JEFFREY H. BOY
ASHOK KUMAR
U.S. Army Corps of Engineers
Champaign, Illinois

Lead is a naturally occurring element that is toxic to humans of all ages when ingested or inhaled in excessive quantities. Lead is most hazardous to children under the age of six, whose developing central nervous systems are particularly vulnerable to the metal and whose normal play activities expose them to harmful pathways such as lead-contaminated dust and lead-contaminated bare soil. High levels of lead in the blood of young children can produce permanent damage to the nervous system. Recent research, although not conclusive, indicates that relatively low blood lead levels can produce significant nervous system effects, such as reduction in intelligence and attention span, reading and learning disabilities, and behavior problems (1). Numerous studies have failed to detect these effects at low blood levels when other confounding factors have been taken into account (2) (see LEAD-BASED PAINT TESTING).

Exposure to lead can come from many sources. Some of the more common sources are potable water that passes through lead pipes or plumbing that contains lead or lead solder; lead-contaminated dust; lead in soils from combustion of leaded gasoline; industrial waste emissions; deterioration or improper abatement of lead-based paint; glazed china and pottery; lead introduced into child's environment by hobbies such as stained glass, lead casting, and ceramics; and lead taken home from occupational sources. Though there are many sources of exposure, this article deals only with lead-based paint. While the use of lead-based paint on residential surfaces was banned in 1978 in the United States by the Consumer Products Safety Commission, buildings constructed before that time are likely to have multiple coats of lead-based paint beneath newer coats of lead-free paint. Lead-based paints were used extensively before the 1950s on both interior and exterior surfaces of buildings because the slow chalking produced self-cleaning surfaces. The use of lead house paints then declined until the 1978 ban by the Consumer Products Safety Commission (3). Primers containing red lead oxide (Pb_3O_4) are still allowed for use on steel in industrial facilities such as bridges and water storage tanks, though that use is declining. Steel structures coated with lead-based paint include highway bridges, railroad bridges and trestles, antenna towers, water storage tanks, and aircraft hangars. There are 300,000 steel highway bridges in United States, many coated with lead-based paint and/or lead-based paint primers.

Of the 20 million children under age six in the United States, an estimated 840,000 (4.4%) have blood lead levels at or above the level of concern established by the Centers for Disease Control and Prevention (CDC) of 10 μg/dL of whole blood (4). Common sources of childhood lead poisoning include lead-contaminated dust, dust from lead-contaminated soil, paint chips, and drinking water. Due to the small particle size, lead-contaminated dust may not be visible to the naked eye and is difficult to clean up. There is a developing body of data that suggests that where there is no visible dust due to frequent and efficient cleaning, there is also a low chance of lead in the household dust being a source of significant contamination. Most commonly, lead-contaminated dust gets on children's hands and toys and then into their bodies through normal hand to mouth activity. Potential sources of lead contaminated dust in a home due to lead-based paint include deterioration (chipping or peeling); generation by friction or impact surfaces; and disturbances during repainting, remodeling, or abatement projects. The other significant pathway of child lead exposure is dust from lead-contaminated bare soil. Soil contamination can be traced to past widespread use of leaded gasoline, to neglect of or improper treatment or abatement of deteriorating exterior paint (on houses, bridges, and industrial facilities), and in some areas to industrial lead sources (1) (see LEAD LEVELS IN SOIL).

Painters, maintenance and renovation workers, and abatement contractors who use unsafe paint removal practices can cause contamination of dust or soils and may, themselves, be exposed to extremely high levels of lead. These workers are at risk of dangerous blood lead levels, although adults are less susceptible than children to some of the damaging effects of lead. Workers who take home lead-contaminated dust on their clothing can also contaminate their homes and expose their own young children.

Lead-based paint also poses a problem during the demolition and disposal of old buildings. Traditionally, the procedure for disposing of most building materials was either to simply demolish the building and dispose of the debris in a landfill or to burn the building down. These acts are now known to release large quantities of lead into the air and soil. There are now restrictions by the U.S. Environmental Protection Agency (EPA) on the demolition and disposal of buildings containing lead-based paint. The debris disposal is regulated by the Resource Conservation and Recovery Act (RCRA). The regulations for disposal of architectural components are under review, and latest guidance from state and federal regulators should be obtained before demolition of older buildings that may contain lead-based paint.

DEFINITION OF LEAD-BASED PAINT HAZARD

There are separate standards for liquid lead-based paint in the container and for dried paint applied to a surface. As a liquid paint in the container, lead-based paint is defined by the U.S. Consumer Product Safety Commission (CPSC) as paint containing more than 0.06% (600 ppm) lead by dry weight. In 1978 the CPSC banned the sale of all lead-based paint for residential use. It also banned products bearing lead-based paint produced or distributed for use by consumers. U.S. lead standards are listed at the end of this article.

The Residential Lead-Based Paint Hazards Reduction Act is often referred to as Title X because it was enacted as Title X of the Housing and Community Development Act of 1992 (U.S. Public Law 102-550). The term *lead-based paint* as used in Title X is paint or other dry surface coatings that contain lead $\geq$1.0 mg/cm^2 or 0.5% by weight (5000 μg/g or 5000 ppm by weight). Title X further defines *lead-based paint hazard* as "any condition that causes exposure to lead from lead-contaminated dust, lead-contaminated soil, [or] lead-contaminated paint that is deteriorated or present in accessible surfaces, friction surfaces or impact surface that would result in adverse human health effects." This definition differs from the absolute standard of lead-based paint and considers other factors such as the paint condition and location. Thus, under this definition, intact lead-based paint on most walls and ceilings is not considered to be a hazard. However, the condition of such paint should be monitored and maintained to ensure that it does not deteriorate.

The approach for lead-based paint hazard control depends on whether the structure is a occupied building or a steel structure. For buildings, especially residences occupied by children, pregnant women, or day care centers, the goal is to prevent exposure to lead-based paint hazards. Lead-based paint may be managed in place through interim controls or abated (removed, covered or encapsulated) to control or eliminate the source of the lead-based paint hazard. However, for steel structures such as bridges, lead-based paint is usually maintained until the paint deterioration warrants removal. For steel structures, the goal is to minimize the exposure of workers to lead during paint removal. Removal of lead-based paint from steel structures increases the cost by as much as 10 times over the cost of removal of paint that does not contain lead.

When lead-based paint and/or lead-based paint hazards have been identified in a building or on a steel structure, several elements must be considered to select the most effective interim control or abatement solution. The potential risk must be assessed; if the lead-based paint is in good condition, it may be managed-in-place. When the lead-based paint hazard can no longer be managed-in-place, partial or full abatement of the lead-based paint must be planned. This may include removal and replacement of painted components.

Many different technologies are commercially available for the removal of lead-based paint from a variety of substrates. Each of these technologies has its own inherent advantages and disadvantages that must be considered when selecting a lead-based paint abatement method (see LEAD-BASED PAINT: TESTING).

REGULATORY REQUIREMENTS

The removal of lead-based paint is regulated by federal, state, and local governments. The principal federal legislation is Title X. In response to this legislation, the U.S. Department of Housing and Urban Development (HUD), has developed implementing guidelines for the evaluation and control of lead-based paint hazards in housing. Other federal legislation that impacts lead-based paint removal in the United States includes RCRA, which regulates the disposal of hazardous waste; the Clean Air Act (CAA), which establishes ambient air quality standards; the Clean Water Act (CWA), which establishes discharge procedures into the nation's waterways; the Toxic Substance Control Act (TSCA), which regulates toxic substances; the Hazardous Materials Transportation Act, which regulates the transportation of hazardous materials; and Occupational Safety and Health Administration (OSHA) worker protection requirements.

Air

Ambient air quality standards in the United States are derived from the CAA and the resulting National Ambient Air Quality Standards (NAAQS) (5). The ambient air quality standard for lead and its compounds measured as elemental lead is 1.5 $\mu g/m^3$ maximum arithmetic mean averaged over a calendar quarter (90 days). The assessment is based on an analysis of the total suspended particulates in air (6).

Dust is defined as particulate materials with aerodynamic particle diameter $\leq 10\ \mu m$. The NAAQS standard for dust is 150 $\mu g/m^3$ for a 24-h average concentration (6). The particulate requirement is based on a 24-h criterion, and the limit for lead is based on a 90-day criterion. The thrust of these regulations are for the long-term continuous monitoring of air quality in a large region.

Facility owners or local air quality regulators may require that surveillance by instrument monitoring be conducted during a lead removal project. Since most paint removal operations are not normally conducted for 90 days, the Steel Structures Painting Council (SSPC) has developed suggestions for establishing daily criteria for monitoring. However, the appropriate air quality officials should be contacted before implementation (7). The guide suggests that a particulate standard of 450 $\mu g/m^3$ over an 8-h period may be a reasonable method for applying the EPA criteria, assuming no emissions during the other 16 h. It is recommended that background levels be established by monitoring before beginning work.

The SSPC guide also provides suggestions on converting the lead criteria to a daily value by determining the number of days work will be performed out of 90 and the number of hours in the day that lead abatement would occur. This would yield an adjusted daily allowance (ADA) for lead emissions. For example, working 45 days out of a quarter for 8 h per day would yield an average daily allowance of 9 $\mu g/m^3$ per 8-h workday, as shown in equation 1:

$$\begin{aligned} \text{ADA} &= \left(\frac{90 \text{ days}}{45 \text{ work days}}\right) \times 1.5\ \gamma g/m^3 \times \left(\frac{24 \text{ h}}{8 \text{ work hours}}\right) \\ &= 9\ \mu g/m^3 \end{aligned} \tag{1}$$

Water

In the United States the EPA has established water quality standards for freshwater (8). Additional regulations require a permit for the discharge of a debris into U.S. waters (9). These permits are generally not issued for lead-based paint debris discharge. Therefore, lead-based paint debris must be collected, contained, properly treated, and disposed. Activities during lead-based paint abatement must be conducted to prevent water runoff that could reach water or storm sewers and eventually pollute our nation's streams and rivers.

Soil

Activities during lead-based paint abatement should also be conducted to prevent additional contamination of the soil. In addition, lead-contaminated soil is a concern both as a direct source of exposure through inadvertent ingestion due to children's normal hand to mouth activities and as a contributor to indoor dust lead levels when tracked inside. Common sources of lead in soil include deteriorating lead-based paint and airborne deposition from point source emissions or leaded gasoline. The background concentration of lead in soil generally ranges from 5 to 50 ppm of lead. Soil lead concentrations vary widely, from <1 ppm to tens of thousands parts per million. The EPA recommends that for bare soils with lead concentrations exceeding 400 ppm evaluation be conducted and physical exposure be reduced, commensurate with the expected degree of risk (10).

Waste Disposal

The primary federal statute that regulates hazardous waste disposal in the United States is RCRA. RCRA defines the criteria of nonhazardous and hazardous waste and regulates all solid waste, which includes liquid, solid, and some gaseous wastes. Most waste generated in lead-based paint abatement activities is solid waste, which can be nonhazardous or hazardous. In addition, solid waste is generally regulated on the state and local levels.

RCRA defines the generator as the facility or location whose act produces a hazardous waste. In the case of lead-based paint abatement, both the property owner and contractor can be considered to be the generator. The generator must determine if the material is defined as hazardous by RCRA. The waste can either be a listed waste or have characteristics of a hazardous waste. Generators must test their waste or use existing knowledge to determine if it exhibits hazardous characteristics. Wastes exhibiting one or more of the following characteristics are considered hazardous: toxicity, corrosivity, ignitability, and reactivity.

Waste toxicity is measured by using the EPA's Toxicity Characteristic Leaching Procedure (TCLP). The waste is granulated and subjected to an acid leach for 18 h. The resulting extract is analyzed for lead (or other elements); and if the concentration is above the regulatory limit, the waste is classified as hazardous. The regulatory limit for leachable lead is 5 mg/L (ppm) (11).

Waste is classified as hazardous due to corrosivity if it is highly acidic, with a pH ≤ 2, or highly basic, with a pH >12.5. Unneutralized caustic chemical paint strippers may be corrosive. Ignitable wastes include liquids with a flash point of $<140°C$, flammable solids, compressed gases, and oxidizers. Certain solvents from organic paint strippers may be ignitable. Reactive wastes are wastes that are capable of generating explosive or toxic gases and are not likely to result from lead-based paint removal processes (12). RCRA requires that all hazardous wastes meet strict requirements before they can be disposed in a landfill. In general, these requirements mandate treatment before disposal.

Containment

Lead-based paint control activities generate lead-contaminated dust, paint chips, and other materials. Containment systems are used to protected the environment during lead-based removal activities. Such a system can range from the use of disposable plastic dropcloths to the complete enclosure of the work area. The type of containment depends on a number of consideration such as the type of the hazard and the size of the work area. Generally, lead-based paint control activities should be conducted in vacant units. Residents and occupants should never be permitted in the work area while work is under way. Resident reentry should be permitted only after the area has been cleaned and passed clearance. HUD guidelines provide additional descriptions of site preparation and containment requirements for lead-based paint control activities in residential housing.

The containment system includes the containment structure itself (containment wall, ceiling, floor, supporting structure and entryways), a ventilation system (consisting of forced or natural air input, and natural or mechanical exhaust), and in many cases a dust collection system. Inside the containment structure, airborne lead levels can be extremely high, requiring the most stringent worker protection.

The containment materials can consist of either rigid materials, such as plywood, aluminum, rigid metal, or plastic, or flexible materials, such as tarpaulins, wind screens, plastic sheeting, or cloth drapes. When dust-producing processes such as abrasive blasting are used, the containment material must be impenetrable to dust and wind. Impenetrable materials must be used with negative air pressure systems. Wind screen materials, such as open-mesh fabrics, which retain larger particles but do allow some dust to escape, should not be used when 100% containment is required (13).

The containment structure should be designed and constructed to support the material, equipment, and operators who will work inside. Scaffolding and flooring should be capable of safely supporting this combined weight. In the case of abrasive blasting this may also include the weight of the blasting materials and the waste (14).

Joints between containment materials must also be sealed. The result can be either partial or complete seals. Sealing materials can include, tape, caulk, urethane foam, Velcro, or other material or a combination of these.

Access to the containment structure must also be provided. It may consist of fully sealed air locks, resealable doorways, or the use of overlapping tarps or sheeting. Fully sealed doors with air locks require entryways that may be opened and then completely resealed. This may consist of a series of chambers with separating doorways. Inside one of the chambers, washing or vacuuming equipment may be present to allow the decontamination of the workers and equipment to prevent the transmission of lead hazards from inside the containment structure to the environment.

Containment structures must also be designed to permit the transmission of debris from the work area to a storage or reclamation area. Debris may be removed via a funnel or hopper system, vacuum systems, conveyor system, or manual systems such as buckets or wheelbarrows. The workers transporting the debris are subjected to the same requirements for worker protection.

The ventilation system depends on the type of abatement being performed. OSHA has published rulings on ventilation system requirements. Ventilation systems include the intake hood or device, duct work, air filtering system, fan, and exhaust stack. Ventilation systems are often designed to provide negative air pressure. This is an air pressure below atmospheric pressure, so that in the event of a leak, the natural tendency is for air to flow from outside into the containment structure which prevents the hazardous lead dust from escaping (15).

Air movement inside the containment structure should be sufficient to maintain the visibility of the workers. The SSPC recommends a cross-draft velocity of 2.83 m^3/min (100 cf/min) and a down draft velocity of 1.70 m^3/min (60 cf/min) (16). Air movement should also be sufficient to prevent heat exhaustion of the workers.

Dust collection is required to remove dust from the air before exhausting it to the environment. Dust collection typically involves wet or dry collection systems, such as bag houses or fabric filters. The fan must be properly sized to provide the required air movement.

Worker Protection

OSHA in 1993 issued standards for lead exposure in the construction industry, which are the minimum requirements that employers must follow to protect workers during lead hazard control projects (17). Employers should refer directly to the construction lead standard for the complete requirements.

The standards define a personnel exposure limit (PEL) as 50 $\mu g/m^3$, which is expressed as an 8-h time-weighted average. Thus if an individual works longer than 8 h, the PEL is reduced by a time-weighted average; for example, a 10-h shift has a PEL of 40 μg/m^3. Various provisions of this standard are triggered by an action level of 30 μg/m^3.

The standards require an initial assessment of airborne lead to determine the appropriate worker protection equipment that shall be used. The assessment is usually accomplished through air monitoring. Until the initial exposure assessment is completed, OSHA standards require the use of interim protective provisions, including respiratory protection, protective clothing and equipment, change areas, hand washing facilities, training, and initial blood sampling and analysis. The employer can discontinue the interim protection for an employee after it has been documented that the employee's exposure is below the PEL.

OSHA standards require the use of respirators when an employee's exposure exceeds the PEL and as interim protection for tasks specified in the task-related triggers. It also requires respirators whenever an employee requests their use. Because there are recognized health effects at blood levels below what is allowed by OSHA, employees may request respirators even when their exposure is below the PEL. Generally, in the absence of other contaminants, half-mask, air-purifying respirators with high efficiency particulate air (HEPA) filters should be adequate for atmospheres with up to 500 μg/m^3 lead. The OSHA standard lists the criteria for the proper selection of respirators (18).

OSHA standards further require that employers provide protective clothing whenever employees are exposed to airborne lead levels above the PEL. Hard hat, goggles, safety shoes, hearing protection, and other personnel protection equipment may also be required by other OSHA standards.

To minimize additional employee exposure to lead, OSHA requires that employers provide hygiene facilities and ensure that good hygiene practices are used. This includes providing change areas, showers, eating areas, and hand washing areas. Good hygiene practices are intended to prevent employee exposure form ingestion or inhalation. They are also intended to prevent the contamination of worker's vehicles or home. Many symptoms of lead exposure are nonspecific and thus worker education in the signs and symptoms of lead exposure should be an important component of worker training.

Workers must undergo both initial and periodic medical surveillance, depending on the level and duration of the airborne lead exposure. This includes the determination of blood lead levels. The OSHA construction industry standard requires that an employee be removed from airborne exposure of lead at full pay until the lead blood level is $<$40 μg/dL (19).

LEAD-BASED PAINT HAZARDS IN RESIDENTIAL HOUSING

More than one-half of the entire housing stock in the United States is believed to contain some lead-based paint; this includes the approximately 57 million housing units built before the 1978 ban on lead in residential paints. In 1990, The U.S. Department of Housing and Urban Development estimated that 13.8 million housing units contain chipping or peeling lead-based paint (20). The study indicated that 6.2 million housing units had elevated lead-contaminated dust levels. Of the 20 million units projected to contain lead-based paint hazards, 3.8 million were estimated to house children under the age of 7. Young children are more vulnerable to adverse medical consequences from lead ingestion.

Historical structures present an additional challenge to lead-based paint abatement. Historic preservation requirements restrict the use of removal methods that may damage the substrate, such as abrasive blasting of wood surfaces. Furthermore, other options such as the use of encapsulants or replacement of components (windows or doors) are not compatible with the requirements for the preservation of historic structures.

LEAD-BASED PAINT HAZARD MITIGATION PROCESSES

Abatement is defined as the removal of either the building component or the paint itself, or the near permanent (25-year) enclosure of the lead-based paint hazards. Proper abatement is the desired response to lead hazards. Abatement strategies include removal of lead-based paint, enclosure of lead-based paint, encapsulation of lead-based paint, and replacement of building components (21).

Removal of lead-based paint involves the greatest degree of disturbance and dust generation. Therefore, onsite removal of lead-based paint from a substrate should be avoided if possible. Enclosure, encapsulation, and building component replacement are the least invasive and most preferred hazard mitigation methods.

Abatement methods prohibited by HUD in residential settings include the following:

- Open flame burning or torching.
- Machine sanding or grinding without high efficiency particulate air (HEPA) attached vacuum systems.
- Uncontained water blasting or high pressure water wash.
- Abrasive blasting or sandblasting without HEPA attached vacuum systems.
- Heat guns operating above 593°C (1100°F).

Emerging Technologies

Emerging technologies for the removal of lead-based paints include lasers, thermal spray vitrification, and microwaves.

Lasers. Laser paint removal process are currently under development and have been identified for use on historic structures. Abrasive blasting, with or without a chemical stabilizer, is often too aggressive for use on fragile historic structures. Chemical paint stripping is often the most appropriate method for removing lead-based paint from wood surfaces in historic structures. However, chemical stripping is expensive due to the large volume of liquid hazardous waste produced and the requirements for worker protection.

Laser paint removal systems have been designed and built for use on fragile historic wood structures. The system contains a CO_2 pulse laser-and-beam delivery system. The laser units also have a self-contained waste collection system that is designed to contain the hazardous lead-based paint debris that is produced and reduce worker protection requirements. The laser is controlled through a multispectral camera and computer program. The spectrum of each paint color to be removed is stored in the computer memory. The computer instructs the laser to fire when these paint colors are seen by the spectrograph; thus the laser can be instructed to remove all coats of paint or to remove only the top coats(s), leaving the primer intact. The system applies the laser to a 1 cm^2 target area for 20 to 30 μs and then moves onto the next target area. The laser continues row by row through a 12 × 12-cm frame and then returns to the start of the frame and repeats the process. This rastoring process allows each area to cool before being processed again, thus reducing thermal damage to the substrate. Once the frame is devoid of the specific color, the system is moved to a new area. The system is designed to remove paint from both interior and exterior wood surfaces.

Evaluation of the laser paint removal system by U.S. Army Construction Engineering Research Laboratories (USACERL) showed potential as a paint removal technology for use on historic wood structures as lead-based paint was successfully removed. The advantages included no containment costs, no requirements for worker protection, and reduction of hazardous waste compared to chemical paint strippers. However, the laser paint removal system far exceeded any present abatement removal techniques in terms of cost and low production rates. The slow production rates brings the cost above any technologies currently used to remove lead-based paint from historic structures. Although the theory and technology of laser based system were demonstrated, further engineering enhancement will be necessary to make the process cost-effective (22). Advantage and disadvantages are shown in Table 1.

Table 1. Laser Paint Removal

Advantages	Disadvantages
Less hazardous waste	Higher cost
Containment costs reduced	Low production rate
Worker protection not required	

Thermal Spray Vitrification. The thermal spray vitrification (TSV) process to remove hazardous lead-based paint was developed and patented by the USACERL (23). The process consists of thermal spraying a molten glass onto the painted surface from which the deteriorating paint is to be removed. Dried glass powder is melted in the high temperature flame of the thermal spray torch. When the glass strikes the substrate, it is molten and reacts with the paint on the substrate. The organic components of the paint are pyrolyzed, while the lead ions are trapped within the silicate tetrahedra network of the glass. The leaded glass and carbon vitrifies, immobilizing the lead ions within the glass network, thereby preventing leaching. The coefficient of thermal expansion difference across the glass-substrate interface and the quenching stresses in the glass cause the glass to crack and spall off the substrate. Remelting the residue results in a nonhazardous waste, and the lead oxide is immobilized in the glassy iron silicate matrix (24).

The TSV process is dust free, eliminating the need for containment. The advantages of the thermal spray vitrification process are the significant savings from not requiring containment structures and the reduced cost of worker protection. The formation of a nonhazardous waste also decreases waste disposal costs. The TSV process can be used to remove lead-based paint from steel.

A simulated nuclear waste glass composition developed by the Savannah River Laboratory for treatment of nuclear waste was modified to serve as the glass waste form for the vitrification of lead, chromium, and cadmium. The iron borosilicate glass composition was considered because of its ability to accommodate a wide variety of hazardous species, its outstanding long-term chemical durability, and its corrosion resistance to a wide range of environmental conditions.

To determine the stabilization capacity of the glass, crucible melts were prepared with increasing amounts of PbO (0–40 wt %) added to the iron silicate glass. TCLP was used to determine durability of the glass for lead. The TCLP results from the lead-loading experiment revealed that the waste form is capable of safely storing up to 25 wt % PbO without further modification.

The iron borosilicate glass powder was sprayed and fused using an oxyacetylene flame spray system onto steel panels coated with lead-based paint. The glass fragments of TSV process were collected from the multiple applications and remelted. The resulting glass was found to be a nonhazardous waste as determined by the EPA's toxicity characteristic leaching procedure (TCLP) test. Lead removal was determined by X-ray florescence (XRF) testing. The lead concentration on the panel steeply decreased with the number of spray applications. The results showed that thermal spray vitrification can remove lead from the substrate to <1.0 mg/cm^2 in two to three passes. The accuracy of XRF measurements below 1.0 mg/cm^2 are questionable and supplementary testing using wipe samples is recommended before making worker protection decisions.

During laboratory testing, the certified results of the airborne lead concentration were below the EPA NAAQS limit of 1.5 μg/m^3 (90-day average), with an adjusted daily allowance of 27 μg/m^3. The certified results of personal exposure monitoring of lead were less than the OSHA action level of 30 μg/m^3 (8-h exposure).

The TSV process creates less hazardous dust and waste than does abrasive blasting and reduces the need, and the associated costs, for containment structure, environmental monitoring, worker health protection, and hazardous waste disposal. The potential benefits from this process include reduced costs of environmental compliance and worker protection associated with the removal of hazardous lead-based paint from steel structures (Table 2).

Microwaves. The microwave-assisted removal (MAR) of paint process is under development at the USACERL and has been patented (25). The MAR process has potential applicability on lead-based paint removal form historic structures and from delicate composite materials. In the MAR process, microwave coupling compounds are applied as a waterborne slurry or as a polymer binder paste to the painted surface. Microwaves have the unique ability to rapidly and selectively heat the coated surface. Compounds such as graphite or silicon carbide can reach temperatures up to 1000°C (1832°F) in <1 min when exposed to microwaves (800 W). The microwave applicator uses standard 2.45 GHz magnetron tubes similar

Table 2. Thermal Spray Vitrification Process

Advantages	Disadvantages
Nonhazardous waste form	Multiple application of the process required for complete removal
Dust-free process	
Reduced containment	
Reduced worker protection	

to those used in household microwave ovens. The applicator is designed to focus microwave energy at the interface where concentrated microwave energy can be used effectively. The paint is debonded from the substrate by heat from the microwaves and is removed easily. A microwave shield is provided for worker protection. In addition, safety switches as well as feedback monitoring and control of the microwave energy make the system safe for workers to operate. No extra clothing or suits are required for the operators. Since the airborne lead levels are below the EPA and OSHA threshold requirements, containment structures, environmental monitoring, and worker health monitoring are not required in this process. Demonstrations of microwave-assisted paint removal conducted in the laboratory successfully removed lead-based paint from wood substrates.

A portable microwave assisted paint removal system has been developed. The unit consist of two parts, the applicator and power supply. The power supply contains the microwave generation components, control panel, and safety interlock circuitry. The control panel contains forward and reflected power meters, which are used to monitor the microwave power delivered to the painted surface. The power supply is connected to the applicator by a flexible high power coaxial cable. The applicator contains and directs the microwave energy onto the painted surface and provides safety switches and shielding to prevent microwave leakage. The aperture is a specially designed microwave window that couples the microwave energy from the waveguide to the painted surface. Susceptor coatings are brushed or sprayed onto the paint to facilitate microwave energy absorption. The hand-held applicator is moved slowly across the painted surface to heat and debond the paint. The operator then scrapes the debonded paint and collects the residue for proper disposal. Exhaust ventilation must be provided as smoke and other fumes are given off when the paint is heated. Table 3 lists the advantages and disadvantages of the MAR process.

Table 3. Microwave Removal Process

Advantages	Disadvantages
Dust-free process	Potential substrate damage
Reduced containment	
Reduced worker protection	

SUMMARY

In the United States, there are about 57 million homes with some lead-based paint. The greatest risk from lead-based paint hazards is to the children below the age of 6. The risk arises from deteriorated lead-based paint, lead-contaminated dust, and lead-contaminated soil. A risk assessment should be conducted according to HUD guidelines; and then, based on the results, a lead hazard control plan should be developed. The lead hazard control plan may include interim controls such as encapsulants and enclosures. Component removal or paint removal should be considered for long-term hazard abatement. There is no one removal technique that is applicable for all substrates. Abrasive blasting, power tools, heat guns, and chemical strippers are the most commonly used technologies for paint removal. Abrasive blasting needs a containment structure and worker protection and cannot be used in the interior of buildings because it creates a lead contaminated dust hazard. Chemical strippers create a large amount of rinse water waste, which is usually hazardous. Power tools are rather slow and labor intensive. Heat guns must operate at <315°C, otherwise the organic binder will degrade and cause a lead-contaminated dust hazard. Emerging technologies include thermal spray vitrification and laser for steel and microwave-assisted paint stripping for wooden substrates. A life cycle cost analysis should be conducted to take advantage of the local conditions, and then proper interim controls and abatement technologies should be chosen.

APPENDIX A: SUMMARY OF LEAD STANDARDS

Lead in paint: Consumer Products Safety Commission standards
- 1.0 mg/cm^2 or 5000 μg/g

Lead in dust levels for risk assessment (by wipe sampling): EPA standards
- Floors: 1075 μg/m^2 (100 μg/ft^2)
- Interior windowsills: 5379 μg/m^2 (500 μg/ft^2)
- Window troughs and wells: 8607 μg/m^2 (800μg/ft^2)

Dust levels for lead hazard screen only: EPA standards
- Floors: 537 μg/m^2 (50 μg/ft^2)
- Window troughs and wells: 4303 μg/m^2 (400 μg/ft^2)

Leaded dust clearance levels by wipe sampling: HUD standards
- Floors: 1075 μg/m^2 (100 μ/ft^2)
- Interior windowsills: 5379 μg/m^2 (500 μg/ft^2)
- Window troughs and wells: 8607 μg/m^2 (800 μg/ft^2)
- Exterior concrete surfaces: 8607 μg/m^2 (800 μg/ft^2)

Bare residential soil: EPA standards
- Painting or removal criteria: 5000 μg/g
- Building perimeter and yard: 2000 μg/g
- Play areas and high contact areas for children: 400 μg/g

Airborne particulates: OSHA standards
- Action level (8-h time-weighted average): 30 μg/m^3
- PEL (8-h time-weighted average): 50 μg/m^3

Elevated blood lead levels for children age 1–5: CDC standards
- Intervention level for children: 20 μ/dL (or 15–19 μg/dL in two consecutive samples take several months apart)

Hazardous waste: EPA standards
- By TCLP: 5 mg/L (ppm)

ACKNOWLEDGMENTS

We would like to acknowledge the assistance of S. Drozdz of the USACERL, Victoria Belfit of the U.S. Army Center for Health Promotion and Preventive Medicine, and B. Nix of the office of the U.S. Army Assistant Chief of Staff, Installation Management (ACSIM) for providing invaluable review and comments on the manuscript. Funding was provided by the Strategic Environmental Research and Development Program (SERDP). J. H. Boy would like to acknowledge that the work was supported in part by an appointment to the

Research Participation Program at USACERL administered by Oak Ridge Institute for Science and Education through an interagency agreement between the U.S. Dept. of Energy and USACERL.

BIBLIOGRAPHY

1. Lead Based Paint Hazard Reduction and Financing Task Force, *Putting the Pieces Together: Controlling Lead Hazards in the Nation's Housing,* U.S. Department of Housing and Urban Development, Washington, D.C., July 1995.
2. W. L. Roper, *Preventing Lead Poisining in Young Children,* Centers for Disease Control and Prevention, Atlanta, Ga., Oct. 1991.
3. 16 *CFR* 1303.
4. *Morbid. Mortal. Weekly Rep.* (Feb. 21, 1997).
5. 40 *CFR* 50.
6. 40 *CFR* 50.12.
7. Steel Structures Painting Council, "Guide for Containing Debris Generated during Paint Removal Operations," *Guide 6 (CON),* SSPC, Pittsburgh, Pa., 1995.
8. *Fed. Reg.* **50,** 30784 (July 29, 1985).
9. 40 *CFR* 122.1.
10. L. R. Goldman, "Guidance on Residential Lead-Based Paint, Lead Contaminated Dust, and Lead Contaminated Soil," *PB94-963284,* U.S. EPA, Washington, D.C., July 1994.
11. 40 *CFR* 50.6.
12. 40 *CFR* 261.20–23, subpart C.
13. K. A. Trimber, "Industrial Lead Paint Removal Handbook, 2nd ed., *SSPC93-02,* KTA-Tator, Pittsburgh, Pa., 1993.
14. 29 *CFR* 1926.451.
15. 29 *CFR* 1920.94.
16. Steel Structures Painting Council, "Guide for Containing Debris Generated during Paint Removal Operations," *SSPC Guide (Con),* SSPC, Pittsburgh, Pa., Mar. 1993.
17. 29 *CFR* 1920.
18. 29 *CFR* 1920.103.
19. 29 CFR 1920.62.
20. U.S. Department of Housing and Urban Development, "Comprehensive Workable Plan for the Abatement of Lead-Based Paint in Privately Owned Housing," *Report to Congress,* Office of Policy Development and Research, Washington, D.C., Dec. 7, 1990.
21. U.S. Department of Housing and Urban Development, *Guidelines for the Evaluation and Control of Lead-Based Paint Hazards in Housing,* Washington, D.C., June 1995.
22. V. F. Hock, C. M. Gustafson, D. M. Cropek, and S. A. Drozdz, "Demonstration of Lead-Based Paint Removal and Chemical Stabilizations Using Blastox," *FEAP TR 96/20,* U.S. Army Construction Engineers Research Laboratories, Champaign, Ill., Oct. 1996.
23. U.S. Pat. 5,292,375 (Mar. 8, 1994), A. Kumar and J. P. Petreanu (to Secretary of the Army).
24. S. W. Covey, J. L. Lattimore, and A. Kumar, paper presented at the *8th National Thermal Spray Conference,* 1995.
25. U.S. Pat. 5, 268,548 (Dec. 1993), A. Kumar (to Secretary of the Army).

LEAD-BASED PAINT: TESTING

JOHN SCHWEMBERGER
Consultant
Rockville, Maryland

In recent years, lead-based paint testing has taken on importance as public policy in the United States has emphasized preventing of lead poisoning instead of reacting to lead poisoning (see LEAD-BASED PAINT: HAZARD MITIGATION). Deteriorated lead-based paint, lead-based paint on friction, impact, or mouthable surfaces, and lead-based paint on surfaces disturbed by renovation or remodeling are regarded as primary sources of lead dust in housing. This dust may be ingested by children, with potentially harmful effects. Testing of paint can be accomplished by laboratory analysis of paint samples, by measurements of paint in place with a portable X-ray fluorescence instrument, or by a combination of the two

Laboratory analysis is usually carried out by atomic spectroscopic analysis, regarded as the most accurate way to test for lead in paint. Residential testing is important because preschool-age children are the group most vulnerable to lead exposure and spend most of their time in their homes. Testing of nonresidential structures has increased because of heightened awareness of lead hazards, with the goals of reducing worker exposures to lead and releases of lead to the environment and the ecology and reducing lead exposure for the population as a whole.

RESIDENTIAL STRUCTURES

Lead in Painted Housing Components

Paint and Its Constituents. Paint consists of a solid material suspended in a liquid. Paint is usually applied to surfaces in housing units to create a protective covering or to provide decoration. The largest portion of the solid material in a paint is usually referred to as the pigment, and the liquid is known as the vehicle or binder. The vehicle dries after the paint is applied, and thereby, the pigment adheres to the painted surface. In addition to the pigment, other solid materials referred to as additives may be included in the manufacturing process to modify the characteristics of the paint. By definition, additives are a small proportion of the total solid material. Most paints will contain one or more pigments and a number of additives as solid materials.

In the past, lead compounds have been used in pigments or additives for house paints in the United States. Lead compounds were used as primary or white pigments, as color pigments, as corrosion-inhibiting pigments, and as driers, a type of additive which accelerates the drying of paint after it is applied to a surface. The largest use of lead was in the primary or white pigments. The primary pigments containing lead were basic lead carbonate, lead sulfate pigments, lead silicate pigments, and leaded zinc oxide. It is generally recognized that most of the mass of lead on painted housing components is attributable to basic lead carbonate. Basic lead carbonate, also known as basic carbonate white lead or white lead, is usually represented by the chemical formula $2\ PbCO_3{\cdot}Pb\ (OH)_2$, and its main constituent is lead carbonate ($PbCO_3$). Basic lead carbonate was used in formulating single pigment paints and paints containing more than one pigment. Lead sulfate compounds were constituents of other leaded pigments, as the paint industry used lead sulfate ($PbSO_4$) and monobasic lead sulfate ($PbSO_4{\cdot}PbO$), and combinations of the two, in pigments. A common lead primary pigment was basic sulfate white lead, also known as sublimed white lead. It is represented by the chemical formula $2\ PbSO_4{\cdot}PbO$. Lead silicate pigments have a long history in the paint industry.

The lead silicate compounds in pigments were different forms of the monobasic lead silicate, whose chemical formula is approximated by 3 $PbO \cdot 2 SiO_2 \cdot H_2O$. Lead silicate compounds were generally used in mixed pigment paints. Leaded zinc oxide, originally known as zinc white lead, was a combination of zinc oxide (ZnO) and basic sulfate white lead.

Color pigments containing lead compounds include so-called chrome greens, chrome yellows and oranges, green shade chrome yellows, and molybdate oranges and reds. Chrome greens contained lead sulfate. Chrome yellows and oranges included lead chromate ($PbCrO_4$) as a constituent, and usually included either lead sulfate or lead monoxide (PbO). Green shade chrome yellows contain lead chromate and lead sulfate. Molybdate oranges and reds include a mixture of lead chromate, lead sulfate, and lead molybdate ($PbMoO_4$). Corrosion-inhibiting pigments were not commonly found in house paints as the primary or color pigments, and corrosion-inhibiting pigments still have application in industrial or commercial settings to reduce or prevent corrosion on metal surfaces. There are a number of lead compounds in this class of pigments: red lead or trilead tetroxide (Pb_3O_4), dibasic lead phosphite (2 $PbO \cdot PbHPO_3 \cdot 1/2H_2O$), calcium plumbate (2 $CaO \cdot PbO_2$), basic lead chromate ($PbCrO_4 \cdot PbO$), blue basic lead sulfate (also known as sublimed blue lead), and basic lead silicochromate. Blue basic lead sulfate contains lead sulfate, lead monoxide, lead sulfide (PbS), and lead sulfite ($PbSO_3$). Basic lead silicochromate consists of basic lead chromate bonded to silica. Although rare, lead metallic flakes were added to some special types of paints. A number of other lead compounds, such as lead salicylate, lead cyanamide, lead chloride, lead fluoride, and others were rarely used. Lead monoxide and lead metallic flakes were included in formulating of some varnishes (1–6).

The History of Lead in Paint. Paint was not commonplace in colonial America. However, by 1795, this had changed, and painting of public buildings and housing gained acceptance. A factory for production of white lead was established in Philadelphia in 1804, and white lead, red lead, and chrome yellow pigments were available in the early 1800s (7). The use of white lead in paint continued in the United States during the nineteenth century, and, by 1890, there were approximately 40 factories in the United States producing white lead. In Australia, in the latter part of the nineteenth century and the early part of the twentieth century, lead in paint was causing lead poisoning of children, and steps to reduce lead exposure from paint were taken in the state of Queensland. By the mid-1920s, medical journals in the United States were linking lead poisoning in children to lead-based paint. In the 1920s, a number of European countries adopted restrictions on the use of lead in paint (8). The United States, however, did not take any action based on either the Australian findings or the medical findings in the United States nor did it ratify the restrictions adopted by some of the European governments. The use of lead in paint in the United States continued in the first half of the twentieth century, with the highest concentrations of lead in paint likely from 1900 to 1920 or 1930. Beginning in the 1940s, the use of lead in paint started to decline. Two developments that began earlier, the use of other primary pigments, such as lithopone and titanium dioxide, and the increasing popularity of latex paints, reduced the amount of lead in paint. Starting in the 1920s, lithopone was used in interior paints, and titanium dioxide began to replace lead-containing pigments in the 1940s. Lead compounds were usually not used in latex paints because of the instability of the emulsion. In addition, by the 1950s, city governments in the United States had begun to pass legislation restricting the use of leaded paint in residences because of the linkage of leaded paint to lead poisoning of children. For example, in 1951, Baltimore banned further use of leaded paint in the interiors of residences because of 83 deaths of children in Baltimore from 1941 to 1951 attributed to lead poisoning (9). In the mid-1950s, the painting industry voluntarily adopted a standard for the maximum amount of lead in interior paint, set at 1% lead by mass (10,000 μg/g) of the non volatile content of the paint.

The 1970s were a period of extensive activity by the federal government. In 1971, the federal Lead-Based Paint Poisoning Prevention Act (LBPPPA) defined lead-based paint as paint with more than 1% lead. In 1972, the U.S. Department of Housing and Urban Development (HUD) banned the use of lead-based paint in housing constructed by the federal government or assisted by the federal government. In 1973, federal legislation amending the LBPPPA reduced the allowable amount of lead in paint for housing with federal construction or assistance to no more than 0.5% (5000 μg/g) and to no more than 0.06% (600 μg/g) after December 31, 1974 unless the Consumer Product Safety Commission (CPSC) declared that a level greater than 0.06% was safe. In 1974 CPSC considered 0.5% a safe level, and so the level for the maximum allowable lead content remained at 0.5%. Additional amendments to the LBPPPA in 1976 reduced the allowable amount of lead in paint to 0.06% unless CPSC decided that a limit above 0.06% and less than 0.5% was safe. In 1977 CPSC found that a limit above 0.06% is unsafe. Following the provisions of the amendments, as of June 23, 1977, the maximum allowable amount of lead in paint for housing assisted or constructed by the federal government was set at 0.06% (600 μg/g). In 1978, the Consumer Product Safety Commission banned the use of paint exceeding 0.06% lead by mass of the nonvolatile content or by mass of the dried paint film in residences or other areas, where consumers had direct access to painted building components, on furniture and toys, and banned the sale of such paint to consumers. When the ban was promulgated, CPSC estimated that 90% of the latex paints and 70% of the oil paints were already below the 0.06% limit (10).

The Extent of Lead-Based Paint in Housing in the United States. The most systematic and consistent information on the extent of lead-based paint in the United States is derived from a national survey of housing conducted by HUD in 1989 and 1990. Although there have been studies of the prevalence of lead-based paint in some cities in the United States, the HUD survey is the only national survey of lead-based paint in housing. The survey was restricted to housing built before 1980. The restriction to pre-1980 housing was made on the assumption that housing built after 1979 was unlikely to contain lead in paint above 0.06%. The survey was restricted to the continental United States and, by design, omitted housing devoted to the elderly, military housing, institutional housing, and vacation homes. There were 77 million private housing units built before 1980 were included in the scope of this survey.

Lead in paint was measured in units of milligrams of lead per square centimeter (mg/cm^2). Lead-based paint was defined as paint at or above 1.0 mg/cm^2 (0.0021 lbs avdp/ft^2) of lead. A rough rule of thumb is that one milligram of lead per square centimeter is approximately 1% lead by weight (11,12). Under this definition, it was estimated that 64 million housing units, or 83% of the pre-1980 privately owned housing units in the scope of the study, have lead-based paint somewhere on the interior or exterior. However, the square footage coated with lead-based paint for units with lead-based paint was relatively small on average: about 56 m^2 (600 ft^2) in the interior and about 81 m^2 (870 ft^2) on the exterior. For the set of houses with lead-based paint on the interior, lead-based paint coated 12% of the interior painted surfaces, by square footage. For the set of houses with lead-based paint on exterior surfaces, lead-based paint covered 44% of the exterior painted surfaces, again by square footage.

It was estimated that 14 million private housing units built before 1980, or 19% of the 77 million units in the scope of the survey, contain nonintact lead-based paint on more than 0.5 m^2 (five ft^2). Where such nonintact paint was found, dust and soil levels were more likely to exceed guideline thresholds for lead in dust and soil. Of the houses with nonintact lead-based paint on the interior, 34% had dust levels exceeding guideline levels. For houses with nonintact paint on the exterior, 53% had dust levels exceeding guideline levels. By comparison, 12 to 18% of those houses with no lead-based paint, intact interior lead-based paint, or intact exterior lead-based paint had dust levels above guideline levels. The case for exterior lead-based paint and soil was similar. The survey found that 56% of the houses with non intact lead-based paint on the exterior had soil levels above guideline levels. Of the houses with intact exterior lead-based paint, 20% had soil levels above guideline levels. Of the houses with no exterior lead-based paint, 9% had soil levels above the guidelines. The survey also estimated that of the 64 million private housing units with lead-based paint, 12 million were inhabited by children under the age of seven. In 3.3 million of these 12 million housing units, nonintact lead-based paint was present on at least 0.5 m^2 (five ft^2) of painted surface area. In 1.7 million of the 12 million units, lead-based paint was present and lead dust was above guideline levels.

The distribution of lead levels observed in the sample of pre-1980 private housing was skewed, with many observations less than 1.0 mg/cm^2. The median lead level (or 50th percentile) for interiors was 0.19 mg/cm^2 (0.00039 lbs avdp/ft^2), and the corresponding arithmetic mean was 0.81 mg/cm^2 (0.0017 lbs/ft^2). The 75th percentile was 0.60 mg/cm^2 (0.0012 lbs/ft^2), and the 90th percentile was 1.66 mg/cm^2 (0.00340 lbs/ft^2). This indicates that between 10% and 25% of the survey's interior measurements made in private housing units met the survey's definition of lead-based paint. Lead measurements for exteriors and for the common areas in multifamily housing units were higher than those for interiors. The median for exterior surfaces was 0.42 mg/cm^2 (0.00086 lbs/ft^2), with an arithmetic mean of 2.07 mg/cm^2 (0.00424 lbs/ft^2). The 75th percentile for exteriors was 1.85 mg/cm^2 (0.00379 lbs/ft^2), and the 90th percentile was 5.81 mg/cm^2 (0.0118 lbs/ft^2). This means that between 25% and 50% of the exterior measurements in private housing units met the definition of lead-based paint. For common areas in multifamily housing, the median was 0.70 mg/cm^2 (0.0014 lbs/ft^2), the arithmetic mean was 2.10 mg/cm^2 (0.00430 lbs/ft^2), the 75th percentile was 2.20 mg/cm^2 (0.00451 lbs/ft^2), and the 90th percentile was 5.54 mg/cm^2 (0.0113 lbs/ft^2). As with exterior surfaces, between 25% and 50% of the survey measurements in the common areas in private housing met the definition of lead-based paint.

Table 1 displays the percentage of painted square footage coated with lead-based paint, in pre-1980 housing units with interior lead-based paint, by component and by substrate. The substrate is the material beneath the paint. Table 1 also includes the average square footage with lead-based paint for components and substrates in housing that has some lead-based paint in the interior. Only 9% of the paint on walls, ceilings, and floors was classified as lead-based, but this category accounted for over half the square footage with lead-based paint. Nonmetal components and the Shelves and other category were more likely to have lead-based paint and accounted for most of the remaining interior square footage with lead-based paint. Wood was more likely to have lead-based paint than any other interior substrate. Wood, plaster, and drywall substrates dominated the square footage of interior components with lead-based paint. Table 2 displays similar information for exterior components and substrates. Exterior walls and exterior nonmetal components were most likely to have lead-based paint, and these two components accounted for almost all the exterior square footage with

Table 1. Painted Area (avg. m^2) with Lead-Based Paint in Privately Owned Housing[a,b]

Component/Substrate Components	Painted Area with Lead-Based Paint, % (Interior)	Average m^2 (ft^2) with Lead-Based Paint (Interior)
Walls, ceilings, and floors	9	34.5 (371)
Metal components	4	0.2 (2)
Nonmetal components	24	13.6 (146)
Shelves and other	36	7.6 (82)
Total	*12*	*55.8 (601)*
Substrates		
Wood	26	22.1 (238)
Metal	4	0.3 (3)
Drywall and plaster	9	32.5 (350)
Concrete	3	0.5 (5)
Undetermined	5	0.6 (6)
Total	*12*	*55.8 (601)*

[a] In pre-1980 housing with lead-based paint on the interior.
[b] Ref. 15.

Table 2. Painted Area (avg. m^2) with Lead-Based Paint in Privately Owned Housing[a,b]

Component/Substrate Component	Painted Area with Lead-Based Paint, % (Exterior)	Average m^2 (ft^2) with Lead-Based Paint (Exterior)
Walls	49	63.3 (681)
Metal components	8	0.7 (7)
Nonmetal components	41	15.7 (169)
Porches and other	12	1.2 (13)
Total	*44*	*80.7 (869)*
Substrate		
Wood	46	50.8 (547)
Metal	33	9.0 (97)
Drywall and plaster	46	3.3 (35)
Concrete	40	12.2 (131)
Undetermined	60	5.4 (58)
Total	*44*	*80.7 (869)*

[a] In pre-1980 housing with lead-based paint on the exterior.
[b] Ref. 15.

lead-based paint. By far, wood substrates accounted for most of the exterior square footage with lead-based paint.

The survey divided the private housing stock into three categories by age of housing: those built before 1940, those built between 1940 and 1959, and those built between 1960 and 1979. There were 21 million homes in the first group, 20 million in the second group, and 36 million in the third group, respectively. The housing units built before 1940 had the highest average lead levels in paint, had more square footage with lead-based paint on both the interior and exterior than the other two groups, were the most likely to have deteriorated paint, were the most likely to have lead levels above guideline levels in dust and soil, and had exceptionally high levels of lead in paint on the exterior. The survey report also suggested a number of regional differences: the Northeast, Midwest, South, and Western regions of the United States all had large percentages of private housing with lead-based paint, but the Western region had the lowest percentage; nonintact lead-based paint was more likely in the Northeast and Midwest; levels of lead in dust above guideline levels were far more common in the Northeast than in other regions; and levels of lead in soil were more likely to be above guideline levels in the Northeast and the Midwest.

The HUD National Survey also tested public housing units in developments built before 1980 that could be occupied by families. There are about 910,000 public housing units of this type in the United States. The survey found that 86% of the public housing units available to families had lead-based paint. In contrast to private housing units, the majority of the square footage of lead-based paint was in the interior. On average, units with interior lead-based paint had 34 m^2 (370 ft^2) of lead-based paint in the interior, and units with exterior lead-based paint had 12 m^2 (130 ft^2) of lead-based paint on the exterior. These square footages corresponded to approximately 12% of the interior painted surfaces and 7% of the exterior painted surfaces for family public housing units that had lead-based paint in the interior or on the exterior. The distribution of observations in public housing was skewed, although not to the extent of those in private housing. For interiors in public housing, the survey median was 0.21, the arithmetic mean was 0.58, the 75th percentile was 0.68, and the 90th percentile was 1.74 mg/cm^2 (0.00043, 0.0012, 0.0014, and 0.00356 lbs/ft^2, respectively). Readings on exteriors and common areas were lower than for their counterparts in private housing. The survey median for exteriors was 0.14, the arithmetic mean was 1.22, the 75th percentile was 0.72, and the 90th percentile was 3.44 mg/cm^2 (0.00029, 0.00250, 0.00147, and 0.00705 lbs/ft^2, respectively). The survey median for common areas in multifamily units was 0.31, the arithmetic mean was 1.17, the 75th percentile was 1.08, and the 90th percentile was 2.42 mg/cm^2 (0.00063, 0.00240, 0.00221, and 0.00496 lbs/ft^2, respectively). For both interior and exterior observations, lead-based paint was found in between 10 and 25% of the observations. Approximately 25% of the observations in common areas were classified as lead-based paint using the survey's definition. As in private housing, exterior lead levels were high for the oldest category of housing reported, in this case housing built before 1950 (13–15).

One important limitation of the HUD National Survey is related to the methodology for measuring lead in paint. Lead measurements were obtained from portable X-ray fluorescence (XRF) analyzers brought to the sites in the survey. These analyzers are not as accurate as laboratory analysis of paint samples. Adjustments to the readings from the XRF analyzer instruments were made, based on quality control measurements, to reduce instrumental bias and the variability between instruments. Portable XRF instruments were chosen as the methodology for paint testing so that there would be no destructive paint testing. Performing destructive paint testing was considered, but excluded because it would have reduced cooperation in the survey, which was voluntary.

The HUD National Survey did not include any measures of the lead levels of children who occupied the housing units in the survey. The third National Health and Nutrition Examination Survey (NHANES III), conducted in 1988 through 1992, provided information on the nationwide levels of lead in blood. Although the survey did not provide any data on lead-based paint in housing, its statistics on the levels of blood lead in the national population are worth comparing to the results of the HUD National Survey. A total of 1.7 million children between ages one and five years had blood lead levels above or equal to 10 mg/dL (0.48 μmol/L), the CDC minimum level of concern as of August 1996. This was about 9% of all children in the United States in the 1–5-year age group. Analysis of the survey data indicates that blood lead levels were higher for central city residents than for noncentral city residents, and higher blood lead levels were correlated with low income, low educational background, and residence in the Northeast region of the United States (16,17).

Comparisons of the NHANES III and the HUD National Survey results suggest that the number of housing units with adversely affected children is probably smaller than the number of housing units with both lead-based paint and children present. A joint federal and private sector task force estimated that 5 to 15 million housing units in the United States contain lead hazards (18). The task force recommended that housing units built before 1950 be given priority status in addressing lead hazards. The task force based its recommendation on data from the HUD National Survey, data from the U.S. Bureau of Mines which showed that 92 percent of the tonnage of lead carbonate added to paint was used before 1950, and preliminary data from NHANES III which indicated that children, aged one to five years, living in housing built before 1946 were two times more likely to have blood lead levels above or equal to 10 μg/dL than their counterparts in housing built between 1946 and 1973 (19).

NONRESIDENTIAL STRUCTURES

Why Test for Lead in Paint

The expansion of the federal government's program for lead-based paint and lead-based paint hazards has had an impact on testing of nonresidential structures. There is increasing awareness of the hazards and potential hazards associated with lead-based paint. Testing of nonresidential structures is motivated by concerns over exposures of children, exposures of workers, exposures of the general population, and releases of lead to the environment and the ecology. Of these, exposures of children are the most important, especially of children less than two years of age.

The 0.06% limitation on the amount of lead in paint, effective in 1978, applies to paint and similar coatings in residences, on furniture, on toys, in schools, hospitals, parks, playgrounds, public buildings and other buildings where consumers have direct access to painted surfaces. The following are exclusions or exemptions from the 0.06% limitation:

- Paints and coatings for automobiles and boats
- Agricultural and industrial equipment refinish coatings
- Industrial and commercial building and equipment maintenance coatings, including traffic and safety marking coatings
- Graphic arts coatings (that is, products marketed solely for application on billboards, road signs, and similar uses, and for identification marking in industrial buildings)
- Touch-up coatings for agricultural equipment, lawn and garden equipment, and appliances
- Catalyzed coatings marketed solely for use on radio-controlled model power aircraft
- Lead-containing backing paint in mirrors that are part of furniture
- Artists' paints and related materials
- Metal furniture (excluding metal children's furniture) bearing factory applied lead coatings (20)

Thus, the 1978 standard that applies to residential housing units also applies to schools, hospitals, and public buildings, but does not apply to industrial and commercial buildings or to road markings, street signs, or highway billboards. Lead above 0.06% is allowed in paint for these applications and the others listed above.

Where and How to Test

Nonresidential structures have not received the same amount of attention as residential units, because of the emphasis on reducing childhood lead poisoning. Nonresidential structures can be divided into several categories. The first category includes places where children are likely to spend time. Examples of these are day care centers; nursery schools; kindergartens; elementary schools; middle, junior high, and high schools; and playgrounds. The second category is comprised of places where the general public has free access. This category includes museums, hotels, airports, and parks. The third category is buildings where people work. These include office buildings, retail stores, factories, and warehouses. The fourth category is outdoor structures. These include bridges, water towers, street signs, billboards, and pavement markings. The 1978 standard applies to the first and second category and to those buildings in the third category to which consumers have access, but it does not apply to the structures in the fourth category.

Many of the principles for lead paint testing in residential housing apply to nonresidential structures in the first three categories. In general, procedures on how to test carry over. The one obvious difference between residential structures and nonresidential structures is that, in nonresidential structures, rooms, exterior sides, and room equivalents are likely to be larger than those in residential structures, and there may be large components in nonresidential structures that are painted a single color and appear to be homogeneous throughout. Two approaches can be used to deal with the large room equivalents and the large components. First, large rooms or large components can be divided into smaller portions by using typical sizes in residential housing. The resulting room or component portions can, then, be tested as separate entities, following the approach in a residential unit. A second approach is to take three samples on a large component, spread over the component. If the results of the samples are within the bounds of sampling and measurement error, then, regard the component as homogeneous, and the three samples as sufficient to characterize its lead level. As noted above, paint sample collection and laboratory measurement alone are have a coefficient of variation of 12 to 17%. Differences within this range are caused by sampling and analysis; differences above this range indicate real spatial differences in the component tested. The first approach is the more conservative. For structures in the fourth category above, the concept of a room or even a side of a house may be difficult to apply. The structures in this category are seemingly homogeneous. A reasonable approach is to test several samples spread over the structure (21).

DISCLAIMER

This article does not necessarily reflect the views and policies of any federal agency or department mentioned.

BIBLIOGRAPHY

1. D. H. Parker, *Principles of Surface Coating Technology,* Interscience Publishers, New York, 1965.
2. V. C. Vesce and L. W. Ryan, "Classification of Pigments;" C. H. Rose, "The Lead Pigments;" H. A. Nelson, "Zinc Pigments;" D. O. Noel, "Bronze Powder;" and H. R. Harlan, "Flaked Metallic Lead;" in J. J. Mattiello, ed., Volume II, *Raw Materials: Pigments, Metallic Powders, and Metallic Soaps of Protective and Decorative Coatings,* John Wiley & Sons, Inc., New York, 1942, pp. 3–57 (Chapt. 1); pp. 337–368 (Chapt. 13); pp. 369–388 (Chapt. 14); pp. 601–602 (Chapt. 26); pp. 603–610 (Chapt. 27).
3. A. G. Roberts, *Organic Coatings: Properties, Selection, and Use,* National Bureau of Standards (now the National Institute of Standards and Technology), Washington, D.C., 1968, pp. 5–14.
4. G. Wormald, "Chromate Pigments," and E. J. Dunn, Jr., "Lead Pigments," in R. R. Myers and J. S. Long, eds., Vol. 3, *Treatise on Coatings,* Part I, *Pigments,* Marcel Dekker, New York, 1975, pp. 305–332 (Chapt. 6) and pp. 333–427 (Chapt. 7).
5. D. Heiserman, *Exploring Chemical Elements and Their Compounds,* TAB Books, Blue Summit Ridge, Penn., 1992, p. 287.
6. P. Goetz, Editor-in-Chief, *The New Encyclopedia Britannica,* 15th ed., Vols. 7, 9, and 21, Encyclopedia Britannica, Inc., Chicago, Ill., 1986, pp. 218–219 (vol. 7), p. 66 (Vol. 9), and pp. 404–406 (vol. 21).
7. W. G. Vanney, "Exterior Finishes," in J. J. Mattiello, ed., vol. III, *Manufacture and Uses: Colloids, Oleoresinous Vehicles and Paints, Water and Emulsion Paints, Lacquers, Printing Inks, Luminescent Paints, and Stains of Protective and Decorative Coatings,* John Wiley & Sons, Inc., New York, 1943, Chapt. 10A, pp. 269–271.
8. D. E. Jacobs, "Lead-Based Paint As A Major Source of Childhood Lead Poisoning: A Review of the Evidence", in M. E. Beard and S. D. A. Iske, eds., *Lead in Paint, Soil, and Dust: Health Risks, Exposure Studies, Control Measures, and Quality Assurance, ASTM STP 1226,* American Society for Testing and Materials, Philadelphia, Pa., 1995, pp. 175–187.
9. National Lead Information Center, "Lead Inform", vol. 1, Washington, D.C., 1995, p. 7, citing *Industrial Health Monthly* (Nov. 1951).
10. U.S. Dept. of Housing and Urban Development, *Comprehensive and Workable Plan for Abatement of Lead-Based Paint in Privately Owned Housing, Report to Congress,* U.S. Dept. of Housing and Urban Development, Washington, D.C., 1990, pp. 1–1 to 1–4.
11. U.S. Environmental Protection Agency, *A Field Test of Lead-Based Paint Testing Technologies: Summary Report,* EPA 747-R-95-002a, U.S. Environmental Protection Agency, Washington, D.C., 1995, pp. 15–16.
12. U.S. Environmental Protection Agency, *A Field Test of Lead-Based Paint Testing Technologies: Technical Report,* EPA 747-R-95-002b, U.S. Environmental Protection Agency, Washington, D.C., 1995, pp. 4–31 to 4–34.
13. U.S. Environmental Protection Agency, *Report on the National Survey of Lead-Based Paint in Housing, Base Report,* U.S. Environmental Protection Agency, Washington, D.C., 1995.
14. U.S. Environmental Protection Agency, *Report on the National Survey of Lead-Based Paint in Housing, Appendix I: Methodology,* U.S. Environmental Protection Agency, Washington, D.C., 1995.
15. U.S. Environmental Protection Agency, *Report on the National Survey of Lead-Based Paint in Housing, Appendix II: Analysis,* U.S. Environmental Protection Agency, Washington, D.C., 1995.
16. D. J. Brody and co-workers, *JAMA* **272**(4), 277–283 (1995).
17. J. L. Pirkle and co-workers, *JAMA* **272**(4), 284–291 (1995).
18. Lead-Based Paint Hazard Reduction and Financing Task Force, *Putting the Pieces Together: Controlling Lead Hazards in the Nation's Housing, Summary,* U.S. Dept. of Housing and Urban Development, Washington, D.C., 1995, p. 4.
19. Lead-Based Paint Hazard Reduction and Financing Task Force, *Putting the Pieces Together: Controlling Lead Hazards in the Nation's Housing, Report,* U.S. Dept. of Housing and Urban Development, Washington, D.C., 1995, pp. 69–70.
20. "Ban of Lead-Containing Paint and Certain Consumer Products Bearing Lead-Containing Paint," *Code of Federal Regulations, Title 16, Part 1303,* Office of the Federal Register, Washington, D.C., 1995, pp. 345–349.
21. ASTM Committee E-6 on Performance of Buildings, *ASTM Standards on Lead-Based Paint Abatement in Buildings,* 03-506194-10, American Society for Testing and Materials, Philadelphia, Pa., 1994.

LEAD LEVELS IN SOIL

B. E. DAVIES
Clemson University
Clemson, South Carolina

Whenever lead and its compounds are heated, dissolved, or pulverized they become mobile and escape to the environment as contaminants. They follow normal biogeochemical pathways until they enter and are trapped in some environmental sink. Soils and sediments are important sinks. Lead has been used for millennia, because it is easily worked and its ores have been found close to the surface in many parts of the world. There has, therefore, been ample opportunity for it to be dispersed and lead concentrations in soils and sediments appear to have increased over the years. Elevated soil concentrations are now found near mines and smelters, in most urban soils, alongside roads and in industrial areas, and in many agricultural soils as a result of using sewage sludge contaminated by industrial wastewaters. Lead has a long residence time in soil due to its low solubility and relative freedom from microbial degradation, and it will persist in soil until far into the future.

Lead is neither an essential nor a beneficial element for plants or animals. It is well known, however, for being poisonous. Although human body burdens are generally well below those that cause clinical symptoms of toxicity there are fears that in some lead-rich environments they may be high enough to cause ill health and mental impairment in young children. Soil and dusts are important sources of lead for children and lead in blood can be related directly to lead in soil or dust. Many investigations have been carried out on lead in environmental materials, including soil, over the recent decades, and there is now a good understanding of its environmental chemistry and health significance.

Although the primary concern is with soil, dust is often not distinguished from soil in many health studies and work on dust will also be reviewed. The essential difference between the two materials is that dust is dirt that has settled on to a surface but is not necessarily formed from the fabric of the surface, whereas soil is the product of chemical and biological weathering of the underlying rock with which it is intimately connected.

A conceptual difference is drawn between contamination and pollution. Contamination is understood to be the addition of lead to a soil by some human activity. A polluted soil is one in which the lead concentration is high enough to cause biological injury. There is a voluminous literature on the clinical and subclinical effects of lead poisoning on adults and children.

CHEMISTRY AND USE OF LEAD

Lead is a member of Group IVB of the Periodic Table of the elements. Its atomic weight is 207.2, and its atomic number is 82. There are four stable isotopes: ^{204}Pb, ^{206}Pb, ^{207}Pb, and ^{208}Pb; the latter three are the ultimate decay products of ^{238}U, ^{235}U, and ^{232}Th, respectively. Because of this association with radioactive decay, lead ores derived from different source rocks at different geological times can show differing isotope abundances. This property has been used to recognize and apportion sources in contamination studies.

Elemental lead is a dense blue-gray metal with a density of 11.3 kg/m^3. It melts at 327°C and boils at 1744°C. Because of the low melting point, even early technological societies were able to smelt, melt, and work lead. The metal is also soft and creeps or flows under sustained pressure and is, therefore, readily cut and shaped. It has long been used to waterproof roofs and to make water pipes. Metallic lead is relatively opaque to ionizing radiation, and it is a valuable shield material in x-ray and radioisotope work.

Lead readily alloys with other metals: the Pb–Sb alloy is used to make battery plates in the lead–acid accumulator and Pb–Sn alloys are often used as solder. In historic times pewter (four parts tin to one part lead with a little antimony) was commonly used for drinking vessels, although these are now mostly ornamental.

Two oxidation states (Pb(II) and Pb(IV)) are stable, but the environmental chemistry of the element is dominated by the plumbous ion Pb^{2+}. Inorganic compounds are widely used, eg, yellow chromate is used in road markings, and paints have in the past contained lead oxides or lead soaps to promote polymerization. Many countries now require no-lead or low-lead paints to be used for interior decoration (see LEAD-BASED PAINT). There is an extensive organic chemistry of Pb(IV) compounds, especially tetra-alkyl and tetra-aryl compounds, which have been used as gasoline additives for most of this century.

LEAD CONTENTS OF ROCKS AND SOIL

The chemical composition of soil depends in the first instance on that of the parent material, ie, the geochemical nature of the subjacent rock. During the course of soil formation, the soil mantle becomes differentiated into horizons. Mobile constituents are leached down the profile and are either captured in subjacent horizons or lost to the groundwater. Humus accumulates in surface layers as biological debris undergoes degradation and transformation. Anthropogenic influences cause trace element enhancement through contamination. Thus over time soil composition changes away from that of the original parent rock.

Reference books quote the crustal abundance of lead as approximately 13 μg/g. The ionic radius of Pb^{+2} is 124 pm, and it isomorphously replaces K^+ (133 pm) in silicate lattices or Ca^{+2} (106 pm) in carbonates and apatites. There is generally an increase in the lead content of igneous rocks in the sequence ultramatic to silicic and the lead content of granitic intrusives is mainly controlled by their potassium feldspar content. The average basalt contains 5 μg Pb/g and the average granite, 20 μg Pb/g.

Although 95% of crustal rocks are of igneous origin, sedimentary rocks account for 75% of surface exposures and are, therefore, the most widespread soil parent material. The most common sedimentary rocks are shales and mudstones (80%), which have an average Pb content of 23 μg/g. Sandstones constitute 15% of sedimentary rocks and contain an average of 10 μg Pb/g, while limestones and dolomites (5% of sedimentary rocks) contain about 71 μg Pb/g. Lead has a strong affinity for sulfur (ie, it is chalcophilic) and it concentrates in sulfide phases in rocks; the major ore mineral is galena (PbS). Black shales are rich in organic matter and iron sulfide and tend to have higher lead contents than other shales.

Reliable estimates of soil lead on a regional or national basis have become available only in recent years. Until the advent of atomic absorption spectrometry in the 1960s the analysis of lead was slow and difficult and was limited to experimental work or occasional soil analyses for agronomic purposes; large surveys were impractical. One of the most comprehensive compilations of published data calculates 29.2 μg Pb/g as the overall mean of 4970 collated values with a range of <1–888 μg/g (1). Systematic surveys have been published over recent years. In the United States, data are available from a collection of 963 surface (0–20 cm) soils (2). For the whole conterminous United States the (geometric) mean was 16 μg/g; west of the 96th meridian the mean was 17 μg/g and east of it, 20 μg/g. The range was <10–700 μg/g. In a comprehensive survey of agricultural soils in England and Wales, the geometric mean was 45.6 μg Pb/g, and the range was 3.0–16338 μg/g (3). In Scotland, for 3944 samples from 896 Scottish soil profiles, the geometric mean content of all mineral samples was 13 μg Pb/g and of all organic samples was 30 μg Pb/g (4).

There is a paucity of data for remote, unsettled areas of the world. One study (5) reported 72.6 μg/g for continental Chile; <0.5 μg/g for the Antarctic peninsula; and 7.1 μg/g for Torres del Paine, a wilderness area located in an inaccessible part of the Cordillera des Andes. It is probable that in remote or recently settled areas, soil lead contents are <20 μg/g; but elsewhere general, low level contamination from industry, ore processing, and the use of leaded gasoline has raised concentrations generally to 30–100 μg/g.

Lead in Soil Profiles

Lead accumulates naturally in surface horizons of soil. In Canada, lead showed the greatest proportional accumulation of a range of trace metals in surface horizons of soils (6). One study (7) used the ratio of lead concentration in topsoil (0–15 cm) to that in subsoil (30–45 cm) as an index of surface contamination. The index was called the relative topsoil enhancement (RTE). In normal agricultural areas of Great Britain, RTE ranged from 1.2 to 2.0, whereas in locations affected by mining and smelting operations, values between 4 and 20 were common.

There is little evidence that lead is readily lost from soil profiles by leaching, and it remains in an insoluble or stable form in surface layers. Soils have rather large capacities for the immobilization of lead and the organic fraction is mainly responsible for the observed fixation of the metal (8). Experimental evidence confirms this. When 11 soils were treated with a natural leachate spiked with lead and other trace elements, lead was immobile in all soils except for one classed as an ultisol, with a loamy sand texture and a low cation retention capacity (9).

Identification of Contaminated Soils

There is no simple, unequivocal way of recognizing when a soil has been contaminated by lead, since all soils contain it at some concentration. It is assumed that for an uncontaminated soil the observed content will be somewhere within a range clustering around the mean. How to establish this range is still a matter for argument. Geochemists now generally accept that the trace element contents of igneous rocks conform to a log-normal model, ie, the frequency distribution for a given element in a given rock type can be represented by the Gaussian, normal curve if the element contents are transformed to natural or common logs. Whether this log-normal model is applicable to soils is still arguable but most researchers assume it is. A property of the normal distribution is that 95% of the area underlying the curve is bounded by the range ± 1.96 times the standard deviation (SD). For convenience the range is usually established using 2 SD and for the log-normal model the range becomes the mean/2 SD to the mean times 2 SD. For the conterminous United States, the 95% probability range for soil lead is 4.3–60 μg/g (2). In Scotland the range is 2.6–83 μg Pb/g (4). Experience suggests that in the case of most mineral soils, the upper limit is 150–250 μg/g.

A high value does not necessarily mean the soil is contaminated in the sense of having been affected by some human activity. Natural lead poisoning has been described in six localities in Norway. This arose because mineralization by galena (disseminated in quartzitic sandstone) occurred close to the surface. Ground quite barren of vegetation contained 24,500 μg Pb/g compared with a local background level of 57 μg/g (10).

In the great majority of cases an unusually high soil lead content is the result of contamination. Nonetheless, such a judgment should be made only by following a set protocol.

There is much public concern about environmental lead, especially in the United States. Remediating an apparently lead-contaminated environment can be costly. It is, therefore, essential that a problem area be carefully and systematically characterized using a protocol similar to that above.

ANTHROPOGENIC SOURCES OF LEAD

There are as many sources of lead for soils as there are uses of the metal or its compounds. Some are comparatively common yet strictly local. Two examples will suffice to illustrate sporadic local lead contamination.

The effects of water running off roofs containing lead can be exemplified by some Scandinavian data. In Sweden, samples of soil that had received the runoff from lead roofs of village churches for more than eight centuries were analyzed. Soil samples from below a depth of 100 cm contained 2–120 μg/g, whereas the upper 20 cm contained concentrations as high as 12,400 μg/g (11). Contamination of soil close to any building will be discussed below.

In Great Britain, November 5 is celebrated as Guy Fawkes Night when bonfires are lit and fireworks let off. The fires are made from any combustible material gathered over the preceding weeks. An examination was made of the site of such a fire in a suburban backyard where an old painted door and a roll of linoleum had been burned. The garden soil background level was 928 μg Pb/g, but the soil under the fire contained 2203 μg Pb/g and the residual ash contained 1958 μg/g. The researcher in soil contamination must always be aware of the possibility of these localized unusual concentrations of lead. A single spot sample does not characterize a whole yard.

There are several well-recognized major sources of lead that are of global interest: contamination from vehicle exhausts, sewage sludge and other materials used in agriculture, and ore mining and smelting.

Lead Derived from Vehicle Exhausts

As the internal combustion engine developed during the early decades of this century there was a growing demand for gasoline of higher octane ratings to avoid uneven combustion in the engine cylinders, ie, knocking. In the early 1920s, it was discovered that the lead alkyls (tetraethyl and tetramethyl lead), when added to gasoline, helped overcome the problem. The first leaded gasoline was sold in 1923 in the United States, and its use rapidly became standard both here and throughout the world. By the 1970s, the global consumption of lead for gasoline additives was some 375,000 t, and <70% of this was used in the United States.

It was not until the 1960s that much attention was paid to the possible dangers of inhaling lead-rich aerosols emitted from vehicle exhausts, despite earlier adverse reports. The introduction of lead-free fuel and the stepwise reduction of lead contents in gasoline in the United States in the 1970s was primarily because the introduction of platinum-based catalytic converters (to reduce photochemical smog in Los Angeles and other cities) required the use of unleaded fuel to avoid poisoning the platinum catalyst (12).

It was in British Columbia, Canada, that it was first discovered that soil and vegetation samples collected near roads contained unusually high lead contents (13), and it was noted that the contributions of lead from gasoline fumes merited attention. Subsequently, it was shown in Denver, Colorado, that soil and grass were contaminated by lead within 152 m downwind of roads. The source of the lead was vehicle exhaust fumes; and the relationship between lead content and distance was exponential, being highest nearest the road edge (14). This important source of lead for soil has now been recognized in many countries. There have been numerous detailed studies of the influence of factors such as particle size, volume of traffic, speed of traffic, grade of hill, etc on the deposition of lead. Depositional trends have been modeled mathematically (15). The literature suggests that there is always a zone of 15–20 m wide on either side of most roads in which the concentration of lead exceeds local background and highest levels are found within the first meter from the edge of the highway. Contamination of the roadside environment is now worldwide as a consequence of the use of leaded gasoline.

The effect of deposition of lead-rich aerosols from exhausts is most marked along roads, but vehicle exhausts have also caused environmental contamination over wide areas. Cores of ice taken from the Greenland ice cap show a steady increase in lead concentrations over the last 1000 years and a remarkable increase since the introduction of leaded gas.

The reduction in consumption of leaded gas has been matched by a decline in airborne lead levels and deposition to the land surface. For example, in Minnesota annual lead fluxes to the ground surface through precipitation decreased from 1979 to 1983: 2000 to 370 ng/cm^2 at an urban site and 430 to 100 ng/cm^2 at a rural site. In Europe, the mandatory reduction of lead in gasoline took place in 1986. From 1984 to 1986 airborne lead at monitoring sites in western Britain decreased by 52–61% (16). Before that date, some 8000 t lead per year is estimated as having escaped from the immediate vicinity of British roads and half of that was deposited over the British land surface (15).

One study, using published data for air concentrations of lead, calculated deposition of the metal from air on to land in various remote regions as 0.4 g/ha/year at the South Pole, 7.2 g/ha/year in northwest Canada, and 6.3 g/ha/year in northern Michigan (17). In rural, industrialized and metropolitan Europe, calculations of deposited lead ranged from 87 to 536 g/ha/year (median 189) and in North America from 71 to 20,498 g/ha/year (median 4,257). The calculated high enrichment in lead was attributed to a "substantial contribution from vehicle exhaust" (17). Normalizing air concentration of metals to those of aluminium (derived from a natural source), it was shown that lead has high enrichment factors (>100) in remote, rural, and metropolitan regions of Europe and North America, probably due to anthropogenic inputs.

The lead contents of surface soils in Norway decrease from an average of <120 μg Pb/g in the south of the country to <10 μg Pb/g in the far north; the lower values are due to remoteness from western European industries (18). When lead concentrations in soil collected at the Rothamsted Experimental Station, England, in 1972 were compared with samples collected in the late nineteenth century, increases ranged from 17 to 46%. Direct roadside contamination of the samples was unlikely, and the rise in levels reflected a general environmental contamination by atmospheric lead during the twentieth century (19).

The most convincing direct evidence for car exhausts as a source of lead is derived from studies of lead isotope ratios. Lead isotope evidence has shown that vehicle exhausts are the main soil contaminant near Adelaide, Australia (20). In southeastern Missouri (where much of the U.S. lead ore and ingot lead is produced), contaminating lead may be derived from either leaded gasoline or from milling and smelting operations. The $^{206/204}$Pb isotope ratios are different for the two sources. In four samples of gasoline the mean ratio was 18.49 whereas for 131 samples of the local ores, the mean ratio was 20.81. The ratios for three soil samples taken from alongside roads were 18.43, 18.58, and 18.82. Contaminating lead close to highways, therefore, appeared to have originated primarily from vehicle exhausts (21).

LEAD POISONING

Frank lead poisoning is normally seen only in industrial exposure or after the accidental swallowing of metal-rich substances (metal salts, leaded paint chips, etc). Early symptoms are evening tiredness and, later, aching in the joints and limbs. The subsequent abdominal colic has long been known, eg, in Devon, England, it was called Devonshire colic and was traced to excess lead in cider. The presses for crushing the apples were often sealed with lead before the danger was recognized. Acute lead poisoning will be associated with vomiting, and chronic lead poisoning with damage to the nerves associated, perhaps, with damage to the wrist extensors ("wrist drop"). Long-term exposure may also cause a dark line in the gums. Acute lead poisoning may lead to death.

Concern over the last 30 years has been expressed over the consequences of environmental exposure to lead at concentrations below those likely to give even the early symptoms of frank lead poisoning. The greatest attention has been paid to children. Children, especially those <3 years, are more susceptible to lead poisoning than are adults, and toddlers spend much of their time playing on the floor and sucking their fingers or licking toys and other objects. In some cases, children develop a marked tendency to ingest soil or dust, a condition known as pica. Imported, cheap toys are occasionally found to contain lead in the paint. Lead in food can sometimes be a significant source, eg, where people rely heavily on home grown vegetables in contaminated areas or where lead has leached from some glazes or from solder used in some tinned cans. Modern cans and glazes are free of this problem.

Air, water, paint, and soil or dust are now generally considered as the most important sources of lead in the urban environment. This article is principally concerned with the implications of soil (dust) contamination, it has been shown how widespread are the sources of contamination. However, it would be incomplete not to mention briefly exposure from air and water, as most individuals are exposed simultaneously to all sources, albeit to differing extents.

CONCLUSION

Over the post-World War II decades, lead has emerged from obscurity in environmental research to arguably the most intensively researched of all the heavy metals. Strong evidence has emerged that nonindustrial exposure to the metal can harm human health, especially young children. We are all likely to be exposed to more than one source, especially lead in air, water, soil, or dust. The problem of exposure to airborne lead compounds is diminishing in all countries where lead-free motor fuel is now mandatory. Similarly, the hazard associated with lead pipes in soft-water areas is decreasing as old plumbing is replaced, and more people live in newer homes with plastic pipes. The problem of lead in soil and dust remains serious, because of its long residence time in these materials.

BIBLIOGRAPHY

1. A. M. Ure and M. L. Berrow, in H. J. M. Bowen, ed., *Environmental Chemistry,* Vol. 2, Royal Society of Chemistry, London, 1982, pp. 94–204.
2. H. T. Schacklette and J. G. Boerngen, "Element Concentrations in Soils and Other Superficial Materials of the Conterminous United States," *U.S. Geological Survey Professional Paper 1270,* U.S. Government Printing Office, Washington, D.C., 1984.
3. S. P. McGrath and P. J. Loveland, *The Soil Geochemical Atlas of England and Wales,* Blackie Academic and Professional, London, 1992.
4. G. A. Reaves and M. L. Berrow, *Geoderma,* 32, 1–8 (1984).
5. A. M. Carrasco and M. Prendez, *Water Air Soil Pollut.,* **57–58,** 713–722 (1991).
6. J. L. Wright, R. Levick, and H. J. Atkinson, *Soil Sci. Soc. Am. Proc.,* **19,** 340–344 (1955).

7. P. Colbourn and I. Thornton, *J. Soil Sci.,* **29,** 513–526 (1978).
8. R. L. Zimdahl and R. K. Skogerboe, *Environ. Sci. Technol.,* **11,** 1202–1207 (1977).
9. N. E. Korte, J. Skopp, E. E. Niebla, and W. H. Fuller, *Water Air Soil Pollut.,* **5,** 149–156 (1975).
10. B. Bolviken and J. Lag, *Trans. Inst. Mining Metal.,* **86(B),** 173–180 (1977).
11. S. S. Jorgensen and M. Willems, *Ambio,* **16**(1), 16–19 (1987).
12. J. O. Nriagu, *Sci. Total Environ.,* **92,** 13–28 (1990).
13. H. V. Warren and R. E. Delavault, *Trans. R. Soc. Can.,* **54,** 11–20 (1960).
14. H. L. Cannon and J. M. Bowles, *Science,* **137,** 765–766 (1962).
15. A. C. Chamberlain, M. J. Heard, P. Little, and R. D. Wiffen, *Phil. Trans. R. Soc. London A,* **290,** 577–589 (1979).
16. R. A. Page, P. A. Cause, and S. J. Baker, *Sci. Total Environ.,* **68,** 71–77 (1988).
17. G. Sposito and A. L. Page, in H. Sigel, *Metal Ions in Biological Systems,* Vol. 18, Marcel Dekker, New York, 1984, pp. 287 – 332.
18. E. Steinnes, J. E. Hanssen, J. P. Rambaek, and N. B. Vogt, *Water Air Soil Pollut.,* **74,** 121–140 (1994).
19. C. Williams, *J. Agric. Sci.,* **82,** 189–192 (1974).
20. B. L. Gulson, K. G. Tiller, K. J. Mizon, and R. M. Meery, *Environ. Sci. Technol.,* **15,** 691–696 (1981).
21. M. B. Rabinowitz and G. W. Wetherill, *Environ. Sci. Technol.,* **8,** 705–709 (1972).

LIBS. See LASER-INDUCED BREAKDOWN SPECTROSCOPY.

LIF SPECTROSCOPY. See LASER-INDUCED FLUORESCENCE SPECTROSCOPY.

LUFTs. See OIL SPILL CONTAINMENT: TERRESTRIAL.